Vegetation Stress

Vegetation Stress

Based on the International Symposium on Vegetation Stress
Munich-Neuherberg, June 19-21, 1995

Edited by Hartmut K. Lichtenthaler

465 figures and 134 tables

Gustav Fischer Stuttgart · Jena · New York · 1996

Prof. Dr. Hartmut K. Lichtenthaler
Botanisches Institut II der Universität Karlsruhe
Kaiserstraße 12
D-76128 Karlsruhe
Germany

Special Edition of Journal of Plant Physiologie
Vol. 148, No. 1–5, 1996.

Die Deutsche Bibliothek – CIP-Einheitsaufnahme

Journal of plant physiology: biochemistry, physiology and mol. biology of plants. – Stuttgart ; New York : G. Fischer.
Erscheint jährlich in 2 Bd. – Aufnahme nach Vol. 147, No. 5 (1996)
Darin aufgegangen: Biochemie und Physiologie der Pflanzen. – Bis Bd. 114, H. 5 (1984) u. d. T.: Zeitschrift für Pflanzenphysiologie
ISSN 0176-1617.

Vol. 148, No. 1/5 = Special ed. Vegetation stress. – 1996.

Vegetation stress: based on the International Symposium on Vegetation Stress, Munich-Neuherberg, June 19–22, 1995; 134 tables / ed. by Hartmut K. Lichtenthaler. – Stuttgart; Jena; New York: G. Fischer, 1996
(Journal of plant physiology ; Vol. 148, No. 1/5 = Special ed.)
ISBN 3-437-20544-7 (Stuttgart ...)
ISBN 1-56081-444-6 (New York)
NE: Lichtenthaler, Hartmut K. [Hrsg.]; International Symposium on Vegetation Stress ⟨1995, München⟩

All business correspondence should be made with:
Gustav Fischer Verlag GmbH & Co KG, Wollgrasweg 49, D-70577 Stuttgart (Register HR Nr. B 4976). Deutsche Bank Stuttgart 80 20 000 (BLZ 600 700 70), Stuttgarter Bank 45 290 (BLZ 600 901 00), Commerzbank Stuttgart, 87 99 900 (BLZ 600 400 71), Landesgirokasse Stuttgart 11 205 49 (BLZ 600 501 01), Postgiroamt Stuttgart 135 56-709 (BLZ 600 100 70).

For USA and Canada:
VCH Publisher Inc., 303 N. W. 12th Avenue, Deerfield Beach, Florida 33442-1705, USA

Type setting: Calwer Druckzentrum, D-75365 Calw
Printed: Calwer Druckzentrum, D-75365 Calw
Bound: Röck, D-74189 Weinsberg
Printed in Germany

Contents

AIR POLLUTION, ACID RAIN, FOREST DECLINE, AND INCREASED CO_2 LEVEL

CHILLING SENSITIVITY AND FROST TOLERANCE

GENERAL PHYSIOLOGY AND STRESS TOPICS

REFLECTANCE AND FLUORESCENCE TECHNIQUES IN STRESS DETECTION

Reflectance Signatures

Fluorescence Signatures

Fluorescence Imaging

J. Plant Physiol. Vol. 148. pp. 1–3 (1996)

Preface of the Editor

This special issue of the Journal of Plant Physiology, which will also appear as a book with the title «Vegetation Stress», contains a large sample of the papers presented at the «First International Symposium on Vegetation Stress». This symposium was held at the GSF Research Center Neuherberg in Munich, Germany, from June 19 through 21, 1995 and had been organized and sponsored by Botany II, Karlsruhe (H. K. Lichtenthaler), the NASA Greenbelt, USA (E. Chappelle) and the GSF Research Center, Munich-Neuherberg (H. Sandermann). It brought together participants from more than 20 countries from all over the world. The idea to organize a vegetation stress symposium found an overwhelming positive response among the colleagues working in different fields of plant stress, and showed the great need for a discussion forum for the many different aspects of vegetation stress. The encouragement by many colleagues was a strong impetus for me to establish this symposium in a joint effort with the colleagues E. Chappelle, E. Middleton, H. Sandermann, C. Langebartels and C. Lütz within a few months.

Although the symposium was open to all forms of vegetation stress, main emphasis was given to stress effects induced by UV-B radiation, air pollution (ozone and SO_2), global change, nutrient deficiency, heavy metals, and forest decline. In addition, a major section was devoted to optical techniques and instrumentations (chlorophyll fluorescence, bluegreen fluorescence, reflectance) for near and also for possibly far distance stress detection in plants and their photosynthetic apparatus, in which the new laser-induced high resolution fluorescence imaging system sets a new standard in the stress detection of plants.

All manuscripts were reviewed by at least two reviewers in the usual evaluation procedure of the Journal of Plant Physiology (JPP). Only revised papers were accepted into this special JPP volume on vegetation stress; in fact, several manuscripts had to be unfortunately rejected. Some papers represent reviews, but the majority consists of original papers. The first contribution represents an introduction to the present stress concept in plants and the definition of plant stress, followed by contributions to the stress topics mentioned above. Some papers describe completely new results and come up with promising conclusions, which will stimulate future research. Other papers demonstrate the state of the art in stress research and give further evidence for already established causal relationships. Another group of papers addresses the application of additional methods and the development of new techniques for early stress assessment in plants. In summary, the approximately 90 contributions of experts in their field yield an excellent overview on major research lines in the present vegetation stress research. Hopefully, this vegetation stress book can and will serve as a reference book for experts as well as for newcomers and students, and it will stimulate further research for gaining a better understanding of stress effects, the mode of action of stressors and the stress coping mechanisms in plants.

I am very grateful to Dr. Wulf von Lucius, G. Fischer Verlag, Stuttgart, for his offer to print this special vegetation stress volume of the Journal of Plant Physiology and to the JPP Editor-in-Chief, Martin Bopp, Heidelberg, for his encouragement. Finally, I wish to thank the many colleagues who helped in the fast reviewing of the manuscripts, to Dr. Anja Maier, Karlsruhe, for essential help in the editorial work, to my secretaries, Ms. Inge Jansche and Gabrielle Johnson, and to my co-workers for valuable assistance and last but not least to my wife Regine for her patience and generous understanding of this time-consuming editorial work.

Karlsruhe, November 1995

Hartmut K. Lichtenthaler

Some participants

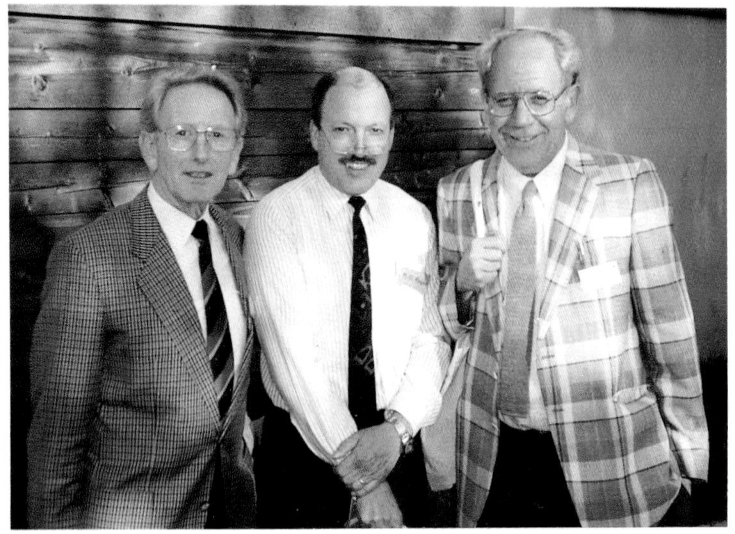

Hartmut Lichtenthaler, Hans-Dieter Payer, Ya'acov Leshem

Nicola D'Ambrosio, Zoltan Tuba

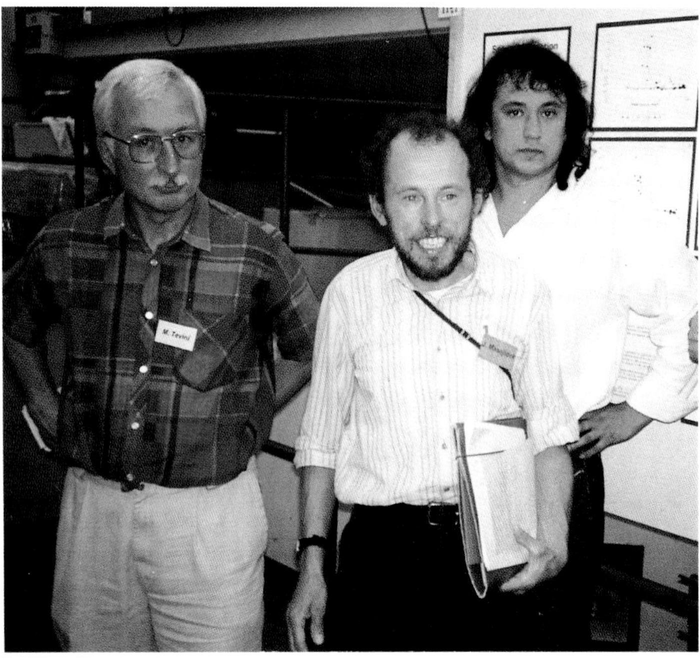

Manfred Tevini, Jiri Masojidek, Stephan Thiel

Reto Strasser, Pierre Dizengremel

Some chairpersons and organizers:
Front row: James Schepers, Jiri Masojidek, Christian Langebartels, Cornelius Lütz, Betsy Middleton, Hartmut K. Lichtenthaler, Anatoly Gitelson, Heinrich Sandermann. *Second row:* Claus Buschmann, Zoltan Tuba, Charles Mulchi, Jim McMurtrey, Erich Elstner, Ya'acov Leshem, Heinz Rennenberg (*from left to right*).

Hanna Wegner, Cornelius Lütz, Margrita Moser, Christian Langebartels

Hubert Ziegler, Heinrich Sandermann

J. Plant Physiol. Vol. 148. pp. 4–14 (1996)

Vegetation Stress: an Introduction to the Stress Concept in Plants

Hartmut K. Lichtenthaler

Botanisches Institut, Lehrstuhl II, University of Karlsruhe, Kaiserstr. 12, 76128 Karlsruhe, Germany

Received September 5, 1995 · Accepted October 12, 1995

Summary

This is a presentation of the essentials of the present stress concept in plants, which has been well developed in the past 60 years. Any unfavorable condition or substance that affects or blocks a plant's metabolism, growth or development, is to be regarded as stress. Plant and vegetation stress can be induced by various natural and anthropogenic stress factors. One has to differentiate between short-term and long-term stress effects as well as between low stress events, which can be partially compensated for by acclimation, adaptation and repair mechanisms, and strong stress or chronic stress events causing considerable damage that may eventually lead to cell and plant death. The different stress syndrome responses of plants are summarized in a scheme. The major abiotic, biotic and anthropogenic stressors are listed. Some stress tolerance mechanisms are mentioned.

Stress conditions and stress-induced damage in plants can be detected using the classical ecophysiological methods. In recent years various non-invasive methods sensing different parameters of the chlorophyll fluorescence have been developed to biomonitor stress constraints in plants and damage to their photosynthetic apparatus. These fluorescence methods can be applied repeatedly to the same leaf and plant, e.g. before and after stress events or during recovery. A new dimension in early stress detection in plants has been achieved by the novel high resolution fluorescence imaging analysis of plants, which not only senses the chlorophyll fluorescence, but also the bluegreen fluorescence emanating from epidermis cell walls which can change under stress induced strain. This powerful new technique opens new possibilities for stress detection in plants.

Key words: Bluegreen fluorescence, chlorophyll fluorescence, damage, resistance, long-term stress, strain, stress-factors.

Introduction

In the past ten years the number of scientific publications, found in journals of botany, plant physiology, ecophysiology and plant biochemistry dealing with plant stress and plant stress detection increased enormously. This process is still continuing and will proceed in an even more enhanced way in the future. Several books, e.g. *Stress and Stress Coping in Cultivated Plants* (McKersie and Leshem, 1994), *Plant Adaptation to Environmental Stress* (Fowden et al., 1993) and proceedings of symposia (Alscher and Cumming, 1990) or plant stress reviews (Larcher, 1987; Lichtenthaler, 1988) have appeared, which describe either plant stress concepts or particular aspects of plant stress, and also name various stressors and stress constraints.

The term «plant stress» is used by most authors in a very broad sense which justifies the establishment of a unifying concept of plant stress. This is fully correct since a multitude of stressors with different modes of action can induce, besides very specific effects, the same or at least similar overall responses in the plant. Plants do not have many response possibilities to stress, but respond, besides specific acclimation, in general with either a high-light type or a low-light type growth or adaptation response (Lichtenthaler, 1984).

From the literature it appears, however, that many authors regard almost every little modification and change of metabolic pathways, growth responses and development pattern of plants as stress responses and stress effects. In this respect the terms «plant stress and stress responses» are «over stressed». The term «stress» should not be applied to mere and fast readjustments of metabolic fluxes, e.g. of photosynthetic rates or respiration and transpiration rates as induced by changes in the photon flux density (sunlight ⇔ clouds), a decrease in temperature or an increase in air humidity. The plants are acclimatized and respond flexibly to such steadily re-occurring switches of cell metabolism and physiological activities as a response to changing environmental conditions. Moreover, diurnal changes in metabolic activities, growth pattern and cell division activities, which are regularly found at the day/night changes in the evening or at the night/day changes in the morning, do not represent stress effects, but can only be regarded as a reorientation of metabolic and growth activities according to the preferential day or night occurrence of certain metabolic processes. In addition, the plants can respond to environmental changes not only by fast acclimations, but also by particular long-term adaptations, e.g. of leaf size and thickness, stomata density, structure and function of chloroplasts as well as enzyme levels to either high-light or low-light growth conditions. Depending on their type and nature, these adaptations may take place within 1 or 2 days or in one week latest. With such adaptation responses plants can avoid stress constraints and adapt in an optimal way to new and changing outdoor growth conditions (Lichtenthaler et al., 1981; Lichtenthaler and Meier, 1984; Meier and Lichtenthaler, 1981).

Despite their capacity for fast acclimation of metabolic fluxes and the somewhat slower adaptation responses as well as certain stress tolerance mechanisms, plants are often exposed to sudden short-term or long-term stress events which reduce cell activity and plant growth to a minimum. This can lead to a severe damage eventually causing cell death if the stress coping mechanisms or repair mechanisms of plants are overworked. There exist many either natural or anthropogenic stress factors, which, depending on their intensity and duration, can cause damage to plants. These stresses can also be characterized as abiotic or biotic stresses.

In order to better differentiate between regular acclimation and adaptation responses of plants on one hand, and stress effects, stressors and stress constraints on the other hand, one needs a unifying general stress concept of plants. In the past 60 years the stress concept (Seleye, 1936) had successively been developed for plants by various authors (Stocker, 1932, 1947; Larcher, 1987; Levitt, 1980; Lichtenthaler, 1988; McKersie and Leshem, 1994). This concept seems to be hardly known to the botanical community although the term stress is presently being used in many publications. During the first international «Vegetation Stress Conference» in Munich, June 1995, the present stress concept of plants was therefore presented as an introduction speech, and is also exposed here at the beginning of this vegetation stress volume. This review not only surveys the different stress approaches, but also gives some examples for stress detection in plants, e.g. by the non-invasive chlorophyll fluorescence techniques, and very recently also by including the bluegreen fluorescence of plants.

Definition of plant stress

The original general stress concept for living organisms was developed by H. Selye (1936, 1956) and can be summarized in the following two sentences: «*All agents can act as stressors, producing both stress and specifc action*» and «*There exist stressor specific responses and non-specific general responses*». J. Levitt (1980) defined stress as: «*Any environmental factor potentially unfavorable to living organisms*».

On the basis of various observations in plants, and also under inclusion of the original concept on drought resistance of the botanist Stocker (1932 and 1947), the plant ecophysiologist Larcher (1987) summarized the stress concept of plants, and he stated that «*Every organism experiences stress, although the way in which it is expressed differs according to its level of organization*». From the botanist's point of view he described stress as a «state in which increasing demands made upon a plant lead to an initial destabilization of functions, followed by normalization and improved resistance» and also «*If the limits of tolerance are exceeded and the adaptive capacity is overworked, the result may be permanent damage or even death*». Larcher (1987) also stated that «stress contains both destructive and constructive elements and that stress is a selection factor as well as a driving force for improved resistance and adaptive evolution».

Eu-stress and dis-stress

Lichtenthaler (1988), who took up a proposal of W. Larcher given in a personal discussion, extended the stress concept of plants by differentiating between eu-stress and dis-stress, in which case eu-stress is an activating, stimulating stress and a positive element for plant development, whereas dis-stress (as seen in the English word distress) is a severe and a real stress that causes damage, and thus has a negative effect on the plant and its development. As formulated by Lichtenthaler (1988): «*A mild stress may activate cell metabolism, increase the physiological activity of a plant, and does not cause any damaging effects even at a long duration. Such mild stimulating stress is favorable for the plant*». In any case one has to consider that stress is a dose-dependent matter. At fairly low concentrations a stressor, e.g. a herbicide, can stimulate plant metabolism and plant growth, as has been observed in the case of various herbicides and plant growth regulators. Thus, very low doses of a stressor and a xenobiotic can, in fact, have the opposite effect than higher doses. Whether this applies to all stressors has yet to be proved. However, at a concentration 10 or 100 times higher the same xenobiotics will cause damage to the plant and induce early senescence finally leading to death if the stressor is not removed. Such damaging stressor concentrations and all other stress constraints at higher doses are negative for the physiology and development of plants, and thus represent a definite stress in the sense of a dis-stress. Within this concept, real stress shows up when a certain threshold of a stressor, which can no longer be compensated for.by the plant, is exceeded. When this threshold of stress-tolerance or stress-resistance has been passed, a short-term high level stress can principally induce the same damage as a long-term low level stress. The applicability of the «stressor dose – stress effect relationship» seems to be obvious, but has

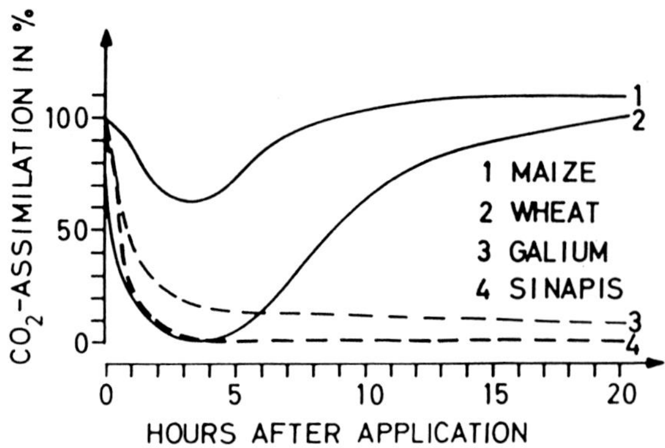

Fig. 1: Inhibition of photosynthetic CO_2 assimilation in the crop plants maize and wheat and the weeds *Galium* and *Sinapis* after spraying of leaves with the herbicide bentazon at doses equivalent to $1\,kg\,ha^{-1}$ (from Lichtenthaler et al., 1982).

not been proved so far in all cases and thus, more research is required in this field.

One should keep in mind that the transition between eu-stress and dis-stress is fluent. The relative position of the stress tolerance threshold depends not only on the plant species, but also on the type of stressors applied and on the predisposition of the plant, i.e. the growth condition and vitality before the stressor starts to act. Plants also differ in their stress coping capacity. This can be illustrated with the example of the application of herbicides in agricultural crops in order to kill weeds. Many crop plants possess the capacity to detoxify herbicides by introduction of a hydroxyl group to the aromatic ring of the herbicide which is then glycosylated to an inactive compound that can no longer bind to its target protein (Devine et at., 1993; Hock et at., 1995). However, this detoxifying capacity is often not present in the weeds to be controlled and the latter will eventually die off. An example is shown in Figure 1 with the application of the herbicide bentazon which blocks the photosynthetic electron transport by binding to the Q_B-binding protein of photosystem II instead of Q_B. After bentazon application the photosynthetic rates initially decline in the crop plants wheat and maize as well as in the weeds *Galium* and *Sinapis* (Lichtenthaler et al., 1982). After several hours, the photosynthetic capacity of maize and wheat is, however, restored since both crop plants possess the ability to hydroxylate, glycosylate and detoxify bentazon. Consequently, the weeds possess a much lower stress tolerance than the crop plants wheat and maize, which exhibit the herbicide detoxifying metabolism.

President Clinton 1995 and Plant Stress Research

In a public speech President B. Clinton regarded a 1 million US$ government-financed «study of plant stress» as an example of wasted funds, since he mistook plant stress as meaning emotional stress of plants. The topic of concern was a governmental grant to Texas Tech University to run a laboratory for the «development of drought-resistant wheat and pasture grasses». Animal or human stress on one hand and plant stress on the other hand, however, are different matters. This should be better emphasized by plant physiologists and made clear to the public. In his letter and response to B. Clinton's remarks, James N. Siedow (1995), the present president of the American Society of Plant Physiology, also gave a clear definition of stress in plants: «*Plant stress refers to a wide range of biological and environmental stresses that crops and other plants are subjected to daily. These include drought, cold and heat, weeds, insects and a host of diseases including those caused by viral, fungal and bacterial pathogens.*»

Stress Concept in Physics and Botany

The stress concept has also been developed in physics, and there the terms stress, strain and damage are well defined. This stress concept can also be applied to plants (see Lichtenthaler, 1988). According to this, these stress terms used in physics mean:

Stress:	is a state of the plant under the condition of a force applied.
Strain:	is the response to the stress and to the force applied to the plant (i.e. the expression of stress before damage occurs)
Damage:	is the result of too high a stress, which can no longer be compensated for.

In botany and plant physiology the term «strain» is rarely used and often not known. Strain is usually replaced by stress responses. Based on the stress concepts in physics it is clear that there can be stress and strain in plants, and that a damage does not necessarily occur even when the plant is under long-term stress and continuous strain. With specific strain (and limited vitality) the plant can survive also under continuous stress constraints although at much reduced metabolic activities and growth rates. An example may be given here. In the Northern Black Forest at Herrenalb a 170 year old pine (*Pinus silvestris* L.) grows on the portal and walls of a former Romanic monastery church, ca. 4 m above ground, but its roots are only found above ground in the stones of this wall and they have no access to soil and water. Thus, under continuous stress (primarily water stress) and strain this pine managed to survive in this unfavorable location and to grow within 170 years to a ca. 9 m high tree which visually appears fully intact and healthy (Fig. 2). The growth limitations set by this location are, however, documented by much less needles per needle year, as well as much shorter and thinner twigs as compared to pines growing in locations with more optimal growth conditions. Reducing the leaf or needle area, i.e. the area for transpiration, is one of the major water stress-coping mechanisms found in broad-leaf and conifer trees.

The different phases induced by stress

Based on the original stress concept of Selye (1936, 1956) and taking into account the results of Stocker (1932, 1956) one has to differentiate among the plant's stress responses three phases (Larcher, 1987) to which a fourth has been added by Lichtenthaler (1988). Before stress exposure the plants are in a certain standard situation of physiology which

Fig. 2: Pine (*Pinus silvestris* L.) having grown for ca. 170 years under continuous water stress on the sandstone portal of a former Romanic monastery church at Herrenalb, Black Forest. The pine roots have no contact to soil and ground water, but end in the wall 2 m above ground (Height of the pine: ca. 9 m).

is an optimum within the limits set by the growth, light and mineral supply conditions of the location. Stressors or complex stress events will then lead to the three stress response phases, and later to the regeneration phase after removal of the stressors if the damage had not been too severe. These are the consecutive four phases:

1. Response Phase: (beginning of stress) — **alarm reaction**
- deviation of the functional norm
- decline of vitality
- catabolic processes exceed anabolism

2. Restitution Phase: (continuing stress) — **stage of resistance**
- adaptation processes
- repair processes
- hardening (reactivation)

3. End Phase: (long-term stress) — **stage of exhaustion**
- stress intensity too high
- overcharge of the adaptation capacity
- chronic disease or death

4. Regeneration Phase: **partial or full regeneration** of the physiological function when the stressor is removed and the damage was not too high.

At the beginning of stress the plants react with a decline of one or several physiological functions, such as the performance of photosynthesis, transport or accumulation of metabolites and/or uptake and translocation of ions. Due to this decrease in metabolic activities, the plants deviate from their normal physiological standard and their vitality declines. Acute damage will occur fast in those plants which possess no

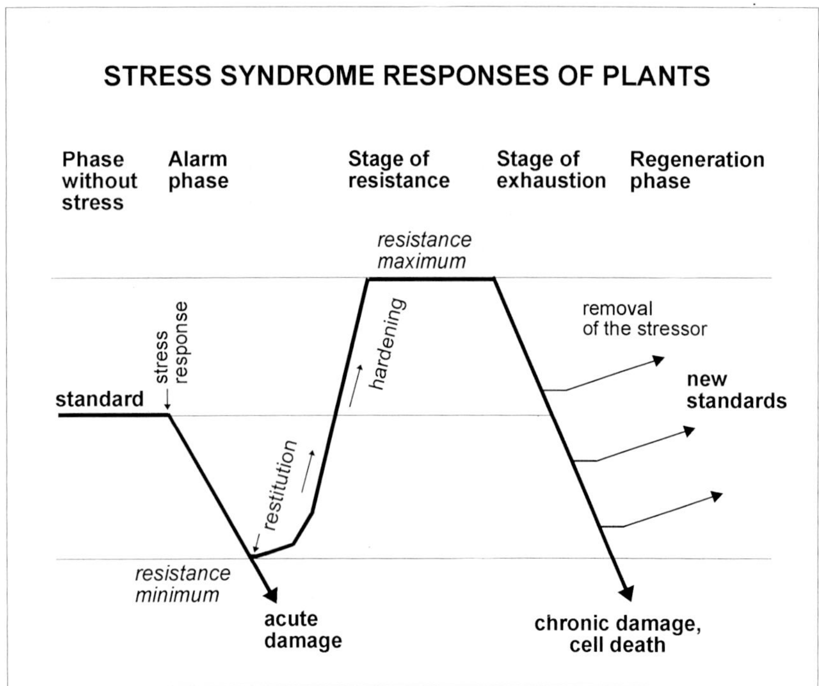

Fig. 3: General concept of the phase sequences and responses induced in plants by stress exposure. Plants growing at a physiological standard condition will respond to and cope with stress. After removal of the stressor(s), new standards of physiology can be reached depending on the time of stressor removal as well as the duration and intensity of the stress.

or only low stress tolerance mechanisms, and thus have a low resistance minimum (Fig. 3). During this alarm phase most plants will, however, activate their stress coping mechanisms by fast acclimations of their metabolic fluxes as well as activating repair processes and long-term metabolic and morphological adaptations. This is also called the general alarm syndrome GAS (McKersie and Leshem, 1994). GAS may also stand for general acclimation syndrome or general adaptation syndrome. Repair processes and adaptations will not only lead to a restitution of the previous physiological functions, but also to a hardening of plants by establishing a new physiological standard, which is an optimum stage of physiology under the changed environmental conditions and which corresponds to the plants' *resistance maximum* (Fig. 3). At long-term stress and a stress-dose overloading the plants' stress coping mechanisms, the stage of exhaustion (end phase) shows up in which physiology and vitality are progressively lost. This causes damage and finally cell death. However, when the stressors are removed in the right time before the senescence processes become dominant, the plants will regenerate and move to new physiological standards (regeneration phase). The time and stage of exhaustion at which the stressors are removed defines to which new physiological standard within the resistance minimum and maximum the plants will move (Fig. 3).

How long the plant will stay at the new physiological standard depends on external and internal factors. In field plants this is certainly not too long. Endogenous changes in the development program of plants have always been associated with changes in their physiology program and activity, and have resulted again in a new physiological standard. Furthermore, the next stress events will show up soon, and these again require a re-orientation of the plant's physiology standard to a new «optimum» within the limited possibilities set by the stress constraints. One should keep in mind that stress exposure of plants is not a rare event, but can occur daily, since there exist many different stressors which often act simultaneously. Therefore, stress and strain are routine events in a plant's life. Continuous stress and strain does, however, not mean that a damage must necessarily occur in a plant. If intensity and duration of stress are not too high and long, the plants will orient themselves within the range set by the resistance minimum and maximum (Fig. 3), and in such cases damage symptoms are not detectable. With respect to such findings one has to differentiate between the detection of stress and strain on one hand and the detection of clear damage symptoms on the other hand. Both processes may require different methods of detection, since the methods for damage detection may not allow to screen stress or strain of plants. If one wants to take countermeasures against stress and strain in order to avoid damage and to guarantee an optimum growth and harvest of plants, one should not wait until damage symptoms are visually detectable but respond much earlier. And this requires an early and efficient stress and strain detection in plants.

Stress constraints and stressors

There exist many stress events and a multitude of stressors in the life cycle of plants. The different kinds of stress factors (stressors) acting on land plants are listed in Table 1 under the grouping of natural stress factors (I) and anthropogenic stress factors (II). One can also list the various kinds of stressors under biotic and abiotic stress factors which is as valid as the grouping given in Table 1. With respect to the new large scale tree and forest decline detected in 1983 in the Northern hemisphere (Europe, USA, Russia, China) (Lichtenthaler and Buschmann, 1984; Rennenberg et al., 1996; Wellburn, 1994) and which is still progressing, it was desired to contrast the

Table 1: List of natural and anthropogenic stress factors acting on terrestrial vegetation.

I. Natural stress factors:

- **high irradiance** (photoinhibition, photooxidation),
- **heat** (increased temperature),
- **low temperatures** (chilling),
- **sudden and late frost,**
- **water shortage** (desiccation problems),
- **natural mineral deficiency** (e.g. nitrogen shortage),
- **long rainy periods,**
- **insects,**
- **viral, fungal and bacterial pathogens.**

II. Anthropogenic stress factors:

- **herbicides, pesticides, fungicides,**
- **air pollutants,** e.g. SO_2, NO, NO_2, NOx,
- **ozone** (O_3) and **photochemical smog,**
- **formation of highly reactive oxygen species** ($^1O_2$, radicals $O_2^{\cdot -}$ and $OH^{\cdot}$, H_2O_2)
- **photooxidants** (e.g. peroxyacylnitrates),
- **acid rain, acid fog, acid morning dew,**
- **acid pH** of soil and water,
- **mineral deficiency** of the soil, often induced by acid rain (shortage of the basic cations K, Mg, Ca, often Mn and sometimes Zn),
- **over-supply of nitrogen** (dry and wet NO_3-deposits),
- **heavy metal load** (lead, cadmium, etc.),
- **overproduction of NH_4^+** in breeding stations (uncoupling of electron transport),
- **increased UV-radiation** (UV-B and UV-A),
- **increased CO_2 level and global climate change.**

potential anthropogenic stress factors (most of which showed up only in the past 40 years) against the many natural abiotic and biotic stress factors to which the trees had been exposed to for a very long time.

One has to consider that the stressors listed in Table 1 rarely act individually and separately on the plant. Usually, several stress factors act simultaneously on the plant, such as the frequently combined heat, water and high-light stress at dry, sunny and warm summer periods. In addition, on plants there often act primary stressors or stress events, which considerably reduce the plants' vitality, such as air pollution followed by secondary stressors, such as bark beetles or particular fungi, which further lower the plant's vitality and will eventually lead to the dying-off of the tree.

Light adaptation and stress tolerance

Plants can adapt their leaf morphology as well as the structure and function of their photosynthetic apparatus to the incident light intensity. This adaptation response is best visualized in the formation of sun and shade leaves of trees which possess not only a different morphology and chemical composition, but also different rates of photosynthesis (Lichtenthaler, 1984). High-light plants and sun leaves exhibit a smaller leaf area (to reduce the transpiration rate) and are thicker (e.g. longer palisade parenchyma cells, or even two rows of palisade cells as in beech) than shade leaves or leaves of low-light plants. They possess sun-type (high-light) chloroplasts with higher rates of photosynthetic quantum conversion and net CO_2 assimilation, and also a higher light compensation point and a higher light saturation point of the overall photosynthetic process (Lichtenthaler et al., 1981; Lichtenthaler and Meier, 1984; Meier and Lichtenthaler, 1981). High-light or sun-type chloroplasts possess much lower amounts of the light-harvesting chlorophyll *a/b* proteins, the LHCPs, a lower degree of stacking of thylakoids, less thylakoids per chloroplast, but more photosynthetic electron transport chains and photosynthetic reaction centers per total chlorophyll compared to low-light or shade-type chloroplasts. The latter, in turn, exhibit much higher and wider grana stacks and have invested in a large light-harvesting antenna to overcome the light shortage at their shade or low-light location (Lichtenthaler and Meier, 1984; Meier and Lichtenthaler, 1981).

The sun-type (high-light) or shade-type (low-light) modification of leaves and chloroplasts is a true adaptation response of plants. The sun and shade leaf modification can only be expressed during leaf growth, but the adaptation of chloroplast ultrastructure and photosynthetic function to high-light (sun) or low-light (shade) growth conditions is possible throughout the vegetation period, and takes about one week in order to fully convert shade-chloroplasts into sun-chloroplasts *or vice versa*. These adaptations to either high or low irradiance make sense from a physiological point of view. High-light plants and sun leaves are better adapted to high-light exposure than low-light plants and shade leaves. Sun leaves with an extreme full sun light exposure can reduce their LHCPs to very low amounts in order to avoid absorption of excess light which cannot be used in photosynthetic quantum conversion.

In other words, high-light plants and sun leaves are much better protected against high-light stress than low-light plants or shade leaves. By a thicker cuticula, more flavonols in their epidermis, etc. they are also better protected against UV-A and UV-B stress or damage as is being shown via fluorescence excitation spectra (Schweiger et al., 1996). This indicates that a high-light adaptation of leaves and chloroplasts is also associated with a higher stress tolerance. The light adaptation capacity of plants is genetically fixed. Many plants are so-called «low-light plants» which grow in the shade of others or in locations with low irradiance. Their adaptation capacity is relatively low, and they cannot grow or do not survive at full sun light (plant group 1 in Fig. 4 A). Other plants, e.g. most of our crop plants, are light plants, which need a high irradiance to yield a reasonable growth and grain yield, but their adaptation capacity is also fairly narrow (plant group 3 in Fig. 4 A). In addition, there exist plants possessing a very wide adaptation range, such as beech (plant group 2) with its sun and shade leaves and their extreme modification capacity of chloroplast ultrastructure and function.

The wider the range of the adaptation capacity of the plant the better they are protected against various stress factors. The «light plants» (group 3 in Fig. 4 A) will be under stress when the irradiance falls below their genetically possible adaptation range. Low light plants (group 1 in Fig. 4 A), in turn, are under stress when the irradiance of their location exceeds their light adaptation capacity. However, plants with a wide range of adaptation capacity (group 2) can respond very flexibly to changes in irradiance, and are thus much better protected against high light stress and photoinhibition. This basic principle of adaptation capacity range and relative stress

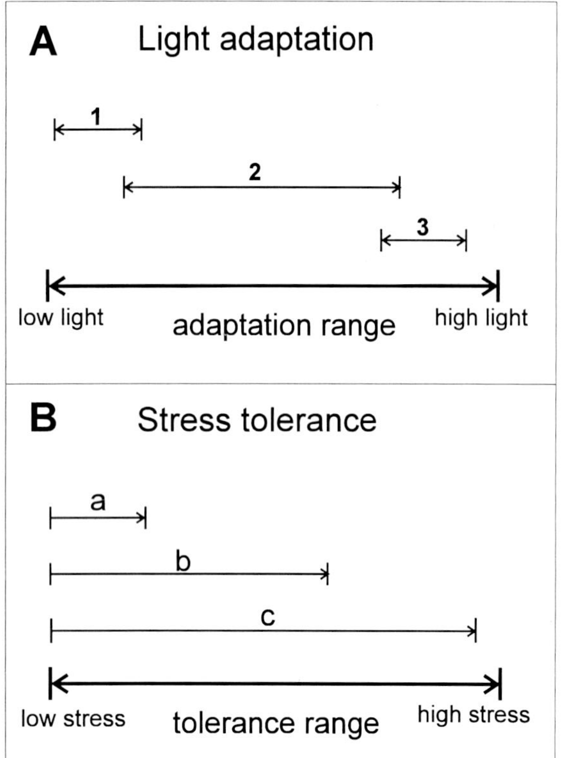

Fig. 4: Light adaptation and stress tolerance range of plants. **A)** The adaptation capacity of leaves and chloroplasts to high-light or low-light growth conditions is low for the plant groups 1 and 3, and high for the plant group 2. **B)** The plants a, b and c possess a low, medium or high stress tolerance, respectively.

tolerance exposed here for light adaptation also applies to all the other adaptation responses plants may possess.

Like the adaptation capacity of plants, also the tolerance capacity of plants is genetically fixed. Some plants possess a low, medium or high stress tolerance as shown in Fig. 4 B. Although light adaptation processes are one essential factor of a relative stress tolerance of a plant, there exist many more factors which determine the overall stress tolerance of plants, such as stress-coping mechanisms and the capacity of repair processes. Flexibility of cell metabolism and its fast acclimation to changes in environmental conditions as found e.g. in the books by Alscher and Wellburn (1994) and Smirnoff (1995), is a first essential step in stress avoidance. In plants with a low stress tolerance the capacity of the different stress-coping mechanisms is very low, and some of the mechanisms may not exist at all. Thus, such plants reach very fast an acute stage of damage, since their stress resistance minimum has fallen short already at a low stress threshold.

Stress coping mechanisms

There exist many stress-coping mechanisms which show up depending on the type and strength of stress, such as proline accumulation during drought and salinity, polyol accumulation (e.g. mannitol, sorbitol) at water stress conditions, formation of heat shock proteins, formation of radical scavenging compounds (ascorbate, glutathione, α-tocopherol), increase of the level of superoxide dismutase, formation of UV-A and UV-B absorbing pigments in the epidermis layer which protect the photosynthetic apparatus in the leaf mesophyll against damaging UV-radiation (Schweiger et al., 1995), or within the thylakoids the fast photoreduction (within minutes) of the carotenoid violaxanthin to zeaxanthin functioning at high-light conditions in the photoprotection of the photosynthetic apparatus (Lichtenthaler and Schindler, 1992; Schindler, Demming and Adams II, 1993; Schindler and Lichtenthaler, 1994). Those plants that are particularly tolerant to photoinhibition, such as the tobacco aurea mutant Su/su, even double their zeaxanthin amounts by *de novo* biosynthesis within a 5 h high light exposure (Schindler and Lichtenthaler, 1994). The exact mechanism of the photoprotective action of zeaxanthin is not yet known, it has, however, in some cases an indirect influence on the quenching of chlorophyll fluorescence. But many chlorophyll fluorescence quenching processes proceed independently of zeaxanthin. A non-enzymic oxidation of zeaxanthin to violaxanthin by either the highly reactive oxygen species ($^1O_2$, $O_2^{\cdot -}$, $OH^{\cdot}$) formed at high-light conditions and/or by detoxifying epoxy groups being formed at double bounds of thylakoid lipids has also been proposed a possible photoprotection mechanism of zeaxanthin (Lichtenthaler and Schindler, 1992; Schindler and Lichtenthaler, 1996).

One essential mechanism in the preservation of reasonable, though reduced photosynthetic rates at an excess of high-light, is the partial inactivation of photosystem II centers by photoinhibition (destruction of the D1-protein) (Godde et al., 1996; Krause and Weis, 1991; Schindler and Lichtenthaler, 1994), a process which protects the remaining photosystem II centers from photodestruction. Partial photoinhibition thus guarantees to still maintain sufficient photosynthetic net CO_2 assimilation rates at high-light conditions in order to allow plant growth and development.

Stress detection by chlorophyll fluorescence

Stress conditions and stress-induced damage in plants can be detected using the classical ecophysiological methods of measuring the rates of photosynthesis, respiration and transpiration, the stomata conductance and water potential, as well as content and ratios of the photosynthetic pigments (chlorophylls and carotenoids) or the concentration of stress metabolites. Most stress factors, even if they do not directly affect the composition of the photosynthetic apparatus or its functions, will affect the photosynthetic process in the long run. At physiological conditions about 80 to 90 % of the absorbed light energy will be dissipated from excited chlorophyll *a* (Chl*) via photosynthetic quantum conversion, whereas de-excitation by heat emission (ca. 5 to 15 %) and the red + far-red chlorophyll fluorescence (0.5 to 2 %) are much lower (Fig. 5 A). Under stress, the photosynthetic quantum conversion declines, however, and correspondingly heat emission and chlorophyll fluorescence increase considerably (Fig. 5 B).

Thus, stress-induced changes in the photosynthetic quantum conversion as well as damages to the photosynthetic apparatus can easily be detected via the non-invasive method of measuring the chlorophyll fluorescence induction kinetics (Kautsky effect), or particular chlorophyll fluorescence ratios (Rfd-values as vitality index, ratio F690/F735, Fo/Fm, Fv/Fo) (Lichtenthaler, 1988, 1990; Lichtenthaler and Rinderle, 1988) as well as different quenching coefficients such as qP and qN (Schreiber et al., 1986; Krause and Weis, 1991). Kautsky and Hirsch (1931) already described the inverse relationship between photosynthetic quantum conversion and chlorophyll fluorescence (see review Lichtenthaler, 1992). The chlorophyll fluorescence shows maxima in the red region near 685 to 690 nm and the far-red region near 730 to 740 nm (Fig. 6). The Rfd-values are best measured at both regions (Rfd 690 and Rfd 730), in which case the values and amplitude changes induced by stress are higher in the 690 nm than in the 740 nm region. From both Rfd-values one can also determine the stress adaptation index (Lichtenthaler and Rinderle, 1988), which provides information on the stress exposure and the strain on the photosynthetic apparatus. The height of the Rfd-values is a measure of the potential photosynthetic capacity of a leaf and is correlated to the photosynthetic net CO_2-assimilation (Tuba et al., 1994; Babani et al., 1996).

From the chlorophyll fluorescence induction kinetics one is able to directly determine the ratio of the chlorophyll fluorescence emission in the red (F690) and far-red fluorescence maximum (F740) and this ratio F690/F740 is an indicator of the *in vivo* chlorophyll content (Hák et al., 1990; Lichtenthaler et al., 1990; D'Ambrosio et al., 1992). Although the various kinds of chlorophyll fluorescence measurements provide valuable information on the state of health or stress exposure of the photosynthetic apparatus, they have an essential disadvantage. It is a fact that the fluorescence data can be collected only from one leaf point per measurement, and one needs to determine the fluorescence signals of various leaf points in order to obtain an approximate realistic picture of the functional state or damage of the photosynthetic apparatus and its function.

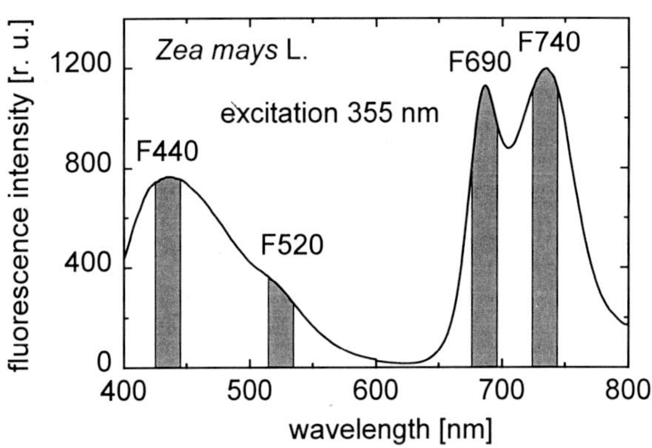

Fig. 6: UV-laser induced fluorescence emission spectrum of a plant leaf with maxima or shoulders in the blue (F440), green (F520), red (F690) and far-red (F740) spectral region. The four fluorescence bands applied in fluorescence imaging and stress detection of plants are indicated.

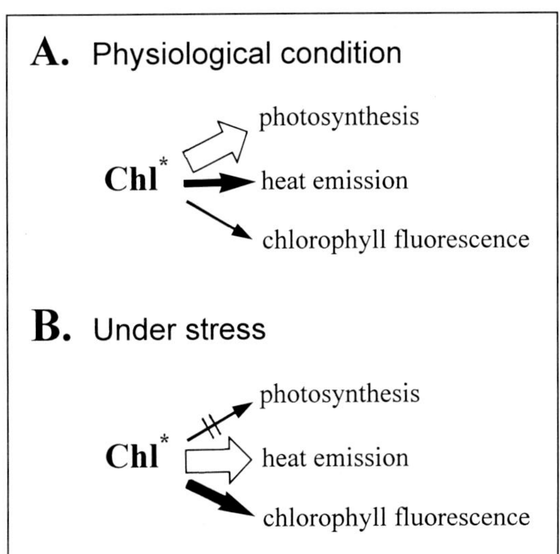

Fig. 5: De-excitation of the excited states of chlorophyll *a* by photosynthetic quantum conversion (photosynthesis), heat emission and red and far-red chlorophyll *a* fluorescence **A)** under physiological and **B)** under stress conditions. The thickness of the arrow indicates the relative proportions of the three de-excitaton processes.

Stress detection by fluorescence imaging

In order to overcome this problem we developed, in a close cooperation with physicists, the first high resolution laser-induced fluorescence imaging system (LIF imaging method), which allows to simultaneously image the chlorophyll fluorescence signatures F690 and F740 at all points of a leaf (Lang et al., 1994, 1996; Lichtenthaler et al., 1995, 1996). Since plants also possess a blue (F440) and green fluorescence emission (F520), when excited with UV-A radiation (Fig. 6), we included the blue and green fluorescence in the fluorescence imaging of leaves. In contrast to chlorophyll fluorescence, the blue and green fluorescences do not show any variable but are constant during the light-induced fluorescence induction kinetics (Stober and Lichtenthaler, 1993 a). The blue and green fluorescence is primarily emitted from diverse plant phenolics (e.g. hydroxy cinnamic acids, flavonols) of the plant epidermis cell walls (Lang et al., 1991; Lichtenthaler et al., 1992; Stober and Lichtenthaler, 1993 b, Stober et al., 1994). The blue-green fluorescence emission is not evenly distributed over the whole leaf surface but is particularly high in the leaf vein regions (Lang et al., 1994; Lichtenthaler, 1996). The red + far-red chlorophyll flurescence emission, in turn, is low in the leaf vein region and high in the intercostal, vein-free leaf regions. This can also be seen in the fluorescence images of a variegated leaf of *Codiaeum* where the chlorophyll is unevenly distributed over the leaf surface (Fig. 7).

Small local differences in bluegreen and red + far-red fluorescence emission as well as fluorescence gradients over the leaf surface can easily be sensed via the high resolution fluorescence imaging system. Fluorescence imaging in the four fluorescence bands sets a new standard in the detection of stress effects in plants. We found that the fluorescence ratios blue/red (F440/F690), blue/far-red (F440/F740) and red/far-red (F690/F740) are very sensitive to any changes in envi-

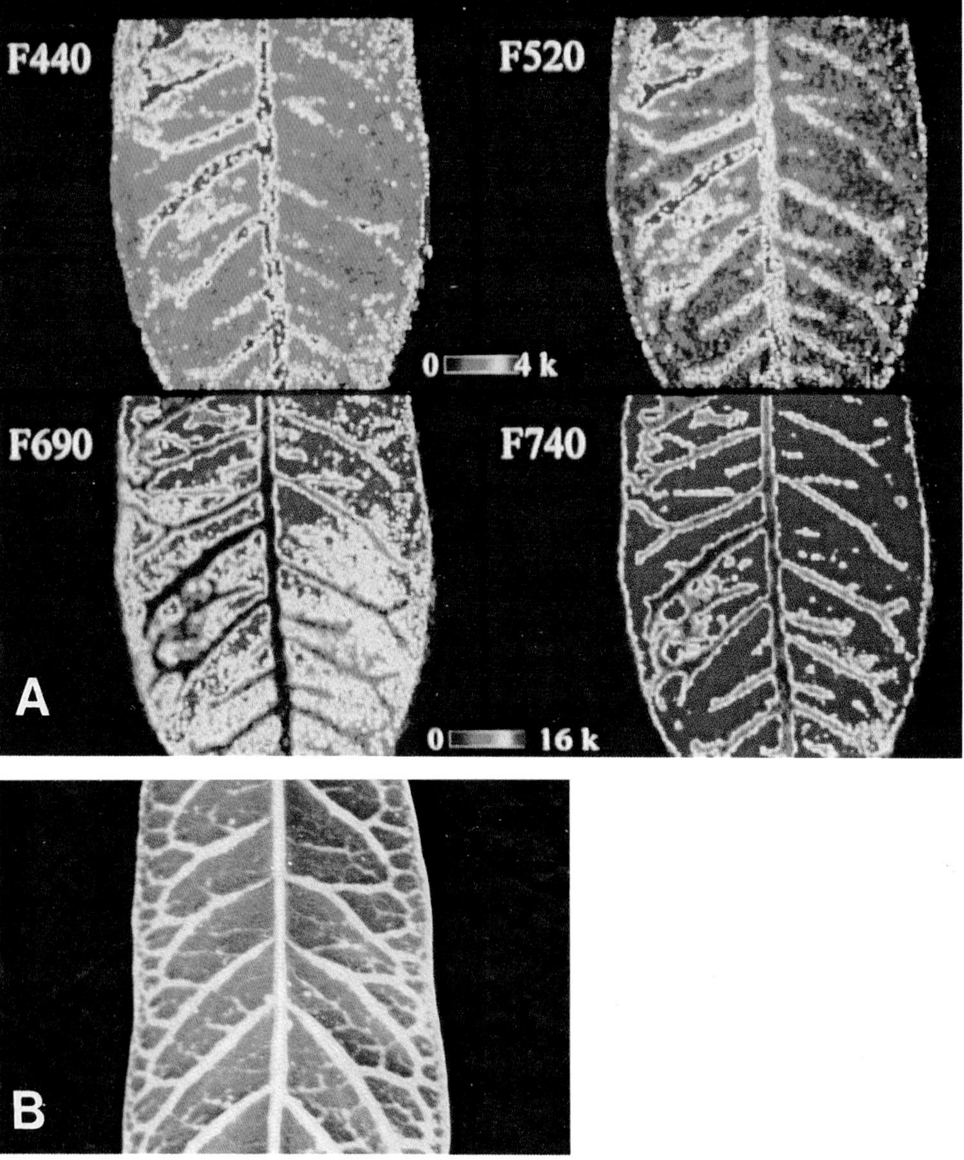

Fig. 7: A. False colour fluorescence images of the blue (F440), green (F520), red (F690) and far-red (F740) fluorescence emission in a young green leaf of *Codiaeum variegatum* L. Fluorescence intensities increase from dark blue via green, yellow to red (see color scale in the figure). **B.** Color photograph of a leaf of *Codiaeum.*

ronment and growth conditions of plants. These ratios are thus early indicators of stress and strain-induced changes in the photosynthetic function of leaves (Lang et al., 1996; Lichtenthaler et al., 1996). Laser-induced fluorescence imaging thus provides a very early and much better and precise stress and damage diagnosis than the point measurements of chlorophyll fluorescence applied so far. This new LIF method can thus be recommended to all those interested in stress detection in plants.

Photon energy flow in plant leaves

Plant stress modifies in multiple ways the energy flow of photons (sun light) through the leaf with the result that the absorption, reflectance and transmittance properties of leaves are changed. Stress also changes the relative proportions of absorbed ligth energy, which are used for photosynthetic quantum conversion, chlorophyll fluorescence, blue-green fluorescence or heat emission as is shown in Fig. 8. This is why red + far-red chlorophyll fluorescence and blue-green fluorescence kinetics and images can successfully be applied in stress detection of plants.

Acknowledgements

I wish to thank Ms. Inge Jansche, Ms. Gabrielle Johnson and Dr. Michael Lang for their excellent assistance in the preparation of the manuscript.

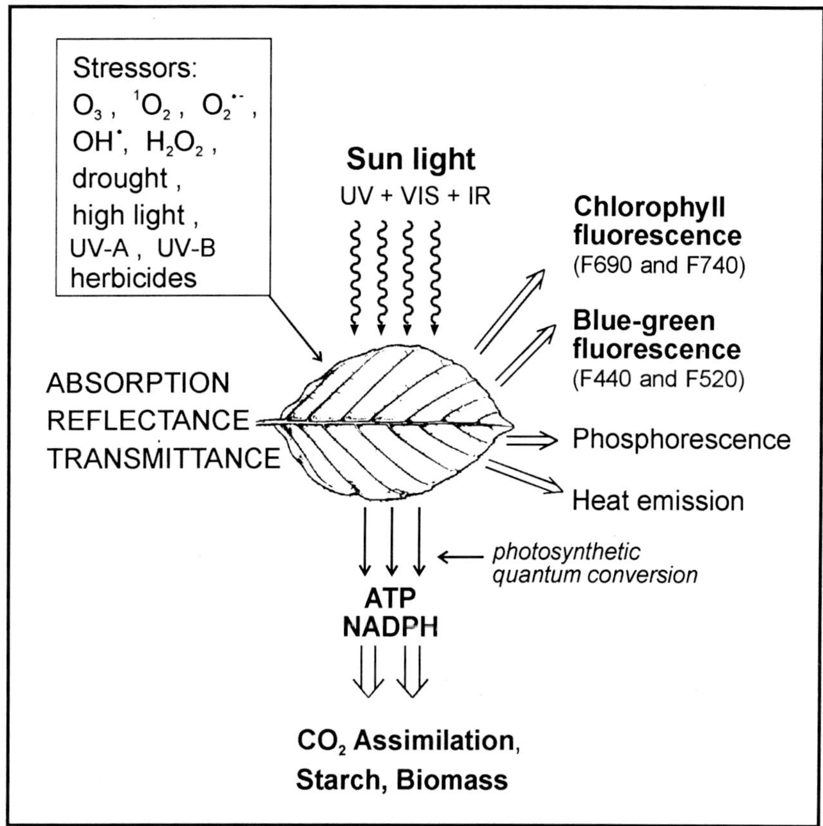

Fig. 8: Scheme of photon energy flow and dissipation in plant leaves which is modified (either blocked or enhanced) by a multitude of natural and anthropogenic stressors, such as highly reactive oxygen species, herbicides, UV-A and UV-B radiation or high-light stress or drought. A stress-induced decline in leaf physiology and photosynthetic quantum conversion can be monitored by non-invasive measurements of the red + far-red chlorophyll fluorescence and the blue-green fluorescence.

References

ALSCHER, R. G. and J. R. CUMMING (eds.): Stress Responses in Plants: Adaptation and Acclimation as Mechanism. J. Wiley-Liss, New York, 1990.

ALSCHER, R. G. and A. R. WELLBURN: Plant Responses to the Gaseous Environment. Chapman & Hall, London, 1994.

BABANI, F., P. RICHTER, and H. K. LICHTENTHALER: Changes in different chlorophyll fluorescence signatures during greening of etiolated barley seedlings as measured with the CCD-OMA fluorometer. J. Plant Physiol. *148,* 471–477 (1996).

DEMMING-ADAMS, B. and W. W. ADAMS III: Photoprotection and other responses of plants to high light stress. Annu. Rev. Plant Physiol. Plant Mol. Biol. *43,* 599–626 (1993).

DEVINE, M. D., S. O. DUKE, and C. FEDTKE: Biochemistry and Physiology of Herbicide Action, pp. 95–112. Springer Verlag, Berlin, 1993.

FOWDEN, L., T. MANSFIELD, and J. STODDART: Plant Adaptation to Environmental Stress. Chapman & Hall, London, 1993.

KONOPKA, C., R. HOLLINDERBÄUMER, V. ELBERT, H. WIETOSKA, and D. GODDE: Imbalances of D1 protein turnover during stress induced chlorosis of a declining spruce tree. J. Plant Physiol. *148* (1996).

HÁK, R., H. K. LICHTENTHALER, and U. RINDERLE: Decrease of the fluorescence ratio F690/F730 during greening and development of leaves. Radiat. Environ. Biophys. *29,* 329–336 (1990).

HOCK, B., C. FEDTKE, and R. R. SCHMIDT: Herbizide. G. Thieme Verlag, Stuttgart, 1995.

JACKSON, M. B. and C. BLACK: Interacting Stresses on Plants in a Changing Climate. NATO Advanced Research Workshop, Springer Verlag, Berlin, 1993.

KAUTSKY, H. and A. HIRSCH: Neue Versuche zur Kohlensäureassimilation. Naturwiss. *19,* 964 (1931).

KRAUSE, G. H. and E. WEIS: Chlorophyll fluorescence and photosynthesis: The basics. Annu. Rev. Plant Physiol. Plant Mol. Biol. *42,* 313–349 (1991).

LANG, M., F. STOBER, and H. K. LICHTENTHALER: Fluorescence emission spectra of plant leaves and plant constituents. Radiat. Environ. Biophys. *30,* 333–347 (1991).

LANG, M., H. K. LICHTENTHALER, M. SOWINSKA, P. SUMM, and F. HEISEL: Blue, green and red fluorescence signatures and images of tobacco leaves. Bot. Acta *107,* 230–236 (1994).

LANG, M., H. K. LICHTENTHALER, M. SOWINSKA, F. HEISEL, H. A. MIEHE, and F. TOMASINI: Fluorescence imaging of water and temperature stress in plant leaves. J. Plant Physiol. *148,* 613–621 (1996).

LARCHER, W.: Streß bei Pflanzen. Naturwissenschaften *74,* 158–167 (1987).

LEVITT, J.: Responses of Plants to Environmental Stresses. Vol. 1, Academic Press, New York, 1980.

LICHTENTHALER, H. K.: Differences in morphology and chemical composition of leaves grown at different light intensities and qualitites. In: BAKER, N. R., W. J. DAVIES, and K. C. ONG (eds.): Control of Leaf Growth, pp. 201–222. Cambridge University Press, Cambridge, 1984.

– *In vivo* chlorophyll fluorescence as a tool for stress detection in plants. In: Lichtenthaler, H. K. (ed.): Applications of Chlorophyll Fluorescence, pp. 129–142. Kluwer Academic Publishers, Dordrecht, 1988.

– Applications of chlorophyll fluorescence in stress physiology and Remote sensing. In: Steven, M. and J. A. Clark (eds.): Applications of Remote Sensing in Agriculture, pp. 287–305. Butterworths Scientific Ltd., London, 1990.

– The Kautsky Effect: 60 years of chlorophyll fluorescenceinduction kinetics. Photosynthetica *27,* 45–55 (1992).

Lichtenthaler, H. K. and D. Meier: Regulation of chloroplast photomorphogenesis by light intensity and light quality. In: Ellis, H. (ed.): Chloroplast Biogenesis, pp. 261–281. Cambridge University Press, Cambridge, 1984.

Lichtenthaler, H. K. and C. Buschmann: Das Waldsterben aus botanischer Sicht. G. Braun Verlag, Karlsruhe, 1984.

Lichtenthaler, H. K. and U. Rinderle: The role of chlorophyll fluorescence in the detection of stress conditions in plants. CRC Critical Reviews in Analytical Chemistry *19,* Suppl. I, 29–85 (1988).

Lichtenthaler, H. K. and C. Schindler: Studies on the photoprotective function of zeaxanthin at high-light conditions. In: Murata, N. (ed.): Research in Photosynthesis, Vol. IV, pp. 517–520. Kluwer Academic Publishers, Dordrecht, 1992.

Lichtenthaler, H. K., C. Buschmann, M. Döll, H.-J. Fietz, T. Bach, U. Kozel, D. Meier, and U. Rahmsdorf: Photosynthetic activity, chloroplast ultrastructure, and leaf characteristics of high-light and low-light plants and of sun and shade leaves. Photosynthesis Research *2,* 115–141 (1981).

Lichtenthaler, H. K., C. Buschmann, U. Rinderle, and G. Schmuck: Applications of chlorophyll fluorescence in ecophysiology. Radiat. Environ. Biophys. *25,* 297–308 (1986).

Lichtenthaler, H. K., R. Hák, and U. Rinderle: The chlorohyll fluorescence ratio F690/F730 in leaves of different chlorophyll content. Photosynth. Res. *25,* 295–298 (1990).

Lichtenthaler, H. K., M. Lang, F. Stober, C. Schindler, H. Edner, H. Johansson, S. Svanberg, and L. O. Björn: Remote multi-colour fluorescence imaging of selected broad-leaf plants. EARSeL Advances in Remote Sensing 3 (Part 3), 2–14 (1995).

Lichtenthaler, H. K., M. Lang, M. Sowinska, F. Heisel, and J. A. Miehe: Detection of vegetation stress via a new high resolution fluorescence imaging system. J. Plant Physiol. *106,* 1127–133 (1996).

McKersie, B. D. and Y. Y. Leshem: Stress and Stress Coping in Cultivated Plants, pp. 1–256. Kluwer Academic Publishers, Dordrecht, 1994.

Meier, D. and H. K. Lichtenthaler: Ultrastructural development of chloroplasts in radish seedlings grown at high and low light conditions and in the presence of the herbicide bentazon. Protoplasma *107,* 195–207 (1981).

Rennenberg, H., C. Herschbach, and A. Polle: Consequences of air pollution on shoot-root interactions. J. Plant Physiol. *148,* 537–547 (1996).

Schindler, C. and H. K. Lichtenthaler: Is there a correlation between light-induced zeaxanthin accumulation and quenching of variable chlorophyll *a* fluorescence? Plant Physiol. Biochem. *32,* 813–823 (1994).

Schindler, C. and H. K. Lichtenthaler: Photosynthetic CO_2-assimilation, chlorophyll fluorescence and zeaxanthin accumulation in fied grown maple trees in the course of a sunny and a cloudy day. J. Plant Physiol. *148,* in press (1996).

Schreiber, U., U. Schliwa, and W. Bilger: Continuous recording of photochemical and non-photochemical chlorophyll fluorescence quenching with a new type of modulation fluorometer. Photosynth. Res. *10,* 51–62 (1986).

Schweiger, J., M. Lang, and H. K. Lichtenthaler: Differences in fluorescence excitation spectra of leaves between stressed and non-stressed plants. J. Plant Physiol. *148,* 537–547 (1996).

Selye, H.: A syndrome produced by various nocuous agents. Nature *138,* 32–34 (1936).

– The stress of Life. McGraw Hill, New York, 1956.

Siedow, J. N.: Public affairs. ASPP Newsletter (American Soc. Plant Physiologists) *22,* No. 2, 6–9 (1995).

Smirnoff, N.: Environment and Plant Metabolism: Flexibility and Acclimation. BIOS Scientific Publishers Ltd., Oxford, 1995.

Stober, F. and H. K. Lichtenthaler: Studies on the constancy of the blue and green fluorescence yield during the chlorophyll fluorescence induction kinetics (Kautsky effect). Radiat. Environ. Biophys. *32,* 357–365 (1993 a).

Stober, F. and H. K. Lichtenthaler: Studies on the localisation and spectral characteristics of the fluorescence emission of differently pigmented wheat leaves. Bot. Acta *106,* 365–370 (1993 b).

Stober, F., M. Lang, and H. K. Lichtenthaler: Blue, green and red fluorescence emission signatures of green, etiolated, and white leaves. Remote Sens. Environ. *47,* 65–71 (1994).

Stocker, O.: Probleme der pflanzlichen Dürreresistenz. Naturwissenschaften *34,* 362–371 (1947).

– Transpiration und Wasserhaushalt in verschiedenen Klimazonen. I. Untersuchungen an der arktischen Baumgrenze in Schwedisch-Lappland. Jahrb. wiss. Botanik *75,* 494 (1932).

Tuba, Z., H. K. Lichtenthaler, Z. Czintalan, Z. Nagy, and K. Szente: Reconstitution of chlorophylls and photosynthetic CO_2 assimilation in the desiccated poikilochlorophyllous plant *Xerophyta scabrida* upon rehydration. Planta *192,* 414–420 (1994).

Wellburn, A.: Air Pollution and Climate change: The Biological Impact. 2nd edition. Longman Scientific & Technical, New York, 1994.

J. Plant Physiol. Vol. 148. pp. 15–20 (1996)

Photomorphogenic Effects of UV-B Radiation on Plants: Consequences for Light Competition

Paul W. Barnes[1], Carlos L. Ballaré[2], and Martyn M. Caldwell[3]

[1] Department of Biology, Southwest Texas State University, San Marcos, Texas 78666-4616 USA

[2] IFEVA-Dept. de Ecología, Facultad de Agronomía, Universidad de, Buenos Aires, Av. San Martín 4453, Buenos Aires, Argentina

[3] Department of Rangeland Resources, Utah State University, Logan, Utah 84322-5230 USA

Received July 24, 1995 · Accepted October 27, 1995

Summary

A combination of field and laboratory studies were conducted to explore the nature of photomorphogenic effects of ultraviolet-B radiation (UV-B; 280–320 nm) on plant morphology and to evaluate the ecological consequences of these alterations in morphology for interspecific competition. Under laboratory conditions, seedlings of cucumber (*Cucumis sativus* L.) and tomato (*Lycopersicon esculentum* Mill.) exhibited appreciable (ca. 50 %) and rapid (<3 h) inhibition in hypocotyl elongation in response to UV-B exposure. In cucumber, this inhibition was reversible, occurred without any associated changes in dry matter production and was caused by UV-B incident on the cotyledons and not the stem or growing tip. Inhibition of stem elongation in etiolated tomato seedlings occurred at least 3 h prior to the onset of accumulation of UV-absorbing pigments and monochromatic UV supplied against a background of visible radiation revealed maximum effectiveness in inhibition around 300 nm. Collectively, these findings suggest that a specific, but yet unidentified, UV-B photoreceptor is involved in mediating certain morphological responses to UV-B. For mixtures of wheat (*Triticum aestivum* L.) and wild oat (*Avena fatua* L.), a common weedy competitor, supplemental UV-B irradiation in the field differentially altered shoot morphology which resulted in changes in canopy structure, light interception and calculated stand photosynthesis. It is argued that, because of its asymmetrical nature, competition for light can potentially amplify the effects of UV-B on shoot morphology and may, therefore, be an important mechanism by which changes in the solar UV-B spectrum associated with stratospheric ozone reduction could alter the composition and character of terrestrial vegetation.

Key words: Avena fatua L., Cucumis sativus L., Lycopersicon esculentum Mill., Triticum aestivum L., canopy structure, competition, morphology, photomorphogenesis, stratospheric ozone depletion, ultraviolet-B radiation, UV-B photoreceptor.

Abbreviations: A_{can} = net canopy photosynthesis; LAI = leaf area index ($m^2 m^{-2}$); PFD = photon flux density, 400–700 nm; PFD_{int} = canopy intercepted PFD; UV-A = ultraviolet-A radiation (320–400 nm); UV-B = ultraviolet-B radiation (280–320 nm).

Introduction

Exposure to ultraviolet-B radiation (UV-B; 280–320 nm) is known to cause cellular damage (Strid et al., 1994; Teramura and Sullivan, 1994) as well to elicit a number of photomorphogenic responses (Wellmann, 1983; Ensminger, 1993) in higher plants. Examples of photomorphogenic effects of UV-B include, 1) the induction of UV-absorbing pigments (phenolic compounds), which serve to protect sensitive tissue from UV-B exposure (Caldwell et al., 1983; Li et al., 1993), and 2) the inhibition of stem and leaf elongation which causes alterations in shoot morphology (Barnes et al., 1990; Ballaré et al., 1991). In contrast to many damaging effects of UV-B (e.g., DNA damage and photosynthesis inhibition), photomorphogenic responses appear most sensitive to UV wavelengths that exhibit the greatest relative change with stratospheric ozone reduction (i.e., $\lambda \approx 300$ nm; Ensminger, 1993) and are, therefore, likely to be especially sensitive to the ongoing and future changes in the solar UV-B climate (Madronich, 1993). Moreover, these photomorphogenic effects on individual plants have the potential to significantly affect the structure and function of terrestrial ecosystems by altering species interactions and biogeochemical cycles (Barnes et al., 1988; McCloud and Berenbaum, 1994; Gehrke et al., 1995).

In this paper, we review results from our studies examining photomorphogenic effects of UV-B on plant morphology, and illustrate how UV-B-induced changes in morphology can potentially influence competition for light.

Photomorphogenic Effects of UV-B on Plant Morphology

Numerous studies have shown that exposure to UV-B radiation can result in the altered morphology of higher plants. These morphological alterations involve both the inhibition and stimulation of growth in various plant organs (Table 1). Under conditions of high UV-B and low PFD, as generally occur in growth chambers and greenhouses, the morphological changes can be associated with reductions in photosynthesis and/or total plant production (e.g., Dai et al., 1992). In these instances, reduced growth of leaves and stems may be due to a general reduction in growth capacity resulting from partial inhibition of photosynthesis and other forms of UV damage. However, plants grown in the field or under laboratory conditions with realistic and balanced levels of UV-B, UV-A, and visible radiation often exhibit little or no evidence

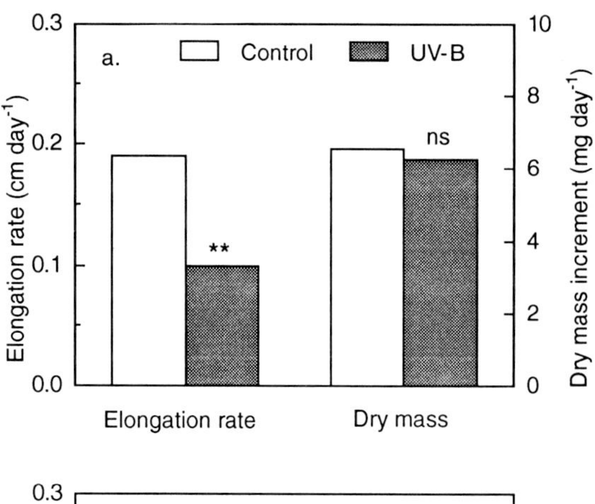

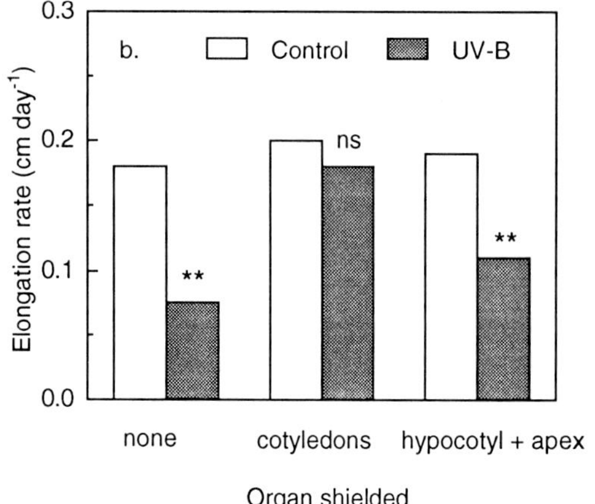

Fig. 1: Influence of UV-B radiation on hypocotyl elongation in greenhouse-grown cucumber seedlings. Control plants received no UV-B while UV-B treated plants received 9.6 kJ m^{-2} d^{-1} weighted UV-B (Caldwell 1971) for 6 h as described by Ballaré et al. (1991). Results in panel b were obtained by shielding plant organs with UV-B-excluding clear polyester. Significant treatment differences at $P<0.01$ as determined by Student t-tests are denoted by "**". Data are from Ballaré et al. (1991).

Table 1: Summary of morphological responses of plants to UV-B radiation.

Morphological Trait	Direction of Effect	Reference[1]
Shoot height	–	a, b, c, d
Stem or leaf elongation	–	e, f, g, h
Leaf size or area	–	a, b, d, i
Shoot : root mass	±	i, j, k
Leaf mass/area	+	b, i, l
Number of leaves	+	k, l, m
Branching or tillering	+	i, k, l
Cotyledon curling	+	n

[1] References are not an exhaustive list but include only recent and/or representative studies conducted under laboratory and field conditions (including ambient and enhanced UV-B). [a] Barnes et al. (1988); [b] Searles et al. (1995); [c] Sullivan et al. (1992); [d] Tevini et al. (1991); [e] Ballaré et al. (1991); [f] Ballaré et al. (1995 a); [g] Sisson and Caldwell (1976); [h] Steinmetz and Wellmann (1986); [i] Sullivan et al. (1994); [j] Barnes et al. (1993); [k] Ziska et al. (1993); [l] Barnes et al. (1990); [m] Larson et al. (1990); [n] Wilson and Greenberg (1993).

of UV-B-induced damage (e.g., Caldwell et al., 1994). Yet, UV-B-induced alterations in morphology are consistently observed in response to ambient or enhanced levels of solar UV-B, even in the absence of any detectable reduction in photosynthetic carbon assimilation (Beyschlag et al., 1988; Ziska et al., 1993), chlorophyll fluorescence (Sullivan et al., 1994; Searles et al., 1995), or total biomass production (Barnes et al., 1990; Tevini et al., 1991).

While some of the morphological alterations described above may simply represent shifts in allocation resulting from UV-B acclimation (e.g., altered shoot : root ratios and increased leaf mass/area; Bornman and Teramura, 1993), others appear to represent true photomorphogenic responses to UV-B. For example, Ballaré et al. (1991) demonstrated that short-term UV-B exposure (6 hours) equivalent to ambient UV-B in the tropics produced rapid and appreciable inhibition (ca. 50 %) in hypocotyl (stem) elongation in light-grown cucumber seedlings in a greenhouse (Fig. 1 a). This inhibition was reversible, occurred without any associated reductions in cotyledon area expansion or total plant biomass accumulation, and was apparently due to UV-B perceived by the cotyledons and not by the stem or growing apex (Fig. 1 b). Thus, in this case, the rapid UV-B-induced reduction in hypocotyl elongation was not due to carbon limitations or to direct damage to the growing cells; a specific photomorphogenic effect of UV-B on stem elongation was therefore implicated.

Results from subsequent studies with etiolated tomato seedlings revealed that maximum inhibition of stem elongation occurred at 297 nm (Fig. 2 a) and elimination of the UV-B-induced elongation inhibition could be achieved by pretreating the seedlings with chemicals (KI, NaN_3 and phenylacetic acid) that interfere with normal flavin (and possibly pterin) photochemistry (Ballaré et al. (1995 a). Few differences in UV-B elongation responses have been observed between mutants and isogenic wild types of cucumber, tomato and *Arabidopsis* that differ in phytochrome content which

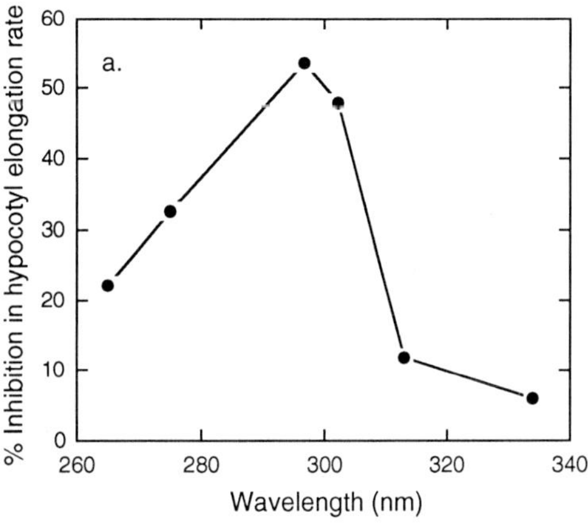

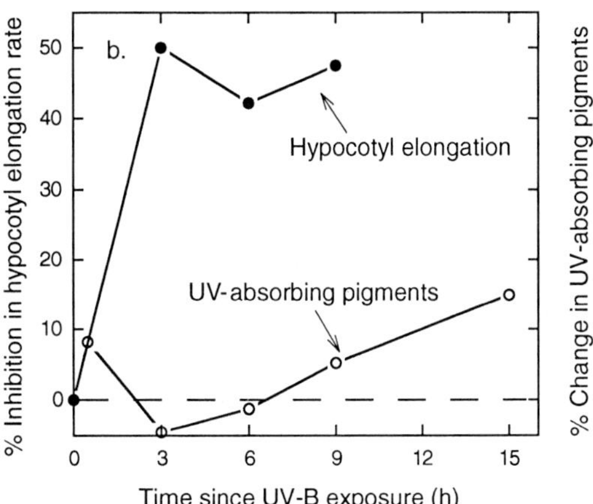

Fig. 2: Wavelength sensitivity (a) and time-course (b) of inhibition of hypocotyl elongation in etiolated tomato seedlings. In panel a, elongation rates were measured after a 4-h exposure to 1 $\mu mol\ m^{-2}\ s^{-1}$ monochromatic radiation applied against a background PFD of 63 $\mu mol\ m^{-2}\ s^{-1}$ as described by Ballaré et al. (1995 a). In panel b, data are expressed as percents of UV-B exposed plants (unweighted broad-band UV-B = 3.1 $\mu mol\ m^{-2}\ s^{-1}$) relative to plants that received no UV-B. UV-absorbing pigment accumulation was measured as crude extract absorbance at 305 nm. Data are from Ballaré et al. (1995 b).

suggests that phytochrome is not required for the expression of this particular effect of UV-B on elongation (Ballaré et al., 1991; 1995 a; Goto et al., 1993). Collectively, these findings and others (Steinmetz and Wellmann, 1986; Ensminger, 1993) suggest that a specific UV-B photoreceptor is involved in mediating these elongation responses.

While the adaptive significance of photomorphogenic effects on pigmentation may be well established, such is not the case for the UV-B-induced changes in morphology. Ballaré et al. (1995 b) proposed that short-term inhibitions in stem elongation in seedlings could serve to delay emergence from the soil until sufficient UV-protective mechanisms are in place. Wellmann (1983) suggested that growth reduction occurring under realistic UV-B irradiation should be viewed as a mechanism by which plants delay cell division and thus, reduce the impact of UV-B effects on DNA integrity. Normal cell division and growth would then proceed, following the UV-B exposure and/or acclimation to UV-B (i.e., accumulation of UV-absorbing pigments). Indeed, for etiolated tomato seedlings in the laboratory, it does appear that UV-B inhibition of stem elongation occurs prior to the accumulation of UV-absorbing pigments, as determined by whole-leaf extracts (Fig. 2 b). In leaves of etiolated rye (*Secale cereale* L.) seedlings, however, Braun and Tevini (1993) found rapid (within 30 min) UV-B-induced increases in flavonoids and shifts of trans-cinnamic esters to the cis-form in epidermal tissue and these changes appeared sufficient to adequately protect these plants from UV-B damage (Tevini et al., 1991). The relationship between changes in phenylpropanoid biosynthesis and leaf elongation, however, was not examined in these studies. Whether temporal differences in morphological and pigmentation responses exist for emerging seedlings in the field has yet to examined.

Effects of UV-B on Plant Competition

Competition is generally considered to be a reciprocal negative interaction between two individuals (Connell, 1990). In plants, competition can occur via direct interference (i.e., allelopathy), though it more commonly involves the mutual utilization of shared resources (e.g., light or nutrients) and is thereby an indirect interaction. In this latter type of competition, termed exploitative or resource competition, one individual reduces the quantity and/or quality of a shared resource and a second, neighboring individual perceives and responds in a negative way to this altered resource condition (i.e., the competitive «effect» and «response»; Goldberg, 1990). Because of the largely unidirectional supply and acquisition of light, competition for this resource is thought to be inherently asymmetrical between neighboring plants (i.e., taller individuals capture disproportionately more of the light that shorter individuals; Weiner, 1990). Thus, taller plants strongly suppress the growth of shorter plants and small initial differences in stature and size among competing individuals can become greatly magnified over time (Schmitt et al., 1987).

Exposure to both ambient (Bogenrieder and Klein, 1982; Gold and Caldwell, 1983) and enhanced solar UV-B (Fox and Caldwell, 1978; Gold and Caldwell, 1983; Barnes et al.,

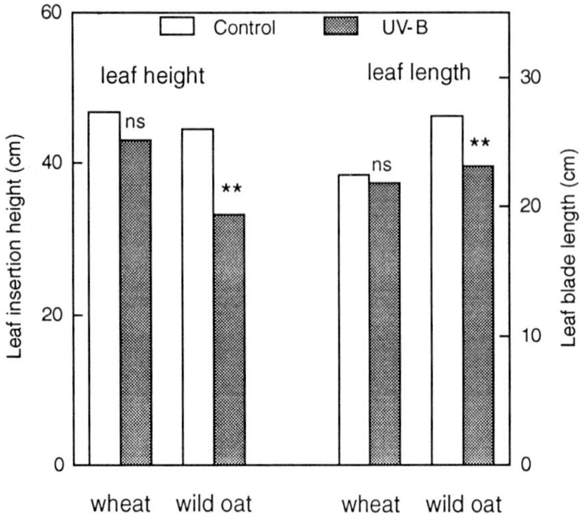

Fig. 3: Insertion height and blade length of midcanopy leaves of wheat and wild oat exposed to control (= ambient UV-B) and enhanced UV-B (simulating 20 % ozone reduction for Logan, Utah, USA) under field conditions. Significant treatment differences at $P < 0.01$ as determined by Duncan's Multiple Range test are denoted by «**». Data are from Barnes et al. (1988).

1988) has been shown to alter the balance of competition in a number of species pairs. Predicting which plant associations are susceptible to UV-B-induced changes in competitive balance and under which conditions these changes would occur requires an understanding of the underlying mechanisms responsible for these competitive shifts. In previous studies with wheat (*Triticum aestivum*) and a common weedy competitor, wild oat (*Avena fatua*), UV-B-induced shifts in competitive balance were associated with differential effects of the UV-B on shoot morphology (Fig. 3) which led to shifts in the relative positioning of leaf area for the two species (Fig. 4). These changes in canopy structure were computed to be sufficient to alter light interception and canopy photosynthesis for these species under greenhouse (Ryel et al., 1990) and field conditions (Figs. 4, 5). Thus, it was hypothesized that UV-B enhancement shifted the balance of competitive indirectly by inducing changes in shoot morphology which, in turn, altered the competition for light between these two species.

At the present time, it is not known whether the hypothesis of morphologically-mediated UV-B-induced shifts in competitive balance is a general mechanism that applies to other species mixtures or plant associations. A previous sur-

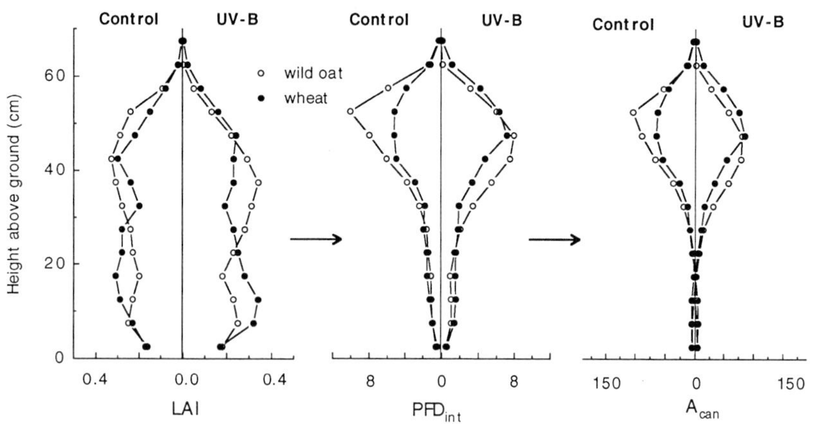

Fig. 4: Height profiles of measured LAI, and calculated intercepted PFD (PFD_{int}; mol quanta m^{-2} ground area d^{-1}) and net canopy photosynthesis (A_{can}; mmol CO_2 m^{-2} ground area d^{-1}) for wheat and wild oat in mixtures under control (= ambient (UV-B) and enhanced UV-B treatments (simulating 20 % ozone reduction for Logan, Utah, USA) under clear-sky conditions. Data are from Barnes et al. (1995).

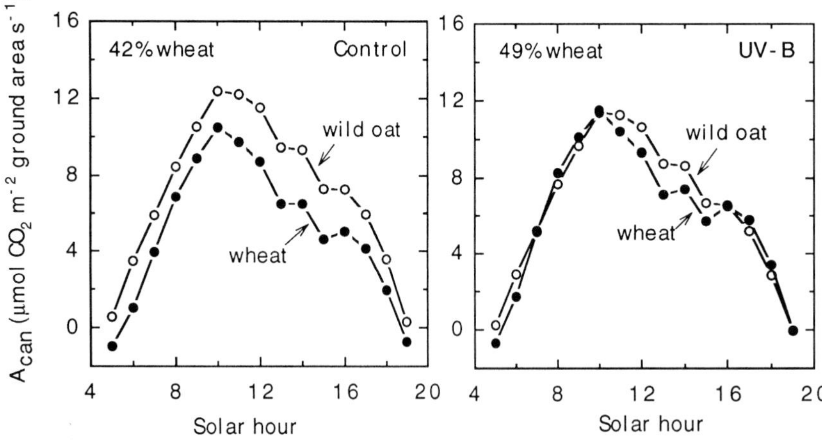

Fig. 5: Simulated midseason net canopy photosynthesis (A_{can}) for wheat and wild oat in mixtures under control (= ambient) and enhanced UV-B treatments (simulating 20 % ozone reduction for Logan, Utah, USA) under clear-sky conditions. Units as in Fig. 4. % wheat = percent of total daily mixture A_{can} contributed by wheat. Data are from Barnes et al. (1995).

vey of several annual species grown in dense monocultures in a greenhouse showed grasses to be generally more responsive, morphologically, to UV-B than dicots (Barnes et al., 1990). Based on these findings, it was proposed that mixtures of grasses might be more susceptible to UV-B-induced competitive changes than mixtures of only dicots. This hypothesis has yet to be tested. Also, previous studies have shown that UV-B responsiveness typically decreases under conditions of limited soil moisture (Sullivan and Teramura, 1990) and nutrients (Murali and Teramura, 1985). Thus, we further hypothesized that UV-B-induced shifts in competitive balance would be most prevalent in moist, fertile habitats where stand density is high, plants are generally similar in stature, and light is a limiting resource for growth and productivity (Barnes et al., 1995). Whether UV-B can alter belowground competition via differential effects on shoot : root allocation has yet to be explored.

Conclusions

Exposure to UV-B at ambient or enhanced levels induces alterations in shoot morphology that are unrelated to general damaging effects of UV-B on photosynthesis and growth. In laboratory case studies involving seedlings of two dicot species, a photomorphogenic effect of UV-B on hypocotyl elongation has been described that is rapid, reversible, involves signal transduction from cotyledons to hypocotyl, and exhibits maximal effectiveness in the middle of the UV-B waveband. At the present time the specific photoreceptor responsible for mediating these morphological responses is unknown, and the adaptive significance of these responses remains uncertain. Also, what relationship, if any, exists between these rapid elongation responses of seedlings grown under laboratory conditions and the morphological alterations displayed by mature, field-grown plants has yet to be established. It is clear, however, that variation in morphological sensitivity to UV-B exists among plant species and these differential effects, though often subtle, have the potential to be greatly magnified through the process of interspecific competition for light. Further tests of the generality of the hypothesis of UV-B-induced alteration in competition for light will aid in increasing our understanding of the conditions under which UV-B-induced shifts in competitive balance are to be expected. Direct extrapolation of findings from simple two-species mixtures to multi-species plant associations in nature, however, may be limited due to the likelihood of higher-order direct and indirect species interactions (both negative and positive) which could have unanticipated but significant consequences in more complex species assemblages (Wooton, 1994).

Acknowledgements

This research was supported by the United States Department of Agriculture (CSRS/NRICG grants 92-37100-7763 to Southwest Texas State University and 92-37100-7630 to Utah State University), the United States Environmental Protection Agency (Cooperative Agreement EPA CR 813125 with Utah State University), and a SWTSU Faculty Research Enhancement Grant. Although the research described in this article has been funded in part by the previously mentioned US EPA cooperative agreement, it has not been subjected to the Agency's required peer and policy review and therefore does not necessarily reflect the views of the Agency, and no official endorsement should be inferred.

References

Ballaré, C. L., P. W. Barnes, and R. E. Kendrick: Photomorphogenic effects of UV-B radiation on hypocotyl elongation in wild type and stable-phytochrome-deficient mutant seedlings of cucumber. Physiol. Plant. *83,* 652–658 (1991).

Ballaré, C. L., P. W. Barnes, and S. D. Flint: Inhibition of hypocotyl elongation by ultraviolet-B radiation in de-etiolating tomato seedlings. I. The photoreceptor. Physiol. Plant. *93,* 584–592 (1995 a).

Ballaré, C. L., P. W. Barnes, S. D. Flint, and S. Price: Inhibition of hypocotyl elongation by ultraviolet-B radiation in de-etiolating tomato seedlings. II. Time-course, comparison with flavonoid responses and adaptive significance. Physiol. Plant. *93,* 593–601 (1995 b).

Barnes, P. W., P. W. Jordan, W. G. Gold, S. D. Flint, and M. M. Caldwell: Competition, morphology and canopy structure in wheat (*Triticum aestivum* L.) and wild oat (*Avena fatua* L.) exposed to enhanced ultraviolet-B radiation. Funct. Ecol. *2,* 319–330 (1988).

Barnes, P. W., S. D. Flint, and M. M. Caldwell: Morphological responses of crop and weed species of different growth forms to ultraviolet-B radiation. Amer. J. Bot. *77,* 1354–1360 (1990).

– – – Early-season effects of supplemented solar UV-B radiation on seedling emergence, canopy structure, simulated stand photosynthesis and competition for light. Global Change Biol. *1,* 43–53 (1995).

Barnes, P. W., S. Maggard, S. R. Holman, and B. S. Vergara: Intraspecific variation in sensitivity to UV-B radiation in rice. Crop Sci. *33,* 1041–1046 (1993).

Beyschlag, W., P. W. Barnes, S. D. Flint, and M. M. Caldwell: Enhanced UV-B irradiation has no effect on photosynthetic characteristics of wheat (*Triticum aestivum* L.) and wild oat (*Avena fatua* L.) under greenhouse and field conditions. Photosynthetica *22,* 516–525 (1988).

Bogenrieder, A. and R. Klein: Does solar UV influence the competitive relationship in higher plants? In: Calkins, J. (ed.): The role of solar ultraviolet in marine ecosystems, pp. 641–649. Plenum Press, New York, NY (1982).

Bornman, J. F. and A. H. Teramura: Effects of ultraviolet-B radiation on terrestrial plants. In: Young, A. R., L. O. Björn, J. Moan, and W. Nultsch (eds.): Environmental UV photobiology, pp. 427–471. Plenum Press, New York (1993).

Braun, J. and M. Tevini: Regulation of UV-protective pigment synthesis in the epidermal layer of rye seedlings (*Secale cereale* L. cv. Kustro). Photochem. Photobiol. *57,* 318–323 (1993).

Caldwell, M. M.: Solar UV irradiation and the growth and development of higher plants. In: Giese, A. C. (ed.): Photophysiology Vol. 6, pp. 131–177. Academic Press, New York, NY (1971).

Caldwell, M. M., S. D. Flint, and P. S. Searles: Spectral balance and UV-B sensitivity of soybean: a field experiment. Plant Cell Environ. *17,* 267–276 (1994).

Caldwell, M. M., R. Robberecht, and S. D. Flint: Internal filters: Prospects for UV-acclimation in higher plants. Physiol. Plant. *58,* 445–450 (1983).

Caldwell, M. M., A. H. Teramura, and M. Tevini: The changing solar ultraviolet climate and the ecological consequences for higher plants. Trends Ecol. Evol. *4,* 363–367 (1989).

Connell, J. H.: Apparent versus «real» competition in plants. In: Grace, J. B. and D. Tilman (eds.): Perspectives on plant competition, pp. 9–26. Academic Press, Inc., San Diego, CA (1990).

Dai, Q., V. P. Coronel, B. S. Vergara, P. W. Barnes, and A. T. Quintos: Ultraviolet-B radiation effects on growth and physiology of four rice cultivars. Crop Sci. *32*, 1269–1274 (1992).

Ensminger, P. A.: Control of development in plants and fungi by far-UV radiation. Physiol. Plant. *88*, 501–508 (1993).

Fox, F. M. and M. M. Caldwell: Competitive interaction in plant populations exposed to supplementary ultraviolet-B radiation. Oecologia *36*, 173–190 (1978).

Gehrke, C., U. Johanson, T. V. Callaghan, D. Chadwick, and C. H. Robinson: The impact of enhanced ultraviolet-B radiation on litter quality and decomposition processes in Vaccinium leaves from the Subarctic. Oikos *72*, 213–222 (1995).

Gold, W. G. and M. M. Caldwell: The effects of ultraviolet-B radiation on plant competition in terrestrial ecosystems. Physiol. Plant. *58*, 435–444 (1983).

Goldberg, D. E.: Components of resource competition in plant communities. In: Grace, J. B. and D. Tilman (eds.): Perspectives on plant competition, pp. 27–49. Academic Press, Inc., San Diego, CA (1990).

Goto, N., K. T. Yamamoto, and M. Watanabe: Action spectra for inhibition of hypocotyl growth of wild-type plants and of the *hy2* long-hypocotyl mutant of *Arabidopsis thaliana* L. Photochem. Photobiol. *57*, 867–871 (1993).

Larson, R. A., W. J. Garrison, and R. W. Carlson: Differential responses of alpine and non-alpine *Aquilegia* species to increased ultraviolet-B radiation. Plant, Cell Environ. *13*, 983–987 (1990).

Li, J., T. Ou-Lee, R. Raba, R. G. Amundson, and R. L. Last: *Arabidopsis* flavonoid mutants are hypersensitive to UV-B irradiation. Plant Cell *5*, 171–179 (1993).

Madronich, S.: UV radiation in the natural and perturbed atmosphere. In: Tevini, M. (ed.): UV-B radiation and ozone depletion. Effects on humans, animals, plants, microorganisms, and materials. pp. 17–69. Lewis Publishers, Boca Raton, USA (1993).

McCloud, E. S. and M. R. Berenbaum: Stratospheric ozone depletion and plant-insect interactions: effects of UVB radiation on foliage quality of *Citrus jambhiri* for *Trichoplusia ni*. J. Chem. Ecol. *20*, 525–539 (1994).

Murali, N. S. and A. H. Teramura: Effects of ultraviolet-B irradiance on soybean. Vll. Biomass and concentration and uptake of nutrients at varying P supply. J. Plant Nutr. *8*, 177–192 (1985).

Ryel, R. J., P. W. Barnes, W. Beyschlag, M. M. Caldwell, and S. D. Flint: Plant competition for light analyzed with a multispecies canopy model. I. Model development and influence of enhanced UV-B conditions on photosynthesis in mixed wheat and wild oat canopies. Oecologia *82*, 304–310 (1990).

Schmitt, J., J. Eccleston, and D. W. Ehrhardt: Dominance and suppression, size-dependent growth and self-thinning in a natural *Impatiens capensis* population. J. Ecol. *75*, 651–665 (1987).

Searles, P. S., M. M. Caldwell, and K. Winter: The response of five tropical plant species to solar ultraviolet-B radiation. Amer. J. Bot. *82*, 445–453 (1995).

Sisson, W. B. and M. M. Caldwell: Photosynthesis, dark respiration, and growth of *Rumex patientia* L. exposed to ultraviolet irradiance (288 to 315 Nanometers) simulating a reduced atmospheric ozone column. Plant Physiol. *58*, 563–568 (1976).

Steinmetz, V. and E. Wellmann: The role of solar UV-B in growth regulation of Cress (*Lepidium sativum* L.) seedlings. Photochem. Photobiol. *43*, 189–193 (1986).

Strid, Å., W. S. Chow, and J. M. Anderson: UV-B damage and protection at the molecular level in plants. Photosyn. Res. *39*, 475–489 (1994).

Sullivan, J. H. and A. H. Teramura: Field study of the interaction between solar ultraviolet-B radiation and drought on photosynthesis and growth in soybean. Plant Physiol. *92*, 141–146 (1990).

Sullivan, J. H., A. H. Teramura, and L. R. Dillenberg: Growth and photosynthetic responses of field-grown sweetgum (*Liquidambar styraciflua;* Hamamelidaceae) seedlings to UV-B radiation. Amer. J. Bot. *81*, 826–832 (1994).

Sullivan, J. H., A. H. Teramura, and L. H. Ziska: Variation in UV-B sensitivity in plants from a 3,000-m elevational gradient in Hawaii. Amer. J. Bot. *79*, 737–743 (1992).

Teramura, A. H. and J. H. Sullivan: Effects of UV-B radiation on photosynthesis and growth of terrestrial plants. Photosyn. Res. *39*, 463–473 (1994).

Tevini, M., U. Mark, and M. Saile-Mark: Effects of enhanced solar UV-B radiation on growth and function of crop plant seedlings. Curr. Topics Plant Biochem. Physiol. *10*, 13–31 (1991).

Tevini, M., J. Braun, and G. Fieser: The protective function of the epidermal layer of rye seedlings against ultraviolet-B radiation. Photochem. Photobiol. *53*, 329–333 (1991).

Weiner, J.: Asymmetric competition in plant populations. Trends Ecol. Evol. *5*, 360–364 (1990).

Wellmann, E.: UV radiation in photomorphogensis. In: Shropshire, W. J., Jr. and H. Mohr (eds.): Encyclopedia of plant physiology. Volume 116 B (New Series). Photomorphogenesis. pp. 745–756. Springer-Verlag, Berlin (1983).

Wilson, M. I. and B. M. Greenberg: Specificity and photomorphogenic nature of ultraviolet-B-induced cotyledon curling in *Brassica napus* L. Plant Physiol. *102*, 671–677 (1993).

Wootton, J. T.: The nature and consequences of indirect effects in ecological communities. Annu. Rev. Ecol. Syst. *25*, 443–466 (1994).

Ziska, L. H., A. H. Teramura, J. H. Sullivan, and A. McCoy: Influence of ultraviolet-B (UV-B) radiation on photosynthetic and growth characteristics in field-grown cassava (*Manihot esculentum* Crantz). Plant Cell Environ. *16*, 73–79 (1993).

J. Plant Physiol. Vol. 148. pp. 21–25 (1996)

Sensitivity to UV-B of Plants Growing in Different Altitudes in the Alps

Werner Rau and Helga Hofmann

Botanisches Institut der Universität München, Menzinger Str. 67, D-80638 München, Germany

Received July 17, 1995 · Accepted October 27, 1995

Summary

The purpose of our investigations was to elucidate whether different species of the same genus growing preferentially in different altitudes have a genetically fixed different sensitivity to ultraviolet B (UV-B, 280–320 nm), possibly by selection, and/or whether they have a different ability to adapt to UV-B. Plants from species pairs or triplets of 5 genera were grown from seeds without or with additional UV-B. Sensitivity was tested in a growth chamber with UV-B-irradiation in addition to white light by estimation of visible injury. Content and accumulation of UV-absorbing compounds were measured from leaf extracts. In plants grown without UV-B to a cushion there were marked differences in the sensitivity to UV-B between different genera. Different species of the same genus showed only slight differences in sensitivity but in 2 genera the alpine species exhibited a better adaptation after UV-B pre-irradiation. UV-B enhanced the accumulation of UV-absorbing substances to different amounts. Marked differences in the sensitivity to UV-B between different genera may lead to a change in ecosystems should the natural UV-B-radiation increase.

Key words: Saxifraga, Draba, Arabis, Galium, Dianthus, adaptation to UV-B, altitudinal gradient, sensitivity to UV-B radiation, UV-B-absorbing compounds.

Introduction

Depletion of the stratospheric ozone layer may lead to an increase in UV-B-radiation (280–320 nm) reaching the earth's surface. This situation has led to many investigations on the effect of UV-B on plant growth. From the results of several studies it is well documented that different species of plants – or even different cultivars of the same species – show large differences in sensitivity to UV-B (for Ref. see: Caldwell, 1981; Teramura, 1983; Tevini et al., 1989).

The amount of UV-B reaching the earth's surface increases with altitude above sea-level and with lower latitudes. One might expect, therefore, that species growing in regions with a high natural fluence rate of UV-B (alpine and low latitude regions) are less sensitive to UV-B than those from lowland or higher latitude regions.

In few studies the effect of UV-B on species and populations from different altitudes has been tested. Earlier reports on this topic were summarized and discussed by Caldwell (1968). Larson et al. (1990) compared an alpine (*A. caerulea* E. James) and a non-alpine (*A. canadensis* L.) species of *Aquilegia* in their response to UV-B-radiation. They found that growth in height was more inhibited in the lowland species while the number of leaves increased faster in the alpine species; there was, however, no difference in the effect of UV-B on other parameters, e.g. biomass, net photosynthesis and increase in flavonoids. Seeds of many species were collected along a 3000 m elevational gradient in Hawaii by Sullivan et al. (1992). They were then germinated and grown in a greenhouse with either no or two different irradiances of UV-B, and height and biomass were estimated. A great variation in responses to UV-B-radiation among the species was found, and not all species effects correlated with altitude. However, from the plants grown from seeds collected in 0–499 m elevation 72 % were sensitive, whereas only 43 % of those from 1500–3000 m were sensitive. In a subsequent study (Ziska et al., 1992) four pairs of species or populations of plants from contrasting elevations grown from the same seed collection were compared under the same conditions. Plants collected from lower elevational ranges were more affected by UV-B than plants from higher elevational ranges with regard to certain parameters e.g. biomass or maximum photosynthesis. Plants from low elevations showed an increase in UV-B-absorbing substances under UV-B irradiation but plants from higher elevations did not.

In all these studies the effect of UV-B was determined in plants grown from germination with UV-B versus controls grown without UV-B. It was, therefore, not possible to distinguish whether differences in sensitivity to UV-B in different species of the same genus growing preferentially in contrasting altitudes is genetically fixed – possibly by selection – or whether they have a different ability to adapt to UV-B.

Only in two reports experimental conditions were used which allowed such a distinction. Caldwell (1968) collected seeds of *Oxyria digyna* (L.) Hill growing either in Alaska at sea level or in Colorado at ca. 1820 m elevation. Plants were grown under controlled conditions without UV-B. No difference in sensitivity of tissue destruction was found between these populations when irradiated with UV-B. However, later on Caldwell et al. (1982) reported that in *Oxyria* epidermal sensitivity to UV-B was greater in arctic than in alpine populations. A comparison of species pairs of two genera (*Lupinus* and *Taraxacum*) collected either in Alaska at sea-level or in equatorial alpine regions and of different populations of *Plantago lanceolata* L. and *Rumex acetosella* L. from different altitudes was carried out by Barnes et al. (1987). Plants were grown from seeds in a greenhouse without supplemental UV radiation. They were then transferred to field conditions with or without supplemental UV-B and several parameters measured. In *Lupinus* and *Plantago* the alpine species or population proved to be less sensitive to UV-B, in *Rumex* no difference was noted and in *Taraxacum* the results were not clearly defined.

The purpose of our investigations was to determine whether a genetically fixed different sensitivity to UV-B exists in different species of the same genus. To this end pairs or triplets of species of 5 genera which grow preferentially in different elevations were grown in the greenhouse without supplemental UV-B and then sensitivity was tested in a growth-chamber under different fluence rates of UV-B. In addition the ability for adaptation to UV-B was investigated by pre-irradiation with supplemental UV-B in the greenhouse prior to the sensitivity test.

One of the mechanisms by which plants are able to protect themselves against injury or damage by high fluence rates of UV-B are compounds absorbing in the UV-B-region. The most prominent among these are phenylpropanoids and in particular flavonoids which are accumulated mainly in the epidermis of the leaves, and in some plants their content increases in response to UV-B exposure (for Ref. see Wellmann, 1983; Tevini et al., 1991; Beggs and Wellmann, 1994). Therefore, the content of UV-B-absorbing compounds was measured in the different species in non-adapted plants and in plants after adaptation in the greenhouse.

Material and Methods

Plant material

Pairs or triplets of species of the same genus which grow preferentially in the alpine region (>1200 m above sea level) or in lower mountain ranges (~1000 m) or in lowlands (~500 m) in the European Alps were selected with the kind advice from colleagues of the Herbarium and the Botanical Garden in Munich. Seeds were obtained from Botanical Gardens in Germany, Austria, Switzerland and Italy, and germinated in soil-containing dishes in the greenhouse. After some weeks the plants were transferred to soil-containing pots (diameter 8 cm) and further cultivated in the greenhouse. When they had grown to a cushion (after ca. 9–12 months) they were used for experiments. Plant species used for the experiments are listed in Table 1.

Radiation conditions

For the sensitivity test in the growth chamber (see below), white light was provided by fluorescent tubes (65 W, Osram, Daylight de Luxe) and incandescent-bulbs (60 W, Osram; about 10 % of the total fluence rate) at a fluence rate of about 30 $W \cdot m^{-2}$ (400–700 nm) at plant level.

Supplemental UV-B radiation was provided by UV-B fluorescent tubes (Philips TL 12/40 W) filtered with 0.08 mm cellulose acetate in a frame hanging between tube and plants. The spectral irradiance was determined with a spectroradiometer (Bentham, Gigahertz-Optik, Puchheim, Germany) equipped with a dual holographic grating. Different fluence rates were adjusted by variation of the distance between tube and plant level and by covering the fluorescent tube with a perforated metal tube with different diameter holes.

Test of sensitivity

For the test plants were selected for uniformity. Some of the plants were adapted to UV-B by cultivation in the greenhouse with additional UV-B at a fluence rate of 150 $mW \cdot m^{-2}$ at plant level. The sensitivity was then tested by transferring the non-adapted and adapted plants to a growth chamber (22 ± 1 °C; 70 % relative humidity) There they were irradiated for 15 h per day with white light supplemented with UV-B in the range from 300–1500 $mW \cdot m^{-2}$ in separate compartments.

Sensitivity was measured by estimation of visible injury during 10 days of irradiation. Grades of injury were determined in the following range:

grade 1: slight yellowing or bronze-colouring on single leaves.
grade 2: bronze-colouring or slight yellowing on half of the leaves.
grade 3: bronze-colouring or yellowing of nearly all leaves.
grade 4: yellowing and drying of nearly all leaves (nearly complete damage).

We are well aware that this rating is in part subjective. To overcome this handicap determination was done independently by two persons.

In the first preliminary experiments it was found that a supplemental irradiation with UV-B of fluence rates of 300 $mW \cdot m^{-2}$ in the sensitivity test caused only slight visible injury. Therefore, for a better differentiation in sensitivity between the different species, for the following tests supplemental fluence rates of 600 and 1000 $mW \cdot m^{-2}$ were chosen.

Due to limited space in the greenhouse and in the growth chamber all 5 genera could not be investigated at the same time. In most experiments two genera were used, and each experimental condition included 3–4 plants of one species. For comparison in all sensitivity tests in the growth chamber, controls consisted of plants grown without supplemental UV-B. The experiments were carried out during nearly 3 years at various seasons. Each genus and species was investigated in 2–4 independent experiments. There were no or only very slight differences in the results between experiments carried out with a genus at different seasons.

UV-B-absorbing compounds

For determination of the content of UV-B-absorbing compounds leaf tissues were ground in ethanol using mortar and pestle, the extract was heated up to 60 °C in order to precipitate any proteins, and centrifuged. Absorption was measured in a scanning spectrophotometer (Uvicon 820, Kontron, Germany), and with a computer programme the integral of the absorption at 280–320 nm was calculated; the value on a fresh weight basis was taken as the relative content. Absorption curves in the UV are very similar in the different species of a genus indicating a similar composition of UV-absorbing compounds. This enables a comparison of these species. There are, however, great differences between the genera.

Results

Non-adapted plants

Results with non-adapted plants are compiled in Table 1 (open crosses). The data show marked differences in sensitivity between different genera. This is in agreement with the results of Sullivan et al. (1992), although the experimental conditions were different. Differences in sensitivity between species of different elevations varied from genus to genus. Only in *Dianthus* the alpine species was less sensitive, whereas in *Arabis* and *Galium* such a difference was less pronounced. However, in 2 genera – *Saxifraga* and *Draba* – the alpine species was slightly more sensitive than the montane or lowland ones. The leaves of the different species of the same genus show a very similar morphology. Therefore, differences in sensitivity between the species cannot be attributed to such variations.

Table 1: Grade of visible injury (grade 1–4) of plants at the end of an irradiation period of 10 days in a growth chamber with white light and supplemental UV-B of two different fluence rates (sensitivity test). Before sensitivity test plants were grown in the greenhouse either without (open crosses) or with supplemtnal UV-B of a fluence rate of 150 $mW \cdot m^{-2}$ for 6 weeks (crosses in parentheses). Mean of 2–4 independent experiments with 3–4 plants per each species and condition.

	Fluence rate of UV-B ($mW \cdot m^{-2}$)			
	600		1000	
Saxifraga				
pedemontana All. (alpine)	++++	(++)	++++	(+++)
rosacea Moench (lowland)	+++	(++)	+++	(+++)
Draba				
dubia Suter (alpine)	+++	(+)	++++	(+)
incana L. (montane)	++	(+)	+++	(++)
nemorosa L. (lowland)	+++	(++)	+++	(++)
Arabis				
pumila Jacq. (alpine)	+	(–)	++	(–)
corymbiflora Vest (montane)	++	(+)	+++	(++)
hirsuta (L.) Scop. (lowland)	++	(+)	+++	(+)
Galium				
pseudohelveticum Ehrend. (alpine)	–	(–)	++	(++)
anisophyllum Vill. (lowland)	+	(+)	+++	(+++)
Dianthus				
subacaulis Vill. (alpine)	–	(–)	–	(–)
silvester Wulf. (lowland)	–	(–)	++	(+)

The data also show differences in tolerance towards increasing fluence rates of UV-B between the genera. Whereas in *Saxifraga* and *Draba* irradiation with a fluence rate of 600 $mW \cdot m^{-2}$ already caused a high grade of visible injury compared to the fluence rate of 100 $mW \cdot m^{-2}$, in *Galium* and *Dianthus* the lower fluence rate was tolerated with no or only minor visible injury, and only the higher fluence rate was really effective. This demonstrates the variations in threshold of tolerance between different genera.

Adaptation

Adaptation to UV-B was investigated by pre-irradiation of plants in the greenhouse with supplemental UV-B for 6 weeks prior to the sensitivity test. The test was carried out together with the non-adapted plants in order to allow a complete comparison. The results are also compiled in Table 1 (crosses in parantheses).

No adaptation effect was obvious in *Galium* and only a minor one in *Saxifraga*, whereas the genera *Draba* and *Arabis* exhibited good adaptation. In the latter two genera the alpine species showed a better ability to adapt. The increase in the grade of injury by irradiation with 1000 $mW \cdot m^{-2}$ UV with time is depicted in Fig. 1a for *Draba* and Fig. 1b for *Arabis* in a comparison of non-adapted and adapted plants. The data reveal that besides a decrease in the grade of injury at the end of the irradiation period (after 10 days), the onset of visible injury is delayed. This indicates that adapted plants may tolerate UV-B-stress for a limited period of time. In *Arabis* adaptation in the alpine species reached such a degree that the adapted plants were nearly non-sensitive to our experimental conditions. Whether or not a pre-irradiation with higher fluence rates of UV-B would have lead to a higher degree of adaptation has yet to be shown.

UV-B-absorbing compounds

Changes in the content of UV-B-absorbing substances during adaptation was measured. The results are compiled in Fig. 2. The data show that in non-adapted plants only in two genera – *Draba* and *Dianthus* – the alpine and montane species had a higher content, whereas *Saxifraga* exhibited a reverse situation. Increased accumulation of UV-B-absorbing compounds induced by UV-B was very different in the different genera. It was lowest in *Galium* correlated with the lowest ability to adapt (see Table 1) and in *Arabis* the higher increase in the alpine species correlated with better adaptation.

Discussion

The intention of this study was to determine whether different species of the same genus growing preferentially in different altitudes have a genetically fixed different sensitivity to UV-B or whether they have a different ability to adapt to UV-B. As an adequate experimental design for such an ap-

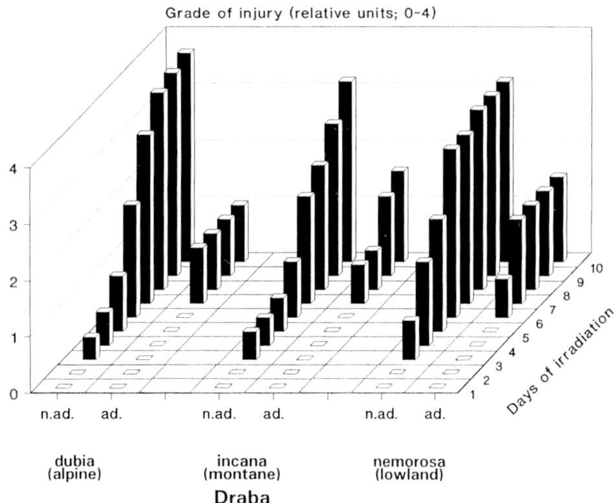

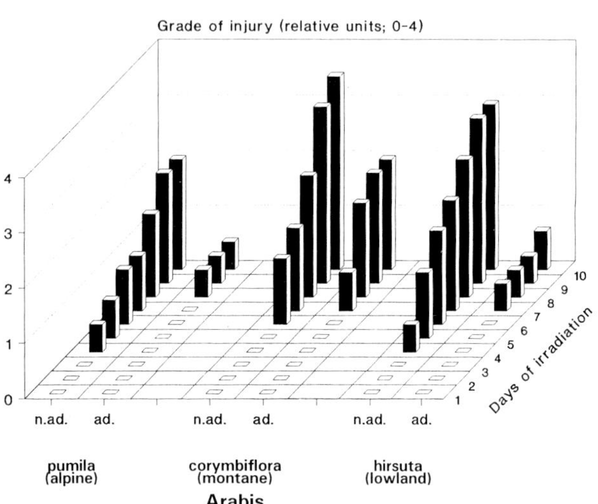

Fig. 1 a+b: Time course of increase in the grade of visible injury by irradiation with supplemental UV-B (1000/mW · m^{-2}) in the growth chamber (sensitivity test). Plants were grown before the sensitivity test in the greenhouse either without (nonadapted; «n.ad.») or with supplemental UV-B of a fluence rate of 150 mW · m^{-2} for 6 weeks (adapted; «ad.»). Mean of 4 independent experiments with 3–4 plants per each species and condition.

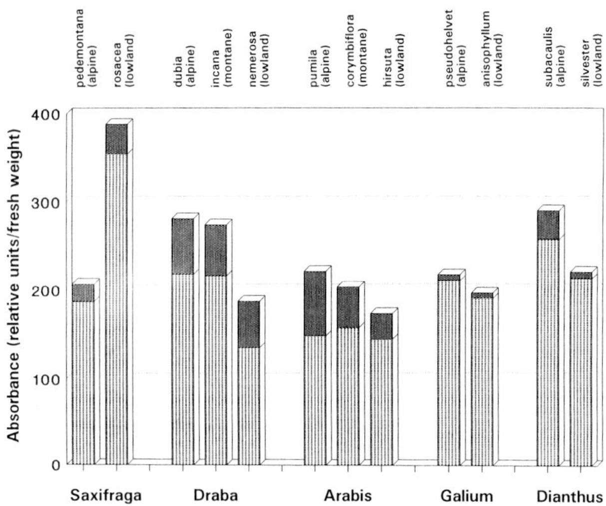

Fig. 2: Content of UV-B-absorbing compounds in plants grown in the greenhouse without supplemental UV-B (light bars) and increase in content after irradiation with supplemental UV-B (150 mW · m^{-2}) in the greenhouse for six weeks (dark bars). Mean of 2–4 independent experiments with 3–4 plants per each species and condition.

proach plants were cultivated in a glass-covered greenhouse so that they were grown without any UV-B. A subsequent sensitivity test in a growth chamber where they were irradiated with different fluence rates of UV-B in addition to white light should show the genetically fixed sensitivity without interference by adaptation.

In the 5 genera investigated, only in three of them the alpine species were, to different extent less sensitive compared to the lowland ones (in agreement with the results of Barnes et al., 1987), but in the other two genera the alpine species were slightly more sensitive. A marked difference between the altitudinal species pairs was only found in *Dianthus*. From these results we conclude that the assumption that alpine species are less sensitive to UV-B by their genetic constitution developed by selection cannot be generalized.

Plants are able to adapt to UV-B radiation. Various protecting mechanisms are known e.g. variations in the epidermal transmittance (for Ref. see Robberecht et al., 1980; Cen and Bornman, 1993) partly caused by reflectance and partly by increase in UV-B-absorbing compounds, mainly flavonoids, induced by UV-B irradiation (for Ref. see Wellmann, 1983; Tevini et al., 1991; Beggs and Wellmann, 1994), and also by other factors (for Ref. see Caldwell et al., 1994). The ability of the different genera and species was investigated by pre-irradiation of the plants in the greenhouse with supplemental UV-B before the sensitivity test. Distinct differences in the extent of the ability were found between the different genera ranging from nearly no (*Galium*) to good (*Draba, Arabis*) adaptation. In the latter two genera the alpine species exhibited a better adaptation compared to the lowland species. We are well aware that the adaptation conditions in our experiments, i.e. fluence rate of supplemental UV-B in the greenhouse and period of time of supplemental UV-B may have been insufficient for some genera or species. However, the data clearly show the differences between the genera and the altitudinal pairs or triplets.

In agreement with the results of Ziska et al. (1992) a marked increase in the content of UV-B-absorbing compounds induced by UV-B irradiation was not found in all genera or species. A lack of increase in *Galium* was correlated with the lowest ability to adapt. In general, however, content of UV-B-absorbing compounds was not clearly correlated with the sensitivity to UV-B.

As a conclusion from our results on distinct differences in sensitivity to UV-B between different genera and in the different ability to adapt it should be pointed out that these differences may lead to a change in natural ecosystems upon an icrease in natural UV-B radiation.

Acknowledgements

This work was supported by the Bavarian Climate Research Programme («BayForklim»). We thank the colleagues of the Herbarium and the Botanical Garden München-Nymphenburg in particular Dr. W. Lippert and Mr. D. Schacht for advice in selection and cultivation of the plants, Mrs. Dagmar Sensburg and Ingrid Duschanek for skillful technical assistance and Mrs. Irmgard Reiber for careful cultivation of the plants.

References

Barnes, P. W., S. D. Flint, and M. M. Caldwell: Photosynthetic damage and protective pigments in plants from a latitudinal arctic/alpine gradient exposed to supplemental UV-B radiation in the field. Arctic and Alpine Research *19,* 21–27 (1987).

Beggs, C. J. and E. Wellmann: Photocontrol of flavonoid biosynthesis. In: Kendrick, R. E. and G. H. M. Kronenberg (eds.): Photomorphogenesis in plants. Vol. 2, pp. 733–750. Kluver Academic, Dordrecht (1994).

Caldwell, M. M.: Solar ultraviolet radiation as an ecological factor for alpine plants. Ecological Monographs *38,* 243–268 (1968).

Plant response to solar ultraviolet radiation. In: Lange, O. L., P. S. Nobel, C. B. Osmond, and H. Ziegler (eds.): Encycl. of Plant Physiol. Vol. 12 A, pp. 169–197. Springer-Verlag, Berlin (1981).

Caldwell, M. M., R. Robberecht, R. S. Nowak, and W. D. Billings: Differential photosynthetic inhibition by ultraviolet radiation in species from the arctic-alpine life zone. Arctic and Alpine Research *14,* 195–202 (1982).

Caldwell, M. M., A. H. Teramura, M. Tevini, J. F. Bornman, L. O. Björn, and G. Kulandaivelu: Effects of increased solar ultraviolet radiation on terrestrial plants. In: Environmental effects of ozone depletion. 1994 Assessment, UNEP (1994).

Cen, Y. P. and J. F. Bornman: The effect of exposure to enhanced UV-B radiation on the penetration of monochromatic and polychromatic UV-B radiation in leaves of *Brassica napus.* Physiol. Plant. *87,* 249–255 (1993).

Larson, R. A., W. J. Garrison, and R. W. Carlson: Differential responses of alpine and non-alpine *Aquilegia* species to increased ultraviolet-B radiation. Plant, Cell and Environment *13,* 983–987 (1990).

Robberecht, R., M. M. Caldwell, and W. D. Billings: Leaf ultraviolet optical properties along a latitudinal gradient in the arctic-alpine life zone. Ecology *61,* 612–619 (1980).

Sullivan, J. H., A. H. Teramura, and L. H. Ziska: Variation in UV-B sensitivity in plants from a 3000-m elevational gradient in Hawaii. Amer. J. of Botany *79,* 737–743 (1992).

Teramura, A. H.: Effects of ultraviolet-B radiation on the growth and yield of crop plants. Physiol. Plant. *58,* 415–427 (1983).

Tevini, M. and A. H. Teramura: UV-B effects on terrestrial plants. Photochem. Photobiol. *50,* 479–487 (1989).

Tevini, M., J. Braun, and G. Fieser: The protective function of the epidermal layer of rye seedlings against ultraviolet-B radiation. Photochem. Photobiol. *53,* 329–333 (1991).

Wellmann, E.: UV radiation: Definitions, characteristics and general effects. In: Encycl. of Plant Physiol. Vol. 16 B, pp. 745–756. Springer-Verlag, Berlin (1983).

Ziska, L. H., A. H. Teramura, and J. H. Sullivan: Physiological sensitivity of plants along an elevational gradient to UV-B radiation Amer. J. of Botany *79,* 863–871 (1992).

J. Plant Physiol. Vol. 148. pp. 26–34 (1996)

UV-B Effects on Crops: Response of the Irrigated Rice Ecosystem

David Olszyk[1], Quijie Dai[2], Paul Teng[2], Hei Leung[3], Yong Luo[4], and Shaobing Peng[2]

[1] U.S. EPA, National Health and Environmental Effects Research Laboratory, Corvallis, Oregon, 97333 USA

[2] International Rice Research Institute, P.O. Box 933, 1099 Manila, Philippines

[3] Department of Plant Pathology, Washington State University, Pullman, Washington 99264-6430, USA

[4] Department of Plant Pathology, University of Arkansas, Fayetteville, Arkansas 72701

Received June 24, 1995 · Accepted October 27, 1995

Summary

Increasing ultraviolet-B (UV-B) radiation resulting from depletion of the stratospheric ozone layer could have damaging effects on crops. This paper reviews recent findings on direct effects of UV-B on rice growth and yield as well as indirect effects via impacts on other organisms in the rice (*Oryza sativa*) agroecosystem. The findings are based on research by scientists at the International Rice Research Institute (IRRI) in Los Baños, the Philippines, and their collaborators in China and the United States; with comparison to research by scientists in other countries. Current results indicate that while enhanced UV-B directly impacts many aspects of rice growth, physiology, and biochemistry under controlled phytotron conditions; in general rice growth and yield are not affected under natural field conditions. The difference in response may be related both to the levels of UV-B exposure used in phytotron vs. field studies and the lower ratio of UV-A to UV-B in the phytotron compared to field. In terms of indirect effects on rice blast disease, enhanced UV-B affected both the fungus itself (*Pyricularia grisea*) and the susceptibility of the rice plant to the fungus. Based on these data, simulation models estimated potential impacts of higher UV-B levels on blast severity and rice yield in different countries of southeast and east Asia. Ultimately, results from rice studies can be used to identify strategies to minimize any negative effects of UV-B on rice productivity.

Key words: Rice, Oryza sativa, UV-B, CO_2, methane, temperature, rice blast, Pyricularia grisea.

Abbreviations: EPA = United States Environmental Protection Agency; IRRI = International Rice Research Institute; UV = ultraviolet radiation; UV-B = ultraviolet-B radiation; UV-B_{BE} = biological effective UV-B; UV-A = ultraviolet-A radiation; SOD = superoxide dismutase; CA = catalase; AP = ascorbate peroxidase; MDA = malondialdehyde; ASA = ascorbic acid; GSH = glutathion; ABA = abscissic acid; IAA = indolacetic acid.

Disclaimer

The information in this document has been funded wholly or in part by the U.S. Environmental Protection Agency under cooperative agreement number 817425 to the International Rice Research Institute. It has been subject to the agency's peer and administrative review. It has been approved for publication as an EPA document. Mention of trade names or commercial products does not constitute endorsement or recommendation for use.

Introduction

Stratospheric Ozone Depletion, UV-B, and Effects on Crops

Scientists and policy makers are concerned that increasing UV-B radiation resulting from a depletion of the stratospheric ozone (O_3) layer may have potential impacts on the earth. Some atmospheric trace gases produced by human activities, especially chlorofluorocarbons, have photodegradation products which destroy O_3 in the stratosphere (Rowland, 1989).

Ozone strongly absorbs UV-B, thereby filtering out much of this UV component before it reaches the earth's surface (UNEP, 1994). Thus decreasing stratospheric ozone levels should result in a substantial increase in UV-B (290–320 nm) radiation reaching the Earth's surface, with the amount dependent on atmospheric and geographic factors (UNEP, 1994). Enhanced UV-B at the surface may have potentially damaging effects on humans, plants, and animals in terrestrial and aquatic ecosystems (SCOPE, 1992; UNEP, 1994).

Potential effects of enhanced UV-B on crops are of concern as serious societal problems could result if global food production decreased. Enhanced UV-B has damaging effects on plant dry matter production and physiological processes such as photosynthesis, respiration, and ion transport (Teramura, 1983). This is due in part to UV-B radiation absorbed by molecules such as those in Photosystem II, the electron transport system, nucleic acids, enzymes, pigments, and growth regulators (Caldwell and Flint, 1994 a).

Many reviews have summarized available information on UV-B responses of crops. For example, Teramura (1990) described UV-B studies through approximately mid-1989, highlighting field studies and effects of UV-B on crop yields – especially for soybean. Of 22 crops evaluated: 5 had definite reductions in yield, 6 had mixed responses depending the study, 9 had no (or negligible) response to enhanced UV-B, and 2 had uncertain responses. More recently Caldwell and Flint (1994 a) concluded that «The data base of appropriate UV-B studies is very limited when compared with other fields of environmental research such as with tropospheric air pollutants.» They particularly were concerned with lack of data for tropical crops in areas of currently high UV-B and interactions of UV-B with other environmental factors.

Rice and UV-B

With respect to grain cereals, rice (*Oryza sativa*) is the most important staple food for direct human consumption (IRRI, 1993), particularly in Asia. Thus rice must be included in any evaluation of the agricultural effects from enhanced UV-B. However, scientific knowledge of the response of the rice plant to UV-B is meager.

At the time of Teramura's review (1990) there was little information on the response of rice to UV-B. Van et al. (1976) reported that rice was «moderately sensitive» to UV-B in comparison to some other crop species based on growth and photosynthetic responses in greenhouse and growth chamber studies. Biggs et al. (H. Biggs, personal communication) found no statistically significant effect of elevated UV-B on rice yield in the field, even though there was a trend towards a reduction in seed yield with some simulated ozone depletions for Florida. Basiouny (1986) indicated that rice was «sensitive» to UV-B based on reductions in height; fresh, dry, and ash weights; and leaf necrosis for enhanced UV-B exposed compared to check plants in a growth chamber. However, of these studies were conducted using U.S. cultivars and exposure conditions and not cultivars and conditions representative of realistic UV-B exposures in tropical regions.

The EPA/IRRI Research Program

Thus, due to the importance of rice and general lack of information on it's response to UV-B, the EPA in collaboration with IRRI have been conducting a major research program on the effects of UV-B on rice (Olszyk and Ingram, 1991). The focus is on irrigated rice – the most productive cropping system. The program concerns not only effects of UV-B on rice, but also effects of elevated CO_2 and climatic change (increased temperature) on rice, as well as methane emissions from rice fields. The program includes effects on rice from an ecosystem viewpoint, i.e. interactions among rice plants, other organisms in rice fields, and the plant-soil system. Ecosystem mediated changes in rice productivity especially could occur through changes in individual species productivity, morphology, and secondary chemistry which could alter crop-weed competition, disease susceptibility, insect herbivory, and decomposition rates (Caldwell and Flint, 1994 b).

This paper reviews major findings on the response of rice to UV-B from the EPA/IRRI studies. The focus is on results from field and phytotron studies at IRRI, with limited additional information from rice-disease (blast), rice-insect, and rice-weed studies; and reference to other key studies with UV-B and rice. The results have either been published previously elsewhere, or are in preparation for publication elsewhere and are included here only to illustrate ongoing research.

Materials and Methods

The following are general procedures applicable across UV-B experiments at IRRI and by their collaborator at Washington State University (WSU). An early study also was conducted in a greenhouse at Corvallis, Oregon (Barnes et al., 1993).

Phytotron

General procedures were described by Dai et al. (1995), Dai et al. (1994 a, b), Liu et al. (1994), Zhang et al. (1994), Dai et al. (1992), and IRRI, personal communciations[1]. Plant culture generally is according to normal experimental procedures used in the IRRI phytotron. Ten-day old rice seedlings were transplanted into pots which were used for three to four week exposures. Due to the short experimental period, all plant samples are taken during vegetative growth stages. A wide variety of biomass, morphological, growth, biochemical, and physiological parameters have been measured.

The UV-B exposures are obtained using individual lamp frames suspended over 1.8×1.2 m carts; each frame with four UV-B emitting fluorescent bulbs for most UV-B treatments except six bulbs for 19.1 kJ m^{-2} d^{-1} $UV\text{-}B_{BE}$ treatment ($UV\text{-}B_{BE}$ based on generalized plant action spectrum of Caldwell, 1971, normalized to unity at 300 nm). Bulbs are surrounded by plastic filters – cellulose acetate to allow transmission of UV-B in wavelengths of 290–320 nm for simulating ozone depletion, and mylar to block transmission of UV-B

[1] Most of IRRI personal communication information published in Olszyk, D. M., D. Bachlet, J. Kern, K. Fischer, K. L. Heong, H.-U. Neue, M. Kropff, S. Peng, P. Teng, L. Ziska, H. Leung, W. H. Patrick, C. W. Lindau, R. D. DeLaune, and J. Bolte: Effects of UV-B and Global Climate Change on Rice. Fourth Annual Program Progress Report. NHEERL-COR-845.

as a check. For most studies the target UV-B_{BE} levels are approximately 0 and 13 kJ m^{-2} d^{-1} for the enhanced UV-B and check treatments, respectively. When a series of UV-B levels are used in exposure-response studies, UV-B_{BE} levels are approximately 0, 6, 13, and 19.1 kJ m^{j-2} d^{-1} to better encompass the range of UV-B exposures occurring in the field and phytotron. Enhanced UV-B treatments use a square-wave pattern of exposure; bulbs come on and off at precisely defined intervals (i.e. on at 9 am and off at 3 pm) and are at a constant intensity during the on period regardless of ambient light conditions.

Field

Experiments are conducted IRRI using procedures described in Dai et al. (1995), and IRRI personal communications. Plots are 4.0 × 1.2 m and general crop management methods are typical for IRRI field studies. Approximately 15-day old seedlings are transplanted from a greenhouse to field plots. Plant samples are obtained at different growth stages for growth and biomass production measurements and UV-screening pigment determination. Yield components, and final growth and biomass measurements are made at final harvest. Biochemical measurements were made primarily in recent studies to evaluate potential mechanisms for differences in UV-B susceptibility in the field vs. phytotron.

Each plot is equipped with an individual lamp frame containing 12 UV-B emitting fluorescent bulbs equipped with filters as in the Phytotron. Most experiments use two treatments, enhanced UV-B (about 20 % ozone depletion for the Philippines) and an ambient UV-B check, with six replicate plots per treatment. The main UV-B treatment for each experiment uses a square-wave pattern of exposure: bulbs come on and off at precisely defined intervals (i.e. on at 9 am and off at 3 pm) and are at a constant intensity during the on period irrespective of ambient light conditions. A supplementary treatment uses a modulated exposure system on a more limited basis, whereby bulb UV-B is altered over the day in direct proportion to the ambient light intensity. Field UV-B exposures levels will be calculated as UV-BE_{BE} based on spectral measurements described below.

Rice Blast

Initial rice-blast studies were conducted in the IRRI phytotron using procedures similar to those for rice cultivar studies (Finckh et al., 1995). Subsequent experiments with rice blast are conducted at WSU using general procedures described in Leung et al. (1995) and H. Leung, personal communications. Laboratory experiments with the fungus use a small frame of UV-B bulbs suspended over cultures. The UV-B treatments are obtained by placing cellulose acetate or mylar filters directly over the culture plates. Rice-fungal disease cycle studies are conducted in a growth chamber or laboratory area equipped with UV-B bulbs and plastic filters for the different treatments.

Spectral Analysis

Complete spectral analyses of UV-B levels in the different experiments are carried out periodically with an Optronics 752 spectroradiometer at IRRI and with an Optronics 742 spectroradiometer at WSU (Dai et al., 1995; Leung et al., 1995). Checks are made routinely to insure that bulbs are on and emitting a relatively constant level of UV radiation using a UV-X radiometer with a UV-X 31 sensor at both IRRI and WSU. Ambient UV-B is measured continuously with a Robertson-Berger (R-B) meter in the field at IRRI. Both the UV-X and R-B meters are compared periodically to the Optronics instruments. Phytotron and laboratory UV-B levels are based solely on spectroradiometer measurements under the lamp frames as the contribution of ambient sunlight is essentially zero. For field studies, both the contribution of UV-B from the bulbs and sunlight under the lamp frames will be considered along with the data from continuous UV monitoring with the R-B meter to estimate total UV-B exposure during different growing seasons. In 1994 both UV-A (321–400 nm) and UV-B (280–320 nm) were measured in the glasshouse and in the field.

Results and Discussion

Rice Cultivars Differ in Response to Enhanced UV-B

A total of 188 rice cultivars representing major rice genotypes and ecosystems were screened for UV-B susceptibility in the IRRI phytotron (Dai et al., 1994 a, b). Seedling plants (up to approximately 4 weeks old) were used so that a large number of cultivars could be tested and several replicate plants could be tested per cultivar. Key growth parameters (height, tiller number, leaf area, and shoot dry weight) were measured and used to calculate responses based on the differences in growth for enhanced UV-B (≈13.0 kJ m^{-2} d^{-1}) vs. check plants (≈0 kJ m^{-2} d^{-1}).

Rice cultivars differed widely in response to elevated UV-B as described in detail by Dai et al. (1995, 1994 a, b). In terms of total shoot dry weight, the response across cultivars followed an ≈ normal distribution with a mean weight reduction of 6 %, with extremes ranging from 35 % (cv. *Dubovskiy*) to 32 % increase in weight (cultivar *Guze*) (Fig. 1 A). Leaf area and height showed similar reductions with enhanced UV-B compared to checks, while tiller number generally increased in response to enhanced UV-B.

There was no relationship between the response of rice cultivars to UV-B and approximate ambient clear sky UV-B intensities for their countries and/or regions of origin (Dai et al., 1994 a; Olszyk, 1994). Cultivars originating from countries at lower latitudes with higher ambient UV-B had similar responses to UV-B exposure as cultivars from countries at higher latitudes with lower UV-B. However, six rice cultivars that had significantly increased biomass with enhanced UV-B exposure originated from high ambient UV-B areas, and may be useful in a UV-B resistance breeding program (Dai et al., 1994 a). Staff at IRRI have crossed six UV-B tolerant and six UV-B-sensitive cultivars (36 separate crosses) and is testing the F1 hybrids in an outdoor experiment for UV-B susceptibility and heritability of tolerance.

The range in response to UV-B among rice cultivars at IRRI was similar to the range found in other studies. For 16 rice cultivars, Teramura et al. (1991) reported a 41 % decrease to 23 % increase in total plant dry weight with enhanced UV-B compared to check plants (Fig. 1 B). Six cultivars had statistically significant decreases in weight and one cultivar had a significant increase in weight with enhanced UV-B compared to check plants. The average response was a 13 % decrease in plant dry weight, which was slightly larger than the average decrease in shoot dry weight found by Dai et al. (1995). Though they did not carry the plants to yield; Teramura et al. reported that the range in panicle dry weight responses to enhanced UV-B closely followed the biomass responses. This research was conducted in a greenhouse at the University of Maryland with UV-B levels of approximately 15.7 kJ m^{-2} d^{-1}

$UV-B_{BE}$ for enhanced compared to ≈0 $UV-B_{BE}$ for check plants, which were similar to the $UV-B_{BE}$ levels used in the IRRI phytotron.

Barnes et al. (1993) reported a wide range in variability in response of rice plants to UV-B based on a greenhouse study in Oregon, USA. However, the extremes in response were less than in the Dai and Teramura studies: a 13 % decrease to 4 % increase, with an average 4 % decrease in total dry weight across 22 cultivars with enhanced UV-B compared to check plants (Fig. 1 C). This reduced response compared to the IRRI and Maryland responses may be related to the lower UV-B exposures used in Oregon, i.e. ≈10.3 kJ m^{-2} d^{-1} $UV-B_{BE}$ for enhanced UV-B compared to ≈0 $UV-B_{BE}$ for check plants.

Sato and Kumagai (1993) reported a wide range in response to enhanced ultraviolet (UV) radiation among 198 cultivars of rice representing a range of Indian (Bengal) and Indonesian ecotype cultivars and Japanese group cultivars. They noted general decreases in shoot fresh weight, plant height, and leaf chlorophyll content for enhanced UV compared to check plants, especially with higher UV exposures. They also noted increases in parameters for some cultivars with enhanced UV, especially at a low level. However, since plants were exposed to UV-C as well as UV-B in Sato and Kumagai's study, the results can not be directly compared to the other studies where plants were exposed only to UV-B.

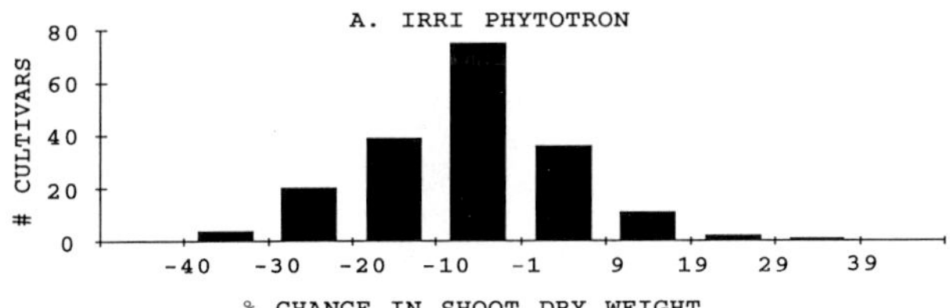

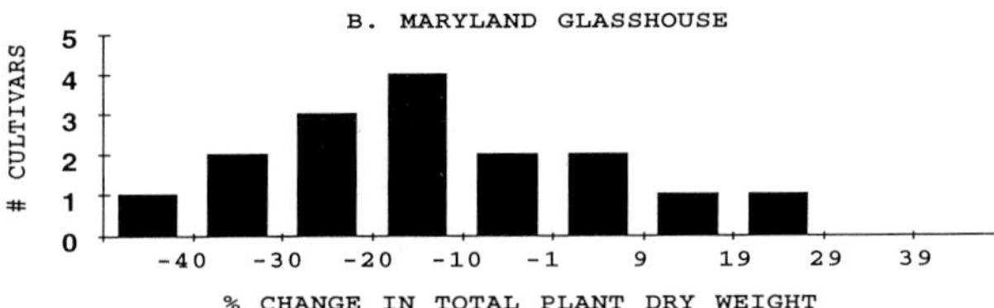

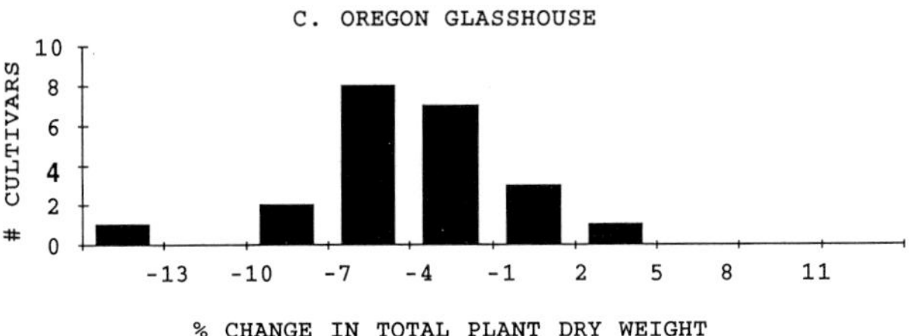

Fig. 1: Relative growth response of rice cultivars exposed to enhanced UV-B vs. check plants for studies in glasshouses at different locations. (A) Philippines data: total shoot weight response of 188 cultivars (Dai et al., 1995). (B) Maryland, USA, data: total plant dry weight response for 16 cultivars (Teramura et al., 1991). (C) Oregon, USA, data: total dry weight for 22 cultivars (Barnes et al., 1993). Number of cultivars in bar includes value to lower right of bar, e.g. range is −9 to 0 for middle IRRI bar.

Table 1: Effects of enhanced UV-B radiation on yield of rice in the field at Los Baños, the Philippines[a].

Year	Season	Cultivar	Treatment	Yield (T/ha)	CV (%)	UV-B/ Check
1992	Dry	IR64 (T)	Check	7.2	3.8	
			Enhanced UV-B	7.0	1.9	0.97
		IR74 (S)	Check	7.2	2.0	
			Enhanced UV-B	6.3	3.2	0.88
	Wet	IR72 (T)	Check	5.2	7.1	
			Enhanced UV-B	5.7	4.3	1.10
		IR74 (S)	Check	4.5	11.8	
			Enhanced UV-B	4.8	6.3	1.07
1993	Dry	IR72 (T)	Check	8.4	4.8	
			Enhanced UV-B	8.5	10.1	1.01
		IR74 (S)	Control	7.6	5.4	
			Enhanced UV-B	7.7	5.1	1.01
	Wet	IR72 (T)	Check	4.3	3.5	
			Enhanced UV-B	4.2	6.8	0.98
		IR74 (S)	Check	4.7	9.2	
			Enhanced UV-B	4.4	3.6	0.94
Mean (n = 4)			Tolerant			1.02
			Susceptible			0.98

[a] Preliminary data on effects of enhanced UV-B radiation on yield of rice in the field at Los Baños, the Philippines, based on data from Dai et al. (1995). Data are for five plots/treatment for 1992 dry season; six plots/treatment for other seasons. The average coefficient of variation was 5.6 across all yields. There were no statistically significant differences between control and enhanced UV-B treatment plants for any season as determined with least significant difference (l.s.d.) tests. The UV-B susceptible cultivar was IR74; UV-B tolerant cultivar was IR72 (except IR64 in 1992 dry season) based on phytotron studies. Check was bulbs with mylar films; enhanced UV-B represented 20 % ozone depletion at Los Baños. However, actual $UV-B_{BE}$ levels are being calculated based on spectroradiometric measurements under the lamp frames and ambient UV-B monitoring and modeling. The 1992 dry season enhanced UV-B level likely was less than level for other season due to methodological problems early in the study; data included for comparison purposes only.

Rice Productivity is Not Directly Affected by UV-B in the Field

Under realistic growing conditions for tropical rice, enhanced UV-B had no overall effect on yield in the field (Dai et al., 1995). Over four seasons (1992–93), a UV-B susceptible rice cultivar (IR74) had only a 2 % decrease and UV-B tolerant cultivars (IR72 and IR64) had a 2 % increase in yield with enhanced UV-B compared to check plants (Table 1). These small changes in average yield with UV-B were less the mean coefficient of variability across yields. A similar pattern (lack of UV-B response) was observed in the 1994 dry season (IRRI, personal communications, data not shown).

Rice did not respond to UV-B either during wet seasons with somewhat lower ambient light intensities, and, hence background UV-B radiation; or during dry seasons when light intensities and UV-B radiation were higher. A general lack of a UV-B response also was consistent across different growth stages (middle tillering, panicle initiation, flowering, and maturity), including early seedling growth stages when effects were seen in the phytotron (as described earlier). Lack of any differential response between the UV-B susceptible and tolerant cultivars further reinforced the conclusion that, overall, rice was not susceptible to yield reductions in response to enhanced UV-B in these studies.

In addition to lack of total yield response was a lack of significant UV-B effects on other specific responses at any harvest (Dai et al., 1995; IRRI personal communications). Parameters measured were standard agronomic indicators of plant productivity: i.e. yield components (panicle weight, panicle numbers, % filled grains/panicle, 1000 grain weight), growth and morphology (plant height, tiller number, leaf area index, specific leaf weight), and biomass (leaf, stem and root dry weights).

The question remains, whether or not the lack of UV-B response on rice in the field was due to use of relatively tolerant cultivars (despite the previous phytotron testing). Thus, recent field studies at IRRI have focused on the relative yield responses of a wider range of 12 cultivars. Results are still being processed, but preliminary data from one test season showed a range in total shoot dry weight from a 10 % decrease to an 8 % increase, with an average 1 % increase for UV-B exposed, compared to check plants (IRRI, personal communications) (Fig. 2 A), but with no statistically significant effects for any cultivar. Yield had a similar response pattern (IRRI, personal communications, data not shown).

The range in response among cultivars in the field and/or for yield at IRRI was very similar to that in recent studies for other locations. In Japan, Nouchi et al. (1995) found a range in total dry weight response among 17 cultivars ranging from 12 % decrease to 7 % increase, with an average 2 % decrease with enhanced UV-B compared to check plants (Fig. 2 B). However, there were no statistically significant effects of UV-B for any cultivar. Because the study was conducted late in

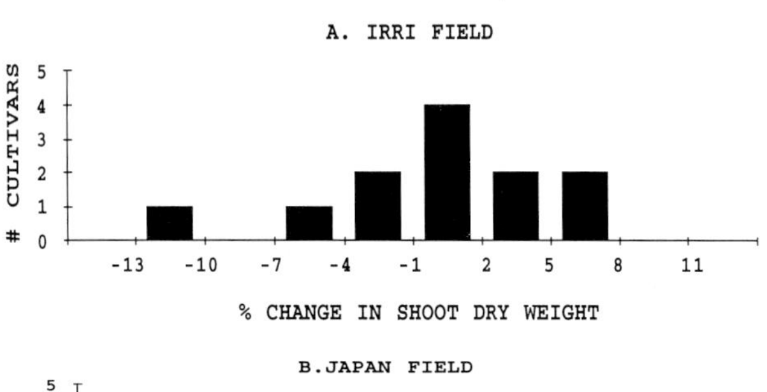

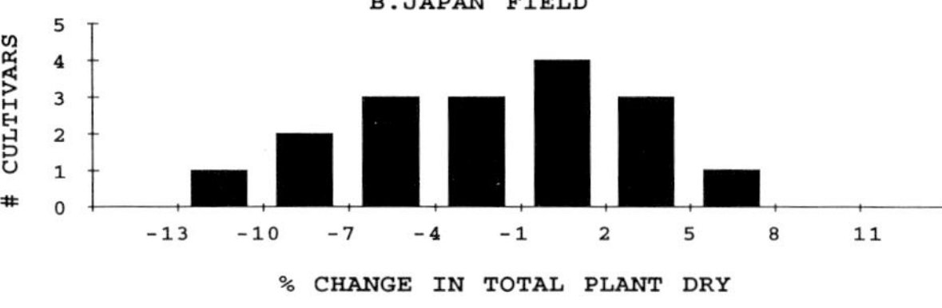

Fig. 2: Relative growth responses for rice cultivars exposed to enhanced UV-B in the field at different locations. Data are as % of check plants. (A) Data from the Philippines: total shoot weight response of 12 cultivars (Dai et al., personal communication, Olszyk et al., 1995). (B) Data from Japan: total plant dry weight response of 17 cultivars (Nouchi and Kobayashi, 1995). Number of cultivars in bar includes value to lower right of bar, e.g. range is 0 to 2 for middle IRRI bar.

the season, they did not report actual yield data, only panicle dry weights which were highly variable among cultivars. This study was conducted in the field with UV-B exposures corresponding to approximately 0.50 to 15.5 kJ m^{-2} d^{-1} UV-B_{BE} for enhanced UV-B compared to 0.23 to 6.2 kJ m^{-2} d^{-1} UV-B_{BE} for check plants. Exposures were with a modulated system which provided for enhanced UV-B output over the day in proportion to ambient UV-B levels.

Similar to the results in the field at IRRI and in Japan, Teramura et al. (1990) reported no effect of UV-B on yield of an Asian rice cultivar, ‹IR36,› based on studies in greenhouses. However, UV-B significantly reduced the growth enhancement due to elevated CO_2 when plants were exposed to both enhanced UV-B and elevated CO_2 compared to ambient UV-B and elevated CO_2.

Reasons for Difference in UV-B Response Between Field and Phytotron Plants

Lower impact of enhanced UV-B on rice plants in the field compared to phytotron may be related to factors including: (1) a lower increment of enhanced UV-B exposure, (2) older plants at a less UV-B susceptible growth stage, and (3) more light, both in terms of UV-A (which stimulates UV-B damage repair mechanisms) and possibly visible light.

In terms of exposure, there is a smaller net UV-B enhancement for field, compared to phytotron plants. For example, during dry seasons, field effects are based on an increase of $\approx$6 kJ m^{-2} d^{-1} UV-B_{BE} for enhanced UV-B compared to check plants, based on averages of $\approx$13 and 7 kJ m^{-2} d^{-1} UV-B_{BE} for enhanced UV-B and check plants, respectively. In contrast, phytotron effects are based on an increase of $\approx$13 kJ m^{-2} d^{-1} UV-B_{BE}, based on averages of $\approx$13 kJ m^{-2} d^{-1} UV-B_{BE} and 0 UV-B_{BE} for enhanced UV-B and check plants, respectively.

In terms of plant age, field plants are older than phytotron plants during most of their UV-B exposure period. Field plants are exposed to UV-B for $\approx$100 days from 15-to-115 days after seeding. In contrast, most phytotron responses are based on data from plants exposed to UV-B only during early vegetative growth. Phytotron plants normally are exposed to UV-B for 21 days, from 10-to-31 days after seeding. In the field, the only noticeable UV-B effects (increased UV-B absorbing pigments and decreased height) occurred during the early vegetative stage similar to the age of the plants in the phytotron.

In terms of light conditions, field plants have much higher levels of UV-A exposure than phytotron (glasshouse) plants, which has been suggested to be important in triggering UV-B repair mechanisms in plants (UNEP, 1994). Readings in the 290–303 nm portion of the spectrum (UV-B) were greater in the phytotron than in the field, but readings above 303 nm - ($\approx$UV-A) were greater in the field than in the phytotron (Fig. 3). The sum of UV-A readings was 7.5 times greater in the field than in the phytotron. The UV-A/UV-B_{BE} and UV-A/UV-B ratios in the field were 62 and 12, respectively, while in the glasshouse, the ratios were 8 and 3, respectively. Following this study, UV-A was monitored together with UV-B routinely in the field (IRRI, personal communication, data not shown).

Field plants also likely had higher visible radiation compared to phytotron plants. Visible radiation also has been as-

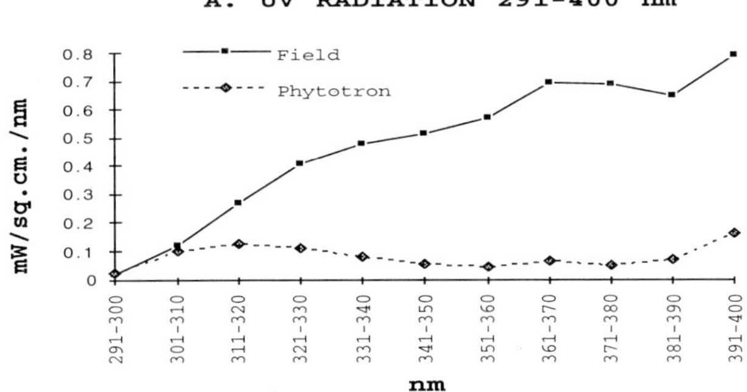

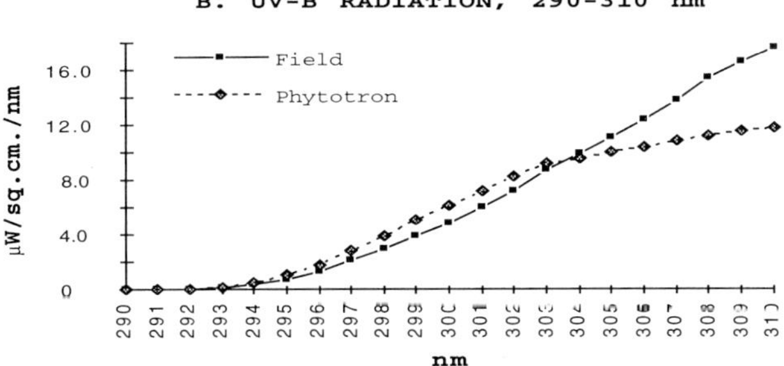

Fig. 3: Example of ambient UV radiation in field vs. phytotron (A) 291–400 nm, (B) 290–310 nm. UV-B_{BE} is 7.764 W cm^{-2} nm^{-1} in the field and 7.766 W cm^{-2} nm^{-1} in the phytotron for 290–320 nm.

sociated with UV-B repair mechanisms and decreased reduced severity of UV-B effects (UNEP, 1994). Kumagai and Sato (1992) demonstrated this for rice showing lower reductions in plant height, leaf number, biomass fresh and dry weights, and total chlorophyll content with enhanced UV-B and UV-C radiation, and higher compared to lower levels of visible light.

Biochemical and physiological responses of rice to UV-B

To better understand the basis for growth responses observed with enhanced UV-B in the phytotron studies, IRRI staff have measured of a wide range of anatomical, physiological, and biochemical parameters. Some of these parameters were studied to evaluate possible UV-B protection and adaptive processes (UNEP, 1994). A few may indicate general responses to stress which may or may not be specific to UV-B (e.g. net photosynthesis, leaf soluble protein). Still others may be involved with indirect effects of UV-B on rice *vis-a-vis* interactions with other organisms (Caldwell and Flint, 1994b). In the phytotron most parameters showed decreases in response with enhanced UV-B relative to the checks, but in the field fewer parameters showed changes in response to enhanced UV-B (IRRI, personal communications, data not shown).

Though rice plants in general did not show any yield or growth changes with enhanced UV-B in the field at IRRI, they apparently did detect and respond to UV-B as shown by the relatively consistent increase in flavonoid compounds (measured by general UV-B absorbance) with enhanced UV-B vs. check plants (Dai et al., 1995; IRRI, personal communications, data not shown). For example, during the 1993 wet season flavonoid concentration was 14 % higher for enhanced UV-B across mid-tillering and panicle initiation stages for both cultivars. There was a general trend towards greater increases in flavonoid concentrations with enhanced UV-B for UV-B susceptible compared to UV-B tolerant cultivars for several harvests and growth stages. This suggested at least an association between UV-B absorbing pigments and UV-B resistance in rice as reported by Nouchi et al. (1995) and He et al. (1993, 1994).

Recently there has been a focus on metabolites involved with free radical production and detoxification as a possible mechanism for less UV-B effect in the field compared to phytotron. Preliminary results from the field indicated changes in indicators of free radical metabolism. There was a reduction of SOD, CA, and AP activities at panicle initiation for both the UV-B tolerant cultivar IR72 and UV-B susceptible cultivar IR74 except that AP activity of IR72 was increased by 19.2 % (IRRI, personal communications, data not shown). At the flowering stage, UV-B radiation did not affect SOD and CA activities significantly, but decreased AP activity by 32.3 % for IR74. A significant accumulation of $O_2^{-\cdot}$ in IR74 leaves was observed at panicle initiation after UVB radiation, while MDA and H_2O_2 contents were increased only by 15.0 % and 8.4 %, respectively. The UV-B treatment also increased activated oxygen scavenger levels by 14.3 % for ASA and 34.4 % for GSH.

Phytotron studies also indicated a wide range of responses in indicators of free radical metabolism. The MDA content for the leaves of two rice cultivars (UV-B susceptible ‹IR74› and ‹Naisarsail›) over 20 days of exposure to 13.0 kJ m^{-2} d^{-1} UV-B_{BE} compared to ≈ 0 kJ m^{-2} d^{-1} UV-B_{BE} check plants (Liu et al., 1994). Enhanced UV-B also was associated with an increased SOD content in leaves during the first 15–20 days of the exposure, but not at the end of the exposure. The changes in MDA and SOD content varied somewhat between cultivars. A range of enhanced UV-B levels increased $O_2^{-\cdot}$ generation rate, and MDA and H_2O_2 contents in the leaves of rice cultivars (UV-B susceptible ‹IR74› and ‹Dular›) after 28 days of UV-B exposure (IRRI, personal communications, data not shown). Change was greatest for $O_2^{-\cdot}$ generation rate, followed by MDA and H_2O_2. Activities of SOD and CA were reduced, but the decline in SOD was lower than in CA under high UV-B (13.0 and 19.1 kJ m^{-2} d^{-1} UV-B_{BE}). High UV-B decreased ASA but increased GSH levels simultaneously. Cultivar IR74 had higher levels of SOD, CA, ASA, and GSH and lower levels of $O_2^{-\cdot}$, MDA and H_2O_2 compared with Dular. Low UV-B (6.0 kJ m^{-2} d^{-1} UV-B_{BE}) did not affect these parameters compared to ≈ 0 UV-B_{BE}.

In the phytotron, ABA and IAA contents in rice leaves were studied as possible links between UV-B exposure and growth effects (Zhang et. al., 1994). For ABA, two weeks of exposure to all levels of UV had no effect on the UV-B susceptible cultivar ‹IR74›, except for an increase with 19.1 kJ m^{-2} d^{-1} UV-B_{BE} compared to 0 UV-B_{BE}. Another UV-B susceptible cultivar ‹Naizersail› showed a slight increase in ABA content with 13.0 and 19.1 kJ m^{-2} d^{-1} UV-B_{BE}. Four weeks of exposure increased ABA contents for both cultivars at all enhanced UV-B levels. For IAA, two weeks of exposure resulted in a decrease in content for IR74 to a similar extent with 6, 13, or 19.1 kJ $m^{-2} d^{-1}$ UV-B_{BE} compared to 0 UV-B_{BE}. Naizersail had a significant decrease in IAA only with 19.1 kJ $m^{-2} d^{-1}$ UV-B_{BE} compared to 0 UV-B_{BE}. Four weeks of exposure resulted in a decrease in IAA contents for both cultivars at all enhanced UV-B levels.

Effects of UV-B on rice diseases, insects, and weeds

Since the rice plant *per se* shows little direct response to enhanced UV-B the field, the principle potential route for yield effects may be through interactions with other organisms. The primary indirect effect of investigated in the IRRI/EPA project concerns impacts of enhanced UV-B on the rice-blast disease complex caused by the fungus *Pyricularia grisea.* Rice blast was chosen because (1) it is important globally, (2) host resistance currently protects rice from losses and any alterations in this resistance could adversely affect rice production, (3) modeling work exists which can be used to evaluate the impacts of UV-B on rice blast on a regional basis, and (4) there was one report of research which indicated that rice blast could be affected by UV-B (Honda and Nemoto, 1985).

The IRRI UV-B and blast research has the following components: (1) preliminary studies on the effects of UV-B on disease severity using different cultivars of rice and isolines of *P. grisea,* (2) determination of the effects of UV-B on the fungus itself including sporulation, colony size, frequency of mutations, and isoline susceptibility, (3) determination of the effects of UV-B on multiple cycles of the disease, and (4) development and testing of a model for estimating and spatial analysis of potential effects of UV-B on disease severity in several Asian countries.

Initial studies with rice blast began at IRRI (Finckh et al., 1995) to determine the effect of UV-B exposure (≈ 0 or 17 kJ $m^{-2} d^{-1}$ UV-B_{BE}) on the response of rice to subsequent blast inoculation. The UV-B response of 36 combinations of 17 rice cultivars and 4 blast isolates was determined by evaluating blast lesion number and size. The results were variable with 18 combinations having higher numbers of lesions with UV-B. However, there were only two statistically significant increases for blast susceptible cultivar-isolate combinations, and one cultivar-isolate combination had a significant reduction in lesion size.

UV-B preinolculation studies used two rice cultivars (‹IR30›, ‹IR72›), one blast isolate, and eight UV-B levels (≈ 0 to 17 kJ $m^{-2} d^{-1}$) (Finckh et al., 1995). For IR30 lesion size increased with increased UV-B, while for IR72, the number of lesions per plant and per area decreased with increased UV-B. Furthermore, for IR30 UV-B damage recovery was delayed, and for IR72 disease effects of UV-B on dry weight and leaf area were enhanced.

In terms of effects of UV-B on the blast fungus itself, most results are still preliminary and subject to verification. Initial findings Leung et al. (1995) indicated that fungal sporulation and spore viability were affected adversely by daily UV-B levels assumed for the tropics. The effects differed among fungal isolates. For example, in isolate Guy11, UV-B exposure to 5.6 kJ m^{-2} UV-B_{BE} (2 hour exposure @ 2.8 KJ $m^{-2} h^{-1}$) resulted in a significantly reduced germination rate, but not final germination percentage; and exposure to 8.4 kJ m^{-2} UV-B_{BE} (3 hour exposure) significantly reduced spore viability. Isolate Guy11 was more susceptible to UV-B than Po6-6.

A wide range of sensitivity was observed among field isolates from the Philippines (Leung et al., 1995). Twenty-two isolates from 15 clonal lineages were evaluated for spore viability and sporulation following exposure from 2.8 to 11.2 kJ m^{-2} UV-B_{BE} (2, 4, 6 or 8 hours exposure to 1.4 kJ $m^{-2} h^{-1}$). Sporulation generally was reduced by UV-B exposure, but due to variability in sporulation capacity of the isolates, no statistically significant differences among isolates in UV-B response were detected.

Ultraviolet-C radiation at wavelengths shorter than 280 nm (UV-C) is well-known as being harmful to microorganisms: thus experiments have been conducted on the relative susceptibility of fungal isolates to UV-C and UV-B. The experiments indicated no apparent relationship between susceptibility to UV-B and UV-C in blast isolates (Leung et al., 1995); suggesting that the mechanisms of tolerance to UV-B and UV-C (to 0.38 kJ/m^2) are different.

Preliminary results also indicated that UV-B could enhance the mutation rate at a nonlethal irradiance level (Leung et al., 1995). Exposure to UV-B increased the frequency of genetic mutations at the BUF locus (a pigment gene) among survivors of Guy11 exposed to UV-B. Because of these effects and the likely low exposure of fungus itself in the field, mutagenic, rather than lethal, effects of UV-B may present the greatest impact on the fungal pathogen population. In order to generalize the mutagenic effects of UV-B on the blast fungus, mutation in genes directly affecting host specificity are being studied.

Current experiments focus on factors which could affect the response of blast and rice to blast under conditions more representative of the field (H. Leung, personal communication, data not shown). For example, photoreactivation of UV-B repair mechanisms by longer wavelengths of light may moderate the mutagenic effect of UV-B. Preliminary studies conducted with polychromatic light and UV-B suggested no detectable influence of polychromatic light on the lethal effect of UV-B on the blast fungus.

Existing data indicate that the rice blast fungus shows growth inhibition (spore viability and sporulation) and enhanced mutations in response to UV-B exposure, and that these responses could affect the disease cycle. However, given the size of the pathogen population it is unknown if UV-B impacts on viability and sporulation would affect the epidemiology of the disease. This is being tested in growth chambers. In preliminary experiments (H. Leung, personal communication, data not shown) seedlings of blast-susceptible rice line Co39 were exposed to 11.2 kJ m^{-2} UV-B_{BE} (1.5 kJ $m^{-2} h^{-1}$ for 7.5 hours) and subsequently inoculated with isolate Po6-6. Disease lesions were than classified as grayish – susceptible (able to sporulate) or brown – resistant (no or limited spores produced). Preliminary results showed that the number of brown lesions were higher under UV-B exposure than the check.

To test computer simulation methodology and spatial analysis and illustrate resulting potential impacts of UV-B, the blast model BLASTSIM (Calvero and Teng, 1991) was linked to the rice model CERES (Crop Environment Resource Synthesis)-RICE (Godwin et al., 1990) to produce a model simulating the impacts of both UV-B and blast on rice yield and disease incidence (Luo et al., 1994). Rice parameters (net assimilation rate; leaf area; leaf, sheath, culm, and root dry weights; grain weight, and length of longest root) were modified for UV-B responses of cultivars IR30 and IR74 based on data from short-term phytotron studies (Dai et al., 1992). Initially, all plant components were considered in more complex coupling analysis, however, single coupling whereby only net assimilation rate was affected by UV-B was best for explaining UV-B effects on rice growth (Luo et al., 1994). It was assumed that UV-B produced a 9–10 % reduction in net assimilation rate. The blast parameters (lesion expansion rate, lesion number, and sporulation rate) were modified based on data blast parameters for isolate Po6-6 on cultivars IR30 and IR72 from Finckh et al. (1994) and Leung et al. (1995). Both the rice plant and blast effects were based on responses from plants or blast receiving a high level of UV-B, e.g., ≈ 13 kJ $m^{-2} d^{-1}$ UV-B_{BE} vs. plants receiving 0 UV-B_{BE}.

Preliminary estimates of impacts of UV-B and blast on rice yield and disease severity have been made for five countries in Asia. The results are based on actual climatic data over 30 years and are part of a larger data set where impacts were evaluated with a series of scenarios reflecting changes in average temperatures (Luo et al., 1994; IRRI, personal communication, data not shown). Over all 53 sites in 5 countries, yield loss and average maximum disease severity were greatest with the combined UV-B and blast stresses. The data suggest a potential interaction between UV-B and blast, i.e. the joint effects of UV-B and blast produced a 13 ± 2 % (average ± standard deviation) yield loss, which was slightly greater than the additive effects of UV-B (10 ± 0.2 %) and blast (1 ± 2 %) alone. However, statistical analysis is necessary to determine if interactions are, in fact, greater than additive or «synergistic». For average disease severity, the joint effects of UV-B and blast produced an average index of 19 ± 7 compared to 9 ± 3 for blast alone.

Luo et al. (1994) evaluated the overall methodology and results for the Philippines in detail. Yield loss caused by UV-B alone was 9–10 % (same as assumed reduction in net assimilation rate), and yield loss caused by blast and UV-B combined was 15–20 % across a range of climatic temperature conditions. Measures of disease severity were more than doubled with blast and UV-B combined. Luo et al. (1994) concluded that to further test the model, data are needed on UV-B responses for additional parameters and rice cultivars, and on blast epidemics under enhanced UV-B. Since such data are difficult to obtain they also concluded that «predictions of global change effects [including UV-B] on blast must remain tentative».

Studies on the effects of UV-B on an insect pest of rice (rice leaffolder, *Cnaphalocrocis medinalis*) on rice plants have been carried out at IRRI through a collaborative project with Dr. Malcolm Whitecross of the Australian National University. Insects were either placed on intact, 60-day-old plants or were fed leaves taken from 22-day old plants exposed to enhanced UV-B (15 kJ m^2) in the IRRI phytotron. Preliminary results (Caasi-Lit et al., personal communication) indicate that leaffolder larvae fed leaves from UV-B exposed plants had a significant reduction in fifth instar larval weight and leaf area consumed compared to insects fed leaves from check plants. Studies with intact plants indicated longer larval periods, lower larval weights, and reduced pupal period lengths compared to check plants. The adverse effects on rice leaffolders were associated with the higher phenolic content of UV-B exposed host rice plants. Thus, the rice leaffolders had a preference for UV-B susceptible rice cultivars which had lower phenolic contents than UV-B tolerant cultivars.

Four weed species found in the lowland ecosystem (*Leptochloa chinensis, Sphenochlea zeylanica, Cyperus difformis,* and *Cyperis iria*) are being grown in an outdoor experiment under ambient and enhanced UV-B conditions (approximately 6 and 13 kJ $m^{-2} d^{-1}$ UV-B_{BE}, respectively). Key growth parameters are being measured: leaf, stem, root, and panicle dry weights; leaf area; and plant height. These parameters will be used to parameterize INTERCOM (rice-weed competition) model for UV-B responses. INTERCOM is based on a simple model which relates crop loss to relative leaf area of weeds within several weeks of crop emergence (Kropff and van Laar, 1993).

Conclusions

Based on field studies at IRRI and other locations, rice yields likely will not be directly affected by enhanced UV-B at levels used in those studies. This conclusion is based on data for direct effects on rice plants. Indirect impacts of enhanced UV-B on rice *vis-a-vis* interactions among rice and other ecosystem components are possible, but data on such effects are preliminary and subject to much uncertainty in terms of applicability to field conditions.

Phytotron studies indicate that a wide array of biochemical, physiological, anatomical, and growth effects can occur with high levels of UV-B. However, few of these responses are found in the field, possibly due to differences in UV-B exposure in the field compared to the phytotron, e.g. lack of UV-A in the phytotron.

Possible UV-B effects still could occur to rice plants under particular circumstances, e.g. if a highly UV-B susceptible rice cultivar was used in a region with high UV-B exposures, or UV-B altered rice susceptibility to factors such as diseases. However, based on the wide range in resistance found in the phytotron studies, even if effects were to occur, breeding-for resistance to UV-B and or other factors such as blast should be able to alleviate any impacts.

Based on this information, policy makers, and others interested in rice and UV-B, will have the tools to begin to assess mitigation and adaptation options to maintain rice productivity if significant damage were to occur. The rice/blast modeling effort illustrates a process which could be expanded to include effects of UV-B and other components of the rice ecosystem to carry out a risk assessment of the impacts of UV-B on a regional basis for key rice producing areas of Asia, by integrating experimental data, ecosystem modeling, and spatial analysis (Olszyk, 1994).

References

Barnes, P. W., S. Maggard, S. R. Holman, and B. S. Vergara: Intraspecific variation in sensitivity to UV-B radiation in rice. Crop Sci. *33,* 1041–1046 (1993).

Basiouny, F. M.: Sensitivity of corn, oats, peanuts, rye, sorghum, soybean and tobacco to UV-B radiation under growth chamber conditions. J. Agron. Crop Sci. *77,* 1354–1360 (1986).

Caldwell, M. M.: Solar UV radiation and the growth and development of higherplants. In: Geise, A. C. (ed.): Photophysiology Vol. *6,* Academic Press, New York (1971).

Caldwell, M. M. and S. D. Flint: Solar ultraviolet radiation and ozone layer change: implications for crop plants. In: Boote, K. J., J. M. Bennett, T. R. Sinclair, and G. M. Paulson (eds.): Physiology and Determination of Crop Yield, pp. 487–507. American Society of Agronomy, Madison, WI. (1994 a).

– – Atmospheric ozone reduction, solar UV-B radiation and terrestrial ecosystems. Climate Change *28,* 375–394 (1994 b).

Calvero, S. B. and P. S. Teng: Computer simulation of tropical rice-leaf blast pathosystem using BLASTSIM.2. In: American Society of Plant Pathology, Annual Meeting (1991).

Dai, Q., B. S. Vergara, A. Q. Chavez, and S. Peng: Response of rice plants from different regions to ultraviolet-B radiation. Intl. Rice Res. Notes *19,* 15–16 (1994 a).

Dai, Q., S. Peng, A. Q. Chavez, and B. S. Vergara: Intraspecific response of 188 rice cultivars to enhanced ultraviolet-B radiation. Environ. Exp. Bot. *34,* 433–442 (1994 b).

Dai, Q., S. Peng, A. Q. Chavez, and B. S. Vergara: Effect of enhanced UV-B radiation on growth and production of rice under greenhouse and field conditions. In: Peng, S., K. T. Ingram, H.-U. Neue, and L. H. Ziska, (eds.): Climate Change and Rice, pp. 242–257. Springer-Verlag, Berlin (1995, in press).

Dai, Q., V. Coronel, B. Vergara, P. Barnes, and A. Quintos: The influence of ultraviolet-B radiation on the growth and physiology of four rice cultivars. Crop Sci. *32,* 1269–1274 (1992).

Finckh, M. R., A. Chavez, Q. Dai, and P. S. Teng: Effects of enhanced UV-B radiation on the growth of rice and its susceptibility to rice blast under glasshouse conditions. Agric. Ecosys. Environ. *52,* 223–233 (1995).

Godwin, D. C., U. Singh, R. J. Buresh, and S. K. de Datta: Modeling of nitrogen dynamics in relation to rice growth and yield. In: 14th International Congress Soil Science Transactions, Kyoto, Japan, Aug. 1990, International Society of Soil Science Vol. IV, pp. 320–325 (1990).

He, J., L.-K. Huang, W. S. Chow, M. I. Whitecross, and J. M. Anderson: Effects of supplementary ultraviolet-B radiation on rice and pea plants. Aust. J. Plant Physiol. *20,* 129–142 (1993).

– – – – – Responses of rice and pea plants to hardening with low doses of ultraviolet-B radiation. Aust. J. Plant Physiol. *21,* 563–567 (1994).

Hondo, Y. and M. Nemoto: Control of seedling blast of rice with ultraviolet absorbing vinyl film. Plant Dis. *69,* 596–598 (1985).

IRRI: IRRI Rice Almanac. International Rice Research Institute (IRRI), Los Baños, the Philippines. 142 pp. (1993).

Kropff, M. J. and H. H. van Laar (eds.): Modeling Crop-Weed Interactions. CAB International, Wallingford, Oxon, UK, The International Rice Research Institute, Los Baños, The Philippines, 274 pp. (1993).

Kumagai, T. and T. Sato: Inhibitory effects of increase in near-UV radiation on the growth of Japanese rice cultivars (*Oryza sativa* L.) in a phytotron and recovery by exposure to visible radiation. Jpn. J. Breed. *42,* 545–552 (1992).

Leung, H., D. Christian, P. Loomis, and N. Bandian: Effects of ultraviolet-B radiation on spore viability, sporulation, and mutation of the rice-blast fungus. In: Peng, S., K. T. Ingram, H.-U. Neue, and L. H. Ziska (eds.): Climate Change and Rice, pp. 204–216. Springer-Verlag, Berlin (1995, in press).

Liu, X., Q. Dai, S. Peng, and B. S. Vergara: Lipid peroxidation and superoxide dismutase activity in rice leaves as affected by ultraviolet-B radiation. Intl. Rice Rese. Notes *19,* 54–55 (1994).

Luo, Y., D. O. Tebeest, P. S. Teng, and N. G. Fabellar: Risk analysis of rice leaf blast epidemics associated with effects of enhanced ultraviolet-B and temperature changes in the Philippines. Intl. Rice Res. Notes *19,* 54–55 (1994).

Nouchi, I. and K. Kobayashi: Effects of enhanced ultraviolet-B radiation with a modulated lamp control system on growth of 17 rice cultivars in the field. J. Agric. Meteor. *51,* 11–20 (1995).

Olszyk, D. M.: UV-B effects on terrestrial ecosystems. In: Kodama, Y. and S. D. Lee (eds.): Proceedings of the 13th UOEH International Symposium & Pan Pacific Cooperative Symposium on Impact of Increased UV-B Exposure on Human Health and Ecosystem, pp. 67–77. U. Occupational and Environmental Health. Kitakyshu, Japan (1994).

Olszyk, D. M. and K. T. Ingram: Effects of UV-B and global climate change on rice production: the EPA/IRRI cooperative research plan. In: Ilyas, M. (ed.): Ozone Depletion, Implications for the Tropics, pp. 234–253. U. Science Malaysia and U.N. Environment Programme, Penang, Malaysia (1991).

Rowland, F. S.: Chlorofluorocarbons and the depletion of stratospheric ozone. Amer. Sci. *77,* 36–45 (1989).

Sato, T. and T. Kumagai: Cultivar differences in resistance to the inhibitory effects of near-UV radiation among Asian ecotype and Japanese lowland and upland cultivars of rice (*Oryza sativa* L.). Jpn. J. Breed. *43,*61–68 (1993).

SCOPE (Scientific Committee on Problems of the Environment: Effects of Increased Ultraviolet Radiation on Biological Systems. SCOPE, Paris, France (1992).

Teramura, A. H.: Effects of ultraviolet radiation on the growth and yield of crop plants. Physiol. Plant. *58,* 415–427 (1983).

– Implications of stratospheric ozone depletion upon plant production. HortScience *25,* 1557–1560 (1990).

Teramura, A. H., L. H. Ziska, and A. Ester Sztein: Changes in growth and photosynthetic capacity of rice with increased UV-B radiation. Physiol. Plant. *83,* 373–380 (1991).

Teramura, A. H., J. H. Sullivan, and L. H. Ziska: Interaction of elevated ultraviolet-B radiation and CO_2 on productivity and photosynthetic characteristics in wheat, rice, and soybean. Plant Physiol. *94,* 470–475 (1990).

UNEP. Environmental Effects of Ozone Depletion: 1994 Assessment. United Nations Environment Programme (UNEP) (1994).

Van, T. K., L. A. Garrard, and S. H. West: Effects of UV-B radiation on net photosynthesis of some crop plants. Crop Sci. *16,* 715–718 (1976).

Zhang, J., S. Huang, Q. Dai, S. Peng, and B. S. Vergara: Effect of elevated ultraviolet-B radiation on abscisic acid and indoleacetic acid content of rice leaves. Intl. Rice Res. Notes *19,* 56–57 (1994).

J. Plant Physiol. Vol. 148. pp. 35–41 (1996)

The Potential Sensitivity of Tropical Plants to Increased Ultraviolet-B Radiation

LEWIS H. ZISKA[1,*]

[1] International Rice Research Institute, P.O. Box 933, 1099 Manila, Philippines

Received June 24, 1995 · Accepted October 27, 1995

Summary

Little is known concerning the impact of stratospheric ozone depletion and increasing ultraviolet (UV)-B radiation on the phenology and growth of tropical plants. This is because, ostensibly, tropical plants are already exposed to relatively high levels of UV-B radiation (relative to a temperate environment) and should, therefore, possess a greater degree of tolerance to increased UV-B radiation. In this brief review I hope to show that, potentially, direct and indirect effects on photosynthesis, assimilate partitioning, phenology and biomass could occur in both tropical crops (e.g. cassava, rice) and native species (e.g. *Cecropia obtusifolia* (Bertol. Fl)., *Tetramolopium humile* (Gray), *Nana sandwicensis* L.). However, it should be noted that differences in sensitivity to UV-B radiation can be related to experimental conditions, and care should be taken to ensure that the quantity and quality of background solar radiation remains at near ambient conditions. Nevertheless, by integrating current and past studies on the impact of UV-B radiation on tropical species, I hope to be able to demonstrate that photosynthesis, morphology and growth in tropical plants could be directly affected by UV-B radiation and that UV-B radiation may be a factor in species and community dynamics in natural plant populations in the tropics.

Key words: Ultraviolet (UV)-B radiation, stratospheric ozone, photosynthesis, ecotypic differentiation, assimilate partitioning, photosynthetic flux density.

Introduction

Stratospheric Ozone Depletion and UV-B Effects on Plant Function

Plants, by necessity, require light for photosynthesis and growth and therefore are exposed to varying amounts of UV-B radiation (i.e. radiation between 280–320 nm). The recent increase in atmospheric chlorofluorocarbons (CFCs), methane (CH_4) and nitrous oxides (N_xO) may lead to significant reductions in global stratospheric ozone. A decrease in stratospheric ozone would, in turn, result in a significant increase in UV-B radiation (relative to current ambient levels) experienced by plants.

Although only a small fraction of the total electromagnetic spectrum, UV-B radiation is sufficiently actinic to evoke a range of photobiological effects (Caldwell 1971). This is due in part to absorption of UV-B radiation by macromolecules such as nucleic acids, proteins, pigments, (and to a lesser extent) lipids.

Most studies which have examined the impact of supplemental UV-B radiation have done so using agronomically important crops, principally from temperate regions (e.g. soybean, cucumber). In these studies, a number of principle effects of UV-B radiation on plant function have been observed. These include direct damage to the photosynthetic apparatus as evidenced by decreases in the Hill reaction and subsequent declines in Photosystem II activity (Wilson and Greenberg 1993, Greenberg et al. 1989, Renger et al. 1989, Noorudeen and Kulandaivelu 1982, Vu et al. 1981) as well as reductions in Ribulose 1,5-bisphosphate carboxylase/oxygenase (Rubisco) (Strid et al. 1990, Vu et al. 1984). Reductions

* New address:
Dr. Lewis H. Ziska, Plant Physiologist, USDA-ARS, Climate Stress Laboratory, Bldg. 046A, 10300 Baltimore Avenue, Beltsville, MD 20705, U.S.A.

in photosynthetic capacity may also directly affect plant growth and development. It is fairly common to observe stunting of seedling growth following irradiation with UV-B radiation (Teramura 1983). Overall, plant growth (as determined by final dry weight) is often significantly reduced by UV-B radiation (Teramura 1983). In addition, alteration of secondary chemistry, specifically in the production of flavonoid compounds (which absorb in the UV-B region) has also been observed (cf. Caldwell et al. 1983 a, Tevini et al. 1991). It has been proposed that flavonoids and other phenolics could effectively screen the amount of UV-B radiation reaching internal plant tissue (Tevini et al. 1991).

Tropical Systems and UV-B radiation

Given the implementation of the Montreal Protocol (which phases out CFC production in industrialized countries), ozone depletion should, eventually, stabilize. However, because CFCs have such a long atmospheric lifetime (ca. 100 years) and because CFC use is increasing in countries not party to the protocol, it is premature to consider the threat of continued ozone depletion to be ended. In fact, recent measurements have indicated a marked reduction in ozone reduction within 15° of the equator from 1979 to 1992 (Madronich and de Gruijl 1993). It is conceivable therefore, that significant increases in UV-B radiation at tropical latitudes are a possibility.

Tropical plant environments are unique for several reasons. They permit a high degree of species diversity (Erwin 1983), are instrumental in reducing soil erosion and maintaining hydrological cycles (Salati and Vose 1984) and serve as an economic base for timber and mineral extraction (Bodley and Benson 1979). These environments include some of the most fertile ecosystems on the planet and account for an estimated 46 % of the living terrestrial carbon pool (Brown and Lugo 1982).

Even if no ozone depletion occurs, plants growing in the tropics (i.e. 20° N to 20° S) already experience high ambient levels of UV-B radiation compared to temperate regions. Ostensibly, since tropical plants reside in an environment which possesses an already naturally high amount of UV-B radiation (because of the shorter solar angle over the tropics), tropical plants are thought to have an inherent resistance to UV-B radiation. However, tropical plants are exposed to a wide range of ambient UV-B radiation as a result of light gaps or changes in elevation (Caldwell et al. 1980). Because of the large number of plant species, alterations in light quality with increased UV-B could, potentially, have a greater impact on plant productivity and species diversity. Recent UV-B exclusion experiments for tropical tree species suggest that subtle changes in growth and morphology could influence competitiveness in a range of microenvironments (e.g. light gaps) (Searles et al. 1995). Other work on tropical crops and native species does suggest that tropical plants and plant systems may be vulnerable to continued increases in UV-B radiation associated with stratospheric ozone depletion (Ziska et al. 1993, Ziska et al. 1992, Teramura et al. 1991). Clearly, the degree of vulnerability will vary among tropical plant communities; yet, we know very little with respect to how tropical plants will respond to UV-B radiation. To date, the majority of UV-B radiation research has focused on temperate crop species.

Methodological Considerations

In many experiments on UV-B radiation tolerance, additional UV-B radiation is provided by placing plants under racks of UV-B lights. The amount of supplemental UV-B is then controlled by adjusting the height of the UV-B lamps at a fixed distance above the tops of the plants during the experiment. The spectral irradiance reaching the leaf surface is weighted with the generalized plant response action spectrum (Caldwell 1971), integrated over wavelength and normalized to 300 nm to obtain the daily biological effective fluence ($UV\text{-}B_{BE}$). The timing of UV-B radiation is usually as a two-step square wave, that is, UV lights are turned on and off each day at preset times in the morning and afternoon. Control racks are also provided but in this instance the lamps are wrapped in mylar to prevent the transmittance of UV-B radiation. Models of simulated UV-B radiation (e.g. Green et al. 1980, Björn and Murphy 1985) are then used to estimate the treatment dose. Such models do not always predict total UV-B irradiation accurately (see Sullivan et al. 1994 for a discussion).

This square-wave type of experimental UV-B irradiation does not precisely match solar spectral characteristics, and can lead to significant increases in UV-B radiation relative to solar background, especially on cloudy days. Many growth chamber and glasshouse studies are derived from square wave experiments using low levels of photosynthetic flux density (PFD 400–700 nm) in combination with unrealistically high levels of UV-B radiation. This can lead to substantial errors in regard to estimating the sensitivity of a particular variable to increased UV-B. In cassava for example, no effect of UV-B radiation is observed for total biomass under field conditions even at very high UV-B radiation levels, while reductions in growth chamber studies for this and other variables are quite dramatic (Table 1). The photosynthetic and growth response to UV-B radiation can be greatly altered by background solar radiation, particularly by UV-A radiation and blue light, both

Table 1: Comparison of growth chamber and field trials with cassava (*Manihot esculenta*) with and without supplemental UV-B radiation. Control and UV-B treatments for the growth chamber and field trials were 0 and 13.5 kJ m^{-2} and 8.4 and 13.9 kJ m^{-2}, respectively. PFD within the growth chamber however was <50 % that of natural sunlight. Harvest data were taken at 77 and 81 days after planting for the growth chamber and field trials, respectively. Additional details for the field trials are given in Ziska et al. 1993.

Variable	Growth Chamber		Field Trial	
	Control	UV-B	Control	UV-B
Photosynthesis ($\mu mol\ m^{-1}\ s^{-1}$)	8.6	5.6*	11.9	11.1
Leaf Area (cm^2)	967	223*	9,481	10,902
Leaf Dry Wt. (gms)	5.3	1.6*	42.4	44.9
Total Biomass (gms)	11.6	3.8*	125.7	118.9

* Indicates significant differences according to the Students paired t-test ($P \leq 0.05$, n = 12 plants) for either experiment.

of which may be necessary for photorepair of UV-B induced damage (cf. Middleton and Teramura 1994, Adamse and Britz 1992).

Alternative systems have been developed which supply UV-B radiation as a proportional supplement to ambient solar conditions (Caldwell et al. 1983b, Yu et al. 1991). These modulated systems can account for both daily and seasonal changes in UV-B fluence. In addition, because ambient UV-B continuously controls how much additional UV-B is added in these systems, modelled predictions of supplemental UV-B are de-emphasized. Comparisons between modulated and square-wave systems illustrate that the total amount of UV-B exposure and subsequent change in plant responses can vary depending on the system used (Sullivan et al. 1994). However, cost and ease of use still favor the utilization of square wave systems in many experiments.

As an alternative, other researchers have focused on comparing plants grown at ambient levels of UV-B radiation with those in which all UV-B radiation has been excluded. Exclusion or reduction of ambient UV-B radiation can take place through the use of selected plastics (e.g. Searles et al. 1995) or by simulating a tropospheric ozone layer (e.g. Tevini et al. 1988). These type of studies can be less costly to initiate and can be useful as an initial analysis of physiological and morphological effects. However, as with greenhouse studies, control plants may not receive any UV-B radiation at all, which is unrealistic for plants grown outdoors under a natural solar spectrum.

Differences in methodology alone should not be the basis for ignoring all previous data concerning plant sensitivity to UV-B radiation. Although previous experiments using square wave technology are not always useful for predicting gross changes in plant productivity as a consequence of stratospheric ozone depletion, this research is still relevant to identifying key processes which could be affected by UV-B radiation and can be used as a starting point for initial assessments. Nevertheless, it is clear that, whenever possible, field experiments should be emphasized. Although preliminary UV-B exclusion experiments would provide worthwhile data, long-term field experiments using modulated UV-B systems to determine the response of tropical plants and communities to increased UV-B are highly desirable.

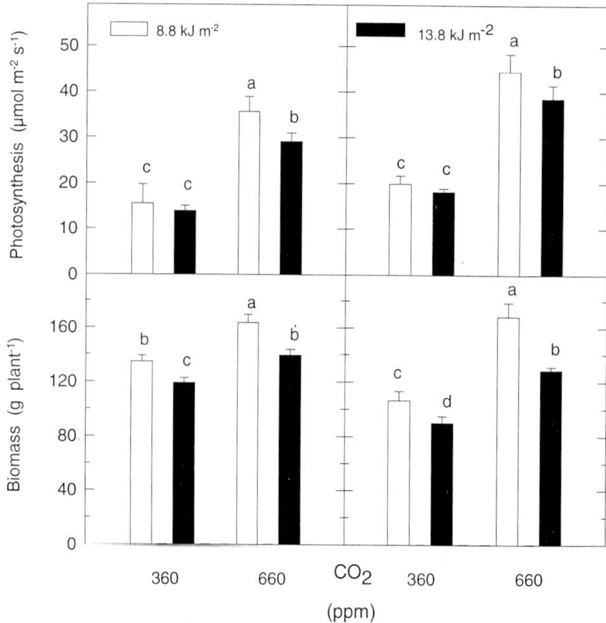

Fig. 1: Percent change in total plant dry biomass for the UV-B treatment (15.7 kJ m^2) compared to the no UV-B radiation control. *indicates significant difference relative to the 0 UV-B control (n = 10). Additional details are given by Teramura et al. 1991.

Case Studies of UV-B and Tropical Plants

Crops

Among tropical plants, important agronomic species include rice, sugar cane, maize and cassava. Rice is especially important, supplying approximately 1 billion people with 70 % of their caloric intake (De Datta 1981). Although rice is grown extensively in tropical areas, it is cultivated under a wide range of environments between latitudes 45° N and 40° S. It is possible, therefore, that rice growing at different latitudes may have developed different sensitivities to UV-B radiation. Knowledge of the range of sensitivities would be invaluable in selecting rice for UV-B tolerance.

In initial greenhouse trials, approximately one third of the 16 rice cultivars examined for UV-B sensitivity showed a significant reduction in total biomass with a simulated 20 % increase in UV-B radiation at the equator (15 kJ m^{-2}) (Fig. 1, Teramura et al. 1991). For these sensitive cultivars, leaf area and tiller number were reduced (Teramura et al. 1991). In these initial studies there was some indication that seeds which originated in already high UV-B areas (i.e. near the equator and/or at high elevations) could have a greater inherent tolerance to UV-B radiation. Later greenhouse studies which screened a large number of rice cultivars to high levels of UV-B radiation (13.0 kJ m^{-2}) also showed a significant reduction in total biomass for approximately one third of the cultivars examined (61 out of 188 cultivars) with 143/188 cultivars exhibiting reductions in plant height (Dai et al. 1994). In these studies, no correlation between sensitivity of plant biomass to UV-B and cultivar origin was observed (Dai et al. 1994). For both studies however, a number of rice cultivars were identified which are tolerant to very high levels of UV-B radiation.

In addition to CFCs, atmospheric carbon dioxide (CO_2) is also increasing. Because CO_2 is the sole source of carbon in plants, increases in photosynthesis and growth are anticipated for both temperate and tropical plants (cf. Ziska et al. 1991). Rice is one of the few plants species in which simultaneous increases in UV-B and CO_2 have been examined (Teramura et al. 1990, Ziska and Teramura 1992). Data obtained for two rice cultivars at the Duke University Phytotron indicate that a doubling of atmospheric CO_2 would result in significant increases in photosynthesis and plant biomass (Fig. 2). However, these increases were either eliminated or reduced when supplemental UV-B radiation corresponding to a 20 % deple-

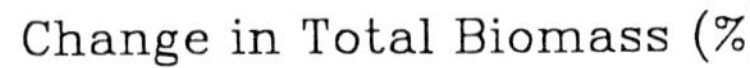

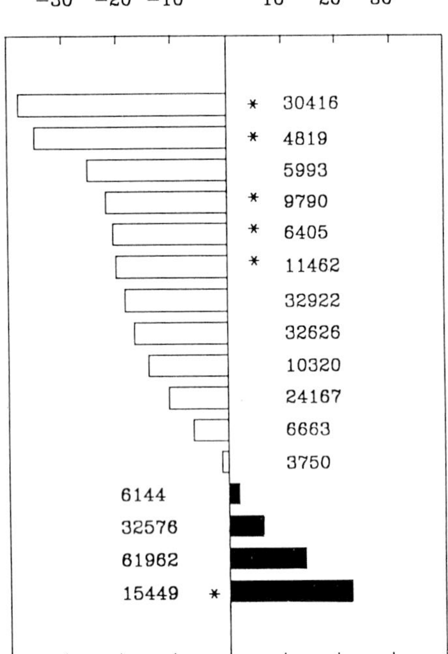

Fig. 2: Changes in photosynthesis and biomass for two rice cultivars, IR-36 (Traditional semi-dwarf indica and Fujisaka-5, Japonica type rice grown at two CO_2 concentrations (360 (ambient) and 660 ppm) and two levels of supplemental UV-B radiation (8.8 and 13.8 kJ m^{-2}). Different letters indicate significant differences separated by the Student-Newman-Keuls multiple range test at $P \leq 0.05$, n=10 for biomass, n=4 for photosynthesis.

Table 2: Changes in growth parameters (per plant) for cassava (*Manihot esculenta*) with and without supplemental increases in UV-B radiation (8.4 and 13.9 kJ m^{-2}, for the control and UV-B treatments, respectively). The UV-B supplement simulates a 15% depletion in stratospheric ozone at the equator using the Green et al. model (1980).

Variable	Control	UV-B
Plant height (m)	0.75	0.64
Total Leaf Area (m^2)	1.13	1.35*
Leaf Biomass (g)	51.9	56.1
Stem Biomass (g)	44.5	43.0
Root Biomass (g)	52.0	35.8*
Total Biomass (g)	148.4	134.9
Shoot/Root Ratio	2.3	3.2*

* Indicates significance at the $P \leq 0.05$ level according to the Student-Newman-Keuls multiple range test, n = 24. Additional details are given in the Materials and Methods section of Ziska et al. 1993.

tion in stratospheric ozone was added (i.e. 8.8 vs. 13.8 kJ m^{-2}). Clearly, additional information on how combinations of factors anticipated with environmental change could affect tropical plant productivity is desired.

Among tropical crops, cassava is one of the most important food crops, ranking fourth as a source of calories (Cock 1985). It is adapted to a wide range of climatic conditions and is frequently grown as a subsistence crop if other crops fail. Cassava grown in the field at temperate latitudes for a 3 month period and exposed to a 15% depletion in stratospheric ozone (as determined at the equator) showed no direct effect on photosynthesis as measured either by CO_2 exchange or by O_2 evolution (Ziska et al. 1993). However, differential partitioning of biomass was observed with a significant increase in shoot/root ratio (ca. 32%, Table 2). This increase could have been due in part, to the UV-B stimulation of axillary leaf production. Presumably, additional carbohydrate used in leaf development was supplied by the roots with a subsequent reduction in root weight with UV-B exposure. Since root production determines the harvestable portion of cassava, these data suggest that increases in UV-B radiation may still impact economic yield even if photosynthesis and total plant biomass are unaffected. In contrast, UV-B exclusion studies for cassava showed no alteration of root/shoot ratios (Searles et al. 1995).

Native Plants

There is significant natural variation in the daily effective UV-B irradiance reaching the earth's surface. This variation is the result of a natural latitudinal gradient in total atmospheric ozone column thickness, prevailing solar angles at different latitudes, elevation above sea level and an optical amplification effect (Caldwell et al. 1980). Adaptation to UV-B radiation may be best developed in species that occur in high UV-B irradiance environments such as low latitude, high elevation locations. In temperate regions it has been demonstrated that UV-B radiation can alter the competitive interactions among agricultural species and weedy competitors in crop systems by altering morphology and/or growth rate (Barnes et al. 1990, Gold and Caldwell 1983).

Studies which have compared the response of non-agronomic plants from tropical environments are rare. To date, only a single study has examined the impact of UV-B exclusion on native tree species in a tropical environment (Searles et al. 1995). In this study, 3 rainforest tree species (*Cecropia obtusifolia* [Bertol. Fl.], *Tetragastris panamensis* [Gaerth.], *Calophyllum longifolium* [Willd.] and 2 economically important species, (*Swietenia macrophylla* [King], *Manihot esculenta* [Crantz]) were grown at near ambient and zero UV-B radia-

Table 3: Percent reduction in total plant biomass for 5 tropical species exposed *in situ* to near ambient UV-B radiation relative to a no UV-B control (Searles et al. 1995). Study was conducted at Barro Colorado Island, Panama (9°N). Significant reductions in *Cecropia obtusifolia* were determined at the $P \leq 0.10$ level as determined by ANOVA (SAS PC version 6.04, SAS Institute, Inc. Cary NC, USA). Additional details can be found in the materials and methods section of Searles et al. 1995.

Species	% Reduction
Cecropia obtusifolia	−22%*
Calophyllum longifolium	−19%
Tetragastris panamensis	−17%
Manihot esculenta	−14%
Swietenia macrophylla	−3%

Table 4: Changes in growth parameters for plants of the same genus (*Tetramolopium, rockii, humile*) and species (*Oenothera stricta*) found at high and low elevations and exposed to different levels of UV-B radiation. Values are means of 6–8 plants. Data were obtained from Figure 1 and Table 1 of Ziska et al. 1992. Data were re-analyzed using the Duncans Multiple Range Test (P ≤ 0.05). Additional details can be found in the Materials and Methods section of Ziska et al. 1992.

Species/Elevation	UV-B kJ m^{-2}	Shoot	Root	Flower	Total
		g			
Oenothera stricta (1,321 m)	0.0	5.53 a	0.95	0.71	7.19 a
	15.5	5.54 a	1.07	0.99	7.53 a
	23.1	3.64 b	0.91	0.40	4.97 b
Oneothera stricta (2,790 m)	0.0	4.22	1.49	0.08 b	5.80 b
	15.5	5.42	1.48	0.64 a	7.54 a
	23.1	5.08	0.95	1.07 a	7.09 a
Tetramolopium rockii (0–100 m)	0.0	14.65	2.32	1.59 a	18.57
	15.5	13.23	2.37	1.31 a	16.91
	23.1	13.49	2.03	0.67 b	16.19
Tetramolopium humile (2,959 m)	0.0	3.03	0.20	0.46	3.73 b
	15.5	3.39	0.29	0.48	3.76 b
	23.1	4.40	0.35	0.53	4.83 a

Table 5: Changes in photosynthetic characteristics for plants of the same genus (*Tetramolopium, rockii; humile,*) and species (*Oenothera stricta*) found at high and low elevations and exposed to different levels of UV-B radiation. Values are means of 6–8 plants. Data were obtained from Figure 3 and Table 3 of Ziska et al. 1992. Data were re-analyzed using the Duncans Multiple Range Test (P ≤ 0.05). Additional details can be found in the Materials and Methods section of Ziska et al. 1992.

Species/Elevation	UV-B kJ m^{-2}	Photosynthesis µmol O_2 m^{-2} s^{-1}	AQE µmol CO_2 µmol PPF^{-1}
Oenothera stricta (1321 m)	0.0	28.3 a	0.018 a
	15.5	26.7 ab	0.016 ab
	23.1	24.8 b	0.013 b
Oneothera stricta (2,790 m)	0.0	29.7	0.020
	15.5	29.4	0.017
	23.1	29.1	0.017
Tetramolopium rockii (0–100 m)	0.0	33.8 a	0.031 a
	15.5	25.1 b	0.021 ab
	23.1	23.8 b	0.018 b
Tetramolopium humile (2,959 m)	0.0	15.1	0.023
	15.5	15.2	0.024
	23.1	15.2	0.024

tion through the use of selected filters. For this trial, reduced plant height at ambient relative to zero UV-B radiation seemed to be the most consistent response, although a significant reduction in plant biomass was observed for *C. obtusifolia* after 61 days (Table 3).

If ambient UV-B radiation is a function of elevation, then the same genus or plant species growing at different elevations may have different sensitivities to UV-B irradiance. To determine this, 33 plant species grown from seed collected from populations along an elevational gradient in Haleakala Crater National Park in Hawaii (20° N) were grown for 12 weeks in unshaded greenhouses at the University of Maryland at UV-B fluences of 0, 15.5 or 23.1 kJ m^2. Reductions in total plant biomass with increased UV-B radiation was used as an indication of UV-B sensitivity. Although the responses varied among species, the general trend was for UV-B tolerance to increase with elevation. Native Hawaiian species which demonstrated reductions in biomass included *Lycium sandwicense* and *Nana sandwicensis* (Sullivan et al. 1992).

In a follow-up study, the growth and photosynthetic response to the same levels of UV-B radiation were examined in populations of the same genera or species from different elevations (Ziska et al. 1992). Populations of *Oenothera stricta* (Ledeb. ex Link). and *Tetramolopium rockii* (Sherff.) from lower elevations demonstrated reductions in shoot, floral and/or total plant biomass (Table 4) as well as significant reductions in maximum photosynthesis (as determined by an O_2 electrode) and apparent quantum efficiency (AQE) (Table 5). Interestingly, for plant populations from higher elevations, maximum growth was correlated with increasing UV-B radiation and in one case (*Oenothera stricta*) supplemental UV-B radiation was required for flowering to occur (Table 4). No reductions in either maximum photosynthesis or AQE were noted for high elevation populations with increasing UV-B radiation (Table 5). Increases in UV-B absorbing compounds (i.e. flavonoids) were noted for low elevation plants; however, plants from high elevations produced a consistently higher amount of these compounds even in the absence of UV-B radiation (Ziska et al. 1992). Elevational studies involving different plant populations suggest that plants from high ambient UV-B environments may have developed or maintained mechanisms which allow maximum growth reproductive and photosynthetic characteristics.

How Could UV-B Radiation Influence Tropical Ecosystems?

At present there is sufficient data to suggest that differential sensitivity to UV-B radiation could have occurred in response to natural UV-B radiation gradients for tropical plants. Consequently, the potential exists for a significant alteration in natural ecosystems if stratospheric ozone depletion continues. Given the paucity of available data, it is still difficult to generalize from individual plant studies to the tropical system as a whole; however, an attempt to speculate on potential scenarios within tropical communities should UV-B radiation increase has been made.

The creation and timing of light gaps within the forest canopy is recognized as a strong force for plant growth, species regeneration and community structure (Pickett 1983). In studies of the forest canopy in Costa Rica for example, ca. 75 % of the canopy species examined required a light gap for maturation and/or seed dispersal (Hartschorn 1980). When a light gap occurs, daily PFD can increase dramatically, from 1 to over 30 mol m^{-2} day^{-1}. Large increases in PFD will have important consequences for growth and development, but the concomitant increase in UV-B radiation could adversely affect growth and morphology. For example in the Searles study, interspecific differences in plant height were observed at UV-B radiation levels associated with light gaps (Searles et al. 1995). It is possible that other physiological processes including leaf development, photosynthetic capacity, reproductive effort could be altered by increased UV-B radiation with subsequent consequences for plant competition.

The occurrence and concentration of secondary chemicals appears to be higher in tropical than temperate plant environments, presumably because herbivore and pathogen selection pressures are greater in these systems (Lee et al. 1979). In studies with tropical mangrove, for example, it was shown that the phenolic content of leaves could vary as a function of UV-A and UV-B radiation (Lovelock et al. 1992). Phenolics and other UV-B absorbing compounds may have a myriad of uses as floral fragrances, seed toxins, chemical defenses etc. For deep shade species, anthocyanin on the undersurface of leaves may improve photosynthetic efficiency by back-scattering incoming light through photosynthetic tissue (Lee et al. 1979). Changes in secondary compounds with UV-B radiation in tropical systems could significantly alter the photosynthetic capacity, growth, phenology, morphology and herbivory of tropical plants with subsequent affects on competition and productivity.

Perhaps one of the best known characteristics of tropical systems is the extent of species diversification (Erwin 1983). Although the biotic and/or abiotic factors responsible for diversity are unknown, the response of temperate species to UV-B radiation does suggest a range of inter and intra-specific responses. This variation in the response to UV-B radiation can lead to changes in the competitive balance even in simple agro-ecosystems (Gold and Caldwell 1983). Different responses of tropical plants to UV-B radiation suggests potential changes in growth, reproduction and productivity with subsequent changes in the degree of plant diversification.

Future Research Priorities

Clearly, UV-B could play a role in both the physiology and ecology of tropical plants. UV-B may play a role in light gap formation, the occurrence and concentration of secondary compounds and even as an abiotic factor influencing species diversification.

Limited data from available studies suggests that photosynthesis, plant height and growth could be both directly and indirectly affected by UV-B radiation and that a range of potential adaptations to increased UV-B radiation may have already developed in tropical environments for different ambient UV-B levels associated with elevation. However, many of these studies have been conducted in greenhouses with no UV-B radiation as a control or with limited amounts of UV-A or blue light. It is unclear in rice for example, if the same degree of sensitivity to UV-B radiation would be observed in the field. Unfortunately, I am aware of no published reports which have examined the response of rice to UV-B radiation *in situ*.

Clearly, much more needs to be known in order to accurately assess the effects of UV-B radiation on tropical plant productivity and to make intelligent scientific and policy decisions as climate change continues. Fundamental information related to single leaves and whole plant processes will contribute greatly to filling gaps in our knowledge base. Information on the impact of increased UV-B radiation on ecosystem level processes such as nutrient recycling, production and distribution of secondary compounds, species distribution and plant competition are also required.

It also important to emphasize that future anthropogenic change will not be confined solely to increases in UV-B radiation, but will also include increases in CO_2 and potential increases in temperature. To date, almost no data exists on the interactions between UV-B radiation and temperature and while some initial work indicates that increased CO_2 could reduce the extent of UV-B induced damage in rice (Teramura et al. 1990, Ziska and Teramura 1992) and Jack Pine (*Pinus banksian* Lamb.) (Stewart and Hoddinott 1993), the mechanistic basis for this response is almost completely unknown.

Lastly, additional improvements in methodology should be made in regard to UV-B assessment studies in order to gain insight into plant response at ambient or near ambient levels of UV-A, UV-B and blue light. New and innovative field approaches are being tried including the use of modulated systems (e.g. Yu et al. 1991) and UV-B exclusion studies (Searles et al. 1995). However, the most important research needs can only be met by establishing *in situ* experiments with modulated supplemental UV-B radiation under tropical conditions, preferably on a long-term basis.

Acknowledgements

The author would like to express his appreciation to Drs. Joe Sullivan and Alan Teramura for their technical assistance and valuable insight in the writing of this review.

References

Adamse, P. and S. J. Britz: Amelioration of UV-B damage under high irradiance. I: Role of photosynthesis. Photochem. Photobiol. *56*, 645–650 (1992).

Barnes, P. W., S. D. Flint, and M. M. Caldwell: Morphological responses of crop and weed species of different growth forms to ultraviolet-B radiation. Amer. J. Bot. *77*, 1354–1360 (1990).

Björn, L. O. and T. M. Murphy: Computer calculation of solar ultraviolet radiation at ground level. Physiologie Vegetale *23*, 555–561 (1985).

Bodley, J. H. and F. C. Benson: Cultural ecology of Amazonian palms. Washington State University Laboratory of Anthropology, Reports of Investigations no. 56, Pullman, WA. (1979).

Brown, S. and A. E. Lugo: The storage and production of organic matter in tropical forests and their role in the global carbon cycle. Biotropica *14*, 161–187 (1982).

Caldwell, M. M.: Solar UV irradiation and the growth and development of higher plants. In: Geise, A. C. (ed): Photophysiology. Vol *6*, pp. 131–177. Academic Press, New York (1971).

Caldwell, M. M., M. R. Robberecht, and W. D. Billings: A steep latitudinal gradient of solar ultraviolet-B radiation in the arctic-alpine life zone. Ecology *6*, 600–611 (1980).

Caldwell, M. M., M. R. Robberecht, and S. D. Flint: Internal filters: prospects for UV acclimation in higher plants. Physiol. Plant. *58*, 445–450 (1983a).

Caldwell, M. M., W. G. Gold, G. Harris, and C. W. Ashorst: A modulated lamp system for solar UV-B (280–320) supplementation studies in the field. Photochem. and Photobiol. *37*, 479–485 (1983).

COCK, J. H.: Cassava: The plant and its importance. In: COCK, J. H. (ed.): Cassava: New Potential for a Neglected Crop, pp. 3–12, Westview Press, Colorado (1985).

DAI, Q., S. B. PENG, A. Q. CHAVEZ, and B. S. VERGARA: Intraspecific responses of 188 rice cultivars to enhanced UVB radiation. Environ. Exp. Bot. *4,* 433–442 (1994).

DEDATTA, S. K.: Rice in perspective. In: DEDATTA, S. K. (ed.): Principles and Practices of Rice Production, pp. 1–9. Wiley and Sons, New York (1981).

ERWIN, T.: Beetles and other insects of tropical forest canopies at Mahaus, Brazil, sampled by insecticidal fogging. In: SUTTON, S. L., T. C. WHITMORE, and A. C. CHADWICK (eds.): Tropical Rain Forest: Ecology and Management, pp. 59–75, Blackwell Scientific, Oxford (1983).

GOLD, W. G. and M. M. CALDWELL: The effects of ultraviolet-B radiation on plant competition in terrestrial ecosystems. Physiol. Plant. *58,* 435–444 (1983).

GREEN, A. E. S., K. R. CROSS, and L. A. SMITH: Improved analytical characterization of ultraviolet skylight. Photochem. Photobiol. *31,* 59–65 (1980).

GREENBERG, B. M., V. GABA, O. CANAANI, S. MALKIN, A. K. MATTOO, and M. EDELMAN: Separate photosensitizers mediate degradation of the 32-kDa photosystem II reaction center protein in the visible and UV spectral regions. P.N.A.S. *86,* 6617–6620 (1989).

HARTSCHORN, G. S.: Neotropical forest dynamics. Biotropica *12,* 23–30 (1980).

LEE, D. W., J. B. LOWRY, and B. C. STONE: Abaxial anthocyanin layer in leaves of tropical rain forest plants: Enhancer of light capture in deep shade. Biotropica *11,* 70–77 (1979).

LOVELOCK, C. E., B. F. CLOUGH, and I. E. WOODROW: Distribution and accumulation of ultraviolet radiation absorbing compounds in leaves of tropical mangroves. Planta *188,* 143–154 (1992).

MADRONICH, S. and F. R. DE GRUIJL: Skin cancer and UV radiation. Nature *366,* 23 (1993).

MIDDLETON, E. M. and A. H. TERAMURA: The role of flavonol glycosides and carotenoids in protecting soybean from ultraviolet-B damage. Plant Physiol. *103,* 741–752 (1994).

NOORUDEEN, A. M. and G. KULANDAIVELU: On the possible site of inhibition of photosynthetic electron transport by ultraviolet (UV-B) radiation. Physiol. Plant. *55,* 161–166 (1982).

PICKETT, S. T. A.: Differential adaptation of tropical tree species to canopy gaps and its role in community dynamics. Trop. Ecol. *24,* 64–84 (1983).

RENGER, G., M. VOLKER, H. J. ECKERT, R. FROMME, S. HOHM-VEIT, and P. GRABER: On the mechanism of photosystem II deterioration by UV-B irradiation. Photochem. Photobiol. *49,* 97–105 (1989).

SALATI, E. and P. B. VOSE: Amazon basin: A system in equilibrium. Science *225,* 129–137 (1984).

SEARLES, P. S., M. M. CALDWELL, and K. WINTER: The response of five tropical dicotyledon species to solar ultraviolet-B radiation. Amer. J. Bot. *82,* 445–453 (1995).

STEWART, J. D. and J. HODDINOTT: Photosynthetic acclimation to elevated carbon dioxide and UV irradiation in *Pinus banksiana.* Physiol. Plant. *88,* 493–500 (1993).

STRID, A., W. S. CHOW, and J. ANDERSON: Effects of supplementary ultraviolet-B radiation on photosynthesis in *Pisum sativum.* Biochim. Biophysica Acta *1020,* 260–268 (1990).

SULLIVAN, J. H., A. H. TERAMURA, and L. H. ZISKA: Variation in UV-B sensitivity in plants from a 3,000-m elevational gradient in Hawaii. Amer. J. Bot. *79,* 737–743 (1992).

SULLIVAN, J. H., A. H. TERAMURA, P. ADAMSE, G. F. KRAMER, A. UPADHYAYA, S. J. BRITZ, D. T. KRIZEK, and R. M. MIRECKI: Comparison of the response of soybean to supplemental UV-B radiation supplied by either square wave or modulated irradiation systems. In: BIGGS, R. H. and M. E. B. JOYNER (eds.): Stratospheric Ozone Depletion, pp. 211–220, Springer-Verlag, Berlin, Germany.

TERAMURA, A. H.: Effects of ultraviolet radiation on the growth and yield of crop plants. Phyisol. Plant. *58,* 415–427 (1983).

TERAMURA, A. H., J. H. SULLIVAN, and L. H. ZISKA: Interaction of elevated ultraviolet-B radiation and CO_2 on productivity and photosynthetic characteristics in wheat, rice and soybean. Plant Physiol. *94,* 470–475 (1990).

TERAMURA, A. H., L. H. ZISKA, and A. E. SZTEIN: Changes in growth and photosynthetic capacity of rice with increased UV-B radiation. Physiol. Planta. *83,* 373–380 (1991).

TEVINI, M. J., P. GRUSEMANN, and G. FIESER: Assessment of UV-B stress by chlorophyll fluorescence analysis. In: LICHTENTHALER, H. K. (ed.): Applications of Chlorophyll Fluorescence), pp. 229–238, Kluwer Press, Dordrecht, Germany (1988).

TEVINI, M. J., J. BRAUN, and G. FIESER: The protective function of the epidermal layer of rye seedlings against UV-B radiation. Photo. Photobio. *50,* 479–487 (1991).

VU, C. V., L. H. ALLEN Jr., and L. A. GARRARD: Effects of supplemental UV-B radiation on growth and leaf photosynthetic reactions of soybean (*Glycine max*). Physiol. Plant. *52,* 353–362 (1981).

– – – Effects of UV-B radiation (280–320 nm) on ribulose 1,5-bisphosphate carboxylase in pea and soybean. Env. Exp. Bot. *24,* 131–143 (1984).

WILSON, M. I. and B. M. GREENBERG: Protection of the D1 photosystem II reaction center protein from degradation in ultraviolet radiation following adaptation of *Brassica napus* L. to growth in ultraviolet-B. Photochem. Photobiol. *57,* 556–563 (1993).

YU, W., J. H. SULLIVAN, and A. H. TERAMURA: The YMT Ultraviolet-B irradiation system: Manual of operation. Final Report to US EPA, Environmental Research Laboratory, Corvallis, OR.

ZISKA, L. H., K. P. HOGAN, A. P. SMITH, and B. G. DRAKE: Growth and photosynthetic response of nine tropical species with long-term exposure to elevated carbon dioxide. Oecologia *86,* 383–389 (1991).

ZISKA, L. H. and A. H. TERAMURA: CO_2 enhancement of growth and photosynthesis in rice (*Oryza sativa*): Modification by increased ultraviolet-B radiation. Plant Physiol. *99,* 473–481 (1992).

ZISKA, L. H., A. H. TERAMURA, and J. H. SULLIVAN: Physiological sensitivity of plants along an elevational gradient to UV-B radiation. Amer. J. Bot. *79,* 863–871 (1992).

ZISKA, L. H., A. H. TERAMURA, J. H. SULLIVAN, and A. MCCOY: Influence of ultraviolet-B (UV-B) radiation on photosynthetic and growth characteristics in field-grown cassava (*Manihot esculenta* Crantz). Plant, Cell & Environ. *16,* 73–79 (1993).

J. Plant Physiol. Vol. 148. pp. 42–48 (1996)

Screening of Freshwater Algae (*Chlorophyta, Chromophyta*) for Ultraviolet-B Sensitivity of the Photosynthetic Apparatus

FUSHENG XIONG[1,2,3], FILIP LEDERER[1,2], JAROMÍR LUKAVSKÝ[2], and LADISLAV NEDBAL[1,4]

[1] Institute of Microbiology, National Research Centre Global Climate Change and Photosynthesis, Opatovický mlýn, CZ-37981 Třeboň, Czech Republic

[2] University of South Bohemia, Faculty of Biological Sciences, Branišovská 31, Č. Budějovice, Czech Republic

[3] on leave from: Yangzhou University, People's Republic of China

[4] to whom the correspondence should be sent

Received June 26, 1995 · Accepted October 24, 1995

Summary

Sixty-seven species of algae (*Chlorophyta, Chromophyta – Xanthophyceae*) were used in a screening experiment in which the UV-B sensitivity of the photosynthetic apparatus was assessed. The species were selected to represent largely different natural environments ranging from mountain snow and high elevation lakes on one extreme and shaded thermal springs and soil on another. The study was aimed at the identification of algal species whose photosynthetic apparatus was either extremely sensitive or resistant to enhanced UV-B radiation.

Samples of algae were exposed to $2\,W{\cdot}m^{-2}$ of artificial UV-B radiation for 2 hours. The response of the algae to the UV-B exposure was monitored by following changes in photosynthetic oxygen evolution rates and changes in the PAM fluorescence emission. The most UV-B susceptible algal species lost 30–50 % of their oxygen evolving capacity during the UV-B exposure. On the other extreme, UV-B exposure stimulated the oxygen evolving capacity of some algal species by as much as 20 %. The emission of chlorophyll fluorescence from algae was also modified by the UV-B exposure with F_m declining and F_o rising. The quantitative correlation between the UV-B-induced changes in oxygen evolving capacity and the fluorescence parameters was found to be weak for F_o and for photochemical quenching (q_P), while, in most species, the correlation was strong for F_v/F_m and Φ_e. The impact of the UV-B exposure on the capacity of the algae to develop non-photochemical quenching was largely species-dependent.

Most UV-B-resistant algae have large cell cross-sections or are growing in large-sized colonies or cenobia that provide effective shading of the internal structures. Alternatively, UV-B-sensitive species tended to be small sized or filamentous. A high fraction of UV-B-resistant species was found among algae originally isolated from mountainous, sun-exposed locations. They have frequently a solid cell wall containing sporopollenin.

Key words: Algae, Chl fluorescence, photosynthesis, UV-B radiation, sporopollenin.

Abbreviations: Chl *a/b* = chlorophyll *a/b*; F_m = maximal fluorescence emitted with the primary quinone acceptor of Photosystem II reduced and quenching mechanisms ineffective; F_o = fluorescence emitted with the primary quinone acceptor of Photosystem II oxidised; F_v = variable fluorescence equals $F_m - F_o$; PSII = Photosystem II; Q_A = primary quinone acceptor of Photosystem II; q_P = photochemical quenching of fluorescence; q_{NP} = non-photochemical quenching of fluorescence; UV-B = ultraviolet-B radiation (280–320 nm); Φ_e = quantum efficiency of linear electron transport estimated from fluorescence emission.

Introduction

Anthropogenic depletion of stratospheric ozone layer results in increased levels of UV-B radiation reaching both terrestrial and aquatic ecosystems (Frederick and Lubin, 1988; Stolarski, 1988; Caldwell et al., 1989, Häder and Worrest, 1991; Cullen and Neale, 1994). Assessment of the present impact of elevated UV-B radiation on autotrophic organisms as well as the prediction of future effects on ecosystems are the focus of numerous studies (for reviews see e.g. Tevini and Teramura, 1989; Häder, 1993; Karentz, 1994). Elevated UV-B radiation was found to affect various fundamental processes of plant cells e.g. photosynthesis (Bornman, 1989; Neale et al., 1993; Döhler and Haas et al., 1995), cell cycle (Behrenfeld et al., 1992), growth and development (e.g. Bothwell et al., 1993) and nitrogen assimilation (Döhler et al., 1991). During evolution, plants have developed several mechanisms that provide protection against the relatively low ambient flux of UV-B photons occurring under the ozone layer (Bornman, 1989, Häder, 1993; Karentz, 1994). The damage caused by elevated UV-B exposure depends therefore on a complex interplay between the UV-B dose, the effectiveness of plant protection mechanisms and the sensitivity of the plant to UV-B. Tevini and Teramura (1989) reviewed screening of about 300 species of annual terrestrial plants for UV-B sensitivity and found a large amount of variability with about half of the species susceptible and half resistant. A wide range of interspecific UV-B sensitivity was also found in marine diatoms (Calkins and Thordardottir, 1980; Karentz et al., 1991; Lesser et al., 1994;) and sea macroalgae and seagrasses (Larkum and Wood, 1993).

Microscopic algae play an important role in terrestrial as well as in freshwater and marine ecosystems. Parallel to their function in nature, algal biotechnology is of increasing importance. The impact of elevated UV-B radiation on the capacity of microscopic freshwater algae to maintain these functions is difficult to predict because of the scarcity of data on the UV-B susceptibility of these organisms. The available information concerning algae is mainly focused on marine organisms and ecosystems (Smith et al., 1992; Cullen and Neale, 1994; Karentz, 1994, Williamson and Zagarese, 1994). Taking into account the large variability in UV-B responses found in other organisms, screening of a large set of algal species is expected to provide information useful both in algal ecophysiology and biotechnology. Equally important, results of such a screening can enhance the potential of microscopic freshwater algae as relatively simple model organisms in the research of UV-B damage and photoprotection.

We have chosen a two-step approach: In the first phase of the experiment, which is reported here, we used a large number of algal species subjected to a relatively high dose of UV-B radiation ($2\ W \cdot m^{-2}$ administered for 2 hours). In order to limit the complexity of the problem, the present study focused only on the response of the photosynthetic apparatus of the algae to a relatively high dose of UV-B radiation. The results of this study allow us to continue the project with a second phase of experiments, in which, the impact of realistic, long term UV-B exposure will be investigated using sets of resistant and susceptible species identified here.

Materials and Methods

Algae

The 67 algal species used in the screening were obtained from the Culture Collection of Autotrophic Organisms of the Institute of Botany, Třeboň (Lukavský et al., 1992). The unialgal cultures listed in Table 1 were collected from agar slants and transferred into 50 mL of growth medium after Zehnder (see Lukavský et al., 1992 for composition of individual media). Cultures were then exposed to $10\ W \cdot m^{-2}$ photosynthetically active radiation (PAR). After 10–15 days, the preinoculum cultures were transferred from Ehrlenmayer flasks to cylindrical tubes with 50 mL of fresh media. The algal suspension was bubbled with air enriched with 2% CO_2, kept at 26 °C and exposed to $10\ W \cdot m^{-2}$ of PAR. After another 2–3 days, 100 mL of fresh media were added and the irradiance was increased to $25\ W \cdot m^{-2}$. After 3–4 additional days, the cultures were in the exponential phase of growth.

UV-B exposure

Exponentially growing cultures were diluted to 8–15 µM Chl and placed into Petri dishes (50 mL, 7×1.5 cm). The algal suspension was initially pretreated for 15 minutes at constant temperature (26 ± 2 °C) and irradiance ($25\ W \cdot m^{-2}$ PAR). After this pretreatment, UV-B radiation was added, which was produced by a single fluorescent tube (Philips TL20W/12) placed above algal suspension. The UV-C radiation was blocked by aged cellulose acetate filter (3/1000 inch thickness) that was regularly replaced after 20 hours of service. The ability of cellulose acetate filters to block the UV-C radiation was checked using a Shimadzu UV-3000 spectrophotometer. The unweighted UV-B irradiance incident at the surface of the algal suspension was measured by International Light SED 240 detector and adjusted to $2\ W \cdot m^{-2}$ by changing the distance between the UV-B light and algal suspension. In order to avoid confusion of the UV-B response and UV-A or visible light induced inhibition, control experiments were done with Mylar filters. The Mylar filter blocks both UV-C and UV-B and is transparent to visible and UV-A photons. No deteriorating effects were seen under Mylar protection when UV-B sensitive strains were exposed.

Photosynthetic oxygen evolution and Chl fluorescence

Aliquots of suspension (6 mL) were taken during the UV-B exposure to measure oxygen evolution and Chl fluorescence. The sample was placed in a temperature-controlled cuvette (28 °C) and HCO_3^- was added to final concentration of 20 mM. After a dark adaptation of 10 minutes, Chl fluorescence parameters F_o and F_m were measured with a PAM fluorimeter (Walz, Germany, Schreiber et al., 1986) based on a protocol similar to that of Ting and Owens (1992) and Hofstraat et al. (1994). Using the same sample, the CO_2-dependent oxygen evolution was measured (YSI Clark-type electrode) during 5 min of exposure of the algae to $150\ W \cdot m^{-2}$ of light from a Xenon lamp filtered by 2% $CuSO_4$ solution. During this actinic light exposure, nearly steady-state fluorescence emission was established and the F_s, F_m' and F_o' parameters were determined for calculation of the non-photochemical (q_{NP}) and photochemical (q_P) quenching (Schreiber et al., 1986) and of the Φ_e: the quantum efficiency of electron flow through PSII (Genty et al., 1989).

The UV-B exposure as well as the accompanying oxygen and fluorescence measurements were repeated 3–5 times for each algal species and means as well as standard deviations were calculated for each of the measured parameters.

Table 1: The list of screened algal species*.

No.	Strain identification	Note	UV-B Response
Chlorophyta, Chlamydophyceae, Chlamydomonadales			
1	*Chlamydomonas chlorococcoides,* strain Schwarz 1975	Yugoslavia, soil	–
2	*C. debaryana,* strain Ettl 1960/4	CS, mts. forest soil	–
3	*C. macropyrenoidosa,* strain Hübel 1964/182	D, forest soil	–
4	*C. peterfii,* strain Holubcová 1959/3	CS, mts., meadow	+
Chlorophyta, Chlamydophyceae, Chlorococcales			
1	*Characium sieboldii,* strain Hindák 1963/70	CS, mts. snow	–
2	*C. starrii,* strain Starr 1951/UTEX, 112 (strain-)	South Africa, soil	–
3	*C. terrestre,* strain Trainor et Bold 1953/Gött. 209-2	USA, Georgia, soil	+
4	*Chlorococcum echinozygotum,* strain Starr 1965/118	Phillipines, soil	–––
5	*C. elbense,* strain Hindák 1969/81	CS, lake	–
6	*C. ellipsoideum,* strain Kováčik 1977/11	CS, therm. spring	+++
7	*C. hypnosporum,* strain Pringsheim 1940/Camb. 237-1	U.K., soil	–
8	*C. minutum,* stain Bold/Camb. 213-7	India, soil	+
9	*C. scabellum,* strain Hindák 1969/125	CS, soil	–––
10	*C. vacuolatum,* strain Starr 1952/UTEX 110	South Africa, soil	–
11	*Chlorolunula* sp. strain Kováčik 1988/5	CS, mts., pool	–
12	*Chlorosarcinopsis aggregata,* strain ARCE/UTEX 779	Cuba, soil	–
13	*C. minuta,* strain Lukešová 1987/4	CS, meadow soil	–
Chlorophyta, Chlorophyceae, Chlorellales			
1	*Ankistrodesmus spiralis,* strain Christensen 1948/4879	DK, freshwater	–––
2	*Chlorella* cf. *minutissima,* strain Cassie/Camb. 211-52	N. Zealand, water	–
3	*C. sorokiniana,* strain Prát et Basler, Praha AC.A14	CS, therm. spring	–
4	*Choricystis* sp. strain Lukešová 1988/8	CS, soil	–
5	*Coelastrella multistriata,* strain Kalina 1967/9	CS, mts., peat bog	–
6	*C. multistriata,* strain Trenkwalder 1975/Inns.T88	Italy, mts. soil	+
7	*Dictyosphaerium pulchellum,* strain Kováčik 1983/7	CS, mts., lake	–
8	*D.* cf. *tetrachotomum,* strain Fott 1959/1	CS, mts.	–
9	*Diplosphaera* cf. *chodatii,* strain Lukešová 1988/6	CS, forest, soil	–
10	*Enallax coelastroides,* strain Kalina 1966/1	CS, mts. meadow	+++
11	*E.* sp., strain Kováčik 1984/	CS, mts., peat bog	+++
12	*Monoraphidium convolutum,* strain Komárek 1964/28	Cuba, pool	+
13	*M. convolutum,* strain Komárek 1964/110	Cuba, soil/dry pool	–––
14	*M. tortile,* strain Hindák 1963/104	CS, mts., soil	–
15	*Pseudococcomyxa.* sp., strain Kováčik 1977/4	CS, therm. spring	+
16	*P.* sp., strain Wydrzycka 1981/GC-10	Costa Rica, soil	–––
17	*Raphidocelis inclinata,* strain Pringsheim 1939/Gött. 243-1	U.K., soil	–
18	*R. valida,* strain George/Camb. 243-2	U. K. freshwater	–
19	*Scenedesmus* cf. *corallinus,* strain Kováčik 1988/4	CS, mts. valley	–
20	*S.* sp., strain NEČAS 1965/N-508, GREI./15, UV mutant	UV selected.	+++
21	*Scotiella chlorelloidea,* strain Komárek 1961/2	U. K., subaerophyt	+++
22	*Scotiellopsis rubescens,* strain Vinatzer/Inns. V	CH, mts. soil	+
23	*S. terrestris,* strain Hindák 1963/59	CS, mts. snow	+
24	*Sphaerocystis bilobata,* strain Lukešová 1987/3	CS, field soil	+++
25	*Tetraedron minimum,* strain Hindák 1964/24	CS, mts., wett stone wall	–
26	*Willea* sp. strain Kováčik 1987/12	CS, mts. lake	–
Chlorophyta, Chlorophyceae, Protosiphonales			
1	*Spongiochloris excentrica,* strain Bold/UTEX 108	USA, soil	–
2	*S. excentrica,* strain Lukešová 1988/5	CS, field soil	+
3	*S. spongiosa,* strain Vischer 1942/318	CH, soil	+
Chlorophyta, Chlorophyceae, Chaetophonales			
1	*Fritschiella tuberosa,* strain Andrews/112.80	USA, field soil	–
2	*Stigeoclonium* sp., strain Gardavský 1985/13	CS, fishpond	–
Chlorophyta, Pleurastrophyceae, Chlorosarchinales			
1	*Myrmecia bisecta,* strain Lukešová 1987/10	CS, meadow soil	–––
2	*Neochloris bilobata,* strain Trenkwald. 1975/Inns.T58	Italy, mts. soil	–
3	*Pleurastrum sarcinoideum,* strain Lukešová 1986/11	CS, meadow soil	–

Table 1: Continued.

No.	Strain identification	Note	UV-B Response
Chlorophyta, Ulvophyceae, Ulotrichales			
1	*Interfilum paradoxum,* strain Pringsheim/UTEX 177	U. K., soil	---
2	*Rhexinema errumpens,* strain Lukešová 1987/8	CS, forest soil	–
Chlorophyta, Ulvophyceae, Neochloridales			
1	*Chlorotetraedron bitridens,* strain starr 1952/120	USA, soil	+
2	*C. polymorphum,* strain DEAN/42	Australia, soil	–
Chlorophyta, Charophyceae, Klebsormidiales			
1	*Chlorhormidium flacciolum,* strain Hindák 1965/96	Cuba, garden soil	---
2	*C.* sp., strain Kováčik 1983/6	CS, mts. lake, benthos	+
3	*Stichococcus exiguus,* strain Komárek 1962/1	CS, mts. snow	+
Chromophyta, Xanthophyceae (Heterokontae), Miscococcales			
1	*Botrydiopsis alpina,* strain Vischer 1949/232	CH, soil	–
2	*B. alpina,* strain Vinatzer 1975/Inns.V181	Italy, mts., soil	–
3	*Chloridella neglecta,* strain Vischer 1940/216	CH, soil, meadow	+
4	*C. simplex,* strain Hindák 1962/16	CS, mts., snow	–
5	*Chlorobotrys regularis,* strain Flint 1964/2	Antarctis, soil	–
6	*Heterococcus brevicellularis,* strain Vischer 1945/351	CH, soil, forest	---
Chromophyta, Xanthophyceae (Heterokontae), Tribonematales			
1	*Bumilleriopsis filiformis,* strain Vischer 1943/360	CH, soil	–
2	*Tribonema aequale,* strain Pringsheim/Cambr. 880-1	CS, soil	---
3	*Xanthonema bristolianum,* strain Hindák 1966/38	CS, mts. snow	---

* UV-B response is identified based on the relative change of photosynthetic oxygen evolution capacity caused by 90-min UV-B exposure. For Resistant: +++: stimulated more than 10 %; +: change between – 10 % and + 10 %; For Sensitive: ---: reduced more than – 40 %; –: reduced between – 40 % and – 10 %. CS = Czechoslovakia.

Results

Oxygen evolving capacity was affected by UV-B exposure in most of the algal species tested (Fig. 1). The response varied from reduction by about 50 % to stimulation of the oxygen evolution rate that was in some species 20 % higher than the pre-exposure value. There is no clear-cut correlation between taxonomic affiliation of the species and UV-B sensitivity. However, in general, relative sensitivity tends to be higher in *Chromophyta* as compared to *Chlorophyta.*

Table 2 presents statistics of the UV-B response of the screened algae. Among the screened algal species 70 % were found to be sensitive to UV-B with suppression of oxygen evolving capacity by more than 10 % following the exposure. Out of the resistant species (30 % of total), 11 species were stimulated by UV-B exposure when the final oxygen evolving capacity was compared to the pre-UV-B exposure value. A relatively even proportion of sensitive to resistant species (9 : 7) was found among the algae originally collected from lakes or snow detritus in mountains. A high proportion of sensitive species was found in algae isolated from soil in mountains (4 : 1), soil in lowlands (27 : 7) and from lowlands plankton (5 : 2). The statistics of the algae from thermal springs is not conclusive due to the low number of species in the screening.

Table 2: UV-B responses among the tested algal species.

Response	Inhibition	Frequency	Location/number of species				
			mts. lake or snow	mts. soil	soil	plankton	thermal spring
sensitive	>10 %	70 %	9	4	27	5	1
resistant	<10 %	30 %	7	1	7	2	2

Among the most sensitive species are: *Ankistrodesmus spiralis* (no. 1, *Chlorellales,* Fig. 1), *Heterococcus brevicellularis* (no. 6, *Mischococcales*), *Myrmecia bisecta* (no. 1, *Chlorosarcinales*), *Klebshormidium bisectum* (no. 1, *Klebsormidiales*), *Pseudococcomyxa* sp. (no. 16, *Chlorellales*), and *Xanthonema bristolianum* (no. 3, *Tribonematales*).

Among the species exhibiting the greatest stimulation by UV-B are: *Chloridella neglecta* (no. 3, *Mischococcales*), *Scenedesmus* (no. 19, 20, *Chlorellales*), *Spongiochloris spongiosa* (no. 3, *Protosiphonales*), *Enallax coelastroides* (no. 10, *Chlorellales*) and *Chlorococcum ellipsoideum* (no. 6, *Chlorococcales*).

Fig. 2 presents a comparison of the relative changes in fluorescence parameters and in oxygen evolution capacity of all screened algae resulting from 90 minutes of UV-B exposure. No correlation was seen between the nearly universal UV-B induced rise in F_0 and the change in the oxygen evolution capacity (Fig. 2 A). Relatively small species to species variation in F_0 response indicate that the underlying molecular mechanism is common for a vast majority of the screened algae. The capacity of the PSII reaction centers of the algae

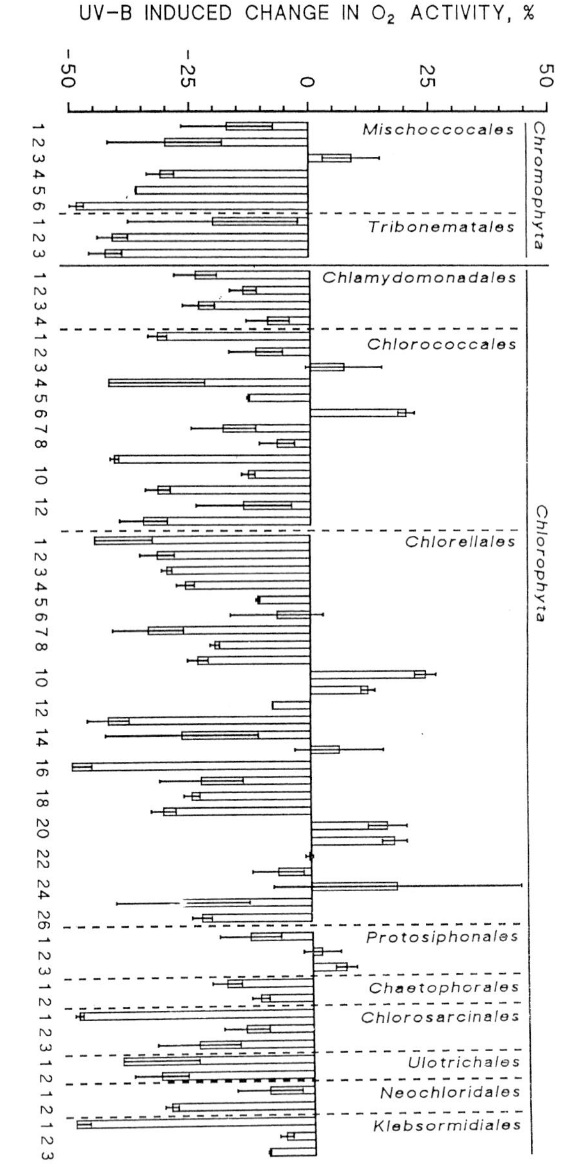

Fig. 1: The relative change in the photosynthetic oxygen evolution capacity of the algae induced by 90 minutes UV-B exposure. Each bar represents change in CO_2-dependent O_2 evolution capacity in an individual species of algae with error bar indicating standard deviation found in 3–5 independent experiments. The ordering of the algal species as well as numbering in individual families (bottom of the figure) is identical to that in Table 1.

for development of photochemical quenching (q_P) was nearly unaffected by UV-B exposure (Fig. 2 D). Limited correlation can be recognised between the UV-B induced change in the capacity of the algae to develop non-photochemical quenching (q_{NP}) and the change in the oxygen evolution capacity (Fig. 2 D). Large variations in the relationship indicate a species-specific character of the underlying molecular mechanism. A relatively high correlation with the UV-B induced changes in the oxygen evolution capacity was found for the F_v/F_m (Fig. 2 B) and for Φ_e, the quantum yield of PSII electron transfer (Fig. 2 C).

Discussion

Freshwater algae exhibit species-dependent sensitivity to UV-B exposure that ranges from reduction to stimulation of the oxygen evolving capacity. Similar variability was reported earlier in higher plants (Teramura et al., 1991, Sullivan et al., 1992) and with CO_2 fixation in algae (Smith et al., 1992). The mechanism underlying such a widely variable response of plants to elevated UV-B remains obscured by the complexity of damaging processes to several targets in the cell including PSII damage and by the complexity of the protection mechanisms (Häder and Worrest, 1991; Willimoson and Zealter, 1994).

The results of this investigation lead to the identification of algal species that are extremely sensitive to UV-B and algal species that are resistant to or even stimulated by UV-B radiation. Analysis of common features in the two groups resulted in the hypothesis that *Chromophyta* tend to be more sensitive than *Chlorophyta.* The minimal cross-cell or cross-colony/cenobium distance is larger in the resistant species (8, 3×4, 30, 2×7, 10 µm) compared to the sensitive species (1, 6, 3, 6, 3 and 4 µm). We propose that resistance to UV-B exposure is more likely to be observed in algae with a large minimal cross-cell distance or with cells organised in colonies or cenobia that can provide shading of the internal structures. In addition, the resistant algae have frequently a solid cell wall containing sporopollenin. The spectroscopical analysis of the cell wall fraction isolated from resistant species is underway to test the hypothesis that the UV-B absorbing or scattering structures are localized in the cell wall of these algae. The group of algae sensitive to UV-B is more heterogeneous. In addition to the small size of cell, characteristics that could lead to UV-B sensitivity include the occurrence of naked zoospores as a part of the cell cycle.

Many of the algal species that were resistant to UV-B radiation were originally isolated from sunlight-exposed locations of high elevation (cf. Larson et al., 1990). Also one of the resistant species, *Scenedesmus* sp., strain NEČAS 1965/N-508, GREI./15, UV (no. 20 *Chlorellales,* Fig. 1) was isolated as a UV resistant strain from *Scenedesmus quadricauda* culture 30 years ago during UV-assisted induction of random mutations. These two examples demonstrate the persistence of the genetically encoded basis for UV-B resistance. On the other extreme, the incidence of sensitive species is found to be high among algae isolated from shaded soil (Table 2). The presence of several resistant species among the soil algae is more difficult to interpret because of lack of information about the depths and character of soil from which the organism was isolated.

The comparison of fluorescence and oxygen evolution data obtained during the progressing UV-B exposure shows that fluorescence can serve as a secondary indicator of the UV-B induced changes in photosynthetic activity. The highest degree of reliability in the correlation with the oxygen evolution data was found with the Φ_e and F_v/F_m (Fig. 2 B and Fig. 2 C). The nearly uniform observation that UV-B induced an increase in F_o (Fig. 2 A) can be tentatively interpreted as a

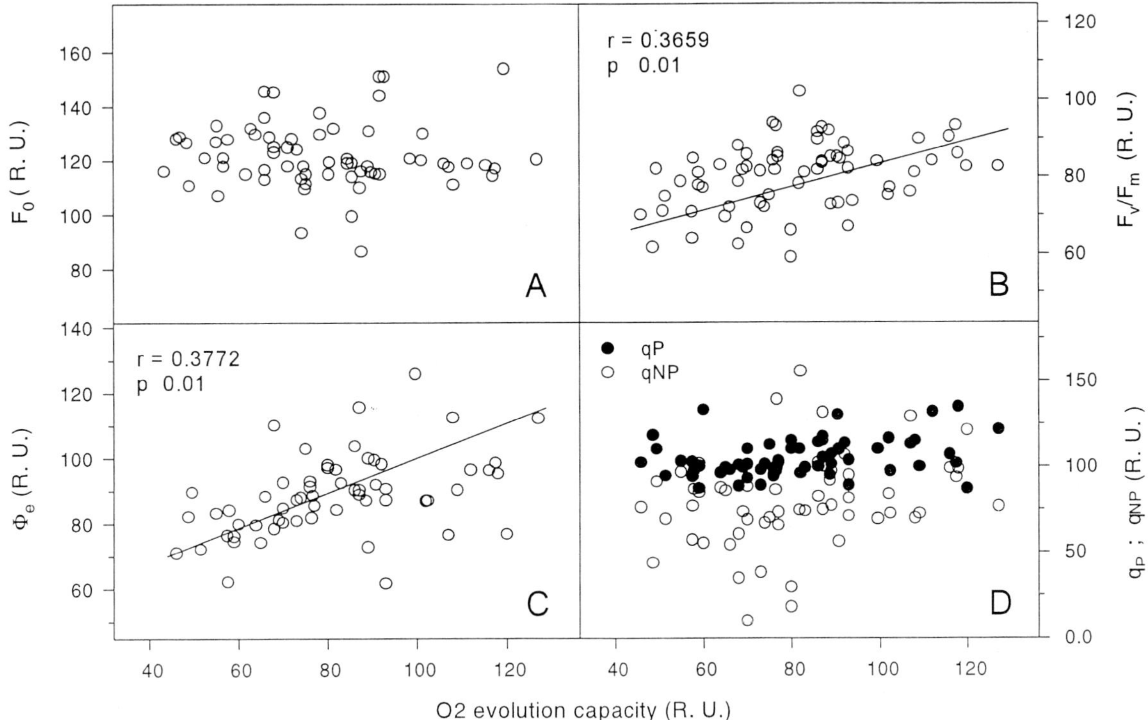

Fig. 2: Comparison of the relative changes in the oxygen evolution capacity of the screened algae with the accompanying changes in the fluorescence parameters F_o (A), F_v/F_m (B), Φ_e (C), and q_P and q_{NP} (D) as response to enhanced UV-B exposure. For more detail, see Materials and Methods. r-coefficients of correlations for probability $p<0.01$.

process occurring in the proximal antenna of PSII in almost all screened algae or as a process which changes the equilibrium of Q_A/Q_A^- in favour of the reduced form. To test these alternatives, future experiments will be done with added quinones. The largely species-dependent decline in the capacity of the algae to develop non-photochemical quenching which accompanies UV-B induced suppression of oxygen evolution capacity can be tentatively ascribed to UV-B-induced damage to the de-epoxidase of the carotenoid cycle (Pfündel et al., 1992). To test this hypothesis, the capacity of the algae to perform the light-induced carotenoid conversion will be compared with the oxygen evolution activity during UV-B exposure of sensitive algal species.

In the next phase of the experiments, the hypothesis will be tested that algal species resistant to UV-B exposure on a short time scale will also be resistant to a more realistic long-term UV-B exposure. It is anticipated that some of the strains susceptible to the short time/high UV-B exposure will be severely affected during more realistic UV-B exposure. A comparative study on the impact of enhanced UV-B exposure on pigment composition of susceptible and resistant algal species will be presented elsewhere. This comparison of susceptible and resistant algae is expected to yield more insight into the molecular mechanisms of protection used by the algae against UV-B stress.

Acknowledgements

The project was supported by contracts of Grant Agency of Czech Republic 206-93-0663 and 202-94-0457. Authors also appreciate assistance of Mrs. Marie Kašpárková and Mrs. Helena Vondrková from Culture Collection of Autotrophic Organisms at Třeboň and critical reading of the manuscript by Dr. Mary Poulson.

References

BEHRENFELD, M. J., J. T. HARDY, and H. LEE II: Chronic effects of ultraviolet-B radiation on growth and cell volume of *Phaeodactylum tricornutum* (*Bacillariophyceae*). J. Phycol. *28,* 757–760 (1992).

BORNMAN, J. F.: Target sites of UV-B radiation in photosynthesis of higher plants. J. Photochem. Photobiol. B. Biology *4,* 145–158 (1989).

BOTHWELL, M. L., D. SHERBOT, A. C. ROBERGE, and R. J. DALEY: Influence of natural ultraviolet radiation on periphytic diatom community growth, biomass and species composition: short-term versus long-term effects. J. Phycol. *29,* 24–35 (1993).

CALDWELL, M. M., A. H. TERAMURA, and M. TEVINI: The change solar ultraviolet climate and the ecological consequences for higher plants. Trends in Ecology and Evolution *4,* 363–366 (1989).

CALKINS, J. and T. THORDARDOTTIR: The ecological significance of solar UV radiation on aquatic organisms. Nature. *283,* 563–566 (1980).

CULLEN, J. J. and P. J. NEALE: Ultraviolet radiation, ozone depletion, and marine photosynthesis. Photosynthesis Res. *39,* 303–320 (1994).

DÖHLER, G., E. HAGMEIER, E. GRIGOLEIT, and K. D. KRAUSE: Impact of solar UV radiation on uptake of ^{15}N-ammonia and ^{15}N-nitrate by marine diatoms and natural phytoplankton. Biochem. Physiol. Pflanzen. *187,* 293–303 (1991).

Döhler, G., E. Hagmeier, and F. T. Haas: UV effects on chlorophylls and carotenoids of the haptophyceaen alga *Pavlova*. Photosynthetica *31*, 157–160 (1995).

Frederick, J. E. and D. Lubin: Possible long-term changes in biologically active ultraviolet radiation reaching the ground. Photochem. Photobiol. *47*, 571–578 (1988).

Genty, B., J. M. Briantais, and N. R. Baker: The relationship between the quantum yield of photosynthetic electron transport and quenching of chlorophyll fluorescence. Biochim. Biophys. Acta. *990*, 87–92 (1989).

Häder, D. P. and R. C. Worrest: Effects of enhanced solar ultraviolet radiation on aquatic ecosystems. Photochem. Photobiol. *53*, 717–725 (1991).

– – Risks of enhanced solar ultraviolet radiation for aquatic ecosystems. In: Round, F. E. and D. J. Chapman (eds.): Progress in Physiological Research. pp. 1–38. vol. *9*, Biopress Ltd. (1993).

Hofstraat, J. W., J. C. H. Peeters, J. F. H. Snel, and C. Geel: Simple determination of photosynthetic efficiency and photoinhibition of *Dunaliella tertiolecta* by saturating pulse fluorescence measurements. Marine Ecology Progress Series. *103*, 187–196 (1994).

Karentz, D., J. E. Cleaver, and D. L. Mitchell: Cell survival characteristics and molecular responses of Antarctic phytoplankton to ultraviolet-B radiation. J. Phycol. *27*, 326–341 (1991).

– – – Ultraviolet tolerance mechanisms in Antarctic marine organisms. In: Weiler, C. S. and P. A. Penhale (eds): Ultraviolet radiation in Antarctica: Measurements and biological effects. Antarctic Research Series, American Geophysical union, Washington, D.C. *62*, 93–110 (1994).

Larkum, A. W. D. and W. F. Wood: The effect of UV-B radiation on photosynthesis and respiration of phytoplankton, benthic macroalge and seagrasses. Photosynth. Res. *36*, 17–23 (1993).

Larson, R. A., W. J. Garrison, and R. W. Carlson: Differential responses of alpine and non-alpine *Aquilegia* species to increased ultraviolet-B. Plant Cell Environ. *13*, 983–987 (1990).

Lesser, M. P., J. J., and P. J. Neale: Carbon uptake in a marine diatom during acute exposure to ultraviolet B radiation: Relative importance of damage and repair. J. Phycol. *30*, 183–192 (1994).

Lukavský, J., V. Cepák, J. Komárek, M. Kašpárková, and M. Takáčová: Catalogue of algal and cyanobacterial strains of Culture Collection of Autotrophic Organisms at Třeboň. Arch. Hydrobiol./Algological Studies *63*, 59–119 (1992).

Neale, P. J., J. J. Cullen, M. P. Lesser, and A. Melis: Physiological bases for detecting and predicting photoinhibition of aquatic photosynthesis by PAR and UV radiation. In: Yamamoto, H. Y. and C. M. Smith (eds.): Photosynthetic Responses to the Environment. pp. 61–77. American Society of Plant Physiologists, Rockville, Maryland (1993).

Pfündel, E. E., R.-S. Pan, and R. A. Dilley: Inhibition of violaxanthin deepoxidation by ultraviolet-B radiation in isolated chloroplasts and intact leaves. Plant Physiol. *98*, 1372–1380 (1992).

Schreiber, U., W. Schliwa, and U. Bilger: Continuous recording of photochemical and non-photochemical chlorophyll fluorescence quenching with a new type of modulation fluorimeter. Photosynth. Research. *10*, 51–62 (1986).

Smith, R. C., B. B. Prézelin, K. S. Baker, R. R. Bidigare, N. P. Boucher, T. Coley, D. Karentz, S. MacIntyre, H. A. Matlick, D. Menzies, M. Ondrusek, Z. Wan, and K. J. Waters: Ozone depletion: Ultraviolet radiation and phytoplankton biology in Antarctic Waters. Science *255*, 952–959 (1992).

Stolarski, R. S.: The Antarctic ozone hole. Sci. Amer. *258*, 30–36 (1988).

Sullivan, J. H., A. H. Teramura, and L. H. Ziska: Variation in UV-B sensitivity in plants from a 3000 m elevation gradient in Hawaii. Amer. J. Bot. *79*, 737–743 (1992).

Tevini, M. and A. H. Teramura: UV-B effects on terrestrial plants. Photochem. Photobiol. *50*, 479–487 (1989).

Teramura, A. H., H. L. Ziska, and E. Sztein: Changes in growth and photosynthesis capacity of rice with increased UV-B radiation. Physiologia Plantarum. *83*, 373–380 (1991).

Ting, C. S. and T. G. Owens: Limitations of the pulse-modulated technique for measuring the fluoresence characteristics of algae. Plant Physiol. *100*, 367–373 (1992).

Williamson, C. B. and H. E. Zagarese: The impact of UV-B radiation on pelagic freshwater ecosystems. Advances in Limnology *43*, 9–11 (1994).

J. Plant Physiol. Vol. 148. pp. 49–56 (1996)

Combination Effects of UV-B Radiation and Temperature on Sunflower (*Helianthus annuus* L., cv. Polstar) and Maize (*Zea mays* L., cv. Zenit 2000) Seedlings

Uwe Mark and Manfred Tevini

University of Karlsruhe, Botanical Institute II, Kaiserstr. 12, D-76128 Karlsruhe, Germany

Received July 16, 1995 · Accepted October 27, 1995

Summary

Sunflower (*Helianthus annuus* L., cv. Polstar) and maize (*Zea mays* L., cv. Zenit 2000) seedlings were grown for 23 days in climate controlled growth chambers under solar radiation. Using UV-transmitting filters as covers (ozone filter technique) a difference in UV-B irradiance of 20 to 29 % was simulated. This difference corresponds to a decrease in stratospheric ozone of about 12 % based on ambient ozone levels at Lisbon (Portugal). During the experimental season in 1990 the daily course of temperature ranged from T_{min} = 13.5 °C at night to T_{max} = 28 °C around midday, and in 1991 from T_{min} = 17.5 to T_{max} = 32 °C simulating an average diurnal temperature increase, which might be expected in the 21st century as a result of the greenhouse effect.

Generally, sunflower and maize seedlings grown under T_{max} = 32 °C had increased growth parameters compared to seedlings grown under T_{max} = 28 °C. In both plant species fresh and dry weight as well as plant size and leaf area were increased as a result of reduced UV-B radiation at T_{max} = 28 °C, but, with very few exceptions, not at T_{max} = 32 °C. Thus, the increase in temperature by 4 °C mainly compensated the reduced growth caused by increased UV-B radiation.

The net photosynthetic rates of sunflower and maize seedlings based on leaf area exhibited higher or similar rates under enhanced UV-B radiation in comparison to the control plants. However, on a plant basis significant UV-B dependent decreases were observed. Increased growing temperature resulted in higher net photosynthetic rates per plant.

Neither UV-B nor temperature had an influence on the chlorophyll content per plant. Respiration and transpiration of sunflower and maize plants related to the entire plant did not show any UV-B dependent differences. Increased temperature resulted in higher respiration rates in maize only, whereas transpiration increased in both species.

Key words: Helianthus annuus, Zea mays, enhanced solar UV-B, growth, net photosynthesis, respiration, transpiration.

Abbreviations: CFC = chlorofluorocarbon; UV-B = ultraviolet-B radiation (280–320 nm); UV-A = ultraviolet-A radiation (320–400 nm); PAR = photosynthetically active radiation (400–700 nm); rH = relative humidity; TKS = ready-to-use growing medium; RGR = relative growth rate; SE = stem elongation rate; IAA = indole acetic acid; d = days; chl = chlorophyll; NP = net photosynthesis; DR = dark respiration; TR = transpiration.

Introduction

There is now substantial evidence for atmospheric changes, mainly the depletion of the ozone layer as a result of anthropogenic emissions, such as CFCs and the global climate change (Molina and Rowland 1974, Watson et al. 1990, Hofmann et al. 1992). The burning of fossil fuels and the deforestation of tropical rain forests are contributing to rising CO_2

concentrations in the atmosphere (Houghton and Woodwell 1989). These concentrations might rise from ~350 µL/L to about 700 µL/L between the years 2030 and 2100 (Schneider 1989, Watson et al. 1990).

Increasing amounts of CO_2 and other trace gases, such as methane, dinitrogen oxide and CFCs contribute to the greenhouse effect with an expected increase of global temperature in the range of 1.5 °C to 5.5 °C (Watson et al. 1990).

Experiments on temperature effects show that plants are able to photosynthesize over a large temperature range (Long 1983, Baker et al. 1988), and that temperature also effects other plant processes, such as leaf and fruit expansion (Auld et al. 1978, Johnson and Thornley 1985, Khayat et al. 1985, Marcelis and Bann Hofman-Eijer 1993). However, the effects of temperature are dependent on species and environmental conditions, as well as on the environmental conditions at the time of measurement. Therefore, temperature optima are different for plants from colder climates compared to those from warmer climates, which is apparent in C_3 and C_4 plant species. Physiological differences between these two groups are the reason for higher temperature optima in C_4 plants. Usually photosynthesis increases with temperature to a certain point when the destruction of the photosystems and denaturation of enzymes occurs. The concomitant increase of respiratory CO_2 loss manifests in higher photorespiration as the ratio of oxygenase relative to carboxylation increases with rising temperature (Jordan and Ogren 1984, Long 1991). Due to O_2 competition only little stimulation of net CO_2 fixation by increased temperature is seen in C_3 plants. Hence, the temperature effect is nearly balanced by increased respiration and photorespiration and often results in a broad temperature response curve between 15 °C and 30 °C. Since photorespiration in C_4 plants is of little significance, they show optima between 30 °C and 40 °C in net photosynthesis. Limited ribulose bisphosphate carboxylase as a result of temperature induced reductions of ATP and NADPH production has also been found in C_3 species.

However, experiments also suggest that plants can adapt to different temperature conditions and shift their temperature optimum with respect to photosynthetic activity (Berry and Björkman 1980) and also to environmental changes, such as CO_2 increase (Long 1991). Once species-dependent extreme temperatures are exceeded plants suffer severely from temperature stress, that affects membrane integrity and enzyme denaturation followed by cessation of photosynthesis (Baker et al. 1988).

Other environmental factors, such as UV-B radiation interact with photosynthesis as well. Numerous studies of UV-B effects on terrestrial plants have been documented (Caldwell et al. 1989, Tevini and Teramura 1989, Bornman and Teramura 1993, Tevini 1994) reporting impaired growth and photosynthesis in many UV-B sensitive plant species, such as sunflower and maize. Very important for the expression of the UV-B effect are the levels of UV-A radiation and visible light which can influence photorepair mechanisms (Langer and Wellmann 1990, Pang and Hays 1991). The stimulation of the protective pigments can also quantitatively alternate UV-B radiation responses of plants (Cen and Bornman 1990, Tevini et al. 1991). Results obtained in growth chambers and greenhouses, where low white light (shade conditions) is applied in addition to an unnaturally high irradiance of artificial UV-B (and in several cases UV-C), cannot be transferred to real nature. Under such artificial UV-stress damaging UV-radiation effects are often much stronger than the inductions of UV-adaptation responses, such as the induction of UV-screening pigments, photorepair, biochemical responses against oxydative stress as well as morphological responses, such as leaf area and growth reduction. Thus, results in growth chambers and greenhouses under artificial UV-B radiation sources are often overestimating the UV-B effects. However, in nature, stress responses as mentioned before must co-act to produce adaptation to stress or to avoid stress using the entire spectrum of the solar radiation as a polychromatic signal. Therefore, in the following experiment growth conditions were chosen applying only solar radiation with UV-B attenuated by ozone in a closed cuvette (Tevini et al. 1990) positioned on top of growth cabinets. Presently, this technique results in the most natural UV-B conditions ever used in UV-B experiments. The aim of this experiment was to investigate the combination effects of two different levels of UV-B radiation and elevated temperature on a C_3 and C_4 species under natural UV-B radiation conditions simulating a 12 % ozone depletion and increases of 4 °C of the diurnal temperature course.

Materials and Methods

Plant and growing conditions

The experiments were carried out over a period of two years, from 1990–1991, at the research station Quinta de São Pedro in Portugal, about 10 km south of Lisbon (38.65° N; 9.18° W), where high levels of ambient UV-B radiation are present during the summer season.

Seeds of sunflower (*Helianthus annuus* L., cv. Polstar) and maize (*Zea mays* L., cv. Zenit 2000) were sown in trays (15 × 15 cm) on standard substrate TKS 1, each tray containing 25 seeds.

The trays were placed in growth chambers covered with an ozone UV-B filter system developed by Tevini et al. (1990). This system, constructed as an UV-transmissible plexiglass cuvette, was placed on top of the growth chamber. The ambient UV-B radiation reaching the plants beneath was reduced by passing ozone through the filter on top, thus, only UV-B levels similar to those found at more northern latitudes were present in this growth chamber. Under these conditions the «control» data (–UV-B) were obtained. Plants in the second growth chamber were subject to ambient UV-B levels (+UV-B) of Portugal when the filter on top contained only air. The difference between control and ambient UV-B levels was about 20 % in 1990 and 30 % in 1991 (Table 1). Thus, ambient radiation which is higher than the control UV-B, is considered as enhanced UV-B radiation compared to lower UV-B radiation ambient at a more northern latitude. 10–12 series of plants were grown from June through August each year. Combination effects of enhanced temperature and UV-B were determined in 1990, where diurnal fluctuations of temperature and relative humidity (rH) were simulated, with a minimum temperature of 17.5 °C (79 % rH) at night and a maximum temperature of 32 °C (35 % rH) during the day. In 1991 the diurnal temperature was lowered by 4 °C with T_{max} of 28 °C. The relative humidity remained the same as in the previous year. Growth and gas exchange parameters were determined after 13, 18 and 23 days in order to observe possible developmental or fluence dependent changes in the response to UV-B radiation under both temperature conditions. Since only one ozone UV-B filter was available each year, the comparison of control data (–UV-B) under «normal» and elevated temperatures had to be made with respect to global radiation differences between the consecutive experimental years.

In tables 2 and 3 results for 1990 were given in columns C and D, for 1991 in columns A and B.

Irradiation conditions

A temperature controlled spectroradiometer Optronics 742 was used for the radiation measurements. The ambient UV-B radiation was measured on all cloudless and cloudy days, but not on rainy days during the experimental period. In 1990 it had rained at 6 days, in 1991 at 10 days.

Mean daily UV-B, UV-A and PAR fluences between June and August in 1990 and 1991 are listed in table 1. UV-A and PAR measurements were taken outside and inside the growth chambers. In all growth chambers, irrespective of the time of year, a UV-A and PAR reduction of about 12 % occurred due to the absorption of the plastic material of the ozone cuvettes on top.

Growth analysis

Plant size, fresh and dry weight were determined in all plants as well as the leaf area of which the total leaf area per plant had been measured. Only plant material above ground was considered for fresh and dry weight, the latter involved the drying of the material at 70 °C. Calculations of the specific growth parameters SE (stem elongation rate) and RGR (relative growth rate) followed the equation by Harper (1977). Measurements of growth parameters in sunflower and maize were taken after 13 days, 18 days and 23 days.

Pigment analysis

Acetone was used to extract chlorophyll, and quantitative calculations were made according to Lichtenthaler (1987). The chlorophyll content was based on leaf area and total plant.

Gas exchange measurements

Net photosynthesis was determined based on three different parameters: leaf area, chlorophyll content, and plant, as well as respiration and transpiration, the latter based on leaf area and plant. Measurements at a constant temperature of 20 °C and relative humidity of 55 % rH were carried out at each developmental stage.

Determination of respiration was carried out at the beginning of each measurement in the dark. Plants were then irradiated with a 75 W cold-light lamp with 2800 $\mu mol/m^2 s$ (PAR), which was sufficient for both species to reach their light saturation point. Actual measurements were made during the steady state of each plant after approximately 1 h of irradiation with CO_2 concentration of 340 ppm during each test. In 13-day old sunflower seedlings the gas exchange measurements could be performed with the whole plant (cotyledones and primary leaves), whereas in maize seedlings it could only be performed with secondary leaves. For further measurements secondary leaves of sunflower and secondary (18-days) or tertiary leaves (23 days) of maize were used.

Gas exchange parameters were calculated following the models by v. Caemmerer and Farquhar (1981).

Statistical analysis

Each experiment involved several repetitions so that the total number of plants used for estimating one parameter ranged between 40 and 75. The following results represent mean values and their respective standard deviations. A two-factor analysis of variance (Model I) was applied to qualify significant ($\alpha = 5$ %) differences of the treatments (UV-B, temperature). For the quantification of significant differences between two means the Fisher's LSD-test was used. Prior to each test a Kolmogoroff-Smirnov-Adjustment test (KSA) was made in order to determine if an unknown, given distribution F(x) would correspond with a normal distribution $N(\mu,\sigma)$. For all series a correspondence with the normal distribution could be seen; the independence of the series resulted from the experimental design.

In figures a significant difference between plants grown under enhanced UV-B compared to control plants (–UV-B, 28 °C) is indicated by an asterisk. In tables different letters within the same row indicate significant differences.

Results

The effects of enhanced solar UV-B radiation on sunflower and maize under two different diurnal temperature courses with T_{max} of 28 °C and 32 °C respectively were investigated in terms of growth, chlorophyll content and gas exchange. The mean daily UV-B fluences (Table 1) between June and August at the experimental location (38.7° N) were about 15 % lower in 1991 than in 1990, which correlates with a reduction in photosynthetically active radiation (PAR) fluences of about 14 %. UV-A was about 17 % lower in 1991 than in 1990 (Table 1).

In both years parallel experiments with the two temperature regimes under lower UV-B have been performed due to possible changes in white light levels in the two years. However, significant differences neither in shoot biomass and size nor in the gas exchange parameters have been observed in the consecutive year (data not shown, Mark 1992). Only small increases in leaf area of sunflower and maize have been seen in 1990 due to higher white light which had no influence on the UV-B effect normally reducing the leaf area. Statistical analyses of the single experiments repeated 10-fold during the year showed the same statistical trend in all experiments.

Growth

Both species grown under higher UV-B and T_{max} of 32 °C exhibited an increase of all growth parameters, such as weight, size and leaf area compared to conditions of +UV-B and T_{max} of 28 °C within 18 d of development (Table 2 and 3). For example, 13-day old sunflower seedlings at 28 °C were 6.3 cm in shoot size compared to 8.9 cm at 32 °C. During further development the differences became smaller for all other growth parameters except plant size.

Fresh and dry weight of sunflowers (Table 2) under higher UV-B radiation (+UV-B) and T_{max} of 28 °C were reduced by

Table 1: Mean daily fluences of UV-B, UV-A and PAR measured within (UV-B) and outside (UV-A, PAR) the growth chamber from June to August in 1990 and 1991.

Year	+UV-B (kJ/m^2)	–UV-B (kJ/m^2)	Δ%	UV-A (MJ/m^2)	PAR (MJ/m^2)
1990	54.39±12.47	45.26±10.45	+20.2	1.33±0.14	14.3±2.15
1991	47.14±10.02	36.54±11.51	+29.0	1.10±0.15	12.3±0.95

about 18 % during the whole duration of the experiment compared to the reduced radiation level (–UV-B, T_{max} 28 °C). An increase of temperature by 4 °C resulted in a compensation of the UV-B effect, that is, plants grown at higher UV-B were similar in weight and size to the control plants grown at the reduced UV-B level at 28 °C.

In young sunflower seedlings (13 days) even an increase of weight at reduced UV-B and T_{max} of 32 °C was observed.

Table 2: Growth- and gas exchange parameters of sunflower leaves with respect to plant age (d = days from sowing) and growth conditions.

Parameter	Age (d)	A –UV-B 28 °C	B +UV-B 28 °C	C –UV-B 32 °C	D +UV-B 32 °C
shoot fresh weight (g)	13	1.77 a*	1.45 b	1.94 c	1.87 ac
	18	2.57 a	2.36 b	2.64 a	2.68 a
	23	2.93 a	2.61 b	3.11 a	3.06 a
shoot dry weight (g)	13	0.14 a	0.12 b	0.16 c	0.15 ac
	18	0.28 a	0.24 b	0.33 c	0.32 c
	23	0.45 a	0.41 b	0.45 a	0.47 a
shoot size (cm)	13	8.64 a	6.26 b	9.83 c	8.88 a
	18	16.77 a	13.23 b	17.78 a	16.94 a
	23	21.87 a	19.20 b	23.53 c	22.39 ac
leaf area (cm^2/plant)	13	29.78 a	24.67 b	36.74 c	34.95 c
	18	34.34 a	31.00 b	45.32 c	41.88 d
	23	43.02 a	37.63 b	52.16 c	48.83 cd
SE (cm/day)	13	0.66 a	0.48 b	0.76 c	0.68 a
	18	1.63 a	1.39 b	1.59 a	1.62 a
	23	1.00 a	1.17 a	1.14 a	1.08 a
RGR (mg/day)	13	0.38 ac	0.37 a	0.39 b	0.38 bc
	18	0.14 a	0.14 a	0.14 a	0.15 a
	23	0.09 a	0.11 a	0.07 a	0.08 a
chl (µg/cm^2)	13	27.96 a	30.89 b	30.72 b	29.97 ab
	18	23.12 a	27.47 b	24.31 a	26.12 b
	23	20.56 a	23.43 b	20.40 a	21.55 ab
chl (mg/plant)	13	0.79 a	0.72 a	1.11 b	1.08 b
	18	0.80 a	0.84 a	1.07 b	1.11 b
	23	0.85 a	0.85 a	1.05 b	1.04 b
NP (µmol CO_2/cm^2 s)	13	19.17 ab	17.86 a	19.80 b	18.94 ab
	18	15.46 a	17.32 b	14.31 a	14.72 a
	23	12.66 a	13.33 a	15.13 b	12.99 a
NP (nmol CO_2/plant s)	13	55.01 a	41.78 b	68.32 c	65.99 c
	18	50.98 a	46.06 b	56.53 c	55.16 ac
	23	57.12 a	47.00 b	85.39 c	62.29 d
NP (µmol CO_2/Chl s)	13	62.87 a	54.29 b	65.33 a	64.19 a
	18	71.98 a	57.93 b	59.53 b	57.61 b
	23	61.00 a	52.41 b	65.09 c	60.67 a
DR (µmol CO_2/m^2 s)	13	1.38 a	1.52 a	1.00 b	1.12 b
	18	0.94 a	1.08 b	0.69 c	0.78 c
	23	0.81 a	0.90 b	0.69 c	0.71 c
DR (nmol CO_2/plant s)	13	4.17 a	3.55 a	3.50 a	3.94 a
	18	3.15 a	2.95 a	2.62 a	2.93 a
	23	3.66 a	3.31 a	3.88 a	3.42 a
TR (µmol H_2O/plant s)	13	7.46 a	6.23 a	10.57 b	10.33 b
	18	7.65 a	7.15 a	8.91 b	9.05 b
	23	9.91 ac	8.10 a	13.74 b	11.91 bc

* Different letters within the same row indicate a significant difference at the 0.05 significance level.

Table 3: Growth- and gas exchange parameters of maize leaves with respect to plant age (d = days from sowing) and growth conditions.

Parameter	d	A –UV-B 28 °C	B +UV-B 28 °C	C –UV-B 32 °C	D +UV-B 32 °C
shoot fresh weight (g)	13	1.42 a*	0.97 b	2.09 c	1.90 c
	18	2.54 a	2.03 b	3.28 c	3.14 c
	23	3.75 a	2.85 b	4.88 c	4.35 ac
shoot dry weight (g)	13	0.16 a	0.10 b	0.23 c	0.21 c
	18	0.29 a	0.22 a	0.45 b	0.40 b
	23	0.47 a	0.33 b	1.19 c	0.72 d
shoot size (cm)	13	22.66 a	17.60 b	28.65 c	26.38 d
	18	33.51 a	28.30 b	38.34 c	36.96 c
	23	39.64 a	34.10 b	44.01 c	41.73 ac
leaf area (cm^2/plant)	13	47.82 a	33.94 b	72.16 c	64.84 d
	18	83.42 a	65.01 b	111.11 c	105.87 c
	23	116.1 a	89.52 b	152.49 c	139.38 ac
SE (cm/day)	13	1.74 a	1.35 b	2.20 c	2.03 d
	18	2.80 a	2.11 b	1.96 b	2.12 b
	23	1.23 a	1.15 ab	1.02 b	0.90 b
RGR (mg/day)	13	0.39 a	0.35 b	0.41 c	0.41 c
	18	0.29 a	0.25 b	0.12 c	0.12 c
	23	0.09 a	0.08 ab	0.11 a	0.05 b
chl (µg/cm^2)	13	18.87 a	24.37 b	18.27 a	18.22 a
	18	14.82 a	17.13 b	11.32 c	11.67 c
	23	10.39 a	12.43 b	8.52 c	8.86 ac
chl (mg/plant)	13	0.91 a	0.86 a	1.34 b	1.15 b
	18	1.05 a	1.06 a	1.33 b	1.31 b
	23	1.02 a	0.91 a	1.38 b	1.19 b
NP (µmol CO_2/cm^2s)	13	14.15 a	15.24 a	11.96 b	11.13 b
	18	9.44 a	8.98 a	7.03 b	6.40 b
	23	5.84 a	5.42 a	3.90 b	4.14 b
NP (nmol CO_2/plant s)	13	53.85 a	46.57 b	85.56 c	70.54 d
	18	72.29 a	57.33 b	94.89 c	73.30 a
	23	51.35 a	43.84 b	64.31 c	52.54 b
NP (µmol CO_2/Chl s)	13	65.42 a	60.18 b	67.03 a	63.66 ab
	18	70.96 a	50.45 b	60.42 c	63.82 c
	23	57.44 a	49.48 b	55.20 ac	50.75 bc
DR (µmol CO_2/m^2s)	13	0.76 a	0.79 a	0.57 b	0.56 b
	18	0.58 a	0.66 a	0.45 b	0.42 b
	23	0.28 a	0.30 a	0.40 a	0.37 a
DR (nmol CO_2/plant s)	13	2.92 a	2.55 a	4.18 b	3.71 b
	18	4.63 a	4.15 a	6.61 b	5.15 ab
	23	2.83 a	2.50 a	7.01 b	4.81 c
TR (µmol H_2O/plant s)	13	4.92 a	4.28 a	8.68 b	7.54 b
	18	7.78 ab	6.22 a	11.24 b	10.06 ab
	23	6.38 a	5.98 a	9.14 a	8.36 a

* Different letters within the same row indicate a significant difference at the 0.05 significance level.

Relative growth rates (RGR) and stem elongation rates (SE) did not differ as much, and only at the age of 13 days a temperature induced increase in SE occurred in sunflower seedlings.

In maize seedlings reductions of fresh and dry weight as well as RGR (Table 3) were observed as a result of higher UV-B radiation. Elevated temperature caused plants to be heavier by 20 % on average. RGR of maize plants older than 18 days was lower under elevated temperature conditions.

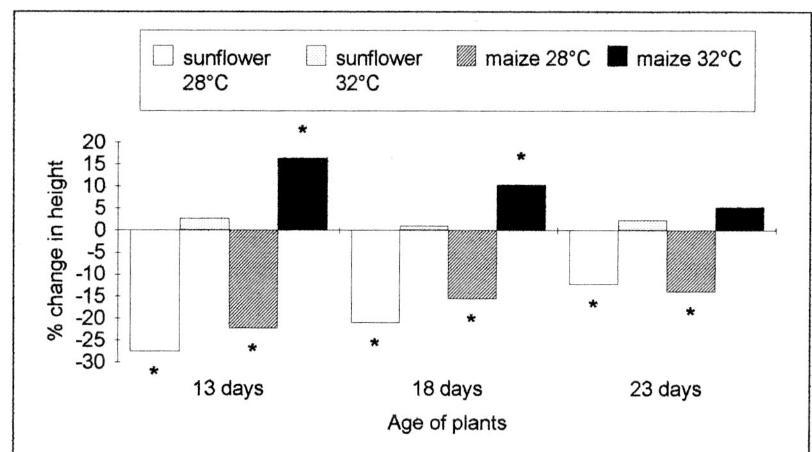

Fig. 1: Percent change in height (cm) of sunflower and maize between −UV-B (28 °C) and +UV-B at 28 °C and 32 °C. Stars indicate significant differences (α = 0.05).

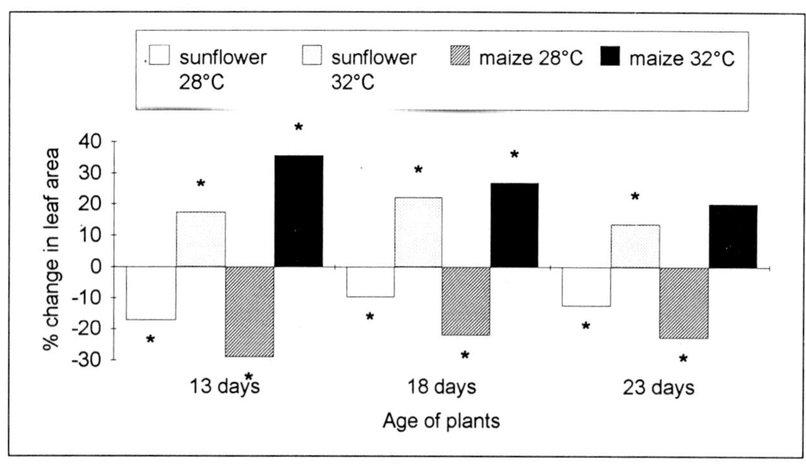

Fig. 2: Percent change in leaf area (cm^2) of sunflower and maize between −UV-B (28 °C) and +UV-B at 28 °C and 32 °C. Stars indicate significant differences (α = 0.05).

Plant height of sunflower seedlings (Table 2) was lowest at elevated UV-B radiation when grown at T_{max} 28 °C. Increased temperature resulted in taller plants and here size reducing UV-B effects were not as obvious as at 28 °C (Fig. 1). Stem elongation rates were seen to be similarly affected up to 18 days of growth, whereas RGRs showed no differences between the treatments.

In maize seedlings grown at 28 °C, enhanced UV-B resulted in decreased height of about 22 % after 13 days (Table 3). With prolonged irradiation this difference became somewhat smaller. Growing seedlings at T_{max} of 32 °C induced a significant increase in height of about 14 %. Although higher UV-B also reduced the size, maize plants grown at T_{max} = 32 °C were still bigger than UV-B plants grown at 28 °C. Thus, after 23 days (Fig. 1) the temperature effect overcompensated for the UV-B effect by 18 %.

Independent of irradiation duration and temperature, leaf area of both species decreased under elevated UV-B radiation (Tables 2 and 3). However, plants grown at higher temperature had smaller leaf area reductions with respect to higher UV-B conditions. A temperature increase of 4 °C generally increased the leaf area (Fig. 2) to such a high extent that even +UV-B treated plants had larger leaf areas than plants grown under reduced UV-B and temperature.

Increased chlorophyll content of approximately 14 % per leaf area was observed in sunflower, independent of plant age as a result of elevated UV-B (Table 2). Under enhanced temperature a UV-B-dependent increase of 7.4 % was noticed after 18 days only. Chlorophyll content per plant did not appear to be affected by UV-B but was generally enhanced under higher temperature in both plant species (Table 2).

Maize plants responded in a similar way with respect to chlorophyll content per leaf area grown at T_{max} of 28 °C, but at T_{max} of 32 °C no UV-B induced chlorophyll reduction was observed (Table 3).

Gas exchange

Sunflower seedlings grown under +UV-B and T_{max} of 28 °C exhibited similar (13, 23 days) or higher (12 % after 18 days) rates of net photosynthesis based on leaf area compared to control plants grown at the lower UV-B level (Table 2). However, net photosynthesis based on chlorophyll and total plant (Table 2) was lower during growth at the higher UV-B level. An increase in temperature did not alter the net photosynthesis of sunflower leaves based on leaf area but increased it drastically on a plant basis. There was no remarkable UV-B effect at T_{max} = 32 °C plants, and thus a temperature increase

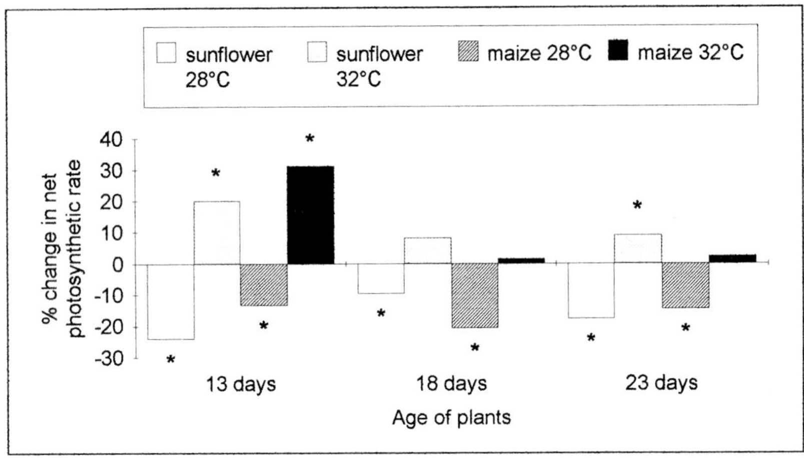

Fig. 3: Percentage change in net photosynthetic rate (nmol CO_2/plant s) of sunflower and maize between –UV-B (28 °C) and +UV-B at 28 °C and 32 °C. Stars indicate significant differences ($\alpha=0.05$).

compensates net photosynthesis reductions measured at T_{max} = 28 °C. In maize plants significant reductions of net photosynthesis based on plant were seen under enhanced UV-B and both temperature regimes, with a maximum reduction of 20.7% (T_{max} 28 °C) and 22.8% (T_{max} 32 °C) after 18 days (Table 3). Effects of temperature generally were a decreased net photosynthetic rate based on leaf area, but an increased net photosynthetic rate per plant compared to plants at T_{max} of 28 °C (Fig. 3).

Respiration in sunflower seedlings based on leaf area increased significantly after 18 days only as a result of higher UV-B radiation at T_{max} 28 °C (Table 2). Plants grown at T_{max} of 32 °C respired less at the time of measurement compared to those grown under 28 °C, and no UV-B effect was observed. Generally, sunflower seedlings grown under higher UV-B radiation and T_{max} of 28 °C respired most. Based on total plant no effects could be observed.

A UV-B dependent increase in respiration based on leaf area was not seen in maize plants (Table 3). An increase of 4 °C in temperature resulted only after 23 days in an increase of respiration per plant of about 90%.

Differences in transpiration did not occur in sunflower and maize plants under elevated UV-B radiation. However, higher temperature resulted in pronounced increases of transpiration per plant, especially in older plants (Tables 2 and 3).

Discussion

The effects of solar UV-B radiation on sunflower and maize seedlings grown at different diurnal temperature courses with T_{max} of 28 °C, and 32 °C respectively, are qualitatively similar in general but quantitatively different in respect to growth, chlorophyll content and gas exchange.

Growth

Sunflower and maize seedlings grown for up to 23 d under higher UV-B radiation at T_{max} of 28 °C, and also under T_{max} of 32 °C exhibited significant reductions in size and leaf area. Growth reductions are often found in UV-sensitive plant species under enhanced UV-B radiation in greenhouses when artificial UV-B sources were used, or in the field when artificial UV-B radiation supplemented solar UV-B (Sullivan and Teramura 1988, 1990, Barnes et al. 1988, Teramura et al. 1990 a, b). The mechanism for the reduction in stem elongation might be due to changes in the phytohormon levels, especially of IAA which is responsible for elongation. In sunflower seedlings grown under artificial UV-B radiation a photooxidation of IAA was found (Ros 1990, Ros and Tevini 1995). Additionally, an activation of peroxidases that dismantle IAA can take place. It is supposed that the same mechanism takes place also under solar UV-B radiation. In contrast to this, the reduction in leaf area is caused by a reduction in cell size (Wolf 1988) and/or a change in leaf structure to smaller but thicker leaves (Tevini et al. 1983).

In both species the temperature increase of 4 °C resulted in higher absolute values of growth parameters regardless of UV-B effects. However, in maize seedlings the temperature effect was more pronounced than in sunflower seedlings. For example, the size of 13 d old maize seedlings increased from 22.7 cm to 28.7 cm representing a 26 % increase, whereas sunflower seedlings gained only 13 % increase at T_{max} of 32 °C compared to a 4 °C lower diurnal course. Growth increases by increasing temperature up to the optimum for photosynthesis are well known in many species (Auld et al. 1978, Johnson and Thornley 1985, Tollenaar 1989). Generally, this growth stimulation is being discussed as an effect of increased photosynthetic rates (Q_{10}-rule) which could also be demonstrated in this study.

Growth differences in sunflower and maize are also well known and are due to physiological changes seen in C_3 and C_4-species.

The biomass (fresh and dry weight over ground) was less affected by enhanced UV-B. The mechanisms for biomass reductions are not yet fully understood. On one hand, decreases in overall photosynthesis were observed (Brandle et al. 1977, Sisson and Caldwell 1976) which is supported by our results, on the other hand, no effects on photosynthetic characteristics were found, e.g. in wheat (Beyschlag et al. 1988). In that case, morphological changes in the canopy structure with different light interception had been discussed for biomass changes. In these sunflower and maize seedlings investigated here, the net photosynthesis per plant was substantially reduced under enhanced UV-B, which is virtually due to a lower total leaf area, since the net photosynthesis per leaf area had not significantly decreased.

The UV-B effects on growth are more pronounced at lower than at the higher temperature. It is possible that the increased temperature was a type of stress for the plants, at least for sunflower. It has been demonstrated that different types of stress resulted in stimulated biosynthesis of flavonoids, which can diminish UV-B effects by their UV-absorbing properties (Murali and Teramura 1987, Tevini et al. 1991). Furthermore, repair mechanisms may also be temperature-dependent and thus prevent UV-B damage at higher temperatures to a higher degree.

Although the leaf area of sunflower and maize was reduced under enhanced UV-B the total chlorophyll content per plant had not been influenced. Therefore the plants showed a higher chlorophyll content per leaf area. The same effect in combination with a thickening of the leaves was observed in different crop plants (Tevini et al. 1983). The influence of temperature on the chlorophyll content was mainly an increase in chlorophyll content per plant. This is probably due to different growth reactions as well as the different chlorophyll contents (per leaf area).

Gas exchange

One important result of this work is that the interpretation of UV-B effects on the gas exchange is dependent on the parameter to which the data are related. Regarding only the net photosynthetic rate based on leaf area the conclusion would be that a stimulation of the photosynthesis for sunflower, and no effect for maize. This would correspond to various studies cited in the literature (Beyschlag et al. 1988, Teramura et al. 1990 a, b). In comparison to this the photosynthetic rate related to chlorophyll content and plant clearly indicates that there is a major impact of UV-B on photosynthesis resulting in a significant decrease. This obvious contradiction is probably produced by the different structure of the leaves from plants grown under normal and enhanced UV-B (Tevini et al. 1983), whereas the chlorophyll content per plant remains unaffected.

The UV-B dependent decrease of photosynthesis can be attributed to lower enzyme activities for ribulose bisphosphate carboxylase and phosphoenolpyruvat carboxylase and/or lower photosystem II activity. Both were found in different plant species (Vu et al. 1982, Bornman et al. 1984, Tevini and Pfister 1985) and especially in sunflower and maize under enhanced artificial UV-B (Mark and Tevini, in preparation). An indirect effect of UV-B on transpiration and therefore on internal CO_2 concentration could not be found during our research.

Concerning the effect of increased temperature it was found that the net photosynthetic rates were mainly higher in sunflower and maize, which may be caused by a general stimulation of photosynthesis (Q_{10}-rule) but is yet not fully investigated. Nevertheless, in both plant species the UV-B effect was compensated. This could be due also to the photorepair mechanism (Teramura 1980) and an increased accumulation of flavonoids which are able to screen out UV-B radiation (Tevini et al. 1991).

Major effects of UV-B and temperature on respiration and transpiration were not found. In correspondence with previous results (Teramura et al. 1980, Mirecki and Teramura 1984) it is suggested that respiration and transpiration of sunflower and maize are insensitive to either enhanced UV-B radiation and moderate temperature increase.

Acknowledgements

This work was supported by the Bundesministerium für Bildung, Wissenschaft, Forschung und Technologie (07 UVB 036).

The authors want to thank the Institut für Internationale Zusammenarbeit in Bildung und Forschung (IZBF) in Hamburg for the logistic support in Portugal as well as for the preparation of the research station Quinta de São Pedro near Lisbon. Special thanks to the manager of the Institute, Mr. A. Pircher, for organisational assistance and the kind reception. The authors thank Mr. E. Heene for technical assistence.

References

Auld, B. A., M. D. Dennett, and J. Elston: The effect of temperature changes on the expansion of individual leaves of *Vicia faba* L. Ann. Bot. *42,* 877–888 (1978).

Baker, N. R., S. P. Long, and D. R. Ort: Photosynthesis and temperature, with particular reference to effects on quantum yield. In: Long, S. P. and F. J. Woodward (eds.): Plants and Temperature. Symp. Soc. Exp. Biol. *42,* pp. 347–375. Company of Biologists, Cambridge (1988).

Barnes, P. W., P. W. Jordan, W. G. Gold, S. D. Flint, and M. M. Caldwell: Competition morphology and canopy structure in wheat (*Triticum aestivum* L.) and wild oat (*Avena fatua* L.) exposed to enhanced ultraviolet-B radiation. Functional Ecology *2,* 319–330 (1988).

Berry, J. A. and O. Björkman: Photosynthetic response and adaption to temperature in higher plants. Annu. Rev. Plant Physiol. *31,* 491–543 (1980).

Beyschlag, W., P. W. Barnes, S. D. Flint, and M. M. Caldwell: Enhanced UV-B irradiation has no effect on photosynthetic characteristics of wheat (*Triticum aestivum* L.) and wild oat (*Avena fatua* L.) under greenhouse and field conditions. Photosynthetica *22,* 516–525 (1988).

Bornman, J. F. and A. H. Teramura: Effects of ultraviolet-B radiation on terrestrial plants. In: Young, A. R., L. O. Björn, J. Moan, and W. Nultsch (eds.): Environmental UV Photobiology, pp. 427–471. Plenum Press, New York (1993).

Bornman, J. F., L. O. Björn, and H. E. Akerlund: Action spectrum for inhibition by ultraviolet radiation of photosystem II activity in spinach thylakoids. Photobiochem. Photobiophys. *8,* 305–313 (1984).

Brandle, J. R., W. F. Campbell, W. B. Sisson, and M. M. Caldwell: Net photosynthesis, electron transport capacity, and ultrastructure of *Pisum sativum* L. exposed to ultraviolet-B radiation. Plant Physiol. *60,* 165–169 (1977).

Caldwell, M. M., S. Madronich, L. O. Björn, and M. Ilyas: Ozone reduction and increased solar ultraviolet radiation. In: Environmental Effects Panel Report, UNEP, Nairobi, pp. 1–10 (1989).

Cen, Y. P. and J. F. Bornman: The Response of Bean Plants to UV-B Radiation under Different Irradiances of Background Visible Light. Exp. Bot. *41,* 1489–1495 (1990).

Harper, J. C.: Population biology of plants. Academic Press, New York (1977).

Hofmann, D. J., S. J. Oltmans, J. M. Harris, S. Solomon, T. Deshler, and B. J. Johnson: Observation and possible causes of new ozone depletion in Antarctica in 1991. Nature *359,* 283–287 (1992).

Houghton, R. A. and G. M. Woodwell: Global climatic change. Sci. Amer. *260,* 18–26 (1989).

Johnson, I. R. and J. H. M. Thornley: Temperature dependence of plant and crop processes. Ann. Bot. *55,* 1–24 (1985).

Jordan, D. B. and W. L. Ogren: The CO_2/O_2 specificity of ribulose-1,5-bisphosphate concentration, pH and temperature. Planta *161,* 308–313 (1984).

Khayate, E., D. Rravad, and N. Zieslin: The effect of various night temperature regimes on the vegetative growth and fruit production of tomato plants. Sci. Hortic. *27,* 9–13 (1985).

Langer, B. and E. Wellmann: Phytochrom induction of photoreactivating enzyme in *Phaseolus vulgaris* L. seedlings. Photochem. Photobiol. *52,* 861–863 (1990).

Lichtenthaler, H. K.: Chlorophylls and carotenoids: Pigments of photosynthetic biomembranes. Methods of Enzymology *148,* 350–382 (1987).

Long, S. P.: Photosynthesis in 4-carbon pathway plants at low temperature. In: Marcelle, R., H. Clijsters, and M. van Poucke (eds.): Advances in Agricultural Biotechnology: Effects of Stress on Photosynthesis, pp. 237–244. Martinus Nijhoof, Dr. W. Junk Publishers, The Hague (1983).

Long, S. P.: Modification of the response of photosynthetic productivity to rising temperature by atmospheric CO_2 concentrations: Has its importance been underestimated? Plant, Cell and Env. *14,* 729–739 (1991).

Marcelis, L. F. M. and L. R. Bann Hofman-Eijer: Effect of temperature on the growth of individual cucumber fruits. Physiol. Plant. *87,* 321–328 (1993).

Mark, U.: Zur Wirkung erhöhter artifizieller und solarer UV-B-Strahlung in Kombination mit erhöhter Temperatur und Kohlendioxidkonzentration auf das Wachstum und den Gaswechsel von ausgewählten Nutzpflanzen. In: Tevini, M. (ed.): Karlsr. Beitr. Entw. Ökophysiol. *11,* 1–220 (1992).

Mirecki, R. M. and A. H. Teramura: Effects of ultraviolet-B irradiance on soybean. V. The dependence of plant sensitivity on the photosynthetic photon flux density during and after leaf expansion. Plant Physiol. *74,* 475–480 (1984).

Molina, M. J. and F. S. Rowland: Stratospheric sink for chlorofluoro-methanes: chlorine atom-catalyzed destruction of ozone. Nature *249,* 810–812 (1974).

Murali, N. S. and A. H. Teramura: Insensitivity of soybean photosynthesis to ultraviolet-B radiation under phosphorus defiency. J. Plant Nutr. *10,* 501–515 (1987).

Pang, D. W. and J. B. Hays: UV-B Inducible and Temperature-sensitive Photoreactivation of Cyclobutane Pyrimidine Dimers in *Arabidopsis thaliana.* Plant Physiol. *95,* 536–543 (1991).

Ros, J.: Zur Wirkung von UV-Strahlung auf das Streckungswachstum von Sonnenblumenkeimlingen (*Helianthus annuus* L.). In: Tevini, M. (ed.): Karlsr. Beitr. Entw. Ökophysiol. *8,* 1–157 (1990).

Ros, J. and M. Tevini: UV-radiation and Indole-3-acetic acid: Interactions during growth of seedlings and hypocotyl segments of sunflower. J. Plant Physiol. *146,* 295–305 (1995).

Schneider, S. H.: The greenhouse effect: science and policy. Science *243,* 771–781 (1989).

Sisson, W. B. and M. M. Caldwell: Photosynthesis, dark respiration and growth of *Rumex patientia* L. exposed to ultraviolet irradiance (280 to 315 nanometers) simulating a reduced atmospheric ozone column. Plant Physiol. *58,* 563–568 (1976).

Sullivan, J. H. and A. H. Teramura: Effects of ultraviolet-B irradiation on seedling growth in the *Pinaceae.* Amer. J. Bot. *72,* 225–230 (1988).

– – Field study of the interaction between solar ultraviolet-B radiation and drought on photosynthesis and growth in soybean. Plant Physiol. *92,* 141–146 (1990).

Teramura, A. H.: Effects of ultraviolet-B irradiance on soybean. I. Importance of photosynthetically active radiation in evaluating ultraviolet-B irradiance effects on soybean and wheat growth. Physiol. Plant. *48,* 333–339 (1980).

Teramura, A. H., R. H. Biggs, and S. Kossuth: Effects of ultraviolet-B irradiances on soybean. II. Interaction between ultraviolet-B and photosynthetically active radiation on net photosynthesis, dark respiration and transpiration. Plant Physiol. *65,* 483–488 (1980).

Teramura, A. H., J. H. Sullivan, and L. Lydon: Effects of UV-B radiation on soybean yield and seed quality: a 6-year field study. Physiol. Plant. *80,* 5–11 (1990 a).

Teramura, A. H., J. H. Sullivan, and L. H. Ziska: Interaction of elevated CO_2 on productivity and photosynthetic characteristics in wheat, rice, and soybean. Plant Physiol. *94,* 470–475 (1990 b).

Tevini, M.: UV-B Effects on terrestrial plants and aquatic organisms. Progress in Botany *55,* 174–189 (1994).

Tevini, M. and K. Pfister: Inhibition of photosystem II by UV-B radiation. Z. Naturforsch. *40c,* 129–133 (1985).

Tevini, M. and A. H. Teramura: UV-B effects on terrestrial plants. Photochem. Photobiol. *50,* 479–487 (1989).

Tevini, M., U. Thoma, and W. Iwanzik: Effects of enhanced UV-B radiation on germination, seedling growth, leaf anatomy and pigments of some crop plants. Z. Pflanzenphysiol. *109,* 435–448 (1983).

Tevini, M., U. Mark, and M. Saile: Plant experiments in growth chambers illuminated with natural sunlight. In: Payer, H. D., T. Pfirrmann, and P. Mathy (eds.): Environmental Research with Plants in Closed Chambers, pp. 240–251. Air pollution Res. Rep. *26,* Commission of the European Communities, Brussels, Belgium (1990).

Tevini, M., J. Braun, and G. Fieser: The protective function of the epidermal layer of rye seedlings against ultraviolet-B radiation. Photochem. Photobiol. *53,* 329–333 (1991).

Tollenaar, M.: Response to dry matter accumulation in maize to temperature. II. Leaf photosynthesis. Crop Sci. *29,* 1275–1279 (1989).

Von Caemmerer, S. and G. D. Farquhar: Some relationships between the biochemistry of photosynthesis and the gas exchange of leaves. Planta *153,* 376–387 (1981).

Watson, R. T., H. Rodhe, H. Oescheger, and U. Siegenthaler: Greenhouse gases and aerosols. In: Houghton, J. T., G. J. Jenkinsand, and J. J. Ephraums (eds.): Climate Change: The IPCC Scientific Assessment, pp. 1–40. Cambridge University Press, Cambridge (1990).

Wolf, U.: Morphologische Untersuchungen zur Wirkung von UV-Strahlung auf das Streckungswachstum von Sonnenblumenkeimlingen. Staatsexamensarbeit. Universität Karlsruhe (1988).

Vu, C. V., L. H. Allen, and L. A. Garrard: Effects of supplemental UV-B radiation on primary photosynthetic carboxylating enzymes and soluble proteins in leaves of C_3 and C_4 crop plants. Physiol. Plant. *55,* 11–16 (1982).

J. Plant Physiol. Vol. 148. pp. 57–62 (1996)

Rapid Fluence-Dependent Responses to Ultraviolet-B Radiation in Cucumber Leaves: The Role of UV-Absorbing Pigments in Damage Protection

Paulien Adamse and Steven J. Britz

U.S. Dept. of Agriculture, Agricultural Research Service, Climate Stress Laboratory, Beltsville, MD 20705-2350 USA

Received June 24, 1995 · Accepted October 20, 1995

Summary

The role of foliar UV-absorbing pigments (UVAP) as optical screening agents in the resistance of *Cucumis sativus* L. to UV-B radiation was investigated by exposing young leaves at a defined developmental stage from sensitive (cv Poinsett) and insensitive (cv Ashley) lines to brief UV-B treatments varying between 4 and 10 h. The amount of blue light (BL) or UV-A radiation during UV-B exposure was also varied. Rapid increases in UVAP immediately following UV-B were compared to damage in the same tissue (increased specific leaf weight or chlorosis) determined 72 h after the start of UV-B. Poinsett was more sensitive to both forms of UV-B damage than Ashley under conditions where the response to UV-B was not saturated. Although UVAP increased rapidly in response to UV-B, it is unlikely that optical screening by these compounds was responsible for genetic differences in sensitivity to UV-B for the following reasons: 1) the kinetics of UVAP increase were similar to that for induction of damage; 2) increases in UVAP in the UV-sensitive line (Poinsett) were similar to those in the resistant line (Ashley); and 3) BL and UV-A radiation significantly reduced damage by UV-B in cv Poinsett when given simultaneously but had relatively small stimulatory effects on rapid UVAP accumulation. These results do not rule out a general role for optical screening by UVAP nor do they exclude the possibility that qualitative differences in UVAP (e.g., as antioxidants) are the basis for cultivar differences.

Key words: Cucumis sativus L., blue light, cucumber, flavonoids, ultraviolet, chlorosis, specific leaf weight.

Abbreviations: BL = blue light 400–500 nm; cv(s) = cultivar(s); LPS = low pressure sodium; PAR = photosynthetically-active radiation 400–700 nm; UV-A = ultraviolet-A 315–400 nm; UV-B = ultraviolet-B 280–315 nm; UVAP = UV-absorbing pigments.

Introduction

Genetic variability in higher plants in response to UV-B is well-documented (Murali and Teramura, 1986; Reed et al., 1992). Determination of mechanisms underlying these differences should contribute to the development of stress tolerant crops and improve our ability to assess the impact of UV-B on plant growth. However, except for some known mutations (Li et al., 1993; Britt et al., 1993), the bases for most genetic differences are not well understood.

Intraspecific differences in UV-B sensitivity in cucumber have been correlated with levels of UVAP (Murali and Teramura, 1986). These compounds (presumptive flavonoids; McClure, 1975) could serve as optical screens, absorbing UV while transmitting PAR (Caldwell et al., 1983). Significantly fewer pyrimidine dimers were formed by UV-B radiation in algal cultures with higher levels of flavonoids (Takahashi et al., 1991). UVAP increase in response to UV-B exposure in many plants, including curcurbits (Sisson, 1981). Although increased flavonoids have been suggested as an indicator of DNA damage (Beggs et al., 1985), data from cucumber correlate accumulation of UVAP in leaves with reduced damage (Adamse et al., 1994). UV-B exposure induced the accumulation of UVAP on a leaf area basis in two cultivars of cucumber, with approximately 50 % greater response after a 3 day treatment in cv Ashley (more resistant to damage) than in cv Poinsett (less resistant). Starting values of UVAP were similar for the two lines.

BL and UV-A often enhance the UV-induced synthesis of phenylpropanoid compounds (Ohl et al., 1989) and can be used to dissect the role of UVAP levels in resistance to UV-B. Supplemental BL during UV-B treatment ameliorated damage and caused both increases in total UVAP per leaf as well as a shift in the relative amounts of different UVAP (Adamse et al., 1994). BL effects were fluence-dependent and not saturated where BL constituted up to 7 % of daily photosynthetic photon flux. Interpretation, however, was complicated by simultaneous effects of BL on leaf growth and pigment content not related to UV-B exposure. Furthermore, damage to the leaf by UV-B was probably induced during the first 24 h of exposure. UVAP on a leaf area basis increased significantly within this time, but the sensitive and resistant lines did not differ in this early pigment response and no effect of BL was detected.

More detailed kinetic analysis of UV-B response during short term exposure is required to test the possibility that rapid increases in foliar UVAP content during the first hours of UV-B exposure contributed to amelioration of damage. The objectives of the current study therefore are to evaluate the role of bulk increases of UVAP in prevention of UV-B damage (e.g., through an optical screening mechanism) by comparing the kinetics of UVAP accumulation induced by UV-B in relation to the time course for induction of damage in cultivars differing in sensitivity to UV-B.

Supplemental BL and UV-A were used to modify damage and UVAP accumulation simultaneously in the same tissue. Sensitivity to UV-B is altered by background levels of PAR (Cen and Bornman, 1990), spectral interactions (Middleton and Teramura, 1994), and atmospheric CO_2 (Adamse and Britz, 1992 b). To minimize the possibility that BL or UV-A act indirectly through photosynthesis or phytochrome, background high PAR from amber LPS lamps was used to maintain high rates of photosynthesis during treatment as well as high levels of phytochrome phototransformation to the far-red-absorbing form (Adamse and Britz, 1992 b; Britz and Sager, 1989).

Developing third leaves immediately subsequent to opening, and without prior UV-B exposure, were irradiated for 10 h or less with UV-B in the presence of high PAR from background LPS lamps ± supplemental BL or UV-A. The third leaf was selected because the prime period of UV-B sensitivity does not overlap with that of other leaves (data not shown) and because there is little direct effect of UV-B on the first leaf which serves as a source of photosynthate (Adamse and Britz, 1992 b; Britz and Adamse, 1994).

Some plants were returned to white light growth conditions without UV-B for quantification of UV-B-induced leaf damage (i.e., chlorosis or increased SLW) 72 h after the start of treatment. This approach allowed us to concentrate on responses induced rapidly by brief UV exposure, while minimizing possible differences in the kinetics of expression. In other plants, young leaves were harvested and extracted immediately after UV-B exposure to determine whether UVAP increased rapidly enough to influence damage by UV-B. Note that increased SLW is interpreted as a sensitive indicator of UV-B damage, since the largest changes in SLW in the third leaf were associated with the greatest inhibition of leaf growth by UV-B (Adamse et al., 1994). SLW is highly conserved and it is possible to see significant changes when differences in leaf area and dry weight are no longer discernible.

Materials and Methods

Plant material

Cucumber (*Cucumis sativus* L.) cvs Poinsett and Ashley were grown in 1.5 l pots with vermiculite, flushed daily with a modified half-strength Hoagland's solution, in a growth chamber at 27 °C, ca. 50 % relative humidity, and ca. 450 µmol mol^{-1} CO_2 (ambient level in the building). Plants were illuminated 14 h per day with a 1 : 1 mixture of 400 W high pressure sodium and metal halide lamps. PAR was stepped diurnally, starting with 1 h at 270 µmol m^{-2} s^{-1}, continuing with 12 h at 840 µmol m^{-2} s^{-1} and concluding with 1 h at 270 µmol m^{-2} s^{-1}. After 11 days, plants were selected for uniformity and transferred to a UV exposure chamber approximately 110 min after the onset of white light. This time corresponded to the unfolding of the third leaf.

Light treatments

The UV exposure chamber was maintained at the same environmental conditions as the growth chamber except for light quality and was divided into two compartments by means of a UV-B absorbing polyester film. An adjustable bank of UV-B lamps (F40WT12-6A/UVB-313, Q-panel Co., Cleveland, OH, USA) was suspended underneath the main lamp canopy. On one side, UVB-313 lamps were wrapped with cellulose diacetate film (0.08 mm) to provide UV-B (+UVB). On the other side, the lamps were wrapped with polyester film (0.13 mm) to remove UV-B while transmitting longer wavelengths (−UVB). Most PAR in the second chamber was provided by LPS lamps (SOX 180 W, Philips North America, Bloomfield, NJ) with or without supplemental BL or UV-A fluorescent lamps (respectively, F40T12/247 and F40T12/BLB GTE Sylvania, Danvers, MA). For description of light levels see Table 1. Note that the low levels of BL (Hg lines at 405 and 436 nm) and UV-A (mainly below 350 but including the Hg line at 365 nm) were emitted by the UV-B source lamps (supplemental lamps off). Wrapping the lamps in polyester substantially reduced the shorter wavelength component of UV-A. Background LPS and BL lamps produced no UV-B. UV-B sources, UV-B absorbing or transmitting filters and the resulting spectra are described in more detail elsewhere (Adamse and Britz, 1992 a).

Two series of experiments were conducted. In the first, plants were treated with UV-B in the presence or absence of supplemental BL (Table 1). The second series used supplemental UV-A instead of BL (Table 1). The two cultivars were treated simultaneously, but only one level of supplemental BL or UV-A could be used in a given experiment. Consequently, +BL and −BL treatments were alternated during the first series, while +UV-A and −UV-A were alternated during the second series.

Specific leaf weight and chlorosis

Plants were irradiated with UV (±UVB) for 4, 8 or sometimes 10 h and returned immediately to white light growth conditions. Zero time controls were subjected to a mock exposure and returned to white light. The plants were harvested 72 h after the start of the UV treatment. Specific leaf weight was determined after measuring

Table 1: Light conditions during UV treatments. Photon fluence rate (μmol m^{-2} s^{-1}) of photosynthetically-active radiation (PAR; 400–700 nm), blue light (BL; 400–500 nm), UV-A (315–400 nm), UV-B (290–315 nm). Biologically-effective UV-B (UV-B_{BE} normalized to 1.0 at 300 nm; Caldwell, 1971) for a maximal 10 h total exposure were 20.8 and 0.3 kJ m^{-2} for +UVB and −UVB, respectively.

Supplement		BL	UV-A	UV-B	PAR
+BL	+UVB	59	9	2.80	826
	−UVB	58	6	0.04	825
−BL	+UVB	6	9	2.80	778
	−UVB	5	6	0.00	777
+UV-A	+UVB	6	52	2.91	531
	−UVB	5	49	0.16	530
−UV-A	+UVB	5	9	2.80	530
	−UVB	4	6	0.05	529

dry weight and area (LiCor Model LI-3000, Lincoln, NE, USA) of the third leaf. Chlorosis was estimated visually. Separate studies evaluating subjective scoring with a leaf scanning system indicated the two methods were comparable. The scanning system consisted of a video camera (Model 652, Dage-MTI, Michigan City, IN) and Ultricon video tube (Burle Industries, Lancaster, PA), connected to a Macintosh PC. Leaf videos were digitized (New Image Technology, Seabrook, MD) and stored as bit-mapped images (Macintosh 512 K, Apple Computer, Cupertino, CA). By adjusting the contrast, an image of the leaf was obtained either with or without chlorotic lesions visible. Pixel number for both images was compared after scanning and the percent damaged area was estimated.

UV-absorbing pigments

Third leaves were harvested after 0, 4, 8 or sometime 10 hours of UV treatment and stored in liquid nitrogen. UVAP were extracted from whole leaves after incubation (24 h in darkness at room temperature) in acidified methanol (methanol : HCl : water : : 40 : 1 : 59) as described previously (Adamse and Britz, 1992 b). Absorption spectra (Model 219, Cary, Varian, Palo Alto, CA, USA) were comparable to those published elsewhere (Cen and Bornman, 1990) and revealed two peaks at approximately 270 and 325 nm. Pigments were quantified at 325 nm in order to maximize sensitivity.

Statistics

Indicated values are averages of 8 to 16 samples, obtained in two or three replicate experiments. Significance of difference between means was determined based on one-sided t-tests at $P<0.05$.

Results

UV-Absorbing Pigments

UV-B treatment caused similar, rapid 20 % increases in UVAP in third leaves of cultivars Poinsett and Ashley extracted immediately following UV-B exposure (−BL and −UV-A; Fig. 1). The response appeared to be slower in the second series (±UV-A), possibly as a result of slight differences in leaf age or background PAR. Although it is not possible to determine unambiguously whether the UV-A or BL components of the UV-B lamps affected the response to UV-B, supplemental UV-A or BL did not affect UVAP accumulation in the absence of UV-B (data not shown) and had very little effect on the UV-B response in Poinsett. However, supplemental UV-A or BL stimulated rapid UVAP accumulation in Ashley, with significant UV-B effects detected after only 4 h of treatment (i.e., 8.3 kJ m^{-2} of UV-B_{BE} total fluence).

Chlorosis

Induction of chlorotic lesions was observed in cv Poinsett (Fig. 2) but not in cv Ashley (data not shown), confirming

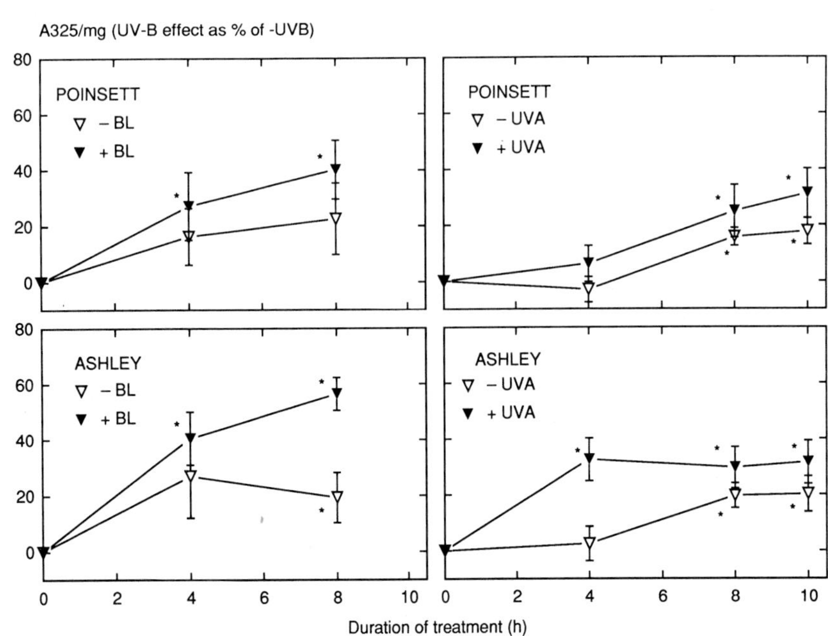

Fig. 1: Effect of blue light (BL) or UV-A radiation on UV-B-induced accumulation of UV-absorbing pigments in the third leaf of cucumber cvs Poinsett and Ashley. White-light-grown plants were treated with UV-B (+UVB) for up to 10 h starting 11 days after planting. Controls (−UVB) were exposed to polyester-wrapped UV-B lamps for comparable intervals and did not receive UV-B radiation. Treatment took place in a growth chamber illuminated with high PAR from LPS lamps with or without supplemental radiation (±BL or ±UV-A). Irradiances are described in Table 1. All treatments began at the same time of day (110 min after lights-on). Leaves were harvested immediately after treatment (±UVB), frozen in liquid nitrogen and extracted in acidic methanol. Values are the average of two (±BL) or three (±UV-A) replicate experiments and are expressed as absorbance change at 325 nm per mg dry weight (+UVB minus −UVB) normalized with respect to −UVB controls. Bars indicate one standard error of the mean; * indicates a significant UV-B effect (P<0.05).

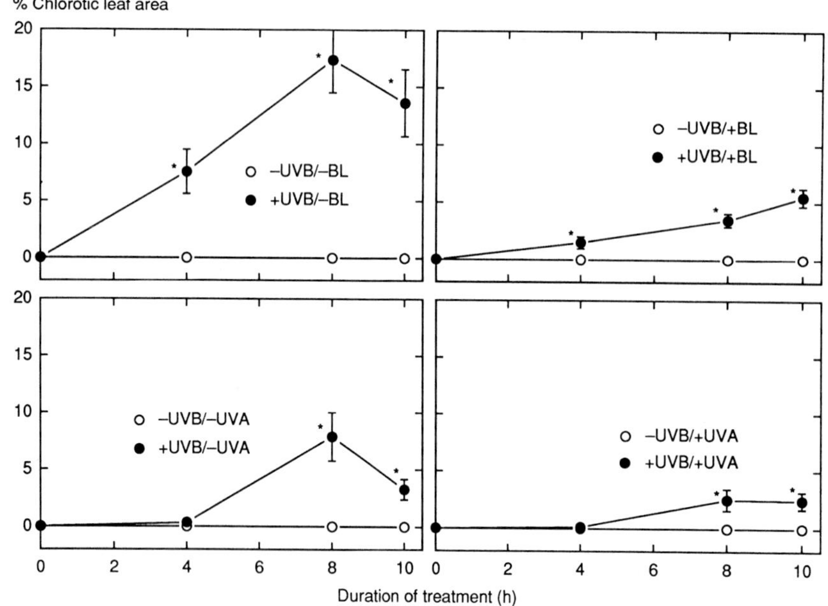

Fig. 2: Effect of blue light (BL) or UV-A radiation on expression of UV-B-induced chlorotic lesions in the third leaf of cucumber cv Poinsett. Plants were returned to the white light chamber immediately after exposure. Damage, expressed as a percentage of the total leaf area, was assayed 72 h after the start of the UV-B treatment. Leaf three expanded from ca. 1 cm^2 to 26 cm^2 in white light controls during this time. Other details are as in Fig. 1.

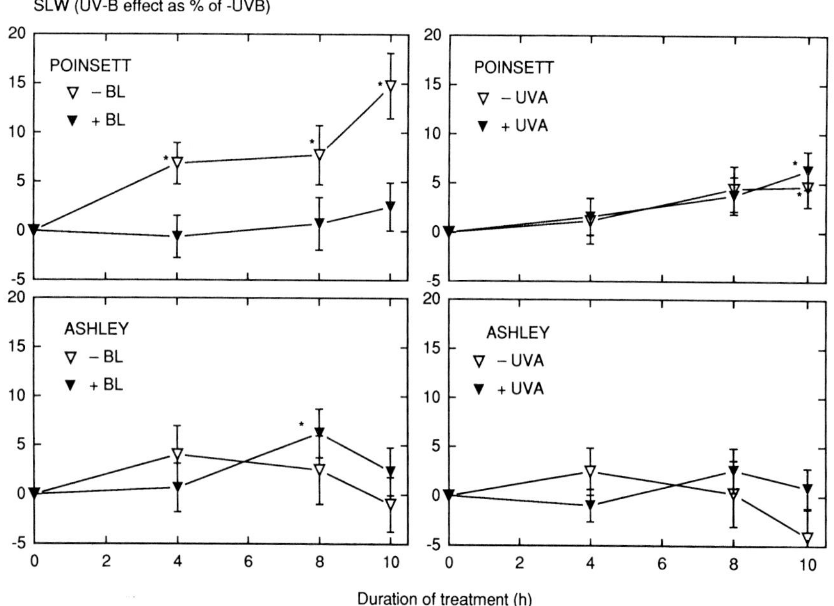

Fig. 3: Effect of blue light (BL) or UV-A radiation on UV-B-induced increase in Specific Leaf Weight (SLW) in the third leaf of cucumber cvs Poinsett and Ashley. Plants were transferred from ±UVB back to white light and leaves were harvested 72 h after the start of treatment. Values are the average of two (±BL) or three (±UV-A) replicate experiments and are expressed as the change in SLW (+UVB minus –UVB) normalized with respect to the –UVB controls. Other details are as in Figs. 1 and 2.

the differential sensitivity of the cultivars to UV-B reported previously (Adamse and Britz, 1992 b; Adamse et al., 1994). Significant chlorosis was induced by 4 h of UV-B in Poinsett in the absence of BL and approached 20 % of the leaf area for an 8 h exposure. Maximal injury was less than 10 % of leaf area during the UV-A interaction trials and longer exposures were required to see a significant effect. In spite of these differences, supplemental BL or UV-A during UV-B treatment reduced damage to a similar extent during an 8 h exposure (about 80 % for BL and 70 % for UV-A). Note that UV-A did not induce chlorosis when given by itself, even at much greater photon fluence rate.

Specific Leaf Weight

SLW in Ashley was not affected by UV-B in either series of experiments (Fig. 3). In Poinsett, UV-B exposure caused significant increases in SLW in both series of experiments, although sensitivity was greatly reduced during the second series (±UV-A), again possibly as a result of slight differences in leaf age or background PAR during UV-B exposure. Supplemental BL dramatically reduced the UV-B-induced increase in SLW, consistent with interpretation of the changes in SLW as a damage response. Although both supplemental BL and UV-A prevented one form of UV-B damage to leaves (i.e. chlorosis), UV-A did not prevent a small but significant increase in SLW following a 10 h exposure to UV-B. Note that comparable increases in SLW following 4 h of UV-B in the first series were clearly reversed by BL. These data confirm earlier results (Adamse et al., 1994) that leaf growth in Poinsett is more sensitive to UV-B than in Ashley when responses induced during the first 24 h of exposure are compared. Moreover, the differences between cultivars are not caused by a more rapid response in Poinsett.

Discussion

Optical screening by UVAP may provide protection from UV-B exposure, but most studies correlating UVAP content with altered susceptibility to UV-B damage are evaluated after long-term UV exposure (days or weeks). It is therefore difficult to assess the significance of quantitative or qualitative differences in UVAP because the time of maximal sensitivity to UV-B damage is generally not known in relation to the kinetics of pigment accumulation. For example, less damaged plants may coincidently accumulate more UVAP. The current study avoids this problem by concentrating on rapid responses under conditions where UV-B effects are not saturated and clear differences between sensitive and insensitive cultivars can be demonstrated. Furthermore, differential damage is not the result of more rapid expression in the sensitive cultivar.

The data obtained do not support a role for differences in UVAP accumulation as the basis for genetic differences in susceptibility to damage from UV-B radiation in cucumber. While UVAP accumulate rapidly in leaves of a sensitive cultivar (Poinsett) during UV exposure, the response curves for simultaneous induction of two forms of damage in the same tissue (chlorotic lesion and increased SLW) are similar to that for pigment accumulation. Thus, it is unlikely that UVAP provide additional screening before damage is induced. Two cultivars, Ashley and Poinsett, that differ in susceptibility to damage had similar UVAP accumulation in the absence of supplemental BL or UV-A radiation. Finally, both supplemental UV-A and BL ameliorated damage in cv Poinsett but produced only a small stimulation of UVAP accumulation induced by UV-B. These results are consistent with earlier studies (Adamse et al., 1994) showing that BL stimulation of large-scale UVAP accumulation was expressed between 24 and 72 h after the start of treatment. Note that UVAP were determined at 325 nm, the absorbance maximum of the extracts, in order to optimize the discrimination of small changes. Absorbance changes, hence changes in screening of incident radiation, would be reduced proportionately at other wavelengths (e.g., 300 nm) where extinction coefficients are lower.

BL and UV-A had qualitatively similar effects on responses to UV-B, but quantitative comparisons are difficult to make for several reasons. The sensitivity of the third leaf to UV-B decreased during the latter part of the study (i.e, ±UV-A), possibly as a result of slight (ca. 12 h) increases in leaf physiological age. PAR was also lower during UV-B treatment (but not during white light growth before or after) because the UV-A lamps shaded the LPS lamps. Nonetheless, differences in leaf sensitivity or PAR do not affect the basic conclusions concerning UVAP involvement, since results within each series are internally consistent.

Comparisons between UV-A and BL are also complicated because the UV-B lamps emitted significant amounts of longer wavelength radiation (Table 1) and because plants received UV-A and BL during normal white light growth chamber conditions. Thus, it is not clear to what extent responses to UV-B were actually the result of long wavelength contamination from the UV-B lamps or interactions in the growth chamber. It is possible that all or part of the apparent differential sensitivity between cultivars is actually the result of differences in damage prevention or repair dependent on exposure to UV-A and BL provided concomitantly by the UV-B lamps. Pure sources of UV-B are required for more precise interpretation of spectral interaction.

Although bulk changes in UVAP do not appear to be involved in the differential sensitivity between cultivars, it should be emphasized that Ashley and Poinsett have different complements of UVAP as detected by HPLC (Adamse et al., 1994). BL treatment differentially affected the relative distribution of these compounds in the two cultivars. Qualitative differences in flavonoid content could be related to biochemical function. For example, flavonoids may act as antioxidants (Husain et al., 1987), protecting against oxidative stresses such as UV-B. Thus, it is possible that genetic differences are related to biochemical differences in flavonoid function.

It is also possible that differences between cultivars as well as the effects of supplemental BL and UV-A are related to DNA damage and repair. The effects of BL and UV-A are consistent with rapid repair of pyrimidine dimers formed during UV irradiation (i.e., photoreactivation; Pang and Hays, 1991; Karentz et al., 1991; Quaite et al., 1992), suggesting that DNA may be a target for damaging physiological action of UV-B exposure. It has been shown that BL increases photoreactivation (e.g., Tanada and Hendricks, 1953; Saito and Werbin, 1969; Beggs et al., 1985, Pang and Hays, 1991). Differential responses to BL and UV-A could be explained by different photoreactivating enzymes with different spectral sensitivity (e.g., Pang and Hays, 1991).

Acknowledgements

This work was supported in part by U.S. Department of Agriculture grant CSRS 89-37280-4799. The authors thank Drs. J. Sullivan and E. Middleton for helpful comments.

References

Adamse, P. and S. J. Britz: Spectral quality of two fluorescent UV sources during long-term use. Photochem. Photobiol. *56*, 641–644 (1992a).

– – Amelioration of UV-B damage under high irradiance. I: Role of photosynthesis. Photochem. Photobiol. *56*, 645–650 (1992b).

Adamse, P., S. J. Britz, and C. R. Caldwell: Amelioration of UV-B damage under high irradiance. II: Role of blue light photoreceptors. Photochem. Photobiol. *64*, 110–115 (1994).

Beggs, C. J., A. Stolzer-Jehle, and E. Wellmann: Isoflavonoid formation as an indicator of UV stress in bean (*Phaseolus vulgaris* L.) leaves. The significance of photorepair in assessing potential damage by increased solar UV-B radiation. Plant Physiol. *79*, 630–634 (1985).

Britt, A. B., J.-J. Chen, D. Wykoff, and D. Mitchell: A UV-sensitive mutant of *Arabidopsis* defective in the repair of pyrimidine-pyrimidinone (6-4) dimers. Science *261,* 1571–1574 (1993).

Britz, S. J. and P. Adamse: UV-B-induced increase in specific leaf weight of cucumber as a consequence of increased starch content. Photochem. Photobiol. *64,* 116–119 (1994).

Britz, S. J. and J. C. Sager: Photomorphogenesis and photoassimilation in soybean and sorghum grown under broad spectrum or blue-deficient light sources. Plant Physiol. *94,* 448–454 (1990).

Caldwell, M. M.: Solar UV irradiation and the growth and development of higher plants. In: Giese, A. C. (ed.), Photophysiology, Vol. *6,* pp.131–177. Academic Press, New York, NY (1971).

Caldwell, M. M., R. Robberecht, and S. D. Flint: Internal filters: prospects for UV-acclimation in higher plants. Physiol. Plant. *58,* 445–450 (1983).

Cen, Y.-P. and J. F. Bornman: The response of bean plants to UV-B radiation under different irradiances of background visible light. J. Exp. Bot. *41,* 1489–1495 (1990).

Husain, S. R., J. Cillard, and P. Cillard: Hydroxyl radical scavenging activity of flavonoids. Phytochemistry *26,* 2489–2491 (1987).

Li, J., T. Ou-Lee, R. Raba, R. G. Amundson, and R. L. Last: *Arabidopsis flavonoid* mutants are hypersensitive to UV-B radiation. The Plant Cell *5,* 171–179 (1993).

Karentz, D., J. E. Cleaver, and D. L. Mitchell: Cell survival characteristics and molecular responses of antarctic phytoplankton to ultraviolet-B radiation. J. Phycol. *27,* 326–341 (1991).

McClure, J. W.: Physiology and function of flavonoids. In: Harborne, B., T. J. Marby, and H. Marby (eds.): The Flavonoids, Part *2,* pp. 970–1055. Academic Press, New York, NY (1975).

Middleton, E. M. and A. H. Teramura: Understanding photosynthesis, pigment and growth response induced by UV-B and UV-A irradiances. Photochem. Photobiol. *60,* 38–45 (1994).

Murali, N. S. and A. H. Teramura: Intraspecific differences in *Cucumis sativus* sensitivity to ultraviolet-B radiation. Physiol. Plant. *68,* 673–677 (1986).

Ohl, S., K. Hahlbrock, and E. Schaefer: A atable blue light-derived signal modulates UV light-induced chalcone synthase gene activation in cultured parsley cells. Planta *177,* 228–236 (1989).

Pang, Q. and J. B. Hays: UV-B-inducible and temperature-sensitive photoreactivation of cyclobutane pyrimidine dimers in *Arabidopsis thaliana.* Plant Physiol. *65,* 536–543 (1991).

Quaite, F. E., B. M. Sutherland, and J. C. Sutherland: Action spectrum for DNA damage in alfalfa lowers predicted impact of ozone depletion. Nature *358,* 576–578 (1992).

Reed, H. E., A. H. Teramura, and W. L. Kenworthy: Ancestral U.S. soybean cultivars characterized for tolerance to ultraviolet-B radiation. Crop Sci. *32,* 1214–1219 (1992).

Saito, N. and H. Werbin: Action spectrum for a DNA-photoreactivating enzyme isolated from higher plants. Radiation Botany *9,* 421–424 (1969).

Sisson, W. B.: Photosynthesis, growth and ultraviolet irradiance absorbance of *Cucurbita pepo* L. leaves exposed to ultraviolet-B radiation (280–315 nm). Plant Physiol. *37,* 120–124 (1981).

Takahashi, A., K. Takeda, and T. Oshini: Light-induced anthocyanin reduces the extent of damage to DNA in UV-irradiated *Centaurea cyanus* cells in culture. Plant Cell Physiol. *32,* 541–547 (1991).

Tanada, T. and S. B. Hendricks: Photoreversal of ultraviolet effects in soybean leaves. Amer. J. Bot. *40,* 634–637 (1953).

J. Plant Physiol. Vol. 148. pp. 63–68 (1996)

Response of Soybean Bulk Leaf Water Relations to Ultraviolet-B Irradiation

Edwin L. Fiscus[1], Fitzgerald L. Booker[2], and Joseph E. Miller[1]

[1] U.S. Department of Agriculture, Agricultural Research Service, and Department of Crop Science, North Carolina State University, 1509 Varsity Drive, Raleigh NC 27606

[2] Air Quality Research Program, Department of Crop Science, North Carolina State University, Raleigh NC 27695

Received June 29, 1995 · Accepted November 12, 1995

Summary

Erosion of the stratospheric ozone layer by anthropogenic emission of halogenated compounds may lead to increased UV-B radiation at ground level. In 1990 soybeans (*Glycine max* (L.) Merr. cv. Essex) were grown in open-top field chambers with 3 levels of UV-B replicated 3 times. UV-B treatments corresponded to changes in the total column ozone thickness of +15 % (low), −20 % (medium) and −35 % (high). Leaves were sampled during four intervals of the growing season and subjected to pressure-volume (P-V) analysis to determine symplastic volume (V_o), maximum turgor pressure (P_{MAX}), symplastic solute content (N_s), tissue elasticity coefficient (z), and the potential at turgor loss (TLP). Leaf conductance, average specific leaf weight (SLW), and area per leaf were measured several times during the season. During the second sampling period at 58 days after planting (DAP), V_o was significantly decreased at the highest UV-B level but was not affected by UV-B treatments again. Also during the second sampling period the elasticity coefficient, z, in the medium and high UV-B treatments was significantly less than in the low UV-B treatment. In the fourth sampling period (100 DAP), z again was significantly affected by UV-B treatment; in this case, however, while z was lower in the high than the medium UV-B treatment, it did not differ from the low UV-B treatment. The relative symplastic volume at the turgor loss potential (RSV_{TLP}) was the only other parameter to show a significant UV-B effect, but only at the highest treatment level and only near the end of the season. Generally, the significant changes in all P-V parameters with plant age were much larger than any treatment effects. There were no significant differences in leaf conductance, leaf area or SLW, indicating that, although the leaf P-V relationships were noticeably altered, these sporadic treatment effects had little real influence on leaf water balance.

Key words: Glycine max (L.) Merr., Conductance, Elasticity, Leaf ontology, Pressure-Volume Analysis, Soybean, Ultraviolet-B, Leaf Water Relations.

Abbreviations: DAP = days after planting; N_s = symplastic solute content; P_{MAX} = turgor potential at full hydration; P-V = pressure-volume; R-B = Robertson-Berger; RSV_{TLP} = relative symplastic volume at the turgor loss potential; SLW = specific leaf weight; TLP = water potential at turgor loss (incipient plasmolysis); UV-B = ultraviolet-B; $UV\text{-}B_{BE}$ = biologically effective UV-B; V_o = symplastic volume at full turgor; z = tissue elasticity coefficient; ε = tissue volumetric elastic modulus; ε_{max} = tissue volumetric elastic modulus at full hydration; $\psi_{\pi 100}$ = symplastic osmotic potential at full hydration; ψ_{lx} = leaf xylem pressure potential.

Introduction

Decreases in the stratospheric ozone layer create the potential for increased ultraviolet-B (UV-B) radiation (280–320 nm) at ground level. Increases in UV-B may pose a hazard to natural aquatic and terrestrial ecosystems as well as suppress worldwide crop production, although the latter is increasingly in doubt (Fiscus et al., 1994; Fiscus and Booker, 1995).

Stomata, which control the plant interface with the gaseous environment, have a special role in plant responses to environmental stresses. Stomatal responsiveness to perturbations of leaf water balance are well known. Reports of stomatal responses to increased UV-B radiation, however, are not consistent and range from no effect (Brandle et al., 1977; Murali and Teramura, 1986; Sisson and Caldwell, 1976; Teramura et al., 1983) to both conductance decreases (Bennett, 1981; Björn, 1989; Murali and Teramura, 1986; Teramura et al., 1983) and conductance increases (Ogoeva, 1974; Teramura et al., 1983). Teramura et al. (1984) found no significant responses of leaf water potentials or leaf conductance in soybean to supplemental UV-B radiation unless the plants had been preconditioned with water stress. Decreases in fresh weight without accompanying decreases in dry weight in response to UV-B exposure (Krizek, 1981; Tevini et al., 1981) also suggest possible perturbations in the water relations of leaves subjected to elevated UV-B radiation. These latter observations, taken together with the reports of disturbed stomatal function, might be symptomatic of changes in bulk leaf water relations. It is surprising that there is little information available on the underlying leaf water relationships that may lead to, or reflect, these responses. The purpose of our experiment, therefore, was to examine the osmotic and turgor relations of fully expanded soybean leaves, through pressure-volume analysis, to determine if changes in the bulk leaf water relations are consistent with reported perturbations in stomatal function. For this experiment we used a soybean cultivar, Essex, that has been reported to be susceptible to UV-B damage (Krizek, 1981; Teramura et al., 1990).

Materials and Methods

In 1990, soybean (*Glycine max* (L.) Merr. cv. Essex) was grown in 15-l pots in a 2 : 1 : 1 (by volume) mixture of soil, sand and Metro-Mix 220 (W. R. Grace Co., Cambridge, MA)[1] in open-top field chambers (Heagle et al., 1979) at Raleigh, NC, USA. Plants were watered daily and fertilized biweekly with «Peters Blossom Booster» (10-30-20 : N-P-K) (Grace-Sierra Horticultural Products Co., Milpitas, CA)[1] and three times during the season with «Peters STEM» soluble trace elements and micronutrient mix (Grace-Sierra Horticultural Products Co., Milpitas, CA)[1].

Low, medium, and high UV-B supplements were provided by banks of fluorescent lamps (model UVB-313, Q-Panel Co., Cleveland, OH)[1] suspended in the open-top chambers as previously described (Booker et al., 1992). Lamps in the low treatment were wrapped in polyester film to filter out radiation less than 315 nm. Lamp irradiance in the medium and high UV-B treatments was filtered with cellulose diacetate (0.13 mm thickness) to remove radiation below 290 nm. Lamp banks were kept at a constant height of 0.4 m above the canopy, and the irradiance was varied with fluorescent dimmer controls on the lamp ballasts.

Broad band erythemal meters (model 2D, Solar Light Co., Philadelphia, PA)[1] with a spectral response similar to that of the Robertson-Berger (R-B) meter were used to set the lamp bank irradiance levels each day, and a Robertson-Berger meter (Berger, 1976) was used to monitor solar UV continuously. Ambient and chamber irradiances also were checked with a UV-visible spectroradiometer (model 742, Optronics Laboratories, Inc., Orlando, FL)[1] equipped with a 3.7-m long quartz fiber optics cable and Teflon diffuser head. The spectroradiometer was calibrated with an NIST-traceable 200 W tungsten-halogen lamp standard of spectral irradiance (model 220A, Optronics Laboratories, Inc., Orlando, FL)[1] driven by a current regulated power source (model 65, Optronics Laboratories, Inc., Orlando, FL)[1]. Wavelength calibration was checked periodically by comparison with Hg emission lines from a UVB-313 lamp. The broadband erythemal meters were calibrated against the spectroradiometer as previously described (Booker et al., 1992) and biologically effective UV-B irradiance (UV-B_{BE}) was calculated by applying Caldwell's generalized plant action spectrum (Caldwell, 1971), normalized to 300 nm, to the spectroradiometer scans.

Stratospheric ozone losses corresponding to the supplemental UV-B_{BE} radiation levels were calculated from the radiative transfer model derived from Green (1983) by Björn and Murphy (1985). The model was extensively modified[2] for ease of use and flexibility and additional functions added for Caldwell's plant action spectrum and the CIE erythemal spectral sensitivity curve (McKinley and Diffey, 1987). The model predicted a clear sky total daily UV-B_{BE} irradiance for 21 June of 5.45 kJ m^{-2}. An average daily solar UV-B_{BE} of 4.88 ± 0.12 (s.d.) kJ m^{-2} measured by the Robertson-Berger meter for the 30 d surrounding 21 June indicated acceptable agreement with the model. Also on 21 June, the daily supplemental UV-B_{BE} irradiance in the three treatments was set to 0, 4.43, and 8.13 kJ m^{-2} for the low (control), medium, and high UV-B treatments. Due to an average UV-B shading effect of the open-top chambers of 24 %, the total daily UV-B_{BE} irradiance for the three treatments was 4.14, 8.57, and 12.27 kJ m^{-2}. The treatments thus corresponded to an increase in the O_3 column thickness of 15 % for the low treatment and decreases of 20 % and 35 % for the medium and high treatments. Supplemental UV-B treatments were administered as a constant daily addition over a 6 h period (0900–1500 EST), with the lamp output levels being set each day, from sowing to harvest. To compensate for seasonal changes in photoperiod and solar UV-B irradiance, the supplemental irradiance levels were adjusted biweekly and thus maintained relatively constant ozone column depletion simulations (Booker et al., 1992). At the end of the season, the actual exposure figures were refined according to the ground-based measurements provided by the R-B meter. Thus, over the entire experimental period, the mean daily UV-B_{BE} irradiances were 3.02, 6.24 and 8.98 kJ m^{-2}. On overcast days the treatments were discontinued if the UV-B_{BE} irradiance fell below 20 % of the maximum calculated for our location on 21 June for more than 30 minutes. Solar UV-B_{BE} was evaluated every 2 h afterward, and treatments resumed if the overcast cleared.

Periodically throughout the season, leaf conductances were obtained in association with carbon exchange rates measured with a LI-COR 6000 portable photosynthesis system (LI-COR, Lincoln, NE)[1]. Growth data were also collected by sequential harvest to obtain, among other things, total leaf area, mean area per leaf and specific leaf weight (SLW) during the season. Additional details of the experimental design, plant cultural conditions, growth analysis, gas exchange, yield and exposure methodologies may be found in Miller et al. (1994).

Leaves for pressure-volume (P-V) analysis were taken from the fourth node from the apex of the main stem within 1.5 h after sunrise to assure a high water potential and to avoid the problems associated with rehydration (Meinzer et al., 1986). Leaves were enclosed in a plastic bag containing a moist paper towel and severed from the plant. Leaves were placed in a pressure chamber (Soil Moisture Equipment Corp, Santa Barbara, CA)[1], which was lined with moist paper toweling, within 5 minutes after cutting. After the initial bal-

[1] Mention of a product or company name does not constitute an endorsement or recommendation by the United States Department of Agriculture.

[2] Copies of the executable files on a computer disk may be obtained from EF at no charge.

ance pressure was determined, the leaf was over-pressurized for 15 to 20 minutes. Pressure was then brought below the previous balance pressure and the expressed sap, which had been collected on absorbent material in contact with the cut petiole surface, was weighed. The new equilibrium balance pressure then was obtained. This sequence was repeated a sufficient number of times to obtain a complete P-V curve as determined by on-line data plots. The Balance Pressure (BP)-Expressed Volume (V_E) data pairs were then analyzed according to the following model:

$$\psi_{lx} = -BP = -\frac{RTN_S}{V_O - V_E} + P_{MAX} \exp(-zV_E) \qquad (1)$$

Where ψ_{lx} is the leaf xylem pressure potential; R, the gas constant; T, the temperature in Kelvins; N_s, the number of osmols of solute in the symplast; V_o, the original symplastic volume; P_{MAX}, the turgor pressure at V_o; and z, the tissue elasticity coefficient. The first term on the right side of the equation describes the osmotic relations of the tissue and was derived by Tyree and Hammel (1972). The second term was first used by Hellkvist et al. (1974) to describe the relationship between turgor and symplastic volume.

Data were fitted by a nonlinear least squares program as described by Fiscus et al. (1995) to obtain the coefficients presented in this paper. The procedure generates values of P_{MAX} and V_o that represent the fully hydrated condition of the leaf.

Following determination of the four coefficients in the equation, additional calculations were performed according to Fiscus et al. (1995) to find the so-called turgor loss potential (TLP) and the relative symplastic volume at that potential (RSV_{TLP}).

Tissue volumetric elastic modulus is a more widely used elasticity term than z and is usually defined as $\varepsilon = V(dP/dV)$ (Hellkvist et al., 1974). According to Sinclair and Venables (1983) and Fiscus et al. (1995), ε can be expressed in terms of equation 1 as

$$\varepsilon = (V_O - V_E)\, z\, P_{MAX} \exp(-z\, V_E) \qquad (2)$$

With the tissue at full hydration ($V_E = 0$), equation 2 simplifies to $\varepsilon_{max} = -V_o z P_{MAX}$ or $\varepsilon_{max} = -zRTN_s$. Further discussion of the relationship between z and ε may be found in Fiscus et al. (1995) and we mention it here only to remind the reader that such physical expressions as the volumetric elastic modulus can be influenced by both osmotic and turgor parameters.

The experiment was a completely randomized design of 3 replicates and 3 levels of UV-B (Low, Medium, High) for a total of 9 open-top chambers. Chambers were sampled four times throughout the experiment by sampling one plant from each of the 9 chambers during each of the four sampling periods. Due to the time required for the pressure-volume processing, each sampling period lasted 12 d. The sampling sequence among the treatments was randomized for each period to minimize temporal bias in the data. Individual plants were sampled only once. All parameters within each sampling period were compared by analysis of variance and by the Student-Newman-Keuls pairwise comparison. Differences were deemed not significant if $P > 0.05$.

Results and Discussion

Developmental Changes

The effects of plant age on the parameters for leaf water relations generally were greater than the effects of the UV-B treatments (Table 1). The integrated effect of normal ontological changes in parameters of leaf water relations is shown in figure 1 where potential-volume lines were calculated from the fitting coefficients in table 1. Seasonal variations in the

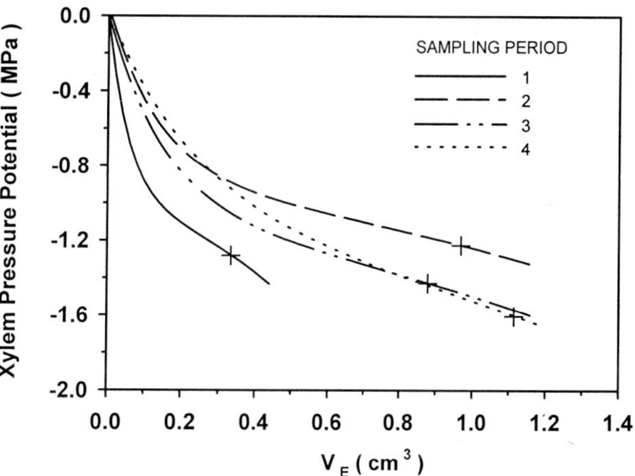

Fig. 1: Normal ontological changes in leaf pressure-volume relationships over the course of the experiments. Each line was calculated from equation 1 using the relevant mean values from table 1. Lines are terminated at $V_E \simeq V_o/3$. TLPs, calculated from the mean coefficients in table 1, are indicated by the crosses on the lines.

pressure-volume curves (Fig. 1) appear to be dominated by z and ψ_π, but predicting the shape of the P-V curve from these two parameters is not an obvious process since ψ_π clearly influences the expression of z. For example, as a result of the large change in z between periods 1 and 2 that is accompanied by little change in $\psi_{\pi 100}$ ($-P_{MAX}$) (Table 1), the initial slope of the curve, in the region dominated by turgor changes, is less steep in period 2 than in period 1 (Fig. 1), and little change occurs in TLP. In contrast, the further increase in z between periods 2 and 3 is accompanied by a decrease in $\psi_{\pi 100}$. This combination of changes results in an initial slope of period 2 that appears less steep than period 3, even though z is lower in period 2. TLP, however, decreases in this instance. The effects of further increases in z and decreases in $\psi_{\pi 100}$ in the fourth period result in the least steep initial slope and a line that converges with the period 3 line at high levels of water loss.

Another especially noticeable feature in figure 1 is that the large increase in absolute quantities of water loss necessary to reach TLP can be attributed directly to larger leaf size and increased elasticity even though there is a seasonal trend toward decreasing TLP. Further discussion of the normal ontological changes may be found in Fiscus et al. (1995).

Supplemental Ultraviolet-B

Among the parameters in table 1, a statistically significant response to the UV-B treatments occurred only for V_o, z and RSV_{TLP}. During period 2 V_o was significantly reduced by the highest UV treatment, while during the same period z was significantly less in the medium and high UV treatments than in the low treatment. Again, in the fourth period, z was significantly less in the high treatment than the medium but did not differ from the low treatment. The only other significant difference occurred during the fourth period when RSV_{TLP} was larger in the high UV treatment. In this case, the differences (Table 1) are due to normal ontological declines

Table 1: Means comparison of soybean leaf parameters. Numbers in parentheses in sampling period column are the mean days after planting (DAP_0 = day 151 of 1990) for that period. SLWs are the means for all the leaves on the plant (n=4 plants) and Leaf Area is the mean area per leaf of all the leaves on the plant (n=4 plants). For all the osmotic and turgor parameters, n=3. UV-B treatment numbers correspond to low (1), medium (2) and high (3) exposure levels. Numbers in a column in the same sampling period are not different at the P<0.05 level if followed by the same letter.

Sampling Period	UV-B Treatment	V_o cm^3	z cm^{-3}	P_{MAX} MPa	N_S mosmol	TLP MPa	RSV_{TLP}	SLW mg cm^{-2}	Leaf Area cm^2
1 (38)	1	1.34 a	−18.1 a	0.96 a	0.52 a	−1.28 a	0.751 a	3.52 a	57.0 a
	2	1.81 a	−12.6 a	0.94 a	0.70 a	−1.26 a	0.755 a	3.36 a	51.0 a
	3	1.54 a	−15.2 a	0.98 a	0.64 a	−1.34 a	0.763 a	3.56 a	57.5 a
2 (58)	1	3.51 a	−7.00 a	0.93 a	1.32 a	−1.22 a	0.767 a	3.48 a	125.7 a
	2	2.91 ab	−9.90 b	0.98 a	1.14 a	−1.24 a	0.790 a	3.27 a	145.2 a
	3	1.81 b	−10.19 b	1.07 a	0.80 a	−1.46 a	0.729 a	3.38 a	133.2 a
3 (72)	1	3.54 a	−6.28 a	1.08 a	1.54 a	−1.43 a	0.752 a	3.95 a	118.2 a
	2	3.26 a	−7.33 a	1.00 a	1.32 a	−1.33 a	0.756 a	3.84 a	120.3 a
	3	3.84 a	−6.57 a	1.03 a	1.61 a	−1.34 a	0.771 a	3.99 a	108.5 a
4 (100)	1	3.62 a	−3.90 ab	1.13 a	1.63 a	−1.62 a	0.694 a	4.30 a	117.1 a
	2	4.62 a	−3.23 a	1.23 a	2.28 a	−1.74 a	0.703 a	4.26 a	130.6 a
	3	3.46 a	−6.24 b	1.21 a	1.71 a	−1.62 a	0.751 b	4.49 a	118.8 a

that did not occur in the high UV treatments. Changes in the P-V parameters generally were small compared to the normal ontological changes recorded over the four sampling periods. Specifically, while the difference in V_o between the medium and the high UV treatment was 48 % in period 2, the seasonal change was 2.7-fold, and corresponding values for z were 31 % and 4. 6-fold.

Integration of all the parametric changes are shown in figures 2–5. The curves in figure 2 (period 1; about mid-vegetative phase) differ primarily because of the non-significant difference in z, although there is little difference in the absolute quantities of water loss to TLP. During period 2 (Fig. 3; early reproductive phase), however, the decreasing values of both V_o and z with increasing UV-B affect the shape of the curves, which indicate a substantial decrease in the total

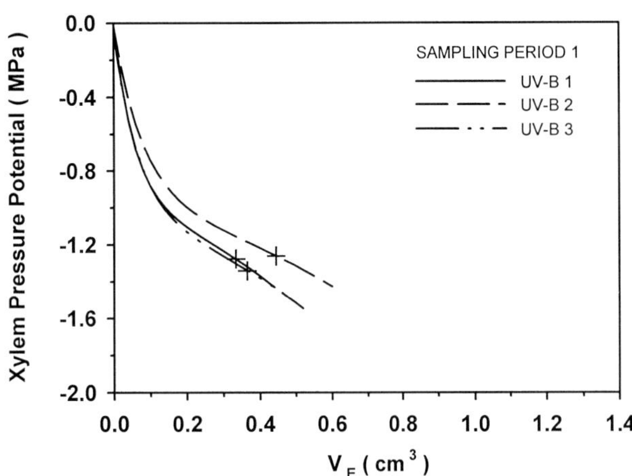

Fig. 2: Changes induced in the leaf pressure-volume relationships by UV-B treatment at sampling period 1. Each line was calculated from equation 1 using the relevant mean values from table 1. Lines are terminated at $V_E \simeq V_o/3$. TLPs, calculated from the mean coefficients in table 1, are indicated by the crosses on the lines.

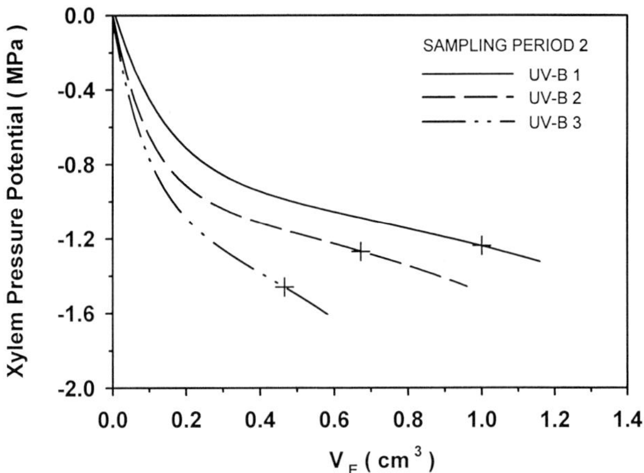

Fig. 3: Changes induced in the leaf pressure-volume relationships by UV-B treatment at sampling period 2. Each line was calculated from equation 1 using the relevant mean values from table 1. Lines are terminated at $V_E \simeq V_o/3$. TLPs, calculated from the mean coefficients in table 1, are indicated by the crosses on the lines.

water loss necessary to reach TLP at the medium and high UV-B levels. Period 3, which occurred during the mid-reproductive stage (Fig. 4), showed no significant effects of the UV-B treatments. Finally, the shape of the curves in period 4 (Fig. 5; late reproductive phase) reflect the lower value of z and the higher RSV_{TLP} under high UV-B as indicated in table 1.

Although measurements were not available for the first sampling period, there was no significant effect of UV-B on stomatal conductance over the remainder of the season starting at a time corresponding to period 2 (Fig. 6). Neither was there a detectable effect on carbon exchange (Miller et al., 1994; Fiscus et al., 1994), nor were there any differences in total leaf area (Miller et al., 1994), mean area per leaf (Table 1) or SLW (Table 1).

The lack of significant changes in the present experiment concurs with the conclusions of Teramura (1983) that SLW is not a reliable indicator of UV-B stress. While Teramura (1983) also stated that leaf area generally is reduced by UV-B treatments, our results do not show such an effect (see also Miller et al., 1994). Changes in V_o, however, may be interpreted as changes in leaf size, and we did observe a statistically significant transient reduction of about 48 % (Table 1) during period 2 at the highest UV-B exposure. Since the symplastic solution constitutes the bulk of the leaf fresh weight and in the absence of any change in SLW or leaf area, the decrease in V_0 during period 2 is consistent with the observations of Krizek (1981) and Tevini et al. (1981).

Although a review of previous experiments by Fiscus and Booker (1995) has shown a consistent occurrence of UV-B induced phenylpropanoid metabolism and synthesis of UV-B absorbing compounds, it seems unlikely that changes in V_0 and z are related to the production of soluble phenolics. In the absence of changes in SLW, it also is unlikely that the differences in elasticity would be due to deposition of additional lignin. Furthermore, there is no evidence that UV-B induces lignin synthesis. It is possible, however, that elasticity differences were due to changes in conformation and cross linking of cell wall phenolics and proteins induced by UV-B (see also Caldwell et al., 1989).

In conclusion, UV-B irradiation treatments resulted in only sporadic changes in V_o and z that noticeably affected the overall leaf P-V relationships. However, these effects were small compared to normal ontological changes and were not sufficient to disturb the leaf water balance in such a way as to affect conductance, gas exchange or eventual yield, despite the severity of the treatments.

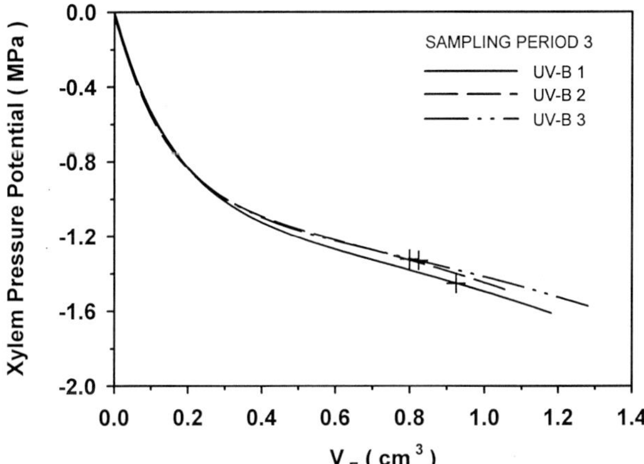

Fig. 4: Changes induced in the leaf pressure-volume relationships by UV-B treatment at sampling period 3. Each line was calculated from equation 1 using the relevant mean values from table 1. Lines are terminated at $V_E \simeq V_o/3$. TLPS, calculated from the mean coefficients in table 1, are indicated by the crosses on the lines.

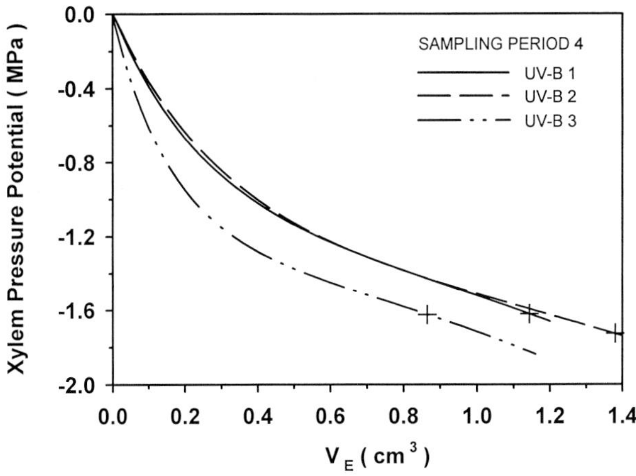

Fig. 5: Changes induced in the leaf pressure-volume relationships by UV-B treatment at sampling period 4 Each line was calculated from equation 1 using the relevant mean values from table 1. Lines are terminated at $V_E \simeq V_o/3$. TLPs, calculated from the mean coefficients in table 1, are indicated by the crosses on the lines.

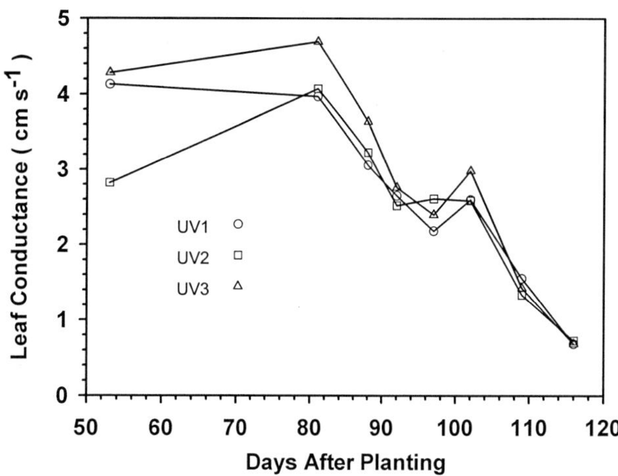

Fig. 6: Leaf conductance measured during carbon exchange measurements during the season. None of these data show any differences at the $P < 0.05$ level of significance.

References

Bennett, J. H.: Photosynthesis and gas diffusion in leaves of selected crop plants exposed to ultraviolet-B radiation. J. Environ. Qual. *10,* 271–275 (1981).

Berger, D.: The sunburning ultraviolet meter: design and performance. Photochem. Photobiol. *24,* 587–593 (1976).

Björn, L. O. and T. M. Murphy: Computer calculation of solar ultraviolet radiation at ground level. Physiol. veg. *23,* 555–561 (1985).

Björn, L. O.: Consequences of decreased atmospheric ozone: Effects of ultraviolet radiation on plants. In: Schneider, T., S. D. Lee, G. J. R. Wolters, and L. D. Grant (eds.): Atmospheric Ozone Research and its Policy Implications, pp. 261–267. Elsevier Scientific Publications B.V., Amsterdam (1989).

Booker, F. L., E. L. Fiscus, R. B. Philbeck, A. S. Heagle, J. E. Miller, and W. W. Heck: A supplemental ultraviolet-B radiation system for open-top field chambers. J. Environ. Qual. *21,* 56–61 (1992).

Brandle, J. R., W. F. Campbell, W. B. Sisson, and M. M. Caldwell: Net photosynthesis, electron transport capacity, and ultrastructure of *Pisum sativum* L. exposed to ultraviolet-B radiation. Plant Physiol. *70,* 165–170 (1977).

Caldwell, M. M.: Solar ultraviolet radiation and the growth and development of higher plants. In: Giese, A. C. (ed.): Photophysiology, pp. 131–177. Academic Press, New York (1971).

Fiscus, E. L., J. E. Miller, and F. L. Booker: Is UV-B a hazard to soybean photosynthesis and yield? Results of an ozone/UV-B interaction study and model predictions. In: Biggs, R. H. and M. E. B. Joyner (eds.): Stratospheric Ozone Depletion/UV-B Radiation in the Biosphere, NATO ASI Series, Vol. I 18, pp. 135–147. Springer-Verlag, Berlin, Heidelberg (1994).

Fiscus, E. L., F. L. Booker, J. E. Miller, and C. D. Reid: Response of soybean leaf water relations to tropospheric ozone. Can. J. Bot. *73,* 517–526 (1995).

Fiscus, E. L. and F. L. Booker: Is increased UV-B a threat to crop photosynthesis and productivity? Photosynth. Res. *43,* 81–92 (1995).

Green, A. E. S.: The penetration of ultraviolet radiation to the ground. Physiol. Plant. *58,* 351–359 (1983).

Heagle, A. S., R. B. Philbeck, H. H. Rogers, and M. B. Letchworth: Dispensing and monitoring ozone in open-top field chambers for plant effects studies. Phytopath. *69,* 15–20 (1979).

Hellkvist, J., G. P. Richards, and P. G. Jarvis: Vertical gradients of water potential and tissue water relations in Sitka spruce trees measured with the pressure chamber. J. Appl. Ecol. *11,* 637–667 (1974).

Krizek, D. T.: Plant physiology: ultraviolet radiation. In: Parker, S. B. (ed.): 1981 Yearbook of Science and Technology, pp. 306–309. McGraw-Hill, New York (1981).

McKinlay, A. F. and B. L. Diffey: A reference action spectrum for ultra-violet induced erythema in human skin. In: Passchier, W. F. and B. F. M. Bosnjakovic (eds.): Human Exposure to Ultraviolet Radiation: Risks and Regulations, pp. 83–87. ELsevier Science Publishers B.V., Amsterdam (1987).

Meinzer, F. C., P. W. Rundel, M. R. Sharifi, and E. T. Nilsen: Turgor and osmotic relations of the desert shrub *Larrea tridentata.* Plant Cell Environ. *9,* 467–475 (1986).

Miller, J. E., F. L. Booker, E. L. Fiscus, A. S. Heagle, W. A. Pursley, S. F. Vozzo, and W. W. Heck: Ultraviolet-B radiation and ozone effects on growth, yield, and photosynthesis of soybean. J. Environ. Qual. *23,* 83–91 (1994).

Murali, N. S. and A. H. Teramura: Effectiveness of UV-B radiation on the growth and physiology of field-grown soybean modified by water stress. Photochem. Photobiol. *44,* 215–219 (1986).

Ogoeva, K.: Reaction of the stomatal apparatus of barley leaves to the effect of natural UV radiation. Izvestiya Akademii Nauk Tadzhikskoi S.S.R., Otdelenie Biologicheskikh Nauk *4,* 48 (1974).

Sinclair, R. and W. N. Venables: An alternative method for analysing pressure-volume curves produced with the pressure chamber. Plant Cell Environ. *6,* 211–217 (1983).

Sisson, W. B. and M. M. Caldwell: Photosynthesis, dark respiration, and growth of *Rumex patientia* L. exposed to ultraviolet irradiance (288–315 nanometers) simulating a reduced atmospheric ozone column. Plant Physiol. *58,* 563–568 (1976).

Teramura, A. H.: Effects of ultraviolet-B radiation on the growth and yield of crop plants. Physiol. Plant. *58,* 415–427 (1983).

Teramura, A. H., M. Tevini, and W. Iwanzik: Effects of ultraviolet-B irradiation on plants during mild water stress. 1. Effects on diurnal stomatal resistance. Physiol. Plant. *57,* 175–180 (1983).

Teramura, A. H., I. N. Forseth, and J. Lydon: Effects of ultraviolet-B radiation on plants during mild water stress. IV. The insensitivity of soybean internal water relations to ultraviolet-B radiation. Physiol. Plant. *62,* 384–389 (1984).

Teramura, A. H., J. H. Sullivan, and J.Lydon: Effects of UV-B radiation on soybean yield and seed quality: a 6-year field study. Physiol. Plant. *80,* 5–11 (1990).

Tevini, M., W. Iwanzik, and U. Thoma: Some effects of enhanced UV-B irradiation on the growth and composition of plants. Planta *153,* 388–394 (1981).

Tyree, M. T. and H. T. Hammel: The measurement of the turgor pressure and the water relations of plants by the pressure-bomb technique. J. Exp. Bot. *23,* 267–282 (1972).

J. Plant Physiol. Vol. 148. pp. 69–77 (1996)

Initial Assessment of Physiological Response to UV-B Irradiation Using Fluorescence Measurements

Elizabeth M. Middleton[1], Emmett W. Chappelle[1], Takisha A. Cannon[2], Paulien Adamse[3], and Steven J. Britz[3]

[1] Laboratory for Terrestrial Physics, National Aeronautics and Space Administration, Goddard Space Flight Center, Greenbelt, MD 20771

[2] Department of Biology, University of Virginia, Charlottesville, VA 22903

[3] Climate Stress Laboratory, US Agricultural Research Service, Beltsville, MD 20705

Received September 21, 1995 · Accepted November 11, 1995

Summary

Fluorescence emissions obtained by excitation at 280 and 340 nm (280EX 300–520 nm; 340EX 360–800 nm) were used to discriminate physiological change induced by ultraviolet B (UV-B) irradiation in two cucumber (*Cucumis sativus* L.) cultivars, Poinsett (UV-B sensitive) and Ashley (insensitive). Plants were grown in chambers with controlled spectral irradiation, including biologically effective UV-B irradiation (21 or 0.3 $kJ\,m^{-2}\,d^{-1}$) provided for 5 days with photosynthetically active radiation (~38 $mol\,m^{-2}\,d^{-1}$). Differentiating UV-B induced effects and cultivar differences proved more successful with a dimethyl sulfoxide (DMSO) leaf extract than with freshly excised, intact leaves. Poinsett exhibited significantly lower ($P \leq 0.01$) fluorescence for most wavelengths or spectral ratios, whether excited at 280 or 340 nm. The single dominant UV-A fluorescence peak observed in all 280EX emission spectra (330–350 nm) was shifted in DMSO from 340 to 350 nm in UV-B irradiated plants (with a significantly higher F350/F475 ratio, $P \leq 0.001$). This could indicate that UV-B irradiation altered the relative amounts of soluble protein invested in enzymes for photosynthesis (e.g., rubisco) versus UV-B protective compounds. In 340EX spectra, UV-B exposed plants also had higher blue/far-red ratios, possibly due to enhanced production of an antioxidant, blue fluorescing compound known to accumulate after UV-B induced degradation of rubisco. In DMSO, this ratio (F450/F730) was linearly related to the total carotenoid/Chl pigment ratio, with qualitatively different responses for the two cultivars. For 340EX spectra, UV-B effects were most successfully discriminated by the far-red peak, alone or included in a ratio with either red or blue fluorescence. UV-B irradiated plants exhibited a significantly lower ($P \leq 0.001$) far-red peak in DMSO and lower far-red/red fluorescence ratios in both media, indicating loss of chlorophyll. The F730/F680 ratio for DMSO was log-linearly dependent on total chlorophyll content ($r^2 = 0.58$; $P \leq 0.001$). Chlorophyll (Chl) *a* and *b* were reduced ($P \leq 0.01$) and the Chl *a*/*b* ($P \leq 0.05$) and carotenoid/Chl ($P \leq 0.001$) ratios were higher under UV-B irradiation. This experiment provides further evidence that UV-B induced damage includes degradation of Chl and photosynthetic function, while photoprotection involves the antioxidant defense system.

Key words: Cucumis sativus, UV-B, fluorescence, cucumber, plant stress.

Abbreviations: Chl = chlorophyll; DMSO = dimethyl sulfoxide; F = fluorescence; N-FK = N-formylkynurenine; PAR = photosynthetically active radiation (400–700 nm); RFI = relative fluorescence emission intensity; rubisco = ribulose 1,5-bisphosphate carboxylase; UV = ultraviolet radiation (290–400 nm); UV-A = ultraviolet-A radiation (320–400 nm); UV-B = ultraviolet-B radiation (290–320 nm); UVAP = UV absorbing pigments; 280 EX = fluorescence emission spectra obtained between 300–520 nm, produced by excitation at 280 nm; 340 EX = fluorescence emission spectra obtained between 360–800 nm, produced by excitation at 340 nm.

Introduction

Crops growing in the mid-latitudes are increasingly at risk of exposure to elevated ultraviolet-B (UV-B, 290–315 nm) radiation resulting from stratospheric ozone depletion (Gleason et al., 1993). Since many UV-B induced reductions in plant growth or productivity are linked to altered leaf physiology and function (Caldwell et al., 1989; Tevini and Teramura, 1989; Tevini et al., 1991), elevated levels of UV-B irradiation can be considered a significant potential environmental stress in natural ecosystems and agriculture, as defined by Lichtenthaler (this issue). UV-B studies are usually conducted at UV-B levels higher than ambient (e.g., comparable to 10–20 % ozone reductions), either to simulate projected future conditions or to augment plant responses. However, the effect of UV-B radiation on whole organism and physiological processes has been difficult to assess, in part because UV-B damage can be partially or wholly ameliorated when UV-A (320–400 nm) and blue (400–500 nm) radiation is provided (Middleton and Teramura, 1993, 1994; Adamse and Britz, this issue). Because the seasonally and diurnally variable solar radiation regime in natural environments (with or without artificially supplemented UV-B irradiation) is superimposed on plant phenological changes and other environmental factors, the relevant physiological processes are difficult to identify and quantify. For example, whereas UV-B radiation directly damages DNA (e.g., Pang and Hays, 1991; Britt et al., 1993), the chain of events involved in its photorepair by the enzyme photolyase is initiated by UV-A/blue radiation, for which a radical and novel mechanism was recently described (Park et al., 1995). Furthermore, both UV-B and UV-A/blue irradiation stimulate the production of UV absorbing pigments (UVAP, e.g., flavonoids) (Beggs et al., 1985; Beggs and Wellmann, 1994; Li et al., 1994), a general category of screening compounds widely believed to provide UV-B photoprotection (although the mechanism has not been adequately explained).

Detection of physiologic stress in plants using fluorescence measurements has been successfully demonstrated (Chappelle et al., 1984 a, b; Lichtenthaler and Rinderle, 1988; Lang et al., 1991, 1992; Stober and Lichtenthaler, 1992, 1993; Stober et al., 1994; Valentini et al., 1994). Therefore, fluorescence measurements might provide a valuable tool for understanding UV-B induced physiological processes, especially in controlled systems. Our study was undertaken to initiate assessments of fluorescence measurements in UV-B radiation studies of agricultural species, beginning with an experiment conducted on two cucumber (*Cucumis sativus* L.) cultivars previously designated (Britz and Adamse, 1994) as UV-B sensitive (Poinsett) and insensitive (Ashley), according to several growth and physiological parameters (Adamse et al., 1994). We examined RFI spectra resulting from excitation at 340 and 280 nm, and photosynthetic pigments for both cultivars, in an artificial UV-B irradiation experiment where background spectral quality and quantity, and other environmental conditions, were controlled in growth chambers.

Material and Methods

Plant material

Cucumber (*Cucumis sativus* L.) cvs. Poinsett and Ashley were selected, based on their previous performance as «UV-B sensitive» and «UV-B insensitive» according to several leaf variables including photosynthetic rate, growth, and flavonoid accumulation. 1.5 l pots, with plants grown individually in vermiculite and flushed daily with a modified half-strength Hoagland's solution, were placed in a growth chamber for 10 days at 27 °C, ~ 50 % relative humidity, and ~450 μmol mol^{-1} CO_2. Photosynthetically active radiation (PAR, 400–700 nm) was provided by an equal mix of 400 W high pressure sodium and metal halide lamps for 14 h per day, with the first and last hour at 270 μmol m^{-2} s^{-1}, and 12 h at 840 μmol m^{-2} s^{-1}.

Experimental Treatments

When the third leaf was unfolding (day 11), 24 plants selected for uniformity were moved to a second growth chamber for UV treatment at 27 °C and ~50 % relative humidity. CO_2 was increased to 750 μmol mol^{-1} to enhance growth. PAR «background radiation» was provided for 14 h per day by low pressure sodium lamps (SOX 180 W, Philips North America, Bloomfield, NJ, USA) and supplemented with blue (400–500 nm) fluorescent lamps (F40T12/247/RS; Sylvania, Danvers, MA, USA). ⅓ of these lamps were turned on for the first and last hour, and all lamps were turned on for the intervening 12 h. The chamber was divided into two compartments by a vertical sheet of UV-B absorbing polyester film. On each side, UV radiation was provided over the middle 10 (of 14) h by fluorescent sun lamps (UV-B 313, F40WT12-6A, Q-panel Co., Cleveland, OH, USA) placed in a horizontal lamp bank suspended below the PAR/blue lamps. The UV lamps in the «+UV-B» compartment were wrapped with cellulose diacetate film (0.08 mm) to provide UV-B and UV-A radiation; those in the «−UV-B» compartment were wrapped with polyester film (0.13 mm) to provide UV-A radiation. All lamps together provided daily irradiation of 35 mol m^{-2} PAR, 2.68 mol m^{-2} supplemental blue, and either 0.101 or 0.001 mol m^{-2} (21 or 0.3 kJ m^{-2} biologically effective UV-B, at 300 nm; Caldwell, 1971), respectively, for «+UV-B» and «−UV-B (Control)» treatments.

Measurements

Steady-state fluorescence emission spectra were obtained on whole leaves and on extracts of leaf material using a Fluorolog II spectrofluorometer (SPEX Industries, Inc., Edison, NJ, USA) which is capable of producing high intensity continuous wavelength monochromatic light excitation (see McMurtrey et al., this issue). Two emission spectra per sample were acquired with excitation wavelengths of 340 and 280 nm, respectively, referred to below as 340 EX (360–800 nm) and 280 EX (300–520 nm). A cut-off filter was used with 340 EX to eliminate harmonic doubling.

On day 15, after 5 days of experimental UV irradiation, the third leaf was harvested from each plant. 340 EX and 280 EX spectra were obtained on individual, whole leaves mounted in the SPEX chamber, irradiated through the adaxial surface near the leaf tip, with the signal measured in the transmitted beam. From the leaf tip region, where UV-B induced chlorosis was most pronounced, four leaf disks (1 cm^2) were removed ½ cm from the leaf edge and placed in polystyrene cuvettes with 4 mL dimethyl sulfoxide (DMSO) for extraction of leaf organic compounds, including pigments. The cuvettes were stored in the dark at room temperature (~25 °C) for two days before 3 mL of extract per sample, diluted 1 : 1 (to bring emissions well below the saturation value), were transferred to a clear quartz cuvette for acquisition of fluorescence measurements. Each sample was then transferred to a standard quartz cuvette for determination of absorbance at 470, 648, 664, and 750 nm (Lichtenthaler, 1987) with a computerized dual beam spectrophotometer (Perkin-Elmer Lambda 3, Perkin-Elmer, Norwalk, CN, USA). Statistical analyses were performed for each plant media (whole leaf, DMSO) using Systat 5.01 (SYSTAT, Inc., Evanston, IL, USA) based on the experimental factors, UV-B (+,–) and Cultivar (Ashley, Poinsett), with separate consideration by excitation wavelength (340, 280 nm).

Results

Fluorescence Emission Spectra Obtained by Excitation at 340 nm

In the 340EX spectra from both the DMSO extract and leaves (Fig. 1a, b), blue through green (<600 nm) RFI for the «less-sensitive» cultivar Ashley exceeded that of Poinsett. In both media, higher RFI observed for Ashley at individual wavelengths was highly significant ($P<0.001$) at 450 and 525 nm, and for DMSO at 730 nm. In DMSO (Fig. 1a), RFI for UV-B plants exceeded that of controls in blue wavelengths (450 nm, $P = 0.012$), but was significantly lower ($P \leq 0.016$) in the far-red (730 nm) and red (680 nm) chlorophyll peaks. In fresh leaves (Fig. 1b), large variability limited the ability to separate UV-B treatments.

In UV-B irradiated plants, significantly lower far-red/red fluorescence ratios were seen in both media (Fig. 2a, b). The lower F730/F680 ratios observed in DMSO were similar for both cultivars, and UV-B treatment differences were highly significant ($P<0.001$). In fresh leaves, the influence of cultivar differences dominated (Ashley > Poinsett, $P=0.013$), although reduction of the F740/F685 ratio due to UV-B exposure was apparent when the cultivars were combined ($P=0.042$). Conversely, UV-B irradiation significantly increased the blue/far-red fluorescence ratios in both media (Fig. 3a, b). The increase in this ratio was highly significant for DMSO (F450/F730, $P<0.001$) with UV-B exposure. In fresh leaves, the higher F460/F740 ratio associated with UV-B exposure was also significant when cultivars were combined ($P=0.043$).

Photosynthetic Pigment Content

In general, UV-B irradiation significantly reduced Chl *a*, Chl *b*, total Chl, and total photosynthetic pigment (Chl + carotenoids), although total carotenoid content was unaffected

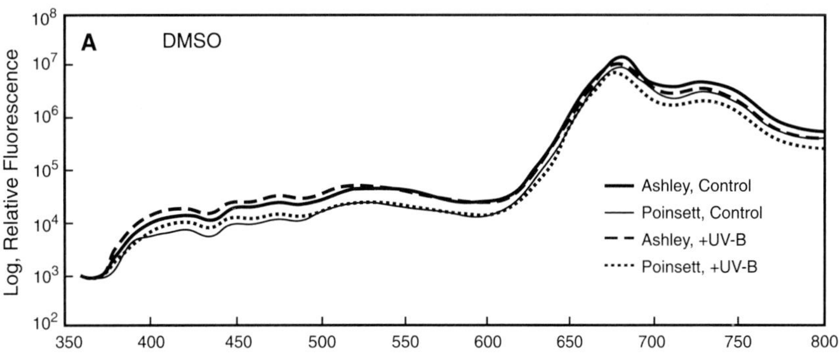

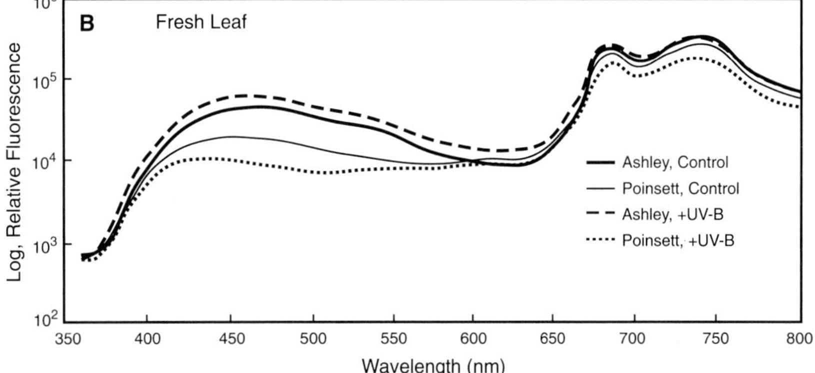

Fig. 1: The average fluorescence spectra (370–800 nm, log relative units) obtained from excitation at 340 nm [referred to as 340EX below] are shown for the four treatment groups (2 Cultivars and 2 UV treatments) in the two media: A) DMSO extract; and B) whole, fresh leaves. Control groups are shown in solid lines and UV-B irradiatiated groups in dashed lines, separately for cultivars Ashley and Poinsett.

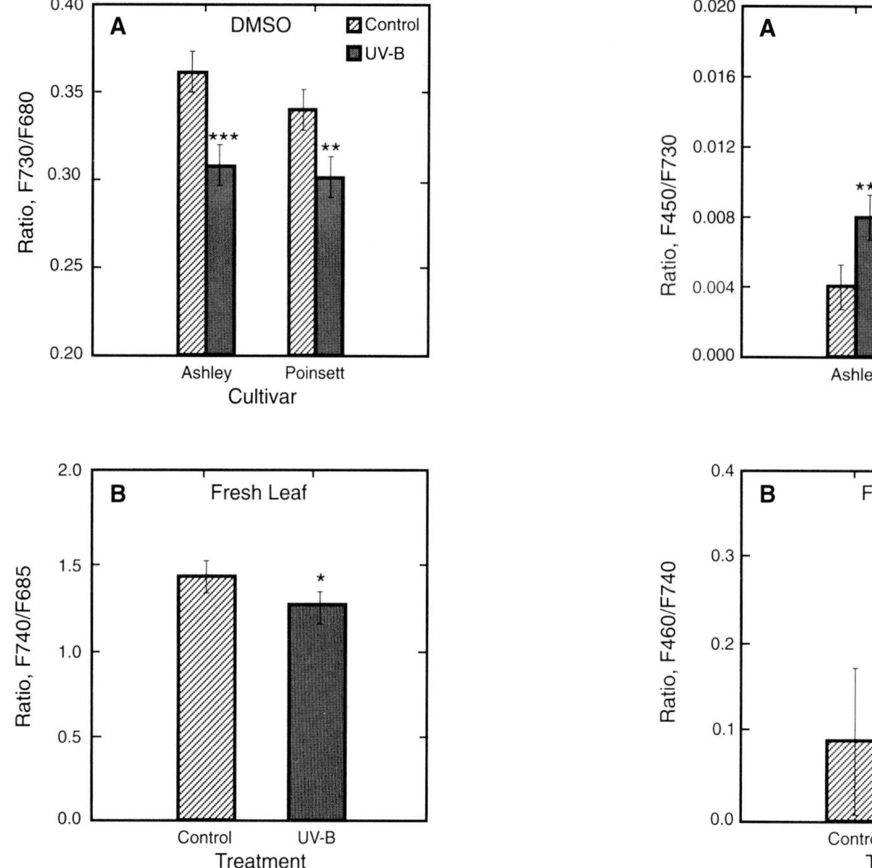

Fig. 2: The reduction in the far-red/red fluorescence ratio (340EX) due to UV-B irradiation is shown in the two media: A) DMSO extract; and B) whole, fresh leaves. For A, the F730/F680 ratio is given separately for cultivars Ashley and Poinsett, where similar and highly significant ($P<0.001$; $r^2=0.61$, UV treatment effect) reductions in the ratio were observed in both cultivars. For leaves, the reduction in the F740/F685 ratio was significant ($P=0.042$) when cultivars were combined.

Fig. 3: The increase in the blue/far-red fluorescence ratio (340EX) due to UV-B irradiation is shown in the two media: A) DMSO extract; and B) whole, fresh leaves. For A (F450/F730), the appropriate fluorescence ratio is given separately for cultivars Ashley and Poinsett. Increases in the ratio were highly significant ($P<0.001$; $r^2=0.72$, UV treatment effect) in both cultivars, but the increase was relatively greater for Ashley. For leaves, the increase in the F460/F740 ratio was significant ($P=0.043$, nonparametric Kruskal-Wallis test) when cultivars were combined.

Table 1: Photosynthetic Pigment Content: Means and Standard Errors (SE) by Treatment Group, as determined from DMSO extracts. Also given are the significance levels[1] for comparisons of means between the control and corresponding UV-B treatment groups.

Group	Treatment	Chl *a* µg cm^{-2}	Chl *b* µg cm^{-2}	Total Chl µg cm^{-2}	Total Carotenoid µg cm^{-2}	Total Pigment µg cm^{-2}	Chl *a*/*b* Ratio	Carotenoid/ Chl Ratio
A. Separate Cultivars (n = 6)								
Ashley	Control	17.76^{a,b}	5.86^{c}	23.46^{b}	4.18^{a}	27.64^{b}	3.00^{a}	0.182^{a}
	UV-B	13.20^{a,2}	3.60^{a} ***	16.80^{a} **	4.24^{a}	21.02^{a} *	3.68^{b} *	0.254^{b} ***
Poinsett	Control	19.76^{b}	5.62^{b,c}	25.20^{b}	5.32^{b}	30.70^{b}	3.52^{a,b}	0.210^{a,b}
	UV-B	16.62^{a,b}	4.60^{a,b}	21.22^{a,b}	4.86^{a,b}	28.08^{a,b}	3.62^{b}	0.230^{b}
	SE	0.68	0.26	0.92	0.14	1.01	0.16	0.014
B. Combined Cultivars (n = 12)								
	Control	18.68^{b}	5.74^{b}	24.42^{b}	4.74^{a}	29.16^{b}	3.26^{a}	0.196^{a}
	UV-B	14.76^{a} **	4.06^{a} ***	18.80^{a} **	4.52^{a}	23.32^{a} **	3.65^{b} *	0.243^{b} ***
	SE	0.28	0.16	0.31	0.12	0.33	0.13	0.030

[1] Statistically different means are indicated by different letter superscripts. The statistical significance levels for the means comparisons for the controls versus UV-B treatment are also indicated: a) ***, $P\leq0.001$; b) **, $P\leq0.01$; c) *, $P\leq0.05$, and d) ^{2}, $P\leq0.10$.

(~4.6 µg cm^{-2}), as presented in Table 1 B. Additionally, both the Chl *a/b* and carotenoid/Chl ratios were significantly increased by UV-B irradiation. These trends were observed in each cultivar, but were significant in Ashley only (Table 1 A), and for the combined data (Table 1 B). For controls, Ashley and Poinsett had equivalent levels of chlorophylls (Chl *a*, ~18 µg cm^{-2}, Chl *b*, ~6 µg cm^{-2}; and total Chl, ~24 µg cm^{-2}), although Poinsett controls had a higher total carotenoid level than Ashley (5.3 vs. 4.2 µg cm^{-2}, P = 0.002). With UV-B exposure, significant differences between the two cultivars for Chl could not be determined (Chl *a*, ~15 µg cm^{-2}; Chl *b*, ~4 µg cm^{-2}; and total Chl, ~19 µg cm^{-2}).

Fluorescence ratios from 340EX DMSO spectra were further investigated with respect to photosynthetic pigments, and two ratios were found log-linearly related to pigment variables (Fig. 4 a, b). The F730/F680 ratio was dependent on total Chl content (r^2 = 0.58, P < 0.001) over all data. Also, the F450/F730 ratio was related to the carotenoid/Chl ratio, with different responses by the two cultivars: for Ashley, the regression relationship was more strongly expressed (r^2 = 0.91) than for Poinsett (r^2 = 0.54).

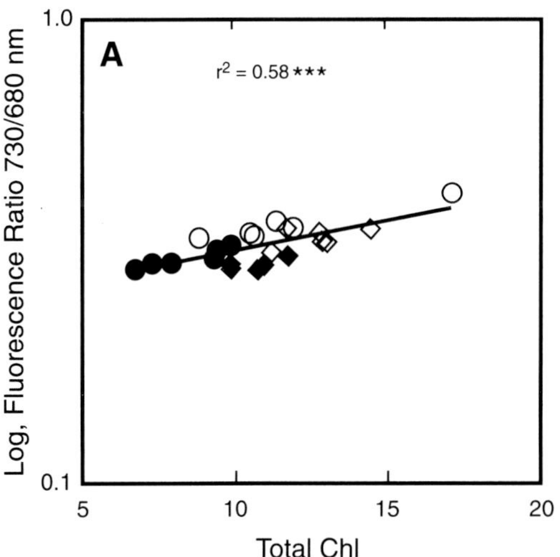

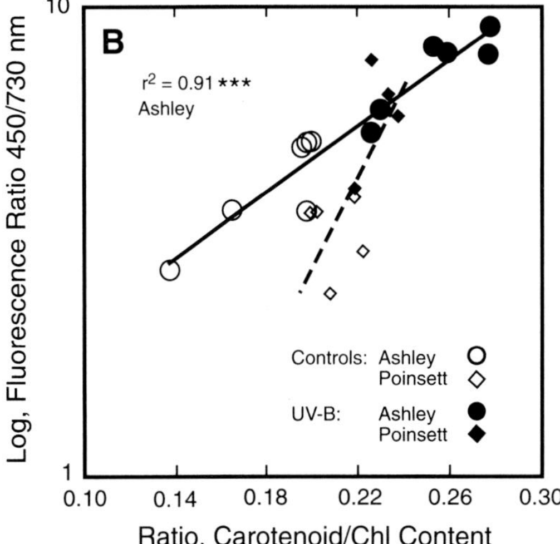

Fluorescence Emission Spectra Obtained with Excitation at 280 nm

In 280EX spectra (Fig. 5 a, b), a dominant near UV-A peak occurred in both media. In fresh leaves, the spectra were smooth and the primary peak occurred at ~330 nm, but no treatment differences could be discerned. However, UV-B exposure clearly produced a higher UV-A peak RFI in DMSO xxx(P < 0.001), which was shifted 10 nm (340 to 350 nm). Cultivar differences were also exhibited (P ≤ 0.01) in DMSO, with RFI for Ashley greater than for Poinsett, up through the minimum located at 435 nm. Secondary peaks were observed at 315, 335, 420 and 450 nm. Cultivar differences were eliminated by normalizing the UV-A peak to the value at 475 nm (i.e., F330/F475 ratio, fresh leaves; F350/F475 ratio, DMSO) and the DMSO ratio was significantly higher (P < 0.001) in the UV-B irradiated plants (Fig. 6 a, b).

Relating Fluorescence Emissions Obtained from 340 and 280 nm Excitations

The RFI measured at 330 nm from 280EX spectra was well correlated in both media to that measured at 680 (685 for leaves) from 340EX spectra (Fig. 7 a, b). A single positive log-linear correlation (i.e., no dependent variable could be designated) was exhibited for fresh leaves, all data combined (r = 0.81). UV-B irradiated plants (r = 0.93) could be separated from Controls (r = 0.88) using the DMSO extraction (Fig. 7 a); in addition, UV-B exposure shifted the response to a higher RFI at 330 nm but lower RFI at 680 nm. Similar, but somewhat lower, correlations were obtained when 730 was substituted for 680 nm, and 350 for 330 nm (data not shown).

Fig. 4: The increase in two fluorescence ratios (340EX) measured as a function of photosynthetic pigment variables is shown: A) F730/F680 ratio as a log-linear function of total chlorophyll content (µg cm^{-2}) for all data [F730/F680 ratio = 0.090 + 0.101*log (total Chl); r^2 = 0.58, P < 0.001]; and B) F450/F730 ratio as a log-linear function of the Carotenoid/Chl pigment ratio. In B, the regression line for Ashley (large symbols) is shown with a solid line [F450/F730 ratio = 19.872 + 9.065*log (Carotenoid/Chl); r^2 = 0.91, P < 0.001]; the small symbols are Poinsett, with dashed regression line [F450/F730 ratio = 37.484 + 21.669*log (Carotenoid/Chl); r^2 = 0.54, P < 0.05]. Open symbols = controls; filled symbols = UV-B exposed plants.

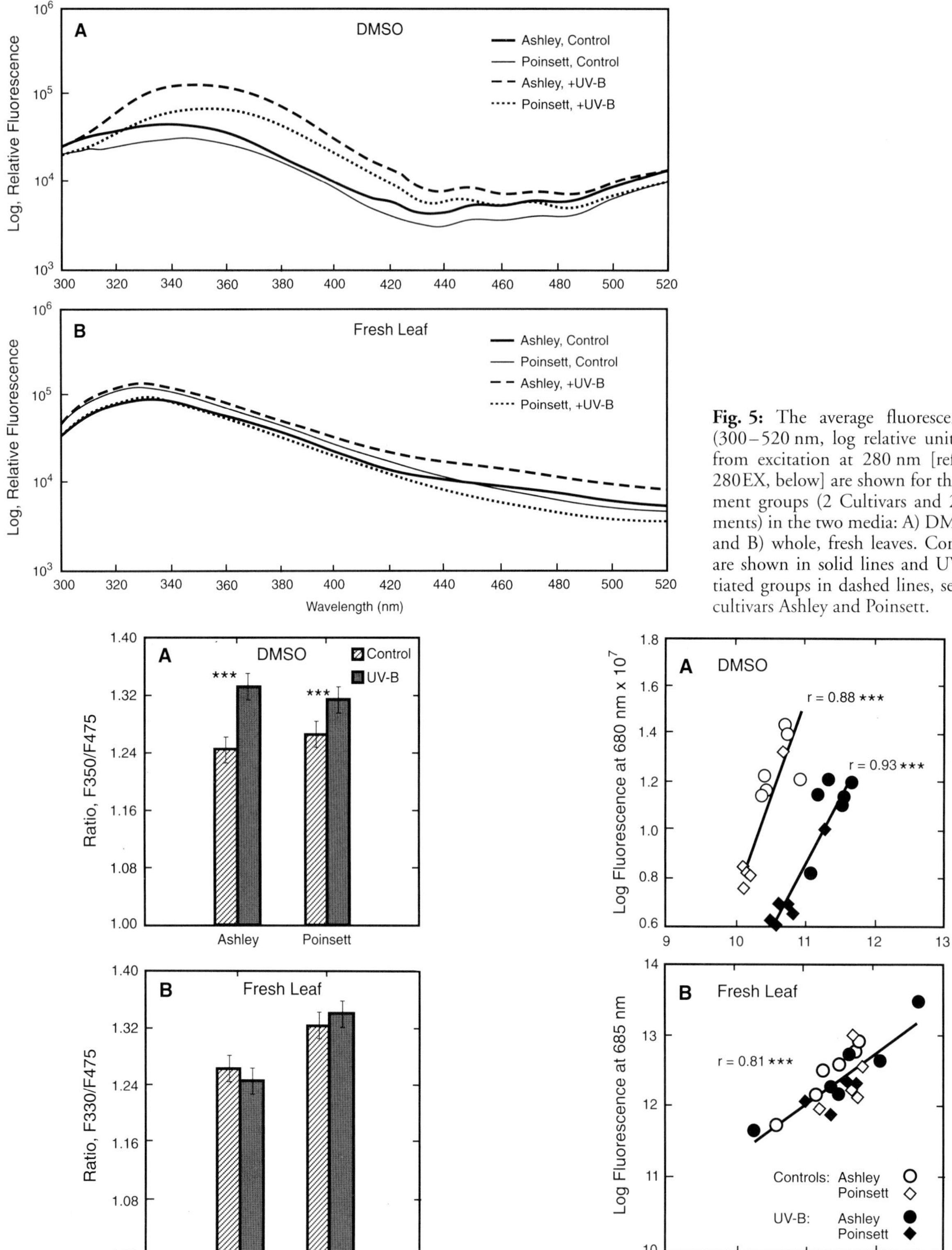

Fig. 5: The average fluorescence spectra (300–520 nm, log relative units) obtained from excitation at 280 nm [referred to as 280EX, below] are shown for the four treatment groups (2 Cultivars and 2 UV treatments) in the two media: A) DMSO extract; and B) whole, fresh leaves. Control groups are shown in solid lines and UV-B irradiatiated groups in dashed lines, separately for cultivars Ashley and Poinsett.

Fig. 6: The increase in the UV-A/blue fluorescence ratio from 280EX with UV-B irradiation is shown for the two media: A) DMSO extract; and B) whole, fresh leaves. For A, the F350/F475 ratio is given separately for cultivars Ashley and Poinsett, where similar and highly significant ($P \leq 0.01$; $r^2 = 0.82$, UV treatment effect) increases were observed in both cultivars. For leaves, no change in the F330/F475 ratio was observed.

Fig. 7: The correlations between fluorescence measured at 680 nm (340EX) and that measured at 330 nm (280EX) are shown for log-log relationships: A) DMSO extract; and B) whole, fresh leaves. A single positive relationship for all data ($r = 0.81$, $P \leq 0.000$) resulted for fresh leaves. Separate, highly significant ($P < 0.001$) correlations resulted in DMSO for UV-B plants ($r = 0.93$) and controls ($r = 0.88$).

Discussion

Similar results were obtained for fresh leaves and the DMSO extract, for either spectrum type (280EX or 340EX). However, whole leaf responses were too variable to determine statistically significant differences, in most cases. This was primarily due to the actual variability of the chlorotic spots along the leaf edges in UV-B exposed plants, coupled with the greater structural complexity of whole leaves, as compared with the uniform response in the extract. Consequently, the extracts were preferable to leaves in separating UV-B induced effects and cultivar differences, due to less total variability, no interference from surface or cellular structures, and concentration of the influential fluorescent compounds.

For 280EX spectra, the UV-A peak (330–350 nm) was clearly higher in the extracts for UV-B exposed plants of both cultivars, and significantly higher for DMSO when expressed as a normalized value (a UV-A/blue ratio, F350/F475). The fluorescence in this region is due primarily to aromatic amino acids and proteins (Fujimori, 1981; Pigault and Gerard, 1984; Caldwell, 1987). In green plants, the most abundant protein is the photosynthetic enzyme, ribulose 1,5-bisphosphate carboxylase (rubisco) which comprises 50–70 % of total soluble leaf protein (Dalling, 1987) and which is expected to dominate the total fluorescence for proteins in the 330–350 nm spectral range (Caldwell, 1993; Corp et al., 1994). The activity and/or content of rubisco is known to be reduced after UV-B exposure (Vu et al., 1984; Strid et al., 1990). Several studies of pure compounds have typically shown that UV-B exposure induced similar reductions in total soluble protein content, rubisco, and tryptophan residues (Vu et al., 1984; Valenzeno, 1987; Jordan et al., 1992; Caldwell, 1993), although total soluble protein content may also increase in UV-B treatments (e.g., Tevini et al., 1981). While protein solubility in DMSO is probably low, fluorescence treatment differences can still be observed. The observed 10 nm shift in peak fluorescence (340–350 nm) coupled with greater relative 350 RFI observed for DMSO could be explained as the net result of decreased rubisco content (maximal fluorescence, ~340 nm) and increased content and/or altered composition of other protein-related products. These products are most likely storage proteins and enzymes associated with the precursor biosynthesis pathway of compounds involved in UV-B protection that fluoresce at slightly longer wavelengths.

In this study, we observed fluorescence changes associated with proteins (280EX) as well as increased production of blue fluorescing compounds (340EX) after UV-B exposure in both cultivars. These observations are consistent with the suggestion by Jordan et al. (1992) that degradation of rubisco enabled redistribution of carbon and nitrogen for augmented synthesis of compounds involved in UV-B defense. Furthermore, a plausible physiological mechanism which is also consistent with our results, and which links the degradation of proteins including rubisco to production of blue fluorescent compounds, has been provided by Caldwell (1993). His *in vitro* study used partially purified rubisco from one of the same cucumber cultivars (Poinsett) used in our study; he showed that UV-B induced photolysis of tryptophan residues from rubisco led to formation of a blue fluorescent photo-oxidation product, tentatively identified as N-formyl-kynurenine (N-FK). Under this scenario, UV-B irradiation disruption of protein synthesis would lead to increased extractable UV-A fluorescing amino acids and protein precursors and/or blue fluorescing degradation products, of which N-FK is an example. If Caldwell's hypothesis is correct, the greater resistance to UV-B exposure consistently observed for Ashley in UV-B experiments and generally higher blue fluorescence (340EX) might be attributed to a higher native production of N-FK or related photo-oxidative compounds. Alternatively, Ashley may be more efficient at converting the degradation products from rubisco into enzymes involved in photorepair and/or antioxidants. Unfortunately, the biochemical measurements to resolve these issues were not acquired in this study. However, the involvement of anti-oxidants in UV-B protection is given credance by the recent more general finding (Willekens et al., 1994) that both UV-B irradiation and ozone fumigation similarly induced the production of mRNA encoding for two catalases and glutathione peroxidase in the antioxidant biochemical pathway.

The possibility is not discounted of flavonoid involvement in photoprotection; increased production of flavonoid pathway enzymes (e.g., phenylalanine ammonia-lyase), may contribute to the 350 fluorescence peak (280EX spectra). However, fluorescence by the UVAP end-products at all wavelengths, including the blue/green region (340EX), is too low to contribute significantly to the observed blue fluorescence treatment differences, and can be discounted as an explanation here. Adamse and Britz (this issue) also concluded it unlikely that UVAP were responsible for the genetic differences in UV-B sensitivity between Poinsett and Ashley, although UVAP increased in response to UV-B irradiation.

The UV-B induced chlorophyll loss was successfully captured in lower far-red/red fluorescence ratios in 340EX spectra, for both media. This result is consistent with those of other investigators (e.g., Lichtenthaler et al., 1990). Presently, we have no satisfactory explanation for the greater Chl degradation which occurred in the UV-B «insensitive» cultivar (Ashley). The 450 RFI peak may also reflect an increase in specific carotenoids in the carotenoid pool, as opposed to the total pool size measured here, since UV-B exposure induced an increase in both the blue/far-red fluorescence ratio and the carotenoid/Chl ratio. The qualitatively different relationship between these two ratios expressed by the two cultivars, with a relatively greater change observed in Poinsett, may indicate a different photoprotective response per cultivar in response to UV-B irradiation. Since the alteration in the carotenoid/Chl ratio was due primarily to lower Chl *b*, damage to the light harvesting complex of Photosystem II is indicated. The blue/far-red fluorescence ratio (340EX) has previously been identified as an indicator of a general plant stress response (Stober and Lichtenthaler, 1992; Lang et al., 1992), and recently found useful in identifying plant response to UV-B irradiation for whole leaves in soybean (Subhash et al., 1995).

The strong correlation in both media between fluorescence emissions at 330 nm (from 280EX) and 680 nm (from 340EX) is especially intriguing and reveals that the same overall process-photosynthetic function/integrity, was captured at these two wavelengths. The DMSO results clearly show that UV-B irradiation altered this relationship in a manner consistent with the explanations given above: although greater UV-A fluorescence (i.e., higher aromatic amino acid levels) was observed, a lower investment was made in photosynthetic constituents (i.e., lower red fluorescence) with UV-B exposure. This general result suggests that the active rubisco fraction in bulk leaf protein is in balance, as expected, with electron transfer efficiency determined by the total intact chlorophyll pool. Therefore, it appears likely that matter and energy are diverted from rubisco synthesis into that of other enzymes which facilitate production of protective compounds, when plants are exposed to elevated UV-B irradiation.

Acknowledgements

We thank Mr. Lawrence Corp (SSAI, Lanham, MD, USA) for technical assistance.

References

Adamse, P. and S. J. Britz: Rapid fluence-dependent responses to ultraviolet-B radiation in cucumber leaves: interaction with blue and ultraviolet-A radiation. J. Plant Physiol., this issue (1996).

Adamse, P., S. J. Britz, and C. R. Caldwell: Amelioration of UV-B damage under high irradiance. II. Role of blue light photoreceptors. Photochem. Photobiol. *64,* 110–115 (1994).

Beggs, C. J., A. Stolzer-Jehle, and E. Wellmann: Isoflavonoid formation as an indicator of UV stress in bean (*Phaseolus vulgaris* L.) leaves: the significance of photorepair in assessing potential damage by increased solar UV-B radiation. Plant Physiol. *79,* 630–634 (1985).

Beggs, C. J. and E. Wellmann: Photocontrol of flavonoid biosynthesis. In: Kendrick, R. E. and G. H. M. Kronenberg (eds.): Photomorphogenesis in Plants, pp. 733–751, Kluwer Academic Publishers, Dordrecht (1994).

Britt, A. B., J.-J. Chen, D. Wykoff, and D. Mitchell: A UV sensitive mutant of *Arabidopsis* defective in the repair of pyrimidine-pyrimidinone (6-4) dimers. Science *261,* 1571–1574 (1993).

Britz, S. J. and P. Adamse: UV-B-induced increase in specific leaf weight of cucumber as a consequence of increased starch content. Photochem. Photobiol. *64,* 116–119 (1994).

Caldwell, C. R.: Temperature-induced protein conformational changes in barley root plasma membrane-enriched microsomes. II. Intrinsic protein fluorescence. Plant Physiol. *84,* 924–929 (1987).

Caldwell, C. R.: Ultraviolet-induced photodegradation of cucumber (*Cucumis sativus* L.) microsomal and soluble protein typtophanyl residues *in vitro.* Plant Physiol. *101,* 947–953 (1993).

Caldwell, M. M.: Solar UV irradiation and the growth and development of higher plants. In: Giese, A. C. (ed.): Photophysiology, Vol. *6,* pp. 131–177, Academic Press, New York, NY (1971).

Caldwell, M. M., A. H. Teramura, and M. Tevini: The changing solar ultraviolet climate and the ecological consequences for higher plants. Trends Evol. Ecol. *4,* 363–367 (1989).

Chappelle, E. W., J. E. McMurtrey, F. M. Wood, and W. W. Newcomb: Laser induced fluorescence of green plants. 2. LIF changes caused by nutrient deficiencies in corn. Appl. Opt. *23,* 139–144 (1984b).

Chappelle, E. W., F. M. Wood, J. E. McMurtrey, and W. W. Newcomb: Laser induced fluorescence of green plants. 1. A technique for the remote detection of plant stress and species differentiation. Appl. Opt. *23,* 134–138 (1984).

Corp, L. A., E. W. Chappelle, J. E. McMurtrey, and M. S. Kim: A new fluorescence band obtained by the excitation of plants at 280 nm and its implications to the remote assessment of vegetation. Proceedings, 1994 International Geoscience and Remote Sensing Symposium (IGARSS'94), Vol. II, Pasedena, CA, pp. 986–989 (1994).

Dalling, M. J.: Proteolytic enzymes and leaf senescence. In: Thomson, W. W., E. A. Nothnagel, and R. C. Huffaker (eds.): Plant Senescence. Biochemistry and Physiology, pp. 54–70, Amer. Soc. Plant Physiol., Rockville, MD (1987).

Fujimori, E.: Blue fluorescence and crosslinking of photooxidized proteins. FEBS Letters *135,* 257–260 (1981).

Gleason, J. F., P. K. Bhartia, J. R. Herman, R. McPeters, P. Newman, R. S. Stolarski, L. Flynn, G. Labow, D. Larko, C. Seftor, C. Wellenmeyer, W. D. Komhyr, A. J. Miller, and W. Planet: Record low global ozone in 1992. Science *260,* 523–526 (1993).

Krause, G. H. and E. Weis: Chlorophyll fluorescence and photosynthesis: the basics. Annu. Rev. Plant Physiol. Plant. Biol. *42,* 313–349 (1991).

Jordan, B. R., J. He, W. S. Chow, and J. M. Anderson: Changes in mRNA levels and polypeptide subunits of ribulose 1,5-bisphosphate carboxylase in response to supplementary ultraviolet-B radiation. Plant, Cell, Environ. *15,* 91–98 (1992).

Lang, M., P. Siffel, Z. Braunova, and H. K. Lichtenthaler: Investigations of the blue-green fluorescence enussion of plant leaves. Bot. Acta *105,* 435–439 (1992).

Lang, M., F. Stober, and H. K. Lichtenthaler: Fluorescence emission spectra of plant leaves and plant constituents. Radiat. Environ. Biophys. *30,* 333–347 (1991).

Li, J., T. Ou-Lee, R. Raba, R. G. Amundson, and R. L. Last: *Arabidopsis* flavonoid mutants are hypersensitive to UV-B radiation. The Plant Cell *5,* 171–179 (1994).

Lichtenthaler, H. K.: Chlorophylls and carotenoids: pigments of photosynthesis. Methods Enzymol. *148,* 350–352 (1987).

– Vegetation stress: An introduction to the stress concept in plants. J. Plant Physiol. *148,* 4–14 (1996).

Lichtenthaler, H. K., R. Hak, and U. Rinderle: The chlorophyll fluorescence ratio F690/F730 in leaves of different chlorophyll content. Photosyn. Res. *25,* 295–298 (1990).

Lichtenthaler, H. K. and U. Rinderle: The role of chlorophyll fluorescence in the detection of stress conditions in plants. CRC Critical Rev. in Anal. Chem. *19,* 29–85 (1988).

McMurtrey III, J. E., E. W. Chappelle, M. S. Kim, and L. A. Corp: Fluorescence and reflectance of field corn (*Zea mays* L.) grain grown under eight different nitrogen fertilizer levels. J. Plant Physiol., this volume (1996).

Middleton, E. M. and A. H. Teramura: The role of flavonol glycosides and carotenoids in protecting soybean from ultraviolet-B damage. Plant Physiol. *103,* 741–752 (1993).

– – Understanding photosynthesis, pigment and growth responses induced by UV-B and UV-A irradiances. Photochem. Photobiol. *60,* 38–45 (1994).

Pang, Q. and J. B. Hays: UV-B inducible and temperature-sensitive photoreactivation of cyclobutanepyrimidine dimers in *Arabidopsis thaliana.* Plant Physiol. *95,* 536–543 (1991).

Park, H.-W., S.-T. Kim, A. Sancar, and J. Deisenhofer: Crystal structure of DNA photolyase from *Escherichia coli.* Science *268,* 1866–1872 (1995).

Pigault, C. and D. Gerard: Influence of the location of tryptophanyl residues in proteins on their photosensitivity, Photochem. Photobiol. *40,* 291–296 (1984).

Strid, A., W. S. Chow, and J. M. Anderson: Effects of supplementary UV-B radiation on photosynthesis in *Pisum sativum.* Biophys. Biochem. Acta *1020,* 260–268 (1990).

Stober, F., M. Lang, and H. K. Lichtenthaler: Blue, green, and red fluorescence emission signatures of green, etiolated, and white leaves. Remote Sensing Environ. *47,* 65–71 (1994).

Stober, F. and H. K. Lichtenthaler: Changes of the laser-induced blue, green and red fluorescence signatures during greening of etiolated leaves of wheat. J. Plant Physiol. *140,* 673–680 (1992).

– – Characterization of the laser-induced blue, green and red fluorescence signatures of leaves of wheat and soybean grown under different irradiance. Physiol. Plant. *88,* 696–704 (1993).

Subhash, N., P. Mazzinghi, F. Fusi, G. Agati, and B. Lercari: Analysis of laser-induced fluorescence line-shape of intact leaves: application to UV stress detection. Photochem. Photobiol. *62,* 711–718 (1995).

Tevini, M., W. Iwanzik, and U. Thoma: Some effects of enhanced UV-B irradiation on the growth and composition of plants. Planta *153,* 388–394 (1981).

Tevini, M., J. Braun, and G. Fieser: The protective function of the epidermal layer of rye seedlings against ultraviolet-B radiation. Photochem. Photobiol. *53,* 329–333 (1991).

Tevini, M. and A. H. Teramura: UV-B effects on terrestrial plants. Photochem. Photobiol. *50,* 479–487 (1989).

Valentini, R., G. Cecchi, P. Mazzinghi, G. Scarascia-Mugnozza, G. Agati, M. Bazzani, P. De Angelis, F. Fusi, G. Matteucci, and V. Raimondi: Remote sensing of chlorophyll *a* fluorescence on vegetation canopies. Remote Sensing Environ. *47,* 29–35 (1994).

Valenzeno, D. P.: Photomodification of biological membranes with emphasis on singlet oxygen mechanisms. Photochem. Photobiol. *46,* 147–160 (1987).

Vu, C. V., L. H. Allen, and L. A. Garrard: Effects of enhanced UV-B radiation (280–329 nm) on ribulose-1,5-bisphosphate carboxylase in pea and soybean. Environ. Exp. Bot. *24,* 131–143 (1984).

Willekens, H., W. V. Camp, M. van Montagu, D. Inzé, C. Langebartels, and H. Sandermann Jr.: Ozone, sulfur dioxide, and ultraviolet B have similar effects on mRNA accumulation of antioxidant genes in *Nicotiana plumbaginifolia* L. Plant Physiol. *106,* 1007–1014 (1994).

J. Plant Physiol. Vol. 148. pp. 78–85 (1996)

Morphological and Physiological Responses of *Brassica napus* to Ultraviolet-B Radiation: Photomodification of Ribulose-1,5-bisphosphate Carboxylase/Oxygenase and Potential Acclimation Processes

Bruce M. Greenberg*, Michael I. Wilson, Karen E. Gerhardt, and Kenneth E. Wilson

Department of Biology, University of Waterloo, Waterloo, ONT N2L 3G1, Canada

Received June 24, 1995 · Accepted October 27, 1995

Summary

As the stratospheric ozone layer is depleted, the biosphere will be exposed to higher levels of ultraviolet-B (UV-B) radiation (290–320 nm). Using laboratory light sources that simulate the spectral quality of sunlight, we are examining some of the mechanisms involved in plant responses to UV-B. It was found that exposure of ribulose-1,5-bisphosphate carboxylase/oxygenase (Rubisco) from *Brassica napus* to UV-B *in vivo* or *in vitro* resulted in production of a high molecular weight (HMW) variant of the large subunit. Coincident with formation of the HMW product *in vitro* was a loss in tryptophan fluorescence. To protect against damage, plants can acclimate to UV-B. To this end, we have studied cotyledon curling in *B. napus;* a photomorphogenic response specific to UV-B. To characterize the photoreceptor for curling, inhibitors of photochemical signaling were employed. A quencher of flavin excitation, and inhibitors of Ca^{++} and cyclic nucleotide signaling diminished curling. Biosynthesis of flavonoids and other UV-absorbing pigments also occurred in *B. napus* exposed to the levels of UV-B that caused curling. To determine which flavonoids and other UV-absorbing compounds were UV-B specific, HPLC analysis was carried out. Approximately 20 distinct UV-absorbing pigments were produced in response to UV-B radiation. Thus, using *B. napus* we were able to follow UV-B induced damage and acclimation.

Key words: Environmental stress, Flavonoids, Photomorphogenesis, Photoreceptor, Photosynthesis, Rubisco.

Abbreviations: A23187 = the Ca^{++} ionophore calcimycin; PAA = phenylacetic acid; HMW = high molecular weight; PAR = photosynthetically active radiation (400–700 nm); PSII = photosystem II; Rubisco = Ribulose-1,5-bisphosphate Carboxylase/Oxygenase; SSR = simulated solar radiation; SDS-PAGE = Sodium dodecyl sulfate – polyacrylamide gel electrophoresis; UV-A = Ultraviolet-A (320–400 nm); UV-B = Ultraviolet-B (290–320 nm).

Introduction

The stratospheric ozone layer, which attenuates solar ultraviolet-B (UV-B) radiation (290–320 nm), is being depleted by pollutants such as chlorofluorocarbons (Frederick, 1990; Kerr and McElroy, 1993). As a result, UV-B levels at the surface of the earth will likely increase, resulting in negative impacts on biological organisms (Tevini et al., 1989). Terrestrial plants are especially vulnerable due to their obligatory requirement for sunlight for photosynthesis (For recent reviews see Tevini et al., 1989; Tevini, 1993; Greenberg et al., 1995a).

* Correspondence.

The sites of UV-B damage are at the molecular level with potential lesions to DNA, proteins, membrane bilayers, photosynthetic pigments and phytohormones (Greenberg et al., 1989; Tevini et al., 1989; Kramer et al., 1991; Quaite et al., 1992; Ros and Tevini, 1995). In particular, the photosynthetic apparatus is vulnerable to UV-B radiation (Bornman et al., 1984; Vu et al., 1984; Greenberg et al., 1989; Strid et al., 1990; Tevini et al., 1991; Jordan et al., 1992; Wilson et al., 1995). Both photosystem II (PSII) and ribulose-1,5-bisphosphate carboxylase/oxygenase (Rubisco) are known targets of UV-B.

UV-B radiation has always been present in the environment, and adaptive mechanisms which diminish the damaging effects of UV-B have evolved in plants (Tevini et al., 1989; Tevini, 1993; Greenberg et al., 1995 a). One of the potential acclimation mechanisms is alterations in leaf transmittance properties, which results in attenuation of UV-B in the epidermis before it reaches the interior of the leaf (Tevini et al., 1989; Tevini et al., 1991; Cen and Bornman, 1993; Wilson and Greenberg, 1993 a). The action of oxygen detoxification enzymes, such as superoxide dismutase combined with glutathione reductase as well as various peroxidase activities which could have elevated levels after UV-B exposure, is another potential protection process (Murali et al., 1988; Kramer et al., 1991; Middleton and Teramura, 1993; Strid, 1993). Also, DNA repair via photoreactivation and excision repair are likely mechanisms that can ameliorate UV-B induced damage (Langer and Wellmann, 1990; Pang and Hays, 1991; Quaite et al., 1992). Activation of most acclimation processes is envisaged to occur via a specific UV-B photoreceptor(s) as well as phytochrome and the blue light/UV-A receptor (Ballaré et al., 1991; Ensminger and Schäfer, 1992).

In this study, laboratory light sources that provided environmentally relevant levels of UV-B radiation were used to study mechanisms of UV-B responses in plants. First, we examined UV-B induced structural changes to Rubisco. When Rubisco is exposed to UV-B the large subunit (55 kD) is photomodified such that it migrates at 66 kD in SDS-PAGE (Wilson et al., 1995). Formation of this photoproduct *in vitro* using extracts from *Brassica napus* was found along with loss of Trp fluorescence from the protein preparation. As well, to investigate protection against protein damage of this nature, morphological changes to *B. napus* were probed. It was found that *B. napus* cotyledons curl upwards in response to UV-B potentially via activation of a flavin-type photoreceptor. In addition, the plants actively synthesize specific flavonoids and other UV-absorbing pigments in response to UV-B radiation.

Materials and Methods

Plant growth and exposure to UV-B radiation

Seeds of *Brassica napus* L. cv Topas (Canola) were sown in Pro-Mix potting media (Premier Brands, Rivière-du-Loup, Quebec, Canada) and placed under artificial lighting in growth chambers at a temperature of 22 °C. The light sources are described below. The spectral photon distributions and fluence rates of all light sources used in this study were measured at the level of the leaves using a calibrated spectroradiometer (Oriel Inc., Stratford, Connecticut).

For generation of the Rubisco photoproduct *in vivo*, plants were grown under 70 µmol $m^{-2} s^{-1}$ of photosynthetically active radiation (PAR, 400–700 nm) with a 16 h light/8 h dark photoperiod for 21 d. The plants were then placed in PAR (65 µmol $m^{-2} s^{-1}$) plus UV-B (1.5 µmol $m^{-2} s^{-1}$) and UV-A (1.7 µmol $m^{-2} s^{-1}$), generated with fluorescent lamps (F48T12-CW for PAR and FS40 for UV-B and UV-A, National Biological Co., Twinsburg, Ohio) filtered to screen out UV-C (<290 nm) (Fig. 1 B) (Wilson et al., 1995). The biologically effective irradiance of UV-B (UV-B_{BE}) used was equivalent to 260 mW m^{-2} using Caldwell's general plant damage action spectrum normalized at 300 nm (Caldwell et al., 1980).

For the flavonoid biosynthesis studies, 21 d old plants that had been grown in PAR (350 µmol $m^{-2} s^{-1}$) were placed in simulated solar radiation (SSR) for 6 d. The SSR source contains cool-white fluorescent lamps, UV-A fluorescent lamps (National Biological Co., Twinsburg, Ohio) and UV-B fluorescent lamps (National Biological Co., Twinsburg, Ohio). The number of lamps was balanced to give a PAR : UV-A : UV-B ratio of 100 : 10 : 2, and a fluence rate of 350 µmol $m^{-2} s^{-1}$ (Fig. 1 C). The radiation was filtered through 0.08 mm cellulose diacetate to remove all of the incident UV-C (200 to 290 nm). The biologically effective irradiance UV-B was approximately 1200 mW m^{-2} (according to Caldwell et al., 1980). The control plants were grown under a mylar screen to remove the UV-B. The PAR photoperiod was 16 h light/8 h dark and the UV-B and UV-A was applied during the middle 8 h of the PAR photoperiod.

Protein Isolation and Analysis

To analyze for the Rubisco photoproduct generated *in vivo*, leaf disks (0.4 cm^2) were taken from the oldest fully expanded leaves of individual plants after a 16 h treatment period. Soluble proteins were isolated from leaf tissue, separated by SDS-PAGE, and detected with a Rubisco holoenzyme anti-serum as previously described (Wilson et al., 1995).

Rubisco Isolation and In Vitro UV-B Irradiation

A crude soluble chloroplast extract containing Rubisco was obtained following the procedure of Salvucci et al. (1986). Rubisco was partially purified from this chloroplast extract via ultrafiltration with a 300 kD cutoff filter in a 150 mL Omegacell according to manufacturer's specifications (Filtron Technology Corporation, Northborough, Massachusetts). The material retained by the filter (retentate) contained primarily Rubisco (approx. 80 %) and a chaperonin (HSP60) complex (approx. 10 %) (data not shown). The concentration of proteins in the retentate was about 0.5 mg mL^{-1} in 50 mM tricine (pH 8), 150 mM NaCl, 0.04 % (w/v) sodium azide.

The partially purified Rubisco (200 µL) was irradiated at 300 nm (2 µmol $m^{-2} s^{-1}$) for 60 min at room temperature. The irradiations were carried out in a cuvette in a scanning spectrofluorometer (Photon Technology Inc., London, Ontario) using the excitation beam as the source of 300 nm radiation. At various time points during the incubation the preparation was analyzed fluorometrically. The excitation wavelength was 300 nm (2 µmol $m^{-2} s^{-1}$) and emission spectra were collected from 310 to 500 nm. At the end of the 60 min treatment period the protein preparation was subjected to SDS-PAGE and analyzed immunologically with the Rubisco anti-serum (Wilson et al., 1995).

UV-B Induced Cotyledon Curling

B. napus seeds were germinated for 5 d in 3 × 3 cell packs containing moistened Pro-Mix potting media (Premier Brands) in growth chambers at 22 °C under PAR (100 µmol $m^{-2} s^{-1}$) given in 16 h light/8 h dark photoperiods. To induce curling (Wilson and Greenberg, 1993 b), the seedlings were exposed for 60 min to UV-B

($2.5\,\mu mol\,m^{-2}\,s^{-1}$) generated with a Xe arc lamp and a 290 nm interference filter (10 nm bandwidth). Following the UV-B treatment, the seedlings were incubated in PAR ($100\,\mu mol\,m^{-2}\,s^{-1}$) for 24-h to allow curling to develop. Curling was quantified by determining the curling angle between the midvein, the cotyledon margin and a line parallel to the soil surface (Wilson and Greenberg, 1993 b). Potential inhibitors of curling were painted onto the cotyledon surfaces 1 h prior to the UV-B treatment. The inhibitors (all purchased from Sigma Chemical Co., St. Louis, Missouri) and their concentrations (in 0.05 % v/v triton X-100) are listed in Table 1. The controls were painted with 0.05 % v/v triton X-100.

Analysis of Flavonoids

B. napus leaf disks were cut from control and UV-B treated plants. The disks (100 mg) were incubated in 80 % methanol, 20 % filter purified water for 24 h at 4 °C. The extracted flavonoids were analyzed by reversed phase HPLC with a photodiode array detector. A 150 µL sample was injected onto a 25 × 4.6 cm C-18 column (Supelco, Oakville, Ontario, Canada) with a mobile phase of 91 % (v/v) 1 mM phosphoric acid (pH 3), 9 % (v/v) acetonitrile. This was followed by a non-linear elution gradient to 100 % acetonitrile as described in the figure legend. Total flavonoids from whole leaves and upper epidermis were also analyzed spectrophotometrically following extraction into 80 % methanol (Wilson and Greenberg, 1993 a).

Results

Laboratory light sources for the study of UV-B effects

At the surface of the earth, the molar ratio of PAR : UV-A : UV-B is about 100 : 10 : 1 (Fig. 1 A), and so we have chosen to work with UV-B levels at about 1 % to 2 % of PAR on a photon basis (Fig. 1 B). *B. napus* can be grown to maturity under this ratio of PAR : UV-B radiation even at total fluence rates as low as $50\,\mu mol\,m^{-2}\,s^{-1}$. If the PAR : UV-B ratio is held at 100 : 1, no adverse signs of stress are observed (Greenberg et al., 1993), and the plants demonstrate adaptive responses (Wilson and Greenberg, 1993 a).

It has been shown that for some plant species high fluence rate PAR is important for optimal adaptation to UV-B (Tevini et al., 1989; Cen and Bornman, 1990). To achieve this, field studies have been performed where natural sunlight (approximately $2000\,\mu mol\,m^{-2}\,s^{-1}$ of PAR) is supplemented with UV-B to mimic ozone losses. However, field conditions are not available to all investigators and high irradiance lighting conditions can be difficult to achieve in the laboratory. A balanced lighting system (i.e. one containing UV-B, UV-A and PAR) might compensate for the need for high fluence rates, because this spectral region triggers several morphogenic responses and activates DNA repair via photolyase (Langer and Wellmann, 1990; Adamse and Britz, 1992).

To generate a balanced lighting system, PAR, UV-B and UV-A can be readily combined. We use a light source that mimics sunlight with respect to the relative amounts of PAR and UV (the spectrum shown in Fig. 1 C has a PAR : UV-A : UV-B ratio of about 100 : 10 : 2) (Greenberg et al., 1995 b). This simulated solar radiation (SSR) source contains cool-white fluorescent lamps, UV-A fluorescent lamps and UV-B fluorescent lamps available from various manufacturers. The

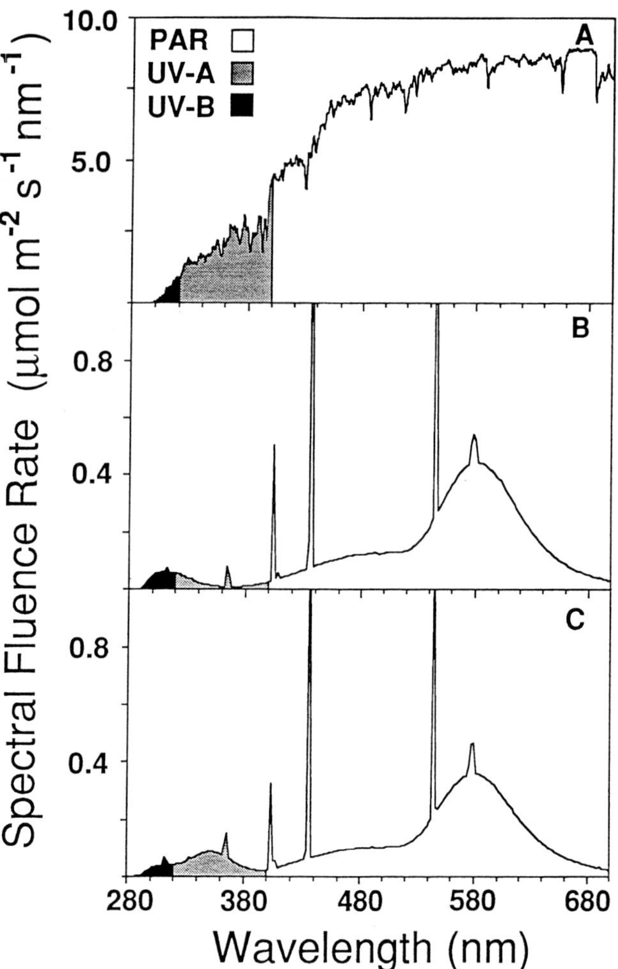

Fig. 1: Spectral distribution of sunlight, a PAR plus UV-B source and a simulated solar radiation (SSR) source. Panel A: Sunlight measured on a cloudless day 38 km north of Cleveland, Ohio (12 July 1994). Panel B: Emission spectrum of a PAR plus UV-B source filtered through cellulose diacetate. Panel C: SSR source filtered through polystyrene. All measurements were made with a calibrated photodiode array based spectroradiometer. See text for further descriptions of the artificial light sources and radiometric measurements.

light is filtered through cellulose diacetate or polystyrene to remove all of the incident UV-C (200 to 290 nm). While the spectral output shown in Fig. 1 C does not precisely match sunlight, the ratio of PAR : UV-A : UV-B is similar to that of terrestrial sunlight in the 290 to 700 nm wavelength span for midsummer in temperate latitudes corresponding to southern Canada and northern USA. Further, the UV-B fluence rate can be lowered by covering the UV-B lamps with cheese cloth. Thus, this source has environmental relevance and is excellent for studies on the mechanisms of UV-B responses in plants. *B. napus* and *Lemna gibba* have been found to grow well under this source, showing no overt signs of UV-B stress (Greenberg et al., 1993).

Detrimental effects of UV-B on plants

The negative effects of UV-B on plants include inhibition of photosynthesis, protein damage, membrane damage and DNA lesions (Greenberg et al., 1989; Tevini et al., 1989; Kramer et al., 1991; Quaite et al., 1992; Caldwell, 1993). It is currently unknown which of these impacts are the most detrimental, although photosynthesis is thought to be an important point of damage. Specific targets in photosynthesis for UV-B radiation are PSII and Rubisco (Bornman et al., 1984; Vu et al., 1984; Greenberg et al., 1989; Strid et al., 1990; Tevini et al., 1991; Jordan et al., 1992; Wilson et al., 1995). *In vivo* exposure of Rubisco to UV-B was recently shown to result in photomodification of the large subunit (LSU, 55 kD) to a high molecular weight (HMW) variant of 66 kDa (Wilson et al., 1995).

We have extended this work using a partially purified preparation of the holoenzyme of Rubisco. It was found that exposure of Rubisco *in vitro* to UV-B resulted in formation of the HMW product that was observed *in vivo* (Fig. 2). In addition, when one analyzes this preparation spectrofluorometrically (Fig. 2), a typical Trp emission spectrum is observed (c.f. Caldwell, 1993). The tryptophan fluorescence signal was diminished when the protein preparation was irradiated under the same *in vitro* UV-B conditions that induced *in vitro* formation of the HMW Rubisco photoproduct (Fig. 2). This suggests that a Trp residue(s) in Rubisco may be a target for UV-B radiation.

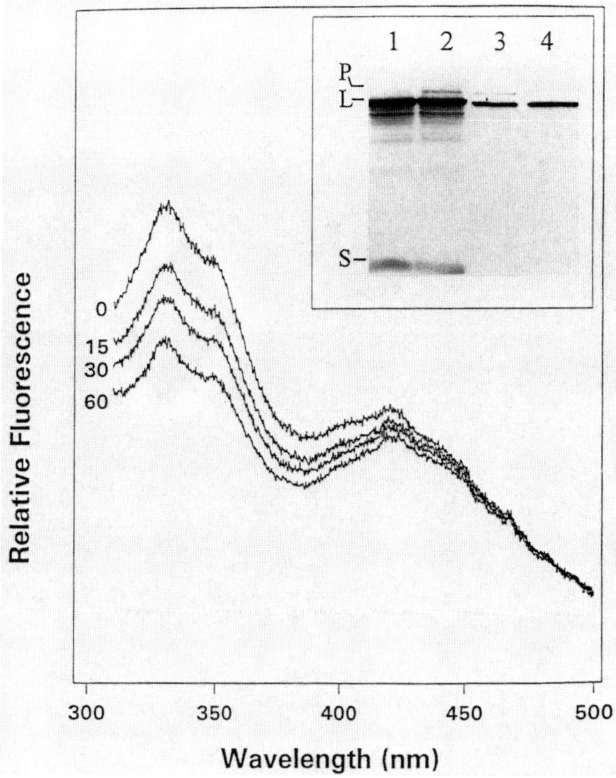

Fig. 2: Treatment of Rubisco with UV-B radiation. A partially purified preparation of Rubisco (approximately 80 % Rubisco, 0.5 mg mL^{-1}) was placed in a cuvette in a spectrofluorometer. The sample was exposed *in situ* to actinic radiation of 300 nm (2 $\mu mol\, m^{-2}\, s^{-1}$) at 23 °C. At 0, 15, 30 and 60 min an emission spectrum from 310 nm to 500 nm was recorded using the 300 nm actinic beam as the excitation wavelength. Inset: Analysis of the protein preparation for UV-B induced production of the HMW Rubisco photoproduct. Proteins were separated by SDS-PAGE and detected immunologically with a Rubisco holoenzyme antiserum. Lane 1: soluble proteins isolated from *B. napus* leaves treated *in vivo* with PAR minus UV-B. Lane 2: soluble proteins isolated from *B. napus* leaves treated *in vivo* with PAR plus UV-B. Lane 3: partially purified Rubisco treated for 60 min in the spectrofluorometer with 300 nm radiation. Lane 4: partially purified Rubisco incubated in the dark at 23 °C for 60 min. Positions of the HMW photoproduct (P), and the large (L) and small (S) subunits of rubisco are indicated.

UV-B induced cotyledon curling

Because UV-B has always been present in the environment, acclimation mechanisms have evolved in plants which diminish the damaging effects of the radiation (Tevini et al., 1989; Barnes et al., 1990). These mechanisms are thought to be triggered by a specific UV-B photoreceptor(s) (Barnes et al., 1990; Ballaré et al., 1991; Hashimoto et al., 1991; Li et al., 1993; Wilson and Greenberg, 1993 b). To examine UV-B photoreceptor(s) it is crucial to have simple and rapid photomorphogenic assays. To this end, we have used upward curling of *Brassica napus* cotyledons. In a previous study we demonstrated that curling has the attributes of a photomorphogenic response and that it is specific for UV-B irradiation (Wilson and Greenberg, 1993 b). The degree of curling showed a log-linear dependence on UV-B fluence, and reciprocity with respect to length of exposure and fluence rate, indicating a single photoreceptor species. Furthermore, curling is triggered by relatively short pulses of UV-B radiation, so the response should be useful for probing the mechanism of UV-B detection in plants. Flavonoids also accumulate during the 24 h following the 60 min pulse of UV-B (Wilson and Greenberg, manuscript in preparation).

In a preliminary study to characterize the photoreceptor for curling, inhibitors of photomorphogenic signaling were used (Table 1). Phenylacetic acid (PAA), which can quench flavin and pterin excited states (Vierstra and Poff, 1981), inhibited UV-B enhanced upward curling, implying a potential chromophore for the photoreceptor. The Ca^{++} ionophore, A23187 (Tretyn et al., 1991), caused the same degree of positive curling in PAR and UV-B, removing the UV-B specificity of the response. Additionally, the cyclic nucleotide phosphodiesterase activator imidazole (Tretyn et al., 1991) diminished curling in UV-B. Conversely, triacontanol, an inhibitor of auxin induced cell expansion (Morré et al., 1991), had only a modest impact on UV-B specific upward curling, and taxol, a stabilizer of microtubules (Chu et al., 1993), did not affect the curling process. The lack of a strong response to triacontanol also argues against pleiotropic effects of PAA on curling by it acting as an exogenous auxin.

UV-B induction of flavonoid biosynthesis in B. napus

One mechanism that could protect the photosynthetic apparatus is alterations in leaf transmittance properties. Epi-

Table 1: The effects of signal transduction inhibitors on cotyledon curling in *Brassica napus.*

Compound[2]		Curling Angle (degrees)[1]	
		PAR[3]	UV-B[3]
Control		−13.2±2.3	24.5±2.8
PAA	(10 mM)	−26.2±1.6	−16.3±3.3
Imidazole	(10 mM)	−23.8±2.1	−14.1±4.4
A23187	(10 µM)	12.5±1.8	6.9±6.8
Taxol	(10 µM)	−19.9±1.9	25.0±3.1
Triacontanol	(10 µM)	−5.2±3.9	13.6±5.8

[1] The curling angle (in degrees) is the angle between the midvein, the cotyledon margin and a line parallel to the soil surface (see Wilson and Greenberg, 1993 a). A positive angle is upward curling. The angle is given ± the standard error of the mean (n = 18).

[2] Prior to UV-B or PAR treatment, the entire cotyledon surface was painted with the indicated concentrations of the compounds in 0.05 % (v/v) triton X-100. The control treatment was 0.05 % (v/v) triton X-100 with no other additions.

[3] Plants were exposed to either PAR (100 µmol m^{-2} s^{-1}) or UV-B (1 µmol m^{-2} s^{-1}) for 100 min and then incubated in PAR (100 µmol m^{-2} s^{-1}) for 24 h, after which the curling angle was measured (see Wilson and Greenberg, 1993 a).

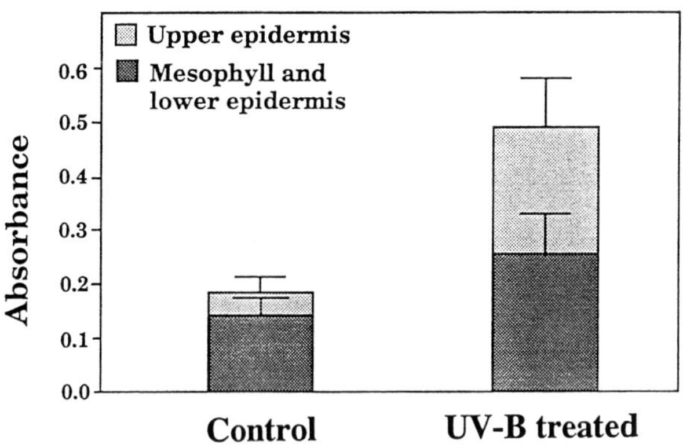

Fig. 3: Distribution of flavonoids in *B. napus.* Plants were adapted to growth in UV-B plus PAR (UV-B treated) or PAR alone (control) as described in materials and methods. The mesophyll was removed from the upper epidermis and the flavonoid content was determined in both fractions (Wilson and Greenberg, 1993 a). The content of flavonoid is reported as relative absorbance in the epidermis and mesophyll of the leaves. Error bars represent standard error of the mean (n = 5). Combining both parts of a histogram gives the total flavonoid content of a leaf.

dermal flavonoids are synthesized in most plant species in response to UV-B (Tevini et al., 1989; Hashimoto et al., 1991; Li et al., 1993, Wilson and Greenberg, 1993 a). They are an efficient screen because they strongly absorb from 280 to 350 nm, but do not absorb PAR, which would diminish the level of photosynthetically active radiation. The potential of flavonoids to act as a UV-B screen has been demonstrated by examining protection of the photosynthetic apparatus from UV-B damage (Tevini et al., 1991; Wilson and Greenberg, 1993 a) and by measuring transmission through epidermal layers (Cen and Bornman, 1993). Consistent with these findings, in *B. napus* the vast majority of synthesis of flavonoids and other UV-absorbing pigments in response to UV-B occurs in the epidermal cell layer (Fig. 3).

We are interested in determining if flavonoid biosynthesis is triggered by the same photoreceptor(s) that initiates cotyledon curling. To begin this work, the production of specific flavonoids in response to UV-B was examined (Fig. 4). Plants were exposed to UV-B levels in an SSR source (Fig. 1 C) similar to the UV-B conditions that cause cotyledon curling during germination (Wilson and Greenberg, 1993 b). Under this SSR source, UV-B induced synthesis of flavonoids and other UV-absorbing compounds was detected after about 1.5 d and saturated after 4 d with a 2 fold increase in total pigment levels (data not shown). HPLC analysis revealed that several flavonoids and other UV-absorbing compounds are synthesized in large amounts in response to UV-B (Fig. 4). Three peaks (peaks 4, 7 and 12) that increase in response to UV-B are indicated on Fig. 4. Two of these peaks (peaks 4 and 12) have been tentatively identified as kaempferol glycosides based on their absorbance spectra in the presence and absence of shift reagents (data not shown). Many of the other peaks induced

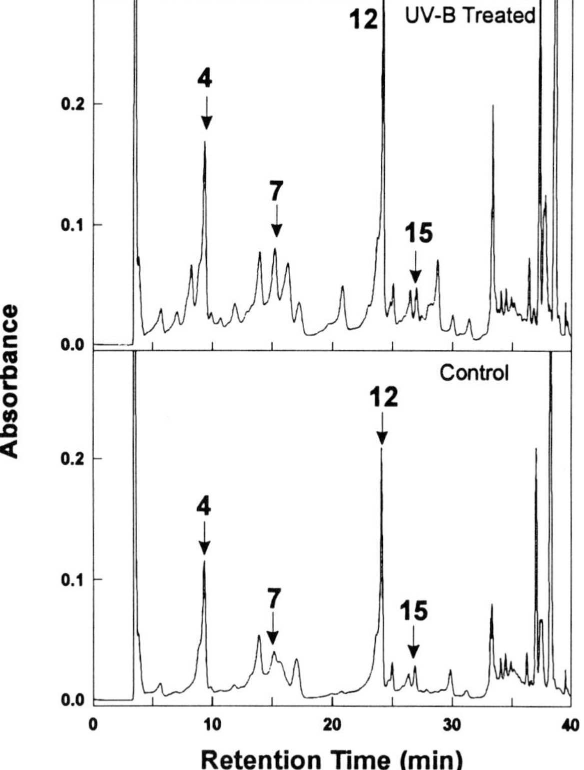

Fig. 4: HPLC analysis of flavonoids from *B. napus.* Methanol (80 %) extracts of leaf disks from UV-B treated and control plants were subjected to reversed phase HPLC and monitored at 340 nm with a diode array detector. The flavonoids were loaded onto the column with a mobile phase of 91 % 1 mM phosphoric acid (pH 3), 9 % acetonitrile. A non-linear gradient to 100 % acetonitrile was then run to elute the flavonoids (3 min at 10 % acetonitrile, 5 min at 11.5 % acetonitrile, 9 min at 14 % acetonitrile, 2 min at 19 % acetonitrile, 9 min at 22 % acetonitrile, 6 min at 100 % acetonitrile). Peaks 4, 7, and 12 increase due to UV-B. Peak 15 remains relatively constant in UV-B.

by UV-B have absorbance spectra characteristic of flavonoids. Other peaks do not increase or increase only modestly in response to UV-B (for instance peak 15 in Fig. 4). Therefore, specific flavonoids are generated in response to UV-B.

Discussion

B. napus is an excellent system to study the mechanism of UV-B effects on plants. Damage to Rubisco (Fig. 2) and PSII (Wilson and Greenberg, 1993 a) can both be used as measures of stress. The cotyledon curling assay (Table 1) is a rapid measure of UV-B induced photomorphogenesis. Finally, epidermal flavonoid synthesis is a strong response to UV-B in *B. napus* (Figs. 3 and 4) and flavonoids have been shown to protect PSII (Wilson and Greenberg, 1993 a).

Proteins are sensitive to UV-B because aromatic amino acids and many protein co-factors (e.g. flavins and quinones) absorb UV-B radiation (Grossweiner, 1984; Greenberg et al., 1989; Caldwell, 1993). Trp is especially sensitive to UV-B because its absorbance tail extends past 300 nm. Indeed, when Rubisco was exposed to 300 nm radiation *in vitro*, a decrease in Trp fluorescence was observed (Fig. 2). Moreover, since Trp is the only amino acid capable of absorbing a 300 nm photon (Grossweiner, 1984), one may conclude that Trp absorbance is required to generate the HMW photoproduct. It will now be interesting to determine if the loss in Trp fluorescence is closely connected to production of the HMW product. In contrast, PSII is damaged by UV-B via absorbance by a quinone or the Tyr that is involved in electron transport (Bornman et al., 1984; Greenberg et al., 1989). This implies that the damage observed in Rubisco is a more general phenomenon as almost all proteins have Trps, and many of these Trp residues will be in orientations that can initiate protein oxidation following absorbance of a UV-B photon.

The ability of plants to adapt to UV-B will depend, in part, on their capacity to detect elevated levels of the radiation. Presumably, acclimation is induced by a photoreceptor(s) that detects UV-B photons and triggers changes in development. Three general classes of morphogenic photoreceptors have been identified in higher plants: phytochrome, a blue light/UV-A photoreceptor(s) and a UV-B photoreceptor(s) (Kendrick and Kronenberg, 1994). Considerably more is known about phytochrome and the blue light/UV-A photoreceptor(s) than the UV-B photoreceptor(s).

To better characterize the UV-B photoreceptor(s) and signal transduction pathway(s), the *B. napus* cotyledon curling response has promise. We have demonstrated the response to be photomorphogenic and developmental in nature, not indiscriminate cellular damage that erroneously appears as a morphogenic process (Wilson and Greenberg, 1993 b). Curling is also rapid and readily quantifiable. We have begun to use the curling assay to probe the nature of a UV-B photoreceptor. It was previously suggested that flavins and/or pterins bound to proteins may be involved in UV-B detection by plants (Ensminger and Schäfer, 1992; Ballaré et al., 1995). Our findings are consistent with this proposal as phenylacetic acid, a quencher of flavin excited states (Vierstra and Poff, 1981), inhibited curling (Table 1). In addition, we found the Ca^{++} ionophore A23187 (Tretyn et al., 1991) prevented UV-B specific positive curling. Although non-specific upward curling was observed (Table 1), a requirement for normal cellular levels of Ca^{++} was implied. Furthermore, the cyclic nucleotide phosphodiesterase activator imidazole (Tretyn et al., 1991) prevented curling indicating that cyclic nucleotides are also involved in signal transduction. This is striking since Ca^{++} and cyclic nucleotides often work in series in signal transduction pathways. Interestingly, Ros and Tevini (1995) reported that degradation of indole acetic acid (auxin) by UV-B leads to inhibition of hypocotyl elongation in sunflower seedlings, suggesting that IAA is a photoreceptor affecting photomorphogenesis. In another study, Braun and Tevini (1993) reported that UV-B induced *trans* to *cis* isomerization of cinnamic acid leads to activation of phenylalanine ammonia lyase and thus increased synthesis of flavonoids, implying that *trans*-cinnamic acid is also a UV-B photoreceptor. Thus, it is reasonable to assume that there are multiple UV-B photoreceptor species.

One of the key acclimation/protection processes activated by UV-B is the synthesis of epidermal flavonoids (Cen and Bornman, 1993; Wilson and Greenberg, 1993 a). It is clear from the results presented here that specific flavonoids and other UV-absorbing pigments are selectively generated in response to UV-B. It will now be important to determine which of these compounds are synthesized in tolerant and sensitive biotypes of *B. napus*. Furthermore, it will be interesting to assess if any of the UV-B-induced flavonoids arise via signaling from the photoreceptor that triggers other UV-B responses like cotyledon curling. Finally, we would like to know which of the UV-B specific flavonoids are localized in the epidermis, where they would provide screening, compared to localization in the mesophyll, where they might act as anti-oxidants.

It is clear that UV-B can have major impacts on plants. At current UV-B levels (approximately 1 % of PAR), plants are able to acclimate to varying degrees. However, as the UV-B : PAR ratio increases, the ability of plants to adapt diminishes and the amount of damage rises. Thus, a major objective of future research on UV-B effects on plants will be to elucidate the most effective UV-B adaptation mechanisms, determine how they are triggered and define which plants are particularly well suited to cope with elevated UV-B levels.

Acknowledgements

We wish to thank Drs. T. S. Babu, C. Duxbury, M. Edelman, E. B. Dumbroff, J. E. Thompson, and A. K. Mattoo for many fruitful discussions. This work was supported by Collaborative, Research and Equipment Grants from NSERC to BMG, and an Ontario Graduate Scholarship to MIW.

References

Adamse, P. and S. J. Britz: Amelioration of UV-B damage under high irradiance. I. Role of photosynthesis. Photochem. Photobiol. *56,* 645–650 (1992).

Ballaré, C. L., P. W. Barnes, and R. E. Kendrick: Photomorphogenic effects of UV-B radiation on hypocotyl elongation in wild-type and stable-phytochrome-deficient mutant seedlings of cucumber. Physiol. Plant. *83,* 652–658 (1991).

Ballaré, C. L., P. W. Barnes, and S. D. Flint: Inhibition of hypocotyl elongation by ultraviolet-B radiation in de-etiolating tomato seedlings. I. The photoreceptor. Physiol. Plant. *93,* 584–592 (1995).

Barnes, P. W., S. D. Flint, and M. M. Caldwell: Morphological responses of crop and weed species of different growth forms to ultraviolet-B radiation. Amer. J. Bot. *77,* 1354–1360 (1990).

Bornman, J. F., L. O. Bjorn, and H. E. Akerlund: Action spectra for photoinhibition by ultraviolet radiation of photosystem II activity in spinach thylakoids. Photobiochem. Photobiophys. *8,* 305–313 (1984).

Braun, J. and M. Tevini: Regulation of UV-protective pigment synthesis in the epidermal layer of rye seedlings. Photochem. Photobiol. *57,* 318–323 (1993).

Caldwell, C. R.: Ultraviolet-induced photodegradation of cucumber (*Cucumis sativus* L.) microsomal and soluble protein tryptophanyl residues *in vitro.* Plant Physiol. *101,* 947–953 (1993).

Caldwell, M. M., R. Robberecht, and W. D. Billings: A steep latitudinal gradient of solar ultraviolet-B radiation in the arctic-alpine life zone. Ecology *61,* 600–611 (1980).

Cen, Y.-P. and J. Bornman: The response of bean plants to UV-B radiation under different irradiances of background visible light. J. Exp. Bot. *41,* 1489–1495 (1990).

– – The effect of exposure to enhanced UV-B radiation on the penetration of monochromatic and polychromatic UV-B radiation in the leaves of *Brassica napus.* Physiol. Plant. *87,* 249–255 (1993).

Chu, B., G. P. Kerr, and J. V. Carter: Stabilizing microtubules with taxol increases microfilament stability during freezing of root tips. Plant Cell Environ. *16,* 883–889 (1993).

Ensminger, P. A. and E. Schäfer: Blue and ultraviolet-B light photoreceptors in parsley cells. Photochem. Photobiol. *55,* 437–447 (1992).

Frederick, J. E.: Trends in atmospheric ozone and ultraviolet radiation: mechanisms and observations for the northern hemisphere. Photochem. Photobiol. *51,* 757–763 (1990).

Greenberg, B. M., V. Gaba, O. Canaani, S. Malkin, A. K. Mattoo, and M. Edelman: Separate photosensitizers mediate degradation of the 32-kDa photosystem II reaction center protein in the visible and UV spectral regions. Proc. Natl. Acad. Sci., U.S.A. *86,* 6617–6620 (1989).

Greenberg, B. M., M. I. Wilson, V. Gaba, L. Ren, and X.-D. Huang: Responses of *Brassica napus* (oilseed rape) to ultraviolet-B. Life Sci. Adv.-Plant Physiol. *12,* 167–176 (1993).

Greenberg, B. M., M. I. Wilson, X.-D. Huang, K. E. Gerhardt, C. L. Duxbury, and B. Gensemer: The effects of ultraviolet-B radiation on higher plants. In: Wang, W., W. Lower, and J. Gorsuch (eds.): Plants for Environmental Studies, In Press. Lewis Publishers, Boca Raton, FL (1995 a).

Greenberg, B. M., D. G. Dixon, M. I. Wilson, X.-D. Huang, B. J. McConkey, C. L. Duxbury, K. E. Gerhardt, and B. Gensemer: Use of artificial lighting in environmental assessment studies. In: LaPoint, T. W., F. T. Price, and E. Little (eds.): Environmental Toxicology and Risk Assessment, 4th Volume, ASTM STP 1262, In Press. American Society for Testing and Materials, Philadelphia, PA (1995 b).

Grossweiner, L. I.: Photochemistry of proteins: a review. Curr. Eye Res. *3,* 137–144 (1984).

Hashimoto, T., C. Shichijo, and H. Yatsuhashi: Ultraviolet action spectra for the induction and inhibition of anthocyanin synthesis in broom sorghum seedlings. J. Photochem. Photobiol. *11,* 353–363 (1991).

Jordan, B. R., J. He, W. S. Chow, and J. M. Anderson: Changes in mRNA levels and polypeptide subunits of ribulose 1,5-bisphosphate carboxylase in response to supplementary ultraviolet-B radiation. Plant Cell Environ. *15,* 91–98 (1992).

Kendrick, R. E. and G. H. M. Kronenberg: Photomorphgenesis in Plants. Kluwer Academic Press, Dordrecht, Holland (1994).

Kerr, J. B. and C. T. McElroy: Evidence for large upward trends of ultraviolet-B radiation linked to ozone depletion. Science *262,* 1032–1034 (1993).

Kramer, G. F., H. A. Norman, D. T. Krizek, and R. M. Mirecki: Influence of UV-B radiation on polyamines, lipid peroxidation and membrane lipids in cucumber. J. Phytochem. *30,* 2101–2108 (1991).

Langer, B. and E. Wellmann: Phytochrome induction of photoreactivating enzyme in *Phaseolus vulgaris* L. seedlings. Photochem. Photobiol. *52,* 861–863 (1990).

Li, J., T.-M. Ou-lee, R. Raba, R. G. Amundson, and R. A. Last: *Arabidopsis* flavonoid mutants are hypersensitive to UV-B irradiation. Plant Cell *5,* 171–179 (1993).

Middleton, E. M. and A. H. Teramura: The role of flavonol glycosides and carotenoids in protecting soybean from ultraviolet-B damage. Plant Physiol. *103,* 741–752 (1993).

Morré, D. J., G. Sellden, X. Z. Zhu, and A. Brightman: Triacontanol stimulates NADH oxidase of soybean hypocotyl plasma membrane. Plant Sci. *79,* 31–36 (1991).

Murali, N. S., A. H. Teramura, and S. K. Randall: Response differences between two soybean cultivars with contrasting UV-B radiation sensitivities. Photochem. Photobiol. *48,* 653–657 (1988).

Pang, Q. and J. B. Hays: UV-B-inducible and temperature sensitive photoreactivation of cyclobutane pyrimidine dimers in *Arabidopsis thaliana.* Plant Physiol. *95,* 536–543 (1991).

Quaite, F. E., B. M. Sutherland, and J. C. Sutherland: Action spectrum for DNA damage in alfalfa lowers predicted impact of ozone depletion. Nature *358,* 576–578 (1992).

Ros, J. and M. Tevini: Interaction of UV-radiation and IAA during growth of seedlings and hypocotyl segments of sunflower. J. Plant Physiol. *146,* 295–302 (1995).

Salvucci, M. E., A. R. Portis Jr., and W. L. Ogren: Purification of ribulose 1,5-bisphosphate carboxylase/oxygenase with high specific activity by fast protein liquid chromatography. Anal. Biochem. *153,* 97–101 (1986).

Strid, A.: Alteration of expression of defense genes in *Pisum sativum* after exposure to supplementary ultraviolet radiation. Plant Cell Physiol. *34,* 949–953 (1993).

Strid, A., W. S. Chow, and J. M. Anderson: Effects of supplementary ultraviolet-B radiation on photosynthesis in *Pisum sativum.* Biochim. Biophys. Acta. *1020,* 260–268 (1990).

Tevini, M.: Effects of enhanced UV-B radiation on terrestrial plants. In: Tevini, M. (ed.): UV-B Radiation and Ozone Depletion: Effects on Humans, Animals, Plants, Microorganisms, and Materials, pp. 125–153. Lewis Publishers, Boca Raton, FL (1993).

Tevini, M., A. H. Teramura, G. Kulandaivelu, M. M. Caldwell, and L. O. Bjorn: Terrestrial plants. In: van der Luen, J. C., M. Tevini, and J. C. Worrest (eds.): Environmental Effects Panel Report Pursuant to Article 6 of the Montreal Protocol on Substances that Deplete the Ozone Layer, pp. 25–37. United Nations Environment Programme, United Nations, N.Y. (1989).

Tevini, M., J. Braun, and G. Fieser: The protective function of the epidermal layer of rye seedlings against ultraviolet-B radiation. Photochem. Photobiol. *53,* 329–333 (1991).

Tretyn, A., R. E. Kendrick, and G. Wagner: The role(s) of calcium ions in phytochrome action. Photochem. Photobiol. *54,* 1135–1155 (1991).

Vierstra, R. D. and K. L Poff: Mechanisms of specific inhibition of phototropism by phenylacetic acid in corn seedling. Plant Physiol. *670,* 1011–1015 (1981).

VU, C. V., L. H. ALLEN Jr., and L. A. GARRARD: Effects of enhanced UV-B radiation (280–320 nm) on ribulose 1,5-bisphosphate carboxylase in pea and soybean. Environ. Exp. Bot. *24,* 131–143 (1984).

WILSON, M. I. and B. M. GREENBERG: Protection of the D1 photosystem II reaction center protein from degradation in ultraviolet radiation following adaptation of *Brassica napus* L. to growth in ultraviolet-B. Photochem. Photobiol. *57,* 556–563 (1993 a).

WILSON, M. I. and B. M. GREENBERG: Specificity and photomorphogenic nature of ultraviolet-B induced cotyledon curling in *Brassica napus* L. Plant Physiol. *102,* 671–677 (1993 b).

WILSON, M. I., S. GHOSH, K. E. GERHARDT, N. HOLLAND, T. S. BABU, M. EDELMAN, E. B. DUMBROFF, and B. M. GREENBERG: *In vivo* photomodification of ribulose-1,5-bisphosphate carboxylase/oxygenase holoenzyme by ultraviolet-B radiation: formation of a 66 kDa variant of the large subunit. Plant Physiol. *109,* 221–229 (1995).

J. Plant Physiol. Vol. 148. pp. 86–91 (1996)

Effect of UV-B on Enzymes of Nitrogen Metabolism in the Cyanobacterium *Nostoc calcicola*

Ashok Kumar[1,2], Rajeshwar P. Sinha[1,2], and Donat-P. Häder[1]

[1] Institut für Botanik und Pharmazeutische Biologie, Friedrich-Alexander-Universität, Staudtstr. 5, D-91058 Erlangen, Germany

[2] School of Biotechnology, Banaras Hindu University, Varanasi-221005, India

Received June 24, 1995 · Accepted October 27, 1995

Summary

The effects of ultraviolet-B (UV-B; 280–315 nm) irradiation on nitrogenase and nitrate reductase (NR) activity have been studied in the filamentous and heterocystous N_2-fixing cyanobacterium *Nostoc calcicola.* Exposure of cultures to UV-B (5 W/ m^2) for as little as 30 min caused complete inactivation of nitrogenase activity whereas nitrate reductase activity was stimulated twofold in comparison to one exposed to fluorescent white light. GS activity was also inhibited by UV-B treatment, but there was no total loss of activity even after 4 h. NR activity showed a gradual stimulation up to 4 h and thereafter it became constant. Stimulation was also obtained in reductant deficient cultures (12 h incubation in the dark) suggesting independence of NR of PS-II under UV-B. NR activity was also unaffected in the presence of DCMU, a known inhibitor of PS-II. However, both O_2 evolution and $^{14}CO_2$ uptake were completely abolished following 30 min of UV-B treatment. Addition of the protein synthesis inhibitor chloramphenicol (25 µg/mL) to cultures did not show any inhibitory effect on NR activity. SDS-PAGE analysis of UV-B treated cultures elicited gradual loss of protein bands with increasing duration of exposure. Our findings suggest that UV-B irradiance has differential effects on the enzymes of the nitrogen metabolism in the cyanobacterium *Nostoc calcicola.* Further studies are needed to reveal the exact mechanism involved in the stimulation of NR activity by UV-B. Whether UV-B has a direct effect on NO_2^- accumulation in the cells needs detailed investigation.

Key words: $^{14}CO_2$ uptake, cyanobacteria, glutamine synthetase, nitrate reductase, nitrogenase, Nostoc, ultraviolet-B.

Abbreviations: DCMU = 3-(3,4-dichlorophenyl)-1,1-dimethylurea; GS = Glutamine synthetase; NED = α-(N-1)-napthylenediamine dihydrochloride; NR = nitrate reductase; SDS-PAGE = sodium dodecyl sulphate-polyacrylamide gel electrophoresis; UV-B = ultraviolet-B (280–315 nm).

Introduction

Cyanobacteria are a primitive group of gram-negative prokaryotes which occupy an important place in both aquatic as well as terrestrial ecosystems (Stewart, 1980). All the cyanobacterial species utilize inorganic combined nitrogen viz., NO_3^- or NH_4^+ as their sole source of nitrogen but many are capable to grow diazotrophically (Stanier and Cohen-Bazire, 1977). Nitrogen-fixing cyanobacteria possess a central position in the nutrient cycling largely due to their inherent capacity to fix atmospheric nitrogen directly into ammonium, with the help of the enzyme nitrogenase; thus making it available for use by higher plants (Kumar and Kumar, 1988). The estimates vary as to the contribution of cyanobacterial nitrogen fixation to the global nitrogen cycle; according to one assumption cyanobacteria fix over 35 million tons of nitrogen annually (Häder et al., 1989). The contribution of nitrogen-fixing cyanobacteria as a natural biofertilizer in increasing the fertility of diverse soils especially of rice paddy fields has been demonstrated by several researchers (Stewart, 1980; Sinha and Kumar, 1992).

Cyanobacteria possess nitrate and nitrite reductase which convert nitrate to nitrite and nitrite to ammonia (Guerrero et al., 1981). NR of cyanobacteria is a membrane bound enzyme and uses reduced ferredoxin rather than a reduced pyridine nucleotide as its electron donor (Guerrero et al., 1981). Both nitrogenase and NR are molybdoenzymes and share some common properties (Kumar and Kumar, 1980). NH_3, the end product of N_2 or NO_3^- reduction is assimilated via glutamine synthetase and glutamate synthetase in almost all the species studied so far (Florencio and Ramos, 1985).

Enhanced levels of UV-B radiation (280–315 nm) have been demonstrated to cause damaging effects in a variety of living organisms (Häder et al., 1989; Häder and Worrest, 1991; Vincent and Roy, 1993). Several workers have reported the deleterious effects of UV-B on growth, pigmentation, carbon fixation, motility and a few other processes in different classes of algae as well as cyanobacteria. Severe inhibition of N_2 fixation has been reported in a few species of cyanobacteria (Newton et al., 1979; Tyagi et al., 1992; Sinha et al., 1995 a). Inhibition of $^{14}CO_2$, $^{15}NO_3^-$ and $^{15}NH_4^+$ uptake has been reported in diatoms (Döhler, 1985; Döhler and Biermann, 1987). However little, if any, study has been made on some other important enzymes of nitrogen metabolism operative especially in blue-green algae (cyanobacteria).

In the present study we have investigated the impacts of UV-B on the activity of a few key enzymes of nitrogen metabolism in an N_2-fixing cyanobacterium *Nostoc* sp.

Material and Methods

Organism and growth conditions

The filamentous and heterocystous N_2-fixing cyanobacterium *Nostoc calcicola* was isolated from a rice paddy field near Varanasi, India. Based on morphological characteristics, this strain was identified with the help of standard taxonomic keys and monographs (Desikachary, 1959). Purification of the strain was done following standard microbiological techniques (Sinha et al., 1995 b). This species showed best growth in a medium described by Safferman and Morris (1964). The culture was routinely grown in an air-conditioned culture room at 27 ± 1 °C and illuminated with fluorescent white light (14.4 W/ m^2). For performing experiments on NR activity cultures were grown in a medium containing 2.5 mM KNO_3. Unless stated otherwise all experiments were done with log phase cultures having a dry weight of approximately 0.1 mg/mL.

Mode and source of UV-B irradiation

Diazotrophically or nitrate-grown cultures were transferred to sterile Petri dishes (25 mm in diameter) and exposed to artificial UV-B produced from a transilluminator (Bachofer, Reutlingen, Germany) or a UV-B lamp (Vilber Lourmat, VL – 15 M, France), with their main output at 312 nm. The suspension was gently agitated by a magnetic stirrer during irradiation to facilitate uniform exposure. A predefined amount of culture was removed from the Petri dishes at indicated time interval following UV-B exposure. The irradiance of UV-B was fixed at about 5 W/ m^2 for all experiments.The irradiance of UV-B was measured by a double monochromator spectroradiometer (Type 752, Optronic Laboratories, Orlando, FL, USA).

Nitrate reductase (NR) activity

In vivo NR activity was estimated by the method of Camm and Stein (1974). The nitrite formed was determined by the diazocoupling method of Lowe and Evans (1964). During UV-B irradiation, 1-mL aliquots were withdrawn at desired intervals and mixed with sulfanilamide. After an interval of 10 min NED was added and the absorbance of the pink color was estimated at 540 nm in a DU 70 spectrophotometer (Beckman, Palo Alto, CA, USA).

Nitrogenase activity

Nitrogenase activity was measured by the acetylene reduction technique (Stewart et al., 1968). After exposure to UV-B for a known time interval 2 mL culture was directly transferred to a vacutainer tube (Becton Dickinson, Rutherford, NJ, USA) and acetylene was injected by a hypodermic syringe so as to attain a final concentration of 10 %. The tubes were incubated in fluorescent light (14.4 W/ m^2) at 27 ± 1 °C. Ethylene formed was measured by removing 0.2 mL gas phase from the tube and injecting it in a gas liquid chromatograph (CIC, Baroda) fitted with a Porapak R column and a flame ionization detector.

Glutamine synthetase (GS) activity

In vivo glutamine synthetase (transferase) activity was measured by the method of Sampaio et al. (1979). GS activity was calculated on the basis of γ-glutamyl hydroxamate formed from a known volume of cultures after a fixed time interval.

$NaH^{14}CO_3$ uptake

Uptake of $^{14}CO_2$ was determined by the method of Kumar et al. (1982). 10 mL of a liquid culture was spiked with 50 µL $NaH^{14}CO_3$ (specific activity 962 Bq/mL) and exposed to UV-B or fluorescent light. 1-mL aliquots were removed at desired time intervals and transferred directly into scintillation vials containing 0.2 mL acetic acid (50 %). After processing, 10 mL scintillation cocktail (Bray's solution) were added in each vial and the samples were counted in an LKB 1209 Rack Beta liquid scintillation counter.

Gel electrophoresis

SDS-PAGE was carried out in a vertical system (2001, Pharmacia LKB, Uppsala, Sweden) with gels of 155 × 130 mm, 1.5 mm thick, using the method described by Laemmli (1970) with a gradient (5–15 % T) in the resolving gel. Gels were stained with Coomassie brilliant blue R 250 and dried in a gel dryer (Bio-Rad, Richmond, CA).

Chlorophyll determination

Chl *a* content was measured after Mackinney (1941).

Chemicals

Chloramphenicol, NED, sulphanilamide and all other biochemicals were purchased from Sigma Chemical Company, St. Louis, MO, USA. $NaH^{14}CO_3$ was obtained from Isotope Division, Bhabha Atomic Research Centre, Bombay, India.

Results

The test organism was exposed to irradiances of 2.5, 5.0, and 10.0 W/ m^2 of UV-B for known time intervals with a view to select a mid-inhibitory survival dose. UV-B exposure at 2.5 W/ m^2 for 30 min did not cause a significant effect on survival, whereas 10.0 W/ m^2 elicited almost complete (100 %) killing. There was only about 50 % survival at 5 W/ m^2 following 30 min exposure (data not shown) and therefore, this intensity was employed in all the experiments. The fact that growth and survival of many algae including N_2-fixing cyanobacteria is severely affected by enhanced levels of UV-B prompted us to study similar effects on a few enzymes of nitrogen metabolism including nitrogenase activity of an actively growing culture of *Nostoc calcicola,* isolated from an Indian rice paddy field. It is evident from Figure 1 that UV-B exposure of as little as 10 min resulted in a drastic decrease (about 50 %) in the activity of the nitrogenase enzyme, and the same was completely abolished at 30 min. On the other hand, such a drastic inhibitory effect was not noticed on nitrate reductase. In fact, the activity of NR was stimulated during the exposure to UV-B (Fig. 2). From the data of Figure 2 it is clear that the activity of NR shows stimulation up to 4 h following UV-B treatment and thereafter it became constant. It is pertinent to mention that severe inhibition of growth and survival of this strain occurred after 120 min of continuous UV-B treatment (data not shown). Knowing that UV-B treatment causes differential effects on nitrogenase and NR, we became interested in the possible factors involved in eliciting the above effects.

Since GS is the main ammonia assimilating enzyme and it is involved in the regulation of nitrogenase and NR activity, it was desirable to test the effect of UV-B on this enzyme. Figure 3 shows that the exposure of cultures to UV-B also caused inhibition of GS activity, however unlike nitrogenase activity there was no complete loss of activity even after 2 h of continuous treatment. There was about 50 % loss in activity after 2 h treatment.

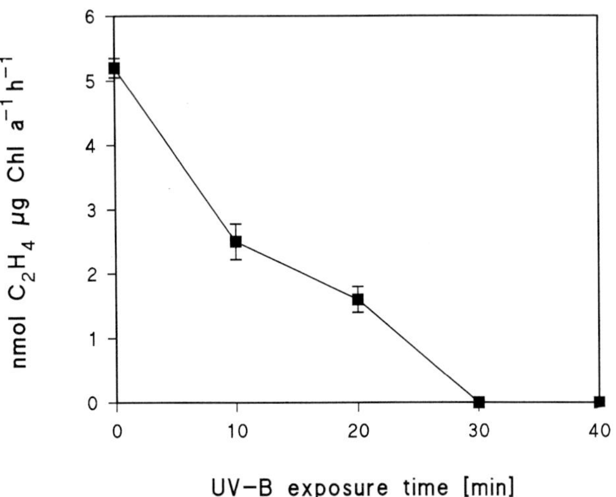

Fig. 1: Effect of UV-B radiation on nitrogenase activity of *Nostoc calcicola.* An actively N_2-fixing culture was exposed to UV-B (5 W/ m^2) for the indicated time period and thereafter a C_2H_2 reduction assay was performed under fluorescent light. The untreated control culture showed an activity of 5.2 nmol C_2H_4 µg Chl a^{-1} h^{-1} throughout the period of assay. Results are based on three separate but identical experiments. S.D. was consistently less than 10 % of means.

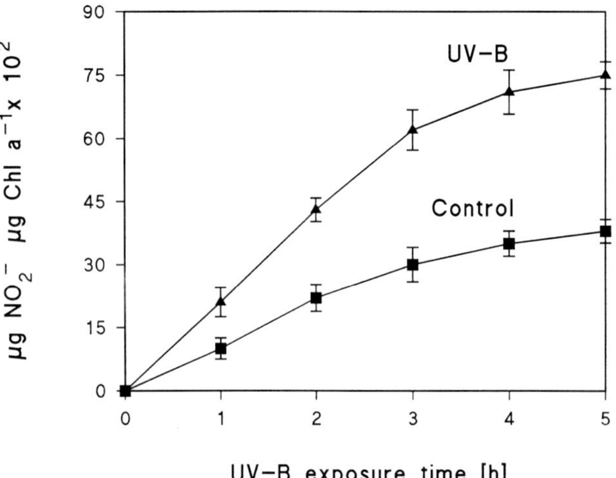

Fig. 2: Response of NR activity of *Nostoc calcicola* towards UV-B. Cultures grown in a medium supplemented with KNO_3 (2.5 mM) were exposed to UV-B, and NR activity was measured at different time periods. Other conditions similar to Figure 1.

Rapid inhibition of nitrogenase activity following UV-B treatment might be due to a) the loss of reductant and ATP; b) complete inactivation of the nitrogenase polypeptide; c) some conformational changes in the enzyme complex and; d) the loss of the oxygen protection in the heterocysts. To answer the above questions we tested the impact of UV-B on photosynthetic processes especially involved in the generation of reductant and ATP. Exposure of cultures to UV-B caused a loss of $^{14}CO_2$ uptake, and the rate became almost negligible after 30 min (data not shown). Similarly, oxygen evolution was stopped after UV-B treatment for as little as 10 min. Like nitrogenase, NR activity is also dependent on a continuous supply of reductant and ATP, however the observed stimulation by UV-B seems intriguing. The data in Table 1 show that the reductant and ATP is probably not directly involved in driving nitrogenase and NR activity during UV-B exposure and that the photosynthetic electron transport is not implicated in this process and respiratory production of reductant and ATP may partly or wholly supply all the needs of N assimilation. Nitrogenase activity was completely lost either after UV-B treatment or 12 h of dark pretreatment; NR activity was again stimulated under UV-B (2 h) and more than 50 % activity remained in dark-pretreated cultures (12 h). Furthermore, UV-B treatment in the dark-pretreated cultures also showed a twofold stimulation in NR activity. Secondly, DCMU, an inhibitor of PS-II did not cause a complete loss of either nitrogenase or NR activity in fluorescent white light in 2 h.

Involvement of *de novo* protein synthesis in the stimulation of NR activity was tested employing a protein synthesis inhibitor, chloramphenicol. Addition of chloramphenicol (25 µg/ mL) to cultures showed a significant decrease in nitrogenase and NR activity under fluorescent light whereas NR activity

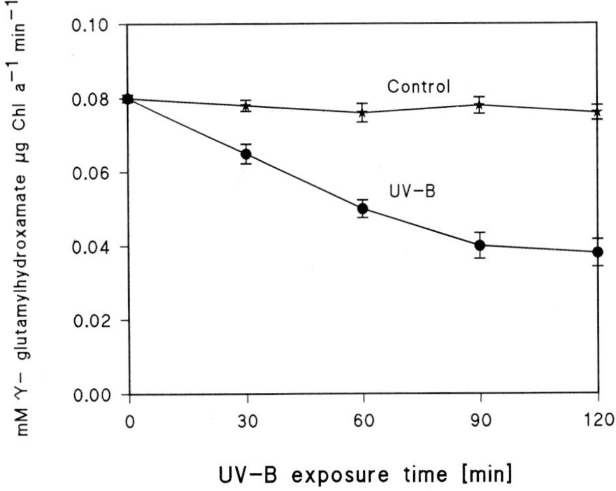

Fig. 3: Effect of UV-B on GS activity of *Nostoc calcicola.* Other conditions similar to Figure 1.

Table 1: Nitrogenase and nitrate reductase activity in reductant deficient cultures of *Nostoc calcicola* under UV-B.

Conditions	Nitrogenase activity (nmol C_2H_4 µg Chl a^{-1} h^{-1})	NR activity (µg NO_2^- µg Chl a^{-1})
Control (FL)[a]	5.40±0.03	127.00±0.12
UV-B treated (2 h)	0.0	251.00±0.08
DCMU (20 µM)+FL (2 h)	2.00±0.04	85.00±0.10
DCMU + UV-B (2 h)	0.0	242.00±0.08
Dark pretreatment (12 h)[b]	0.0	72.00±0.08
Dark pretreatment (12 h) + UV-B (2 h)	0.0	160.00±0.08

[a] FL – Fluorescent light.
[b] Nitrogenase activity was completely lost after 8 h dark pretreatment. NR activity was not lost completely even after 24 h dark incubation.
The values represent means ± S.D.

was stimulated at least up to 3 h under UV-B in chloramphenicol treated cultures (Table 2). Nitrogenase activity did not appear even after 12 h in chloramphenicol treated cultures receiving UV-B irradiance. Synthesis of new protein(s) including nitrogenase and NR polypeptides was also not apparent from the electrophoretic pattern of UV-B treated cultures (Fig. 4). Drastic loss in protein bands including number and intensity was detected with increasing UV-B exposure time.

Discussion

Our findings clearly demonstrate that UV-B radiation has a differential effect on the key enzymes of nitrogen metabolism in the cyanobacterium *Nostoc calcicola.* It causes complete inhibition of nitrogenase, partially inhibits GS and stimulates NR activity. Inhibition of nitrogenase activity by UV-B treatment has also been reported in *Anabaena flos-aquae* and *Nostoc spongiaeforme* (Newton et al., 1979; Tyagi et al., 1992). Damaging effects of UV-B are known on a number of physiological processes of cyanobacteria such as pigmentation, photosynthesis and others (Häder and Worrest, 1991). It has been reported that UV-B exposure has a deleterious effect on the photosynthetic apparatus leading to the reduction in the supply of ATP and $NADPH_2$ (Kulandaivelu and Noorudeen, 1983). As such, disruption of cell membrane and/or alteration in thylakoid integrity as a result of UV-B radiation may partly or wholly destroy the components required for photosynthesis and thus may affect the rate of CO_2 assimilation (Vu et al., 1981). There is a possibility that the inhibition of the nitrogenase activity might be due to reduced supply of reductant and ATP following UV-B treatment. However, rapid inhibition followed by complete loss of nitrogenase activity (within 30 min) seems surprising because survival of the organism was not affected significantly by 30 min of UV-B treatment. It is also pertinent to mention that the loss of reductant and ATP does not occur immediately in cyanobacteria since many N_2-fixing species are capable to drive nitrogenase activity at the cost of an endogenous pool (Stewart, 1980; Kumar et al., 1982). Such a fact is also evident from our data of DCMU-treated cultures (Table 1) where nitrogenase activity was detected up to 2 h even in the absence of a functional PS-II. This rules out the possibility of direct involvement of reductant and ATP supply in causing instant inhibition of nitrogenase activity. Most probably UV-B treatment causes complete inactivation/denaturation of the nitrogenase and thus the cultures become devoid of nitrogenase activity. This conclusion is corroborated from the fact that chloramphenicol treatment did not cause complete loss of nitrogenase activity even after 12 h in fluorescent light but there was a total loss of activity under UV-B + chloramphenicol treated cultures (Table 2). UV-B is known to damage proteins and enzymes, especially those rich in aromatic amino acids (Vincent and Roy, 1993). Loss of the proteins that carry the photoreceptor chromophores of the phytoflagellate *Euglena* and the phycobiliproteins of cyanobacteria and certain

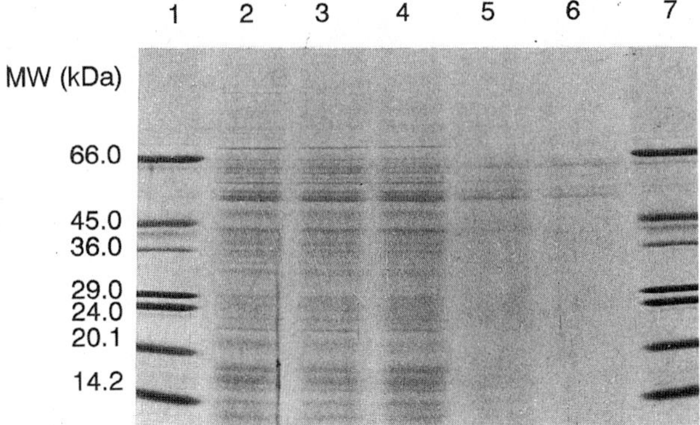

Fig. 4: Vertical SDS-PAGE (gradient 5–15 % T) protein profile of *Nostoc calcicola* following increasing exposure to UV-B radiation. Lanes 1 and 7: marker proteins (MWS – 877 L, Sigma), lane 2: control, lane 3: 30 min, lane 4: 60 min, lane 5: 90 min and lane 6: 120 min UV-B treated cells. Equal amount of cells were withdrawn at indicated time period and subjected to SDS-PAGE analysis.

Table 2: Effects of UV-B on nitrogenase and NR activity of *Nostoc calcicola* in the presence of protein synthesis inhibitor chloramphenicol (25 µg/mL).

Treatment	Nitrogenase activity (nmol C_2H_4 µg Chl a^{-1} h^{-1})					NR activity (µg NO_2^- µg Chl a^{-1})				
Time [h]	0	3	6	9	12	0	1	2	3	4
Control (FL)[a]	5.20 ±0.03	–	–	–	–	12.70 ±0.08	13.20 ±0.06	25.60 ±0.06	33.00 ±0.06	45.00 ±0.08
Chloramphenicol (FL)	5.20 ±0.03	4.80 ±0.04	4.50 ±0.04	4.00 ±0.04	2.00 ±0.04	12.70 ±0.08	20.20 ±0.06	26.00 ±0.08	31.00 ±0.06	32.00 ±0.06
Chloramphenicol (UV-B)	5.20 ±0.03	0.0	0.0	0.0	0.0	12.70 ±0.08	19.00 ±0.06	28.04 ±0.06	34.00 ±0.08	45.50 ±0.08

[a] FL – Fluorescent light. The values represent means ± S.D.

phytoflagellates is well known (Häder and Worrest, 1991; Sinha et al., 1995c).

The observed stimulation of NR activity following UV-B exposure can not yet be explained. In general, UV-B exposure of cyanobacterial cultures is known to cause deleterious effects on almost all metabolic processes studied so far (Häder et al., 1994). However, stimulation of NR activity following UV-B treatment has been reported in an angiospermic plant, *Crotalaria juncea* (Saralabai et al., 1989). From our findings, it is evident that stimulation of NR activity is not at all dependent on PS-II activity. We have observed that UV-B causes complete loss of $^{14}CO_2$ uptake and O_2 evolution within 30 min of exposure. Furthermore, cultures treated with DCMU and then exposed to UV-B also showed twofold stimulation of NR activity. Similarly, dark pretreated cultures (deficient in reductant) also showed enhanced levels of NR upon exposure to UV-B. These findings suggest that NR activity may operate under UV-B even in the absence of functional PS-II. A possible role of new protein including synthesis of the NR polypeptide in the stimulation of NR activity is also ruled out from the data of chloramphenicol treatment as well as SDS-PAGE analysis. NR activity was found at an enhanced level with chloramphenicol treatment. Our findings suggest that the effect of UV-B on nitrogen metabolism of cyanobacteria may not be generalized. Certain enzymes are extremely sensitive whereas a few are tolerant to UV-B. A direct role of UV-B on the accumulation of NO_2^- in the cells may not be ruled out since the observed stimulation of NR activity appears unique. Although, similar to the UV-B effect, stimulation of NR activity has also been reported by exposure of *Oscillatoria princeps* to blue light (Kumar et al., 1986). Further studies are needed to reveal the exact mechanism of the UV-B mediated stimulation of NR activity.

We conclude that any substantial increase in the solar UV-B radiation might adversely affect the cyanobacterial populations in terms of nitrogen fixation which in turn will affect the productivity of higher plants especially rice in paddy fields. N_2-fixing cyanobacteria are being considered as an alternate natural source of nitrogenous fertilizers for rice and other crops and any adverse effects on these ecologically and economically important organisms may stress the agricultural economy of developing countries.

Acknowledgements

A. Kumar thankfully acknowledges the DAAD for awarding a short term study visit. This work was financially supported by the DAAD and the Ministry of H. R. D., Govt. of India to R. P. Sinha and by the Bundesminister für Forschung und Technologie (project KBF 57) and the European community (EV5V-CT91-0026) to D.-P. Häder.

References

Camm, E. L. and J. R. Stein: Some aspects of the nitrogen metabolism of *Nodularia spumigena* (cyanophyceae). Can. J. Bot. *52,* 719–726 (1974).

Desikachary, T. V.: Cyanophyta. Indian Council of Agricultural Research, New Delhi, India (1959).

Döhler, G.: Effects of UV-B radiation (290–320 nm) on the nitrogen metabolism of several marine diatoms. J. Plant Physiol. *118,* 391–400 (1985).

Döhler, G. and I. Biermann: Effect of UV-B irradiation on the response of ^{15}N-nitrate uptake of *Lauderia annulata* and *Synedra planktonica.* J. Plankton. Res. *9,* 881–890 (1987).

Florencio, F. J. and J. L. Ramos: Purification and characterization of glutamine synthetase from the unicellular cyanobacterium *Anacystis nidulans.* Biochim. Biophys. Acta *838,* 39–48 (1985).

Guerrero, M. G., J. M. Vega, and M. Losada: The assimilatory nitrate-reducing system and its regulation. Annu. Rev. Plant Physiol. *32,* 169–204 (1981).

Häder, D.-P., R. C. Worrest, and H. D. Kumar: Aquatic ecosystems. UNEP Environmental Effects Panel Report, 39–48 (1989).

Häder, D.-P. and R. C. Worrest: Effects of enhanced solar-ultraviolet radiation on aquatic ecosystems. Photochem. Photobiol. *53,* 717–725 (1991).

Häder, D.-P., R. C. Worrest, H. D. Kumar, and R. C. Smith: Effects of increased solar ultraviolet radiation on aquatic ecosystems. UNEP Environmental Effects Pannel Report, 65–77 (1994).

Kulandaivelu, G. and A. M. Noorudeen: Comparative study of the action of ultraviolet-C and ultraviolet-B radiation on photosynthetic electron transport. Physiol. Plant. *58,* 389–394 (1983).

Kumar, A. and H. D. Kumar: Tungsten-induced inactivation of molybdoenzymes in *Anabaena.* Biochim. Biophys. Acta *613,* 244–248 (1980).

Kumar, A., F. R. Tabita, and C. van Baalen: Isolation and characterization of heterocysts from *Anabaena* sp. strain CA. Arch. Microbiol. *133,* 103–109 (1982).

KUMAR, A. and H. D. KUMAR: Nitrogen-fixation by blue-green algae. In: SEN, S. P. (eds.): Proceedings of the Plant Physiology Research, Society for Plant Physiology and Biochemistry, 1st International Congress of Plant Physiology, New Delhi, India, pp. 85–103 (1988).

KUMAR, H. D., M. JHA, and A. KUMAR: Stimulation of nitrate reductase activity by blue light in a thermophilic cyanobacterium *Oscillatoria princeps*. Br. Phycol. J. *21,* 165–168 (1986).

LAEMMLI, U. K.: Cleavage of structural proteins during the assembly of the head of bacteriophage T4. Nature *227,* 680–685 (1970).

LOWE, R. H. and H. J. EVANS: Preparation and some properties of a soluble nitrate reductase from *Rhizobium japonicum*. Biochem. Biophys. Acta *85,* 377–389 (1964).

MACKINNEY, G.: Absorption of light by chlorophyll solution. J. Biol. Chem. *140,* 315–322 (1941).

NEWTON, J. W., D. D. TYLER, and M. E. SLODKI: Effect of ultraviolet-B (280–320) radiation on blue-green algae (Cyanobacteria), possible biological indicators of stratospheric ozone depletion. Appl. Environ. Microbiol. *37,* 1137–1141 (1979).

SAFFERMAN, R. S. and M. E. MORRIS: Growth characteristics of the blue-green algal virus LPP-1. J. Bacteriol. *88,* 771–773 (1964).

SAMPAIO, M. J. A. M., P. ROWELL, and W. D. P. STEWART: Purification and some properties of glutamine synthetase from the nitrogen-fixing cyanobacteria *Anabaena cylindrica* and a *Nostoc* sp. J. Gen. Microbiol. *111,* 181–191 (1979).

SARALABAI, V. C., P. THAMIZHCHELVAN, and K. SANTHAGURU: Influence of ultraviolet-B radiation on fixation and assimilation of nitrogen in *Crotalaria juncea* Linn. Ind. J. Plant Physiol. *32,* 65–67 (1989).

SINHA, R. P. and A. KUMAR: Screening of blue-green algae for biofertilizer. In: PATIL, P. S. (eds.): Proceedings of the National Seminar on Organic Farming, Pune, India, pp. 95–97 (1992).

SINHA, R. P., N. SINGH, A. KUMAR, H. D. KUMAR, M. HÄDER, and D.-P. HÄDER: Impacts of UV-B irradiation on certain physiological and biochemical processes in cyanobacteria. J. Photochem. Photobiol. B: Biol. (1995 a). In press.

SINHA, R. P., H. D. KUMAR, A. KUMAR, and D.-P. HÄDER: Effects of UV-B irradiation on growth, survival, pigmentation and nitrogen metabolism enzymes in cyanobacteria. Acta Protozool. *34,* 187–192 (1995 b).

SINHA, R. P., M. LEBERT, A. KUMAR, H. D. KUMAR, and D.-P. HÄDER: Spectroscopic and biochemical analyses of UV effects on phycobiliproteins of *Anabaena* sp. and *Nostoc carmium*. Bot. Acta *108,* 87–92 (1995 c).

STANIER, R. Y. and G. COHEN-BAZIRE: Phototrophic prokaryotes: the cyanobacteria. Annu. Rev. Microbiol. *31,* 225–274 (1977).

STEWART, W. D. P., G. P. FITZGERALD, and R. H. BURRIS: Acetylene reduction by nitrogen-fixing blue-green algae. Arch. Microbiol. *62,* 336–348 (1968).

STEWART, W. D. P.: Some aspects of structure and function in N_2-fixing cyanobacteria. Annu. Rev. Microbiol. *34,* 497–536 (1980).

TYAGI, R., G. SRINIVAS, D. VYAS, A. KUMAR, and H. D. KUMAR: Differential effect of ultraviolet-B radiation on certain metabolic processes in a chromatically adapting *Nostoc*. Photochem. Photobiol. *55,* 401–407 (1992).

VINCENT, W. F. and S. ROY: Solar ultraviolet-B radiation and aquatic primary production: damage, protection, and recovery. Environ. Rev. *1,* 1–12 (1993).

VU, C. V., L. H. ALLEN Jr., and L. A. GARRARD: Effects of supplemented UV-B radiation on net photosynthetic reactions of Soybean (*Glycine max*). Physiol. Plant. *52,* 353–362 (1981).

J. Plant Physiol. Vol. 148. pp. 92–99 (1996)

Involvement and Non-Involvement of Pyrimidine Dimer Formation in UV-B Effects on *Sorghum bicolor* Moench Seedlings

MEGUMI HADA[1], SEIJI TSURUMI[2], MITSUHIRO SUZUKI[1,6], ECKARD WELLMANN[3], and TOHRU HASHIMOTO[4,5]

[1] Department of Biology, Kobe University, Rokkodai, Kobe, 657, Japan

[2] Graduate School of Science and Technology, Kobe University, Rokkodai, Kobe, 657, Japan

[3] Biological Institute II, Freiburg University, Schänzlestr. 1, 79104 Freiburg, Germany

[4] Department of Life Science, Kobe Women's University, Higashisuma, Kobe, 654, Japan

Received June 24, 1995 · Accepted October 27, 1995

Summary

In plants ultraviolet B (UV-B) radiation causes not only DNA lesion and other deleterious effects, but also assumingly contributes to normal photomorphogenesis mediated by a putative UV-B photoreceptor(s). In order to examine which UV-B effects do or do not involve DNA lesion, cyclobutane pyrimidine dimers (CPDs), thymine dimer (CTD), and pyrimidine (6-4) pyrimidinone photoproducts (6-4PP) were determined by enzyme-linked immunosorbent assay or radioimmunoassay with regard to UV-B-induced anthocyanin synthesis and inhibition as well as coil formation (abnormal growth) of the first internode of *Sorghum bicolor* Moench. For anthocyanin, dark-grown 3-day-old seedlings were irradiated at various fluences with each of the three UV radiations having bands of (A) 295–345, (B) 300–350, and (C) 325–385 nm at ¼ peak height, followed by far-red light (FR) to exclude phytochrome actions. For coil formation, a radiation similar to A was used. The variation of CPD level versus the wavelengths and fluences of UV-B did not correlate with that of anthocyanin synthesis, but did so with the inhibition of anthocyanin synthesis above a critical level of CPDs. Coil formation was also correlated with CPDs (measured as CTD) above a critical level, which was observed not only when CPD level was raised by increasing the fluence of UV-B, but also when it was lowered by UV-A photorepair. The results supported the view that anthocyanin induction does not involve DNA damage, whereas anthocyanin inhibition may be associated with it. The coil formation was indicated to be caused by it. Pyrimidine (6-4) pyrimidinone photoproducts were formed in proportion to CPDs, and hence, it was impossible to determine which of the two kinds of dimers was responsible. The presence of a fast elimination of CPDs (half-life of ca. 30 min) in the dark was suggested.

Key words: Sorghum bicolor, anthocyanin synthesis, coil formation (abnormal growth), ELISA, photoreactivation, pyrimidine dimers, pyrimidine (6-4) pyrimidinone photoproducts, RIA, thymine dimer, UV-B effects.

Abbreviations: BEI = biological effective irradiance; CPD = cyclobutane pyrimidine dimer; CTD = cyclobutane thymine dimer; 6-4PP = pyrimidine (6-4) pyrimidinone photoproduct; ELISA = enzyme-linked immunosorbent assay; FR = far-red light; RIA = radioimmunoassay; UV-A, UV-B, and UV-C = ultraviolet light of 320 to 400 nm, 280 to 320 nm, and 200 to 280 nm, respectively.

[5] Corresponding author.

[6] Present address: Department of Biology, Kohnan University, Okamoto, Kobe, 658, Japan.

Introduction

In plants UV-B causes not only DNA lesion and other deleterious effects, but also is believed to contribute to normal photomorphogenesis (Ballaré et al., 1991; Hashimoto, 1994). The former include the formation of cyclobutane pyrimidine dimers and pyrimidine (6-4) pyrimidinone photoproducts (Britt et al., 1993; Chen et al., 1994), leakage of K^+ ion, abnormal growth, arrest of growth, tissue collapse (Cen and Bornman, 1990), and degeneration of some organelles (He et al., 1994; Santos et al., 1993). As photomorphogenic effects UV-B induces the synthesis of anthocyanin and other flavonoids (Beggs et al., 1986), and promotes the synthesis of chlorophylls and carotenoids as well as the expansion of the cotyledons and photomorphogenic growth inhibition of the hypocotyl (Ballaré et al., 1991, 1995; Staxen and Bornman, 1994). Also, it retards the senescence of leaves (For references on these effects, refer to a review by Hashimoto, 1994).

As the mechanisms of the deleterious UV-B actions, DNA lesion (Britt et al., 1993, Mitchell et al., 1989) and cell membrane damage (Murphy et al., 1993) are considered, and for the photomorphogenic effects the mediation of UV-B photoreceptor(s) are proposed based mainly on action spectral data (Ballaré et al., 1995; Beggs et al., 1986; Hashimoto and Yatsuhashi, 1984; Hashimoto et al., 1991; Takeda and Abe, 1992; Yatsuhashi et al., 1982). Action spectra in the UV-B region are variable. This is probably due to a screening by UV-absorbing cell constitutents (Quaite et al., 1992), but it is also possible that the action peak is due to coaction of DNA lesion and UV absorption by another photoreceptor, and is shifted to wavelengths other than absorption peaks of DNA and the photoreceptor. Thus an action spectrum which does not agree with the absorption of DNA could provide no evidence for rejecting the possibility of DNA lesion being involved. On the other hand, few papers are available to show the relevance of DNA lesion to the actual physiological phenomena of plants caused by UV-B, an exception being the work by Chen et al. (1994). Thus more studies are required to show relationship of DNA lesion and UV-induced physiological phenomena.

Etiolated *Sorghum bicolor* Moench seedlings used in the present studies are sensitive to UV as well as red light, and action spectra for some UV-B actions have been determined. The action peak at 290 nm for anthocyanin induction together with the absence of far-red light (FR) reversion of the UV-B action and the presence of coaction of UV-B and red light led to the proposal that the UV-B action is mediated by a putative UV-B photoreceptor (Hashimoto et al., 1991; Yatsuhashi et al., 1982). At high fluences, on the other hand, UV-B inhibits anthocyanin synthesis, and its action spectrum was rather similar to that for the *in vitro* formation of CPDs and 6-4PPs (Hashimoto et al., 1991; Matsunaga et al., 1991). Another similar action spectrum has been obtained for an abnormal growth (coil formation) in the first internode caused by high-fluence UV-B (Hashimoto et al., 1984). Thus, the latter two high fluence UV-B actions have been assumed to be caused via DNA lesion.

In the present study DNA is extracted, and CPDs and 6-4PPs are determined by immunological methods, and their possible involvement or non-involvement in these three UV-B-induced physiological phenomena is examined by the presence or absence of quantitative correlations.

Materials and Methods

Plant material

Seeds of *Sorghum bicolor* Moench, cv. Acme Broomcorn, 1991 crop from Kobe University Farm at Kasai, were soaked for 24 h in a water bath of 24 °C, in which temperature-adjusted tap water being supplemented and circulated, sown in vermiculite in pots, 8×8×7 (height) cm^3. Seedlings were grown for 3 days at 24 °C until they reached a height of ca. 7 cm at the time of irradiation.

Light sources

In experiments for anthocyanin synthesis K-295, K-300, and BLB-325 as well as FR were used. K-295 was UV-B radiation obtained from four 20 W UV-B fluorescent tubes (FL20S-E, Toshiba, Tokyo) through a quartz filter (U340, Hoya Corp., Akishima, Japan). K-300 was obtained by filtering K-295 with a sheet of polyvinyl chloride, «Nangoku» (Mitsubishi Kasei Chemicals, Tokyo). BLB-325 was supplied from four 20 W black light fluorescent tubes (FL20S-BLB, Toshiba, Tokyo) without filter. The spectral photon distributions of these three kinds of UV are depicted in Fig. 1. Far-red light (FR) was the light having λ_{max} 760 nm, $\lambda½$ 736–808 nm, provided from far-red fluorescent tubes (FL20S-FR-74) through a sheet of polyacrylic resin (Delaglass A900, Asahi Kasei Chemicals, Tokyo) (cf. Shichijo et al., 1993). The UV-B source for experiments of coil formation was a fluorescent tube (TL 40W/12, Philips), which emitted radiation having λ_{max} 310 nm, $\lambda½$ 40 nm. The light source for photoreactivation was a UV-A fluorescent tube (TL 36W/73, Osram) emitting light having λ_{max} 367nm, $\lambda½$ 20nm.

The biologically effective irradiances (BEI) of K-295, K-300, and BLB-325 can be calculated by multiplying irradiances in $W m^{-2}$ corresponding to photon fluences given in Results by the BEI conversion factors, 17.9×10^{-3}, 6.52×10^{-3} and 6.05×10^{-5} for K-295, K-300, and BLB-325, respectively. Using the data on sunlight by De Fabo et al. (1990) it was estimated that BEls of K-295, K-300, and BLB-325, respectively, at 20, 20, and 60 $\mu mol\ m^{-2} s^{-1}$ correspond to 2700, 980, and 8 % of that of the sunlight at noon in July at 40°N latitude.

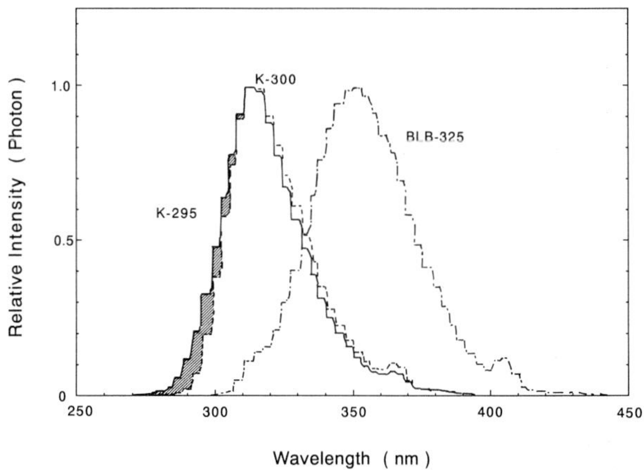

Fig. 1: Spectral photon density distribution of the UV sources. Shaded portions represent the difference in the short half waveband between K-295 and K-300.

The BEI conversion factors were calculated as follows: The photon fluence distribution curves in Fig. 1 were converted into spectral irradiances in W m^{-2} using thus-obtained spectral irradiances and the weighting factors in Table 1 in Smith et al. (1980), which were taken from Setlow's (1974) ‹average action spectrum for affecting DNA› normalized to the action at 265 nm, the BEI conversion factors were calculated by the equation $\int I_\lambda E_\lambda d_\lambda / \int I_\lambda d_\lambda$, where I_λ and E_λ are irradiance and weighting factor at a given wavelength, respectively.

Photon fluence rates for anthocyanin induction were determined with a photon density meter (HK-I, Riken, Wako-shi, Japan), and the irradiances for coil formation were measured with a thermopile (CA-1, Kipp and Zonen, Delft, Holland), and converted into photon fluences. Spectral photon fluence rate curves were obtained with a Spectrophotondensitometer (HK-II-2, Riken, Wako-shi, Japan).

Determination of DNA lesion

In the experiments to examine the relation with anthocyanin synthesis, DNA lesion was measured as follows. Immediately after seedlings were irradiated unilaterally, a 30-mm segment was excised from an irradiated or a non-irradiated seedling, leaving the 15-mm top part of the first internode, and 20 of them (ca. 400 mg) were collected and frozen. They were homogenized with 600 µL of 100 mM Tris-HCl buffer (pH 8.0) containing 50 mM EDTA, 500 mM NaCl, and 10 µL mercaptoethanol, then supplemented with 80 µL of 10 % SDS and heated at 65 °C for 10 min. After adding 200 µL of 5 M potassium acetate containing 11.5 % acetic acid, and keeping on ice for 20 min, the homogenate was centrifuged to collect the supernatant, which was then filtered with Miracloth (Calbiochem). From the filtrate, DNA/RNA mixture was precipitated with the addition of 0.7 volume of isopropyl alcohol on ice, and centrifuged. The resulting pellet was dissolved in 400 µL of 0.5 M Tris-HCl buffer (pH 8.0) containing 0.25 M EDTA, shaken with 400 µL of phenol/chloroform (lower phase) saturated with 10 mM Tris-HCl (pH 8.0) and centrifuged. The upper phase was subjected to precipitation by the addition of 0.6 volume of 20 % polyethylene glycol 6000 containing 2.5 M NaCl for 10 min and centrifuged. The pellet was washed with cold 70 % ethanol twice, and then with 99 % ethanol to desiccate *in vacuo*. Finally DNA was collected from the DNA/RNA mixture in TE by RNase treatment.

Cyclobutane pyrimidine dimers (CPDs) and pyrimidine (6-4) pyrimidinone photoproducts (6-4PPs) were quantified by peroxidase-linked immunosorbent assay (ELISA) according to Mori et al. (1991) with Nikaido's antibodies TDM-2 and 64M-2, respectively. The former antibody has been shown to recognize specifically CPDs in the decreasing order of affinity to TT-, CT-, CC- and TC-dimers, and the latter, 6-4PPs in the order of TT-, TC-, CC-, and CT-dimers (Mori et al., 1991).

In the experiments for coil formation, seedlings were irradiated unilaterally, a 5-mm segment was excised from the apical part (elongation zone) of the first internode. Twenty five segments for each test were collected, frozen and subjected to DNA extraction. Frozen segments were homogenized with 1 mL of 50 mM Tris-HCl buffer (pH 8.0) containing 10 mM EDTA, 0.1 M NaCl, and 2 % SDS. The homogenate was heated at 60 °C for 15 min, and then centrifuged. The supernatant was mixed with an equal volume of phenol-chloroform (1 : 1, v/v) saturated with 0.01 M Tris-HCl buffer (pH 7. 5), and centrifuged. The supernatant was washed with an equal volume of chloroform. Adding to the upper phase 1 mL of ethanol and 50 µL of 3 M sodium acetate (pH 4.8), DNA (containing RNA) was precipitated and collected by centrifugation. The DNA was hydrolyzed in 1.8 mL of 90 % aqueous formic acid in a sealed glass tube at 170–180 °C for 75 min, and subjected to radioimmunoassay (RIA) with cyclobutane thymine dimer-specific antibody (Klocker et al., 1984).

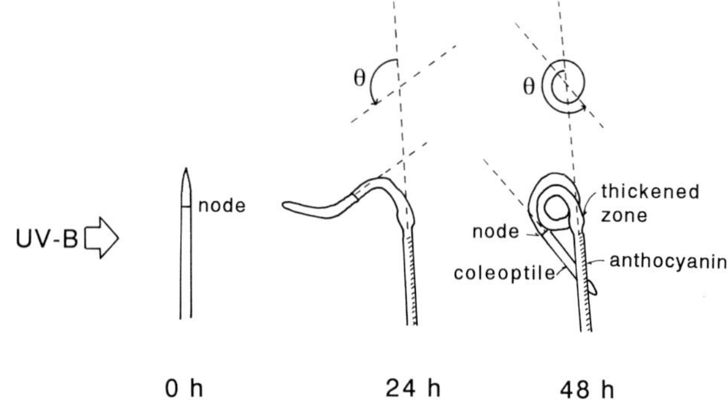

Fig. 2: Coil formation of the *Sorghum* first internode and its quantification by the coil angle θ. When unilaterally irradiated with UV-B, the top ca. 4 mm part (elongation zone) of the internode formed a coil. The transition zone between the coil and the lower straight part thickened, serving as a marker of the end of the coil. Anthocyanin was formed on the shaded flank instead of the irradiated flank, because the intense UV-B inhibited anthocyanin synthesis on this flank.

Data points of the same symbols in the figures on DNA lesion are replications determined with the same DNA samples.

Quantification of coil formation

After unilateral irradiation a coil is formed by unequal elongations of the irradiated and shaded flanks of the internode. The size of a coil was determined by the angle, θ, formed by the straight parts below and above the coiled part of the internode (Fig. 2). After a lag period of less than one h the coil formation appeared and the coil angle increased linearly with time up to 72 h or more after irradiation (Hashimoto et al., 1984). Since the angle includes that of the phototropic curvature, which does not continue to increase beyond some hours, the increment of the angle between 24 h and 48 h was taken as size of a coil.

Quantification of anthocyanin

Anthocyanin is formed at the internode part of ca. 35 mm (beginning 5 mm below the coleoptilar node at the time of irradiation) which almost completed elongation. Twenty 40-mm segments including the 35-mm part were excised, extracted with 6 mL of 1 % hydrochloric acid methanol (v/v), and the differential absorbance at 528 nm and 650 nm was determined as anthocyanin content according to Shichijo et al. (1993).

Results

Induction and inhibition of anthocyanin synthesis and DNA damage

Anthocyanin synthesis was induced by K-290 and K-300 at the same rate relative to their fluences, whereas the inhibition of anthocyanin synthesis by K-295 started at 2 mmol m^{-2} and was more severe than that by K-300, which started at 6 mmol m^{-2} (Fig. 3 A). With BLB-325 no inhibition appeared up to 70 mmol m^{-2}, although more anthocyanin was synthesized than the maximum levels produced by the former two UV-B irradiations. The results of determination of CPD and 6-4PP are shown in Fig. 3 B, C. The production of CPD differed greatly in K-295 from K-300, and no appreciable amount of CPD was produced by BLB-325 at a fluence range tested. Similar tendencies of 6-4PP production were observed. Comparison of the curves for anthocyanin synthesis and CPD generation revealed that with K-295 as well as K-300 the inhibition of anthocyanin synthesis started to occur at the same level of CPD, which was given by K-295 at $2\,mmol\,m^{-2}$ and by K-300 at $6\,mmol\,m^{-2}$. Similar situation of 6-4PP versus anthocyanin inhibition was noted (Fig. 3C).

The CPDs and 6-4PPs generated in the internodes decreased fast in the dark, exponentially against time (Figs. 4, 5). When plotted on a log scale, the data in the figures gave straight lines. The half-life was ca. 30 min for CPDs and ca. 60 min for 6-4PPs. An hour later, even the high level of CPDs generated by $12\,mmol\,m^{-2}$ (of K-295 declined below the dimer level produced at $2\,mmol\,m^{-2}$, where no inhibition of anthocyanin synthesis occurred (cf. Fig. 3A, B).

These data clearly show that CPDs and 6-4PPs are not responsible for anthocyanin induction, but may be involved in anthocyanin inhibition.

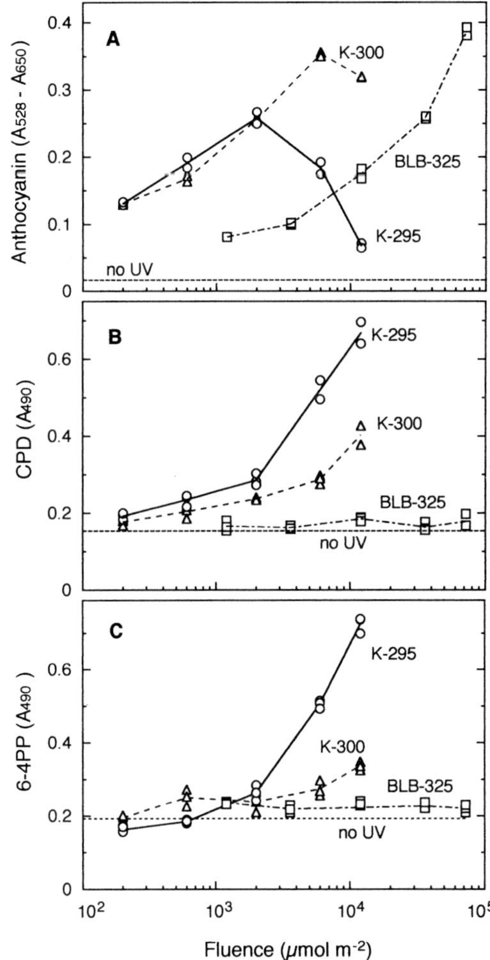

Fig. 3: Effects of different bands of UV radiation on the induction and inhibition of anthocyanin synthesis (A), formation of cyclobutane pyrimidine dimers (CPDs) (B), and pyrimidine (6-4) pyrimidinone photoproducts (6-4PPs) (C) in *Sorghum* first internodes. The fluence was changed by the irradiation period of each UV radiation; K-295 of $20\,\mu mol\,m^{-2}\,s^{-1}$ ($7.62\,W\,m^{-2}$), K-300 of $20\,\mu mol\,m^{-2}\,s^{-1}$ ($7.51\,W\,m^{-2}$), and BLB-325 of $60\,\mu mol\,m^{-2}\,s^{-1}$ ($20.5\,W\,m^{-2}$). All UV irradiations were followed by far-red light (FR) ($80\,\mu mol\,m^{-2}\,s^{-1}$, 3 min), and «no UV» control received FR alone (horizontal broken line). ELISA: (B) 40 ng DNA $well^{-1}$, the antibody TDM-2; (C) 300 ng DNA $well^{-1}$, the antibody 64M-2.

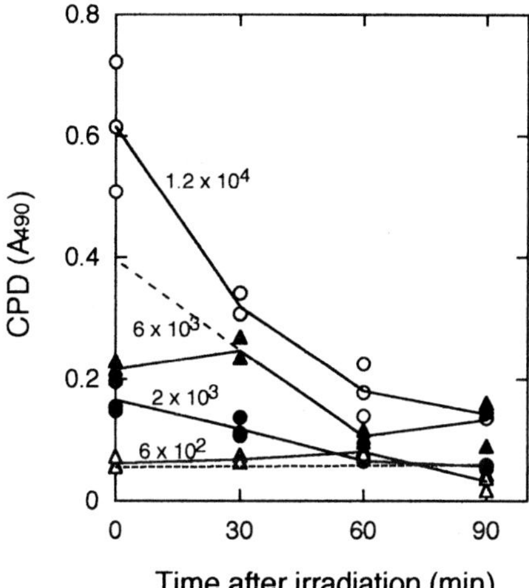

Fig. 4: Decay in the dark of CPDs in *Sorghum* first internodes after UV-B. Seedlings were irradiated with UV-B (K-295) at the indicated fluences followed by FR, and then kept in the dark at 24 °C. ELISA: 40 ng DNA $well^{-1}$, the antibody TDM-2. Horizontal broken line, background with DNA obtained from no-UV control; the extension by a broken line for $6\,mmol\,m^{-2}$, based on repeated experiments. ELISA, the same as in Fig. 3.

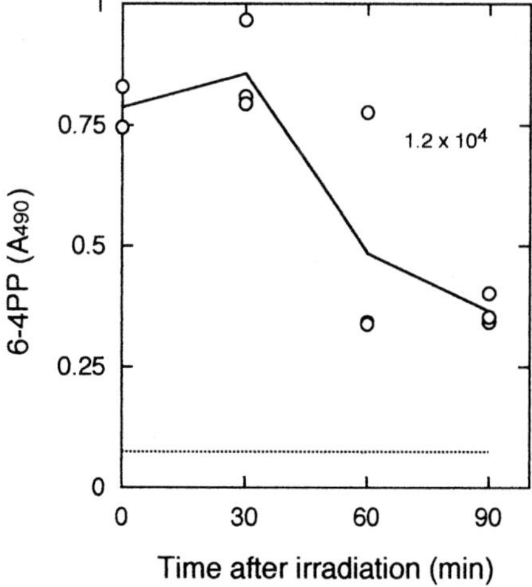

Fig. 5: Decay in the dark of 6-4PPs in *Sorghum* first internodes after UV-B. UV-B, $12\,mmol\,m^{-2}$ of K-295. Other explanations are the same as in Fig. 3.

Coil formation and DNA lesion

The coil angle observed between 24 h and 48 h after UV-B increased up to ca. 300° with the increase in fluence of UV-B. Further rise in fluence did not increase the angle, but resulted in a decrease in the coil radius (Fig. 6). The amount of cyclobutane thymine dimer (CTD), by contrast, increased proportionally with the fluence of UV-B up to 30.6 mmol m^{-2} (for 30 min at 17 µmol $m^{-2} s^{-1}$ or 6.4 W m^{-2}).

Next, photoreactivation was tested. A 10-min pulse of UV-B was given, and followed by varied periods of exposure to UV-A at 40 µmol $m^{-2} s^{-1}$ (13.6 W m^{-2}). Both dimer content and coil angle decreased with the increasing fluence of UV-A. On exposure to UV-A for more than 20 min, no coiling appeared, although some amount of dimer remained non-photorepaired (Fig. 7). Since UV-A also activates phytochrome, a possible involvement of phytochrome was excluded by FR (17 µmol $m^{-2} s^{-1}$ for 3 min) following UV-A, but no appreciable effect of FR on the recovery was observed (data not shown). It is to be added that interestingly the dimer formation was linear to UV-B fluence, whereas the photoreactivation, linear to the log of UV-A fluences (Figs. 6, 7).

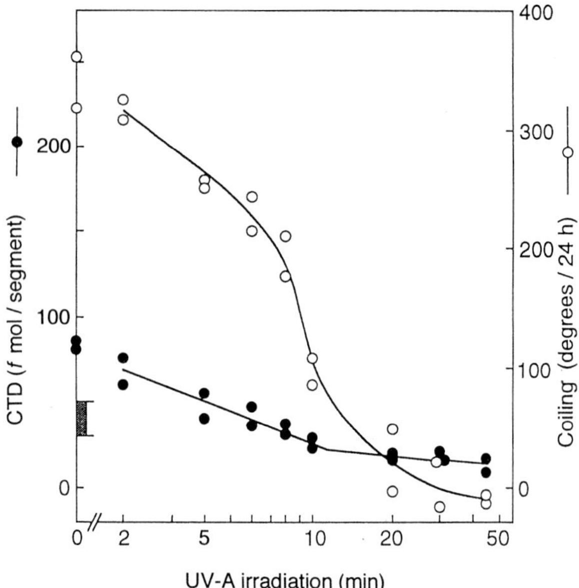

Fig. 7: Effects of UV-A photoreactivation on coil formation (○) and CTD level (●) in *Sorghum* first internodes. Unilateral UV-B irradiation for 10 min at 17 µmol $m^{-2} s^{-1}$ (6.4 W m^{-2}) was followed by vertical UV-A irradiation at 40 µmol $m^{-2} s^{-1}$ (13.6 W m^{-2}) for the indicated periods. Other explanations are the same as in Fig. 6.

Repeated induction and photoreactivation of dimer and coil formation

What happens to dimer content and coil formation when alternate induction and photoreactivation are repeated aroused our interest. The pulses of UV-B and UV-A was given as shown in Fig. 8 without a pause of more than a minute between the pulses, finally followed by FR. Seedlings for dimer determination were harvested immediately, and coiling was determined after 48 h. The coil angles thus-determined may include those of curvature, but they are valid as an index of coil formation, because the curvature is less than 120° (Hashimoto et al., 1984). When the irradiation ended with UV-B before FR, CTD of ca. 70 fmol/segment was generated and coil formation was greater than 400°, whereas, when it ended with UV-A, CTD was ca. 20 fmol/segment, and no coiling occurred. With the increase in number of the pulses the coil angles tended to slightly decline probably due to some growth inhibition of the coiling part by UV-B or UV-A.

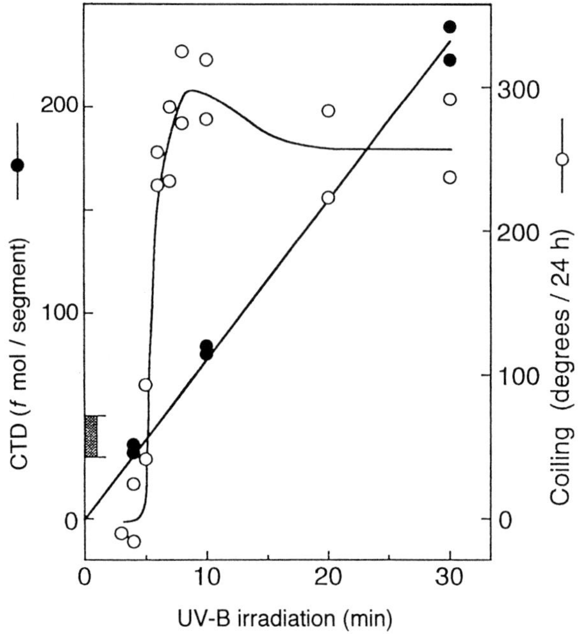

Fig. 6: : Effects of UV-B on coil formation (○) and cyclobutane thymine dimer (CTD) levels (●) in *Sorghum* first internodes. Coiling on ordinate represents increase of coil angle from 24 h to 48 h after UV-B (17 µmol $m^{-2} s^{-1}$ or 6.4 W m^{-2} for the indicated periods). The shaded bar on ordinate indicates the assumed critical level of CTD for coil formation (see Discussion). CTD was determined by RIA.

On leaving UV-B-irradiated seedlings on the laboratory bench for one day or so after determining coil angles, we noted a white line running in the middle along the concave flank of coiled internodes against the pink-coloured background due to room light-induced anthocyanin. A microscopic observation revealed that the white line was a crack due to a collapse of the epidermis and a few cell layers beneath. The crack was always found when coiling occurred. Under near-saturated high humidity air conditions coil formation was hard to occur. In ordinary humidity similar coiling occurred without UV-B, when an artificial lesion was given by peeling with forceps a small strip of the epidermal layers of the internode.

Discussion

In the present paper the physiological relevance of UV-B-induced DNA lesion was examined in *Sorghum bicolor* seedlings which have been well analysed for some physiological UV-B actions. Action spectra showed that anthocyanin synthesis induction occurs maximally at 290 nm, while the inhibition of anthocyanin synthesis is greater as wavelength comes down toward 257 nm (Hashimoto et al., 1991). Action

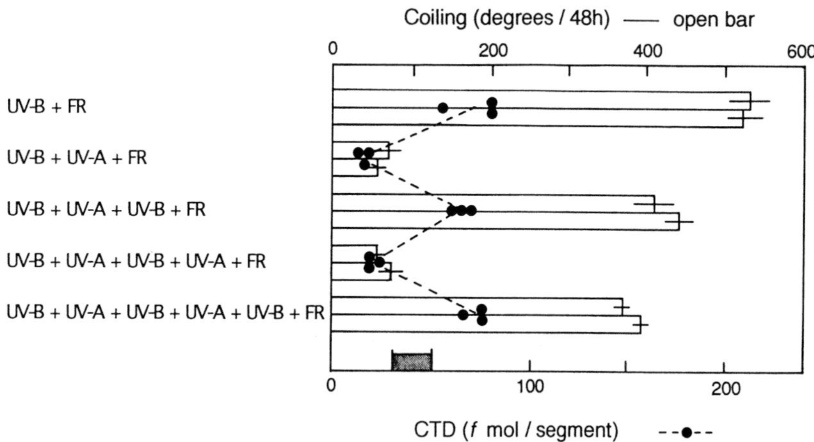

Fig. 8: Effects of alternate irradiations with UV-B and UV-A on coiling growth (open bars) and CTD level (●). The two open bars for each treatment represent coil angles in two independent experiments, and coil angles are the values attained during 48 h after irradiation. UV-B was given unilaterally at $17 \mu mol\, m^{-2} s^{-1}$ ($6.4\, W\, m^{-2}$) for 8 min for the first pulse, and for 5 min for second and third pulses; UV-A, vertically at $40 \mu mol\, m^{-2} s^{-1}$ ($13.6\, W\, m^{-2}$) for 20 min for every pulse; FR, vertically at $17 \mu mol\, m^{-2} s^{-1}$ ($2.8\, W\, m^{-2}$) for 3 min. The shaded bar on ordinate indicates the assumed critical CTD level for coil formation (see Discussion).

spectra for the generation of CPDs and 6-4PPs had a peak at 260 nm (Matsunaga et al., 1991; Rosenstein and Mitchell, 1987). Taking account of these findings and the emission spectra of the three kinds of UV radiation (Fig. 1), the findings with anthocyanin synthesis and its inhibition as well as with the generation of CPDs and 6-4PPs in Fig. 3 may readily be understood. The great differences of the generation of the DNA photoproducts among K-295, K-300, and BLB agreed with the differences of the biologically effective irradiances of the three kinds of radiation (See Materials and Methods). The absence of correlation between the induction of anthocyanin synthesis and the levels of CPDs and 6-4PPs (Fig. 3) allows us to conclude that neither of the DNA photoproducts are responsible for anthocyanin induction. The results also exclude the possibility that the DNA photoproducts might coact with a UV-B photoreceptor in anthocyanin synthesis, because if so, anthocyanin synthesis by K-295 and K-300 should differ from each other in the anthocyanin induction phase (at fluences lower than $2\, mmol\, m^{-2}$) where the levels of CPD and 6-4PP differed between K-295 and K-300 (Fig. 3 A). This conclusion supports the view that that the induction of anthocyanins and other flavonoids is mediated by a UV-B photoreceptor, but not through DNA lesion (Hashimoto and Yatsuhashi, 1984; Wellmann, 1974; Yatsuhashi et al., 1982).

The inhibition of anthocyanin synthesis is, by contrast, closely correlated with the trends of DNA lesion observed (Fig. 3). Although K-295 and K-300 generated different levels of DNA photoproducts, no difference in anthocyanin induction at fluences lower than $2\, mmol\, m^{-2}$ was resulted. It is assumed that CPDs and/or 6-4PPs above the critical levels may be responsible for anthocyanin inhibition, whereas below them the DNA lesion may not be effective. However, to obtain a causal connection between DNA lesion and anthocyanin inhibition, more evidence is needed. Although a photoreactivation experiment seems useful, it is not conclusive, because photoreactivation light enhances anthocyanin synthesis by coaction with phytochrome and/or a UV-B photoreceptor (Drumm-Herrel and Mohr, 1981; Duell-Pfaff and Wellmann, 1982).

Coil formation of the first internode occurred when CTD exceeded a critical level, 30–50 fmol per a top 5 mm of internode (Fig. 6). When CTD was produced once, and then photorepaired down below the critical level, coil formation failed to occur (Fig. 7). Whether coil formation occurs or not depends on a small difference in CTD level (Fig. 8). The results may allow us to conclude that CTD above a critical level causes coil formation. No photoreactivation has yet been discovered with damages other than CPD generation, and hence we conclude that CPDs are responsible for coil formation. As the mechanism of coil formation, we assume that CPDs may cause a collapse of epidermal cells, which results in the crack formation of the epidermal cell layers. In ordinary humidity this allows unilateral water deficit and, in turn, an unequal growth inhibition at the irradiated and shaded sides, resulting in the coil formation of the internode (cf. Results).

It is interesting that the thresholds of DNA photoproducts were found with both anthocyanin inhibition and coil formation. No strict comparison is possible whether the thresholds are the same for anthocyanin inhibition and coil formation because of the different chemical species were determined between the two. However, previous experiments with monochromatic 281 nm and 283 nm radiation (Hashimoto et al., 1984, 1991) showed that both responses started to appear between fluences of 200 and $500 \mu mol\, m^{-2}$, and hence the critical levels for the two responses are considered to be close.

Cyclobutane pyrimidine dimers and 6-4PPs decayed rather fast in the dark, the half-life being ca. 30 min for CPDs, and ca. 60 min for 6-4PPs (Figs. 4, 5). A similar fast dark elimination of 6-4PPs has been reported in the progenitor line of a UV-B sensitive mutant *uvr1* of *Arabidopsis,* while the elimination in *uvr1* is very slow. By contrast, the elimination of CPDs in the dark are slow: half-life is more than 20 h for both of the above-stated *Arabidopsis* lines (Britt et al., 1993), 8–17 h for *Arabidopsis thaliana* ecotype Columbia (Pang and Hays, 1991), and as an example of a faster decay, 2 h for alfalfa (Quaite et al., 1994). The fast dark elimination of CPDs in *Sorghum* is unique. Since it is conceivable that some conformational change of the DNA which may interfere with the attachment of the antibody and hence the detection of the CPDs may occur in a short time in the vicinity of CPDs, confirmation of the fast elimination may be needed by another approach. If the fast repair is really the case, it is tempting to consider how the signal of the DNA photoproducts is transmitted and manifested.

In coil formation the repeated UV-B pulses followed by UV-A (photorepair light) did not accumulate the action of UV-B in terms of either CTD generation or the magnitude of coiling (Fig. 8), indicating that the reversal of the action of each UV-B pulse was complete during 20 min of a UV-A pulse. A case of similarly fast photorepair is reported with *Arabidopsis* (Britt et al., 1993; Chen et al., 1994), while in alfalfa it takes more than 60 min to photorepair 80 % of the CPDs formed (Quaite et al., 1994). In some seeds and dark-grown plants photolyase is scarce and induced by light, either white or red light (Chen et al., 1994; Langer and Wellmann, 1990). Dark-grown *Sorghum* seedlings are assumed to have the enzyme in a sufficient amount without light exposure.

In the present paper, DNA photoproducts determined were CPDs and 6-4PPs for anthocyanin inhibition, and CTD for coil formation. Both CPDs and 6-4PPs were formed and decayed in similar trends (Figs. 4, 5), although the absolute amounts may be different. Hence it was impossible to determine which photoproduct was responsible for anthocyanin inhibition and coil formation.

In conclusion, quantitative comparison of UV-B-induced DNA lesion with UV-B-induced anthocyanin synthesis and its inhibition as well as with coil formation in the *sorghum bicolor* Moench first internode has revealed that anthocyanin synthesis induction does not involve DNA lesion, whereas coil formation is caused by DNA lesion. Very close similarities of the trends of DNA lesion for anthocyanin inhibition to the case for coil formation have strongly suggested that anthocyanin inhibition also involves DNA lesion. It has been indicated that there are thresholds above which DNA lesion is effective. The dark repair of CPDs in this dark-grown plant is uniquely/fast, providing a good subject for future studies.

Acknowledgements

We thank Prof. Dr. Osamu Nikaido, Kanazawa University, Kanazawa for the gifts of the antibodies, and Mr. G. Buchholz and Ms. Lilo Riegger, Freiburg University, Freiburg for their assistance with RIA. The financial supports were a grant (No. 03045031, to T. H.) under the Monbusho International Scientific Research Program, a grant (to T. H.) of the Global Environmental Research from the Environment Agency of Japan, and finances (for T. H. and S. T.) by the German Academic Exchange Service.

References

BALLARÉ, C. L., P. W. BARNES, and R. E. KENDRICK: Photomorphogenic effects of UV-B radiation on hypocotyl elongation in wild type and stable-phytochrome-deficient mutant seedlings of cucumber. Physiol. Plant. *83,*, 652–658 (1991).

BALLARÉ, C. L., P. W. BARNES, and S. D. FLINT: Inhibition of hypocotyl elongation by ultraviolet-B radiation in de-etiolating tomato seedlings. I. The photoreceptor. Physiol. Plant. *93,* 584–592 (1995).

BEGGS, C. J., E. WELLMANN, and H. GRISEBACH: Photocontrol of flavonoid biosynthesis. In: KENDRICK, R. E. and G. H. M. KRONENBERG (eds.): Photomorphogenesis in Plants. pp. 467–499. Martinus Nijhoff Publishers, Dordrecht (1986).

BRITT, A. B., J.-J. CHEN, D. WYKOFF, and D. MITCHELL: A UV-sensitive mutant of *Arabidopsis* defective in the repair of pyrimidine-pyrimidinone (6-4) dimer. Science *261,* 1571–1574 (1993).

CHEN, Y.-P. and J. F. BORNMAN: The response of bean plants to UV-B radiation under different irradiances of background visible light. J. Exp. Bot. *41,* 1489–1495 (1990).

CHEN, J.-J., D. L. MITCHELL, and A. B. BRITT: A light-dependent pathway for the elimination of UV-induced pyrimidine (6-4) pyrimidinone photoproducts in *Arabidopsis.* Plant Cell *6,* 1311–1317 (1994).

DE FABO, E. C., F. P. NOONAN, and J. E. FREDERICK: Biologically effective doses of sunlight for immune suppression at various latitudes and their relationship to changes in stratospheric ozone. Photochem. Photobiol. *52,* 811–817 (1990).

DRUMM-HERREL, H. and H. MOHR: A novel effect of UV-B in a higher plant (*Sorghum vulgare*). Photochem. Photobiol. *33,* 391–398 (1981).

DUELL-PFAFF, N. and E. WELLMANN: Involvement of phytochrome and a blue light photoreceptor in UV-B induced flavonoid synthesis in parsley (*Petroselinum hortense* Hoffm.) cell suspension culture. Planta *156,* 213–217 (1982).

HASHIMOTO, T.: UV-B effects on plants – The present knowledge and prospect for future studies. In: Proceedings, Tsukuba Ozone Workshop – Global Environment Tsukuba '94, pp. 134–145, Center for Global Environmental Research, Tsukuba (1994).

HASHIMOTO, T. and H. YATSUHASHI: Ultraviolet photoreceptors and their interaction in broom sorghum – analysis of action spectra and fluence-response curves. In: SENGER, H. (ed.): Blue Light Effects in Biological Systems, pp. 125–136. Springer Verlag, Berlin (1984).

HASHIMOTO, T., S. ITO, and H. YATSUHASHI: Ultraviolet light-induced coiling and curvature of broom sorghum first internodes. Physiol. Plant. *61,* 1–7 (1984).

HASHIMOTO, T., C. SHICHIJO, and H. YATSUHASHI: Ultraviolet action spectra for the induction and inhibition of anthocyanin synthesis in broom sorghum seedlings. J. Photochem. Photobiol. *B: Biol. 11,* 353–363 (1991).

HE, J., L.-K. HUANG, and M. I. WHITECROSS: Chloroplast ultrastructure changes in *Pisum sativum* associated with supplementary ultraviolet (UV-B) radiation. Plant, Cell and Environment *17,* 771–775 (1994).

KLOCKER, H., B. AUER, H. J. BURTSCHER, M. HIRSCH-KAUFFMANN, and M. SCHWEIGER: A synthetic haptene for induction of thymine-dimer-specific antibodies. Eur. J. Biochem. *142,* 313–316 (1984).

LANGER, B. and E. WELLMANN: Phytochrome induction of photoreactivating enzyme in *Phaseolus vulgaris* L. seedlings. Photochem. Photobiol. *52,* 861–863 (1990).

MATSUNAGA, T., K. HIEDA, and O. NIKAIDO: Wavelength dependent formation of thymine dimers and (6-4)photoproducts in DNA by monochromatic ultraviolet light ranging from 150 to 365 nm. Photochem. Photobiol. *54,* 403–410 (1991).

MITCHELL, D. L., J. E. VAUGHAN, and R. S. NAIRN: Inhibition of transient gene expression in chinese hamster ovary cells by cyclobutane dimers and (6-4) photoproducts in transfected ultraviolet-irradiated plasmid DNA. Plasmid *21,* 21–30 (1989).

MORI, T., M. NAKANE, T. HATTORI, T. MATSUNAGA, M. IHARA, and O. NIKAIDO: Simultaneous establishment of monoclonal antibodies specific for either cyclobutane pyrimidine dimer or (6-4) photoproduct from the same mouse immunized with ultraviolet-irradiated DNA. Photochem. Photobiol. *54,* 225–232 (1991).

MURPHY, T. M., Y. C. QIAN, C. K. AUH, and C. VERHOEVEN: UV-induced events at plant plasma membranes. In: SHIMA, A. et al. (ed.): Frontiers of Photobiology, pp. 555–560. Excerpta Medica, Amsterdam (1993).

PANG, Q. and J. B. HAYS: UV-B-inducible and temperature-sensitive photoreactivation of cyclobutane pyrimidine dimers in *Arabidopsis thaliana.* Plant Physiol. *95,* 536–543 (1991).

QUAITE, F. E., B. M. SUTHERLAND, and J. C. SUTHERLAND: Action spectrum for DNA damage in alfalfa lowers predicted impact of ozone depletion. Nature *358,* 576–578 (1992).

QUAITE, F. E., S. TAKAYANAGI, J. RUFFINI, J. C. SUTHERLAND, and B. M. SUTHERLAND: DNA damage levels determine cyclobutyl pyrimidine dimer repair mechanism in alfalfa seedlings. Plant Cell *6,* 1635–1641 (1994).

ROSENSTEIN, B. S. and D. L. MITCHELL: Action spectra for the induction of pyrimidine(6-4)pyrimidone photoproducts and cyclobutane pyrimidine dimers in normal human skin fibroblasts. Photochem. Photobiol. *45,* 775–780 (1987).

SANTOS, I., J. M. ALMEIDA, and R. SALEMA: Plants of *Zea may* L. developed under enhanced UV-B radiation. I. Some ultrastructural and biochemical aspects. J. Plant Physiol. *141,* 450–456 (1993).

SETLOW, R. B.: The wavelengths in sunlight effective in producing skin cancer: a theoretical analysis. Proc. Nat. Acad. Sci. USA *71,* 3363–3366 (1974).

SHICHIJO, C., T. HAMADA, M. HIRAOKA, C. B. JOHNSON, and T. HASHIMOTO: Enhancement of red-light-induced anthocyanin synthesis in sorghum first internodes by moderate low temperature given in the pre-irradiation culture period. Planta *191,* 238–245 (1993).

SMITH, R. C., K. S. BAKER, O. HOLM-HANSEN, and R. OLSON: Photoinhibition of photosynthesis in natural waters. Photochem. Photobiol. *31,* 585–592 (1980).

STAXEN, I. and J. F. BORNMAN: A morphological and cytological study of *Petunia hybrida* exposed to UV-B radiation. Physiol. Plant. *91,* 735–740 (1994).

TAKEDA, J. and S. ABE: Light-induced synthesis of anthocyanin in carrot cells in suspension. IV. The action spectrum. Photochem. Photobiol. *56,* 69–74 (1992).

WELLMANN, E.: Regulation der Flavonoidbiosynthese durch ultraviolettes Licht und Phytochrom in Zellkulturen und Keimlingen von Petersilie (*Petroselinum hortense* Hoffm.). Ber. Deutsch. Bot. Ges. *87,* 267–273 (1974).

YATSUHASHI, H., T. HASHIMOTO, and S. SHIMIZU: Ultraviolet action spectrum for anthocyanin formation in broom sorghum first internodes. Plant Physiol. *70,* 735–741 (1982).

J. Plant Physiol. Vol. 148. pp. 100–106 (1996)

Amelioration of Ultraviolet-B-Induced Down-Regulation of mRNA Levels for Chloroplast Proteins, by High Irradiance, is Mediated by Photosynthesis

SOHEILA A.-H. MACKERNESS, P. JOE BUTT, BRIAN R. JORDAN*, and BRIAN THOMAS

Horticulture Research International, Wellesbourne, Warwick, CV35 9EF, UK

Received June 24, 1995 · Accepted October 5, 1995

Summary

The mechanism by which increasing photosynthetically active radiation (PAR) reduces the sensitivity of RNA transcripts to UV-B radiation was studied in pea (*Pisum sativum* L.). mRNA transcript levels for *rbc S, rbc L, cab* and *psb A* were measured over an 8 d experimental period in pea plants supplemented with UV-B radiation under a range of conditions. Under low light (150 µmol $m^{-2} s^{-1}$), UV-B resulted in a significant decline in the levels of transcripts for all four genes which was prevented by increasing the background irradiance to 350 µmol $m^{-2} s^{-1}$ (high light) with white light from fluorescent lamps. Increasing CO_2 levels to give photosynthesis rates equivalent to the high light treatment partially protected *rbc S* and *cab* transcripts and fully protected *rbc L* transcripts but did not prevent visible injury. Increasing light with low pressure sodium lamps, which increase photosynthesis but are not effective for activation of the DNA repair enzyme, photolyase, gave results which were not significantly different from white fluorescent high light treatments. Protection by high light was lost in the presence of the photosynthesis inhibitors CCCP and DCMU. The UV-B induced increase in the expression of chalcone synthase (*chs*) genes was delayed by the treatments which increased photosynthesis rates and conferred protection. The results indicate that photosynthesis plays a key role in the amelioration of UV-B induced decline in mRNA levels for proteins. The minimal role of DNA repair by photolyase indicates that reduction in photosynthesis gene transcripts in response to UV-B represents a specific regulation rather than being a consequence of DNA damage.

Key words: Pisum sativum, Photosynthesis, high PAR, RNA transcripts, UV-B stress.

Abbreviations: UV-B = Ultraviolet-B radiation; PAR = Photosynthetically Active Radiation; RUBISCO = ribulose 1,5-bisphosphate carboxylase/oxygenase; LL = Low light; HL_W = high broad spectrum light; HL_{Na} = high light using sodium lamps; H_{CO_2} = high CO_2 (LL); LPS = low pressure sodium lamps; CCCP = carbonyl cyanide m-chlorophenylhydrazone; DCMU = 3-(3,4 dichlorophenyl) dimethylurea.

Introduction

Reduction in stratospheric ozone has raised concern over the increases in the levels of ultraviolet-B radiation (UV-B: 280–320 nm) reaching the earth's surface (Blumthaler and Ambach, 1990). Exposure of plants to UV-B most notably leads to reduction in photosynthesis resulting in biomass reduction; anatomical changes such as decreases in root to shoot ratio and development of smaller, thicker leaves; (Teramura, 1980; Krizeck, 1975) and increases in the UV-B absorbing flavonoids (Tevini et al., 1991; Robberecht and Caldwell, 1986; Warner and Caldwell, 1983). Recently molecular studies have shown that exposure to UV-B leads to the down regulation of several key photosynthetic genes re-

* New Zealand Institute for Crop and Food Research Limited, Levin Research Centre, Kimberley Road, Private Bag 4005, Levin, New Zealand.

sulting in the subsequent decline in the proteins. These studies have been able to explain, to some extent, the mechanism by which UV-B inhibits photosynthesis (Jordan, 1995).

It has been recognised for some time that longer wavelength radiation (PAR: 400–700 nm) can minimise UV-B induced damage. This protection occurs at the physiological, biochemical and also molecular level (Cen and Bornman, 1990; Flint et al., 1985; Mirecki and Teramura, 1984; Jordan et al., 1992). Plants grown under high PAR have thicker leaves and increased levels of flavonoids as compared to those growth under low levels (Teramura, 1980; Cen and Bornman, 1990; Tevini et al., 1991) which tends to lessen the sensitivity of plants to UV-B (Warner and Caldwell, 1983; Murali and Teramura, 1985). These changes, however, cannot explain why plants grown under low PAR and exposed to UV-B under high PAR are less sensitive to UV-B than plants grown under low PAR and then irradiated with UV-B under low PAR (Kramer, et al., 1992; Mirecki and Teramura, 1984). The effect of high PAR, under these conditions, has been attributed to activation of photorepair mechanisms (Fernbach and Mohr, 1992; Mirecki and Teramura, 1984) which ameliorates the UV-B stress induced damage. Photorepair is thought to involve the activation of the DNA repair enzyme, photolyase (Pang and Hays, 1991). This enzyme uses energy in the range of 300 to 500 nm to reverse UV-B induced photoproducts, cyclobutane pyrimidine dimers (Sutherland, 1981). However, higher levels of visible light may also contribute to protection by providing additional substrate through increases in photosynthesis for the repair or replacement of damaged organelles or tissues (Adamse and Britz, 1992).

Jordan et al. (1992) noted that increased PAR can reduce the UV-B induced down regulation of photosynthetic genes. We have used this system to study the role of photosynthesis in the protection against UV-B damage at the molecular level. In order to separate the potential contribution of photosynthesis and non-photosynthetic perception mechanisms to this protection, photosynthesis rates were manipulated by altering CO_2 levels and PAR levels using low pressure sodium lamps (LPS) which are effective for photosynthesis (Withers et al., 1991) but not for activation of photorepair enzymes. In addition, the effect of lowering of photosynthesis rates on transcript levels, under high light conditions, using inhibitors, was also studied.

Material and Methods

Plant material and experimental conditions

Pea (*Pisum sativum* L. cv Feltham first) seedlings were grown in a controlled environment cabinet, partitioned in two halves, with 12 h light (22 °C), 12 h dark (16 °C) cycles, for 17 d. Incident radiation was provided by Philips warm white fluorescent tubes giving an irradiance of 150 µmol $m^{-2} s^{-1}$ PAR. Half of the plants were then given supplementary UV-B radiation from two UV Lamps (Philips TL 40) during the 12 h photoperiod under the four experimental conditions. To exclude UV radiation below 290 nm, the UV-B lamps were wrapped in cellulose acetate sheets which were changed daily. The other half of the plants (controls) were subjected to the same treatments but in the absence of UV-B radiation. For low light experiments (LL), irradiation was maintained at 150 µmol $m^{-2} s^{-1}$ for the duration of the UV-B treatment. In high light experiments (HL), the illumination was increased to 350 µmol $m^{-2} s^{-1}$ using either white fluorescent tubes (HL_W) or low pressure sodium lamps (LPS) (HL_{Na}). High CO_2 (H_{CO_2}) experiments were carried out under LL conditions and at 1500 ppm CO_2. All experiments were carried out in Weiss cabinets except for HL_{Na}. which were carried out in Richmond cabinets as LPS lamps could be accommodated more easily in these cabinets. The spectral irradiance between 260–320 nm was determined by using an Optronics 740A spectroradiometer (Optronics Laboratories, Inc., Orlando, Fl, USA). The levels of UV-B were 4.22 and 16.7 mW m^{-2} nm^{-1} at 297 nm and 313 nm respectively. These UV-B levels represent 8.5 % at 297 nm and 7.5 % at 313 nm of UV-B radiation used in a comparable study (Jordan et al., 1992). The levels of UV-B in the control half of the cabinet were negligible. The third pair of leaflets was routinely used for experiments unless otherwise stated.

Photosynthesis

Photosynthesis rates of the third pair of leaflets, at incident irradiation, were determined using a Combined Infrared Gas Analysis System (CIRAS-1, PP Systems, Hitchin, UK). The concentration of CO_2 in H_{CO_2} experiments and irradiance levels in HL_{Na} experiments were chosen to mimic the photosynthesis levels at the start of the HL_W experiments. Values are quoted as the mean from 5 separate measurements ± standard error of the mean.

Analysis of RNA

Total RNA was extracted from leaf tissue and analysed by Northern blotting as described previously (Jordan et al., 1991). Specific RNAs were detected by hybridisation with ^{32}P labelled homologous cDNA probes. Relative amounts of hybridisation to specific bands were quantified by excising the appropriate area of the filter and determining the amount of bound radioactive probe by liquid scintillation counting in Omnisafe scintillation fluid (Life Technologies, UK). The quantitative data is presented as percentage transcripts from UV-B treated plants as compared to transcripts from control plants which were treated under the same conditions but in the absence of UV-B irradiation.

The *rbc L* and *rbc S* cDNA sequences for the large and the small subunit of RUBISCO respectively are described in Jordan et al. (1992). The *psbA* sequence for the D1 polypeptide of Photosystem II is described in Jordan et al. (1991). The *cab* sequence for chlorophyll *a/b* binding protein and the *chs 1* sequence for chalcone synthase are described in Jordan et al. (1994).

Photosynthesis inhibitor studies

The effect of the photosynthesis inhibitors on the transcript levels under HL_W conditions were determined by using leaf discs. Ten leaf discs (1 cm^2) were floated adaxial side up in 0.1 % Tween containing either 5 µM CCCP (Sigma, UK), 25 µM DCMU (Sigma, UK) or dH_2O (control). The discs were then incubated under control or under supplemental UV-B radiation. After appropriate incubation periods, the discs were washed three times in distilled water and then frozen immediately in liquid nitrogen. The samples were prepared for Northern blot analysis as described above and quantified by liquid scintillation counting.

Statistical analysis

An analysis of variance (ANOVA) was performed on quantitative data obtained from duplicate Northern analysis, for each treatment, and least significant difference (LSD) at the 5% level determined.

Results

Visible damage

In all treatments, bronzing of the leaf tissue was the first visible sign of UV-B stress and was apparent after 2 d UV-B treatment under low light (LL) conditions. This became more pronounced throughout the treatment. By 4 d the leaves had acquired a glazed appearance and by 8 d signs of chlorosis were visible. Under all other treatments, the progressive damage was less pronounced, with plants grown under high light (HL_W) and high sodium light (HL_{Na}) having reduced signs of chlorosis and tissue death after 8 d UV-B treatment. Under high CO_2 conditions (H_{CO_2}) although initially plants appeared healthier than those under ambient CO_2 levels (LL), strong bronzing became visible between day 5 and day 8 of the treatment, and their appearance was comparable to those grown under low light (LL; ambient CO_2) conditions.

Effect of supplemental UV-B on transcript levels under LL, HL_W, HL_{Na} and H_{CO_2} conditions

Total RNA was extracted from the third pair of leaflets from the base of the pea seedlings, grown in the presence and absence of supplemental UV-B, under all 4 experimental conditions. The integrity of the RNA was determined using Northern blot analysis. Discrete bands were observed for all probes used in hybridisations (data not shown). The steady state RNA levels were quantified by counting of the radioactive probe bound to the specific transcript after hybridisation with ^{32}P labelled *rbc S, cab, rbc L* and *psb A* cDNA probes and are shown in Figure 1. Under all test conditions the nuclear encoded genes, *rbc S* and *cab* (Fig. 1a, b) were more sensitive to UV-B treatment than chloroplast encoded genes, *rbc L* and *psb A* (Fig. 1c, d). Under LL conditions, the decline in transcript levels for all 4 genes studied in the Richmond cabinets was not significantly different from values obtained in Weiss cabinets and, therefore, the data was combined, and values represented in Figure 1 and Table 1 are average values from the 2 experiments.

Photosynthetic rates of the third pair of leaflets were measured under LL conditions and were 2.5 (± 0.2) mg CO_2 $dm^2 h^{-1}$ at the beginning of the experiment. Under LL conditions the nuclear encoded *cab* and *rbc S* transcripts were reduced dramatically by the second day of UV-B treatment, continuing to decline up to 8 d where transcripts were present only in trace amounts (Fig. 1a, b). The effect of UV-B on the chloroplast genes was not as dramatic. Transcripts for *rbc L* were not substantially reduced after 2 d of UV-B treatment (Fig. 1c). The levels began to decline by the third day of treatment and continued to fall until the last sampling day. In contrast, the *psb A* transcript levels were not substantially affected by UV-B supplementation for up to 4 d of treatment (Fig. 1d). The percentage of RNA transcripts remaining on

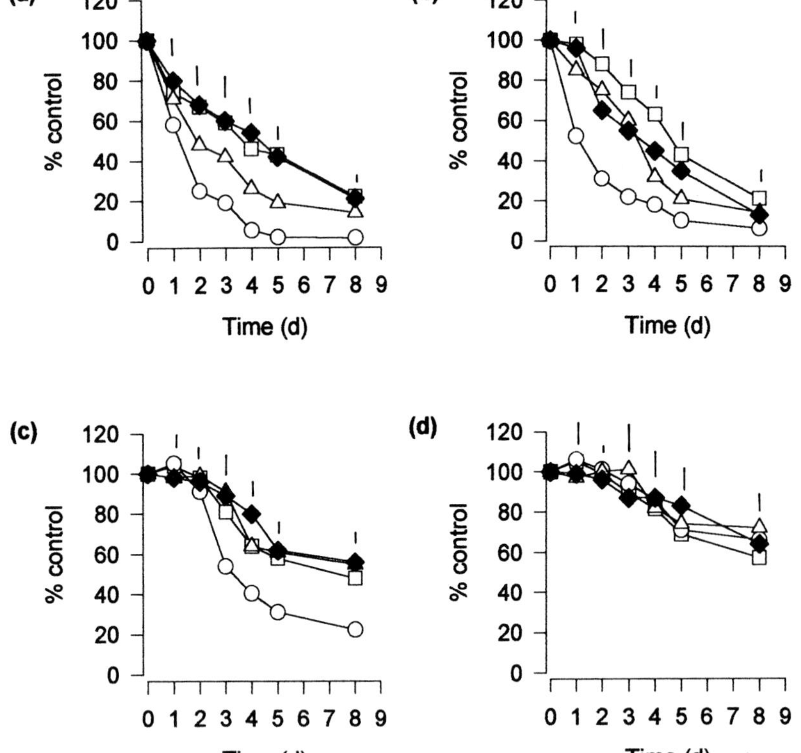

Fig. 1 a, b, c, d: Steady state levels of leaf mRNA, under the four treatment conditions, after exposure to UV-B. mRNA Northern blots, probed with cDNA probes, were quantified by excides of appropriate area of the filter and counting in a liquid scintillation counter. The effect is defined as the percentage counts for UV treatments with respect to the counts from control plants, treated under the same conditions but in the absence of UV-B radiation, on equivalent days. Each point represents the mean value obtained from duplicate blots. Error bars indicate the least significant difference (LSD) taken at the 5% level. Probes used were **a** *rbc S,* **b** *cab,* **c** *rbc L,* **d***psb A.* Symbols: LL (○), HL_W (□), H_{CO_2} (△), HL_{Na} (◆).

Table 1: Steady state levels of mRNA, under the four treatments, after exposure to UV-B for 3 d (values taken from Figure 1 and are given as the percentage counts for UV treatments with respect to the counts from control plants on equivalent days, treated under the same conditions but in the absence of UV-B radiation). Values from the third day of treatment were chosen to illustrate effects of UV-B on plants during the four treatments as differences in treatments was most pronounced on this day. Means in the same row marked by * are not significantly different at the 5 % level as determined by ANOVA.

Genes	Treatments				LSD
	LL	HL_W	H_{CO_2}	H_{Na}	
rbcS	24	59*	42	60*	10.4
cab	19	74	60*	55*	8.6
rbc L	54	81*	91*	89*	10.2
psb A	90*	94*	100*	87*	13.3

the third day of each treatment from plants supplimented with UV-B as compared to control transcript levels, from plants treated under the same conditions but in the absence of UV-B radiation, are shown for all genes in Table 1.

Increasing the background irradiance from 150 to 350 $\mu mol\ m^{-2}\ s^{-1}$ resulted in an increase in photosynthetic rates from 2.5 to 8.9 (± 0.3) mg CO_2 $dm^2\ h^{-1}$, at the start of the experiment. The profile of RNA levels for *rbc S, cab* and *rbc L* under HL_W conditions was similar to LL conditions but transcript levels were not as dramatically reduced (Fig. 1 a, b, c). HL_W treatment had no significant effect on transcript levels for *psb A.* These transcripts were, however, not significantly affected by UV-B treatment under LL conditions (Fig. 1 d). The protection of *rbc S* and *cab* transcripts was maintained throughout the 8 d UV-B treatment (Fig. 1 a and b). Increased illumination under HL_W conditions resulted in a significant protection of *rbc L* transcripts after the third day of UV-B treatment, at the point where levels sharply decline under LL levels (Fig. 1 c). Table 1 clearly illustrates the extent of protection against UV-B reduction in transcript levels on the third day of UV-B treatment.

Photosynthesis rates were raised by increasing CO_2 levels, under LL conditions (HL_{CO_2}), to 8.8 (± 0.5) mg CO_2 $dm^2\ h^{-1}$. This photosynthesis rate was similar to that under HL_W. This treatment resulted in protection of the transcripts, but only for *rbc L* transcripts to levels comparable to HL_W (Fig. 1 c). However, under H_{CO_2} conditions, the *rbc S* and *cab* mRNA levels were significantly higher than under LL (ambient CO_2) conditions for the duration of the UV-B treatment (Fig. 1 a, b). The protection of *rbc L* transcripts under H_{CO_2} was complete, resulting in maintenance of levels at HL_W values after 3 d UV-B treatment (Fig. 1 c). Due to the insensitivity of *psb A* transcripts towards UV-B, there was little change in their levels under these conditions (Fig. 1 d).

Low pressure sodium lights were used to increase photosynthesis to 9.0 (± 0.2) mg CO_2 $dm^2\ h^{-1}$ (HL_{Na}), a rate again comparable to that under HL_W conditions (Fig. 1). The profile of mRNA levels obtained under the HL_{Na}. conditions was not significantly different from values obtained under HL_W for all the genes (Fig. 1 a, c, d) with the exception of *cab* (Fig. 1 b) where levels were slightly lower on the second to fourth day of UV-B treatment. However, the levels were still significantly higher than LL and H_{CO_2} levels (Fig. 1 b, Table 1).

Effect of photosynthesis inhibitors on HL_W protection

Increasing photosynthesis rates result in protection of the transcript levels for the 4 genes being studied (Fig. 1). In order to determine whether a decrease in photosynthesis results in loss of this protection under HL_W conditions, 2 photosynthesis inhibitors, DCMU and CCCP, were used in leaf disc experiments. DCMU is an inhibitor of photosynthetic electron transport and CCCP is an uncoupler of photophosphorylation. The presence of both inhibitors lead to the loss of protection under HL_W conditions for all 4 genes (Fig. 2).

Changes in chs transcript levels under the four treatments

UV-B irradiation leads to transient increases in *chs* expressions under all 4 treatments. Maximum transcript levels were reached after different time periods depending on the treatment but levels were comparable in all treatments (Fig. 3). Under LL conditions the peak value was reached on the second day in both Weiss and Richmond cabinets (Fig. 3). This peak was shifted to the third day under H_{CO_2} conditions and to the fifth day under both HL_W and HL_{Na} conditions (Fig. 3).

Discussion

In this study, by manipulating photosynthesis rates without increasing the activity of photolyase, it was possible to determine the contribution of photosynthesis to the protection of transcript levels for three key photosynthetic proteins, RUBISCO, chlorophyll *a/b* binding protein and D1 polypeptide of PSII, under high irradiance conditions. Jordan et al. (1994) indicated that effects of UV-B on transcript levels are detectable long before any physiological differences are measurable and, therefore, this approach allows the determination of subtle effects more easily and over shorter time periods.

Increased photosynthesis protects UV-B induced decreases in transcript levels

Under all conditions, supplemental UV-B resulted in the decrease in the transcript levels for all 4 genes studied. Nuclear encoded transcripts were more sensitive to UV-B radiation than chloroplast encoded transcripts (Fig. 1). These observations are consistent with previous studies on the effect of UV-B on the mRNA levels for photosynthetic proteins (Jordan et al., 1991; 1992).

High irradiance during UV-B treatment reduced the UV-B induced down-regulation of the photosynthetic genes *rbc S, cab* and *rbc L.* Similar protection for *rbc S* (Jordan et al., 1992) and *cab* (Jordan et al., 1991) has been noted previously. Due to the relative insensitivity of *psb A* transcript levels to UV-B, there was little change in levels under these conditions (Fig. 1 d; Table 1). The decline in transcript levels after day 4

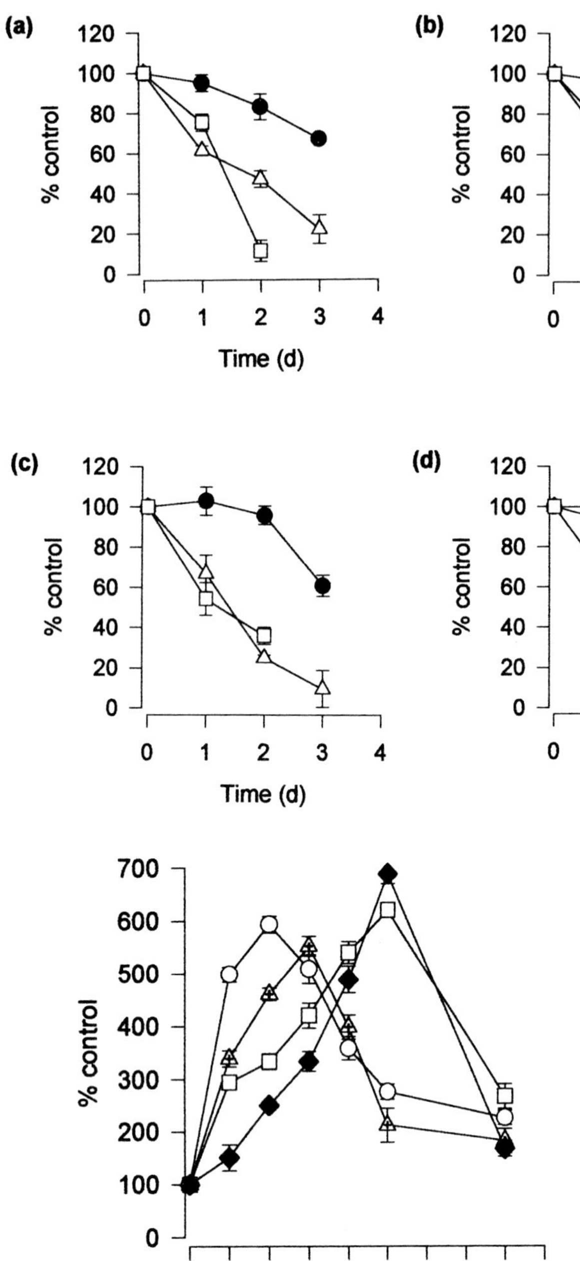

Fig. 2 a, b, c, d: Steady state mRNA levels of leaf discs, supplimented with UV-B under HL_W conditions, after addition of water (●), CCCP (□) and DCMU (△). The effect is defined as the percentage counts for UV treatments with respect to the counts from leaf discs treated under the same conditions, but in the absence of UV-B radiation, on equivalent days. Each point represents the mean value obtained from duplicate experiments. Error bars indicate the variance in the duplicate experiments. Probes used were **a** *rbc S*, **b** *cab*, **c** *rbc L*, **d** *psb A*.

Fig. 3: Steady state levels of leaf mRNA under the four treatments, after exposure to UV-B. Northern blots, probed with *chs* cDNA, were quantified by excision of appropriate area of the filter and counting in a liquid scintillation counter. The effect is defined as the percentage counts for UV treated plants with respect to the counts from control plants, treated under the same conditions but in the absence of UV-B radiation, on equivelant days. Each point represents the mean value obtained from duplicate blots. Error bars indicate the variation in duplicates. Symbols: LL (○), HL_W (□), H_{CO_2} (△), HL_{Na} (◆).

of UV-B treatment is most likely due to premature senesence of the leaves as a results of UV-B exposure.

Increasing photosynthesis levels to HL_W levels by using high CO_2 and sodium lamps allows the separation of the potential contribution of photosynthesis and non-photophotosynthetic perception mechanisms in protection under HL_W conditions. LPS lamps emit at a narrow waveband at 589 nm and do not activate DNA repair enzymes, photolyases, which require light between 300 and 500 nm (Pang and Hays, 1991). Sensitivity of transcripts to supplemental UV-B was reduced under both H_{CO_2} and HL_{Na} conditions for *rbc S*, *cab* and *rbc L*. High CO_2 levels were, however, not as effective as HL_W or HL_{Na} treatments in the protection of steady state RNA levels for the nuclear encoded genes *rbc S* and *cab* (Fig. 1 a, b) as compared to the chloroplast encoded gene, *rbc L* (Fig. 1 c). Due to the relative insensitivity of *psb A* transcript levels to UV-B, again, there was little change in levels under these conditions (Fig. 1 d; Table 1).

Elevated CO_2 concentrations, leading to increased photosynthetic rates, resulted in a significant reduction in the decline in *rbc S* and *cab* transcripts as compared to LL, ambient CO_2 conditions, throughout the duration of the experiment (Fig. 1 a, b). Previous studies have looked at the interaction of CO_2 and UV-B at the physiological level but the responses observed were inconsistent. For example, UV-B was found to reduce any CO_2 induced increases in biomass in rice and wheat (Teramura et al., 1990). In contrast, under identical conditions, no effect of UV-B was recorded in soybean (Teramura et al., 1990). Very few molecular studies have been carried out on plants grown in high CO_2 and none under both elevated CO_2 and UV-B conditions. In tomato, the abundance of transcripts for the nuclear-encoded chloroplast proteins were reduced in response to high CO_2 but chloroplast-encoded transcripts were not significantly affected under the same conditions (Van Oosten et al., 1994; Van Oosten and Besford, 1994). A similar pattern was observed in the present study with pea (data not shown). Therefore, the differential effect of CO_2 on the protection of nuclear and chloroplast encoded genes towards supplemental UV-B, is likely to be due to the differences in the sensitivity of these genes to CO_2. Transcripts of *rbc L* are insensitive to high levels of CO_2 and, therefore, the protective effect of higher photosynthesis under these conditions was not masked by any CO_2 effect. In contrast, *rbc S* and *cab* are sensitive to CO_2 and this may explain the apparently inefficient protection of transcripts. Complete protection of all genes under increased illumination by LPS lamps, a system which also does not activate photorepair mechanisms, further implies that CO_2 effects are the likely cause of the decreased protection efficiency under H_{CO_2} conditions.

These observations strongly suggest a significant role of photosynthesis in the protection of these transcripts under high irradiance and a limited contribution of photorepair mechanism. This was supported by the loss of protection when photosynthesis levels were reduced under HL_W conditions using photosynthesis inhibitors, further indicating that light *per se* is not the major contributor to protection. The presence of photoreactivating enzyme activity has been demonstrated in many plant species, for example in water plants (Degani et al., 1980) and wheat (Taylor et al., 1995) but could not been demonstrated in several other plant species (Steinmitz and Wellman, 1986). Therefore, it is possible that in pea, photoreactivation, by photolyase, plays a negligible part in protection against short term UV-B damage or that pea does not have an active photolyase.

UV-B stimulation of chs expression

Chs transcripts, under all experimental conditions, were increased to a peak value in response to UV-B and then declined again. The time taken for peak values to be reached was shifted from day 2 under LL conditions to day 3 under H_{CO_2} and day 4 under HL_W and HL_{Na} (Fig. 3). The *chs* genes encode the enzyme chalcone synthase, a key enzyme in the synthesis of flavonoids. Flavonoids are UV-B screening pigments produced in plants in response to UV-B radiation (Caldwell, 1981; Wellman, 1983). Increases in *chs* expression, in response to supplemental UV-B have been previously reported (Strid, 1993; Takeda et al., 1993). The peak shift noted in this study may indicate that stimulation of *chs* expression is controlled by the level of damage inflicted by UV-B and that the shift represents the protection of damage by HL_W, HL_{Na} and H_{CO_2} leading to a delay in full expression of this gene. Protection was most effective under HL_W and HL_{Na} and the delay in maximal expression of *chs* was longest in these treatments. However, more detailed studies will be needed to confirm this.

Flavonoid concentrations under these experimental conditions have been measured and a correlation between the level of gene expression and the concentration of flavonoids during the 4 treatments has been observed (Mackerness, Jordan, Thomas, unpublished data). Maximal levels were reached at day 5 for HL_W and HL_{Na} treatments, and days 2 and 3 for LL and H_{CO_2} respectively. However, flavonoids did not accumulate to higher levels under any one treatment (Mackerness, Jordan, Thomas, unpublished data). This observation is not surprising since flavonoid accumulation is a direct result of incident UV-B radiation (Jordan, 1993; Jordan et al., 1994) and as background lighting did not contain any UV-B, and high levels of CO_2 do not activate flavonoid synthesis (Adamse and Britz, 1992; Ziska and Teramura, 1992), the increase in flavonoids would mainly be a result of supplemental UV-B, which was constant for all treatments.

In conclusion, we have shown that photosynthesis or a process linked to photosynthesis plays a key role in the amelioration of UV-B induced decline in pea mRNA levels for photosynthetic proteins. In the absence of photolyase activation, increases in photosynthesis were sufficient to account for protection of transcript levels against UV-B irradiation. Conversely, application of inhibitors of photosynthesis resulted in loss of this protection. The seemingly limited role of photolyase in this protection mechanism suggests that there is a selective and specific regulation of gene expression by UV-B radiation, corroborating previous studies (Jordan et al., 1994; Strid, 1993).

Acknowledgements

This research was supported by the Biotechnology and Biological Sciences Research Council. We are grateful to Ron Pierce for running the Weiss cabinets, Rodney Edmondson for helping with the statistical analysis and to Diana Wilkins for her help with the CIRAS measurements.

References

Adamse, P. and S. J. Britz: Amelioration of UV-B damage under high irradiance: Role of photosynthesis. Photochem. Photobiol. *56,* 645–650 (1992).

Blumthaler, M. and W. Ambach: Indication of increasing solar ultraviolet-B radiation flux in alpine regions. Exper. Bot. *42,* 547–554 (1990).

Caldwell, M. M.: Plant responses to solar UV-B radiation. Encycl. Plant Physiol. *41,* 169–197 (1981).

Cen, Y. P. and J. F. Bornman: The response of bean plants to UV-B radiation under different irradiances of background visible light. J. Exp. Bot. *41,* 1489–1495 (1990).

Degani, N., E. Ben-Hur, and E. Riklis: DNA damage and repair: induction and removal of thymine dimers in ultraviolet light irradiated intact water plants. Photochem. Photobiol. *31,* 31–36 (1980).

Fernbach, E. and H. Mohr: Photoreactivation of UV light effects on growth of Scots pine (*Pinus sylvestris* L.) seedlings. Trees *6,* 232–235 (1992).

Flint, S. D., P. W. Jordan, and M. M. Caldwell: Plant protective response to enhanced UV-B radiation under field conditions: Leaf optical properties and photosynthesis. Photochem. Photobiol. *41,* 95–99 (1985).

Jordan, B. R.: The molecular biology of plants exposed to ultraviolet-B radiation and the interaction with other stresses. In: Jackson, M. B. (ed.): Interacting stresses on plants in a changing climate. NATO ASI Series, pp. 153–170. Springer-Verlag, Berlin (1993).

– The effect of ultraviolet-B radiation on plants: a molecular prospective. In: Callow, J. A. (ed.): Advances in Botanical Research. Academic Press Ltd., in press (1995).

Jordan, B. R., H. J. Chow, and J. M. Anderson: Changes in mRNA levels and polypeptide subunits of ribulose 1,5-bisphosphate carboxylase in response to supplementary ultraviolet-B radiation. Plant Cell Environ. *15,* 91–98 (1992).

Jordan, B. R., P. E. James, A. Strid, and R. G. Anthony: The effect of ultraviolet-B radiation on gene expression and pigment composition in etiolated and green pea tissue: UV-B induced changes in gene expression are gene specific and dependent upon tissue development. Plant Cell Environ. *15,* 91–98 (1994).

Jordan, B. R., W. S. Chow, A. Strid, and J. M. Anderman: Reduction in *cab* and *psb A* RNA transcripts in response to supplementary ultraviolet-B radiation. FEBS Lett. *248,* 5–8 (1991).

Kramer, G. F., D. T. Krizek, and R. M. Mirecki: Influence of photosynthetically active radiation and spectral quality on UV-B induced polyamine accumulation in soybean. Phytochem. *31,* 1119–1125 (1992).

Krizeck, D. T.: Influence of ultraviolet radiation on germination and early seedling growth. Physiol. Plant. *34,* 182–186 (1975).

Mirecki, R. M. and A. H. Teramura: Effects of ultraviolet-B irradiance on soybean. Plant Physiol. *74,* 475–480 (1984).

Murali, N. S. and A. H. Teramura: Effects of ultraviolet-B irradiance on soybean. Influence of phosphorus nutrition on growth and flavonoid content. Plant Physiol. *63,* 413–416 (1985).

Pang, Q. and J. B. Hays: UV-B-inducible and temperature-sensitive photoreactivation of cyclobutane pyrimidine dimers in *Arabidopsis thaliana.* Plant Physiol. *95,* 536–543 (1991).

Robberecht, R. and M. M. Caldwell: Leaf optical properties of *Rumex patientia* L. and *Rumex obtusifolius* L. in regard to a protective mechanism against solar UV-B radiation injury. In: Worrest, R. C. and M. M. Caldwell (eds.): Stratospheric ozone reduction, solar ultraviolet radiation and plant life, pp. 251–259. Springer-Verlag, Berlin (1986).

Steinmitz, V. and E. Wellmann: The role of solar UV-B in growth regulation of cress (*Leidium sativum* L.) seedlings. Photochem. Photobiol. *43,* 193–198 (1986).

Strid, A.: Alteration in the expression of defence genes in *Pisum sativum* after exposure to supplementary UV-B radiation. Plant Cell Physiol. *34,* 949–953 (1993).

Sutherland, B. M.: Photoreactivating enzymes. In: Boyer, P. D. (ed.): The Enzymes. Academic Press, New York (1981).

Takeda, J., S. Abe, Y. Hirose, and Y. Ozeki: Effects of light and 2,4-dichlorophenoxyacetic acid on the level of mRNA for phenylalanine ammonia-lyase and chalcone synthase in carrot cell culture in suspension. Physiol. Plant. *89,* 4–10 (1993).

Taylor, R. M., O. Nikaid, B. R. Jordan, J. Rosamond, C. M. Bray, and A. K. Tobin: Ultraviolet-B-induced DNA lesions and their removal in wheat (*Triticum aestivum* L.) leaves. Plant, Cell Environ., in press (1995).

Teramura, A. H.: Effects of ultraviolet-B radiation on soybean. Physiol. Plant. *48,* 333–339 (1980).

Teramura, A. H., J. H. Sullivan, and L. H. Ziska: Interaction of elevated ultraviolet-B radiation and CO_2 on productivity and photosynthetic characteristics in wheat, rice and soybean. Plant Physiol. *94,* 470–475 (1990).

Tevini, M., J. Braun, and G. Gieser: The protective function of the epidermal layer of rye seedlings against ultraviolet-B radiation. Photochem. Photobiol. *53,* 329–333 (1991).

Van Oosten, J. J., D. Wilkins, and R. T. Besford: Regulation of the expression of photosynthetic nuclear genes by CO_2 is mimicked by regulation by carbohydrates: a mechanism for the acclimation of photosynthesis to high CO_2? Plant, Cell Environ. *11,* 913–923 (1994).

Van Oosten, J. J. and R. T. Besford: Sugar feeding mimics effect of acclimation to high CO_2-rapid down regulation of RUBISCO small subunit transcripts but not of the large subunit transcripts. J. Plant Physiol. *143,* 306–312 (1994).

Warner, C. W. and M. M. Caldwell: Influence of photon flux density in the 400–700 nm waveband on inhibition of photosynthesis by UV-B (280–320 nm) irradiation in soybean leaves: Separation of indirect and immediate effects. Photochem. Photobiol. *38,* 341–346 (1983).

Wellman, E.: UV radiation: definition, characteristics and general effects. In: Mohr, H. and W. Shropshire (eds.): Photomorphogenesis 16, pp. 745–756 (1983).

Withers, A. C., D. Vince-Prue, and B. Thomas: Identity of the photoreceptors for the perception of irradiance in photosynthetic light acclimation in tomato. Photochem. Photobiol. *54,* 451–457 (1991).

Ziska, L. H. and A. H. Teramura: CO_2 enhancement of growth and photosynthesis in rice (*Oryza sativa*). Plant Physiol. *99,* 473–481 (1992).

J. Plant Physiol. Vol. 148. pp. 107–114 (1996)

Scaling Plant Ultraviolet Spectral Responses from Laboratory Action Spectra to Field Spectral Weighting Factors

STEPHAN D. FLINT and MARTYN M. CALDWELL

Department of Rangeland Resources and the Ecology Center, Utah State University, Logan, UT 84322-5230 USA

Received July 15, 1995 · Accepted October 20, 1995

Summary

Biological spectral weighting functions (BSWF) play a key role in calculating the increase of biologically effective solar ultraviolet-B radiation (UV-B_{BE}) due to ozone reduction, assessing current latitudinal gradients of UV-B_{BE}, and comparing solar UV-B_{BE} with that from lamps and filters in plant experiments. Plant UV action spectra (usually determined with monochromatic radiation in the laboratory with exposure periods on the order of hours) are often used as BSWF. The realism of such spectra for plants growing day after day in polychromatic solar radiation in the field is questionable. We tested the widely used generalized plant action spectrum since preliminary data from an action spectrum being developed with monochromatic radiation for a cultivated oat variety indicate reasonable agreement with the generalized spectrum. These tests involved exposing plants to polychromatic radiation either from a high-pressure xenon arc lamp in growth chambers or in the field under solar radiation with supplemental UV-B lamps. Different broad-spectrum combinations were achieved by truncating the spectrum at successively longer UV wavelengths with various filters. In the growth chamber experiments, the generalized plant spectrum appeared to predict plant growth responses at short (<310 nm) wavelengths but not at longer wavelengths. The field experiment reinforced these conclusions, showing (in addition to the expected direct UV-B effects) both direct UV-A effects and UV-A mitigation of UV-B effects.

Key words: Avena sativa, Action spectrum, Biological spectral weighting function, spectral response, UV-A, UV-B.

Abbreviations: BSWF = biological spectral weighting function; O_3 = ozone; PFD = photosynthetic flux density (400–700 nm); RAF = radiation amplification factor; UV-A = radiation in the 320–400 nm waveband (sometimes considered to begin at 315 nm); UV-B = radiation in the 280–320 nm waveband; UV-B_{BE} = biologically effective UV (UV radiation weighted with a BSWF; may extend into the UV-A); UV-C = radiation <280 nm.

Introduction

Depletion of stratospheric ozone, and the consequent increase of solar ultraviolet-B (280–320 nm) radiation reaching the ground surface, continues to be a major concern for the biosphere. The Antarctic «ozone hole» has become larger and the depletion of ozone more severe (Herman et al., 1995). A general erosion of ozone at lower latitudes has also been well established (Gleason et al., 1993). The biological effects of increased solar UV-B are only significant if two conditions are met: Organisms must be sensitive to the increase in UV-B flux rates received, and the spectral sensitivity of these effects (i.e., the relative response to the different wavelengths) must have rather specific characteristics. Most research addressing plant effects deals with absolute sensitivity and very little with spectral response (Tevini and Teramura, 1989; Caldwell and Flint, 1994 a, b; Runeckles and Krupa, 1994).

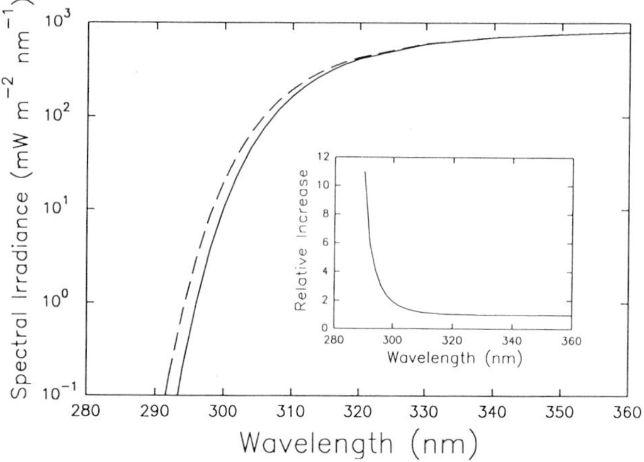

Fig. 1: Solar spectral irradiance (direct beam + diffuse) at noon at a temperate latitude (40° N) location in summer with normal (continuous line) and a 20 % reduction of the ozone column (dashed line). In the inset is the factor for relative increase of spectral irradiance at each wavelength due to the ozone column reduction. From Caldwell and Flint (1995).

Spectral responses are pivotal in assessing the biological consequences of ozone reduction. The absorption coefficient of ozone increases by orders of magnitude with decreasing wavelength (Molina and Molina, 1986). A reduction of the ozone layer results in a very specific increase in solar UV-B and only in a waveband of about 30 nm – between 290 and 320 nm. The relative enhancement of UV-B is also highly wavelength specific (Fig. 1). The quantity of additional radiation resulting from even an appreciable ozone depletion is trivial when considered against the background of total solar UV. Thus, this additional radiation is only important if the biological responses are much more sensitive to the shorter than to longer wavelength radiation. To incorporate the biological importance in expressing UV-B, a biological weighting function is convoluted with spectral irradiance at each wavelength as

$$UV\text{-}B_{BE} = \int I_\lambda \, E_\lambda \, d\lambda \qquad (1)$$

where UV-B_{BE} is the integrated biologically effective UV-B irradiance, I_λ is the spectral irradiance, and E_λ are biological spectral weighting functions (BSWF), representing the manner in which the plant UV response is influenced by wavelength (i.e., the spectral response – usually taken from an action spectrum). The UV-B_{BE} may include some longer wavelength radiation if E_λ extends beyond the UV-B. The increment of UV-B_{BE} resulting from ozone reduction, i.e., ΔUV-$B_{BE}/\Delta O_3$, is called the radiation amplification factor (RAF).

Because of the very specific manner in which solar UV-B changes with wavelength, BSWF in Eq 1 must have particular characteristics in order to result in an appreciable RAF. Only BSWF that decrease very sharply with increasing wavelength will result in an appreciable RAF with ozone reduction. Therefore, biological photoreactions that do not result in BSWF of this character should receive much less attention in the assessment of ozone reduction consequences. While the foregoing is well known and has been discussed in reviews (e.g., Caldwell et al., 1986, 1989; Coohill, 1991, 1992), it is not always fully appreciated.

The BSWF are not only important in assessing the phenomenon of ozone reduction directly, but are involved in several other aspects of assessing the relevance of plant response to UV-B. The magnitude of the natural latitudinal gradient, low at the poles and high at the equator (Caldwell et al., 1980), is totally dependent on the BSWF used to portray it (Caldwell and Flint, 1994 a). Similarly, BSWF are essential in experiments utilizing UV-B lamps since the lamps do not perfectly match sunlight (Caldwell et al., 1986; Caldwell and Flint, 1995).

The BSWF used to evaluate ozone depletion traditionally have been taken directly from analytical action spectra (as termed by Coohill, 1991) which are usually employed to elucidate photobiological mechanisms, and specifically, to identify potential chromophores. They are commonly developed by evaluating biological responsiveness to monochromatic irradiation. In order to identify potential chromophores, there has been an emphasis on the fine structure of action spectra and little attention has been directed to the tails of these spectra.

For use in the ozone reduction problem, the tails of BSWF at longer wavelengths can become very important if they do not decline markedly with increasing wavelength since sunlight increases by orders of magnitude at longer wavelengths. Apart from uncertainties about the nature of action spectra tails, analytical action spectra which are usually derived using monochromatic radiation in the laboratory over periods of hours may not predict the spectral response of plants exposed day after day to the full solar spectrum in the field. Interactions between UV-B and PFD (photosynthetic photon flux density, i.e., total photon flux in the 400–700 nm waveband) have been investigated over several years (Caldwell and Flint, 1994 a). Most of this work shows that at greater ratios of PFD/UV-B there is much less damage or effect of the UV-B than at lower PFD/UV-B ratios. Less study has been directed to the interaction of UV-B and UV-A, or all three wavebands, but there is also evidence that UV-A can ameliorate the effects of UV-B (Middleton and Teramura, 1994; Caldwell et al., 1994). While certain effects of these longer wavelengths are known, there is not a complete mechanistic understanding of how radiation at different wavelengths and at different flux densities interacts to cause biological effects, especially over periods of weeks and months. Thus, this must be addressed empirically.

To evaluate the appropriateness of the analytical action spectra to the ozone reduction problem, they must be linked to more realistic field conditions. This process, also known as scaling, uses experiments under a variety of conditions with polychromatic radiation. Plant response under these conditions is compared to the response that would be expected if the analytical action spectra were used to weight the irradiation in the experiment. When the weighted irradiance does not correspond with the actual plant response, the spectra are modified both with further laboratory study conducted with a combination of monochromatic and polychromatic radiation and again tested under field conditions. The ultimate goal is to determine BSWF that, under reasonable limitations, can aid in assessing the ozone reduction problem.

Here we present two experiments conducted in growth chambers with xenon arc irradiation where filters were used to progressively truncate the UV wavelengths from the lamp-emission spectrum. An additional UV-B sensitivity experiment was conducted in the field, with sunlight filtered in part of the experiment to remove UV-A. We used growth and morphology as characteristics of particular ecological interest, since such parameters are more consistently affected in UV-B experiments than are photosynthetic processes (Caldwell and Flint, 1994 b). Development of an action spectrum using traditional monochromatic radiation experimentation is also in progress and some preliminary data are shown for whole-plant responses.

Materials and Methods

Plant culture

Cultivated oat (*Avena sativa* L. cv. otana) was selected for these experiments because previous work showed this variety to be morphologically responsive to UV-B (Barnes et al., 1990). As in these previous experiments, seedlings were grown in 0.15-L conically shaped containers (Ray Leach Nursery, Canby OR) at an interplant spacing of 4.5 cm, producing a density of 494 plants m^{-2} for all experiments. Oats and many plant species typically occur in stands rather than in isolated conditions and the resulting canopy light climate has a profound influence on development and response to radiation (e.g., Smith, 1981), therefore, we have taken this approach to the culture and experimentation with these plants. Seed were planted directly into the potting medium and allowed to germinate in the light (growth chamber control conditions or ambient sunlight in the field experiment). More containers were planted than needed, so that uniform seedlings (<1 cm in height) could be selected and randomly assigned to the different treatments. Thus, these experiments only examined UV-B effects on seedling and young plant development, not potential UV-B effects on germination and emergence.

Spectral irradiance measurement

Spectral irradiance of the UV waveband in both the growth chamber and the field experiments was measured with a double-monochromator spectroradiometer (Optronic model 742, Orlando FL). Calibration procedures and modification of the instrument for field use are described in Caldwell et al. (1994).

The generalized plant action spectrum (Caldwell, 1971 as formulated by Green et al., 1974, normalized to 300 nm) was used as a provisional BSWF since it is widely used and preliminary data from an action spectrum using the same oat plants appears to correspond reasonably well with this spectrum. In the growth chamber experiments, this weighted spectral irradiance was used to predict the magnitude of the morphological changes that would be seen at the intermediate UV level. This was done using the measured effects seen in the treatments with shorter or longer wavelength cutoffs. In the field experiment, they were used to adjust the lamp output as described below.

Growth chamber trimming experiments

Growth chambers (Mallory Co., Salt Lake City UT) illuminated by water-cooled 6000-W xenon arc lamps (Atlas Electric, Chicago, IL) and filtered by different plastic films (Table 1) provided spectra trimmed of various shortwave portions of the UV-B or UV-B and UV-A. A 12-h photoperiod at 700 $\mu mol\ m^{-2}\ s^{-1}$ PFD produced a total daily irradiance of 30.2 mol m^{-2}. Plant position was rotated daily, and every second day, plants and appropriate filters were moved to a different chamber and PFD levels were checked and adjusted. Temperature was gradually changed from a maximum of 25 C during the photoperiod to a minimum of 12 C during the dark period. Humidity was not controlled.

Experiments were run for 14 d (short-wavelength trimming experiment) and 16 d (long-wavelength trimming experiment). At this time plants were harvested, morphological parameters were measured, and shoot material was oven dried. There were between 76 and 80 plants per treatment.

Table 1: Treatments, weighted UV-B, radiation sources and filters used in the different experiments.

Treatment	Weighted UV-B kJ m^{-2} d^{-1}	Radiation sources and filters	
		Short wavelength trimming experiment	
		Filtering xenon arc lamp	
I[3]	5.6	1 sheet 0.13-mm cellulose diacetate	
II	3.2	2 sheets 0.13-mm cellulose diacetate	
III	0	1 sheet 0.13-mm clear polyester	
		Long wavelength trimming experiment	
		Filtering xenon arc lamp	
IV	5.2	1 sheet 0.13-mm cellulose diacetate	
V	0.04	1 sheet 0.025-mm clear polyester	
VI	0	2 sheets 0.05-mm Llumar[1] plus 0.025-mm clear polyester	
		Field trimming experiment	
		Filtering fluorescent lamps	Filtering sunlight
+UV-B +UV-A	6.8	0.13-mm cellulose triacetate	0.13-mm clear polyester
−UV-B +UV-A	0.1	0.13-mm clear polyester	0.13-mm clear polyester
+UV-B −UV-A	6.8	0.13-mm cellulose triacetate	0.13-mm Llumar[2]
−UV-B −UV-A	0.1	0.13-mm clear polyester	0.13-mm Llumar[2]

[1] Llumar, type SCL SR PS2, Martin Energy Products, Martinsville, VA.
[2] Llumar weatherized polyester, Norton Performance Plastics, Wayne, NJ.
[3] Corresponding spectral irradiances are shown in Fig. 3.

Field trimming experiment

This experiment evaluated plant response to UV-B under two contrasting UV-A environments: one with near-ambient solar UV-A, and the other with much of the background of solar UV-A removed. It was run in the field from 13 August to 8 September 1994 near Logan, Utah, USA (41.5° N latitude, 1460 m elevation). Racks of 6 UV-B fluorescent lamps (Model UV-B-313, Q Panel Co, Cleveland OH) were operated by an irradiation control system that tracks ambient UV-B radiation and provides proportional lamp output corresponding with current ambient solar radiation conditions (Caldwell et al., 1983). Lamp output was adjusted to provide the equivalent of a 30 % ozone depletion in the +UV-B treatment for this time of year (as described in Caldwell et al. (1994)).

Lamps were filtered to remove UV-C and short wavelength UV-B and to provide treatment and control conditions (Table 1). In addition, a filter covering the whole plot was used to remove UV-A from one of the treatments (Table 1). To avoid creating different microclimates, it was necessary to have a similar cover over the +UV-A plots. Because of this, all UV-B in this experiment was provided by the lamps.

Four plots were used for this experiment, and each was divided in half with a polyester barrier so that half the plot could be used for +UV-B treatment and half for control (no UV-B) conditions. Thus, there were two subplots (each with 48 plants) for each set of experimental conditions shown in Table 1. All plants were rotated within subplots every two days and a portion were also rotated between the two equivalent subplots.

Morphological measurements were made throughout the experiment. At the final harvest the shoot material was oven dried.

Results

Action spectrum

The generalized plant action spectrum has been used as a provisional BSWF to weight spectral irradiance in the trimming experiments reported here. Data from a preliminary action spectrum for inhibition of growth in oat seedlings (Fig. 2) correspond rather well with the generalized spectrum. This action spectrum is currently being developed in our laboratory using a large double monochromator with water-filled quartz-window prisms (modified from Fluke and Setlow, 1954, Ballaré et al., 1995) to irradiate light-grown seedlings. The inhibition of the insertion height of the first leaf, an index for internode elongation (10 d after UV exposure) was the parameter used in development of this action spectrum. Further work, especially at longer wavelengths, is needed.

Growth chamber experiments

The short-wavelength trimming experiment showed a close agreement between the observed UV-B effects on growth and morphological parameters with predictions from the generalized plant action spectrum for these polychromatic radiation conditions (Fig. 3, top). Thus, the generalized plant action spectrum reasonably describes the plant responses to the shorter wavelengths (i.e., <310 nm) used in this experiment. Note that not all the responses should be necessarily interpreted as damage; there is a UV-B enhancement of leaf number in both growth chamber trimming experiments.

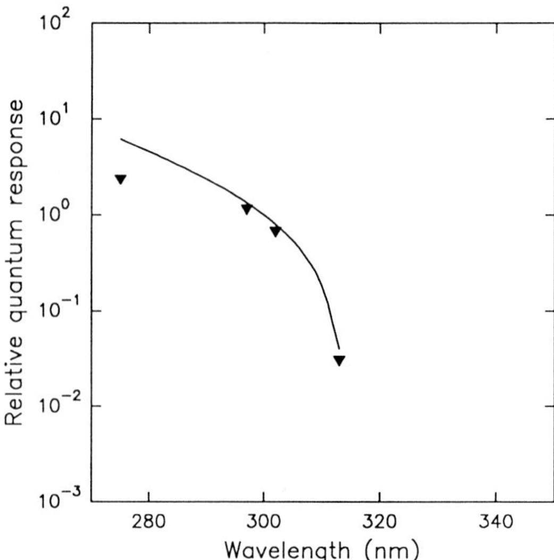

Fig. 2: The widely-used generalized plant action spectrum (Caldwell, 1971) (continuous line) and preliminary data from an action spectrum for the inhibition of elongation of the first internode of oat seedlings (triangles).

The long-wavelength trimming experiment failed to show such good agreement between this action spectrum and observed plant responses (Fig. 3, bottom). The generalized plant action spectrum predicts little response to the radiation between 310 and 350 nm, yet appreciable plant response to radiation in this region was observed.

In addition to the parameters shown in Fig. 3, total main-shoot leaf length decreased with increasing UV. This decrease showed a pattern similar to the other morphological data in Fig. 3: experimental values were close to those predicted by the generalized plant spectrum in the short-wavelength trimming experiment but not in the long-wavelength trimming experiment (data not shown).

Shoot mass did not respond in a consistent pattern in these two experiments. Total shoot mass showed a significant decrease with UV in the short-wavelength trimming experiment but not the long-wavelength trimming experiment (data not shown).

Field experiment

The field experiment was designed to provide a substantial UV-B treatment (simulating a 30 % ozone depletion) both with near-ambient solar UV-A present (polyester cover over the lamp rack) or with much of the UV-A removed (Llumar film over the lamp rack). Thus, there was both a UV-B treatment and a control (no UV-B) in each of the two UV-A conditions (Fig. 4).

The oat seedlings showed an early response to the treatments, and responded to both UV-B and UV-A. A typical morphological response is shown in Fig. 5 for the insertion height of the first leaf. (This is the same parameter used to develop the preliminary action spectrum in Fig. 2.) Both UV-B and UV-A had significant effects and there was a signifi-

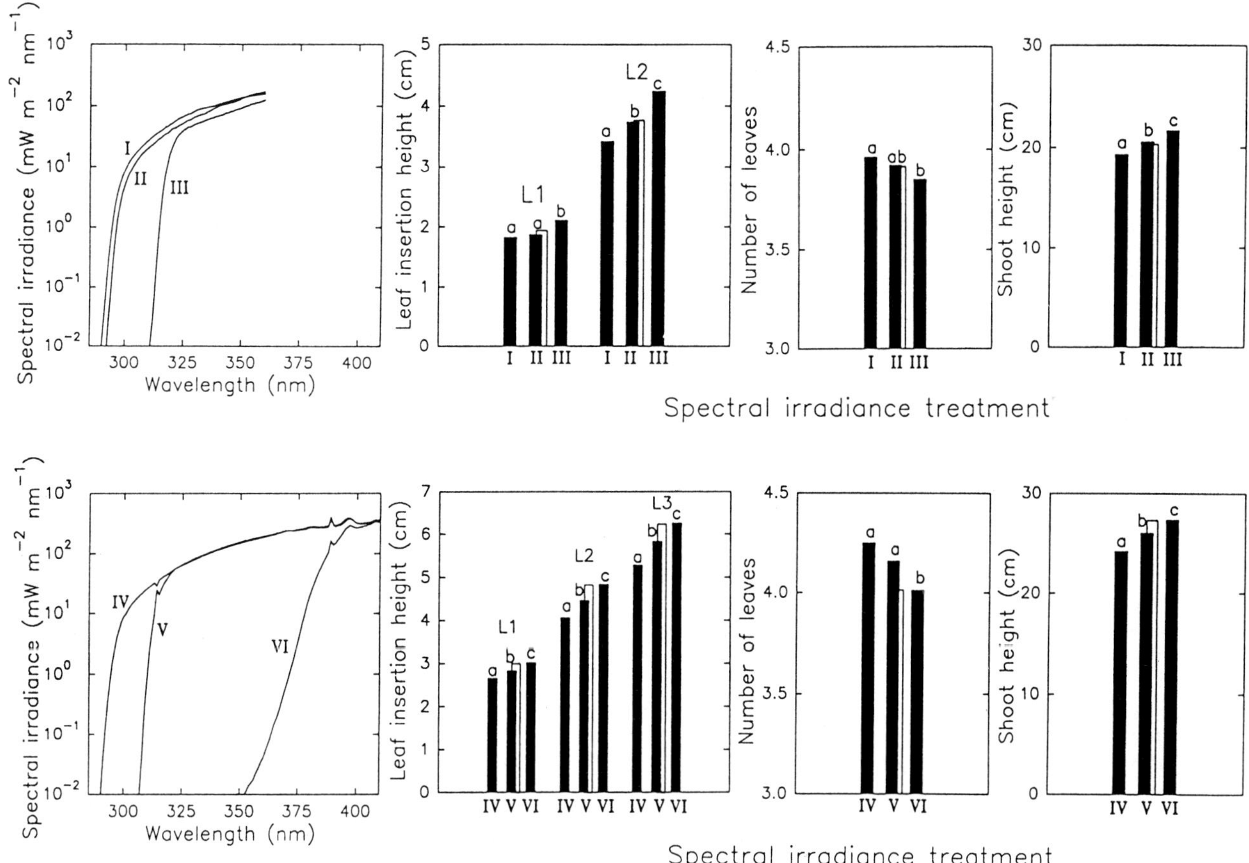

Fig. 3: Morphological responses in the growth chamber trimming experiments. The UV spectral irradiance conditions are portrayed at the left side of the figure. Internode length, (represented as height of leaf insertion; L1, L2, and L3 are the heights of leaves 1, 2, and 3, respectively), number of leaves per plant, and total shoot height are presented to the right. Roman numerals indicate the spectral irradiance treatment. Solid bars indicate the measured data, and the open bar represents the value that would be expected in the intermediate UV level based on weighting the different spectral combinations with the generalized plant action spectrum (see Methods). For each measured parameter, solid bars marked with a different letter are significantly different at $P<0.05$ in a Duncan's new multiple range test. Each solid bar represents the mean of 76–80 plants.

cant UV-A X UV-B interaction. The UV-A strongly ameliorated the UV-B reduction of internode elongation and yet UV-A had a strong depressing effect on internode length in the absence of UV-B. This type of response, with the significant interaction between UV-B and UV-A and with at least one significant main effect, was seen for other measurements such as early (9 d) measurement of leaf 1 insertion height, shoot height (at 5 d and 9 d) and leaf lengths (at 13 d and 18 d). During the first part of the experiment, UV-A or UV-B main effects were seldom seen without a significant interaction.

Another common response, seen generally at the final harvest (26 d) showed both significant UV-A and UV-B effects but without a significant interaction. Both UV-B and UV-A reduced leaf area in an additive manner (Fig. 6). This was also seen with mass of individual leaves and whole-plant leaf mass and in the insertion height of the third leaf (at 18 d). However, total shoot mass exhibited no significant change due to either UV-B or UV-A (data not shown).

Discussion

Development of BSWF that reflect plant spectral responses under realistic field conditions is essential to an assessment of the ozone reduction problem. We are taking the approach of scaling action spectra developed with monochromatic radiation to more realistic radiation conditions and time periods.

The growth chamber trimming experiments at shorter wavelengths (Fig. 3 top), show that the provisional action spectrum (the generalized plant spectrum) reasonably predicts plant response under these particular conditions. However, the chamber trimming experiments at longer wavelengths (Fig. 3 bottom), indicated that wavelengths longer than ca. 310 nm are more effective than would be predicted by the provisional spectrum. When this generalized spectrum was originally compiled from available plant action spectra, most of these spectra indicated no response beyond 313 nm (an emission line of mercury lamps commonly used in action spectrum development). Thus, this generalized spectrum abruptly truncates at this point which in reality is unlikely. Both the long-wavelength chamber trimming experiment and the field experiment indicate response beyond this point. Preliminary data from the action spectrum for oat seedling internode elongation (Fig. 2) appear to fit the generalized plant action spectrum, but further development at longer wavelengths may indicate a tail in this region. (Another commonly used composite action spectrum in the ozone reduction problem, the DNA damage spectrum of Setlow (1974), declines more steeply with increasing wavelength than the generalized plant spectrum and would be less successful in fitting the response from the growth chamber and field experiments.)

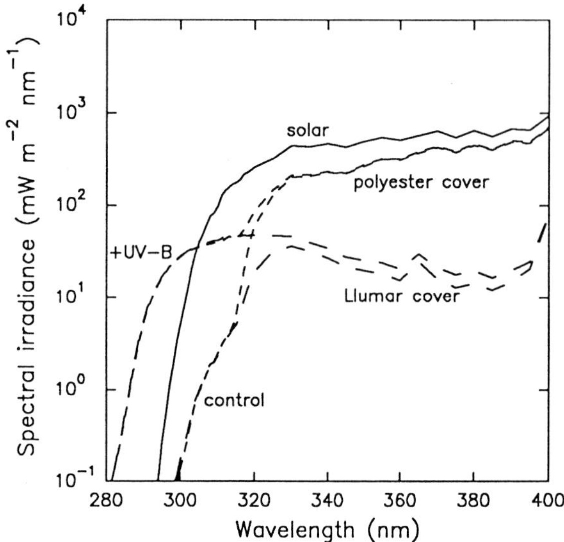

Fig. 4: Noontime clear-sky spectral UV irradiance in the field experiment. The four different UV-B and UV-A combinations are represented along with ambient solar UV.

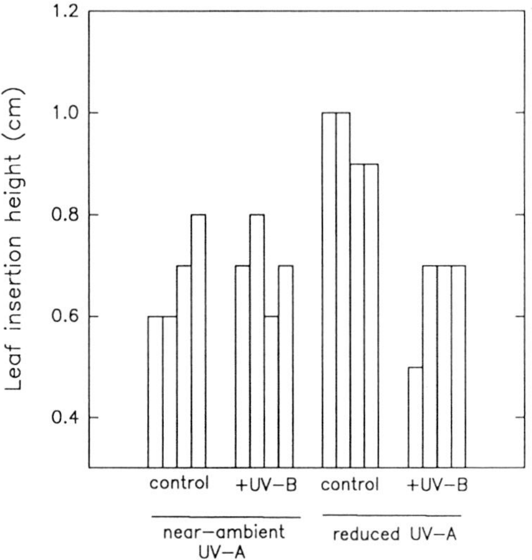

Fig. 5: Internode length, represented as height of insertion for first leaf, under the different field radiation combinations shown in Fig. 4. Each bar represents an individual replicate plant group (a rack with 24 plants per group) within a particular radiation combination. Using a two-factor ANOVA, there was a significant UV-B X UV-A treatment interaction (P = 0.01) along with significant main effects UV-A, P = 0.02; UV-B, P = 0.004). Plants had been treated 13 days under these UV conditions.

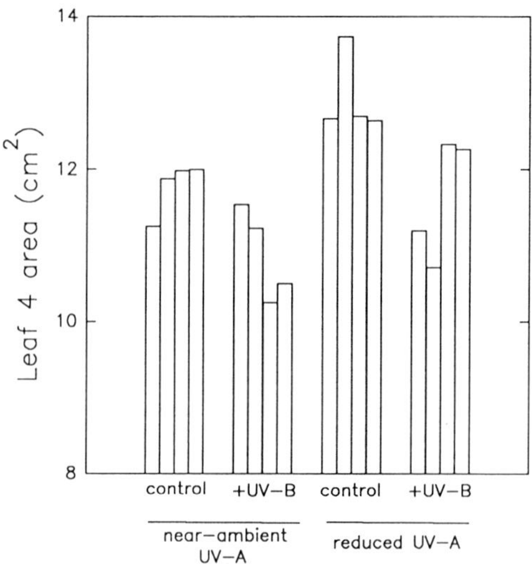

Fig. 6: Leaf 4 area under the different field radiation combinations shown in Fig. 4. Each bar represents an individual replicate plant group (a rack with 24 plants per group) within a particular radiation combination. Using a two-factor ANOVA, there were significant main effects (UV-A, P = 0.008; UV-B, P = 0.003) but no significant UV-A X UV-B interaction (P = 0.51). Plants had been treated 26 days under these UV conditions.

Action spectra developed with monochromatic radiation may not accurately predict the spectral response for plants in polychromatic radiation such as sunlight. Radiation absorbed by chromophores at longer wavelengths may not only diminish the response to UV-B, but also change the relative spectral response. This is a central theme being tested in our studies. Our field experiment (Fig. 5, 6) indicated both a direct effect of UV-A and a modification of UV-B effects by the UV-A environment. The direct effects of UV-A on plant growth suggest that the appropriate BSWF would have a more pronounced tail at longer wavelengths than the generalized plant spectrum (as was also suggested by the growth chamber trimming experiment at longer wavelengths). On the other hand, the amelioration of UV-B effects by UV-A may also influence the steepness of the BSWF. If UV-A reduces the effectiveness of UV-B, it may steepen the BSWF over at least a portion of the wavelength range. The net effect of the direct response to UV-A and the mitigation of UV-B by UV-A and how these affect the slope of the BSWF are being explored in continuing studies.

Action spectrum development with polychromatic radiation (a background of longer wavelength radiation with the sequential addition of increments of shorter wavelength radiation) increases realism but has some drawbacks. Several possible spectra can emerge from the same data set in the deconvolution process, and the tails of the spectra at longer wavelengths can be highly variable (Rundel, 1983). Furthermore, these are still usually developed over short periods of time with high flux densities of radiation. Attempts to develop action spectra in the field environment using various lamp and filter combinations are apparently meeting with some success (Holmes, personal communication).

Studies with marine and freshwater phytoplankton have concentrated on photosynthetic inhibition normally assayed as ^{14}C incorporation. While these indicate that UV-B has greater photon effectiveness than UV-A, both wavebands are responsible for inhibition (e.g., Smith, 1993; Kim and Watanabe, 1993). However, in addition to a UV-A inhibition of photosynthesis, Smith et al. (1992) found that UV-A as well as PFD were effective in mitigating UV-B photosynthetic inhibition. Action spectra for marine phytoplankton photosynthetic inhibition developed with polychromatic radiation indicate tails into the UV-A waveband (e.g., Smith, 1993; Cullen and Neale, 1994). For higher plants, UV inhibition of photosynthesis has been shown under laboratory and glasshouse conditions, but it is seldom evident in the field (Caldwell and Flint, 1994b). Action spectra have been recently developed using DNA cyclobutane dimer production (Quaite et al., 1992) and ultraweak luminescence (an indicator of free radical formation) (Cen and Björn, 1994). However, the link between these molecular and physiological indicators and whole-plant performance in the field has not been established. Thus, in the present study we have concentrated on morphological changes in response to UV. Shoot yield was not influenced by UV in our field experiment with oat which corresponds with other studies for several species indicating morphological response to UV-B to be more apparent than effects on yield (Barnes et al., 1990, 1993).

Although plankton photosynthesis is readily depressed by UV-A, and higher plants show some responses (Caldwell, 1984; Kartusch and Mittendorfer, 1990; Panagopoulos et al., 1990; Fernbach and Mohr, 1992), they generally have not been observed in full unobstructed sunlight. There are also several reports that UV-A in addition to PFD can mitigate effects of UV-B (e.g., Iwanzik, 1986; Caldwell et al., 1994), but the net effect of different wavebands of UV-A on realistic BSWF remains to be resolved.

Refinement of BSWF is also needed at shorter wavelengths in the UV-B waveband (<305 nm) since the steep rise with decreasing wavelength has a sizeable effect on how BSWF translate into radiation amplification factors and in comparing sunlight with lamp UV. The effective bandwidth employed in the development of many action spectra (especially with monochromatic radiation) is often rather broad (e.g., 10 nm); thus, the spectral resolution reported in an action spectrum appears to be greater than is justified. Shortwave trimming experiments with small waveband increments between successive filters (as in Fig. 3) should help to resolve the appropriate BSWF.

Resolving both the steepness of BSWF within the UV-B waveband and the nature of the tails into the UV-A are important goals in the scaling exercise. Because of the effort involved, these studies must concentrate on a single species and a limited suite of characteristics. While it would be unreasonable to expect that even a realistic BSWF developed for one species and one plant property be applicable to other species or processes, some simplifying rules may emerge from the present work that will allow translation of BSWF from one species or process to another.

Acknowledgements

Financial support of this research was provided by the United States Department of Agriculture (CSRS/NRICG grants 92-37100-7630 and 95-37100-1612) and the Utah Agricultural Experiment Station. We gratefully acknowledge the research assistance of Tim Hudelson, Pisal Sujcha, Javad Torabinejad and Jason West. Charles Ashurst provided electronic engineering support for this work. Peter Searles assisted in data analysis and provided helpful comments on an earlier draft.

References

Ballaré, C. L., P. W. Barnes, and S. D. Flint: Inhibition of hypocotyl elongation by ultraviolet-B radiation in de-etiolating tomato seedlings. I. The photoreceptor. Physiologia Plantarum *93*, 584–592 (1995).

Barnes, P. W., S. D. Flint, and M. M. Caldwell: Morphological responses of crop and weed species of different growth forms to ultraviolet-B radiation. American J. Botany *77*, 1354–1360 (1990).

Barnes, P. W., S. Maggard, S. R. Holman, and B. S. Vergara: Intraspecific variation in sensitivity to UV-B radiation in rice. Crop Science *33*, 1041–1046 (1993).

Caldwell, M. M.: Solar ultraviolet radiation and the growth and development of higher plants. In: Giese, A. C. (ed.), Photophysiology, pp. 131–177. Volume *6*, Academic Press, New York (1971).

– Effects of UV radiation on plants in the transition region to blue light. In: Senger, H. (ed.): Blue light effects in biological systems, pp. 20–28. Springer-Verlag, Berlin (1984).

Caldwell, M. M., L. B. Camp, C. W. Warner, and S. D. Flint: Action spectra and their key role in assessing biological consequences of solar UV-B radiation change. In: Worrest, R. C. and M. M. Caldwell (eds.): Stratospheric ozone reduction, solar ultraviolet radiation and plant life, pp. 87–111. Springer, Berlin (1986).

Caldwell, M. M. and S. D. Flint: Solar ultraviolet radiation and ozone layer change: Implications for crop plants. In: Boote, K., J. M. Bennett, T. R. Sinclair, and G. M. Paulsen (eds.): Physiology and determination of crop yield, pp. 487–507. ASA-CSSA-SSSA, Madison, WI (1994a).

– – Stratospheric ozone reduction, solar UV-B radiation and terrestrial ecosystems. Climatic Change *28*, 375–394 (1994b).

– – Lighting considerations in controlled environments for nonphotosynthetic plant responses to blue and ultraviolet radiation. In: Tibbitts, T. W. (ed.): Light in controlled environments. National Aeronautics and Space Administration Contractor Report, pp. 113–124. NASA, Ames Research Station, Moffett Field, CA (1995).

Caldwell, M. M., S. D. Flint, and P. S. Searles: Spectral balance and UV-B sensitivity of soybean: a field experiment. Plant, Cell and Environment *17*, 267–276 (1994).

Caldwell, M. M., W. G. Gold, G. Harris, and C. W. Ashurst: A modulated lamp system for solar UV-B (280–320 nm) supplementation studies in the field. Photochemistry and Photobiology *37*, 479–485 (1983).

Caldwell, M. M., R. Robberecht, and W. D. Billings: A steep latitudinal gradient of solar ultraviolet-B radiation in the arctic-alpine life zone. Ecology *61*, 600–611 (1980).

Caldwell, M. M., A. H. Teramura, and M. Tevini: The changing solar ultraviolet climate and the ecological consequences for higher plants. Trends in Ecology and Evolution *4,* 363–367 (1989).

Cen, Y. P. and L. O. Björn: Action spectra for enhancement of ultraweak luminescence by UV radiation (270–340 nm) in leaves of *Brassica napus.* Journal of Photochemistry and Photobiology B: Biology *22,* 125–129 (1994).

Coohill, T. P.: Action spectra again? Photochemistry and Photobiology *54,* 859–870 (1991).

– Action spectra revisited. Journal of Photochemistry and Photobiology B: Biology *13,* 95–100 (1992).

Cullen, J. J. and P. J. Neale: Ultraviolet radiation, ozone depletion, and marine photosynthesis. Photosynthesis Research *39,* 303–320 (1994).

Fernbach, E. and H. Mohr: Photoreactivation of the UV light effects on growth of Scots pine (*Pinus sylvestris* L.) seedlings. Trees *6,* 232–235 (1992).

Fluke, D. J. and R. B. Setlow: Water-prism ultraviolet monochromators. Journal of the Optical Society of America *44,* 327–330 (1954).

Gleason, J. F., P. K. Bhartia, J. R. Herman, R. McPeters, P. Newman, R. S. Stolarski, L. Flynn, G. Labow, D. Larko, C. Seftor, C. Wellemeyer, W. D. Komhyr, A. J. Miller, and W. Planet: Record low global ozone in 1992. Science *260,* 523–526 (1993).

Green, A. E. S., T. Sawada, and E. P. Shettle: The middle ultraviolet reaching the ground. Photochemistry and Photobiology *19,* 251–259 (1974).

Herman, J. R., P. A. Newman, R. McPeters, A. J. Krueger, P. K. Bhartia, C. J. Seftor, O. Torres, G. Jaross, R. P. Cebula, D. Larko, and C. Wellemeyer: Meteor 3/total ozone mapping spectrometer observations of the 1993 ozone hole. Journal of Geophysical Research D *100,* 2973–2983 (1995).

Iwanzik, W.: Interaction of UV-A, UV-B and visible radiation on growth, composition, and photosynthetic activity in radish seedlings. In: Worrest, R. C. and M. M. Caldwell (eds.): Stratospheric ozone reduction, solar ultraviolet radiation and plant life, pp. 287–301. Springer-Verlag, Berlin (1986).

Kartusch, R. and B. Mittendorfer: Ultraviolet radiation increases nicotine production in *Nicotiana* callus cultures. J. Plant Physiology *136,* 110–114 (1990).

Kim, D. S. and Y. Watanabe: The effect of long wave ultraviolet radiation (UV-A) on the photosynthetic activity of natural population of freshwater phytoplankton. Ecological Research *8,* 225–234 (1993).

Middleton, E. M. and A. H. Teramura: Understanding photosynthesis, pigment and growth responses induced by UV-B and UV-A irradiances. Photochemistry and Photobiology *60,* 38–45 (1994).

Molina, L. T. and M. J. Molina: Absolute absorption cross sections of ozone in the 185- to 350-nm wavelength range. Journal of Geophysical Research D *91,* 14, 501–14, 508 (1986).

Panagopoulos, I., J. F. Bornman, and L. O. Björn: Effects of ultraviolet radiation and visible light on growth, fluorescence induction, ultraweak luminescence and peroxidase activity in sugar beet plants. Journal of Photochemistry and Photobiology B: Biology *8,* 73–87 (1990).

Quaite, F. E., B. M. Sutherland, and J. C. Sutherland: Action spectrum for DNA damage in alfalfa lowers predicted impact of ozone depletion. Nature *358,* 576–578 (1992).

Rundel, R. D.: Action spectra and estimation of biologically effective UV radiation. Physiologia Plantarum *58,* 360–366 (1983).

Runeckles, V. C. and S. V. Krupa: The impact of UV-B radiation and ozone on terrestrial vegetation. Environmental Pollution *83,* 191–213 (1994).

Setlow, R. B.: The wavelengths in sunlight effective in producing skin cancer: a theoretical analysis. Proceedings of the National Academy of Science USA *71,* 3363–3366 (1974).

Smith, H. (ed.): Plants and the daylight spectrum. Academic Press, London (1981).

Smith, R. C., B. B. Prezelin, K. S. Baker, R. R. Bidigare, N. P. Boucher, T. Coley, D. Karentz, S. Macintyre, H. A. Matlick, D. Menzies, M. Ondrusek, Z. Wan, and K. J. Waters: Ozone depletion: ultraviolet radiation and phytoplankton biology in Antarctic waters. Science *255,* 952–959 (1992).

Smith, R. C.: Implications of increased solar UV-B for aquatic ecosystems. In: Chanin, M.-L. (ed.): The role of the stratosphere in global change, pp. 473–493. Springer, Berlin (1993).

Tevini, M. and A. H. Teramura: UV-B effects on terrestrial plants. Photochemistry and Photobiology *50,* 479–487 (1989).

J. Plant Physiol. Vol. 148. pp. 115–119 (1996)

Spectral Shaping of Artificial UV-B Irradiation for Vegetation Stress Research

Thorsten Döhring, Matthias Köfferlein, Stephan Thiel, and Harald K. Seidlitz

GSF-Forschungszentrum für Umwelt und Gesundheit GmbH, Expositionskammern, D-85758 Oberschleissheim, Germany

Received July 20, 1995 · Accepted October 20, 1995

Summary

Ecological plant experiments using artificial light sources require careful shaping of the spectral irradiance. This includes the steep UV-absorption characteristics resulting from the filtering of solar radiation by atmospheric ozone. Borosilicate and soda-lime glass filters screen radiation very similarly to ozone. They have a high mechanical stability and are available in large filter sheets and are, therefore, suited for the simulation of future scenarios of enhanced solar UV-B radiation in large scale vegetation stress experiments. Although such filters meet many requirements of light engineering, there are limitations due to the slope of the UV-edge and due to solarisation effects. Thus, the interpretation of the artifical radiation spectra and their comparison to UV scenarios of decreasing stratospheric ozone need careful discussion. Different methods to classify spectra of artificial UV-radiation are presented, and a new classification by a cut-off wavelength of the UV-edge and its slope is introduced.

Key words: Ultraviolet-B, filter materials, global radiation, global change, vegetation stress, sun simulator, ozone absorption.

Abbreviations: DU = Dobson units; PAR = photosynthetic active radiation (400–700 nm); UV-A = ultraviolet-A radiation (320–400 nm); UV-B = ultraviolet-B radiation (280–320 nm); UV-C = ultraviolet-C radiation (200–280 nm).

Introduction

Stratospheric ozone attenuates the solar spectral irradiance in the UV-B range by several orders of magnitude within a narrow wavelength range. Measurements during the last few years indicate that a depletion of the stratospheric ozone layer occurred (Ambach and Blumthaler 1994). A concomitant increase of UV-B irradiation can be expected, which may cause radiation stress to plants, animals, and human beings.

The response of plants to UV-B radiation has been the subject of a large number of studies, many of them are carried out using artificial UV-sources. In case these experiments are aimed to be of ecological relevance, careful shaping of the spectral irradiation according to natural solar radiation is necessary. Critical points are a realistic balance of UV-B, UV-A, and PAR (Caldwell et al. 1994), and the steep UV cut-off caused by stratospheric ozone. A flexible variation of the cut-off wavelength is especially necessary for UV-B stress research. There are different experimental approaches to establish enhanced UV-B scenarios for vegetation stress studies:

- Naturally enhanced global radiation in comparison to Central European lowland conditions (control), is available at lower latitudes or at high elevation. Controls can be obtained by suitable filtering of the excess radiation (Tevini et al. 1989).
- In various field experiments global UV-B radiation is artificially increased by supplementing radiation from fluorescent lamps (Caldwell et al. 1994).
- In closed chamber systems using artificial plant lighting UV-B can be supplied independently from visible light (Seckmeyer and Payer 1993).

Special measures have to be taken in these experiments to simulate high levels of UV-B irradiance without allowing the UV-B absorption edge of the spectrum to shift to wavelengths below the limits set by the absorption of stratospheric ozone. The employment of such filters in large-scale phytotrons imposes specific technical requirements:

(i) A steep cut-off at the desired UV-B wavelength and low losses in the remaining range of the spectrum, i.e. in the UV-A, visible and infrared spectral region.
(ii) A high temporal stability of the spectral transmission characteristics.
(iii) A high mechanical stability.
(iv) An availability of large filter sheets, if several square meters of experimental area are to be covered.

The requirements mentioned in (ii) and (iii) ask for special attention, as the filter material is exposed to a nearly extraterrestrial UV-radiation environment. The filters have to withstand a high-level UV-B irradiation of approx. 30 W m^{-2} within a phytotron.

It is the aim of this paper to discuss the physical and technical requirements of UV-filtering for ecological plant research and to provide new criteria for evaluating the spectral quality of the UV-edge of artificial lighting systems.

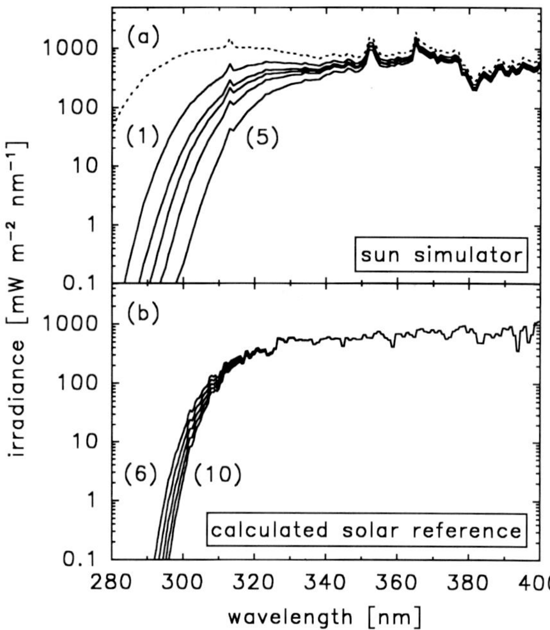

Fig. 1 (a): UV-spectra of the small GSF sun simulator using different combinations of glass filters. (1) 1 layer of Tempax batch 2, (2) 2 layers of Sanalux, (3) 2 layers of Tempax batch 2, (4) 2 layers of Tempax batch 3, (5) 3 layers of Tempax batch 3, (Dashed line: spectrum of sun simulator without filtering).

Fig. 1 (b): Calculated reference spectra of global irradiation for different values of the total stratospheric ozone column, (6) 240 DU, (7) 280 DU, (8) 320 DU, (9) 360 DU, (10) 400 DU.

Materials and Methods

The GSF-Research Center of Environment and Health operates a phytotron consisting of seven sun simulators of various size (Thiel et al. 1996, this volume). Effects on plants due to irradiation, climate, air pollutant stress, and the combination of these parameters can be studied. The technical arrangements of these closed chamber systems guarantee reproducible experimental conditions within a wide range. A combination of different lamps simulates the visible light, and UV-B fluorescent lamps provide the UV-B part of the radiation (Seckmeyer and Payer 1993, Payer et al. 1994). The spectral composition of the irradiation within the sun simulators is determined by spectroradiometric measurements using a double monochromator system (Bentham, Reading, U.K.). The details of the measurement system are described by Thiel et al. (1996, this volume).

In the sun simulators different batches of borosilicate glass sheets (5 mm Tempax®, manufactured by Schott, Mainz, Germany) and soda-lime glass sheets (4 mm Sanalux®, manufactured by DESAG, Grünenplan, Germany) are used as UV-blocking filters. Ageing of filter materials was studied by exposing samples in the lamphouse of a sun simulator with a UV-B irradiation of approx. 30 W m^{-2}. The spectral irradiance during exposure is shown in figure 1 (a) (dashed line). The normal incidence transmittance of the filters was measured after different exposure times by use of a spectrophotometer (Biochrom 4060, Pharmacia) at a wavelength resolution of 1 nm.

Results

Borosilicate and soda-lime glass sheets provide a steep cut-off at approximately 300 nm. The filter transmission varies from batch to batch (data not shown). Therefore, the filters have to be selected for each experiment. Figure 1 (a) presents the spectral irradiance obtained by various combinations of glass sheets. The maximum variation of radiation spectra ranges between the unfiltered lamp spectrum (dashed line) and a total blocking of ultraviolet radiation below 400 nm.

The experimental area of a sun simulator can be divided in up to four compartments equipped with different filters, which allow to provide different UV-B irradiation scenarios simultaneously. Each sector is covered with the same number of filter layers, as each additional layer reduces the radiation intensity in the whole spectral range by approx. 8 % due to reflection losses. We usually employ a combination of two filter layers.

The engineering quality of the atmospheric UV absorption by this glass filter technique is sufficient for most plant experiments. However, this method has limitations. The strong UV exposure cause rapid ageing of borosilicate glass filters as seen in Fig. 2. This is due to a physico-chemical effect known as solarisation and originates in UV-induced changes in the oxidation state of iron contaminations present in the glass matrix (Vogel 1992). The oxidation of Fe^{2+} to Fe^{3+} is accompanied by an absorption shift to longer wavelengths within the UV range (Fig. 2).

The temporal progress of the decrease of the glass transmittance at 300 nm due to solarisation is shown in figure 3. A big change occurs in fresh glasses within the first few hours, followed by a slow long-term decrease. Hence, transmission changes during the plant experiment can mostly be avoided by pre-ageing of new filters. In addition, there is a gradient of ageing with the depth of the glass (data not shown).

Soda-lime glass (Sanalux®) exhibits a very much reduced ageing compared to borosilicate glass (Tempax®). The solarisation of the soda-lime glass ceases after a few hours of UV treatment. This may be due to the lower content of iron con-

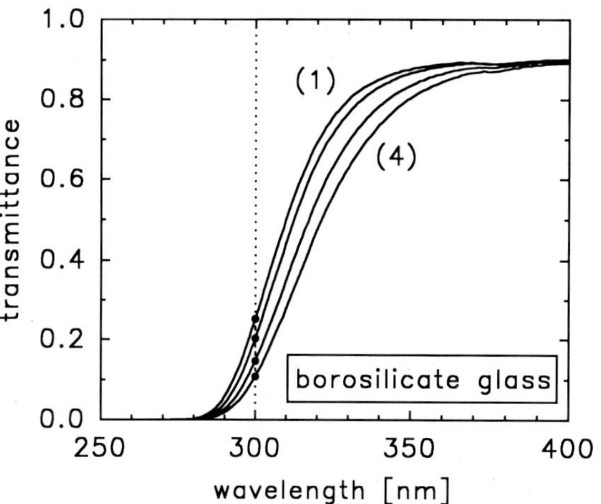

Fig. 2: Temporal change in spectral transmittance of borosilicate glass (Tempax batch 1) during UV-B treatment after (1) 0 h, (2) 10 h, (3) 130 h, (4) 440 h duration of exposure.

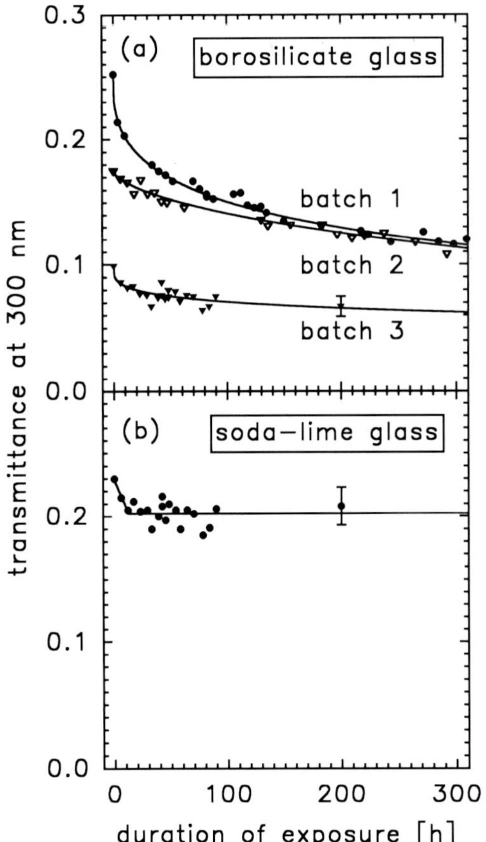

Fig. 3: Temporal change of transmittance at 300 nm during UV-B treatment (solid lines: fitted curves) (a) borosilicate glass of different batches. (b) soda-lime glass.

taminants. At present, this type of soda-lime glass is available only with a thickness of 4 mm. The ageing speed of a borosilicate filter layer can be drastically reduced by using a protecting soda-lime glass sheet on top. Spectral shifts caused by ageing of the material during an experiment are considered when evaluating the biological data.

Discussion

Various materials are currently used for UV-B filtering, but their application for plant experiments is often restricted by technical limitations. As stated in the introduction, the filters require (i) a steep cut-off, (ii) a high optical stability, (iii) a high mechanical stability, and (iv) must be available in large sheets. For choosing the optimum UV-filter of a sun simulator, the characteristics of polymer films, colour glass filters, borosilicate glass, and soda-lime glass are compared.

Polymers like cellulose acetate foils, mylar foils, acrylic glass or others are often employed for UV-filtering (Caldwell et al. 1994). They comply quite well with requirements (i) and (iv) but not with (ii) and (iii). Under strong exposure polymer foils can only be used for a few hours or days. At extreme conditions of nearly «extraterrestrial» UV radiation inside the lamp house of a phytotron these materials deteriorate very rapidly and must be changed frequently. In addition, the stability of polymer filters is reduced by mechanical, chemical, and temperature stress. Polymers exhibit strong absorption bands in the infrared range and the use of different filter materials may, therefore, cause differences in the experiment.

UV colour glass filters with specified UV-B transmittance, e.g. Schott cut-off filters, fulfil the requirements (i) to (iii). However, they are available only in small dimensions and it is no economic approach to cover large experimental areas with this material. Furthermore, the graduation of the UV-B edge is restricted to a few available UV-filter types and can be varied by filter thickness only within a limited range.

Borosilicate or soda-lime glass filters fulfil the requirements (i) to (iv), provided that the remaining solarisation of borosilicate glass filters can be handled during the experiment. By using filter combinations of soda-lime and borosilicate glasses a realistic simulation of the UV-part of global radiation can be achieved. This is demonstrated in figure 4, where the simulated spectrum is compared to a solar reference.

The reference spectrum of global radiation was created by use of the radiation transfer model TERRA I (Seckmeyer et al. 1994). This model considers a sun elevation of 60° and an ozone column of 320 DU, representing a typical global radiation at noon on a clear summer day in Central Europe. The calculation was also performed for other UV-scenarios by changing the parameter of the ozone columns between 240 DU and 400 DU (Fig. 1 (b)).

The spectral irradiance obtained by filtering seems to agree quite well with natural global radiation on a linear scale, however, the differences are made apparent on the logarithmic plot (Fig. 1, Fig. 4). Therefore, a careful discussion of the spectra is necessary, if experiments with simulated ultraviolet radiation are evaluated. Especially the different steepness of simulated and natural radiation spectra requires the description and classification of the artificial sunlight.

Full information on the radiation conditions is only given by the complete spectral data. Therefore, spectroradiometric measurements of the radiation are necessary for the interpretation of plant experiments. For practical purposes it is desirable to reduce the spectral data set. Therefore, irradiation data of biological experiments are often given by integration of distinct spectral bands, weighted with different action spectra e.g. DNA damage, plant damage, or erythema (Caldwell et al. 1986). These classifications do not take into account the exact form of the spectral distribution of the UV-B, which might lead to wrong conclusions on the response of the biological system. Furthermore, the action spectra known so far are mostly obtained by monochromatic action spectroscopy and do not take into account interactions with other spectral plant responses.

We, therefore, suggest a spectral classification of the UV absorption edge by defining a UV cut-off wavelength and the slope of the UV-edge. The cut-off wavelength is defined as the wavelength position at an intensity limit of 0.1 mW m^{-2} nm^{-1}. This definition depends only on the physical characteristics of the irradiance and not on biological weighting. We have chosen this intensity as the unequivocally detectable limit of state-of-the-art spectroradiometers. The definition implies a signal-to-noise ratio of approx. 100 (Seckmeyer et al. 1994, Ambach and Blumthaler 1994). In order to classify the steepness of the UV edge, the slope is defined as increase per nm between the values of 0.1 and 1 mW m^{-2} nm^{-1}. This definition represents mainly the absorbance characteristics of the filter material. For most filter materials the slope is linear to wavelength on a logarithmic plot. Cut-off wavelength and slope of the UV-B spectra are marked in figure 4. In table 1 the integrated and weighted values for the different methods of spectral classification are listed in order to compare the spectra of the GSF sun simulator to calculated solar references. The numerical data correspond to the spectra plotted in figure 1.

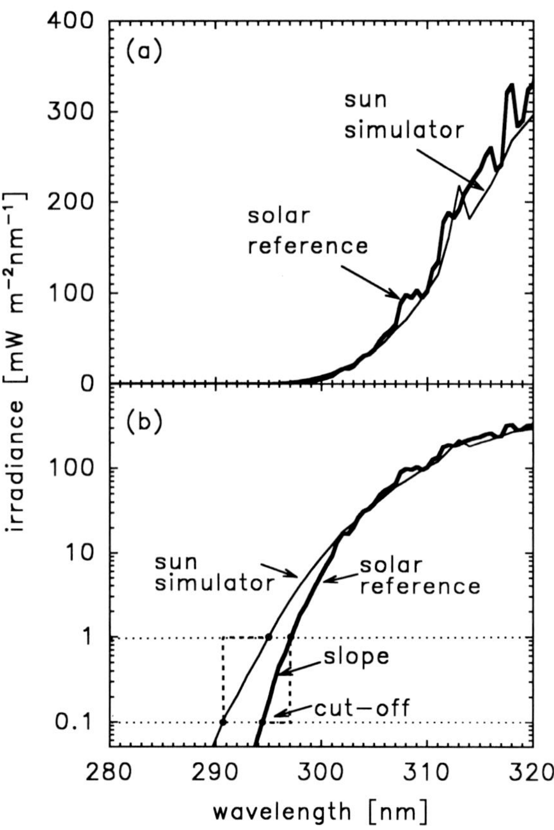

Fig. 4: UV-B spectra of the small GSF sun simulator and calculated solar reference (60°, 320 DU) (a) Linear plot. (b) Logarithmic plot. Cut-off wavelength and slope of the UV-edge are marked.

Table 1: Integrated UV-B irradiance data of the sun simulator and calculated solar reference spectra corresponding to Fig. 1 for different methods of classification (see text).

	spectrum no.				
	(1)	(2)	(3)	(4)	(5)
sun simulator					
UV-B [W/m²]	6.7	3.5	2.5	1.6	0.6
plant damage [mW/m²]	1369	433	217	104	21
DNA-damage [mW/m²]	1160	270	105	48	12
erythemal dose rate [MED/h]	14	5.0	3.0	1.7	0.7
cut-off wavelength [nm]	284	288	291	294	297
slope [nm^{-1}]	2.7	2.4	2.3	2.1	1.9
	spectrum no.				
	(6)	(7)	(8)	(9)	(10)
calculated natural global radiation, model TERRA I					
UV-B [W/m²]	3.4	3.1	2.8	2.6	2.4
plant damage [mW/m²]	402	313	249	201	164
DNA-damage [mW/m²]	197	139	102	78	61
erythemal dose rate [MED/h]	4.6	3.7	3.2	2.7	2.4
cut-off wavelength [nm]	292	293	294	295	296
slope [nm^{-1}]	3.4	3.8	3.6	3.7	3.6

Conclusion

In this paper we have discussed the requirements of UV-filtering for ecological plant experiments. Particularly the spectral shaping of the UV edge in order to simulate the absorption of stratospheric ozone calls for special attention. Different filter materials are compared. Soda-lime glass reduces the problem of ageing caused by intense UV irradiation due to its low iron content.

The graduation of different spectra according to enhanced UV scenarios can be achieved by the combination of different glass filter materials (borosilicate glass and soda-lime glass). The different filters have different ageing features, reaching different saturation levels, but this problem can be handled in the sun simulators. During various experiments in the GSF phytotron the glass filter technique was successfully applied to simulate scenarios of both enhanced and reduced UV-B. The simulation quality is assessed by means of different methods of classification and by considering the cut-off wavelength and the slope of the UV-edge.

Gaseous Ozone is an efficient way to increase the steepness of the UV absorption edge in technical filters (Tevini et al. 1989) and should be considered for further development. When employing an ozone filter in a closed chamber system with artificial irradiation, technical problems related to the toxicity, the chemical aggressiveness, and the instability of ozone ougth to be taken into account to provide operational safety. An ozone filter for deployment at the GSF phytotron facilities is presently under development.

Acknowledgements

We thank Dr. Payer (GSF Research Center) for his support, and Prof. Tevini (Universität Karlsruhe) for stimulating discussions.

References

AMBACH, W. and M. BLUMTHALER: Characteristics of solar UV irradiance. Meterol. Zeitschrift *3,* 211–220 (1994).

CALDWELL, M. M., L. B. CAMP, C. W. WARNER, and S. D. FLINT: Action spectra and their role in assessing biological consequences of solar UV-B radiation change. In: WORREST, R. C. and M. M. CALDWELL (eds.): Stratospheric ozone reduction, solar ultraviolet radiation and plant life, pp. 87–96. Springer-Verlag, Heidelberg (1986).

CALDWELL, M. M., S. D. FLINT, and P. S. SEARLES: Spectral balance and UV-B sensitivity of soybeans: a field experiment. Plant, Cell and Environment *17,* 267–276 (1994).

PAYER, H. D., P. BLODOW, M. KÖFFERLEIN, M. LIPPERT, W. SCHMOLKE, G. SECKMEYER, H. K. SEIDLITZ, D. STRUBE, and S. THIEL: Controlled environment chambers for experimental studies on plant responses to CO_2 and interactions with pollutants. In: SCHULZE, E. D. and H. A. MOONEY (eds.): Design and Execution of Experiments with CO_2 enrichment, pp. 127–165. Commission of the European Communities, Brussels (1993).

SECKMEYER, G. and H. D. PAYER: A new sunlight simulator for ecological research on plants. J. Photochem. Photobiol. *B 21,* 175–181 (1993).

SECKMEYER, G., S. THIEL, M. BLUMTHALER, P. FABIAN, S. GERBER, A. GUGG-HELMINGER, D.-P. HÄDER, M. HUBER, C. KETTNER, U. KÖHLER, P. KÖPKE, H. MAIER, J. SCHÄFER, P. SUPPAN, E. TAMM, and E. THOMALLA: Intercomparison of spectral-UV-radiation measurement systems. Applied Optics *33,* 7805–7812 (1994).

TEVINI, M., U. MARK, and M. SAILE: Plant experiments in growth chambers illuminated with natural sunlight. In: PAYER, H. D., T. PFIRRMANN, AND P. MATHY (eds.): Environmental research with plants in closed chambers, pp. 240–251. Commission of the European Communities, Brussels (1989).

THIEL, S., T. DÖHRING, M. KÖFFERLEIN, A. KOSAK, P. MARTIN, and H. K. SEIDLITZ: A phytotron for plant stress research: How far can artifical lighting compare to natural sunlight? J. Plant Physiol. *148,* 456–463 (1996).

VOGEL, W.: Glaschemie. Springer-Verlag, Berlin, 1992.

J. Plant Physiol. Vol. 148. pp. 120–128 (1996)

Avoidance of Photoinhibition and Examples of Photodestruction in High Alpine *Eriophorum*

CORNELIUS LÜTZ

GSF-Forschungszentrum für Umwelt und Gesundheit, Expositionskammern, Ingolstädter Landstr. 1, D-85764 Oberschleißheim, Germany

Received July 17, 1995 · Accepted October 20, 1995

Summary

The ecological adaptation of plant development and of the photosynthetic apparatus to contrasting temperature regimes and strong irradiation is studied with plants growing at high alpine locations. Higher plants, surviving between 2000 m and 4000 m altitude in the Alps have developed mechanisms which avoid freezing damage as well as photodestruction.

The following studies were made with *Eriophorum angustifolium* plants which develop «etiolated» leaves already under snow and allow to examine simultaneously the whole sequence of greening of plants growing in a distance of a few meters. We investigated plant development and ecophysiology of photosynthesis as it occurs in the field under high irradiance at cloudless days, which can easily change to low light conditions, often under a snow cover.

In green leaves of *Eriophorum* a sudden frost period with snow fall followed by strong irradiation leads to yellow coloured leaf areas, which regreen very slowly. Most of the thylakoids disintegrate and form clumped aggregates, indicating the breakdown of mainly chlorophylls and their binding proteins. Even high amounts of zeaxanthin in the affected leaf parts cannot avoid the damage. While hydrophilic antioxidants (glutathione, ascorbate) seem not to be directly involved in the destruction, β-carotene and α-tocopherol decrease.

Development of photosynthesis in some high alpine plants such as *Eriophorum* is retarded until favourable temperature conditions for use of photosynthetic products are present. In contrast, occasional weather changes, as common in alpine environments, can cause strong destructions in leaves and give an idea about the consequences if the observed adaptive retardation would not have been developed by adaptation.

Key words: Eriophorum, alpine plants, chilling, greening, pigments, photoinhibition, ultrastructure.

Introduction

Plant life in the Central Alps at altitudes between 2000 m and 4000 m is exposed to often contrasting temperatures, light intensities and sudden weather changes with occasionally snow falls even in summer. High alpine plants must have developed mechanisms which prevent damage by freezing or by photodestruction. Furthermore, these organisms grow fast, because a vegetation period of about three months only is available. Ecophysiological studies on alpine plant photosynthesis have shown that temperature response of photosynthesis is well acclimated to these special conditions, often starting from subzero temperatures and maximal at slightly lower temperatures compared to European lowland species, but with a much broader maximum (Moser 1973, Wagner and Larcher 1981, Larcher 1983, Bergweiler 1987, Körner and Diemer 1987, Larcher 1994) to cope with short-time changes in the environment. Furthermore, most alpine species manage leaf growth and leaf expansion near 0 °C, but this development is slow (Körner and Larcher 1988). Billings (1974) characterized the special features of alpine plants versus arctic individuals, pointing out that alpine species may react more flexible to environmental changes.

There are no indications that typical high mountain plants suffer from photoinhibition in the field, even not in full sunshine. However, mountain climate can easily change to cold conditions, which potentially result in photoinhibition or photodestruction (Huner 1985, Somersalo and Krause 1990). Many observations and measurements of alpine vegetation we made during the last years showed that the combination of cold and light was not a main stressor (Bergweiler 1987). This also holds for young, developing tissue close to snow banks. In addition, with increasing altitude the leaves contain more antioxidants to protect against radical activity (Wildi and Lütz 1996).

In the laboratory, most dark-grown seedlings of angiosperms form partially functional thylakoids several hours after transfer into light. However, some alpine plants do not start a comparable development even under high irradiance, if the temperature in the soil or under a snow cover ranges between 0 °C and approx. 10 °C. High contents of carotenoids protect this sensitive stage of development against photodestruction. Some photosynthetic activities start already in this «etiolated» system, while dark respiration remains unchanged.

Here we want to show a) the adaptation of plastid development to the high altitude environmental situation, b) the typical photodestruction of photosynthetically active tissue under the combined stress of high irradiation and cold in extreme situations.

Materials and Methods

Plant material and collecting

The sampling sites, acid peat bogs near Obergurgl/Tirol (Austria), range in altitudes between 2200 m and 2400 m. These areas are wettened by melting snow water until late summer or for the whole snow-free period (usually from June to September). They mostly contain *Eriophorum angustifolium.* The described effects were found in different *Eriophorum* species; their pigment composition, ultrastructure, and photosynthesis do not differ significantly. Here the data raised on *Eriophorum angustifolium* are presented. The plants were collected in different years always at the end of June on clear days and non-chilling air temperatures.

Fully greened plants (Fig. 1) grew on a soil which had been found dried at the surface thus warmed up by solar irradiation occasionally over 23 °C (in −3 cm below soil surface). Yellow appearing plants («etiolated») were found either in running ice-cold melting water (Fig. 2), or were found under a snow cover of 5–30 cm (Fig. 3).

Plants with yellow spots at the snow surface in otherwise green tissues grew in flat areas which were covered with new snow in June by 10–15 cm (Fig. 5); these sites were visited again after the snow had disappeared (Fig. 6).

Pigment assays

Pigments were either extracted in Dimethylformamide at the sampling sites according to Bergweiler and Lütz (1986) with the possibility to determine leaf area after extraction, or leaf parts were frozen in liquid nitrogen at the sites immediately after photography and fresh weight determination (portable balance, ± 1 mg accuracy). Further extraction, separation, and quantification by diode-array HPLC is as was described by Siefermann-Harms (1988) and by Wildi and Lütz (1996).

Assay of antioxidants

The antioxidants ascorbate, glutathione and α-tocopherol were extracted, separated, and quantified by HPLC after testing for recovery rates essentially as described by Wildi and Lütz (1996). Contents of β-carotene were assayed in the pigment system.

Measurement of photosynthetic oxygen evolution

The photosynthetic oxygen evolution was determined by a portable Hansatech (Kings Lynn, Norfolk, UK) leaf oxygen electrode at the respective sampling sites in 2300–2400 m altitude. This system had been adapted by us from the original construction (Delieu and Walker 1981) for the use in extended field campaigns: The electrode chamber was equipped with a thermoelement in the gas phase which is in contact with the lower surface of the analysed leaf; cooling was accomplished by means of a small 6 V water pump using water from springs or melting snow warmed up by sunshine in 30–40 min in a reservoir flask. The required temperature (20 ± 2 °C) was held by adding cold water or snow to the reservoir. The electrode signals were transformed online by an interface card in a notebook computer and the rate of oxygen evolution/consumption was calculated by a self-written Turbo-Pascal programme which incorporates the actual air pressure (obtained from a barometer). The programme also calculates temperatures deviating from the preset value, as measured by the interface connected thermoelement, to cope with changes in gas volume in the cuvette. This was a necessary add-on, because in the field, especially in the wind-exposed mountains, temperature with deviations less than 1 °C can hardly be maintained constant by instruments limited in weight and energy supply.

The whole system including computer and pump is supplied by two 12 V/2.6 Ah batteries, which are re-loaded by means of a solar panel supplying 37 Watts in full sun. This setup allows photosynthesis measurements for one day in the field and has been used now for seven years with only slight modifications in our high alpine ecophysiological research.

During measurements, the irradiation ranged from 1700 μMol/m^2 sec to 2100 μMol/m^2 sec.

Chlorophyll fluorescence

Fast *in vivo* chlorophyll fluorescence (Kautsky effect) was recorded using the Hansatech Plant Efficiency Analyser (PEA) after predarkening of the leaves by clips for 30 min (five measurements per developmental stage). The interpretation of fluorescence records followed Strasser et al. (1995); in this work the ratio Fv/Fm and Fo will be discussed.

Electron microscopy

Samples were prepared at the field sites and fixed in 2.5 % buffered glutaraldehyde, followed by 1 % osmic acid in the cold. Further processing was performed as described by Lütz and Moser (1977) and Lütz (1987). The ultrathin sections were studied in a Zeiss EM 10 CR electron microscope.

Results

Normal development

New leaves of *Eriophorum* plants grow from rhizomes already under snow, with 2–4 leaves developing first. It is not

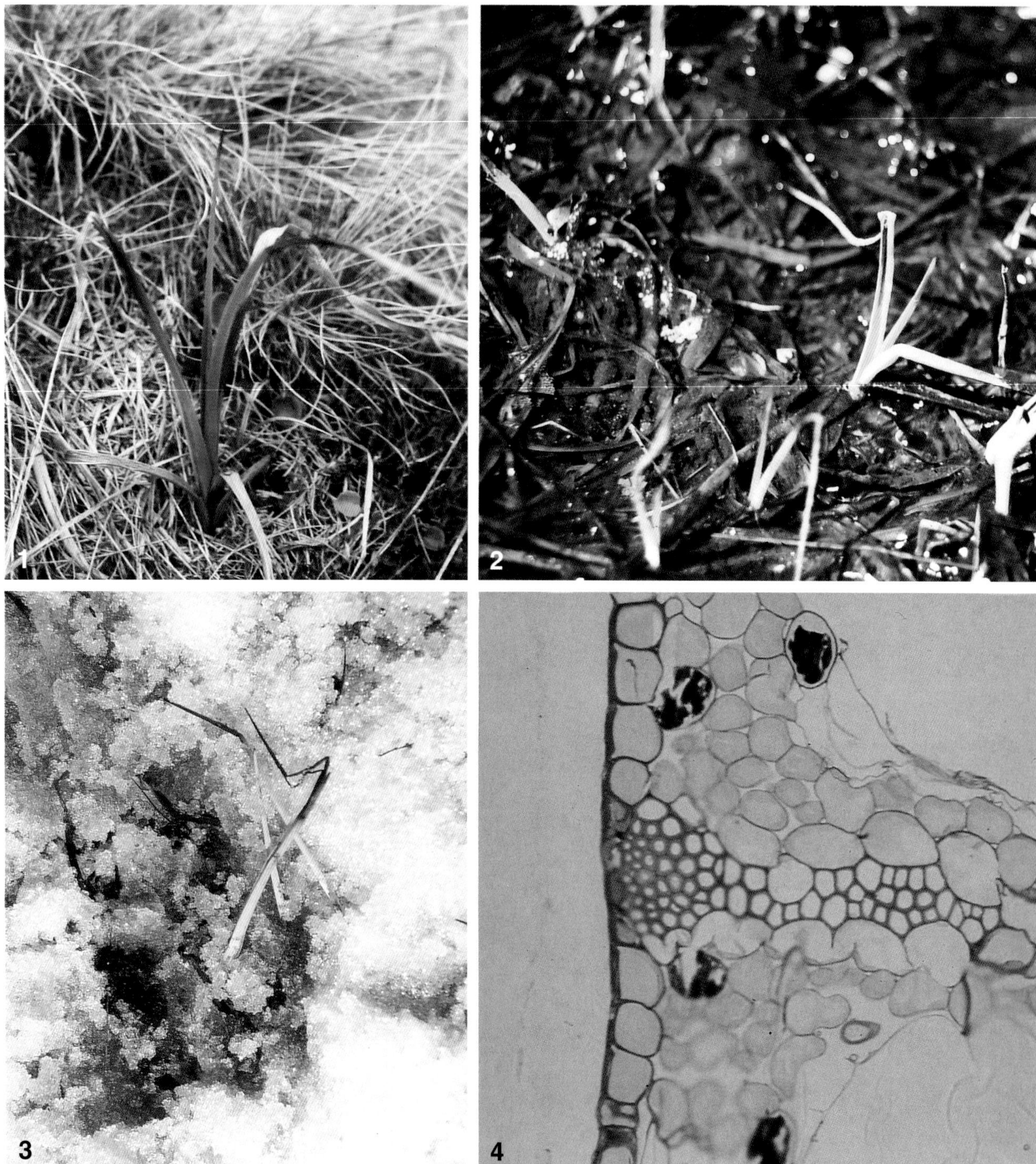

Figs. 1–4: Different developmental stages of *Eriophorum angustifolium* plants in the field, end of June, at an elevation of 2300 m. Fig. 1: fully greened plant before flower formation. Fig. 2: Plants growing in cold melting water but without snow protection. Fig. 3: Plants taken from a 20 cm cover of snow, macroscopically similar to those shown in Fig. 2. Fig. 4: Part of a green *Eriophorum* leaf, showing sclerenchyme bundles, mesophyll and parts of the aerenchyme. Magn.: 800×.

clear which snow depth or light transmission starts this process. Table 1 indicates that snow covers of approx. 50 cm may allow sufficient light access for tissue development and even for greening. Leaf growth and colouration by carotenoids, not by chlorophylls, is characteristic (Fig. 3), also for the plants which grow up in full sun from cold soil and melting water (Fig. 2). Greening occurs when soil and/or water temperatures reach approx. 10 °C or more (Fig. 1, Table 2). Only

Table 1: Attenuation of solar irradiation by snow, measured by three different sensors (LiCor) in a snowbank at 2300 m altitude on a clear day end of June at 11:00 MEZ. The snow was lightly covered with dust.

sensor	snow surface	−20 cm	−45 cm	−85 cm	−155 cm
$\mu Mol/m^{-2}$ sec	2 350	70	17	0.2	0.005
W/m^{-2}	1 100	13	0.5	0.005	0.001
Lux	125 000	5 000	100	14	3

Table 2: Typical temperature gradient in the brownish-black coloured soil or the melting snow water, respectively, in an *Eriophorum* stand at 2300 m altitude. The temperatures were measured at 3 cm depth in the soil or in the running water on a cloudless day at the end of June between 11 h and 12 h.

a) plants fully greened, flowering, on humid soil: 22 °C
b) plants fully greened, flowering, on humid, but hay-covered soil: 19 °C
c) plants fully greened, flowering, in water: 19 °C
d) plants with greening leaves, 30 cm from the rim of snow, in water over dark soil: 13 °C
e) plants with yellow leaves, 10 cm from the rim of snow, in water: 4°C
f) plants with yellow leaves, under snow and in melting water: 1 °C
g) plants with yellow spots on green leaves, 50 cm from the rim of snow, in water: 18 °C

the fully greened plants start flowering; in this stage the soil temperatures can rise easily to 22 °C or more by irradiation despite humid or wet conditions (Table 2).

This development is mirrored in plastid ultrastructure: Fig. 7 shows «etioplasts», which can be characterized by prolamellar bodies and a few thylakoids, in both samples with yellow leaves. The snow-free plants develop more small grana, but still retain prolamellar bodies, in general slightly smaller compared to the samples taken from the snow. However, these differences are small. The fully greened leaves contain normal chloroplasts (Figs. 8, 9). Near the leaf base, the meristematic zone, occasionally small prolamellar body like structures appear connected to thylakoids as a sign of development (Fig. 8). The mesophyll cells group together around the sclerenchyme and vascular strands which stiffen the aerenchyme of the leaf (Fig. 4).

The presence of etioplasts and prolamellar bodies even in strong light found in plants growing in cold soil/water conditions, but an otherwise normal leaf development indicates that not light, rather temperature controls plastid formation and subsequent onset of photosynthetic activities. Photosynthetic functioning was recorded as variable to maximum chlorophyll fluorescence (Fv/Fm) and leaf oxygen development. These measurements were performed at the growth sites of the plants with portable equipment, because the transfer of plants to lower altitudes or laboratory conditions would have changed too many parameters.

As a measure for quantum efficiency in photosystem II the ratio Fv/Fm ranges between 0.26 and 0.33 in non-green plants, while in the fully greened leaves the ratio is near 0.78, a normal value on sunny days. The low values in the yellowish leaves show that despite the occurrence of some small grana (see above) the light harvesting system of thylakoids is still under formation. The initial fluorescence Fo (approx. 0.07 vs. 0.38) also indicates the small amount of active chlorophyll in these young stages.

Table 3: Photosynthesis and dark respiration expressed as oxygen development or consumption per leaf area and chlorophyll content, respectively. Values for «net photosynthesis» were not corrected for respiration losses, as in «gross photosynthesis». Means of five determinations plus standard error.

	respiration/ cm^2	net photo-synthesis/cm^2	gross photo-synthesis/chl
+ snow, yellow	−3.2 (0.1)	−1.1 (0.2)	+33.1 (5.6)
– snow, yellow	−3.5 (0.2)	+0.5 (0.1)	+15.4 (4.2)
normal, green	−3.1 (0.2)	+6.4 (0.4)	+2.9 (0.3)

Whole leaf photosynthesis changes strongly according to the development (Table 3): leaves which are taken from the snow consume oxygen, but less than in dark respiration. However, the carbon budget is negative. The yellowish plants which develop close to the snow bank and are already exposed to the sun for several days may just cope with respiratory losses. High photosynthetic activities allow the green leaves to use the short vegetation period for the required carbon fixation.

The calculation basis «cm^2» is used because the leaves grow from a basal meristem which means the leaf segments under study are fully differentiated and may change in freshweight or dry matter whereas leaf area is a stable basis. In comparison, the described effects for photosynthesis based on fresh weight give similar results, but with higher deviations of values. If photosynthesis is corrected for respiration losses and is related to chlorophyll content, the high efficiency of the young photosynthetic units appears: the «+snow, yellow» samples seem to develop a more than 10 times higher activity compared to the fully developed stage.

Dark respiration is found with similar activities at least under the conditions supplied with the oxygen electrode, which means no drastic difference in temperature acclimation of respiration between the developmental stages. Also on an ultrastructural level no differences in mitochondria structure or number could be found in the mesophyll cells.

Drastic changes between developmental stages occur in plastid pigments (Table 4). They reflect i) the assembly of photosynthetic units and b) the role of shielding pigments under strong irradiation.

Leaves taken from −20 cm snow contain mostly carotenoids showing ratios of chl/car = 0.09, with the ratios increase in the yellow samples (shown in Fig. 2) to 0.1 and in the green stage to 3.2. β-carotene and neoxanthin (data not shown) developed from 15 % over 25 % to the final 100 % in the green leaf, while lutein is already present in the snowfree yellow samples like in green leaves. A comparable high amount of xanthophyll cycle pigments (37 % of final) is present in the «etioplasts», raising to a final of 44 % in the second (snow free) stage; here, due to the unfiltered irradiation, mostly zeaxanthin is present. It is interesting, that chloro-

Table 4: Concentration of individual leaf pigments in the three developmental stages, based on leaf surface (mg/cm^2). Chl/car: ratio of total chlorophylls to total carotenoids. X-cycle: sum of the xanthophyll cycle pigments violaxanthin, antheraxanthin and zeaxanthin. Means of five determinations plus stadard error.

	+ snow, yellow	– snow, yellow	normal, green
carotenoids	0.35 (0.09)	0.87 (0.03)	1.33 (0.08)
chlorophylls	0.03 (0.01)	0.08 (0.01)	4.20 (0.02)
chl/car	0.08	0.09	3.2
X-cycle	0.12 (0.03)	0.14 (0.02)	0.32 (0.03)
violaxanthin	0.08 (0.03)	0.02 (0.01)	0.22 (0.02)
zeaxanthin	0.02 (0.01)	0.09 (0.02)	0.08 (0.01)
lutein	0.21 (0.02)	0.66 (0.04)	0.63 (0.07)
β-carotene	0.02 (0.01)	0.03 (0.01)	0.12 (0.01)

plasts contain more violaxanthin and antheraxanthin (data not shown), but not more zeaxanthin than etiochloroplasts.

Pigment composition the young plastids is mainly characterized by the retarded chlorophyll development and by a well adapted set of xanthophylls which will help to avoid photodestruction of thylakoids in this sensitive developmental stage.

Symptoms of photodestruction

Occasionally fully greened plants may be exposed to new snow even in summer in altitudes above timberline. If snow fall does not cover plants completely, parts of the leaves will be exposed to high irradiation at sudden weather changes. This situation is explained by Fig. 5, followed by an appearance shown in Fig. 6 after snow has melted away. Yellow spots develop in the leaves near to the snow surface where the tissue is exposed to cold plus high irradiation enhanced with reflected irradiation. Continuous observation of plants has shown that regreening of these parts takes 2–3 weeks.

By light microscopy (Fig. 4) no difference appears between green and yellow parts of a leaf. Ultrastructural observation demonstrates that only chloroplasts are affected: in comparison to the normal plastids of the green part (Fig. 9), plastid thylakoids from yellow tissue (Fig. 10) are been destroyed nearly completely. Together with remnants of thylakoids (often dilated) and a few grana most of the membraneous material clogged together in non-resolvable structures like lipid aggregations and in plastoglobuli. However, envelope membranes seem to be intact.

These photodestructed parts do not develop any photosynthetic activity, which result is supported by the low chlorophyll fluorescence of Fv/Fm of 0.34. The fate of the individual thylakoid pigments is described in Table 5: carotenoids remain at 64 %, while chlorophylls are broken down to 23 % (preferentially chl *b*, as is shown by the increase in the ratio chl *a/b*). In case of xanthophyll cycle pigments the reduction of violaxanthin by almost 90 % reflects the formation of zeaxanthin (approx. 40 % more zeaxanthin), but also a degradation: the sum of x-cycle pigments is reduced by about 35 %. Stronger reductions were found for neoxanthin and β-carotene contents.

Table 5: Comparison of pigments and antioxidants in the green and in the yellow photodestructed part of individual *Eriophorum* leaves. Abbreviations as in table 4. Values are given in mg/cm^2 from 5 determinations plus standard error.

	green part	yellow part
carotenoids	1.13 (0.15)	0.72 (0.03)
chlorophylls	3.63 (0.41)	0.84 (0.01)
X-Cycle	0.54 (0.08)	0.36 (0.06)
violaxanthin	0.32 (0.10)	0.04 (0.01)
zeaxanthin	0.12 (0.01)	0.19 (0.04)
lutein	0.45 (0.12)	0.29 (0.04)
neoxanthin	0.06 (0.02)	0.01 (0.01)
β-carotene	0.13 (0.04)	0.06 (0.01)
chl a/b	4.2	8.7
chl/car	3.2	1.2
ascorbate	0.078 (0.01)	0.083 (0.01)
glutathione	0.185 (0.02)	0.216 (0.02)
α-tocopherol	0.225 (0.02)	0.171 (0.01)

In order to study the mechanisms of these destructive processes, the changes in two hydrophilic antioxidants (glutathione, ascorbate) and two lipophilic antioxidants (β-carotene, α-tocopherol) were followed (Table 5). Only β-carotene (see above) and tocopherol seem to be consumed in the destruction process, while the contents of the two hydrophilic antioxidants increase slightly as a possible secondary stress reaction.

These observations meet the ultrastructural results: the combined action of strong light and cold affects first the lipid domains of the thylakoids, as is indicated by the losses of β-carotene and tocopherol, if chlorophylls degrade, their stabilizing proteins will also be removed. In consequence, the remaining structure has a much higher lipid/protein ratio, which leads to prolamellar body like structures as they are found in the electron microscope. Because other cell membranes or compartments obviously are not concerned, this reaction must be regarded as special for chloroplasts in full photosynthetic activity.

Discussion

The climatic environments of high alpine ecosystems can be characterized by extreme regimes e.g. in temperatures and irradiation. The short vegetation period and the individual microclimate of plants from high elevations result in numerous adapations (Bergweiler 1987, Körner and Larcher 1988, Larcher 1994, Moser 1973, Wildi and Lütz, 1996).

In order to extend the growing season at high altitudes, the plant development of most alpine species starts already below snow (as an example: –20 cm snow), at temperatures between –5 °C to 0 °C (Bergweiler 1987) and when light attenuation allows weak to moderate light conditions of approx. 70 µMol/m^2 sec (see Table 1). For plants like oat or bean, this irradiation would be sufficient for chloroplast formation, completed at normal temperatures within 10–20 hours (Krol and Huner 1989, Eskins 1993, Rüdiger 1993). In the *Eriophorum* example, chloroplasts develop only very slowly as long as the

Fig. 5: *Eriophorum* plants, partially covered with new snow (approx. 10 cm) in June. Near to the snow surface the tissue changed from green to yellow due to photodestruction in thylakoids.

Fig. 6: The same plot as in Fig. 5, but after snow melting. The yellow photodestructed leaf parts regreen slowly.

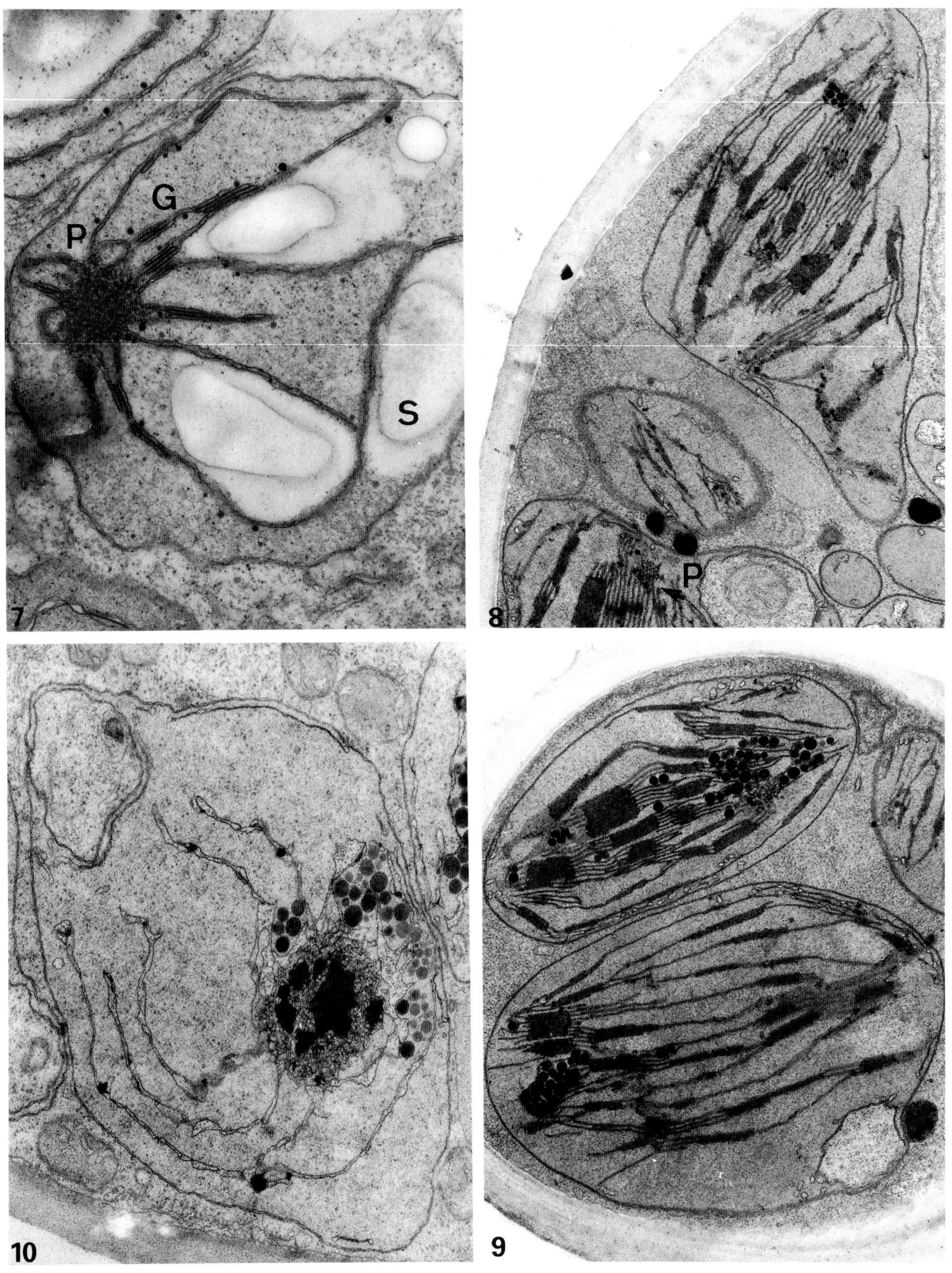
G
P
S
7
8
P
10
9

microenvironment of the plants is cold (Figs. 2, 3). The regulating step seems to be the esterification of chlorophyllide (Rüdiger 1993). Despit the presence of prolamellar bodies, HPLC analysis could not find protochlorophyllide, but small amounts of chlorophyllide together with some chlorophylls. We measured that chlorophyllide esterification is accelerated in some alpine species, including *Eriophorum,* at temperatures above 12 °C (Blank-Huber 1986, Rüdiger 1993). A second regulative step may be the formation of delta-aminolevulinic acid (ALA) as the precursor of porphyrins. Hodgins and v. Huystee (1986) described that ALA synthesis is low at temperatures below 12 °C in *Zea mays.* As soon as soil and air temperatures are more favourable for metabolic enzyme reactions, *Eriophorum* plants green completely in 1–2 days (Fig. 1). The yellow leaves (Figs. 2, 3) developed as normal adaptation to alpine conditions. Their sensitive developmental stages are well protected against photodestruction by a set of carotenoids and antioxidants. *Eriophorum* contains comparably high amounts of antioxidants similar to other high alpine plants (Wildi and Lütz 1996).

The xanthophyll cycle pigments as well as lutein accumulated in high amounts in the yellow leaves. As zeaxanthin synthesis occurs already to a considerable extent in the lower light under 10–20 cm of snow and at low temperatures, its enzymatic formation (Siefermann-Harms 1990, Pfündel und Bilger 1994) seems to be active also at lower temperatures. The sum of the xanthophyll cycle pigments is similar in the two «yellow» and developing samples and is doubled in the green leaves. The histological development of the leaves is nearly completed in all samples (Fig. 4).

Photosynthesis was measured as oxygen release/uptake at the growth sites of *Eriophorum.* Even in the leaves which were taken from the snow and immediately measured (however at approx. 20 °C) small activities exist. A similar, even more retarded development under the snow is reported by Hamerlynck and Smith (1994) for the snowbank geophyte *Erythronium grandiflorum* and by Bergweiler (1987) for *Taraxacum alpinum* and *Ranunculus glacialis.* Their etiochloroplasts indicate that some photosynthetic units are already working in leaves still covered by snow or emerging from cold, running melting water. The development occurs in similar steps as described by Krol and Huner (1989) for winter rye under intermittent light. The existence of prolamellar bodies in the light in the presence of chlorophyllide instead of protochlorophyllide indicates that most of the lipids and carotenoids of *Eriophorum* are synthesized for an «expected» fast greening at favourable conditions. Prolamellar bodies are formed in developing plastids by prolonged darkness or by otherwise inhibited thylakoid formation resulting in an overproduction of lipidic compounds (Lütz 1986).

Leaf dark respiration is nearly equal in all samples. As soon as the storage tissues (rhizomes) in the ground warm up and may supply energy to the developing tissues, the plants green fast. They preform a metabolic condition, ready for a fast development, if environmental conditions improve and the danger of photodestruction due to overreduction of the electron transport chain is reduced. *Eriophorum* appears well equipped for sustaining a broad range of contrasting growth conditions.

Yellow coloured leaves do not only appear during dark development of plants due to inhibited chlorophyll biosynthesis because light is missing (Bergweiler et al. 1984, Lütz 1981) as is discussed above. Yellowing also developes occasionally as a consequence of very unfavourable environmental conditions, which lead to destruction of chlorophyll protein complexes, mostly leaving back certain carotenoids and membrane lipids, which partially assemble in plastoglobuli (Lichtenthaler 1969). The retarded development during the cold microclimate is seen as a mechanism to prevent more radicals to be formed than would be removed by the tissue scavangers. For instance, porphyrin accumulation results in photodynamic damage of chloroplasts (Chakraborty and Tripathy 1992). Therefore photoinhibition as a result of freezing and light stress occurs mainly in fully developed chloroplasts (Somersalo and Krause 1990). Also the *Eriophorum* example as presented in Figs. 5 and 6 supports the effects of an overreduced electron chain, followed by radical formation resulting in photodestructed thylakoid membranes. Acclimated leaf tissue can cope with such a stress combination (Huner 1985), but the much higher light intensities at the snow surface even break the higher antioxidant pool as found in acclimated alpine plants (Wildi and Lütz 1996). The pigment analyses have shown that β-carotene and α-tocopherol were mostly affected by the stressors, followed by chl *b* and neoxanthin. The two former compounds are well known as radical quenchers in all living systems (Ong and Packer 1992, Tinkler et al. 1994). Their capacity could not avoid destruction of most of the chlorophylls. Chlorophyll binding proteins are then destabilized and will be degraded. The resulting ultrastructure shows dilated thylakoids together with clogged material, which is very similar in structure to prolamellar bodies treated with proteolytic enzymes (Lütz and Tönissen 1985). Therefore, the breakdown products formed should be enriched in remaining thylakoid lipids and peripheral proteins together with carotenoids – visible as yellow spots on the leaves. A repair requires intensive protein and pigment biosynthesis.

Selected high alpine plants allow numerous ecophysiological studies, down to the molecular level, on the response of metabolic activities and adaptation processes to microenvi-

Fig. 7: Etioplast from a yellow coloured leaf without snow cover, as in Fig. 2. P: prolamellar body; G: grana, they occur less frequent in otherwise similar etioplasts from snow covered samples. S: starch grain. Magn.: 30,000×.

Fig. 8: Chloroplasts from the lower part of a green *Eriophorum* leaf. Occasionally small prolamellar bodies(P) are visible. Magn.: 20,000×.

Fig. 9: Typical chloroplasts from fully developed leaf tissue. They contain many black coloured plastoglobuli. Magn.: 20,000×.

Fig. 10: Chloroplast in the yellow, photodestructed area as is shown in Fig. 6. Remaining thylakoids are dilated, most membrane material clumped together. Numerous plastoglobuli are formed, often greyish coloured. Magn.: 25,000×.

ronments. Structural and physiological answers are stronger than found in most other plants. Therefore, insights in different important processes in plant life can be easily obtained for instance from *Eriophorum*. This contributes to understand from an ecophysiological point of view, what we see in the field: a well adapted, beautiful flora, mostly resistant to the harsh conditions of the High Alps.

Acknowledgements

I thank Dr. Karl-Heinz Pettinger for writing the Turbo-Pascal-programme for the oxygen electrode, the interface setup and configuring and his help in many field measurements. A part of this work was only possible by using the laboratory at the Obergurgl station of the Institut für Hochgebirgsforschung (Prof. G. Patzelt, M. Strobl). I thank Prof. W. Larcher for his comments and Mrs. M. Moser for correcting the manuscript.

References

Bergweiler, P., U. Röper, and C. Lütz: The development and ageing of membranes from etioplasts of *Avena sativa.* Physiol. Plantar. *60,* 395–400 (1984).

Bergweiler, P.: Charakterisierung von Bau und Funktion der Photosynthesemembran ausgewählter Pflanzen unter den Extrembedingungen des Hochgebirges. Ph. D. Thesis, Universität Köln (1987).

Bergweiler, P. and C. Lütz: Determination of leaf pigments by HPLC after extraction with N,N-dimethylformamide: ecophysiological applications. Env. Exptl. Botany *26,* 207–210 (1986).

Billings, W. D.: Adaptations and origins of alpine plants. Arctic and Alpine Research *6,* 129–142 (1974).

Blank-Huber, M.: Untersuchungen zur Chlorophyll-Biosynthese. Solubilisierung und Eigenschaften der Chlorophyll-Synthetase. Ph. D. Thesis, University of Munich (1986).

Chakraborty, N. and B. Tripathy: Involvement of singlet oxygen in 5-aminolevulinic acid induced photodynamic damage of cucumer (*Cucumis sativus* L.) chloroplasts. Plant Physiol. *98,* 7–11 (1992).

Delieu, T. and D. A. Walker: Polarographic measurements of photosynthetic oxygen evolution by leaf discs. New Phytologist *89,* 165–178 (1981).

Eskins, K.: Light and temperature regulation of chloroplast development. In: Sundqvist, Ch. and M. Ryberg (eds.): Pigment protein complexes in plastids. Academic Press, London, pp. 63–90 (1993).

Hamerlynck, E. and W. Smith: Subnivean and emergent microclimate, photosynthesis, and growth in *Erythronium grandiflorum* Pursh, a snowbank geophyte. Arctic and Alpine Research *26,* 21–28 (1994).

Hodgins, R. R. and R. B. v. Huystee: Delta-aminolevulinic acid metabolism in chill stressed maize (*Zea mays* L.). Plant Physiol. *126,* 257–268 (1986).

Huner, N. P. A.: Acclimation of winter rye to cold-hardening temperatures results in an increased capacity for photosynthetic electron transport. Can. J. Botany *63,* 506–511 (1985).

Körner, Ch. and M. Diemer: *In situ* photosynthetic responses to light, temperature and carbon dioxide in herbaceous plants from low and high altitude. Funct. Ecology *1,* 179–194 (1987).

Körner, Ch. and W. Larcher: Plant life in cold climates. In: Long, S. and F. Woodward (eds.): Plants and temperature. Sympos. Soc. Exp. Biol. *42,* 25–52 (1988).

Krol, M. and N. P. Huner: Low temperature development under intermittend light conditions results in the formation of etiochloroplasts. J. Plant Physiol. *134,* 623–628 (1989).

Larcher, W.: Ökophysiologische Konstitutionseigenschaften von Gebirgspflanzen. Ber. Deutsch. Bot. Ges. *96,* 73–85 (1983).

– Hochgebirge: An den Grenzen des Wachstums. In: Ökologische Grundwerte in Österreich. Biosystematics and Ecology Series (ed. Österr. Acad. Wiss.) Wien, 304–343 (1994).

Lichtenthaler, H. K.: Die Plastoglobuli von Spinat, ihre Größe und Zusammensetzung während der Chloroplastendegeneration. Protoplasma *68,* 315–326 (1969).

Lütz, C. and W. Moser: Beiträge zur Cytologie hochalpiner Pflanzen. I. Untersuchungen zur Ultrastruktur von *Ranunculus glacialis* (L.). Flora *166,* 21–34 (1977).

Lütz, C.: On the significance of prolamellar bodies in membrane development of etioplasts. Protoplasma *108,* 99–115 (1981).

Lütz, C. and H. Tönissen: Effects of enzymatic cleavage on prolamellar bodies and prothylakoids prepared from etioplasts. Israel. J. Botany *33,* 195–209 (1985).

Lütz, C.: Prolamellar bodies. In: Arntzen, C. and A. Staehelin (eds.): Photosynthetic membranes, Encyclopedia of Plant Physiol. New Ser. *19,* Springer, Berlin, pp. 683–692 (1986).

– Cytology of high alpine plants. II. Microbody activity in leaves of *Ranunculus glacialis* (L.) Cytologia (Tokyo) *52,* 679–686 (1987).

Moser, W.: Licht, Temperatur und Photosynthese an der Station «Hoher Nebelkogel» (3184 m). In: Ellenberg, H. (ed.): Ökosystemforschung, Springer, Berlin, 203–223 (1973).

Ong, A. S. H. and L. Packer (eds.): Lipid-soluble antioxidants: biochemistry and clinical applications. Birkhäuser, Basel (1992).

Pfündel, E. and W. Bilger: Regulation and possible function of the violaxanthin cycle. Photosynthesis Res. *42,* 89–109 (1994).

Rüdiger, W.: Esterification of chlorophyllide and its implication for thylakoid development. In: Sundqvist, Ch. and M. Ryberg (eds.): Pigment protein complexes in plastids. Academic Press, London, pp. 219–240 (1993).

Siefermann-Harms, D.: High-performance liquid chromatography of chloroplast pigments: One step separation of carotene and xanthophyll isomers, chlorophylls and pheophytins. J. Chromatogr. *448,* 411–416 (1988).

– Chlorophylls, carotenoids and the activity of the xanthophyll cycle. Environm. Pollution *68,* 293–303 (1990).

Somersalo, S. and G. H. Krause: Photoinhibition at chilling temperatures and effects of freezing stress on cold acclimated spinach leaves in the field. Physiol. Plant. *79,* 617–622 (1990).

Strasser, R., A. Srivastava, and Govindjee: Polyphasic chlorophyll *a* fluorescence transient in plants and cyanobacteria. Photochem. Photobiol. *61,* 32–42 (1995).

Tinkler, J., F. Böhm, W. Schalch, and T. Truscott: Dietary carotenoids protect human cells from damage. J. Photochem. Photobiol. *26,* 283–285 (1994).

Wagner, J. and W. Larcher: Dependence of CO_2 gas exchange and acid metabolism of the alpine CAM plant *Sempervivum montanum* on temperature and light. Oecologia *50,* 88–93 (1981).

Wildi, B. and C. Lütz: Antioxidant composition of selected high alpine plant species from different altitudes. Plant, Cell, Environm. *19,* in press (1996).

J. Plant Physiol. Vol. 148. pp. 129–134 (1996)

Photoinhibition *in Situ* in Norway Spruce

Tereza Schulzová

Department of Ecological Physiology of Forest Trees, Institute of Landscape Ecology, Academy of Sciences, Květná 8, Brno, 603 00, Czech Republic

Received June 24, 1995 · Accepted October 31, 1995

Summary

Photoinhibition, expressed as the decay of the Fv/Fm ratio during sunny days, has been found *in situ* in a canopy of Norway spruce (*Picea abies* (L.) Karst). For the investigation of photoinhibition the chlorophyll fluorescence method was used combined with parallel continuous measurements of photosynthetically active radiation inside and above the canopy and temperature inside the canopy. We have found differences in daily course of Fv/Fm ratio between sunny and cloudy days and between the upper and lower part of the canopy. Decay of the Fv/Fm ratio did not only occur in the months with an extremely high photon flux density but this phenomenon has been present throughout the whole growing season.

Key words: Picea abies, chlorophyll fluorescence, photoinhibition.

Abbreviations: PAR = photosynthetically active radiation; PS II = photosystem II; Fo = grown level of chlorophyll *a* fluorescence, all reaction centers of PS II are open; Fm = maximal level of chlorophyll *a* fluorescence, all reaction centers of PS II are closed; Fv/Fm = ratio of variable and maximal chlorophyll *a* fluorescence.

Introduction

Photosynthetic activity of forest ecosystems is influenced especially by the state and function of the assimilation apparatus and by microclimatic conditions inside the stand. The fluctuation of microclimatic elements from the growing optimum can appear as stress (Rosenqvist et al. 1991). The most significant elements of microclimate during the growing season are for example high intensities of PAR, extreme temperatures, lack of water (Kalina et al. 1994). The synergetic influence increases the negative effect on the photochemistry of the photosynthesis (Powles 1984). In droughted *Sorghum,* high light together with high temperature can cause more damage to photosynthesis, than high temperature alone (Ludlow 1987). On a site, stress factors have daily and seasonal patterns. The level of stress induced can be defined by the time needed to complete recovery (Bolhar-Nordenkampf et al. 1994).

The basic role in the modification of a forest microclimate is played by solar radiation, which is a directive component for other microclimatic elements and at the same time is the sole energetic input for photosynthetic processes. Photosynthetically active radiation is the essential condition for photosynthesis on the one hand, but the high intensities of PAR can reduce the conversion efficiency of solar radiation in the processes of assimilation on the other hand (Ögren and Rosenqvist 1992).

There are numerous studies based on the investigation of the negative effects (lowered rate of net CO_2 uptake, lowered net O_2 evolution, lowered rate of electron transport) of the high level of irradiance incident on the surface of the photosynthetic organisms, for which is common be used the term «photoinhibition». Photoinhibition of photosynthesis had originally been defined as a decrease in the rate of photosynthesis, particularly in the efficiency of photosynthetic energy conversion (Demmig-Adams and Adams III 1992).

Inhibition of photosynthesis by excess of excitation energy is initiated in the reaction centres of photosystem II (Krause 1988). The actual level of photoinhibition depends therefore on the photoinactivation process, the synthesis of D1 protein and the activity of protective mechanisms (Demmig-Adams and Adams III 1990), however studies with sun and shade leaves clearly demonstrated that D1 protein turnover was high in sun leaves and low in shade leaves and showing no correlation with photoinhibition (Bolhar-Nordenkampf et al. 1994).

Some research investigated photoinhibition induced by natural conditions of light and other environmental factors (Bolhar-Nordenkampf et al. 1989, Hanelt et al. 1993), others turned their attention to the mechanism of photoinhibition under precise laboratory conditions (Demmig-Adams and Adams III 1993, Ögren and Rosenqvist 1992, Öquist et al. 1992).

A number of articles have documented photoinhibition under low temperature (Hanelt et al. 1993) and photoinhibition investigated with the inhibitors of various photoprotective mechanisms (Demmig-Adams and Adams III 1993, Greer et al. 1991, Krause et al. 1990, Öquist et al. 1992, Richter et al. 1992, Samuelsson et al. 1987).

Decrease in CO_2 uptake and O_2 evolution have often been correlated with the decrease in variable fluorescence of PS II (usually parameterized as Fv/Fm) *in vivo*, which implies decreased photochemical conversion efficiency of PS II (Demmig and Björkman 1987, Ögren and Rosenqvist 1992). Modifications in photochemical efficiency (Fv/Fm), which often shows the earliest signs of stress, were determined on a daily as well as an annual basis (Bolhar-Nordenkampf et al. 1994).

From the photosynthetic organisms, a wide range of the species was found to be sensitive to photoinhibitory treatment (Powles 1984). A higher sensitivity to photoinhibition appears in shaded plants or plants (or leaves) which developed in low light and then were exposed to high light (Herbert and Waaland 1988, Ögren 1988, Öquist et al. 1992). Under natural low light conditions the differences of the photochemical capacity between shaded and sun-exposed leaves were negligible in all three (*Phaseolus vulgaris, Zea mays* and *Helianthus annuus*) species. On a day with full sunlight a decline of Fv/Fm was observable at noon-time in the sun-exposed leaves (Bolhar-Nordenkampf et al. 1991).

In the field studies or in the laboratory studies simulating field conditions photoinhibition is typically found (Hanelt et al. 1993, Špunda et al. 1993). At noon the leaves of sun-exposed plants exhibited a well developed reduction in photochemical efficiency, which recovered to differing degrees during the afternoon (Bolhar-Nordenkampf et al. 1994). For willow leaves was found that received full daylight caused a 15% decline in the Fv/Fp ratio in the afternoon on cloudless days in July (Fp is the fluorescence at the peak of the induction curve). Together with a similar decline in the maximum quantum yield of O_2 evolution it is presumed influence of high light to photochemical efficiency of PS II (Ögren 1988).

The forest stands are exposed to high values of PAR during the vegetation season (Hansen 1992). In our study we would like to present the effect of photoinhibition under natural conditions in forest ecosystems. We have focused our experiments on investigating three aspects of the influence of high level of PAR on forest trees: i) daily course of photoinhibition expressed as a decay in photochemical efficiency of PS II, ii) photoinhibition throughout the whole season (from May till September) and iii) differences in daily courses of Fv/Fm ratio for needles from sun-exposed and shaded parts of the tree. In all above mentioned aspects we compared changes in chlorophyll *a* fluorescence on sunny and cloudy days.

Materials and Methods

Study site

All the measurements were carried out in the Experimental Ecological Study Site (EESS) of Bílý Kříž (lat. 49° 30′ 17″ N, long. 18° 32′ 28″ E) located in the Moravskoslezské Beskydy Mts. at an altitude of 943 m a.s.l. throughout the growing season 1994. It is a slightly cold, humid area, rich in precipitation. The mean annual temperature is 4.9 °C, the mean annual amount of precipitation 1100 mm, mean annual relative air humidity 80%.

The stand under study consisted of 15-year-old spruce (*Picea abies* (L.) Karst) monoculture, located on a mild slope of southerly exposure, with density of 2.776 trees/ha.

Study of photoinhibition

Our attention was devoted to different responses of the upper and lower part of the canopy during sunny and cloudy days of all months during the growing season, respectively. From the chlorophyll fluorescence measurements were determined the relative daily changes in Fv/Fm ratio. Qualitatively different responses of the sun-exposed and shaded needles to excessive light are assumed.

Modulated chlorophyll a fluorescence (fast part of Kautsky effect)

For the investigation of all fluorescence characteristics a PAM Fluorometer (H. Walz, Effeltrich, Germany) was used. Samples containing five one-year-old needles were taken from sunny and shaded parts of trees and predarkened for 30 minutes. After illumination (upper parts of needles) by actinic light (120 $\mu mol\, m^{-2} s^{-1}$), the characteristic fast fluorescence induction curves were recorded for two seconds and afterwards a white saturation pulse (2500 $\mu mol\, m^{-2} s^{-1}$, 1s duration) was applied.

Study of the microclimate of the stand

During the growing season the values of photosynthetically active radiation (PAR) incident upon the stand and temperature were measured. The downward PAR above stand (i.e. open space) was measured using LI 190S quantum sensors (LI-COR, USA).

The temperature data (sensors KTY 81) were taken from a meteorological station located about 200 meters from the forest stand under study. Values of PAR and temperature were recorded automatically on a logger (Delta-T, Great Britain) at 30-minute intervals.

For detailed investigation of penetrating PAR into tree canopy self-made instruments was used. These measurements were performed in order to characterise the differences in light conditions of sun-exposed and shaded parts of trees. The instrument is based on a 1 m-long solid metal bar carrying 20 self-made sensors based on BPW 21 photodiodes (with a range 400–700 nm and maximal sensitivity 550 nm) (Siemens, EC). Every hour the measurements of penetrating PAR were carried out in 8 world sites. Mean values from all 160 values (or less when the width of the tree canopy was closer than 1 m) were taken to determine the daily courses of penetrating PAR into sun-exposed and shaded parts of trees (Fig. 1).

Results

We have found different levels of photochemical efficiency of PS II for sunny and cloudy days and for sun-exposed and shaded parts of the spruce crown throughout the whole growing season (Figs. 2, 3). These findings are in agreement with the results from study with field-grown plants (*Phaseolus vul-*

garis, Zea mays and *Helianthus annuus*) reflecting photoinhibitory effect under natural conditions.

The highest value for Fv/Fm ratio was found on cloudy days of July, about 0.76. This is a relative low value, which could be due to the experimental methods of fluorescence measurements. The value of Fm was determined after short illumination with intensity of light about 120 µmol $m^{-2} s^{-1}$, which might have influenced the reduction state of PQ and probably also Q_A and the degree of non-photochemical quenching and might have lowered the value of Fv/Fm ratio. For the clarification of that influence we compared data from measurements of Fv/Fm ratio before and after short illumination by low light (120 µmol $m^{-2} s^{-1}$) after predarkening the samples for 30 minutes. We have also compared data from fluorescence measurements of needles collected in dark (4:00 am, when we expected the highest level of relaxation of photosystem II after photoinhibition throughout the day) and data from noon (12:00, when the expected photoinhibition is highest). The results you can see in Table 1. The highest value was reached for samples collected early in the morning without short illumination before the value of Fm was determined. The lowest one was observed for needles collected in noon and after short illumination before the value of Fm was determined. So we should not neglect the influences of short illumination. However, some other mechanisms for seasonal decline in Fv/Fm ratio are supposed, from which we prefer influence of drought throughout the season. We have decided for this presumption on the basis of data

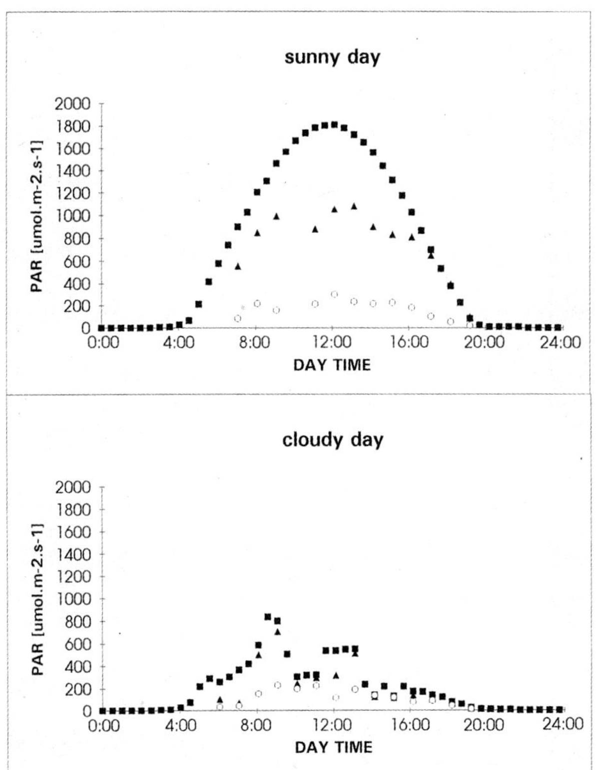

Fig. 1: Daily courses of incident PAR ■ above the tree canopy and penetrating PAR into two different (sun-exposed ▲ and shaded ○) parts of the tree canopy.

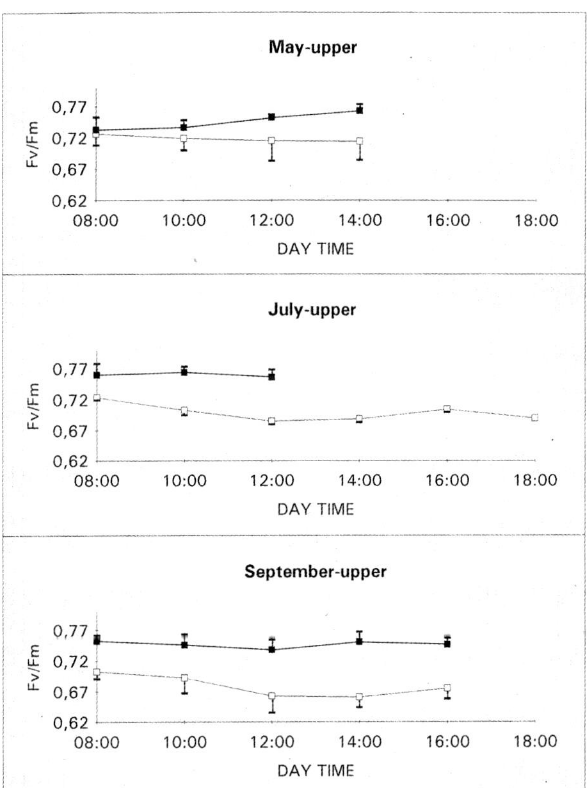

Fig. 2: The daily courses of Fv/Fm ratio in three chosen months for needles from upper (sun-exposed) parts of tree crown; black points – cloudy days, white points – sunny days (SD, n = 12).

measured in the season 1995, not loaded by drought. All values of Fv/Fm ratio from this season were higher than values from the preceding season.

An other reason for lowered Fv/Fm ratio is probably insufficient relaxation from photoinhibitory treatment. The slower relaxation in dark was found to comparison with relaxation in low light (data not shown). However, for that reason higher values of Fv/Fm ratio could be expected for cloudy days and for shaded parts of the tree crown, which are not affected by high light. It is clear that if this mechanism is present, it does not affect the Fv/Fm ratio after 30 minutes alone, but together with other effects, for example the above mentioned methodological aspects or prolonged photoinhibition provoked by natural environmental conditions (many sunny days throughout the spring and the whole growing season, extremely dry season etc.).

In Table 2 are expressed the percentual changes of mean values of Fv/Fm ratio from upper (sun-exposed) and lower (shaded) parts of the spruce crown. From the three aspects of investigation we have found: i) During the sunny days decay was observed in the Fv/Fm ratio (Figs. 2, 3). This decrease is significant at midday (maximal values of incident PAR) and reaches about 5 % compared with the morning value for the sunny days and for sun-exposed needles. The Fv/Fm ratio has a dynamic course, in the evening and in the morning the Fv/Fm ratio reaches the maximum. ii) The seasonal investigations reveal, that photoinhibition does not occur on particular days only, when the photoinhibition processes are extremely fraught with high intensity of incident PAR together with other stresses (drought) but this phenomenon exists in forest ecosystems generally. iii) The daily course in Fv/Fm ratio is typical for sun-exposed parts of tree crown. The shaded parts have a steady level of Fv/Fm through the whole day. In shaded needles and for cloudy days, PS II quantum efficiency was stable during the whole day. The differences in Fv/Fm ratio between sunny and cloudy days at noon were tested by t-test(Microsoft Excel). All differences were found to be signifficant ($P > 0.005$ for data from July and September, $P > 0.025$ for data from May).

From the seasonal viewpoint, on cloudy days the Fv/Fm ratio from the morning measurements exhibits the lowest value in May (for both sun-exposed and shaded needles). It could be due to climatic conditions, low temperature during nights and high intensities of PAR during the days before our investigations (data not shown). The similar results were found by Godde (1992). In his study Fv/Fm was somewhat lower in spring during the development of the new flush.

The lower shaded parts of trees are sensitive to light on sunny days too, but the differences are not so signifficant as for the upper sun-exposed tree parts.

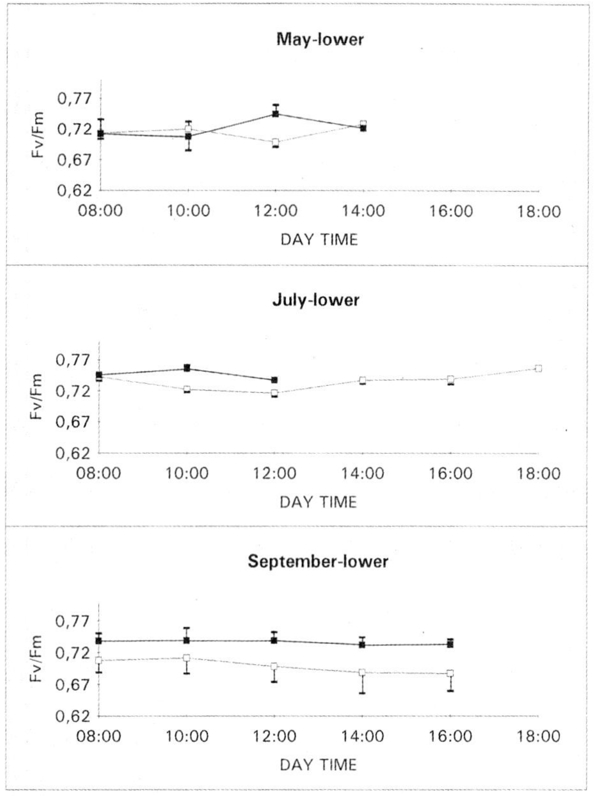

Fig. 3: The daily courses of Fv/Fm ratio in three chosen months for needles from lower (shaded) parts of tree crown, black points – cloudy days, white points – sunny days (SD, n = 12).

Table 1: The comparison of Fv/Fm ratio for different photochemical states of photosystem II (night and noon) and for various methods of measurement (the determination of Fm after short illumination by actinic light (120 µmol m^{-2} s^{-1}) and without short illumination), (SD, n = 12).

	Fv/Fm ± SD
night + illumination	0.792 ± 0.004
night	0.804 ± 0.013
noon + illumination	0.760 ± 0.013
noon	0.777 ± 0.015

Table 2: Percentage expression of decay in Fv/Fm ratio at 12 : 00 against 8 : 00 (100 %); PAR (mol m^{-2} d^{-1}) – mean value of sum of photosynthetically active radiation, Temp. (°C) – mean value of maximal daily temperature in measured days, Upper – sun-exposed part of tree crown (third whorl), Lower – shaded part of tree crown (seventh whorl).

Month	PAR [mol m^{-2} d^{-1}]	Temp. [°C]	Fv/Fm Upper %	Fv/Fm Lower %
Sunny days				
May	47.8	27	97.8	97.9
July	56.7	23	95.3	96.8
September	25.7	17	94.3	98.2
Cloudy days				
May	7.1	6	100	100
July	19.3	15	99.3	99.2
September	16.7	8	98.1	100

Discussion

There were described many investigations of Fv/Fm ratio to obtain more information about changes in photochemical efficiency of PS II. The fluorescence measurements of photosystem II unaffected by high light show the values of Fv/Fm ratio above 0.80 (Demmig-Adams and Adams III 1992, Ögren and Rosenqvist 1992, Špunda et al. 1993). From the cloudy days we can presume the maximal photochemical efficiency of PS II expressed by the Fv/Fm ratio. The registered values do not reach this level probably due to the causes mentioned above (see Results).

The radiation regime in spruce crowns is characterised by horizontal and vertical distribution of solar radiation. The crown creates a heterogenous optical system, in which certain parts of the canopy are exposed to high levels of PAR at different times of the days and season. For a little better characterization of heterogeneity of PAR for sun-exposed and shaded parts of tree crown we carried out observations with a self-made solarimeter carrying 20 sensors (see Materials and Methods) for sunny and cloudy days. It is clear that values from the upper part (third whorl) are below real ones (because the registered values represent the mean), the real values are near that of incident PAR. So the maximal differences of PAR between sun-exposed and shaded parts is about 83 % for sunny days, so that the new shoots develop under very different radiation conditions. On cloudy days the differences are not so significant.

Some authors suggest mechanistic differences in photoinhibition of sun and shaded plants. In sun plants, an active repair cycle of PS II replaces photoinhibited reaction centres with photochemically active ones, thereby conferring partial protection against photoinhibition. In shade plants, this repair cycle is less important for protection against photoinhibition, instead, photoinhibited PS II reaction centres may give, as they accumulate, increased protection to the remaining connected, functional PS II centres by controlled, nonphotochemical dissipation of excess excitation energy (Öquist et al. 1992). In our case it is questionable lower Fv/Fm ratio for shaded needles on sunny days than on cloudy days. It could be explained by higher sensitivity to increased PAR and quantitatively and qualitatively different responses to light (Powles 1984). The shaded needles have larger light-absorbing antennae complexes than the sun-exposed needles, so the protection could be seen in the disconnection between antennae complexes and reaction centers. For our site under study the high values of PAR are determined by the values above 750 $\mu mol\, m^{-2}\, s^{-1}$, it is the value of PAR, when the rate of photosynthesis is saturated (data not shown). On sunny days the values of radiation above saturating PAR are present approximately from 7:00 to 17:00 hours (depending on the month of the season), that means the photoinhibition influences the photochemical efficiency of PS II for the majority of sunny days. Lower saturating PAR could be expected for shaded than for sun-exposed needles.

We should not forget about the effect of temperature on photoinhibition. The temperature 30–40 °C can lead to lowered photochemical efficiency if any other stress is present (water stress) (Ludlow 1987). The temperature 42–48 °C can influence the photoinhibitory behaviours without any other stress, just by presence of high light. On sunny days the temperature on the needle surface can reach temperature near 40 °C, thereby we could not eliminate the influences of temperature on the level of photoinhibiton.

In some studies the Fv/Fm ratio increased during summer, probably as an adaptation to higher irradiances (Godde 1992). The low Fv/Fm ratio for sunny days at the end of the season does not indicate the seasonal acclimation, however there can exist the acclimation between spring and summer.

Conclusion

We have found the decrease of photochemical efficiency of PS II during the sunny days of the growing season to be permanent. The effect of the photoinhibition is not considerable, so the photoprotective mechanisms are sufficient to remove all of the excessive radiation. The photoinhibition effect under natural conditions is a regulatory mechanism and is under the control of the photosynthetic apparatus. The decay in Fv/Fm ratio does not cause any irreversible damage.

References

Bolhar-Nordenkampf, H. R., M. Hofer, and E. G. Lechner: Analysis of light-induced reduction of the photochemical capacity in field-grown plants. Evidence for photoinhibition? Photosynthesis Research *27,* 31–39 (1991).

Bolhar-Nordenkampf, H. R., S. P. Long, N. R. Baker, G. Öquist, U. Schreiber, and E. G. Lechner: Chlorophyll fluorescence as a probe of the photosynthetic competence of leaves in the field: a review of current instrumentation. Functional Ecology *3,* 497–514 (1989).

Demmig-Adams, B. and W. W. Adams III: Photoprotection and other responses of plants to high light stress. Annu. Rev. Plant Physiol. Plant Mol. Biol. *43,* 599–626 (1992).

– – The carotenoid zeaxanthin and high-energy-state quenching of chlorophyll fluorescence. Photosynth. Res. *25,* 187–197 (1990).

– – The xanthophyll cycle, protein turnover, and the high light tolerance of sun-acclimated leaves. Plant Physiol. *103,* 1413–1420 (1993).

Demmig, B. and O. Björkman: Comparison of the effect of excessive light on chlorophyll fluorescence (77K) and photon yield of O_2 evolution in leaves of higherplants. Planta *171,* 171–184 (1987).

Godde, D.: Photosynthesis in relation to modern forest decline; content and turnover of the D-1 reaction centre polypeptide of photosystem 2 in spruce trees *(Pices abies).* Photosynthetica *27 (1–2),* 217–230 (1992).

Greer, D. H., C. Ottander, and G. Öquist: Photoinhibition and recovery of photosynthesis in intact barley leaves at 5 and 20 °C. Physiologia Plantarum *81,* 203–210 (1991).

Hanelt, D., K. Huppertz, and W. Nultsch: Daily course of photosynthesis and photoinhibition in marine macroalgae investigated in the laboratory and field. Mar. Ecol. Prog. Ser. *97,* 31–37 (1993).

Hansen, U.: Photoinhibitory fluorescence quenching in spruce growing in a montane forest. Photosynthetica *27 (1–2),* 253–256 (1992).

Herbert, S. K. and J. R. Waaland: Photoinhibition of photosynthesis in a sun and a shade species of the red algal genus *Porphyra.* Marine Biology *97,* 1–7 (1988).

Kalina, J., M. V. Marek, and V. Špunda: Combined effects of irradiance and first autumn frost on CO_2 assimilation and selected parameters of chlorophyll *a* fluorescence in Norway spruce shoots. Photosynthetica *30* (2) 233–242 (1994).

Krause, G. H.: Photoinhibition of photosynthesis. An evaluation of damaging and protective mechanisms. Physiologia plantarum *74,* 566–574 (1988).

Krause, G. H., S. Somersalo, E. Zumbusch, B. Weyers, and H. Laasch: On the mechanism of photoinhibition in chloroplasts. Relationship between changes in fluorescence and activity of photosystem II. J. Plant Physiol. *136,* 472–479 (1990).

Ludlow, M. M.: Light stress at high temperature. Photoinhibition, edited by D. J. Kyle, C. B. Osmond, and C. J. Arntzen, Elsevier Science Publishers B. V., Biomedical Division, 89–109 (1987).

Ögren, E.: Photoinhibition of photosynthesis in willow leaves under field conditions. Planta *175,* 229–236 (1988).

Ögren, E. and E. Rosenqvist: On the significance of photoinhibition of the field and its generality among species. Photosynthesis Research *33,* 63–71 (1992).

Öquist, G., J. M. Anderson, S. McCaffery, and W. S. Chow: Mechanistic differences in photoinhibition of sun and shade plants. Planta *188,* 422–431 (1992).

Powles, S. B.: Photoinhibition of photosynthesis induced by visible light. Ann. Rev. Plant Physiol. *35,* 15–44 (1984).

Richter, M., B. Böthin, and A. Wild: Changes of the quantum yield of oxygen evolution and the electron transport capacity of isolated spinach thylakoids during photoinhibition. J. Plant Physiol. *140,* 244–246 (1992).

Rosenqvist, E., G. Wingsle, and E. Ögren: Photoinhibition of photosynthesis in intact willow leaves in response to moderate changes in light and temperature. Physiologia Plantarum *83,* 390–396 (1991).

Samuelsson, G., A. Lönneborg, P. Gustafsson, and G. Öquist: The susceptibility of photosynthesis to photoinhibition and the capacity of recovery in high and low light grown cyanobacteria, *Anacystis nidulans.* Plant Physiol. *83,* 438–441 (1987).

Smillie, R. M., S. E. Hetherington, J. He, and R. Nott: Photoinhibition at chilling temperatures. Aust. J. Plant Physiol. *15,* 207–222 (1988).

Špunda, V., J. Kalina, R. Kuropatwa, M. Mašláň, and M. Marek: Responses of photosystem 2 photochemistry and pigment composition in needles of Norway spruce samplings to increased radiation level. Photosynthetica *28* (3), 401–413 (1993).

J. Plant Physiol. Vol. 148. pp. 135–141 (1996)

Changes in Chlorophyll *a* Fluorescence Parameters in Leaves and Stems of the CAM Plant *Cissus quinquangularis* Chiov. Exposed to High Irradiance

NICOLA D'AMBROSIO[1,2], CHRISTIANE SCHINDLER[1], AMALIA VIRZO DE SANTO[2], and HARTMUT K. LICHTENTHALER[1]

[1] Botanisches Institut II der Universität, Kaiserstraße 12, D-76128 Karlsruhe, Germany

[2] Dipartimento di Biologia Vegetale, Universitá di Napoli «Federico II», Via Foria 223, I-80139 Napoli, Italia

Received September 5, 1995 · Accepted October 27, 1995

Summary

The effects of high irradiance (700 and 1400 µmol photons $m^{-2} s^{-1}$) on the regulation of photochemistry in leaves and stems of the CAM plant *Cissus quinquangularis* Chiov. were investigated by chlorophyll *a* fluorescence parameters. Significant decreases in F_o, F_m and F_v were observed in the leaf and in the stem tissues with increasing light intensity and exposure time. F_v/F_m, the PSII photochemical efficiency, decreased at an higher extent in stems and leaves at 1400 µmol photons $m^{-2} s^{-1}$. Non-photochemical quenching increased with the intensity and duration of light exposure, being higher in leaves than in stems. Upon high-light stress, a slighter reduction in quantum yield of the PSII electron transport was observed in the stem than in the leaf tissue. A significant decrease of photochemical quenching was observed in leaves but not in stems, thus suggesting a different strategy in the regulation of photochemistry in leaves and stems of *C. quinquangularis* plants exposed to high-light stress.

Key words: Cissus quinquangularis, CAM, chlorophyll fluorescence, light stress, fluorescence quenching, photoinhibition, photosynthesis.

Abbreviations: F_o, or F_m = minimal, maximal fluorescence when all PSII reaction centres are open or closed in the dark-adapted state; F'_o or F'_m = minimal, maximal fluorescence when all PSII reaction centres are open or closed in the light-adapted state; F_v = variable fluorescence (i.e. $F_v = F_m - F_o$); F = fluorescence in light-adapted state; F_v/F_m = PSII photochemical efficiency in the dark-adapted state; F'_v/F'_m = PSII photochemical efficiency or efficiency of excitation energy capture by open PSII reaction centres in the light-adapted state; ϕ_e = quantum yield of PSII electron transport; q_P = photochemical fluorescence quenching coefficient; q_N = non-photochemical fluorescence quenching coefficient; Q_A = the primary stable electron acceptor of PSII; Q_{red}/Q_{tot} = fraction of reduced or closed PSII centres.

Introduction

Light energy absorbed by photosynthetic pigments can be utilized in photochemistry, re-emitted radiatively as chlorophyll fluorescence and dissipated non-radiatively as thermal energy. Measurements of light-induced *in-vivo* chlorophyll *a* fluorescence in higher plants provide a powerful non-intrusive tool in the study on integrity and functionality of the photosynthetic apparatus as well as on mechanisms whereby the utilization of absorbed light energy is regulated within the photosystems (Krause and Weis, 1991; Horton and Ruban, 1992; Schreiber and Bilger, 1993).

Light energy absorbed by photosynthetic pigments beyond the amount utilized through the photosynthetic electron transport has to be dissipated safely to prevent an accumulation of excitation energy potentially harmful to the photosynthetic apparatus (Osmond, 1981; Björkman, 1987). Changes in partitioning of absorbed light energy between photochemistry and dissipative pathways lower the chlorophyll fluorescence yield. Any environmental stress reducing the photosynthetic activity brings about changes in the chlorophyll *a* fluorescence parameters. Different chlorophyll *a* fluorescence parameters and fluorescence ratios have been introduced to define the functionality of the photosynthetic apparatus (Schreiber et al, 1986; Lichtenthaler and Rinderle, 1988; Adams III et al., 1990). Moreover, the light energy partitioning within the photosynthetic apparatus can be detected by resolving the fluorescence quenching into photochemical (q_P) and non-photochemical (q_N) quenching coefficients.

Chlorophyll *a* fluorescence techniques are widely used to examine the regulation of photochemistry and the process of photoinhibition in plants exposed to light stress. There are few information on the partitioning of excitation energy within the photosystems in CAM plants. Moreover it has been suggested that Crassulacean acid metabolism as a CO_2-concentrating mechanism could prevent photoinhibitory damage when light energy is absorbed beyond the amount utilized in photochemistry (Osmond et al., 1980; Gil, 1986; Adams III and Osmond, 1988).

The purpose of this study was to assess how leaves and stems of CAM plant *Cissus quinquangularis* Chiov. regulate photochemistry when exposed to high light (700 and 1400 µmol photons m^{-2} s^{-1}) by monitoring changes in chlorophyll *a* fluorescence parameters during illumination and subsequent recovery.

Material and Methods

Plant material

Cissus quinquangularis Chiov. (Chiovenda, 1932), the CAM plant used in this work, is originated from Southern Somaliland (Africa). Both leaves and stems of this species are photosynthetically active and exhibit diurnal fluctuation of malic acid. Plants of *C. quinquangularis* were grown in pots, containing common soil, compost and sand, kept outdoors in full sunlight at the Botanical Garden of Karlsruhe University in summer. The stem shows a green layer of photosynthetic parenchyma up to ca. 1.5 mm under the epidermis and an inner light green, mainly water storing parenchyma that contains photosynthetic pigments in traces. The leaves are caducous, fleshy, glabrous and ca. 2 mm thick (Chiovenda, 1932).

High-light treatment

Plants of *C. quinquangularis* with fully developed leaves and stems were exposed at 700 and 1400 µmol photons m^{-2} s^{-1} for 6 hours. Before starting the light treatments plants were kept in darkness for 12 hours (generally correspondent to the night time). After the light treatments plants were kept in darkness in order to examine the recovery process after a potential photoinhibitory damage. The light source during illumination consisted of two 500 W HQIE lamps. By using one or two of these lamps the two irradiances chosen were obtained at the sample surface. A water layer of 80 mm was placed between plants and lamps to avoid heating up.

Chlorophyll a fluorescence measurements

Chlorophyll *a* fluorescence was measured at room-temperature using a pulse amplitude modulation fluorometer (PAM 101, Walz,

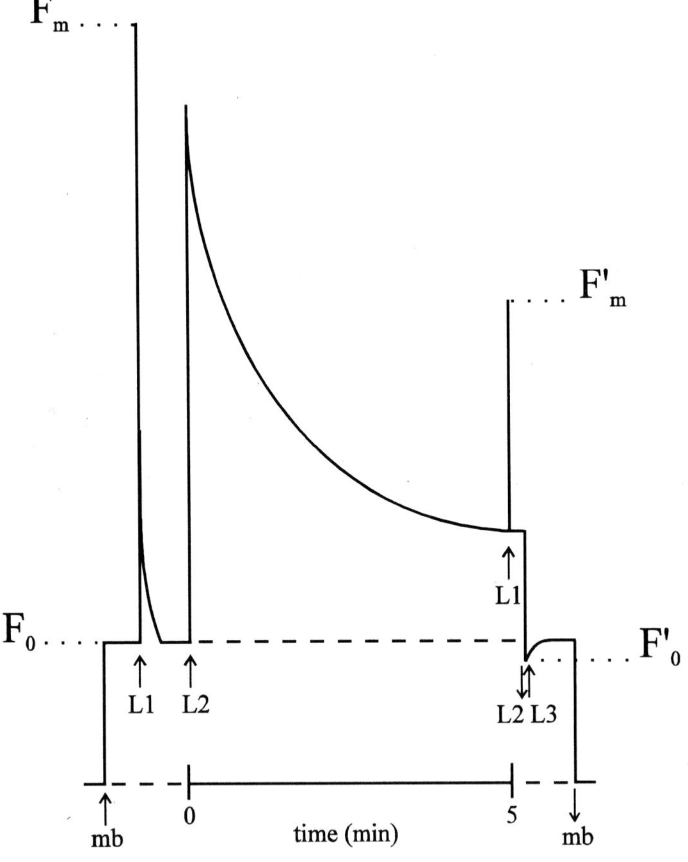

Fig. 1: Experimental protocol showing the procedure used to measure the fluorescence parameters. L1 = white light saturating 1 s pulse (3000 µmol m^{-2} s^{-1}); L2 = white actinic light (200 µmol m^{-2} s^{-1}); L3 = far-red illumination; mb = measuring beam of weak far-red light (1–2 µmol photons m^{-2} s^{-1}); ↑ = light on; ↓ = light off.

Effeltrich, Germany) as described by Schreiber et al. (1986). Measurements of chlorophyll fluorescence emission, from the upper surface of leaves and from the more-exposed side of stems, were made on attached leaves and stems. Illumination and fluorescence measurements were made via a fiber optic. To ensure a complete and constant illumination on all samples the fiber optic was located directly on the photosynthetic tissue surface.

After 12 h of darkness (corresponding to night time) fluorescence was excited with a measuring beam of weak far-red light (1–2 µmol photons $m^{-2} s^{-1}$) from a pulse light-emitted diode to obtain the minimal fluorescence yield when all PSII reaction centres are open (F_o). A saturating white light pulse (1 s) (L1) provided by a Schott KL 1500 lamp (Schott AG, Mainz, Germany) was then applied to temporarily close all photosystem reaction centers reducing the acceptor Q, and therefore to determine the maximum fluorescence yield (F_m) of the dark-adapted sample. The green tissue samples were then illuminated for 5 min with actinic light (200 µmol photons $m^{-2} s^{-1}$) (L2) provided by a slide projector with a halogen lamp (250 W) fitted with a KG1 heat filter. After 5 min illumination the maximal fluorescence yield (F'_m) was determined by applying a saturating white light pulse (1 s) (L1). Then the actinic light (L2) was turned off and a far-red illumination (L3) was quickly turned on to obtain complete reoxidation of the primary electron acceptor of PSII (Q_A) and to determine the minimal fluorescence yield when all PSII reaction centres are open in the steady-state light (F'_o). The protocol for these chlorophyll fluorescence measurements is illustrated in Fig. 1. The weak modulated measuring beam was used at 1.6 kHz in the dark and increased to 100 kHz during illumination. Saturating pulses from the Schott 1500 lamp, the intensity (3000 µmol photons $m^{-2} s^{-1}$) of which was adjusted to saturate the fluorescence yield as well as to avoid photoinhibition during the pulse, were controlled from a PAM 103 Trigger Control Unit. Resolution of photochemical (q_P) and non-photochemical (q_N) quenching was made by the 1 s saturating pulse given after 5 min illumination with actinic light. The quantum yield of PSII electron transport (ϕ_e) was calculated as $(F'_m\text{-}F)/F'_m$ according to Genty et al. (1989).

The initial fluorescence parameters measured after a 12 h dark period were considered as starting values. Subsequently chlorophyll fluorescence measurements were made at 0.5, 1, 3 and 6 hours during the high light treatment and at 14, 17–19 and 24 hours during the recovery at low light. Therefore all the fluorescence parameters measured upon high-light and dark exposure were expressed as percentage of the starting values. Prior to the fluorescence measurements during high-light treatments all plants were predarkened for 20 min at room temperature.

The nomenclature and calculation of the chlorophyll fluorescence followed the recommendations of van Kooten and Snel (1990).

Statistical analysis

Analysis of variance (ANOVA) was used to test significant differences between different high-light treatments and between the fluorescence parameters of leaves and stems. In order to determine whether significant differences existed, the LSD-test (Least Significant Difference) was used based on a significance level of $p \leq 0.05$. The data presented are at least the mean ± standard deviation for two measurements from three different leaf samples. In all figures where error bars are indicated, the standard deviation is shown.

Results and Discussion

Chlorophyll *a* fluorescence measurements were performed in leaves and stems of *C. quinquangularis* Chiov. during illumination at 700 (L1) and 1400 (L2) µmol photons $m^{-2} s^{-1}$ and a subsequent dark or low light phase. Significant decreases ($p < 0.01$–0.001) in F_o, F_m and F_v, were observed in the leaf and in the stem tissues with increasing exposure time (Fig. 2 A–F) and these were more pronounced at L2 than L1 light treatments. A lowered yield of F_v in response to an excessive light intensity has been related to photoinhibition of photosynthesis (Krause, 1988; Krause and Weis, 1991). The changes observed in these fluorescence parameters suggest that the two light intensities to which *C. quinquangularis* plants have been exposed in these experiments represent a light intensity stress. Moreover in both photosynthetic tissues at L1 and L2 treatment F_v showed a higher decrease than F_o.

According to Butler's model (Butler, 1978; Butler and Kitajima, 1975) a decline in F_v and F_o indicates an increase of non-radiative excitation energy dissipation. In contrast to the pronounced quenching of F_m and F_v, the F_v/F_m ratio, defining the PSII photochemical efficiency, was accompanied by only a slighter significant decrease ($p < 0.01$) in stems and leaves of *C. quinquangularis* after L1 and L2 treatment (Fig. 3 A–B). After 6 h exposure to 700 µmol photons $m^{-2} s^{-1}$ the non-radiative energy dissipation was able to prevent a significant reduction of the PSII photochemical efficiency in stems

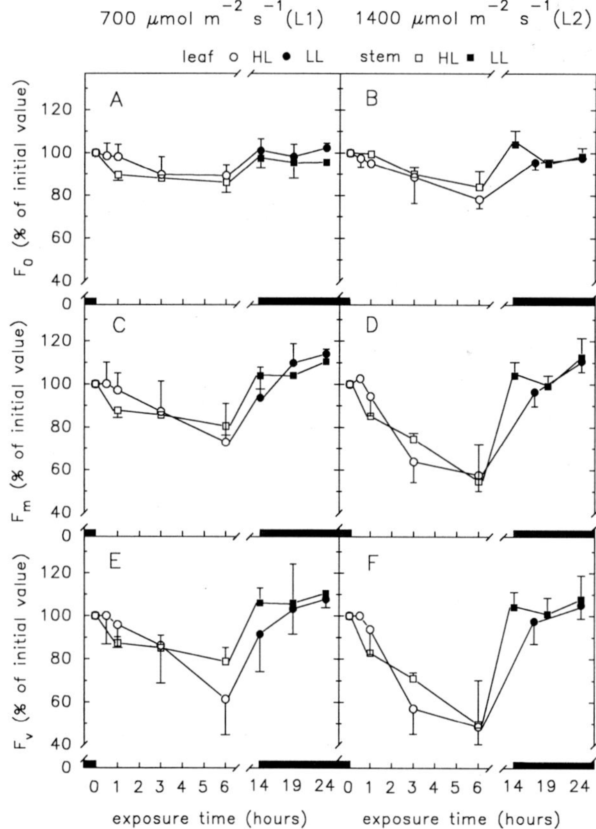

Fig. 2: Changes in minimal fluorescence (F_o), maximum fluorescence (F_m) and variable fluorescence (F_v) in leaves (○, ●) and stems (□, ■) of *C. quinquangularis* plants during illumination (open symbols) and during subsequent dark phase (closed symbols). Plants were exposed to 700 and 1400 µmol $m^{-2} s^{-1}$ (left and right side respectively). Values represent the percentage of the initial value before the light treatment.

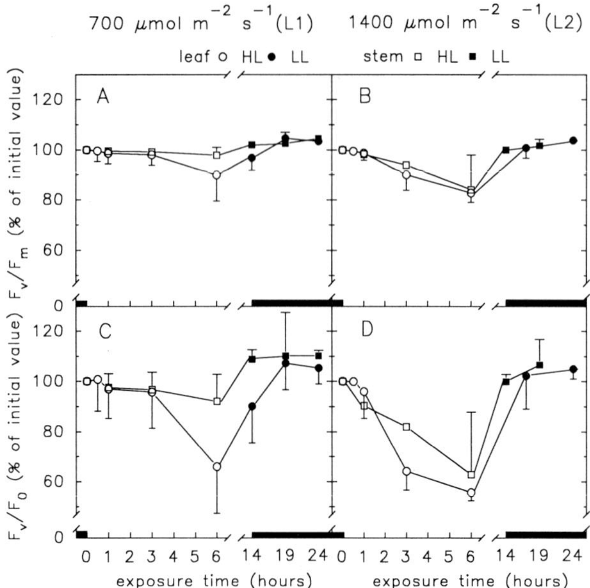

Fig. 3: Changes in the PSII photochemical efficiency expressed as F_v/F_m and F_v/F_o in leaves (○, ●) and stems (□, ■) of *C. quinquangularis* induced by light treatment (open symbols) and dark recovery (closed symbols). Plants were exposed to 700 and 1400 µmol m^{-2} s^{-1} (left and right side respectively). Values are expressed in percent of the initial value.

but not in leaves. In fact, in stems F_v/F_m was reduced only by 2.1 %, wheras in leaves the reduction amounted to 10.1 %. After exposing to 1400 µmol photons m^{-2} s^{-1}, in spite of the higher quenchings in F_m and F_o (Fig. 2 B, D, E), F_v/F_m decreased to an higher extent in stem and in leaf tissues (17.3 % and 16.0 % respectively) than after L1 treatment (Fig. 3 B). The photochemical efficiency can be detected also by the F_v/F_o ratio which exhibits a greater sensibility being able to reflect any decrease of photochemical efficiency better than F_v/F_m (Fig. 3 C–D) (Lichtenthaler et al., 1992). It is evident from these data that after a 6 h exposure to 1400 µmol photons m^{-2} s^{-1} photoinhibition occurs in stem as well as in the leaf tissue; however this photoinhibition is completely recovered after 14–17 h darkness (Fig. 3 A–B).

The rate of energy dissipation has been quantified by a non-photochemical quenching coefficient (q_N) (Bradbury and Baker, 1981, 1984; Bilger and Schreiber, 1986; Schreiber et al., 1986). Upon 6 h exposure to 700 µmol m^{-2} s^{-1} q_N was almost constant in the stem tissue (109.4 %) while it increased up to 170.2 % in the leaf tissue (Fig. 4 A). The increase in q_N was higher in both photosynthetic tissues at L2 compared to L1 treatment and after 6 h exposure q_N amounted to 121.3 % in stems and to 200.3 % in leaves (Fig. 4 B). These results suggest that the non-photochemical energy dissipation increases with the intensity and duration of light exposure and that it was higher in leaves than in stems.

Recently, it has been demonstrated that a decrease in photosynthetic energy conversion (photoinhibition) can be

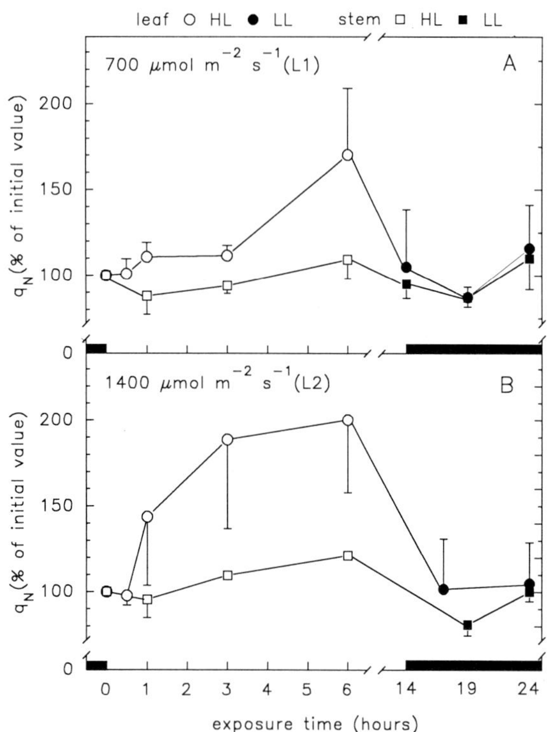

Fig. 4: Non-photochemical quenching (q_N) in leaves (○, ●) and stems (□, ■) of *C. quinquangularis* upon illumination (open symbols) and recovery in darkness (closed symbols). Values represent the percentage of the initial value before the light treatment.

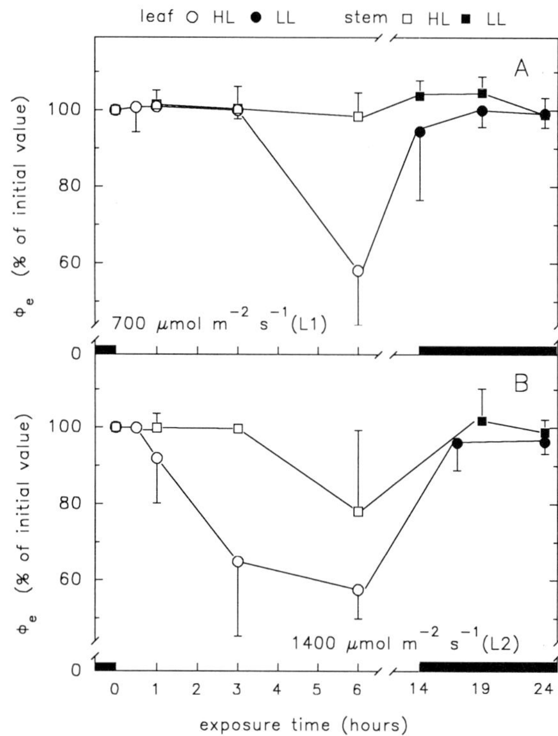

Fig. 5: Changes in the quantum yield of PSII electron transport (ϕ_e) in leaves (○, ●) and stems (□, ■) of *C. quinquangularis* induced by light treatments (open symbols) and dark recovery (closed symbols). Values are expressed in percent of the initial value.

caused not only by a damage to PSII but also by photoprotective processes, such as the thermal energy dissipation that represents an early response before a damage occurs to the photosynthetic apparatus (Demming-Adams and Adam III, 1992). It has been demonstrated that increases in q_N by increasing light intensity represent a «down regulation» of photosynthesis (Baker, 1991). This means that any increase of non-radiative excitation energy dissipation can be an early process of a overall regulatory mechanism to minimize or prevent damage to PSII reaction centers and consequently to allow a photochemical efficiency be kept upon a light stress. In *C. quinquangularis* plants exposed to light stress the energy dissipation is an important photoprotective process in both photosynthetic tissues; however, the leaves exhibited higher rates of energy dissipation that seem to be related to a more intensive light stress of leaves as compared to the stems.

In order to obtain information on the regulation of photochemistry in *C. quinquangularis* plants exposed to high-light treatments the quantum yield of PSII electron transport (ϕ_e) was also measured. When plants were exposed to 700 µmol photons $m^{-2} s^{-1}$ no change in ϕ_e was observed throughout the exposure time in the stems, whereas a significant decrease ($p<0.001$) to 58.1% only occurred in leaves after a 6 h exposure (Fig. 5 A). Also after exposing to 1400 µmol photons $m^{-2} s^{-1}$ higher values of ϕ_e were always found in the stems; however a decrease of ϕ_e was apparent in the stems

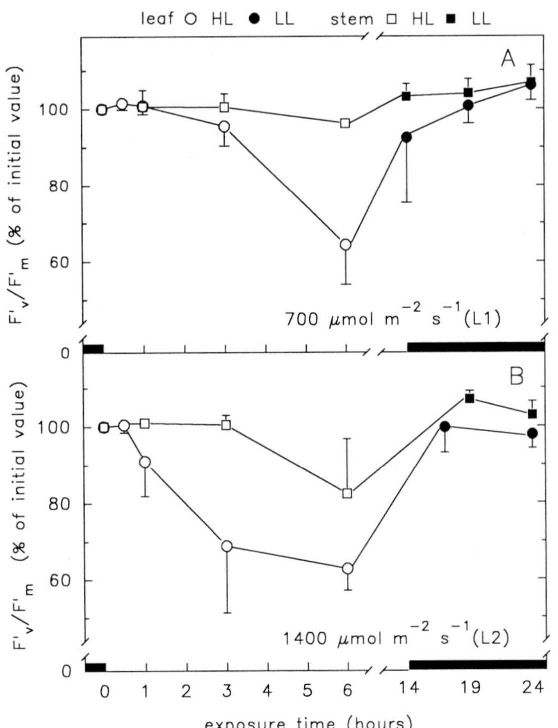

Fig. 6: Changes of the efficiency of excitation energy capture by open PSII reaction centres (F'_v/F'_m) in leaves (○, ●) and stems (□, ■) of *C. quinquangularis* plants exposed to light exposures (open symbols) and during subsequent recovery in darkness (closed symbols). Values represent the percentage of the initial value before the light treatment.

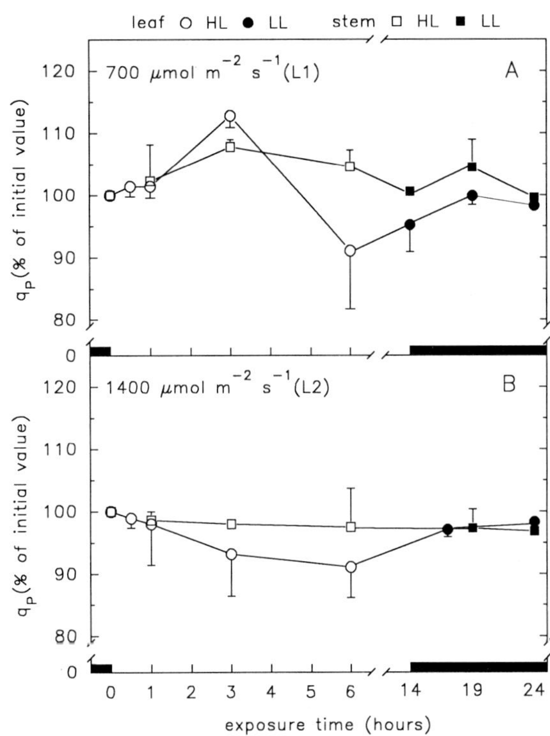

Fig. 7: Effects of high irradiances (open symbols) with subsequent recovery in darkness (closed symbols) on photochemical quenching (q_P) in leaves (○, ●) and stems (□, ■) of *C. quinquangularis* plants. Values are expressed in percent of the initial value.

(78.1%) only after the 6 h high-light exposure, whereas in leaves such a decrease up to 57.5% was significant ($p<0.001$) and continuous throughout the light exposure (Fig. 5 B). These data are evidence to a greater capacity of the stem as compared to the leaf tissue to avoid, upon a high light stress, a reduction in quantum yield of the PSII electron transport.

As described by Genty et al. (1989) the quantum yield of PSII electron transport (ϕ_e) is the product between the efficiency of excitation energy capture by open PSII reaction centres in a light-adapted state (F'_v/F'_m) and photochemical quenching (q_P). In these experiments it was also demonstrated that the leaf tissue was able to avoid a reduction in ϕ_e only until 3 h of exposure to 700 µmol photons $m^{-2} s^{-1}$ (Fig. 5 A). In fact, at this point it was observed that ϕ_e remained constant, in spite of a slight F'_v/F'_m reduction (Fig. 6 A), due to an increase in q_P (Fig. 7 A). Such an increase in q_P, a measure of available excitation energy that is used in photochemistry, could represent a mechanism to prevent a ϕ_e reduction only up to this light stress intensity. By increasing light and time exposure the leaf tissue was not able to avoid reduction of quantum yield of PSII electron transport. Moreover, during light treatments in stems ϕ_e was influenced only by F'_v/F'_m being q_P always constant, while in leaves reductions of ϕ_e values was accompanied by decreases in F'_v/F'_m and q_P (Fig. 6 A–B, 7 A–B).

During light exposure the reduction state of Q_A was also determined, the primary stable electron acceptor of PSII, expressed as fraction of reduced or closed PSII centres

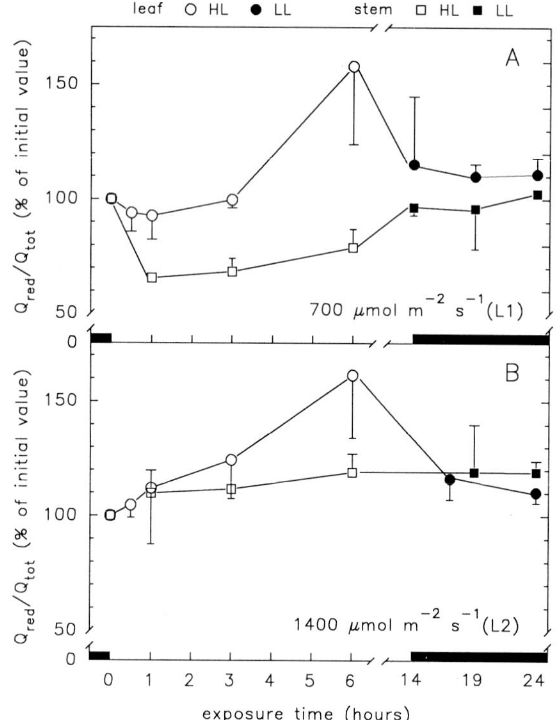

Fig. 8: Changes in the fraction of reduced or closed PSII centres (Q_{red}/Q_{tot}) upon light treatments (open symbols) with subsequent recovery in darkness (closed symbols) in leaves (○, ●) and stems (□, ■) of *C. quinquangularis* plants. Values represent the percentage of the initial value before light treatment.

(Q_{red}/Q_{tot}). Upon exposure to 700 µmol photons m^{-2} s^{-1} in the stem tissue Q_{red}/Q_{tot} was always lower than the starting value, whereas in the leaf tissue it increased significantly ($p<0.001$) up to 158.1% at a 6 h light exposure (Fig. 8 A). At an exposure to 1400 µmol photons m^{-2} s^{-1} Q_{red}/Q_{tot} increased continuously in both of the photosynthetic tissues, but the increase was lower in the stem (up to 118.9%) than in leaf tissue (up to 161.5%) (Fig. 8 B). It had been observed that photoinhibition correlates positively with the reduction state of PSII centres (Horton and Hague, 1988: Ögren, 1991). In the experiments, described here, a significant increase in the reduction state of Q_A was observed only in the leaf tissue. In contrast, during high-light exposure, the photochemistry rate was not significantly influenced in the stem tissue.

Our data evidenced that, in spite of the same degree of photoinhibition, leaves and stems of CAM plant *C. quinquangularis* exposed to high-light stress showed a different strategy in the regulation of photochemistry. The stem tissue exhibited a greater capacity to keep the photochemistry at a higher rate and the non-photochemical energy dissipation process was more limited than in the leaf tissue. However, even if the leaves seemed to experience a higher degree of light stress, as evidenced by a lowered photochemistry rate, all chlorophyll fluorescence parameters were completely recovered after 14 to 17 h in darkness in the leaf as well as in the stem tissue. The question of the relative contribution of leaves and stems of *C. quinquangularis* to the overall carbon gain by these two different strategies of regulation of photochemistry under high-light stress conditions remains to be investigated.

The leaf's strategy of regulation of photochemistry, evidenced by these experiments in the CAM plant *C. quinquangularis* Chiov. exposed to high-light stress, support the view, observed in several C_3 and C_4 plants, that photoinhibition has to be considered also as a long-term down-regulation of PSII under high-light stress (Öquist et al., 1992 a, 1992 b; Baker, 1991).

Acknowledgements

A grant of DAAD (Deutscher Akademischer Austauschdienst), Bonn to N. D. is gratefully acknowledged. We would like to thank the Botanical Garden of the University of Naples for providing *C. quinquangularis* plants and Mr. Bernhard Dill and coworkers of the Botanical Garden of the University of Karlsruhe for their skillful assistance in propagating and growing the plants.

References

Adams, W. W. III and C. B. Osmond: Internal CO_2 supply during photosynthesis of sun and shade grown CAM plants in relation to photoinhibition. Plant Physiol. *86,* 117–123 (1988).

Adams, W. W. III, B. Demming-Adams, K. Winter, and U. Schreiber: The ratio of variable to maximum chlorophyll fluorescence from photosystem II, measured in leaves at ambient temperature and at 77K, as an indicator of the photon yield of photosynthesis. Planta *180,* 166–174 (1990).

Baker, N. R.: A possible role for photosystem II in environmental perturbations of photosynthesis. Physiol. Plant. *81,* 563–570 (1991).

Bilger, W. and U. Schreiber: Energy-dependent quenching of dark level chlorophyll fluorescence in intact leaves. Photosynth. Res. *10,* 303–308 (1986).

Björkman, O.: Low-temperature chlorophyll fluorescence in leaves and its relationship to photon yield of photosynthesis in photoinhibition. In: Kyle, D. J., C. B. Osmond, and C. J. Arntzen (eds.): Photoinhibition, pp. 123–144. Elsevier Science Publishers, Amsterdam (1987).

Bradbury, M. and N. R. Baker: Analysis of the slow phases of the *in vivo* chlorophyll fluorescence induction curve. Changes in the redox state of photosystems II electron acceptors and fluorescence emission from photosystems I and II. Biochim. Biophys. Acta *635,* 542–551 (1981).

– – A quantitative determination of photochemical and non-photochemical quenching during the slow phase of the chlorophyll fluorescence induction curve of bean leaves. Biochim. Biophys. Acta *765,* 275–290 (1984).

Butler, W. L.: Energy distribution in the photochemical apparatus of photosynthesis. Annu. Rev. Plant Physiol. *29,* 345–378 (1978).

Butler, W. L. and M. Kitajima: Fluorescence quenching in PSII of chloroplasts. Biochim. Biophys. Acta *376,* 116–125 (1975).

Chiovenda, E.: Flora somala II. p. 141. Pubblicazione del Ministero delle Colonie, Modena (Real Orto Botanico), Italy (1932).

Demming-Adams, B. and W. W. Adams III: Photoprotection and other responses of plants to high light stress. Annu. Rev. Plant Physiol. Plant Mol. Biol. *43,* 599–626 (1992).

Genty, B., J.-M. Briantais, and N. R. Baker: The relationship between the quantum yield of photosynthetic electron transport and quenching of chlorophyll fluorescence. Biochim. Biophys. Acta *990,* 87–92 (1989).

GIL, F.: Origin of CAM as an alternative photosynthetic carbon fixation pathway. Photosynthetica *20,* 494–507 (1986).

HORTON, P. and A. HAGUE: Studies of the induction of chlorophyll fluorescence in isolated barley protoplasts. IV. Resolution of non-photochemical quenching. Biochim. Biophys. Acta *932,* 107–115 (1988).

HORTON, P. and A. V. RUBAN: Regulation of photosystem II. Photosynth. Res. *34,* 375–385 (1992).

KRAUSE, G. H.: Photoinhibition of photosynthesis. An evaluation of damaging and protective mechanisms. Physiol. Plant. *74,* 566–574 (1988).

KRAUSE, G. H. and E. WEIS: Chlorophyll fluorescence and photosynthesis: the basics. Annu. Rev. Plant Physiol. Plant Mol. Biol. *42,* 313–349 (1991).

LICHTENTHALER, H. K. and U. RINDERLE: The role of chlorophyll fluorescence in the detection of stress conditions in plants. CRC Crit. Rev. Anal. Chem. *19* (Suppl.), S 29–S 85 (1988).

LICHTENTHALER, H. K., S. BURKART, C. SCHINDLER, and F. STOBER: Changes in photosynthetic pigments and *in vivo* chlorophyll fluorescence parameters under photoinhibitory growth conditions. Photosynthetica *27,* 343–353 (1992).

ÖGREN, E.: Prediction of photoinhibition of photosynthesis from measurements of fluorescence quenching components. Planta *184,* 538–544 (1991).

ÖQUIST, G., W. S. CHOW, and J. A. ANDERSON: Photoinhibition of photosynthesis represents a mechanism for the long-term regulation of photosystem II. Planta *186,* 450–460 (1992 a).

ÖQUIST, G., J. A. ANDERSON, S. MCCAFERRY, and W. S. CHOW: Mechanistic differences in photoinhibition of sun and shade plant. Planta *188,* 422–431 (1992 b).

OSMOND, C. B.: Photorespiration and photoinhibition: some implications for the energetics of photosynthesis. Biochimica Biophysica Acta *639,* 77–98 (1981).

OSMOND, C. B., K. WINTER, and S. POWLES: Adaptative significance of carbon dioxide recycling during photosynthesis in water-stressed plants. In: TURNER, N. C. and P. J. KRAMER (eds.): Adaptation of plants to water and high temperature stress, pp. 139–154. John Wiley & Sons, New York (1980).

SCHREIBER, U. and W. BILGER: Progress in chlorophyll fluorescence research: major developments during the past years in retrospect. Progr. Bot. *54,* 161–173 (1993).

SCHREIBER, U., U. SCHLIWA, and W. BILGER: Continuous recording of photochemical and non-photochemical chlorophyll fluorescence quenching with a new type of modulation fluorometer. Photosynth. Res. *10,* 51–62 (1986).

VAN KOOTEN, O. and J. F. H. SNEL: The use of chlorophyll fluorescence nomenclature in plant stress physiology. Photosynth. Res. *25,* 147–150 (1990).

J. Plant Physiol. Vol. 148. pp. 142–147 (1996)

Carbon Metabolism Enzyme Activities and Carbon Partitioning in *Pinus halepensis* Mill. exposed to Mild Drought and Ozone

D. Gerant[1], M. Podor[2], P. Grieu[2], D. Afif[1], S. Cornu[1], D. Morabito[1], J. Banvoy[1], C. Robin[2], and P. Dizengremel[1]

[1] Laboratoire de Biologie Forestière, associé INRA, équipe Ecophysiologie Cellulaire, Université Henri Poincaré, Nancy I, BP 239, 54506 Vandoeuvre les Nancy Cedex, France

[2] Laboratoire Agronomie & Environment ENSAIA-INRA, BP 172, 54 505 Vandoeuvre les Nancy Cedex, France

Received June 26, 1995 · Accepted October 10, 1995

Summary

Since several years, accelerated decline of Aleppo pine (*Pinus halepensis*) forests has been observed in mediterranean areas. In fact, the combination of various environmental factors (photochemical oxidants, drought, high light, ...) was suspected to cause this decline. In this study, three year-old Aleppo pines were exposed during 3 months to ozone fumigation (100 ppb) combined or not with mild drought to study the effects of these combined factors on some sequences linked to carbon partitioning and primary carbon metabolism within the tree. After a cumulative ozone exposure of 132 ppm·h, ozone induced a significant decrease in specific activity of the whole-plant (−38 %) combined with a disequilibrium of the carbon transfer between root and shoot in favour of the shoots (non significant). Moreover, while the same cumulative dose of ozone had no effect on total Rubisco activity in one year-old needles, mitochondrial NAD malic enzyme activity increased significantly (+32 %). By combining ozone with mild drought, the ozone-induced responses of all the parameters were significantly amplified and Rubisco activity was significantly decreased (by 44 %). These results allowed us to conclude that at 132 ppm·h, ozone alone led to an increase in dark respiration. Moreover, by the combination of ozone and mild drought, a decrease carbon fixation capacity was associated to a decrease of the carbon transfered to the roots, leading to a reduced root growth. Thus, there are indications that high levels of ozone during the summer months may impair the ability of *Pinus halepensis* to withstand severe water stress in its natural environment.

Key words: Pinus halepensis Mill., carbon partitioning, growth chambers, mild drought, mitochondrial NAD malic enzyme, ozone fumigation, Rubisco.

Abbreviations: DC = droughted control trees; DO = droughted and ozone-treated trees; DW = dry weight; HC = humid control trees; HO = humid ozone-treated trees; MBq = Mega Becquerel; ψ_b = predawn needle water potential; NAD-ME = NAD-malic enzyme; PVC = polyvinyl chloride; Rubisco = ribulose bisphosphate carboxylase oxygenase; SA = specific activity; TA = total activity.

Introduction

The aim of the EC project (EV5V CT93 0263) in which this present work is integrated is to determine the fundamental effects of ozone on Aleppo pine (*Pinus halepensis* Mill.) against a background of other environmental factors and to understand the physiological and biochemical mechanisms that enable this species to be successful in a mediterranean environment. Since several years, coinciding with the rapid increased traffic, accelerated decline of *Pinus halepensis* forests has been observed in mediterranean areas. As *Pinus halepensis* is of major ecological significance in the Mediterranean region (particularly Greece and Spain), the responses of *Pinus halepensis* to the seasonal and diurnal variations in its environment, either with or without the added stress of ozone have to be known. A previous EC project work (STEP CT91 0017) has confirmed that severe chlorosis exhibited by Aleppo pines is not primarily nutritional but is due to a photo-oxidation of the chlorophylls under drought and high temperature (Velissariou et al., 1992; Gimeno et al., 1992). Whether this chlorosis has any relationship with ozone is still not understood and we explored the hypothesis that ozone pre-disposes the pine to summer stresses and more particularly in this paper to drought.

The carbon budget is of importance in regard to the growth processes in higher plants. Numerous studies have shown that the impact of environmental factors (atmospheric pollutants, soil water deficit, ...) on plant growth and carbon budget is considered as a major target for these stresses within the plant (Chen et al., 1994). Among the different processus concerned, net photosynthesis/dark respiration (Reich, 1983; Wallin et al., 1990; Heath, 1994 a) and carbon allocation (Mc Laughin et al., 1982; Spence et al., 1990; Friend and Tomlinson, 1992) are particularly affected. By contrast, few is known about the biochemical responses at the cellular level.

Early in the spring, in Greece and Spain, ozone pollution and mild drought naturally occur and may affect *Pinus halepensis.* To reproduce natural conditions occurring in the mediterranean areas at this period of the year, experiments were carried out in growth chambers. Three year-old Aleppo pines were fumigated with ozone alone or combined with mild drought. The aim of this investigation was to determine if the ozone dose used increases drought sensitivity of *Pinus halepensis.* To answer this question we have chosen to integrate studies at different levels. At the cellular level, the responses of the activities of two enzymes considered as markers of the main primary carbon sequences (Rubisco for photosynthesis and NAD malic enzyme (EC 1.1.1.39) specifically located in the mitochondria for respiration) will be confronted, at the whole-plant level, to the behaviour of the pool of carbon newly available within *Pinus halepensis* tree.

Material and Methods

Plant material and growth conditions

Three year-old Aleppo pine (*Pinus halepensis* Mill.) seedlings were grown in greenhouse in 4 kg containers filled with a potting mixture, sand and perlite (50 %, 30 %, 20 %), with one plant per container. A slow-released fertilizer was added: 2 g of Nutricot 100 (N/P/K: 16/9/9) per pot. Before the experiment, 144 pines were transfered into 4 growth chambers during two months for acclimation under the following day/night conditions: 14/10 h of photoperiod, relative humidity 70/90 %, air temperature 22/15 °C, CO_2 molar fraction in the air 400 ± 20 $\mu mol \cdot m^{-2} \cdot s^{-1}$ and photon flux density at the top of the plants 500 $\mu mol \cdot m^{-2} \cdot s^{-1}$. No discrimination in watering occurred until the beginning of the experiment: trees were irrigated every two days at the field capacity by weighing the individual pots.

Ozone exposure and water supply

Half of pots were exposed to 100 ppb (200 $\mu g/m^3$) of ozone in 2 chambers in a complete randomized design during 3 months. Fumigations were performed 14 h/day during the photoperiod. The remaining pines were continuously exposed to charcoal-filtered air during the same period in the two other chambers. The environmental conditions were those of the acclimation period. Ozone was generated from pure oxygen with an ozone generator (Ozone generator model 500M, Fischer™). The flow through the chamber corresponded to 1.2 air changes per minute. Ambient air was analysed continuously by an ozone analyser ($O_3$41M, Environment S.A.). At the sampling date presented here (104 days), cumulative ozone exposure was estimated to 132 ppm·h.

Two different irrigation regimes were applied in each growth chamber: (i) control, irrigated at the field capacity every two days, (ii) water deficit irrigated with an amount of water equivalent in a first part to 75 % of their evapotranspiration rate and to 50 % in the last part of the experiment. Predawn needle water potentiel (ψ_b) was used as an index of water deficit. Predawn needle water potentiel was determined on a single needle with a Scholander pressure chamber (Scholander et al., 1965). ψ_b was measured on one needle per tree, on 10 trees per treatment.

Carbon fixation and transfer

C-14 exposure. After 104 days of ozone fumigation, 6 pines of each treatment were exposed to a labelling with $^{14}CO_2$. For each labelling, a chamber consisting of altuglass (130 × 72 × 55 cm^3) supported on a PVC frame was placed over the pines to be exposed. At the bottom of the chamber a PVC support was laid flat around it and a join of about 15 cm of water provides an effective seal. At the top a join of water protected the trees from a warming of the atmosphere due to the infra red radiations. The environmental conditions were the same as in the growth chambers. Three fans located inside the labelling chamber insured a good mixing of the air and a good repartition of ^{14}C during the labelling. A source of 1.48 MBq (40 microcuries) from $NaH^{14}CO_3$ (Amersham, specific activity: 2.04 GBq/mmol) per plant is used in a small vial fixed on the wall inside the labelling chamber. At the beginning of the labelling, the specific activity of the atmosphere was 148.29 MBq/g of carbon (C). The whole plants were exposed to $^{14}CO_2$ during a 3 h labelling period. After this period the atmosphere inside the labelling chamber was renewed and the remaining $^{14}CO_2$ was fixed on soda lime for 21 h (chase period). Pines were then transfered in a growth chamber for 15 hours. After this period pines were harvested and separated in 5 compartments: current-year needles, previous-years needles, current-year stems, previous-years stems and roots. The samples were immediately frozen at –180 °C in liquid nitrogen, freeze-dried for 3 days and weighted.

Carbon content analysis – Counting

The carbon content of small subsamples (1 to 5 mg DW) was determined after combustion and gas chromatography in a Nitrogen/Carbon autoanalyser (NA 1500 CARLO ERBA). The CO_2 carried by helium was then trapped by phenethylamine of a scintillator cocktail (Carbo Max Plus, Packard) for ^{14}C counting in liquid scintillation.

The ^{14}C activity of each sample was counted by a Packard TRI-CARB 460 D liquid scintillation spectrometer. Two parameters were measured: (i) the specific activity in the whole plant for each treatment: SA (Bq/g of dry mass), this parameter represents the measurement of the amount of CO_2 fixed in the plant during the labelling experiment (ii) the radioactivity in the various plant fractions expressed as % of total activity in the plant (% TA) calculated from:

$$\% \text{ TA} = \frac{\text{Bq of the organ}}{\text{Bq of the whole plant}}$$

Enzyme extraction

Fresh needles (500 mg) were homogenized in 10 mL of extraction medium (Dizengremel et al., 1994). The extract was centrifuged at 20,000×g for 45 min. The crude supernatant was desalted through PD-10 columns (Sephadex G25, Pharmacia, Piscataway, NJ, U.S.A.) with elution medium (extraction medium without protective agents). All the extraction and desalting steps were performed at 4 °C.

Enzyme activities

Rubisco activities were analyzed spectrophotometrically (UVIKON 930, Kontron U.K., Ltd.) according to Lilley and Walker (1974) with some modifications (Dizengremel et al., 1994). The assays were performed at 25 °C and Rubisco total activity was measured after a 15 min incubation period with $MgCl_2$ 20 mM and $NaHCO_3$ 25 mM before adding 0.5 mM ribulose 1,5-bisphosphate (RuBP). NAD-malic enzyme activity was measured as described by Davis and Patil (1975), the reaction was initiated by the addition of $MnCl_2$ 2.5 mM. Enzyme activities were expressed as nanomoles NADH oxidized (Rubisco) or NAD reduced (NAD-ME) per second (nanokatal) in relation to total soluble proteins unit.

Total chlorophyll content of needles was determined in 80 % acetone according to Lichtenthaler (1987). Total soluble proteins in the extracts were determined by the Coomassie Blue G 250 method of Bradford (1976) with bovine serum albumin as standard.

Table 1: Effects of ozone alone or combined with mild drought on total soluble proteins (mg/g DW) and total chlorophylls (mg chl/g DW) contents in one year-old needles of *Pinus halepensis* Mill. Standard errors were given for n=3. The letters a, b and c represent significant differences between treatments after ANOVA analysis (p<0.05).

Treatments	Total soluble proteins	chlorophyll a+b
HC	10.12±1.30 c	1.39±0.20 b
DC	3.66±0.36 a	0.77±0.19 b
HO	6.60±0.27 b	1.05±0.26 b
DO	3.91±0.21 a	0.53±0.18 a

Table 2: Effects of ozone alone or combined with mild drought on total Rubisco and NAD malic enzyme (NAD ME) activities (nanokat/mg protein) of one year-old *Pinus halepensis* needles. Values in parentheses are percent of control (HC). Standard errors were given for n=3. The letters a, b and c represent significant differences between treatments after ANOVA analysis (p<0.05).

Treatments	Total Rubisco activity		NAD ME activity	
HC	9.07±0.78 b	(100)	0.57±0.01 a	(100)
DC	3.99±0.89 a	(44)	1.09±0.04 c	(191)
HO	8.80±0.88 b	(97)	0.75±0.01 b	(132)
DO	5.07±0.34 a	(56)	1.12±0.12 c	(197)

Results

Specific activity of the whole-plant (SA) as a carbon gain index

After 104 days of experiment (Fig. 1), the specific activity of the whole plant (in Bq/g of carbon) for well-watered ozone-treated pines (HO) was significantly reduced by 38 % compared to well-watered control pines (HC). For the same period, this paramater was strongly decreased for droughted control (DC) and droughted ozone-treated pines (DO) (respectively −84 and −95 % compared to well-watered control, HC). At the end of the experiment, predawn needle water potential was about −0.8 MPa.which corresponds to a mild drought.

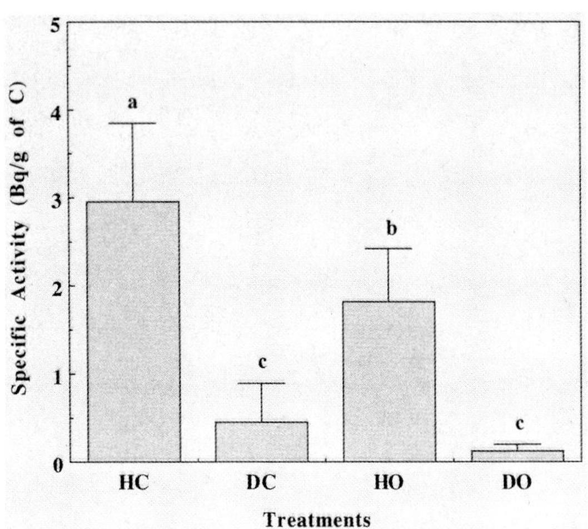

Fig. 1: Effects of ozone alone or combined with mild drought on specific activity of ^{14}C incorporation of the whole-plant in *Pinus halepensis* Mill. Values are the mean from six independant determinations with vertical bars indicating standard deviations.The letters a, b and c represent significant differences between treatments after ANOVA analysis (p <0.05). HC = humid control trees, DC = drought control trees, HO = humid and ozone-treated trees, DO = drought and ozone-treated trees.

Effect of ozone associated or not with mild drought on total soluble proteins and total chlorophyll contents

The chlorophyll *a+ b* content of the one year-old needles is presented in Table 1. No significant difference was observed when the chlorophyll content was expressed on a dry weight basis between DC or HO needles vs HC needles. By contrast, the chlorophyll content in DO needles was significantly decreased.

Total soluble protein content expressed on a dry weight basis is presented in Table 1. Protein content of HO one year-old needles was lower (up to 35 %) than that of control (HC) needles and significant decreases in total soluble proteins content were noticed for DC (36 % of the control HC) and for DO (39 % of the control).

Effect of ozone associated or not with mild drought on enzyme activities

Rubisco activities of the one year-old needles, measured at the end of the experiment (104 days), are shown on Table 2. Ozone exposure (HO vs HC) had no effect on total Rubisco

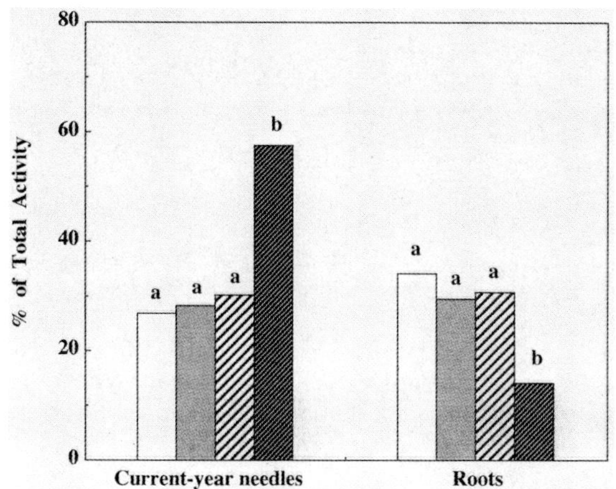

Fig. 2: Effects of ozone alone or combined with mild drought on carbon partitioning in the roots and the current-year needles in *Pinus halepensis* Mill. Values are the mean from six independant determinations.The letters a, b and c represent significant differences between treatments after ANOVA analysis ($p<0.05$). (□): HC = humid control trees, (▒): DC = drought control trees, (▨): HO = humid and ozone-treated trees, (■): DO = drought and ozone-treated trees.

activity whereas mild drought decreased significantly Rubisco activities up to 55 % (DC vs HC) in the same needles. By combining ozone and mild drought (DO vs HC), Rubisco was decreased to a similar level as for drought alone.

A significant increase in activity was observed for NAD ME (about 32 %) in the HO needles and the stimulation was higher in DC and DO needles where NAD ME activity was two times higher than in HC needles.

Carbon transfer

Only the results concerning the roots and the current-year needles are presented in this paper as no significant variations were noticed for the other compartments (previous-years needles, current and previous-year stems). At 104 days (Fig. 2) there was a slight increase in carbon transfer to current-year needles (+11.4 % compared to the well-watered control) due to ozone. This was amplified and became significant with the combination of ozone with drought treatment (+53.5 %). Ozone tended to reduce carbon transfer to the roots (by 10 %). Combined with mild water deficit carbon transfer to the roots was significantly decreased by 58.4 % compared to well-watered control (Fig. 2).

Discussion

A reduction of the specific activity of ^{14}C-incorporation in the whole-plant (Fig. 1) was observed in ozone-treated pines (HO). This result suggests that CO_2 exchange rates were different for ozone-treated seedlings (HO) and control seedlings (HC). However, in *Pinus halepensis* exposed to a cumulative dose of ozone of 44 ppm·h, we found no modification in net CO_2 assimilation (Grieu and Robin, 1993). This previous result is reinforced in the present study by no significant modification of fully activated Rubisco capacity in the ozone-treated pines (Table 2) even at 132 ppm·h. In the literature, a reduction in net photosynthesis rates due to ozone has been observed in conifers (Barnes, 1972; Tseng et al., 1988). However, the effect of ozone on photosynthetic processes depends on ozone concentration and on the shoot age (Wallin et al., 1990). In *Pinus taeda* saplings exposed to ozone (Richardson et al., 1992 a, b; Dizengremel et al., 1994), the significant decrease in needle photosynthetic rates was proportional to the cumulative ozone exposure, decreasing by 50 % after 350 ppm·h. Moreover, in loblolly pine grown under elevated ozone (Dizengremel et al., 1994), there was a positive correlation between the decrease in photosynthetic rate and Rubisco capacity and quantity. Under controlled ozone exposure, a decrease in Rubisco activity as well as Rubisco quantity was also observed in spruce (*Picea abies* L. Karst.) needles but at lower ozone concentrations (Dizengremel et al., 1993) indicating an higher sensibility of this species to ozone compared to pine species. In our experiment, in one year-old needles of *Pinus halepensis* exposed to 132 ppm·h of ozone, Rubisco capacity was not yet modified. This result gives new arguments for the existence of a threshold value of cumulative ozone exposure (up to 100 ppm·h for *Pinus taeda*) beneath which photosynthetic processes (net photosynthesis, Rubisco activity) were unchanged.

The decrease in specific activity noticed for ozone-treated Aleppo pines associated with no modification in net assimilation led to consider the processes at the whole-plant level whereby carbon loss might occur: dark respiration, rhizodeposition and needle senescence. At the end of the treatment, the cumulative dose of ozone that *Pinus halepensis* seedlings received was not sufficient to induce important visible injury symptoms and there was no needle loss. To our knowledge, there is no reference about rhizodeposition in condition of ozone pollution and/or water deficit on trees. However, in our experiment, the strong decrease of carbon transfer to the roots (Fig. 2) eliminated the hypothesis of increased rhizodeposition in response to ozone fumigation. Thus, we focused on the hypothesis that ozone treatment induced an increase in dark respiration rates as observed in ozone-stressed *Pinus stobus* (McLaughin et al., 1982), in *Pinus sylvestris* (Skärby et al., 1987). To get better knowledge about the functioning of the respiratory processes in *Pinus halepensis,* we measured, in one year-old needles, the activity of NAD malic enzyme, choosen as specific marker of mitochondrial activity. This enzyme was shown to be stimulated by ozone treatment (Table 2). An increase in the activity of some mitochondrial enzymes was also reported in spruce trees exposed to elevated ozone (Gérant et al., 1988) although the magnitude of the response varied depending on the clone. The activity of G6PDH, the first enzyme of the pentose phosphate pathway, was also shown to increase in plants exposed to pollution (Koziol et al., 1988; Dizengremel and Petrini, 1993; Dizengremel et al., 1994). These reactions in the foliar cells could explain the increase in dark respiration rates leading to a decrease in net carbon fixation generally observed in ozone-stressed plants. Moreover, respiration processes largely participate to the production of energy in the plant cell. Indeed, catabolic pathways provide the energy to drive detoxification and repair processes which are activated by ozone-stress (Friend and Tomlinson, 1992; Rapport de l'Académie des Sciences, 1993).

The decrease in specific activity in response to ozone, was reinforced in drought x ozone fumigation conditions (DO) by the fact that drought by itself induced a decrease in carbon fixation capacity: i) decrease in net photosynthesis linked to stomatal closure (Grieu and Robin, 1993), decrease in Rubisco capacity (Table 2), decrease in chlorophyll content (Table 1) ii) an higher NAD malic enzyme activity stimulation (Table 2). Moreover, ozone associated with mild drought led to an increase in carbon transfer to the current-year needles and in parallel to a decrease in carbon transfer to the roots (Fig. 2). Our results could thus be consistant with those of Roberts (1990) who found that ozone and drought act together additively to alter physiological processes. On the other hand, these results are in contrast with the hypothesis that drought by inducing stomatal closure reduces damages due to ozone (Tingey and Hogsett, 1985). Ozone by itself in well watered trees induced increase (Pearson and Mansfield, 1993), no effects or even decrease (Eamus et al., 1990) in stomatal resistance depending on the growth and measurement conditions (Barnes et al., 1990; Heath, 1994 b). The response of stomatal resistance to ozone during water stress is thus of importance. In addition, ozone is given to cause a desequilibrium of the carbon transfer favouring shoots. For Spence et al. (1990) in *Pinus taeda,* total carbon transport to the roots was decreased (−40 %) and carbon was accumulated in the stems. For Friend and Tomlinson (1992), a short term carbon retention coupled with a diversion of carbon from storage compounds are foliar mechanisms to repair damages due to ozone. Wellburn and Wellburn (1994), in the same EC Project, showed an extensive accumulation of starch and crushing of the phloem sieve cells at mid-summer in Aleppo pine fumigated with episodes of ozone. They found a significant decrease in the root/shoot ratio in high level ozone-treated trees by the end of two summer seasons.

The combination of ozone and mild drought induced a decrease of the carbon fixation capacity associated to a decrease of the carbon transfered to the roots, leading probably to a reduced root growth in Aleppo pine. These results give arguments that in natural conditions, episodes of ozone probably affect the normal ability of *Pinus halepensis* to withstand moderate water deficit. The response of *Pinus halepensis* to summer conditions (severe water deficit, high episodes of ozone) should thus be particularly worrying. This may account for their reduced vitality in Mediterranean regions in condition of ozone pollution.

Acknowledgements

This work is supported by the Commission of the European Communities (STEP EV5V CT 93 0263: «Interactions between ozone, climatic and nutritional factors on coniferous tree physiology»).

References

Barnes, R. D.: Effects of chronic exposure to ozone on photosynthesis and dark respiration of pines. Environ. Pollut. *3,* 133–136 (1972).

Barnes, J. D., D. Eamus, and K. J. Brown: The influence of ozone, acid mist and soil nutrient status on Norway spruce (*Picea abies* (L.) Karst.). II. Photosynthesis, dark respiration and soluble carbohydrates of tree during late autumn. New Phytol. *115,* 149–156 (1990).

Bradford, M. M.: A rapid and sensitive method for the quantitation of microgram quantities of proteins utilizing the principle of protein-dye binding. Anal. Biochem. *72,* 248–254 (1976).

Chappelka, A. H. and P. H. Freer-Smith: Predisposition of trees by air pollutants to low temperatures and moisture stress. Environmental Pollution *87,* 105–117 (1995).

Chen, C. W., W. T. Tsai, and L. E. Gomez: Modeling responses of Ponderosa pine to interacting stresses of ozone and drought. Forest Science. *40* (2), 267–288 (1994).

Davis, D. D. and K. D. Patil: The control of NAD specific malic enzymes from cauliflower bud mitochondria by metabolites. Planta *126,* 197–211 (1975).

Dizengremel, P. and M. Petrini: Effects of air pollutants on the pathways of carbohydrate breakdown. In: Alscher, R. G. and A. R. Wellburn (eds.): Plant Responses to the Gaseous Environment. Molecular, Metabolic and Physiological Aspects, pp. 255–278. Chapman and Hall, London (1993).

Dizengremel, P., D. Afif, M. Petrini, C. Namysl, and O. Queiroz: Expérience de simulation de Montardon: effets des polluants atmosphériques sur le métabolisme de l'épicéa. In: Bonneau, M. and G. Landmann (eds.): Pollution atmosphérique et dépérissement des forêts dans les montagnes françaises, pp. 207–212. INRA, Nancy (1993).

Dizengremel, P., T. W. Sasek, K. J. Brown, and C. J. Richardson: Ozone-induced changes in primary carbon metabolism enzymes of loblolly pine needles. J. Plant Physiol. *144,* 300–306 (1994).

Eamus, D., J. D. Barnes, L. Mortensen, H. RoPoulsen, and A. W. Davison: Persistent stimulation of CO_2 assimilation and stomatal conductance by summer ozone fumigation in Norway spruce. Env. Pollution *63,* 365–379 (1990).

Friend, A. L. and P. T. Tomlinson: Mild ozone exposure alters ^{14}C dynamics in foliage of *Pinus taeda* L. Tree Physiology. *11,* 215–227 (1992).

Gérant, D., A. Citerne, C. Namysl, P. Dizengremel, and M. Pierre: Study of some respiratory enzymes in foliar organs and root systems of spruce and oak trees. Relation with forest decline. In: Bervaes, J., P. Mathy, and P. Evers (eds.): Air Pollution Research Report, 109–118. E. Guyot S.A., Bruxelles (1988).

Gimeno, B. S., D. Velissariou, J. D. Barnes, R. Inclan, J. Pena, and A. W. Davison: Efectos del ozono sobre aciculas de *Pinus halepensis* en Grecia y Espana. Ecologia *6,* 131–134 (1992).

Grieu, P. and C. Robin: Water status, gas exchange rate and carbon partitioning of Aleppo pine (*Pinus halepensis* Mill.) under water stress and ozone pollution – Final report of European Project STEP CT 910117 DTEE (1993).

Heath, R. L.: Alterations of plant metabolism by ozone exposure. In: Alsher, R. and A. R. Wellburn (eds.): Plant Responses to the gaseous environment, Molecular, Metabolic and Physiological Aspects, pp. 121–145. Chapman and Hall (1994 a).

– Possible mechanisms for the inhibition of photosynthesis by ozone. Photosynth. Res. *39,* 439–451 (1994 b).

Koziol, M. J., F. R. Whatley, and J. D. Shelvey: An integrated view of the effects of gaseous air pollutants on plant carbohydrate metabolism. In: Schulte-Hostede, S., N. M. Darrall, L. W. Blank, and A. R. Wellburn (eds.): Air Pollution and Plant Metabolism, pp. 148–168. Elsevier Applied Science, London (1988).

Lichtenthaler, H. K.: Chlorophylls and carotenoids: Pigments of Photosynthetic Biomembranes. Methods in Enzymology. *148,* 350–382 (1987).

LILLEY, R. C. and D. A. WALKER: An improved spectrophotometric assay for ribulose bisphosphate carboxylase. Biochim. Biophys. Acta *358,* 226–229 (1974).

McLAUGHIN, S. B., R. K. McCONATHY, D. DUVICK, and L. K. MANN: Effects of chronic air pollution stress on photosynthesis, carbon allocation, and growth of pine trees. Forest Sci. *28,* 60–70 (1982).

PEARSON, M. and T. A. MANSFIELD: Interacting effects of ozone and water stress on the stomatal resistance of beech (*Fagus sylvatica* L.). New Phytol. *123,* 351–358 (1993).

RAPPORT DE L'ACADÉMIE DES SCIENCES: Ozone et propriétés oxydantes de la troposphere, Essai d'évaluation scientifique. In: Technique et Documentation, *30.* Lavoisier, Paris (1993).

REICH, P. B.: Effects of low concentrations of O_3 on net photosynthesis, dark respiration, and chlorophyll contents in aging hybrid Poplar leaves. Plant Physiol. *73,* 291–296 (1983).

RICHARDSON, C. J., T. W. SASEK, and E. A. FENDICK: Implications of physiological responses to chronic air pollution for forest decline in the southeastern U.S.A. Environm. Toxicol. Chem. *11,* 1105–1114 (1992 a).

RICHARDSON, C. J., T. W. SASEK, E. A. FENDICK, and L. W. KRESS: Ozone exposure-response relationships for photosynthesis in genetic strains of loblolly pine seedlings. Forest Ecol. Management *51,* 163–178 (1992 b).

ROBERTS, B. R.: Physiological response of yellow-poplar seedlings to stimulate acid rain, ozone fumigation, and drought. Forest Ecology and Management. *31,* 215–224 (1990).

SCHÖLANDER, P. F., H. T. HAMMEL, E. D. BRADESTREES, and E. A. HEMMINGSEN: Sap pressure in vascular plant. Science. *148* (3668), 339–346 (1965).

SKÄRBY, L., E. TROENG, and C. A. BOSTROM: Ozone uptake and effects on transpiration, net photosynthesis, and dark respiration in Scots pine. Forest Science *33,* 801–808 (1987).

SPENCE, R. D., E. J. RYKIEL Jr., and P. J. H. SHARPE: Ozone alters carbon allocation in loblolly pine: assessment with carbon-11 labeling. Environm. Pollution *64,* 93–106 (1990).

TINGEY, R. D. and W. E. HOGSETT: Water stress reduces ozone injuries via a stomatal mechanism. Plant Physiol. *77,* 944–947 (1985).

TSENG, E. C., J. R. SEILER, and B. I. CHEVONE: Effects of ozone and water stress on greenhouse-grown Fraser fir seedling growth and physiology. Environ. Exp. Bot. *28,* 37–41 (1988).

VELISSARIOU, D., A. W. DAVISON, J. D. BARNES, T. PFIRRMANN, and C. D. HOLEVAS: Effects of air pollution on *Pinus halepensis* Mill. Pollution levels in Attica, Greece. Atmospheric Environ. *26 (3),* 373–380 (1992).

WALLIN, G., L. SKÄRBY, and G. SELLDÉN: Long-term exposure of Norway spruce, *Picea abies* (L.) Karst., to ozone in open-top chambers. I. Effects on the capacity of net photosynthesis, dark respiration and leaf conductance of shoots of different ages. New Phytol. *115,* 335–344 (1990).

WELLBURN, F. A. M. and A. R. WELLBURN: Atmospheric ozone affects carbohydrate allocation and winter hardiness of *Pinus halepensis* Mill. J. Exp. Bot. *45,* 607–614 (1994).

J. Plant Physiol. Vol. 148. pp. 148–154 (1996)

Reflectance Measurements of Leaves for Detecting Visible and Non-visible Ozone Damage to Crops

MARTIN KRAFT[1], HANS-JOACHIM WEIGEL[2], GERD-JÜRGEN MEJER[2], and FRANK BRANDES[1]

[1] Institut für Biosystemtechnik, Bundesforschungsanstalt für Landwirtschaft (FAL), Bundesallee 50, D-38116 Braunschweig, Germany

[2] Institut für Produktions- und Ökotoxikologie, Bundesforschungsanstalt für Landwirtschaft (FAL), Bundesallee 50, D-38116 Braunschweig, Germany

Received June 24, 1995 · Accepted October 20, 1995

Summary

Spring wheat (*Triticum aestivum* cv. Turbo), white clover (*Trifolium repens* cv. Karina) and maize (*Zea mays* cv. Bonny) plants were exposed for 20–30 days in open top chambers to charcoal-filtered air (CF, control) and CF air supplied with O_3 for 8–12 h/per day in the concentration range of 180–240 µg O_3/m^3 (8–12 h/day treatment mean). At the end of the O_3 treatment spectral reflectance measurements were made on single leaves of all 3 species and on canopies of wheat and clover using a CCD (Charged Coupled Device) camera and wavelength filters with 11 wavelength bands ranging from 450 nm to 950 nm. Different vegetation indices such as the normalized difference vegetation index (NDVI) and the «main inflection point» (MIP) were calculated. Based on these results it was shown that visible O_3 damages were correlated to the spectral reflectance changes: Both leaves and canopies showed an increased reflectance of visible light after ozone treatment. While clover and maize leaves as well as clover and wheat canopies showed a decreased near infrared (NIR) reflectance, the NIR reflectance of wheat leaves did not change, even if the leaves had visible symptoms. A decreased infrared reflectance was detectable for all clover leaves after O_3 treatment although for part of the leaves no visible foliar damage symptoms could be observed.

Key words: Triticum aestivum cv. Turbo, Trifolium repens cv. Karina, Zea mays cv. Bonny, blue shift, clover, maize, ozone, red edge, spectral reflectance, vegetation index, wheat.

Abbreviations: ND, NDVI = normalized difference (vegetation index); MIP = main inflection point; R_{wl} = reflectance at wavelength wl; vis = visible wavelength(s); NIR, nir = near infrared wavelength(s) (used NIR wavelengths are 750 nm to 950 nm); IHS = intensity/hue/saturation color model; I = intensity; H = hue.

Introduction

Tropospheric ozone (O_3) has become one of the most phytotoxic air pollutants affecting agricultural and forest plants and other types of vegetation. At the biochemical and physiological level O_3 is causing membrane damage, changes in plant cellular structures, altered carbon allocation and impairment of photosynthesis. Depending on the exposure concentration and duration as well as on the plant type, general responses to O_3 include visible foliar injury symptoms, early senescence and reduced growth and biomass production (Lefohn, 1992; Bender and Weigel, 1995). Plant growth can also be affected before visible foliar damage can be noticed. Recent European experiments have shown yield reductions of some sensitive crop species under current ambient O_3 concentrations (Fuhrer and Achermann, 1994). However, the spatial extent to which O_3 must be regarded as an additional stress factor for agricultural plants across Europe remains largely unknown.

Stress induced physiological damage in plants is manifested in an altered reflectance spectrum, and several spectral indices using the near-infrared region were proposed as stress indicators (Hunt et al., 1987; Ripple, 1986). However, in comparison to the various other techniques to assess O_3 effects on plants, leaf spectral reflectance as a non-destructive and rapid method to detect O_3 stress has hardly been used for European crop species (Runeckles and Resh, 1975; Schutt et al., 1984; Williams and Ashenden, 1992; Carter et al., 1992).

Spectral index values are important tools for the interpretation of spectral plant signatures. Especially in the area of remote sensing, indices have been used for several years, e.g. for nitrogen nutrition estimation in grassland (Büker et al., 1991), monitoring of the growth state of several crops (Lagouarde et al., 1986) and determination of the chlorophyll content in leaves (Guyot et al., 1988; Buschmann and Nagel, 1993; Gitelson and Merzylak, 1994). Williams and Ashenden (1992) were able to show that O_3 treatment of white clover led to a decreased vegetation index of the plant canopies (using a simple vegetation index NIR/R with R: 600 nm to 700 nm, and NIR: 760 nm to 1100 nm). For an overwiev of previously used vegetation indices see Buschmann (1993).

In the present report different types of indices are applied to the single leaf reflectance of wheat, clover and maize in order to find out, which of the indices is suitable to detect and quantify O_3 effects on different crop species and especially, whether the technique is suitable to assess «previsual» or «invisible injuries» due to an O_3 stress (Williams and Ashenden, 1992).

Materials and Methods

Plant culture and O_3 treatment

White clover (*Trifolium repens* cv. Karina), spring wheat (*Triticum aestivum* cv. Turbo) and maize (*Zea mays* cv. Bonny) were grown in 5 litre pots on a standard prefertilized mix of soil and peat (ED 73) and watered during the experiments according to their water demand.

Exposure of plants took place during the growing season 1993 in open-top chambers equipped with dispensing and monitoring systems for O_3. Details of the exposure facility are described by Weigel and Jäger (1988). Treatments consisted of charcoal filtered air (control; CF) and CF air supplemented with O_3 generated from electrical discharge in pure oxygen. O_3 was supplied to the chambers daily from 9.00 a.m. to 9.00 p.m. (wheat, clover, 3–23 August) and 12.00 a.m. to 8.00 p.m. (maize, 27 August–13 October). Mean O_3 concentrations in the chambers supplemented with O_3 were 243 $\mu g/m^3$ and 197 $\mu g/m^3$, respectively. Average O_3 concentrations in the CF chambers were always $<20\,\mu g/m^3$.

Plant selection for reflectance measurements

Reflectance measurements were carried out on 23 August (wheat, clover) and 13 October (maize). Leaves were randomly selected from each treatment. While clover leaves were not selected for leaf age, for wheat and maize only the second youngest leaves were used. The leaves were categorized into 3 (wheat, clover) or 2 (maize) groups: «healthy» (cf. Figs. 2–6) = leaves from the CF treatment (wheat n=5; clover n=10; maize n=6); «O_3, no visible discolouration» (cf. Figs. 2–6) = leaves from the O_3 treatments without visible changes (wheat n=3; clover n=10; maize n=6); «O_3, visibly discoloured» (cf. Figs. 2–6) = leaves from the O_3 treatments with visible O_3 symptoms (wheat n=9; clover n=15).

Equipment and procedure of leaf spectral reflectance measurement

As a spectroradiometer with high wavelength resolution in the visible, near, and middle infrared range of the light (e.g. Büker et al., 1991; Gitelson and Merzylak, 1994) was not available, a camera system, including a CCD sensor with 512×512 pixels (image points) was used. This «panchromatic» camera system is sensitive to visible and near infrared light. Bandpass filters ranging from 450 nm up to 950 nm in steps of 50 nm were used, the bandwidth of which is about 30 to 40 nm.

The grey value of each object (e.g. a leaf) was calculated as the average of the image pixels covered by the object. A calibration which was needed in order to map the grey value into a reflectance factor was accomplished by reflectance standards (spectralon; reflectance factors used were 2 %, 5 %, 10 %, 20 %, 40 %, and 50 %) which were placed close to the object and measured simultaneously. (Reflectance means the ratio reflected light/incident light.)

Images were taken in a «lighting tower» with black walls (Fig. 1). On the top a white cotton fabric was fixed on a plexiglass plate and illuminated by halogene lamps. This lighting arrangement was a compromise between a single point light source and a homogeneous hemispherical lighting.

For single leaf reflectance measurements excised leaves were fixed on a black varnished plate by weights. Images were taken only from adaxial leaf sides within 60 minutes after the excision starting with the filter with the shortest wavelength.

For canopy reflectance measurements whole pots were placed into the «lighting tower» together with the reflectance standards. A rectangular part of the canopy of each pot was selected for the measurements.

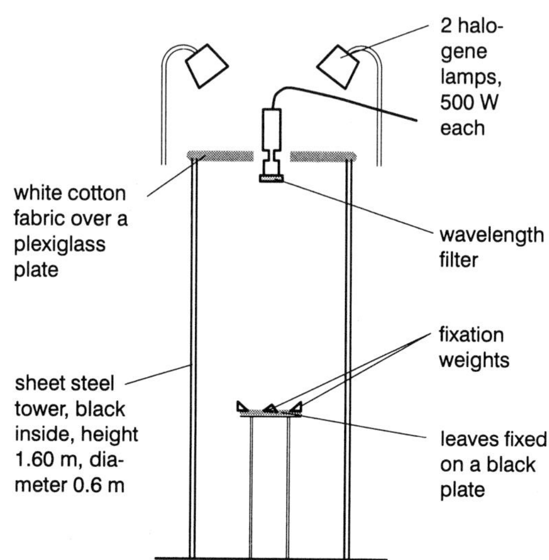

Fig. 1: Schematic presentation of the illumination and image acquisition system for reflectance measurements of excised plant leaves.

Calculation of index values

While in remote sensing techniques index formulas are usually applied to (large) canopy's signatures, in the present experiments several indices were calculated using the results from the single leaf measurements.

Normalized difference vegetation indices (NDVI, Rouse et al., 1974) are scalar numbers calculated from two reflectance values, one in the visible range of light (R_{vis}) and one infrared reflectance value (R_{nir}):

$$\mathrm{NDVI} = (R_{nir} - R_{vis})/(R_{nir} + R_{vis}) \quad (1).$$

Four different choices of visible resp. NIR wavelengths or wavelength ranges were investigated: ND1 corresponding to vis = 700 nm and nir = 750 nm according to Gitelson and Merzylak (1994). These authors showed that this index is highly correlated to the chlorophyll *a* content for e.g. maple and chestnut leaves. ND2 corresponding to vis = 550 nm and nir = 850 nm (den Dulk, 1989). ND3 uses broad band combinations of visible and infrared reflectance, similar to the spectral bands of the satellite sensor system NOAA-AVHRR (Lagouarde et al., 1986). R_{vis} is calculated here as the average of the reflectance at 600 nm and 650 nm, R_{nir} is the average of the reflectance at 750 nm, 800 nm, 850 nm, and 900 nm. ND4 uses vis = 650 nm and nir = 850 nm (Schellberg and Kühbauch, 1991).

According to Collins (1978), Guyot and Baret (1988), Guyot et al. (1988), Büker et al. (1991), and Carter et al. (1992) the red edge inflection point or «main inflection point» (MIP) can also be used as a vegetation index. The MIP is the wavelength position of the «red edge», i.e. the steep slope from the red to the infrared reflectance. The «blue shift» of the red edge, i.e. the shift towards shorter wavelengths has been observed under different stress conditions (Carter et al., 1992). Collins (1978) showed that the blue shift is induced by a decreasing chlorophyll *a* content. In the present experiments the MIP was approximated using the lambda 1 formula from Guyot and Baret (1988) and the wavelength set 1: 650 nm, 2: 700 nm, 3: 750 nm, 4: 800 nm; this subset of the measured wavelengths is closest to the wavelengths proposed by Guyot and Baret (1988).

The intensity (I) and hue (H) values of the IHS colour model correspond to the visible brightness and colour of an object. They were calculated here using the wavelengths red = 650 nm, green = 550 nm, and blue = 450 nm:

$$I = (R_{red} + R_{green} + R_{blue})/3 \quad (2)$$

and

$$H = \mathrm{atan2}\left(\frac{(R_{red} - R_{green})}{\sqrt{2}}, \frac{-R_{red} - R_{green} + 2 \cdot R_{blue}}{\sqrt{6}}\right) \quad (3)$$

with

$$\mathrm{atan2}\,(x, y) = \begin{cases} \arctan\,(x/y) & \text{if } x>0 \text{ and } y>0; \\ n \cdot PI + \arctan\,(x/y) & \text{else, with n such that } x/y = \tan\,(\mathrm{atan2}\,(x,y)). \end{cases}$$

Results and Discussion

Plant growth and development

O_3 caused visible damage symptoms on wheat leaves, i.e. lamina (Adaros et al., 1991; Bender et al., 1994). On an average O_3-treated plants were smaller and leaves were shorter and narrower than the controls. However, O_3-treated wheat plants also included leaves (especially the younger ones), which did not show any visible symptoms at all.

O_3 symptoms on clover included typical chlorotic spots (oxidant stipple; Ommen et al., 1995), larger necrotic areas across the leaf and bleaching of the whole leaf area. The plant density of clover «canopies» was more closed in control treatments than in O_3 stressed plants. Selection of leaves whithout visible discolouration resulted in a collection of leaves which on an average were darker than randomly selected leaves from the control plants. All O_3 treated leaves were smaller than control leaves.

The O_3 treated maize plants showed only little visible damage: size, shape and colour of the younger leaves did not differ from the control plants. O_3 symptoms (chlorophyll bleaching) were only observed at the tip of oldest leaves.

Reflectance measurements

Wheat

In Fig. 2 the mean reflectance values for each of the three classes are shown, together with the calculated standard deviation of reflectance values of single leaves within each class, assuming a normal distribution. The following reflectance properties are obvious:

1. In the visible wavelengths reflectance values of visibly damaged leaves vary significantly (e.g. up to 4.1 % (absolute) at 600 nm). In accordance with the visible symptoms reflectance of O_3 treated leaves is higher than that of healthy and O_3 treated leaves without visible damage. However, changes in the reflectance due to the relatively high O_3 treatment concentrations used here are small e.g. in comparison to O_3 induced reflectance changes of beans under chronic low level O_3 exposure (Runeckles and Resh, 1975).

2. In the same wavelength region the variation within the classes «healthy» and «O_3, no visible discolouration» is very small. Leaves from O_3 treated plants reflect about 1 % (absolute) more visible light than control («healthy») leaves. However, due to the small sample size this can only be regarded as an assumption.

3. In the near infrared wavelength region mean reflectance values of all three classes fall within a narrow band of 0.8 % (absolute) and thus are not distinguishable from measurement error. The variation of infrared reflectance within visibly damaged leaves (about 1 % to 1.5 % absolute) is significantly lower than in the visible wavelengths.

The reflectance signatures of wheat canopies are shown in Fig. 3. The near infrared reflectance of O_3 treated wheat is significantly lower than the reflectance of control plants. Since single leaves do not show such a pronounced difference, this effect can probably be explained by a smaller leaf area index in O_3 treated canopies compared to control plants.

Clover

As shown in Fig. 4 the following reflectance properties can

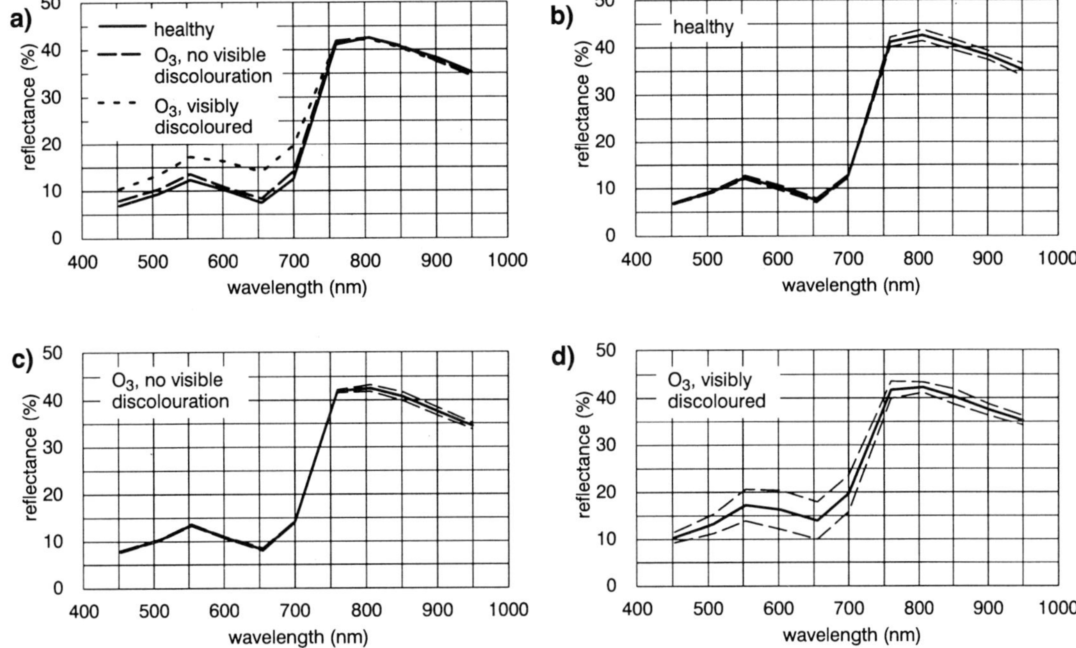

Fig. 2: Reflectance spectra of excised wheat leaves (*Triticum aestivum* cv. Turbo) of plants grown in open-top chambers with filtered air (= healthy) or filtered air supplemented with ozone (O_3). a = averages for three different damage classes. b–d = averages and (single) standard deviations for each of the damage classes.

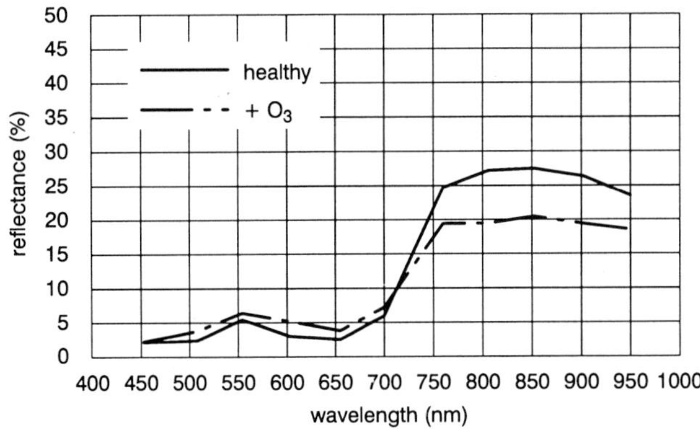

Fig. 3: Reflectance spectrum of model wheat canopies (*Triticum aestivum* cv. Turbo, potted plants) after growth in open-top chambers with filtered air (= healthy) or filtered air supplemented with ozone (O_3).

1. In the visible wavelength range control leaves as well as leaves without visible discolourations absorb a high percentage of light, as the respective reflectance values are lower than for wheat or maize leaves.

2. In the visible wavelength range the reflectance of control leaves (up to 1.5 % absolute) varies more than for wheat or maize which is in accordance with the visible impression. This might be correlated to the fact that leaves were not selected for age classes.

3. In the visible wavelength range O_3 treated leaves with visible discolourations clearly reflect more light than control leaves, and the standard deviation within the sample of this group of leaves is high (up to 2.7 % absolut, at 650 nm). Since the selected leaves without visible discolouration were darker than the controls, a randomly selected sample from the damaged canopy would on an average differ less from the control leaves, but the standard deviation would increase significantly.

4. In the near infrared wavelength region damaged leaves reflect significantly less (distance >3 % absolute at 800 nm), which is different from wheat or maize (see below). Interestingly, both classes of O_3 treated leaves on an average showed the same infrared reflectance, independent from their appearance in the visible wavelength range. With standard deviations between 1.2 % and 2.3 % (absolute) the variation of reflectance within all three classes is similar. Changes of the NIR reflectance have been attributed to changes of the mesophyll structure, which in some plant stress experiments lead to an increase, in other experiments to a decrease of NIR reflectance (Gausman and Quisenberry, 1990).

The signatures of the clover canopies are shown in fig. 5. In the visible wavelength range O_3 treated canopies reflect slightly more than controls. In the near infrared range damaged canopies reflect considerably less than controls, as was also observed by Williams and Ashenden (1992) for the clover cultivar *Grasslands Huia* in combination with acid mist. These authors discussed the thinning of clover canopies as a reason for these changes, which, in addition to the lower leaf reflectance, might also be true in the present case.

Maize

The signatures of maize leaves without visible O_3 symptoms (Fig. 6) can be interpreted as follows:

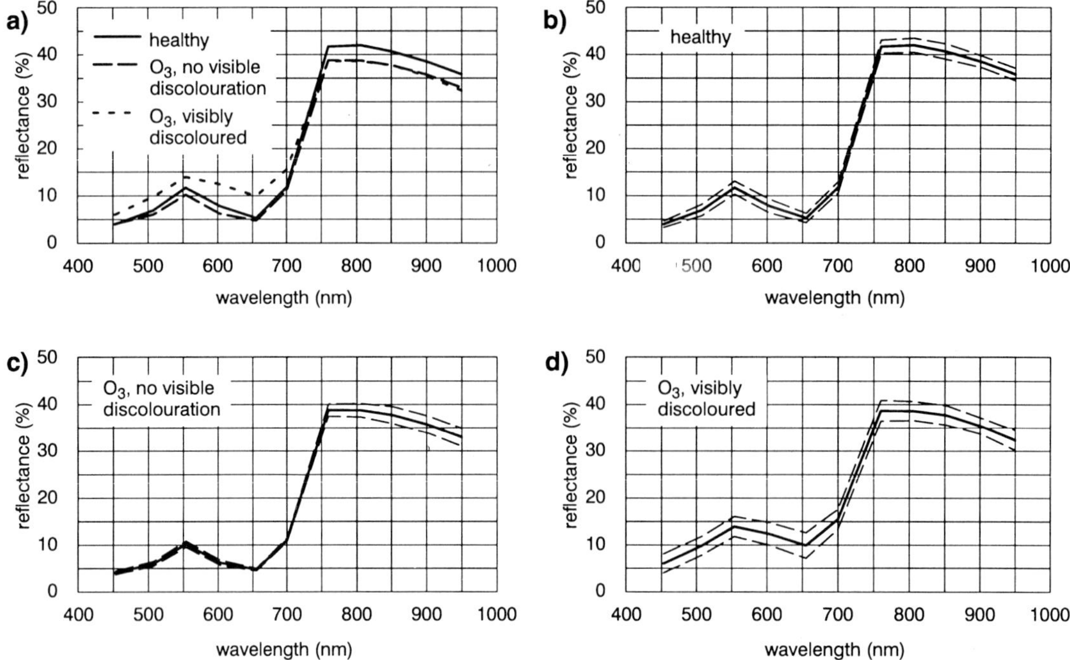

Fig. 4: Reflectance spectra of excised clover leaves (*Trifolium repens* cv. Karina) of plants grown in open-top chambers with filtered air (= healthy) or filtered air supplemented with ozone (O_3). a = averages for three different damage classes. b–d = averages and (single) standard deviations for each of the damage classes.

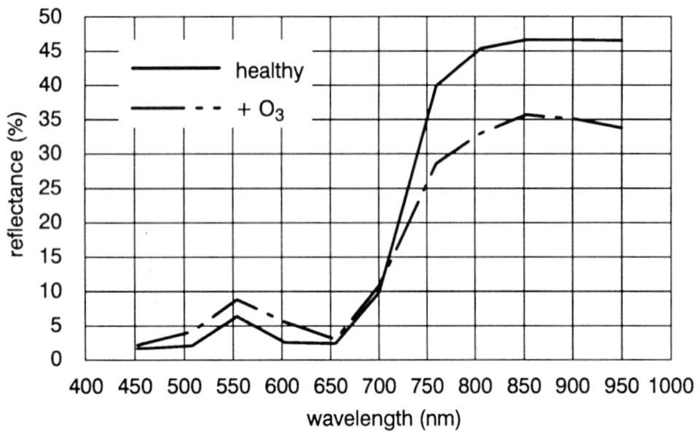

Fig. 5: Reflectance spectrum of model clover canopies (*Trifolium repens* cv. Karina, potted plants) after growth in open-top chambers with filtered air (= healthy) or filtered air supplemented with ozone (O_3).

Table 1: Index values for leaves from healthy plants, for undiscoloured leaves from O_3 damaged crop plants, and for visibly discoloured leaves from O_3 damaged crop plants. ND1 through ND4 are normalized difference vegetation indices, MIP is the main inflection point, I and H are two coordinates of the IHS model. For detailed explanations see text.

index	wheat leaves			clover leaves			maize leaves	
	healthy	O_3, no visible discol.	O_3, visibly discol.	healthy	O_3, no visible discol.	O_3, visibly discol.	healthy	O_3, no visible discol.
ND1	0.534	0.495	0.359	0.560	0.558	0.428	0.526	0.490
ND2	0.534	0.501	0.400	0.554	0.575	0.460	0.479	0.449
ND3	0.646	0.621	0.456	0.724	0.749	0.543	0.701	0.676
ND4	0.692	0.663	0.486	0.771	0.779	0.585	0.766	0.747
MIP	722.02	720.31	719.11	719.82	719.25	718.89	718.52	717.47
I	8.83	9.88	13.85	6.94	6.27	9.93	9.05	9.68
H	−2.25	−2.16	−2.64	−2.19	−2.20	−2.60	−2.18	−2.20

1. In the visible wavelength range O_3 treated leaves reflect slightly more (up to 1.2 % at 550 nm) than control ones, but the standard deviation within the latter group (1.1 % at 350 nm) is almost as high as the difference of the mean signatures of the two classes.
2. In the infrared wavelength range both classes reflect equally; no influence of O_3 treatment can be seen.

Index values for single leaves of all species

The results of the calculated index values for all species investigated are shown in table 1.

As expected, leaves with visible O_3 induced discolouration have lower NDVI values than those from the control treatment. High NDVI values of leaves without visible discolouration show, that O_3 effects in the visible wavelength range dominate the NDVI.

Especially for clover the normalized differences do not reflect the effects in the infrared wavelength region, but only the visible properties. Gitelson and Merzylak (1994) were able to show that for maple and chestnut leaves, ND1 is highly correlated to the chlorophyll *a* content. If this holds true for the species investigated here, then O_3 treated wheat and maize leaves, which were classified as to have no visible

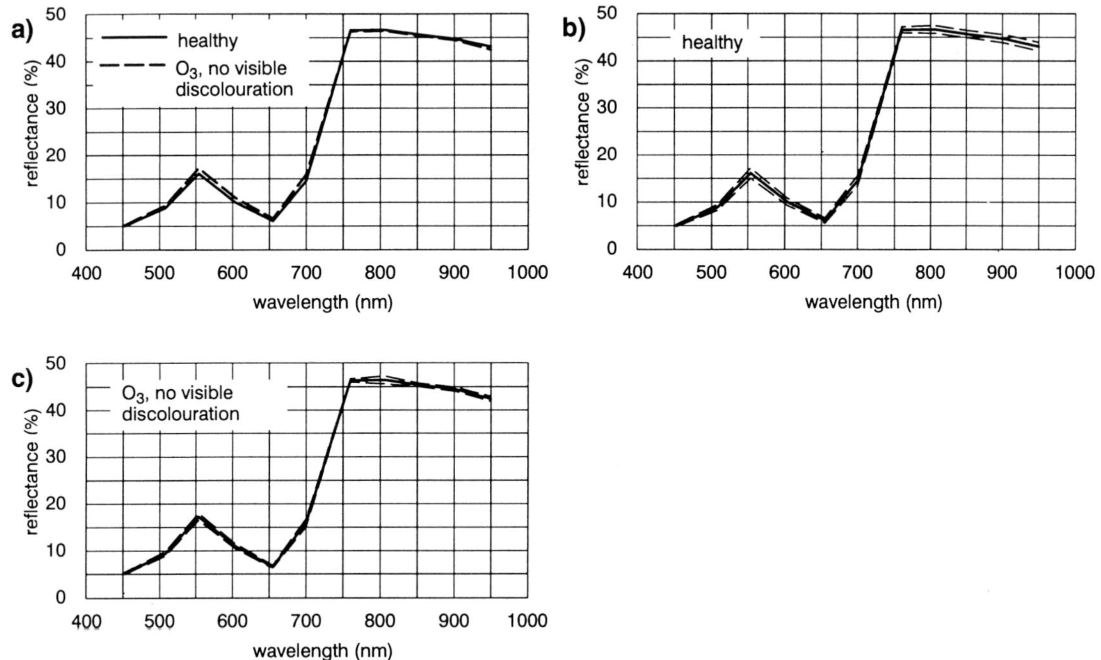

Fig. 6: Reflectance spectra of excised maize leaves (*Zea mays* cv. Bonny) of plants grown in open-top chambers with filtered air (= healthy) or filtered air supplemented with ozone (O_3). The measured leaves of the O_3 treated plants showed no visible discolouration. a = averages for two different damage classes. b–c = averages and (single) standard deviations for each of the damage classes.

discolouration, are clearly indicated to be chlorotic by ND1, whereas O_3 treated clover leaves without visible discolourations should have the same chlorophyll *a* content as leaves from the control treatment. Buschmann and Nagel (1993) propose log (R_{800}/R_{550}) as a vegetation index with high correlation to chlorophyll content. The values of this index are for clover: 0.557 (healthy leaves), 0.581 (O_3 treated leaves without visible discolouration), and 0.443 (O_3 treated leaves with visible discolourations). Thus this index even suggests a slightly higher chlorophyll content for O_3 treated leaves without visible discolouration compared to healthy clover leaves. Since chlorophyll was not measured in this investigation, this has to be clarified by further experiments.

The main inflection point (MIP) is shifted towards shorter wavelengths by the O_3 induced damages, as was already reported for loblolly pine needles by Carter et al. (1992). Leaves without visible discolourations from O_3 treated plants have inflection points ranging between those of visibly discoloured and control leaves. Thus, the MIP seems to be an appropriate index for the degree of O_3 damages. Especially O_3 treated clover leaves without visible discolouration but with a slightly darker colour can clearly be indexed as being affected by this pollutant.

As expected, intensity (I) and hue (H) correspond (Table 1) to the visible impression of brightness and colour of leaves. This type of visible damage may thus be quantified using the intensity or hue values as indices. O_3 treated leaves without visible discolouration cannot be separated from healthy leaves of the control treatment by these two indices.

Conclusion

The present report is the first to describe spectral reflectance changes of German crop species after growth in an atmosphere with increased O_3 concentration. All species showed increases of leaf reflectance of visible light similar to earlier reports for other crops. The NIR reflectance of wheat leaves did not change, even if chlorotic damages were clearly visible. On the other hand, O_3 led to a significantly decreased NIR reflectance of clover leaves, even before injuries become visible. Thus, O_3 induced different NIR reflectance changes for different plant species.

Among the vegetation indices investigated here only the MIP appeared to be sensitive even to «invisible injuries» induced by O_3. As this parameter is also sensitive to other environmental factors, it has to be found out whether the O_3 induced changes of the MIP can be separated from these parameters. The present results, which must be regarded as preliminary, did not correlate reflectance changes to physiological or biochemical alterations or growth changes due to the O_3 impact. Also, the O_3 concentrations supplied to the plants were relatively high compared to the situation in the field. In order to fully exploit the potential of spectral reflectance changes as an early indicator of an O_3 stress, further experiments under more realistic exposure conditions should be carried out.

Acknowledgements

The authors thank Wolfgang Paul for the discussion of novel vegetation index formulas, Axel Munack for multiple discussions of the results and their presentation, Henrik M. Nielsen for his help in the English translation of this paper, and Karin Tamoschat for her help with the figures.

References

Adaros, G., H. J. Weigel, and H. J. Jäger: Impact of ozone on growth and yield parameters of two spring wheat cultivars (*Triticum aestivum* L.) J. Plant Diseases and Protection *98,* 113–124 (1991).

Bender, J. and H. J. Weigel: Zur Gefährdung landwirtschaftlicher Kulturpflanzen durch troposphärische Ozonkonzentrationen. Berichte über Landwirtschaft *73,* 136–156 (1995).

Bender, J., H. J. Weigel, U. Wegner, and H. J. Jäger: Responses of cellular antioxidants to ozone in wheat flag leaves at different stages of plant development. Environ. Pollut. *84,* 15–21 (1994).

Büker, C., J. G. P. W. Clevers, and W. Kühbauch: Die Position des Rot-Infrarot-Anstiegs in Reflexionsspektren von Grünland – ein Maß für die Höhe der Stickstoffversorgung. Fernerkundung in der Land- und Forstwirtschaft, Berichte der GIL *1,* 45–54 (1991).

Buschmann, C.: Fernerkundung von Pflanzen: Ausbreitung, Gesundheitszustand und Produktivität. Naturwissenschaften *80,* 439–453 (1993).

Buschmann, C. and E. Nagel: *In vivo* spectroscopy and internal optics of leaves as basis for remote sensing of vegetation. Int. J. Remote Sensing *14,* 711–722 (1993).

Carter, G. A., R. J. Mitchell, A. H. Chappelka, and C. H. Brewer: Responses of leaf spectral reflectance in loblolly pine to increased atmospheric ozone and precipitation acidity. J. Exp. Bot. *43,* 577–584 (1992).

Collins, W.: Remote sensing of crop type and maturity. Photogramm. Engin. and Rem. Sens. *44,* 43–55 (1978).

den Dulk, J. A.: The interpretation of remote sensing, a feasibility study. Proefschrift Wageningen (1989).

Fuhrer, J. and B. Achermann (eds.): Critical levels for ozone. UN-ECE workshop in Bern, 1–4 November 1993, Schriftenreihe der FAC Liebefeld–Bern (1994).

Gausman, H. W. and J. E. Quisenberry: Spectrophotometric detection of plant leaf stress. In: Katterman, F. (ed.): Environmental Injury to Plants, pp. 257–280. Academic Press, San Diego (1990).

Gitelson, A. and M. N. Merzylak: Quantitative estimation of chlorophyll *a* using reflectance spectra: experiments with autumn chestnut and maple leaves. J. Photochem. Photobiol. B: Biol. *22,* 247–252 (1994).

Guyot, G., F. Baret, and D. J. Major: High spectral resolution: determination of spectral shifts between the red and near infrared. Int. Arch. of Photogrammetry and Remote Sensing *27* (11), 750–760 (1988).

Guyot, G. and F. Baret: Utilisation de la haute résolution spectrale pour suivre l'état des couverts végétaux. Proc. 4th Int. Coll. on Spectral Signatures of Objects in Remote Sensing, Aussois, France, 18–22 January 1988 (ESA SP-287, April 1988), pp. 279–286 (1988).

Hunt, E. R., B. R. Rock, and P. S. Nobel: Measurement of leaf relative water content by infrared reflectance. Remote Sens. Environ. *22,* 429–435 (1987).

Lagouarde, J.-P., B. Seguin, S. Clinet, S. Gandia, and Y. Kerr: Intérêt de l'indice de végétation des satellites NOAA-AVHRR pur le suivi des cultures (résultats d'une étude dans la basse vallée du Rhône). Agronomie *6* (10), 949–954 (1986).

Lefohn, A. S.: Surface level ozone exposures and their effects on vegetation. Lewis publishers, Chelsea, USA. (1992).

Ommen, O. E., J. Bender, and H. J. Weigel: Effects of ozone on white clover: protection of ozone damage by ethylenediurea. J. Applied Botany *68,* 146–150 (1994).

Ripple, W. J.: Spectral reflectance relationship to leaf water stress. Photogramm. Eng. Remote Sens. *52,* 1669–1675 (1986).

Rouse, J. W., R. H. Haas, J. A. Schell, D. W. Deering, and J. C. Harlan (eds.): Monitoring the vernal advancement of retrogradation of natural vegetation. NASA/GSFC, Type III, Final Report, Greenbelt, MD (1974).

Runeckles, V. C. and H. M. Resh: The assessment of chronic ozone injury to leaves by reflectance spectrophotometry. Atmospheric environment *9,* 447–452 (1975).

Schellberg, J. and W. Kühbauch: Reflexionsmessungen von Weizenbeständen zum Zwecke der fernerkundlichen Zustandsbeschreibung und Ertragsschätzung. Fernerkundung in der Land- und Forstwirtschaft, Berichte der GIL *1,* 61–68 (1991).

Schutt, J. B., R. A. Rowland, and H. E. Heggestad: Identification of injury resulting from atmospheric pollutants using reflectance measurements. J. Environ. Qual. *13,* 605–608 (1984).

Weigel, H. J. and H. J. Jäger: Zur Ökotoxikologie von Luftschadstoffen: Aufbau und Funktionsweise einer Expositionsanlage aus open-top Kammern zur Untersuchung von Immissionswirkungen auf Pflanzen. Landbauforschung Völkenrode *38,* 182–195 (1988).

Williams, J. H. and T. W. Ashenden: Differences in the spectral characteristics of white clover exposed to gaseous pollutants and acid mist. New Phytol. *120,* 69–75 (1992).

J. Plant Physiol. Vol. 148. pp. 155–159 (1996)

Manganese Deficiency Enhances Ozone Toxicity in Bush Beans (*Phaseolus vulgaris* L. cv. Saxa)

HORST MEHLHORN and ANDREA A. WENZEL

FB 9.4, Institut für Angewandte Botanik, Henri-Dunant-Str. 65, Universität-GH-Essen, D-45117 Essen, Germany

Received June 20, 1995 · Accepted October 10, 1995

Summary

Manganese deficiency was found to enhance ozone toxicity in bush beans (*Phaseolus vulgaris* L. cv. Saxa). This difference in plant sensitivity to ozone could not be explained by increased stomatal uptake nor decreased abilities to detoxify superoxide anions or hydrogen peroxide. Thus, manganese deficiency decreased transpiration, had no effect on ascorbate-dependent peroxidase or *Cu/Zn*-superoxide dismutase (SOD) activity and even increased levels of *Mn*-SOD activity. By contrast, manganese deficiency reduced ascorbic acid content in leaves and enhanced ozone-induced stress ethylene formation. This supports the hypothesis that environmental stress increases ozone sensitivity of plants because of increased stress ethylene formation. Moreover, these results suggest that MnSOD contributes little to make plants more tolerant when they are exposed to ozone.

Key words: Phaseolus vulgaris, ethylene, ozone, ascorbate peroxidase, superoxide dismutase, manganese deficiency.

Abbreviations: SOD = superoxide dismutase.

Introduction

Reactive oxygen species such as hydrogen peroxide, singlet oxygen, hydroxyl radicals or superoxide radicals are thought to be responsible for the phytotoxicity of ozone (Hoigné and Bader, 1975; Grimes et al., 1983; Hewitt et al., 1990; Mehlhorn et al., 1990; Kanofsky and Sima, 1991). Plants possess powerful antioxidants to limit cellular injury when they are exposed to such oxidative stress (Alscher, 1989). Thus, ascorbic acid (Freebairn, 1957; Wenzel et al., 1995), ascorbate peroxidase (Mehlhorn, 1990) and glutathione reductase (Guri, 1983) have been shown to be important for plant resistance to ozone. However, it is less clear if superoxide dismutases (SOD) are also important (Lee and Bennett, 1982; McKersie et al., 1982; Chanway and Runeckles, 1984; Pitcher et al., 1991; Bowler et al., 1992; van Camp et al., 1994).

Most higher plants contain two types of SOD that have been classified according to their metal cofactor, Cu/ZnSOD and MnSOD. Of these, Cu/ZnSODs are the major isozymes in plants but inside the mitochondrial matrix superoxide anions are detoxified by MnSOD (Scandalios, 1993). Zinc deficiency has been shown to reduce levels of Cu/ZnSOD activity (Vaughan et al., 1982; Cakmak and Marschner, 1987). Recently, this was shown to increase plant sensitivity to ozone (Wenzel and Mehlhorn, 1995). Therefore, similar studies were also carried out with manganese-deficient plants. Thereby, it was investigated if MnSOD may also contribute to detoxification of superoxide anions when plants are exposed to ozone.

Materials and Methods

Plant growth

Seeds of bush beans (*Phaseolus vulgaris* L., cv. Saxa) were sown in a mixture of unfertilized potting compost, sand and perlite (1:1:1). After germination under greenhouse conditions the seedlings were transferred to 500 mL plastic flascs filled with continously aerated nutrient solutions. The composition of the nutrient solutions was as follows: K_2SO_4 (0.75 mM); KCl (0.1 mM); $Ca(NO_3)_2$ (2 mM); $MgSO_4$ (0.65 mM); KH_2PO_4 (0.5 mM); H_3BO_3 (10 µM); Fe-EDTA (0.1 mM); $ZnSO_4$ (50 nM); $CuSO_4$ (50 nM);

$(NH_4)_6Mo_7O_{24}$ (5 nM). Manganese was supplied as $MnSO_4$ at concentrations of 100 nM for sufficiently supplied plants (+Mn). For establishment of manganese deficiency, nutrient solutions contained 20 nM $MnSO_4$ during the first week of cultivation. Subsequently, plants were grown in the absence of manganese until harvest (−Mn). Nutrient solutions were changed twice per week.

Ozone exposures

Two weeks after germination, plants were exposed to ozone for 10 days in controlled environment fumigation cabinets described previously (Wenzel et al., 1995). Chambers were supplied with ambient air that had been purified by passage through a series of filtration modules containing either high efficiency penetration air filter material, activated charcoal or Purafil to remove dust particles and ambient air pollutants. Air (16 °C) passing into the cabinets had been humidified to a relative humidity of 90 % to maintain the relative humidity inside the cabinets at 60 %. The temperature inside the cabinets was 23 °C. The chambers were illuminated by 24 metal halide lamps (OSRAM PowerStars HQI TS 1000W/D-235, Tonding, Essen, Germany) which was reflected off the cabinet walls, as the back and two side walls had been lined on the outside with aluminium foil to give an even light distribution of approximately 350 µmol m^{-2} s^{-1}. Plants were fumigated with 100 nL L^{-1} ozone daily from 11:00 until 18:00 h local time. Ozone was generated from oxygen (technical grade) to avoid formation of N_2O_5 (Brown and Roberts, 1988).

During an ozone episode in the summer, plants were also exposed to ambient ozone concentrations in the greenhouse where daily maximum ozone concentrations reached similar levels as in the controlled fumigation experiment, but where average ozone concentrations were only 63 nL L^{-1} over the same 7 h period.

Plant analysis

Visible leaf injury (% necrosis of total leaf area), manganese content, levels of enzyme activities, soluble protein (Wenzel and Mehlhorn, 1995), ascorbic acid (Lucas et al., 1993) and transpiration (Wenzel et al., 1995) were determined as described previously. Rates of ethylene emissions were estimated after incubating whole leaves for 24 h in the light together with 3.5 mL 10 mM sodium phosphate buffer (pH 7.0) inside glass vessels (N 20-5, Macherey & Nagel, Düren, Germany) closed air-tight and agitated on a water bath held at room temperature. From these bottles, 1-mL air samples were then taken from the headspace. Ethylene in these samples was then measured isothermically (60 °C) on a gas chromatograph (Intersmat GC16, Delsi Instruments GmbH, Mettmann, Germany) equipped with a flame ionization detector and a 1.83 m × 3 mm Porapak N column (Macherey & Nagel, Düren, Germany). Nitrogen was used as carrier gas at a flow rate of 40 mL min^{-1}.

Statistics

All experiments were repeated at least twice and were carried out in duplicate chambers. Data are given as means ± s.e.m. and represent at least five replicates per treatment and chamber resulting in a minimum of 10 plants per chamber and a total of at least 40 observations for each experiment. Treatment effects were analysed by two-way analysis of variance using raw data from individual experiments where data from replicate chambers were pooled as no chamber effects were detected. Due to large variability and unequal variances, foliar injury and ethylene formation was analysed after arcsine transformation of the data. In addition, the two ozone treatments were also compared using Student's *t*-test. All statistical analyses were carried out using the MINITAB computer package and results were considered to be significant at the 5 % significance level.

Results

Effects of manganese deficiency on foliar manganese concentrations, and cellular antioxidants

When bushbeans were grown on nutrient solutions in the presence or absence of manganese for three weeks, primary leaves contained less than one third of the amount of manganese than plants sufficiently supplied with this nutrient (Table 1). However, manganese deficiency had no significant effects on growth and plants also did not develop manganese deficiency symptoms. Similarly, levels of Cu/ZnSOD or ascorbate peroxidase activity were not affected by manganese deficiency (Table 1). By contrast, manganese deficiency doubled levels of MnSOD activity while content of ascorbic acid and proteins in leaves were reduced in manganese-deficient plants (Table 1).

Effects of manganese deficiency on plant sensitivity to ozone

After 10 days of fumigation with ozone (100 nL L^{-1}, 7 h d^{-1}), almost all plants which had been fully supplied with manganese developed no visible leaf injury. By contrast, ozone caused severe foliar injury on plants deficient in manganese (Table 2). Similar results were also observed when plants were exposed to ambient concentrations of ozone under greenhouse conditions (Table 2). This increased sensitivity to ozone of manganese-deficient plants could not be attributed to a higher uptake of ozone as manganese deficiency and ozone both slightly reduced rates of transpiration (Table 1). Treatment with ozone also did not affect levels of Cu/ZnSOD, MnSOD, ascorbate peroxidase activity, ascorbic acid or protein content (Table 1). However, ozone significantly increased stress ethylene formation (Fig. 1). Moreover, there was a significant treatment interaction between manganese deficiency and ozone. After two days of fumigation,

Table 1: Manganese concentrations, levels of MnSOD, Cu/ZnSOD and ascorbate peroxidase activities, soluble proteins, ascorbic acid content and transpiration in primary leaves of four week-old manganese-sufficient (+Mn) and manganese-deficient (−Mn) bush beans after 10 days of fumigation with ozone (100 nL L^{-1}, 7 h d^{-1}). Significant treatment differences are indicated by different letters ($P \leq 0.05$).

	Control		Ozone	
	+Mn	−Mn	+Mn	−Mn
Manganese (µg g^{-1} dry weight)	34.3±3.7 a	13.1±1.5 b	33.5± 3.5 a	7.4± 1.6 c
MnSOD (units g^{-1} fresh weight)	9.9±2.4 a	18.3±4.1 b	12.2± 5.2 a	16.0± 7.3 b
Cu/ZnSOD (units g^{-1} fresh weight)	89.0±9.2 a	92.1±2.0 a	95.7±16.1 a	103.7±19.0 a
Ascorbate peroxidase (µmol g^{-1} fresh weight h^{-1})	11.5±1.5 a	11.5±0.8 a	13.4± 1.6 a	13.2± 1.2 a
Proteins (mg g^{-1} fresh weight)	11.5±1.1 a	9.1±0.4 b	11.6± 0.6 a	10.0± 0.7 b
Ascorbic acid (µmol g^{-1} fresh weight)	5.1±1.5 a	2.9±1.1 b	5.0± 1.0 a	3.7± 1.9 b
Transpiration (g H_2O $plant^{-1}$ day^{-1})	29.9±1.6 a	23.8±1.7 b	23.9± 1.9 b	18.8± 1.0 c

Table 2: Visible leaf injury (%) on primary leaves of four week-old manganese-sufficient (+Mn) and manganese-deficient (−Mn) bush beans after 10 days of exposure to ozone either in controlled environment fumigation experiments or after exposure to ozone in the greenhouse under ambient conditions. For each experiment consisting of four and two treatments, respectively, significant treatment differences are indicated by different letters as tested by analysis of variance and Student's *t*-test ($P \leq 0.05$). Analysis of variance indicated a significant treatment interaction ($P \leq 0.01$).

	Climatic Chamber 100 nL L^{-1} (7 h d^{-1})		Ambient Air 63 nL L^{-1} (7 h mean)
	Control	Ozone	
+Mn	0 a	2.3±0.9 b	2.3±0.4 b
−Mn	0 a	33.8±6.5 c	41.5±6.2 c

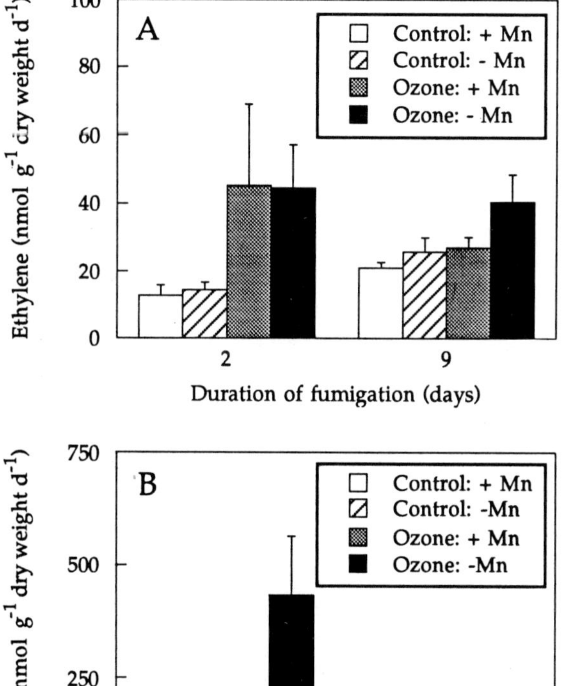

Fig. 1: Ethylene formation in primary leaves (A) and first trifoliate leaves (B) of 4-week-old manganese-sufficient (+Mn) and manganese-deficient (−Mn) bush beans after 2 d and 9 d of fumigation with ozone (100 nL L^{-1}, 7 h d^{-1}).

ozone increased rates of ethylene formation in primary leaves regardless of manganese nutrition but after nine days of fumigation with ozone stress ethylene formation was only observed in manganese-deficient plants (Fig. 1 A). In the first trifoliate leaves, ozone only increased rates of ethylene formation in manganese-deficient plants after two days of fumigation while ozone reduced rates of ethylene formation after nine days of fumigation both in plants that were supplied with all nutrients and in manganese-deficient plants (Fig. 1 B).

Discussion

Recent investigations with zinc-deficient plants demonstrated that trace nutrient deficiency may increase plant sensitivity to ozone (Wenzel and Mehlhorn, 1995). Similar results were also obtained in the present study when bush beans suffered from manganese deficiency (Table 2), i.e. where concentrations of manganese in the leaves were less than 20 p.p.m. (Baumeister and Ernst, 1978; Bergmann, 1988; Pearson and Rengel, 1995). This may contribute to explain foliar injury symptoms in field grown Norway spruce trees which were characterized as deficient in manganese (Polle et al., 1992).

In the experiments with zinc-deficient plants, increased ozone sensitivity could be correlated with reduced levels of CuZnSOD. Thereby, it could be demonstrated that CuZnSOD plays a role in detoxificaton processes when plants are exposed to ozone (Wenzel and Mehlhorn, 1995). For MnSOD, this seems less likely. Although levels of MnSOD activity were increased in manganese-deficient plants, these plants were more sensitive to ozone than plants fully supplied with this trace nutrient (Table 2). Experiments with transgenic plants also demonstrated that overexpression of MnSOD in mitochondria does not increase plant tolerance to ozone (van Camp et al., 1994). Moreover, ozone does not increase transcript levels of MnSOD (Willekens et al., 1994) whereas cytosolic (Sharma and Davis, 1994) and chloroplastic (Willekens et al., 1994) Cu/ZnSODs are induced by ozone. Histological studies also indicate that ozone only affects mitochondria after effects are seen in the cytosol and in chloroplasts (Lee, 1968; Ojanperä et al., 1992). While it is unclear why levels of MnSOD activity were increased in manganese deficient plants, enhanced levels of SOD activity in nutrient deficient plants have been observed before (Cakmak and Marschner, 1992; Iturbe-Ormaetxe et al., 1995). Together, these results suggest that formation of toxic oxygen species in mitochondria is not a primary event during ozone toxicity.

Differences in ozone sensitivity also could not be explained by effects of manganese deficiency on foliar uptake of the pollutant. Thus, manganese deficiency decreased rates of transpiration and, thereby, reduced uptake of ozone (Table 1). This should have reduced ozone toxicity but the opposite was observed (Table 2). Similarly, manganese deficiency also had no effect on activities of Cu/ZnSOD or ascorbate peroxidase. Thus, cellular ability to detoxify superoxide anions or hydrogen peroxide was not impaired (Table 1). Tissue injury from increased formation of hydrogen peroxide also seems unlikely. Thus, while levels of MnSOD were increased in manganese-deficient plants (Table 1), no foliar injury was observed in manganese-deficient plants that had not been fumigated with ozone (Table 2).

By contrast, manganese deficiency enhanced ozone-induced rates of ethylene formation. Thus, in plants sufficiently supplied with manganese, stress ethylene formation was only observed in primary leaves after two days of fumigation with ozone while ethylene formation was still enhanced after nine days in manganese-deficient plants (Fig. 1 A). Moreover, in first trifoliate leaves, stress ethylene formation was only observed in manganese-deficient plants (Fig. 1 B). Stress ethylene formation has been shown to determine plant sensitivity to ozone (Mehlhorn and Wellburn, 1987; Mehlhorn et al., 1991; Reddy et al., 1993; Wenzel et al., 1995). Therefore, effects of manganese deficiency on rates of ethylene formation may explain why manganese-deficient plants were more sensitive to ozone than plants fully supplied with this nutrient (Table 2). Similar effects of nitrogen oxides on ozone toxicity already have been reported (Mehlhorn and Wellburn, 1987).

Manganese deficiency also reduced ascorbic acid content in leaves (Table 1). Like ethylene, ascorbic acid also has been shown to affect plant sensitivity to ozone (Freebairn, 1960; Siegel, 1962; Dass and Weaver, 1968). Thus, it was suggested that ascorbic acid may detoxify ozone in the apoplast (Chameides, 1989). However, ozone does not deplete ascorbic acid in the apoplast (Polle et al., 1995) and uptake of ozone far exceeds oxidation of ascorbic acid (Luwe et al., 1994). By contrast, ascorbic acid may reduce ethylene biosynthesis and, thereby, reduce plant sensitivity to ozone (Wenzel et al., 1995). This may explain why, after nine days of fumigation, ozone-induced stress ethylene formation in primary leaves was only observed in manganese-deficient plants which had reduced levels of ascorbic acid (Table 1, Fig. 1 A). Similarly, this may explain why foliar levels of ascorbic acid are increasing with altitude in needles of Norway spruce (Polle et al., 1995).

Alternatively, manganese deficiency may have increased ozone toxicity because of its ability to modulate enzyme activities. Thus, decarboxylases, dehydrogenases, phosphokinases or phosphotransferases are activated *in vitro* by manganese (Marschner, 1986). However, it is less clear if this is also important *in vivo* as magnesium can replace manganese and is present at much higher concentrations (Clarkson and Hanson, 1980). Manganese also is involved in photosynthetic oxygen formation and, thereby, may impair the first step of the electron transport chain of the light reaction. This may have negative effects on subsequent reactions such as photophosphorylation and carbon fixation (Marschner, 1986). However, no reduction in growth was observed in manganese-deficient plants. This indicates that photosynthesis was not impaired. Accumulation of nitrite is another possibility (Marschner, 1986) but no foliar injury was observed in non-fumigated plants (Table 2).

In conclusion, manganese deficiency was found to enhance ozone toxicity. This is attributed to effects on ozone-induced stress ethylene formation, possibly, because of effects of manganese deficiency on ascorbic acid content. In addition, it is proposed that formation of toxic oxygen species in mitochondria is not a primary event during ozone toxicity and that MnSOD contributes little to make plants less sensitive to ozone.

Acknowledgements

The authors are grateful to Prof. Dr. R. Guderian for making available the experimental facilities for this work, H. Hennern, H. Schlautmann, C. Kosch for technical assistance and the Commission of the European Communities (grant EV5V-CT92-0092) for financial support.

References

Alscher, R. G.: Biosynthesis and antioxidant function of glutathione in plants. Physiol. Plant. *77*, 457–464 (1989).

Baumeister, W. and W. Ernst: Mineralstoffe und Pflanzenwachstum. Gustav Fischer Verlag, Stuttgart (1978).

Bergmann, W.: Ernährungsstörungen bei Kulturpflanzen (2nd ed.). Gustav Fischer Verlag, Stuttgart (1988).

Bowler, C., M. van Montagu, and D. Inzé: Superoxide dismutase and stress tolerance. Annu. Rev. Plant Physiol. Molec. Biol. *43*, 83–116 (1992).

Cakmak, I. and H. Marschner: Mechanism of phosphorus-induced zinc deficiency in cotton. III. Changes in physiological availability of zinc in plants. Physiol. Plant. *70*, 13–20 (1987).

– – Magnesium deficiency and high light intensity enhance acitivities of superoxide dismutase, ascorbate peroxidase, and glutathione reductase in bean leaves. Plant Physiol. *98*, 1222–1227 (1992).

Chameides, W. L.: The chemistry of ozone deposition to plant leaves: Role of ascorbic acid. Environ. Sci. Technol. *23*, 595–600 (1989).

Chanway, C. P. and V. C. Runeckles: The role of superoxide dismutase in the susceptibility of bean leaves to ozone injury. Can. J. Bot. *62*, 236–240 (1984).

Clarkson, D. T. and J. B. Hanson: The mineral nutrition of higher plants. Annu. Rev. Plant Physiol. *31*, 239–298 (1980).

Dass, H. C. and G. M. Weaver: Modification of ozone damage to *Phaseolus vulgaris* by antioxidants, thiols and sulfhydryl reagents. Can. J. Plant Sci. *48*, 569–574 (1968).

Freebairn, H. T.: Reversal of inhibitory effects of ozone on oxygen uptake of mitochondria. Science *126*, 303–304 (1957).

– The prevention of air pollution damage to plants by use of vitamin C sprays. J. Air Pollut. Contr. Assoc. *10*, 314–317 (1960).

Grimes, H. D., K. K. Perkins, and W. F. Boss: Ozone degrades into hydroxyl radical under physiological conditions. A spin trapping study. Plant Physiol. *72*, 1016–1020 (1983).

Guri, A.: Variation in glutathione and ascorbic acid content among selected cultivars of *Phaseolus vulgaris* prior to and after exposure to ozone. Can. J. Plant Sci. *63*, 733–737 (1983).

Hewitt, C. N., G. L. Kok, and R. Fall: Hydroperoxides in plants exposed to ozone mediate air pollution damage to alkene emitters. Nature *344*, 56–58 (1990).

Hoigné, J. and H. Bader: Ozonation of water: role of hydroxyl radicals as oxidizing intermediates. Science *190*, 782–784 (1975).

Iturbe-Ormaetxe, I., J. F. Moran, C. Arrese-Igor, Y. Gogorcena, R. V. Klucas, and M. Becana: Activated oxygen and antioxidant defences in iron-deficient pea plants. Plant, Cell Environ. *18*, 421–429 (1995).

Kanofsky, J. R. and P. Sima: Singlet oxygen production from the reaction of ozone with biological molecules. J. Biol. Chem. *266*, 10478–10481 (1991).

Lee, E. H. and J. H. Bennett: Superoxide dismutase. A possible protective enzyme against ozone injury in snap beans (*Phaseolus vulgaris* L.). Plant Physiol. *69*, 1444–1449 (1982).

Lee, T. T.: Effect of ozone on swelling of tobacco mitochondria. Plant Physiol. *43*, 133–139 (1968).

Lucas, P. W., L. Rantanen, and H. Mehlhorn: Needle chlorosis in Sitka spruce following a three-year exposure to low concentrations of ozone: changes in mineral content, pigmentation and ascorbic acid. New Phytol. *124*, 265–275 (1993).

LUWE, M. W. F., U. TAKAHAMA, and U. HEBER: Role of ascorbate in detoxifying ozone in the apoplast of spinach (*Spinacia oleracea* L.) leaves. Plant Physiol. *101*, 969–976 (1993).

MARSCHNER, H.: Mineral nutrition in higher plants. Academic Press, London (1986).

MCKERSIE, B. D., W. D. BEVERSDORF, and P. HUCL: The relationship between ozone insensitivity, lipid-soluble antioxidants, and superoxide dismutases in *Phaseolus vulgaris*. Can. J. Bot. *60*, 2686–2691 (1982).

MEHLHORN, H.: Ethylene-promoted ascorbate peroxidase activity protects plants against hydrogen peroxide, ozone and paraquat. Plant, Cell Environ. *13*, 971–976 (1990).

MEHLHORN, H. and A. R. WELLBURN: Stress ethylene formation determines plant sensitivity to ozone. Nature *327*, 417–418 (1987).

MEHLHORN, H., B. J. TABNER, and A. R. WELLBURN: Electron spin resonance evidence for the formation of free radicals in plants exposed to ozone. Physiol. Plant. *79*, 377–383 (1990).

MEHLHORN, H., J. M. O'SHEA, and A. R. WELLBURN: Atmospheric ozone interacts with stress ethylene formation by plants to cause visible leaf injury. J. Exp. Bot. *42*, 17–24 (1991).

OJANPERÄ, K., S. SUTINEN, H. PLEIJEL, and G. SELLDEN: Exposure of spring wheat, *Triticum aestivum* L., cv. Drabant, to different concentrations of ozone in open-top chambers: effects on the ultrastructure of flag leaf cells. New Phytol. *120*, 39–48 (1992).

PEARSON, J. N. and Z. RENGEL: Uptake and distribution of ^{65}Zn and ^{54}Mn in wheat grown at sufficient and deficient levels of Zn and Mn II. During grain development. J. Exp. Bot. *46*, 841–845 (1995).

PITCHER, L. H., E. BRENNAN, A. HURLEY, P. DUNSMUIR, J. M. TEPPERMAN, and B. A. ZILINSKAS: Overproduction of petunia chloroplastic copper/zinc superoxide dismutase does not confer tolerance in transgenic tobacco. Plant Physiol. *97*, 452–455 (1991).

POLLE, A., K. CHAKRABARTI, S. CHAKRABARTI, F. SEIFERT, P. SCHRAMEL, and H. RENNENBERG: Antioxidants and manganese deficiency in needles of Norway spruce (*Picea abies* L.) trees. Plant Physiol. *99*, 1084–1089 (1992).

POLLE, A., G. WIESNER, and W. M. HAVRANEK: Quantification of ozone influx and apoplastic ascorbate content in needles of Norway spruce trees (*Picea abies* L., Karst.) at high altitude. Plant, Cell Environ. *18*, 681–688 (1995).

REDDY, G. N., R. N. ARTECA, Y. R. DAI, H. E. FLORES, F. B. NEGM, and E. J. PELL: Changes in ethylene and polyamines in relation to mRNA levels of the large and small subunits of ribulose bisphosphate carboxylase/oxygenase in ozone stressed potato foliage. Plant, Cell Environ. *16*, 819–826 (1993).

SCANDALIOS, J. G.: Oxygen stress and superoxide dismutases. Plant Physiol. *101*, 7–12 (1993).

SHARMA, Y. K. and K. R. DAVIS: Ozone-induced expression of stress-related genes in *Arabidopsis thaliana*. Plant Physiol. *105*, 1089–1096 (1994).

SIEGEL, S. M.: Protection of plants against airborne oxidants: cucumber seedlings at extreme ozone levels. Plant Physiol. *37*, 261–266 (1962).

VAN CAMP, W., H. WILLEKENS, C. BOWLER, M. VAN MONTAGU, D. INZÉ, P. REUPOLD-POPP, H. SANDERMANN Jr., and C. LANGEBARTELS: Elevated levels of superoxide dismutase protect transgenic plants against ozone damage. Bio/Technology *12*, 165–168 (1994).

VAUGHAN, D., P. C. DEKOCK, and B. G. ORD: The nature and localization of superoxide dismutase in fronds of *Lemma gibba* L. and the effect of copper and zinc deficiency on its activity. Physiol. Plant. *54*, 253–257 (1982).

WENZEL, A. A. and H. MEHLHORN: Zinc deficiency enhances ozone toxicity in bush beans (*Phaseolus vulgaris* L. cv. Saxa). J. Exp. Bot. *46*, 867–872 (1995).

WENZEL, A. A., H. SCHLAUTMANN, C. A. JONES, K. KÜPPERS, and H. MEHLHORN: Aminoethoxyvinylglycine, cobalt and ascorbic acid all reduce ozone toxicity in mung beans by inhibition of different steps during ethylene biosynthesis. Physiol. Plant. *93*, 286–290 (1995).

WILLEKENS, H., W. VAN CAMP, M. VAN MONTAGU, D. INZÉ, C. LANGEBARTELS, and H. SANDERMANN Jr.: Ozone, sulfur dioxide, and ultraviolet B have similar effects on mRNA accumulation of antioxidant genes in *Nicotiana plumbaginifolia* L. Plant Physiol. *106*, 1007–1014 (1994).

J. Plant Physiol. Vol. 148. pp. 160–165 (1996)

The Assessment of Ozone Stress by Recording Chromosomal Aberrations in Root Tips of Spruce Trees [*Picea abies* (L.) Karst]

MARIA MÜLLER[1], BÄRBEL KÖHLER[1], MICHAEL TAUSZ[1], DIETER GRILL[1], and CORNELIUS LÜTZ[2]

[1] Institute of Plant Physiology, University of Graz, Schubertstraße 51, A-8010 Graz, Austria

[2] GSF Forschungszentrum für Umwelt und Gesundheit, Expositionskammern, Neuherberg, D-85764 Oberschleissheim, Germany

Received June 24, 1995 · Accepted October 15, 1995

Summary

Spruce plants were exposed in environmental chambers to different levels of ozone for six weeks to study the influence of increased ozone on root tip chromosomes of spruce trees. No visible symptoms resulted from these treatments. The classification of chromosomal defects was used to characterise the influences on the root tips caused by ozone. Directly after the fumigation had ceased the fumigated variant showed a significantly increased number of chromosomal aberrations in comparison to the control. Five further investigations of both variants of this experiment up to two years after the fumigation had ended showed a long-term after-effect to ozone on the genetic material of spruce trees.

The observed chromosomal aberrations in all variants of the experiment consisted of chromosomal stickiness, chromosomal breakage, and fragmentation. The main type of the observed chromosomal abnormalities was chromosome stickiness leading to cell death.

Key words: Picea abies (L.) Karst., chromosomal defects, ozone.

Introduction

Today there can be no doubt that concentrations of ozone in air at ground levels are capable of reducing growth and causing visible injury to vegetation. There has also been a concern about the role of ozone as a contributory factor to forest decline, the widespread decrease in tree vitality observed over North America and Europe (Cape et al. 1994). Ambient concentrations of ozone in Europe are generally below the levels that cause direct visible injuries. In recent years, the concentration of ozone in the lower atmosphere in Europe has been increasing (Penkett 1984), and short peak-ozone episodes can cause a direct damage to sensitive plants. Prolonged exposure to ozone results in suppression of photosynthesis and plant growth and in changes of assimilate partitioning (Darral 1989). At low ozone levels a large proportion of assimilate is usually allocated to leaves and stems at the expense of roots and crowns (Heck et al. 1983). Many of these effects were noted before any visible injury occurred (Spence et al. 1990), thus showing the need for a sensitive method for monitoring ozone effects on plants.

It is still of interest to find new methods and modify old test systems for an early detection of environmental influences. One of these test systems is the classification of chromosomal aberrations, that comprise a sensitive method for the assessment of genotoxic effects caused by chemical treatments. The cytological studies are relatively simple to perform and give valuable information to effects of cell division and chromosomes (Levan 1938, Fiskesjö 1985). In our search for an easy and reliable assay for the detection of genotoxic substances on forest trees we have carried out a test system by the classification of chromosomal aberrations with spruce trees. Some other test systems with forestry trees still exist. Fiskesjö (1989) examined the aluminium toxicity in root tips of spruce trees, beech and oak and found disturbances in dividing cells, as c-mitotic effects as well as growth restrictions and browning of the roots and so-called ‹Al-structures›. Another test system showed that spruce trees are mitotically highly active throughout the whole year and therefore they may be affected by pollutants also during their morphological state of dormancy (Matschke et al. 1994). Schubert and Rieger (1994) worked out a genotoxicity test, based on the evaluation of sister chromatid exchange frequencies and suggest their test as a suitable and reliable tool to observe genotoxic effects in spruce plants. According to the basic work with spruce trees of a Slovenian working group (Druskovic 1988), in 1989 we performed our plant test system by the classification of chromosomal aberrations with the spruce tree as bioindicator plant. We used this method at natural sites (Grill et al. 1993, Müller et al. 1991, 1992, 1994) and under defined conditions in climate chambers (Müller and Grill 1994), greenhouses (Müller et al., submitted) and open-top chambers (Müller et al., in press) for the assessment and evaluation of genetic risks of trees from environmental influences. These studies showed that the classification of chromosomal aberrations in spruce is a sensitive measure to investigate various environmental effects on trees.

In the present paper the classification of chromosomal aberrations as a test system was applied to spruce trees fumigated in climate chambers. The aims of this study are: (1) to investigate the influences of ozone on the genetic material for an immediate effect and for further two years for a long-term after effect and (2) to discuss possible mechanisms leading to chromosomal aberrations in the root tips of spruce trees after ozone-treatment.

Materials and Methods

Plants and soils

Clonal spruce trees (*Picea abies* (L.) Karst., clone 4611) were potted in acidic, sandy forest soil and kept outdoors for 2 years before the experiment. The plants were irrigated regularly, fertilised twice a year, and protected from pathogens.

Environmental conditions

Climate chamber studies were undertaken in the phytotron at the GSF-Munich (Germany).

1. Immediate effect. The spruce plants were transferred to the environmental chambers and grown in four independent growth chambers of a large phytotron under completely controlled conditions. During a 42-day experimental period in 1993, the plants were exposed to a climatic programme simulating variations of temperature and day length, as observed at higher elevations in the Bavarian forest. The lack of higher light intensities was compensated for by extended noon-time illumination resulting in daily mean energy values comparable to field conditions. The relative humidity within the chambers varied in the following range: maximum rh 98–100 %; mean maximum rh 78 %; mean minimum rh 63 % (Blank and Lütz 1990). Each experimental group consisted of eight potted five-year-old trees. The pots (12-cm-diameter) were installed in temperature-controlled root boxes to provide root temperatures comparable to field conditions and connected to an automatic irrigation system to provide water at a soil water tension of 150 hPa.

The treatments were 20 nL L^{-1} ozone and 100 nL L^{-1} ozone. Two chambers (each with 4 trees) were used for each treatment. Fumigation was provided for 24 h per day for 6 weeks (February–April). Background concentrations of about 8–10 nL L^{-1} SO_2 and NO_x were present in the chambers.

2. Long-term after-effect. After the fumigation had ceased the spruce plants were transferred to Graz first to climate chambers at the Institute of Plant Physiology. The temperatures were 19 °C and 17 °C (± 1 to 2 °C) for day and night, respectively. The relative humidity within the chambers varied between 74 and 89 %. and the day – night rhythm was 14 to 10 hours. After three month under these conditions the spruce plants were transferred to a field near the institute and exposed to normal air and weather conditions. Samples were taken after 88 days (in summer 1993), 196 days (in autumn 1993), 447 days (in summer 1994), 554 days (in autumn 1994) and 770 days (in spring 1995).

Sample treatment

The root tips were cut, treated with 1-bromonaphthalene, washed and fixed in ethanol : glacial acetic acid (3 : 1; v/v). After fixation the root tips were hydrolysed in 3N HCl for 3 min at 63 °C, stained in freshly prepared Schiff's reagent and squashed in a few drops of carmine acetic acid. (Treatment according to Müller et al. 1991).

Evaluation

The cells in metaphase were classified in the following categories: metaphase, break, fragment, ring, connection, clumping, amorphous chromatin mass. Connection, clumping and amorphous chromatin masse were scored and classified in one category (= stickiness), also break, fragment and ring. 100 to 200 metaphases (from cells with the whole cell wall around) per replicate tree of the two experimental groups were examined for chromosomal aberrations. The percentages of abnormalities in total metaphases were calculated. All slides were coded and examined blind.

Statistics

Statistical evaluations were completed using Statistica® (StatSoft, USA, 1994) software package. Effects of ozone exposure and sampling were calculated by two-way analysis of variance (ANOVA). Post-hoc comparisons of means were carried out by Scheffé's test following significant ANOVA results.

Results

Observations of visible damage were made on a qualitative basis during and at the end of ozone fumigation and during 25 months after the fumigation had ceased. No visible evidence of injuries or of stunted growth due to the effect of increased ozone was observed in any of the plants at the end of the fumigation and also 2 years later. Differences in the results between the trees of the replicated chambers could not be observed. The differences in the number of chromosomal aberrations between the variants of the experiments are shown in Table 1 and Fig. 1. Increased ozone (100 nL L^{-1}) had significant effects on the genetic material, as indicated in an in-

Table 1: Effects on Norway spruce root cells treated with ozone for 6 weeks directly after fumigation had ceased and for further two years. The chromosomal aberrations were scored in metaphase. For each variant 8 trees were used, and if possible 100 to 200 cells per tree were examined for chromosomal aberrations. Numbers in brackets are the standard deviation.

days after the fumigation period		$20\ nL\ L^{-1}$	$100\ nL\ L^{-1}$
0	metaphases	1023	860
	aberrations %	2.1 (0.2)	3.9 (0.7)***
	stickiness	1.8 (0.4)	3.7 (0.8)
	others	0.3 (0.4)	0.2 (0.2)
88	metaphases	1507	942
	aberrations %	2.3 (0.4)	5.1 (0.8)***
	stickiness	2.0 (0.6)	4.8 (1.0)
	others	0.3 (0.3)	0.3 (0.4)
196	metaphases	866	656
	aberrations %	4.3 (0.6)	5.6 (0.6)*
	stickiness	2.7 (0.7)	4.3 (1.0)
	others	1.6 (0.5)	1.3 (0.7)
447	metaphases	888	926
	aberrations %	4.5 (0.2)	6.0 (0.4)ns
	stickiness	2.4 (0.6)	4.2 (0.6)
	others	2.1 (0.6)	1.8 (0.9)
554	metaphases	653	705
	aberrations %	4.5 (0.3)	5.8 (0.6)*
	stickiness	2.5 (0.8)	3.3 (0.8)
	others	2.0 (0.8)	2.5 (0.9)
770	metaphases	778	844
	aberrations %	5.2 (0.2)	6.9 (0.3)***
	stickiness	2.8 (1.6)	3.8 (2.0)
	others	2.4 (1.6)	3.1 (1.9)

The symbols indicate significant differences between the two variants of the experiment; ns = not significant; * $p<0.01$; *** $p<0.001$.

creased amount of chromosomal abnormalities (3.9 % ± 0.7) compared to the control (Table 1). The root meristems of the plants from the control chamber ($20\ nL\ L^{-1}$) represented the lowest aberration rate with 2.1 % ± 0.2 (= mean ± SD). 88 days later, the plants grew in the meantime in climate chambers in Graz, the amount of chromosomal defects of the former fumigated variant was increased (5.1 % ± 0.8) compared to the ozone-treated series of the experiment directly after the fumigation had ended ($p = 0.08$). The control variant showed no increase of chromosomal aberrations (2.3 %±0.4).

To this time in June 1993 the spruce trees were transferred to a field near the institute. Four further investigations of the spruce trees during the next two years (in October 1993, in June 1994, in October 1994 and in May 1995) were carried out and gave clear evidence of a long-term after-effect of ozone on the genetic material of spruce trees. The former ozone fumigated variant showed under environmental conditions 5.6 % to 6.9 % of chromosomal aberrations, thus causing a statistically significant increase of chromosomal defects ($p<0.001$) compared to the results of the climate chambers (Fig. 1). The former control variant responded under normal air and weather conditions with 4.3 % to 5.2 % of chromosomal abnormalities, showing an elevation of the aberration rate at p-level <0.001 in relation to the climate chamber studies (Fig. 1).

The differences of the investigations of the former ozone-treated variant to the control were statistically highly significant at the p-level <0.01 and at $p<0.001$ (Table 1). Only the investigation after 447 days showed a lower significance ($p = 0.06$) between the two variants of the experiment (Table 1).

In both variants the main type of chromosomal aberrations was chromosome stickiness, resulting in connections, clumped metaphases and amorphous chromatin masses. The relation between the single chromosome defects (break, fragment, ring) and the defects of the whole metaphase (stickiness) changed from the climate chamber studies to the field studies. In the climate chambers up to 92 % of all chromosomal abnormalities resulted in chromosome stickiness, whereas under field conditions around 61 % of the observed chromosomal defects were chromosome stickiness (Table 1).

Discussion

Plant responses to ozone exposure have been subject of many investigations in several laboratories (Runeckles and Chevone 1992). Many studies have been performed with agricultural crops, whereas forest trees received much less attention. The results presented in this paper show that the spruce clone is not sensitive to the ozone treatment of $100\ nL\ L^{-1}$ for 6 weeks. The symptoms of ozone injury including bleaching, chlorotic mottling, changes in pigmentation and necrosis could not be seen (Kress et al. 1982, Williams 1986, Polle et al. 1993). Visible damage had not been found in other experiments in which concentrations of about 50–70 nL ozone L^{-1} were used (Alscher et al. 1989, Wallin et al. 1990). It has been suggested, that acute short-term exposure to high ozone concentrations (>$200\ nL\ L^{-1}$) generally results in visible damage, whereas long-term, chronic exposure to lower ozone concentrations generally leads to reduced growth without visible foliar damage (cp. Heagle 1989).

The results of the chromosomal analysis directly after fumigation had ceased showed a significantly increased number of chromosomal aberrations in the root tip meristems of the former fumigated variant, thus indicating the genetic material of the root tips of spruce trees is a sensitive measure for oxidative stress. Although ozone-induced plant damage has been studied quite extensively, the mechanism of ozone injury is still unknown (Kangasjärvi et al. 1994). Examples of the ozone-induced physiological changes are reduced photosynthesis, increased respiration rate, membrane lipid peroxidation, enhanced rate of senescence, reduced transpiration due to stomatal closing (Runeckles and Chevone 1992) and disturbances in the genetic material (Müller and Grill 1994). It is assumed that ozone enters the mesophyll via stomata, where it is rapidly dissolved in water and converted into reactive oxygen species such as superoxide anions, hydroxyl radicals, and hydrogen peroxide (Grimes et al. 1983, Mehlhorn et al. 1990, Kanofsky and Sima 1991). The formation of these active free radicals in ozone-exposed plants before symptoms appear has been demonstrated using electron spin resonance

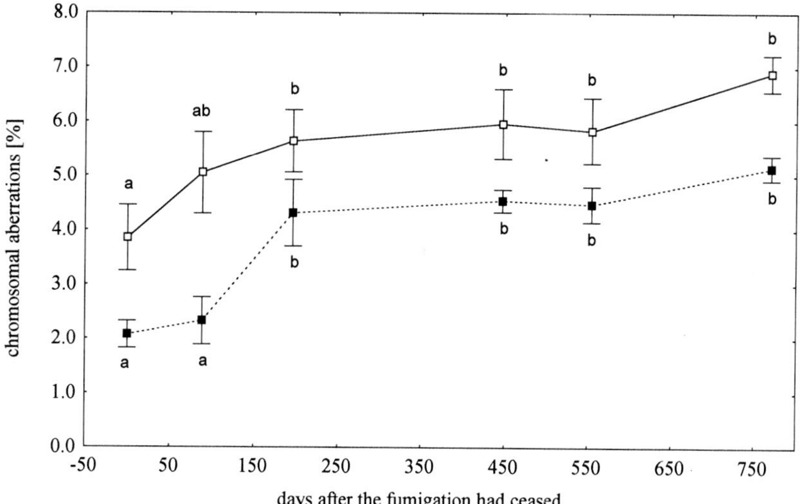

Fig. 1: Effects on Norway spruce root cells treated with ozone for 6 weeks directly after fumigation had ceased and for further two years. The chromosomal aberrations were scored in metaphase. For each variant 8 trees were used, and if possible 100 to 200 cells per tree were examined for chromosomal aberrations. Significant differences ($p<0.001$) within the variants of the experiment are indicated by different letters (a, b); ---- = control variant; —— = former ozone fumigated variant.

spectroscopy (Mehlhorn et al. 1990). The active oxygen species are most probably responsible for the plasma membrane lipid peroxidation that has been demonstrated to occur after ozone treatment (Pauls and Thompson 1980, Heath 1987, Chevrier et al. 1990). In several reviews the role of free oxygen radicals in disease and genetic damage has been reported (Slater 1984, Comporti 1985, Halliwell and Gutteridge 1984, 1985, 1986). Furthermore, reports have provided evidence for a relationship between lipid peroxidation processes and genotoxicity (Ames et al. 1982, Ames 1983, Kensler et al. 1984, Vaca et al. 1988). One of the earliest studies showing the interaction between lipid peroxides and DNA in plants were presented by Swaminathan and Natarajan (1956, 1959). Today there is increasing evidence that also aldehydes generated endogenously during the process of lipid peroxidation are causally involved in some of the physiological effects associated with oxidative stress in cells and tissues. Lipid peroxidation is an amplifier for initial free radicals and the reactive aldehydes generated in this process may well act as second toxic messengers of the complex chain reactions that are initiated if polyunsaturated fatty acids of the membrane bilayer are converted to lipid hydroperoxides (Esterbauer et al. 1991). Compared with free radicals, aldehydes are rather long lived and can diffuse from the membranes (i.e. the side of their origin) and reach and attack targets intracellularly or extracellularly which are distant from the initial free event. Furthermore, lipid peroxidation and products formed in this process could inhibit certain DNA repair systems (Krokan et al. 1985, Cox et al. 1988, Curren et al. 1988), thus indirectly increasing the rate of spontaneous mutation (Esterbauer et al. 1991). So far a particularly sensitive test system for genotoxicity of aldehydes is primary cultures of rat hepatocytes. With this cell type it was shown that aldehydes are genotoxic and lead to significant increase of micronuclei, sister chromatid exchange and chromosomal aberrations (Eckl and Esterbauer 1989, Esterbauer et al. 1990, Eckl et al. 1993). The role of aldehydes with chromosomal damage in plants is still unknown. The possibility exists that endogenously formed aldehydic lipid peroxidation products react with the genetic material in the root tips and induce mutations. Up to now in all investigations the aldehydes were added to the respective test system and the effects produced in response to the exogenous aldehyde were then determined after a certain time (Esterbauer et al. 1991). In this connection the distance from the side of ozone influence (the top of a tree) to the root meristems must be taken into consideration. The rather long lived aldehydes must be translocated in the phloem to the root tips, in which the toxic peroxidation products may damage the phloem. Phloem damage interrupts translocation of assimilates, which correlates with observed reductions in root growth (Spence et al. 1990). Various studies have noted accumulations of soluble sugars and starch in shoots, that were best explained by reduced or delayed phloem translocation (Tingey et al. 1976, Vogels et al. 1986, Gorissen and Van Veen 1988, Spence et al. 1990, Willenbrink and Schatten 1993). If ozone reduces translocation of sugars to roots, the long-term after effect may not be evident for years. A major difficulty in studying ozone effects on plants is to determine whether biophysical, biochemical or physiological changes are occurring within the plant when no visible, external response is observed (Spence et al. 1990).

Investigations with an ozone-sensitive clone (Lippert 1992, Polle et al. 1993), showing effects on the genetic material of spruce trees 21 months after fumigation had ceased are previously described (Müller et al. 1994). The results of the present experiment under normal air and weather conditions showing a significantly increased amount of chromosomal aberrations in the former ozone-treated spruce trees proved a long-term after-effect of ozone on the genetic material of spruce plants. From our work at natural sites we can suggest that the genetic material of spruce trees responds to changing environment with an increase or a decrease in the number of chromosomal aberrations (Müller et al. 1992). This explains the elevation of the aberration rate from the control variant under field conditions compared to this variant under charcoal filtered air. Approximately 4 to 5 % of chromosomal aberrations can be found in meristems of spruce trees grown in natural areas (Müller et al. 1991, 1992). If no long-term effects were manifest in the genetic material of the former fumigated variant both variants of the experiment would show approximately the same frequency of chromosomal defects under field conditions. Post-fumigation long-term effects of ozone are well known and among other things they were observed for plant growth and vitality in loblolly pine (Spence et al. 1990), for secondary metabolites and antioxidants as catechin (Langebartels et al. 1990) and for pigments in spruce needles (Lütz 1992).

The main type of chromosomal defects in all investigations was chromosome stickiness. This type of aberration may result from improper folding of the chromosome fiber into single chromatids, that chromosomes become attached to each other by subchromatid bridges (McGill et al. 1974, Klasterska et al. 1976). Stephen (1979) also suggested that stickiness is a type of physical adhesion that involves mainly the proteinaceous matrix of the chromatin material. Chromosome stickiness reflects highly toxic effects, usually of an irreversible type, probably leading to cell death (Liu et al. 1993). The further, but less frequent, defects found in this series of experiment are breaks, fragments and rings, which are genotoxic effects and they may transfer the damage to the following generations of cells (Fiskesjö 1994). Under conditions of constant ozonetreatment in the climate chambers a significantly increased number of sticky chromosomes was observed. Under field conditions also an increased number of chromosome stickiness, but considerable more breaks and fragments occurred. It is difficult to decide wether certain influences or combinations lead to specific chromosomal abnormalities. Druskovic (1988) suggested acute influences creating abnormalities on single chromosomes, as breaks and fragments, and constant long-lasting influences causing stickiness.

In conclusion, the results of this study combined with those of others, point toward immediate and long-term effects of exposure to moderate concentrations of ozone on growth and vitality in trees even when no visible damage occurs. The classification of chromosomal aberrations is a sensitive measure for environmental influences on spruce trees, for non-accumulating compounds as ozone, too. Despite the widespread occurrence of ozone injuries, the mechanisms of the damaging process and the plant defence systems against ozone attacks are still poorly understood. Considering the importance of our forest tree species it is essential to characterize their short- and long-term responses to a variety of environmental and pollution impacts.

Acknowledgements

This study was supported by the ‹Jubiläumsfonds der österreichischen Nationalbank›. We are grateful to Anita Geiszinger for technical assistance. We thank the ‹GSF Forschungszentrum für Umwelt und Gesundheit, Expositionskammern, Neuherberg (Germany)› for the fumigation experiment in the climate chambers and the ‹Forstliche Bundesversuchsanstalt Vienna› for plant material.

References

Alscher, R. G., R. G. Amundson, J. R. Cumming, J. Fincher, G. Rubin, P. van Leuken, and L. H. Weinstein: Seasonal changes in the pigments, carbohydrates and growth of red spruce as affected by ozone. New Phytol. *113*, 211–223 (1989).

Ames, B. N.: Dietary carcinogens and anticarcinogens, oxygen radicals and degenerative diseases. Science *221*, 1256–1264 (1983).

Ames, B. N., M. C. Hollstein, and R. Cathart: Lipid peroxidation and oxidative damage to DNA. In: Yagi, K. (ed.): Lipid Peroxides in Biology and Medicine, pp. 339–351. Academic Press, New York (1982).

Blank, L. and C. Lütz: Tree exposure experiment in closed chambers. Environ. Poll., Special issue, *64*, 189–399 (1990).

Cape, J. N., R. I. Smith, and D. Fowler: The influence of ozone chemistry and meteorology on plant exposure to photo-oxidants. Proc. Roy. Soc. Edinb. *102 B*, 11–31 (1994).

Chevrier, N., Y. S. Chung, and F. Sarhan: Oxidative damages and repair in *Euglena gracilis* exposed to ozone II. Membrane permeability and uptake of metabolites. Plant Cell Physiol. *31*, 987–992 (1990).

Comporti, M.: Biology of disease. Lipid peroxidation and cellular damage in toxic liver injury. Lab. Invest. *53*, 599–623 (1985).

Cox, R., S. Goorha, and C. C. Irving: Inhibition of DNA-methylase activity by acrolein. Carcinogenesis *9*, 463–465 (1988).

Curren, R. D., L. L. Yang, P. M. Coklin, R. C. Grafström, and C. C. Harris: Mutagenesis of Xeroderma pigmentosum fibroblasts by acrolein. Mut. Res. *209*, 17–22 (1988).

Darral, N. M.: The effect of air pollutants on physiological processes in plants. Plant, Cell, Environ. *12*, 1–30 (1989).

Druskovic, B.: Cytogenetic bioindication – Utilization of cytogenetic analysis for detection of the effects of genotoxic pollutants on forest trees (Material and method) (in Slovenian). Biol. Vestn. *36*, 1–18 (1988).

Eckl, P. and H. Esterbauer: Genotoxic effects of 4-hydroxyalkenals. Adv. Biosci. *76*, 141–157 (1989).

Eckl, P. M., A. Ortner, and H. Esterbauer: Genotoxic properties of 4-hydroalkenals and analogous aldehydes. Mut. Res. *290*, 183–192 (1993).

Esterbauer, H., P. Eckl, and A. Ortner: Possible mutagenes derived from lipids and lipid precursors. Mut. Res. *238*, 223–233 (1990).

Esterbauer, H., R. J. Schaur, and H. Zollner: Chemistry and biochemistry of 4-hydroxynonenal, malonaldehyde and related aldehydes. Free Rad. Biol. Med. *11*, 81–128 (1991).

Fiskejö, G.: The *Allium* test as a standard in environmental monitoring. Hereditas *102*, 99–112 (1985).

– Aluminium toxicity in root tips of *Picea abies* (L.) Karst., *Fagus sylvatica* L., and *Quercus robur* L. Hereditas *111*, 149–157 (1989).

– *Allium* test II: Assessment of a chemical's genotoxic potential by recording aberrations in chromosomes and cell divisions in root tips of *Allium cepa* L. Environ. Tox. Water Qual. *9*, 235–241 (1994).

Gorissen, A. and J. A. van Veen: Temporary disturbance of translocation of assimilates in Douglas firs caused by low levels of ozone and sulfur dioxide. Plant Physiol. *88*, 559–563 (1988).

Grill, D., G. Zellnig, E. Stabentheiner, and M. Müller: Strukturelle Veränderungen in Abhängigkeit verschiedener Luftschadstoffe. Forstwiss. Cbl. *112*, 2–11 (1993).

Grimes, H. D., K. K. Perkins, and W. F. Boss: Ozone degrades into hydroxyl radical under physiological conditions. A spin trapping study. Plant Physiol. *72*, 1016–1020 (1983).

Halliwell, B. and J. M. C. Gutteridge: Oxygen toxicity, oxygen radicals, transition metals and disease. Biochem. J. *219*, 1–14 (1984).

– – Free Radicals in Biology and Medicine. Oxford Science Publications, Clarendon Press, Belfast, 1985.

– – Oxygen free radicals and ironin relation to biology and medicine: Some problems and concepts. Arch. Biochem. Biophys. *246*, 501–514 (1986).

Heagle, A. S.: Ozone and crop yield. Ann. Rev. Phytopathol. *27*, 397–423 (1989).

HEATH, R. L.: The biochemistry of ozone attack on the plasma membrane of the plant cells. In: SAUNDERS, J. A., L. KOSAK-CHANNING, and E. E. CONN (eds.): Recent Advances in Phytochemistry. Pytochemical Effects of Environmental Compounds, pp. 29–54. Plenum Press, New York (1987).

HECK, W. W., E. U. BLUM, R. A. REINERT, and A. S. HEAGLE: Effects of air pollution on crop production. In: MEUD, W. J. (ed.): Strategies of Plant Reproduction, pp. 333–350. Osmum Publishers, Granada (1983).

KANGASJÄRVI, J., J. TAVINEN, M. UTRIAINEN, and R. KARJALAINEN: Plant defence systems induced by ozone. Plant Cell Environ. *17*, 783–794 (1994).

KANOFSKY, J. R. and P. SIMA: Singlet oxygen production from the reactions of ozone with biological molecules. J. Biol. Chem. *266*, 9039–9042 (1991).

KENSLER, T. W. and M. A. TRUSH: Role of oxygen radicals in tumor promotion. Environ. Mutagen. *6*, 593–616 (1984).

KLASTERSKA, I., A. T. NATARAJAN, and C. RAMEL: An interpretation of the origin of subchromatid aberrations and chromosome stikkiness as a category of chromatid aberrations. Hereditas *83*, 153–162 (1976).

KRESS, L. W., J. M. SKELLY, and K. H. HINKELMAN: Growth impact of ozone, NO_2 and/or SO_2 on *Pinus taeda* L. Environ. Mon. Assess. *1*, 229–239 (1982).

KROKAN, H., R. C. GRAFSTRÖM, D. SUNDQVIST, H. ESTERBAUER, and C. C. HARRIS: Cytotoxicity, thiol depletion and inhibition of o-6-methylguanine-DNA methyltransferase by various aldehydes in cultured human bronchial fibroblasts. Carcinogenesis *6*, 755–759 (1985).

LANGEBARTELS, C., W. HELLER, K. KERNER, S. LEONARDI, D. ROSEMANN, M. SCHRAUDNER, M. TROST, and H. SANDERMANN: Ozone induced defence reactions in plants. In: PAYER, H. D., T. PFIRRMANN, and P. MATHY (eds.): Environmental Research with Plants in Closed Chambers. CEC – Air Pollution Report no. 26, pp. 358–368 (1990).

LEVAN, A.: The effect of colchicine on root mitosis in *Allium*. Hereditas *24*, 471–486 (1938).

LIPPERT, M.: Multifaktorieller Ansatz zur Analyse der Langzeitwirkung erhöhter CO_2- und Ozonkonzentrationen: Gaswechselmessungen an jungen Fichten und Buchen in Expositionskammern. Ph. D. Thesis, University of Würzburg 1992.

LIU, D., J. WUSHENG, and L. DESHEN: Effects of aluminium ion on root growth, cell division, and nucleoli of garlic (*Allium sativum* L.). Environ. Poll. *82*, 295–299 (1993).

LÜTZ, C.: Reaktionen von Fichten bei unterschiedlichen Immissionsbelastungen. In: GUTTENBERGER, H., E. BERMADINGER, and D. GRILL (eds.): Pflanze, Umwelt, Stoffwechsel, pp. 87–98 (1992).

MATSCHKE, J., S. HINNAH, F. ALBERS, and R. AMENDA: Schädigungen der Gehölze durch Zelldefekte in den Meristemen. Angew. Bot. *68*, 71–78 (1994).

MCGILL, M., S. PATHAK, and T. C. HSU: Effects of ethidium bromide on mitosis and chromosomes: a possible material basis for chromosome stickiness. Chromosoma *47*, 157–167 (1974).

MEHLHORN, H., B. TABNER, and A. R. WELLBURN: Electron spin resonance evidence for the formation of free readicals in plants exposed to ozone. Physiol. Plant. *79*, 377–383 (1990).

MÜLLER, M. and D. GRILL: Induction of chromosomal damage by ozone in the root meristems of Norway spruce. Proc. Roy. Soc. Edinb. *102 B*, 49–54 (1994).

MÜLLER, M., H. GUTTENBERGER, D. GRILL, B. DRUSKOVIC, and J. PARADIZ: A cytogenetic method for exmining the vitality of spruces. Phyton (Horn, Austria) *31*, 143–155 (1991).

MÜLLER, M., H. GUTTENBERGER, E. BERMADINGER-STABENTHEINER, and D. GRILL: Die praktischen Erfahrungen mit der cytogenetischen Bioindikation zur Früherkennung von Vegetationsschäden. Allg. Forst- u. J.-Ztg. *163*, 164–168 (1992).

MÜLLER, M., B. KÖHLER, H. GUTTENBERGER, D. GRILL, and C. LÜTZ: Effects of various soils, different provenances and air pollution on root tip chromosomes in Norway spruce. Trees *9*, 73–79 (1994).

MÜLLER, M., A. GEISZINGER, and D. GRILL: Occurrence and frequency of chromosomal defects in root tip meristems of spruce trees under conditions of air pollutants. Proceedings of the 16th International Meeting for Specialists in Air Pollution Effect on Forest Ecosystems (submitted).

MÜLLER, M., A. FANGMEIER, M. TAUSZ, D. GRILL, and H.-J. JÄGER: The genotoxic effects of ozone on the chromosomes of root tip meristems of Norway spruce (*Picea abies* (L.) Karst.). Angew. Bot. *69*, 125–129 (1995).

PAULS, K. P. and J. E. THOMPSON: *In vitro* simulation of senescenserelated membrane damage by ozone-induced lipid peroxidation. Nature *283*, 504–506 (1980).

PENKETT, S. A.: Ozone increases in ground-level European air. Nature *311*, 14–15 (1984).

POLLE, A., T. PFIRRMANN, S. CHARKRABARTI, and H. RENNENBERG: The effects of enhanced ozone and enhanced carbon dioxide concentrations on biomass, pigments and antioxidative enzymes in spruce needles (*Picea abies* L.). Plant Cell Environ. *16*, 311–316 (1993).

RUNECKLES, V. C. and B. I. CHEVONE: Crop responses to ozone. In: LEFOHN, A. S. (ed.): Surface Level Ozone Exposures and Their Effects on Vegetation, pp. 189–252. Lewis Publishers, Chelsea, UK (1992).

SCHUBERT, I. and R. RIEGER: Sister-chromatid exchanges in *Picea abies* – a test for genotoxicity in forest trees. Mut. Res. *323*, 137–142 (1994).

SLATER, T. F.: Free radical mechanisms in tissue injury. Biochem. J. *222*, 1–15 (1984).

SPENCE, R. D., E. J. RYKIEL, and P. J. H. SHARPE: Ozone Alters Carbon Allocation in Loblolly Pine: Assessment with Carbon-11 Labeling. Environ. Poll. *64*, 93–106 (1990).

STEPHEN, J.: Cytological causes of spontaeous fruit abortion in Haemanthus katherinae Baker. Cytologia *44*, 805–812 (1979).

SWAMINATHAN, M. S. and A. T. NATARAJAN: Chromosome breakage induced by vegetable oils and edible fats. Curr. Sci. *25*, 382–384 (1956).

– – Cytological and genetic changes induced by vegetable oils in triticum. J. Hered. *4*, 177–187 (1959).

TINGEY, D. T., R. G. WILHOUR, and C. STANDLEY: The effect of chronic ozone exposures on the metabolic content of Ponderosa pine seedlings. For. Sci. *22*, 234–241 (1976).

VACA, C. E., J. WILHELM, and M. HARMS-RINGDAHL: Interaction of lipid peroxidation products with DNA. A review. Mut. Res. *195*, 137–149 (1988).

VOGELS, K., R. GUDERIAN, and G. MASUCH: Studies on Norway spruce (*Picea abies* Karst.) in damaged forest stands and in climate chamber experiments. In: SCHNEIDER, T. (ed.): Acidification and Its Policy Implications, pp. 171–186. Elsevier, Amsterdam (1986).

WALLIN, G., L. SKÄRBY, and G. SELLDÉN: Long term exposure of Norway spruce *Picea abies* (L.) Karst., to ozone in open top chambers 1990. I. The effects on the capacity of net photosynthesis, dark respiration and leaf conductance of shoots of different ages. New Phytol. *115*, 335–344 (1990).

WILLENBRINK, J. and TH. SCHATTEN: CO_2-Fixierung und Assimilatverteilung in Fichten unter Langzeitbegasung mit Ozon. Forstw. Cbl. *112*, 50–56 (1993).

WILLIAMS, W. T.: Effect of oxidant air pollution on needle health and annual-ring width in a Ponderosa pine forest. Environ. Cons. *13*, 229–234 (1986).

J. Plant Physiol. Vol. 148. pp. 166–171 (1996)

The Impact of Ozone and Drought on the Water Relations of Ash Trees (*Fraxinus excelsior* L.)

S. Reiner[1], J. J. J. Wiltshire[2], C. J. Wright[1], and J. J. Colls[3]

[1] Department of Agriculture and Horticulture, Faculty of Agricultural and Food Sciences, The University of Nottingham, Sutton Bonington, Loughborough, Leicestershire, LE12, 5RD, UK

[2] ADAS Terrington, Terrington St. Clement, King's Lynn, Norfolk, PE34 4PW, UK

[3] Department of Physiology and Environmental Science, Faculty, of Agricultural and Food Sciences, The University of Nottingham, Sutton Bonington, Loughborough, Leicestershire, LE12 5RD, UK

Received June 24, 1995 · Accepted October 10, 1995

Summary

Two year old potted saplings of ash (*Fraxinus excelsior* L.) were exposed to ozone episodes in open-top chambers. Ozone exposure concentrations were 0 ppb and 150 ppb, and there were 22 exposures of 8 h, beginning in May 1994. Additionally half of the plants of each pollution treatment were subjected to three drought cycles of 7–14 days, while the rest of the plants were well-watered.

Stomatal conductance and radial increment at the stembase and at the base of the new shoot growth were investigated. Ringwidth and cell number per ringwidth of the annual ring produced in 1994 were determined microscopically, using samples taken from the stembase.

The ozone episodes were shown to have an impact on the stomatal responsiveness of the plants, restricting stomatal aperture of the droughted ozone-exposed treatment after drought cycles when all plants were maintained at a high soil moisture. The drought cycles alone, however, left the functioning of the stomata unimpaired. The restriction of stomatal aperture, which was found for the droughted, ozone-exposed treatment, was concomitant with significantly decreased radial growth at the stembase of those plants. Microscopical analysis of the annual rings showed that this was caused by a reduction of cell numbers in xylem tissue. Radial increment at the base of the new shoot growth was less affected by the pollution treatment, but the drought treatment alone caused a significant, but smaller growth reduction. This response to ozone could lead to less water uptake, and thus water supply to the crown might become limiting.

Key words: Fraxinus excelsior L., ozone, drought, stomatal conductance, growth, annual rings.

Introduction

According to Hegi (1975), ash (*Fraxinus excelsior* L.) has one of the highest transpiration rates of European tree species. Comparison of the data presented in the review of Körner et al. (1979) confirms this. Therefore, ash is naturally found in moist localities, both as regards soil and atmosphere (Hegi, 1975).

In recent years ash trees in England have experienced a die-back apparently related to dry summers (Hull and Gibbs, 1991). Since dry summer weather often coincides with increased ozone concentrations, the latter may be an additional factor contributing to the decline.

Like some other air pollutants, ozone is known to influence stomatal responsiveness of many plant species (Koch and Maier-Maerker, 1992; Pearson and Mansfield, 1993; Reich and Lassoie, 1984) and this implies an impact on plant water relations and water use. Growth (Cooley and Manning, 1987; Edwards et al., 1992) and biomass allocation (Edwards et al., 1994; Spence et al., 1994) can be affected in polluted plants, and this can also influence water relations.

Generally, plant responses to elevated ozone concentrations below that which causes visible injury are found to vary with species, as well as with actual pollutant uptake; the latter is highly dependent on stomatal conductance (Darrall, 1989).

For ash, susceptibility to foliar chlorosis induced by ozone has already been reported by Ashmore et al. (1985). Regarding the increased ash dieback during dry summers, the interactions between water relations, elevated pollutant concentrations and drought, are of particular interest. To investigate these interactions, stomatal conductance, radial growth and the morphology of annual rings of drought stressed and well watered plants from different pollution treatments were assessed.

Materials and Methods

At The University of Nottingham, Sutton Bonington Campus, UK (latitude 52° 50′ N) trees of *Fraxinus excelsior* L., were treated with air of different qualities during the field season of 1994. The 64 two year old, seed grown saplings were transplanted into pots (12 L) four weeks before the exposure to ozone. The pots contained sieved soil (mesh size: 10 mm) from the experimental site. For thermal insulation the pots were wrapped in a double layer of bubble polythene and a covering layer of white polythene.

The treatments were imposed in 8 open top chambers, 3.8 m high and 3.5 m in diameter. The sides of the chambers could be covered or uncovered rapidly, allowing the trees to be uncovered, when treatments were not being applied, thereby decreasing the cumulative thermal effects of the chambers (details have been published by Wiltshire et al., 1992). When the chambers were covered, the air entering the chambers was passed through Purafill and activated charcoal filters. Four chambers received filtered air (CF), while the other four had 150 ppb of ozone, generated by electrical discharge in oxygen, added to filtered air ($CF+O_3$). The sides of the chambers were covered and the chambers operated on days when the plants were exposed to ozone and on days when the ambient O_3 concentrations exceeded 50 ppb in order to ensure that the trees in the CF treatment were never exposed to ozone levels greater than 50 ppb. At all other times the chambers were uncovered and the plants in both treatments received ambient air. On exposure days the plants were fumigated for 8 hours (10.00–18.00 GMT), and exposure episodes lasted between 1–4 days in succession. The course of the exposures during the 1994 season is illustrated in figure 1. Exposure dates depended on the ambient conditions and preference was given to sunny days with low windspeeds and above-average temperatures. Altogether there were 22 exposure days, with two-thirds of exposures between mid-June and mid-July.

The two year old potted plants were randomly distributed to the chambers (8 pots per chamber). Two litres of NPK fertilizer (NPK 10 : 10 : 27; Phostrogen, Phostrogen Ltd., Corwen, Clwyd, Wales, UK) were applied to the pots monthly. The application dates are indicated in figure 1. Four of the eight potted plants per chamber were randomly selected for the drought treatment. The «well watered» plants were irrigated regularly throughout the season: water was applied on evenings when the soil surface appeared dry and the pots were watered until water was observed running from the drainage holes. The droughted plants were subjected to three drought episodes of 7–14 days (Fig. 1). After each drought episode the droughted plants were irrigated regularly for two weeks, to allow recovery from drought stress.

Thus the experimental treatments were:

Exposed to ozone, well watered:	$CF+O_3$-w
Exposed to ozone, droughted:	$CF+O_3$-d
Protected from ozone, well watered:	CF-w
Protected from ozone, droughted:	CF-d

Measurement of stomatal conductance were taken with a diffusion porometer (AP4, Delta-T Devices, Burwell, Cambridge, UK). For each assessment up to three (in some cases only one) fully developed leaves were chosen randomly. Stomatal conductance was mainly assessed immediately before, during and shortly after ozone exposure episodes. Except for those measurement made during a fumigation, all the other assessments were undertaken between 7.00 and 10.00 GMT. During fumigations, measurements were taken between 10.30 and 13.30 GMT and on some days additionally between 14.00 and 17.00 GMT. Because of the large weather-related variability in stomatal conductance values between days, these data are presented as the difference between a control treatment (CF-w) and other treatments, giving a ‹relative stomatal conductance›. Linear regression analysis was used to fit lines to the data.

Radial increment at the stembase and at the base of the new apical shoot growth were assessed by measuring stem diameter every two weeks throughout the season. Relative growth rates for the first and the second half of the season, as well as for the whole season were calculated.

Plants were destructively harvested on 4 October 1994 and stem discs were taken from the stembase. Samples were fixed in glutaraldehyde solution (3 %, at 5 °C) and embedded in LR White resin (medium grade). A minimum of 4 sections of 3 µm were prepared from each sample. Toluidin blue was used for staining.

From these sections ringwith and cell number of the latest annual ring were measured. Split plot analysis of variance with the main plots, CF and $CF+O_3$ and the subplots, CF-d, CF-w, $CF+O_3$-d

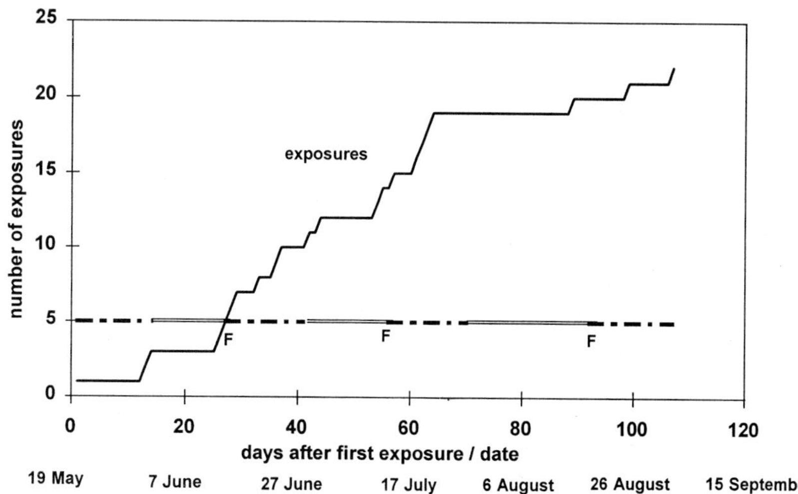

Fig. 1: Ozone exposure regime and drought treatment imposed on 2 year old ash trees in 1994; —— ozone exposures; ═══ watering of all plants; - - - - watering of well watered treatment only; F = fertilizer application. Gravimetric soil water content [%] as measured on the last day of each drought episode: 14 June: droughted: 10.05 ± 0.721; well-watered: 19.595 ± 2.18. 13 July: droughted: 6.97 ± 1.02; well watered: 19.84 ± 1.911. 17 August: droughted: 7.00 ± 1.09; well watered: 18.37 ± 1.97

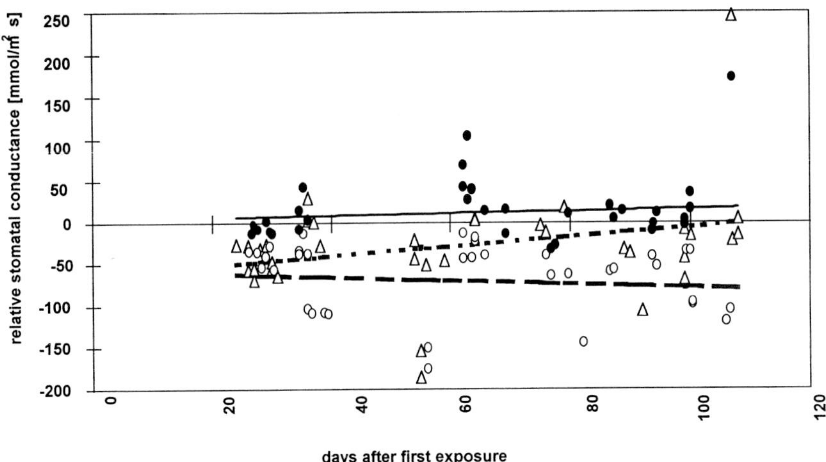

Fig. 2: Differences of mean stomatal conductance between well watered CF plants and the three other drought pollution interactions: ΔCFw-CFd; ○ CFw-CF+O_3d; ● CFw-CF+O_3w; - - - - CFw-CFd; — — — CFw-CF+O_3d; —— CFw-CF+O_3w.

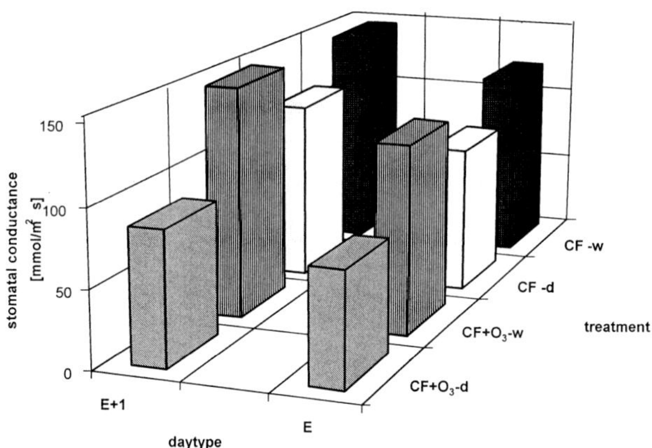

Fig. 3: Treatment combination means for exposure days (E) and first days following an exposure (E+1). Daytype means were calculated from six replications. Split plot analysis of the data showed a highly significant effect ($P<0.001$) of drought and a significant interactions of ozone and drought ($P=0.024$).

and CF+O_3-w and simple analysis of variance (statistical package: GENSTAT 5) were performed. Data were blocked, using the location of the plots on the experimental site as the blocking criterion.

Results

Between 11 June and 3 September 1994, 36 assessments of stomatal conductance were made. Figure 2 shows the calculated relative stomatal conductances plotted against time, with linear trends fitted to the data. The differences of stomatal conductance mainly showed an impact of the drought treatment: CF+O_3 and CF plants reacted in a similar way, with a lower conductance for the water stressed plants during the drought cycles. After rewatering, the CF-d treatment recovered more quickly than the CF+O_3-d treatment to a similar stomatal conductance to the well watered plants. The fitted trends indicate that at the beginning of the growing season both droughted treatments showed similar differences to the control. However, as the season progressed the relative stomatal conductance in the CF+O_3-d treatment continued to decrease so that 23 days after the first exposure they were significantly lower. In contrast the CF-d treatment initially showed a decrease in relative stomatal conductance, but this was not maintained as the season progressed so that by the end of the measurement period there was no significant difference between it and the well watered treatments. Split-plot analysis of the data revealed a highly significant effect of drought on stomatal conductance ($P<0.001$) as well as a significant interaction of the pollution treatment with drought ($P=0.024$). Figure 3 shows the mean stomatal conductance for the different treatments for exposure days (E) and first days following an exposure (E+1). A series of assessments with three measurements per plant was used for the calculation of the means and for statistical analysis, with six replicates of the respective daytypes. There was similar reaction of both pollution treatments to drought: stomatal conductance of the droughted plants was decreased ($P<0.001$). Additionally a drought-pollution treatment interaction ($P=0.024$) became evident from the comparison of the two droughted pollution treatments, manifested in the much lower conductance values of the ozone treated plants.

The relative growth rates of radial increment at the stembase and radial increment at the base of the new shoot growth are given in figure 4 as the half-seasonal and the seasonal means of the drought pollution combinations, obtained from split plot analysis. From the seasonal means (m) for the stembase (Fig. 4 a) a drought effect ($P=0.005$) as well as an interaction of drought with the pollution treatment $P=0.03$) was evident. The latter showed in the much lower relative growth rate of the CF+O_3-d plants, as compared to the other three treatments. Thus the ozone-drought interaction caused a decrease in the relative growth rate of 32.8 % (CF+O_3-d), whereas drought stress alone (CF-d) decreased the relative growth rate by 5.2 % only. However during the first half of the season (1) there was a significant effect only of drought on radial increment ($P=0.018$), while for the second half of the season (2), there was a suggestion of a drought pollution interaction ($P=0.06$).

Relative growth rates at the base of the new shoot growth (Fig. 4 b) were generally higher than those for the stembase. There was no evidence for a drought pollution interaction on radial increment at the base of the new shoot growth. During

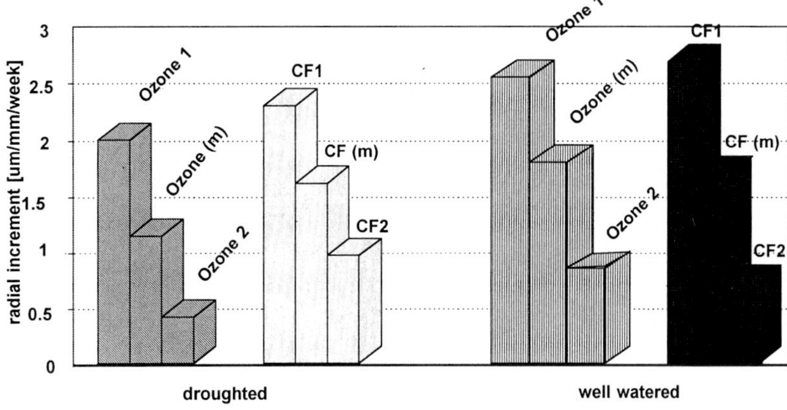

Fig. 4 a: Mean relative growth rates: stembase. 1: drought effect: P = 0.018. m: drought effect: P = 0.005, ozone-drought interaction: P = 0.03. 2: ozone-drought interaction: P = 0.06. Relative growth rates for the first (1) and second (2) half of the season, as well as the seasonal mean growth rate (m) for assessment of the radial increment at the stembase.

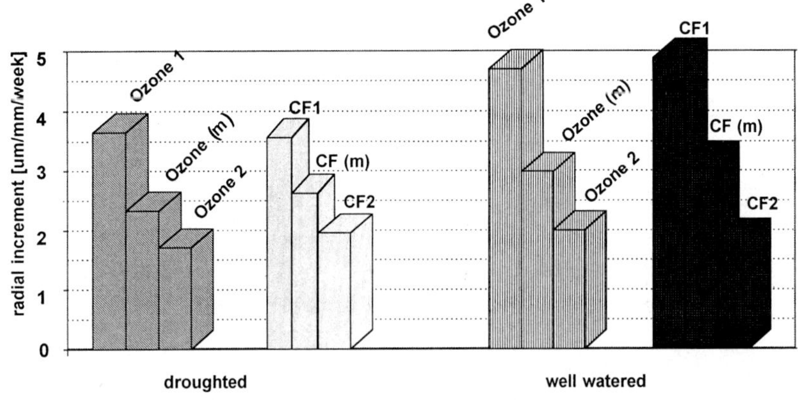

Fig. 4 b: Mean relative growth rates: base of the new shoot growth. 1: drought effect: P=0.007. m: drought effect: P=0.017. 2: –

the first half of the season (1), however, radial growth of the new shoot was decreased by the drought treatment (P=0.007), and this effect was also apparent from the seasonal means (m; P=0.017).

The results of the microscopical analysis of stem sections taken from the stembase are given in figure 5. Mean width of the latest annual ring clearly showed treatment effects on xylem development and the results showed similar trends to those found by the measurements of relative growth rates at the stembase (reflecting the combined effects of phloem and xylem development). A highly significant effect of drought (P <0.001), as well as a highly significant interaction of drought with the pollution treatment – (P<0.001) on annual ring width were revealed. The cell numbers, printed into the columns in figure 6, mirror the results obtained for the width of the annual ring.

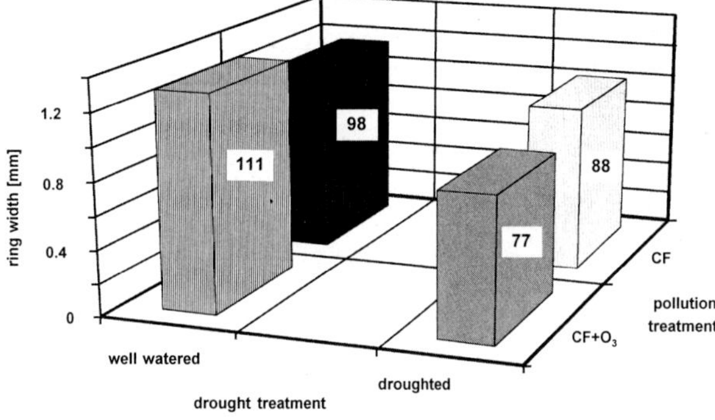

Fig. 5: Mean width of the annual ring 1994. Drought effect: P <0.001; ozone-drought interaction: P <0.001. Numbers in columns indicate the mean cell number per ringwidth.

Discussion

In this experiment pot grown ash trees showed a significant effect of the interaction of ozone and drought on stomatal conductance, shown by reduction of stomatal aperture in the high ozone treatment when droughted only. This result corresponds well with findings for other species, which show that drought protects from direct pollutant injury (Beyers et al., 1992; Dobson et al., 1990; Wieser and Havranek, 1993). However, Pearson and Mansfield (1993) in a similar experiment on beech reported conflicting results with increased stomatal resistance for the well watered ozone stressed plants and a smaller increase in stomatal resistance as the water stress developed.

Comparison of the data from the droughted and the well-watered pollution treatment demonstrates the importance of the drought protection, since well-watered CF+O_3 plants exhibited slightly increased conductances, although not significant, in response to the exposures. Besides increasing ozone uptake of the plants, such stomatal opening could be expected to enhance water loss during periods of good water availability and accordingly to decrease water use efficiency (Reich and Lassoie, 1984). Further it could lead to faster depletion of soil water and consequently to earlier and longer lasting periods of drought. Since ash is adapted to a habitat with good soil water availability and naturally shows rather high values of stomatal conductance, as can be seen from comparison to other woodland trees (Körner et al., 1979), ozone uptake can be assumed to reach high levels as well. Therefore ash is at comparatively high risk from ozone injury, and a significantly increased amount of foliar chlorosis in ambient air has already been reported for a site in south eastern England (Ashmore et al., 1985).

Across-season analysis of the data showed for the ozone and drought stressed plants (CF+O_3-d) that stomatal conductance decreased over the season, as compared to the control (CF-w). This supports the results of Wiltshire et al. (1994) for four year old ash trees, who reported a larger decrease of mean daily water use over the season, when plants had been exposed to ozone, although in that work there was no control of plant water status. Similarly Olszyk et al. (1991) detected small but consistently decreasing stomatal conductance and transpiration rates for ‹Valencia› orange in ambient air, as compared to filtered air. Such a decline in leaf water conductance can possibly be attributed to decreasing responsiveness of the stomata, as has been described for many other species (Keller and Häsler, 1984; Reich and Lassoie, 1984; Skärby et al., 1987). In contrast the drought treatment alone (CF-d) while showing a marked decrease in relative stomatal conductance early in the growing season when drought treatments were first applied apparently acclimated to these drought episodes so that by the end of the experiment, when all plants were being well watered, there was no difference between the CF-d and the CF-w treatments. This suggests that the responsiveness of stomata was unimpaired by drought cycles alone, and that the decrease in stomatal conductance shown by the CF+O_3 plants must be attributed to the interactions of drought and ozone.

There were similar differences between the levels of conductance values for the two different daytypes for all drought-pollutant combinations. These may be due to differences in atmospheric conditions at the time of the assessments. Water vapour pressure deficit, which is accepted as one of the main determinants of stomatal conductance, even at decreased soil water content (Schulze et al., 1975), is assumed to have differed for the respective daytypes. According to Wieser and Havranek (1993) water vapour pressure deficit is the climatic factor most positively correlated with O_3 concentrations. However, since high leaf-air deficits lead to stomatal closure, this factor reduces potential O_3 stress.

The results show that the effect of drought reduced ozone uptake and consequently mitigated the direct injury by the pollutant. However the impaired stomatal responsiveness and the small stomatal aperture of the ozone and drought stressed plants can reduce carbon assimilation and growth. The determination of relative growth rates of the radial increment indicated that growth of the CF+O_3-d plants was strongly decreased. In contrast radial increment at the base of the CF+O_3-w plants was slightly increased, again corresponding with the results for stomatal conductance. Similarly, an increase in stem diameter as a response to O_3 was described for *Pinus halapensis* by Wellburn and Wellburn (1994). However, decreases in radial growth have more commonly been reported (Edwards et al., 1992; Temple and Miller, 1994).

Unlike at the stembase, radial growth at the base of the new shoot growth did not seem to be affected by the pollution treatment; the drought treatment, however, caused a decrease in radial growth particularly during the first half of the season. This discrepancy between the effects on the two different parts of the shoot may be due to differences in sink strength, as well as to an ozone induced alteration of carbon partitioning. Thus in the CF+O_3-d treated plants partitioning may have been affected such that allocation to the lower parts of the stem and the roots was suppressed. Similar findings have been reported for *Quercus rubra* L. (Edwards et al., 1994), *Picea rubens* Sarg. (Thornton et al., 1992), *Phaseolus vulgaris* (Okano et al., 1984) and many other species. According to Spence et al. (1990) such changes in carbon allocation can be attributed to impaired phloem loading.

The analysis of annual rings yielded results that were consistent with the findings for radial growth and showed that xylem growth was affected. Reductions in xylem tissue growth, when plants were exposed to ozone, were also found by Landolt et al. (1993). From the assessment of cell numbers per ringwidth it can be deduced, that a reduction of cell number, rather than cell size caused the observed differences. For *Populus tremula* Matyssek et al. (1993) stated that suppressed latewood formation was the cause of an observed reduction in the weight/length ratio.

In conclusion, ozone episodes can impair stomatal responsiveness of ash, thus affecting growth. During periods of good water availability the high stomatal aperture could lead to earlier depletion of soil water reserves. In dry years this could expose plants to longer periods of drought, when highly decreased stomatal conductance would then impair carbon assimilation.

Acknowledgements

This study was made possible by the financial support of the U.K. Department of the Environment and the German Academic Exchange Service (DAAD). We thank Mr. G. Landon for technical assistance and operation of the open-top chambers, and Mr. J. Craigon for advice concerning data analysis.

References

Ashmore, M., N. Bell, and J. Rutter: The role of ozone in forest damage in West Germany. Ambio *14,* 81–87 (1985).

Beyers, J. L., G. H. Riechers, and P. J. Temple: Effects of long term ozone exposure and drought on the photosynthetic capacity of ponderosa pine (*Pinus ponderosa* Laws.). New Phytologist *122,* 81–90 (1992).

COOLEY, D. R. and W. J. MANNING: The impact of ozone on assimilate partitioning in plants: a review. Environmental Pollution *47,* 95–113 (1987).

DARRALL, N. M.: The effect of air pollutants on physiological processes in plants. Plant, Cell and Environment *12,* 1–30 (1989).

DOBSON, M. C., G. TEST TAYLOR, and P. H. FREER-SMITH: The control of ozone uptake by *Picea abies* (L.) and *P. sitchensis* (Bong.) Carr. during drought and interacting effects on shoot water relations. New Phytologist *116,* 465–474 (1990).

EDWARDS, G. S., S. D. WULLSCHLEGER, and J. M. KELLY: Growth and physiology of northern red oak: preliminary comparisons of mature tree and seedling responses to ozone. Environmental Pollution *83,* 215–221 (1994).

EDWARDS, N. T., G. L. EDWARDS, M. KELLY, and G. E. TAYLOR: Three-year growth responses of *Pinus taeda* L. to simulated rain chemistry, soil magnesium status, and ozone. Water, Air and Soil Pollution *63,* 105–118 (1992).

HEGI, G.: Illustrierte Flora von Mitteleuropa, Vol. V, Part 3. Verlag Paul Parey, Hamburg, pp. 1926–1934 (1975).

HULL, S. K. and J. N. GIBBS: Forestry Commission bulletin 93. Ash dieback – a survey of non woodland trees. London: HMSO (1991).

KELLER, T. and R. HÄSLER: The influence of a fall fumigation with ozone on the stomatal behaviour of spruce and fir. Oecologia *64,* 284–286 (1984).

KOCH, W. and U. MAIER-MAERCKER: Die Bedeutung des Wasserhaushalts für die Beurteilung von Waldschäden. Allgemeine Forstzeitung *8,* 394–400 (1992).

KÖRNER, CH., J. A. SCHEEL, and H. BAUER: Maximum Leaf Diffusive Conductance in Vascular Plants. Photosynthetica *13 (1),* 45–82 (1979).

LANDOLT, W., M. GÜNTHARDT-GEORG, I. PFENNINGER, and C. SCHEIDEGGER: Ozone-induced microscopical changes and quantitative carbohydrate contents of hybrid poplar (*Populus* × *euramericana*). Trees *8,* 183–190 (1994).

MATYSSEK, R., T. KELLER, and T. KOIKE: Branch growth and leaf gas exchange of *Populus tremula* exposed to low ozone concentrations throughout two growing seasons. Environmental Pollution *79,* 1–7 (1993).

OKANO, K., O. ITO, G. TAKEBA, A. SHIMIZU, and T. TOTSUKA: Alteration of ^{13}C-assimilate partitioning in plants of *Phaseolus vulgaris* exposed to ozone. New Phytologist *97,* 155–163 (1984).

OLSZYK, D. M., D. K. TAKEMOTO, and M. POE: Leaf photosynthetic and water relations responses for ‹Valencia› orange trees exposed to oxidant air pollution. Environmental and Experimental Botany *91,* 427–436 (1991).

PEARSON, M. and T. A. MANSFIELD: Interacting effects of ozone and water stress on the stomatal resistance of beech (*Fagus sylvatica*). New Phytol. *123,* 351–358 (1993).

REICH, P. B. and J. P. LASSOIE: Effects of low level O_3 exposure on leaf diffusive conductance and WUE in hybrid poplar. Plant, Cell and Environment *7,* 661–668 (1984).

SCHULZE, E.-D., O. L. LANGE, L. KAPPEN, M. EVENARI, and U. BUSCHBOM: The role of air humidity and leaf temperature in controlling stomatal resistance of *Prunus armeniaca* L. under desert conditions. II. The significance of leaf water status and internal carbon dioxide concentration. Oecologia *18,* 219–233 (1975).

SKÄRBY, L., E. TROENG, and C. BOSTROM: Ozone uptake and effects on transpiration, net photosynthesis and dark respiration in Scots pine. Forest Science *33,* 801–808 (1987).

SPENCE, R. D., E. J. RYKIEL, and P. J. H. SHARPE: Ozone alters carbon allocation in Loblolly Pine: Assessment with carbon-11 labeling. Environmental Pollution *64,* 93–106 (1990).

TEMPLE, P. J. and P. R. MILLER: Foliar injury and radial growth of ponderosa pine. Canadian Journal of Forest Research *24,* 1877–1882 (1994).

THORNTON, F. C., P. A. PIER, and C. MCDUFFIE: Red spruce response to ozone and cloudwater after three years exposure. Journal of Environmental Quality *21,* 196–202 (1992).

WELLBURN, F. A. M. and A. R. WELLBURN: Atmospheric ozone affects carbohydrate allocation and winter hardiness of *Pinus halapensis* (Mill.). Journal of Experimental Botany, Vol. 45, No. *274,* 607–614 (1994).

WIESER, G. and W. M. HAVRANEK: Ozone uptake in the sun and shade crown of spruce: quantifying the physiological effects of ozone exposure. Trees *7,* 227–232 (1993).

WILTSHIRE, J. J. J., C. J. WRIGHT, and M. H. UNSWORTH: A new method for exposing mature trees to ozone episodes. Forest Ecology and Management *51,* 115–120 (1992).

WILTSHIRE, J. J. J., M. H. UNSWORTH, and C. J. WRIGHT: Seasonal changes in water use of ash trees exposed to ozone episodes. New Phytologist *127,* 349–354 (1994).

J. Plant Physiol. Vol. 148. pp. 172–178 (1996)

Direct Effects of Ozone on the Reproductive Development of *Brassica* Species

C. A. STEWART[1], V. J. BLACK[1], C. R. BLACK[2], and J. A. ROBERTS[2]

[1] Department of Geography, Loughborough University of Technology, Loughborough, Leics., LE11 3TU, UK

[2] Department of Physiology and Environmental Science, University of Nottingham, Sutton Bonington Campus, Loughborough, Leics., LE12 5RD, UK

Received June 24, 1995 · Accepted October 10, 1995

Summary

A novel controlled environment system was constructed to allow the reproductive structures of small annual plants to be exposed to gaseous atmospheric pollutants independently of the vegetative component. The design and experimental application of this system, which permits up to 12 plants to be exposed simultaneously, are described. A single exposure of flowering racemes of *Brassica campestris* L. to 100 $nL\,L^{-1}$ ozone for 6 h had no significant effect on the numbers of reproductive sites produced or aborted. This contrasts sharply with the related species, *B. napus* L., in which a single exposure to 100 $nL\,L^{-1}$ ozone induced significant losses of reproductive sites. However, multiple exposures of *B. campestris* to ozone produced significant effects on seed abortion and mature seed number per pod at final harvest, the extent of which depended on the developmental stage of the reproductive organ at the time of exposure. Despite these effects, seed number per plant, mean seed weight and total seed weight per plant at maturity were not significantly affected, indicating a high degree of compensation during reproductive development. The origins and ecological significance of these responses are discussed.

Key words: Brassica, ozone, reproductive development.

Introduction

Reproductive development is a critical phase in the life of annual plant species since any impairment of pollination, fertilisation or seed development has potentially serious implications for seed quality and quantity in agricultural crops and the competitive ability of species growing in natural ecosystems. The susceptibility of reproductive processes to environmental stresses such as temperature and drought is well established (e.g. Stirling and Black, 1991; Rao et al., 1992), but, with the exception of pollen, the adverse effects of atmospheric pollutants on reproductive development have not been recognised until relatively recently (Dubay and Murday, 1983 a, b; Cox, 1983; Kasana, 1991). In most previous studies, the vegetative and reproductive structures have been exposed simultaneously to pollutants (Reiling and Davison, 1992 a; Fernandez-Bayon et al., 1993; PORG, 1993), making it impossible to separate direct effects on reproductive development from indirect effects induced by injury to the vegetative organs.

Recent studies of oilseed rape (*Brassica napus* L.) have demonstrated that reproductive development may be affected *directly* by pollutant exposure. Bosac et al. (1993) demonstrated that a single 6 h exposure of the reproductive organs to 100 $nL\,L^{-1}$ O_3 or 100 $nL\,L^{-1}$ O_3 + 30 $nL\,L^{-1}$ SO_2 produced both short and long-term effects on reproductive development. These included reductions in pollen germinability and pollen tube growth, both of which are essential for successful fertilisation and seed production. Significant losses of reproductive sites were also observed 2–5 d after exposure to 200 $nL\,L^{-1}$ SO_2, 100 $nL\,L^{-1}$ O_3 or 100 $nL\,L^{-1}$ O_3 + 30 $nL\,L^{-1}$ SO_2 (Bosac et al., 1994). Different racemes exhibited varying sensitivity to pollutant exposure, with the second and third laterals being most affected (Black et al., 1993). Intervarietal differences were also observed, as has previously been reported for vegetative processes in other species (Reiling and Davison, 1992 b; PORG, 1993).

The extensive initial losses of reproductive sites observed in oilseed rape 2–5 d after pollutant exposure decreased relative to the clean air controls by 25 d after exposure (Bosac et al., 1994) because its indeterminate growth habit confers considerable compensatory flexibility during reproductive growth. Thus, plants experiencing a single pollution event were able to replace at least some of the lost reproductive sites. Significant effects on pod length, grain yield and seed oil content were nevertheless observed at final harvest (Bosac et al., 1993), and seed germination and seedling vigour were also depressed (Bosac, 1992).

These studies with oilseed rape demonstrate that exposure to ozone alone, or in combination with SO_2, may exert important *direct* effects on reproductive development that are both persistent and of potential agricultural and ecological significance. It is important to establish whether similar effects occur not only in crops, whose harvestable component often comprises grain, fruit or pods, but also in native species whose long-term ecological viability may depend on their reproductive success. This work seeks to establish whether the sensitivity of *B. napus* to ozone and its subsequent ability to compensate is also observed in *B. campestris,* a related species with a more determinate pattern of reproductive development. The results obtained will provide a greater understanding as to whether the reproductive strategy adopted by particular species is an important factor influencing sensitivity to ozone exposure during reproductive development.

Materials and Methods

Plant material

Brassica campestris L. was chosen for study because it is a compact, rapidly cycling species which can be grown from seed to maturity within five weeks (Tompkins and Williams, 1990). Although related to *Brassica napus, B. campestris* is more determinate in its reproductive development, producing only 20–30 flowers in one or two racemes as opposed to many hundred in 7–8 racemes. Most of the flowers are borne on the terminal raceme c 14 days after germination and appear in a regular and predictable order, with one flower opening every 8 hours. The flowers are large, have high female fertility and are predominantly self-incompatible, facilitating studies of pollen × stigma interactions. The final fruit number is established 1–2 weeks after anthesis. These features, and in particular the ordered predictability of development, are ideally suited to detailed studies of the impact of ozone on embryogenesis, reproductive development and seed production.

Brassica campestris L. seeds (MacIntyre, Mottingham Garden Centre, London) were germinated in Fisons Levington multipurpose compost in a G3600THTL Fisons Fitotron growth cabinet. Continuous fluorescent and tungsten lighting provided 230 µmol $m^{-2} s^{-1}$ PAR (400–700 nm) 60 cm below the light source. Temperature was maintained at a constant 24 °C and relative humidity at 70 %. After three days seedlings were transplanted individually into 10 cm pots containing multipurpose compost and then returned to the growth cabinet and watered when required. Newly opened flowers were present from day 15 onwards and these were cross-pollinated at 24 hour intervals using pollen from untreated plants until day 30, by which time all the flowers on the terminal raceme had opened. Watering was then discontinued to allow the plants to undergo a ripening-off period and the seeds were harvested around day 50. Similar timecourses of development to those described by Tompkins and Williams (1990) were observed in plants grown under these experimental conditions (Fig. 1).

Exposure

When plants reached day 20 of their life cycle, i.e. when all stages of floral development from buds to pods were present on the terminal raceme, their inflorescences were enclosed in a specially designed exposure system. Sets of 10 plants were exposed either to filtered air (see below) or to a single 6 h exposure to a target ozone concentration of 100 nL L^{-1} on day 20 or to four consecutive 6 h

Fig. 1: Stages of plant development in *B. campestris.* Note the rapid succession of developmental stages and the presence of pods, open flowers and buds at day 21 after sowing.

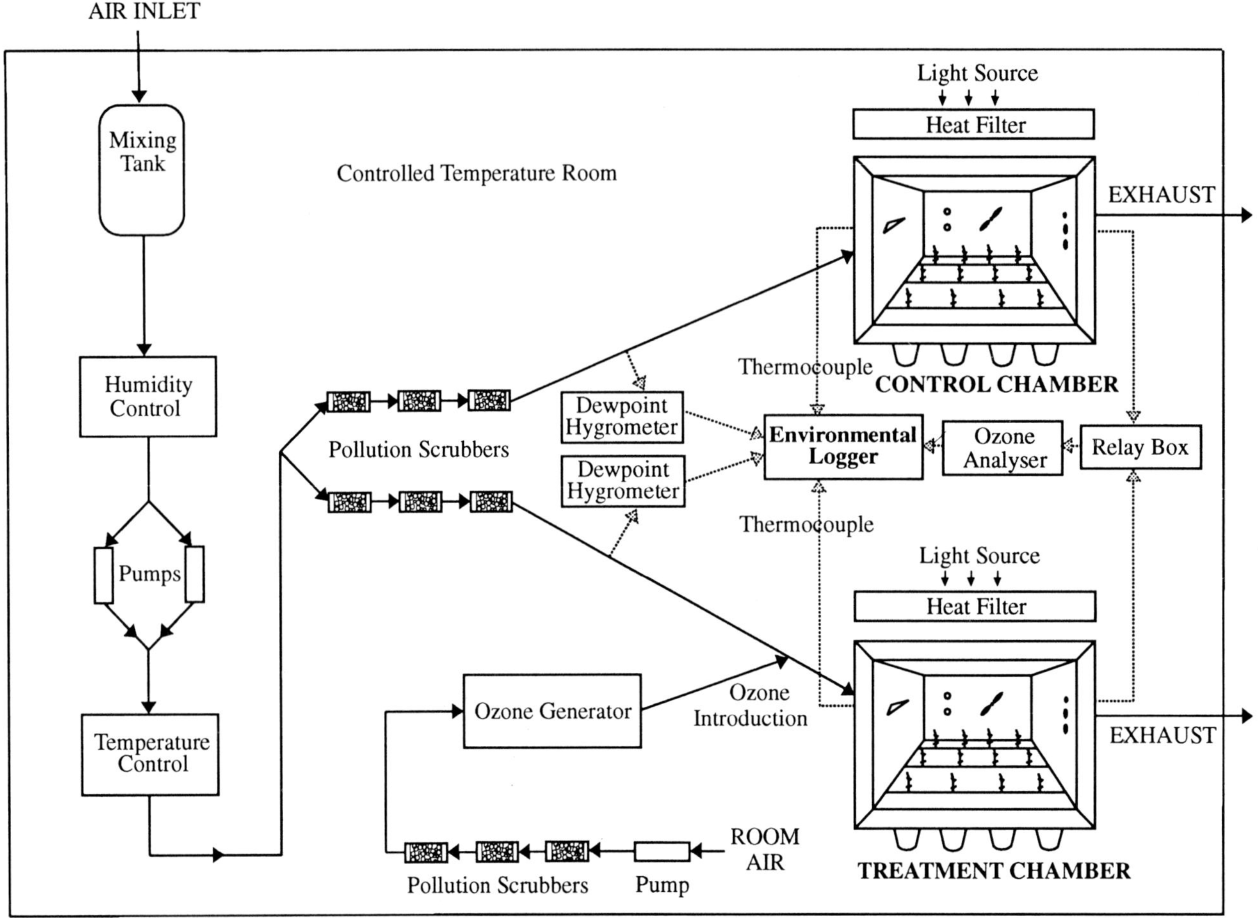

Fig. 2: Schematic representation of the exposure system.

exposures to filtered air or 100 nL L^{-1} O_3 on days 20–23. These exposure regimes were selected to simulate naturally occurring ozone episodes in spring and early summer in the UK (PORG, 1993). The vegetative parts of plants in both treatments were maintained in clean air. The plants were kept within the exposure system throughout the exposure period and subsequently returned to the growth cabinets.

Exposure System

A schematic diagram of the exposure system is shown in Fig. 2. Two perspex chambers (75 L) (one treatment and one control) were designed specifically to permit the inflorescences of up to 12 plants per chamber to be fumigated separately from the vegetative parts. The base of each chamber comprised five removable perspex strips which provided twelve regularly spaced 1 cm diameter holes when in place. These were foam-lined and permitted the introduction of the reproductive structures into the chamber. Once the plants had been placed in position, the perspex base-pieces were brought together to form a seal around the plant stem. A removable chamber front allowed access for pollination and measurements.

Light was provided by a single Osram Powerstar HQI-E 400W/D lamp and reflector positioned above each chamber, and heat from the light sources was filtered out by waterbaths (Fig. 2). The air supply for the chambers was drawn from outside the building, via a 46 L plastic mixing tank, using two diaphram pumps (Charles Austin Pumps Ltd., Weybridge). Because of the known interaction between humidity and plant reponses to ozone (Mortensen, 1989), the air stream was passed through a simple humidity control system before entering the chambers. This involved bubbling the air through water at a temperature of 10 °C to maintain the relative humidity of air entering the chambers between 50 and 70 %. Temperature control within the chambers was achieved by passing the air stream through a copper cooling coil immersed in a water bath.

The conditioned air was then divided into two streams, each of which was passed through a series of three pollution scrubbers (80 cm^3), two containing activated charcoal separated by one containing Purafil. Air entered the chambers through holes positioned on the upper part of the back wall and exited through holes in the lower part of the right hand wall (Fig. 2). The air flow rates to each chamber were 16 L min^{-1}, resulting in approximately one air change every five minutes. An internal fan ensured adequate mixing of air within the chamber. Ozone generated by a purpose-built UV ozone generator (Bosac, 1992) was introduced into the airline supplying the treatment chamber.

Air temperature within each chamber was measured continuously using copper/constantan thermocouples inserted into the chamber and the relative humidity of the air entering each chamber was measured by dewpoint hygrometers (Models 1100DP and 1100AD General Eastern Systems, Watertown, Massachusetts). Mean temperature and humidity values within the control and treatment chambers during the single and multiple exposure experiments were 22.2 ± 0.7 °C and 54.9 ± 1.6 % and 21.9 ± 0.7 °C and

56.4 ± 1.7% respectively. PAR fluxes at inflorescence height within the chambers and at leaf height below the chamber were 200 and 150 $\mu mol\ m^{-2}\ s^{-1}$ respectively. Ozone concentrations within the chambers were measured at 10 min intervals using an ozone analyser (Model 8810 Monitor Labs. Inc., San Diego, California). Average ozone concentrations of 109.9 + 10.3 $nL\,L^{-1}$ and 100.1+0.02 $nL\,L^{-1}$ were recorded during the single and multiple exposures respectively. Concentrations in the control chamber were invariably below the detection level of the analyser. Environmental data and ozone concentrations were recorded on a 21XL Micrologger (Campbell Scientific, Logan, Utah).

Plant Measurements

The numbers of reproductive sites and stages of development (pods, flowers, buds) were recorded daily prior to, during and after exposure. The number of aborted sites was also recorded. When the plants had reached maturity (day 45–50), the pods from each plant were sub-divided into the following fractions: i) sites exposed as developing pods only; ii) sites exposed as open flowers during a single exposure or as open flowers at any time during multiple exposures; and iii) sites exposed as buds only. The numbers of mature and aborted seeds per pod were counted for each fraction. Mature seed was bulked for each fraction of individual plants before being weighed to determine individual seed weight and total seed weight per plant. The untransformed data for the number of reproductive sites present, number of sites aborted, and seed weights for treatment and control plants were compared using paired t-tests.

Results

A single 6 h exposure of *B. campestris* to 100 $nL\,L^{-1}$ O_3 had no detectable effect on the numbers of reproductive sites retained, lost through abortion or percentage losses (Table 1). This is in sharp contrast to previous research using the related species, oilseed rape (*B. napus* cv. Tapidor), in which significant losses of reproductive sites were observed 5 d after exposure to 100 $nL\,L^{-1}$ O_3 for 6 h (Table 1; after Bosac, 1992; Bosac et al., 1994). The much larger numbers of reproductive sites retained and lost in *B. napus* are also apparent from Table 1. Even after four similar exposures on consecutive days, no significant effect on the retention or losses of reproductive sites was observed in *B. campestris* (Table 1).

Table 1: Number of fertile reproductive sites retained or lost and percentage losses following single and multiple exposures of inflorescences of *B. campestris* or *B. napus* cv. Tapidor to filtered air (control) or air containing 100 nL L^{-1} O_3 for 6 h. ** Denotes significance at $P<0.01$; NS indicates no significant effect.

Species	Number of Sites	Control	O_3	Significance
B. campestris	Retained	28.9	30.0	NS
Single exposure	Lost	1.2	1.7	NS
(8 d)[1]	% Losses	4.2	5.7	NS
B. campestris	Retained	29.2	30.4	NS
4 day exposure	Lost	5.8	6.4	NS
(5 d)[1]	% Losses	19.9	21.1	NS
B. napus	Retained	451.6	464.5	NS
Single exposure	Lost	173.0	262.4	**
(5 d)[1]	% Losses	38.3	56.5	**

[1] Numbers in parenthesis indicate the number of days after exposure when site losses were recorded.

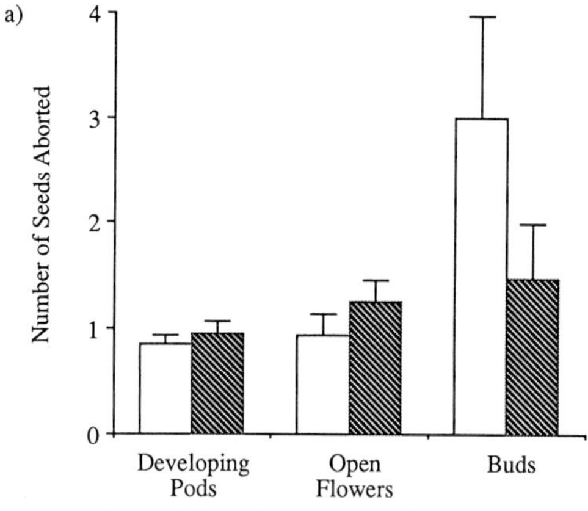

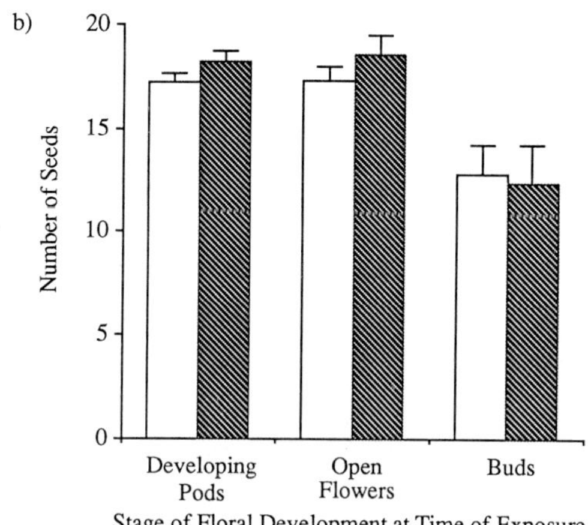

Fig. 3: Mean numbers of (a) aborted and (b) mature seeds per pod in *B. campestris* 8 days after a single exposure to 100 $nL\,L^{-1}$ O_3 or clean air for sites which were exposed as developing pods, open flowers or buds. □, ozone; ▧, control. Single standard errors are shown; n = 10.

The numbers of seeds which were aborted or developed to maturity within the pods of *B. campestris* were also unaffected by a single exposure to ozone (Fig. 3). Seed number was lower in pods which developed from reproductive sites present as buds at the time of exposure than in those which developed from open flowers or pods already present (Fig. 3 b). Seed abortion was greater in pods produced from reproductive sites which were exposed to ozone as buds (Fig. 3 a), although this effect was not significant following a single exposure. Exposure to ozone on four consecutive days also had no significant effect on the timing, rate or extent of reproductive site losses in *B. campestris* (Fig. 4), although the numbers of sites involved were substantially greater in both control and treated plants than after a single exposure (Table 1). This forms a sharp contrast to *B. napus,* which exhibited significant reproductive site losses after a single exposure.

Although multiple exposures to ozone had no discernable effect on reproductive site losses in *B. campestris,* there were

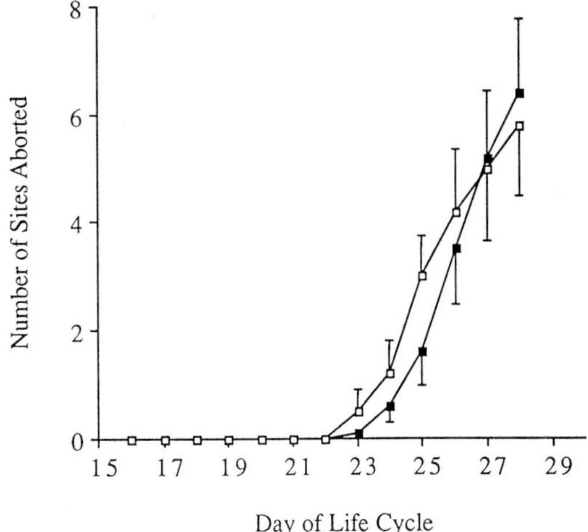

Fig. 4: Timecourses of the mean number of reproductive sites aborted on the terminal raceme of *B. campestris* when exposed to 100 nL L^{-1} O_3 or clean air for 6 h on four consecutive days commencing on day 20. ■, ozone; □, control. Single standard errors are shown; n = 10.

Table 2: Seed yield from *B. campestris* exposed to 100 nL L^{-1} O_3 for 6 h on four consecutive days commencing on day 20 of its life cycle. Standard errors are shown; n = 10.

	Control	O_3
Total number of seeds per plant	402.7 ±27.7	425.1 ±30.1
Individual seed weight (mg)	0.88± 0.04	0.87± 0.04
Total seed weight per plant (g)	0.35± 0.03	0.36± 0.02

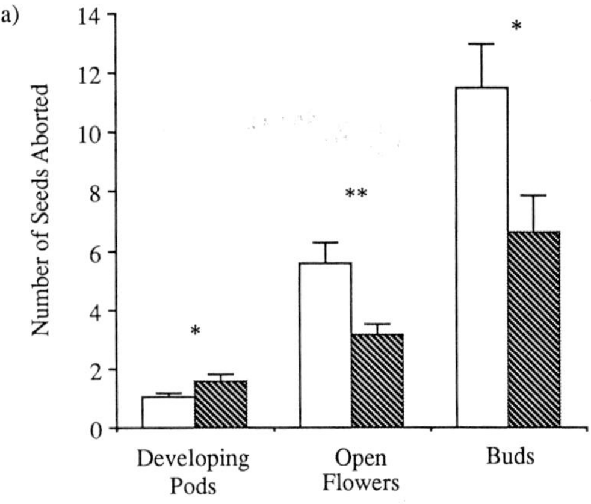

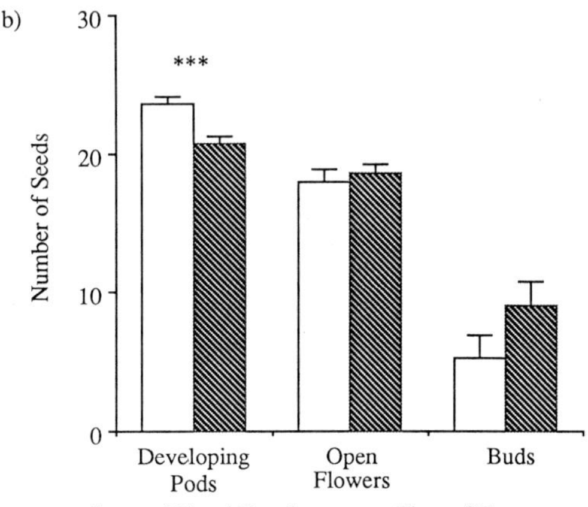

Fig. 5: Mean numbers of (a) aborted and (b) mature seeds per pod in *B. campestris* 5 days after the last of four daily 6 h exposures to 100 nL L^{-1} O_3 or clean air for sites exposed as developing pods, open flowers or buds. □, ozone; ▨, control. Single standard errors are shown. *, ** and *** indicate significance at $P<0.05$, $P<0.01$ and $P<0.001$; n = 10.

significant effects on seed abortion in all categories of pod (Fig. 5 a), and also in the number of mature seed per pod at harvest in the developing pods category (Fig. 5 b). Seed number was significantly increased in the developing pods category of ozone-treated plants, but marginally reduced in the open flowers and buds categories.

Despite the significant effects of ozone on seed retention and abortion, there was no detectable effect on seed number per plant, mean seed weight or total seed weight per plant at maturity (Table 2). Thus, although seed abortion was significantly greater in pods formed from sites exposed as open flowers or buds, this was offset by lower abortion and greater seed retention to maturity in pods present at the time of exposure. This may reflect a redirection of assimilates away from the developing seeds and pods produced from open flowers and buds towards reproductive sites already present at the time of exposure.

Discussion

The exposure system described here permits groups of up to 12 plants to be exposed simultaneously to well-defined pollution events under highly reproducible environmental conditions. The ability to fumigate the reproductive structures separately from the vegetative component permits the *direct* effects of pollutant exposure to be ascertained, an approach which has been attempted in few previous studies. However, the flexibility of the exposure system also permits whole plants to be exposed, thereby enabling the additional impact of *indirect* effects mediated through the vegetative organs to be assessed.

We have previously shown that a 6 h exposure of the inflorescences of *B. napus* to 100 nL L^{-1} ozone significantly affects the abscission and abortion of flower buds (Bosac et al., 1994), but that the final number of fertile sites was not affected. Detailed analyses revealed that ozone-treated plants compensated for the initial losses of reproductive sites by increasing the length and number of raceme branches produced (Bosac et al., 1994). Effects on grain yield and seed quality characters such as oil, protein and carbohydrate content were nevertheless observed at final harvest (Bosac, 1992). Although the structure of the inflorescences of *B. napus* was well-suited for our previous studies, its extended duration of development (approximately 70 days from flower bud to mature pod) and extreme plasticity of raceme growth made it

Table 3: Effect of O_3 on various stages of reproductive development in *Brassica napus* and *Brassica campestris*, + = affected by O_3; × = not affected by O_3.

Aspect of Reproductive Development	*Brassica napus* cv. Tapidor	*Brassica campestris*
Bud abortion	+	×
Bud abscission	+	×
Seed abortion	×	+
Seeds per pod	×	+
Seeds per plant	×	×

difficult to achieve the high level of reproducibility required to establish with certainty the detailed impact of ozone on reproductive development. Our current studies have therefore focussed on the floral biology of *B. campestris,* a species which offers many of the advantageous features of *B. napus,* but also has a much shorter life-cycle and exhibits less developmental flexibility. Moreover, the plants are self-incompatible, allowing the impact of ozone on pollination and fertilisation to be examined in terms of effects on both the male and female reproductive processes.

It is apparent that bud abortion and flower abscission in *B. campestris* were not affected by either a single exposure to 100 nL L^{-1} ozone or repeated episodes over four consecutive days. However, significant effects of repeated exposures on seed abortion were apparent. Intriguingly, the number of mature seeds developing from flowers that had undergone fertilisation prior to ozone treatment was significantly higher than in control plants, whereas the abortion of seeds in pods that developed from exposed flowers or buds was increased. Consequently, the total number of seeds per plant was unaffected by the ozone treatment. These observations suggest that, like *B. napus, B. campestris* can compensate for losses of reproductive sites promoted by ozone, although the nature of the loss and the mechanism by which this compensation is achieved differ between the two species (Table 3). *B. campestris* exhibits a more limited flexibilty of raceme development, both temporally and spatially, than *B. napus* since the duration of the reproductive phase is restricted to approximately 14 days. As a result, there is little opportunity for a compensatory mechanism operating through the development of additional racemes on the inflorescence. Suporting evidence for this view comes from the fact that over 600 fertile reproductive sites may be produced by a single *B. napus* plant, whereas the equivalent value for *B. campestris* is approximately 30.

Two hypotheses may be advanced to explain these observations. The first is that ozone directly affects seed development in *B. campestris,* with maturation being promoted in flowers fertilised prior to exposure (developing pods category) and reduced in those fertilised either at the time of exposure or after treatment (flowers and buds categories). Alternatively, an element of the impact of ozone may be mediated via an indirect route. In view of our previous data from *B. napus,* we favour the second proposal and suggest that the impact of ozone results from the inhibition of seed development in the flowers and buds categories, and that this indirectly increases the success of those seeds which have already begun the grain-filling and maturation process. In support of this hypothesis, previous studies with maize suggest that the numbers of ears per plant and kernals per ear at maturity are particularly sensitive to environmental stresses (Fischer and Palmer, 1983), and that the primary effect of drought or shade imposed immediately after flowering is on the number of aborting kernals (Kiniry and Ritchie, 1985; Schussler and Westgate, 1991). Indeed, Lafitte and Edmeades (1995) have suggested that stress-induced reductions in the number of florets per ear permit more uniform development of the remaining reproductive sites, thereby reducing the abortion of apical kernals within the cob. This may reflect a more general response to abiotic stresses which may be extended to other species. The implication of this hypothesis is that plants of *B. campestris* have a finite capacity for mature seed production and that this is regulated at the level of seed development rather than in terms of sites where fertilisation may take place, as in *B. napus.*

This proposed reproductive strategy for *B. campestris* requires the existence of a signalling mechanism between mature and developing seeds which limits the growth of the latter once the capacity to support developing seed to maturity has been reached. This capacity might well be determined by the production and partitioning of assimilates by the plant. If this were so, it would be anticipated that exposure of entire *B. campestris* plants to ozone might reduce total seed yield due to additional effects on assimilate production and distribution by the vegetative parts. Certainly, it has been well documented that ozone has deleterious effects on the photosynthetic capacity and partitioning of assimilates in a range of plant species (TERG, 1988; Wolfenden and Mansfield, 1991; TERG, 1993; PORG, 1993). It is therefore essential to establish the impact of stresses such as ozone on both the vegetative and reproductive stages of development in plants since this may enable the precise biochemical and physiological mechanisms by which ozone affects plant growth to be ascertained.

Acknowledgements

This work was supported by the U.K. Natural Environment Research Council.

References

Black, V. J., J. J. Colls, and C. R. Black: Ozone/sulphur dioxide interactions in temperate arable crops. In: Jackson, M. B. and C. R. Black (eds.): Interacting stresses on plants in a changing climate, NATO ISI series, Vol. *16,* pp. 83–110. Springer-Verlag, Berlin (1993).

Bosac, C.: The impact of ozone and sulphur dioxide on reproductive development in oilseed rape. PhD Thesis, University of Nottingham, U.K. (1992).

Bosac, C., V. J. Black, C. R. Black, J. A. Roberts, and F. M. Lockwood: Impact of O_3 and SO_2 on reproductive development in oilseed rape (*Brassica napus* L.). I. Pollen germination and pollen tube growth. New Phytol. *124,* 439–446 (1993).

Bosac, C., J. A. Roberts, V. J. Black, and C. R. Black: Impact of O_3 and SO_2 on reproductive development in oilseed rape (*Brassica napus* L.). II. Reproductive site losses. New Phytol. *126,* 71–79 (1994).

Cox, R. M.: Sensitivity of forest plant reproduction to long range transported air pollutants: *in vitro* and *in vivo* sensitivity of *Oenothera parviflora* L. pollen to simulated acid rain. New Phytol. *97,* 63–70 (1983).

Dubay, D. T. and W. H. Murdy: The impact of sulphur dioxide on plant sexual reproduction: *In vitro* and *in vivo* effects compared. J. Env. Qual. *12,* 147–149 (1983a).

– – Direct adverse effects of SO_2 on seed set in *Geranium carolinianum* L.: a consequence of reduced pollen germination on the stigma. Bot. Gaz. *144,* 376–381 (1983b).

Fernandez-Bayon, J. M., J. D. Barnes, J. H. Ollerenshaw, and A. W. Davison: Physical effects of ozone on cultivars of watermelon (*Citrullus lanatus*) and muskmelon (*Cucumis melo*) widely grown in Spain. Environ. Pollut. *81,* 199–206 (1993).

Fischer, K. S. and A. F. E. Palmer: Maize. In: IRRI (ed.): Proc. symp. on potential productivity of field crops under different environments. Los Baños, Philippines. pp. 155–180. IRRI, Los Baños, Laguna, Philippines (1980).

Kasana, M. S.: Sensitivity of three leguminous crops to O_3 as influenced by different stages of growth and development. Environ. Pollut. *69,* 131–149 (1991).

Kiniry, J. R. and J. T. Ritchie: Shade-sensitive interval of kernal number in maize. Agron. J. *77,* 711–715 (1985).

Lafitte, H. R. and G. O. Edmeades: Stress tolerance in tropical maize is linked to constitutive changes in ear growth characteristics. Crop Sci. *35,* 820–826 (1995).

Mortensen, L. M.: Review: Effects of ozone on plants in relation to other environmental conditions. Meded. Norsk. Inst. Skogforsk. *42,* 57–66 (1989).

Rao, G. U., A. Jain, and K. R. Shivanna: Effects of high temperature stress on *Brassica* pollen: viability, germination and ability to set fruits and seeds. Ann. Bot. *68,* 193–198 (1991).

Reiling, K. and A. W. Davison: Effects of a short ozone exposure given at different stages in the development of *Plantago major* L. New Phytol. *121,* 643–647 (1992a).

Reiling, K. and A. W. Davison: Spatial variation in ozone resistance of British populations of *Plantago major* L. New Phytol. *122,* 699–708 (1992b).

Schussler, J. R. and M. E. Westgate: Maize kernal set at low water potential: I. Sensitivity to reduced assimilates during early kernal growth. Crop Sci. *31,* 1189–1195 (1991).

Stirling, C. M. and C. R. Black: Stages of reproductive development in groundnut (*Arachis hypogaea*) most susceptible to environmental stress. Trop. Agric. *68,* 296–300 (1991).

Tompkins, S. T. and P. H. Williams: Fast plants for finer science. An introduction to the biology of rapid cycling *Brassica campestris* (*rapa*) L. J. Biol. Educ. *24,* 239–249 (1990).

United Kingdom Photochemical Oxidants Review Group: Ozone in the United Kingdom. United Kingdom Photochemical Oxidants Review Group Third Report, Air Quality Division, Department of the Environment, London, U.K. pp. 1–169. (1993).

United Kingdom Terrestrial Effects Review Group: The Effects of Acid Deposition on the Terrestrial Environment in the United Kingdom, First Report, HMSO, London, U.K. pp. 1–120 (1988).

– Air Pollution and Tree Health in the United Kingdom, HMSO, London, U.K. pp. 1–88 (1993).

Wolfenden, J. and T. A. Mansfield: Physiological disturbances in plants caused by air pollutants. Proc. Roy. Soc. Edin. B *97,* 117–138 (1991).

J. Plant Physiol. Vol. 148. pp. 179–188 (1996)

Ozone Affects Birch (*Betula pendula* Roth) Phenylpropanoid, Polyamine and Active Oxygen Detoxifying Pathways at Biochemical and Gene Expression Level

J. TUOMAINEN[1], R. PELLINEN[1], S. ROY[2], M. KIISKINEN[1], T. ELORANTA[3], R. KARJALAINEN[4], and J. KANGASJÄRVI[1,*]

University of Kuopio, [1] Department of Ecology and Environmental Science, [2] Department of Physiology, [3] Department of Biochemistry and Biotechnology, Box 1627, SF-70211 Kuopio, Finland

[4] University of Helsinki, Department of Plant Pathology, Finland

Received June 24, 1995 · Accepted October 31, 1995

Summary

We have studied ozone-induced reactions at biochemical and mRNA level in two birch clones that differ in their ozone-sensitivity. When exposed to a single 8 hour ozone pulse (150 ppb), first visible injuries appeared in 24 hours in the sensitive clone and lead eventually to partial tissue chlorosis and necrosis, while the insensitive clone was unaffected. Cell plasma membrane damage was measured by vital stain Evan's Blue permeability. After ozone-exposure, the relative number of vital stain permeable cells increased equally in both clones reaching maximum at 24 hours and decreasing thereafter. The damaged cells were randomly distributed, but in some leaves of the sensitive clone cell death spread forming necrotic lesions. The total cellular activities of superoxide dismutase, peroxidase and glutathione reductase increased following the change in the relative Evan's Blue permeability. The enzyme activity increase was considerably higher in the sensitive clone suggesting that it is somehow related to the cell damage. PAL, that controls the phenylpropanoid biosynthesis, is a good indicator of the coordinated plant defense responses. Gene encoding PAL was induced rapidly but transiently in both clones during the ozone exposure. This indicates that defense responses were induced in both clones, when only the sensitive clone showed widespread cell death. This suggests that the ozone-induced defense reactions and cell death in these birch clones are two separately controlled processes. Increased putrescine levels have often been suggested to be involved in plant ozone tolerance. On the contrary to the induction of putrescine accumulation usually detected in ozone tolerant plants, the accumulation of free putrescine occurred in the ozone sensitive birch clone.

Key words: Betula pendula, ozone, cell death, SOD, peroxidase, glutathione reductase, PAL, polyamines, defense reactions, oxygen radicals.

Abbreviations: [1]AdoMet = adenosyl methionine; AOS = active oxygen species; GR = glutathione reductase; GSH = reduced glutathione; PAL = phenylalanine ammonia-lyase; POX = peroxidase; rbcS = Rubisco small subunit; SOD = superoxide dismutase.

* Correspondence.

[1] The nucleotide sequence data reported in this paper will appear in the EMBL, GenBank and DDBJ Nucleotide Sequence Databases under the accession numbers X76077 (PAL) and X92858 (Rubisco).

Introduction

When atmospheric ozone enters the leaf intercellular airspace, oxygen radicals (active oxygen species, AOS) are formed in the apoplast by reaction of ozone with the apoplastic structures. The role of these AOS in the plant-ozone interactions is not yet fully resolved. In plant-pathogen interactions, however, oxygen radicals are known to play an important role in defense response activation (Dixon et al. 1994) and triggering of programmed cell death (Levine et al. 1994) in the hypersensitive tissue necrosis.

Changes in several biochemical parameters have been detected as a response to ozone in O_3 exposed plants (Kangasjärvi et al. 1994). For example, the activities of enzymes like superoxide dismutase (SOD), peroxidases (POX) and glutathione reductase (GR) that remove the AOS formed in the cell during normal metabolism and under stress change often, but not always (Bowler et al. 1992). Similar changes in enzyme activities have been observed during the hypersensitive tissue necrosis in plant-pathogen interactions (Croft et al. 1990). Increased tolerance to several stress situations have been correlated with increased activities of SOD, GR and POX, but whether their role in plant ozone tolerance is protective, or a sign of developing tissue damage is still unresolved (Madamanchi et al. 1992, Foyer et al. 1994, Kangasjärvi et al. 1994, Willekens et al. 1994, Broadbent et al. 1995).

Polyamine concentrations respond to ozone-exposure and other free radical generating conditions (Bors et al. 1989, Rowland-Bamford et al. 1989, Langebartels et al. 1991, Reddy et al. 1993); the cell wall-localized hydroxycinnamic acid conjugates of polyamines can act as scavengers of free radicals (Bors et al. 1989). It has been suggested that stress-ethylene synthesis (Mehlhorn and Wellburn 1987) and induction of polyamine metabolism in response to ozone exposure (Bors et al. 1989, Rowland-Bamford et al. 1989, Langebartels et al. 1991) are important factors in determining plant's sensitivity to ozone. Both pathways share a common precursor and have negative regulatory interactions with each other (Apelbaum et al. 1985). Generally, in ozone insensitive plants polyamine levels have been higher than in sensitive plants, or increased as a response to ozone exposure (Langebartels et al. 1991). Barley plants became sensitive to ozone when polyamine synthesis was prevented (Rowland-Bamford et al. 1989). It has been concluded (Langebartels et al. 1991) that based on the differential responses in antioxidant or hormone levels, other biochemical or physiological reactions may either promote or inhibit lesion formation by the oxygen radicals.

Phenylalanine ammonia-lyase (PAL) is the key enzyme in the biosynthesis of phenylpropanoids in higher plants (Hahlbrock and Scheel 1989). Phenylpropanoid derivatives serve as pigments, phytoalexins, UV-protectants, insect repellents, signal molecules and polymeric constituents of surface and support structures. PAL gene expression is induced by several environmental stimuli, and also by ozone (Rosemann et al. 1991, Eckey-Kaltenbach et al. 1994, Sharma and Davis 1994). Since PAL regulates an important pathway that is induced during the general, coordinated plant defense responses (Hahlbrock and Scheel 1989), it is widely used as an indicator of their activation.

Several experiments with ozone-induced cellular responses have been performed with annual herbaceous plants, like tobacco, parsley, *Arabidopsis*, soybean and potato, but so far only a few experiments concerning ozone-induced changes in the biochemistry and gene expression with deciduous trees have been performed. We have analyzed the activities of active oxygen detoxifying enzymes and studied the expression of photosynthetic and defense-related genes in two ozone-exposed birch clones that differ in their sensitivity to ozone. In this report we show that ozone-damaged individual cells are present in equal abundancy in both clones, but only the sensitive clone displays visible tissue damage, and that the changes in AOS-detoxifying enzyme activities and putrescine accumulation correlate with the extend and progress of tissue damage, not ozone resistance of the clones. We are also interested in ozone-regulation of defense systems and show that they are activated in both clones.

Material and Methods

Plant material

Tissue culture-derived birch (*Betula pendula* Roth) clones KL-5-M and KL-2-M were obtained from a commercial supplier (Enso-Gutzeit Oy, Imatra, Finland). Previous experiments (Pääkkönen et al. 1993) have shown that the clone KL-5-M is sensitive and clone KL-2-M insensitive to ozone. Saplings were planted before the bud burst in 2 L. Plastic pots filled with medium-fertilized peat : sand (4 : 1) and kept in plant fumigation chambers in ozone-free air until ozone-exposure.

Ozone exposures

Plants were kept under 22 : 2 h photoperiod (light : dark) with daytime temperature 19 °C and night temperature 12 °C and relative humidity of 55/80 % (day/night). Light intensities at pot-level were 300 $\mu mol \cdot m^{-2} \cdot s^{-2}$. Ozone was generated from pure O_2 with Fischer 500 ozone generator. Ozone was mixed to filtered air and the ozone-concentration inside the fumigation chambers was measured continuously with model 1008-RS ozone analyzer (Dasibi Environmental Corp.) and regulated with a computer-controlled system. Plants were exposed to a single eight-hour ozone-pulse of 150 ppb (± 5 ppb) four weeks after bud burst.

Leaf samples were collected one day before exposure (−24), 0, 2, 8, 12, 24, 48 and 72/96 hours after the beginning of the 8-hour ozone-exposure. Only fully expanded middle-aged leaves from four individuals of each clone at every time point were collected and frozen immediately in liquid nitrogen. Control samples were collected similarly from unexposed plants. Frozen leaves were kept at −70 °C. Enzyme activities, polyamine concentrations and gene expression were all determined from same samples collected from four individual plants for each time point and treatment.

Cell viability

Leaf disks (1 cm diameter) were cut with a cork borer from the non-damaged area of ten randomly selected leaves per plant and vacuum infiltrated with 0.25 % Evan's Blue solution as described in Greenberg and Ausubel (1993). Evan's Blue dye is normally excluded from cells that have intact membranes. Penetration of the dye to cells was determined under a microscope with 100× magnification from cells underlaying the intersections of a 17×17 lines grid in the eyepiece of the microscope. A total number of 1445 cells was scored from each leaf disk from five random locations. The viability of the cells was calculated as percentage of Evan's Blue-permeable cells from total number of cells scored.

Enzyme activities

Supernatant from the frozen plant tissue was prepared as described previously (Roy et al. 1992). Spectrophotometric analyses were performed with a UV/Vis LAMBDA 2 double beam spectrophotometer (Perkin Elmer Co.; USA). All the enzyme activities were assayed from the same supernatant. Peroxidase (POX, EC 1.11.1.7) activity was assayed spectrophotometrically in a mixture containing the enzyme source, 3.4 mM guaiacol as substrate, 0.9 mM H_2O_2 and 50 mM sodium phosphate buffer (pH 6.0). The rate of increase in absorbance was measured at +25 °C and 470 nm (Putter 1975). The enzyme activity was calculated using an extinction coefficient of 26.6 mM cm^{-1} at 470 nm for tetraguaiacol. Glutathione reductase (GR, EC 1.6.4.2) was assayed by monitoring the decrease in absorbance of NADPH at 340 nm. One unit of enzyme oxidized 1 µmol of NADPH min^{-1} at 25 °C (Carlberg and Mannervik 1985). The total cellular superoxide dismutase (SOD, EC 1.1.15.1) activity was assayed spectrophotometrically by monitoring reduction of NBT (Nitroblue tetrazolium) in a xanthine oxidase system at 560 nm. One unit of the enzyme activity inhibited the rate of reduction of NBT by 50 % (Beauchamp and Fridovich 1971). The protein content of the enzyme source was determined with Lowry-assay using crystalline bovine serum albumin as standard.

Polyamines

Concentrations of leaf putrescine, spermidine and spermine were determined according to Eloranta et al. (1990). 0.15 mg of leaf material was homogenized under liquid nitrogen and dissolved in 1.05 mL of 5 % sulphosalicylic acid. Insoluble material was removed with centrifugation for 5 minutes at 15,000 × g in a microcentrifuge. 500 µL of the supernatant was hydrolyzed for total polyamine determination with equal volume of HCl for 16 to 18 hours in 100 °C, evaporated under nitrogen and redissolved in 500 µL of 5 % sulphosalicylic acid. Rest of the supernatants and the hydrolyzed samples were kept at −20 °C until analysis.

After centrifugation and filtering through a 0.2 µm Acrodisc LC13 membrane (Gelman Sci.), a 20 µL sample was injected to an HP 1090 Liquid Chromatograph (Hewlett Packard) and polyamines were separated in a reversed-phase column (Hewlett Packard Hypersil ODS 5 µm, 100 × 2.1 mm). Elution, post-column derivatization with ortophthalaldehyde and fluorometric detection were performed as described earlier (Eloranta et al. 1990). Known amounts of standard polyamines (putrescine from Calbiochem; spermidine and spermine from Sigma) were added to the samples and compared to the chromatograms from samples to test the reliability of the concentration measurement by means of peak area. A 10 µL aliquot of a 20 µM polyamine mixture (putrescine, spermidine and spermine) in 5 % sulphosalicylic acid was injected in every ten samples as a standard for the calculation of the polyamine concentrations.

Duplicate samples were prepared for the HPLC analysis from four individual plants for each time point. The means of the sample peak areas were calculated for putrescine, spermidine and spermine and the final concentrations were calculated with comparison to the means of the standard peak areas.

Statistical analysis

The statistical significanse between the control and treated plants in enzyme activities and polyamine concentrations were analyzed with student's t-test using SPSS-statistical package.

RNA and DNA isolation

RNA was isolated from 0.5–1 g of frozen leaves with the procedure of Friemann et al. (1991). Following extraction, 0.4 g of CsCl/mL was added to the supernatant, which was layered ontop of a 1.5 mL CsCl pad (5.7 M). Tubes were spun overnight at 265,600 × g in a TH-641 (Sorvall) rotor. RNA pellet was dissolved in H_2O, extracted with acid phenol (pH 4.0), phenol : chloroform (pH 4.0) and chloroform. RNA was precipitated with ethanol, resuspended into H_2O and kept frozen until further use.

DNA was isolated from leaves with a modified procedure of Möller et al. (1992). Pulverized leaves (30–60 mg) were first extracted with methanol and pellet was dried. Instead of SEVAG, suspension was extracted with chloroform.

Gene expression

Probes for birch *pal* and Rubisco small subunit (*rbcS*) were amplified with PCR using genomic DNA as a template. Probe for birch *pal* was amplified with degenerate nucleotide primers (5′ ACGTGGATCCCA(TC)GGNGGNAA(TC)TT(TC)CA(AG)GG 3′ and 5′ACGTGGATCCAC(AG)TC(TC)TG(AG)TT(AG)TG(TC)TG(TC)TC 3′) corresponding to conserved amino acid sequences (AGLVDQNHQE and RGQHGGNFGG). *pal* primers were kind gifts from Y. Helariutta (Biotechnology Institute, University of Helsinki). Birch *rbcS* probe was amplified with primers 5′ GGCTTGTAGGCAATGAAACTGATGCACTGGACTTGACG 3′ and 5′ GACATTACTTCCATCACAAGCAACGGAGGAAG 3′. PCR conditions were: 95 °C, 1 min; 55 °C, 1 min; 72 °C, 1 min; 30 cycles using Taq DNA polymerase (Promega) with supplied buffer. Amplified *rbcS* fragment was cloned into BlueScript KS vector and *pal* fragment into pUC19 vector. Clones were verified with sequencing of both strands. Tubulin probe was a kind gift from H. Kokko (Department of Biochemistry and Biotechnology, University of Kuopio). RNA was separated on 1.0 % agarose gel in MOPS buffer (Sambrook et al. 1989) and transferred by capillary blotting onto MSI Magnagraph membrane. Filters were baked at 80 °C for an hour, prehybridized at 42 °C for 2–4 hours in 50 % deionized formamid, 5× SSC, 2× Denhardt's solution, 1 % SDS and 25 mg/mL denatured herring sperm DNA. Probes were labelled with nick translation to the specific activity of $1–3 \times 10^8$ cpm/µg and added to the prehybridization solution. Hybridization was performed at 42 °C overnight. Following washes [2× 5 minutes (2× SSC) at room temperature, 30 minutes at 65 °C (2× SSC, 1 % SDS)], hybridization was detected at −70 °C with X-ray film.

Results

A single 8 hour pulse of ozone (150 ppb) was sufficient to induce tissue damage and necrosis on the palisade parenchyma of the fully expanded leaves in the ozone sensitive birch clone (Fig. 1). First visual ozone-damage was detectable in the sensitive clone 24 hours after the beginning of the exposure as watery lesions visible when examined against light (Fig. 1 b). Purple spots, brown necrotic flecks (not shown), and larger chlorotic and necrotic sections (Fig. 1 c) were visible in the palisade of few leaves of the sensitive clone by 48 hours while the insensitive clone was unaffected (Fig. 1 a). Tissue-damage did not proceed any further than shown in Figure 1 c. If the plants were exposed to two ozone peaks during consecutive days, tissue damage in the sensitive clone was

Fig. 1: a–c: Birch leaves from exposed plants after a single 8 hour pulse of ozone (150 ppb). (**a**) insensitive clone 48 hours, (**b**) sensitive clone 24 hours and (**c**) 48 hours after the beginning of the exposure. **d–i.** Microscope pictures of Evan's Blue-stained cells from the birch leaves: clean air (**d** and **g**), 24 hours (**e** and **h**) and 48 hours (**f** and **i**) after the beginning of the ozone exposure. Evan's Blue stain can enter the cells that have damaged plasma membranes, thus damaged cells remain blue after the removal of excess stain. The brown, presumably dead cells at 48 h are indicated with an arrow in **i**. d, e, and f: 400× magnification; g, h, and i: 1600× magnification.

much more excessive than shown in Figure 1 c. In every experiment only leaves of the sensitive clone that had just expanded to their final size showed visible ozone-damage, the young leaves and several weeks old mature leaves did not show any visible symptoms.

Cell viability and tissue damage

Once ozone enters plant leaf it will form active oxygen species (AOS) at the vicinity of the plasma membrane (Mehlhorn et al. 1990). These AOS can damage the plasma membrane and alter its composition and permeability (Heath 1987). We measured the ability of the plasma membrane to exclude Evan's Blue dye as an indicator of the membrane intactness (Fig. 1d–i). Intact cells exclude the dye. In the ozone-sensitive clone the leaf discs were cut from the visually non-damaged areas of the leaves. The number of the Evan's Blue-permeable cells in the healthy-looking areas of randomly selected leaves of the experimental plants increased from less than one percent to 10–12 percent of the total number of cells in both clones during the 24 hours from the beginning of the exposure and decreased to almost zero by 72 hours (data not shown). In the leaves collected 48–72 hours from the beginning of the O_3 exposure, the presumably damaged cells excluded the dye and were detectable as brown, light-impermeable structures (Fig. 1f and i). The variability in the leaf cell permeability to the dye between leaves collected from different parts of the same plant was excessive ranging from 0% to 70% at 24 hours, but no differences in the means were detectable between the clones. This was reflected also as high SD (not shown) of the Evan's Blue permeable cells since leaf disks were cut from randomly selected leaves of variable developmental stage and ozone-damage formed only in leaves of particular developmental stage. Though individual leaky cells were visible in the visually non-damaged leaves of the both clones equally, large visibly damaged areas were present only in the leaves of the sensitive clone.

Enzyme activities

We measured the total cellular activities of superoxide dismutase (SOD), quaiacol peroxidase (POX) and glutathione reductase (GR) in the ozone-exposed birch leaves (Fig. 2). During the eight-hour ozone exposure no increase in the total SOD or POX activities was apparent. By twelve hours from the beginning of the exposure, SOD activity had increased in both clones significantly (Fig. 2). The SOD-activity reached its maximum 24 hours after the beginning of the exposure and was higher in the sensitive clone. By 48 hours the SOD activity had decreased to control levels. Hydrogen peroxide generated by SOD is removed by, e.g., actions of various peroxidases. Slight increase in the quaiacol peroxidase activity was evident by twelve hours and the activity reached maximum at 24 hours (Fig. 2). Decrease of POX activity was slower than in SOD and activity increased more in the sensitive clone.

Glutathione reductase regenerates reduced glutathione (GSH) which is used in several cellular redox reactions where GSH is used as electron donor. The ozone-sensitive birch clone had lower base level GR activity than the insensitive clone (Fig. 2). Ozone exposure increased the activity of GR in both clones with maximal activity at 24 hours. The increase in activity was significantly higher in the sensitive clone and it took place earlier and remained significantly higher still at the end of the experiment. In the ozone-sensitive birch clone the magnitude of the enzyme activity increase due to ozone exposure was significantly higher for all the enzymes assayed.

Polyamine concentrations

Polyamines have previously been shown to play an important role in determining plant ozone sensitivity (Rowland-Bamford et al. 1989, Langebartels et al. 1991). We measured concentrations of putrescine, spermine and spermidine from the experimental plants. Ozone-exposure increased free putrescine levels in the sensitive clone by 24 hours (Fig. 3) and caused a slight decrease in free spermine in the ozone-exposed insensitive clone (data not shown). All polyamine concentrations of control plants were similar between the clones during the experiment. No significant differences were detected between the total and free polyamine content indicating that almost all the polyamines were in free form in the plants (data not shown).

Ozone-induced gene expression

PCR- amplified gene fragments of birch *pal* and *rbcS* were cloned and used as probes in gene expression analyses. Rubisco SSU transcript steady state levels decreased in the ozone-exposed plants during the exposure when compared to control plants (Fig. 4). This decrease lasted past the duration of the exposure and was more prominent in the insensitive clone. In the ozone sensitive clone *rbcS* transcript levels increased right after the ozone exposure. The steady-state *pal* transcript levels, however, were almost undetectable in fully expanded leaves, but were rapidly and transiently increased during the exposure in both clones. *pal* transcript levels were highest at 8 hours, decreased to almost unobservable levels by 12 hours and increased again at 24 hours. No differences were detectable between the clones.

Discussion

Ozone is known to induce several responses in plants at both biochemical and gene expression level (Kangasjärvi et al. 1994), but these studies have been performed mostly with annual herbaceous plants. In tree species where ozone has more severe impact to the growth of the plant due to the annually cumulative effect, the few biochemical and gene expression studies of ozone-induced reactions have been limited to coniferous species (Rosemann et al. 1991, Galliano et al. 1993) though broad-leave trees are generally more sensitive to ozone (Darrall 1989). We have analyzed ozone-induced changes in the metabolism and gene expression of European white birch by exposing ozone-sensitive and insensitive clones to a single ozone pulse.

It has been shown in several experiments (Rich et al. 1970) that plants can close their stomata in high ozone concentrations. Gas exchange measurements with the birch clones used have shown negligible changes in the gas exchange parameters under ozone exposure (Pääkkönen and Kärenlampi, in preparation). Thus the factors determining the difference in the ozone sensitivity of these clones must lay on the biochemical differences between them.

It has often been suggested that the AOS detoxifying enzymes, like SOD, POX and GR could protect plants from ozone damage. In our experiments the eight-hour ozone exposure increased activities of AOS detoxifying, mostly organellar localized SOD, GR and POX at 24 hours (Fig. 2). According to Foyer et al. (Foyer et al. 1994), there is no evidence that the antioxidative defense systems would anticipate the potential damage caused by the AOS formed in various

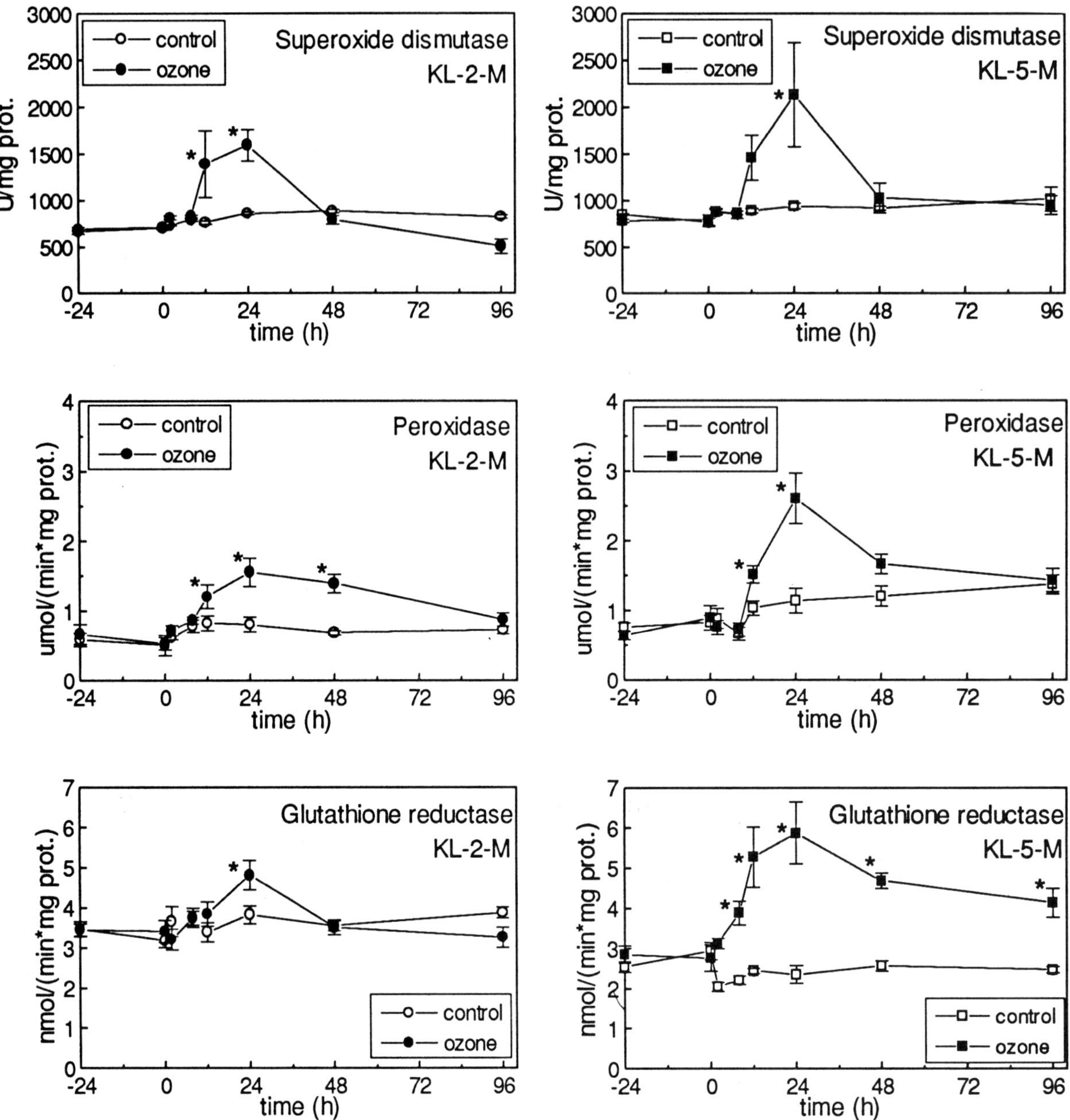

Fig. 2: Effect of ozone exposure on the activity of active oxygen species detoxifying enzymes superoxide dismutase, peroxidase, and glutathione reductase in ozone insensitive (KL-2-M), and ozone sensitive (KL-5-M) birch clones. The enzyme activities were determined from samples collected from four plants of control material kept in clean air and four plants that were given a single 8 hour ozone pulse of 150 ppb. (mean values ± SD, n=4, statistically significant (p=0.05) differences between treatments at each time point are indicated with a *).

stress situations, but instead react only as a response to actual damage. This can be seen in the timing of the increase in the antioxidative enzyme activities measured in this study. The maximal increase in the activities took place concomitantly to the increase in the membrane permeability, 16 hours after the end of the ozone exposure when also the appearing tissue damage could first be seen (Fig. 1 b) and when the relative abundance of the Evan's Blue permeable cells (Fig. 1 e and h) was at its maximum. Interestingly, in a study of incompatible interactions between French bean and its avirulent pathogen that causes hypersensitive tissue necrosis (Croft et al. 1990), the membrane permeability and POX and SOD activities increased during the hypersensitive cell collapse between 12 and 24 hours after infiltration of the leaves with the avirulent pathogen, i.e., at the same time than these changes took place in ozone exposed birch (see Fig. 2).

The peak in the AOS detoxifying enzyme activities at 24 hours suggests presence of AOS in the tissue well after the ozone exposure. The increase in the enzyme activity after ozone exposure implies also that their activity might be more

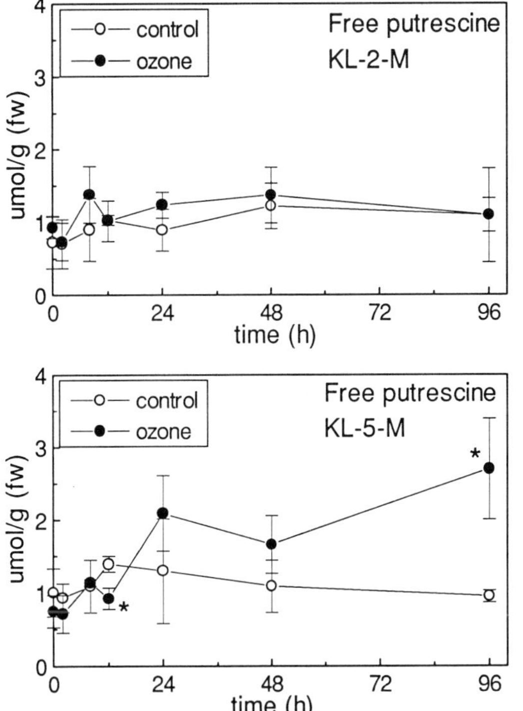

Fig. 3: Effect of ozone treatment on free putrescine levels of the ozone insensitive (KL-2-M) and ozone sensitive (KL-5-M) birch clones. Samples were from the same plants as in Fig. 2. Free polyamines were extracted from leaves and separated by HPLC. Concentrations were calculated from peak areas by comparison to standards (mean values ± SD, n = 4, statistically significant (p = 0.05) differences between treatments at each time point are indicated with a *).

tissue damage-related than protective from the oxygen radicals regenerated directly from ozone. The higher magnitude in the enzyme activity increase in the sensitive birch clone that also displayed tissue damage (Fig. 1 b–c) further strengthens this notion. In experiments with tobacco, it has similarly been seen that changes in SOD and ascorbate peroxidase transcript levels take place only with actual tissue damage (Willekens et al. 1994), and with bean (Chanway and Runeckles 1984), changes in SOD activity correlated with visible injury. In similar studies with pea (Madamanchi et al. 1992), however, no change in GR activity was observed though the higher ozone level used, 150 ppb, did cause necrotic lesions in the experimental plants. On the other hand, Edvards et al. (1994) did see an increase in the GR activities in ozone-exposed pea-plants when no visible damage was apparent. In those experiments, though, the plants were exposed to ozone for several days and enzyme activity was measured at the end of the treatment. Thus, the correlation between progress of tissue damage and increase in the AOS scavenging enzyme activities can not be regarded as general response, though such connection has been observed in several studies.

Ozone exposure increased cell plasma membrane permeability in the non-damaged areas of the leaves equally in both clones. The dark brown individual cells detectable in green leaf tissue of both clones 48–72 hours after the ozone exposure (Fig. 1 f and i) had similar spatial distribution as the Evan's Blue permeable cells at 24 hours (Fig. 1 e and h). These brown cells presumably are the cells that allowed the vital stain to enter 24 hours earlier. Only in the sensitive clone cell death spread further in some leaves usually encompassing the area encircled by leaf veins (Fig. 1 c). Recently several *Arabidopsis* mutants have been described (Greenberg and Ausubel 1993, Dietrich et al. 1994, Greenberg et al. 1994), where the plants develop uncontrollable spread of necrotic lesions during the hypersensitive reaction and display continuous induction of defense genes like PAL, GST and PR proteins. This suggests that the defense gene induction is a part of coordinated, preprogrammed responses. In addition to the cell death (Fig. 1), rapid induction of genes encoding PAL (Fig. 4), PR proteins (Ernst et al. 1992, Schraudner et al. 1992, Eckey-Kaltenbach et al. 1994) and GST (Sharma and Davis 1994) has been observed in ozone exposed plants.

Greenberg et al. (1994) proposed that active oxygen formed by the cells into the apoplast might be the signal that triggers the programmed cell death. Oxygen radicals are formed within minutes in the cell wall in plant-pathogen interactions (Apostol et al. 1989) and this oxidative burst is closely related to the subsequent events in the programmed cell death during the hypersensitive reaction (Levine et al. 1994). Oxygen radicals formed in ozone-exposed plants (Mehlhorn et al. 1990) could thus have similar programmed cell death triggering effect in the sensitive plants.

The phenylpropanoid pathway is an important part of the plant defense mechanism against pathogen attack (Hahlbrock and Scheel 1989) and it is a well recognized example of the coordinated plant defense response induction. Continuous *pal* expression was also observed in the accelerated cell death mutants of *Arabidopsis* (Greenberg et al. 1994). The *pal* induction and hypersensitive cell death are not, though, linked processes, they are regulated separately (Jakobek and Lindgren 1993). In our experiments ozone caused changes in PAL mRNA steady state levels in both birch clones (Fig. 4), but only one clone showed wide spread cell death (Fig. 1 c). This suggests that also the ozone-induced defense reactions and cell death are independently controlled processes. The observed changes in *pal* steady state mRNA levels are consistent with *pal* response to ozone-exposure in herbaceous species (Eckey-Kaltenbach et al. 1994, Sharma and Davis 1994). The timing of the increase in *pal* transcript levels in ozone exposed birch plants followed also quite closely the corresponding *pal* induction in birch by wounding of the leaves (Pellinen and Kangasjärvi, unpublished) and pathogen exposure or elicitor treatment in bean (Jakobek and Lindgren 1993).

Decrease in photosynthesis has been observed frequently in ozone-exposed plants and several mechanisms for this inhibition of photosynthesis has been proposed (Heath 1994). In a recent study (Schlagnhaufer et al. 1995), connection with gene-induction for the ethylene-synthesis controlling ACC synthase and decrease in *rbcS* transcript levels was observed. We also observed decreased *rbcS* transcript levels in both birch clones used (Fig. 4) and have seen increases in the mRNA levels for the second ethylene synthesis controlling enzyme, ACC oxidase at the very beginning of the ozone exposure (Kiiskinen, Tuomainen and Kangasjärvi, unpublished). Whether a similar mechanistic connection between

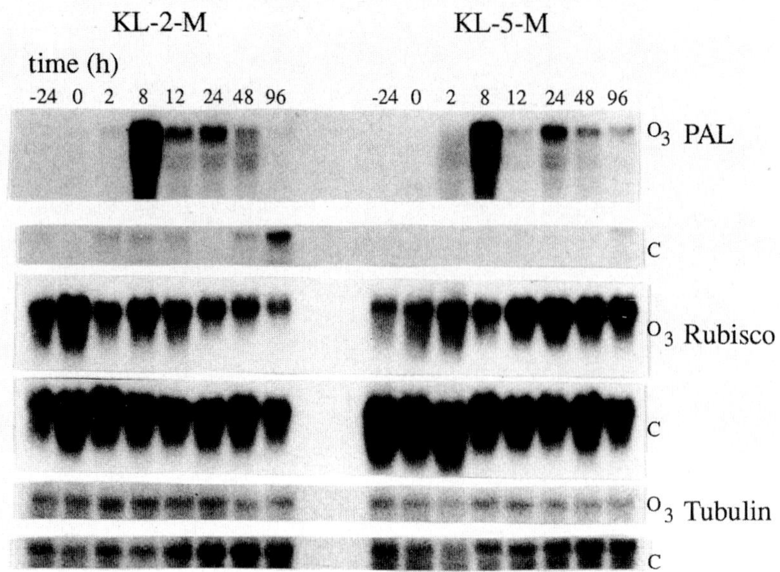

Fig. 4: The effect of ozone exposure on the levels of *rbcS* and *pal* transcripts in ozone insensitive KL-2-M and ozone sensitive KL-5-M birch clones. Northern blots of total RNA (10 μg per lane) from birch leaves collected as described in Fig. 2. Equal loading of RNA on the blots was tested with tubulin probe.

induction of ethylene synthesis and decrease in *rbcS* transcript levels exists in birch as was shown for potato (Schlagnhaufer et al. 1995) needs to be tested in further experiments.

In connection to ethylene, polyamines are thought to play an important role in determining plant ozone sensitivity and their concentrations have increased in ozone insensitive plants (Rowland-Bamford et al. 1989, Langebartels et al. 1991). The extracellular conjugated polyamines have been suggested to act as antioxidants in the cell wall and scavenge ozone-derived radicals in the apoplast (Bors et al. 1989, Rowland-Bamford et al. 1989, Langebartels et al. 1991). Birch clones behaved quite differently in our experimental conditions (Fig. 3). Ozone-exposure increased putrescine concentration of the sensitive birch clone by 24 hours when tissue damage became detectable. This is opposite to the similar experiments with tobacco, where the polyamine concentrations increased in the ozone-insensitive Bel B tobacco (Langebartels et al. 1991). In birch the increased putrescine levels seem to have no connection with protection from ozone as in tobacco or barley. One explanation for this might be that all the putrescine in birch was in soluble form. It has been shown that only the conjugated polyamines, not the free forms, act as radical scavengers. The increase in the putrescine concentrations could result from either increased production or decreased turnover or transport. Increased activity of arginine decarboxylase, that has often been observed in ozone-exposed plants (Rowland-Bamford et al. 1989, Langebartels et al. 1991, Reddy et al. 1993) would result in increased polyamine production. On the other hand, decreased diamine oxidase activity, like observed in ozone-exposed beans (Peters et al. 1989), would have the same effect.

In the present study an extensive approach to unravel the metabolic reasons for the ozone-sensitivity of a broad-leaved tree was taken. In all the parameters studied, differences in the responses to ozone-exposure between the sensitive and insensitive birch clones were of quantitative rather than of qualitative nature. Furthermore, all the differences recorded between the clones in the metabolism of fully expanded middle-aged leaves seem to result from, rather than to be the cause of the damages created by ozone. Thus, in these birch clones, the biochemical processes that were assayed seem not to be responsible for the differences in their ozone-sensitivity. Additional comparative studies with the leaf populations at the developmental stage vulnerable to ozone injuries are required.

Acknowledgements

This work was supported by grants from the Academy of Finland (no. 3148) and Ministry of Forestry and Agriculture to JK. We want to thank Ms. Eija Korhonen for help in the polyamine analyses, Ms. Maarit Valjakka for help during the cell viability determinations and Mr. Timo Oksanen for ozone exposures. Photographs 1 a–c are courtesy of Dr. Jarmo Holopainen.

References

Apelbaum, A., A. Goldlust, and I. Icekson: Control by ethylene of arginine decarboxylase activity in pea seedlings and its implication for hormonal regulation of plant growth. Plant Physiol. *79,* 635–640 (1985).

Apostol, I., P. F. Heinstein, and P. S. Low: Rapid stimulation of an oxidative burst during elicitation of cultured plant cells. Role in defense and signal transduction. Plant Physiol. *90,* 109–116 (1989).

Beauchamp, C. and I. Fridovich: Superoxide dismutase: Improved assay and an assay applicable to acrylamide gels. Anal. Biochem. *44,* 276–287 (1971).

Bors, W., C. Langebartels, C. Michel, and H. Sandermann Jr.: Polyamines as radical scavengers and protectants against ozone damage. Phytochem. *28,* 1589–1595 (1989).

Bowler, C., M. van Montagu, and D. Inzé: Superoxide dismutase and stress tolerance. Annu. Rev. Plant Physiol. Plant Mol. Biol. *43,* 83–116 (1992).

BROADBENT, P., G. P. CREISSEN, B. KULAR, A. R. WELLBURN, and P. M. MULLINEAUX: Oxidative stress responses in transgenic tobacco containing altered levels of glutathione reductase activity. Plant J. *8,* 247–255 (1995).

CARLBERG, I. and B. MANNERVIK: Glutathione reductase. Meth. Enzymol. *113,* 484–490 (1985).

CHANWAY, C. P. and V. C. RUNECKLES: The role of superoxide dismutase in the susceptibility of bean leaves to ozone injury. Can. J. Bot. *62,* 236–240 (1984).

CROFT, K. P. C., C. R. VOISEY, and A. J. SLUSARENKO: Mechanism of hypersensitive cell collapse: correlation of increased lipoxygenase activity with membrane damage in leaves of *Phaseolus vulgaris* (L.) inoculated with an avirulent race of *Pseudomonas syringae* pv. *phaseolicola.* Physiological and Molecular Plant Pathology *36,* 49–62 (1990).

DARRALL, N. M.: The effect of air pollutants on physiological processes in plants. Plant, Cell Environ. *12,* 1–30 (1989).

DIETRICH, R. A., T. P. DELANEY, S. J. UKNES, E. R. WARD, J. A. RYALS, and J. L. DANGL: *Arabidopsis* mutants simulating disease resistance response. Cell *77,* 565–577 (1994).

DIXON, R. A., M. J. HARRISON, and C. J. LAMB: Early events in the activation of plant defense responses. Annu. Rev. Phytopathol. *32,* 479–501 (1994).

ECKEY-KALTENBACH, H., D. ERNST, W. HELLER, and H. SANDERMANN Jr.: Biochemical plant responses to ozone. IV. Cross-induction of defensive pathways in parsley plants. Plant Physiol. *104,* 67–74 (1994).

EDWARDS, E. A., C. ENARD, G. P. CREISSEN, and P. M. MULLINEAUX: Synthesis and properties of glutathione reductase in stressed peas. Planta *192,* 137–143 (1994).

ELORANTA, T. O., A. R. KHOMUTOV, R. M. KHOMUTOV, and T. HYVÖNEN: Aminooxy analogues of spermidine as inhibitors of spermine synthase and substrates of hepatic polyamine acetylating activity. J. Biochem. *108,* 593–598 (1990).

ERNST, D., M. SCHRAUDNER, C. LANGEBARTELS, and H. SANDERMANN Jr.: Ozone-induced changes of mRNA levels of β-1,3-glucanase, chitinase and ‹pathogenesis-related› protein 1b in tobacco plants. Plant Mol. Biol. *20,* 673–682 (1992).

FOYER, C. H., M. LELANDAIS, and K. J. KUNERT: Photooxidative stress in plants. Physiol. Plant. *92,* 696–717 (1994).

GALLIANO, H., M. CABANÉ, C. ECKERSKORN, F. LOTTSPEICH, H. SANDERMANN Jr., and D. ERNST: Molecular cloning, sequence analysis and elicitor-/ozone-induced accumulation of cinnamyl alcohol dehydrogenase from Norway spruce (*Picea abies* L.). Plant Mol. Biol. *23,* 145–156 (1993).

GREENBERG, J. T. and F. M. AUSUBEL: *Arabidopsis* mutants compromised for the control of cellular damage during pathogenesis and aging. Plant J. *4,* 327–341 (1993).

GREENBERG, J. T., A. GUO, D. F. KLESSIG, and F. M. AUSUBEL: Programmed cell death in plants: A pathogen-triggered response activated coordinately with multiple defense functions. Cell *77,* 551–563 (1994).

HAHLBROCK, K. and D. SCHEEL: Physiology and molecular biology of phenylpropanoid metabolism. Annu. Rev. Plant Physiol. Plant Mol. Biol. *40,* 347–369 (1989).

HEATH, R. L.: The biochemistry of ozone attack on the plasma membrane of the plant cells. In: SAUNDERS, J. A., L. KOSAK-CHANNING, and E. E. CONN (eds.): Recent Advances in Phytochemistry. Phytochemical Effects of Environmental Compounds, Vol. *21,* pp. 29–54. Plenum Press, New York (1987).

– Possible mechanisms for the inhibition of photosynthesis by ozone. Photosyn. Res. *39,* 439–451 (1994).

JAKOBEK, J. L. and P. B. LINDGREN: Generalized induction of defense responses in bean is not correlated with the induction of the hypersensitive reaction. Plant Cell *5,* 49–56 (1993).

KANGASJÄRVI, J., J. TALVINEN, M. UTRIAINEN, and R. KARJALAINEN: Plant defense systems induced by ozone. Plant, Cell Environ. *17,* 783–794 (1994).

LANGEBARTELS, C., K. KERNER, S. LEONARDI, M. SCHRAUDNER, M. TROST, W. HELLER, and H. SANDERMANN Jr.: Biochemical plant responses to ozone I. Differential induction of polyamine and ethylene biosynthesis in tobacco. Plant Physiol. *95,* 882–889 (1991).

LEVINE, A., R. TENHAKEN, R. DIXON, and C. LAMB: H_2O_2 from the oxidative burst orchestrates the plant hypersensitive disease resistance response. Cell *79,* 583–593 (1994).

MADAMANCHI, N. R., J. V. ANDERSON, R. G. ALSCHER, C. L. CRAMER, and J. L. HESS: Purification of multiple forms of glutathione reductase from pea (*Pisum sativum* L.) seedlings and enzyme levels in ozone-fumigated pea leaves. Plant Physiol. *100,* 138–145 (1992).

MEHLHORN, H. and A. R. WELLBURN: Stress ethylene formation determines plant sensitivity to ozone. Nature *327,* 417–418 (1987).

MEHLHORN, H., B. TABNER, and A. R. WELLBURN: Electron spin resonance evidence for the formation of free radicals in plants exposed to ozone. Physiol. Plant. *79,* 377–383 (1990).

MÖLLER, E. M., G. BAHNWEG, H. SANDERMANN Jr., and H. H. GEIGER: A simple and efficient protocol for isolation of high molecular weight DNA from filamentous fungi, fruit bodies, and infected plant tissues. Nucl. Acid Res. *20,* 6115–6116 (1992).

PÄÄKKÖNEN, E., S. PAASISALO, T. HOLOPAINEN, and L. KÄRENLAMPI: Growth and stomatal responses of birch (*Betula pendula* Roth) clones to ozone in open-air and chamber fumigations. New Phytol. *125,* 615–623 (1993).

PETERS, J. L., F. J. CASTILLO, and R. L. HEATH: Alteration of extracellular enzymes in pinto bean leaves upon exposure to air pollutants, ozone and sulfur dioxide. Plant Physiol. *89,* 159–164 (1989).

PUTTER, J.: Peroxides. In: BERGMEYER, H. U. (ed.): Methods of enzymatic analysis, pp. 685–690. Verlag Chemie Weinheim Academic Press Inc., New York (1975).

REDDY, G. N., R. N. ARTECA, Y.-R. DAI, H. E. FLORES, F. B. NEGM, and E. J. PELL: Changes in ethylene and polyamines in relation to mRNA levels of the large and small subunits of ribulose bisphosphate carboxylase/oxygenase in ozone-stressed potato foliage. Plant, Cell Environ. *16,* 819–826 (1993).

RICH, W., P. E. WAGONER, and H. TOMLINSON: Ozone uptake by bean leaves. Science *169,* 79–80 (1970).

ROSEMANN, D., W. HELLER, and H. SANDERMANN Jr.: Biochemical plant responses to ozone II. Induction of stilbene biosynthesis in scots pine (*Pinus sylvestris* L.) seedlings. Plant Physiol. *97,* 1280–1286 (1991).

ROWLAND-BAMFORD, A. J., A. M. BORLAND, P. J. LEA, and T. A. MANSFIELD: The role of arginine decarboxylase in modulating the sensitivity of barley to ozone. Environ. Poll. *61,* 95–106 (1989).

ROY, S., R. IHANTOLA, and O. HÄNNINEN: Peroxidase activity in lake macrophytes and its relation to pollution tolerance. Environ. Exp. Bot. *32,* 457–464 (1992).

SAMBROOK, J., T. MANIATIS, and E. F. FRITSCH: Molecular Cloning, Ed. 2. Cold Spring Harbor Laboratory Press, Cold Spring Harbor, NY (1989).

SCHLAGNHAUFER, C. D., R. E. GLICK, R. N. ARTECA, and E. J. PELL: Molecular cloning of an ozone-induced 1-aminocyclopropane-1-carboxylate synthase cDNA and its relationship with a loss of *rbcS* in potato (*Solanum tuberosum* L.) plants. Plant Mol. Biol. *28,* 93–103 (1995).

Schraudner, M., D. Ernst, C. Langebartels, and H. Sandermann Jr.: Biochemical plant responses to ozone III. Activation of the defense-related proteins β-1,3-glucanase and chitinase in tobacco leaves. Plant Physiol. *99,* 1321–1328 (1992).

Sharma, Y. and K. Davis: Ozone-induced expression of stress-related genes in *Arabidopsis thaliana.* Plant Physiol. *105,* 1089–1096 (1994).

Willekens, H., W. van Camp, M. van Montagu, D. Inzé, C. Langebartels, and H. Sandermann Jr.: Ozone, sulfur dioxide, and ultraviolet B have similar effects on mRNA accumulation of antioxidant genes in *Nicotiana plumbaginifolia* (L.). Plant Physiol. *106,* 1007–1014 (1994).

J. Plant Physiol. Vol. 148. pp. 189–194 (1996)

Evaluation of Ozone Impact on Mature Spruce and Larch in the Field

GERHARD WIESER and WILHELM M. HAVRANEK

Forstliche Bundesversuchsanstalt, Abt. Forstpflanzenphysiologie, Rennweg 1, A-6020 Innsbruck, Austria

Received June 24, 1995 · Accepted October 10, 1995

Summary

During the growing seasons 1986 to 1993 we examined the effects of ozone (O_3) on the gas exchange of mature Norway spruce (*Picea abies*) and European larch (*Larix decidua*) trees under field conditions at a low and a high elevation site. Twigs were enclosed in chambers and exposed to different O_3 concentrations for one or two seasons tracking ambient climatic conditions. After one and two fumigation periods, only mean O_3 concentrations higher than 100 ppb caused a pronounced decline in gas exchange, both in spruce and larch. The observed lack in symptom expression at mean O_3 concentrations lower than 100 ppb can be attributed to modifications in the amount of O_3 entering the needles. At both study sites O_3 uptake (F_{O_3}) was effectively controlled by stomatal conductance (gH_2O) and therefore by factors such as light, humidity and water status, controlling gH_2O. Water vapour pressure deficit (VPD) was the climatic factor most closely correlated with ambient O_3 concentration. Thus, when O_3 concentrations were highest, F_{O_3} tended to be restricted by stomatal narrowing. Mitigation of potential O_3 stress by stomatal narrowing was more pronounced at the low elevation site where soil water stress and VPD were greater than at high altitude. On the other hand, the capability to detoxify oxygen radicals is greater in plants growing at high altitude. Therefore, we conclude that ambient O_3 concentration presently does not constitute an acute danger for spruce and larch trees.

Key words: Larix decidua, Picea abies, altitudinal gradient, forest decline, gas exchange, ozone, ozone uptake, stomatal conductance.

Abbreviations: A = ambient air; CF = charcoal-filtered; F_{O_3} = flux or ozone uptake; gH_2O and gO_3 = stomatal conductance for water vapour and ozone; O_3 = ozone; Pn = net CO_2 assimilation rate; VPD = vapour pressure deficit; VPD_{LA} = leaf-air-vapour pressure difference.

Introduction

Since the early eighties ozone was believed to be involved in the new kind of forest decline phenomena (Arndt et al., 1982; Prinz, 1984). In Austria this hypothesis was supported by the fact that damage of spruce trees at timberline increased in parallel with ambient O_3 concentration (cf. Amt der Tiroler Landesregierung, 1985–1995). However, unambiguous evidence of O_3 injury did not exist. Not even at timberline, where average O_3 concentrations exceeded values, which clearly caused injury in young conifers under controlled conditions in chamber experiments (cf. Reich, 1987). Such results should not be used without reservation to predict a response of adult trees in the field, because of morphological and physiological differences between juvenile and mature tissues (Rebbeck et al., 1992, 1993). Furthermore, seedlings in chambers also do not experience such a variety of interacting stress as will commonly occure in the field (Reich, 1987).

For a better understanding of chronic O_3 effects on forest trees long-term experiments seemed to be necessary. Furthermore, little was known about O_3 effects on adult trees in the field, where environmental conditions such as light, temperature, humidity, soil moisture and nutrition exert a pronounced influence on the physiological status of a tree, and therefore on its response to O_3 (Tingey and Taylor, 1982). Therefore, it was the goal of our studies to examine if ambient and above ambient O_3 concentrations can cause any visible injury like necrosis or chlorosis of the needle surface and affect physiological parameters like the gas exchange in mature forest trees at different altitudes after a fumigation period of one or two growing seasons.

Material and Methods

Study site and pollution load

The experiments were conducted on 33 to 65-year-old spruce (*Picea abies* (L.) Karst.) and larch (*Larix decidua* Mill.) trees at two sites in the Tyrol, Austria: namely at a low elevation site at 1000 m a.s.l. with a montane climate (Zillertal) and at timberline on Mount Patscherkofel in 1950 m a.s.l. with a cool subalpine climate (Table 1). At both study sites O_3 was the dominant air pollutant, whereas the concentrations of SO_2, NO and NO_x were neglibile low. The high elevation site was exposed to higher mean O_3 concentrations than the low elevation site (Table 1). Half-hour-maxima were up to 120 ppb at both sites. At the low elevation site, the photochemical formation and destruction of O_3 led to a pronounced diurnal pattern of O_3 with rising values during daylight hours and a decline during the night, whereas at timberline the diurnal variations were only small (Fig. 1).

Experimental design and O_3 treatments

At both sites an aluminium platform provided access to branches in the upper shade and sun crown. There, twigs were enclosed in fumigation chambers made from thin perspex (the sun exposed ones were climatised) as described by Havranek and Wieser (1990, 1994). The O_3 exposure regimes, tracking ambient fluctuations, were charcoal-filtered (CF) air, non filtered ambient air (A) and A enriched with 30, 60 or 90 ppb O_3, with four to eight replicates per treatment. During the vegetation periods 1986–1993 twigs were fumigated continuously for 6 to 23 weeks (Table 2). When twigs were fumigated for two periods the chambers were removed in autumn and the twigs were kept in ambient air over winter.

Measurements

The effects of O_3 on gas exchange of twigs were determined *in situ* under comparable conditions using fully climatised chambers (Walz, Effeltrich, FRG). Additionally, continuous climatic, O_3 and gas exchange measurements under ambient conditions were used to calculate O_3 uptake according to the flux equation:

$$F_{O_3} = [O_3] * gO_3$$

where F_{O_3} is the flux or uptake rate of O_3, $[O_3]$ the O_3 concentration of the ambient air, and gO_3 the stomatal conductance for O_3. The latter was calculated by multiplying the conductance of water vapour (gH_2O) by 0.613, the ratio of diffusivities of water vapour and O_3. To faciliate calculating F_{O_3}, O_3 concentration inside the needles was assumed to be zero (cf. Tingey and Taylor, 1982; Laisk et al., 1989). All gas exchange parameters were calculated according to Von Caemmerer and Farquhar (1981) and related to total needle surface area estimated with glass-beads (Thompson and Leyton, 1971).

Table 1: Macroclimate in the Central Tyrolean Alps at 1000 and 2000 m above sea level (after Fliri, 1975), as well as mean O_3 concentration during the vegetation periods (May–October) 1986–1993. Summer refers to June, July and August.

Altitude	1000 m	2000 m
Global radiation, percent of 1000 m	100	107
Annual average air temperature, °C	7.9	2.5
Average air temperature in summer, °C	17.4	11.0
Average VPD in summer, hPa	5.8	2.6
Annual sum of precipitation, mm	760	980
Precipitation in summer, mm	290	360
Average wind speed in summer, m s^{-1}	1.3	3.4
O_3, ppb*	41	63

* Own measurements and from Amt der Tiroler Landesregierung (1985–1994).

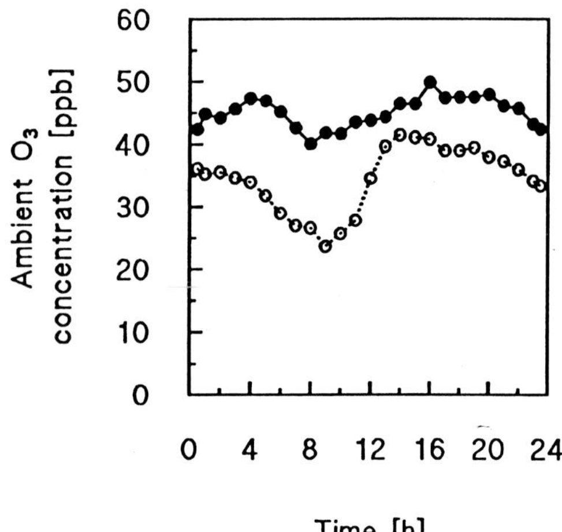

Fig. 1: Average diurnal course of ambient O_3 concentration at the low altitude study site during June 19 and August 15, 1991 (open symbols) and at timberline during the period August 5–31, 1992 (closed symbols).

Table 2: Duration of the fumigation periods for spurce and larch and corresponding mean ambient O_3 concentrations.

Site	Year	Fumigation (weeks)	Mean $[O_3]$ (ppb)	Tree and crown position
1950 m	1986	12	65	spruce shade crown
1950 m	1987	12	64	spruce shade crown
1000 m	1988	23	47	spruce shade crown
1000 m	1989	16	37	spruce shade crown
1000 m	1990	16	25	spruce shade crown
		10	30	spruce sun crown
1000 m	1991	9	34	larch sun crown
1950 m	1993	6	45	larch sun crown

Results and Discussion

In our exposure experiments at both study sites even the highest O_3 concentration (A+90 ppb) did neither cause any visible injury like chlorosis, chlorotic mottling or necrotic spots nor affect the 100-needle dry weight, specific leaf area and chlorophyll content of spruce and larch needles when compared to CF-controls.

After one fumigation period no clear treatment effects on net photosynthesis and gH_2O were observed in current-year spruce needles at average O_3 concentrations below 100 ppb (Havranek et al., 1989; Wieser et al., 1991; Wieser and Havranek, 1994). Only O_3 concentrations higher than 100 ppb caused a pronounced decline in Pn which, however, was statistically significant only at the low altitude site (Fig. 2). One- and two-year-old spruce needles reacted barely to increased O_3.

Twigs fumigated for a second period, following ambient climatic conditions during winter, did not show any chlorotic injury or needle loss in the following spring and the new flush was not affected in growth. However, O_3 effects became more severe and were more pronounced in the current flush than in older needle age classes (Fig. 2).

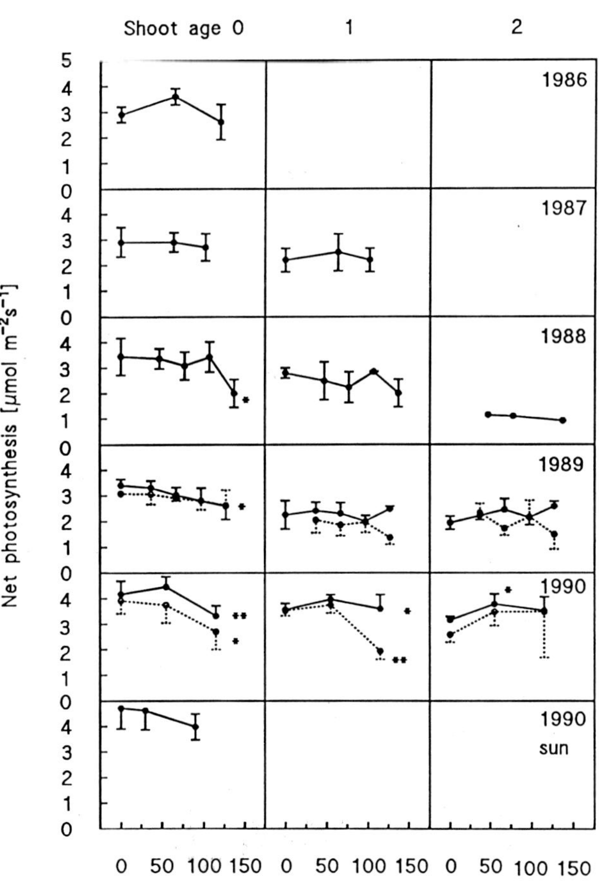

Fig. 2: Net photosynthesis of Norway spruce shoots of different age and different O_3 treatments after one (solid lines) and two (dotted lines) fumigation periods at timberline (1986, 1987) and at the low altitude site (1988–1990). Measurements were taken after 6 or 23 weeks of fumigation in the shade crown (1986–1990) and in the sun crown in 1990 (see Table 2). Statistically significant to control (0 ppb O_3): $^*P<0.05$, $^{**}P<0.01$.

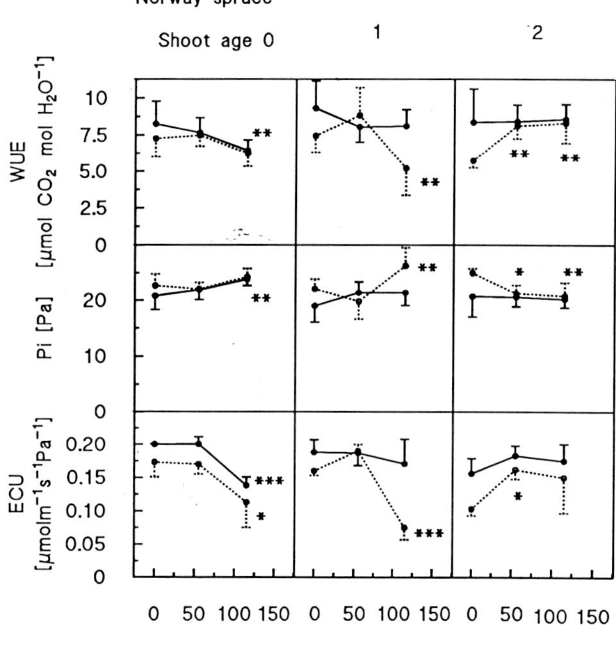

Fig. 3: Water use efficiency (WUE; top), internal partial pressure of CO_2 (Pi; middle) and efficiency of carbon dioxide uptake (ECU; bottom) of Norway spruce shoots of different age and different O_3 treatments after one (solid lines) and two (dotted lines) fumigation periods at the low altitude study site investigated in the shade crown in 1990. Statistically significant to control (0 ppb O_3): $^*P<0.05$, $^{**}P<0.01$, $^{***}P<0.001$.

Furthermore, observed losses in Pn of current and one-year-old spruce needles continuously exposed to O_3 concentrations higher than 100 ppb were greater than reductions in gH_2O. A gas exchange response reflecting also a decline in the water-use efficiency (Fig. 3) and indicating that stomatal narrowing alone did not limit CO_2 uptake (Reich and Amundson, 1987; Matyssek et al., 1991; Pell et al., 1992; Wallin et al., 1992). Stomatal narrowing which occurred after fumigation with the highest O_3 concentrations was probably a function of increased intercellular CO_2 partial pressure resulting from a reduced CO_2 fixation (Fig. 3). However, the sequence of events leading to Pn inhibition is poorly understood (Chapellka and Chevone, 1992).

Similar results as in the O_3 tolerant spruce were also obtained in larch, a species classified as sensitive (VDI, 1989; Davis and Wilhour, 1976) or also intermediate (Chappelka and Chevone, 1992) to O_3. In exposure experiments at the low elevation site even two-fold ambient O_3 concentrations given for two months did not effect Pn and gH_2O of larch needles compared to needles exposed to CF air (Havranek and Wieser, 1993). However, by adding 150 to 200 ppb O_3 to CF air for three to six weeks at timberline O_3 reduced Pn and gH_2O (Volgger, 1995).

In order to elucidate the response of gas exchange to changing environmental conditions Pn and gH_2O of twig pairs were measured in parallel with two chambers under a standardized diurnal time course of increasing and decreasing light, temperature and humidity (cf. Havranek and Wieser, 1993, 1994). Under the same chamber conditions there were no differences in the regulatory capacity of stomata between CF twigs and twigs fumigated with either A or two-fold ambient O_3 concentration, neither in spruce nor in larch. However, regardless to the O_3 pretreatment gas exchange was strongly dependent on microclimatic conditions outside the chambers (Wieser et al., 1991) and on the water status of the whole tree (Havranek and Wieser, 1993), both having a much greater influence on Pn and gH_2O than possible small O_3 effects.

Compared to chamber experiments (cf. Reich, 1987; Pye, 1988; Darrall, 1989) several factors limited our ability to detect O_3 effects at long-term concentrations below 100 ppb. This observed lack in symptom expression can, for the most part, be attributed to modifications in the amount of O_3 entering the plant. Variations in O_3 uptake reflect variation in both O_3 concentration and gH_2O. In general high O_3 concentrations occurred during periods of high irradiance and high temperature. At both study sites O_3 concentration was best correlated with VPD_{LA} which, similar to O_3, is dependent on irradiance and temperature. This correlation was stronger at the low elevation site (Fig. 4), but seems to be a general feature (cf. Grünhage and Jäger, 1994).

Both, in spruce and larch stomatal aperture decreased with increasing VPD_{LA} and with soil water stress and thus avoiding a high F_{O_3} at peak ambient O_3 concentrations (Havranek and Wieser, 1993; Wieser and Havranek, 1993a; 1995). At a low VPD_{LA}, however, stomata were wide open, which resulted in a high F_{O_3} into the needles even at low ambient O_3 concentrations (Fig. 4). Compared to spruce, larch displayed a higher maximum gH_2O, but stomatal narrowing with increasing VPD_{LA} was also more pronounced. Evidence that drought stress protected young conifers from O_3 injury through its influence on stomatal narrowing has also been obtained in chamber experiments (Freer-Smith et al., 1989; Dobson et al., 1990; Fincher and Alscher, 1992; Wallin and Skärby, 1992; Temple et al., 1992; Roberts and Cannon, 1992; Kronfuß et al., 1996).

Beside drought, different climatic conditions at different habitats, or even in the sun and shade side of a tree, as well as nutrition, all influence leaf morphology and physiology of stomata, and by this F_{O_3}. Under similar exposure to light maximum gH_2O of spruce and larch was higher in sun needles than in shade needles (Wieser and Havranek, 1994; 1995). Needles of fertilized spruce trees showed a more pronounced stomatal reaction to changing VPD_{LA} and irradiance than needles with nutrient deficiency (Wieser and Havranek, 1993b). Furthermore, fertilization of spruce trees diminished needle yellowing, even without any change in O_3 exposure (Wieser and Havranek, 1993b).

In general, F_{O_3} was higher at timberline, where diurnal variation of O_3 is only small but O_3 concentration is continuously higher than in valleys. Furthermore, in alpine regions precipitation and soil water reserves tend to increase, whereas VPD decreases with increasing altitude (cf. Table 1). Because of the lower evaporative demand at high altitude trees at timberline are rarely forced to restrict water loss (Tranquillini,

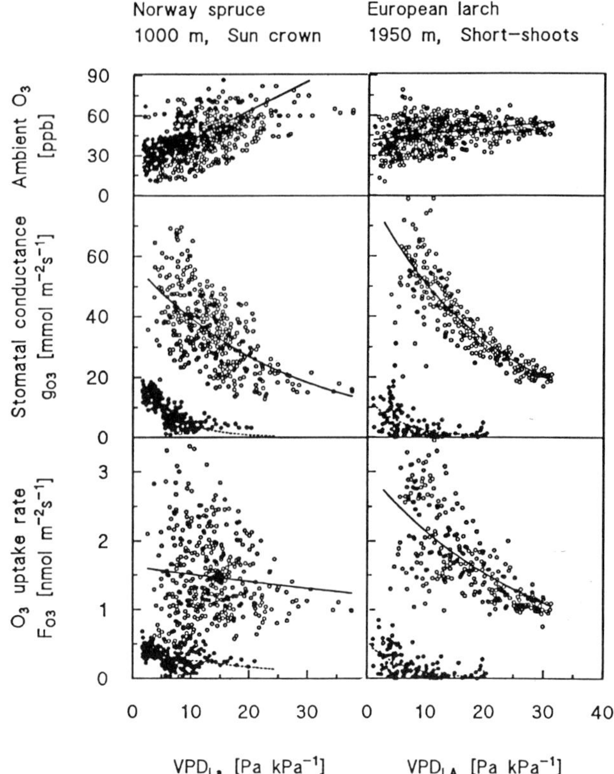

Fig. 4: Means of ambient O_3 concentration (top), of stomatal conductance for O_3 (gO_3, middle) and O_3 uptake (F_{O_3}, bottom) in relation to leaf-air vapour pressure difference (VPD_{LA}) for sun needles of Norway spruce at 1000 m a.s.l and for short-shoot needles of European Larch at 1950 m a.s.l. o photon flux density >= 600, o <= 100 $\mu mol\ m^{-2}\ s^{-1}$.

1979), whereas in valleys conductance is often reduced. Comparative measurements in Norway spruce similar in age indicated, that maximum gH_2O increases with altitude and that stomatal narrowing at midday and in the afternoon as a consequence of increasing VPD_{LA} is less pronouced at high altitude (Fig. 5). On the contrary, long term measurements of Häsler et al. (1991) showed that average gH_2O of a spruce tree near timberline was approximately 30% lower than that of a tree at low altitude. Perhaps the higher age and its diminishing effect on gH_2O of the tree at high altitude could be one reason for the different results (Yoder et al., 1994).

However, there is also evidence, that the capability to detoxify oxigen radicals is greater in plants growing at high altitude (Rennenberg, 1988), where plants are generally exposed to chronically high oxidative stress including low temperature, high radiation and high O_3 concentration (see Table 1). The apoplastic ascorbate concentration in spruce needles at timberline was found to be 25% higher than in spruce needles from trees grown at 1000 m a.s.l. (Polle et al., 1995). Furthermore, even within a species there may be genetically differences in O_3 sensitivity. In chamber experiments low-altitude provenances of young spruce (Havranek et al., 1990) and larch seedlings (unpublished) seemed to be more sensi-

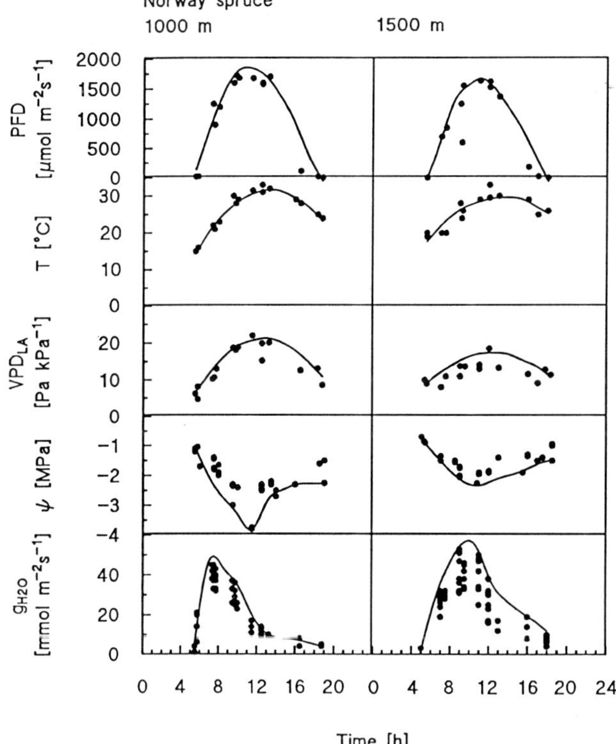

Fig. 5: Diurnal courses of photon flux density (PFD), air temperature (T), leaf-air vapour pressure difference (VPD_{LA}), as well as boundary lines for needle water potential (ψ) and stomatal conductance (gH_2O) of current-year's Norway spruce needles at a high- and a low elevation site. Measurements were made during sunny warm weather conditions in August 1992.

tive to enhanced O_3 concentrations than high-altitude provenances. This observed differences in symptom expression might be an evolutionary adaptation based on the higher ambient O_3 levels at high altitudes.

In conclusion, the above results do not support the hypothesis that tropospheric O_3 is the causal factor in the observed new kind of forest decline phenomena because chronic O_3 levels lower than 100 ppb did not cause any visible injury or even affect gas exchange of mature spruce and larch trees at low and high altitudes. In both, spruce and larch gH_2O provides an important principal limiting factor for F_{O_3} and hence for the «physiological threshold dose». Even if the quantity of O_3 uptake is known, O_3 effects are also dependent on the detoxification capacity, which varies in time (Esterbauer et al., 1980; Schupp and Rennenberg, 1988) and can be influenced by several factors (Manderscheid et al., 1991; Polle et al., 1992; Kronfuß et al., 1996). However, knowledge is limited about detoxification of O_3 in combination with other stress factors under field conditions and further research is needed to establish mechanisms influencing F_{O_3} and detoxification systems in trees. Both should also be considered in discussion about critical levels of O_3 for forest trees (Fuhrer and Achermann, 1994).

References

Amt der Tiroler Landesregierung: Zustand der Tiroler Wälder. Untersuchungen über den Waldzustand und die Immissionsbelastung. Amt der Tiroler Landesregierung – Landesforstdirektion, Innsbruck (1985–1995).

Arndt, U., G. Seufert, and W. Nobel: Die Beteiligung von Ozon an der Komplex-Krankheit der Tanne (*Abiea alba* Mill.) – eine prüfenswerte Hypothese. Staub-Reinhaltung der Luft *42,* 243–247 (1982).

Chapellka, A. H. and B. I. Chevone: Tree response to ozone. In: Lefohn, A. S. (ed.): Surface level ozone exposures and their effects on vegetation, pp. 271–342. Lewis, Chelsea, MI (1992).

Darrall, N. M.: The effect of air pollutants on physiological processes in plants. Plant Cell Environm. *12,* 1–30 (1989).

Davis, D. D. and R. Wilhour: Susceptibility of woody plants to sulphur dioxide and photochemical oxidants. A literature review. EPA 600/3-76-102, Envir. Res. Lab. Corvallis, Oregon (1976).

Dobson, M. C., G. Taylor, and P. H. Freer-Smith: The control of ozone uptake by *Picea abies* (L.) Karst. and *P. stichensis* (Bong.) Carr. during drought and interacting effects on shoot water relations. New Phytol. *116,* 465–474 (1990).

Esterbauer, H., D. Grill, and R. Welt: Der jahreszeitliche Rhythmus des Ascorbinsäuresystems in Nadeln von *Picea abies.* Z. Pflanzenphysiol. *98,* 392–402 (1980).

Fincher, J. and R. G. Alscher: The effect of long-term ozone exposure on injury of seedlings of red spruce (*Picea rubens* Sarg.) New Phytol. *120,* 49–59 (1992).

Fliri, F.: Das Klima der Alpen im Raume von Tirol. Monographien zur Landeskunde Tirols I. Universitätsverlag Wagner, Innsbruck (1975).

Freer-Smith, P. H., M. Dobson, and G. Taylor: Factors controlling the rates of O_3 uptake by spruce and beech. In: Bucher, J. B. and I. Bucher-Wallin (eds.): Air Pollution and Forest Decline. (Proceedings of the 14th international meeting for specialists in air pollution effects on forest ecosystems, IUFRO P2.05, Interlaken, Switzerland, 2–8 October 1988.) pp. 407–409. Eidgenössische Anstalt für das Forstliche Versuchswesen, Birmensdorf (1989).

Fuhrer, J. and B. Achermann: Critical levels for ozone. A UN-ECE workshop report. Schriftenreihe der Eidgenössischen Forschungsanstalt für Agrikulturchemie und Umwelthygiene 16, Liebefeld, Bern (1994).

Grünhage, L. and H. L. Jäger: Influence of the atmospheric conductivity on the ozone exposure of plants under ambient conditions: considerations for establishing ozone standards to protect vegetation. Environ Pollut. *85,* 125–129 (1994).

Häsler, R., C. Savi, and K. Herzog: Photosynthese und stomatäre Leitfähigkeit der Fichte unter dem Einfluß von Witterung und Luftschadstoffen. In: Stark, M. (ed.): Luftschadstoffe und Wald, pp. 143–168. Verlag der Fachvereine, Zürich (1991).

Havranek, W. M., P. Pfeifhofer, and D. Grill: Pigmentgehalte und Gaswechsel von Tief- und Hochlagenfichten nach chronischer Ozonbelastung. Forstw. Cbl. *109,* 200–209 (1990).

Havranek, W. M. and G. Wieser: Research design to measure ozone uptake and its effects on gas-exchange of spruce in the field. In: Payer, H. D., T. Pfirrmann, and P. Mathy (eds.): Environmental research with plants in closed chambers. Air pollution research report *26.* pp. 148–152. Commission of the European Communities, Brussels (1990).

– – Zur Ozontoleranz der Lärche (*Larix decidua* Mill.). Forstw. Centralbl. *112,* 56–64 (1993).

– – Design and testing of twig chambers for ozone fumigation and gas exchange measurements in mature trees. Proc. Roy. Soc. Edinburgh Sect. B. *102,* 541–546 (1994).

Havranek, W. M., G. Wieser, and M. Bodner: Ozone fumigation of Norway spruce at timberline. Ann. Sci. For. *46* suppl., 581s–585s (1989).

Kronfuss, G., A. Polle, G. Wieser, and W. M. Havranek: Effects of ozone and mild drought stress on total and apoplastic guaiacol peroxidase and lipid peroxidation in needles of young Norway spruce (*Picea abies* (L.) Karst. J. Plant Physiol. *148,* 203–206 (1996).

Laisk, A., O. Kull, and H. Moldau: Ozone concentration in leaf intercellular air spaces is close to zero. Plant Physiol. *90,* 1163–1167 (1989).

Manderscheid, R., H. J. Jäger, and M. M. Schoenberger: Dose-response relatioships of ozone effects on foliar levels of antioxidants, soluble polyamids and peroxidase activity of *Pinus taeda* (L.): assessment of the usefullness as early ozone indicators. Angew. Bot. *65,* 373–386 (1991).

Matyssek, R., M. S. Günthardt-Goerg, T. Keller, and C. Scheidegger: Impairment of gas exchange and structure in birch leaves (*Betula pendula*) caused by low ozone concentrations. Trees *5,* 5–13 (1991).

Pell, E. J., N. Eckhardt, and A. J. Enyedi: Timing of ozone stress and resulting status of ribulose bisphosphate carboxylase/oxygenase and associated net photosynthesis. New Phytol. *120,* 397–405 (1992).

Polle, A., K. Chakrabarti, S. Chakrabarti, F. Seifert, P. Schramel, and H. Rennenberg: Antioxidants and manganese deficiency in needles of Norway spruce (*Picea abies* L.) trees. Plant Physiol. *99,* 1084–1089 (1992).

Polle, A., G. Wieser, and W. M. Havranek: Quantification of ozone influx and apoplastic ascorbate content in needles of Norway spruce trees (*Picea abies* (L., Karst.) at high altitude. Plant Cell Environm. *18,* 681–688 (1995).

Prinz, B.: Woran sterben unsere Wälder. Umschau *18,* 544–549 (1984).

Pye, J. M.: Impact of ozone on the growth and yield of trees: a review. J. Environ. Qual. *17,* 347–360 (1988).

Rebbeck, J., K. F. Jensen, and M. S. Greenwood: Ozone effects on the growth of grafted mature and juvenile red spruce. Can. J. For. Res. *22,* 756–760 (1992).

– – – Ozone effects on grafted mature and juvenile red spruce: photosynthesis, gH_2O and chlorophyll concentration. Can. J. For. Res. *23,* 450–456 (1993).

Reich, P. B. and R. G. Amundson: Ambient levels of ozone reduce photosynthesis in tree and crop species. Science *230,* 566–570 (1987).

Reich, P. B.: Quantifying plant response to ozone: aunifying theory. Tree Physiol. *3,* 63–91 (1987).

Rennenberg, H.: Wirkung von Photooxidantien auf Pflanzen. In: Schulte-Hostede, S., M. Kirchner, and M. Reuter (eds.): Internationales Symposium «Verteilung und Wirkung von Photooxidantien im Alpenraum» Garmisch-Partenkirchen 11–15 April 1988. pp. 360–370. GSF Bericht *17,* München (1988).

Roberts, B. R. and W. J. Cannon Jr.: Growth and water relationship of red spruce seedlings exposed to atmospheric deposition and drought. Can. J. For. Res. *22,* 193–197 (1992).

Schupp, R. and H. Rennenberg: Diurnal changes in the glutathione content of spruce needles (*Picea abies* L.). Plant Sci. *57,* 113–117 (1988).

Temple, P. J., G. H. Riechers, P. R. Miller, and R. W. Lennox: Growth response of ponderosa pine to long-term exposure to ozone, wet and dry acidic deposition, and drought. Can. J. For. Res. *23,* 59–66 (1992).

Thompson, F. B. and L. Leyton: Method for measuring the leaf surface area of complex shoots. Nature *299,* 572 (1971).

Tingey, D. T. and G. E. Taylor Jr.: Variation in plant response to ozone: a conceptual model of physiological events. In: Unsworth, H. M. and D. D. Ormond (eds.): Effects on gaseous air pollution in agriculture and horticulture. pp. 113–138. Butterworth, London (1982).

Tranquillini, W.: Physiological ecology of the alpine timberline. Ecological Studies 31. Springer, Berlin (1992).

VDI: Maximale Immissions-Konzentrationen für Ozon. VDI Handbuch Reinhaltung der Luft. Beuth, Berlin, Blatt 6 (1989).

Volgger, E.: Zur Ozonempfindlichkeit der europäischen Lärche (*Larix decidua* Mill.) an der Waldgrenze. Diplomarbeit Universität Innsbruck (1995).

Von Caemmerer, S. and G. D. Farquhar: Some relationships between the biochemistry of photosynthesis and the gas exchange of leaves. Planta *153,* 376–387 (1981).

Wallin, G., S. Ottosson, and G. Sellden: Long-term exposure of Norway spruce, *Picea abies* (L.) Karst, to ozone in open top chambers. In: Effects on the stomatal and non-stomatal limitation of photosynthesis and on carboxylation efficiency. New Phytol. *121,* 395–401 (1992).

Wallin, G. and L. Skärby: The influence of ozone on the stomatal and non-stomatal limitation of photosynthesis in Norway spruce, *Picea abies* (L.) Karst., exposed to soil moisture deficit. Trees *6,* 128–136 (1992).

Wieser, G. and W. M. Havranek: Ozone uptake in the sun and shade crown of spruce: quantifying the physiological effects of ozone exposure. Trees *7,* 227–232 (1993 a).

– – Einfluß der Nährstoffversorgung auf den Gaswechsel von Fichte. Centralbl. Ges. Forstwes. *110,* 135–149 (1993 b).

– – Exposure of mature Norway spruce to ozone in twig-chambers: effects on gas exchange. Proc. Roy. Soc. Edinburgh Sect. B *102,* 119–125 (1994).

– – Environmental control of ozone uptake in *Larix decidua* Mill.: a comparison between different altitudes. Tree Physiol. *15,* 253–258 (1995).

Wieser, G., M. Weih, and W. M. Havranek: Ozone fumigation in the sun crown of Norway spruce. In: Reuter, M., M. Kirchner, E. Kirchinger, H. Reiter, K. Rösel, and U. Pfeifer (eds.): Waldschadensforschung im östlichen Mitteleuropa und in Bayern. Proc. Expertentagung Schloß Neuburg/Inn bei Passau 13–15 November 1990, pp. 567–573. GSF-Bericht *24,* München (1991).

Yoder, B. J., M. G. Ryan, R. H. Waring, A. W. Schoettle, and M. R. Kaufmann: Evidence of reduced photosynthetic rates in old trees. Forest Science *40,* 513–527 (1994).

J. Plant Physiol. Vol. 148. pp. 195–202 (1996)

Destabilization of the Antenna Complex LHC II During Needle Yellowing of a Mg-Deficient Spruce Tree Exposed to Ozone Pollution – Comparison With Other Types of Yellowing

DOROTHEA SIEFERMANN-HARMS

Universität Karlsruhe, Botanisches Institut I, D-76128 Karlsruhe und Forschungszentrum Karlsruhe, Institut für Toxikologie, Postfach 3640, D-76021 Karlsruhe

Received July 14, 1995 · Accepted October 31, 1995

Summary

The stability of the peripheral antenna complex of Photosystem II, LHC II, has been examined in the course of different stress-induced and naturally occurring yellowing processes. In needles of a Mg-deficient spruce tree (*Picea abies* (L.) Karst.) growing under ozone exposure at a Black Forest stand the LHC II became highly unstable at an early stage of yellowing. No destabilization of LHC II was observed when Mg-deficient spruce seedlings turned yellow or when healthy spruce trees went through natural yellowing in winter. Also the senescence-dependent yellowing of snowdrop (*Galanthus nivalis* L.) leaves was not accompanied by LHC II destabilization. – In declining spruce trees at Black Forest stands the yellowing process was characterized by a rapid loss of chlorophyll rather than a slow decrease as observed for stress conditions dominated by Mg deficiency. – It is suggested that the needle yellowing of Mg-deficient spruce trees is accelerated under ozone pollution and the yellowing mechanism is modified to include destabilization of the LHC II.

Key words: Spruce, Picea abies, ozone, Mg deficiency, yellowing mechanism, forest decline, light-harvesting Chl-a/b-protein complex.

Abbreviations: AC = age class (AC 1: needle age 0–11 months, AC 2, needle age 12–23 months, etc); Chl = chlorophyll; DC = damage class (DC 0: without symptoms of decline; DC 2: with moderate symptoms of decline); DM = dry matter; FM = fresh matter; HPLC = high pressure liquid chromatography; LHC II = peripheral light-harvesting Chl-*a/b*-protein complex of Photosystem II; PAG = polyacrylamide gel; V+A+Z = violaxanthin+antheraxanthin+zeaxanthin.

Introduction

Needle yellowing of spruce trees at altitudes over 700 m a.s.l. is a widespread symptom of forest decline in the subalpine ranges of Central Europe. The chlorotic trees usually suffer from nutritional deficiencies (e.g. Zech and Popp, 1983; Mies and Zöttl, 1985; Köstner et al., 1990; Polle et al., 1992; Evers, 1994). Air pollution imposes additional stress on the plants (Guderian, 1977; Arndt et al., 1983; Wentzel, 1985). In the mountain ranges of Southern Germany the predominant gaseous pollutant is ozone. At higher altitudes of the Black Forest the mean monthly ozone concentrations in summer exceed 100 $\mu g/m^3$ (Baumbach and Baumann, 1992). These ozone concentrations are not high enough to induce visible symptoms in spruce. In the Black Forest yellowing is generally accompanied by Mg deficiency.

The long-term fumigation experiments at Hohenheim (Arndt, 1990; Seufert et al., 1990) and Kettwig (Krause et al., 1991, Landesanstalt für Immissionsschutz von Nordrhein-Westfalen, unpublished) revealed that spruce trees growing under sufficient mineral supply also develop yellowing symptoms if they are exposed to realistic levels of polluted air (Siefermann-Harms et al., 1995 b). The artificial atmospheres of these experiments contained ozone at concentrations comparable to Black Forest conditions, whereas the SO_2 (and NO_2) concentrations in winter were similar to those of German urban areas. The experiments suggest that the air pollution at Black Forest sites is close to the critical level inducing damage *per se*. Therefore one has to ask whether the yellowing symptoms occurring under Black Forest-type stress conditions are solely due to Mg deficiency or reflect the combined action of insufficient Mg supply and ozone pollution.

The peripheral antenna complex of Photosystem II, LHC II, is the major pigment-binding protein of the photosynthetic apparatus. When isolated from healthy plants, this complex is remarkably stable under strong light plus aerobiosis or at low pH: The LHC-bound pigments are hardly photooxidised and efficiently protected from proton attack (Siefermann-Harms, 1990 a). At our research site at Schöllkopf (Northern Black Forest) we examined the stability of LHC II in needles of healthy and declining spruce. In case of yellowing needles, the LHC-bound pigments were photooxidised more rapidly and the portion of LHC-bound Chl-*a* which could be attacked by protons, was increased (Siefermann-Harms, 1992). Therefore, LHC II destabilization was suggested to participate in the yellowing mechanism.

Here we compare the behaviour of LHC II in the course of different stress-induced and natural yellowing processes. A destabilisation of LHC II was only detected during the rapid yellowing processes observed at Black Forest sites. The presented data suggest that in case of Black Forest-type pollution the needle yellowing of Mg deficient spruce trees is accelerated and destabilization of the LHC II is involved in the complex reaction pattern of yellowing.

Materials and Methods

Plant material

Needles of Norway spruce (*Picea abies* (L.) Karst.) from the field stand Schöllkopf, spruce seedlings of a hydroculture experiment and leaves of snowdrop (*Galanthus nivalis* L.) from a public garden were examined:

Schöllkopf site: This forest stand is located at the plateau of Schöllkopf south-west of the town of Freudenstadt (Northern Black Forest), 830 to 840 m a.s.l. Thirthy to 60-year-old spruce trees with no visible symptoms of decline, and trees with yellow needles and heavy needle loss grow in close neighbourhood. Low Mg supply and high ozone levels are typical for this site. In the years 1989–1991 the Freudenstadt-Schöllkopf site has been investigated in an interdisciplinary approach by scientists working in the fields of forestry, soil chemistry, air pollution, plant biochemistry, genetics and physiology (for review see Siefermann-Harms et al., 1995 b). The data on natural chlorosis in fall and winter (Fig. 5) have been collected in the course of this interdisciplinary study (healthy spruce trees No. 10 and 16, sun-exposed needles from the seventh whirl). The behaviour of a declining spruce tree (spruce No. 33) and a healthy control (spruce No. K3) were examined in 1993 in short intervalls (Fig. 3; sun-exposed needles of the tenth whirl). Spruce No. 33 (DC 2) behaves like the other declining spruce trees at Schöllkopf site: Needle yellowing and subsequent shedding take place in the second needle age class during some time in summer.

Hydroculture experiment: Seeds of autochthonous Norway spruce from district 840-08 (Freudenstadt) were obtained from Staatsklenge Baden-Württemberg (Forstamt Nagold). After germination and cultivation on agar (1.4 % DIFCO Bacto-agar in water, pH 5.5) for 4 weeks the seedlings were transferred to specially designed hydroculture vessels (Steegborn, 1991) and kept in a greenhouse for 8 months at a daily regime of 15 h light (150–250 $\mu mol\, m^{-2} s^{-1}$ PAR), 21 °C and 9 h dark, 15 °C. Nutrient solutions (14 L/vessel) were continuously aerated and exchanged every week. Two nutrient solutions were compared, one with and one without Mg. They contained the following major elements (mg/L): 105 Cl, 15 Mg, 20 S (or, when Mg was excluded: 70 Cl, 0 Mg, 15.5 S, respectively), 25 N as NH_4^+, 25 N as NO_3^-, 10 P, 50 K, 40 Ca, 1 Fe as Fe(III)-EDTA, 0.95 Na, the following minor elements (mg/L): 0.57 Si, 0.17 Mn, 0.17 B, 0.02 Zn, 0.02 Cu, 0.003 Mo, and the following trace elements (µg/L): 0.18 Al, 0.4 Br, 0.4 J, 0.18 Co, 0.09 Li, 0.27 Ni, 0.4 Ti and 0.27 Sn. The pH of the solutions was adjusted with HCl to pH 5.5. For further details see Steegborn (1991).

Public garden at Tübingen: Leaves of snowdrop growing wild in the public garden at Eberhardshöhe, 440 m a.s.l., were analysed in March through May 1987.

Methods

Isolation of thylakoid membranes: Spruce twigs were harvested, placed in water and transferred to the laboratory, where they were kept at room temperature under dim light. Thylakoid membranes were isolated within two days after sampling. All isolation steps were performed at 0–4 °C. Needles (2 g) in 25 mL half frozen extraction medium (250 mM sorbitol; 42 mM HEPES buffer pH 7.5; 15 mM $MgCl_2$; 10 mM Na ascorbate; 8 mM Na Cl; 2 % wt/wt Polyvinylpyrrolidone 15, Serva 33 422) were homogenized in the 35-mL vessel of a Waring blendor for 5 s at low and for 20 s at high speed. The homogenate was mixed with 10 mL of icecold hexane and immediately filtered through wet Nylon cloth (mesh width 35 µm) which retained the colourless hexane phase and needle fragments. The hexane treatment improved the membrane yield but did not affect the membrane properties studied here. The filtrate was centrifuged for 3 min at 500 ×g, the supernatant for 5 min at 12,000 ×g, the 12,000 ×g pellet was resuspended in washing medium (300 mM sorbitol; 50 mM HEPES buffer pH 7.5; 10 mM Na Cl) and centrifuged again at 12,000 ×g for 5 min. The pellet was osmotically shocked in distilled water, centrifuged for 10 min at 40,000 ×g, and resuspended in distilled water. Portions of 350 µL containing 300 µg Chl were frozen and stored in liquid nitrogen for further use.

Membrane solubilization and isoelectric focussing on PAG plates: The thawed membranes were mixed with Triton X-100 (Chl/Triton X-100 = 1/30, wt/wt; 2 % Triton X-100), stirred on ice for 5 min and centrifuged for 10 min at 39,000 ×g. The supernatant which contained more then 90 % of the Chl was immediately placed on digitonin-containing PAG plates and subjected to isoelectric focussing according to Siefermann-Harms and Ninnemann (1982). The gel plates used in this study contained 0.5 % (wt/wt) water-soluble digitonin (Serva 901 119). When focussing was complete the gel plates were scanned at 675 nm using a home-made device in combination with the SLM spectrophotometer DW 2000. The Chl-protein complexes were eluted from the gels with TRICINE buffer (50 mM; pH 7.8) as described before (Siefermann-Harms and Ninnemann, 1979).

Spectroscopic tests: All spectroscopic tests were performed at room temperature using SLM spectrophotometer DW 2000. P-700 and the cytochromes were assayed as in Siefermann-Harms and Ninnemann (1979, 1981). Acid stability of LHC II was monitored in the DUAL mode at 675 versus 720 nm. The presented data are mean values obtained from two LHC II preparations per membrane sample. For further details see Fig. 2.

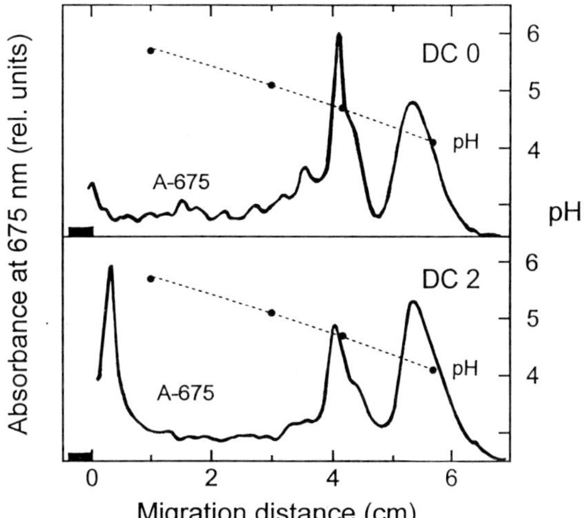

Fig. 1: Gel scan of solubilized Chl-protein complexes of spruce after separation by isoelectric focussing. Needles of age class 1992 of a healthy (DC 0) and a declining (DC 2) spruce tree sampled at Schöllkopf in fall 1993 are compared. The black bar represents the sample application trough.

Chl determination: Two 100-mg aliquots of fresh needles or leaves were quantitatively extracted in 100 % acetone (Siefermann-Harms, 1988), and two 100-mg aliquots were dried at 105 °C for 24 h. The Chl content of the extracts was determined in 80 % acetone according to Lichtenthaler and Wellburn (1983). For the respective parallels, the Chl content per FM and the FM/DM ratio differed by less then 5 %. Mean values were used for calculations.

Pigment analysis: The pigments were separated by HPLC and quantified according to Siefermann-Harms (1988, 1990 b). Twenty-μL samples of extracted pigments in 100 % acetone or of LHC II in TRICINE buffer at pH 7.8 (A-675 = 0.23–0.26 cm^{-1}) were injected.

Results

Properties of the light-harvesting Chl-a/b-protein complex LHC II of a healthy and a declining spruce tree at the Schöllkopf site

When solubilized Chl-protein complexes of spruce thylakoids are separated by our isoelectric focussing procedure (Fig. 1) a single strong band focusses at pH 4.1–4.3, a strong double band at pH 4.6–4.8, two week bands at pH 4.8–5.0 and several week bands at pH 5.0–5.8. The band focussing at pH 4.1–4.3 represents the peripheral light-harvesting Chl-*a/b*-protein complex of Photosystem II, LHC II, as judged from its isoelectric point, absorbance spectrum (absorbance maxima at 675 and 650 nm), Chl content (35–45 % of total Chl, depending on sampling date), high Chl-*b* and xanthophyll to Chl-*a* ratios (Table 1) and lack of P-700 and cytochromes.

The bands focussing at higher pH are characterized here only briefly (data not shown). They contain, relative to Chl-*a,* small amounts of Chl-*b* and xanthophylls but high amounts of α- and β-carotene. The strong band focussing at pH 4.6–4.8 contains both, cytochrome b-559, a component associated with reaction center II and P-700, the pigment of reaction center I, indicating the presence of both photosystems. The bands focussing at pH >4.8 contain P-700 and increasing amounts of the cytochrome b-563/f-complex. The fact that in case of spruce thylakoids our isoelectric focussing procedure does not yield pure Photosystem II complexes is in contrast to observations with *Phaseolus vulgaris* (Siefermann-Harms and Ninnemann, 1979, 1981), *Lactuca sativa* (Siefermann-Harms and Ninnemann, 1982) and *Spinacea oleracia* (Siefermann-Harms, 1984). A reasonable explanation would be that reaction-center complexes of different species can differ in their isoelectric points.

The Chl-protein pattern of thylakoids from green needles of a healthy (DC 0) and yellow needles of a declining (DC 2) spruce agree only in part (Fig. 1). The isoelectric points of the respective bands were comparable but a band remaining near the starting trough was much stronger in case of yellow then green needles. The red absorbance maximum of this band (673–674 nm) indicates that most of its Chl-*a* is protein-bound. The band contains P-700 and cytochromes and the Chl-*a/b* ratio is higher than that of thylakoids. Therefore it is assumed to consist of poorly solubilized membrane material enriched in one or both photosystems.

The pigment pattern of LHC II of green spruce needles (Table 1, DC 0) agrees with that of other species (Siefermann-Harms, 1985; Lichtenthaler, 1987). With Chl-*a/b* ratios of 1.25–1.35 (as compared to *a/b*-ratios of the needles of 3.5 ± 0.1) this complex binds most of the Chl-*b*. It also binds most of the lutein and neoxanthin (35–36 lutein and 16–17 neoxanthin/100 Chl-*a* for LHC II as compared to values of 15–17 and 7–8, respectively, for the needles), whereas the

Table 1: Chl-*a* content of last year's spruce needles and properties of the Chl-*a/b*-protein complex LHC II isolated from the needles. A healthy (DC 0) and a declining (DC 2) spruce tree form the Freudenstadt–Schöllkopf site (Northern Black Forest, 840 m a.s.l.) are compared.

Damage class (DC)	Sampling date (1994, * 1993)	Chl-*a* content of the needles (mg/g DM)	H^+-labile Chl-*a* in LHC II (%)	Ratio of LHC II bound pigments (mol/100 mol Chl-*a*)				
				Chl-b	Lutein	Neoxanthin	V+A+Z	α+β-Carotene
DC 0	May 5	1.61	17.7	73.5	35.6	15.9	8.1	2.7
	August 16	3.14	16.1	78.7	33.5	16.5	9.5	2.6
	September 28	3.14	14.5	81.3	35.9	16.6	7.6	2.1
DC 2	May 5	1.82	23.9	73.0	37.5	18.4	7.6	3.5
	August 16	1.32	31.9	64.9	34.9	15.8	9.2	2.9
	September 20*	1.33	29.4	62.5	37.7	19.0	7.7	2.6

pigments of the xanthophyll cycle, V+A+Z (7.5–9.5 pigments/100 Chl-*a* for LHC II as compared to 8–11 for needles), are not enriched in LHC II and the carotenes (2–3 carotenes/100 Chl-*a* for LHC II as compared to 12–14 for needles) are preferentially bound to other complexes.

In needles of the declining spruce trees at Schöllkopf, the ratios of Chl-*b*, neoxanthin and carotene to Chl-*a* do not change during yellowing whereas the ratios of lutein and V+A+Z to Chl-*a* increase (Siefermann-Harms, 1992). In contrast to these observations, none of the LHC II-bound carotenoids changed its ratio to Chl-*a* significantly in yellow needles (Table 2, DC 2, sampling dates in August and September). However, Chl-*b* was decreased in LHC II of yellow as compared to green needles, yielding Chl-*a*/*b* ratios of 1.5–1.7. This observation holds for all LHC II preparations of yellow needles examined so far.

As observed for other species (Siefermann-Harms and Ninnemann, 1982, 1983), LHC II preparations isolated from healthy spruce bind most of their Chl-*a* at protein sites not accessible to protons. The kinetics of pheophytin formation from LHC II-bound Chl-*a* at low pH, monitored as absorbance decrease at 675 nm, illustrate this property (Fig. 2). While ‹free› Chl-*a* (e.g. in aqueous detergent solution) at low pH is completely converted into pheophytin within two minutes (Siefermann-Harms, unpublished results), most of the LHC-bound Chl-*a* remains unchanged under such conditions. Only 13.4 % of LHC-bound Chl-*a* were transferred in case of the healthy spruce (Fig. 2, DC 0) and the conversion of this ‹H^+-labile Chl-*a*› proceeded rather slowly. About twice as much H^+-labile Chl-*a* was detected in LHC II of the declining spruce (Fig. 2, DC 2) but also in this case the major part of the LHC-bound Chl-*a* remained stable at low pH throughout the measuring period. In the following, the H^+-labile portion of the LHC II-bound Chl-*a* was used to quantify changes in LHC II stability.

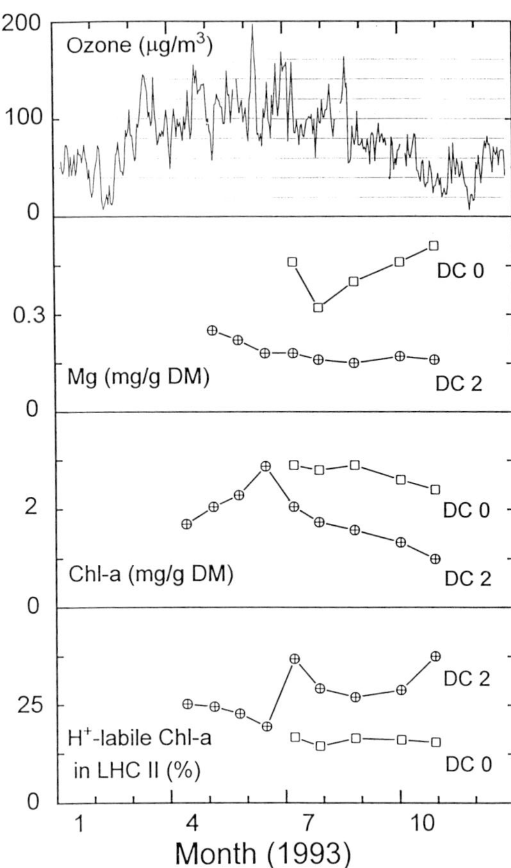

Fig. 3: Ozone concentration (daily means) at Schöllkopf site, Mg- and Chl-*a* content of spruce needles from this site, and H^+-labile Chl-*a* in the isolated LHC II. Needles of age class 1992 of a healthy (DC 0) and a declining (DC 2) spruce tree are compared. Source of the ozone data: UMEG, «Landesweites Meßnetz». Determination of Mg: Forstliche Versuchsanstalt Baden-Württemberg.

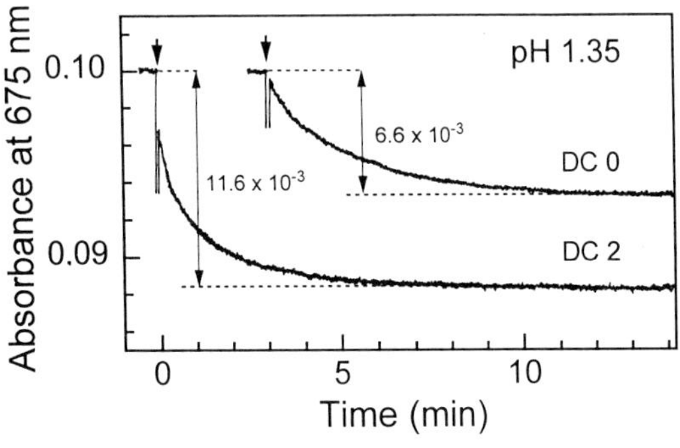

Fig. 2: Partial conversion of Chl-*a* into pheophytin-a in acid-treated LHC II. The LHC II bands focussing at pH 4.1–4.3 (Fig. 1, DC 0 and DC 2) were examined. Samples in aqueous solution (A-675 = 0.1 cm^{-1}, pH 7.8) were acidified (arrows indicate rapid mixing of the 1.5-mL samples with 20 µL concentrated HCl) and the total absorbance decrease at 675 nm was monitored. By correcting for sample dilution the acid-induced absorbance change ΔA-675 was obtained. Based on ΔA-675 and the observation that a 100 % conversion of Chl-*a* into pheophytin corresponds to ΔA-675 = 0.04, the LHC II preparations contained 13.4 % and 26.1 % of H^+-labile Chl-*a*, respectively.

Fig. 3 shows the annual time course of the ozone concentration at Schöllkopf. It further compares the Mg and Chl content and the LHC II stability of spruce needles from a healthy and a declining tree growing at that site. Between March and April the ozone level increased from daily means of 20–60 µg/m^3 in winter to values of 100–140 µg/m^3 in summer. The needles of the declining spruce contained only about one third of the Mg found in needles of the healthy spruce. The Mg values of the declining tree (0.16–0.20 mg/g DM) were within the critical range at which yellowing symptoms become evident at the Schöllkopf site (Siefermann-Harms et al., 1995b). In spite of the low Mg level the needles of the declining spruce went through the normal greening process observed in spring. Between April and middle of June the Chl content almost doubled reaching values similar to those of the healthy tree while the new shoot developed to its final length. End of June the needles suddenly turned yellow. They lost about 40 % of the Chl within three weeks while maintaining the initial Chl-*a*/*b* ratio. At the same time the stability of LHC II decreased strongly. In August and September the yellowing process slowed down and LHC II stabilized to some degree.

At Schöllkopf the sudden needle yellowing and the decrease of LHC II stability occurred together. In the following it is asked whether LHC II destabilization is a general feature of yellowing processes or develops only under special stress conditions.

Behaviour of LHC II during yellowing of spruce seedlings under Mg deficiency and healthy spruce trees in winter

Mg deficiency is a widespread cause for the yellowing of plants (Evers, 1994). Therefore we examined spruce seedlings grown in hydroculture in the presence or absence of Mg (Steegborn, 1991). Under both conditions the seedlings developed similarly for about 4 months after germination indicating a sufficient Mg supply from the seeds. Also the Chl content of the primary needles was similar (Fig. 4, top). In the following months the growth rate of the seedlings cultivated without Mg slowed down, and the needles turned yellow. A 50 % decrease of the Chl content was reached after 3 months, whereas the Chl-*a*/*b* ratio was hardly affected. As shown in Fig. 4 (bottom), the yellowing process was not accompanied by changes in LHC II stability.

The Chl content of healthy spruce trees goes through annual changes with minimum in late winter and maximum in summer (Köster et al., 1990; Siefermann-Harms, 1994). At the Schöllkopf site we observed an almost 50 % decrease of the Chl-*a* content between August and February (Fig. 5, top) together with a slight increase of the Chl-*a*/*b* ratio (Siefermann-Harms, 1995). This natural yellowing process occurred without LHC II destabilization. Instead, LHC II stability increased during fall and winter, as observed for both, the new flush and older needles (Fig. 5, bottom).

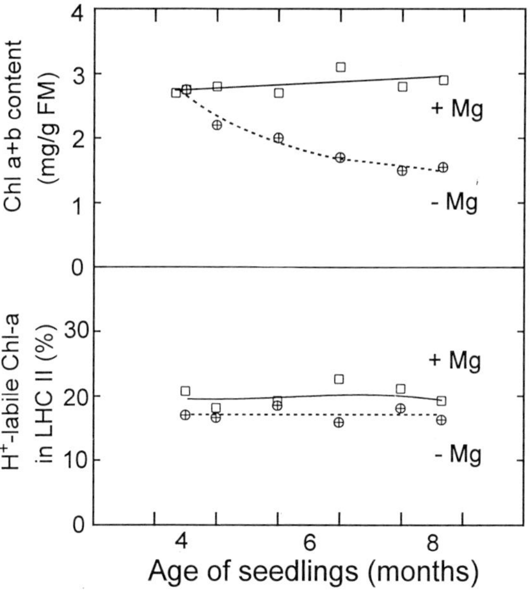

Fig. 4: Chl content and amount of H^+-labile Chl-*a* in the LHC II of spruce seedlings cultivated in hydroculture with optimal Mg supply (+Mg) and under conditions of complete Mg exclusion (−Mg). From Steegborn (1991), modified.

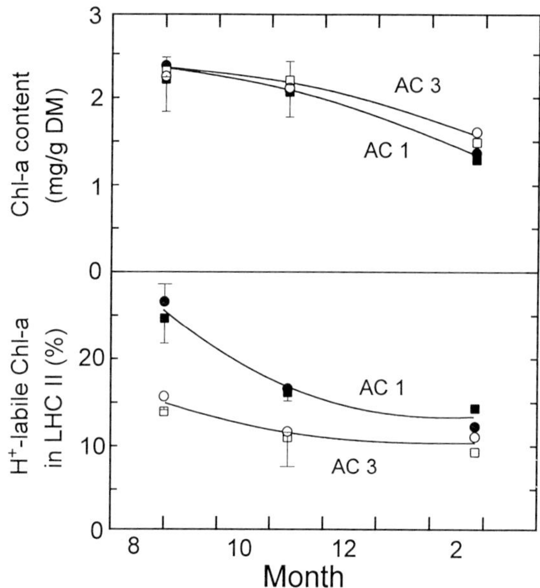

Fig. 5: Chl-*a* content of spruce needles and amount of H^+-labile Chl-*a* in their LHC II during needle yellowing in winter. Needles of age class 1 (AC 1) and 3 (AC 3) from healthy spruce trees of the Schöllkopf site (squares: spruce No. 10; circles: spruce No. 16) were sampled in the winters 1988/89 to 1990/91.

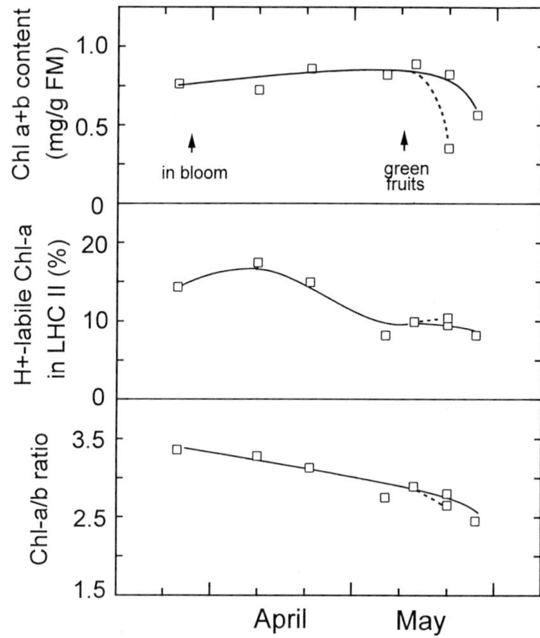

Fig. 6: Changes of the Chl content of snowdrop (*Galanthus nivalus* L.) leaves, the amount of H^+-labile Chl-*a* in their LHC II, and the Chl-*a*/*b* ratio of the leaves during snowdrop development from blooming to leaf yellowing.

Stability changes of LHC II during senescence-dependent yellowing of snowdrop leaves (Galanthus nivalus L.)

Forest decline symptoms have been discussed in relation to accelerated senescence (e.g. Hampp et al., 1990). Therefore we asked whether LHC II destabilization is associated with senescence-dependent yellowing processes. This question was examined using the short-lived leaves of snowdrop (*Galanthus nivalus* L.) rather than the long-lived and slowly senescent needles of spruce. As shown in Fig. 6 (top), the Chl content of snowdrop leaves did not change significantly while the plants flowered and developed fruits. In May a sudden yellowing was observed spreading from leaf tip to base and the leaves died. LHC II was highly stable throughout this yellowing process (Fig. 6, middle). The Chl-*a*/*b*-ratio of the leaves decreased steadily between March and May (Fig. 6, bottom) suggesting that the amount of LHC II increased relative to the complexes preferably binding Chl-*a*. A slightly accelerated decrease of the Chl-*a*/*b*-ratio was observed during yellowing in May which points at a preferential disappearance of Chl-*a* binding proteins.

Discussion

At the Schöllkopf site declining spruce trees are exposed to at least two types of stress, severe Mg deficiency and enhanced levels of ozone (Fig. 3, DC 2). The ozone concentrations at Schöllkopf are not high enough to cause yellowing if the trees are supplied with sufficient Mg (Fig. 3, DC 0). Therefore Mg deficiency is the key factor when the spruce trees turn yellow. In addition, ozone (or some accompanying compound) acts on the trees. As discussed in the following, ozone affects the mechanism of yellowing and enhances the velocity of the yellowing process.

The needle yellowing of declining spruce trees at Schöllkopf occurs together with a loss of LHC II stability (Fig. 3, bottom). This is not a general feature of yellowing processes. Yellowing of spruce needles in winter (Fig. 5) and of snowdrop leaves during senescence (Fig. 6) take place without destabilisation of LHC II. Obviously the yellowing of declining spruce is not comparable with normal needle chlorosis under wintry conditions and appears not to be due to accelerated senescence. Also the yellowing of spruce seedlings cultivated without Mg is not accompanied by changes in LHC II stability (Fig. 4). Therefore, the yellowing process at the Schöllkopf site cannot be explained solely in terms of Mg deficiency.

The kinetics of yellowing support this view. Yellowing events at Schöllkopf proceed rather rapidly (Fig. 3 and Siefermann-Harms, 1994). Chl losses of 40 to 50 % may occur within 20 to 30 days at some time in summer. This behaviour differs from the observations of Köstner et al. (1990) who investigated needle yellowing of chlorotic spruce trees in a forest decline area of the Fichtelgebirge. Similar to the trees at Schöllkopf, the chlorotic trees of the Fichtelgebirge area suffered from severe Mg deficiency (Lange et al., 1989). In contrast to the Schöllkopf site, the ozone level – with mean monthly concentrations in May through August not exceeding 58 µg/m^3 (Köstner et al., 1990) – was rather low. The chlorotic trees growing under these conditions accumulated less Chl in early summer than healthy trees did, but rapid losses of half the Chl were not observed. Köstner et al. (1990) suggest that needle yellowing in the Fichtelgebirge area was mainly due to nutritional deficiencies rather than air pollutants. They further assume that Chl formation ceased under those conditions.

Direct evidence to support this view comes from the experiments of Mehne-Jakobs (1994), who induced Mg deficiency in young spruce trees. One-year-old needles of the spruce grown under moderate Mg deficiency accumulated less Chl in spring than the controls, but a fast decrease of the Chl did not occur. Under severe Mg deficiency, Chl formation slowed down already in the new flush, but again, no rapid breakdown of the Chl content was observed.

In case of the spruce seedlings grown on hydroculture (Fig. 4), a $^{14}CO_2$ pulse/chase experiment allowed to show that under Mg deficiency less $^{14}CO_2$ was incorporated into the apoproteins and the Chl of the major Chl-protein complexes, while the breakdown of the complexes was not enhanced (Steegborn, 1991; Siefermann-Harms and Steegborn, 1992). These data together with the field observations of Köstner et al. (1990) and the experimental results of Mehne-Jakobs (1994) show that Mg deficiency – single or as dominant stress factor – acts by inhibiting the synthesis of Chl and other thylakoid components rather than accelerating the natural Chl breakdown.

Rapid Chl losses have first been detected by Mies and Zöttl (1985) at Kälbelescheuer, a research site in the Southern Black Forest. While examining a collective of 4 highly chlorotic trees, the authors observed that between end of July and end of August the Chl content of last year's needles decreased from 2 to 1 mg/g DM. The trees suffered from severe Mg deficiency and they grew under significantly higher ozone levels than the trees at the Fichtelgebirge area (Anonymus, 1987: Mean monthly ozone conzentrations at Kälbelescheuer in 1985, May through August, 73–106 µg/m^3). Also at Kälbelescheuer we obtained the first evidence for LHC II destabilization in yellowing spruce needles (Siefermann-Harms and Ninnemann, 1987). These data suggest that at the two Black Forest sites the Chl content of the needles decreases faster than at the Fichtelgebirge site and that the mechanism of rapid yellowing includes LHC II destabilization.

Recently we showed that the Chls and carotenoids of the stable LHC II are protected from photooxidation by an O_2 barrier located in the protein (Siefermann-Harms and Angerhofer, 1995). The O_2-barrier as well as the acid stability of LHC II are lowered by hydrophobic compounds, by long-chain fatty acids (Siefermann-Harms, 1990 a, c) and corresponding lysophosphatidic acids (unpublished results) but not by natural galactolipids (Siefermann-Harms and Yamamoto, 1982, unpublished results). We assume that the former compounds loosen the trimer structure of LHC II without affecting the efficient energy transfer from carotenoid to Chl-*a* within the monomers. The observation that destabilized LHC II from spruce contains less Chl-*b* while the other pigments are not affected (Table 1) might support this assumption. According to the LHC II model of Kühlbrandt et al. (1994) Chl-*b* is less deeply burried in the LHC monomers than Chl-*a* and the carotenoids. If the trimer structure of LHC II is loosened one might therefore expect Chl-*b* to be less firmly bound to the protein *in situ* or to be more easily released during LHC II isolation.

Based on our studies with LHC II (Siefermann-Harms and Angerhofer, 1995) the following scenario might take place during rapid needle yellowing: When Mg-deficient spruce trees are exposed to enhanced levels of ozone some compound is formed able to destabilize a few LHC II molecules. Through their action the O_2-barrier in the LHC II molecules becomes leaky, the pigments can be oxidised under light, energy transfer from triplet-state Chl to ground-state oxygen yields harmful singlet oxygen, proteases act on the LHC II apoprotein, and surrounding membrane lipids are peroxidised or attacked by lipases. In the course of this breakdown new LHC II-destabilizing components can be generated supporting a light-driven chain reaction.

The suggested mechanism of LHC II destruction does not explain the fact that during rapid yellowing the photosynthetic apparatus decreases as a whole while the surviving part remains photosynthetically active (Siefermann-Harms et al., 1995 b). To get insight into the overall process of yellowing, many other parameters of the spruce trees presented in Fig. 3 have been investigated. Changes of their fluorescence behaviour, Chl, P-700 and carotenoid contents, size of thylakoids and starch grains (Siefermann-Harms et al., 1995 a), content and synthesis of the D1 protein of Photosystem II (Godde, 1995) have been studied at short sampling intervalls. None of the parameters changed significantly before yellowing began. While yellowing became evident, the synthesis of the D1 protein was enhanced, and the LHC II lost stability. The available data show that D1 turnover and LHC II stability are affected at an early stage of the yellowing process; possible interactions between both changes remain to be established in further studies.

Acknowledgements

The author thanks Ms. G. Eisenhardt, Ms. K. Lisiecki, Ms. S. Grafe and Ms. M. Setzer for expert technical assistence during different periods of this work. I am highly endebted to Ms. G. Trefz-Malcher, Forstliche Versuchsanstalt Baden-Württemberg at Freiburg, for analysing the element content of the needles. I further thank the Gesellschaft für Umweltmessungen und Umwelterhebungen mbH (UMEG), Karlsruhe, for making their ozone data available to me. Financial support of the Projekt Europäisches Forschungszentrum für Maßnahmen zur Luftreinhaltung (PEF), Karlsruhe, is greatfully acknowledged.

References

Anonymus: Monatswerte der Immissionskonzentrationen, LFU Baden-Württemberg 1985. In: Siefermann-Harms, D. and E. Kilz (eds.): Interdisziplinärer PEF-Forschungsschwerpunkt Kälbelescheuer/Südschwarzwald – Quelldaten 1985. Report Forschungszentrum Karlsruhe KfK-PEF *10,* 195–200 (1987).

Arndt, U.: The Hohenheim Long-term Experiment: Synoptic discussion of methods and results. Environ. Pollut. *68,* 435–451 (1990).

Arndt, U., W. Obländer, E. König, and W. Maier: Auftreten und Wirkung gasförmiger Luftverunreinigungen in Waldgebieten Baden-Württembergs. VDI-Ber. *500,* 249–256 (1983).

Baumbach, G. and K. Baumann: Schadgasaufkommen und -eintrag in Waldbestände. Selbstverlag des Institutes für Verfahrenstechnik und Dampfkesselwesen der Universität Stuttgart, Bericht Nr. 25–1992 (1992).

Evers, F. H.: Magnesiummangel, eine verbreitete Erscheinung in Waldbeständen – Symptome und analytische Schwellenwerte. Mitt. Ver. Forstl. Standortskunde u. Forstpflanzenzüchtung *37,* 7–16 (1994).

Godde, D.: Funktion und Reparatur des photosynthetischen Apparates in mangelernährten Fichten bei verschiedenen Lichtintensitäten – Molekularer Mechanismus der streßinduzierten Vergilbung. Report Forschungszentrum Karlsruhe FZKA-PEF (before 1995: KfK-PEF) *130,* 191–199 (1995).

Guderian, R. (ed.): Air pollution. Phytotoxicity of acidic gases and its significance in air pollution control (Ecological studies 22). Springer, Berlin–Heidelberg–New York (1977).

Hampp, R., W. Einig, and B. Egger: The Hohenheim Long-term Experiment: Energy and redox status, and carbon allocation in one- to three-year-old needles. Environ. Pollut. *68,* 305–318 (1990).

Köstner, B., F.-C. Czygan, and O. L. Lange: An analysis of needle yellowing in healthy and chlorotic Norway spruce (*Picea abies*) in a forest decline area of the Fichtelgebirge (N.E. Bavaria). I. Annual time-course changes in chloroplast pigments for five different needle age classes. Trees *4,* 55–67 (1990).

Kühlbrandt, W., D. N. Wang, and Y. Fujiyoshi: Atomic model of plant light-harvesting complex by electron crystallography. Nature *367,* 614–621 (1994).

Lange, O. L., R. M. Weikert, M. Wedler, J. Gebel, and U. Heber: Photosynthese und Nährstoffversorgung von Fichten aus einem Waldschadensgebiet auf basenarmem Untergrund. Allg. Forst Z. 3/1989, 55–64 (1989).

Lichtenthaler, H. K.: Chlorophylls and carotenoids, the pigments of the photosynthetic biomembranes. In: Douce, R. and L. Packer (eds.): Methods in Enzymology Vol. *148,* pp. 350–382, Academic Press, New York (1987).

Lichtenthaler, H. K. and A. R. Wellburn: Determination of total carotenoids and chlorophylls *a* and *b* of leaf extracts in different solvents. Biochem. Soc. Transactions *603,* 591–592 (1983).

Mehne-Jakobs, B.: Auswirkungen eines experimentell induzierten Magnesiummangels auf die Zusammensetzung der Chloroplastenpigmente bei Fichte (*Picea abies* (L.) Karst.). Allg. Forst- u. J.-Ztg. *165,* 221–227 (1994).

Mies, E. and H. W. Zöttl: Zeitliche Änderung der Chlorophyll- und Elementgehalte in den Nadeln eines gelb-chlorotischen Fichtenbestandes. Forstw. Cbl. *104,* 1–8 (1985).

Polle, A., M. Mössnang, A. v. Schönborn, R. Sladkovic, and H. Rennenberg: Field studies on Norway spruce trees at high altitudes. I. Mineral, pigment and soluble protein contents of needles as affected by climate and pollution. New Phytol. *121,* 89–99 (1992).

Seufert, G., V. Hoyer, H. Wöllmer, and U. Arndt: The Hohenheim Long-term Experiment: General Methods and Materials. Environ. Pollut. *68,* 253–273 (1990).

Siefermann-Harms, D.: Evidence for a heterogenous organization of violaxanthin in thylakoid membranes. Photochem. Photobiol. *40,* 205–229 (1984).

– Carotenoids in Photosynthesis. I. Location in photosynthetic membranes and light-harvesting function. Biochim. Biophys. Acta *811,* 325–355 (1985).

– High-performance liquid chromatography of chloroplast pigments. One-step separation of carotene and xanthophyll isomers, chlorophylls and pheophytins. J. Chromatogr. *448,* 411–416 (1988).

– Protective function of the apoprotein of the light-harvesting chlorophyll-*a/b*-protein complex in pigment photooxidation. J. Photochem. Photobiol. *4B,* 283–295 (1990 a).

– The Hohenheim Long-term Experiment: Chlorophyll, carotenoids and the activity of the xanthophyll cycle. Environ. Pollut. *68,* 293–303 (1990 b).

– Photooxidation and pheophytin formation of chlorophyll in the light-harvesting Chl-*a/b*-protein complex exposed to fatty acids: Protective role of the intact apoprotein. In: Baltscheffsky, M. (ed.): Current Research in Photosynthesis. Vol. *2,* pp. 245–248. Kluwer Academic Publ., Dordrecht–Boston–London (1990 c).

– The yellowing of spruce in polluted atmospheres. Photosynthetica *27,* 323–341 (1992).

– Light and temperature control of season-dependent changes in the α- and β-carotene content of spruce needles. J. Plant Physiol. *143,* 488–494 (1994).

– Untersuchungen zum Vergilbungsvorgang von Fichten am Standort Freudenstadt-Schöllkopf: Pigmentgehaltsänderungen, Lichtbedarf der Vergilbung und Vorgänge im Chlorophyll-Protein-Komplex LHC II. In: Bittlingmaier, L., W. Reinhardt, and D. Siefermann-Harms (eds.): Waldschäden im Schwarzwald. Ecomed, Landsberg am Lech (1995).

Siefermann-Harms, D. and A. Angerhofer: An O_2-barrier in the light-harvesting Chl-*a/b*-protein complex LHC II protects chlorophylls and carotenoids from photooxidation. In: Mathis, P. (ed.): Photosynthesis: from Light to Biosphere, Vol. *4,* pp. 71–74. Kluwer Academic Publishers, Dordrecht, The Netherlands (1995).

Siefermann-Harms, D. and H. Ninnemann: The separation of photochemically active PS-I and PS-II containing chlorophyll-protein-complexes by isoelectric focussing of bean thylakoids on polyacrylamide gel plates. FEBS Lett. *104,* 71–77 (1979).

– – The chlorophyll-protein complexes of flashed bean leaves. In: Akoyunoglou, G. (ed.): Photosynthesis, Vol. *3,* pp. 635–645. Balaban Internatl. Sciences Services, Philadelphia, Pa. (1981).

– – Pigment organization in the light-harvesting chlorophyll-*a/b*-protein complex of lettuce chloroplasts. Evidence obtained from protection of the chlorophylls against proton attack and from excitation energy transfer. Photochem. Photobiol. *35,* 719–731 (1982).

– – Differences in acid stability of the chlorophyll-protein complexes in intact thylakoids. Photobiochem. Photobiophys. *6,* 85–91 (1983).

– – Chlorophyll- und Carotinoidgehalt der Nadeln und Verhalten des Lichtsammelnden Chlorophyll-*a/b*-Proteins in saurem Milieu. In: Siefermann-Hamrs, D. and E. Kilz (eds.): Interdisziplinärer PEF-Forschungsschwerpunkt Kälbelescheuer/Südschwarzwald – Quelldaten 1985. Report Forschungszentrum Karlsruhe KfK-PEF *10,* 65–71 (1987).

Siefermann-Harms, D. and J. Steegborn: Physiologische und biochemische Untersuchungen zur Montanen Vergilbung von Fichten. – Teil V: Wirkungen von kombiniertem Standortstreß und von einzelnen Streßfaktoren auf Bestandteile des Photosyntheseapparates. Report Forschungszentrum Karlsruhe KfK-PEF *94,* 21–34 (1992).

Siefermann-Harms, D., Ch. Weinmann, Th. Hofer, C. Boxler-Baldoma, H.-G. Heumann, and A. Seidel: Weitere Untersuchungen zum Vergilbungsvorgang – keine kurzfristige Ergrünung nach Mg-Sulfat Düngung. Report FZKA-PEF Forschungszentrum Karlsruhe (before 1995: KfK-PEF) *130,* 175–189 (1995 a).

Siefermann-Harms, D. and 32 co-authors: Ergebnisse und Folgerungen der interdisziplinären Studie zur Vergilbung von Fichten in montanen Lagen des Schwarzwaldes (Freudenstadt–Schöllkopf 1989–1991). In: Bittlingmaier, L., W. Reinhardt, and D. Siefermann-Harms (eds.): Waldschäden im Schwarzwald. Ecomed, Landsberg am Lech (1995 b).

Steegborn, J.: Einfluß von Magnesium auf den Turnover von Pigment-Protein-Komplexen in Fichtenkeimlingen aus definierten Anzuchtbedingungen. PhD Thesis, University of Freiburg (1991).

Wentzel, K. F.: Smogbelastung im Wald am höchsten. Hessischer Waldbesitzerverband *33,* 33–34 (1985).

Zech, W. and E. Popp: Magnesiummangel, einer der Gründe für das Fichten- und Tannensterben in NO-Bayern. Forstw. Cbl. *102,* 50–55 (1983).

J. Plant Physiol. Vol. 148. pp. 203–206 (1996)

Effects of Ozone and Mild Drought Stress on Total and Apoplastic Guaiacol Peroxidase and Lipid Peroxidation in Current-Year Needles of Young Norway Spruce (*Picea abies* L., Karst.)

G. KRONFUSS[1], G. WIESER[2], W. M. HAVRANEK[2], and A. POLLE[3]

[1] Universität Innsbruck, Institut für Botanik, Sternwartestraße 15, A-6020 Innsbruck, Austria

[2] Forstliche Bundesversuchsanstalt, Institut 6, Abteilung Baumphysiologie, Rennweg 1, A-6020 Innsbruck, Austria

[3] Albert-Ludwigs-Universität Freiburg, Institut für Forstbotanik und Baumphysiologie, Professur für Baumphysiologie, Am Flughafen 17, D-79085 Freiburg, Germany

Received July 10, 1995 · Accepted September 30, 1995

Summary

Drought stress has been shown to decrease O_3-uptake and is thought to mitigate O_3 stress in trees. To test this hypothesis we investigated the effects of ozone fumigation (100 nL/L, 17 weeks) singly and in combination with mild drought stress on guaiacol peroxidase (POD) activity (EC 1.11.1.7) and lipid peroxidation in Norway spruce needles (*Picea abies* L.). After 10 weeks of fumigation a subset of the trees was subjected to mild drought. Activity of apoplastic guaiacol peroxidase was decreased by ozone in needles of well-watered trees. Mild drought caused a reduction in cumulative ozone-uptake by about 17% as compared to well-watered trees. In spite of this reduction in ozone-uptake, activity of apoplastic guaiacol peroxidase was also low in drought-stressed ozone-fumigated trees. Ozone exposure caused a slight increase of total guaiacol peroxidase activity whereas mild drought led to a decrease that was more pronounced in combination with ozone. We found enhanced contents of malondialdehyde, an indicator of lipid peroxidation, in drought subjected trees but not as a consequence of ozone exposure.

Key words: Picea abies, apoplast, cumulative ozone-uptake, drought stress, guaiacol peroxidase, malondialdehyde.

Abbreviations: EWF = extracellular washing fluid; CF = ozone free control; CF+D = control with mild drought stress; O_3 = ozone fumigated; O_3+D = ozone-fumigated and drought-stressed trees; POD = guaiacol peroxidase.

Introduction

The extracellular matrix is the first compartment of the needles ozone reaches after passing through the stomata. Ozone is solubilized in the aqueous phase of the extracellular matrix, generating harmful decomposition products such as superoxide anion, hydroxyl radicals and hydrogen peroxide (Hewitt et al., 1990). If ozone or secondary products can react with the sensitive plasmamembrane damage through lipid peroxidation will occur (Sakaki et al., 1983; Price et al., 1990). Therefore, a protective mechanism against activated oxygen must exist in the apoplastic space to protect the plasmamembrane from injury. Guaiacol peroxidase (POD) can reduce hydrogen peroxide by using low-molecular phenolic compounds like guaiacol or coniferyl alcohol as cosubstrates (Gaspar et al., 1982; Polle et al., 1994). It is supposed that peroxidases play a role in the defence against hydrogen peroxide (Matschke et al., 1988; Polle and Junkermann, 1993) and, therefore, may contribute to the protection against ozone mediated injury. Several short-term fumigation experiments with high ozone concentrations were performed to investigate mechanism of acute damage and defence reactions in herbaceous plants (Castillo et al., 1984; Peters et al., 1989) and in Norway spruce (Castillo et al., 1987). A general conclusion from these studies was, that POD activity increased by ozone and that the observed increase was more pronounced in the apoplastic than in the symplastic space. Long-term fumigation experiments, representing a chronic exposure to activated oxygen, resulted also in an increase of total POD activity in the leaves from herbaceous plants (Bender et al., 1990; Fangmeier et al., 1994) and in the needles of loblolly pine (Manderscheid et al., 1991). In young needles of Norway spruce, however, a decrease in the apoplastic POD was reported by Ogier et al. (1991). To our knowledge no study has been dealing with a combined stress of drought and ozone on Norway spruce. The aim of our study was to examine effects of moderate long-term ozone exposure and mild drought on POD under particular consideration of the apoplastic space. A further objective was to investigate the often discussed relation between ozone-uptake and ozone effects in plants. Atmospheric drought conditions (Wieser and Havranek, 1993) and soil water deficits cause stomatal narrowing and, therefore, are the major factors limiting ozone-uptake. We measured ozone-uptake and examined if mild drought can reduce ozone effects on POD or if interacting effects occur between these two stress factors. In addition we used analysis of malondialdehyde as indicator of membrane damage.

Materials and Methods

Experimental conditions

Potted four-year-old Norway spruce seedlings (*Picea abies*) were obtained from the Landesforstpflanzgarten Stams (Tyrol, Austria). The provenance used was from middle altitude (1450 m). The trees were continuously exposed to charcoal-filtered ozone free air (CF) or to air with 100 nL/L ozone for 17 weeks from 26th June to 28th October in two climatised fumigation chambers. Each chamber contained 20 trees. The photoperiod was 14 hours and the average photon flux density was about 360 $\mu mol\ m^{-2}\ s^{-1}$ at the height from which the samples were harvested. Day and night temperatures were kept at 20 °C and 15 °C, respectively. The dew point was 13 °C at both day and night. After 10 weeks a subset of the trees was subjected to soil-drying by withholding water for further 38 days. Samples were taken at the end of the experiment in October.

Calculation of cumulative ozone-uptake

We calculated the ozone-uptake rate into needles as described previously (Wieser and Havranek, 1995) by using porometer measurements of the leaf conductance. 6 trees per day were measured at 10.30–11.00 and 22.30–23.00. Cumulative O_3-uptake was calculated as the uptake rate integrated over the measurement period.

Preparation of EWF and total needle extracts

Extracellular washing fluids (EWF) of the needles were obtained using a vacuum infiltration method as described by Polle et al. (1990). Needles were cut from the twigs and needle tips were removed. After washing the needles with distilled water they were vacuum infiltrated in 50 mL of a buffer containing 50 mM MES, 40mM KCl and 2 mM $CaCl_2$ (KOH pH 6). Subsequently, the needles were put into a centrifuge tube with a perforated bottom, placed over an Eppendorf tube and centrifuged for 15 min at 800 g. The collected EWFs were frozen in liquid nitrogen and stored at −80 °C until further analysis.

To extract enzymes from total needles, 100 mg of the samples were ground in liquid nitrogen and added to 10 mL 0.1 M phosphate buffer (KH_2PO_4/K_2HPO_4 (pH 7.8), 0.5 % Triton X-100, 100 mg PVP). The mixture was centrifuged for 20 min at 20,000 g. The supernatant was removed from the pellet and centrifuged again for 20 min at 20,000 g. Before the measurement of enzyme activities, EWFs and total needle extracts were desalted over Sephadex G-25 columns (PD-10 or NAP-5, Pharmacia). For the equilibration of the columns phosphate buffer (0.1 M KH_2PO_4/K_2HPO_4, pH 7.8) and 50 mM MES (KOH pH 6) was used for total needle extracts and EWFs, respectively.

Measurements of peroxidase activity

50 µL of different dilutions of the enzyme extract were added to an assay containing 500 µL 0.1 M phosphate buffer (KH_2PO_4/K_2HPO_4 pH 5.25) and 400 µL 0.1 M guaiacol (Polle et al., 1990). The reaction was started by addition of 50 µL 0.2 M H_2O_2. Changes in absorbtion were recorded at a wavelength of 436 nm. POD activities were calculated using an extinction coefficient of 25.5 $mM^{-1}\ cm^{-1}$ and a stoichiometry of 4.

Determination of lipid peroxidation

Malondialdehyde was determined according to Peever and Higgins (1989). Aliquots of the desalted enzyme extracts from total needles from 300 to 500 µL were filled up to a volume of 1 mL with a solution containing 20 % (v/v) trichloroacetic acid and 0.5 % (v/v) thiobarbituric acid. The mixture was incubated for 30 min at 95 °C. After centrifugation, thiobarbituric acid reaction products were determined from the difference in absorbance at 532 and 600 nm, using an extinction coefficient of 1.5 $M^{-1}\ cm^{-1}$.

Results

Mild drought stress caused stomatal narrowing and a decrease in the predawn needle water potential to a minimum of −0.5 MPa in needles of CF and O_3 treated trees after 38 days (results not shown). The reduction in stomatal conductance diminished cumulative ozone-uptake by about 17 % at the end of the drought period as compared with well-watered trees (Fig. 1).

In both ozone treatments apoplastic POD activity was reduced as compared to CF controls (Fig. 2, A). However, on the basis of the 5 samples analysed, the differences were not significant.

POD of total needle extracts (Fig. 2, B) tended to increase by O_3 in well watered trees. Drought alone caused a decline in POD activity which was much more pronounced in combination with ozone. Similar results were obtained when POD activity was based on protein content (data not shown).

The amount of malondialdehyde (Fig. 3) in the needles was increased by drought but not by ozone.

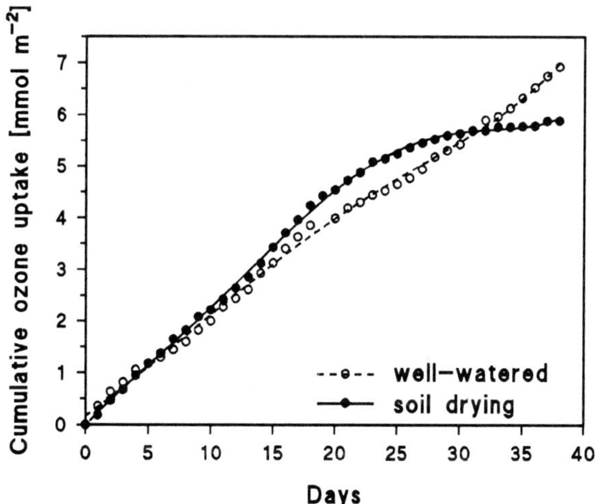

Fig. 1: Effect of drought on cumulative ozone-uptake into spruce needles during 38 days of soil drying and fumigation with 100 nL L^{-1} O_3. One curve represents average cumulative O_3-uptake of 10 trees. Legend: (○) = well watered and (●) = drought stressed trees.

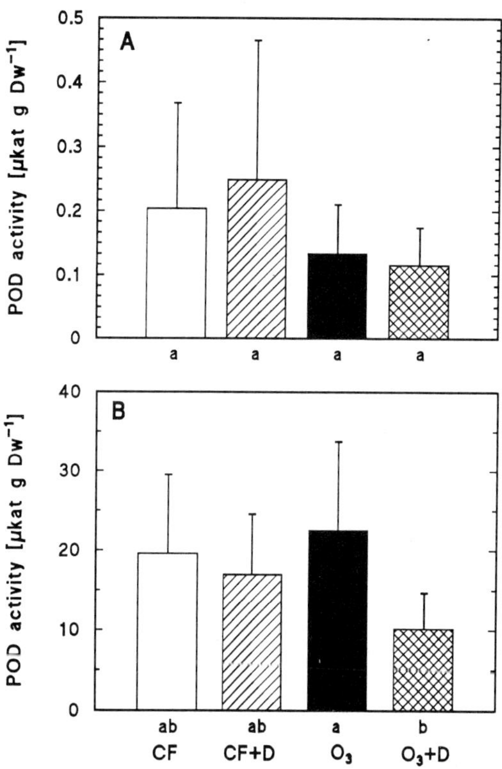

Fig. 2: Guaiacol peroxidase activity from (A) EWF (n = 5) and (B) total needle extracts (n = 7) of Norway spruce after different treatments (means ± SD). Legend: open bars (CF) represent means of ozone-free controls, hatched bars (CF+D) drought subjected ozone-free controls, solid bars (O_3) ozone-exposed and squared bars (O_3+D) ozone and drought stressed trees. Different letters denote significant differences at $p \leq 0.05$ (t-test). Dw = dry weight.

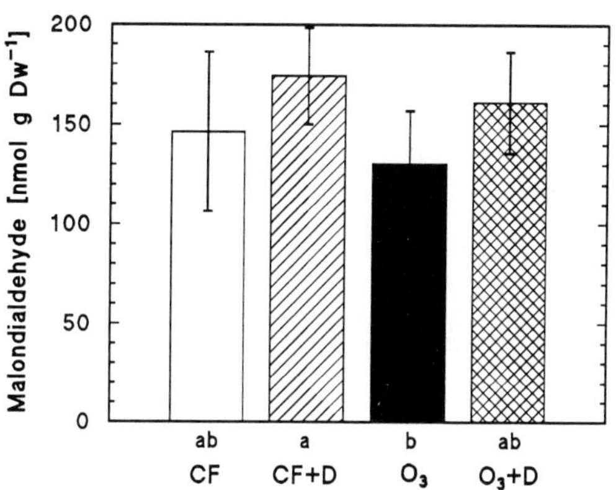

Fig. 3: Amounts of malondialdehyde in needle extracts of Norway spruce (n = 7) after different treatments (means ± SD). For further details see legend to Figure 2.

Discussion

Ozone fumigation affected POD differently in the needle apoplast and in the symplast. We observed that POD activity in the needle apoplast was reduced due to ozone fumigation (Fig. 2). A similar effect was described by Ogier et al. (1991) who observed a drastic reduction of apoplastic POD activity after 9 week of ozone (100 nL/L) exposure of Norway spruce. It is known that ozone and ozone reaction products can destroy proteins and consequently also enzymes. Therefore, the observed decline in apoplastic POD activity by ozone may support the idea (Polle and Rennenberg, 1993) that this enzyme was injured due to ozone exposure. In contrast, in the symplast we hardly observed an effect caused by ozone alone.

Although mild drought led to stomatal narrowing and thereby clearly reduced ozone-uptake apoplastic POD activity of trees subjected to ozone plus drought remained low. If POD is really important in the defence against ozone a decline in apoplastic POD activity will mean a loss in detoxification capacity. However, ozone exposure did not cause an enhanced lipid peroxidation within the needles. These results suggest that probably also other more efficient pathways exist to protect the plasmamembrane from ozone mediated injury and that it is not a main function of POD to protect from ozone damage.

We found enhanced lipid peroxidation due to mild drought stress in both, ozone free and ozone fumigated trees. Thus, mild drought affected membranes more than ozone. Total POD activity seemed to decrease due to mild drought in the needles of CF trees, and combined treatment with drought and ozone caused a significant reduction in POD activity. Thus, it seems that total POD has no direct function in protection from active oxygen species which possibly occur during drought stress. We found no explanation for the significant decrease of total POD in trees treated with ozone plus drought. However, it seems possible that a strong disturbance of the normal POD activity may release several metabolic processes which may be important in the defence against oxidative stress. Thereby, POD may act like a stress indicator for the plant itself.

Acknowledgements

We wish to thank Thomas Gigele for excellent technical assistance, Mr. Braunegger (Landesforstpflanzgarten, Stams) and all researchers who work in the «Institut für Baumphysiologie» in Freiburg for their helpfulness.

References

Bender, J., H. J. Weigel, and H. J. Jäger: Regression analysis to describe yield and metabolic responses of beans (*Phaseolus vulgaris*) to chronic ozone stress. Angew. Botanik *64,* 329–343 (1990).

Castillo, F. J., P. R. Miller, and H. Greppin: «Waldsterben» Extracellular biochemical markers of photochemical oxidant air pollution damage to Norway spruce. Experientia *43,* 111–115 (1987).

Castillo, F. J., C. Penel, and H. Greppin: Peroxidase release induced by ozone in *Sedum album* leaves. Plant Physiol. *74,* 846–851 (1984).

Fangmeier, A., S. Brunschön, and H. J. Jäger: Time course of oxidant stress biomarkers in flag leaves of wheat exposed to ozone and drought stress. New Phytol. *126,* 63–69 (1994).

Gaspar, T., C. Penel, T. Thorpe, and H. Greppin: Peroxidases 1970–1980. A survey of their biochemical and physiological roles in higher plants. In: Lobarzewski, J., H. Greppin, C. Penel, and T. Gaspar (eds.): pp. 124–129. University Press, Switzerland (1982).

Hewitt, C. N., G. L. Kok, and R. Fall: Hydroperoxides in plants exposed to ozone mediate air pollution damage to alkene emitters. Nature *344,* 56–58 (1990).

Manderscheid, R., H. J. Jäger, and M. M. Schöneberger: Dose-response relationships of ozone effects on foliar levels of antioxidants, soluble polyamines and peroxidase activity of *Pinus taeda* (L.): Assessment of the usefulness as early ozone indicators. Angew. Botanik *65,* 373–386 (1991).

Matschke, J., H. Hertel, C. Ewald, and E. Nöring: Vitalitätsbestimmung immissionsbeeinflußter Koniferen durch die Analyse von Defensivenzymen. Beitr. Forstwissenschaft *22,* 3, 125–133 (1988).

Ogier, G., H. Greppin, and F. Castillo: Ascorbate and guaiacol peroxidase capacities from apoplastic and cell material extracts of Norway spruce needles after long-term ozone exposure. In: Lobarzewski, J., H. Greppin, C. Penel, and T. Gaspar (eds.): Biochemical, molecular, and physiological aspects of plant peroxidases. pp. 391–400. University Press, Geneva, Switzerland (1991).

Peever, T. L. and V. J. Higgins: Electrolyte leakage, lipoxygenase, and lipid peroxidation induced in tomato leaf tissue by specific and nonspecific elicitors from *Cladosporium fluvum.* Plant Physiol. *90,* 867–875 (1989).

Peters, J. L., F. J. Castillo, and R. L. Heath: Alteration of extracellular enzymes in pinto bean leaves upon exposure to air pollutants, ozone and sulphur dioxide. Plant Physiol. *89,* 159–164 (1989).

Polle, A., K. Chakrabarti, W. Schürmann, and H. Rennenberg: Composition and properties of hydrogen peroxide decomposing systems in extracellular and total extracts from needles of Norway spruce. Plant Physiol. *94,* 312–319 (1990).

Polle, A. and W. Junkermann: Does atmospheric hydrogen peroxide contribute to damage to forest trees? Environ. Sci. Tecnol. *28,* 812–815 (1994).

Polle, A., T. Otter, and F. Seifert: Apoplastic peroxidases and lignification in needles of Norway spruce (*Picea abies* L.). Plant Physiol. *106,* 53–60 (1994).

Polle, A. and H. Rennenberg: Significance of antioxidants in plant adaptation to environmental stress. Fowden, L., T. Mansfield, and J. Stoddart (eds.): Plant adaptation to environmental stress. pp. 263–273. Chapman and Hall (1993).

Price, A., P. W. Lucas, and P. J. Lea: Age dependent damage and glutathione metabolism in ozone fumigated barley. Journal of Experimental Botany *41,* 1309–1317 (1990).

Sakaki, T., N. Kondo, and K. Sugahara: Break down of photosynthetic pigments and lipids in spinach leaves with ozone fumigation: Role of active oxygen. Physiol Plant. *59,* 28–34 (1983).

Wieser, G. and W. M. Havranek: Ozone uptake in the sun and shade crown of spruce: Quantifying the physiological effects of ozone exposure. Trees *7,* 227–232 (1993).

– – Environmental control of ozone uptake in *Larix decidua* Mill.: a comparison between different altitudes. Tree Physiology *15,* 253–258 (1995).

J. Plant Physiol. Vol. 148. pp. 207–214 (1996)

Different Responses to Ozone of Tobacco, Poplar, Birch, and Alder

MADELEINE S. GÜNTHARDT-GOERG

Swiss Federal Institute for Forest, Snow and Landscape Research, Zürcherstrasse 111, CH-8903 Birmensdorf, Switzerland

Received July 19, 1995 · Accepted October 10, 1995

Summary

Plants of an ozone sensitive tobacco clone (*Nicotiana tabacum* L. var. BelW3), cuttings of poplar (*Populus × euramericana* var. Dorskamp) and birch (*Betula pendula* Roth), and seedlings of alder (*Alnus glutinosa* (L.) Gaertn.) were grown in the field or in the field fumigation chambers during one growing season. Twenty chambers (each with one plant per species) were used for four treatments with either filtered air, or filtered air with added ozone (75 ppb) from 07:00 to 19:00, or from 19:00 to 07:00, or continuously.

Tobacco did not respond to ozone applied during the nighttime, but leaf injury symptoms of plants grown in the field under shade appeared at an ozone dose similar to that applied during the daytime. The leaves of the deciduous trees showed injury symptoms in all ozone treatments. In the tree species the stomatal pores were open at 06:00 (closed in tobacco), but they were narrowed at 10:00 under ozone as compared to filtered air. The time span until ozone-induced leaf injury symptoms or premature leaf loss occurred, was influenced by the species, the season, and the ozone regime. Similarities and differences, in particular in the starch metabolism or the formation of cell wall exudates, could be traced from early symtoms to cell collapse in all species.

In contrast to agricultural plants, ozone impact on the leaves of deciduous trees is as important during dusk, night and dawn as during daytime.

Key words: Alnus glutinosa, Betula pendula, Nicotiana tabacum, Populus × euramericana, microscopy, nighttime exposure, ozone, phenology.

Introduction

For European countries the phytotoxicity of ozone causes the major crop losses attributable to air pollution (UK PORG 1993). Data for yield-reduction in wheat related to exposure (Fuhrer 1994) were used to establish recommendations for critical ozone levels. An AOT40-value (accumulated exposure over a threshold of 40 ppb, calculated for daylight-hours and three months) of 5.3 ppm-h was defined for crops in the ‹workshop on critical levels› (Ashmore and Fuhrer 1994) in relation to a 10 % yield reduction. For forest trees, 10 ppm-h for 24 hours a day during a 6-month period were very provisionally proposed to cover the highest sensitivity of the receptor, because it was based on a very small data base (Matyssek et al. 1992, Braun and Flückiger 1994, Küppers et al. 1994). On sunny days the 40 ppb threshold was surpassed e.g. on a site in the Swiss Mittelland during 11 hours/day, at an alpine site during 9.5 hours (Fig. 1). To reach the critical level for crops (5.3 ppm-h) or for forest trees (10 ppm-h), 30 or 57 such sunny days would be enough on the Mittelland site, and 54 or 102 days on the alpine site. Although ozone concentration during nighttime is higher on the alpine than on the Mittelland site, it rarely surpasses 40 ppb and is similar to that on cloudy days (Fig. 1). This experiment was therefore designed to establish the relative ozone sensitivity of a particularly ozone sensitive tobacco clone compared to three deciduous tree species in relation to the daytime and the season of the exposure to ozone. Visual symptoms and cell injury

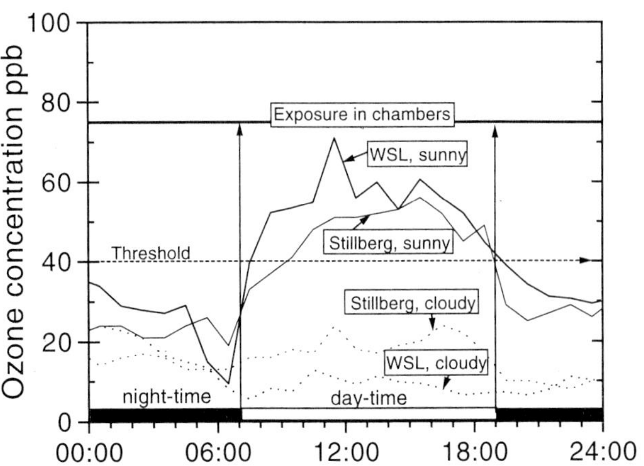

Fig. 1: Daily time course of ozone concentrations (hourly means) at a Swiss Mittelland (WSL, Birmensdorf) and an alpine site (Stillberg, Davos), each on a sunny or a cloudy day relative to the concentration in the exposure chambers and to the AOT40 (accumulated exposure over a threshold of 40 ppb).

were examined with respect to further experiments in the field, where yield and particularly underground biomass is difficult to quantify.

Materials and Methods

Tobacco (*Nicotiana tabacum* L. var. BelW3, which is particularly ozone (O_3) sensitive) was grown from seeds and exposed in the Birmensdorf field fumigation chambers from June 25 until August 28. Poplar (hybrid *Populus* × *euramericana* var. ‹Dorskamp›, originally from J. Mooi, Wageningen, IPO The Netherlands) was grown from cuttings and exposed from May 17 until July 13. Cuttings from local birch (*Betula pendula* Roth, clone WSL 31-2) and seedlings from local alder (*Alnus glutinosa* (L.) Gaertn.) were exposed from May 17 until October 1. The plants were harvested at a height of over 2 m. Twenty chambers (one plant/10-L pot, four species/chamber) were used for four treatments with either charcoal-filtered air (ozone concentration <3 ppb) or filtered air with added O_3 (75 ppb) from 07:00 to 19:00 (daytime, D), or from 19:00 to 07:00 (nighttime, N, including dawn and dusk), or continuously (ND, relation to ambient O_3 concentration see Fig. 1). The plants were well watered and supplied with nutrients. O_3 was generated from pure oxygen and continuously monitored with a Monitor labs 8810 instrument in the chambers and in the field. A shadow roof covered the open-top chambers whenever the photon flux density surpassed 600 µmol $m^{-2} s^{-1}$ to prevent over-heating in the chambers. Five tobacco and birch plants each were exposed in the field near the fumigation chambers in such shaded conditions, and five plants per species in a sunny place.

The leaves were numbered, classified by visual symptoms (Günthardt-Goerg et al. 1993) and examined for the time-span until ozone-induced early visual symptoms appeared (white dots in tobacco, light-green leaf tip or leaf rim in poplar, light-green dots or bronze discoloration in birch, and brownish discoloration between 2nd order veins in alder), or established symptoms (necrotic points in all species), or premature leaf loss.

For microscopy, discs of 8 mm diameter were excised (10:00–11:00) between 2nd-order veins (in the middle of the right and left halves of the lamina) in leaves selected by their phenology (same visual injury class) from all species and regimes. Discs excised were immediately immersed in methanol (bleaching and conservation, several changes) and later dyed with KI-I (2 g KI and 1 g I in 100 mL distilled water) to determine the starch patterns. Other discs were excised and immersed in 2.5 % phosphate buffered glutaraldehyde, pH 7.4, evacuated, cut into strips within the fixation medium, and stored at 4 °C. For light microscopy (LM, Leica microscope Leitz DM/RB, with micrographsystem Wild MPS 48/52). The specimens were dehydrated (Feder and O'Brien 1968) and embedded in Technovit 7100 (Kulzer Histo-Technik). Sections of 2.5 µm were stained for pectins with Ruthenium Red (Gerlach 1984), or Coriphosphine (Weis et al. 1988), or with Aniline Blue (Gerlach 1984) for callose.

For transmission electron microscopy (TEM) one mm^2 pieces from fixed birch leaf strips were rinsed twice in phosphate buffer and shipped submerged in buffer to Delaware OH (USA) Forest Service Laboratory for further processing. For low temperature scanning electron microscopy (LTSEM) leaf discs were excised to a drop of 1 % methylcellulose in water on aluminium stubs, immediately plunge-frozen in and stored under liquid nitrogen (at −96 °C) until further processing in the WSL SEM-laboratory (methods see Scheidegger et al. 1991, Matyssek et al. 1991, Günthardt-Goerg et al. 1993). Because the number of leaves formed on the main stems did not differ between the treatments (Fig. 4), discs for observation of the stomatal aperture were taken from the 13th leaf (age 25 d) from the top of the poplar trees on 10:00, July 9, and from the 6th (6:00) and 7th (10:00, leaf age 20 d) leaf on September 26 from the alder trees. The different fumigation treatments were sampled against each other. Stomatal aperture was measured in SEM micrographs and related to the length of the guard cells to match the different cell size within the four species. In tobacco leaves the pore width of the guard cells was shaded by overarching cuticular ledges. This complication, however, was irrelevant, because the micrograph information

Fig. 2: A: Tobacco, left leaf exposed in the field in a sunny, right leaf with early O_3 symptoms (arrow) in a shaded place. B: Left poplar leaf grown under filtered-air (control), right with necrotic spots (arrow) under the daytime O_3 treatment. C: Birch, left control, right leaf with light-green dots (dashed arrow) and necrotic spots (arrow) grown under nighttime O_3. D: Alder, left control, right leaf grown under nighttime O_3, leaf with bronze necrotic dots (arrow) between the 2nd order leaf veins. E–H: Left leaf = filtered-air control treatment, right = leaf from O_3 regime. Magnification bar for E = 52 µm, for F, G, and H = 8 µm. E: Starch patterns in an alder leaf lower surface; left control leaf with even starch distribution in the mesophyll cells (m), few starch grains in stomata (st), right leaf without visual injury symptoms but groups of mesophyll cells without starch (m), increased starch grains in the stomata (st) and in bundle sheath cells along veinlets (v). F: Poplar leaf cross sections, stained with Aniline Blue for callose and viewed under ultraviolet light; left control with starch grains, but hardly visible cell walls, right leaf with visual injury symptoms (as Fig. 2 B) with less starch, but cell walls with exudates (arrow), which did not show the intensive white/yellow colour of callose. G: The same poplar leaves, but stained with Ruthenium Red for pectins (arrow = cell wall exudates). H: Mesophyll cells in birch leaf cross sections stained with Coriphosphine for pectin-specific fluorescence at 630 nm (here shown for better contrast with excitation 450–490 nm), starch grains in the left micrograph appear greenish yellow, cell wall exudates in the right micrograph are indicated by arrows.

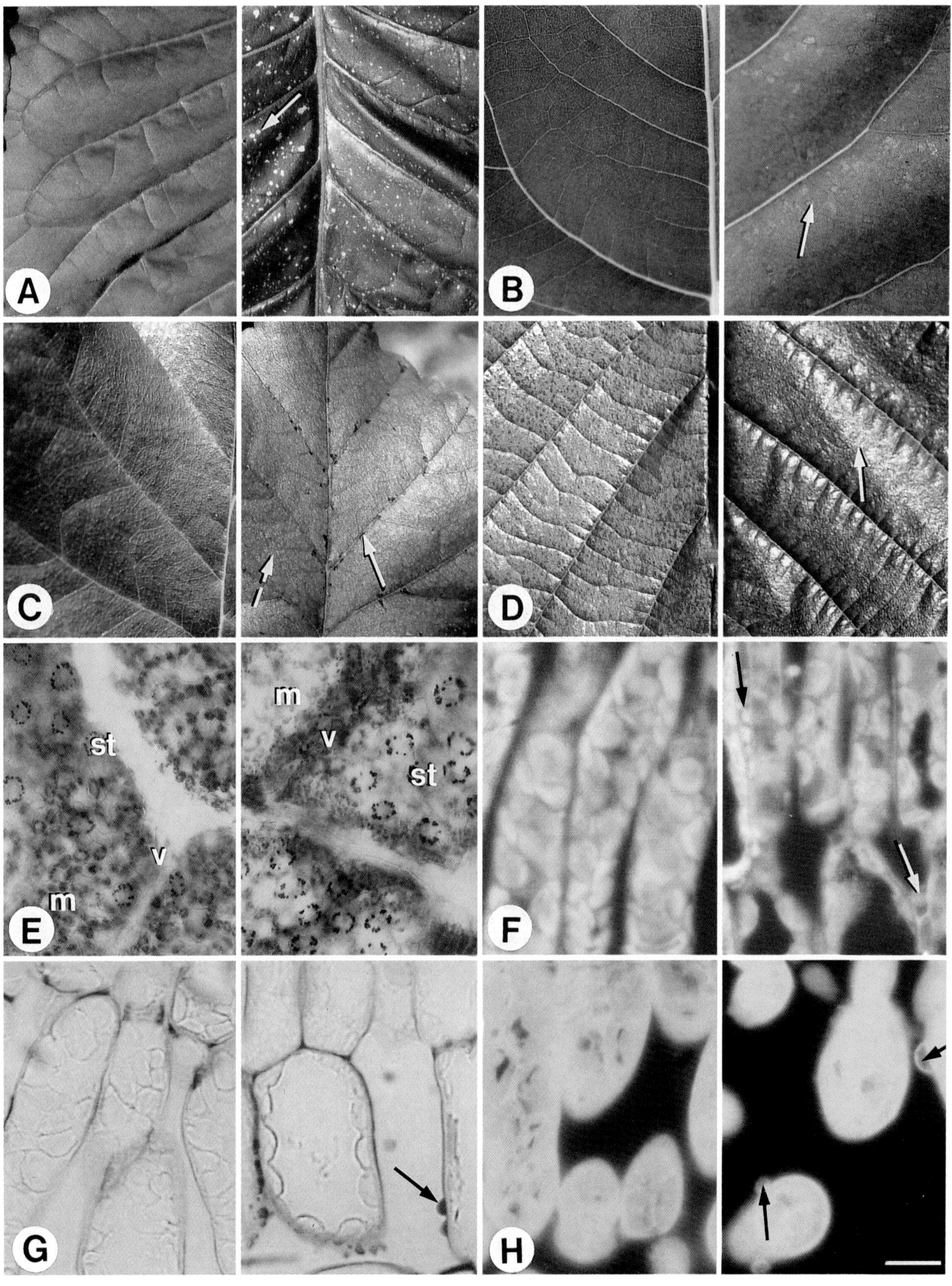
A
B
C
D
st
v
m
m
v
st
E
F
G
H

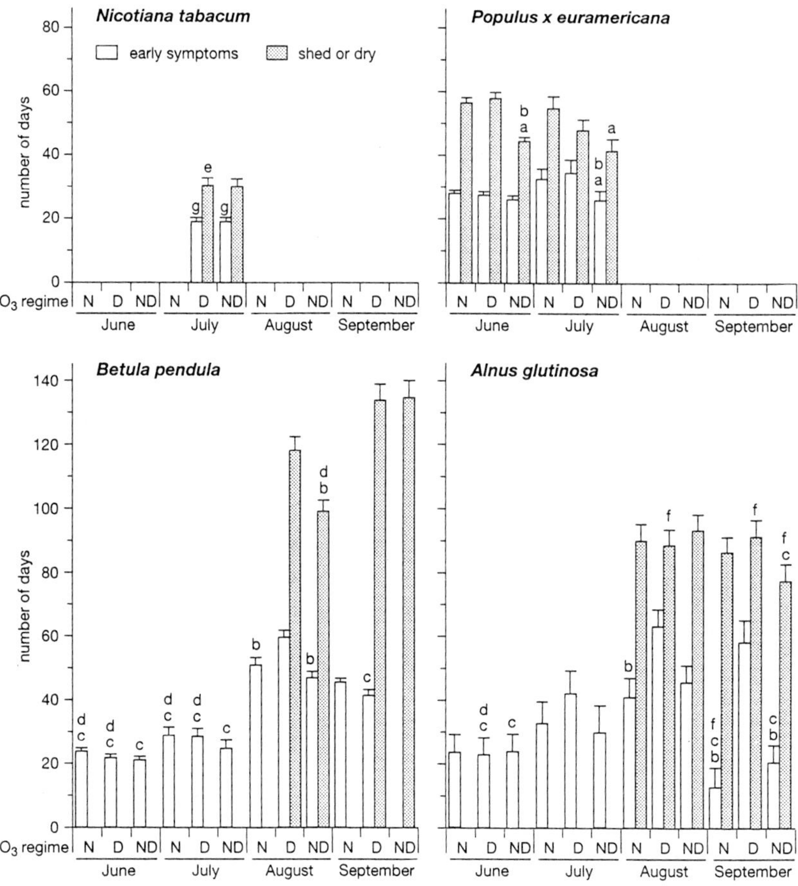

Fig. 3: Number of days until O_3-induced early visual leaf symptoms occurred, or the leaves were shed (= totally dry in tobacco) in relation to the fumigation regime, and the season (June–September) for the four species. Fumigation regimes: N = filtered air with added O_3 (75 ppb) during nighttime from 07:00 to 19:00, D = added O_3 during daytime from 19:00 to 07:00, ND = O_3 added continuously. Lower case characters above the columns (mean values with standard error bars) indicate that according to ANOVA (LSD 99 %) these means are significantly smaller than corresponding means of:

- the same month and species, but in another fumigation regime: a = nighttime O_3 fumigation regime, b = daytime
- the same regime and species, but in another month: c = August, d = September
- the same regime and month, but in another species: e = poplar, f = birch, g = alder.

of closed stomata at 06:00, but opened at 10:00 was sufficient. In alder the pore width was also sometimes narrowed compared to that of the cuticular ledges, whereas both, the width of guard cells and that of the cuticular ledges coincided in poplar. Aperture width of the cuticular ledges was selected as a common parameter for alder and poplar stomata as shown in Fig. 6.

Results

Tobacco did not respond to ozone (O_3) applied during nighttime (N). Early leaf injury symptoms (white dots, photograph Fig. 2 A) appeared after the same time-span in the daytime (D) as in the continuous (ND) exposure to O_3 (Fig. 3). In the leaves of the deciduous trees (Fig. 3) early injury symptoms had developed after a similar time span in June as in July, without significant differences between the exposure to O_3 at nighttime-, daytime-, or continuously (except poplar with enhanced appearance under continuous O_3 in July). In August (particularly when O_3 was applied during daytime) birch and alder were less sensitive in terms of days until early visual leaf symptoms appeared, compared to June or July. In September newly formed alder leaves were most sensitive when exposed to O_3 during nighttime.

The same visual symptoms developed under the different fumigation regimes, namely light-green leaf colour in poplar, light-green dots in birch, and bronze discoloration between 2nd order veins in alder. These early symptoms later became necrotic spots in all species (photograph Fig. 2 B–D). Poplar leaves in June and July, and birch leaves in August, were shed earlier when the trees were continuously exposed to O_3 than when exposed during day- or nighttime (Fig. 3). However, birch exposed during nighttime did not shed the leaves with necrotic spots until November in contrast to the exposure during daytime or continuously (Fig. 3).

Tobacco and poplar exposed in the field in a sunny place did not develop visual leaf symptoms, in contrast to birch and alder, which after 51 (birch) or 54 (alder) days, had developed similar early injury symptoms as those in the fumigation chambers. Under shade birch leaf symptoms appeared after 39 days, tobacco leaf symptoms after 34 days. The life span of the leaves grown in the field (days until shed) lasted until autumnal leaf fall.

The daytime of O_3 application had no significant influence on the total number of leaves formed the end of the vegetation period, but a small influence on the phenology of the whole tree foliage (Fig. 4). Fewer leaves were shed at the end of the vegetation period in the nighttime, compared to the daytime and the continuous regime, whereas few green leaves were left in the continuous regime. The three tree species followed different strategies to cope with leaf injury. Poplars coped with the prominent leaf loss through a fast growth of the terminal shoot (reason for removing the trees from the chambers in July, no side twigs) and by this turnover had

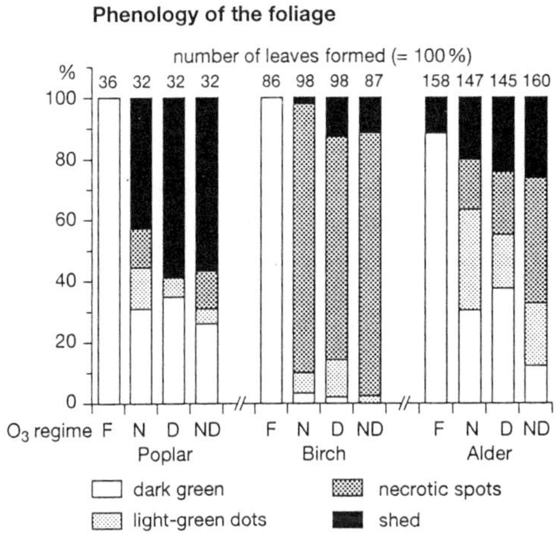

Fig. 4: Number of leaves formed and phenology of the foliage on (main + side) twigs of poplar (exposed 57 days), birch and alder (exposed 125 days). O_3 fumigation regimes as Fig. 3 with F = filtered-air control.

only a small number of injured leaves. Birch foliage remained longer on the twigs and thus was mainly constituted of leaves with necrotic spots. Alder was intermediate.

Corresponding to the occurrence of visual O_3 injury symptoms, stomatal pores were closed at 06:00 in tobacco leaves, but not in the leaves of the tree species (Fig. 5 A, B and Fig. 6). In a snap-shot of the pore apertures by LTSEM (measured in micrographs) the aperture in alder leaves did not significantly differ between the O_3 regimes at 06:00 a.m. The stomatal pores were widened at 10:00 (compared to 06:00) in the leaves from the filtered-air control, but did not change in the fumigated treatments. O_3 added only during daytime may have induced the further narrowing of the stomatal pores in poplar leaves at 10:00.

The starch content in the guard cells of all species increased with narrowing of the pore and thus with the applied ozone dose. Less starch was found in the leaf mesophyll cells within O_3-induced light-green dots or discoloration. In addition to the decreased starch formation, the starch remained accumulated in the bundle sheath cells of veinlets (particularly in birch, but less conspicuous in tobacco and alder, Fig. 2 E). With increasing visual O_3 symptoms amylopectin granules were formed in epidermal cells of tobacco, poplar, and birch leaves, but were not observed in alder and were missing in the control leaves under filtered air.

In mesophyll cells with reduced starch granules, adjacent to collapsed cells or in leaves with beginning necrotic spots (visible as tiny black points in a dissecting microscope), the cell walls thickened and formed exudates, where they faced the intercellular space. These exudates had an irregular form in poplar cells (Fig. 2 F and Fig. 2 G), were droplet-like in birch (Fig. 2 H) and tobacco (Fig. 5 C), but small and less frequent in alder (Fig. 5 D). The thickened outermost cell wall layer including these exudates was visible under ultraviolet light (Fig. 2 F), but did not appear yellow, as callose did. Staining reactions for pectins were positive (Fig. 2 G, H). Both thickened cell walls and exudates originating from the outermost pectinaceous cell layer were observed under TEM (Fig. 5 E = control with smooth thin cell wall, 5 F = leaf with early injury, cell with thickened wall and early exudates, 5 G and 5 H = cells with exudates).

Discussion

The daily AOT40-value was 420 ppb-h when O_3 was applied during daytime and nighttime, but 840 ppb-h when applied continuously. The critical level for crops (5.3 ppm-h during daylight hours only) as established on a 10 % wheat yield reduction would have been reached in our treatment after 12.6/6.3 days (daytime/continuous exposure), that for forest trees (10 ppm-h during 24 hours) after 23.8/11.9 days (daytime or nighttime/continuous exposure). The first visual leaf injury symptoms actually appeared in June and July on tobacco leaves after a mean of 18.9 ± 2.3 days (daytime exposure AOT40 = $18.9\,d \times 12\,h \times 35\,ppb$ = 7.9 ppm-h/24 h exposure = 15.8 ppm-h), and on the deciduous tree leaves after 27.0 ± 2.0 days (daytime exposure AOT40 = 11.3 ppm-h/24 h exposure = 22.7 ppm-h), both later than expected according to the critical level value. Whereas the appearance of visual symptoms under continuous fumigation (Günthardt-Goerg et al. 1993) was dose dependent, the present results report that:

a) visual symptoms appeared after a certain threshold of the applied O_3,
b) the halfday exposure to O_3 (with reduced acclimation but recover possibilities during halfday exposure to filtered air) had induced visual symptoms after the same time as the 24 h exposure,
c) the O_3 applied during nighttime was as effective as when applied during daytime.

Birch and alder trees exposed in the sunny field had developed early symptoms in July after 51 or 54 days respectively, but not poplar (Landolt et al. 1994 b) or tobacco plants. This time-span was close to the 57 days calculated by the AOT40-value of sunny days (s. introduction and Fig. 1). In the birch exposed in the field, but under shade, O_3-induced leaf symptoms appeared earlier (after 39 days) than in the sun, and thus close to that of the sensitive tobacco clone (34 days). Similar sensitivity has been reported at several areas of Europe for beans and clover (Sanders et al. 1993). Later in the season no symptoms appeared on the newly formed leaves of trees exposed in the field: in August most probably because the stomata were closed during the hot and dry daytime when the O_3-concentration was high, and in September, because the ambient O_3 concentrations were low.

Narrowing, but not closure of the stomata during nighttime has been measured by gas exchange in birch leaves (Matyssek et al. 1995) and larch needles (Wieser and Havranek 1995). Through such continuous measurements O_3 flux into the leaves could be calculated. The species-dependent influence of temperature, light and water conditions on the stomatal conductance and therefore on O_3 influx together with different acclimation and repair facilities (Morgan 1990) complicate the prediction of the occurrence of injury symptoms.

Biomass reduction measured at the end of the season was mainly related to premature leaf loss in summer in birch

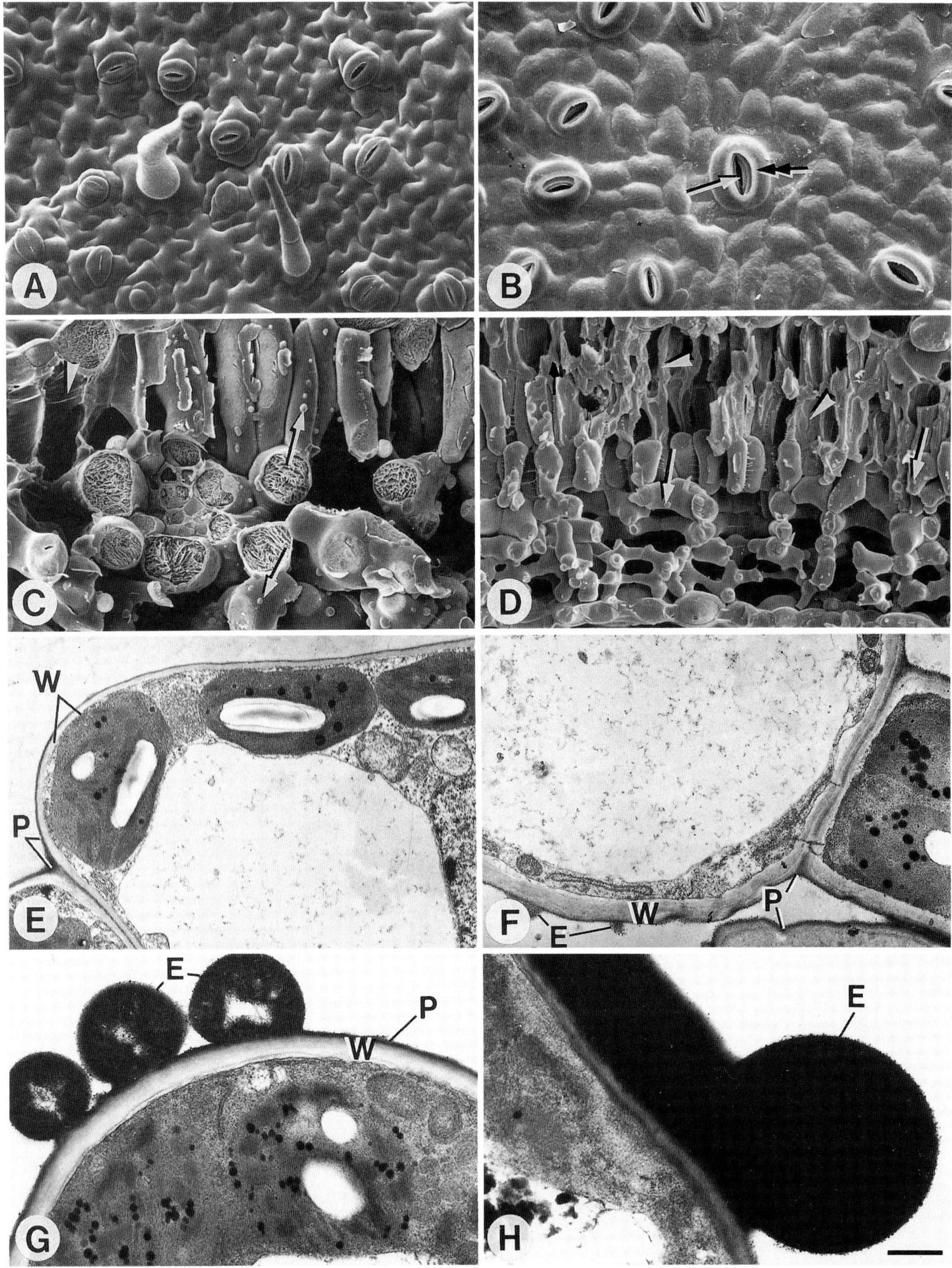
A
B
C
D
W
P
E
F
E
W
P
E
P
W
G
E
H

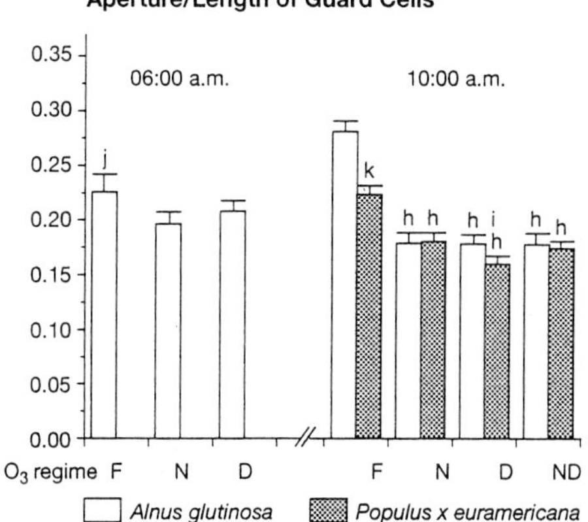

Fig. 6: Stomatal aperture (μm pore width of cuticular ledges in guard cells/μm length of the guard cells) in alder and poplar under leaf surfaces at 06:00 and 10:00 a.m., as measured in SEM micrographs. Fumigation regimes as in Fig. 4. Lower case characters above the columns (mean values with standard error bars) indicate that according to ANOVA (LSD = 99 %) these means are significantly smaller than corresponding means of:

- the same species and time, but in the filtered-air control (h), or the continuous O_3 fumigation regime (i);
- the same species and regime, but at 10:00 (j);
- the same regime and time, but in alder (k).

(Matyssek et al. 1992) and poplar (Matyssek et al. 1993). In the present investigation a biomass reduction on the base of leaf loss would therefore be expected for tobacco, but not for the trees exposed in the field (without premature leaf loss). For the trees exposed in the chambers the probability of biomass reduction would be tobacco > poplar > birch and alder (Fig. 3). However, biomass was only analysed in birch trees (Matyssek et al. 1995). Although the leaves with necrotic spots were not shed in the birches exposed to O_3 during nighttime, biomass was similarly reduced as in the trees exposed during daytime. The biomass (in particular root) reduction and the reduced root/shoot ratio were correlated with the applied O_3 dose (Matyssek et al. 1995, similar results were also found in beech by Braun and Flückiger 1995). The necrotic leaves in birch exposed to O_3 during nighttime, therefore, must have had a similar inhibitory effect on carbon allocation to the roots as shed leaves. Because the seasonal appearance and the insertion height of the leaves with necrotic spots is important, both visual injury without (red maple) or with biomass reduction (black cherry) has been found by Samuelson (1994). Visual injury, however, only has been considered for short-term critical levels referring to acute injury in beans and clover (AOT40 of 700 ppb-h accumulated during three consecutive days during daylight hours, Ashmore and Fuhrer 1994).

Changed starch patterns in the leaves before or when visual injury symptoms occur (Fig. 2 E, Günthardt-Goerg et al. 1991, and 1993, Landolt et al. 1994 a) may indicate disturbed carbon allocation to the roots, in particular in the leaves of the lowest third of the stem, which export photosynthates to the roots in trees with undetermined growth (Dickson and Larson 1981). According to the intensity of accumulated starch along leaf veinlets, birch leaves (in the lowest third of the stem) with necrotic spots would have a stronger inhibitory effect on carbohydrate allocation to the roots than similarly injured alder leaves, but a smaller effect than poplar leaves, which were prematurely shed. O_3, which enters the leaf through the stomatal openings, but which is rapidly decomposed in the apoplast (Laisk et al. 1988), may easily induce cell wall reactions in leaves with large intercellular air spaces as compared to compact leaves such as wheat or conifer needles. The occurrence of cell wall exudates (in birch and tobacco > poplar > alder, Fig. 2 F–G, Fig. 5 C and D) confirmed this hypothesis. Pectinaceous exudates, which in leaves with O_3 injury preceded cell collapse, have a ‹repair-function› in graft unions (Jeffree and Yeoman 1983, Miller and Barnett 1993), where they establish a mechanical union between the cell surfaces.

In general the present investigations show that ozone impact on the leaves of deciduous trees is as important during dusk, night and dawn as during daytime, contrasting wheat and other crops (Fuhrer 1994, UK PORG 1993), but also that the selected critical levels are just below the observed values which induced visual injury. In particular the present investigation reminds us, that leaf injury symptoms depend on the species (with larger variation in seedlings than in cloned cuttings, Günthardt-Goerg et al. 1994), on the climatic conditions of the season and the year including the occurrence or application of O_3, on shade (Tjoelker and Reich 1993), and on acclimation during onthogenesis (Günthardt-Goerg 1993).

Acknowledgements

The author thanks A. Burkhart and U. Bühlmann for tending the plants, C. Rhiner, T. Koller and J. Bolliger for technical assistance, W. Landolt and P. Bleuler for the fumigation management and the O_3 measurements at the WSL, R. Häsler for the O_3 data from Stillberg, Davos, J. Bucher and S. Maurer for critical reading of the manuscript, and M. J. Sieber for the stylistic editing of the English text. Collaboration with the WSL REM-laboratory (Fig. 5 A–D by C. Scheidegger and P. Hatvani), and the TEM-laboratory (Fig. 5

Fig. 5: SEM and TEM micrographs, bar for A = 49 μm, for B, C, and D = 24 μm, for E and F = 1.4 μm, for G and H = 0.74 μm. A: SEM micrograph of a tobacco lower leaf surface with closed stomatal pores and B: alder lower leaf surface with partially opened stomatal pores at 6:00 a.m., the white arrow indicates the pore width confined by the guard cells, the black arrow indicates the cuticular ledges of the guard cells. C: Freeze fracture of a tobacco leaf (with visual white dots as Fig. 2 A) showing mesophyll cells with cell wall exudates (arrows) facing the intercellular space and collapsed cells (arrow head). D: Fracture through an alder leaf, arrows as C. E: TEM micrograph of a birch mesophyll cell (control) with thin cell walls (W) lined with a black electron-dense outermost layer (P for pectinaceous, according to the stains in Fig. 2 F–H). F– H: Pectinaceous cell wall exudates (E) of different size in mesophyll cells of a birch leaf with necrotic spots (as Fig. 2 C).

E–H by C. McQuattie) of the USDA forest service Delaware, Ohio is gratefully acknowledged.

References

* Detailed reference see Ashmore and Fuhrer (1994).

Ashmore, M. and J. Fuhrer: Workshop summary. In: Fuhrer, J. and B. Achermann (eds.): Critical levels for ozone, a UN-ECE workshop report. Schriftenreihe der Eidgen. Forschungsanstalt für Agrikulturchemie und Umwelthygiene (FAC) *16,* CH-Liebefeld-Bern, pp. 4–12 (1994).

Braun, S. and W. Flückiger: Critical levels of ambient ozone for growth of tree seedlings.* Schriftenreihe der FAC *16,* CH-Liebefeld–Bern, pp. 88–97 (1994).

– – Effects of ambient ozone on seedlings of *Fagus sylvatica* L. and *Picea abies* (L.) Karst. New Phytol. *129,* 33–44 (1995).

Dickson, R. E. and P. R. Larson: ^{14}C fixation, metabolic labelling pattern, and translocation profiles during leaf development in *Populus deltoides.* Planta *152,* 461–470 (1981).

Feder, N. and T. P. O'Brien: Plant microtechnique: Some principles and new methods. Amer. J. Bot. *55,* 123–142 (1968).

Fuhrer, J.: The critical level for ozone to protect agricultural crops. An assessment of data from European open-top chamber experiments.* Schriftenreihe der FAC *16,* CH-Liebefeld–Bern, pp. 42–57 (1994).

Gerlach, D.: Botanische Mikrotechnik. 3. Auflage, Thieme-Verlag, Stuttgart (1984).

Günthardt-Goerg, M. S., R. Matyssek, C. Scheidegger, and T. Keller: Bioindikationen in Birkenblättern unter niedrigen Ozonkonzentrationen. VDI Berichte *901,* 631–647 (1991).

– – – – Differentiation and structural decline in the leaves and bark of birch (*Betula pendula*) under low ozone concentrations. Trees *7,* 104–114 (1993).

Günthardt-Goerg, M. S., W. Landolt, R. Matyssek, and J. B. Bucher: Ozone sensitivity of deciduous trees.* Schriftenreihe der FAC *16,* CH-Liebefeld–Bern, pp. 306–309 (1994).

Jeffree, C. E. and M. M. Yeoman: Development of intercellular connections between opposing cells in a graft union. New Phytol. *93,* 491–509 (1983).

Küppers, K., J. Boomers, C. Hestermann, S. Hanstein, and R. Guderian: Reaction of forest trees to different exposure profiles of ozone-dominated air pollution mixtures.* Schriftenreihe der FAC 16, CH-Liebefeld–Bern, pp. 98–110 (1994).

Laisk, A., O. Kulli, and H. Moldau: Ozone concentration in intercellular air spaces is close to zero. Plant Physiol. *90,* 1163–1167 (1988).

Landolt, W., M. S. Günthardt-Goerg, I. Pfenninger, and C. Scheidegger: Ozone induced microscopical changes and quantitative carbohydrate contents of hybrid poplar (*Populus × euramericana*). Trees *8,* 183–190 (1994 a).

Landolt, W., B. Lüthy-Krause, M. S. Günthardt-Goerg, R. Matyssek, and J. Bucher: Environmental factors modify plant response to ozone.* Schriftenreihe FAC, *16,* CH-Liebefeld–Bern, pp. 310–312 (1994 b).

Matyssek, R., M. S. Günthardt-Goerg, T. Keller, and C. Scheidegger: Impairment of gas exchange and structure in birch leaves (*Betula pendula*) caused by low ozone concentrations. Trees *5,* 5–13 (1991).

Matyssek, R., M. S. Günthardt-Goerg, M. Saurer, and T. Keller: Seasonal growth, $\delta^{13}C$ in leaves and stem, and phloem structure of birch (*Betula pendula*) under low ozone concentrations. Trees *6,* 69–76 (1992).

Matyssek, R., M. S. Günthardt-Goerg, W. Landolt, and T. Keller: Whole-plant growth and leaf formation in ozonated hybrid poplar(*Populus × euramericana.* Environ. Pollution *81,* 207–212 (1993).

Matyssek, R., M. S. Günthardt-Goerg, S. Maurer, and T. Keller: Nighttime exposure to ozone reduces whole-plant production in *Betula pendula.* Tree Physiology *15,* 159–165 (1995).

Miller, H. and J. R. Barnett: The structure and composition of bead-like projections on sitka spruce callus cells formed during grafting and in culture. Ann. Bot. *72,* 441–448 (1993).

Morgen, P. W.: Effects of abiotic stresses on plant hormone systems. In: Alscher, R. G. and J. R. Cummings (eds.): Stress responses in plants: adaptation and acclimation mechanisms, pp. 113–146. Wiley-Liss, Inc., New York, (1990).

Samuelson, L. J.: Ozone-exposure responses of black cherry and red maple seedlings. Environ. Experim. Bot. *34,* 355–362 (1994).

Sanders, G., G. Balls, and C. Booth: Ozone critical levels for agricultural crops. Analysis and interpretation of results from the UN-ECE International Cooperative Programme for Crops.* Schriftenreihe FAC, *16,* CH-Liebefeld–Bern, pp. 58–72 (1994).

Scheidegger, C., M. S. Günthardt-Goerg, R. Matyssek, and P. Hatvani: Low-temperature scanning electron microscopy of birch leaves after exposure to ozone. J. Microscopy *161,* 85–95 (1991).

Tjoelker, M. G. and P. B. Reich: Forest canopy response to ozone: light environment and photosynthesis in a sugar maple stand.* Schriftenreihe FAC, *16,* CH-Liebefeld–Bern, pp. 112–120 (1994).

UK PORG: Third report of the United Kingdom Photooxidants Review Group. Department of the environment, London, pp. 83–107 (1993).

Weis, K. G., V. S. Polito, and J. M. Labavitch: Microfluorometry of pectic material in the dehiscence zone of almond (*Prunus dulcis* [Mill.] DA Webb) fruits. J. Histochem. Cytochem. *36,* 1037–1041 (1988).

Wieser, G. and W. M. Havranek: Environmental control of ozone uptake in *Larix decidua* Mill.: a comparison between different altitudes. Tree Physiol. *15,* 243–258 (1995).

J. Plant Physiol. Vol. 148. pp. 215–221 (1996)

β-1,3-Glucanase mRNA is Locally, but not Systemically Induced in *Nicotiana Tabacum* L. cv. BEL W3 after Ozone Fumigation

DIETER ERNST[1,*], ACHIM BODEMANN[1], ELMON SCHMELZER[2], CHRISTIAN LANGEBARTELS[1], and HEINRICH SANDERMANN Jr.[1]

[1] GSF-Institut für Biochemische Pflanzenpathologie, D-85758 Oberschleißheim, Germany

[2] MPI für Züchtungsforschung, D-50829 Köln, Germany

Received October 13, 1995 · Accepted October 31, 1995

Summary

Ozone fumigation of tobacco caused a strong induction of basic β-1,3-glucanase mRNA. *In situ* hybridization revealed a uniform labeling of the whole tissue section, except for the necrotic areas. Ozone-treated leaf parts showed an increased transcript accumulation, whereas no transcripts were detectable in non-treated leaf parts. Treatment of a single leaf increased β-1,3-glucanase mRNA level only in this leaf and not in other leaves. Conversely, protection of a single leaf resulted in transcript accumulation only in the fumigated leaves. These results indicate that ozone-induced β-1,3-glucanase mRNA accumulation is restricted to the ozone-treated area, and that ozone does not lead to systemic induction of this ‹pathogenesis-related› (PR) protein.

Key words: Nicotiana tabacum L. cv Bel W3, β-1,3-glucanase, mRNA localization, ozone, systemic response.

Abbreviations: ACC = 1-aminocyclopropane-1-carboxylic acid; CLEO = cuvette for local exposure of ozone; PR = pathogenesis-related; SAR = systemic acquired resistance; SSC = 150 mM NaCl, 15 mM sodium citrate.

Introduction

Ozone is an important photochemical air pollutant potentially reducing growth and yield of crop plants (Heck et al., 1988; Fuhrer and Achermann, 1993) Exposure to subacute ozone levels can result in the induction or inhibition of biochemical and physiological processes in the plant. The induced reactions generally are part of the plant defense against oxidative stress (reviews Sandermann et al., 1989; Kangasjärvi et al., 1994). Recent results have shown that at the molecular level, transcripts for photosynthesis genes were reduced (Reddy et al., 1993; Bahl and Kahl, 1995). In contrast, expression of genes which are involved in plant defense against pathogens was increased upon ozone fumigation. The ozone induction of genes of cell wall biosynthesis (Galliano et al., 1993; Schneiderbauer et al., 1995), of the phenyl-propanoid pathway (Eckey-Kaltenbach et al., 1994 a), of the antioxidant system (Sharma and Davis, 1994; Willekens et al., 1994), of ethylene biosynthesis (Schlagnhaufer et al., 1995) and of ‹pathogenesis-related› (PR) proteins (Ernst et al., 1992; Eckey-Kaltenbach et al., 1994 b) has been recognized.

The PR proteins of tobacco have recently been classified into 11 major groups (review van Loon et al., 1994). The basic PR proteins, including β-1,3-glucanase, are mainly localized in the vacuole, whereas the acidic forms are mainly localized extracellularly (review Bol et al., 1990; Keefe et al., 1990). Basic β-1,3-glucanase mRNA is strongly induced by various biotic and abiotic stress factors (review Bol et al., 1990), that include ozone (Ernst et al., 1992; Schraudner et

* Corresponding author.

al., 1992). *In situ* β-1,3-glucanase mRNA localization studies in potato have revealed a more or less uniform labeling of the whole tissue section, except for the necrotic areas colonized by the inducing fungus employed (Schröder et al., 1992). Similarly, in transgenic tobacco containing the β-1,3-glucanase promotor fused to the β-glucuronidase (GUS) reporter gene, a specific staining reaction was restricted to the cells surrounding the hypersensitive lesions (Castresana et al., 1990).

Systemic acquired resistance (SAR) is an important component of plant defense against pathogen infection (review Ryals et al., 1994). The most thoroughly characterized example of SAR exists for tobacco (Ward et al., 1991 b). Most genes for acidic PR proteins were systemically induced in the uninfected upper leaves, whereas genes encoding the basic PR proteins were not systemically induced (Brederode et al., 1991). In contrast, the mRNA for the class I glucanase (encoding basic glucanase) was only weakly induced in systemic tissue of an infected tobacco plant (Ward et al., 1991 a). Infection of potato leaves with the fungal pathogen, *Phytophthora infestans,* caused a strong increase of a 1.4 kb transcript in the infected leaf areas and a slightly delayed increase in the non-infected leaf parts, after RNA hybridization with basic β-1,3-glucanase cDNA from tobacco (Schröder et al., 1992).

To study the response of β-1,3-glucanase mRNA towards ozone fumigation of tobacco plants in more detail, we analyzed transcript localization by *in situ* hybridization near the ozone-induced necrotic areas. In addition, Northern blotting of total RNA from fumigated and non-fumigated leaf areas, as well as of whole leaves, was carried out.

Material and Methods

Growth conditions and fumigations

Ozone-sensitive (*Nicotiana tabacum* L. cv Bel W3) tobacco plants were cultivated for 12 weeks under a 12 h day/night cycle in an environmental growth chamber (Langebartels et al., 1991). Local leaf exposures were performed in a small cuvette (CLEO I) with a total volume of 0.217 l at adaxial and abaxial sides of the leaf. The covered leaf area was 16 cm^2 (8×2 cm) and the air exchange rate was 0.6 L/min. The gasket of this cuvette was covered with rubber to avoid leaf damage. Single leaf fumigations were carried out in cuvettes (CLEO II) having a volume of 5.3 L and an air exchange rate of 12 L/min. A fan in the chamber provided an even air circulation. Ozone was generated by electrical discharge in dry oxygen (Langebartels et al., 1991). The desired ozone concentration (0.3 up to 0.5 µL/L, 5 h) was obtained by computer control. Temperature (25 ± 2 °C) and humidity (75 ± 5 %) were as in in the growth chamber, while photosynthetic photon fluence rate (400–700 nm) was approximately 120 µmol $m^{-2} s^{-1}$ in the leaf chamber.

Harvesting plant tissue

After removing the mid vein, as well as top and bottom of the leaves, leaf samples were frozen in liquid nitrogen and stored at −80 °C until use for RNA isolation and 1-aminocyclopropane-1-carboxylic acid (ACC) determination, respectively. In local leaf exposure experiments the tissue was divided as follows: ozone fumigation area (1), gasket of the cuvette (2), adjacent 1 cm zone around (3), residual leaf area (4) (Fig. 4 A).

Determination of ACC levels

The ACC content was assayed according to Langebartels et al. (1991), ethylene being determined by gas chromatography.

RNA isolation and Northern blot analysis

0.25 g of frozen leaf tissue was ground to a fine powder using mortar and pestle, and total RNA was isolated as described (Ernst et al., 1992). For Northern blotting, total RNA (5 µg) was fractionated in 1.5 % agarose-formaldehyde gels and then electroblotted onto Hybond membranes (Amersham) as described (Ernst et al., 1992). After UV cross-linking (Stratagene), the membranes were first prehybridized and then hybridized in the presence of a specific β-1,3-glucanase clone (pGL 43) (Mohnen et al., 1985) as described (Ernst et al., 1992). Filters were washed at room temperature for 10 min in 2× SSC, followed by 2× SSC, 0.1 % SDS at 65 °C for 30 min, and then by 0.1× SSC at 65 °C for 30 min. Blots were autoradiographed at −80 °C on pre-flashed Fuji X-ray films.

Tissue fixation, embedding and sectioning

Tobacco leaves were cut into pieces of about 4 mm^2 and fixed in 1 % glutaraldehyde, 100 mM sodium phosphate buffer, pH 7.0 according to Schmelzer et al. (1989). The fixed tissue was dehydrated in a series of aqueous ethanol solutions: 30 %, 50 %, 70 %, 80 %, 90 %, 100 % (v/v), then infiltrated with t-butanol and paraplast, and the cooled paraffin block was sectioned at 10 µm thickness (Schmelzer et al., 1989). The sections were transferred to poly-L-lysine coated slides, and deparaffinized in xylol/ethanol (Schmelzer et al., 1989).

In situ hybridization

Prior to *in situ* hybridization sections were pronase treated and post-fixed (Somssich et al., 1988). For *in vitro* transcription, the β-1,3-glucanase cDNA was subcloned in pT7/T3α-19 (GIBCO BRL) and *in vitro* transcripts were partially hydrolyzed, resulting in fragments with an average length of 50–150 bp (Somssich et al., 1988). Conditions of *in situ* hybridization with ^{35}S-labeled antisense and sense RNA transcripts and microautoradiography were as described by Cuypers and Hahlbrock (1988). Photomicrographs were taken under a Zeiss Axioplan microscope equipped with dark-field and phase-difference optics, using Kodak Tmax-100 films.

Results

In situ hybridization

Tobacco plants were exposed to ozone concentrations of 0.3 µL/L for 5 h, which led to approximately 40 % necrotic area after 24 h and resulted in β-1,3-glucanase mRNA accumulation (Ernst et al., 1992). The necrotic area showed the typical blue autofluorescence after UV light excitation (Schraudner et al., 1992). Phase-difference microscopy revealed no difference in leaf morphology of ozone-treated and control plants outside of the necrotic area (Fig. 1 A, C). However, in necrotic areas of ozone-treated plants, both epidermal cell layers were close together, caused by lysis of parenchymatic cells (Fig. 2 A). *In situ* hybridization was carried out in sections from ozone-treated leaves, in order to analyze the localization of β-1,3-glucanase mRNA in and around the necrotic area. Hybridization was performed with sections of pa-

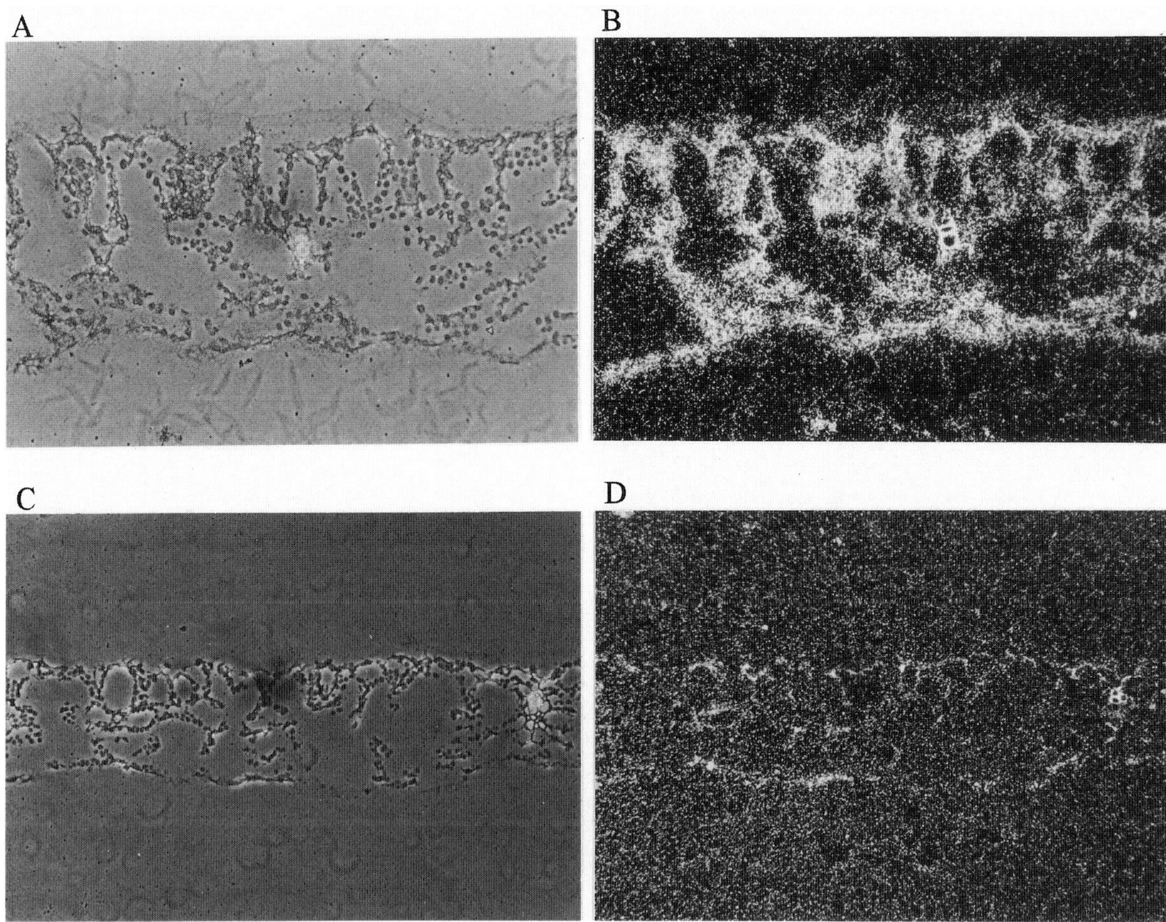

Fig. 1: *In situ* localization of β-1,3-glucanase mRNA in cross-sections of tobacco leaves. Leaves were treated with 0.3 µL/L ozone (A, B) or filtered air (C, D) for 5 h and further incubated in filtered air for 10 h. Sections were hybridized with antisense β-1,3-glucanase mRNA derived from β-1,3-glucanase cDNA (pGL 43). Phase-difference microscopy (A, C), dark-field conditions (B, D). Magnification: ×60 (A, B); ×80 (C, D).

raffin-embedded tissue, and antisense β-1,3-glucanase mRNA. 14 h after the onset of fumigation β-1,3-glucanase mRNA accumulated throughout the leaf, as can be seen in the uniform distribution of silver grains (Fig. 1 B). In sections of exposed control plants, no hybridization signals were visible (Fig. 1 D). Similarly, no signals above background were seen when sense β-1,3-glucanase mRNA transcripts were used as controls for non-specific binding (Fig. 2 C). Tissue sections of necrotic areas, as well as cells directly surrounding this area showed little or no hybridization signals, in contrast to the non-destroyed cell types (Fig. 2 B).

Accumulation of β-1,3-glucanase mRNA is not a result of systemic gene activation

The accumulation of β-1,3-glucanase transcripts around the necrotic area prompted us to analyze a possible systemic response of the plant. The middle-aged leaf no. 3 from the top was fumigated with 0.5 µL/L ozone in whole leaf cuvettes (CLEO II) for 5 h and total RNA was isolated after 14 h from leaves no. 1–6. A strong increase of the β-1,3-glucanase mRNA level could only be observed in the fumigated leaf no. 3, whereas transcript levels did not change in the other leaves, exposed to ozone-free air (Fig. 3 B). Small amounts of β-1,3-glucanase mRNA in leaf no. 5 and 6 of the fumigated, as well as in leaf no. 6 of the control plants, are caused by an inherent increase of β-1,3-glucanase in older leaves (Schraudner et al., 1992). In complementary experiments leaf no. 3 was protected and received pollutant-free air, while the rest of the tobacco plant was fumigated with ozone (0.5 µL/L, 5 h). As can be seen in Fig. 3 C, transcript accumulation was strong in leaves no. 4–6 and weak in leaf no. 2 and 1. The latter effect was probably caused by the lower stomatal conductance of young tobacco leaves (Schraudner et al., 1992). No β-1,3-glucanase transcript accumulation was evident for the protected leaf no. 3. These results demonstrate that the accumulation of basic β-1,3-glucanase mRNA is restricted to ozone-treated leaves.

Using a cuvette for local ozone exposure (CLEO I) of tobacco leaf parts, β-1,3-glucanase accumulation was studied within exposed as well as in the surrounding leaf area. 14 h after onset of fumigation with 0.5 µL/L ozone for 5 h, leaf parts

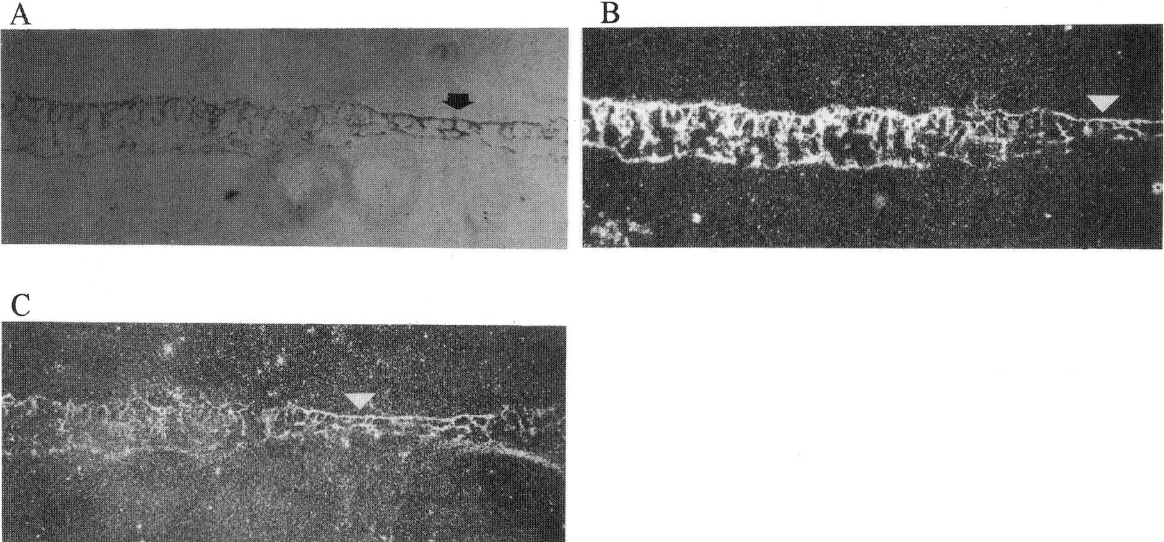

Fig. 2: *In situ* localization of β-1,3-glucanase transcripts near a necrotic area. Leaves were treated with ozone (0.3 µL/L, 5 h) and further incubated in filtered air for 10 h. Phase-difference microscopy (A); hybridization with antisense transcripts, dark-field conditions (B); hybridization with sense transcripts, dark-field conditions (C). Arrows indicate the necrotic area. Magnification: ×40.

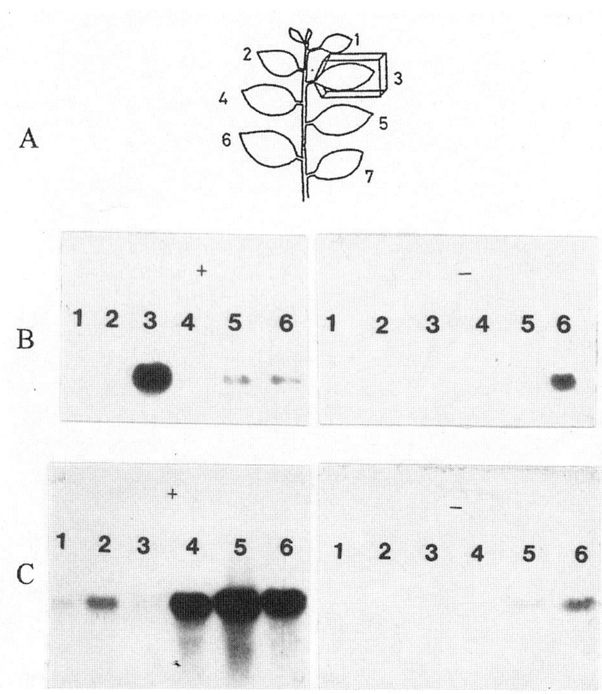

Fig. 3: Northern blot analysis of β-1,3-glucanase mRNA in ozone-treated (+) and control (–) tobacco. Ozone exposure scheme (A). Leaf no. 3 was exposed to ozone (0.5 µL/L, 5 h) or filtered air (B). Leaf no. 3 was ozone-protected and the residual leaves were ozone-exposed (C). RNA was isolated after 14 h and 5 µg were electrophoresed in denaturating agarose gels, blotted onto nylon membranes and hybridized with a β-1,3-glucanase probe (pGL 43). Numbers correspond to the leaf numbers, taking the youngest leaf greater than 8 cm in length as no. 1.

were separated, as indicated in Fig. 4 A, and analyzed for mRNA accumulation. No β-1,3-glucanase mRNA was detectable in the filtered-air treated leaf area of leaf no. 4, as well as in the gasket area, excluding wounding in this area as a local trigger (Fig. 4 B). When middle-aged leaves were locally exposed to ozone, a dramatic increase in the level of β-1,3-glucanase mRNA could be observed in the fumigated leaf part (area 1) (Fig. 4 B). Reduced levels were seen in the gasket area of the cuvette (area 2), as well as in a distance of 1 cm around to the cuvette (area 3). No β-1,3-glucanase transcripts could be detected in the residual leaf (area 4) and in the adjacent upper and lower leaves (Fig. 4 B). The content of the ethylene precursor, ACC, increased in the fumigated leaf area from near zero to 4 nmol/g fresh weight.

Discussion

In situ localization

Ozone can cause visible injury and a variety of physiological, biochemical and molecular responses in susceptible plants (reviews Sandermann et al., 1989; Kangasjärvi et al., 1994). The results of this paper extend our earlier observations of the induction of β-1,3-glucanase mRNA and enzyme activity upon ozone fumigation (Ernst et al., 1992; Schraudner et al., 1992). Tissue localization of β-1,3-glucanase mRNA in leaves exposed to ozone-free air revealed no signals over background (Fig. 1 D). Ozone exposure led to a more or less uniform distribution of β-1,3-glucanase mRNA over the whole cross-section of tobacco leaves (Fig. 1 B). No preference for individual cell types, like epidermal, palisade- or spongy-parenchyma cells was evident. Similar results have

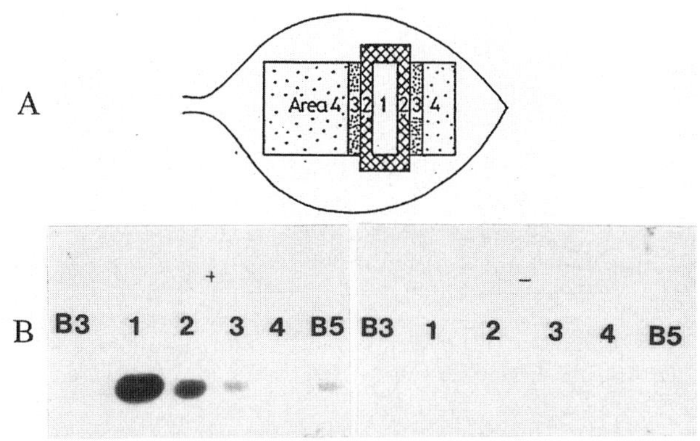

Fig. 4: Northern blot analysis of β-1,3-glucanase mRNA after local ozone exposure of leaf no. 4. Ozone exposure scheme (A); 1: ozone fumigation area (0.5 μL/L, 5 h), 2: gasket of the exposure cuvette, 3: adjacent 1 cm zone around, 4: residual leaf area. RNA was isolated after 14 h and 5 μg were separated on denaturating agarose gels, electroblotted onto nylon membranes and hybridized with a β-1,3-glucanase probe (pGL 43). B3: leaf no. 3, B5: leaf no. 5, numbers correspond to the leaf areas.

been obtained by Schröder et al. (1992) in potato leaves infected with *Phytophthora infestans*. The expression was not observed in necrotic lesions of ozone-treated plants. However, cells surrounding this area in a distance of about 1 mm showed a strong transcript accumulation in all cell types (Fig. 2 B). Again, such results for β-1,3-glucanase mRNA have also be found in fungus-infected potato leaves (Schröder et al., 1992). Similarly, in transgenic tobacco the expression of β-1,3-glucanase, as well as of PR la protein has been shown to be localized around the infected area upon pathogen infection (Castresana et al., 1990; Ohshima et al., 1990). The spatial induction of PR proteins upon ozone fumigation, as well as pathogen infection suggests a common mediator for this transcript accumulation. Formation of ethylene, as well as its precursor, ACC, has been shown to increase in ozone-treated tobacco (Langebartels et al., 1991) and potato leaves (Reddy et al., 1993). Furthermore, ethylene treatment of tobacco leaves induced β-1,3-glucanase to an equal extend in all leaf cells (Keefe et al., 1990). These findings support the idea of ethylene as a mediator of ozone effects on β-1,3-glucanase (Ernst et al., 1992; Schraudner et al., 1992). Additional support is provided by the ozone-induced increase of ACC synthase activity in tobacco (Betz et al., 1995), as well as the accumulation of ACC synthase mRNA in potato upon ozone fumigation (Schlagnhaufer et al., 1995).

Basic β-1,3-glucanase mRNA is non-systemically expressed in tobacco leaves

The similarity of leaf injury between ozone-treated and pathogen-infected tobacco plants, as well as the *in situ* hybridization results prompted us to analyze a possible β-1,3-glucanase mRNA induction in untreated leaves. Although the ozone-exposed leaves clearly showed visible ozone injuries and strong transcript accumulation, no response was detectable in non-exposed tobacco leaves of the same plant (Fig. 3). In contrast, a systemically induced increase of basic β-1,3-glucanase has been previously reported for potato leaves, infected with *Phytophthora infestans* (Schröder et al., 1992). This difference might be explained by the different plant material used. Virus infection of tobacco cv Xanthi led to a weak systemic induction of basic class I β-1,3-glucanase (Ward et al., 1991 a). However, in tobacco cv. Samsun, basic PR gene expression was not systemically induced in uninfected upper leaves (Brederode et al., 1991). A thorough analysis of the expression of several PR genes showed no accumulation of basic β-1,3-glucanase mRNA in uninfected tobacco tissue (Ward et al., 1991 b). These results are in agreement with our finding that ozone did not systemically induce basic β-1,3-glucanase mRNA.

Wounding of leaves, on the other hand has resulted in transcript accumulation of several acidic, as well as basic PR proteins (Brederode et al., 1991) and systemically signalling has also been found (review Bowles, 1990). However, to detect this systemical transcript accumulation, large areas of leaves were repeatedly wounded (Peña-Cortes et al., 1988; Parsons et al., 1989; Neale et al., 1990). In order to extend the ozone-damaged leaf area we protected a single leaf in a leaf cuvette (CLEO II) from ozone during fumigation. The ozone-exposed leaves showed a drastic increase of the basic β-1,3-glucanase mRNA level, whereas this mRNA was not detectable in the protected leaf (Fig. 3 C). This was in contrast to the systemic increase of basic β-1,3-glucanase mRNA (FB7-5) observed in the control leaf of wounded tobacco plants (Neale et al., 1990). It therefore seems unlikely that the signal evoked by wounding and ozone, respectively, are mediated in the same way.

The observation, that *in situ* hybridization showed an ozone-induced increase of β-1,3-glucanase mRNA around the necrotic region (Fig. 2 B), prompted us to analyze an ozone-exposed leaf area, as well as the non-exposed adjacent tissue. While high mRNA levels were found in the ozone-treated area, there was a drastic decline in the gasket of the cuvette and in an adjacent 1 cm zone around (Fig. 4 B). This mRNA accumulation in the non-exposed leaf area could be a direct ozone effect or an effect mediated by other induced signals, because virtually no ozone concentrations could be measured outside of the cuvette: (1) after reaching the intercellular leaf space ozone may diffuse into the adjacent non-fumigated leaf tissue. However, this seems to be unlikely, since Laisk et al. (1989) have shown that ozone concentrations inside of leaf tissues are close to zero and that ozone is rapidly decomposed in the cell wall and plasmalemma. (2) it might be possible that ozone-increases ACC and ethylene contents locally (Langebartels et al., 1991). Both products may spread out for about 2 cm from the ozone-exposed area, where they lead to increased transcript levels. This mechanism is supported by results obtained with tomato (Spanu and Boller, 1989). They have shown that ACC synthase activity was highest in necrotic tissue and the ACC content remained elevated at a greater distance in tomato plants infected by *Phytophthora infestans*. It was speculated that the ACC formed diffused into the surrounding tissue and there caused ethylene production (Spanu and Boller, 1989). Therefore ACC or ethylene might be the mediator in signalling the ozone effect over a short distance, but this remains to be established experimentally.

In conclusion, ozone treatment of tobacco leaves resulted in transcript accumulation of basic β-1,3-glucanase, surrounding the necrotic areas. No systemic induction of basic β-1,3-glucanase mRNA took place upon ozone fumigation of tobacco leaves.

Acknowledgements

The authors would like to thank E. Kiefer and L. Gößl for their excellent technical asssistance, P. Bader for his great help in the development of the exposure cuvettes, and Dr. W. Heller for critical reading of the manuscript. The β-1,3-glucanase clone (pGL 43) was kindly provided by Prof. F. Meins (Friedrich Miescher-Institut, Basel).

References

Bahl, A. and G. Kahl: Air pollutant stress changes the steady-state transcript levels of three photosynthesis genes. Environ. Pollut. *88,* 57–65 (1995).

Betz, C., J. Tuomainen, J. Kangasjärvi, D. Ernst, Z.-H. Yin, C. Langebartels, and H. Sandermann: Ozone activation of ethylene biosynthesis in tomato. Conference on Vegetation Stress, p. 13. GSF, München/Neuherberg, Germany (1995).

Bol, J. F., H. J. M. Linthorst, and B. J. C. Cornelissen: Plant pathogenesis-related proteins induced by virus infection. Annu. Rev. Phytopathol. *28,* 113–138 (1990).

Bowles, D.: Signals in the wounded plant. Nature *343,* 314–315 (1990).

Brederode, F. Th., H. J. M. Linthorst, and J. F. Bol: Differential induction of acquired resistance and PR gene expression in tobacco by virus infection, ethephon treatment, UV light and wounding. Plant Mol. Biol. *17,* 1117–1125 (1991).

Castresana, C., F. de Carvalho, G. Gheysen, M. Habets, D. Inzé, and M. van Montagu: Tissue-specific and pathogen-induced regulation of a *Nicotiana plumbaginifolia* β-1,3-glucanase gene. Plant Cell *2,* 1131–1143 (1990).

Cuypers, B. and K. Hahlbrock: Immunohistochemical studies of compatible and incompatible interactions of potato leaves with *Phytophthora infestans* and of the nonhost response to *Phytophthora megasperma.* Can. J. Bot. *66,* 700–705 (1988).

Eckey-Kaltenbach, H., D. Ernst, W. Heller, and H. Sandermann: Biochemical plant responses to ozone. IV. Cross-induction of defensive pathways in parsley (*Petroselinum crispum* L.) plants. Plant Physiol. *104,* 67–74 (1994a).

Eckey-Kaltenbach, H., E. Grosskopf, H. Sandermann, and D. Ernst: Induction of pathogen defence genes in parsley (*Petroselinum crispum* L.) plants by ozone. Proc. Royal Soc. Edinburgh *102 B,* 63–74 (1994b).

Ernst, D., M. Schraudner, C. Langebartels, and H. Sandermann: Ozone-induced changes of mRNA levels of β-1,3-glucanase, chitinase and ‹pathogenesis-related› protein 1b in tobacco plants. Plant Mol. Biol. *20,* 673–682 (1992).

Fuhrer, J. and B. Achermann: Workshop on Critical Levels for Ozone. Swiss Federal Office of Environment and Swiss Federal Research Station for Agricultural Chemistry and Environmental Hygiene, Bern (1993).

Galliano, H., M. Cabané, C. Eckerskorn, F. Lottspeich, H. Sandermann, and D. Ernst: Molecular cloning, sequence analysis and elicitor-/ozone-induced accumulation of cinnamyl alcohol dehydrogenase from Norway spruce (*Picea abies* L.). Plant Mol. Biol. *23,* 145–156 (1993).

Heck, W. W., O. C. Taylor, and D. T. Tingey: Assessment of Crop Loss from Air Pollutants. Elsevier, London (1988).

Kangasjärvi, J., J. Talvinen, M. Utriainen, and R. Karjalainen: Plant defence systems induced by ozone. Plant Cell Environ. *17,* 783–794 (1994).

Keefe, D., U. Hinz, and F. Meins: The effect of ethylene on the cell-type-specific and intracellular localization of β-1,3-glucanase and chitinase in tobacco leaves. Planta *182,* 43–51 (1990).

Laisk, A., O. Kull, and H. Moldau: Ozone concentration in leaf intercellular air spaces is close to zero. Plant Physiol. *90,* 1163–1167 (1989).

Langebartels, C., K. Kerner, S. Leonardi, M. Schraudner, M. Trost, W. Heller, and H. Sandermann: Biochemical plant responses to ozone. I. Differential induction of polyamine and ethylene biosynthesis in tobacco. Plant Physiol. *95,* 882–889 (1991).

Mohnen, D., H. Shinshi, G. Felix, and F. Meins: Hormonal regulation of β-1,3-glucanase messenger RNA levels in cultured tobacco tissues. EMBO J. *4,* 1631–1635 (1985).

Neale, A. D., J. A. Wahleithner, M. Lund, H. T. Bonnett, A. Kelly, D. R. Meeks-Wagner, W. J. Peacock, and E. S. Dennis: Chitinase, β-1,3-glucanase, osmotin, and extensin are expressed in tobacco explants during flower formation. Plant Cell *2,* 673–684 (1990).

Ohshima, M., H. Itoh, M. Matsuoka, T. Murakami, and Y. Ohashi: Analysis of stress-induced or salicylic acid-induced expression of the pathogenesis-related la protein in transgenic tobacco. Plant Cell *2,* 95–106 (1990).

Parsons, T. J., H. D. Bradshaw, and M. P. Gordon: Systemic accumulation of specific mRNAs in response to wounding in poplar trees. Proc. Natl. Acad. Sci. USA *86,* 7895–7899 (1989).

Peña-Cortes, H., J. Sanchez-Serrano, M. Rocha-Sosa, and L. Willmitzer: Systemic induction of proteinase-inhibitor-II gene expression in potato plants by wounding. Planta *174,* 84–89 (1988).

Reddy, G. N., R. N. Arteca, Y.-R. Dai, H. E. Flores, F. B. Negm, and E. J. Pell: Changes in ethylene and polyamines in relation to mRNA levels of the large and small subunits of ribulose bisphosphate carboxylase/oxygenase in ozone-stressed potato foliage. Plant Cell Environ *16,* 819–826 (1993).

Ryals, J., S. Uknes, and E. Ward: Systemic acquired resistance. Plant Physiol. *104,* 1109–1112 (1994).

Sandermann, H., R. Schmitt, W. Heller, D. Rosemann, and C. Langebartels: Ozone-induced early biochemical reactions in conifers. In: Longhurst, J. W. S. (ed.): Acidic Deposition. Sources, Effects and Controls, pp. 243–254. British Library, London (1989).

Schlagnhaufer, C. D., R. E. Glick, R. N. Arteca, and E. J. Pell: Molecular cloning of an ozone-induced 1-aminocyclopropane-1-carboxylate synthase cDNA and its relationship with a loss of *rbcS* in potato (*Solanum tuberosum* L.) plants. Plant Mol. Biol. *28,* 93–103 (1995).

Schmelzer, E., S. Krüger-Lebus, and K. Hahlbrock: Temporal and spatial patterns of gene expression around sites of attempted fungal infection in parsley leaves. Plant Cell *1,* 993–1001 (1989).

Schneiderbauer, A., E. Back, H. Sandermann, and D. Ernst: Ozone induction of extension mRNA in Scots pine, Norway spruce and European beech. New Phytol. *130,* 225–230 (1995).

Schraudner, M., D. Ernst, C. Langebartels, and H. Sandermann: Biochemical plant responses to ozone. III. Activation of the defense-related proteins β-1,3-glucanase and chitinase in tobacco leaves. Plant Physiol. *99,* 1321–1328 (1992).

Schröder, M., K. Hahlbrock, and E. Kombrink: Temporal and spatial patterns of 1,3-β-glucanase and chitinase induction in potato leaves infected by *Phytophthora infestans.* Plant J. *2,* 161–172 (1992).

Sharma, Y. K. and K. R. Davis: Ozone-induced expression of stress-related genes in *Arabidopsis thaliana.* Plant Physiol. *105,* 1089–1096 (1994).

Somssich, I. E., E. Schmelzer, P. Kawalleck, and K. Hahlbrock: Gene structure and *in situ* transcript localization of pathogenesis-related protein 1 in parsley. Mol. Gen. Genet. *213,* 93–98 (1988).

Spanu, P. and T. Boller: Ethylene biosynthesis in tomato plants infected by *Phytophthora infestans.* J. Plant Physiol. *134,* 533–537 (1989).

Van Loon, L. C., W. S. Pierpoint, Th. Boller, and V. Conejero: Recommendations for naming plant pathogenesis-related proteins. Plant Mol. Biol. Rep. *12,* 245–264 (1994).

Ward, E. R., G. B. Payne, M. B. Moyer, S. C. Williams, S. S. Dincher, K. C. Sharkey, J. J. Beck, H. T. Taylor, P. Ahl-Goy, F. Meins, and J. A. Ryals: Differential regulation of β-1,3-glucanase messenger RNAs in response to pathogen infection. Plant Physiol. *96,* 390–397 (1991 a).

Ward, E. R., S. J. Uknes, S. C. Williams, S. S. Dincher, D. L. Wiederhold, D. C. Alexander, P. Ahl-Goy, J.-P. Métraux, and J. A. Ryals: Coordinate gene activity in response to agents that induce systemic acquired resistance. Plant Cell *3,* 1085–1094 (1991 b).

Willekens, H., W. van Camp, M. van Montagu, D. Ínze, C. Langebartels, and H. Sandermann: Ozone, sulfur dioxide, and ultraviolet B have similar effects on mRNA accumulation of antioxidant genes in *Nicotiana plumbaginifolia* L. Plant Physiol. *106,* 1007–1014 (1994).

J. Plant Physiol. Vol. 148. pp. 222–228 (1996)

Ozone and Ultraviolet B Effects on the Defense-related Proteins β-1,3-Glucanase and Chitinase in Tobacco

MICHAELA THALMAIR[1], GUY BAUW[3], STEPHAN THIEL[2], THORSTEN DÖHRING[2], CHRISTIAN LANGEBARTELS[1]*, and HEINRICH SANDERMANN Jr.[1]

[1] GSF – Forschungszentrum für Umwelt und Gesundheit GmbH, Institut für Biochemische Pflanzenpathologie and
[2] Arbeitsgruppe Expositionskammern, Neuherberg, D-85764 Oberschleißheim, Germany

[3] Laboratorium voor Genetika, via the Department of Genetics, affiliated to the Flanders Interuniversity Institute for Biotechnology, Universiteit Gent, B-9000 Gent, Belgium

Received October 13, 1995 · Accepted November 15, 1995

Summary

The air pollutant ozone is a potent abiotic inducer of defense-related enzymes such as pathogenesis-related proteins. Here we report on the accumulation of β-1,3-glucanase and chitinase in *Nicotiana tabacum* L. treated with ozone and ultraviolet B radiation, singly and in combination, under a simulated sunlight spectrum. Ozone ($0.16\,\mu L \cdot L^{-1}$, 2×5 h) induced the basic isoforms of β-1,3-glucanase in both, ozone-sensitive (Bel W3) and -tolerant (Bel B) cultivars, while chitinase was only affected in cv. Bel W3. Ultraviolet B radiation (7.5 MED) alone did not lead to β-1,3-glucanase or chitinase induction. In combined treatments ultraviolet B increased the ozone-dependent lesion formation and reduced chitinase accumulation in the sensitive cv. Bel W3. Analysis of the intercellular washing fluid of ozone-treated plants revealed the accumulation of a major ozone-related protein (O_3R-1) of 28 kDa within 32 h. Microsequence analysis of two tryptic peptides showed 100 % homology to acidic chitinase PR-3b. These results indicate that basic β-1,3-glucanase and chitinase are distinctly regulated in ozone and ultraviolet B treated tobacco, and that ultraviolet B radiation with a similar UV edge as the solar spectrum does not lead to an accumulation of basic pathogenesis-related proteins.

Key words: Nicotiana tabacum L., apoplast, β-1,3-glucanase, chitinase, ozone, ultraviolet B, intercellular washing fluid.

Abbreviations: IWF = intercellular washing fluid; O_3R = ozone-related; PPFR = photosynthetic photon fluence rate; PR = pathogenesis-related; TMV = tobacco mosaic virus.

Introduction

Ozone and ultraviolet B radiation (UV-B, 280–320 nm) are important features of the environment in which plant species evolved. Adverse effects of both factors on plants have been demonstrated in controlled experiments when certain threshold levels were surpassed (Runeckles and Krupa, 1994). On the other hand, constitutive as well as inducible antioxidants and UV protectants are known, e.g. antioxidant molecules and proteins as well as UV-B absorbing secondary compounds (Tevini, 1993; Sandermann et al., 1994). Ample information exists on ozone and UV-B radiation as single factors, but information about the effects of simultaneous exposure to both stressors is very scarce (Runeckles and Krupa, 1994).

Pathogenesis-related (PR) proteins are known to be induced by biotic factors, but also in response to a variety of abiotic stimuli, including organic chemicals, heavy metals and air pollutants (Bol et al., 1990; Simmons, 1994). Functional analysis of PR proteins has revealed that they mainly

* Correspondence.

exhibit β-1,3-glucanase (PR-2, EC 3.2.1.39) and chitinase (PR-3, 8 and 11, EC 3.2.1.14) activities. Thaumatin-like proteins (PR-5), proteinase inhibitors (PR-6) and peroxidases (PR-9) have also been included in this group (van Loon et al., 1994). The photochemical air pollutant ozone has been characterized as an important abiotic elicitor of PR proteins (Schraudner et al., 1992; 1994). Ozone exposure of the bioindicator plants for ozone, tobacco Bel W3 and Bel B (Heggestad, 1991), increased the levels of mRNAs for β-1,3-glucanase, chitinase and PR protein 1b in this temporal sequence (Ernst et al., 1992). Similarly, UV-C radiation (200–280 nm) is known to induce the accumulation of PR proteins, and increased the resistance of tobacco plants to TMV infection (Brederode et al., 1991; Yalpani et al., 1994). The UV-B induced accumulation of acidic PR-1 proteins was recently reported by Green and Fluhr (1995). These results indicate that the abiotic stressors ozone and UV-B radiation may lead to similar inductions of PR proteins.

In this study we have compared the response of β-1,3-glucanase and of chitinase to ozone and ultraviolet B applied singly and in combination under carefully controlled climatic conditions. Plants were treated under a radiation spectrum close to natural solar radiation (Payer et al., 1994; Thiel et al., 1996). Besides total leaf extracts, we have analysed proteins of the intercellular washing fluid of tobacco leaves and have characterized by microsequencing a major ozone-responsive protein as an acidic chitinase.

Materials and Methods

Plant Material and Treatment Conditions

Nicotiana tabacum L. cv. Bel W3 and Bel B plants were grown in pollutant-free air as described by Langebartels et al. (1991). Ten-week-old plants were transferred into walk-in chambers of the GSF phytotron (Payer et al., 1994) and were acclimated for 6 days in pollutant-free air to higher light intensities without UV-B radiation. Climatic conditions included a 14 h light (6 a.m. to 8 p.m., $PPFR_{400-700\,nm}$ 1100 ± 50 μmol $m^{-2} s^{-1}$, 25 ± 1 °C) – 10 h dark cycle (20 ± 1 °C) and relative humidities between 60 and 70 %. PPFR increased continuously between 6 and 8 a.m. and decreased between 6 and 8 p.m. All other conditions were identical to those described by Willekens et al. (1994).

Plants were exposed to ozone, UV-B and ozone plus UV-B for 5 h/d (between 9 a.m. and 2 p.m.), respectively, on two successive days (Fig. 1). Control plants were held in pollutant-free air without UV-B radiation. Ozone was generated by electrical discharge in pure oxygen and was administered through mass flow controllers into the air stream entering the growth chambers (Langebartels et al., 1991). Maximum ozone concentrations were 0.16 ± 0.01 $\mu L \cdot L^{-1}$. The chamber atmosphere was sampled using teflon lines at the canopy level at eight locations per chamber. Concentrations of ozone were measured using a UV-type ozone analyzer (CSI 3100; Columbia Scientific Instruments, Austin, TX).

UV-B radiation was provided between 9 a.m. and 2 p.m. by UV-B fluorescent lamps (TL 12; Philips, Eindhoven, The Netherlands). UV-C was excluded by two layers of borosilicate glass (5 mm Tempax, 6.5 mm Pyran; Schott, Mainz, Germany). UV-B spectral irradiance during the experiments was measured by a double-monochromator with chopper and lock-in amplifier (Thiel et al., 1996). The $UV\text{-}B_{BE}$ irradiance was calculated using the generalized plant action spectrum of Caldwell (1971), normalized to 300 nm, and was approx. 55 $mW \cdot m^{-2}$. A Robertson-Berger meter (Biometer Model 501, Solar light, Philadelphia, USA) was used for continuous monitoring of UV-B radiation (Fig. 1 a). Weighted UV-B irradiance was expressed in minimal erytheme effective dose (MED) units and was approx. 0.67 MED/h during the exposure phase (9 a.m. to 2 p.m.) or 7.5 MED as total dose (6 a.m. to 8 p.m.). Control plants were held under ordinary float glass and received <0.5 MED total dose or $UV\text{-}B_{BE}$ below 0.1 $mW \cdot m^{-2}$.

Experiments were conducted in a completely randomized design with three replicates for each treatment. Duncan's multiple range test ($P<0.05$) was used to separate treatment means.

Determination of β-1,3-glucanase activities

Leaves were numbered from the top of the plant (leaf 1 larger than 8 cm; Langebartels et al., 1991). Leaves 1 to 7 were harvested separately excluding the mid-veins, as well as top and bottom of the leaves, and were ground in liquid nitrogen. β-1,3-Glucanase activity was assayed fluorimetrically with laminarin (Sigma, Deisenhofen, Germany) as substrate (Schraudner et al., 1992). β-1,3-Glucanase activity is expressed in katal per gram fresh weight on the basis of glucose reducing equivalents released from laminarin.

Protein extraction and analysis

Proteins were extracted in three volumes of 200 mM Tris-HCl (pH 8.0) containing 1 mM PMSF, 5 mM DTT, 0.25 mM EDTA. The resulting extract was clarified before analysis (8800 g, 15 min at room temperature). β-1,3-Glucanase and chitinase proteins were measured by a two-antibody ELISA according to Sticher et al. (1992) in microtiter plates (Costar, Biorad). Western blot analysis was performed according to Towbin et al. (1979). The antibodies against basic and acidic β-1,3-glucanase and basic-type chitinase were a kind gift of Dr. F. Meins, Basel, Switzerland.

Intercellular washing fluid (IWF) was prepared from longitudinal leaf halves without the mid veins by vacuum-infiltration of 50 mM sodium acetate buffer (pH 5.0), 100 mM KCl, 5 mM DTE. IWF was collected by centrifuging the rolled leaf halves at 1000 g for 10 min. Contamination with intracellular markers was less than 1 % (Langebartels et al., 1991).

A volume of 1.5 mL IWF (corresponding to 4.5 g leaf tissue) was concentrated by freeze-drying, and the residue was then dissolved in 0.3 mL bidistilled water. After addition of 30 μL 0.15 % (w/v) sodium desoxycholate and incubation in an ice bath for 30 min, the protein was precipitated with TCA (1 h at 0 °C). The pellet obtained by centrifugation (5000 g, 10 min) was dissolved in 90 μL SDS sample buffer (Laemmli, 1970) and 10 μL 1 M Tris base, and was heated at 80 °C for 10 min. SDS-PAGE was carried out using a 5 % acrylamide stacking gel and a 12 to 15 % gradient separation gel. Proteins were blotted on PVDF-membranes (Millipore, Eschborn, Germany) and stained with Amido black prior to microsequence analysis (Bauw et al., 1989). The PVDF-bound proteins were subjected to an *in situ* digestion with trypsin, and the eluted peptides were purified by reversed-phase HPLC. The main peptides were analysed in a protein sequencing apparatus (470A, Applied Biosystems; Bauw et al., 1989).

Results

Effects of ozone and ultraviolet B on tobacco plants

Tobacco plants cv. Bel W3 and Bel B were acclimated to high light intensities (1100 ± 50 μmol $m^{-2} \cdot s^{-1}$) for 6 days, and were then treated with two pulses of ozone, UV-B radiation and ozone plus UV-B, respectively, for 5 h on two conse-

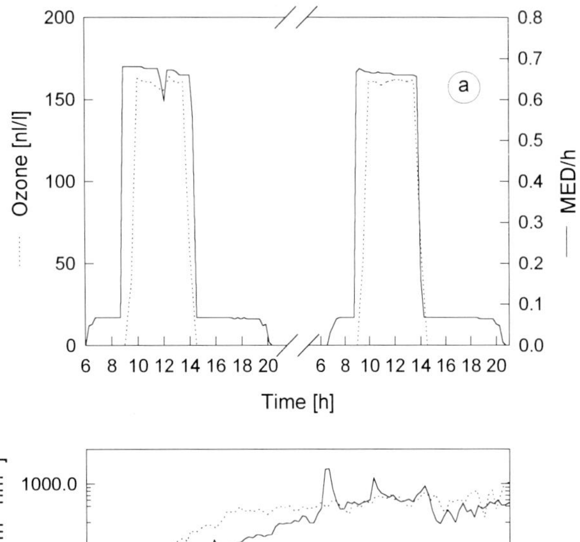

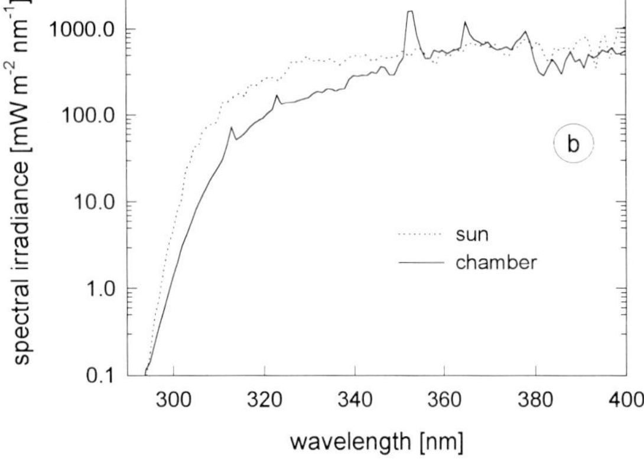

Fig. 1: Exposure profiles for ozone and UV-B radiation. (a) Time course of ozone (dotted line) and of minimal erytheme effective dose (MED; full line) during two consecutive days of exposure. (b) Mean spectral irradiance for the experimental area measured at plant canopy (chamber) in comparison to the solar spectrum (60° sun elevation, 298 Dobson units total ozone column, <2 Oktas cloud cover, low turbidity, albedo 0.05).

cutive days (Fig. 1 a). UV-B radiation was approx. 0.67 $MED \cdot h^{-1}$ (7.5 MED total dose), while biologically effective irradiance for the general plant response ($UV\text{-}B_{BE}$; Caldwell, 1971) amounted to $55\,mW \cdot m^{-2}$. The spectral distribution of irradiance at the plant canopy level was similar to the solar spectrum (60° sun elevation) with cut-off wavelengths at $0.1\,mWm^{-2} \cdot nm^{-1}$ (Döhring et al., 1996) of 293 and 294 nm, respectively (Fig. 1 b). Ozone exposure alone ($0.16 \pm 0.01\,\mu L \cdot L^{-1}$; 2×5 h) led to typical pergament-like lesions on the middle-aged leaves of the ozone-sensitive tobacco cv. Bel W3 (Fig. 2). Ozone exposure plus ultraviolet B radiation led to significantly higher percentages of necrotic leaf area on leaves 5 and 6 (Figs. 2, 3). As expected, the ozone tolerant cv. Bel B did not show any symptoms following ozone or ozone plus UV-B treatment (data not shown). UV-B radiation alone did not lead to any visible injury in both cultivars.

Effect of ozone and UV-B radiation on β-1,3-glucanase and chitinase

β-1,3-Glucanase activity in the middle-aged and older leaves 4 to 7 increased already 8 h after the beginning of the first ozone pulse and was 6- to 8-fold higher than in control plants after two subsequent pulses (32 h; Fig. 4). Similar results were found with β-1,3-glucanase protein content where the levels were elevated up to 70-fold. The combined treatment of ozone and UV-B did not lead to major changes in the ozone-dependent induction of β-1,3-glucanase. UV-B alone led to a weak induction in the older leaves 6 and 7 where β-1,3-glucanase activity inherently increases under developmental regulation (Schraudner et al., 1992; Ernst et al., 1996). Western blotting revealed the ozone-dependent accumulation of a basic isoform of β-1,3-glucanase at 33 kDa, but acidic isoforms were induced neither by ozone nor by UV-B (data not shown). As can be seen in Fig. 4, β-1,3-glucanase was similarly, but less drastically induced at the activity and

Fig. 2: Visible symptoms on leaves of tobacco plants cv. Bel W3. Middle aged-leaves of ozone- (left) and ozone plus UV-B-(right) treated plants are shown 32 h after the onset of exposure.

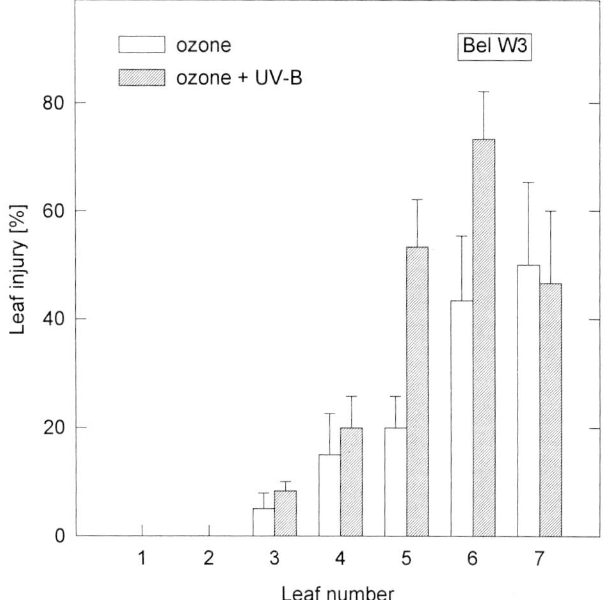

Fig. 3: Leaf injury in tobacco cv. Bel W3 in percent necrotic leaf area, 32 h after the onset of ozone exposure (0.16 ± 0.01 μL·L^{-1}, 2×5 h) with (shaded columns) and without UV-B radiation (open columns, 7.5 MED total dose, 2×5 h). Leaves are counted from the top of the plant, leaf 1 being larger than 8 cm. Means ± S.E. (n=3).

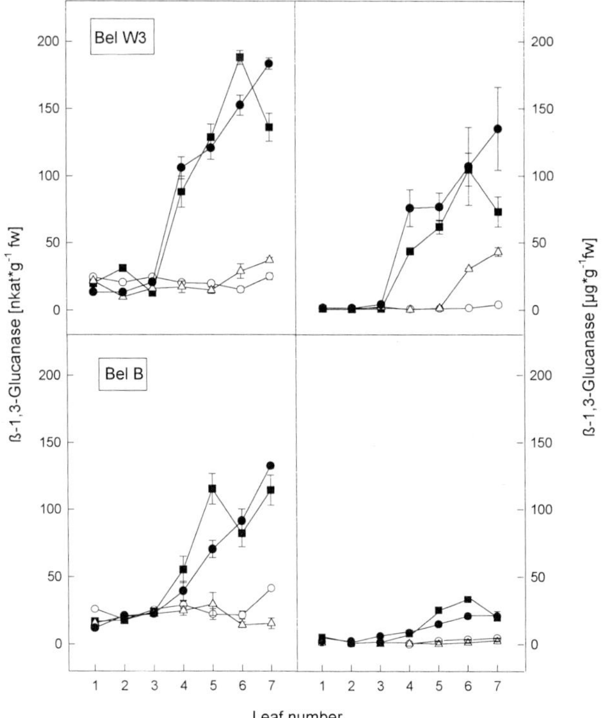

Fig. 4: β-1,3-Glucanase enzyme activity (left) and protein content (right) of O_3 and UV-B treated tobacco plants (top, cv. Bel W3; bottom, cv. Bel B). Plants were exposed to ozone (●), ozone plus UV-B (■), UV-B (△) or to pollutant-free air without UV-B (○) for 5 h on two consecutive days. Leaves 1 to 7 were harvested after 32 h. Means ± S.E. (n=3).

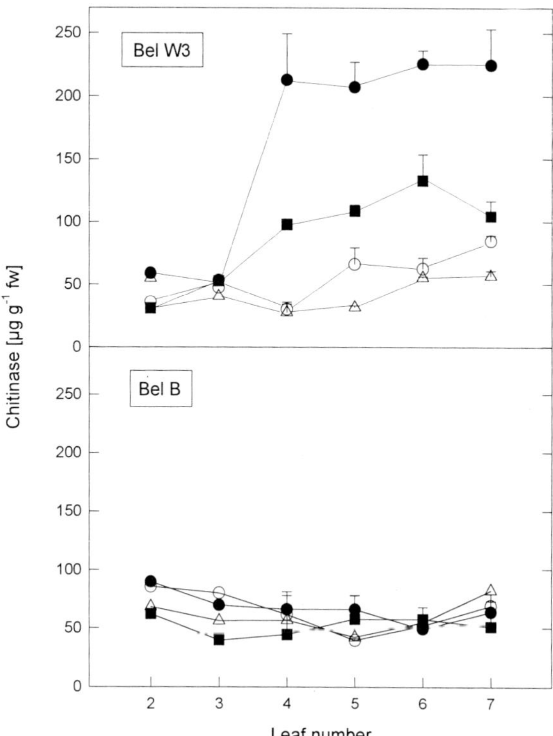

Fig. 5: Chitinase protein content in O_3 and UV-B treated tobacco plants (top, cv. Bel W3; bottom, cv. Bel B). Plants were exposed to ozone (●), ozone plus UV-B (■), UV-B (△) or to pollutant-free air without UV-B (○) for 5 h on two consecutive days. Leaves 1 to 7 were harvested after 32 h. Means ± S.E. (n=3).

protein level in the ozone-tolerant cv. Bel B. The combination of ozone and UV-B radiation led to slightly higher increases of β-1,3-glucanase in this cultivar (significant only for leaf 5); all these reactions occurred in the absence of visible symptoms.

As shown in Fig. 5, chitinase protein levels were elevated in ozone-treated tobacco cv. Bel W3 only. In contrast to β-1,3-glucanase, chitinase was significantly ($P<0.05$) reduced in the combined treatment. No responses of chitinase protein levels to ozone, UV-B and ozone plus UV-B, respectively, were evident in cv. Bel B.

Identification of an ozone-related apoplastic protein

Treatments of tobacco plants with a single pulse of ozone predominantly led to the induction of intracellular β-1,3-glucanase and chitinase activities (Schraudner et al., 1992). When the IWF obtained by vacuum infiltration of tobacco leaves harvested after 24 and 32 h was analysed by SDS-PAGE, several ozone-induced or reduced protein bands were found (M. Thalmair, G. Bauw, C. Langebartels, unpublished results). A major protein band (O_3R-1) induced in ozone-treated tobacco Bel W3 occurred at 28 kDa. This protein was analysed by microsequence analysis of tryptic peptides generated by in *in situ* digestion of the PVDF-bound proteins (Bauw et al., 1989). The sequences of two tryptic peptide fragments are given in Fig. 6. Data bank research revealed 100 % identity of both sequences with class II tobacco acidic chitinase PR-3b (PR-Q; Payne et al., 1990).

		136																152
TOB O_3R-1	..	A	G	T	A	I	G	Q	E	L	V	N	N	P	D	L	V	A
TOB PR-3b (II)	..	A	G	T	A	I	G	Q	E	L	V	N	N	P	D	L	V	A
TOB PR-3a (II)	..	A	G	N	A	I	R	Q	D	L	V	N	N	P	D	L	V	A
TOB PR-3c (I)	..	C	G	R	A	I	G	V	D	L	L	N	N	P	D	L	V	A
TOB PR-3d (I)	..	C	G	R	A	I	G	V	D	L	L	N	N	P	D	L	V	A
TOB PR-3e (I)	..	C	G	R	A	I	G	Q	N	L	L	N	N	P	D	L	V	A

		186								195	
TOB O_3R-1	..	P	S	A	A	D	Q	A	A	N	R	..
TOB PR-3b (II)	..	P	S	A	A	D	Q	A	A	N	R	..
TOB PR-3a (II)	..	P	S	A	A	D	Q	S	A	N	R	..
TOB PR-3c (I)	..	P	S	A	G	D	R	A	A	N	R	..
TOB PR-3d (I)	..	P	S	S	A	D	R	A	A	N	R	..
TOB PR-3e (I)	..	P	S	A	A	D	R	A	A	N	R	..

Fig. 6: Comparison of amino acid sequences of tryptic peptides of the ozone-related protein O_3R-1 and tobacco chitinases. Source of data: TOB PR-3b (class II) and TOB PR-3a (II), tobacco PR-P and PR-Q cDNA clones (Payne et al., 1990); TOB PR-3c (I), TOB PR-3d (I), TOB PR-3e (I), tobacco genomic clones λCHN 17, λCHN 200 and λCHN 14, respectively (Shinshi et al., 1990; van Buuren et al., 1992).

Discussion

In recent years the use of plants exhibiting differential ozone tolerance has led to rapid advances in our understanding of plant responses to oxidative stress. It was observed that several ozone responses resemble those of plant-pathogen interactions (Sandermann et al., 1994). Studies on ozone effects on plant gene expression have shown that the steady-state levels of mRNAs for PR proteins (Ernst et al., 1992), antioxidant enzymes (Willekens et al., 1994), ethylene biosynthetic enzymes (Schlagnhaufer et al., 1995) and other stress-related proteins (Eckey-Kaltenbach et al., 1994; Sharma and Davis, 1994) increase in ozone-treated plants. The biomonitor plants for ozone, tobacco Bel W3 and Bel B (Heggestad, 1991), have been characterized biochemically in detail (Sandermann et al., 1994) and have also been used in the present study to monitor the responses of the PR proteins β-1,3-glucanase and chitinase to ozone and UV-B radiation singly and in combination.

β-1,3-Glucanase induction by ozone occurred at the enzyme and protein levels in both cultivars, but was higher in the ozone-sensitive tobacco cv. Bel W3. This induction was confined to middle-aged and older leaves which exhibit the highest stomatal conductance (Langebartels et al., 1991) and thereby experience the highest internal ozone doses. Chitinase protein levels, on the other hand, were only increased in the ozone-sensitive cultivar Bel W3. Similar results have been obtained by Ernst et al. (1992) where steady-state mRNA levels for basic β-1,3-glucanase rose as early as one hour after the onset of exposure, while basic chitinase (3 h), acidic chitinase (14 h) and PR-1b (14 h) mRNAs accumulated later. Schraudner et al. (1992) showed that ozone induction factors in tobacco were between 35 and 70-fold for β-1,3-glucanase, but only 3- to 4-fold for chitinase activity. A pronounced coordination of basic isoforms of β-1,3-glucanase and chitinase has been found in TMV-infected tobacco (Vögeli-Lange et al., 1988) ethylene-treated bean (Vögeli et al., 1988) and potato leaves infected by *Phytophthora infestans* (Schröder et al., 1992). On the other hand, independent regulation has also been reported with other interactions such as tobacco infected with *Pseudomonas tabaci* (Meins and Ahl, 1989) or treated with salicylic acid (Bol, 1988). We conclude that basic β-1,3-glucanase and chitinase are not coordinately regulated in ozone- and UV-B treated tobacco, but their induction may be mediated by related signal transduction pathways.

It was shown in a preceding study that ozone predominantly induced the basic (intracellular) isoforms of β-1,3-glucanase and chitinase (Schraudner et al., 1992). Here we analysed for the first time the protein profile of the intercellular washing fluid in ozone-treated and control tobacco leaves. Besides other so far unidentified ozone-related (O_3R) proteins, a major band at 28 kDa was identified by microsequence as class II acidic chitinase PR-3b (nomenclature according to van Loon et al., 1994). This chitinase has been described by Payne et al. (1990) in virus-infected tobacco, and it occurs also in the apoplastic fluid of ozone-treated plants as shown in the present paper. Sequence homology of the peptides to acidic PR-3a (PR-P; class II) was 85 %, while the basic chitinases PR-3c, 3d and 3e (class I; Meins et al., 1992) were identical between 74 and 81 %. Accumulation of this acidic chitinase was, however, only observed after two subsequent ozone treatments in the sensitive cultivar Bel W3. As visible leaf injury occurred at the same time (Fig. 3), acidic chitinase is considered as a secondary response to ozone which waits for tissue damage to occur.

UV-B radiation alone led to slight inductions of β-1,3-glucanase activity and protein in older leaves of cv. Bel W3, while no reactions were found in cv. Bel B. When plants were exposed to ozone and UV-B radiation simultaneously, ozone-induced β-1,3-glucanase activity and protein levels were virtually not affected, while chitinase protein levels in cv. Bel W3 decreased by 50 %. This result again clearly points to differential regulation of PR proteins. Irradiation with shorter wavelengths (UV-C) has previously been demonstrated to cause induction of all PR protein families (Brederode et al., 1991). This induction correlated with drastic tissue damage which may be the cause of the observed responses (Brederode et al., 1991). Yalpani et al. (1994) showed a similar accumulation of PR-1 proteins by UV-C light and by ozone. However, the stratospheric ozone layer totally excludes UV-C light from reaching the earth's surface so UV-C radiation is unsuitable to demonstrate UV-B responses in plants.

Recently, Green and Fluhr (1995) reported the accumulation of acidic-type PR-1 proteins in UV-B treated tobacco plants cv. Samsun NN. Other PR proteins were only weakly affected. It thus seems that (acidic) PR-1 proteins are regulated differently from the (basic) PR-2 and PR-3 investigated in this study. It is also possible that the overall light regime applied during pre-cultivation and exposure is of importance for PR protein accumulation. Unfortunately, no spectral data were provided by Green and Fluhr (1995). In the present study, the plants were acclimated and exposed under spectral characteristics close to the solar spectrum (Fig. 1 b). This included high light intensities in the visible range (PPFR above $1000\,\mu mol \cdot m^{-2} \cdot s^{-1}$), balanced UV-B/PPFR ratios and a cut-off wavelength (Döhring et al., 1996) at 293 nm which is very close to that of the solar spectrum (294 nm). Preliminary experiments suggest that cut-off wavelengths below the solar UV edge (294 nm) lead to an induction of PR proteins while cut-off at longer wavelenghts do not affect basic β-1,3-glucanase and chitinase (M. Thalmair, C. Langebartels, unpublished results).

The signalling pathways for PR protein induction in ozone- and UV-B-exposed plants are not understood as yet, although evidence for a role of ethylene and/or activated oxygen species is accumulating. Ozone increased ethylene biosynthesis primarily in the sensitive cultivar Bel W3 (Langebartels et al., 1991; Yin et al., 1994). Exogeneously applied ethylene is known to affect mainly the basic isoforms of PR proteins (Boller, 1993) which are also increased by ozone. Other ozone-induced genes and proteins such as phenylalanine ammonia-lyase, peroxidase, glutathione S-transferase are also ethylene-inducible (Kangasjärvi et al., 1994). The recent findings of ethylene-responsive elements in the promoter region of PR proteins (Eyal et al., 1993) add to the importance of ethylene in PR protein regulation. Up to now, UV-B related effects on ethylene biosynthesis have only been described by Ros (1990). This author has shown that ethylene emission from sunflower seedlings was elevated by 50 and 80 % with increasing irradiance of UV-B at cut-off wavelengths of 295 and 280 nm, respectively.

Under experimental conditions comparable to those used in the present study, UV-B radiation and ozone led to an increase in the steady-state transcript levels of distinct antioxidant genes (Willekens et al., 1994). It was shown that one of three catalase isoforms (Cat 2) as well as glutathione peroxidase responded to both stresses with similar rapid increases (starting at 3 h). Catalase is involved in H_2O_2 degradation, while glutathione peroxidase may repair lipid-derived peroxides (Willekens et al., 1994). Gene expression of several antioxidant enzymes also increased in ozone-exposed *Arabidopsis thaliana* with glutathione *S*-transferase transcripts responding after only 30 minutes (Conklin and Last, 1995). Treatments with ozone or UV-B radiation in parallel experiments for eight days led to approx. 50 % increases in antioxidant compounds and enzyme activites in the same species (Rao and Ormrod, 1995). In addition, ozone and UV-B radiation increase the levels of activated oxygen species and lipid peroxidation in plants as reviewed by Foyer et al. (1994). Activated oxygen species may therefore act as signal molecules in ozone (Schraudner et al., 1992) and UV-B treated plants (Green and Fluhr, 1995) as proposed in plant-pathogen studies (Dixon et al., 1994).

In conclusion, we propose that the induction of intracellular basic isoforms of β-1,3-glucanase, and possibly chitinase, is mediated by ozone-induced stress ethylene. Other responses, such as the accumulation of acidic PR 1 proteins (Yalpani et al., 1994; Green and Fluhr, 1995) may be mediated via activated oxygen species, but these reactions seem to occur rather late in ozone-treated plants or at wavelengths below the solar UV edge.

Acknowledgements

We thank Lucia Göβl and Josef Hintermair as well as the staff of the GSF phytotron for their excellent technical assistance. The authors are grateful to Dr. F. Meins, FMI Basel, for introducing one of us (M.T.) to ELISA techniques and for the kind gift of β-1,3-glucanase and chitinase antibodies. This work was supported by grants from the Bundesministerium für Forschung und Technologie and the Fonds der Chemischen Industrie.

References

Bauw, G., J. van Damme, M. Puype, J. Vandekerckhove, B. Gesser, G. P. Ratz, J. B. Lauridsen, and J. E. Celis: Protein-electroblotting and -microsequencing strategies in generating protein data bases from two-dimensional gels. Proc. Natl. Acad. Sci. USA *86,* 7701–7705 (1989).

Bol, J. F., J. M. Linthorst, and B. J. C. Cornelissen: Plant pathogenesis-related proteins induced by virus infection. Annu. Rev. Phytopathol. *28,* 113–138 (1990).

Bol, J. F.: Structure and expression of plant genes encoding pathogenesis-related proteins. In: Verma, D. P. S. and R. B. Goldberg (eds.): Temporal and Spatial Regulation of Plant Genes, pp. 201–221. Springer, Wien (1988).

Boller, T.: Antimicrobial functions of the plant hydrolases, chitinase and β-1,3-glucanase. In: Fritig, B. and M. Legrand (eds.): Mechanisms of Plant Defense Responses, pp. 391–400. Kluwer, Dordrecht (1993).

Brederode, F. T., H. J. M. Linthorst, and J. F. Bol: Differential induction of acquired resistance and PR gene expression in tobacco by virus infection, ethylene treatment, UV light and wounding. Plant Mol. Biol. *17,* 1117–1125 (1991).

Caldwell, M. M.: Solar UV irradiation and the growth and development of higher plants. In: Giese, C. (ed.): Photophysiology, Vol. VI, pp. 131–268. Academic Press, New York (1971).

Conklin, P. L. and R. L. Last: Differential accumulation of antioxidant mRNAs in *Arabidopsis thaliana* exposed to ozone. Plant Physiol. *109,* 203–212 (1995).

Dixon, R. A., M. J. Harrison, and C. J. Lamb: Early events in the activation of plant defense responses. Annu. Rev. Phytopathol. *32,* 479–501 (1994).

Döhring, T., M. Köfferlein, S. Thiel, and H. K. Seidlitz: Spectral shaping of artificial UV-B irradiation for vegetation stress research. J. Plant Physiol. *148,* 115–119 (1996).

Eckey-Kaltenbach, H., D. Ernst, W. Heller, and H. Sandermann: Biochemical plant responses to ozone: IV. Cross-induction of defensive pathways in parsley (*Petroselinum crispum* L.) plants. Plant Physiol. *104,* 67–74 (1994).

Ernst, D., A. Bodemann, E. Schmelzer, C. Langebartels, and H. Sandermann: β-1,3-Glucanase mRNA is locally, but not systemically induced in *Nicotiana tabacum* L. cv. Bel W3 after ozone fumigation. J. Plant Physiol. *148,* 215–221 (1996).

Ernst, D., M. Schraudner, C. Langebartels, and H. Sandermann: Ozone-induced changes of mRNA levels of β-1,3-glucanase, chitinase and pathogenesis-related protein 1b in tobacco plants. Plant Mol. Biol. *20,* 673–682 (1992).

Eyal, Y., Y. Meller, S. Lev-Yadun, and R. Fluhr: A basic-type PR-1 promoter directs ethylene responsiveness, vascular and abscission zone-specific expression. Plant J. *4,* 225–234 (1993).

Foyer, C. H., M. Lelandais, and K. J. Kunert: Photooxidative stress in plants. Physiol. Plant. *92,* 696–717 (1994).

Green, R. and R. Fluhr: UV-B-induced PR-1 accumulation is mediated by active oxygen species. Plant Cell *7,* 203–212 (1995).

Heggestad, H. E.: Origin of Bel-W3, Bel-C and Bel-B tobacco varieties and their use as indicators of ozone. Environ. Pollut. *74,* 264–291 (1991).

Laemmli, U. K.: Cleavage of structural proteins during the assembly of the head of bacteriophage T4. Nature *227,* 680–685 (1970).

Langebartels, C., K. Kerner, S. Leonardi, M. Schraudner, M. Trost, W. Heller, and H. Sandermann: Biochemical plant responses to ozone. I. Differential induction of polyamine and ethylene biosynthesis in tobacco. Plant Physiol. *95,* 882–889 (1991).

Kangasjärvi, J., J. Talvinen, M. Utriainen, and R. Karjalainen: Plant defence systems induced by ozone. Plant Cell Environ. *17,* 783–794 (1994).

Meins, F. and P. Ahl: Induction of chitinase and β-1,3-glucanase in tobacco plants infected with *Pseudomonas tabaci* and *Phytophthora parasitica* var. *nicotianae.* Plant Sci. *61,* 155–161 (1989).

Meins, F., J.-M. Neuhaus, C. Sperisen, and J. Ryals: The primary structure of plant pathogenesis-related glucanohydrolases and their genes. In: Boller, T. and F. Meins (eds.): Genes involved in Plant Defense, pp. 245–282. Springer-Verlag, Wien (1992).

Payer, H. D., P. Blodow, M. Köfferlein, M. Lippert, W. Schmolke, G. Seckmeyer, H. Seidlitz, D. Strube, and S. Thiel: Controlled environment chambers for experimental studies on plant responses to CO_2 and interactions with pollutants. In: Schulze, E.-D. and H. A. Mooney (eds.): Design and Execution of Experiments on CO_2 Enrichment. pp. 127–145. Commission of the European Communities, Brussels (1994).

Payne, G., P. Ahl, M. Moyer, A. Harper, J. Beck, F. Meins, and J. Ryals: Isolation of complementary DNA clones encoding pathogenesis-related proteins P and Q, two acidic chitinases from tobacco. Proc. Natl. Acad. Sci. USA *87,* 98–102 (1990).

Rao, M. V. and D. P. Ormrod: Impact of UVB and O_3 on the oxygen free radical scavenging system in *Arabidopsis thaliana* genotypes differing in flavonoid biosynthesis. Photochem. Photobiol. *62,* 719–726 (1995).

Ros, J.: Zur Wirkung von UV-Strahlung auf das Streckungswachstum von Sonnenblumenkeimlingen (*Helianthus annuus* L.). Karls. Beitr. Entw. Ökophysiol. *8,* 1–157 (1990).

Runeckles, V. C. and S. V. Krupa: The impact of UV-B radiation and ozone on terrestrial vegetation. Environ. Pollut. *83,* 191–213 (1994).

Sandermann, H., D. Ernst, W. Heller, and C. Langebartels: Biochemical markers for stress detection and ecophysiology. In: Schulze, E.-D. and H. A. Mooney (eds.): Design and Execution of Experiments on CO_2 Enrichment, pp. 45–51. Ecosystems Research Report 6, Commission of the European Communities, Brussels (1994).

Schlagnhaufer, C. D., R. E. Glick, R. N. Arteca, and E. J. Pell: Molecular cloning of an ozone-induced 1-aminocyclopropane-1-carboxylate synthase cDNA and its relationship with a loss of rbcS in potato (*Solanum tuberosum* L.) plants. Plant Mol. Biol. *28,* 93–103 (1995).

Schraudner, M., D. Ernst, C. Langebartels, and H. Sandermann: Biochemical plant responses to ozone. III. Activation of the defense-related proteins β-1,3-glucanase and chitinase in tobacco leaves. Plant Physiol. *99,* 1321–1328 (1992).

Schraudner, M., U. Graf, C. Langebartels, and H. Sandermann: Ambient ozone can induce plant defense reactions in tobacco. Proc. Royal Soc. Edinburgh *102 B,* 55–61 (1994).

Schröder, M., K. Hahlbrock, and E. Kombrinck: Temporal and spatial patterns of 1,3-β-glucanase and chitinase induction in potato leaves infected by *Phytophthora infestans.* Plant J. *2,* 161–172 (1992).

Sharma, Y. K. and K. R. Davis: Ozone-induced expression of stress-related genes in *Arabidopsis thaliana.* Plant Physiol. *105,* 1089–1096 (1994).

Shinshi, H., J.-M. Neuhaus, J. Ryals, and F. Meins: Structure of a tobacco endochitinase gene: evidence that different chitinases genes can arise by transposition of sequences encoding a cysteine-rich domain. Plant Mol. Biol. *14,* 357–368 (1990).

Simmons, C. R.: The physiology and molecular biology of plant 1,3-β-1,3-D-glucanases and 1,3;1,4-β-D-glucanases. Crit. Rev. Plant Sci. *13,* 325–387 (1994).

Sticher, L., U. Hinz, A. D. Meyer, and F. Meins: Intracellular transport and processing of a tobacco vacuolar β-1,3-glucanase. Planta *188,* 559–565 (1992).

Tevini, M.: Effects of enhanced UV-B radiation on terrestrial plants. In: Tevini, M. (ed.): UV-B radiation and ozone depletion. pp. 125–153. Lewis Publishers, Boca Raton (1993).

Thiel, S., T. Döhring, M. Köfferlein, A. Kosak, P. Martin, and H. K. Seidlitz: A phytotron for plant stress research: How far can artificial lighting compare to natural sunlight? J. Plant Physiol. *148,* 456–463 (1996).

Towbin, H., T. Staehelin, and J. Gordon: Electrophoretic transfer of proteins from polyacrylamide gels to nitrocellulose sheets. Procedure and some applications. Proc. Natl. Acad. Sci. USA *76,* 4350–4354 (1979).

Van Buuren, M., J.-M. Neuhaus, H. Shinshi, J. Ryals, and F. Meins: The structure and regulation of homeologous tobacco endochitinase genes of *Nicotiana sylvestris* and *N. tomentosiformis* origin. Mol. Gen. Genet. *232,* 460–469 (1992).

Van Loon, L. C., W. S. Pierpoint, T. Boller, and V. Conejero: Recommendations for naming plant pathogenesis-related proteins. Plant Mol. Biol. Rep. *12,* 245–264 (1994).

Vögeli, U., F. Meins, and T. Boller: Co-ordinated regulation of chitinase and β-1,3-glucanase in bean leaves. Planta *174,* 364–372 (1988).

Vögeli-Lange, R., A. Hansen-Gehri, T. Boller, and F. Meins: Induction of the defense-related glucanohydrolases, β-1,3-glucanase and chitinase, by tobacco mosaic virus infection of tobacco leaves. Plant Sci. *54,* 171–176 (1988).

Willekens, H., W. van Camp, M. van Montagu, D. Inze, C. Langebartels, and H. Sandermann: Ozone, sulfur dioxide, and ultraviolet B have similar effects on mRNA accumulation of antioxidant genes in *Nicotiana plumbaginifolia* L. Plant Physiol. *106,* 1007–1014 (1994).

Yalpani, N., A. J. Enyedi, J. Leon, and I. Raskin: Ultraviolet light and ozone stimulate accumulation of salicylic acid, pathogenesis-related proteins and virus resistance in tobacco. Planta *193,* 372–376 (1994).

Yin, Z.-H., C. Langebartels, and H. Sandermann: Specific induction of ethylene biosynthesis in tobacco plants by the air pollutant, ozone. Proc. Royal Soc. Edinburgh *102 B,* 127–130 (1994).

J. Plant Physiol. Vol. 148. pp. 229–236 (1996)

Scots Pines after Exposure to Elevated Ozone and Carbon Dioxide Probed by Reflectance Spectra and Chlorophyll *a* Fluorescence Transients

OUTI MEINANDER[1], SUSANNE SOMERSALO[2], TOINI HOLOPAINEN[3], and RETO J. STRASSER[4]

[1] Finnish Meteorological Institute, Air Quality Department, Sahaajankatu 20 E, FIN-00810 Helsinki

[2] University of Helsinki, Department of Crop Production, FIN-00014 University of Helsinki

[3] University of Kuopio, Department of Ecology and Environmental Science, P.O. Box 1627, FIN-70211 Kuopio

[4] University of Geneva, Bioenergetics Laboratory, CH-1254 Jussy-Geneva

Received June 24, 1995 · Accepted November 25, 1995

Summary

Natural Scots pines have been exposed to filtered air, ambient air and air with elevated O_3 or/and CO_2 in open top chambers. The trees showed no differences in their optical responses prior to the fumigations. After the fumigation period of three months the plants were in good health. The position of the maximum derivative of the green light reflectance in carbon dioxide fumigated pines was shifted from the control pines inflection point, by approximately 4 nm towards shorter wavelengths. The position of the red edge derivative maximum showed no significant changes. By fluorescence techniques (as OJIP – fast fluorescence transients) nearly no change was found in the quantum yield for electron transport (φ_o or excitation energy trapping φ_{Po}. However, the estimated activities as absorption, trapping or electron transport per cross-section increased considerably for all samples with elevated O_3 or CO_2. This increased activity seems to be due to an increased antenna size in O_3 treated samples. At elevated CO_2 the antenna size is decreased whereas the density of reaction centers per cross-section increased. This means that two different stress-adaptation mechanisms can lead to a similar macroscopic phenomenon like e.g. an increased metabolic activity.

Introduction

Solar radiation reflected from leaves has a characteristic spectrum and each tree species has its own spectral properties. The spectral reflectance is an indicator of pigment composition and architecture of the sample. Chlorophyll fluorescence, on the other hand, is a process competing with photosynthetic electron transport, and can be employed to study the potential photosynthetic capacity and detect damages of the photosynthetic apparatus. Many previous studies have shown that measurements on reflectance or chlorophyll fluorescence can provide a powerful tool in the detection of forest damage (Lichtenthaler and Rinderle 1988, Rock et al. 1988, Ruth and Weisel 1993). Ruth et al. (1991) have reported that the combination of reflectance and chlorophyll fluorescence methods can be used to distinguish between trees of different damage classes. The fluorescence measurements can be mainly used to determine activities of the photosynthetic apparatus such as the fluxes of photons absorbed, energy trapped or electrons transported. Therefore, fluorescence measurements can be used as a tool for early diagnosis of stress, not only before any changes in the visual appearance of the samples have occured, but even when no or only slight changes in their chemical composition can be detected.

Elevated ozone concentrations in the atmosphere from today and elevated carbon dioxide concentrations in the near future will have some influences on the global vegetation. The concentrations however are such that dramatic changes within a short time cannot be expected. However, early diagnosis of damages is necessary to influence the legislation about environmental protection. In open top chambers it is possible to simulate an expected future climate with elevated CO_2, elevated O_3 or both. Non destructive methods like bio-spectroscopy and imaging of canopy, single plants or even parts of leaves, are today available.

Here we report how Scots pines react after exposure to elevated ozone or/and carbon dioxide.

Materials and Methods

Fumigation

The reflectance and chlorophyll fluorescence of Scots Pine (*Pinus sylvestris* L.) needles were measured from pines exposed to ozone (O_3) or/and carbon dioxide (CO_2). Five different treatments were used: filtered air, air containing ambient O_3, elevated O_3, elevated CO_2, and elevated O_3 and CO_2. An automatic fumigation control at the Mekrijärvi Research Station, University of Joensuu, Finland, was used. The intended values of ozone concentration (40–70 ppb, depending on the time of growth period) and carbon dioxide (350–700 ppm) were about twice as much as those of their ambient concentrations in Finland. The total cumulative doses and the critical doses above 40 ppb (1-hour average concentration exceeding 40 ppb) of ozone received by the pines are shown in Table 1.

In the measurements, twenty natural Scots pine saplings, approximately two meters high, were used. Each of the saplings was surrounded by a plastic open top chamber. Five pine saplings growing in a natural forest environment (no surrounding chamber) were also measured.

Reflectance measurements

The reflectance of shoots from the 25 pines was measured both prior to the fumigations and after three months of fumigation. The reflectance was measured at 512 channels between 320–1050 nm with the Personal Spectrometer™ (Analytical Spectral Devices, USA). The reflectance of horizontally oriented shoots was measured using eastern branches (except for one tree where southern branches were measured) at a height of 1.5 m with a fixed 15 degrees field of view (FOV) at 10 cm distance. A black background with constant reflectance properties was placed behind each measured shoot. The incoming light was calibrated by the reference reflectance standard Spectralon™. Prior to the fumigations, the reflectance of over-wintered first-year Scots pine shoots was measured in a forest environment on June 10th 1994, under stable fully clouded dry weather conditions. At the end of the growing season in 1994, i.e. after three months exposures, the reflectance of over-wintered second year Scots pine shoots was measured on September 14th, under fully clouded weather conditions after a rain shower. The measured shoots were of the same age group as those measured prior to the fumigations.

Table 1: Average cumulative total and critical (above 40 ppb) ozone doses received by the experimental seedlings during the growth period of 1994.

Treatment	cumulative dose ppb×h	
	>0	>40
O_3 exposure	109651	16499
Ambient air	80977	1274
Filtered air	59243	203

Data analysis was carried out by estimating the spectral position of the maximum derivative for the red edge (680–750 nm) and the green light (505–575 nm). This kind of analysis can be considered to be invariant with the illumination or the amount of background within the FOV of the spectroradiometer (Curran et al. 1990). The derivative vector dy was calculated by fitting a least squares line to every point of the vector y. The least squares line was calculated using four points before and four after the middle point. The aim was to minimize the amount of points around the middle point without strengthening the noise. The statistical testing on these wavelength positions was made using the non-parametric Mann-Whitney U-test.

Chlorophyll fluorescence

Chlorophyll fluorescence was measured with a Plant Efficiency Analyser (Hansatech, UK). The fluorescence transients of dark-adapted needles were measured as described by Strasser and Govindjee (1992) and Strasser et al. (1995). The fluorescence signal was measured in the far red (above 720 nm) with a time resolution of 10 µs and a safe first measurement 40 µs after the onset of illumination. The needles measured were of the same needle year class as those used for the reflectance measurements. Prior to the fumigations, fluorescence signals were measured on June 9th. At the end of the growing season, the measurements were made on the same day as those of reflectance. The fluorescence curve of a 1 second illumination was digitized into 1198 datapoints (Fig. 1). From each fluorescence transient of the type OJIP (as shown in Fig. 1 a and b), nine values were selected as indicated in Tables 2 a and 2 b.

By a built in routine in the PEA instrument an initial fluorescence value ($F_{0\,calc}$) is calculated. This value is very close to the value measured at 50 µs. Therefore, we use for all calculations $F_{50\,\mu s}$ as the initial fluorescence. The fluorescence data were used, according to Strasser and Strasser (1995), to calculate the biophysical fluxes: absorption (ABS), max. trapping (TR_0), max. electron transport (ET_0). These fluxes were all normalised over the number of reaction centers of PSII (RC) or the cross-section (CS) of the sample. Fig. 2 shows a model which contains these expressions and Table 3 presents the equations which link the measured signals to the biophysical expressions, as presented elsewhere (Strasser and Strasser 1995). The cross link of the different expressions is presented in Fig. 3.

Results

Reflectance signals prior to the fumigations

In the case of the red edge derivative maximum, no statistical differences were found between pines representing the different groups (2-sided Mann-Whitney U-test). Secondly, the positions of the green light derivative maximum of the pines were compared. None of these comparisons showed any difference (2-sided Mann-Whitney U-test, Table 4).

These results justified the comparison of the spectral responses after the fumigations.

Fluorescence signals prior to the fumigations

There were no differences in the fluorescence characteristics of the needles grown in chambers planned for filtered or ambient air or in the natural forest conditions, all these samples together were control-values. No difference in the fluorescence signals was found between the control plants and the plants to be treated (1-sided t-test).

Table 2: Reduction of a whole fluorescence transient to nine distinct values. t_{Fmax} = time to reach the maximal fuorescence intensity; Area = Summation of all values $(F_M - F_t)$ for $t = 50\,\mu s$ to $t = t_{Fmax}$; $F_{0\ calc}$ = Initial fluorescence intensity, calculated by the instrument; F_M = Maximal measured fluorescence intensity; F_1 to F_5 = Fluorescence intensities measured at the corresponding time. 2a shows the raw data of the trees in free air and in the chambers with filtered and ambient air. 2b shows the raw data of the fumigated trees in the open top chambers. The values of the experimental ratio F_0/F_M are used for the statistical calculations.

File no.	t_{Fmax} msec	area msec	F_0 calc.	F_M	F_1 50 µs	F_2 100 µs	F_3 300 µs	F_4 2 ms	F_5 30 ms	F_0/F_M
2a										
free air control										
2	1600.00	14300	78	405	82	89	110	178	320	0.202
3	402.00	14300	102	507	108	118	149	237	387	0.213
29	697.00	31300	132	888	146	160	215	407	704	0.164
30	766.00	22800	103	623	113	123	165	315	499	0.181
39	691.00	31100	120	750	135	150	212	401	588	0.180
40	723.00	5250	21	115	23	24	29	49	85	0.200
49	738.00	30400	127	843	144	160	228	463	667	0.171
50	540.00	21200	116	860	134	149	219	463	701	0.156
avg	769.63	21331	100	624	111	122	166	314	494	0.177
std	356.98	9537.2	36.1	269.9	41.4	46.5	69.0	148.9	218.7	0.020
cov	0.464	0.447	0.362	0.433	0.375	0.382	0.416	0.474	0.443	0.113
chamber air filtered										
7	426.00	18300	144	806	156	173	226	365	649	0.194
27	618.00	26100	137	798	150	165	219	380	623	0.188
28	703.00	20400	124	656	134	147	192	323	517	0.204
33	826.00	21800	94	593	104	115	160	304	475	0.175
34	766.00	6040	24	120	26	27	32	52	86	0.217
45	755.00	31300	114	804	126	138	186	388	626	0.157
46	711.00	18600	108	681	124	139	205	412	571	0.182
43	675.00	38600	142	930	159	176	246	466	731	0.171
avg	685.00	22643	111	674	122	135	183	336	535	0.182
std	121.90	9696.3	39.2	247.6	43.1	48.2	66.6	125.4	197.9	0.019
cov	0.178	0.428	0.354	0.368	0.352	0.357	0.363	0.373	0.370	0.105
chamber air (ambient ozone)										
5	548.00	14600	110	640	123	137	190	310	526	0.192
9	443.00	15800	107	569	115	127	164	266	448	0.202
25	683.00	25200	131	816	145	159	214	391	653	0.178
31	749.00	27300	120	772	134	146	201	408	614	0.174
avg	605.75	20725	117	699	129	142	192	344	560	0.185
std	137.00	6455.7	10.9	114.6	13.1	13.6	21.2	67.2	91.8	0.013
cov	0.226	0.311	0.093	0.164	0.101	0.096	0.110	0.195	0.164	0.071
2b										
chamber air+ozone										
10	641.00	41700	167	1124	185	204	278	495	861	0.165
11	590.00	33400	163	1062	185	208	295	511	853	0.174
14	531.00	28100	125	802	134	146	190	336	595	0.167
15	578.00	20500	100	601	108	118	154	263	453	0.180
16	622.00	12400	78	425	87	97	132	218	346	0.205
17	686.00	21100	112	719	129	145	209	351	584	0.179
18	973.00	16700	96	577	105	117	160	274	469	0.182
53	695.00	30900	111	747	122	132	175	357	567	0.163
avg	664.50	25600	119	757	132	146	199	351	591	0.174
std	136.16	9666.6	31.5	238.5	36.0	40.4	58.9	105.5	183.7	0.013
cov	0.205	0.378	0.265	0.315	0.273	0.277	0.296	0.301	0.311	0.077
chamber air+carbon dioxide										
12	608.00	28500	134	890	148	164	223	420	697	0.166
13	566.00	31000	139	935	150	163	209	385	701	0.160
24	708.00	25300	126	813	141	157	221	435	658	0.173
35	650.00	24800	112	786	128	141	201	409	607	0.163
36	752.00	34600	130	888	140	151	192	378	643	0.158
44	634.00	33200	146	1018	164	181	258	563	828	0.161
52	698.00	33400	140	968	156	170	237	530	795	0.161
avg	659.43	30114	132	900	147	161	220	446	704	0.163
std	63.93	3984.3	11.2	82.2	11.7	13.0	22.4	72.2	80.6	0.005
cov	0.097	0.132	0.085	0.091	0.080	0.081	0.102	0.162	0.114	0.032

Table 2: Continued.

File no.	t_{Fmax} msec	area msec	F_0 calc.	F_M	F_1 50 µs	F_2 100 µs	F_3 300 µs	F_4 2 ms	F_5 30 ms	F_0/F_M
chamber air + ozone + carbon dioxide										
19	741.00	22700	117	703	130	142	193	353	556	0.185
20	693.00	14900	115	546	128	142	192	302	453	0.234
21	696.00	13200	121	577	134	147	197	311	475	0.232
22	683.00	14800	112	597	131	147	212	335	493	0.219
37	650.00	32900	141	1018	158	174	243	530	808	0.155
41	610.00	19600	98	601	111	123	176	353	485	0.185
42	718.00	13000	79	473	89	99	140	291	394	0.188
47	638.00	31200	155	1100	184	207	321	663	927	0.167
48	680.00	30800	149	1009	166	183	254	528	808	0.165
avg	678.78	21456	121	736	137	152	214	407	600	0.186
std	40.54	8255.0	24.4	238.7	28.8	32.3	52.5	132.3	193.6	0.030
cov	0.060	0.385	0.202	0.324	0.211	0.213	0.245	0.325	0.323	0.159

Table 3: Definitions of the used symbols in Fig. 3 and their link to the fluorescence signals. V_J = relative variable fluorescence at 2 ms; Mo = slope of the origin of the fluorescence transient normalised to the maximal variable fluorescence ($F_M - F_0$); S_m = normalised area (see Table 2).

$$\frac{RC}{CS} = \frac{\Phi_{Po}}{\Phi_{Fo}} \cdot \frac{Fo}{Mo} \cdot V_J$$

$$V_J = \frac{F_{2ms} - F_{50\mu s}}{F_M - F_{50\mu s}}$$

$$Mo = \frac{dV}{dto} = \frac{F_{300\mu s} - F_{50\mu s}}{F_M - F_{50\mu s}} \cdot 4$$

$$\Phi_{Po} = 1 - \frac{F_{50\mu s}}{F_M}$$

$$p_o = 1 - \frac{1}{N}$$

$$N = \frac{Mo}{V_J} \cdot S_m$$

$$S_m = \frac{Area}{F_M - F_{50\mu s}}$$

Table 4: The wavelength position of the green light and red edge maxima of the reflectance derivative from Scots pine shoots, prior to fumigations. The number of measured shoots n = 25.

Treatment	n	Green light derivative max [nm] +	Red edge derivative max [nm] +
Control (chamber, filtered air)	4	522.9±2.7	711.7±7.6
Control (chamber, ambient air)	4	523.6±1.8	705.6±5.7
Control (free air)	5	521.6±1.2	701.1±5.9
O_3	4	521.5±1.8	712.0±7.4
CO_2	4	522.3±3.8	710.2±6.8
CO_2+O_3	4	521.5±1.3	712.7±6.0

+ Data are presented as mean ± 1 standard deviation.

Table 5: The wavelength position of the green light and red edge maxima of the reflectance derivative from Scots pine shoots, fumigated for three months with ozone or/and carbon dioxide, and the controls. The number of measured shoots n = 25.

Treatment	n	Green light derivative max [nm] +	Red edge derivative max [nm] +
Control (chamber, filtered air)	4	527.0±1.3	708.4±8.0
Control (chamber, ambient air)	4	525.6±2.4	707.7±5.1
Control (free air)	5	525.4±2.2	713.8±6.0
O_3	4	524.6±2.0	712.0±7.4
CO_2	4	523.1±2.0	712.0±6.7
CO_2+O_3	4	524.6±1.2	704.1±8.7

+ Data are presented as mean ± 1 standard deviation.

Reflectance signals after three months fumigations

The derivative maximum of the green light reflectance feature of the carbon dioxide fumigated pines was shifted, by approximately 4 nm towards shorter wavelengths (blue shift), away from the inflection point of the pines fumigated with filtered air (2-sided Mann-Whitney U-test, p=0.03, Table 5). Ozone exposure alone or in combination with carbon dioxide was not found to cause any significant change. In the case of the red edge derivative maximum, no statistical differences were found (2-sided Mann-Whitney U-test).

Physiological information of fluorescence signals after fumigations

The measured fluorescence data are listed in the Tables 2 a and 2 b. From these values several expressions have been calculated according to Table 3 and listed in Table 6. The deviation (in %) from the samples grown under filtered air (100 %) is indicated for each treatment. *Specific* activities are expressed as absorption, trapping or electron transport fluxes *per reaction center* of PSII, whereas the *phenomenological* activities are expressed as absorption, trapping or electron trans-

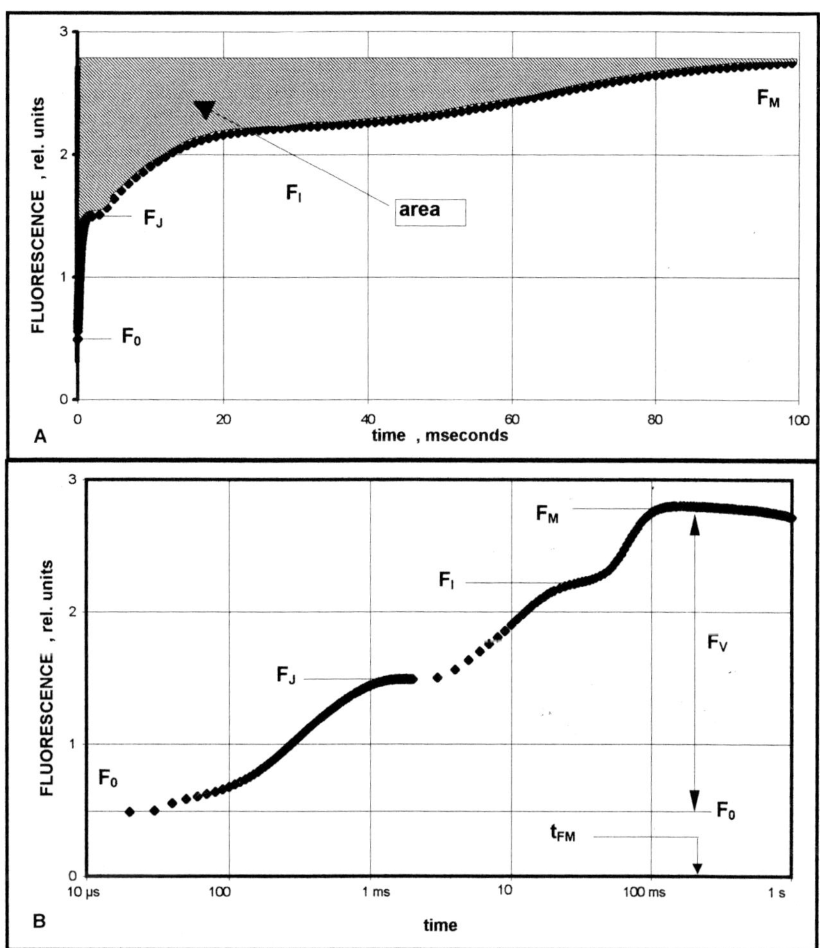

Fig. 1: Typical fluorescence induction curve of green leaves and needles. A: First 100 milliseconds on a linear time scale. B: The same curve in logarithmic time scale.

F
ABS
Chl*
Ne⁻
TR_t
Qa (RC) Qa⁻
(N-1)e⁻
ET_t

$$\frac{ABS}{RC} \quad \frac{TR_t}{RC} \quad \frac{ET_t}{RC}$$

$$\frac{TR_t}{ABS} = \varphi_{pt} \qquad \frac{ET_t}{TR_t} = p_o$$

$$\frac{ET_t}{ABS} = \varphi_t = \varphi_{pt} * p_o$$

Fig. 2: Energy-flux model of PSII. ABS Absorption – flux of photons. TR_t Energy – flux trapped by the reaction center at time t. ET_t Electron transport – flux at time t. RC Reaction center of PSII. Chl* Excited chlorophyll of the antenna of PSII. φ_{Pt} Quantum yield for excitation energy trapping at time t. φ_t Quantum yield for electron transport at time t. p_0 Probability that a trapped exciton is moving an electron beyond Qa at time zero.

port fluxes *per cross section* of the sample. The ratio of fluxes or yields φ_0 is defined as electron transport flux per absorption flux which can be compared with the photosynthetic activity per chlorophyll of PSII. The correlation of the specific activities to the phenomenological activities is given by the density of reaction centers of PSII per cross section of the sample. Fig. 4 shows the relative values (relative to the samples grown in filtered air) of the specific activities per reaction center (left) and of the phenomenological activity per cross-section (right). The quantum yield for electron transport φ_0 and for excitation energy trapping φ_{P_0}, as well as the density of the reaction centers of PSII are shown in the middle. In this way it can be shown that a similar behavior (stimulation of phenomenological activities) after two different treatments ($+CO_2$ or $+O_3$) can be due to different mechanistic causes as revealed by the specific activities.

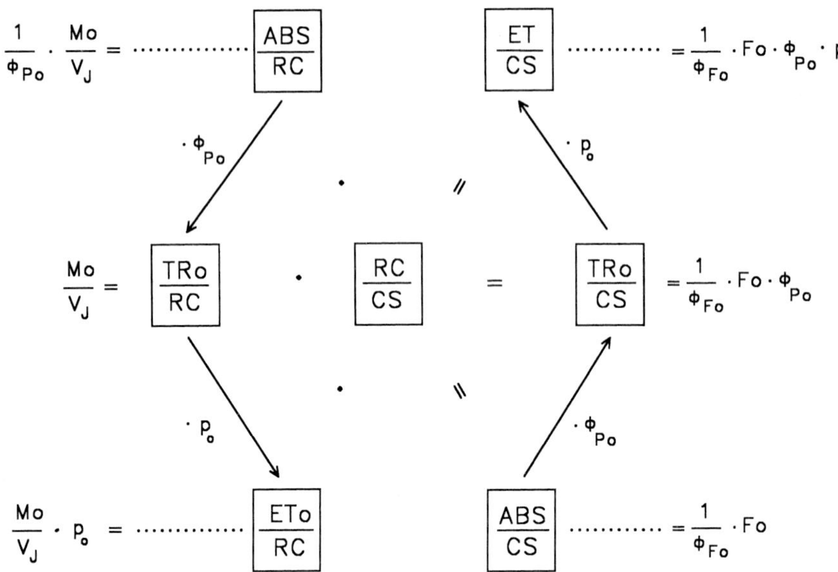

Fig. 3: Correlations of the specific activities to the phenomenological activities and their link to fluorescence signals. For definitions of Mo, V_J, φ_{Po}, see Table 3.

Table 6: Calculated expressions as defined in Fig. 2 and Fig. 3 and Table 3. All numbers are in relative units except for the yields. The deviation, in %, from the filtered air control (100%) is indicated as well. Light absorption per reaction center is proportional to the antenna size ABS/RC = AS.

	free air control	open top chamber				
		air filtered	air	O_3	CO_2	O_3+CO_2
Flux per RC						
ABS/RC = AS	1.322−4	1.378±0	1.43 +4	1.474+7	1.182−14	1.422+3
TR_0/RC	1.087−4	1.128±0	1.166+3	1.217+8	0.989−12	1.158+3
ET_0/RC	1.063−4	1.104±0	1.138+3	1.193+8	0.964−13	1.13 +2
Flux-Ratio = Yield						
$TR_0/ABS = \varphi_{p0}$	0.823+1	0.818±0	0.815±0	0.826+1	0.837+2	0.814±0
$ET_0/TR_0 = p_0$	0.978±0	0.978±0	0.976±0	0.980±0	0.975±0	0.976±0
‖						
$ET_0/ABS = \varphi_0$	0.804±0	0.801±0	0.796−1	0.809+1	0.816+2	0.795−1
Density of RC = D_0						
RC/CS = (Chl/CS):(Chl/RC)						
$D_0 = F_0/AS$	83.7−6	88.8±0	90.4+2	89.5+1	124.2+40	96.2+8
Activities						
$ABS/CS = F_0$	111−9	122±0	129+6	132+8	147+20	137+12
$TR_0/CS = F_0 \cdot \varphi_{p0}$	91−9	100±0	105±5	109+9	123+23	111+11
$ET_0/CS = F_0 \cdot \varphi_0$	89−9	98±0	103+5	107+9	120+22	109+11

Discussion

The reflectance data suggested that CO_2 fumigation shifts the green light derivative maximum of current year shoots towards shorter wavelengths.

These results are supported by the preliminary results of Palomäki et al. (1995) studying the photosynthesis and needle ultrastructure of the same pines. It seems that the green light reflectance data could correlate with the size of chloroplasts to some degree. The size of chloroplasts increased in all treatments and correspondingly the reflectance derivative maximum showed shifts of some extent towards shorter wavelengths, although this change was not always statistically significant. Earlier, Ustin and Curtiss (1990) had studied the spectral characteristics of ozone-treated conifers. According to them, changes in the proportions and species of carotenoid and xanthophyll pigments may possibly contribute to variances in the 440–590 nm region.

From the reflectance data we conclude that all samples showed no or only very slight chemical and structural

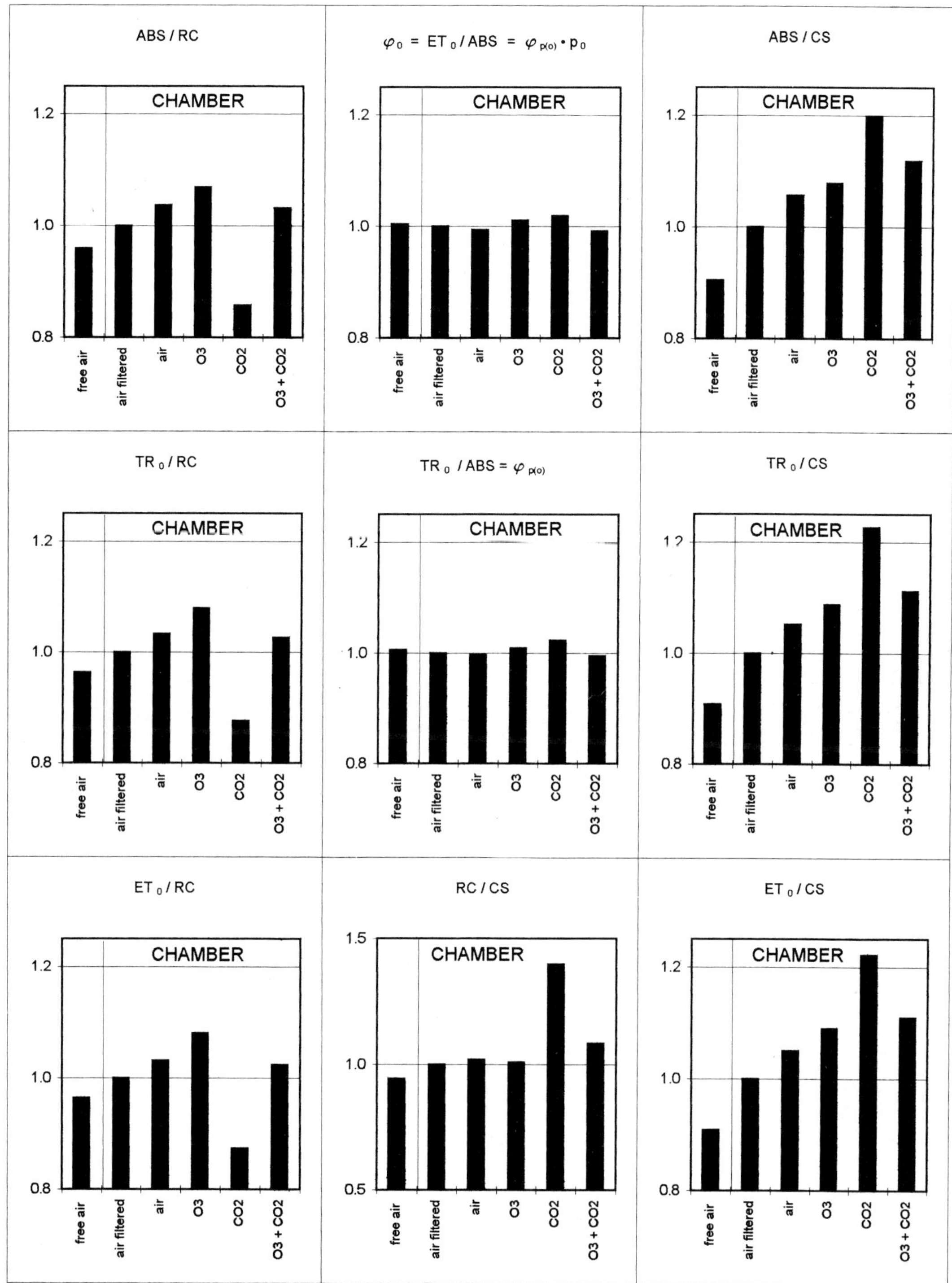

Fig. 4: Presentation of the expressions defined in Table 3 and Fig. 3 relatively to the values of the control with filtered air. left: Specific activities such as Absorption or Trapping or Electron Transport, per reaction center. right: Phenomenological activities such as Absorption or Trapping or Electron Transport, per cross-section of the sample. middle: – Quantum yield for electron transport (top). – Maximum quantum yield for exciton-energy trapping (center). – Density of reaction centers in the sample (bottom). Five treatments were applied in the open top chambers. One free air control was measured outside the chamber.

changes due to the fumigation during the period of the experiments. Functional and dynamic variation can therefore be considered as early detections of stress on basically intact samples.

Fluorescence signals can show nearly no or very typical changes upon fumigation according to the expressions analysed. The most popular index measured in the field is φ_{Po} (= $1 - F_0/F_M = F_V/F_M$), which indicates the maximal efficiency that an absorbed photon is trapped by a reaction center of PSII. This yield, as well as the yield for electron transport, remained nearly constant for all treatments. The general practice shows that chemical and structural changes can be observed when the yield φ_{Po} is altered. As a consequence, all tests which can reveal changes in a sample before φ_{Po} is affected, can be considered as early diagnosis of stress.

All fluorescence data show some differences between the control plants inside or outside the open top chambers. Variations in the microclimate (wind, humidity, temperature) may be the reason.

The phenomenological activities increased in the presence of O_3 (ambient or added) and in elevated CO_2. However, an antagonism is observed when both gases (O_3 and CO_2) are simultaneously applied. Under these conditions the activities per cross-section are lower than when O_3 is added alone and higher than when CO_2 is added alone. It has to be emphasized here that the activities refer only to PSII. Oxidations (due to activated oxygen or O_3) which may happen between PSII and PSI would account for an activity of PSII but not of PSI or CO_2 fixation. The specific activities show that the antenna size of a PSII increases in the presence of O_3. Therefore, the trapping flux and the electron transport increase. Quantum yields and the density of reaction centers are nearly unchanged. Therefore, the increased activity caused by the ozone treatment is triggered by an increased antenna size. Quite different is the situation under elevated CO_2. Antenna size and specific activities decrease significantly but, however, they are compensated and even over-compensated by an increased density of reaction centers per cross-section. Even though the yields φ_{Po} and φ_0 remained nearly unchanged, the phenomenological activities (e.g. as electron transport per cross-section) increase under elevated CO_2. This has been observed in many similar experiments and reported by several authors. The fast fluorescence rise, within the measuring time of only one second, is a powerful tool to screen in a short time many samples under *in vivo in situ* conditions. According to the proposed data handling routines (JIP-test) some estimations about structural and functional changes in the photosynthetic apparatus can be calculated.

The fumigations will continue during the year 1995. The results of these studies will show the effect of over-wintering and prolonged exposures on the reflectance and fluorescence signals.

Acknowledgements

We wish to thank Mr. Marko Meinander for writing the program for derivative calculations, and Dr. P. Eggenberg for establishing some tables and figures. Financial support from the Academy of Finland to O. Meinander and T. Holopainen is gratefully acknowledged.

References

CURRAN, P. J., J. L. DUNCAN, and H. GHOLZ: Exploring the relationship between reflectance red edge and chlorophyll content in slash pine. Tree Physiology *7,* 33–48 (1990).

LICHTENTHALER, H. K. and U. RINDERLE: The role of chlorophyll fluorescence in the detection of stress conditions in plants. CRC *19* (1), 29–85 (1988).

PALOMÄKI, V., K. LAITINEN, S. KELLOMÄKI, and T. HOLOPAINEN: Effects of changing carbon dioxide and ozone concentrations on photosynthesis and needle ultrastructure of Scots pine saplings. An abstract submitted to the 5th International Conference on Acidic Deposition, Science-Policy, Gothenburg, Sweden, 26–30 June, 1995.

ROCK, B., T. HOSHIZAKI, and J. MILLER: Comparison of *In situ* and Airborne Spectral Measurements of the Blue Shift Associated with Forest Decline. Remote Sensing of Environment *24,* 109–127 (1988).

RUNECLES, V. and H. RESH: The Assesment of chronic ozone injury to leaves by reflectance spectrometry. Atmospheric Environment *9,* 447–452 (1975).

RUTH, B. and B. WEISEL: Investigations on the photosynthetic system of spruce affected by forest decline and ozone fumugations in closed chambers. Environ. Pollut. *79,* 31–35 (1993).

RUTH, B., E. HOQUE, B. WEISEL, and J. S. HUTZLER: Reflectance and Fluorescence Parameters of Needles of Norway Spruce Affected by Forest Decline. Remote Sensing of Environment *38,* 35–44 (1991).

STRASSER, R. J. and GOVINDJEE: The Fo and the O-J-I-P fluorescence rise in higher plants and algae. In: ARGYROUDI-AKOYUNOGLOU, J. H. (ed.): «Regulation of Chloroplast Biogenesis». Plenum Press, New York, pp. 423–426 (1992).

– – On the O-J-I-P fluorescence transient in leaves and D1 mutants of Chlamydomonas reinhardtii. Proceedings of the IX International Congress on Photosynthesis, Nagoya, Japan. In: MURATA, N. (ed.): «Research in Photosynthesis». Kluwer Academic Publishers, Dordrecht, Vol. II, pp. 29–32 (1992).

STRASSER, R. J., A. SRIVASTAVA, and GOVINDJEE: Polyphasic chlorophyll *a* fluorescence transient in plants and cyanobacteria. Photochemistry and Photobiology, Vol. *61,* No. 1, pp. 32—42 (1995).

STRASSER, B. J. and R. J. STRASSER: Measuring fast fluorescence transients to address environmental questions: The JIP-test. Proceedings of the X International Congress on Photosynthesis. Kluwer Academic Publishers (in press).

USTIN, S. L. and B. CURTISS: Spectral characteristics of ozone-treated conifers. Environmental and Experimental Botany *30* (3), 293–308 (1990).

J. Plant Physiol. Vol. 148. pp. 237–242 (1996)

Dark Respiration Under Low Pressure and Increased Ethylene

B. G. Ageev[1], T. P. Astafurova[2], Yu. N. Ponomarev[1], V. A. Sapozhnikova[1], T. A. Zaitseva[2], and A. P. Zotikova[2]

[1] Institute for Atmospheric Optics, Siberian Branch of the Russian Academy of Sciences, Academicheskii ave., 1, 634055, Russia

[2] Research Institute of Biology and Biophysics, Tomsk State University, Lenin Ave., 36, 634050, Tomsk, Russia

Received June 9, 1995 · Accepted October 27, 1995

Summary

Responses of plants seedlings to hypobaria (reduced pressure of air) exposure (up to 8 kPa) and exogenously applied ethylene (10^{-4}, 10^{-5}, 10^{-6} kPa) was studied using a gasometric system based on CO_2-laser photoacoustic spectrometer. The increase of CO_2-evolution (in the dark) by pea (*Pisum sativum* L.), wheat (*Triticum aestivum* L.), maize (*Zea mays* L.), barley (*Hordeum vulgare* L.) and pine seedlings (*Pinus sylvestris* L.) was revealed. Some differences in dynamics of CO_2-evolution and signal magnitudes are probably stipulated by the objects specificity. The alcohol-dehydrogenase, isocitrate dehydrogenase, glucose-6-phosphate dehydrogenase, phosphoenolpyruvate carboxylase activity and contents of malate, pyruvate and lactate were studied in pea leaves. It was shown that an anaerobic metabolism increases with time and also when the air pressure is decreased. An increase of CO_2-evolution from ethylene-treated pea seedlings and alcohol-dehydrogenase activation were observed.

Key words: Hordeum vulgare L., Pisum sativum L., Pinus sylvestris L., Triticum aestivum L., Zea mays L., gas-exchange, hypobaria, metabolism, ethylene, respiration.

Abbreviations: ADH = alcohol dehydrogenase; GPDH = glucose-6-phosphate dehydrogenase; IDH = isocitrate dehydrogenase; PEPC = phosphoenolpyruvate carboxylase; PA = photoacoustic.

Introduction

The study of the physiological and biochemical state of plants under modified atmospheric conditions has become increasingly important in recent years. Atmospheric changes disturb the balance of physiologically active gases and influence the main energy-transforming process in autophytes. The properties of the ambient air vary both under variations of gaseous components of natural and anthropogenic origin and under reduced pressure. Particular attention has been given to ethylene, the level of which increases in regions of gas and oil production and chemical processing.

The effects of pressure variation of gaseous medium on plants are poorly understood up to now (Hochachka and Somero, 1988), because under natural conditions of the plants growing they are not subjected to strong pressure variations. Nevertheless, they may undergo the pressure variations in man-made conditions. Some data are known on the increased pressure influence on growth and respiration of roots (Sarquis and Jordan, 1992). The development of space researches and populating of Alpine areas stimulated such kind of investigations. At present the data are available on the effects of reduced pressure on photosynthetic reactions and respiratory metabolism (Gale, 1972; Musgrave et al., 1988; Astafurova et al., 1990, 1993). Some papers are known devoted to plant growing in vacuum (Costes and Vartapetian, 1978). The results of these investigations may be used in space engineering on creation of vitality maintenance systems, growing the plants under artificial atmospheric conditions as well as in biotechnology.

Ethylene becames of particular interest as the air pollutant that falls into the group of gaseous stressors, even small concentrations of which significantly influence vital activity processes of plants (Smith and Hall, 1984; Taylor and Gunderson, 1986). Ethylene is as well the plant hormon, the formation of which increases under stressors exposure, for example under air pressure variation.

Investigation of plant responses to variations of pressure or gaseous contents of the air is of great interest to specialists studying the potential of plants and mechanisms of their adaptation to stress.

Results of investigations of various respiratory metabolism responses in leaves of plants under reduced pressure conditions are presented in this paper together with fragmentary data on intensity of respiratory CO_2-exchange of plants exposed to exogenic ethylene. A gasometric system, based on a photo-acoustic spectrometer with a tunable laser, is described and some methodological aspects of the measurements are considered.

Materials and Methods

Plant materials and growth conditions

The objects of our investigation were the 8-day-old pea (*Pisum sativum* L.), wheat (*Triticum aestivum* L.) and barley seedlings (*Hordeum vulgare* L.) as well as 14-day-old maize (*Zea mays* L.) and pine (*Pinus sylvestris* L.) seedlings grown in conditions of soil culture under luminescent lamps of 40 W/m^2 intensity, at air temperatures of 22 to 24 °C, 12-hour photoperiod and normal atmospheric pressure of 101 kPa. Before the experiment, the above-ground parts of the plants were cut off and put inside wet material.

Exposure system

The control and testing groups of the plants were simultaneously put into exposure chambers of 10^{-3} m^3 volume. The hypobaria was reached by pumping with a vacuum pump up to the necessary pressure. The measurements were conducted at three values of pressure inside the exposure chambers: 54, 29 and 8 kPa (what equivalent approximately to heights of $5.1 \cdot 10^3$, $10 \cdot 10^3$ and $20 \cdot 10^3$ m above the sea level). When studying the influence of exogenous ethylene, the total pressure was 101 kPa and the partial pressures of C_2H_4 were approximately 10^{-6}, 10^{-5} and 10^{-4} kPa. Ethylene-treated plants, after exposure time, were put into the chambers with air without ethylene. The control plants were under normal aeration conditions. The temperature of 22–24 °C remained constant in all cases. To eliminate the photosynthesis process during the experiment, the plants were kept in the dark. The duration of the plants' exposure to stressors varied from 2 to 48 hours.

Gas-exchange measurements

The dark respiration intensity of the seedlings inside the exposure chambers was estimated by CO_2 evolution using photoacoustic (PA) spectrometers with a tunable CW CO_2 laser (Fig. 1) (Ageev et al., 1994). The radiation source was selected due to the fact that some products of the plants vital activity (CO_2, C_2H_4, NH_3) have intense absorption lines within the CO_2-laser generation band. A modified CW CO_2-laser ILGN-705 (manufactured in Russia) was used in the experiments. For wavelength tuning the laser output mirror was replaced by a combination of a diffraction grating of 100 lines/mm and a plane 100 % reflected mirror. The radiation wavelength was selected by the plane mirror tuning. The resonator construction enabled us to obtain the generation of twelve discrete lines in (10.5–

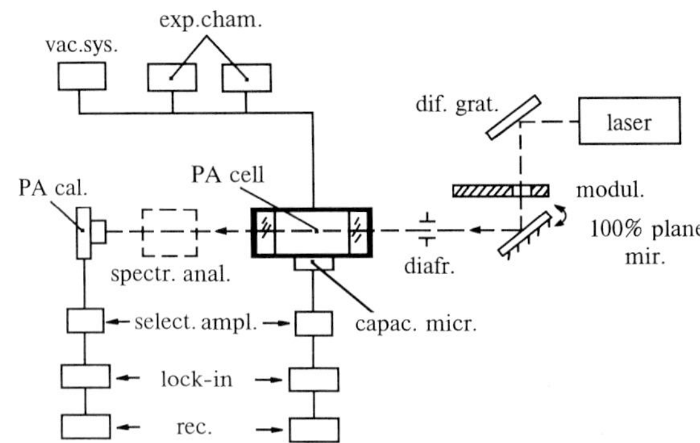

Fig. 1: Block-diagram of set-up.

10.7) µm region. Identification of the laser generation lines was carried out with a noncommercial panoramic spectrum analyzer, which was used on occasion for wavelength measurements. Amplitudely modulated radiation was directed with the plane mirror through a diaphragm of 3 mm in diameter into the cell of the PA detector of our design (100 mm in length and 10 mm in diameter, with BaF_2 windows). The diaphragm was used to screen the cell walls from laser radiation. PA signal arising inside the cell was measured with a plane capacitor microphone of our own construction, mounted into the cell wall. An electric signal from the microphone was preamplified and then recorded by the recording system: a selective amplifier, lock-in amplifier and recorder (all the devices manufactured in Russia). The reference signal from the modulator reaches the input of the lock-in amplifiers. Similar channels were used to recorded a signal from a homemade PA calorimeter which was located behind the PA cell. The calorimeter is a nonselective PA detector which measures the radiation power passing through the cell.

Reference (control) and experimental groups of plants were placed in the exposure chambers connected to the vacuum system and PA-cell.

In the course of the experiment the gas samples from the exposure chambers were successively put into the evacuated measuring PA-cell. When radiation is absorbed, at a given wavelength, the amplitude of the electric signal from the PA-detector U is directly proportional to the concentration C of the absorbing gas (or a gas mixture). The ratio A, characterizing the absorbtivity of the gas or gaseous mixture, was determined during the experiment as following;

$$A = U/W = \alpha C,$$

where W is the incident laser power and α is the sensitivity of the PA detector. When measuring gaseous mixtures with small absorption coefficients and using a short PA cell, the value of incident radiation power is equal to the value of the radiation power which has passed through the cell. α is the function of the total gas pressure P in the cell (Antipov et al., 1984). For the used PA-detector $\alpha = \alpha_{max}$ at P ~8 kPa, therefore, all the measurements were carried out at the same pressure. All experiments on C_2H_4 influence were carried out at the pressure of 101 kPa.

For absolute measurements of CO_2 concentrations the calibration of the PA – detector has been made: the magnitude of α was estimated from A measurements with etalon gas mixtures (PO Analitpribor, Russia).

It is known that the plants during their vital activity evolve not only the main gaseous products (CO_2, O_2, H_2O), but ethylene (C_2H_4) participating in hormonal balance, ammonia (NH_3) characterizing the albumin exchange, and a number of volatile metabolites as well. To identify the gases being involved in gas exchange we observed variations of A with time for the two CO_2 laser wavelength $\lambda_1 = 10.591\,\mu m$ (IP(20)) and $\lambda_2 = 10.532\,\mu m$ (IP(14)). Those wavelengths were chosen because carbon dioxide (at λ_1) and ethylene (at λ_2) are the main contributors to absorption (Hurren, 1990). C_2H_4 absorptivity on the IP(20) line is known to be low (Meyer and Sigrist, 1990). To estimate its influence on absorption the gas samples have been passed through the CO_2 chemical absorber of ascarite.

Biochemical analysis

The plant material was analyzed just after the experiment completion or after its fixation. Different steps of the respiratory metabolism functioning were estimated by the due, to enzymes activity: oxidating pentose phosphate pathway – glucose-6-phosphate dehydrogenase («EC» 1.1.1.49); Krebs cycle NAD-dependent isocitrate dehydrogenase («EC» 1.1.1.41); ethanol fermentation-alcohol dehydrogenase («EC» 1.1.1.1). Carboxylating activity was inferred from phosphoenol pyruvate carboxylase («EC» 4.1.1.31) functioning. The leaf tissue (1 g) was frozen and ground in liquid N_2 extracted in 10 mL of 0.1 M tris-HCl, pH 7.8 containing 2 mM EDTA, 10 mM $MgCl_2$ and centrifuged for 15 min at 12,000×g (Verkhoturova and Astafurova, 1983). The resultat supernatant was analyzed for enzymes activity.

The measurements were conducted spectrophotometrically in supernatant by the methods described earlier (Chapman and Osmond, 1974; Gavrilenko et al., 1975; Möller et al., 1977) which were then modified with regard to the objects peculiarities (Verkhoturova and Astafurova, 1983). The malate, pyruvate and lactate concentrations were determined enzymatically after the leaves fixation by perchloric (Astafurova et al., 1994). The experiment had 2–3 fold repetition in 3 biological recurrings. The arithmetic means and standard errors calculated from the data of all experiments are presented in Tables and Figures. Differences from control values were significant at $p<0.01$ and $p<0.05$.

Results and Discussion

An integral index of the plants functional activity is the gas-exchange characteristic. It is used to estimate the plants resistance to soil and atmosphere pollution, drought, frost and other stressors (Nitchiporovitch, 1990). We have conducted 2 runs of independent experiments to study the plants gas-exchange dynamics under low pressure of the ambient air and increased concentration of exogenous ethylene. The measurement results (Fig. 2) show CO_2 evolution by pea seedlings under reduced pressure to exceed markedly the control results and increase with the pressure lowering. At the pressure of 8 kPa an explicit two-peak curve is seen; at 54 kPa the two peaks are smoothed and shifted to later time. In 24 hours since the exposure's start the maximum CO_2 evolution is marked and then it declines. The first CO_2 evolution under the hypobaria conditions is possible to attribute to rate increase of its diffusion from intercellular spaces (Gale, 1973; Musgrave et al., 1988). The second peak can be explained by intensification of its intercellular formation due to activation of endogenous substrate decarboxylation under such conditions (Astafurova et al., 1990). Accumulation of CO_2 inside cells and tissues takes place due to the processes of spirit fermentation, sugars oxidation by pentose phosphate pathway, tricarboxylic acid cycle functioning, and other decarboxylating reactions.

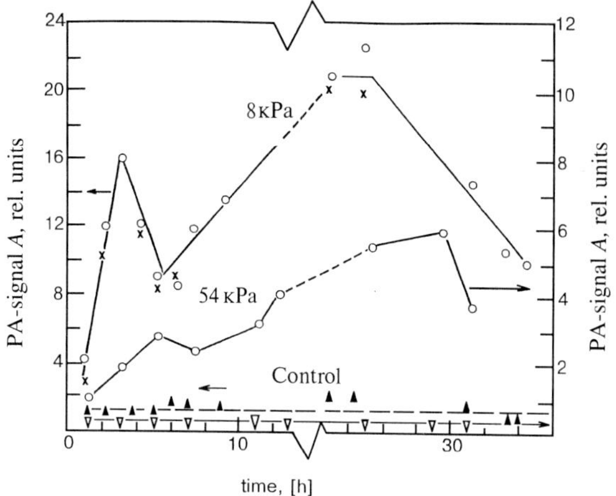

Fig. 2: Photoacoustic signal A vs. time for pea seedling at 8 and 54 kPa (▲, ∇ – A for control plants at P = 101 kPa). - - - - the night-time interval free of measurements.

Table 1: Alcohol dehydrogenase activity in pea leaves at rising hypobaria. Exposure time is 24 h.

Treatments	Air pressure, kPa	Alcohol dehydrogenase: μmol NADH mg^{-1} protein min^{-1}	%
Control	101	1.53±0.09	100
Hypobaria	54	2.76±0.14	181
	29	4.98±0.18	326
	8	14.29±0.41	932

Data represent mean±standard deviation. Differences from control values were significant at $p<0.01$.

Table 1 indicates a pronounced activation of alcohol dehydrogenase in pea leaves at a pressure lower than to 54 kPa. Under an increase of hypobaria (to 29 or 8 kPa) and the exposure time (to 24 h) the enzyme activity increases markedly (Fig. 3). For GPDH and PEP-carboxylase 6-hour activation is observed. But after 24 hour exposure an inhibition of the enzymes is noted. The lowering of isocitrate dehydrogenase activity noted at 6-hour exposure remains after 24-hour experimenting and testifies to deceleration of Krebs cycle tricarbon step. The study of respiratory substrates content has shown a decrease of pyruvate and malate quantity in pea leaves and an increase of lactate level at a pressure lowering to 54 kPa. As is seen from Table 2, at 29 kPa pressure the quantitative increase of all metabolites mentioned above is observed. The further air rarefication up to 8 kPa results in a significant increase of pyruvate and malate contents and a decrease of the lactate level.

It may be inferred from the results obtained that the exposure of the plant leaves to integrated stressors arising inside the closed volume of the exposure chambers under air rarefication (low pressure, O_2, CO_2 and other gases concentration lowering) leads to oxygen deficiency as evidenced by the increase of lactate, activation of ADH and GPDH, and inhibition of IDH. These changes are known to be the indices of

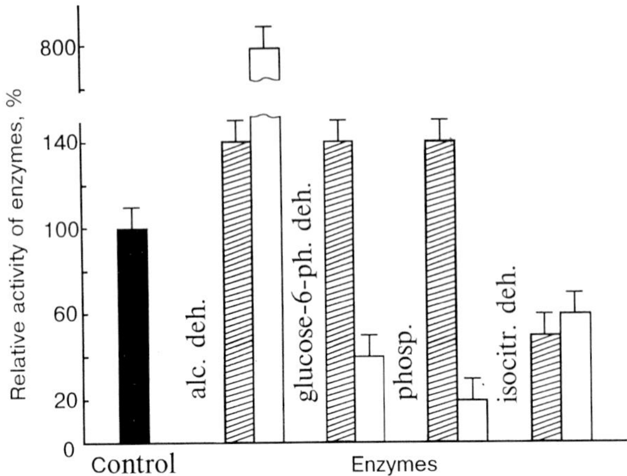

Fig. 3: Relative activity of enzymes (%): alcohol dehydrogenase, glucose-6-phosphate dehydrogenase, phosphoenolpyruvate-carboxylase, isocitrate dehydrogenase in pea leaves at P = 8.3 kPa. Exposure time: ▨ – 6 h, □ – 24 h.

Table 2: Organic acids content in pea leaves at rising hypobaria. Exposure time is 24 h.

Treatments	Air pressure, kPa	Pyruvate	Malate	Lactate
		μmol fresh weight g^{-1}		
Control	101	2.08 ±0.24	0.90 ±0.01	1.83 ±0.16
Hypobaria	54	1.08*±0.04	0.49 ±0.01	2.34 ±0.16
	29	3.47 ±0.12	1.47*±0.12	2.58 ±0.21
	8	4.55 ±0.13	2.64 ±0.11	1.04*±0.12

Data represent mean ± standard deviation. Differences from control values were significant at $p<0.01$. The asterisk represents a significance at $p<0.05$.

anaerobic metabolism since they indicate both oxidative processes in tricarboxylic acids cycle and domination of reducing reactions (Kennedy et al., 1992; Ricard et al., 1994). Several authors believe that the lactate formation takes place predominantly at early stages of anaerobiosis and more extended or intense action leads to its decrease (Hoffman et al., 1986; Chirkova, 1988). Such succession of the reactions was revealed in pea leaves under hypobaria growing (Tables 1, 2), which agrees with the dynamics of CO_2 evolution (Fig. 2). Judging from the physiological response, the pea leaves are free from acute deficit of O_2 under reduced pressure. The activation of PEP-carboxylase leads to an increase of malate content. An extended exposure of pea leaves to hypobaria tends to decrease carboxylating activity. Conceivably, it points to saturation of a cell with CO_2, the excess of which diffuses outside. The enhancement of CO_2 evolution from animal organisms under exposure to low pressure inside an exposure chamber for a long time has also been noted (Agadzhanyan, 1972; Petrukhin et al., 1975).

The response of different organisms to the stressors is known to depend on species, age and individual typological features. Some of them die quickly, others are able to resist stressors for a long time. An analysis of the dynamics of CO_2 evolutions by plants of various systematic groups has shown an enhancement of CO_2 evolution by all species exposed to the low pressure of 54 kPa (Fig. 4) for a long time. Nevertheless, some difference in the shape of the curves, signal magnitudes, and peaks shifting in time is probably stipulated by the objects specificity. For example, the activity of CO_2 evolution by pea seedlings relative to control plants exceeds the same by other species. The periodic type of responses is characteristic of peas (Fig. 2). Only one explicit maximum is inherent to wheat, barley, and pine seedlings. CO_2 evolution by maize seedlings increases initially with time and then remains constant. The same physiological response may be attributed to structurally-functional peculiarities of the plants. The maize accumulates a significant part of CO_2 in the organic acid form which decarboxylates easily via principal and alternative pathways. On the other hand, high carboxylating ability of maize allows intercellular CO_2 to take an active part in biosynthetical cycles. The existance of the functional mechanism of CO_2 concentration and anatomic features of leaves are favorable for CO_2 accumulation inside C_4 plants. This fact is likely to explain the lower evolution of CO_2 by maize seedlings as compared to other species. The reactions of substrates decarboxylation under hypobaric hypoxia dominate in pea over carboxylating activity (Astafurova et al., 1993). Developed spongy parenchyma, large intercellulars, and weak cuticle facilitate the gases diffusion from the pea leaves surface. So, at reduction of the air pressure the hypobaric hypoxia occurs in the plant tissues. This fact is confirmed by findings of other researchers (Musgrave et al., 1988). Under low pressure

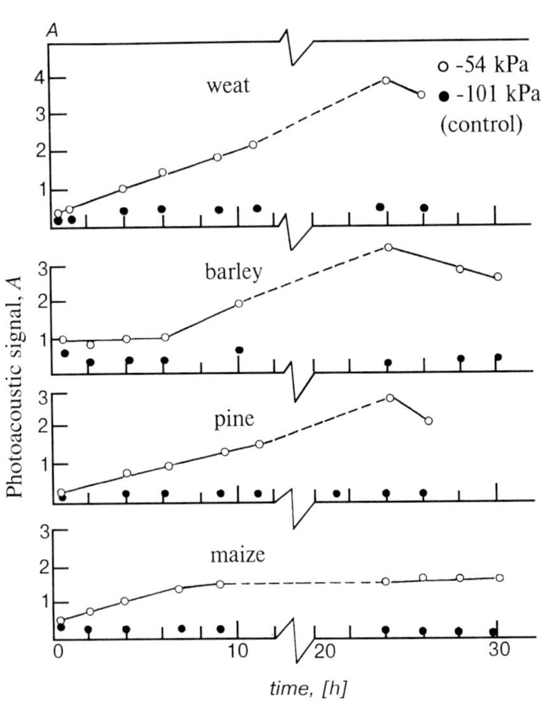

Fig. 4: Temporal dependence of *A* for weat, barley, pine, maize seedlings. ● – P = 101 kPa, ○ – P = 54 kPa. Dotted line denotes the night-time free of experiments.

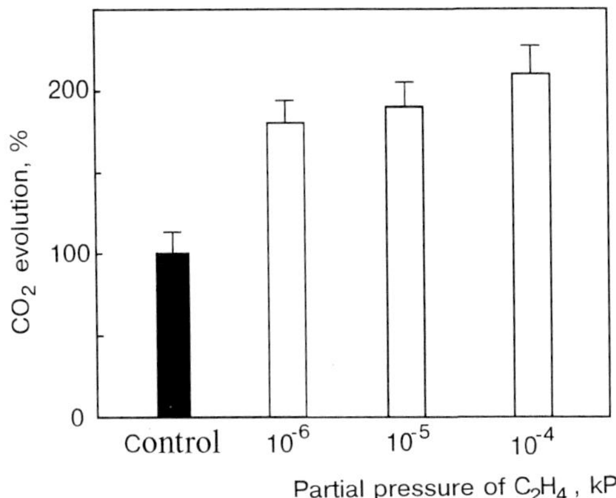

Fig. 5: Influence of exogenously applied ethylene on CO_2 evolution by pea seedling (%).

Table 3: Alcohol dehydrogenase activity in ethylene-treated pea leaves. Exposure time is 48 h.

Experimental variants	C_2H_4, partial pressure, kPA	Alcohol dehydrogenase μmol NADH mg^{-1} protein min^{-1}	Alcohol dehydrogenase %
Control	–	2.03±0.10	100
C_2H_4-treated	10^{-6}	2.52±0.18	124
	10^{-5}	8.26±0.55	407
	10^{-4}	4.08±0.30	201

Data represent mean ± standard deviation. Differences from control values were significant at $p<0.05$.

in plant leaves the rate and direction of various respiratory reactions vary.

Our experiments have shown that a 48-hour exposure of pea seedlings to C_2H_4 of various concentrations increases CO_2 evolution relative to control plants (Fig. 5). But a further increase of C_2H_4 concentration does not lead to significant enhancement in CO_2 evolution. Exogenous ethylene influences stronger the alcohol dehydrogenase activity at a 10^{-5} kPa concentration (Table 3). Reference data are available on enhancement of respiratory activity of plants exposed to ethylene (Warman and Theophanes, 1988), nevertheless its role in alcoholdehydrogenase activation is still not clear (Morrell and Greenway, 1989).

We are planning supplementary investigations to study the mechanisms of the respiratory metabolism responses to exogenous ethylene exposure. So, the results obtained show the variation of pressure and exogenous ethylene concentration to affect markedly the physiological processes in plants. The plant reactions depend on the type, intensity, and duration of the stressors exposure as well as on species peculiarities of the objects.

The developed instrumentation allows us to investigate gas-exchange in plants both at various physical states of environment (pressure, temperature, light) and at various gas content. A study of plant reactions to polluting gases exposure, particularly, of industrial pollutants, may be of great importance for solution of ecological problems.

Acknowledgements

This work was particularly supported by State Committee of Russian Federation on High Education (project No. 93-4-3).

References

Agadzhanyan, N. A.: Organism and gaseous environment. Meditsina, Moskva (1972) (In Russian).

Ageev, B. G., T. P. Astafurova, Yu. N. Ponomarev, V. A. Sapozhnicova, and K. L. Kositsyn: Use of a CO_2-laser-based opto-acoustic spectrometer in studying gas exchange of vegetation. Atmos. Oceanic Opt. *7*, 528–530 (1994) (In Russian).

Antipov, A. B., V. A. Kapitanov, Yu. N. Ponomarev, and V. A. Sapozhnicova: Opto-acoustic method in laser spectroscopy of molecular gases. Nauka, Novosibirsk (1984) (In Russian).

Astafurova, T. P., O. B. Vaishlya, G. S. Verkhoturova, T. A. Zaitseva, and T. V. Chirkova: Photosynthetic and respiratory metabolism in plants as affected by hypobaric hypoxia. Fiziol. Rast. *37*, 690–696 (1990) (In Russian).

Astafurova, T. P., O. B. Vaishlya, G. V. Lapina, A. V. Rakitin, and I. M. Magomedov: Activity of anaerobic dehydrogenases and carboxylic enzymes in Pea leaves under complex effects of environmental factors. Vestnik St-Peterburgh University *3*, 73–78 (1993) (In Russian).

Astafurova, T. P., O. B. Vaishlya, T. A. Zaitseva, G. V. Lapina, and A. V. Rakitin: Activity of photosynthetic apparatus of pea leaves under rarefied atmosphere. Fiziol. and Biochem. Cult. Rast. *26*, 450–455 (1994) (In Russian).

Chapman, E. A. and S. B. Osmond: The effect of light on the tricarboxylic acid cycle in green leaves. III. A comparison between some C_3 and C_4 plants. Plant Physiol. *53*, 893–898 (1974).

Chirkova, T. V.: On pathways of plant adaptation of hypoxia and anoxia. Fiziol. Rast. *35*, 393–411 (1988) (In Russian).

Costes, C. and B. B. Vartapetian: Plant growth in a vacuum: the ultrastructure and functions of mitochondria. Plant Sci. Lett. *11*, 115–119 (1978).

Gale, I.: Availability of carbon dioxide for photosynthesis at high altitudes: theoretical considerations. Ecology *53*, 494–497 (1972).

Gavrilenko, V. F., M. E. Ladugina, and L. M. Khandobina: The large practice on plant Physiology. Vysshaya Shola, Moskva (1975) (In Russian).

Hochachka, P. W. and G. N. Somero: Biochemical Adaptation. Mir, Mockva (1988) (In Russian).

Hoffman, N. E., A. F. Bent, and A. D. Hanson: Induction of lactate dehydrogenase isozymes by oxygen deficit in barley root tissue. Plant Physiol. *82*, 658–663 (1986).

Hurren, E. J. M., F. G. C. Bejnen, J. Reuss, L. A. C. J. Voesenek, and C. W. P. M. Blom: Sensitive Intracavity Photoacoustic Measurements with CO_2 Waveguide Laser. Appl. Phys. *50*, 137–144 (1990).

Kennedy, R. A., M. E. Rumpho, and T. C. Fox: Anaerobic metabolism in plants. Plant Physiol. *100*, 1–6 (1992).

Meyer, P. L. and M. W. Sigrist: Atmospheric pollution monitoring using CO_2-laser photoacoustic spectroscopy and other techniques. Rev. Sci. Instrum. *61*, 1779–1807 (1990).

Möller, G., P. Stamp, and G. Geisler: Fotometrische Messung der PEP-Carboxylase-Aktivität in Maisblättern unter Berücksichtigung des Entwicklungszustandes der Pflanze. Z. Pflanzenernähr. und Bodenkunde *140*, 481–490 (1977).

MORRELL, S. and H. GREENWAY: Evidence does not support ethylene as a cue for synthesis of alcohol dehydrogenase and pyruvate decarboxylase during exposure to hypoxia. Aust. J. Plant Physiol. *16,* 469–475 (1989).

MUSGRAVE, M. E., W. A. GERTH, H. W. SCHELD, and B. R. STRAIN: Growth and Mitochondrial respiration of mungbeans (*Phaseolus aureus* Roxb.) germinated at low pressure. Plant Physiol. *86,* 19–22 (1988).

NICHIPOROVICH, A. A. (ed.): Infrared gas analyzers for investigations of gas exchange in plants. Nauka, Moskva (1990) (In Russian).

RICARD, B., I. COUEE, P. RAYMOND, P. H. SAGLIO, V. SAINT-GES, and A. PRADET: Plant metabolism under hypoxia and anoxia. Plant Physiol. Biochem. *32,* 1–10 (1994).

SARQUIS, J. I., W. R. JORDAN, and P. W. MORGAN: Effect of atmospheric pressure on maize roots growth and ethylene production. Plant Physiol. *100,* 2106–2109 (1992).

SMITH, A. R. and M. A. HALL: Mechanism of ethylene action. Plant Growth Regul. *2,* 151–165 (1984).

TAYLOR, G. E. Jr. and C. A. GUNDERSON: The response of foliar gas exchange to exogenously applied ethylene. Plant Physiol. *82,* 653–657 (1986).

VERKHOTUROVA, G. S. and T. P. ASTAFUROVA: On the direction of some Krebs cycle reactions in green leaves in the light. Fiziol. Rast. *30,* 580–586 (1983) (In Russian).

WARMAN, T. V. and S. THEOPHANES: Ethylene production and action during foliage senescence in *Hedera helix* L. J. Exp. Bot. *39,* 685–694 (1988).

J. Plant Physiol. Vol. 148. pp. 243–248 (1996)

Leaf Contamination by Atmospheric Pollutants as Assessed by Elemental Analysis of Leaf Tissue, Leaf Surface Deposit and Soil

Anna Alfani, Giulia Maisto, Paola Iovieno, Flora A. Rutigliano, and Giovanni Bartoli

Dipartimento di Biologia Vegetale – Università «Federico II» di Napoli, via Foria, 223-80139 Napoli, Italy

Received August 10, 1995 · Accepted October 27, 1995

Summary

In order to evaluate the influence of air pollutants influx on leaf elemental composition, the concentration of N, S, Cu, Fe and Pb were analyzed in the surface deposit and tissue of *Quercus ilex* L. leaves from 8 sites of the urban area of Naples. The soil from the trunk base area of *Q. ilex* trees in the same sites was also analyzed for total contents of N and S and for available contents of Cu, Fe and Pb.

In the leaf surface deposit S content was high though significantly ($P<0.001$) lower than in the leaf tissue, whilst N was not detectable. Cu, Pb and Fe contents in leaf surface deposit were conspicuous. The Pb content was higher in the leaf surface deposit than in the leaf tissue. No correlation between leaf tissue and surface deposit contents was found for S or for Fe. By contrast, positive and significant correlations ($P<0.01$) were found between leaf deposit and leaf tissue for both Cu and Pb. N and S contents in the leaves were not correlated to the respective contents in the soil and the same was also found for Cu and Fe. In contrast with the presence of limiting concentrations in the soil, N, S and Fe leaf contents were significantly higher than in the leaves from remote sites. The data suggest that direct uptake of airborne pollutants, in addition to root absorption, may influence leaf elemental composition of *Q. ilex* L. leaves.

Key words: Quercus ilex L., Air pollution, Leaf contamination, Nitrogen, Sulphur, Trace metals.

Abbreviations: TEA = triethanolamine; DTPA = diethylenetriaminopentacetic acid; PVC = polyvinyl chloride.

Introduction

Deposition of airborne particulates and gases on vegetation and soil in polluted areas may influence leaf and soil elemental composition by influx of trace metals as well as nitrogen and sulphur oxides. Higher concentrations of N, S and trace elements have been found in leaves from urban and industrial areas than from remote areas (Little, 1973; Buchauer, 1973; Dolske, 1988; Rovinsky et al., 1993; Houdijk and Roelofs, 1993; Meng et al., 1995). Uptake and accumulation of elements in plants may follow two different paths, i.e. the root system and the foliar surface. The relative importance of these routes for pollutant flux towards the leaves is not clear. Moreover, although the uptake of airborne pollutants has been extensively analyzed (Okano et al., 1988; Spink and Parsons, 1995), only a few studies have been performed in the field (Houdijk and Roelofs, 1993; Meng et al., 1995).

The aim of this research is an evaluation of the two uptake routes in city trees subjected to different load of air pollutants and different availability of nutrients and metals in the soil.

Naples, a densely populated city with major industrial plants within the urban area and a heavy load of vehicular traffic, is characterized by a conspicuous trace metals and nitrogen oxides air pollution. Associated with air pollution is a massive contamination of soil and plants, notably along roads with high traffic flow (Alfani et al., 1996). In particular, a positive correlation between traffic flow and trace metal

(Pb, V, Cu, Fe, Zn, Ni) contents has been found in unwashed leaves of *Quercus ilex* L. trees from several sites in the city (Alfani et al., 1989). Moreover a removable dry deposit on the leaf surface was found to be positively correlated to traffic flow; a positive correlation was also found between the amount of leaf deposit and its Pb, Cu, Fe and S content (Alfani et al., 1995).

In this work we have studied the distribution of N, S and three trace metals (Pb, Cu and Fe) in the leaf surface deposit, leaf tissue and soil in order to evaluate leaf contamination by pollutant influx. Analyses have been carried out on leaves of Holly oak (*Quercus ilex* L.) and samples of the underlying soil from several sites in the urban area of Naples.

Material and Methods

Quercus ilex L. is an evergreen mediterranean oak very common as an ornamental tree in Naples. In January 1993, 6 branches of the last generation were collected from around the lower foliage of 2 or 3 at least 50-year-old trees in 8 sites of the urban area. The sites were in parks (CE, OB), squares (PC, PT), roads (CC, VC, VM, ST); the roads support intensive traffic flow, i.e. an average of 30,158; 48,004; 48,000 and 76,365 vehicles/day, respectively (Comune di Napoli, unpublished data). The sampling was carried out after a prolonged rainless period.

For each site, 2 comparable samples of 30 leaves were selected by taking away 2 adiacent leaves in good condition from the collected branches. A sample was washed by shaking in about 500 mL of tap water for 30 minutes in order to remove the leaf surface deposit, whilst the other was left unwashed for comparison.

Soil samples were collected on May 1993 and January 1994 from the same sites as for leaf sampling. In each site, 3 soil samples were collected from the surface horizon (0–10 cm) at the trunk base with a PVC cylinder and a plexiglass trowel and stored in polythene bags. In laboratory the soil samples were passed through a 2 mm mesh sieve.

Samples of soil and leaves were oven-dried at 75 °C to constant weight and grounded to a fine powder by an agate pocket in a Fritsh pulverisette.

N and S analyses of soil and leaves were carried out on 3 subsamples by a Nitrogen Carbon Sulphur Analyzer (Carlo Erba NA 1500). Pb, Cu and Fe contents of soil and leaves were determined by atomic absorption spectrometry (SpectrAA 20 Varian). Leaf samples (250 mg) were digested in a Mileston (mls 1200) Microwave Laboratory Systems with a mixture of nitric and hydrofluoric acid (HF 50%: HNO_3 65% = 2:4). The available amount of Pb, Cu and Fe in the soil were estimated in extracts obtained by shaking soil samples for 2 hours in Lindsay and Norwell reactive (DTPA, $CaCl_2$ and TEA at pH 7.3) in the ratio 1:2 (Lindsay and Norwell, 1978). Analyses have been carried out on 2 or 3 leaf subsamples and on 3 soil subsamples.

The elemental concentration of the leaf surface deposit was calculated by difference between unwashed and washed leaves. This procedure is supported by experimental evidence showing that careful rinse in water removes a very large percentage of pollutants burden from the leaf surface (Little, 1973). For *Glycine max* L. (Dolske, 1988) and oak leaves (Lindberg et al., 1979) an effective remotion of surface deposit and a complete dissolution of NO_3^- and SO_4^- from surface deposits were observed in the first 2 minutes of the washing treatment, followed by a slower leaching of these compounds from within the leaf. Accordingly, in this report the values of elemental concentration in washed leaves are referred to as leaf tissue values. The soil contents of analyzed elements are reported as means values of the May and January samplings.

The significance of differences of elemental content between leaf and deposit, as well as between sampling sites was checked by one-way ANOVA test. Relationships between leaf tissue and leaf deposit elemental contents as well as between leaf and soil values were tested by the Spearman's coefficient of rank correlation (r_s).

Results

Nitrogen and sulphur

N and S contents of *Quercus ilex* L. leaf tissue ranged from 16 to 21 $mg\,g^{-1}$ d.w. and from 1.1 to 1.5 $mg\,g^{-1}$ d.w. respectively with significant differences among sites (Figs. 1, 2).

N in the leaf deposit was not detectable; by contrast S content of leaf deposit was in the range 0.1–0.9 $mg\,g^{-1}$ d.w., with significant differences among sites (Fig. 2). The S content in leaf tissue was always significantly higher ($P<0.001$) than in leaf deposit, and no correlation was found between values in the leaf tissue and leaf deposit.

Total N and S contents in the soil were in the range 2.1–7.2 and 0.5–1.1 $mg\,g^{-1}$ d.w. respectively (Figs. 1, 2). No correlation was found between leaves and soil N and S contents. The ratio between leaf and soil N and S contents was always above 1, besides conspicuous differences were found between sites (Fig. 3).

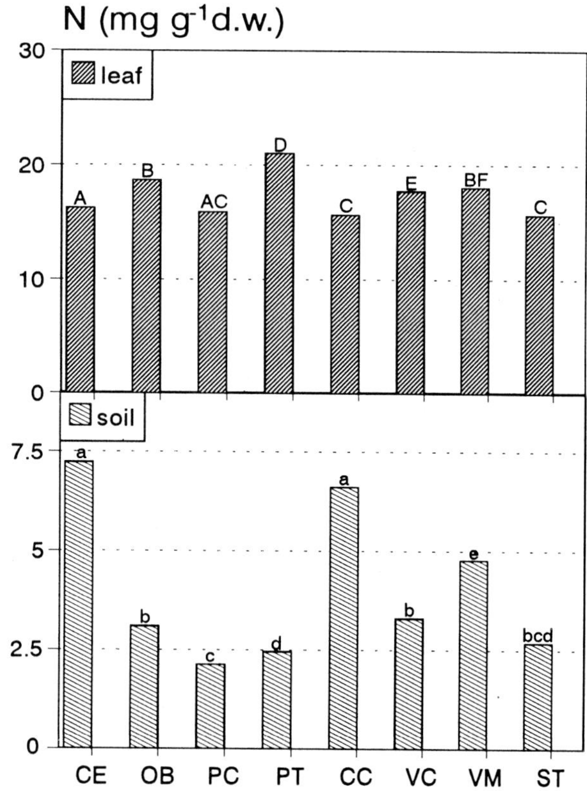

Fig. 1: Differences among eight sites in the urban area of Naples for N concentration in *Quercus ilex* leaf tissue (above) and in soil (below). Different letters indicate significant ($P<0.05$) differences between sites.

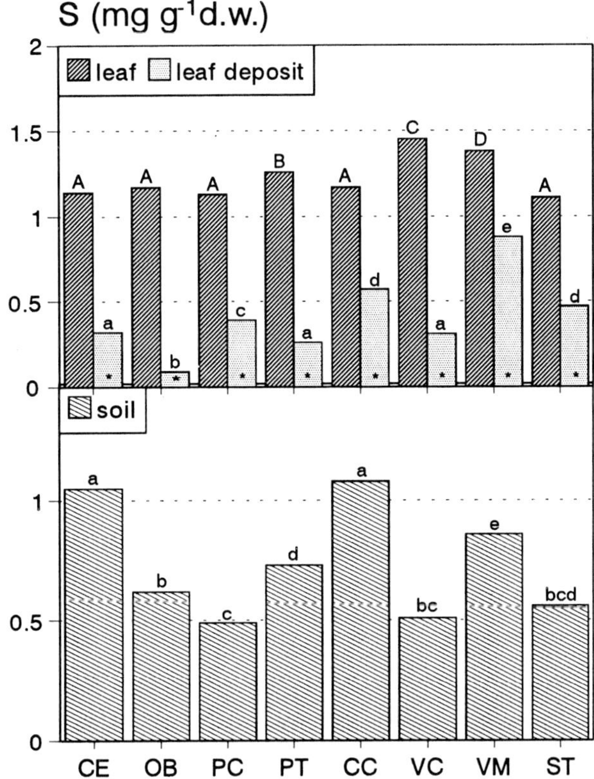

Fig. 2: Differences among eight sites in the urban area of Naples for S concentration in *Quercus ilex* leaf tissue and leaf deposit (above), and in soil (below). Different letters indicate significant ($P<0.05$) differences between sites: for leaf tissue (capital letters), for leaf deposit and soil (small letters). The asterisk indicates significant differences ($P<0.05$) between leaf deposit and leaf tissue content.

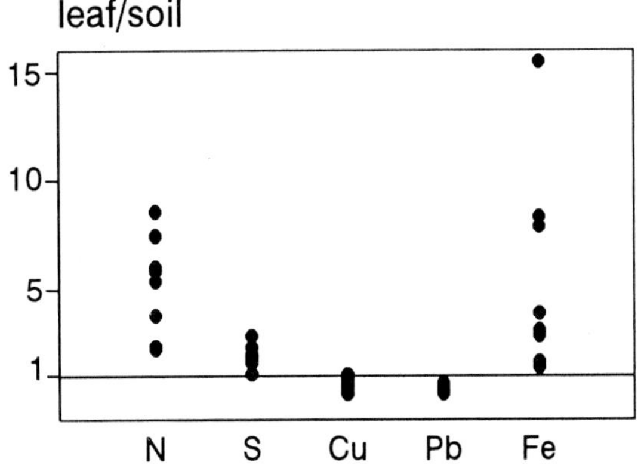

Fig. 3: Ratio between leaf and soil contents of N, S, Cu, Fe and Pb, in samples from the eight studied sites.

Organic carbon in the soil varied in the range 1.8–8.5 % (Table 1). Soils relatively rich in organic matter (CE, CC) showed higher N and S contents (Tab. 1, Figs. 1, 2) compared to poorer soils (PT, PC, ST). The values of N and S contents in the soil were positively correlated with each other ($r_s = 0.844$; $P<0.001$) as well as with soil organic carbon content ($r_s = 0.897$; $P<0.001$ and $r_s = 0.618$; $P<0.05$ respectively).

Table 1: Water holding capacity (WHC), cationic exchange capacity (CEC), pH and content of organic carbon in soils from some sites in the urban area of Naples.

Sites	WHC (H_2O % d.w.)	CEC (meq/100 g d.w.)	pH	C_{org} (%)
CE	75	45.7	7.25	8.5
OB	59	25.0	7.15	4.3
PC	55	13.4	6.09	2.1
PT	50	23.7	5.34	1.8
CC	65	21.5	4.77	6.6
VC	69	21.0	6.73	4.5
VM	52	23.0	7.81	4.7
ST	49	11.1	6.88	3.3

Copper

The values of Cu content in the leaf tissue and leaf deposit were in the range 7–31 $\mu g\,g^{-1}$ d.w. and 1–80 $\mu g\,g^{-1}$ d.w., respectively, with strong differences among sites for both values (Fig. 4). The Cu content in the leaf tissue was significantly higher ($P<0.01$) than in the leaf deposit with the exception of leaves collected from VM and ST, where Cu content in the deposit was significantly higher ($P<0.01$) than in the tissue (Fig. 4). A significant correlation ($r_s = 0.857$; $P<0.01$) was found between Cu content in the leaf tissue and leaf deposit.

The available Cu content in the soil, ranging from 13.8 to 66.5 $\mu g\,g^{-1}$ d.w. (Fig. 4), was higher than in leaves as shown in Fig. 3. No correlation was found between Cu contents in the soil and leaf tissue.

Iron

The Fe contents in the leaf tissue and leaf deposit ranged from 197 to 914 $\mu g\,g^{-1}$ d.w. and from 140 to 1459 $\mu g\,g^{-1}$ d.w., respectively (Fig. 5), with strong differences among sites for both values. The highest concentrations were found in leaves from trees along roads with high traffic flow. Differences in the Fe content in the leaf tissue and leaf deposit were not always significant, nevertheless the Fe content was sometimes higher in the leaf tissue, sometimes in the leaf deposit (Fig. 5). The Fe contents in the leaves and deposit were not correlated.

The available Fe content in the soil was in the range 58–243 $\mu g\,g^{-1}$ d.w. (Fig. 5) and showed no correlation with Fe content in the leaves. The concentration of available Fe in the soil was usually lower than in the leaves (Fig. 3), the range of the leaf/soil ratio of Fe content was very wide.

Lead

The Pb content was in the range 2–21 $\mu g\,g^{-1}$ d.w. in the leaf tissue and 1–24 $\mu g\,g^{-1}$ d.w. in the leaf deposit (Fig. 6), with strong differences among sites for both. The values were higher in the deposit than in the tissue (Fig. 6), even if significant differences were found in some samples only. A positive

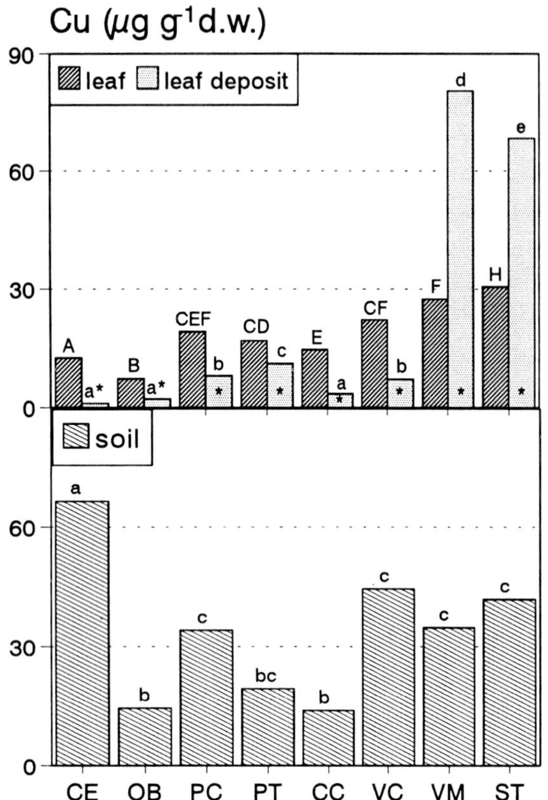

Fig. 4: Differences among eight sites in the urban area of Naples for Cu concentration in *Q. ilex* leaf tissue and leaf deposit (above), and soil (below). Letters and symbols as in Fig. 2.

and significant correlation ($r_s = 0.929$; $P < 0.001$) was found between the values of Pb content in the leaf tissue and in the surface deposit.

The available Pb content in the soil, in the range 11.3–46.3 $\mu g\,g^{-1}$ d.w. (Fig. 6), was higher than leaf content (Fig. 3). The Pb contents in the leaf tissue were positively correlated to soil Pb contents ($r_s = 0.714$; $P < 0.05$). The highest values of Pb in leaves and soil were found in the same site (ST), on a road (Fig. 6).

Discussion

All the elements considered in this study showed marked variation in their concentrations in the soil, in leaf tissue and in leaf surface deposit, with major differences among sites. This situation provides a favourable model for estimating the relative contribution of elemental soil availability and pollutant deposition to leaf contamination.

Nitrogen and Sulphur

In the soils of some sites tested in this study the values of total N were lower than figures (5–15 $mg\,g^{-1}$ d.w.) reported by Allen (1989) for organic soils. Most likely this is a consequence of removal of the litter by wind and/or garbage collectors, since both N and S in soil are strictly correlated to organic carbon concentration. In several plants such as *Betula, Alnus, Eucalyptus* and conifers, nutrient concentration in seedlings was found to be proportional to the relative rates of supply to the root system (Ingestad, 1982; Cromer et al., 1984). N concentration in needles of *Pseudotsuga menziesii* Mirb., *Pinus nigra* (var. maritima (Ait.) Melville) and *Pinus sylvestris* L. showed a regional trend that was related to deposition and N soil concentration (Houdijk and Roelofs, 1993). The absence of correlation between soil and leaf tissue content of either N and S strongly suggests that in *Q. ilex* foliar uptake from the air, besides root uptake, is an important source for N and S. The high values of N and S content in leaves of *Quercus ilex* plants from soils poor in N and S might be accounted for, in accordance with Cadle et al. (1991), by postulating some N and S transfer from the leaf deposit to the underlying leaf tissue and/or transcuticular and stomatal uptake of airborne nitrogen and sulphur oxides. This hypothesis fits well with previous observations showing that the N content in needles of *Abies alba* Mill. collected from plants growing in Naples increased with aging whilst remained unchanged in plants from a remote site with a higher N level in the soil (Alfani et al., 1994).

The S content in the leaf deposit was remarkably high whilst, unexpectedly, the N content was not detectable. This may indicates that airborne N pollutants are not incorporated in the leaf deposit or, more likely, that they are transferred from the deposit to the leaf tissue more quickly than S com-

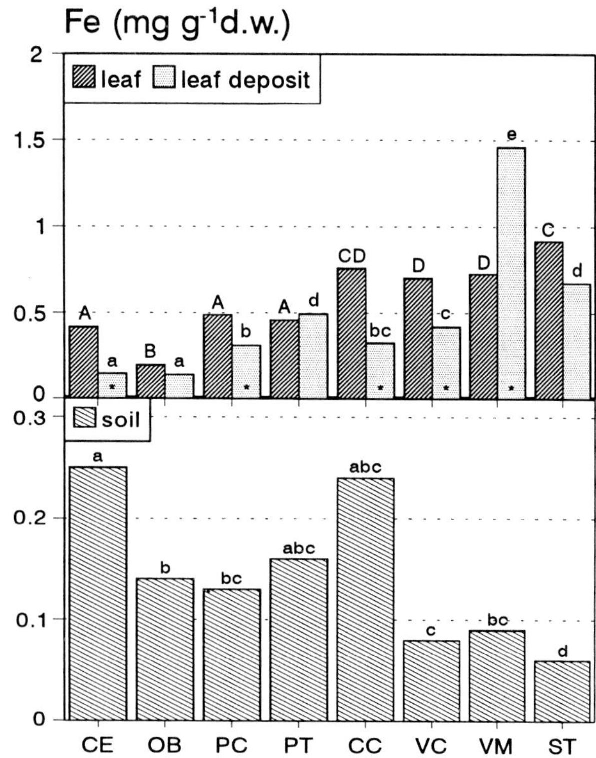

Fig. 5: Differences among eight sites in the urban area of Naples for Fe concentration in *Q. ilex* leaf tissue and leaf deposit (above), and soil (below). Letters and symbols as in Fig. 2.

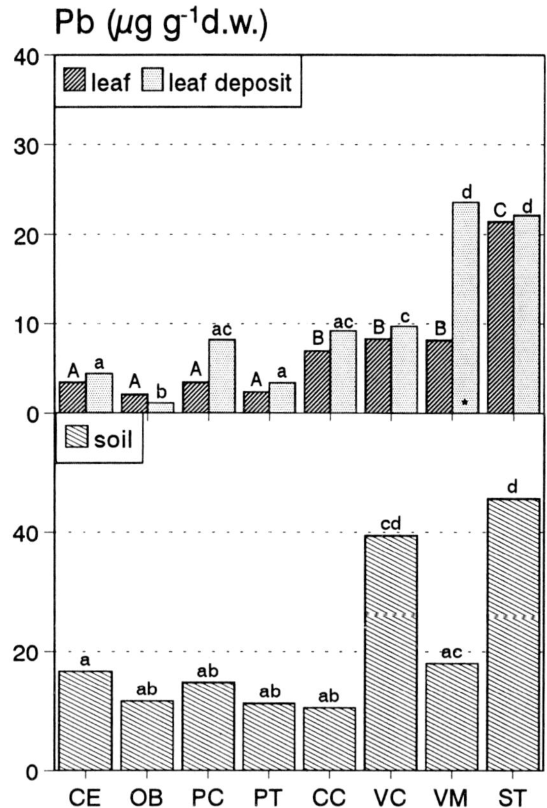

Fig. 6: Differences among eight sites in the urban area of Naples for Pb concentration in *Q. ilex* leaf tissue and leaf deposit (above), and soil (below). Letters and symbols as in Fig. 2.

pounds. In accordance with the latter hypothesis, Meyers and Hicks (1988) found that SO_2 transfer across the leaf cuticle and/or through stomata into the mesophyll tissue is limited by a surface uptake resistance, whilst the transfer of HNO_3 is mostly affected by atmospheric turbulence and near-surface diffusion.

N and S concentrations in *Q. ilex* leaves in the urban area of Naples were higher than in leaves from a remote area (Alfani, unpublished data) or reported in literature for trees in natural woods (Lossaint and Rapp, 1978). The observation that every year in the Spring, the plants of *Q. ilex* in Naples are massively infestated by aphids, notably at site PT that showed the highest N content, is in line with the report by Bolsinger and Fluckinger (1989) showing a good relationship between aphid population growth and aminoacids accumulation in plants exposed to NO_x.

Copper, Iron and Lead

Our results provide evidence of a massive deposition of airborne heavy metals on the leaf surface and of a remarkable accumulation in leaf tissue. Among factors involved in this process are probably important the stellate trichomes investing the abaxial surface of *Q. ilex* leaves, that increase the surface area effective in metal interception (Dolske, 1988). The existence of a positive correlation between the concentration of Pb and Cu in the leaf deposit and leaf tissue suggests that the deposition of airborne components directly influences leaf elemental composition.

Soil samples from sites along roads exhibited higher concentrations of both total and available Cu and Pb. In contrast, the available Fe was low in the sites where the total content was very high (Alfani, 1996). In some sites the concentration of available Fe in the soils underlying *Q. ilex* trees was close to the minimum value ($50\,\mu g\,g^{-1}$ d.w.) reported by Allen et al. (1989) for organic soils.

Soil concentrations of Fe were about tenfold lower than leaf concentrations, whilst Cu and Pb contents of soil, were about in the same range as in the leaves. Cu and Fe contents in the leaf tissue were higher than in the leaf deposit, except for the more contaminated sites where particularly high Cu and Fe contents were found in the deposit. These data are suggestive of Cu and Fe transfer from the surface deposit to the leaf tissue. Direct foliar uptake is likely to be particularly important for Fe in consideration of the low availability of this element in the soil as opposed to the high levels found in the leaf tissue and deposit.

Lead appears to be equally accumulated in the leaf deposit and leaf tissue. This is likely to reflect a passive diffusion mechanism from the deposit to the tissue, since Pb root-uptake is known to be very scarce and translocation from roots to leaves negligibile (Kabata-Pendias et al., 1993). The superficial deposit removed by washing includes soluble lead fractions that may penetrate the leaf tissue through the guard cells and cuticle (Bukovac and Vittwer, 1957; Franke, 1967) and insoluble fractions that may be incorporated in the leaf tissue at a much lower rate (Little, 1973). Studies carried out in background areas have revealed the absence of a direct correlation between trace metal contents in the soil and in the associated vegetation (Rovinsky et al., 1993). The correlation observed in the present study between Pb content in the soil and in the leaf tissue probably reflects accumulation on all exposed surfaces, soil as well as leaves, proportional to the degree of air Pb contamination at different sites.

In conclusion, our data indicate that in spite of limiting concentrations in the soil, N, S and Fe contents in leaves of *Q. ilex* in the urban area of Naples are high, thereby suggesting foliar uptake of airborne pollutants. Our data, however, do not permit to establish whether the uptake involves active or passive mechanisms nor to assess the relative contribution of root and leaf route.

Acknowledgements

This work was financially supported by Consiglio Nazionale delle Ricerche and Ministero dell'Università e della Ricerca Scientifica e Tecnologica.

References

Alfani, A., G. Bartoli, A. Virzo De Santo, M. Lombardi, F. Rutigliano, A. Fioretto, and E. Gargiulo: Leaf elemental composition of *Quercus ilex* L. in the urban area of Naples. I. Trace elements. Conference on: «Man and the environment» The plant components in anthropic systems, pp. 121–135. ENEL, Società Botanica Italiana (1989).

Alfani, A., G. Bartoli, and A. Virzo De Santo: L'azoto nelle foglie delle piante in ambiente urbano. In: Ferrari, C., F. Manes, and E. Biondi (eds.): Alterazioni ambientali ed effetti sulle piante, pp. 24–35. Edagricole, Bologna (1994).

Alfani, A., G. Bartioli, P. Caserta, and G. Andolfi: Amount and elemental composition of dry deposition to leaf surfiace of *Quercus ilex* in the urban area of Naples. Agr. Med. Special Volume, 194–199 (1995).

Alfani, A., G. Bartoli, F. A. Rutigliano, G. Maisto, and A. Virzo De Santo: Trace metals in the soil and the leaves of *Quercus ilex* in the urban area of Naples. Biological Trace Metals Research *51* (1996) in press.

Allen, S. E.: Chemical analysis of ecological materials. Blackwell scientific publications, Oxford London (1989).

Bolsinger, M. and W. Flückinger: Ambient air pollution induced changes in amino acid pattern of phloem sap in host plants – Relevance to aphid infestation: Environ. Pollut. *56,* 209–216 (1989).

Buchauer, M. J.: Contamination of soil and vegetation near a zinc smelter by zinc, cadmium copper and lead. Environ. Sci. Technol. *7,* 131–135 (1973).

Bukovac, M. J. and S. H. Vittwer: Absorption and mobility of foliar application of nutrients. Plant Physiol. *32,* 428–435 (1957).

Cadle, S. H., J. D. Marshall, and P. A. Mulawa: A laboratory investigation of the routes of HNO_3, dry deposition to coniferous seedlings. Environ. Pollut. *72,* 287–305 (1991).

Cromer, R. N., A. M. Wheeler, and N. J. Barr: Mineral nutrition and growth of Eucalyptus. New Zeland Journal of Forest Science *14,* 229–239 (1984).

Dolske, D. A.: Dry deposition of airborne sulfate and nitrate to soybeans. Environ. pollut. *53,* 1–12 (1988).

Franke, W.: Foliar penetration of solutions. Annu. Rev. Plant Physiol *18,* 281–300 (1967).

Houdijk, A. L. F. M. and J. G. M. Roelofs: The effects of atmospheric nitrogen deposition and soil chemistry on the nutritional status of *Pseudotsuga menziesii, Pinus nigra* and *Pinus sylvestris.* Environ. Pollut. *80,* 79–84 (1993).

Ingestad, T.: Relative addition rate and external concentration; driving variables in plant nutrition research. Plant, Cell and Environ. *5,* 443–453 (1982).

Kabata-Pendias, A., M. Piotrowska, and S. Dudka: Trace metals in legumes and monocotyledons and their suitability for the assessment of soil contamination. In: Markert, B. (ed.): Plants as biomonitors. Indicators for heavy metals in the terrestrian environment, pp. 485–494 (1993).

Lindberg, S. E., R. C. Harriss, R. R. Turner, D. S. Shriner, and D. D. Huff: Mechanisms and rates of atmospheric deposition of selected trace elements and sulfate to a deciduous forest watershed. Report ORNL/TM-6674. Oak Ridge National Laboratory. Oak Ridge, Tennessee, 550 (1979).

Lindsay, W. L. and A. Norwell: Development of a DTPA soil test for zinc, iron, manganese and copper. Soil Sci. Soc. Am. J. *42,* 421–428 (1978).

Little, P.: A study of heavy metal contamination of leaf surfaces. Environ. Pollut. *5,* 159–172 (1973).

Loissant, P. and M. Rapp: La forêt méditerranéenne de chênes verts (*Quercus ilex* L.). In: Lamotte, M. and F. Bourlière (eds.): Problèmes d'écologie: Structure et fonctionnement des écosystèmes terrestres, pp. 129–185. Masson, Paris (1978).

Meng, F. R., C. P. A. Bourque, R. F. Belczewski, N. J. Whitney, and P. A. Arp: Foliage responses of spruce trees to long-term lowgrade sulfur dioxide deposition. Environ. Pollut. *2,* 143–152 (1995).

Meyers, T. P. and B. B. Hicks: Dry deposition of O_3, SO_2, and HNO_3 to different vegetation in the same exposure environment. Environ. Pollut. *53,* 13–25 (1988).

Okano, K., T. Machida, and T. Totsuka: Absorption of atmospheric NO_2 by several herbaceous species: estimation by the ^{15}N dilution method. New Phytol. *109,* 203–210 (1988).

Rovinsky, F. Y., L. V. Burtseva, and T. B. Chicheva: Heavy metal in the vegetation as indicators for the environmental pollution in the area of the former USSR. In: Markert, B. (ed.): Plants as biomonitors. Indicators for heavy metals in the terrestrian environment, pp. 507–514 (1993).

Spink, A. J. and A. N. Parsons: An experimental investigation of the effects of nitrogen deposition to *Narthecium ossifragum.* Environ. Pollut. *2,* 191–198 (1995).

J. Plant Physiol. Vol. 148. pp. 249–257 (1996)

Mechanisms of Oxygen Activation during Plant Stress: Biochemical Effects of Air Pollutants

SUSANNE HIPPELI and ERICH F. ELSTNER

Lehrstuhl für Phytopathologie, Technische Universität München, 85350 Weihenstephan, Germany

Received June 6, 1995 · Accepted September 15, 1995

Summary

Green plants can adapt to drought, temperature, high and low light, infections, air pollution and soil contamination. Dependent on the strength of these impact(s), several symptoms indicate the deviation from normal, steady- state-metabolism. Most of these visible or measurable symptoms are connected with oxygen activation where principally a transition from heterolytic (two electron transitions) to increased homolytic (one electron transitions) reactions is observed. Homolytic reactions create free radicals, which are generally counteracted by a parallel increase of radical scavenging processes or compounds thus warranting metabolic control within certain limits. At advanced states of stress control may be gradually lost and chaotic radical processes dominate. Finally, cellular decompartmentalizations allow lytic and necrotic processes. Every episode during this cascade is characterizable by the balance between pro- and antioxidative capacities. Photosynthetic processes which are under metabolic and oxygen-detoxifying control are converted into photodynamic reactions which are only controlled by light and scavenger- and/or quencher-availabilities. Photoinhibition may represent the threshold between these two light driven events. This (more or less theoretical) sequence of events is not yet fully understood and in most cases can only punctually be characterized and followed by indicator reactions such as ESR.

In this communication we shall concentrate on basic redox-mechanisms during oxidative stress situations involving primary air pollutants. Chemical interactions between different gaseous pollutants and generally mechanisms of the finally toxic effects on target molecules will be shown.

Key words: Air pollution, photooxidations, plant stress.

Abbreviations: DCMU = Dichlorophenyl-dimethylurea; DOPA = 3,4-Dihydroxyphenylalanine; ESR = electron spin resonance; PA = primary receptor; PAN = peroxyacetyl nitrate; SH = sulphhydryl; VOC = volatile organic compounds.

1. Induction of plant stress

The term stress stems from physics and has been uncritically introduced in medicine and botany. In general mechanics it is clearly defined as the point or degree of bending of an elastic system just between symptomless reversibility and irreversible deformation or break. In medicine and botany, stress is supposed to indicate all situations beyond normal, defined by the observer i.e. man.

All organs of higher plants (with some exceptions) perform aerobic metabolism and are thus subject to activated oxygen species, oxygen-«oversaturation» and thus possibly oxygen «stress». This may occur under high light intensities over longer time periods, during drought periods, after poisoning such as soil contamination or air pollution or under mineral deficiency. A well-balanced antioxidative strategy has been elaborated during some 3 to 4 billion years. In this review basic reactions operating during oxidative stress situations will be discussed where certain prooxidative situations in plant life, air pollution effects and antioxidative processes will be dealt with.

Plant stress underlies sincerely a different definition as compared to what humans may define for themselves. It is difficult to exactly define the term stress, however, since it is

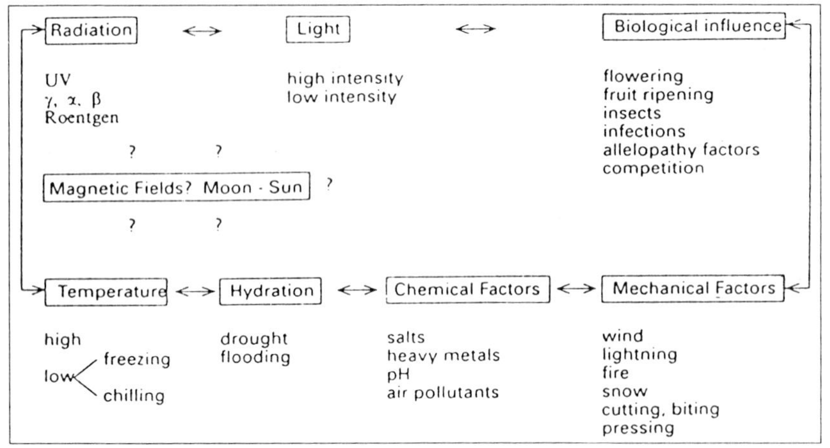

Fig. 1: The seven «stressfields» for plants (from Elstner et al., 1988; Elstner and Osswald, 1994).

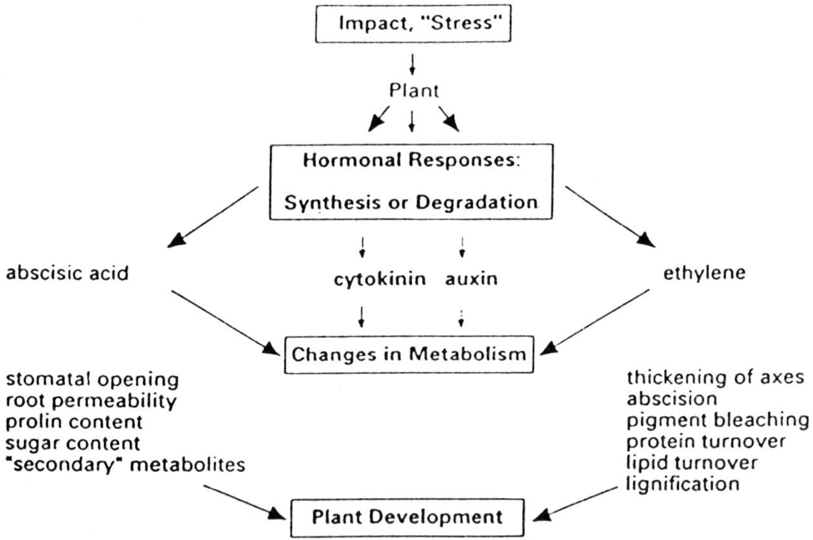

Fig. 2: Stress induced signal transduction and plant development (from Elstner et al., 1988; Elstner and Osswald, 1994).

used for the description of processes and syptoms during which the individuum is supposed to be in a more or less critical situation or to suffer, undoubtedly coupled with message feedback between different organs. The individual platforms indicating the start depends on the genetic outfits and the amount of impact necessary. It thus is almost impossible to exactly define fixed stress points i.e. the thresholds between normal metabolism and stress metabolism. In the following an attempt is made to more closely look at the most characteristic metabolic features underlying principle routes and reactions operating close to the above mentioned threshold. Seven «stressfields» have been defined for plants (Fig. 1).

2. Fundamental metabolic conditions governing stress situations

As outlined and reviewed by several authors and editors (Elstner et al., 1988; Pell and Steffen, 1991) ecological factors and preconditions, developmental processes, stress and oxygen activation are tightly connected with each other in several, up to date only partially understood ways. Wounding or other mechanical influences such as bending, increased transport system resistance induce oxidative processes such as fatty acid peroxidation through partial or complete decompartmentalization and induction or inactivation of detoxification pathways. Similarly, ionizing and UV radiation, drought, flooding, osmotic impacts (high salt concentrations, desiccation), deficiency in macro- or micronutrients, dramatic temperature changes such as heating, chilling or freezing as well as poisoning (air pollution, water and soil pollution, herbicide treatment) cause changes in pro- and antioxidative potentials. These are responded by hormone synthesis or release from endogenous stores inducing increased resistance, stability, avoidance, tolerance or repair (Fig. 2).

All the above mentioned influences on plants underly feedback to the chloroplast: independent on the site of transport- or metabolic «block», be it in the roots (salt, drought, mineral

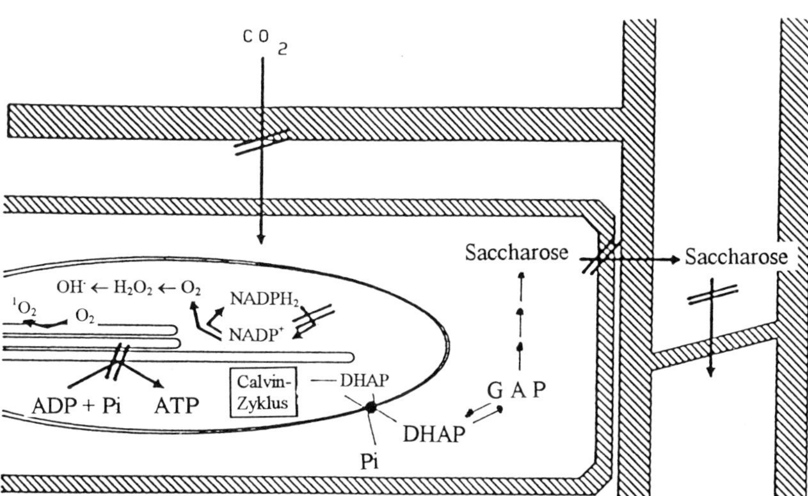

Fig. 3: Basic feedback events finally blocking photosynthetic electron transport (from Elstner et al., 1988; Elstner and Osswald, 1994).

deficiency), in the transport system (xylem or phloem blocks by infections) in phloem loading (photooxidants, infections) limitations in CO_2 fixation (stomatal closure, lack of Calvin cycle activities due to enzyme inhibition) or in photosynthetic electron transport, photosynthetic oxygen activation and/or photodynamic processes are provoked, leading to a chain of reactions outlined in a following section. A summary of the mentioned feedback system is given in Fig. 3.

The situations indicated in Fig. 3 has been simulated with isolated chloroplasts in the presence or absence of the natural electron acceptor, ferredoxin + NADP and the electron transport blocker, DCMU. Bleaching of photosynthetic pigments and formation of ethane from peroxidized linolenic acid (Konze and Elstner, 1978; Elstner and Pils, 1979, Elstner and Oßwald, 1980) are effects of «electron transport limitations».

Several lines of evidence lead us to conclude (Elstner et al., 1988) that electron flow through photosystem II under production of hydrogen peroxide is not initiating pigment bleaching but electron block after photosystem II initiates bleaching via lipid peroxidation by singlet oxygen. This also lead us to assume that block of the electron transport is the final «stressing» condition for green plants. Before this occurs, a series of metabolic and regulatory activities are passed. Massive production of singlet oxygen in the photosystems is thus a function of impaired metabolism allowing a definition of plant stress and metabolic events at the threshold between fully reversible (controlled) and only partially reversible (chaotic) reaction sequences from a «radical view».

From the recent literature it is evident that continuing stress leads to changes in metabolism (see above), regulation and morphological symptoms, indicating changes in the limits of anabolic, catabolic, amphibolic and anaplerotic pathways. In the course of stress «development» a maximum of «steering» activities is achieved, concerning the establishment of tolerance, avoidance, and detoxification by opening the stress- and rescue boxes. In the following, oxidative processes yield an aging of the most strongly stressed compartments where parts of the regulatory functions are lost. At the end of this line, aldehydes and lipofuscins are chemically produced from peroxidized lipids and protein amino groups. These products contribute to cell necrotization in addition to phenoloxidase and peroxidase products (i.e. o-quinones from o-dihydroxy phenylpropanes such as DOPA and caffeic acid) responsible for the browning. A summary of these events is outlined in Fig. 4.

Oxygen is also activated under «normal», metabolic steady state conditions but especially in plant disease development (Tzeng and DeVay, 1993). Thus, reactive oxygen species have come into the focus of investigations on pathological and toxicological situations both in human medicine and phytomedicine since approximately two decades. (To mention just a few: Michelson et al., 1977; Bannister and Bannister, 1980; Ciba Foundation Symposium, 1979; Bannister and Hill, 1980; Cohen and Greenwald, 1983; Greenwald and Cohen, 1983; Elstner, 1982, 1987, 1990, 1991, 1991 a; Gilbert, 1981; Halliwell and Gutteridge, 1989; Hayaishi, 1974; Pell and Steffen, 1991; Pryor, 1975–1982; Scandalios, 1992; Sies, 1985, 1991).

3. Air pollutants as inducers of oxidative stress

The apparent local and global increases of certain atmospheric trace gases have become a world wide focus of discussion since all living organisms and ecosystems are concerned by these atmospheric and stratospheric «global changes» (Manning and v. Tiedemann, 1995). The role of oxidative processes in air pollution has already been reported in the last century where the production of acid rain was recognized as an oxidation by atmospheric hydrogen peroxide of sulfur dioxide stemming from coal combustion (Holle, 1892). Similarly, chlorophyll bleaching was described as a process catalyzed by oxidizing enzymes (Woods, 1899). Nevertheless we were greatly surprised in our days by the overwhelming number of recent findings that oxygen toxicity is the dominating factor in general toxicology and pathology (Elstner, 1990).

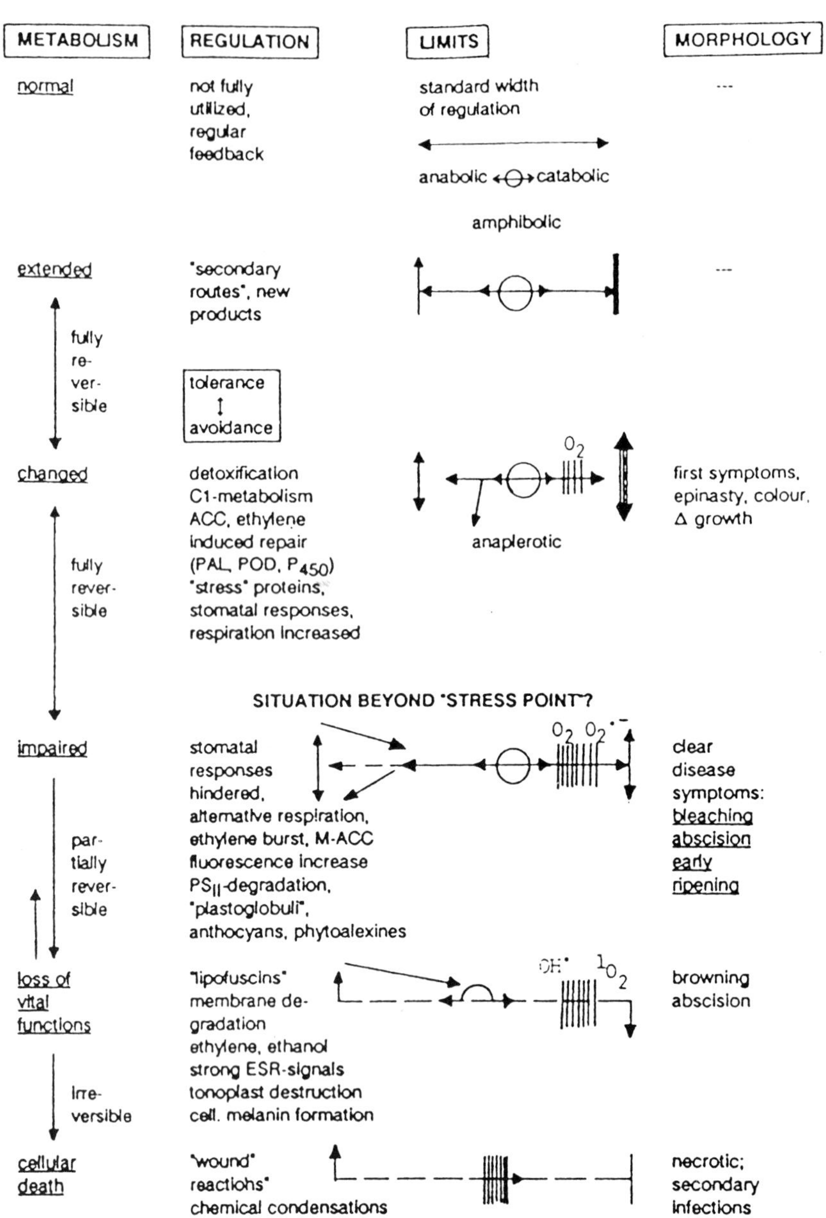

Fig. 4: Metabolic, regulatory and morphological changes during stress development (from Elstner et al., 1988; Elstner and Osswald, 1994).

4. General Biochemical Aspect of Reactions with Air Pollutants

Basic chemical reactions of the most prominent potentially toxic trace gases will be discussed. Biologically it is almost impossible to differentiate between the individual influences under natural conditions since there is no single influence of just one component of air pollutants on aerobic cells. Nevertheless we will separately discuss the influences of SO_2, NO_x ozone and peroxyacetyl nitrate (PAN) on important target molecules.

4.1 Dark Reactions

4.1.1 SO_2 and Derivatives

SO_2 is produced during combustion of organic materials especially of coal. SO_2 and HSO_3^- are involved in radical chain processes. HSO_3^- is a reductant which, however, can accelerate oxidative processes by reducing peroxidic bonds, thus producing anions and radicals. Principally, SO_2 or HSO_3^- preferentially reacts with the following types of molecules:

(a) aldehydes and ketones, where hydroxysulfonates are formed, which in turn are inhibitors of several enzymes;
(b) olefines, where sulfonic acids are formed and the double bond is lost;
(c) pyrimidines under formation of dihydrosulfonates, which may have mutagenic effects;
(d) disulfides under formation of S-sulfonates, where the S-S-bridge is split.
(e) superoxide ($O_2^{\cdot -}$) under formation of the very reactive disulfite radical:

$$O_2^{\cdot -} + HSO_3^- + 2H^+ \longleftrightarrow HSO_3\cdot + H_2O_2$$

In this reaction, superoxide may be substituted by transition metal ions:

$$M^{n+} + SO_3^{2-} \rightarrow M^{(n-1)+} + SO_3\cdot^-$$

$HSO_3^{\cdot}$ is a powerful initiator of several radical chain processes such as lipid peroxidation. Under aerobic conditions, the sulfate radical is generated from the sulfite radical:

$$\begin{aligned} SO_3^{\cdot -} + O_2 &\rightarrow SO_5^{\cdot -} \\ SO_5^{\cdot -} + SO_3^{2-} &\rightarrow SO_4^{\cdot -} + SO_4^{2-} \\ SO_4^{\cdot -} + SO_3^{\cdot -} &\rightarrow SO_4^{2-} + SO_3^{\cdot -} \end{aligned}$$

Both radical anions may be involved in destructive reactions initiated by SO_2 expositions.

Most of the physiological effects can be induced by hydrogen abstraction as the primary reaction. None of the above mentioned reactions, however, seems to be solely responsible for the observed toxic effects or symptoms. Of special importance is certainly the reaction with preformed hydroperoxides (see below).

SO_2 is a highly water soluble gas which is rapidly hydrated forming sulfite (Peiser and Yang, 1978):

$$SO_2\,(g) \rightleftharpoons SO_2\,(aq) \xrightarrow[pK=1.76]{H_2O\ H^+} HSO_3^- \xrightarrow[pK=7.20]{H^+} SO_3^{2-}$$

Sulfite modifies the cellular energy metabolism in mammalian tissues by decreasing the cellular ATP pool; this may be due to changing the function of pyridine and flavin nucleotides (McManus et al., 1989).

If sulfite is not metabolized via sulfite oxidase, it may be oxidized via radical chain mechanisms. In the presence of oxygen, a series of SO_x- and oxygen radicals are generated (see below). This reactions induce lipidperoxidation and numerous connected reactions.

In clean air, SO_2 is present in concentrations lower than 20 µg/m³. In heavily polluted areas close to coal power plants concentrations up to 1200 µg/m³ may be observed. Conversion factors for SO_2: 1 ppm = 2670 µg/m³; ppb = µg/m³ × 0.375; µg/m³ = ppb × 2.67.

4.1.2 NO_x

NO and NO_2 are free radicals. Due to their unpaired electrons several one electron reactions can be initiated where the attack of olefinic structures (double bonds) is one characteristic. The first reaction is most probably the abstraction of a hydrogen atom.

$$NO_2 + \text{—CH=CH—CH}_2\text{—} \rightarrow HNO_2 + \text{—}\dot{C}H\text{—}\dot{C}H\text{—}\dot{C}H\text{—} \xrightarrow{(O_2)} \text{chain reaction}$$

Addition reactions can also be observed:

$$NO_2 + \text{—C=C—} \rightarrow NO_2\text{—C—C—}^{\cdot}$$

$$NO_2\text{—C—C—}^{\cdot} + O_2 \rightarrow NO_2\text{—C—C—OO}^{\cdot} \rightarrow \text{chain reaction}$$

The primary product of combustion is NO. Since NO and NO_2 are tightly connected with each other in atmospheric chemistry via a very complicated chain of reactions (see Hippeli and Elstner, 1992) involving ozone and a vast amount of volatile organic molecules (VOC) calculating with activities or concentrations of only NO or NO_2 is not very meaningful, however.

Conversion factors for NO_2: 1 ppm = 1910 µg/m³; ppb = µg × 0.523; µg/m³ = ppb × 1.91.

4.1.3 *Ozone and peroxides (PAN)*

Ozone is formed from oxygen and NO_x catalyzed by VOC in a chain of atmospheric events. Ozone ($(-)O-O(+)=O$) with a standard redox potential of about 2 volts is a strong oxidant and a very reactive molecule due to its ionic resonance structure. Characteristic chemical reactions with biochemical relevance are the splitting of olefinic bonds and the reactions with thiols. Ozonisation of olefines yields intermediary ozonides, which in turn decompose into ketones and other products probably via the so-called «Criegee zwitterions». The resonance structures allow additions of carboxylic acids, alcohols and water, finally yielding different peroxidic products:

$$\text{>C=C<} + O_3 \rightarrow \underset{\text{ozonide}}{\text{>C(O–O–O)C<}} \rightarrow \left[\text{—C}^{(+)}\text{–O–O}^{(-)} \leftrightarrow \text{>C=}\overset{+}{O}\text{–}\overset{-}{O}\right] + \text{>C=O}$$

additions to the Criegee zwitterion

$$\text{>C}^{(+)}\text{–O–O}^{(-)} + ROH \rightarrow \text{—C(O–OH)—OR}$$

$$+ RCOOH \rightarrow \text{—C(O–OH)—O—C(=O)—R}$$

$$+ H_2O \rightarrow \text{>C=O} + H_2O_2$$

peroxidic products

Thiol oxidation leads to sulfonic acids or disulfides:

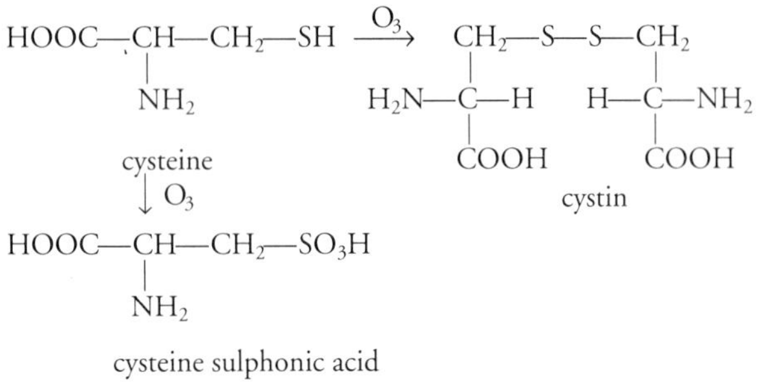

$$\mathrm{HOOC{-}CH(NH_2){-}CH_2{-}SH} \xrightarrow{O_3} \mathrm{CH_2{-}S{-}S{-}CH_2}\ \text{with}\ \mathrm{H_2N{-}C{-}H\ (COOH)}\ \text{and}\ \mathrm{H{-}C{-}NH_2\ (COOH)}$$

cysteine → cystin

$\downarrow O_3$

$$\mathrm{HOOC{-}CH(NH_2){-}CH_2{-}SO_3H}$$

cysteine sulphonic acid

Similarly, glutathione and proteinic SH-groups are oxidized thus drastically changing the antioxidative power and cellular redox balances together with enzymic functions. The oxidation of SH-groups is responsible for inhibition of the lipid synthesis in mitochondria and microsomes: Acyl-CoA-thioesterases, -thiokinases and acyltransferases are blocked, detectable as inhibition of the glycerol-3-phosphate acylation (Peters and Mudd, 1982). Glycerinaldehyd-3-phosphate dehydrogenase (GAPDH), a key enzyme for glycolysis, is also inactivated, where the oxidation of SH-groups in the active center is responsible for the rapid loss of function. In addition, peripheric SH-groups, tryptophan, histidine and methionine groups are oxidatively damaged (Knight and Mudd, 1984) thus provoking proteolytic degradation. Ozone inactivates α1-proteinase inhibitor via methionine oxidation (Mohsenin and Gee, 1989), and lysozyme via the conversion of tryptophan into N-formyl kynurenine (Dooley and Mudd, 1982). Due to its ionic resonance structure ozone is more water soluble than oxygen. Therefore hydrophilic compartments may be the main site of its reaction. Under acidic conditions ozone is quite stable, under basic conditions it decomposes into molecular oxygen. At 22 °C hundred ppm ozone on a water surface yield micromolar solutions. One may calculate that up to 10^{-6} moles of ozone are dissolved in 55 moles H_2O. In the presence of aromatic compounds O_3 is decomposed into OH· radicaltype oxidants. Dependent on the concentrations of NO, aromatic compounds and high light intensities, O_3 can increase strongly where up to 400–500 µg/m^3 can be reached on hot and sunny days. Average concentrations in rural areas during sommer days may be between 20–80 µg per m^3, depending on daytime and altitude. In sommer 1994 in Germany episodes with 150–250 µg/m^3 O_3 have frequently been observed.

Conversion factors for O_3: 1 ppm = 2000 µg/m^3, ppb = µg/m$^3 \times 0.501$, µg/m^3 = ppb$\times$2,00.

Different peroxides are present in polluted air; in this respect peroxyacetyl nitrate (PAN) and H_2O_2 are of major importance. In centers of «photosmog» air pollution such as the Los Angeles region PAN may yield up to 5–20% of the corresponding ozone concentrations. Besides PAN other peroxides such as peroxypropionyl-, peroxybutyryl- and also peroxybenzyl nitrate have to be considered as plant-toxic species in polluted air. PAN at physiological pH (7.2) has a half live of 4.4 min. It decays base-catalyzed essentially yielding acetate and nitrite:

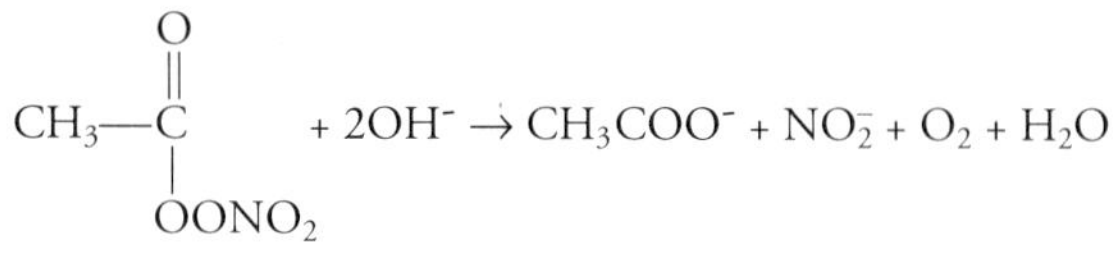

$$\mathrm{CH_3{-}C({=}O){-}OONO_2} + 2OH^- \rightarrow CH_3COO^- + NO_2^- + O_2 + H_2O$$

PAN may react with the following chemicals and biochemicals:

(a) thiol groups, especially in enzymes, such as the SH-groups of glucose-6-phosphate dehydrogenase; the substrates, Glu-6-P and NAD may partially prevent SH-oxidation. Smaller S-containing molecules are also oxidized: methionine is converted into methionine sulfoxide and liponic acid and coenzyme A are converted into the corresponding disulfides. With glutathione one mole PAN converts 3 moles of SH-groups.

(b) PAN reacts with NADPH, where the oxidized compounds (NAD, NADP) do not seem to further react with PAN.

(c) PAN reacts with olefines under formation of epoxides and further products of lipidperoxidation:

$$\text{PAN} + \text{olefine} \rightarrow CH_3COO^{\cdot} + NO_2 + \text{epoxide(s)}$$
$$CH_3COO^{\cdot} \rightarrow CO_2 + CH_3^{\cdot}$$
$$CH_3^{\cdot} + LH \rightarrow L^{\cdot} + CH_4$$

(d) Reactions with amines result in their acetylations:

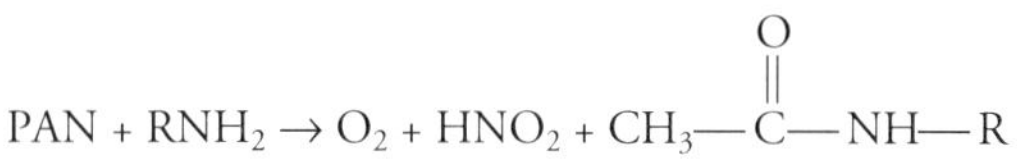

$$\text{PAN} + RNH_2 \rightarrow O_2 + HNO_2 + \mathrm{CH_3{-}C({=}O){-}NH{-}R}$$

Conversion factors for PAN: 1 ppm PAN = 4370 µg/m^3; ppb = µg/m$^3\times$0.223; µg/m^3 = ppb$\times$4.37.

4.2 Light-driven Reactions

Like humans and animals, plants under several conditions may suffer from «stress», although these conditions by definition may be completely different from animal systems (Elstner and Osswald, 1994). Besides drought, flooding, frost, heat, soil chemicals and many others, air pollution represents one major stressor. As not outlined in detail at this point, most of the mentioned conditions under continuing light influence yield an overreduction of the photosynthetic electron transport chain due to limited $NADPH_2$ reoxidation.

Under conditions with limited $NADP^+$ in the stroma of the chloroplasts (i.e. strong light, transport limitations, ATP-uncoupling, inhibition of Calvin-cycle activities and others) reductive and photodynamic oxygen activation are measurable. Especially inhibited CO_2-fixation due to stomatal closure or inhibition of the Calvin-cycle due to a blocked translocation of photosynthetic products lead to this limited reoxidation of $NADPH_2$. These conditions result in the mentioned «over»-reduction of the electron transport chain and a lack of the potential of charge separation after chlorophyll activation in the light. Thus, inhibited energy transfer supports photodynamic reactions of type II generating singlet oxygen (Elstner et al., 1985):

$$\text{Chl} \rightarrow \text{Chl}^* \quad (\text{Chl}^* = \text{activated chlorophyll})$$
$$\text{Chl}^* \rightarrow {}^3\text{Chl} \quad ({}^3\text{Chl} = \text{triplet state chlorophyll})$$
$${}^3\text{Chl} + O_2 \rightarrow \text{Chl} + {}^1O_2$$

The electrophilic nature of singlet oxygen results in chemical reactions with double bonds or other electron rich functionality. Singlet oxygen may eventuelly be quenched by carotenoides or α-tocopherol in the thylakoid membranes or directly react with unsaturated fatty acids (LH) forming hydroperoxides (LOOH) (Knox and Dodge, 1985):

$$LH + {}^1O_2 \rightarrow LOOH$$

Membrane disruption due to lipid peroxidation is a common feature of photodynamic actions. Valenzo (1987) reviewed different photomodifications of biological membranes with emphasis on singlet oxygen mechanisms. He concluded that effective membrane sensitization usually involves an association of the sentiziser with the membrane. This is the fact especially in thylakoid membranes of the chloroplasts.

Limited $NADP^+$-availability results also in the formation of $O_2^{\cdot -}$ and H_2O_2 via «over»-reduction of electrontransport system and subsequent electron channelling to oxygen. Oxygen may be reduced by either the primary acceptor (PA) of photosystem I, by reduced ferredoxin (fd_{red}), or by both (Osswald and Elstner, 1986):

$$\begin{aligned} fd_{red} + O_2 &\rightarrow fd_{ox} + O_2^{\cdot -} \\ fd_{red} + O_2^{\cdot -} + 2H^+ &\rightarrow fd_{ox} + H_2O_2 \\ PA_{red} + O_2 &\rightarrow PA_{ox} + O_2^{\cdot -} \end{aligned}$$

H_2O_2 may be further reduced by reduced ferredoxin forming oxygen species with properties similar to the hydroxylradical (OH·) (Elstner et al., 1985):

$$fd_{red} + H_2O_2 \rightarrow fd_{ox} + OH^- + (OH\cdot)$$

Light dependent generation of reactive oxygen species in the thylakoid membrane is summerized in fig. 5.

Reductively and photodynamically generated reactive oxygen species degradate fatty acids in the membranes. The formed hydroperoxides of fatty acids decay metal catalyzed forming alkoxylradicals (LO·), which in turn cooxidize chlorophyll and other pigments, initiating their bleaching (Elstner et al., 1985):

$$\begin{aligned} LOOH + Me &\rightarrow LO\cdot + OH - + Me_{ox} \\ LO\cdot + Chl_{red} &\rightarrow LOH + Chl_{ox} \end{aligned}$$

Reactive oxygen species can be detoxified in the chloroplasts. Superoxide dismutation is catalyzed by copper zinc SOD, which is found in the stroma as well as bound to the thylakoids. Since chloroplasts are devoid of catalase, H_2O_2 detoxification is brought about the ascorbate ($AscH_2$/Asc) and glutathione (2GSH/GSSG) redox cycle at the expense of $NADPH_2$ («Beck-Halliwell-Asada cycle», c.f. Elstner et al., 1985):

$H_2O_2 + AscH_2$	Ascorbate peroxidase	→	$Asc + 2H_2O$
$Asc + 2GSH$		→	$AscH_2 + GSSG$
$GSSG + NADPH_2$	Glutathione reductase	→	$2GSH + NADP$

The whole process not only detoxifies H_2O_2 but also decreases the $NADPH_2$/ATP ratio. Thus, lipid peroxidation and pigment cooxidation due to the generation of reactive oxygen species only take place when the defence system of the chloroplast is overcharged. Furthermore accumulation of H_2O_2 in the stroma inhibits certain Calvin-cycle enzymes (generally the phosphatases).

Uncoupling of oxidative phosphorylation results in insufficient reoxidation of $NADPH_2$ and thus yields inhibition of photosynthesis in the light due to a lack of ATP.

4.2.1 SO_2 and derivatives

Plant damage, caused by SO_2 is oxygen and light dependent. Typical SO_2-effects are the rapid inhibition of photosynthesis before direct symptoms such as bleaching of chlorophyll can be observed. SO_2 or its aqueous solution, sulfite (HSO_3^-) can be oxidized in the presence of oxygen via a radical chain mechanism (Peiser and Yang, 1978):

Initiation:	$SO_3^{2-} + O_2$	→	$SO_3^{\cdot -} + O_2^{\cdot -}$
Progapation:	$SO_3^{\cdot -} + O_2$	→	$SO_5^{\cdot -}$
	$SO_5^{\cdot -} + SO_3^{2-}$	→	$SO_4^{\cdot -} + SO_4^{2-}$
	$SO_4^{\cdot -} + SO_3^{2-}$	→	$SO_4^{2-} + SO_3^{\cdot -}$
	$SO_4^{\cdot -} + OH^-$	→	$SO_4^{2-} + OH^{\cdot}$
	$OH\cdot + SO_3^{2-}$	→	$OH^- + SO_3^{\cdot -}$
Termination:	$OH\cdot + SO_3^{\cdot -}$	→	$OH^- + SO_3$
	$SO_3^{\cdot -} + O_2^{\cdot -} + 2H^+$	→	$SO^3 + H_2O_2$
	$O_2^{\cdot -} + O_2^{\cdot -} + 2H^+$	→	$H_2O_2 + O_2$

The initiation can be started by metal ions, UV-irradiation or enzymatic reactions. Free radicals generated during sulfite autoxidation attack cellular membranes via lipid peroxidation and subsequent chlorophyll bleaching. In addition lipid peroxidation is initiated by sulfite radicals generated during the reaction between sulfite and superoxide (Peiser and Yang, 1978):

$$\begin{aligned} SO_3^{2-} + O_2^{\cdot -} + 2H^+ &\rightarrow SO_3^{\cdot -} + H_2O_2 \\ SO_3^{\cdot -} + LH &\rightarrow HSO_3^- + L^{\cdot} \\ L^{\cdot} + O_2 &\rightarrow LOO^{\cdot} \\ LOO^{\cdot} + LH &\rightarrow LOOH + L^{\cdot} \end{aligned}$$

The bleaching of chlorophyll by lipid peroxidation is only observed, if HSO_3^- is simultaneously present. Neither LOOH nor HSO_3^- alone can destroy the green color of chlorophyll .

The generated H_2O_2 is thought to be responsible for the inactivation of ascorbate peroxidase, glutathione reductase and enzymes of the Calvin-cycle, such as phosphatases. SO_2 inhibits the translocation of photosynthetic products via a disturbed phloem loading.

4.2.2 Ozone

A disturbed phloem loading is also observed after ozone exposure of higher plants and is associated with an inactivation of the plasmalemma bound ATPase (Dominy and Heath, 1985). Secondary events lead to an increased starch accumulation and finally bleaching of the photosynthetic pigments.

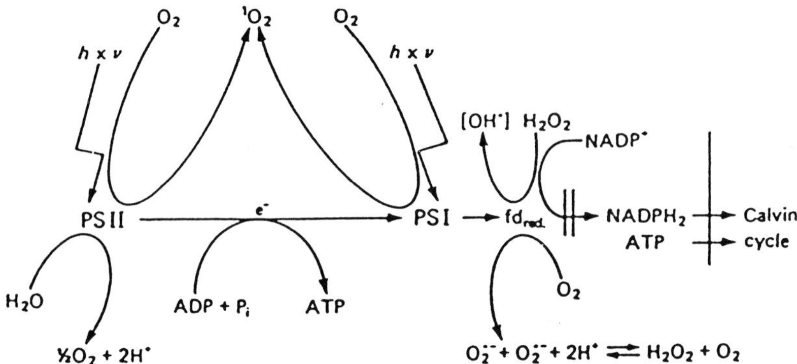

Fig. 5: Light-dependent generation of reactive oxygen species in the thylakoid membrane (after Osswald and Elstner, 1986).

Ozone further acts as an effective uncoupler of the photophosphorylation (Robinson and Wellburn, 1983). Studies with isolated chloroplasts showed, that ozone blockes the energy transfer from the primary light acceptors to the reaction centers, initiating photodynamic reactions of type II as mentioned above. The significance of this observations remains to be established *in vivo,* since ozone can probably not penetrate the outer chloroplast membrane.

Aben and coworkers (1990) showed that low level ozone (120 µg/m^3 for 8 hours per day over two weeks) can have a direct effect on the stomata and photosynthetic system, causing decreased photosynthesis and stomatal conductance.

References

ABEN, J. M. M., M. JANSSEN-JURKUVICOVA, and E. H. ADEMA: Effects of low-level ozone exposure under ambient conditions on photosynthesis and stomatal control of *Vicia faba* L. Plant Cell Environ. *13,* 463–469 (1990).

BANNISTER, J. V. and H. A. O. HILL (eds.): Chemical and biochemical aspects of superoxide and superoxide dismutase. Developments in Biochemistry, Vol. 11 A, Elsevier/North Holland, New York (1980).

BANNISTER, W. H. and J. V. BANNISTER (eds.): Biological and clinical aspects of superoxide and superoxide dismutase. Developments in Biochemistry, Vol. 11 B., Elsevier/North Holland, New York (1980).

Ciba Foundation Symposium 65 (New Series) Oxygen free radicals and tissue damage. Excerpta Medica, Amsterdam (1979).

COHEN, G. and R. A. GREENWALD (eds.): Oxy Radicals and their Scavenger Systems. Vol. I, Molecular Aspects Elsevier Biomedical, New York (1983).

DOMINY, P. J. and R. L. HEATH: Inhibition of the K^+-stimulated ATPase of the plasmalemma of pinto bean leaves by ozone. Plant Physiol. *77,* 43–45 (1985).

DOOLEY, M. M. and J. B. MUDD: Reaction of ozone with lysozyme under different exposure conditions. Arch. Biochem. Biophys. *218,* 459–471 (1982).

ELSTNER, E. F., W. F. OSSWALD, and R. J. YOUNGMAN: Basic mechanisms of pigment bleaching and loss of structural resistance in spruce (*Picea abies*) needles: advances in phytomedical diagnostics. Experientia *41,* 591–597 (1985).

ELSTNER, E. F. and W. OSSWALD: Chlorophyll photobleaching and ethane production in dichlorophenyldimethylurea (DCMU)- or paraquat-treated *Euglena gracilis* cells. Z. Naturforsch. *35 c,* 129–135 (1980).

ELSTNER, E. F. and I. PILS: Ethane formation and photobleaching in DCMU-treated *Euglena gracilis* cells and isolated spinach chloroplasts. Z. Naturforsch. *34 c,* 1040–1043 (1979).

ELSTNER, E. F.: Comparism of «inflammation» in pine needles and humans. In: BORS, W., M. SARAN, and D. TAIT (eds.): Oxygen radicals in chemistry and biology. Walter de Gruyter & Co Berlin, pp. 967–981 (1984).

– Oxygen activation and oxygen toxicity. Annu. Rev. Plant Physiol. *33,* 73–96 (1982).

– Metabolism of activated oxygen species. In: DAVIES, P. P. (ed.): The Biochemistry of Plants, Vol. 11, Academic Press, London, pp. 253–315 (1987).

– Der Sauerstoff – Biochemie, Biologie, Medizin, BI-Wissenschaftsverlag Mannheim (1990).

– Oxygen radicals – biochemical basis for their efficacy. Klin. Wochenschr. *69,* 949–956 (1991).

– Mechanisms of oxygen activation in different compartments of plant cells. In: PELL, E. and K. STEFFEN (eds.): Current Topics in Plant Physiology, American Soc. of Plant Physiologists, pp. 13–25 (1991 a).

ELSTNER, E. F., G. A. WAGNER, and W. SCHÜTZ: Activated oxygen in green plants in relation to stress situations. In: RANDALL, D. D., D. G. BLEVINS, and W. H. CAMPBELL (eds.): Current topics in plant biochemistry and physiology, Vol 7, University of Missouri–Columbia Press, pp. 159–187 (1988).

ELSTNER, E. F. and W. OSSWALD: Mechanisms of oxygen activation during plant stress. Proceedings of the Royal Soc. of Edinburgh *102 B,* 131–154 (1994).

GILBERT, D. L. (ed.): Oxygen and living processes. An interdisciplinary approach, Springer Verlag, New York, Heidelberg, Berlin (1981).

HALLIWELL, B. and J. M. C. GUTTERIDGE (eds.): Free radicals in biology and medicine, Clarendon Press, Oxford (1989).

HAYAISHI, O. (ed.): Molecular oxygen in biology. Topics in molecular oxygen research, North-Holland/American Elsevier, Amsterdam (1974).

HIPPELI, S. and E. F. ELSTNER: Oxygen radicals and air pollution. In: SIES, H. (ed.): Oxidative stress – oxidants and antioxidants. Academic Press, London, pp. 3–56 (1991).

HOLLE, G.: Vergiftete Nadelhölzer. In: Die Gartenlaube, *48,* 795–797 (1892).

KNIGHT, K. L. and J. B. MUDD: The reaction of ozone with Glyceraldehyde-3-phosphate dehydrogenase. Arch. Biochem. Biophys. *229* (1), 259–269 (1984).

KNOX, J. P. and A. D. DODGE: Review article number 7: Singlet oxygen and plants. Phytochem. *24,* 889–896 (1985).

KONZE, J. R. and E. F. ELSTNER: Ethane and ethylene formation by mitochondria as indication of aerobic lipid degradation in reponse to wounding of plant tissue. Biochim. Biophys. Acta *528,* 213–221 (1978).

MANNING, W. J. and A. v. TIEDEMANN: Climate changes: Potential effects of increased atmospheric carbon dioxide (CO_2), ozone (O_3) and ultraviolet (UV-B) radiation on plant diseases. Environmental Pollution *88,* 219–245 (1995).

MCMANUS, M. S., L. C. ALTMAN, J. Q. KOENIG, D. L. LUCHTEL, D. S. COVERT, F. S. VIRANT, and C. BAKER: Human nasal epithelium: characterization and effects of *in vitro* exposure to sulfur dioxide. Exp. Lung Res. *15,* 849–865 (1989).

MICHELSON, A. M., J. M. MCCORD, and I. FRIDOVICH (eds.): Superoxide and superoxide dismutases. Academic Press, London (1977).

MOHSENIN, V. and B. L. GEE: Oxidation of α1-protease inhibitor: role of lipid peroxidation products. J. Appl. Physiol. *66,* 2211–2215 (1989).

OSSWALD, W. F. and E. F. ELSTNER: Fichtenerkrankungen in den Hochlagen der Bayerischen Mittelgebirge. Ber. Deutsch. Bot. Ges. *99,* 313–339 (1986).

PEISER, G. D. and S. F. YANG: Chlorophyll destruction in the presence of bisulfite and linolenic acid hydroperoxide. Phytochem. *17,* 79–84 (1978).

PELL, E. J. and K. L. STEFFEN (eds.): Active oxygen/oxidative stress and plant metabolism. In: Current topics in plant physiology, Vol. 6., Amer. Soc. Plant Physiol. Rockville (1991).

PETERS, R. E. and J. B. MUDD: Inhibition by ozone of t of glycerol 3-phosphate in mitochondria and micr rat lung. Arch. Biochem. Biophys. *216,* 34–41 (1982).

PRYOR, W. A. (ed.): Free radicals in biology. Vols. I–V, Academic Press, New York (1975–1982).

ROBINSON, D. C. and A. R. WELLBURN: Light-induced changes in the quenching of 9-aminoacridine fluorescence by photosynthetic membranes due to atmospheric pollutants and their products. Environ. Pollution *32,* 109–120 (1983).

SCANDALIOS, J. G. (ed.): Molecular biology of free radical systems. Cold Spring Harbor Laboratory Press, New York (1992).

SIES, H. (ed.): Oxidative Stress – Oxidants and Antioxidants. Academic Press (1985 and 1991).

TZENG, D. D. and J. E. DEVAY: Role of oxygen radicals in plant disease development. In: ANDREWS, J. H. and I. C. TOMMERUP (eds.): Advances in plant pathology. Academic Press, London, pp. 1–34 (1993).

VALENZO, D. P.: Photomodification of biological membranes with emphasis on singlet oxygen mechanisms. Photochem. Photobiol. *46,* 147–160 (1987).

J. Plant Physiol. Vol. 148. pp. 258–263 (1996)

The Characterization and Contrasting Effects of the Nitric Oxide Free Radical in Vegetative Stress and Senescence of *Pisum sativum* Linn. Foliage

YA'ACOV Y. LESHEM and ESTHER HARAMATY

Department of Life Sciences, Bar-Ilan Univeristy, Ramat Gan 52900, Israel

Received June 24, 1995 · Accepted September 8, 1995

Summary

As monitored respectively by a nitric oxide (NO)-specific probe and gas chromatography, emission of both NO and ethylene from pea foliage increases with duration of stress and/or senescence promoting conditions. Depending on concentration, NO appears to exert a stress-coping or, contrariwise, an inhibitory effect on leaf growth. Stress coping is evidenced by the marked deceleration of stress ethylene by application of the NO releasing compound S-nitroso-N-acetylpenicillamine (SNAP) and possibly also by significant increment of leaf disc expansion upon treatment with low concentrations of chemically generated NO gas. At higher NO concentrations, leaf expansion is inhibited by NO treatment as well as by three different NO releasing compounds: SNAP, 3-morpholinosydnonimine, and N-*tert*-butyl-α-phenylnitrone. Further evidence to the assumption of endogenous involvement of NO metabolism is lent by the inhibition reversal induced by the nitric oxide synthase inhibitor N^G-methyl-*L*-arginine. Reconciliation of the opposite effects is discussed in terms of translocation and cellular binding sites of the free radical gas.

Key words: Pisum sativum Linn., Ethylene, nitric oxide, senescence, stress.

Abbreviations: ACC = 1-aminocyclopropane-1-carboxylic acid; EPPS = N-[2-hydroxyethyl]-piperazine-N′-[3-propane-sulphonic acid]; L.S.D. = least significant difference; MES = 2-(N-morpholino) ethanesulphonic acid; N-ARG = N^G-methyl-*L*-arginine; NO = nitric oxide; NOS = nitric oxide synthase; PBN = N-*tert*-butyl-α-phenylnitrone; Sin-1 = 3-morpholinosydnonimine; SNAP = S-nitroso-N-acetylpenicillamine.

Introduction

In the past few years, NO, a potentially toxic, relatively unstable, small diatomic free radical gas has become one of the most intensely investigated and fascinating entities in biological chemistry. It is highly diffusible from its site of synthesis making control of its synthesis the key to regulation of its activity. The overall formation of NO is as follows:

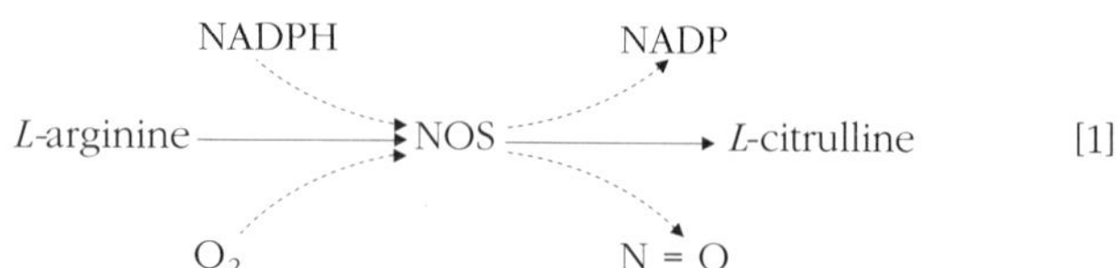

[1]

In mammalian systems promotory processes under its control include neurotransmission (Snyder, 1992; Madison, 1993), vasodilation (Konorev et al., 1995), coronary relaxation (Moneada et al., 1991), long-term memory potentiation (Fin et al., 1995) and penile erection (Andersson and Wagner, 1995). Moreover, NO is a means whereby macrophages exert their tumoricidal and bactericidal action (Snyder, 1992). These effects are on the ‹credit› side. On the ‹debit› side, NO participates in neurological disease (Koprowski et al., 1993), anaphylaxis (Mitsuhata et al., 1994), stroke episodes (Prince and Gunson, 1993) and migraine headaches (Oleson et al., 1994). NO moreover impairs the ability of hemoglobin to bind O^2: in its negative action current opinion holds that NO can damage cells by attacking Fe-S clusters in pivotal enzymes such as guanyl cyclase (Feldman et al., 1993).

Of particular interest in terms of phospholipid membrane components is the ability of NO to react with the superoxide radical, O_2^-, to form peroxynitrite $ONOO^-$, which together with its protonated form ONOOH, have the propensity to harm cells by oxidizing lipids, lipo-proteins or thiols (Radi et al., 1991; Fukuhori et al., 1995).

Concerning vegetative stress, it is well known that the pollution-produced NO may inhibit plant growth (Anderson and Mansfield, 1979; Neighbour et al., 1990). However as opposed to mammalian systems, to the best of our knowledge, as yet nothing is known concerning endogenous occurrence of NO in plants, barring, indirectly, as an intermediary step in nitrification, denitrification and nitrate reductase action (Beevers, 1976; Payne, 1983; Van Spanning et al., 1995). In view of the above, and within a framework of research on senescence and vegetation stress-associated processes, the aim of the present investigation was to ascertain if plants produce the NO free radical gas and, if so, to elucidate, if only partially, its mode of action in plant growth regulation.

Materials and Methods

Plant material

All experiments were performed on 21-day-old pea plants (*Pisum sativum* Linn. cv. P.F. 70A, USA) grown on a vermiculite medium (detailed by Leshem et al., 1993) in a growth chamber at 22 °C and in relative humidity of 60 per cent. Light source from GRO-LUX fluorescent tubes provided a 125 $\mu mol\, m^{-2} s^{-1}$ (PAR) intensity, the photoperiod being 12 h.

NO Determination

NO was directly measured by employing an ISO-NO Nitric Oxide Sensor (WPI – Florida, USA) equipped with a 2 mm NO specific probe. Experiments usually employed 4 replicates, each replicate consisting of three plants severed from their roots and placed in 25 mL vials containing 2 mM buffer EPPS pH 8.5, 10^{-5} M $CaCl_2$ either with or without 2 mM ACC which is the ethylene precursor. The vials containing the plants were then placed under 0.5 L plastic bell jars and sealed with rubber serum caps. Bell jars were also equipped with an adaptor inlet through which the NO probe was inserted and, aided by an O-ring held in place directly over the pea foliage, the purpose being to monitor NO emission, if any, immediately after its release from the plants, and thus to obviate as a sensor misreading the conversion of NO to NO_2, the half life of NO being ca. 5 s (Snyder, 1992). Preliminary trials carried out in bell jars purged with N_2 provided essentially similar results thus indicating registering only of NO and not of NO_2 to which the probe also possesses a limited degree of sensitivity.

Pea Leaf Expansion Trials

As a measure of possible NO effect on plant growth, the Pea Leaf Expansion Assay (Elzenga and van Volkenburg, 1994) was employed. This parameter expresses control over cell wall extensibility and turgor that in the final outcome regulate leaf expansion, the final result of which is construction of sites for photosynthesis and surface area for evaporative cooling (van Volkenburg, 1992).

With the aid of a 0.5 cm diameter cork borer, leaf discs were excised from pea foliage. Each disc was weighed and subsequently placed in Petri dishes, each dish containing a double layer of Whatman No. 2 filter paper to which 4 mL of 0.01 M buffer MES pH 5.5 also containing 5 mM sucrose, 10^{-5} M $CaCl_2$, 40 units of calmodulin, 0.03 µM NADPH and 0.01 M*L*-arginine, were added. The latter four ingredients were added in order to optimize any potential NOS activity. In the experiments testing the effects of the NO generating compounds SNAP, Sin-1 and PBN and of effects of the NOS inhibitor, N-ARG, these were added to the incubation medium in concentrations as mentioned in the captions to the figures. The Petri dishes were subsequently incubated for 22 h at 22 °C in the light of the intensity as outlined above after which the discs were removed, gently placed between two sheets of filter paper to remove excess moisture, subsequently weighed and weight increase calculated. Each experiment was conducted employing 4 replicates, each comprised of 10 discs. Mode of statistical analyses are mentioned in each case in captions to the pertinent figures or to Table 1.

Table 1: Inhibition of Pea Leaf Disc Expansion by NO Releasing Compounds. Results are 4 replicate means. Statistical computation of L.S.D. was by «Analysis of Variance». See text for experimental details. The 42 percent increase of leaf disc fresh weight observed in the non-treated controls after a 22 hr incubation period served as a basis of comparison.

Treatment	Leaf Disc Expansion (relative)
Control	100
Sin-1	83
PBN	68
SNAP	12
L.S.D. $p>0.05$	17

Direct Exposure of Leaf Discs to NO

In WPI – ISO NO Nitric Oxide Sensor calibration procedure, NO may be generated by employing two solutions: A. 0.1 M KI + 0.1 M H_2SO_4; B. 50 µM KNO_2. Upon mixing the two solutions, the following reaction takes place:

$$2KNO_2 + 2KI + 2H_2SO_4 \rightarrow 2NO + I_2 + 2H_2O + 2K_2SO_4 \quad [2]$$

Because the reaction goes to completion, equation [2] states that the ratio between KNO_2 and NO is 1:1. Therefore, the amount of NO generated in the solution will be equal to the amount of KNO_2 added. The final concentration of NO will be equal to the diluted concentration of KNO_2 in the solution.

Fig. 1 indicates how, in the present research, plant discs were exposed to NO: Solution A was entered into a 50 mL infiltration flask positioned over a magnetic stirrer, and whose sidearm was stoppered with a rubber serum cap through which later, with the aid of a syringe at given periods, spikes of solution B were injected to obtain final NO concentrations as presented in Fig. 3. Over the mouth of the infiltration flask, the lower section of a Ø 5 cm Kontes glass column equipped with a stopcock was firmly placed. A stiff partition of 55 µM mesh Miracloth, serving as a gas penetrable support for the pea foliage discs, was placed over the upper surface of Kontes column base. This attachment was then covered with a closely fitting hourglass which contained two gas inlets, one which served for purging the system with N_2 gas and the other for an air outlet during the purging. Purging with N_2 was performed in order to obviate the complication which may stem from conversion of NO to NO_2 which occurs in free air. The hourglass and Kontes column base were carefully sealed with a double layer of Parafilm to prevent gas leakage.

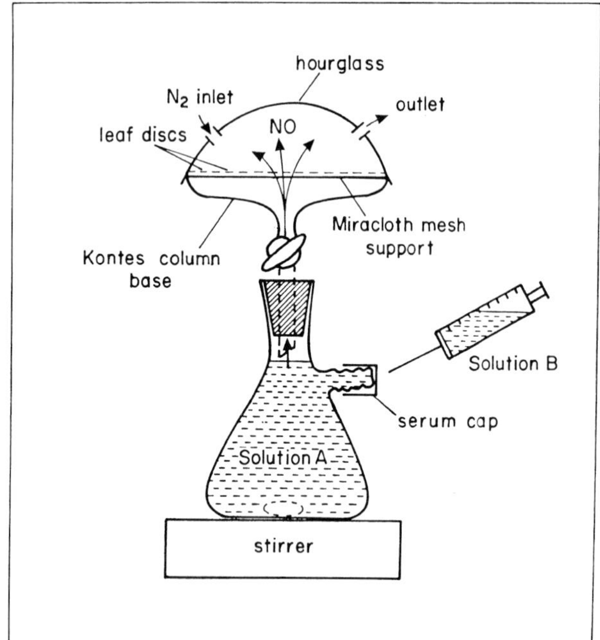

Fig. 1: *Apparatus employed for generating NO and NO treatment of leaf discs.* Solution A = 0.1 M Kl + 0.1 M H_2SO_4; Solution B = 50 µM KNO_2; Experimental procedure is outlined in the text.

In each of the quadruplicated treatments, 10 weighed leaf discs were placed on the Miracloth support, care being taken to place the lower leaf faces containing the stomates, facing downwards to later meet the upward stream of NO. The base support was then covered with the hourglass, sealed with Parafilm and with the stopcock open, the whole system, including the remaining air space above solution B in the infiltration flask, was purged for 2 min with N_2 after which the N_2 inlet and outlet were stoppered.

Ethylene Determination

At given periods, employing a hypodermic syringe, 1 mL gas samples were withdrawn through the rubber serum caps of the bell jars containing the pea foliage to assess ethylene production. To prevent moisture accumulation, an open vial containing 5 g anhydrous $CaCl_2$ was also placed in each bell jar. Ethylene measurement was performed on a Varian Model 3400 FID gas chromatograph equipped with an alumina column and a Varian 4290 integrator. Injection temperature was 150 °C and column temperature 110 °C. Readings presented are four replicate means.

Endogenous NO regulation

SNAP, Sin-1 and PBN release NO under physiological conditions (Hery et al., 1989; Shaffer et al., 1992; Chamulitrat, 1993) while on the other hand NOS activity may be inhibited by the *L*-arginine analog N-ARG (Hibbs et al., 1988). The experimental rationale here employed was that application of the above compounds to plants and observation of physiological effects would shed farther light on the role of NO and NOS activity in growth regulation. Their mode of employment is described in the Experiments and Results section.

Experiments and Results

NO Production under Stress

Emission of NO was measured in three groups of plants which were severed from their roots and thereupon a) immediately placed in the EPPS buffer solution under the bell jar as outlined in the previous section; b) allowed to wilt at 24 °C for 1 h in dry air before placing in the buffer medium; c) allowed to wilt for 2 h.

Results presented in Fig. 2 clearly indicate NO evolution by all three groups of stressed plants and furthermore that with increase of duration of wilting stress, NO production is intensified. These results lead to the obvious question – ‹is NO production a stress-coping or a stress-inducing strategy?› An endeavor to address this issue was made by the following experiments.

Effect of Direct Exposure to NO on Pea Leaf Expansion

As graphically depicted in Fig. 1 and described in the ‹Materials and Methods› section, direct NO treatment to pea foliage was applied to weighed leaf discs by spiking solution A with solution B by injection through the sidearm of the infiltration flask. Each spike was 1 mL of solution B administered with the stopcock open during injection time. An ascending series of NO concentrations was attained by measuring the number of spikes/min from 1–6. Controls were injected with like volumes of H_2O. Immediately before injections, the stopcock was closed and reopened after briefly stirring the two solutions. Discs were held in the NO atmosphere for 1 h and then transferred to Petri dishes containing the leaf expansion media as outlined previously. After 22 hr, leaf discs were again weighed and weight increments calculated.

Results presented in Fig. 3 indicate that as compared to expansion of controls (which on a relative scale was 100 %)

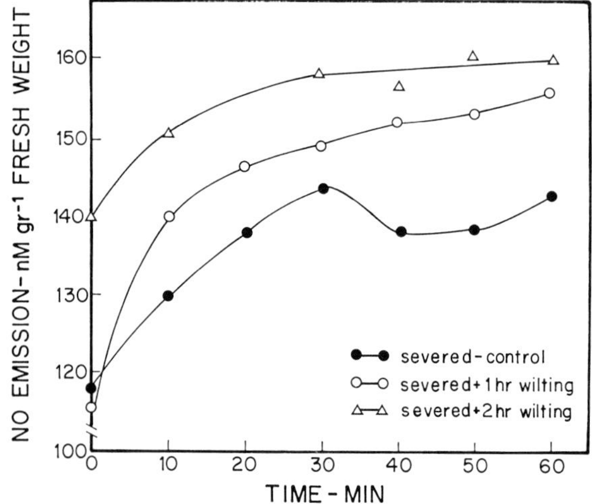

Fig. 2: *Effect of water stress on NO emission in Pisum sativum* Linn. *foliage.* Results presented are of one of three typical experiments. In all three treatments, standard deviations did not exceed 8 % of given data points.

with increase of NO concentration up to 5×10^{-6} M, NO considerably increased leaf growth (ca. 40 per cent more than controls) after which at higher concentrations the promotive effect tapers off.

Interlinkage of NO and Ethylene Emission in Senescing Pea Foliage

In the experiment whose results are seen in Fig. 4, both NO and C_2H_4 in the presence and absence of the ethylene precursor ACC were monitored simultaneously. The short-term experimental duration was based upon an earlier report that upon severing pea plants from their roots, rate of C_2H_4 emission was most pronounced during the initial 60 min after severance (Leshem et al., 1993). Results of this experiment presented in Fig. 4 clearly indicate presence of NO where emission proceeds simultaneously with that of ethylene. This figure also shows that not only is C_2H_4 emission decreased, as expected, by lack of precursor ACC, but also that of NO. Since both co-ordinate axes (NO and C_2H_4) are on an equimolar scale, it is moreover noteworthy that NO emission exceeds that of ethylene. The overall implications of this observation if found as a general plant phenomena and not limited to protein-rich legumes, are far reaching.

The rationale underlying the next experiment was that if indeed NO and C_2H_4 are metabolically interlinked, then addition of an NO releasing compound to the growth buffered solution should effect C_2H_4 emission by the plants. This was achieved by adding 2×10^{-3} M SNAP to the buffer medium, preliminary experimentation having shown that this is the optimal SNAP concentration for ethylene enhancement. This experiment was continued for a lengthier period than the former one in order to allow purported physiologically released NO from the SNAP to interact with ethylene metabolism. Results seen in Fig. 5 indicate that C_2H_4 production was markedly decreased by NO. Implications of this observation will be dealt with in the Discussion.

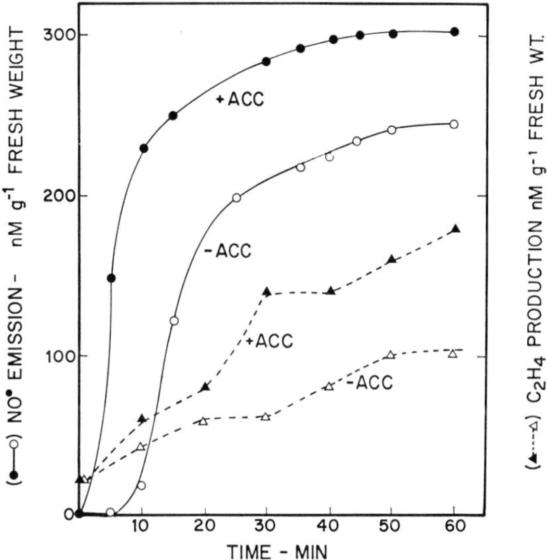

Fig. 4: *Interlinkage of NO and C_2H_4 in senescing Pisum sativum* Linn. *foliage.* 2 mM ACC was present in or absent from the buffer medium. This experiment was repeated 3 times, each producing similar results; the above figure represents those of one typical trial. In all four treatments, standard deviations did not exceed 11.5 percent of given data points.

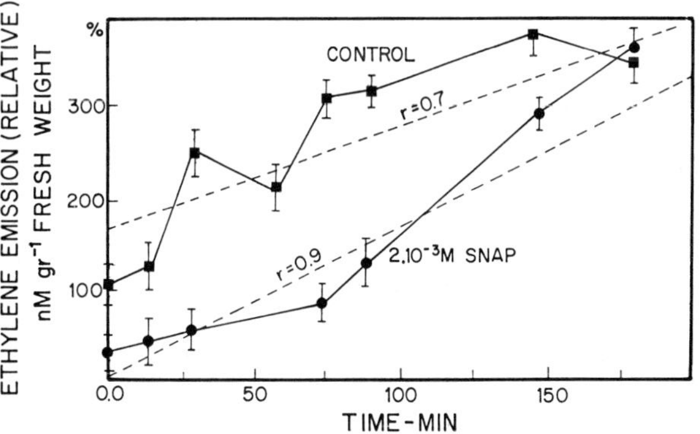

Fig. 5: *Effect of the NO releasing compound SNAP on C_2H_4 emission in senescing Pisum sativum* Linn. *foliage.* See ‹Materials and Methods› for experimental details. Results presented are 4 replicate means. - - - - indicates regression lines.

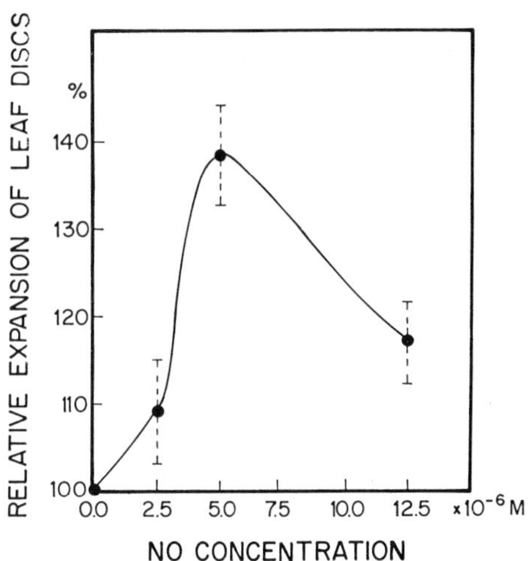

Fig. 3: *Pea leaf expansion as effected by direct exposure to NO.* Four replicate means. Concentrations of NO seen in the figure were generated by KNO_2 spiking of a solution of KI and H_2SO_4. See details in ‹Materials and Methods›. Basis of relative comparison as outlined in the caption to Table 1.

Endogenous NO Regulation Trials

a. NO releasing compounds: The former experiment indicated that while lower NO concentrations had a promoting effect on pea leaf expansion, higher ones indicate a decreasing tendency (Fig. 5). A further series of trials was therefore undertaken, employing a somewhat different approach, i.e., in addition to SNAP, the application of two further NO releasing compounds – Sin-1 and PBN, at a considerably higher concentration,

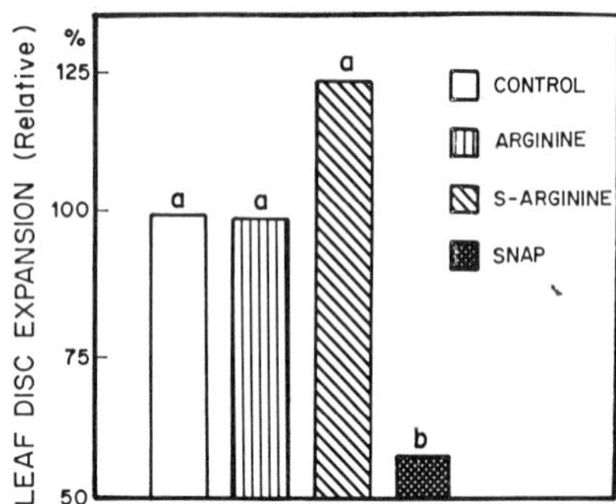

Fig. 6: *Pisum sativum* Linn. *leaf expansion as effected by N-arginine and SNAP.* Four replicate means. See text for experimental details. Different lower case letters at the heads of each histogram column indicate statistically significant differences at the $p > 0.05$ level as computed by ‹Analysis of Variance›.

2×10^{-3} M, which as indicated in preliminary experiments is within the inhibitory range.

Results of this experiment indicate that all three compounds significantly inhibited leaf disc expansion and that in this respect the most effective was SNAP (Table 1).

b. Regulation of NOS-like activity: The present experiment was conducted in order to further pinpoint mode of leaf disc inhibition as obtained above, the rationale being that if the inhibitory response is attributed to the NO released by SNAP, concomitant decrease in endogenous NO levels should reverse the inhibition. The latter aim could be attained by treatment with N-arg, an inhibitor of the NO producing enzyme NOS (Hibbs et al., 1988). In this experiment, N-arg concentration was 10^{-3} M, and in addition to an untreated control, a further one, regular *L*-arginine of the same molarity as of the N-arg, was also included.

The histogram of experimental results (Fig. 6) shows that the two controls produced essentially identical results while, as expected, SNAP markedly inhibited disc expansion (43 per cent inhibition as compared to controls). Moreover, the experimental hypothesis apparently is borne out by the contrasting, promotory effect of N-arg (23 % promotion).

Discussion

Besides demonstrating the presence of plant NO and possibly of NOS-like activity, the results of this research present contrasting and even opposite – both promotory and inhibitory – effects of NO in growth regulation of pea foliage. In a stress/senescence situation induced by partial severance from roots and subsequent partial wilting, NO emission from plantlets increases with time (Fig. 2). Moreover, direct exposure of leaf tissue to NO indicates that with increase of concentration to the 5×10^{-6} M range, there is a concomitant increase of leaf disc expansion (Fig. 3). These observations regarded together with those showing a marked dependence of emission of both NO and C_2H_4 upon the availability of C_2H_4 precursor ACC (Fig. 4), and the significant reduction of C_2H_4 production induced by application of NO-releasing SNAP (Fig. 5) suggest a stress coping role of the NO free radical gas. In some as yet unelucidated manner, the latter may serve to counteract emission of ethylene which is well documented to be both stress- and senescence-inducing (Leshem et al., 1986). This promotory function is possibly in keeping with disease and stress combatting effects of NO encountered in mammalian systems (Snyder, 1992).

As opposed to the above essentially promotive results, invariably obtained with low NO concentrations, higher concentrations of this gas produced opposite results. At concentrations exceeding the optimal for expansion, a marked growth decreasing trend is observed upon directly exposing leaf discs to NO. At a higher molar concentration simulating that of directly applied NO gas, three different NO generating compounds, Sin-1, PBN and SNAP, manifest a statistically significant inhibition of leaf growth expansion the order of effectivity being SNAP > PBN > Sin-1 (Table 1). Further credence to high concentration NO-associated inhibitory effect may be obtained from the reversal by N-arg, of the SNAP-induced inhibition (Fig. 6), this observation also suggesting NOS-like activity in pea foliage.

Concerning the differential effect of NO – low concentrations promoting, higher ones inhibiting, one possible and as yet tentative speculation may be offered. The high degree of NO mobility in biological tissues (Madison, 1993) which in plants, from source of synthesis, probably is initially via the apoplast, its immediate contact is with cell wall constituents whose contact with this gaseous free radical may exert a loosening and, hence, a growth promoting effect by means of permitting inherent turgor to increase cell expansion. Upon buildup of NO concentration, it may subsequently come into contact with membrane bilayer phospholipid components exert a fluidizing effect on the plasmalemma and with the same outcome. Upon further buildup of NO concentration, by reacting with the superoxide free radical, peroxynitrite may be formed (Pryor, 1994), and cause membrane leakage. Finally, at even higher concentrations, NO could subsequently diffuse into the cytosol and attack key enzymes as outlined above. To our knowledge, this is the first demonstration of NO-associated regulatory action in higher plants, and offers a new strategy for plant growth, stress and senescence control.

Acknowledgements

The authors wish to thank Prof. E. B. Dumbroff and Mrs. Pat Dumbroff of the Faculty of Agriculture, The Hebrew Universitv of Jerusalem. Rehovoth, Israel, for their expert aid and advice in operation of the NO sensor, and also Shlomit Klopman and Yael Friedman for their aid in the ethylene trials. Part of this research was conducted by E.H. within the framework of requirements for a graduate thesis.

References

Anderson, L. S. and T. A. Mansfield: The effect of nitric oxide pollution on growth of tomato. Environ. Pollut. *20,* 113–121 (1979).

Andersson, K. E. and Y. Wagner: Physiology of penile erection. Physiol. Rev. *75,* 191–236 (1995).

Beevers, L.: Nitrogen Metabolism in Plants. Elsevier, New York, pp. 18–19 (1976).

CHAMULITRAT, W., S. J. JORDAN, R. P. MASON, K. SAITO, and R. G. CULTER: Nitric oxide formation during light-induced decomposition of phenyl-N-tert butylnitrone. J. Biol. Chem. *268,* 11520–11527 (1993).

ELZENGA, J. T. M. and E. VAN VOLKENBURG: Characterization of ion channels in the plasma membrane of epidermal cells of expanding pea (*Pisum sativum*) leaves. Membr. Biol. *137,* 227–235 (1994).

FELDMAN, P. L., O. W. GRIFFITH, and D. J. STUEHR: The surprising life of nitric oxide. Chem. Eng. News. *71,* 26–38 (1993).

FIN, C., C. DE CUNHA, C. BROMBERG, P. K. SMITZ, A. BIANCHIN, J. H. MEDINA, and I. IZQUIERDO: Experiments suggesting a role for nitric oxide in the hippocampus in memory processes. Neurobiol. Learn Memory, 113–115 (1995).

FUKUHORI, M., K. ICHIMORI, H. NAKAGAWA, and H. OKNO: Nitric oxide reversibly suppresses xanthine oxidase activity. Free Rad. Res. *21,* 203–212 (1995).

HERY, P. J., J. D. HOROWITZ, and W. J. LOUIS: Nitroglycerin-induced tolerance affects multiple sites in the organic nitrate bioconversion cascade. J. Pharmacol. Exp. Ther. *248,* 762–768 (1989).

HIBBS, J. B., R. R. TUMTOR, Z. VAVRIN, and E. M. RACHLIN: NO is a highly reactive endogenous chemical produced by activated macrophages and serves as a mediator for expressing activity. Biochem. Biophys. Res. Comm. *157,* 87–94 (1988).

KONOREV, E. A., M. M. TARPEY, J. JOSEPH, J. E. BAKNER, and B. KALYANARAMAN: Nitronyl oxides as probes to study the mechanism of vasodilatory action. Free Rad. Biol. Med. *18,* 169–177 (1995).

KOPROWSKI, H., Y. M. ZHENG, E. HEBER-KATZ, N. FRAZER, L. RORKE, Z. F. FU, C. HANLON, and B. DIETZSCHOLD: *In vivo* expression of inducible nitric oxide synthase in experimentally induced neurological diseases. Proc. Natl. Acad. Sci. USA *90,* 3024–3027 (1993).

LESHEM, Y. Y., A. H. HALEVY, and C. FRENKEL: Processes and Control of Plant Senescence. Elsevier Science Publishers, Amsterdam, p. 49 (1986).

LESHEM, Y. Y., D. RAPAPORT, A. A. FRIMER, G. STRUL, U. ASAF, and I. FELNER: Buckministerfullerene (C-60 Carbon Allotrope) inhibits ethylene evolution from 1-aminocyclopropane-1-carboxylic acid (ACC)- treated shoots of pea (*Pisum sativum*), broad-bean (*Vicia faba*) and flowers of carnation (*Dianthus caryophyllus*). Ann. Bot. *72,* 457–461 (1993).

MADISON, D. V.: Pass the nitric oxide. Proc. Natl. Acad. Sci. USA *90,* 4329–4331 (1993).

MITSUHATA, H., J. SAITOH, H. TAKEUCHI, Y. HARIGUCHI, and R. SHIMIZU: Production of nitric oxide in anaphylaxis in rabbits. Shock *2,* 381–384 (1994).

MONEADA, S., R. M. J. PALMER, and F. A. HIGGS: Nitric oxide: physiology, pathophysiology and pharmacology. Pharmacol. Rev. *43,* 109–142 (1991).

NEIGHBOUR, E. A., M. PEARSON, and H. MEHLHORN: Purafil filtration prevents the development of ozone-induced frost injury: a potential role for nitric oxide. Atmos. Environ. *24 A,* 711–715 (1990).

OLESON, J., L. LYKKE, and H. THOMPSON-IVERSEN: Nitric oxide is a key molecule in migraine and other vascular headaches. Trend. Pharmacol. Sci. *15,* 149–153 (1994).

PAYNE, W. J.: Denitrification. Wiley Interscience Publishers, New York, pp. 90–97 (1983).

PRINCE, R. C. and D. E. GUNSON: Rising interest in nitric oxide synthase. Trend Biol. Sci. *18,* 35–36 (1993).

PRYOR, W. A.: Nitric oxide and peroxynitrite. Soc. Free Rad. Res. Newslet. *4,* 12 (1994).

RADI, R., J. S. BECKMAN, K. M. BASH, and R. A. FREEMAN: Peroxylnitrite induced membrane lipid peroxidation: cytotoxic potential of superoxide and nitric oxide. Arch. Biochem. Biophys. *228,* 481–487 (1991).

SHAFFER, J. E., H. BA-JIN, W. H. CHERN, and F. W. LEE: Lack of tolerance to a 24-hour infusion of SNAP in conscious rabbits. J. Pharmacol. Exp. Ther. *260,* 286–293 (1992).

SNYDER, S. H.: Nitric oxide: first in a new class of neurotransmitters. Science *257,* 494–496 (1992).

VAN SPANNING, R. J. M., P. N. ANTHONIUS, W. N. M. DE BOER, S. REIJNDERS, H. V. WESTERHOFF, H. STOUTHAMER, and J. VAN DER OOST: Nitric acid and nitric oxide reduction in *Paracoccus denitrificans.* FEBS Lett. *360,* 151–154 (1995).

VAN VOLKENBURG, E.: Leaf cell expansion. In: BAKER, N. R. and H. THOMAS (eds.): Crop Photosynthesis: Spatial and Temporal Determinants, pp. 235–252. Elsevier Science Publishers, Amsterdam (1992).

J. Plant Physiol. Vol. 148. pp. 264–270 (1996)

The Effect of «Acid Rains» and Mineral Fertilization on the Development of Biometrical Features of *Fagus sylvatica* L. Seedlings

STANISŁAW MAŁEK and JAN GRESZTA

Department of Forest Ecology, Forest Faculty, Agricultural Academy of Kraków, Al. 29-go Listopada 46, PL-31425 Kraków, Poland

Received June 26, 1995 · Accepted October 8, 1995

Summary

The purpose of the present experiment was to describe the influence of various methods of mineral fertilization: NK into leaves (Florovit), NPK onto soil (Fruktus 2) and liming (dolomite magnesium) on the development of *Fagus sylvatica* L. seedlings subject to the influence of simulated acid rains of pH 2.5, 3.0, 3.5, 4.0, 4.5.

As a result of the carried out analysis of the correlation coefficient a reciprocal high correlation was found between particular parameters of growth of *Fagus sylvatica* L. seedlings i.e. of the diameter of root nect, root mass, total mass of: seedlings, leaves, main shoots, lateral shoots, ligneous parts, overground parts, as well as of total length of lateral shoots. Whereas no high correlation was observed between the above mentioned parameters and a total heigth of the seedling.

The obtained results of the analysis of changes in biometric characteristics of *Fagus sylvatica* L. seedlings showed, that the best development was achieved at pH 4.5, whereas their weakest development was observed at 2.5. The analysis showed the decrease in values of analysed characteristic together with the increase in acidity of aqueous solutions.

The first and second and third year of the experiment for *Fagus sylvatica* L. showed that kind of fertilization did not effect positively the increase in values of the analyzed biometrical feature – total height of seedling comparing with those of control plots. The significant improvement of total height of seedling was noticed on control plot with pH 3.5 in relation to pH 2.5 on control plot in first year. The best increment of total height of seedling was noticed for pH 3.5 for each year.

The third year of the experiment showed that using fertilization: Ca + Mg onto soil, NK into leaves and NPK onto soil it is possible to restrict the unfavourable influence of «acid rains» on total mass of *Fagus sylvatica* L. seedling. The best significant increment of this biometrical feature was observed on pH 4.0. When one wants to restrict the unfavourable influence of «acid rains» on analyzed total mass of seedlings within pH 2.5–4.5 it is possible to choose optimaly one from three applied fertilizing variants.

Key words: Acid rain, fertilizers, biometrical features, tree seedlings, Fagus sylvatica L.

Abbreviations: pH = pH of simulated acid rains.

Introduction

It is claimed that one of the main causes of the new forest dieback is the changes in soil chemistry provoked by acid rain. «Acid rain» causes marked disturbance in the uptake of nutrients by plants and at the same time limits their growth (Altshuller & Linthurst 1976, Ulrich 1983, Ulrich 1986, Abrahamsen et al. 1987, Krause 1988, Białobok 1989, Greszta at al. 1990, Huettl 1990, Gruszka 1991). This process occurs expecially intensively on barren grounds e.g. on sands for:

Table 1: Chemical characteristics of water used for preparation of aqueous solution of sulphuric acid.

	unit	pH in H_2O	Na	K	Ca	Mg	Fe	Mn	Zn	Cu	Pb	Cd	Ni	Cr	Cl
Water used	mg/L	7.8	21.48	1.0	31.4	6.2	0.35	6.2	0.175	<0.005	0.03	<0.005	0.07	<0.01	0.97
Added with water	kg/ha/y	7	279.2	13.0	408.2	80.6	4.55	80.6	1.99	–	0.39	–	0.81	–	4.55

Table 2: Chemical characteristics of the soil used for experiment.

	unit	pH in H_2O	pH in KCl	N	C	P	Na	K	Ca	Mg	Fe	Mn	Zn	Cu	Pb	Cd	Ni	Cr
Soil	µg/g	7.3	7.1	420	6700	–	130	380	910	410	3270	241	22	14	18.2	1.4	6.3	5.5
exchangeable elements	µg/g			–	–	27.1	–	18.5	11.5	–	–	–	–	–	–	–	–	–

Pinus sylvestris L., *Pinus strobus* L., *Larix decidua* Mill., *Abies alba* L., and *Fagus sylvatica* L. (Gruszka 1991). To some extent is counteracted by supplying soil with nutrients and liming (Rugge 1978, Huettl 1988, 1989, Zoettl et al. 1989, Huettl et al. 1990, Huettl & Fink 1991). It is believed that the negative effects of acid rain may be limited by supplying the soil with fertilizers in the form of easily assimilative macroelements i.e. NPK, dolomite (Charitonov 1970, Huettl & Wiśniwski 1987, Huettl 1988, 1989, Huettl et al. 1990, Huettl & Fink 1991).

Mineral fertilization and liming has become one of the main operations to reestabilize the forest ecosystem. In this case fertilization and liming is an important agent of prevention and counteraction negative effects of air pollution and in particular of acid rain on bio-, pedo- and hydrosphere (Huettl 1986, Zoettl & Huettl 1986, Brocksen et al. 1988, Huettl & Wiśniwski 1987).

In our investigation the main aim has been to detect the acid rain effects on the productivity of *Fagus sylvatica* L. seedlings treated also with fertilizers and lime.

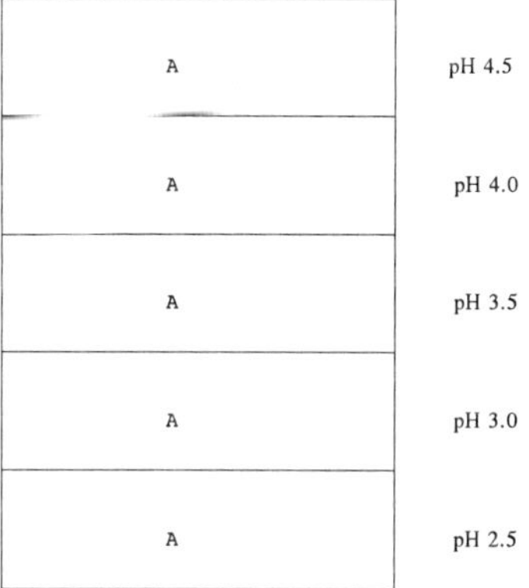

Block A

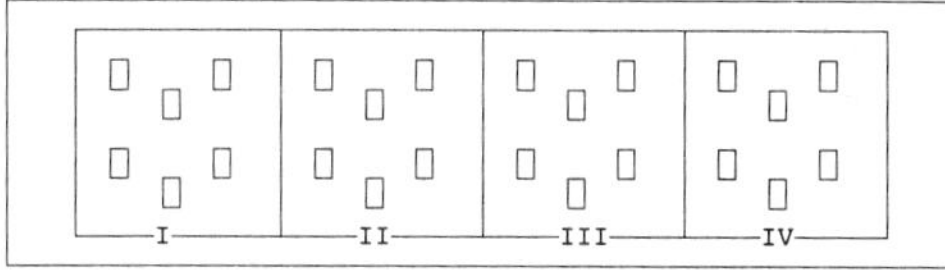

Fig. 1: Diagram of field experiment.
pH 2.5, pH 3.0, pH 3.5, pH 4.0, pH 4.5 – pH of simulated acid rain; A – Block A; I – control (without fertilization); II – fertilizer: Ca+Mg onto soil; III fertilizer: NK into leaves; IV – fertilizer: NPK onto soil; □ – single *Fagus sylvatica* L. seedling.

Material and Methods

Artificial acid rain was obtained from a water solution of H_2SO_4 to reach the following pH in KCl values: 2.5, 3.0, 3.5, 4.0, 4.5 under a plastic roof. The average yearly rainfall in nearest area is about 600 mm, but in order to compensate for an increased transpiration under the roof the seedlings were exposed to higher rainfall, 1300 mm per year. Concentrated H_2SO_4 was used in preparing the acid solutions. After mixing the solutions were checked and corrected with H_2SO_4 and NaOH to obtain required pH-values. The water was taken from a deep water well, and the water was filtered to absorb eventual particles. The chemical composition of the water is given in Table 1.

The seedlings were prevented from the influence of precipitation by a roof without front and side walls, making other atmospheric factors such as temperature, humidity, fog, dew and others not much different from natural conditions.

One year old seedlings of *Fagus sylvatica* L. were used, identical as far as their quality and health were concerned – I class acc. to estimation used in PGL (BN-76/9212-02). *Fagus sylvatica* L. seeds were from the Brzesko Forestry Inspectorate (habitat type of forestry Lósw, 60–80 years). The seeds were germinated in an open greenhouse.

The seedlings were planted in the open greenhouse but under similar conditions on mixed loamy sand (Table 2) using separated plots measuring 80×80 cm. The plots were separated from one another by a concrete slab 60 cm deep in the ground, in order to avoid the overgrowth of the roots into neighbouring plots and a horizontal dislocation of water solutions of different pH. Each plot contained 6 seedlings. A general plan of the experiment is shown in Fig. 1.

The experiment was established in the spring of 1989. The seedlings were planted in March and watered with original ground water for 2 months to facilitate root stabilizing and to eliminate losses caused by the transplantation stress.

From mid-June on, the seedlings were watered with artificial acid rain each third day. The control plots of the experiment were sprayed only with acid rain of different pH: 2.5, 3.0, 3.5, 4.0, 4.5 each year.

Table 3: Multiple correlation matrix of the analyzed biometric characteristics for *Fagus sylvatica* L.

	RND	RM	NM	LSM	MSM	LPM	APM	SM	SH	LSL
RND	1.000	0.838	0.834	0.767	0.854	0.858	0.860	0.872	0.848	0.842
RM		1.000	0.931	0.853	0.859	0.896	0.928	0.974	0.730	0.869
NM			1.000	0.842	0.864	0.894	0.949	0.958	0.763	0.907
LSM				1.000	0.820	0.932	0.925	0.912	0.628	0.843
MSM					1.000	0.972	0.960	0.936	0.843	0.857
LPM						1.000	0.989	0.967	0.794	0.891
APM							1.000	0.988	0.803	0.917
SM								1.000	0.787	0.914
SH									1.000	0.784
LSL										1.000

Names of Variables: RND – diameter of all roots neck (cm); RM – mass of a root (g); LM – total mass of leaves (g); LSM – total mass of lateral shoots (g); MSM – total mass of main shoots (g); LPM – total mass of ligneous part (g); APM – total mass of aboveground part (g); SM – total mass of seedlings (g); SH – total height of seedlings (cm); LSL – total length of lateral shoots (cm).

A second serie of plots was sprayed with acid rain like the control plots, and was also fertilized from April to September with dolomite on the soil surface (Ca+Mg on soil): 5 % MgO, 70 % CaO each year. Eight g was given on each plot and each month, which means: 37.5 kg MgO $ha^{-1} \cdot y^{-1}$ and 525 kg CaO $ha^{-1} \cdot y^{-1}$.

A third serie of plots was sprayed with acid rain like the control part, and was fertilized from April to September once a month with the mineral fertilizer NK «Florovit» onto the needles (NK onto needles) each year. These plots received fertilizers containing: 3.0 % – N, 2.4 % – K_2O, 0.75 % – Ca, 0.3 % – S, 0.1 % – MgO, 400 mg/dm^3 – Fe, 150 mg/dm^3 – Zn, 150 mg/dm^3 – Mn, 70 mg/dm^3 – Cu, 30 mg/dm^3 – B, 20 mg/dm^3 – Mo. Of the fertilizer material was taken 25 mL per 5 L of water from April to September meaning: 12 kg N, 9.6 kg K_2O, 3.0 kg Ca, 1.2 kg S, 0.4 kg MgO, 0.16 kg Fe, 0.06 kg Zn, 0.028 kg Cu, 0.018 kg B, 0.008 kg Mo per ha, per year.

A fourth serie of plots was sprayed with acid rain like the control part, and was fertilized once a month with a solid mineral fertilizer «Fruktus 2» (NPK onto soil) each year. These plots received fertilizers containing: 13.5 % N, 5 % P_2O_5, 20 % K_2O, 5 % MgO, 0.2 % – Cu, 0.2 % – Zn, 0.2 % – Mn, 0.2 % – Br, 0.01 % – Mo, in the proportion of 38 g per one plot – which means 175.5 kg N, 65 kg P_2O_5, 260 kg K_2O, 65 kg MgO, 2.6 kg Cu, 2.6 kg Zn, 2.6 kg Mn, 2.6 kg Br, 0.13 kg Mo per ha, per year.

The experiment was carried out in the nursery at Jodłówka for three years, close to Brzesko – 60 km eastwards of Kraków, on the permises of the Brzesko Forest Inspectorate belonging to Kraków District Head Office for State Forests.

After three vegetation seasons, Oct. 1991, the total biomass of leaves, shoots and roots was taken for analyses of *Fagus sylvatica* L.. The plant material was separated as follows: main shoots grown in the years 1988, 1989, 1990, 1991, lateral shoots from years: 1988, 1989, 1990, 1991, leaves and roots from the whole seedling; assimilation system for beech (1990, 1991). For determinin a mean mass of 1000 leaves the material was collected from the whole plant.

The following biometric characteristics were analysed for plants:

1. total height of a seedling before beginning and in 1, 2, 3, year of the experiment,
2. diameter of the root neck,
3. total mass of: roots, seedling, lateral shoots, main shoots, lingneous and aboveground parts, after the third year of the experiment,
4. mean total lenght of lateral shoots after the third year of the experiment,
5. mean mass of 1000 leaves in the 2, 3, year of the experiment,
6. mean length of lateral shoots in the 1-st, 2nd and 3rd year of the experiment.

The mass of the analyzed vegetable material was determined by means of the laboratory balance of Medicat Ltd. 1600C type with an accurracy up to ± 0.01 g, and the length and height by the ruler of Skala type with an accurracy of ± 0.1 cm.

Results

Correlation coefficients for the biometric variables between each characteristic were calculated. The results are given in Table 3.

On the basis of analysis of correlation coefficient diameter of the root neck, total dry weight of: leaves, lateral shoots, ligneous and above ground parts of seedlings, seedling at the end of the experiment, total length of lateral shoots – for these variables the correlation coefficient is higher than 0.850, that is to say, the above mentioned characteristics are highly correlated between each other. For further analysis the total dry weight of seedlings was taken as a representative value for this group. The total height of the seedling, which is not highly correlated with remaining variables (the correlation coefficient is lower than 0.850). This variable should be separately analyzed features. Thus two characteristic features were chosen for further analysis: the total height of a seedling and its total mass (Oktaba 1971). The results after three years of the experiment on chosen biometrical features are given in Table 4.

The average of total mass of seedlings, on controlled plots in the 3-rd year of the experiment, changed together with the change of pH of the aqueous solution; ranging from the least 17.62 g value at pH 2.5 it systematically increased up to pH 4.0, reaching the largest mass of 112.04 g. After exceeding pH 4.0, the mass value decreased and at pH 4.5 it was 88.06 g (Table 4).

The average of the total height of seedlings after 3rd year of the experiment on the control plots varied together with the increase of the pH from the least 22.88 cm value at pH 2.5, to the highest one of 86.64 cm at pH 4.0. Above this pH the height decreased to 84.77 cm at pH 4.5 (Table 4). The same tendency to decreasing the increment of the main shoot has been observed in particular years. At pH 2.5 a total

Table 4: Average values of chosen biometric characteristics of *Fagus sylvatica* L. at the end of the experiment of different pH treatments ± standard deviation).

pH	total height of seedling in cm on control plots after						total mass of seedlings on control in g in 1991		total height of seedling in cm on Ca+Mg plots after						total mass of seedlings on Ca+Mg plots in g in 1991	
	1989		1990		1991				1989		1990		1991			
	av.	±	av.	±	av.	±	av.	±	av.	±	av.	±	av.	±	av.	±
2.5	9.35	7.86	15.20	13.51	22.88	22.16	17.62	21.85	16.45	10.23	33.55	19.31	52.77	28.21	62.51	39.51
3.0	15.30	4.78	22.48	5.56	31.27	10.39	21.17	15.60	20.92	6.35	37.15	9.91	60.07	18.30	73.14	54.95
3.5	22.15	9.80	39.28	22.96	59.23	29.69	72.54	76.05	28.05	12.58	51.15	23.35	78.47	36.96	80.51	61.59
4.0	26.55	5.74	49.13	19.15	86.72	16.52	112.04	41.40	23.67	9.27	49.10	15.43	71.15	21.29	98.93	92.67
4.5	33.15	15.42	53.85	10.95	84.77	27.68	88.06	77.45	24.15	4.89	41.47	9.42	65.23	16.64	92.67	42.12

pH	total height of seedling in cm on NK plots after						total mass of seedlings on NK plots in g in 1991		total height of seedling in cm on NPK plots after						total mass of seedlings on NPK plots in g in 1991	
	1989		1990		1991				1989		1990		1991			
	av.	±	av.	±	av.	±	av.	±	av.	±	av.	±	av.	±	av.	±
2.5	30.18	19.41	43.07	27.16	60.15	35.79	81.82	56.65	21.88	5.56	37.93	16.95	56.67	19.20	65.20	28.88
3.0	23.68	8.52	46.35	17.26	66.42	22.45	79.32	39.34	21.95	8.15	39.08	17.57	63.80	24.66	66.14	45.63
3.5	28.05	6.49	46.45	12.64	61.35	15.43	96.34	52.71	22.53	4.59	42.52	16.56	70.02	22.17	106.29	61.46
4.0	25.33	7.74	45.97	15.90	75.30	20.47	112.67	71.08	24.10	8.48	46.67	21.89	81.03	30.17	109.57	66.17
4.5	29.42	9.68	47.43	14.08	70.88	16.98	147.45	127.21	30.98	12.84	47.95	15.15	81.93	21.52	114.35	33.51

height of seedlings in 1990 was 15.20 cm (the least value), then it systematically increased up to 53.85 cm at pH 4.5 achieving the largest value (Table 4). In 1989 the total height of the seedlings was changing from the least value of 9.35 cm at pH 2.5 it increased up to 33.15 cm at pH 4.5 achieving the highest value (Table 4).

The average of total mass of seedlings, on Ca + Mg plots in the 3-rd year of the experiment, changed together with the change of pH of the aqueous solution; from the least amont of 62.51 g value at pH 2.5 it systematically increased up to pH 3.5, reaching the largest mass of 98.93 g. After exceeding pH 4.0, the mass decreased and at pH 4.5 it was 92.67 g (Table 4).

The average of the total height of seedlings on the Ca + Mg plots in the 3-rd year of the experiment growed together with the increase of the pH from the least value 52.77 cm value at pH 2.5, to the greatest one of 78.47 cm at pH 3.5. Above this pH value, the height decreased to 65.23 cm at pH 4.5 (Table 4). The same tendency to decreasing the increment of the main shoot has been observed in individual years year. At pH 2.5 a total height of seedlings in 1990 was 33.55 cm (the least value), while at pH 3.5 it was 51.15 cm (the highest value), above this pH the height of seedlings decreased up to 41.47 cm at pH 4.5 (Table 4). In 1989 the total height of the seedlings was changing from the least value of 16.45 cm at pH 2.5, it increased achieving the largest value of 28.05 cm at pH 3.5, and then this height decreased to 24.15 cm at pH 4.5 (Table 4).

According to pH of acid rain the average of total mass of seedlings, on NK plots in the 3-rd year of the experiment, changed together with the change of pH of the aqueous solution; from the 81.82 g value at pH 2.5 the mass decreased up to pH 3.0, reaching the least value of 79.32 g, and then it incereased achieving the largest value 147.75 g at pH 4.5 (Table 4).

The average of the total height of seedlings on the NK plots in the 3-rd year of the experiment varied together with the change of the pH from the lesst value 60.15 cm at pH 2.5, it increased to 66.42 cm at pH 3.0. Above this pH value followed the height decrease to 61.75 cm at pH 3.5, and then the height increased to the largest value of 75.30 cm at pH 4.0, but at pH 4.5 it again decreased to 70.88 cm (Table 4). The same tendency has been observed for particular years. At pH 2.5 a total height of seedlings in 1990 was 43.07 cm achieving the least value, then this value increased achieving the largest value 46.45 cm at pH 3.5, and above this pH the decrease in the seedlings height ranged 45.97 cm for pH 4.0 and afterwards increased to 47.43 cm at pH 4.5 (Table 4). In 1989 the total height of the seedlings changed in the following way: from 30.18 cm the largest value at pH 2.5, the value decreased up to 23.68 cm at pH 3.0 achieving the least value, and then this height increased to 28.05 cm at pH 3.5, and again it decreased up to 25.33 cm at pH 4.0, and next it increased to 29.42 cm at pH 4.5 (Table 4).

According to pH of acid rain the average of total mass of seedlings, on NPK plots, changed together with the change of pH of the aqueous solution; from the least value of 65.20 g at pH 2.5 the mass increased systematically up to pH 4.5 reaching the largest value 114.35 g (Table 4).

The average of the total height of seedlings on the NPK plots in the 3-rd year of the experiment varied together with the change of the pH from the least value 56.67 cm at pH 2.5, than it increased systematically achieving the largest value 81.93 cm at pH 4.5 (Table 4). The same tendency has been

Table 5: Two way Analysis of variance indicats significant effect (P = 0.05) of: pH of acid rain treatment, kind of fertilization, interaction between pH and fertilization for two biometrical features of *Fagus sylvatica* L. seedlings treated with different pH of artificial acid rain and different kind of fertilization for taken homogeneous scheme.

biometrical feature	total height of seedling												total mass of seedling after 1991		
	in 1988			in 1989			in 1990			in 1991					
scheme: kind of fertilization/significant factor	pH	kind of fertilization	interaction pH-fertilization	pH	kind of fertilization	interaction pH-fertilization	pH	kind of fertilization	interaction pH-fertilization	pH	kind of fertilization	interaction pH-fertilization	pH	kind of fertilization	interaction pH-fertilization
homogenous scheme for pH	2.5, 3.0, 3.5, 4.0, 4.5			2.5, 3.5, 4.5			2.5, 3.5, 4.5			2.5, 3.5, 4.5			2.5, 3.0, 3.5, 4.0, 4.5		
Ca+Mg onto soil-control				*			*			*			*	*	
NK into leaves-control										*			*	*	
NPK onto soil-control				*		*	*			*			*	*	
homogenous scheme for pH	2.5, 3.0, 3.5, 4.0, 4.5			2.5, 3.5, 4.5			2.5, 3.5, 4.5			2.5, 3.5, 4.5			2.5, 3.0, 3.5, 4.0, 4.5		
Ca+Mg onto soil – NK into leaves															
Ca+Mg onto soil – NPK onto soil															
NK into leaves – NPK onto soil															

observed for particular years. At pH 2.5 a total height of seedlings in 1990 was 37.93 cm (the least value), then this value increased achieving the largest value of 47.93 cm at pH 4.5 (Table 4). In 1989 the tottal height of the seedlings changed in the following way: fron the least value 21.88 cm at pH 2.5, the value increased to the largest one of 30.98 cm at pH 4.5 (Table4).

Significant factor effect

The plan of the experiment described in material and methods considered two different opposing principles:

1. using of fertilization or not:
   Ca + Mg on soil-control, NK on leaves-control, NPK on soil-control
2. fertilizer application:
   Ca + Mg on soil – NK on leaves, Ca + Mg on soil – NPK on soil, NK on leaves – NPK on soil.

Inhomogenously of data for a certain treatment indicate that other factors might influence on results. Being a factorial experiment the results were confirmed by analyses of variance (two way Anova factor analysis – Oktaba 1971). With pH as one factor and fertilizers as the other.

Statistical analyse showed that the total height of seedlings before experiment (1988) was equal and we can draw a conclusion that the experiment was established in a correct way (Table5).

Statistical analyse for total height of seedling (homogenous for pH: 2.5, 3.5, 4.5) after first year indicate significant differences for pH in: «Ca + Mg onto soil-control» and «NPK onto soil-control», and interaction between pH and fertilization in: «NPK onto soil-control» on biometrical feature changing the value in first case: between pH: 2.5 and 3.5 about 12.20 cm, 2.5 and 4.5 about 15.75 cm – the best increment was noticed for pH 4.5 the worst for pH 2.5, in a second case: changing the value between pH: 2.5–3.5 about 12.98 cm. Interaction between pH and fertilization caused significant effect on biometrical feature on control and NPK plots – for control plot with pH 2.5 in relation to different pH of control plots an average increased as follow: to pH 3.5 about 23.80 cm. The worst development was noticed on control plot with pH 2.5, the best on control plot with pH 3.5.

Statistical analyse for total height of seedling (homogenous for pH: 2.5, 3.5, 4.5) after second year indicate significant differences for pH in: «Ca + Mg onto soil-control» and «NPK onto soil-control» (Table 5) on biometrical feature changing the value in first case: between pH: 2.5 and 3.5 about 20.84 cm, 2.5 and 4.5 about 20.93 cm – the best increment was noticed for pH 4.5 the worst for pH 2.5, for second case: between pH: 2.5–3.5 about 21.33 cm. The worst development was noticed for pH 2.5, the best for pH 3.5.

Statistical analyse for total height of seedling (homogenous for pH: 2.5, 3.5, 4.5) after third year indicate significant differences for pH in: «Ca + Mg onto soil-control», «NK into leaves-control» and «NPK onto soil-control» (Table 5) on biometrical feature changing the value in first case: between pH: 2.5 and 3.5 about 31.09 cm, 2.5 and 4.5 about 37.26 cm, for second scheme: 2.5 and 3.5 about 36.31 cm, for third scheme: 2.5 and 3.5 about 43.13 cm. The best increment was noticed for pH 3.5 the worst for pH 2.5.

Statistical analyse for total mass of seedling (homogenous for pH: 2.5, 3.0, 3.5, 4.0, 4.5) after third year indicate significant differences for fertilization and pH for: «Ca + Mg onto soil-control», «NK into leaves-control» and «NPK onto soil-control» (Table 5) – on biometrical feature changing the value using fertilization-increasing the value in first scheme: about 19.20 g, in second scheme about 41.24 g and in third scheme about 30.00 g in relation to control plots. Significant effect of pH on biometrical feature changed the value between pH for first scheme: 2.5 and 4.0 about 65.52 g, 2.5 and 4.5 about 50.70 g, 4.0 and 4.0 about 58.58 g, for second scheme: 2.5 and 4.0 about 62.64 g, 2.5 and 4.5 about 68.34 g, 3.0 and 4.0 about 62.36 g, for third scheme: 2.5 and 4.0 about 69.40, 2.5 and 4.5 about 60.09 g, 3.0 and 4.0 about 67.40 g, 3.0 and 4.5 about 58.10 g. The best increment was noticed for pH 4.0 the worst for pH 2.5.

Discussion

The obtained results of the analysis of changes of biometric characteristic of *Fagus sylvatica* L. seedlings, growing on plots: control, fertilized with: Ca + Mg onto soil, NK into leaves (Florovit), NPK onto soil (Fruktus 2) showed at pH: 2.5, 3.0, 3.5, 4.0, 4.5 of simulated acid rains that seedlings achieved the best development at pH 4.5, whereas the weakest one at pH 2.5. This confirms the hypothesis presented at the Conference of United Nations (1971) and also Dahl and Skre (1971) that the low reaction is not suitable to beech (Baule and Fricker 1971, Gruszka 1991) and that the optimum growth of *Fagus sylvatica* L. is achieved at pH of soils within the range of pH in H_2O: 5.5–6.5, and maximum 7.1 (Ivanov 1970, Fielder et al. 1973 – referred to Kowalkowski 1982).

It was found that the coefficient a reciprocal high correlation was found between particular parameters of growth of *Fagus sylvatica* L., seedlings i.e. of the diameter of root nect, root mass, total mass of: seedlings, leaves, main shoots, lateral shoots, ligneous parts, overground parts and lenght of lateral shoots. Whereas no high correlation was observed between the above mentioned parameters and a total heigth of the seedling.

In the event of introduction of higher acidification into soil, the beech reacted by decrease in all values of growth parameters. This confirms the results obtained by Gruszka (1991).

Fagus sylvatica L. with *Quercus alba* L., *Carya ovata* Mill., *Fagus grandifolia* Ehnh. belong to a group of trees hardy to air pollution. The limitation of growth and damades of the assimilation system were found only at pH 2.5 of acid rains. But, in general at pH 2.5 overcolouring and necrosis as well as weaker growth of seedlings may be observed (Foy 1974, Jensen and Dochinger 1989).

The first and second and third year of the experiment for *Fagus sylvatica* L. showed that kind of fertilization did not effect positively the increase in values of the analyzed biometrical feature – total height of seedling comparing with those of control plots. The significant improvement of total height of seedling was noticed on control plot with pH 3.5 in relation to pH 2.5 on control plot in first year. The best increment of total height of seedling was noticed for pH 3.5 for each year.

The third year of the experiment showed that used fertilization: Ca + Mg onto soil, NK into leaves and NPK onto soil it is possible to restrict the unfavourable influence of «acid rains» on total mass of *Fagus sylvatica* L. seedling. The best significant increment of this biometrical feature was observed on pH 4.0. When one wants to restrict the unfavourable influence of «acid rains» on analyzed total mass of seedlings within pH 2.5–4.5 it is possible to choose optimaly one from three applied fertilizing variants.

The obtained results confirm that *Fagus sylvatica* L. develops best on limestone soils without strong acidification (Peninningsfeld 1964, Charitonov 1970, Baule and Fricker 1971, Kowalkowaki 1982). The satisfactory results after liming may be expected at an sufficient supply with remaining nutrients. The fertilization experiments carried out in nurseries and on stand areas at simultaneous supplying with various combinations of: N, P, K, Ca show a diversified influence of fertilization the increment of the cross-section sufrace and of the height of a tree. The favourable influence was observed by: Trillmich (1969), Kern and Moll (1971), Pasternak et al. (1978). No reaction was observed by Hausser (1971), whereas the negative results were found by: Mitscherlich and Wittich (1963), Kenel and Wehrmann (1967), Moller et al. (1969). In comparision with other species the beech settles more fertile sites, well supplied with nutrients. Therefore satisfactory results after fertilization of stands are are obtained on exceptionally poor site (Fober 1990).

References

ABRAHAMSEN, G. B., B. TVEITE, and A. O. STUANES: Wet acid deposition on soil properties in relation to forest growth. Experimental results. Paper given at the IUFRO Conference: Woody Plant Growth in a Chenging Physical and Chemical environment, Vancouver July 27–31 (1987).

ALTSCHULLER, A. P. and R. A. LINTHURST: The acidic deposition phenomenon and its effects: Critical assessment review papers. Volume II. Effects Sciences. PA-600-8-83 016A, U.S. Environmental Protection Agency, Corvallis, OR. (1976).

BAULE, H. and C. FRICKER: Nawożenie drzew leśnych. PWRiL. Warszawa (1971).

BIAŁOBOK, S.:' Wpływ kwaśnych opadów atmosferycznych na drzewa i lasy. W: Życie drzew w skażonym środowisku. PWN – Instytut Dendrologii *21,* 171–193 (1989).

BN-76/9212-02. Material sadzeniowy. Sadzonki drzew i krzewów do upraw leśnych, plantacji i zadrzewieff. Wydawnictwa Normalizacyjne – Warszawa.

BROCKSEN, R. W., H. W. ZOETL, D. B. PORCELLA, R. F. HUETTL, K. FEGER, and J. WIŚNIEWSKI: Experimental liming of watersheds: An international cooperative effort between the United States and West Germany Water, Air, and Soil Pollution *41,* 455–471 (1988).

CHARITONOV, G.: Znacenie izviestkowania i gipsovanija pri vyrascivani lesnych kultur v Karpatach. Izviestija Vyssich Ucebnych Zavedenij, Lesnoj Żurnal *13,* (4), 5–7 (1970).

Conference of United Nations: Royal Ministry for Foreigen Affairs and Royal Ministry of Agriculture: Air pollution across national boundaries. The impact on the environment of sulfur in air and precipitation. (Sweded's case study for the United Nations Conference on the Human environment. Stockholm. Conference Proceedlings), (1971).

DAHLE, E. and O. SKRE: Konferens om avsvavling, Stockholm 11, 11, 1969. Conference roceedlings 1. (1971).

FIEDLER, H. J., W. NEBE, and F. HOFFMAN: Forstliche Pflanzenernährung und Düngung, Stuttgart, 481 pp. (1973).

FOBER, H.: Mineralne żywienie. W: Buk zwyczajny 1990, PWN. Instytut Dendrologii *10,* 143–157 (1990).

FOY, C. D.: Effects of aluminium on plant growth. In: CARSON, E. W. (ed.): The plant root and its environment. University Press of Virginia, Charlottesville, 601–642 (1974).

GRESZTA, J., A. GRUSZKA, and T. WĄCHALEWSKI: Humus degradation under the influence of simulated acid rains. IUFRO Montreal, Canada. Division 2, 419–443 (1990).

GRUSZKA, A.: The impact of simulated acid rains on seedlings of selected forest tree species. Scientific Papers of Kraków Agricultural University, *257,* 5–101 (1991).

HAUSSER, K.: Düngungsversuche zu 70- bis 90jährigen Buchenbeständen auf der Schwäbischen Alp. Allg. Forst- u. Jagdztg. *142,* (819), 225–233 (1971).

HUETTL, R. F.: Forest fertilization: results from Germany, France and the Nordic countries. The Fertiliser Society, 1–40 (1986).

Huettl, R. F. and J. Wiśniewski: Fertilization as a tool minitage forest decline associated with nutrient deficiencies. Water, Air and Soil Poll. *33,* 265–273 (1987).

– New type Forest declines and restabilization/revitalization strategies. A pragmatic focus. Water, Air, and Soil Poll. *41,* 95–111 (1988).

– Liming and fertilization as mitigation tools in declining forest ecosystems. Water, Air, and Soil Poll. *44,* 93–118 (1989).

– Fertilization in multi-purpose forestry and its role in mitigating the decline of the world's forest resources. Plenary paper: 10th World Fertilizer Congress of CIEC, 21–27 October 1990, Micosia, Cyprus, *1,* 1–32 (1990).

Huettl, R. F., S. Fink, H. J. Lutz, M. Poth, and J. Wiśniewski: Forest decline, nutrient supply and diagnostic fertilization in southwestern Germany and southern California. For. Ecol. Manage. *30,* 341–350 (1990).

Huettl, R. F. and S. Fink: Pollution, nutrition and plant Function. In: Porter, J. R. and D. W. Lawlor (eds.): Plant growth: interaction with nutrition and environment. Soc. For Experimental Biology Seminar, Series *43,* 207–226 (1991).

Ivanov, A. F.: Rost drewiesnych rastenij i kislotnost poczw. Nauka i Technika, Mińsk (1970).

Jensen, K. F. and L. S. Dochinger: Response of eastern hardwood species to ozone, sulfur dioxide and acid precipitation. JAPCA Journal, *39,* 852–855 (1989).

Kennel, R. and J. Wehrmann: Ergebnis eines Düngungsversuches mit extrem hohen Stickstoffgaben in einem Kiefernbestand geringer Bonität. XIV IUFRO-Kongress VI, 216–231 (1967).

Kern, K. G. and W. Moll: Zur Düngung von Kifern- Buchen-Kulturen. Allg. Forst- u. Jagdztg. *142,* 127–139 (1971).

Kowalkowski, A.: Nawożenie mineralne drzewostanów. SGGW-AR, Warszawa, 120 pp. (1982).

Krause, G. H. M.: Inpact of Air Pollutants on Above – Ground Plant Parts on Forest Trees. In: Mathy, P. (ed.): Air Pollution and Ecosystems. Proc. Internat. Symp., Grenoble, France 18–22 05 1987, pp. 168–2160. Reidel Publ. Comp., Dordrecht, 168–216 (1988).

Mitscherlich, G. and W. Wittich: Fertilizer trials in older stands in the Lutter a. B. forest district. Aus dem Walde, Hannover 6, 5. Forestry Abstr. *24,* No. 4978 (1963).

Moller, C. M., O. Scharff, and J. R. Dragsted: 10 years' fertilizing experiments in Norway spruce and beech representing the main variations in growth conditions in Denmark. Forstl. Forsogsv. Danm. *31,* (2), 85–276 (1969).

Oktaba, W.: Metody statystyki matematycznej w doświadczalnictwie. PWN-Warszawa, 320 pp. (1971).

Pasternak, P. S., V. P. Stefurak, G. G. Zadorowa, S. N. Slobodjan, and V. E. Sav'juk: Effektivnost' mineral'nych udobrenij v el'nikach i bucinach severnych megasklonov Ukrainskich Karpat. Lesovodstvo i Agrolesomelioracija *51,* 34–38 (1978).

Penningsfeld, F.: Nährstoffmangelerscheinungen bei Baumschulgehölzen. Die Phosphorsäure *24,* (3/4), 199–212 (1964).

Rugge, U.: Physiologische Schäden durch Umweltfaktoren. In: Bäume in der Stadt, pp. 121–166. Ulmer Verlag, Stuttgart (1978).

Trillmich, H. D.: Düngung von Mischbeständen in einem Rauchschadengebiet des Erzgebirges. Wiss. Z. Tech. Univ. Dresden *18,* 807–816 (1969).

Ulrich, B.: Soil acidity and its relations to acid deposition. In: Ulrich, B. and J. Pankrath (eds.): Effects of Accumulation of Air pollutants in Forest Ecosystems, pp. 233–243. Reidel, Publishing, Company, Dordrecht (1983).

– Die Rolle der Bodenversäuerung beim Waldsterben: Langfristige Konsequenzen und forstliche Möglichkeiten. Forstw. Zbl. *105,* 421–435 (1986).

Zoettl, H. W. and R. F. Huettl: Nutrient supply and forest decline in southwest-Germany. Water, Air, and Soil Pollut. *31,* 449–461 (1986).

Zoettl, H. W., R. F. Hiettl, S. Fink, G. H. Tomlinson, and J. Wiśniewski: Nutritional disturbance and histological changes in declining forest. Water, Air, and Soil Pollut. *48,* 87–109 (1989).

J. Plant Physiol. Vol. 148. pp. 271–275 (1996)

The Effect of Simulated Acid Rain on Chlorophyll Fluorescence Spectra of Spruce Seedlings (*Picea abies* L. Karst.)

PAVEL ŠIFFEL[1], ZUZANA BRAUNOVÁ[1], EVA ŠINDELKOVÁ[1], and PAVEL CUDLÍN[2]

[1] Department of Photosynthesis, Institute of Plant Molecular Biology, Academy of Sciences of the Czech Republic, Branišovská 31, České Budějovice, 370 05, Czech Republic

[2] Department of Forest Ecology, Institute of Landscape Ecology, Academy of Sciences of the Czech Republic, Na sádkách 7, České Budějovice, 370 05, Czech Republic

Received June 12, 1995 · Accepted September 12, 1995

Summary

Three-year old Norway spruce seedlings (*Picea abies* L. Karst), were grown in peat soil in a greenhouse and sprayed either with simulated acid mist (a mixture of H_2SO_4 and HNO_3, pH 2.7) or with distilled water of pH 5.6 for three months. During spraying, the soil was covered with plastic foil to avoid acidification of the soil. The photosynthetic pigment apparatus of spruce needles was analyzed using SDS-polyacrylamide gel electrophoresis, spectrophotometric determination of pigment content and low temperature chlorophyll fluorescence spectroscopy.

The chlorophyll(Chl)-protein composition of spruce thylakoids was not affected by the acid treatment. Chl *a/b* ratio as well as the content of Chl (*a* + *b*) and total carotenoids tended to decrease in the acidified spruce seedlings but these changes were not significant.

The acidification of spruce foliage resulted in a small but highly significant increase in the intensity of chlorophyll fluorescence emission of the light-harvesting complex (LHC). This increase indicates structural changes of the photosynthetic pigment apparatus resulting from direct foliage-mediated action of acid rain.

Key words: Acid mist, Picea abies, pigment content, chlorophyll fluorescence spectra, chlorophyll-proteins.

Abbreviations: Chl = chlorophyll; CP1 = chlorophyll-protein of photosystem I; CPa = chlorophyll-protein of photosystem II; F_{681}/F_{685}= ratio of fluorescence intensities at emission maxima of light-harvesting complex and photosystem II; F_{685}/F_{735} = ratio of fluorescence intensities at emission maxima of photosystem II and photosystem I; LHC = light-harvesting complex; PS I = photosystem I; PS II = photosystem II; PSs = photosystems; SDS = sodium dodecylsulphate.

Introduction

Woody plants are subject to atmospheric and soil stress conditions which reduce growth and vigor, increase susceptibility to high-light stress, insects and pathogens, and cause premature death (Kramer, 1987). Among the atmospheric stress factors that cause the decline of coniferous forests, acid precipitation is potentially one of great importance (Schulze, 1989). Long-term exposure to acid rain can cause vegetation injury over large areas (McLaughlin, 1988; Esher et al., 1992).

Acid rain/mist can affect trees both through contact with the cuticula and via soil acidification. Disturbance of nutrition of plants has been recognized as the major factor influencing its physiological responses in areas exposed to increased acid depositions (Schultze, 1989). Besides, direct contact of acid rain with foliage can cause significant impairment of assimilative organs (for review see Hodgan, 1992): first, it can alter the cuticular surface (Rinallo et al., 1986; Schmitt et al., 1987; Barnes and Brown, 1990) and functioning of stomata. Consecutively, variations in cation leaching from foliage (Hogan, 1992), photosynthesis, water relations (Barnes et al., 1990 a; 1990 b; Flückiger et al., 1988), and in carbon metabolism (for review see Hampp, 1992) can be found as well. Second, input of excess protons into stroma can depress the rate of enzymic reactions of the Calvin cycle (Werdan et al., 1975; Woodrow at al., 1984), thus increasing sensitivity of the photosynthetic apparatus to photodamage (Osmond, 1981). Increased acidity in chloroplasts can be harmful also for Chl-proteins, particularly those involved in PS II (Siefermann-Harms, 1992; Lebedev et al., 1986).

In contrast with the relatively well documented influence of acid depositions on the needle surface, studies on photosynthesis and pigment apparatus lead to contradictory results (McLaughlin, 1988; Hogan, 1992). The influence of acid rain on the pigment apparatus has usually been characterized by analyzing the pigment composition. These studies indicate that acid rain has positive as well as negative effects on Chl content and chl *a*/*b* ratio, or sometimes, has no effect at all (Darrall and Jäger, 1984; Barnes et al., 1990 b; Eamus and Fowler, 1990). Some carotenoid ratios have also been tested as indicators of acid rain impact to photosynthetic apparatus (Wolfenden et al., 1988). Increase in Chl content has been related to nitrogen fertilization by acid rain (Reich et al., 1987; N'soukpoe-Kossi et al., 1992).

In this study, we present the effect of spraying spruce needles with simulated acid rain of pH 2.7 on low temperature chlorophyll fluorescence emission spectra indicating acid rain induced damage to photosynthetic apparatus. We also determined the effect on pigment content and Chl-proteins content of spruce needles.

Material and Methods

Plant material and acid rain treatment

One-year-old seedlings of Norway spruce (*Picea abies* L. Karst.), grown from seeds originating from the Jizerské hory Mts. (50° 51′ 50″ N, 15° 11′ 30″ E; altitude 850 m) under cold greenhouse conditions were transferred into 10 × 10 cm (1 liter volume) plastic pots with peat-bark-perlite (3 : 3 : 2 v/v) substrate with the addition of 0.5 kg · m^{-3} NPK fertilizer. The seedlings were maintained outdoors for two years under low nutrient supply. Then, the seedlings were transported into a greenhouse which was open at both ends and after a short acclimation the simulated acidification was begun. An acid solution of pH 2.7 was prepared with distilled water in equilibrium with atmospheric CO_2 pressure and a mixture of sulphuric and nitric acids (96 % H_2SO_4: 65 % HNO_3 in 2.8 : 1 v/v ratio). Ten seedlings were sprayed five times a week on their foliage and stems with 20 mL of acid solution, and another 10 seedlings with distilled water (pH 5.6). During spraying of the shoots, the soil was covered with polyethylene foil to avoid the acidification of the roots and soil. The soil was watered with 30 mL of distilled water after each shoot spraying. The current year needles were sampled and analyzed twice, within 85 and 95 days after the start of the acidification process. The total input of SO_4^{-2} and NO_3 ions per one seedling was 120 mg and 27.6 mg, respectively.

Isolation of chloroplasts and Chl-protein complexes

The needles collected from one spruce seedling were homogenized for 1 min in an isolation medium containing 0.33 M sorbitol, 1 mM $MgCl_2$, 1 mM $MnCl_2$, 0.5 % bovine serum albumin, 50 mM HEPES (pH 7.5), 1 % polyethyleneglycol 4000, 2 mM EDTA, 8 % polyvinylpyrrolidon 350,000 and 3 mM sodium ascorbate. The chloroplasts were sedimented at 4000 × g for 10 min. The pellet was resuspended in a medium containing 0.33 M sorbitol, 1 mM $MgCl_2$, 1 mM $MnCl_2$, 2 mM EDTA, 50 mM HEPES (pH 7.5). Chl-proteins were isolated by a SDS polyacrylamide gel electrophoresis using the gel and buffer system described in Anderson (1980). The SDS/Chl ratio was 15 : 1. The gels were scanned at 676 nm. For details of isolation and scanning see Šiffel et al. (1992).

Chlorophyll content

The amounts of Chls *a* and *b* and total carotenoids were determined in 80 % acetone extracts with a PU 8800 spectrophotometer (Philips-Pye Unicam, U.K.) according to Lichtenthaler (1987).

Chlorophyll fluorescence emission spectra

Emission spectra were measured with needle homogenates prepared as follows: twenty current year needles were put into a precooled glass homogenizer with 5 mL of isolation medium (see Chloroplast isolation) and homogenized. The homogenate was soaked into the filter paper and immediately frozen in liquid nitrogen. The Chl concentration in the samples was ca. 0.5 µg Chl (*a*+ *b*) cm^{-2}. The time interval between the start of homogenization and the freezing of the sample was shorter than 40 s. The chlorophyll fluorescence emission spectra of chloroplasts or thylakoids were measured using a Fluorolog spectrofluorometer (SPEX, U.S.A.), using a 450 W Xe-lamp, two double-grating monochromators and a cooled photomultiplier. All spectra were measured at 77 K with the spectral bandwidth of emission monochromator of 2 nm. The excitation wavelength was 490 nm.

Dry weight

Dry weight of needles, shoots and roots were determined with plant organs dried at 105 °C for 24 h.

Statistics

Standard errors and statistical significance of the found differences were calculated using the Student T-test.

Results

The low temperature chlorophyll fluorescence emission spectrum of isolated chloroplasts consists of three main bands. Two of them belong to photosystem (PS) II (maxima at 685 and 695 nm); the third (with the maximum at 735 nm) originates from PS I (e.g. Satoh and Butler, 1978; Satoh, 1982; Mimuro et al., 1987). The second derivative spectrum also reveals the presence of a band of the light-harvesting complex (LHC, maximum at 681 nm, see e.g. Šiffel et al., 1991). The relative intensity of these bands depends on the content of the particular pigment-protein complexes in thylakoid membranes, and on their interaction.

Figure 1 compares the chlorophyll emission spectra of needle homogenates prepared from seedlings sprayed with acid rain with those sprayed with distilled water. The seedlings treated with acid rain showed a slight increase of chlorophyll fluorescence intensity between 675 and 690 nm (Fig. 1). Second derivatives revealed that this rise was mainly related to an increase of the emission band with maximum at 681 nm (Fig. 1), i.e. the emission band originating from LHC. The ratio of emission spectra (acid rain/distilled water sprayed seedlings) showed the sharp peak at 680–681 nm (Fig. 1, top), thereby confirming the finding of derivative spectroscopy. The mean value of the F_{681}/F_{685} ratio calculated from 20 spectra (recorded for 10 spruce seedlings) of seedlings sprayed with acid rain was significantly (P=0.02) higher than this ratio determined for seedlings treated with distilled water (Fig. 2, left). A nonsignificant increase of the F_{685}/F_{735} ratio, found for acid treated seedlings, (Fig. 2, right) was likely the consequence of the increase of LHC emission intensity and of overlapping of PS II and LHC emission bands.

The contents of Chl-proteins of PS I (CP1), PS II (CPa) and LHC isolated by SDS electrophoresis were not significantly changed by acid rain treatment (Table 1). Chl (*a+b*) and total carotenoid contents, based on the dry weight of needles, decreased in response to simulated acid rain (Fig. 3). A similar trend was also found for the Chl *a/b* ratio. These changes, however, were not significant.

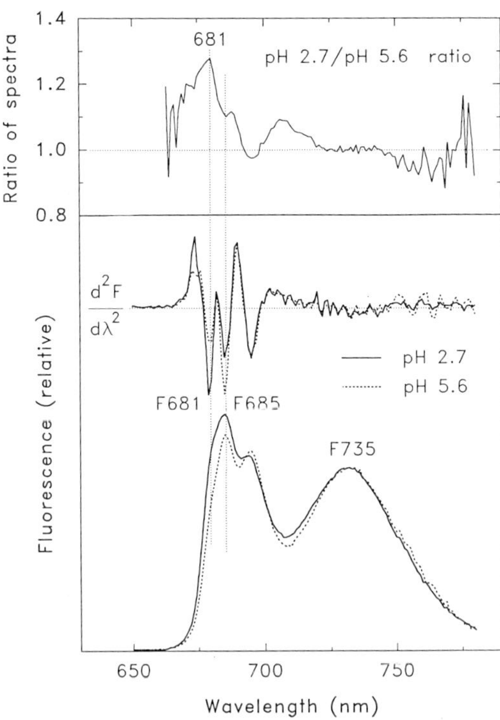

Fig. 1: Chlorophyll fluorescence emission spectra of the needle homogenates prepared from current year needles of spruce seedling sprayed with distilled water (·····) and acid solution (——). The second derivatives of the spectra are shown at the top. Spectra were measured at 77 K. Spectra were normalized at 735 nm.

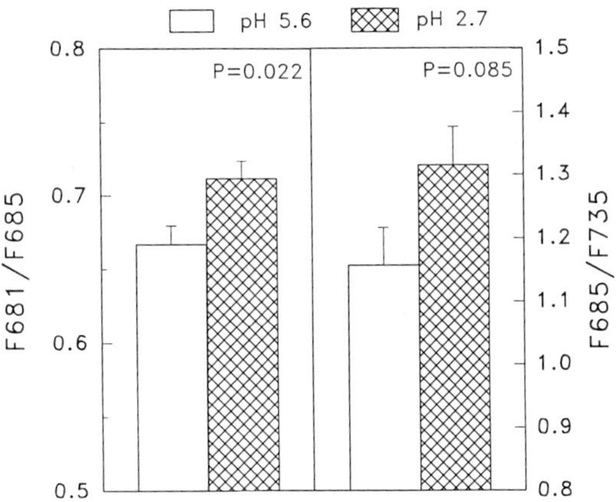

Fig. 2: Effect of simulated acid rain on the ratios of chlorophyll fluorescence intensities at: LHC and PS II emission maxima (i.e. F_{681}/F_{685})-left, and PS II and PS I emission maxima (i.e. F_{685}/F_{735})-right. Two treatments were different at a probability level P indicated at the top of each graph.

Table 1: Amounts of Chl-protein complexes separated by SDS electrophoresis from the needles of spruce seedlings sprayed with acid rain (pH 2.7) and distilled water (pH 5.6) (percent of total). CP1-protein of PS I, CPa-Chl-protein of PS II, LHC-light-harvesting complex, LHC_{olig}-oligomer, LHC_{mono}-monomer, FP-free pigments.

	CP 1	LHC_{mono}	CPa	LHC_{olig}	FP
pH 5.6	12.0±3.0	4.8±1.8	3.0±0.8	47.2±4.7	33.0±3.7
pH 2.7	10.7±2.5	4.0±1.1	2.2±0.7	49.4±6.2	33.7±4.3

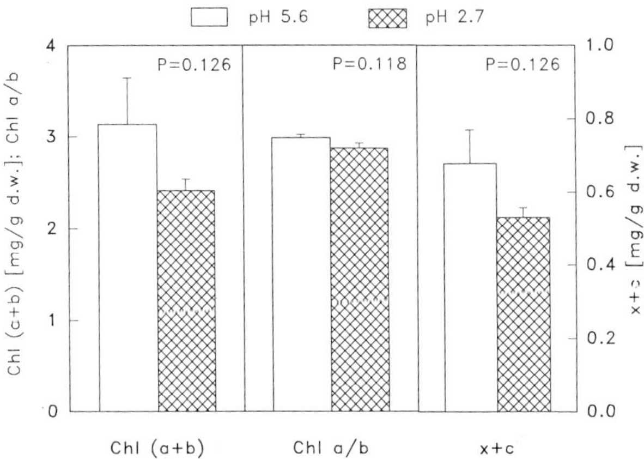

Fig. 3: Effect of simulated acid rain on Chl and total carotenoids contents. Two treatments were different at a probability level P indicated at the top of each graph.

Discussion

In this study, the effects of simulated acid rain on three characteristics of pigment apparatus, namely Chl-protein content, pigment content and chlorophyll fluorescence spectra, were determined. A significant difference between seedlings sprayed with acid rain and of those sprayed with distilled water was found only for chlorophyll fluorescence spectra showing an increase in LHC emission intensity.

Most of the light energy absorbed in LHC is transferred to PSs and only a small portion (≈ 1 %) is emitted as fluorescence. The increase in LHC emission intensity indicated a partial interruption of energy transfer between LHC and PS II and/or PS I. Fluorescence quantum yield at 77 K of the detergent-solubilized LHC is about 20 to 30 times higher than the quantum yield of thylakoid bound LHC (see decay times of the LHC in Mimuro et al., 1987 and Il'ina et al., 1981 and fluorescence intensities of the LHC in Satoh, 1982 and Shubin et al., 1976). Thus, even a very small decrease in the efficiency of energy transfer from LHC to PSs can bring about a detectable increase of LHC emission.

The changes observed in emission spectra (Fig. 1.) could be caused by the direct action of the acid rain on the thylakoid membrane or could result from photoinhibitory damage brought about by acid-induced inhibition of electron transport in chloroplasts. An increase in emission intensity of the LHC was induced by treating isolated chloroplasts with acids (Lebedev et al., 1986). Similar changes in the emission spectrum were also found for tobacco plants cultivated under photoinhibitory conditions (in a CO_2 deficient atmosphere) for several days (Šantrůček et al., 1991). Light-stimulation of Chl breakdown in senescent needles, or in needles exposed to pollutants, was described by Lichtenthaler and Buschmann (1984) and Siefermann-Harms (1992). Chl breakdown occurring in yellowing leaves has been shown to result in relative increase in LHC emission for several plant species (Šiffel et al., 1991).

Regardless of the mechanism of acid rain action on spruce needles, the results presented demonstrated that acid precipitation can lead to structural changes of the photosynthetic pigment apparatus.

References

Anderson, J. M.: P-700 content and polypeptide profile of chlorophyll-protein complexes of spinach and barley thylakoids. Biochim. Biophys. Acta *591,* 113–126 (1980).

Barnes, J. D. and K. A. Brown: The influence of ozone and acid mist on the amount and wettability of the surface waxes in Norway spruce [*Picea abies* (L.) Karst.]. New Phytol. *114,* 531–535 (1990).

Barnes, J. D., D. Eamus, and K. A. Brown: The influence of ozone acid mist and soil nutrient status on Norway spruce [*Picea abies* (L.) Karst.]. I. Plant-water relations. New Phytol. *114,* 713–720 (1990 a).

Barnes, J. D., D. Eamus, and K. A. Brown: The influence of ozone acid mist and soil nutrient status on Norway spruce [*Picea abies* (L.) Karst.]. II. Photosynthesis, dark respiration and soluble carbohydrates of trees during late autumn. New Phytol. *115,* 149–156 (1990 b).

Darrall, N. M. and H. J. Jäger: Biochemical diagnostic tests for the effect of air pollution on plants. In: Koziol, M. J. and F. R. Whatley (eds.): Gaseous Air Pollutants and Plant Metabolism, pp. 333–351. Butterworths, London (1984).

Eamus, D. and D. Fowler: Photosynthetic and stomatal conductance responses to acid mist of red spruce seedlings. Plant Cell Environ. *13,* 349–357 (1990).

Esher, R. J., D. H. Marx, S. J. Ursic, R. J. Baker, L. R. Brown, and D. C. Coleman: Simulated acid rain impact on fine roots, endomycorrhizae, microorganisms, and invertebrates in pine forests of the southern United States. Water, Air and Soil Poll. *61,* 269–278 (1992).

Flückiger, W., S. Leonardi, and S. Braun: Air pollutant effects on foliar leaching. In: Cape, J. N. and P. Mathy (eds.): Scientific Basis of Forest Decline Symptomatology, pp. 160–169. Report of No. 15, COST/CEC Brussels, Belgium (1988).

Hampp, R.: Comparative evaluation of the effects of gaseous pollutants, acidic deposition, and mineral deficiencies on the carbohydrate metabolism of trees. Agriculture, Ecosystems and Environ. *42,* 333–364 (1992).

Hogan, G. D.: Physiological effects of direct impact of acidic deposition on foliage. Agriculture, Ecosystems and Environ. *42,* 307–319 (1992).

Il'ina, M. D., E. A. Kotova, and A. Yu. Borisov: The detergent and salt effect on the light-harvesting chlorophyll *a/b* complex from green plants. Biochim. Biophys. Acta *636,* 193–200 (1981).

Kramer, P. J.: The role of water stress in tree growth. J. Arboric. *13,* 33–38 (1987).

Lebedev, N. N., P. Šiffel, E. V. Pakshina, and A. A. Krasnovskii: The effect of acidification on absorption and fluorescence spectra of French bean chloroplasts and kinetics of pheophytin formation. Photosynthetica *20,* 124–130 (1986).

Lichtenthaler, H. K.: Chlorophylls and carotenoids: Pigments of photosynthetic biomembranes. Methods in Enzymology. *148,* 350–382 (1987).

Lichtenthaler, H. K. and C. Buschmann: Photooxidative changes in pigment composition and photosynthetic activity of air-polluted spruce needles (*Picea abies* L.). In: Sybesma, C. (ed.): Advances in Photosynthesis Research. Vol. *IV,* pp. 245–250. Martinus Nijhoff/Dr. W. Junk Publ., The Hague, Boston, Lancaster (1984).

McLaughlin, S. B.: Whole tree physiology and air pollution effects on forest trees. In: Bervaes, J., P. Mathy, and P. Evers (eds.): Relationship between Above and Below Ground Influences of Air Pollutants on Forest Tree, pp. 8–25. Rep. No. 16, COST/CEC Brussels, Belgium (1988).

Mimuro, M., N. Tamai, T. Yamazaki, and I. Yamazaki: Excitation energy transfer in spinach chloroplasts. Analysis by the time-resolved fluorescence spectrum at –196 °C in the picosecond range. FEBS Lett. *213,* 119–122 (1987).

N'Soukpoé-Kossi, C. N., C. Trottier, C. A. Achi, D. Charlebois, and R. M. Leblanc: Effect of acid watering of the soil on the photosynthetic activity, growth, and foliar pigments of sugar maple seedlings. J. Environ. Sci. Health *A27,* 863–877 (1992).

Osmond, C. B.: Photorespiration and photoinhibition. Some implications for the energetics of photosynthesis. Biochim. Biophys. Acta *639,* 77–89 (1981).

Reich, P. B., A. W. Schoettle, H. F. Stroo, J. Troiano, and R. G. Amundson: Effects of acid rain on white pine (*Pinus strobus*) seedlings grown in five soils. I. Net photosynthesis and growth. Can. J. Bot. *65,* 977–987 (1987).

Rinallo, C., P. Raddi, R. Gellini, and V. Di. Lonardo: Effect of simulated acid deposition on the surface structure of Norway spruce and silver fir needles. Sonderdruck Eur. J. Forest Pathol. *16,* 440–446 (1986).

Šantrůček, J., P. Šiffel, M. Polišenská, and H. Pešková: Species specific response of photosynthetic apparatus to closed *in vitro* conditions. Photosynthetica *25,* 203–209 (1991).

Satoh, K.: Fractionation of thylakoid-bound chlorophyll-protein complexes by isoelectric focussing. In: Edelman, M., R. B. Hallick, and N. H. Chua (eds.): Methods in Chloroplast Molecular Biology. pp. 845–856, Elsevier Biomedical Press, Amsterdam, New York, Oxford (1982).

Satoh, K. and W. L. Butler: Low temperature spectral properties of subchloroplast fractions purified from spinach. Plant Physiol. *61,* 373–379 (1978).

Schmitt, U., M. Ruetze, and W. Liese: Rasterelectronen-mikroskopische Untersuchungen an Stomata von Fichten- und Tannennadeln nach Begasung und saurer Beregnung. Eur. J. Forest Pathol. *17,* 118–124 (1987).

Schulze, E. D.: Air pollution and forest decline in a spruce (*Picea abies*) forest. Science *244,* 776–783 (1989).

Shubin, V. V., V. A. Sinestichekov, and F. F. Litvin: Peremennaya fluorestzentziya nativnykh form chlorofilla pri –196 °C. [Variable fluorescence of native forms of chlorophyll at –196 °C.]. Biofizika *21,* 760–762 (1976).

Siefermann-Harms, D.: The yellowing of spruce in polluted atmosphere. Photosynthetica *27,* 323–341 (1992).

Šiffel, P., J. Kutík, and N. N. Lebedev: Spectroscopically analyzed degradation of chlorophyll-protein complexes and chloroplast ultrastructure during yellowing of leaves. Photosynthetica *23,* 395–407 (1991).

Šiffel, P., E. Šindelková, M. Durchan, and M. Zajícová: Photosynthetic characteristics of *Solanum tuberosum* L. plants transformed by genes for synthesis of phytohormones. 2. Pigment apparatus. Photosynthetica *27,* 441–447 (1992).

Werdan, K., H. W. Helt, and M. Milovancev: The role of pH in the regulation of carbon fixation in the chloroplast stroma. Studies on CO_2 fixation in the light and in the dark. Biochim. Biophys. Acta *396,* 276–292 (1975).

Wolfenden, J., D. C. Robinson, J. N. Cape, I. S. Paterson, B. J. Francis, H. Nehlhorn, and A. R. Wellbrun: Use of carotenoid ratios, ethylene emissions and buffer capacities for early diagnosis of forest decline. New Phytol. *109,* 85–95 (1988).

Woodrow, I. E., D. J. Murphy, and E. Latzko: Regulation of stromal sedoheptulose-1-7-bisphosphatase activity by pH and Mg^{2+} concentration. J. Biol. Chem. *259,* 3791–3795 (1984).

J. Plant Physiol. Vol. 148. pp. 276–286 (1996)

Chronic SO_2- and NO_x-Pollution Interferes with the K^+ and Mg^{2+} Budget of Norway Spruce Trees

STEFAN SLOVIK[1]

[1] Julius-von-Sachs-Institut für Biowissenschaften mit Botanischem Garten der Universität Würzburg, Lehrstuhl für Botanik I, Mittlerer Dallenbergweg 64, D-97082 Würzburg, Germany

Received June 24, 1995 · Accepted October 30, 1995

Summary

Quantitative data concerning the impact of chronic SO_2 and NO_x (= NO_2 + NO) pollution at ambient concentrations on the nitrogen (N_{org}), sulphur (S_{org}), K^+ and Mg^{2+} budget of Norway spruce trees growing in synchronized monocultures are summarized. (i) Stomatal SO_2 and NO_x uptake rates, (ii) epicuticular SO_2 and NO_x deposition rates (including aerosol deposition), and (iii) sulphur and nitrogen deposition rates via acid precipitation are considered. Atmospheric SO_2 and NO_2 concentrations required to supply the total S- and N-demand of growing spruce trees are deduced. Based on the total S_{org} and N_{org} demand (i) of whole spruce trees or alternatively, (ii) of harvested trunk wood per monoculture turnover, critical load rates for sulphur and nitrogen are deduced and compared with stomatal, epicuticular and ombrogenic sulphur and nitrogen deposition rates in spruce forests of central Europe. Reductive and oxidative SO_2 detoxification pathways in spruce needles are quantified. Measured SO_2- and NO_x-dependent cation throughfall rates from spruce canopies are compared with the additional cation demand rates of spruce stands after stomatal uptake of SO_2 (for SO_4^{2-} neutralization purposes) or of NO_x (for avoidance of [K^+ or Mg^{2+}]: N_{org} dilution). Chronic SO_2-pollution is 2.0 to 2.6 times more phytotoxic than equally high NO_2 concentrations in air. The here presented own data, which are available for different site elevations and tree age classes, are discussed within the context of «spruce decline».

Key words: Air pollution (SO_2, NO_2, NO), Cation competition (K^+, Mg^{2+}), Critical load (SO_4^{2-}, NO_3^-), Chronical deposition (aerosol, cuticle, stomata, precipitation), Detoxification (trace gases), Forest decline (Norway spruce, Picea abies [L.] Karst., Pinaceae), Tree nutrition (nutrient cycling).

Abbreviations: A_{spec} = Specific needle surface area per kg needle dry matter [$m^2\,kg^{-1}$ DW]; Cat = Any cation species (K^+, Mg^{2+} etc.); G_{H_2O} = Stomatal H_2O conductance of mean needles [mmol H_2O m^{-2} needle surface s^{-1}]; GP = Trunk growth period (= duration of the annual «vegetation period») [$d\,a^{-1}$]; h = Site elevation above the sea level [m a.s.l.]; J'_{gas} = Specific stomatal trace gas uptake rate [$\mu mol\,m^{-2}$ needles d^{-1} (GP) (nPa $Pa^{-1})^{-1}$]; J_{NO_x} = Stomatal (NO_2 + NO) flux per unit needle surface [$\mu mol\,m^{-2}\,d^{-1}$ (GP)]; J_{SO_2} = Stomatal SO_2 flux per unit needle surface [$\mu mol\,m^{-2}$ needle surface d^{-1} (GP)]; L_{ion} = Measured canopy throughfall rate [mmol kg^{-1} needle DW a^{-1}]; L_o = Background leaching rate in clean air [mmol kg^{-1} needle DW a^{-1}]; L'_{ion} = Specific ion leaching rate per needle dry matter [mmol kg^{-1} DW a^{-1} (nPa $Pa^{-1})^{-1}$]; N_{org} = Organic nitrogen compounds; $[NO_2]_a$ = Annual mean of the ambient NO_2 concentration in air [nPa Pa^{-1} = ppb]; r = Pearson correlation coefficient [no dimension]; $[SO_2]_a$ = Annual mean of the ambient SO_2 concentration in air [nPa Pa^{-1} = ppb]; S_{org} = Organic sulphur compounds; S_{rel} = Relative fraction of the annual stomatal SO_2 dose that is oxidized to accumulating SO_4^{2-} within needles [mol SO_4^{2-} mol^{-1} SO_2]; ΣCat = Stoichiometric sum of cation equivalents [Eq]; t = Stand (tree) age of synchronized spruce monocultures in years [a]; TP_{rel} = Relative contribution of a regarded cation Cat to the charge balance of SO_4^{2-}; z = Stoichiometric factor for different cation species (Cat_zSO_4).

Introduction

Norway spruce (*Picea abies* [L.] Karst.) is one of the economically most important tree species of the northern hemisphere (Schmidt-Vogt, 1986). For almost two decades, symptoms of canopy thinning, early needle senescence, «unusual» trunk increment growth rates (and dynamics) and at extreme sites even a local decline of spruce trees became apparent in central Europe. Possible natural and anthropogenic reasons for these observations are discussed (e.g. Zech and Popp, 1983; Wentzel, 1985; Zöttl, 1986; Schulze, 1989; Hüttl, 1991; Kandler, 1994). In this communication, field data obtained for stomatal, epicuticular and ombrogenic (= via precipitation) sulphur and nitrogen deposition rates are compared to the growth demand for sulphur (S_{org}), nitrogen (N_{org}), K^+ and Mg^{2+} of initially healthy spruce monocultures (i.e. before development of «decline» symptoms). Focussing for example on stomatal SO_2 uptake (cf. Fig. 1), generated (bi)sulfite anions may cause acute damage symptoms, which are out of the scope of this discussion, since observed SO_2 concentrations are usually too small in the field (exceptions admitted). Still, chronic SO_2 uptake by needles forces

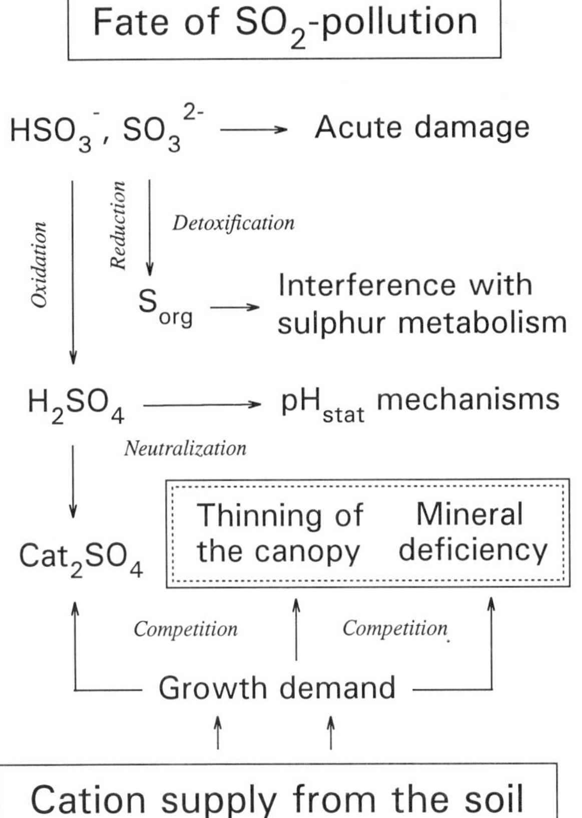

Fig. 1: Overview of chronically activated detoxification ($SO_2 \rightarrow S_{org}$, or $SO_2 \rightarrow SO_4^{2-}$) and compensation pathways in spruce trees after long lasting SO_2 pollution at ambient SO_2 concentrations in the field. Concerning NO_x (= NO_2 + NO), only reductive detoxification ($NO_x \rightarrow N_{org}$) is observed in the field at ambient NO_x concentrations.

reduction ($SO_2 \rightarrow S_{org}$) or oxidation ($SO_2 \rightarrow H_2SO_4$) of generated (bi)sulfite in the longterm (cf. Ziegler, 1975). This metabolic «bifurcation» of SO_2 detoxification pathways is absent at ambient annual means of NO_2 pollution in spruce forests (ca. 8 to 16 nPa NO_2 Pa^{-1} = ppb; cf. Slovik et al., 1996 a). There is NO_2-dependent NO_3^- accumulation in spruce needles only after fumigation with much higher NO_2 concentrations (200 to 500 nPa NO_2 Pa^{-1} = ppb) than those which occur in the field (Tischner et al., 1988; Kaiser et al., 1993). Generation of H_2SO_4 in the needle mesophyll activates pH-stat mechanisms, which mobilize base (OH^-) and cations (Cat) as counterions of SO_4^{2-}. In the field, pH-stat-mechanisms are not overburdened (Börtitz, 1969, Kaiser et al., 1991). Cations, which are sequestered into needle vacuoles together with accumulating SO_4^{2-} (mainly K^+; cf. Slovik et al., 1996 c), are unavailable for growth. There is little SO_4^{2-} retranslocation before needle abscission. Thus, cations immobilized in the vacuole (e.g. K_2SO_4) represent an additional cation loss in shed spruce needles, which would be absent in clean air. Similarly, there is additional cation leaching after absorption of SO_2 in spruce canopies (Slovik et al., 1996 b). Thus, chronic SO_2 pollution causes competition for cations, which is harmful mainly (i) at unfertilized or unlimed, poor stands if water soluble sulphate salts further leak from the soil matrix into the ground water, or (ii) if shed or leached cations, e.g. K^+, are taken up by competitive plants or soil microorganisms. This lack of local nutrient recycling is dominant mainly at stands growing on acidic soils (case i), or if there is concomitant nitrogen deposition, which acts as a fertilizer of soil organisms (destruents) and of fast growing herbaceous plants (case ii). The by these plants incorporated cations become unavailable to spruce trees. Thus, cation deficiency or – alternatively – thinning of the canopy structure must be the consequence in the long-term. In contrast to SO_2, NO_2 is quantitatively assimilated ($NO_2 \rightarrow N_{org}$) in spruce needles (acute cell damage by NO_2 is omitted here; cf. Wellburn, 1990). Stomatal NO_2 uptake either induces additional tissue growth (if a raising N_{org} content in spruce tissues is to be avoided), or it leads to an increase of the N_{org} content in spruce tissues (if stimulation of growth is absent). In the first case, there is an additional cation demand for tissue growth if dilution of cations (→ cation deficiency) is to be avoided in the longterm (cf. Schulze, 1989). Also in the second case, there is an additional cation demand if the Cat: N_{org} ratio is to be kept constant as it was in healthy spruce trees before onset of NO_2 pollution, since a high and stable K^+: N_{org} ratio is an important parameter of pathogen tolerance and frost hardiness in the field. Thus, chronic SO_2 and NO_x pollution both induce an additional cation demand, which would be absent in clean air, since stomatal trace gas uptake is – of course – not accompanied by an equivalent stomatal cation uptake. Concerning SO_2, this cation equivalency is defined (i) by the relative sulphate formation percentage S_{rel} per unit of stomatal SO_2 uptake flux [mol SO_4^{2-} mol^{-1} SO_2] and (ii) by the relative contribution of different cations to the charge balance of vacuolar sulphate TP_{rel} [Eq Cat Eq^{-1} SO_4^{2-}]; TP stands for «tonoplast symport». Concerning NO_2 (incl. NO), its cation equivalency is defined by the Cat: N_{org} demand ratio for the growth of healthy spruce trees. In this communication, the as yet mentioned expectations are quantified and discussed in the context of «spruce decline», focussing only on chronic effects which develop after a couple of years.

Materials and Methods

A set of results of the author is presented here in the manner of a short review. Thus, the applied approaches are only briefly summarized here (cf. citations):

Stomatal SO_2 and NO_x uptake

Annual trace gas doses at six spruce stands in Hessen (Germany) were numerically integrated from 1984 to 1992 with high time resolution (30 min; 17,520 integration steps per year). Meteorological field data, which served as the data basis for modelling stomatal conductances after Körner et al. (1995), and pollution data of SO_2, NO_2 and NO were measured within spruce forests by the Hessische Landesanstalt für Umwelt (Wiesbaden, FRG). Obtained annual doses from all available sites and years (= «site·years») were first standardized on a unit length of the trunk growth period GP [$d\,a^{-1}$], since this «annual vegetation period» is the predominant time period with open stomata per year. Standardized quotients were then correlated versus the corresponding annual means of trace gas concentrations of all site·years. Thus, statistical functions became available, which allow the estimation of annual stomatal trace gas doses of spruce needles in the field if only (i) annual means of trace gas concentrations and (ii) the length of the growth period GP are available; cf. Slovik et al. (1995) and Slovik et al. (1996 a) for details: substomatal concentrations, canopy effects, data validation etc.

Epicuticular SO_2 and NO_x absorption (incl. aerosol deposition)

Annual canopy throughfall and precipitation rates of SO_4^{2-} and NO_3^-, and of cations (NH_4^+, K^+, Mg^{2+} etc.), measured at the same sites in Hessen where stomatal trace gas fluxes were integrated (cf. above), are available after Balázs (1991). Throughfall data, that originate from canopy surfaces themselves, were first corrected for expected artificial concentration increases after evaporation of intercepted precipitation water and then correlated with concomitantly measured annual means of SO_2 or NO_2 concentrations in air. Linear correlation and regression analysis was performed also for cations vs. (SO_2 or NO_2) in order to identify the «mean» ionic composition pattern of trace gas-dependent spruce canopy leachates in the field; cf. Slovik et al. (1996 b).

Nutrient analysis data

After pressure ashing of samples, analysis of metal cations (K^+, Mg^{2+}, u. a.) and of total sulphur was performed by inductively coupled plasma (ICP) atomic emission spectrophotometry (Model JY 70 PLUS, ISA Jobin-Yvon, France) with an automatic sample injector (Gilson model 222, Villiers le Bel, France). The carbon and nitrogen content of spruce tissues was determined by element analysis (CHN-O-Rapid, Heraeus, Hanau, Germany). After hot water extraction of spruce tissues, anions (SO_4^{2-}, NO_3^-) were detected by isocratic anion chromatography (HPLC; Ionenchromatograph IC 100, Biotronik, Maintal, FRG). Tissue analysis data from healthy spruce trees growing in Würzburg (Botanical garden) were taken as references.

Tissue growth rates in ageing spruce monocultures

As references, typical growth and turnover rates of all important tissues and organs (needles, branch wood, branch bark, trunk wood, trunk bark, root wood, root bark, fine roots etc.) of healthy spruce trees growing in ageing monocultures at different site elevations were reconstructed on the basis of historical trunk yield tables after «Wiedemann» (cf. Schober, 1987), growth data after Vanselow (1951; cf. Schmidt-Vogt, 1986), morphometric data after Mette and Korell (1989) and supplementary morphometric field data (cf. Slovik, 1996 for derivation of tissue growth scenaria [kg tissue DW $ha^{-1}\,a^{-1}$] and tests of data consistency).

Results

Stomatal trace gas uptake

The specific stomatal uptake dose of trace gases, given in µmol (SO_2 or NO_x) m^{-2} needle surface day^{-1} (growth period GP) ($nPa\,Pa^{-1}$)$^{-1}$ annual mean of concentration in the air, depends on the annual trace gas pollution pattern measured in spruce forests («mean» data in Fig. 2) and on the annual kinetics of the stomatal spruce canopy conductance in the field («mean» data in Fig. 3). Separate integration of stomatal trace gas fluxes per m^2 needle surface after Slovik et al. (1995, 1996 a) for all available n = 45 site·years in Hessen yields 45 annual doses for SO_2 and NO_x (= NO_2 + NO), which are standardized in Fig. 4 per observed number of annual days of the trunk growth (= «vegetation») period GP [$d\,a^{-1}$], since stomatal aperture is maximum in the summer months (cf. Fig. 3). Results are plotted versus the observed annual means of trace gas concentrations in air (cf. data points in Fig. 4). Stomatal SO_2 and NO_x fluxes yield linear regression functions, which are summarized in the legend of Fig. 4. Concerning SO_2, below ca. 3 nPa SO_2 Pa^{-1} (= ppb) the regression function must be extrapolated to zero (not to an intercept at ca. 0.647 µmol SO_2 $m^{-2}\,d^{-1}$). Concerning NO_x, there is an apparent NO_2 compensation point at ca. 7 nPa NO_2 Pa^{-1}, which is about two times higher than the observed compensation point of NO_2 itself (cf. Slovik, 1996 a), since Fig. 4 presents NO_x (NO_2 + NO) flux data as plotted versus annual means of NO_2 concentrations in air. Pollution kinetics of NO_2 and NO are largely parallel (cf. Fig. 2). Fig. 4 allows the

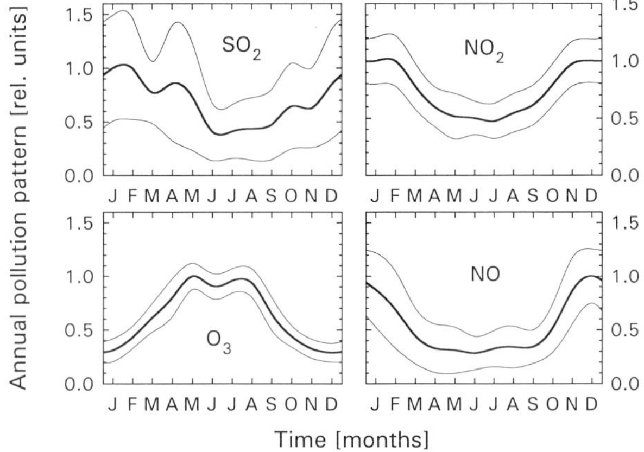

Fig. 2: Kinetics of the annual SO_2, NO_2, NO and O_3 pollution pattern in Norway spruce forests (relative units), measured in central Europe (pooled data from 6 sites in Hessen, FRG). Bold lines represent monthly means of n = 45 site·years. Thin lines denote standard deviations (SD) of these field data (after Slovik et al., 1996 a).

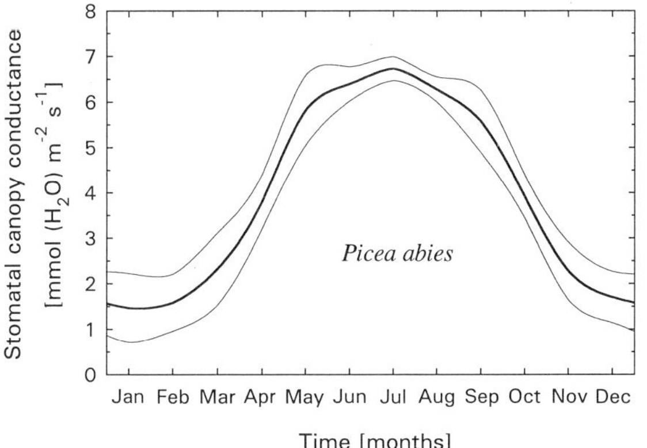

Fig. 3: Typical annual kinetics of the stomatal canopy conductance [mmol H_2O m^{-2} total needle surface s^{-1}] of Norway spruce canopies in central Europe (Hessen, FRG). Bold lines indicate monthly means (± SD, thin lines) of results from n = 45 pooled site·years. The given data already consider that the stomatal conductance of mean needles of spruce canopies is only ca. 0.32 times the needle conductance of exposed needles (canopy effects!). The stomatal SO_2 conductance is (0.51 to 0.53) times, and the stomatal NO_2 conductance is (0.58 to 0.63) times the H_2O conductance; modelling of stomatal conductances after Körner et al. (1995) was based on meteorological field data in Hessen (after Slovik, 1996; cf. also Slovik et al., 1995, 1996 a).

estimation of stomatal SO_2 and NO_x (NO_2 + NO) uptake fluxes in the field if (i) annual means of SO_2 and NO_2 pollution, and (ii) the annual length of the growth period GP [$d\,a^{-1}$] are known. The latter depends on the geographical site elevation and on the site latitude of northern and central Europe (cf. Wiersma, 1963).

Epicuticular trace gas and aerosol deposition

In Fig. 4 stomatal uptake rates of SO_2 and NO_x are compared with spruce canopy throughfall rates of SO_4^{2-} and NO_3^- as depending on annual means of SO_2 or NO_2 concentrations in air, measured at the same sites in Hessen (cf. Slovik et al., 1996 b). Compared with stomatal uptake functions, «epicuticular» sulphur and nitrogen deposition functions (including aerosol deposition) show a congruent shape of trace gas dependency (cf. Fig. 4). Both, stomatal and «epicuticular» nitrogen deposition, show a «compensation point» at (5 to 7) nPa NO_2 Pa^{-1} (cf. arrows in Fig. 4). Comparison of obtained regression slopes in Fig. 4 (including NH_4^+ throughfall data versus NO_2 pollution data; not shown here) yields ca. 2.3 times higher trace gas dependent throughfall rates of SO_4^{2-} or NO_3^- (plus NH_4^+) per unit of trace gas pollution than corresponding specific stomatal SO_2 or NO_x uptake rates (cf. Slovik et al., 1996 b). Integration of the area below the «typical» stomatal conductance curve in central Europe after Fig. 3 allows the quantification of the ratio of total annual hours (8760 $h\,a^{-1}$) per annual hours with «open» stomata. This ratio equals ca. 2.3 $h\,h^{-1}$ for SO_2 and for NO_2 (cf. Slovik et al., 1996 a, b). Thus, there seems to be a concomitant SO_2 and NO_x deposition into spruce canopies, which by 70 % (= 100 %·2.3/3.3) occurs all the year at day and night (= «epicuticular» deposition), and which by 30 % (100 %·1.0/3.3) is possible only if stomata are open mainly (not only) in the summer at day hours (= stomatal uptake of SO_2 and NO_x). Since both, stomatal trace gas uptake and «epicuticular» trace gas deposition, share (i) similar deposition resistances (unstirred boundary layers, «canopy effects», impact of wind velocities), and (ii) a common water dissolution step (in cell wall water within the needle mesophyll, on wet canopy surfaces or first in fog water before aerosol deposition), similar statistical deposition functions are not only expected in the field, but indeed observed (cf. Fig. 4). Fig. 4 allows the estimation of «typical» SO_4^{2-} and NO_3^- throughfall rates per unit of needle dry matter if only annual means of SO_2 and NO_2 pollution are known.

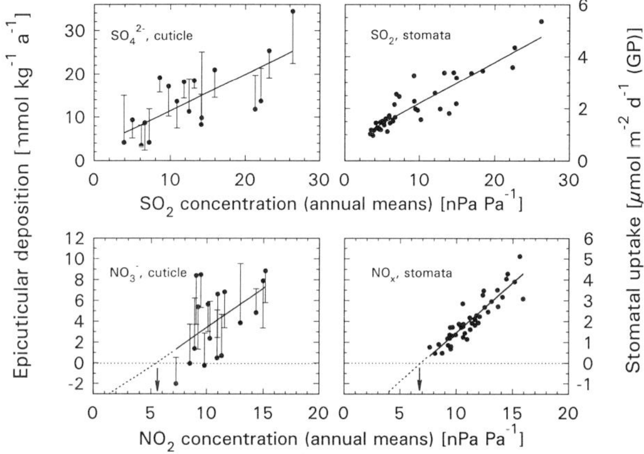

Fig. 4: Stomatal SO_2 and NO_x (= NO_2 + NO) uptake rates by, and epicuticular SO_2 (→ SO_4^{2-}) and NO_x (→ NO_3^-) deposition rates in Norway spruce canopies as depending on the SO_2 or NO_2 concentration in air (pooled available data since 1984 from 6 spruce sites in Hessen, FRG). Epicuticular deposition rates after Balázs (1991; cf. Slovik et al., 1996 b) are given in mmol (SO_4^{2-} or NO_3^-) kg^{-1} needle DW $year^{-1}$. Error bars denote standard deviations (SD for n = 3 or 4 plots per site). Parameters of epicuticular trace gas deposition rates (L_{SO_4}, L_{NO_3}) are r = Pearson correlation coefficient [no dimension], L_o = intercept [mmol kg^{-1} DW a^{-1}] and L'_{ion} = slope [mmol kg^{-1} needle DW a^{-1} (nPa Pa^{-1})$^{-1}$ annual mean of trace gas pollution]:

$L_{SO_4}=f([SO_2]_a)$ r = 0.701*** L_o = 3.11 ± 3.09 L'_{SO_4} = 1.682 ± 0.428
$L_{NO_3}=f([NO_2]_a)$ r = 0.489* L_o = −4.10 ± 3.85 L'_{NO_3} = 0.753 ± 0.347

Data of annual stomatal SO_2 and NO_x uptake rates in µmol SO_2 or NO_x (= NO_2 + NO uptake) per m^2 total needle surface, and standardized per day (d) of the annual trunk growth period GP [$d\,a^{-1}$], after Slovik et al. (1995, 1996 a). Parameters of stomatal trace gas uptake rates (J_{SO_2}, J_{NO_x}) are: r = Pearson correlation coefficient [no dimension], b = intercept [µmol m^{-2} needle surface d^{-1} growth period] and J'_{gas} = slope [µmol m^{-2} needle surface d^{-1} growth period (nPa Pa^{-1})$^{-1}$ annual mean of trace gas pollution]:

$J_{SO_2}=f([SO_2]_a)$ r = 0.909**** b = 0.647 ± 0.114 J'_{SO_2} = 0.157 ± 0.011
$J_{NO_x}=f([NO_2]_a)$ r = 0.903**** b = −3.309 ± 0.401 J'_{NO_x} = 0.477 ± 0.035

The specific needle surface A_{spec} of spruce needles equals (11 to 13) m^2 total surface kg^{-1} needle DW. The dependency of the annual growth period GP [$d\,a^{-1}$] on the site elevation in central Europe (latitude ≈ 50° northern hemisphere) is given in Table 2 (cf. Wiersma, 1963).

Data basis of annual nutrient demand rates

As a reference, total nutrient contents (N_{org}, S_{org}, K^+, Mg^{2+}) of different tissues from healthy spruce trees growing in Würzburg (cf. Table 1) were interrelated with «typical» site- and stand age-dependent canopy, trunk and root tissue growth rates of healthy spruce monocultures in the field (cf. Slovik, 1996) in order to obtain total annual nutrient demand rates of entire spruce trees (i) either on a hectare basis, or (ii) per unit stock of needle dry matter (= interface to stomatal SO_2 or NO_x uptake rates after Fig. 4). Table 2 exemplifies results obtained for «typical» spruce growth rates focussing on the CO_2 balance (annual net photosynthesis, annual respiration [incl. e.g. root leaching], CO_2 fixation in dry matter) of spruce monocultures growing at different site elevations. Shown are annual CO_2 balance rates of entire spruce trees, but given on a needle dry matter basis. Results of the spruce growth model after Slovik (1996) are consistent with independent investigations.

Mere atmospheric nitrogen and sulphur supply

Atmospheric SO_2 and NO_x may serve as S- and N-sources for plant growth. In Table 3 NO_2 and SO_2 concentrations in air (annual means) have been deduced, which would be sufficient to meet the total N_{org} and S_{org} demand of healthy spruce monocultures growing at two selected site elevations (ca. 10 and ca. 1000 m a.s.l.). Necessary SO_2 concentrations range from ca. 28 nPa SO_2 Pa^{-1} (= ppb SO_2) in very young spruce stands (ca. 5 years old) to only 17 or 11 nPa SO_2 Pa^{-1} in ca. 120 years old spruce stands. At lower elevations more SO_2 is «necessary» to formally supply the total S_{org} demand of growing spruce stands than in the mountains. Deduced SO_2 concentration equivalents do occur in the field (up to more than 30 nPa SO_2 Pa^{-1}; cf. Slovik et al., 1996 a), but required annual means of NO_2 concentration in air for the entire N_{org} supply via stomata largely exceed measured annual means of NO_2 pollution in the field of (8 to 16) nPa NO_2 Pa^{-1} (= ppb NO_2; cf. Slovik et al., 1996a). Thus, ambient NO_2 concentrations supply only a few percent of the total annual N_{org} demand of spruce trees growing in the field. These small stomatal NO_x doses can be easily assimilated to N_{org} compounds.

Table 1: Organic nitrogen (N_{org}), sulphur (S_{org}), K^+ and Mg^{2+} contents of different spruce tissues (*Picea abies*) analyzed in Würzburg (Botanical Garden) in 1991 (± SD; n = 50).

Tissue or tree organ *Picea abies*	Org. nitrogen N_{org} [mmol kg^{-1}]	Org. sulphur S_{org} [mmol kg^{-1}]	Potassium K^+ [mmol kg^{-1}]	Magnesium Mg^{2+} [mmol kg^{-1}]
Green needles	652±121	25.2±9.9	122.9±32.2	31.1±6.8
Brown needles	438± 92	16.8±2.9	82.0±14.1	37.6±5.1
Retranslocation	214± 44	8.4±3.0	40.9±35.2	−6.5±8.5
Branch bast	537±119	14.9±7.4	99.0±18.5	37.3±6.6
Branch wood	133± 27	3.9±0.9	20.4± 4.2	6.4±1.6
Root bast	458± 63	11.4±1.6	139.1±21.4	43.9±9.5
Root wood	273± 22	5.6±2.3	77.5±33.5	15.2±1.4

Table 2: Annual CO_2 fixation (dry matter formation), net photosynthesis and respiration (incl. root leaching) of Norway spruce trees growing at different site altitudes. Data of whole trees are given on a needle dry matter basis [mol CO_2 kg^{-1} needle DW a^{-1}]. The mean length of the annual trunk growth period is estimated after Wiersma (1963). Percentage data of respiration and CO_2 fixation are given on a net photosynthesis basis = 100%; data were calculated after Slovik (1996).

Site altitude	Growth period	CO_2 fixation	net photo-synthesis	respi-ration	respi-ration	CO_2 fixation
[m]	[d a^{-1}]	[mol C kg^{-1} a^{-1}]			[%]	[%]
10	222	41.0±4.2	118	77	65	35
400	200	40.2±4.3	106	66	62	38
800	176	39.6±4.4	93	53	57	43
1200	154	38.5±4.4	82	44	54	46
1600	130	36.5±4.3	69	32	46	54
2000	108	33.6±4.1	57	23	40	60

Table 3: «Necessary» NO_2 and SO_2 pollution (annual means) that formally supplies – after stomatal uptake – the entire N_{org} or S_{org} demand of ageing Norway spruce trees growing in the field. Shown are amplitudes from young trees (t ≈ 5 years; first value) up to old trees (t ≈ 120 years; second value). Data were calculated after Slovik (1996).

Site elevation [m a. s. l.]	SO_2 [nPa Pa^{-1} = ppb]	NO_2 [nPa Pa^{-1} = ppb]
10 m	≥28–17	≥340–230
1000 m	≥28–11	≥340–180

Critical load rates for nitrogen and sulphur

In natural spruce ecosystems growing in clean air, there are only small nitrogen losses via e.g. NH_3 or NO_x emission into the air, or via e.g. NO_3^- leaching into the ground water. Only these nitrogen losses must be compensated in a balanced ecosystem (atmospheric input, N_2-fixation e.g. by microorganisms). After harvesting of biomass in managed spruce forests, there is loss of carbon and nitrogen. This carbon loss should reduce the microbial soil activity and hence cause reduced N_2-fixation. Additionally, there is loss of nitrogen. Is this nitrogen loss in central Europe compensated by presently measured total nitrogen (= NH_4^+ + NO_3^-) deposition rates via precipitation water? Is there a widespread excess supply of nitrogen in central Europe that forces spruce trees to «grow to death», i.e. to run into another – e.g. Mg^{2+} – shortcut (cf. Schulze, 1989)? Necessary annual wet deposition rates of nitrogen and sulphur in order to resupply the mean annual N_{org} and S_{org} loss by harvesting spruce trunk wood at stands growing at varying site elevations are given in Table 4. Shown are mean annual values per a 120 years lasting monoculture turnover period; cf. Schober (1987). The mean annual nitrogen load (NO_3^- and NH_4^+) contained in precipitation water at six spruce sites in central Germany (State of Hessen) since 1984 equals only (421 ± 106) mol N ha^{-1} a^{-1} (pooled data after Balázs, 1991). Thus, on an ecosystem level, there is a higher mean annual nitrogen loss by trunk wood harvesting

Table 4: Dependency of critical wet deposition rates (precipitation flux) [mol (or kg) ha^{-1} a^{-1}] of nitrogen (NO_3^- + NH_4^+) and sulphur (SO_4^{2-}) on the site elevation [m a. s. l.]. Data are calculated from the cumulative N_{org} and S_{org} fixation in harvested trunk wood (= nutrient loss) per spruce monoculture turnover (cf. Slovik, 1996) using the nutrient contents given in Tab. 1; SO_2 and NO_2-dependent epicuticular and stomatal SO_2 and NO_2 deposition rates (cf. Fig. 4) are omitted here (but cf. Tab. 5). Observed field data after Balázs (1991).

Site altitude	Critical load of nitrogen			Critical load of sulphur		
[m a. s. l.]	[mol ha^{-1} a^{-1}]		[kg ha^{-1} a^{-1}]	[mol ha^{-1} a^{-1}]		[kg ha^{-1} a^{-1}]
10 m	921	=	12.9	27.3	=	0.87
400 m	823	=	11.5	24.4	=	0.78
800 m	724	=	10.1	21.5	=	0.69
1200 m	626	=	8.8	18.5	=	0.59
1600 m	527	=	7.4	15.6	=	0.50
2000 m	428	=	6.0	12.7	=	0.41
Observed	421±106	=	5.9±1.5	180±32	=	5.7±1.0

Table 5: Total nitrogen (NO_3^-, NH_4^+, NO_2, NO) and sulphur (SO_4^{2-}, SO_2) wet deposition (precipitation rates) and dry deposition (stomatal uptake, epicuticular deposition) in spruce forests growing in Hessen (central Germany; pooled data from Biebergemünd in the Spessart mountains, Frankenberg at the Sauerland mountains, Fürth in the Odenwald mountains, Grebenau at the Vogelsberg mountain, Königstein in the Taunus mountains and Witzenhausen close to Kassel) as compared with the N_{org} and S_{org} demand of these spruce stands. Given is the range of available annual means (summary of data after Balázs, 1991; Slovik et al., 1996a, b).

Deposition component	Nitrogen compounds [mol ha^{-1} a^{-1}]	Sulphur compounds [mol ha^{-1} a^{-1}]
Stomatal uptake flux	14 – 178	32– 153
Epicuticular deposition	0 – 404	58– 485
Annual precipitation	286 – 547	136– 218
Total nutrient input	300 –1129	226– 856
Total growth demand	2211 –2680	67– 81
Percentage of supply [%]	13.6– 42.1%	337–1057%
Trunk wood growth demand	724 – 823	22– 24
Percentage of supply [%]	41.4– 72.9%	1027–3567%

than there is concomitant resupply of nitrogen by wet deposition of NO_3^- plus NH_4^+ (cf. Table 4). Consequently, chronic wet deposition of nitrogen alone presently still not even compensates the N_{org} demand for harvested trunk wood in central Germany. The N_{org} demand of trunk bark – if it is also harvested (not decomposing in the field) – is even as yet disregarded. Thus, considering time ranges up to centuries, there is no excess nitrogen burden via «acid rain» on managed spruce ecosystems, but instead a nitrogen deficit if epicuticular and stomatal NO_2 deposition (= «direct» NO_2 effects; cf. Fig. 4) would be disregarded. Only the wet deposition of sulphur exceeds the S_{org} demand for trunk growth by roughly about one order of magnitude (Table 4). Table 5 balances the total nitrogen and sulphur uptake of spruce ecosystems (stomatal uptake, epicuticular deposition, precipitation) at six sites in central Germany. Results are compared (i) with the total annual N_{org} and S_{org} demand of all growing tissues in spruce stands and (ii) with the N_{org} and S_{org} loss via harvesting of trunk wood. The total nitrogen deposition rate at six spruce stands in central Germany supplies only (13.6 to 42.1) % of the total N_{org} demand of whole spruce trees, and (41.4 to 72.9) % of the N_{org} demand only of trunk wood growth (Table 5). If compared with absolute clean air (and N-free precipitation water), then this additional supply percentage would correspond to an excess nitrogen supply into unmanaged forests (trunk harvesting absent). But this conclusion is usually not sound, since the correct nitrogen balance of managed spruce ecosystems must at first compare the aboveground input and output of nitrogen in order to get a remainder that burdens the soil below. Similar to Table 4, in Table 5 there is no massive nitrogen, but only a massive sulphur burden of managed spruce ecosystems, which exceeds the annual S_{org} growth demand of entire spruce trees by (337 to 1057) % – or by (1027 to 3567) % = (10 to 35-fold) if compared with the S_{org} loss via harvesting of trunk wood.

SO_2 detoxification pathways in spruce needles

As yet, nitrogen and sulphur balances of native spruce stands have been presented (ecological aspect). Now, the nitrogen and sulphur budget of spruce needles after stomatal NO_x or SO_2 uptake must be considered (physiological aspect). At ambient concentrations NO_2 is essentially assimilated to N_{org} compounds, but not oxidized to accumulating NO_3^- (cf. Tischner et al., 1988; cf. Kaiser et al., 1993). The situation is different concerning chronic SO_2-pollution at ambient concentrations. After stomatal uptake and hydration of SO_2, sulfurous acid (H_2SO_3) and finally (bi)sulfite anions are formed in the needle mesophyll. In the longterm, (bi)sulfite is either assimilated to reduced sulphur compounds, or it is oxidized to sulphate anions yielding sulphuric acid. Even in the Erzgebirge (ca. 30 nPa SO_2 Pa^{-1}) only 1% of the SO_2 flux taken up via open stomata is balanced by re-emission of H_2S (Kindermann et al., 1995 a, b). Other reduced volatile sulphur compounds (CS_2, COS etc.) are not detectable products of reductive SO_2 detoxification in spruce needles. Also the reductive assimilation of SO_2 to organic sulphur compounds S_{org} is an unimportant SO_2 detoxification pathway: Fig. 5 indicates that there is only a weak and small positive correlation of the total S_{org} content of spruce needles (harvested in the Erzgebirge) with the sulphate content of these needles (cf. Slovik et al., 1996a). The intercept of the shown linear regression (Fig. 5) is close to the S_{org} content of needles harvested at sites with rather clean air and small SO_4^{2-} contents in needles (25.2 mmol S_{org} kg^{-1} needle DW in Würzburg; cf. Table 1). The sulphate content of spruce needles correlates with observed ambient SO_2 concentrations (Kaiser et al., 1993; Hüve et al., 1995; Slovik et al., 1995). The slope of the linear regression function in Fig. 5 equals (0.103 ± 0.058) mol S_{org} mol^{-1} SO_4^{2-}. Thus, 9.3 % (= 100 %·0.103/1.103) of the total accumulating sulphur is reduced sulphur S_{org} and 90.7% (= 100 %·1.000/1.103) of the total accumulating sulphur is sulphate. Clearly, SO_4^{2-} is the

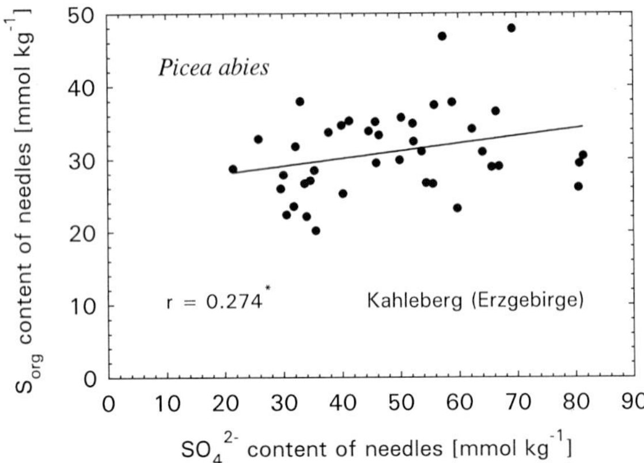

Fig. 5: Correlation of the organic sulphur content S_{org} in ca. 3.3 year old spruce needles (= oldest available needles) versus the sulphate content of these needles from massively damaged spruce trees growing on the summit of the Kahleberg (Erzgebirge, East Saxony, FRG); n = 41 trees; the regression equation equals $S_{org} = (0.103 \pm 0.058) \cdot SO_4^{2-} + (26.0 \pm 3.0)$; The weak correlation is significant (error probability $P<5\%$; r = 0.274*); data after Slovik et al. (1996c).

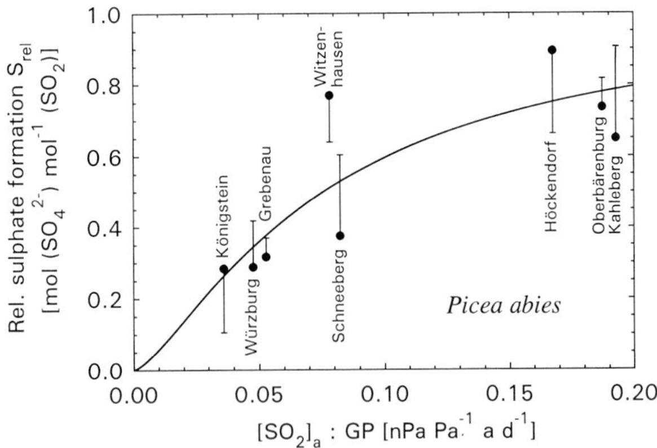

Fig. 6: The relative sulphate formation S_{rel} [mol (SO_4^{2-}) mol^{-1} (SO_2)] depends on the ratio $[SO_2]_a$: GP, since a high SO_2 pollution and a short trunk growth period GP [d a^{-1}] synergistically promotes sulphate accumulation, vice versa; data points after Slovik et al. (1995); error bars are obtained using the Gauß law of error propagation. The indicated eight sites are located in central Germany (Hessen: Königstein, Grebenau, Witzenhausen; Northern Bavaria: Würzburg, Schneeberg; Eastern Saxony: Höckendorf, Oberbärenburg, Kahleberg); cf. detailed site characterization after Slovik et al. (1995). The solid line is a least square fit to shown data.

dominant detoxification product in spruce needles after chronic SO_2-pollution if acute needle damage by SO_2 is still largely absent. Massive S_{org} accumulation is evident only in visually SO_2-damaged spruce needles (cf. Grill and Esterbauer, 1973; Grill et al., 1979; Grill et al., 1980). In Fig. 6 the percentage S_{rel}, [mol SO_4^{2-} mol^{-1} SO_2] of in spruce needles accumulating SO_4^{2-} [mmol SO_4^{2-} kg^{-1} needle DW a^{-1}] per annual stomatal SO_2 dose [mmol SO_2 kg^{-1} needle DW a^{-1}] is plotted versus the ratio of the observed ambient SO_2 concentration $[SO_2]_a$ in air [nPa SO_2 Pa^{-1}] to the length of the growth period GP [d a^{-1}] (cf. Slovik et al., 1995). At sites with a low ratio $[SO_2]_a$: GP, i.e. at sites with clean air and long growth periods at lower elevations (= high tree growth rates and S_{org} demand rates) in Königstein, Würzburg and Grebenau (all in central Germany), only S_{rel}, ≈ 0.3 mol SO_4^{2-} mol^{-1} SO_2 (≈ 30 % of the low stomatal SO_2 uptake rate) accumulate as sulphate, while – apparently – 70 % of the annual (small) SO_2 dose yields sulphur that can be exported via the phloem sap. In spruce needles growing in the mountains at high ambient SO_2 concentrations in the Erzgebirge and high $[SO_2]_a$: GP ratios in Höckendorf, Oberbärenburg and Kahleberg, (70 to 90) % of the stomatal SO_2 dose accumulates as SO_4^{2-} in spruce needles (cf. Slovik et al., 1995). Under these conditions, there is an apparent limitation of phloem loading with sulphur compounds that forces the accumulation of SO_4^{2-} in spruce needle vacuoles. In the mountains with short growth periods GP [d a^{-1}] there is *grosso modo* only a restricted number of days per year with open stomata and efficiently working bast (phloem) elements. The function of phloem cells depends on species-dependent minimum temperatures (cf. Gamalei et al., 1994), that must be exceeded within the time interval GP. Wiersma (1963) reports a minimum temperature of +6 °C (daily mean) as the best ecological representative for the quantification of annual (trunk) growth periods GP (i.e. with largely open stomata and with dominant assimilate transport flow in the phloem sap).

SO_2-dependent vacuolar cation sequestration

Gymnospermal needles accumulate more than 99 % of the analyzed sulphate not in the mesophyll apoplast (cf. Polle et al., 1994), but in vacuoles (cf. Kaiser et al., 1989). Since there is no pH-decrease of needle homogenates even in spruce needles originating from the SO_2-polluted Erzgebirge (Börtitz, 1969; Kaiser et al., 1991), generated vacuolar H_2SO_4 (cf. Kaiser et al., 1993) must have been neutralized in the field by $H^+ \Leftrightarrow Cat^+$ exchange at the tonoplast level. Correlation analysis data (Cat^+ vs. SO_4^{2-} contents of needles) indicate that potassium is the main vacuolar cation that neutralizes sulphate in Würzburg and in the Erzgebirge (Table 6). The relative stoichiometric importance of potassium ranges from $TP_{rel}(K^+)$ ≈ 0.59 Eq K^+ Eq^{-1} SO_4^{2-} in Würzburg to $TP_{rel}(K^+)$ ≈ 0.82 Eq K^+ Eq^{-1} SO_4^{2-} in the Erzgebirge (Kahleberg; cf. Table 6). Other significantly correlating cations (Mg^{2+} and Zn^{2+} or Mn^{2+} and Al^{3+}) play a minor role. Interestingly, there is no correlation between the Ca^{2+} and the SO_4^{2-} content in spruce needles at both sites (cf. Table 6). This has physicochemical reasons, since the oxalate content of spruce needle vacuoles is too high at the vacuolar pH value and the ionic strength of the cell sap for the coexistence of soluble $CaSO_4$ and extractable oxalic acid without vacuolar precipitation of Ca-oxalate (cf. Slovik et al., 1996c). Fig. 7 shows the theoretical maximum contribution $TP_{rel}(Ca^{2+})$ [Eq Ca^{2+} Eq^{-1} SO_4^{2-}] to the vacuolar SO_4^{2-} neutralization in spruce needle vacuoles. Even at SO_2 polluted sites in the Erzgebirge there is no significant contribution of Ca^{2+} to the neutralization of observed SO_2-dependent SO_4^{2-} accumulation (cf. Fig. 6). Still, there is concomitant accumulation of Ca^{2+} and SO_4^{2-} in ageing spruce needles, but both accumulation rates are not causally interrelated:

Table 6: Vacuolar cation sequestration stoichiometrics TP_{rel} [Eq Cat Eq^{-1} SO_4^{2-}] for different cation species as observed in the field in Würzburg (Botanical Garden, Lower Franconia, h = 200 m a. s. l., 5 to 10 nPa SO_2 Pa^{-1}) or on the summit of the Kahleberg (Erzgebirge, Ore mountains, h = 905 m a. s. l., ca. 30 nPa SO_2 Pa^{-1}). Blanks (–) indicate that this cation species did not correlate with the SO_4^{2-} content of needles at this site. There is no SO_4^{2-}-dependent Ca^{2+}, Na^+, Fe^{2+}, or Cu^{2+} accumulation at both sites (but a needle age-dependent SO_4^{2-} and Ca^{2+} accumulation); data after Slovik et al. (1996 c).

Cation	Würzburg (Bavaria) [Eq Cat Eq^{-1} SO_4^{2-}]	Kahleberg (Saxony) [Eq Cat Eq^{-1} SO_4^{2-}]
K^+	0.589±0.260	0.818±0.166
Mg^{2+}	0.360±0.096	–
Zn^{2+}	0.051±0.005	–
Mn^{2+}	–	0.105±0.053
Al^{3+}	–	0.077±0.024
Ca^{2+}	–	–
Σ Cat^+	1.000±0.311	1.000±0.176

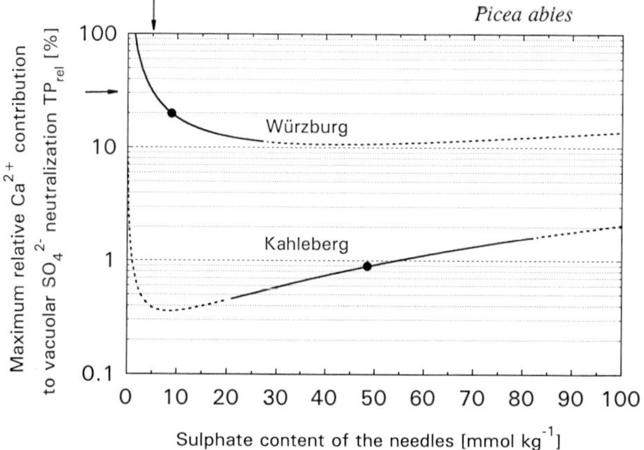

Fig. 7: Maximal relative contribution of Ca^{2+} to the vacuolar SO_4^{2-} neutralization that is possible without vacuolar precipitation of insoluble Ca-oxalate crystals. This theoretical maximum percentage TP_{rel} (Ca^{2+}) [mol Ca^{2+} mol SO_4^{2-}] was calculated based on (i) the pK_a values and amount of soluble oxalic acid in needles, (ii) on the pH value of needle homogenates (≈ pH of vacuolar sap) and (iii) on the solubility product $K_{sol} = 2.56 \cdot 10^{-9}$ $(kmol\, m^{-3})^2$ of Ca-oxalate. Using the Debye-Hückel-Onsager theory, K_{sol} was recalculated for ionic strengths of needle extracts; cf. Slovik et al. (1996 c). Solid lines indicate the interval of occurring SO_4^{2-} concentrations in needle samples from Würzburg and the Kahleberg (Erzgebirge). Dashed lines are extrapolations. Filled symbols indicate the mean sulphate contents in needles at both sites. Arrows indicate that at needle sulphate contents below 5 mmol SO_4^{2-} kg^{-1} needle DW TP_{rel} (Ca^{2+}) rapidly exceeds 30 % and reaches 100 % at ca. 2 mmol SO_4^{2-} kg^{-1} needle DW. Only below ca. 5 mmol SO_4^{2-} kg^{-1} needle DW it is possible to consume Ca^{2+} for SO_4^{2-} neutralization purposes in needle vacuoles without precipitation of Ca-oxalate.

There is no correlation of Ca^{2+} vs. SO_4^{2-} if pseudocorrelations, caused by pooled analysis data from all needle age classes (cf. Hüve et al., 1995), are thoroughly avoided (cf. Slovik et al., 1996 c).

Trace gas-dependent cation leaching from spruce canopies

The relative composition of counterions of SO_4^{2-}, that leak from spruce canopies in the field as statistically depending on ambient SO_2 concentrations in air, is given in Table 7. In contrast to vacuolar cation data given in Table 6, Ca^{2+} compensates ca. 40 % (eq.) of SO_4^{2-}, and is therefore the dominant cation that is «extracted» from spruce canopies after chronic SO_2 pollution («epicuticular» SO_2 absorption). Thus, ca. 40 % of the SO_2-dependent SO_4^{2-} concentration increment in the spruce canopy throughfall is gypsum ($CaSO_4$). Potassium contributes only by about 26 % to the stoichiometric cation balance of leaching SO_4^{2-} from spruce canopies (Table 7). Al^{3+}, Mg^{2+} and Fe^{3+} play only a minor role. Since H_3O^+ concentrations in the canopy throughfall did not correlate with ambient SO_2 (or NO_2) concentrations in air, deposited SO_2 must have been fully neutralized by $Cat^+ \Leftrightarrow H_3O^+$ exchange at spruce canopy surfaces or by dissolution of alkaline dust, ash etc. particles deposited onto spruce crown surfaces (Slovik et al., 1996 b). Concerning NO_2, the situation is different (cf. Table 7). With increasing NO_2 concentration in air, leaching of K^+ from spruce canopies is strongly reduced (consumption by nitrogen-fertilized epiphytic mosses, lichens etc.?). The NH_4^+ throughfall rates parallel increasing NO_2 concentrations, but there was no correlation between NH_4^+ leaching rates and ambient SO_2 concentrations, i.e. no codeposition of $(NH_4)_2SO_4$ was identified (Slovik et al., 1996 b). It may be assumed that increased throughfall rates of NH_4^+ at elevated NO_2 concentrations in

Table 7: Results of linear correlation and regression analysis of cation leaching rates L_{ion} from Norway spruce canopies in the field (pooled data since 1984 from six stands in Hessen, FRG) vs. concomitantly measured annual means of SO_2 or NO_2 concentrations in air [nPa Pa^{-1} = ppb]. Only significant correlations are shown. The regression intercept L_o [mmol Cat kg^{-1} needle DW a^{-1}] estimates the background leaching rate in «clean» air (extrapolation to zero). The regression slope L'_{ion} [mEq Cat kg^{-1} needle DW a^{-1} (nPa gas $Pa^{-1})^{-1}$] is the specific trace gas-dependent canopy leaching increment. Slopes and intercepts are given ± standard errors (SE). Calculated percentage data (± SE) balance the ion stoichiometry of L'_{ion} and therefore estimate the «mean» ionic composition of trace gas-dependent cation leachates from spruce canopies in the field; data after Slovik et al. (1996 b; cf. Balázs, 1991).

Ion [–]	Correlation coefficient	Intercept L_o [mmol kg^{-1} a^{-1}]	Slope L'_{ion} [mmol kg^{-1} a^{-1} ppb^{-1}]	Percentage [%]
Sulphur dioxide				
Ca^{2+}	0.435*	3.52± 3.96	1.058±0.547	39.6± 20.5
K^+	0.497*	18.85± 4.30	0.680±0.297	25.5± 11.1
Al^{3+}	0.764****	0.05± 0.53	0.525±0.111	19.7± 4.2
Mg^{2+}	0.404*	0.25± 1.26	0.308±0.174	11.5± 6.5
Fe^{3+}	0.629**	0.04± 0.15	0.099±0.031	3.7± 1.2
ΣCat^+	–	–	2.670±0.784	100.0± 29.4
Nitrogen dioxide				
K^+	0.475*	48.69±10.05	−1.895±0.906	−200.7± 96.0
NH_4^+	0.642**	−13.38± 7.00	1.711±0.616	181.2± 65.2
Mn^{2+}	0.484*	−1.03± 2.78	1.074±0.501	113.8± 53.1
Zn^{2+}	0.562*	−0.35± 0.14	0.054±0.024	5.7± 2.5
ΣCat^+	–	–	0.944±1.205	100.0±127.6

air are caused by the reduction of at first generated HNO_2 (or HNO_3) to NH_4^+ (NO_x → $HNO_{2(3)}$ → NH_4^+ by epiphytic plants?), but this question still remains open.

NO_x-dependent cation demand

Stomatal uptake of NO_x (NO_2 and NO) causes an additional cation demand if both, (i) dilution of cations (NO_x consumed for growth), and (ii) increase of the $N_{org}:Cat^+$ ratio (NO_x not consumed for growth) are to be avoided in the longterm. There is an additional cation demand of about 0.2 mol K^+ mol^{-1} NO_x and ca. 0.06 mol Mg^{2+} mol^{-1} NO_x, which is only slightly dependent on the site elevation (cf. Table 8). Data are given on a whole tree basis.

Trace gas-dependent potassium and magnesium depletion

The NO_x-caused cation demand (Table 8) can be compared with the SO_2-caused cation demand (Table 6) regarding also the relative SO_4^{2-} formation percentage S_{rel} per SO_2-uptake flux (cf. Fig. 6). Results of this comparison are given in Table 9, which summarizes the «necessary» NO_2 and SO_2 concentrations in air (annual means) that cause – after stomatal uptake – the complete consumption of the entire annual K^+ or Mg^{2+} budget of spruce trees. Our data (Table 9) indicate that (70 to ca. 140) nPa SO_2 Pa^{-1} (= ppb) are necessary for a complete SO_2-dependent K^+ consumption (vacuolar K_2SO_4 formation) if K^+ contents of spruce tissues are similar to the Würzburg situation (cf. Table 1), but about (180 to ca. 340) nPa NO_2 Pa^{-1} are «necessary» for a full K^+ depletion. SO_2 and NO_2 equivalents for the complete depletion of the annual Mg^{2+} uptake from the soil range from 90 to ca. 200 nPa SO_2 Pa^{-1}, respectively 180 to ca. 320 nPa NO_2 Pa^{-1} (Table 9). Thus, the chronic impact of SO_2 on the cation budget of spruce trees is higher than that of chronic NO_2 pollution (acute damage omitted).

Table 8: NO_x-dependent K^+ and Mg^{2+} demand for growth of entire spruce trees [mol Cat mol^{-1} NO_x] after complete assimilation of the stomatal NO_2 and NO uptake in the needle mesophyll ($NO_x \rightarrow N_{org}$). Both, (i) cation dilution after growth stimulation by NO_x, and (ii) an ascending N_{org}: Cat ratio in spruce tissues (if growth stimulation is absent) is to be avoided in the longterm. Data are amplitudes from young trees (t ≈ 5 years; first value) up to old trees (t ≈ 120 years; second value). Results are based on growth scenaria (Slovik, 1996) and tissue contents of healthy spruce trees (Tab. 1).

Site elevation [m a.s.l.]	Potassium [mol K^+ mol^{-1} NO_x]	Magnesium [mol Mg^{2+} mol^{-1} NO_x]
10 m	0.200–0.195	0.057–0.065
1000 m	0.201–0.198	0.056–0.063

Table 9: «Necessary» NO_2 and SO_2 concentrations in air (annual means) that cause – after stomatal uptake – a 100 % competition with the annual K^+ or Mg^{2+} uptake from the soil if stimulation of cation uptake – e. g. after fertilization or liming – is absent in the field. Data are amplitudes from young trees (t ≈ 5 years; first value) up to old trees (t ≈ 120 years; second value) if the nutritional status of spruce trees approximates the tissue contents shown in Tab. 1 (= Würzburg situation).

Site elevation [m a.s.l.]	SO_2 [nPa Pa^{-1} = ppb]	NO_2 [nPa Pa^{-1} = ppb]
Potassium		
10 m	≥140– 95	≥340–220
1000 m	≥140– 70	≥340–180
Magnesium		
10 m	≥200–130	≥320–220
1000 m	≥200– 90	≥320–180

Table 10: The necessary $NO_2:SO_2$ pollution ratio [nPa Pa^{-1} (nPa $Pa^{-1})^{-1}$] for the entire K^+ or Mg^{2+} depletion of ageing spruce trees (after Tab. 9) is a relative measure of the $SO_2:NO_2$ toxicity ratio at equal ambient concentrations in air (common target: annual K^+ or Mg^{2+} budget of spruce trees). Data are amplitudes from young trees (t ≈ 5 years; first value) up to old trees (t ≈ 120 years; second value).

Site elevation [m a.s.l.]	Potassium, K^+	Magnesium, Mg^{2+}
	[nPa Pa^{-1} (nPa $Pa^{-1})^{-1}$]	
10 m	2.00–2.40	1.60–1.65
1000 m	2.10–2.60	1.75–1.90

Relative chronic phytotoxicity of SO_2 and NO_2

Based on the impact of SO_2 and NO_2 on the K^+ and Mg^{2+} budget of entire spruce trees, the relative chronic phytotoxicity of these two trace gases can be compared (Table 9). Ratios of ambient $NO_2:SO_2$ pollution can be calculated, which would cause the same absolute impact on the cation budget of spruce trees (Table 10). For the K^+ budget, the relative phytotoxicity of SO_2 is 2.0 to 2.6 times higher than an equally high NO_2 concentration in air. Concerning Mg^{2+}, the SO_2 toxicity is 1.6 to 1.9 times higher than equally high NO_2 concentrations in air. Interestingly, chronic NO_2-pollution affects the Mg^{2+} budget relatively more than the K^+ budget, but SO_2 is chronically more phytotoxic than NO_2 at all site elevations [m a.s.l.] and in any spruce tree age class.

Discussion

Stomatal uptake of the trace gases SO_2 or NO_x (NO_2 + NO) is of course not accompanied by an «equivalent» stomatal cation flux. Thus, irrespective of reductive or oxidative SO_2 or NO_x detoxification in needle mesophyll cells, the stomatal SO_2 or NO_x supply must induce an additional cation demand that finally must be supplied from the soil. The quantification of this additional cation demand depends on the growth rate of spruce tissues in the field (Slovik, 1996), on their nitrogen, sulphur and cation contents (Table 1) and on the relative participation of oxidative (SO_2 → SO_4^{2-}) versus reductive (SO_2 → S_{org}) SO_2 detoxification pathways (cf. Fig. 5 and 6). In the absence of cation fertilization or liming, this additional cation demand can be supplied only from fertile soils, but not at cation-depleted stands. Hence, chronic SO_2 and NO_2 pollution can cause mineral deficiency symptoms and they may induce the depression of canopy and root growth rates, since root bast and needles contain the highest absolute K^+ and Mg^{2+} contents (cf. Table 1). Consequently, reduction of root growth (→ mineral and water deficiency) and needle growth (→ thinning of the canopy structure) can save cations e.g. for vacuolar sulphate neutralization demands. Still, the rationale of «spruce decline» proposed here is a tentative attempt to explain observed symptoms in the field (cf. Anonymus, 1989; Anonymus, 1993). NO_2 usually occurs in too low concentrations in spruce forests (only 8 to 16 nPa NO_2 Pa^{-1}; cf. Slovik et al., 1996 a) for a massive induction of the here proposed causal rationale. The expected impact of stomatal NO_2 uptake on the N_{org} budget of spruce trees lies within the natural scattering of N_{org} analysis data (cf. Table 1). In contrast, the occurring SO_2-pollution is high enough at some exceptional sites (e.g. in the Erzgebirge; ca. 30 nPa SO_2 Pa^{-1} = ppb) for a massive competition mainly with the K^+ budget of spruce. But even then there must be a combination of infavourable environmental conditions for the induction of «canopy thinning» symptoms that exceed the natural dynamics of ca. ± 20 % of the stock of needles in healthy canopies. Mainly the cumulative combination of (i) high SO_2 concentrations in air, (ii) short growth periods in the mountains, (iii) poor K^+ or Mg^{2+} depleted soils and (iv) concomitant acid precipitation (→ K^+ or Mg^{2+} leaching into the ground water) will synergistically enhance reversible canopy thinning symptoms. The here proposed rationale can explain moderate canopy thinning symptoms, but spruce decline is not a generally occuring phenomenon in central Europe. The role of wet deposited nitrogen as a dominant cause of «spruce decline» is obscure (cf. Table 4), since observed annual nitrogen wet deposition rates usually do not exceed nitrogen loss rates via harvesting of trunk wood. Exceptions, e.g. in The Netherlands, may be admitted, but usually spruce trees will not «grow to death» after excess nitrogen supply (cf. Schulze, 1989). Nevertheless, «epicuticular» NO_x-deposition in natural spruce ecosystems may trigger harmful consequences if (i) herbaceous plants and microorganisms (incl. pathogens) stimulate their growth and (ii) protein-rich (phytophage) insects, deers etc. increase their population densities (kg ha^{-1}). All these competitive organisms consume K^+ and Mg^{2+}, which – at least transiently – will become unavailable for spruce trees. Most affected by chronic K^+ and Mg^{2+} depletion is the growth and turnover of those spruce tissues, which show high K^+- and Mg^{2+} contents (roots, needles, cf. Table 1). Thus, reduced needle and root growth rates are expected, while trunk wood growth may be largely unaffected, since its K^+ and Mg^{2+} content is low (cf. Table 1). Cation deficiency and canopy thinning symptoms of less «fit» ecosystem components – here of spruce trees – may be the consequence if there is a soil-defined shortage of K^+ or Mg^{2+} on a hectare basis. It is a well known phenomenon that N-fertilized grass-lands change their plant species composition (→ ecosystem drift). Additionally, biodiversity will be reduced. It may be expected that similar longterm effects must occur also in N-burdened unmanaged spruce forests, but investigation of such time-consuming processes on an ecosystem level is a difficult task.

Acknowledgements

I am grateful to Prof. Dr. U. Heber and Prof. W. M. Kaiser (Würzburg) for stimulating discussions and advice. This work has been performed within the research program of the Sonderforschungsbereich SFB 251 (TP B1) of the University of Würzburg. Support by the Deutsche Forschungsgemeinschaft is gratefully acknowledged.

References

Anonymus: Dritter Bericht des Forschungsbeirats Waldschäden/ Luftverunreinigungen, FBW. Karlsruhe: Kernforschungszentrum (ISSN 0931 – 7805), 611 pp. (1989).

– Waldzustandsbericht der Bundesregierung – Ergebnisse der Waldschadenserhebung – 1993. Bonn (Germany): Referat Öffentlichkeitsarbeit und Besucherdienst of the Government of the Federal Republic of Germany (1993).

Balázs, Á.: Niederschlagsdeposition in Waldgebieten des Landes Hessen. Ergebnisse von den Meß-Stationen der «Waldökosystemstudie Hessen». Forschungsberichte Hessische Forstliche Versuchsanstalt, Vol. *11,* 168 pp, Hann. Münden: Hessisches Ministerium für Landesentwicklung, Wohnen, Landwirtschaft, Forsten und Naturschutz (1991).

Börtitz, S.: Physiologische und biochemische Beiträge zur Rauchschadenforschung. 11. Mitteilung: Analysen einiger Nadelinhaltsstoffe an Fichten unterschiedlicher individueller Rauchhärte aus einem Schadgebiet. Archiv Forstwesen *18* (2), 123–131 (1969).

Gamalei, Y. V., A. J. E. van Bel, M. V. Pakhomova, and A. V. Sjutkina: Effects of temperature on the conformation of the endoplasmic reticulum and on starch accumulation in leaves with the symplasmic minor-vein configuration. Planta *194,* 443–453 (1994).

Grill, D. and H. Esterbauer: Cystein und Glutathion in gesunden und SO_2-geschädigten Fichtennadeln. Eur. J. Forest Pathol. *3,* 65–71 (1973).

Grill, D., H. Esterbauer, and U. Klösch: Effect of sulfur dioxide on glutathione in leaves of plants. Environ. Pollution *13,* 87–194 (1979).

Grill, D., H. Esterbauer, M. Scharner, and Ch. Felgitsch: Effect of sulfur-dioxide on protein-SH in needles of *Picea abies.* Eur. J. Forest Pathology *10,* 263–267 (1980).

Hüttl, R. F.: Die Nährelementversorgung geschädigter Wälder in Europa und Nordamerika. Habilitation thesis, Freiburger Bodenkundliche Abhandlungen (ISSN 0344-2691) *28,* 440 pp. (1991).

Hüve, K., A. Dittrich, G. Kindermann, S. Slovik, and U. Heber: Detoxification of SO_2 in conifers differing in SO_2-tolerance. A comparison of *Picea abies, Picea pungens* and *Pinus sylvestris.* Planta *195,* 578–585 (1994).

Kaiser, W. M., E. Martinoia, G. Schröppel-Meier, and U. Heber: Active transport of sulfate into the vacuole of plant cells provides halotolerance and can detoxify SO_2. J. Plant Physiol. *133,* 756–763 (1989).

Kaiser, W. M., A. P. M. Dittrich, and U. Heber: Sulfatakkumulation in Fichtennadeln als Folge von SO_2-Belastung. In: Reuther, M., M. Kirchner, and K. Rösel (eds.): 2. Statusseminar der Projektgruppe Bayern zur Erforschung der Wirkung von Umweltschadstoffen (PBWU), München-Neuherberg, 4.–6. Febr. 1991, GSF-Bericht *26/91,* 425–438 (1991).

– – – Sulfate concentrations in Norway spruce needles in relation to atmospheric SO_2: a comparison of trees from various forests in Germany with trees fumigated with SO_2 in growth chambers. Tree Physiology *12,* 1–13 (1993).

Kandler, O.: Vierzehn Jahre Waldschadensdiskussion. Szenarien und Fakten. Naturwiss. Rundschau *11/94,* 419–425 (1994).

KINDERMANN, G., K. HÜVE, S. SLOVIK, H. LUX, and H. RENNENBERG: Emission of Hydrogen Sulfide by twigs of conifers – a comparison of Norway spruce (*Picea abies* (L.) Karst.), Scots pine (*Pinus sylvestris* L.) and Blue spruce (*Picea pungens* Engelm.). Plant and Soil *168/169*, 421–423 (1995a).

– – – – – Is Emission of Hydrogen Sulfide a dominant factor of SO_2 detoxification? A comparison of Norway spruce (*Picea abies* (L.) Karst.), Scots pine (*Pinus sylvestris* L.) and Blue spruce (*Picea pungens* Engelm.) in the Ore mountains. Phyton (Austria), *35*(2), 255–267 (1995b).

KÖRNER, CH., J. PERTERER, CH. ALTRICHTER, A. MEUSBURGER, S. SLOVIK, and M. ZÖSCHG: Ein einfaches empirisches Modell zur Berechnung der jährlichen Schadgasaufnahme von Fichten- und Kiefernadeln. Allgemeine Forst- und Jagdzeitung *166/1*, 1–9 (1995).

METTE, H.-J. and U. KORELL: Richtzahlen und Tabellen für die Forstwirtschaft. Berlin: Deutscher Landwirtschaftsverlag (1989).

POLLE, A., M. EIBLMEIER, and H. RENNENBERG: Sulphate and antioxidants in needles of Scots pine (*Pinus sylvestris* L.) from three SO_2-polluted field sites in eastern Germany. New Phytologist *127*, 571–577 (1994).

SCHMIDT-VOGT, H.: Die Fichte. Ein Handbuch in zwei Bänden. Vol. *II/1* Wachstum, Züchtung, Boden, Umwelt, Holz. Paul Parey, Hamburg und Berlin, 563 p. (1986).

SCHOBER, R.: Ertragstafeln wichtiger Baumarten. 3rd edn., Frankfurt am Main: J. Sauerländer Verlag, 166 p. (1987).

SCHULZE, E.-D.: Air pollution and forest decline in a spruce (*Picea abies*) forest. Science *244*, 776–783 (1989).

SLOVIK, S.: Tree Physiology. In: HÜTTL, R. F. and W. SCHAAF (eds.): Nutrition of Ecosystems. Magnesium deficiency in forest ecosystems, ca. 120 pp. Kluwer Academic Publishers, Dordrecht (The Netherlands), in press (1996).

SLOVIK, S., A. SIEGMUND, G. KINDERMANN, R. RIEBELING, and Á. BALÁZS: Stomatal SO_2 uptake and sulfate accumulation in needles of Norway Spruce stands (*Picea abies*) in Central Europe. Plant and Soil *168/169*, 405–419 (1995).

SLOVIK, S., A. SIEGMUND, H. W. FÜHRER, and U. HEBER: Stomatal uptake of SO_2, NO_x and O_3 by spruce crowns (*Picea abies*) and canopy damage in Central Europe. New Phytologist, 132, in press (1996a).

SLOVIK, S., Á. BALÁZS, and A. SIEGMUND: Canopy throughfall of *Picea abies* (L.) Karst. as depending on trace gas concentrations. Plant and Soil *178*(2), 255–267 (1996b).

SLOVIK, S., K. HÜVE, G. KINDERMANN, and W. M. KAISER: SO_2-dependent cation competition and compartmentation in Norway spruce needles. Plant, Cell and Environment, in press (1996c).

TISCHNER, R., A. PEUKE, D. L. GODBOLD, R. FEIG, G. MERG, and A. HÜTTERMANN: The Effect of NO_2 fumigation on Aseptically Grown Spruce Seedlings. J. Plant Physiol. *133*, 243–246 (1988).

VANSELOW, K.: Krone und Zuwachs der Fichte in gleichaltrigen Reinbeständen. Forstwiss. Cbl. *70*, 705–719 (1951).

WELLBURN, A. R.: Why are atmospheric oxides of nitrogen usually phytotoxic and not alternative fertilizers? Tansley Review No. 24, New Phytologist *115*, 395–429 (1990).

WENTZEL, K.-F.: Hypothesen und Theorien zum Waldsterben. Forstarchiv *56*, 51–56 (1985).

WIERSMA, J. H.: A new method of dealing with results of provenance tests. Silvae Genet. *12*, 200–205 (1963).

ZECH, W. and E. POPP: Magnesiummangel, einer der Gründe für das Fichten- und Tannensterben in NO-Bayern. Forstwiss. Cbl. *102*, 50–55 (1983).

ZIEGLER, I.: The effect of SO_2 pollution on plant metabolism. Residue Rev. *56*, 79–105 (1975).

ZÖTTL, H.: Possible causes of forest damage in Germany. CONCAWE Report *86/61*, 55–70 (1986).

[illegible] Pollution Stress on Forest Ecosystems [illegible] the Northern Tyrolean Limestone Alps

[illegible]tute of Air Pollution Research and Forest Chemistry, Seckendorff-Gudent-Weg 8, A-1130 Vienna, Austria

Received June 24, 1995 · Accepted September 25, 1995

Summary

Forest ecosystems, especially forest ecosystems in mountain regions, are under the influence of complex stressors. Apart from the natural, mostly climatic factors, they include human influences, such as air pollution, tourism, accessibility of forests, and various types of management for silvicultural purposes, wildlife, and grazing.

Air pollution stresses to the forest ecosystems of the Northern Tyrolean Limestone Alps are caused by local emitters and long-distance transport. Apart from the gaseous components SO_2, NO_x, and O_3, they contain predominantly protons, nitrogen compounds, and heavy metals.

The air pollution stress of the Northern Tyrolean Limestone Alps was described by using the results of all-Austrian investigations (Austrian Bioindicator Grid), air monitoring data from the Federal Province of the Tyrol, model calculations, and surveys from the area of Achenkirch. The monitoring data were evaluated in accordance with effect-related limiting values or Critical Loads.

The evaluation of the air monitoring data for the area under investigation showed that the stress patterns differed significantly depending on the absolute altitude above sea level or the relative altitude above the valley floor. In valleys, effect-related limiting values for SO_2 and NO_2, which were established to protect sensitive plant species, are exceeded in conglomerations, and ozone has been a permanent stressor especially at higher altitudes. The influence of accumulative pollutants (e.g. heavy metals) on the soil and the fact that these pollutants have increased and reached the sensitive zone of the timber line is particularly important in respect of soil-biological processes. While acid inputs can easily be buffered by the bedrock of the Northern Tyrolean Limestone Alps, the determined nitrogen inputs constitute a potential risk to forest ecosystems.

Key words: Northern Tyrolean Limestone Alps, ozone, acid depositions, nitrogen depositions, heavy metals.

Abbreviations: F_o = basic fluorescence; F_m = maximum fluorescence; $F_v = F_m - F_o$; NTLA = Northern Tyrolean Limestone Alps; VOC's = volatile organic compounds.

Introduction

Forest ecosystems are influenced by manifold stressors, which, in consequence of the differences in the landscapes of the Alpine area, cause very different effects. The Alps are stressed by climatic or natural influences, avalanches, wildlife, and grazing animals (damage by browsing, rubbing, stepping and bark peeling) as well as by human influences (accessibility of forests for tourism and silviculture etc.). Moreover, also gaseous pollutants (SO_2, NO_x, ozone, VOC's) and wet depositions (nitrogen or acid inputs; Herman and Stefan, 1992; Smidt and Gabler, 1994; Smidt et al., 1995 a) cause stress. Altogether, the negative impacts produce stress patterns of daily, seasonal, and altitudinal dynamics (Bolhar-Nordenkampf, 1989). The degrees of crown defoliation, which were determined in the course of the crown condition surveys,

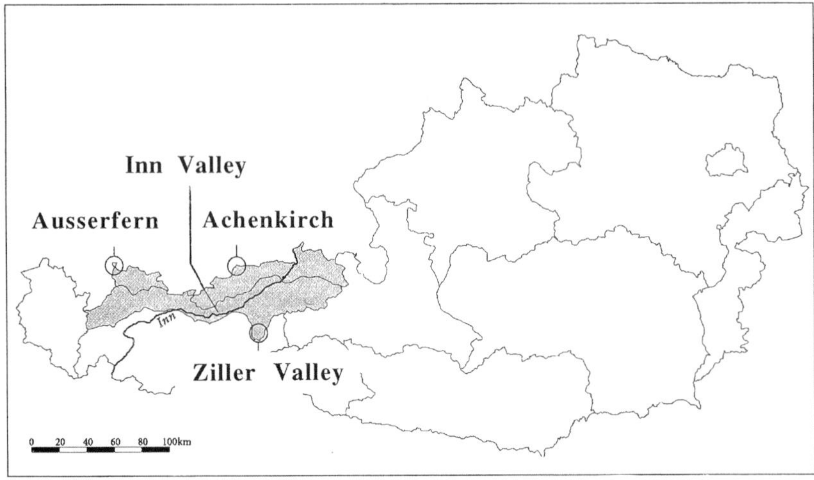

Fig. 1: The area of the Northern Tyrolean Limestone Alps (Austria) and the locations of the investigation areas.

showed that the NTLA (Northern Tyrolean Limestone Alps) are particularly seriously affected in this respect (Office of the Provincial Government of the Tyrol, 1995; Fig. 1).

To be able to protect Alpine forest ecosystems one must know the relevant stressors, as only exact knowledge of the stressors allows to work out useful measures. In the following, the air pollution stresses of the NTLA are to be characterized (growth areas 2.1 and 4.1 according to Kilian et al., 1994) and to weigh the individual air pollutants in respect of their risk. Stress-physiological and ecophysiological aspects of air pollution will equally be considered.

Methods

To assess the risk connected with the air pollution to which the NTLA are exposed, extensive data is required which must be linked and interpreted on an interdisciplinary basis. In this connection, it is important to distinguish between results from large-scale investigations and those from special plots (several hundred km^2) or small-scale investigations (not exceeding few km^2). The results of the investigations that are relevant in respect of air pollution were evaluated according to criteria, but also model approaches were applied.

Large-scale investigations and the Tyrolean air monitoring grid

For statements about a growth area, results from integrated monitoring, including those from sample grids, are available. However, such results inform only about visible characteristics: The Austrian Forest Condition Survey («Waldzustandsinventur») and the Forest Damage Monitoring System («Waldschaden-Beobachtungssystem») are restricted to the determination of the degrees of crown defoliation and yellowing; the Forest Inventory («Waldinventur») is spezialized on mechanical damage, and the Forest Development Plan («Waldentwicklungsplan») to again other criteria (damage of protective areas due to wildlife). The Bioindicator Grid applies foliar analyses to determine particulary the impact of SO_2 and nutrient deficiencies/imbalances, and the Forest Soil Monitoring System («Waldboden-Zustandsinventur») uses soil analyses to investigate the «condition» of the soils. Data from the Tyrolean air monitoring grid provide additional information. With the help of model approaches some of the results of these investigations can also be used for conclusions in respect of larger areas (Table 1).

Investigations in parts of the Tyrol and small-scale investigations

Research approaches carried out in the course of interdisciplinary forest research (the projects «Zillertal Altitude Profile» and «Achenkirch Altitude Profiles») provide further data which, due to the complex procedures involved, cannot be determined in the frame of large-scale investigations. These approaches include special investigations of larger areas (Tyrolean Transit Study; remote sensing by means of IR false-colour photography). Results from physiological bioindication and from measurements of the photosynthesis activity, for instance, allow conclusions about physiological («hidden») injuries (e.g. Smidt and Herman, 1994; Smidt et al., 1994 a; Table 2) and therefore are the basis of an early diagnosis of tree stress (Bolhar-Nordenkampf and Lechner, 1989).

Evaluation criteria for plants

Concentrations of gaseous air pollutants are evaluated by means of effect-related limiting values for SO_2, NO_2 and O_3; for ozone, also provisional critical doses were defined (Table 3). It must be noted, however, that an evaluation using effect-related limiting values of pollutants does not allow any conclusion in respect of the plant-physiologically relevant (effective) doses that have actually been taken up. For the time being, the significance of VOC's and heavy metal input cannot be clearly evaluated, as there exist no effect-related limiting values for plants.

Impact of pollutants

The impact of SO_2 is determined by measuring the sulphur content of spruce needles, the heavy metal impacts (e.g. of Pb and Cd) by measuring the concentrations in spruce needles, spruce barks, mosses and fungi (Table 3).

Pollutant inputs

H, S and N inputs can be assessed by using the UN-ECE Critical Loads (1988). For model calculations of Critical Loads of nitrogen, various forest soil maps (Knoflacher and Piechl, 1992; Knoflacher and Loibl, 1993) were used in addition; for that of the depositions, the evaluation and area-related interpolation of results (Kovar and Puxbaum, 1991; Kovar et al., 1991). In scanning plots the sensitivity of receptors to certain pollutants is determined and compared to the input of agents. With the help of models the ion concentration in the soil water can be estimated, taking into account the ion deposi-

Table 1: Large-scale investigations* and other research activities for the monitoring of pollutant stress and their significance in respect of forest damage.

Investigations	Significance in respect of forest damage	References
Large scale investigations		
Austrian Bioindicator-Grid*	Sulfur-impact; content/deficiency of main nutrients in Norway spruce needles and nutrient imbalances resp.	Stefan, 1995
Austrian Forest Condition Survey (1983–1989)*	Crown condition	Federal Forest Research Centre, 1991
Austrian Forest Damage Monitoring System (since 1989)*	Crown condition, wood increment, integrating pollutant measurements	Federal Forest Research Centre, 1992 a
Austrian Forest Soil Monitoring System*	Soil analyses; sensibility against proton input, anthropogenous heavy metal input	Federal Forest Research Centre, 1992 b
Other Research		
Air pollutant and deposition measurements in the Tyrol, Tyrolean Transit Study	SO_2, NO_x, O_3, dust; wet field depositions	Provincial Government of the Tyrol (1991, 1987–1995); Smidt and Gabler, 1994; Smidt, 1994
Pesticides in rainwater		Lorbeer et al., 1994
Chlorocarbons in spruce needles and in the ambient air		Plümacher and Schröder, 1994, Schröder, 1994
Aerial photography Ausserfern	Crown condition (crown thinning)	Gärtner, 1993 a, 1993 b
Large scale investigations (not pollutant-related)		
Austrian Forest Investory*	Damage by wildlife (bark peeling, browsing etc.), logging damage; regeneration, endangering of protected forests	Schadauer, 1995
Forest Development Plan*	Leading functions and endangering factors (damage by deer; no respect to air pollutants)	Austrian Federal Ministry of Agriculture and Forestry, 1991

tion, the weathering, the erosion, and the withdrawal by the vegetation (UN-ECE, 1991; Table 3).

Plant-physiological parameters

Also the evaluation of the stress impacts by means of plant-physiological parameters requires criteria for their interpretation in respect of stress and pollution impacts. The established criteria are, among other things, based on long-term surveys from «clean air areas» of the Tyrol («Zillertal Altitude Profile», «Achenkirch Altitude Profile»). Altogether, they allow concrete conclusions about the total stress, the effects being clear even before there are visible symptoms («early stress diagnosis»). However, for an interpretation of these data one must know the natural ranges to be able to differentiate between air-pollution-related and natural effects of stress (Table 4).

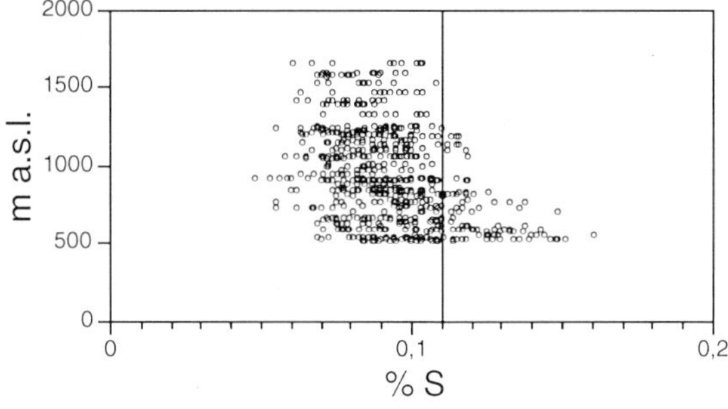

Fig. 2: S-content in spruce needles of the Austrian Bioindicator Grid (needle year 1, 1983–1993; n = 66; limiting value (Austrian Federal Law Gazette, 1984: 0.11% S).

Results and Discussion

Results of large-scale investigations

Sulphur content in spruce needles (Austrian Bioindicator Grid, 1983–1992): Up to an altitude of 1,200 m a.s.l. the sulphur content of the one-year-old spruce needles examined on the sample plots (n = 66) of the NTLA exceeded the limiting values established under the Second Regulation against Forest-Damaging Air Pollutants, which indicates air pollution impacts by SO_2 (Stefan, 1995; Fig. 2). As, contrary to the emission situation in Austria, the sulphur concentrations of the spruce needles of the NTLA has not significantly decreased since 1983, these values might also be a result of large-scale sulphur transports or imports from abroad (Federal Environmental Agency, 1994); partly also the uptake of sulphate from the ground may be an explanation of that phenomenon.

Heavy-metal content of the soils (Austrian Forest Soil Monitoring System, Inventory 1992): The Pb and Cd concentrations of the soil that were determined in the course of the Forest Soil Monitoring System (Federal Forest Research Centre, 1992 a) showed amounts between 50 and 120 mg of

Table 2: Methods of investigation and their significance in respect of the stress of forest trees (forest stands)* indicates pollution.

Method/Parameters	Significance in respect of the stress	References
Measurements of air pollutants*		
Concentrations of SO_2, NO_2, O_3 and organic compounds in the ambient air; concentrations of ions and pesticides, H-, S- and N-input; basic meteorological data	Excesses of effect-related limiting values (Critical Levels) and/or Critical Loads	Office of the Provincial Government of the Tyrol, 1990–1993; Smidt et al., 1995 a; Lorbeer et al., 1994; Kovar and Puxbaum, 1991; Smidt, 1994 b; Fliri, 1975
Accumulative bioindication*		
Content of pollutants in spruce needles, heavy metals in bark, lichens, mosses and fungi	Impact of SO_2 and S-input (needle analyses), proton input (lichen mapping), heavy metal input	Herman, 1991; Hofmann, 1992 and 1994; Stefan, 1995; Zechmeister, 1995
Reactive bioindication* (whole organism level)		
Assessment of crown thinnings	Unspecific reference to injury (crown thinning)	Federal Forest Research Centre, 1991
Aerial photography	Unspecific reference to injury (crown thinning)	Gärtner, 1993 a, 1993 b
Reactive, biochemical and physiological bioindication*		
Content of antioxidative compounds in spruce needles and pigments, lipoids and enzymes, proteins; measurement of the photosynthetic capacity	Alterations of the concentrations of antioxidative compounds: increased influence of photooxidants, deviation from «normal» values of lipoids and the chlorophyll fluorescence: reference to physiological condition	Bermadinger-Stabentheiner, 1994; Puchinger and Stachelberger, 1994; Plümacher and Schröder, 1994; Bolhar-Nordenkampf and Götzl, 1992; Lütz and Dodell, 1994
Reactive, cytogenetic bioindication*		
Chromosome aberrations of fine root tips	Various stress factors (e. g. pollutant input)	Müller, 1995
Forest-pathological investigations	Specific biotic and abiotic stress factors	Cech and Tomiczek, 1995
Other investigations		
Soil- and soil water measurements*	Nutrient supply, heavy metal input, nitrate transfer into the groundwater	Mutsch, 1992, 1995; Berger, 1994
Microbial soil investigations*: microbial activity, mycorrhiza	Various stress factors (e. g. pollutant input)	Insam and Rangger, 1995; Rangger and Insam, 1995

Pb/kg of soil and 3 mg of Cd/kg of soil. The accumulation in the upper soil and the increasing concentrations of these two elements with altitude indicate that the soils are effective heavy metal sinks (Mutsch, 1995).

Crown conditions: In the Tyrol the area of over 60-year old stands affected by crown defoliation (total of heavily + moderately + slightly defoliated) and registered in the framework of the Austrian Federal Forest Condition Monitoring, rose by approximately 30 % to more than 4 % until 1988; since then it remained almost constantly at 40 %. Compared to 1984, the damage determined in the NTLA significantly increased in 1993; on the other hand, the situation improved in the Inn Valley, and in the Central Alps the damage was much less serious (26 %). (In the Northern Alps of the Tyrol 48 % of the forested area showed defoliation symptoms in 1993, while the Inn Valley and the Central Alps were significantly less affected [36 % and 26 %, respectively]; Office of the Provincial Government of the Tyrol, 1995).

Concentrations of SO_2, NO_2, and ozone (values from 1990–1993): In consequence of the reduction of the air pollution during the past decade the annual SO_2 averages are relatively low in the Tyrol; the maximum annual average of 25 μg/m³ set by IUFRO (1975/1978) was not at all exceeded and the maximum daily and half-hour averages were exceeded in congested areas (Fig. 2) only. Maximum annual averages, half-hour averages, and daily averages of NO_2 were exceeded in valleys with heavy traffic (particularly in the Inn Valley) between approx. 500 m and 600 m a.s.l. (Smidt et al., 1995 a). The NO_2 concentrations, especially the maximum daily and half-hour averages, decrease relatively heavily with the altitude above the valley floor (Fig. 3).

In the Alpine area the long-term ozone concentrations have proved to significantly increase with altitude until the timber line (Smidt and Gabler, 1994). As the vegetation period limiting value (60 μg/m³) may be significantly exceeded even in the case of annual ozone means <60 μg/m³, permanently exceeded values and, consequently, a long-term risk for sensitive plants are likely not only at altitudes above 1,000 m. Also if the daily means (reference value again 60 μg/m³), which reached 180 μg/m³ in the NTLA (the maximum daily means also increase significantly up to the timber line), are taken as a basis, a risk must be assumed. The maximum half-hour mean, however, rarely exceeds 220 μg/m³ and therefore is clearly below the effect-related limiting value. A two-dimensional representation of the ozone distribution, prepared by means of a model calculation, showed that dur-

Table 3: Selected evaluation criteria for plant (forest ecosystem) stress.

Parameter	Limiting values/Critical Levels, Critical Loads	Source/Legal regulations*/Criteria**
Pollutant concentrations (-doses)		
SO_2-concentrations (µg/m^3)	Year: 25 April–October: 50 (24 h)/70 (0.5 h) November–March: 100 (24h)/150 (0.5 h)	IUFRO (1975/1978)** Austrian Federal Law Gazette (1984)*, Austrian Academy of Sciences (1975)**
NO_2-concentrations (µg/m^3)	Year: 30, 24 h: 80, 0.5 h: 200	Austrian Academy of Sciences (1987)**
O_3-concentrations (µg/m^3)	Vegetation period (9.00–16.00, April–October): 60; 8 h: 60; 0.5 h: 300	Austrian Academy of Sciences (1989)*
O_3-doses	1 h mean values >40 ppb (preliminary): 10 ppm·h	UN-ECE (1994)**
H-, S- and N-input		
Proton input	>2 kg/ha.a (carbonate)	UN-ECE (1988)**, Mutsch and Smidt (1994)**
Equivalent sulfur input	>32 kg/ha·a (carbonate)	UN-ECE (1988)**
Nitrogen input	10–12$^{1)}$/>20$^{2)}$ kg/ha·a (coniferous forests) <15 kg/ha·a (deciduous forests)	UN-ECE (1988)**
Pollutant content in spruce needles and spruce bark (pollutant input)		
S-content in spruce needles	0.11% (needle year 1) 0.14% (needle year 2)	Austrian Federal Law Gazette (1984)*
Pb in spruce needles	>4 mg Pb/kg	Knabe (1984)
Pb and Cd in spruce bark	>8 mg Pb/kg/>0.60 mg Cd/kg	Herman (1991)

1 Nutrient imbalances depending on the Mg- and Ca-concentration and on the nitrification rate of the soil.
2 Shifting of species of the herbaceous layer and shrubs depending on the uptake by the trees and the base saturation of the soil.

Table 4: Plant-physiological parameters (Herman and Smidt, 1994).

Paramters	Ranges	References
Spruce needles		
Photosynthetic capacity (Fv/Fm (Fo)	Fv/Fm (Fo) «Normal»: 0.85 (0.18) Lower limit of the natural variation: 0.72 (0.14) Strong, reversible disturbances: >0.60 (0.28) Strong, also structural disturbances >0.30 (0.10)	Bolhar-Nordenkampf and Götzl, 1992
Thioles (needle year 1)	0.25–0.7 µmol/g d.w.	Bermadinger-Stabentheiner, 1994
Ascorbic acid (needle year 1)	1.1–3.9 mg/g d.w.	
Peroxidase activity (needle year 1)	5–50 units/g d.w.	
Glutathione reductase (needle year 1)	0.6–1.5 units/g d.w.	
Total chlorophyll (needle year 1)	60–1500 µg/g d.w.	
Glutathion-S-transferase	0.1–1.0 nkat/mg protein	Plümacher and Schröder, 1994
Lipoids	Upper/lower limit (900 m–1000 m a.s.l.)	Puchinger and Stachelberger, 1994
Fatty acids (total, % d.w.)	<0.6/>0.8	
Unsaturated fatty acids (%)	<50/>60	
Linoleic acid (%)	<25/>30	
Oleic acid (%)	<8/>8	
Steroles (% d.w.)	<0.25/>0.35	
Triglycerides (% d.w.)	<0.15/>0.25	
Fine roots of spruce	Cytogenetic site index (CSI)	Müller, 1995
Cytogenetic bioindication of chromosomal aberrations	>1.4	

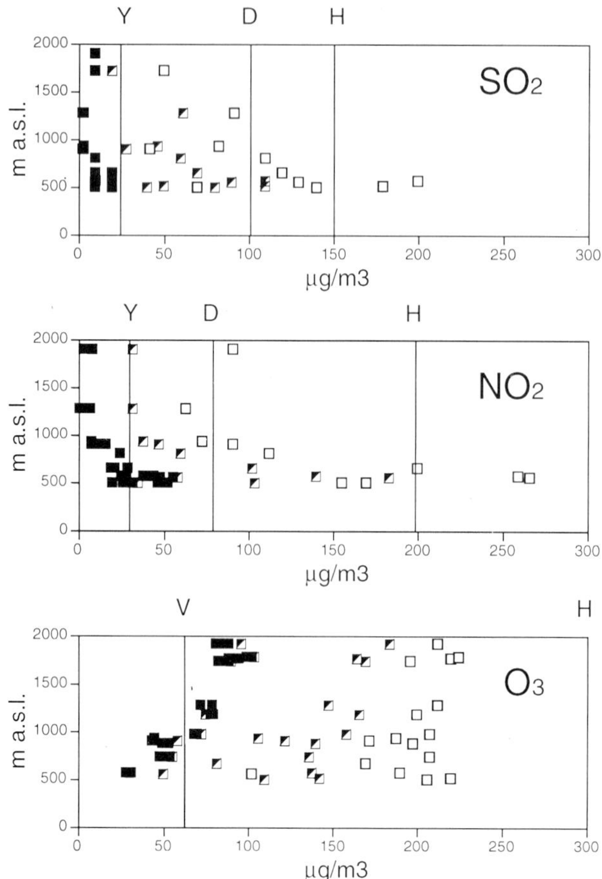

Fig. 3: Annual mean values, maximum daily mean values and maximum half-hour mean values in the NTLA (1990–1993). ■ Annual mean values, ◩ Maximum daily mean values, □ Maximum half-hour mean values*; Y: Limiting value (annual mean value), D: Critical Level (daily mean value), H: Critical Level (half-hour mean value), V: Critical Level (vegetation period mean value).
The absolute half-hour maximum of 660 µg/m³ SO_2 was registered in Brixlegg (Inn Valley, station Innweg).

ing an ozone episode of summer 1991 the Tyrol and Vorarlberg were the federal provinces with the highest concentrations (Loibl et al., 1991).

H, S, and N inputs (1985–1993): The proton inputs amount to up to 0.55 kg/ha.a in the NTLA (three monitoring stations in Northern Tyrol [Office of the Provincial Government of the Tyrol]; four stations in the area of Achenkirch [Smidt, 1994 b]); they do not degrade the soil, as the latter are in most cases influenced by carbonate. This applies also to the protons deposited with the sulfate (the S inputs in the field are equally far below the Critical Loads of 32 kg/ha.a for calcareous soils; UN-ECE, 1988).

The nitrogen inputs may, however, constitute a potential risk: The Critical Loads of the growth areas of the NTLA are between 5 and 20 kg N/ha.a, just as the inputs measured in the field (11 stations, including three in Loisachtal/Bavaria). As, according to Glatzel (1990), the demand of the forest ecosystems for N input is also in other parts of Austria exceeded by about 7 kg/ha.a, it can be assumed that the available amounts of nitrogen exceed by far the need and cause a number of unfavorable effects (nutrient imbalances, eutrophication, increased pollution of the groundwater with nitrate). As dry and occult deposition (fog; Winkler and Pahl, 1993) have to be added to the above values, the latter are minimum approximate values of the actual input.

Results from parts of the Tyrol

Inventories by aerial photographs (Ausserfern, 1990)

Since 1989 inventories by aerial photographs with false colour photography (IR) have been carried out in more than 30 areas of Austria (Gärtner, 1993 a). A comprehensive catalogue of characteristics is used to define the crown conditions. From the Ausserfern (district of Reutte), an area in the northwest of the Tyrol with a large share of protection forest, results from 560 km² are available. The evaluations showed that 43 % of the forested area there (a little less than total area of the NTLA; s.a.) must be considered affected. In this area, forests stocking on dolomite are much more seriously damaged than those on limestone sites; and protection forests are clearly more seriously affected than commercial forests. Damage in this area increases with altitude.

Tyrolean Transit Study (1991)

Investigation of specific air pollutants: Apart from SO_2, NOx, and O_3, also exhaust gases from cars and a number of organic components and heavy metals were investigated along the transit route between the Inn Valley and the Brennero; the traffic-related air pollution stress was described (Office of the Provincial Government of the Tyrol, 1991); along the highway, the stress through NO_x, heavy metals, and de-icing road salts proved to be very high and the ozone limiting values were very often exceeded. The investigation of *lichen communities* from various Tyrolean valleys and transects (Hall/Schwaz; Brixlegg/Wörgl; Matrei/Brennero, Zell am Ziller/Mayrhofen, and Achenkirch) shows that especially the areas of the valleys are stressed by acid components and that also the exposure plays a role in this connection, while ozone does not have any influence on the lichen flora (Office of the Provincial Government of the Tyrol, 1991; Hofmann, 1992 and 1994). Both the evaluation of the air analyses and lichen mappings and of other investigations carried out in that area (regarding climate, VOC's, soil) have shown that the area of the transit routes is particularly heavily polluted, particularly at sites located below, or not higher than, about 200 m above the valley floor, where the situation is even more critical due to frequent inversions.

Moss analyses

In the framework of investigations of mosses which were used as accumulation indicators in an all-European Grid, several Tyrolean sites and 5 altitude profiles (2 of them in the Tyrol) were sampled and the depositions of heavy metals were determined through the chemical analysis of 5 moss species. Deposition rates were found to increase with altitude (Zechmeister, 1995), which is in accordance with the results from the Forest Soil Monitoring System.

Results of small-scale investigations («Achenkirch Altitude Profiles»)

Based on long-term, interdisciplinary, stress-physiological investigations in the Ziller Valley (Bolhar-Nordenkampf, 1989; Smidt and Herman, 1992), which is located south of the Inn Valley in Achenkirch, the work has since 1990 been continued at two differently exposed altitude profiles close to the German border (920–1,690 m a.s.l. and 930–1,420 m a.s.l.) (Smidt et al., 1994). Apart from comprehensive investigations in view of an exact characterization of the respective areas and their situation in respect of pollution (Smidt and Gabler, 1994), several methods of physiological bioindication have been applied (Table 2).

Some of the below methods allow clear statements about the effects of pollution (accumulation indicators, soil analyses). Others, such as the cytogenetic bioindication, lipid analyses, and measurements of the photosynthetic activity, provide rather unspecific information about stresses. In many cases, however, a synopsis allows the interpretation of the results in respect of air pollution effects.

The soil analyses carried out under the Forest Soil Monitoring System include both altitude profiles of the project area. At both altitude profiles man-made heavy-metal emissions (Pb, Cd) were found to have enriched. The altitude-related heavy-metal accumulation in mosses, which were equally found in the project area, corresponded to the results of other altitude profiles.

In the Tyrol, results of stand deposition measurements are available only for the area of Achenkirch (Office of the Provincial Government of the Tyrol, 1989; Berger, 1995). Compared to other Austrian plots, the throughfall of the Christlum Profile (2 plots) showed the lowest nitrogen inputs (12 kg/ha.a).

Methods of bioindication

* *Heavy-metal analyses of fungi*

Compared to data from European (Pb) or Alpine (Cd) investigation areas both the medians of Pb concentrations in the collected fungi (10 µg/g d.w.) and that of the Cd content (4.8 mg/g d.w.) were very high. *Cortinarius* species are particularly well suited as Cd, Zn, and Cu accumulation indicators (Peintner, 1995).

* *Cytogenetic bioindication*

Chromosome aberrations in fine roots of mature and juvenile spruce can be due to air pollution stress (Müller, 1995). The highest impact in Achenkirch was found between 1,000 m and 1,400 m; at lower altitudes the results corresponded well with those of other bioindication methods (e.g. determination of the sulphur content ot needles).

* *Physiological bioindication*

Components of the antioxidative system counteract the formation of radicals and are produced even in cases of increased stress. «Normal» modifications with higher altitude, which are also connected with increasing pollution and photooxidative stress, are increased concentrations of thiols and ascorbic acid as well as an increased peroxidase activity. Synthetic pigments and α/β-carotene, however, decrease with altitude (Bermadinger-Stabentheiner, 1994).

Lipoid components fulfil vital tasks in the metabolism and in the building of membranes. The amounts in which they occur (lipid patterns) allow conclusions in respect of the total stress of spruce; in some cases even conclusions about the type of stress are possible. The total content of fatty acids increases with the nitrogen content of the soil and decreases as a result of dry stress; combined occurrence of SO_2 and NO_2 causes the degradation of shorter-chain fatty acids (e.g. lauric acid); the influence oxidants have on the pattern of fatty acids is, however, not yet clear (Puchinger and Stachelberger, 1994; Puchinger, pers. comm.).

Information about influences to the photosynthetic apparatus can be drawn from investigations of the measurements of the chlorophyll fluorescence, of the oxygen production through photosynthesis, and from the analysis of the thylakoid proteins (Lütz and Dodell, 1994); in the area of Achenkirch no significant impairment was found. Also the long-term field investigations carried out in the Ziller Valley indicate only a slight impairment of the photosynthesis, but neither in spruce (Wieser and Havranek, 1993) nor in larch (Havranek and Wieser, 1993) a significant reduction of the photosynthesic capacity through ambient ozone was found.

Organic components of spruce needles and depositions

Investigation of depositions

The volatile chlorocarbons found in the spruce needles of Achenkirch contained approximately 1 µg/kg of volatile compounds (VOC's) and up to 7.5 mg/kg of trichloroacetic acid, which proves the ubiquitarian presence of this family of agents even in «clean-air-areas» (Plümacher and Schröder, 1994). In Achenkirch the amounts of pesticides in wet depositions (atrazine, lindane, etc.) were relatively high, but can still be considered unrisky (Smidt et al., 1995 b). According to the present state of knowledge the importance of VOC's will mostly lie in its indirect effects; if any, direct effects are found in congested areas only (Smidt, 1994 a).

Conclusion

The air pollution stress in the NTLA can be described by using large-scale monitoring methods and data from the air monitoring grid. For more exact data about particularly seriously endangered or affected forest stands, detailed investigations (e.g. evaluation of aerial photos from IR colour films) provide valuable additional information. Forest damage research on an interdisciplinary basis is realistic in rather small areas only; altitude profiles, however, allow conclusions about corresponding altitudinal ranges.

When evaluating air pollution, one must take into consideration that, just as the characteristic physiological values, pollutants are subject to seasonal changes and that the stress pattern of forest trees is therefore under constant change. Moreover, also altitude and site influence the stress pattern significantly; the deposition of protons, sulphur and nitrogen is, however, differentiated.

Gaseous air pollutants: Compared to the applicable limiting values, primarily the valleys of the NTLA within inversion zones are stressed by SO_2 and NO_2; they become less critical with increasing altitude, while the importance of ozone increases with altitude.

In the case of the wet depositions, nitrogen has caused the most significant problems: The amounts of nitrogen registered in the two growth areas of the NTLA may have a negative effect on forest sites as the latter are adjusted to low nitrogen reserves. The Critical Loads are likely to be exceeded in large areas of the NTLA, which is problematic not so much because of the proton inputs (the bedrock contains carbonate and can therefore easily buffer protons), but because of the risk of an «imbalanced fertilization» and, consequently, nutrient imbalances, which have already been determined in the framework of the Austrian Bioindicator Grid (Stefan and Herman, 1995). The fact that also in the NTLA the supply with nitrogen (determined in the spruce needles of the Bioindicator Grid) is predominantly poor, does not contradict the above, as it is probably less a result of the lack of nitrogen than of a reduced uptake.

The *heavy metal inputs* into forest ecosystems, especially those that do not have a function as trace elements, must be considered a potential risk. This is true even if they are not taken up by the roots, like the Pb, as they can negatively influence the mycorrhiza and the microflora of the soil and can also accumulate via the metabolism, for instance through fungi. However, no certain results about concrete effects of the heavy-metal inputs into forest ecosystems of the NTLA are available yet.

The methods of the *physiological bioindication* have contributed significantly to the early diagnosis of stress (Bolhar-Nordenkampf, 1989). At the example of the components of the antioxidative system these methods were used to prove altitude-related stress patterns in altitude profiles. With the help of air monitoring and other results (e.g. electron microscopy, lichen mapping) these patterns were interpreted as the results of the different SO_2 and ozone stress (Bermadinger-Stabentheiner, 1994). Also chemical investigations of lipids indicate that the «physiological condition» may be explained with air pollution stress (Puchinger and Stachelberger, 1994).

Apart from the methods mentioned in the present contribution, also others are suited to determine the «tree-physiological condition» (e.g. tree ring analysis, mycorrhiza investigations; biometric methods, wax structure and composition of wax; measurement of water absorbing capacity (cf. Bolhar-Nordenkampf and Lechner, 1989); however, if used by themselves, these methods do not allow clear conclusions as to the effects of air pollutants and therefore have to be interpreted synoptically.

Altogether, one can say that in connection with air pollution the most important stressors of the forest ecosystems of the NTLA are the combined effects of SO_2 and NO_x (especially in valleys), ozone and heavy-metal depositions (increasing with increasing altitude), and nitrogen inputs. The stress patterns, which may be visible (e.g. defoliation of tree tops, yellowing) or cause physiological damage (modification of certain, stress-indicating foliar agents), differ greatly according to location and altitude and are influenced also by other factors.

References

Austrian Academy of Sciences: Luftqualitätskriterien Schwefeldioxid. Federal Ministry of Environment, Youth and Family (ed.) (1975).

– Luftqualitätskriterien Stickstoffdioxid. Federal Ministry of Environment, Youth and Family (ed. (1987).

– Luftqualitätskriterien Ozon. Federal Ministry of Environment, Youth and Family (ed.) (1989).

Austrian Federal Law Gazette (Bundesgesetzblatt) no. 199/1984, 89th copy. Second regulation against air pollutants causing damage to forests (1984).

Austrian Federal Ministry of Agriculture and Forestry (ed.): Waldentwicklungsplan (1991).

Berger, T.: Eintrag und Umsatz langzeitwirksamer Luftschadstoffe in Waldökosystemen der Nordtiroler Kalkalpen. FBVA-Berichte (Federal Forest Research Centre Vienna, Austria, ed.) *87,* 133–144 (1995).

Bermadinger-Stabentheiner, E.: Stress-physiological investigations on spruce trees from the «Achenkirch Altitude Profiles»: Phyton *34,* 97–112 (1994).

Bolhar-Nordenkampf, H. R. (ed.): Streßphysiologische Ökosystemforschung Höhenprofil Zillertal. Phyton *29* (1989).

Bolhar-Nordenkampf, H. and E. Lechner: Synopse streßbedingter Modifikationen der Anatomie und Physiologie von Nadeln als Frühdiagnose einer Disposition zur Schadensentwicklung bei Fichte. Phyton *29,* 255–302 (1989).

Bolhar-Nordenkampf, H. R. and M. Götzl: Chlorophyllfluoreszenz als Indikator der mit der Höhenlage zunehmenden Streßbelastung von Fichtennadeln. FBVA-Berichte (Federal Forest Research Centre, ed.) *67,* 119–132 (1992).

Cech, T. and C. Tomiczek (eds.): Forstpathologische Erhebungen im Gebiet Achental. FBVA-Berichte (Federal Forest Research Centre Vienna, Austria, ed.) *86,* (1995).

Federal Environmental Agency (ed.): Umwelt in Österreich, Daten und Trends 1994 (1994).

Federal Forest Research Centre (ed.): Zusammenfassende Darstellung der Waldzustandsinventur. Mitteilungen der Forstlichen Bundesversuchsanstalt *166* (1991).

– Österreichische Waldbodenzustandsinventur. Mitteilungen der Forstlichen Bundesversuchsanstalt *168/1* and *168/2* (1992 a).

– Österreichisches Waldschadenbeobachtungssystem. FBVA-Berichte (Federal Forest Research Centre Vienna, Austria, ed.) *71* (1992 b).

Fliri, F.: Das Klima der Alpen im Raume von Tirol. Monographien zur Landeskunde, Folge 1.

Gärtner, M.: Luftbildinventur an der FBVA. Österr. Forstztg. 1, *49* (1993 a).

– Außerfern 1990 – Untersuchungsbericht der Luftbildinventur. Österr. Forstztg. *9,* 61 (1993 b).

Glatzel, G.: The nitrogen status of Austrian forest ecosystems as influenced by atmospheric deposition, biomass harvesting and lateral organomass exchange. Plant and Soil *128,* 67–74 (1990).

Havranek, W. M. and G. Wieser: Zur Ozontoleranz der europäischen Lärche (*Larix decidua* Mill.). Forstw. Cbl. *112,* 56–64 (1993).

Herman, F.: Schwermetallgehalte von Fichtenborken als Indikator für anthropogene Luftverunreinigungen. VDI-Berichte *901,* 375–390 (1991).

Herman, F. and K. Stefan: Zusammenschau. FBVA-Berichte (Federal Forest Research Centre, Vienna, Austria, ed.) *67,* 139–147 (1992).

Herman, F. and S. Smidt: Results from the fields of Integrated Monitoring, Bioindicators and Indicator Values for the characterization of the physiological condition of trees. Phyton *34,* 169–192 (1994).

Hofmann, P.: Immissionsbezogene Flechtenkartierung im Zillertal. FBVA-Berichte (Federal Forest Research Centre Vienna, Austria, ed.) *67,* 119–132 (1992).

Hofmann, P.: Pollutant-related mapping of lichens in the area of Achenkirch, Phyton *34*, 71–84 (1994).

Insam, H., H. Rangger, and F. Göbl: Bodenmikrobielle Untersuchungen auf beweideten und nicht beweideten Flächen am Schulterberg. FBVA-Berichte (Federal Forest Research Centre Vienna, Austria, ed.) *87*, 215–220 (1995).

IUFRO (International Union of Forestry Research Organizations) 1975–78: Luftqualitätskriterien.

Kilian, W., F. Müller, and F. Starlinger: Die forstlichen Wuchsgebiete Österreichs. FBVA-Reports (Federal Forest Research Centre Vienna, Austria, ed.) *82* (1994).

Knabe, W.: Merkblatt zur Entnahme von Blatt- and Nadelproben für chemische Analysen. Allg. Forstztg. *39* (1984).

Knoflacher, H. M. and T. Piechl: Critical Loads – Level 2: Flächenbezogene Berechnung und Kartendarstellung von kritischen Belastungsgrenzen und tatsächlichen Belastungen in Österreich. Austrian Research Centre Seibersdorf, ÖFZS–A–2187 (1992).

Knoflacher, H. M. and W. Loibl: Mapping of Critical Loads of nitrogen for Austria – Preliminary results. Austrian Research Centre Seibersdorf, ÖFZS–A–2521 (1993).

Kovar, A., A. Kasper, H. Puxbaum, G. Fuchs, M. Kalina, and M. Gregori: Kartierung der Deposition von SOx, NOx, NHx und basischen Kationen in Österreich. Vienna Technical University, Institute of Analytical Chemistry, Report no. 9/91 (1991).

Kovar, A. and H. Puxbaum: Nasse Deposition im Ostalpenraum. Vienna Technical University, Institute of Analytical Chemistry, Report 14/91 (1991).

Loibl, W., J. Züger, and A. Kopsca: Flächenhafte Ozonverteilung in Österreich für ausgewählte Ozonepisoden 1991. Federal Environmental Agency/Federal Ministry of Environment, Youth and Family, Report UBA-93-071 (1991).

Lorbeer, G., W. Hartl, and R. Kohlert: Determination of trichloroacetic acid in rainwater from Achenkirch and other Austrian sites. Phyton *34*, 57–62 (1995).

Lütz, C. and B. Dodell: Photosynthetic performance of spruce trees from selected research areas in the mountain forests near Achenkirch. Phyton *34*, 127–140 (1994).

Müller, M.: Cytogenetische Bioindikationen an den «Höhenprofilen Achenkirch» – Untersuchungen von Chromosomenaberrationen im Wurzelmeristem der Fichte. FBVA-Berichte (Federal Forest Research Centre Vienna, Austria, ed.) *87*, 169–176 (1995).

Mutsch, F.: Österreichische Waldbodenzustandsinventur. Ergebnisse Waldbodenbericht. Teil 6: Schwermetalle Mitteilungen der Forstlichen Bundesversuchsanstalt *168/2*, 145–188 (1992).

Mutsch, F. and S. Smidt: Durch Protoneneintrag gefährdete Waldgebiete in Österreich. CBl. f. d. ges. Forstwesen *111*, 57–66 (1994).

Mutsch, F.: Schwermetalle im Boden als Immissionsindikatoren auf einem Prallhang im Raum Achenkirch. FBVA-Berichte (Federal Forest Research Centre Vienna, Austria, ed.) *87*, 153–160 (1995).

Office of the Provincial Government of the Tyrol: Zustand der Tiroler Wälder, 1989.

Office of the Provincial Government of the Tyrol: Luftmeßberichte (1990–1993).

Office of the Provincial Government of the Tyrol (ed.): Auswirkungen des Straßenverkehrs auf die Umwelt. Teilbericht Luft, Vegetation, Boden. Bericht an den Tiroler Landtag (1991).

– Zustand der Tiroler Wälder. Bericht an den Tiroler Landtag (1995).

Peintner, U.: Schwermetallgehalte von Basidiomyceten und deren Eignung als Bioindikatoren. FBVA-Berichte (Federal Forest Research Centre Vienna, Austria, ed.) *87*, 161–169 (1995).

Plümacher, J. and P. Schröder: Accumulation of C_1-/C_2-chlorocarbons and trichloroacetic acid and a possible correlation with glutathione S-transferases in conifer needles. Phyton *34*, 141–154 (1994).

Puchinger, L. and H. Stachelberger: Determination of lipids in spruce needles as stress indicators in the «Achenkirch Altitude Profiles». Phyton *34*, 113–126 (1994).

Rangger, A. and H. Insam: Mikrobielle Aktivitäten und Biomasse entlang eines Höhenprofiles in den Nordtiroler Kalkalpen. FBVA-Berichte (Federal Forest Research Centre Vienna, Austria, ed.) *87*, 95–104 (1995).

Schadauer, K.: Beschreibung der Nordtiroler Kalkalpen anhand der Parameter der Österreichischen Waldinventur unter besonderer Berücksichtigung der Gefährdungsfaktoren im Schutzwald. FBVA-Berichte (Federal Forest Research Centre Vienna, Austria, ed.) *87*, 221–230 (1995).

Schröder, P.: Immission, Aufnahme und Entgiftung von Xenobiota in Fichten am Schwerpunktstandort Achenkirch. Interim Report, Fraunhofer Institut and GSF Forschungszentrum für Umwelt und Gesundheit, Munich (1994).

Smidt, S. and F. Herman: Ecosystem studies at different elevations in an Alpine valley. Phyton *32*, 177–200 (1992).

Smidt, S.: Gefährdung von Waldökosystemen durch flüchtige organische Verbindungen. Z. Pfl. Krkh. Pfl.schutz *101*, 423–445 (1994a).

Smidt, S.: Measurements of field depositions in the area of Achenkirch. Phyton *34*, 45–56 (1994b).

– Die Beurteilung von Ozonmeßdaten anhand von Critical Levels der UN-ECE. Bericht 3/1994, Federal Forest Research Centre, Institute of Air Pollution Research and Forest Chemistry (1994c).

Smidt, S. and K. Gabler: SO_2, NO_x and ozone records along the «Achenkirch Altutude Profiles». Phyton *34*, 33–44 (1994).

Smidt, S. and F. Herman: Waldökosystemforschung in inneralpinen Tälern Tirols. Z. Umweltchem. Ökotox. *6*, 203–208 (1994).

Smidt, S., F. Herman, D. Grill, and H. Guttenberger: Studies of ecosystems in the Limestone Alps – «Achenkirch Altitude Profiles». Phyton *34* (1994).

Smidt, S., M. H. Knoflacher, F. Mutsch, K. Stefan, and F. Herman: Belastung der Nordtiroler Kalkalpen durch Schadstoffdepositionen. FBVA-Berichte (Federal Forest Research Centre Vienna, Austria, ed.) *87*, 245–262 (1995a).

Smidt, S., R. Womastek, and G. Lorbeer: Pestizideinträge durch nasse Depositionen. FBVA-Berichte (Federal Forest Research Centre Vienna, Austria, ed.) *87*, 145–152 (1995b).

Stefan, K.: Österreichisches Bioindikatornetz. Ergebnisse der Schwefelanalysen der Probenahme 1993 und Vergleich der Resultate der von 1983–1993 und von 1985–1993 bearbeiteten Probepunkte. Federal Forest Research Centre, Institute of Air Pollution Research and Forest Chemistry, Report BIN-S 106 (1995).

Stefan, K. and F. Herman: Ergebnisse chemischer Nadelanalysen der Jahre 1983–1992 von Fichten aus dem Tiroler Kalkalpin. FBVA-Berichte (Federal Forest Research Centre Vienna, Austria, ed.) *87*, 231–244 (1995).

UN-ECE: Critical Loads Workshop, 19–24 March 1988, Skokloster, Sweden (1988).

– Mapping Critical Loads for Europe. CCE Technical Report No. 1 (1991).

– Critical Levels for ozone. A UN-ECE Workshop Report (J. Fuhrer and B. Achermann, eds.). Schriftenreihe der FAC Liebefeld *16* (1994).

Wieser, G. and W. M. Havranek: Ozone uptake in the sun and shade crown of spruce: quantifying the physiological effects of ozone exposure. Trees *7*, 227–232 (1993).

Winkler, P. and S. Pahl: Eintrag von Spurenstoffen durch Nebel auf Wälder. Proceedings of the Int. Symposium «Stoffeinträge aus der Atmosphäre und Waldbodenbelastung in den Ländern der ARGE ALP und ALPEN ADRIA», Berchtesgaden, 27.–29. April 1993, GSF-Report 39/93, 126–134 (1993).

Zechmeister, H.: Correlation between altitude and heavy metal deposition in the Alps. Environmental Pollution *89*, 73–80 (1995).

J. Plant Physiol. Vol. 148. pp. 296–301 (1996)

Consequences of Air Pollution on Shoot-Root Interactions

Heinz Rennenberg, Cornelia Herschbach, and Andrea Polle

Albert-Ludwigs-Universität Freiburg, Institut für Forstbotanik und Baumphysiologie, Am Flughafen 17, D-79085 Freiburg i. Br., Germany

Received September 24, 1995 · Accepted November 10, 1995

Summary

The impact of SO_2, NO_2 and O_3 on physiological processes in plants and their consequences at the whole-plant level are discussed in the present paper. Ozon interacts with carbon allocation most likely by inhibiting sucrose export. This causes an accumulation of carbohydrates and starch in leaves and results in a reduction of photosynthesis. Thus, O_3-exposure can diminish the availability of photosynthetate for growth and development and result in an increased shoot to root ratio and an overall reduction in biomass. By contrast, SO_2 and NO_2 can act as nutrients. SO_2 affects the sulfate and the organic sulfur pools of the leaves and will cause an enhanced export of sulfur. As a consequence, plants fumigated with SO_2 contain enhanced amounts of reduced sulfur, mainly glutathione, in the roots. Glutathione acts as a signal to control sulfate uptake from the soil and inhibits the process of xylem loading. Apparently, sulfur from atmospheric pollution can interact with the sulfur nutrition of plants. NO_2 may interact with the nitrogen nutrition of plants in a similar way. The absorbed NO_2 is used to synthesize amino acids which are translocated in the phloem to the roots. Since amino acids transported in the phloem can decrease nitrate uptake by roots, it is feasible that nitrogen taken up *via* the leaves can interact with whole nitrogen nutrition of plants as described for sulfur. The significance of SO_2, NO_2 and O_3 in affecting root-shoot interactions, will depend on the availability of defence systems, the size of internal storage pools and the actual growth rate of the plant.

Key words: Ozone, sulfur dioxide, nitrogen dioxide, shoot-to-root interaction, whole plant processes.

Introduction

Tropospheric sulfur dioxide, nitrogen dioxide, and ozone are major air pollutants of national and international concern. Chemically, these air pollutants or their degradation products in plants may directly attack organic materials, or may initiate the formation of radicals, thereby, destroying lipids, proteins, and other constituents of living tissues (Elstner, 1984). These primary reactions of sulfur dioxide, nitrogen dioxide, and ozone occur in above-grounds parts of plants and initiate a sequence of secondary reactions causing profound effects on whole-plant physiology. In addition, sulfur dioxide and nitrogen dioxide are converted to sulfate and nitrate, respectively, enter the xylem-born pools of these macro-nutrients in the leaves, and thus interact with whole-plant sulfur and nitrogen nutrition (Wellburn, 1990; Rennenberg and Polle, 1994). To understand the action of sulfur dioxide, nitrogen dioxide, and ozone on whole-plant physiology, it is essential to trace reactions leading at primary sites of injury eventually to changes in carbohydrate and nutrient allocation. Since it is beyond the scope of this article to review the entire sulfur dioxide, nitrogen dioxide, and ozone literature, we will consider only some recent studies on important intrinsic and environmental factors that modulate responses of plants to these pollutants and illustrate effects at the whole-plant level. Further aspects of physiological effects of air pollutants on plants have recently been reviewed (e.g. Bytnerowicz and Grulke, 1992; De Kok, 1990; Matyssek et al., 1995; Pearson and Stewart, 1993; Sonnewald and Willmitzer, 1992; Wellburn, 1990).

Influx of air pollutants and primary reactions

The diffusive flux of sulfur dioxide, nitrogen dioxide, and ozone into leaves is determined by the concentration gradient between the external gas-phase and the intercellular air spaces of the leaves. The uptake of pollutants depends on a series of resistances, such as the presence of boundary layers, stomatal aperture and density, and leaf internal (mesophyll) resistances (Wellburn, 1990; Runeckles, 1992; Rennenberg and Polle, 1994). The influx of these air pollutants through the cuticle is at least 10^4-times lower than stomatal uptake (Kerstiens and Lendzian, 1989) and, thus, negligible. The mesophyll resistance to ozone influx into leaves is insignificant (Laisk et al., 1989); internal resistances may, however, reduce the diffusive flux of sulfur dioxide and nitrogen dioxide, respectively (Pfanz et al., 1987; Thoene et al., 1991). Still the influx of ozone as well as sulfur dioxide and nitrogen dioxide can mainly be described as a function of stomatal conductance of the leaves and external pollutant concentration (Wellburn, 1990; Runeckles, 1992; Rennenberg and Polle, 1994).

The physico-chemical rules, which govern pollutant uptake by plants, have several important consequences: (1) Plants with high stomatal conductance for water vapor, i.e. many herbaceous plants and deciduous trees such as poplar (200 to 1000 mmol H_2O m^{-2} s^{-1}) may be exposed to up to 20-fold higher effective doses of pollutants than plants with low stomatal conductance such as conifers (50 mmol H_2O m^{-2} s^{-1}) at the same external pollutant concentration (Sen Gupta et al., 1991; Wieser and Havranek, 1993). (2) During hot and dry weather plants will close their stomata to limit water loss (Dobson et al., 1990) and, thereby, may have a reduced risk to suffer from injury by pollutant influx. (3) Plants, which maintain relatively wide stomatal openings at night, are exposed to higher pollutant doses than plants which close their stomata in the dark. It must be emphasized, however, that sulfur dioxide, nitrogen dioxide, and ozone also affect stomatal conductance. It seems that stomata initially tend to reduce gas exchange in response to these air pollutants and at later stages of pollutant impact seem to suffer from disturbed regulation (Black, 1985; Darrell, 1989; Wellburn, 1990).

Once an air pollutant has entered the leaf, it must get dissolved in the aqueous apoplastic matrix and diffuse across the cell wall before itself or its secondary products reach sensitive structures such as the plasma membrane or membrane-localized proteins (Polle and Rennenberg, 1993; Rennenberg and Polle, 1994). Because of its reactivity ozone taken up by the leaf is oxidized entirely, nitrogen dioxide and sulfur dioxide at least partially in the apoplastic matrix (Laisk et al., 1989; Ramge et al., 1993; Rennenberg and Polle, 1994). It has been shown for example that exposure to moderate ozone concentrations induces a range of defence reactions in the apoplastic space, including pathogen-related proteins (Schraudner et al., 1992; Kangasjärvi et al., 1994) and various antioxidative compounds (Bors et al., 1989; Castillo and Greppin, 1988; Polle et al., 1995; Luwe and Heber, 1995). The efficiency of apoplastic defences varies greatly in different plants species and was high in beech and spruce (Luwe and Heber, 1995; Polle et al., 1995) and low in spinach, *Sedum album* and broad bean leaves (Luwe and Heber, 1995; Luwe et al., 1993; Castillo and Greppin, 1988). Apparently, such species-inherent features may play a role in mediating tolerance to air pollutants.

Interaction of ozone with carbohydrate allocation

Though it is unlikely that ozone molecules reach intracellular targets, a common range of physiological responses to moderate ozone concentrations has been reported including (1) accumulation of carbohydrates or starch and (2) a decrease in photosynthesis related to reduction in ribulose-1,5-bisphosphate carboxylase/oxygenase (Rubisco) activity (Darrall, 1989; Farage et al., 1991; Günthard-Goerg et al., 1993; Heath, 1994; Matyssek et al., 1992; Willenbrink and Schatten, 1993; Städler and Ziegler, 1993; Pell and Nagi Reddy, 1991). Data of Eckhardt et al. (1991) suggest that reactive oxy-products of ozone might diffuse to chloroplasts and oxidize SH-groups on Rubisco, thereby marking this protein for proteolytic degradation. On the other hand, there is also evidence that ozone initially injures membrane proteins (Castillo and Heath, 1990). This may affect transport processes resulting in an inhibition of sucrose export and an accumulation of carbohydrates and starch in mesophyll cell and along the veins (Fig. 1) (Matyssek et al., 1992; Günthard-Goerg et al., 1993; Städler and Ziegler, 1993; Wellburn and Wellburn, 1994; Willenbrink and Schatten, 1993). It is now well-known that accumulation of carbohydrate in leaves results in a decreased quantum yield efficiency of photosystem II, a down regulation of photosynthetic electron flux, a decrease in Rubisco activity and bleaching of the affected leaves (Sonnewald and Willmitzer, 1992).

At the whole-plant level a reduced supply with photosynthate has negative effects on the production of storage compounds and on biomass accumulation. Under the influence of ozone a reduction in biomass accumulation has frequently been reported (Darrell, 1989). The formation of side-branches is reduced and lesions in the stem can be observed resulting in changes in plant architecture (Günthard-Goerg et al., 1993; Matyssek et al., 1993). Root respiration decreased (Gorissen and van Veen, 1988) and root growth is generally diminished resulting in an increased shoot-to-root ratio in ozone-treated plants (Winner and Atckinson, 1986). Less rootstocks on one hand, and disturbed regulation of the stomata on the other hand, can render plants more susceptible to drought stress as observed by Maier-Mercker and Koch (1991).

Interaction of sulfur dioxide with whole-plant sulfur nutrition

In contrast to ozone, sulfur dioxide taken up by the leaves is considered to be a nutrient that may contribute significantly to the plant's sulfur requirement in protein synthesis, especially if the sulfur supply in the soil is scarce (Leone and Brennan, 1972; Klein et al., 1978; De Kok, 1990). This view is based on the observation that, as a consequence of apoplastic and symplastic oxidation, exposure of shoots to sulfur dioxide results in an accumulation of sulfate. The absolute

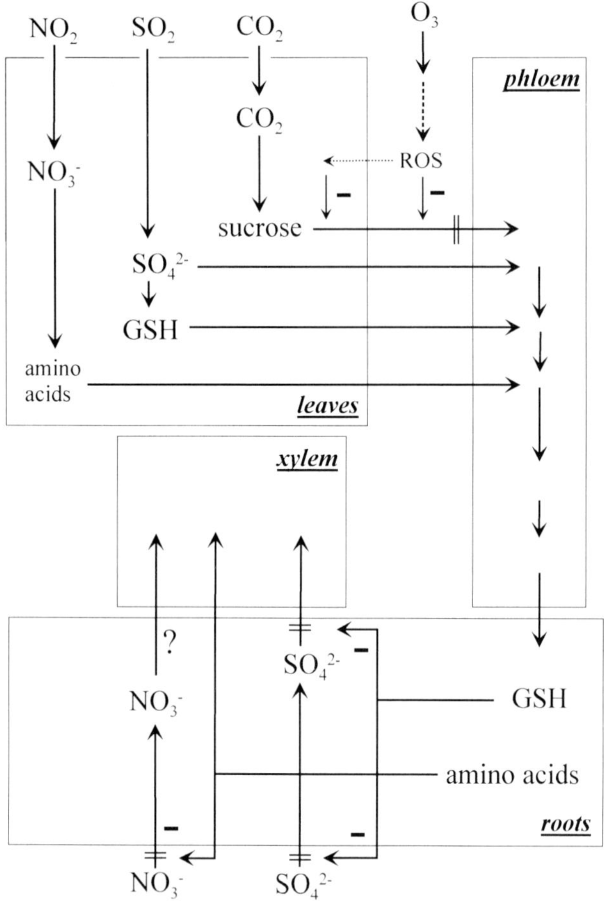

Fig. 1: The influence of atmospheric pollution by O_3, SO_2 and NO_2 on shoot-root interactions in plants. Reactive oxygen species (ROS) released by ozone impact inhibit sucrose export from leaves. Absorbed NO_2 or SO_2 are used to synthesize amino acids or glutathione (GSH), respectively, which are translocated in the phloem to the roots. Amino acids may affect nitrate uptake; glutathione reduces sulfate uptake and the xylem loading process in roots.

amount of sulfate accumulated in the shoot is dependent on the sulfur dioxide concentration applied, the duration of exposure, environmental factors such as temperature, the species analyzed, and its sulfur nutritional status (De Kok, 1990). Intracellular sulfate derived from atmospheric sulfur dioxide may enter a metabolically inactive storage pool, i.e. the vacuole, and a metabolically active pool in the cytoplasm as well (Rennenberg and Polle, 1994). A small but distinct fraction of the cytoplasmic sulfate generated from sulfur dioxide is reduced and incorporated into organic sulfur compounds such as cysteine and glutathione (De Kok, 1990).

Part of the sulfur from the sulfur dioxide absorbed by the shoot is translocated to the root and found in various sulfur fractions of the root (Faller et al., 1970; Faller, 1972 a, b; Garsed and Read, 1977 a, b; Herschbach et al., 1995 a). Transport of sulfur to the root may proceed in reduced and oxidized form, i.e. glutathione and sulfate, which are mobile in the phloem (Rennenberg, 1984; Rennenberg and Lamoureux, 1990; Rennenberg and Herschbach, 1995). In herbaceous plants and in conifers such as spruce, sulfate and glutathione are readily exchanged between phloem and xylem (Biddulph, 1956; Schupp et al., 1992; Schneider et al., 1994). It has, therefore, been suggested that leaves exposed to sulfur dioxide will feed sulfur dioxide derived sulfate and glutathione into circulating sulfur pools of the vascular system and, thereby, will interact with whole-plant sulfur nutrition (Rennenberg and Polle, 1994). The roots may act as a sink of sulfur dioxide derived sulfate and glutathione fed into the vascular system. Apparently, removal of sulfate from a circulating pool by the roots of plants exposed to sulfur dioxide is too small to result in sulfate accumulation in the roots (Rennenberg and Polle, 1994). However, the recent finding of elevated thiol contents in the roots of spinach plants fumigated with sulfur dioxide may be an indication for an enhanced removal of glutathione from a circulating pool by the roots (Herschbach et al., 1995 a). This finding is of particular significance, since glutathione can inhibit external sulfate uptake by the roots as well as its allocation to the shoot at the level of xylem loading (Fig. 1; Herschbach and Rennenberg, 1991; 1994).

Recent fumigation experiments with spinach and tobacco plants provide evidence that fumigation of shoots with sulfur dioxide interacts with whole-plant sulfur nutrition (Herschbach et al., 1995 a, b). In spinach plants fumigated with sulfur dioxide, allocation of sulfate from the roots to the shoot was inhibited at the level of xylem loading irrespective of whether the plants were grown with low, normal, or high sulfur supply (Herschbach et al., 1995 a). Sulfate uptake by the roots was also inhibited upon sulfur dioxide exposure, but only in plants grown with low sulfate supply. Similar results were obtained with tobacco plants (Herschbach et al., 1995 b). It can therefore be assumed that sulfur dioxide interacts with the demand-driven control by the shoot of sulfate nutrition *via* the root. Apparently, xylem loading of sulfate is more susceptible to this mechanism of shoot-to-root control than sulfate uptake by the roots. However, these findings with spinach and tobacco plants cannot be generalized, since phloem-to-xylem exchange of sulfur is not a general phenomenon in plants. For example, in beech, both sulfate and glutathione are transported in phloem and xylem, but a phloem-to-xylem exchange along the stem does not take place (Herschbach and Rennenberg, 1995). Apparently, circulating pools of sulfate and glutathione do not exist in beech trees.

Interaction of nitrogen dioxide with whole-plant nitrogen nutrition

For several reasons it is tempting to assume that – similar to the interaction of sulfur dioxide absorption by the shoot with sulfate uptake and allocation by the root – nitrogen dioxide absorbed by the shoot may interact with nitrate uptake by the root. (1) Nitrogen dioxide absorbed by the leaves undergoes disproportionation to nitrate and nitrite (Lee and Schwartz, 1981) and/or reacts with ascorbate to yield nitrite in the aqueous phase of the apoplastic space (Ramge et al., 1993). The products of these reactions are rapidly transported into the cytoplasm and are incorporated into amino compounds by the cells' general nitrogen metabolism (Nußbaum et al., 1993; Weber et al., 1995). (2) Fumigation experiments with ^{15}N-nitrogen dioxide have shown that amino compounds synthesized from the absorbed nitrogen dioxide are translocated in the phloem to the roots (Weber et al., 1995). (3) Phloem translocated amino compounds were found to inhibit nitrate uptake (Muller and Touraine, 1992). This inhibition is thought to be part of a mechanism that regulates nitrate uptake at the whole plant level *via* a circulating organic nitrogen pool (Touraine et al., 1994).

Still the significance of sulfur dioxide and nitrogen dioxide absorption by leaves for plant nutrition cannot directly be compared. Nitrogen is required by plants for growth and development in amounts approx. two orders of magnitude higher than those of sulfur (Marschner et al., 1991). As a consequence, nitrogen dioxide absorption by the leaves can contribute only at a minor extent to the nitrogen metabolism of fast growing species, when realistic atmospheric gas mixing ratios are applied (Wellburn, 1990). An influence of nitrogen dioxide absorption by the leaves on nitrate uptake by the roots cannot be expected under these conditions. This may considered to be different for slowly growing tree species like conifers. However, growth of trees is more dependent on the size of storage pools and mobilization from these pools than on the actual nitrogen supply (Kozlowski et al., 1991). Therefore, interactions between nitrogen dioxide absorption by the leaves and nitrate uptake by the roots may be difficult to analyze with tree species (Fig. 1).

This dilemma has recently been overcome by the use of spruce seedlings that exhibit a rate of growth that is higher than that of adult trees, but still low as compared to fast growing herbaceous plants (Muller et al., 1995). The growth of these seedlings will only shortly depend on the relatively small amount of nitrogen stored in the seeds and, subsequently, will depend entirely on the actual nitrogen supply. When spruce seedlings grown at different levels of nitrogen supply were fumigated with 100 ppb nitrogen dioxide for 48 h, absorption of nitrogen dioxide accounted for 20 to 40 % of nitrate uptake. In seedlings grown with nitrate plus ammonium, fumigation with nitrogen dioxide decreased nitrate uptake by the roots, while this process was not affected in plants grown with nitrate as the sole source of nitrogen and in plants kept under nitrogen deficiency (Muller et al., 1995). When fumigation with 100 ppb nitrogen dioxide was performed continuously over a period of 7 weeks, nitrogen dioxide absorption by the spruce seedlings accounted for 10 to 15 % of the total nitrogen budget of the plants. Root uptake of nitrate and ammonium as well as nitrate reduction in the roots declined upon nitrogen dioxide fumigation, while organic nitrogen transport in the phloem readily increased by a factor of about 3 as compared to control plants (Muller et al., 1995). These findings are consistent with the hypothesis that nitrogen uptake by the roots is regulated by the demand of the plant for nitrogen *via* downward transport of amino acids (Touraine et al., 1994). Apparently, atmospheric nitrogen dioxide can interact with this regulatory process in spruce seedlings.

Conclusions

Recent experiments show that ozone, sulfur dioxide, and nitrogen dioxide affect physiological processes at the whole-plant level. Reaction products of ozone interact with carbon allocation in plants most likely by inhibiting sucrose export, thereby, causing an accumulation of carbohydrate and starch in the leaves. Sulfur dioxide and nitrogen dioxide apparently interact with shoot-to-root signalling in sulfur and nitrogen nutrition, respectively, thereby inhibiting uptake and allocation of sulfate and nitrate. Because of the regulatory coupling of sulfur and nitrogen metabolism (Brunold, 1993), effects of sulfur dioxide on nitrogen nutrition and effects of nitrogen dioxide on sulfur nutrition can be expected, but have not been studied so far. With the data presently available only a partial description of the interaction of air pollutants with whole-plant processes has been achieved. Differences between species in this interaction have to be expected because of differences in carbohydrate allocation as well as sulfur and nitrogen nutrition, but have not been analyzed. Thus, intensive research in the field of air pollution interactions with whole-plant processes is required, to understand these interactions at the physiological, biochemical and molecular level.

Acknowledgements

Work performed in the authors' laboratory has been financially supported by the German National Science Foundation (DFG), the Bundesminister für Bildung, Forschung und Technologie (BMBF) through the EUREKA program EUROSILVA, and the Bayerisches Staatsministerium für Landesentwicklung und Umweltfragen (BayStLMU).

References

BIDDULPH, S. F.: Visual indications of S^{35} and P^{32} translocation in the phloem. Am. J. Bot. *43,* 143–148 (1956).

BLACK, V. J.: SO_2 effects on stomatal behavior. In: WINNER, W. E., H. A. MOONEY, and R. A. GOLDSTEIN (eds.): Sulfur Dioxide and Vegetation, pp. 96–117, Stanford University Press, Stanford (1985).

BORS, W., C. LANGEBARTELS, C. MICHEL, and H. SANDERMANN: Polyamines as radical scavengers and protectants against ozone damage. Phytochem. *28,* 1589–1595 (1989).

BRUNOLD, CH.: Regulatory interactions between sulfate and nitrate assimilation. In: DE KOK, L. J., I. STULEN, H. RENNENBERG, C. BRUNOLD, and W. E. RAUSER (eds.): Sulfur nutrition and assimilation in higher plants, pp. 61–75, SPB Academic Publishing bv, The Hague, The Netherlands (1993).

BYTNEROWICZ, A. and N. E. GRULKE: Physiological effects of air pollutants on western trees. In: OLSON, R. K., D. BINKLEY, and M. BÖHM (eds.): The response of western forests to air pollution, pp. 183–233, Springer Verlag, Berlin (1992).

CASTILLO, F. and H. GREPPIN: Extracellular ascorbic acid and enzyme activities related to ascorbic acid metabolism in *Sedum album* leaves after ozone exposure. Exp. Environ. Bot. *28,* 231–238 (1988).

CASTILLO, F. and R. HEATH: Ca^{2+}-transport in membrane vesicles from pinto bean leaves and its alteration after ozone exposure. Plant Physiol. *94,* 788–795 (1990).

DARRELL, N. M.: The effect of air pollutants on physiological processes in plants. Plant Cell Environ. *12,* 1–30 (1989).

DE KOK, L. J.: Sulfur metabolism in plants exposed to atmospheric sulfur. In: RENNENBERG, H., CH. BRUNOLD, L. J. DE KOK, and I. STULEN (eds.): Sulfur nutrition and sulfur assimilation in higher plants – fundamental environmental and agricultural aspects, pp. 111–130, SPB Academic Publishing, The Hague, The Netherlands (1990).

Dobson, M. C., G. Taylor, and P. H. Freer-Smith: The control of ozone uptake by *Picea abies* (L.) Karst. and *P. sitchensis* (Bong.) Carr. during drought and interacting effects on shoot water relations. New Phytol. *116*, 465–474 (1990).

Eckhardt, N. A., L. G. Landry, and E. J. Pell: Ozone-induced changes in RuBPCase from potato and hybrid poplar. In: Pell, E. and K. Steffen (eds.): Active Oxygen/Oxydative Stress and Plant Metablism, pp. 233–237, Curr. Top. Plant Physiol. Vol. *6* (1991).

Elstner, E. F.: Schadstoffe, die über die Luft zugeführt werden. In: Hock, B. and E. F. Elstner (eds.): Pflanzentoxikologie – Der Einfluß von Schadstoffen und Schadwirkungen auf Pflanzen, pp. 67–94, Bibliographisches Institut Mannheim/Wien/Zürich, Wissenschaftsverlag (1984).

Faller, N., K. Herwig, and H. Kühn: Die Aufnahme von Schwefeldioxyd ($^{35}SO_2$) aus der Luft. II Aufnahme, Umbau und Verteilung in der Pflanze. Plant and Soil *33*, 283–295 (1970).

Faller, N.: Schwefeldioxid, Schwefelwasserstoff, nitrose Gase und Ammoniak als ausschließliche S- bzw. N-Quellen der höheren Pflanze. Z. Pflanzenernähr. Düngg. Bodenkde. *131*, 120–130 (1972a).

– Absorption of sulfur dioxide by tobacco plants differently supplied with sulfate. In: Isotopes and radiation in soil-plant relationships including forestry, pp. 51–57, Vienna: Int. Atom. Energ. Agenc. (1972b).

Farage, P., S. Long, E. Lechner, and N. Baker: The sequence of changes within the photosynthetic apparatus of wheat following short-term exposure to ozone. Plant Physiol. *95*, 529–535 (1991).

Garsed, S. G. and D. J. Read: Sulfur dioxide metabolism in soybean, *Glycine max* var. biloxi. I. The effects of light and dark on the uptake and translocation of $^{35}SO_2$. New Phytol. *78*, 111–119 (1977a).

– – Sulfur dioxide metabolism in soy-bean, *Glycine max* var. biloxi. II. Biochemical distribution of ^{35}S products. New Phytol. *79*, 583–592 (1977b).

Gorisson, A. and J. van Veen: Temporary disturbance of translocation of assimilates in douglas firs caused by low levels of ozone and sulfur dioxide. Plant Physiol. *88*, 559–563 (1988).

Günthard-Goerg, M. S., R. Matyssek, C. Scheidegger, and T. Keller: Differentiation and structural decline in the leaves and bark of birch (*Betula pendula*) under low ozone concentration. Trees *7*, 104–114 (1993).

Heath, R.: Possible mechanisms for the inhibition of photosynthesis by ozone. Photosyn. Res. *39*, 439–451 (1994).

Herschbach, C. and H. Rennenberg: Influence of glutathione (GSH) on sulfate influx, xylem loading and exudation in excised tobacco roots. J. Exp. Bot. *42*, 1021–1029 (1991).

– – Influence of glutathione (GSH) on net uptake of sulfate and and sulfate transport in tobacco plants. J. Exp. Bot. *45*, 1069–1076 (1994).

– – Long-distance transport of ^{35}S-sulphur in three-year-old beech trees (*Fagus sylvatica* L.). Physiol. Plant. *95*, 379–386 (1995).

Herschbach, C., L. J. De Kok, and H. Rennenberg: Net uptake of sulfate and its transport to the shoot in spinach plants fumigated with H_2S or SO_2: Does atmospheric sulfur affect the ‹inter-organ› regulation of sulfur nutrition? Bot. Acta *108*, 41–46 (1995a).

– – – Net uptake of sulfate and its transport to the shoot in tobacco plants fumigated with H_2S or SO_2. Plant and Soil *175*, 75–84 (1995b).

Kangasjärvi, J., J. Talvinen, M. Utriainen, and R. Karjalainen: Plant defence systems induced by ozone. Plant Cell Environ. *17*, 783–794 (1994).

Kerstiens, G. and K. Lendzian: Interactions between ozone and plant cuticles. I. Ozone deposition and permeability. New Phytol. *112*, 13–19 (1989).

Klein, H., H.-J. Jäger, W. Domes, and C. H. Wong: Mechanism contributing to differential sensitivities of plants to SO_2. Oecologia *33*, 203–208 (1978).

Kozlowski, T. T., P. J. Kramer, and S. G. Pallardy: The physiological ecology of woody plants. Academic Press, Inc., San Diego, California, USA.

Laisk, A., O. Kull, and H. Moldau: Ozone concentration in leaf intercellular air spaces is close to zero. Plant Physiol. *90*, 163–1167 (1989).

Lee, Y. N. and S. E. Schwarz: Reaction kinetics of nitrogen dioxide with liquid water at low partial pressure. J. Phys. Chem. *85*, 840–848 (1981).

Leone, I. A. and E. Brennan: Modification of sulfur dioxide injury to tobacco and tomato by varying nitrogen and sulfur nutrition. J. Air Pollut. Contr. Ass. *22*, 544–547 (1972).

Luwe, M. and U. Heber: Ozone detoxification in the apoplast and symplast of spinach, broad bean and beech leaves at ambient and elevated concentrations of ozone in air. Planta *197*, 448–455 (1995).

Luwe, M., U. Takahama, and U. Heber: Role of ascorbate in detoxifying ozone in the apoplast of spinach (*Spinacia oleracea*) leaves. Plant Physiol. *101*, 969–976 (1993).

Maier-Mercker, U. and W. Koch: Experiments on the control capacity of stomata of *Picea abies* (L.) Karst. after fumigation with ozone and in environmentally damaged material. Plant Cell Environ. *14*, 175–184 (1991).

Marschner, H., M. Häussling, and E. George: Ammonium and nitrate uptake rates and rhizosphere pH in non-mycorrhizal roots of Norway spruce [*Picea abies* (L.) Karst.]. Trees *5*, 14–21 (1991).

Matyssek, R., M. Günthardt-Goerg, M. Saurer, and T. Keller: Seasonal growth, $\delta^{13}C$ in leaves and stem, and phloem structure of birch (*Betula pendula*) under low ozone concentrations. Trees *6*, 69–76 (1992).

Matyssek, R., T. Keller, and T. Koike: Branch growth and leaf gas exchange of *Populus tremula* exposed to low ozone concentrations throughout two growing seasons. Environ. Poll. *79*, 1–7 (1993).

Matyssek, R., P. Reich, R. Oren, and W. E. Winner: Response mechanisms of conifers to air pollutants. In: Smith, W. K. and T. M. Hinckley (eds.): Ecophysiology of coniferous forests. pp. 255–308. Academic Press, New York (1995).

Muller, B. and B. Touraine: Inhibition of NO_3^- uptake by various phloem-translocated amino acids in soybean seedlings. J. Exp. Bot. *43*, 617–623 (1992).

Muller, B., B. Touraine, and H. Rennenberg: Interaction between atmospheric and pedospheric nitrogen nutrition in spruce (*Picea abies* L. Karst.) seedlings. Plant Cell Environ. in press (1996).

Nussbaum, S., P. von Ballmos, H. Gfeller, U. P. Schlunegger, J. Fuhrer, D. Rhodes, and Ch. Brunold: Incorporation of atmospheric $^{15}NO_2$-nitrogen into free amino acids by Norway spruce *Picea abies* (L.) Karst. Oecologia *94*, 408–414 (1993).

Pearson, J. and G. R. Stewart: The deposition of atmospheric ammonia and its effects on plants. New Phytol. *125*, 283–305 (1993).

Pell, E. J. and G. Nagi Reddy: Oxidative stress and its role in air pollution toxicity. In: Pell, E. and K. Steffen (eds.): Active Oxygen/Oxydative Stress and Plant Metablism, pp. 67–75, Curr. Top. Plant Physiol. Vol. *6*, (1991).

Pfanz, H., E. Martinoia, O.-L. Lange, and U. Heber: Mesophyll resistance to SO_2 fluxes into leaves. Plant Physiol. *85*, 922–927 (1987).

POLLE, A. and H. RENNENBERG: Significance of antioxidants in plant adaptation to environmental stress. In: FOWDEN, L., T. MANSFIELD, and J. STODDART (eds.): Plant Adaptation to Environmental Stress, pp. 263–273, Chapman and Hall, New York (1993).

POLLE, A., G. WIESER, and W. M. HAVRANEK: Quantification of ozone influx and apoplastic ascorbate content in needles of Norway spruce trees (*Picea abies* L., Karst.) at high altitude. Plant Cell Environ. *18,* 681–688 (1995).

RAMGE, P., F.-W. BADECK, M. PLÖCHL, and G. H. KOHLMAIER: Apoplastic antioxidants as decisive elimination factors within the uptake process of nitrogen dioxide into leaf tissues. New Phytol. *125,* 771–785 (1993).

RENNENBERG, H.: The fate of excess sulfur in higher plants. Ann. Rev. Plant Physiol. *35,* 121–153. (1984).

RENNENBERG, H. and C. HERSCHBACH: Responses of plants to atmospheric sulphur. In: WILEY, J. and SONS (eds.): Plant Responses to Air Pollution, in press (1996).

RENNENBERG, H. and G. L. LAMOUREUX: Physiological processes that modulate the concentration of glutathione in plant cells. In: RENNENBERG, H., CH. BRUNOLD, L. J. DE KOK, and I. STULEN (eds.): Sulfur nutrition and sulfur assimilation in higher plants – fundamental environmental and agricultural aspects, pp. 111–130, SPB Academic Publishing, The Hague, The Netherlands (1990).

RENNENBERG, H. and A. POLLE: Metabolic consequences of atmospheric sulphur influx into plants. In: ALSCHER, R. G. and A. R. WELLBURN (eds.): Plant Responses to the Gaseous Environment, pp. 165–180, Chapman and Hall, London (1994).

RUNECKLES, V. C.: Uptake of ozone by vegetation. In: LEFOHN, A. S. (ed.): Surface-level ozone exposures and their effects on vegetation, pp. 157–188, Lewis Publishers, Chelsea (1992).

SCHNEIDER, A., T. SCHATTEN, and H. RENNENBERG: Exchange between phloem and xylem during long distance transport of glutathione in spruce trees (*Picea abies* [Karst.] L.). J. Exp. Bot. *45,* 457–462 (1994).

SCHRAUDNER, M., D. ERNST, C. LANGEBARTELS, and H. SANDERMANN: Biochemical plant responses to ozone: III. Activation of the defense-related protein β-1,3-glucanase and chitinase in tobacco leaves. Plant Physiol. *99,* 1321–1328 (1992).

SCHUPP, R., T. SCHATTEN, J. WILLENBRINK, and H. RENNENBERG: Long-distance transport of reduced sulphur in spruce (*Picea abies* L.). J. Exp. Bot. *43,* 1243–1250 (1992).

SEN GUPTA, A., R. G. ALSCHER, and D. MCCUNE: Response of photosynthesis and cellular antioxidants to ozone in *Populus* leaves. Plant Physiol. *96,* 650–655 (1991).

SONNEWALD, U. and L. WILLMITZER: Molecular approaches to sink-source interactions. Plant Physiol. *99,* 1267–1270 (1992).

STÄDLER, S. and H. ZIEGLER: Illustration of the genetic differences in ozone sensitivity between the varieties *Nicotiana tabacum* Bel W3 and Bel B using various plant systems. Bot. Acta *106,* 265–276 (1993).

THOENE, B., P. SCHRÖDER, H. PAPEN, A. EGGER, and H. RENNENBERG: Absorption of atmospheric NO_2 by spruce (*Picea abies* L.) trees. I. NO_2 influx and its correlation with nitrate reduction. New Phytol. *117,* 575–585 (1991).

TOURAINE, B., D. T. CLARKSON, and B. MULLER: Regulation of NO_3^- uptake at the whole plant level. In: ROY, J. and E. GARNIER (eds.): A whole plant perspective on carbon-nitrogen interactions, pp. 11–30, SPB Academic Publishing, The Hagen, The Netherland (1994).

WEBER, P., A. NUSSBAUM, J. FUHRER, H. GFELLER, U. P. SCHLUNEGGER, CH. BRUNOLD, and H. RENNENBERG: Uptake of atmospheric $^{15}NO_2$ and its incorporation into free amino acids in wheat (*Triticum aestivum*). Physiol. Plant. *94,* 71–77 (1995).

WELLBURN, A. M. and A. R. WELLBURN: Atmospheric ozone affects carbohydrate allocation and winter hardiness of *Pinus halepensis* (Mill.). J. Exp. Bot. *45,* 607–614 (1994).

WELLBURN, A. R.: Why are atmospheric oxides of nitrogen usually phytotoxic and not alternative fertilizers? New Phytol. *115,* 395–429 (1990).

WIESER, G. and W. M. HAVRANEK: Ozone uptake in sun and shade crown of spruce: quantifying the physiological effcts of ozone exposure. Trees *7,* 227–232 (1993).

WILLENBRINK, J. and T. SCHATTEN: CO_2-Fixierung und Assimilatverteilung in Fichten unter Langzeitbegasung mit Ozon. Forstw. Cbl. *112,* 50–56 (1993).

WINNER, W. E. and C. J. ATKINSON: Absorption of air pollutants by plants and consequences for growth. Trends Ecol. Evol. *1,* 15–18 (1986).

J. Plant Physiol. Vol. 148. pp. 302–308 (1996)

Microscopical and Mycological Investigations on Wood of Penduncluate Oak (*Quercus robur* L.) Relative to the Occurrence of Oak Decline

ANDREA KAUS, VOLKER SCHMITT, ANDREAS SIMON, and ALOYSIUS WILD

Institut für Allgemeine Botanik, Johannes Gutenberg-Universität Mainz, D-55099 Mainz, Germany

Received June 28, 1995 · Accepted September 21, 1995

Summary

Microscopical studies exhibited great differences between healthy and damaged penduncluate oak trees (*Quercus robur* L.) relative to the state of sapwood vessels and the degree of fungal infection. In the sapwood of damaged trees deposits, discolorations and early tylosis formations were found which frequently occurred in combination with fungal infections. These results lead to the assumption that the defence reaction of the trees probably induced by fungal infection changes the structure of the tree vessels to such an extent that their water transfer function is disturbed. Some fungal genera playing an important role in the discussion of the causes of oak decline could be isolated. Fungal hyphae in the vessels and tylosis formation in the vessels cause the breakdown of the water economy in the trees.

These observations, however, are not evidence for causal predisposition of tree damage by fungal infections.

Key words: Oak decline, tyloses, fungal infection, water stress.

Introduction

The massive oak decline which has been reported from Eastern Europe since the middle of the 70s (Marcu, 1987; Prpic and Raus, 1987; Oleksyn and Przybyl, 1987) and from Western Europe since the middle of the 80s (Skadow and Traute, 1986; Hämmerli and Stadler, 1988; Hartmann et al., 1989; Balder, 1993) is attributed to multiple causes, which have not been explained satisfactorily yet. Oak decline seems to bee almost independent of tree age, forest community, habitat conditions as well as forest management (Balder, 1989 and 1993).

In the course of the decline, the crown becomes increasingly transparent due to abscission of twigs and dieback of complete branches. The leaves are small and arranged in bunches and frequently show necroses and chlorosis shortly after budbreak (Balder, 1989). As the decline advances brown bark necroses may occur in combination with black slime flux on the trunk (Balder and Dujesiefken, 1991). Sometimes bark and cambium die axially in stripes. The twigs also exhibit histological changes (Seehann and Liese, 1991). In particular, brownish necroses at the stem base may be associated with *Armillaria mellea* and other fungi (Balder, 1989; Hartmann et al., 1989). In case of such bark damage the sapwood shows frequently black discolorations. Weather conditions (e.g. frost, drought), insect infestations (e.g. *Agrilus* sp.), fungal infections and anthropogenic influences (e.g. high nitrogen depositions) are regarded as possible reasons for oak decline.

To test these assumptions, morphological, dendrochronological, microscopical and mycological studies on penduncluate oaks of different damage classes were made. This paper focuses on microscopical and mycological aspects and their possible influence on the water status of trees.

Material and Methods

The studies presented here were carried out on the wood of six penduncluate oaks (*Quercus robur* L.) of two adjacent locations. Both locations are situated in the Westerwald (Forestry office Hachenburg-Süd, forest district Mündersbach, sections R12b and

R17b) in 335 to 340 m (R12b) and 325 to 345 m (R17b) above sea level. In 1993, the average age of the two oaks stands was 152 years (R12b) and 156 years (R17b).

In November 1993 and March 1994, six oaks, i.e. three oaks of each location, were felled. Two of them were apparently healthy (tree 1 and 2), two were damaged (tree 3 and 4) and two were seriously damaged (tree 5 and 6). In November 1993, forest experts classified the tree vitality according to the crown structure (Roloff, 1989). Trunk discs were taken from a trunk height of 1.30 m and from the base of the crown. From the seriously damaged tree 5 (R17b), an additional disc was removed at 4 m trunk height, and from tree 3 (same location) additionally at 17 m trunk height. The discs were processed the same day.

Dendrochronology

The surface of the trunk discs cut at 1.30 m trunk height were planed by means of a ribbon grinder. The annual diameter growth rate of every disc was measured along four rectangular radii by digital positions meter according to Johann.

Microscopical techniques

For light microscopical studies of wood from the trunk, samples of about 2–3 cm-cubes were cut and immediately fixed in an ethanol-formol-acetic acid solution. Sections were cut on a sliding microtome and evaluated by means of a Zeiss Axioplan light microscope.

For scanning electron microscopical (SEM) investigation, small cubes were cut from the trunk disc and planed by means of a razor blade. The specimen were then frozen in liquid nitrogen and freeze-dried for 24 hours in a Leybold-Heraeus Lyovac GT2 freeze drying equipment. The samples were mounted on specimen stubs by means of electrically conductive glue (Elecolith 340) and twice sputtered with gold (Polaron Sputter coater) for three minutes. A Cambridge Stereoscan 250 MK 2 was used for the investigations. The scanning electron micrographs were taken at 10 kV.

Mycological techniques

For mycological studies, endophytic fungi were isolated from the trunk discs. For this purpose, small wood splinters were sterilised in 70 % ethanol and laid out on 2 % malt Agar containing streptomycin (50 mg/L). Growing mycelium or spores were transferred to fresh malt Agar, then the different genera were isolated and identified.

Results

Visual state and growth of the sample trees

Location R17b. Tree 1 (damage class 0) was located at the periphery of clearing with NE exposure, had a broad crown with numerous buds and only a few scars of abscised short shoots. The width of the annual rings varied considerably. On the average, however, the annual radial growth rate was high and without any substantial decrease in vitality (Fig. 1).

Tree 3 (damage class 2) was located in the interior of the stand. Almost half of the crown was in a good condition. The tree exhibited a reduced length of annual shoots and number of buds, and an increased number of scars of abscised short shoots. In general, the radial growth rate of this tree was very low and the annual ring width was more uniform than in tree 1 (Fig. 1).

Tree 5 (damage class 2–3) was located at the periphery of a clearing with W exposure. The crown consisted of only one branch and numerous shoots from preventitous buds covered trunk. Buds were missing and the number of scars of shed short shoots was fairly high. In general, the radial stem growth of this tree was higher than for tree 3, however, with declining tendency. In 1993, there was an abrupt declining in annual ring width (Fig. 1).

Location R12b. Tree 2 (damage class 0) was located in the interior of the stand and was characterised by a broad crown. The twigs, however, had a lot of scars and only a few buds. The dendrochronological analysis showed great fluctuations of the annual growth with a tendency of decreasing annual ring widths (Fig. 2).

Tree 4 (damage class 1–2) was also located inside the stand. Compared to tree 3, the crown of this tree showed a higher state of vitality. The twigs produced many more buds and the number of scars was considerably lower than in tree 3. The diameter growth rate of tree 4 was similar to that of a healthy tree, i.e. large growth fluctuations and high overall growth rate (Fig. 2).

Tree 6 (damage class 2–3) was situated at the periphery of a clearing NE exposure. The major part of the crown was withered. Slime flux was observed at numerous points on the trunk, the twigs bore no buds and exhibited a great number

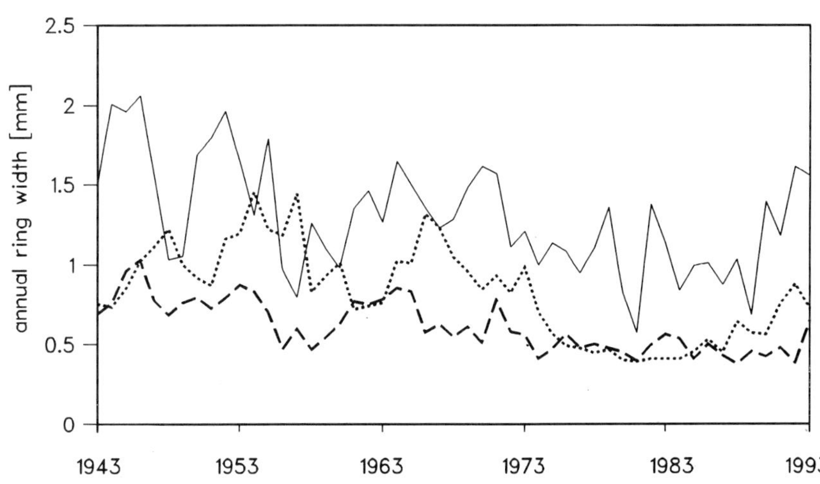

Fig. 1: Annual ring widths [mm] in 1.30 m trunk height of the oak trees 1 (healthy), 3 (damaged) and 5 (severely damaged) of location R17b in the course of the years 1943–1993. ——tree1; - - - -tree5; · · · · ·tree 3.

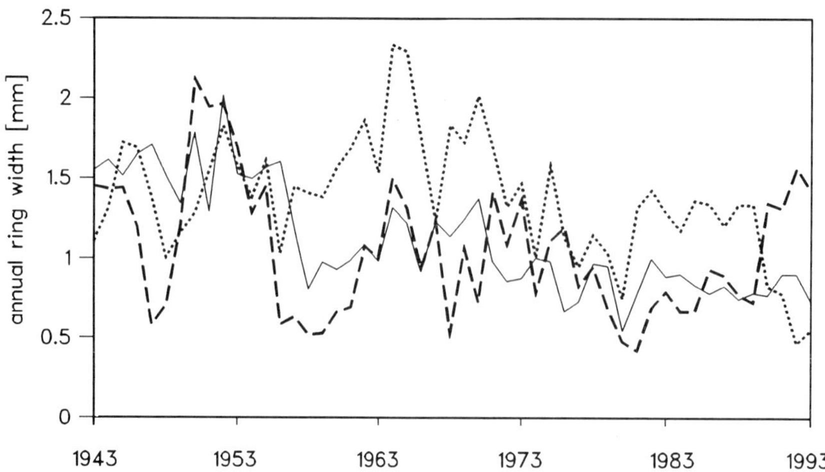

Fig. 2: Annual ring widths [mm] in 1.30 m trunk height of the oak trees 2 (healthy), 4 (damaged) and 6 (severely damaged) of location R12b in the course of the years 1943–1993. —— tree 2; - - - - tree 4; · · · · · tree 6.

of scars. Until 1990, the annual diameter growth rate was very high and showed great fluctuations comparable to that of healthy trees. Since 1990, however, there was a dramatic decrease in annual ring growth (Fig. 2).

Microscopical studies

The microscopical studies (LM and SEM) showed that the tylosis formation increased with the degree of damage, in earlywood vessels as well as latewood vessels. Whereas the heartwood vessels were generally sealed by tyloses, as was expected, the degree of tylosis formation in the sapwood was considerably different. The healthy trees 1 and 2 showed a low degree of tylosis formation, i.e. at least the vessels of the last five (tree 2) or six (tree 1) annual rings were free of tyloses. The seriously damaged trees 5 and 6 exhibited a high degree of tylosis formation particularly in the two youngest annual rings. In general, the degree of vessels sealed by tyloses increased from the stem base (1.30 m trunk height) to the base of the crown (see Tables 1 and 2).

The frequency of observed fungal hyphae and discolorations also increased with an increasing degree of tree damage. In the vessels of the both healthy trees 1 and 2 only a few hyphae could be observed in the sapwood, while the heartwood was apparently sterile. The damaged trees 3 and 4, however, exhibited hyphae also in the heartwood. The highest infection rates were noticed in the severely damaged trees 5 and 6 (see Tables 1 and 2 and Fig. 3). In a single case (tree 3, decayed heartwood at 17.00 m trunk height, S exposure), even conidiophores with conidia (presumably of *Penicillium* sp.) were found in the vessels.

Furthermore, by means of SEM studies, deposits on the walls of early-wood vessels were observed in different areas (in tree 1 and 2, sapwood at 1.30 m and 20.00 m trunk height, S exposure).

As far as starch deposits were concerned, the situation was just the opposite. Whereas the trees of the damage class 0 and 1–2 showed numerous starch deposits in the parenchyma cells of the sapwood, no starch was found in the severely damaged trees 5 and 6. Close to areas of decayed wood, tylosis formation, discoloration, fungal infections and a dissolution of starch deposits were observed (see Tables 1 and 2).

Table 1: Microscopical (LM, SEM) examinations; oak stand R17b.

Tree	Trunk height	Area	Aspect	Hyphae	Starch	Discoloration	Tylosis formation
1	1.30 m	sapwood	N	+	+++	–	from 7. a.r.
		heartwood	N	–	–	–	+++
	20.00 m	sapwood	S	–	n.d.	–	from 6. a.r.
		heartwood	S	–	–	–	+++
3	1.30 m	sapwood	N	+	+++	–	from 4./5. a.r.
		sapwood	E	+	+++	–	from 5./6. a.r.
		sapwood	W	+	+++	–	from 4./5. a.r.
		heartwood	SE	++	–	–	+++
	12.00 m	sapwood	E	+	+++	–	from 5./6. a.r.
		sapwood	N	+++	+++	–	from 3.–5. a.r.
		heartwood[1]	N	+++	–	+++	+++
	17.00 m	sapwood[1]	N	++	+++[1]	+++	++
		sapwood[2]	S	++	+++[1]	+++	from 2. a.r.
		heartwood	S	+++	–	+	+++
5	1.30 m	sapwood	S	+++	+	+++	++[1]
		sapwood	N	+++	–	+++	++[1]
		heartwood	N	+	–	++	+++
	4.00 m	sapwood	E	+++	–	+++	+++
		sapwood	S	+++	–	+++	+++
		sapwood	NW	+++	+	+++	+++
		heartwood[3]	SE	++	–	+++	+++
	11.00 m	sapwood	N	+++	+	+++	from 3. a.r.
		sapwood	W	+++	–	++	++[1]
		heartwood[1]	W	++	–	++	+++

Area: [1] = rotten; [2] = near wood-pecker hole; [3] = deformed.
Aspect: N = North; S = South; W = West; E = East.
Hyphae: – = absent; + = occasional; ++ = frequent; +++ = abundant.
Starch: – = no starch deposition; +, ++, +++ = increasing degree of starch deposition; [1] = no starch deposition close to rotten areas; n.d. = not determined.
Discoloration: – = no discoloration; +, ++, +++ = increasing degree of discoloration.
Tylosis formation: a.r.= annual ring; ++ = nearly all vessels are sealed; +++ = all vessels sealed; [1] = particularly vessels of 2 youngest annual rings sealed.

Mycological studies

In the course of the mycogical studies, different genera of fungi were isolated from the wood of the sample trees. To some of these fungi a possible pathological importance is at-

Table 2: Microscopical (LM, SEM) examinations; oak stand R12 b.

Tree	Trunk height	Area	Aspect	Hyphae	Starch	Discolor-ation	Tylosis formation
2	1.30 m	sapwood	S	+	+++	–	from 6. a.r.
		sapwood	N	+	+++	–	from 6. a.r.
		heartwood	W	–	+	–	+++
		heartwood	SE	–	+	+	+++
	20.00 m	sapwood	NE	+	+++	–	from 4./5. a.r.
		sapwood	SE	+	+++	–	from 4./5. a.r.
		heartwood	NE	+	+	–	+++
4	1.30 m	sapwood	W	+	+++	–	from 6. a.r.
		sapwood	S	+	+++	–	from 5. a.r.
		sapwood	N	+	+++	–	from 6. a.r.
		heartwood	N	+	+	+	+++
	21.00 m	sapwood	W	+	+++	–	from 3./4. a.r.
		sapwood	SE	+	+++	–	from 5. a.r.
		heartwood	NW	++	+	++	+++
6	1.30 m	sapwood	W	+++	–	+++	+++
		sapwood	E	+++	–	++	+++
		sapwood	S	+++	–	+++	+++
		sapwood	N	+++	–	+++	+++
		heartwood	E	+++	–	+++	+++
	21.00 m	sapwood	W	+++	–	++	++
		sapwood	S	+++	–	++	++
		sapwood	N	+	–	+	+++
		heartwood	central	–	–	+++	+++
		heartwood[3]	S	+	–	+++	+++

Area: [1] = rotten; [2] = near wood-pecker hole; [3] = deformed.
Aspect: N = North; S = South; W = West; E = East.
Hyphae: – = absent; + = occasional; ++ = frequent; +++ = abundant.
Starch: – = no starch deposition; +, ++, +++ = increasing degree of starch deposition; [1] = no starch deposition close to rotten areas; n.d. = not determined.
Discoloration: – = no discoloration; +, ++, +++ = increasing degree of discoloration.
Tylosis formation: a.r.= annual ring; ++ = nearly all vessels are sealed; +++ = all vessels sealed; [1] = particularly vessels of 2 youngest annual rings sealed.

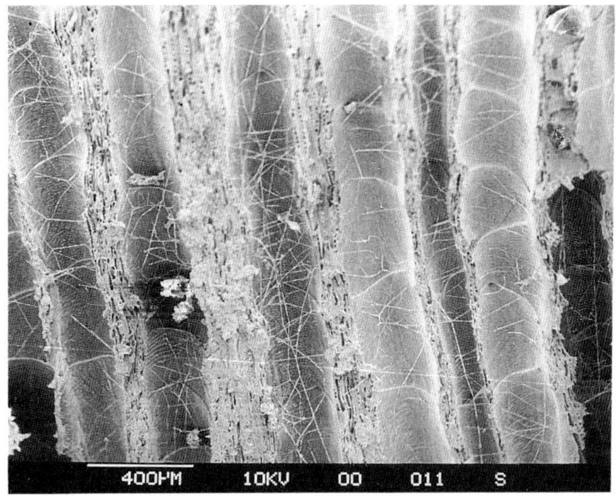

Fig. 3: Longitudinal section through early-wood vessels colonised by fungal hyphae from the sapwood of oak tree 5 (location R17b). Magn. ×1,000.

tached. For example, a representative of the genus *Graphium, Ophiostoma querci,* was isolated from the trees of location R17b (see Tables 3 and 4). The extent of fungal infection was correlated to increasing tree damage. The greatest number of different fungal genera was isolated from the seriously damaged trees 5 and 6.

Discussion

Microscopical observations

The microscopical studies pointed out that the occurrence of hyphae is accompanied by tylosis formation. This leads to the assumption that the sealing of the vessels represents a defence reaction against the fungal infection. By sealing the vessels the axial spread of the fungi *via* the vascular system may be hindered. In SEM specimens hyphae emerging from the vessel pits or penetrating them were frequently observed. The pits allow the hyphae to spread to the adjacent tissues (Fig. 4). The presence of fungal hyphae in pits seems to be a potential way of fungal propagation in wood (Wozny and Siwecki, 1990).

It is difficult to say, whether the depositions on the vessels walls can be related to the isolated fungi (Fig. 5). In ultrastructural studies, Wozny and Siwecki (1990) noticed altered surface structures of the vessels of damaged trees. They also describe gelatinous substances in the interior of the vessels, which are sometimes so extensive that the lumina of the vessels are completely sealed. These depositions result from the dissolution of the pit membranes, which is caused by fungal pectinolytic activity (Cooper and Wood, 1980).

Particularly at injured or rotten areas, the adjacent tissue is characterised by a higher degree of tylosis formation. The cells were probably dead and formed a protective layer between the healthy and the necrotic tissue. No starch was produced and deposited any more. Similar observations were also reported by Seehann and Liese (1990). The discolorations of the reaction zones are mainly caused by suberin depositions (Pearce and Holloway, 1984). Suberin is a constructive biopolymer acting as a protective barrier against pathogenic micro-organisms. It prevents both the germination of fungal spores and the penetration of mycelium, and thus development of pathogens within the plant (Grosser, 1987).

Almost no hyphae were found in the healthy tree 2 at location R12b at 1.30 m trunk height, neither in LM nor in SEM. The sapwood was not sealed by tyloses at least in the youngest six annual rings. However, in the SEM depositions were found in the South-exposed sapwood vessels. This might be an indication for the presence of endophytic fungi. At 20.00 m trunk height tylosis formation spread to younger annual rings, as expected, because the degree of tylosis formation rises with increasing trunk height (Babos, 1993). The reason for this early sealing of vessels might be the presence of hyphae noticed in LM and SEM observations. As a consequence of the early sealing of the sapwood vessels, the water transfer capacity of the tree is probably disturbed in this area. The tree which had been classified as healthy exhibited a decrease in vitality, as had already been proved in crown structure and dendrochronology. In general, it seems that the

Table 3: Isolated fungi; oak stand R17b.

Tree	Trunk height	Area	Aspect	Growth Dynamic	Fungal Genera
1	1.30 m	sapwood	N	2	*Penicillium* sp. *Graphium* sp. (*Ophiostoma*) n.i. I
		heartwood	N	3	*Penicillium* sp.
	20.00 m	sapwood	S	2	*Basidiomycetes gen.* sp.
		heartwood	S	–	sterile
3	1.30 m	sapwood	N	2	*Penicillium* sp. n.i. II
		sapwood	E	2	*Penicillium* sp. n.i. II
		sapwood	W	2	*Penicillium* sp. n.i. II
		heartwood	SE	1	*Penicillium* sp.
	12.00 m	sapwood	E	1	*Penicillium* sp. n.i. II
		sapwood	N	1	*Penicillium* sp. n.i. II
		heartwood[1]	N	2	*Penicillium* sp. *Acremonium* sp.
	17.00 m	sapwood[1]	N	1	*Basidiomycetes gen.* sp. *Graphium* sp. (*Ophiostoma*) n.i. III
		sapwood[2]	S	1	*Penicillium* sp.
		heartwood[1]	S	1	*Penicillium* sp. n.i. III
5	1.30 m	sapwood	S	1	*Penicillium* sp. *Graphium* sp. (*Ophiostoma*) n.i. IV
		sapwood	N	1	*Penicillium* sp. *Gliocladium* sp. *Trichoderma* sp. *Paecilomyces* sp. *Mucorales gen.* sp. n.i. IV
		heartwood	N	1	*Penicillium* sp. n.i. V
	4.00 m	sapwood	E	1	*Penicillium* sp. *Cephalosporium* sp.
		sapwood	S	1	*Geniculosporium* sp.
		sapwood	NW	1	*Botrytis cinerea*
		heartwood[3]	SE	1	*Penicillium* sp.
	11.00 m	sapwood	N	1	*Paccilomyces* sp. *Geniculosporium* sp.
		sapwood	W	1	*Basidiomycetes gen.* sp.
		heartwood[1]	W	1	*Penicillium* sp. *Trichoderma* sp. *Cephalosporium* sp.

Area: [1] = rotten; [2] = near wood-pecker hole; [3] = deformed.
Aspect: N = North; S = South; W = West; E = East.
Growth Dynamik: 1, 2, 3 = First mycelial appearance at first, second, or third growth control.
Fungal Genera: n.i. = not identified.

Table 4: Isolated fungi; oak stand R12b.

Tree	Trunk height	Area	Aspect	Growth Dynamic	Fungal Genera
2	1.30 m	sapwood	S	1	*Penicillium* sp.
		sapwood	N	1	*Penicillium* sp.
		heartwood	W	–	sterile
		heartwood	SE	1	*Penicillium* sp.
	20.00 m	sapwood	NE	1	*Mucorales gen.* sp.
		sapwood	SE	1	*Penicillium* sp.
		heartwood	NE	–	sterile
4	1.30 m	sapwood	W	1	*Penicillium* sp.
		sapwood	S	1	*Penicillium* sp.
		sapwood	N	1	*Penicillium* sp.
		heartwood	N	1	*Penicillium* sp.
	21.00 m	sapwood	W	3	*Penicillium* sp.
		sapwood	SE	1	*Penicillium* sp.
		heartwood	NW	–	sterile
6	1.30 m	sapwood	W	1	*Penicillium* sp. *Verticillium* sp. *Graphium* sp. *Coniella* sp.
		sapwood	E	1	*Basidiomycetes gen.* sp.
		sapwood	E	1	*Verticillium* sp.
		sapwood	N	1	*Penicillium* sp. *Verticillium* sp.
		heartwood	E	1	*Cephalosporium* sp.
	21.00 m	sapwood	W	1	*Penicillium* sp. *Cytospora intermedium*
		sapwood	S	1	*Penicillium* sp. *Graphium* sp. (*Sporothrix*) *Coniella* sp.
		sapwood	N	1	*Cytospora* sp.
		heartwood	central	1	*Penicillium* sp. *Verticillium* sp.
		heartwood[3]	S	2	*Graphium* sp. (*Sporothrix*)

Area: [1] = rotten; [2] = near wood-pecker hole; [3] = deformed.
Aspect: N = North; S = South; W = West; E = East.
Growth Dynamik: 1, 2, 3 = First mycelial appearance at first, second, or third growth control.
Fungal Genera: n.i. = not identified.

crown of the tree is damaged first. This assumption could be confirmed by mycological studies.

The tree 5 and 6 which had been evaluated as severely damaged owing to their crown structure, showed extensive fungal infections in both LM and SEM studies. The starch deposits were dissolved or had not been produced and especially the youngest annual rings were heavily sealed. This defence mechanism of the tree against penetration of pathogens, however, also results in sealing of the vessels and finally the breakdown of the water flow. In particular, the severely damaged tree 6 illustrates how rapidly oak decline may involve lethal effects even on trees with high diameter growth rates (Fig. 1).

In general, it can be stated that the vessels of the damaged trees were colonised by fungal hyphae, sometimes to such an extent that water transfer was probably seriously impaired. The defence reactions of the tree, i.e. sealing of the sapwood vessels to isolate the infected trunk areas, reduces the water transfer (→water stress). Therefore it can be assumed that the reason for oak decline is closely related to the disturbances in the water status (see also Wozny and Siwecki, 1990).

Significance of the isolated fungi

The influence of fungi in oak decline has not been proven yet and is currently studied by several laboratories. The occurrence of American oak wilt, which was suspected initially, could not be proved. The pathogen *Ceratocystis fagacearum* (Bretz) Hunt has not been isolated yet in any case. Particularly in Eastern Europe, however, some species of the genus *Ceratocystis* which are made responsible for a mycosis similar to American oak wilt, were isolated (Kryukowa and Balder, 1991). Among the genus *Ceratocystis,* however, there are also

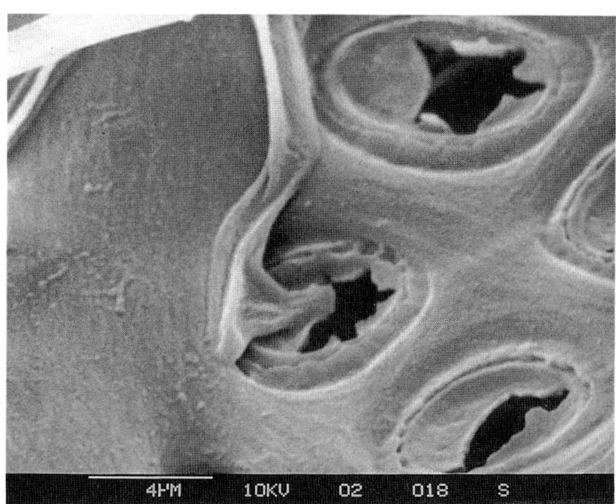

Fig. 4: Fungal hypha penetrating an early-wood vessel pit of an oak tree. Magn. ×11,400.

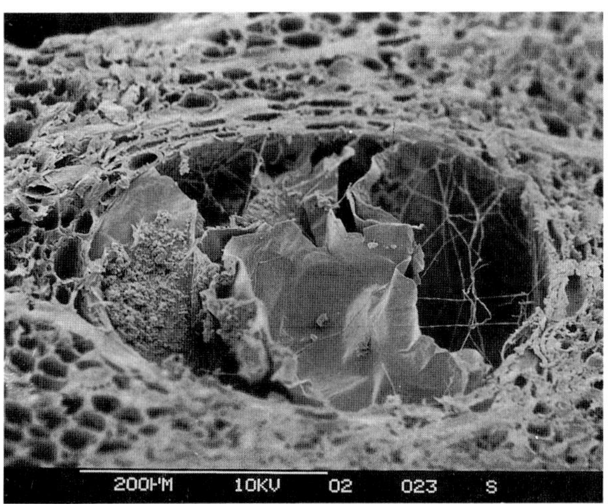

Fig. 5: Transversal section of sapwood of an oak tree showing tyloses formation, deposits and presence of hyphae in an early-wood vessel of tree 5 (location R17b). Magn. ×2,400.

non-aggressive species which only cause a bluish discoloration of wood (Butin, 1989).

Vascular mycosis-like symptoms in oaks have also been described in Germany (Balder, 1992). Besides cambium and phloem necroses, Balder (1992) also observed discolorations of the adjacent tissue. These areas contained larger amounts of lignin and water. Furthermore, the vessels were clearly blocked by tyloses. Similar observations were made in the damage sample trees of location R17b and R12b.

Especially from location R17b, an anamorph of *Ceratocystis* sp., *Graphium* sp. (= *Ophiostoma querci*) could be isolated from all trees. In tree 1, which was evaluated as healthy, *Graphium* was isolated at 1.30 m trunk height although no macroscopically or microscopically visible damage symptoms were found. The retarded growth indicates that the mycelium *in situ* had not developed very far. This could either mean that the infection was still very recent or that the defence mechanism of the tree prevented the spread of fungi. The latter would mean that a predisposition by other stress factors is necessary to weaken a tree so that its defence system is not able to suppress the pathogens. From location R12b, *Graphium* was only isolated from tree 6 which was seriously damaged. Kehr and Wulf (1993) assume that the low frequency of *Ceratocystis* and *Ophiostoma* which could be isolated from fresh necroses shows that fungi cannot be primarily related to oak decline. The fungi they isolated in the course of their studies were classified as potential weak parasites, which colonised predisposed tissue. The existence of such secondary parasites is understood by Kehr and Wulf (1993) as an indicator of stress.

In the present study, the common presence of the genus *Penicillium* is significant. It can be assumed that at least a great proportion of the isolated *Penicillium* cultures was produced by spores which penetrated into the vessels during cutting. On the other hand, Niethammer found as early as 1949, that there are interactions between *Penicillium* and different kinds of wood-decomposing fungi. Therefore, it is possible that the genus *Penicillium* plays a certain role in oak decline.

Verticillium is another genus which could be isolated from the wood of the sample trees and which leads to disturbance in the vascular system. The external symptoms of a *Verticillium* wilt are withering of leaves and shoot tips. In the sapwood there are greenish-brown discolorations (spots, dots or ring-shaped discolorations). The vessels of the early-wood are obstructed by numerous hyphae and in combination with the secretion of wilt toxins, the water status is disturbed. The infection is mainly induced by injuries (Butin, 1989) or *via* roots (Peg, 1974). Only from tree 6 of location R12b, representatives of this species could be isolated, what indicates that *Verticillium* is a secondary parasite. This is also true for the genus *Cephalosporium* (Synonym: *Acremonium*) which also could be isolated from the heavily damaged tree 6.

Other species with pathogenic potential were isolated from tree 5 of location R17b. *Botrytis cinerea* is a parasite and causes grey rot in wood. Again, this parasite could only be isolated from seriously damaged wood and therefore cannot contribute to explain the oak decline phenomenon. This also applies to the genus *Geniculosporium* which causes white rot. Because of their capability to decompose cellulose and lignin, it is assumed that they accelerate decomposition in vital plant parts and thus senescence (Petrini and Müller, 1986). In the present study, *Geniculosporium* most probably plays the role of a weak parasite.

Hartmann and Blank (1992) regularly succeeded in isolating *Cytospora intermedium* from necrotic border layers, as almost the only fungi species. It has been reported that this species is able to cause necroses in the phloem of elder oaks (Hartmann et al., 1989). However, this ability is still being discussed (Kowalski, 1991). As a weak parasite, *Cytospora intermedium* participates in the formation of vast phloem necroses, however, the tree has to be predisposed (Hartmann and Blank, 1992). In the present study, *Cytospora intermedium* was found in the sapwood of the seriously damaged tree 6 at 21.00 m trunk height.

The isolated saprophytic fungi of genera *Trichoderma, Gliocladium* and *Paecilomyces* play hardly any role in the discussion of oak decline. Representatives of *Mucorales* which could be isolated, are of no special importance either. In this study it was not possible to determine *Basidiomycetes,* which had been identified microscopically by means of their clamp connections. Many *Basidiomycetes* which occur in oaks are pathogen causing brown cubical rot and white rot. It is interesting that *Basidiomycetes* could be isolated in healthy as well as in medium and severely damaged trees.

References

BABOS, K.: Tyloses formation and the state of health of *Quercus petrara* trees in Hungary. IAWA Journal *14,* 239–243 (1993).

BALDER, H. and D. DUJESIEFKEN: Stand der Untersuchungen zum Eichensterben in Westberlin. Allg. Forst. Z. *32,* 845–848 (1989).

– – Vascular mycosis-like symptoms in oaks. In: Oak decline in Europe. Proceedings of the International Symposium, Kórnik, Poland, pp. 321–323 (1990).

BALDER, H.: Pathogene Aspekte der Eichenerkrankung. Arbeitsmittel Berliner Forsten, *2* (1992).

BUTIN, H.: Krankheiten der Wald- und Parkbäume. Diagnose – Biologie – Bekämpfung. G. Thieme Verlag, Stuttgart (1989).

COOPER, R. M. and R. K. S. WOOD: The role of cell wall degrading enzymes in vascular wilt diseases. III. Possible envolvement of endopectin lyase. Physiol. Plant. Pathol. *16,* 285–300 (1980).

GROSSER, D.: Chemische Abwehrstoffe der Pflanze. Biol. Rundschau *25,* 225–237 (1987).

HÄMMERLI, F. and B. STADLER: Eichenschäden. Eine Übersicht zur Situation in Europa und in der Schweiz. Phytosanitärer Beobachtungs- und Meldedienst. Bulletin Nr. *3.* Birmensdorf (1988).

HARTMANN, G. and R. BLANK: Winterfrost, Kahlfraß und Prachtkäferbefall als Faktoren im Ursachenkomplex des Eichensterbens in Norddeutschland. Forst und Holz *47,* 443–452 (1992).

HARTMANN, G., R. BLANK, and S. LEWARK: Eichensterben in Norddeutschland. Verbreitung, Schadbilder und mögliche Ursachen. Forst und Holz *44,* 475–487 (1989).

KEHR, R. D. and A. WULF: Fungi associated with aboveground portions of declining oaks (*Quercus robur*) in Germany. Eur. J. For. Path. *23,* 18–27 (1993).

KOWALSKI, T.: Oak decline. I. Fungi associated with various disease symptoms on overground portions of middle-age and old oaks (*Quercus robur* L.). Eur. J. For. Path. *21,* 136–151 (1991).

KRYUKOWA, E. A.: Genus *Ceratocystis* fungi as a cause of oak wilt disease in the south of the USSR. In: Oak decline in Europe. Proceedings of the International Symposium, Kórnik, Poland, pp. 123–128 (1990).

MARCU, G.: Ursachen des Eichensterbens in Rumänien und Gegenmaßnahmen. Österr. Forst. Z. *98,* 53–54 (1987).

NIETHAMMER, A.: Die Gattung *Penicillium* Link. Merkmale, Verbreitung, Leistungen, Antibiotika, Arten. Eugen Ulmer Verlag, Stuttgart (1949).

OLEKSYN, J. and K. PRZYBYL: Oak decline in the Soviet Union – scale and hypotheses. Eur. J. For. Path. *17,* 321–336 (1987).

PEARCE, J. E. and P. J. HOLLOWAY: Suberin in the sapwood of oak (*Quercus robur* L.): its composition from a compartimentalisation barrier and occurrence in tyloses in undecayed wood. Physiol. Plant. *24,* 71–81 (1984).

PEG, G. F.: *Verticillium* diseases. Rev. Plant. Pathol. *53,* 157–182 (1974).

PETRINI, L. E. and E. MUELLER: Telomorphs and anamorphs of European species of *Hypoxylon* (*Xylariaceae, Sphaeriales*) and allied genera. Mycol. Helv., 501–622 (1986).

PRPIC, B. and D. RAUS: Stieleichensterben in Kroatien im Licht ökologischer und vegetationskundlicher Untersuchungen. Österr. Forst. Z. *98,* 55–57 (1987).

ROLOFF, A.: Kronenentwicklung und Vitalitätsbeurteilung ausgewählter Baumarten der gemäßigten Breiten. Schr. Forst. Fak. Göttingen *93* (1989).

SEEHANN, G. and W. LIESE: Histological observations on Eastern European oak decline symptoms in branches. In: Oak decline in Europe. Proceedings of the International Symposium, Kórnik, Poland, pp. 325–332 (1990).

SKADOW, K. and H. TRAUTE: Untersuchungsergebnisse zum Vorkommen einer Eichenerkrankung im nordöstlichen Harzvorland. Beitr. Forstwirtschaft *20,* 64–74 (1986).

WOZNY, A. and R. SIWECKI: Ultrastructural studies of oak disease. In: Oak decline in Europe. Proceedings of the International Symposium, Kórnik, Poland, pp. 315–320 (1990).

J. Plant Physiol. Vol. 148. pp. 309–316 (1996)

Physiological, Biochemical and Molecular Effects of Sulfur Dioxide

CAMELLIA MOSES OKPODU, RUTH GRENE ALSCHER, ELIZABETH A. GRABAU, and CAROLE L. CRAMER

Dept. of Plant Pathology, Physiology and Weed Science, Virginia Tech, Blacksburg, VA 24061

Received July 6, 1995 · Accepted November 3, 1995

Summary

Damage of leaves due to air pollutants such as sulfur dioxide is mediated through the production of reactive oxygen species. The site of action of sulfur dioxide is the chloroplast and deleterious effects on foliar tissue depend on light and photosynthetic electron transport. Protection may be afforded, in part, by components of the antioxidant (photo)scavenging cycle. Relative resistance to sulfur dioxide and cross-resistance to other oxidative stresses which originate in the chloroplast have been correlated, in many cases, with elevated levels of various antioxidant proteins and/or substrates. Recent studies utilizing differentially sensitive cultivars, antioxidant enzyme analyses which differentiate between specific isoforms at the gene and protein levels, and plants genetically engineered to alter the expression of specific antioxidant isozymes, have provided new insights into the mechanisms of resistance to sulfur dioxide and other stresses. These data suggest that complex regulatory mechanisms function at both the gene and protein level to coordinate antioxidant responses and that a critical role is played by organellar localization and inter-compartment coordination. An involvement of a strong developmental component in resistance is indicated.

Key words: Review, sulfur dioxide, radicals, antioxidants, SOD, glutathione reductase, ascorbate peroxidase.

I. Introduction

At high concentrations in air, sulfur dioxide exerts its deleterious action on foliar tissue through the generation of toxic oxygen species (Kenyon and Duke, 1985; Dodge, 1971). One initial product, the superoxide radical (O_2^-), upon further reaction within the cell, can form more reactive oxygen products such as hydroxyl radicals and singlet oxygen (Ehleringer and Björkman, 1977; Youngman, 1984).

The enzymes ascorbate peroxidase (AP; EC 1.11.1.7), glutathione reductase (GR; EC 1.6.4.2) and superoxide dismutase (SOD; EC 1.15.1.1), among other enzymes, are involved in the regeneration of antioxidants that are important in detoxification of these reactive oxygen species (Fig. 1; for reviews see Foyer et al., 1991; Beyer et al., 1991; Foyer and Mullineaux, 1994). These enzymes are involved in a metabolic cycle, (the hydrogen peroxide-scavenging cycle) that successively oxidizes and re-reduces the antioxidant substrates glutathione and ascorbate. The rapid operation of this cycle is important since several intermediate enzymes (AP and SOD) may be inhibited by the product, H_2O_2. The scavenging cycle may be an important component in the antioxidant defense mechanism conferring resistance against stress in plant cells, at least when the original site of stress is in the chloroplast.

We have focused our work on two pea cultivars whose sensitivity to sulfur dioxide differs with respect to apparent photosynthesis. Studies of responses to sulfur dioxide and paraquat at the whole plant and the protoplast level have been carried out. Apparent photosynthesis at the whole plant level was more resistant in cv. Progress and more sensitive in cv. Nugget (Alscher et al., 1987; Madamanchi and Alscher, 1991; Madamanchi et al., 1992, 1994 a, b). Evidence from our work and from those of others is summarized here which supports the hypothesis that antioxidant scavenging processes in plants are critical for tolerance to a broad range of oxidant stresses, including paraquat and sulfur dioxide.

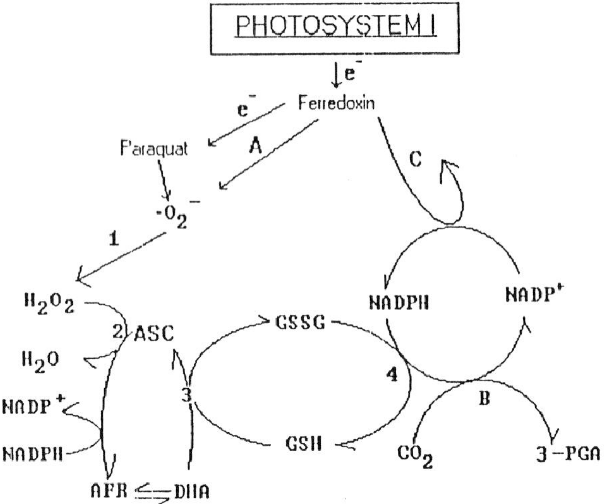

Fig. 1: Antioxidant mechanism for the scavenging of oxy-free radicals in plant chloroplasts. A = Mehler reaction. B = CO_2 fixation. C = Reduction of $NADP^+$ by reduced ferredoxin. 1 = Superoxide dismutase (SOD); 2 = Ascorbate peroxidase (AP); 3 = Dehydroascorbate reductase (DHAR); and, 4 = Glutathione reductase (GR). (Abbreviations: GSSG, oxidized glutathione; GSH, reduced glutathione; ASC, ascorbate; AFR, ascorbate free radical; DHA, dehydroascorbate; and, 3-PGA, 3-phosphoglycerate).

II. Physiological Effects of Sulfur Dioxide

It is widely accepted that the primary site of sulfur dioxide action is in the chloroplast (Schreiber and Neubauer, 1990; Ziegler, 1975; Sutinen, 1988; Schiffgens-Gruber and Lutz, 1992). Upon entering the leaf cell, and the chloroplast in particular, sulfur dioxide as sulfite can be oxidized to sulfate. The oxidation pathway is an aerobic process which involves light and the photosynthetic electron transport chain (Asada, 1980). A chain reaction involving the production of oxyradicals occurs which is terminated through the action of superoxide dismutase (SOD). The product from the SOD reaction is H_2O_2, which itself can inhibit photosynthesis at low levels through oxidation of thiol-mediated stromal enzymes. Therefore, because carbon assimilation is inhibited by H_2O_2 (IC_{50} = 10 µM), the photosynthetic capabilities of the plant will be impaired if H_2O_2 is not efficiently removed.

III. The Ascorbate-Glutathione Scavenging Cycle

a. Antioxidant substrates

Both ascorbate and glutathione are important substrates in the ascorbate-glutathione scavenging cycle. Ascorbate occurs at relatively high levels (ca. 25 mM) in the chloroplast (Foyer, 1993). It functions as a major antioxidant, engaging in nonenzymatic as well as enzymatic processes within the chloroplast and elsewhere throughout the plant cell. Ascorbate is thought to be synthesized in the cytosol, although the exact nature of the biosynthetic pathway is not yet known. A specific ascorbate translocator functions to transport ascorbate across the chloroplast envelope into the stroma (Beck et al., 1983). Glutathione is a thiol-containing tripeptide which is synthesized in the chloroplast and which exerts its antioxidant function in the reduced form (reviewed by Hausladen and Alscher, 1993). Like ascorbate, it is present in the chloroplast at high levels (ca. 1–5 mM) and also has a multiplicity of antioxidant functions. In unstressed chloroplasts it exists 90–95 % in the reduced form and functions to maintain the redox state of the stroma as well as to detoxify H_2O_2 produced as a result of the action of SOD. A common response to oxidative stress is an increase in the foliar glutathione pool (reviewed in Alscher, 1989).

The proposed mechanism for the scavenging of reactive oxygen is outlined in Fig. 1. A major site of superoxide production is the univalent reduction of molecular oxygen on the reducing side of PSI. The O_2^- is dismutated by SOD to H_2O_2. One site of dismutase activity is associated with the thylakoid. Those molecules of O_2^- which are not reduced at the thylakoid may be scavenged by the stromal form of SOD. H_2O_2 is removed in a subsequent reaction through the action of ascorbate peroxidase (Fig. 1, Step 2). Ascorbate, abundantly present in the chloroplast, is oxidized in two sequential steps, first producing monodehydroascorbate (MDA) and, subsequently, dehydroascorbate (DHA) (Loewus, 1988). The MDA can be either reduced by reduced ferredoxin at the thylakoid membrane before diffusion into the stroma, or reduced to ascorbic acid by the stromal MDA reductase. If MDA is not reduced by either reduced ferredoxin or MDA reductase, DHA can be converted to ascorbic acid by DHA reductase that uses reduced glutathione as an electron donor (Neubauer and Schreiber, 1989).

b. Antioxidant enzymes

Ascorbate peroxidase (AP)

AP plays a major role in the detoxification of hydrogen peroxide in the chloroplast and also in the cytosol. AP isoforms present in the cytosol (1 form, characterized in pea) and the chloroplast (2 forms, characterized in spinach) have been studied at the biochemical level (Chen and Asada, 1989; Nakano and Asada, 1981; Mittler and Zilinskas, 1991). Each isoform has distinct biochemical properties (Asada, 1994). Within the chloroplast, a stromal (sAP) and a thylakoid-bound (tAP) exist. tAP, which was only recently described (Miyake and Asada, 1992), is proposed to function both to remove superoxide anions produced as a result of the univalent reduction of molecular oxygen by the photosynthetic electron transport chain and to participate in the regeneration of membrane-bound antioxidants such as xanthophyll and α-tocopherol. The nature of the association between tAP and the thylakoid membrane has not been fully defined. The stromal form of the enzyme, which has received more attention, is thought to function primarily in the stromal ascorbate-glutathione scavenging cycle (Nakano and Asada, 1991).

Superoxide dismutase

SODs have been subclassified according to their metal cofactor. Three types of SOD are known to exist, FeSOD, MnSOD and Cu/Zn SOD. *In situ* staining procedures allow detection of enzyme activity in native gels (Beaucamp and

Fridovich, 1971). Isoforms can be further distinguished based on distinct biochemical properties.

FeSOD was first isolated from *Escherichia coli* (*E. coli*) (Fridovich, 1986). It is resistant to cyanide and is inhibited by H_2O_2 (Bannister et al., 1987; Fridovich, 1986). In plants, cDNA sequences for FeSOD have been isolated from *Glycine max, Nicotiana plumbaginifolia* and *Arabidopsis thaliana* (Crowell and Amasino, 1991; Van Camp et al., 1990), facilitating analyses of FeSOD regulation. The FeSOD is localized to the chloroplasts.

Biochemical studies have revealed the existence of FeSOD in several taxonomically distant plant species only. This apparently «random» distribution of FeSOD may be due to problems in assay methods and/or lability of the protein. Alternatively, FeSOD expression may be responsive to environmental conditions. In fact, Tsang et al. (1991) observed siginificant increases in steady state levels of FeSOD mRNA in *Nicotiana plumbanigifolia* in response to paraquat stress. They also observed a decrease in FeSOD mRNA levels when light-grown plants were placed in the dark for three days. Preliminary data suggest that FeSOD activities in pea may be under developmental control (Donahue and Alscher, unpublished data).

MnSOD is found in the mitochondrial matrix of all species of plants analyzed (Baum and Scandalios, 1981; Bowler et al., 1994). There are other forms associated with glyoxysomes and peroxisomes (Sandalio and del Rio, 1987). Both the MnSOD and the FeSOD have a subunit molecular weight of 23 kDa with one atom of metal per subunit (Fridovich, 1986). The molecular weight varies among different organisms, ranging from 23 to 92-kDa (Salin, 1988). MnSODs have 2–4 subunits of equal size (Salin, 1988). MnSODs are frequently dimeric, although the enzyme has been reported to be tetrameric in bacteria and in mitochondria (Fridovich, 1973). However, the ability to form dimeric and tetrameric complexes may be a reason why the molecular weight has varied among species.

Cu/ZnSOD has been isolated and characterized from numerous sources (review by Van Camp et al., 1994). The Cu/ZnSOD form is a dimer of 33 kDa and contains one Cu^{2+} and one Zn^{2+} per subunit (Misra and Fridovich, 1972). Cu/ZnSOD activity is inhibited by cyanide and inactivated by H_2O_2. In plants, Cu/ZnSOD it is the most abundant of the three metalloproteins. The enzyme has been localized to the chloroplast and cytosol and a thylakoid-associated form has also been reported (reviewed by Asada, 1994).

Using the pea system, our data showed that antiserum against the chloroplast Cu/ZnSOD recognized one protein with an approximate subunit molecular weight of 17.9 kDa (Madamanchi et al., 1994 a). These results are in close agreement with *in vitro* translation data for a cDNA of the chloroplastic Cu/ZnSOD with a predicted subunit molecular weight of 17.4 kDa. Antibodies directed toward the cytoplasmic form cross-reacted with two proteins in pea leaf extracts with approximate molecular weight of 17.9/16.8 kDa. The polypeptide with a molecular weight of 17.9 kDa was thought to be the chloroplastic form.

Glutathione reductase

Six to eight isoforms of GR, separable by isoelectric focusing (IEF), have been observed in leaf extracts of pea and tobacco (Madamanchi et al., 1992; Edwards et al., 1990; Foyer et al., 1991). 70–85 % of the total GR activity is associated with the chloroplast, with the remainder distributed among the cytosol and the mitochondrion. Our data show that two of the six IEF-separable isoforms are associated with the extraplastidic compartment and the remainder with the chloroplast (Madamanchi et al., 1992).

The various isoforms of GR which are separable on IEF gels are not apparent in protein preparations obtained from pea and purified by chromatographic methods. Our data (Madamanchi et al., 1992) and those of Edwards et al. (1990) revealed that only one GR protein with native molecular mass of 114 kDa was present in pea chloroplast preparations and also in pea leaf extracts. The protein has a subunit molecular weight of 55 kDa, suggesting that in its native form it is a homodimer. The relationship of the IEF data to those obtained with the purified protein remains to be elucidated. Edwards et al. (1990) ruled out phosphorylation and glycosylation as possible GR post-translational modification mechanisms.

Clones of GR from a number of plant species (*Arabidopsis*, soybean, pea) have been isolated (reviewed by Creissen et al., 1994). The cDNA sequences are all highly homologous to each other (ca. 70 % similarity at the deduced amino acid level) and most likely all encode a plastid form of the protein. Identification of additional GR-encoding clones will be instrumental in addressing the source and molecular basis of multiple GRs in these systems. At the protein level, three distinct GR proteins have been isolated from red spruce (Hausladen and Alscher, 1994 a, b).

IV. Proposed Involvement of Antioxidant Processes in Resistance to Oxidative Stress

The many documented cases in which a species, biotype or cultivar showing cross-resistance to two or more stresses (e.g., paraquat, high light, sulfur dioxide) have been used to propose a common basis for resistance (Gressel and Galun, 1994; Malan et al., 1990). In the case of *Conyza,* for example, resistance to paraquat was correlated with tolerance to high light conditions (Malan et al., 1990). In the case of our pea system, resistance to sulfur dioxide was correlated with resistance to paraquat (Madamanchi et al., 1994 b). It is possible that cross-resistance occurs between stresses which originate in the same subcellular compartment (e.g., the chloroplast in the case of paraquat and sulfur dioxide) but not between stresses which have disparate sites of action within the plant leaf. In tobacco, for example, ozone tolerance was not correlated with resistance to paraquat (Aono et al., 1991). (The site of ozone action is thought to be the extracellular space, Mehlhorn and Wellburn, 1994).

Genetic analyses of paraquat resistance and susceptible biotypes of Conyza showed that paraquat resistance is conferred by a single dominant locus (Shaaltiel et al., 1988). Activities of plastid SOD, AP and GR co-segregated with resistance in both the F_1 and F_2 generations. These data suggest that common regulatory factors mediate the coordinate expression of these plastid antioxidant proteins.

Table 1: Properties of Ascorbate-Hydrogen Peroxide Scavenging Cycle.

Name	Native (kDa)	Subunit (kDa)			Other Features
1. Ascorbate Peroxidase[a]			Inhibitors	K_m Ascorbate (mM)	
thylakoid (tAP)	–	40	DTT	10	bound to stroma of thylakoids; homologous at N-terminus to sAP
stroma (sAP)	–	30	DTT	10	stroma in a soluble form homologous at N-terminus to tAP not cAP
cytosol (cAP)	–	30–57	KCN, p-chloromercuric sulfonic acid	0.5	liable in the absence of ascorbate; possible several isoforms
2. Glutathione Reductase				K_m NADPH (μM)	
Source					
pea[b]	114	55	–	4.8	5 isoforms; plastidic
pea[c]	135	–	GSH (5 mM), NEM (100 μM)	2.7	inactive at pH 6 and high temperatures; broad pH optimum 6.9–8.2; plastidic
pea[d]	135	–	GSH, $ZnCl_2$, $CuSO_4$	2.3	pH optimum 7.7; thermostable; stable at pH 6; roots
Spinach[e]	145	72.5	–	2.8	pH optimum 8.5–9.0; K_m GSSG 200 μM
Maize[f]	190	65/63	NEM, GSH, β-ME	1.7 μM	pH optimum 8; heterotetramer; NADPH does not inactivate
Red Spruce[g]			K_m GSSG (μM)	K_m NADPH (μM)	
GR1NH (Nonhardened)	–	52/30	72.5	4.0	K_m NADPH 5-fold higher for harden needles at low temperature versus non-harden; location unknown
GR2NH	–	55	132	2.0	
GR1H (Hardened)	–	54/60/30	56	5.8	
GR2H	–	57/58	139	2.7	
3. Superoxide Dismutase[h, i]			Inhibitors		
Cu/Zn	–	32	H_2O_2, cyanide		Contains one Cu/Zn per subunit; very stable
Mn	–	23–96	–		Mitochondrion; 2–4 subunits
Fe	–	23	H_2O_2		one atom of Fe/subunit

[a] Asada, 1994; [b] Madamanchi et al., 1992; [c] Bielawski and Joy, 1986; [d] Bielawski and Joy, 1986; [e] Halliwell and Foyer, 1978; [f] Mahan and Burke, 1987; [g] Hausladen and Alscher, 1994 (a, b); [h] Fridovich, 1986; and [i] Salin, 1988.

Data obtained with the pea system suggest that significant developmental differences in sensitivity to oxidative stress also exist. We have observed this developmental control both within and between the two cultivars which have been studied. Responses to paraquat were studied in the first three leaves of Progress and Nugget. Paraquat resistance, expressed as leaf injury, antioxidant enzyme activities (GR, AP, SOD) and transcript levels of GR, plastid Cu/ZnSOD and cytosolic AP were measured in pea leaves which had been exposed to 10^{-5} M paraquat. The data suggest that developmental stage and/or leaf position is more important for resistance than the relatively slight changes in antioxidants (AP, GR, and SOD) activities brought about by actual exposure to the stress. No significant increase in RNA abundance was observed with paraquat exposure for plastid GR and plastid SOD, although there was good correlation between the amount of transcript present in the three leaf types and their relative resistances to paraquat (Donahue and Alscher, unpublished data).

In protoplasts, differential sensitivity was also observed in two cultivars of pea when sulfite was used as a stress (Okpodu and Alscher, unpublished data). CO_2-dependent O_2-evolution was inhibited by 50 % in ‹Progress› after 1 hour and continued to decline after 2 hours of exposure to sulfite (90 % inhibition). In ‹Nugget› protoplasts, exposure to 10 mM sulfite for 1 hour decreased photosynthesis by 60 %. However, photosynthesis subsequently returned to control levels by 1.5 hours. The activity of GR was unchanged in ‹Progress› over the entire time course of the experiment. In ‹Nugget›, the GR activity increased by 30 % after 1 hour, followed by a decline to pre-exposure rates. In both cultivars, only a single polypeptide corresponding to GR activity was detected on one dimensional gels. An increase in GR activity in ‹Nugget› protoplasts was observed on activity gels at 30 minutes, after which time the GR activity returned to control levels. When the protein extracts from protoplasts isolated from both cultivars were resolved on native analytical isoelectric focusing gels, six distinct isoforms were detected using a GR activity stain (pI's ranging from 4.1 to 5.6). In ‹Nugget› protoplast extracts, the polypeptide with a pI of 4.6 increased as compared with the control protoplasts after 30 minutes exposure to 10 mM sulfite. New forms with pIs of 4.2 and 4.3 appeared in treated protoplasts obtained from the leaves of ‹Nugget›. Preliminary data obtained from chloroplasts isolated from 10 mM sulfite treated (30 minutes) ‹Nugget› protoplasts showed an increase in the pI 4.6 isoform. These data suggest that the chloroplast isoform with a pI 4.6 may be related to processes conferring resistance to sulfite stress. Future studies will determine if this polypeptide with a pI of 4.6 is due to a different gene product, or if it is the result of post-transcriptional or post-translational modifications of a single gene product.

a. SOD may play several roles

The contribution of extraplastidic scavenging processes to foliar sulfur dioxide and paraquat resistance has not been adequately assessed. It is entirely possible that the cytosolic compartment is involved in antioxidant responses to paraquat and to sulfur dioxide exposure. Perl-Treves and Galun (1991) have demonstrated a paraquat-mediated increase in tomato cytosolic Cu/ZnSOD at the level of steady state mRNA. This increase was detected in a shorter time and at a lower paraquat dose than that needed to bring about increases in steady state levels of mRNA encoding plastid Cu/ZnSOD. Perl et al. (1993) showed that transgenic potatoes overexpressing either the cytosolic or the chloroplast Cu/ZnSOD were more resistant to paraquat. In our pea system (Madamanchi et al., 1994), a 115 % increase in cytosolic CuZn/SOD protein and ca. a 90 % increase in plastid Cu/ZnSOD was observed in ‹Progress› with no increase occurring in Nugget when both cultivars were exposed to 0.8 ppm sulfur dioxide for 3.5 hours (Madamanchi et al., 1994 a). Interestingly, no comparable sulfur dioxide-mediated change in steady state mRNA levels encoding, plastid Cu/ZnSOD was detected. Thus, post-transcriptional processes must also be involved in antioxidant resistance responses to sulfur dioxide stress. Relatively low levels of transcript remained in the leaves of the susceptible cultivar by the end of the exposure period. Since mRNA transcript abundance decreased in the leaves of Nugget as a result of sulfur dioxide exposure, transcript stability may also play an important role in resistance to stress.

The stress-mediated increases in SOD detected by Perl-Treves and Galun (1991) and Madamanchi et al. (1994 a) were relatively slow compared to other observed changes in antioxidants. Total glutathione and glutathione reductase activities increased in the first 40 minutes of the exposure. This is in contrast to the responses of SOD where increases in protein were not detected until the 180 minute time point and changes in enzyme activities were detected 30 minutes later. The physiological significance of this lag in the response of SOD to stress remains to be assessed.

Transgenic tobacco plants which overexpressed a pea plastid Cu/ZnSOD at a level four-times higher than wild-type showed increased resistance to high light and chilling (Sen Gupta et al., 1993 a). Thus SOD activities can be important for stress resistance. However, transgenic plants over-expressing the same plastid Cu/ZnSOD 50-fold were actually more sensitive to paraquat stress than untransformed wild-type plant. (Tepperman and Dunsmuir, 1991). Since Cu/ZnSOD activity is inhibited by the reaction product, H_2O_2, a massive increase in H_2O_2 such as would have occurred in the transformed plants with 50-fold over-expression of SOD may have lead to inhibition of enzyme activity. This result suggests that rates of H_2O_2 removal and replenishment of antioxidant substrates are crucial steps in antioxidant resistance processes in plant leaves. The observation that glutathione levels increased relatively rapidly in stressed pea leaves, by comparison with SOD, can be interpreted as evidence of the importance of other components of the scavenging cycle for resistance.

Increases in SOD activities occur in response to stress under other circumstances also. Toumainen et al. (this volume) have observed increases in SOD activities in response to ozone stress in the leaves of sensitive but not of resistant birch clones. This increase preceded the appearance of injury symptoms in the sensitive clone. In our laboratory, we have also observed increases in cytosolic SOD activities in bleached protoplasts which have been exposed to paraquat (Doulis and Alscher, unpublished data). More work is needed to establish and define the different roles of SOD in cellular responses to oxidative stress.

In an effort to avoid the biochemical problem posed by the inhibition of Cu/ZnSOD by its reaction product, transgenic plants have been produced in which mitochondrial MnSOD has been introduced into chloroplasts. MnSOD activity is unaffected by H_2O_2 and thus should continue to function even in an environment with elevated levels of H_2O_2. A mitochondrial transit peptide was replaced by a chloroplast signal sequence and the resulting transgenic tobacco showed chloroplast MnSOD activity and increased resistance to oxidative stress imposed by paraquat and ozone (Bowler et al., 1991; Trolinder and Allen, 1994). These results suggest a possible means by which crop plants could be engineered for greater resistance to oxidative stresses which originate in the chloroplasts. However, the signaling process through which antioxidant mechanisms including changes in SOD are marshaled in response to stress imposition *in vivo* remain to be elucidated.

b. Glutathione reductase is involved in antioxidant resistance mechanisms and is responsive to environmental cues. Multiple forms of GR

Three forms of GR were recently identified in red spruce needles (Hausladen and Alscher, 1994 a, b). The appearance of one of these forms was correlated with the onset of cold tolerance and was concluded to be a separate gene product, based on a comparison of its N-terminal amino acid sequence with that of the form which was present in the needles throughout the year.

Foyer et al. (1991) transformed tobacco plants with an *E. coli* gene (*gor*) encoding GR. The protein, from the GR construct which was under the regulation of the 35S CaMV promoter was expressed in the cytosol. Transgenic plants expressed statistically significantly higher GR activities (2–10 fold) than the non-transformed plants. Overexpression of the cytosolic form of GR alone did not improve tolerance to paraquat. Foyer et al. (1991) concluded that overproduction of cytosolic GR, or a single component in the antioxidant system, was not sufficient to confer resistance. In another report, transgenic tobacco plants expressing the *E. coli* GR in the cytosol showed lower susceptibility to paraquat in the light (Aono et al., 1991).

Aono et al. (1991), showed that transgenic tobacco plants which expressed the *E. coli* GR protein in the chloroplast stroma were less susceptible to sulfur dioxide and paraquat than control plants but showed no differential susceptibility to ozone. In another study, Aono et al. (1995) introduced antisense constructs to GR (AGR) into tobaccco and demonstrated a suppression of GR activity. Plants with suppressed GR activity were more sensitive to paraquat stress. Seventeen AGR plants were obtained with a 50–70 % decrease in GR activity in comparison to the wild-type plants. The plants with the lowest GR activity were self-fertilized to obtain the F_1 generation. Although these plants showed extremely low GR activity, they behaved like the wild-type plants when they were grown under non-stressed conditions. Some amount of scavenging activity is necessary to maintain physiological function, even under optimal conditions. This result could be an expression of the relatively low levels of GR needed to maintain photosynthetically active plant cells under unstressed conditions.

The impact of the GR antisense on specific GR isoforms was not directly assessed. There is evidence for the existence of multiple isoforms of GR which may be differentially expressed (Okpodu and Alscher, see above). Edwards et al. (1994) showed that changes in the composition of GR isoenzymes occurs when peas are subjected to four different oxidative stress conditions.

We have also found that leaf position and/or a type of developmental regulation of glutathione reductase activity is apparent in maturing pea leaves (Okpodu and Alscher, unpublished data). Thus the combined evidence to date suggests that multiple forms of GR exist and that some subset of these forms appear in response to environmental and developmental cues, including stress imposition. Work is currently underway to identify a GR multi-gene family, to understand its relationship to the various GR proteins present in leaves, and to define the roles played by its various members. In cases where sequence data for more than one form is available, antisense strategies which target specific isoforms may provide powerful tools for elucidating isoform function.

c. The Expression of Ascorbate Peroxidase may be linked to SOD Action

Several cDNAs encoding cAPs have been isolated and characterized in plants (Asada, 1994). cDNAs which encode the chloroplast AP (APchl) have not been identified. Sen Gupta et al. (1993) reported that 3-fold increases in AP activity and mRNA occurs in transgenic tobacco plants which overexpress chloroplast Cu/ZnSOD. This induction could be mimicked in non-transformed tobacco leaf discs with the addition of H_2O_2. These results suggest that increasing the level of gene product in one part of the pathway affects other enzymes in the same pathway. This type of co-regulation may be critical in interpreting results using antisense plants where the cytosolic Cu/ZnSOD is suppressed. These data also suggest that H_2O_2 plays a role in a stress-responsive signal transduction pathway.

V. Conclusion and Future Directions

The production of active oxygen species is inevitable in photosynthetically active plant cells even under the most favorable of environmental conditions. Accumulated molecular, biochemical and physiological evidence suggests that antioxidant mechanisms are involved in protecting the cell against oxidative stress. The use of transgenic plants to investigate the molecular basis of stress tolerance are important tools in elucidating the pathway to resistance (Rennenberg and Polle, 1994). It is clear from the evidence cited in this paper that developmental controls of stress tolerance play a larger role than previously expected. The identification and characterization of mutants in the antioxidant pathway may give deeper insights on how to genetically engineer stress-resistance into important plant species.

Acknowledgements

This work was supported by BARD (United States-Israel Binational Agricultural Research and Development Fund) grant # US 2043-91.

References

ALSCHER, R. G.: Biosynthesis and antioxidant function of glutathione in plants. Physiol. Plant. *77,* 459–464 (1989).

ALSCHER, R. G., J. L. BOWER, and W. ZIPFEL: The basis for different sensitivities of photosynthesis to SO_2 in two cultivars of pea. J. Exp. Bot. *38,* 99–108 (1987).

AONO, M., A. KUBO, H. SAJI, K. TANAKA, and N. KOMDO: Enhanced tolerance to photooxidative stress of transgenic *Nicotiana tabacum* with high chloroplastic glutathione reductase activity. Plant Cell Physiol. *34,* 129–135 (1993).

AONO, M., A. KUBO, H. SAJI, T. NATORI, K. TANAKA, and N. KONDO: Resistance to active oxygen toxicity of transgenic *Nicotiana tabacum* that expresses the gene for glutathione reductase from *Escherichia coli.* Plant Cell Physiol. *32,* 691–697 (1991).

AONO, M., H. SAJI, K. FUJIYAMA, M. SUGITA, M. KONDO, and K. TANAKA: Decrease in activity of glutathione reductase enhances paraquat sensitivity in transgenic *Nicotiana tabacum.* Plant Physiol. *107,* 645–648 (1995).

ASADA, K.: Formation and scavenging of superoxide in chloroplasts, with relation to injury by sulfur dioxide. Res. Rep. Natl. Inst. Environ. Stud. *11,* 165–179 (1980).

– Production and action of active oxygen species in photosynthetic tissues. In: FOYER, C. H. and P. M. MULLINEAUX (eds.): Causes of Photooxidative Stress and Amelioration of Defense Systems in Plants, pp. 77–104. CRC Press, Boca Raton, Florida (1994).

BANNISTER, J. V., W. H. BANNISTER, and G. ROTILIO: Aspects of the structure, function and application of superoxide dismutase. In: FASMAN, G. D. (ed.): Critical Reviews in Biochemistry, pp. 111–180. CRC Press, Inc., Boca Raton, Florida (1987).

BAUM, J. A. and J. G. SCANDALIOS: Isolation and characterization of the cytosolic and mitochondrial superoxide dismutases of maize. Arch. Biochem. Biophys. *206,* 249–264 (1981).

BEAUCHAMP, C. O. and I. FRIDOVICH: Superoxide dismutase: improved assays and an assay applicable to acrylamide gel. Anal. Biochem. *44,* 276–287 (1971).

BECK, A., A. BURKERT, and M. HOFMANN: Uptake of L-ascorbate by intact spinach chloroplasts. Plant Physiol. *73,* 41–45 (1983).

BEYER, W., J. IMLAY, and I. FRIDOVICH: Superoxide dismutases. Prog. Nucleic Acids Res. *40,* 221–253 (1991).

BIELAWSKI, W. and K. W. JOY: Properties of glutathione reductase from chloroplasts and roots of pea. Phyotochem. *25,* 2261–2265 (1986).

BOWLER, C., W. VAN CAMP, M. VAN MONTAGU, and D. INZE: Superoxide dismutase in plants. Critical Review in Plant Sciences *13,* 199–218 (1994).

BOWLER, C., L. SLOOTEN, S. VANDENBRANDER, R. D. RYCKE, J. BOTTERMAN, C. SYBESMA, M. VAN MONTAGU, and D. INZE: Manganese superoxide dismutase can reduce cellular damage mediated by oxygen radicals in transgenic plants. EMBO J. *10,* 1723–1732 (1991).

CHEN, G. X. and K. ASADA: Ascorbate peroxidase in tea leaves: occurrence of two isozymes and the differences in their enzymatic and molecular properties. Plant Cell Physiol. *30,* 987–998 (1989).

CREISSEN, G., A. EDWARDS, and P. MULLINEAUX: Glutathione reductase and ascorbate peroxidase. In: FOYER, C. H. and P. MULLINEAUX (eds.): Causes of photooxidative stress and amelioration of defense systems in plants. pp. CRC Press, Inc., Boca Raton, Florida (1994).

CROWELL, D. N. and R. M. AMASINO: Nucleotide sequence of an iron superoxide dismutase complementary DNA from soybean. Plant Physiol. *96,* 1393–1394 (1991).

DODGE, A. D.: The mode of action of bipyridylium herbicides, paraquat and diquat. Endeavour *30,* 130–135 (1971).

EDWARDS, E. A., S. RAWSTHORNE, and P. M. MULLINEAUX: Subcellular distribution of multiple forms of glutathione reductase in leaves of peas (*Pisum sativum* L.). Planta *180,* 278–284 (1990).

EDWARDS, E. A., C. ENARD, G. P. CREISSEN, and P. M. MULLINEAUX: Synthesis and properties of glutathione reductase in stressed peas. Planta *192,* 137–143 (1994).

EHLERINGER, J. and O. BJÖRKMAN: Quantum yields for CO_2 uptake in C_3 and C_4 plants: dependence on temperature, CO_2 and O_2 concentrations. Plant Physiol. *59,* 86–90 (1977).

FOYER, C. H. and P. M. MULLINEAUX (eds.): Causes of photooxidative stress and amelioration of defense systems in plants. CRC Press, Inc., Boca Raton, Florida (1994).

FOYER, C. H. and M. LELANDAIS: The role of ascorbate in the regulation of photosynthesis. Current Topics in Plant Physiol. *8,* 89–101 (1993).

FOYER, C. H., M. LELANDAIS, E. A. EDWARDS, and P. M. MULLINEAUX: The role of ascorbate in plants, interactions with photosynthesis and regulatory significance. Current Topics in Plant Physiol. *6,* 131–145 (1991).

FOYER, C. H., M. LELANDAIS, C. GALAP, and K. J. KUNERT: Effects of elevated cytosolic glutathione reductase activity on the cellular glutathione pool and photosynthesis in leaves under normal and stress conditions. Plant Physiol. *97,* 863–872 (1991).

FRIDOVICH, I.: Superoxide dismutases. Adv. Enzymol. *58,* 61–97 (1986).

GRESSEL, J. and E. GALUN: Genetic controls of photooxidant tolerance. In: FOYER, C. and P. MULLINEAUX (eds.): Photooxidative stresses in plants, pp. 237–275. CRC Press, Boca Raton, Florida (1994).

HALLIWELL, B. and C. H. FOYER: Properties and physiological function of a glutathione reductase purified from spinach leaves by affinity chromatography. Planta *139,* 9–17 (1978).

HAUSLADEN, A. and R. G. ALSCHER: Cold-hardiness-specific glutathione reductase isozymes in red spruce. Plant Physiol. *105,* 215–223 (1994b).

– – Purification and characterization of glutathione reductase isozymes specific for the state of cold hardiness of red spruce. Plant Physiol. *105,* 205–213 (1994a).

– – Glutathione. In: ALSCHER, R. G. and J. L. HESS (eds.): Antioxidants in Higher Plants, pp. 1–30. CRC Press, Boca Raton, Florida (1993).

KENYLON, W. H. and S. D. DUKE: Effects of acifluorfen on endogenous antioxidants and protective enzymes in cucumber. Plant Physiol. *79,* 862–866 (1985).

LOEWUS, F. A.: Ascorbic acid and its metabolic products. In: PREISS, J. (ed.): The Biochemistry of Plants, pp. 85–107. Academic Press, New York (1988).

MADAMANCHI, N. R., J. V. ANDERSON, R. G. ALSCHER, C. L. CRAMER, and J. L. HESS: Purification of multiple forms of glutathione reductase from pea (*Pisum sativum* L.) seedlings and enzyme levels in ozone-fumigated pea leaves. Plant Physiol. *100,* 138–145 (1992).

MADAMANCHI, N. R., X. YU, A. DOULIS, R. G. ALSCHER, K. K. HATZIOS, and C. L. CRAMER: Acquired resistance to herbicides in pea cultivars through pretreatment with sulfur dioxide. Pest. Biochem. and Physiol. *48,* 31–40 (1994a).

MADAMANCHI, M. R., J. L. DONAHUE, C. L. CRAMER, R. G. ALSCHER, and K. PEDERSEN: Differential response of Cu,Zn superoxide dismutases in two pea cultivars during a short-term exposure to sulfur dioxide. Plant Mol. Biol. *26,* 95–103 (1994b).

MADAMANCHI, N. R. and R. G. ALSCHER: Metabolic bases for differences in sensitivity of two pea cultivars to sulfur dioxide. Plant Physiol. *97,* 88–93 (1991).

MAHAN, J. R. and J. J. BURKE: Purification and characterization of glutathione reductase from corn mesophyll chloroplasts. Physiol. Plant. *71,* 352–358 (1987).

MALAN, C., M. M. GREYLING, and J. GRESSEL: Correlation between CuZn superoxide dismutase and glutathione reductase and environmental and xenoboitic stress tolerance in maize inbreeds. Plant Sci. *69,* 157–166 (1990).

MEHLHORN, H. and A. R. WELLBURN: Man-induced causes of free radical damage: O_3 and other gaseous pollutants. In: FOYER, C. and P. MULLINEAUX (eds.): Photooxidative stresses in plants, pp. 155–176. CRC Press, Boca Raton, Florida (1994).

MISRA, H. P. and I. FRIDOVICH: Purification and properties of superoxide dismutase from a red alga, *Porphyridium.* J. Biol. Chem. *252,* 6421–6423 (1977).

MITTLER, R. and B. A. ZILINSKAS: Purification and characterization of pea cytosolic ascorbate peroxidase. Plant Physiol. *97,* 962–968 (1991).

MIYAKE, C. and K. ASADA: Ferrodoxin-dependent photoreduction of the monodehydroascorbate radical in spinach thylakoids. Plant Cell Physiol. *35,* 539–549 (1994).

NAKANO, Y. and Y. ASADA: Purification of ascorbate peroxidase from spinach chloroplasts: its inactivation in ascorbate-depleted medium and reactivation by monodehydroascorbate radical. Plant Cell Physiol. *28,* 131–140 (1987).

NEUBAUER, C. and U. SCHREIBER: Photochemical and non-photochemical quenching of chlorophyll flourescence induced by hydrogen peroxide. Z. Naturforsch *44 C,* 262–270 (1989).

PERL, A., R. PERL-TREVES, S. GALILI, D. AVIV, E. SHALGI, S. MALKIN, and E. GALUN: Enhanced oxidative-stress defense in transgenic potato expressing tomato Cu,Zn superoxide dismutases. Theor. Appl. Genet. *85,* 568–576 (1993).

PERL-TREVES, R. and E. GALUN: Developmental and light-induced regulation and stress-enhanced expression of two Cu,Zn superoxide dismutase genes in tomato. Plant Mol. Biol. *17,* 745–760 (1991).

RENNENBERG, H. and A. POLLE: Protection from oxidative stress in transgenic plants. Biochem. Soc.Trans. *22,* 936–939 (1994).

SALIN, M. L.: Plant superoxide dismutases: A means of coping with oxygen radicals. Curr. Top. Plant Biochem. Physiol. *7,* 188–200 (1988).

SANDALIO, L. M. and L. A. DEL RIO: Localization of superoxide dismutase in glyoxysomes from *Citrullus vulgaris.* Functional implications in cellular metabolism. J. Plant Physiol. *127,* 395–409 (1987).

SCHIFFGENS-GRUBER, A. and C. LUTZ: Ultrastructure of mesophyll cell chloroplasts of spruce needles exposed to O_3, SO_2, and NO_2 alone and in combination. Environ. Expt. Botany *32,* 243–254 (1992).

SCHREIBER, U. and C. NEUBAUER: O_2-dependent electron flow, membrane energisation and the mechanism of non-photochemical quenching of chlorophyll fluorescence. Photosynthesis Res. *25,* 279–293 (1990).

SEN GUPTA, A., J. L. HEINEN, A. S. HOLADAY, J. J. BURKE, and R. D. ALLEN: Increased resistance to oxidative stress intransgenic plants that overexpress chloroplasts Cu/Zn superoxide dismutase. Proc. Natl. Acad. Sci. USA *90,* 1629–1633 (1993).

SHAALTIEL, Y., N.-H. CHUA, S. GEPSTEIN, and J. GRESSEL: Dominant pleiotropy controls enzymes co-segregating with paraquat resistance in *Conyza bonariensis.* Theor. Appl. Genet. *75,* 850–856 (1988).

SUTINEN, S.: Ultrastructure of mesophyll cells in and near necrotic spots on otherwise normal needles of Norway spruce. Eur. J. For. Pathol. *85,* 379–384 (1988).

TEPPERMAN, J. M. and P. DUMSMUIR: Transformed plants with elevated levels of chloroplastic SOD are not more resistant to superoxide toxicity. Plant Mol. Biol. *14,* 501–511 (1990).

TROLINDER, N. L. and R. D. ALLEN: Expression of chloroplast localized Mn SOD in transgenic cotton. J. Cell Biochem. 18A, *97* (1994).

TSANG, E. W. T., C. BOWLER, D. HEROUART, W. VAN CAMP, R. VILLARROEL, C. GENETELLO, M. VAN MONTAGU, and D. INZE: Differential regulation of superoxide dismutases in plants exposed to environmental stress. Plant Cell *3,* 783–792 (1991).

VAN CAMP, W., C. BOWLER, R. VILLARROEL, E. W. T. TSANG, M. VAN MONTAGU, and D. INZE: Characterization of iron superoxide dismutase cDNAs from plants obtained by genetic complementation in *Escherichia coli.* Proc. Natl. Acad. Sci. USA *87,* 9903–9907 (1990).

YOUNGMAN, R. J.: Oxygen activation: is the hydroxyl radical always biologically relevant? Trends Biochem. Sci. *9,* 280–285 (1984).

ZIEGLER, I.: The effect of SO_2-pollution on plant metabolism. Residue Rev. *56,* 79–105 (1975).

J. Plant Physiol. Vol. 148. pp. 317–323 (1996)

Effects of Root Damage on the Nutritional Status and Structure of Scots Pine Needles

ANNE JOKELA[1], VIRPI PALOMÄKI[2], SATU HUTTUNEN[1], and RISTO JALKANEN[3]

[1] University of Oulu, Department of Botany, FIN-90570 Oulu, Finland

[2] University of Kuopio, Department of Ecology and Environmental Science, P.O. Box 1627, FIN-70211 Kuopio, Finland

[3] The Finnish Forest Research Institute, Rovaniemi Research Station, P.O. Box 16, FIN-96301 Rovaniemi, Finland

Received June 24, 1995 · Accepted September 25, 1995

Summary

In this experiment root damage was induced artificially in 35-year-old Scots pine trees growing on a nutrient-poor pine heath in northern Finland by cutting their surface roots at the beginning of the growing season. Nutrient concentrations in the needles were measured during the next growing season and the development of changes in the inner structure of the needles was observed. The needles were examined by light microscopy and at the ultrastructural level to reveal early symptoms of possible nutrient imbalances. Needle nutrient concentrations were lowered after root cutting, and changes in needle microscopic structure caused by nutrient imbalance were detected. Root reduction induced shorter needle length growth, and morphometric measurements with an image analyzer showed that the needle cross-sectional area, areas of various tissues and the sizes of certain cell types were reduced. Injury to the phloem tissue in the vascular bundle was indicative of a nutritional imbalance in the needles. The injuries to the mitochondrial structure and abnormally arranged chloroplast thylakoids at ultrastructural level pointed to P deficiency and the translucency of the chloroplast stroma to N deficiency. Vacuolization of the cytoplasm was induced, possibly indicating cold damage.

Key words: Pinus sylvestris L., needle morphology, needle ultrastructure, root damage.

Abbreviations: LRR = light root reduction; SRR = severe root reduction.

Introduction

The premature needle loss observed among Scots pines (*Pinus sylvestris* L.) in northern Finland in 1987 raised the question whether their nutrient uptake had been affected by root injury. The oldest needles of Scots pine started to turn yellow at the height of the growing season, in July 1987, and they had been shed within one month, well before the end of the growing season in southern part of northern Finland. This exceptional needle loss was most severe on the poorest dry pine heaths, characterized by a fine-textured sandy soil, a thin humus layer and a very effectively browsed *Cladonia* spp. layer (Jalkanen, 1988). From one to five needle year classes were lost, normal needle retention being 4 to 7 classes. The height growth of the pines was typically retarded quite considerable in the same summer in which the needle loss occurred. The shortened shoots of peculiar appearance had dwarfed needles. The trees began to recover in 1988 (Jalkanen, 1990).

It was shown experimentally that the needle loss was caused by frost affecting the roots during the cold, snowless winter (Jalkanen et al., 1995), although air pollutants, and especially ozone were also suspected as the main causal agents at the early stages of the research (Sutinen and Pesonen, 1990, see also Tikkanen and Raitio, 1990). As there was a contradiction between the ultrastructural observations on Scots pine needles and the experimental results regarding the effects of cold stress on the roots, there arose a need for more careful investigations into the relations between root reduction and tree health.

We have now used microscopic methods to evaluate the possible role of nutrients in needle loss. Microscopy is a useful method for investigating nutritional disturbances (Holopainen et al., 1992) as cell structure injuries can be detected before the occurrence of any visible injuries to trees, such as needle discoloration. The specific reactions of microscopically viewed target cells to the actions of pathogenic factors seem to offer a more differentiated diagnostic approach than the quantitative biochemical analysis of parameters in tissue homogenates (Fink, 1993). Experimentally induced nutritional disturbances have been observed to alter the cell structure of Scots pine needles, and specific symptoms of nutrient deficiencies have been described (Palomäki and Holopainen, 1994; Jokela et al., 1995 a and b; Palomäki and Holopainen, 1995).

The aim of the present work was to find out what nutritional disturbances, if any, are caused by artificial root reduction, what injuries are visible in the needle morphology and cell structure and what interactions may take place between nutrient deficiency and microscopic injuries.

Materials and Methods

Experimental sites

The first artificial root reduction experiment (experiment 1) was established on a dry pine heath with fine-textured sandy soil in late May 1990 in the Kivalo experimental area of the Finnish Forest Research Institute in northern Finland (66° 20′ N, 26° 45′ E) and the second one (experiment 2) in the late May 1992. In experiment 1 four randomly selected trees were studied per each treatment on the 20 sample plots (Jalkanen et al., 1995). In experiment 2, three evenly sized sample trees were chosen on each of the ten sample plots, located randomly but at least 10 m apart in the stand. The trees were on average 2.9 m and 2.2 m in height in experiments 1 and 2, respectively. One pine on each sample plot for each experiment served as a control tree and other two were subjected to either light root reduction (LRR) or severe root reduction (SRR). In SRR all the surface roots to a depth of 20 cm were cut with a spade at a distance of 1.0 m from the stem, while in the case of LRR the roots were cut only around the northern half of the circle in experiment 1 and the western half in experiment 2.

Samples for microscopy

The needle samples required for microscopy in experiment 1 were obtained in winter, on Feb. 18, 1991, nine months after root reduction. In experiment 2, needles were collected twice, on June 30 and Sept. 21, 1992, one and four months after root reduction. The needles were collected from the middle of the shoot directly into test tubes filled with 2 % glutaraldehyde prefixative in phosphate buffer (pH 7.0, 0.1 M) in experiment 1 and 1.5 % paraformaldehyde-glutaraldehyde in cacodylate buffer (pH 7) of molarity 0.05 M on 30. 6. 1992 and 0.075 M on 21. 9. 1992 in experiment 2. The needles were cut into 0.5 mm cross-sections and these were postfixed with 1 % OsO_4 for 6 h at +4 °C, infiltrated with propylenoxide and embedded in Ladd's epon. Needle cross-sections were cut with an ultramicrotome. Semi-thin sections of 1–3 µm for light microscopy were stained with toluidine blue, and ultra-thin sections of 50–70 nm with lead citrate and uranyl acetate.

Observations by light microscopy and needle length measurements

Observations were made by light microscopy only in experiment 2 employing samples collected both in early summer and at the end of growing season, in order to follow the development of the needles after root cutting. The samples were examined under a Nikon OPTIPHOT-2 light microscope, and morphometrical measurements were performed using a digital image analyser (Microscale TM/TC, Digithurst Ltd., England) and a video camera (Hitachi KP-C571 CCD color camera). Compression and shrinkage of the tissues during embedding in plastic and sectioning were assumed to be minimal and approximately equal in all the samples (Toth, 1982). The variables measured (Fig. 1) were the thickness and width of the needle, the thicknesses of the mesophyll and epidermis+hypodermis (measured on the adaxial and abaxial sides), the diameter of the central cylinder (including bundle sheath cells, transfusion parenchyma cells and sclerenchyma tissue and measured at the vascular bundle and at the centre of the cylinder), the size of the resin ducts, the size of the bundle sheath cells (adaxial and abaxial), the thickness of the sclerenchyma cell walls, and the areas of the whole needle, the central cylinder, the phloem and the xylem. The number of resin ducts was also calculated.

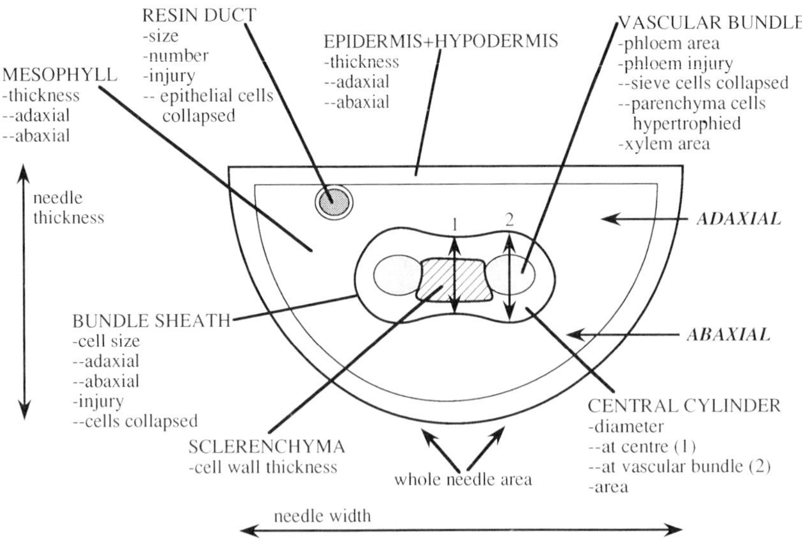

Fig. 1: Cross-section of a Scots pine needle in light microscopy with morphometric variables measured with an image analyzer and tissue condition classifications.

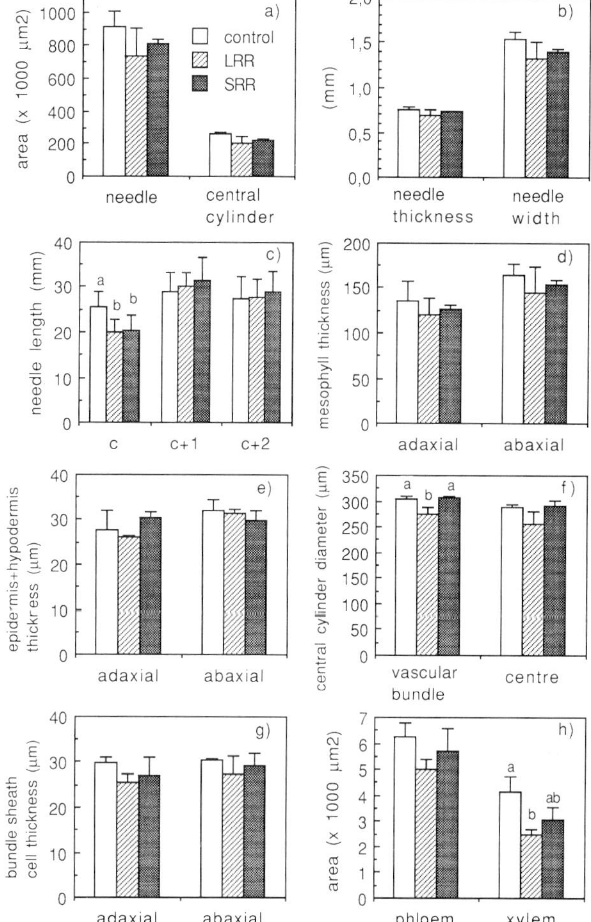

Fig. 2: Results of morphometric measurements of Scots pine needle cross-section performed with an image analyzer and measurements of needle length in control, LRR and SRR, n=3 trees per treatment. The different letters indicate statistically significant differences between the treatments ($p<0.05$). a. Needle and central cylinder area. b. Needle thickness and width. c. Needle length in current (c), current+1 (c+1) and current+2 (c+2) needle years. (The current-year needles had developed after root reduction, and c+1 and c+2 needle years before it). d. Mesophyll thickness measured on the adaxial and abaxial sides of the needle. e. Epidermis+hypodermis thickness measured at the corresponding points to mesophyll thickness. f. Central cylinder diameter measured at the vascular bundle and at the centre of the central cylinder. g. Bundle sheath cell thickness measured on the adaxial and abaxial sides of the needle. h. Phloem and xylem area.

The thicknesses of the mesophyll, epidermis+hypodermis and central cylinder in relation to needle thickness and the areas of the central cylinder, phloem and xylem in relation to the whole needle area were calculated.

The various tissues, the phloem in the vascular bundle, bundle sheath cells and resin duct epithelium, were classified in terms of condition (Fig. 1). A total of 51 needles were examined (25 collected on 30.6. and 26 collected on 21.9.), representing 2–3 needles per age class and 3 trees per treatment.

The branches of the 4th whorl were used for needle length measurements, taking 25 short shoots per needle age class.

Ultrastructural observations

At the ultrastructural level the condition of the chloroplast, mitochondria, cytoplasm, plasma membrane and tonoplast was assessed in current-year needles. A total of 12 and 9 trees, respectively, were studied in experiments 1 (needles collected on Feb. 18) and 2 (needles collected on Sept. 21). 1–2 needles per tree and 4–13 cells per needle cross-section were examined in each experiments.

Nutrient analysis

The needles on the sixth branch whorl were sampled for nutrient analysis. These had developed in 1987–90 in experiment 1, sampled on 7th September 1990, and in 1989–92 in experiment 2, sampled on 28th July 1992. Nitrogen was analysed by the Kjeldahl method and P, K, Ca and Mg by plasma emission spectrometry (Halonen et al., 1983).

Statistics

The morphometric and tissue injury classifications and ultrastructural data were analysed using the SAS (1988) and nutrient analysis data with SYSTAT (1992). The differences in morphometric data between treatments were tested by an analysis of variance and Tukey's studentized range (HSD) test using the means of two to three replicate trees. The distribution of the variables into tissue and cell injury classes was assessed with the likelihood ratio chi-square test.

Results

Observations by light microscopy and needle length

Light microscopy was used in experiment 2 to ascertain how root reduction influenced the development of the needles during the growing season. Needle thickness and width and the areas of the whole needle and the central cylinder, the size of bundle sheath cells and the thickness of mesophyll at the end of the growing season (Sept. 21) were all greatest in the controls and smallest in the LRR samples (Fig. 2 a, b, d and g). Needle length (Fig. 2 c) was greatest in the controls, but nearly the same after LRR and SRR in the current-year needles ($p=0.000$), while no differences in the length of the older needles (current+1 and current+2) was found between the treatments. The thickness of the epi+hypodermis was greatest after SRR on the adaxial side of the needle (Fig. 2 e).

The diameter of the central cylinder was smallest after LRR (Fig. 2 f), but almost the same in the control and SRR needles when measured at the vascular bundle ($p=0.013$) and at the centre of the cylinder. No differences in sclerenchyma cell wall thickness were noted between the treatments (data not shown). The resin ducts were indicatively larger in the controls (117 ± 4 µm), than in the LRR (100 ± 13 µm) and SRR needles (102 ± 5 µm). No significant changes in the number of resin ducts were found after root reduction (control 6.8 ± 0.8, LRR 5.8 ± 0.4 and SRR 6.3 ± 1.3).

The areas of phloem and xylem were smallest in the LRR samples and largest in the controls, the differences in xylem area being statistically significant ($p=0.012$, Fig. 2 h). A slight increase in phloem injuries could be observed after SRR (Fig. 3 a), these being exhibited by 33 % of the needles after this treatment but in no cases after the other treatments. The resin ducts and bundle sheath cells did not show any increase in injuries as a consequence of root reduction.

The thicknesses of the mesophyll, epidermis+hypodermis and central cylinder in relation to needle thickness and the areas of the central cylinder, phloem and xylem in relation to the whole needle area showed the same trend in differences between treatments as the original variables (thicknesses of mesophyll, epidermis+hypodermis and central cylinder, and areas of central cylinder, phloem and xylem).

The morphometric measurements at the beginning of the growing season (June 30) showed that the control needles were already larger than the LRR and the SRR needles, but there were smaller differences between the LRR and SRR samples than at the end of the growing season (Sept. 21). This was especially clear in whole needle and central cylinder areas, needle thickness and width and resin duct size. The phloem cells were still in good condition at the beginning of growing season after all the treatments.

Ultrastructural observations

In experiment 1 (Table 1) the needles collected in February showed swelling of the mitochondria in the mesophyll as the clearest symptom after LRR treatment (Fig. 3b), while SRR treatment had led to swelling of the mitochondria, a decrease in the number of cristae in them, an accumulation of large lipid bodies in cytoplasm and vacuolization of the cytoplasm (Fig. 3c). In addition, translucency of the chloroplast stroma (Fig. 3c) was noted in 86% of the cells and a decrease in the number of thylakoids in the chloroplasts in 42% after SRR, whereas these changes were observed in under 10% of cases after LRR and in the controls the differences being statistically significant.

Curling membraneous structures with black globules were frequently observed in the mesophyll cells after LRR (Table 1) in experiment 2 (needles collected in September), and also in the control and SRR needles (Figs 3d and e), but not so frequently. Translucent cytoplasm was a common symptom in the control (Fig. 3d) and LRR samples (Table 1). Abnormally arranged chloroplast thylakoids were seen after LRR, and the plasma membrane in the mesophyll cells was separated from the cell wall. Both small and large lipid accumulations were observed after SRR. No increase in the changes in mitochondrial structure were detected as a consequence of root reduction. Tonoplast proliferation (Fig. 3f), cytoplasmic vacuolization and swelling of the chloroplast thylakoids were observed in all treatments in experiment 2.

Table 1: Ultrastructural observations of changes in cell structure. Percentages of observed cells possessing the injury. The number of observed cells per treatment was 14–18 in experiment 1 and 27–30 in experiment 2. The needles in experiment 1 were collected in February and those in experiment 2 in September. The differences between the treatments were assessed with the likelihood ratio chi-square test ($p<0.05^{*}$, $p<0.01^{**}$, $p<0.001^{***}$). LRR = light root reduction, SRR = severe root reduction.

Variable	experiment	treatment			p
		control	LRR	SRR	
Chloroplast					
swelling of chloroplast	1	0	0	7	NS
thylakoids	2	17	7	13	NS
abnormal arrangement of	1	29	28	64	NS
chloroplast thylakoids[1]	2	37	61	35	*
Cytoplasm					
curling structures	1	21	22	21	NS
	2	23	61	13	***
small lipid accumulations	1	100	94	100	NS
	2	43	25	74	***
large lipid accumulations	1	21	39	100	***
	2	3	0	16	*
cytoplasmic vacuolization	1	50	44	93	***
	2	33	39	55	NS
translucent cytoplasm	1	7	11	21	NS
	2	40	46	10	***
Membranes					
proliferation of tonoplast	1	36	11	7	NS
	2	10	25	19	NS
plasma membrane separated	1	29	6	7	NS
from cell wall	2	27	68	36	***
Mitochondria					
swelling of mitochondria[1]	1	0	100	100	***
	2	83	48	50	*
swollen cristae[1]	1	0	22	7	*
	2	22	20	11	NS
decrease in number of cristae[1]	1	0	39	57	***
	2	17	16	17	NS

[1] See Palomäki and Holopainen (1994).

Nutrient analysis

Root reduction altered the nutritional status of the pine needles in both experiments (Table 2), reducing the N, P, K, Ca and Mg contents in the current-year needles. The contents of Ca and Mg in the current-year needles were significantly lowered after LRR and SRR in both experiments, and

Fig. 3: Light and transmission electron micrographs. a. Light microscopy level. Phloem injury in a needle that had developed after SRR (experiment 2). The parenchyma cells are hypertrophied (arrow) and the sieve cells (s) have collapsed. b = bundle sheath cell, sc = sclerenchyma cells, x=xylem. Bar = 0.1 mm. b.–f. Mesophyll cells at ultrastructural level. b. Swelling of mitochondria (m) and lipid accumulations (l) in a mesophyll cell after LRR (experiment 1). Bar = 1 µm. c. Swelling of mitochondria with a decrease in the number of cristae (m). There is also vacuolization (v) of the cytoplasm (experiment 1) and the chloroplast (c) stroma is translucent after LRR (experiment 1). Bar = 1 µm. d. Translucent cytoplasm in a mesophyll cell of a control needle. There are numerous curling membraneous structures (arrows) in the cytoplasm (experiment 2). Bar = 2 µm. e. Curling structure (arrow) in a mesophyll cell after SRR (experiment 2). Bar = 1 µm. f. Heavily proliferated tonoplast (arrow) in a SRR mesophyll cell (experiment 2). n = nucleus, Bar = 1 µm. Abbreviations: c = chloroplast, v = vacuole in the cytoplasm.

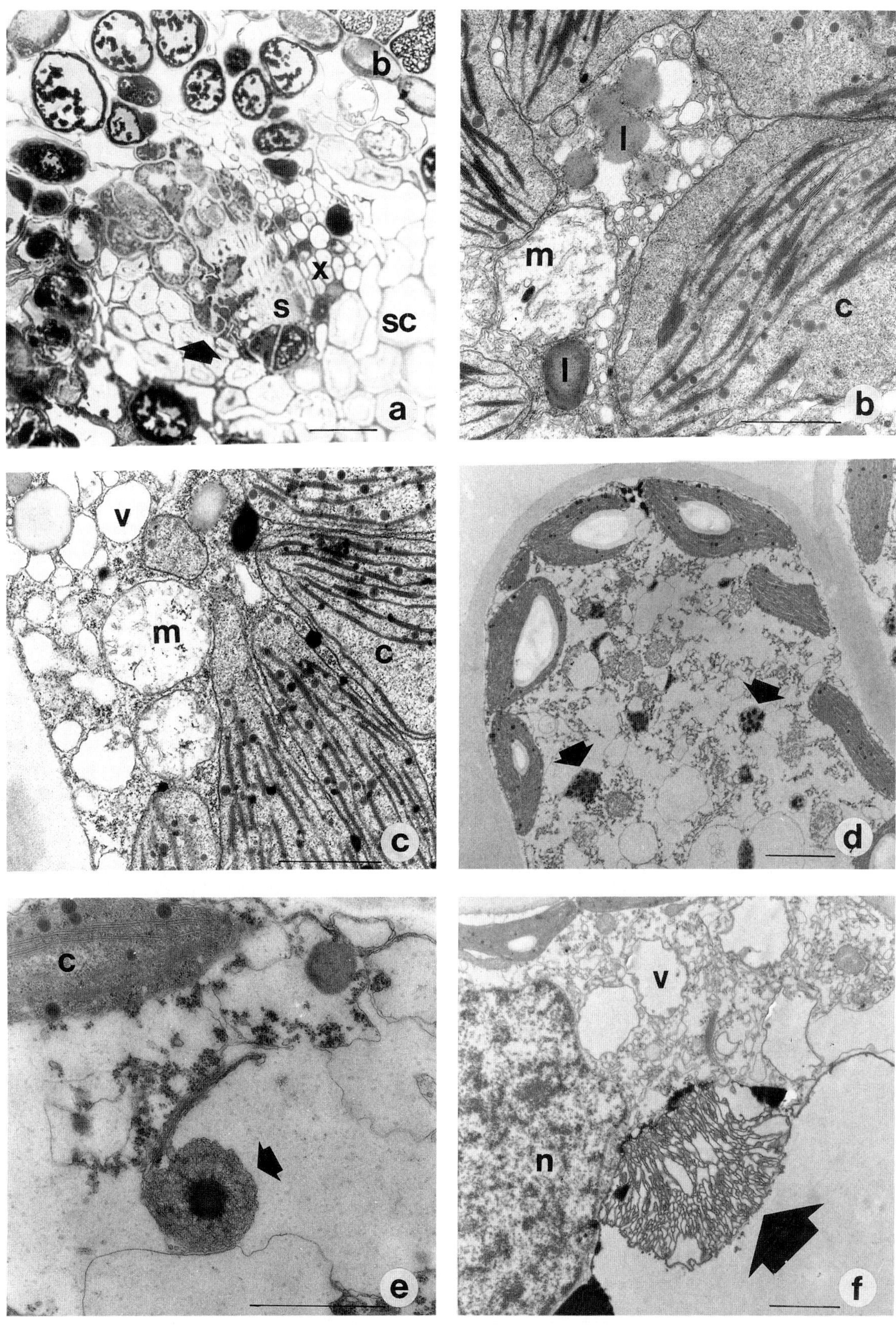
b
x
s
sc
a
l
m
l
c
b
v
m
c
c
d
c
e
v
n
f

Table 2: Needle nutrient contents in current, current+1, current+2 and current+3 needle years in experiment 1 (7th September, 1990) and experiment 2 (28th July, 1992), n=number of trees, analysis of variance: $p<0.05^*$, $p<0.01^{**}$, $p<0.001^{***}$.

nutrient	experi-ment	treatment						p
		control		LRR		SRR		
		mean	STD	mean	STD	mean	STD	
current								
N, % DW^{-1}	1	1.06	0.08	1.04	0.12	0.98	0.11	NS
	2	1.04	0.07	0.87	0.10	0.80	0.05	***
P, g/kg	1	1.27	0.14	1.24	0.09	1.20	0.11	NS
	2	1.43	0.16	1.32	0.14	1.22	0.08	*
K, g/kg	1	4.13	0.38	4.11	0.40	4.03	0.58	NS
	2	5.90	0.70	5.88	1.45	4.67	0.53	*
Ca, g/kg	1	2.60	0.34	2.00	0.39	1.82	0.55	***
	2	2.00	0.35	1.15	0.20	0.95	0.13	**
Mg, g/kg	1	1.12	0.14	1.06	0.16	0.97	0.14	*
	2	0.92	0.12	0.78	0.12	0.67	0.10	***
n	1	15		18		15		
n	2	10		9		8		
current+1								
N, % DW^{-1}	1	1.02	0.14	0.97	0.11	0.93	0.11	NS
	2	0.88	0.06	0.75	0.06	0.72	0.07	***
P, g/kg	1	1.15	0.12	1.11	0.11	1.10	0.11	NS
	2	0.83	0.04	0.78	0.05	0.71	0.08	***
K, g/kg	1	3.28	0.36	3.20	0.33	3.12	0.51	NS
	2	2.94	0.43	2.56	0.45	2.28	0.26	**
Ca, g/kg	1	3.50	0.63	2.91	0.53	2.94	0.52	**
	2	3.10	0.50	2.27	0.25	2.34	0.28	***
Mg, g/kg	1	0.92	0.21	0.87	0.18	0.84	0.12	NS
	2	0.74	0.13	0.69	0.10	0.63	0.08	NS
n	1	19		20		19		
n	2	10		9		8		
current+2								
N, % DW^{-1}	1	0.95	0.12	0.93	0.10	0.88	0.09	NS
	2	0.88	0.07	0.75	0.06	0.73	0.06	***
P, g/kg	1	1.19	0.09	1.17	0.16	1.12	0.14	NS
	2	0.85	0.05	0.72	0.09	0.72	0.05	***
K, g/kg	1	3.15	0.38	3.13	0.31	3.08	0.53	NS
	2	2.60	0.45	2.28	0.41	2.13	0.21	*
Ca, g/kg	1	3.96	0.90	3.54	0.81	3.65	0.86	NS
	2	4.71	0.94	3.50	0.52	3.70	0.36	**
Mg, g/kg	1	0.81	0.21	0.76	0.16	0.77	0.14	NS
	2	0.63	0.16	0.60	0.10	0.55	0.10	NS
n	1	20		20		20		
n	2	10		8		8		
current+3								
N, % DW^{-1}	1	0.92	0.08	0.87	0.10	0.83	0.09	NS
	2	0.85	0.06	0.74	0.05	0.71	0.07	***
P, g/kg	1	1.12	0.11	1.12	0.13	0.88	0.21	*
	2	0.87	0.06	0.74	0.06	0.73	0.08	***
K, g/kg	1	2.80	0.21	2.85	0.24	2.30	0.79	NS
	2	2.48	0.43	2.17	0.39	1.96	0.20	*
Ca, g/kg	1	4.05	0.90	3.62	0.69	3.40	1.31	NS
	2	5.23	1.17	4.20	0.76	4.22	0.42	*
Mg, g/kg	1	0.66	0.20	0.53	0.14	0.48	0.14	NS
	2	0.54	0.18	0.53	0.11	0.45	0.12	NS
n	1	7		9		4		
n	2	10		9		8		

the N content was lowered after LRR and SRR in experiment 2 as were P and K after SRR. The N content of the needles was quite low in all treatments in both experiments. The nutrient contents were also highest in the controls and lowest in the SRR material in the older needle years, and this was expecially clear with regard to N and P in experiment 2.

Discussion

Root damage in the form of artificial root reduction clearly altered the nutritional status of the pine needles (e.g. Raitio, 1990), and the fact that the differences between treatments were clearer in experiment 2 than in experiment 1 may be caused by the different sampling season (e.g. Palomäki and Raitio, 1995). The shorter needles with a smaller cross-sectional area seen after root reduction may indicate increased xeromorphism (Stover, 1944), but the limited nutrient resources seem to be efficiently used for needle thickness and width growth after SRR, as the needle cross-sectional dimensions were larger after SRR than after LRR. The thickness of the central cylinder has been observed earlier to be increased at a low needle K level (Jokela et al., 1995 b), but a thick central cylinder was not a specific symptom of nutrient imbalance in this material.

Although nutritional imbalances have been reported earlier to cause thinning of the cell walls in the sclerenchyma (Raitio, 1981; Jokela et al., 1995 a), the sclerenchyma cell walls studied here did not respond to root reduction. The injuries to the phloem structure of conifer needles observed at the end of the present experiment are closely related to nutrient imbalances (Fink, 1989, 1991 and 1993; Jokela et al., 1995 a). The worsening of the nutritional status during the growing season may have caused these phloem injuries.

Swelling of the mitochondria, a decrease in the number of cristae and an abnormal arrangement of chloroplast thylakoids in pine needles, observed in the present experiments, have earlier been attributed to a phosphorus deficiency in similar P concentrations (Palomäki and Holopainen, 1994). Changes in mitochondrial structure have been linked with a decrease in ATP synthesis in maize leaves (Ciamporova and Mistrik, 1993). The clearer response of the mitochondrial structure to root reduction in experiment 1 was probably caused by the longer exposure time. The translucency of the chloroplast stroma connected earlier with N deficiency (Palomäki and Holopainen, 1995) may result from a decrease in ribulose-1,5-bisphosphate (RuBP) carboxylase. The synthesis of photosynthetic enzymes in the stroma is closely controlled by the extent of N influx into the leaf (Makino et al., 1984). Translucency of cytoplasm, which was frequent in all treatments, may have been caused by the naturally low N (Palomäki and Holopainen, 1995). The vacuolization of the cytoplasm frequently observed here in wintertime needles after SRR has been connected earlier with a lowering in cold tolerance induced by acid precipitation (Bäck et al., 1994).

Tonoplast proliferations have earlier been regarded as symptoms of potassium deficiency (Holopainen et al., 1992) or excess N (Soikkeli and Kärenlampi, 1984; Jokela et al., 1995 a). In the present experiments, this effect, and also the lipid accumulations in the cytoplasm and separation of the plasma membrane from the cell wall, may be general stress symptoms (Holopainen et al., 1992). The curling membraneous structures with black globules have earlier been found in conifers affected by air pollutants (Soikkeli, 1981), but their causal agent is still unknown.

The injuries detected by light microscopy and at the ultrastructural level reflect a nutrient imbalance caused by root reduction. Symptoms related to deficiencies in major nutrients (N and P, Palomäki and Holopainen, 1994 and 1995) and to general nutrient imbalance (Fink, 1991; Jokela et al., 1995 a and b) were observed.

Acknowledgements

We thank Mrs. Tellervo Siltakoski, Eija Kukkola, M. Sc. and Mr. Tarmo Aalto, for technical assistance, Dr. Toini Holopainen and Dr. Jaana Bäck for valuable comments, and Mr. Malcolm Hicks for revising the language of the manuscript. This research was financed by the Kone Foundation and the Academy of Finland.

References

Bäck, J., S. Neuvonen, and S. Huttunen: Pine needle growth and fine structure after prolonged acid rain treatment in the subarctic. Plant, Cell Environ. *17,* 1009–1021 (1994).

Ciamporova, M. and I. Mistrik: The ultrastructural response of root cells to stressful conditions. Environ. Exp. Bot. *33,* 11–26 (1993).

Fink, S.: Pathological anatomy of conifer needles subjected to gaseous air pollutants or mineral deficiencies. Aquilo Ser. Bot. *27,* 1–6 (1989).

– Structural changes in conifer needles due to Mg and K deficiency. Fertilizer Research *27,* 23–27 (1991).

– Microscopic criteria for the diagnosis of abiotic injuries to conifer needles. In: Hüttl, R. F. and D. Müller-Dombois (eds.): Forest Decline in Atlantic and Pacific Region, pp. 175–188. Springer-Verlag, Berlin (1993).

Halonen, O., H. Tulkki, and J. Derome: Nutrient analysis methods. Metsäntutkimuslaitoksen tiedonantoja *121,* 28 p. (1983).

Holopainen, T., S. Anttonen, A. Wulff, V. Palomäki, and L. Kärenlampi: Comparative evaluation of the effects of gaseous pollutants, acidic deposition and mineral deficiencies: structural changes in the cells of forest plants. Agric. Ecosystems Environ. *42,* 365–398 (1992).

Jalkanen, R.: Faller även de sista barren av träden i norra Finland? Norra Finlands hårt provade träd. Skogsaktuellt *1,* 8–11 (1988).

– Root cold stress causing a premature yellowing of oldest Scots pine needles. In: Merril, W. and M. E. Ostry (eds.): Recent research on foliage diseases, proc. IUFRO WP S2.06.04 Carlisle, Pennsylvania, USA, 29 May–2 June 1989. USDA For. Serv., Gen. Tech. Rep. *WO-56,* 34–37 (1990).

Jalkanen, R., T. Aalto, K. Derome, K. Niska, and A. Ritari: Lapin neulaskato. Männyn neulaskatoon 1987 johtaneet tekijät Pohjois-Suomessa. Loppuraportti. Summary: Needle loss in Lapland. Factors leading to needle loss of Scots pine in 1987 in northern Finland. Final report. Metsäntutkimuslaitoksen tiedonantoja *544,* 75 p. (1995).

Jokela, A., J. Bäck, S. Huttunen, and R. Jalkanen: Excess nitrogen fertilization and the structure of Scots pine needles. Eur. J. For. Path. *25,* 109–124 (1995 a).

Jokela, A., T. Sarjala, S. Huttunen, and S. Kaunisto: Microscopic structure and polyamines of Scots pine needles with varying potassium levels. Tree Physiol. accepted (1995 b).

Makino, A., T. Mae, and K. Ohira: Relation between nitrogen and ribulose-1,5-bisphosphate carboxylase in rice leaves from emergence through senescence. Plant Cell Physiol. *25,* 429–437 (1984).

Palomäki, V. and T. Holopainen: Effects of phosphorus deficiency and recovery fertilization on growth, mineral concentration, and ultrastructure of Scots pine needles. Can. J. For. Res. *24,* 2459–2468 (1994).

– – Effects of nitrogen deficiency and recovery fertilization on ultrastructure, growth and mineral concentrations of Scots pine needles. Can. J. For. Res. *25,* 198–207 (1995).

Palomäki, V. and H. Raitio: Chemical composition and ultrastructural changes in Scots pine needles in a forest decline area in southwestern Finland. Trees *9,* 311–317 (1995)

Raitio, H.: Effects of macronutrient fertilization on the structure and nutrient content of pine needles on a drained short sedge bog. Folia For. *465,* 1–9 (1981).

– The foliar chemical composition of young pines (*Pinus sylvestris* L.) with or without decline. In: Kauppi, P., P. Anttila, and K. Kenttämies (eds.): Acidification in Finland, pp. 699–713. Springer-Verlag Berlin (1990).

Sas: SAS Procedures guide. Release 6.03 edition. SAS Institute, Cary, North Carolina, USA (1988).

Soikkeli, S.: The effects of chronic urban pollution on the inner structure of Norway spruce needles. Savonia *4,* 1–19 (1981).

Soikkeli, S. and L. Kärenlampi: The effects of nitrogen fertilization on the ultrastructure of mesophyll cells of conifer needles in northern Finland. Eur. J. For. Path. *14,* 129–136 (1984).

Stover, E. L.: Varying structure of conifer leaves in different habitats. Bot. Gaz. *106,* 12–25 (1944).

Sutinen, S. and R. Pesonen: Männyn ja kuusen neulasten soluvaurioista Lapissa. In: Varmola, M. and P. Palviainen (eds.): Lapin metsien terveys. Metsäntutkimuspäivät Rovaniemellä 1989. Metsäntutkimuslaitoksen tiedonantoja *347,* 69–75 (1990).

Systat: SYSTAT for Windows: Statistic, Version 5th Edition. Evanston, IL: Systat Inc. (1992).

Tikkanen, E. and H. Raitio: Nutrient stress in young Scots pines suffering from needle loss in a dry heath forest. Water Air Soil Pollut *54,* 281–293 (1990).

Toth, R.: An introduction to morphometric cytology and its application to botanical research. Amer. J. Bot. *69* (10), 1694–1706 (1982).

J. Plant Physiol. Vol. 148. pp. 324–331 (1996)

Imbalances of D1 Protein Turnover during Stress Induced Chlorosis of a Declining Spruce Tree

CLAUDIA KONOPKA, REINHARD HOLLINDERBÄUMER, VOLKER EBBERT, HELGA WIETOSKA, and DORIS GODDE

Lehrstuhl für Biochemie der Pflanzen, Ruhr-Universität, D-44780 Germany

Received June 24, 1995 · Accepted September 30, 1995

Summary

Development of chlorosis and loss of PSII activity were compared in a declining spruce from the Schöllkopf site (Freudenstadt, Black Forest, Germany) during two vegetation periods. As a control tree, an apparently healthy, green tree was used. Loss of leaf chl content in the declining tree occurred during the development of the new flush in the second year needles. It preceded clearly any permanent functional inhibition of PSII as detected by changes in the chl fluorescence parameter F_v/F_m. The degradation of the photosynthetic apparatus was due to changes in thylakoid protein turnover, especially of the PSII D1 protein. Together with the loss of chl, an increase in D1 protein synthesis per chl determined by radioactive labelling was observed. It was up to 5 times higher than measured before the onset of chlorosis and more than 2 times higher than in the green control. Immunological determination by Western-blotting revealed that D1 protein content per chl decreased to 30 % of the green control. Thus, D1 protein degradation was even more stimulated than its synthesis. The loss of the D1 protein was accompanied by a comparable loss of CP 47 showing that polypeptides of both the PSII reaction centre and its inner antennae were lost by degradation. This process probably prevented a large accumulation of inactivated PSII. However, together with other factors, it might initiate the breakdown of the other polypeptides, especially of LHCII, resulting in the observed chlorosis.

Key words: Chlorosis, D1 protein turnover, Forest Decline, Photoinhibition, Photosystem II.

Abbreviations: chl = chlorophyll; DM = dry matter; Fo = yield of intrinsic fluorescence when all PSII centres are open in the dark; F_m = yield of maximal fluorescence when all reaction centres are closed; F_m' = fluorescence yield when all reaction centres are closed (after a saturating flash) under steady-state conditions; F_v = yield of variable fluorescence, (difference between F_o and F_m); ΔF = yield of variable fluorescence under steady state conditions, difference between F_m' and F_t, the fluorescence yield at steady state conditions; LHC = light harvesting complex; PFD = photon flux density; PS = photosystem; Q_A = primary quinone acceptor of PSII; Q_B = secondary quinone acceptor of PSII; q_p = photochemical quenching; q_n = non-photochemical quenching.

Introduction

Plants exposed to stress conditions like mineral deficiencies or high loads of air pollutants develop a chlorosis of their leaves indicating that the size of their photosynthetic apparatus has been reduced (Marschner and Cakmak 1989, Lange et al. 1989). A similar chlorosis is shown by stress damaged spruce trees (Lange et al. 1989, Siefermann-Harms 1992, Siefermann-Harms et al. 1993). Together with needle loss, needle yellowing is therefore used to determine the degree of forest decline. The development of stress induced chlorosis both in herbaceous plants (Marschner and Cakmak 1989) and in trees is a light dependent process (Siefermann-Harms 1992). In this respect, stress induced yellowing is similar to the chlorosis that can be observed, when plants are exposed to photoinhibitory conditions. Photoinhibition of photosynthesis is located at the level of PSII, which looses its ability to initiate photosynthetic electron transport (Aro et al. 1993, Ohad et al. 1994). In photoinactivated PSII centres oxygen radicals are formed leading to the photooxidation of both pigments and proteins (Telfer and Barber 1994). Since stress conditions are known to enhance the probability of photoinhibition (Godde and Hefer 1994), it was concluded that stress induced chlorosis is a consequence of photoinhibition and consists in the photooxidative breakdown of already nonfunctional components of the photosynthetic machinery (Demmig-Adams 1992, Wild 1988).

Recently, evidence has been provided, that in mineral deficient spinach where the allocation of assimilates is affected the development of stress induced chlorosis precedes inhibition of PSII (Godde and Dannehl 1994). Also in plants with disturbances in their source-sink relations like transgenic plants with increased activity of invertase (von Schaewen et al. 1991) or glucose supplied cell cultures (Schäfer et al. 1992) yellowing proceeds without a permanent inhibition of photosynthesis. In these cases, the degradation of the photosynthetic apparatus is not a mere photooxidative breakdown, but is well regulated process where the size of the photosynthetic apparatus is reduced while its function is retained. Such a process is very likely controlled by changes in protein turnover. A mechanistic key role seems to be played by the D1 protein from the reaction centre of PSII (Godde and Dannehl 1994) which is characterized by its rapid light dependent turnover (Mattoo et al. 1994). Even under optimal conditions for photosynthesis the D1 protein is rapidly degraded and replaced by resynthesis. The turnover of the D1 protein is normally associated with the process of photoinhibition (Ohad et al. 1994) and is known to play an essential role in the repair of photoinactivated PSII centres (Aro et al. 1993, Ohad et al. 1994). When the degradation of the D1 protein becomes limiting, which is the case under high light conditions, inactivated PSII centres accumulate leading to the phenomenon of photoinhibition (Aro et al. 1994, Kim et al. 1993). On the other hand, when synthesis of the D1 protein is limiting, the degraded D1 protein cannot be replaced. As a consequence, also the other components of PSII are lost by proteolysis (Schuster et al. 1988). In spinach under mineral stress, synthesis of the D1 protein, although stimulated, cannot keep up with its even more enhanced degradation (Godde and Dannehl 1994). This prevents the accumulation of inactivated PSII centres, however, it leads to specific loss of PSII proteins. Under these conditions, also the LHCII which has bound between 40–50% of the chl, was found to be degraded. Its loss was seen as chlorosis. A stimulation of D1 protein turnover by stress conditions has also been observed in ozone fumigated (Godde and Buchhold 1992) and in yellowing trees from outdoors (Godde 1992).

To get insight into the molecular mechanism of needle yellowing, we compared the development of the chl content, already published by Siefermann-Harms et al. (1994, 1995), with the changes in PSII function and in the synthesis and content of thylakoid proteins in a yellowing spruce tree from the stand of Schöllkopf near Freudenstadt (Germany) over two vegetation periods. As control, one of the apparently undamaged trees was used. On the Schöllkopf, the trees suffer under a complex stress situation (Siefermann-Harms et al. 1993). Due to a severe acidification of the soil, the supply of the trees with cationic nutrients, especially with magnesium, is extremely critical. In addition, the trees are exposed to a high load of ozone during summer. Besides these two factors, the watersupply of the soil shows extreme variations, so that highly chlorotic trees stand side to side with apparently healthy looking, green trees.

Material and Methods

Plant material

In 1993 and 1994 every three weeks twigs were taken from the top (8th whorl) of a 45–50 year old spruce (*Picea abies* L.). The twigs were kept in the dark for about 20 h until the next day when all measurements were performed.

Chl determination

Pigment analysis was performed by Siefermann-Harms as part of a joint project. The data are already published (Siefermann-Harms et al. 1994, 1995).

Preparation of thylakoids

For immunological determination of thylakoid proteins, thylakoids from spruce needles were prepared as described in Godde (1992).

Pulse modulated chl fluorescence induction

The yield of intrinsic fluorescence when all PSII centres are open in the dark, F_o, and the yield of maximum fluorescence when all reaction centres are closed, F_m, were determined by pulse modulated chl fluorescence induction according to Schreiber et al. (1986) in leaf discs using a PAM fluorometer (Walz, Effeltrich, Germany). The configuration consisted of the basic module PAM 101, the emitter/detector unit ED-101, the module 102 with a PAM 102 l LED light source and the PAM 103 module with the high-intensity light source Fl-103. The fluorometer and its measuring principles have been described previously (Schreiber et al. 1986). The PFD of the actinic light was 80 $\mu mol\, m^{-2} \cdot s^{-1}$ and the 500 ms saturating flashes had a PFD of 4000 $\mu mol\, m^{-2} \cdot s^{-1}$. Prior to the determination of chlorophyll fluorescence, the needles taken from a dark chamber were again dark adapted for 5 min. No substantial increase of F_v/F_m could be observed after further dark adaptation. The value F_o was determined by applying only the measuring light (peak wavelength 710 nm). F_m was estimated from the fluorescence yield achieved on the addition of a saturating 500 ms-pulse.

Immunological determination of thylakoid proteins

The relative contents of certain thylakoid proteins per mg chl were determined immunologically by Western-blotting. Thylakoids were first solubilized in 5% SDS, 15% glycerine, 50 mM Tris (pH 6.8) and 2% mercaptoethanol at room temperature for 30 min. After sonication with the microtip of a sonicator (Branson, Schwäbisch-Gmünd, Germany), starch was centrifuged-off. The polypeptides were separated by denaturing polyacrylamide gel (15%) electrophoresis according to Schägger et al. (1985). Proteins were then transferred onto nitrocellulose by electroblotting using a Bio-Rad

(München, Germany) Transblot chamber for 3 h at 0.4 A in a cold room at −28 °C. After saturation with 3 % gelatine in Tris buffer (pH 7.5), the first antibody was allowed to react overnight at room temperature in 1 % gelatine. After washing with Tris and 0.05 % Tween-20, the second antibody (horseradish peroxidase; Bio-Rad, München, Germany) was allowed to react in 1 % gelatine for 2 h. For detection of the D1 protein we used an antibody (Johanningmeier 1987) against part of the D1 protein (aminoacids 167–353) and β-galactosidase. The antibody against CP 47 was a kind gift from R. Berzborn (Ruhr-Universität, Bochum, Germany).

Pulse labelling of the D1 protein

20 h after harvesting, needles (0.7 g) were cut into 1 cm^2 pieces. An aliquot of 10 mL of a 1 mM solution of [^{14}C] leucine with a specific radioactivity of 9.25 MBq was applied by vacuum infiltration (3·20 s). Synthesis of D1 protein was determined after 15 min incubation in the light (500 µmol photons·m^{-2}·s^{-1}) according to Godde et al. (1991). The densitometry of Western blots and autoradiograms was performed with an Ultroscan XL Laser Densitometer (LKB, Bromma, Sweden).

Results and Discussion

To test whether stress induced chlorosis in outdoor spruce is a consequence of photoinhibition, the development of needle chl content was compared with changes of the chl fluorescence parameter F_v/F_m, which reflects the ability of PSII to reduce its primary quinone acceptor Q_A and to initiate photosynthetic electron transport (Krause and Weis 1991) (Fig. 1A and B). The degradation of the photosynthetic apparatus as it is reflected by the decrease in needle chl content took place in both years during the transitions from the first needle year to the second one and seems to be coupled with the development of the new flush. Similar results have been obtained by earlier measuring campaigns on this site (Siefermann-Harms et al. 1993), but also from other sites (Lange et al. 1989). In 1993, chl content first rose from 2,44 mg/g DM in early March (week 10) to 3,55 mg/g DM in mid June (week 25) before it declined (Fig. 1A). In 1994, the chl content of the yellowing tree declined without a prior increase from 1.89 mg/g DM determined in week 14 (April) to 0.879 mg/g DM in early July, week 27 (Fig. 1 B). The needle chl content of the undamaged control tree went up from 1.63 mg/g DM in April, week 14, to 3.08 mg/g DM in early July 1994, week 27.

In both vegetation periods, chlorosis precedes the loss of PSII function. In the year 1993, almost no loss of F_v/F_m could be observed (Fig. 1 a). Only at the end of summer, a slight accumulation of inactivated PSII centres was detected by chl fluorescence. In 1994, substantial loss of PSII function was observed during July, week 27, six weeks later than chl content started to decrease (Fig. 1 b). No substantial loss of PSII function was detected in the control needles at this time (see also Godde 1995). The results are basically congruent to the ones obtained with spinach under mineral stress (Godde and Dannehl 1994). This confirms that stress induced chlorosis is not initiated by photoinhibition, but instead is a well regulated process where the size of the photosynthetic apparatus is reduced while its functions remain. This does

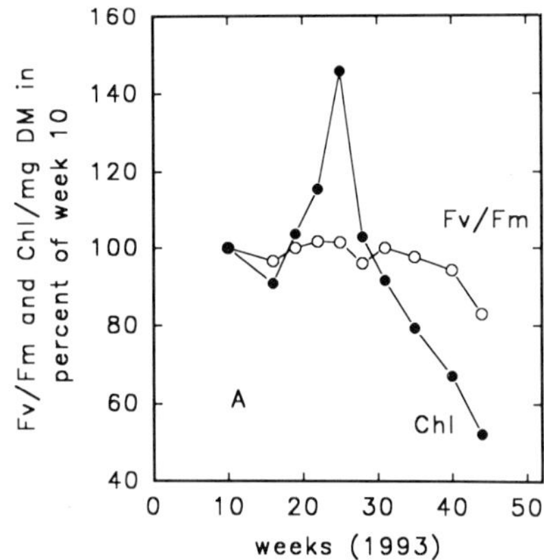

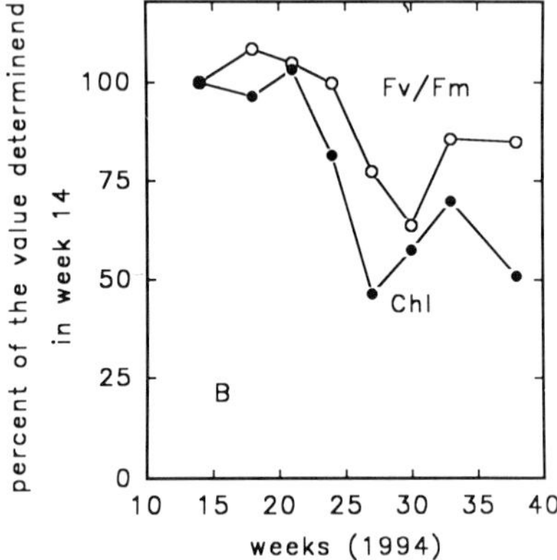

Fig. 1: Chlorosis preceds loss of PSII function in a stress damaged tree from the Schöllkopf site near Freudenstadt (Black Forest, Germany). PSII function was determined as decrease in F_v/F_m. The data represent the average of at least 4 different measurements. Standard deviation was less than 5 %. Chl content of the needles were taken from (Siefermann-Harms et al. 1994, 1995). Fig. 1A compares the development of both parameters during the vegetation period 1993, Fig. 1B during 1994. In week 10 1993 F_v/F_m was 0.784 ± 0.011, Chl content was 2,439 mg/gDM. In week 14 1994, F_v/F_m was 0.766 ± 0.013, chl content was 1.89 ± 0.069 mg/gDM. Development of the new flush took started in both years in late May and was finished in early July. (●) Chl content, (○) F_v/F_m.

not mean, that photoinhibition in chlorotic plants never occurs. The chance for photoinhibition under stress conditions is even enhanced (Godde and Hefer, 1994). However, photoinhibition is a late event during the course of chlorosis. In mineral deficient spinach a 10 % loss of F_v/F_m was observed when leaf chlorophyll content was alredy reduced to 50 %. Comparable results have been obtained from outdoor spruce during a measuring campagne on the Schöllkopf (Siefermann-Harms et al. 1993).

To test whether the degradation of the photosynthetic apparatus was due to changes in protein turnover, light dependent thylakoid protein synthesis was followed by vacuum-infiltration of needles with ^{14}C leucine and subsequent illumination for 15 min with 500 µmoles m^{-2} s^{-1}. The twigs had been dark adapted for 20 h after harvesting. As shown in Fig. 2, only one polypeptide was substantially labelled during the short light period. This protein was identified by Western-blotting as the D1 reaction centre protein of PSII. Some minor incorporation of ^{14}C leucine was seen in the higher molecular region, but no labelling could be detected in the molecular region smaller than the 32 kDa D1 protein.

To get information about the turnover of the D1 protein, D1 protein synthesis was determined on a chl basis and then compared to its content determined immunologically in the same needles. As shown in Fig. 3, D1 protein synthesis per chl showed strong seasonal fluctuations in 1994. Similar changes in D1 protein synthesis were observed in 1993 (Fig. 3a). After a maximum in early April 1994 around the 14th week, D1 protein synthesis decreased during late April and May, both in the declining and in the control tree. In the control tree, D1 protein synthesis per chl increased again during the greening process starting from early June, week 21, and reached a second maximum in early July, at the 27 week. At this time it was almost 3 times higher than in early June. In August and September, it declined to a very low level. In this tree, D1 protein content per chl was constant all over the vegetation period (Fig. 5c). As a consequence, not only D1 protein synthesis but also D1 protein degradation was stimulated during the greening process. Obviously, not all of the D1 protein synthesized was stable incorporated into the photosynthetic membrane. To account for both the increased D1 protein turnover and for the enlargement of the photosynthetic apparatus, D1 protein synthesis when related on a dry matter basis increased during the greening process about a factor of 4.9 (Fig. 4).

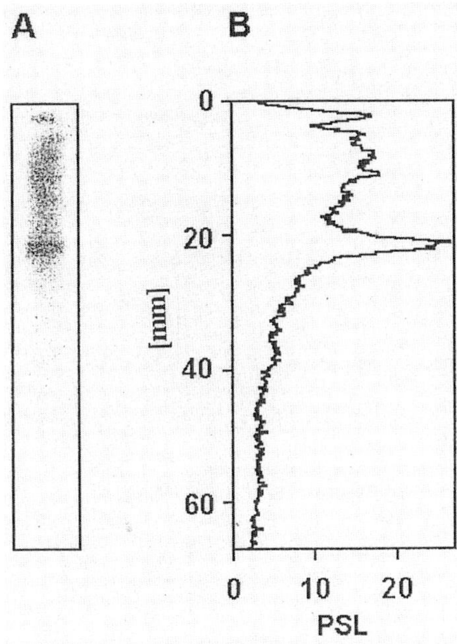

Fig. 2: The D1 reaction centre polypeptide of PSII is the most rapid synthesized polypeptide in a yellowing spruce from the Schöllkopf site in June 1994. Fig. 2A shows the phosphoimage of ^{14}C leucine incorpoaration into thylakoid proteins, Fig. 2B the corresponding the vertical scan. Labelling with ^{14}C leucine, isolation of thylakoids and separation of their polypeptides were performed as Materials and methods.

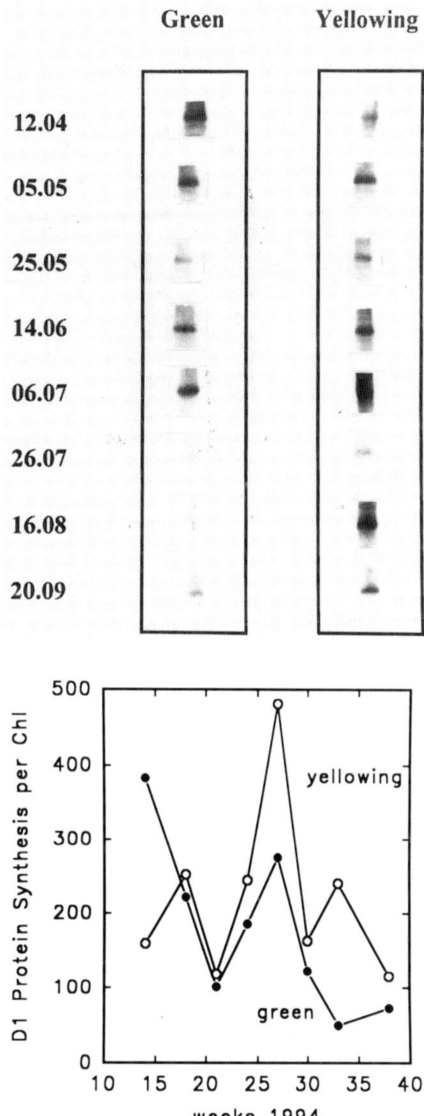

Fig. 3: D1 protein synthesis when related to a chl basis is stimulated in the yellowing tree when compared to the green control spruce from the same site. Relative D1 protein synthesis was measured as described in Materials and methods. Fig. 3A shows the incorporation of ^{14}C leucine into the D1 protein of the yellowing and control needles during the vegetation period 1993 as determined by autoradiography. Fig. 3B shows the D1 protein synthesis in 1994 as determined and evaluated by phosphoimaging. (●) green control, (○) yellowing tree.

In the yellowing tree, D1 protein synthesis per chl also increased in early June, week 21, in the now 2 year old needles although their chl content went down (Fig. 3). In early July, week 27, D1 protein synthesis per chl was stimulated almost by a factor of 5. At this time, it was twice as high as in the control tree. Then it decreased, but stayed still up to 4 times higher than in the non-yellowing control. In the yellowing needles, the immunological determined content of the D1 protein per chl decreased when compared to the control both in 1993 and 1994 (Fig. 5 A and B). In 1993, the loss of D1 protein was mostly pronounced in July when the degradation of the photosynthetic apparatus took place. In 1994, the content of the D1 protein per chl of the declining tree was diminished by 20 % when compared to the control already in early June. During the process of chlorosis it declined further and at the 27th week, it was only 30 % of the value of the control. This confirms results of Wild and coworkers that the content of the D1 protein although determined *via* its binding properties of herbicides is reduced in yellowing trees from Schöllkopf. Thus, D1 protein turnover in the yellowing needles was more than 5 fold higher than in the control. An almost complete loss of PSII core polypeptides has recently been reported from spruce during the winterly yellowing of *Pinus sylvestris* in Scandinavia (Ottander et al. 1995).

In vitro experiments have shown that D1 protein degradation is induced by a damage to PSII caused by disturbances on the acceptor or donor side of PSII (Aro et al. 1993, Ohad et al. 1994). Whether such a damage is also prerequisite for D1 protein turnover *in vivo* is still under debate. In mineral deficient spinach a stimulation of D1 protein turnover was observed (Godde and Dannehl 1994), when ΔpH over the thylakoid membrane did not relax and down-regulated PSII activity in the light (Krieger et al. 1993). The increase in membrane energetisation was caused by disturbances in assimilate allocation (Dietz and Heilos 1990, Godde and Hefer 1994) which have been shown to inhibit CO_2-fixation (Foyer 1988). As a consequence more ATP and NADPH were formed by the photosynthetic electron transport system than could be used in the Calvin-cycle. An increase membrane energetisation as indicated by a high non-photochemical quench (Zimmer-Rinderle and Lichtenthaler 1992, Siefermann-Harms et al. 1993) and in the ATP/ADP ratio (Einig et al. 1992, Siefermann-Harms et al. 1993) has also been observed in other yellowing trees from the same site. This gives rise to the assumption that also in stress damaged trees the allocation of assimilates is disturbed, even more since the declining trees accumulate more starch than healthy trees during late summer (Einig et al. 1991). Like in the Mg/S deficient spinach, the disturbances of assimilate allocation will lead to an increased turnover of the D1 protein. However, at this time it cannot be decided which stress factor is mainly responsible for the observed damage since both ozone and magnesium deficiency, the two dominating stress factors at Schöllkopf, are known to interfere with the allocation of assimilates (Cakmak et al. 1994, Willenbrink and Schatten 1993).

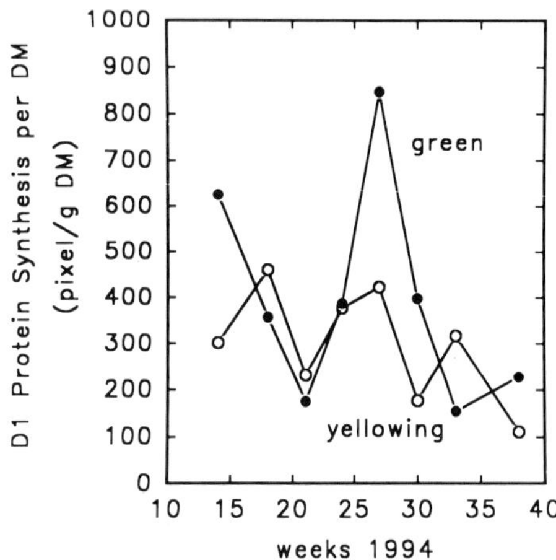

Fig. 4: Capacity for D1 protein synthesis is reduced in the yellowing tree when compared to the green control. Relative D1 protein synthesis was determined per chl as shown in Fig. 3 B. Data were then related to dry matter (DM) by multiplying with the chl content per dry matter shown in Fig. 1 B. (●) grenn control, (○) yellowing tree.

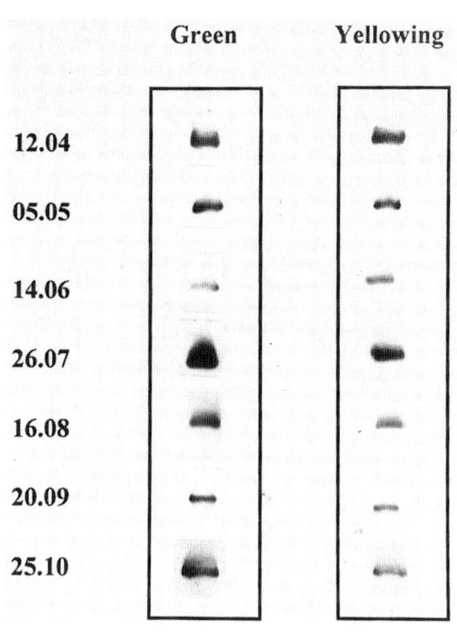

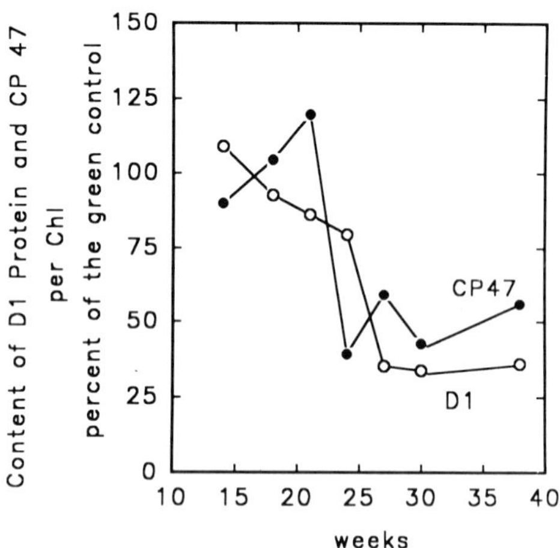

Fig. 5: The content of PSII polypeptides like the D1 protein and CP 47 is reduced in the yellowing needles in comparison to the green control both when related to chl. Fig. 5 A shows the original Western-blots against the D1 protein which was based on the identical chl content. Samples were taken in 1993, Fig. 5 B shows the laserdensitometrical evaluation of Western-blots from samples taken in 1994. The immunological determination of the PSII polypeptides has been carried out 3 times in 3 different experiments. In the green control the content of D1 protein and CP 47 was stable all over the vegetation period (Fig. 5 c). (●) D1 protein, (○) CP 47.

In vivo, the degree of photoinhibition is determined by the rate of photoinactivation and by the rate of D1 protein degradation. Thus, the rapid degradation of D1 protein in the yellowing needles might prevent an accumulation of inactivated PSII centres. Only in summer, when D1 protein turnover has declined, accumulation of inactivated PSII centres was observed as decrease in F_v/F_m. This shows that D1 protein turnover is regulating both the amount and the activity of PSII. The strong seasonal variations in D1 protein turnover indicate that plants have the possibility for its regulation according to the physiological conditions. Thus, although photoinactivation and damage to PSII might be necessary for D1 protein degradation, they are not the only factors that determine D1 protein turnover, at least not in higher plants. In contrast, the changes in D1 protein turnover seem to be independent of the actual amounts of inactivated PSII centres indicating that D1 protein degradation *in vivo* is no simple autocatalytic event intrinsic to the reaction centre complex (Telfer and Barber 1994).

The decrease of the D1 protein content per chl shows clearly that D1 protein synthesis in the yellowing tree, although stimulated on a chl basis, was insufficient to outbalance the enhanced degradation. On a dry matter basis, the loss of D1 protein became even more obvious. At the 27th week, the yellowing needles contained only 15 % of the D1 protein when compared to the green control (Fig. 6). As shown in Fig. 4, D1 protein synthesis when related to dry matter declined steadily in the yellowing tree throughout the vegetation with some minor fluctuations. Obviously, the capacity of the yellowing tree to synthesize D1 protein was so restricted that not all of the degraded D1 protein could be replaced. No excess D1 protein could be synthesized which would have been necessary for an enlargement of the photosynthetic apparatus.

The loss of the D1 protein per chl was accompanied by a comparable loss of CP 47 from the inner antennae of PSII (Fig. 5 b, Fig. 6). When related to dry matter, the content of D1 protein and CP 47 was reduced by about 80 % when compared to the green control (Fig. 6). Studies with mineral deficient spinach have shown that also the D2 protein and cyt b_{559} of PSII were degraded under conditions when D1 protein cannot be replaced sufficiently. The ratio of LHCII per chl remained constant. This has to be expected, since LHCII binds more than 40 % of the total chl. The specific decrease of PSII components in relation to chl implies that under chronic stress conditions the degradation of the PSII core polypeptides is faster than the loss of the light-harvesting system. This is in contrast to high light conditions, where the degradation of the photosystems follows the reduction of the antenna size. Here, the degradation of LHCII is due to an induction of a proteolytic activity in the stroma (Lindahl et al. 1995). However, also the LHCII has to undergo certain conformational changes before it can be attacked by the protease. These changes might be related to phosphorylation. We assume that under stress conditions the loss of PSII destabilises the other electron transport components especially, the LHCII, which is now in excess to PSII. The loss of PSII might be necessary, but not sufficient for LHCII degradation. One can imagine that also here phosphorylation, which is significantly altered under stress conditions (Dannehl et al. 1995, Godde 1993), plays an important role in signaling LHCII degradation.

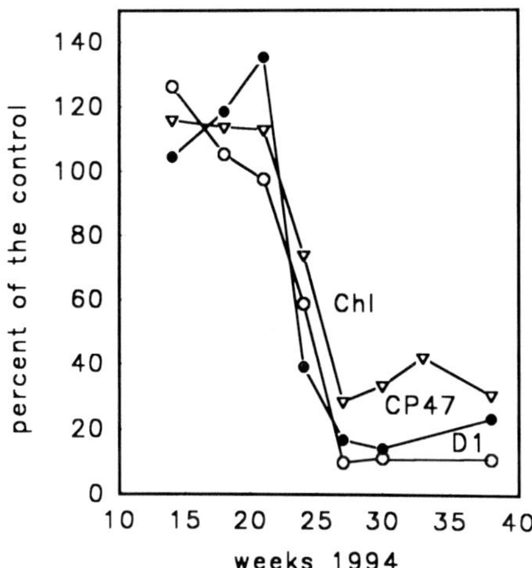

Fig. 6: The content of chl and PSII polypeptides is reduced in the yellowing needles when based on dry matter (DM). The data of the original measurements shown in Fig. 5 B were related to dry matter (DM) by multiplying with the chl content. (○) D1 protein, (●) CP 47, (∇) Chl.

Conclusion

Stress induced chlorosis in declining spruce trees is not caused by an irreversible functional inhibition of the photosynthetic apparatus. In contrast, the photosynthetic apparatus is degraded, while the functionality of its remaining components remains unaffected. Only at later stages of chlorosis, an accumulation of inactivated PSII centres can occur. The degradation of the photosynthetic apparatus is induced by changes in protein turnover, especially of the D1 protein from the reaction centre of PSII. Under stress conditions, this protein is rapidly synthesized, but even more rapidly degraded. The imbalance between its synthesis and degradation rate results in a loss of PSII centres. This loss seems to destabilize the other electron transport components and the LHCII. Its degradation becomes visible as chlorosis. At the beginning of the yellowing process, the degradation of the D1 protein is so fast that inactivated PSII centres do not accumulate. However, when D1 protein turnover becomes limiting later in the vegetation period, accumulation of inactivated PSII centres occurs giving rive to the observed decrease in the chl fluorescence parameter F_v/F_m. The changes in D1 protein turnover are observed under stress conditions that interfere with the allocation of assimilates like high loads of ozone or certain disturbances in the supply with mineral nutrient like magnesium deficiency.

Acknowledgements

This work was made possible by a grant from the Projekt Europäisches Forschungszentrum in cooperation with Dr. D. Siefermann-Harms, KFK-Karlsruhe, Karlsruhe, Germany.

References

Aro, E.-M., I. Virgin, and B. Andersson: Photoinhibition of photosystem II. Inactivation, protein damage and turnover. Biochim. Biophys. Acta *1143,* 113–134 (1993).

Maro, E.-M., S. McCaffery, and J.-M. Anderson: Recovery from photoinhibition in peas (*Pisum sativum* L.) acclimated to varying growth irradiances. Plant Physiol. *104,* 1033–1041 (1994).

Cakmak, I., C. Hengeler, and H. Marschner: Changes in Phloem export of sucrose in leaves in response to phosphorous, potassium and magnesium deficiency in bean plants. J. Exp. Bot. *45,* 1251–1257 (1994).

Demmig-Adams, B. and W. W. Adams III: Photoprotection and other responses of plants to high light stress. Annu. Rev. Plant Physiol. Plant Mol. Biol. *43,* 599–626 (1992).

Dietz, K. J. and L. Heilos: Carbon metabolism in spinach leaves as affected by leaf age and phosphorous und sulfur nutrition. Plant Physiol. *93,* 1219–1225 (1990).

Einig, W., C. Groshupp, U. Zimmer-Rinderle, B. Egger, R. Hampp, and H. K. Lichtenthaler: Gaswechsel und Verbindungen des Kohlenhydrat- und Intermediärstoffwechsels von Nadeln unterschiedlich geschädigter Fichten in Abhängigkeit vom Tagesgang. In: 8. Statuskolloquium des PEF vom 17. bis 19. März 1992 im Kernforschungszentrum Karlsruhe, KfK-PEF 94, pp. 35–46 (1992).

Foyer, C. H.: Feedback inhibition of photosynthesis through source-sink regulation in leaves. Plant. Physiol. Biochem. *26,* 483–492 (1988).

Genty, B., J. M. Briantais, and N. R. Baker: The relationship between the quantum yield of photosynthetic electron transport and quenching of chlorophyll fluorescence. Biochim. Biophys. Acta *990,* 87–92 (1989).

Godde, D.: Photosynthesis in relation to modern forest decline; content and turnover of the D-1 reaction centre polypeptide of photosystem 2 in spruce trees (*Picea abies*). Photosynthetica *27,* 217–230 (1992).

– Funktion und Reparatur des photosynthetischen Apparates in mangelernährten Fichten bei verschiedenen Lichtintensitäten. In: 10. Statuskolloquium des PEF vom 15.–17. März 1994, Projekt Europäisches Forschungszentrum für Maßnahmen zur Luftreinhaltung (PEF) ed., pp. 181–190, 1994.

Godde, D.: Funktion und Reparatur des photosynthetischen Apparates in magelernährten Fichten bei verschiedenen Lichtintensitäten – Molekularer Mechanismus der streßinduzierten Vergilbung. In: 11. Statuskolloquium des PEF vom 14. bis 16. März im Forschungszentrum Karlsruhe, Projekt Europäisches Forschungszentrum für Maßnahmen zur Luftreinhaltung (PEF) ed. FZKA-PEF *130,* pp. 191–200 (1995).

Godde, D. and J. Buchhold: Effect of long term fumigation on the turnover of the D-1 reaction center polypeptide in spruce (*Picea abies*). Physiol. Plant. *86,* 568–574 (1992).

Godde, D. and H. Dannehl: Stress induced chlorosis and increase in D1 protein turnover precedes photoinhibition in spinach suffering under combined magnesium and sulphur deficiency. Planta *195,* 291–300 (1994).

Godde, D. and M. Hefer: Photoinhibition and light-dependent turnover of the D1 reaction centre polypeptide of photosystem II are enhanced by mineral stress conditions. Planta *193,* 290–299 (1994).

Godde, D., H. Schmitz, and M. Weidner: Turnover of the D-1 reaction center polypeptide from photosystem II in intact spruce needles and spinach leaves. Z. Naturforsch *46c,* 245–251 (1991).

Johanningmeier, U.: Expression of psbA gene in *E. coli.* Z. Naturforsch. *42c,* 755–757 (1987).

Kim, J. H., J. A. Nemson, and A. Melis: Photosystem II reaction center damage and repair in *Dunaliella salina* (green alga). Plant Physiol. *103,* 181–189 (1993).

Krause, G. H. and E. Weis: Chlorophyll fluorescence and photosynthesis: The basics. Annu. Rev. Plant Physiol. Plant Mol. Biol. *42,* 313–349 (1991).

Krieger, A., E. Weis, and S. Demeter: Low-pH-induced Ca^{2+}ion release in the water-splitting system is accompanied by a shift in the midpoint redox potential of the primary quinone acceptor Q_A. Biochim. Biophys. Acta *1144,* 411–418 (1993).

Lange, O. L., U. Heber, E.-D. Schulze, and H. Ziegler: Atmospheric pollutants and plant metabolism. In: Schulze, E.-D., O. L. Lange, and R. Oren (eds.): Forest decline and air pollution, pp. 238–276. Springer, Berlin (1989).

Lindahl, M., D. H. Yang, and B. Andersson: Regulatory proteolysis of the major light-harvesting chlorophyll *a/b* protein of photosystem II by a light-induced membrane-associated enzymic system. Eur. J. Biochem. *231,* 503–509 (1995).

Marschner, H. and I. Cakmak: High light intensity enhances chlorosis and necrosis in leaves of zinc, potassium and magnesium deficient bean (*Phaseolus vulgaris*) plants. J. Plant Physiol. *134,* 308–315 (1989).

Mattoo, A. K., H. Hoffmann-Falk, J. B. Marder, and M. Edelmann: Regulation of protein metabolism: coupling of photosynthetic electron transport to *in vivo* degradation of the rapidly metabolised 32-kilodalton protein of the chloroplast membrane. Proc. Natl. Acad. Sci. USA *81,* 1380–1384 (1984).

Ohad, I., N. Keren, H. Zer, H. Gong, T. S. Mor, A. Gal, S. Tal, and Y. Domovich: Light-induced degradation of the photosystem II reaction centre D1 protein *in vivo:* an integrative approach. In: Baker, N. R. and J. R. Bowyer (eds.): Photoinhibition of photosynthesis from molecular mechanisms to the field, BIOS, Oxford, pp. 161–177 (1994).

Ottander, C., D. Campbell, and G. Öquist: Seasonal changes in photosystem II organisation and pigment composition in *Pinus sylvestris.* Planta, submitted (1995).

Schäfer, C., H. Simper, and B. Hofmann: Glucose feeding results in coordinated chanes of chlorophyll content, ribulose-1,5-bisphosphate carboxylase-oxygenase activity and photosynthetic potential in photoautotrophic suspension cultured cells of *Chenopodium rubrum.* Plant, Cell and Environment *15,* 343–350 (1992).

Schägger, H., U. Borchert, H. Aquila, T. A. Link, and G. Jagow: Isolation and aminoacid sequence of the smallest subunit of beef heart bc1 complex. FEBS Lett. *190,* 89–94 (1985).

Schreiber, U., U. Schliwa, and W. Bilger: Continous recording of photochemical and non-photochemical fluorescence quenching with a new type of modulation fluorometer. Photosynth. Res. *10,* 51–62 (1986).

Schuster, G., R. Timberg, and I. Ohad: Turnover of thylakoid photosystem II proteins during photoinhibition of *Chlamydomonas reinhardtii.* Eur. J. Biochem. *177,* 403–410 (1988).

Siefermann-Harms, D.: The yellowing of spruce in polluted atmospheres. Photosynthetica *27,* 323–341 (1992).

Siefermann-Harms, D., P. Bourgeois, K. Baumann, G. Baumbach, C. Buschmann, W. Einig, S. Fink, F. Franke, R. Hampp, E. E. Hildebrand, E. Hochstein, H.-J. Jäger, M. Konnert, I. Kottke, H.-K. Lichtenthaler, R. Manderscheid, W. Schmidt, H. Schneckenburger, D., Seemann, H. Spiecker, W. Urfer, A. V. Wild, K. Wilpert, and U. Zimmer-Rinderle: Ergebnisse und Schlußfolgerungen der koordinierten Forschung zur montanen Vergilbung in Hochlagen des Schwarzwaldes (Standort Schöllkopf, Freudenstadt). In 9. Statuskolloquium des PEF vom 9.–11. März 1993 im Kernforschungszentrum Karlsruhe, pp. 79–126 (1993).

SIEFERMANN-HARMS, D., CH. WEINMANN, H. SCHNECKENBURGER, H.-G. HEUMANN, and A. SEIDEL: Der Vergilbungsvorgang und seine Aufhebung durch künstliche Beschattung. In: 10. Statuskolloquium des PEF vom 15.–17. März 1994 (1994).

SIEFERMANN-HARMS, D., CH. WEINMANN, TH. HOFER, C. BOXLER-BALDOMA, H.-G. HEUMANN, and A. SEIDEL: Weitere Untersuchungen zum Vergilbungsvorgang – keine kurzfristige Ergrünung nach Mg-Sulfat Düngung. In: 11. Statuskolloquium des PEF vom 14.–16. März 1995, Projekt Europäisches Forschungszentrum für Maßnahmen zur Luftreinhaltung (ed.), FZKA-PEF *130,* pp. 175–190 (1995).

TELFER, A. and J. BARBER: Elucidating the molecular mechanisms of photoinhibition by studying isolated photosystem II reaction centres. In: BAKER, N. R. and J. R. BOWYER (eds.): Photoinhibition of photosynthesis – from molecular mechanisms to the field, Bios Scientific Publishers, Oxford, pp. 25–50 (1994).

VAN KOOTEN, O. and J. H. F. SNEL: The use of chlorophyll fluorescence nomenclature in plant stress physiology. Photosynth. Res. *25,* 147–150 (1990).

VAN WIJK, K. J. and P. R. VAN HASSELT: Photoinhibition of photosystem II is preceded by down-regulation through light-induced acidification of the lumen: Consequences for the mechanism of photoinhibition *in vivo.* Planta *189,* 359–368 (1993).

VON SCHAEWEN, A., M. STITT, R. SCHMIDT, U. SONNEWALD, and L. WILLMITZER: Expression of a yeast-drived invertase in the cell wall of tobacco and *Arabidopsis* plants leads to accumulation of carbohydrate and inhibition of photosynthesis and strongly influences growth and phenotype of transgenic tobacco plants. EMBO J. *9,* 3033–3044 (1990).

WILD, A.: Licht als Streßfaktor bei Waldbäumen. Naturwiss. Rundsch. *41,* 93–96 (1988).

WILD, A. and S. TIETZ-SIEMER: Physiologische und cytomorphologische Untersuchungen an ungeschädigten und geschädigten Fichten im Nordschwarzwald (Freudenstadt). In: 8. Statuskolloquium des PEF vom 17. bis 19. März 1992 im Kernforschungszentrum Karlsruhe, PEF-Berichte 94, pp. 47–60 (1992).

WILLENBRINK, J. and T. SCHATTEN: CO_2-Fixierung und Assimilatverteilung in Fichten unter Langzeitbegasung mit Ozon. Forstwiss. Cbl. *112,* 50–56 (1993).

ZIMMER-RINDERLE, U. and H. K. LICHTENTHALER: Jahresverlauf der Photosyntheseaktivität geschädigter Fichten im Nordschwarzwald. In: 8. Statuskolloquium des PEF vom 17. bis 19. März 1992 im Kernforschungszentrum Karlsruhe, PEF-Berichte 94, pp. 81–92 (1992).

J. Plant Physiol. Vol. 148. pp. 332–338 (1996)

Chamber Effects and Responses of Trees in the Experiment using Open Top Chambers

DALIBOR JANOUŠ, VÍTĚSLAV DVOŘÁK, MAGDA OPLUŠTILOVÁ, and JIŘÍ KALINA

Institute of Landscape Ecology, Academy of Sciences of the Czech Republic, Květná 8, 603 00 Brno, Czech Republic

Received June 28, 1995 · Accepted September 30, 1995

Summary

Chamber effects are unintentional concomitants of experiments using open-top field chambers (OTCs). Cylindrical OTCs were used in a mountain Norway spruce stand. Chamber effects were investigated as well comparing samples in OTCs with ambient air and samples not enclosed in OTCs (control). The OTCs were 2.5 m in diameter at the base, 6.0 m high and the volume was 20 m^3. The iron frames were covered by transparent 0.2 mm PVC film.

Changes of growth conditions as follows: Transmission of the chambers covered by a new clear film was 91 ± 2 % of the solar photosynthetically active radiation (84 ± 4 % at the end of the growing season). Mean air temperatures inside OTCs were 1.3 °C higher than outside OTCs on sunny days. Inside there were decreases in relative humidity during the sunny hours by as much as 10 %.

Responses of the trees as follows: Continual increase of chlorophyll content (*a* + *b*) was recorded during the cultivation of the trees in the OTCs in comparison with the control (76 %). A similar situation was observed for the carotenoids content. We recorded seasonal increases of F_v/F_{max} values and the $t_{1/2}$ parameter in the OTCs in comparison with the control treatment. At the beginning of the seasons (1993 and 1994) stem respiration was greater in the control than in the OTC (70 % and 85 % of the control) but at the end of the seasons the situation was opposite (115 % and 230 % of the control). No differences were found in the development of apical shoots. Differences were found only in length increases of whorl branches, in the season 1994 (p=0.01). The length increase in the OTCs was greater by 38 %.

Key words: Picea abies, chamber effects, pigment constants, chlorophyll fluorescence, stem respiration, growth analyses.

Abbreviations: *a* + *b* = chlorophyll *a* and *b* content; Carx + c = total carotenoides content; E = wapour tension; F_{max} = maximum fluorescence of chlorophyll; F_o = ground fluorescence of chlorophyll; F_s = lowest level of fluorescence after being illuminated for 2 seconds by actinic light; F_v/F_{max} = ratios of the variable chlorophyll fluorescence; OTC = open top chamber; PhAR = photosynthetically active radiation; PSII = photosystem II; Q_{10} = calculated coefficient of respiration rate change by 10 °C; R_{20} = respiration rates re-counted for 20 °C; $t_{1/2}$ = half time measurement of the maximum fluorescence level after using actinic light.

Introduction

The effects of global changes to the atmosphere on temperate forests and predicting the influence of these changes after 50 years (Eamus and Jarvis, 1989) is an urgent task for environmental plant physiologists.

A suitable field chamber is in many cases necessary for the controlled simulation of growth conditions or any of their factors. The use of a chamber is also necessary for the application of an accetable amount (from the poin of view of expenses) of added matter (in our case CO_2) and energy. However, changes of controlled factors in the chamber are always accompanied by changes in other factors, these factors provide so called «chamber effects». The main chamber effects are as follows (see Allen et al., 1992): (1) reduction and change in quality of solar radiation environment (especially UV-B radiation), (2) unnatural wind flow, changes in the microclimate, and finally (3) disturbed soil water patterns. Total elimination of chamber effects is impossible, it is only possible to reduce certain factors using some supplementary device.

Open top chambers have been commonly used for field studies in the influence of air pollution on assimillatory apparatus over the last 20 years (Heagle et al., 1973; Olszyck et al., 1980). OTC's are very useful tools for the simulation of global climatic changes (Allen et al., 1992; Ceulemans and Mousseau, 1994).

The design of the OTC follows the findings of a compromise between an open environment (f.e. used FACE systems – Hendrey, 1992) and a closed growth chamber. The size of the chamber opening (together with the effect of the lateral walls) represents this compromise. Cylindrical, open-top field chambers described by Heagle et al. (1973) or square-walled, open-top chambers used by Nakayama and Kimball (1988) – both types are without frustum – on the one side of the existing gamut of possible OTC designs. Field exposure chambers with rain exclusion devices (rain cap) of Heagle et al. (1989) represent the opposite side of this gamut.

This paper presents chamber effects of our OTCs to asses changes of growth conditions and the effects on 15 year old Norway spruce individuals growing in a forest stand in the Beskydy Mts.

Materials and Methods

The research project was located on the study site Bily Kriz (coordinates 49° 30′ North, 18° 32′ East, at an altitude 943 m above sea level), in the Moravskoslezske Beskydy Mts., North Moravia, in the Czech Republic. The study site Bily Kriz belongs to a moderately cold and wet region with an average annual air temperature 5 °C (−4 °C in January, and 14 °C in July), the average annual amount of precipitation is 1400 mm (80 mm in January, and 150 mm in July) and the annual average number of days with snow cover being 160 (Janous, 1990).

Eight open top chambers were constructed around trees *in situ*, with one tree per chamber in a Norway spruce stand (*Picea abies* (L.) Karst.) planted in the year 1982 (average height of trees was 3.4 m, and width of crown 2.2 m in 1993/1994). Three groups of four sample trees were investigated in the following treatments: i) samples in four OTCs with elevated concentrations of CO_2 (not of interest in this paper), ii) samples in four OTCs with ambient concentrations, and iii) samples for comparison not enclosed in OTCs. The experiment was conducted from 1992 and was interrupted during winter months due to high snow cover.

Chamber Design

The special design of the chambers used (as the result of an optimization process for the study site conditions) was determined by certain factors which were as follows: i) to minimize the entry of ambient air into the chamber on the extremly windy study site, ii) not to reduce interception of vertical precitipation (however, plants growing in OTC's are virtually starved of horizontal precipitation), iii) to minimize changes of irradiation of sample trees.

The chamber was cylindrical with an added double frustum and a rain collection device – funnel. An iron frame was covered by transparent, 0.2 mm polyvinyl chloride film containing UV inhibitors (non degradable). The chamber was 2.5 m in diameter at the base, 6.0 m high and the volume was 20 m^3. The upper hoop of the funnel was the same diam as the chamber base. Top openings of frustum were 25 % of the base (1.25 m in diam). Frustum angles were 23° from the vertical. The frustum was equiped with a collar for dripping the rain to minimize running on the chamber wall. The overall shape of the chamber was similar to the «flask» chamber described by Heagle et al. (1989), however, the function of the upper part of their chamber was rain exclusion, contrary to ours (rain collection).

Air handling system of the chamber was supplied by a fan (500 W, Janka, CR) located on the north facing side of the chamber and by a 0.3 m diameter perforated polyethylene duct located on the inner side of the chamber base. The volume of air in the «growth space» of the chamber (see below) was exchanged once per minute.

Growth Conditions

The daily course of photosynthetic photon flux density (PPFD, at tree top level) was measured using self-made quantum sensors based on a diode (BPW 21, Siemens, E.C.) calibrated with a Quantum Sensor (LI-COR 190S, Lincoln, USA) in each chamber and outside.

Air temperature (at a height of 2.0 m, within crown) and soil temperature (0.1 m deep below the projection of crown) were measured continuously using thermocouples (KTY 122-81, Siemens, E.C.) for each of the sample trees.

Relative humidity was measured inside and outside OTCs in crown spaces, height 2.0 m, and on the open site by electronic modification of Assman's psychrometer based on ventilated thermocouples, mentioned above, calibrated using a mercury one.

The data was collected by data logger (Delta T, England).

The transmission of the PVC film was measured in the laboratory with Specord M400 (Carl Zeiss Jena, Germany).

Pigment Content

The one-year old needles were removed bewteen 8:00 a.m. and 10:00 a.m. from the S/SW-exposed middle crown parts of the individual tree (four ambient and four control trees). Samples for the determination of pigment content of the individual tree were prepared as a mixture of needles from different parts of the investigated tree. The content of pigments was calculated on needle area basis. The projected needle area was estimated using a planimeter (LI-3000A, LI-Cor. USA). Each mixed sample was divided into three subsamples. The needle area of each subsample was estimated and the subsamples were ground in 80 % acetone in darkness at room temperature with a small amount of $MgCO_3$. The content of chlorophylls and total carotenoids was determined spectrophotometrically using a spectral resolution of 2 nm (Specord M400, Carl Zeiss Jena, Germany) according to Lichtenthaler (1987).

Modulated Chlorophyll a Fluorescence

Chlorophyll *a* fluorescence was measured using a chlorophyll fluorometer (PAM 101, 102, 103, Heinz Walz, Effeltrich, Germany). Samples containing five one-year-old needles were excised on four occasions (April and September 1993, April and October 1994) and predarkened for 60 minutes. After illumination by actinic light (120 $\mu mol\ m^{-2}\ s^{-1}$), the characteristic fast fluorescence induction curves were recorded for two seconds and afterwards a white saturation pulse was applied. The parameters F_o – all reaction centers of PSII are open, F_{max} – all reaction centers of PSII are closed, F_s – lowest level of fluorescence after being illuminated for 2 seconds by actinic light and $t_{1/2}$ – half time measurement of the maximum fluorescence level after using actinic light. The photochemical efficiency of PSII (F_v/F_{max}) was determined according to Havaux et al. (1991) as follows:

$$F_v/F_{max} = (F_{max} - F_o)/F_{max}$$

Parameters of the fluorometer PAM 101, 102, 103 were adjusted in an identical manner for all measurements.

Respiration

Respiration of woody parts (stems and branches) was measured during the growing seasons utilising a portable photosynthesis system (LI-6200, Li-Cor, USA) and respiratory chambers fixed to the stem and branch of the sample trees. To compare respiration rates they were re-counted for 20 °C (R_{20}), using calculated coefficient Q_{10} (respiration rate change by 10 °C, Svensson, 1989; Sprugel and Benecke, 1990).

The surface temperature of woody parts was measured by thermocouple thermometer (HH-25KC, ONE OMEGA DRIVE, USA).

Growth and Phenological Analysis

We have observed on each tree:

1. Flushing stages and development of tree terminal bud.
2. Flushing stages and development of apical buds of a branch on each tree at the first, third and fifth whorl counted from the top of the tree. The branches were selected in the southern part of the trees.
3. Diameter and length change of these branches. The diameter was measured in two directions from a marked spot at the branch base.
4. Diameter change of stems were observed and measured from a marked spot on every stem as close to the ground as possible (ca. 20 cm).

The length increase and diameter increase were measured using a digital caliper Mitutoyo, England.

The data was tested by t-Test using the Analysis tools from Microsoft Excel version 4.0.

The beginning of the growing seasons was defined in our case, as the first day in spring after a period of five days with an average daily temperature higher than 5 °C and the end of growing season as a day in autumn after period of five days with daily average temperature lower than 5 °C (Krecmer, 1980).

Results

Growth Conditions

Transmission of the used PVC film was:

- in the spectral range 400–900 nm constantly 92 %,
- in the range 330–360 nm 30 %,
- in the range 250–295 nm 1 %.

PhAR transmission of the OTC covered by a new clear film was 91 ± 2 % (mean of 2350 measurements from May).

The degradation of the film during the season had the effect of decreasing transmission: 84 ± 4 % (mean of 2400 measurements from September).

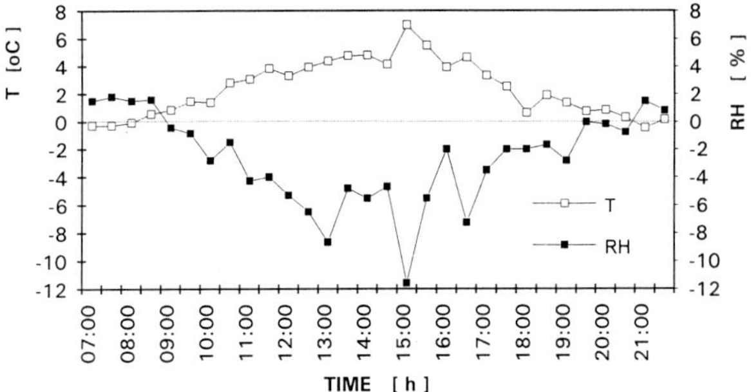

Fig. 1: Day record of differences between temperatures (T) and relative humidities (RH) inside and outside an open top chamber from a sunny day in July 1994.

Mean temperatures of air inside the OTCs were 1.3 °C above outside OTCs on sunny days. Night differences were not found but mid day increases in air temperature were 4.9 °C above ambient on sunny days (mid day irradiation 900 $W \cdot m^{-2}$, measured by pyranometer Kipp-Zonen, Delft, NL). Differences in soil temperatures were not found.

Our measurements recorded a decrease in the relative humidity during the sunny days by as much as 10 % (Fig. 1).

Pigment Content

Continual increases of chlorophyll content (*a* + *b*) were recorded during cultivation of the trees in the OTCs compared with the control (Table 1). Differences between the treatments increased from 16 % in April 1993 to 76 % in October 1994. We observed a similar situation in carotenoides content (Carx + c, from 16 % to 57 %). The changes of both the chlorophyll and carotenoid ratio are not so expressive and we did not find any trend.

Table 1: Total chlorophyll content (a + b), total caroteinods (x + c) expressed on a projected needle area basis, chlorophyll a to chlorophyll b ratio (a/b) and total chlorophyll to total carotenoids ratio (a + b/x + c) for control trees (control) and trees grown in open-top chamber (OTC). Each value is the mean of 12 measurements and standard error is shown (SE). Statistically significant differences between control and OTC treatment for each measurement month at P = 0.01 and P = 0.05 is indicated by two and one asterix (ns – not significant difference).

Sample	a+b [g m^{-2}]	SE	x+c [g m^{-2}]	SE	a/b	SE	a+b/ x+c	SE
control-April-93	0.20	0.007	0.05	0.002	3.17	0.072	4.15	0.063
OTC-April-93	0.23	0.009	0.06	0.002	3.11	0.077	4.10	0.088
	**		**		ns		ns	
control-October-93	0.30	0.018	0.06	0.004	2.82	0.034	4.78	0.076
OTC-October-93	0.39	0.017	0.08	0.003	2.91	0.022	5.00	0.065
	**		**		**		**	
control-April-94	0.27	0.014	0.07	0.003	2.96	0.045	3.83	0.099
OTC-April-94	0.42	0.042	0.10	0.008	2.79	0.012	4.00	0.090
	**		**		**		ns	
control-October-94	0.41	0.021	0.09	0.004	2.89	0.032	5.12	0.059
OTC-October-94	0.72	0.071	0.14	0.012	2.89	0.016	5.22	0.026
	**		**		*		ns	

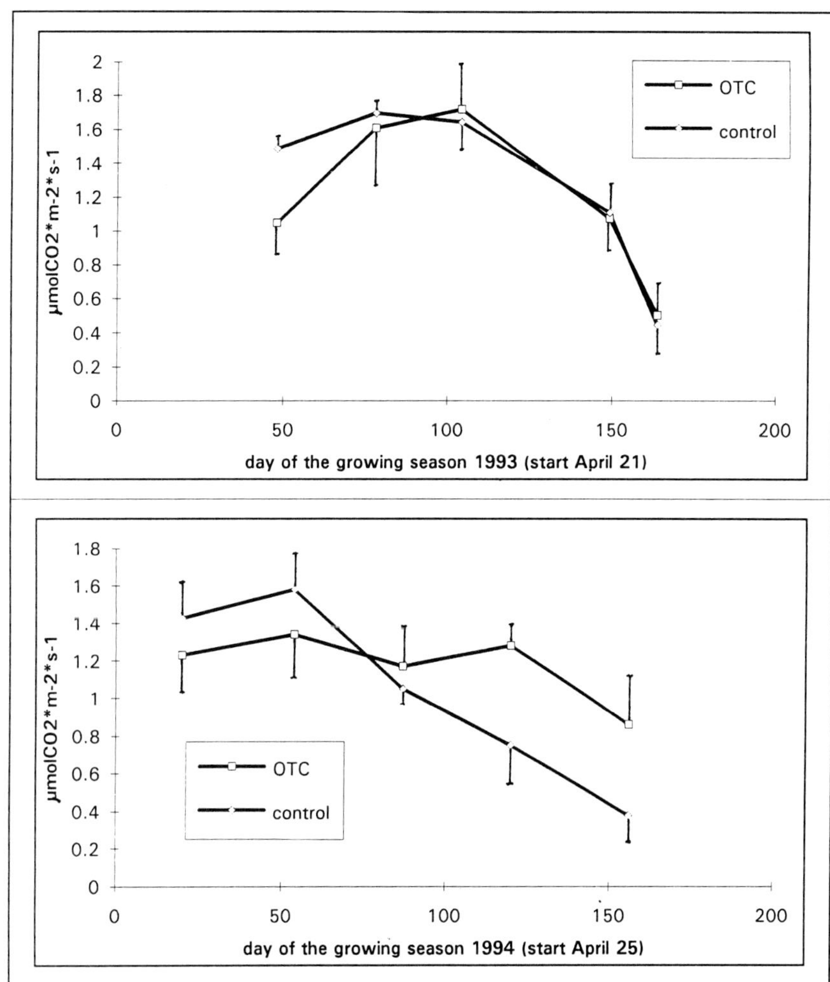

Fig. 2: Respiration rate R_{20} of the Norway spruce stems in open top chambers (OTC) and in open air (control) in the growing seasons 1993 and 1994.

Modulated Chlorophyll a Fluorescence

The increase in photochemical yield of photosystem II due to chamber effects was demonstrated by a seasonal increase of F_v/F_{max} values compared to trees grown side out of the OTCs. We observed significant ($P = 0.01$, $n = 12$) by 7 % higher F_v/F_{max} for trees grown in the OTCs after both seasons (Table 2). The parameter $t_{1/2}$ was also influenced, we have recorded increases of this parameter during the cultivation in the OTCs compared with the control treatment (about 37 % at October 1994).

Respiration

Seasonal course of respiratory rate of the stems is given in Fig. 2. At the beginning of the seasons (1993 and 1994) respiration was greater in control than in OTC (70 % and 85 % of the control) but at the end of the seasons the situation was opposite (115 % and 230 % of the control). Maximal rates of stem respiration R_{20} were:

in the OTC:

1.72 ± 0.21 (1.63 ± 0.20) $\mu mol \cdot m^{-2} \cdot s^{-1}$ on July 1993 (June 1994),

Table 2: The half time measurement of the maximum fluorescence level after using actinic light ($t_{1/2}$) and photochemical efficiency of PS II (F_v/F_m) for control trees (control) and trees grown in open-top chamber (OTC). Each value is the mean of 12 measurements and standard error is shown (SE). Statistically significant differences between control and OTC treatment for each measurement month at P = 0.01 and P = 0.05 is indicated by two and one asterix (ns – not significant difference).

Sample	$t_{1/2}$	SE	F_v/F_m (Fm-Fo')/Fm	SE
control-April-93	0.686	0.004	0.631	0.012
OTC-April-93	0.661	0.007	0.659	0.010
	**		*	
control-October-93	0.326	0.006	0.705	0.008
OTC-October-93	0.335	0.007	0.735	0.005
	ns		**	
control-April-94	0.278	0.008	0.569	0.015
OTC-April-94	0.284	0.005	0.594	0.018
	ns		ns	
control-October-94	0.334	0.010	0.710	0.013
OTC-October-94	0.440	0.012	0.750	0.003
	**		**	

in the control:
1.70 ± 0.11 (1.90 ± 0.32) $\mu mol \cdot m^{-2} \cdot s^{-1}$ on June 1993 (June 1994).
Calculated coefficient $Q_{10} = 2.2$ was used.

Relative Increases of Tree Heights

There were no differences to be found between treatments at the time of flushing of the terminal buds. Percentage changes in the total length of terminals in seasons 1993 and 1994 were compared with the average length of whorls (100 %), which was counted as an average value for whorl lengths for the seven years prior to 1992, when the experiment began. There were no significant differences.

Relative Increases of Stem Diameters

Fig. 3 shows the percentage seasonal increase in stem diameter for 1993 and 1994 compared with the stem diameter at the beginng of each season (100 %). The differences are not statistically significant.

Relative Diameter Increase of Branches

The percentage change in the branch base diameter was calculated in the same way as the tree terminal increase – 100 % of the Y-axis presents the average annual increase of branch base diameter in years leading up to 1992, when the experiment began.
The differences between treatments are not statistically significant.

Relative Length Increases of Branches

Length increases of branches in 1993 and 1994 are compared with the average annual length increase before 1992 (100 %) when the trees were enclosed in the OTC's.
Influence of chamber effect was found in length increases of whorl branches in season 1994 (p=0.01) – bigger increases in the OTCs (Fig. 4).

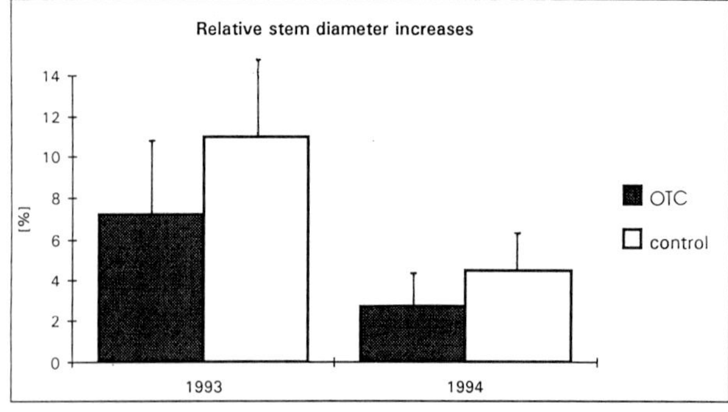

Fig. 3: Relative increases of stem diameters in open top chambers (OTC) and in open air (control) in growing seasons 1993 and 1994. The stem diameter at the begining of each season is 100 %.

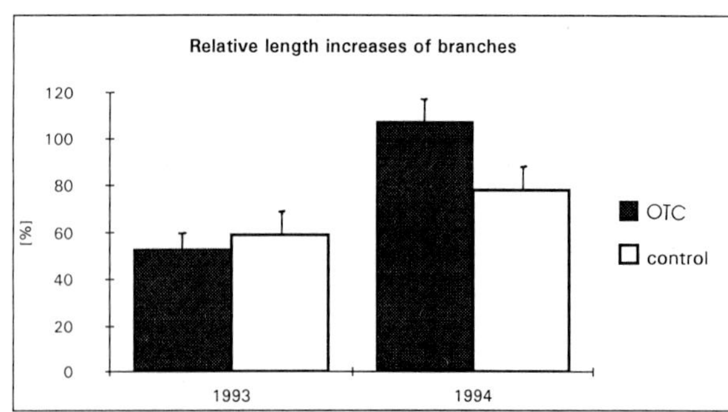

Fig. 4: Relative length increases of whorl branches in open top chambers (OTC) and in open air (control) in growing seasons 1993 and 1994. The average annual length increase before 1992, when the experiment began, was 100 %.

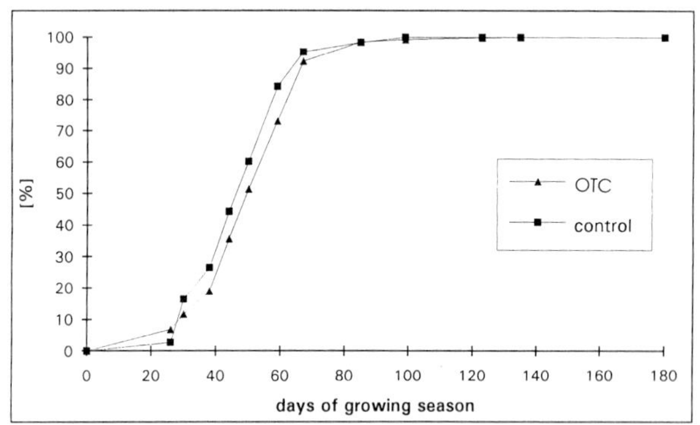

Fig. 5: Growth rate of apical shoots in open top chambers (OTC) and in open air (control) during growing season 1994. Start of the growing season was on April 25. All differences were not statistically significant.

Growth Rate of Apical Shoots

Growth rate was recorded during both seasons at weekly intervals. The beginning of the growing season was on April 21st and April 25th, respectively for 1993 and 1994 and the end of the season was on October 17th and October 24th for the years 1993 and 1994, respectively.
The growth of apical shoots in the OTC was in its middle stage (50 % of total length increase) after 43 days of the growing season compared to 38 days of the control treatment for 1993. A similar situation existed in 1994 (Fig. 5), when the shoots in the OTC reached one half of their final length after 49 days, control after 46 days. The differences are not statistically significant.

Discussion

It is suitable to characterize two different areas within the OTCs in a vertical direction: (A) growth space and (B) protective space. The growth space is determined by the space between the uppermost part of the plant and the ground. The protective space is from the upper part of the plant up to the frustum of the OTC. Efficiency of the protective air layer is dependent on both the thickness of the layer and the flow rate of the air in this layer.

Protective properties of the frustum (valued by Baldocchi et al., 1989) were multiplied by using the double frustum and the rain collection device-funnel.

Transmission of the PVC film used in the spectral range of photosynthetically active radiation (PhAR) was constant. The PVC film (and used UV stabilizers) blocked transmission bellow PhAR as well as other materials (Drake et al., 1989). Heagle et al. (1989) reported transmission of the PVC film about 90 % and Drake et al. (1989) reported transmission of the polyester film from 85 % at 400 nm to 90 % at 700 nm. PhAR transmission of our chamber covered by a new clear film was 91 %. The iron frame of the chamber has no significant shading influence. Transmission decreases with the decreasing angle of the sun's rays on the film. Therefore irradiation within the OTC is minimal close to the side of the chamber in view of the position of the sun. Irradiation within the chamber is fully homogenated a little further from the film (0.1 m). Heagle et al. (1989) reported transmission of an OTC with rain cap 85 % (design of the upper part of our OTC is similar to this one), while Drake et al. (1989) reported transmission of an OTC as approximatelly 90 %, that is higher than the transmission of the polyester film used (size of the opening of this OTC is larger than ours and a diffuse role of the films was not tested). The degradation of our film during the season had the effect of decreasing transmission caused in the first place by the deposition of impurities on the film.

Mean temperatures of air within the canopy inside OTCs were higher than outside OTCs on sunny days. These mean values (1.2 °C) were not so different from the chamber air temperatures reviewed by Allen et al. (1992). Differences between inner and ambient temperatures were strongly dependent on solar radiation. Soil temperatures under the canopy were dependent on air temperatures but there were no significant differences in daily means inside and outside OTC.

Olszyk et al. (1980) reported higher relative humidity within the chambers than within the canopy of plants outside the chambers. Allen et al. (1992) reviewed besides others, that relative humidity increased by day and decreased by night. Heagle et al. (1973) reported that their measurements of relative humidity indicated no differences between the chambers and ambient air (OTC without frustum). Our measurements recorded a decrease in the relative humidity during the day. There is a negative correlation with temperature differences inside and outside OTCs (results from three sunny days in July: $r^2 = 0.81$, $P < 0.001$, $n = 90$). With increasing temperature, the value of saturated vapour tension (E) is increasing, relative humidity is decreasing.

Pigment contents between investigated characterictics were found to be the most sensitive to chamber effects. In OTCs we recorded continual increases of chlorophyll and carotenoids content during the growth of the trees. Differences between OTC and control treatments increased during both growing seasons. It probably shows the influence of changes in radiation intensity inside the OTC (Björgman, 1981; Senser et al., 1975). Because there was no change in the quality of photosynthetically active radiation we did not investigate the chamber effect on pigment ratio (Anderson et al., 1988).

The results from chlorophyll *a* fluorescence obtained during the cultivation in OTCs showed greater photochemical efficiency in the capture of radiant energy for PSII in comparison with the control. Also in parameter $t_{1/2}$, a simple indicator of the pool size and the transfer of electrons from the reaction centers to the quinone pool, had bigger values in the OTCs. The whole reduction of this parameter during the growing season 1993 could be assumed to result from environmental stress (warm summer and precipitation deficiency compared with long-term yearly averages). Also chlorophyll *a* fluorescence significantly detected environmental changes (Lichtenthaler and Rinderle, 1988; Long et al., 1993).

Regarding the increase in the chlorophyll content the light harvesting complex of PSII were increasing (Zuber, 1985). Therefore, photochemical efficiency for the capture of photosynthetically active radiation and its utilization for the photochemical processes were higher.

Respiration rates corresponded with the phenological phase of the individual tree especially at the start of the season when the maximal variance between samples becomes evident. The significant influence of chamber effects is shown in Fig. 2, and it was also found when comparing the development of respiration in years 1993 and 1994. The whole stem respiration in OTCs in the year 1994 was 122 % in comparison to the control treatment, in the year 1993 respiratory losses were not different. This development shows that in the OTC respiration is increasing in comparison with control during one growth season and throughout the whole application of the chamber devices, although any greater biomass increment of the stems was not found. The results are consistent with the following: The maintenance respiration is temperature-dependent (Linder and Troeng, 1980). The growth respiration is proportional to the amount of new produced biomass. Its energetic losses produced per biomass unit are independent of the temperature at which the biomass was produced (Penning de Wries, 1975; Landsberg, 1986).

Growth and phenological analysis in the seasons 1993 and 1994 did not show any clear trends, which appear in Norway spruce trees growing in the influence of the OTCs. With the exception of length increases of whorl branches between ambient and control in 1994, we found no statistically significant differences. Variability within the trees in our experiment removed any statistical differences between treatments. Growth is a final part in a sequence of physiological processes and responses of the trees in growth would be damped and delayed (see Fritts, 1976).

Conclusion

Chamber effects of the open top chambers for field studies are not insignificant. Chamber effects had the largest influence on the quality of the photosynthetical apparatus, especially the amount of chlorophyll content. We recorded an increase in the amount of pigments, their ratio however was unchanged. We found an increase in the stem respiration in the OTCs during the growing seasons. Growth of the trees was not influenced as much. We found only a significant effect on the length of whorl branches, other parameters including phenology were not significantly changed during the investigated years.

Acknowledgements

Research was supported by the EC R&D Programme (ECO-CRAFT Contract No. EV5V-CT92-0127) and by the Ministry of Environmental Protection of the CR (Contract No. GA/1194/93).

References

Allen, L. H. Jr., B. G. Drake, H. H. Rogers, and J. H. Shinn: Field techniques for exposure of plants and ecosystem to elevated CO_2 and other trace gases. Crit. Rev. Plant Sci. *11,* 85–119 (1992).

Anderson, J. M., W. S. Chow, and D. J. Goodchild: Thylakoid membrane organisation in soon/shade acclimation. Australian J. Plant Physiol. *15,* 11–26 (1988).

Björgman, O.: Responses to different quantum densities. In: Lange, O. L., P. S. Nobel, and H. Ziegler (eds.): Physiological plant ecology. I. Response to the physical environment, pp. 57–108. Springer Verlag, Berlin–Heidelberg–New York (1981).

Ceulemans, R. and M. Mousseau: Effects of elevated atmospheric CO_2 on woody plants. New Phytol. *127,* 425–446 (1994).

Drake, B. G., P. W. Leadley, W. J. Arp, D. Nassiry, and P. S. Curtis: An open top chamber for field studies of elevated atmospheric CO_2 concentration on saltmarsh vegetation. Functional Ecol. *3,* 363–371 (1989).

Eamus, D. and P. G. Jarvis: The direct effect of increase in the global atmospheric CO_2 concentration on natural and commercial temperate trees and forests. Adv. Ecol. Res. *19,* 1–55 (1989).

Fritts, H. C.: Tree Rings and Climate. Academic Press, London–New York–San Francisco (1976).

Havaux, M., R. J. Strasser, and H. Greppin: A theoretical and experimental analysis of the q_P and q_N coefficients of chlorophyll fluorescence quenching and their relation to photochemical and nonphotochemical events. Photosynth. Res. *27,* 41–55 (1991).

Heagle, A. S., D. E. Body, and W. W.: Heck: An open top field chamber to assess the impact of air pollution on plants. J. Environ. Qual. *2,* 365–368 (1973).

Heagle, A. S., R. B. Philbeck, R. E. Ferrell, and W. W. Heck: Design and performance of a large, field exposure chamber to measure effect of air quality on plants. J. Environ. Qual. *18,* 361–368 (1989).

Hendrey, G. R. (ed.): FACE: Free-Air CO_2 Enrichment for plant research in the field. Chelsea, Michigan: C.K. Smoley Publish (1992).

Janous, D.: Cambial activity of Norway spruce (*Picea abies* (L.) Karst.). Ph. D. diss. Agricultural Univ., Brno (1990).

Krecmer, V. (ed.): Bioclimatological Dictionary. Academia, Prague. p. 106 (1980) (in Czech).

Landsberg, J. J.: Physiological Ecology of Forest Production. Academic Press, London (1986).

Lichtenthaler, H. K.: Chlorophylls and Carotenoids: Pigments of Photosynthetic Biomembranes. Methods in Enzymology *148,* 350–382 (1987).

Lichtenthaler, H. K. and U. Rinderle: The role of chlorophyll fluorescence in the detection of stress conditions in plants. CRC Crit. Rev. Anal. Chem. *19,* 29–85 (1988).

Linder, S. and E. Troeng: Photosynthesis and transpiration of 20-year-old Scots pine. In: Persson, T. (ed.): Structure and Function of Northern Coniferous Forest – An Ecosystem Study. Ecol. Bull. (Stockholm) *32,* 165–181 (1980).

Long, S. P., N. R. Baker, and C. A. Raines: Analysing the responses of photosynthetic CO_2 assimilation to the long-term elevation of atmospheric CO_2 concentration. Vegetatio *104/105,* 33–45 (1993).

Olszyk, D. M., T. W. Tibbitts, and W. M. Hertsberg: Environment in open-top field chambers utilized forair pollution studies. J. Environ. Qual. *9,* 610–615 (1980).

Penning de Vries, F. W. T.: Use of assimilates in higher plants. In: Cooper, J. G. (ed.): Photosynthesis and Productivity in Different Environments, pp. 459–480. Cambridge Univ. Press, Cambridge (1975).

Nakayama, F. S. and B. A. Kimball: Soil carbon dioxide distribution and flux within the open-top chamber. Agron J. *80,* 394–398 (1988).

Senser, M., S. Schötz, and E. Back: Seasonal changes in structure and function of spruce chloroplasts. Planta *126,* 1–10 (1975).

Sprugel, D. G. and U. Benecke: Woody-tissue respiration and photosynthesis. In: Lassoie, J. P. and T. M. Hinckley (eds.): Methods and Approaches in Tree Ecophysiology. CRC Press, Boca Raton (1990).

Svensson, J.: Stem respiration in young stands of *Eucalyptus globbulus* Labill. A thesis for the degree Master of Science in Forestry. Uppsala (1989).

Zuber, H.: Structure and function of light-harvesting complexes and and their polypeptides. Photochem. and Photobiol. *42,* 821–844 (1985).

J. Plant Physiol. Vol. 148. pp. 339–344 (1996)

Influence of Altitude, Sampling Year and Needle Age Class on Stress-Physiological Reactions of Spruce Needles Investigated on an Alpine Altitude Profile

EDITH BERMADINGER-STABENTHEINER

Institut für Pflanzenphysiologie, Schubertstraße 51, A-8010 Graz, Austria

Received July 9, 1995 · Accepted September 25, 1995

Summary

The aim of the presented study was to characterize the physiological state of spruce needles from an alpine profile depending on altitude (970–1420 m above sea level) and needle age class (current year needles, one-year-old needles) during two consecutive years by investigating components of the antioxidative scavenging system (water extractable thiols, ascorbic acid, total peroxidase). The chlorophyll content was measured as a non-specific indicator of the general plant condition. With the exception of thiols the one-year-old needles always had higher needle contens of antioxidants and chlorophylls and the means of thiols and chlorophylls were significantly higher in the second sampling year. A more or less continous increase of antioxidants and a decrease of chlorophylls with increasing altitude was observed from the middle elevations (1140–1240 m a.s.l.) up to the highest site near the timberline (1420 m a.s.l.) and was interpreted as a response of the needles to increased oxidative stress due to the more uncomfortable environmental conditions at these elevations. However, high levels of antioxidants near the valley bottom (970–1050 m a.s.l.) also indicated an enhancement of the detoxification system and rather low chlorophyll levels (means of one-year-old needles <1100 µg/g dw) probably reflected high environmental impact on the trees at these sites.

Key words: Spruce, Picea abies (L.) Karst., environmental stress, ascorbate, water extractable thiols, peroxidase activity, chlorophyll, altitude profile.

Introduction

Forest ecosystems are exposed to a broad variety of different environmental parameters. The Achenkirch project (Tyrol, Austria) was organized by the Austrian Federal Forest Research Centre with the major goal to describe and assess the importance of the most diverse correlations between site conditions, natural stressors and air pollutants and to characterize the response of spruce trees to this surrounding (Herman and Smidt, 1994). Within the scope of this project stress-physiological investigations on needles were of particular interest.

Activated oxygen species and free radicals are involved in stress-symptom development under a great number of different conditions (Elstner and Osswald, 1994). The antioxidative scavenging system of plants therefore turned out to be a valuable tool for an evaluation of the response of plants to various stressors (Alscher and Amthor, 1988; Rennenberg and Polle, 1989; Bermadinger et al., 1990; Smith et al., 1990; Madamanchi et al., 1991; Tausz et al., 1994). In field studies, however, the physiological response of the needles is always influenced by a great many of different environmental factors. It therefore is very difficult – if at all possible – to identify a special cause of stress (Schulze et al., 1989). A field-approach for stress-physiological indication will only be successful in a well documented research area, where a detailed description of the investigated area helps to detect and to rate the influence of several environmental factors, including air pollutants, on the forest ecosystem. These requirements are given in the Achenkirch area (Herman and Smidt, 1994).

As a first step towards a better understanding of the physiological reactions of needles in an alpine environment it is necessary to know the general dependence of the investigated parameters on altitude and needle age class. The physiological fluctuations within consecutive sampling years are also of fundamental interest. Only then will it be possible to determine the connections between climatic and site conditions, man-made stressors and the physiological response of the trees (Herman and Smidt, 1994).

Therefore the aim of the presented study was to characterize a part of the antioxidative scavenging system (water extractable thiols, ascorbic acid, activity of total peroxidase) of spruce needles from an alpine profile depending on altitude and needle age class during two consecutive sampling years. The chlorophyll content was measured as a non-specific indicator of the general plant condition (Oren et al., 1993; Tausz et al., 1994).

Material and Methods

The sampling sites are situated in the Northern Calcareous Alps, west of Achental, a valley that connects the Inn valley with South Bavaria/Germany. The Christlum profile includes six sampling sites from the valley bottom (970 m above sea level; site 1) to the timberline (1420 m a.s.l.; site 6; compare Table 1). The predominant soil types are Rendzina and Terra fusca, developed from dolomite and various forms of limestone. Following the assessment criteria of the Forest Condition Inventory (WZI) and the Forest Damage Monitoring System (WBS) little or medium crown thinning could be found (Herman and Smidt, 1994). According to Smidt and Gabler (1994) the limits for SO_2 (limit for the annual mean: 25 µg SO_2/m^3) and NO_2 (limit for the annual mean: 30 µg NO_2/m^3) were not exceeded in the research area during the investigation period. The ozone limits (limits for 8-hour average and for the vegetation period: 60 µg O_3/m^3), however, were frequently exceeded (Smidt and Gabler, 1994).

Table 1: Altitude (m above sea level), exposition and number of trees (*Picea abies* (L.) Karst.) of the Christlumprofile investigated in August 1991 and 1992.

Site	Altitude (m)	Exposition	Number of trees
6	1420	north	4
5	1320	north	4
4	1240	north-east	5
3	1140	north-east	5
2	1050	north-east	5
1	970	east	5

For details of sampling of the spruce needles and conservation procedure compare Bermadinger-Stabentheiner (1994 a).

Soluble thiols, that consist of 95–100 % of glutathione (Grill et al., 1982), were determined photometrically with 5,5′-dithiobis(2-nitrobenzoic acid) (Ellman reagent) (Grill et al., 1982). The content of *ascorbic acid* was determined by an isocratic high-performance-liquid-chromatography (HPLC)-method according to Kneifel and Sommer (1985). Peroxidase activity was measured according to Keller and Schwager (1971). The chloroplast pigments were analysed using a gradient HPLC method according to Pfeifhofer (1989). Chlorophylls, carotenes and xanthophylls are highly correlated (chlorophyll/xanthophyll: $r=0.94$, $p<<0.001$, $n=210$; chlorophyll/carotene: $r=0.69$, $p<<0.001$, $n=210$; Bermadinger-Stabentheiner, 1994 a) and only the results for the chlorophylls (sum of chlorophyll *a* and *b*) are presented. All data are presented on dry weight basis (per g dw).

Statistical evaluations were completed using Statistica© (StatSoft, USA, 1994) software package. Differences between sites, needle age and sampling date were calculated by analysis of variance (ANOVA). Small violations of the normality assumption did occur, but such small deviations from this assumption usually do not adversely bias the F-test in ANOVA procedure. However, since site means and standard deviations correlated, a fact that actually may lead to misleading ANOVA-results, significant main effects had to be verified by non-parametric methods (Kruskall-Wallis test) (Tausz et al., 1996).

Results

The needle contents of water extractable thiols, ascorbic acid, peroxidase activity, and total chlorophylls of the current year and one-year-old needles sampled in August 1991 and 1992 are shown in Figs. 1–4 in dependence on altitude (mean; box = mean ± standard error of the mean; whisker = mean ± standard deviation). The results of analysis of variance (ANOVA) with altitude, sampling date and needle age as main effects (three-way ANOVA) are shown in Table 2 and the results of the corresponding Kruskall-Wallis test in Table 3.

Table 3: Non-parametric verification of ANOVA (Table 2) with Kruskall-Wallis test; n – number of samples; * – $p=0.01–0.05$; ** – $p=0.001–0.01$; *** – $p<0.001$; n. s. – non significant.

	n	Altitude	Sampling year	Needle age class
Thiols	106	n. s.	***	n. s.
Peroxidase	110	**	n. s.	***
Ascorbic acid	109	n. s.	n. s.	***
Chlorophyll	91	**	***	***

Table 2: Results of analysis of variance (ANOVA) with altitude (970–1420 m a. s. l.), sampling year (1991 and 1992) and needle age class (current year needles and one-year-old needles) as main effects; * – $p=0.01–0.05$; ** – $p=0.001–0.01$; *** – $p<0.001$; n. s. – non significant.

	df Error	Altitude (1)	Sampling year (2)	Needle age class (3)	(1)×(2)	(1)×(3)	(2)×(3)	(1)×(2)×(3)
Thiols	82	*	***	*	n. s.	n. s.	n. s.	n. s.
Peroxidase	86	*	n. s.	***	n. s.	n. s.	**	n. s.
Ascorbic acid	85	***	n. s.	***	*	n. s.	n. s.	n. s.
Chlorophyll	67	***	***	***	n. s.	*	n. s.	n. s.

Water extractable thiols

Mean responses of water extractable thiols were affected only by the sampling year (Table 2) with distinctly higher thiol contents ($p < 0.001$) observed in the second year of investigation. Significant ANOVA-results for altitude and needle age could not be verified by Kruskall-Wallis test (Table 3). The highest thiol contents of the needles were always observed near the timberline (1320–1420 m) and near the valley bottom (970–1050 m; Fig. 1). Wheras the current year needles did not show a distinct altitudinal dependence the one-year-old needles revealed a characteristic altitudinal pattern with high values of thiols at the valley bottom and near the timberline and lower values at middle elevations (1240 m). This pattern, however, was more pronounced in the first sampling year.

Ascorbic acid

In the first sampling year the altitudinal pattern of the needle content of ascorbic acid resembled very much that of the thiols (Fig. 2). Maximum values of ascorbic acid could be observed near the timberline (1420 m) and at the valley bottom (970 m), wheras the needles from the middle elevations (1140–1240 m) where characterized by lower needle contents (current year needles as well as one-year-old needles). In the second sampling year the ascorbic acid content of the current year needles did not show a clear altitude dependence and was thus also comparable to the thiols. Mean responses of ascorbic acid were affected only by the needle age class (Table 2). The one-year-old needles always had higher levels of ascorbic acid than the current year needles ($p < 0.001$). Significant results of ANOVA with altitude as main effect could not be verified by Kruskall-Wallis test (Table 3).

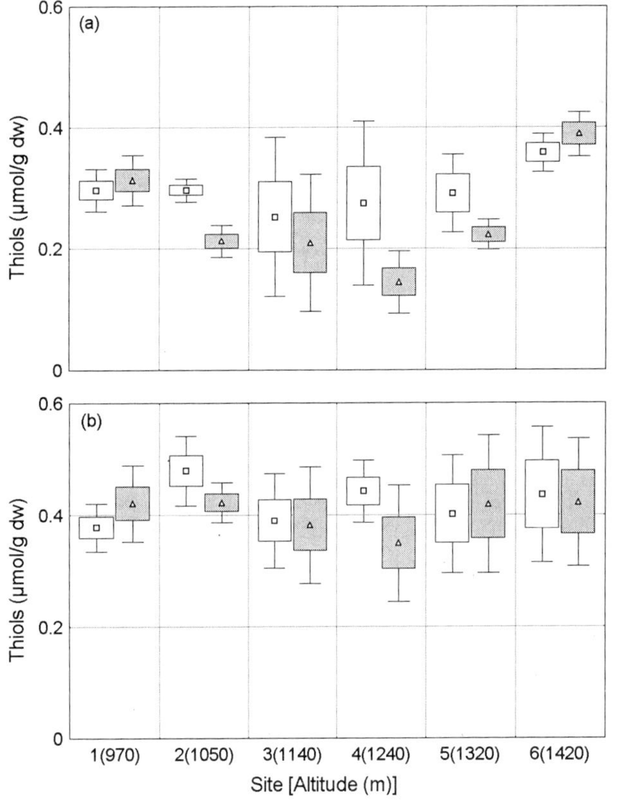

Fig. 1: Water extractable thiols (µmol/g dw) in dependence on altitude (970–1420 m a.s.l.; site 1–6) of the current year needles (open squares, open boxes) and one-year-old needles (open triangles, dotted boxes); (a) first sampling year (August 1991); (b) second sampling year (August 1992); open squares and triangles represent the mean, box = mean ± standard error of mean, whisker = mean ± standard deviation.

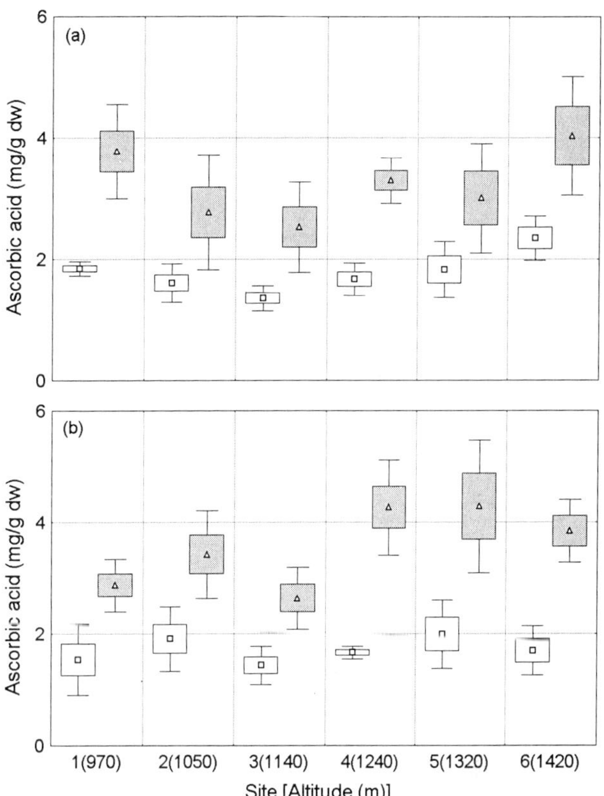

Fig. 2: Ascorbic acid (mg/g dw) in dependence on altitude (970–1420 m a.s.l.); for further detail compare Fig. 1.

Activity of total peroxidase

Mean responses of the activity of total peroxidase were affected by altitude as well as by needle age class (Tables 2, 3). The one-year-old needles often revealed higher enzyme activities ($p < 0.001$) than the current year needles. The peroxidase activity of the current year needles was characterized by a more or less continous increase with increasing altitude (1991 as well as 1992; Fig. 3). In both investigation years the highest enzyme activities of the one-year-old needles were observed at 1240–1320 m.

Chlorophylls

Mean responses of chlorophylls were affected by altitude, needle age class and the sampling year (Tables 2, 3). Comparable to the thiols the chlorophyll values of the needles were significantly higher in the second sampling year ($p<0.001$) wheras corresponding to the needle contents of ascorbic acid and to the peroxidase activity the one-year-old needles were characterized by higher contents than the current year needles ($p<0.001$). In both sampling years the chlorophyll content of the current year needles slightely increased with increasing altitude (Fig. 4). In the first sampling year the one-year-old needles did not show an altitude dependence with generally rather low chlorophyll contents at all sites (means <1200 µg/g dw). The altitudinal pattern of the chlorophyll values of the needles observed in the second sampling year was characterized by very low levels at the valley bottom (means <1000 µg/g dw), a continuous decrease from 1050 to 1320 m and higher levels again near the timberline (1420 m; mean >1600 µg/g dw).

Discussion

The aim of the presented study was to evaluate the influence of altitude, sampling year and needle age class on content of ascorbic acid, water extractable thiols, chlorophyll and activity of total peroxidase of spruce needles from an alpine altitude profile.

Plants at higher elevations have to cope with lower temperature, higher irradiation and higher concentrations of naturally occurring photooxidants (Bolhar-Nordenkampf and Lechner, 1989). These harsh environmental conditions are known to generate oxidative stress to the cells by raising the cellular concentrations of peroxides and free radicals and as a response metabolic antioxidant quenching mechanisms are activated (Alscher and Amthor, 1988; Smith et al., 1990; Elstner and Osswald, 1994).

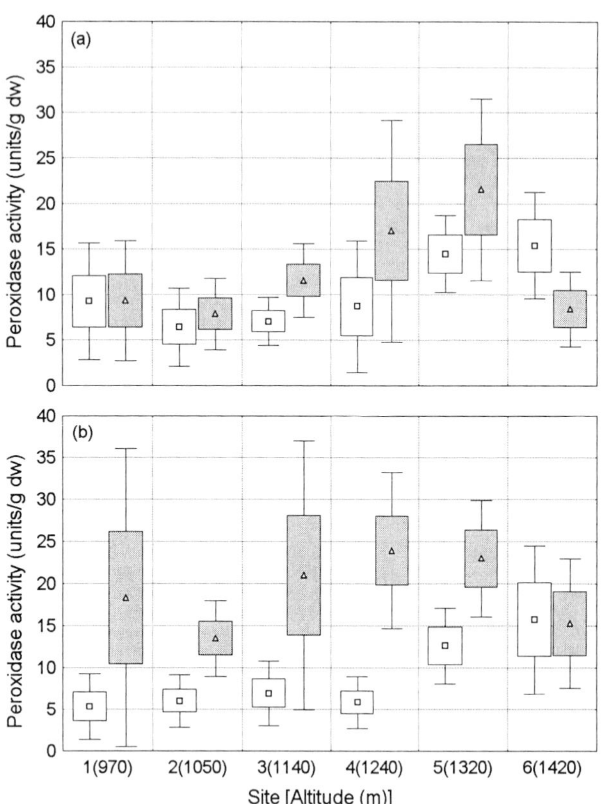

Fig. 3: Activity of total peroxidase (units/g dw) in dependence on altitude (970–1420 m a.s.l.); for further detail compare Fig. 1.

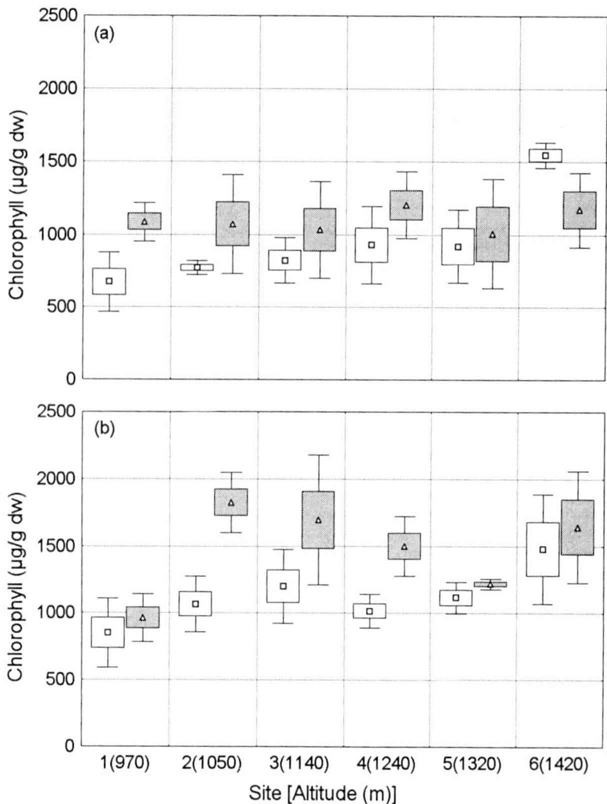

Fig. 4: Chlorophyll (µg/g dw) in dependence on altitude (970–1420 m a.s.l.); for further detail compare Fig. 1.

In the here presented study from the Achenkirch research area the highest levels of antioxidants could always be observed in needles from the timberline (1420 m) thus indicating an enhanced detoxification capacity of the plant cells as a response to the prevailing, uncomfortable environmental conditions at these elevations. Furthermore, the altitudinal course of these antioxidants was characterized by a more or less continous increase from middle elevations (1140 m) up to the timberline. Since a comparable enhancement of the antioxidative scavenging system with increasing altitude is confirmed by other field studies with *Picea abies* and *Picea rubens* from different research plots in Austria, Germany and North America (Grill et al., 1988; Bermadinger et al., 1989; Madamanchi et al., 1991; Bermadinger-Stabentheiner et al., 1991; Polle and Rennenberg, 1992; Bermadinger-Stabentheiner, 1994 a), this increase with increasing altitude can be designated as a general response of the antioxidative scavenging system of spruce trees in alpine areas. However, ozone concentrations increased during the last decades (Smidt et al., 1994) and ozone limits are frequently exceeded at the higher elevations of the Achenkirch area (Smidt and Gabler, 1994). It is well known, that ozone exposure results in an enhancement of the scavenging system (Alscher and Amthor, 1988; Elstner and Osswald, 1994). Therefore it is highly probable that the high ozone concentrations in Achenkirch also participate in generating oxidative stress to the plants (Smidt et al., 1994).

However, the needles from sites near the valley bottom (970–1050 m) were also characterized by higher needle contents of ascorbic acid and thiols comparable to those at the timberline thus resulting in an altitudinal course with high levels of these antioxidants near the valley bottom (970 m) and the timberline (1420 m) and lower values at the middle elevations (1140–1240 m). It is known that a lot of various impacts (e.g., salt, drought, mineral deficiency, infections, air pollutants) lead to oxidative stress and to an increase in the cellular antioxidant level (Elstner and Osswald, 1994). Results of a lichen mapping (Hofmann, 1994) and investigations of epicuticular wax structures (Bermadinger-Stabentheiner, 1994 b) classified the sites near the valley bottom as moderately polluted. Low concentrations of sulfur dioxide and nitrogen oxides (Smidt and Gabler, 1994), however, give evidence that pollutants can not only be responsible for these physiological reactions. It should be considered that ambient concentrations of common air pollutants can modify plant responses to a number of environmental stressors and vice versa (Davison and Barnes, 1993). The rather low chlorophyll levels near the valley bottom (means of one-year-old needles <1100 µg/g dry weight) also reflected a high environmental impact on the trees at these sites (Oren et al., 1993; Siefermann-Harms, 1994). Similar altitudinal patterns of antioxidants with high levels near the timberline (1700 m) and the valley bottom (870 m) and lower values at the middle elevations (1270 m) were also observed at the near-by Wank (Bavarian Calcareous Alps), but were interpreted as a decrease in the scavenging capacity indicating a weakening of the spruce trees (Polle and Rennenberg, 1992).

The mean responses of the needle content of ascorbic acid as well as of peroxidase-activity were not affected when comparing two consecutive investigation years. Water extractable thiols and chlorophylls, however, were significantly increased in the second sampling year. There are first indications for the assumption that thiols and chlorophylls correlate with daily mean temperature the months before sampling whilst this was not the case with ascorbic acid and peroxidase (Bermadinger-Stabentheiner, unpublished). Such dependencies of the physiological condition of the needles on climatic variations must be studied in further detail. With the exception of thiols the one-year-old needles always had higher needle contents of antioxidants and chlorophylls and are therefore better suited as bioindicators, since the differences between sites are more pronounced.

Summarizing, the needle contents of antioxidants and chlorophylls depend on altitude, needle age and also on the sampling date. Deviations from expected altitudinal patterns (increase of antioxidants and decrease of chlorophylls with increasing altitude) indicate different and additional environmental impacts the plants have to cope with. Starting from these detailed description of the physiological response of spruce needles from the Achenkirch research area the next step will be to detect and to rate further connections between the nutrient situation, the climatic conditions, manmade stressors and the physiological response of the trees under field conditions and on an interdisziplinary basis (Herman and Smidt, 1994).

Acknowledgements

This study was financially supported by the Austrian Federal Ministry of Agriculture and Forestry. I am also very grateful to the whole team of the Federal Forest Research Centre for their support, and to the team of our laboratory for their technical assistance.

References

Alscher, R. G. and J. S. Amthor: The physiology of free-radical scavenging: maintenance and repair processes. In: Schulte-Hostede, P., N. M. Darrall, L. W. Blank, and R A. Wellburn (eds.): Air pollution and plant metabolism, pp. 94–115. Elsevier, New York (1988).

Bermadinger, E., D. Grill, and H. Guttenberger: Thiole, Ascorbinsäure, Pigmente und Epikutikularwachse in Fichtennadeln aus dem Höhenprofil «Zillertal». Phyton (Austria) *29,* 163–185 (1989).

Bermadinger, E., H. Guttenberger, and D. Grill: Physiology of young Norway spruce. Environ. Pollut. *64,* 319–330 (1990).

Bermadinger-Stabentheiner, E.: Stress-physiological investigations on spruce trees (*Picea abies* (L.) Karst.) from the «Achenkirch Altitude Profiles». Phyton (Austria) *34,* 97–112 (1994 a).

– Epikutikularwachse von Fichtennadeln am Höhenprofil Achenkirch. FBVA-Reports (Federal Forest Research Centre, ed.) *78,* 117–120 (1994 b).

Bermadinger-Stabentheiner, E., D. Grill, and T. Kern: Physiologisch-biochemische Streßindikation an Fichten aus verschiedenen Höhenlagen. VDI-Ber. *901,* 391–406 (1991).

Bolhar-Nordenkampf, H. R. and E. G. Lechner: Synopse streßbedingter Modifikationen der Anatomie und Physiologie der Nadeln als Frühdiagnose einer Disposition zur Schadensentwicklung bei Fichte. Phyton (Austria) *29,* 255–301 (1989).

Davison, A. W. and J. D. Barnes: Predisposition to stress following exposure to air pollution. In: Jackson, M. B. and C. R. Black (eds.): Interacting stresses on plants in a changing climate, NATO ASI Series, Vol. *I 16,* pp. 111–123. Springer Verlag, Berlin, Heidelberg (1993).

Elstner, E. F. and W. Osswald: Mechanisms of oxygen activation during plant stress. Proceedings of the Royal Society of Edinburgh *102 B,* 131–154 (1994).

Grill, D., H. Esterbauer, and K. Hellig: Further studies on the effect of SO_2-pollution on the sulfhydryl-systems of plants. Phytopath. Z. *104,* 264–271 (1982).

Grill, D., H. Pfeifhofer, A. Tschulik, K. Hellig, and K. Holzer: Thiol content of spruce needles at forest limits. Oecologia *76,* 294–297 (1988).

Herman, F. and S. Smidt: Ecosystem studies in a limestone area – «Achenkirch altitude profiles». Phyton (Austria) *34,* 9–24 (1994).

Hofmann, P.: Pollutant-related mapping of lichens in the area of Achenkirch. Phyton (Austria) *34,* 71–84 (1994).

Keller, T. and H. Schwager: Der Nachweis unsichtbarer («physiologischer») Fluor-Immissionsschädigungen an Waldbäumen durch eine einfache kolorimetrische Bestimmung der Peroxidase-Aktivität. Eur. J. For. Path. *1,* 6–18 (1971).

Kneifel, W. and R. Sommer: HPLC-Methode zur Bestimmung von Vitamin C in Milch, Molke und Molkegetränken. Z. Lebensm. Unters. Forsch. *181*, 107–110 (1985).

Madamanchi, N. R., A. Hausladen, R. G. Alscher, R. G. Amundson, and S. Fellows: Seasonal changes in antioxidants in red spruce (*Picea rubens* Sarg.) from three field sites in the northeastern United States. New Phytol. *118*, 331–338 (1991).

Oren, R., K. S. Werk, N. Buchmann, and R. Zimmermann: Chlorphyll-nutrient relationships identify nutritionally caused decline in *Picea abies* stands. Can. J. For. Res. *23*, 1187–1195 (1993).

Pfeifhofer, H.: Evidence of chlorophyll *b* and lack of lutein in *Neottia nidus-avis* plastids. Biochem. Physiol. Pflanzen *184*, 55–61 (1989).

Polle, A. and H. Rennenberg: Field studies on Norway spruce trees at high altitudes: II. Defence systems against oxidative stress in needles. New Phytol. *121*, 635–642 (1992).

Rennenberg, H. and A. Polle: Effects of photooxidants on plants. In: Georgii, H. W. (ed.): Mechanisms and effects of pollutant transfer into forests, pp. 251–258. Kluwer Academic Publishers, Dordrecht (1989).

Schulze, E. D., R. Oren, and O. L. Lange: Processes leading to forest decline: a synthesis. In: Schulze, E. D., O. L. Lange, and R. Oren (eds.): Ecological Studies, Vol. *77*, pp. 459–468. Springer Verlag, Berlin, Heidelberg (1989).

Siefermann-Harms, D.: Light and temperature control of season-dependent changes in α- and β-carotene content of spruce needles. J. Plant Physiol. *143*, 488–494 (1994).

Smidt, S. and K. Gabler: SO_2, NO_x and ozone records along the «Achenkirch altitude profiles». Phyton (Austria) *34*, 33–44 (1994).

Smidt, S., E. Bermadinger-Stabentheiner, and F. Herman: Altitude-dependent ozone concentrations and changes of ozone related plant-physiological parameters in the needles of Norway spruce. Proceedings of the Royal Society of Edinburgh *102 B*, 113–117 (1994).

Smith, I. K., A. Polle, and H. Rennenberg: Glutathione. In: Alscher, R. and J. Cumming (eds.): Stress responses in plants: Adaptation and acclimation mechanisms, pp. 201–215. Wiley-Liss., Inc., New York (1990).

Tausz, M., M. Müller, E. Bermadinger-Stabentheiner, and D. Grill: Stress-physiological investigations and chromosomal analysis on Norway spruce (*Picea abies* (L.) Karst.) – A field study. Phyton (Austria) *34*, 291–308 (1994).

Tausz, M., L. J. De Kok, I. Stulen, and D. Grill: Physiological responses of Norway spruce trees to elevated CO_2 and SO_2. J. Plant Physiol. *148*, 362–367 (1966).

J. Plant Physiol. Vol. 148. pp. 345–350 (1996)

Reconstitution of Photosynthesis Upon Rehydration in the Desiccated Leaves of the Poikilochlorophyllous Shrub *Xerophyta scabrida* at Elevated CO_2

ZSOLT CSINTALAN[1], ZOLTÁN TUBA[1,3], HARTMUT K. LICHTENTHALER[2], and JOHN GRACE[3]

[1] Plant Physiological Section, Department of Botany and Plant Physiology, Agricultural University of Gödöllő, H-2103 Gödöllő, Hungary

[2] Botanical Institute II (Plant Physiology and Plant Biochemistry), University of Karlsruhe, Kaiserstraße 12, D-76128 Karlsruhe, Germany

[3] Institute of Ecology and Resource Management, University of Edinburgh, Darwin Building, Mayfield Road, Edinburgh EH9 3JU, UK

Received June 24, 1995 · Accepted September 25, 1995

Summary

We report the resynthesis of the photosynthetic apparatus and the restoration of its function in the monocotyledonous C_3 shrub *Xerophyta scabrida* (Pax) Th. Dur. et Schinz (Velloziaceae) following a period of 5 years in the air-dried state. Detached leaves were rehydrated at present (350 µmol mol^{-1}) and at elevated CO_2 (700 µmol mol^{-1}). Elevated CO_2 concentration had no effect on the rate of rehydration, nor on the *de novo* resynthesis pattern of the chlorophylls and carotenoids or the development of photochemical activity in the reviving desiccated leaves. The time required to fully reconstitute the photosynthetic apparatus and its function in the air-dried achlorophyllous leaves on rehydration did not differ at the two CO_2 concentrations. However, respiratory activity during rehydration was more intensive and of longer duration at high CO_2 and net CO_2 assimilation first became apparent 12 h later than in the leaves rehydrated at present CO_2. After reconstitution of the photosynthetic apparatus, the net CO_2 assimilation rate was higher in the high CO_2 leaves, however it rapidly declined to a value lower than that in the present CO_2 plants due to acclimation. This acclimation to elevated CO_2 occurred only after complete reconstitution of the photosynthetic apparatus. The downward acclimation of photosynthesis was accompanied by a decrease in content of photosynthetic pigments (chlorophyll *a*+ *b* and carotenoids x+c) and stomatal conductance. The initial slope of the A/c_i curve for the high CO_2 leaves was much lower and net CO_2 assimilation rates were lower at all c_i's than in the present CO_2 plants. The rate of respiration also decreased and the C-balance of the high CO_2 leaves therefore remained similar to that of leaves in present CO_2.

Key words: Acclimation, A/c_i function, carotenoids, elevated CO_2, net CO_2 assimilation, respiration, stomatal conductance, variable chlorophyll fluorescence decrease ratio (Rfd value).

Abbreviations: A = CO_2 assimilation rate; a+b = chlorophyll *a* and *b;* c_i = intercellular CO_2 concentration; g_s = stomatal conductance; HDT = homoiochlorophyllous desiccation tolerant; PDT = poikilochlorophyllous desiccation tolerant; Rfd 690 = variable chlorophyll fluorescence ratio at 690 nm; RuBisCO = RuBP carboxylase/oxyenase; SLA = specific leaf area; x+c = total carotenoids (xanthophylls x and carotenes c).

Introduction

There have been many reports on the impacts of long-term high CO_2 on plants (see eg. recent reviews of Jarvis, 1993; Ceulemans and Mousseau, 1994). Dahlman (1993) has reviewed the combined effects of elevated CO_2 and other environmental factors. The interaction of high CO_2 with water stress (drought) has been especially intensively investigated (Morison, 1993; Rogers and Dahlman, 1993). However, the effect of elevated CO_2 on desiccation tolerant plants has not been investigated. These species can survive years of desiccation.

Desiccation tolerance is relatively infrequent in plants and is rare in angiosperms (Gaff, 1977; Gaff, 1989). However, it is important in certain vegetation zones, such as deserts and semi-deserts in the dry tropics, in dry-wet tropical vegetation formations, in temperate continental climates and in arid boreal areas. Any analysis of the impact of high CO_2 on terrestrial vegetation guarantees investigation of this rather neglected group, the desiccation tolerant (DT) plants.

DT plants form two groups (Hambler, 1961; Bewley, 1979; Gaff, 1989; Tuba et al., 1993 a): a) those, which preserve their chlorophyll content and photosynthetic apparatus during desiccation, are called homoiochlorophyllous DT (HDT), and b) those, which lose their chlorophylls and thylakoids, are termed poikilochlorophyllous DT (PDT). In the latter the photosynthetic apparatus with its pigments and thylakoids is resynthesised on rehydration (Gaff, 1989; Tuba et al., 1994 a).

Elevated CO_2 affects photosynthesis (Sage et al., 1989; Arp and Drake, 1991; Jarvis, 1993), and the PDT mechanism itself causes profound changes in the photosynthetic apparatus during desiccation and rehydration. Therefore, it was hypothesised that high atmospheric CO_2 would influence both the reconstitution of the photosynthesis and the functional revival of the desiccated PDT plants on rehydration. Hence, we examined the reconstitution of the photosynthetic apparatus and its function including CO_2 assimilation in *Xerophyta scabrida,* a PDT monocot C_3 shrub, at rehydration of detached leaves at present (350 μmol mol^{-1}) and at elevated (700 μmol mol^{-1}) CO_2 following a period of 5 years in the air-dried state. The development of pigment resynthesis, thylakoid function, photosynthetic CO_2 assimilation and respiration was followed from the beginning of rehydration and revival through the acclimation of photosynthesis to elevated CO_2. This is the first report on the reconstitution of photosynthesis in a desiccated (PDT) plant at elevated CO_2.

Materials and Methods

Species examined

Xerophyta scabrida (Pax) Th. Dur. et Schinz, is a member of the Velloziaceae family (Velloziales, related to Bromeliales). It is a C_3 (stable carbon isotope ratio $\delta^{13}C = -27.3$ ‰) PDT pseudoshrub of about 40–90 cm in height, with perennial leaves, lacking secondary thickening (Tuba et al., 1993 b). On the top of cliffs *X. scabrida* forms an almost semi-desertlike bush vegetation on biotite migmatite and hornblende gneiss rocks with a long dry season of 5–6 months (Pócs, 1976). Air-dried leaves were collected in Tanzania (Uluguru Mts., Mindu Hill, SSW of Morogoro town at 650 m altitude) by the end of the dry season in 1988 and were stored in airtight polythene bags.

Rehydration procedure

Desiccated leaves were allowed to rehydrate as described earlier (Tuba et al., 1994 a) in 45 L plexiglass chambers, which provided the control of CO_2, air humidity, light and temperature. Glass jars containing the leaves (fixed in a vertical position) with one half of their laminae immersed in distilled water were placed in the chambers. The measurements were made on the middle portion of upper halves of the leaves which were rehydrated and revived in air.

The chambers were illuminated by a halogen light source (Osram Powerstar Mercury HQIE lamps) to provide 1000 μmol $m^{-2} s^{-1}$ of photon flux density. A 5 cm thick filter of streaming water was applied between the light source and the chambers. Temperature was kept at 25 °C. The air CO_2 concentrations were maintained at 350 μmol mol^{-1} in the control and at 700 μmol mol^{-1} in the elevated CO_2 chambers. An electromagnetic valve controlled by an infrared gas analyser (Tuba et al., 1994 b) was used to maintain the elevated CO_2 concentration and axial ventilators provided the mixing of the entering CO_2 with air. For the ambient chambers air cylinders supplied the 350 μmol mol^{-1} CO_2 air at a constant rate. Identical concentrations of CO_2 were continuously passed through the distilled water in which the leaves were immersed and an air pump also ensured the further aeration of the water. The water was kept at 23 °C and frequently changed.

CO_2 gas exchange measurements

The rates of (rehydration and dark) respiration and CO_2 exchange in the light were measured using an LCA2-type IRGA system (ADC Co. Ltd., Hoddesdon, U.K.), operated in differential mode and Parkinson LC-N leaf chamber, as previously described (Tuba et al., 1994 a). CO_2 dependence of light saturated net CO_2 assimilation rates (A/c_i curves) were measured using the same system (Tuba et al., 1994 b). Ambient CO_2 concentrations of 100, 350, 500, 700 and 1000 μmol mol^{-1} were produced by a gas diluter (GD 600, ADC Co. Ltd., Hoddesdon, U.K.). Gas-exchange parameters (A, c_i,) were calculated according to von Caemmerer and Farquhar (1981). Stomatal conductance (g_s) was measured with an AP4 type mass flow porometer (Delta-T Devices Cambridge, U.K.). Leaf area was measured by a leaf area meter (LAM 001, Delta T Devices, Cambridge, U.K.). The Mitscherlich function (Thornley and Johnson, 1990) was used to fit curves to the data on A/c_i.

Other methods

Leaf water and photosynthetic pigment (chlorophyll *a*+ *b* and carotenoid x+c) content and the variable chlorophyll fluorescence decrease ratio (Rfd) were measured as described earlier (Tuba et al., 1993 b, 1994 a and 1994 b). Rehydration, with regreening was repeated four times with 150 leaves for both CO_2 concentration. Measurements were made in at least five replicate samples.

Results

Water content and specific leaf area

On placing the air-dried leaves of *X. scabrida* in water, their water uptake began instantly and they rapidly unfolded (1–1.5 h). The leaf water content increased reaching over 90 % of its maximum in 8 h (Fig. 1 A) with no further increase after 12 h. The specific leaf area (SLA) increased in a similar manner from 52.6 $cm^{-2} g^{-1}$ dry weight at the beginning to 163.5 $cm^{-2} g^{-1}$ after 12 h (Fig. 1 B).

Resynthesis of photosynthetic pigments

Of the photosynthetic pigments only some carotenoids x+c were present in the desiccated leaves of *X. scabrida,* as had been reported for other PDT's, too (Gaff, 1989). The *de novo* resynthesis of chlorophyll *a+b* and carotenoid x+c began 12 h after the start of rehydration (Fig. 2 A and B). There were no statistically significant differences in chlorophyll *a+b* and carotenoid x+c content between leaves reviving at ambient and high CO_2, although in the latter, the amount of both pigments was less at 20 h and remained increasingly so for 72 h. The amount of chlorophyll *a+b* and carotenoid x+c did not change in the present CO_2 leaves after 72 h. In contrast, following the maximum at 72 h, there was a slight decrease of chlorophyll *a+b* and carotenoid x+c content in the high CO_2 leaves which was apparent at 120 h (Fig. 2).

Photochemical activity

Photochemical activity, measured as variable decrease ratio at 690 nm (Rfd690), in the rehydrating leaves was first detected at 12 h, coincident with the beginning of chlorophyll *a+b* resynthesis (Fig. 3). This was followed by a gradual increase in the Rfd690 values for 72 h. At 72 h leaves at both

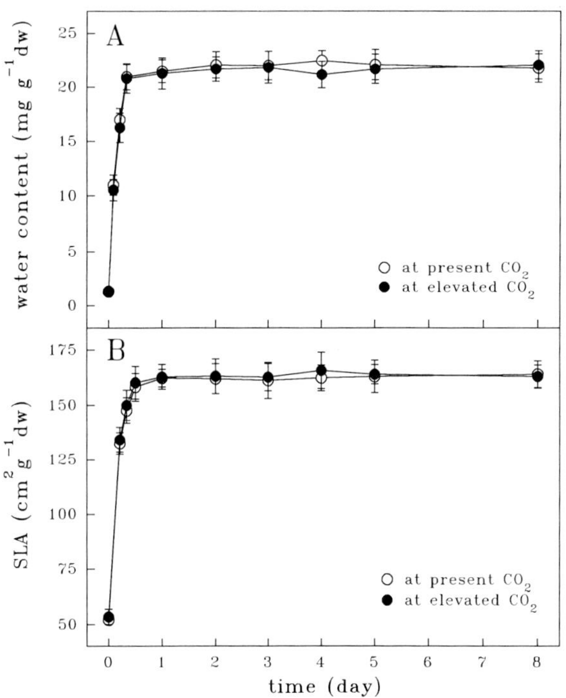

Fig. 1: Changes in water content **(A)** and in specific leaf area, SLA, **(B)** in air-dried achlorophyllous leaves of *X. scabrida* during the rehydration process at present (350 µmol mol^{-1}) and elevated CO_2 (700 µmol mol^{-1}) concentration. Dark bars represent standard deviations.

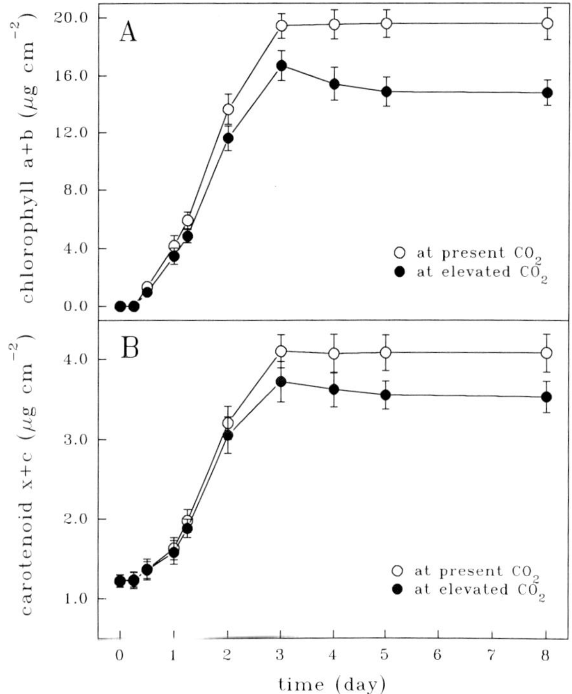

Fig. 2: Resynthesis and accumulation of chlorophyll *a+b* **(A)** and carotenoids x+c **(B)** in air-dried, achlorophyllous leaves of *X. scabrida* during rehydration and revival at present (350 µmol mol^{-1}) and elevated CO_2 (700 µmol mol^{-1}) concentration. Dark bars represent standard deviation.

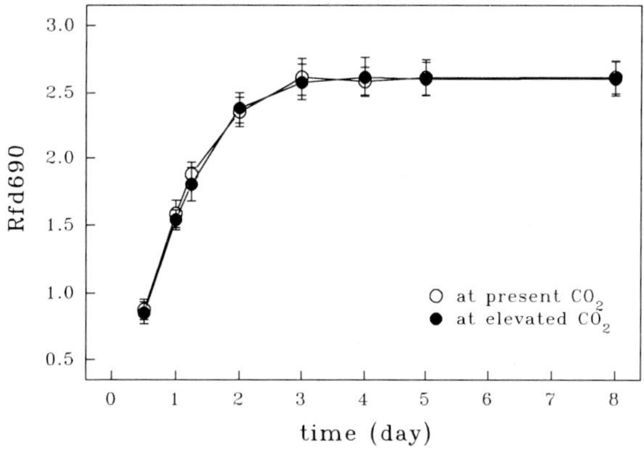

Fig. 3: Development of the variable chlorophyll fluorescence decrease ratio (Rfd690) in air-dried, achlorophyllous leaves of *X. scabrida* during revival and regreening at present (350 µmol mol^{-1}) and elevated CO_2 (700 µmol mol^{-1}) concentration. Dark bars represent standard deviation.

CO_2 concentrations exceeded a value of 2.5, which is commonly found for photosynthetically fully active leaves (Lichtenthaler and Rinderle, 1988).

CO_2 gas exchange

Respiration in the rehydrating leaves of *X. scabrida* was detectable 30 min after the start of rehydration (Fig. 4 A). CO_2

evolution increased sharply and reached its maximum after 2 h. The rehydration respiration had a higher peak in the high CO_2 leaves than in the present CO_2 ones, and it required longer (48 h) for them to decrease to a stable value. For both, the present and the high CO_2 leaves, the CO_2 exchange rates measured in the dark were identical to those in the light for 18 h after rehydration (compare Fig. 4 A with Fig. 4 B). After this the CO_2 production in the light decreased gradually due to the start of CO_2 assimilation.

In the leaves rehydrating at present CO_2, net CO_2 assimilation was first measurable 24 h after rehydration but not until 36 h for the high CO_2 leaves (Fig. 4 B). Maximum rates of light saturated net CO_2 assimilation were reached at 72 h at both CO_2 concentrations. Subsequently, the rate fell in the high CO_2 leaves but was maintained in the present CO_2 leaves.

8 days after rehydration the A/c_i curve for the revived leaves did not reach saturation in the high CO_2 leaves, while it was saturated at 400 µmol mol^{-1} CO_2 in the present CO_2 plant leaves (Fig. 5 A). The initial slope of A/c_i curve for the high CO_2 leaves was much lower than for the present CO_2 ones, and net CO_2 assimilation rates in the high CO_2 leaves were lower at all c_i's than in the leaves at present CO_2.

The increase in stomatal conductance after rehydration followed a similar course at both CO_2 concentrations but was finally lower in the leaves of high CO_2 plants than in the present CO_2 leaves (Fig. 5 B).

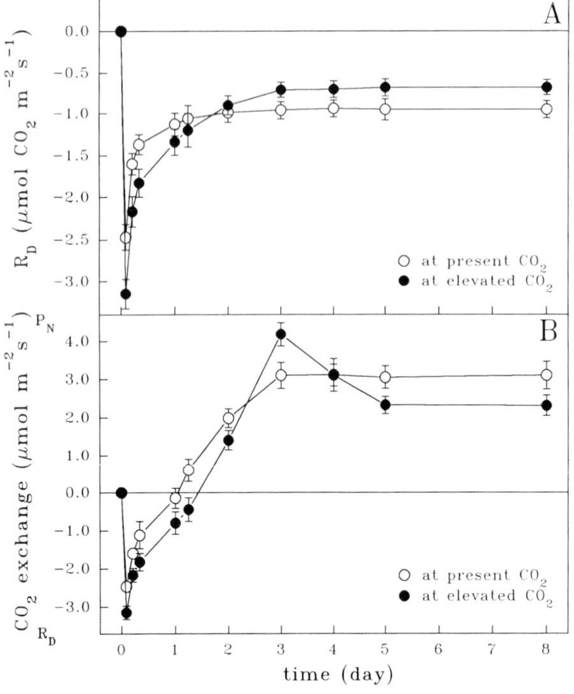

Fig. 4: Changes in the respiration rate, R_D **(A)** and kinetics of the CO_2 gas exchange in illuminated **(B)** air-dried, achlorophyllous leaves of *X. scabrida* during revival and regreening at present (350 µmol mol^{-1}) and elevated CO_2 (700 µmol mol^{-1}) concentration. CO_2 gas-exchange in light was measured at saturating (800 µmol m^{-2} s^{-1}) photon flux density. Relative humidity and temperature were kept constant during the measurements at 85 % and 20 ± 0.5 °C, respectively. Dark bars represent standard deviation.

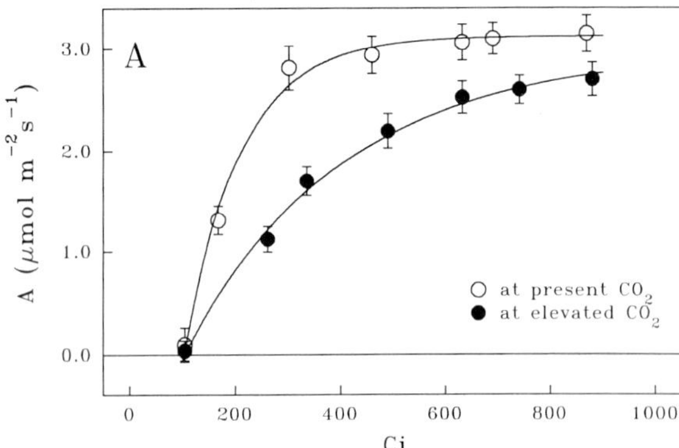

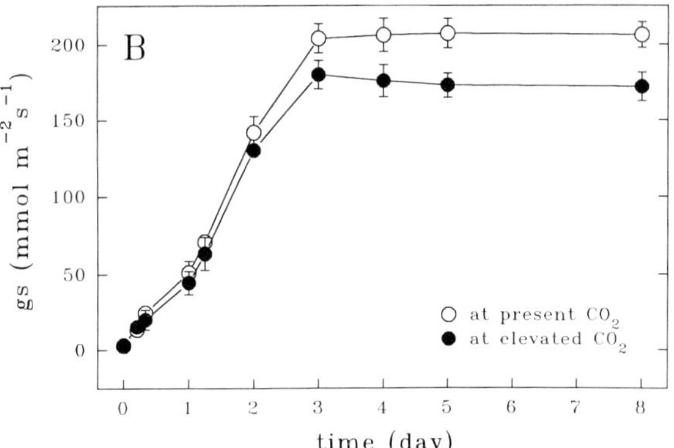

Fig. 5: **(A)** Net CO_2 assimilation rate, A, as a function of intercellular CO_2 concentration, c_i (µmol mol^{-1}), in the leaves of *X. scabrida* at 8 days after the rehydration and revival at present (350 µmol mol^{-1}) and elevated CO_2 (700 µmol mol^{-1}) concentration. CO_2 assimilation was measured at 800 µmol·m^{-2}·s^{-1} photon flux density and at 20 ± 0.5 °C leaf temperature. **(B)** Changes in stomatal conductance (g_s) of air-dried, achlorophyllous leaves of *X. scabrida* in the course of rehydration and regreening at present (350 µmol mol^{-1}) and elevated (700 µmol mol^{-1}) CO_2 concentration. Dark bars represent standard deviation.

Discussion

The photosynthetic apparatus in the air-dried leaves of the PDT *X. scabrida,* which were collected in its natural habitat, was found to be completely destroyed and the desiccoplasts (former chloroplasts) (Tuba et al., 1993 b) did not contain chlorophylls. Water uptake prompted an immediate response in the leaves and signalled the end of the anabiotic state and the beginning of revival processes.

Water relations and SLA

CO_2 concentration had no effect on the rate of water uptake and water content in the leaves, and water relations were restored as in a previous study (Tuba et al., 1994 a). The development of SLA was also unaffected by CO_2 concentration. The over three-fold increase in SLA at both CO_2 concentrations ensured a favourable leaf area to mass ratio for gas exchange in the reviving leaves (Tuba et al., 1994 a).

Respiration

Respiration began rapidly at both CO_2 concentrations and became fully operational after 6 h of rehydration before the cells had reached full turgidity (Tuba et al., 1994 a). This suggests that most of the respiratory enzymes were preserved at desiccation and during long-term air-dried anabiosis in the PDT *X. scabrida* (Tuba et al., 1994 a), in a similar way to that reported elsewhere for HDT's (Bewley, 1979; Harten and Eickmeier, 1986; Schwab et al., 1989). The duration of rehydration respiration was longer in the high CO_2 leaves than in the present CO_2 ones. Most reports on long-term exposure to high CO_2 indicate a decrease in respiration (Amthor, 1991; Bunce, 1994). However, those reports refer to studies made on photosynthesising leaves. Rehydration respiration is a special type of high intensity respiration which occurs in both HDT and PDT plants on rehydration preceding CO_2 assimilation (Smith and Molesworth, 1973; Proctor, 1990; Meenks et al., 1991; Tuba et al., 1994 a). The rate of the later appearing ‹normal› respiration in the fully revived leaves of *X. scabrida* was consistent with the usual pattern and was lower at high CO_2 than at the present ones.

Photosynthetic apparatus

On reviving, PDT plants need to rebuild a functional photosynthetic apparatus prior to the commencement of CO_2 assimilation. The rebuilding of the photosynthetic apparatus in PDTs involves the resynthesis of chlorophyll *a+ b* (Tuba et al., 1993 a) and the reassembly of the thylakoids (Tuba et al., 1993 b). The amount of chlorophyll *a+ b* resynthesised in the leaves fully revived at high CO_2 concentration was lower than that in the present CO_2 leaves and is in agreement with earlier reports, which suggested that high CO_2 causes a decrease in chlorophyll *a+ b* content (Eamus and Jarvis, 1989; Houpis et al., 1988). The resynthesis of functional thylakoids were reflected by the developing of the photochemical activity (Fig. 3).

Photochemical activity and CO_2 assimilation

Slow chlorophyll fluorescence kinetics (Kautsky effect) and variable fluorescence as expressed by Rfd values indicate the potential photochemical activity (Lichtenthaler, 1988; Lichtenthaler and Rinderle, 1988). Rfd values in this study indicated that the resynthesised chlorophylls and thylakoids in the rehydrating leaves became functional after 12 h and reached full capacity in 72 h. Thylakoid activity resumed in a similar manner with no difference in its intensity at both CO_2 concentrations. Over the first 2 days of rehydration there is yet not net CO_2 assimilation which only appeared after a threshold Rfd ratio of 1.5 was reached.

Acclimation of photosynthesis to an elevated CO_2 concentration

The synchronous appearance of maximum net photosynthesis in the leaves at both CO_2 concentrations indicated that the resynthesis of the photosynthetic apparatus and its full activation required identical time, independent of atmospheric CO_2 concentration. Acclimation to high CO_2 (Jarvis, 1993) occurred only after the rebuilding of the photosynthetic apparatus was complete as was indicated by a drop in net photosynthesis rate, chlorophyll *a+ b* content and stomatal conductance between 72 h and 120 h.

The differences in the characteristics of the photosynthetic apparatus in the leaves revived at the two different CO_2 concentrations are illustrated in Fig. 5 A. At high CO_2 the reduced initial slope of the A/c_i curve indicates a reduced Rubisco capacity (Caemmerer and Farquhar, 1984; Sharkey, 1985; Sage et al., 1989).

The decrease in stomatal conductance at high CO_2 in this study (Fig. 5 B) is yet another case where CO_2 assimilation is affected to a large extent by reduced stomatal conductance in long-term high CO_2 exposure studies (Rogers et al., 1983; DeLucia et al., 1985; Jarvis, 1993). Stomatal conductance decreases as c_i increases (Mott, 1988), and in plants grown at high CO_2 concentrations it may be more pronounced than in those grown at present CO_2 (Eamus and Jarvis, 1989; Tuba et al., 1994 b).

Conclusions

Photosynthesis in photosynthetically active leaves of *X. scabrida,* which were revived and maintained at high CO_2, showed a downward acclimation (sensu Jarvis, 1993) of the net CO_2 assimilation rate as compared to leaves of plants kept at present CO_2. It is suggested that the lower net CO_2 assimilation was caused by reduced Rubisco capacity. The rate of respiration also decreased in high CO_2 plants, and the C-balance of the leaves therefore remained similar to that in the present CO_2 plants. The revival of the leaves is of great importance in the natural habitat of *X. scabrida* at the beginning of the rainy season. The leaves revive first and resume activity; the initiation of adventitious root begins only when the revival of the leaves has been completed (Tuba et al., 1993 b). Therefore the findings of in this study are of relevance for modelling the revival of *X. scabrida* and possibly other PDT's at their natural habitat, especially in view of a future higher CO_2 environment.

Acknowledgements

The present work was funded by the ECOCRAFT Environment R&D Programme (EC Brussels), also supported by the Hungarian Scientific Research Foundation (OTKA I/3.1545, I/4 F5359, and CO294). Financial support by Phare/Accord and Soros Foundation (Budapest) is also acknowledged (ZT). We wish to thank Professor T. Pócs (Eger, Hungary – Morogoro, Tanzania) for collecting the plant material and Dr. N. Smirnoff (Exeter) for his valuable suggestions on the manuscript.

References

Amthor, J. S.: Respiration in a future, higher-CO_2 world. Plant Cell Environ. *14,* 13–20 (1991).

Arp, W. J. and B. G. Drake: Increased photosynthetic capacity of *Scirpus alnei* after four years of exposure to elevated CO_2. Plant, Cell Environ. *14,* 1003–1006 (1991).

Bewley, D. J.: Physiological aspects of desiccation tolerance. Ann. Rev. Plant Physiol. *30,* 195–238 (1979).

Bunce, J. A.: Responses of respiration to increasing atmospheric carbon dioxide concentrations. Physiologia Plantarum *90,* 427–430 (1994).

Caemmerer, S. von and G. D. Farquhar: Some relationships between the biochemistry of photosynthesis and the gas exchange of leaves. Planta *153,* 376–387 (1981).

– – Effects of partial defoliation, changes of irradiance during growth, short term water stress and growth at enhanced $p(CO_2)$ on the photosynthetic capacity of leaves of *Phaseolus vulgaris* L. Planta *160,* 320–329 (1984).

Ceulemans, R. and M. Mousseau: Effects of elevated atmospheric CO_2 on woody plants. New Phytol. *127,* 425–446 (1994).

Dahlman, R. C.: CO_2 and plants: revisited. Vegetatio *104/105,* 339–355 (1993).

DeLucia, E. H., T. W. Sasek, and B. R. Strain: Photosynthetic inhibition after long-term exposure to elevated levels of atmospheric carbon dioxid. Photos. Res. *7,* 175–184 (1985).

Eamus, D. and P. G. Jarvis: Direct effects of increase in the global atmospheric CO_2 concentration on natural and commercial trees and forests. Advences in Ecol. Res. *19,* 1–55 (1989).

Gaff, D. F.: Desiccation tolerant vascular plants of Southern Africa. Oecologia (Berlin) *31,* 95–109 (1977).

Gaff, D. F.: Responses of desiccation tolerant ‹resurrection› plants to water stress. In: Kreeb, K. H., H. Richter, and T. M. Hinckley (eds.): Structural and functional responses to environmental stresses, pp. 255–268. SPB Academic Publishing bv, The Hague (1989).

Hambler, D. J.: A poikilohydrous, poikilochlorophyllous angiosperm from Africa. Nature *191,* 1415–1416 (1961).

Harten, J. B. and W. G. Eickmeier: Enzyme dynamics of the resurrection plant *Selaginella lepidophylla* (Hook. & Grev.) Spring during rehydration. Plant Physiol. *82,* 61–64 (1986).

Houpis, J. L. J., K. A. Surano, S. Cowles, and J. H. Shinn: Chlorophyll and carotenoid concentrations in two varieties of *Pinus ponderosa* seedlings subjected to long-term elevated carbon dioxide. Tree Physiology *4,* 187–193 (1988).

Jarvis, P. G.: Global change and plant water relations. In: Borghetti, M., J. Grace, and A. Raschi (eds.): Water Transport in Plants under Climatic Stress, pp. 1–13. Cambridge University Press, Cambridge (1993).

Lichtenthaler, H. K.: *In vivo* chlorophyll fluorescence as a tool for stress detection in plants. In: Lichtenthaler, H. K. (ed.): Applications of Chlorophyll Fluorescence, pp. 129–142. Kluwer Academic Publisher, Dordrecht (1988).

Lichtenthaler, H. K. and U. Rinderle: The role of chlorophyll fluorescence in the detection of stress conditions in plants. CRC Critical Reviews in Analytical Chemistry, *19,* Suppl. 1. 29–85 (1988).

Meenks, J. D. L., Z. Tuba, and Zs. Csintalan: Ecophysiological responses of *Tortula ruralis* upon transplantation around a power plant in West Hungary. Journ. Hattori Bot. Lab. *69,* 21–35 (1991).

Morison, J. I. L.: Response of plants to CO_2 under water limited conditions. Vegetatio *104/105,* 193–209 (1993).

Mott, K. A.: Do stomata respond to CO_2 concentrations other than intercellular? Plant Physiol. *86,* 200–203 (1988).

Pócs, T.: Vegetation mapping in the Uluguru Mountains (Tanzania, East Africa). Boissiera *24 b,* 477–498 + 1 map (1976).

Proctor, M. C. F.: The physiological basis of bryophyte production. Bot. J. Linnean Soc. *104,* 61–77 (1990).

Rogers, H. H. and R. C. Dahlman: Crop responses to CO_2 enrichment. Vegetatio *104/105,* 117–131 (1993).

Rogers, H. H., J. F. Thomas, and G. E. Bingham: Response of agronomic and forest species to elevated CO_2. Science *220,* 428–429 (1983).

Sage, R. F., T. D. Sharkey, and J. R. Seeman: Acclimation of photosynthesis to elevated CO_2 in five C_3 species. Plant Physiol. *89,* 590–596 (1989).

Schwab, K. B., U. Schreiber, and U. Heber: Response of photosynthesis and respiration of resurrection plants to desiccation and rehydration. Planta *177,* 217–227 (1989).

Sharkey, T. D.: Photosynthesis of intact leaves of C_3 plants: physics, physiology and rate limitations. Bot. Rev. *51,* 53–105 (1985).

Smith, D. C. and S. Molesworth: Lichen physiology. XIII. Effect of rewetting dry lichens. New Phytol. *722,* 525–533 (1973).

Thornley, J. H. M. and I. R. Johnson: Plant and Crop Modelling, Clarendon Press Oxford, Oxford, pp. 669 (1990).

Tuba, Z., H. K. Lichtenthaler, Zs. Csintalan, and T. Pócs: Regreening of desiccated leaves of the poikilochlorophyllous *Xerophyta scabrida* upon rehydration. J. Plant Physiol. *142,* 103–108 (1993 a).

Tuba, Z., H. K. Lichtenthaler, I. Maróti, and Zs. Csintalan: Resynthesis of thylakoids and chloroplast ultrastructure in the desiccated leaves of the poikilochlorophyllous plant *Xerophyta scabrida* upon rehydration. J. Plant Physiol. *142,* 742–748 (1993 b).

Tuba, Z., H. K. Lichtenthaler, Zs. Csintalan, Z. Nagy, and K. Szente: Reconstitution of chlorophylls and photosynthetic CO_2 assimilation upon rehydration in the desiccated poikilochlorophyllous plant *Xerophyta scabrida* (Pax) Th. Dur. et Schinz. Planta *192,* 414–420 (1994 a).

Tuba, Z., K. Szente, and J. Koch: Response of photosynthesis. stomatal conductance, water use efficiency and production to long-term elevated CO_2 in winter wheat. J. Plant Physiol. *144,* 661–668 (1994 b).

J. Plant Physiol. Vol. 148. pp. 351–355 (1996)

Interactive Effects of Elevated CO_2, Ozone and Drought Stress on the Activities of Antioxidative Enzymes in Needles of Norway Spruce Trees (*Picea abies,* [L.] Karsten) Grown with Luxurious N-Supply

PETER SCHWANZ[1], KARL-HEINZ HÄBERLE[2], and ANDREA POLLE[1,*]

[1] Albert-Ludwigs-Universität Freiburg, Baumphysiologie, Am Flughafen 17, D-79085 Freiburg, Germany

[2] GSF-Forschungszentrum für Umwelt und Gesundheit, Expositionskammern, Neuherberg, Postfach 11 29, D-85764 Oberschleißheim, Germany

Received June 24, 1995 · Accepted September 25, 1995

Summary

The aim of the present study was to address the complex interactions of environmental constraints, ozone and drought stress, with elevated atmospheric CO_2 on the activities of antioxidative enzymes and soluble protein contents in needles of Norway spruce trees (*Picea abies* L.). Five-year-old spruce trees were kept from bud break in June until January of the following year in phytochambers under climatic conditions similar to those of a natural site in the Bavarian forest. The trees were well-supplied with nitrogen and exposed to either elevated CO_2 (ambient + 200 $\mu L L^{-1}$), elevated ozone (80 $nL L^{-1}$, from June to October) or to a combination of both factors. Controls were grown with 20 $nL L^{-1}$ O_3 and ambient CO_2 levels. In each chamber, a subset of trees was subjected to episodical drought stress in summer. Needles from controls investigated in October (summer conditions) and January (winter conditions) showed little seasonal variation of superoxide dismutase (SOD), an approximately 2-fold reduction in catalase (CAT), and a 2-fold increase in guaiacol peroxidase (POD) activity. Exposure to elevated CO_2 did not affect the activities of any of these enzymes in October and January, respectively, but caused a significant reduction in soluble protein. Ozone had no significant effect. Drought stress caused memory effects. In January, needles from trees drought-stressed in summer contained higher activities of defence enzymes and soluble protein contents than needles from well-watered trees. Three weeks after the end of a drought episode in summer, needles from spruce trees grown at elevated CO_2 contained increased CAT and POD activities as compared to needles from trees grown at ambient CO_2. This response was increased, if elevated ozone was present as an additional stress factor. These observations suggest that Norway spruce trees grown under elevated atmospheric CO_2 concentrations might better be able to compensate environmental stresses than trees grown at ambient atmospheric CO_2 concentrations.

Key words: Picea abies, superoxide dismutase, peroxidase, catalase, seasonal changes, stress.

Abbreviations: ANOVA = multifactorial analysis of variance; CAT = catalase; POD = peroxidase; SOD = superoxide dismutase; SD = standard deviation; DW = Dry weight; FW = fresh weight.

Introduction

The current atmospheric CO_2 concentration of about 355 $\mu L L^{-1}$ is expected to double by the end of the next century (Strain, 1987). Since photosynthesis of C_3 plants is not saturated at the present ambient CO_2 concentration, rising CO_2 levels can stimulate growth and biomass production, if other environmental factors are not limiting (Mousseau and Saugier, 1992). However, CO_2 is also a greenhouse gas, i.e., it contributes to global warming. Climate models predict that

* To whom correspondence should be addressed.

further increases in CO_2 will drive increases in temperature which in turn will cause shifts in precipitation patterns and result in drought periods in northern mid-latitudes in summer (Roeckner, 1992). Tropospheric ozone concentrations are also anticipated to rise, especially in industrialised regions (Chameides et al., 1994). Thus, in a future scenario it can be expected that on one hand plants might have an increased capacity for photosynthesis but on the other hand might suffer more frequently from stresses such as drought and ozone.

At the cellular level both drought and ozone cause an increased oxidative stress (Smirnoff, 1993; Heath, 1995). To prevent injury, plants are protected by a defence system composed of antioxidative enzymes and substrates (Foyer et al., 1994). In this system, SODs (EC 1.15.1.1) catalyse the removal of $O_2^{\cdot -}$. The product of this reaction, H_2O_2, is detoxified by PODs (EC 1.11.1.7) and CATs (EC 1.1 1.1.6). It has previously been observed that growth of Norway spruce for one season at an elevated CO_2 concentration resulted in decreased activities of SOD and CAT in current-year needles (Polle et al., 1993). CAT activity was also reduced in tobacco grown at elevated CO_2 concentration (Havir and McHale, 1989). Therefore, it has been suggested that rising CO_2 levels can potentially decrease the oxidative stress plant tissues are normally exposed to (Polle, 1996).

It is still an open question whether a decrease in antioxidant capacity is a general response to CO_2 enhancement and whether such plants differ in their ability to compensate oxidative stresses. It has been suggested that plants grown with elevated CO_2 might be better protected from environmental stresses by an increased capacity for detoxification and repair (Carlson and Bazzaz, 1982). In radish grown with elevated CO_2, ozone-induced reduction in photosynthesis was delayed as compared with seedlings grown at an ambient CO_2 concentration (Barnes and Pfirrmann, 1992). However, in Norway spruce, ozone-induced needle injury was not prevented in plants grown at elevated CO_2 concentrations (Polle et al., 1993). The aim of the present study was to address the effects of multiple stresses (ozone and drought) on antioxidative defences (SOD, POD, CAT) and soluble protein content in needles of Norway spruce trees (*Picea abies* L.) grown at ambient and elevated CO_2 concentrations under summer and winter conditions.

Materials and Methods

Four 3-year-old clonal spruce seedlings (clon 3663, Staatliche Samenklenge und Baumschule, Laufen, Germany) were potted into 40 L pots containing acidic brown forest soil from the Bavarian forest (Germany). The experimental set-up used for the present investigation was a subset of a greater study (Lippert et al., 1966) and consisted of 32 pots. Before the experimental exposures, the plants were kept outdoors for two years. During this adaption phase the trees were fertilised regularly with micro- and macronutrients including four additional doses of 750 mg NH_4NO_3 per pot (Häberle, 1995). The trees were protected from pathogens by application of insecticides and acaricides described elsewhere (Häberle, 1995). Shortly before bud break in June 1992, the trees were transferred into walk-in phytochambers at the GSF (cf. Payer et al., 1993). Each of the chambers was equipped with 8 pots containing 4 trees each. During the experimental period until January 1993, the plants were kept under climatic conditions similar to those observed in the Bavarian forest. «Summer» climate (16th June to 31st October) was characterized by alternating combinations of air temperatures/relative air humidities of 18 °C/60 %, 20 °C/50 %, 24 °C/40 %. At night, air temperature and relative humidity were 12 °C and 90 %, respectively. During «fall» (1st November to 7th December) the air temperature was gradually decreased from 18 °C to 7 °C in the light period and from 12 °C to 0 °C in the dark period. In «winter» (8th December to 12th January 1993), the day and night temperatures were 3 °C and −3 °C, respectively. Occasionally, the plants were exposed to night temperatures of −7 °C in «winter». The day lengths were varied as follows: 14 hours in «summer», 9 hours in «fall», and 8 hours in «winter». The photon flux density was 750 μmol quanta $m^{-2} s^{-1}$ at tree height. During the experimental period, the plants were fertilised with micro- and macronutrients and received additionally 10 applications of 750 mg NH_4NO_3 per pot (Häberle, 1995). The plants were subjected to the following experimental treatments in four chambers: «control» (filtered ambient air with a mean CO_2 concentration of about 400 $\mu L\,L^{-1}$ + 20 $nL\,L^{-1}$ ozone continuously), «elevated ozone» (filtered ambient air + 80 $nL\,L^{-1}$ ozone continuously for 11 hour a day during summer + occasional peaks of 140 $nL\,L^{-1}$ ozone, 20 $nL\,L^{-1}$ ozone residual time of day), «elevated CO_2» (filtered ambient air + 20 $nL\,L^{-1}$ ozone continuously + 200 $\mu L\,L^{-1}$ CO_2), and «elevated ozone + CO_2» (same treatment as for «ozone» + 200 $\mu L\,L^{-1}$ CO_2). The pots were connected with an automatic irrigation system which supplied the trees with water according to their demand; i.e., during most time of the growth phase the soil water potential was between 80 and 120 hPa. To generate drought stress, 4 pots per chamber were disconnected from the automatic irrigation system and not supplied with water from 17th July to 31st July and from 7th September to 5th October. In the second drought period, the trees were rewatered when the predawn water potential had dropped to about −2.2 MPa.

Current-year needles were harvested under summer (30th October 1992) and winter (7th January 1993) conditions, respectively. Needles from two trees per pot were pooled, frozen in liquid nitrogen and stored at −80 °C until extraction of enzymes and soluble protein. Generally, 6 replicates per treatment were analysed individually. Needle extracts were prepared in 100 mM KH_2PO_4/K_2HPO_4, pH 7.8, 0.5 % Triton X-100, 2.5 % insoluble polyvinylpyrrolidone and gel-filtered over Sephadex G-25 (Pharmacia, Freiburg, Germany) as described previously (Polle et al., 1989). Enzymatic activities were determined as described elsewhere: SOD (Polle et al., 1989), CAT (Aebi, 1983), and guaiacol POD (Polle et al., 1990). Soluble protein was determined in gel-filtered extracts with bicinchoninic acid reagent (Pierce, München, Germany). Bovine serum albumin was used as standard. The dry weight of needles was taken after 72 hours at 80 °C. The data were subjected to multiple analysis of variance and the significance of treatment effects was tested by employing a multiple range test (method of least significant differences, LSD) at $p < 0.05$ (Statgraphics, STSC Inc. Rockville, MD, USA).

Results and Discussion

In the present study clonal spruce trees were subjected to a complex experimental treatment including elevated CO_2 (about 600 $\mu L\,L^{-1}$ vs. an ambient concentration of about 400 $\mu L\,L^{-1}$), elevated ozone (80 $nL\,L^{-1}$ vs. 20 $nL\,L^{-1}$ vs. 20 $nL\,L^{-1}$) and episodical drought stress. The effects of acute drought stress were not addressed. Needles were investigated three weeks (October) and three months (January), respectively, after the drought treatment had been terminated. Inspite of these long recovery phases, in some of the treatments

Table 1: Soluble protein content and DW-to-FW-ratio of current-year needles of Norway spruce trees exposed to a combination of stress factors (drought, ozone) in the presence of ambient or elevated CO_2. Experimental conditions were as described in Materials and Methods. Figures indicate means (n = 5 to 6, ±SD). Different letters in columns refer to significant differences at p≤0.05 (LSD), ND = not determined.

Harvest	CO_2 $\mu L L^{-1}$	O_3 $nL L^{-1}$	Drought	Protein (mg g^{-1} DW)	DW-to-FW ratio
October	400	20	–	39.2± 6.7 b, c	0.520±0.014 b, c
	400	20	+	46.8± 6.3 c, d	0.491±0.018 a, b
	400	80	–	36.6± 9.4 a, b, c	0.511±0.011 a, b, c
	400	80	+	52.5± 7.0 d	0.500±0.021 a, b, c
	600	20	–	34.1± 6.2 a, b	0.521±0.045 c
	600	20	+	46.4±10.2 c, d	0.479±0.029 a
	600	80	–	28.4± 6.6 a	0.508±0.016 a, b, c
	600	80	+	42.2± 6.7 b, c, d	0.512±0.014 a, b, c
January	400	20	–	37.8± 4.5 α, β	0.487±0.014 α, β
	400	20	+	46.2± 4.3 α	0.464±0.013 α
	600	20	–	38.7± 5.8 α	0.490±0.025 β
	600	20	+	43.5± 5.1 α, β	0.481±0.024 α, β

Table 2: Summary of a multifactorial analysis of variance of the effects of elevated CO_2, ozone, drought and season on soluble protein content (protein), DW-to-FW-ratio, and antioxidative enzyme activities (SOD, CAT, POD) in needles of Norway spruce. P indicates calculated probability (n = 5 or 6 replicates per treatment). Experimental conditions were as described in Materials and Methods.

Variable	p				
	Protein	DW-to-FW-ratio	SOD	CAT	POD
Season	0.9243	*0.0039*	*0.0035*	*0.0000*	*0.0000*
Ozone (October)	0.5117	0.6365	*0.0139*	0.9969	0.0507
Elevated CO_2 (October)	*0.0334*	0.7925	0.5847	*0.0088*	*0.0301*
Drought (October)	*0.0000*	*0.0063*	*0.0566*	0.5582	0.0408
Elevated CO_2 (January)	0.3860	0.7448	0.8770	0.0686	0.8258
Drought (January)	*0.0001*	*0.0008*	*0.0093*	0.0993	*0.0028*

significant drought-induced increases in the soluble protein content and decreases in the DW-to-FW ratio of the needles were observed (Table 1). An ANOVA on all experimental data revealed that the drought-induced effects on dry matter and soluble protein were highly significant (Table 2). The decrease in dry matter of the drought-stressed needles which was particularly pronounced in the presence enhanced CO_2, points to limitations in biomass production. In fact, periodical water deficiency resulted in 10 % less biomass of current-year shoots than in those of well-watered trees regardless of whether elevated CO_2 was present or not (Lippert et al., 1996). Ozone had neither a significant effect on the soluble protein content nor on the DW-to-FW-ratio of the needles (Tables 1, 2). The soluble protein content of needles from spruce trees grown at 600 $\mu L L^{-1}$ CO_2 tended to decrease as compared with that of needles from trees grown at ambient CO_2 (Table 1). It is known from previous studies that growth under elevated concentrations of CO_2 can result in reductions in the foliar nitrogen concentration affecting the concentration of foliar proteins (Petterson and McDonald, 1994). However, the spruce trees used in the present investigation were highly fertilized with nitrogen to prevent such limitations. The N-fertilisation regime resulted in mean N-contents of 20.6 ± 3.5 mg N g^{-1} DW regardless of whether the plants were grown with elevated CO_2 or not but caused, especially in the presence of elevated CO_2, imbalances in other nutrients such as P, K, and Mg (Lippert et al., 1996; Häberle, 1995).

The antioxidative enzymes SOD, CAT, and POD responded differentially to the environmental variables employed in the present investigation (CO_2, ozone, drought-stress, Figs. 1–3). A overall statistical analysis by ANOVA revealed that antioxidative enzymes showed significant differences between October and January harvests (Table 2). These fluctuations corresponded to seasonal changes observed in the

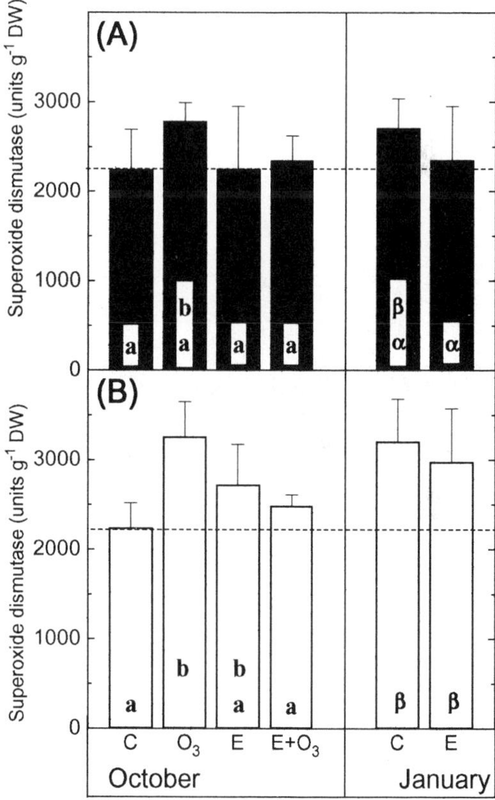

Fig. 1: Superoxide dismutase activity in current-year needles of Norway spruce trees exposed from bud break in June 1992 to January 1993 to ambient conditions (C), elevated ozone (O_3), elevated CO_2 (E) and a combination of elevated ozone and CO_2 (E+O_3). (A) Well-watered trees, (B) episodical drought periods in summer. Samples were collected in October (summer conditions) and January (winter conditions). Details of the experimental treatments are described in Materials and Methods. Bars indicate means (n=5 to 6, ± SD). Different letters in bars refer to significant differences at p<0.05 (LSD-test). Broken line indicates the level of the unstressed ambient control.

field (SOD: Kröniger et al., 1995; POD: Polle et al., 1994) or were also caused by experimental temperature decreases (CAT: Schittenhelm et al., 1994).

An increased CO_2 concentration had no significant effect on the activity of SOD of current-year needles neither in October nor in January (Fig. 1 A). In the light of a previous study (Polle et al., 1993), the lack of responses of defence enzymes to CO_2 enhancement was unexpected. However, it should be noted that several experimental conditions differed between the two studies. First, the degree by which CO_2 was increased was smaller than previously (+200 $\mu L\ L^{-1}$ vs. +400 $\mu L\ L^{-1}$). Second, the spruce trees in the present study were pre-adapted to and grown during the experimental exposures with a luxurious N-supply which caused limitations in other nutrients. Finally, in each of these experiments only one spruce clone was used. It is possible that spruce clones differ in their responsiveness to CO_2 enhancement.

Ozone had no effect on the SOD activity of spruce trees grown at an enhanced CO_2 concentration (Fig. 1). It has been shown that the response of SOD activity to ozone fumigation is quite variable (Polle and Rennenberg, 1991 and references therein). The present data show that SOD activity was significantly increased in plants exposed to both drought-

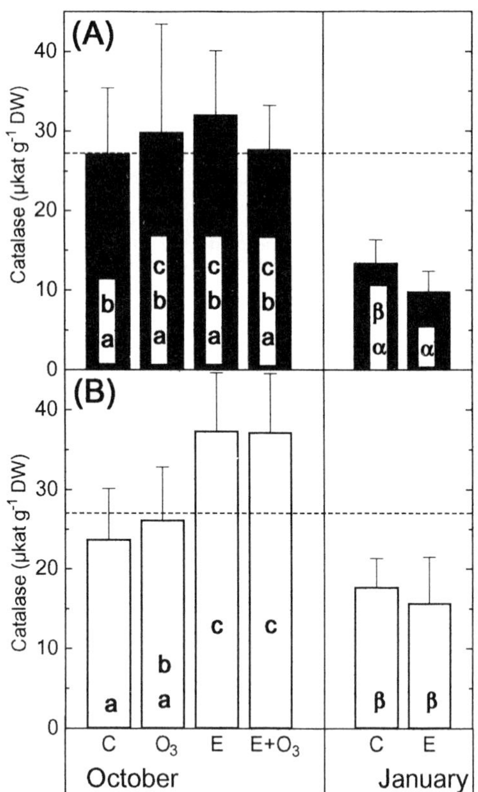

Fig. 2: Catalase activity in current-year needles of Norway spruce trees exposed from bud break in June 1992 to January 1993 to ambient conditions (C), elevated ozone (O_3), elevated CO_2 (E) and a combination of elevated ozone and CO_2 (E+O_3). (A) Well-watered trees, (B) episodical drought periods in summer. Further conditions as in Fig. 1. Broken line indicates the level of the unstressed ambient control.

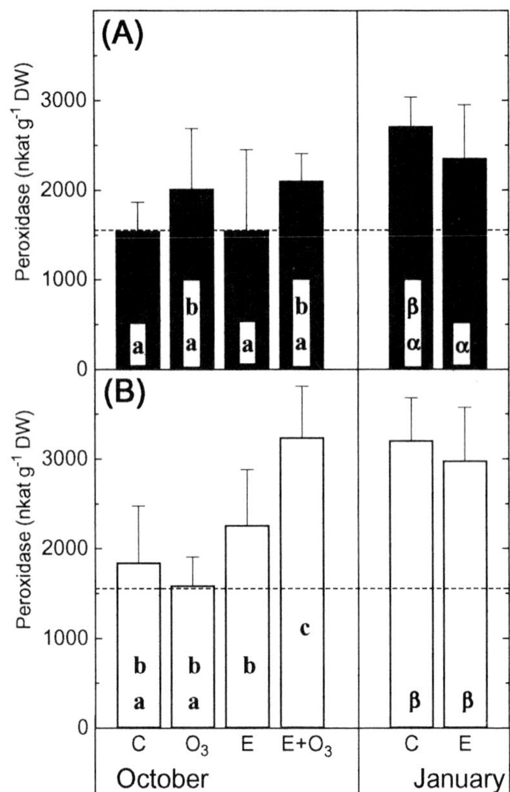

Fig. 3: Guaiacol peroxidase activity in current-year needles of Norway spruce trees exposed from bud break in June 1992 to January 1993 to ambient conditions (C), elevated ozone (O_3), elevated CO_2 (E) and a combination of elevated ozone and CO_2 (E+O_3). (A) Well-watered trees, (B) episodical drought periods in summer. Further conditions as in Fig. 1. Broken line indicates the level of the unstressed ambient control.

stress and ozone at elevated CO_2 concentrations but not at ambient CO_2 (Fig. 1 B). Surprisingly, in January needles from spruce trees, that had been subjected to drought stress in summer, showed higher SOD activities than those from well-watered plants (Fig. 1 B). To our knowledge, such a drought-induced «memory effect» has not been reported before.

Neither ozone, nor elevated CO_2 or a combination of both factors resulted in significant changes of CAT or POD activities (Figs. 2 A, 3 A). However, if the plants were additionally drought-stressed, CAT activity tended to decline in needles from spruce trees grown at ambient CO_2 concentrations, whereas CAT activity increased in needles from plants grown with elevated CO_2 concentrations (Fig. 2 B). Reductions in CAT activity in response to ozone or acute drought stress have been observed previously (Zhang and Kirkam, 1994; Polle et al., 1993). POD activity of drought-stressed needles showed a significant increase in needles from trees grown with elevated CO_2 concentrations but not in needles from trees grown at ambient CO_2 (Fig. 3 B). This increase was even more pronounced, if the trees were additionally ozone-stressed (Fig. 3 B). In January, needles from trees previously subjected to drought-stress still contained elevated CAT and POD activities (Fig. 2 B, 3 B). These observations suggest that acclimation to environmental stresses may be improved in spruce trees grown at elevated than ambient CO_2 concentrations.

Acknowledgements

We are grateful to M. Eiblmeier for excellent technical assistence and to Prof. Dr. H. Rennenberg (Universität Freiburg) for critical reading of the manuscript. Parts of this study were financially supported by the EC (EV5V-CT92-0093 and STEP-CT90-0035).

References

AEBI, H.: Catalase. In: BERGMEIER, H. (ed.): Methods of Enzymatic Analysis, pp. 121–126. Verlag Chemie, Weinheim (1983).

BARNES, J. and T. PFIRRMANN: The influence of CO_2 and O_3, singly and in combination, on gas exchange, growth, and nutrient status of radish (*Raphanus sativus* L.) New Phytol. *121,* 403–412 (1992).

CARLSON, R. W. and F. A. BAZZAZ: Photosynthetic and growth response to fumigation with SO_2 at elevated CO_2 in C_3 and C_4 plants. Oecologia *54,* 50–54 (1982).

CHAMEIDES, W. L., P. S. KASIBHATLA, J. YIENGER, and H. LEVY II: Growth of continental-scale metro-agro-plexes, regional ozone pollution and world food production. Sci. *264,* 74–77 (1994).

FOYER, C., M. LELANDAIS, and K. J. KUNERT: Photooxidative stress in plants. Physiol. Plant. *92,* 696–717 (1994).

HÄBERLE, K.-H.: Wachstumsverhalten und Wasserhaushalt eines Fichtenklones (*Picea abies* [L.] Karst.) unter erhöhten CO_2 und O_3-Gehalten der Luft bei variierter Stickstoff- und Wasserversorgung, Thesis, LMU München, 1995.

HAVIR, E. A. and N. A. MCHALE: Regulation of catalase activity in leaves of *Nicotiana sylvestris* by high CO_2. Plant Physiol. *89,* 952–957 (1989).

HEATH, R. L.: Possible mechanisms for the inhibition of photosynthesis by ozone. Photosynth. Res. *39,* 439–451 (1994).

KRÖNIGER, W., H. RENNENBERG, and A. POLLE: Developmental changes of CuZn- and Mn-superoxide dismutase isozymes in seedlings and needles of Norway spruce (*Picea abies* L.). Plant Cell Physiol. *34,* 1145–1149 (1993).

LIPPERT, M., K.-H. HÄBERLE, K. STEINER, H.-D. PAYER, and K. E. REHFUESS: Interactive effects of elevated CO_2 and O_3 on photosynthesis and biomass production of Norway spruce seedlings (*Picea abies* [L.] Karst.) under different nitrogen nutrition and irrigation treatments. Trees, in the press (1996).

MOUSSEAU, M. and B. SAUGIER: The direct effect of increased CO_2 on gas exchange and growth of forest tree species. J. Exp. Bot. *43,* 1121–1130 (1992).

PAYER, H.-D., M. KÖFFERLEIN, G. SECKMEYER, H. SEIDLITZ, D. STRUBE, and S. THIEL: Controlled environmental chambers for experimental studies on plant responses to CO_2 and interactions with pollutants. In: SCHULZE, ED. and H. A. MOONEY (eds.): Design and execution of experiments on CO_2 enrichment. Ecosystem Research Report, pp. 127–146, CEC, Brussels-Luxembourg (1993).

PETTERSON, R. and J. S. MCDONALD: Effects of nitrogen supply on the acclimation of photosynthesis to elevated CO_2. Photosynth. Res. *39,* 389–400 (1994).

POLLE, A. and H. RENNENBERG: Superoxide dismutase activity in needles of Scots pine and Norway spruce under field and chamber conditions: lack of ozone effects. New Phytol. *117,* 335–343 (1991).

POLLE, A., B. KRINGS, and H. RENNENBERG: Superoxide dismutase activity in needles of Norwegian spruce trees (*Picea abies* L.). Plant Physiol. *90,* 1310–1315 (1989).

POLLE, A., F. SEIFERT, and T. OTTER: Apoplastic peroxidase and lignification in needles of Norway spruce (*Picea abies* L.). Plant Physiol. *106,* 53–60 (1994).

POLLE, A., K. CHAKRABARTI, W. SCHÜRMANN, and H. RENNENBERG: Composition and properties of hydrogen peroxide decomposing systems in extracellular and total extracts from needles of Norway spruce (*Picea abies* L., Karst.) Plant Physiol. *94,* 312–319 (1990).

POLLE, A., T. PFIRRMANN, S. CHAKRABARTI, and H. RENNENBERG: The effects of enhanced ozone and enhanced carbon dioxide concentrations on biomass, pigments and antioxidative enzymes in spruce needles (*Picea abies* L.). Plant Cell Environmen. *16,* 311–316 (1993).

POLLE, A.: Protection from oxidative stress in trees as affected by elevated CO_2 and environmental stress. In: MOONEY, H. and G. KOCH (eds.): Terrestrial Ecosystem Response to Elevated CO_2. Physiological Ecology series, pp. 299–315, Academic Press, New York (1996).

ROECKNER, E.: Past, present and future levels of greenhouse gases in the atmosphere and model projections of related climatic changes. J. Exp. Bot. *43,* 1097–1109 (1992).

SCHITTENHELM, J., S. TODER, S. FATH, S. WESTPHAL, and E. WAGNER: Photoinactivation of catalase in needles of Norway spruce. Physiol. Plant. *90,* 600–606 (1994).

SMIRNOFF, N.: The role of active oxygen in the response of plants to water deficit and desiccation. New Phytol. *125,* 27–58 (1993).

STRAIN, B.: Direct effects of increasing CO_2 on plants and ecosystems. Trends Ecol. Evol. *2,* 18–21 (1987).

ZHANG, J. and M. B. KIRKHAM: Drought stress-induced changes in activities of superoxide dismutase, catalase and peroxidase in wheat species. Plant Cell Physiol. *35,* 785–791 (1994).

J. Plant Physiol. Vol. 148. pp. 356–361 (1996)

Responses of CO_2 Assimilation, Transpiration and Water Use Efficiency to Long-Term Elevated CO_2 in Perennial C_3 Xeric Loess Steppe Species*

Zoltán Tuba[1,2], Kálmán Szente[1], Zoltán Nagy[1], Zsolt Csintalan[1], and Judit Koch[1]

[1] Plant Physiological Section, Department of Botany and Plant Physiology, Agricultural University of Gödöllő, H-2103 Gödöllő, Hungary

[2] Institute of Ecology and Resource Management, University of Edinburgh, Darwin Building, King's Buildings, Mayfield Road, Edinburgh EH9 3JU, UK

Received June 24, 1995 · Accepted September 25, 1995

Summary

CO_2 assimilation (A), transpiration (E), water use efficiency (WUE), leaf-nitrogen and carbohydrate responses to 11 months elevated (700 µmol mol^{-1}) CO_2 exposure in four perennial C_3 species (*Festuca rupicola, Dactylis glomerata, Filipendula vulgaris, Salvia nemorosa*) from a xeric temperate loess steppe are reported. The responses in the species varied greatly owing to their differing acclimation. The acclimation of photosynthesis was somewhat downward in *F. rupicola,* fully downward in *D. glomerata,* and upward in *S. nemorosa* and *F. vulgaris.* The reduction in the initial slope of the A/c_i response curve in *F. rupicola* and *D. glomerata* suggested a decrease in Rubisco capacity. Net CO_2 assimilation at 700 µmol mol^{-1} CO_2 c_a in the high CO_2 *F. rupicola* was higher than in those grown at present (350 µmol mol^{-1}) CO_2; there was no difference in *D. glomerata.* The initial slope of the A/c_i curve indicated an increased Rubisco capacity in high CO_2 *F. vulgaris* and *S. nemorosa.* Their net CO_2 assimilation was higher in the plants grown in the high CO_2 treatment at c_i's over 200 µmol mol^{-1} than that in the plants grown at present CO_2. The A/c_i response curves, which were saturated in all species grown at present CO_2, did not reach saturation in the plants grown at elevated CO_2, reflecting that the Pi limitation of CO_2 assimilation was alleviated in the plants grown at high CO_2. Transpiration decreased with an increase in c_i in both the present and elevated CO_2 *F. rupicola* and *D. glomerata.* In *F. vulgaris,* an increase in c_i caused a reduction in transpiration in the plants grown at high CO_2 only. Transpiration rate in both the present and elevated CO_2 *S. nemorosa* was not affected by any change in c_i. It is suggested then that long-term exposure to high CO_2 causes a similar acclimation of stomatal regulation and transpiration to that of photosynthesis. High CO_2 caused a significant decrease in protein-nitrogen content only in *D. glomerata.* Starch increased in *F. rupicola* and *D. glomerata* and soluble sugar content was higher in all species grown at high CO_2 than at ambient. Instantaneous WUE significantly increased in all species grown at elevated CO_2.

Key words: Acclimation, A/c_i function, downward regulation, E/c_i function, net CO_2 assimilation, protein-nitrogen, starch, sugars, transpiration, upward regulation.

Abbreviations: A = net CO_2 assimilation rate; c_a = ambient air CO_2 concentration; c_i = intercellular CO_2 concentration; E = transpiration rate; Pi = orthophosphate; RuBP = riblose 1,5-biphosphate; Rubisco = RuBP carboxylase/oxygenase; WUE = water use efficiency of photosynthesis.

* Dedicated to academician Prof. Gábor Fekete on the occasion of his 65th birthday.

Introduction

In C_3 plants exposed to long-term high CO_2, the rate of CO_2 assimilation at high CO_2 does not always increase, but in many cases it is unchanged or may indeed decrease in comparison with that in plants grown at present CO_2 (DeLucia et al., 1985; Tissue and Oechel, 1987; Sage et al., 1989; Jarvis, 1993). An ‹upward regulation› of photosynthesis will result in an increase in net CO_2 assimilation (Arp and Drake, 1991; Campbell et al., 1988) while, as being reported in an increasing number of publications, a ‹downward› regulation will cause lower photosynthetic rates (Jarvis, 1993; Ceulemans and Mousseau, 1994). Downward regulation can be partly attributed to a decrease in Rubisco capacity (Sage et al., 1989). A decrease in CO_2 assimilation may also be caused by an end-product (starch) feedback inhibition which occurs when the amount of available sucrose formed at an elevated CO_2 concentration exceeds sink demand (Azcon-Bieto, 1983). Pi-regeneration capacity increases relative to RuBP regeneration capacity during acclimation to high CO_2 (Sage et al., 1989).

Protein-nitrogen content is often lower in plants grown at an elevated CO_2 concentration than at present CO_2 ones (see Jarvis, 1993). Stomatal aperture and stomatal conductance are also often reduced in long term high CO_2 treated plants (Rogers et al., 1983) which in turn results in lower intensities of transpiration (Kimball and Idso, 1983). One of the most widely reported effects of elevated CO_2 is that photosynthetic water use efficiency (WUE) increases (Rogers et al., 1983; Morison and Gifford, 1984).

Very little is known about the long-term responses of natural vegetation including herbaceous species (e.g. Tissue and Oechel 1987; Curtis et al., 1989; Ziska et al., 1990; Newton, 1991) to elevated atmospheric CO_2 concentrations; there are only four studies which have lasted for more than two seasons (see Körner and Miglietta, 1994). Data on herbaceous perennials are particularly sparse.

In this paper we present the net CO_2 assimilation, transpiration and WUE responses to 700 μmol mol^{-1} CO_2 concentration in four perennial C_3 species from a xeric temperate loess grassland after a 11-month period of exposure. The changes in carbohydrate and protein-nitrogen contents are also reported.

Materials and Methods

The studied vegetation is a xeric temperate loess steppe (Salvio-Festucetum rupicolae pannonicum) (Zólyomi and Fekete, 1994) situated at the border of the Hungarian Great Plain (Albertirsa, Monor-Irsa hills, 48 km south east of Budapest) at 160 m altitude a.s.l. The parent rock is sandy loess and loess with thick humus- and nutrient-rich A layer (humus layer: 100 cm, humus content: 6.1 % and available nitrogen: 14.2 g m^{-2}). The original grassland is made up of more than 90 species. The open air CO_2 fumigation experiment was carried out in the Botanical Garden of the Agricultural University at Gödöllő (28 km east of Budapest). The climate of the two locations did not differ: temperate continental with hot dry summers; mean annual precipitation 500 mm or less; annual mean temperature of 11 °C; and large annual amplitude of mean temperature changes (22 °C).

50 cm × 50 cm × 30 cm monoliths were extracted from the grassland and transplanted into the open top chambers described by Tuba et al. (1994 a). The soil in the chambers was removed and it was replaced by soil from the profiles where the monoliths had been collected from. Four weeks after the transplantation of the monoliths the grass was cut. Following a two-month adaptation period, the monoliths were exposed gradually, over a 4 weeks period, to 700 μmol mol^{-1} CO_2. The elevated CO_2 treatments were maintained at 700 μmol mol^{-1} and the present CO_2 treatments at 350 μmol mol^{-1} after 01 August 1992. The air CO_2 concentrations in the elevated and present CO_2 chambers were maintained as described earlier (Tuba et al., 1994 a). The monoliths were occasionally watered.

The measurements were made on *Festuca rupicola* Heuff., *Dactylis glomerata* L., *Filipendula vulgaris* Mönch. and *Salvia nemorosa* L. after 11 months exposure to 700 μmol mol^{-1} and 350 μmol mol^{-1} CO_2. *F. rupicola,* the dominant species of the grassland, has sclerophyllous erect leaves of waxy surface, while the others are frequent characteristic species of the grassland with different leaf characteristics: *Dactylis* has flat blades; *Filipendula* has soft, large incised leaves; and *Salvia* has broad, entirely waxy, abaxially hairy rosette and stem-leaves. All species are perennial and have C_3 photosynthesis.

Measurement of net CO_2 assimilation

The rates of light saturated (1200 μmol photons m^{-2} s^{-1} at 23 °C leaf temperature) net CO_2 assimilation were measured using an LCA2-type IRGA system (ADC Co. Ltd., Hoddesdon, U.K.), operated in differential mode and Parkinson LC-N and LC-B leaf chambers, as previously described (Tuba et al., 1994; Nagy et al., 1995). Intercellular CO_2 dependence of light saturated net CO_2 assimilation rates (A/c_i curves) was measured using the same system (Tuba et al., 1994). Ambient CO_2 concentrations of 100, 350, 500, 700 and 1000 μmol mol^{-1} were produced by a gas diluter (GD 600, ADC Co. Ltd., Hoddesdon, U.K.). Measurements were carried out in the morning hours. Prior to measurements leaves were acclimated to the experimental conditions for at least 20 min. Gas-exchange parameters were calculated according to von Caemmerer and Farquhar (1981). Water use efficiency of photosynthesis (WUE) is given as the ratio of net CO_2 assimilation rate to transpiration rate. Leaf areas were measured by a leaf area meter (LAM 001, Delta T Devices, Cambridge, U.K.). The Mitscherlich function (Thornley and Johnson, 1990) was used to fit curves to the data on A/c_i.

Other methods

Carbohydrate (starch and soluble sugars) and leaf protein-nitrogen content were measured as described earlier (Tuba et al., 1994). The same leaves were used for gas-exchange, carbohydrate and leaf protein-nitrogen content determination. The measurements were done in at least 5 replicates with different leaves. Statistical analyses (t-statistics, linear regression analysis) were performed using GB-STAT programs (Friedmann, 1988).

Results

CO_2-dependence of net CO_2 assimilation rate

The initial slope of the A/c_i response curve was lower in the high *F. rupicola* and *D. glomerata* than in the ones grown at present CO_2 (Fig. 1). The A/c_i curves indicated that photosynthesis in the present CO_2 treatments reached saturation in both species. At a c_i corresponding to present 350 μmol mol^{-1} c_a, net CO_2 assimilation was lower in the high CO_2 *F. rupicola* than in the present ones. Conversely, when net CO_2 assimilation was measured at a c_a of 700 μmol mol^{-1}, the plants grown at high CO_2 assimilated more CO_2 than ones grown at

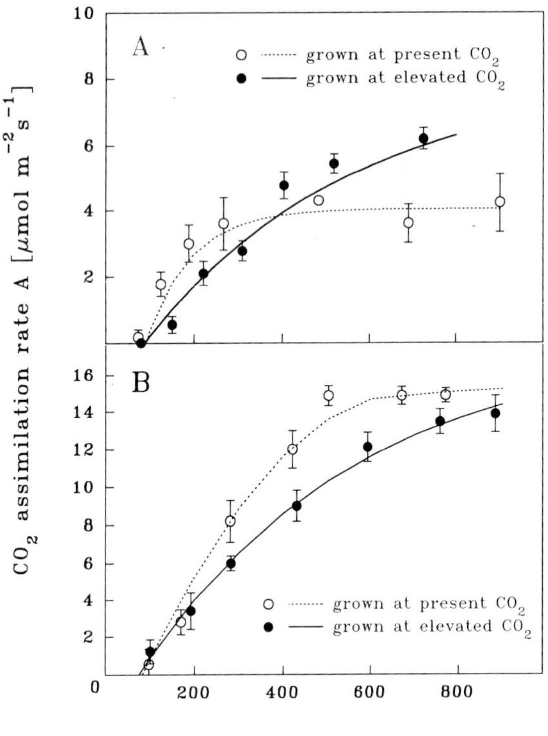

Fig. 1: (A) Net CO_2 assimilation rate as a function of intercellular CO_2 concentration (c_i) in *F. rupicola* and *D. glomerata* (B) grown at present (350 µmol mol^{-1}) and elevated (700 µmol mol^{-1}) CO_2 concentration. CO_2 assimilation was measured at 1200 µmol photons m^{-2} s^{-1} saturation irradiance and at 23 ± 0.5 °C leaf temperature after 11 months of exposure. Bars represent standard deviation.

present CO_2. In *D. glomerata,* there was no significant difference between the treatments at a c_i corresponding to 350 µmol mol^{-1} c_a. Plants grown at present CO_2 had a higher net CO_2 assimilation rate at a c_i corresponding to 700 µmol mol^{-1} c_a.

Net CO_2 assimilation reached saturation in the present CO_2 *S. nemorosa* and *F. vulgaris* (Fig. 2). In contrast, saturation was not reached in either species at elevated CO_2 and their net CO_2 assimilation was higher than that in the plants grown at present CO_2 at c_i's over 200 µmol mol^{-1} CO_2 with the difference ever increasing at higher c_i's. In the high CO_2 *S. nemorosa* and *F. vulgaris* the initial slope of the A/c_i response curve was higher than in the present ambient ones (Fig. 2).

Transpiration rate and water use efficiency of photosynthesis

In all four species there was a linear relationship between the rate of transpiration and c_i. With increasing c_i transpiration rate decreased in both the present and high CO_2 *F. rupicola* and *D. glomerata* (Fig. 3). The values for transpiration in the high CO_2 *F. rupicola* and *D. glomerata* were significantly lower than in the present CO_2 ones. In *F. vulgaris,* an increase in c_i also caused a reduction in transpiration in the plants grown at high CO_2 and did not cause any decrease in the ones grown at present CO_2 (Fig. 4 A). Transpiration rate in both the present and elevated CO_2 *S. nemorosa* was not affected by any change in c_i (Fig. 4 B). There were no significant differences in the rate of transpiration between the present and the high CO_2 *S. nemorosa.*

Values of instantaneous WUE at a c_a of 700 µmol mol^{-1} CO_2 were higher in all elevated CO_2 plants than the ones grown at present 350 µmol mol^{-1} CO_2 (Table 1). In *D. glome-*

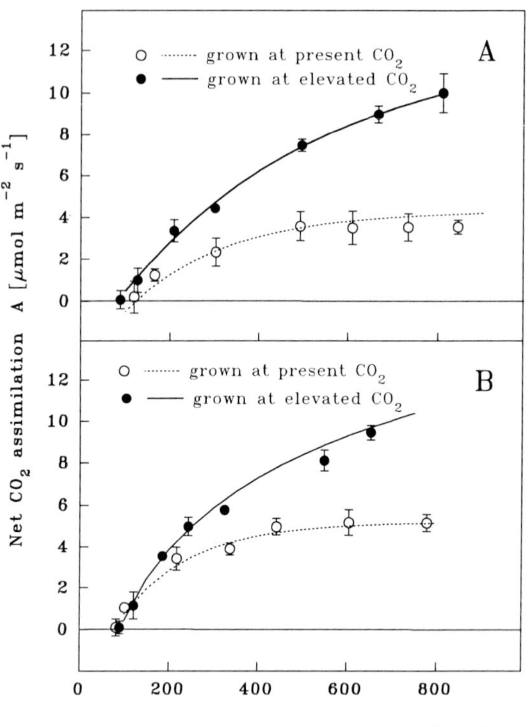

Fig. 2: (A) Net CO_2 assimilation rate as a function of intercellular CO_2 concentration (c_i) in *F. vulgaris* and *S. nemorosa* (B) grown at present (350 µmol mol^{-1}) and elevated (700 µmol mol^{-1}) CO_2 concentration. CO_2 assimilation was measured at 1200 µmol photons m^{-2} s^{-1} saturation irradiance and at 23 ± 0.5 °C leaf temperature after 11 months of exposure. Bars represent standard deviation.

Table 1: Water use efficiency (WUE) in *F. rupicola, D. glomerata, F. vulgaris* and *S. nemorosa* grown at present (350 µmol mol^{-1}) and elevated (700 µmol mol^{-1}) CO_2 concentration measured both at present and elevated CO_2. Conditions during the measurements were same as in Figs. 1 and 2. (Means ± standard deviations; * indicates significant difference at P≤0.05. Comparisons were made between the treatments (elevated vs present CO_2 concentration).

	WUE measured at	
	350 µmol mol^{-1} CO_2	700 µmol mol^{-1} CO_2
F. rupicola (at present)	0.67±0.13	0.83±0.28
F. rupicola (at elevated)	1.20±0.35*	2.58±0.56*
D. glomerata (at presente)	1.25±0.19	1.66±0.27
D. glomerate (at elevated)	0.75±0.08	1.78±0.37*
F. vulgaris (at present)	1.13±0.17	1.98±0.31
F. vulgaris (at elevated)	1.53±0.34	4.71±0.87*
S. nemorosa (at present)	2.43±0.61	3.24±0.19
S. nemorosa (at elevated)	2.62±0.66	4.77±0.63

rata, WUE at 350 µmol mol^{-1} was higher in the ambient plants than in the ones grown at elevated CO_2.

Protein-nitrogen and corbohydrate content

Although protein-nitrogen content was reduced in all species in the high CO_2 treatments, it was significantly lower only in *D. glomerata* (Table 2). Starch content was higher in *F. rupicola* and *D. glomerata* in the high CO_2 treatment while

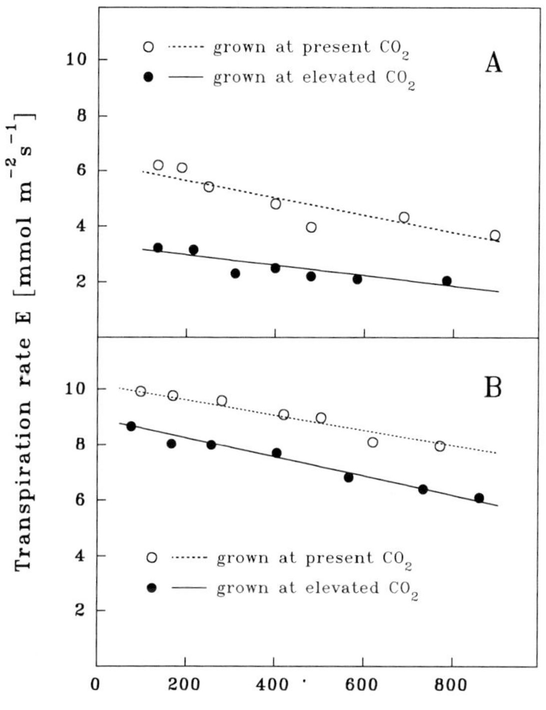

Fig. 3: (A) Transpiration rate as a function of intercellular CO_2 concentration (c_i) in *F. rupicola* and *D. glomerata* (B) grown at present (350 µmol mol^{-1}) and elevated (700 µmol mol^{-1}) CO_2 concentration. Conditions during the measurements were same as in Fig. 1.

Table 2: Protein-nitrogen, starch and soluble sugar content in the leaves of *F. rupicola, D. glomerata, F. vulgris* and *S. nemorosa* grown at present (350 µmol mol^{-1}) and elevated (700 µmol mol^{-1}) CO_2 concentration. (Means ± standard deviations; * indicates significant difference at P≤0.05. Comparisons were made between the treatments (elevated vs present CO_2 concentration).

	protein-nitrogen ($g\ m^{-2}$)	starch ($mg\ m^{-2}$)	soluble sugars ($mg\ m^{-2}$)
F. rupicola (at present)	1.63±0.35	8.29±0.35	11.45±0.72
F. rupicola (at elevated)	1.50±0.18	9.72±0.43*	16.65±0.58*
D. glomerata (at present)	2.48±0.04	10.44±1.41	16.96±1.24
D. glomerata (at elevated)	2.15±0.15*	13.27±2.17*	23.64±0.7*
F. vulgaris (at present)	3.53±0.12	40.01±0.31	32.49±2.19
F. vulgaris (at elevated)	3.41±0.15	26.50±0.32	36.21±1.18*
S. nemorosa (at present)	2.89±0.16	24.78±2.41	15.44±0.62
S. nemorosa (at elevated)	2.70±0.15	23.16±0.81	28.34±1.8*

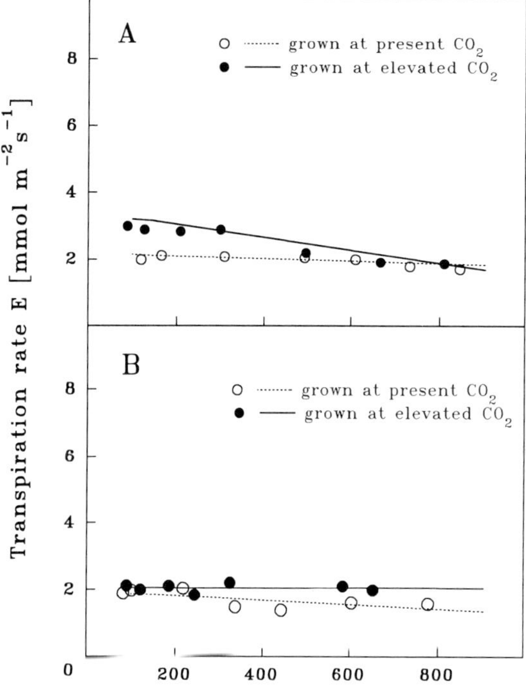

Fig. 4: (A) Transpiration rate as a function of intercellular CO_2 concentration (c_i) in *F. vulgaris* and *S. nemorosa* (B) grown at present (350 µmol mol^{-1}) and elevated (700 µmol mol^{-1}) CO_2 concentration. Conditions during the measurements were same as in Fig. 2.

it was unchanged in *F. vulgaris* and *S. nemorosa* (Table 2). The soluble sugar content was increased at high CO_2 in all species (Table 2).

Discussion

The physiological responses in the xeric perennial C_3 loess grassland species varied greatly owing to their differing acclimation to an elevated CO_2 concentration. To characterise the type and degree of acclimation (‹downward› or ‹upward›) the net CO_2 assimilation versus c_i response curve has been widely used (Sage et al., 1989; Jarvis, 1993). The reduction in the initial slope of the A/c_i response curve in *F. rupicola* and *D. glomerata* can be interpreted as a decrease in Rubisco capacity. The reduction in Rubisco capacity on exposure to high CO_2 has been reported (e.g. Caemmerer and Farquhar, 1984; Sage et al., 1989) and has been explained by a decrease in enzyme activation state or in the amount of Rubisco or both (Sharkey, 1985). As a large proportion of leaf protein-nitrogen of the C_3 plants is found in Rubisco (Evans, 1989), the reduced leaf nitrogen content in *F. rupicola* and in the *D. glomerata* might have been an indication of a decrease in the amount of Rubisco. It appears that in *F. rupicola* and in *D. glomerata* there was a downward acclimation of photosynthesis. In *D. glomerata* for example, net CO_2 assimilation in the high CO_2 treatment was lower at all c_a's than in the present CO_2 treatment (Fig. 2). This was different in *F. rupicola* in the high CO_2 treatment where the initial slope of the A/c_i response curve was reduced, yet it photosynthesised above c_i's of 400 μmol mol^{-1} at a higher rate than plants grown at present CO_2. In contrast, in *S. nemorosa* and *F. vulgaris,* the high CO_2 environment did not reduce the Rubisco capacity (steeper initial slope of their A/c_i response curve) and they assimilated much more CO_2 than their controls grown at present CO_2 with ever increasing differences at higher c_i's. These were clear indications of an upward regulation of photosynthesis (Arp and Drake, 1991; Campbell et al., 1988) to elevated CO_2.

If the sink demand cannot accommodate to the enhanced level of photosynthates this can lead to accumulation of starch, as it was observed in high CO_2 grown plants (Cave et al., 1981; Peet et al., 1986). In this study, only the high CO_2 *F. rupicola* and *D. glomerata* accumulated starch while all four species had higher soluble sugar content. This suggests that the ‹non-starch accumulating› two other species (with upward regulation of photosynthesis) still had an active sink to ensure the transport of photosynthates from the chloroplasts after eleven months of exposure to high CO_2.

The non-saturated A/c_i response curve in all of the four high CO_2 species reflects that the limitation of the Pi-regeneration capacity (Caemmerer and Farquhar, 1981) was decreased in the high CO_2 treatments. In some species the decrease in limitation of the Pi-regeneration capacity increased the starch accumulation (Azcon-Bieto, 1983). However the changes in starch content of the investiged loess species indicate that the decrease in limitation of the Pi-regeneration capacity is not accompanied by starch accumulation in every cases (Sage et al., 1989). Furthermore the starch accumulation does not correlate with the degree of the limitation of the Pi-regeneration capacity.

It has long been established that transpiration is directly related to stomatal conductance (Meidner and Mansfield, 1965). Stomatal conductance and transpiration are usually reduced by long-term CO_2 exposure (Rogers et al., 1983; Jarvis, 1993). However in some cases no change or even an increase was reported (Eamus and Jarvis, 1989). Significant reduction in transpiration in this study only occurred in the *F. rupicola* and the *D. glomerata* which showed a downward regulation of photosynthesis as opposed to the two other species with upward regulation where there was not significant change. Therefore it is suggested that long-term exposure to high CO_2 causes a similar acclimation of stomatal and transpiration regulation to that of photosynthesis. The contrasting transpiration and stomatal acclimation may be explained by the varying sensitivity of species to high CO_2 (Eamus and Jarvis, 1989) which is not fully understood as yet.

The marked increase in instantaneous WUE on long-term exposure to high CO_2 has long been known (Rogers et al., 1983). In this study, this was also observed in all species. In the high CO_2 *F. rupicola* and *D. glomerata,* the increase in WUE occurred with downward acclimation of photosynthesis and a large reduction in transpiration. In the elevated CO_2 *S. nemorosa* and *F. vulgaris,* WUE increased because the plants maintained low transpiration and greatly increased their net CO_2 assimilation. The higher WUE in high CO_2 than in ambient can increase drought tolerance (Morison, 1993) in all four species which may confer an advantage on the plants in the sometimes extremely dry conditions in the grassland (Nagy et al., 1994).

The acclimation of photosynthesis to long-term high CO_2 is influenced by pot size also (Arp, 1991). It has been found that small pot size resulted in a downward acclimation, while large pots caused an upward regulation. However both types of acclimation occurred in our work which was carried out without any pot size limitation (without pots) conflicts with Arp's (1991) results.

The differences in acclimation of photosynthesis (upward in dicots and downward in mononcots) may be considered as differences in the species' responses with implications in sink-source regulation and/or it can also be interpreted as dicot and monocot responses.

The present results may have important ecological implications. For example the downward acclimation in *F. rupicola* in this study indicates that its dominance in a future high CO_2 climate may be reduced and/or replaced by other, now less abundant species due to their upward acclimation.

Acknowledgements

Present work was funded by the ECOCRAFT Environment R&D Programme (EC Brussels) also supported by the Hungarian Scientific Research Foundation (OTKA I/3.1545, I/4 F. 5358, F5360 and CO294). We thank Dr. N. Smirnoff for his valuable suggestions on the manuscript.

References

Arp, W. J.: Effects of source-sink relations on photosynthetic acclimation to elevated CO_2. Plant, Cell and Environ. *14,* 869–875 (1991).

Arp, W. J. and B. G. Drake: Increased photosynthetic capacity of *Scirpus alnei* after four years of exposure to elevated CO_2 Plant, Cell Environ. *14,* 1003–1006 (1991).

Azcon-Bieto, J.: Inhibition of photosynthesis by carbohydrates in wheat leaves. Plant Physiol. *73,* 681–686 (1983).

Caemmerer, S. von and G. D. Farquhar: Some relationships between the biochemistry of photosynthesis and the gas exchange of leaves. Planta *153,* 376–387 (1981).

– – Effects of partial defoliation, changes of irradiance during growth, short term water stress and growth at enhanced p(CO_2) on the photosynthetic capacity of leaves of *Phaseolus vulgaris* L. Planta *160,* 320–329 (1984).

Campbell, W. J., L. H. J. R. Allen, and G. Bowes: Effects of CO_2 concentration on rubisco activity, amount and photosynthesis in soybeen leaves. Plant Physiology *88,* 1310–1316 (1988).

Cave, G., L. C. Tolley, and B. R. Strain: Effect of carbon dioxid enrichment on chlorophyll content, starch content and starch grain structure in *Trifolium subterraneum leaves.* Physiol. Plant. *51,* 171–174 (1981).

Ceulemans, R. and M. Mousseau: Effects of elevated atmospheric CO_2 on woody plants. New Phytol. *127,* 425–446 (1994).

Curtis, P. S., B. G. Drake, P. W. Leadley, W. J. Arp, and D. F. Whigham: Growth and senescence in plant communities exposed to elevated CO_2 concentrations on an estuarine marsh. Oecologia *78,* 20–26 (1989).

Delucia, E. H., T. W. Sasek, and B. R. Strain: Photosynthetic inhibition after long-term exposure to elevated levels of atmospheric carbon dioxid. Photos. Res. *7,* 175–184 (1985).

EAMUS, D. and P. G. JARVIS: Direct effects of increase in the global atmospheric CO_2 concentration on natural and commercial trees and forests. Advences in Ecol. Res. *19,* 1–55 (1989).

EVANS, J. R.: Photosynthesis and nitrogen relationships in leaves of C_3 plants. Oecologia *78,* 9–19 (1989).

FRIEDMAN, P.: GB-STAT Professional Statistics and Graphics (Version 1.5), Silver Spring, MD (301), 384–2754 (1988).

JARVIS, P. G.: Global change and plant water relations. In: BORGHETTI, M., J. GRACE, and A. RASCHI (eds.): Water Transport in Plants under Climatic Stress. Cambridge University Press, pp. 1–13 (1993).

KIMBALL, B. A. and S. B. IDSO: Increasing atmospheric CO_2: effects on crop yield, water use and climate. Agricultiral Water Management *7,* 55–72 (1983).

KÖRNER, C. and F. MIGLIETTA: Long term effects of naturally elevated CO_2 on mediterranean grassland and forest trees. Oecologia *99,* 343–351(1994).

MEIDNER, H. and T. A. MANSFIELD: Physiology of Stomata. McGraw-Hill, London, pp. 179 (1965).

MORISON, J. I. L.: Response of plants to CO_2 under water limited conditions. Vegetatio *104/105,* 193–209 (1993).

MORISON, J. I. L. and R. M. GIFFORD: Plant growth and water use with limited water supply in high CO_2 concentrations. I. Leaf area, water use and transpiration. Aust. J. Plant Physiol. *11,* 361–384 (1984).

NAGY, Z., Z. TUBA, K. SZENTE, and G. FEKETE: Photosynthesis and water use efficiency during degradation of semiarid loess steppe. Photosynthetica *30,* 307–311 (1994).

NAGY, Z., Z. TUBA, F. ZSOLDOS, and L. ERDEL: CO_2 exchange and water relation responses of maize and sorghum during water and salt stress. J. Plant Physiol. *145,* 539–544 (1995).

NEWTON, P. C. D.: Direct effects of increasing carbon dioxide on pasture plants and communities. New Zealand Journal of Agricultural Research *34,* 1–24 (1991).

PEET, M. M., S. C. HUBER, and D. T. PATTERSON: Acclimation to high CO_2 in monoccious cucumbers. II. Carbon exchange rates, enzyme activities, and starch and nutrient concentrations. Plant Physiol. *80,* 63–67 (1986).

ROGERS, H. H., J. F. THOMAS, and G. E. BINGHAM: Response of agronomic and forest species to elevated CO_2. Science *220,* 428–429 (1983).

SAGE, R. F., T. D. SHARKEY, and J. R. SEEMAN: Acclimation of photosynthesis to elevated CO_2 in five C_3 species. Plant Physiol. *89,* 590–596 (1989).

SHARKEY, T. D.: Photosynthesis of intact leaves of C_3 plants: physics, physiology and rate limitations. Bot. Rev. *51,* 53–105 (1985).

THORNLEY, J. H. M. and I. R. JOHNSON: Plant and Crop Modelling, Clarendon Press Oxford, Oxford, pp. 669 (1990).

TISSUE, D. T. and OECHEL, W. C.: Response of *Eriophorum vaginatum* to elevated CO_2 and temperature in the Alaskan tundra. Ecology *68,* 401–410 (1987).

TUBA, Z., K. SZENTE, and J. KOCH: Response of photosynthesis, stomatal conductance, water use efficiency and production to long-term elevated CO_2 in winter wheat. J. Plant Physiol. *144,* 661–668 (1994).

ZISKA, L. H., B. G. DRAKE, and S. CHAMBERLAIN: Long-term photosynthetic response in single leaves of A C_3 and C_4 salt marsh species grown at elevated atmospheric CO_2 *in situ.* Oecologia *83,* 469–472 (1990).

ZÓLYOMI, B. and G. FEKETE: The Pannonian loess steppe: differentiation in space and time. Abst. Bot. *18,* 29–41 (1994).

J. Plant Physiol. Vol. 148. pp. 362–367 (1996)

Physiological Responses of Norway Spruce Trees to Elevated CO_2 and SO_2

MICHAEL TAUSZ[1], LUIT J. DE KOK[2], INEKE STULEN[2], and DIETER GRILL[1]

[1] Institute of Plant Physiology, Karl-Franzens-University of Graz, Schubertstraße 51, A-8010 Graz, Austria

[2] Department of Plant Biology, University of Groningen, P.O. Box 14, 9750 AA Haren, The Netherlands

Received July 7, 1995 · Accepted September 25, 1995

Summary

Young Norway spruce (*Picea abies* (L.) Karst.) trees were exposed to elevated CO_2 (0.8 mL L^{-1}), SO_2 (0.06 µL L^{-1}), and elevated CO_2 and SO_2 (0.8 mL L^{-1} and 0.06 µL L^{-1}, respectively) for three months. Exposure to elevated CO_2 resulted in an increased biomass production of the needles, while the pigment content was decreased. Exposure to SO_2 hardly affected growth and pigment contents. Chlorophyll/carotenoid- and chlorophyll *a*/chlorophyll *b*-ratios were not affected by either CO_2 or SO_2. The epoxidation state of the xanthophyll-cycle was changed upon SO_2-exposure, due to a higher zeaxanthin and a lower violaxanthin content. Chlorophyll fluorescence measurements showed F_v/F_m-ratios of 0.7 or higher for all needles, which indicated a healthy photosynthetic apparatus. Exposure to SO_2 resulted in increased foliar contents of sulfate, total glutathione (reduced and oxidized form), cyst(e)ine, and a slightly more reduced redox state of glutathione. Exposure to elevated CO_2 resulted in a slight decrease in glutathione contents, but it did not affect sulfate or cyst(e)ine contents of the spruce needles. Neither ascorbic acid content nor its redox state were affected by CO_2 or SO_2. The effects of CO_2 and SO_2 were independent from each other, since significant interactions $CO_2 \times SO_2$ were not observed.

Key words: Picea abies, antioxidants, CO_2, pigments, SO_2, sulfur metabolism.

Introduction

Atmospheric concentration of CO_2 is rising dramatically and a doubling of the present concentration would occur during the last half of the next century. Presently, much attention is paid to the effects of increasing CO_2 on physiological processes in plants and ecosystems (Graham et al., 1990). Exposure of plants to elevated CO_2 generally resulted in positive effects on biomass production due to an increased photosynthesis (reviewed by Ceulemans and Mousseau, 1994). Allocation of assimilates and initial growth stimulation by CO_2 has been shown to be strongly dependent on the level of mineral nutrition (Stulen et al., 1993).

The impact of air pollutants on plants is still subject of discussion. Despite the extensive literature the processes underlying the phytotoxicity of the ‹classical› air pollutant SO_2 are still unresolved. Harmful effects of this pollutant have been ascribed to: (1) The reactions of sulfite ions formed by hydratation of SO_2 with various cellular compounds (Jäger et al., 1986; De Kok, 1990; Rennenberg and Polle, 1994). (2) The oxidation of sulfite to sulfate which may result in the formation of cytotoxic radicals, causing oxidative stress (Tanaka et al., 1982; Miszalski and Ziegler, 1992; Elstner and Osswald, 1994). Antioxidative defense systems may protect the cells against injury caused by radical reactions (Alscher and Amthor, 1988; Miszalski and Ziegler, 1992).

The interactive effects of elevated CO_2 and air pollutant stress on physiological functioning of plants are of great interest. It has been observed that elevated CO_2 may protect plants from air pollutant injury. Two possible mechanisms have been suggested: (1) The reduction of pollutant uptake due to a decreased stomatal conductance (Allen, 1990). (2) The stimulation of the internal cellular detoxification mechanisms (Barnes and Pfirrmann, 1992). For instance, it has been shown that the synthesis of ascorbic acid, a major antioxidant, depended on carbohydrate availability (Loewus et al., 1990). It has been suggested that carbon assimilation stimulated by elevated CO_2 enhanced the antioxidative detoxification capacity of the cells enabling the plant to cope with SO_2 (Rao and De Kok, 1994). However, data on these interactions between elevated CO_2 and air pollutants are still inconsistent. Experiments with combined exposure of plants to elevated CO_2 and O_3 resulted in additive effects (Heagle et al., 1993), while other studies revealed that elevated CO_2 might even reduce tolerance against oxidative stress (Polle et al., 1993).

In the present paper we investigated the physiological responses of young clonal Norway spruce (*Picea abies* (L.) Karst.) trees to long-term exposure (3 months) to SO_2 in combination with elevated CO_2. Special interest was paid to the interactions of CO_2 impact with partly already known effects of SO_2 exposure and to the redox status of ascorbic acid and glutathione.

Materials and Methods

Plant material

One year old clonal Norway spruce plants (*Picea abies* (L.) Karst. from Boomkwekerij Zundert B. V., Meirseweg 45, 4881 MJ Zundert, The Netherlands) were received on January 26, 1993 and planted in ‹forest soil› in 9 cm diameter plastic pots. Plant height was 20 cm. Soil characteristics were as follows: Total N 5.47 g kg^{-1} soil dry weight (dw), total C 118 g kg^{-1} dw, NH_4^+ 0.15 g kg^{-1} dw, NO_3^- 26.9 mg kg^{-1} dw, P 5.4 mg kg^{-1} dw, K 25.6 mg kg^{-1} dw, Mg 134 mg kg^{-1} dw, pH 7.2. Trees were grown in the greenhouse at a day/night temperature of 22/18 °C (relative humidity was 70 ± 10 %). Trees were repotted on July 6, 1993 and were randomly designed for the fumigation cabinets (15 pots per cabinet). Plant height was 25 cm. For biochemical investigations plant material from 9 trees of each cabinet was harvested on October 11, 1993. Needles were cut from the branches, shock-frozen in liquid nitrogen immediately and lyophilized afterwards. Biomass measurements, chlorophyll fluorescence measurements, and sulphate determination were carried out on 6 trees per chamber harvested October 22, 1993.

Fumigation conditions

Four treatment variants were carried out: Ambient air (‹Control›), SO_2 (‹SO_2›), elevated CO_2 (‹CO_2›), and a combination of SO_2 and elevated CO_2 (‹SO_2+CO_2›). Plants were fumigated in 150 L cylindrical stainless steel cabinets with polycarbonate tops. Pressurized CO_2 and/or SO_2 diluted with nitrogen were injected into the incoming air stream at the desired concentrations by ASM electronic mass flow controllers (Bilthoven, The Netherlands). Air exchange rate in the cabinets was 50 L min^{-1} and the air inside the cabinets was circulated by a ventilator (with an air movement capacity of 20 L s^{-1}) to reduce the boundary layer surrounding the leaves. The air temperature was 21 ± 2 °C and the relative humidity amounted to 40 ± 8 %. Photoperiod was set at 14 h day^{-1} and the light intensity at plant height was 160–300 µmol m^{-2} s^{-1} (within the 400–700 nm range). A Philips HPI-T 400 W lamp was used as light scource. CO_2 concentrations in the air stream were measured periodically with an IRGA (ADC Model 225, MK 3 Hoddesdon, UK) and the SO_2 concentrations were monitored by a Lear Siegler Sulfur Dioxide Analysator (ML 9850, Englewood, Colorado, USA). Mean day/night CO_2 concentrations in the cabinets were 365/400 µL L^{-1} (ambient) and 775/825 (elevated). Mean SO_2 inlet concentrations were 0.002 (ambient) and 0.063 µL L^{-1} (SO_2). Mean day/night outlet concentrations of SO_2 were 0.040/0.042 µL L^{-1} with the SO_2 treatment and 0.015/0.029 µL L^{-1} with the SO_2+CO_2 treatment. The trees were irrigated with rain water daily.

Experimental methods

Biomass:

Shoots were separated in needles and stems and weight (fresh weight = fw) was determined. They were dried at 70 °C for 36 h and weighed again (dry weight = dw).

Pigments:

Pigments were separated and measured using a HPLC gradient-method as described by Pfeifhofer (1989): Two Milton Roy Mini-Metric 1721 pumps with Gradient Master, UV/vis detector Milton Roy SpectroMonitor D, Integrator Shimadzu C-R1B, column Spherisorb S5 ODS2 25 × 4.6 mm with precolumn S5 ODS2 5 × 4.6 mm, solvent A: acetonitril : methanol : water = 100 : 10 : 5 (v/v/v), solvent B: acetone : ethylacetate = 2 : 1 (v/v), linear gradient from 90 % (v) A to 30 % (v) A in 17 min, running time = 30 min, flow rate = 1 mL min^{-1}. Absorbance of the column effluent was measured at 440 nm. Acetone extracts of lyophilized needle powder were injected by a cooled (0 °C) Marathon autosampler.

Chlorophyll fluorescence:

Chlorophyll fluorescence ratio (F_v/F_m) was determined from the needle surfaces of 15 min dark adapted needles using a portable chlorophyll fluorometer (PAM-2000, H. Walz, Effeltrich, Germany). The PPFD of modulation light was about 0.2 µmol m^{-2} s^{-1}. After measuring F_o, F_m was measured by exciting the surface with white light of 6000 µmol m^{-2} s^{-1} 24 measurements from 4 needles randomly collected from 6 trees were done per treatment.

Sulfate:

Sulfate content was measured by an isocratic HPLC-method. Extraction was carried out on needle dry matter which was soaked at 60 °C in H_2O for 12 h, homogenized in an Ultra Turrax, the homogenate was filtered through Miracloth and centrifuged at 30,000 × g for 10 min. The supernatant was centrifuged once again and subjected to HPLC analysis. HPLC-system consisted of a Kratos Spectroflow pump, a Ionosphere A anion exchange column (Chrompack 200 × 4.6 mm), and a Knauer Differential-Refractometer. Solvent was 20 mM potassiumhydrogenphtalate containing 0.02 % sodiumazide (pH 4.3). Flow rate was 1 mL min^{-1}.

Thiols:

Glutathione, cysteine, γ-glutamyl-cysteine (oxidized and reduced) were determined as described before (Kranner and Grill, 1996). Thiols were extracted from 0.1 g lyophilized needle powder in 3 mL 0.1 M HCl. 400 µL of this extract was incubated with 600 µL of CHES-buffer (200 mM 2-(N-cyclohexylamino)ethanesulfonic acid, pH 9.3) and 100 µL of 3 mM dithiothreitol (DTT) for sixty minutes at room temperature to reduce thiol-groups. The SH-groups were labeled by monobromobimane by incubating a 1 mL aliquot of the reduced extracts with a 60 µL aliquot of 15 mM monobromobimane at room temperature in the dark for 15 minutes. Derivatization was stopped by adding a 760 µL aliquot of 0.25 % (v/v) methansulfonic acid. For the determination of the content of oxidized thiols 30 µL of 50 mM N-ethylmaleimide (NEM) and 600 µL 200 mM CHES-buffer were added to 400 µL of the thiol extract and the mixture was subsequently incubated at room temperature for 15 min in order to block the SH-groups. The excess of NEM was removed by extraction with toluene and the remaining oxidized thiols were reduced by DTT and derivatized with monobromobimane as described above. Separation and determination of the derivatized thiols were done on a gradient HPLC-system as described in Kranner and Grill (1993): Two Knauer HPLC-pumps 64, Knauer Software/Hardware package, Shimadzu Fluorescence monitor RF-535 (excitation 380 nm, emission 480 nm), and a cooled Marathon autosampler. Column Spherisorb S5 ODS2 25×4.6 mm. Solvent A: 0.25 % (v/v) acetic acid in water containing 5 % methanol, pH 3.9. Solvent B: 90 % (v/v) methanol in water, Gradient: 5 % solvent B to 15 % solvent B in 20 min, 100 % solvent B for 6 min, and 5 % solvent B for another 8 min. Flow rate was 1 mL min^{-1}.

Ascorbic acid:

Ascorbic acid and dehydroascorbic acid were determinated simultaneously following an adapted derivatization procedure with o-phenylenediamine (Tausz et al., 1996). 700 µL of extracts of freeze-dried needle powder in 1.5 % (w/v) metaphosphoric acid (containing 1 mM EDTA) were treated with 450 µL TRIS-buffer (0.2 M Tris(hydroxymethyl)-aminoethane, pH 8.5) and 10 µL 0.1 % (w/v) o-phenylenediamine for 25 minutes in the dark. Reaction was stopped by the addition of orthophosphoric acid (final pH 2). Separation was carried out on an isocratic HPLC system: LKB 2150 pump, Hewlett Packard diode array detector IB with autosampler 1050 and software package. Column: Spherisorb S5 ODS2 25×4.6 mm, precolumn Spherisorb S5 ODS2 5×4.6 mm, solvent: 1 mM hexadecylammoniumbromide and 0.05 % (w/v) NaH_2PO_4 in methanol : water = 3 : 7 (v/v), running time = 20 min, flow rate = 1 mL min^{-1}. Absorbance of the column effluent was measured at 248 and 348 nm for the measurement of ascorbic acid and the dehydroascorbic-acid-derivative, respectively.

Statistics

Because of the complex design of the experiment we had no possibility to repeat the exposure in different fumigation cabinets. We decided to investigate a sufficient number of individuals of one spruce clone instead of fewer individuals from more clones in order to obtain results that could be statistically evaluated.

Statistical evaluations were completed using Statistica® (StatSoft, USA, 1994) software package. If sample numbers were sufficient (n = 9) differences between treatments would be calculated by two-way analysis of variance with SO_2 and CO_2 as main effects. ANOVA assumptions were tested by means of Levene's test for homogeneity of variances and the χ^2-test of normality, respectively. Small violations of the normality assumption did occur, but such small deviations from this assumption usually do not adversely bias the F-test in ANOVA procedure. However, since treatments' means and standard deviations were correlated, a fact that actually may lead to misleading ANOVA-results, significant main effects had to be verified by non-parametric methods (Kruskal-Wallis test). For sulfate data and biomass data due to smaller sample sizes non-parametric Kruskal-Wallis analysis of variance was used exclusively. Post-hoc comparisons of means were carried out by Scheffé's test following significant ANOVA results (Glaser, 1978). After Kruskal-Wallis-test proved significance, cross-comparisons of experimental groups were completed according to Conover's test (Bortz et al., 1990).

Results

Exposure of young spruce trees to elevated CO_2 (0.8 mL L^{-1}) and to a combination of elevated CO_2 (0.8 mL L^{-1}) and SO_2 (0.06 µL L^{-1}) resulted in a slightly higher needle biomass than that of the control (Table 1). On the other hand, exposure to SO_2 (0.06 µL L^{-1}) alone did not affect needle biomass production (Table 1). These results illustrate a significant main effect of CO_2 on needle biomass (Table 2).

Needles from trees exposed to elevated CO_2 or elevated CO_2 and SO_2 contained lower chlorophyll, carotene, and xanthophyll contents than needles from the control, which illustrates a significant effect of CO_2 on pigment content (Table 1 and 2). SO_2-exposure hardly affected the pigment content of spruce needles (Table 1 and 2). Upon exposure to elevated CO_2 the contents of all pigments were lowered to the same extent, which was also illustrated by the ratios of the different pigments (Table 1). However, α-carotene/β-carotene-ratios were significantly lower for elevated CO_2, SO_2, and elevated CO_2 in combination with SO_2 (Table 1 and 2). SO_2-exposure resulted in significant changes in the carotenoid pools participating in the xanthophyll cycle (violaxanthin = V, antheraxanthin = A, and zeaxanthin = Z), while elevated CO_2 concentrations did not affect the status of the xanthophyll cycle (Table 1 and 2). Needles of SO_2-exposed trees showed a higher Z/(V+A+Z) ratio, due to a higher zeaxanthin content in combination with a lower violaxanthin content.

Measurements of chlorophyll fluorescence showed F_v/F_m-ratios of 0.7 and higher for all treatments (Table 1). Slight differences between the treatments were corroberated by significant main effects of CO_2 and SO_2 in ANOVA (Table 2).

Exposure of trees to SO_2 and elevated CO_2 and SO_2 resulted in a significantly enhanced sulfate content of the needles (Table 1). Its content increased from 18 (control, elevated CO_2) to 43 µmol g^{-1} dw (SO_2, elevated CO_2 and SO_2).

Exposure of spruce trees to SO_2 and to elevated CO_2 in combination with SO_2 caused significantly higher thiol contents of the needles than that of the control and elevated CO_2, respectively (Table 1), which illustrates a significantly positive effect of SO_2 on these parameters (Table 2). The content of total glutathione (reduced + oxidized glutathione), the predominant thiol compound present in spruce needles, doubled upon SO_2 exposure. Also the cyst(e)ine content was enhanced. Spruce needles contained only traces of γ-glutamyl-cysteine (data not shown). There was a significant change in the redox state of glutathione upon SO_2-exposure, illustrated by a lower percentage of oxidized glutathione (Table 1). In addition to these SO_2-effects we found that glutathione contents were lower in the needles of the CO_2-exposed trees than in the control trees and significantly decreased in needles of the trees exposed to SO_2 and CO_2 compared to those exposed to SO_2 alone (Table 1). This result illustrates a significantly negative effect of CO_2-exposure on foliar glutathione content (Table 2).

The content of total ascorbate (ascorbate + dehydroascorbate) and the redox state of ascorbate were not significantly affected by SO_2 or CO_2 (Table 1 and 2).

Discussion

The present data on biomass are consistent with those observed for other plant species (Polle et al., 1993; Ceulemans

Table 1: Physiological responses of spruce trees to elevated CO_2 and SO_2. Data represent means of 6 to 24 measurements (± standard deviations). Different letters (*a, b, c*) indicate significant differences ($p<0.050$) according to Scheffé's test following significant ANOVA results or Conover's cross-comparisons following significant Kruskal-Wallis test results, respectively. (dw = needle dry weight, N = neoxanthin, A = antheraxanthin, V = violaxanthin, Z = zeaxanthin, L = lutein, X = xanthophylls, C = carotenes).

	treatment			
	control	CO_2	SO_2	CO_2+SO_2
shoot fresh weight [g $tree^{-1}$]	7.37±0.69	8.15±0.93	6.84±1.20	7.47±0.48
needle fresh weight [g $tree^{-1}$]	4.60±0.61	5.31±0.78	4.31±0.80	4.96±0.20
needle dry weight [g $tree^{-1}$]	1.92±0.30	2.52±0.38	1.95±0.36	2.23±0.10
chlorophyll a+b [mg g^{-1} dw]	2.05±0.91 *a*	1.30±0.54 *b*	1.63±0.77 *ab*	1.24±0.41 *b*
carotenes α+β [μg g^{-1} dw]	85±30 *a*	51±13 *b*	57±23 *ab*	46±9 *b*
xanthophylls N+V+A+Z+L [μg g^{-1} dw]	180±51 *a*	117±30 *b*	142±37 *ab*	116±26 *b*
chlorophyll *a/b*	2.7±0.2	2.6±0.2	2.5±0.4	2.6±0.2
chlorophylls/xanthophylls	8.4±0.7	7.9±1.0	7.4±2.9	7.6±0.9
chlorophylls/carotenes	18.1±1.2	18.0±0.9	19.0±2.9	18.5±0.9
chlorophylls/(X+C)	5.7±0.3	5.4±0.5	5.3±0.7	5.3±0.4
α-carotene/β-carotene	0.85±0.20 *a*	0.53±0.13 *b*	0.57±0.25 *b*	0.47±0.19 *b*
Z/(V+A+Z)	0.49±0.10 *a*	0.50±0.14 *ab*	0.64±0.21 *ab*	0.72±0.16 *b*
chlorophyll fluorescence (F_v/F_m)	0.76±0.04 *a*	0.73±0.04 *ab*	0.73±0.06 *ab*	0.70±0.06 *b*
sulfate content [μmol g^{-1} dw]	17.2±2.3 *a*	18.0±4.4 *a*	42.4±5.8 *b*	43.6±7.8 *b*
total glutathione [nmol g^{-1} dw]	717±165 *a*	477±235 *a*	1459±559 *b*	951±339 *c*
total cysteine [nmol g^{-1} dw]	38±16 *a*	58±33 *ab*	114±67 *b*	97±62 *ab*
oxidized glutathione [% of total glutathione]	7.5±3.7 *ab*	11.0±5.5 *b*	4.4±2.6 *a*	6.0±1.3 *a*
total ascorbate [μmol g^{-1} dw]	20.10±3.12	25.00±5.05	23.28±7.55	19.30±4.60
dehydroascorbate [% of total ascorbate]	19±9	30±10	25±10	16±6

Table 2: Physiological responses of spruce trees to elevated CO_2 and SO_2. Probabilities (p) were calculated by two-way analysis of variance (ANOVA) and by Kruskal-Wallis non-parametric analysis of variance, respectively. «Not significant» (ns) indicates $p>0.050$.

	n	ANOVA-effect			Kruskal-Wallis test	
		CO_2	SO_2	$CO_2 \times SO_2$	CO_2	SO_2
shoot fresh weight [g $tree^{-1}$]	24	–	–	–	ns	ns
needle fresh weight [g $tree^{-1}$]	24	–	–	–	p=0.028	ns
needle dry weight [g $tree^{-1}$]	24	–	–	–	p=0.014	ns
chlorophyll a+b [mg g^{-1} dw]	36	p=0.001	p=0.032	ns	p=0.010	ns
carotenes α+β [μg g^{-1} dw]	36	p=0.002	p=0.022	ns	p=0.021	ns
xanthophylls N+V+A+Z+L [μg g^{-1} dw]	36	p=0.001	ns	ns	p=0.005	ns
chlorophyll *a/b*	36	ns	ns	ns	ns	ns
chlorophylls/xanthophylls	36	ns	ns	ns	ns	ns
chlorophylls/carotenes	36	ns	ns	ns	ns	ns
chlorophylls/(X+C)	36	ns	ns	ns	ns	ns
α-carotene/β-carotene	36	p=0.003	p=0.013	ns	p=0.010	p=0.037
Z/(V+A+Z)	36	ns	p=0.002	ns	ns	p=0.004
F_v/F_m	96	p=0.011	p=0.013	ns		
sulfate content [μmol g^{-1} dw]	24				ns	p<0.001
total glutathione [nmol g^{-1} dw]	36	p=0.004	p<0.001	ns	p=0.016	p<0.001
total cysteine [nmol g^{-1} dw]	36	ns	p=0.001	ns	ns	p=0.003
oxidized glutathione [% of total glutathione]	36	p=0.039	p=0.002	ns	ns	p=0.002
total ascorbate [μmol g^{-1} dw]	36	ns	ns	p=0.020	ns	ns
dehydroascorbate [% of total ascorbate]	36	ns	ns	p=0.003	ns	ns

and Mousseau, 1994; Rao and De Kok, 1994). The increase in biomass due to elevated CO_2 may be ascribed to an enhanced carbon assimilation (Ceulemans and Mousseau, 1994). SO_2 is known to be phytotoxic even at low levels (De Kok, 1990; Rao and De Kok, 1994). However, the present SO_2 concentration was apparently too low to induce negative growth effects.

An elevated CO_2-induced decrease in pigment content, mainly in that of chlorophyll, has also been observed in other studies (Surano et al., 1986; Houpis et al., 1988; Heagle et al., 1993), but results are not always consistent for conifers (Polle et al., 1993). It has been reported that SO_2-impact may result in a loss of pigments of conifers, even at lower levels (Pérez-Soba et al., 1994). However, impact of SO_2 in the present experiment was obviously too low to cause these effects. The de-epoxidation of violaxanthin via antheraxanthin to zeaxanthin is an important process in the dissipation of excess energy and protection of the photosynthetic membranes from injury (Siefermann-Harms, 1977). An increase in zeaxanthin production in leaves upon SO_2-exposure has also been observed by Veljovic-Jovanovic et al. (1993), who explained it as an effect of acidification of the thylakoid systems by sulfate.

Chlorophyll fluorescence data differed slightly among treatments, F_v/F_m-ratios were 0.7 or higher for all needles, which indicated a healthy photosynthetic apparatus, regardless of the different exposure conditions (Lechner and Bolhàr-Nordenkampff, 1989).

Sulfate accumulation upon exposure to SO_2 is a general phenomenon in leaves of plants (De Kok, 1990; Polle et al., 1994). It has been suggested that elevated CO_2 may counteract the injurious effects of air pollution by a decreased pollutant uptake through a reduction in stomatal aperture (Allen, 1990). Apparently, in our experiments the SO_2 uptake was not affected by the CO_2 concentration.

Cysteine and glutathione are central metabolites in the sulfur assimilatory pathway (Bergmann and Rennenberg, 1993). If sulfur uptake by the roots is by-passed and sulfur is directly supplied to foliar tissue, it generally results in an increased cysteine and glutathione content, as the consequence of an enhanced and uncontrolled sulfur assimilation rate (Grill et al., 1979 a, 1982; De Kok and Stulen, 1993). In our experiment, this effect was obviously mediated by CO_2 influence, which had a negative effect on thiol contents. As shown by the sulfate contents, this was not the consequence of less uptake and oxidation but rather of a lower assimilation rate. However, the mechanisms of this CO_2 effect must remain speculative.

Results in literature on the ascorbic acid system under SO_2 pollution are contradictory: Ascorbic acid content was reported to either decrease (Grill et al., 1979 b), remain unaffected (Rao and De Kok, 1994), or increase (Mehlhorn et al., 1986) upon SO_2-exposure. Ascorbic acid is a major cellular antioxidative compound and is necessary as a reductant for the violaxanthin de-epoxidation (Siefermann-Harms, 1977). Although its metabolism is closely connected to carbohydrate assimilation (Loewus et al., 1990) we could not observe a significant stimulation of its formation by CO_2 enrichment.

It is important to note that all effects of SO_2 and CO_2 were independent from each other, i.e. significant interactions $SO_2 \times CO_2$ were not observed. The responses of the spruce trees to elevated CO_2 in combination with SO_2 could be explained as an addition of the single effects.

With respect to the main question addressed in this study, wether the influence of elevated CO_2 could mediate the effect of SO_2 pollution, the following facts are to be pointed out: (1) Our results do not confirm the theory that CO_2 decreases pollution uptake via reduced stomatal conductance. Apparently, SO_2 uptake and oxidation to sulfate remained unchanged by CO_2. (2) A certain CO_2 mediated reduction of the sulfur assimilation illustrated by lower thiol concentrations was found. Our data do not allow conclusions if this means a protective effect or even an enhanced sensibilisation. (3) CO_2 might enhance the resistance against air pollutants by stimulating the cellular defense capacity (Barnes and Pfirrmann, 1992). Previous studies reporting an absence of negative effects of SO_2 in the presence of elevated CO_2 explained this effect by a more reduced redox state of ascorbate and glutathione (Rao and De Kok, 1994). However, changes in the redox status that occurred in the present experiment were probably too small to alter antioxidative capacities (compare Rao and De Kok, 1994) and, on the other hand, there were no injurious effects of SO_2 on the spruce trees detectable in the present experiment.

Acknowledgements

We thank A. Wonisch and D. Reicher for their technical support in pigment analysis and C. Brunold, Bern, for the supply of the soil and the data of soil characteristics.

References

Allen, L. H.: Plant response to rising carbon dioxide and potential interactions with air pollutants. J. Environ. Quality *19,* 15–34 (1990).

Alscher, R. G. and J. S: Amthor: The physiology of free radical scavenging: Maintainance and repair processes. In: Schulte-Hostede, S., N. M. Darrall, L. W. Blank, and A. R. Wellburn (eds.): Air pollution and Plant Metabolism, pp. 94–115. Elsevier, London, New York (1988).

Barnes, J. D. and P. J. Pfirrmann: The influence of CO_2 and O_3, singly and in combination, on gas exchange, growth and nutrient status of radish (*Raphanus sativus* L.). New Phytol. *121,* 403–412 (1992).

Bergmann, L. and H. Rennenberg: Glutathione metabolism in plants. In: De Kok, L. J., I. Stulen, H. Rennenberg, C. Brunold, and W. E. Rauser (eds.): Sulfur nutrition and assimilation in higher plants: regulatory, agricultural, and environmental aspects, pp. 109–123. SPB Academic Publishing, The Hague, The Netherlands (1993).

Bortz, J., G. A. Lienert, and K. Boenke: Verteilungsfreie Methoden in der Biostatistik. Springer Verlag, Berlin, Heidelberg, New York (1990).

Ceulemans, R. and M. Mousseau: Tansley Review No. 71. Effects of elevated atmospheric CO_2 on woody plants. New Phytol. *127,* 425–446 (1994).

De Kok, L. J.: Sulfur metabolism in plants exposed to atmospheric sulfur. In: Rennenberg, H., C. Brunold, L. J. De Kok, and I. Stulen (eds.): Sulfur Nutrition and Sulfur Assimilation in Higher Plants: Fundamental, Environmental and Agricultural Aspects, pp. 125–138. SPB Academic Publishing, The Hague, The Netherlands (1990).

De Kok, L. J. and I. Stulen: Role of glutathione in plants under stress. In: De Kok, L. J., I. Stulen, H. Rennenberg, C. Brunold, and W. E. Rauser (eds.): Sulfur Nutrition and Assimilation in Higher Plants: Regulatory, Agricultural and Environmental Aspects, pp. 125-138. SPB Academic Publishing, The Hague, The Netherlands (1993).

Elstner, E. F. and W. Osswald: Mechanisms of oxygen activation during plant stress. Proceedings of the Royal Society of Edinburgh *102 B,* 131–154 (1994).

Glaser, W. R.: Varianzanalyse. First Edition, Gustav Fischer Verlag, Stuttgart, New York (1978).

Graham, R. L., M. G. Turner, and V. H. Dale: How increasing CO_2 and climate change affect forests. BioScience *40,* 575–587 (1990).

Grill, D., H. Esterbauer, and U. Klösch: Effect of sulphur dioxide on glutathione in leaves of plants. Environ. Poll. *19,* 187–194 (1979 a).

Grill, D., H. Esterbauer, and R. Welt: Einfluß von SO_2 auf das Ascorbinsäuresystem der Fichtennadeln. Phytopath. Z. *96,* 361–368 (1979 b).

Grill, D., H. Esterbauer, and K. Hellig: Further studies on the effect of SO_2-pollution on the sulfhydryl-system of plants. Phytopath. Z. *104,* 264–271 (1982).

Heagle, A. S., J. E. Miller, D. E. Sherrill, and J. O. Rawlings: Effects of ozone and carbon dioxide mixtures on two clones of white clover. New Phytol. *123,* 751–762 (1993).

Houpis, J., K. Surano, S. Cowles, and J. Shinn: Chlorophyll and carotenoid concentrations in two varieties of *Pinus ponderosa* seedlings subjected to long-term elevated carbon dioxide. Tree Physiol. *4,* 187–193 (1988).

Jäger, H. J., H. J. Weigel, and L. Grünhage: Physiologische und biochemische Aspekte der Wirkung von Immissionen auf Waldbäume. Eur. J. For. Path. *16,* 98–109 (1986).

Kranner, I. and D. Grill: Content of low-molecular-weight thiols during the imbibition of pea seeds. Physiol. Plant. *88,* 557–562 (1993).

Kranner, I. and D. Grill: Determination of glutathione disulfide in lichens: A comparison of frequently used methods. Phytochemical Analysis *7,* 24–28 (1996).

Lechner, E. G. and H. R. Bolhàr-Nordenkampff: Seasonal and stress-induced modifications of the photosynthetic capacity of Norway spruce at the altitude profile ‹Zillertal›. A. Characteristics of chlorophyll fluorescence induction. Phyton (Austria) *29* (3), 187–206 (1989).

Loewus, M. W., D. L. Bedgar, K. Saito, and F. A. Loewus: Conversion of L-sorbonose to L-asorbic acid by a NADP dependent dehydrogenase in bean and spinach leaves. Plant Physiol. *94,* 1492–1495 (1990).

Mehlhorn, H., G. Seufert, A. Schmidt, and K. J. Kunert: Effect of SO_2 and O_3 on productions of antioxidants in conifers. Plant Physiol. *82,* 336–338 (1986).

Miszalski, Z. and H. Ziegler: Superoxide dismutase and sulfite oxidation. Z. Naturforsch. *47,* 360–364 (1992).

Peréz-Soba, M., L. van der Eerden, and I. Stulen: Combined effects of gaseous ammonia and sulphur dioxide on the nitrogen metabolism of the needles of Scots pine trees. Plant Physiol. Biochem. *32,* 539–546 (1994).

Pfeifhofer, H.: Evidence of chlorophyll *b* and lack of lutein in *Neottia nidus-avis* plastids. Biochem. Physiol. Pflanzen *184,* 55–61 (1989).

Polle, A., T. Pfirrmann, S. Chakrabarti, and H. Rennenberg: The effects of enhanced ozone and enhanced carbon dioxide concentrations on biomass, pigments and antioxidative enzymes in spruce needles (*Picea abies* L.). Plant Cell Environ. *16,* 311–316 (1993).

Polle, A., M. Eiblmeier, and H. Rennenberg: Sulphate and antioxidants in needles of Scots pine (*Pinus sylvestris* L.) from three SO_2-polluted field sites in eastern Germany. New Phytol. *127,* 571–577 (1994).

Rao, M. V. and L. J. De Kok: Interactive effects of high CO_2 and SO_2 on growth and antioxidant levels in wheat. Phyton (Austria) *34,* 279–290 (1994).

Rennenberg, H. and A. Polle: Metabolic consequences of atmospheric sulfur influx into plants. In: Alscher, R. G. and A. Wellburn (eds.): Plant responses to the gaseous environment, pp. 165–180. Chapman and Hall, London (1994).

Siefermann-Harms, D.: The xanthophyll cycle in higher plants. In: Tevini, M. and H. K. Lichtenthaler (eds.): Lipids and Lipid Polymers in Higher Plants, pp. 218–230. Springer Verlag, Berlin, Heidelberg, New York (1977).

Stulen, I., J. Den Hertog, and C. M. Jansen: The influence of atmospheric CO_2 enrichment on allocation patterns of carbon and nitrogen in plants from natural vegetations. In: Abrol, Y. P., P. Mohanty, and G. Govindjee (eds.): Photosynthesis – Photoreactions to Plant Productivity, pp. 509-524. Oxford & IBH Publishing Co., New Delhi, India (1993).

Surano, K., P. Daley, J. Houpis, J. Shinn, J. Helms, R. Pallassou, and M. Costella: Growth and physiological responses of *Pinus ponderosa* Dougl. ex P. Laws to long-term elevated CO_2 concentrations. Tree Physiol. *2,* 243–259 (1986).

Tanaka, K., N. Kondo, and K. Sughara: Accumulation of hydrogen peroxide in chloroplasts of SO_2-fumigated spinach leaves. Plant Cell Physiol. *23,* 999–1007 (1982).

Tausz, M., I. Kranner, and D. Grill: Simultaneous determination of ascorbic acid and dehydroascorbic acid in plant materials by high-performance liquid chromatography. Phytochemical Analysis, in press (1996).

Veljovic-Jovanovic, S., W. Bilger, and U. Heber: Inhibition of photosynthesis, acidification and stimulation of zeaxanthin formation in leaves by sulfur dioxide and reversal of these effects. Planta *191,* 365–376 (1993).

J. Plant Physiol. Vol. 148. pp. 368–373 (1996)

Effect of Short-term and Long-term Low Temperature Stress on Polyamine Biosynthesis in Wheat Genotypes with Varying Degrees of Frost Tolerance

I. RÁCZ[2], M. KOVÁCS[1], D. LASZTITY[2], O. VEISZ[1], G. SZALAI[1], and E. PÁLDI[1]

[1] Agricultural Research Institute of the Hungarian Academy of Sciences, H-2462 Martonvásár, P.O. Box 19, Hungary

[2] Department of Plant Physiology, Eötvös Loránd University, H-1088 Budapest, Múzeum krt 4, Hungary

Received July 9, 1995 · Accepted October 18, 1995

Summary

Two series of experiments were carried out to examine the short- and long-term effects of low temperature on polyamine biosynthesis in wheat. In the first series, studies were made on the polyamine accumulation in the leaves, crowns and roots of winter wheat varieties with varying degrees of frost tolerance subjected to short-term low temperature stress (6 h, −2 °C). A marked accumulation of Put was observed. Agm accumulation was also examined and found comparable to that of Put. This suggests that Agm, which is an intermediate product of Put synthesis only in higher plants, may play an important role during short-term cold treatment.

The second series of experiments was aimed at discovering the effect of wheat chromosomes 5A and 7A, which contain major genes responsible for frost resistance, on the polyamine synthesis taking place in various parts of the seedlings during long periods of cold treatment, and especially on the alternative metabolic pathway present only in higher plants.

Key words: Cold stress, polyamines, frost tolerance, wheat, Triticum aestivum L.

Abbreviations: Put = putrescine; Agm = agmatine; Spd = spermidine; Spm = spermine; PAs = polyamines; TLC = thin layer chromatography; HPLC = high performance liquid chromatography; ADC = arginine decarboxylase; ODC = ornithine decarboxylase; LT = lethal temperature; fw = fresh weight; Mv4 = Martonvásári 4; Mv14 = Martonvásári 14; Bu20 = Bucsányi 20.

Introduction

Polyamines (Put, Spd, Spm) are essential components of living organisms and their implications in the physiological and biochemical processes have long been studied (Slocum et al., 1984; Smith, 1985). They are known to play an important regulatory role in a variety of growth and developmental processes (Bagni, 1970; Galston, 1983; Pegg, 1986; Faure et al., 1991; Páldi et al., 1993), in senescence (Anguillesi et al., 1990), in plant diseases (Walters, 1989; Rajam et al., 1985) and in stress reactions (Bagni et al., 1979; Chu et al., 1992; Galiba. et al., 1993; Kramer et al., 1990). Polyamine accumulation as a result of various stress factors (pH, water deficiency, lack of K, salinity, osmotic stress, low and high temperatures) is well known and Put especially has been associated with stress conditions, but the physiological significance of these changes is not yet understood (Evans et al., 1989). There is an alternative metabolic pathway for synthesising Put (arginine – agmatine – N-carbamoylputrescine – putrescine) which exists only in higher plants. It has been shown that ADC, the enzyme responsible for Agm biosynthesis, is also involved in the stress-induced reactions, but no real attention has been given so far to its product (Flores et al., 1984). Though exposure to low temperature has also long been reported to induce Put accumulation in different species (Guye et al., 1986; McDonald et al., 1986; Kushad et al.,

1987) data concerning the dynamics of polyamine accumulation under cold stress in cereals, especially in wheat, are scarce (Nadeau et al., 1987).

In the present work the research concept consisted of two parts. The first was based on the fact that in intercontinental climatic areas, such as Hungary, the young wheat plants are often exposed to short periods of below-freezing temperatures in autumn, particularly in the second half of October and early November. These short-term effects do not represent the beginning of long-term low temperature hardening, which is essential for the development of frost resistance in wheat plants. These short periods of cold shock may retard the development of the young seedlings to a considerable extent, preventing them from preparing adequately for the longer period of hardening at low, but non-freezing temperatures. In a knowledge of this field experience it was decided to obtain information on how young wheat seedlings respond to such effects and on what changes this induces in the synthesis of the polyamines used as stress markers.

The foundation for the second part of the research concept was that in the temperate zone it is essential for the cultivated wheat varieties to possess adequate frost resistance. This character is genetically determined and develops as the result of a specific period of hardening at low temperature. Varieties with different degrees of frost resistance and substitution lines containing genes responsible for frost resistance provide an excellent opportunity for tracing the changes occurring in the synthesis of polyamines during a long period of cold treatment and for studying the possible correlations.

The first aim of the present study was to examine the changes in the levels of polyamines during a short exposure to low temperature in a number of winter wheat cultivars differing in their sensitivity to freezing stress and to determine the distribution of polyamines in different parts of the plants.

The other was to study the effect of the chromosomes carrying the two genes responsible for frost resistance on the polyamines being synthesised in various plant organs during long-term low temperature treatment, with special regard to the significance of the alternative metabolic pathway present only in higher plants.

Materials and Methods

Plant material

In the first series of experiments four cultivated wheat varieties (Martonvásári 4 and Martonvásári 14 from Hungary, Bucsányi 20 from Slovakia, Vitka from Yugoslavia) significantly differing from each other with respect to frost resistance were used. The sterilised seeds were sown (ten to a pot) in sterilised soil consisting of a mixture of loamy soil, Vegasca (humus-containing additive manufactured by Florasca Ltd., Hungary) and sand (2 : 2 : 1, v : v : v). The plant material was grown in plant growth chambers (PGW-35 model, Conviron, Canada). The sprouts were irrigated five times a week with deionized water and twice a week with nutrient solution containing deionized water and Wuxal (500 : 1, v : v). The plants were kept under a 16 h light photoperiod at 16 °C, under a mixture of fluorescent and incandescent lighting giving an irradiance of 300 $\mu Em^{-2} s^{-1}$ with 80 % relative humidity day and night. After 3 weeks of growth, the wheat plants were treated as follows. The temperature was lowered to 2 °C (temperature gradient: 0.5 °C/h) at an irradiance level of 530 $\mu Em^{-2} s^{-1}$. After 6 hours at 2 °C the temperature was dropped further to −2 °C (temperature gradient: 0.5 °C/h) and the plants were kept at −2 °C for 6 hours. Samples were collected at the end of the 3rd week at 16 °C and after the 6th hour of the 2 °C and −2 °C treatments from the leaves, crowns and roots of the plants.

In the second series of experiments two wheat varieties differing in their degrees of frost resistance (Cheyenne and Chinese Spring) were used together with the substitution lines CS(Ch5A) and CS(Ch7A) from chromosome substitution lines produced by Sears from a cross between Chinese Spring and Cheyenne (Sears, 1988).

Germinated grains were sown in wooden boxes (42 × 31 × 18 cm) in a 4 : 1 mixture of garden soil and sand. During the two weeks of preliminary growth plants were kept at day and night temperatures of 15 °C and 10 °C, respectively, with a 12-hour day. After this the temperature was maintained at 2 °C continuously for 70 days, again with a 12-hour day. During the 70-day hardening period samples were taken every day for the first three days, then every 10 days from the 10th day. Samples were divided into leaf, crown (tillering node) and root sections for the determination of polyamine content.

Estimation of frost tolerance

For the purpose of the frost tests, seedlings of the various wheat varieties were raised for two weeks in an earth-sand mixture (3 : 1). Freezing was carried out in programmed freezing chambers, where the temperature gradient of cooling and warming was 0.5 °C an hour. In order to determine the LT_{50} value the 14-day-old seedlings were first hardened for 5 days at 2 °C, after which the plants were divided into as many parts as there were freezing temperatures (−5, −10, −15, −18, −20 and −25 °C). The seedlings were kept at the relevant freezing temperatures for 48 hours, then raised for a further 14 days at 15 °C (16 h light, 8 h dark, 70 % relative humidity) in a phytotron unit (Conviron, type: PGV-36). The survival percentage was then determined as a function of the freezing temperatures applied. On the basis of the graphs obtained, the critical temperature was taken to be that at which 50 % of the plants survived. The mean results of three biological replications were taken as the basis for determining the LT_{50} values. The LT_{50} values of the varieties examined in the experiments were as follows: Martonvásári 4 (Mv 4): −19.8 ± 3.3 °C; Bucsányi 20 (Bu 20): −13.5 ± 2.9 °C; Martonvásári 14 (Mv 14): −9.7 ± 2.6° Vitka: −5.1 ± 2.4 °C. Significant differences were exhibited in the survival rates in frost resistance tests (Veisz and Sutka, 1989) at −14 °C, ranging from 93.3 ± 2.1 % for the most resistant variety, Cheyenne (Ch) and 13 ± 3.8 % for the most sensitive, Chinese Spring (CS), with intermediary values of 43.3 ± 4.3 % and 36.9 ± 3.6 % for CS(Ch5A) and CS(Ch7A), respectively.

Assay of polyamine content

Polyamine content was determined by the method of Nielsen (1990) with a slight modification. 0.1 g tissue from leaves, crowns or roots was homogenised in 5 mL 5 % (v/v) $HClO_4$. The extracts were left in melting ice for an hour, then centrifuged for 30 minutes at 20,000 g. 0.5 mL dansyl chloride (10 mg/mL, dissolved in acetone) and 0.5 mL of saturated Na_2CO_3 were added to 0.5 mL of supernatant. The mixture was vortexed and incubated in darkness at 60 °C for an hour. The dansylated polyamines were extracted with two volumes of toluol. The polyamines were separated on silica gel TLC plates (Merck, Art. 5554 Kieselgel 60 F 254) with cyclohexane : ethyl acetate (5 : 4 v/v) as a solvent. Spots corresponding to comigrant polyamine standards were visualised under UV light, scraped off and eluted in 2 mL of ethyl acetate, after which their fluorescence was measured with a spectrofluorimeter (excitation: 350 nm, emission: 495 nm). The results were compared with standards dansylated and chromatographed in the same way. Agmatine is not a stable diamine. Before dansylating, the commercial product should be purified twice by high performance liquid chromatography because of decomposed substances.

Results

The changes in polyamine content observed in the leaves, crowns and roots of four winter wheat varieties with varying degrees of frost resistance as the result of short-term low temperature are summarised in Table 1. One aim of the study was to determine whether there was any difference between the quantities of polyamines synthesised in various parts of the seedlings. When comparing the plant organs the largest quantity of polyamines was found in the crown and the lowest quantity in the roots. This was valid for all four varieties. An analysis of the changes occurring due to temperature indicated that no significant differences could be demonstrated between the 2 °C treatment and the control as regards the polyamine contents of the leaves and roots. In the crown, however, which contained the largest quantity of polyamines, there was evidence that low temperature induced a higher level of polyamine synthesis. As the result of treatment at −2 °C the quantity of polyamines rose in all three plant organs in all four varieties. The decisive role in this short-term low temperature effect was played by putrescine, agmatine and spermidine. Spermine had no role in the leaves and roots. The results obtained for putrescine are presented in Table 1, which clearly illustrates the inductive effect of low temperature. There was a positive correlation between the induced quantity of putrescine and the frost resistance of the varieties, characterised by the LT_{50} value. Special attention was paid to the role of agmatine. The results suggest that agmatine plays an important role in the defence response to short-term low temperature stress. As a consequence of this, a significantly greater quantity of agmatine could be demonstrated in the −2 °C treatment than in that carried out at 2 °C. It is interesting to note that, in contrast to putrescine, the quantity of agmatine was in negative correlation with the LT_{50} value of the variety. The same tendency can be established in the case of spermidine.

Studies were also made on the effect of long-term cold treatment on changes in the free polyamine content. Changes in free polyamine content were examined in a frost tolerant wheat cultivar, Cheyenne, and a frost sensitive cultivar Chinese Spring, together with the chromosome substitution lines CS(Ch5A) and CS(Ch7A) in which the 5A and 7A chromosomes of the recipient Chinese Spring cultivar were substituted with the 5A and 7A chromosomes (responsible for frost tolerance) of the donor Cheyenne cultivar. During the 70-day cold hardening period characteristic changes could be observed in the polyamine level and composition in all four wheat genotypes. Marked increases were measured in the polyamine content, especially in the putrescine and agmatine, during the first few days of hardening followed by a slow decrease and then a slow increase after the 40th day (Fig. 1). The initial polyamine levels were higher in the roots and leaves of Cheyenne than in those of Chinese Spring, but the values changed to a lesser extent during the first few days of cold stress than in Chinese Spring. By contrast, the lower initial values in the crown of Cheyenne increased to a greater extent during the first few days than Chinese Spring. In other words, the greatest changes could be observed in the crown part of Cheyenne, which is the surviving organ of this extremely frost resistant variety. By the end of cold treatment higher polyamine levels developed in Cheyenne than in Chinese Spring. The polyamine levels in chromosome substitution lines exhibited transient characteristics, resembling the recipient variety to a greater extent. Among the polyamines, the putrescine and agmatine contents exhibited significant changes, while much smaller alterations, with less characteristic tendencies, could be observed in the amounts of spermidine and spermine.

Discussion

One objective of this paper was to examine the changes in polyamine content during a short exposure to low temperature in four wheat cultivars differing in their sensitivity to freezing, and to gain new information on the distribution of polyamines in the leaves, crowns and roots of the plants. Information regarding the effect of short-term cold stress on the regulation of the polyamine levels in wheat is scarce. The present results are in agreement with previous observations that cold stress induces major changes in the polyamine levels (Guye et al., 1986; McDonald et al., 1986; Kushad et al., 1987). The data clearly demonstrate that the cultivars responded to low temperature with a significant increase in their polyamine levels (Table 1). The most pronounced changes were found in the level of Put; Spd showed a less dramatic, but sustained increase in concentration, while Spm levels appeared to be the least responsive to short-term low temperature stress. Similar findings have already been reported by Nadeau et al. (1987). It is a well-known fact that in higher plants there is an alternative metabolic pathway for the synthesis of putrescine, starting from arginine, with agmatine as the intermediary product (Flores et al., 1984).

The most important observation in this study is that very high levels of Agm were found in two of the cultivars studied. The dynamics and magnitude of the Agm accumulation appeared to be similar to those of Put. It has earlier been reported that plants subjected to various stress conditions exhibited greater ADC activity (Flores et al., 1984; Galston et al., 1990), but no data have been presented concerning the implication of its product, Agm, in the stress reactions. In the light of the fact that the accumulation of Agm appeared comparable to that of Put, the possibility of Agm playing a major role in cold-induced reactions seems obvious and merits further investigation mainly as a precursor for Put. Its accumulation is consistent with the dynamics of Put accumulation. This has implications for the regulation of Put by ADC during short-term cold stress. It can also be stated that leaves and roots showed different polyamine accumulations at low temperature. The roots generally exhibited the lowest polyamine levels. This could reflect diversity in cold sensitivity between the various plant parts, or the temperature buffering of the soil.

Interesting findings were recorded with respect to correlations between the frost resistance of the varieties and the in-

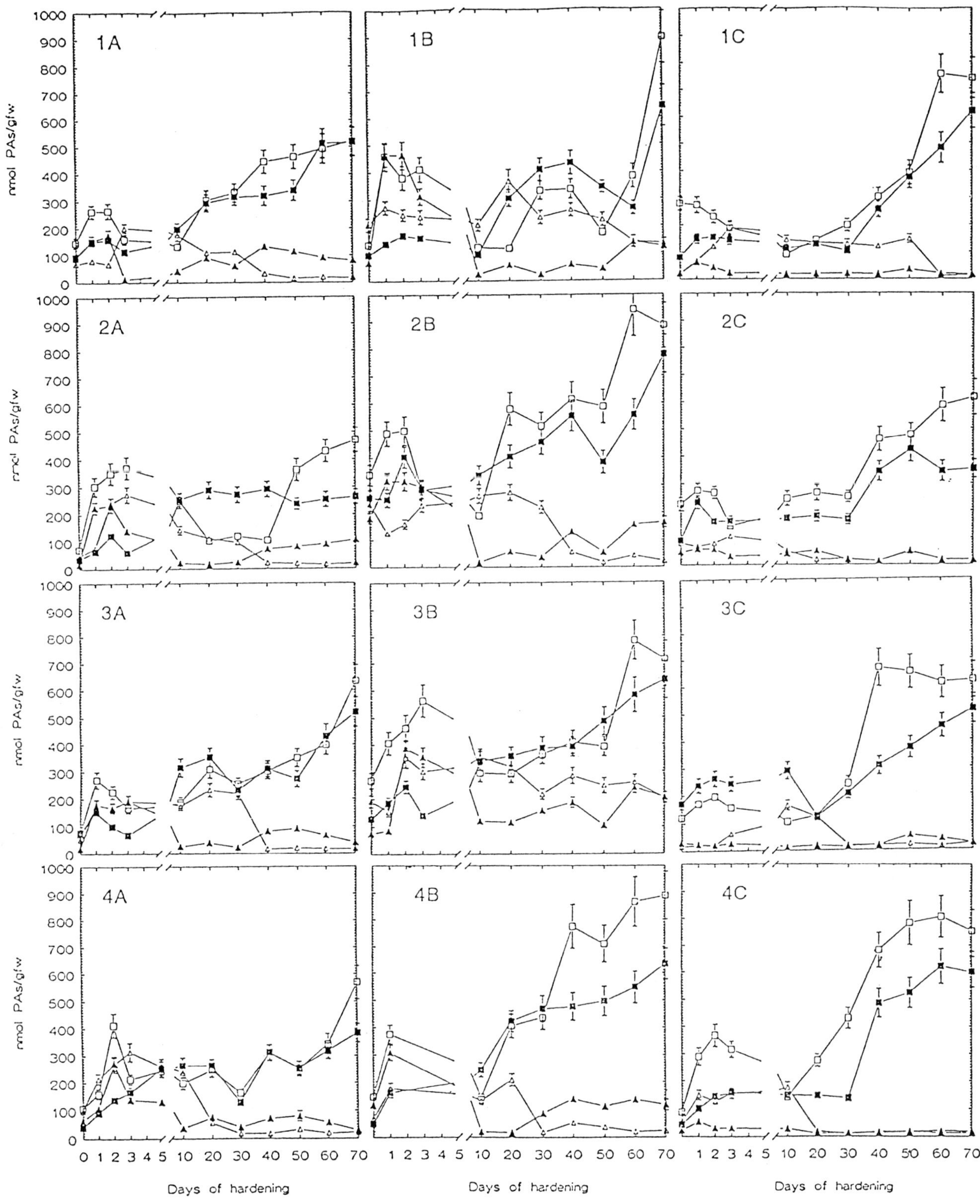

Fig. 1: Changes in polyamine contents in the leaves (A), crowns (B) and roots (C) of winter wheat cultivar Cheyenne (1), spring wheat cultivar Chinese Spring (2) and the chromosome substitution lines CS/Ch5A (3) and CS/Ch7A (4) during long-term cold treatment. Symbols: □: Put, ■: Agm, △: Spd, ▲: Spm. Each point is the mean of four individual measurements ± SD.

Table 1: Changes in polyamine levels in the roots, crowns and leaves of winter wheat (*Triticum aestivum* L.) cultivars with varying degrees of frost tolerance after short-term cold stress measured at 2 °C and –2 °C. LT_{50} values are as follows: Mv-4 (–19.8 °C), Bu-20 (–13.5 °C), Mv-14 (–9.7 °C) and Vitka (–5.1 °C). Duration of cold stress at 2 °C and –2 °C was 6 hr. Measurements at 16 °C were taken as the control. Values are means ± SD (n = 5). (nf: not found).

Plant organ	Type of PAs	Mv-4 µg PA/g fw			Bu-20 µg PA/g fw			Mv-14 µg PA/g fw			Vitka µg PA/g fw		
		16 °C	2 °C	–2 °C	16 °C	2 °C	–2 °C	16 °C	2 °C	–2 °C	16 °C	2 °C	–2 °C
Root	Agm	103±14	101±13	172±16	106±14	108±13	161±12	95±10	96±18	152±16	88±11	120±14	134±12
	Put	101±18	110±19	192±21	81±10	110±11	149±15	123±13	129±15	178±19	95±12	98±10	208±21
	Spd	nf	nf	20±12	nf	17±11	18±10	nf	nf	nf	nf	nf	nf
	Spm	58±19	59±10	63±12	37±18	38±17	45±11	58±17	74±18	91±11	28±17	31±18	43±17
Crown	Agm	110±11	122±15	208±19	71±18	121±13	167±15	124±11	132±19	158±14	103±10	106±19	132±14
	Put	145±12	209±18	513±49	205±23	324±27	723±52	344±39	455±38	906±53	269±21	336±26	468±55
	Spd	162±17	220±19	441±45	95±15	135±10	145±13	36±16	121±11	255±21	224±25	287±26	448±39
	Spm	114±18	124±10	137±14	117±19	145±14	151±14	139±13	198±25	256±22	21±18	66±11	87±10
Leaf	Agm	117±15	135±16	194±23	120±12	145±17	217±23	131±15	151±14	183±16	112±19	120±13	149±14
	Put	130±11	144±12	165±24	126±18	111±10	233±31	142±11	104±15	135±12	144±18	111±12	284±33
	Spd	89±19	92±10	313±33	104±11	128±11	269±33	83±16	137±12	181±12	70±11	79±13	97±11
	Spm	32±18	52±14	73±14	29±12	33±13	42±14	39±13	49±16	51±14	nf	nf	42±13

duced polyamine synthesis. The quantity of putrescine increased as the result of treatment at –2 °C and this tendency was in positive correlation with the frost resistance of the varieties. With the exception of a single case, there was an increase in putrescine induction parallel to an increase in frost resistance.

The effect of long-term (70-day) cold treatment on polyamine biosynthesis was studied on special wheat genotypes. Answers were sought to two questions: a) What effect was exerted by chromosomes bearing genes responsible for frost resistance on the quantities and mutual ratios of the polyamines being synthesised in various parts of the seedlings (leaves, crown, roots)? b) Could the decisive role of the alternative metabolic pathway present only in higher plants, and of its main product, agmatine, be proved also in the case of long-term cold treatment? The plant material used in the experiment consisted of the varieties Cheyenne, which possesses excellent frost resistance, and Chinese Spring, which has extremely poor frost resistance, together with two substitution lines of the latter, in which chromosomes 5A and 7A were replaced by those of Cheyenne. It can be seen from the results summarised in Fig. 1 that in all four genotypes the extent of polyamine synthesis was far the greatest in the crown, followed by the leaves and roots. If the role of the individual polyamines is examined, it is obvious that the quantities of putrescine and agmatine synthesised are much the highest, both in the various plant organs and in different genotypes. The quantities of these two polyamines rise during the first few days, then, after a brief pause, show an unambiguous increase until the 60th day, after which they stagnate, indicating that cold has no further inductive effect on these polyamines. If spermidine and spermine are considered, it can be observed that these only have a significant role on the crown. The quantity of spermidine and spermine synthesised in the leaves is far less than that recorded in the crown, while in the roots they play hardly any role at all. By contrast to putrescine and agmatine, the quantity of spermidine exhibits a constant decline, while the quantity of spermidine is hardly detectable towards the end of the treatment. If the quantity of polyamine synthesised is considered as a function of genotype, it is clear that this quantity is highest in Cheyenne, which has excellent frost resistance. A comparison of Chinese Spring and the substitution lines reveals no great differences, with the exception of the roots. In the latter case, the quantity of putrescine and agmatine synthesised in the two substitution lines is significantly greater than that in Chinese Spring. If the effects of short-term and long-term cold treatments are compared, it is quite clear that the alternative pathway has a significant role in the polyamine biosynthesis taking place at low temperature. The key enzyme in the alternative pathway is ADC, which is responsible for the synthesis of agmatine and was found by Galston and Sawhney (1990) to be a stress enzyme which can be induced readily by low temperature, particularly in cereals. ADC is induced to a greater extent than ODC, since the activity of the ODC enzyme is in any case greater than that of ADC under non-stress conditions.

The connection of polyamines with nucleic acids, membrane stability and protein synthesis is well known. The results obtained in the present work confirm the findings of investigations on the effect of low temperature in wheat at the nucleic acid (Dévay and Páldi, 1977; Páldi and Dévay, 1977, 1983; Páldi et al., 1994) and more recently at the protein synthesis level (Lasztity et al., 1994).

Acknowledgements

The authors wish to thank Dr. L. Láng (Department of Wheat Breeding, Agricultural Research Institute, Martonvásár) for providing the winter wheat varieties and Dr. T. Kremmer (Dept. of Biochem. Inst. of Oncopathol., Budapest) for a gift of dansylated agmatine. We are grateful for generous support from the National Scientific Research Fund of Hungary (OTKA, No. I/2 1122 and I/3 140 and 150).

References

Anguilessi, C. M., I. Grilli, R. Tazziolo, and C. Flores: Polyamine accumulation in aged wheat seeds. Biol. Plant. *32,* 189–197 (1990).

Bagni, N.: Metabolic changes of polyamines during the germination of *Phaseolus vulgaris.* New Phytol. *69,* 159–164 (1970).

Bagni, N. and D. Serafini-Fracassini: Polyamines and plant tumor. J. Biochem. *28,* 392–394 (1979).

Chu, C. and T. M. Lee: Regulation of chilling tolerance in rice seedlings by plant hormones. Korean J. of Crop Sci. *37,* 188–198 (1992).

Dévay, M. and E. Páldi: Cold-induced rRNA synthesis in wheat cultivars during the hardening period. Plant Sci. Lett. *8,* 191–195 (1977).

Evans, P. T. and R. L. Malmberg: Do polyamines have roles in plant development? Annu. Rev. of Plant Physiol. and Mol. Biol. *40,* 235–269 (1989).

Faure, O., M. Mengoli, A. Nougarede, and N. Bagni: Polyamine pattern and biosynthesis in zygotic and somatic embryo stages of *Vitis vinifera.* J. Plant Physiol. *138,* 545–549 (1991).

Flores, H. E., N. D. Young, and A. W. Galston: Polyamine metabolism and plant stress. In: Key, L. and T. Kosuge (eds): Cellular and Molecular Biology of Plant Stress, 93–114. Allen R. Liss Publ., New York (1984).

Galiba, G., G. Kocsy, R. Kaur-Sawhney, J. Sutka, and A. W. Galston: Chromosomal localization of osmotic and salt stress-induced differential alterations in polyamine content in wheat. Plant Sci. *92,* 203–211 (1993).

Galston, A. W.: Polyamines as modulators of the plant development. Bioscience *33,* 249–261 (1986).

Galston, A. W. and R. K. Sawhney: Polyamines in Plant Physiology. Plant Physiol. *94,* 406–410 (1990).

Guye, M. G., L. Vigh, and J. M. Wilson: Polyamine titre in relation to chilling sensitivity in *Phaseolus* species. J. Exp. Bot. *37,* 1036–1043 (1986).

Kramer, G. F. and C. Y. Wang: Effects of chilling and temperature preconditioning on the activity of polyamine biosynthetic enzymes in zucchini squash. J. Plant Physiol. *136,* 115–119 (1990).

Kushad, M. M. and G. Yelenosky: Evaluation of polyamine and proline levels during low temperature acclimation of Citrus. Plant Physiol. *84,* 692–695 (1987).

Lasztity, D., I. Rácz, and E. Páldi: Activity of protein synthesising system in vernalised wheat seedlings at low temperature. In: Dörffling, K., B. Brettschneider, H. Tantau, and K. Pithan (eds.): Agriculture. Crop adaptation to cool climates, pp. 349–354. ECSP-EEC-EAEC, Brussels, Luxembourg (1994).

McDonald, R. E. and M. M. Kushad: Accumulation of putrescine during chilling injury of fruits. Plant Physiol. *82,* 324–326 (1986).

Nadeau, S., S. Delaney, and J. Chouinard: Effect of cold hardening on the regulation of polyamine levels in Wheat (*Triticum aestivum* L.) and Alfalfa (*Medicago sativa*). Plant Physiol. *84,* 73–77 (1987).

Páldi, E. and M. Dévay: Characteristics of the rRNA synthesis taking place at low temperatures in wheat cultivars with varying degrees of frost hardiness. Phytochemistry *16,* 177–179 (1977).

– – Relationship between the cold-induced rRNA synthesis and the rRNA cistron number in wheat cultivars with varying degrees of frost hardiness. Plant Sci. Lett. *30,* 61–67 (1983).

Páldi, E., T. Kremmer, and D. Lasztity: Specific polyamine synthesis during vernalisation in wheat. Physiol. Plant. *83,* 394 (1993).

Páldi, E., I. Rácz, and D. Lasztity: Studies on the correlation between the quantity of 1.4 MD rRNA precursor and frost resistance. In: Dörffling, K., B. Brettschneider, H. Tantau, and K. Pithan (eds.): Agriculture. Crop adaptation to cool climates, pp. 341–348. ECSP-EEC-EAEC, Brussels, Luxembourg (1994).

Páldi, E., G. Szalai, I. Rácz, and D. Lasztity: Polyamine biosynthesis during vernalisation in winter and spring wheat varieties. In: Dörffling, K., B. Brettschneider, H. Tantau, and K. Pithan (eds.): Agriculture. Crop adaptation to cool climates, pp. 195–204. ECSP-EEC-EAEC, Brussels, Luxembourg (1994).

Pegg, A. E.: Recent advances in the biochemistry of polyamines in eukaryotes. Biochem. J. *234,* 249–262 (1986).

Pukacka, S., Z. Szczotka, and M. Zymanczyk: Arginine decarboxylase, ornithine decarboxylase and polyamines under cold and warm stratification of Norway maple (*Acer platanoides* L.) seeds. Acta Physiol. Plant. *13,* 247–252 (1991).

Rajam, M. V., L. H. Weinstein, and A. W. Galston: Prevention of a plant disease by specific inhibition of fungal polyamine biosynthesis. Proc. Natl. Acad. Sci. USA *82,* 6874–6878 (1985).

Slocum, R. D., R. Kaur-Sawhney, and A. W. Galston: The physiology and biochemistry of polyamines in plants. Arch. Biochem. Biophys. *235,* 283–303 (1984).

Smith, T. A.: Polyamines. Annu. Rev. Plant Physiol. *36,* 117–143 (1985).

Walters, D.: Polyamines and plant disease. Plants Today *1,* 22–26 (1989).

J. Plant Physiol. Vol. 148. pp. 374–377 (1996)

Effect of Low Temperature on the rRNA Processing in Wheat (*Triticum aestivum* L.)

E. Páldi[1], I. Rácz[2], and D. Lasztity[2]

[1] Agricultural Research Institute of the Hungarian Academy of Sciences, H-2462 Martonvásár, P.O. Box 19, Hungary

[2] Eötvös Loránd University, Department of Plant Physiology, H-1088 Budapest, Múzeum krt 4/a, Hungary

Received July 9, 1995 · Accepted October 18, 1995

Summary

In the course of the experiments, studies were made on the effect of low temperature on rRNA processing. The two wheat lines in this work differed from each other only as regards frost resistance. It was proved that in the case of the weakly frost resistant line quantitative and qualitative changes took place in the rRNA maturation process as the result of low temperature. During cold treatment the last precursors (1.4 and 0.9 MD) of the two stable cytoplasmic rRNAs (1.3 and 0.7 MD) accumulated at the expense of the stable rRNA fractions. This accumulation increased as cold treatment proceeded. In the line with good frost resistance this change could not be demonstrated. The results indicate that in the poorly resistant line the cold treatment had an inhibitory effect on the last step in the maturation process, i.e. at low temperature this process is unable to proceed to completion.

Key words: Triticum aestivum, frost tolerance, low temperature, RNA precursor, RNA processing, wheat.

Abbreviations: rRNA = ribosomal ribonucleic acid; MD = million dalton; LT = lethal temperature; Mv = Martonvásári; PAGE = polyacrylamide gel electrophoresis.

Introduction

The physiological and biological processes which lead to cold tolerance or the adaptation of plants to low temperature are extremely complex. The molecular and genetic mechanisms which regulate these processes are not sufficiently well known. In recent years, however, more and more information has been reported in this field (Guy et al., 1985; Mohapatra et al., 1987; Perras and Sarhan, 1989). Perras and Sarhan studied the interaction between cold effect and protein synthesis in detail and proved that, due to the cold effect, changes occurred in the soluble protein composition. These changes were correlated primarily with the frost resistance of the variety (Perras and Sarhan, 1989; Sarhan and Perras, 1987). In the course of later work these authors identified an mRNA group which only functioned during the cold effect and coded proteins synthesised only in frost resistant varieties. This suggests the existence of a positive correlation between the quantity of proteins synthesised due to the cold effect and the frost resistance of the varieties (Chauvin et al., 1993).

The effect of low temperature on the nucleic acid metabolism of wheat has long been the subject of research in Martonvásár. In the course of this work it has been demonstrated that cold-induced rRNA synthesis takes place in seedlings as the result of low temperature during the first few days of cold treatment (Dévay and Páldi, 1977; Páldi and Dévay, 1977). It was established that the extent of cold-induced rRNA synthesis is closely correlated to the rRNA cistron number (Páldi and Dévay, 1983). Further examinations proved the existence of a close correlation between cold-induced synthesis and the frost resistance of wheat, though the cold treatment only exerted an inductive effect in winter wheat (Dévay and Páldi, 1977). Results obtained in other fields showed that low temperature leads to changes in the minor base composition of the cytoplasmic RNAs and even in the isoacceptor spectrum of certain tRNAs. The aim of the present research was to determine whether low temperature has any effect on rRNA processing and if so, how this is expressed.

Materials and Methods

Plant material

The two lines were selected by scientists from the Wheat Breeding Department of the institute from a population of the variety Martonvásári 6, derived from a cross between Bezostaya 1, which has excellent frost resistance, and the French variety Moisson, which is poorly resistant to frost. For the purposes of biochemical analyses, Dévay and co-workers further strengthened or weakened the frost resistance characters of the two lines by selecting the weakly frost-resistant line Mv 13–74 for frost sensitivity and the frost-resistant line Mv 11–75 for frost resistance over a number of years.

Growth and labelling of wheat seedlings

The seeds were germinated under sterile conditions on 1 % agar containing 2 % sucrose at room temperature for 48 h in darkness. After cold treatment (1, 3, 5 and 8 days) the intact seedlings were incubated with their root tips resting in 5 mL 1000-fold diluted Knop solution per 50 seedlings, containing 0.5 mCi $^{32}P_i$, for 14 h at 3 °C.

Preparation of RNA

The wheat seedlings were homogenised in buffered sucrose medium (Loening, 1965) and the homogenate was centrifuged at 1000 g for 5 min. This sedimented the cell debris and nuclear material and left the bulk of the cytoplasmic fractions in suspension. The debris fraction was resuspended in the same medium. Sodium dodecyl sulphate (0.5 %) and sodium 4-aminosalicylate (5 %, w/v) were added to both fractions (Kirby, 1965) and the suspensions were shaken with phenol containing 0.1 % (w/v) 8-hydroxyquinoline at 0–5 °C. The phases were separated by centrifugation at 1000 g for 10 min and the phenol extraction was repeated once or twice. The RNA was precipitated from the final supernatant by the addition of 2 % (w/v) sodium acetate and 2.5 vol. of ethanol at –20 °C. The final RNA precipitate was reprecipitated from twice 0.3 M sodium acetate, washed once with ethanol and partially dried for a few minutes *in vacuo*. It was then dissolved in electrophoresis buffer containing 5 % (w/v) sucrose to give a final concentration between 0.5 and 2.0 mg/mL. The purified RNA was fractionated by electrophoresis on 2.4 % polyacrylamide gel (50 V for 3.5 h). The distribution of radioactivity was determined by freezing the gels in dry ice to scanned length prior to cutting 0.5 mm slices, which were dried on filter paper. The slices were counted in a liquid scintillation spectrometer (Loening, 1967; Loening and Ingle, 1967; Loening, 1969).

Digestion of DNA

20 µg/mL of electrophoretically purified DNAse and 40 µL/mL of 1 M $MgCl_2$ were added to the nucleic acid solution (100 µg/mL in 0.05 M Tris-HCl, pH 7.2) and incubated at 20 °C for 20 min. After incubation the solution was deproteinized by shaking for 15 min with an equal volume of a 24 : 1 mixture of $CHCl_3$ and 3-methylbutan-1-ol. The top aqueous layer was removed after centrifugation (2500 g for 15 min, 20 °C) and further deproteinized as described above. The DNA-free RNA was precipitated by the addition of two vol. of EtOH (Wells and Ingle, 1970).

Estimation of frost tolerance

For the purpose of the frost tests, seedlings of the various wheat varieties were raised for two weeks in an earth-sand mixture (3 : 1). Freezing was carried out in programmed freezing chambers, where the temperature gradient of cooling and warming was 0.5 °C an hour. In order to determine the LT_{50} value the 14-day-old seedlings were first hardened for 5 days at 2 °C, after which the plants were divided into as many parts as there were freezing temperatures (–5, –10, –15, –18, –20 and –25 °C). The seedlings were kept at the relevant freezing temperatures for 48 hours, then raised for a further 14 days at 15 °C (16 h light, 8 h dark, 70 % relative humidity) in a phytotron unit (Conviron, type: PGV-36). The survival percentage was then determined as a function of the freezing temperatures applied. On the basis of the graphs obtained, the critical temperature was taken to be that at which 50 % of the plants survived. The mean results of three biological replications were taken as the basis for determining the LT_{50} values.

Results

In order to study the rRNA maturation process, two wheat lines were chosen which differed, as the result of selection, chiefly as regards the degree of frost resistance. These lines were Mv 11–75, which has good frost resistance (LT_{50}: –16.8 °C), and Mv 13–74, which has poor resistance (LT_{50}: –3.1 °C). In the course of the experiments, studies were made on the effect of cold induction on the relative proportions of rRNA precursors. Etiolated 72-hour-old seedlings were exposed to cold treatment at 3 °C for various lengths of time (1, 3, 5 and 8 days). After the incorporation of $^{32}P_i$ through the roots the rRNAs were purified and then separated by means of PAGE. Their radioactivity was then determined using the liquid scintillation method.

It could be seen from the data that, as the result of cold treatment (for example after 5 days), the proportion of 2.9 MD/2.3 MD high molecular weight rRNA precursors in the line Mv 13–74 shifted in favour of the latter fraction, while the proportion of 1.4 MD/1.3 MD and 0.9 MD/0.7 MD rRNA precursors shifted in favour of the former fractions as treatment continued (Table 1). In the frost-resistant line Mv 11–75 the 1.4 MD and 0.9 MD rRNA precursors could not be demonstrated during the treatment. The 2.9 MD/2.3 MD ratio was practically constant at a value above 2 (Table 2). The change in the relative proportion of precursors as cold treatment proceeded in the poorly frost-resistant line suggests the retardation of the rRNA maturation process. By contrast, in the frost-resistant line the absence of 1.4 MD and 0.9 MD precursors indicates the uninterrupted synthesis of stable 1.3 MD and 0.7 MD cytoplasmic rRNAs.

Discussion

The results mentioned in the introduction suggest that low temperature must also affect rRNA processing.

Contrary to our normal practice, varieties were not used in this experiment, since they always consist of a mixture of several lines and are thus genetically heterogeneous. In order to pinpoint the correlation between the changes assumed to take place due to the cold effect and the frost resistance of the

Table 1: Effect of low temperature on rRNA processing in the wheat line Mv 13-74 ($LT_{50} = -3.1\,°C$).

Lenght of cold treatment (day)	Incorporation of $^{32}P_i$ into the cytoplasmic rRNAs and their precursors					
	2.9	2.3	1.4 (million dalton)	1.3	0.9	0.7
Control	13,167	5,981	–	24,173	–	16,822
3	11,052	9,168	4,636	21,002	3,451	14,143
5	7,871	9,095	14,825	12,964	13,794	6,075
8	7,659	9,877	20,464	5,598	14,268	3,699
$LSD_{5\%}$	850	442	1,200	949	1,092	813
$LSD_{1\%}$	1,192	619	1,682	1,331	1,531	1,140

1) Molecular weights of stable cytoplasmic rRNAs are 1.3×10^6 and 0.7×10^6 dalton.
2) Incorporation of $^{32}P_i$ into the different rRNA precursors at 3 °C for 24 h. 50 µg total rRNA was separated by PAGE (2.4 %, 50 V, 3.5 h).
3) LT_{50} = critical temperature characteristic of the actual frost hardiness of the wheat line.
4) Data are the average of five experiments.

Table 2: Effect of low temperature on rRNA processing in the wheat line Mv 11-75 ($LT_{50} = -16.8\,°C$).

Lenght of cold treatment (day)	Incorporation of $^{32}P_i$ into the cytoplasmic rRNAs and their precursors					
	2.9	2.3	1.4 (million dalton)	1.3	0.9	0.7
Control	13,908	6,558	–	26,115	–	15,644
3	12,956	5,766	–	22,958	–	14,173
5	12,214	5,658	–	21,814	–	15,006
8	11,482	5,059	–	20,988	–	14,319
$LSD_{5\%}$	1,228	675	–	2,367	–	1,232
$LSD_{1\%}$	1,722	947	–	3,318	–	1,728

1) Molecular weights of stable cytoplasmic rRNAs are 1.3×10^6 and 0.7×10^6 dalton.
2) Incorporation of $^{32}P_i$ into the different rRNA precursors at 3 °C for 24 h. 50 µg total rRNA was separated by PAGE (2.4 %, 50 V, 3.5 h).
3) LT_{50} = critical temperature characteristic of the actual frost hardiness of the wheat line.
4) Data are the average of five experiments.

given wheat genotype characterised by the LT_{50} value, the rRNA maturation process was studied in two lines, chosen because they differed, as the result of selection, chiefly as regards the degree of frost resistance. These lines were Mv 11–75, which has good frost resistance (LT_{50} value: −16.8 °C), and Mv 13–74, which has poor resistance (LT_{50} value: −3.1 °C). The results indicate that as the result of low temperature both quantitative and qualitative changes occur in the wheat line with poor frost resistance. This is obvious not only from the mutual ratio of the two high molecular weight precursors, but more importantly from the appearance in ever larger quantities of the last precursors of the two stable cytoplasmic rRNA fractions (1.3 and 0.7×10^6 dalton). These precursors cannot be demonstrated in the line with good frost resistance.

The results suggest that a reduction in frost resistance disturbs the rRNA maturation process, as a consequence of which there is an increase in the last precursors (1.4 and 0.9 MD) of the two stable cytoplasmic rRNAs. It should also be noted that the demonstration of the 0.9 MD precursor has only been successful in a few plant species (e.g. onion, blue algae, etc.) due to its high rate of decomposition. These results make a small contribution to the better understanding of the mechanism of the low temperature effect. Further studies will be required to shed light on the finer details of how rRNA processing is retarded in wheat genotypes with weak frost resistance.

The connection of nucleic acids with polyamines and protein synthesis is well known. The results obtained in the present work confirm the findings of investigations on the effect of low temperature in wheat at the polyamine (Páldi et al., 1994) and more recently at the protein synthesis level (Lasztity et al., 1994).

Acknowledgements

We are grateful for generous support from the National Scientific Research Fund of Hungary (OTKA, I/2 1122) to E. Páldi.

References

Danyluk, J. and F. Sarhan: Differential mRNA transcription during the induction of freezing tolerance in spring and winter wheat. Plant Cell Physiol. *31* (5), 609–619 (1990).

Dévay, M. and E. Páldi: Cold-induced rRNA synthesis in wheat cultivars during the hardening period. Plant Sci. Lett. *8*, 191–195 (1977).

Guy, C. L., K. J. Niemi, and R. Brambl: Altered gene expression during cold acclimation of spinach. Proc. Natl. Acad. Sci. USA *82*, 3673–3677 (1985).

Kirby, K. S.: Isolation of nucleic acids with phenolic solvents. In: Grossman, L. and K. Moldave (eds.): Methods in Ezymology, Vol. *XIIB*, pp. 87–89. Academic Press, New York and London (1968).

Lasztity, D., I. Rácz, and E. Páldi: Activity of protein synthesising system in vernalised wheat seedlings at low temperature. In: Dörffling, K., B. Brettschneider, H. Tantau, and K. Pithan (eds.): Agriculture. Crop adaptation to cool climates, pp. 349–354. ECSP-EEC-EAEC, Brussels, Luxembourg (1994).

Loening, U. E.: The fractionation of high molecular weight ribonucleic acid by polyacrylamide gel electrophoresis. Biochem. J. *102*, 251–257 (1967).

Loening, U. E. and J. Ingle: Diversity of RNA components in green plant tissues. Nature *215*, 363–367 (1967).

Loening, U. E.: The determination of the molecular weight of ribonucleic acid by polyacrylamide gel electrophoresis. The effects of changes in conformation. Biochem. J. *113*, 131–138 (1969).

Mohapatra, S. S., R. J. Poole, and R. S. Dhindsa: Changes in protein patterns and translatable messenger RNA populations during cold acclimation of Alfalfa. Plant Physiol. *84*, 1172–1176 (1987).

Páldi, E. and M. Dévay: Characteristics of the rRNA synthesis taking place at low temperatures in wheat cultivars with varying degrees of frost hardiness. Phytochemistry *16*, 177–179 (1977).

– – Relationship between the cold-induced rRNA synthesis and the rRNA cistron number in wheat cultivars with varying degrees of frost hardiness. Plant Sci. Lett. *30,* 61–67 (1983).

PÁLDI, E., I. RÁCZ, and D. LASZTITY: Studies on the correlation between the quantity of 1.4 MD rRNA precursor and frost resistance. In: DÖRFFLING, K., B. BRETTSCHNEIDER, H. TANTAU, and K. PITHAN (eds.): Agriculture. Crop adaptation to cool climates, pp. 341–348. ECSP-EEC-EAEC, Brussels, Luxembourg (1994).

PÁLDI, E., G. SZALAI, I. RÁCZ, and D. LASZTITY: Polyamine biosynthesis during vernalisation in winter and spring wheat varieties. In: DÖRFFLING, K., B. BRETTSCHNEIDER, H. TANTAU, and K. PITHAN (eds.): Agriculture. Crop adaptation to cool climates, pp. 195–204. ECSP-EEC-EAEC, Brussels, Luxembourg (1994).

PERRAS, M. and F. SARHAN: Synthesis of freezing tolerance proteins in leaves, crown and roots during cold acclimation of wheat. Plant Physiol. *89,* 577–585 (1989).

WELLS, R. and J. INGLE: The constancy of the buoyant density of chloroplast and mitochondrial deoxyribonucleic acids in a range of higher plants. Plant Physiol. *46,* 178–179 (1970).

J. Plant Physiol. Vol. 148. pp. 378–383 (1996)

Role of Light in the Development of Post-chilling Symptoms in Maize

G. SZALAI[1], T. JANDA[1], E. PÁLDI[1], and Z. SZIGETI[2]

[1] Agricultural Research Institute of the Hungarian Academy of Sciences, H-2462 Martonvásár, P.O.B. 19, Hungary

[2] Department of Plant Physiology, Eötvös Loránd University, H-1445 Budapest, P.O.B. 330, Hungary

Received June 24, 1995 · Accepted October 5, 1995

Summary

The effect of light (250 µmol m^{-2} s^{-1}) on the appearance of post-chilling symptoms was investigated in cold-treated maize (*Zea mays* L. hybrid Furio) seedlings using electrolyte leakage and chlorophyll fluorescence induction measurements as indicators. The longer the cold pretreatment (0.5 °C) in complete darkness, the more rapid the decrease in F_v/F_m and the increase in electrolyte leakage during cold treatment in the light. The most important difference in the changes in these two parameters is that the changes in F_v/F_m occur much earlier if the cold treatment is carried out in the light. These findings suggest that chilling stress in cold sensitive maize plants led to an increased susceptibility to photoinhibition at low temperatures. F_v/F_m and electrolyte leakage changed not only at low temperatures but also after a certain cold pretreatment period at normal temperature. When the seedlings were returned to 25 °C after various chilling periods in the dark both parameters showed that post-chilling symptoms appeared much more rapidly in the light than in the dark. By contrast to the change in F_v/F_m, where plants chilled for only two days exhibited differences in post-chilling changes in the light and dark, a substantial increase in electrolyte leakage was only observed after four days of cold pretreatment. These results suggest that photoinhibition has a role not only during the chilling period, but also in the appearance of post-chilling symptoms.

Key words: Zea mays L., chilling, ion leakage, photoinhibition, post-chilling symptoms.

Abbreviations: F_v and F_m = variable and maximum chlorophyll fluorescence in the dark adapted state, respectively; F_o = initial chlorophyll fluorescence in the dark adapted state; PS II = photosystem II.

Introduction

Chilling-sensitive plants, such as maize, may suffer severe damage when exposed to low temperature (Öquist and Martin, 1986; Raison and Lyons, 1986). It was suggested that at low temperature membrane lipids undergo a phase change from fluid to solid phase, which results in conformational changes and higher activation energy in membrane-bound enzymes and in dysfunction of the membranes (Lyons, 1973; Raison, 1974). This phase transition of lipids was considered the primary event of chilling injury (Lyons et al., 1979). Chilling injury may cause an increase in the rate of respiration and electrolyte leakage (Creencia and Bramlage, 1971).

Low temperature also has an effect on the photosynthetic apparatus (Öquist, 1983) in cold sensitive plants. There is evidence that the chilling-induced depression of photosynthesis has an important role in the limiting of early growth in maize (Miedema, 1982; Ort and Baker, 1988; Stirling et al., 1993). A decrease in the photosynthetic activity of maize plants was observed in the field when the temperature was below 12 °C (Farage and Long, 1987). After low temperature treatment a decrease in the apparent quantum yield of CO_2 assimilation (Long et al., 1983) and a decrease in the photosynthetic oxygen evolution (Baker et al., 1983) were observed. In another study it was shown that in maize growing below 13 °C there is an alteration in pigment content and arrangement, and an inhibition of the electron flow (Csapó et al., 1991).

These adaptation processes depend not only on the ambient temperature but also on the irradiance (Long, 1983; Hetherington et al., 1989). In a previous paper we reported that the higher the irradiance and the lower the temperature the more pronounced changes occurred in certain chlorophyll fluorescence parameters (Janda et al., 1994).

It is well known that the exposure of plants to irradiance higher than that required for the saturation of photosynthesis results in different responses by the photosynthetic apparatus. There is evidence that the major site of the response is located in PS II (Kirilovsky et al., 1990; Kyle et al., 1984), probably either at the Q_A (Allakhverdiev et al., 1987; Styring et al., 1990; Vass et al., 1988, 1992) or at the Q_B level of the acceptor site (Kyle, 1987; Ohad et al., 1990; Janda et al., 1992) (for review see Aro et al., 1993). This phenomenon is known as photoinhibition.

It is well established that all factors which cause a decrease in the photosynthesizing ability (low temperature, drought, etc.) increase the level of photoinhibition (Öquist, 1987). Especially in the case of thermophilic plants (for example maize, which originates from the subtropical region) even a relatively moderate light intensity may lead to photoinhibitory damage at chilling temperatures, while certain types of damage do not appear in the dark. Some of the damage symptoms occurring in maize plants due to low temperature appear only after the plants are returned to a non-chilling temperature (Hetherington and Öquist, 1988). It is not known, however, whether photoinhibition plays any role in the appearance of post-chilling symptoms.

The aim of the present study was to examine the role of light, not only during the chilling period but also after the low temperature treatment, in the appearance of post-chilling symptoms under controlled growth conditions using two separate methods, the chlorophyll fluorescence induction method and the electrolyte leakage measurement technique.

Materials and Methods

Plant material

Sterilized seeds of maize (*Zea mays* L. hybrid Furio) were sown (one in a pot) in sterilized soil consisting of a mixture of loamy soil, Vegasca (humus-containing additive, manufactured by Florasca) and sand (2:2:1, v:v:v). The third fully expanded leaves of young (2–3 week old) maize plants were used in the experiments. The seedlings were grown in a Conviron PGV-36 chamber in the phytotron of the Agricultural Research Institute of the Hungarian Academy of Sciences, Martonvásár. The temperature was 25/23 °C with a 16/8-hour light/dark periodicity, the irradiance at the level of the leaves was a constant 250 µmol $m^{-2} s^{-1}$ using metal halide lamps, and the relative humidity was 75 %. The cold treatment was carried out in a chamber of the same type.

Cold treatment

Various cold treatments and preparatory treatments were applied in order to study the role of light in chilling and post-chilling symptoms.

Cold treatment 1. Effect of light during chilling

The plants were kept in the dark at 0.5 °C for 0, 2 or 4 days, after which F_v/F_m and electrolyte leakage were measured in the light at 0.5 °C immediately and after 24 h.

Cold treatment 2. Effect of light on the development of post-chilling symptoms

2/A. After preliminary treatment at 0.5 °C in the dark for 2 or 4 days the plants were transferred to 25 °C, some in the light and some in the dark to discover where and when the post-chilling symptoms were most pronounced.

2/B. The plants were kept in the dark at 0.5 °C for 2 or 4 days, after which they were exposed to a temperature of 25 °C for 1, 2 or 3 days, still in the dark. Following this preliminary treatment the plants were exposed to light, still at 25 °C, in order to determine whether post-chilling symptoms developed, and to what extent.

Chlorophyll fluorescence measurements

The chlorophyll fluorescence induction parameters of the leaves were determined at room temperature using a pulse amplitude modulated (PAM) fluorometer (PAM-2000, Walz, Effeltrich, Germany). The saturating light intensity was 2500 µmol $m^{-2} s^{-1}$. Before the measurements, the plants were dark-adapted for 30 min. For the chlorophyll fluorescence induction parameters the nomenclature of van Kooten & Snel (1990) was used. The results are the means of 6–8 replications, and were statistically evaluated using the standard deviation and T-test methods.

Electrolyte leakage measurements

Leaf disks cut to a size of 9 mm in diameter were placed two to a vial in 1.5 mL ultrapure water (distilled water was purified using a Milli-Q 50 water purification system) and were shaken for 1 h in the dark. Conductivity was measured using an Automatic Seed Analyzer (ASA610, Agro Sciences, Inc. USA). The results were the means of 20 replications, and were statistically evaluated using the standard deviation and T-test methods.

Results

Effects of chilling injury on the sensitivity of chilling-induced photoinhibition in maize

Young maize plants were chill-treated at 0.5 °C in the dark and in the light. To determine changes in the PS II efficiency, the F_v/F_m parameter was measured, which is the most reliable marker for photoinhibition (Krause, 1988). The decrease in this parameter was much more pronounced when the cold

Table 1: Changes in the variable fluorescence ratio (F_v/F_m) in 2-week-old maize seedlings during cold treatment (0.5 °C) in the light (250 µmol $m^{-2} s^{-1}$) carried out after a cold pretreatment in the dark (cold treatment 1) (n = 6–8; *: significant at $p \leq 0.001$).

Cold pre-treatment in the dark (days)	F_v/F_m values Cold treatment in the light 0 h	 7 h	 24 h
0	0.798 (±0.010)	0.489 (±0.029)*	0.240 (±0.095)*
2	0.784 (±0.009)	0.385 (±0.018)*	0.048 (±0.023)*
4	0.725 (±0.024)	0.186 (±0.033)*	0.028 (±0.009)*

treatment was carried out in the light (Table 1) than in the dark and was due to a slight increase in the F_o and a severe decrease in the F_m. Although the F_v/F_m parameter was hardly affected by low temperature (0.5 °C) in the dark even after several days, exposing the pre-chilled plants to low temperature in the light accelerated the effect of photoinhibition induced by low temperature. It can be seen from Table 1 that the longer the cold pretreatment in complete dark, the greater the extent of the decrease in F_v/F_m in the light.

Electrolyte leakage measurement is a useful tool for the characterization of the state of membranes (Creencia and Bramlage, 1971). Certain stresses may cause changes in the structure of the membranes, which may in turn lead to the leakage of ions from the cells. Table 2 shows that a 24 h cold treatment in the light without cold pretreatment did not cause an increase in the electrolyte leakage (indeed there was a slight decrease), but after various periods of dark cold pretreatment during which this parameter hardly changed even after 4 days, the electrolyte leakage increased after exposing the plants to light cold conditions. As in the case of F_v/F_m the longer the cold pretreatment in the dark, the more pronounced was the change occurring in this parameter. The most important difference in the changes in these two parameters is that the changes in F_v/F_m occur much earlier if the cold treatment is carried out only in the light.

Role of irradiance in the appearance of post-chilling symptoms

Figure 1 shows the changes in the electrolyte leakage of young maize plants during dark cold treatment (0.5 °C) and after returning them to a warm, non-chilling temperature (25 °C) in the dark and in the light (cold treatment 2/A). After one day at low temperature in the dark there was a slight decrease in electrolyte leakage, after which it started to increase slowly. A sudden increase occurred only after the 4th day. When the plants were returned to a non-chilling temperature after the 2nd day, changes in this parameter were halted, independently of the irradiance. However, if the plants were returned to the warm after the 4th day of dark cold treatment, an increase in electrolyte leakage occurred. The level of this increase depended on the irradiance: it was much more pronounced in the light. In some cases an increase in the anthocyanin content was observed during cold treatment. There was no change in the anthocyanin content during dark cold treatment carried out at 0.5 °C even after 4 days, and also none in the case of plants treated first to dark cold and then to dark warm (25 °C) conditions. However,

Table 2: Changes in the electrolyte leakage in 2-week-old maize seedlings during cold treatment (0.5 °C) in the light (250 µmol m^{-2} s^{-1}) carried out after a cold pretreatment in the dark (cold treatment 1) (n = 20; **: significant at p≤0.001).

Cold pretreatment in the dark (days)	Electrolyte leakage (relative units)	
	Cold treatment in the light	
	0 h	24 h
0	36.5 (±4.2)	28.8 (± 3.5)**
2	29.9 (±3.0)	43.0 (± 5.4)**
4	35.1 (±4.9)	97.6 (±16.2)**

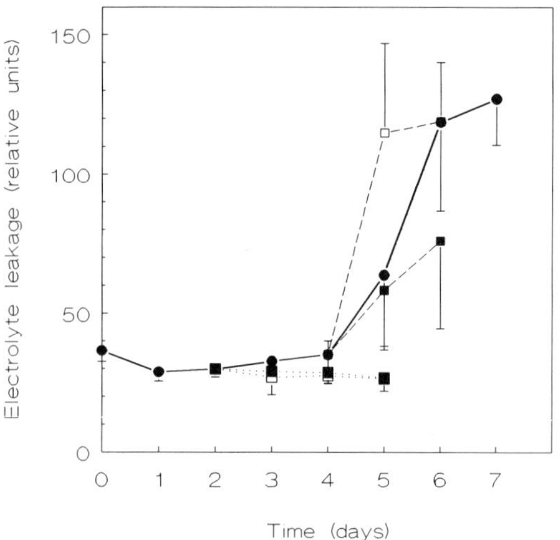

Fig. 1: Changes in electrolyte leakage during cold treatment (0.5 °C) in the dark in two-week-old maize seedlings. The plants were kept for seven days at 0.5 °C in the dark (solid line) or shifted back to 25 °C after a 2 day (·····) or 4 day (- - - -) chilling period and then kept in the dark (■) or in the light (250 µmol m^{-2} s^{-1}) (□) (cold treatment 2/A).

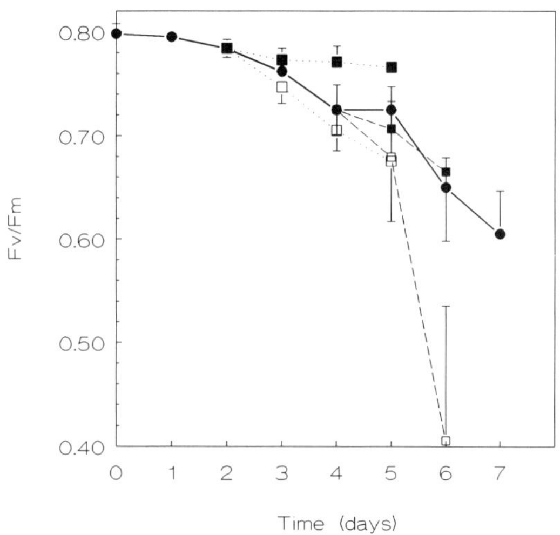

Fig. 2: Changes in the variable chlorophyll fluorescence ratio (F_v/F_m) during cold treatment (0.5 °C) in the dark in two-week-old maize seedlings. The plants were kept for seven days at 0.5 °C in the dark (solid line) or shifted back to 25 °C after a 2 day (·····) or 4 day (- - - -) chilling period and then kept in the dark (■) or in the light (250 µmol m^{-2} s^{-1}) (□) (cold treatment 2/A).

when the plants were returned to a non-chilling temperature in the light after 2 days in cold dark conditions, an increase in the anthocyanin content occurred. This increase was very moderate after one day, but dramatic after the 2nd day. There was also a slight increase in plants treated to 2 days cold, 1 day warm in the dark and then 1 day warm in the light.

Figure 2 shows that under dark cold conditions a sudden decrease in the F_v/F_m parameter only occurred after the 5th day, but this decrease was less than that which occurred within 7 h at 0.5 °C in the light (Table 1). When the plants

Table 3: Changes in electrolyte leakage values in 2-week-old maize seedlings after returning maize seedlings pretreated in different ways to a higher temperature in the light (250 µmol m^{-2} s^{-1}) (cold treatment 2/B) [D = dark; L = light; W = warm (25 °C); C = cold (0.5 °C) n = 20; ns: not significant at $p \leq 0.05$; *: significant at $p \leq 0.001$].

Pretreatment	Electrolyte leakage (relative units)	
	0 h LW	24 h LW
2 d DC+1 d DW	29.0 (± 2.7)	27.4 (± 3.0)^{ns}
2 d DC+2 d DW	28.6 (± 3.7)	28.0 (± 2.9)^{ns}
2 d DC+3 d DW	26.8 (± 4.8)	60.7 (±29.3)*
4 d DC+1 d DW	58.1 (±20.2)	116.0 (±24.4)*

Table 4: Changes in the variable fluorescence ratio (F_v/F_m) values in 2-week-old maize seedlings after returning maize pretreated in different ways to a higher temperature in the light (250 µmol m^{-2} s^{-1}) (cold treatment 2/B) [D = dark; L = light; W = warm (25 °C); C = cold (0.5 °C); n = 6–8; *: significant at $p \leq 0.05$; **: significant at $p \leq 0.001$].

Pretreatment	F_v/F_m values	
	0 h LW	24 h LW
2 d DC+1 d DW	0.773 (±0.011)	0.719 (±0.014)**
2 d DC+2 d DW	0.771 (±0.015)	0.731 (±0.009)**
2 d DC+3 d DW	0.766 (±0.007)	0.512 (±0.252)*
4 d DC+1 d DW	0.707 (±0.023)	0.361 (±0.296)*

were returned to a non-chilling temperature (cold treatment 2/A) after the 2nd day of dark cold treatment the decrease in the F_v/F_m parameter halted only in the dark. In the light this decrease was faster than the decrease under dark cold conditions. After a longer dark cold pretreatment this difference was greater. When returned to warm temperature in the light after 6 days at 0.5 °C, F_v/F_m decreased to zero within 7 h, while in the dark this value was 0.487. It can be concluded from these data that post-chilling symptoms appear more slowly in the dark.

Next, the effect of light on changes in the electrolyte leakage and F_v/F_m was examined in maize plants pretreated to dark cold and dark warm conditions (cold treatment 2/B). It can be seen from Table 3 that when maize plants cold treated for two days in the dark were kept in the dark at 25 °C for 1 or 2 days no change was then observed in the electrolyte leakage after they were returned to the light at 25 °C. However, there was an increase in electrolyte leakage after 3 days under dark warm conditions. As was shown above, when maize chilled for 4 days was returned to 25 °C the appearance of post-chilling symptoms was less dramatic in the dark. However, if the plants are returned to the light after 1 day of dark pretreatment, the light may cause damage symptoms to appear (Table 3).

Similar results were obtained from measurements of the F_v/F_m parameter, with the difference that there was also a decrease in this parameter in the case of maize plants chilled for 2 days when they were returned to the light after 1–3 days in the dark at 25 °C (Table 4).

Discussion

Chilling injury has an important role in limiting the production of maize over a wider area. It is well known that at low temperature in the case of cold-sensitive plants photoinhibition plays an important role in the damage (Yakir et al., 1986; Hetherington and Smillie, 1989; Janda et al., 1994). The cultivation of cold-resistant plants at low, but not freezing temperatures leads to an increased resistance to photoinhibition, as was demonstrated for spinach (Boese and Huner, 1990; Somersalo and Krause, 1989) and winter rye (Öquist and Huner, 1991). However, as was demonstrated in this study, in the case of maize growing at low temperature (0.5 °C) the susceptibility of the plant to low temperature-induced photoinhibition increased.

It is known that some damage symptoms cannot be seen at low temperature, but only when the plants are returned to a non-chilling temperature after the cold treatment. Compared with non-chilled seedlings, low temperature-treated plants showed depressed rates of growth, depending on the chilling period, and there was a decrease in the maximal rate of rise of the induced chlorophyll fluorescence (F_R) (Smillie et al., 1987; Hetherington and Öquist, 1988).

The aim of the present work was to examine whether photoinhibition had any role in the appearance of post-chilling symptoms. The F_v/F_m ratio indicates the photochemical efficiency of PS II, and a decrease in this parameter is the most reliable sign of photoinhibition (Krause, 1988). In the case of cold-treated plants, a return to 25 °C caused a decrease in the F_v/F_m ratio in spite of the fact that this parameter hardly changed during cold treatment in complete darkness. This decline depended on the ambient light: it was much more pronounced in the light than in the dark. This suggests that the cold-treated cold-sensitive maize plants became more light-sensitive. The longer the cold treatment in the dark, the more pronounced was the difference between plants returned to a non-chilling temperature in the light and in the dark as regards the appearance of post-chilling symptoms. Although the primary effect of photoinhibition at low temperatures seems to be a reduction in the efficiency of PS II, the damage is not confined to the photosynthetic apparatus. Another rapidly occurring effect is increased leakage of ions from chilled tissue, presumably due to increased membrane permeability of the tissues (Creencia and Bramlage, 1971). By contrast to the change in F_v/F_m, where plants chilled for only two days exhibited differences in post chilling changes in the light and dark, a substantial increase in ion leakage was only observed after four days of cold pretreatment. Similarly to F_v/F_m, the changes in the electrolyte leakage after cold pretreatment were much more pronounced in the light than in the dark.

Although during cold treatment 2/A plants kept throughout in the dark exhibited a reduction in F_v/F_m and an increase in electrolyte leakage, no visible damage to the plants was observed. Cold treatment 2/B was devised to discover whether these apparently healthy plants were capable of further growth, for which light was required. It was found that plants kept at 0.5 °C in the dark for only 2 days were capable of compensating for cold-induced damage. After cold dark treatment for 4 days, when the plants were transferred to 25 °C, those kept in the dark exhibited a much smaller reduction in F_v/F_m and increase in ion leakage than those kept in the light, but when later exposed to light these too died within a short time. In other words, cold treatment at 0.5 °C for 4 days caused permanent damage to the plants which could not be compensated for. If the cold treatment was carried out not in the dark, but in the light, the plants died within 2 days.

Another surprising phenomenon, observed only when plants cold-treated for 2 days were exposed to light immediately on transfer to 25 °C or within a day, was the appearance of anthocyanin pigmentation. In agreement with data presented by Christie et al. (1994), no change could be detected in the anthocyanin level at low temperature (0.5 °C); however, a dramatic increase in the pigment content was found after shifting to 25 °C if the low temperature treatment was carried out in complete darkness. These results suggest that the impairment of the transcriptional and translational processes important for anthocyanin biosynthesis is due not only to chilling, but also to light stress.

These results suggest that photoinhibition has a role not only during the chilling period, but also in the appearance of post-chilling symptoms.

Acknowledgements

We are gratefully indebted to Edit Kövesdi and Zsuzsa Kóti for their technical assistance and to Dr. C. Marton for providing the maize seeds. We wish to thank Prof. H. K. Lichtenthaler for his critical reading of the manuscript. Thanks are also due to Barbara Harasztos for revising the manuscript linguistically. This work was supported by grants from the Hungarian National Scientific Research Foundation (OTKA 140, OTKA 2667, F 016001, W 015469), which are gratefully acknowledged.

References

Allakhverdiev, S., E. Setlikova, V. V. Klimov, and J. Setlik: In photoinhibited photosystem II particles pheophytin photoreduction remains unimpaired. FEBS Letters *226,* 186–190 (1987).

Aro, E.-M., I. Virgin, and B. Andersson: Photoinhibition of Photosystem II. Inactivation, protein damage and turnover. Biochim. Biophys. Acta *1143,* 113–134 (1993).

Baker, N. R., T. M. East, and S. P. Long: Chilling damage to photosynthesis in young *Zea mays.* J. Exp. Bot. *34,* 189–197 (1983).

Boese, S. R. and N. P. A. Huner: Effects of growth temperature and temperature shifts on spinach leaf morphology and photosynthesis. Plant Physiol. *94,* 1830–1836 (1990).

Christie, P. J., M. R. Alfenito, and V. Walbot: Impact of low temperature stress on general phenylpropanoid and anthocyanin pathways: Enhancement of transcript abundance and anthocyanin pigmentation in maize seedlings. Planta *194,* 541–549 (1994).

Creencia, R. P. and W. J. Bramlage: Reversibility of chilling injury to corn seedlings. Plant Physiol. *47,* 389–392 (1971).

Csapó, B., J. Kovács, E. Páldi, and Z. Szigeti: Fluorescence induction characteristics of maize inbred lines after long-term chilling treatment during the early phase of development. Photosynthetica *25,* 575–582 (1991).

Farage, P. K. and S. P. Long: Damage to maize photosynthesis in the field during periods when chilling is combined with high photon fluxes. In: Biggins, J. (ed.): Progress in Photosynthesis Research, Vol. IV. pp. 139–142. Martinus Nijhoff, Dordrecht (1987).

Hetherington, S. E. and G. Öquist: Monitoring chilling injury: A comparison of chlorophyll fluorescence measurements, post-chilling growth and visible symptoms of injury in *Zea mays.* Physiol. Plantarum *72,* 241–247 (1988).

Hetherington, S. E., J. He, and R. M. Smillie: Photoinhibition at low temperature in chilling-sensitive and -resistant plants. Plant Physiol. *90,* 1609–1615 (1989).

Janda, T., G. Szalai, J. Kissimon, E. Páldi, C. Marton, and Z. Szigeti: Role of light intensity in the chilling injury of young maize plants studied by chlorophyll fluorescence induction measurements. Photosynthetica *30,* 293–299 (1994).

Janda, T., W. Wiessner, E. Páldi, D. Mende, and S. Demeter: Thermoluminescence investigation of photoinhibition in the green alga *Chlamydobotrys stellata* and in *Pisum sativum* L. leaves. Z. Naturforsch. *47c,* 585–590 (1992).

Kirilovsky, D., J.-M. Ducruet, and A.-L. Etienne: Primary events occurring in photoinhibition in *Synechocystis* 6714 wild type and atrazine-resistant mutant. Biochim. Biophys. Acta *1020,* 87–93 (1990).

Krause, G. H.: Photoinhibition of photosynthesis. An evaluation of damaging and protective mechanisms. Physiol. Plantarum *74,* 566–574 (1988).

Kyle, D. J.: The biochemical basis for photoinhibition of photosystem II. In: Kyle, D. J., C. B. Osmond, and C. J. Arntzen (eds.): Photoinhibition, pp. 197–226. Elsevier, Amsterdam (1987).

Kyle, D. J., I. Ohad, and C. J. Arntzen: Membrane protein damage and repair: selective loss of a quinone-protein function in chloroplast membranes. PNAS, USA *81,* 4070–4074 (1984).

Long, S. P.: C_4 photosynthesis at low temperature. Plant Cell Env. *6,* 345–363 (1983).

Long, S. P., T. East, and N. R. Baker: Chilling damage to photosynthesis in young *Zea mays* I. Effects of light and temperature variation on photosynthetic CO_2 assimilation. J. Exp. Bot. *34,* 177–188 (1983).

Lyons, J. M.: Chilling injury in plants. Ann. Rev. Plant Physiol. *24,* 445–466 (1973).

Lyons, J. M., J. K. Raison, and P. L. Steponkus: The plant membrane in response to low temperature: an overview. In: Lyons, J. M., D. Graham, and J. K. Raison (eds.): Low Temperature Stress in Crop Plants, pp. 1–24. Academic Press, New York (1979).

Miedema, P.: The effects of low temperature on *Zea mays.* Adv. in Agronomy *35,* 93–128 (1982).

Ohad, I., N. Adir, H. Koike, D. J. Kyle, and Y. Inoue: Mechanism of photoinhibition *in vivo.* J. Biol. Chem. *265,* 1972–1979 (1990).

Ort, D. R. and N. R. Baker: Consideration of photosynthetic efficiency at low light as a major determinant of crop photosynthetic performance. Plant Physiol. Biochem. *26,* 555–565 (1988).

Öquist, G.: Effects of low temperature on photosynthesis. Plant Cell Env. *6,* 281–300 (1983).

– Light stress at low temperature. In: Kyle, D. J., C. B. Osmond, and C. J. Arntzen (eds.): Photoinhibition, pp. 67–87. Elsevier, Amsterdam (1987).

Öquist, G. and N. P. A. Huner: Cold-hardening-induced resistance to photoinhibition of photosynthesis in winter rye is dependent upon an increased capacity for photosynthesis. Planta *189,* 150–156 (1993).

Öquist, G. and B. Martin: Photosynthesis in cold climates. In: Baker, N. R. and S. P. Long (eds.): Topics in Photosynthesis Vol. *7,* pp. 237–293. Elsevier, Amsterdam (1986).

Raison, J. K.: A biochemical explanation of low-temperature stress in tropical and sub-tropical plants. In: Bielski, R. L., A. R. Ferguson, and M. M. Cresswell (eds.): Mechanisms of Regulation of Plant Growth, Bulletin *12,* pp. 487–497. The Royal Society of New Zealand, Wellington (1974).

RAISON, J. K. and J. M. LYONS: Chilling injury: a plea for uniform terminology. Plant Cell Env. *9,* 685–686 (1986).

SMILLIE, R. M., R. NOTT, S. HETHERINGTON, and G. ÖQUIST: Chilling injury and recovery in detached and attached leaves measured by chlorophyll fluorescence. Physiol. Plantarum *69,* 419–428 (1987).

SOMERSALO, S. and G. H. KRAUSE: Photoinhibition at chilling temperature. Fluorescence characteristics of unhardened and cold acclimated spinach leaves. Planta *177,* 409–416 (1989).

STIRLING, C. M., V. H. RODRIGO, and J. EMBERRU: Chilling and photosynthetic productivity of field grown maize (*Zea mays*); changes in the parameters of the light-response curve, canopy leaf CO_2 assimilation rate and crop radiation-use efficiency. Photosynth. Res. *38,* 125–133 (1993).

STYRING, S., J. VIRGIN, A. EHRENBERG, and B. ANDERSSON: Strong light photoinhibition of electron transport in Photosystem II. Impairment of the function of the first quinone acceptor, Q_A. Biochim. Biophys. Acta *1015,* 269–278 (1990).

VAN KOOTEN, O. and J. H. F. SNEL: The use of chlorophyll fluorescence nomenclature in plant stress physiology. Photosynth. Res. *25,* 147–150 (1990).

VASS, I., N. MOHANTY, and S. DEMETER: Photoinhibition of electron transport activity of Photosystem II in isolated thylakoids studied by thermoluminescence and delayed luminescence. Z. Naturforsch. *43 c,* 871–876 (1988).

VASS, I., S. STYRING, T. HUNDALL, A. KOIVUNIEMI, E. ARO, and B. ANDERSSON: Reversible and irreversible intermediates during photoinhibition of photosystem II: Stable reduced Q_A species promote chlorophyll triplet formation. Proceed. Nation. Acad. Sciences, USA *89,* 1408–1412 (1992).

YAKIR, D., J. RUDICH, B. A. BRAVDO, and S. MALKIN: Prolonged chilling under moderated light: effect on photosynthetic activity measured with the photoacoustic method. Plant Cell Env. *9,* 581–588 (1986).

J. Plant Physiol. Vol. 148. pp. 384–390 (1996)

The F685/F730 Chlorophyll Fluorescence Ratio as Indicator of Chilling Stress in Plants

GIOVANNI AGATI[1], PIERO MAZZINGHI[1], MICHELE LIPUCCI DI PAOLA[2], FRANCO FUSI[1], and GIOVANNA CECCHI[3]

[1] Istituto di Elettronica Quantistica – Consiglio Nazionale delle Ricerche, Via Panciatichi, 56/30, 50127 Firenze, Italy

[2] Dipartimento di Biologia delle Piante Agrarie, Università di Pisa, Viale delle Piagge, 23, 56124 Pisa, Italy

[3] Istituto di Ricerca sulle Onde Elettromagnetiche – Consiglio Nazionale delle Ricerche, Via Panciatichi, 64, 50127 Firenze, Italy

Received June 24, 1995 · Accepted October 10, 1995

Summary

The response of chlorophyll fluorescence to chilling temperatures was evaluated by two different experiments. In the first, the F685/F730 and the F_v/F_m chlorophyll fluorescence ratios were measured in *Phaseolus vulgaris* L., cv. Mondragone plants under chilling stress at 4 °C and moderate light (100 µmol $m^{-2} s^{-1}$) up to 72 hours. F_v/F_m decreased linearly with chilling time indicating a photoinhibitory effect (no change was observed in the dark under the same conditions). F685/F730 underwent a rapid exponential decay followed by a linear slow decline. In a second experiment, the F685/F730 ratio, the total chlorophyll fluorescence, F685+F730, and the leaf temperature were monitored on a single leaf in a climate chamber as the temperature was decreased from 20 to 4 °C. The experiment was run simultaneously on the chilling-sensitive *Phaseolus vulgaris* and on the chilling-tolerant *Pisum sativum* L. (cv. Shuttle) plants. For both species two phases related to the leaf temperature can be distinguished: the first 4-hour period during which the leaf temperature decreased from 24 to 4 °C, and a second period during which the leaf temperature slightly oscillated around 4 °C. The behaviour of F685/F730 for the bean was quite different from that of the pea plant. During the first phase, it decreased markedly for the chilling-sensitive bean while a slight increase was observed for the chilling-resistant pea. In the following period, the F685/F730 values for the pea remained constant while those for the bean were found still to decrease. On the basis of our results, the use of the chlorophyll fluorescence ratio as indicator of plant chilling sensitivity can be envisaged.

Key words: Phaseolus vulgaris, Pisum sativum, chilling temperature, chlorophyll fluorescence, continuous monitoring, environmental factors, photoinhibition, stress detection in plants.

Abbreviations: F_o = initial fluorescence; F_m = maximum fluorescence; F_v = variable fluorescence (F_m-F_o; Chl = chlorophyll; ChlFR = ratio between the red and near-ir bands of the chlorophyll fluorescence spectrum; F685 and F730 = chlorophyll fluorescence intensity at 685 and 730 nm respectively; PPFD = photosynthetic photon flux density; LEAF = Laser Excited Automatic Fluorometer.

Introduction

Temperature affects the photosynthetic capacity of plants by a series of complex biochemical and biophysical processes (Öquist et al., 1983). Plants adapted to high-temperature environments undergo inhibition of photosynthesis when exposed to low temperatures. The extent of inhibition is dependent on the species, age of leaves, chilling temperature regime and intensity of the environmental light (Öquist et al., 1987; Hetherington et al., 1989; Hodgson et al., 1989). A decrease in the photosynthetic rate was observed in many species of higher plants under chilling, no-freezing, temperatures in the dark. A more severe reduction occurs when chilling is induced under high light, since the susceptibility to photoinhibition is highly increased at low temperatures.

The mechanisms regulating the inhibition of photosynthesis under chilling are not completely understood. It seems definitely accepted that the primary target for chilling stress in the light is at the reaction center of PSII. On the other hand, storage at 0 °C in the dark appears to inhibit specifically the electron transport from water to PSII (Smillie and Nott, 1979).

Measuring the fluorescence ratio F_v/F_m of the variable chlorophyll (Chl) fluorescence is a widely used method to monitor reversible or irreversible damage to PSII activity. In fact, this parameter equals the quantum yield for PSII photochemistry (Butler, 1978). F_v/F_m is remarkably constant among non-stressed plants, regardless of the species (F_v/F_m equal to about 0.8). On the other hand, it decreases drastically in plants kept few hours at chilling temperatures even under moderate light intensities (Hetherington et al., 1989; Somersalo and Krause, 1989).

The chilling induced reduction of F_v/F_m, measured at room temperature (Neuner and Larcher, 1991; Sthapit et al., 1995) or at 77 K (Hetherington et al., 1989; Huang et al., 1989), was suggested as a useful parameter to identify the plant sensitivity to low temperatures. Other Chl fluorescence parameters have been also indicated as plant chilling tolerance indexes. Hetherington et al. (1983) introduced the measurement of F_r, the maximal rate of the induced rise in Chl fluorescence, that markedly decreases in chilling-sensitive plants kept at 0 °C. Relative chilling tolerance is defined as the time taken for F_r to decrease by 50 % in leaves chilled at 0 °C. More recently, Walker et al. (1990) evaluated the chilling tolerance in different tomato lines using the ratio of the initial to the peak fluorescence of the induction kinetics after 72 h at 2 °C. Neuner and Larcher (1990) used the vitality index, Rfd = $(f_m-f_s)/f_s$, the photochemical quenching coefficient and the initial slope of the non-photochemical quenching coefficient as a monitoring method of screening soybean varieties for chilling susceptibility.

Recently, the chlorophyll fluorescence ratio (ChlFR), F685/F730, between the red and the near-infrared bands of the fluorescence spectrum has been proposed as stress indicator (Rinderle and Lichtenthaler, 1988). For example, ChlFR was observed to increase in plants under water stress or after mechanical injuries, with respect to controls (Lichtenthaler and Rinderle, 1988). Chilling stress in beans, on the contrary, was seen to induce a decrease of ChlFR (Lipucci di Paola et al., 1992).

The use of the ChlFR in the detection of chilling stresses is attractive since it does not need predarkening of the leaf and therefore it would result more useful and advantageous with respect to Chl fluorescence induction measurements. In theory, it can also be used in remote sensing monitoring of vegetation (Valentini et al., 1994). This work is an extension of our previous observation (Lipucci di Paola et al., 1992) in order to accurately investigate the changes in the F685/F730 ratio induced by chilling temperatures under moderate light on a chilling sensitive plant. These data were compared with the fluorescence parameter F_v/F_m of the fluorescence induction kinetics under the same treatment. The potential use of the ChlFR as screening parameter of plant chilling tolerance was also evaluated by measuring the F685/F730 ratio in the chilling-sensitive bean and the chilling-resistant pea species.

Materials and Methods

Plant material and growth conditions

Experiments were performed on 20-d old seedlings of bean (*Phaseolus vulgaris* L., cv. Mondragone) and pea (*Pisum sativum* L., cv. Shuttle) at the stage of fully expanded primary leaves. Seeds were germinated in perlite and transplanted to perforated plastic pots (8 cm diameter) filled with expanded clay and placed in tanks (61 capacity, 24 plants per tank) with a continuously aerated nutrient solution. Seedlings were grown in a climate chamber under continuous light (100 µmol $m^{-2} s^{-1}$). The chamber temperature was set at 23 ± 1 °C and the relative humidity was 80 %.

Chilling treatment

Chilling treatment was performed in the growth chamber at 4 ± 1 °C and at a constant relative humidity of 80 %. Experiments were done at low light intensity (100 µmol $m^{-2} s^{-1}$) to minimise photoinhibitory effects. A separate batch of plants was used as control, maintained at 23 ± 1 °C, relative humidity of 80 %, in the same light conditions as stressed plants. Chilling stress was followed up to 3 days. All the experiments were conducted at atmospheric CO_2 concentration.

Fluorescence measurements and data analysis

The Chl fluorescence ratio, F685/F730, was measured by the portable two-wavelengths fluorometer LEAF® (Laser Excited Automatic Fluorometer) (Loto, Florence, Italy) described previously (Lipucci di Paola et al., 1992). The instrument excites Chl fluorescence by a He-Ne laser and detects both the 685 and 730 nm fluorescence peaks by a couple of photomultiplier detectors preceded by suitable interference filters. An improved version of the LEAF instrument that uses a diode laser at 635 nm to excite fluorescence and a couple of photodiodes as detection system was also employed.

Parameters F_o, F_m and F_v/F_m of the fluorescence induction kinetics were obtained using a pulse-amplitude modulations fluorometer (PAM-101; H. Walz, Effeltrich, Germany) as reported previously (Schreiber, 1986).

At various time intervals of the chilling treatment, fluorescence was measured on each leaf by the LEAF device on both controls and stressed plants. The rapidity of the LEAF method allows for the collection of a large number of data on each leaf. Consequently, the F685/F730 value for each leaf was the average of about 30 measurements. The same leaves were then dark-adapted for 15 min and measured by the PAM instrument. Values of the different fluorescence parameters were the average on 6 leaves from 6 different plants.

All data were collected from the upper side of fully expanded primary leaves attached to the plant.

In a separate experiment, one plant of each species was continuously monitored while the chamber temperature was decreased from 23 to 4 °C under 100 µmol $m^{-2} s^{-1}$ and 80 % of relative humidity. The ChlFR and the total Chl fluorescence, F685+F730, were measured in automatic mode at regular 1 min time intervals from the beginning of the chilling treatment contemporary on both bean and pea leaves by using two fiber optics LEAF fluorometers. The fiber optics tips of the two LEAF instruments were positioned at about 45° with respect to the leaf surface at a distance of about

2 mm in order to avoid shading of the leaf area under measurement. At the same time the temperature of both leaves was also measured by two infrared thermometers (Land Cyclops Compac 3, Minolta) at a frequency of 1/60 Hz. This experiment was aimed to accurately investigate the initial period during which the plant acclimates to 4 °C starting from room temperature. The measurements was carried out up to 20 hours maintaining the chamber temperature at 4 °C.

The variation of F685/F730 and F_v/F_m versus the chilling time was analysed by fitting the mean values.

Results

F685/730 and F_v/F_m comparison

The F685/F730 and F_v/F_m fluorescence parameters as function of the time of chilling treatment in bean are presented in Fig. 1. Values are relative to controls. It is evident that F685/F730 decreases quickly in the first few hours of the treatment and then more slowly.

A quite different behaviour is showed for F_v/F_m that decreases linearly with treatment. The decrement of $F_v/F_m = (F_m-F_o)/F_m$ is mostly due to the decrease of F_m rather than to a variation of F_o (data not shown).

Curve fitting of the average values of the above experimental data results in the parameters reported in Table 1. Values of F_v/F_m are well fitted by a single linear curve with negative slope of −0.008. The best fit of F685/F730 consists of a rapid exponential decay followed by a slow linear decrement. In order to carefully evaluate the significance of the linear variation of F685/F730, we analysed the data by the one-way analysis of variance. It results that the change of F685/F730 at time of the chilling treatment longer than 6.5 h is significant ($0.025<P<0.05$).

The Chl content, as measured previously (Lipucci et al., 1992), was found to remain constant during the chilling treatment.

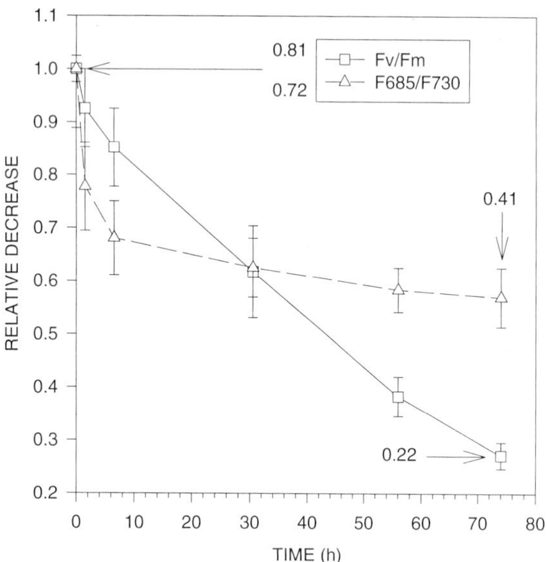

Fig. 1: Time course of the F685/F730 (△) and F_v/F_m (□) chlorophyll fluorescence ratios measured in *Phaseolus vulgaris* L., cv. Mondragone during chilling at 4 °C and 100 µmol $m^{-2} s^{-1}$. Values are relative to controls and are the average ± standard deviation of six different measurements.

Table 1: Curve fitting parameters for F685/F730 and F_v/F_m versus time of chilling treatment in *Phaseolus vulgaris.*

	exponential fitting curve		linear fitting curve		r^2
	amplitude	time constant (h)	slope ($\times 10^3$)	intercept	
F685/F730 vs t	0.10±0.01	1.27±0.23	−0.9±0.1	0.65±0.01	0.998
F_v/F_m vs t	–	–	−7.7±0.5	0.76±0.02	0.984

Fluorescence and temperature continuous monitoring

The results of the fluorescence continuous monitoring, as the environmental temperature was decreased and maintained at 4 °C up to 20 hours, are shown in Figs. 2 a and 2 b for the bean and pea plants respectively. The ChlFR and the total F685+F730 Chl fluorescence are reported along with the leaf temperature as function of the chilling time. For both bean and pea, two phases related to the leaf temperature can be distinguished: 1) a first 4-hour period during which the leaf temperature exponentially decreased from 25 to 4 °C, and 2) a second period during which the leaf temperature slightly oscillated around 4 °C. The behaviour of Chl fluorescence is instead different for the two species. During the first phase, F685/F730 decreases markedly for the chilling-sensitive bean while a slight increase is observed for the chilling-resistant pea. In the following period, the ChlFR for the pea remained constant while those for the bean are found still to decrease. The total Chl fluorescence, F685+F730, increases in both pea and bean plants with decreasing leaf temperature to a maximum and then decreases to a plateau. However, in bean the fluorescence intensity increase is much smaller (50 % vs. 130 %) and shorter (2.5 vs. 5 h) than that observed in pea.

The relationship between ChlFR and the leaf temperature for the chilling-sensitive and chilling-tolerant species is better shown in Fig. 3 during the first phase. It is evident that the slope of the decline in F685/F730 observed in bean increases when the leaf temperature becomes lower than 7 °C. In pea, instead, F685/F730 is inversely correlated to the leaf temperature with a single regression line over the temperature range reported.

Discussion

Comparison between F685/F730 and F_v/F_m

The effect of chilling temperatures on F685/F730 and F_v/F_m under moderate light depends on the time of treatment. For chilling times longer than about 6 hours the behaviour of the two chlorophyll fluorescence parameters is similar. In *Phaseolus vulgaris,* both factors decrease linearly. The decrease of F_v/F_m with chilling can be explained as due to a photoin-

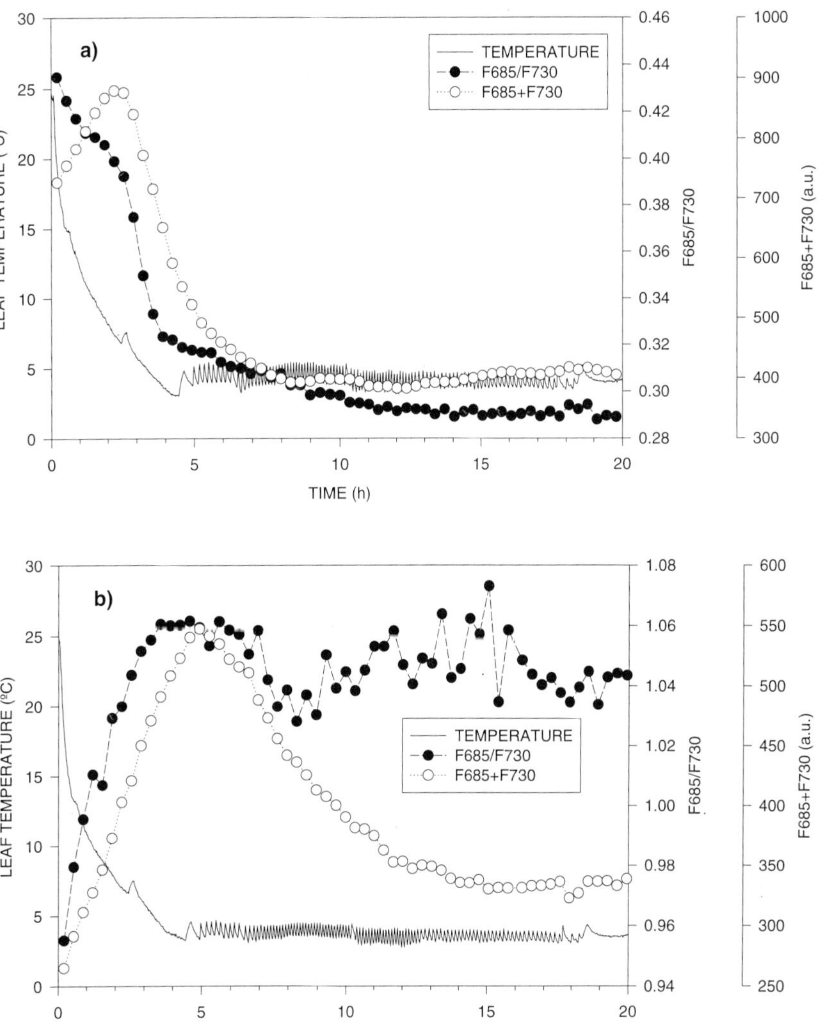

Fig. 2: Variation of the chlorophyll fluorescence ratio, F685/F730 (●), total chlorophyll fluorescence, F685+F730 (○); and leaf temperature (solid line) during a continuous monitoring in a climate chamber of *Phaseolus vulgaris* (a) and *Pisum sativum* (b) attached leaves. Light intensity was 100 µmol $m^{-2} s^{-1}$ and relative humidity was 80%.

hibitory effect. It is well known that chilling temperatures increase the plant sensitivity to photoinhibition so that also relatively low PPFD can inhibit the photosynthetic activity (Hetherington et al., 1989; Ottander et al., 1992; Somersalo and Krause, 1989; Powles et al., 1983).

Photoinhibition consists of complex mechanisms aimed to protect the photosynthetic apparatus from excess light energy. It can be seen as the result of concomitant damaging and repairing processes. Greer et al. (1986) already observed photoinhibition in *Phaseolus vulgaris* at 5 °C and 140 µmol $m^{-2} s^{-1}$ and suggested that the inhibition was favoured by low temperatures by inactivation of the repair mechanism.

The dissipating mechanisms of photoinhibition are in competition with the radiative relaxation of chlorophyll. Consequently, photoinhibition is accompanied by a reduction in the chlorophyll fluorescence. The ratio F_v/F_m was shown to be an useful indicator of photosynthesis inhibition under high-light intensity and physiological temperatures (Björkman, 1987), or at moderate light intensity and chilling temperatures (Somersalo and Krause, 1989; Hetherington et al., 1989). In our work, the presence of photoinhibition in beans at 4 °C and 100 µmol $m^{-2} s^{-1}$ was than proved by the decrease of F_v/F_m with the treatment. Moreover, in a test experiment performed in the dark we found that no change in F_v/F_m was induced by chilling alone (data not shown).

Interpretation of the change with chilling of the F685/F730 ratio is complicated because of the scarce knowledge available on the physiological meaning of this fluorescence parameter. The main contribution to chlorophyll fluorescence in the red-far-red spectral region comes from PSII. The photosystem PSI seems to provide significant fluorescence, peaked at about 740 nm, only at low temperatures (Krause and Weis, 1991), so that the ratio between the shorter and the longer wavelength emission bands at 77 K is used to monitor changes in the energy distribution between PSI and PSII. Several experimental evidences, however, suggest that at physiological temperatures the contribution of PSI to the 740 nm fluorescence is not negligible with respect to PSII fluores-

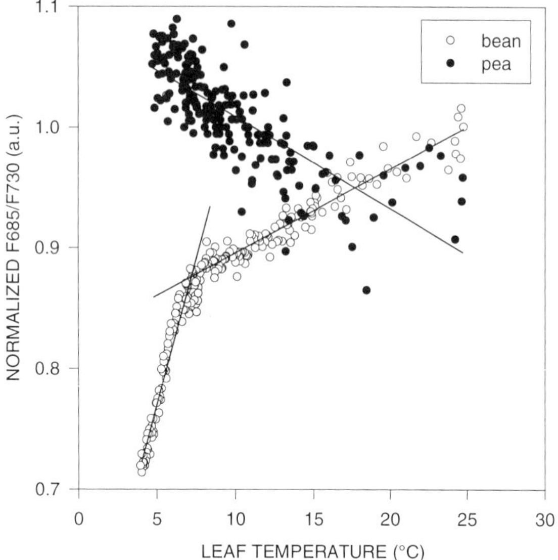

Fig. 3: The chlorophyll fluorescence ratio, F685/F730 as function of the leaf temperature under moderate light intensity (100 µmol $m^{-2} s^{-1}$) in *Phaseolus vulgaris* (○) and *Pisum sativum* (●). Solid lines are regression curves.

cence (Gently et al., 1990; Pfündel, 1995) and that variations in the ChlFR may reflect changes in the photosynthetic activity of the two photosystems (Agati et al., 1995).

What is the main target of photoinhibition is still under debate and it seems to depend markedly on the environmental conditions. It was observed that *in vivo* PSI is more resistant than PSII to strong light stress (Havaux and Eyletters, 1991). Under moderate high-light stress (420 W/m^2, 1 h) of sugar maple leaves, the energy storage and the photosynthetic activity in PSI are augmented (Charland et al., 1992), while PSII is specifically inhibited. At higher light intensities both PSII and PSI are inhibited, but the later at a lower extent.

Recently, it was observed that at low temperatures a preferential inactivation of PSI occurs in both chilling-resistant (Havaux and Davaud, 1994) and chilling-sensitive plants (Terashima et al., 1994). The mechanism proposed for the PSI photoinhibition involves the inactivation of the electron acceptor side without destruction of the P-700 complex (Sonoike and Terashima, 1994). Consequently, being the PSI Chl fluorescence yield independent of the redox state of P-700 (Butler, 1978), no change in the Chl fluorescence at 730 nm can be ascribed to the potential chilling-induced photoinhibition of PSI.

In our experimental conditions, photoinhibition of PSII was proved to occur by the reduction of F_v/F_m. Consequently, a decrease in the steady-state fluorescence emitted by PSII at both fluorescence maxima with chilling treatment is expected. As indicated above, the contribution of PSI to the longer-wavelength Chl fluorescence should remain constant under chilling. The larger quenching on the red fluorescence peak with respect to the far-red one under photoinhibition is confirmed by fluorescence measurements at 77 K (Ögren and Öquist, 1984; Demmig and Björkman, 1987; Somersalo and Krause, 1989).

The decrease of F685/F730 we observed in bean plants under chilling and moderate light treatment can then be explained as due to the inhibition of PSII.

The variation in F685/F730 induced by chilling stress is opposite to that reported for all other stress conditions in which the Chl concentration is reduced (Rinderle and Lichtenthaler, 1988; Lichtenthaler and Rinderle, 1988). In these cases, in fact, an increase in ChlFR is observed as due to a decrease in the fluorescence reabsorption at 690 nm (Agati et al., 1993). In our experiment, the extent of the stress is too short to lead to a Chl degradation, as confirmed by pigment extractions (Lipucci di Paola et al., 1992). Variations in the absorbance and scattering properties of the different Chl-protein complexes, going from room to chilling temperature, can not be ruled out. These effects could change the fluorescence reabsorption on the short-wavelength band (Agati et al., 1993). However, in our experiment, we verified that the leaf transmittance spectra were the same before and after the chilling treatment.

Fluorescence and temperature continuous monitoring

The short-term effect of a decrease in temperature on the Chl fluorescence has been investigated simultaneously on the chilling-resistant pea and the chilling-sensitive bean. The decrement in F685/F730 observed in bean after 6.5 h (32 %) is in accordance with that measured at the same time of chilling in the previous above experiment. It is evident from Fig. 3 that the effect of temperature on ChlFR is markedly different for the two species. In the chilling-sensitive bean, F685/F730 is directly correlated to the leaf temperature and there is a critical temperature, around 7 °C, below which the slope of the regression line drastically increases. This is not the case for the chilling-tolerant pea in which a single inverse linear correlation between F685/F730 and leaf temperature is found. Also the behaviour of the total Chl fluorescence with leaf temperature is species dependent. In bean F685+F730 starts to decrease at about 7 °C, while in pea it increases until the leaf temperature decreases to 4 °C. The initial behaviour of ChlFR and F685+F730 in bean is analogue to that previously seen on *Ficus benjamini* in the 14–25 °C temperature range (Agati et al., 1995). Explanation of these results is difficult because a complete characterisation of the temperature profiles of the Chl fluorescence bands at the steady-state is lacking. In fact, most of the studies have been dedicated to the detection of the Chl fluorescence induction kinetics parameters at different temperatures. It was found that the steady-state Chl fluorescence in soybean increased with progressive cooling from 20 to 0 °C (Neuner and Larcher, 1990). This result was correlated to a large reduction of the photochemical quenching at lower temperatures while the non-photochemical quenching at the steady-state was only slightly affected by temperature. Analogue dependencies of the photochemical and non-photochemical fluorescence quenching on leaf temperature were reported by Koroleva et al. (1994) for a chilling tolerant species. Furthermore, decreasing leaf temperature induces a reduction of the thylakoid membrane fluidity which inhibits the reoxidation of plastoquinones leading to an increase in the Chl fluorescence yield (Havaux and Gruszecki, 1993).

Understanding the change with temperature of F685/F730 is even more complicate and requires further investigations. It is however evident that different mechanisms operate on the two species during the chilling. In bean, at the initial decrease in ChlFR that is luckily due to a temperature effect only, a photoinhibitory effect adds below 7 °C inducing a quenching of the total fluorescence and a more rapid decrease in F685/F730. In pea, the temperature effect on F685/F730 is modest but opposite to that of bean. The reason for this difference probably relies on diverse mechanisms of energy dissipation, including PSII → PSI transfer, developed by the two species and on their dependence on leaf temperature (Holaday et al., 1992; Koroleva et al., 1994; Adams et al., 1995).

F685/F730 as screening parameter for chilling tolerance

Large attention have been devoted to the development of screening methods to identify the plant sensitivity to low temperatures (Hetherington et al., 1983; Walker et al., 1990; Neuner and Larcher, 1990, Hetherington et al., 1989; Sthapit et al., 1995). At present, all the procedures used are based on the measurement of Chl fluorescence induction parameters and, therefore, their practical application is limited by the requirement of predarkening of the leaf.

Our data on plants with different chilling susceptibility indicate a new, rapid method for the screening of the plant chilling tolerance.

We observed (Figs. 2 and 3) that the temperature effect on F685/F730 is species dependent, being opposite in the chilling-resistant pea with respect to the chilling-sensitive bean. Although more experimental work is needed to understand the physiological meaning of the change of F685/F730 with temperature in relation to the plant chilling sensitivity, the usefulness of this parameter can be envisaged. We demonstrated that the F685/F730 ratio between the two Chl fluorescence peaks can be used to monitor chilling stresses under light. The measurement of F685/F730 by two wavelengths fluorometers, as the LEAF instrument, presents the advantage, with respect to fluorescence kinetics measurements, of no needs for predarkening of the plant. A large number of leaves can then be sampled in a short time improving the statistics of data. Moreover, this method can result useful, in principle, in remote sensing of vegetation. The reliability of the method must be checked on a large number of different species and the suitable parameter, such as the slope of the F685/F730 vs. leaf temperature curve, to be used as index of chilling sensitivity must be found.

References

Adams, W. W. III, A. Hoehn, and B. Demmig-Adams: Chilling temperatures and the xanthophyll cycle. A comparison of warm-grown and overwintering spinach. Aust. J. Plant Physiol. *22,* 75–85 (1995).

Agati, G., F. Fusi, P. Mazzinghi, and M. Lipucci di Paola: A simple approach to the evaluation of the reabsorption of chlorophyll fluorescence spectra in intact leaves. J. Photochem. Photobiol. B: Biol. *17,* 163–171 (1993).

Agati, G., P. Mazzinghi, F. Fusi, and I. Ambrosini: The F685/F730 chlorophyll fluorescence ratio as a tool in plant physiology: response to physiological and environmental factors. J. Plant. Physiol. *145,* 228–238 (1995).

Björkman, O.: Low-temperature chlorophyll fluorescence in leaves and its relationship to photon yield of photosynthesis in photoinhibition. In: Kyle, D. J., C. B. Osmond, and C. J. Arntzen (eds.): Photoinhibition, pp. 123–144. Elsevier Science Publishers B.V., Amsterdam (1987).

Butler, W. L.: Energy distribution in the photochemical apparatus of photosynthesis. Ann. Rev. Plant Physiol. *29,* 345–378 (1978).

Charland, M., K. Veeranjaneyulu, and R. M. Leblanc: Simultaneous determination of *in vivo* activities of PSI and PSII following high-light stress. In: Murata, N. (ed.): Research in Photosynthesis, pp. 459–462. Kluwer Academic Publishers, Dordrecht (1992).

Demmig, B. and O. Björkman: Comparison of the effect of excessive light on chlorophyll fluorescence (77K) and photon yield of O_2 evolution in leaves of higher plants. Planta *171,* 171–184 (1987).

Gently, B., J. Wonders, and N. R. Baker: Non-photochemical quenching of F_0 in leaves is emission wavelength dependent: consequences for quenching analysis and its interpretation. Photosynth. Res. *26,* 133–139 (1990).

Greer, D. H., J. A. Berry, and O. Björkman: Photoinhibition of photosynthesis in intact bean leaves: role of the light and temperature, and requirement for chloroplast-protein synthesis during recovery. Planta *168,* 253–260 (1986).

Havaux, M. and A. Davaud: Photoinhibition of photosynthesis in chilled potato leaves is not correlated with a loss of Photosystem-II activity. Photosynth. Res. *40,* 75–92 (1994).

Havaux, M. and M. Eyletters: Is the *in vivo* Photosystem I function resistant to photoinhibition? An answer from photoacoustic and far-red absorbance measurements in intact leaves. Z. Naturforsch. *46c,* 1038–1044 (1991).

Havaux, M. and W. I. Gruszecki: Heat- and light-induced chlorophyll *a* fluorescence changes in potato leaves containing high or low levels of the carotenoid zeaxanthin: indications of a regulatory effect of zeaxanthin on thylakoid membrane fluidity. Photochem. Photobiol. *58,* 607–614 (1993).

Hetherington, S. E., J. He, and R. M. Smillie: Photoinhibition at low temperature in chilling-sensitive and -resistant plants. Plant Physiol. *90,* 1609–1615 (1989).

Hetherington, S. E., R. M. Smillie, A. K. Hardacre, and H. A. Eagles: Using chlorophyll fluorescence *in vivo* to measure the chilling tolerances of different populations of maize. Aust. J. Plant Physiol. *10,* 247–256 (1983).

Hodgson, R. A. J. and J. K. Raison: Inhibition of photosynthesis by chilling in moderate light: a comparison of plants sensitive and insensitive to chilling. Planta *178,* 545–552 (1989).

Holaday, A. S., W. Martindale, R. Alred, A. L. Brooks, and R. C. Leegood: Changes in activities of enzymes of carbon metabolism in leaves during exposure of plants to low temperature. Plant Physiol. *98,* 1105–1114 (1992).

Huang, L.-K., C. B. Osmond, and I. Terashima: Chilling injury in mature leaves of rice. II. Varietal differences in the response to interactions between low temperature and light measured by chlorophyll fluorescence at 77 K and the quantum yield of photosynthesis. Aust. J. Plant Physiol. *16,* 339–352 (1989).

Koroleva, O. Y., W. Brüggemann, and G. H. Krause: Photoinhibition, xanthophyll cycle and *in vivo* chlorophyll fluorescence quenching of chilling-tolerant *Oxyria digyna* and chilling-sensitive *Zea mays.* Physiol. Plant *92,* 577–584 (1994).

Krause, G. H. and E. Weis: Chlorophyll fluorescence and photosynthesis: the basics. Ann. Rev. Plant Physiol. Plant. Mol. Biol. *42,* 313–349 (1991).

Lichtenthaler, H. K. and U. Rinderle: The role of chlorophyll fluorescence in the detection of stress conditions in plants. CRC Crit. Rev. Anal. Chem. *19* (Suppl. 1), S 29–S 85 (1988).

Lipucci di Paola, M., P. Mazzinghi, A. Pardossi, and P. Vernieri: Vegetation monitoring of chilling stress by chlorophyll fluorescence ratio. EARSeL Adv. Rem. Sens. *2,* 2–6 (1992).

Neuner, G. and W. Larcher: Determination of differences in chilling susceptibility of two soybean varieties by means of *in vivo* chlorophyll fluorescence measurements. J. Agron. Crop Sci. *164,* 73–80 (1990).

– – The effect of light, during and subsequent to chilling, on the photosynthetic activity of two soybean cultivars, measured by *in vivo* chlorophyll fluorescence. Photosynthetica *25,* 257–266 (1991).

Ögren, E. and G. Öquist: Photoinhibition of photosynthesis in *Lemna gibba* as induced by the interaction between light and temperature. III. Chlorophyll fluorescence at 77 K. Physiol. Plant. *62,* 193–200 (1984).

Öquist, G.: Effects of low temperature on photosynthesis. Plant Cell Environ. *6,* 281–300 (1983).

Ottander, C., T. Hundal, B. Andersson, N. P. A. Huner, and G. Öquist: On the susceptibility of photosynthesis to photoinhibition at low temperatures. In: Murata, N. (ed.): Research in Photosynthesis, pp. 455–458. Kluwer Academic Publishers, Dordrecht (1992).

Pfündel, E.: PSI fluorescence at room temperature: possible effects on quenching coefficients. In: Mathis, P. (ed.): Proceedings of the 10th International Congress on Photosynthesis. Kluwer Academic Publishers, Dordrecht (in press).

Powles, S. B., J. A. Berry, and O. Björkman: Interaction between light and chilling temperature on the inhibition of the photosynthesis in chilling-sensitive plants. Plant Cell Environ. *6,* 117–123 (1983).

Rinderle, U. and H. K. Lichtenthaler: The chlorophyll fluorescence ratio F690–F735 as a possible stress indicator. In: Lichtenthaler, H. K. (ed.): Applications of Chlorophyll Fluorescence, pp. 189–196. Kluwer Academic Publishers, Dordrecht (1988).

Schreiber, U.: Detection of rapid induction kinetics with a new type of high-frequency modulated chlorophyll fluorometer. Photosynth. Res. *9,* 261–272 (1986).

Sonoike, K. and I. Terashima: Mechanism of photosystem-I photoinhibition in leaves of *Cucumis sativus* L.. Planta *194,* 287–293 (1994).

Smillie, R. M. and R. Nott: Assay of chilling injury in wild and domestic tomatoes based on photosystem activity of the chilled leaves. Plant Physiol. *63,* 796–801 (1979).

Somersalo, S. and G. H. Krause: Photoinhibition at chilling temperature. Planta *177,* 409–416 (1989).

Sthapit, B. R., J. R. Witcombe, and J. M. Wilson: Methods of selection for chilling tolerance in Nepalese rice by chlorophyll fluorescence analysis. Crop Sci. *35,* 90–94 (1995).

Terashima, I., S. Funayama, and K. Sonoike: The site of photoinhibition in leaves of *Cucumis sativus* L. at low temperatures is photosystem I, not photosystem II. Planta *193,* 300–306 (1994).

Valentini, R., G. Cecchi, P. Mazzinghi, G. Scarascia Mugnozza, G. Agati, M. Bazzani, P. de Angelis, F. Fusi, G. Matteucci, and V. Raimondi: Remote sensing of Chlorophyll *a* fluorescence of vegetation canopies: 2. physiological significance of fluorescence signal in response to environmental stresses. Remote Sens. Environ. *47,* 29–35 (1994).

Walker, M. A., D. M. Smith, K. P. Pauls, and B. D. McKersie: A chlorophyll fluorescence screening test to evaluate chilling tolerance in tomato. Hort. Sci. *25,* 334–339 (1990).

J. Plant Physiol. Vol. 148. pp. 391–398 (1996)

A Comparison of the Relative Rates of Transport of Ascorbate and Glucose Across the Thylakoid, Chloroplast and Plasmalemma Membranes of Pea Leaf Mesophyll Cells

CHRISTINE H. FOYER[1] and MAUD LELANDAIS[2]

[1] Department of Environmental Biology, IGER, Plas Gogerddan, Aberystwyth, Dyfed SY23 3EB, UK

[2] Laboratoire du Métabolisme, INRA, Route de Saint-Cyr, 78026 Versailles Cedex, France

Received June 28, 1995 · Accepted September 25, 1995

Summary

Mesophyll protoplasts, chloroplasts and thylakoids were prepared from the leaves of 10–12 day old pea seedlings. They contained 1618 ± 351, 197 ± 30 and 25 ± 7 nmol ascorbate per milligram of chlorophyll respectively equivalent to average concentrations of 8.4, 9.8 and 3.8 mM ascorbate in the intact protoplasts, intact chloroplasts and thylakoid membranes respectively. Transport of [^{14}C]-ascorbate and [^{14}C]-glucose across the plasmalemma, chloroplast envelope and thylakoid membranes was measured in these fractions. Saturation kinetics for [^{14}C]-ascorbate and [^{14}C]-glucose uptake suggest that the plasmalemma and chloroplast envelope membranes contain carriers for both ascorbate and glucose. In contrast, the thylakoid membranes appear to have no transport system for ascorbate, permeability resulting from diffusion alone. Dehydroascorbate inhibited ascorbate transport but not glucose transport across the chloroplast envelope membrane. Ascorbate stimulated [^{14}C]-glucose transport across the plasmalemma and chloroplast envelope membranes. Similarly, glucose had a positive effect on [^{14}C]-ascorbate transport across these membranes. These observations suggest the presence of independent carriers for ascorbate and glucose into both protoplasts and chloroplasts. Unlike animal membranes where the hexose transporters have been found to transport dehydroascorbate the plant counterparts does not appear to do so, either in the chloroplast envelope or the plasmalemma membranes. The ascorbate transporters on these membranes are hence discrete entities not involved with general hexose transport. However, the observed synergistic effects of either glucose or ascorbate on the transport of the other metabolite suggest co-ordination of the two processes.

Key words: Ascorbate transporters, glucose transporters, oxidative stress, photosynthesis, pea.

Abbreviations: B.S.A. = Bovine Serum Albumin; Hepes = 4-(2-hydroxyethyl)-1-piperazine-ethane sulfonic acid; MES = Morpholinoethane sulfonic acid.

Introduction

Ascorbate is a versatile and important antioxidant in plants and animals. It has many functions in leaves both within the intracellular compartments and in the apoplastic space. It has been known for many years that ascorbate is an essential antioxidant of the stromal compartment (Foyer, 1993; Foyer et al., 1994). It is involved in the protection and regulation of photosynthesis. It is, for example, an essential co-factor for the synthesis of the energy quencher, Zeaxanthin, in the thylakoid lumen (Hager, 1969; Pfündel and Bilger, 1994). The conversion of violaxanthin to Zeaxanthin is catalysed by the water soluble enzyme violaxanthin de-epoxidase that is located in the lumen of the thylakoid membranes (Hager, 1969; Hager and Holocher, 1994). The permeability of the thylakoid membrane for ascorbate is important in determining the lumen concentration of ascorbate and hence the activity of the enzyme (Neubauer and Yamamoto, 1994).

More recently monodehydroascorbate-dependent electron transport by enzymes on the plasmamembrane has been implicated in growth and other processes (Arrigoni, 1994; Navas et al., 1994; Morré et al., 1988; Citterio et al., 1994). Similarly, ascorbate has been shown to be an important component of the apoplastic space in leaves (Luwe et al., 1993; Takahama and Oniki, 1992; Castillo and Greppin, 1988; *Vigna:* Takahama, 1993 a, b; Castillo et al., 1987; Polle et al., 1990, 1995). The measured concentrations of ascorbate in the apoplast range from below 1 mM up to 5 mM. In the apoplast, ascorbate is involved in the regulation of cell-wall associated enzymes, for example, ascorbate inhibits the peroxide-mediated oxidation of coniferyl alcohol by reducing the radical products of the reaction (Takahama and Oniki, 1992). Ascorbate itself is oxidised in this process and the initial substrate for peroxidase is regenerated (Takahama, 1993 a, b) such that net oxidation of coniferyl alcohol was observed only after the oxidation of all the ascorbate present. In addition to these functions, extracellular ascorbate could also be part of the antioxidant defences of the apoplast protecting against harmful oxidants such as pollutants (Luwe et al., 1993; Luwe and Heber, 1995; Polle et al., 1995). An understanding of the mechanisms involved in the translocation of ascorbate across the plasmalemma membrane into the apoplastic space is important because of the roles of ascorbate cell extension, lignin biosynthesis and in the extracellular defences (Takahama and Oniki, 1992; Takahama, 1993; Luwe et al., 1993; Luwe and Heber, 1995; Polle et al., 1990). Ozone is taken up through open stomata and penetrates the intracellular air spaces of the leaves. It must be destroyed in the apoplastic space to prevent rapid absorption by the mesophyll tissues or destructive reaction with the plasmalemma. Ascorbate intercepts ozone in the apoplast and is oxidised to monodehydroascorbate and dehydroascorbate. Since the apoplast has no enzymic system for recycling dehydroascorbate, it has to be returned to the cytosol to regenerate reduced ascorbate. In this study the nature of the systems responsible for the transport of ascorbate and dehydroascorbate across the plasmalemma, chloroplast envelope and thylakoid membranes were studied in order to determine capacity of the cell to respond to demand for reduced ascorbate.

The biosynthetic pathway of ascorbate biosynthesis in higher plants is uncertain and its intracellular localisation is unknown. The pathway of ascorbic acid synthesis in plants appears not to be directly comparable to that found in animals. A branch point occurs in the reaction pathway when D-glucose is oxidised to the level of sugar acid (Loewus, 1980; Loewus et al., 1990). Higher plants primarily use a pathway which conserves the carbon chain in the same order and carbon sequence as that found in D-glucose. The alkaloid lycorine, which is a potent inhibitor of ascorbic acid biosynthesis, has been found to inhibit L-galactono-γ-lactone dehydrogenase (De Gara et al., 1994) and higher plants readily convert L-galactono-γ-lactone to ascorbic acid. This substrate, however, does not seem to be a natural constituent of plants which do not convert D-glucose to L-galactono-1,4-lactone. The presence of L-galactono-γ-lactone dehydrogenase is also incompatible with the ‹direct› reaction scheme. It is possible that there is a natural substrate for this enzyme, as yet unknown, that arises from a non-inversion pathway. Early indications suggested that ascorbate biosynthesis may occur in the cytosol (Loewus, 1980). If this is the case, ascorbate must then be transported into the chloroplasts. Whatever the location of ascorbate biosynthesis in the cell, this essential vitamin must be transported from the site of synthesis across cellular membranes to all the other compartments where it is required. Ascorbate does not readily cross biological membranes except with the aid of specific carrier proteins. The chloroplast envelope contains an ascorbate translocator (Anderson et al., 1983 a; Beck et al., 1983). In the case of spinach chloroplasts transport was found to occur by a process of facilitated diffusion (Beck et al., 1983). Similarly, ascorbate and dehydroascorbate were found to be transported into barley leaf protoplasts across the plasmalemma by a carrier whose activity was driven by the electrochemical gradient across the membrane (Rautenkraaz et al., 1994). In contrast, transport across the tonoplast membrane into the vacuole was found to occur by diffusion alone since no active transport could be demonstrated (Rautenkraaz et al., 1994).

Carrier-mediated transport of ascorbic acid and dehydroascorbate has been found in membranes isolated from animal tissues as well as in those from plants. In the case of mammalian cells the transport of ascorbate has been linked with the hexose transport systems. Experiments with human neutrophils and a *Xenopus lavis* oocyte expression system revealed that mammalian facilitative hexose transporters are efficient transporters of dehydroascorbate but not ascorbate (Vera et al., 1993). These authors suggested that a single transport protein with two functional sites with different affinities for dehydroascorbate and glucose was present and suggest that the glucose transporters are used to transport dehydroascorbate and hence accumulate ascorbate in human cells (Vera et al., 1994). One of the objectives of the present study was to determine whether a similar system could operate in plant cell membranes, since this would mean that glucose and ascorbate compete for the same transport proteins.

Materials and Methods

Plant material and protoplast isolation

Pea plants (*Pisum sativum* L. cv. Frisson (INRA, France) were grown in pots in a Sanyo Fitotron growth chamber at 400 μmol $m^{-2} s^{-1}$ irradiance and a 16 h photoperiod, 20 °C day/17 °C night, 75 % humidity. The pots were watered with complete nutrient solution (Coic and Lesaint, 1975). Pea leaf mesophyll protoplasts were isolated essentially as described by Mills and Joy (1980) in 450 mM sorbitol, 10 mM $MgCl_2$ and 1 mM K H_2PO_4 medium. The leaf pieces floated on a digestion medium consisting of 450 mM sorbitol, 20 mM Mes/KOH buffer (pH 5.5), 1 mM $CaCl_2$, 2 % cellulase (Onozuka 3S), and 0.5% pectinase (macerozyme). Cellulase and macerozyme were obtained from Yakult Biochemicals, Nishinomiya, Japan. The enzymes were allowed to digest the pea cell-walls for a period of 1 h under low light and the resultant protoplast suspensions were filtered, resuspended in medium A and pelleted by centrifugation at 100 g for 5 min. The isolated protoplasts were purified on a 28 % percoll step gradient as described previously (Mills and Joy, 1980) and re-suspended in 330 mM sorbitol, 5 mM and 50 mM Hepes/KOH buffer (pH 7.6) and stored on ice. The metabolic function and integrity of the isolated protoplasts was determined by their capacity for CO_2-dependent O_2 evolution in the light.

CO_2-dependent O_2 evolution

CO_2 fixation was estimated by the addition of protoplasts at a final concentration of 50 µg chlorophyll mL^{-1} to a medium containing 330 mM sorbitol, 5 mM $CaCl_2$, 50 mM Hepes-KOH buffer (pH 7.6) and 5 mM NaH CO_2. CO_2-dependent O_2 evolution in the above reaction medium was measured polarographically at 20 °C in Hansatech electrodes illuminated by white light at an irradiance of 500 µmol m^{-2} s^{-1}. Only protoplasts with a rate of CO_2-dependent O_2 evolution greater than 95 µmol O_2 mg^{-1} chlorophyll h^{-1} were used in the studies of ascorbate and glucose transport.

Chloroplast preparation

Intact chloroplasts were prepared mechanically from pea leaves according to the method of Mills and Joy (1980). The final chloroplast pellet was resuspended in medium containing 0.33 M sorbitol, 1 mM $MgCl_2$, 1 mM $MnCl_2$, 2 mM EDTA, 0.1 % (w/v) BSA and 100 mM Hepes-KOH buffer (pH 7.6) and the intactness of the chloroplasts was determined by the ferricyanide method (Lilley et al., 1975). All chloroplast preparations used in the following studies were over 90 % intact with a CO_2-dependent O_2 evolution rate of between 70–80 µmol O_2 h^{-1} Chl as determined in the oxygen electrode according to the method of Walker (1987).

Thylakoid preparation

Thylakoids were prepared from intact chloroplasts by resuspension in a medium containing 5 mM $MgCl_2$ and 10 mM dithiothreitol 1 mM EDTA, 10 mM Hepes-KOH buffer (pH 7.6) in distilled water. After 1 min the thylakoid membranes were pelleted by centrifugation and resuspended in 0.33 M sorbitol, 1 mM EDTA, 10 mM KCl, 50 mM Hepes-KOH buffer (pH 7.9) and 10 mM $MgCl_2$ and stored on ice.

Ascorbate and glucose uptake measurements

All compartment volumes were calculated as described by Heldt (1980). Uptake of L-[carboxyl-^{14}C]ascorbate or [^{14}C]-glucose (Amersham International plc, France) in the medium described for protoplasts, chloroplasts or thylakoids was determined essentially as described by Heldt (1980). Reaction mixtures also contained protoplasts, chloroplasts or thylakoids (100 µg chlorophyll (Chl)) and either 5, 10 or 20 mM [^{14}C]ascorbate or [^{14}C]glucose (1.8, 5 MBq $mmol^{-1}$) in a volume of 1 mL. These were incubated at 20 °C in darkness unless otherwise stated in the text. Samples (100 µL) were withdrawn at intervals for up to 5 min and the protoplasts, chloroplasts or thylakoids rapidly separated from the incubation medium by centrifugation through 100 µL silicone oil. The silicone oil composition was as follows:

Protoplasts: 15 % (v/v) of silicone fluid AR20 in silicone fluid AR200.

Chloroplasts: 5 % silicone fluid AR20 in silicone fluid AR200.

Thylakoids: 25 % silicone fluid AR20 in silicone fluid AR200.

The silicone oils were obtained from Wacker Chemie S.A., Lyon, France.

The effect of dehydroascorbate, ascorbate or glucose on the uptake of [^{14}C]-compounds was estimated by inclusion of the appropriate concentrations of these in the incubation media.

Total ascorbate was determined by the method of Foyer et al. (1983). Chlorophyll was measured in 80 % acetone extracts by the method of Arnon (1949).

Results

Intact mesophyll protoplasts and chloroplasts were isolated from pea leaves. Their functional integrity was measured *via* CO_2-dependent O_2 evolution and their total internal volumes and ascorbate contents measured (Table 1). The chloroplasts and protoplasts preparations used in the following studies were over 90 % intact. Minimal rates of CO_2-dependent O_2 evolution were 70 µmol mg^{-1} chlorophyll h^{-1} and 95 µmol mg^{-1} chlorophyll h^{-1} respectively. The thylakoid membrane fraction was also isolated from the pea leaf chloroplasts and the ascorbate contents measured. The ascorbate content of the pea protoplasts and chloroplasts was consistent with previous measurements on spinach (Foyer et al., 1983). The chloroplasts contained approximately 10 mM ascorbate while the thylakoid membranes had about 4 mM ascorbate (Table 1). By subtracting the ascorbate values for the chloroplasts from those of the intact protoplasts we can obtain estimates of the ascorbate contents of the rest (cytosol plus vacuole) of the leaf mesophyll cells. They are in the order of 1500 nmol mg^{-1} Chl and demonstrate that other cellular compartments as well as the chloroplasts contain high concentrations of ascorbate as has been found previously (Franke and Heber, 1964; Foyer et al., 1983; Rautenkranz et al., 1994). The values in Table 1 indicate the base levels of ascorbate in the individual leaf cell compartments before the following studies on ascorbate, dehydroascorbate and glucose were performed.

The rates of uptake of [^{14}C]-ascorbate into the thylakoid lumen were very slow and did not show saturation kinetics (Fig. 1). [^{14}C]-ascorbate uptake was measured for periods up to 30 minutes but only the initial 30 s is shown in Figs. 1 and 2 to illustrate the apparent absence of concentration-dependent ascorbate transport. Kinetic differences in ascorbate uptake by thylakoids (Figs. 1 and 2) and intact chloroplasts (Fig. 3) were observed. Since the initial concentration of ascorbate in the thylakoid membranes was approximately 3.8 mM (Table 1) an influx of ascorbate would be expected at external ascorbate concentrations of 5 or 10 mM. This was not observed (Fig. 2) suggesting that ascorbate (the base form) is not the diffusing species. Hence, it is likely that only the uncharged

Table 1: Pea leaf mesophyll protoplast, chloroplast and thylakoid volumes and ascorbate contents. The numbers in brackets indicate the number of separate preparations that were used to make the estimations.

	Volume (µL mg^{-1} Chl)	Ascorbate content (nmol mg^{-1} Chl)	Average concentration (mM)
Pea leaf:			
protoplasts	193±31 (8)	1618±351	8.4
chloroplasts	20±4 (11)	197±30	9.8
thylakoids	6.5±3.2 (18)	25±7	3.8

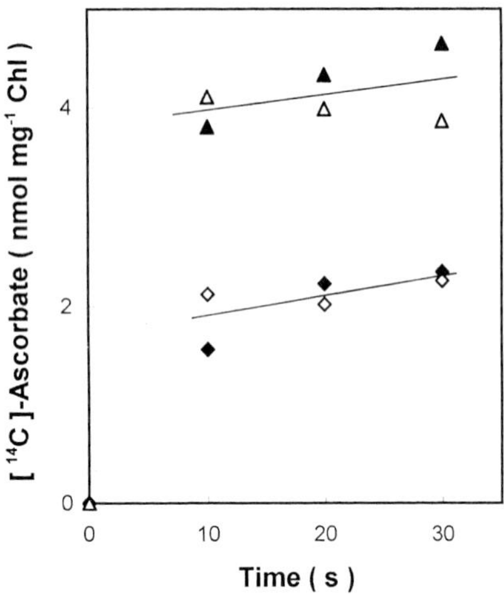

Fig. 1: Uptake of [^{14}C]-ascorbate by isolated pea leaf thylakoid membranes. Isolated thylakoid membranes were incubated with either 0.5 mM (diamonds) or 1 mM (triangles) ascorbate in the absence (filled symbols) or presence (open symbols) of 1 mM dehydroascorbate.

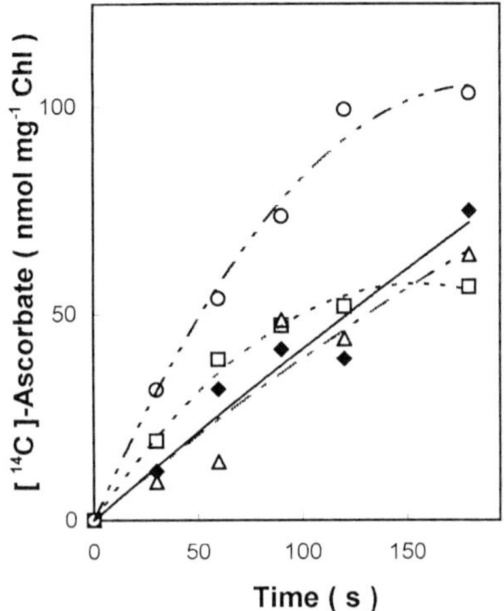

Fig. 3: The effect of glucose on [^{14}C]-ascorbate uptake by intact pea leaf chloroplasts. Intact chloroplasts were incubated with 10 mM [^{14}C]-ascorbate in the absence (◆) or presence of 5 mM (□), 10 mM (△) or 20 mM (○) glucose.

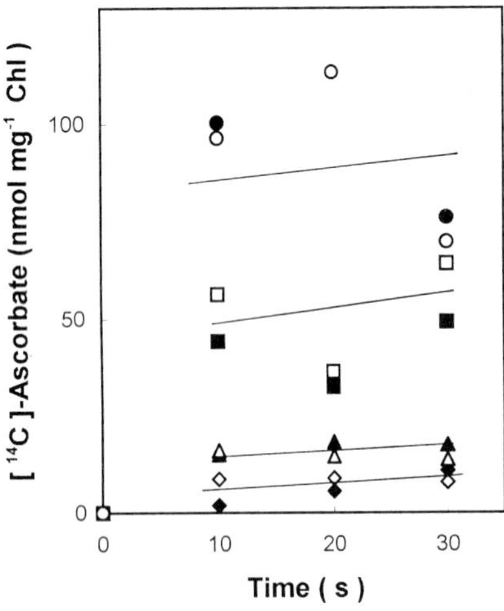

Fig. 2: The effect of ascorbate concentration on the uptake of [^{14}C]-ascorbate by pea leaf thylakoid membranes in the absence and presence of dehydroascorbate. Thyalkoid membranes were incubated with either 1 mM (◆,◇), 2.5 mM (▲,△), 5 mM (■,□) and 10 mM (●,○) ascorbate in either the absence (filled symbols) or presence (open symbols) of 2 mM dehydroascorbate.

species, ascorbic acid, can diffuse across the membrane. At pH 7.9 and 10 mM ascorbate the concentration of ascorbic acid is small (1.9 μM; pKa ascorbic acid = 4.18). The lumenal ascorbic acid concentration at 3.8 mM ascorbate (pH 7.9) would be about 0.7 μM.

The difference in [^{14}C]-ascorbate content of the membranes between the zero time point and the 10 s time point does not reflect uptake but is most probably due binding of ascorbate to the membranes (Figs. 1 and 2). A wide range of ascorbate concentrations (Fig. 2 shows data with up to 10 mM ascorbate) was studied (0.2–20 mM) but no measurable differences in the rate of uptake were found (data not shown). Dehydroascorbate had no effect on ascorbate uptake (Figs. 1 and 2). Rates of ascorbate translocation across the thylakoid membranes were similar in the dark and in the light and not affected by ATP (data not shown). Taken together, these data indicate the absence of carrier-mediated translocation of ascorbate into the thylakoid lumen. The values of ascorbate uptake into the thylakoid lumen are low and an order of magnitude lower than those obtained for ascorbate transport across the plasmalemma and chloroplast envelope membranes (Table 2).

Uptake of [^{14}C]-ascorbate into intact chloroplasts followed saturation kinetics (Fig. 3) suggesting that the transport process was carrier-mediated. Glucose concentrations in the range of 5–20 stimulated [^{14}C]-ascorbate uptake (Fig. 3). In contrast, dehydroascorbate concentrations of the same order inhibited the transport of ascorbate (Fig. 4). [^{14}C]-glucose was also rapidly transported into the pea chloroplasts (Fig. 5). The transport of glucose was unaffected by the presence of dehydroascorbate (Fig. 5), indicating that dehydroascorbate was not transported by the same carrier that was used to transport glucose across the chloroplast envelope.

The rate of uptake of [^{14}C]-ascorbate into pea leaf protoplasts was concentration-dependent and followed saturation

Table 2: Rates of [^{14}C]-ascorbate and [^{14}C]-glucose uptake into intact protoplasts and chloroplasts isolated from pea leaves.

A	Ascorbate uptake into:	Rate of uptake (nmol/min) of 10 mM [C^{14}]-ascorbate
(a)	Chloroplasts	36.64±1.52
(b)	Protoplasts	33.76±5.85
B	Glucose uptake into:	Rate of uptake (nmol/min) of 10 mM [C^{14}]-glucose
(a)	Chloroplasts	50.02±4.0
(b)	Protoplasts	60.1 ±25.7

kinetics suggesting carrier-mediated transport across the plasmalemma (Fig. 6). The transport of [14C]-ascorbate was not inhibited by glucose and indeed glucose appeared to stimulate [14C]-ascorbate transport into the protoplasts (Fig. 7). Similarly, dehydroascorbate did not inhibit [14C]-ascorbate uptake by the protoplasts (Fig. 8). [14C]-glucose uptake into the protoplasts followed saturation kinetics and was not inhibited by the presence of ascorbate (Fig. 9). As was found with the effect of glucose on [14C]-ascorbate, uptake ascorbate has a small but positive effect on [14C]-glucose uptake into the protoplasts (Fig. 9). A comparison of the data in Figures 7 and 9 suggests that there is some synergy between glucose and ascorbate transport across the plasmelemma membrane. In contrast to ascorbate, dehydroascorbate had no effect on [14C]-glucose transport into the protoplasts (Fig. 10).

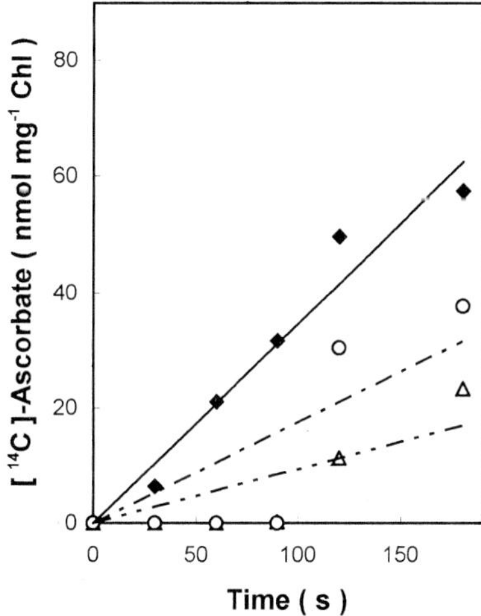

Fig. 4: The effect of dehydroascorbate on [14C]-ascorbate transport into intact pea leaf chloroplasts. [14C]-ascorbate was supplied at a concentration of 5 mM and dehydroascorbate concentrations were either 0 (◆), 10 mM (○) or 20 mM (◇).

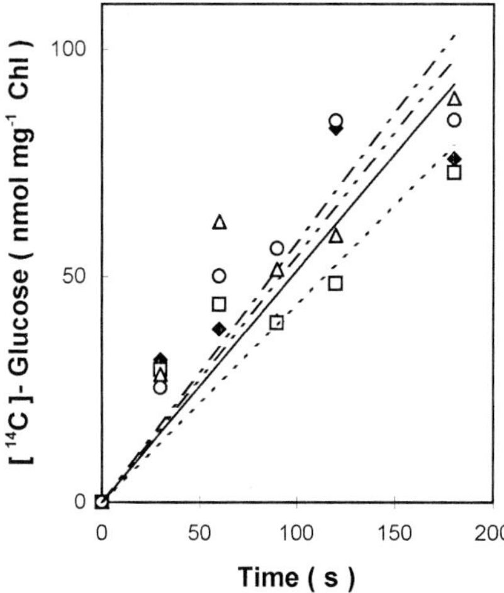

Fig. 5: Uptake of [14C]-glucose by pea leaf chloroplasts. The glucose concentration was 10 mM and dehydroascorbate concentrations were either 0 (◆), 5 mM (□), 10 mM (△) or 20 mM (○).

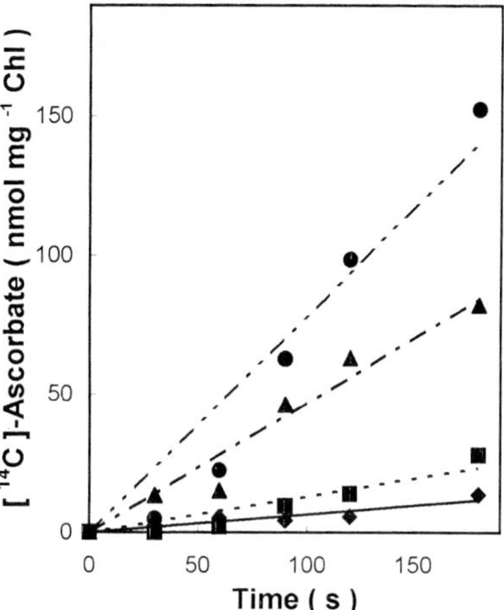

Fig. 6: Uptake of [14C]-ascorbate by intact pea leaf protoplasts. The concentrations of [14C]-ascorbate provided were 2 mM (◆), 4 mM (■), 6 mM (▲) and 10 mM (●).

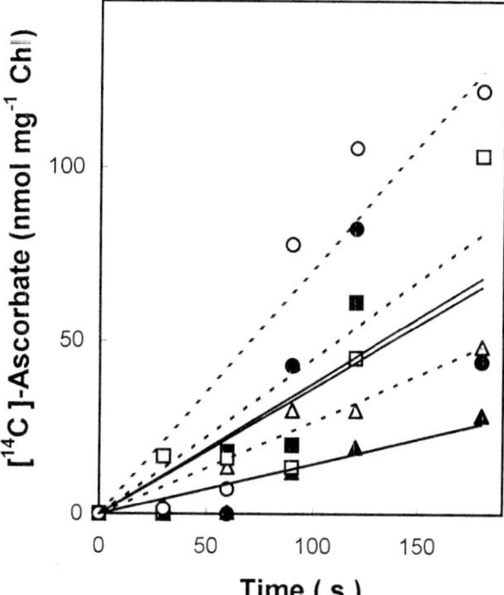

Fig. 7: The effect of glucose on [14C]-ascorbate uptake by pea leaf protoplasts. The [14C]-ascorbate concentrations used were 4 mM (triangles), 6 mM (squares) and 10 mM (circles). In each case the [14C]-ascorbate was supplied in the absence (filled symbols) or presence (open symbols) of 5 mM glucose.

Rapid rates of transport of ascorbate and glucose into intact protoplasts and intact chloroplasts were found (Table 2).

Discussion

The ascorbate concentrations of the pea leaf protoplasts, chloroplasts and thylakoid membranes are compatible with previous measurements (Foyer, 1993; Foyer et al., 1983; Franke and Heber, 1964; Rautenkranz et al., 1994). The chloroplasts contained 10 mM ascorbate, 4 mM of which was associated with the thylakoid membranes. In the present study only 12 % of the ascorbate in the mesophyll cells was localised in the chloroplasts. This value agrees with data obtained by Rautenkranz et al. (1994) but is somewhat lower than previous values of between 30–50% (Foyer et al., 1983; Franke and Heber, 1964). This discrepancy probably arises from differences in the methods of chloroplast isolation as well as species differences and differences in growth irradiance. Nevertheless, all studies agree that a substantial amount of the leaf mesophyll ascorbate pool is found in cellular compartments other than the chloroplast. If this ascorbate was evenly distributed between the other cellular compartments a value of about 8.2 mM ascorbate for the cell minus chloroplasts can be calculated. Rautenkranz et al. (1994) found that relatively little of the ascorbate measured in protoplasts prepared from barley seedlings was present in the vacuole. In barley seedlings grown under similar irradiances to those under which the pea plants were grown in the present study about 26% of the cellular ascorbate was found in the isolated vacuoles (Rautenkranz et al., 1994). If a similar situation existed in the present case and assuming that the vacuole occupies 70% of the protoplast volume then the ascorbate content of the cytosol would be about 20 mM. The proportion of ascorbate in the apoplastic compartment has not been measured in the present study. Transport of ascorbate across the plasmalemma membrane is facilitated by the presence of a carrier. If the apoplastic ascorbate content is below 5 mM as is the case in other studies (Castillo and Greppin, 1988; Polle et al., 1990, 1995; Takahama and Oniki, 1992; Luwe et al., 1993), then transport from the cytoplasm to the apoplast will follow a relatively steep gradient that allows the concentrations of apoplastic ascorbate to maintain levels required for antioxidative protection as well as other functions. Oxidised ascorbate may be recycled *via* NADH-Monodehydroascorbate reductase enzymes present on the plasmalemma (Navas et al., 1994).

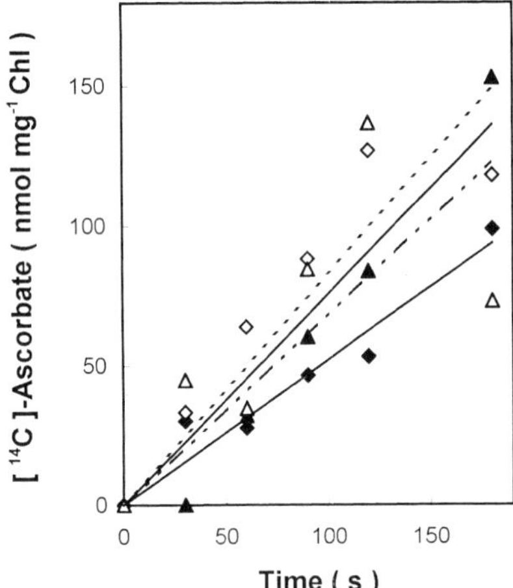

Fig. 8: The effect of dehydroascorbate on [^{14}C]-ascorbate uptake by intact pea leaf protoplasts. The [^{14}C]-ascorbate concentrations supplied were 6 mM (diamonds) or 10 mM (triangles) and uptake was measured either in the absence (filled symbols) or presence (open symbols) of 5 mM dehydroascorbate.

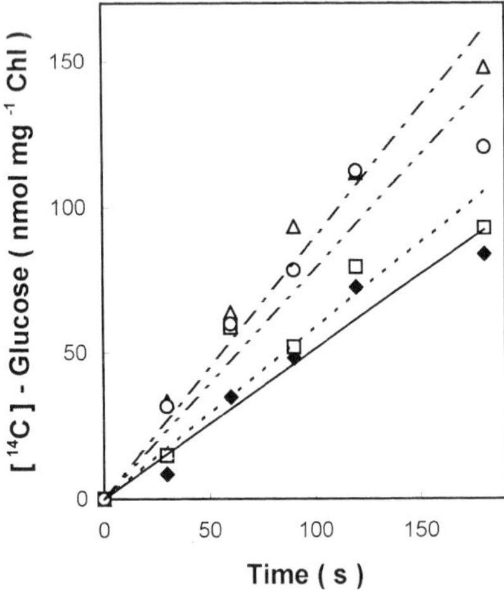

Fig. 9: The effect of ascorbate on [^{14}C]-glucose uptake by intact pea leaf protoplasts. [^{14}C]-glucose was supplied to intact protoplasts at a concentration of 5 mM. [^{14}C]-glucose uptake was measured in either the absence (◆) or presence of 5 mM (□), 10 mM (△) or 20 mM (○) ascorbate.

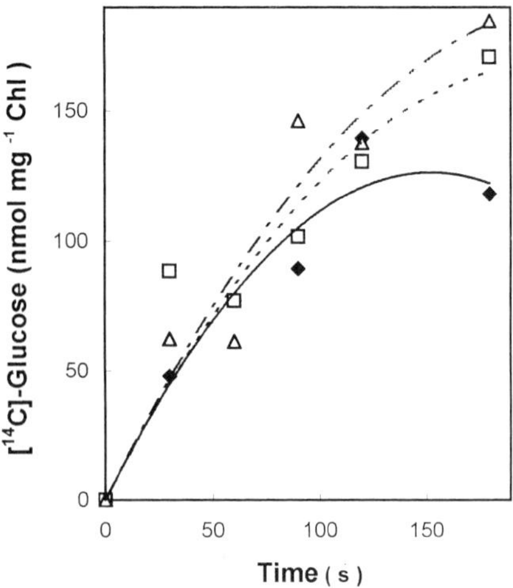

Fig. 10: The effect of dehydroascorbate on [^{14}C]-glucose uptake by intact pea leaf protoplasts. The [^{14}C]-glucose concentration was 10 mM and this was supplied to the protoplasts in either the absence (◆) or presence of dehydroascorbate at concentrations of 5 mM (□) or 10 mM (△).

The thylakoid membranes do not appear to facilitate transport of ascorbate into the thylakoid lumen. No mechanism of ascorbate transport, other than diffusion, could be detected. This will have important consequences. At a constant concentration of external ascorbate the uptake rate should be strongly pH-dependent since only the uncharged species ascorbic acid can diffuse across the membrane. Similarly, only the ascorbic acid concentration would equilibrate between the lumen and stroma, the total ascorbate content (acid+base forms) being different in the stromal and lumenal compartments as long as a pH-gradient existed. At pH-equilibrium the concentration of ascorbate should be similar in both compartments. The absence of ascorbate transport to the lumen is surprising considering the important function of ascorbate in Zeaxanthin synthesis. Neubauer and Yamamoto (1994) pointed out that permeability of the thylakoid membrane for ascorbate would limit the activity of violaxanthin de-epoxidase. The lumen pH is also important in the activation and mobility of this enzyme (Pfündel and Dilley, 1993; Pfündel and Bilger, 1994; Bratt et al., 1995). As the pH of the thylakoid lumen decreases, the violaxanthin de-epoxidase binds to the lumenal side of the thylakoid membrane and becomes essentially immobile. The Km of violaxanthin de-epoxidase for ascorbate is also highly dependent on pH. The true substrate for the enzyme is ascorbic acid rather than the base form. Bratt et al. (1995) calculate a Km of 0.1 ± 0.02 mM for ascorbic acid. Hence, as the lumen pH falls and Zeaxanthin formation proceeds, ascorbic acid will move into the lumen. It is possible that this is sufficient to satisfy the affinity of the enzyme for ascorbic acid but modulation of the availability of ascorbic acid *via* limitation of transport may have regulatory significance.

Ascorbate is taken up by intact protoplasts and chloroplasts by a system that displays both saturation and accumulation. This indicates that in both cases ascorbate uptake is mediated by specific transporters at the cell membrane and the chloroplast envelope. In contrast to animal cells where ascorbate transport is linked to the hexose transport system glucose uptake was not inhibited by ascorbate or dehydroascorbate or ascorbate uptake inhibited by glucose in either the intact protoplasts or chloroplasts. This suggests that the ascorbate transporters are discrete entities and not part of the general hexose transport system. The observed small but apparent stimulation of ascorbate accumulation caused by the presence of glucose and *vice versa* by the intact protoplasts is intriguing. The mechanism of this synergistic effect is, at present, unknown. Rapid translocation of ascorbate across the cell membrane and chloroplast envelope ensures the provision of adequate levels of ascorbate at appropriate sites. Inhibition of [^{14}C]-ascorbate transport by dehydroascorbate at the chloroplast envelope membrane suggests that dehydroascorbate competes for the same transport carrier. The absence of any dehydroascorbate-induced inhibition of [^{14}C]-ascorbate uptake by intact protoplasts suggests that the ascorbate transporter on this membrane does not recognise dehydroascorbate. This is intriguing since oxidised ascorbate must be recycled. One possibility is that a separate transporter for dehydroascorbate exists on the plasmalemma. Alternatively, the plasmalemma-based enzymes using oxidised ascorbate as a substrate may quickly and effectively regenerate reduced ascorbate (Navas et al., 1994). Transport of oxidised ascorbate back to the cytosol may introduce an element of futile competition in this case.

References

Anderson, J. W., C. H. Foyer, and D. A. Walker: Light-dependent reduction of dehydro-ascorbate and uptake of exogenous ascorbate by spinach chloroplasts. Planta *158*, 442–450 (1983 a).

Arnon, D. I.: Copper enzymes in isolated chloroplasts. Polyphenol oxidase in *Beta vulgaris* L. Plant Physiol. *24*, 1–15 (1949).

Arrigoni, O.: Ascorbate system in plant development. J. Bioenergetics and Biomembranes *26*, 407–419 (1994).

Beck, E., A. Burkert, and M. Hofmann: Uptake of L-ascorbate by intact chloroplasts. Plant Physiol. *73*, 41–45 (1983).

Bratt, C. E., P.-O. Arvidsson, M. Carlson, and H.-E. Åkerlund: Regulation of violaxanthin de-epoxidase activity by pH and ascorbate concentration. Photosyn. Res. *45*, 169–175 (1995).

Castillo, F. J. and H. Greppin: Apoplastic ascorbic acid and enzyme activities related to acid metabolism in *Sedum album* L. leaves after ozone exposure. Environ. Exp. Bot. *28*, 231–238 (1988).

Castillo, F. J., P. R. Miller, and H. Greppin: «Waldsterben», Part IV: Extracellular biochemical markers of photochemical oxidant air pollution damage to Norway spruce. Experientia *43*, 111–115 (1987).

Citterio, S., S. Sgorbati, S. Scippa, and E. Sparvoli: Ascorbic acid effect on the onset of cell proliferation in pea root. Physiol. Plant. *92*, 601–607 (1994).

Coic, Y. and C. Lesaint: La nutrition mineralé et en eau des plantes en horticulture acancée. Document Technique de SCPA *23*, 1–22 (1975).

De Gara, L., C. Paciolla, F. Tommasi, R. Liso, and O. Arrigoni: *In vivo* inhibition of galactono-γ-lactone conversion to ascorbate by lycorine. J. Plant Physiol. *144*, 649 (1994).

Foyer, C. H.: Ascorbic acid. In: Alscher, R. G. and J. L. Hess (eds.): Antioxidants in higher plants. CRC Press Inc., Boca Raton, Florida, USA, 31–58 (1993).

Foyer, C. H., J. Rowell, and D. A. Walker: Measurement of the ascorbate content of spinach leaf protoplasts and chloroplasts during illumination. Planta *157*, 239–244 (1983).

Foyer, C. H., M. Lelandais, and K. J. Kunert: Photooxidative stress in plants. Physiol. Plant. *92*, 616–717 (1994).

Franke, W. und U. Heber: Über die quantitative Verteilung der Ascorbinsäure innerhalb der Pflanzenzelle. Z. Naturforsch. *196*, 1146–1149 (1964).

Hager, A.: Lichtbedingte pH-Erniedrigung in einem Chloroplasten-Kompartiment als Ursache der enzymatischen Violaxanthin-Zeaxanthin-Umwandlung, Beziehungen zur Photophosphorylierung. Planta *89*, 224–243 (1969).

Hager, A. and K. Holocher: Localization of the xanthophyll-cycle enzyme violaxanthin de-epoxidase within the thylakoid lumen and abolition of its mobility by a (light-dependent) pH decrease. Planta *192*, 581–589 (1994).

Heldt, H. W.: Measurement of metabolite movement across the envelope and of the pH in the stroma and the thylakoid space in intact chloroplasts. Methods Enzymol. *69*, 604–613 (1980).

Luwe, M. W. F., U. Takahama, and U. Heber: Role of ascorbate in detoxifying ozone in the apoplast of spinach (*Spinacia oleracea* L.) leaves. Plant Physiol. *101*, 969–976 (1993).

Lilley, R. MCC., M. P. Fitzgerald, K. G. Rienits, and D. A. Walker: Criteria of intactness and photosynthetic activity of spinach chloroplast preparations. New Phytol. *75*, 1-10 (1975).

Loewus, F. A.: L-ascorbic acid: metabolism, biosynthesis, function. In: Preiss, J. (ed.): The Biochemistry of Plants, Vol. *3*. Academic Press, San Diego, CA, pp. 77–99 (1980).

Loewus, M. W., D. L. Bedgar, K. Saito, and F. A. Loewus: Conversion of L-ascorbic acid by a NADP-dependent dehydrogenase in bean and spinach leaves. Plant Physiol. *94*, 1492 (1990).

Luwe, M. W. F., U. Takahama, and U. Heber: Role of ascorbate in detoxifying ozone in the apoplast of spinach (*Spinacia oleracea* L.) leaves. Plant Physiol. *101*, 969–976 (1993).

Luwe, M. and U. Heber: Ozone detoxification in the apoplast and symplast of spinach, broad bean and beech leaves at ambient and elevated concentrations of ozone in air. Planta (in press) (1995).

Mills, W. R. and K. W. Joy: A rapid method for isolation of purified, physiologically active chloroplasts used to study the intracellular distribution of amino acids in pea leaves. Planta *148*, 75–83 (1980).

Morré, D. J., A. O. Brightman, L.-Y. Wu, R. Barr, B. Leak, and F. L. Crane: Role of plasmamembrane redox activities in elongation growth in plants. Physiol. Plant. *73,* 187–193 (1988).

Navas, P., J. M. Villalba, and F. Cordoba: Ascorbate function at the plasmamembrane. Biochim. Biophys. Acta *1197,* 1–13 (1994).

Neubauer, C. and H. Y. Yamamoto: Membrane barriers and Mehler-peroxidase reaction limit the ascorbate available for violaxanthin de-epoxidase activity in intact chloroplasts. Photosynth. Res. *39,* 137–147 (1994).

Pfündel, E. E. and B. Bilger: Regulation and possible function of the violaxanthin cycle. Photosyn. Res. *42,* 89–109 (1994).

Pfündel, E. E. and R. A. Dilley: The pH dependence of violaxanthin de-epoxidation in isolated pea chloroplasts. Plant Physiol. *101,* 65–71 (1993).

Polle, A., K. Chakrabarti, W. Schücrmann, and H. Rennenberg: Composition and properties of hydrogen peroxide decomposing systems in apoplastic and total extracts from needles of Norway spruce (*Picea abies* L. Karst.). Plant Physiol. *94,* 312–319 (1990).

Polle, A., G. Wieser, and W. M. Harranek: Quantification of ozone influx and apoplastic ascorbate content in needles of Norway spruce trees (*Picea abies* L. Karst.) at high altitude. Plant, Cell and Environ. *18,* 681–688 (1995).

Rautenkranz, A. A. F., L. Li, F. Machler, E. Martinoia, and J. J. Oertli: Transport of ascorbic and dehydroascorbic acids across protoplast and vacuole membranes isolated from barley (*Hordeum vulgare* L., cv. Gerbel) leaves. Plant Physiol. *106,* 187–193 (1994).

Takahama, U.: Redox state of ascorbic acid in the apoplast of stems of *Kalanchoe daigremontiana.* Physiol. Plant. *89,* 791–798 (1993).

Takahama, U.: Regulation of peroxidase-dependent oxidation of phenolics by ascorbic acid – different effects of ascorbic acid on the oxidation of coniferyl alcohol by the apoplastic soluble and cell wall-bound peroxidases from *Vigna angularis.* Plant Cell Physiol. *34,* 809–817 (1993a).

– Redox state of ascorbic acid in the apoplast of stems of *Kalanchoë daigremontiana.* Physiol. Plant. *89,* 791–798 (1993b).

Takahama, U. and T. Oniki: Regulation of peroxidase-dependent oxidation of phenolics in the apoplast of spinach leaves by ascorbate. Plant Cell Physiol. *33,* 379–387 (1992).

Vera, J. C., C. I. Rivas, J. Fischbarg, and D. W. Golde: Mammalian facilitative hexose transporters mediate the transport of dehydroascorbic acid. Nature *364,* 79–82 (1993).

Walker, D. A.: Experiment 3. CO_2 and PGA-dependent O_2 evolution. In: The use of the oxygen electrode and fluorescence probes in simple measurements of photosynthesis. Oxygraphic Ltd., obtained from Hansatech Instruments Ltd., Kings Lynn, UK, 129 (1987).

J. Plant Physiol. Vol. 148. pp. 399–412 (1996)

Photosynthetic CO_2-Assimilation, Chlorophyll Fluorescence and Zeaxanthin Accumulation in Field Grown Maple Trees in the Course of a Sunny and a Cloudy Day*

Christiane Schindler and Hartmut K. Lichtenthaler

Botanisches Institut, Lehrstuhl II, University of Karlsruhe, Kaiserstr. 12, D-76128 Karlsruhe, Germany

Received October 5, 1995 · Accepted December 5, 1995

Summary

Changes in photosynthetic CO_2 fixation, variable chlorophyll fluorescence ratios, carotenoid composition, zeaxanthin accumulation and the de-epoxidation state DEPS of the zeaxanthin cycle were simultaneously determined in leaves of field-grown maple trees (*Acer platanoides* L.) on a sunny and a cloudy day in August. In the course of the sunny day the photosynthetic net CO_2 assimilation rates P_N rose with the PPFD in the morning and were then predominantly determined by the stomata opening (partial closure at noon and further closure in the afternoon as seen from the decrease in gH_2O-values). When the rate of CO_2-fixation began to decrease at a high sun irradiance (high PPFD) at about 10:00 am (partial closure of stomata), large zeaxanthin amounts were accumulated in a biphasic process: 1) by a fast phototransformation of existing violaxanthin into zeaxanthin and 2) by a slow but continuous increase of the zeaxanthin level of 50 to 60 % *via de novo* biosynthesis. A strong zeaxanthin accumulation depends on a high proton gradient and only proceeded when P_N was light-saturated and was reduced by a partial closure of stomata. The zeaxanthin accumulation in the morning followed the rise in PPFD with some delay as well as the zeaxanthin retransformation to violaxanthin at the PPFD decrease in the late afternoon and evening.

On the sunny day the photochemical and non-photochemical quenching coefficients of the chlorophyll fluorescence, qP and qN, showed an inverse relationship. With increasing PPFD qP declined to very low values (below 0.1) and qN increased to high values >0.9. The variable chlorophyll fluorescence ratios Fv′/Fm′ and ΔF/Fm′, indicators of the quantum efficiency of photosynthetic electron transport of photosystem II, declined parallel to the high PPFD-induced decline of qP and the rise in qN. There was, however, no clear correlation between the changes in chlorophyll fluorescence parameters and the rise and fall of zeaxanthin in the course of the sunny day. In fact, in the afternoon the zeaxanthin level decreased much earlier than the decline in qN and the rise and regeneration of the initial values of qP, Fv′/Fm′ or ΔF/Fm′. Despite the strong increase in qN and the large decrease in qP, Fv′/Fm′ and ΔF/Fm′, the net CO_2 assimilation rates P_N of maple leaves only declined by 22 % as compared to the maximum P_N values at 10 : 15 am indicating that the photosynthetic apparatus was only marginally photoinhibited. It is assumed that only a small fraction of chloroplasts at the upper leaf side became photoinhibited and functioned as a light filter for the other chloroplasts in the lower mesophyll parts which remained photosynthetically active. The results demonstrate that the changes of the chlorophyll fluorescence parameters qP and qN as well as of the ratios Fv′/Fm′ or ΔF/Fm′, measured with the PAM fluorometer at the upper leaf side, are not representative of the total pool of leaf chloroplasts. For this reason it is suggested to measure the chlorophyll fluorescence parameters not only at the upper but also at the lower leaf side in order to obtain more representative information on the chloroplasts of the whole leaf. In addition, chlorophyll fluorescence measurements should always be complemented by net CO_2 fixation measurements in order to investigate to which degree a presumed photoinhibition really exists at the whole leaf level.

On the cloudy day up to 3 pm the PPFD was very low, yet net photosynthesis (P_N) was performed at good rates. There occurred, however, hardly any increase in zeaxanthin content, except for a short sunny

* Dedicated to Prof. Dr. Guy Ourisson, Strasbourg, on the occasion of his 70th birthday.

period in the early afternoon. An accumulation of zeaxanthin by *de novo* biosynthesis did, however, not occur. Significant changes in the quenching coefficients qP and qN or in the variable fluorescence ratios ΔF/Fm′ and Fv′/Fm′ were not detectable in the course of the cloudy day.

The results suggest that zeaxanthin seems to play an essential role in photoprotection of the photosynthetic apparatus against high-light exposure in plants growing at field conditions in direct sunlight. The major part of high light induced chlorophyll fluorescence quenching, as seen in the decline of qP and the fluorescence ratios Fv′/Fm′ and ΔF′/Fm′ and in the rise of qN, is apparently not dependent on zeaxanthin. The possible function of zeaxanthin (detoxifying highly reactive oxygen species, removing of expoy groups from lipids or as spacer between LHCII and the reaction center of PSII) is being discussed, yet the exact mode of the photoprotective action of zeaxanthin remains to be clarified.

Key words: Acer platanoides L., carotenoids, chlorophyll fluorescence, daily cycle, photoprotection, photosynthesis, zeaxanthin, zeaxanthin cycle.

Abbreviations: an = antheraxanthin; *c* = α+β-carotene; chl *a* = chlorophyll *a;* chl *b* = chlorophyll *b;* c_i = internal CO_2-concentration; DEPS = deepoxidation state of the zeaxanthin cycle; Fv′/Fm′, ΔF/Fm′ = ratios of the variable chlorophyll fluorescence under illuminated conditions; gH_2O = stomatal conductance; *l* = lutein; *n* = neoxanthin; PPFD = photosynthetic photon flux density; P_N = net photosynthesis rate; PSII = photosynthetic photosystem II; qP and qN = photochemical and non-photochemical quenching coefficients; *v* = violaxanthin; *v*+*an*+*z* = sum of zeaxanthin cycle carotenoids; *z* = zeaxanthin.

Introduction

It has been established by several research groups that among the photosynthetic carotenoids, which are present in the photosynthetic membranes, three xanthophylls undergo light-dependent interconversions (Hager, 1975; Siefermann-Harms, 1977; Yamamoto, 1979; Schindler and Lichtenthaler, 1994). This light-driven zeaxanthin cycle involves the enzymic conversion (photoreduction) of the di-epoxide violaxanthin to the epoxide-free zeaxanthin *via* the intermediate antheraxanthin. This process proceeds predominantly at high-light conditions when photosynthetic quantum conversion and CO_2 assimilation are light-saturated. In low-light or darkness the back-reaction occurs, i.e. the oxygenation of zeaxanthin to its di-epoxide violaxanthin. The high-light induced accumulation of zeaxanthin is often a biphasic process, with a fast photoconversion of violaxanthin into zeaxanthin which is followed by a slow increase in the total amount of zeaxanthin cycle carotenoids by a *de novo*-biosynthesis of zeaxanthin (Lichtenthaler and Schindler, 1992; Schindler and Lichtenthaler, 1994). This double mechanism of zeaxanthin accumulation provides a fast acclimation of the photosynthetic apparatus to high-light conditions. There is agreement that zeaxanthin itself or together with antheraxanthin plays a photoprotective role against photoinhibitory irradiances (Demmig-Adams and Adams III, 1992; Lichtenthaler und Schindler, 1992; Gilmore and Yamamoto, 1993; Pfündel und Bilger, 1994; Schindler and Lichtenthaler, 1994: Mohanty et al., 1995; Mohanty et al., 1995).

Plants are subjected to various kinds of natural and anthropogenic stressors (Lichtenthaler, 1996), whereby outdoor high-light stress at locations with strong sun irradiance is a frequent matter to which the plants have to respond by acquisition of adaptive and photoprotective mechanisms. Other environmental stressors, such as heat and water shortage, can enhance the damage of excessive light energy. Whenever an environmental stress factor causes a decrease in the rate of photosynthesis, the prevailing PPFD to which the plants are exposed can become excessive in situations where such PPFDs were not excessive before. At such conditions light is being absorbed by chlorophylls and carotenoids in the light-harvesting antenna, but cannot be utilized in photosynthesis, since the photosynthetic quantum conversion and CO_2 fixation processes are reduced, e.g. by photoinhibition (Krause and Weis, 1991) or photooxidative damage. This is then seen in a reduction of either photosynthetic electron transport and carbon assimilation or in a decrease of the values of the ratios of the variable chlorophyll fluorescence Fv/Fm and ΔF/Fm′ as well as in an increase of the non-photochemical quenching coefficient qN (Schreiber et al., 1986; Lichtenthaler und Rinderle, 1988; Genty et al., 1989; Havaux et al., 1991 a and b). Photoinhibition of photosynthesis is generally manifest as a sustained decrease in the efficiency of photosynthetic quantum conversion.

Whereas the formation of zeaxanthin by photoreduction of violaxanthin has extensively been studied at laboratory conditions, there have been only a few determinations of the diurnal sun-induced changes in zeaxanthin cycle carotenoids of field-grown higher plants. Apart from this there are some indications that leaves growing in natural sunlight show decreases in several variable chlorophyll fluorescence parameters during the course of the day which parallel the increase in incident PPFD (Adams et al., 1987, 1988, 1989; Einig et al., 1991). Although high levels of zeaxanthin in full sunlight and changes in the epoxidation state of the zeaxanthin cycle carotenoids in the course of a day were described for several field grown species (Adams and Demmig-Adams, 1992), there is very littly information about the daily course of the components of the zeaxanthin cycle in response to variation in the net photosynthetic CO_2 fixation rates. In fact, simultaneous kinetics of zeaxanthin accumulation and the rates of photosynthetic CO_2 fixation have not yet been determined. Although chlorophyll fluorescence coefficients and ratios can be taken as indicators of the photosynthetic processes, they cannot replace the measurements of net CO_2 fixation rates.

For this reasons we have determined the kinetics of photosynthetic CO_2 fixation and chlorophyll fluorescence parameters and contrasted these to changes in zeaxanthin levels in leaves of field-grown trees of *Acer platanoides* L. in the course of a sunny and a cloudy day. The kinetics were recorded in response to the diurnal changes in the incident PPFD in leaves of young maple trees growing outdoors in the Botanical Garden.

Materials and Methods

Plants

Four year old maple plants (*Acer platanoides* L.) were cultivated in 10 L pots under field conditions on a humus rich soil. The trees with stem and side branches reached a height of about 1.5 meters. The leaves were arranged on the stem with two leaves on opposite sides of each whorl. Gas exchange parameters, changes in chlorophyll fluorescence and contents of the individual photosynthetic pigments were determined in parallel measurements on fully expanded dark-green leaves. The fully developed leaves were of similar size and showed an average leaf area of about 120 cm^2. The daily watered maple trees were set up on a shadeless place in the Botanical Garden. The leaves to be investigated were oriented towards the sun in such a way that they received full sunlight during the whole day. Most measurements were performed from 8:00 a.m. to 10:00 p.m. Due to the «daylight savings time» in Germany in the summer these hours correspond, however, to 7:00 am through 9:00 pm of Central European time.

Gas exchange measurements

For the gas exchange measurements with the CO_2/H_2O-porometer (Walz, Effeltrich, Germany) one leaf of a maple whorl was placed into the measuring cuvette of the porometer, which was always adjusted towards the sun during the day, so that the whole leaf surface was exposed to full sunlight. The parameters of net photosynthesis rate (P_N), transpiration, stomatal conductance (gH_2O), internal CO_2-concentration (c_i) and photon flux density (PPFD) were determined every 15 min throughout the day.

Chlorophyll fluorescence measurements

The chlorophyll fluorescence kinetics were measuremed using the PAM-fluorometer (Walz, Effeltrich, Germany). These measurements were performed with the parallel leaf on the same whorl of the maple tree. The leaf was fixed in such a way that the measuring fiber optic of the PAM-fluorometer was as close as possible to the leaf surface and did not shade the leaf surface. The red actinic light of the PAM fluorometer together with the sun light functioned as actinic light source. A simultaneous saturating pulse of white light of 3000 $\mu mol\ m^{-2}\ s^{-1}$ applied using the PAM-fluorometer allowed to determine the fluorescence parameters Fm′ and Fv′ in the illuminated state (Fig. 1). The fluorescence ratios Fv′/Fm′ and ΔF/Fm′, as well as the quenching coefficients qP and qN were calculated according to Schreiber et al. (1986) and Lichtenthaler and Rinderle (1988) using the nomenclature of van Kooten and Snel (1990). The ratio ΔF/Fm′ is defined as ratio of (Fm′-F)/Fm′ (see Fig. 1). The related chlorophyll fluorescence ratio Fv′/Fm′ is also measured in the illumination state of leaves (confer Fig. 1). The photochemical quenching coefficient qP is defined as (Fm-F)/Fv′ = ΔF/Fv′, whereas the non-photochemical quenching coefficient qN is given as the difference 1-(Fv′/Fv).

Pigment analysis

Maple leaves of the whorl next to the leaves used for gas exchange and chlorophyll fluorescence measurements were selected to quantify the level of chlorophylls *a* and *b* and individual carotenoids during the course of the day using a high-pressure liquid chromatography system (HPLC) as described before (Schindler and Lichtenthaler, 1994). We had checked before that the chlorophyll and carotenoid composition of the green maple leaves at two neighbouring whorles were the same within a narrow range of variation. During the investigations in the course of a full day the leaf samples were collected continuously as leaf discs (9 mm diameter) from the vein-free regions of the leaves and stored in liquid nitrogen. Four comparable leaf samples were taken for pigment determination at each measuring point. Pigment analysis was performed on the following day. After extraction of leaf discs with 100 % acetone (quartz sand, mortar) the pigments were separated and quantitatively determined by reversed phase HPLC applying a Nucleosil C_{18} column (particle

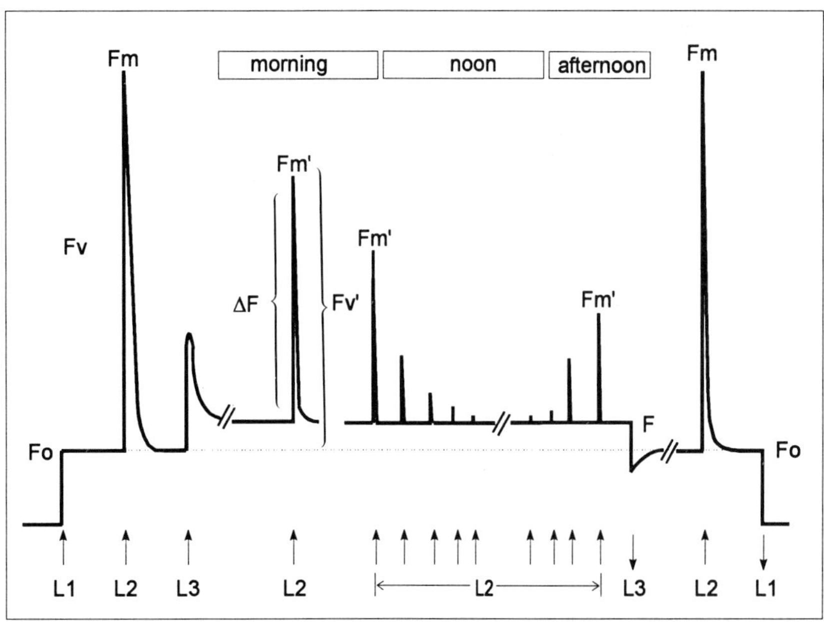

Fig. 1: Sequence of the determination of various chlorophyll fluorescence parameters in maple leaves in the course of the sunny day using the Walz PAM fluorometer applying the following light sources: L1 = modulated red light (0.1 $\mu mol\ m^{-2}\ s^{-1}$); L2 = white saturating light pulse of 1s (3000 $\mu mol\ m^{-2}\ s^{-1}$) and L3 = continuous red actinic light (60 $\mu mol\ m^{-2}\ s^{-1}$) plus natural sunlight of low PPFD in the morning and evening and of high PPFD between ca. 10 am to ca. 3.45 pm. These kinetics allowed to measure the actual quantum yield of photosystem II in the illuminated state (ratios ΔF/Fm′ and Fv′/Fm′) as well as the photochemical und non-photochemical quenching coefficients qP and qN. (↑ indicates light on and ↓ light off for L1 and L3; in the case of L2 ↑ indicates the 1s light pulse).

size 4 µm). The solvent systems for the combined isocratic and gradient separation were acetonitrile-methanol-0.2 M Tris/HCl buffer and methanol-hexane as described (Lichtenthaler et al., 1992). For comparative reasons and fast information the levels of total chlorophyll *a+b* and total carotenoid *x+c* were determined spectrophotometrically in the 100% acetone extract solution using the new, redetermined absorption coefficients and equations (Lichtenthaler, 1987). These allow to calculate the levels of chlorophylls *a* and *b* and total carotenoids from one extract solution.

The *de-epoxidation state DEPS* of the zeaxanthin cycle was determined as the ratio of $(z+0.5an)/(v+an+z)$. Zeaxanthin is free of epoxides, whereas antheraxanthin contains one epoxy group in one of the two β-ionone rings. This is why antheraxanthin (*an*) is only considered by the factor 0.5 in the numerator of the above ratio.

Results

1. Sunny day

Changes in irradiance and photosynthesis

The green maple leaves investigated here exhibited a light saturation of photosynthesis at 800 to 900 µmol photons m^{-2} s^{-1}. On a very clear and warm day in August with exceptionally bright sunshine the photosynthetic photon flux density PPFD measured in the leaf sample cuvette and the ambient air temperature increased during the course of the day (Fig. 2A.). The PPFD showed a maximum at 1 pm, which was higher on this unusual bright day than on other sunny days. Some sun light reflection contributed to this very high PPFD. The PPFD strongly declined after 3.45 pm, when some clouds showed up. The temperature rose from 20 °C in the morning to a broad maximum at or above 35 °C from 3 to 7 pm (Fig. 2A.) and then declined to 26 °C at 10 pm. The net photosynthesis CO_2 fixation P_N (Fig. 2B.) started with good rates even at a low PPFD and then increased fast from 8:30 am to a maximum value of 11.3 µmol CO_2 $m^{-2}s^{-1}$ at 10:00 am, when the actual PPFD was 2600 µmol m^{-2} s^{-1}. Thereafter the net photosynthetic rates decreased by about 22%, stabilized at P_N values of 8.8 µmol CO_2 m^{-2} s^{-1} and only further declined with the decrease in PPFD in the later afternoon at and after 4 pm. The short PPFD-increase towards 6 pm caused a parallel short increase of the net photosynthetic rate. With the beginning of darkness at 10:00 pm the rate of net photosynthesis was zero. The increase in photosynthetic net CO_2 fixation after 10:00 am as well as the decline thereafter correlated very well with the stomata conductance of water vapour gH_2O (Fig. 2C.), thus showing that the decline in photosynthesis was primarily caused by a partial closure of the stomata and the corresponding lack of CO_2. The rate of transpiration exhibited a somewhat different kinetic (Fig. 2B.). The internal CO_2 concentration c_i, in turn, showed a similar trend to P_N and gH_2O, except for the late afternoon after 6 pm when PPFD and photosynthetic CO_2 fixation declined (Fig. 2C.).

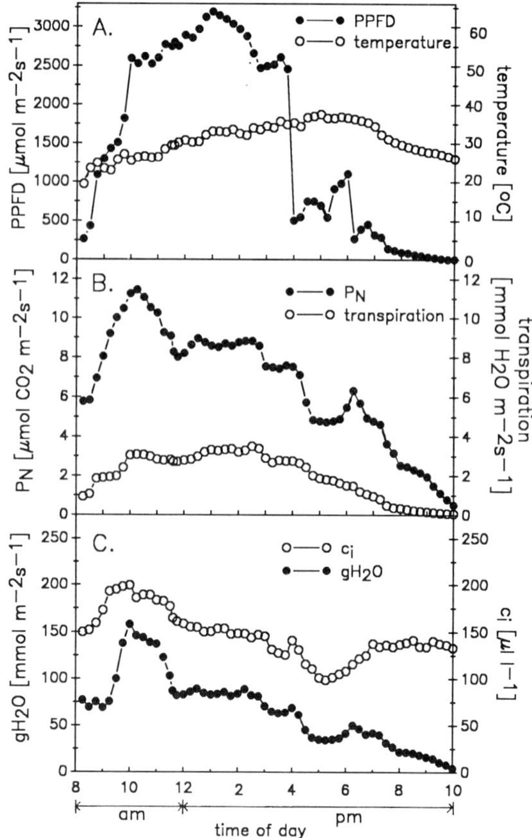

Fig. 2: Course of air temperature and photosynthetic photon flux density (PPFD) (**A.**), net photosynthesis rate (P_N) and transpiration (**B.**) as well as stomatal conductance (gH_2O) and internal CO_2-concentration (c_i) (C.) in *Acer platanoides* leaves during the course of a sunny day.

Changes in pigments and zeaxanthin cycle carotenoids

In the course of the sunny day the level of leaf chlorophyll *a* and total carotenoids *x+c* showed a slight tendency for higher pigment levels at the end of the day, whereas the content of chlorophyll *b* was little affected (Fig. 3A.). The levels of the major individual carotenoids lutein and β-carotene showed a tendency for an increase of about 20% as compared to the morning values, which appeared to be significant ($p<0.05$). The maple leaves also contained some minor amounts of α-carotene which significantly rose during the day. The level of neoxanthin did, however, not exhibit any change (Fig. 3B.).

In contrast, the zeaxanthin cycle xanthophylls showed big changes during the course of the sunny and warm day. The increase in PPFD during the morning was paralleled by a decrease in violaxanthin content and a concomitant increase of its de-epoxidation product zeaxanthin. The zeaxanthin level increased from a value of 0.16 µg cm^{-2} in the early morning to a maximum value of 1.66 µg cm^{-2} leaf area at noon time (Fig. 4A.). The accumulation of zeaxanthin always proceeded with a clear delay to the increase in PPFD. The major photoconversion of violaxanthin into zeaxanthin started when the linear rise in P_N rates (Fig. 5) stopped at 9:45 am, and the P_N rates began to decline. A stable maximum content of zeaxanthin was reached when the maximum P_N rate had declined by about 22% at noon.

The further PPFD rise to 3200 µmol m^{-2} s^{-1} (measured in the cuvette) from noon to 1 pm was, however, not accompanied by a further increase in the zeaxanthin level. The sharp

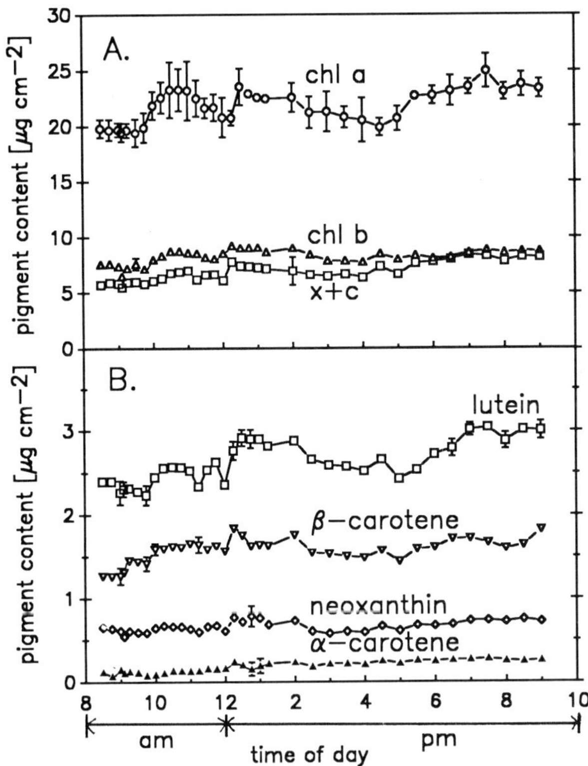

Fig. 3: Daily cycle of chlorophylls *a* and *b* and total carotenoid content (*x*+*c*) (**A.**) and α-carotene and β-carotene as well as xanthophylls lutein and neoxanthin (**B.**) in leaves of *Acer platanoides* in the course of a sunny day. Values are means (n=4) with standard deviation.

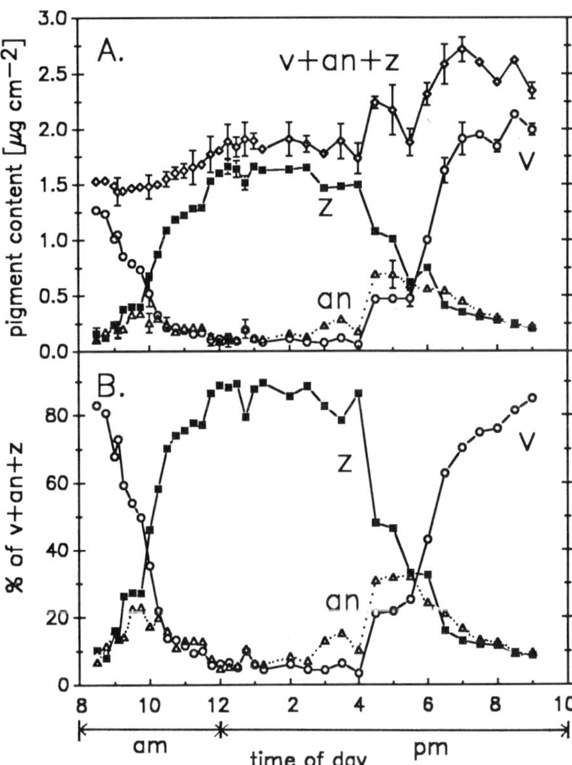

Fig. 4: Changes in the zeaxanthin cycle xanthophylls zeaxanthin (*z*), antheraxanthin (*an*), violaxanthin (*v*) and the sum of *v*+*an*+*z* in green maple leaves during the course of the sunny day. **A**) Absolute amounts per leaf area and **B**) percentage amounts per sum of *v*+*an*+*z* (mean of 4 determinations).

decrease in PPFD around 4:00 pm in the afternoon was accompanied by reverse pigment conversions, a decrease in zeaxanthin and an increase in violaxanthin content. There also occurred a small, transient increase in the level of the intermediate antheraxanthin at the beginning of zeaxanthin formation in the morning, and another one in the late afternoon, when the PPFD decreased after 4:00 pm and zeaxanthin was reconverted into violaxanthin.

The accumulation of zeaxanthin on the expense of violaxanthin was completed at about 11:00 am, yet the zeaxanthin content further increased by *de novo* biosynthesis of zeaxanthin which also augmented the total xanthophyll cycle carotenoid content *v*+*an*+*z* by about 15% from 12:00 am to 4:00 pm (Fig. 4A.). A further increase in the level of zeaxanthin cycle carotenoids of about 50% by *de novo* biosynthesis took place during the PPFD decrease in the afternoon between 4:00 and 7:00 pm. During this late afternoon period with decreasing PPFD zeaxanthin was almost completely reoxidized to its di-epoxide violaxanthin (Fig. 4B.). During the high PPFD sunny period of the day zeaxanthin made up more than 80% of the xanthophyll cycle carotenoids *v*+*an*+*z* and decreased to a low percentage (<10%) in the late afternoon (Fig. 4B.). Only the short sunny period reappearing at 6:00 pm interrupted the zeaxanthin decrease for a short time.

The course of the changes in zeaxanthin content and that of the incident PPFD are contrasted in Fig. 6A. Noteworthy

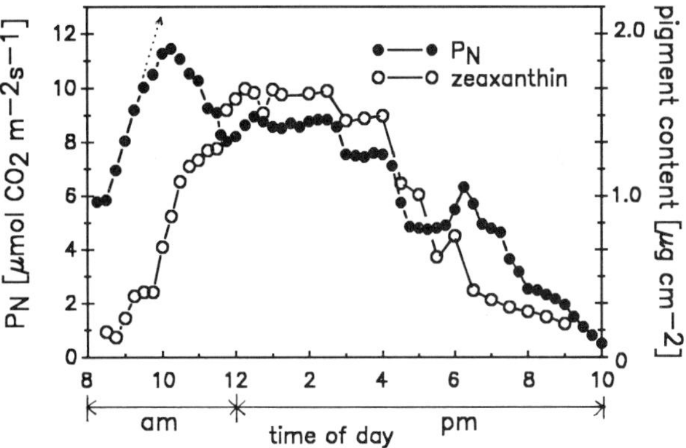

Fig. 5: Development of the net photosynthetic rate P_N and the zeaxanthin content of maple leaves during the course of the sunny day (with some clouds in the afternoon at about 4 pm).

is that the violaxanthin de-epoxidation to zeaxanthin in the morning and the epoxidation of zeaxanthin to violaxanthin in the evening proceeded with a delay as compared to the PPFD increase and decrease. This is also the case when the de-epoxydation state DEPS is plotted against PPFD (Fig. 6B.). The DEPS includes the additional changes in the content of antheraxanthin which contains one epoxy group in

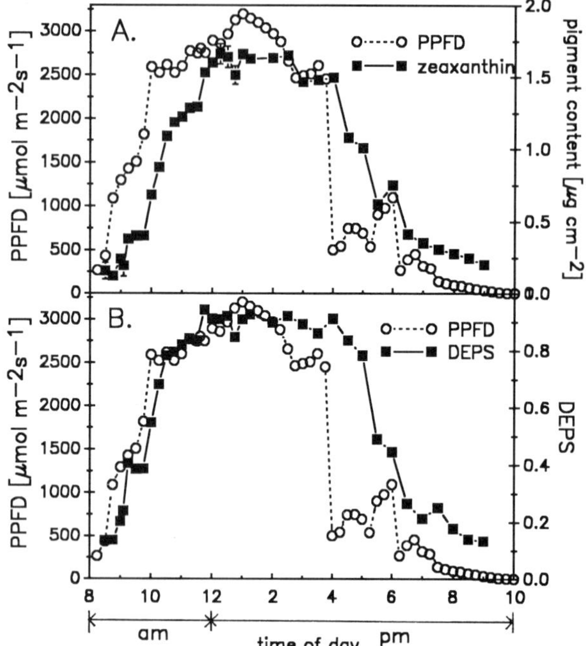

Fig. 6: Changes in photosynthetic photon flux density PPFD plotted against light-induced changes in zeaxanthin content (**A.**) and against the de-epoxidation state DEPS of the zeaxanthin cycle carotenoids (**B.**) in the course of the bright sunny day. The DEPS is defined as the ratio of $(z+0.5\,an)/(v+an+z)$.

the one β-ionone ring, whereas the other β-ionone ring is free of epoxide. The decrease of the DEPS in the afternoon is even more delayed with respect to the decrease in PPFD than that of zeaxanthin, since antheraxanthin (the intermediate of the conversion) is temporarily increased (Fig. 4 B.).

Changes in chlorophyll fluorescence parameters

Chlorophyll fluorescence ratios Fv′/Fm′ and ΔF/Fm′ are considered indicators of the actual quantum yield of the photosynthetic photosystem II of illuminated leaves and reflect the efficiency of photosynthetic quantum conversion in the illuminated state of leaves. The variable chlorophyll fluorescence Fv′/Fm′ decreased in the morning from a value of 0.76 at 8.15 am to a value of 0.16 at 1 : 30 pm and then remained nearly unchanged until 4 : 00 pm. Only when the PPFD decreased from 4 pm rose the ratio Fv′/Fm′ again and reached the starting value of 0.77 around 9 pm. The course of the ratio ΔF/Fm′ (Fig. 7 A.) showed a similar kinetic to that of Fv′/Fm′. At 12 : 30 am this fluorescence ratio reached a very low minimum of 0.06 (10.3 % of starting value). The regeneration of ΔF/Fm′ took place in two phases: a slow phase from 2 : 00 to 5 : 15 pm (increase to 32 % of the starting value) was followed by a faster phase, after which the starting value of 0.63 was regenerated at 9 : 00 pm.

Changes in quenching coefficients

The photochemical quenching coefficient qP decreased from 0.83 in the morning to an extremely low minimum of 0.03 at 2 : 00 pm (Fig. 7 B.). With the appearance of some clouds in the afternoon qP increased again and reached gradually a value of 0.91 at 9 : 00 pm. During this sunny day the non-photochemical quenching coefficient qN exhibited an opposite kinetic (Fig. 7 B.). From a starting point of 0.05 the coefficient qN increased to a maximum value of 0.83 at 0 : 30 pm. With the appearance of some clouds in the afternoon qN decreased again to a final value of 0.09 in the evening. The short PPFD increase at nearly 6 : 00 pm did not influence the values of qP and qN. The non-photochemical quenching coefficient qN rose faster than the zeaxanthin content (Fig. 8 A.), whereas the rise in the DEPS degree was at first delayed but then proceeded earlier than the PPFD. In the afternoon qN declined much later than the zeaxanthin content and the de-epoxidation state DEPS (Fig. 8 A. and B.).

2. Cloudy day

Photosynthesis parameters

The changes of zeaxanthin cycle xanthophylls and the kinetics of photosynthetic CO_2 fixation were also determined in maple leaves on a cloudy day, which showed up after 3 days of bright sunshine. On the cloudy day the temperature showed values between 19 and 21 °C. Most of the day the PPFD exhibited values in the range of 100 to 300 µmol m^{-2} s^{-1} except for some clearing up at 1 : 45 pm and 3 : 30 pm. The stomata were not fully open as seen from the gH_2O-values which ranged from 25 to 50 between 10 : 00 am and 5 : 00 pm. Net photosynthesis increased in the morning despite a thick cloud cover and low PPFD, and yielded good

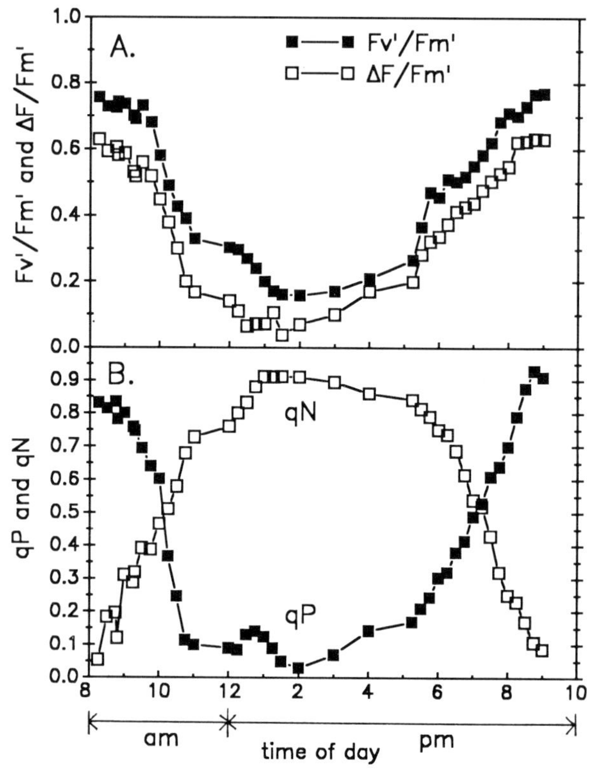

Fig. 7: Daily cycle of the variable chlorophyll fluorescence ratios Fv′/Fm′ and ΔF/Fm′ (**A.**) and the quenching coefficients qP and qN (**B.**) in maple leaves on the sunny day.

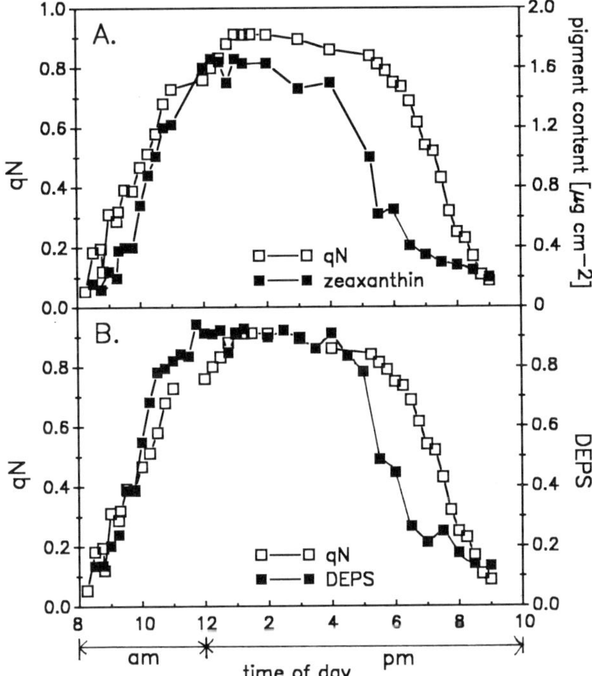

Fig. 8: Changes in the non-photochemical quenching coefficient qN plotted against the zeaxanthin content (**A.**) and the de-epoxidation state DEPS of the zeaxanthin cycle carotenoids (**B.**) in the course of the sunny day.

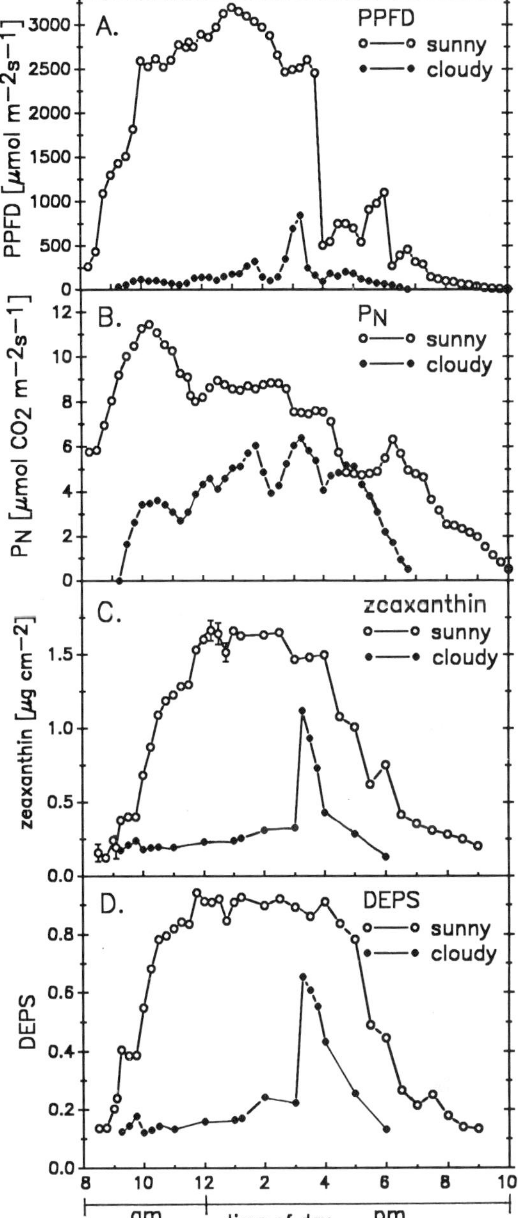

Fig. 9: Comparison of several parameters measured in maple leaves on a sunny and a cloudy day. **A)** photosynthetic photon flux density (PPFD), **B)** net photosynthesis rate (P_N), **C)** zeaxanthin content and **D)** de-epoxidation state DEPS of the zeaxanthin cycle carotenoids.

P_N values from 10:00 am to 5:00 pm (Fig. 9 A. and B). The maximum P_N values on the cloudy day were determined between 6 to 6.5 µmol CO_2 $m^{-2} s^{-1}$, and were reached during two short phases of clearings at 2:00 pm and from 3:00 to 3:30 pm. At the low PPFD after 5:00 pm P_N was gradually reduced to 0.3 µmol CO_2 $m^{-2} s^{-1}$ at 6:45 pm. Although the PPFD level on the cloudy day almost reached for a short moment (3:30 pm) the light saturation level of photosynthetic CO_2 fixation (800–900 µmol CO_2 $m^{-2} s^{-1}$) of the maple leaves investigated, the corresponding P_N rate of 6 µmol CO_2 $m^{-2} s^{-1}$ was far below the maximum P_N rate of 11.3 µmol CO_2 $m^{-2} s^{-1}$ reached on the sunny day. This demonstrates that the PPFD must be maintained for a longer time period before maximum P_N rates can be reached.

Chlorophyll fluorescence parameters

In the course of the cloudy day the chlorophyll fluorescence ratios Fv′/Fm′ and ΔF/Fm′, which reflect the actual quantum yield and the photochemical efficiency of photosystem II, remained almost at the same high level (Fig. 10 A.). The only exception was a small decrease of Fv′/Fm′ and ΔF/Fm′ during the short clearing period at 3:15 pm, but the values were regenerated fast when the cloud cover reappeared and the PPFD fell back to the previous light conditions. The photochemical quenching coefficient qP, in turn, remained on an unchanged high level of near 0.92 ± 0.02 during the whole day (Fig. 10 B.). The non-photochemical quenching coefficient qN exhibited very low values below 0.1 and responded only slightly to the short PPFD increases at 2:00 and 3:15 pm (Fig. 10 B.).

Zeaxanthin cycle xanthophylls

At the low-light conditions of the cloudy day the zeaxanthin cycle xanthophylls were predominantly present in the oxidized form as violaxanthin (80.3 % ± 1.74 % of *v+an+z*) (Fig. 11 A. and B.). The remaining 19.7 % consisted of zeaxanthin and antheraxanthin in fairly equal amounts. The to-

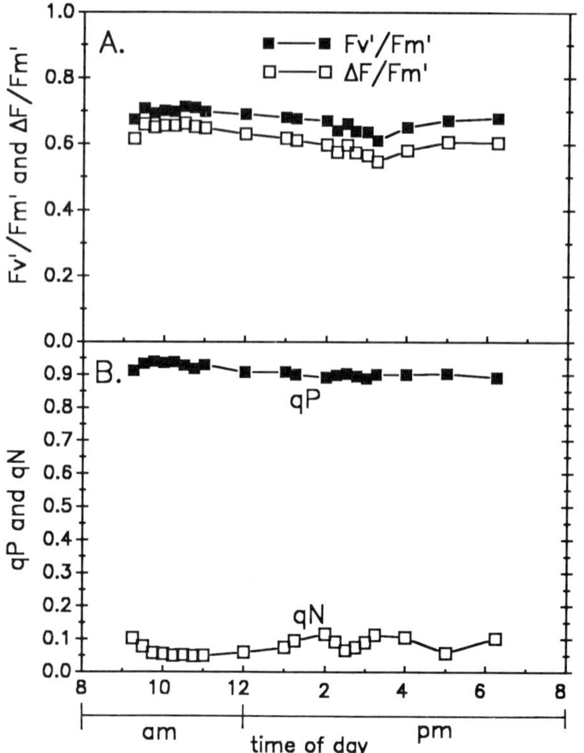

Fig. 10: Development of the chlorophyll fluorescence ratios Fv′/Fm′ and ΔF/Fm′ (**A.**) and the quenching coefficients qP and qN (**B.**) determined in maple leaves on the cloudy day.

tal amounts of *v+an+z,* which had been increased from 1.5 to 2.5 μg cm^{-2} on the sunny day (Fig. 4 A.), decreased to about 1.65 μg cm^{-2} in the course of the cloudy day (Fig. 11 A.). Also on the cloudy day a fast de-epoxidation of violaxanthin to zeaxanthin was detected, when at 3 : 15 pm the PPFD increased to a value of over 840 μmol m^{-2} s^{-1}. This zeaxanthin accumulation to values of 0.33 μg cm^{-2} took place without delay after the rise in PPFD (Fig. 9 C.). At the end of this short clearing up period zeaxanthin was again epoxidized to violaxanthin, but this epoxidation process proceeded with some delay to the PPFD decrease as on the sunny day (Fig. 9 A. and B.). This retardation of violaxanthin formation is seen more clearly in the DEPS values (Fig. 9 D.). It needs to be said that zeaxanthin formation on the cloudy day proceeded at a PPFD of ca. 840 μmol m^{-2} s^{-1} which corresponded to a light saturation point of photosynthetic net CO_2 fixation P_N. Except for the decrease in total zeaxanthin cycle carotenoids *v+an+z,* no changes in other carotenoids or in the chlorophylls were observed on the cloudy day.

Due to the significant increase ($p<0.01$) in zeaxanthin cycle carotenoids *v+an+z* on the sunny day (Fig. 4 A.) and the significant decline ($p<0.05$) in the absolute zeaxanthin cycle xanthophyll content on the cloudy day (Fig. 11 A.), the relative proportions of zeaxanthin cycle xanthophylls *v+an+z* of the total carotenoid content of maple leaves were modified from morning to evening of the cloudy and sunny day (Fig. 12). On the sunny day the relative proportions of *v+an+z* significantly increased ($p<0.05$) from 25.6 to 28.8 % as well as that of carotenes from 23.3 to 25.5 %. At the same time the relative proportions of lutein and neoxanthin, which are bound to the light-harvesting protein of photosystem II (LHCII), significantly declined ($p<0.02$) to 36.9 % and 8.8 % respectively (Fig. 12). On the cloudy day the relative proportions of the zeaxanthin cycle xanthophylls significantly ($p<0.05$) shifted in the reverse direction with a clear increase ($p<0.05$) in the proportions of carotenes, whereas the relative levels of lutein increased and the relative level of neoxanthin remained the same (Fig. 12).

Discussion

Performance of photosynthesis

The results of this investigation with green leaves of maple trees at an outdoor location showed that during the course of a sunny day there occur not only changes (increases and decreases) in photosynthetic activity, and chlorophyll fluorescence parameters, but also in the level of zeaxanthin cycle xanthophylls *v+an+z,* zeaxanthin content and the de-epoxidation state DEPS which are induced by the increase of the photon flux density PPFD in the morning and its decrease in the afternoon. On the cloudy day with a low PPFD these changes did not show up, except for some variations in the photosynthetic net CO_2 fixation rates.

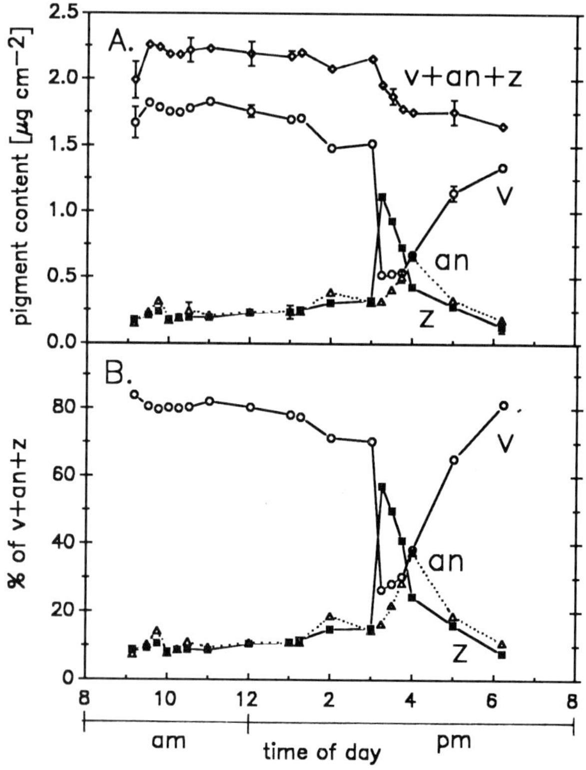

Fig. 11: Changes in the absolute and relative content of zeaxanthin cycle pigments in *Acer platanoides* leaves in the course of the cloudy day. **A.**) absolute amounts per leaf area and **B.**) percentage amounts of total *v+an+z.* Values are means (n=4) with standard deviation.

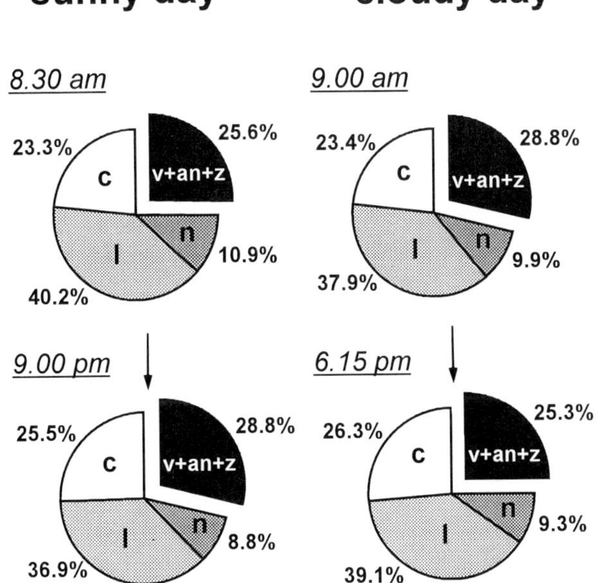

Fig. 12: Changes in the percentage of the individual carotenoids determined on the starting point in the morning and on the last measuring point in the evening of both, a sunny and a cloudy day in *Acer platanoides* leaves.

The CO_2/H_2O gas exchange measurements on the sunny day demonstrated that the P_N rates of the fully developed, attached leaves primarly depended on the opening state of the stomata, i.e. the stomata resistance towards CO_2 and water vapour as judged from the gH_2O values (Fig. 2). As expected, the stomata opening followed the increase in PPFD (increase in gH_2O values). In fact, the highest P_N rates and gH_2O values were obtained at 10:00 am; thereafter the gH_2O values decreased and the concomitant P_N was reduced. With the partial closure of the stomata and their adjustment to a new, fairly stable level from 11:30 am to 2:45 pm the leaf and its photosynthetic apparatus was entering a different physiological state. The decreased availability of CO_2 reduced the rate of CO_2 fixation, yet the light absorption continued. In this experiment the photon flux density even increased until 1 pm. The excess light caused an over-reduction of the photosynthetic electron transport chain and an accumulation of excess ATP and NADPH, which could no longer be consumed in photosynthetic CO_2 fixation. Such high-light stress requires physiological adaptations and adjustments of cell metabolism and chloroplast metabolism. There exist various other chloroplast pathways, which can consume ATP and NADPH, such as nitrite reduction, sulfate and sulfite reduction and *de novo* fatty acid biosynthesis (Golz et al., 1994). The capacity of these biochemical processes seems, however, not to be sufficiently large to prevent the over-reduction of the photosynthetic apparatus at high irradiances.

In order to protect the photosynthetic apparatus against photoinhibition and concomitant photooxidative degradation of photosynthetic pigments as well as peroxidations of thylakoid lipids, additional processes must be activated. An essential one seems to be the deactivation of excited chlorophyll states *via* an enhanced emission of heat and red + far-red chlorophyll fluorescence (Schreiber et al., 1986; Lichtenthaler and Rinderle, 1988; Krause and Weis, 1988 and 1991; Burkart, 1994). One mechanism to guarantee such a non-photochemical energy-dissipation appears to be the detachment of the light-harvesting complex LHCII consisting of the chlorophyll *a/b*-lutein-neoxanthin proteins (LHCPs) (e.g. Lichtenthaler et al., 1982; Lichtenthaler, 1987) from the reaction center of photosystem II, which has been investigated in detail by Horton and co-workers (Horton et al., 1991; Ruban and Horton, 1995). Another photoprotective mechanism is the high-light induced accumulation of zeaxanthin by phototransformation of violaxanthin which has been observed under artificial light (Hager, 1975; Yamamoto, 1979; Lichtenthaler et al., 1992; Schindler and Lichtenthaler 1994) and also outdoors in sun light (Adams and Demmig-Adams, 1992).

In sun-exposed maple leaves the changes in incident PPFD were followed by subsequent changes in the epoxidation state of the zeaxanthin cycle. The zeaxanthin content rose in the morning of the sunny day with a delay to the increase of PPFD, and in the afternoon a decrease of these parameters was observed (Fig. 6). Similar diurnal patterns of changes in the zeaxanthin cycle components at natural growth conditions were also found in the crop species sunflower, pumpkin and cucumber (Adams and Demmig-Adams, 1992). Thus, major changes in the de-epoxidation state of the zeaxanthin cycle, finely tuned by the diurnal changes in irradiance, occur daily in plants growing in natural sunlight. On the cloudy day with low irradiance the rise and decline of the zeaxanthin levels did not occur in the maple leaves investigated here, except during a short clearing up in the afternoon when the photon flux density had briefly reached the light saturation point of photosynthetic CO_2 fixation (Fig. 11).

Accumulation of zeaxanthin

The enzyme responsible for the light-induced zeaxanthin accumulation, the violaxanthin de-epoxidase, becomes active only at a sufficiently acid pH in the thylakoid lumen of about 5.5 (Hager, 1975; Hager and Holocher, 1994) usually achieved at irradiances which are higher than the light saturation point of photosynthetic CO_2 fixation (Lichtenthaler et al., 1992; Schindler and Lichtenthaler, 1994). Zeaxanthin can also accumulate at medium irradiances, e.g. at the onset of photosynthesis upon irradiation of predarkened leaves, when the two photosystems are impaired, which causes an acidification of the thylakoid lumen for some minutes. The latter then induces a transitory accumulation of zeaxanthin, which declines after several minutes by its reconversion to violaxanthin (Schindler and Lichtenthaler, 1994), when the photosynthetic light and electron transport reactions (state 1/state 2 transitions) and the dark reactions have been readjusted and reach a stationary balance reducing the intially high pH gradient. A stable zeaxanthin accumulation at medium irradiances can, however, also occur at lower temperatures (Schindler, 1995), which slow down the rate of CO_2 fixation (Calvin cycle performance), and also induce a higher pH-gradient and acidification of the thylakoid lumen.

Also, in the maple leaves investigated the accumulation of zeaxanthin can be understood and explained on the basis of light-induced changes in the pH of the thylakoid lumen. Although some zeaxanthin accumulation occurred in the morning parallel to the rise in PPFD and in P_N rates, the strong transformation of violaxanthin into zeaxanthin only proceeded at and after 10:00 am when the linear rise in P_N rates (dotted line in Fig. 5) creased and the net CO_2 fixation rates eventually declined due to a partial closure of the stomata (decline of gH_2O values). At this time (10:00 am) the PPFD ($2600\,\mu mol\,m^{-2}\,s^{-1}$) had already far exceeded the light saturation point of photosynthetic CO_2 fixation (ca. 800 to $900\,\mu mol\,m^{-2}\,s^{-1}$). This caused a stronger acidification of the thylakoid lumen, and activated the de-epoxidase to form zeaxanthin on the expense of violaxanthin.

Another phenomenon, which is not generally observed in plants cultivated in growth chambers or glasshouses, is the presence of a distinct level of zeaxanthin in leaves of plants growing in the field prior to sunrise and even after an entire night. This phenomenon was also found in other plants (Adams and Demmig-Adams, 1992; Demmig-Adams and Adams, 1992). In the course of the sunny day we observed a significant rise in the sum of zeaxanthin cycle pigments $v+an+z$ by *de novo* biosynthesis of about 50–60 %. This high-light induced enlargement of the zeaxanthin cycle pool represents a relatively fast adaptation mechanism of the photosynthetic apparatus to the high-light conditions in the course of the sunny day. It had been shown before in laboratory experiments that an efficient biphasic accumulation of zeaxanthin a) by photoconversion of violaxanthin and b) by *de novo* biosynthesis mediates a photoprotection of the photosynthetic apparatus against photoinhibition (Lichtenthaler und Schindler, 1992; Schindler und Lichtenthaler, 1994). From the high capacity for zeaxanthin accumulation of the field-grown maple leaves it can be concluded that zeaxanthin, also at high sun irradiances and field conditions, functions in photoprotection of the photosynthetic apparatus.

Function of zeaxanthin

It had been shown in different plant species that a quenching of the variable chlorophyll fluorescence (Fv or the ratio Fv/Fm) occurred at increasing irradiance with artificial light or at a high sun irradiance (Adams, 1988; Adams et al., 1987, 1989; Lichtenthaler and Schindler, 1992; Schindler and Lichtenthaler, 1994). On the basis of some preliminary measurements it was postulated that zeaxanthin formed by the high-light induced transformation of violaxanthin into zeaxanthin might at least be partially responsible for the decrease in variable chlorophyll fluorescence through a zeaxanthin induced dissipation of absorbed light energy (Demmig et al., 1987; Demmig-Adams et al., 1989). In detailed kinetic measurements the two processes zeaxanthin accumulation and decline in maximal and variable chlorophyll fluorescence could, however, not be correlated (Lichtenthaler and Schindler, 1992; Lichtenthaler et al., 1992; Schindler et al., 1994). In the meantime various zeaxanthin-dependent and non-dependent chlorophyll fluorescence quenching effects have been described by other authors (Adams et al., 1990; Havaux et al., 1991 a; Gilmore and Yamamoto, 1993; Richter et al., 1994). More recently antheraxanthin has been intriduced as an additional xanthophyll to be involved in chlorophyll fluorescence quenching (Gilmore and Yamamoto, 1993).

The steep rise of the non-photochemical quenching coefficient qN in green maple leaves on the sunny day clearly precedes the accumulation of zeaxanthin (Fig. 8 A.), which would exclude a causal relationship between both processes. When, however, the de-epoxidation state DEPS of the zeaxanthin cycle, which includes antheraxanthin, is applied, the situation is different. The DEPS initially rises parallel to qN, and in the time after 10 am even precedes the rise in qN. This would be the only time where a presumable interrelationship could be considered. However, the decrease in zeaxanthin and in the DEPS in the afternoon of the sunny day is much faster than the decline in qN. This, in turn, clearly contradicts a causal relationship of both processes.

The matter is that in cases of a presumable zeaxanthin-dependent non-photochemical quenching this quenching occurs in the presence of zeaxanthin. Whether this quenching is really zeaxanthin-dependent or whether the quenching and zeaxanthin formation are independent but might both be induced by the same mechanism, such as the build-up of a pH gradient (acidification of the thylakoid lumen), has not exactly been differentiated and is difficult to decide. It is not disputed (Demmig-Adams and Adams, 1992; Schindler and Lichtenthaler, 1994) that zeaxanthin plays a photoprotective role in thylakoids. An aurea tobacco mutant, which at high-light exposure is able to double the zeaxanthin amounts by *de novo* biosynthesis, is much better protected against photoinhibition than the normal green wild-type tobacco (Schindler and Lichtenthaler, 1994) in which the total level of zeaxanthin cycle carotenoids is hardly increased. Also the biphasic increase in zeaxanthin content in maple leaves on the sunny day by violaxanthin de-epoxidation and *de novo* biosynthesis is in favour of such a photoprotective role of zeaxanthin. This function is further supported by the decline in total zeaxanthin cycle carotenoids on the cloudy day when this photoprotective function is not needed.

What then is the mode of action of zeaxanthin? *One possible mechanism* could be that zeaxanthin reacts in a non-enzymic way with the various highly reactive oxygen species, which are formed in chloroplasts at prolonged high light conditions (Elstner, 1990; Asada and Takahashi, 1987; Asada, 1992; Foyer and Harbinson, 1994; Halliwell and Gutteridge, 1989; Price et al., 1989) to yield its oxidized product violaxanthin. This possible role of zeaxanthin, proposed by Lichtenthaler and Schindler (1992), and Lichtenthaler (1993), is presented in Fig. 13. The function of the high light driven xanthophyll cycle would then be to de-epoxidize this violaxanthin in order to restore the zeaxanthin pool.

As a *second mechanism,* we propose that zeaxanthin might not directly react with the reactive oxygen species but in an indirect way by removing epoxy groups from the fatty acids of structural thylakoid lipids which had been introduced by the reactive oxygen species formed at high PPFD. Also Havaux et al. (1991 a) assume that zeaxanthin protect thylakoids against lipid peroxidation. According to both mechanisms presented here, zeaxanthin would be oxidized to violaxanthin, which, in turn, would be de-epoxidized by the light-driven zeaxanthin cycle to regenerate the photoprotective acceptor molecule zeaxanthin (Fig. 13). Such a high irradiance in-zeaxanthin —O_2*→ violaxanthin

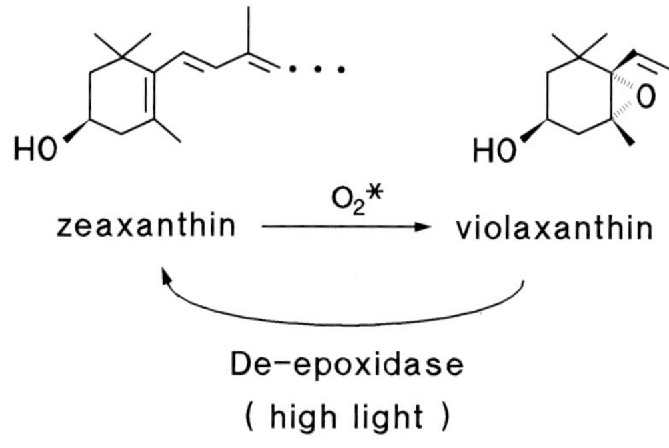

De-epoxidase
(high light)

O_2* = highly reactive oxygen species:

$^1O_2$	singlet oxygen
$O_2^{\bullet -}$	superoxide radical anion
$OH^\bullet$	hydroxyl radical
H_2O_2	hydrogen peroxide

Fig. 13: Scheme of the possible function of the violaxanthin/zeaxanthin cycle of photochemically active thylakoids under high photon flux densities (high light) in the dissipation of highly reactive oxygen species O_2*. Zeaxanthin might detoxify O_2* by becoming oxidized in a non-enzymic way to violaxanthin, which then is de-epoxidized to zeaxanthin by the light-activated de-epoxidase.

duced zeaxanthin cycle performance (forward reaction) would also make sense from a physiological point of view.

It is in favour of these mechanisms that violaxanthin (plus zeaxanthin) is not specifically enriched like the other carotenoids in a particular chlorophyll-carotenoid protein of thylakoids but seems to be distributed all over the thylakoid membrane (Lichtenthaler et al., 1982; Lichtenthaler, 1987). After PAGE the major part of violaxanthin is found in the free pigment fraction, even after application of very mild detergents (Lichtenthaler et al., 1982). The presence of some violaxanthin in the LHCII (Peter and Thornber, 1991; Ruban and Horton, 1992; Lee and Thornber, 1995) after disruption of thylakoids does not necessarily mean that it is localized in the LHCII or that the zeaxanthin cycle would be performed in the LHCII. One should always compare the relative violaxanthin (plus zeaxanthin) content of an isolated pigment-protein with the relative violaxanthin content of whole chloroplasts before the isolation of pigment-proteins. Only when an enrichment of violaxanthin (plus zeaxanthin) as compared to chloroplasts would be found, could one think of a partial localization of these xanthophylls in a particular pigment protein. So far this has neither yet been shown for violaxanthin, antheraxanthin or zeaxanthin.

Horton et al. (1991); Ruban and Horton (1995) assume that the aggregation and dissociation of light harvesting chlorophyll proteins from the reaction center of PSII are a major source of the high light induced fluorescence quenching. This concept opens a *third mechanism* and possibility for zeaxanthin function. It could act as a spacer between the aggregated LHCII and the reaction centers of PSII, thus preventing the transfer of excitation energy from LHCII to the PSII reaction centers and indirectly reduce the variable chlorophyll fluorescence as well as cause an increase of the non-photochemical energy dissipation as seen here in the observed rise in qN. The spontaneous response of the vioaxanthin/zeaxanthin cycle to yield zeaxanthin at an acidification of the thylakoid lumen due to a light saturated photosynthesis and/or a decline of the photosynthetic performance, would be a fast acclimation reaction to avoid photoinhibitory and photooxidative degradation. This concept of zeaxanthin function would also imply that violaxanthin with ist two epoxide groups cannot replace zeaxanthin as a spacer. A regulatory effect of zeaxanthin on thylakoid membrane fluidity is also assumed (Gruszecki and Strzalka, 1991; Havaux and Gruszecki, 1993). In any case much further research is required to finally understand the exact mode of photoprotective actions of zeaxanthin.

Chlorophyll fluorescence parameters

The combination of a strong sun irradiance and high temperatures on the sunny day led to distinct changes in several chlorophyll fluorescence parameters of the maple leaves. It is well known that at a high irradiance high temperatures can cause or amplify a photoinhibition of the photosynthetic apparatus (Ludlow, 1987). The continuous decrease in the fluorescence ratios Fv′/Fm′ and ΔF/Fm′ and in the photochemical quenching coefficient qP to low values from morning to noon with rising PPFD and temperature on the sunny day demonstrated that the efficiency of photochemical quantum conversion was strongly reduced in maple leaves. This was further emphasized by the steep rise of the non-photochemical quenching coefficient qN to extremely high values of >0.9. A decline in quantum efficiency of PSII and of photosynthetic quantum conversion at high irradiances, once the photosynthesic processes are light saturated, is a quite normal process.

These changes in the chlorophyll fluorescence ratios and the quenching coefficients do, however, not necessarily mean that in the course of the sunny day the photosynthetic apparatus was photoinhibited by the high irradiance in combination with the high temperature of $>35\,°C$. The two related chlorophyll fluorescence ratios ΔF/Fm′ and Fv′/Fm′ are indicators of the actual quantum yield of the photosystem II in illuminated leaves at steady state conditions of fluorescence. They show the efficiency of photosynthetic quantum capture of photosystem II in the light-adapted state at partially or fully closed reaction centers II. At the low PPFD in the morning and evening the ratio Fv′/Fm′ exhibited values of 0.75 to 0.78 and the ratio ΔF/Fm′ values of 0.6 to 0.63. Per definition (confer Fig. 1) these two fluorescence ratios always possess lower values than the variable fluorescence ratio Fv/Fm which is the efficiency of excitation capture (quantum yield) of photosystem II in the dark-adapted state with fully open reaction centers. As in other plants the variable fluorescence ratio Fv/Fm exhibited values of 0.85 to 0.9 also in maple leaves, which were determined in leaves predarkened for 15 min in the morning and in the evening of the sunny and cloudy day.

The decline in the variable chlorophyll fluorescence ratios Fv′/Fm′ and ΔF/Fm′ in the course of the sunny day indicates that the reaction centers of PSII became increasingly closed due to the rise in PPFD. This also demonstrates that the linear photosynthetic electron transport of PSII is light-saturated. Therefore, at a high irradiance, as on a sunny day, the pulses of saturating light applied to the sun-illuminated leaves yielded a Fm′ level which hardly exceeded the F level, with the result that Fv′ and ΔF became very low (cf. Fig. 1), and as a consequence also the values of Fv′/Fm′ and ΔF/Fm′. In other words, the two ratios Fv′/Fm′ and ΔF/Fm′ are not suited to differentiate between a high-light induced closure of the PSII reaction centers on one hand, and a photoinhibition of photosystem II on the other hand, since both processes result in low values of Fv′ and ΔF.

At the very high irradiance and high temperature of the bright sunny day one has, however, to assume a partial photoinhibition of the photosynthetic apparatus. If the decline in the ratios Fv′/Fm′ and ΔF/Fm′ were solely due to a closure in photosystem II reaction centers by the high PPFD, the values of both fluorescence ratios should have rapidly increased when the PPFD sharply declined due to the appearance of clouds in the afternoon. This was, however, not the case. The regeneration of the values of Fv′/Fm′ and ΔF/Fm′ proceeded very slowly at decreasing PPFD and responded with a large delay of several hours. This is a clear indication that the photosystem II of the maple leaves was partially photoinhibited on that sunny day.

Further insight into the process of photosynthetic quantum conversion and the efficiency of photosynthetic energy capture was obtained by determination of the photochemical and non-photochemical chlorophyll fluorescence quenching coefficients qP and qN. At low and medium irradiances the photochemical quenching coefficient qP, defined as ΔF/Fv′, is considered to be proportional to the quantum yield of noncyclic electron transport (Genty et al., 1989). In maple leaves qP exhibited high values of 0.83 to 0.93. Similar values were found in other plants (Krause and Weis, 1988; Lichtenthaler and Rinderle, 1988; Schreiber et al. 1986; Havaux et al., 1991 b). In contrast, the non-photochemical quenching coefficient, qN, defined as 1-Fv′/Fv, exhibits at low irradiances low values of <0.2. The coefficient qN is composed of several components: qE = high energy quench, which is related to the formation of the pH gradient, qT = a quench related to state transitions of the photosynthetic apparatus and phosphorylation of LHCP (Bennet, 1984) and qI = photoinhibitory quench indicating a photoinactivation of PSII. The three components can be determined by dark relaxation kinetics (Burkart, 1994) with qE as the fast relaxing component ($t_{1/2}$<1 min) and qT as the medium relaxing component ($t_{1/2}$ = 8–12 min), whereas qI as the slowly relaxing component ($t_{1/2}$>30 min) is due to photoinhibitory damage of photosystem II and requires repair mechanisms, e.g. synthesis and integration of the D1 protein, which are slow.

The decline in the photochemical qP and the rise in the non-photochemical quenching coefficient qN on the sunny day indicated that not only the process of photosynthetic quantum conversion was light-saturated, but that also the non-photochemical de-excitation of absorbed light energy was considerably enhanced in order to dissipate the excess light energy. Since at 1:00 pm the PPFD had reached a value which was about 4 times higher than the light saturation point of photosynthesis (800–900 μmol m^{-2} s^{-1} PAR), this development was to be expected. The decline in qP is understandable since ΔF (the spikes obtained by saturating light pulses) became very low at noon of the sunny day and was hardly above the actual fluorescence F of the kinetic (cf. Fig. 1), The steep rise of qN from values <0.1 in the morning to values of 0.93 at 1:00 pm is caused by the extreme decline of Fv′ as compared to the variable fluorescence Fv of dark-adapted leaves. The regeneration of the morning values of qP and qN proceeded with a long delay with respect to the sharp decline in PPFD after 3.45 pm (Figs. 2 and 7). This indicates that the contribution of the photoinhibitory component qI to the overall non-photochemical quenching coefficient qN was relatively high on that sunny day. Although the relative amounts of qI, qT and qE of the maple leaves could not be determined for technical reasons, we determined the relative proportions of the three non-photochemical quenching components on another day with leaves from young maple trees and soybean plants (Burkart, 1994). These results indicated that the photoinhibitory quench qI made up more than 50% of the overall qN after several hours of irradiance with excessive light (2 to 3 times above the light saturation point of P_N). From this we conclude that on the sunny day qI contributed to a considerable extent to the overall qN of that proportion of chloroplasts, which can be reached by fluorescence measurements at the upper leaf side.

In contrast to the sunny day, there was no excess light available on the cloudy day and qP remained high (values around 0.9). qN exhibited very low values at or below 0.1. As expected at a low PPFD, the two fluorescence ratios Fv′/Fm′ and ΔF/Fm′ exhibited no decrease in the course of the day, since the irradiance (actinic light + daylight) was not high enough to close the reaction centers of photosystem II. Thus Fv′ and ΔF as well as Fm′ gave high signals resulting in regular values of Fv′/Fm′ and ΔF/Fm′.

The results of this investigation show that the chlorophyll fluorescence ratios Fv′/Fm′ (the maximum photochemical efficiency of PSII) and ΔF/Fm′ (the actual photochemical efficiency of PSII) are very suitable, together with the quenching coefficients qP and qN, in order to describe photochemical and non-photochemical dissipation processes of absorbed light energy in the photosynthetic apparatus. A decline in Fv′/Fm′, ΔF/Fm′ and qP and the concomitant increase in qN at high irradiance above the light saturation point of photosynthesis do, however, not allow to differentiate between a high-light induced closure of the reaction centers of PSII and a photoinhibitory damage, since both processes decrease Fv′ and ΔF, and thus cause the changes observed in all four chlorophyll fluorescence parameters determined here. A photoinhibitory damage of photosystem II can only be deducted from a low regeneration of these chlorophyll fluorescence parameters, when the plants are brought back to a low light level, and even more exactly by the performance of dark relaxation kinetics.

It should also be emphasized that the changes in the chlorophyll fluorescence ratios Fv′/Fm′ and ΔF/Fm′ and quenching coefficients qP and qN did not at all reflect the net photosynthetic CO_2 fixation rates P_N of the maple leaves on that sunny day (cf. Figs. 5 and 7). P_N was primarily determined by the stomata opening as seen from the gH_2O values and declined only by 22% between 11:30 am and 2:45 pm, whereas the two fluorescence ratios and quenching coefficients had decreased more than 75% as compared to morning values. This situation can be explained by two assumptions: 1) either the large changes in the determined chlorophyll fluorescence parameters are only marginally attributed due to photoinhibition of the photosynthetic apparatus of maple leaves or 2) the chlorophyll fluorescence signals excited and sensed from the upper sun-exposed leaf side were not representative of the whole leaf chloroplasts but only of a small pool of chloroplasts at the upper leaf side, which may have been photoinhibited to a high degree. The second assumption would imply that the partially «shaded» chloroplasts in the lower leaf half as well as those of the upper leaf half, which are located below the first layer of sun light exposed chloroplasts, performed an undisturbed photosynthetic net CO_2 assimilation which was then only determined by the stomata conductance and the availability of CO_2. The fact that the chlorophyll fluorescence signatures predominantly reflect the characteristics of the chloroplasts of that leaf side (either upper or lower leaf side), where the fluorescence was excited and recorded (Lichtenthaler and Rinderle, 1988; Lichtenthaler et al., 1986; Schweiger et al., 1996), is in favour of this second assumption. Thus, it appears that the first layer of partially or fully photoinhibited chloroplasts may function as a screen and filter for the chloroplasts of the subsequent lower layers of the leaf mesophyll, which can perform normal net photosynthetic rates P_N. This is an interesting acclimation and photoprotective mechanism of leaves against too high an irradiance.

These observations emphasize that the chlorophyll fluorescence signals measured of the upper leaf side are not representative of the chloroplasts of the whole leaf. This statement particularly applies to the PAM fluorometer, in which the excitation wavelength consists of red light of about 650 nm, which in dark green leaves is readily absorbed by the first chloroplast layer and therefore only a small portion of excitation light penetrates further into the leaf and to the chloroplasts of the lower leaf half. For this reason the chlorophyll fluorescence signatures measured of dark-green leaves mainly derive from the first chloroplast layer of the mesophyll near to the upper epidermis layer. Therefore it is recommended to record chlorophyll fluorescence signatures from the upper and lower leaf side, if one wants to have a reliable information on the physiology of chloroplasts of a whole leaf. In addition, we recommend to rather use excitation light of the green spectral region, which penetrates much deeper into the leaf, than the red light applied so far. Also, blue excitation light should be avoided, if one wants to have fluorescence information on the whole leaf, since much of the blue light is readily absorbed by chlorophylls and carotenoids in the first chloroplast layer of the leaf mesophyll. This especially applies to green and dark green leaves with a high chloroplast and chlorophyll content. Only in cases of very low chlorophyll content can blue or red excitation light yield chlorophyll fluorescence signals which are representative of the chloroplasts of the whole leaf level.

Our results clearly indicate that the non-invasive chlorophyll fluorescence measurements taken at the upper leaf side cannot replace the net CO_2 fixation measurements using a CO_2/H_2O porometer. Non-invasive chlorophyll fluorescence measurements allow a first fast screen of a possible decline or damage to the photosynthetic apparatus, but should always be done simultaneously at the upper and lower leaf sides. Such chlorophyll fluorescence measurements need, however, to be complemented by P_N measurements in order to decide whether a photoinhibition or damage (as presumed from fluorescence measurements) is representative of the chloroplasts of the whole leaf and whether the net photosynthetic rates had really declined or not.

Acknowledgements

Part of this work was supported by a grant of the BMFT, Bonn, within the EUREKA research program LASFLEUR (EU380). We wish to thank the members of the Botanical Garden of the University of Karlsruhe, in particular Karlheinz Knoch, for growing the maple trees and to Gabrielle Johnson for checking the English text.

References

ADAMS, W. W. III: Photosynthetic acclimation and photoinhibition of terrestrial and epiphytic CAM tissues growing in full sunlight and deep shade. Aust. J. Plant Physiol. *15*, 123–134 (1988).

ADAMS, W. W. III. and B. DEMMIG-ADAMS: Operation of the xanthophyll cycle in higher plants in response to diurnal changes in incident sunlight. Planta *186*, 390–398 (1992).

ADAMS, W. W. III., S. D. SMITH, and C. B. OSMOND: Photoinhibition of the CAM succulent *Opuntia basilaris* growing in Death Valley: Evidence from 77K fluorescence and quantum yield. Oecologia *71*, 221–228 (1987).

ADAMS, W. W. III., I. TERASHIMA, E. BRUGNOLI, and B. DEMMIG: Comparisons of photosynthesis and photoinhibition in the CAM vine *Hoya australis* and several C_3 vines growing on the coast of eastern Australia. Plant Cell Environ. *11*, 173–181 (1988).

ADAMS, W. W. III., M. DIAZ, and K. WINTER: Diurnal changes in photochemical efficiency, the reduction state of Q, radiationless energy dissipation, and non-photochemical fluorescence quenching in cacti exposed to natural sunlight in northern Venezuela. Oecologia *80*, 553–561 (1989).

ADAMS, W. W. III., B. DEMMIG-ADAMS, and K. WINTER: Relative contributions of zeaxanthin-related and zeaxanthin-unrelated types of «High-Energy-State» quenching of chlorophyll fluorescence in spinach leaves exposed to various environmental conditions. Plant Physiol. *92*, 302–309 (1990).

ASADA, K.: Production and scavenging of active oxygen in chloroplasts. In: SCANDALIOS, J. G. (ed.): Molecular Biology of Free Radical Scavenging Systems, pp. 173–192. Cold Spring Habor Laboratory Press, Plainview, New York (1992).

ASADA, K. and M. TAKAHASHI: Production and scavenging of active oxygen in photosynthesis. In: KYLE, D. J., C. B. OSMOND, and C. J. ARNTZEN (eds.): Photoinhibition, p. 227. Elsevier Science Publishers, Amsterdam (1987).

BENNET, J.: Chloroplast protein phosphorylation and the regulation of photosynthesis. Physiol. Plant. *60*, 583–590 (1984).

BURKART, S.: Investigations on the relaxation kinetics of non-photochemical quenching of the chlorophyll fluorescence of leaves under photoinhibitory conditions. Karlsr. Contr. Plant Physiol. *27*, 1–136 (1994).

DEMMIG-ADAMS, B. and W. W. ADAMS III: Photoprotection and other responses of plants to high light stress. Annu. Rev. Plant Physiol. Plant Mol. Biol. *43*, 599–626 (1992).

DEMMIG, B., G. WINTER, A. KRÜGER, and F.-C. CZYGAN: Photoinhibition and zeaxanthin formation in intact leaves. Plant Physiol. *84*, 218–224 (1987).

DEMMIG-ADAMS, B., W. W. ADAMS III, A. WINTER, A. KRÜGER, and F.-C. CZYGAN: Light stress and photoprotection related to the carotenoid zeaxanthin in higher plants. In: BRIGGS, W. R. (ed.): Photosynthesis, pp. 375–391. Alan R. Liss Inc., New York (1989).

ELSTNER, E. F.: Der Sauerstoff (Biochemie, Biologie, Medizin), BI-Wissenschaftsverlag, Mannheim (1990).

EINIG, W., P. WEIDMANN, R. HAMPP, C. HAGG, U. RINDERLE, and H. K. LICHTENTHALER: Tagesgänge der Photosynthese-Aktivität in Fichtennadeln: Gaswechsel, Energiehaushalt und Regulation des Kohlensäurestoffwechsels. In: 7. Statuskolloquium des PEF, KfK-PEF Report *80*, pp. 81–91. Kernforschungszentrum, Karlsruhe (1991).

Foyer, C. H. and J. Harbinson: Oxygen metabolism and the regulation of photosynthetic electron transport. In: Foyer, C. H. and P. M. Mullineaux (eds.): Causes of Photooxidative Stress and Amelioration of Defense Systems, pp. 1–41. CRC Press, Boca Raton, Florida (1994).

Genty, B., J.-M. Briantais; and N. R. Baker: The relationship between the quantum yield of photosynthetic electron transport and quenching of chlorophyll fluorescence. Biochem. Biophys. Acta *990,* 87–92 (1989).

Gilmore, A. M. and H. Y. Yamamoto: Linear models relating xanthophylls and lumen acidity to non-photochemical fluorescence quenching. Evidence that antheraxanthin explains zeaxanthin-independent quenching. Photosynth. Res. *35,* 67–78 (1993).

Golz, A., M. Focke, and H. K. Lichtenthaler: Inhibitors of *de novo* fatty acid biosynthesis in higher plants. J. Plant Physiol. *143,* 426–443 (1994).

Gruszecki, W. I. and K. Strzalka: Does the xanthophyll cycle take part in the regulation of fluidity of the thylakoid membrane? Biochim. Biophys. Acta *1060,* 310–314 (1991).

Hager, A.: Die reversiblen, lichtabhängigen Xanthophyllumwandlungen im Chloroplasten. Ber. Deutsch. Bot. Ges. *88,* 27–44 (1975).

Hager, A. and K. Holocher: Localization of the xanthophyll-cycle enzyme violaxanthin de-epoxidase within the thylakoid lumen and abolotion of its mobility by a (light dependent) pH decrease. Planta *192,* 581–589 (1994).

Halliwell, B. and J. M. C. Gutteridge: Protection against radical damages: Systems with problems. In: Balliwell, B. and J. M. C. Gutteridge (eds.): Free Radicals in Biology and Medicine, pp. 277–298. Clarendon Press, Oxford (1989).

Havaux, M. and W. I. Gruszecki: Heat and light-induced chl *a* fluorescence changes in potato leaves containing high or low levels of the carotenoid zeaxanthin, indications of a regulatory effect of zeaxanthin on thylakoid membrane fluidity. Photochem. Photobiol. *58,* 607–614 (1993).

Havaux, M., W. I. Gruszecki, I. DuPont, and R. M. Leblanc: Increased heat emission and ist relationship to the xanthophyll cycle in pea leaves exposed to strong light stress. J. Photochem. Photobiol.B: Biol. *8,* 361–370 (1991a).

Havaux, M., R. J. Strasser, and H. Greppin: A theoretical and experimental analysis of the qP and qN coefficients of chlorophyll fluorescence quenching and their relation to photochemical and non-photochemical events. Photosynth. Res. *27,* 41–44 (1991b).

Horton, P., A. V. Ruban, D. Rees, A. A. Pascal, G. Noctor, and A. J. Young: Control of the light-harvesting function of chloroplast membranes by aggregation of the LHCII chlorophyll-protein complex. FEBS Lett. *292,* 1–4 (1991).

Krause, G. H. and E. Weis: The photosynthetic apparatus and chlorophyll fluorescence. An introduction. In: Lichtenthaler, H. K. (ed.): Applications of chlorophyll fluorescence, pp. 3–11. Kluwer Academic Publishers, Dordrecht (1988).

– – Chlorophyll fluorescence and photosynthesis: The basics. Annu. Rev. Plant Rev. Plant Physiol. Plant Mol. Biol. *42,* 313–349 (1991).

Lee, A. I. and J. P. Thornber: Analysis of the pigment stoichiometry of pigment-protein complexes from barley (*Hordeum vulgare*). Plant Physiol. *107,* 565–574 (1995).

Lichtenthaler, H. K.: Chlorophylls and carotenoids: Pigments of photosynthetic biomembranes. Methods Enzymol. *148,* 350–382 (1987).

– The plant prenyllipids including carotenoids, chlorophylls and prenylquinones. In: Moore, T. S. (ed.): Lipid Metabolism in Plants, pp. 427–470. CRC Press Inc., Boca Raton, Florida (1993).

– Vegetation stress: an introduction to the stress concept in plants. J. Plant Physiol. *148,* 4–14 (1996).

Lichtenthaler, H. K. and U. Rinderle: The role of chlorophyll fluorescence in the detection of stress conditions of plants. CRC Crit. Rev. Anal. Chem. *19* (Suppl.), pp. 29–85 (1988).

Lichtenthaler, H. K. and C. Schindler: Studies on the photoprotective function of zeaxanthin at high-light conditions. In: Murata, N. (ed.): Research in Photosynthesis, Vol. *IV,* pp. 517–520. Kluwer Academic Publishers, Dordrecht (1992).

Lichtenthaler, H. K., U. Prenzel, and G. Kuhn: Carotenoid composition of chlorophyll-carotenoid-proteins from radish chloroplasts. Z. Naturforsch. *37c,* 10–12 (1982).

Lichtenthaler, H. K., C. Buschmann, U. Rinderle, and G. Schmuck: Application of chlorophyll fluorescence in ecophysiology. Radiat. Environ. Biophys. *25,* 297–308 (1986).

Lichtenthaler, H. K., S. Burkart, C. Schindler, and F. Stober: Changes in photosynthetic pigments and *in vivo* chlorophyll fluorescence parameters under photoinhibitory growth conditions. Photosynthetica *27,* 343–353 (1992).

Ludlow, M. M.: Light stress at high temperature. In: Kyle, D. J., C. B. Osmond, and C. J. Arntzen (eds.): Photoinhibition, pp. 89–110. Elsevier, Amsterdam (1987).

Mohanty, N. and H. Y. Yamamoto: Mechanism of non-photochemical chlorophyll fluorescence quenching. I. The role of de-epoxidised xanthophylls and sequestered thylakoid membrane protons as probed by dibucaine. Aust. J. Plant Physiol. *22,* 231–238 (1995).

Mohanty, N., A. M. Gilmore, and H. Y. Yamamoto: Mechanism of non-photochemical chlorophyll fluorescence quenching. II. Resolution of rapidly reversible absorbance changes at 530 nm and fluorescence quenching by the effects of antimycin, dibucaine and cation exchanger, A23187. Aust. J. Plant Physiol. *22,* 239–247 (1995).

Peter, G. F. and J. P. Thornber: Biochemical composition and organization of higher plant photosystem II light-harvesting pigment-proteins. J. Biol. Chem. *266,* 16745–16754 (1991).

Pfündel, E. and W. Bilger: Regulation and possible function of the violaxanthin cycle. Photosynth. Res. *42,* 89–109 (1994).

Price, A. H., N. M. Atherton, and G. A. F. Hendry: Plants and drought stress generate activated oxygen. Free Rad. Res. Commun. *8,* 61–66 (1989).

Richter, M., R. Goss, B. Böthin, and A. Wild: Zeaxanthin dependent and zeaxanthin independent changes in non-photochemicalenergy dissipation. J. Plant Physiol. *143,* 495–499 (1994).

Ruban, A. V. and P. Horton: Mechanism of ΔpH-dependent dissipation of absorbed excitation energy by photosynthetic membranes. I. Spectroscopic analysis of isolated light-harvesting complexes. Biochim. Biophys. Acta *1102,* 30–38 (1992).

– – Regulation of non-photochemical quenching of chlorophyll fluorescence in plants. Aust. J. Plant Physiol. *22,* 221–230 (1995).

Schindler, C.: Investigations on the zeaxanthin accumulation in plants. Karlsr. Contr. Plant Physiol. *32,* 1–203 (1995).

Schindler, C. and H. K. Lichtenthaler: Is there a correlation between light-induced zeaxanthin accumulation and quenching of variable chlorophyll *a* fluorescence? Plant Physiol. Biochem. *32,* 813–823 (1994).

Schreiber, U., U. Schliwa, and W. Bilger: Continuous recording of photochemical and non-photochemical chlorophyll fluorescence quenching with a new type of modulation fluorometer. Photosynth. Res. *10,* 51–62 (1986).

Schweiger, J., M. Lang, and H. K. Lichtenthaler: Differences in fluorescence excitation spectra of leaves between stressed and non-stressed plants. J. Plant Physiol. *148* (in press) (1996).

Siefermann-Harms, D.: The xanthophyll cycle in higher plants. In: Tevini, M. and H. K. Lichtenthaler (eds.): Lipids and Lipid Polymers in Higher Plants, pp. 218–230. Springer, Berlin (1977).

Van Kooten, O. and J. F. H. Snel: The use of chlorophyll fluorescence nomenclature in plant stress physiology. Photosynth. Res. *25,* 147–150 (1990).

Yamamoto, H. Y.: Biochemistry of the violaxanthin cycle in higher plants. Pure Appl. Chem. *51,* 639–648 (1979).

J. Plant Physiol. Vol. 148. pp. 413–417 (1996)

Preparation of Carotenoid Enriched Vesicles of Monogalactosyldiacylglycerol and Digalactosyldiacylglycerol for Use in Raman Studies to Investigate Leaf Canopy Damage

BERTHA ALLEN[1], HELEN A. NORMAN[2], JAMES S. VINCENT[1], and EMMETT W. CHAPPELLE[3]

[1] University of Maryland, Baltimore County, Catonsville, Maryland 21228, USA

[2] Weed Science Laboratory, United States Department of Agriculture, Agricultural Research Service, Beltsville, Maryland 20705, USA

[3] National Aeronautics and Space Administration, Goddard Space Flight Center, Laboratory of Terrestrial Physics, Greenbelt, Maryland 20771, USA

Received July 7, 1995 · Accepted October 15, 1995

Summary

New methods were developed for the semi-preparative isolation of galactolipids (GL) monogalactosyldiacylglycerol (MGDG), and digalactosyldiacylglycerol (DGDG) from *Glycine max* L. Merr. (soybean) leaf tissue. Model membranes composed of isolated MGDG and DGDG were enriched with the natural carotenoid, lutein and subjected to vibrational spectroscopic analysis as a function of temperature. The carotenoid was introduced to provide a resonance Raman (rR) indicator that would be sensitive to the macroscopic environment of the lipid assembly. The preliminary Infra-red (IR), Raman and rR spectra in the ordered gel phase and in the relatively disordered liquid crystalline (lc) phase of the phospholipid (PL) model membrane system distearoylphosphatidylcholine (DSPC) and spectra of the GL model membrane systems, that span the temperature range of –30 °C to 0 °C, indicate that vesicles prepared by this method provide relevant model membrane systems.

Key words: Glycine max L. Merr., bilayers, carotenoids, galactolipids, model membrane systems, order/disorder, thermotropic profiles.

Abbreviations: DGDG = digalactosyldiacylglycerol; DSPC = disteroylphosphatidylcholine; GL = galactolipid; IR = Infra-Red; MGDG = monogalactosyldiacylglycerol; MLV = multilamellar vesicles; PL = phospholipid; rR = resonance Raman. In the shorthand numbering system used for fatty acids e.g. 16:0 the integer preceding the colon represents the number of carbon atoms in the fatty acid while that following the colon indicates the number of double bonds present.

Introduction

The photosynthetic machinery of plants exists in the chloroplast thylakoid membranes. Carotenoids are bound to chlorophylls in the chlorophyll-carotenoid-protein complexes of the thylakoid membranes of non-stressed plants. The partially damaged chloroplasts of stressed plants exhibit chlorophyll degradation resulting in an accumulation of carotenoids in the osmiophilic plastoglobuli (Lichtenthaler, 1968). The osmiophilic plastoglobuli are believed to be storage reservoirs for lamellar lipids prior to thylakoid formation and are also found to increase in size and number in damaged and senescing chloroplasts (Dodge, 1970; Mühlethaler, 1971).

Carotenoids have been previously used as both rR and absorption spectral probes of phospholipid (PL) membrane structure (Mendelsohn and Van Holten, 1979). Non-invasive techniques for determining the macroscopic structure in membranes include infra-red (IR), Raman, and resonance Raman (rR) spectroscopy (Mendelsohn et al., 1989; Casal and McElhaney, 1990; Vincent and Levin, 1991). The rR of carotenoids in GL is a potentially valuable tool for monitoring structural changes in GL that could be a model for changes resulting from plant stress.

The initial study is an investigation of the spontaneous Raman of the well characterized PL, DSPC, and then progresses to the rR of a carotenoid immersed in DSPC. Later studies will involve more complex model GL systems that resemble natural chloroplast membranes. To execute these studies, it was first necessary to develop new methods for semi-preparative purification of leaf GL. Resonance Raman and Raman thermotropic profiles of unsaturated MGDG and DGDG will provide a basis for understanding the role of fatty acid unsaturation in describing conformational order and membrane fluidity.

Materials and Methods

Materials

Soybean plants (DeKalb CX360) were grown in field plots at USDA-ARS, Beltsville, Maryland, USA. Trifoliate leaves were removed, weighed, lyophilized and stored at −80 °C. DSPC in chloroform was purchased from Avanti Polar Lipids, Inc. (Alabaster, Ala., USA). Both boron trifluoride in methanol and xanthophyll, containing greater than 70 % lutein, were purchased from Sigma Chemical Co. (St. Louis, MO., USA).

Lipid purification

Total lipids were extracted based on the procedure of Bligh and Dyer (1959). Lipid fractions were initially separated on the basis of polarity, into neutral lipid, GL, and PL classes, using silica Sep-Pak cartridges (Millipore/Waters, Milford, Mass, USA). Chloroform, chloroform : methanol, (85 : 15), and methanol were used respectively, to elute the different lipid fractions (Norman and St. John, 1986). The GL fraction was loaded onto a set of 3 Sep-Paks connected in series and washed with chloroform, chloroform : methanol, (85 : 15), and methanol.

TLC on silica gel G, using an acetone : acetic acid : water, (100 : 2 : 1) (v/v/v), developing solvent system, indicated the presence of a mixture of carotenoids that had to be removed before the GL could be isolated. To achieve this, the sep-pak elutions were repeated, followed by elution of the GL fraction through activated charcoal. Purity of the GL fractions was confirmed by TLC on silica gel G. To estimate yield, MGDG and DGDG were quantified based on galactose content (Rouglan and Batt, 1968). Recoveries were confirmed using radioactive ^{14}C. Lipid phosphorus was determined spectrophotometrically (Norman and St. John, 1986).

MGDG and DGDG were resolved by normal phase HPLC using a mobile phase composed of hexane with 1 % tetrahydrofuran, isopropyl alcohol, and water (49.0/46.2/4.8) (v/v/v) delivered at 0.8 mL min^{-1}. The column was a MAXSIL 5 µ SI (Phenomenex, Torrance, CA., USA) (150 × 10.00 mm). The HPLC analysis was achieved using a Hitachi (Model L-6200A) pump and a Linear (UVIS 200) variable wavelength detector operated at 208 nm to detect the peaks. The presence of lipid was verified by monitoring the CH_2 vibrational stretching mode region using a Nicolet FTIR (Model 60SX), which was purged of carbon dioxide. Fatty acid methyl esters were prepared directly with boron trifluoride/methanol and analyzed by GLC using a Hewlett Packard Model 5890 Series II gas chromatograph (Norman and St. John, 1986).

Raman sample preparation – Aqueous Dispersions of PL and GL –

Both PL and GL model systems were dissolved in chloroform solvent, subjected to a stream of N_2 gas and then put under high vacuum for 24 h to remove all traces of organic solvents. Multilamellar vesicles (MLV), of DSPC formed readily after a few freeze-thaw cycles. However, many freeze-thaw cycles of sonicating at 45 °C, freezing, vortexing with ice crystals and sonicating again were required to prepare MLV's of DGDG. After the seventh freeze-thaw cycle, the samples were heated to 80 °C and immediately concentrated under a stream of N_2, transferred to evacuated capillary tubes and sealed under nitrogen pressure.

Aqueous dispersions of MGDG were even more difficult to achieve. After completing the procedure outlined above for DGDG, it was necessary to store the sample frozen overnight. The next day two more freeze-thaw cycles were necessary before the samples were fully hydrated. The difficulty in preparing MLV's of MGDG may be related to the formation of non-lamellar structures (hexagonal II phase) in unsaturated systems whereas DGDG forms lamellar bilayers (Shipley, 1973; Sen, 1983). These non-lamellar structures of MGDG have been suggested to facilitate insertion and stabilization of large protein complexes of the photosystems and may provide a structural role in forming the curved edges of the appressed areas of thylakoid membranes (Curatolo, 1986).

Physical studies

Raman scattering studies of the aqueous dispersions were achieved using a modified Spex triplemate spectrometer with the excitation radiation provided by a 514.5 nm coherent Innova 100 Argon ion laser with powers that did not exceed 100 mW. Raman scattering was detected with a diode array consisting of 650 pixels that spanned the spectral range of about 900 cm^{-1} to 1600 cm^{-1} with a 1200 line/mm grating. Spectral frequencies were calibrated with the atomic emission lines of a neon lamp. Raman line shape changes of the CH_2 stretching region of DSPC, MGDG and DGDG as a function of temperature were used to describe the phase behavior (Vincent and Levin, 1991). In addition, the resonance Raman carbon-carbon stretching modes of lutein enriched aqueous dispersions of DSPC and DGDG were analyzed from prepared samples (Mendelsohn and Van Holten, 1979).

Results and Discussion

Purification of MGDG and DGDG

GL fractions have been typically recovered by silicic acid or acid treated-florisil column chromatography and further resolved by TLC (Christie, 1992; Gunstone et al., 1994). To replace these procedures, several laboratories have described HPLC techniques for the direct isolation of lipid classes. Moreau et al. (1990) were successful in resolving microgram quantities of lipid extracts from different plant tissues by normal phase analytical HPLC with an isooctane : isopropanol : water ternary gradient system. Rezanka and Podojil (1989) separated a crude polar lipid extract of *Chlorella kessleri* by HPLC using a preparative silica gel column. In this case mg quantities of lipid were recovered. Semi-preparative isolation of phosphatidylcholine from soybean leaf tissue included isocratic normal phase HPLC of a PL fraction recov-

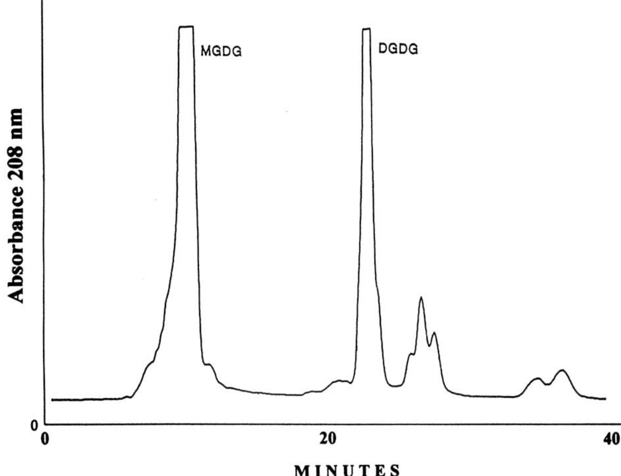

Fig. 1: Normal phase HPLC separation of MGDG and DGDG using an isocratic mobile phase composed of hexane with 1% tetrahydrofuran, isopropyl alcohol, and water (49.0/46.2/4.8) (v/v/v).

ered from a silica Sep-Pak (Glass, 1990, 1991). This method made use of an isocratic mobile phase composed of isooctane, isopropanol and water (40/41/9) (v/v/v), modified from conditions previously described (Moreau et al., 1990).

Semi-preparative isolation of GL classes (i.e. MGDG and DGDG) from leaf tissue is complicated by the fact that these lipids elute relatively early on a normal phase HPLC column, and the MGDG peak in particular is easily contaminated by carotenoids when mg samples of crude lipid extracts are injected. We were unsucessful in resolving MGDG and DGDG from contaminating lipids when a total lipid fraction was injected. It was first necessary to recover a GL fraction from crude extracts by 2 silica Sep-Pak (Millipore/Waters, Milford, Mass., USA) elutions followed by active charcoal to remove contaminating carotenoids. The resulting GL fraction was then resolved by normal phase HPLC using an isocratic mobile phase composed of hexane with 1% tetrahydrofuran, isopropyl alcohol, and water (49.0/46.2/4.8) (v/v/v). The retention times achieved for MGDG and DGDG were 12 and 25 min, respectively (Fig. 1). Aliquots of the recovered MGDG and DGDG peaks each gave a single spot on a TLC plate and no lipid phosphorus was detectable. We were successful in isolating milligram quantities of MGDG and DGDG from soybean leaf tissue by this purification scheme. The resultant yields of MGDG and DGDG were 8.7 mg/10 g and 4.6 mg/10 g fresh weight of leaf tissue, respectively. GLC analysis of methyl ester of DGDG fatty acids revealed the following composition (wt%): 15.0% 16:0, 7.2% 18:0, 3.0% 18:2 and 74.8% 18:3. The equivalent MGDG composition was 1.9% 16:0, 0.9% 18:0, 0.6% 18:1, 3.8% 18:2, and 92.8% 18:3. This relatively high level of unsaturation is typical of leaf tissue (Kates and Harwood, 1994).

Raman Spectroscopic Data

Two spectral regions, the C-H stretching region (~2800–3100 cm^{-1}) and the ~1000–1600 cm^{-1} region in which the C-C stretching mode, the C-H deformation mode and the CH_2 twisting mode occur in the spontaneous Raman of the acyl chain lipids and the resonance enhanced C-C and C=C stretching modes of the carotenoids have served to indicate macroscopic organization in lipid assemblies (Mendelsohn and Van Holten, 1979; Levin, 1984). The line shape and frequency of those vibrational modes vary as the the assemblies change with temperature. To be certain that the behavior of DSPC as a function of temperature is not inordinately pertubed by the inclusion of lutein, the Raman spectra of both lutein enriched and pure DSPC MLV's were observed as a function of temperature. The peak height intensity ratio of the 2850 cm^{-1} methylene symmetric stretching mode divided by the 2880 cm^{-1} methylene assymetric stretching mode is one parameter that has proved useful in describing the transition from the well ordered gel state at low termperatures to the liquid crystalline phase above the main lipid phase transition temperatures. This peak height intensity ratio I_{2850}/I_{2880} describes intermolecular chain disorder (Mendelsohn and Van Holten, 1979). The band at ~2850 cm^{-1} represents the methylene symmetric stretching modes of the DSPC acyl chain and the band at 2880 cm^{-1} represents the well ordered all trans gel phase acyl chain methylene asymmetric stretching modes (Levin, 1984). The phase transition temperatures of 52 ± 1.5 °C for pure DSPC MLV's and 51.5 ± 0.5 °C for lutein enriched DSPC MLV's were determined and found to be consistent with transition temperatures of 52 ± 2 °C and 52 ± 2.6 °C presented by Mendelsohn and Van Holten (1979) and Verma et al. (1980), respectively. Our results indicate that there is little or no pertubation of the lipid structure by lutein at normal concentrations of 0.4 mole percent or less.

The resonance enhanced C-C and C=C stretching mode region of the carotenoid has been used to probe the environment around the carotenoid. These vibrational modes of lutein enriched DSPC were monitored as a function of temperature. Figure 2 describes the C-C stretching modes of lutein enriched DSPC above and below the transition temperature.

The C-C and C=C stretching modes were also monitored as a function of temperature (Fig. 3) in DGDG and MGDG model systems. Since the band at ~1130 cm^{-1} is associated with chains in the all trans conformation and the band at ~1090 cm^{-1} is associated with acyl chains with gauche conformers; the intensity ratio I_{1090}/I_{1130} describes the number of

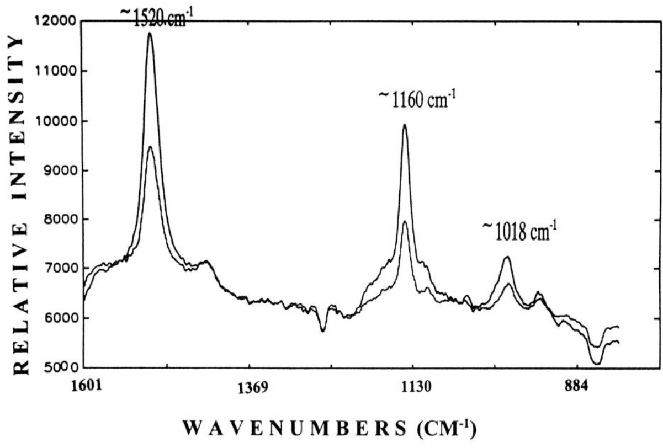

Fig. 2: Resonance enhanced C-C (carbon-carbon) stretching modes of lutein enriched DSPC aqueous dispersions above (upper curve) and below (lower curve) the phase transition temperature.

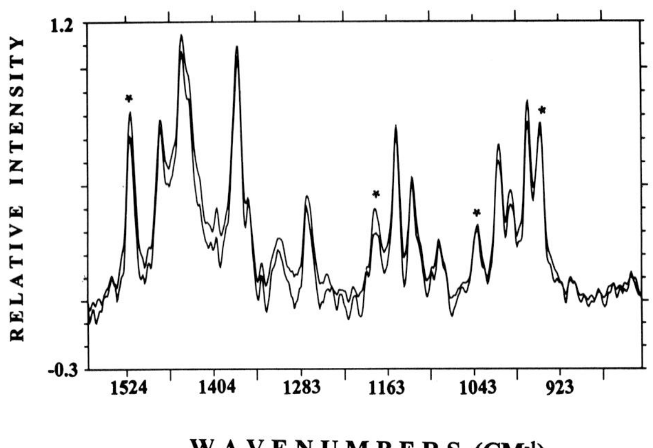

Fig. 3: Resonance Raman enhanced C-C (carbon-carbon) stretching modes of lutein enriched DGDG aqueous dispersions at −30 °C (lower curve) and 0 °C (upper curve).

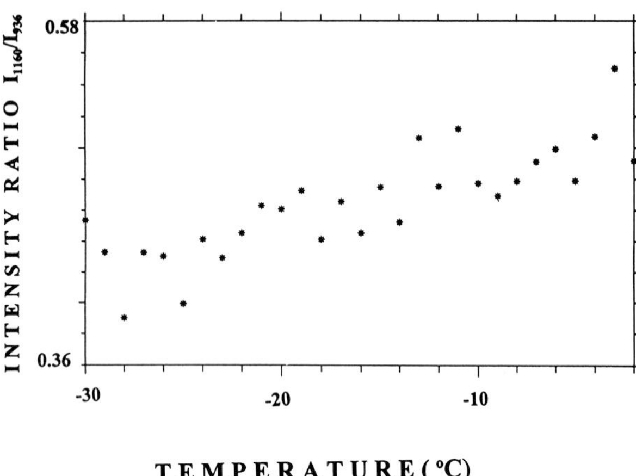

Fig. 4: Raman spectral order parameter of intensity ratio 1160/936-vs-Temperature (°C) of DGDG MLV's.

acyl chains in the all trans conformation (Mendelsohn and Van Holten, 1979; Levin, 1984). In addition, the spectral order parameter I_{1160}/I_{936} was monitored as a function of temperature, where the resonance enhanced carotenoid band at 1160 cm^{-1} and the band at 936 cm^{-1} is associated with the internal $NaClO_4$ standard. The results revealed an increase in disorder on increasing temperature and therefore support using the resonance enhanced modes of the carotenoid probe to describe orientational and organizational order of the GL (Fig. 4).

Acknowledgements

The authors would like to thank Ruth Mangum for her work on helping to produce appropriate developing solvent systems. We would also like to thank Dr. J. George Bhuta for the use of the Nicolet FTIR. In addition, we offer our gratitude to Drs. Jeffrey Mason and Curt Meuse for many helpful discussions on the preparation of aqueous dispersions of GL and raman spectroscopic techniques and applications, respectively. We would also like to thank Dr. Ira W. Levin for providing the use of Raman spectroscopic instrumentation in his laboratory at the National Institutes of Health, Bethesda, Md. USA.

References

Bligh, E. G. and W. J. Dyer: A Rapid Method of Total Lipid Extraction and Purification. Can. J. Biochem. Physiol. *37,* 911–917 (1959).

Casal, H. L. and R. N. McElhaney: Quantitative Determination of Hydrocarbon Chain Conformational Order in Bilayers of Saturated Phosphatidylcholines of Various Chain Lengths by Fourier Transform Infrared Spectroscopy. Biochemistry *29,* 5423–5427 (1990).

Christe, W. W.: Detectors for High-Performance Liquid Chromatography of Lipids with Spectral Reference to Evaporative Light Scattering Detector. In: Christie, W. W. (ed.): Advances in Lipid Methodology *7,* pp. 239–271. The Oily Press Ltd., Scotland (1992).

Curatolo, W.: Glycolipid Function. Biochim. Biophys. Acta *906,* 137–160 (1987).

Devlin, M. T. and I. W. Levin: Acyl Chain Packing Properties of Deuterated Lipid Bilayer Dispersions: Vibrational Raman Spectral Parameters. J. Raman Spectrosc. *21,* 441–451 (1990).

Dodge, Y. D.: Changes in Chloroplast Fine Structure during Autumnal Senscence of Betula Leaves. Ann. Bot. *34,* 817–824 (1970).

Glass, R. L.: Semipreparative High-Performance Liquid Chromatographic Separation of Phosphatidylcholine Molecular Species from Soybean Leaves. J. Liquid Chromat. *14,* 339–349 (1991).

Kates, M. and J. L. Harwood: Separation and Isolation Procedures. In: Gunstone, F. D., M. Kates, and J. L. Harwood (eds.): The Lipid Handbook, Second Edition, pp. 229–234. Chapman and Hall Chemical Database, London (1994).

Levin, I. W.: Vibrational Spectroscopy of Membrane Assemblies. In: Clark, R. J. H. and R. E. Hester (eds.): Advances in Infrared and Raman Spectroscopy *2,* pp. 1–47. Wiley Heyden Publishers (1984).

Lichtenthaler, H. K.: Plastoglobuli and Fine Structure of Plastids. Endeavour *27,* 144–149 (1968).

Mendelsohn, R. and M. A. Davies: Quantitative Determination of Conformational Disorder in the Acyl Chains of Phospholipid Bilayers by Infrared Spectroscopy. Biochemistry *28,* 8934–8939 (1989).

Mendelsohn, R. and R. W. Van Holten: Zeaxanthin ([3R,3′R]-beta-beta-Carotene-3-3′-Diol) as a Resonance Raman and Visible Absorption Probe of Membrane Structure. Biochemistry *28,* 8934–8939 (1979).

Moreau, R. A., P. T. Asmann, and H. A. Norman: Analysis of Major Classes of Plant Lipids by High-Performance Liquid Chromatography with Flame Ionization Detection. Phytochemistry *29,* 2461–2466 (1990).

Muhlethaler, K.: The Ultrastructure of Plastids. In: Gibbs, M. (ed.): Structure and Function of Chloroplasts, pp. 12–13. Springer-Verlag, Berlin (1971).

Norman, H. A. and J. B. St. John: Metabolism of Unsaturated Monogalactosyldiacylglycerol Molecular Species in *Arabidopis thaliana* Reveal Different Sites and Substrates for Linolenic Acid Synthesis. Plant Physiol. *81,* 731–736 (1986).

Rezanka, T. and M. Podojil: Preparative Separation of Algal Polar Lipids and of Individual Molecular Species by High-Performance Liquid Chromatography and Their Identification by Gas Chromatography-Mass Spectrometry. J. Chromat. *464,* 397–408 (1989).

Rouglan, P. G. and R. D. Batt: Quantitative Analysis of Sulfolipid (Sulfoquinovosy diglycerides) and Galactolipids (Monogalactosyl and digalactosyl diglycerides) in plant tissues. Anal. Biochem. *22,* 74–88 (1968).

Sen, A., W. P. Williams, and P. J. Quinn: The Structure and Thermotropic Properties of Pure 1,2-Diacylgalactosylglycerols in Aqueous Systems. Biochim. Biophys. Acta *663*, 380–389 (1981).

Shipley, G. C., J. P. Green, and B. W. Nichols: The Phase Behavior of Monogalactosyl, Digalactosyl, and Sulphoquinovosyl Diglycerides. Biochim. Biophys. Acta *311*, 531–544 (1973).

Snyder, R. G. and M. W: Poore: Conformational Structure of Polyetheylene Chains from Infrared Spectrum of the Partially Deuterated Polymer. Macromolecules *6*, 708–715 (1973).

Verma, S. P., D. F. H. Wallach, and J. D. Sakura: Raman Analysis of the Thermotropic Behavior of Lecithin-Fatty Acid Systems and of their Interactions with Proteolipid Apoprotein. Biochemistry *19*, 574–579 (1980).

Vincent, J. S. and I. W. Levin: Raman Spectroscopic Studies of Dimyristoylphosphatidic Acid and Its Interaction with Ferricytochrome C in Cationic Binary and Ternary Lipid-Protein Complexes. Biophys. J. *59*, 1007–1021 (1991).

J. Plant Physiol. Vol. 148. pp. 418–424 (1996)

Soil Salinity Effects on Crop Growth and Yield – Illustration of an Analysis and Mapping Methodology for Sugarcane

Craig Wiegand, Gerry Anderson, Sarah Lingle, and David Escobar

U.S. Department of Agriculture, Agricultural Research Service, Subtropical Agricultural Research Laboratory, 2413 E. Bus. Hwy 83, Weslaco, Texas 78596, U.S.A.

Received June 24, 1995 · Accepted October 10, 1995

Summary

The effects of soil salinity on growth and yield of sugarcane (*Saccharum* spp. hybrids) are used to illustrate how soil and plant samples («ground truth»), digital videographic or SPOT HRV spectral observations, and image analysis by unsupervised classification can be used jointly to quantify and map variations in weighted electrical conductivity (WEC, dS m^{-1}) of the root zone and YIELD (metric tons of millable stalks ha^{-1}). The combined data for the 1992 and 1993 growing seasons of the study showed that each dS m^{-1} increase in WEC reduced stalk population by 0.6 stalks m^{-2}, stalk weight by 0.14 kg, and stalk yields by 13.7 metric tons ha^{-1}. Sugarcane growth and yield were not affected by root zone salinities less than about 2 dS m^{-1}, but no millable stalks were produced at salinities in excess of 10 dS m^{-1}. The 25 pixels ha^{-1} of SPOT is a good scale for mapping salt stress patterns and taking site-specific ameliorative actions. The combination of satellite or aerial spectral observations, ground truth, and image classification procedures demonstrated in this study is readily applicable to other vegetation stresses.

Key words: Salinity, soil salinity, soil electrical conductivity, vegetation stress, remote sensing, image analysis, yield, yield components, sugarcane, Saccharum spp., videography, SPOT, SPOT HRV, spectral components analysis, ground truth, site specific management, mapping, crop growth, cost : benefit analysis, reclamation, economic yield.

Abbreviations: HRV = high resolution visible radiometer; SPOT = French polar orbiting satellite; WEC = weighted electrical conductivity of the root zone; YIELD = metric tons of millable stalks ha^{-1}; DC = digital counts; pixel = picture element; NDVI = normalized difference vegetation index; NIR = near-infrared wavelength interval, or band; Red = visible red band; YG = yellow-green visible band; SCA = spectral components analysis; ECe = electrical conductivity of water extracted from saturated soil, dS m^{-1}; band = electromagnetic energy wavelength interval, or waveband, over which a sensor operates.

Introduction

Saline soil is an example of a widespread condition that causes vegetative stress. Soils are considered saline or salt-affected when the electrical conductivity of water extracted (ECe) from water-saturated soil samples from the root zone exceeds 4 dS m^{-1} (Richards, 1954). The dissolved salts create an osmotic stress that adds to any existing matric water stress (Thomas and Wiegand, 1970), so that plants that grow are stunted. At salinities greater than about 16 dS m^{-1} the seeds of even salt tolerant crop plants do not germinate. Consequently, areas within fields more saline than that are barren.

Szabolics (1989) estimated that salt-affected soils comprise 19 % of the 2.8 billion hectares of arable land on Earth. They occur naturally in arid and semi-arid climates from weathering of indigenous minerals (Tanzi, 1990) but are most important economically where irrigation is practiced to produce crops (Carter, 1975). The soluble salts in the applied irrigation water accumulate in the root-zone as plants extract water unless deep percolation of rainfall and water applied for leaching, or man-made drainage, maintains a favorable salt balance.

Sugarcane (*Saccharum* spp.) is widely grown in tropical and subtropical climates under supplemental irrigation. Maas and Hoffman (1977) and Maas (1990) rated it moderately sensitive to soil salinity, and reported a threshold ECe of 1.7 dS m^{-1} above which stalk and sugar yields decline. Joshi and Naik (1964) reported 50 % reductions in yield in India at a root zone ECe of 3.8 dS m^{-1}, while Robinson and Worker (1965) reported a 50 % reduction in yield in California at 4.2 dS m^{-1}. Based on Jamaican experience on clay soils, Shaw (1982) classified soils for sugarcane as nonsaline (ECe <2 dS m^{-1}), low in salinity (ECe 2 to 3 dS m^{-1}), and high in salinity (ECe >4 dS m^{-1}).

Lichtenthaler (1990) and Rinderle and Lichtenthaler (1989) used chlorophyll flourescence to examine trees for stresses including forest decline, whereas Buschmann et al. (1991) have detected stress in coniferous trees spectrophotometrically.

The purpose of this paper is to use soil salinity as the vegetation stress and sugarcane as the crop to illustrate a methodology for mapping the extent and severity of vegetation stress that combines ground observations, image analysis, and remote spectral digital observations (aerial photography or videography, or satellite multi-spectral imagery). The paper builds on previous experience with cotton (*Gossypium hirsutum* L.) (Wiegand et al., 1992, 1994 a) and the 1992 season's results for sugarcane (Wiegand et al., 1994 b).

Methods

1. Test Field, and Soil and Plant Samples

A 59 ha (372 m long by 1586 m wide) commercial field of first ratoon sugarcane (cv. CP 70–321) at lat. 26° N and long. 98° W was chosen and studied during the 1992 and 1993 crop seasons. In 1992, soil and plant sample sites were positioned at 40 m intervals down the crop rows on six transects across the 1039-row-wide field. Transect pairs were 40 m (25 crop rows) apart and there were nine sample sites per transect for a total of 54 sample sites. In 1993, 70 sample sites were located at 20 m intervals along five paired and two unpaired transect segments positioned to cross areas where 1992 soil samples and imagery indicated the sugarcane was salt-affected. Sample transects were moved three rows from their 1992 positions to avoid any carryover effects of foot traffic and trampling. Figure 1 displays the sampling pattern in the field both years. Sample sites were staked and flagged before soil and plant sampling began.

Soil samples were taken on 23 April 1992 and on 13 April 1993 by 0–30, 30–60, and 60–90 cm depth increments using a tractor-

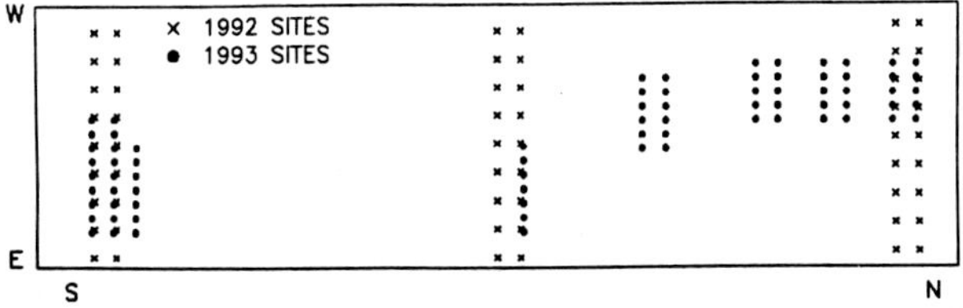

Fig. 1: Soil and plant sample locations by years within the 59-ha sugarcane field studied. x- - - -1992 sites; ●- - - -1993 sites.

mounted Giddings* hydraulic sampler. The samples were air-dried, ground to pass a 2 mm sieve, and salinity was measured as electrical conductivity of saturated soil extracts (ECe, dS m^{-1}) according to the procedures of Rhoades (1982). A weighted electrical conductivity of the root zone defined by

$$WEC = 0.5EC1 + 0.4EC2 + 0.1EC3, \quad [1]$$

where EC1, EC2, and EC3 refer to electrical conductivities of the 0–30, 30–60, and 60–90 cm soil depths, respectively, was also calculated for each site.

Counts of millable stalks per 10 m^2 were made 20 October 1992 and 27 August 1993 at all sample sites. Fifteen-stalk samples were acquired at 26 sample sites 9 December 1992 and at 30 sites 27 October 1993 to determine average stalk weight. The stalks were topped at the uppermost dewlap (collar) and stripped of green leaf blades (lamina) and sheaths. Fresh weight was determined and recorded. Stalk yield (YIELD, metric tons ha^{-1}) was calculated from the yield components, plant population and average stalk weight.

2. Spectral Observations

A high resolution multispectral video system, described by Everitt et al. (1991), was used to acquire videography in midday from an altitude of 3200 m on 18 September, 1992 and 29 September, 1993. The three camera system acquires observations in the yellow-green (YG, 543–552 nm), red (Red, 644–656 nm), and near-infrared (NIR, 845–857 nm) wavelength intervals. Ground resolution of the video data was 3.3 m. SPOT HRV (Chevrel et al., 1981) which has 20 m ground resolution (25 samples ha^{-1}) digital data were acquired on two dates in 1993 (4 June and 4 October) but on no dates in 1992 because cloud cover on overpass dates prevented it.

The digital images of the other two bands of both the video and SPOT HRV data were registered to the respective NIR band. Then 8-bit digital line printer maps of the field for each band were scaled to the dimensions of the field, and the digital counts for the sample sites were manually extracted.

3. Statistical and Image Analyses

The NIR, Red, and YG digital counts (DC) and the normalized difference vegetation index (NDVI) defined by

$$NDVI = (NIR - Red)/(NIR + Red) \quad [2]$$

were the spectral observations used for all analyses. The weighted electrical conductivity of the root zone (WEC), the number of millable stalks m^{-2}, average stalk weight, and stalk yields (YIELD, metric tons ha^{-1}) were the ground truth variables. Statistical analyses consisted of linear regressions of stalk population, stalk weight, and YIELD on WEC and NDVI, and multiple linear regression estimates of WEC and YIELD from the NIR, Red, and YG observations.

The high altitude videography acquired 18 September 1992 was submitted to a 16-category unsupervised spectral classification using the K CLUSTER software of the PCI* system. Each class was assigned a color and the classification map of the field was produced and displayed. The SPOT data for 4 October 1993 were subjected to 6- and 15-category unsupervised classifications on a SUN* work station using GRASS* software. A report that gave the number of

* Trade names are included for the benefit of the reader and do not infer endorsement of, nor preference for, the mentioned products by the U.S. Department of Agriculture.

pixels in each class and mean DC and its standard deviation for each band for each spectral class was also generated by the software.

To relate the spectral categories from the unsupervised classifications to salinity within the field, WEC for the sample sites was estimated from the three video (or SPOT) bands by multiple linear regression. Then the band means for each of the spectral classes were inserted into the WEC estimation equation to obtain the salinity class means that corresponded to the spectral class means. Upper and lower limits of salinity classes were set as the average of adjacent salinity class means.

Next the salinity of every pixel in the field was calculated, assigned to a class, color-coded, and mapped. The number of pixels in each salinity class was counted. From the report of those numbers and the total number of pixels in the field, the percentage of the pixels in each class for the field was calculated. The joint use of the unsupervised classifications to determine the spectrally different crop condition classes and then the insertion of the class mean digital counts into the regression equations from the sample sites to define the salinity is a crucial step in the procedure. For clarity, a simple chart of the procedures is included as Figure 2.

Results

1. Spectral-agronomic Relations

The combined data for 1992 and 1993 on millable stalks m^{-2}, stalk weight, and metric tons of sugarcane stalks ha^{-1} (YIELD) are summarized versus both root zone salinity (WEC) and the normalized difference vegetation index (NDVI) in Figure 3. The regression equations show that millable stalks estimated at $9.1\ m^{-2}$ under a nominal $1\ dS\ m^{-1}$ nonsaline soil condition decreased by 0.56 stalks m^{-2} for each $1\ dS\ m^{1-}$ increase in root zone salinity. Likewise, stalk weight, estimated at 1.41 kg under nonsaline conditions decreased by 0.14 kg for each $1\ dS\ m^{-1}$ increase in salinity. The soil salinity effects on those two components of yield (Figs. 3 A and 3 B) translate to a 13.7 metric ton per hectare decrease in stalk yield per unit increase in root zone salinity (Fig. 3 C).

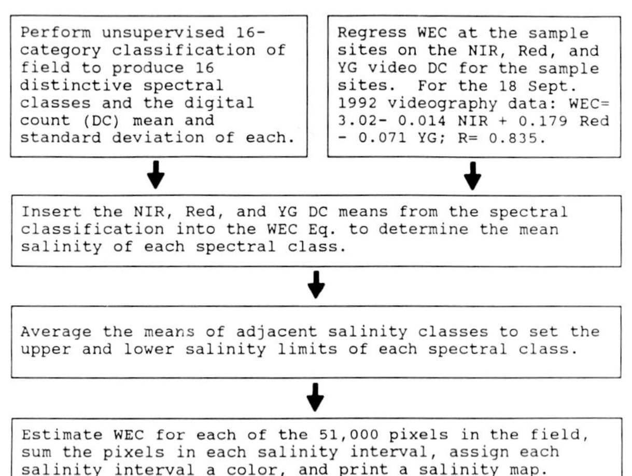

Fig. 2: Flow diagram of major steps used to produce soil salinity maps of salt-affected sugarcane. Numerical data in the flow diagram apply specifically to the 18 September 1992 videography data of Figure 4 A.

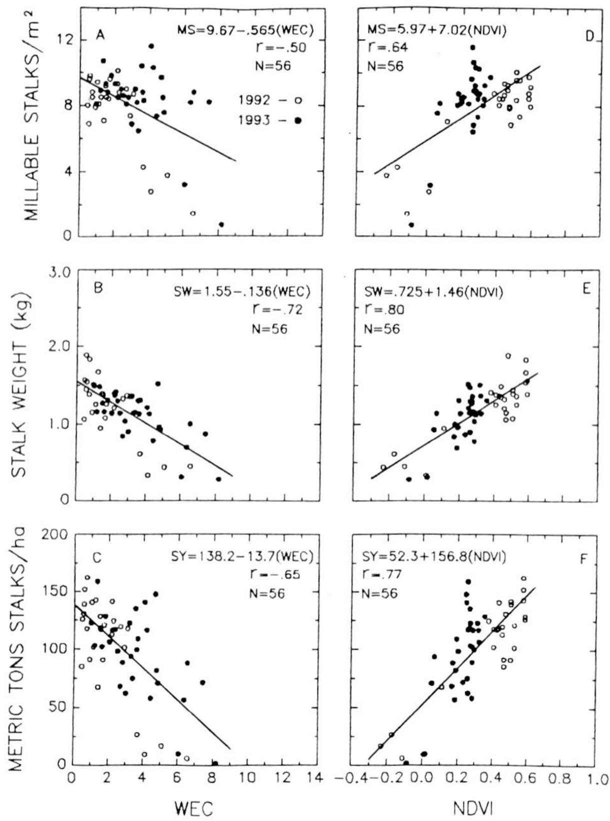

Fig. 3: Weighted root zone salinity (WEC) and the normalized difference vegetation index (NDVI) versus crop YIELD (metric tons stalks ha^{-1}) and its components, millable plants m^{-2} and stalk weight in kg.

The relation between NDVI from the September 1992 and 1993 videography observations and stalk yield and its components are displayed in parts D, E, and F of Figure 3. The range in NDVI is from −0.4 to 0.6 where values in the approximate range −0.4 to −0.2 correspond to bare soil, and values more positive than −0.2 correspond to increasing plant cover. Thus, similar to the WEC estimation equations, for a value of 0.0 NDVI, the regression equation of Figure 3 F predicts a yield of 52 metric tons stalks ha^{-1}. At the maximum value of NDVI of 0.55 the equation in Figure 3 F predicts 138 metric tons stalks ha^{-1}, the same as the equation in part C predicts at $WEC = 0\ dS\ m^{-1}$.

All parts of Figure 3 demonstrate that salinity's effects on crop production can be better estimated from spectral observations of the canopies than from measurements of the cause of the surpressed growth, salinity. This is partly due to problems of sampling, but, in our experience, a better explanation is that crop yields relate more closely to the amount of photosynthetically active tissue in the canopies (Wiegand et al., 1991) achieved than to an abiotic factor of the environment. It is well established that plants integrate the effects of growing conditions and express their responses through the canopies produced (Wiegand and Richardson, 1984, 1987).

Table 1 summarizes the equations developed from the sampling for estimating root zone salinity (WEC) and YIELD

Table 1: Equations by data sets for estimating root zone salinity (WEC) and YIELD in metric tons sugarcane stalks ha^{-1}, from spectral observations expressed both as digital counts of individual bands and as NDVI.

Data Source; Date	n[1]	Equation	R^2
VIDEO; 18 Sept '92	26	WEC = 1.29 – 0.0139 NIR + 0.1139 Red – 0.0236 YG	.800
		YIELD = 25.80 + 1.961 NIR – 1.400 Red – 0.1801 YG	.921
	54	WEC = 3.02 – 0.0136 NIR + 0.1790 Red – 0.0709 YG	.835
		WEC = 5.387 – 8.073 (NDVI)	–.811
VIDEO; 24 Sept '93	30	WEC = 2.11 – 0.0229 NIR + 0.0447 Red – 0.0096 YG	.621
		YIELD = 10.23 + 0.792 NIR – 0.570 Red – 0.139 YG	.729
	74	WEC = 3.40 – 0.016 NIR + 0.0052 Red – 0.0702 YG	.744
		WEC = 7.12 – 16.533 (NDVI)	–.753
SPOT; 4 June '93	30	WEC = 49.86 – 0.233 NIR + 0.022 Red – 0.138 YG	.722
		YIELD = –487.83 + 3.957 NIR + 1.189 Red – 1.200 YG	.831
	74	WEC = 53.11 – 0.2612 NIR + 0.0515 Red – 0.1827 YG	.799
		WEC = 12.188 – 18.19 (NDVI)	–.700
SPOT; 4 Oct '93	30	WEC = –23.96 + 0.002 NIR + 0.468 Red + 0.128 YG	.679
		YIELD = –272.2 + 2.860 NIR + 7.007 Red – 4.425 YG	.736
	74	WEC = –19.55 – 0.033 NIR + 0.4656 Red + 0.1268 YG	.748
		WEC = 14.80 – 23.01 (NDVI)	–.752

[1] Number of observations.
[2] Multiple correlation coefficients (R) or simple product moment correlation coefficients (r) for significance at the 0.05 level for 26, 30, 54 and 74 observations (n) are 0.374, .349, .264, and .226 for one independent variable and .523, .476, .357, and 323 for three independent variables. (R values, not R^2 values are tabulated).

from the spectral data. The following conclusions can be reached from the contents of the table: yield is estimated better from the spectral data than is salinity; in both 1992 and 1993, the data from all sample sites estimated WEC better than did the sites where plant samples were taken; the video data for 1992 estimated the salinity and yields better than either the videography or SPOT did in 1993; and, the video and SPOT data for 1993 estimated yield of sugarcane and salinity of the soil about equally well.

The equations for the sample sites calibrate the remote spectral observations to ground conditions and permit maps of the whole population of pixels in the field to be produced.

2. *Image Analysis Results*

The purpose of unsupervised classification is to objectively produce statistically distinctive spectral categories for further interpretation. The number of categories desired is an option in the program. For salt-tolerant crops, the categories 0–4, 4–8, 8–12, 12–16, and >16 dS m^{-1} have proven useful. But, sugarcane experiences a growth and yield reduction at 1.7 dS m^{-1} (Maas and Hoffman, 1977; Maas, 1990) and the equation in Figure 3 C predicts zero yield of sugarcane at 10.1 dS m^{-1}. Therefore, salinity categories such as <1.7, 1.7–3.0, 3.0–4.5, 4.5–6.0, 6.0–8.0, 8.0–10.0, and >10 dS m^{-1} are more appropriate for sugarcane.

The statistically different spectral categories are usually color-coded and displayed on the CRT. However, they remain spectral categories unless there is a way to relate them meaningfully to soil salinity. For the data sets of this study, that was done by inserting the category means by band into the sample site WEC estimation equations of Table 1. If yields are of interest, then the spectral means of the categories from the computer-generated classification report can be entered into the regression equations for yield, such as those in Table 1, for the respective data sets.

Figure 4 A is a salinity map of the test sugarcane field from the 18 September 1992 videography data. For this map a 16-category unsupervised classification was performed and the salinity categories corresponding to the 16 spectral classes were determined using the regression equation included in the methods flow chart of Figure 2 and in Table 1 for 54 sample sites. Nine of the 16 categories, representing only 3.7% of the image area, were for perimeter pixels consisting of turnrows, a drain ditch and spoil banks along the west edge of the field, and roads and ditches on the south and east sides of the field. Because these «land uses» were predominantly barren, the regression equations estimated salinities >8.61 dS m^{-1} for all of them.

In the image shown in Figure 4 A, the salinities are reported to three decimal places. This exceeds the precision justified by the sampling and analysis procedures. However, there were 51,000 videography pixels in the image and calculations were carried to three decimal places rounded to keep pixels on the borderline between salinity categories from being undefined. That is, the salinity categories for calculation were ≤1.095, 1.096–1.400, 1.401–2.860, etc. Once pixel assignments have been made the practical salinity categories can be stated as <1.1, 1.1–1.4, 1.4–2.9, 2.9–5.6, 5.6–8.6, and >8.6 dS m^{-1}. These are within the meaningful range for sugarcane suggested earlier.

One can also request more categories in the unsupervised classifications, then combine adjacent classes. However, there has to be a spectral response in the crop corresponding to the salinity intervals desired for this to be a viable alternative. Our experience indicates a minimum of five and a maximum

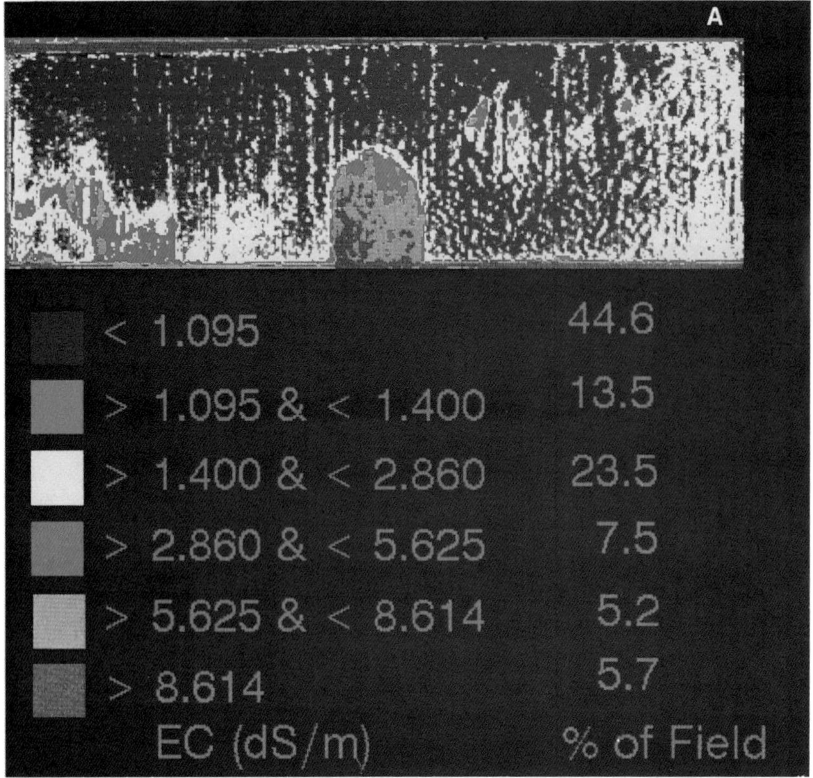

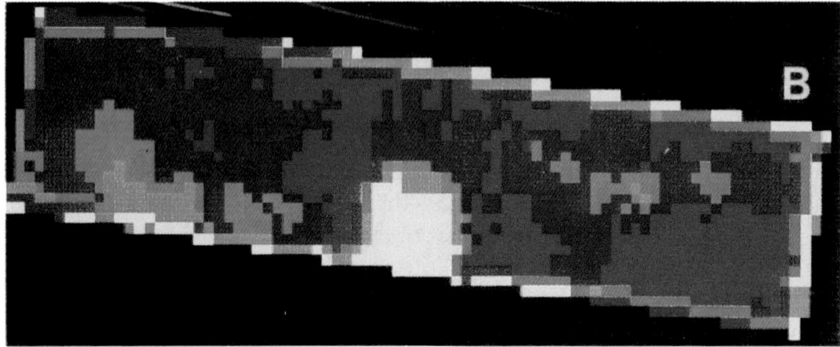

Fig. 4: Salinity map (part A) of 18 September 1992 videography processed according to Figure 2 procedures, and spectral category map (part B) of 4 October 1993 SPOT observations directly from unsupervised image classification procedure.

of seven salinity categories are useful. Although 16 categories were specified for the image of Figure 4 A, the pixels outside the planted sugarcane area constituted 9 of them and the CRT background was another. The peripheral pixels mainly help orient the field for the human observer.

Figure 4 B is a six category color coded spectral classification map for the 4 October 1993 satellite image. There were 1636 pixels in the polygon of the field, so the image is blocky, compared with the videographic image of Figure 4 A.

Table 2 summarizes the salinity intervals and fractional area of the sugarcane field of Figure 4 for the 18 September 1992 videography and the 4 October 1993 SPOT imagery. The salinity intervals are those obtained by entering the digital count means for the spectral categories into the WEC estimation equations of Table 1, then averaging the WEC means of adjacent classes to obtain the upper and lower bounds of the intervals. Six intervals are produced in each case, and in each case two are nonsaline categories, two have been labeled moderately saline and two very saline in keeping with sugarcane's sensitivity to soil salinity. The intervals are practical and realistic for this crop. The percentage of the field in the more saline categories is higher for the October 1993 than the September 1992 image in keeping with the lower salinity of the most saline classes and the higher proportion of boundary pixels in the SPOT image (Fig. 4 B) than in the video image (Fig. 4 A).

Discussion

Our results demonstrate how spectral observations of crop canopies, ground observations, and unsupervised image classifications can be combined to help quantify soil salinity and map its distribution. Unsupervised image classification produces spectrally different categories within the field completely independently of other information. However, one

Table 2: Salinity intervals, and percent of field in each salinity interval for two spectral observation dates.

Date	Salinity Range (dS/m)	Percent of Field and Salinity Rating (%)		
18 Sept 1992 (VIDEO)	< 1.10	44.6	58.1	Nonsaline
	1.10–1.40	13.5		
	1.40–2.86	23.5	31.0	Moderately Saline
	2.86–5.62	7.5		
	5.62–8.61	5.2	10.9	Very Saline
	> 8.61	5.7		
		100.0		
4 Oct 1993 (SPOT)	< 1.50	14.2	47.6	Nonsaline
	1.50–2.00	33.4		
	2.00–3.08	19.4	27.5	Moderately Saline
	3.08–4.94	8.1		
	4.94–6.68	11.8	24.9	Very Saline
	> 6.68	13.1*		
		100.0		

* Part of these pixels are field roads and turnrows that were bare soil and were classed as saline.

usually wants to know what the categories «mean» or what caused them. The ground samples of the soil and plants enable the categories to be further interpreted.

In this study we developed multiple linear regression equations to relate the responses in the three spectral bands to both WEC and YIELD at plant and soil sampling sites. Nix (1981) pointed out that regression models have useful predictive value in an environment where one factor exerts overwhelming dominance over crop responses. Salinity is such a factor; in many salt-affected fields, portions of the field are not affected at all while other portions of the field are so salty no plants grow. For crops, other individual factors that can be dominant are drought (soil water availability), herbicide damage, heavy metal toxicities, freeze damage, and foliar or root diseases. The pattern of occurrence of damage within and among fields often provides clues to the cause of damage that can be verified by ground inspection and sampling.

Among fields there are additional variations in growth due to tillage type and timeliness, past and current fertilizer applications, crop rotations employed, cultivars, etc. There are also variations in soil texture and topography within fields. These introduce uncontrolled variations in plant growth that are sensed by the vegetation indices. Therefore, yield variations among and within fields are usually proportional to the vegetation indices.

Yield maps such as those that can be produced using the equations of Table 1 provide the economic data producers are interested in. The yield maps can also be useful in conjunction with geographic information systems (GIS), (Burrough, 1986) and global positioning systems for site specific management of subareas of fields. Likewise, reclamation specialists, drainage engineers, and loan agencies need the YIELD versus WEC functional relations for cost: benefit analyses and for amortizing reclamation costs.

Acknowledgements

We thank Rene Davis for helping acquire the videography data; Romeo Rodriguez, Paul Thompson, and Jean Ann Pearcy for helping acquire and process the field data; Ricky Villarreal for computer programming and analysis assistance; Romeo Rodriguez and Stephen Neck for statistical analyses; Romeo Rodriguez and Michelle Noriega for figure preparation; and, Angie Cardoza for manuscript preparation.

References

Burrough, P. A.: Principles of Geographic Information Systems for Land Resources Assessment. Clarendon Press, Oxford, 194 p. (1986).

Buschmann, C., U. Rinderle, and H. K. Lichtenthaler: Detection of stress in coniferous forest trees with the VIRAF spectrometer. IEEE Trans. Geosci. and Remote Sens. *29*, 96–100 (1991).

Carter, D. L.: Problems of salinity in agriculture. In: Poljakoff-Mayber, A. and J. Gales (eds.): Ecological Studies, Analysis, and Synthesis, Vol. *15*, pp. 25–35, Plants in Saline Environments. Springer-Verlag, New York (1975).

Chevrel, M., M. Courtois, and G. Weill: The SPOT satellite remote sensing mission. Photogramm. Engin. and Remote Sens. *47*, 1163–1171 (1981).

Everitt, J. H., D. E. Escobar, and J. R. Noriega: A high resolution multispectral video system. Geocarto Int. *6*, 45–51 (1991).

Joshi, G. V. and G. R. Naik: Salinity effect on growth and photosynthetic productivity in sugarcane var. Co 740. Indian Sugar *27*, 329–332 (1964).

Lichtenthaler, H. K.: Applications of chlorophyll fluorescence in stress physiology and remote sensing. In: Steven, M. and J. A. Clark (eds.): Applications of Remote Sensing in Agriculuture, pp. 287–305. Butterworths Scientific Ltd., London (1990).

Maas, E. V.: Crop salt tolerance. In: Tanji, K.. K. (ed.): Agriculture Salinity Assessment and Management, ASCE Manuals and Reports on Engineering Practice No. 71, pp. 262–304. Amer. Soc. Civil Engineers, New York (1990).

Mass, E. V. and G. J. Hoffman: Crop salt tolerance – current assessment. J. Irrig. and Drainage *103* (IR2), 115–134 (1977).

Nix, Henry: Simplified simulation models based on specified minimum data sets: The CROPEVAL concept. In: Berg, A. (ed.): Application of Remote Sensing to Agricultural Production Forecasting, pp. 151–170. A. A. Balkema, Rotterdam (1981).

Rhoades, J. D.: Soluble salts. In: Page, A. L. (ed.): Methods of Soil Analysis, Part 2, 2nd ed. Agronomy 9, pp. 167–178. Amer. Soc. of Agronomy, Madison, Wisconsin (1982).

Richards, L. A. (ed.): Diagnosis and Improvement of Saline and Alkali Soils. U.S. Dept. of Agriculture Handbook No. 60. U.S. Government Printing Office, Washington, D.C. (1954).

Rinderle, U. and H. K. Lichtenthaler: The various chlorophyll fluorescence signatures as a basis for physiological ground truth control in remote sensing of forest decline. In: Intl. Geosc. and Remote Sens. Sympos., IGARSS '89, Vancouver, Vol. *2*, pp. 674–677 (1989).

Robinson, F. E. and G. F. Worker: Growth of sugarcane in areas irrigated with Colorado River water. Calif. Agric. *19*, 2–3 (1965).

Shaw, M. E. A.: Aspects of the management of salinity on swelling clay soils in Jamaica. Trop. Agric. (Trinidad) *59*, 167–172 (1982).

Szabolics, I.: Salt-affected soils. The Chemical Rubber Company Press, Inc. Boca Raton, Florida (1989).

Tanzi, K. K.: The nature and extent of agricultural salinity problems. In: Tanzi, K. K. (ed.): Agriculture Salinity Assessment and Management, ASCE Manual and Reports on Engineering Practice No. 71, pp. 1–17. Amer. Soc. Civil Engineers, New York (1990).

Thomas, J. R. and C. L. Wiegand: Osmotic and matric suction effects on relative turgidity, temperature, and the growth of cotton leaves. Soil Science *109,* 85–92 (1970).

Wiegand, C. L. and A. J. Richardson: Leaf area, light interception and yield estimates from spectral components analysis. Agron J. *76,* 543–548 (1984).

– – Spectral components analysis. Rationale and results for three crops. Intl. J. Remote Sens. *8,* 1011–1032 (1987).

Wiegand, C. L., A. J. Richardson, D. E. Escobar, and A. H. Gerbermann: Vegetation indices in crop assessments. Remote Sens. Environ. *35,* 105–119 (1991).

Wiegand, C. L., J. H. Everitt, and A. J. Richardson: Comparison of multispectral video and SPOT-1 HRV observations for cotton affected by soil salinity. Int. J. Remote Sens. *13,* 1511–1525 (1992).

Wiegand, C. L., J. D. Rhoades, D. E. Escobar, and J. H. Everitt: Photographic and videographic observations for determining and mapping the response of cotton to soil salinity. Remote Sens. Environ. *49,* 212–223 (1994 a).

Wiegand, C. L., D. E. Escobar, and S. E. Lingle: Detecting growth variation and salt stress in sugarcane using videography. Proc. 14th Biennial Workshop, Color Aerial Photography and Videography for Resource Monitoring, pp. 185–200. Amer. Soc. Photogramm. and Remote Sens. Bethesda, Maryland (1994 b).

J. Plant Physiol. Vol. 148. pp. 425–433 (1996)

Salt Stress Responses of Higher Plants: The Role of Proton Pumps and Na^+/H^+-Antiporters

THOMAS RAUSCH, MATTHIAS KIRSCH, RAINER LÖW, ANGELIKA LEHR, RUTH VIERECK, and AN ZHIGANG

Botanisches Institut, Im Neuenheimer Feld 360, D-69120-Heidelberg, Germany

Received September 5, 1995 · Accepted October 30, 1995

Summary

In salt-stressed higher plants NaCl may either be excluded from the cells or sequestered into the vacuole. Different pathways may dominate in different plants and different organs of the same plant. The proteins involved in salt transport across the plasma membrane and the tonoplast i.e. proton pumps and Na^+/ H^+-antiporters have been identified. Progress in cloning of the P-type H^+-ATPase, the V-type H^+-ATPase, and the vacuolar H^+-PP_iase has provided important tools for the study of the molecular mechanisms involved in ion sequestration. However, not a single plant has as yet been studied in sufficient detail to allow a comprehensive evaluation of the relative importance of individual transport processes for the salt tolerance of an intact plant. This review summarizes our present as yet limited knowledge and identifies promising areas for future research.

Key words: P-type H^+-ATPase, V-type H^+-ATPase, vacuolar H^+ PP_iase, Na^+/H^+ antiporter(s), salt stress, salt compartmentation, higher plants.

Introduction

The stress caused by high soil salinity is a major challenge to non-halophytic higher plants. It poses serious limitations to agriculture in many areas around the world, particularly on irrigated farmlands. However, little is yet known about the genetic basis for the observed diversity in salt-sensitivity between different crops (Epstein et al., 1980; Sumaryati et al., 1992). Plant breeding has confirmed that salt tolerance is not conferred by a single trait but is the consequence of complex gene interactions (Cheeseman, 1988; Bartels and Nelson, 1994). As a result progress in understanding the network of molecular mechanisms leading to salt tolerance has been slow (Flowers et al., 1977; Greenway and Munns, 1980; Cheeseman, 1988; Munns, 1993). Dogmatic views on some of the factors assumed to limit plant growth in saline soils i.e. NaCl effects on i) turgor, ii) photosynthesis, and iii) production of particular metabolites, are not substantiated by sufficient experimental proof and have delayed progress (Munns, 1993).

A biphasic model has been proposed for the growth responses of plants to salinity: During the initial phase, which may comprise days or weeks, water deficit dominates, resulting in reduced leaf growth. With the beginning of the secondary phase toxic ion effects (Na^+, Cl^-) initiate early leaf senescence (Munns, 1993). As salinity is first perceived in the root, it is likely that a root-derived signal, presumably abscisic acid, is formed, which directly or indirectly down-regulates leaf expansion rate (Passioura, 1988; Wolf et al., 1990; Zhao et al., 1991). Only the external cell layers of the root are exposed to the soil concentration of NaCl, whereas in the transpiration stream the salt level is considerably lower (Munns, 1993). Therefore, it is generally assumed that toxic concentrations of NaCl build up first in the fully expanded leaves. Here, NaCl is compartmentalized in the vacuoles. Only after their loading capacity is surpassed, the cytosolic and apoplasmic concentrations will reach toxic levels, resulting in enzyme inhibition and turgor loss, respectively. Thus salt exclusion in the root and/or salt sequestration in the leaf cell vacuoles are likely to be critical determinants for salt tolerance.

Recently, molecular tools became available to study some of the proteins involved in transport of Na^+ across the plasma membrane and the tonoplast. Thus cloning, expression analysis and studies on structure-function relationships of the primary proton pumps, including the plasma membrane P-type H^+-ATPase (Sussman, 1994; Michelet and Boutry, 1995), the vacuolar H^+-ATPase (Sze et al., 1992), and the vacuolar H^+-PP_iase (Sarafian et al., 1992; Tanaka et al., 1993; Leigh et al., 1994), have dramatically increased our knowledge about these proteins, which energize secondary ion transport. Conversely, our understanding of the not yet cloned Na^+/H^+-antiporters in both membranes is still limited (DuPont, 1992). It should be noted that a survey on different plant species indicated that Na^+/H^+ antiporters are not ubiquitously present, although developmental regulation of their expression could easily obscur their presence (Mennen et al., 1990).

A serious drawback of many investigations on the effect(s) of salt-stress on ion transport comes from the fact that often only a single aspect, like steady state transcript level for a particular proton pump, is addressed. However, it is obvious that a comparison of mRNA level, protein level, protein modification and protein activity in membrane vesicles is required to evaluate the physiological significance of any salt-induced changes (Lüttge et al., 1995). Also, at present direct proof for an essential role of the abovementioned membrane proteins in conferring salt tolerance is still lacking. Clearly experiments with transgenic plants over- or underexpressing the protein of interest are urgently needed (Munns, 1993; Bartels and Nelson, 1994). However, it should be kept in mind that the essential housekeeping functions of the primary proton pumps for cell metabolism will make it difficult to carve out their specific contribution to salt tolerance in transgenic plants.

This review will mainly focus on the involvement of proton pumps and Na^+/H^+-antiporters in the salt stress response (Fig. 1) and outline promising areas for future research. As Na^+-ions are assumed to be toxic to many cytosolic enzymes, emphasis will be on energized Na^+-transport, although salinity stress comprises exposure to Na^+ and Cl^-. It is important to note that enzymes are sensitive to the Na^+/K^+-ratio (Greenway and Munns, 1980). Therefore cytosolic toxicity of Na^+ may be countered by decreasing the Na^+ level, increasing the K^+ level, or both. Other important salt-induced metabolic changes, like the formation of compatible solutes, and the induction of specific stress proteins will not be included (Arakawa and Timasheff, 1985; Bray, 1993; Tarcynski et al., 1993; Bartels and Nelson, 1994; Bohnert et al., 1995).

The molecular biology of plant proton pumps, «work horses» for cellular transport processes

Cloning of higher plant P-type H^+-ATPases has revealed the existence of numerous isoforms most of which are expressed in all tissues, though to different degrees, whereas some show narrowly defined spatial expression patterns (Sussman, 1994; Michelet and Boutry, 1995). Their multiple housekeeping functions for cell metabolism involve i) energization of secondary ion and metabolite transport, ii) regulation of turgor and cell wall extension, and iii) intracellular pH regulation (Michelet and Boutry, 1995). The functional unit of the P-type H^+-ATPase is a homodimer, and the single polypeptide of about 100 kDa is predicted to be embedded in the plasma membrane *via* 10 membrane-spanning domains. The catalytic cycle involves a phosphorylated intermediate, but additional phosphorylation involved in pump regulation has been implicated. The enzyme has an autoinhibitory domain at its C-terminal end which might be involved in post-translational enzyme regulation (Palmgren et al., 1991; Michelet and Boutry, 1995). Despite the recent progress in the analysis of structure-function relationships of this important plant membrane protein, it has not yet been established by which mechanisms NaCl stress may affect the expression and regulation of this primary proton pump.

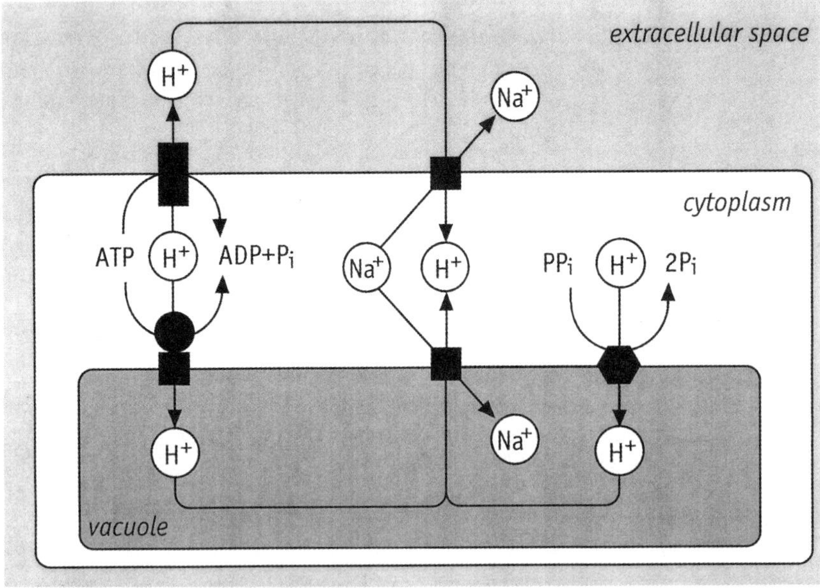

Fig. 1: Localization of proton pumps and Na^+/H^+ antiporters in plant cells. Note that antiporters may be absent, present at the tonoplast and/or plasma membrane, inducible or constitutively expressed, depending on the plant species and cell type. Conversely, the proton pumps, the P-type H^+-ATPase, the V-type H^+-ATPase, and the vacuolar H^+-PP_iase are ubiquitously present in all plants, although the latter may be absent in older tissues.

The V-type H^+-ATPase is localized in the tonoplast and Golgi membranes of higher plant cells (Sze et al., 1992). This proton pump is a multimeric protein with a complex structure, composed of a membrane-integral complex, V_0 and a peripheral complex, V_1. The V_0-complex comprises at least 4 different subunits, one of which, subunit c (the proteolipid), is present in six copies per holoenzyme. The six subunit c polypeptides provide the proton channels. The peripheral V_1-complex, which is exposed to the cytosolic side, includes a hexamer of three A-(70-kDa) and three B-(60 kDa) subunits, which is linked to V_0 by a stalk like structure composed of the subunits C, D and E (Sze et al., 1992; Lüttge et al., 1995). The V_1-complex may dissociate from the V_0-complex, and it has been proposed that this represents a mechanism for pump regulation (Sze et al., 1992; Ward and Sze, 1992).

In 1988 Zimniak et al. and Manolson et al. cloned the first higher plant cDNAs for subunits A and B, respectively, and Struve et al. (1990) provided the first promoter characterization of a V-type H^+-ATPase gene (subunit A). Right at the beginning the first cDNA sequences revealed that the V-type H^+-ATPase is an evolutionarily highly conserved protein complex (Nelson and Taiz, 1989). The extensive homology between V-type H^+-ATPase genes of different eukaryotic organisms has facilitated the rapid sequence acquisition for many of its subunits, particulary in yeast and mammels (Lüttge et al., 1995 and literature cited therein). However, till recently cloning of higher plant V-type H^+-ATPase cDNAs was limited to subunits A, B and c (Narasimhan et al., 1991; Löw et al., 1994; Hörtensteiner et al., 1994; Löw et al., 1995). Like for the P-type H^+-ATPase the existence of differentially expressed isoforms was soon established for several subunits of the V-type H^+-ATPase (Bernasconi et al., 1990; Lai et al., 1991; Berkelman et al., 1994; Kirsch, Viereck, and Rausch, unpublished results). Little is yet known about the factors regulating expression of the isoforms. Based on results obtained with plants transformed with a subunit A-antisense construct, Gogarten et al. (1992) proposed the existence of specific vacuolar and Golgi isoforms, an interpretation which has not found widespread acceptance. Whether specific salt-responsive isoforms exist is not yet known. Interestingly, the promoter of the carrot subunit A-gene contains a putative abscisic acid responsive cis-element (Guiltinan et al., 1990), but its function has not been tested (Struve et al., 1990). However, Binzel and Dunlap (1994) provided evidence against a function of abscisic acid in NaCl-induced up-regulation of subunit A expression.

The vacuolar H^+-PP_iase of higher plants has only recently been cloned (Sarafian et al., 1992; Tanaka et al., 1993; Leigh et al., 1994). The H^+-PP_iase consists of a single polypeptide (M_r ranging from 64 to 81 kDa), however, the size of the functional unit is not yet clear (monomer, dimer, or tetramer?). As this proton pump resides on the same membrane as the V-type H^+-ATPase its physiological function has been a matter of speculation. Recently, it could be shown that the H^+-PP_iase catalyzes K^+/H^+-cotransport into the vacuole (Davies et al., 1992; Leigh et al., 1994). As the expression of this pump is high in growing tissues but considerably reduced in older tissues a participation in vacuolar sequestration of Na^+ seems unlikely (Lüttge et al., 1995), except in cell cultures (Colombo and Cerana, 1993).

Salt exclusion at the plasma membrane via a Na^+/H^+-antiporter driven by the P-type H^+-ATPase

The entry of NaCl into the root cells, its symplasmic transport past the Casparian band (Peterson, 1988), and its transfer into the transpiration stream, are the primary steps of salt accumulation in the plant. The pathways for uptake of Na^+ into the root symplasmic space have not yet been established in much detail. It has been generally assumed that non-selective (K^+-)ion channels allow entry into the cells (Schachtman et al., 1991), however, more recent results suggest that Na^+ is taken up with K^+ by a high affinity K^+-uptake carrier (Schachtman and Schoeder, 1994; Schachtman et al., 1995). Exclusion of Na^+ from the root cells would therefore either be dependent on changed characteristics of these non-specific and/or specific uptake routes, or, alternatively, on an active extrusion mechanism.

The operation of an Na^+/H^+ antiporter in the plasma membrane of root cells, powered by the P-type H^+-ATPase, would provide an active mechanism for Na^+-extrusion. Root plasma membrane vesicles from the halophyte *Atriplex nummularia* Lindl. (Braun et al., 1986) grown on 400 mM NaCl showed a significantly increased vanadate-sensitive ATP-driven SCN^--accumulation, as well as an increased H^+-pumping rate, when compared to control plants. For comparison, plasma membrane vesicles of control and salt-grown *Gossypium hirsutum* L., a glycophytic species, were analyzed, but here no differences were found (Hassidim et al., 1986). However, as the authors point out correctly the halophyte *A. nummularia* grows more vigorously under saline condition. Therefore, it could be argued that the salt-free control-treatment is in fact a stress-treatment. The same authors presented evidence for the presence of a Na^+/H^+-antiporter in plasma membrane vesicles of *A. nummularia* (Braun et al., 1988; Hassidim et al., 1990), which was not saturable between 20 and 180 mM Na^+. Despite the obvious presence of the basic machinery for active Na^+-extrusion in root cells, *Atriplex* species accumulate considerable amounts of NaCl in the leaves (Glenn et al., 1994). This may indicate that the extrusion of Na^+ across the plasma membrane *via* the Na^+/H^+-antiporter helps to keep the Na^+-concentration in the root cells below the threshold level of toxicity, while allowing sufficient NaCl to pass to the transpiration stream.

Recently, Niu et al. (1993 a, b) have cloned a cDNA encoding the P-type H^+-ATPase of *A. nummularia*. In suspension-cultured cells the steady state transcript levels increased only when deadapted cells (previously exposed to NaCl) were re-exposed to NaCl, and this induction was dependent on the developmental stage of the cell culture (Niu et al., 1993 a). Unadapted cells were not responsive to salt. In whole plants treatment with 400 mM NaCl induced an up-regulation of P-type H^+-ATPase mRNA in roots and leaves, but not in stem (Niu et al., 1993 b). In roots the effect was most pronounced in the elongation zone as shown by *in situ*-hybridization. It is noteworthy that the up-regulation of mRNA levels in the root over the first 24 hours in the salt-treated plants was much less pronounced than the down-regulation in the salt-free control plants during the same period. Although the authors do not comment on this important observation, it may indicate that *A. nummularia* ‹needs› salt for vigorous growth (see above). It again confirms that for a ‹real› halophyte the salt-free condition is no control-treatment, and low mRNA levels for the P-type H^+-ATPase may merely reflect non-optimal growth conditions. Interestingly, in old leaves, where the basic transcript level is low, salt-treatment led to a massive induction after 24 hours, whereas in young leaves with its high mRNA level NaCl did not lead to a further increase. This may indicate that in non-growing leaves the requirement for active ion transport at the plasma membrane is low (Niu et al., 1993 b). The results also show that expression of the P-type H^+-ATPase is developmentally regulated. Therefore, care has to be taken to separate *bona fide* salt-effects on gene expression from indirect effects caused by salt-induced changes of plant growth and development (see below). Molecular studies on gene expression tend to neglect this important issue.

In a comprehensive study Brüggemann and Jenisch (1987; 1988; 1989) have compared the activities of the P-type H^+-ATPase in salt-tolerant and salt-sensitive *Plantago* species. Highly purified outside-out plasma membrane vesicles were prepared and their vanadate-sensitive H^+-ATPases characterized. No differences were found between the different species, and in *P. maritima* L. and *P. crassifolia* L. the enzyme showed identical properties when isolated from control or salt-treated plants. The authors conclude that in the halophytic *Plantago* species the P-type H^+-ATPase, despite its basic role in energization of secondary ion transport, is not involved in the *regulation* of ion transport. It is noteworthy, that in roots of *P. maritima* a Na^+/H^+ antiporter was induced after salt-treatment, however this activity was confined to the tonoplast and absent in the plasma membrane (Staal et al., 1991). Conversely, no antiporter could be detected in the roots of control and salt-grown *P. media* L.

Summarizing the above and other more recent results (Wilson and Channon, 1995) on the P-type H^+-ATPase and the Na^+/H^+-antiporter in the plasma membrane in relation to salt tolerance of higher plants two major conclusions may be drawn: 1) comparisons between unrelated species, even within the functional group of halophytes, are of limited value, and 2) for all systems studied so far the data are incomplete and do not allow a proper assessment of the roles of both membrane enzymes during plant development. Detailed studies on the time course of changes in P-type H^+-ATPase mRNA, protein and enzyme activity in relation to Na^+/H^+-antiporter activity during plant development and in response to salt are not available for a single species. It appears that among halophytic species studies are most advanced for *Atriplex* and *Plantago.* Thus it would be advantageous if future work would focus on species of these genera, rather than adding incomplete data sets on other species to our fragmentary knowledge. From the experimental point of view the combination of detailed molecular analysis with non-invasive techniques like nuclear magnetic resonance (Spickett et al., 1993) holds particular promise.

Salt accumulation into the vacuole via a Na^+/H^+-antiporter driven by the V-type H^+-ATPase and/or the vacuolar H^+-PP_iase

Once Na^+-ions have entered the transpiration stream they are transported into the leaves where the only possibility to counter their potential toxicity for cytosolic enzymes is the vacuolar sequestration. In fact, the concentration of Na^+ in the vacuole may exceed 2- to 5-fold the cytoplasmic concentration (Yeo, 1981; Matoh et al., 1987; Binzel et al., 1988). However, already in the root cells Na^+ may be sequestered into the vacuole. Therefore, the tonoplast and its proton pumps as well as tonoplast Na^+/H^+ antiporters have been studied in great detail in relation to their putative role(s) for salt tolerance. Salt-induced changes in tonoplast proteins have been most thoroughly analyzed in *Beta vulgaris* L. (red beet and sugar beet), *Mesembryanthemum crystallinum* L., a facultative halophyte with inducible Crassulacean acid metabolism (CAM), and barley (*Hordeum vulgare* L.). The evolutionary conservation of V-type H^+-ATPase structure (Nelson and Taiz, 1989; Sze et al., 1992) has considerably facilitated the molecular analysis of its gene expression in different plant species after salt exposure (Narasimhan et al., 1991; Löw et al., 1992; Löw et al., 1994; Binzel and Dunlap, 1994; Löw et al., 1996; Lüttge et al., 1995; Kirsch and Rausch, unpublished results), as homologous gene probes became readily available using the RT-PCR technique (Löw et al., 1992, 1994; Löw and Rausch, 1994). Conversely, the molecular analysis of the vacuolar Na^+/H^+ antiporter has proven to be more difficult. All attempts to clone a plant antiporter have as yet not been successful. Recently, bacterial and mammalian antiporters have been cloned, revealing only moderate sequence conservation, even among isoforms of a single species (Orlowski et al., 1992; Pinner et al., 1992; Padan et al, 1994). In a study on ion channels of the leaf cell tonoplast of *Suaeda maritima* L. Maathuis et al. (1992) did not detect any special adaptation in this halophyte, as compared to glycophytic species (Leach et al., 1990). Thus it is unlikely, that ion channels contribute significantly to salt tolerance.

The moderately salt-tolerant *Beta vulgaris* exhibits an inducible Na^+/H^+ antiporter with an apparent $K_{m(Na^+)}$ of 26 mM (Blumwald and Poole, 1985; 1987). After photolabeling tonoplast proteins with an amiloride analogue ([$_3$H]5-(N-methyl-N-isobutyl)-amiloride; Blumwald et al., 1987; Barkla et al., 1990), several polypeptides showed specific binding. Interestingly, one of the labelled proteins with an apparent size of 170 kDa corresponded to a protein which could be induced several-fold by amiloride treatment in suspension-cultured sugar beet cells (Barkla and Blumwald, 1991). A polyclonal antiserum raised against the denatured 170 kDa-protein inhibited antiport activity *in vitro,* indicating that the 170 kDa protein is identical with or part of the Na^+/H^+-antiporter.

Recently, we could show that in leaves of salt-treated sugar beet plants the expression of the V-type H^+-ATPase is increased as judged from Western blot analysis with an antiserum against the holoenzyme. Furthermore, mRNA levels for two V-type H^+-ATPase subunits, i.e. the catalytic subunit A and the proton channel-forming subunit c, were increased about 5-fold in leaves of salt-treated plants (Table 1; Kirsch, An Zhigang and Rausch, unpublished results). The effect became apparent after 48 hours and was found in plants of different age. It is not yet known whether the delayed response reflects the slow build-up of leaf Na^+-content. It is noteworthy that in young leaves and roots the transcript levels for both subunits were higher than in expanded leaves, suggest-

Table 1: Salt-induced changes in leaf mRNA levels for the V-type H^+-ATPase subunits A and c in adult plants of different species. Transcript levels were assessed by densitometric quantitation of Northern blots (Löw and Rausch, 1994; Löw et al., 1996; Kirsch and Rausch, unpublished results). Note that plants were stressed with different NaCl concentrations. A-mRNA and c-mRNA indicate transcript amounts for subunits A and c, respectively.

Plant species	mM NaCl	c-mRNA: NaCl/control	A-mRNA: NaCl/control	Hours of treatment
Beta vulgaris	400	5–10	>5	48
Daucus carota	100	2	≥2	24/48
Zea mays	50/200	ca. 1	ca. 1	48
Mesembryanthemum crystallinum	400	2	2	8/24

ing that expression of V-type H^+-ATPase is also dependent on plant development as earlier mentioned for the P-type H^+-ATPase (see above). Again, careful studies are needed to differentiate between direct salt effects on pump expression as compared to indirect effects *via* growth retardation.

DuPont and co-workers (Garbarino and DuPont, 1988, 1989; DuPont, 1992) have demonstrated the presence of a tonoplast Na^+/H^+-antiporter in barley roots with an apparent $K_{m(Na^+)}$ of 9 mM, which was induced with a half-time of only 15 min (Garbarino and DuPont, 1989). The induction was not dependent on protein synthesis and activation by post-translational modification has been suggested. It has been previously shown, that mammalian Na^+/H^+-antiporters may be regulated by phosphorylation (Sardet et al., 1990; Bianchini et al., 1991), but as yet no evidence of such regulatory mechanism has been provided for plants.

In addition to the rapid salt-induced up-regulation of antiporter activity the apparent H^+/ATP-stoichiometry of the V-type H^+-ATPase approximately doubled, although this change became apparent only after 2 days of growth in the presence of 100 mM NaCl (DuPont, 1992). The interpretation of the apparent change in pump stoichiometry is certainly not straightforward, as at present all results have been obtained only with sucrose gradient-purified membrane vesicle preparations (DuPont, 1992). Therefore, further work is needed to show that an intrinsic change of the pump characteristics does indeed take place. Interestingly, in salt-treated suspension-cultured cells of *Nicotiana tabacum* L. (Reuveni et al., 1990) and *Daucus carota* L. (Löw and Rausch, unpublished results) similar increases of the H^+/ATP ratio have been observed. A possible explanation could be the reversible inactivation of the pump by oxidation as has been recently shown for the V-type H^+-ATPase of coated vesicles from bovine brain (Feng and Forgac, 1992; 1994). Assuming that H^+-pumping is more sensitive to oxidation of cysteines 254 and 532 of the catalytic subunit A than is ATP hydrolysis, the inactivation of a significant portion of the pumps in control cells by reversible oxidation would lead to a lower apparent H^+/ATP ratio. The apparent stoichiometry could be increased after salt exposure by fully reducing all available pumps. It is noteworthy that in coated vesicles about 50% of the pumps are in the oxidized, inactive state (Feng and Forgac, 1992; 1994). Recently, we could show for carrot suspension-cultured cells that only in membrane vesicles of control cells the addition of glutathione (GSH) activated H^+-pumping whereas in membrane vesicles of NaCl-treated cells (100 mM NaCl, 3 hours) H^+-pumping rates were higher as compared to control cells but could not be further stimulated by the addition of GSH (Table 2; Verstappen and Rausch, unpublished results). The results indicate that in barley roots the post-translational regulation of pump and antiporter characteristics may be more important than changes in gene expression, emphasizing the urgent need for more detailed studies at the protein level.

The multiple effects of salt-treatment on the structure and function of the vacuolar H^+-ATPase of the facultative halophyte *Mesembryanthemum crystallinum,* a species with inducible CAM (Lüttge, 1993; Ratajczak et al., 1994; Lüttge et al., 1995), have been studied in much detail. Conversely, the Na^+/H^+-antiporter of *M. crystallinum* has as yet been neglected although in salt-grown plants the leaves accumulate considerable amounts of Na^+. The interpretation of salt effects on tonoplast transport functions in a plant with salt-inducible CAM poses specific problems as a clear differentiation between salt responses *per se* and changes due to the transition from C_3-photosynthesis to CAM is sometimes difficult. Again the crucial question arises: what is the proper control treatment?

Recently, we observed a doubling of transcript levels for three V-type H^+-ATPase subunits (A, B and c; Löw et al., 1994; 1996) within 8 hours after treating 4 week-old plants in the C_3-state with 400 mM NaCl. This rapid effect was confined to the roots and young leaves, whereas in fully expanded leaves only subunit c-mRNA increased, and this increase was transient. As full induction of the CAM-state takes several days we interprete this up-regulation as a rapid and direct effect of salt. In addition to the observed salt-effects expression of the V-type H^+-ATPase genes also showed organ-specific differences, again confirming the dependence of the stress response on plant development (Löw et al., 1994; 1996). After prolonged salt exposure (1 week) Ratajczak et al. (1994) found a 2.5-fold increase of V-type H^+-ATPase protein in the leaves. After withdrawl of NaCl from the growth medium the salt-induced up-regulation proved to be completely reversible, reaching control values within a period of 2 days. This indicates a considerable turnover of the V-type H^+-ATPase in *M. crystallinum.*

Table 2: Effect of 10 mM glutathione (GSH) on *in vitro* H^+-pumping activity ($\Delta F/min/mg_{protein}$) of vacuolar H^+-ATPase in tonoplast-enriched membrane vesicles isolated from suspension-cultured *Daucus carota* cells. Six days after transfer to fresh medium cells were exposed to 100 mM NaCl for 1 or 3 hours. Membranes were isolated in the presence of 20 mM ascorbic acid and 1.5 mM dithiothreitol (Verstappen and Rausch, unpublished results). Values in brackets represent percent of control without GSH. The standard error of the mean (3 independent membrane isolations) was ≤25%.

cell treatment	no GSH $\Delta F/min/mg_{protein}$	10 mM GSH $\Delta F/min/mg_{protein}$
control	0.031±0.006 (100)	0.049±0.009 (158)
100 mM NaCl, 1 h	0.057±0.013 (183)	0.051±0.011 (165)
100 mM NaCl, 3 h	0.047±0.010 (152)	0.050±0.012 (161)

Image analysis of the V_1-complex has recently revealed pentameric structures, in addition to the expected hexameric structures thought to represent three copies each of subunits A and B (Kramer et al., 1995). The former were interpreted as transition states during assembly and/or disassembly of the V_1-complex. Further evidence for structural changes of the V-type H^+-ATPase during salt-induced C_3-CAM transition comes from the appearance of additional polypeptides of 32 and 28 kDa, which may be co-precipitated with the holoenzyme by an antiserum directed against subunit A (Ratajczak et al., 1994). As an antiserum directed against the 32 kDa polypeptide excised from the gel cross-reacts with subunits A and B of the tobacco holoenzyme, it is likely that polypeptide(s) of this size are degradation products of the larger subunits (An Zhigang, Ratajczak and Rausch, unpublished results), albeit tightly bound to the holoenzyme. Thus, further studies are needed to positively confirm the existence of stress-induced *bona fide* additional subunits. Whether the observed partial proteolysis has a regulatory function or merely reflects the presence of increased protease activity in the CAM-state is not yet known, but a protease activity was associated with tonoplast membranes from CAM plants which was absent in membranes of plants kept in the C_3-state. It is noteworthy, that in *Citrus sinensis* L. Osbeck plants salt-stress induces the formation of a 35 kDa splitting product of subunit A, which, however, dissociates from the holoenzyme (Banuls et al., 1995). The splitting product retains ATP hydrolytic activity. In summary the observations suggests that salt-induced changes in V-type H^+-ATPase stability and turnover may be a more general phenomenon.

Additional salt-induced changes have been found for the membrane integral V_0-complex (Rockel et al., 1994). First, salt-treatment leads to an apparent relative increase of subunit c protein as compared to other V-type H^+-ATPase subunits. Furthermore, the abundance and apparent diameter of intramembrane particles exposed by freeze fracture analysis increase in salt-stressed plants. These particles have been shown to represent V_0-complexes (Rockel et al., 1994). Although the implications of the size-increase of V_0-complexes for pump function are not yet clear, this observation certainly merits further study.

A mechanistic link between Na^+-transport and Ca^{2+}-homeostasis

Many studies have confirmed that NaCl stress may be partially alleviated by increased Ca^{2+}-supply to the growth medium (Läuchli, 1990 and literature cited therein). As, depending on the concentration ratio, Na^+ and Ca^{2+} may displace each other from the plasma membrane, it is obvious that Na^+ may affect cellular Ca^{2+}-homeostasis, whereas Ca^{2+} may reduce Na^+-toxicity. Also, Ca^{2+} seems to affect K^+/Na^+ selectivity at the plasma membrane (Cramer et al., 1987). Furthermore, vacuolar sequestration of Na^+ may affect Ca^{2+}-sequestration in the vacuole. The latter is catalyzed by a H^+/Ca^{2+} exchanger (Schumaker and Sze, 1990). A salt-induced disturbance of cytosolic Ca^{2+}-homeostasis may be the reason for the observed up-regulation of a Ca^{2+}-ATPase by NaCl (Perez-Prat et al., 1992; Wimmers et al., 1992). Thus, as the transport of H^+ and Na^+ across the tonoplast and the plasma membrane may affect the cellular distribution of Ca^{2+}, an important link exists between Na^+-transport and the distribution of an important second messenger.

Relevance of Na^+ membrane transport processes for salt tolerance

Recently, molecular biologists have developed new approaches to study proteins involved in salt tolerance of plants. *Saccharomyces cerevisiae* (Gaxiola et al., 1992; Mendoza et al., 1994), *Schizosaccharomyces pombe* (Jia et al., 1992) and *Chlamydomonas reinhardii* (Prieto et al., 1995) have been used as eukaryotic model organisms to identify proteins involved in salt tolerance. Mutants with reduced salt tolerance were used to isolate the corresponding genes by functional complementation. With this approach a soluble protein could be identified which by a yet unknown mechanism affects the cytosolic Na^+/K^+ ratio in yeast (Gaxiola et al., 1992). The protein phosphatase calcineurin was shown to be essential for salt tolerance, indicating that NaCl adaptation in yeast is dependent on signal transduction involving Ca^{2+} and protein phosphorylation/dephosphorylation (Mendoza et al., 1994). In fission yeast gene amplification of a locus encoding a putative Na^+/H^+ antiporter resulted in sodium tolerance (Jia et al., 1992). Certainly, these and similar studies will increase our understanding of the basic cellular mechanisms contributing to salt tolerance.

From a practical point of view one of the major goals of research on salt tolerance should be to provide guidelines for breeding salt tolerant crops and/or identify specific genes to be used for plant transformation (Bartels and Nelson, 1994). Therefore, in our opinion the following major points need emphasizing: 1) a higher plant is more complex than a unicellular eukaryote; 2) salt tolerance may depend on a very subtle regulatory network of organ specific transport process, making the search for specific salt tolerance genes a risky approach; 3) the primary proton pumps at the tonoplast and the plasma membrane, which energize salt sequestration and/or exclusion, have important housekeeping functions which cannot be easily distinguished from their putative role for salt tolerance.

As demonstrated for the P-type H^+-ATPase of the halophyte *A. nummularia* and the V-type H^+-ATPase of *M. crystallinum,* a comprehensive analysis of salt-induced changes requires a combination of different experimental approaches, extending from membrane physiology to gene expression studies and even ultrastructural analysis (Lüttge et al., 1995). Furthermore, as the stress response is superimposed on organ-specific developmental changes, extreme caution is required when experimental data from one plant species albeit grown under different conditions are compared. It is absolutely essential, that molecular biologists are aware of the ecophysiology of salt-stressed plants. Otherwise we will witness the accumulation of gene expression data with little or no relevance to the salt-stress response (Radin, 1993). Although specific stress proteins may be formed upon acute salt-induced damage to cytosolic proteins (Bray, 1993 and literature cited therein), it is likely that the transport proteins excluding salt from the symplasmic space or guiding salt to the sites of intracellular disposal, i.e. the vacuoles, will be more important for salt tolerance of crop plants under field conditions (Radin, 1993). Elucidation of the functional integration of these proteins within cellular metabolism, including regulation at different levels, is one of the major challenges for plant molecular ecophysiology in the coming years.

Acknowledgements

Our research is supported by the Deutsche Forschungsgemeinschaft (SFB 199). We gratefully acknowledge stimulating discussions with P. M. Hasegawa (West Lafayette, USA), M. Binzel (El Paso, USA), M. Tsantis (Cambridge, U.K.), U. Lüttge and R. Ratajczak (Darmstadt, F.R.G.).

References

ARAKAWA, T. and S. N. TIMASHEFF: The stabilization of proteins by osmolytes. Biophys. J. *47,* 411–416 (1985).

BANULS, J., R. RATAJCZAK, and U. LÜTTGE: NaCl-stress enhances proteolytic turnover of tonoplast H^+-ATPase of *Citrus sinensis:* Appearance of a 35 kDa polypeptide still exhibiting ATP hydrolysis activity. Plant, Cell Environm., in press (1995).

BARKLA, B. J., H. M. CHARUK, E. J. CRAGOE, and E. BLUMWALD: Photolabeling of tonoplast from sugar beet cell suspensions by $[^3H]$5-(N-Methyl-N-Isobotyl)-Amiloride, an inhibitor of the vacuolar Na^+/H^+ antiport. Plant Physiol. *93,* 924–930 (1990).

BARKLA, B. J. and E. BLUMWALD: Identification of a 170-kDa protein associated with the vacuolar Na^+/H^+ antiport of *Beta vulgaris.* Proc. Natl. Acad. Sci. USA, *88,* 11177–11181 (1991).

BARTELS, D. and D. NELSON: Approaches to improve stress tolerance using molecular genetics. Plant Cell Environm. *17,* 659–667 (1994).

BERKELMAN, T., K. A. HOUTCHENS, and F. M. DUPONT: Two cDNA clones encoding isoforms of the B subunit of the vacuolar ATPase from barley roots. Plant Physiol. *104,* 287–288 (1994).

BERNASCONI, P., T. RAUSCH, I. STRUVE, L. MORGAN, and L. TAIZ: An mRNA from human brain encodes an isoform of the B subunit of the vacuolar H^+-ATPase. J. Biol. Chem. *265,* 17428–17431 (1990).

BIANCHINI, L., M. WOODSIDE, C. SARDET, J. POUYSSÉGUR, A. TAKAI, and S. GRINSTEIN: Ocadaic acid, a phosphatase inhibitor, induces activation and phosphorylation of the Na^+/H^+ antiport. J. Biol. Chem. *266,* 15406–15413 (1991).

BINZEL, M. L., F. D. HESS, R. A. BRESSAN, and P. M. HASEGAWA: Intracellular compartmentation of ions in salt adapted tobacco cells. Plant Physiol. *86,* 607–614 (1988).

BINZEL, M. and J. R. DUNLAP: Does abscisic acid function in sodium chloride-induced expression of the tonoplast H^+-ATPase. Plant Physiol. *105* (suppl.), 68 (1994).

BLUMWALD, E. and R. J. POOLE: Na^+/H^+ antiport in isolated tonoplast vesicles from storage tissue of *Beta vulgaris.* Plant Physiol. *78,* 163–167 (1985).

– – Salt tolerance in suspension cultured cells of sugar beet. Induction of Na^+/H^+ antiport activity at the tonoplast by growth in salt. Plant Physiol. *83,* 884–887 (1987).

BLUMWALD, E., E. J. CRAGOE, and R. J. POOLE: Inhibition of Na^+/H^+ antiport acitivity by analogs of amiloride. Plant Physiol. *85,* 30–33 (1987).

BOHNERT, H. J., D. E. NELSON, and R. G. JENSEN: Adaptations to environmental stresses. Plant Cell *7,* 1099–1111 (1995).

BRAUN, Y., M. HASSIDIM, H. R. LERNER, and E. REINHOLD: Evidence for a Na^+/H^+ antiporter in membrane vesicles isolated from roots of the halophyte *Atriplex nummularia.* Plant Physiol. *87,* 104–108 (1988).

BRAUN, Y., M. HASSIDIM, H. R. LERNER, and L. REINHOLD: Studies on H^+-translocating ATPases in plants of varying resistance to salinity. Plant Physiol. *81,* 1050–1056 (1986).

BRAY, E. A.: Molecular responses to water deficit. Plant Physiol. *103,* 1035–1040 (1993).

BRÜGGEMANN, W. and P. JANIESCH: Characterization of plasma membrane H^+-ATPase from salt-tolerant and salt-sensitive *Plantago* species. J. Plant Physiol. *130,* 395–411 (1987).

– – Comparison of plasma membrane ATPase from salt-treated and SALT-FREE GROWN *Plantago maritima* L. J. Plant Physiol. *134,* 20–25 (1989).

– – Properties of native and solubilized plasma membrane ATPase from the halophyte *Plantago crassifolia* grown under saline and non-saline conditions. Physiol. Plant. *74,* 615–622 (1988).

CHEESEMAN, J. M.: Mechanisms of salinity tolerance in plants. Plant Physiol. *87,* 547–550 (1988).

COLOMBO, R. and R. CERANA: Enhanced activity of tonoplast pyrophosphatase in NaCl-grown cells of *Daucus carota.* J. Plant Physiol. *142,* 226–229 (1993).

CRAMER, G. R., J. LYNCH, A. LÄUCHLI, and E. EPSTEIN: Influx of Na^+, K^+, and Ca^{2+} into roots of salt-stressed cotton seedlings. Plant Physiol. *83,* 510–516 (1987).

DAVIES, J., R. J. POOLE, P. A. REA, and D. SANDERS: Potassium transport into plant vacuoles energized directly by a proton-pumping inorganic pyrophosphatase. Proc. Natl. Acad. Sci. USA *89,* 11701–11705 (1992).

DUPONT, F. M.: Salt-induced changes in ion transport: Regulation of primary pumps and secondary transporters. In: COOKE, D. T. and D. T. CLARKSON (eds.): Transport and receptor proteins of plant membranes, pp. 91–100. Plenum Press, New York (1992).

EPSTEIN, E., J. D. NORLYN, D. W. RUSH, R. W. KINGSBURY, D. B. KELLEY, G. A. CUNINGHAM, and A. S. WRONA: Saline culture of crops, a genetic approach. Science *210,* 399–403 (1980).

FENG, Y. and M. FORGAC: A novel mechanism for regulation for regulation of vacuolar acidification. J. Biol. Chem. *267,* 19769–19772 (1992).

– – Inhibition of vacuolar H^+-ATPase by disulfide bond formation between cysteine 254 and cysteine 532 in subunit A. J. Biol. Chem. *269,* 13224–13230 (1994).

FLOWERS, T. J., P. F. TROKE, and A. R. YEO: The mechanism of salt tolerance in halophytes. Annu. Rev. Plant Physiol. *28,* 89–121 (1977).

GARBARINO, J. and F. M. DUPONT: NaCl induces a Na^+/H^+ antiport in tonoplast vesicles from barley roots. Plant Physiol. *86,* 231–236 (1988).

– – Rapid induction of Na^+/H^+ exchange activity in barley root tonoplast. Plant Physiol. *89,* 1–4 (1989).

GAXIOLA, R., I. F. DE LARRINOA, J. M. VILLALBA, and R. SERRANO: A novel and conserved salt-induced protein is an important determinant of salt tolerance in yeast. EMBO J. *11,* 3157–3164 (1992).

GLENN, E. P., M. OLSEN, R. FRYE, D. MOORE, and S. MIYAMOTO: How much sodium accumulation is necessary for salt tolerance in subspecies of the halophyte *Atriplex canescens* ? Plant Cell Environm. *17,* 711–719 (1994).

GOGARTEN, J. P., J. FICHMANN, Y. BRAUN, L. MORGAN, P. STYLES, L. S. TAIZ, K. DELAPP, and L. TAIZ: The use of antisense mRNA to inhibit the tonoplast H^+-ATPase in carrot. Plant Cell *4,* 851–864 (1992).

GREENWAY, H. and R. MUNNS: Mechanisms of salt tolerance in nonhalophytes. Annu. Rev. Plant Physiol. *31,* 149–190 (1980).

GUILTINAN, M. J., W. R. MARCOTTE, and R. S. QUATRANO: A plant leucine zipper protein that recognizes an abscisic acid response element. Science *250,* 267–271 (1990).

Hassidim, M. J., Y. Braun, H. R. Lerner, and L. Reinhold: Na^+/$H^+$and K^+/H^+ antiport in root membrane vesicles isolated from the halophyte *Atriplex* and the glycophyte cotton. Plant Physiol. *94,* 1795–1801 (1990).

– – – – Studies on H^+-translocating ATPases in plants of varying resistance to salinity. Plant Physiol. *81,* 1057–1061 (1986).

Hörtensteiner, S., E. Martinoia, and N. Amrhein: Factors affecting the reformation of vacuoles in evacuolated protoplasts and the expression of two vacuolar pumps. Planta *192,* 395–403 (1994).

Jia, Z.-P., N. McCullough, R. Martel, S. Hemmingsen, and P. G. Young: Gene amplification at a locus encoding a putative Na^+/H^+ antiporter confers sodium and lithium tolerance to fission yeast. EMBO J. *11,* 1631–1640 (1992).

Kramer, D., B. Mangold, A. Hille, I. Emig, A. Hess, R. Ratajczak, and U. Lüttge: The head structure of a higher plant V-type H^+-ATPase is not always a hexamer but also a pentamer. J. Exp. Bot., in press (1995).

Lai, S., J. C. Watson, J. N. Hansen, and H. Sze: Molecular cloning and sequencing of cDNAs encoding the proteolipid subunit of the vacuolar H^+-ATPase from a higher plant. J. Biol. Chem. *266,* 16078–16084 (1991).

Läuchli, A.: Calcium, salinity and the plasma membrane. In: Leonard, R. T. and P. K. Hepler (eds.): Calcium in Plant Growth and Development. The American Society of Plant Physiologists Symposium Series, pp. 26–35, Vol. *4* (1990).

Leach, R. P., K. P. Wheeler, T. J. Flowers, and A. R. Yeo: Molecular markers for ion compartmentation in cells of higher plants. II. Lipid composition of the tonoplast of the halophyte *Suaeda maritima* (L.) Dum. J. Exp. Bot. *41,* 1089–1094 (1990).

Leigh, R. A., R. Gordon-Weeks, S. H. Steele, and V. D. Korenkov: The H^+-pumping inorganic pyrophosphatase of the vacuolar membrane of higher plants. In: Blatt, M. R., R. A. Leigh, and D. Sanders (eds.): Membrane transport in plants and fungi: molecular mechanisms and control. Soc. Exp. Bot., The Company of Biologists Ltd., Cambridge, pp. 61–75 (1994).

Löw, R. and T. Rausch: Sensitive non-radioactive Northern blots using alkaline transfer of total RNA and PCR-amplified biotinylated probes. BioTechniques *17,* 1026–1030 (1994).

Löw, R., A. Lehr, M. Kirsch, L. Taiz, and T. Rausch: The carrot V-type H^+-ATPase: Towards an understanding of the coordinate expression of its genes. Plant Physiol. *99* (suppl.), 71 (1992).

Löw, R., B. Rockel, M. Kirsch, An Zhigang, R. Ratajczak, U. Lüttge, and T. Rausch: Expression of V-Type H^+-ATPase genes in response to salt stress: Comparison of transcript levels for the subunits A, B, and c in a glycophyte and a halophyte. Plant Physiol. *105* (suppl.), 39 (1994).

Löw, R., B. Rockel, M. Kirsch, R. Ratajczak, S. Hörtensteiner, E. Martinoia, U. Lüttge, and T. Rausch: Early salt stress effects on the differential expression of vacuolar H^+-ATPase genes in roots and leaves of *Mesembryanthemum crystallinum.* Plant Physiol. *110,* 259–265 (1996).

Lüttge, U., R. Ratajczak, T. Rausch, and B. Rockel: Stress responses of tonoplast proteins: an example for molecular ecophysiology and the search for eco-enzymes. Acta Bot. Neerl. *44,* 343–362 (1995).

Lüttge, U.: The role of crassulacean acid metabolism (CAM) in the adaptation of plants to salinity. New Phytol. *125,* 59–71 (1993).

Maathuis, F. J. M., T. J. Flowers, and A. R. Yeo: Sodium chloride compartmentation in leaf vacuoles of the halophyte *Suaeda maritima* (L.) Dum. and its relation to tonoplast permeability. J. Exp. Bot. *43,* 1219–1223 (1992).

Manolson, M. F., B. F. F. Quellette, M. Filion, and R. J. Poole: cDNA sequence and homologies of the 57-kDa nucleotide-binding subunit of the vacuolar ATPase from *Arabidopsis.* J. Biol. Chem. *263,* 17987–17994 (1988).

Matoh, T., J. Watanabe, and E. Takahashi: Sodium, potassium, chloride, and betaine concentrations in isolated vacuoles from salt-grown *Atriplex gmelini* leaves. Plant Physiol. *84,* 173–177 (1987).

Mendoza, I., F. Rubio, A. Rodriguez-Navarro, and J. M. Pardo: The protein phosphatase calcineurin is essential for NaCl tolerance of *Saccharomyces cerevisiae.* J. Biol. Chem. *269,* 8792–8796 (1994).

Mennen, H., B. Jacoby, and H. Maschner: Is sodium proton antiport ubiquitous in plant cells? J. Plant Physiol. *137,* 180–183 (1990).

Michelet, B. and M. Boutry: The plasma membrane H^+-ATPase. Plant Physiol. *108,* 1–6 (1995).

Munns, R.: Physiological processes limiting plant growth in saline soils: some dogmas and hypotheses. Plant Cell Environm. *16,* 15–24 (1993).

Narasimhan, M. L., M. L. Binzel, E. Perez-Prat, Z. Chen, D. E. Nelson, N. K. Singh, R. A. Bressan, and P. M. Hasegawa: NaCl regulation of tonoplast ATPase 70-kD subunit mRNA in tobacco cells. Plant Physiol. *97,* 562–568 (1991).

Nelson, N. and L. Taiz: The evolution of H^+-ATPases. TIBS *14,* 113–116 (1989).

Niu, X., J.-K. Zhu, M. L. Narasimhan, R. A. Bressan, and P. M. Hasegawa: Plasma membrane H^+-ATPase gene expression is regulated by NaCl in the cells of the halophyte *Atriplex nummularia* L. Planta *190,* 433–438 (1993 a).

Niu, X., M. L. Narasimhan, R. A. Salzman, R. A. Bressan, and P. M. Hasegawa: NaCl regulation of plasma membrane H^+-ATPase gene expression in a glycophyte and a halophyte. Plant Physiol. *103,* 713–718 (1993 b).

Orlowski, J., R. A. Kandasamy, and G. E. Shull: Molecular cloning of putative members of the Na/H exchanger gene family. J. Biol. Chem. *267,* 9331–9339 (1992).

Padan, E. and S. Schuldiner: Molecular physiology of the Na^+/H^+ antiporter in *Escherichia coli.* J. Exp. Biol. *196,* 443–456 (1994).

Palmgren, M. G., M. Sommarin, R. Serrano, and C. Larsson: Identification of an autoinhibitory domain in the C-terminal region of the plant plasma membrane H^+-ATPase. J. Biol. Chem. *266,* 20470–20475 (1991).

Passioura, J. B.: Root signals control leaf expansion in wheat seedlings growing in drying soil. Australian J. Plant Physiol. *15,* 687–693 (1988).

Perez-Prat, E., M. L. Narasimhan, M. L. Binzel, M. A. Botella, Z. Chen, V. Valpuesta, R. A. Bressan, and P. M. Hasegawa: Induction of a putative Ca^{2+}-ATPase mRNA in NaCl-adapted cells. Plant Physiol. *100,* 1471–1478 (1992).

Peterson, C. A.: Exodermal Casparian bands: their significance for ion uptake by roots. Physiol. Plant. *72,* 204–208 (1988).

Pinner, E., E. Padan, and S. Schuldiner: Cloning, sequencing, and expression of the nhaB gene, encoding a Na^+/H^+ antiporter in *Escherichia coli.* J. Biol. Chem. *267,* 11064–11068 (1992).

Prieto, R., R. A. Bressan, and P. M. Hasegawa: Genes involved in NaCl tolerance in *Chlamydomonas reinhardii.* Plant Physiol. *108* (suppl.), 59 (1995).

Radin, J. W.: Water relations of cotton in controlled environments and the field. In: Close, T. J. and E. A. Bray (eds.): Plant responses to cellular dehydration during environmental stress. American Soc. Plant Physiol. Vol. *10,* pp. 67–78 (1993).

Ratajczak, R., J. Richter, and U. Lüttge: Adaptation of the tonoplast V-type H^+-ATPase of *Mesembryanthemum crystallinum* to salt stress, C_3-CAM transition and plant age. Plant Cell Environm. *17,* 1101–1112 (1994).

Reuveni, M., A. B. Bennett, R. A. Bressan, and P. M. Hasegawa: Enhanced H^+-transport capacity and ATP hydrolysis activity of tonoplast H^+-ATPase after NaCl adaptation. Plant Physiol. *94,* 524–530 (1990).

Rockel, B., R. Ratajczak, A. Becker, and U. Lüttge: Changed densities and diameters of intra-membrane particles in tonoplast vesicles of *Mesembryanthemum crystallinum* in correlation with NaCl-induced CAM. J. Plant Physiol. *143,* 318–324 (1994).

Sarafian, V., Y. Kim, R. J. Poole, and P. A. Rea: Molecular cloning and sequencing of cDNA endoding the pyrophosphate-energized vacuolar membrane proton pump of *Arabidopsis thaliana.* Proc. Natl. Acad. Sci. USA *89,* 1775–1779 (1992).

Sardet, C., L. Counillon, A. Franchi, and J. Pouysségur: Growth factors induce phosphorylation of the Na^+/H^+ antiporter, a glycoprotein of 110 kD. Science *247,* 723–726 (1990).

Schachtman, D. P. and J. I. Schroeder: Structure and transport mechanism of a high-affinity potassium uptake transporter from higher plants. Nature *370,* 655–658 (1994).

Schachtman, D. P., A. Fairhead, and O. Babourina: A pathway for sodium into plant cells? Plant Physiol. *108* (suppl.), 38 (1995).

Schachtman, D. P., S. T. Tyerman, and B. R. Terry: The K^+/Na^+ selectivity of a cation channel in the plasma membrane of root cells does not differ in salt-tolerant and salt-sensitive wheat species. Plant Physiol. *97,* 598–605 (1991).

Schumaker, K. S. and H. Sze: Solubilization and reconstitution of the oat root vacuolar H^+/Ca^{2+} exchanger. Plant Physiol. *92,* 340–345 (1990).

Spickett, C. M., N. Smirnoff, and R. G. Ratcliffe: An *in vivo* magnetic resonance investigation of ion transport in maize (*Zea mays*) and *Spartina anglica* roots during exposure to high salt concentrations. Plant Physiol. *102,* 629–638 (1993).

Staal, M., F. J. M. Maathuis, J. T. M. Elzenga, J. H. M. Overbeek, and H. B. A. Prins: Na^+/H^+ antiport activity in tonoplast vesicles from roots of the salt-tolerant *Plantago maritima* and salt sensitive *Plantago media.* Physiol. Plant. *82,* 179–184 (1991).

Struve, I., T. Rausch, P. Bernasconi, and L. Taiz: Structure and function of the promotor of the carrot V-type H^+-ATPase catalytic subunit gene. J. Biol. Chem. *265,* 7927–7932 (1990).

Sumaryati, S., I. Negrutiu, and M. Jacobs: Characterization and regeneration of salt- and water-stress mutants from protoplast culture of *Nicotiana plumbaginifolia* (Viviani). Theor. Appl. Gen. *83,* 613–619 (1992).

Sussman, M. R.: Molecular analysis of proteins in the plasma membrane. Annu. Rev. Plant Physiol. Plant Mol. Biol. *45,* 211–234 (1994).

Sze, H., J. M. Ward, and S. Lai: Vacuolar H^+-translocating ATPases from plants: Structure, function, and isoforms. J. Bioenerg. Biomembr. *24,* 371–381 (1992).

Tanaka, Y., K. Chiba, M. Maeda, and M. Maeshima: Molecular cloning of cDNA for vacuolar membrane proton-translocating inorganic pyrophosphatase in *Hordeum vulgare.* Biochem. Biophys. Res. Commun. *190,* 1110–1114 (1993).

Tarcynski, M. C., R. G. Jensen, and H. J. Bohnert: Stress protection of transgenic tobacco by production of the osmolyte mannitol. Science *259,* 508–510 (1993).

Ward, J. M., A. Renders, H. T. Hsu, and H. Sze: Dissociation and reassembly of the vacuolar H^+-ATPase complex from oat root. Plant Physiol. *99,* 161–169 (1992).

Wilson, C. and M. C. Shannon: Salt-induced Na^+/H^+ antiport in root plasma membrane of a glycophytic and halophytic species of tomato. Plant Science *107,* 147–157 (1995).

Wimmers, L. E., N. N. Ewing, and A. B. Bennett: Higher plant Ca^{2+}-ATPase: Primary structure and regulation of mRNA abundance by salt. Proc. Natl. Acad. Sci. USA *89,* 9205–9209 (1992).

Wolf, O., W. D. Jeschke, and W. Hartung: Long distance transport of abscisic acid in NaCl-treated intact plants of *Lupinus albus.* J. Exp. Bot. *41,* 593–600 (1990).

Yeo, A. R.: Salt tolerance in the halophyte *Suaeda maritima* L. Dum.: intracellular compartmentation of ions. J. Exp. Bot. *32,* 487–497 (1981).

Zhao, K., R. Munns, and R. W. King: Abscisic acid synthesis in NaCl-treated barley, cotton and saltbush. Australian J. Plant Physiol. *18,* 17–24 (1991).

Zimniak, L., P. Dittrich, J. P. Gogarten, H. Kibak, and L. Taiz: The cDNA sequence of the 69-kDa subunit of the carrot vacuolar H^+-ATPase. J. Biol. Chem. *263,* 9102–9112 (1988).

J. Plant Physiol. Vol. 148. pp. 434–439 (1996)

Effects of Pb and Cd on Cucumber Depending on the Fe-Complex in the Culture Solution

FERENC FODOR, ÉVA SÁRVÁRI, FERENC LÁNG, ZOLTÁN SZIGETI, and EDIT CSEH

Department of Plant Physiology, Eötvös University, Budapest P.O.B. 330, H-1445 Hungary

Received July 7, 1995 · Accepted October 10, 1995

Summary

Modifying effects of EDTA and citrate as Fe(III)-complexes on Pb- and Cd-toxicity were investigated with cucumber (*Cucumis sativus* L.) grown in culture solution. Pb and Cd were applied in 10 µM concentration. Pb and Cd inhibited the growth of plants with Fe-citrate but with Fe-EDTA only Cd did. Chlorophyll content of Cd-treated plants – independently on the Fe-complex – was very low similarly to the iron deficient plants. Pb is proved to inhibit the chlorophyll accumulation only in plants supplied with Fe-citrate. Photosynthetic activity of 6-week-old plants was characterized by *in vivo* CO_2 fixation. The highest activities were observed in the middle (third-fourth) leaf storeys, and decreased in the lower and higher leaf storeys. Cd caused more than 50 % inhibition of the photosynthetic activity when applied with Fe-EDTA. With Fe-citrate, the inhibition exceeded 90 %. The photosynthetic activities in the Pb-treated plants were not significantly different from the control plants. The amount of chlorophyll containing complexes, especially that of PSI was highly affected by Cd particularly in the lower leaves. The results correlate with ^{59}Fe translocation into the shoot. Fe translocation was stimulated by Pb in the presence of Fe-EDTA but it was inhibited in the presence of Fe-citrate. Cd completely inhibited Fe translocation from the root to the shoot with both chelators. It is concluded that strong iron deficiency has a great but not exclusive role in the observed symptoms, and that the quality of the chelator strongly influences the effects of these polluting metals.

Key words: Cucumis sativus, heavy metal toxicity, chlorophyll-proteins, CO_2 fixation, Fe-complexes, Fe-uptake and -translocation.

Abbreviations: Chl = chlorophyll; CP = chlorophyll-protein; EDTA = ethylenediaminotetraacetic acid; LHC = light-harvesting complex; PS = photosystem.

Introduction

Pb and Cd are the most abundant heavy metals polluting the environment. They are taken up by plants mostly through the root system (Cutler and Rains, 1974; Foy, 1978; Vojtechova and Leblová, 1991) and partly – in much smaller amounts – through their leaves (Greger et al., 1993). It was also proved that the composition of the culture solution can considerably modify the uptake and even the effect of these ions (Hardyman and Jacoby, 1984; Wozny et al., 1990).

The most apparent symptom of Cd effect is the chlorosis of the leaves. This is accompanied by the decrease in net photosynthetic activity, PSII activity, the overall inhibition of the dark reactions of photosynthesis and serious damages in chloroplast ultrastructure (for review see Wozny et al., 1990; Malik et al., 1992). Pb is less effective in damaging the photosynthetic apparatus (Ahmed and Tajmir-Riahi, 1993) unless it is applied in extremely high concentrations.

However, the toxic heavy metals can affect plant metabolism and development at several levels. Increasing levels of Pb in the nutrient solution were reported to reduce root growth and the concentration of other elements in the needles of *Picea abies* seedlings (Godbold and Kettner, 1991). Cd was also found to inhibit K uptake into oat roots (Keck, 1978), and reduce growth and iron content in different plants (Wong et al., 1984; Smith et al., 1985). However, no clear effect can be found if one compares lower and higher concentrations of the applied heavy metal treatment (Godbold and Kettner, 1991), different plant species (Kahn and Kahn, 1983) and the concentration of other elements in the root and shoot (Wong et al., 1984).

The aim of the present work was to investigate the *in vivo* effects of Pb and Cd on the growth, photosynthetic activity and iron uptake of cucumber in more details on the same sample, and to analyze the modifying effect of EDTA and citrate as iron(III) chelating agents supplied to the culture solution on the impact of these heavy metals.

Materials and Methods

The experiments were conducted with cucumber (*Cucumis sativus* L.) grown hydroponically in modified Hoagland solution of ¼ strength (Fodor and Cseh, 1993). Iron was supplied as Fe-EDTA or Fe-citrate in 4 µM concentration. Heavy metals ($Pb(NO_3)_2$, $Cd(NO_3)_2$) were added to the culture solution in 10 µM concentration from the 7th day after germination. The pH of the nutrient solution containing EDTA was 5.2–5.4 and that of the nutrient solution containing citrate was 4.7–4.8. The culture solution was renewed every second day. The plants were irradiated with 75 W m^2 light intensity and with 12 h night/day photoperiod. The temperature was 20/25 °C night/day and the relative humidity of the air was 55–60 %.

6-week-old plants bearing 7 leaves were harvested and fresh and dry weight, chlorophyll (Chl) content, *in vivo* CO_2 fixation, Chl-protein (CP)/polypeptide patterns were determined for each leaf.

Chl concentrations of the leaves were determined photometrically in 80 % acetone using the equations of Porra et al. (1989).

The rate of light-induced CO_2 fixation was studied in detached leaves, in an atmosphere containing $^{14}CO_2$ in a sealed glass chamber with a Hg-blockade according to the method of Láng et al. (1985). The radioactivity of the samples was determined by liquid scintillation technique (Beckman LS5000TD).

Isolation of thylakoids and CP complexes by Deriphat polyacrylamide gel electrophoresis was performed as described by Sárvári and Nyitrai (1994). Polypeptide patterns of thylakoids or CPs were determined by denaturing gel electrophoresis according to Laemmli (1970).

Iron uptake and translocation was measured with 2-week-old cucumber plants grown in the same conditions described above but without Fe, Cd and Pb in the medium. Before labelling, the roots of plants were rinsed in ion exchanged distilled water in order to remove the loosely adsorbed ions of the culture solution from the surface. Then they were transferred to complete, quarter strength Hoagland solution containing Fe-EDTA or Fe-citrate labelled with ^{59}Fe and Pb or Cd, respectively. All these heavy metals were applied in 10 µM concentration. After 48 hours, the roots were rinsed again in 100 mL of ion exchanged distilled water. Radioactivity of the roots, hypocotyls, cotyledons, epicotyls and leaves were measured separately in test tubes using a GAMMA-NK 350 counter. Iron uptake was calculated on the basis of the specific activity of the radiolabelled culture solution.

Results and Discussion

The growth retardation of plants by Cd and especially of the roots by Pb is a well-known phenomenon (Wozny and Jerczynska, 1991; Kahle, 1993). We found that the effect of Pb and Cd on the growth of cucumber was greatly influenced by the chelating agent in which the Fe(III) was supplied. Pb seemed to slightly stimulate the growth of both roots and leaves of cucumber in the presence of Fe-EDTA but growth is inhibited if Fe-citrate was supplied. Cd drastically reduced growth at every treatment (Fig. 1).

The Chl content of the second leaves of cucumber was severely decreased by Cd applied together with Fe-EDTA and Fe-citrate, respectively. The magnitude of the decrease was found to be much larger than it was reported earlier (Burzynski and Buczek, 1989). Pb reduced the Chl content to 70 % of the control in case of Fe-citrate but it was much less affected with Fe-EDTA (Table 1).

Photosynthetic activity measured as CO_2 fixation of 6-week-old plants bearing 7 leaves was the highest in the middle (third-fifth) leaf storeys, and decreased in the lower and higher leaf storeys. Cd inhibition was very strong in the presence of both Fe-complexes (Fig. 2).

In order to test the effect of Pb and Cd on the activity of the photosynthetic electron transport chain we measured fluorescence induction parameters. Preliminary results show that photosynthetic electron transport between Q_A and Q_B was slower in Cd treated plants.

Some authors emphasized the inhibiting effect of Cd on CO_2 assimilation, particularly at the level of some Calvin-cycle enzymes (Sheoran et al., 1990; Malik et al., 1992). The reduced demand for ATP and NADPH after Calvin-cycle inhi-

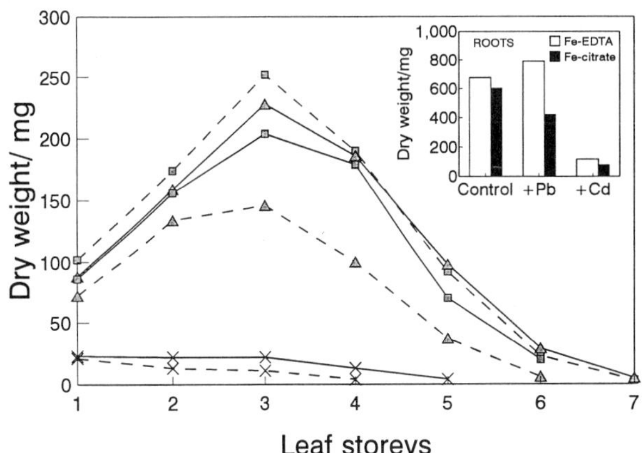

Fig. 1: Effect of Pb and Cd on the growth of cucumber grown on Fe-EDTA (——) and Fe-citrate (- - - -), respectively. Control (□), Pb-treated (△), Cd-treated (×). Data are presented as means of 5 replicates. SE <10 % of the values.

Table 1: Chlorophyll concentration of the second leaves of cucumber treated with Pb and Cd. Values are presented as means of 5 replicates ±SE.

Treatment	Fe-EDTA	Fe-citrate
	µg Chl/g fresh weight	
control	2266±113	1864±142*
+Pb	1888±119	1277±172
+Cd	405± 18	228± 22

* Third leaf.

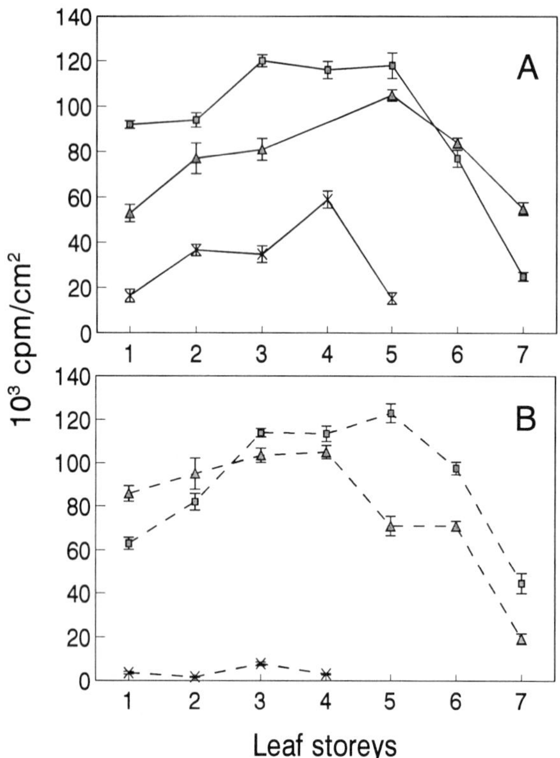

Fig. 2: Effect of Pb and Cd on the $^{14}CO_2$ fixation of cucumber grown on Fe-EDTA (**A**) and Fe-citrate (**B**), respectively. Control (□), Pb-treated (△), Cd-treated (×). Presented values are means of 6 replicates ± SE.

The total amounts of PSI, PSII and LHCII did not decrease in Pb-treated plants if iron was supplied as Fe-EDTA but the amounts of PSI and PSII were reduced if Fe-citrate was in the culture solution (Fig. 4). In contrast, Cd strongly reduced the amounts of all complexes in the order of PSI > PSII > LHCII. The effect was more pronounced in the plants grown on Fe-citrate.

Similar changes can be seen on denatured polypeptide patterns of thylakoids isolated from Cd-treated plants (Fig. 5). The decrease of the core complexes (polypeptides in the molecular mass range of 30–70 kDa) was more pronounced than that of LHCs (22–27 kDa).

The changes in the thylakoid organization of Cd-treated plants showed some similarities to those of iron deficient ones, i.e. the amount of PSI particles extremely decreased in both cases (Fodor et al., 1995). This is in agreement with the data referring to insufficient iron supply of shoots in Cd-treated plants (Siedlacka and Baszynski, 1993). However, the amount of PSII was much more affected by Cd-treatment than in case of iron deficiency. Some structural parts of PSII seem to be sensitive to Cd treatment (Maksymiec and Baszynski, 1988). Moreover the ratio of LHCII to PSII was higher in the Cd-treated plants compared to the iron deficient ones. Thus, the Chl *a*/*b* ratio decreased in the Cd-treated plants but did not change or increased in the iron deficient plants. Cd-induced changes in the trans-Δ^3-hexadecenoic fatty acid content were reported to influence the oligo-

bition causes a down-regulation of PSII photochemistry and of the yield of linear electron transport (Krupa et al., 1993). This can be reflected in the slower rate of electron transport between Q_A and Q_B. According to Siedlacka and Baszynski (1993), the decrease in electron transport around PSI in Cd-treated maize plants correlated with a low ferredoxin content which, in turn, was correlated with a low iron concentration suggesting Cd-induced iron deficiency. These results are consistent with our observations.

Structural differences between control plants and the ones treated with heavy metals were observed as changes in the CP and polypeptide pattern of treated thylakoids. Chl-containing bands (Fig. 3) were identified according to their Chl *a*/*b* ratio and polypeptide pattern (not shown) after Sárvári and Nyitrai (1994). The Chl *a*/*b* ratio of bands 1 and 2 was around 6–8 while only Chl *a* was present in bands 3 and 4. All of them contained PSI polypeptides. However, the 22–24 kDa LHCI polypeptides were absent in bands 3 and 4. Chl *a*-containing fractions of PSII were found in bands 5 (PSII CC) and 6 (CP47 and PSII RC) and in band 10 (CP43), respectively. Fractions of the PSII connecting antenna were found in bands 7 and 11 (CP29, Chl *a*/*b* was around 3.5), 12 (CP26, CP24, Chl *a*/*b* was around 2) in complexed and monomer forms, respectively. Some solubilized LHCI polypeptides were also present in band 7 and 12. LHCII (Chl *a*/*b* ratio was around 1.6) was detected in band 9.

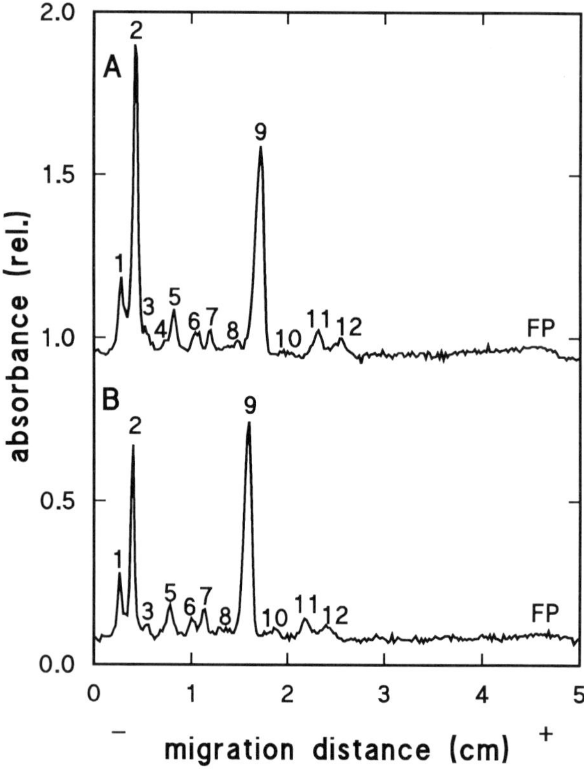

Fig. 3: Representative densitograms of the chlorophyll-protein patterns of thylakoids isolated from the control (**A**) and cadmium-treated (**B**) leaves of cucumber plants supplied with Fe-EDTA. FP = free pigment.

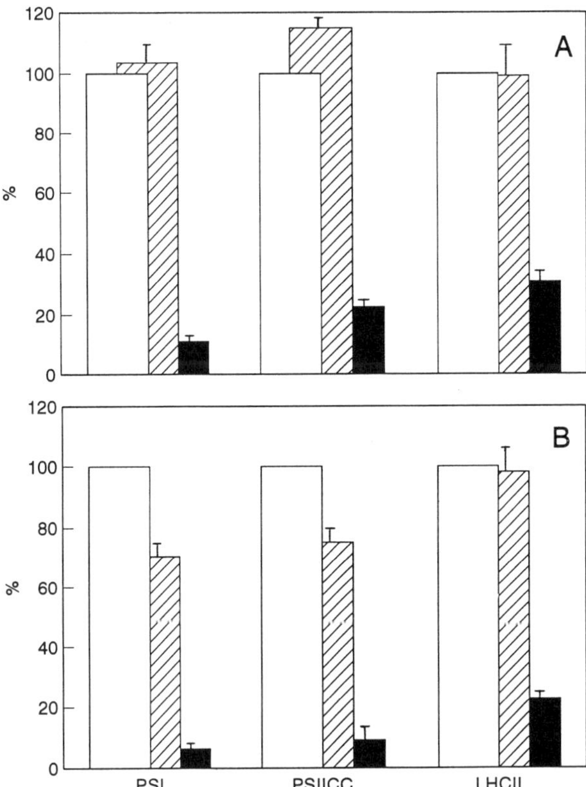

Fig. 4: The absolute amounts of Chl-containing complexes expressed as % of the control originating from plants supplied with Fe-EDTA (**A**) or Fe-citrate (**B**). The amount of a given complex shown here was calculated by summing up densitogram bands corresponding to the complex. Then the Chl content of leaves was shared among the complexes according to their relative proportions. Empty = control, hatched = Pb-treated, filled = Cd-treated. Bars represent the means of 3–4 replicates ± SE.

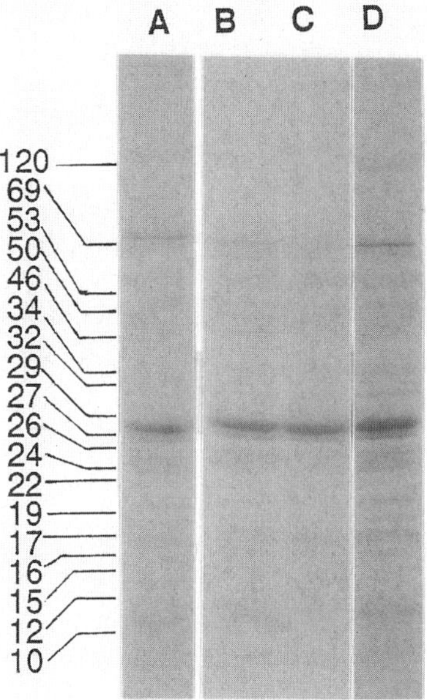

Fig. 5: Polypeptide patterns of Fe-EDTA control (**A**), Cd-treated (**B**) and Fe-citrate Cd-treated (**C**), control (**D**) plants, respectively. Numbers are molecular masses in kDa.

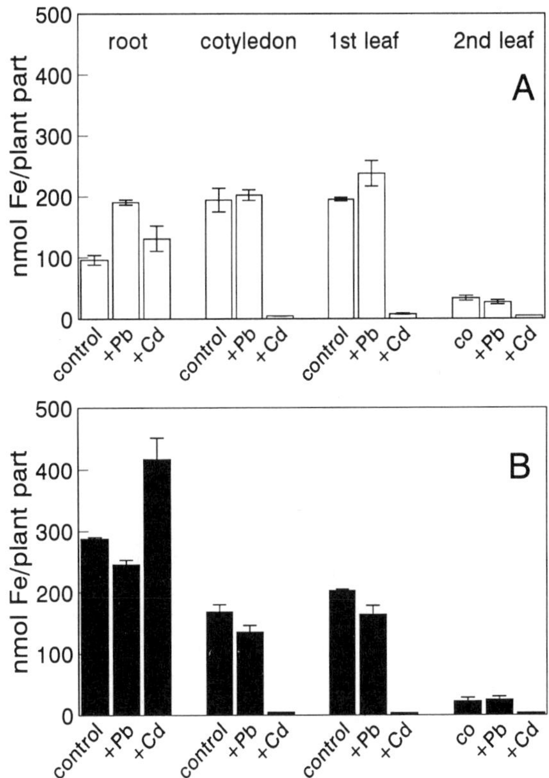

Fig. 6: Effect of Pb and Cd on the uptake and translocation of Fe in cucumber grown on Fe-EDTA (**A**) and Fe-citrate (**B**), respectively. Bars represent the means of 5 replicates ± SE.

merization of LHCII (Krupa, 1988), while Ahmed and Tajmir-Riahi (1993) found conformational changes of LHCII subunits binding heavy metals as Pb and Cd *in vitro*. Conformational changes of the components may cause the incomplete assembly followed by degradation. These observations may refer to the multi-level effect of Cd.

To find out the possible connections between the Cd-effect and iron deficiency we examined the uptake and translocation of iron. The iron uptake in the roots of plants supplied with Fe-citrate was significantly higher than in the ones supplied with Fe-EDTA (Fig. 6). This is the most apparent in the control and in the Cd-treated plants. However, most of the iron was undoubtedly non-specifically adsorbed on the binding sites of the root cell wall (Römheld and Marschner, 1986; Cseh et al., 1994). In the plants supplied with Fe-EDTA, both Pb and Cd had a stimulatory effect on the iron uptake of the roots. On the other hand iron translocation was only moderately affected (stimulated!) by Pb in the presence of Fe-EDTA compared to the control plants but it was inhibited in the presence of Fe-citrate. Cd completely inhibited iron translocation from the root to the shoot with both Fe-complexes, therefore the Fe accumulated in the root is inaccessible for plant metabolism. These data support the earlier results of Burzynski and Buczek (1989).

According to literature data Cd exerts a strong inhibition on the root Fe(III) reductase in 5 μM concentration while Pb in 100 μM concentration was ineffective. Cd may inactivate the root Fe(III) reductase thereby inhibiting iron uptake (Alcántara et al., 1994). This may be due to the interference of Cd with the SH-groups of these enzymes (Hendrix and Higinbotham, 1974; Wozny et al., 1990). Our preliminary data show that 10 μM Pb does not exert any effect on either the standard or the enhanced («turbo») electron transport functioning in the plasmalemma. Cadmium strongly inhibits the «turbo» electron transport and the proton efflux. The impairment of both the reductive capacity and the generation of the acidic pH in the rhizosphere, that are the conditions of the normal iron supply of most dicotyledonous plants, may also contribute to the development of iron deficiency in the presence of Cd.

In conclusion: Although iron deficiency seems to play an important role we believe that a series of other reactions concerning different levels of metabolism are also involved in the formation of Cd-toxicity symptoms. The most important ones may be the inhibition of the Calvin cycle followed by the down-regulation of PSII as well as the direct effects of Cd on PSI and, especially, on PSII functions. Moreover, the incomplete assembly of conformationally changed components and their subsequent degradation may also be substantial factors in the toxicity. Cucumber appears to be more tolerant to Pb especially if it was applied with Fe-EDTA. Its toxic effect would probably be more apparent at higher concentrations. Considering the modifying effect of chelating agents, citrate makes the plants more susceptible to heavy metal toxicity.

Acknowledgements

This work was supported by grant EEC CIPA-CT 93-0202 which is gratefully acknowledged.

References

Ahmed, A. and H. A. Tajmir-Riahi: Interaction of toxic metal ions Cd^{2+}, Hg^{2+}, and Pb^{2+} with light-harvesting proteins of chloroplast thylakoid membranes. An FTIR spectroscopic study. J. Inorg. Biochem. *50*, 235–243 (1993).

Alcantara, E., F. J. Romera, M. Canete, and M. D. De la Guardia: Effects of heavy metals on both induction and function of root Fe(III) reductase in Fe-deficient cucumber (*Cucumis sativus* L.) plants. J. Exp. Bot. *45*, 1893–1898 (1994).

Burzynski, M. and J. Buczek: Interaction between cadmium and molybdenum affecting the chlorophyll content and accumulation of some heavy metals in the second leaf of *Cucumis sativus* L. Acta Phys. Plant. *11*, 137–145 (1989).

Cseh, E., Gy. Váradi, and F. Fodor: Effect of Fe-complexes and N-forms on the Fe absorption, uptake and translocation of cucumber plants. Bot. Közl. *81*, 47–55 (1994).

Cutler, J. M. and D. W. Rains: Characterization of cadmium uptake by plant tissue. Plant Physiol. *54*, 64–71 (1974).

Fodor, F., B. Böddi, É. Sárvári, Gy. Záray, E. Cseh, and F. Láng: Correlation of iron content, spectral forms of chlorophyll and chlorophyll-proteins in iron deficient cucumber. Physiol. Plant. *93*, 750–756 (1995).

Fodor, F. and E. Cseh: Effect of different nitrogen-forms and iron-chelates on the development of stinging nettle. J. Plant Nutr. *16*, 2239–2253 (1993).

Foy, C. D., R. L. Chaney, and M. C. White: The physiology of metal toxicity in plants. Annu. Rev. Plant Physiol. *29*, 511–566 (1978).

Godbold, D. L. and C. Kettner: Lead influences root growth and mineral nutrition of *Picea abies* seedlings. J. Plant Physiol. *139*, 95–99 (1991).

Greger, M., M. Johansson, A. Stihl, and K. Hamza: Foliar uptake of Cd by pea (*Pisum sativum*) and sugar beet (*Beta vulgaris*). Physiol. Plant. *88*, 563–570 (1993).

Hardiman, R. T. and B. Jacoby: Absorption and translocation of Cd in bush beans (*Phaseolus vulgaris*). Physiol. Plant. *61*, 670–674 (1984).

Hendrix, D. L. and N. Higinbotham: Heavy metals and sulphhydryl reagents as probes of ion uptake in pea stem. In: Zimmermann, U. and J. Dainty (eds.): Membrane Transport in Plants, pp. 412–416. Springer-Verlag, Berlin, Heidelberg, New York (1974).

Kahle, H.: Response of roots of trees to heavy metals. Env. Exp. Bot. *33*, 99–119 (1993).

Kahn, S. and N. N. Kahn: Influence of lead and cadmium on the growth and nutrient concentration of tomato (*Lycopersicon esculentum*) and egg-plant (*Solanum melongena*). Plant and Soil *74*, 387–394 (1983).

Keck, R. W.: Cadmium alteration of root physiology and potassium ion fluxes. Plant Physiol. *62*, 94–96 (1978).

Krupa, Z.: Cadmium-induced changes in the composition and structure of light-harvesting chlorophyll *a/b* protein complex 11 in radish cotyledons. Physiol. Plant. *73*, 518–524 (1988).

Krupa, Z., G. Öquist, and N. P. A. Huner: The effects of cadmium on photosynthesis of *Phaseolus vulgaris* – a fluorescence analysis. Physiol. Plant. *88*, 626–630 (1993).

Laemmli, U. K.: Cleavage of structural proteins during assembly of the head of bacteriophase T4. Nature *227*, 680–685 (1970).

Láng, F., É. Sárvári, and Z. Szigeti: Apparatus and method for rapid determination of photosynthetic CO_2 fixation of leaves. Biochem. Physiol. Pflanzen *180*, 333–336 (1985).

Maksymiec, W. and T. Baszynski: The effect of Cd^{2+} on the release of proteins from thylakoid membranes of tomato leaves. Acta Soc. Bot. Pol. *57*, 465–474 (1988).

Malik, D., I. S. Sheoran, and R. Singh: Carbon metabolism in leaves of cadmium treated wheat seedlings. Plant Physiol. Biochem. *30*, 223–229 (1992).

Porra, R. J., W. A. Thompson, and P. E. Kriedemann: Determination of accurate extinction coefficients and simultaneous equations for assaying chlorophyll *a* and *b* extracted with four different solvents: verification of the concentration of chlorophyll standards by atomic absorption spectroscopy. Biochim. Biophys. Acta *975*, 384–394 (1989).

Römheld, V. and H. Marschner: Mobilization of iron in the rhizosphere of different plant species. Adv. Plant Nutr. *2*, 155–204 (1986).

Sárvári, É. and P. Nyitrai: Separation of chlorophyll-protein complexes by Deriphat polyacrylamide gradient gel electrophoresis. Electrophoresis *15*, 1068–1071 (1994).

Sheoran, I. S., H. R. Singal, and R. Singh: Effect of cadmium and nickel on photosynthesis and the enzymes of the photosynthetic carbon reduction cycle in pigeonpea (*Cajanus cajan* L.). Photosynth. Res. *25*, 345–351 (1990).

Siedlecka, A. and T. Baszynski: Inhibition of electron flow around photosystem I in chloroplasts of Cd-treated maize plants is due to Cd-induced iron deficiency. Physiol. Plant. *87*, 199–202 (1993).

Smith, G. C., E. G. Brennan, and B. J. Greenhalgh: Cadmium sensitivity of soybean related to efficiency in iron utilization. Env. Exp. Bot. *25*, 99–106 (1985).

VOJTECHOVA, M. and S. LEBLOVÁ: Uptake of lead and cadmium by maize seedlings and the effect of heavy metals on the activity of phosphoenolpyruvate carboxylase isolated from maize. Biol. Plant. *33,* 386–394 (1991).

WONG, M. K., G. K. CHUAH, L. L. KOH, K. P. ANG, and C. S. HEW: The uptake of cadmium by *Brassica chinensis* and its effect on plant zinc and iron distribution. Env. Exp. Bot. *24,* 189–195 (1984).

WOZNY, A., A. STRONSKI, and E. GWOZDZ: Plant cell responses to cadmium. Seria Biologia No. 44. pp. 29 Uniwersitet Im. Adama Mickiewicza W Poznaniu, Poznan (1990).

WOZNY, A. and E. JERCZYNSKA: The effect of lead on early stages of *Phaseolus vulgaris* L. growth *in vitro* conditions. Biol. Plant *33,* 32–39 (1991).

J. Plant Physiol. Vol. 148. pp. 440–444 (1996)

Aerial Photography to Detect Nitrogen Stress in Corn

TRACY M. BLACKMER and JAMES S. SCHEPERS

Soil Scientist, USDA-Agricultural Research Service, 241 Keim Hall, University of Nebraska, Lincoln, NE 68583-0915

Received June 24, 1995 · Accepted October 10, 1995

Summary

Economic and environmental benefits can result from improved nitrogen (N) management in corn (*Zea mays* L.) production. This research project was conducted to determine the utility of aerial photographs to detect N deficiency in a crop canopy caused by natural soil variability and variable fertilizer N application rates. Chlorophyll meter readings and digitized aerial photograph data were compared to corn grain yield and stalk nitrate concentrations from a 60-ha field in Central Nebraska. In addition to natural field variability, 30.4-m long treatments were imposed through the center of the field by applying fertilizer at rates of 0, 56, 112, 168, and 224 kg N ha^{-1}. Grain yield, chlorophyll meter readings, photographic brightness and stalk nitrate samples were collected. Despite a significant effect from fertilizer treatments, chlorophyll meter readings did not correlate well with grain yields. The brightness of the red component in a digitized color photograph showed a significant inverse relationship ($r^2 = 0.42$) with grain yield. Stalk nitrate concentrations at harvest, when compared to red brightness from the digitized photograph, provided a better confirmation of N status than grain yield. Aerial photographs appear to be capable of detecting management induced variability as well as reduced yield portions of a field that result from natural variability.

Key words: Nitrogen, chlorophyll meter, Zea mays L., remote sensing, stalk nitrate, aerial photographs.

Abbreviations: N = nitrogen; GPS = global positioning system.

Introduction

Economic and environmental benefits from improved N management in corn can frequently be enhanced by using real-time methods to evaluate management practices. Tissue testing is one technique that can provide valuable information on the crop's N status.

Current tissue tests for N used in production agriculture require sampling specific plant parts and making interpretations according to generalized criteria. It follows that tissue samples can be no more representative of a field than the plants that are sampled. Therefore, if soil properties or other conditions (climate, disease, insects, etc.) change spatially or temporally, it may be difficult to collect and analyze the appropriate number of samples that permit identification of N stressed areas in a field. This limitation is because of the labor requirements, time constraints, and analytical expenses.

Chlorophyll meters have recently been introduced as a tool to quantify N status. Application of this technology extends to monitoring the dynamics of N transformations in soil and assessing N availability to crops. Researchers have shown that chlorophyll meters are capable of detecting N stress in corn (Piekielek and Fox, 1992; Schepers et al., 1992; Wood et al., 1992; Blackmer and Schepers, 1995). Chlorophyll meters measure light transmittance through a leaf and provide rapid results without the expenses of laboratory analysis. However, interpretation and application of chlorophyll meter data are hindered by the problem of obtaining a representative sample from a potentially variable field.

Canopy reflectance measurements integrate many factors influencing the general health of the crop. Reflectance measurements can range from those that deal with individual leaves to those that encompass an entire field, depending on the configuration of the sensor and distance from the crop. In any case, some ground truthing and analytical results are required to interpret reflectance images, whether they be scans from a spectroradiometer or an aerial photograph. Nonetheless, reflectance measurements potentially offer a more rapid and less expensive assessment of growing conditions than is possible with traditional chemical analysis using leaf tissue or plant sap.

Canopy reflectance in the visible and near infrared wavelengths has been shown to detect N deficiency in corn (Walburg et al., 1982; Blackmer et al., 1995). This permits rapid evaluation of larger areas than is realistically possible with chlorophyll meters or traditional tissue testing procedures. Another advantage is that reflectance techniques can be automated for use on high clearance vehicles.

Aerial photography is one of several techniques to record canopy reflectance. Researchers have been able to detect various types of stress in different crops by use of aerial photography (Colwell, 1956; Wildman, 1982; Jackson, 1986). More specifically, research conducted on N response trials has demonstrated that aerial photography can be a good indicator of deficiency in corn (Blackmer, 1995). The benefit of an aerial photograph is that an entire field can be monitored with high resolution and minimal concern for changing light conditions during the time when measurements are taken. Photographs permit identification of potential problems for even small areas within a field. When coupled with GPS technology, these small areas could be located and treated.

The objective of this study is to evaluate aerial photography as a technique to detect N deficiency in a crop canopy caused by natural or fertilizer induced variability.

Materials and Methods

Corn was planted in late April, 1993 in a 60-ha field near Shelton, Nebraska at a rate of 70,000 kernels ha^{-1}. Nitrogen fertilizer rates (0, 56, 112, 168, and 224 kg N ha^{-1}) were applied as anhydrous ammonia to randomized end-to-end plots that were 12.2-m wide (16 rows at 76-cm spacing) by 30.4-m long and replicated four times. The crop was approximately 30-cm tall at the time of N application. An adjacent strip 12.2-m wide received a constant rate of 106 kg N ha^{-1} (recommended amount based on soil analysis) as anhydrous ammonia.

Chlorophyll meter readings were collected at the R3 (milk) growth stage (Ritchie et al., 1986) by taking two sets of 30 readings using a Minolta SPAD-502 chlorophyll meter[1]. A color aerial photograph (Kodak Gold ASA 400 film) of the entire 60-ha field was taken from an airplane (altitude of ~1500 m) on a clear day when the corn was at the R4 (dough) growth stage. The image was digitized and then rectified using coordinates of 9 landmarks determined with a GPS and ERDAS software. Pixel size was adjusted to one square meter. Average red brightness of the center group of seven by seven pixels was used to represent each plot. Red brightness on the film is determined by light exposure on the film from wavelengths around 600–700 nm.

Grain yields were determined by hand harvesting 15.2 m segments from two rows in each plot. Grain yields were adjusted to 155 g kg^{-1} moisture. Stalk samples consisting of a 15-cm section immediately above the upper crown roots of 12 plants per plot were collected after harvest. Stalk segments were coarsely ground dried, finely ground (1.0-mm sieve), and analyzed for nitrate-N concentration using automated wet chemistry procedures.

[1] Mention of trade names or proprietary products does not indicate endorsement by USDA, and does not imply its approval to the exclusion of other products that may also be suitable.

Results and Discussion

Differences in grain yield were measured for plots that received varying amounts of N fertilizer (Fig. 1). Variability in yield response to each N treatment made it difficult to identify the optimal N rate. Reasons for the differential responses to fertilizer are attributed to variability in the quantity of available N supplied by other potential sources and the influence of various other yield-limiting factors. In that replications were as far as 400 m apart, differential response to N fertilizer is expected. Variability in crop response to N fertilizer was even observed within the long plots. Rows 4 and 13 of the 16-row wide plots used for hand harvesting had an average yield difference between the two segments within each plot of 0.75 Mg ha^{-1}. Differences between subsample yields were greater than expected considering that each subsample represented an area larger than is typically measured for hand harvest.

Chlorophyll meters have been shown to be an effective research tool to detect an N deficiency in corn (Piekielek and Fox, 1992; Schepers et al., 1992; Wood et al., 1992; Blackmer and Schepers, 1995). The above studies generally showed a strong correlation between chlorophyll meter readings and grain yield. These findings based on plot research are in contrast to the field scale results of this study that showed chlorophyll meter readings and grain yield were poorly correlated (Fig. 2). If the yield variation in response to N fertilizer was a result of differences in N availability from sources other than fertilizer, then the chlorophyll meter readings should have detected a corresponding difference in crop N status. One would normally expect a close correlation between relative crop N status at the R3 growth stage and relative yield because of the short time between these events (<30 days). The poor relationship between chlorophyll meter readings and grain yield suggests that either the sampling methods were

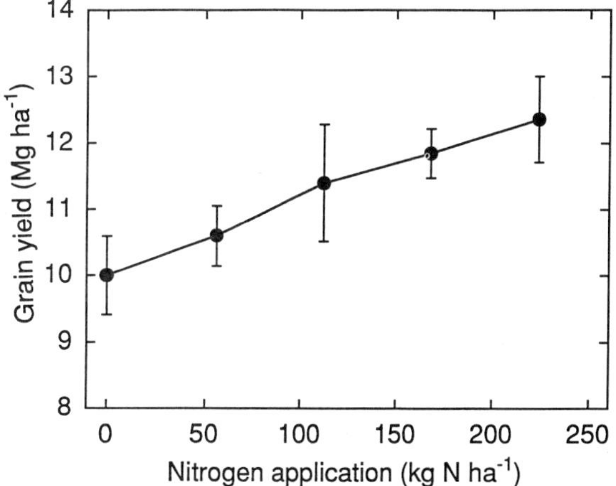

Fig. 1: Corn grain yields of hand harvested plots receiving different amounts of N fertilizer.

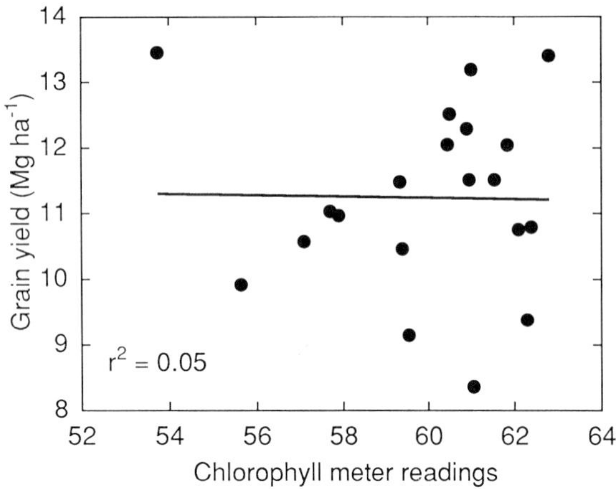

Fig. 2: Relationship between chlorophyll meter readings at the R3 growth stage and grain yield.

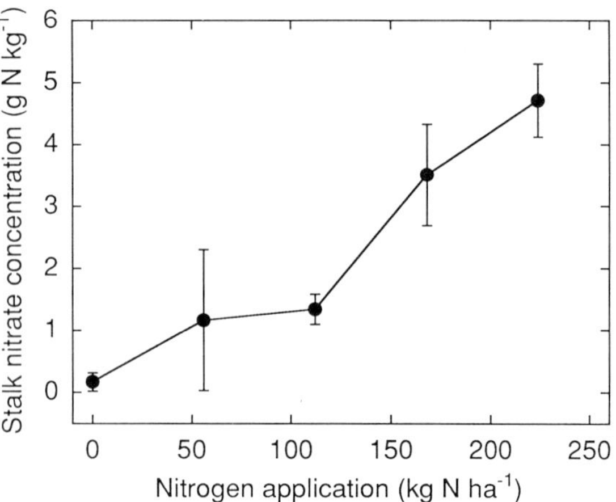

Fig. 3: Stalk nitrate-N concentrations for corn plots receiving different amounts of N fertilizer.

not adequate for either yield or chlorophyll content. Other possibilities are that factors such as disease, N loss, nonuniform N fertilizer application, or plant population affected grain yield. The wide range in grain yield for chlorophyll meter readings above 59 indicates that many factors other than N affected yields (some yields were higher than expected as well).

When available N is present in excess of crop requirement, corn often exhibits luxury consumption by continuing to take up N even though no response occurs. Because chlorophyll meter readings do not generally detect luxury consumption, a late season test of crop N status would be helpful to evaluate N management practices. Determining stalk nitrate-N concentration at harvest is a test that has been used to quantify crop N status of corn at the end of the growing season (Binford et al., 1990). The test is based on the concept that excess N availability in corn is likely to be associated with luxury consumption during the growing season. Extensive evaluation of the late season stalk nitrate test has shown that it is generally able to differentiate between where crop N is adequate and where it is excessive (i.e., luxury consumption) (Binford et al., 1990). The implication is that stalk nitrate-N concentrations above an adequate range are probably caused by excess fertilization, which is known to contribute to nitrate contamination of ground water.

Stalk nitrate-N concentrations in this study showed differences in N status between treatments. Even within a given N treatment, considerable variation in stalk nitrate-N concentration was noted, especially in treatments receiving >112 kg N ha^{-1}. Nitrate-N concentrations above 1800 mg kg^{-1} (Binford et al., 1992) typically indicate that yield was not limited by N availability (Fig. 3). Therefore, high concentrations of stalk nitrate-N minimize the possibility of N deficiency causing the fluctuations in yield noted in Figure 1.

Another method used to detect stress in crops is through analysis of aerial photographs. Photographs can be digitized so that the brightness of each pixel within the photograph is qualitatively indicative of the light reflecting from a given area on the landscape. When combined, the brightness of these pixels generates a digital image that mimics the photograph. This process makes it feasible to compare the brightness of an individual pixel or group of pixels with a designated ground-based measurement.

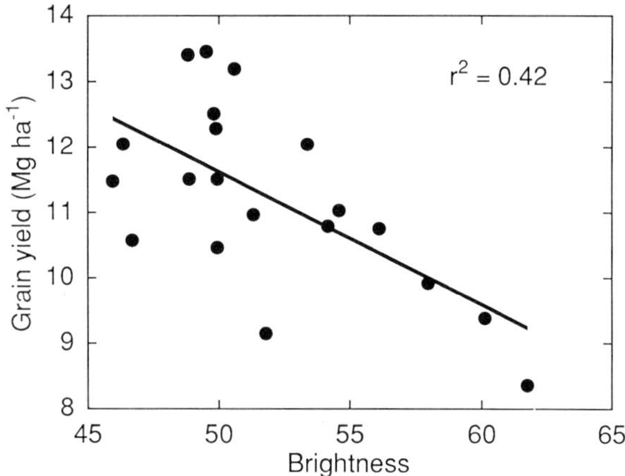

Fig. 4: The relationship between red brightness from a digitized photograph and grain yield.

Digital analysis of the aerial photograph taken at the R4 growth stage showed a significant relationship between red brightness and grain yield (Fig. 4). Because this relationship uses the same yield data as used in Figure 2, it raises questions about the reliability of the chlorophyll meter data. Perhaps the hypothetical problem with the chlorophyll meter data is associated with where on the plant the two types of data are collected. In the case of the chlorophyll meter, the ear leaf was monitored. In contrast, the digitized field image primarily represented the upper portion of the canopy. Perhaps a better relationship with the photograph could have been achieved if the yield measurements were made on the same 7 by 7-m area digitized from the photograph. Aerial photographs have the advantage over ground sampling methods in that the number of pixels represent the number of observations. Automated image analysis permits the user to design the shape and size of the field areas to be compared.

Another technique for examining the ability of the photograph to detect N deficiency is to compare the brightness of reflected light from a plot for a given N treatment with that for an adjacent plot receiving a constant N rate (assumed to be adequate for maximum yield). In this case, the comparison was a field strip receiving a constant rate of $106\,kg\,N\,ha^{-1}$ (based on soil testing data) (Fig. 5). The brightness for each N rate was adjusted by comparing the brightness value to the adjacent area, which makes this data specific for this field. Brightness differences observed for the 0 and $56\,kg\,N\,ha^{-1}$ treatments and not the higher treatments is consistent with the initial fertilizer recommendations of $106\,kg\,N\,ha^{-1}$ and the stalk nitrate data (Fig. 3). Initial soil recommendations, stalk nitrate concentrations and image brightness values all agree with each other, but do not agree with the yield data. This raises further questions about the value of the hand harvested yield estimates.

In an attempt to identify and understand the reasons for the spatial variability shown in digitized images, an area identified as deficient in the photograph was hand harvested. In the photograph, the selected location appeared to have a recurring pattern in the severity of deficiency at 16 row intervals. Field sampling showed that both grain yield and stalk nitrate-N concentration at harvest demonstrated a somewhat similar pattern (Fig. 6). The N fertilization application was made with an eight row applicator. The applicator had 9 knives but the distribution manifold had eight outlets. Fertilizer from one port was split between the two outside knives. Direction of travel during application was reversed so any variation in application rate is expected to be a mirror image of the adjacent eight rows. The pattern could be somewhat confounded by planting and other field operations which were performed with 12-row equipment.

Conclusion

Aerial photographs make it possible to identify atypical or possibly N deficient areas within a field. This study showed considerable variability associated with yield response to N

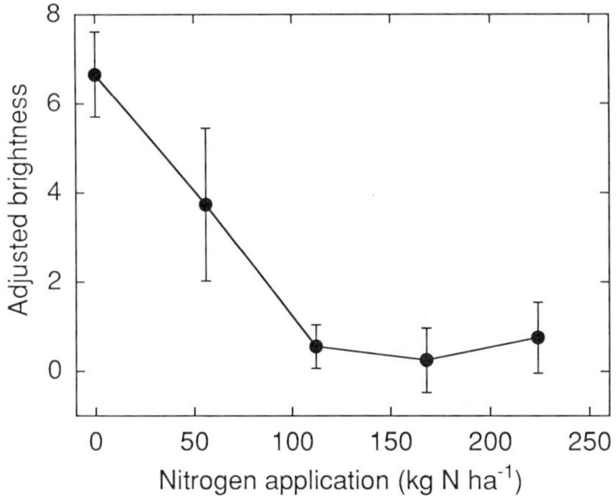

Fig. 5: Difference in red brightness between an adequately fertilized reference area and plots receiving various rates of N fertilizer.

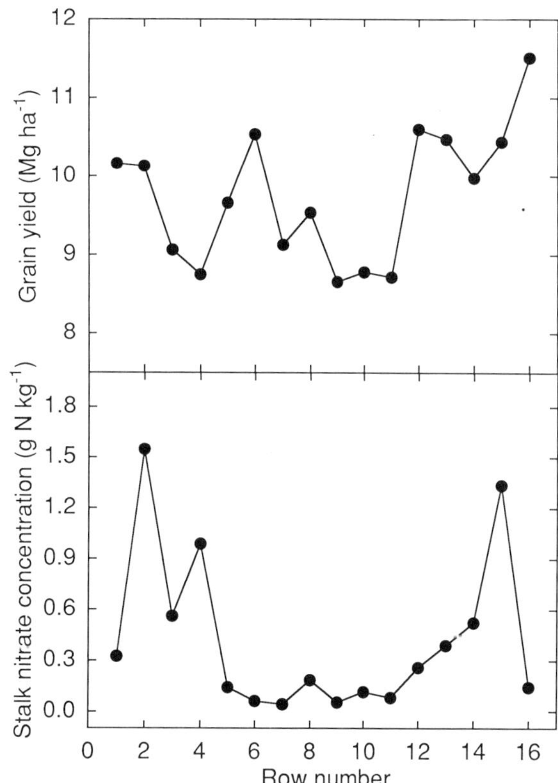

Fig. 6: Stalk nitrate-N concentration and grain yield from consecutive corn rows in a portion of the field identified as N deficient in the photograph.

fertilizer treatments which complicated photographic verification. Stalk nitrate-N data helped confirm the photographic capabilities of identifying N deficient areas. Chlorophyll meter readings taken at the R3 growth stage were poorly correlated with grain yield in this study. Photographic data collected at the R4 growth stage were more highly correlated with grain yield than chlorophyll meter data. These findings may be the result of more intensive sampling and evaluation permitted with the photographic technique. Overall, the photograph provided the best indication of crop N deficiency and permitted evaluation of the entire field at relatively little cost.

References

Binford, G. D., A. M. Blackmer, and B. G. Meese: Optimal concentration of nitrate in cornstalks at maturity. Agron. J. *84,* 881–887 (1992).

Binford, G. D., A. M. Blackmer, and N. M. El-Hout: Tissue test for excess nitrogen during corn production. Agron. J. *82,* 124–129 (1990).

Blackmer, T. M.: Remote sensing techniques to detect nitrogen stress in corn. Ph. D. Dissertation. University of Nebraska, Lincoln (1995).

Blackmer, T. M. and J. S. Schepers: Use of a chlorophyll meter to monitor N status and schedule fertigation of corn. J. Prod. Agric. *8,* 56–60 (1995).

Blackmer, T. M., J. S. Schepers, G. E. Varvel, and E. A. Walter-Shea: Nitrogen deficiency detection capabilities of reflected shortwave radiation from irrigated corn canopies. Agron. J. (In Press) (1995).

Colwell, R. N.: Determining the prevalence of certain cereal crop diseases by means of aerial photography. Hilgardia *26,* 223–86 (1956).

Jackson, R. D.: Remote sensing of biotic and abiotic plant stress. Ann. Rev. Phytopathol. *24,* 265–87 (1986).

Ritchie, S. W., J. J. Hanway, and G. O. Benson: How a corn plant develops. Special Report No. 48 revised. Iowa State Univ., Coop. Ext. Serv., Ames, IA.

Piekielek, W. P. and R. H. Fox: Use of a chlorophyll meter to predict sidedress nitrogen requirements for maize. Agron. J. *84,* 59–65 (1992).

Schepers, J. S., D. D. Francis, M. Vigil, and F. E. Below: Comparison of corn leaf nitrogen concentration and chlorophyll meter readings. Comm. Soil Plant Anal. *23,* 2173–2187 (1992).

Walburg, G., M. E. Bauer, C. S. T. Daughtry, and T. L. Housley: Effects of nitrogen nutrition on the growth, yield, and reflectance characteristics of corn canopies. Agron. J. *74,* 677–683 (1982).

Wildman, W. E.: Detection and management of soil, irrigation, and drainage problems, p. 387–401. In: Johannsen, C. J. and J. L. Sanders (eds.): Remote Sensing for Resource Management. Soil Conserv. Soc. Am., Ankeny, Iowa (1982).

Wood, C. W., D. W. Reeves, R. R. Duffield, and K. L. Edmisten: Field chlorophyll measurements for evaluation of corn nitrogen status. J. Plant Nutr. *15,* 487–500 (1992).

J. Plant Physiol. Vol. 148. pp. 445–455 (1996)

Stress and Stress Management of Land Plants During a Regular Day

ALAKA SRIVASTAVA and RETO J. STRASSER

Laboratory of Bioenergetics, University of Geneva, Jussy, CH-1254, Geneva, Switzerland

Received June 27, 1995 · Accepted September 18, 1995

Summary

In vivo photochemical activity of photosystem II (PSII) was measured by Chlorophyll *a* fluorescence intensity of pea leaves exposed to different types of stresses of daily time course. Pea plants grown at 22 °C show a dramatical decrease in the activity of PSII when exposed to higher temperatures. High light exacerbates the damage of PSII when imposed together with high temperature. But the low light acts as an efficient protector of the photochemical activity against its inactivation by heat. The thermal tolerance of PSII was also triggered by exposing the leaves to moderately elevated temperature and low light (30 °C and 30 Wm^{-2}) before exposing them to higher temperatures. The increase in thermo-tolerance was more than 5 to 10 °C. Pre-heat treatment causes an upward shift in the photoinhibitory light. The stability of PSII against heat was increased strongly in water stressed leaf discs. The experimental results show the existence of an adaptive mechanism of land plants to protect themselves against heat and strong light, the usual changes during the diurnal cycle of the day. A comparable behavior has not been found in unicellular organisms. Our results demonstrate the existence of an antagonism between different stresses (heat, light, water deficit) which are associated to the time course of the day.

Key words: Chlorophyll a fluorescence, heat, light, water-stress, daily time course.

Abbreviations: Chl *a* = chlorophyll *a;* F_o^d and F_m^d = minimal and maximal Chl *a* fluorescence yield in dark adapted leaf; F_s^l = steady state Chl *a* florescence in light; F_m^l and F_o^l = maximal and minimal Chl *a* fluorescence yield under steady state conditions; I_L, I_C, I_H and I_S = low, control, high and stressed light intensities respectively; Mod L = modulated light; AL = actinic light; SL = saturating light; PS (I,II) = photosystem (I, II); ΦPo = maximum quantum yield efficiency of PSII in steady state conditions ($\Phi Po = 1-(F_o^l/F_m^l)$); ΦP = actual quantum yield efficiency of PSII in steady state conditions ($\Phi P = 1-(F_s^l/F_m^l)$); RC = reaction center; T_L, T_C, T_H, T_S = low, control, high and stressed temperature respectively; V_s = relative variable Chl *a* fluorescence in steady state conditions under any given light intensity ($V_s = (F_s^l-F_o^l)/(F_m^l-F_o^l)$); W_s = water stress.

Introduction

The productivity of a plant is directly related to the rate of photosynthetic carbon assimilation which, in turn, is determined by a complex interplay between the plants photosynthetic apparatus and the environment. Although the nature of the photosynthetic machinery within the plants will determine the maximum capacity for the photosynthesis, the extent to which capacity is achieved depends on the external environmental parameters. During the time course of the day, plants must cope with changes in heat (Lu et al., 1994), water deficit and several orders of magnitude in the quantity of light (Pearcy, 1990; Srivastava and Zeiger, 1995). Light is the essential prerequisite for the plant life, but excessive light can photoinhibit photosynthesis and may lead to photooxidative destruction of the photosynthetic apparatus. Plants have adapted to tolerate the variations in the light quantity and quality *via* several dissipative energy processes (Anderson et. al, 1988; Falkowski and LaRoche, 1991; Adams and Adams, 1992; Long et al., 1994; Dau, 1994; Schindler and Lichtenthaler, 1994).

When plants are exposed to temperature above or below the normal physiological range, photosynthesis gets inactivated. The PSII complex is the most susceptible to heat among various components of the photosynthetic apparatus (Berry and Björkman, 1980; Mamedov et al., 1993). It is believed that increasing temperature leads first to a blockage of PSII reaction centers and then to a dissociation of the antenna pigment protein complexes from the central core of the PSII light harvesting complex (Armond et.al., 1980; Gounaris et al., 1984). The oxygen evolving process is also very heat sensitive (Mamedov et.al., 1993). Nash et al. (1985) demonstrated that the heat inactivation of oxygen evolution is caused by the release of functional manganese ions from the PSII complex. Therefore, the heat stability of oxygen evolution should determine the overall heat tolerance of the photosynthetic process.

The heat stability of PSII has been reported to vary substantially due to the influence of various environmental factors. In general, bright light strongly magnify the heat damage of PSII, as it does also for injuries from other physicochemical stresses such as chilling, freezing or drought (for reviews, see Kyle et al., 1987). This effect has been attributed to the decreased rate of photosynthesis caused by these various stresses, which predisposes stressed leaves to photoinhibition. On the other hand low light has been shown to markedly reduce damage to PSII during heat stress (Havaux and Strasser, 1990; Havaux et al., 1991). The heat protection by light depends on the intensity and spectral characteristics of the light (Havaux and Strasser, 1990, 1992). Other physico-chemical stresses such as high salinity (Larcher et al., 1990), hypertonic stress (Kaiser, 1984) or leaf water potential (Seemann et al., 1986; Havaux, 1992) have also been reported to increase heat tolerance of PSII in intact leaves. These observations suggest the existence of antagonistic interactions between environmental stresses, with one stress enhancing the tolerance of photosynthesis towards another.

During the time course of the day heat stress is often combined with strong light and water deficit. In the present work, we have examined the effect of combinations of different stress factors on PSII activity of the plants, by measuring Chl *a* fluorescence as a signal. Chlorophyll *a* fluorescence signal has been used to probe the fate of excitation energy within the photosynthetic apparatus and to provide insight into the mechanism and regulation of photosynthesis *in vivo* (Papageorgiou, 1975; Govindjee et al., 1986; Lichtenthaler and Rinderle, 1988; Lichtenthaler, 1988, 1992). Our results show that the phenomenon of heat protection by low light is an unique feature of the land plants. Training the plants with moderate temperature can also protect the plants from strong light and high temperature. Furthermore, leaf desiccation also enhances the resistance of PSII to high-temperature stress.

Materials and Methods

Plant materials

The land plants (*Pisum sativum, Lycopersicum esculentum, Camellia* sp.) were grown in a glass house (day/night temperature approximately 22/18 °C) under natural sunlight conditions. The swimming plants (*Salvinia natans* and *Pistia stratiotes*) and submerged plants (*Elodea* sp. and *Potamogeton* sp.) were obtained from the Botanical Garden of the University of Geneva, Geneva. Unicellular organisms (*Chlorella* sp., *Chlorella rubrum* and *Scenedesmus* sp. in modified Detmer medium (Pringsheim, 1951) as described by Calderon (1991)) and *Chlamydomonas reinhardtii* (in tris-acetate phosphate medium (Gorman and Levine, 1965)) were grown in liquid medium with constant shaking in a growth chamber at 25 °C and illuminated for 12 hours by a fluorescent lamp (white light).

Stress Treatment

To apply thermal stress, leaf discs were placed on a moist filter paper on a metal block. The temperature of the block was modified by circulating water from a thermostated water bath (Colora Meßtechnik, Lorch/Württ., Germany). Temperature was monitored with a Digi-Sense 8528-20 thermocouple thermometer (Cole-Parmer Instruments, Chicago, IL, USA) stuck on the lower surface of the leaf. Submerged plants were heated in distilled water and the unicellular organisms were heat stressed in their respective medium in DW-1 Hansatech cuvette (Hansatech Ltd., King's Lynn, Norfolk, England). Blue light (through broad band blue light, 400 to 600 nm, Corning filter, CS 4-96)) or the white light was supplied by a halogen lamp (KL1500 light source; Schott, Mainz, Germany) via fiber optics.

For water stress the leaf samples were dehydrated in air by placing them on a filter paper in an open Petridish in the dark at room temperature; control leaf samples were kept on moist filter paper under the same conditions.

Chlorophyll a Fluorescence Measurement
Fast Fluorescence Kinetics

After each treatment, the leaf discs were dark adapted for 20 minutes at room temperature. The fast Chl *a* fluorescence induction curve was monitored by a Plant Efficiency Analyzer (PEA, manufactured by Hansatech Ltd., King's Lynn, Norfolk, England) with 600 Wm^{-2} of light intensity. Illumination was provided by an array of 6 light emitting diodes (peak, 650 nm), focused on the sample surface to provide a homogeneously illuminated spot of about 4 mm in diameter as described in detail by Strasser et al. (1995).

Slow Fluorescence Kinetics in Steady State

The slow Chl *a* fluorescence emission was measured at room temperature at 690 nm by using modulated Chl *a* fluorescence system (Hansatech Ltd.). Chl *a* fluorescence was excited with modulated yellow light (maximum transmittance at 585 nm; modulated frequency, 4.8 KHz). A typical example of different levels of modulated Chl *a* fluorescence signals are shown in Fig. 1. The yellow modulated beam (Mod L) was turned on and the low fluorescence level was measured when it reached a constant value and is referred to as F_o^d (d stands for dark adapted leaf). Then, a short pulse (1 s) of saturating blue light, SL (Schott 1500 lamp with a Corning filter, CS 4-96 and a built in shutter system) was given to measure the maximum level of modulated Chl *a* fluorescence (F_m^d) in the previously dark adapted sample. The variable Chl *a* fluorescence was induced by an actinic continuous blue light (AL) which was adjusted from 0–280 Wm^{-2} using an electronically controlled light source (Schott, KL1500 Electronic lamp with a corning filter, CS 4-96). After few minutes, the Chl *a* fluorescence yield reached a stable and low steady state level (F_s^l). The maximum fluorescence level (F_m^l, where l refers to the fact that the samples had been under continuous illumination with actinic light) at any steady state condition was obtained by giving a short pulse of saturating light (SL). Immedi-

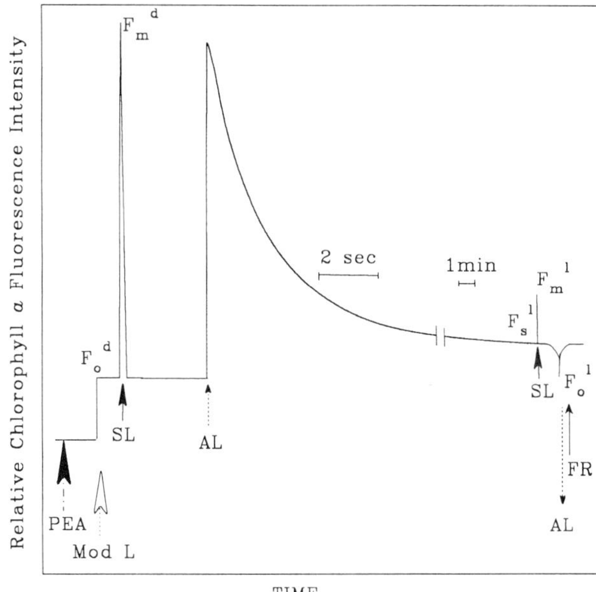

Fig. 1: A typical experimental protocol used to measure different Chl *a* fluorescence parameters of a pea leaf disc. After dark adaptation, the leaf disc was first excited with Plant Efficiency Analyzer (PEA) for 1 s to measure the fast Chl *a* fluorescence induction curve. Later on, the sample was excited with the measuring beam of a weak modulated light, Mod L to obtain minimal level of fluorescence (F_o^d). During a 1 sec pulse of saturating light, SL, (300 Wm^{-2}), the maximum fluorescence emission (F_m^d) was measured. The samples were then irradiated with actinic light (AL). After a prolonged illumination with AL, the Chl *a* fluorescence reached a low steady state level (F_s^l). The maximal (F_m^l) level in the steady state condition was obtained by applying a pulse of saturating light, SL. The minimal (F_o^l) fluorescence level was obtained by simultaneous switching off AL and turning on the far red (FR) light.

ately after that the F_o level (F_o^l) was measured by simultaneous turning off the actinic (AL) and turning on the far red (FR) light.

Results

A Common Optimum Behavior of the Plant

It has been reported earlier (Srivastava and Strasser, 1995 a) that every sample exhibits an optimum curve for the yield of steady state Chl *a* fluorescence versus light intensity. As shown in Fig. 2, with increasing light intensity the fraction of closed reaction centers of PSII (V) which is probed by the normalized relative variable fluorescence ($V_s = (F_s^l - F_o^l)/(F_m^l - F_o^l)$) increased. However, the maximum quantum efficiency ($\Phi Po = 1-(F_o^l/F_m^l)$) as well as the actual quantum efficiency ($\Phi P = 1-(F_s^l/F_m^l)$) decreased (Fig. 2). The flux of energy trapped per unit time which is a measure for the electron transport ET through PSII ($ET = \Phi P^*I$), versus light intensity shows an optimal curve. The maximum yield of the ET can be expected from these samples by exposing them to 80–120 Wm^{-2}. This value decreases by further increasing the light intensity or, under the same light intensity but by changing other environmental factors e.g. temperature or water stress etc. Any type of variations in the environmental conditions drive the system into suboptimality and, according to the stress concept of Strasser (1988) can provoke changes in the optimum behavior of the sample which creates a driving force for a state change, which is sensed by the system as a stress.

Antagonistic and synergistic behavior of light on heat stressed leaves

In vivo modulated Chl *a* fluorescence was measured separately on pea leaves which were exposed to different temperatures (Fig. 3). A typical effect of heat on Chl *a* signals is presented in Fig. 3. As previously reported (Havaux and Strasser, 1990, 1992), heat stress in dark has a marked effect on the PSII of pea leaf, causing a drastic loss of the variable Chl *a* fluorescence. During the heat stress in the dark the F_m^l was sharply decreased whereas the F_o^l was not affected. The fluorescence ratios which are the estimate of the maximum quantum yield (ΦPo), actual quantum yield (ΦP) and photochemical quenching qP = (1-V) also decreased at higher temperatures (Fig. 4). It has been shown earlier that PSII is more sensitive to heat stress in comparison to PSI (Berry and Björkman, 1980; Quinn and Williams, 1985). An early symptom of the heat injury is the loss of O_2 evolution activity and has been correlated with the deactivation of PSII due to denaturation of certain functional proteins (Thompson et al., 1989), dissociation of light harvesting pigments and the PSII complex (Armond et al., 1978, 1980; Gounaris et al., 1984; Sundby et al., 1986), and release of the functional Mn from the PSII (Nash et al., 1985).

But when the leaves were irradiated with very low light (30 Wm^{-2}) during the heat treatment, no changes in F_m^l up

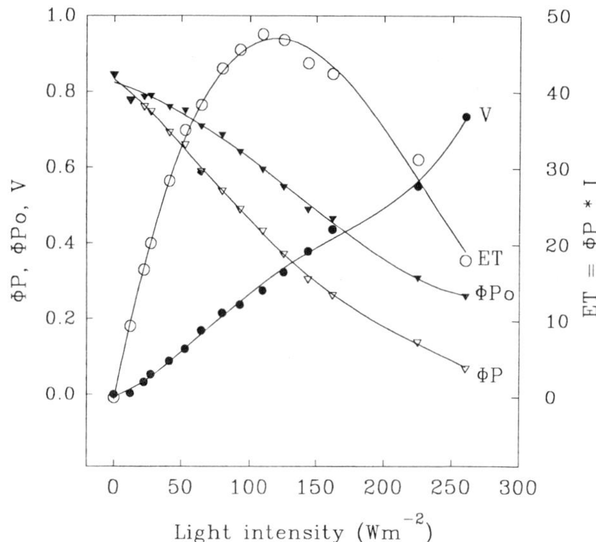

Fig. 2: Light intensity dependence of actual quantum efficiency ($\Phi P = 1-(F_s^l/F_m^l)$), maximum quantum efficiency ($\Phi Po = 1-(F_o^l/F_m^l)$), the fraction of closed reaction centers ($V_s = (F_s^l-F_o^l)/(F_m^l-F_o^l)$) and energy flux being trapped by the RC which allows the electron transport ($ET = \Phi P^*I$, where I is the light intensity) of a pea leaf. Experimental details are the same as in Fig. 1.

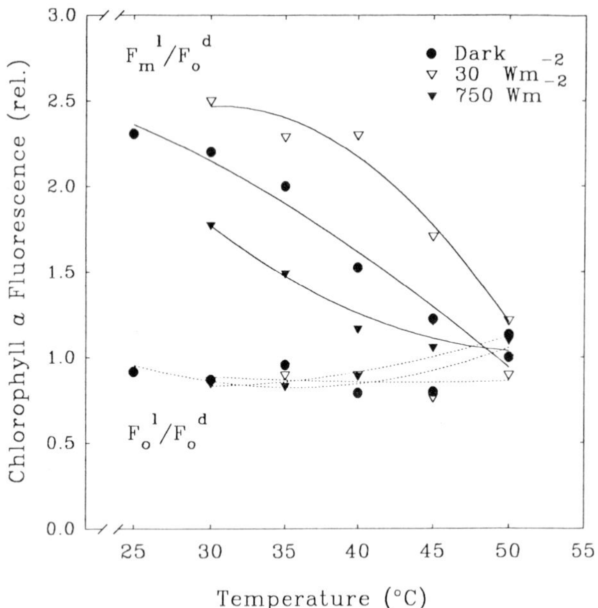

Fig. 3: Effect of low (30 Wm^{-2}) and strong (750 Wm^{-2}) light on temperature induced changes in the pea leaf discs on the relative levels of maximum (F_m^l/F_o^d) and minimum (F_o^l/F_o^d) fluorescence emissions in steady state conditions. Leaf discs were treated in different conditions for 30 min. Fluorescence measurements were performed at room temperature by irradiating them with 100 Wm^{-2} blue light as an actinic light. Other experimental details are the same as in Fig. 1.

to 40 °C was observed. At 45 °C also the F_m^l level was much less affected in comparison to the leaf discs treated at the same temperature in the dark (Fig. 3). Insignificant differences in the ΦPo and ΦP were observed when the leaves were heated up to 40 °C in presence of low light (Fig. 4). This clearly shows the antagonistic effect of low light on heat stressed leaves. A similar type of result was reported earlier (Havaux and Strasser, 1990; Havaux et al., 1991; Srivastava and Strasser, 1995 b). The mechanism by which the low light protects PSII is still unclear. However, Weis (1982 a, b) has suggested the importance of proton and metal cation concentrations of the medium for the heat stability of the isolated chloroplasts. It is possible that light stabilizes the thylakoid membranes by creating a proton gradient and by maintaining ionic equilibrium. Al-Khatib and Wiest (1990) showed that elevated temperature induces conformational changes in the PSII reaction centers resulting in a modification of lipid protein interaction which is believed to play a crucial role in the maintenance of the supra-molecular organization of the photosystem (Webb and Green, 1991). The modified lipid composition of thylakoid membranes can change the PSII thermotolerance (Thomas et al., 1986; Hugly et al., 1989). The heat induced alterations in the PSII reactions were considerably alleviated by low light indicating that light efficiently protected the photochemical apparatus and photosynthesis against heat inactivation.

In contrast to low light when the leaves were heated in presence of high light intensity (750 Wm^{-2}) a sharp decrease in the F_m^l was observed (Fig. 3). The temperature dependence curves of ΦP and ΦPo also shifted towards the lower temperatures. For instance, at 40 °C, ΦPo and ΦP were reduced by 56 % and 75 % when the leaf discs were exposed to 750 Wm^{-2} light intensity in comparison to 21 % and 46 % respectively in leaf discs treated in the dark. These results confirm that heat pre-disposes leaves to photoinhibition damage (Ludlow, 1987; Havaux, 1994). Higher temperature and strong light have been shown to interact at the level of the thylakoid membranes (Mishra and Singhal, 1993).

As it has been reported earlier (Neubauer and Schreiber, 1987; Strasser et al., 1995) that the data obtained during the fast Chl *a* fluorescence induction kinetics give significant information about the physiological state of the sample, we have measured the fast fluorescence rise starting from 40 µs up to minutes.

Fig. 5 shows the experimental results on a leaf disc which were cold, or heat treated for 30 min in the dark and then Chl *a* fluorescence induction kinetics was measured for 1 min with a light intensity of 600 Wm^{-2}. The samples were illuminated again after different times of dark adaptation at room temperature. The fluorescence transient starts from an initial Fo intensity and increases to a maximum peak P through two

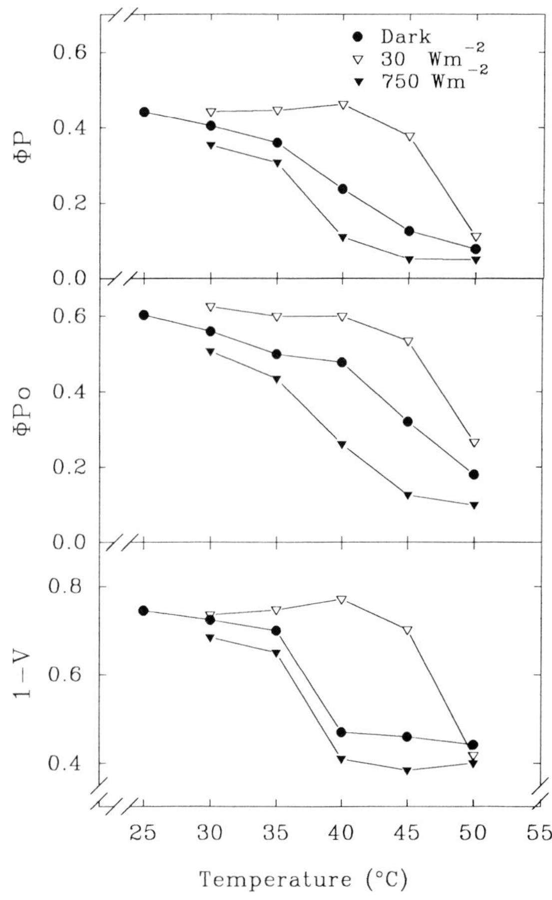

Fig. 4: Effect of low (30 Wm^{-2}) and strong (750 Wm^{-2}) light on temperature induced changes in the ΦP, ΦPo and 1-V. Experimental details are the same as described in Fig. 3.

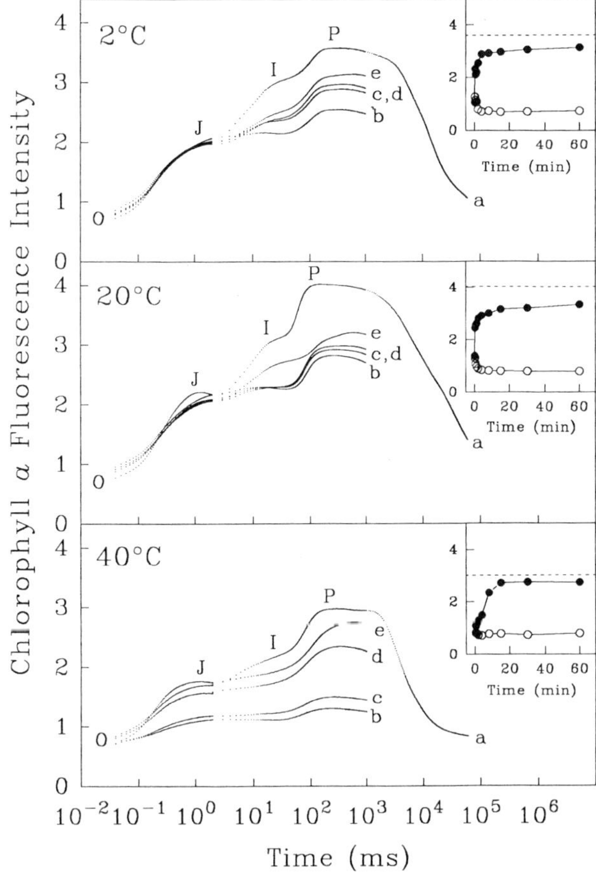

Fig. 5: Effect of dark adaptation time on the Chl *a* fluorescence induction kinetics of the pea leaf discs which were kept at 2 °C, 20 °C or 40 °C in a water bath in the dark for 30 min. Immediately after that Chl *a* fluorescence transients were measured by PEA using 600 Wm^{-2} red actinic light for 1 min (a). Samples were dark adapted for 2 (b), 4 (c), 8 (d) and 60 (e) min and then illuminated further for 1 s with 600 Wm^{-2} red light. Inserts show the recovery in the dark (up to 60 min) of Fo (open circles) and Fp (closed circles). Dotted lines show the Fp level just after the temperature treatment (P of curves a).

intermediate steps between Fo and Fp labelled as J and I (Fig. 5). The first rise from O to J occurs within 2 ms. The second intermediate step I levels in about 20 ms and the final peak in about 200 ms (for more detail see Strasser et. al., 1995). This is followed by a terminal steady state which is very close to Fo. Cooling or heating of the leaf disc in total darkness resulted in a decrease in the variable Chl *a* fluorescence (Fp-Fo). The recovery was very rapid in the samples exposed to cold or at room temperature, with a measured half time t½ of about less than a minute. In contrast, the leaf discs heated at 40 °C recovered very slowly with the t½ of more than 5 min.

Adaptation of PSII thermal tolerance in pea leaves after pre-heat treatment

Figure 6 shows the Chl *a* fluorescence transient of the pea leaf discs heated in dark either directly at 42 °C for 30 min (curve d) or pre-heated with mild temperature and low light (30 °C and 30 Wm^{-2}, for 30 min) before exposing them further for 30 min at 42 °C (curve c). Leaf discs heated with 30 °C in presence of low light show a typical O-J-I-P fluorescence transient (curve b) like control leaf discs (curves a). But when the leaf discs were heated directly at 42 °C a well defined K peak can be seen. As reported earlier (Guisse et al., 1995), after the heat treatment of the leaf the typical O-J-I-P Chl *a* fluorescence transient transformed into O-K-J-I-P transient with a new peak "K" at about 200–400 µs. Several field measurements have shown the appearance of K peak in stressed plants (unpublished results, R. J. Strasser). The K peak becomes the dominant peak among all the steps of Chl *a* fluorescence induction kinetics when leaf discs were treated with higher temperatures. The origin of the K peak has been proposed to be due to the inhibition of the water splitting system and a partial inhibition of the electron transport flow on the acceptor side before Q_A, the primary electron acceptor of PSII. About 67% decrease in variable fluorescence was observed in leaves heated at 42 °C for 30 min in comparison to the control leaves which were kept at room temperature (Fig. 6). If the leaf discs were exposed to 42 °C after a mild preheat treatment (30 °C and 30 Wm^{-2} for 30 min) the variable Chl *a* fluorescence was less reduced and the K peak was also less pronounced. The changes in the amplitude in the K peak after heating the leaf discs at different temperature (1) or for different time at 42 °C (2) are shown in the insert of Fig. 6. It is apparent from the figure that pre-heat adaptation shifts the appearance of K peak towards higher temperature or longer duration of heat treatment. The data obtained by the modulated Chl *a* fluorescence measurement in the steady state conditions also indicate the protective role of pre-heat adaptation on heat stress as it can be seen by the Chl *a* fluorescence ratios, ΦP and ΦPo (Fig. 7). The exact mechanism by which PSII thermostability can rapidly increase in leaves exposed to mild heat-stress conditions remains to be elucidated. But one can speculate that the fast adaptive changes in PSII complex may be either due to conformational changes in PSII or changes in the surroundings of the chloroplastic membranes as suggested by various *in vivo* (Havaux, 1993, 1994) and *in vitro* (Weis, 1982a, b; Seemann et al., 1986) studies.

Adaptation of PSII Photoinhibition tolerance in pea leaves after pre-heat treatment

In contrast to synergistic effect of high light and high temperature (Figs. 3, 4), results presented in Fig. 7 and Fig. 8 show that pre-heat treatment (30 °C together with 30 Wm^{-2} for 30 min) results as well in the protection of the photosynthetic mechanism exposed to the high light. Data obtained from the Chl *a* fluorescence induction curve (Fig. 8) reveal that about 55 % decrease in the variable Chl *a* fluorescence was observed when the leaf discs were exposed directly to 750 Wm^{-2} (curve d) in comparison to 22 % when the leaf discs were exposed to 750 Wm^{-2} after pre-treatment (curve c). Although after the light stress a small bend in the range of K peak was observed, it was not as obvious as in the heat stressed leaf discs. However, it becomes pronounced at higher light intensities (insert in Fig. 8). Again a close correlation was observed between the changes in ΦPo and ΦP during light stress and pre-heat treated leaves (Fig. 7). From these results, one can draw the conclusion that a moderate and brief elevation of leaf temperature markedly stabilizes PSII against photoinhibition. The exact mechanism of heat induced photoprotection of PSII is not known. However, temperature is known to exert a strong influence on the physical properties of the lipid composition (Van der Meer, 1993; Vigh, 1993). Considering the role played by lipid protein interaction in the maintenance of photosystems (Webb and Green, 1991), one can speculate that pre-heat adaptation changes the conformation of PSII which some how result in photoprotection of PSII reaction centers against strong light. Stapel et.al. (1993) observed higher resistance of barley against photoinhibition after pre-heat treatment and they have correlated the results with the accumulation of heat shock proteins which protect the plants from photoinhibition.

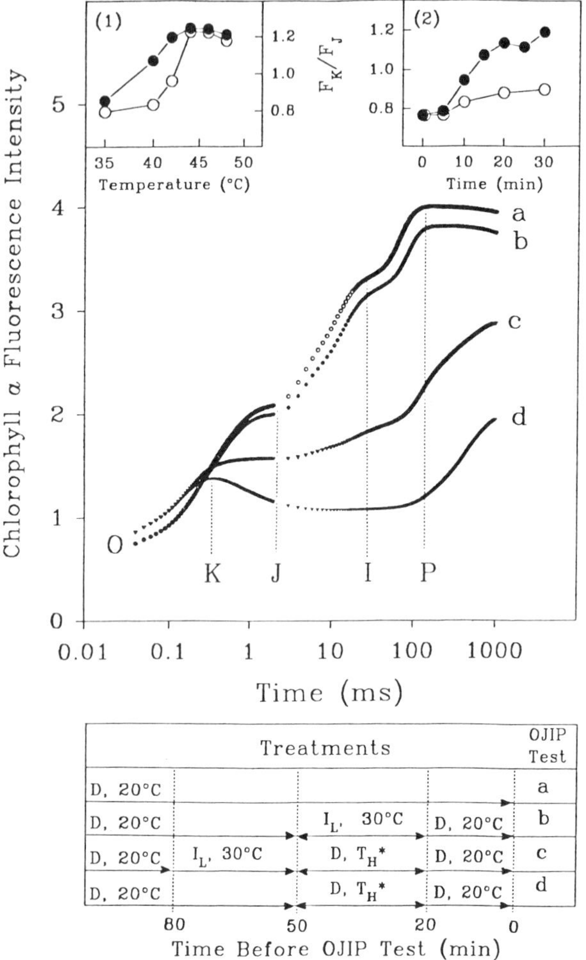

Fig. 6: Effect of pre-treatment on the Chl *a* fluorescence induction curve of pea leaves. The experimental detail of each curve is indicated in the lower panel of the figure. *Representative curves were obtained from the leaf disc treated with 42 °C (T_H), are shown in the figure. I_L is 30 Wm^{-2}. Insert shows appearance of the K peak, expressed as a ratio of F_K/F_J after 30 min exposure to different temperature (1) or for different time duration at 42 °C (2). Leaf discs were either exposed directly (solid circles) or after 30 min of pre-treatment at 30 °C with 30 Wm^{-2} blue light (hollow circles).

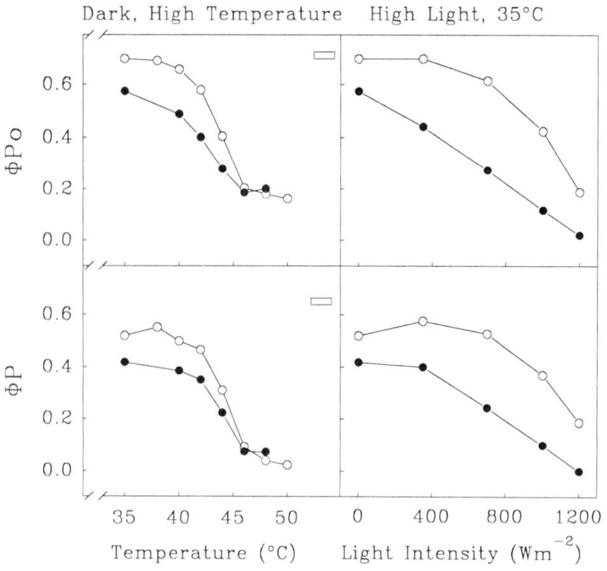

Fig. 7: Effect of pre-treatment (30 °C and 30 Wm^{-2} for 30 min) of pea leaf discs on temperature and light induced changes in the ΦPo and ΦP. Leaf discs were exposed to different temperature and light directly (solid circles) or after pre-treatment (hollow circles). Hollow rectangular in the left panel show the ΦPo (top) and ΦP (lower) values of the leaf discs just after the pre-treatment. All the experimental details are the same as in Fig. 1 but the actinic light intensity was 60 Wm^{-2} light.

Adaptation of PSII Heat Tolerance in Pea Leaves After Rapid Water Stress

Detached pea leaves were subjected to dehydration in air for several hours, resulting in a rapid decrease in the water content. For instance after 5 hours of dehydration in the dark, water content was decreased by 40 %. The effects of this treatment on the PSII function were monitored by measuring the characteristics of chlorophyll *a* fluorescence emission. Although the water content of the leaf discs were very low, the Chl *a* fluorescence induction kinetics (Fig. 9) and the photochemical efficiencies (ΦPo and ΦP, Fig. 10) were only marginally affected. It has been mentioned above that photochemistry sharply decreased after heat treatment. But interestingly water stress antagonizes the effect of heat stress on the leaf discs. In comparison to the wet leaves the water stressed leaf discs placed at 40 °C did not exhibit any changes in the photochemical efficiencies of the PSII (Fig. 10). In the water stressed leaves the heat induced K peak did not get pronounced (insert in Fig. 9) even after 30 min of exposure at 40 °C. Our data on the tolerance of PSII activity to high level of water deficit are in agreement with the previous papers (Sharkey, 1990; Havaux, 1992). Water stress may increase the stability of PSII to heat by strengthening the interaction between PSII proteins and their lipid environment (Ferrari-Iliou et. al, 1984; Prabha et al., 1985).

An Evolutionary Trend of the Protection Against Heat Stress by low Light

In extending the study for the adaptation of the land plants to every day changes in light and temperature, we have measured the low light induced heat protection mechanism in different plants ranging from aquatic unicellular organisms to terrestrial plants. The aquatic plants have several advantages over their terrestrial counterparts. A major advantage of the water plant is that the variations in temperature are not so high as on the land. But free floating members of the aquatic flora are on the surface of the water and most of their parts are exposed to light and heat like the land plants. Table 1 shows that all the land plants have the capacity to protect themselves against heat stress in the presence of low light. The free floating water-plants also showed the same response like the land plants. But the submerged water-plants acted differently. Some of them were able to protect themselves with low light against heat stress but this phenomenon was totally absent in others. The variation in response could be due to the depth of the water tank. In contrast to all these

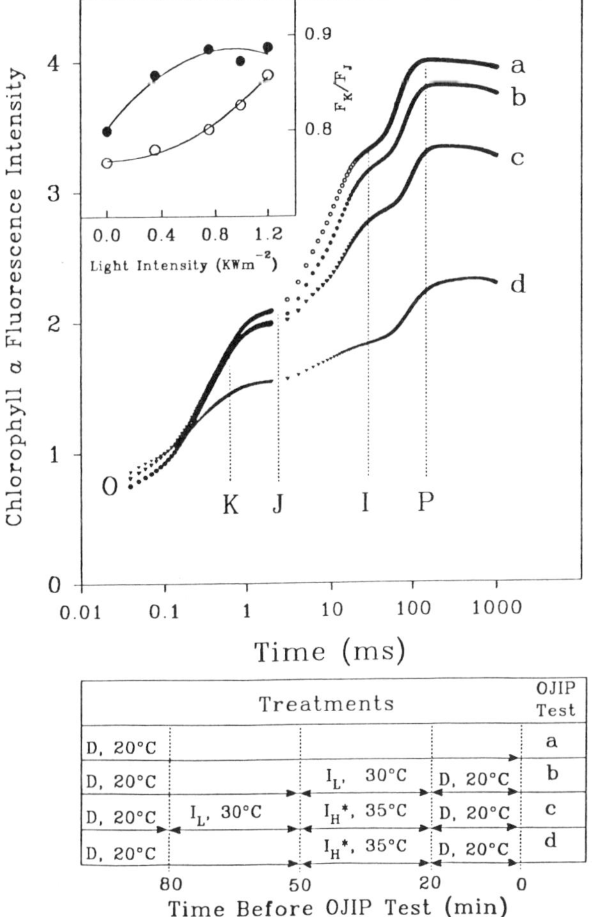

Fig. 8: Effect of pre-treatment on the changes in the Chl *a* fluorescence induction curve of a pea leaf due to high light intensity. All the experimental details about each curve are mentioned below the figure. *Representative curves were obtained from the leaf disc exposed to $750\,Wm^{-2}$ (I_H), are shown in the figure. Low light (I_L) is $30\,Wm^{-2}$. The insert shows the ratio of F_K/F_J after 30 min exposure to different light intensities at 35 °C of pea leaf discs. Leaf discs were either exposed directly (solid circles) or after pre-exposure to 30 °C and $30\,Wm^{-2}$ blue light for 30 min (hollow circles).

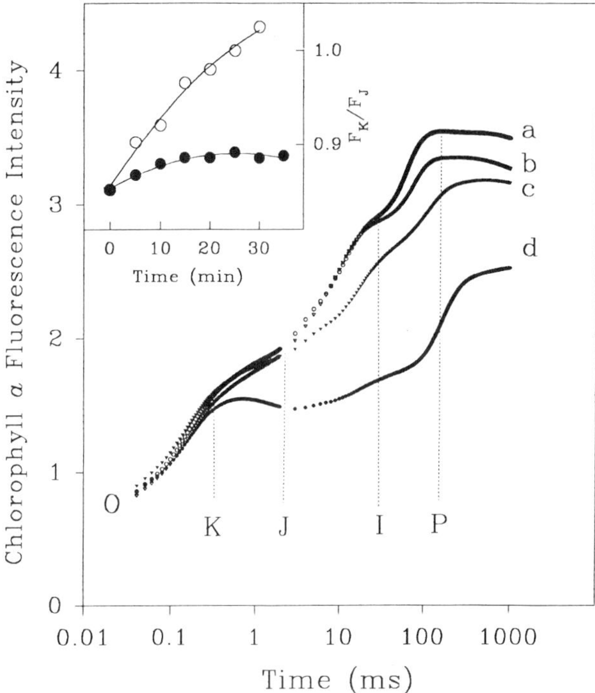

Fig. 9: Effect of temperature on the Chl *a* fluorescence transient of wet and 5 hours dehydrated leaf discs. The curve are: (a) wet, (b) 5 hours dehydrated, (c) 5 hours dehydrated and then exposed to 40 °C for 30 min, (d) directly exposed to 40 °C for 30 min. Insert figure shows the effect of heat treatment (40 °C) for different time duration on the F_K/F_J ratio of dehydrated (solid circles) and wet (hollow circles) pea leaf discs.

terrestrial and aquatic plants, the unicellular organisms totally lack this behavior. Plant evolution had to tread a path between maximizing the light interception for photosynthesis and minimizing the potential for damage arising from the interaction with other environmental factors. In this evolutionary process the terrestrial plants have developed this unique feature for protecting themselves from heat stress. This protecting mechanism remained or got lost when some land plants found their new inhabitant submerged in the water.

Discussion

The present paper shows that, besides these long term processes, the most thermolabile component of the photosynthetic apparatus, PSII, can rapidly adjust the level of its resistance to temperature stress within minutes. Additionally, it was observed that slight changes in leaf temperature (e.g., 30 °C) were enough to trigger some thermal adaptation of PSII. A simplified model is presented in Fig. 11. A moderate high temperature shifts the tolerance towards higher light intensity. In the same way low light treatment shifts the tolerance of plants towards higher temperature. In a way, both factors are interlinked and each of them protect from the other up to a certain extent.

Irrespective of the exact molecular mechanism for the observed changes, this paper confirms the thermal plasticity of PSII *in vivo*. Clearly the reported changes in PSII indicate a rapid process for photosynthetic adaptation to daily temperature variation of the environment. During the time course of the day, the usual increase in temperature can be quite high and the variation can be as high as 20 °C (Lu et al., 1994). One can put the possible scenario during the course of the day as in Fig. 12. Under natural conditions light intensity

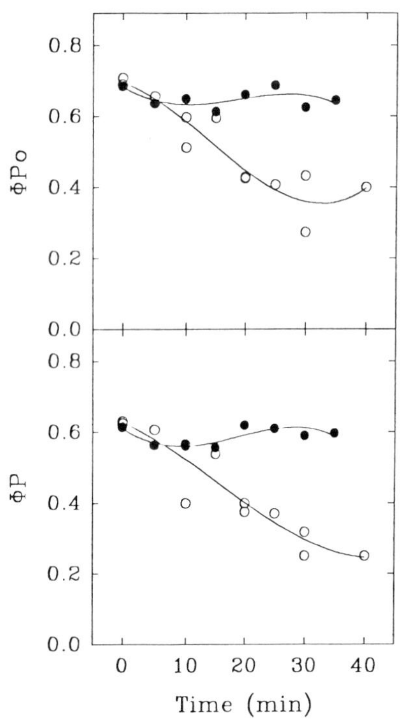

Fig. 10: Effect of heat treatment (40 °C) for different time duration on ΦPo and ΦP of wet (hollow circles) and dehydrated (closed circles) pea leaf discs.

Table 1: Heat protection mechanism by low light in different organisms. For experimental details, see the text.

Land Plants	
Pisum sativum	+ + + +
Lycopersicum esculentum	+ + + +
Camellia sp.	+ + + +
Swimming Plants	
Salvinia natans	+ + + +
Pistia stratiotes	+ + + +
Submerged Plants	
Elodea	+ + – –
Potamogeton sp.	+ – – –
Unicellular organisms	
Chlorella sp.	– – – –
Chlorella rubrum	– – – –
Chlamydomonas reinhardtii	– – – –
Scenedesmus	– – – –

+ Represents the relative heat protection by low light.
– indicates absence of heat protection mechanism by low light.

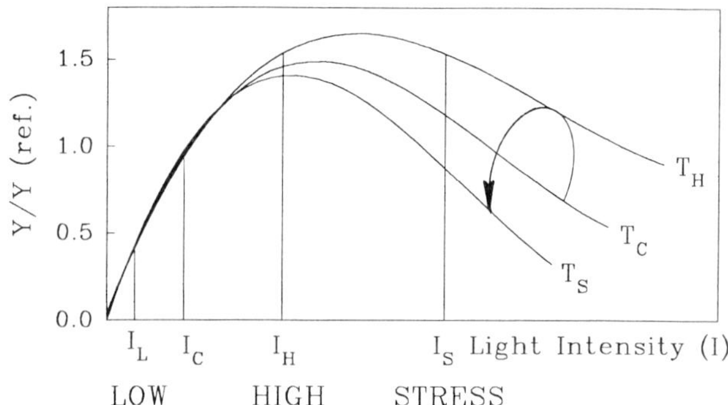

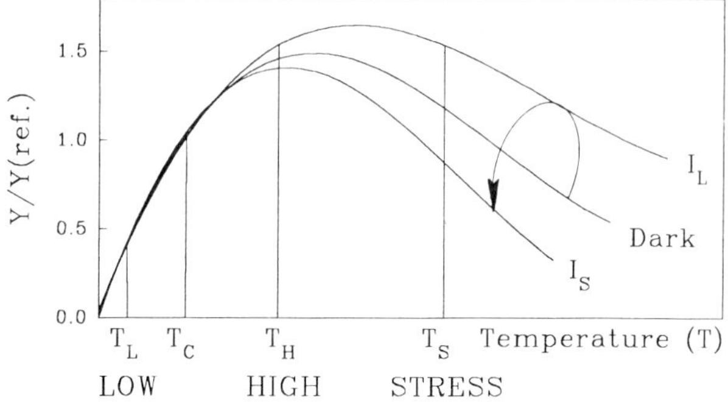

Fig. 11: A model indicating the interactions of different levels of light intensities (I_L, low intensity, I_C control, I_H high intensity, I_S stressed intensity) and temperatures (T_L, low temperature, T_C, control temperature, T_H, high temperature and T_S, stressed temperature) on photosynthetic systems.

and temperature increase steadily over a period of several hours before reaching the maximum. In the morning, plants acclimatize themselves by low temperature for high light, and this protects them during mid-day when the light is very intense. In early afternoon, when the temperature also starts to rise, some plants protect themselves by closing the stomata and by decreasing the rate of photosynthesis like mid-day depression (Ögren, 1988; Wise et al., 1991). In late afternoon, although the temperature is quite high, the plants can protect themselves due to the presence of low light.

In conclusion, the presented data illustrate the complexity of the photosynthetic responses to environmental stresses, with the effect of a given stress being markedly modulated by the other environmental factors. This study demonstrates the antagonism/synergism between the main variables of the day, i.e., temperature, light and water, and shows that a combination of heat and light elicit less injurious effects on the *in vivo* PSII function than heat stress alone. As a result, the *in vivo* PSII activity could be substantially more heat resistant in the field than previously estimated from the laboratory experiments. A mild treatment by low light or moderate temperature and moderate water loss induces an extension of the buffer capacity of the plants against stresses: e.g., light, heat, drought. Due to the temporary environmental changes the system shifts to a suboptimal state which creates the driving force for the state change leading to synergistic adaptation.

DEVELOPMENT OF STRESS TOLERANCE OVER DAILY TIME COURSE

A REAL ADAPTATION PHENOMENON

Environment				Plant System	
Day Time	Light	Temperature	Water	Stress condition	Protection Mechanism
↓	L_L	T_L		No	Induction of High Light Tolerance
	L_H	T_L		No	Being Protected for High Light
					Induction of High Temperature Tolerance
	L_S	T_S	W_S	Strong	Being Protected for High Light and High Temperature
					Maximum Synergistic Protection
	L_L	T_ε	W_3	No	High Temperature Protection also e.g., under the tree
	L_L	T_L		No	

Relaxation and Re-establishment of the State Suited for the Early Morning

Fig. 12: A model showing the changes in light intensity, temperature and water stress and the development of possible stress tolerance by the plants during the time course of the day.

The interaction between stressors suggests that the stress effects on the plants monitored under controlled environmental conditions in the laboratory, where the effects of a defined stress are studied in a "one factor-one response" test, might be quite different from the plant's responses in the field, where several factors usually change simultaneously and interact. The results also indicate that plants have adapted themselves, for their best, for any given environmental condition. In the stress concept by Strasser (1988) this situation has been called: the plant is in the HARMONY with its environment. The thermal protection by low light and by training the plants with pre-mild heat treatment suggests that the present terrestrial plants are well tuned for their natural inhabitant for every combinations of changes in the present daily cycle of the earth, for example heat light and water stress, as discussed here. Geiger and Servaites (1994) have reviewed distinct levels of physiological organizations of the photosynthetic apparatus which provide the flexibility of the plants to acclimate effectively to changing diurnal irradiance. The interaction of different mechanism are complex and provide flexibility for regulation. Out of several mechanisms is the self regulation, which integrate the complex interactions and thus refines regulations. Extreme environmental conditions or drastic manipulation may compromise the plants ability to acclimate. Changes in the global environment, such as increased environmental CO_2 and temperature fluctuations, which are product of human activities may drive regulatory mechanism to instability.

Further work will show how the daily and seasonal changes in the environmental conditions can modify the photosynthetic machinery by extension of their buffer capacity against stress to allow an optimum behavior, optimum growth and an optimum life cycle.

Acknowledgements

This work was supported by National Swiss Foundation grant (#31-33678-92) to RJS.

References

Adams, B. D. and W. W. Adams III: Photoprotection and other responses of plants to high light stress. Ann. Rev. Plant Physiol. Plant Mol. Biol. *43,* 599–626 (1992).

Al-Khatib, K. and S. C. Wiest: Heat-induced reversible and irreversible alterations in the structure of *Phaseolus vulgaris* thylakoid proteins. J. Therm. Biol. *15,* 239–244 (1990).

Anderson, J. M., W. S. Chow, and D. J. Goodchild: Thylakoid membrane organization in sun/shade acclimation. Aust. J. Plant Physiol. *15,* 11–26 (1988).

Armond, P. A., U. Schreiber, and O. Björkman: Photosynthetic acclimation to temperature in the desert shrub, *Larrea divaricata.* Plant Physiol. *61,* 411–415 (1978).

Armond, P. A., O. Björkman, and A. Staehelin: Dissociation of supramolecular complexes in chloroplast membranes. A manifestation of heat damage to the photosynthetic apparatus. Biochim. Biophys. Acta *601,* 433–442 (1980).

Berry, J. and O. Björkman: Photosynthetic response and adaptation to temperature in higher plants. Ann. Rev. Plant Physiol. *31,* 491–543 (1980).

Calderon, C.: Etude, en cultures synchrones autotrophes, mixotrophes et hétérotrophes, de l'évolution énergetique (P) de *Chlorella rubescens,* CHOD. Ph. D. Thesis, Faculte des Science, Université de Genève, Geneva, Switzerland (1991).

Dau, H.: Short-term adaptation of land plants to changing light intensities and its relation to photosystem II photochemistry and fluorescence emission. J. Photochem. Photobiol. *26,* 3–27 (1994).

Falkowski, P. G. and J. LaRoche: Adaptation to spectral irradiance in unicellular algae. J. Phycol. *27,* 8–14 (1991).

Ferrari-Iliou, R., A. T. Thi Pham, and J. Vieira de Silva: Effect of water stress on the lipid and fatty acid composition of cotton (*Gossypium hirsutum*) chloroplasts. Plant Physiol. *62,* 219–224 (1984).

Geiger, D. R. and J. C. Servaites: Diurnal regulation of photosynthetic carbon metabolism in C_3 plants. Ann. Rev. Plant Physiol. Plant Mol. Biol. *45,* 235–256 (1994).

Gorman, D. S. and R. P. Levine: Cytochrome f and plastocyanin: Their sequence in the photosynthetic electron transport chain of *Chlamydomonas reinhardtii.* Proc. Natl. Acad. Sci. USA *54,* 1665–1669 (1965).

Gounaris, K., A. P. R. Brain, P. J. Quinn, and W. P. Williams: Structural reorganization of chloroplast thylakoid membranes in response to heat stress. Biochim. Biophys. Acta *776,* 198–208 (1984).

Govindjee, J. Amez and D. C. Fork (eds.): Light Emission by Plants and Bacteria. Academic Press, Orlando, FL. (1986).

Guissé, B., A. Srivastava, and R. J. Strasser: High temperature induces a new rapid phase in the Chl *a* fluorescence induction kinetics of leaves. Archs Sci. Geneve *48,* 147–160 (1995).

Havaux, M. and R. J. Strasser: Protection of photosystem II by light in heat-stressed pea leaves. Z. Naturforsch. *45c,* 1133–1141 (1990).

Havaux, A., H. Greppin, and R. J. Strasser: Functioning of photosystem I and II in pea leaves exposed to heat stress in the presence and absence of light. Planta *186,* 88–98 (1991).

Havaux, M.: Stress tolerance of photosystem II *in vivo.* Plant Physiol. *100,* 424–432 (1992).

Havaux, M. and R. J. Strasser: Antagonistic effects of red and far-red lights on the stability of photosystem II in pea leaves exposed to heat. Photochem. Photobiol. *55,* 621–624 (1992).

Havaux, M.: Characterization of thermal damage to the photosynthetic electron transport system in potato leaves. Plant Sci. *94,* 19–33 (1993).

– Temperature-dependent modulation of the photoinhibition-sensitivity of photosystem II in *Solanum tuberosum* leaves. Plant Cell Physiol. *35,* 757–766 (1994).

Hugly, S. L., J. Kunst, J. Browse, and C. Somerville: Enhanced thermal tolerance of photosynthesis and altered chloroplast ultrastructure in mutant of *Arabidopsis* deficient in lipid desaturation. Plant Physiol. *90,* 1134–1142 (1989).

Kaiser, W. M.: Response of photosynthesis and dark-CO_2 fixation to light, CO_2 and temperature in leaf slices under osmotic stress. J. Exp. Bot. *35,* 1145–1155 (1984).

Kyle, D. J., C. B. Osmond, and C. J. Arntzen (eds.): Photoinhibition, Elsevier, Amsterdam (1987).

Larcher, W., J. Wagner, and A. Thammathaworn: Effects of superimposed temperature stress on *in vivo* chlorophyll fluorescence of *Vigna unguiculata* under saline stress. J. Plant Physiol. *136,* 92–102 (1990).

Long, S. P. S. Humphries, and P. G. Falkowski: Photoinhibition of photosynthesis in nature. Annu. Rev. Plant Physiol. Plant Mol. Biol. *45,* 633–662 (1994).

Lichtenthaler, H. K. and U. Rinderle: The role of chlorophyll fluorescence in detection of stress conditions in plants. CRS Crit. Rev. Anal. Chem. *19* (Suppl. I), S 29–S 85 (1988).

Lichtenthaler, H. K.: *In vivo* chlorophyll fluorescence as a tool for stress detection in plants. In: Lichtenthaler, H. K. (ed.): Applications of Chlorophyll Fluorescence, pp. 143–149. Kluwer Academic Publisher, Dordrecht (1988).

– The Kautsky effect: 60 years of chlorophyll fluorescence induction kinetics. Photosynthetica *27,* 45–55 (1992).

Lu, Z. M., J. W. Radin, E. L. Turcotte, R. Pearcy, and E. Zeiger: High yield in advanced lines of Pima cotton are associated with higher stomatal conductance, reduced leaf area and lower leaf temperature. Physiol. Plant. *92,* 266–272 (1994).

Ludlow, M. M.: Light stress at high temperature. In: Kyle, D. J., C. B. Osmond, and C. J. Arntzen (eds.): Photoinhibition. Elsevier, Amsterdam, pp. 89–109 (1987).

Mamedov, M., H. Hayashi, and N. Murata: Effects of glycinebetaine and unsaturation of membrane lipids on heat stability of photosynthetic electron-transport and phosphorylation reactions in *Synechocystis* PCC6803. Biochim. Biophys. Acta *1142,* 1–5 (1993).

Mishra, R. K. and G. S. Singhal: Photosynthetic activity and peroxidation of thylakoid lipids during photoinhibition and high temperature treatment of isolated wheat chloroplast. J. Plant Physiol. *141,* 286–292 (1993).

Nash, D., M. Miyao, and N. Murata: Heat inactivation of oxygen evolution in photosystem II particles and its acceleration by chloride depletion and exogenous manganese. Biochim. Biophys. Acta *807,* 127–133 (1985).

Neubauer, C. and U. Schreiber: The polyphasic rise of chlorophyll fluorescence upon onset of strong continuous illumination: 1. Saturation characteristic and partial control by the photosystem II acceptor side. Z. Naturforsch. *42c,* 1246–1254 (1987).

Papageorgiou, G.: Chlorophyll fluorescence: an intrinsic probe of photosynthesis. In: Govindjee (ed.): Bioenergetics of Photosynthesis. Academic Press, New York, pp. 319–371 (1975).

Pearcy, R. W.: Sunflecks and photosynthesis in plant canopies. Annu. Rev. Plant Physiol. Plant Mol. Biol. *41,* 421–453 (1990).

Prabha, C., Y. K. Arora, and D. S. Wagle: Phospholipids of wheat chloroplasts and its membranes under water stress. Plant Sci. *38,* 13–16 (1985).

Pringsheim, G. H.: Methods for the cultivation of algae. In: Smith, G. M. (ed.): Manual of Phycobiology. Waltham, Massachusetts, USA, Chronica Botanica Company, pp. 347–357 (1951).

Ögren, E.: Photoinhibition of photosynthesis in willow leaves under field conditions. Planta *175,* 229–236 (1988).

Quinn, P. J. and W. P. Williams: Environmentally induced changes in chloroplast membranes and their effects on photosynthetic function. In: Barber, J. and N. R. Baker (eds.): Photosynthetic Mechanism and the Environment, Vol. *6,* Elsevier Science Publisher, The Netherlands, pp. 1–47 (1985).

Seemann, J. R., W. J. Y. Downton, and J. A. Berry: Temperature and leaf osmotic potential as factors in the acclimation of photosynthesis to high temperature in desert plants. Plant Physiol. *80,* 926–930 (1986).

Schindler, C. and H. K. Lichtenthaler: Is there a correlation between light-induced zeaxanthin accumulation and quenching of variable chlorophyll *a* fluorescence? Plant Physiol. Biochem. *32,* 813–823 (1994).

Sharkey, T. D.: Water stress effects on photosynthesis. Photosynthetica *24,* 651 (1990).

Srivastava, A. and E. Zeiger: Guard cell zeaxanthin tracks photosynthetic active radiation and stomatal apertures in *Vicia faba* leaves. Plant Cell Environ. *18,* 813–817 (1995).

Srivastava, A. and R. J. Strasser: The steady state chlorophyll *a* fluorescence exhibits *in vivo* an optimum as a function of light intensity which reflects the physiological state of the plants. Plant Cell Physiol. *36,* 839–848 (1995 a).

– – How do land plants response to stress temperature and stress light. Arch. Sci. (Genéve) *48,* 135–146 (1995 b).

Stapel, D., E. Kruse, and K. Kloppstech: The protective effect of heat shock protein against photoinhibition under heat shock in barley. J. Photochem. Photobiol. *21,* 211–218 (1993).

Strasser, R. J.: A concept for stress and its application in remote sensing. In: Lichtenthaler, H. K. (ed.): Applications of Chlorophyll Fluorescence, pp. 333–337. Kluwer Academic Publishers, Dordrecht (1988).

Strasser, R. J., A. Srivastava, and Govindjee: Polyphasic chlorophyll *a* fluorescence transient in plants and cyanobacteria. Photochem. Photobiol. *61,* 32–42 (1995).

Sundby, C., A. Melis, P. Mäenää, and B. Anderson: Temperature-dependent changes in the antenna size of photosystem II. Reversible conversion of photosystem II_α to photosystem II_β. Biochim. Biophys. Acta *851,* 475–483 (1986).

Thomas, P. G., P. J. Dominy, L. Vigh, A. R. Mansourian, P. J. Quinn, and W. P. Williams: Increased thermal stability of pigment-protein complexes of pea thylakoids following catalytic hydrogenation of membrane lipids. Biochim. Biophys. Acta *849,* 131–140 (1986).

Thompson, L. K., R. Blaylock, J. M. Sturtevant, and G. W. Brudvig: Molecular basis of heat denaturation of photosystem II. Biochemistry *28,* 6686–6695 (1989).

Van der Meer, B. W.: Fluidity, dynamics and order. In: Shinitzky, M. (ed.): Biomembranes: Physical Aspects. VCH, Weinheim, Germany, pp. 97–158 (1993).

Vigh, L., D. A. Los, I. Horvath, and N. Murata: The primary signal in the biological perception of temperature: Pd-catalyzed hydrogenation of membrane lipids stimulated the expression of the desA gene in *Synechocystis* PCC6803. Proc. Natl. Acad. Sci. USA *90,* 9090–9094 (1993).

Webb, M. S. and B. R. Green: Biochemical and biophysical properties of thylakoid acyl lipids. Biochim. Biophys. Acta *1060,* 133–158 (1991).

Weis, E.: Influence of light on the heat sensitivity of the photosynthetic apparatus in isolated spinach chloroplasts. Plant Physiol. *70,* 1530–1534 (1982 a).

– Influence of metal cations and pH on the heat sensitivity of photosynthetic oxygen evolution and chlorophyll fluorescence in spinach chloroplasts. Planta *154,* 41–47 (1982 b).

Wise, R. R., D. H. Sparrow, A. Ortiz-Lopez, and D. R. Ort: Biochemical regulation during the mid-day decline of photosynthesis in field-grown sunflower. Plant Sciences *74,* 45–52 (1991).

J. Plant Physiol. Vol. 148. pp. 456–463 (1996)

A Phytotron for Plant Stress Research: How Far Can Artificial Lighting Compare to Natural Sunlight?

STEPHAN THIEL, THORSTEN DÖHRING, MATTHIAS KÖFFERLEIN, ANDRE KOSAK, PETER MARTIN, and HARALD K. SEIDLITZ

GSF-Forschungszentrum für Umwelt und Gesundheit, Expositionskammern, Postfach 1129, D-85758 Oberschleißheim, Germany

Received July 20, 1995 · Accepted October 18, 1995

Summary

Plants have adapted very efficiently to their natural light habitat. Artificial plant illumination, therefore, requires careful design. Not only the quantity of radiation per area or volume (intensity) but also the spectral quality has to match seasonal and diurnal variations of natural global radiation as close as possible. The GSF Research Center has developed a phytotron system especially devoted to plant stress research, where these requirements are of particular importance. The phytotron consists of seven closed chambers (4 walk-in size chambers, two medium and one small sun simulator). Our contribution outlines the basic design of the lighting and presents spectral data.

A good approximation of terrestrial global radiation is achieved if several commercially available lamp types are combined and adequate filters are applied to reject unwanted infrared and harmful ultraviolet radiation. A programmable switch control for the individual lamp banks allows a variation of both spectrum and intensity of the illumination.

Spectroradiometric measurements show that the maximum level of illumination in the small and in the medium size chambers can compete both in spectral distribution and in intensity with outdoor global radiation for solar elevations up to 60°. The maximum light level available inside the large walk-in chambers reaches an irradiance corresponding to solar elevation of 50°. The UV-B : UV-A : PAR ratio, which mirrors the spectral balance of plant lighting, can be adjusted to values following the diurnal variation of natural global radiation.

Key words: Artificial plant lighting, growth chambers, diurnal variation, phytotron, spectral quality of radiation, UV-radiation.

Abbreviations: UV-C = ultraviolet C radiation (100–280 nm); UV-B = ultraviolet B radiation (280–320 nm); UV-A = ultraviolet A radiation (320–400 nm); PAR = photosynthetic active radiation (400–700 nm); PFD = photon flux density (400–700 nm); IL = illuminance; RMS = root-mean-square; DU = Dobson Unit.

Introduction

Light is certainly the most important environmental factor for plant growth and development. It does not only serve as an energy source thus being the main ‹nutrient›, it also provides the information regulating various physiological processes. Light quality, quantity, direction and temporal variation may have tremendous effects on plant life. In their natural habitat, plants are exposed to sunlight. Their pigments and receptors have, therefore, adapted to the unique structure of the sun's spectral output. An artificial plant illumination qualified for ecologically oriented plant stress research must, therefore, consider the photobiological demands and which rely on the natural global radiation. It is, therefore, necessary to simu-

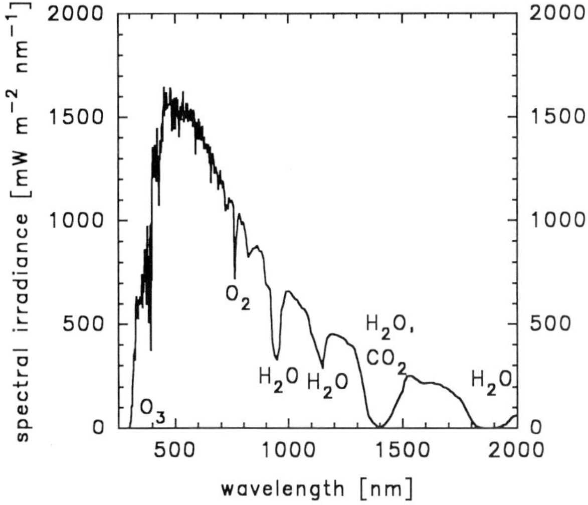

Fig. 1: Calculated global radiation for clear sky condition around midday during summer at moderate latitudes, using a combination of model Terra I (Seckmeyer et al., 1994) and a model of Justus and Paris (1985): Spectral irradiance and some prominent absorption bands.

late global radiation as closely as possible. The spectrum of global radiation, which is the sum of direct sunlight and diffuse scattered radiation, covers approximately the 300 to 4000 nm wavelength range. A typical spectrum of global radiation is depicted in Fig. 1. The spectral irradiance shows a peak at approx. 500 nm and some dominant absorption bands, which are caused by carbon dioxide and atmospheric water vapour. At approx. 300 nm the spectrum exhibits a steep cut-off due to the filtering effect of stratospheric ozone. The spectrum can be, somewhat arbitrarily, divided into several spectral regions. In the infrared part (above 800 nm), no specific receptors are known from higher plants, this radiation is absorbed by water within the plant tissue and determines the temperature of the plant or parts of the plant. The portion between 400 and 700 nm is termed photosynthetic active radiation (PAR) as it is mainly used for photosynthesis. A small region in the red and far-red part of the spectrum between approx. 600 and 750 nm regulates the phytochrome balance (Quail et al., 1983). On the ultraviolet side of the solar spectrum the interest is presently focused on the UV-B part (280–320 nm), as there is major concern that elevated levels, resulting from anthropogenic depletion of the stratospheric ozone, might cause severe damages in plants as well as in animals and man (e.g. Tevini, 1993; Young et al., 1993). The long wave UV-A (320–400 nm) and bluelight initiates photorepair processes and mitigates UV-B damages.

Various experiments have demonstrated that a naturally balanced UV-B : UV-A : PAR ratio is a prerequisite for a realistic ecological plant experiment (Teramura, 1980; Caldwell et al., 1994). However, reducing the complete spectral information to a double ratio is an, admittedly, convenient simplification. It does not take into account the exact shape of the spectrum, which might have far reaching consequences, especially in the UV-B range (Döhring et al., 1996).

Plant lighting has been the subject of many studies and comprehensive reviews exist (see e.g. Bickford and Dunn, 1972; Warrington et al., 1978; Tibbitts et al., 1983; Tibbitts (ed.), 1994). For vegetation stress research, however, a more ‹holistic› approach may be more appropriate, i.e. considering the whole spectrum as one entity. It is the aim of this paper to outline the design and performance of artificial lighting for a phytotron system showing that the demands stated above can be met even for large scale facilities with several square meters of experimental area.

Materials and Methods

The lay out of artificial plant lighting

There is no single artificial light source able to simulate both spectral quality and spectral quantity of global irradiance as shown in Fig. 1. Seckmeyer and Payer (1993) have, therefore, proposed a combination of metal halide lamps, quartz halogen lamps and blue fluorescent tubes in order to simulate the spectrum from the UV-A to the infrared. Excess infrared, mainly caused by the quartz halogen lamps and a few infrared lines emitted by metal halide lamps can be removed by a layer of water. The missing UV-B is supplemented by UV-B fluorescent tubes. Unfortunately the radiation output of these fluorescent tubes extends to well below 290 nm. This portion must be blocked very efficiently for the safety of both the plants and the personel operators. Selected borosilicate and lime glass filters provide a sufficiently steep cut-off at the desired wavelength. Different combinations of these glasses allow a variation of the cut-off wavelength, thus enabling us to simulate various UV-B scenarios. Details of the UV-B filtering technique are described elsewhere (Döhring et al., 1996). All data given below refer to combinations matching as closely as possible spectra of global radiation typically encountered in mid latitudes during summer. Fig. 2 shows the schematical outline of our plant lighting configuration. In addition to the lamps already described the figure shows extra quartz halogen lamps below the water layer. The water layer used, has to be maintained at a minimum thickness of approx. 20 mm in order to achieve homogeneous rheological conditions and spectral absorption over the whole area. It absorbs, however, a little too much of infrared radiation which has to be supplemented by the additional quartz halogen lamps. The individual spectral contributions of each lamp type, as employed in our plant lighting are shown in Fig. 3.

Technical data of GSF-Phytotron facilities

In our phytotron facilities we operate three different types of sun simulators with an lighting configuration as outlined above.

a) Four *walk-in chambers* with an experimental space of $(3.4 \times 2.8 \times 2.5)\ m^3$ (length × width × height). The lighting assembly consists of 96 metal halide lamps (Osram HQI/D 250 W and 400 W), 104 quartz halogen lamps, (Osram Halostar 200 W, 300 W and 500 W) and 38 blue fluorescent lamps (Philips TLD 18, 36 W). 248 UV-B lamps (Philips TL12 20 W and 40 W) cover the walls of the lamp house below the water filter of 20 mm thickness. The UV-B glass filter system is made of two layers of 6.5 mm borosilicate glass (TEMPAX®, PYRAN®, Schott, Mainz, Germany). The total electrical power consumption of the lighting is 96 kW per chamber.
The level of illumination is controlled by a programmable switch control by which individual banks of each lamp type can be turned on and off independently thus allowing a simulation of the diurnal variation of the irradiation both in quantity and in spectral quality. UV-B, UV-A and PAR irradiation levels are con-

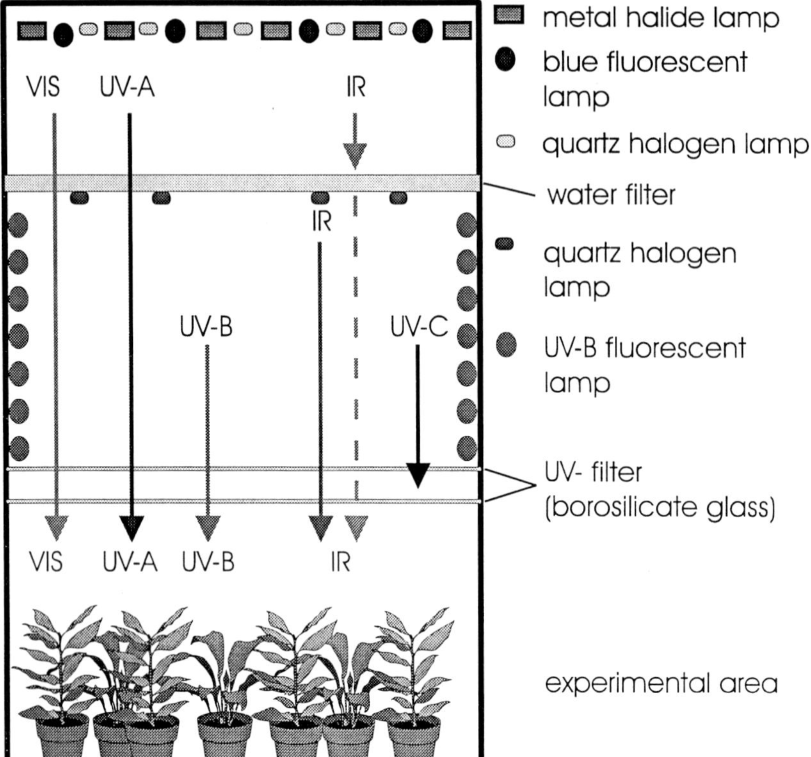

Fig. 2: Schematical outline of the lamp and filter configuration of GSF sun simulators.

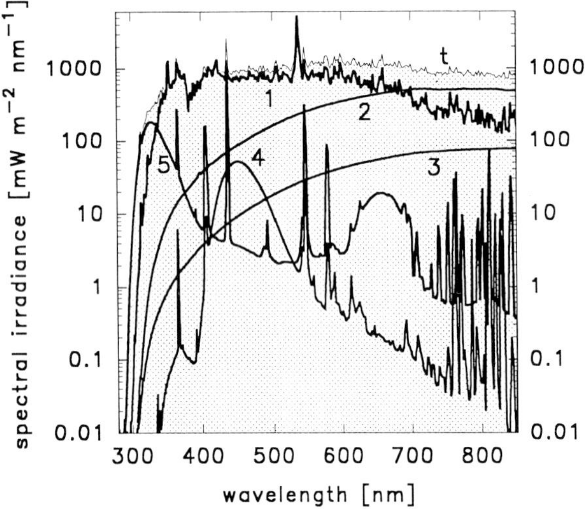

Fig. 3: Spectral contribution of the individual lamp types towards the total irradiance inside a medium size chamber; 1: metal halide lamps; 2: quartz halogen lamps above the water filter; 3: quartz halogen lamps below the water filter; 4: blue fluorescent lamps; 5: UV-B fluorescent lamps; t: total spectral irradiance.

tinuously monitored with appropriate integral sensors. Our walk-in simulators provide full control on remaining environmental factors, such as air temperature (−20 °C to +40 °C) and relative humidity (25 % to 95 %), soil temperature (−15 °C to +25 °C) and moisture (100–800 hPA) and include the possibility for controlled fumigation of gaseous pollutants (Payer et al., 1993). The systems are designed for ecological plant experiments over several months.

b) Two *medium size chambers* with an experimental space of $(1.4 \times 1.4 \times 1.0)\ m^3$ (length × width × height). The basic lay out is the same as described above. Fewer lamps are required due to the smaller size. The lamp ceiling is fitted with 36 metal halide lamps (Osram HQI/D 400 W), 32 quartz halogen lamps (Osram Halostar 300 W and 500 W) and 6 blue fluorescent lamps (Philips TLD 18, 36 W). UV-B irradiation is provided by 96 UV-B lamps (Philips TL12 40 W). The power consumption of a medium size chamber is 36 kW. Programmable lamp control is also available. Climate control is similar to that in a walk-in chamber, however, the range of variability of the parameters is somewhat narrower. The experimental space is fitted with a height adjustable plug-in unit on to which the sample plants are placed. The chambers are mainly devoted to experiments under high light and UV intensity conditions.

c) One *small size chamber* (Seckmeyer and Payer, 1993) with an experimental space of $(1.2 \times 1.2 \times 0.25)\ m^3$ (length × width × height). The lighting assembly is basically the same as in the medium size simulators but the arrangement is more compact. There is a limited control on the climate parameters air temperature and humidity. The small simulator provides the highest light and UV intensities available in our phytotron. The design allows a very flexible handling of UV-filters, such that various enhanced UV-B scenarios with small plants, seedlings or microorganisms can be simulated (Döhring et al., 1996).

Spectroradiometric and radiometric measurements

Spectral data presented in this study were obtained by spectroradiometric measurements. We operate three types of spectroradiometers for different requirements.

a) *Range 250–500 nm:* We use a double monochromator (Bentham M300HR/2, Reading, U.K.) equipped with two gratings of 2400 grooves/mm each, combined with a special UV sensitive photomultiplier (EMI 9558BQ, Ruislip, U.K.) as photodetector. The spectral resolution is 1 nm, the detection limit is 1 $\mu W\,m^{-2}\,nm^{-1}$ at a signal to noise ratio of 2.
b) *Range 280–850 nm:* We use a double monochromator (Bentham TDM300/2, Reading, U.K.) equipped with two pairs of gratings of 2400 grooves/mm and 1200 grooves/mm respectively, combined with a photomultiplier (Bentham DH3, Reading, U.K.) as photodetector. The spectral resolution is 1 nm in the range between 280–400 nm and 2 nm in the range between 400–850 nm, the detection limit is better than 10 $\mu W\,m^{-2}\,nm^{-1}$ at a signal to noise ratio of 2.
c) *Range 400–850 nm:* We use a single monochromator (Bentham M300 Reading, U.K.) equipped with a grating of 1200 grooves/mm, combined with a photomultiplier (EMI 9558BB, Ruislip, U.K.) as photodetector. The spectral resolution is 5 nm, the detection limit is approx. 10×$\mu W\,m^{-2}\,nm^{-1}$ at a signal to noise ratio of 2.

The input optics consist of quartz fiber bundles fitted with a cosine diffuser. Regular calibration is performed with 1000 W and 100 W quartz halogen lamps whose calibration is traceable to the Physikalisch Technische Bundesanstalt, Braunschweig Germany and the National Physics Laboratory, Teddington, U.K.

Our spectral light data are compared with spectroradiometric field data obtained on a nearly cloudless summer day (see table 1 for details) with approx. 60° maximum solar elevation. These measurements are performed with system a) and c) in a temperature stabilized cabinet. Integrated and weighted irradiance values are obtained by numerical integration of the spectral data.

Several factors contribute to the measurement uncertainty of outdoor spectroradiometry. Among these are the calibration error, the cosine error of the diffusor, the diurnal variation of the ratio of direct to diffuse irradiance and the anisotropy of sky radiation. In order to estimate the maximum measurement uncertainty we compare spectral irradiance weighted with the visual response (V_λ) to luxmeter readings obtained simultaneously. For solar elevations above 20° we estimate 20 % and for solar angles below 20° 40 % measurement uncertainty. Due to more stable conditions and less influencing parameters the uncertanties of chamber measurements did not exceed 10 %.

Some additional radiometric measurements are performed with spectral integrating instruments. Total global radiation in the 300–2500 nm range is measured with a pyranometer (Kipp & Zonen CM11, Delft, Holland; accuracy: ± 3–5 %). Illuminance is determined by using a luxmeter (Luxmeter 110, PRC Krochmann, Berlin, Germany; accuracy: ± 3 %). UV-B irradiance (erythemally weighted) is obtained by a Robertson-Berger Meter (Solar Light, Biometer 501, Philadelphia, U.S.A.; accuracy: ± 5 % for daily total).

All radiation measurements are performed at a level which lies above the canopy level in the individual chambers.

Results and Discussion

Integral lighting data

The integrated irradiation values in several spectral regions and data obtained by weighting with selected action spectra (DNA, plant damage, visual response V_λ) are listed in Table 1. The corresponding outdoor data are also presented for comparison.

The homogeneity of illuminance at the canopy level, show that the horizontal distribution of light is basically flat. The root-mean-square (RMS) deviation for maximum light level is 5–10 % (see Table 1). If the light level is reduced, i.e. several lamp banks are switched off, the RMS values double. The UV-B irradiance shows RMS values of approx. 20 % in the walk-in size and medium size chambers and 10 % in the

Table 1: Irradiation data of the GSF-Phytotron facilities and an outdoor spectrum measured at Neuherberg (48.2° N, 11.5° E, altitude 500 m) on July 30, 1992 (solar elevation 60°, mean total ozone column 300 DU, cloud cover <2 Oktas) obtained by spectroradiometric measurements. Plant damage: biologically effective (BE) irradiance using the general plant action spectrum (Caldwell, 1971; Green et al., 1974) normalized to 300 nm.

parameter	small size chamber	medium size chamber	walk-in chamber	sun, 60°, 300 DU	unit
total irradiance[1]	1038	715	670	840	[$W\,m^{-2}$]
UV-C (<280 nm)	$<10^{-7}$	$<10^{-7}$	$<10^{-7}$	–	[$W\,m^{-2}$]
UV-B (280–320 nm)	2.5	1.53	0.70	2.01	[$W\,m^{-2}$]
DNA damage	105	56	25	93	[$mW\,m^{-2}$]$_{DNA}$
plant damage	217	121	52	226	[$mW\,m^{-2}$]$_{BE}$
UV-A (320–400 nm)	53.5	45.2	37	43	[$W\,m^{-2}$]
PAR (400–700 nm)	446	340	274	360	[$W\,m^{-2}$]
PFD (400–700 nm)	2076	1571	1260	1648	[$\mu mol\,m^{-2}\,s^{-1}$]
illuminance (IL)	126	93.3	72	91.4	[klux]
IR (>800 nm)[2]	410	323	290	342	[$W\,m^{-2}$]
UV-B:UV-A:PAR	1:21:179	1:30:222	1:53:391	1:21:179	–
RMS (‹homogeneity›)	5 (IL) 8 (UV-B)	5 (IL) 20 (UV-B)	10 (IL) 20 (UV-B)	–	%
vertical gradient (IL)	–	58[3]	25[4]	–	[$\%\,m^{-1}$]

DNA damage: irradiance weighted with the action spectrum for DNA damage (Setlow, 1974; Green and Miller, 1975) normalized to 300 nm.
RMS: root-mean-square.
[1] Measured with a pyranometer.
[2] Calculated as a difference of total irradiance[1] and integrated data of spectroradiometric measurements.
[3] Measured between 0.5 m and 1.5 m above ground.
[4] Measured between 0.9 m and 1.7 m above ground.

small simulator indicating less homogeneity than the visible radiation.

An estimation of the vertical gradient of the illuminance is obtained from measurements at two different heights (medium size chamber at 0.5 m and 1.5 m, walk-in chamber at 0.9 m and 1.7 m above ground level). The values of illuminance decrease by 60 % per m and 25 % per m within the medium size and the walk-in chamber, respectively. As the experimental space of the small simulator has only a height of 0.25 m no data are given.

Comparison of chamber irradiance and field irradiance

The maximum available spectral irradiance in the small size, medium size and walk-in size chambers is shown in Figs. 4–6 on a linear scale over the 300–850 nm range. A logarithmic plot over the 280–400 nm range is used in order to clarify the steep UV cut-off of the spectral irradiance. Medium size and small size chambers are equipped with similar lighting configurations. The maximum available irradiance is compared to an outdoor spectrum resulting from a solar elevation of 60° and a mean total ozone column of 300 DU.

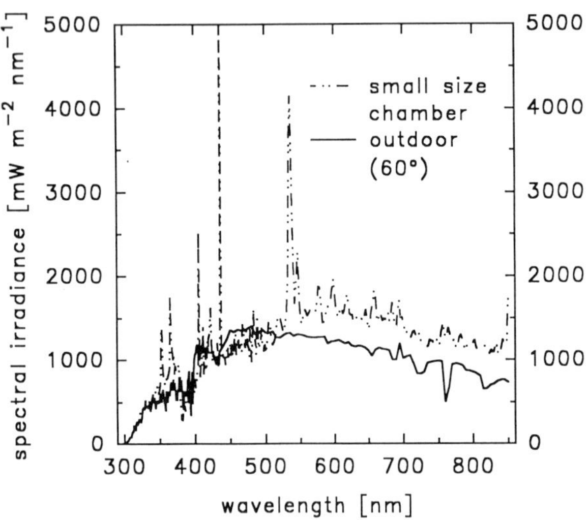

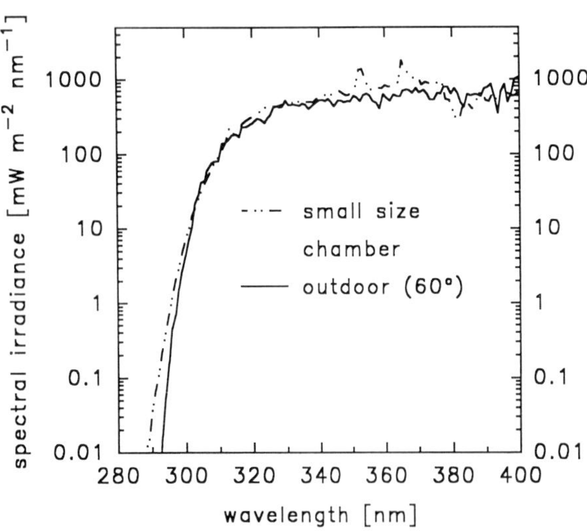

Fig. 4: Maximum available spectral irradiance in the small size chamber and an outdoor spectrum for 60° solar elevation (details see table 1) for comparison. a) linear scale; b) logarithmic scale.

The maximum spectral irradiance of the small chamber exhibits a very good match in the 300–500 nm range (Fig. 4 a). The spectrum shows a few prominent but narrow spectral lines at wavelengths between 350 and 550 nm, which originate from the metal halide lamps[1]. Due to their very narrow spectral width, however, the contribution to the total energy is small (<5 %). The slight excess in the UV-B range below 300 nm (Fig. 4 b) is actually desired for UV-B stress experiments. Excess irradiance in the range above 500 nm represents a reserve for future upgrading and can be omitted by switching off several quartz halogen lamps. The ratios of the spectrally integrated UV-B, UV-A and PAR irradiance, 1 : 21 : 179 for the small-size chamber and for the outdoor spectrum are virtually identical.

The maximum spectral irradiance available in our medium size chambers (Figs. 5 a/5 b) compares very well with an outdoor spectrum (solar elevation 60°) over the range of 300 to 850 nm. The only exception appears in the blue range (400–500 nm) where the relatively poor output of the blue fluorescent tubes causes a deficiency of 30 %. The effect of this deficiency on photosynthetic activity is probably negligible, as its energetic contribution to the total PAR is very small. Its significance, however, for the response of bluelight receptors others than chlorophyll might be considered. The UV-B : UV-A : PAR ratio is 1 : 30 : 222 (outdoor value 1 : 21 : 179), which is still a good approximation.

Figs. 6 a and 6 b show the maximum spectral irradiance available inside the walk-in chambers. Due to different proportions of the dimensions of both the lamphouse and the experimental space and to a somewhat different lamp equipment (less UV-B lamps per usable area, as compared to the smaller chambers) the maximum values are significantly lower than in the smaller chambers. The spectral distribution within this chamber is, therefore, compared to an outdoor spectrum obtained at a solar elevation of 50°. The spectral irradiance above 350 nm shows a good agreement with the reference spectrum, except of the blue portion. The reason is the same as mentioned before. Below 350 nm, especially in the UV-B region (Fig. 6 b) the spectral irradiance at the level of the experiment is quite low. The reason for the UV-B deficiency is the smaller number of UV-B lamps installed (per unit experimental area) and the low UV-reflectance of the stainless steel corrosion resistant cladding used in the experimental area of these chambers. However, it should be noted, that integrated over a 8 h period, a daily UV-B dose of 1.5 kJ_{BE} can be applied. UV-B : UV-A : PAR ratios are 1 : 53 : 391 and 1 : 23 : 207 for the walk-in chamber and the outdoor spectra of 50° sun elevation respectively.

The diurnal variation of the spectral composition

The combination of several lamp types with different spectral output allows a very flexible tuning of both intensity

[1] Medium size and walk-in chambers are equipped with HQI lamps from a new production lot. These lamps emit a line structure slightly different from former lots.

and spectral quality of plant lighting. Using a least square fit algorithm we can correlate the spectra of different lighting levels with outdoor spectra belonging to different angles of solar elevation. We then evaluate the integrated irradiance in the UV-B, UVA and PAR regions for all spectra and compare the UV-B : PAR ratio both of the chamber and the corresponding outdoor spectrum, thus an impression of the quality of the simulation of diurnal varying solar radiation can be obtained.

Fig. 7 shows the UV-B : PAR ratio of outdoor global radiation (open circles) which rises monotonously with increasing elevation angles. The main reason is the reduction of the ozone absorption path length at higher solar elevation. In our plant lighting we can account for this effect by adjusting the amount of active UV-B lamps using the programmable lamp control. Filled diamonds, circles and squares in Fig. 7 show the positions of correlated spectral ratios from a small size, medium size and from a walk-in chamber, respectively. The

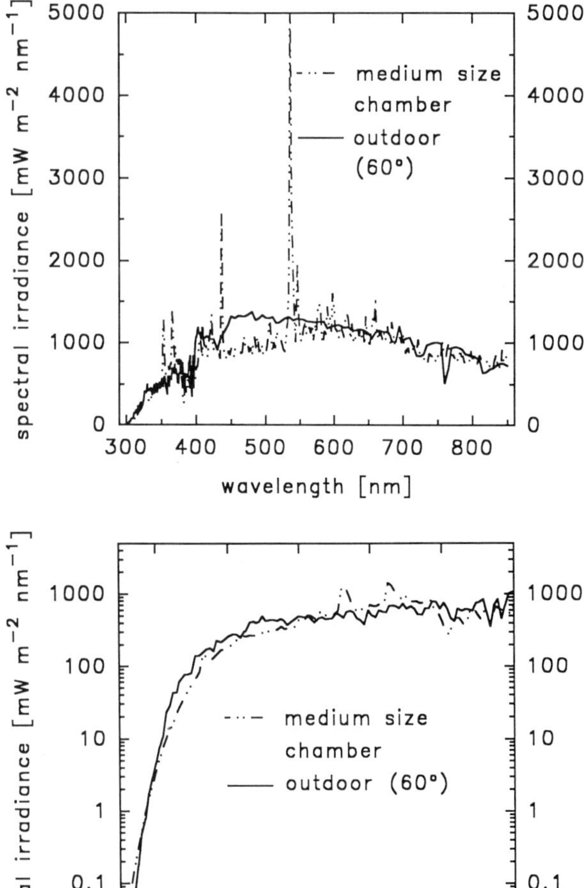

Fig. 5: Maximum available spectral irradiance in the medium size chamber and an outdoor spectrum for 60° solar elevation (details see table 1) for comparison. a) linear scale; b) logarithmic scale.

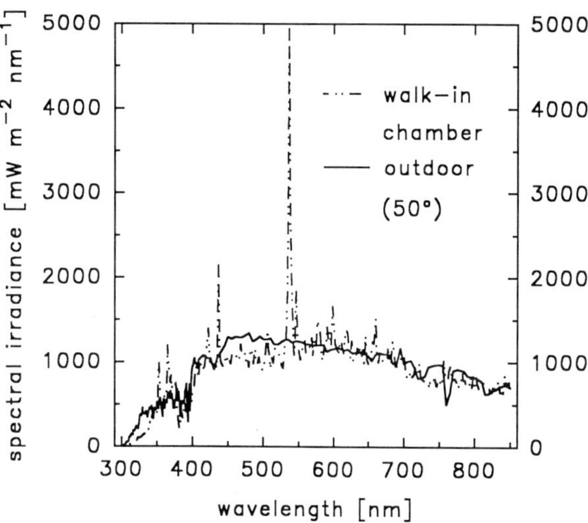

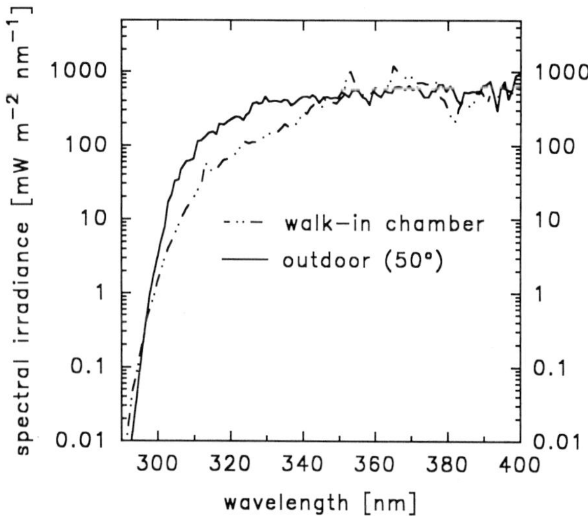

Fig. 6: Maximum available spectral irradiance in the walk-in chamber and an outdoor spectrum for 50° solar elevation (details see table 1) for comparison. a) linear scale; b) logarithmic scale.

UV-B : PAR ratio corresponding to the highest solar elevation (60°) is reached within the small chamber, the respective value from a medium size chamber shows a deficiency of approx. 25 %. All other ratios from those two chambers are in almost perfect agreement with their outdoor counterparts. As already discussed maximum UV-B levels inside walk-in size chambers do not quite reach natural values which results in low UV-B : PAR ratios for high levels of illumination. The UV-A : PAR ratios (approx. 0.1) do not vary strongly for the solar elevations considered in Fig. 7 and are, therefore, not shown.

Conclusion

Controlled environments are a valuable tool in vegetation stress research, provided the environmental parameters are matched as closely as possible to natural conditions. A recent

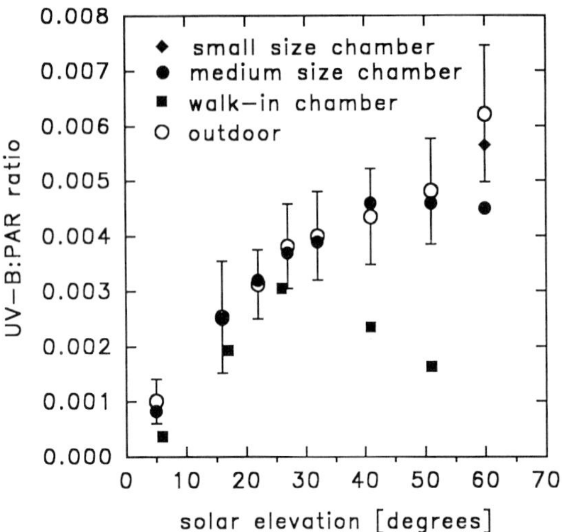

Fig. 7: Simulation of diurnal changes in the spectral balance between UV-B and PAR ranges. Open circles represent the UV-B : PAR ratio obtained from outdoor measurement on clear sky conditions (see Table 1) at different angles of solar elevation. Filled symbols represent the UV-B : PAR ratios for different light and radiation levels within the chamber. Each chamber value corresponds to an outdoor spectrum for a particular solar elevation. Error bars for outdoor data indicate maximum measurement uncertainties. Chamber measurements exhibit much less uncertainties and bars are, therefore, not shown.

essay (Caldwell and Flint, 1994) surveys 33 growth chamber experiments between 1990–1993, based on the UV-B : PAR criterium. In most of those evaluations UV-B : PAR ratios are far from natural data, only two reach solar values for daily doses by adjusting the daylength.

The lighting data of the GSF-Phytotron facilities demonstrate that plant lighting technology has gained a high quality level. Our spectral measurements demonstrate that in chambers up to 2 m^2 experimental area a steep, realistic shape of the UV-B edge, a UV-B : UV-A : PAR ratio close to nature, and high quantum fluxes up to 2000 $\mu mol\, m^{-2} s^{-1}$ can be obtained simultaneously. Large walk-in simulators can provide a well balanced spectrum, too, but due to limitations in the physical dimensions and in power consumption somewhat reduced light levels are available.

Outdoor measurements show that spectral data vary significantly during sun's diurnal course. Artifical lighting qualified for ecological plant research should, therefore, be able to simulate these variations. The data presented demonstrate, that this can be achieved by using a flexible electronic control of a series of individual lighting circuits which feed various lamp types with different spectra of emission.

Acknowledgements

We gratefully acknowledge the extensive support and advice by Dr. H.-D. Payer and we want to thank our technical staff D. Arthofer, H. Egger, P. Kary, W. Kratzl, J. A. Meier and B. Rieger for their efforts to a successful completion and operation of our phytotron facility. One of the authors (S.T.) wants to express his special thanks to Dr. G. Seckmeyer from Fraunhofer Institut für atmosphärische Umweltforschung (IFU) at Garmisch-Partenkirchen, Germany for continuous encouragement during his diploma thesis.

References

Bickford, E. D. and S. Dunn: Lighting for plant growth. The Kent State University Press (1972).

Caldwell, M. M., S. D. Flint, and P. S. Searles: Spectral balance and UV-B sensivity of soybean: a field experiment. Plant, Cell and Environment *17,* 267–276 (1994).

Caldwell, M. M. and S. D. Flint: Stratospheric ozone reduction, solar UV-B radiation and terrestrial ecosystems. Climatic change *28,* 375–394 (1994).

Caldwell, M. M.: Solar ultraviolet radiation and the growth and development of higher plants. In: Giese, A. C. (ed.): Photophysiology Vol. *6,* pp. 131–177. Academic Press, New York (1971).

Döhring, T., M. Köfferlein, S. Thiel, and H. K. Seidlitz: Spectral shaping of artificial UV-B irradiation for vegetation stress research. J. Plant Physiol. *148,* 115–119 (1996).

Green, A. E. S. and J. H. Miller: Measures of biologically effective radiation in the 280–340 nm region. In: Nachtwey, D. S., M. M. Caldwell, and R. H. Biggs (eds.): Impacts of Climatic Change on the Biosphere. CIAP Monograph 5, Part 1 – Ultraviolet Radiation Effects, pp. 2–60 to 2–70. DOT, Washington, D.C. (1975).

Green, A. E. S., T. Sawada, and E. P. Shettle: The middle ultraviolet reaching the ground. Photochem. Photobiol. *19,* 251–259 (1974).

Iqbal, I.: An Introduction to Solar Radiation. Academic Press, Toronto (1983).

Justus, C. G. and M. V. Paris: A Model for Solar Spectral Irradiances and Radiances at the Bottom and Top of a Cloudless Atmosphere. J. Cl. Appl. Met. *24,* 193–205 (1985).

Madronich, S., L. O. Björn, M. Ilyas, and M. M. Caldwell: Changes in biologically active ultraviolet radiation reaching the earth's surface. In: Environmetal effects of ozone depletion, United Nations Environment Program (UNEP), P.O. Box 30552, Nairobi (Kenia), 1991 update, Chapter 1, pp. 1–14 (1991).

Payer, H.-D., P. Blodow, M. Köfferlein, M. Lippert, W. Schmolke, G. Seckmeyer, H. K. Seidlitz, D. Strube, and S. Thiel: Controlled environment chambers for experimental studies on plant responses to CO_2 and interactions with pollutants. In: Schulze, E. D. and H. A. Mooney (eds.): Design and Execution of Experiments on CO_2 Enrichment, Ecosystems Research Report 6, pp. 127–145. ECSC-EEC-EAEC, Bruxelles-Luxembourg (1993).

Quail, P. H., J. T. Colbert, H. P. Hersley, and R. D. Vierstra: Phytochrome: Molecular Properties and Biogenesis. Phil. Trans. R. Soc. Lond. *B 303,* 387–402 (1983).

Seckmeyer, G. and H.-D. Payer: A new sunlight simulator for ecological research on plants. J. Photochem. Photobiol. B: Biol. *21,* 175–181 (1993).

Seckmeyer, G., S. Thiel, M. Blumthaler, P. Fabian, S. Gerber, A. Gugg-Helminger, D.-P. Häder, M. Huber, C. Kettner, U. Köhler, P. Köpke, H. Maier, J. Schäfer, P. Suppan, E. Tamm and E. Thomalla: Intercomparison of spectral-UV-radiation measurement systems. Appl. Opt. *33,* No. 33, 7805–7812 (1994).

Setlow, R. B.: The wavelengths in sunlight effective in causing skin cancer: A theoretical analysis. Proc. Nat. Acad. Sci. *71,* 3363–3366 (1974).

Teramura, A. H.: Effects of ultraviolet-B irradiances on soybean. I. Importance of photosynthetically active radiation in evaluating ultraviolet-B irradiance effects on soybean and wheat growth. Physiologia Plantarum *48,* 333–339 (1980).

Tevini, M. (ed.): UV-B Radiation and Ozone Depletion: Effects on Humans, Animals, Plants, Microorganisms, and Materials. CRC Press, Boca Raton, USA (1993).

Tibbitts, T. W., D. C. Morgan, and I. J. Warrington: Growth of lettuce, spinach, mustard and wheat plants under four combinations of high-pressure sodium, metal halides and tungsten halogen lamps at equal PPFD. J. Am. Soc. Hortic. Sci. *108,* 622–630 (1983).

Tibbitts, T. W. (ed.): Lighting in controlled environments. Proceedings of a workshop in Madison, Wisconssin March 27–30, 1994. NASA Conference Publication CP3309, NASA Kennedy Space Center, Florida (1994).

Warrington, I. J., T. Dixon, R. W. Robotham, and D. A. Rook: Lighting systems in major New Zealand controlled environment facilities. J. Agric. Eng. Res. *23,* 23–36 (1978).

Young, A. R., L. O. Björn, J. Moan, and W. Nultsch: Environmental UV Photobiology. Plenum Press, New York, USA (1993).

J. Plant Physiol. Vol. 148. pp. 464–470 (1996)

Concept of Application of Synthetic Optical Spectra in Photobiological Research of Plants

PETER I. RICHTER[1] and HARTMUT K. LICHTENTHALER[2]*

[1] Department of Atomic Physics, Technical University of Budapest, Budafoki út 8, H-1111 Budapest, Hungary

[2] Botanical Institute, University of Karlsruhe, Kaiserstraße 12, D-76128 Karlsruhe, Germany

Received September 15, 1995 · Accepted October 30, 1995

Summary

A concept is presented for the formation of defined synthetic optical spectra for the investigation of photobiological processes in plants at selected wavelength regions. This new type of a multispectral light source is based on a tunable acousto-optic filter which can electronically be modified by radio frequency signals. There are given some application possibilities of synthetic optical spectra (based on acousto-optic filters) in various fields of photobiological plant research, and in stress detection of plants via fluorescence.

Key words: Acousto-optic filter, fluorescence spectra, radio frequency synthesizer, stress detection, synthetic optical spectra.

Abbreviations: AOTF = acousto-optic tunable filter; LIF = laser-induced fluorescence; PS1 and PS2 = photosynthetic photosystems 1 and 2; RF = radiofrequency.

Introduction

In nature and living organisms photobiological processes are driven by terrestrial solar radiation, i.e. by the irradiance of the sun near the earth surface, which is modified by the transmittance, absorption and scattering of radiation in the atmosphere. The spectral distribution of this radiation (see Fig. 1) depends on the solar zenith angle as well as on changes in concentration of absorbents and scatterers in the atmosphere (Zissis, 1993).

Plants respond to the incident light by two specific light receptor systems: a) the red/far-red reversible photomorphogenetic response system phytochrome, which exists in the inactive form P_{660} as well as in the physiologically active form P_{730} and b) the blue light receptor crytochrome. Both receptor systems control a multitude of photobiological processes in plants, such as the phytochrome control of photomorphogenesis, pigment formation (chlorophyll, carotenoids, anthocyanins), photoperiodism as well as induction or inhibition of flower formation in long-day and short-day plants, whereas the blue light-sensitive cryptochrome regulates the phototropic growth-response of plant shoots towards the light source and can influence many other light modulated processes (e.g. Häder and Tevini, 1987; Mohr and Schopfer, 1995). Plants also show an adaptation of their leaves and chloroplast to high-light or low-light growth conditions as seen in the formation of sun and shade leaves as well as sun-type and shade-type chloroplasts, which are different in chemical composition and their rates of photosynthetic quantum conversion (Lichtenthaler et al., 1981 a, 1982 a, 1982 b). The formation of high-light induced sun-type chloroplasts can be simulated by low irradiance blue light, and the formation of low-light or shade-type chloroplast, in turn, can be monitored by low irradiance red light (Buschmann et al., 1978; Lichtenthaler et al., 1980; Lichtenthaler and Buschmann, 1978). Due to the particular absorption characteristics of chlorophylls and carotenoids as well as their *in vivo* pigment proteins (Lichtenthaler et al., 1981 b; Lichtenthaler, 1987) incident blue light and red light are more readily absorbed by green plants than is green-orange light, and thus they guarantee higher photosynthetic rates than the green-orange photons, which are more readily transmitted and reflected.

There are many more light-modulated processes in plants, and many of these have not yet been analyzed in full detail concerning width and spectral range of the effective wave-

* Correspondance.

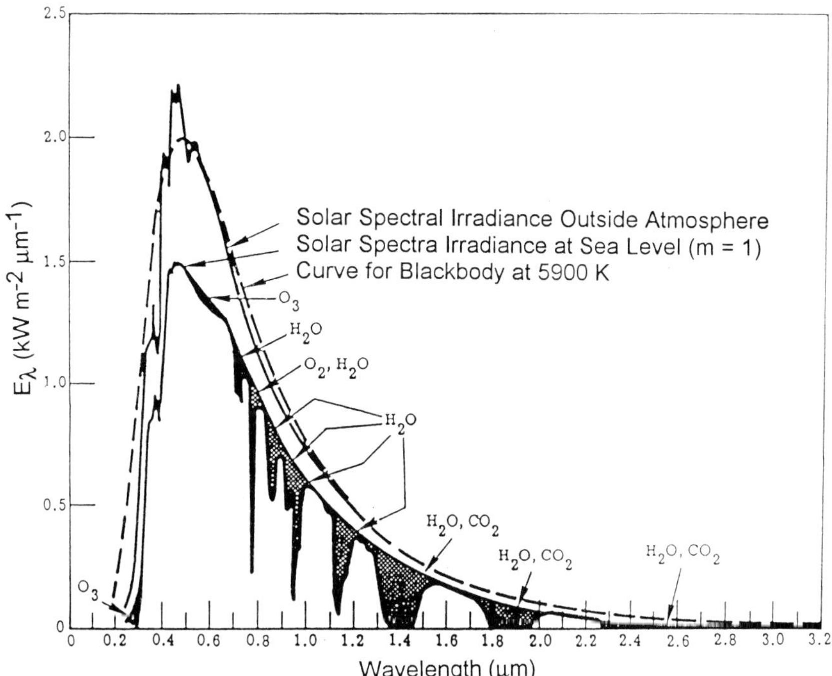

Fig. 1: Spectral radiation power distribution curve of the sun emission. The shaded areas indicate absorption at sea level due to the atmospheric constituents shown (modified after Zissis, 1993).

lengths. Also, the possible overlapping and interaction of different photobiological processes, which compete for the same incident blue and red light, have not yet been clarified in detail. In order to understand photobiological processes they must be studied under a controlled environment in the laboratory. However, in case of artificial light sources the range of available spectral distributions is rather limited and often does not correspond with these particular wavelength regions, which are necessary for clearly demonstrating a specific photobiological process. It is also desirable to study the specific interaction of particular photobiological plant responses by irradiating the plants with those two or three specific wavelength regions required to induce particular photo-responses. Such studies are, however, bound to the existence of defined synthetic light sources providing the two or three wavelength bands required.

The light sources presently available for a specific wavelength region are highly monochromatic, such as lasers that may be tunable as well. There also exist broadband light sources, such as black bodies, arc and discharge sources (high and low pressure) that possess characteristic light emission spectra. Wavelength selection can then be achieved by using different types of monochromators. However, such set-ups are very limited in their application.

It is clear that preparing a light source with a specified spectral distribution in one, two or three selected wavelength ranges is desirable for the study of the plants' photoresponses, their sensitivity to particular wavelength bands and for the study of positive and negative interactions. Following we describe the concept of a new type of light source based on a tunable acousto-optic filter that allows the synthesis of defined optical spectra. Some application possibilities of such new synthesized light sources in photobiological plant research will be indicated.

Technique of synthesizing arbitrary light spectra

In order to produce spectral distributions in a broad range, the application of a broadband radiation source is necessary. Therefore we do not take lasers into account. Wavelength selective monochromators are typically either dispersive (e.g. prims), diffractive (e.g. gratings) or interferometric (interference filters). The first two provide spatial separation of the different spectral components, and a single wavelength can be selected through a slit (Fig. 2). Multilayer interference filters can be constructed that have transmittance at pre-specified wavelengths. However, the spectral characteristics of such filters have to be specified before they are designed, and it is not possible to modify them after their production.

For the purpose of investigating the influence of spectral characteristics of light on photobiological processes optical filters with variable spectral characteristics are most desirable. This can be realized using a tunable acousto-optic filter AOTF (Karpel, 1988). Acousto-optic interaction may occur when an acoustic wave and a light beam are present in an appropriate medium. By launching an acoustic wave (ultrasound) into the medium, it generates a refractive index wave that behaves like an optical grating. The grating constant is inversely proportional to the frequency of the ultrasound,

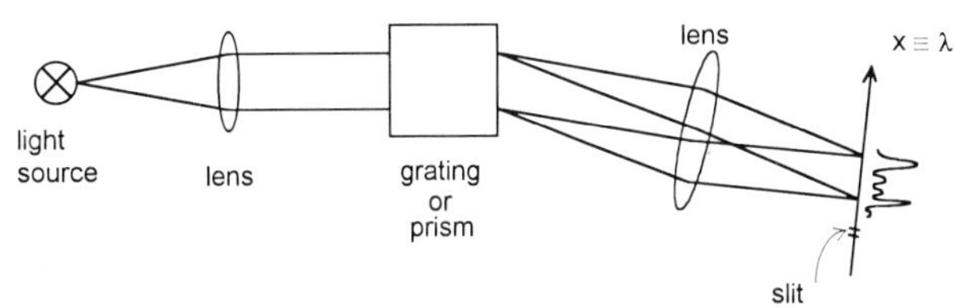

Fig. 2: Dispersive and diffractive monochromators provide spatially resolved spectral distributions.

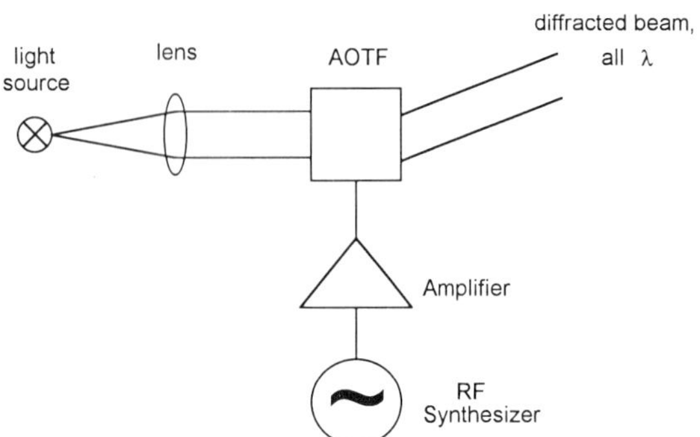

Fig. 3: Acousto-optic tunable filters (AOTF) possess selective wavelength transmission (diffraction) characteristics that are electronically controllable.

and the diffraction efficiency is proportional to its intensity. By appropriate selection of the interaction medium and design the device can be made to behave as an electronically tunable bandpass filter with a fixed optical beam path. A unique feature of the AOTF is the option to superpose acoustic waves with different wavelengths by driving the ultrasound transducer with a composite radiofrequency (RF) signal (Fig. 3). These signals can be generated by radio-frequency synthesizers (Hewlett-Packard, 1995). The latter are capable of synthesizing RF signals with complex spectra that can be transposed to the optical diffraction spectra of the AOTF. In this way complex spectra containing peaks of different heights and widths can be synthesized in the optical range. Filters of this kind can be produced for any wavelength range from 0.2 µm to 12 µm (using different interaction media). The spectral distribution can electronically be set and controlled and varied in time as well, if necessary.

Possible Applications

There certainly is a wide field in plant research for the application of AOTF supplied synthetic optical spectra. Various photosynthetic and photobiological processes are induced and/or controlled by photons of particular wavelength regions. These can be evaluated in more detail with the new technique. A few examples are given:

Photosynthesis

The photosynthetic process in chloroplasts is driven by the two photosystems, PS1 and PS2, which possess differential light absorption characteristics (e.g. Lawlor, 1993). PS2 can make full use of photons only up to 680 nm, whereas PS1 also works with longer photons up to 710 nm. The light-harvesting chlorophyll *a/b* carotenoid proteins (LHCPs = LHC-II) and the chlorophyll *a*/β-carotene proteins CPa (of photosystem 2) as well as CPI and CPIa (with LHC-I) of PS1, which can be separated and isolated by polyacrylamide gel electrophoresis, exhibit specific absorption spectra (Lichtenthaler et al., 1981 b) with a different wavelength position of the maxima and shoulders reflecting the differential pigment composition of the individual pigment proteins (Lichtenthaler et al., 1982 c; Lichtenthaler, 1987).

By means of a selective excitation of the photosynthetic apparatus in the blue region at the *in vivo* absorption bands or maxima, which are different for chlorophyll *a,* chlorophyll *b* and the carotenoids (range 434 to 471 nm), one can study the efficiency of the excitation energy transfer (excitation migration) from carotenoids to chlorophyll *b,* and from the latter to the long wavelength forms of chlorophyll *a* molecules in the reaction centers of photosystems 1 and 2. One can also specifically excite chlorophyll *b* molecules in their red absorption band near 653 nm, which excludes an excitation of the carotenoids absorbing in the blue spectral region only. Furthermore, with red light of ≥700 nm it is possible to exclude PS2-reactions, to oxidize the photosynthetic electron transport chain, and to study solely PS1-dependent reactions, e.g. cyclic photophosphorylation.

All these photochemical and photosynthetic processes have been studied using monochromatic laser light or special filter systems when applying a white light source. The newly developed technique of creating synthetic optical spectra with the help of AOTF provides, however, the possibility to select two, three or even four defined wavelength bands, and to study in a selected way these processes and their resulting interactions by simultaneous and consecutive excitation at different wavelengths and of different pigment forms. One can also synthesize spectra, in which the photon flux density of one wavelength region is being modified, whereas two other spectral bands are kept constant. Such studies with modified synthetic light spectra, permitting illumination with light of different spectral distributions, can provide a better insight into the performance and regulation of the photochemical processes of photosynthetic quantum conversion, and into the relative contribution of different pigment forms of chlorophylls and carotenoids regarding the primary photochemical events.

In fact, photosynthetic investigations with multispectral synthetic light might also permit a better definition of stress-induced damage to the photosynthetic apparatus, as induced by biotic or non-biotic stress constraints. Also the photoinhibition process at high-light conditions including the detection and localization of particular damage sites (Baker and Bowyer, 1994; Krause and Weis, 1991; Lichtenthaler et al., 1992) might be studied in more details.

Photomorphogenesis and other photobiological processes

The morphological development of plants, the formation of photosynthetic pigments (chlorophylls, carotenoids), and in some plants also that of the red anthocyanins is controlled by active phytochrome as well as the light-induced development of functional chloroplasts in dark-grown, etiolated seedlings (Lichtenthaler and Becker, 1972; Lichtenthaler and Kleudgen, 1975). In addition, many other light-dependent processes in plants are regulated or controlled by the active phytochrome, such as photoperiodism responses, flower induction, photonastic responses, in which positive or negative responses, i.e. inductions or inhibitions are possible (see Häder and Tevini, 1987; Mohr and Schopfer, 1995).

The physiologically inactive form of phytochrome P_r or P_{660}, present in dark grown seedlings, exhibits absorption maxima in the blue spectral region and in the near red at 660 nm. By means of a short red light pulse it is transferred to its active form P_{fr} or P_{730}, which exhibits a far-red absorption maximum near 730 nm, but also shows a broad blue absorption band. Using either far-red light of 730 nm or longer dark periods, active phytochrome P_{fr} is transformed to its physiologically inactive form P_r. The active form $P_{fr} \rightarrow$ induces and controls many of the plants' photoresponses.

Plants also possess a blue light receptor, the cryptochrome, which appears to be sensitive to a broad range of radiation from UV-A to blue light. The action spectra obtained for different blue light-dependent processes (e.g. phototropism) show a maximum near 450 nm (see Häder and Tevini, 1982; Mohr and Schopfer, 1995). It has been speculated that particular cellular flavins may be the responsible cryptochrome pigment, but more recently also zeaxanthin has been discussed as the possible blue-light receptor of plants (Quinones, 1994). Whereas phytochrome seems to be an endogenous sensor for light quality in plants, i.e. the spectral composition of the incident light, cryptochrome appears to be the plants' sensor for light quantity. This view is supported by the observation that low intensity blue light has the ability to induce the formation of high-light and sun-type chloroplasts (Lichtenthaler and Buschmann, 1978).

Cryptochrome is not the only blue light receptor in plants, since both forms of phytochrome, P_r and P_{fr}, also possess distinct and partially overlapping absorption bands in the blue spectral region. Depending on the wavelength region blue light can be used to shift the phytochrome balance either towards the P_r or the P_{fr} form. For this reason there may exist many interaction possibilities between cryptochrome induced and active phytochrome-induced photoresponses. In many cases it appears not be clear if a particular blue light-induced photoresponse is solely due to cryptochrome or if it is a co-action of active phytochrome and the blue light receptor. Here the application of the new technique of creating synthetic optical spectra may help. By using a blue light band, activating cryptochrome and a red or far-red band transferring phytochrome either to its active or inactive form, P_{fr} or P_r, it can clearly be differentiated between pure cryptochrome and a possible co-action response with active phytochrome. In this way one could investigate a possible existence of cryptochrome responses that either require the presence of P_{fr} or of its inactive form P_r in order to be implemented.

UV-A and UV-B effects

In recent years the potential damaging effect of an enhanced UV-B radiation (250–320 nm) on the physiology, growth and development of plants, including a decrease in the yield of crop plants, has attracted much attention (Mark and Tevini, 1996). The reason for the exposure of the earth surface to enhanced UV-B radiation is the progressing reduction of the stratospheric ozone layer, which had so far absorbed a major part of the incident solar UV-B radiation. Thus, more solar UV-irradiance can pass through the earth atmosphere unabsorbed (cf. Fig. 1), and this proportion is expected to increase much further in the future.

Numerous UV-B effects have been described including the points of interaction and damage sites with cell metabolism. From the UV-B research results of different groups it appears, however, that various UV-B effects are only found when sufficient amounts of UV-A (320–390 nm) simultaneously act on the plants together with UV-B. This apparent co-action of UV-A and UV-B requires a thorough investigation with defined but varying levels of the relative proportions of UV-A and UV-B. For this type of research the new technique of synthetic optical light spectra using AOTFs provides ample research possibilities. From a broad band light source spectra, which also emits UV-A and UV-B, the AOTFs allow only particular wavelengths in the UV-A, UV-B and visible light regions to pass on to the plant.

The synthetic optical spectra technique also permits to test which wavelength regions of the broad UV-B and UV-A bands are more damaging than others, and which wavelength regions are necessary for an apparent co-action of UV-A and UV-B. At the same time the photon flux density of the visible light (390 to 730 nm) can be modulated with respect to the UV-A and UVB amounts. This opens the opportunity to study and simulate UV-A and UV-B effects, e.g. under high-light and low-light growth conditions of plants. Much of the incident UV-radiation is absorbed by the epidermis of leaves (Caldwell et al., 1983; Schweiger et al., 1996).

The epidermis contains depending on the plant species and growth conditions varying amounts of UV-A and UV-B absorbing flavonols, hydroxycinnamic acids and other secondary plant products, that may serve as an internal protection and filter, e.g. of the photosynthetic apparatus in the mesophyll cells, or against damaging UV-A and UVB radiation. In fact, via UV-induced fluorescence excitation spectra we were able to show that high-light plants and outdoor plants are much better protected against UV-A and UV-B than low-light plants and green house plants (Stober and Lichtenthaler, 1993; Schweiger et al., 1996). Defined synthetic optical spectra can help to find out which UV-radiation regions and which dose is going to induce the enhanced formation of protective plant phenolics and other UV absorbing compounds in epidermis cells and cell walls. In addition, with the application of defined acousto-optic filters, it is also possible to determine if the formation of such protective substances are solely induced by UV-radiation, or if they already show up at high doses of UV-free visible light.

Interaction with phytochrome and cryptochrome

With the application of defined synthetic optical spectra one could test the effects of UV-A and UV-B on plants when the red/far-red reversible photomorphogenic photoreceptor phytochrome is in the physiologically active (P_{fr}) or inactive form (P_r). Similar interactions of UV-A and UV-B with the blue light photoreceptor cryptochrome might exist and can be studied as well.

Application for measuring plant fluorescence spectra and images

Upon excitation with either UV-radiation or blue, green and red light (λ 640 nm) green plants emit the red and far-red chlorophyll *a* fluorescence. It is characterized by maxima in the red spectral region between 684 and 692 nm, and in the far-red range from 730 to 740 nm, which is referred to as F690 and F740 respectively (Fig. 4, right half).

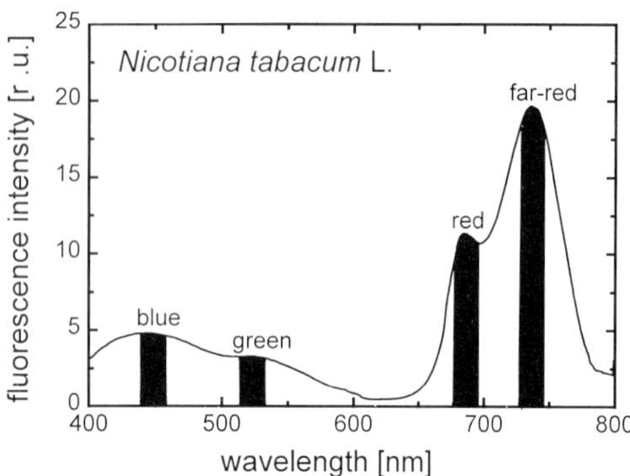

Fig. 4: UV-A induced (355 nm) fluorescence emission spectrum of a green plant leaf with maxima or shoulders in the blue (F440), green (F520), red (F690) and far-red region (F740). The four fluorescence bands applied for high resolution fluorescence imaging of plants (cf. Lichtenthaler et al., 1996) are indicated.

Under normal physiological and photosynthetic conditions the predominant part of the light energy absorbed by leaves is used for photochemical quantum conversion, and only a very small fraction is de-excited via chlorophyll *a* fluorescence emission. Although isolated chlorophyll *b* also exhibits a fluorescence emission (Lichtenthaler and Pfister, 1978), this does not occur *in vivo* in the leaves and chloroplasts, since there the excited states of chlorophyll *b* are transferred 100 % to chlorophyll *a*. The photosynthetic carotenoids are yellow pigments, which absorb light for photosynthetic quantum conversion in the blue region, and transfer their excited states to chlorophyll *a* and *b*. Even though this excitation transfer from carotenoids does not proceed to 100 %, the carotenoids do not show any fluorescence emission, but de-excite apparently only by heat emission.

By using different excitation wavelengths, e.g. in the absorption bands of carotenoids, chlorophyll *b* and chlorophyll *a*, one can study the efficiency of the excitation transfer from the different light absorbing pigments by measuring the red and far-red chlorophyll fluorescence yield. This can be achieved by either measuring the fluorescence bands F690 and F730, or by sensing the complete chlorophyll fluorescence emission spectra, e.g. applying a CCD-OMA fluorometer (charge couple device optical multichannel analyzer) (Buschmann et al., 1996). With the help of synthetic optical spectra filter systems for a strong broad-band light source as excitation source, that type of excitation energy transfer measurements can more easily be performed. It could be proved, if blue light applied in the *in vivo* absorption bands of carotenoids near 470 nm can still increase the chlorophyll *a* fluorescence yield, when the latter had already been excited via the *in vivo* red absorption bands of chlorophyll *b* near 653 nm.

2. *Blue-green fluorescence*

When excited with UV-A radiation plants also possess a blue and green fluorescence emission primarily emitted by various plant phenolics (e.g. hydroxy-cinnamic acids) (Lang et al., 1991; Stober and Lichtenthaler, 1993 a, b), which are covalently bound to the cellulose of the epidermis cells. The blue fluorescence exhibits a maximum near 440 to 450 nm and is referred to as F440, whereas the green fluorescence exhibits a shoulder near 520 to 530 nm, termed F520 (Fig. 4). Under particular growth or stress conditions the green emission can be represented by a maximum that may be higher than that of the blue fluorescence. In contrast to the red and far-red chlorophyll fluorescence, which exhibits a variable component, the blue and green fluorescence emission of plant leaves is constant during illumination (Stober and Lichtenthaler, 1993 b), and can be taken as an internal standard to detect, e.g. stress induced changes in the chlorophyll fluorescence emission. Using synthetic AOTF filters for excitations, the interaction between blue and green fluorescence emission and its possibly different origin could be studied.

3. *Fluorescence imaging*

More recently the laser induced high resolution fluorescence imaging system was established in a joint cooperation between physicists and plant physiologists (Lang et al., 1994 and 1996; Lichtenthaler et al., 1996), and allows to screen the blue, green, red and far-red fluorescence images of whole leaves or even several leaves. This LIF imaging proves to be an excellent technique for early stress detection in plants, and therefore is much superior to the sampling of fluorescence data from punctuated leaf parts (point data measurements) applied so far. LIF images cannot only be obtained from the pure blue, green, red and far-red fluorescence bands, but also from the fluorescence ratios blue/red (F440/F690) and blue/red (F440/F740) as well as images of the red/far-red chlorophyll fluorescence ratio (F690/F740) or the ratio blue/green (F440/F520). It was proved that in particular the fluorescence ratios blue/red and blue/far-red are very sensitive to changes in growth conditions of plants, and even more so in response to various stress constraints (Lang et al., 1996; Lichtenthaler et al., 1996).

At present the LIF images in the four fluorescence bands blue, green, red and far-red are recorded one after the other by manually changing the filters (in a filter wheel). With the new technique of creating synthetic AOTF-filters, the consecutive imaging in the four fluorescence bands could be controlled electronically and performed much faster and cheaper. In addition a 5th wavelength region (the 620 nm range) can be included in the imaging, in order to obtain the information if the blue-green fluorescence declines in the orange region (which is mostly the case) or if it might extend into and overlaps with the red fluorescence emission band F690. If the latter is the case, the F690 chlorophyll value will have to be corrected with the help of the 620 nm fluorescence information.

At present a ND:YAG (λ 355 nm) is applied as excitation source in the LIF imaging device. Yet, such lasers are very expensive and their use might be restricted for financial reasons to an outdoor LIF imaging sensor being used for remote sensing of plants which requires a strong laser. In the case of laboratory LIF instrumentations one could create, however, a much cheaper excitation source by applying particular synthetic optical spectra (AOTFs as filters) which permit only the desired wavelength range from a broad-band light source to pass and to be sent as excitation light to the plant to be investigated. This is the topic of our present research within the cooperation of the Technical University of Budapest and the University of Karlsruhe.

Conclusion

The new technique of synthesizing defined optical spectra, using special acousto-optic filters (AOTFs) and applying different radiowave frequencies, creates many opportunities to study the different photobiological processes in plants under the influence of light with a characteristic spectral composition. Defined synthetic optical spectra will allow:

1) to check the relative importance and contribution of different pigment forms (chlorophylls, carotenoids) for the performance of the photosynthetic light reactions,
2) to optimize the spectral characteristics of particular photobiological processes and to design, on the basis of the optimized spectra, large scale illumination units for photobiological studies,
3) to study the interaction of the blue light receptor cryptochrome with the red/far-red sensitive photoreceptor phytochrome, and
4) to create special transmittance filters for the excitation and recording of the plants' fluorescence emission spectra and in particular for high resolution fluorescence imaging of plants.

More applications of AOTFs in plant research are possible once a particular photoeffect on the formation of plant substances has been detected. Thus, one could think of a multispectral waveband-directed synthesis of particular plant enzymes or secondary plants products when the particular wavelength regions necessary for their accumulation have been established.

Acknowledgements

The concept presented was developed within the scientific cooperation of the Technical University of Budapest, Hungary with the University of Karlsruhe (T.H.), Germany. We wish to thank the German Academic Exchange Service (DAAD) Bonn and the Hungarian counterpart in Budapest for travel support.

References

Baker, N. R. and J. R. Bowyer (eds.): Photoinhibition of Photosynthesis Scientific Publishers, Oxford, 1994.

Buschmann, C., D. Meier, H. K. Kleudgen, and H. K. Lichtenthaler: Regulation of chloroplasts development by red and blue light Photochemistry and Photobiology *27,* 195–198 (1978).

Buschmann, C., J. Schweiger, P. Richter, and H. K. Lichtenthaler: Application of the Karlsruhe CCD-OMA LIDAR-Fluorosensor in stress detection of plants. J. Plant Physiol. *148,* 548–554 (1996).

Caldwell, M., R. Robberecht, and S. D. Flint: Internal filters: prospect for UV-acclimation in higher plants. Physiol. Plantarum *58,* 445–450 (1983).

Häder, D.-P. and M. Tevini: General Photobiology. Pergamon Press, Oxford, 1987.

Hewlett-Packard: 8770A, Arbitrary Waveform Synthesizer. T & M Catalog, 1995.

Karpel, A.: Acousto-Optics. Marcel Dekker, New York, 1988.

Krause, G. H. and E. Weis: Chlorophyll fluorescence and photosynthesis: The basics Ann. Rev. Plant Physiol. Plant Mol. Biol. *42,* 319–349 (1991).

Lang, M., F. Stober, and H. K. Lichtenthaler: Fluorescence emission spectra of plant leaves and plant constituents. Rad. Environ. Biophysics *30,* 333–347 (1991).

Lang, M., H. K. Lichtenthaler, M. Sowinska, P. Summ, and F. Heisel: Blue, green and red fluorescence signatures and images of tobacco leaves. Botanica Acta *107,* 230–236 (1994).

Lang, M., H. K. Lichtenthaler, M. Sowinska, F. Heisel, J. A. Miehé, and F. Tomasini: Fluorescence imaging of water and temperature stress in plant leaves. J. Plant Physiol. *148,* 613–621 (1996).

Lawlor, D. W.: Photosynthesis, Molecular, Physiological and Environmental Processes (2nd edition). Longman Scient. & Technical, Harlow, 1993.

Lichtenthaler, H. K.: Chlorophylls and carotenoids, the pigments of photosynthetic biomembranes. In: Douce, R. and L. Packer (eds.): Methods Enzymol. *148,* 350–382. Academic Press Inc., New York (1987).

– *In vivo* chlorophyll fluorescence as a tool for stress detection in plants In: Lichtenthaler, H. K. (ed.): Applications of Chlorophyll Fluorescence, pp. 143–149. Kluwer Academic Publishers, Dordrecht (1988).

Lichtenthaler, H. K. and K. Becker: Changes of the plastidquinone and carotenoid metabolism associated with the formation of functioning chloroplasts in continuous far-red and white light. In: Proceedings of the 2nd Internat. Congress on Photosynth. Research, Vol. *3,* pp. 2451–2459. Dr. W. Junk, N.V. Publishers, The Hague (1972).

Lichtenthaler, H. K. and C. Buschmann: Control of chloroplast development by red light, blue light and phytohormones. In: Akoyunoglou, G. et al. (eds.): Chloroplast Development, pp. 801–816. Elsevier/North-Holland Biomedical Press, Amsterdam (1978).

Lichtenthaler, H. K. and K. Pfister: Praktikum der Photosynthese. Quelle & Meyer Verlag, Heidelberg, 1978.

Lichtenthaler, H. K. and H. K. Kleudgen: Phytochromsystem und Lipochinonsynthese in den Plastiden etiolierter *Hordeum*-Keimlinge. Z. Naturforsch *30c,* 64–66 (1975).

Lichtenthaler, H. K., C. Buschmann, and U. Rahmsdorf: The importance of blue light for the development of sun-type chloroplasts. In: Senger, H. (ed.): The Blue Light Syndrome, pp. 485–494. Springer-Verlag, Berlin (1980).

Lichtenthaler, H. K., C. Buschmann, M. Döll, H.-J. Fietz, T. Bach, U. Kozel, D. Meier, and U. Rahmsdorf: Photosynthetic activity, chloroplast ultrastructure, and leaf characteristics of high-light and low-light plants and of sun and shade leaves Photosynth. Research *2,* 115–141 (1981a).

Lichtenthaler, H. K., G. Burkard, G. Kuhn, and U. Prenzel: Light-induced accumulation and stability of chlorophylls and chlorophyll-proteins during chloroplast development in radish seedlings. Z. Naturforsch. *36c,* 421–430 (1981b).

Lichtenthaler, H. K., G. Kuhn, U. Prenzel, and D. Meier: Chlorophyll-protein levels and stacking degree of thylakoids in radish chloroplasts from high-light, low-light and bentazon-treated plants. Physiol. Plant. *56,* 183–188 (1982a).

Lichtenthaler, H. K., G. Kuhn, U. Prenzel, C. Buschmann, and D. Meier: Adaptation of chloroplast-ultrastructure and of chlorophyll-protein levels to high-light and low-light growth conditions. Z. Naturforsch *37c,* 464–475 (1982b).

Lichtenthaler, H. K., U. Prenzel, and G. Kuhn: Carotenoid composition of chlorophyll-carotenoid-proteins from radish chloroplasts Z. Naturforsch *37c,* 10–12 (1982c).

Lichtenthaler, H. K., S. Burkart, C. Schindler, and F. Stober: Changes in photosynthetic pigments and *in vivo* chlorophyll fluorescence parameters under photoinhibitory growth conditions. Photosynthetica *27,* 343–353 (1992).

Lichtenthaler, H. K., M. Lang, M. Sowinska, F. Heisel, and J. A. Miehé: Detection of vegetation stress via a new high resolution fluorescence imaging system J. Plant Physiol. *148,* 599–612 (1996).

Mark, U. and M. Tevini: Combination effects of UV-B radiation and temperature on sunflower (*Helianthus annuus* L.) and maize (*Zea mays* L.) seedlings. J. Plant Physiol. *148,* 49–56 (1996).

Mohr, H. and P. Schopfer: Plant Physiology. Springer Verlag, Berlin, 1995.

Quinones, M. A. and E. Zeiger: A putative role of the xanthophyll zeaxanthin in blue light photoreception of corn coleoptiles. Science *264,* 558–561 (1994).

Schweiger, J., M. Lang, and H. K. Lichtenthaler: Differences in fluorescence excitation spectra of leaves between stressed and non-stressed plants. J. Plant Physiol. *148,* 537–547 (1996).

Stober, F. and H. K. Lichtenthaler: Characterisation of the laser-induced blue, green and red fluorescence signatures of leaves of wheat and soybean leaves grown under different irradiance. Physiologia Plantarum *88,* 696–704 (1993a).

Stober, F. and H. K. Lichtenthaler: Studies on the constancy of the blue and green fluorescence yield during the chlorophyll fluorescence induction kinetics (Kautsky effect). Radiat. Environ. Biophysics *32,* 357–365 (1993b).

Zissis, G. J. (ed.): Infrared Handbook, Vol. 1. SPIE, Bellingham, 1993.

J. Plant Physiol. Vol. 148. pp. 471–477 (1996)

Changes of Chlorophyll Fluorescence Signatures during Greening of Etiolated Barley Seedlings as Measured with the CCD-OMA Fluorometer*

FATBARDHA BABANI[1,2], HARTMUT K. LICHTENTHALER[1**], and PETER RICHTER[3]

[1] Botanisches Institut, Universität Karlsruhe, Kaiserstr. 12, D-76128 Karlsruhe, Germany

[2] Institute of Biological Research, Academy of Sciences, Tirana, Albania

[3] Department of Atomic Physics, TU Budapest, Budafoki ut. 8, H-1111 Budapest, Hungary

Received May 20, 1995 · Accepted July 10, 1995

Summary

In greening barley seedlings the chlorophyll fluorescence emmission spectra were recorded by the Karlsruhe CCD-OMA spectrofluorometer during the fast rise (fast component of Kautsky effect) and the slow decline (slow component of Kautsky effect) of the chlorophyll fluorescence kinetics. The relationships between fluorescence signatures and photosynthetic pigments were used to characterize the development of photosynthetic activity in 7 d old etiolated barley seedlings during illumination with continuous white light. The shape of the chlorophyll fluorescence spectra exhibited characteristic changes during the greening of etiolated barley seedlings. At the onset of greening and a very low chlorophyll content only one fluorescence peak near 690 nm was detectable, whereas the second fluorescence peak near 735 nm, initially expressed only as a shoulder, was developed during the chlorophyll accumulation to a separate fluorescence maximum. The time course of the fluorescence intensity near 690 nm and 735 nm at maximum (fm) and steady-state of the chlorophyll fluorescence (fs) can be explained on the basis of a partial reabsorption of the emitted red chlorophyll fluorescence band F690 by the leaf chlorophyll. The chlorophyll fluorescence ratios F690/F735 at fm and fs were determined from the CCD-OMA spectra. The decreasing of the chlorophyll fluorescence ratio F690/F735 with increasing chlorophyll content during greening of the etiolated barley leaves can be expressed by a power function (curvilinear relationship): $y = a x^{-b}$. The variable chlorophyll fluorescence decrease ratios (Rfd-values) as vitality index were calculated from the fluorescence intensities measured at 690 nm and 735 nm of the CCD fluorescence spectra. The Rdf690 and Rfd735 values increased during greening and exhibited the characteristics of a saturation curve. The latter was reached for Rfd690 after 24 h of illumination (chlorophyll content 14 $\mu g\, cm^{-2}$) and for Rfd735 already after 6 h of illumination (chlorophyll content 8 $\mu g\, cm^{-2}$). The Karlsruhe CCD-OMA spectrofluorometer is excellently suited for photosynthetic studies as well as stress detection in plants.

Key words: Carotenoid level, chlorophyll content, chlorophyll fluorescence ratio F690/F735, Hordeum vulgare L., Rfd-values, photosynthetic activity.

Abbreviations: a+b = chlorophyll *a* and *b*; CCD = charge couple device; fm = maximum chlorophyll fluorescence; Fs = steady state chlorophyll fluorescence; fd = fluorescence decrease from fm to fs during the chlorophyll fluorescence induction kinetics; F690 and F735 = chlorophyll fluorescence bands near 690 and 735 nm; F690/F735 = ratio of the two chlorophyll fluorescence emission peaks near 690 nm and 735 nm; OMA = optical multichannel analyzer; Rfd = fluorescence decrease ratio, variable fluorescence ratio; Rfd690 and Rfd735 = variable chlorophyll fluorescence ratios measured at 690 nm and 735 nm; x+c = total carotenoids (xanthophylls, x, and carotenes, c).

* This paper is dedicated to Prof. Ulrich Lüttge, Darmstadt, on the occasion of his 60th birthday.

** Correspondence.

Introduction

During greening of etiolated seedlings in the light the newly synthesized photosynthetic pigments (chlorophylls and carotenoids) are incorporated into the photochemically active thylakoids which are formed in the developing chloroplasts (Lichtenthaler, 1967 and 1969; Stober and Lichtenthaler, 1992). The photosynthetic apparatus, which is built up, results in a gradual development of photosynthetic activity (Boardman and Anderson, 1978; Lichtenthaler, 1969; Sestak, 1985; Sestak and Siffel, 1988).

Chlorophylls and carotenoids absorb radiant energy, which is used in the photosynthetic light and electron transport reactions, to generate NADP and ATP for the reduction of CO_2. A small part of the absorbed light energy is emitted, either as heat or as red chlorophyll fluorescence, which posseses two maxima near 690 nm (F690) and 735 nm (F735) (Virgin, 1954; Lichtenthaler, 1986; Lichtenthaler and Buschmann, 1987). The inverse relationship between *in vivo* chlorophyll fluorescence and photosynthetic activity was first described by Kautsky and Hirsch (1931). The chlorophyll fluorescence can be analyzed by measuring the decay of fluorescence, the kinetics of the fluorescence induction or the complete fluorescence emission spectra in the course of the induction kinetics (Baker and Bradbury, 1981; Buschmann and Lichtenthaler, 1988; Lichtenthaler, 1986; Papageorgiou, 1975; Sestak and Siffel, 1988). The study of the light induced *in vivo* chlorophyll fluorescence provides the basic information on the functioning of the photosynthetic apparatus (Krause and Weis, 1984 and 1991; Lichtenthaler and Rinderle, 1988; Schreiber, 1983). The chlorophyll fluorescence ratio at the two maxima F690/F735 is established as an indicator of the *in vivo* chlorophyll content (Hák et al., 1990; D'Ambrosio et al., 1992; Lichtenthaler and Rinderle, 1988). The complete fluorescence emission spectra at room temperature during the fast fluorescence induction kinetics can be recorded only with a fast optical multichannel analyzer system (OMA) which gives the advantage to carry out spectra measurements within a short time delay (10 to 20 ms range) (Buschmann and Schrey, 1981; Lichtenthaler and Buschmann, 1987). The shape of the complete laser-induced fluorescence emission spectra during induction kinetics can better be determined by the new CCD-OMA spectrofluorometer which exhibits a parallel readout (Szabó et al., 1992). The time resolution of the CCD-OMA is 10 ms. The fluorescence ratio F690/F735 and the variable fluorescence decrease ratio, Rfd values at 690 and 735, can be determined from the emission spectra.

The aim of this work was to study the development of photosynthetically active chloroplasts during illumination of etiolated barley seedlings by following the accumulation of photosynthetic pigments (chlorophylls and carotenoids) and by measuring the changes in several chlorophyll fluorescence parameters using the Karlsruhe CCD-OMA spectrofluorometer.

Materials and Methods

Plants

Barley seedlings (*Hordeum vulgare* L.) were grown for 7 days in the dark on peat with full complement of minerals for growth (TKS2). After 7 days the etiolated plants were exposed to moderate continuous white light (180 $\mu mol\ m^{-2} s^{-1}$) for 32 hours. The greening process was followed by determination of chlorophyll fluorescence emission spectra and kinetics as well as pigment content. The greening process was repeated four times.

Pigment determination

The pigments of the photosynthetic apparatus (chlorophyll *a, b* and total carotenoids x+c) of primary barley leaves were extracted in 100 % acetone and determined in a Shimadzu Spectrophotometer W-2101PC using the redetermined extinction coefficients and equations of Lichtenthaler (1987). In the first hour of illumination only trace amounts of chlorophyll *b* were found in the greening barley leaves which could not be estimated by the spectrophotometric method applied here. Up to 1 h of illumination the level of chlorophyll *a* was determined on the basis of the specific absorption coefficients of chlorophyll *a* (of 92.45 at 661.6 nm) and of total carotenoids x+c at 444 nm using an absorption coefficient of E 1 % = 2500. The pigment extracts were prepared from the middle part of the primary leaves of etiolated and greening barley seedlings between 3 and 5 cm from the top. The values represent the mean of 8 separate extracts made from small leaf discs (0.25 cm^2) of 15 leaves (0 to 4 h of illumination) and 10 leaves after 6 hours of greening.

Fluorescence emission spectra

The complete chlorophyll fluorescence emission spectra were measured of the middle part of 20 min pre-darkened etiolated primary leaves of barley during the greening process using the multichannel CCD-OMA chlorophyll fluorometer as described by Szabó et al. (1992). A 10 mv He/Ne laser (632.8 nm) was used as excitation source. The laser beam illuminated the sample (ca. 0.8 mm^2) after opening a fast electromechanical shutter (1 ms opening time). The emitted chlorophyll fluorescence was imaged into the entrance slit of a spectrograph (Jobin Yvon H25). In front of the spectrograph a red cut-off filter (>645 nm) was applied. Inside the spectrograph a grating (600 lines per mm and 6 nm per mm at the image plan) was used to resolve the complete chlorophyll fluorescence spectrum from 650 to 800 nm. A charge coupled device (CCD) sensor (with 2048 elements) simultaneously detected the complete chlorophyll fluorescence emission spectrum in the 650–800 nm wavelength region. Data acquisition and processing were accomplished by a PC computer. The fluorescence spectra were measured during the fast rise (fast component of the Kautsky effect) and the slow decline of the chlorophyll fluorescence kinetics (show component of the Kautsky effect) to the steady-state (0–300 s) in 20 ms periods at different times during the chlorophyll fluorescence induction kinetics. The chlorophyll fluorescence ratio F690/F735 was determined at maximum fluorescence (fm) and at steady-state fluorescence (fs). The variable fluorescence decrease ratios (Rfd-values at 690 nm and 735 nm), which are defined as the ratio of fluorescence decrease fd (from fm to fs) to the steady state fluorescence (Rfd = (fm-fs)/fs = fd/fs), were calculated from the fluorescence intensities measured at 690 and 735 nm and expressed as Rfd 690 and Rfd 735.

Results

Pigment content and pigment ratio

The level of the chlorophylls and total carotenoids continuously increased during greening of etiolated barley seedlings (Fig. 1 A), yet the pigments exhibited different rates of accumulation. The chlorophyll content (a+b) of 7 d old etiolated barley seedlings started from 0.3 μg cm^{-2} after 10 min of illumination and then consisted solely of chlorophyll *a*. Chlorophyll *b* was clearly detectable after 2 h of illumination. The chlorophyll *a*+*b* content initially rose in a sigmoid pattern (Fig. 1 A). The rate of the accumulation of carotenoids, which had already been formed during the growth of seedlings in darkness, was different. It started from a level of 1.9 μg cm^{-2} and reached 3.8 μg cm^{-2} leaf area after 30 h of illumination. The ratio of chlorophylls *a*/*b* (Fig. 1 B) was high (14.9) after 2 h of illumination and reached constant values of ca. 3 after 8 h of illumination. The ratio of green to yellow pigments, i.e. of chlorophylls to carotenoids (a+b)/(x+c), increased continuously from 0.1 after 10 min of illumination of etiolated leaves to values of 5 (= regular values of green leaves) after 30 h of illumination exhibiting a saturation curve.

Chlorophyll fluorescence emission spectra

The laser-induced chlorophyll fluorescence emission spectra as measured during the rise and decline of the fluorescence in the course of the induction kinetics, showed a fast rise to maximum fluorescence (within 200–240 ms) and then declined to the steady-state fluorescence after 300 s (Fig. 2). This is known as Kautsky effect and shown here for the fluorescence emission spectra. The shape of the chlorophyll fluorescence

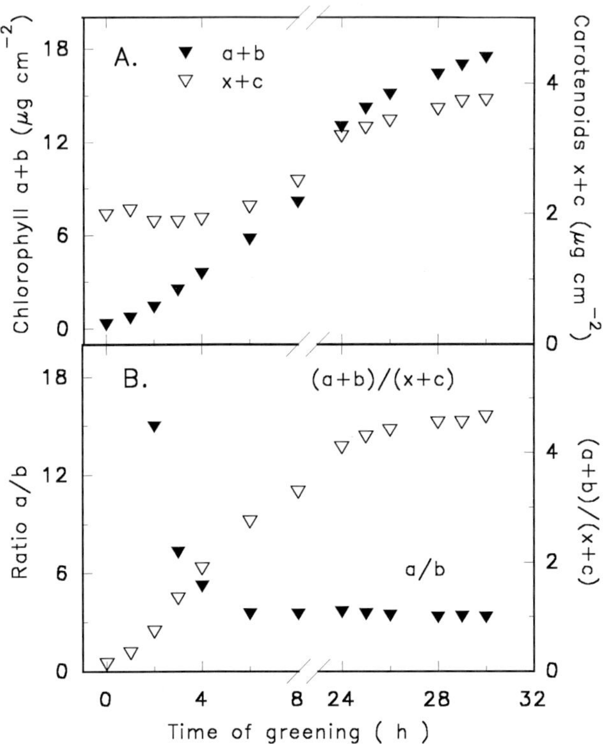

Fig. 1: Level of chlorophylls *a*+*b* and total carotenoids x+c as well as pigment ratios chlorophyll *a*/*b* and chlorophylls to carotenoids (a+b)/(x+c) in etiolated barley leaves during greening. Mean of 8 determinations. Standard deviations ≤10 % from 0 to 4 h and ≤5 % from 6 to 30 h of illumination.

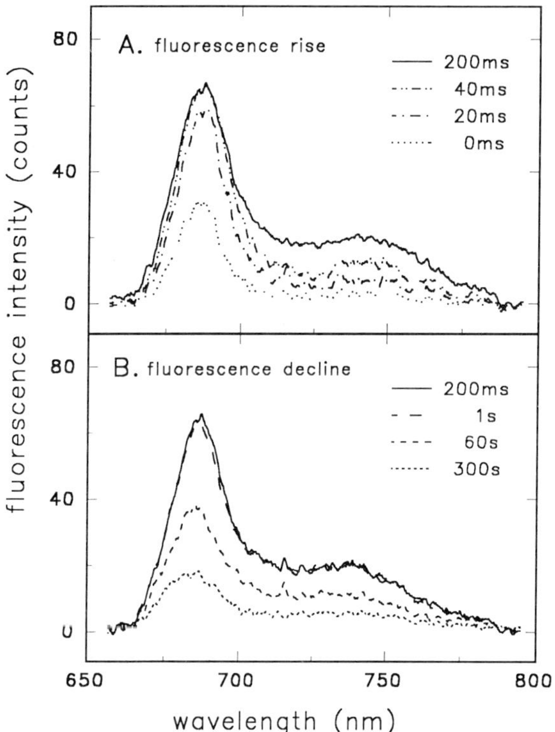

Fig. 2: Chlorophyll fluorescence emission spectra taken during the light induced fluorescence rise (A) and decline (B) and measured in primary leaves of etiolated barley seedlings after 2 h of illumination with white light and a low chlorophyll content.

emission spectra exhibited characteristic changes during the greening of etiolated barley seedlings. At the onset of greening with very low chlorophyll content of leaves only one fluorescence peak near 690 nm was detectable, whereas the second peak near 735 nm was expressed only as a shoulder, but developed further during the chlorophyll accumulation to a separate maximum.

The time course of the fluorescence intensity F690 and that of F735 at maximum and steady-state of the chlorophyll fluorescence induction kinetics showed different characteristics (Fig. 3). The fluorescence intensity at fm of F690 increased during the first hours of illumination until 8 h and then decreased. In contrast the fluorescence intensity F735 continuously increased during the greening of etiolated leaves. This differential change in the intensities of F690 and F735 during greening was also observed at the steady-state fluorescence fs (Fig. 3 B). The decrease of F690 with increasing illumination time is due to a partial reabsorption of the emitted red chlorophyll fluorescence absorption band of chlorophyll *a* (D'Ambrosio et al., 1992; Hák et al., 1990; Lichtenthaler and Rinderle, 1988). The difference in F690 and F735 is equally true if the two fluorescence bands are determined as a narrow or broad band of 1 nm or 10 nm band width (Fig. 4 A, B). The fluorescence intensity F735 at fm continuously increased with increasing of chlorophyll content in barley leaves, whereas F735 at fs showed a saturation curve with increasing chlorophyll content (Fig. 4 A, B).

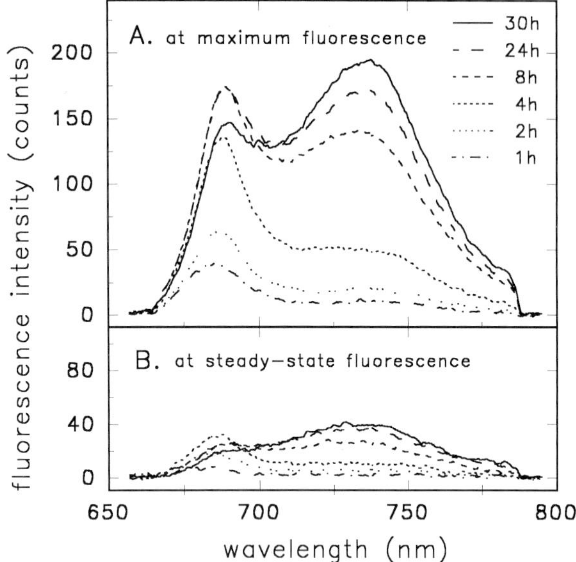

Fig. 3: Shape of the chlorophyll fluorescence emission spectra at maximum fluorescence fm and steady-state fluorescence fs during greening of primary leaves of etiolated barley seedlings. The illumination times in hours are indicated.

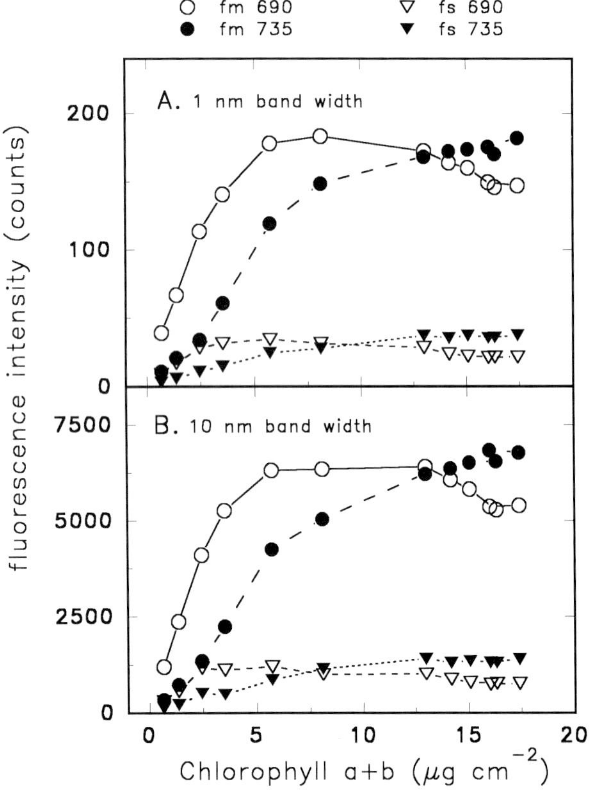

Fig. 4: Development of maximum (fm) and steady-state chlorophyll fluorescence (fs) during greening of primary leaves of etiolated barley seedlings. The fluorescence parameters are given for the 690 nm (○ fm; ▽ fs) and the 735 nm region (● fm; ▼ fs). In A) the fluorescence intensity was measured at a 1 nm band width and in B. at a 10 nm band width. Mean of 6 determinations, standard deviation <5%.

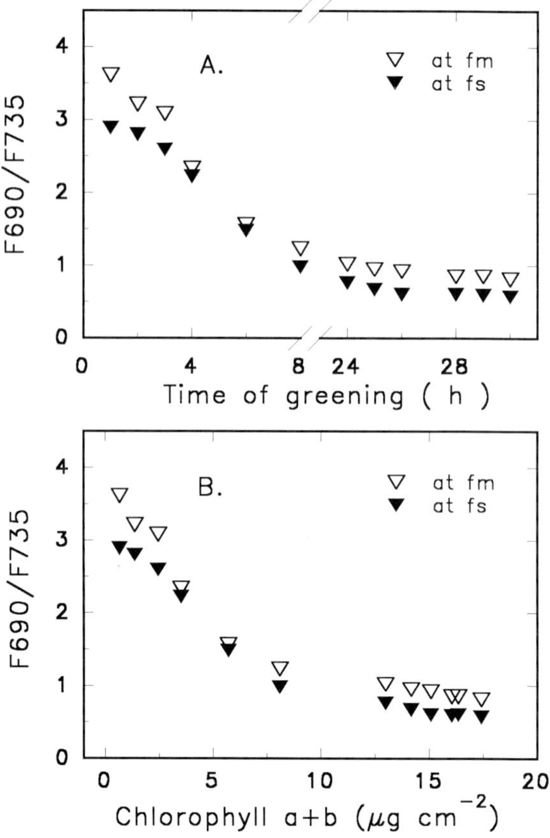

Fig. 5: Decrease in the values of the chlorophyll fluorescence ratio F690/F735 with increasing greening time (A) and chlorophyll *a+b* content (B) of primary barley leaves. The ratio F690/F735 was determined from the fluorescence emission spectra measured in the CCD-OMA fluorometer at maximum and steady state fluorescence fm and fs (band width 1 nm). Standard deviation 6% or less.

Chlorophyll fluorescence ratio F690/F735

In the primary leaves of etiolated barley seedlings the chlorophyll fluorescence ratios F690/F735, determined at maximum (fm) and steady-state (fs) fluorescence, were decreasing with the increasing chlorophyll content during the illumination period of 30 h (Fig. 5). The values of the F690/ F735 ratio at fm (3.7) and fs (2.9) decreased fast within the first 8 h of illumination and up to a chlorophyll *a+b* content of 8 µg cm^{-2}. Thereafter the decrease was much slower with final values of 0.85 at fm and 0.6 at fs. The F690/ F735 ratios were calculated as a sum of the fluorescence measuring points in the 10 nm band width (690 ± 5 nm and 735 ± 5 nm bands) and 1 nm band width in each peak. The values were the same in both cases indicating that it is not required to measure at a broad band width when one wants to determine the fluorescence ratio F690/ F735 or use this ratio for a non-destructive chlorophyll determination. The correlation between the ratio F690/F735 and chlorophyll content can be expressed by a curvilinear function ($y = a x^{-b}$): at fm a = 3.40, b = 0.424 and at fs a = 2.85, b = 0.439 (10 nm band width); at fm a = 3.40, b = 0.427 and at fs a = 2.85, b = 0.438 (1 nm band width). The average values of the fluorescence ratio

F690/F735 are between 20 to 30% higher at maximum fluorescence fm than at steady-state fluorescence fs as had been shown before (Kocsanyi et al., 1988). The decline of F690/F735 values with increasing greening time, due to the increasing reabsorption of the F690 fluorescence, proceeds, however, in both cases, thus reflecting the continously increasing chlorophyll content. Correspondingly, the correlation between the fluorescence ratios F690/F735 at fm and fs can be expressed by a linear function (y = ax+b) with: a = 1.171, b = 0.118 at a 10 nm band width and a = 1.086, b = 0.018 at a 1 nm band width (cf. Fig. 6).

The Rfd values

The fluorescence decrease ratio Rfd = fd/fs is a vitality index of plants and of the photosynthetic apparatus (Lichtenthaler and Rinderle, 1988) and reflects the potential photosynthetic net CO_2 assimilation rate of leaves (Tuba et al., 1994). The variation of the chlorophyll fluorescence decrease ratio (Rfd690 and Rfd735), determined from the CCD-OMA spectra at 690 nm and 735 nm (10 nm band width) during the increase of the chlorophyll content in the primary leaves of etiolated barley seedlings, exhibited specific characteristics (Fig. 7). The Rfd690 values increased during illumination up to a chlorophyll content 15 $\mu g\,cm^{-2}$ at 25 h of illumination. In contrast, the Rfd735 values exhibited the characteristics of a saturation curve (Fig. 7). At the beginning of the greening process the Rfd735 values increased from 2 (1 h light) to 4 (6 h light), but remained fairly constant at and beyond a chlorophyll content of 8 $\mu g\,cm^{-2}$. These results indicate that the Rfd690 values exhibit a much wider amplitude than the Rfd735 values and are thus a better and more correct measure of the increasing photosynthetic CO_2 fixation activity.

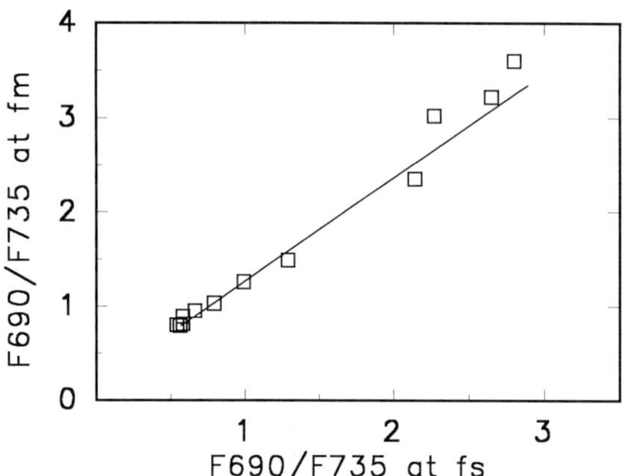

Fig. 6: Linear correlation between the chlorophyll fluorescence ratio F690/F735 determined at maximum fluorescence fm and at steady-state fluorescence fs of greening primary leaves of barley seedlings. The values of the ratio F690/F735, while decrease with increasing chlorophyll content, are based on fluorescence measurements made at a band width of 10 nm

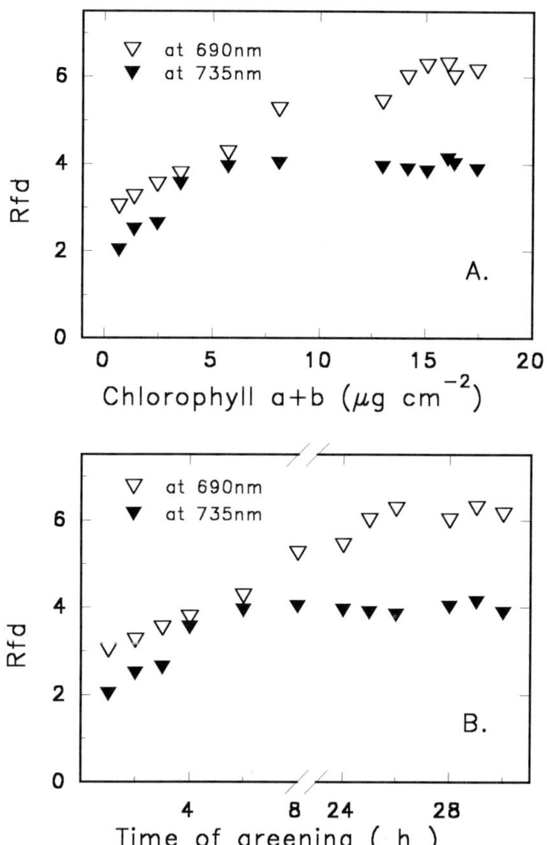

Fig. 7: Increase in the values of the variable chlorophyll fluorescence ratio (Rfd values at 690 and 735 nm) with increasing chlorophyll content (A) and time of greening (B) of primary leaves of barley seedlings. The values are based on fluorescence measurements at a band width of 10 nm. (Standard deviation 5% or less).

Discussion

During greening and development of a functional photosynthetic apparatus in etiolated barley seedlings the level of chlorophyll *a+b* and total carotenoids x+c increases, whereas the pigment ratios a/b and (a+b)/(x+c) decrease (Fig. 1A, B). After 30 h illumination of etiolated barley seedlings the pigment ratios are those of other green leaf tissues. The rate of accumulation of chlorophylls during illumination was higher than that of carotenoids, since the latter had already been accumulated in the dark grown etiolated seedlings. The strong decrease of the chlorophyll *a/b* ratio within 8 h of illumination demonstrates that in the first hours of greening the rate of accumulation of chlorophyll *a* precedes that of chlorophyll *b*. Since the etiolated seedlings already contained carotenoids the pigment ratio (a+b)/(x+c) was initially very low at the beginning of greening (up to 8 h illumination) and then increased to normal values of about 4.8 to 5 which are regularly found in green leaf tissues of many plants. These differential pigment accumulations and changes in pigment ratios during light-induced chloroplast development had been shown before (Lichtenthaler, 1967 and 1968) and are confirmed here.

The complete fluorescence emission spectra measured in the CCD-OMA spectrofluorometer during the fluorescence induction kinetics exhibited changes which can be correlated with the chlorophyll content of the greening etiolated barley seedlings. At the beginning of greening the chlorophyll fluorescence spectra of the primary leaves exhibited a maximum near 690 nm (685–686 nm) and a fairly low shoulder near 735 nm (Fig. 2 A, B). At the ontset of greening the chlorophyll content in the primary leaves was very low and reabsorption of the emitted chlorophyll fluorescence F690 by the leaf chlorophylls can be excluded (Lichtenthaler et al., 1986; Lichtenthaler and Rinderle, 1988; Virgin, 1954). Consequently, at the beginning of greening the chlorophyll fluorescence ratio F690/F735 exhibited high values at the fluorescence maximum (200 ms) and at steady-state of fluorescence (300 s) of 3.7 and 2.9, respectively. The fluorescence intensity of the F690 band at fm and fs increased with the chlorophyll content up to 6–8 $\mu g\,cm^{-2}$ (6–8 h illumination), but decreased thereafter due to a reabsorption of the F690 fluorescence. The fluorescence intensity of the F735 band, in turn, which does not overlap with the chlorophyll absorption bands, increased continuously and initially fast (up to 8 h of illumination) and thereafter at a lower rate (Fig. 4 A, B). During greening the fluorescence emission band F735 developed from a shoulder (0–6 h of illumination) to a distinct fluorescence maximum (after 8 h of illumination) to be seen at fm and also at fs. This can be explained by the fact that the emitted chlorophyll fluorescence F735 is little or not affected by reabsorption processes (Lichtenthaler, 1986; Lichtenthaler and Rinderle, 1988; Stober and Lichtenthaler, 1994). One has also to take into account that at room temperature the photosynthetic photosystem I can show a slight chlorophyll fluorescence emission near 735 nm which can contribute to the overall increase in the F735 fluorescence band.

There exists an inverse curvilinear relationship between the chlorophyll fluorescence ratio F690/F735 at maximum and steady-state of fluorescence and the chlorophyll $a + b$ content of primary leaves of etiolated barley seedlings which can be expressed by a power function ($y = a x^{-b}$) (Fig. 5). The same inverse relationship, which can be used for a non-destructive determination of the *in vivo* chlorophyll content, had been described before in etiolated leaves of wheat (Stober and Lichtenthaler, 1994), green leaves of trees (Hák et al., 1990) and during autumnal chlorophyll breakdown (D'Ambrosio et al., 1992) using a different fluorometer LITWaF. There exists, however, a linear relationship between the chlorophyll fluorescence ratio F690/F735 at maximum fm and at steady-state of fluorescence fs (Fig. 6) as determined here from the CCD-OMA spectra. The same correlation had been established before by measurements with the two wavelength fluorometer LITWaF (D'Ambrosio et al., 1992; Hák et al., 1990; Lichtenthaler and Rinderle, 1988) which only permits to measure the chlorophyll fluorescence kinetics in the 690 and 735 nm region. The advantage of the CCD-OMA, which gives full spectral information and kinetics as well, is evident.

The increase of the Rfd values taken at 690 nm and 735 nm, which function as indicators of the potential photosynthetic CO_2 assimilation rates of leaves (Lichtenthaler and Rinderle, 1988; Tuba et al., 1994), reflect the increasing photosynthetic net CO_2 assimilation during the greening process. The Rfd690 values increased during the illumination period up to 26 hours and are thus a better indicator of the increasing photosynthetic CO_2 assimilation rates, which increased up to about 48 h of illumination (Babani and Lichtenthaler, 1996), than the Rfd735 values which had reached a maximum already after 6 h of illumination.

Our results demonstrate the excellent suitability of the Karlsruhe CCD-OMA fluorosensor for the study of chlorophyll fluorescence emission spectra, for non-invasive determination of the *in vivo* chlorophyll content (F690/F735 ratios) as well as the determination of the fluorescence induction kinetics (Kautsky effect) at the red (F690) and far-red (F735) emission bands and the Rfd values as plant vitality indices. The CCD-OMA chlorophyll fluorometer cannot only be used for studies on chloroplast development, but also in the stress detection of plants, since most stressors affect sooner or later photosynthetic function and cause a decline in chlorophyll content.

Acknowledgements

This work was supported by a grant of BMFT Bonn within the EUREKA research programme LASFLEUR (EU 380) to H. K. Lichtenthaler and by a 3-month fellowship from DAAD, Bonn to F. Babani, which is gratefully acknowledged. We wish to thank Stefan Herzog for technical assistance, Drs. C. Buschmann and K. Szabó for the calibration of the Karlsruhe CCD-OMA fluorometer and Dr. Michael Lang for help during the preparation of the figures.

References

Baker, N. R. and M. Bradbury: Possible applications of chlorophyll fluorescence techniques for studying photosynthesis *in vivo.* In: Smith, H. (ed.): Plants and the Daylight Spectrum, pp. 355–373. Academic Press, London (1981).

Babani, F. and H. K. Lichtenthaler: Light-induced and age-dependent development of chloroplasts in etiolated barley leaves as visualized by determination of photosynthetic pigments, CO_2 assimilation rates and different kinds of chlorophyll fluorescence parameters. J. Plant Physiol. *148,* 555–566 (1996).

Boardman, N. K. and J. M. Anderson: Composition, structure and photochemical activity of developing and mature chloroplasts. In : Akoyunoglou, G. and J. H. Argyroudi-Akoyunoglou (eds.): Chloroplast Development, pp. 1–14. Elsevier, Amsterdam (1978).

Buschmann, C. and H. K. Lichtenthaler: Complete fluorescence spectra determined during the induction kinetics using a diode-array detector. In: Lichtenthaler, H. K. (ed.): Applications of Chlorophyll Fluorescence, pp. 77–84. Kluwer Academic Publishers, Dordrecht (1988).

– – Reflectance and chlorophyll fluorescence signatures of leaves. In: Lichtenthaler, H. K. (ed.): Applications of Chlorophyll Fluorescence, pp. 325–332. Kluwer Academic Publishers, Dordrecht (1988).

Buschmann, C. and H. Schrey: Fluorescence induction kinetics of green and etiolated leaves by recording the complete *in vivo* emission spectra. Photosynth. Res. *1,* 133–241 (1981).

D'Ambrosio, N., K. Szabó, and H. K. Lichtenthaler: Increase of the chlorophyll fluorescence ratio F690/F735 during the autumnal chlorophyll breakdown. Radiat. Environ. Biophys. *31,* 51–62 (1992).

Hák, R., H. K. Lichtenthaler, and U. Rinderle: Decrease of the chlorophyll fluorescence ratio F690/F730 during greening and development of leaves. Radiat. Environ. Biophys. *29,* 329–336 (1990).

Kautsky, H. and A. Hirsch: Neue Versuche zur Kohlenstoffassimilation. Naturwissenschaften *19,* 964 (1931).

Kocsanyi, L., M. Haitz, and H. K. Lichtenthaler: Measurement of laser-induced chlorophyll fluorescence kinetics using a fast acousto-optic device. In: Lichtenthaler, H. K. (ed.): Applications of Chlorophyll Fluorescence, pp. 143–149. Kluwer Academic Publishers, Dordrecht (1988).

Krause, G. H. and E. Weis: Chlorophyll fluorescence as a tool in plant physiology. II. Interpretation of fluorescence signal. Photosynth. Res. *5,* 139–157 (1984).

– – Chlorophyll fluorescence and photosynthesis: the basics. Ann. Rev. Plant. Pysiol. Mol. Biol. *42,* 313–349 (1991).

Lichtenthaler, H. K.: Beziehungen zwischen Zusammensetzung und Struktur der Plastiden in grünen und etiolierten Keimlingen von *Hordeum vulgare* L. Z. Pflanzenphys. *56,* 273–281 (1967).

– Light-stimulated synthesis of plastid quinones and pigments in etiolated barley seedlings. Biochim. Biophys. Acta *184,* 164–172 (1969).

– Laser-induced chlorophyll fluorescence of living plants. In: Proceedings of the International Geoscience and Remote Sensing Symposium (IGARSS), Zürich, Vol. III., pp. 1571–1576. ESA, Scientific and Technical Publications Branch, Noordwijk (1986).

– Chlorophylls and Carotenoids: pigments of photosynthetic biomembranes. Methods Enzymol. *148,* 350–382 (1987).

Lichtenthaler, H. K. and C. Buschmann: Chlorophyll fluorescence spectra of green leaves. J. Plant Physiol. *129,* 137–147 (1987).

Lichtenthaler, H. K. and U. Rinderle: The role of chlorophyll fluorescence in the detection of stress conditions in plants, CRC Critical Reviews in Analytical Chemistry *19,* [Suppl. 1], 29–85 (1988).

Lichtenthaler, H. K. and F. Stober: Laser-induced chlorophyll fluorescence and blue fluorescence of green vegetation. In: Proc 10th EMSeL Symposium Toulouse, pp. 234–241. EMSeL, Boulogne-Billancourt (1990).

Papageorgiou, G.: Chlorophyll fluorescence, an intrinsic probe of photosynthesis. In: Govindjee (ed.): Bioenergetics of Photosynthesis, pp. 319–371. Academic Press, New York (1975).

Rinderle, U. and H. K. Lichtenthaler: The chlorophyll fluorescence ratio F690/F730 as a possible stress indicator. In: Lichtenthaler, H. K. (ed.): Applications of Chlorophyll Fluorescence, pp. 189–196. Kluwer Academic Publishers, Dordrecht (1988).

Schreiber, U.: Chlorophyll fluorescence yield changes as a tool in plant physiology. 1. The measuring system. Photosynth. Res. *4,* 361–372 (1983).

Sestak, Z. and P. Siffel: Changes in chloroplast fluorescence among leaf development. In: Lichtenthaler, H. K. (ed.): Applications of Chlorophyll Fluorescence, pp. 85–91. Kluwer Academic Publishers, Dordrecht (1988).

Sestak, Z.: Photosynthesis during Leaf Development, pp. 76–107. Academia, Prague (1985).

Sprey, B. and H. K. Lichtenthaler: Zur Frage der Beziehungen zwischen Plastoglobuli und Thylakoidgenese in Gerstenkeimlingen. Z. Naturforsch. *21 b,* 697–699 (1966).

Stober, F. and H. K. Lichtenthaler: Changes of the laser-induced blue, green and red fluorescence signatures during greening of etiolated leaves of wheat. J. Plant Physiol. *140,* 673–680 (1992).

Szabó, K., H. K. Lichtenthaler, L. Kocsanyi, and P. Richter: A CCD-OMA device for the measurement of complete chlorophyll fluorescence emission spectra of leaves during the fluorescence induction kinetics. Radiat. Environ. Biophys. *31,* 153–160 (1992).

Tuba, Z., H. K. Lichtenthaler, Z. Csintalan, Z. Nagy, and K. Szente: Reconstitution of chlorophylls and photosynthetic CO_2 assimilation in the dessicated poikilochlorophyllous plant Xerophyta scabida upon rehydration. Planta *192,* 414–420 (1994).

Virgin, H. I.: The distortion of fluorescence spectra in leaves by light scattering and its reduction by infiltration. Physiol. Plant. *7,* 560–570 (1954).

J. Plant Physiol. Vol. 148. pp. 478–482 (1996)

Diurnal Changes in Flavonoids

Markus Veit[1] *, Wolfgang Bilger[1], Thomas Mühlbauer[1], Wolfgang Brummet[1], and Klaus Winter[2]

[1] Julius-von-Sachs-Institut für Biowissenschaften, Universität Würzburg, Mittlerer Dallenbergweg 64, 97082 Würzburg, Germany

[2] Smithsonian Tropical Research Institute, P.O. Box 2072, Balboa, Republic of Panama

Received October 30, 1995 · Accepted December 12, 1995

Summary

Field studies of a tropical tree, *Anacardium excelsum,* and a northern hemisphere high altitude fern, *Cryptogramma crispa,* revealed marked diurnal changes in soluble flavonoid content of leaves and fronds, respectively. The flavonoid content increased during the morning and decreased during the afternoon. In plants of *C. crispa* covered with UV-B absorbing filters, the flavonoid content remained at a constant level throughout the day/night cycle. Upon removal of UV-B absorbing filters (at night), the flavonoid content increased the next morning in a fashion similar to that observed in control plants maintained without filters. Decreases in photosystem II photochemical efficiency upon exposure of *C. crispa* to natural daylight were similar in plants previously covered with UV-B absorbing filters and in control plants, probably owing to the observed ability of plants to rapidly accumulate UV-B protective flavonoids.

Key words: Anacardium excelsum, chlorophyll fluorescence, Cryptogramma crispa, flavonoids, UV-B radiation.

Abbreviations: d.w. = dry weight; F_m = maximal fluorescence yield; F_o = initial fluorescence yield; F_v (= $F_m - F_o$) = variable chlorophyll fluorescence; MeOH = methanol; MET = middle European summer time; PFD = photon flux density (400–700 nm); PSII = photosystem II; UV-A = ultraviolet radiation between 320 and 400 nm; UV-B = ultraviolet radiation between 280 and 320 nm.

Introduction

Depletion of stratospheric ozone concentrations may substantially increase solar UV-B radiation at the tropospheric ground level (Blumthaler and Ambach, 1990; Madronich, 1993). UV-B radiation is known to alter anatomical and morphological plant characteristics (Strid et al., 1994; Teramura and Sullivan, 1994) and can cause damage to nucleic acids, proteins, and membrane lipids (Bornman and Teramura, 1993). PSII appears to be particulary sensitive to UV-B (Bornman et al., 1984; Strid et al., 1990).

Protection against solar UV-B radiation can be accomplished by the accumulation of leaf flavonoids that have strong spectral absorbance in the UV-B region (Robberecht and Caldwell, 1978; Caldwell et al., 1983). These flavonoids are epicuticular, bound to epidermal cell-walls and/or accumulate in the vacuoles of epidermal cells, thereby protecting the photosynthetically active chlorenchyma cells from UV damage (Caldwell et al., 1983; Vogt et al., 1991; Tevini et al., 1991).

Plants from natural ecosystems normally exposed to high UV-B radiation (montane and subalpine species as well as tropical plants from low latitudes) are generally considered to be well adapted to enhanced UV-B radiation because they have high leaf flavonoid contents (Barnes et al., 1987). Nevertheless, the protective function of leaf flavonoids has rarely been studied in wild plants *in situ.* Most UV-B sensitivity studies have been performed with seedlings of crop plants exposed to supplementary UV-B radiation in glass houses and environmental growth chambers (Tevini et al., 1991;

* Author for correspondence.

Bornman and Teramura, 1993; Reuber et al., 1993; Wilson and Greenberg, 1993; Fiscus and Booker, 1995).

Growth of plants under UV-B absorbing filters is known to decrease leaf flavonoid contents (Tevini et al., 1991; Caldwell et al., 1994) and is a useful tool to study the functional significance of flavonoids in UV-B protection under natural conditions. We have used this and other techniques such as measurements of chlorophyll *a* fluorescence to assess the role of leaf flavonoids in the temperate zone, alpine fern, *Cryptogramma crispa,* and in outer-canopy leaves of a tropical tree, *Anacardium excelsum.* We found that the content of soluble leaf flavonoids may undergo marked diurnal changes *in situ.*

Material and Methods

Plant material

Cryptogramma crispa R. Gr. ex Hook. (Pteridaceae) is a small rupestral fern, widely distributed in the montane to subalpine zone of the temperate and boreal regions of Eurasia, and occurring over considerable altitudinal ranges. Plants were studied in a natural alpine stand at 2020 m altitude in Val Calnegia, Ticino, Southern Switzerland (26° 22′ N) in August 1993 and June 1994. Single sterile fronds were collected for flavonoid analysis. *Anacardium excelsum* (Bertero & Balb.) Skeels (Anacardiaceae), a tropical tree, was studied in the seasonally dry forest of Parque Metropolitano, Panama City, Republic of Panama. Disks (1.5 cm diameter each, 6 disks per sample) for flavonoid analysis were taken from sun-exposed outer-canopy leaves of a mature tree on March 18, 1994. A construction crane provided access to the tree canopy.

On June 4, 1993, part of the population of *C. crispa* was covered with either UV-B transmitting Perspex (GS2458, 4 mm, Röhm, Darmstadt, Germany) or UV-B absorbing Perspex (Farblos 233, 4 mm, Röhm, Darmstadt, Germany), respectively. The filters (60×60 cm) were fixed (tilted 5°–30° due to inclination of the terrain) at about 40 cm height using aluminum poles and strings. Average plant height was 16 cm (fertile fronds) and 10 cm (sterile fronds). Natural UV-B radiation was reduced to 4 % under the UV-B absorbing filters. Two filters of each type were installed covering 2 to 3 plants each. Plants not covered with filters were used as controls. All plants grew on a south facing slope with run off water being available for the plants covered by filters. Experiments were conducted in August 1993 and June 1994.

UV-B measurements

UV-B radiation was measured semiquantitatively using a self-built detector. The detector consisted of a UV-sensitive photodiode UVS-1 (Vistek, Seefeld, Germany) which was enclosed in cylindrical brass housing. The photodiode was positioned under in a circular window (0.8 cm diameter) and a white Teflon sheet (1.5 mm thickness). Two optical filters were used to ensure wavelength selectivity (UG11, 1 mm; GG19, 2 mm; Schott, Mainz, Germany). The spectral response of the sensor is shown in Figure 1. The response was recorded with the photodiode and the optical system, but after removal of the Teflon sheet, using a spectrophotometer (Lambda 2, Perkin Elmer, Munich, Germany). The signal of the photodiode was corrected by the emission spectrum of the deuterium lamp of the spectrophotometer. As shown in Figure 1, the sensitivity of the sensor was maximal at 310 nm and reached values close to zero at 280 and 350 nm, respectively. Covering the sensor under full sunlight conditions (about 2000 µmol photons $m^{-2} s^{-1}$) on a horizontal plane with a UV-B and UV-A blocking filter (Balzers, Liechtenstein) caused the signal to decline to 2 %. Under the UV-B absorbing Perspex filters, which transmitted on average 2.8 % of the radiation between 290 and 330 nm (Fig. 1), 4 to 5 % of the signal obtained in natural sunlight without filter was recorded.

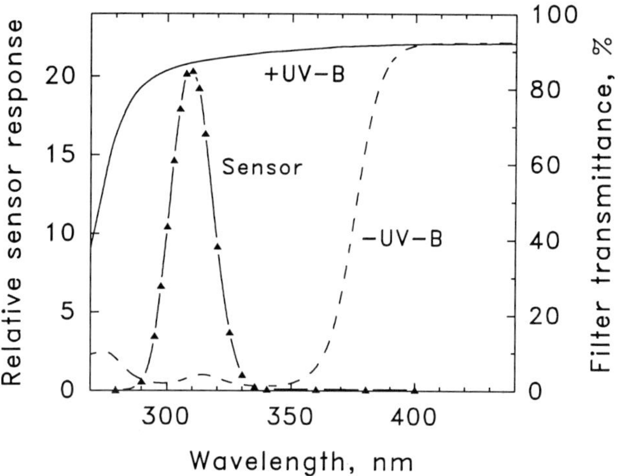

Fig. 1: Spectral sensitivity of a self-built UV-B sensor and transmittance of Perspex filters (UV-B transmitting (+UV-B), GS 2458, Röhm, Darmstadt, Germany; UV-B absorbing (−UV-B), GS 233, Röhm).

Determination of flavonoids

Fronds of *C. crispa* were air-dried in the field. (In a previous experiment, we found no difference in soluble leaf flavonoid content between samples which had been air-dried and samples frozen in liquid nitrogen and freeze-dried). Leaf disks of *A. excelsum* were immediately frozen in liquid nitrogen and freeze dried. About 100 mg of dried plant material was pulverized in a mortar and, after addition of 5 mL of internal standard solution, flavonoids were extracted with MeOH by heating the mixture to the boiling point, filtering the resulting extracts after cooling to room temperature and reextracting the residue twice. The combined extracts were evaporated to near dryness and redissolved in 1 mL 50 % (v/v) MeOH. After centrifugation at 2200 g, the supernatant was used for HPLC analysis. Due to interfering retention times, two different internal standards had to be used: naringenin (1.0 mg mL^{-1}) for *C. crispa* and dimethoxybenzoic acid (0.3 mg mL^{-1}) for *A. excelsum.*

Pumps (type 126), detector (PDA, type 168), autosampler (type 502), integration software (GOLD) of the HPLC system were from Beckman (Munich, Germany). Chromatographic conditions were as follows: column-temperature, 15 °C, thermostated; column, Merck Supersphere RP 18(e); 4 µm; 120·4 mm; guard column, Merck Lichrochart RP 18 4·4 mm (Merck, Darmstadt, Germany); elution profile, A (0.15 % (w/v) aqueous H_3PO_4 : MeOH, 77 : 23, pH = 2) and B (MeOH). Initially isocratic for 3.6 min (100 % A) then with a linear gradient: 3.6 min (100 % A) – 24.0 min (80.5 % A) again isocratic – 30 min than with a linear gradient – 60 min (51.8 % A) and linear – 67.2 min (100 % B); flow rate, 1.0 mL min^{-1}. Due to the distinct spectral properties of the different standards used for the two species, detection and calibration was done at 350 nm for *C. crispa* and at 260 nm for of *A. excelsum.*

Peak purity and identity were checked by comparison of the online UV spectra (photo diode array detector) and cochromatography (TLC, HPLC) with authentic reference substances. The calibration curves were obtained from quercetin-3-*O*-galactoside for *C. crispa* and quercetin-3-*O*-glucoside for analysis of *A. excelsum,* respectively. For unknown peaks only compounds showing typical on-line-UV spectra of flavonoids were calculated. Peak identity in *C. crispa* was proven for quercetin-3-*O*-glucoside, quercetin-3-*O*-galactoside, kaempferol-3-*O*-glucoside, and kaempferol-3-*O*-galactoside. The identity of the flavonoid peaks from *A. excelsum* was not determined. Retention times and spectral characteristics of identified compounds have been published previously (Veit et al., 1995).

Chlorophyll a fluorescence

Pinnae were removed from *C. crispa* at different times during the day (August 3, 1993), and after 15 min dark adaptation PSII quantum efficiency was assessed using a PAM-2000 chlorophyll fluorometer (Walz, Effeltrich, Germany). The PFD of the light pulse to induce F_m was 5000 µmol $m^{-2} s^{-1}$.

Results

In fronds of *C. crispa* the main methanol soluble flavonoids were quercetin-3-*O*-glucoside, quercetin-3-*O*-galactoside, kaempferol-3-*O*-glucoside, and kaempferol-3-*O*-galactoside. In addition, some unidentified flavonole glycosides that contributed less than 8 % to the total leaf flavonoid content were detected. Fronds also contained considerable amounts of hydroxycinnamic esters, mainly chlorogenic acid and 5-*O*-caffeoylshikimic acid. In plants from high altitudes, soluble flavonoids represented up to 10 % of total dry weight of fronds.

Diurnal levels of soluble flavonoid contents were studied in plants that had been covered for over 10 months with UV-B absorbing and UV-B transmitting filters. Experiments were performed with fronds that had developed during this 10 month period. On two consecutive sunny days in June 1994, fully sun-exposed fronds that received natural UV-B radiation showed marked diurnal changes in the content of soluble flavonoids, with increases from dawn to midday, followed by decreases in the course of the afternoon and evening (Fig. 2). In control plants (never covered with filters) the diurnal course of flavonoid content roughly followed the diurnal course in PFD, while in plants under UV-B transmitting filters, the decline in flavonoids in the afternoon was slower than the decline in PFD. Surprisingly, shaded fronds not exposed to direct sunlight and not covered with filters also showed diurnal alterations in soluble flavonoid content, with a magnitude similar to that in sun-exposed fronds.

No diurnal changes in flavonoid content were observed in plants covered with UV-B absorbing filters. At predawn, fronds of all treatments had essentially the same flavonoid content (about 5 % of dry weight). Upon removal of filters (Fig. 2, night of June 29), fronds previously covered with UV-B absorbing filters showed increases in flavonoid content during the morning hours of the next day that were indistinguishable from those in fronds that had always been exposed to natural UV-B radiation (Fig. 2, June 30).

In a similar experiment diurnal changes in F_v/F_m, a measure of photoinhibition of photosynthesis, were determined in fronds previously covered with UV-B absorbing filters and exposed to natural UV-B radiation starting at predawn (Fig. 3). During the first day of exposure to natural UV-B radiation, the decrease in F_v/F_m in these fronds between dawn and midday was not significantly different from fronds which had always been exposed to natural UV-B radiation (Fig. 3).

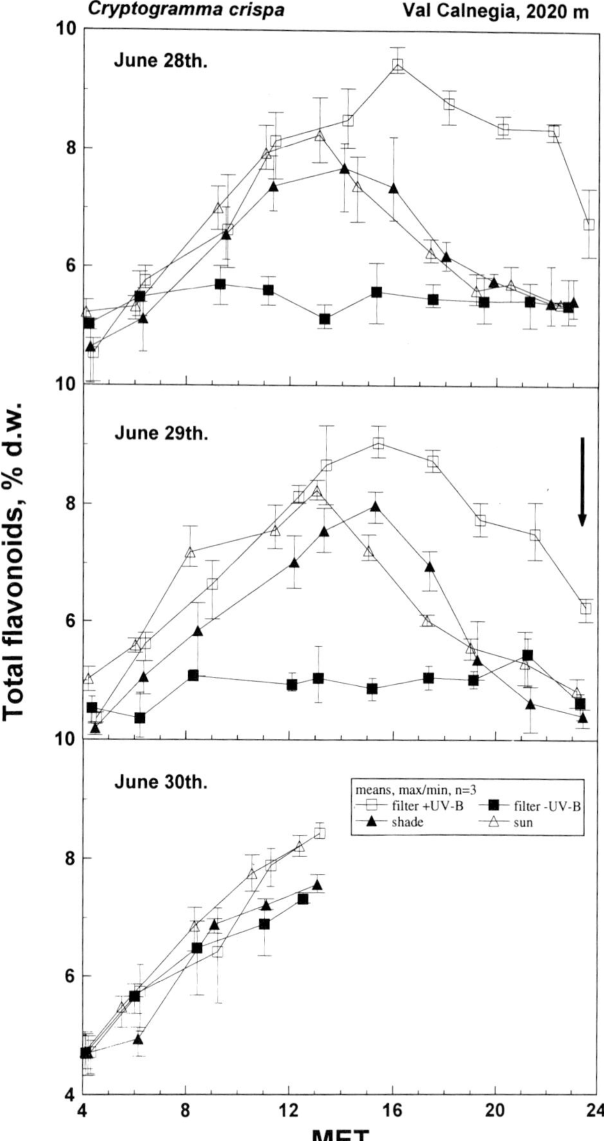

Fig. 2: Content of total methanol soluble leaf flavonoids (calculated with the calibration data of quercetin-3-*O*-galactoside) in fronds of *Cryptogramma crispa* at 2020 m altitude in the Swiss Alps during 3 subsequent days in June 1994. Ambient temperature ranged from 11 °C at night to 23 °C during the day. Clear skies with few clouds in the afternoon prevailed on June 28 and June 29. On June 30, sky was cloudless until 14:00 and a thunderstorm began at 15:00. Squares denote plants that had been grown under UV-B absorbing filters (–UV-B, ■) or UV-B transmitting filters (+UV-B, □) for 10 month prior to experiments. Triangles denote plants without filters, either exposed to full natural sunlight (sun, △) or growing in the shade (shade, ▲). The arrow in the middle panel (June 29, 23:00) indicates removal of filters so that next day (June 30) all plants were exposed to natural sunlight. Vertical bars represent maximal and minimal values (n=3).

Pronounced diurnal increases and decreases in the content of methanol-soluble flavonoids were also observed in leaves of a tropical tree, *Anacardium excelsum,* in its natural habitat in Panama (Fig. 4). Flavonoid contents increased three-fold during the first part of the day, and rapidly decreased to a basal level of approximately 1 % of leaf dry weight when cloud cover in the afternoon led to a reduction in PFD from $1600 \mu mol\, m^{-2} s^{-1}$ to $350 \mu mol\, m^{-2} s^{-1}$. In both *C. crispa* and *A. excelsum,* the diurnal changes in total soluble flavonoid content reflected similar percentage increases of all flavonoid compounds present in the fronds and leaves, respectively.

Discussion

The most important result of this study is the finding that levels of methanol-soluble leaf-flavonoids in these wild plants adapted to high natural levels of UV-B radiation undergo marked diurnal changes, with increases in flavonoid content during the morning hours and decreases during the afternoon. Early work with *Vicia faba* has already suggested that the content of UV-absorbing flavonoids may rapidly increase and decrease in response to alterations in PFD (Lautenschlager-Fleury, 1955). However, this previous study was based on measurements of epidermal transmittance rather than on quantitative determination of leaf flavonoids. Our finding that the changes in flavonoid levels in *C. crispa* and *A. excelsum* roughly followed the natural diurnal course of PFD and that the changes were suppressed in plants covered with UV-B-absorbing filters is consistent with a large body of evidence now available that the key enzymes of flavonoid biosynthesis (phenylalanin ammonia lyase, cinnamoyl-CoA-ligase, chalcone synthase and chalcone-flavanone isomerase) are inducible by UV-radiation (Hahlbrock and Scheel, 1989; Stafford, 1990). At present, however we do not know whether the diurnal changes in flavonoids observed in *C. crispa* and *A. excelsum* are the result of flavonoid synthesis and degradation, or the consequence of interconversions between methanol-soluble and methanol-insoluble (membrane-bound) pools (Schnitzler et al., 1996).

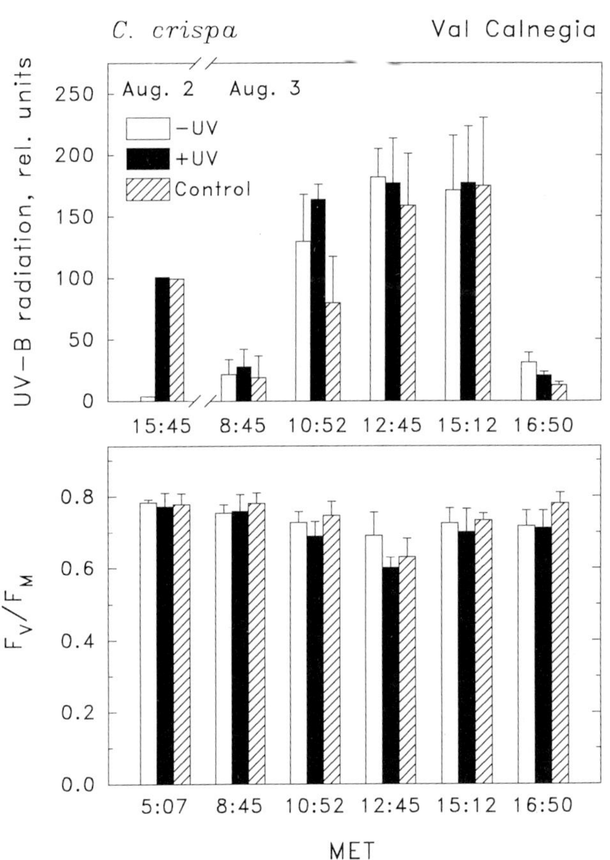

Fig. 3: Effect of exposure to natural sunlight on diurnal changes of F_v/F_m in fronds of *Cryptogramma crispa* previously covered for 2 months with UV-B absorbing and transmitting filters. The upper panel shows the level of UV-B radiation incident on fronds before (Aug. 2, 15:45) and after (Aug. 3) removal of filters. Lower panel shows diurnal changes in PSII photochemical efficiency, F_v/F_m on Aug. 3. i.e. during the first day after removal of filters. F_v/F_m measurements were done with pinnae which had been dark-adapted for 15 min. Vertical bars represent standard deviations (n=5). Control = plants always exposed to natural sunlight; +UV-B = plants kept under UV-B transmitting filters until Aug. 2; –UV-B = plants kept under UV-B absorbing filters until Aug. 2.

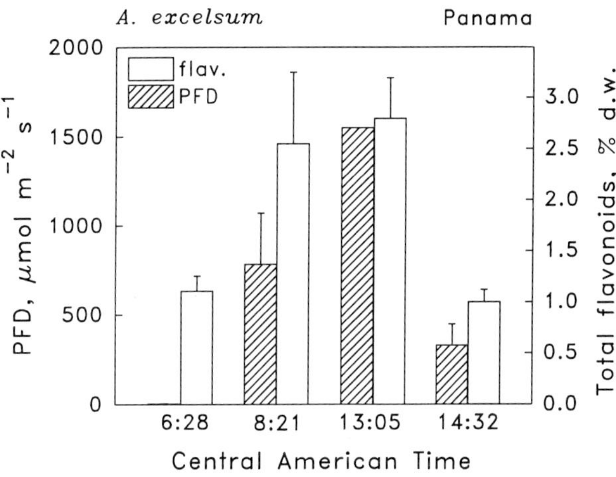

Fig. 4: Content of total methanol-soluble leaf-flavonoids (calculated with the calibration data of Quercetin-3-glucoside) in outer-canopy leaves of the tropical tree *Anacardium excelsum* in Parque Metropolitano, Panama, on March 18, 1994. Fine weather prevailed until noon. Clouds developed at 13:30, followed by rain. Maximum and minimum temperatures averaged 33 and 23 °C, respectively.

Plants previously covered with UV-B absorbing filters retained the ability to rapidly accumulate soluble flavonoids in response to natural radiation, and the putative photoprotection thereby provided may have been the reason that photoinhibition of photosynthesis, measured as decline in F_v/F_m in response to high levels of PFD at midday was not more pronounced than in plants always exposed to natural light.

In most species studied thus far, relatively slow increases in flavonoids in response to UV-B have been reported (e.g. soybean; Wellmann, 1971; Caldwell et al., 1994). Flavonoid accumulation was also found to be stimulated by UV-A, which usually accompanies the increase in UV-B radiation as visible radiation increases (Beggs et al., 1986; Cen and Bornman, 1990). Since the UV-B absorbing filters used in our experiments allowed short wavelength UV-A to penetrate (Fig. 1), diurnal variations could have been triggered by both UV-B and short-wavelength UV-A. Long-wavelength UV-A can be excluded in stimulating the diurnal variations in flavonoids.

We are puzzled by the finding that shaded *C. crispa* showed diurnal fluctuations in flavonoids similar to those of fully sun-exposed plants. Interpretation of data must await future studies which include detailed radiation measurements of shaded study sites. Since scattering of light increases with decreasing wavelength, the reduction in diffuse UV-B radiation in shaded habitats is usually much less pronounced than the reduction in visible light.

Our data demonstrate that photosynthetic organs of two widely differing species of wild plants, adapted to high radiation levels, contain two relatively large pools of soluble flavonoids. One pool seems little affected by light quality and light intensity and does not change during the day/night cycle. A second pool that is superimposed on this constitutive pool, shows strong diurnal variations related to the daily course of natural radiation. These findings have important implications for the short-term regulation of flavonoid accumulation and metabolism on a biochemical-molecular level. Our results also emphasize the importance of studies of intact mature plants in their natural environment to better understand the functional significance of leaf flavonoids in UV protection.

Acknowledgements

This work was supported by grants from Jubiläumsstiftung der Universität Würzburg, Fond der Chemischen Industrie, Deutsche Forschungsgemeinschaft (SFB 251) and by a travel grant from the Smithsonian Tropical Research Institute. We also thank Adriano Brüderli and Andreas Protte for assistance in the field and Michael Kriegmeier for help with the analyses.

References

Barnes P. W., S. D. Flint, and M. M. Caldwell: Photosynthesis damage and protective pigments in plants from a latitudinal arctic/alpine gradient exposed to supplemental UV-B radiation in the field. Arctic and Alpine Research *19,* 21–27 (1987).

Beggs, C. J., U. Schneider-Ziebert, and E. Wellmann: UV-B radiation and adaptive mechanisms in plants. In: Worrest, R. C. and M. M. Caldwell (eds.): Stratospheric Ozone Reduction, Solar Ultraviolet Radiation, and Plant Life, pp. 235-250. Springer Verlag, Berlin, Heidelberg (1986).

Blumthaler, M. and W. Ambach: Indication of increasing solar ultraviolet-B radiation flux in alpine regions. Science *248,* 206–208 (1990).

Bornman, J. F., L. O. Björn, and H. E. Åkerlund: Action spectrum for inhibition by ultraviolet radiation of Photosystem II activity in spinach thylakoids. Photobiochem. Photobiophys. *8,* 305–313 (1984).

Bornman, J. F. and A. H. Teramura: Effects of ultraviolet radiation on terrestrial plants. In: Young, A. R., L. O. Björn, and J. Moan (eds.): Environmental UV Photobiology, pp. 427–471. Plenum Press, New York (1993).

Caldwell, M. M., R. Robberecht, and S. D. Flint: Internal filters: prospects for UV-acclimation in higher plants. Physiol. Plant. *58,* 445–450 (1983).

Caldwell, M. M., S. D. Flint, and P. S. Searles: Spectral balance and UV-B sensitivity of soybean. A field experiment. Plant Cell Environ. *17,* 267–276 (1994).

Cen, Y. B. and J. F. Bornman: The response of bean plants to UV-B radiation under different irradiances of background visible light. J. Exp. Bot. *41,* 1489–1495 (1990).

Fiscus, E. L. and F. L. Booker: Is increased UV-B a threat to crop photosynthesis and productivity? Photosynthesis Res. *43,* 81–92 (1995).

Hahlbrock, K. and D. Scheel: Physiology and molecular biology of phenylpropanoid metabolism. Ann. Rev. Plant Physiol. Plant Mol. Biol. *40,* 347–369 (1989).

Lautenschlager-Fleury, D.: Über die Ultraviolett-Durchlässigkeit von Blattepidermen. Ber. Schweiz. Bot. Ges. *65,* 343–386 (1955).

Madronich, S.: The atmosphere and UV-B radiation at ground level. In: Young, A. R., L. O. Björn, and J. Moan (eds.): Environmental UV Photobiology, pp. 1–39. Plenum Press New York (1993).

Reuber, S., J. Leitsch, G. H. Krause, and G. Weissenböck: Metabolic reduction of phenylpropanoic compounds in primary leaves of rye (*Secale cereale* L.) leads to increased UV-B sensivity of photosynthesis. Z. Naturforsch. *48c,* 749–756 (1993).

Robberecht, R. and M. M. Caldwell: Leaf epidermal transmittance of ultraviolet radiation and its implications for plant sensitivity to ultraviolet radiation induced injury. Oecologia *32,* 277–287 (1978).

Schnitzler, J. P., T. P. Jungblut, W. Heller, M. Köpferlein, P. Hutzler, U. Heinzmann, E. Schmelzer, D. Ernst, C. Langebartels, and H. Sandermann Jr.: Tissue localisation of UV-B screening pigments and of chalcone synthase mRNA in needles of Scots pine (*Pinus sylvestris* L.) seedlings. New Phytologist Bd. *132,* 247–258 (1996).

Stafford, H. H.: Flavonoid metabolism. CRC Press, Boca Raton (1990).

Strid, Å., W. S. Chow, and J. M. Anderson: Effects of supplementary Ultraviolet B radiation on photosynthesis in *Pisum sativum.* Biochim. Biophys. Acta *1020,* 260–268 (1990).

– – – UV-B damage at the molecular level in plants. Photosynth. Res. *39,* 475–489 (1994).

Teramura, A. H. and J. H. Sullivan: Effects of UV-B radiation on photosynthesis and growth of terrestrial plants. Photosynth. Res. *39,* 463–473 (1994).

Tevini, M., J. Braun, and G. Fieser: The protective function of the epidermal layer of rye seedlings against ultraviolet-B radiation. Photochem. Photobiol. *53,* 329–333 (1991).

Veit, M., C. Beckert, C. Höhne, K. Bauer, and H. Geiger: Interspecific and intraspecific variations of phenolics in the genus *Equisetum* subgenus Equisetum. Phytochem. *38,* 881–891 (1995).

Vogt, T., P. G. Gülz, and H. Reznik: UV radiation dependent flavonoid accumulation of *Cistus laurifolius* L. Z. Naturforsch. *46c,* 37–42 (1991).

Wellmann, E.: Phytochrome mediated flavone glycoside synthesis in cell suspension cultures of *Petroselinum hortense* after preirradiation with ultraviolet light. Planta *101,* 283–286 (1971).

Wilson, M. I. and B. M. Greenberg: Protection of the D1 photosystem II reaction center protein from degradation in ultraviolet radiation following adaptation of *Brassica napus* L. to growth in ultraviolet-B. Photochem. Photobiol. *57,* 556–563 (1993).

J. Plant Physiol. Vol. 148. pp. 483–493 (1996)

Non-Destructive Determination of Chlorophyll Content of Leaves of a Green and an Aurea Mutant of Tobacco by Reflectance Measurements

HARTMUT K. LICHTENTHALER[1], ANATOLY GITELSON[2], and MICHAEL LANG[1]

[1] Botanisches Institut, Lehrstuhl II, University of Karlsruhe, Kaiserstr. 12, D-76128 Karlsruhe, Germany

[2] The Remote Sensing Laboratory, J. Blaustein Institute for Desert Research, Ben Gurion University of the Negev, Sede Boker, Campus 84933, Israel

Received July 12, 1995 · Accepted November 30, 1995

Summary

Reflectance spectra from 400 to 800 nm with a spectral resolution of 1 nm and the content of photosynthetic pigments were acquired for leaves of an aurea mutant and a green tobacco (*Nicotiana tabacum* L.) covering a range of chlorophyll *a* content from 7 to 30.8 $\mu g\,cm^{-2}$ and total chlorophyll *a*+ *b* level from 8.4 to 41.4 $\mu g\,cm^{-2}$ leaf area. At a chlorophyll *a* content of more than 7 $\mu g\,cm^{-2}$ leaf area the reflectance near 670 to 680 nm was not sensitive to a variation in chlorophyll content due to saturation of the relationship «absorption vs. chlorophyll content». The wavelength position of the red reflectance minimum (Rmin) in the main red absorption bands of *in vivo* chlorophyll *a* proved to be independent of the chlorophyll content of leaves and thus is not suitable for chlorophyll determination. Maximum sensitivity of leaf reflectance to variation in chlorophyll content was, however, found in a wide spectral range from 530 to 630 nm and near 700 nm. The wavelength of the red edge position of the reflectance spectrum (inflection point IP) in the range of 694 to 706 nm correlated very closely with the leaves' chlorophyll content in a curvilinear manner. The Ip is also closely correlated in a linear manner with the reflectance signals at 550 and 700 nm. Reflectances near 700 nm and 550 nm were found to be sensitive indicators of the red edge position and the chlorophyll content of leaves as well. The ratios of reflectances in the near infra-red range of the spectrum (above 750 nm) to that at 700 nm and at 550 nm, R_{NIR}/R700 (or R750/R700) and R_{NIR}/R550 (or R750/R550), were directly proportional ($r^2 > 0.93$) to the leaves' chlorophyll content. These two novel indices for chlorophyll determination (R750/R700 and R750/R550) allowed the estimation of leaf chlorophyll content with an error of less than 2.1 $\mu g\,cm^{-2}$. The two new vegetation indices, R750/R700 and R750/R550, are linearily correlated to the chlorophyll content. They exhibit a more than 6 times wider dynamic range than the widely used Normalized Difference Vegetation Index NDVI. The new vegetation index ratios are very sensitive to changes in chlorophyll content at low, medium and high chlorophyll content which is not the case for the NDVI. This suggests the application of the new indices R750/R700 and R750/R550 in the near and remote sensing of the chlorophyll content of terrestrial vegetation.

Key words: Chlorophyll content, inflection point IP, red edge, reflectance spectra, red reflectance minimum (Rmin), remote sensing of chlorophyll content, new vegetation index.

Abbreviations: Chl = chlorophyll; Chl *a* and *b* = chlorophyll *a* and *b; c* = carotenes; IP = position of the inflection point at the red edge of the reflectance spectrum; LHCPs = Light-harvesting chlorophyll-proteins of the photosynthetic photosystem II; NDVI = normalized difference vegetation index; NIR = near infra-red range of the spectrum; red edge = rise in the reflectance spectrum of leaves between 680 to 750 nm; Rmin = red reflectance minimum near 680 nm caused by the main red absorption bands of *in vivo* chlorophyll *a;* R550 and R700 = reflectance signals at 550 and 700 nm, respectively; *x*+ *c* = total leaf carotenoids; *x* = xanthophylls.

Introduction

Terrestrial vegetation is exposed to various kinds of natural and anthropogenic stressors (Lichtenthaler, 1996), many of which act simultaneously and will reduce the chlorophyll content of plants by a photooxidative chlorophyll breakdown at a long-term stress exposure. Remote determination of the chlorophyll content from near and far distance by non-destructive methods is therefore a good mean to detect stress conditions in plants (Lichtenthaler, 1989). By non-invasive techniques, such as reflectance measurements, one can routinely follow at the same plant or canopy the successive decline of chlorophylls at continuous stress exposure, and also the regeneration of plants and their photosynthetic pigment apparatus when the stressors are removed.

The increase in leaf reflectance in the visible range of the spectrum at a decrease of the chlorophyll content is very much used in ecophysiology and stress research of plants and has been applied by many authors (Buschmann and Lichtenthaler, 1988; Buschmann et al., 1991; Chappelle et al., 1992; Carter et al., 1993 and 1994; Horler et al., 1983; Lichtenthaler, 1989; Rock et al., 1986; Schmuck et al., 1987; Thomas and Gaussmann, 1987; Yoder and Waring, 1994). For this purpose different reflectance signals at particular wavelengths in the visible and near infra-red (NIR) range of the spectrum were determined, and special ratios of the reflectance signatures are formed which function as indicators of leaf area index, canopy cover and chlorophyll content. A much applied index is the Normalized Difference Vegetation Index NDVI (Rouse et al., 1974; Guyot, 1990; Baret et al., 1992), which is not only sensed in near distance but also from airborne systems including satellites. The NDVI, bound to the reflectance signals at 680 nm (R680, R_{red}) in the red range of the spectrum near the red absorption bands of chlorophyll *a* and in the NIR range above 750 to 800 nm (R800), is defined as the ratio of $(R_{NIR}-R_{red})/(R_{NIR}+R_{red})$ usually expressed as (R800−R680)/(R800+R680). The vegetation index NDVI is, however, not suitable for chlorophyll determination. It is precise only at fairly low chlorophyll levels, but is not sensitive to variations in chlorophyll content at medium and higher chlorophyll levels of the vegetation. For this reason several attempts have been made in recent years in order to increase the accuracy of a reflectance-based chlorophyll determination by inclusion of other or additional reflectance bands and ratios (Buschmann and Lichtenthaler, 1988; Buschmann and Nagel, 1993; Buschmann et al., 1991; Gitelson and Merzlyak 1994 a, 1994 b and 1996; Yoder and Waring, 1994).

The shift of the inflection point IP at the red edge rise of the reflectance spectrum towards shorter wavelengths (blue shift of IP) seems to be a good indicator of a decline in the chlorophyll content (Horler et al., 1983; Lichtenthaler and Buschmann, 1987; Buschmann and Lichtenthaler, 1988; Curran et al., 1991; Baret et al., 1992; Gitelson et al., 1996 b) and has also been used in the remote sensing of forest decline (Rock et al., 1986; Schmuck et al., 1987; Lichtenthaler, 1994). The red reflectance dip Rmin in the main absorption bands of *in vivo* chlorophyll *a* broadens with increasing chlorophyll content. Whether the wavelength position of the reflectance minimum Rmin depends upon the chlorophyll content and whether it exhibits any relation to the wavelength positions of the IP of the red edge has, however, not yet been determined. Chlorophyll *a* is bound *in vivo* in the photosynthetically active thylakoids of chloroplasts to several chlorophyll-carotenoids-proteins which possess differential Chl *a* absorption bands (Lichtenthaler et al., 1981). The different photosynthetic pigment proteins can be separated by polyacryl gel electrophoresis and yield the photosystem I pigment-proteins CPIa and CPI, the photosystem II pigment-protein CPa as well as the light-harvesting chlorophyll-proteins LHCPs, which belong to the light harvesting complex LHCII of photosystem II. In the photosystem I-proteins, CPIa and CPI, the Chl *a* absorption maximum lies at 678 nm, whereas in the photosytem II protein CPa and the Chl *a*/Chl *b*-proteins of the pigment antenna (the LHCPs) the absorption band of the majority of Chl *a* molecules is found at 672 nm (Lichtenthaler et al., 1981). In the reaction centers of both photosystems there exist, however, minor Chl *a* forms which absorb at longer wavelengths such as the Chl *a* dimers P680 (peak at 680 nm) and P700 (peak at 700 nm) (e.g. Butler and Hopkins, 1970; French et al., 1972; Lawlor, 1995). One can assume that the relative proportions of these pigment-proteins in leaves should determine the wavelength position of the red reflectance minimum of leaves.

The reflectance spectra of the tobacco leaves, taken in the visible and near infra-red range of the spectrum, were analysed in order to find spectral bands with a maximum sensitivity to a variation in chlorophyll content and to possibly device new vegetation indices for a better non-invasive remote chlorophyll determination. The research focused on the following spectral features of tobacco leaves which appear to be useful for a remote chlorophyll determination:

(1) high sensitivity of reflectance in the range near 550 nm to variation in Chl *a* content (Thomas and Gaussman, 1977; Tanner and Eller, 1986; Buschmann and Nagel, 1993; Yoder and Waring (1994); Gitelson and Merzlyak, 1994 a and b; 1996; Gitelson et al., 1996 a, b) and to different kinds of plant stress (Carter et al., 1993 and 1994; Schmuck et al., 1987);

(2) high sensitivity of reflectance in quite narrow spectral bands from 690 to 710 nm to variation in Chl *a* content (Gitelson and Merzlyak 1994 a, b and 1996; Gitelson et al., 1996 a, b);

(3) very close correlation between reflectances at 550 nm and 700 nm (Chappelle et al., 1992; Gitelson and Merzlyak 1994 a and b; 1996);

(4) very close correlation between reflectances near 500 nm and 670 nm (Gitelson and Merzlyak 1996; Gitelson et al., 1996 a); and

(5) a close relation between the red edge position and reflectance near 700 nm (Gitelson et al., 1996 b).

In addition, the wavelength position dependence of the inflection point IP of the red edge as well as that of the red reflectance dip (Rmin) in the chlorophyll *a* absorption bands around 680 nm on the chlorophyll content was determined. The aim of the investigation presented here was to develop a better non-destructive technique for the determination of total chlorophyll content by re-investigating different reflectance signatures using a data set of reflectance spectra from leaves with different chlorophyll content of an aurea mutant and a green variety of *Nicotiana tabacum* L.

Material and Methods

Plants

Eight to ten week old tobacco plants (*Nicotiana tabacum* L.), green form su/su and aurea mutant Su/su (Schmidt, 1971; Lichtenthaler et al., 1975; Santrucek et al., 1992), were cultivated on a mineral-containing peat in the greenhouse of the Botanical Garden of the University of Karlsruhe under standard conditions (cf. Lang et al., 1996).

Determination of pigment content

Photosynthetic pigments, chlorophylls and carotenoids, were determined spectrophotometrically (UV-2001 PC, Shimadzu, Duisburg, Germany) from extracts in 100 % acetone using the redetermined extinction coefficients and equations of Lichtenthaler (1987) which allow the simultaneous determination of Chl *a* and *b* and total carotenoids (*x*+ *c*) in the same extract solution.

Reflectance measurements

Reflectance spectra from 350 to 900 nm were recorded from the upper leaf side with a spectrophotometer (UV-2001PC, Shimadzu, Duisburg, Germany) applying an integrating sphere and $BaSO_4$ as reference. The wavelength scanning speed was set to 100 nm/min. For the determination of the reflectance signals at 490, 550, 670, 680, 700 and 750 nm, which were correlated to each other or used for the calculation of reflectance ratios (R750/R550 and R750/R700), a spectral band width (slit width) of 5 nm (position 05 in the UV-2101PC) was applied with the specification that the height was half in order to reduce the stray light.

Data analysis

All calculations including reflectance ratios as well as statistics were performed with Excel (Microsoft, USA). Curve fitting was done with the help of Origin (MicroCal, USA). The *normalized difference vegetation index* NDVI was calculated on the basis of the reflectance signatures at 680 and 800 nm as the ratio (R800–R680)/(R800+R680) (Rouse et al., 1974; Lichtenthaler, 1994).

The position of the *inflection point IP* of the red edge rise in the reflectance spectrum was determined as point of intersection with the zero line of the 2nd derivative of the reflectance spectrum, which was calculated applying the spectrophotometer software of the UV-2001 PC (Shimadzu).

The red *reflectance minimum Rmin* in the red absorption bands of *in vivo* chlorophyll *a* was determined from the intersection with the zero line in the 1st derivative of the reflectance spectrum. It can also be determined with the peak search program present in most of the commercial spectrophotometers.

Table 1: *Aurea tobacco Su/su*: Level of photosynthetic pigments, pigment ratios, reflectance ratios and position of the inflection point (IP) of the red edge and red reflectance minimum (Rmin) in yellowish-green to light-green leaves. The leaves are arranged by increasing chlorophyll content. Pigment values are based on 3 determinations (SD<5 %) and are given in $\mu g \cdot cm^{-2}$ leaf area.

Leaf number	7	6	5	4	3	2	1
Pigment content							
Chlorophyll *a*	7.00	10.80	12.00	12.80	13.50	14.90	15.50
Chlorophyll *b*	1.37	2.27	2.79	3.24	3.51	4.52	4.94
Total chlorophylls *a*+*b*	8.37	13.07	14.79	16.04	17.01	19.42	20.44
Total carotenoids *x*+*c*	3.21	3.95	3.91	3.46	3.34	3.54	3.67
Pigment ratios							
a/b	5.11	4.75	4.30	3.95	3.85	3.30	3.14
(*a*+*b*)/(*x*+*c*)	2.61	3.31	3.78	4.63	5.10	5.49	5.57
Reflectance ratios							
R750/R550	1.39	1.74	1.82	1.83	1.85	1.96	2.10
R750/R700	1.54	1.82	2.16	2.25	2.27	2.52	2.54
Red edge							
position of IP (nm)	694.0	697.0	698.0	698.0	699.0	700.2	700.0
position of Rmin (nm)	678.4	678.4	678.4	678.7	677.5	678.7	678.6

Table 2: *Green tobacco su/su*: Level of photosynthetic pigments, pigment ratios, reflectance ratios and position of the inflection point (IP) of the red edge and red reflectance minimum (Rmin) in green leaves. The leaves are arranged by increasing chlorophyll content. Pigment values are based on 3 determinations (SD<5 %) and are given in $\mu g \cdot cm^{-2}$ leaf area.

Leaf number	6	5	4	3	2	1
Pigment content						
Chlorophyll *a*	21.70	22.50	27.80	27.30	29.00	30.80
Chlorophyll *b*	7.83	7.84	9.92	10.30	10.28	10.61
Total chlorophylls *a*+*b*	29.53	30.34	37.72	37.6	39.28	41.41
Total carotenoids *x*+*c*	5.47	5.67	6.77	8.03	8.16	7.09
Pigment ratios						
a/b	2.77	2.87	2.80	2.65	2.82	2.85
(*a*+*b*)/(*x*+*c*)	5.40	5.33	5.57	4.68	4.81	5.84
Reflectance ratios						
R750/R550	2.44	2.55	2.93	3.50	3.16	3.49
R750/R700	2.82	2.99	3.53	4.10	3.86	4.17
Red edge						
position of IP (nm)	702.0	702.0	703.5	704.5	705.6	706.0
position of Rmin (nm)	678.4	679.0	678.7	679.0	679.0	677.3

Results

Pigment content in the leaves

In the leaves of aurea tobacco, which possesses low chlorophyll levels and a retarded greening (Schindler et al., 1994), the total chlorophyll content (Chl *a*+ *b*) ranged from 8.4 to 20.4 $\mu g\, cm^{-2}$ leaf area. In the leaves of green tobacco the chlorophyll amounts were higher and ranged from 29.5 to 41.4 $\mu g\, cm^{-2}$ leaf area (Tables 1 and 2). The seven yellowish-green to light-green leaves of the aurea tobacco Su/su showed high values for the ratio Chl *a/b,* which declined with increasing chlorophyll content and age. The values of the ratio of green to yellow pigments, chlorophylls/carotenoids (*a*+ *b*)/(*x*+ *c*) were low (2.61) in the young leaves of aurea tobacco, but increased with increasing age and chlorophyll content to a value of 5.57 which is normal for green plant tissue. The six green leaves of green tobacco, which shows a fast and normal greening process response, exhibited values of the pigment ratios, Chl *a/b* (2.65–2.87) and (*a*+ *b*)/(*x*+ *c*) (4.68–5.84) as usually found in green plant tissue (Tables 1 and 2).

Spectral features of reflectance spectra of leaves

The typical reflectance spectra of leaves of aurea and green tobacco with different chlorophyll content (Chl *a:* aurea no. 5 = 12.0 μg cm^{-2}; aurea no. 4 = 15.5 μg cm^{-2}; green no. 5 = 22.5 μg cm^{-2}; green no. 3 = 27.3 μg cm^{-2}) are shown in Fig. 1. They are characterised by a low reflectance in the blue spectral region between 400 to 500 nm, a reflectance maximum in the green region near 550 nm, a reflectance minimum near 680 nm followed by a steep rise in reflectance (known as red edge) to the near infra-red region and a relatively constant maximum reflectance above 750 nm (Fig. 1). The reflectance above 750 nm is not influenced by the red absorption bands of *in vivo* chlorophyll of leaves, and the variation of reflectance in this NIR range was relatively low for all 13 leaves and ranged from 42 to 53 % reflectance. In fact, the coefficient of reflectance variation R750 and R_{NIR} (ratio of standard deviation of reflectance to average reflectance value) was less than 5 % for all 7 aurea and 6 green tobacco leaves studied here.

Reflectance in the blue region

The blue reflectance from 400 to 500 nm was characterised by a fairly low signal (range from 5 to 9 % reflectance) in all measured leaves (Fig. 1). Aurea tobacco leaves with a Chl *a* content of 7 to 12 μg cm^{-2} showed a tendency for a small reflectance dip near 440 nm probably due to the blue chlorophyll absorption band (Soret Band). This reflectance dip was no longer seen at a leaf chlorophyll *a* content above 16 μg cm^{-2}.

Reflectance in the green region

The reflectance spectra showed maxima in the green region near 550 nm that were inversely correlated (hyperbolic relationship $y = ax^{-b}$) to the Chl *a* and Chl *a*+ *b* content of the investigated leaves (Fig. 2). Aurea leaves with a Chl *a* content of 12 μg cm^{-2} (aurea no. 5) exhibited a reflectance near 550 of about 30 % (Fig. 1). With increasing Chl *a* content the 550 nm reflectance decreased to less than 12 % in the green tobacco leaf with a Chl *a* content of 27.3 μg cm^{-2} (green leaf no. 3). The reflectance from 530 to 630 nm showed maximum sensitivity to Chl *a* and decreased to the same extent in all wavelength positions of this range with increasing Chl *a* content from 7 to 30.8 μg cm^{-2}. The curvilinear inverse relationship of Chl *a* and R550 exhibited a determination coefficient of $r^2 = 0.966$ and total Chl *a*+ *b* and R550 of $r^2 = 0.974$.

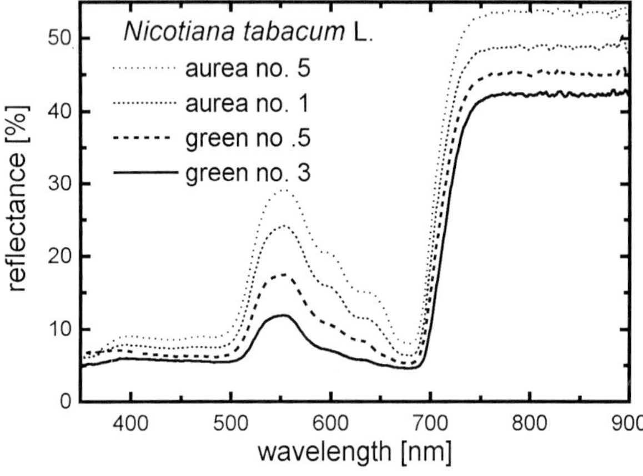

Fig. 1: Reflectance spectra of leaves with increasing chlorophyll content of an aurea mutant (leaves no. 5 and no. 1) and a green variety (leaves no. 5 and no. 3) of *Nicotiana tabacum* L. (see Tables 1 and 2). A minimum sensitivity to changes in chlorophyll content was observed in the blue spectral range (400 to 490 nm) and near 670 nm, whereas maximum sensitivity was found in a wide range of the spectrum from 530 to 630 nm and near 700 nm.

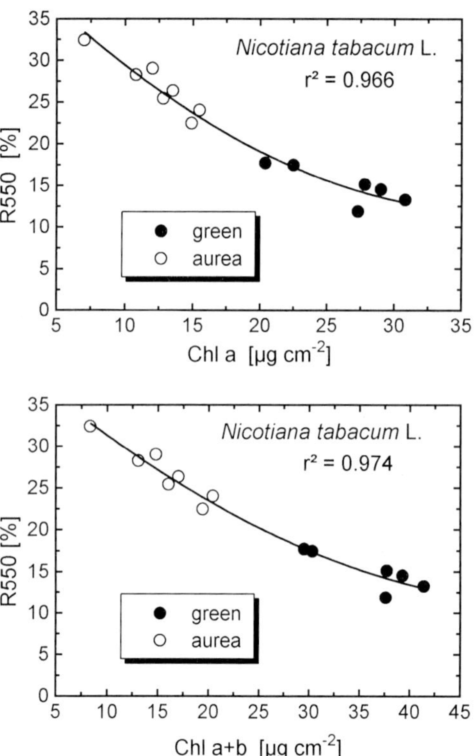

Fig. 2: Hyperbolic relationship ($y = ax^{-b}$) between the green reflectance R550 and chlorophyll *a* and total chlorophyll content (Chl *a*+ *b*) in green and aurea tobacco leaves.

Reflectance in the red region

Reflectance spectra of tobacco leaves exhibited a reflectance dip close to the red absorption bands of Chl *a* near 670 to 680 nm. However, the variation in reflectance in this region at 670 to 680 nm with respect to increasing levels of Chl *a* or Chl *a*+ *b* content was relatively small. An increase in Chl *a* from 7 to 22.5 μg cm^{-2} resulted only in a decrease from approximately 10 % to 5 % in the percentage leaf reflectance in this region, and then the reflectance remained nearly constant when Chl *a* increased to more than 30 μg cm^{-2}. In the blue region near 490 nm and in the red spectral region near 670 nm the variation in reflectance with increasing chlorophyll *a* content was very small, and for a Chl *a* content above 15 μg cm^{-2} the relationship «reflectance vs. Chl-content» was saturated (data not shown). The similar behavior of the reflectances in the blue range and near 670 nm (with small variation and sensitivity to changing Chl *a* and Chl *a*+ *b* content) is documented by a very high correlation (r^2 = 0.990) between the reflectance near 670 nm, R670, and the reflectance near 490 nm, R490, as shown in Fig. 3.

Red edge of leaf reflectance spectra and inflection point

Another spectral range with high sensitivity to changes in chlorophyll content was found in the reflectance near 700 nm in the range of the inflection point of the red edge rise of the reflectance spectrum (Fig. 1). As distinct as the green range of the spectrum, a high spectral sensitivity to changes in the Chl *a* and Chl *a*+ *b* level was found in a relatively narrow spectral band from 690 to 710 nm with a maximum sensitivity at 700 nm. The coefficient of the reflectance variation in this range (R700) was even higher as compared to that for the green reflectance R550 (data not shown). The relationship between the reflectance R700 (Fig. 4), and also that of R550 (Fig. 2), and the Chl *a* or Chl *a*+ *b* content was hyperbolic. In a similar way as for leaves of soybean (Chappelle et al., 1992) as well as maple and chestnut (Gitelson and Merzlyak, 1994a and b; 1996) a very close correlation (r^2 = 0.978) was found between the green and red reflectances, R550 and R700 (Fig. 5).

The rise of reflectance in the near infra-red spectral range, the red edge, can be characterised by the position of its *inflection point IP*. The determination of the inflection point of the red edge is shown for a green tobacco leaf with medium Chl *a* content (22.5 µg cm^{-2}) and for an aurea tobacco leaf with a low Chl *a* content (7 µg cm^{-2}) (Fig. 6). In the 2nd derivative spectra the inflection point of the red edge of the aurea tobacco leaf (IP at 694 nm) was at shorter wavelengths (blue shift) in comparison to the green tobacco leaf (IP at 702 nm) (Fig. 6). The wavelength positions of the inflection point correlate to the Chl *a* and Chl *a*+ *b* content in a curvilinear relationship with a very high correlation coefficient (r^2 = 0.973 for Chl *a* and 0.966 for Chl *a*+ *b*).

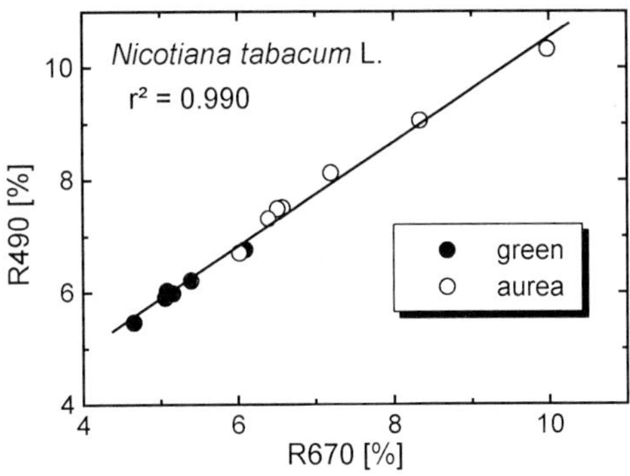

Fig. 3: Linear correlation of the reflectance at 490 nm with that at 670 nm (R490 versus R670) for all studied 13 leaves of an aurea mutant and a green form of *Nicotiana tabacum* L. (chlorophyll *a* content from 7 to 30.8 µg cm^{-2}. The solid line presents a linear relationship with the square of correlation coefficient r^2 = 0.99. The waveband near 670 nm lies in the range of the red absorption bands of chlorophyll *a*, whereas the waveband near 490 nm corresponds to joint absorption bands of chlorophyll *a*, chlorophyll *b* and carotenoids.

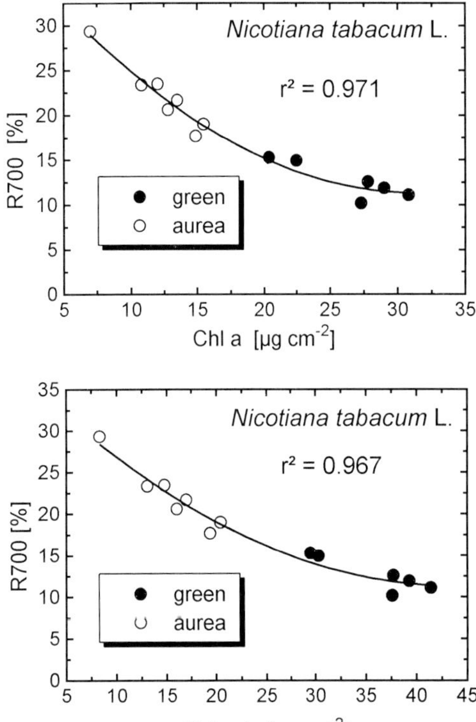

Fig. 4: Correlation of the reflectance at 700 nm (R700) versus the chlorophyll *a* and *a*+ *b* content for 13 leaves of an aurea mutant and green form of *Nicotiana tabacum* L. The lines indicate a hyperbolic relationship between chlorophyll content and R700. The square of the correlation coefficient was 0.97.

Position of the inflection point IP in relation to reflectance at 700 nm

High sensitivity of the position of the IP to chlorophyll content was observed in the spectral range from 690 to 705 nm (Figs. 1 and 7). The high correlation of Chl *a* content, both with R700 and with the IP, suggests a close relation between these characteristics of the red edge of the reflectance spectrum. In fact, a linear relation between R700 and the IP position was found for green and aurea tobacco leaves with r^2 = 0.956 (Fig. 8). Therefore, R700 measurements appear to be a suitable parameter to obtain information on the relative position of the inflection point of the red edge and on the leaves' chlorophyll content. The IP also correlates with the R550 in a linear way (Fig. 8).

Position of the red reflectance minimum Rmin at the red edge

With increasing chlorophyll content the reflectance dip in the red chlorophyll absorption band near 680 nm became broader (Figs. 1 and 6). The wavelength position of the minimum near 680 nm showed, however, similar values in the range of 677.3 to 679 nm for the 13 leaves with different

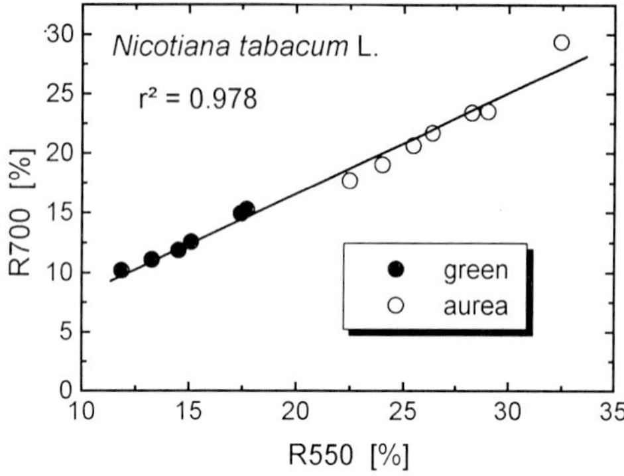

Fig. 5: Linear correlation of the reflectance R700 versus that of R550. Means of aurea and green tobacco with a chlorophyll *a* content from 7 to 30.8 μg cm^{-2}. The square of the correlation coefficient was $r^2 = 0.978$. The two reflectance signatures R700 and R550 were found to exhibit a maximum sensitivity to variations in chlorophyll content of leaves.

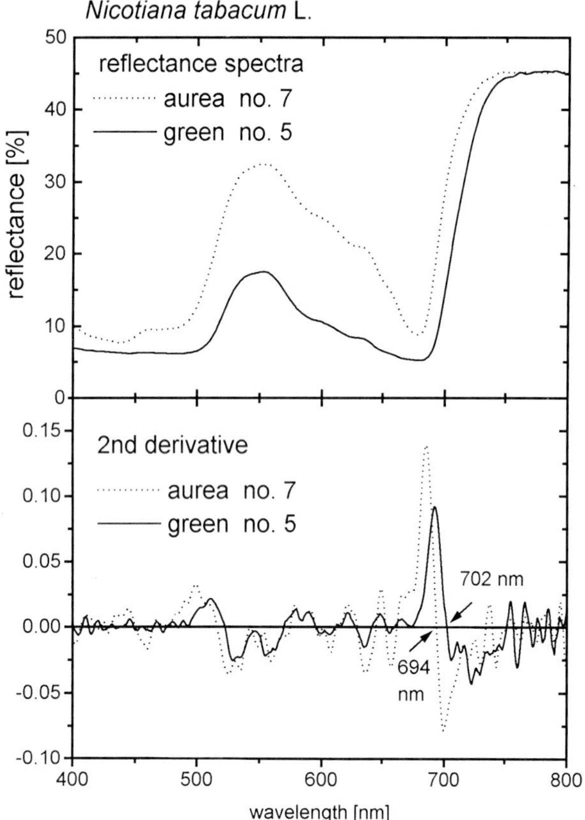

Fig. 6: (**A**) Reflectance spectra of leaves of an aurea mutant (dotted line) and a green form (solid line) of *Nicotiana tabacum* L. Differences in the chlorophyll content between the two leaves can be monitored by the shift of the red edge as estimated *via* the 2nd derivative of the reflectance spectrum. The position of the inflection point IP (as intersection with the zero line) of the red edge in the green tobacco leaf was near 702 nm, whereas that of the aurea leaf was shifted to a shorter wavelength at 694 nm (which is known as blue shift of the IP).

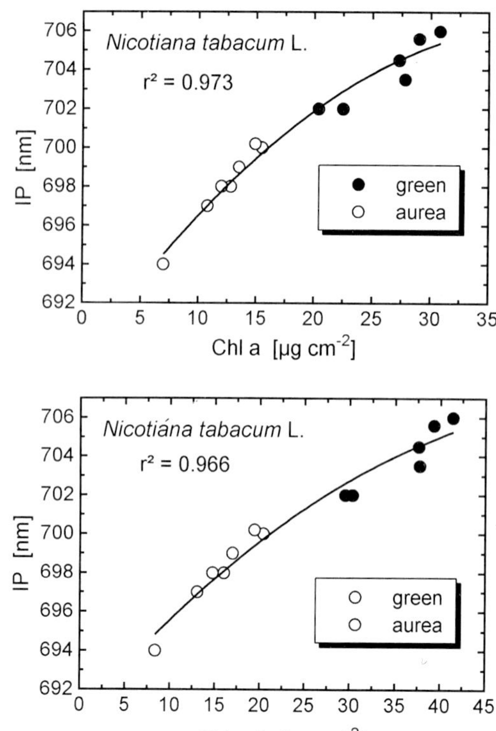

Fig. 7: Position of the inflection point (IP) of the red edge for all 13 leaves of the aurea mutant and green tobacco versus chlorophyll *a* and total chlorophyll content. The position of the inflection point (IP) stringly correlated with the chlorophyll *a* and *a*+ *b* content in an exponential manner (solid lines). Within the group of aurea and green leaves it would be possible to draw a linear relationship, the hyperbolic relationship, however, is the best fit for both groups.

chlorophyll content (Tables 1 and 2). Thus, the wavelength position reflectance minimum Rmin did not show a relationship to the Chl *a* content (Fig. 9) or to the position of the inflection point IP of the red edge (Fig. 10). This excludes Rmin as a mean for *in vivo* chlorophyll determination. The fact, that Rmin does not change with increasing chlorophyll content of leaves, is emphasized by the wavelength position of Rmin being solely determined by the *in vivo* Chl *a* absorption bands of the different chlorophyll/carotenoid-proteins of thylakoids. These absorption bands do not change with increasing or decreasing Chl content, as shown here.

The reflectance ratios R750/R550 and R750/R700 in comparison to the NDVI

With respect to remote sensing of the Chl content of plants *via* non-invasive reflectance measurements one has, however, to consider that the remote determination of the exact wavelength position of the inflection point IP is a relatively critical and difficult matter, and only accurate when high spectral resolution radiometry is used. A remote determination of the exact wavelength positions of IP is, however, not required when one wants to obtain information on the chlorophyll content of leaves and canopies. It is sufficient to concentrate on those reflectance parameters which show great sensitivity to changes in Chl *a* and Chl *a*+ *b* content, such as the reflectance signals R550 and R700. When the latter are related to a «standard reflectance parameter», which shows a low or no sensitivity to changes in Chl content, such as R750, one can establish new reflectance ratios for the non-invasive Chl determination.

Due to the high sensitivity of the reflectance signals near 550 nm and 700 nm and the insensitivity of the NIR reflec-

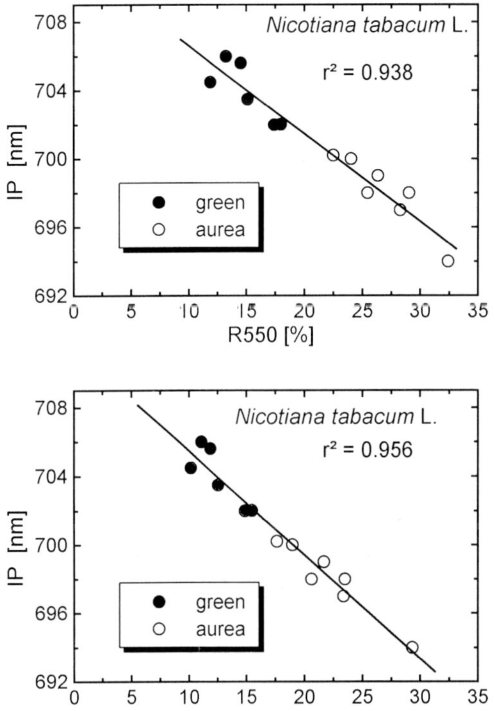

Fig. 8: Linear correlation of the position of the inflection point (IP) with R550 nm and R700 nm for all 13 leaves of an aurea mutant and a green tobacco variety. The correlation coefficient r^2 was somewhat higher for the correlation of the IP with R700 than with R550.

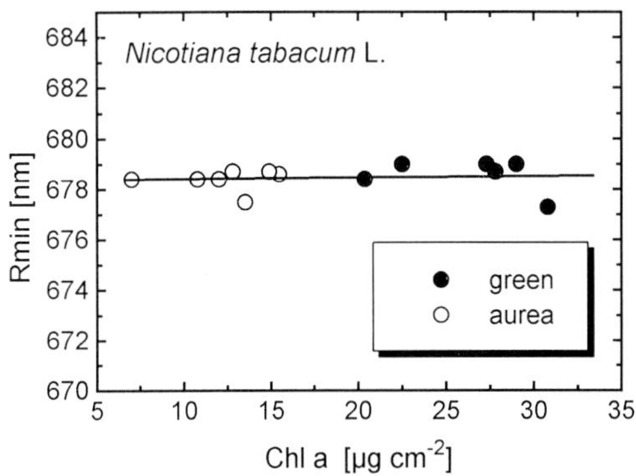

Fig. 9: The wavelength position of the red reflectance minimum Rmin (in nm) is independent of the Chl *a* content of the 13 green and aurea tobacco leaves studied.

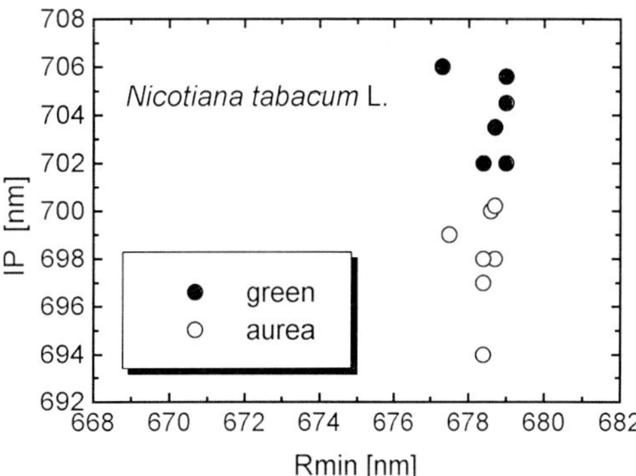

Fig. 10: There is no correlation between the wavelength position of the inflection point IP and that of the red reflectance minimum Rmin.

tance at 750 nm with respect to the chlorophyll content described here and also found in chestnut and maple leaves (Gitelson and Merzlyak, 1994 b), we determined a very high correlation between the two reflectance ratios R750/R550 and R750/R700 and the Chl *a* content with r^2 of 0.960 and 0.962, and the total Chl *a*+ *b* content of r^2 0.936 and 0.927, respectively (Fig. 11). Consequently, a linear relationship was observed between the reflectance ratios R750/R550 and R750/R700 with a high correlation coefficient of $r^2 = 0.984$ (Fig. 12).

The normalized difference vegetation index NDVI is often used for estimation of the chlorophyll content of leaves and terrestrial vegetation as well as green biomass. The great disadvantage of the NDVI for chlorophyll determination is, however, that at medium and higher chlorophyll levels it does not show a good response or correlation to the chlorophyll content. In fact, the NDVI values declined only from 0.81 to 0.75 when the Chl *a* content decreased from 31 to 10 µg cm^{-2}, and the total Chl *a*+ *b* level dropped from 41 to 13 µg cm^{-2} (Fig. 13). This connection is based on the extremely low sensitivity of the R680 reflectance to changes in chlorophyll content at medium or high chlorophyll levels. In contrast, the new reflectance ratios R750/R550 and R750/R700 established here showed a very high linear correlation to the Chl *a* and total Chl *a*+ *b* content for the whole range of chlorophyll levels with $r^2 = 0.956$ and 0.960, respectively (Fig. 13). In the range of Chl *a* content from 7 to 30.8 µg cm^{-2}, the coefficient of NDVI variation was only 4.4 %, whereas for the ratios R750/R550 and R750/R700 values of 28.6 % and 29.5 %, respectively were achieved.

Discussion

Reflectance of plant leaves exhibits signals in the visible domain (VIS) ranging from 400 to 700 nm, in the near-infrared domain (NIR) from 700 to 1300 nm and in the middle-infrared domain (MIR) from 1300 to 2500 nm (Guyot, 1990; Lichtenthaler, 1989 and 1994). The middle-infrared domain contains leaf reflectance minima caused by strong water absorption bands near 1450, 1950 and 2500 nm, but this range is neither of interest nor suitable concerning the determina-

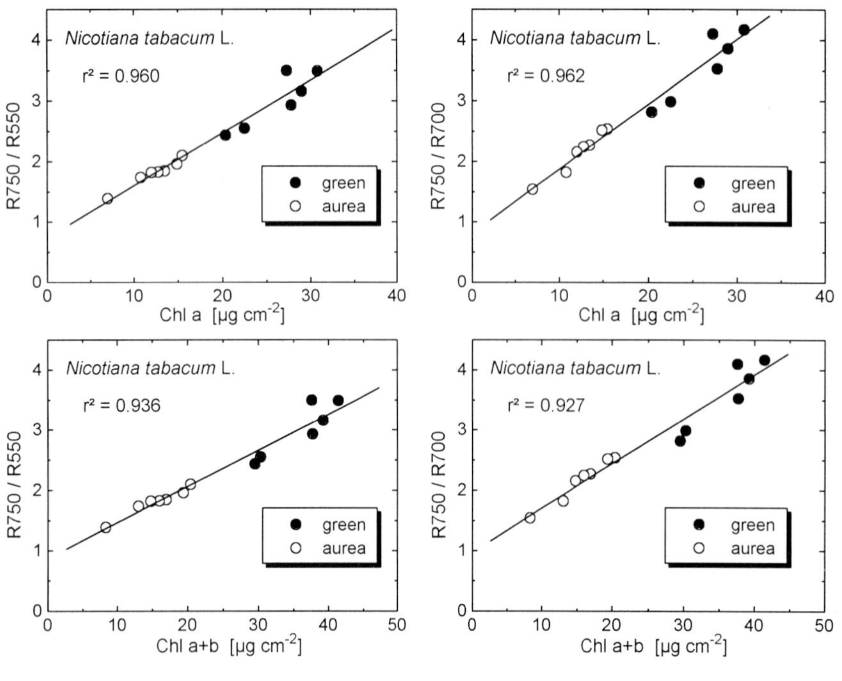

Fig. 11: Linear correlation of the new vegetation indices R750/R550 (**A**) and R750/R700 (**B**) with the chlorophyll *a* and total chlorophyll content for all 13 green and aurea tobacco leaves studied.

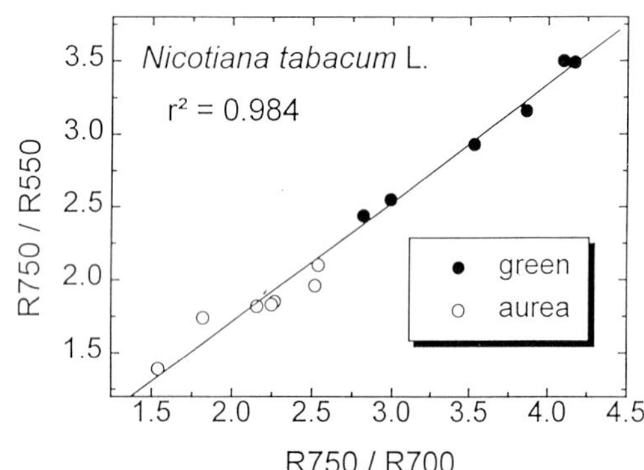

Fig. 12: Linear relationship between the two new vegetation indices R750/R550 and R750/R700 as established with 13 leaves of different chlorophyll content of a green and an aurea variety of tobacco.

tion of photosynthetic pigments. The NIR region includes the chlorophyll-dependent steep rise of reflectance (red edge) as well as the infrared reflectance plateau which depends upon the leaf structure as well as on the size of cells and aerial interspaces in the leaf mesophyll (Lichtenthaler, 1989; Gausman and Quisenberry, 1990). The VIS domain is characterised by low reflectance and low transmittance due to the absorption bands of chlorophylls and carotenoids in the blue and red part of the electromagnetic spectrum. The reflectance maximum in the green region near 550 nm, which is due to an absorption minimum of leaf pigments in this region, is responsible for the green colour of leaves together with leaf transmittance which exhibits similar spectral characteristics as the reflectance (Lichtenthaler, 1994).

Remote sensing of terrestrial vegetation *via* reflectance spectroscopy led to the formation of ratios of reflectance bands (Howard, 1991). The ratio of infrared to red reflec-

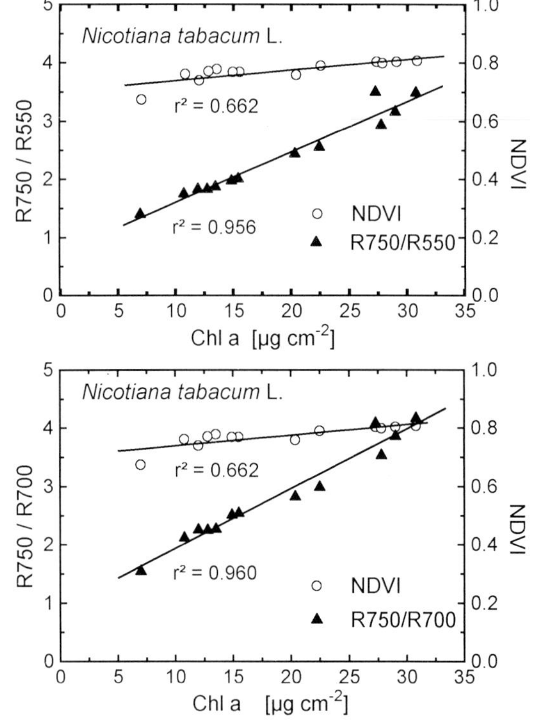

Fig. 13: Comparison of the new vegetation indices R750/R550 and R750/R700 with the old vegetation index NDVI. **A.** Comparison of R750/R550 and versus Chl *a*. The reflectance ratio R750/R550 shows a better correlation to the chlorophyll *a* content ($r^2 = 0.96$) than the vegetation index NDVI ($r^2 = 0.66$). **B.** Comparison of the variation of R750/R700 and NDVI with chlorophyll *a* content. Both new vegetation indices exhibited a better correlation to the chlorophyll *a* content than the old index NDVI. The new vegetation indices R750/R700 and R750/R550 are much more sensitive to changes in chlorophyll content than the old index NDVI.

tance was established as vegetation index in order to estimate green productivity. Biomass productivity was evaluated applying the normalized difference vegetation index NDVI (Tucker et al., 1980), which is based on the red (near 680 nm) and NIR reflectances. Recent studies (Buschmann and Nagel, 1993) demonstrated that there exists no clear correlation between NDVI and Chl *a* content, except for very low chlorophyll levels which are usually not found in leaves and plants. This is due to the fact that the R680 reflectance exhibits only an extremely low sensitivity towards changes in the Chl *a* or total Chl *a+b* content at levels higher than 5 to 7 $\mu g\, cm^{-2}$ (cf. Fig. 1). Thus a decrease in chlorophyll *a+b* content from 41 to 13 $\mu g\, cm^{-2}$ leaf area only caused a decrease in the NDVI values of 9 % (see Fig. 13). An approximate linear relationship and a larger sensitivity of the NDVI values to changing chlorophyll contents was found in aurea tobacco leaves only below a Chl *a* and Chl *a+b* content of 7 and 10 $\mu g\, cm^{-2}$, respectively. Except for aurea mutants, such low chlorophyll levels are only found in highly damaged and stressed plants, e.g. during autumnal leaf senescence.

With respect to chlorophyll determination we examined the new reflectance ratios R750/R550 and R750/R700, which are very accurate for an *in vivo* chlorophyll determination:

(i) The infrared to green reflectance ratio, R750/R550, is composed of the relatively insensitive (with respect to chlorophyll content) infra-red signal R750 and the chlorophyll-content sensitive green reflectance signal R550. This is why the values of the ratio R750/R550 decrease a lot from 3.5 to 1.4 (−60 %) with decreasing Chl *a+b* content from 41 to 8.4 $\mu g\, cm^{-2}$ (−80 %) exhibiting a variation coefficient of 28.6 %. For the two tobacco varieties (green form su/su and aurea mutant Su/su) we found an exact linear correlation between R750/R550 and Chl *a* content ($r^2 = 0.956$) (see Fig. 13). Our results thus confirm former studies of Buschmann and Nagel (1993 and 1995), who found best linear correlation between chlorophyll content and the reflectance ratio infra-red/green, defined as log (R800/R550).

(ii) The second newly established reflectance ratio R750/R700 is based a) on the NIR reflectance signal R750 which is fairly inert and changes very little in tobacco leaves of different age and development, and b) on the reflectance signal R700 which responds with a large amplitude and sensitivity towards changes in the Chl *a* and Chl *a+b* content of leaves. The NIR signal R750 thus serves as an internal reference with respect to reflectance properties caused by leaf structure, whereas the R700 indicates changes going on in the leaf chlorophyll content. In fact, the newly established ratio R750/R700 declined from values of 4.2 to 1.6 (−62 %) with a decline in Chl *a* content of −80 %. The ratio R750/R700 possesses a high linear correlation to the Chl *a* content with $r^2 = 0.960$ (see Fig. 13).

Our results with differently pigmented leaves of green and aurea tobacco are in full agreement with the results of Gitelson and Merzlyak (1994 a and b), who found for maple and chestnut leaves a linear correlation of the two ratios R750/R550 and R750/R705 with the Chl *a* content ($r^2 > 0.97$). This demonstrates the general validity and significance of the two new ratios R750/R550 and R750/R700 as excellent indicators of the *in vivo* chlorophyll content of plants.

The inflection point IP of the red edge also exhibited a very good curvilinear correlation to the chlorophyll content of leaves. Since its blue shift to shorter wavelengths with decreasing chlorophyll content is associated with a considerable increase in the reflectance yield, it is a very exact indicator of changes in the chlorophyll content. From complete leaf reflectance spectra, which can be recorded by high resolution radiometers, the determination of the wavelength position of the IP is possible. In case of remote sensing of the chlorophyll content the determination of the IP is quite difficult, since it would require reflectance sensors of high spectral resolution, which are not available in the existing and future satellite systems. The present reflectance sensors provide the radiance in several quite wide bands in the visible and NIR-range of the spectrum.

The efficiency of the new vegetation indices, as compared to the presently used NDVI, can be estimated using their coefficients of variation. For the same Chl *a* range from 7 to more than 30 $\mu g\, cm^{-2}$, the coefficients' ratio of the variation of the ratios R750/R550 and R750/R700 to that for the NDVI were 6.5 and 6.7, respectively. It means that the newly established indices are more than 6-fold more sensitive to variations in Chl *a* content than the NDVI.

Since the reflectance R700 is a good indicator of the red edge position and also of the chlorophyll content, as shown here, a determination of the exact wavelength position of the IP is not required if one wants to remotely sense the chlorophyll content of terrestrial vegetation. In such cases it is highly sufficient to measure the reflectance in a narrow band near 700 nm or even in somewhat broader bands (near 550 nm and near 750 nm). From these reflectance signals the ratio R750/R700, and alternatively the ratio R750/R550 are formed, which then function as indicators of the chlorophyll content. The reflectance in the NIR-range does not necessarily have to be R750, it could also be R780 or R800, since the reflectance in this NIR region is fairly constant over a longer wavelength range. By selecting the appropriate filters one should, however, stick very closely to the reflectance bands R550 and R700, which are very sensitive to changes in chlorophyll content.

The wavelength position of the red reflectance dip in the red chlorophyll *a* absorption bands, Rmin, was found to be constant at wavelengths between 677.3 and 679 nm. It is therefore independent of the chlorophyll content of the leaves of green or aurea tobacco (see Fig. 9). This demonstrates that the different chlorophyll-proteins with Chl *a* forms showing maxima at various red wavelengths (672, 678 nm) (Lichtenthaler et al., 1981), and the minor Chl *a* forms P680 and P700 and others (e.g. French et al., 1972; Lawlor, 1993) are present in leaves, no matter whether leaves contain a high or low chlorophyll content. The leaves of the aurea tobacco show higher Chl *a/b* values (Table 1) than those of green tobacco (Table 2), which is an indicator that the aurea leaves possess lower amounts of LHCPs than leaves of green tobacco. Although the relative proportions of LHCPs with respect to the other chlorophyll proteins in green and aurea tobacco are somewhat different, this has no influence on the overall wavelength position of the red reflectance minimum Rmin. Small differences in the Rmin from leaf to leaf or even within different points of a leaf may also be due to the fact that Rmin does not only reflect chlorophyll absorption. At low reflectance signals also the internal leaf optics may show a larger influence in this reflectance range as well as the relative contribution of the emitted red chlorophyll fluorescence with a maximum near 684 to 690 nm (F684–F692) (Lichtenthaler and Buschmann, 1987; Buschmann and Lichtenthaler, 1988; Buschmann et al., 1994), which increases with decreasing chlorophyll content. Therefore our findings exclude the application of the wavelength position of the reflectance parameter Rmin near 680 nm in the remote sensing of terrestrial vegetation.

Conclusion

The two new reflectance ratios R750/R550 and R750/R700 are very suitable parameters for a rapid, non-destructive chlorophyll determination of leaves and plants to be used in stress physiology and remote sensing of vegetation. These two new vegetation indices have the great advantage that they match with the chlorophyll content of leaves and plants in a linear way. For this reason they are much better suited for remote sensing of chlorophyll content of terrestrial vegetation than the NDVI, which is only applicable at very low leaf chlorophyll levels, but insensitive and not suitable at a medium and higher chlorophyll content.

Acknowledgements

We wish to thank Gabrielle Johnson for checking the English text.

References

Baret, F., S. Jacquemoud, and G. Guyot: Modeled analysis of the biophysical nature of spectral shift and comparison with information content of broad bands. Remote Sens. Environ. *41,* 133–142 (1992).

Buschmann, C. and H. K. Lichtenthaler: Reflectance and chlorophyll fluorescence signatures of leaves. In: Lichtenthaler, H. K. (ed.): Applications of Chlorophyll Fluorescence, pp. 325–332. Kluwer Academic Publishers, Dordrecht (1988).

Buschmann, C. and E. Nagel: *In vivo* spectroscopy and internal optics of leaves as basis for remote sensing of vegetation. Internat. J. Remote Sens. *14,* 711–722 (1993).

– – Reflexionsspektren von Blättern und Nadeln als Basis für die physiologische Beurteilung von Baumschäden. In: Bittlingmaier, L., W. Reinhardt, and D. Siefermann-Harms (eds.): Waldschäden im Schwarzwald, pp. 288–298. Ecomed, Landsberg (1995).

Buschmann, C., U. Rinderle, and H. K. Lichtenthaler: Detection of stress in coniferous forest trees with the VIRAF spectrometer. IEEE Transactions on Geoscience and Remote Sensing *29,* 96–100 (1991).

Buschmann, C., E. Nagel, K. Szabó, and L. Koscányi: Spectrometer for fast measurements of *in vivo* reflectance, absorptance and fluorescence in the visible and near-infrared. Remote Sens. Environ. *48,* 18–24 (1994).

Butler, W. L. and D. W. Hopkins: Higher derivative analysis of complex absorption spectra. Photochem. Photobiol. *12,* 439–456 (1970).

Carter, G. A.: Responses of leaf spectral reflectance to plant stress. American J. Bot. *80,* 239–243 (1993).

– Ratios of leaf reflectances in narrow wavebands as indicators of plant stress. International J. Remote Sens. *15,* 697–703 (1994).

Chappelle, E. W., M. S. Kim, and J. E. McMurtrey: Ratio analysis of reflectance spectra (RARS): An algorithm for the remote estimation of the concentrations of chlorophyll *a,* chlorophyll *b,* and carotenoids in soybean leaves. Remote Sens. Environ. *39,* 239–247 (1992).

Curran, P. J., J. L. Dungan, B. A. Macler, and S. E. Plummer: The effect of a red leaf pigment on the relationship between red edge and chlorophyll concentration. Remote Sens. Environ. *35,* 69–76 (1991).

French, C. S., Y. S. Brown, and M. C. Lawrence: Four universal forms of chlorophyll *a.* Plant Physiol. *49,* 421–429 (1972).

Gausman, H. W. and J. E. Quisenberry: Spectrophotometric detection of plant leaf stress. In: Katterman, F. (ed.): Environmental Injury to Plants, pp. 257–280. Academic Press, San Diego (1990).

Gitelson, A. and M. N. Merzlyak: Spectral reflectance changes associated with autumn senescence of *Aesculus hippocastanum* L. and *Acer platanoides* L. leaves. Spectral features and relation to chlorophyll estimation. J. Plant Physiol. *143,* 286–292 (1994 a).

– – Quantitative estimation of chlorophyll *a* using reflectance spectra: Experiments with autumn chestnut and maple leaves. J. Photochem. Photobiol. (B) *22,* 247–252 (1994 b).

– – Signature analysis of leaf reflectance spectra: algorithm development for remote sensing of chlorophyll. J. Plant Physiol. *148,* 495–500 (1996).

Gitelson, A., Y. Kaufman, and M. N. Merzlyak: An atmospherically resistant «green» vegetation index (ARGI) for EOS-MODIS. J. Remote Sensing of Environment. (1996 a) in press.

Gitelson, A., M. N. Merzlyak, and H. K. Lichtenthaler: Detection of red edge position and chlorophyll content by reflectance measurements near 700 nm. J. Plant Physiol. *148,* 501–508 (1996 b).

Guyot, G.: Optical properties of vegetation canopies. In: Steven, M. D. and J. A. Clark (eds.): Applications of Remote Sensing in Agriculture, pp. 19–27. Butterworths Scientific Ltd., London (1990).

Horler, D. N., M. Dockray, and J. Barber: The red edge of plant leaf reflectance. International J. Remote Sens. *4,* 273–288 (1983).

Howard, J. A.: Remote Sensing of Forest Resources, pp. 47 ff. Chapman & Hall, London (1991).

Lang, M., H. K. Lichtenthaler, M. Sowinska, F. Heisel, J. A. Miehé, and F. Tomasini: Fluorescence imaging of water and temperature stress in plant leaves. J. Plant Physiol. *148,* 613–621 (1996).

Lawlor, D. W.: Photosynthesis: Molecular, Physiological and Environmental Processes, 2nd ed., Longman Scientific & Technical, Harlow Essex (1993).

Lichtenthaler, H. K.: Chlorophylls and carotenoids: pigments of photosynthetic biomembranes. Methods Enzymol. *148,* 350–382 (1987).

– Possibilities for remote sensing of terrestrial vegetation by a combination of reflectance and laser-induced chlorophyll fluorescence. In: Proceed. Internat. Geoscience and Remote Sensing Symposium, IGARSS'89, Vancouver, Vol. *3,* pp. 1349–1354. Library of Congress, No. 89-84217 (1989).

– Spektroskopische Eigenschaften von Pflanzen und ihre Nutzung zur Fernerkundung der Vegetation. Fridericiana *49,* 25–45 (1994).

– Vegetation stress: an introduction to the stress concept in plants. J. Plant Physiol. *148,* 4–14 (1996).

LICHTENTHALER, H. K. and C. BUSCHMANN: Reflectance and chlorophyll fluorescence signatures of leaves. In: Proceedings of the Remote Sensing Symposium (IGARSS), Michigan, Vol. II, pp. 1201–1206. The University of Michigan, Ann Arbor 1987.

LICHTENTHALER, H. K., V. STRAUB, and K. H. GRUMBACH: Unequal formation of prenyl-lipids in a plant tissue culture and in leaves of *Nicotiana tabacum* L. Plant Science Letters *4*, 61–65 (1975).

LICHTENTHALER, H. K., G. BURKARD, G. KUHN, and U. PRENZEL: Light-induced accumulation and stability of chlorophylls and chlorophyll proteins during chloroplast development in radish seedlings. Z. Naturforsch. *36c*, 421–430 (1981).

ROCK, B. N., T. HOSHIZAKI, H. K. LICHTENTHALER, and G. SCHMUCK: Comparison of *in situ* spectral measurements of forest decline symptoms in Vermont (USA) and the Schwarzwald (F.R.G.). In: Proceedings of the Remote Sensing Symposium (IGARSS), Zürich 1986, Vol. III, pp. 1667-1672. ESA Publications Division, Noordwijk (1986).

ROUSE, J. W., R. H. HAAS, J. A. SCHELL, D. W. DEERING, and J. C. HARLAN: Monitoring the vernal advancement of retrogradation of natural vegetation. NASA/GSFC, Type III, Final Report. Greenbelt, MD, USA (1974).

SANTRUCEK, J., P. ŠIFFEL, M. LANG, H. K. LICHTENTHALER, C. SCHINDLER, H. SYNKOVA, V. KONECNA, and K. SZABO: Photosynthetic activity and chlorophyll fluorescence parameters in aurea and green forms of *Nicotiana tabacum*. Photosynthetica *27*, 529–543 (1992).

SCHINDLER, C., P. REITH, and H. K. LICHTENTHALER: Differential levels of carotenoids and decrease of zeaxanthin cycle performance during leaf development in a green and an aurea variety of tobacco. J. Plant Physiol. *143*, 500–507 (1994).

SCHMUCK, G., H. K. LICHTENTHALER, G. KRITIKOS, V. AMANN, and B. N. ROCK: Comparison of terrestrial and airborne reflection measurements of forest trees. In: Proceedings of the Remote Sensing Symposium (IGARSS), Michigan, Vol. II, pp. 1207–1212. The University of Michigan, Ann Arbor (1987).

THOMAS, J. R. and H. W. GAUSSMAN: Leaf reflectance vs. leaf chlorophyll and carotenoid concentration for eight crops. Agron. J. *69*, 799–802 (1987).

TUCKER, C. J., B. N. HOLBEN, J. H. ELGIN, and J. E. MCMURTREY: Relation of spectral data to grain yield variation. Photogrammetric Engineering and Remote Sensing *46*, 657–666 (1980).

YODER, B. J. and R. H. WARING: The normalised difference vegetation index of small Douglas-fir canopies with varying chlorophyll concentrations. Remote Sens. Environ. *49*, 81–91 (1994).

J. Plant Physiol. Vol. 148. pp. 494–500 (1996)

Signature Analysis of Leaf Reflectance Spectra: Algorithm Development for Remote Sensing of Chlorophyll

ANATOLY A. GITELSON[1] and MARK N. MERZLYAK[2]

[1] J. Blaustein Institute for Desert Research, Ben-Gurion University of the Negev, Sede-Boker Campus 84993, Israel

[2] Department of Cell Physiology and Immunology, Faculty of Biology, Moscow State University, 119899 GSP Moscow W-234, Russia

Received June 24, 1995 · Accepted September 18, 1995

Summary

The goal of the study is to investigate the basic spectral properties of plant leaves to develop spectral indices more sensitive to chlorophyll concentration than the presently widely used Normalized Difference Vegetation Index. These indices can serve as indicators of stress, senescence, and disease in higher plants. The spectral reflectance of senescing leaves of two deciduous species (maple and chestnut) as well as their pigment content were measured. Spectral indices were developed using reflectances corresponding to wavelengths with maximum and minimum sensitivity to variation in pigment concentration. The signature analysis of reflectance spectra indicated that, for a wide range of leaf greenness (completely yellow to dark green leaves), the maximum sensitivity of reflectance coincides with the maximum absorption of chlorophyll *a* at 670 nm. However, for yellow-green to green leaves (minimum chlorophyll *a* as low as 3–5 nmol/cm^2), the reflectance near 670 nm is not sensitive to chlorophyll concentration due to saturation effects. Therefore, it seems inappropriate to use this spectral band for pigment estimation in yellow-green to green vegetation. The spectral bands ranging from 400 to 480 nm and above 730 nm are not sensitive to chlorophyll concentration as found for 670 nm. The reflectances at these wavelengths could be used as references in the vegetation indices. Maximum sensitivity to chlorophyll *a* concentration was found at 550–560 nm and 700–710 nm. Reflectances at 700 nm correlated very well with that at 550 nm for a wide range of chlorophyll concentrations for both plant species studied. The inverse reflectance, $(R_{550})^{-1}$ and $(R_{700})^{-1}$ are proportional to chlorophyll *a* concentration; therefore indices R_{750}/R_{550} and R_{750}/R_{700} are directly proportional (correlation $r^2 > 0.95$) to chlorophyll concentration. These indices were tested for a wide range of chlorophyll *a* concentration, using several independent data sets. The estimation error in the derivation of chlorophyll concentration from the indices is assessed to be less than 1.2 nmol/cm^2.

Key words: Reflectance spectra of leaves, remote sensing, vegetation indices.

Abbreviations: Chl *a* and *b* = chlorophyll *a* and *b;* Car = total carotenoids; STD = standard deviation.

Introduction

Plant senescence, diseases and many long-term stresses result in a loss of chlorophyll content, therefore it is extremely important to develop technique for non-destructive estimation of chlorophyll content of a vegetation. The application of remote sensing technology in the visible part of the electromagnetic spectrum to yield information about the state of vegetation is limited by a complexity of interacting factors involved in the reflectance response (Andrieu and Baret, 1993; Baret and Guyot, 1991; Curran et al., 1991; Horler et al., 1983; Huete et al., 1994). Not enough is yet understood about the peculiarities, specific features and optical properties of a leaf (Horler et al., 1983; Fukshansky, 1981; Vogelmann and Björn, 1986). Nevertheless, several specific bands in the reflectance spectrum were successfully used to monitor chlorophyll and carotenoid contents in intact leaves (Baret et al., 1987; Baret et al., 1992; Buschmann and Nagel, 1993; Chappelle et al., 1992; Curran et al., 1991; Gitelson and Merzlyak, 1994 a, b; Kim et al., 1994; McMurtrey III et al., 1994).

We found in previous experiments with yellowing autumn leaves of two deciduous species that the bands near 550 and 700 nm were the regions with the maximal sensitivity of reflectance to a variation in chlorophyll (Chl) concentration. Several indices utilizing spectral features located outside the main bands of photosynthetic pigment absorption, were developed for estimation of Chl concentration (Gitelson and Merzlyak, 1994 a, b).

The first objective of this research was to understand in more detail the specific spectral features of the plant photosynthetic tissues through examination of the reflectance spectra of both mature and senescing green leaves covering a wide range of pigment concentration. Specific wavelengths sensitive to pigment variation were discovered and the algorithms for Chl assessment at leaf level using reflectances at 550, 700, and longer 750 nm were developed. They were tested by independent data sets for a range of Chl *a* from 0.3 to 44.8 nmol/cm^2 for maple and chestnut leaves.

The second objective was to determine which bands in the reflectance spectrum could be employed for estimation of pigment quantities and other characteristics of vegetation canopies in the presence of non-photosynthetic materials. In addition to the reflectance at 550 and 700 nm, the reflectances at 500 and 670 nm were also found to be highly correlated in yellow-green to dark green leaves. It was suggested to use the indices $[(R_{700}/R_{550})-1]$ and $[(R_{670}/R_{500})-1]$ as indicators of the contribution of non-photosynthetic materials and apply them to remote estimation of pigments in canopies.

Materials and Methods

The experiments were performed in October 1991, 1993 on horse chestnut (*Aesculus hippocastanum* L.), and in October 1992–1993 on Norway maple (*Acer platanoides* L.) leaves. Leaves of both trees were collected in the Botanical Garden of the Moscow State University, as described previously (Gitelson and Merzlyak, 1994 a; Merzlyak and Gitelson, 1995). In addition to the senescing samples, mature green leaves of both species collected in July of 1994 were also examined. This sampling scheme intended to cover a variation of pigment concentrations as high as possible in each experiment. Only leaves having homogeneous dark green, green, green-yellow, yellow-green and yellow color without anthocyanin pigmentation were selected.

Hemispherical reflectance spectra were recorded from 400 to 750 nm for the upper surface of the leaves with a Hitachi 150-20 spectrophotometer equipped with an integrating sphere. The spectral resolution was 2 nm. The reflectance spectra were measured against barium sulfate as a reference standard. A light trap was designed to eliminate the specular reflected component of the radiance, and black velvet was used as a background to absorb the light passing through the leaf (Gitelson and Merzlyak, 1994 a; Merzlyak and Gitelson, 1995). The reflectance was expressed as a ratio of the radiance of the leaf to the radiance of the reference. The spectra were recorded for the sections of the leaves between main veins (maple) or with a removed main vein (chestnut).

Chl *a, b* and total carotenoid (Car) concentrations in the leaves were determined in acetone extracts and calculated using equations and specific extinction coefficients reported by Lichtenthaler (1987). The average mol.wt for Car of 570 was used.

Results

Pigment content in the leaves

Chl *a, b* and Car concentrations in the leaves are presented in Fig. 1 where data are sorted by Chl *a*+ *b* concentration. The dominant pigment, Chl *a,* ranged from 0.3 to 44.8 nmol/cm^2 in maple leaves and from 0.5 to 42.4 nmol/cm^2 in chestnut leaves. The green leaves collected both in summer and autumn (Chl *a*+ *b* > 20 nmol/cm^2) contained approximately equal proportions of pigments. Although the leaves of both species lost Chl *a* and *b* during the progression of autumn senescence, relatively high concentrations of carotenoids were present (Gitelson and Merzlyak, 1994 a; Merzlyak and Gitelson, 1995; Merzlyak and Hendry, 1994). Only trace amounts of chlorophylls were detected in completely yellow leaves.

Reflectance spectral changes in the leaves

The representative reflectance spectra of the maple leaves which changed color from dark green to completely yellow contained decreasing amounts of Chl are shown in Fig. 2. The reflectance spectral features were found to be similar for both species studied. Although the spectra obtained with *A. platanoides* leaves will be considered later, the results being very close to those obtained with *A. hippocastanum* leaves.

The maximum reflectance was found at 750 nm and was independent of pigment concentration and the stage of leaf development. The lowest reflectance was observed in the blue range of the spectrum from 400 to 500 nm. For completely yellow leaves (Chl *a* from 0.3 to 1 nmol/cm^2), the reflectance in the above range was much higher than that of the other

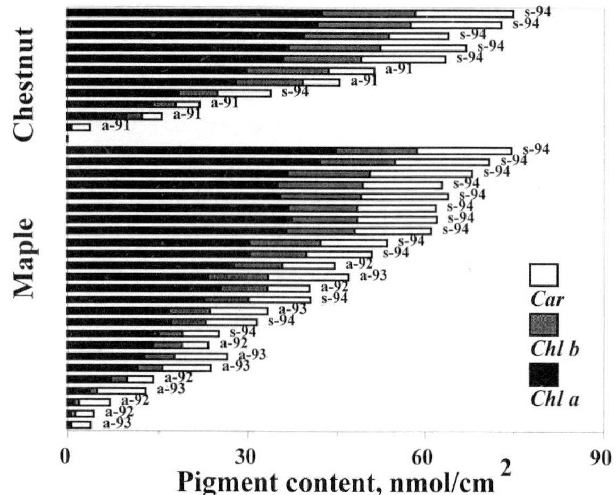

Fig. 1: Pigment concentrations in mature and senescent maple and chestnut leaves. Car is total carotenoids, Chl *a* and Chl *b* are chlorophyll *a* and *b,* respectively. An example of abbreviations: a-91 are autumn leaves, collected in 1991, s-94 are mature leaves collected in summer 1994.

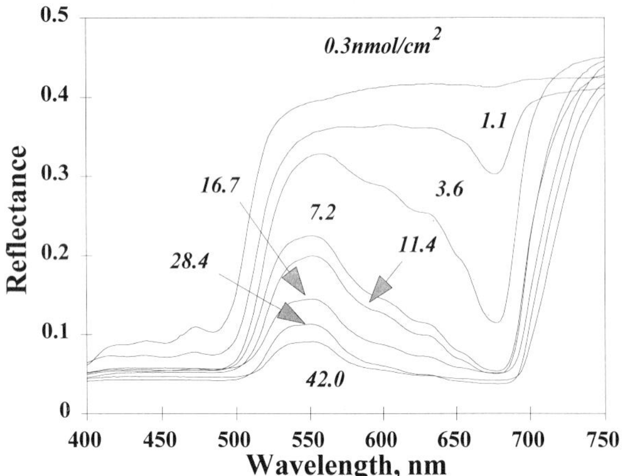

Fig. 2: The representative reflectance spectra of maple leaves containing different concentration of pigments. Chl *a* concentrations in nmol/cm^2 are indicated at reflectance curves.

leaves, and the typical three carotenoid absorption bands were clearly seen (Fig. 2). A very small increase in Chl *a* and *b* concentrations from 0.3 to 1 nmol/cm^2 induced a significant decrease in reflectance and shifted the «green edge» (rise of the reflectance in the range from 500 to 550 nm) at the reflectance spectra toward longer wavelengths. This alteration in the shape of the spectrum occurred at virtually the same carotenoid concentrations. An increase in pigment concentration from 3 to 44.8 nmol/cm^2 did not lead to any variation in the reflectance in this spectral range.

An increase in reflectance occurred near 500 nm for all leaves studied. For a completely yellow leaf (Chl *a* = 0.3 nmol/cm^2), a wide plateau up to 750 nm came after this increase. In the leaves with Chl *a* >3 nmol/cm^2, the prominent maximum of reflectance occurred in the green range of the spectrum. In yellow-green leaves, this peak was wide and reached 30–40 %, while for green-yellow to green leaves (Chl *a* >3 nmol/cm^2), it became narrow and was of decreased magnitude. The reflectance near 550 nm did not exceed 10 % in green leaves (Chl *a* >20 nmol/cm^2).

A decrease in reflectance followed the «green» peak. The reflectance in the range 600–650 nm varied from about 5 % for dark-green leaves up to 40 % for yellow ones. In dark-green to yellow-green leaves (Chl *a* >3 nmol/cm^2), the reflectance near 670 nm (i.e., the red maximum of Chl *a* absorption) was low and remained virtually the same in the range of Chl *a* variation from 5 to more than 44 nmol/cm^2. Only when the Chl *a* dropped to a level of less than 3 nmol/cm^2, was a considerable increase in reflectance observed.

A minimum near 670 nm was followed by a sharp increase in reflectance towards longer wavelengths. The slope of the reflectance increase (the «red edge») varied widely, decreasing with increase in Chl concentration.

To better understand the nature of the observed reflectance changes and to find which spectral bands are maximally sensitive to variation in pigment concentrations, the standard deviation (STD) of the reflectance was studied. STD was calculated for different groups of leaves. The first group contained leaves with a Chl *a* concentration from 0.3 (minimum concentration found in the experiments) to 44.8 nmol/cm^2 (maximum concentration). The second group included leaves with Chl *a* from 0.6 to 44.8 nmol/cm^2. In the third group, the minimum Chl *a* concentration was 1.1 nmol/cm^2 and maximum Chl remained the same (44.8 nmol/cm^2). In other words, the groups had different minimum Chl *a* concentrations, while the maximum concentration remained the same. The first group of leaves can be considered to represent of a wide-ranging process of senescence, when the color of the leaves turn from dark green to completely yellow. The following groups, in order, corresponded to different stages of senescence (or/and stress and disease). This range ended with leaves in the very early stages of stress and senescence, when they were still green, but the suppression of biosynthesis and/or increased degradation of green pigments already had been initiated. In the last group, the summer leaves with natural variation in Chl *a* concentration due to light adaptation (between leaves in the sun and those in shade) were presented. Since the range of variation in Chl concentration was different for each group of the leaves, for their comparison STD of reflectance, STD(R), was normalized to STD of chlorophyll *a*, STD(Chl *a*), and the data are presented in Fig. 3.

The obtained STD(R)/STD(Chl *a*) spectra showed several notable features. They clearly indicated different spectral behavior for groups of leaves containing «completely yellow to green» leaves (samples with minimum Chl *a* <1.1 nmol/cm^2) as compared to «yellow-green to green» (samples with mini-

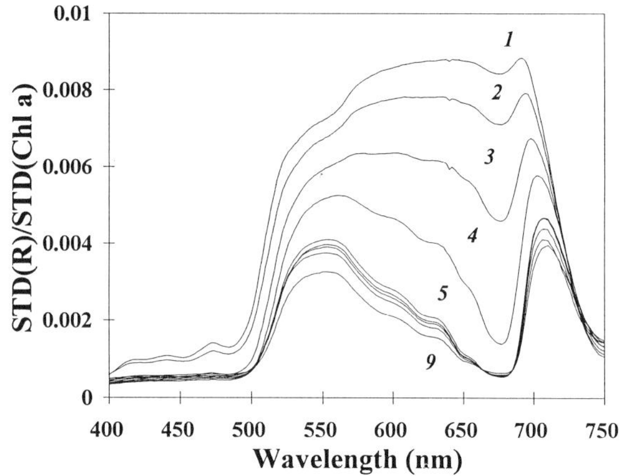

Fig. 3: The ratio of the standard deviation of reflectance, STD(R), normalized to the standard deviation of chlorophyll *a*, STD(Chl *a*), for groups of maple leaves selected from 25 samples, listed in Fig. 1. The groups had different minimal Chl *a* concentrations: 1–0.3 nmol/cm^2; 2–0.6; 3–1.1; 4–3.6; 5–7.1; 6–11.4; 7–12.6; 8–14.0; 9–17.0. Maximum Chl *a* concentration was 44.8 nmol/cm^2 for all leaf groups. The first group of leaves can be considered to represent of a wide-ranging process of senescence, when the color of the leaves turn from dark-green to completely yellow. The following groups, in order, corresponded to different stages of senescence (or/and stress and disease). This range ended with leaves in the very early stages of stress and senescence, when they were still green, but the suppression of biosynthesis and/or increased degradation of green pigments already had been initiated. In the last group with minimal Chl *a* concentration of 17 nmol/cm^2, the summer leaves were presented with natural variation in Chl *a* concentration due to light adaptation (i.e., between leaves in the sun and those in shade).

mum Chl $a > 3$ nmol/cm^2) leaves. For leaves with $44.8 >$ Chl $a > 0.3$ nmol/cm^2, three distinct maxima due to carotenoid absorption near 425, 450 and 490 nm could be seen (Fig. 3). Between 550 and 660 nm, a broad peak of STD existed. Near 670 nm, a small minimum occurred. It was followed by insignificant peak near 690 nm. Then a sharp decrease of STD towards longer wavelengths was recorded. STD was low and almost the same in the blue and the near infra-red ranges of the spectrum. In leaves with $44.8 >$ Chl $a > 1$ nmol/cm^2, the spectral behavior was very different from the group of completely yellow to green leaves. In the blue range, the STD decreased two-three fold and spectral features were not detected. The minimum near 670 nm became prominent. The small peak at 690 nm was transformed to a noticeable maximum which shifted towards the longer wavelengths when the minimal Chl a concentration increased.

When the minimal Chl a in the leaf group was more than 3 nmol/cm^2, the broad maximum between 550 and 650 nm was transformed to a quite small narrow peak centered at 550 nm. The magnitude of the ratio STD(R)/STD(Chl a) at 670 was very low and remained virtually the same for the leaf groups with a minimal Chl a of more than 3 nmol/cm^2. The existence of two spectral bands (one, wide near 550 nm and the other, narrow near 700 nm), where variation of reflectance was found to be much higher than at 670 nm for leaf groups with minimal Chl $a > 1$ nmol/cm^2, was an important finding. It is also interesting to note that the variation of reflectance near 550 and 700 nm remained almost the same for leaf groups with a minimum Chl a concentration of more than 3 nmol/cm^2.

Reflectances in the range of 520 to 650 nm were strongly correlated with reflectance near 700 nm and were much more weakly correlated with reflectances in the blue (400 to 480 nm), in the near infra-red ranges and near 670 nm (data not shown). For all leaves studied an extremely high correlation ($r^2 > 0.99$) was found between the reflectance at 700 nm, R_{700}, and at 550 nm, R_{550} (Fig. 4).

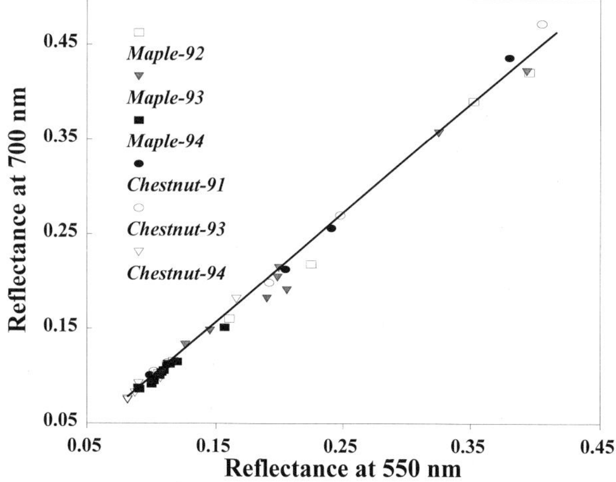

Fig. 4: The reflectance at wavelength 700 nm versus that at 550 nm. For all leaves and both species studied in different years in the range of Chl a concentration from 0.3 to 44.8 nmol/cm^2 the correlation between the reflectances at 550 and 700 nm was extremely high ($r^2 > 0.99$).

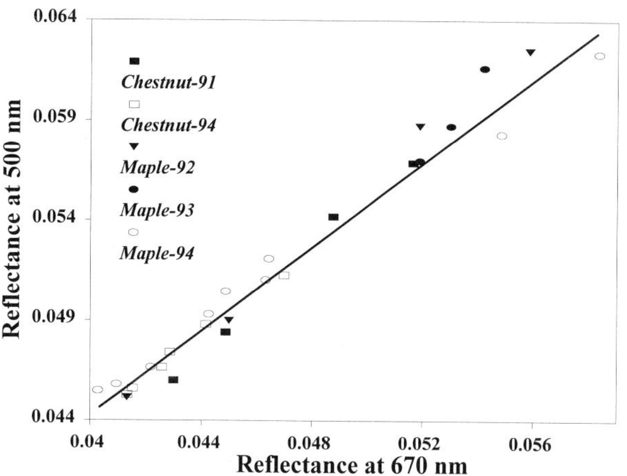

Fig. 5: The reflectance at wavelength 670 nm versus that at 500 nm for a yellow-green to dark-green maple and chestnut leaves with minimum Chl a concentration > 3 nmol/cm^2. The determination coefficient, r^2, for this relationship was more than 0.97.

Another feature of the reflectance spectra was a strong correlation of R_{670} with the reflectances in the range 500–520 nm (Fig. 5). For all leaves studied (Chl $a > 0.3$ nmol/cm^2), the maximum correlation coefficient was at 520 nm (data not shown). For green-yellow to green leaves (Chl $a > 3$ nmol/cm^2), the maximum correlation was R_{670} with R_{500} ($r > 0.98$). This spectral feature of «yellow-green» to «green» vegetation was found also for chestnut leaves.

Algorithms for chlorophyll detection

The index for Chl estimation should be maximum sensitive to Chl and invariant with respect to other factors. Therefore it would be very useful to find the wavelength where only one dominant factor influences the reflectance variation. The ratio STD(R)/STD(Chl a) (Fig. 3) clearly indicates such specific spectral bands with a high sensitivity of reflectance to Chl a variation – near 550 and 700 nm. Fig. 6 demonstrates the relationships between Chl a concentration and reflectances at 550, 700 and 670 nm. There is a hyperbolic relationship between Chl a and R_{550} and R_{700} (Fig. 6). For the linear function Chl a versus $(R_{700})^{-1}$ and $(R_{550})^{-1}$, the correlation was very strong ($r^2 > 0.95$) with an error of Chl a estimation of less than 2.8 nmol/cm^2. Reflectance at 670 nm decreased sharply when Chl a increased up to 3–5 nmol/cm^2; thereafter R_{670} was almost pigment-concentration independent. Therefore, it would be inappropriate to use the reflectance near 670 nm as a sensitive term in the index for Chl estimation. The lowest variation of reflectance took place in the near infra-red (above 750 nm) and in the blue (shorter than 500 nm) ranges of the spectrum.

Thus, the reflectances near 700 nm and in the range from 540 to 600 nm were the only features found to be sensitive to Chl a concentration which could be used in constructing the index. R_{750} can be taken as a term insensitive to Chl a concentration. Taking into consideration that $(R_{700})^{-1}$ and $(R_{550})^{-1}$ are directly correlated to Chl a and that R_{750} virtually did not depend on Chl a, the indices R_{750}/R_{550} and R_{750}/R_{700}

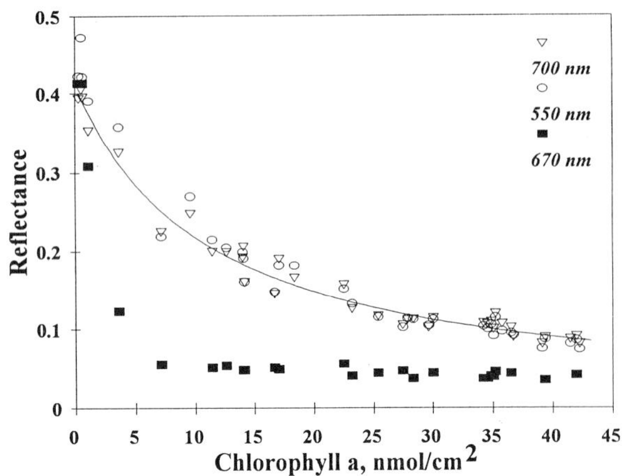

Fig. 6: The reflectance at the wavelengths 550, 670, and 700 nm versus Chl *a* concentrations for maple leaves. Reflectance at 670 nm, R_{670}, decreased sharply when Chl *a* increased up to 3–5 nmol/cm^2; thereafter R_{670} was almost pigment-concentration independent. There is a hyperbolic relationship between Chl *a* and R_{550} and R_{700}. The solid line represents the equation Chl = –7.2 + 4.01 · $(R_{700})^{-1}$ with determination coefficient $r^2>0.95$.

were chosen for Chl *a* assessment. These indices were compared with analytically-measured Chl *a* and Chl *a*+ *b* concentrations. For leaves measured in 1991 and 1992 linear regressions were obtained (Gitelson and Merzlyak, 1994 a, b):

– for maple leaves

$$\text{Chl}\,a = -(8.16 \pm 1.96) + 8.59 \cdot R_{750}/R_{550} \quad (1)$$

with $r^2>0.97$,

– for chestnut leaves

$$\text{Chl}\,a = -(11.52 \pm 1.69) + 12.50 \cdot R_{750}/R_{700} \quad (2)$$

with $r^2>0.98$. For the relationship Chl *a*+ *b* versus the above reflectance ratios, r^2 was higher than 0.96.

The models (1) and (2) were validated by independent data sets obtained in 1993 and 1994. We performed the validation employing Eq. 1 and 2 to calculate the Chl *a* concentration using reflectance data measured in 1993 and 1994. The resulting predicted Chl *a* and Chl *a*+ *b* values were compared to the analytically-measured chlorophyll concentrations in 1993 and 1994 (Fig. 7). The correlation between predicted and measured Chl *a* concentrations was r^2 = 0.99 for both R_{750}/R_{550} and R_{750}/R_{700} ratios, with an error in the Chl *a* estimation of less than 1.26 nmol/cm^2 and 1.14 nmol/cm^2, respectively. A minimal estimation error of Chl *a*+ *b* (less than 1.3 nmol/cm^2) was obtained for R_{750}/R_{700} ratio.

Discussion

Mechanisms responsible for these revealed spectral signatures have to be understood in order to ascertain whether the parameters of above algorithms will be stable for a wide range of pigment concentrations in the leaves and can be applied to a number of plant species. The leaves studied (Fig. 1) covered a very wide range of pigment concentrations (corresponding to leaf color change from completely yellow to dark green). They could be considered as a model of different physiological states of a plant, which revealed the following important relationships.

(1) Near 670 nm, as well as in the blue region of the spectrum (400 to 500 nm), the relationship reflectance versus Chl *a* was saturated for a concentration of the pigment more than 3 nmol/cm^2, whereas near 550 and 700 nm the relation was monotonous and reflectance remained sensitive to pigment concentration for Chl *a*>40 nmol/cm^2 (Figs. 2–3, 6).

(2) For all leaves and both species studied in the range of Chl *a* concentration from 0.3 to 44.8 nmol/cm^2 the correlation between the reflectances at 550 and 700 nm was extremely high ($r^2>0.99$) (Fig. 4).

(3) For all leaves, a strong correlation between R_{670} and reflectance in the range 500–520 nm was found. For green-yellow to green leaves (Chl *a*>3–5 nmol/cm^2) the correlation coefficient peaked at 500 nm and reached r>0.98.

The range near 700 nm belongs to «the red edge» which is a unique feature of green vegetation (e.g., Baret et al., 1992; Curran et al., 1991; Horler et al., 1983; Fukshansky, 1981). It results from two optical properties of plant leaves: high Chl *a* absorption resulting in low reflectance near 670 nm and high internal leaf scattering causing large near infra-red reflectance. Thus, in the wavelength range lower than 700 nm the reflectance is primarily determined by strong Chl *a* absorption, while beyond 700 nm, it is governed less by absorption and much more by scattering (e.g., Horler et al., 1983; Fukshansky, 1981). The reasons for the high sensitivity of this spectral band to Chl *a* concentration is its location which is rather far from the main absorption bands of chloroplast pigments. This prevents the saturation of relationship reflectance versus chlorophyll (Gitelson and Merzlyak, 1994 a). On the

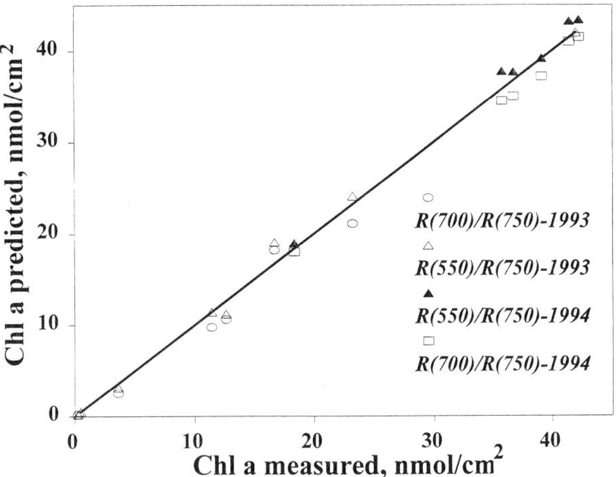

Fig. 7: The results of validation of the models (1) and (2) by independent data sets obtained in 1993 and 1994. Predicted chlorophyll *a* concentrations plotted versus measured in maple and chestnut leaves analytically. Predicted Chl *a* concentrations were calculated from Eq. 1 and 2, with the reflectances of the leaves measured in 1993 and 1994. The correlation between predicted and measured Chl *a* concentrations was r^2 = 0.99 for both R_{750}/R_{550} and R_{750}/R_{700} ratios, with maximum error in the Chl *a* estimation of less than 1.26 nmol/cm^2. Solid line presents equation $\text{Chl}_{pred} = \text{Chl}_{meas}$.

other hand, this band is too far from near infra-red range with high scattering, to provide enough sensitivity to Chl *a* variation. Apparently the reflectance near 700 nm is a fundamental spectral feature that is produced by an equilibrium between two these competing processes. Reflectance near 700 nm was also found to be the transition point at the red edge «between two more-or-less linear phases of the relationship ‹position of inflection point vs. Chl *a*›» (Horler et al., 1983). The high variation of R_{700} with Chl concentration is a result of the shift of the red edge and it is caused by the same physical processes (e.g., Baret et al., 1992; Curraet al., 1991; Gitelson et al., 1996; Horler et al., 1983; Vogelmann et al., 1993).

Reflectance at 550 nm, R_{550}, is located between two wide bands of strong pigment absorption. R_{550} is below the green edge in the reflectance spectrum, around 520 nm, on the long wavelength side of carotenoid as well as the blue Chl *a* and Chl *b* absorption band. The spectral behavior of this edge was found to be very similar to that of the red edge (Horler et al., 1983). The difference between them is that the green edge is primarily determined by Car, Chl *a*, and Chl *b* absorption, while the red edge is governed primarily by Chl *a*. (The contribution of Chl *b* to the absorption at near 700 nm is very low. The maximum for *in vivo* absorption of Chl *b* present in light-harvesting complex is assumed to be located around 650 nm.) At longer wavelengths R_{550} is located before the large range of absorption by chlorophylls. Thus, near 550 nm, two strong absorption processes reach their minimum, producing the monotonous relationship ‹Chl *a* versus R_{550}› with a high sensitivity to Chl *a* concentration.

Reflectances near 700 and 550 nm were found to be of equal sensitivity to variation in Chl *a*. Moreover, R_{550} and R_{700} correlated extremely well in both maple and chestnut leaves from senescent yellow to mature dark green (Fig. 4). The same phenomenon was discovered in soybean (Chappelle et al., 1992; Kim et al., 1994) and for corn leaves (McMurtrey III et al., 1994), when the plants were affected by nitrogen deficiency. This is consistent with the observations of Horler et al. (1983) where, again, a strong correlation between R_{540} and the position of the red edge was found. We also compared the values of R_{550} and R_{700} for young juvenile, mature and yellowing leaves of European beach (Fig. 3 in Tanner and Eller, 1986), and found that they were also very close to each other.

Such similarity between R_{550} and R_{700} could be understood if the absorption at these wavelengths was affected by pigments occurring exactly in the same proportion. Chl *a* is mainly responsible for absorbance near 700 nm. At 550 nm, as well as at longer wavelengths, both Chl *a* and *b* play a major, even dominant, role in light absorption. The contribution of carotenoids is probably much less, as indicated by the reflectance spectra of yellow leaves (Fig. 2). In the presence of trace amounts of both chlorophylls (<0.3 nmol/cm^2) and considerable quantities of carotenoids (>3 nmol/cm^2), no evidence for the contribution of carotenoids to reflectance at 550 nm exist (upper curve in Fig. 2). However, an increase in Chl *a* concentration up to 3 nmol/cm^2 on a background of approximately the same amounts of carotenoids led to a significant decrease in the reflectance near 550 nm.

In solution, pure chlorophylls possessed low, but measurable absorbance between 500 to 600 nm, with a molar extinction coefficient about 3–5 mM^{-1} cm^{-1}. The comparison of absorbance spectra of pure chlorophylls with those of maple leaf extracts in methanol indicated that the contribution of carotenoids at wavelengths near 550 nm and longer was at least 8–10-fold less than that of green pigments (Merzlyak, unpublished). Among the carotenoids present in the chloroplasts of higher plant leaves, only β-carotene exhibited an extremely small absorbance at 550 nm (Lichtenthaler, 1987). Therefore, the perfect covariation ($r^2 > 0.99$) between R_{550} and R_{700} (Fig. 4) may be explained by the fact that Chl *a* covaries reasonably well with Chl *b* ($r^2 > 0.96$) for the chestnut and maple leaves studied (data not shown), while the carotenoids did not contribute significantly to reflectance at 550 nm.

These results have confirmed and quantified our previous findings (Gitelson and Merzlyak, 1994 a, b). The developed indices work for leaves of both species studied, maple and chestnut in a very wide range of Chl concentrations. The indices are directly proportional to Chl *a* allowing precise estimation of Chl *a* and Chl *a*+ *b* concentration at the leaf level. The reflectance near 700 nm is affected mainly by Chl *a*, whereas R_{550} is affected by both Chl *a* and *b*. Since these pigments covary reasonably well, total Chl concentrations can be determined from reflectance measurements at both wavelengths. The sensitivity of reflectance to Chl *a* concentration remained approximately the same in spectral band from 540 to 600 nm. Reflectance in this rather wide range can be used for estimating total Chl, whereas the reflectance near 700 nm is more suitable for Chl *a* assessment.

The index $\log(R_{800}/R_{550})$ was found to be a good indicator of chlorophyll content per leaf area for intact bean leaves (Buschmann and Nagel, 1993). They reported that the ratio R_{800}/R_{550} was also closely correlated with Chl *a* concentration ($r^2 > 0.88$). Carter found high sensitivity of the reflectances near 550 nm and 700 nm to stress of plants (Carter, 1993; 1994). The high sensitivity of R_{700} to Chl *a* concentration was also demonstrated by Chappelle et al. (1992), Kim et al. (1993), and McMurtrey III et al. (1994) for soybean and corn leaves. They constructed the index for Chl *a* estimation, using the ratio R_{700}/R_{670}. Taking into account that R_{670} virtually did not depend on Chl *a* concentration (see Fig. 2 and Table 9 in McMurtrey III et al., 1994), one can conclude that R_{700} was used as a term sensitive to Chl *a* concentration, whereas R_{670} was the insensitive one.

The parameters of the relationships between reflectance and pigment concentration depend on many factors; the primary ones are species, pigment composition, and developmental stage. It is indeed remarkable that the error of Chl *a* estimation in the range 0.3 to 44.8 nmol/cm^2 for two species studied was as low as 1.2 nmol/cm^2 considering the sources of «noise» in the algorithms.

A high correlation between reflectances at 500 and 670 nm was found for the leaves of both plant species with Chl *a* >3 nmol/cm^2 (Fig. 5). This means that absorbance by Chl *a*, *b* and Car at 500 nm and by both chlorophylls at 670 nm was almost similar over a wide range of pigment variation. The correlation of R_{500} and R_{670} was not so high if chlorophyll concentration dropped to less than 3–5 nmol/cm^2, while carotenoid concentration remained relatively high (i.e., for yellow-green and completely yellow leaves). Therefore, the closest correlation between R_{500} and R_{670} took place when a certain proportion of green pigments and carotenoids existed. Apparently this phenomena is unique for yellow-green to green vegetation where a decrease in green pigment during senescence or disease is followed by a proportional decrease in carotenoid concentration.

These spectral features may be useful in construction the indices for the estimation of pigment concentration at the canopy level. The coincidence (for vegetation) of reflectances R_{500} with R_{670} and R_{550} with R_{700} over a wide range of pigment concentrations allows assessment of the effect of background reflectance (e.g. soil). The variation in background reflectance for the same Chl concentration can be recognized in differences between the ratio R_{500}/R_{670} (or/and the R_{700}/R_{550}). The indices $[(R_{700}/R_{550})-1]$ and $[(R_{670}/R_{500})-1]$ could be used to counteract the effects of background reflectance.

References

Andrieu, B. and F. Baret: Indirect methods of estimating crop structure from optical measurements. In: Varlet-Grancher, C., R. Bonhomme, and H. Sinoquet (eds.): Crop Structure and Light Microclimate. Characterization and Applications, pp. 285–322. INRA edition, Paris (1993).

Baret, F., I. Champion, G. Guyot, and A. Podaire: Monitoring wheat canopies with high spectral resolution radiometer. Remote Sens. Environ. *22,* 367–378 (1987).

Baret, F. and G. Guyot: Potential and limits of vegetation indices for LAI and APAR assessment. Remote Sens. Environ. *35,* 161–173 (1991).

Baret, F., S. Jacquemoud, and G. Guyot: Modeled analysis of the biophysical nature of spectral shift and comparison with information content of broad bands. Remote Sens. Environ. *41,* 133–142 (1992).

Buschmann, C. and E. Nagel: *In vivo* spectroscopy and internal optics of leaves as basis for remote sensing of vegetation. Int. J. Remote Sens. *14,* 711–722 (1993).

Carter, G. A.: Responses of leaf spectral reflectance to plant stress. American Journal of Botany *80,* 239–243 (1993).

– Ratios of leaf reflectances in narrow wavebands as indicators of plant stress. Int. J. Remote Sensing *15,* 697–703 (1994).

Chappelle, E. W., M. S. Kim, and J. E. McMurtrey: III. Ratio analysis of reflectance spectra (RARS): An algorithm for the remote estimation of the concentrations of chlorophyll *a,* chlorophyll *b,* and carotenoids in soybean leaves. Remote Sens. Environ. *39,* 239–247 (1992).

Curran, P. J., J. L. Dungan, B. A. Macler, and S. E. Plummer: The effect of a red leaf pigment on the relationship between red edge and chlorophyll concentration. Remote Sens. Environ. *35,* 69–76 (1991).

Fukshansky, L.: Optical properties of plants. In: Smith, H. (ed.): Plants and the Daylight Spectrum, pp. 21–39. Academic Press, London (1981).

Gitelson, A. and M. N. Merzlyak: Spectral reflectance changes associated with autumn senescence of *Aesculus hippocastanum* L. and *Acer platanoides* L. leaves. Spectral features and relation to chlorophyll estimation. J. Plant Physiol. *143,* 286–292 (1994 a).

– – Quantitative estimation of chlorophyll *a* using reflectance spectra: Experiments with autumn chestnut and maple leaves. J. Photochem. Photobiol. (B) *22,* 247–252 (1994 b).

Gitelson, A., M. N. Merzlyak, and H. K. Lichtenthaler: Detection of changes in red edge position and chlorophyll content in leaves by reflectance measurements. J. Plant Physiol. *148,* 501–508 (1996).

Huete, A., C. Justice, and H. Liu: Development of vegetation and soil indices for MODIS-EOS. Rem. Sens. Environ. *49,* 224–234 (1994).

Horler, D. N., M. Dockray, and J. Barber: The red edge of plant leaf reflectance. Int. J. Remote Sens. *4,* 273–288 (1983).

Jacquemoud, S. and F. Baret: Prospect: a model of leaf optical properties spectra. Remote Sens. Environ. *34,* 75–91 (1990).

Kim, M. S., S. T. Daughtry, E. W. Chappelle, J. E. McMurtrey, and C. L. Walthall: The use of high spectral resolution bands for estimating absorbed photosynthetically radiation (A*par*). In: VI Symposium Int. Physical Measurements and Signatures in Remote Sensing. Val d'Isere, France, 17–21 January 1994, pp. 299–306. CNES, Paris (1994).

Lichtenthaler, H. K.: Chlorophyll and carotenoids: Pigments of photosynthetic biomembranes. Meth. Enzym. *148,* 331–382 (1987).

McMurtrey III, J. E., E. W. Chappelle, M. S. Kim, J. J. Meisinger, and L. A. Corp: Distinguishing nitrogen fertilization levels in field corn (*Zea mays* L.) with actively induced fluorescence and passive reflectance measurements. Remote Sens. Environ. *47,* 36–44 (1994).

Merzlyak, M. N. and A. Gitelson: Why and what for the leaves are yellow in autumn? On the interpretation of optical spectra of senescing leaves (*Acer platanoides* L.). J. Plant Physiol. *145,* 315–320 (1995).

Merzlyak, M. N. and G. A. F. Hendry: Free radical metabolism, pigment degradation and lipid peroxidation in leaves during senescence. Proc. Royal Soc. Edinbourgh *102 B,* 459–471 (1994).

Tanner, V. and B. M. Eller: Veränderungen der spektralen Eigenschaften der Blätter der Buche (*Fagus silvatica* L.) von Laubaustrieb bis Laubfall. Allg. Forst- u. J.-Ztg. *157,* 108–117 (1986).

Vogelmann, T. C. and L. O. Björn: Plants as light traps. Physiol. Plant. *68,* 704–708 (1986).

J. Plant Physiol. Vol. 148. pp. 501–508 (1996)

Detection of Red Edge Position and Chlorophyll Content by Reflectance Measurements Near 700 nm

ANATOLY A. GITELSON[1], MARK N. MERZLYAK[2], and HARTMUT K. LICHTENTHALER[3]

[1] The Remote Sensing Laboratory, J. Blaustein Institute for Desert Research, Ben Gurion University of the Negev, Sede Boker Campus 84993, Israel

[2] Department of Cell Physiology and Immunology, Faculty of Biology, Moscow State University, 119899 GSP Moscow W-234, Russia

[3] Botanisches Institut, Lehrstuhl II, University of Karlsruhe, Kaiserstr. 12, D-76128 Karlsruhe, Germany

Received June 24, 1995 · Accepted October 15, 1995

Summary

Pigment contents was determined in and high spectral resolution reflectance measurements were acquired for spring, summer and autumn maple and horse chestnut leaves covering a wide range of chlorophyll content. Consistent and diagnostic differences in the red edge range (680–750 nm) of the reflectance spectrum were obtained for the various leaf samples of both species studied. This included the differences in the wavelength position of the red edge and in the reflectance values in the range of 690 to 710 nm. Both characteristics were found to be dependent on leaf chlorophyll concentration. The first derivative of reflectance spectra showed four peaks at 685–706, 710, 725 and 740 nm that were dependent in different degree on leaf age and pigment concentration in the leaves. The position and the magnitude of the first peak showed a high correlation with the leaf chlorophyll concentration. Reflectance at 700 nm was linearly dependent on the wavelength of the first peak. Variation of inflection point position with change in chlorophyll content was found small for yellow-green to dark green leaves (total chlorophyll in the range above 10 nmol/cm^2). Reflectance near 700 nm was found to be a very sensitive indicator of the red edge position as well as of chlorophyll concentration. The ratio of reflectances at 750 nm to that near 700 nm (R_{750}/R_{700}) was directly proportional (correlation $r^2 > 0.95$) to chlorophyll concentration. The ratio R_{750}/R_{700} as a newly established index for non-invasive *in-vivo* chlorophyll determination was tested by independent data sets in the range of Chl contents from 0.6 to more than 60 nmol/cm^2 of maple and chestnut leaves with an estimation error of Chl of less than 3.7 nmol/cm^2.

Key words: chlorophyll content, reflectance spectra of leaves, red edge position, vegetation indices.

Abbreviations: Chl = chlorophyll; Chl *a* and *b* = chlorophyll *a* and *b;* Car = total carotenoids; red edge = increase in the reflectance spectrum of leaves between 680 and 750 nm; NIR = near infra-red range of the spectrum; λ_{re} = wavelength position of inflection point at red edge of the reflectance spectrum.

Introduction

The quantitative changes in chlorophyll (Chl), the main photosynthetic leaf pigment, have received relatively little attention in remote sensing studies. Several investigators have related the changes in Chl concentration to the shift in the red edge, the inflection point that occurs in the rapid transition between red and near infra-red reflectance (e.g., Horler et al., 1983; Lichtenthaler and Buschmann, 1987; Lichtenthaler, 1989; Curran et al., 1990, 1991). This shift has been associated with plant stress, forest decline and leaf development (e.g., Collins, 1978; Rock et al., 1986 and 1988; Schmuck et al., 1987; Buschmann and Lichtenthaler, 1988; Buschmann et al., 1991).

Few studies have analyzed relationships between chlorophyll concentration and vegetation indices and, especially relations between wavelength of the red edge and reflectance in the red region of the spectrum. Horler et al. (1983) found that the position of inflection point of the red edge was related to Chl concentration in the leaves of dicots and temperate cereals. They also showed that spectral behavior of the 1st derivative of reflectance is complicated and controlled at least two factors: chlorophyll absorbance and scattering by the leaf. Moss and Rock (1991) had observed that the reflectance ratio $R_{734-747}/R_{715-726}$ was excellent for assessing Chl in red spruce. Vogelmann et al. (1993) demonstrated that this ratio to be applicable for total chlorophyll determination in sugar maple. They also found that reflectance ratio $R_{734-747}/R_{715-720}$, and a ratio of the first derivative at 705 nm and 715 nm were highly correlated with total Chl content. Baret et al. (1992) suggested to use three spectral bands (705–715 nm, 732–737 nm, and 772–780 nm) to evaluate the red edge inflection point shift from space observations. Gitelson and Merzlyak (1994 a and b) have observed high sensitivity of reflectance near 700 nm to Chl concentrations in maple and chestnut leaves and found that the relationship between Chl concentration and R_{700}, is hyperbolic. They used this relation and observed low sensitivity of the NIR reflectance to the Chl level to construct the vegetation index R_{750}/R_{700} that was found to be directly proportional to the Chl concentration of leaves.

In this article an attempt is made to analyze different characteristics of the reflectance spectra of plant leaves in the red edge in order to understand the fundamental properties of vegetation in this range of the spectrum. We will particularly focus on the comparison of spectral shift characterized by the wavelength position of the inflection point of the red edge with the reflectance at particular wavelengths in this range of the spectrum. We present a detailed study of the red edge based on laboratory measurements of the reflectance of maple and horse chestnut leaves. We will test two hypothesis, (i) that the reflectance at 700 nm, R_{700}, is primarily controlled by the Chl content and is a measure of the red edge shift, and (ii) that the vegetation indices R_{750}/R_{700} and $R_{NIR}/R_{700-715}$ are more sensitive to changes in Chl content of leaves than the wavelength of inflection point of red edge and that these ratios can be used for a non-invasive remote estimation of the Chl content in leaves with a high degree of accuracy.

Materials and Methods

The experiments were performed in October 1991 and 1993, on horse chestnut leaves (*Aesculus hippocastanum* L.), and in October 1992–1993 on Norway maple leaves (*Acer platanoides* L.). Leaves of both trees were collected in the Botanical Garden of the Moscow State University, as described previously (Gitelson and Merzlyak, 1994 a; Merzlyak and Gitelson, 1995). In addition to the senescing samples, the mature green leaves of both species collected in July of 1994 and 1995 were examined (Gitelson and Merzlyak, 1996). The sampling scheme was intended to cover as high a variation of pigment concentrations as possible. Only leaves having homogeneous dark green, green, greenish-yellow, yellowish-green and yellow color without anthocyanin pigmentation were selected.

Hemispherical reflectance spectra were recorded for the upper surface of the leaves with a Hitachi 150-20 spectrophotometer, equipped with an integrating sphere attachment at the rate of 100 nm/min. The spectra were determined for the sections of the leaves between main veins (maple) or with a removed main vein (chestnut). The reflectance spectra were measured with a spectral resolution of 2 nm against barium sulfate as a reference standard with a light trap to eliminate the specular reflected component of the radiance, and black velvet was used as a background in order to absorb the light passing through the leaf (Gitelson and Merzlyak, 1994 a). Reflectance was expressed as a ratio of the radiance of the leaf to that of the standard. The first derivative spectra were used to determine the wavelength position of the inflection point of the red edge, defined as the wavelength of the maximum of the 1st derivative $dR/d\lambda$. The values of $dR/d\lambda$ were calculated.

The coefficients of variation of the reflectance were calculated as a ratio of standard deviation of the reflectance to average reflectance value.

Chl *a*, *b* and total carotenoid (Car) concentrations in the leaves were determined in acetone extracts and calculated using equations and specific extinction coefficients as reported by Lichtenthaler (1987).

Results and Discussion

Pigment content in the leaves

The dominant pigment was Chl *a* and total Chl contents ranged from 0.64 to 57 nmol/cm^2 in maple leaves and from 0.68 to 62.9 nmol/cm^2 in horse chestnut leaves. The green leaves collected both in summer and autumn (total Chl > 20 nmol/cm^2) contained Chl *a* and Chl *b* in a ratio of 3 and a lower level of carotenoids with a ratio of chlorophylls to carotenoids of 4.5 to 5.5. When the leaves of both species lost Chl *a* and Chl *b* in the progress of autumn senescence, relatively high concentrations of carotenoids were present (see also, Gitelson and Merzlyak, 1994 a; Merzlyak and Gitelson, 1996). In completely yellow leaves only trace amounts of chlorophylls were detected.

Reflectance spectra of the leaves in the red range of the spectrum

Reflectance spectra of maple leaves are shown in Fig. 1. The highest reflectance was in the near infra red range beyond 750 nm. In this range, reflectance remained virtually the same for leaves with a very wide variation in Chl content. The main feature of the reflectance spectra was a sharp increase in leaf reflectance between 680 and 750 nm which is known as red edge. For completely yellow leaves with a very low Chl content (0.6 nmol/cm^2) this red edge may hardly be seen (upper curve in Fig. 1). When the Chl concentration increases, the reflectance decreases reaching minimum values near 675 nm. For completely yellow to dark green maple leaves (Chl = 0.64 to 57 nmol/cm^2), spectral range near 675 nm was sensitive to Chl variation (Fig. 2). However, for yellowish-green to dark green leaves (Chl content from 5 to 57 nmol/cm^2), the reflectance near 675 nm remained approximately the same. This range of the reflectance spectrum showed a minimum sensitivity to variation in Chl content (as can be seen from the coefficient of variation curve in Fig. 2).

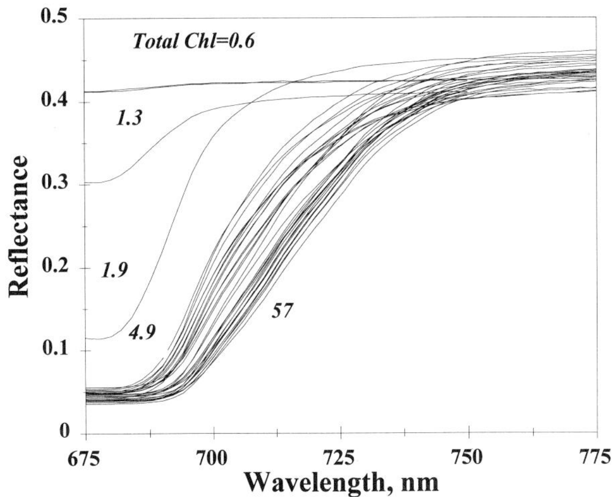

Fig. 1: Reflectance of maple leaves in the red edge range of the spectrum from 675 to 775 nm. The total chlorophyll contents (in nmol/cm^2 leaf area) are indicated for some leaves.

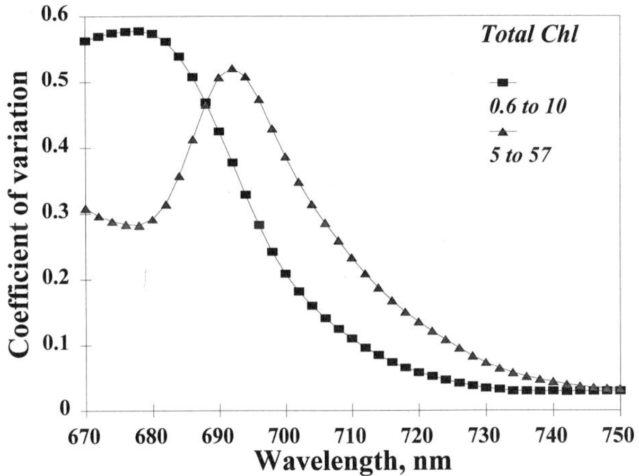

Fig. 2: Coefficient of variation of reflectance for two groups of maple leaves in the red range of the spectrum. The first group contained leaves with a total Chl concentration from 0.6 (the minimum concentration found in our experiments) to 10 nmol/cm^2. The second group included leaves with Chl from 5 to 57 nmol/cm^2 (the maximum concentration found in our experiments). The first group of leaves can be considered to represent a process of senescence or stress, when the color of the leaves turn from completely yellow to light-green. The second group corresponded to various stages of senescence (or/and stress), when color of the leaves vary from greenish-yellow to dark green. The coefficients of variation for these groups have very different spectral behavior. For completely yellow to light-green leaves the variation coefficient shows peak near 675 nm, in the range of the red chlorophyll *a* absorption maximum. For greenish-yellow to dark green leaves the variation coefficient has minimal values near 675 nm, while a pronounced maximum was found between 690 and 700 nm.

In contrast essential differences in reflectance of these green leaves were observed from 690 to 740 nm with a maximum at near 695 nm.

The reflectance near 700 nm, R_{700}, is specific for this region of the spectrum. It was found that the relationship between R_{700} and Chl concentration is hyperbolic with a correlation coefficient of more than 0.95 for both species (Gitelson and Merzlyak, 1996). Apart from the very strong correlation between R_{700} and Chl content of leaves, reflectances in the range between 690 and 730 nm were strongly mutually correlated and were weakly correlated with R_{750}. For Chl concentration of more than 4.9 nmol/cm^2 reflectances in the range from 690 to 730 nm did not correlate with R_{675} (non shown). This suggests that the main control of reflectance in the 690 to 730 nm range for our data set was exerted by Chl concentration rather than by leaf internal structure.

Position of inflection point in relation to Chl concentration

Spectra of the first derivative of reflectance spectra of both maple and chestnut leaves of low to high Chl content showed the presence of four distinguishable peaks at 685–705, near 710, 720 and 740 nm in all leaf samples studied (Fig. 3), demonstrating the presence of universal Chl spectral forms throughout all stages of leaf development. The first peak was dominant in young spring and yellowing autumn leaves of both species, while other three peaks at longer wavelengths were weak and often not completely resolved. In summer leaves all peaks were clearly distinguished.

The magnitude of the first peak increased with Chl increase up to 5 nmol/cm^2, and then dropped sharply to a half of the maximum value for Chl = 57 nmol/cm^2 (Fig. 3 a). For a Chl level of more than 40–45 nmol/cm^2, the 1st derivative was comparatively flat near 700 nm and the peak transformed into a shoulder. The first peak position changed from 685 to near 705 nm with an increase in Chl content from 0.64 to 57 nmol/cm^2 for maple (Fig. 4 a) and from 0.64 to 62.3 nmol/cm^2 for chestnut (Fig. 4 b) leaves.

In quantitative terms, maple and chestnut showed a similar «first peak position vs. Chl concentration» relationships (Figs. 4 A and 4 B). For maple leaves, the best regression was obtained for the first peak position in the form

$$\text{First peak position} = 687\,(\text{Chl})^{0.0056}$$

with the coefficient of determination $r^2 > 0.95$ (solid line in Fig. 4).

The second peak of the 1st derivative of the reflectance spectra was found near 710 nm (Fig. 3). At a low Chl content it was a shoulder near 708 nm, that increased with an increase in Chl content and became the main peak of the 1st derivative curve at high Chl concentration. The wavelength position of this peak increased slightly with the Chl increase, achieving 712 nm for a Chl level >50 nmol/cm^2. For Chl <35 nmol/cm^2 the first peak was dominant, but for Chl >45 nmol/cm^2 the second peak became increasingly evident and finally dominant. For Chl concentration of 35–45 nmol/cm^2, the magnitudes of the first and second peaks became virtually the same and it was hardly to recognize which of them is dominating (Fig. 5). For these Chl concentrations, the main peak changed sharply its position from 701–703 nm (it belonged to the first peak) to 710 nm (as the position of the second peak). Wavelength position of the inflection point

λ_{re} represents whichever peak was dominant, therefore the above mentioned λ_{re} shift involved a jump between two peaks, leading to λ_{re} discontinuity in relationship between λ_{re} and Chl content (Fig. 6 a and 6 b; see also Horler et al., 1983).

The graph «main peak position vs. Chl» (Figs. 6 A and 6 B) had three phases with a transitions at λ_{re} about 698 and 703 nm. The first phase occurred at lower Chl concentrations and λ_{re} values. The increase in Chl concentrations from 0.6 to 8–10 nmol/cm^2 (completely yellow to yellowish-green leaves) gave more-or-less linear increase in λ_{re} from 687 to 698 nm. In the second phase, Chl concentration from 10 to 35 nmol/cm^2 were associated with variation in λ_{re} values from 698 to 704 nm. There was a gap between the λ_{re} limits of the second and third phases. In the third phase, the variation in Chl concentration from 35 to near 60 nmol/cm^2 virtually did not change λ_{re} values. A 1 nm shift of λ_{re} represented a Chl change of slightly less than 1 nmol/cm^2 in phase 1, about 5 nmol/cm^2 in phase 2 and more than 12 nmol/cm^2 in phase 3. Therefore, the apparent greater sensitivity of phase 1 would provide greater accuracy in estimation of low Chl concentration.

The third peak of the 1st derivative located near 725 nm. Its magnitude increased two-fold when Chl varied from 10 to 60 nmol/cm^2. The position of this peak was practically inde-

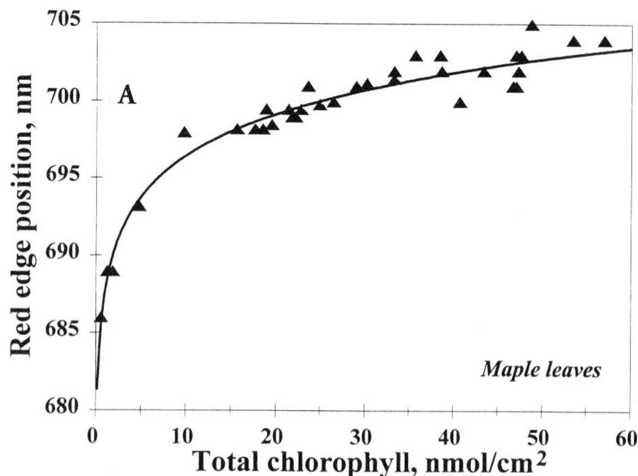

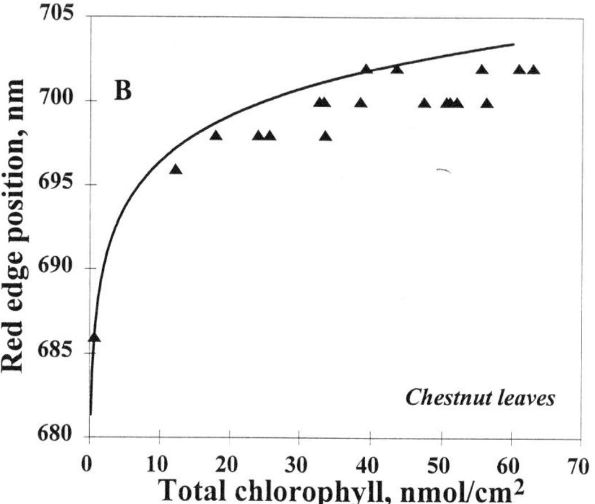

Fig. 4: Short wavelength inflection point position of the red edge (the position of the first peak at the 1st derivative of reflectance spectra) versus total chlorophyll concentration for maple (A) and chestnut (B) leaves. For leaf chlorophyll concentration from 0.6 to 30 nmol/cm^2 the peak in the range 687–703 nm shows up as a main peak at the 1st derivative curve. For chlorophyll concentration of more than 35 nmol/cm^2 this peak is lower in magnitude than the second one located near 710 nm. The position of short wavelength inflection point (range from 687 to 703 nm) correlated highly with total chlorophyll for both species studied. For maple leaves best fit function with ($r^2 > 0.95$) is $\lambda_{re} = 687 * Chl^{0.056}$, where λ_{re} is the position of short wavelength inflection point in nm, Chl is total chlorophyll concentration in nmol/cm^2. Solid line in both (A) and (B) figures represents best fit function for maple leaves. For chestnut leaves with chlorophyll contents of more than 15 nmol/cm^2, red edge position was found at slightly shorter (698–702 nm), than for maple leaves (698–704 nm), wavelengths.

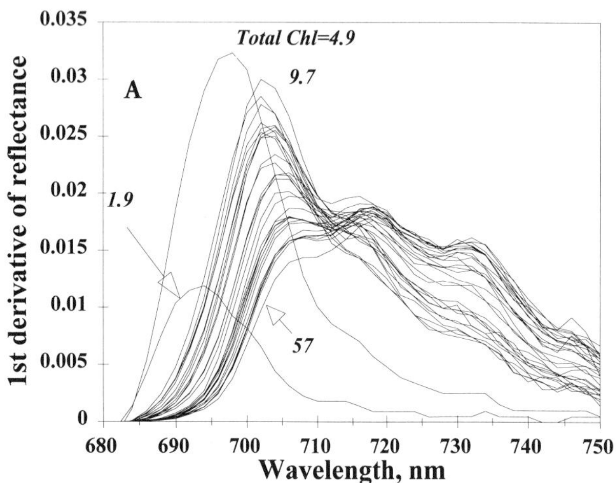

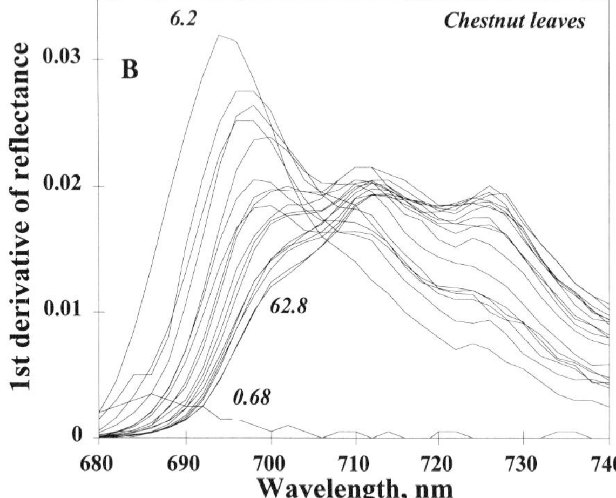

Fig. 3: The first derivative of reflectance spectra of 34 maple leaves (A) and for 19 horse chestnut leaves (B) in the red range of the spectrum. Total chlorophyll contents for some samples are indicated in nmol/cm^2 leaf area. For leaves with chlorophyll contents of more than 4.9 nmol/cm^2, four pronounced peaks can be seen: near 695 nm, 712 nm, 728 nm, and a peak/shoulder around 740 nm.

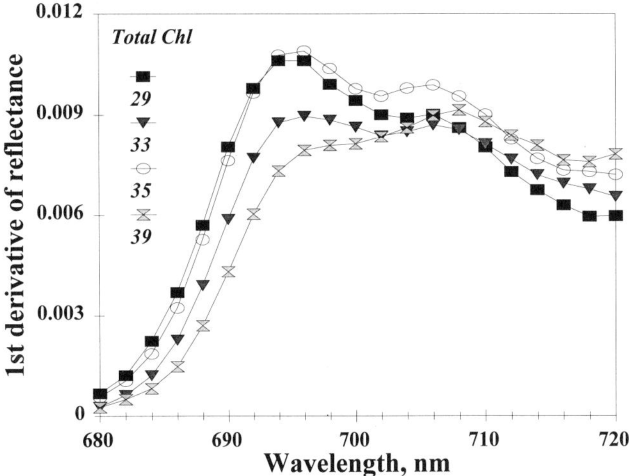

Fig. 5: The first derivative of reflectance spectra for individual maple leaves with a total chlorophyll content from 29 to 39 nmol/cm^2. In this range of Chl content, the 1st derivative has two peaks with approximately the same magnitude. The wavelength position of the first peak is near 698–700 nm and the second one at 710 nm.

pendent of the Chl concentration. The fourth peak was located near 740 nm and its magnitude, as in the case of previous peaks, increased with an increasing in Chl content.

The existence of three phases in the relation «λ_{re} vs. Chl» shows that total chlorophyll content was not the only factor determining λ_{re}. Phases 1 and 2 were associated primarily with the changes in Chl concentration. A peak position at shorter wavelength was strongly correlated with total Chl concentration (Figs. 4 A and 4 B) and was quantitatively in agreement with the results of calculation for spondy mesophyll (N=2.5) derived from PROSPECT model (Baret et al., 1992; Fig. 1 a). Therefore, the position of the shorter wavelength peak in the 1st derivative of reflectance may be considered as the actual red edge position, and its position relates closely to Chl level of leaves. Thus the term λ_{re} we will be referred in the further discussion as the position of the shorter wavelength peak in the 1st derivative of reflectance.

The transition between phases 2 and 3 was not necessarily associated with the changes in Chl concentration. Horler et al. (1983) suggested the reasons for this phenomenon. One of them was a possible variation in Chl *a*/Chl *b* ratio. However, for our data set Chl *a* and Chl *b* were highly correlated; the coefficient of determination (r^2) between them was more than 0.967. Therefore, the ratio Chl *a*/Chl *b* did not correlate at all with λ_{re} although strong correlation between Chl and λ_{re} was observed (Fig. 4).

Position of inflection point in relation to reflectance at 700 nm

High sensitivity of reflectance to Chl concentration was observed in the spectral range from 690 to 705 nm (Fig. 2). High correlation of chlorophyll concentration with R_{700} and λ_{re} suggests a close relation between two these characteristics of the red edge; figures 7 A and 7 B represent this linear relation with high determination coefficient. Therefore, R_{700} is able to serve as a measure of red edge position, that can be measured quite easily.

The spectral resolution required of λ_{re} values can be estimated from the comparison of the variation of R_{700} and λ_{re} values. It was found that a range of λ_{re} values spanning 18 nm corresponded to an R_{700} range of 35 per cent (Fig. 7 a).

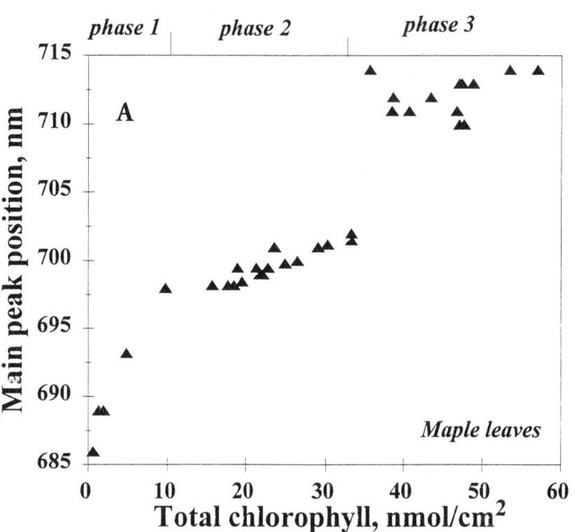

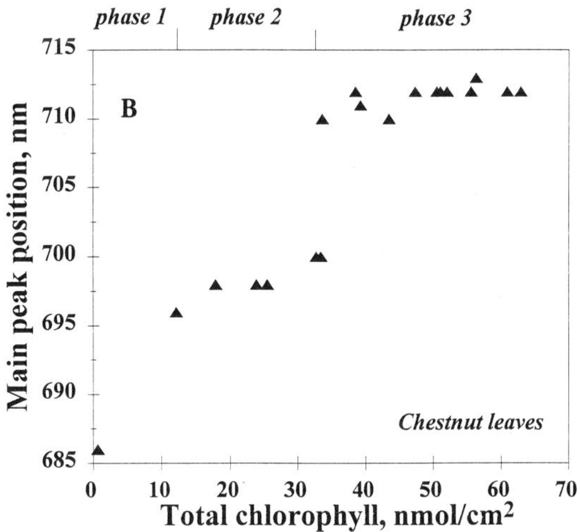

Fig. 6: Position of the main peak in the 1st derivative curve versus total chlorophyll concentration for maple (A) and chestnut (B) leaves. Three main phases characterize this relationships. The first phase occurred at lower Chl concentrations and λ_{re} values. The increase in Chl concentrations from 0.6 to 8–10 nmol/cm^2 gave more-or-less linear increase in λ_{re} from 687 to 698 nm. In the second phase, Chl concentration from 10 to 35 nmol/cm^2 were associated with low variation in λ_{re} values (from 698 to 704 nm). In third phase, the variation in Chl concentration from 35 to near 60 nmol/cm^2 virtually did not change λ_{re} values. In the range of Chl 0.6 to 35 nmol/cm^2 the short wavelength peak is the main peak. For Chl concentrations near 35 nmol/cm^2, the magnitude of this short wavelength peak is approximately equal to the magnitude of the second peak located near 710 nm. In this range it is impossible to recognize exactly which peak is the main and thus to determine the position of inflection point.

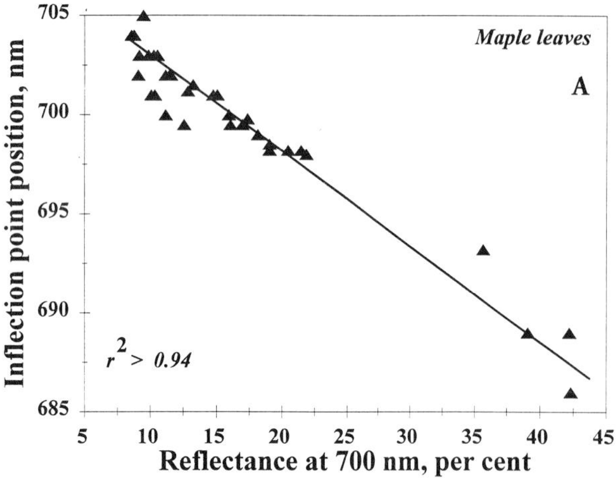

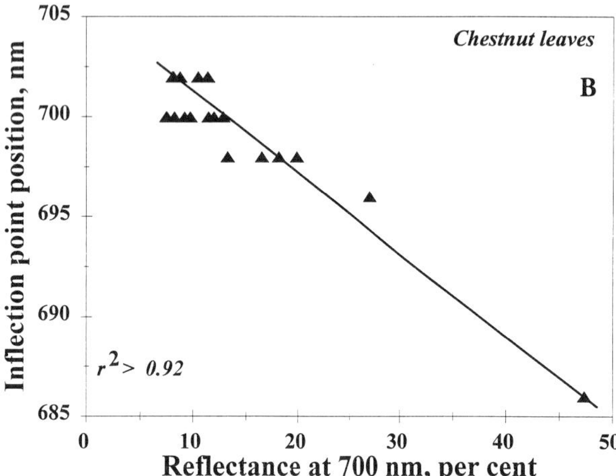

Fig. 7: Wavelength of the inflection point position (the first peak at the 1st derivative curve) versus the reflectance at 700 nm for maple (A) and chestnut (B) leaves. Linear relationship with correlation coefficient $r^2 > 0.94$ (for 34 maple leaves) and $r^2 > 0.92$ (for 19 chestnut leaves) was obtained for very wide range of variation in Chl content from 0.6 to 69 nmol/cm^2.

Therefore, a radiometric resolution of 1 per cent reflectance corresponded to a spectral resolution of near to 0.5 nm for λ_{re}.

Relationship between reflectance ratios and chlorophyll concentrations

The reflectance R_{700} hyperbolically related to Chl concentration (Gitelson and Merzlyak, 1996). On the other hand, reflectance in the NIR was practically independent of chlorophyll levels (Figs. 1 and 2). This suggests that the reflectance ratio R_{NIR}/R_{700} should be linearly proportional to the Chl concentration of leaves. Really, the use of the ratio R_{750}/R_{700} allows to obtain a fairly linear relationship of this index with Chl (Fig. 8). The coefficient of determination for this relationship was more than 0.95. Similar results were consistently found for spring, summer and autumn leaves of both maple and chestnut. The relationship between the ratio R_{750}/R_{700} and Chl content was tested by several independent data sets and an estimation error of less than 3.75 and 2.95 nmol/cm^2 for total Chl and Chl *a* concentration, respectively, was achieved.

A very important feature of the relationship R_{750}/R_{700} vs. Chl is a high sensitivity of the index to variation in Chl content of green leaves. The index R_{750}/R_{700} was equally sensitive to variation in Chl content for completely yellow leaves to dark green ones. It is particularly important because of a rather low sensitivity of λ_{re} to differences between yellow-green and dark green leaves. In the range of leaf color from yellowish-green to dark green (i.e. from 10 to 60 nmol/cm^2 Chl), λ_{re} values varied from 698 to 704 nm (Fig. 4a and 4b). This corresponded to a variation in R_{700} of more than 15 per cent (Fig. 7) and to a more than two-fold variation in the ratio R_{750}/R_{700} (Fig. 8). Thus, the differences between relatively dark green leaves could be measured quantitatively by using the newly established vegetation index R_{750}/R_{700}.

A quite high sensitivity of reflectance to Chl levels of leaves was found in the spectral range from 690 to 705 nm. A maximum sensitivity to Chl was achieved for the reflectance ratio R_{750}/R_{695} and the sensitivity decreased when reflectances at wavelengths longer than 705 nm were used. In Fig. 9 several vegetation indices were plotted versus Chl levels of leaves. The ratio R_{750}/R_{700} showed a maximum determination coefficient and a minimum estimation error of Chl content. A greater sensitivity gave the reflectance ratio R_{750}/R_{695} with an estimation error of Chl content slightly higher than for ratio R_{750}/R_{700} (4.12 nmol/cm^2 for total Chl and 3.2 nmol/cm^2 for Chl *a* content). We also included in this figure the reflectance ratio R_{740}/R_{720} (as suggested recently by Vogelmann et al., 1993). This index provided quite precise estimation of Chl (the estimation error of total Chl for our data set was, however, higher than for the above mentioned indices: 5.9 nmol/cm^2), but the sensitivity to Chl variation was much lower than that for other indices. In addition to rather low sensitivity to Chl content, airborne and space borne measurements of the reflectances R_{740} and R_{720} require a proper atmospheric calibration which is a difficult matter. As another possibility

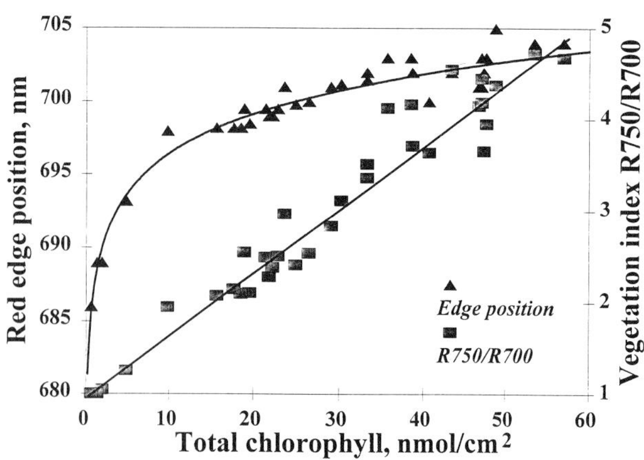

Fig. 8: Reflectance ratio R_{750}/R_{700} and 1st inflection point position of the red edge versus total chlorophyll concentration for maple leaves. Solid lines represent best fit functions for these relationships.

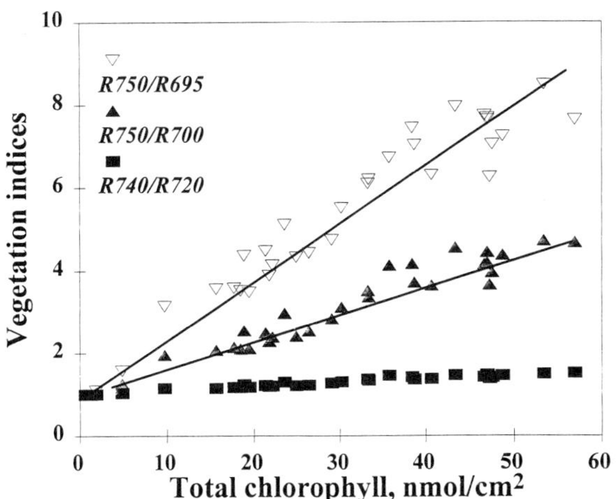

Fig. 9: Vegetation indices versus total chlorophyll concentration in maple leaves. The minimum estimation error of Chl < 3.7 nmol/cm^2 was achieved for the R_{750}/R_{700} ratio. The highest estimation error of 5.9 nmol/cm^2 was obtained using the index R_{740}/R_{720} while possessed a very low sensitivity to the Chl concentration. The index R_{750}/R_{695}, in turn, allows to estimate total Chl concentration with an error of less than 4.12 nmol/cm^2 and exhibited the highest sensitivity.

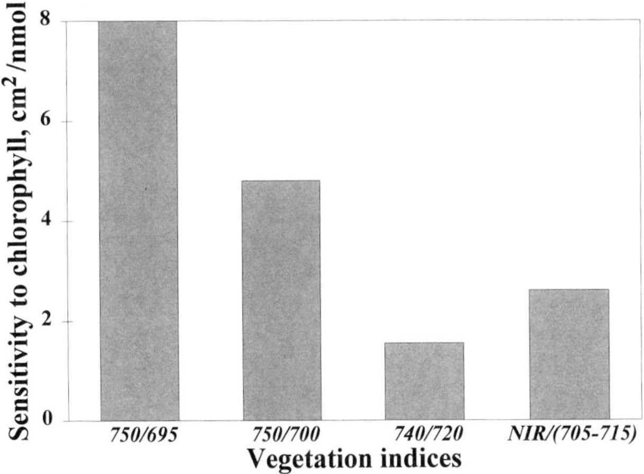

Fig. 10: Sensitivity of different vegetation indices to changes in chlorophyll content, expressed here as a slope of the curves of the vegetation indices versus total chlorophyll content. Maximum sensitivity for leaf chlorophyll can be achieved using index R_{750}/R_{695}, and minimal for the index R_{740}/R_{720}. The index $R_{NIR}/R_{705-715}$, is still quite sensitive to changes in chlorophyll content, thus allowing to work in spectral bands with minimal atmospheric disturbances.

to estimate the Chl content from space borne observation platforms we suggest to use the spectral bands from 705 and 715 nm and NIR. The application of the reflectance ratio $R_{NIR}/R_{705-715}$ allowed us to achieve an estimation error for total Chl content of less than 5.5 nmol/cm^2 and for Chl *a* of less than 4.1 nmol/cm^2, with quite enough sensitivity to the Chl level (Fig. 10).

While the current study demonstrates that the newly established indices are accurate in quantifying the leaf-level Chl content, we can not exactly predict how well these algorithms will hold at the canopy level. The study will also have to be extended to whole canopy spectra, before final conclusions for space application can be made. Our preliminary results on reflectance measurements of different kinds of vegetation (various species of potatoes and tobacco) showed that the new vegetation indices R_{750}/R_{700} and $R_{NIR}/R_{705-715}$ are accurate for assessing the Chl content not only at the leaf level, but also at the canopy level (Gitelson et al., in preparation). Another example of successful application of the vegetation index R_{750}/R_{700} for estimation of early stages of plant stress has been done by Carter (1993; 1994). The range near 700 nm was also found to be sensitive to the leaf area index as indicated by Danson and Plummer (1995; Fig. 2). Nevertheless, we recognize that differences in species, illumination, canopy architecture and other factors may potentially influence and decrease the correlation between the newly established vegetation indices and the Chl content. The relationships between canopy chlorophyll content and proposed vegetation indices need to be examined further.

References

Baret, F., S. Jacquemoud, and G. Guyot: Modeled analysis of the biophysical nature of spectral shift and comparison with information content of broad bands. Remote Sens. Environ. *41,* 133–142 (1992).

Buschmann, C. and H. K. Lichtenthaler: Reflectance and chlorophyll fluorescence signatures of leaves. In: Lichtenthaler, H. K. (ed.): Application of Chlorophyll Fluorescence, pp. 325–332. Kluwer Academic Publishers, Dordrecht (1988).

Buschmann, C. and E. Nagel: *In vivo* spectroscopy and internal optics of leaves as basis for remote sensing of vegetation. International J. Remote Sens. *14,* 711–722 (1993).

Buschmann, C., U. Rinderle, and H. K. Lichtenthaler: Detection of stress in coniferous forest trees with the VIRAF spectrometer. In: Internat. Geoscience and Remote Sensing Symposium, IGARSS '89, Vancouver, Vol. *4,* pp. 2641–2644 (1989) (enlarged version published in: IEEE Transactions on Geoscience and Remote Sensing. *29,* 96–100 (1991)).

Carter, G. A.: Responses of leaf spectral reflectance to plant stress. American Journal of Botany *80,* 239–243 (1993).

– Ratios of leaf reflectances in narrow wavebands as indicators of plant stress. International J. Remote Sensing *15,* 697–703 (1994).

Chappelle, E. W., M. S. Kim, and J. E. McMurtrey III.: Ratio analysis of reflectance spectra (RARS): An algorithm for the remote estimation of the concentrations of chlorophyll *a,* chlorophyll *b,* and carotenoids in soybean leaves. Remote Sens. Environ. *39,* 239–247 (1992).

Collins, W.: Remote Sensing of crop type maturity. Photogrammametric Engineering and Remote Sensing, *44,* 43–55 (1978).

Curran, P. J., J. L. Dungan, and H. L. Gholz: Exploring the relationship between reflectance red-edge and chlorophyll content in slash pine. Tree Physiol. *7,* 33–48 (1990).

Curran, P. J., J. L. Dungan, B. A. Macler, and S. E. Plummer: The effect of a red leaf pigment on the relationship between red edge and chlorophyll concentration. Remote Sens. Environ. *35,* 69–76 (1991).

Danson, F. M. and S. E. Plummer: Red-edge response to forest leaf area index. International J. Remote Sensing *16,* 183–188 (1995).

Gitelson, A. and M. N. Merzlyak: Spectral reflectance changes associated with autumn senescence of *Aesculus hippocastanum* L. and *Acer platanoides* L. leaves. Spectral features and relation to chlorophyll estimation. J. Plant Physiol. *143,* 286–292 (1994a).

– – Quantitative estimation of chlorophyll-*a* using reflectance spectra: Experiments with autumn chestnut and maple leaves. J. Photochem. Photobiol. (B), *22,* 247–252 (1994b).

– – Signature analysis of leaf reflectance spectra: algorithm development for remote sensing of chlorophyll. J. Plant Physiol. *148,* 494–500 (1996).

Horler, D. N., M. Dockray, and J. Barber: The red edge of plant leaf reflectance. International J. Remote Sens. *4,* 273–288 (1983).

Lichtenthaler, H. K.: Chlorophyll and carotenoids: Pigments of photosynthetic biomembranes. Meth. Enzym. *148,* 331–382 (1987).

– Possibilities for remote sensing of terrestrial vegetation by a combination of reflectance and laser-induced chlorophyll fluorescence. In: Internat. Geoscience and Remote Sensing Symposium, IGARSS '89, Vancouver, Vol. *3,* pp. 1349–1354 (1989).

– – Reflectance and chlorophyll fluorescence signatures of leaves. In: Proceedings of the Remote Sensing Symposium (IGARSS), Michigan, Vol. *II,* pp. 1201–1206. The University of Michigan, Ann Arbor 1987.

Moss, D. M. and B. N. Rock: Analysis of red edge spectral characteristics and total chlorophyll values for red spruce (*Picea rubens*) branch segments from Mt. Moosilauke, NH, U.S.A. 11th Annual International Geoscience and Remote Sensing Symposium (IGARSS '91), 3–6 June, 1991, Helsinki, Finland, Vol. *III,* pp. 1529–1532. New York: IEEE (1991).

Rock, B. N., T. Hoshizaki, H. K. Lichtenthaler, and G. Schmuck: Comparison of *in situ* spectral measurements of forest decline symptoms in Vermont (U.S.A.) and the Schwarzwald (F.R.G.). In: Proceedings of the Remote Sensing Symposium (IGARSS), Zürich 1986, Vol. *III,* pp. 1667–1672. ESA Publications Division, Noordwijk (1986).

Rock, B. N., T. Hoshizaki, and J. R. Miller: Comparison of *in situ* and airborne spectral measurements of the blue shift associated with forest decline. Remote Sens. of Environ. *24,* 109–127 (1988).

Schmuck, G., H. K. Lichtenthaler, G. Kritikos, V. Amann, and B. N. Rock: Comparison of *terrestrial and airborne* reflection measurements of forest trees. In: Proceedings of the Remote Sensing Symposium (IGARSS), Michigan, Vol. *II,* pp. 1207–1212. The University of Michigan, Ann Arbor (1987).

Vogelmann, T. C. and L. O. Björn: Plants as light traps. Physiol. Plant. *68,* 704–708 (1986).

J. Plant Physiol. Vol. 148. pp. 509–514 (1996)

Blue-green Fluorescence and Visible-infrared Reflectance of Corn (*Zea mays* L.) Grain for *in situ* Field Detection of Nitrogen Supply

JAMES E. MCMURTREY III[1], EMMETT W. CHAPPELLE[2], MOON S. KIM[3], LAWRENCE A. CORP[3], and CRAIG S. T. DAUGHTRY

[1] USDA/ARS, Remote Sensing Research Laboratory, 10300 Baltimore Avenue, Beltsville, Maryland 20705, USA

[2] NASA/GSFC, Laboratory for Terrestrial Physics, Greenbelt, Maryland 20771

[3] Science Systems and Applications, Inc., 5900 Princess Garden Parkway, Lanham, Maryland 20706

Received June 24, 1995 Accepted October 20, 1995

Summary

The sensing of spectral attributes of corn (*Zea mays* L.) grain from site specific areas of the field during the harvest process may be useful in managing agronomic inputs and production practices on those areas of the field in subsequent growing seasons. Eight levels of nitrogen (N) fertilization were applied to field grown corn at Beltsville, Maryland. These N treatments produced a range of chlorophyll levels, biomass and physiological condition in the live plant canopies. After harvest. spectra were obtained in the laboratory on whole grain samples. Fluorescence emissions were acquired from 400 to 600 nm and percent reflectance were measured in the visible (VIS) near infrared (NIR) and mid-infrared (MIR) regions from 400 nm to 2400 nm. A ultraviolet (UV) excitation band centered at 385 nm was the most effective in producing fluorescence emission differences in the blue-green region of the fluorescence spectrum with maxima centered from 430–470 nm in the blue and with an intense shoulder centered at around 530–560 nm in the green region. Reflectance showed the most spectral differences in the NIR and MIR (970–2330 nm) regions.

Key words: Zea mays, fluorescence, grain, reflectance, vegetation stress.

Introduction

Assessing the potential for measuring and monitoring *in-situ* spectral properties of plant materials in the field or during the harvest process is the goal of many research and development projects. Information gathered from site specific areas of the field during or shortly after harvest could be related to the status of the crop during the previous growing season. These attributes may have a bearing on management decisions for the next growing season. Prescription farming management and application methods may lead to more efficient use of agronomic inputs by analyzing and assessing site specific areas of a field.

The reflectance and fluorescence spectral properties of grain kernels that came from specific N fertilization treatment areas of the field are the subject of this study. The objectives are to determine the potential of the fluorescence and reflectance spectral properties of whole grain as a bioindicator of the live plant growth condition during the prior vegetative growth period. The study was conducted on whole kernel grain harvested from a standing corn canopy.

Numerous plant constituents are involved in photosynthesis and biochemistry of live plant metabolism. The concerted effects of electron transfer, synthesis, formulation, function and metabolism combine to influence the fluorescence of compounds when excited by the proper electromagnetic energy in live plants. Fluorescence has been reported from compounds in live plants in the blue-green area of the spectrum. There are many plant constituents that could contribute to the convoluted fluorescence spectra found in the blue-green (Chappelle et al., 1984 a, 1991 and 1993; Lang and Lichtenthaler, 1991; Lang et al., 1991 and 1992; Lichtenthaler et al., 1991). Chlorophyll species are primarily responsible for the fluorescence at 685 nm in the red and at 740 nm in the near infrared (NIR) (Chappelle et al., 1984 b; Lichtenthaler et al., 1989). Undoubtedly they also influence the synthesis, concentration and function of many of the compounds that fluoresce in other areas of the spectrum. Studies of live vegetation fluorescence in the blue-green-red-NIR areas have been shown to be associated with assessing photosynthetic efficiency and plant vigor. Overall, the potential of fluorescence technique has been demonstrated as a remote sensing method for species identification and as a method for the detection of plant stresses in live intact canopies (Chappelle et al., 1984 a, 1984 b, and 1995; Lang and Lichtenthaler, 1991; Lang et al., 1991; Lichtenthaler et al., 1991; Mazzinghi et al., 1994; McMurtrey et al., 1994 a and 1994 b).

Recent studies have shown that senescent leaves and the resulting dead dry crop residues emit a broad band fluorescence (and possibly other forms of luminescence) that are strong in the blue-green-yellow regions (400 nm – 600 nm) when excited with UV radiation (McMurtrey et al., 1993; Daughtry et al., 1994). These studies concluded that the fluorescence intensity of residues were up to 20 times greater then that of the soils. The fluorescence of soil when compared to alive or dead vegetation is near the instrumentations noise line and is essentially zero. Thus, living plant material and dead plant material (crop residue) can be discriminated from most soils. This technology is being used to develop a crop residue meter as a method to quantify the amount of ground residue cover with fluorescence technology (McMurtrey et al., 1993; Daughtry et al., 1994; Chappelle et al., 1995). Compounds associated with the structural components of cellulose and lignin maybe the primary contributors to the fluorescence of dry dead crop residue biomass. Flavins, pteridines and phenolic substances have been reported to fluoresce under UV excitation (McMurtrey et al., 1993).

Remote sensing of the fluorescence or reflectance of grain as it is harvested from a *in-situ* site specific area of the field may indicate a need for differential fertilization amendments or other changes in the crops management to these segments of the field in subsequent years. Yearly cumulative mean measurements and mean spectral behavior that reliably monitors temporal deviations from «normal» biomass development could be an important long-term management tool. N fertilizer amendments are known to affect the concentration and efficiency of the primary plant pigments. Previous studies showed that there was a possibility of detecting agronomically important over fertilization as well as under-fertilization levels in the live green plant canopy with fluorescence techniques (McMurtrey et al., 1994 a). But, the most cost effective information might be obtained from the harvested product, at harvest time, as the harvesting machinery is making it's trip through the field.

Diffuse reflectance spectra from 400 nm – 2400 nm have been reported to show differences in grain quality. Certain bands that are known to be strong in water absorption (Finney and Norris, 1978) showed differences at 970, 1190, 1450, and 1940 nm. The best spectral differences in the NIR and MIR result with a transformation of log 1/reflectance. Various areas in the MIR at 1200 1455, 1571, 1733, 1931, 2100, and 2330 nm are well known for having strong absorbance features for water, oil, starch, cellulose and protein (Williams and Norris, 1987).

This research was conducted to determine the utility of the information contained in fluorescence spectra compared to the reflectance information of corn grain after harvest for the assessment of inherent soils fertility effects on the previous crops growth condition.

Materials and Methods

Eight different levels of N fertilization were applied to field grown corn at USDA/ARS, Beltsville Agricultural Research Center, Beltsville, Maryland. The field was plowed and prepared with minimum tillage methods. The corn was planted with a no-till planter in 77 cm rows. The crop was planted at a density of 58,000 plants per hectare. Eight N levels were applied to the corn field plots on a Maryland Woodstown sandy loam soil. The field was maintained at the optimal rates for the essential nutrients P and K. Maintenance of optimal rates for N, P, and K are recommended by the local farm extension service and are known to produce a healthy profitable corn crop in an average year in this production area. The pH of the field was maintained at 6.5 with dolomitic lime and supplied the essential minerals of Ca and Mg. Other essential nutrients (such as S, Fe, B, Mn, Cu, and Zn) were supplied by the natural mineralization in the parent soil. The essential nutrient N was applied at the optimal recommended rate, at five progressively lower levels, and at two higher levels as described in Table 1. These treatments produced a range of chlorophyll levels, biomass, and physiological conditions. The corn crop was allowed to become senescent, and die, as is customary with normal harvesting conditions. The grain was taken form the standing plants at harvest time. The canopy was dry and showed brownish coloration. Statistical differences in dry biomass and grain yield are reported in Table 1. Fluorescence and reflectance spectra of grain kernels at harvest and after harvest, from the N treatments, were measured in the laboratory.

Laboratory Spectral Measurements of Grain

A Fluorolog II spectrofluorometer, (SPEX Industries, Inc., Edison, NJ), was used to excite corn grain kernels at a 385 nm wavelength and to measure fluorescence emission spectra in the blue-green-yellow area of the spectrum, respectively. The fluorescence excitation and detection system, used in the laboratory, was configured similar to that described by McMurtrey et al., 1994 b. The SPEX spectrofluorometer utilizes two 0.22 m double monochrometers with resolutions of 1.7 nm per mm slit width. The double monochrometer on the emission side of the instrument is attached to a photon counting photo multiplier tube (PMT) where the fluorescence emission is captured from the sample with data acquisition software. The PMT is sensitive in the range of 290 through 850 nm. Wavelength dependent fluctuations in lamp intensity are corrected by the use of a beam splitter which delivers a portion of the excitation energy to a rhodamine dye cuvette. The fluorescent response of the rhodamine dye is monitored by a silicon photo diode and is proportioned to the excitation energy through the wavelength range of 220 to 600 nm. The 385 nm UV excitation was chosen by comparing excitation maxima of spectra for several peak emission wavelengths. The fluorescence emissions over the 400 nm to 600 nm wavelength range were measured. Generally the longest wavelength absorption maximum (excitation) area will tend to give the strongest fluorescence emission from a given compound (Gilespie, 1985). Since several compounds may have contributing influences to the wavelengths that fluoresce in the blue-green areas, a compromise of an excitation wavelength at 385 nm was used to give a high intensity emission across the major emission areas of the blue-green fluorescence.

An IRIS MARK IV spectrometer made by Geophysical Environmental Research Corporation was used to collect the diffuse reflectance from corn grain kernels. This instrument is a dual beam system that has the ability to look at two targets simultaneously. The spectral range is from 350 nm to 3000 nm. Two detection surfaces are used. A silicon diode for the visible (VIS)-NIR (350 nm–1000 nm) and a Pbs for the MIR (1000 nm–3000 nm) are used with gratings scanned by stepping motors. Spectral bandwidth in the visible-NIR is approximately 2 nm and is 4 nm in the MIR region. The aperture is f/3.5 and the FOV is 6 degrees by 3 degrees and can be varied. A Transportable Hemispheric Illumination Source (Williams and Wood, 1989) with 12, quarts halogen 75w lamps wae used for 6 replications of reflectance measurements from 350 nm to 2500 nm. A halon panel was used as a reference.

Grain samples were collected during the 11/1/94 harvest and stored at ambient room temperature and humidity for later analysis. Analysis was made on the blue-green fluorescence and on the VIS-NIR and MIR reflectance of corn grain that came from site specific N fertilizer treatment areas of the corn field.

Results

The blue-green fluorescence and the VIS-NIR and MIR reflectance of corn grain that came from site specific N fertilizer treatment areas of the field were compared for their ability to detect treatment differences.

During harvest a preliminary set of data was taken form the range of N fertilization levels to determine if the treatment differences were acutely evident in the fluorescence spectra of the freshly harvested grain (Fig. 1). This procedure was repeated twice and gave excellent separation. In this initial data the main fluorescence peak was centered at 430 nm in the blue with a shoulder in the green at 530 nm. The 0 % N treatment had the lowest fluorescence intensity, while the 100 % of the optimal rate for N had the highest fluorescence intensity. Total biomass, biomass without grain, and grain yields for the treatment areas are reported in Table 1. The 1994 growing season for corn had plentiful rainfall. Means could be separated between every treatment for total biomass, with grain yield giving less separation at the upper levels of N fertilization.

The fluorescence properties of the grain samples after 2 months in storage are shown in Table 2. We expected the grain samples to give the same relative fluorescence and reflectance differences between N treatments even after a storage period. The fluorescence characteristics of these samples had a completely different shape and ranking than those taken immediately after harvest (Figs. 1 and 2). The treatment that had been supplied 0 % N during the growing season now had the highest intensity at 440 nm in the blue. The fluorescence features in the green that had previously been a shoulder were now more intense than the blue with distinct green maxima at 560 nm in grain from the 0 %, 12 %, 25 %, 100 % treatments. In general, there was an inverse ranking trend for blue

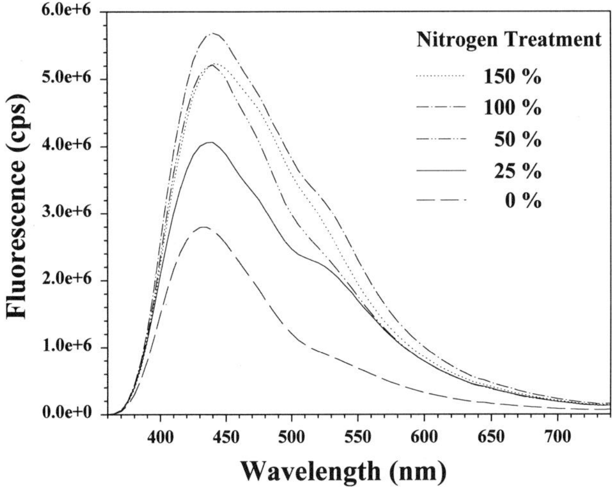

Fig. 1: Fluorescence of field corn *Zea mays* L. grain at harvest time (11/1/94), UV excitation at 385 nm.

Table 1: Percent of optimal N fertilizer treatments and means for dry biomass and yield of field corn in kg/ha.

% of N rate	N Applied	Dry Yield (kg/ha)		
		Vegetation	Grain	Total
150.0	270.0	7958.2 a*	9505.1 a	17463.3 a
125.0	225.0	6678.1 b	8733.7 ab	15411.8 b
100.0	180.0	5819.5 c	7840.9 bc	13660.4 c
75.0	135.0	4953.6 d	7155.7 cd	12109.3 d
50.0	90.0	4274.4 e	6133.6 d	10408.0 e
25.0	45.0	4025.7 e	4430.8 e	8456.5 f
12.5	22.5	2944.7 f	3310.1 f	6254.8 g
0.0	0.0	1996.5 g	2356.8 f	4353.3 h

* For each column, means with the same letter are not significantly different by Student-Newman-Keuls (n=3, α=0.05).

Table 2: Mean fluorescence changes in the blue and green regions of the spectrum on grain from field corn N treatments.

% of N rate	Mean Fluorescence Counts per Second ($\times 10^6$)*				
	11/1/94	1/3/95		4/5/95	
	(n=1)	(n=5)		(n=5)	
	430 nm	440 nm	560 nm	450 nm	470 nm
150.0	5.1	2.8 a	4.2 c	5.3 c	4.5 c
125.0	–	3.7 ab	4.2 c	5.8 c	4.9 c
100.0	5.8	4.3 ab	8.8 bc	7.1 abc	5.9 bc
75.0	–	4.2 ab	5.0 c	6.7 bc	5.6 bc
50.0	5.1	3.8 ab	5.0 c	7.6 abc	6.4 bc
25.0	4.0	2.6 ab	6.2 c	9.5 abc	8.1 abc
12.5	–	3.3 ab	11.4 ab	11.3 a	10.0 a
0.0	2.8	5.7 b	14.3 a	11.0 ab	9.3 ab

* For each column, means with the same letter are not significantly different by Student-Newman-Keuls (n=5, α=0.05).

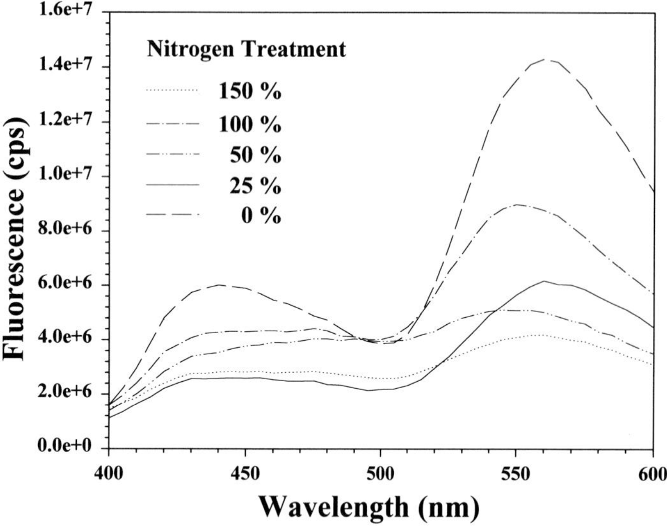

Fig. 2: Mean fluorescence of field corn *Zea mays* L. grain 2 months after harvest (1/3/95), UV excitation at 385 nm.

and the green areas of the spectrum with statistical separation between the upper and lower levels of N fertilization.

Since moisture could play and important role in quenching fluorescence or in increasing the fluorescence of water soluble compounds, samples were tested to determine the effect of drying to zero percent moisture and re-hydrating (Fig. 3). The air dried sample at 12 % moisture was dried to 0 % moisture in a drying oven at 40 °C. After drying the green fluorescence at 560 nm markedly declined and the blue fluorescence reappeared at 460 nm. Upon re-hydrating in a 100 % relative humidity chamber for 72 hours, the green fluorescence reappeared and the blue fluorescence declined. This phenomena could be due to compounds in the grain that behave like excimers or exciplexes (Turro, 1978). When the donor molecule in a certain concentration is in an excited state and the acceptor molecule in a certain concentration is in a ground state they form a metastable complex. This complex has different fluorescence properties than the excited donor molecule. The absorption maximum of the complex differs from that of the individual molecules as does the fluorescence emissions. Apparently in the grain kernel the donor or the acceptor, or perhaps both are water soluble which then causes the green region to fluoresce. The relationship between these molecules is such that the fluorescence emission shifts toward the green depending on the concentration of the compounds and on the hydration state. The relative effect fluorescence quenching by water can be seen in the comparison of the decrease in intensity between 15 % and 12 % moisture sample. The grain was again analyzed for its fluorescence after 6 months (Fig. 4 and Table 2). By this period of time the grain had lost its intense green fluorescence emission and re-hy-

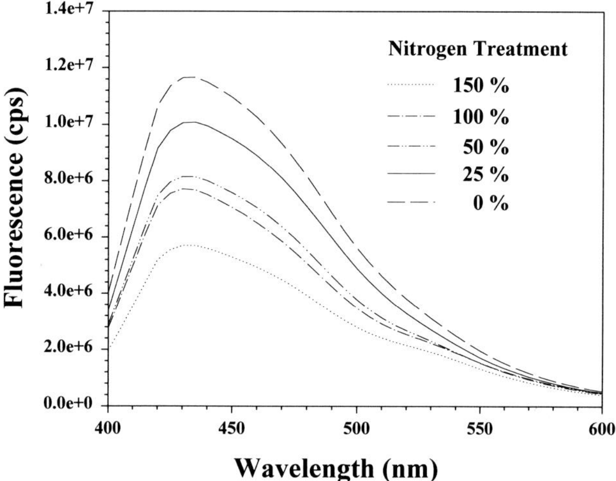

Fig. 4: Mean fluorescence of field corn *Zea mays* L. grain 6 months after harvest (4/5/95), UV excitation at 385 nm.

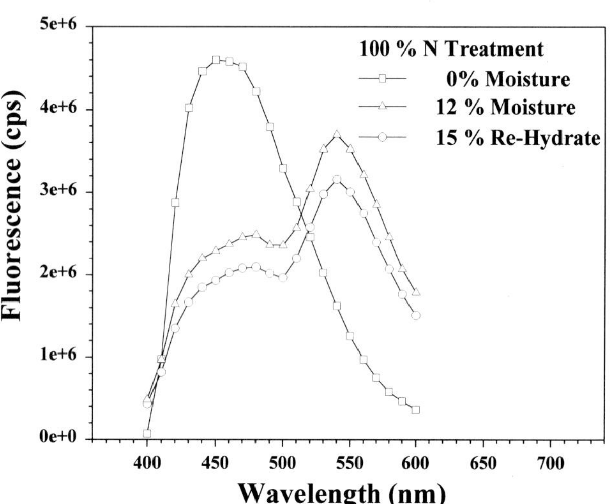

Fig. 3: Fluorescence of field corn *Zea mays* L. grain at ambient relative humidity, after drying, and after re-hydrating, UV excitation at 385 nm.

dration only caused water quenching effects to the blue emission region (data not presented). The green emission area did not return. Presumably the compound(s) responsible are in low concentrations or have degraded into other products.

Diffuse reflectance spectra from 400 nm–2400 nm, were taken on the grain samples after 6 months in storage. There were little spectral reflectance differences in the visible portions of the spectrum. The bands that are known to be strong in water absorption (Finney and Norris, 1978) showed differences at 970, 1190, 1450, and 1940 nm between the zero and low N levels, and the high N fertilization levels. The best spectral differences between treatments occurred with a transformation of log 1/reflectance (Fig. 5 and Table 3). Various areas in the MIR at 1200, 1455, 1571, 1733, 1931, 2100, and 2330 nm are well known for having strong absorbance features for water, oil, starch, cellulose and protein (Williams and Norris, 1987).

The plant ultimately partitions the photosynthate into its reproductive parts at the end of each growth cycle. Many of these compounds become concentrated into the seed grain kernels in corn. At the end of season considerable amounts of carbohydrates and proteins are deposited in the grain (Massie and Norris, 1965; Williams and Norris, 1987). Other degraded products may also contribute to this and other regions of the blue-green-yellow fluorescence of grains and other dry dead plant material.

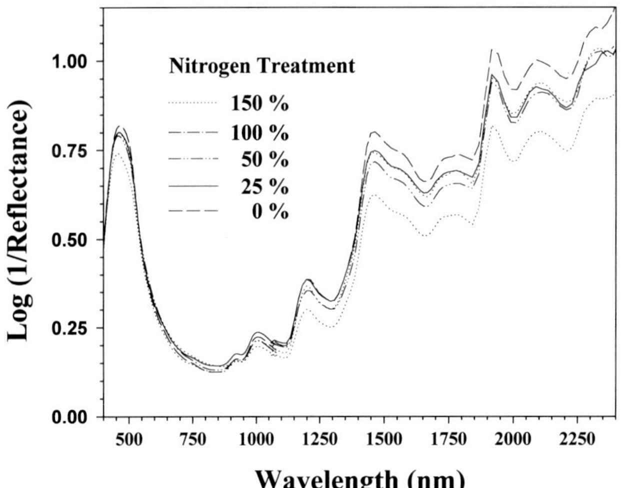

Fig. 5: Mean reflectance transformation log 1/R of field corn *Zea mays* L. grain (4/5/95).

Table 3: Near infrared refelctance transformation $\log^1/_R$ of field corn nitrogen treatments.

% of N Rate	Means (n=6)*						
	1200 nm	1455 nm	1571 nm	1733 nm	1931 nm	2100 nm	2330 nm
150.0	0.37 ab	0.75 ab	0.60 ab	0.68 a	0.95 ab	0.94 ab	1.04 ab
155.0	0.40 a	0.79 a	0.73 a	0.72 a	1.01 a	0.99 ab	1.10 a
100.0	0.39 ab	0.80 a	0.74 a	0.73 a	1.04 a	1.00 a	1.09 a
75.0	0.38 ab	0.76 ab	0.70 a	0.69 a	0.99 ab	0.96 ab	1.05 ab
50.0	0.39 ab	0.75 ab	0.69 ab	0.69 a	0.96 ab	0.92 ab	1.00 ab
25.0	0.35 bc	0.72 ab	0.66 ab	0.65 ab	0.94 ab	0.91 abc	1.03 ab
12.5	0.33 cd	0.67 cb	0.61 bc	0.60 bc	0.88 bc	0.86 bc	0.95 ab
0.0	0.30 d	0.63 c	0.57 c	0.56 c	0.82 c	0.80 c	0.90 b

* For each column, means with the same letter are not significantly different by Student-Newman-Keuls (n=5, α=0.05).

Discussion

Testing of grain samples on a laboratory actively induced fluorescence system established treatment differences between the fluorescence properties of dry corn grain kernels. These treatment differences were similar to the trends for plant biomass and grain yield for plant growth that was derived from varying different N fertilization rates. The blue fluorescence band from 430–470 nm gave separation of the means between the highest and lowest dry biomass and grain yield levels. A high green fluorescence appeared after a couple of months of storage. It's appearance probably is related to an increase in the concentration of certain degradation compounds which may behave as excimers or exciplexs. They are capable of forming a metastable complex that has different fluorescence properties than the excited donor molecule(s). These may be related to changes in the carotinoids in the epidermal surface of the seed coat. The moisture content of the grain at the time of measurement can be a factor in determining the fluorescence properties of grain. The period of time between when crop is first ready for harvest and its actual harvest date may be an important factor in the state and concentration of the compounds that fluoresce in grain.

Previous studies by McMurtrey et al., 1993 and Daughtry et al., 1994 illustrated that the fluorescence technique can be used to distinguish soils from crop residues and to quantify percent ground cover by residue materials. The present study indicates that information from blue-green fluorescence of dry corn grain at harvest time can be significantly associated with the plant growth that results in differences in grain yield. These factors can be used as bio-indicators of crop growth condition during the prior growing period. Use of information from the same site specific areas of the field year after year could lead to better management decisions for agronomic inputs to produce optimal growth for the subsequent growing season.

The physiology of the live corn crop during the previous growing season, in part, can manifest itself in the spectral qualities of the grain. Changes in the management of N fertilizer application could be made within the field in subsequent growing seasons if methods to determine the spectral characteristics of the grain are perfected. More work needs to be done in determining which products are responsible for the blue-green fluorescence and the contributions of the compounds responsible for the NIR and MIR reflectance in grain kernels at harvest time.

References

Chappelle, E. W., F. M. Wood, J. E. McMurtrey, and W. W. Newcomb: Laser induced fluorescence of green plants. 1. A technique for the remote detection of plant stress and species differentiation. Appl. Opt. *23,* 134–138 (1984 a).

Chappelle, E. W., J. E. McMurtrey, F. M. Wood, and W. W. Newcomb: Laser induced fluorescence of green plants. 2. LIF changes caused by nutrient deficiencies in corn. Appl. Opt. *23,* 139–144 (1984 b).

Chappelle, E. W., J. E. McMurtrey III, and M. S. Kim: Identification of the pigment responsible for the blue fluorescence band in the laser induced fluorescence (LIF) spectra of green plants, and the potential use of this band in remotely estimating rates of photosynthesis. Remote Sens. Environ. *36,* 213–218 (1991).

Chappelle, E. W., J. E. McMurtrey, M. S. Kim, and L. A. Corp: The significance of the blue fluorescence band in the laser inducer fluorescence (LIF) spectra of vegetation. Proc. Internat. Geoscience and Remote Sensing Symposium, IGARSS '93, Tokyo, Japan *3,* 1333–1336 (1993).

Chappelle, E. W., C. S. T. Daughtry, and J. E. McMurtrey III: United States Patent, Method for determining surface coverage by materials exhibiting different fluorescent properties, U.S. Patent Number 5, 412, 219, May 2 (1995).

Daughtry, C. S. T., J. E. McMurtrey III, and E. W. Chappelle: Measuring crop residue cover by fluorescence imaging. Proc. Internat. Geoscience and Remote Sensing Symposium, IGARSS '94, Pasadena, CA., pp. 625–654 (1994).

Finney Jr., E. E. and K. H. Norris: «Determination of moisture in corn kernels by near-infrared transmittance measurements». Transaction of the ASAE *21,* 581–584 (1978).

Gilespie Jr., A. M.: «A manual of fluorometric and spectrophotometric experiments». Gordon and Breach Science Publishers. Cooper Station, New York, New York 10276 (1985).

Lang, M. and H. K. Lichtenthaler: Changes in the blue-green and red fluorescence emission spectra of beech leaves during the autumnal chlorophyll breakdown. J. Plant Physiol. *138,* 550–553 (1991).

Lang, M., F. Stober, and H. K. Lichtenthaler: Fluorescence emission spectra of plant leaves and plant constituents. Radiat. Environ. Biophys. *30,* 333–347 (1991).

Lang, M., P. Siffel, Z. Braunova, and H. K. Lichtenthaler: Investigations of the blue-green fluorescence emission of plant leaves. Bot. Acta *105,* 435–439 (1992).

Lichtenthaler, H. K., F. Stober, C. Buschmann, U. Rinderle, and R. Hak: Laser-Induced Chlorophyll Fluorescence and Blue Fluorescence of Plants. Proc. Internat. Geoscience and Remote Sensing Symposium, IGARSS '89, College Park, MD., pp. 1913–1918 (1989).

Lichtenthaler, H. K., M. Lang, and F. Stober: Nature and variation of blue fluorescence spectra of terrestrial plants. Proc. Internat. Geoscience and remote Sensing Symposium IGARSS '91 University of Helsinki, Helsinki, pp. 2283–2286 (1991).

Massie, D. R. and K. H. Norris: Spectral reflectance and transmittance properties of grain in the visible and near infrared. Trans. of the American Society of Agricultural Engineers *8,* 598–600 (1965).

Mazzinghi, P., G. Agati, and F. Fusi: Interpretation and physiological significance of blue-green and red vegetation fluorscence. Proc. Internat. Geoscience and Remote Sensing Symposium, IGARSS '94, Pasadena, CA, pp. 640–642 (1994).

McMurtrey III, J. E., E. W. Chappelle, C. S. T. Daughtry, and M. S. Kim: Fluorescence and reflectance of crop residue and soil. J. Soil Water Conservation, *48,* 207–213 (1993).

McMurtrey III, J. E., E. W. Chappelle, M. S. Kim, and L. A. Corp: Distinguishing nitrogen fertilization levels in field corn with actively induced fluorescence and passive reflectance measurements. Remote Sen. Environ. *47,* 36–44 (1994 a).

McMurtrey, J., E. Chappelle, C. Daughtry, J. Kalshoven, L. Corp, and M. Kim: Field canopy and leaf level fluorescence for distinguishing plant condition differences due to nitrogen fertilization level. Proc. Internat. Geoscience and Remote Sensing Symposium, IGARSS '94, Pasadena, CA, pp. 982–985 (1994 b).

Turro, N. J.: «Modern Molecular Photochemistry». Benjamin/ Cummings Menlo Park, Calif. (1978).

Williams, D. L. and F. M. Wood: A Transportable Hemispherical Illumination System for Making Reflectance Measurements. Remote Sens. Environ. *23,* 131–140 (1987).

Williams, Ph. and K. Norris: «Near-Infrared Technology in the Agricultural and Food Industries». Amer. Assoc. of Cereal Chemists, Inc. 3340 Pilot Knob Road, St. Paul, Minnesota 55121 (1987).

J. Plant Physiol. Vol. 148. pp. 515–522 (1996)

Narrow-band Reflectance Imagery Compared with Thermal Imagery for Early Detection of Plant Stress

GREGORY A. CARTER, WILLIAM G. CIBULA, and RICHARD L. MILLER

NASA, Earth Observation Research Office, Stennis Space Center, MS 39529, USA

Received June 27, 1995 · Accepted October 10, 1995

Summary

A field experiment compared plant stress detection by narrow-band reflectance and ratio images with thermal infrared images. Stress was induced in a mixed stand of 5 year old loblolly pine (*Pinus taeda* L.) and slash pine (*Pinus elliottii* Engelm.) by a soil application of diuron (DCMU) on 22 August followed by bromacil on 19 September, 1994. Herbicide-induced stress was first indicated on 24 and 26 September by significant ($p \leq 0.05$) decreases in photosynthesis and the ratio of variable to maximum fluorescence (F_v/F_m), respectively. Stress was first detected remotely on 5 October by 694 ± 3 nm reflectance imagery and its ratio with reflectance at 760 ± 5 nm ($p \leq 0.05$). This reflectance increase was detected at least 16 days prior to the first visible signs of damage, as quantified by the CIE color coordinate u′, that occurred between 21 and 26 October. Reflectance images at 670 ± 5 nm, 700 ± 5 nm and 760 ± 5 nm first detected stress on 21 October, 12 October and 20 December, respectively. Canopy temperature as indicated by imagery in the 8 to 12 µm band never differed significantly between herbicide-treated and control plots. This resulted from the close coupling of leaf temperatures with air temperature, and the tendency of wind and environmental moisture to equalize temperatures among treatments. The high sensitivity to stress of reflectance imagery at 694 ± 3 nm supports similar conclusions of earlier work, and indicates that imagery in the 690 to 700 nm band is far superior to thermal imagery for the early and pre-visual detection of stress in pine.

Key words: Pinus taeda, Pinus elliottii, plant stress, diuron, bromacil, chlorosis, canopy reflectance, canopy temperature, narrow-band imagery.

Abbreviations: A = net CO_2 assimilation rate; a.i. = active ingredient; ANOVA = analysis of variance; CCD = charge-coupled device; CIE = Commission Internationale de l'Eclairage; F_λ = chlorophyll fluorescence at the subscripted λ; F_t = terminal or steady-state fluorescence; F_v/F_m = variable to maximum fluorescence ratio; λ = wavelength in nanometers; n = number of observations per mean; p = statistical probability; Ψ_w = water potential; r^2 = coefficient of determination; R_λ = reflectance at the subscripted λ; T_{canopy} = imaged canopy temperature; u′ = CIE chromaticity coordinate.

Introduction

When unfavorable growth conditions result in plant physiological stress, leaf chlorophyll content typically begins to decrease (e.g., Waring et al., 1986; Hendry et al., 1987; Lange et al., 1987). Consequently, the reflection of incident radiation from within the leaf interior increases, providing an optical indicator of stress. Reflectance sensitivity analysis (Carter, 1991; Cibula and Carter, 1992) has shown that increased reflectance in the 695 ± 5 nm waveband provides an earlier and more consistent indication of stress than reflectance at other wavelengths in the 400 to 2,500 nm spectrum as a result of the absorption properties of chlorophyll (Carter, 1993; Carter and Miller, 1994). Depending on the severity of stress and the accompanying chlorosis, this reflectance response can be detected prior to damage symptoms apparent to the unaided eye (Cibula and Carter, 1992). The ratio of R_{695} to R_{760} or R_{800} was the most consistent stress indicator among several ratios tested (Carter, 1994; Carter and Miller, 1994) and is quite sensitive to leaf chlorophyll content (Carter et al., 1995). As stress develops and leaf chlorophyll concentrations begin to decrease, R_{695}/R_{760} increases such that stressed plants appear brighter than healthy plants in black-and-white digital images of the ratio (Carter and Miller, 1994).

Similar results have been obtained using other data analysis methods. Reflectance variance analysis indicated a strong relationship of R_{705} and R_{750}/R_{705} with leaf chlorophyll content (Gitelson and Merzlyak, 1994, and this issue). Ratios of spectral reflectance curves indicated the use of R_{675}/R_{700} yielding a precise relationship with chlorophyll content (Chappelle et al., 1992). The technique used most widely in recent years for determining relationships of reflectance with chlorophyll content is determination of the wavelength location of the reflectance curve red edge (Gates et al., 1965 a; Gates, 1980). The inflection point, which generally occurs in the 680–760 nm range, is strongly dependent on chlorophyll content (e.g., Horler et al., 1983; Rock et al., 1988; Curran et al., 1990; Buschmann and Nagel, 1993; Vogelmann et al., 1993; Munden et al., 1994). Other analyses, some of which have not focused on the red to near-infrared transition spectrum, conclude that reflectance near 550 nm or its ratio with the near-infrared provides the closest correlation with leaf chlorophyll content (e.g., Thomas and Gausman, 1977; Tsay et al., 1982; Saxena et al., 1985; Buschmann and Nagel, 1993). The sensitivity of reflectance near 550 nm to chlorophyll may be similar to that near 700 nm (Buschmann and Nagel, 1993; Carter, 1993; Gitelson and Merzlyak, 1994), but reflectance at or near 550 nm has been less reliable as a stress indicator than reflectance near 700 nm (Cibula and Carter, 1992; Carter, 1994; Carter and Miller, 1994).

Increasingly, the measurement of narrow-band reflectance near 700 nm and its ratio with near-infrared reflectance appears to be an optimal basis for the remote sensing of incipient chlorosis and plant stress. However, plant radiative properties other than reflectance may indicate stress at earlier stages. The fluorescence ratio F_{690}/F_{735} increases with a decline in photosynthetic electron transport even prior to a drop in chlorophyll content, and increases further as leaf chlorophyll content begins to decrease (Lichtenthaler and Rinderle, 1988). F_{690}/F_{735}, other fluorescence parameters, and various methods by which they may be measured remotely have been reviewed recently (Schmuck, 1990; Lichtenthaler et al., 1992; Moya et al., 1992; Rosema et al., 1992).

Leaf temperature might also provide an earlier stress indication than reflectance. If plant stress compromises leaf photosynthetic capacity, stomatal conductance would be expected to decrease as a result of decreased demand for atmospheric CO_2 (Farquhar and Sharkey, 1982). This would result in decreased latent heat loss by transpiration and increased leaf temperatures (Nobel, 1991). Thermal radiometry has proven useful in large-scale monitoring of forest canopy temperatures (e.g., Luvall, 1989) and stress detection in crops (e.g., Pinter et al., 1979; Mottram et al., 1983; Berliner et al., 1984). Thus, a change in leaf thermal emittance may indicate stress prior to reflectance near 700 nm.

The purpose of the present study was to test, under field conditions, the hypothesis that thermal infrared emission in the 8 to 12 μm band will provide an earlier indication of plant stress than reflectance in the 695 ± 5 nm range. Also tested was the hypothesis that either measurement would detect stress prior to visibly apparent chlorosis. Stress was induced artificially in slash pine (*Pinus elliottii* Engelm.) and loblolly pine (*Pinus taeda* L.) by soil applications of the photosystem II herbicides diuron (DCMU) and bromacil. Diuron and photosystem II herbicides in general inhibit electron transport by blocking and binding to the Q_B-binding protein (Lichtenthaler and Rinderle, 1988). Photosynthetic rate, fluorescence induction kinetics, and leaf water potential were used as independent, *in contactu* measurements of stress. Radiative and physiological responses to treatment were monitored over a 4-month period.

Materials and Methods

Experimental Design and Treatment Applications

Plant stress was induced by application of herbicides to a mixed planting of 5 year old loblolly and slash pines. The trees had been planted at a density of approximately 40 trees m^{-2} in a 1 m × 20 m outdoor bed located at the Harrison Experimental Forest, 25 km north of Gulfport, MS, USA. When the study began on 22 August, 1994, the trees were 3.3 to 5.5 m in height, 3 to 8 cm in trunk basal diameter, and formed a dense, closed canopy. The 20 m row of trees was divided into 6, 1 m × 3.3 m plots, and herbicide treatment was assigned randomly to plots 1, 2 and 4. Plots 3, 5 and 6 served as non-stressed controls. The herbicide treatment was initiated at 16.00 hours on 22 August when 16 g of 80 % a.i. diuron [(3,4-dichlorophenyl)-1,1-dimethylurea, abbreviated DCMU; Karmex, du Pont de Nemours, Wilmington, DE, USA] was dissolved as a wettable powder in 1 liter of water and sprayed onto the soil in the central 1 m^2 of each treated plot. The area encompassing all 6 plots then was irrigated with approximately 4 cm of tap water. Measurements of plant physiological and radiative properties began on the morning of 23 August, and continued every 1 to 18 days through 20 December. For nearly 1 month after diuron application, physiological measurements indicated little or no stress. Thus, on 19 September, 10 mL of 21.9 % a.i. bromacil [(5-bromo-3-sec-butyl-6-methyluracil); Hyvar X–L, du Pont de Nemours, Wilmington, DE, USA] was injected undiluted as a water-soluble liquid into the upper 6 cm of soil at the base of each tree in the central 1 m^2 of each treated plot. All plots were then watered as above. No additional herbicide was applied during the remainder of the study. All plots received ambient rainfall.

Canopy Imagery and Spectroradiometry

Each plot was imaged to detect treatment effects on pine canopy reflectance, key reflectance ratios and canopy temperature. The reflectance imaging system consisted of a black and white CCD camera (570 horizontal × 485 vertical elements; Model WV-BL202, Panasonic, Secaucus, NJ, USA) connected to a VHS video cassette recorder (Model AG-2400, Panasonic, Secaucus, NJ) and a 9 cm black and white monitor (Model M300, Auto Vision Systems, Miami, FL, USA). The system, mounted on a transportable cart and powered by a 12 V battery, was placed at a marked point approximately 50 m east of the pines. Plots were imaged 2.5 to 3.5 hours prior to solar noon to insure side illumination of the canopies and using a horizontal view angle. When the pines were viewed at this angle, no background objects could be seen through the canopies. Each plot was imaged through a 50 mm telephoto lens and narrow-band interference filters (Andover Corporation, Salem, NH, USA) that were mounted in front of the lens. The telephoto lens was required to allow larger canopy images and force a smaller acceptance angle for scene elements that were imaged furthest from the optical axis of the system. The small acceptance angle is necessary because the spectral transmissivity of interference filters changes as incidence angle deviates from perpendicular to the filter plane. Plots were imaged at 670 ± 5 nm, 694 ± 3 nm, 700 ± 5 nm and 760 ± 5 nm. These filters were selected because leaf and canopy reflectance at 760 nm is not greatly affected while reflectances at 694 and 700 nm are quite sensitive to the onset of chlorosis; reflectance near 670 nm is relatively unresponsive to mild chlorosis (Cibula and Carter, 1992; Buschmann and Nagel, 1993; Carter, 1993; Carter and Miller, 1994; Gitelson and Merzlyak, 1994). Approximately 15 min were required to image the 6 plots and a calibration grey scale through the 4 filters. The grey scale was composed of 5 cards that ranged from white to black. Each card was calibrated to percent reflectance in sunlight using a scanning radiometer and a white reference (see below). These images were used to calibrate the pine images to percent reflectance. Reflectances of the grey cards were approximately 88 %, 28.9 %, 9.8 %, 4.5 % and 2.5 % across the 670 to 760 nm range. All images were recorded on videotape.

Canopy temperatures were imaged in the 8 to 12 µm waveband immediately after reflectance images were acquired using an infrared imaging radiometer with 3× telescope lens (Model 760, Inframetrics, Billerica, MA, USA). The black and white thermal images were recorded on videotape for each plot as above. The radiometer was adjusted such that a digital grey value range of 15 to 255 corresponded linearly with a temperature range of 5 °C. Emissivity of the pine canopies was assumed to approximate 1.0, as is typical of conifer emissivities in the 8 to 12 µm range (Tibbals et al., 1964; Gates et al., 1965 b; Arp and Phinney, 1980). Imaging of canopy temperatures required approximately 5 min.

After canopy temperatures were imaged, canopy spectral reflectances were recorded at 1 nm intervals throughout the 400 to 850 nm spectrum using a portable spectroradiometer (LI-1800UW with 3° field-of-view telescope, LI-COR, Inc., Lincoln, NE, USA). For each plot, reflected solar radiance was measured from a 3 m diameter circular area encompassing the majority of the total plot canopy. The average of 3, 20 s scans was recorded as reflected canopy radiance. Canopy radiance then was multiplied by 100 and divided by radiance reflected from a white reference (Spectralon SRT-99-05, Labsphere, Inc., North Sutton, NH, USA) to compute percent spectral reflectance. To quantify changes in canopy color that occurred with stress, CIE chromaticity coordinates were computed directly from canopy spectral reflectances and CIE color matching functions (Wyszecki and Stiles, 1982; LI-1800UW software).

Physiological Measurements

Approximately 20 min after completion of canopy imaging and spectroradiometry, net CO_2 assimilation rate (A) in the most recently produced needles was measured using a portable gas exchange system (LI-6200 with 0.25 L transparent chamber, LI-COR, Inc., Lincoln, NE, USA). In each plot A was measured for 3 replicate samples comprised of 3 needles each. Because A was to be measured repeatedly during the study and large treatment differences were expected, an average value of 14 ± 0.6 cm^2 was determined for the total surface area of 3 needles placed within the 11 cm length of the sample chamber. This value together with the rates at which the sample depleted CO_2 and increased humidity in the chamber during a 10 s period were used to compute A and stomatal conductance, as described previously (Leuning and Sands, 1989). Measurements were made under quantum fluxes that ranged from 400 µmol $m^{-2} s^{-1}$ on overcast days to 2,200 µmol $m^{-2} s^{-1}$ on clear or partly-cloudy days. Of the 33 sampling dates, 22 were clear, 7 were partly-cloudy and 4 were overcast. To provide an additional stress indicator, fluorescence induction kinetics were measured for 3, 3-needle samples using a portable fluorometer (Model CF-1000, Morgan Instruments, Andover, MA, USA). Three needles were selected from the lower canopy and approximately 0.5 cm^2 of needle projected area was dark-adapted for 15 min in a cuvette. Induction kinetics were measured when the dark-adapted leaf area was exposed to 1000 µmol $m^{-2} s^{-1}$ of actinic light *via* a fiber optic probe. Fluorescence parameters recorded were F_t and the ratio F_v/F_m as a measure of photosynthetic quantum yield (Adams et al., 1990). To monitor plant water stress, Ψ_w were measured for 3 needle fascicles per plot using a pressure chamber (Model 1001, PMS Instrument Company, Corvallis, OR, USA).

Image Processing

Videotape of the pine canopies acquired in the narrow bands and thermal infrared was returned to the laboratory where an image of each plot in each spectral region was captured onto a personal computer. This was accomplished by using a video cassette recorder with single-frame advance capability and a video display and capture board (WIN/TV, Hauppauge Computer Works, Hauppauge, NY, USA). Digital images then were processed using the FIGMENT image processing software (Miller, 1993). Narrow-band images were calibrated to reflectance by regressing the average digital values of the grey cards at each waveband against their known percent reflectances. A quadratic model was used and r^2 values at all wavebands were greater than 0.99. Reflectance ratio images were generated by the quotient of the appropriate reflectance images.

Data Analysis

All data were analyzed for significant treatment effects ($p \leq 0.05$) within each sampling date using ANOVA (Steel and Torrie, 1961). A completely randomized block design was used as the statistical model. For each image, mean canopy reflectance or temperature and the corresponding variance was computed for a rectangular area of approximately 40,000 pixels using FIGMENT. This sample represented the central 1 m of the given plot and was used in the ANOVA to yield a statistical test that was identical to that used for the physiological variables. Stress developed first in plot 4 followed by plot 1, and never developed strongly in plot 2 during the study period. This yielded an unbalanced experimental design such that the data for each treated plot (1, 2 and 4) were compared separately with the combined data from all control plots (3, 5 and 6). Thus, for A, F_v/F_m and Ψ_w, n = 3 and 9 observations per mean for the herbicide and control treatments, respectively. In the following x-y plots of treatment means (see Results), all control means are shown. However, variable means from herbicide-treated plots are shown only where they differ significantly ($p \leq 0.05$) from the corresponding control mean. Treatment means were not compared statistically for spectral reflectance data because only a mean curve per plot was recorded in the field. Instead, these data and plot means of A for the entire study period were used to determine relationships of reflectance or reflectance ratios with A by least-squares regression analysis (Steel and Torrie, 1961).

Results

Photosynthetic rate A and F_v/F_m were the earliest indicators of stress induced by soil applications of diuron followed by bromacil in plots 1, 2 and 4. In plot 4 A and F_v/F_m decreased below control values ($p \leq 0.05$) beginning on 24 and 26 September, respectively (Fig. 1 a, b). Similar decreases began in plot 1 on 5 October and 30 September, respectively. In plot 2 F_v/F_m decreased beginning 15 October but A never de-

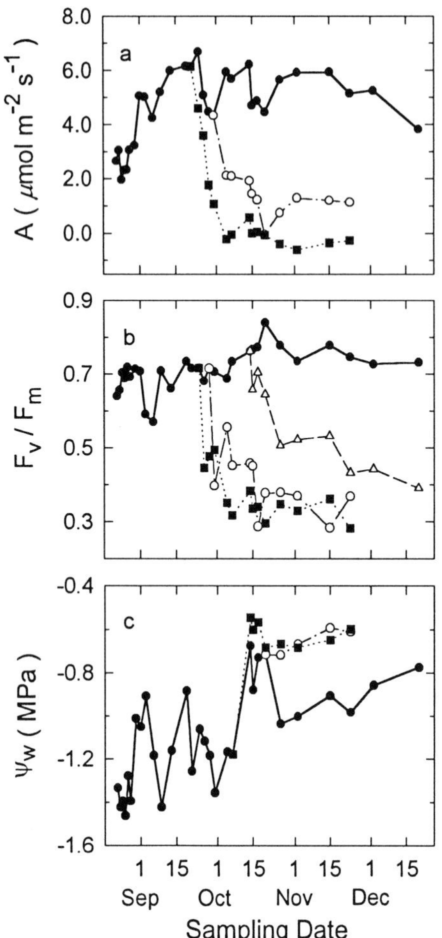

Fig. 1: Net CO_2 assimilation rate A (a), the chlorophyll fluorescence ratio F_v/F_m (b) and leaf water potentials Ψ_w (c) in control pines (●- - - -●) and pines treated with diuron followed by bromacil (○·-·-·○, plot 1; △- - - -△, plot 2; ■·····■, plot 4). Control data represent the mean response of 3 control plots (n = 9). Means for a herbicide-treated plot (n = 3) are shown *only* where they differ significantly ($p \leq 0.05$) from control values according to ANOVA. Photosynthesis first indicated plant stress on 24 September in plot 4 and 5 October in plot 1. In plot 2 A never differed significantly from the controls. F_v/F_m first indicated stress on 26 and 30 September in plots 4 and 1, and 5 October in plot 2, respectively. Ψ_w varied with rainfall, but were consistently greater in plots 4 and 1 beginning 14 and 26 October, respectively. Physiological data were not acquired in plots 1 and 4 after 23 November because needles had turned brown.

creased significantly below control values. The herbicide treatment first decreased stomatal conductance on 5 and 14 October in plots 4 and 1; conductance never decreased below control values in plot 2 (data not shown). F_t was ineffective as a stress indicator. It did not increase significantly until 2 November in plot 1, 2 December in plot 4 and never differred from controls in plot 2 (data not shown). Ψ_w varied with rainfall, but was consistently greater than control values in plots 4 and 1 beginning 14 and 26 October, respectively (Fig. 1 c). The herbicide treatment never affected Ψ_w in plot 2.

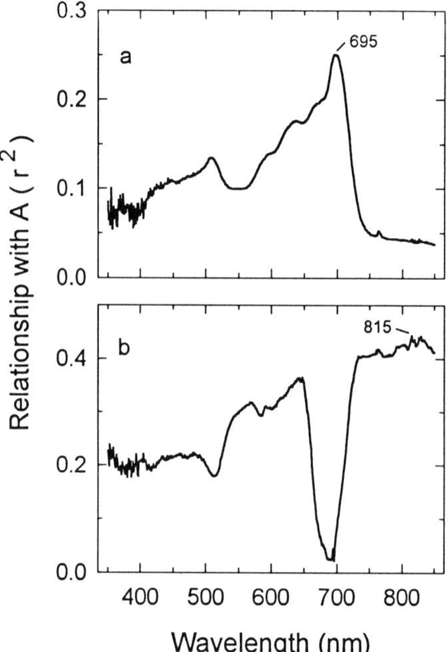

Fig. 2: Coefficients of determination (r^2) for the relationship with net CO_2 assimilation rate A at each 1 nm wavelength interval of reflectance (maximum $r^2 = 0.25$ at 695 nm) (a), and reflectance at 695 nm divided by reflectance at the wavelength indicated by the horizontal axis (maximum $r^2 = 0.44$ for R_{695}/R_{815}, $r^2 = 0.41$ for R_{695}/R_{760}) (b). Regression equations were of the form $y = a + bx + cx^2$ where y is reflectance or a reflectance ratio and x represents mean canopy photosynthesis. All plot means of spectral reflectance and A collected during the 4-month study were included in regression analyses regardless of environmental conditions at the time of measurement.

When all plot means of A and spectral reflectance for the entire study period were combined to include clear, partly-cloudy, overcast, warm and cold days, A was best correlated with canopy reflectance at 695 nm among all wavelengths in the 350 to 850 nm range ($r^2 = 0.25$) (Fig. 2 a). The relationship with A improved substantially when reflectance at 695 nm was divided by reflectance at near-infrared wavelengths (Fig. 2 b). Maximum r^2 was 0.44 for the 695 to 815 nm reflectance ratio. The 695 to 760 nm ratio yielded an r^2 of 0.41.

Images of canopy reflectance at 694 ± 3 nm (R_{694}) and R_{694} divided by reflectance at 760 ± 5 nm (R_{760}) provided the earliest remotely-sensed indications of stress. R_{694} and R_{694}/R_{760} first indicated stress ($p \leq 0.05$) on 5 October, 21 October and 20 December in plots 4, 1 and 2, respectively (Fig. 3 a, b; Fig. 4 a, b). Reflectance images at 670 ± 5 nm (R_{670}) and 700 ± 5 nm (R_{700}) and their ratios with R_{760} first detected stress in plot 4 on 21 and 12 October, respectively; R_{760} did not indicate stress until 20 December (data not shown). Ratios of R_{694} and R_{700} with R_{670} first indicated stress relatively late, on 26 October and 2 November, respectively (data not shown). Thermal imagery never indicated plant stress in any of the treated plots (Figs. 3 c; 4 c).

Among the various CIE color coordinates (Wyszecki and Stiles, 1982) u′ was most sensitive to plant stress. This is un-

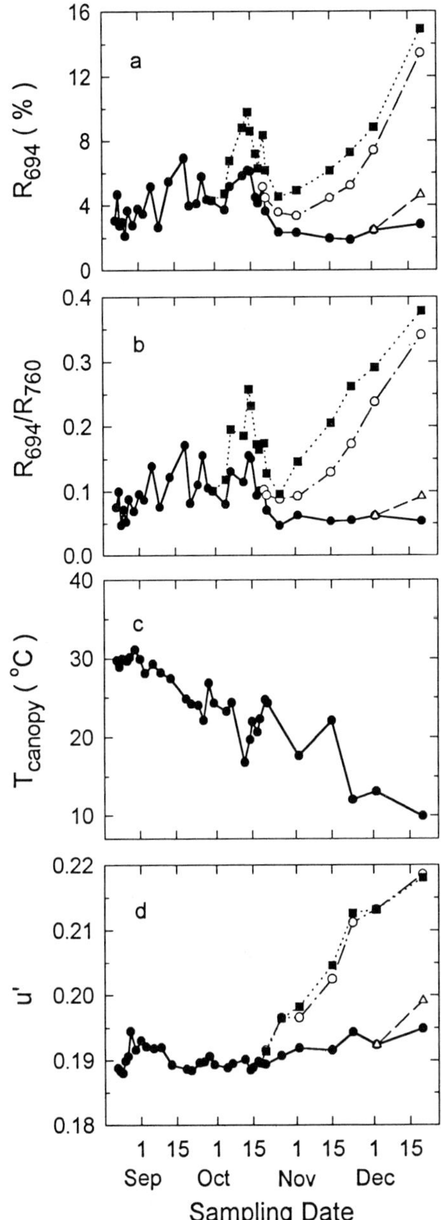

Fig. 3: Canopy reflectance at 694 ± 3 nm (a), its ratio with reflectance at 760 ± 5 nm (b), canopy temperature as determined from thermal imagery (c), and the CIE color coordinate u′ as an indicator of visible color change (d) in control pines (●- - - -●) and pines treated with diuron followed by bromacil (○·- - -·○, plot 1; △- - - -△, plot 2; ■· · · · ■, plot 4). Control data points represent means of 3 control plots (n=9). Means for a herbicide-treated plot (n=3) are shown *only* where they differ significantly ($p \leq 0.05$) from control values according to ANOVA. R_{694} and R_{694}/R_{760} first indicated plant stress on 5 and 21 October in plots 4 and 1, respectively, and on 20 December in plot 2. T_{canopy} never indicated plant stress. Stress was first visually apparent between 21 and 26 October as indicated by u′.

derstandable, since a change in u′ represented a change in hue from green toward yellow. The u′ coordinate indicated that the color of herbicide-treated pine first changed significantly between 21 and 26 October in plot 4. This occurred at least 16 days after stress in plot 4 was indicated by R_{694} and R_{694}/R_{760} (Figs. 3 d; 4 d).

Discussion

The hypothesis that thermal imagery would detect stress in the herbicide-treated pines prior to reflectance at 695 ± 5 nm clearly must be rejected. Imagery of the pine canopies in the 8 to 12 µm range was completely ineffective for stress detection. Although thermal radiometry is quite useful in measuring conifer canopy temperatures (e.g., Luvall, 1989; present results), and has been used to detect stress in crop species (Pinter et al., 1979; Mottram et al., 1983; Berliner et al., 1984), it appears unsuitable for early stress detection in pine. This can be explained by the relatively close coupling of conifer leaf temperatures with atmospheric temperature (Tibbals et al., 1964; Gates et al., 1965 b; Smith and Carter, 1988). In the present case, even mild gusts of wind tended to equalize leaf temperatures among plots and thus mask potential treatment differences. Potential treatment effects were masked also when heavy dew occassionaly persisted on the canopies until late morning.

In contrast to results for the 8 to 12 µm spectrum, the hypothesis that reflectance in the 695 ± 5 nm range would detect stress previsually is supported strongly. R_{694} and R_{694}/R_{760} indicated stress at least 16 days prior to a change in canopy color. This high sensitivity to plant stress was predicted by earlier results (Carter, 1993; 1994; Carter and Miller, 1994). R_{700} also detected stress previsually, as in earlier photography (Cibula and Carter, 1992), but this occurred 7 days later than stress detection by R_{694}. This may have resulted from the inclusion of 700 to 705 nm reflectance in the R_{700} band (see Methods). Reflectance in the 700 to 705 nm region is more variable than in the 690 to 700 nm range for green leaves (Gitelson and Merzlyak, 1994). Thus, a greater reflectance difference between stressed and control leaves must occur at 700–705 nm than at 690–700 nm for the difference to be declared statistically significant. This emphasizes the importance of narrow bandwidth reflectance measurements within the 690 to 700 nm range. Images of increased reflectance near 700 nm are visualizations of the spectral «blue shift» of the red to infrared reflectance transition curve (Cibula and Carter, 1992). This shift occurs frequently with plant stress, and is related closely with leaf chlorophyll content (Horler et al., 1983; Rock et al., 1988; Curran et al., 1990; Demetriades-Shah et al., 1990; Vogelmann et al., 1993; Filella and Peñuelas, 1994) and visible damage symptoms (Ruth et al., 1991; Hoque and Hutzler, 1992).

As described earlier (Carter, 1993; 1994; Carter and Miller, 1994; Carter et al., 1995) the high sensitivity of reflectance in the 690 to 700 nm range to plant stress is explained by the narrowing of the chlorophyll absorption curve that occurs with decreasing leaf chlorophyll content (Gates et al., 1965 a; Gates, 1980). The specific absorption coefficient of chlorophyll *a in vitro* approaches zero near 695 nm compared with approximately 70 near 670 nm (Hoff and Amesz, 1991). Thus, as chlorophyll is lost gradually, leaf reflectance increases at 695 nm while it remains largely unchanged at wave-

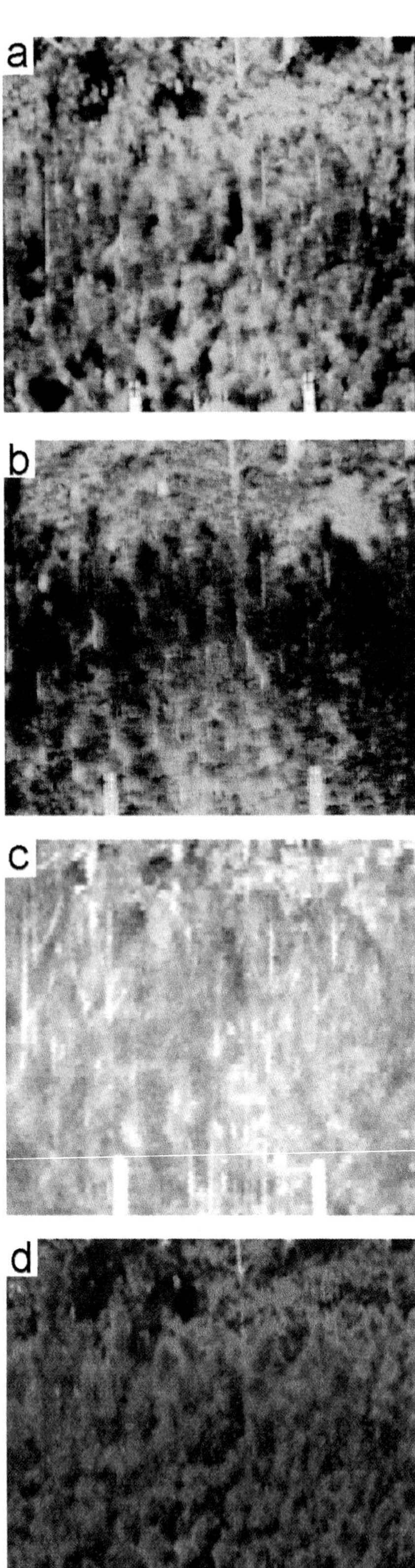

Fig. 5: Digital image of plots 1 through 6 (left to right, foreground) at 700 ± 5 nm. The image was acquired on 27 October from a light aircraft at an altitude of 150 m above the ground. Stress (chlorosis) is indicated by brighter foliage in plots 1 (left) and 4 (mid-frame). The background row of trees was not involved in the study. Automobile in background indicates approximate scale.

lengths of strong absorption, such as 670 nm (e.g., Cibula and Carter, 1992; Chappelle et al., 1992; Carter, 1993). Dividing reflectance at 695 ± 5 nm by near-infrared reflectance, which changes relatively little with early chlorosis (Carter, 1993), may serve to cancel wavelength-independent effects on leaf reflectance caused by multiple reflections inside leaves (Carter and Miller, 1994). Thus, reflectance ratios often relate more closely with leaf chlorophyll content (e.g., Buschmann et al., 1991; Ruth et al., 1991; Chappelle et al., 1992; Buschmann and Nagel, 1993; Vogelmann et al., 1993; Gitelson and Merzlyak, 1994; Carter et al., 1995) and photosynthetic rate (Chappelle et al., 1992; present results) than does reflectance *per se.* Furthermore, reflectance ratios at the canopy and larger scales tend to reduce variability caused by wavelength-independent phenomena such as mutual leaf shading, leaf orientation and solar angle.

With respect to wavelength, a 695 to 760 nm reflectance ratio is similar to the 690 nm to 735 nm fluorescence ratio that also is highly responsive to plant stress (Lichtenthaler and Rinderle, 1988). This fluorescence ratio depends on leaf chlorophyll content as well as stress-induced changes in photochemical energy transfer (Lichtenthaler and Rinderle, 1988). Thus, the high stress-sensitivity of R_{695}/R_{760} is likely

Fig. 4: Digital imagery of reflectance at 694 ± 3 nm (a), reflectance at 694 ± 3 nm divided by reflectance at 760 ± 5 nm (b), canopy temperature determined by emittance in the 8–12 µm band (c) and a color photograph (d) for the diuron and bromacil-treated plot 4. Images were acquired on 5 October, the first day on which R_{694} and R_{694}/R_{760} indicated plant stress. Note left and right plot boundaries at bottom of each frame. Stress (incipient chlorosis) is indicated by brighter foliage in the lower one-third of the plot (a, b) compared with the adjacent control plots (3 and 5). Background trees fill the top of each image.

due in part to the detection of fluorescence as a component of the apparent reflectance signal (Carter and Miller, 1994).

Reflectance at 695 ± 5 nm and its ratio with near-infrared reflectance continues to appear optimal for early detection of chlorosis and plant stress. From a ground-based platform, R_{694} and R_{694}/R_{760} detected incipient chlorosis in pine canopies more than 2 weeks prior to visual symptoms. The same narrow-band approach can be used in light aircraft to detect stress at larger scales (Fig. 5). This narrow-band imaging technique thus should be quite useful in field studies of plant responses to the environment and in plant stress detection for a broad variety of applications. The Space Remote Sensing Center, Institute for Technology Development at Stennis Space Center has recently used R_{694}/R_{840} from light aircraft to detect previsually the stress in *Poa* spp. caused by soil compaction, insufficient irrigation and a bacterial wilt on golf courses (Mr. Tim Gress, personal communication). The use of this technique for early detection of forest damage caused by the Southern Pine Beetle, currently an epidemic in the southeastern USA, is also being explored. In the potential application of this technique from high altitude aircraft and spaceborne platforms, a near-infrared band located between 770 and 800 nm (Wyszecki and Stiles, 1982) would avoid the strong atmospheric O_2 absorption band present at 760 nm.

Acknowledgements

The authors thank Paulette Baham, Toby Feibelman and Virginia Miller for assistance in the field, Larry Lott and the U.S. Forest Service crew at the Harrison Experimental Forest for field assistance and technical support, and Mike Link of du Pont for consultations regarding herbicide applications. This research was supported by the Stennis Space Center Director's Discretionary Fund.

References

ADAMS, W. W., B. DEMMIG-ADAMS, K. WINTER, and U. SCHREIBER: The ratio of variable to maximum chlorophyll fluorescence from photosystem II, measured in leaves at ambient temperatures and at 77K, as an indicator of the photon yield of photosynthesis. Planta *180,* 166–174 (1990).

ARP, G. K. and D. E. PHINNEY: Ecological variations in thermal infrared emissivity of vegetation. Environmental and Experimental Botany *20,* 135–148 (1980).

BERLINER, P., D. M. OOSTERHUIS, and G. C. GREEN: Evaluation of the infrared thermometer as a crop stress detector. Agricultural and Forest Meteorology *31,* 219–230 (1984).

BUSCHMANN, C. and E. NAGEL: *In vivo* spectroscopy and internal optics of leaves as basis for remote sensing of vegetation. Int. J. Remote Sensing *14,* 711–722 (1993).

BUSCHMANN, C., U. RINDERLE, and H. K. LICHTENTHALER: Detection of stress in coniferous forest trees with the VIRAF spectrometer. IEEE Trans. Geosci. Remote Sensing *29,* 96–100 (1991).

CARTER, G. A.: Primary and secondary effects of water content on the spectral reflectance of leaves. Am. J. Bot. *78,* 916–924 (1991).

– Responses of leaf spectral reflectance to plant stress. Am. J. Bot. *80,* 239–243 (1993).

– Ratios of leaf reflectances in narrow wavebands as indicators of plant stress. Int. J. Remote Sensing *15,* 697–703 (1994).

CARTER, G. A. and R. L. MILLER: Early detection of plant stress by digital imaging within narrow stress-sensitive wavebands. Remote Sens. Environ. *50,* 295–302 (1994).

CARTER, G. A., J. REBBECK, and K. E. PERCY: Leaf optical properties in *Liriodendron tulipifera* and *Pinus strobus* as influenced by increased atmospheric ozone and carbon dioxide. Can. J. For. Res. *25,* 407–412 (1995).

CHAPPELLE, E. W., M. S. KIM, and J. E. MCMURTREY: Ratio analysis of reflectance spectra (RARS): an algorithm for the remote estimation of the concentrations of chlorophyll *a,* chlorophyll *b,* and carotenoids in soybean leaves. Remote Sens. Environ. *39,* 239–247 (1992).

CIBULA, W. G. and G. A. CARTER: Identification of a far-red reflectance response to ectomycorrhizae in slash pine. Int. J. Remote Sensing *13,* 925–932 (1992).

CURRAN, P. J., J. L. DUNGAN, and H. L. GHOLTZ: Exploring the relationship between reflectance red edge and chlorophyll content in slash pine. Tree Physiol. *7,* 33–48 (1990).

DEMETRIADES-SHAH, T. H., M. D. STEVEN, and J. A. CLARK: High resolution derivative spectra in remote sensing. Remote Sens. Environ. *33,* 55–64 (1990).

FARQUHAR, G. D. and T. D. SHARKEY: Stomatal conductance and photosynthesis. Annu. Rev. Plant Physiol. *33,* 317–345 (1982).

FILELLA, I. and J. PEÑUELAS: The red edge position and shape as indicators of plant chlorophyll content, biomass and hydric status. Int. J. Remote Sensing *15,* 1459–1470 (1994).

GATES, D. M.: Biophysical ecology. Springer-Verlag, New York (1980).

GATES, D. M., H. J. KEEGAN, J. C. SCHLETER, and V. R. WEIDNER: Spectral properties of plants. Applied Optics *4,* 11–20 (1965 a).

GATES, D. M., E. C. TIBBALS, and F. KREITH: Radiation and convection for Ponderosa pine. Am. J. Bot. *52,* 66–71 (1965 b).

GITELSON, A. and M. N. MERZLYAK: Spectral reflectance changes associated with autumn senescence of *Aesculus hippocastanum* L. and *Acer platanoides* L. leaves. Spectral features and relation to chlorophyll estimation. J. Plant Physiol. *143,* 286–292 (1994).

HENDRY, G. A. F., J. D. HOUGHTON, and S. B. BROWN: Tansley Review No. 11: The degradation of chlorophyll – a biological enigma. New Phytol. *107,* 255–302 (1987).

HOFF, A. J. and J. AMESZ: Visible absorption spectroscopy of chlorophylls. In: SCHEER, H. (ed.): Chlorophylls, p. 726. CRC Press, Boca Raton, FL, USA (1991).

HOQUE, E. and P. J. S. HUTZLER: Spectral blue-shift of red edge monitors damage class of beech trees. Remote Sens. Environ. *39,* 81–84 (1992).

HORLER, D. N. H., M. DOCKRAY, and J. BARBER: The red edge of plant leaf reflectance. Int. J. Remote Sensing *4,* 273–288 (1983).

LANGE, O. L., H. ZELLNER, J. GEBEL, P. SCHRAMEL, B. KÖSTNER, and F.-C. CZYGAN: Photosynthetic capacity, chloroplast pigments, and mineral content of the previous year's spruce needles with and without the new flush: analysis of the forest-decline phenomenon of needle bleaching. Oecologia (Berlin) *73,* 351–357 (1987).

LEUNING, R. and P. SANDS: Theory and practice of a portable photosynthesis instrument. Plant Cell Environ. *12,* 669–678 (1989).

LICHTENTHALER, H. K. and U. RINDERLE: The role of chlorophyll fluorescence in the detection of stress conditions in plants. CRC Crit. Rev. Anal. Chem. *19,* S 29–S 85 (1988).

LICHTENTHALER, H. K., F. STOBER, and M. LANG: The nature of different laser-induced fluorescence signatures of plants. EARSeL Advances in Remote Sensing *1,* 20–32 (1992).

LUVALL, J. C. and H. R. HOLBO: Measurements of short-term thermal responses of coniferous forest canopies using thermal scanner data. Remote Sens. Environ. *27,* 1–10 (1989).

MILLER, R. L.: High resolution image processing on low-cost microcomputers. Int. J. Remote Sensing *14,* 655–667 (1993).

Mottram, R., J. M. DeJager, and J. R. Duckworth: Evaluation of a water stress index for maize using an infra-red thermometer. Crop Production *12,* 26–28 (1983).

Moya, I., G. Guyot, and Y. Goulas: Remotely sensed blue and red fluorescence emission for monitoring vegetation. ISPRS Journal of Photogrammetry and Remote Sensing *47,* 205–231 (1992).

Munden, R., P. J. Curran, and J. A. Catt: The relationship between red edge and chlorophyll concentration in the Broadbalk winter wheat experiment at Rothamsted. Int. J. Remote Sensing *15,* 705–709 (1994).

Nobel, P. S.: Physicochemical and Environmental Plant Physiology. Academic Press, San Diego (1991).

Pinter, P. J., M. E. Stanghellini, R. J. Reginato, S. B. Idso, A. D. Jenkins, and R. D. Jackson: Remote detection of biological stresses in plants with infrared thermometry. Science *205,* 585–587 (1979).

Rock, B. N., T. Hoshizaki, and J. R. Miller: Comparison of *in situ* and airborne spectral measurements of the blue shift associated with forest decline. Remote Sens. Environ. *24,* 109–127 (1988).

Rosema, A., G. Cecchi, L. Pantani, B. Radicatti, M. Romuli, P. Mazzinghi, O. Van Kooten, and C. Kliffen: Monitoring photosynthetic activity and ozone stress by laser induced fluorescence in trees. Int. J. Remote Sensing *13,* 737–751 (1992).

Ruth, B., E. Hoque, B. Weisel, and P. J. S. Hutzler: Reflectance and fluorescence parameters of Norway spruce affected by forest decline. Remote Sens. Environ. *38,* 35–44 (1991).

Saxena, A. K., V. K. Tewari, H. B. Tripathi, and J. S. Singh: Spectro-reflectance characteristics of certain plants of the Kumaun Himalaya and relationship of pigment concentration with leaf reflectance. Proc. Indian Natn. Sci. Acad. *B 51,* 223–234 (1985).

Schmuck, G.: Applications of *in vivo* chlorophyll fluorescence in forest decline research. Int. J. Remote Sensing *11,* 1165–1177 (1990).

Smith, W. K. and G. A. Carter: Shoot structural effects on needle temperatures and photosynthesis in conifers. Am. J. Bot. *75,* 496–500 (1988).

Steel, R. G. D. and J. H. Torrie: Principles and Procedures of Statistics. McGraw-Hill, New York (1961).

Thomas, J. R. and H. W. Gausman: Leaf reflectance vs. leaf chlorophyll and carotenoid concentrations for eight crops. Agron. J. *69,* 799–802 (1977).

Tibbals, E. C., E. K. Carr, D. M. Gates, and F. Kreith: Radiation and convection in conifers. Am. J. Bot. *51,* 529–538 (1964).

Tsay, M. L., D. H. Gjerstad, and G. R. Glover: Tree leaf reflectance: a promising technique to rapidly determine nitrogen and chlorophyll content. Can. J. For. Res. *12,* 788–792 (1982).

Vogelmann, J. E., B. N. Rock, and D. M. Moss: Red edge spectral measurements from sugar maple leaves. Int. J. Remote Sensing *14,* 1563–1575 (1993).

Waring, R. H., J. D. Aber, J. M. Melillo, and B. Moore III: Precursors of change in terrestrial ecosystems. Bioscience *36,* 433–438 (1986).

Wyszecki, G. and W. S. Stiles: Color Science: Concepts and Methods, Quantitative Data and Formulae. John Wiley and Sons, New York (1982).

J. Plant Physiol. Vol. 148. pp. 523–529 (1996)

Transmittance and Reflectance Measurements of Corn Leaves from Plants with Different Nitrogen and Water Supply

J. S. Schepers[1], T. M. Blackmer[2], W. W. Wilhelm[3], and M. Resende[4]

[1] Soil Scientist, USDA-ARS, University of Nebraska Lincoln, NE, 68583-0915

[2] Research Associate, USDA-ARS, University of Nebraska, Lincoln, NE 68583-0915

[3] Plant Physiologist, USDA-ARS, University of Nebraska, Lincoln, NE 68583-0915

[4] Soil Scientist, EMBRAPA, CNP Milho e Sorgo, Sete Lagoas, MG-Brazil

Received July 20, 1995 · Accepted October 30, 1995

Summary

Nitrogen is essential for crop production, but also contributes to eutrophication of surface water and degradation of drinking water quality. Modern corn production requires relatively large quantities of N, which are generally supplied by fertilizers. Over-application of N fertilizers and animal wastes frequently results in nitrate leaching. Synchronizing N availability with crop N need offers the potential to protect the environment without sacrificing production. Tools are needed to rapidly and easily monitor crop N status to make timely decisions regarding fertilizer application. Analytical and optical techniques were evaluated with greenhouse grown corn at silking to evaluate several methods to monitor crop N status. A portable chlorophyll meter was used to measure chlorophyll content of leaves by means of transmittance measurements. Leaf N concentration and chlorophyll meter readings were positively correlated, but were also affected by water stress and hybrid differences. Water stress decreased chlorophyll meter readings but increased leaf N content and diffusive resistance. Nitrogen stress decreased leaf N concentration, chlorophyll meter readings, and diffusive resistance. Both water and N stresses affected crop reflectance measurements. Reflectance values in the green and near IR portions of the spectrum were inversely related to crop N status. Water stress increased reflectance in red, green, and near IR wavelengths. Water stress by N status interactions were significant for chlorophyll meter readings as well as reflectance measurements. Both leaf reflectance and chlorophyll meter measurements provided a good indication of N status for adequately watered plants, but the relationships were poor for plants grown under prolonged water stress.

Key words: Maize, transmittance, reflectance, nitrogen fertilizer, water stress.

Abbreviations: N = nitrogen; C = carbon; W = water stress; VT = stage of corn growth after complete tassel emergence but before silk emergence.

Introduction

Nitrogen (N) fertilizer management in modern corn production is an important area of applied research that has both economic and environmental consequences. Tissue testing is one tool that can aid in fertilizer management by identifying crop requirements for additional N. Measuring leaf optical properties (i.e., reflectance or transmittance) of radiant energy can be considered a form of tissue testing.

Variation in plant pigmentation can be caused by many factors, but nutrient stress is generally a primary consideration. Maas and Dunlap (1989) used reflectance measurements to characterize pigmentation differences between leaves of untreated corn plants, leaves of fluridone-induced albino plants, and leaves from etiolated plants (various lengths of darkness) that had different chlorophyll concentrations. In another study, Al-Abbas et al. (1974) used reflectance measurements to demonstrate that deficiencies of N, P, K, Mg, S, and Zn in corn leaves affected leaf pigmentation and chlorophyll concentrations. Other research has shown that leaf reflectance measured at 550 nm from sweet pepper (*Capsicum annum* L.) leaves had a strong relationship with N concentration (Thomas and Oerther, 1972). They concluded leaf reflectance at 550 nm was a good indicator of leaf chlorophyll content. More recently, others have also shown reflectance measurements can be a good indicator of N stress in corn leaves (McMurtrey et al., 1994; Blackmer et al., 1994).

Chlorophyll meters have been shown to be effective at detecting N stress in corn (*Zea mays* L.) leaves (Schepers et al., 1992 a; Wood et al., 1992; Blackmer et al., 1994). Although transmittance measurements with the Minolta[1] SPAD-502 chlorophyll meter are rapid and easy to make, they represent only a very small portion of a leaf. The closed chamber created when the chlorophyll meter is clamped onto the leaf eliminates interferences by external light sources. On the other hand, an integrating sphere is essentially a larger and more versatile type of chlorophyll meter (also closed system with internal lighting). An integrating sphere not only measures leaf transmittance, but it also measures reflectance at many wavelengths. A major difference between the two instruments is that the Minolta chlorophyll meter measures specific wavelengths (centered at 650 and 940 nm) while the integrating sphere can measure many wavelengths in small increments.

Reflectance measurements can also be made beyond the single leaf confines of an integrating sphere. When canopy reflectance measurements are made, external or natural light sources are used. Variation in external lighting requires special calibration procedures. The advantage of reflectance measurements is that when made from above the canopy, they represent a large area relative to a single leaf. Reflectance measurements made near the crop canopy integrate plant-to-plant variation, while those taken a greater distance from the crop will integrate a larger area. Depending on the field of view of the instrument, measurements taken well above the canopy should make it possible to identify atypical areas in a field.

Leaf pigments (e.g., carotenoids and anthocyanins) absorb various amounts of light in the visible range of the spectrum. These leaf characteristics influence the reflectance signature of crops. Reflection of visible light (400–700 nm) from vegetative tissue, like a corn leaf, is least at wavelengths where chlorophyll absorption is greatest. This phenomena results in characteristically greater reflectance around 550 nm and lower reflectance around 450 and 650 nm. In combination, these absorption/reflection characteristics result in relatively large differences in light reflection compared to relatively small fluctuations in chlorophyll concentrations. It follows that leaf reflectance in the visible portion of the spectrum is indicative of chlorophyll concentration (Benedict and Swidler, 1961; Sinclair et al., 1971) and carotenoid concentration (Thomas and Gausman, 1977).

Water stress can increase reflectance from corn leaves in both the visible and near infrared portions of the spectrum (Wooley, 1971). Previous research has shown that N concentration and chlorophyll content are affected by both N and water stress (Wolfe et al., 1988). They found leaf chlorophyll concentration in corn showed a water stress by N stress interaction, however, leaf N concentration was not affected by the interaction. This is probably because leaf N concentration is related to the N uptake process whereas chlorophyll content is predominantly a metabolic parameter.

The objective of this study was to evaluate the effect of water stress on monitoring crop N status by tissue and optical methods. The approach was to utilize data from a variety of analytical procedures in an attempt to better characterize the interactions between N and water stresses.

Materials and Methods

This greenhouse study was established on 2 August 1993. Three replications of Pioneer brand hybrids[1] 3398 and 3379 were planted into pots 30-cm diameter by 30-cm depth. Artificial lighting was provided to extend the day length. Three levels of N (low, near adequate, and somewhat excessive) were established prior to planting by applying ammonium nitrate and lightly watering into the potting media which consisted of equal parts of sand, soil, peat moss, and vermiculite. After germination, seedlings were thinned to four per pot. Plants in all pots were maintained in a well watered state (minimum of 50 % of water holding capacity) until the V6 growth stage (Ritchie et al., 1992), when two water regimes were established. Plants in one treatment were well watered while plants in the second treatment were allowed to become stressed. Water stress was imposed by reducing the watering frequency. Diffusive resistance was measured to qualitatively determine watering frequency.

At the VT growth stage (21 September), water and N-related measurements were made on the ear leaf of two of the original four plants in each pot. For N-related measurements, thirty chlorophyll meter readings were taken with a Minolta SPAD 502 chlorophyll meter from midway along the length of the ear leaf and midway between the margin and midrib of two plants (Peterson et al., 1993). The same position on these leaves was used to make spectral reflectance measurements using a Li-Cor[1] integrating sphere with a halogen white light source. A $BaSO_4$ reference was used to calibrate all reflectance measurements. Each spectral scan was composed of reflectance from 512 bands between 348 and 1070-nm wavelengths in 1.4-nm increments. Approximately 350 ms were used to complete each spectral scan. Data from selected wavelengths (550, 650, 710, 850, and 940 nm) were used for statistical analysis.

Water related measurements included diffusive resistance, made on the ear leaf using a Li-Cor[1] model 1600 steady state porometer. After chlorophyll meter, reflectance and diffusive resistance measurements were made, the same area of the same leaves was semi-destructively sampled using a 1-cm dia. leaf punch. A total of 40 disks were collected between the margin and midrib from both plants in each pot. Leaf disks were dried, weighed, and finely ground (200 mesh) prior to analysis for C, total N, and ^{15}N using a Carlo Erba NA1500 CNS analyzer coupled to a Europa tracermass mass spectrometer (Schepers et al., 1989). Even though ^{15}N fertilizer was not applied in this study, the analyzer system automatically provided isotopic data.

Analysis of variance was performed on all data using the general linear model procedures of SuperANOVA software (Abacus Concepts, Inc., 1989).

[1] Mention of brand names does not indicate an endorsement by the USDA-ARS.

Results and Discussion

Statistical analysis indicated that leaf N concentration was affected by hybrid, N rate, and water stress (Table 1). These findings are consistent with trends commonly noted in the literature (Schepers et al., 1992 b). Chlorophyll meter readings are typically affected by the same factors that affect leaf N concentration. Differences in chlorophyll meter readings between hybrids were significant at P = 0.055. Both N and water stresses reduced chlorophyll meter readings. Water stress had a considerable effect on chlorophyll meter readings when adequate N was available, but had little effect when N was limiting plant growth (Fig. 1).

Earlier studies (Blackmer et al., 1994; Schepers et al., 1992 a) did not address the influence of water stress on chlorophyll meter readings although there was a strong positive correlation between leaf N concentration and chlorophyll meter readings. Data reported here resulted in only a weak to moderate correlation (r=0.65) between leaf N concentration and chlorophyll meter readings when combined across two different water levels. The relationship between leaf N concentration and chlorophyll meter readings improved significantly by differentiating between levels of crop water status (Fig. 1). Adequate water increased both the chlorophyll meter reading response to N fertilizer and the correlation between meter readings and leaf N concentration.

Table 1: Means and statistical significance of fertilizer N rate, water stress, and cultivar on leaf N, C, and ^{15}N content, chlorophyll meter readings, and diffuse resistance.

Variable	Nitrogen (mg/g)	Carbon (mg/g)	^{15}N atom %	Chlorophyll meter (reading)	Diffuse resistance (s/cm)
Hybrid					
3379	26.7	426	0.378	42.9	348
3398	29.6	419	0.375	44.0	55
N rate					
low	19.6	410	0.384	37.0	9
medium	30.9	426	0.374	44.5	209
high	34.0	431	0.372	48.9	386
Water status					
stressed	29.2	422	0.376	41.1	397
nonstressed	27.1	423	0.377	45.7	6
Probability					
Hybrid	**	*	**	ns	ns
N rate	**	**	**	**	ns
Water stress (W)	**	ns	ns	**	*
Hybrid × N rate	ns	ns	*	ns	ns
N rate × W	ns	ns	ns	**	ns
Hybrid × W	**	ns	ns	ns	ns
Hybrid × N rate × W	ns	ns	*	ns	ns

*, ** Significant at the 0.05 and 0.01 levels, significantly.

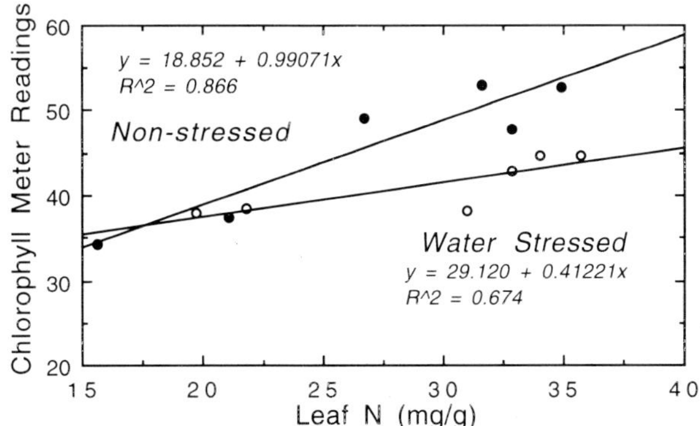

Fig. 1: Relationship between leaf N concentration and chlorophyll meter readings of water stressed and adequately watered corn.

Even though chlorophyll meter readings responded similarly to N for both hybrids in this study, other studies have indicated hybrids may have different chlorophyll meter readings when fertilized at the same N rate (Schepers et al., 1992 b). In this study, at the lowest fertilizer N level Pioneer brand hybrid 3398 contained 22 % greater leaf N concentration than 3379 across both water regimes, but showed only a 2 % difference at the highest N rate (Table 1). The significant hybrid by water stress interaction occurred because leaf N concentration for 3398 was not affected by water stress, while that for 3379 showed an 18 % increase in leaf N concentration with the water stress.

Several scenarios can be proposed to support the above trends in leaf N concentration. First, hybrid 3398 may be more efficient in terms of N uptake than 3379 resulting in greater leaf N concentrations. An alternative is that 3379 might be able to maintain growth at a lower leaf N concentration than 3398 and may partition a greater proportion of N to metabolite production. It is also possible that 3379 is more subject to volatile N losses from its tissue than 3398 or that 3398 is more effective at adsorbing ammonia from the atmosphere than 3379 (Francis et al., 1993). Some support for the volatilization scenario is provided by isotopic N data showing a greater ^{15}N enrichment (i.e., above normal abundance levels) for 3379 than for 3398 (Table 1). For this scenario to be feasible and for 3379 to have a lower leaf N concentration than 3398, it would probably require fractionation (i.e., preference for ^{14}N vs. ^{15}N or vise-versa) of N during volatilization. Fractionation also occurs when using the microdiffusion process to prepare samples for isotopic analysis (Hauck, 1982). As a result, in this case, a great proportion of ^{14}N would be volatilized, thus leaving proportionately more ^{15}N in the plant. This scenario could also account for lower leaf N concentrations for 3379 than for 3398.

Occurrence and fate of N in the environment also involves ammonia absorption from the atmosphere by vegetation, which in some ways can be thought of as the reverse of N volatilization from leaves. Plants under the greatest N stress would have the greatest likelihood of absorbing atmospheric ammonia. If for some reason, the atmosphere was enriched with ^{15}N ammonia or fractionation occurs during absorption, then the most N deprived plants might be expected to become enriched proportionately more than plants having adequate N availability.

Discussion of ^{15}N abundance in the plant tissue is confounded because microbial processes tend to concentrate ^{15}N in soil and manures through fractionation. Increased ^{15}N abundance in leaf tissue can be explained by the fact that commercial N fertilizers tend to have a normal abundance of

^{15}N (i.e., 0.367 atom % ^{15}N). In this case, a shortage of N fertilizer would result in proportionately greater uptake of residual soil N, which could be enriched if the soil contained peat or had a history of manure application. This hypothesis is supported by the strong inverse correlation (r = 0.92) between leaf N concentration and atom % ^{15}N in the leaf. Further study would be required to characterize these processes.

The C content of plant tissue (Table 1) is typically about 400 mg/g, plus or minus 10 %. Data from this study fall within this range, but showed a significant N rate and hybrid effect. No explanation is provided as to why leaf C concentration decreased with N availability other than perhaps it is related to a slight dilution of C content caused by enhanced N uptake or metabolite storage in the leaves. At the time of sampling, the position of the ear leaf was easy to identify and pollination was in progress, but it is unlikely that translocation of metabolites had occurred to a significant extent. This hypothesis is supported by leaf thickness calculations made from leaf disk weight and surface area data, assuming a constant density, which showed a lower leaf thickness for N stressed plants.

Diffusive resistance measurements generally showed more variability than the other measurements. Nonetheless, plant water stress resulted in greater diffusive resistance, which affects exchange of water vapor and gases with the atmosphere, thereby affecting plant metabolism and growth.

Hybrid differences in this study did not affect any of the reflectance values (Table 2). In contrast, nearly every value (i.e., specific wavelengths) was affected by N availability and water status. Reflectance increased with water stress for all wavelengths shown in Table 2. This finding was expected because of the effect water stress has on diffusive resistance and plant metabolism in general. It should be noted that reflectance at 850 and 940 nm decreased with N stress, but concurrently increased at 550 and 710 nm and showed mixed results at 650 nm. These same measurements all showed an N stress by water stress interaction, which prompted a closer examination of the data than is evident from the mean values provided in Table 2. Water stress on plants with adequate N increased reflectance at 550 nm but had little affect under N stressed conditions (Fig. 2). Reflectance at 650 nm was not affected by N status as long as the plants received adequate water, but water stress increased reflectance of plants with adequate N (Fig. 2). Implications of these interactions are that interpretation of leaf reflectance data to evaluate crop N status is likely to be confounded by crop water status.

Since one of the goals of this research was to identify reflectance signatures (i.e., key wavelengths) that could provide similar information as the leaf transmittance measurements collected with the Minolta chlorophyll meter, reflectance data were compared with chlorophyll meter readings. Correlations between meter readings and individual wavelengths (Table 3) were the strongest for 550 nm (green color) and 710 nm (red edge area). Correlations decreased between these two wavelengths where corn plants were absorbing the highest proportion of light in the 650 nm portion of the spectrum (red light) (Table 2).

The lack of any meaningful relationship between chlorophyll meter readings and reflectance at either 850 or 940 nm (near infrared wavelengths) is typical because plants do not

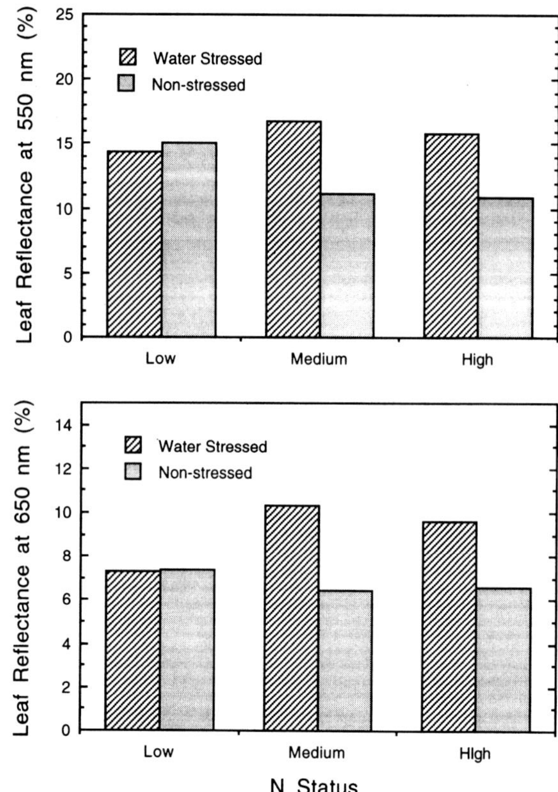

Fig. 2: Ear-leaf reflectance at 550- and 650-nm wavelengths of corn grown at three N levels and under water-stressed and control conditions

Table 2: Mean values and statistical significance of fertilizer N rate, water stress and cultivar on crop reflectance measurements.

Factor	Wavelength (nm)						
	550 (%)	650 (%)	710 (%)	850 (%)	940 (%)	550/850	710/850
Hybrid							
3379	14.2	7.79	18.8	40.8	40.5	0.349	0.462
3398	13.8	8.08	18.1	41.0	40.6	0.336	0.439
N rate							
low	14.8	7.32	19.4	39.6	39.2	0.373	0.490
medium	14.0	8.39	18.5	41.2	40.8	0.337	0.447
high	13.4	8.09	17.5	41.9	41.6	0.318	0.415
Water status							
stressed	15.7	9.06	20.3	42.0	41.4	0.373	0.483
nonstressed	12.4	6.80	16.6	39.8	39.7	0.312	0.418
Probability							
Hybrid	ns	ns	ns	ns	ns	ns	ns
N rate	ns	**	*	**	*	**	**
Water stress	**	**	**	**	*	**	**
Hybrid × N rate	ns	ns	ns	*	ns	ns	ns
N rate × stress	**	**	**	**	*	**	**
Hybrid × stress	ns	ns	ns	ns	ns	ns	ns
Hybrid × N rate × stress	ns	ns	ns	ns	ns	ns	ns

*, ** Significant at the 0.05 and 0.01 levels, significantly.

Table 3: Correlation between chlorophyll meter readings of maize and ear leaf reflectance at several wavelengths.

Wavelength (nm)	Correlation coefficient (r)
550	0.77
650	0.36
710	0.78
850	<0.03
940	0.05
550/850	0.86
650/850	0.46
710/850	0.89
550/940	0.83
650/940	0.46
710/940	0.84

use this type of light (Fig. 3). For this reason, the near infrared portion of the spectrum is frequently used as a reference to normalize data collected at other wavelengths. In essence, infrared wavelengths tend to respond to factors other than those related to photosynthesis. Reflectance in the near infrared portion of the spectrum is influenced by the frequency of intercellular air spaces of leaf tissue, which is influenced by crop water status (Gausman et al., 1974). The effect of water stress on near infrared reflectance is illustrated in Figure 3. Thus, by referencing wavelengths that are responsive to photosynthetic activity and perhaps other unknown factors (i.e., 550, 650, and 710 nm) to nonresponsive wavelengths (i.e., 850 and 940 nm), the resulting ratio should improve the sensitivity of the reflectance measurement. Others have also used this technique to standardize their data (Takebe et al., 1990; Walburg et al., 1982). Correlations between both 550 and 710 nm and chlorophyll meter readings were improved by normalizing with data collected at either 850 or 940 nm (Table 3). The 850-nm reference provided a slightly better correlation with the other wavelengths than did the 940-nm wavelength. This observation could be an artifact because the instrumentation becomes insensitive at about 1050 nm.

Normalized data using the 550/850 nm wavelengths has the advantage over other ratios because the reflectance patterns are broad at these wavelengths (Fig. 3). Data for this reflectance ratio exhibited a highly significant N by water stress interaction (Table 2). Regression analysis showed a strong inverse relationship between leaf N concentration and the 550/850 reflectance ratio for the adequately watered plants (Fig. 4). This is because reflectance values at 550 nm

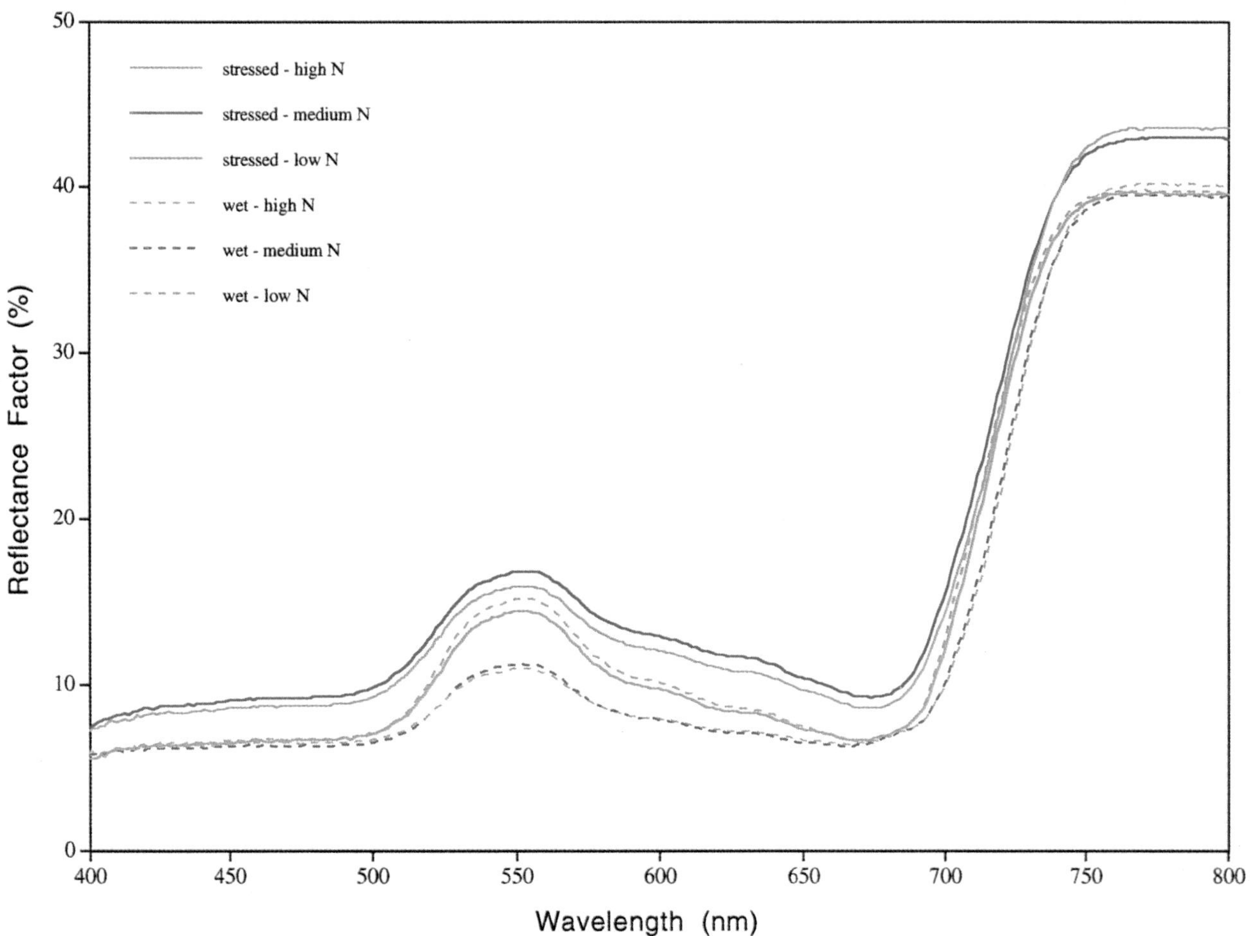

Fig. 3: Reflectance of corn ear-leaves at various wavelengths as affected by N level and water status.

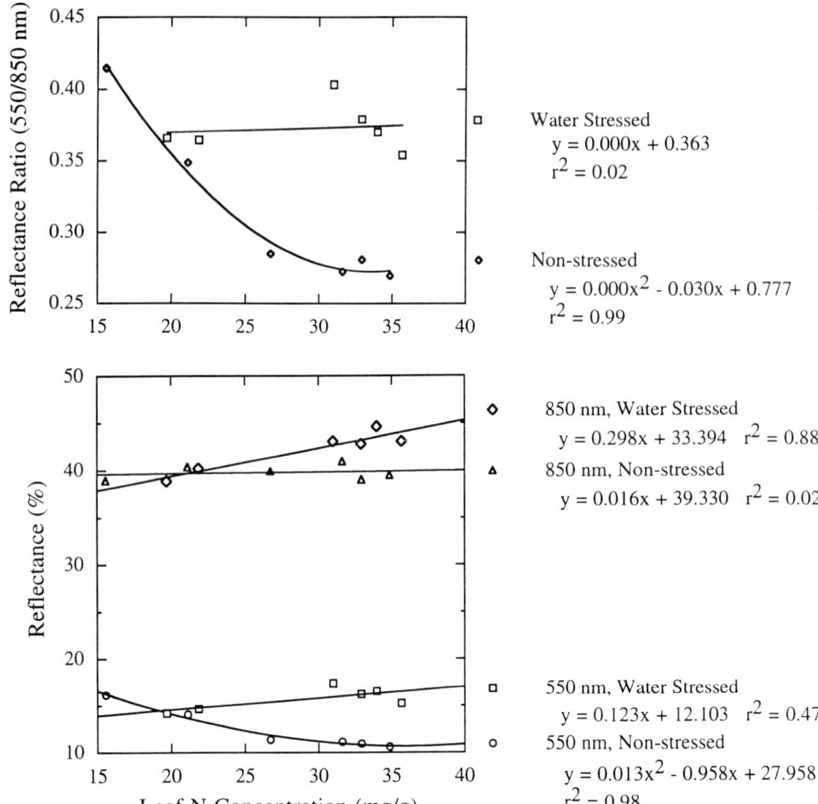

Fig. 4: Ear-leaf reflectance at 550 and 850 nm and the ratio of 550/850 nm values for corn grown under water-stressed and control conditions as related to leaf N concentration.

were highly correlated with leaf N concentration and values at 850 nm were unaffected by leaf N concentration. Water stress rendered reflectance values at 550 nm relatively insensitive to leaf N concentration while the reference values at 850 nm became sensitive to leaf N concentration. These findings raise serious concerns about the reliability of using reflectance measurements from water stressed plants to characterize crop N status. This conclusion may be an artifact of the experimental procedure because the plants used in the study had been under prolonged water stress at the time of measurement. Under more natural field conditions, plants would gradually change water content and would not likely remain under water stress for a prolonged period.

Values for the 550/850 nm reflectance ratio for adequately watered plants showed a strong inverse curvilinear relationship ($r^2 = 0.97$) with chlorophyll meter readings (Fig. 5). In contrast, the similar comparison for water stressed plants was very weak ($r^2 = 0.15$). Questions remain regarding the effect of water stress on chlorophyll meter readings, but the effect of intercellular air spaces must be considered, as noted above. The observation that plants with high leaf N content had a relatively small effect on the value of the 550/850 nm ratio compared to more N deficient plants (Fig. 4) is attributed to a situation where other nutrients or growth factors limit leaf chlorophyll content at high leaf N levels. This effect carries over to the relationship between chlorophyll meter readings and the 550/850 reflectance ratio (Fig. 5).

The 940-nm wavelength was included in this study because the Minolta chlorophyll meter uses this wavelength for

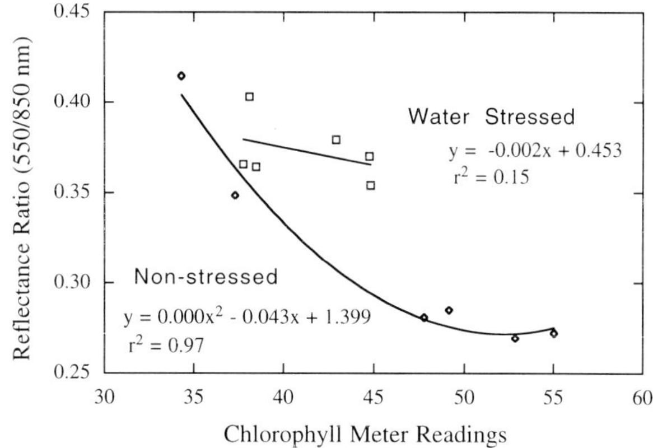

Fig. 5: Ear-leaf reflectance ratio of 550/850 nm values for corn grown under water-stressed and control conditions as related to chlorophyll meter readings.

calibration and normalizing light transmittance at 650 nm. Details of signal handling at these wavelengths within the chlorophyll meter are not published. Data from this study showed a poor relationship between chlorophyll meter readings and the 650/940 ratio ($r = 0.44$). Perhaps this rather poor relationship can be explained because both of these wavelengths were measured as reflected light in our study rather than using transmitted light as with the Minolta chlorophyll meter. Another contributing factor may be that the Minolta meter measures the amount of light transmitted through a 2 by 3 mm segment of the leaf. The rest of the light is either absorbed by the leaf or reflected into the closed chamber around the leaf.

In practical terms, the shape of a reflectance trace as a function of wavelength needs to be considered when evaluating the feasibility of developing a simple and inexpensive sensor system. The peaks at 550 and 650 nm are quite broad and the plateau in the near infrared range (i.e., 850 and 940 nm) make these wavelengths good candidates for further consideration (Fig. 2). Even though reflectance at 710 nm is quite specific for photosynthesis, the narrowness of the band may make it difficult to develop a sensor that would be accurate as well as reliable.

Conclusions

Common laboratory procedures that quantify crop N status were able to detect differences in soil N availability of corn leaves at silking as they responded to water stress. A hand-held chlorophyll meter provided similar information relative to crop N status. Chlorophyll meters make gathering crop N status information faster and easier than traditional laboratory techniques. Reflectance measurements offer the potential to evaluate the N status of a single leaf by using an integrating sphere or an entire crop canopy using a spectroradiometer. Canopy type measurements have many advantages compared to single leaf measurements afforded by the Minolta chlorophyll meter. Even though reflectance measurements taken in this study involved a closed chamber instrument (i.e., integrating sphere) that would be impractical and expensive to use under commercial applications, the results indicate that the concept of measuring leaf reflectance to evaluate crop N status is sound. Normalizing reflectance data at 550 nm to the 850 nm wavelength (i.e., 550/850 ratio), should provide a reasonable measure of crop N status over a range of water regimes.

References

Abacus Concepts, Inc.: SuperANOVA Users Manual. Berkeley, California 94704 (1989).

Al-Abbas, A. H., R. Barr, J. D. Hall, F. L. Crane, and M. F. Baumgardner: Spectra of normal and nutrient-deficient corn leaves. Agron. J. *66*, 16–20 (1974).

Benedict, H. M. and R. Swidler: Nondestructive methods for estimating chlorophyll content of leaves. Science *133*, 2015–2016 (1961).

Blackmer, T. M. and J. S. Schepers: Use of a chlorophyll meter to monitor N status and schedule fertigation of corn. J. Prod. Agric. *8*, 56–60 (1995).

Blackmer, T. M., J. S. Schepers, and G. E. Varvel: Light reflectance compared to other N stress measurements in corn leaves. Agron. J. *86*, 934–938 (1994).

Chappelle, E. W., M. S. Kim, and J. E. McMurtrey: Ratio analysis of reflectance spectra (RARS): an algorithm for the remote estimation of the concentrations of chlorophyll *a*, chlorophyll *b*, and carotenoids in soybean leaves. Remote Sens. Environ. *39*, 239–247 (1992).

Francis, D. D., J. S. Schepers, and M. F. Vigil: Post-anthesis nitrogen loss from corn plants. Agron. J. *85*, 659–663 (1993).

Gausman, H. W., W. A. Allen, and D. E. Escobar: Refractive index of plant cell walls. Applied Optics *13*, 109–111 (1974).

Hauck, R. D.: Nitrogen – Isotopic-Ratio Analysis. In: Page, A. L. (ed.): Methods of Soil Analysis. Number 9, Part 2, pp. 735–779. Am. Soc. Agron., Madison, Wisconsin (1982).

Maas, S. J. and J. R. Dunlap: Reflectance, transmittance, and absorptance of light by normal, etiolated, and albino corn leaves. Agron. J. *81*, 105–110 (1989).

McMurtrey, J. E., III, E. W. Chappelle, M. S. Kim, J. J. Meisinger, and L. A. Corp: Distinguishing nitrogen fertilization levels in field corn [*Zea mays* L.] with actively induced fluorescence and passive reflectance measurements. Remote Sens. Environ. *47*, 36–44 (1994).

Peterson, T. A., T. M. Blackmer, D. D. Francis, and J. S. Schepers: Using a chlorophyll meter to improve N management. Nebguide G93-1171A. Coop. Ext. Serv., Univ. of Nebraska, Lincoln (1993).

Ritchie, S. E., J. J. Hanway, and G. O. Benson: How a corn plant develops. Iowa Cooperative Extension Service Special Report 48, Iowa State University, Ames, Iowa (1992).

Schepers, J. S., D. D. Francis, and M. T. Thompson: Simultaneous determination of total C, total N, and ^{15}N on soil and plant material. Commun. Soil Plant Anal. *20*, 949–960 (1989).

Schepers, J. S., T. M. Blackmer, and D. D. Francis: Predicting N fertilizer needs for corn in humid regions: using chlorophyll meters. In: Bock, B. R. and K. R. Kelley (eds.): Predicting Fertilizer Needs for Corn in Humid Regions. Bull. Y-226. pp. 105–114. National Fertilizer and Environmental Research Center, Tennessee Valley Authority, Muscle Shoals, AL 35660 (1992 a).

Schepers, J. S., D. D. Francis, M. Vigil, and F. E. Below: Comparison of corn leaf nitrogen concentration and chlorophyll meter readings. Comm. Soil Plant Anal. *23*, 2173–2187 (1992 b).

Sinclair, T. R., R. M. Hoffer, and M. M. Schreiber: Reflectance and internal structure of leaves from several crops during a grosing season. Agron. J. *63*, 864–868 (1971).

Takebe, M., T. Yoneyama, K. Inada, and T. Murakam: Spectral reflectance of rice canopy for estimating crop nitrogen status. Plant and Soil. *122*, 295–297 (1990).

Thomas, J. R. and H. W. Gausman: Leaf reflectance vs. leaf chlorophyll and carotenoid concentration for eight crops. Agron. J. *69*, 799–802 (1977).

Thomas, J. R. and G. F. Oerther: Estimating nitrogen content of sweet pepper leaves by reflectance measurements. Agron. J. *64*, 11–13 (1972).

Walburg, G., M. E. Bauer, C. S. T. Daughtry, and T. L. Housley: Effects of nitrogen nutrition on the growth, yield, and reflectance characteristics of corn canopies. Agron. J. *74*, 677–683 (1982).

Wolfe, D. W., D. W. Henderson, T. C. Hsiao, and A. Alvino: Interactive water and nitrogen effects on senescences of corn. II. Photosynthetic decline and longevity of individual leaves. Agron. J. *80*, 865–870 (1988).

Wood, C. W., D. W. Reeves, R. R. Duffield, and K. L. Edmisten: Field chlorophyll measurements for evaluation of corn nitrogen status. J. Plant Nutr. *15*, 487–500 (1992).

Woolley, J. T.: Reflectance and transmittance of light by leaves. Plant Physiol. *47*, 656–662 (1971).

J. Plant Physiol. Vol. 148. pp. 530–535 (1996)

Digital Image Analysis to Estimate Leaf Area

BRENT BAKER[1], DAVID M. OLSZYK[2], and DAVID TINGEY[2]

[1] Ogden Professional Services, and

[2] U.S. Environmental Protection Agency, U.S. EPA National Health and Environmental Effects Research Laboratory, Western Ecology Division, 200 SW 35th St., Corvallis, OR 97333

Received June 24, 1995 · Accepted October 20, 1995

Summary

Digital image processing was evaluated as a nondestructive technique to estimate leaf area of Douglas-fir trees (*Pseudotsuga menziesii*). Four photographs were obtained per tree from side two angles ≈ 90 degrees apart, with separate white and black backgrounds for each angle. Photographs were acquired using a digital camera with computer images (spatial resolution of 1012 by 1524 pixels) recorded directly to a magnetic hard drive, or a standard film camera with photos scanned to produce computer images (spatial resolution 800 by 1200 pixels). Images were separated into red, green, and blue bands with the intensity of the signal digitized into 256 levels (8 bits of accuracy) in each band. Images were processed through a series of operations: choosing of appropriate band for analysis, finding an intensity threshold below which most of the pixels were noise, separating foreground and background pixels, counting the foreground pixels, and converting the number of foreground pixels to a silhouette leaf area (SLA) through comparison with a reference area included in the image. Projected leaf area (PLA) per tree was measured through destructive harvest. Two experiments were carried out: the first to provide an initial function relating SLA to PLA using a small, select, group of trees in pots under diffuse, i.e., scattered, light; and the second using data from trees growing *in situ* to validate the function under field conditions. In the first experiment, r^2 for SLA vs. PLA was high (0.907) and an initial function was calculated. In the second experiment, predicted PLA for individual plants based on image analysis and the initial function, overestimated measured PLA by ≈ 19 %. Thus, a new function for SLA vs. PLA ($r^2 = 0.861$) was calculated based on the larger data set in experiment two. With additional validation, this technique may provide a valuable tool for estimating leaf area nondestructively for studies of environmental stress and vegetation.

Key words: Douglas-fir, Pseudotsuga menziesii, leaf area, digital images.

Abbreviations: PLA = projected leaf area; SLA = silhouette leaf area; SE = standard error.

Disclaimer

The information in this document has been funded by the U.S. Environmental Protection Agency under interagency agreement number GS00K94BHD0107 to the General Services Administration who have a contract with Ogden Government Services. It has been subject to the agency's peer and administrative review. It has been approved for publication as an EPA document. Mention of trade names or commercial products does not constitute endorsement or recommendation for use.

Introduction

Accurate estimation of leaf area is vital for evaluation of plant productivity and physiological processes for a variety of studies on impacts of environmental stress (Kvet and Marshall, 1971). Destructive harvesting of plants followed by sampling of leaves (Evans, 1972) is the most accurate means to measure area, but is not suitable for applications requiring repeated measurements on the same trees or remote measurement of trees. In such cases, the areas of intact leaves can be estimated based on empirically-derived relationships between area and measurements of leaf size and shape (Kvet and Marshall, 1971). However, these measurements can be tedious and time consuming, especially during periods when leaf area is changing rapidly.

Nondestructive measurements using leaf images have the potential for providing reliable estimates of area without harvesting the trees. Leaf images have been obtained through standard film photography and processed using a computer (Lindsey and Bassuk, 1992). Studies have used photographs taken with standard film cameras and computer controlled scanners to digitize the photos to determine leaf area; or used a black and white video camera and monitor to view images and determine area. Images have typically been in black and white for high resolution between leaf and background material. Top-down images have been obtained to estimate ground surface area covered by leaves (Thomas et al., 1988), and side images have been obtained to estimate total leaf surface area (Diebolt and Mudge, 1988; Miller and Lightner, 1987; Snydor et al., 1975). Lindsey and Bassuk (1992) described a side image technique and the close relationship between image area and actual leaf area for broad leaves ($p \leq 0.01$). Diebolt and Mudge (1988) used similar techniques and found a linear relationship between image area and leaf area ($r^2 = 0.83$) for a conifer (*Pinus ponderosa*).

Leaf area based on image analysis may have the additional benefit of being more closely related to key plant physiological processes than area based on destructive sampling. For example, Smith et al. (1991) reported that silhouette leaf area, based on the image of an intact shoot as it intercepts light, was more closely related to net photosynthesis than leaf area based on destructive harvest of needles.

Thus, two experiments were conducted to evaluate the usefulness of image processing methodologies to nondestructively predict leaf area for Douglas-fir (*Pseudotsuga menziesii*) seedlings. Available digital image processing methodology was used with color digital images taken either directly onto a hard drive using a digital camera or digitized from color film photographs, high capacity computer workstations, and image processing software so that more information could be obtained and manipulated than with previous image analysis techniques.

Methods

Tree Selection

Douglas-fir seedlings obtained from Weyerhauser Corporation were used in two experiments. Experiment one provided a general evaluation of the usefulness of image analysis methodology and an initial function relating image leaf areas to measured leaf area. Data were from five trees chosen to represent a range of canopy sizes and shapes, and leaf colors. Trees were grown for 12 months in forest soil in pots.

Experiment two provided a validation of the initial image area function using data from 13 trees with distinctly different sizes providing a range of leaf areas. The trees were from the same population as the trees in experiment one, but were grown for 16 months in field plots along an elevation gradient from the Willamette Valley to the Oregon Cascade mountains. Site altitude and number of trees were: Corvallis–≈122 m and 3 trees, low site–536 m and 4 trees, middle site–951 m and 4 trees, and high site–1219 m and 2 trees.

Image Collection

A digital camera (Kodak DC5200) was used for photographing trees in the first experiment and three trees (Corvallis) from the second experiment. This camera had a standard Nikon 35 mm camera body and lens. The image was captured by a photo sensitive charged coupled device (CCD) and recorded directly to a magnetic hard. The camera separated the image into red, green, and blue bands, with signal intensity digitized into 256 levels, (8 bits of accuracy) in each of band. The spatial resolution of the computer image was 1524 by 1012 pixels. For the other 10 trees in the second experiment (middle, low, high sites), color film photographs were taken with a 35 mm camera using Kodak Ektachome 200 ASA film. The photos were printed as 10.2×15.2 cm ($4'' \times 6''$) prints, scanned at 6200 dots cm^{-2} (40,000 $inch^{-2}$), and had a pixel size of 800 by 1200. A Sharp Model JX-450 color scanner was used with the «ChromaScan» Version 2.1.1 software driver running on an Apple Macintosh®.

Trees were photographed to calculate silhouette leaf area (SLA), i.e, the area which would appear on a flat contrasting surface directly behind a tree viewed from the side. Images were acquired from two angles to overcome errors associated with most of the trees' branches being in a flat plane – the first angle chosen at random on one side; the second taken 90 degrees away on the other side of the tree (Miller and Lightner, 1987). At each angle, images were taken with 2 different backdrops – black velvet (Fig. 1 A) and white cotton (Fig. 1 B) – in order to overcome problems associated with lighting and shadows. To calibrate SLA, a 1 cm^2 piece of contrasting black or white plastic was included in each image in the same plane as the tree.

Digital image processing is sensitive to lighting conditions; i.e. diffuse, uniform, light improves images by reducing shadows. Thus, for the first experiment, trees were placed in a greenhouse with a translucent ceiling and photographs were taken on a cloudy day. However, for experiment two, photographs were acquired under available sunlight conditions as the trees were *in situ* at remote field sites and photographed on prescheduled dates.

Image Processing

Images were down loaded from the hard disk of the camera or from the scanner, through an Apple Macintosh® to a Sun SPARC workstation, and processed using the «Khoros» software developed at the University of New Mexico. Image processing consisted of a series of operations shown in Fig. 2). Use of the reference square made it unnecessary to measure the exact distance from the subject to the camera, and, thus, eliminated the requirement for the complicated calculation involving distance and focal length of the camera lens. The SLA was based on the average of the two black or white images, or the average of the four images from both colors depending on which images are best correlated with PLA. False values in a corner or edge of the image indicating the presence of needles where none actually occurred were clipped away using an image processing function. The analysis software displayed the red, green, and blue bands independently, each as a grey scale image with 8 bits of resolution.

Tree Harvest

After photographs were acquired, trees were harvested and leaves removed. A subsample of leaves was measured for single-sided projected area using a photocell area meter (Lambda Instruments Co. Model 3100). Both the subsample and remainder of leaves were dried for at least 3 days in a 70 C oven and weighed. Total measured projected leaf area (PLA) was calculated based on area and dry weight for a subsample of leaves (Evans, 1972), and total dry weight for all leaves, i.e., PLA = total dry weight * (subsample area/subsample dry weight). For experiment one, all leaves were measured

1 A

1 B

Fig. 1: Original images with (A) black background, and (B) white background.

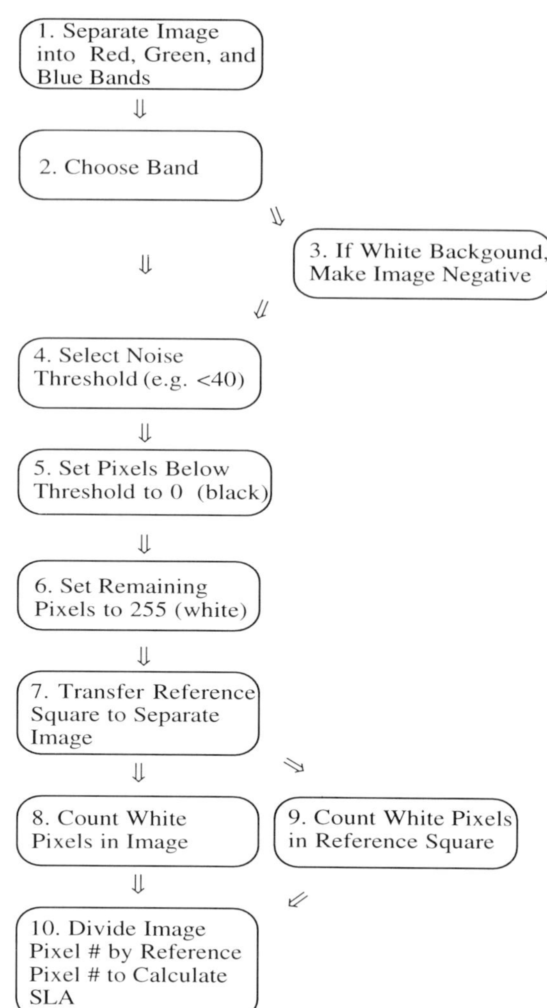

Fig. 2: Flow chart of sequence of image analysis procedures. Step 3 involves inverting the image intensity. SLA = Silhoutte leaf area.

together and not differentiated by age class. For experiment two, leaves were separated and measured by age class, and total leaf area was calculated as the sum of the areas of all age classes. Before harvest, stem diameter was measured to determine the relationship between stem diameter and PLA. Diameter measurements were made just below the cotyledonary swelling using a digital caliper (Mitutoyo Corporation, Digimatic).

Statistical Analysis

Regression analysis software was StatView® Version 4.01 for the Macintosh®. Data from experiment one were used to calculate an initial function (F1) relating SLA to PLA. Data from the second experiment then were used to validate the initial function, i.e., individual plant predicted PLAs were calculated based on image analysis and F1, followed by calculation of the standard error (SE) for the differences between predicted PLAs and measured PLAs. Data from experiment two were then used to calculate a function (F2A) relating SLA to PLA for comparison of regressions from the two experiments, and to calculate a function (F2B) relating SLA to PLA for future validation. Finally, PLA and stem diameter data from experiment two were used to calculate a function relating leaf area to tree growth without image analysis. Regression equations were tested using both untransformed and log (ln) transformed (to equalize variances) data, and both linear and polynomial (second order) equations. For experiment two, data for the three digital camera trees were within the range of the data for the 10 film camera trees so the data were combined for statistical analysis.

Results and Discussion

General Relation Between Image and Measured Leaf Area (Experiment One)

There was a high correlation ($r^2=0.907$) between SLA and PLA (Fig. 3). The linear regression equation using untransformed SLA and the average of white + black background images best fit the data, with a larger r^2 value than either ln transformed or polynomial regression equations and either the individual black or white background (Table 1).

Apparently, averaging the four sample images per tree was sufficient to overcome minor variations in tree size and shape, and lighting during the image collection to provide a close relationship between SLA and PLA. The mean ratio between PLA and SLA per tree was 3.3 (± 0.3 standard deviation, range of 2.8 to 3.7) which was similar to the ratio of about 3.0 found for Douglas-fir by Smith et al. (1991) who determined SLA using detached twigs from mature trees.

Table 1: Coefficients of determination for regressions of image projected leaf area vs. actual leaf area based on destructive harvests.

Experiment	Regression	Transformation	r^2 for actual projected leaf area vs. image silhouette area		
			Black	White	Black + White
One (n=5)	Linear	none	0.587	0.737	0.907
		ln	0.478	0.781	0.885
Two (n=13)	Linear	none	0.004	0.805	0.676
		ln	0.008	0.649	0.523
Two (n=13)	Polynomial[a]	none	0.210	0.861	0.731
		ln	0.264	0.729	0.584

[a] 2nd Order.

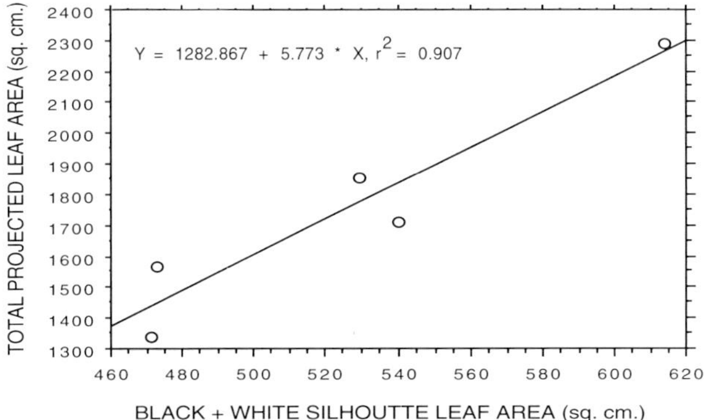

Fig. 3: Relationship between silhouette leaf area based on average of two black and two white background digital images, and projected leaf area based on destructive harvest, first experiment. N=5, standard error (SE) of intercept = 563.524, and SE of slope = 1.067.

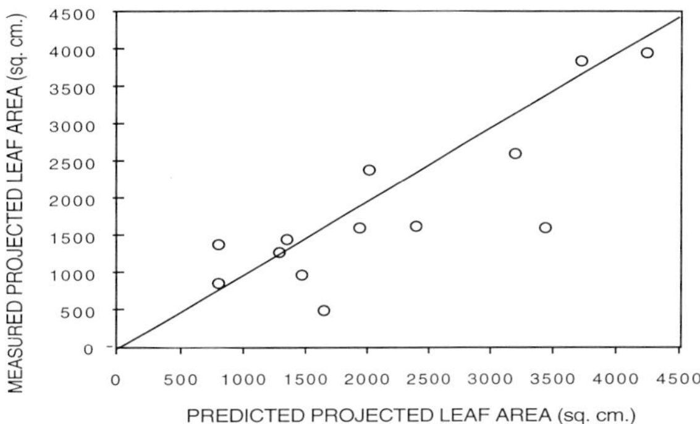

Fig. 4: Relationship between actual measured and actual predicted leaf areas for experiment trees based on image analysis and regression equation F1 from the first experiment, N=13. The solid line indicates a theoretical one-to-one relationship between predicted and measured projected leaf area.

Validation of Image Area Function (Experiment Two)

Data from experiment two did not validate F1 (initial function from experiment one) as there was no close relationship between individual tree predicted PLAs based on F1 and measured PLAs (Fig. 4). The average predicted PLA was 19% greater than the average measured PLA. Differences between predicted and measured PLA were highly variable, with a mean (± standard deviation) difference of 528 ± 497 cm^2.

A comparison of regression equations from experiments one and two using PLA and comparable black and white image data for SLA indicated no likely statistically significant differences between regressions from the two experiments (equation F2A, data not shown). This lack of differences between regressions likely was due more to the small sample sizes and relatively large SEs for the equations, rather than to similarities between trees in the two experiments. In contrast, differences in lighting, tree sizes, leaf separation by age class, and image collection procedures suggested that the data from experiment two should be analyzed separately from experiment one.

Further regression analysis of the experiment two data indicated a high correlation between white background SLA and PLA (Equation F2B, Fig. 5). There was a low correlation between black background images and PLA which likely contributed to the small r^2 between black+white SLA and PLA (Table 1). The ratio between PLA and SLA was 2.6 ± 0.8 standard deviation, with a range of 1.1 to 4.4.

Usefulness of Image Analysis Procedure

Both experiments indicated that SLA can provide a reasonably good prediction of PLA, despite methodological considerations that need further evaluation as described below.

Logistics

The subject tree must be clear of surrounding trees so as to permit an unobstructed view, e.g. on the edge of a plot. There must also be enough clear space next to the tree so that the camera can be positioned far enough away for the image

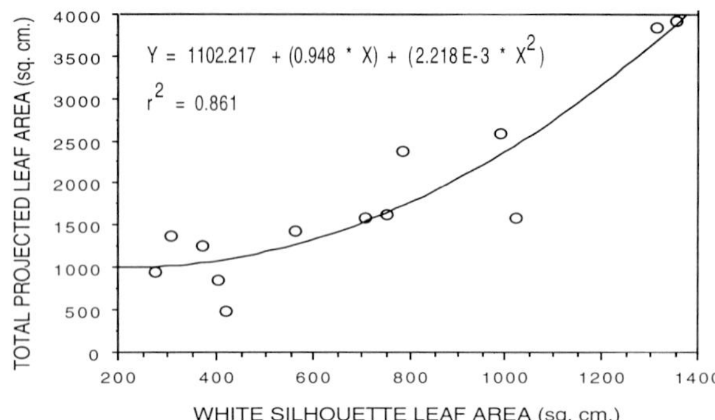

Fig. 5: Relationship between digital image silhouette leaf area and projected leaf area, second experiment. N = 13, standard error (SE) for intercept = 631.894, SE for white area = 1.821, and SE for white area^2 = 1.115E-3.

to be in focus. The black and white backdrops must also be set up, and care taken to minimize shadows and glare. Downloading the image from the camera to computer that is used for image processing takes about 15 minutes per image. The image processing itself takes 5 to 10 minutes per image. This time might be shortened with different image processing software and more powerful hardware.

Lighting

Differences in lighting may have contributed to the larger r^2 values for experiment one where light was more diffuse and images more uniform, compared to experiment two where ambient light was used, and images taken with the white backdrop tended to be underexposed. Light could be made more diffuse in the field by using a portable canopy to scatter the light over trees during image collection, thereby possibly facilitating exposure uniformity.

Color Bands

It was initially assumed that the red vs. green band data would be useful to distinguish leaf area. Surfaces covered with bark (stems and branches) were thought to have a much higher value in the red band, and that leaves would have a higher reflectance in the green band (Yoder and Waring, 1994). In such a case, simple subtraction of red from green values in each pixel would have give a negative value corresponding to bark. The pixels with negative values could then be discarded, leaving only pixels corresponding to leaves.

However, the difference between the different bands turned out to be small in our experiments, and, in most cases, the red from green subtraction process was not very dependable in separating bark from leaves. A solution to using color separation to distinguish needles from bark may involve use of more narrow portions of the spectrum in which the colors of needles and bark do not overlap as indicated by a device such as a reflectance meter. If there is a clear difference in the portions of the visible spectrum reflected by needles and bark, the CCD in the digital camera could be chosen to take advantage of this difference. In the case where photographic film is used and the image later digitized, the color balance of the film would have to be chosen to take advantage of the difference between needles and bark, and the digitizer would need to be sensitive in the same portions of the spectrum.

Fourier Transformation of Data

Fourier transformations were attempted as a method of separating the needles from the background in the image. Because the needles are all a similar size, the assumption was made that they could be filtered out with a fairly narrow band pass filter in Fourier space, and that converting the filtered image back to image space would eliminate the background and perhaps some of the bark. However, because the needles were quite narrow, they were in the same part of the spectrum as the noise. Thus, Fourier processing tended to enhance noise and decrease contrast. In some cases, Fourier processing eliminated most of the bark, but also eliminated a significant portion of the needles.

Destructive Harvest

Although using digital image processing to determine leaf area gives reasonably accurate results, it is still necessary that a destructive harvest occur to obtain data to calculate the SLA vs. PLA function. Trees must be set a side for this purpose and undergo labor intensive separation of needles and analysis of PLA. Periodic harvests (at least annual) likely would be required for long-term studies to check whether the function is valid as trees grow and mature.

Comparison of Leaf Images with Other Predictors of Leaf Area

The ultimate usefulness of the image analyses technique will depend on its relative practicality for routine determination of leaf area compared to other techniques. In contrast to image analysis, simpler measurements may be more efficient to determine leaf area if time and labor are available. For example, stem diameter can be measured rapidly and accurately for usually all trees in a population. Stem diameter has been used as an indicator of tree canopy size, and, hence, leaf area for larger trees; due to the close relationship between basal area (diameter at breast height) and sapwood basal area is closely correlated with leaf area (Waring et al., 1980). Since seedling stems likely contain primarily sapwood, they should have an especially close relationship between stem diameter and leaf area. Based on stem diameter measurements in experiment two, there was a high correlation between d^2 (stem diameter squared) and PLA (r^2 = 0.877) (Fig. 6). Thus, stem diameter could be used to predict leaf area of Douglas-fir seedlings since it was easier to measure repeatedly compared to SLA obtained from leaf images.

Conclusions

Preliminary data indicate that current technology image analysis can be used to measure SLA, and that SLA is corre-

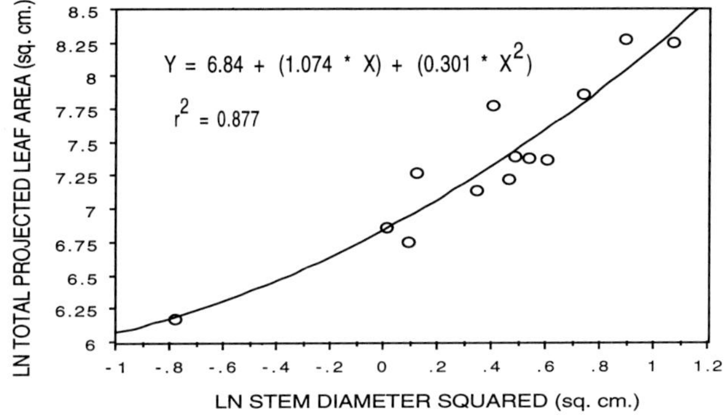

Fig. 6: Relationship between ln of stem diameter squared (d^2) and ln of total projected leaf area, experiment two. N = 13, standard error (SE) for intercept = 0.096, SE for diameter = 0.157, and SE for d^2=0.212.

lated with leaf area. However, additional data would be needed to further validate the new function relating PLA to SLA for Douglas-fir seedlings. If methodological problems can be resolved, especially in terms of lighting and obtaining extra information from color bands, this technique may provide a valuable tool for estimating leaf area nondestructively for studies of environmental stress on vegetation.

References

DIEBOLT, K. S. and K. W. MUDGE: Use of video-imaging system for estimating leaf surface area of *Pinus sylvestris* seedlings. Can. J. For. Res. *18,* 377–390 (1988).

EVANS, G. C.: The Quantitative Analysis of Plant Growth. University of California Press, Berkeley (1972).

KVET, J. and J. K. MARSHALL: Assessment of leaf area and other assimilating plant surfaces. In: SESTAK, Z., J. CATSKY, and P. G. JARVIS (eds.): Plant Photosynthetic Production Manual of Methods, pp. 517–555. Junk, N.V. Pubs., The Hague (1971).

LINDSEY, P. A. and N. L. BASSUK: A nondestructive image analysis technique for estimating whole-tree leaf area. HortTechnology *2,* 66–72 (1992).

MILLER, S. S. and G. W. LIGHTNER: Computer-assisted determination of apple tree canopy volume. HortScience *22,* 393–395 (1987).

SMITH, W. K., A. W. SCHOETTLE, and M. CUI: Importance of the method of leaf area measurement to the interpretation of gas exchange of complex shoots. Tree Physiol. *8,* 121–127 (1991).

SNYDOR, T. D., T. A. FRETZ, D. E. CREAN, and M. R. COBBS: Photographic estimation of plant size. HortScience *10,* 219–220 (1975).

THOMAS, D. L., F. J. K. DASILVA, and W. A. CROMER: Image processing technique for plant canopy cover evaluation. Trans. Amer. Soc. Agric. Eng *31,* 428–434 (1988).

WARING, R. H., W. G. THIES, and D. MUSCATO: Stem growth per unit of leaf area: a measure of tree vigor. For. Sci. *26,* 112–117 (1980).

YODER, B. J. and R. H. WARING: The normalized difference vegetation index of small Douglas-fir canopies with varying chlorophyll concentrations. Remote Sens. Environ. *49,* 81–91 (1994).

J. Plant Physiol. Vol. 148. pp. 536–547 (1996)

Differences in Fluorescence Excitation Spectra of Leaves between Stressed and Non-Stressed Plants*

JOACHIM SCHWEIGER, MICHAEL LANG, and HARTMUT K. LICHTENTHALER

Botanisches Institut, Universität Karlsruhe, Kaiserstr. 12, D-76128 Karlsruhe, Germany

Received August 15, 1995 · Accepted October 18, 1995

Summary

Using a Perkin-Elmer fluorometer the fluorescence excitation spectra of the four main fluorescence bands of plants (blue, green, red and far-red) were determined in the UV-region between 250 to 400 nm. The results provide valuable information on the most suitable wavelength for fluorescence excitation which is different for the blue and green fluorescence from that of the red and far-red chlorophyll fluorescence. The differing fluorescence yield in dependence of the excitation wavelength can be quantified by the fluorescence ratio blue/red (F450/F690) and blue/far-red (F450/F735), as well as red/far-red (F690/F735) and blue/green (F450/F530). The differential values of the fluorescence ratios, as found for the usually applied laser-excitation wavelengths of 308, 337, 355 and 397 nm, are contrasted. The study demonstrates that fluorescence emission spectra of leaves can successfully be applied for stress detection in plants. The fluorescence ratios blue/red (F450/F690) and blue/far-red (F450/F735) proved to be the most sensitive and best suited stress indicators. For simultaneous excitation of the blue-green fluorescence and the red+far-red chlorophyll fluorescence an excitation wavelength between 355 and 390 nm is recommended. Via fluorescence excitation spectra it was also proved that sun leaves, outdoor plants and high-light treated plants are better protected against UV-A and UV-B than shade leaves, greenhouse plants or low-light plants. The transformation in epidermis cells of UV-radiation into blue and green fluorescence, which can be used by mesophyll cells for photosynthetic quantum conversion, is a new protection mechanism of plants and their photosynthetic apparatus against UV-radiation.

Key words: blue-green fluorescence, fluorescence excitation spectra, red chlorophyll fluorescence F690, far-red chlorophyll fluorescence F735, fluorescence ratios blue/red and blue/far-red, stress detection, UV-protection.

Abbreviations: F450 = Blue fluorescence; F530 = green fluorescence; F690 = red chlorophyll fluorescence; F735 = far-red chlorophyll fluorescence; F450/F530 = ratio of the blue to green fluorescence; F450/F690 = fluorescence ratio blue/red; F450/F735 = fluorescence ratio blue/far-red; F690/F735 = chlorophyll fluorescence ratio red/far-red.

Introduction

In their natural location plants are exposed to a multitude of stressors, many of which act simultaneously (Lichtenthaler, 1996). Chlorophyll fluorescence emission spectra and kinetics of plant leaves can be applied in plant physiology to detect stress in plants and in the photosynthetic apparatus (Lichtenthaler and Rinderle, 1988; Lichtenthaler, 1988 and 1990). In the past 30 years steadily increasing numbers of investigations dealt with the red chlorophyll fluorescence of plants, of which Kautsky had shown already that it is inversely correlated to photosynthesis (Kautsky and Hirsch, 1931). This fundamental work of Kautsky and co-workers has been summarized in a recent review (Lichtenthaler, 1992). Excitation of leaves with long-wavelength ultraviolet radiation (UV-A,

* Dedicated to Prof. Dr. Horst Senger on the occasion of his 65th birthday.

320 to 380 nm) results, however, in a fluorescence emission in the blue-green spectral region as well. Thus, a total of four characteristic fluorescence bands can be differentiated in plants: The blue fluorescence near 440 to 450 nm (F450), the green fluorescence near 520 to 530 nm (F530), the red fluorescence near 690 nm (F690) and the far-red fluorescence band near 730 to 740 nm (F735) (Chappelle et al., 1984; Goulas et al., 1990; Lichtenthaler et al., 1991; Stober and Lichtenthaler, 1992, 1993 a, b, c; Stober et al., 1994).

The emitter of the red and far-red fluorescence emission is the protein-bound chlorophyll *a* in the chloroplasts of the mesophyll cells. Due to the physiological role of chlorophyll in photosynthesis, the red chlorophyll fluorescence has been well examined (Govindjee et al., 1986; Lichtenthaler and Rinderle, 1988; Lichtenthaler, 1988 and 1990). The chlorophyll fluorescence ratio F690/F735 is an excellent indicator of the *in vivo* chlorophyll content of leaves. Due to the partial overlapping of the near red 690 nm chlorophyll fluorescence emission band with the 670 to 680 nm absorption bands of chlorophyll (Lichtenthaler, 1987 and 1990), the fluorescence ratio F690/F735 decreases with increasing chlorophyll content (Hák et al., 1990; Lichtenthaler et al., 1990, D'Ambrosio et al., 1992). The chlorophyll fluorescence ratio F690/F735 can also be sensed in a superiour way by the novel high resolution fluorescence imaging over the whole leaf area (Edner et al., 1995; Lang, 1995, 1996; Lichtenthaler et al., 1996), in which case small local spot-size differences in fluorescence emission and gradients from the leaf rim to the center part of the leaf can precisely be detected and used for early stress detection in plants.

In contrast, the blue-green fluorescence primarily emanates form the cell walls of epidermal cells and is emitted by various plant phenolics (Lang et al., 1991; Stober and Lichtenthaler, 1993 a; Harris and Hartley, 1976). The blue and the major part of the green fluorescence are predominantly emitted by epidermis cells and leaf veins (Lang et al., 1992 and 1994; Stober and Lichtenthaler, 1993 c), whereas a smaller part of the green fluorescence also seems to originate from the leaf mesophyll (Lang and Lichtenthaler, 1991; Cerovic et al., 1994; Lang, 1995). The molecular origin of the blue-green fluorescence is not yet completely clear, but blue-green fluorescence appears to be a mixed signal emitted by various plant phenolics, such as hydroxycinnamic acids bound to the cell walls and possibly other phenolic plant products (Goulas et al., 1990; Lang et al., 1991). In contrast to the red and far-red chlorophyll *a* fluorescence which exhibit a transient (Kautsky and Hirsch, 1931), the blue-green fluorescence does not show any fluorescence induction kinetics (Stober and Lichtenthaler, 1993 b). Since the blue-green fluorescence emission of pre-darkened and light-exposed leaves is the same, it can be used as an internal standard in the fluorescence imaging of plants.

The chlorophyll fluorescence ratio F690/F735 has already been established as a non-invasive stress indicator in plant ecophysiology (Lichtenthaler et al., 1987; Rinderle et al., 1991). It significantly increases a) by 15 to 30 % at a sudden impairment of photosynthetic quantum conversion and b) even several times by long-term stress which decreases the chlorophyll content of leaves. The ratio F690/F735 has been applied in the detection of water stress, mineral or nitrogen deficiency, forest decline, and other stresses (Dahn et al., 1992; Rinderle et al., 1991; Hák et al., 1993; Morales et al., 1994). Conventional methods, such as the determination of fluorescence emission spectra of different leaf points (Lang et al., 1992; Chappelle et al., 1985; Chappelle et al., 1986; Stober et al., 1993, 1994), or fluorescence lifetime measurements of leaf pieces (Goulas et al., 1990; Schmuck et al., 1992) have been applied by various authors, as well as the classical chlorophyll fluorescence induction kinetics (Krause and Weis, 1991; Lichtenthaler and Rinderle, 1988; Schreiber et al., 1986; Govindjee, 1995). The fluorescence ratios blue/red (F450/F690) and blue/far-red (F450/F735) are new fluorescence ratios, which have more recently been introduced to stress physiology by Lichtenthaler and co-workers (Lang and Lichtenthaler, 1991; Lichtenthaler et al., 1992, 1993; Stober and Lichtenthaler, 1993 a, b, c; Stober et al., 1994).

The most recent approach is the application of the laser-induced high-resolution fluorescence imaging (Lang et al., 1994, 1995, 1996; Lichtenthaler et al., 1996), which permits the detection of gradients in the fluorescence emission over the leaf area and also local disturbances, which could not be seen using the classical point data fluorescence measurements of individual points of a leaf. Since the four fluorescence bands blue, green, red and far-red considerably change during stress and damage to plants, their ratios can be taken as stress indicators. This is particularly valid since the fluorescence ratios vary very little and much less from leaf to leaf than the absolute fluorescence yield, which may be considerably different from sample to sample depending on the sensing angle and the roughness of the leaf surface. In order to obtain a high fluorescence yield and reproducible results one would like to know which UV-A or UV-B radiation is best to excite the four fluorescence bands F450, F530, F690 and F735. Furthermore, it should be possible to see differences in the fluorescence excitation spectra between healthy, physiologically active plants on one hand and stressed or partially damaged plants on the other.

However, fluorescence excitation spectra of the stressed and the control plants have not yet been performed. For these reasons we determined and compared the fluorescence excitation spectra for the four fluorescence bands in the UV-region of 250 to 400 nm of sun and shade leaves, of greenhouse and field plants and of plants under water, heat and high-light stress. The differences in fluorescence yield due to the excitation wavelength were quantified in the fluorescence ratios blue/red, blue/far-red and red/far-red, which were contrasted for the wavelengths of several lasers used for fluorescence excitation.

Materials and Methods

Plants

Beech leaves *Fagus sylvatica* L. were taken from a 60-year old tree on the campus of the University of Karlsruhe. Maple leaves (*Acer platanoides* L.) came from 4-year old trees grown in pots in the Botanical Garden of the University of Karlsruhe. *Cucurbita pepo* L. and *Phaseolus vulgaris* L. were cultivated in the greenhouse under standard conditions (500 µmol photons $m^{-2} s^{-1}$, 25 °C, 60 % rel. humidity). Maize plants (*Zea mays* L.) were cultivated either in the greenhouse or outdoors in the Botanical Garden.

Fluorescence spectroscopy

Fluorescence excitation spectra were performed on the Perkin-Elmer LS-50 Luminescence Spectrometer (Perkin-Elmer, Überlingen, Germany). The fluorescence yields were recorded for the four major fluorescence bands of plants at 450, 530, 690 and 735 nm with the excitation wavelengths from 250 to 400 nm. The spectral band widths of the slits for excitation and emission monochromators were adjusted to 10 nm. A 430 nm cut-off filter was applied in order to exclude scattered light. The excitation spectra given in the figures have been based on 4 spectra recorded per growth condition. The standard deviation amounted to <7 % for the excitation spectra of the four fluorescence bands and to <4 % for the excitation spectra of the fluorescence ratios.

Results

Sun-exposed leaves

Beech leaves: The fluorescence excitation spectra of sun and shade leaves of beech (*Fagus sylvatica* L.) (Fig. 1) differed for the blue-green and the red/far-red chlorophyll fluorescence emission. The yield of the blue and green fluorescences (F450 and F530) was high in the UV-B region (250 to 320 nm) and declined towards 400 nm (Fig. 1). The fluorescence excitation spectra were similar for sun and shade leaves, however, the fluorescence yield of the shade leaves was ca. 2 times higher than in sun leaves of beech. In contrast, chlorophyll fluorescence excitation spectra showed distinct spectral differences between sun and shade leaves, and the fluorescence yield was very low in sun leaves. In shade beech leaves, the fluorescence increase towards the chlorophyll fluorescence excitation maximum (which is near 435 nm) started at about 330 nm, whereas in sun leaves the red and far-red fluorescence yield increased very little, and only at wavelengths >380 nm. The green fluorescence (F530) excitation spectra of sun and shade beech leaves exhibited much lower fluorescence intensities when compared to the blue fluorescence. The differences in fluorescence excitation spectra can best be demonstrated by forming the fluorescence ratio spectra as shown in Fig. 2. The maxima for the two fluorescence ratios blue/red and blue/far-red were found at the same wavelength, but they were significantly different for shade leaves (315 nm and 317 nm) compared to sun leaves (325 nm and 326 nm for the ratios blue/red and blue/far-red, respectively) (Fig. 2). These differences can be seen in the 1st derivative of the fluorescence ratio excitation spectra as point of intersection with the zero line (Fig. 3). The fluorescence ratio spectra clearly indicated that the height of the fluorescence ratio depended very much upon the excitation wavelength. Differences in the chlorophyll content between sun and shade leaves can also be detected via the excitation spectra of the chlorophyll fluorescence ratio F690/F735 (Fig. 2). The shade leaves of beech, containing about 25 % less chlorophyll than sun leaves, exhibited from 250 to 380 nm – as expected at a lower chlorophyll content – significantly higher values of the chlorophyll fluorescence ratio F690/F735.

Maple leaves: Sun and shade leaves of *Acer platanoides* L. provided similar results as obtained for beech leaves (Fig. 4). In the fluorescence excitation spectra, the intensities of blue and green fluorescence were only slightly higher in shade leaves than in sun leaves. In contrast, the differences in the red and far-red chlorophyll fluorescence between sun and shade leaves were significantly different to beech. In the fluorescence excitation spectra of the fluorescence ratios blue/red and blue/far-red the maxima in sun leaves were shifted by 16 and 17 nm to longer wavelengths, respectively, as compared to shade leaves. This is shown in the first derivative in Fig. 3 as intersection with the zero line. Due to the low red and far-red chlorophyll fluorescence yield in sun leaves, the values of the fluorescence ratios blue/red (F450/F690) and blue/far-red (F450/F735) were higher at most wavelengths than in maple shade leaves (Fig. 5).

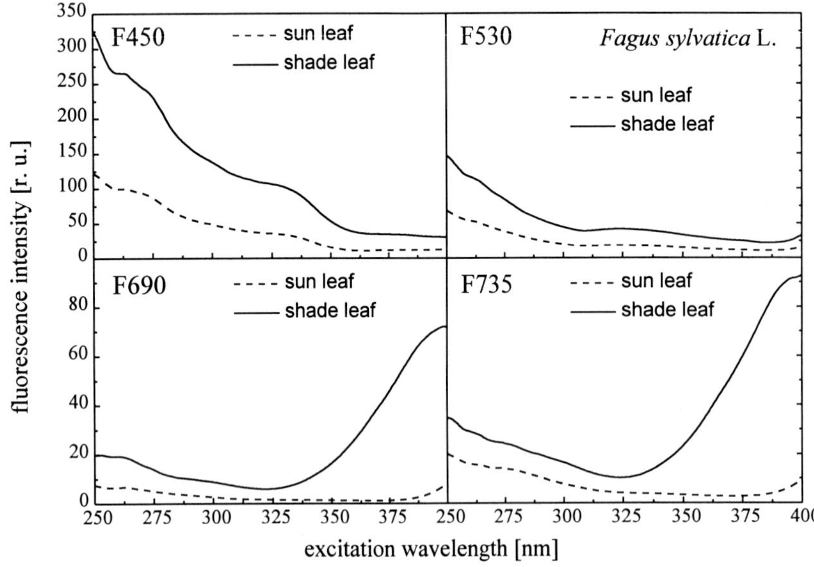

Fig. 1: Fluorescence excitation spectra of sun and shade leaves of *Fagus sylvatica* L. The fluorescence emission was determined at the wavelengths 450 nm, 530 nm, 690 nm and 735 nm.

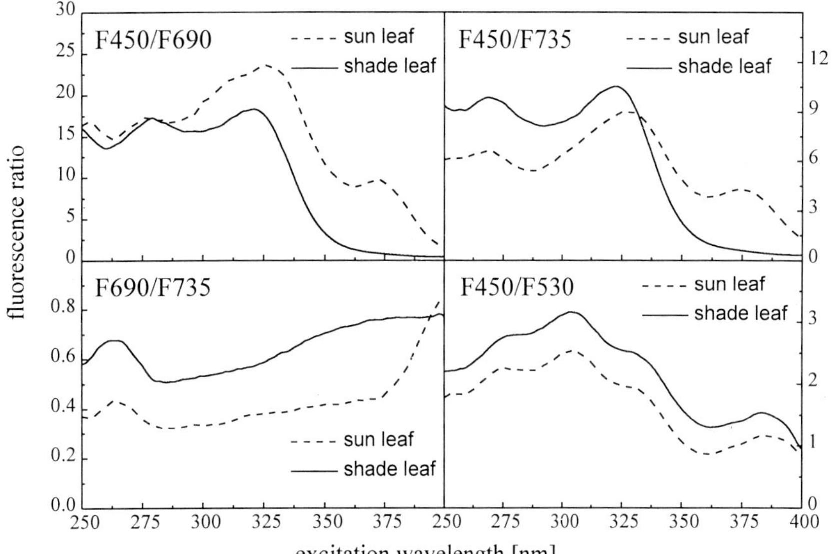

Fig. 2: Fluorescence ratio excitation spectra: Excitation wavelength-dependent variation of the fluorescence ratios blue/red (F450/F690) and blue/far-red (F450/F735) as well as the chlorophyll fluorescence ratio red/far-red (F690/F735) and the fluorescence ratio blue/green (F450/F530) of sun and shade leaves of beech (*Fagus sylvatica* L.).

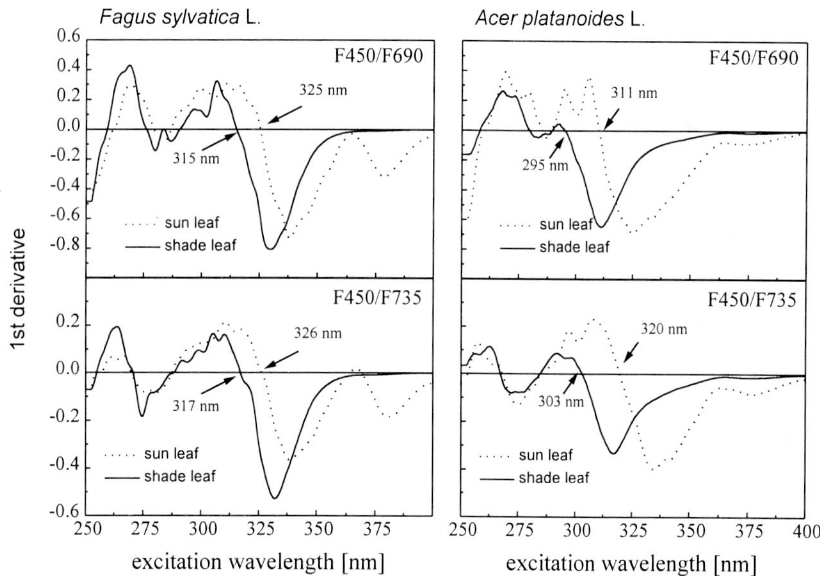

Fig. 3: First derivative of fluorescence ratio excitation spectra of the blue/red (F450/F690) and blue/far-red (F450/F735) in sun and shade leaves of beech (*Fagus sylvatica* L.) and maple (*Acer platanoides* L.).

The differences in the four fluorescence ratios blue/red, blue/far-red, red/far-red and blue/green as determined from the fluorescence ratio excitation spectra of beech and maple (Figs. 2 and 5) are listed in Table 1 for the excitation wavelengths 308, 337, 355 and 397 nm. These are the emitter wavelengths of various lasers which have been applied so far for the excitation of fluorescence signals in plants.

Field plants

The fluorescence excitation spectra of maize plants (*Zea mays* L.) grown in the field and in the greenhouse showed similar results as obtained for sun and shade leaves (Fig. 6). In comparison to greenhouse plants, the blue and green fluorescence emissions ranging from 250 to 400 were about 1.3 to 1.5 times higher in field plants. The excitation maximum in the UV-A region for both blue and green fluorescence was approximately near 330 nm in field plants as well as in greenhouse plants, and at longer wavelengths the blue and green fluorescence yield declined. In shade and sun leaves the rise in the red and far-red chlorophyll fluorescence yield started in greenhouse plants at 325 nm already, but in field plants only at wavelengths >375 nm. These differences in the relative fluorescence yields are more clearly documented in the excitation spectra of the fluorescence ratio (Fig. 7). Maximum values for the ratios blue/red (F450/F690) and blue/far-red (F450/F735) were found at an excitation wavelength around 330 nm. The fluorescence ratios blue/red and blue/far-red showed the largest differences between greenhouse and field plants and appear to be the best indicators to sense differ-

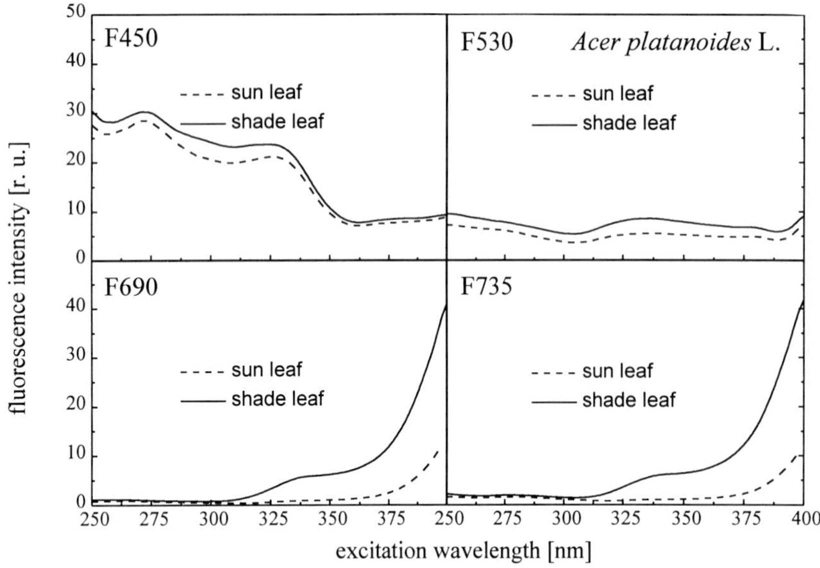

Fig. 4: Fluorescence excitation spectra of sun and shade leaves of *Acer platanoides* L. for the emission wavelengths 450 nm, 530 nm, 690 nm and 735 nm.

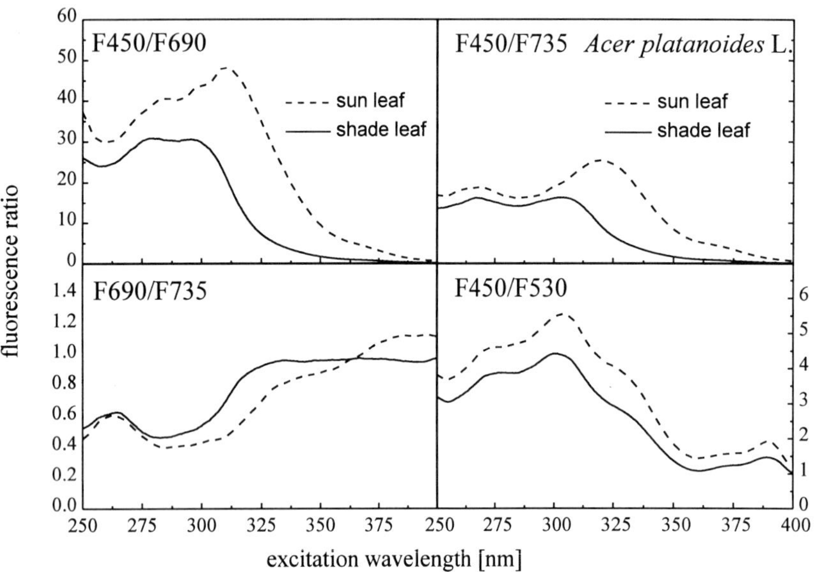

Fig. 5: Fluorescence ratio excitation spectra of shade leaves of *Acer platanoides:* Dependence of the fluorescence ratios blue/red (F450/F690), blue/far-red (F450/F735), red/far-red (F690/F735) and blue/green (F450/F530) on the excitation wavelength.

ences in irradiance between greenhouse and field plants during growth. The chlorophyll fluorescence ratio red/far-red (F690/F735), functioning as an indicator of the *in vivo* chlorophyll content in the range of 250 to 380 nm, was significantly lower in the outdoor plants, which possessed a chlorophyll content 30 to 40 % higher per leaf area unit than in the indoor plants. The fluorescence ratio blue/green (F450/F530) was almost the same in greenhouse and outdoor plants, and exhibited a broad maximum between 290 and 350 nm.

High-light exposure

Bean leaves (*Phaseolus vulgaris* L.) were irradiated for 2 h at 2000 $\mu mol\,m^{-2}\,s^{-1}$. Excitation spectra of the blue (F450) and green (F530) fluorescence showed a 3 to 5 times higher increase of blue and green emission after high-light treatment (Fig. 8). The red and far-red fluorescence excitation yield decreased considerably in the range from 350 to 400 nm under high-light exposure. Thus, the high-light treated leaves exhibited similar chlorophyll fluorescence excitation spectra as those found in sun leaves. Correspondingly, the values of the excitation spectra of the fluorescence ratios changed upon high-light treatment (Fig. 9). The fluorescence ratios blue/red (F450/F690) and blue/far-red (F450/F735) had particularly increased. The chlorophyll fluorescence ratio F690/F735 also increased to much higher values, but this was associated only with a partial decline of the chlorophyll content. This aspect, which may be related to photoinhibition, needs further investigation.

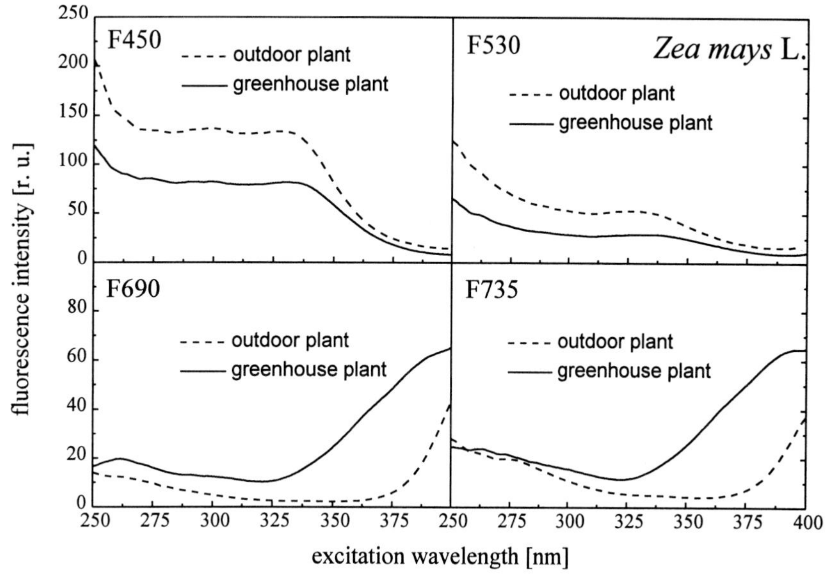

Fig. 6: Fluorescence excitation spectra of *Zea mays* L. leaves grown in the field (outdoor plant) or in the greenhouse for the emission wavelengths 450 nm, 530 nm, 690 nm and 735 nm.

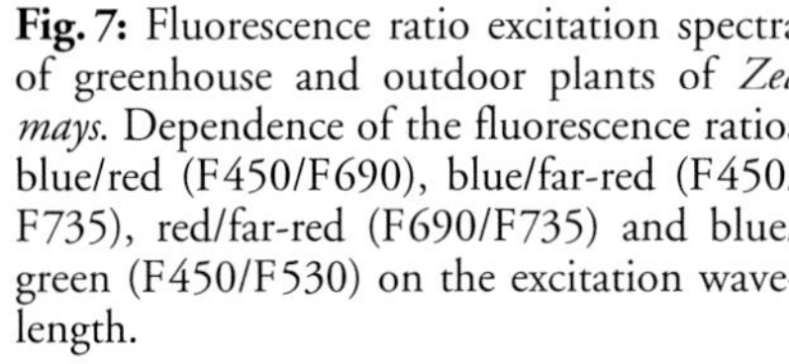

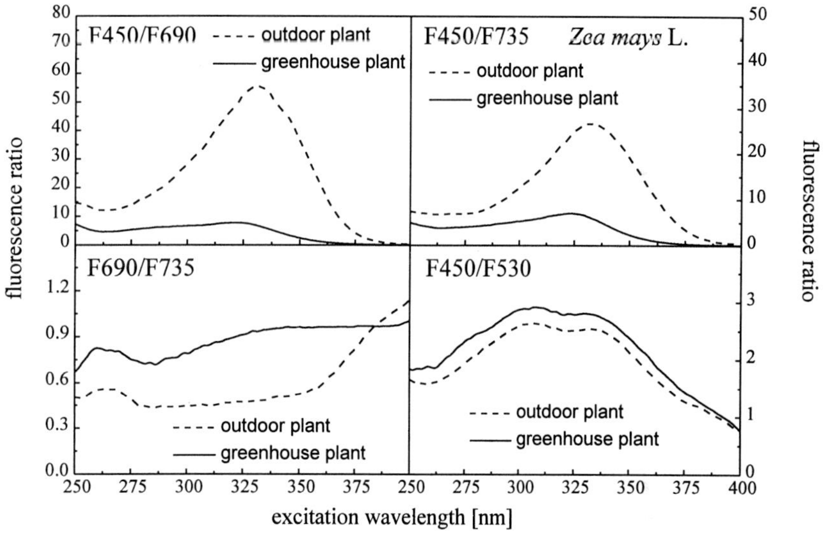

Fig. 7: Fluorescence ratio excitation spectra of greenhouse and outdoor plants of *Zea mays*. Dependence of the fluorescence ratios blue/red (F450/F690), blue/far-red (F450/F735), red/far-red (F690/F735) and blue/green (F450/F530) on the excitation wavelength.

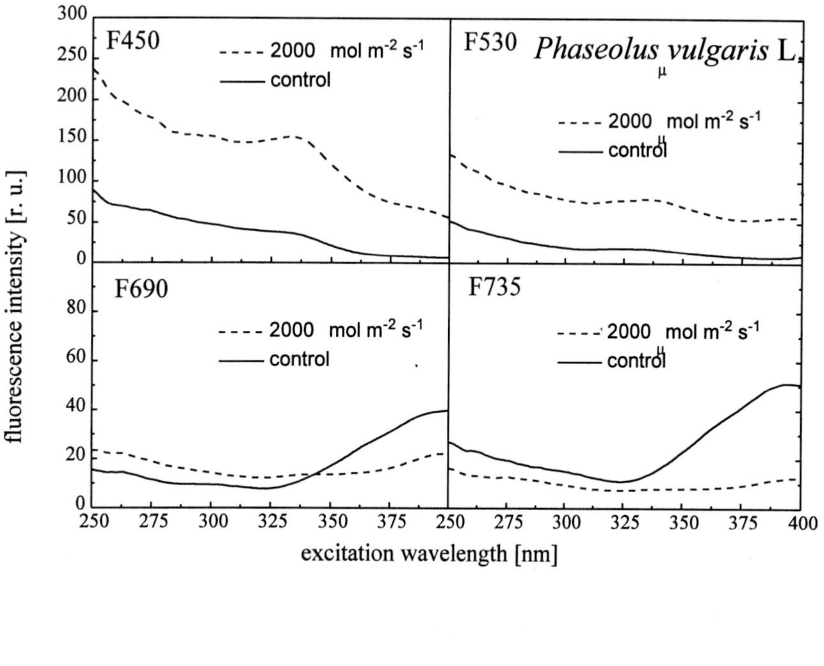

Fig. 8: Fluorescence excitation spectra of high-light stressed bean leaves (*Phaseolus vulgaris* L.) and control leaves. Green bean plants were exposed to a high-light stress of 2000 µmol m^{-2} s^{-1} for 3 h, controls were kept at a medium irradiance of 500 µmol m^{-2} s^{-1}. The spectra are given for the fluorescence emission wavelengths 450 nm (blue), 530 nm (green), 690 nm (red) and 735 nm (far-red).

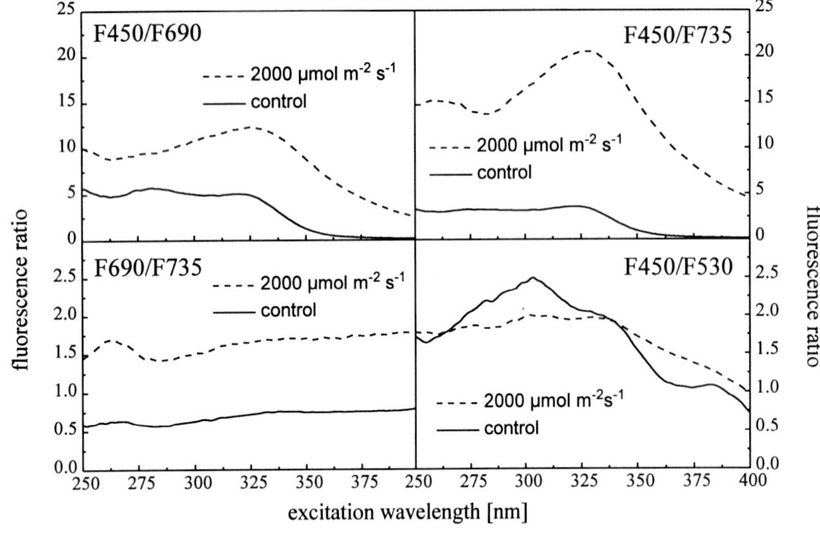

Fig. 9: Fluorescence ratio excitation spectra of leaves of bean plants (*Phaseolus vulgaris*) without (control) and with a 2 h high-light treatment.

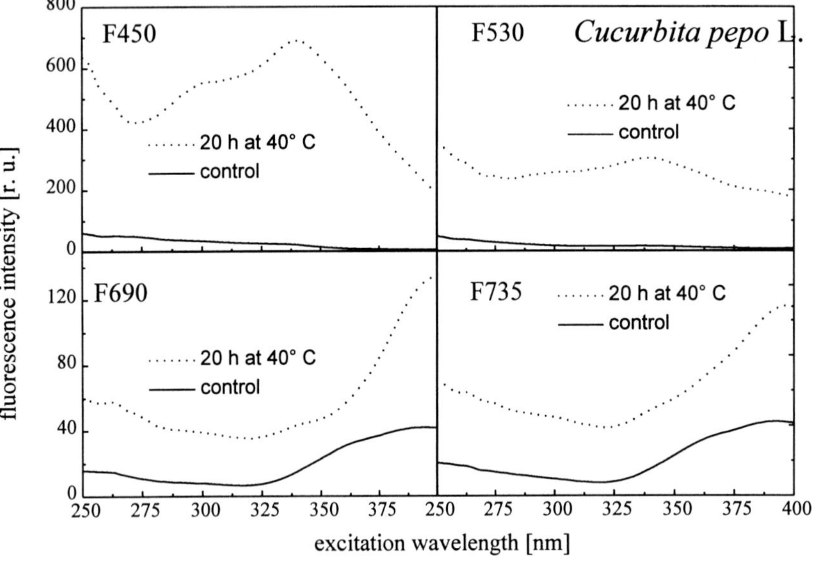

Fig. 10: Fluorescence excitation spectra at the emission wavelengths 450 nm, 530 nm, 690 nm and 735 nm of *Cucurbita pepo* L. plants exposed to heat stress (40 °C for 20 h).

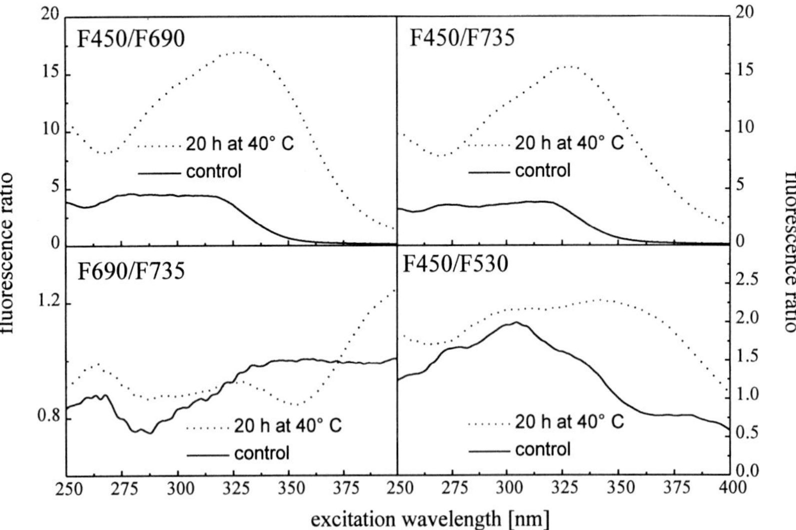

Fig. 11: Fluorescence ratio excitation spectra of leaves of pumpkin seedlings (*Cucurbita pepo*) without (control) and with exposure to a 20 h heat stress.

Table 1: Variation in the fluorescence ratios blue/red (F450/F690), blue/far-red (F450/F735), Red/far-red (F690/F735) and blue/green (F450/F530) of sun and shade leaves of beech (*F. sylvatica*) and maple (*A. platanoides*) and of field plants and greenhouse plants of maize (*Z. mays*) in dependence of the excitation wavelength, as determined from fluorescence excitation spectra. 308 nm corresponds to the emission of XeCl-excimer laser, 337 nm to that of the N_2-laser, 355 nm to that of the 3rd harmonic of Nd:YAG-laser, and 397 nm to the emission of the 1st Stokes of D_2 Raman shifted 3rd harmonic of Nd:YAG-laser.

excitation wavelength	308 nm	337 nm	355 nm	397 nm
Fagus sylvatica L.				
F450/F690				
sun leaf	21.13	20.38	9.90	1.98
shade leaf	16.45	10.42	2.10	0.43
F450/F735				
sun leaf	7.19	7.98	4.13	1.62
shade leaf	9.00	6.70	1.51	0.34
F690/F735				
sun leaf	0.34	0.39	0.42	0.82
shade leaf	0.55	0.64	0.72	0.78
F450/F530				
sun leaf	2.48	1.75	0.92	0.95
shade leaf	3.11	2.32	1.41	1.12
Acer platanoides L.				
F450/F690				
sun leaf	47.86	20.20	6.93	0.75
shade leaf	24.02	3.56	1.31	0.25
F450/F735				
sun leaf	21.73	16.80	6.29	0.85
shade leaf	15.71	3.44	1.27	0.24
F690/F735				
sun leaf	0.45	0.83	0.91	1.13
shade leaf	0.65	0.96	0.97	0.97
F450/F530				
sun leaf	5.41	3.25	1.53	1.38
shade leaf	4.19	2.31	1.15	1.17
Zea mays L.				
F450/F690				
field plant	34.70	52.48	27.95	0.39
greenhouse plant	6.87	5.32	1.65	0.13
F450/F735				
field plant	15.67	25.84	15.32	0.44
greenhouse plant	5.77	5.04	1.58	0.13
F690/F735				
field plant	0.45	0.49	0.55	1.10
greenhouse plant	0.84	0.95	0.96	0.99
F450/F530				
field plant	2.62	2.52	1.95	0.82
greenhouse plant	2.91	2.75	2.19	0.87

Long-term temperature stress

Primary leaves of pumpkin (*Cucurbita pepo* L.) were kept at 40 °C for 20 h. The fluorescence excitation spectra for blue, green, red and far-red fluorescence strongly increased after temperature treatment (Fig. 10). The blue and green fluorescence yield drastically increased by the heat treatment. Also, the red and far-red chlorophyll fluorescence increased considerably, but not as much as the blue-green fluorescence. Correspondingly, the fluorescence ratios blue/red and blue/far-red in heat-stressed plants increased by several times at all excitation wavelengths and exhibited maxima near 330 nm (Fig. 11). The changes in the chlorophyll fluorescence ratio red/far-red depended on the excitation wavelength and did not show any clear trend.

Water stress

In another experiment pumpkin seedlings (*Cucurbita pepo* L.) were exposed to a 6 d water stress. Also under this condition a strong increase in the blue and green fluorescence emission occurred (Fig. 12). In the water stressed plant the red fluorescence F690 had increased, whereas the far-red chlorophyll fluorescence yield F740 showed a decline at wavelengths >340 nm as compared to the control plants. In that case the ratios F450/F690 and F450/F735 were much higher in leaves of water stressed plants with a maximum near 330 nm (Fig. 13).

Discussion

The results of this investigation demonstrated that the excitation spectra of blue and green fluorescence as well as of red and far-red chlorophyll fluorescence are different for sun and shade leaves in both beech and maple trees. In both cases the red and far-red chlorophyll fluorescence yield in sun leaves ranging from 325 to 400 nm was much lower than in shade leaves (Figs. 1 and 4). This means that the sun-exposed leaves accumulate UV-absorbing substances in their epidermis cells preventing the exciting UV-radiation from penetrating through the epidermis into the mesophyll layer oft the leaves where the red and far-red fluorescing chlorophylls are bound to the photosynthetic biomembranes of chloroplasts. These results indicated that sun leaves are better protected against UV radiation than shade leaves. The chlorophyll fluorescence excitation spectra also indicated that maple sun and shade leaves (very low chlorophyll fluorescence yield) are better protected against UV-A and UV-B than sun and shade leaves of beech (Figs. 1 and 4). The best indicators of the different fluorescence yield between sun and shade leaves were the fluorescence ratios blue/red and blue/far-red. These fluorescence ratios yielded maximum values in the range of 320 to 330 nm.

In a similar way as in sun leaves we found a significantly lower chlorophyll fluorescence yield in field plants of maize (as compared to greenhouse plants), and in greenhouse bean plants exposed to a short-term high irradiance of 2000 µmol $m^{-2}s^{-1}$ (Figs. 6 and 8). In both cases the field plants as well as the high-light treated plants exhibited a much higher blue and green fluorescence emission. As a consequence, the excitation spectra of the fluorescence ratios blue/red and blue/far-red showed higher values over the whole UV-A and

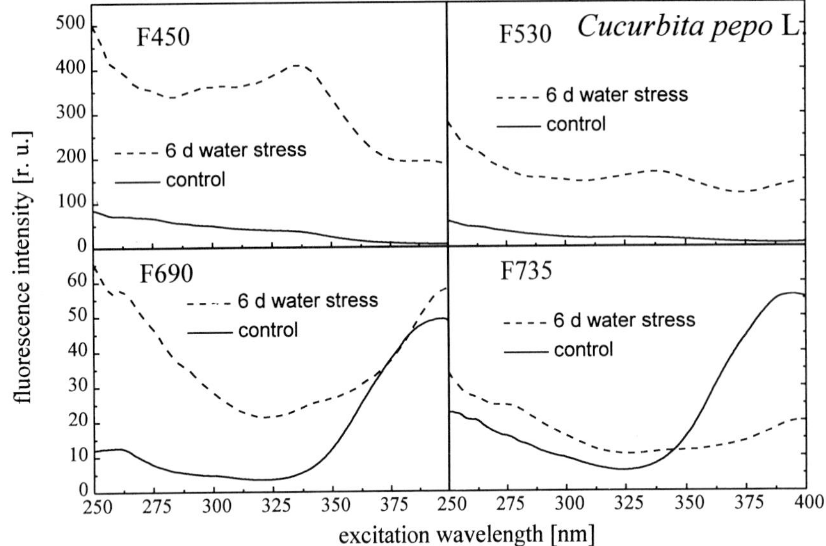

Fig. 12: Fluorescence excitation spectra at the emission wavelengths 450 nm, 530 nm, 690 nm and 735 nm of *Cucurbita pepo* L. control plants and plants which were subjected to a 6 d water stress.

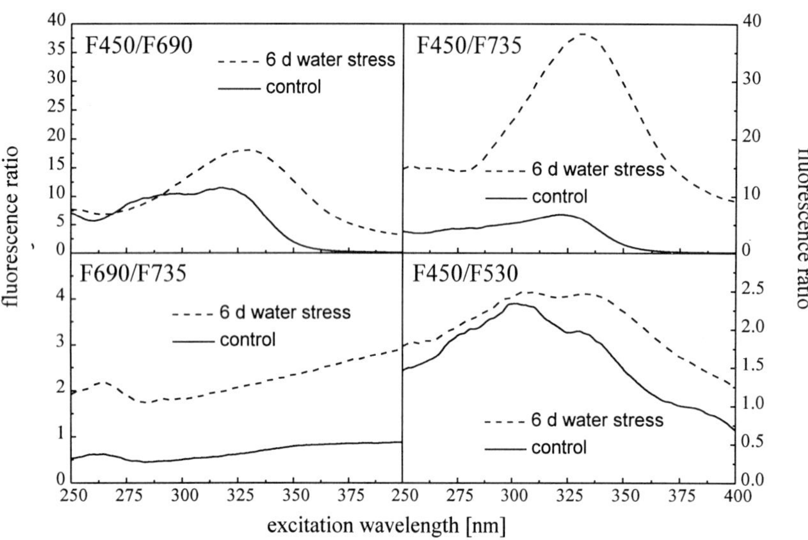

Fig. 13: Fluorescence ratio excitation spectra of leaves of pumpkin seedlings (*Cucurbita pepo*) without (control) and with a 6 d water stress treatment.

UV-B range measured (Figs. 7 and 9). Both fluorescence ratios, F450/F690 and F450/F735, exhibited a maximum near 325 to 330 nm.

Heat treatment (20 h at 40 °C) and water stress treatment in pumpkin seedlings (6 d) resulted in an extreme increase in the blue and green fluorescence yield over the whole fluorescence excitation region from 250 to 400 nm. The red and far-red chlorophyll fluorescence yield changes were not so clear. Both chlorophyll fluorescence bands increased by several times after heat treatment due to the impairment of photosynthetic quantum conversion (cf. Greer et al., 1986), but the increase was less pronounced in the water stressed plants and showed up only at the shorter UV-wavelengths from 250 to about 350 nm. The increases in red and far-red chlorophyll fluorescence yields in heat and water stressed plants were due to a particular mechanism related to the large increase in blue and green fluorescence emission. The blue and green fluorescence, which primarily emanates from the epidermis, is emitted in all directions and had drastically risen after heat and water-stress treatment. It also penetrated into the leaf mesophyll in substantial amounts, and there excited red and far-red chlorophyll fluorescence. In this way the rather damaging UV-A and UV-B radiation were detoxified in the epidermal cells via transformation into visible light (blue-green fluorescence), which would be absorbed by chlorophylls and be used for either chlorophyll fluorescence emission or for photosynthetic quantum conversion. The detoxification of incident UV-A and UV-B light via transformation into visible light (in form of blue and green fluorescence), which can be consumed by the chloroplasts of the mesophyll cells in photosynthetic quantum conversion, is a new photoprotection mechanism of plants against UV-radiation.

In contrast to heat and water stressed pumpkin seedlings, the blue and green fluorescence of sun leaves of beech and maple had not increased as compared to shade leaves. For this reason the sun leaves did not show an increase in chlorophyll fluorescence, which was induced in stressed pumpkin seedlings by the enhanced blue-green fluorescence emission, but they showed a considerable decrease in the chlorophyll fluorescence yield. The latter must have been caused by the accumulation of such UV-absorbing substances in the sun leaves' epidermis, which did not fluoresce since the overall blue-green fluorescence did not increase. These substances, possibly flavonols and other phenolics (Lang et at., 1991), function as UV-filters for the protection of the photosynthetic apparatus against damaging UV-radiation. It had been pointed out before that 80 % or more of the incident UV-radiation is absorbed by the epidermis layer in field plants (Caldwell et at., 1983).

From these results obtained with the fluorescence excitation spectra of stressed and non-stressed plant leaves one can deduct that leaves possess two mechanisms against UV-protection:

1) In their epidermis layer they accumulate UV-absorbing substances which function as UV-filter for the leaf mesophyll, and

2) they can transform potentially damaging UV-radiation into blue and green fluorescence no longer damaging that can be used either for photosynthetic quantum conversion, or in part also be re-emitted as red and far-red chlorophyll fluorescence.

Both UV-protection mechanisms work at the same time. However, it depends on the growth conditions, on the plant species and on the stress applied whether protection mechanism 1 or 2 is expressed to a larger extent. In grass plants (Poaceae), which exhibit a larger blue-green fluorescence emission than dicotyledonous plants (Stober and Lichtenthaler, 1993 b), and also under particular stress constraints (as shown here for field plants, high-light stress, heat and water stress enhancing the blue-green fluorescence emission) the UV-protection mechanism 2 is particularly active. The fluorescence excitation spectra shown here also underline that UV-light is not necessary to induce the two UV-protection mechanisms, but that these can be induced directly by heat stress or water stress treatment and also by high-light stress. This is an essential finding of this investigation.

With respect to the most suitable excitation wavelength for the four fluorescence bands of plants the fluorescence excitation spectra show that emission wavelengths ranging from 300 to 340 nm are most suitable in the case of the blue-green fluorescence. Thus, the XeCl excimer laser (λ emission 308 nm) and the nitrogen laser with an emission wavelength of 337 nm appear to be good excitation sources for the blue-green fluorescence of plants, if one onty wants to know the blue-green fluorescence signatures of plants. Although the blue-green fluorescence yield declines at wavelengths > than 340 nm, the Nd:YAG-laser (λ emission 355 nm) is still a good and useful excitation source.

Concerning the excitation of red and far-red chlorophyll fluorescence the XeCl excimer laser and the nitrogen laser provide, however, a very low chlorophyll fluorescence and are not suitable for its excitation. The Nd:YAG-laser (355 nm) is a good compromise, if one wants to simultaneously excite and measure the blue-green as well as the red and far-red chlorophyll fluorescence. Longer wavelengths ranging from 370 to 390 nm would be more suitable for chlorophyll fluorescence excitation, in particular in field plants, stressed plants and sun leaves, but in that case the blue-green fluorescence would be at a relatively low yield. Thus, for the new fluorescence imaging procedure of monitoring the state of health of plants (Lang et al., 1994, 1996; Lichtenthaler et al., 1996) the recommendation for the proper fluorescence excitation wavelengths can be given as the range from 355 and 390 nm. This wavelength region permits the excitation of the four fluorescence bands blue, green, red and far-red. If only the blue-green fluorescence is of interest, a wavelength region of 300 to 340 nm can be recommended. If only the red and far-red chlorophyll fluorescence should be excited, the blue light range from 420 to 460 nm would be most suitable for excitation, since the red and far-red chlorophyll fluorescence excitation spectra of leaves exhibit a broad maximum in this region. But also a He/Ne laser (λ 632.8 nm) or green light is suitable for the excitation of the chlorophyll fluorescence.

However, monitoring of stress constraints in plants by high resolution fluorescence imaging is not and cannot be based on the absolute fluorescence yield in the four fluorescence bands alone, which can vary from sample to sample due to the sensing angle and roughness of the leaf surface, but is essentially bound to the four fluorescence ratios blue/red, blue/far-red, red/far-red and blue/green. The excitation spectra of these fluorescence ratios (Figs. 2, 5, 7, 9, 11 and 13) demonstrated that the two fluorescence ratios blue/red (F450/F690) and blue/far-red (F450/F735), which increased under the influence of stress, are the most suitable indicators for stress and strain detection in plants. Since the fluorescence ratios blue/red and blue/far-red exhibit maxima in the range from 320 to 340 nm, that wavelength range would be favoured as a fluorescence excitation source. However, also wavelengths longer than 340 nm can detect stress in plants via the fluorescence ratios blue/red and blue/far-red. Since in highly stressed plants hardly any chlorophyll fluorescence was detected at an excitation wavelength of 320 to 340 nm, this favours again an excitation between 355 nm (Nd:YAG-laser) and 390 nm.

References

Caldwell, M. M., R. Robberecht, and S. D. Flint: Internal filters: Prospects for UV-acclimation in higher plants. Physiol. Plant. *58,* 445–450 (1983).

Cerovic, C. G., F. Morales, and I. Moya: Time-resolved spectral studies of blue-green fluorescence of leaves, mesophyll and chloroplasts of sugar beet (*Beta vulgaris* L.). Biochim. Biophys. Acta *1188,* 58–68 (1994).

Chappelle, E. W., F. M. Wood, J. E. McMurtrey, and W. W. Newcomb: Laser-induced fluorescence of green plants. 1: A technique for remote detection of plant stress and species differentiation. Appl. Opt. *23,* 134–138 (1984).

Chappelle, E. W., F. M. Wood, W. W. Newcomb, and J. E. McMurtrey: Laser-induced fluorescence of green plants. 3: LIF spectral signatures of five major plant types. Appl. Opt. *24,* 74–80 (1985).

D'AMBROSIO, N., K. SZABÓ, and H. K. LICHTENTHALER: Increase of the chlorophyll fluorescence ratio F690/F735 during the autumnal chlorophyll breakdown. Radiat. Environ. Biophys. *31,* 51–62 (1992).

DAHN, H. G., K. P. GÜNTHER, and W. LÜDECKER: Characterisation of drought stress of maize and wheat canopies by means of spectral resolved laser induced fluorescence. EARSeL Adv. Remote Sens. *1,* 12–19 (1992).

EDNER, H., J. JOHANSSON, S. SVANBERG, H. K. LICHTENTHALER, M. LANG, F. STOBER, C. SCHINDLER, and L. O. BJÖRN: Remote multi-colour fluorescence imaging of selected broad-leaf plants. EARSeL Advances in Remote Sensing *3,* (Part 3), 2–14 (1995).

GOULAS, Y., I. MOYA, and G. SCHMUCK: Time resolved spectroscopy of the blue fluorescence of spinach leaves. Photosynth. Res. *25,* 299–307 (1990).

GOVINDJEE: Sixty-three years since Kautsky: chlorophyll *a* fluorescence. Austral. J. Plant Physiol. *22,* 131–160 (1995).

GOVINDJEE, J. AMESZ, and D. C. FORK: Light emission by Plants and Bacteria. Academic Press Orlando, Florida, USA (1986).

GREER, D. H., J. BERRY, and O. BJÖRKMAN: Photoinhibition of photosynthesis in intact bean leaves: role of light, temperature and requirement of chloroplast-protein synthesis during recovery. Planta *168,* 253–260 (1986).

HÁK, R., H. K. LICHTENTHALER, and U. RINDERLE: Decrease of the fluorescence ratio F690/F730 during greening and development of leaves. Radiat. Environ. Biophys. *29,* 329–336 (1990).

HÁK, R., U. RINDERLE-ZIMMER, H. K. LICHTENTHALER, and L. NÁTR: Chlorophyll *a* fluorescence signatures of nitrogen-deficient barley leaves. Photosynthetica *28,* 151–159 (1993).

HARRIS, P. J. and R. D. HARTLEY: Detection of bound ferulic acid in cell walls of the gramineae by ultraviolet fluorescence microscopy. Nature *259,* 508–510 (1976).

KAUTSKY, H. and A. HIRSCH: Neue Versuche zur Kohlensäureassimilation. Naturwiss. *19,* 964 (1931).

KRAUSE, G. H. and E. WEIS: Chlorophyll fluorescence and photosynthesis: The basics. Ann. Rev. Plant Physiol. Plant Mol. Biol. *42,* 313–349 (1991).

LANG, M.: Studies on the blue-green and chlorophyll fluorescence of plants and their application for fluorescence imaging of leaves. Karlsruhe Contrib. Plant Physiol. *29,* 1–110 (1995).

LANG, M. and H. K. LICHTENTHALER: Changes in the blue green and red fluorescence emission spectra of beech leaves during the autumnal chlorophyll breakdown. J. Plant Physiol. *138,* 550–553 (1991).

LANG, M., F. STOBER, and H. K. LICHTENTHALER: Fluorescence emission spectra of plant leaves and plant constituents. Radiat. Environ. Biophys. *30,* 333–347 (1991).

LANG, M., P. ŠIFFEL, Z. BRAUNOVA, and H. K. LICHTENTHALER: Investigations on the blue-green fluorescence emission of plant leaves. Bot. Acta *105,* 435–440 (1992).

LANG, M., H. K. LICHTENTHALER, M. SOWINSKA, P. SUMM, and F. HEISEL: Blue, green and red fluorescence signatures and images of tobacco leaves. Bot. Acta *107,* 230–236 (1994).

LANG, M., H. K. LICHTENTHALER, M. SOWINSKA, P. SUMM, F. HEISEL, J. A. MIEHE, and F. TOMASINI: Application of laser-induced fluorescence imaging in the detection of plant stress. In: RICHTER, P. I. and R. C. HERNDON (eds.): Proceed. 2nd Internat. Symposium and Exhibition on Environmental Contamination in Central and Eastern Europe, Budapest 1994, pp. 88–90. Government Institutes, Rockville USA (1995).

LANG, M., H. K. LICHTENTHALER, M. SOWINSKA, F. HEISEL, J. A. MIEHE, and F. TOMASINI: Fluorescence imaging of water and temperature stress in plant leaves. J. Plant Physiol. *148,* 613–621 (1996).

LICHTENTHALER, H. K.: Chlorophyll fluorescence signatures of leaves during autumnal chlorophyll breakdown. J. Plant Physiol. *131,* 101–110 (1987).

– *In vivo* chlorophyll fluorescence as a tool for stress detection in plants. In: LICHTENTHALER, H. K. (ed.): Applications in Chlorophyll Fluorescence, pp. 99–107. Kluwer Acad. Publishers, Dordrecht (1988).

– Applications of chlorophyll fluorescence in stress physiology and remote sensing. In: STEVEN, M. and J. A. CLARK (eds.): Applications of Remote Sensing in Agriculture, pp. 287–305. Butterworths Scientific Ltd., London (1990).

– The Kautsky Effect: 60 years of chlorophyll fluorescence induction kinetics. Photosynthetica *27,* 45–55 (1992).

– Vegetation stress: an introduction to the stress concept in plant. J. Plant Physiol. *148,* 4–14 (1996).

LICHTENTHALER, H. K. and U. RINDERLE: The role of chlorophyll fluorescence in the detection of stress conditions in plants. CRC Crit. Rev. Analyt. Chem. *19* (Suppl. 1), 29–85 (1988).

LICHTENTHALER, H. K., C. BUSCHMANN, U. RINDERLE, and G. SCHMUCK: Applications of chlorophyll fluorescence in ecophysiology. Radiat. Environ. Biophys. *25,* 297–308 (1986).

LICHTENTHALER, H. K., R. HÁK, and U. RINDERLE: The chlorophyll fluorescence ratio F690/F730 in leaves of different chlorophyll content. Photosynth. Res. *25,* 295–298 (1990).

LICHTENTHALER, H. K., M. LANG, and F. STOBER: Nature and variation of blue fluorescence spectra of terrestrial plants. In: Internat. Geoscience and Remote Sensing Symposium IGARSS '91, Espoo (Finland), pp. 2283–2286. Helsinki University of Technology, Espoo (1991).

LICHTENTHALER, H. K., F. STOBER, and M. LANG: The nature of the different laser-induced fluorescence signatures of plants. EARSeL Advances in Remote Sensing *1,* 20–32 (1992).

– – – Laser-induced fluorescence emission signatures and spectral fluorescence ratios of terrestrial vegetation. In: Internat. Geoscience and Remote Sensing Symposium IGARSS '93, Tokyo (Japan), pp. 1317–1320. Kogakuin University, Tokyo (1993).

LICHTENTHALER, H. K., M. LANG, M. SOWINSKA, F. HEISEL, and J. A. MIEHÉ: Detection of vegetation stress via a new high resolution fluorescence imaging system. J. Plant Physiol. *148,* 599–612 (1996).

MORALES, F., Z. G. CEROVIC, and I. MOYA: Characterisation of blue-green fluorescence in the mesophyll of sugar beet (*Beta vulgaris* L.) leaves affected by iron deficiency. Plant Physiol. *106,* 127–133 (1994).

RINDERLE, U., C. SCHINDLER, and H. K. LICHTENTHALER: The laser-induced chlorophyll fluorescence ratio F690/F735 of spruce needles and beech leaves during the course of a year. In: 5th International Colloquium on Physical Measurements and Spectral Signatures in Remote Sensing, Courchevel 1991, pp. 731–734. ESA Publications Division, Nordwijk (1991).

SCHMUCK, G., I. MOYA, A. PEDRINI, D. VAN DER LINDE, H. K. LICHTENTHALER, F. STOBER, C. SCHINDLER, and Y. GOULAS: Chlorophyll fluorescence lifetime determination of waterstressed C3 and C4-plants. Radiat. Environ. Biophys. *31,* 141–151 (1992).

SCHREIBER, U., U. SCHLIWA, and W. BILGER: Continous recording of photochemical and non-photochemical chlorophyll fluorescence quenching with a new type of modulation fluorometer. Photosynth. Res. *10,* 51–62 (1986).

STOBER, F. and H. K. LICHTENTHALER: Changes of the laser-induced blue, green, and red fluorescence signatures during greening of etiolated leaves of wheat. J. Plant Physiol. *140,* 673–680 (1992).

– – Characterisation of the laser-induced blue, green and red fluorescence signatures of leaves of wheat and soybean leaves grown under DIFFERENT IRRADIANCE. PHYSIOL. PLANT. *88,* 696–704 (1993 a).

– – Studies on the constancy of the blue and green fluorescence yield during the chlorophyll fluorescence induction kinetics (Kautsky effect). Radiat. Environ. Biophys. *32,* 357–365 (1993 b).

– – Studies on the localisation and spectral characteristics of the fluorescence emission of differently pigmented wheat leaves. Bot. Acta *106,* 365–370 (1993 c).

STOBER, F., M. LANG, and H. K. LICHTENTHALER: Blue, green and red fluorescence emission signatures of green, etiolated, and white leaves. Remote Sens. Environ. *47,* 65–71 (1994).

J. Plant Physiol. Vol. 148. pp. 548–554 (1996)

Application of the Karlsruhe CCD-OMA LIDAR-Fluorosensor in Stress Detection of Plants

CLAUS BUSCHMANN[1], JOACHIM SCHWEIGER[1], HARTMUT K. LICHTENTHALER[1], and PETER RICHTER[2]

[1] Botanisches Institut II, University of Karlsruhe, Kaiserstr. 12, D-76128 Karlsruhe, Germany

[2] Institute of Physics, Department of Atomic Physics, Technical University of Budapest, Budafoki út 8, H-1111 Budapest, Hungary

Received August 12, 1995 · Accepted October 20, 1995

Summary

The Karlsruhe CCD-OMA LIDAR-Fluorosensor (excitation: cw HeNe laser, 632.8 nm; detection: complete chlorophyll fluorescence spectra between 650–800 nm within 10 ms) was developed in a joint project by the Technical University of Budapest and the University of Karlsruhe for a non-destructive stress detection of plants. The computer-aided fluorosensor permits to measure the full fluorescence spectra at 8 different time intervals during the chlorophyll fluorescence induction kinetics, from which the the variable chlorophyll fluorescence ratio Rfd as plant vitality index can be calculated. The position of the chlorophyll fluorescence emission maxima of green leaves are found in the red region near 690 nm and the far-red region near 735 nm. The wavelength position of the two maxima remained unchanged during the light-induced fluorescence induction kinetics (Kautsky effect). The absolute fluorescence intensity and the ratio of the two chlorophyll fluorescence bands (F690/F735) depended on the chlorophyll content and the photosynthetic activity of a leaf as has been demonstrated here with the CCD-OMA LIDAR fluorosensor.

With increasing chlorophyll content of the leaf the fluorescence intensity of the red band near 690 nm decreased, whereas that of the far-red band near 735 nm remained constant or slightly increased. Consequently, the fluorescence ratio F690/F735 increased with decreasing chlorophyll concentration of leaves (curvi-linear relationship: $y = a \cdot x^{-b}$). Changes in the fluorescence ratio F690/F735 can be taken as an indicator of long-term stress affecting the chlorophyll content of leaves. The inverse relationship between the intensity of the chlorophyll fluorescence and the rate of photosynthesis (Kautsky effect) can be used for detecting, via Rfd-values, a short-term damage to plants, which affects photosynthetic activity, but does not yet decrease the chlorophyll content of the leaf. Damage or stress is indicated by low Rfd-values (ratio fluorescence decrease) or low Ap-values (stress adaptation index). Examples are shown for different types of damage and/or stresses (water stress, forest decline phenomena, application of a herbicide inhibiting photosynthesis, and biological stress, e.g. damage by mites). The Karlsruhe CCD-OMA LIDAR-Fluorosensor proved to be a valuable tool for fast detection of stress to plants in the laboratory, but can also be applied for ground-truth control measurements during remote sensing of the state of health of terrestrial vegetation.

Key words: Laser-induced chlorophyll fluorescence, stress detection of plants, Kautsky effect, ratio F690/F735, Rfd-values.

Abbreviations: a+b = total chlorophyll *a+ b* content; Ap = stress-adaptation index; CCD = charged coupled device; DCMU = 3-(3,4-dichlorophenyl)-1,1-dimethyl urea (diuron); F690/F735 ratio of chlorophyll fluorescence at 690 and 735 nm; fmax = chlorophyll fluorescence maximum of the induction kinetics (Kautsky effect); fs = chlorophyll fluorescence in the steady state of the induction kinetics (Kautsky effect); LIDAR = light-induced detection and ranging; OMA = optical multichannel analyzer; Rfd = ratio of the variable chlorophyll fluorescence (ratio fluorescence decrease); x+c = total carotenoid content (xanthopyhylls+carotenes).

Introduction

All vegetation is subjected to different kinds of natural and anthropogenic stresses. The plant may adapt to low-level and short-term stress, but high-level stress and long-term stress lead to damages (see Lichtenthaler, 1996). Damage effects on plants may also be dangerous to human health. By means of optical methods stress and damage effects to terrestrial vegetation can be monitored remotely. In contrast to reflection measurements, which are widely used today for remote sensing of vegetation, the red and far-red chlorophyll fluorescence emitted by leaves after absorption of light is a specific characteristic of plants. The shape of the chlorophyll fluorescence emission spectra and the variation of fluorescence intensity in the second and minute range during the fluorescence induction kinetics change under stress and can be applied to characterize the state of health of plants (e.g. Lichtenthaler et al., 1986; Rinderle and Lichtenthaler, 1988; Hák et al., 1990; Lichtenthaler et al., 1990; D'Ambrosio et al., 1992). Complete fluorescence emission spectra of leaves have been measured before by means of diode array detectors (Buschmann and Schrey, 1981; Lichtenthaler and Buschmann, 1987; Buschmann and Lichtenthaler, 1988). Diode array detectors have a serial read-out of the data, thus, the signals are not collected at the same time and small spectral changes may occur during the read-out procedure, in particular when the measuring times are short and approach the read-out times. In contrast, CCD detectors (Charged Coupled Device) have the advantage that the read-out of the data proceeds in parallel, and thus occurs simultaneously for all wavelengths. In a joint project between physicists and plant physiologists we developed a CCD-OMA LIDAR-Fluorosensor (OMA = Optical Multichannel Analyzer, LIDAR = light-induced detection and ranging), which is able to record the complete spectrum of the chlorophyll fluorescence between 650 and 800 nm within 10 milliseconds or also longer time periods if desired (Szabó et al., 1992). Here we give several examples for the detection of various stress constraints to plants by means of this laser-equipped CCD-OMA fluorosensor.

Materials and Methods

Plants

Leaves of 4 to 6-week-old plants of tobacco (*Nicotiana tabacum* L.) and bean (*Phaseolus vulgaris* L.), grown in the greenhouse of the Botanical Garden of the University of Karlsruhe were examined. Leaves of healthy bean plants were compared to those of plants affected by mites (*Tetranychus urticae*). Leaves of ivy (*Hedera helix* L.) were taken from a plant growing on the campus of the University of Karlsruhe. On May 30, 1995 we examined young, only several week old needles of this year and one-year-old needles of spruce (*Picea abies* Karst.) from a healthy and a stressed tree at the «Schöllkopf» forest decline test site near Freudenstadt (Black Forest) at 800 m above sea level.

DCMU Treatment

The herbicide diuron, known as DCMU (3-(3,4-dichlorophenyl)-1,1-dimethyl urea), was applied by wetting the lower leaf side of a tobacco leaf with a 0.1 mM solution of DCMU dissolved in a buffer by means of a hair brush.

Pigments

The content of chlorophylls and carotenoids was determined from 100 % acetone extracts of leaves and needles using the redetermined extinction coefficients and equations of Lichtenthaler (1987) which permit simultaneous determinations of both pigment groups in one extract solution. The pigment values given (e.g. Table 1) are based on 2 determinations with a standard deviation of $<5\%$. In the case of spruce 20 needles were extracted from the one-year-old needles, and 80 needles from the fresh this year needles. The pigments in the needles were referred to the projected needle area determined by computer-aided image analysis (Buschmann et al., 1990).

Fluorescence Measurements

The Karlsruhe CCD-OMA fluorescence system (Fig. 1) consists of a 10 mW He/Ne laser (632.8 nm) as light source for excitation of the fluorescence, a spectrograph (grating: 600 lines per mm; entrance slit: 2 mm providing an optical resolution of 10 nm; 6 nm per mm dispersion at the image plane) and a CCD-line as detector (2048 pixel elements). For a more detailed description see Szabó et al. (1992). Before the fluorescence measurements the leaves were adapted to darkness for about 15 min. The 10 mW He/Ne laser provides a high photon flux density of ca. 40,000 $\mu mol\, m^{-2} s^{-1}$ at which the chlorophyll fluorescence emission is saturated. Despite this high irradiance, we did not find any damage in leaves and the measurements could be repeated several times. The leaves were fixed with the upper leaf side facing the excitation light and the detection system. The onset of illumination of the leaf sample was started by opening a fast electromechanical shutter (opening time of 1 ms) about 20 ms after starting the measurement. A personal computer (AT) controled the measurement and the data acquisition from the detector via an interface card developed and constructed in the Budapest laboratory. Each spectrum consisted of 150 measuring points which were detected via a parallel read-out at a minimum integration time of 10 ms. Eight spectra can be consecutively measured at any time during illumination which can freely be chosen by the experimentor. For example 3 spectra before, directly at and shortly after reaching the maximum chlorophyll fluorescence fmax and then 5 spectra during the slow decrease of the chlorophyll fluorescence induction kinetics to the steady state fluorescence fs, which is reached after ca. 5 min of illumination. Each data point represents the sum of four neighbouring detector pixels. The variable fluorescence ratios known as Rfd-values, ratio (fmax – fs)/fs, and the chlorophyll fluorescence ratio F690/F735 were calculated using 10 data points (i.e. a 3 nm range) in the fluorescence emission maxima near 690 and also near 735 nm, thus reducing the arbitrary influence of signal variation due to the measuring noise. Data processing and the display of

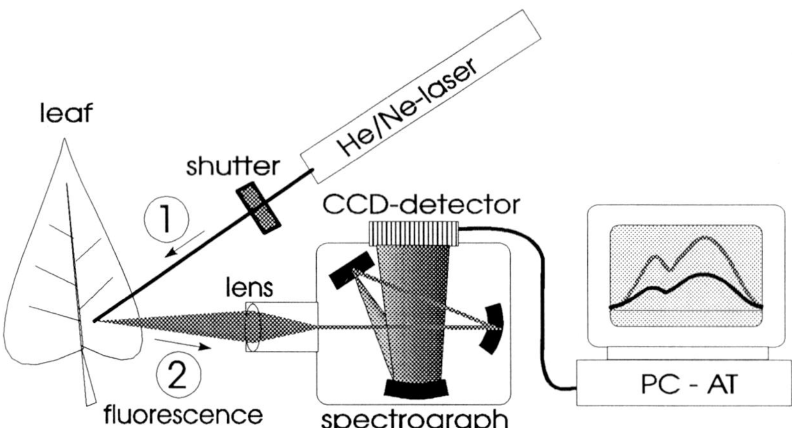

Fig. 1: Scheme of the Karlsruhe laser-eqipped CCD-OMA LIDAR-fluorosensor which enables to measure complete emission spectra of the red and far-red chlorophyll fluorescence (650–800 nm) in the millisecond range. 1) excitation light 632.8 nm of a 10 mW He/Ne laser; 2) fluorescence emitted from the leaf. For further details see Materials and Methods.

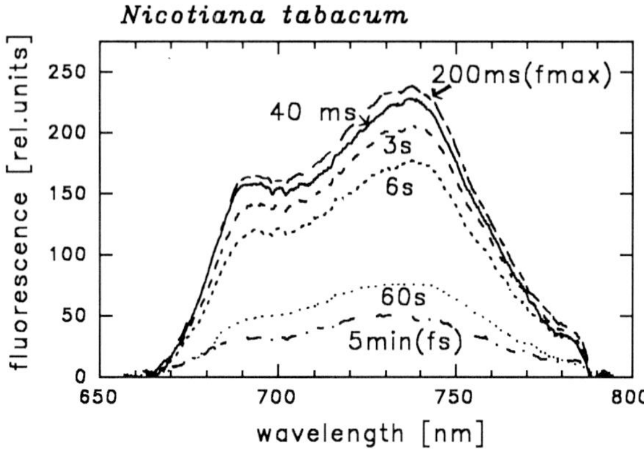

Fig. 2: Example for the fast measurement of complete chlorophyll fluorescence emission spectra of an intact predarkened leaf during the light-induced induction of photosynthesis. The spectra of a healthy dark green tobacco leaf (*Nicotiana tabacum* L.) were taken with an integration time of 10 ms. The measuring time of the fluorescence spectrum after onset of the illumination (from 40 ms to 5 min) is given for each spectrum. Maximum chlorophyll fluorescence fmax was reached 200 ms after onset of illumination, the steady state fluorescence fs was reached 5 min after onset of illumination (chlorophyll content: 38 μg per cm^2 leaf area).

the resulting spectra is carried out by a computer programme written by K. Szabó. The data are transferred into a worksheet programme (LOTUS-123) and then plotted by means of graphic software (SIGMAPLOT).

Results and Discussion

Emission spectrum of the chlorophyll fluorescence of a leaf

The *in vivo* emission spectrum of the chlorophyll fluorescence of a leaf taken at ambient temperature is characterized by two bands with maxima in the regions at 690 and 735 nm (Fig. 2). It has been shown before that the positions of these maxima remain unchanged during the chlorophyll fluorescence induction kinetics regardless of the chlorophyll content and the photosynthetic activity of the leaf (Buschmann and Schrey, 1985; Lichtenthaler and Buschmann, 1987), which is confirmed here with the CCD-OMA fluorosensor characterized by a parallel read-out of data. Changes in the relative intensity of the fluorescence and in the ratio of the two fluorescence bands F690/F735, however, occur during the induction period. In the early chlorophyll fluorescence studies, when the signals were detected through a filter in a more or less wide spectral range, it could not be ruled out with certainty whether changes in the intensity of the fluorescence signal might have been due to a variation of the position of the emission maxima. But this possibility can be excluded completely with the present fast recording, high resolution fluorometers, such as the CCD-OMA fluorosensor.

Ratio F690/F735 as indicator of chlorophyll content of a leaf

With increasing chlorophyll content of the leaf the band at 690 nm decreases, whereas the band at 735 nm increases slightly or does not change. This can be seen by comparing the spectra (Fig. 3) of a bean leaf before and after damage by mites (*Tetranychus urticae*). Mites populate the lower leaf side and suck out the chloroplasts from cells and produce yellow-

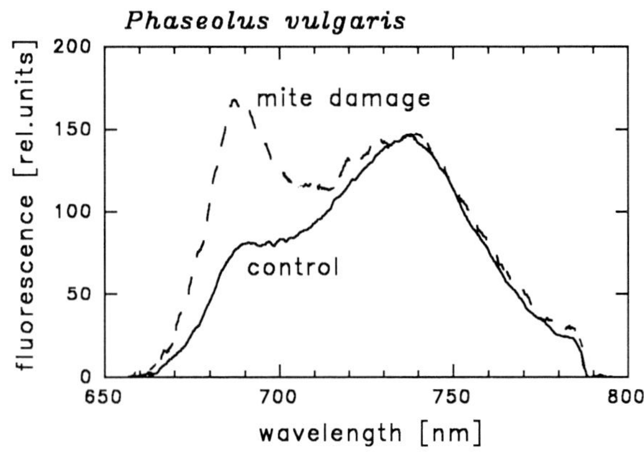

Fig. 3: Chlorophyll fluorescence emission spectra of a healthy bean leaf (*Phaseolus vulgaris* L.) and a bean leaf damaged by mites as measured in the kinetic fluorescence maximum 200 ms after onset of illumination (integration time: 20 ms). Control leaf: 35 μg chlorophyll per cm^2 leaf area; leaf with heavy mite damage: 15 μg chlorophyll per cm^2 leaf area.

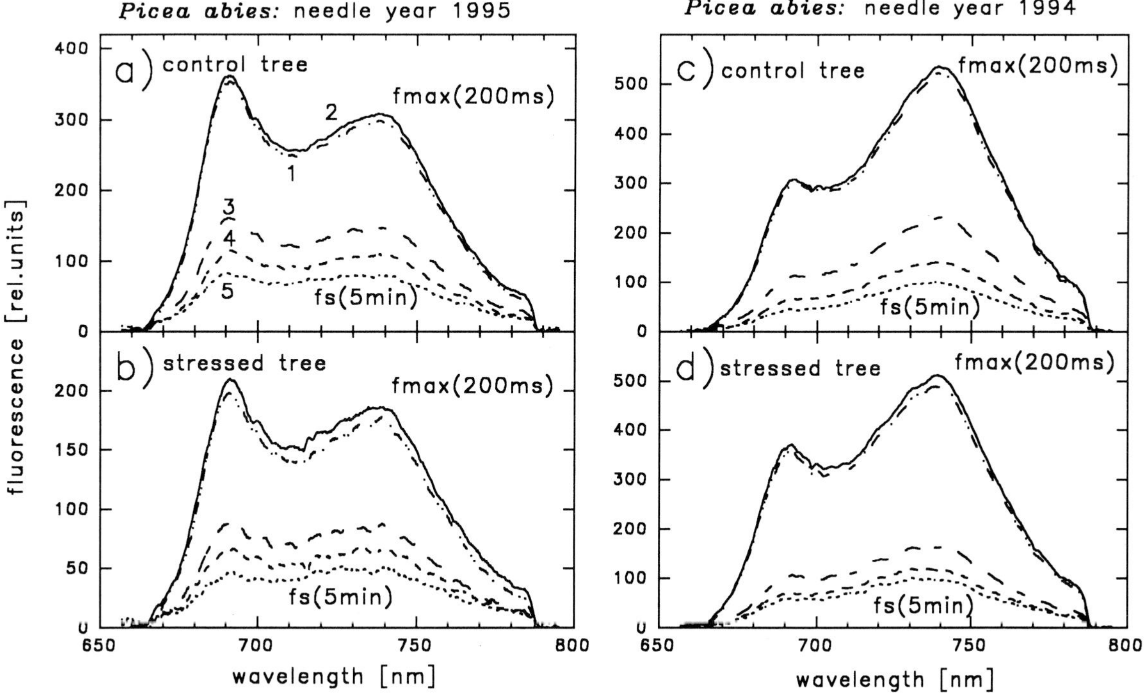

Fig. 4: Chlorophyll fluorescence emission spectra of spruce needles (*Picea abies* Karst.) from a healthy tree (control) and a tree affected by forest decline phenomena at the Schöllkopf test-site close to Freudenstadt (Black Forest). From each tree the youngest light green needles (needle year 1995) were compared with the dark green one-year-old needles (needle year 1994), see Table 1. The spectra were taken with an integration time of 100 ms at different times of the fluorescence induction kinetics: 100 ms, 200 ms (maximum fluorescence fmax), 20 s, 60 s, and 5 min (steady state fluorescence fs) as shown in the spectra curves 1 to 5, respectively.

ish-white spots distributed over the leaves. The chlorophyll fluorescence emission spectra of such mite-affected leaves with a lower chlorophyll content show a higher 690 nm maximum than that of the undamaged controls which possess a higher chlorophyll content. Thus mite attack can be monitored via increases in the fluorescence ratio F690/F735.

Another example is shown in Figure 4 for the comparison of needles from a healthy spruce tree and a spruce affected by forest decline damage symptomes. The young and small needles with a (still) low and almost similar chlorophyll content for both trees (Table 1) show a higher F690/F735 ratio than the larger, dark-green one-year-old needles. For the latter the fluorescence ratio F690/F735, however, is significantly higher in the needles of the stressed tree as compared to those of the healthy tree. This can be explained by the lower chlorophyll content of the needles of the stressed tree (Table 1), which is one stress symptom.

The decrease of the 690 nm maximum in a healthy leaf or needle with increasing chlorophyll content is caused by the overlap of the fluorescence emission spectrum in the 690 nm region with the absorption spectrum of the leaf chlorophyll which leads to a partial reabsorption of the 690 nm region fluorescence inside the leaf tissue (Rinderle and Lichtenthaler, 1988). It has been demonstrated before by several authors (Rinderle and Lichtenthaler, 1988; Hák et al., 1990; D'Ambrosio et al., 1992) that the ratio of the two fluorescence maxima (F690/F735) is decreasing with increasing chlorophyll content in a curvi-linear relationship which can be described by the equation:

$$F690/F735 = k_1 \cdot chl^{-k_2}$$

where chl is the chlorophyll content per leaf area determined from the leaf extract, and k_1 as well as k_2 are constants which vary within narrow limits depending on the plant species or leaf structure. A theoretical treatment of this relationship of the two fluorescence maxima was presented by Strasser et al. (1988). Since long-term stress often leads to a reduced chlorophyll content, the chlorophyll fluorescence ratio F690/F735 can be used for stress detection and chlorophyll decline in terrestrial vegetation. In addition, the detection of the ratio F690/F735 can be applied for a fast, non-destructive determination of the chlorophyll content (C_{chl}) of a leaf. For this purpose the equation given above must be transformed:

$$C_{chl} = \left(\frac{k_1}{F690/F735}\right)^{1/k_2}$$

Fluorescence induction kinetics as indicators of photosynthetic activity

When plants kept in the dark are illuminated the photosynthetic processes are first impaired, and several minutes are required for the induction of full photosynthetic function.

Table 1: Pigment content and various chlorophyll fluorescence characteristics of spruce needles (*Picea abies* Karst.) from a healthy control tree and a stressed tree at the Schöllkopf forest decline test-site close to Freudenstadt (Black Forest). From each tree the small light green needles of the youngest year (1995) are compared with the dark green one-year-old needles (needle year 1994). The emission spectra of the chlorophyll fluorescence of the needles are shown in Figure 4.

	control tree		stressed tree	
	needle year 1995	needle year 1994	needle year 1995	needle year 1994
needle area [mm^2]	3.2	23.8	4.2	14.4
chlorophyll (a+b) [$mg\ m^{-2}$]	140	660	110	420
carotenoid (x+c) [$mg\ m^{-2}$]	35	139	33	94
chlorophyll *a/b*	2.61	3.07	3.72	3.53
(a+b)/(x+c)	4.03	4.72	3.35	4.51
fluorescence signals [rel. untis]				
F690 at fmax	595	510	345	610
F690 at fs	138	75	80	100
F735 at fmax	505	880	310	840
F735 at fs	130	165	80	160
fluorescence ratios				
F690/F735 at fmax	1.18	0.58	1.11	0.73
F690/F735 at fs	1.06	0.45	1.00	0.63
Rfd at 690 nm	3.31	5.80	3.30	5.10
Rfd at 735 nm	2.88	4.33	2.87	4.25
Ap	0.10	0.22	0.10	0.14

The measurement of induction kinetics («Kautsky effect», Kautsky and Hirsch, 1931) shows a fast chlorophyll fluorescence rise within the first second of illumination to a maximum (fmax) and then – parallel to the onset of photosynthesis – a slow decrease within about 5 min to a steady state (fs). By means of the CCD-OMA fluorosensor (Fig. 1) the fluorescence emission spectra can be measured during fluorescence rise and decline.

The decline of the fluorescence during the induction kinetic, sometimes also called variable fluorescence, is related to the photosynthetic electron transport around photosystem II and to state 1/state 2 transitions (for a review see Krause and Weis, 1991). The decrease of fluorescence is ascribed to two components: the photochemical and the non-photochemical quenching (Schreiber et al., 1986). This decline of the chlorophyll fluorescence from the fluorescence maximum fmax to the steady state fs is often expressed as Rfd value (ratio of fluorescence decrease fd, Lichtenthaler et al., 1982; Lichtenthaler and Rinderle, 1988): (fmax−fs)/fs or as fd/fs. The height of the Rfd-values, which can be measured in the 690 and 735 nm region (Rfd690 and Rfd735), is proportional to the leaves' net CO_2 assimilation rate P_N (Tuba et al., 1994; Babani et al., 1996).

The Rfd-values at 690 nm amount to 3.0 to 5 for leaves with a medium to high photosynthetic capacity, and are zero in leaves with fully inhibited photosynthesis. The Rfd735-values are always lower than the Rfd690-values, but completely reflect the differentially photosynthetic activity between leaves. When the photosynthetic electron transport around photosystem II is specifically blocked by the herbicide diuron (DCMU), which binds to the Q_B-binding protein instead of Q_B, the chlorophyll fluorescence does not decline to the steady state of controls. Then fs equals fmax and the Rfd-values become zero (Fig. 5). Short term stress blocking photosynthetic electron transport can also be detected via the fluorescence ratio F690/F735, which increased by about 20 to 30 % at diuron treatment. The smaller decrease of the fluorescence signal indicating a reduced photosynthetic activity is also shown in the chlorophyll fluorescence spectra of the older spruce needles taken from the stressed tree as compared to those of the unstressed tree (Fig. 4 and Table 1). Another example is given in Figure 6 for an ivy leaf subjected to water stress. The chlorophyll fluorescence intensity decreased as well as the Rfd-values which can be calculated from the spectra at fmax and fs.

There are many further examples given in the literature for the detection of photosynthetic activity and the *in vivo* chlorophyll content via Rfd-values, and the fluorescence ratio F690/F735 (e.g. for the development of photosynthetic activity in greening seedlings: Buschmann, 1981; Babani et al., 1996) showing that the changes in the induction kinetics expressed as Rfd-values can be taken as an indicator of short-term stress, which does not necessarily lead to reduced chlorophyll content of the leaf. Since the red 690 nm fluorescence shows a larger amplitude during the induction kinetics than the far-red 735 nm fluorescence the Rfd-values deter-

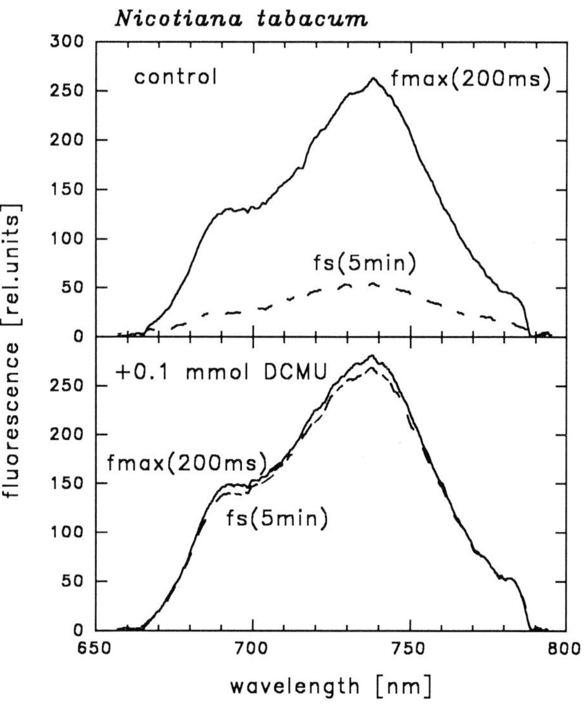

Fig. 5: Chlorophyll fluorescence emission spectra of a tobacco leaf (*Nicotiana tabacum* L.) taken 200 ms (at maximum fluorescence fmax) and 5 min (at steady-state fluorescence fs) after onset of illumination before (control) and after treatment with 0.1 mM DCMU (integration time: 20 ms).

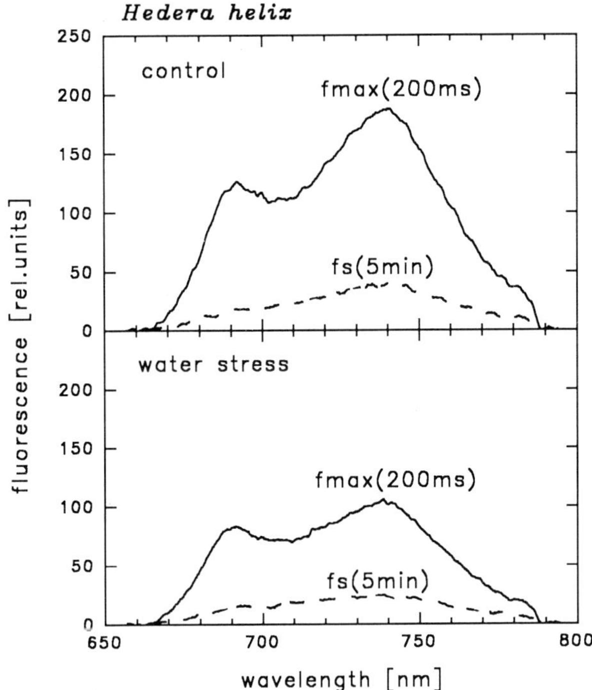

Fig. 6: Chlorophyll fluorescence emission spectrum of an ivy leaf (*Hedera helix* L.) taken 200 ms (at maximum fluorescence fmax) and 5 min after onset of illumination (at steady-state fluorescence fs) before (control) and after 8 hours of water stress induced by detaching the leaf (integration time: 20 ms).

mined at 690 nm are always higher than those at 735 nm (Table 1).

During the induction kinetics of the chlorophyll fluorescence the position of the two fluorescence maxima remained constant, but the fluorescence at 690 nm decreased more strongly than the fluorescence at 735 nm, and thus the ratio of the two maxima F690/F735 decreased (Fig. 2 and Fig. 4). The stronger decrease of the 690 nm fluorescence band during the fluorescence induction kinetics was earlier explained a) by the assumption that the 735 nm fluorescence band may have a higher contribution from deeper cell layers, where the intensity of the excitation light is lower, and thus the photosynthetic induction and the decrease of the fluorescence may be less pronounced than in the upper layers and b) by a possible energy transfer towards chlorophylls emitting at longer wavelengths (Buschmann and Schrey, 1981). Another explanation is the reabsorption of the red 690 nm fluorescence by the leaf chlorophyll which does not apply to the 735 nm fluorescence.

Strasser et al. (1987) found that damage effects of long-term, low-level ozone to beech and poplar showed up earlier when comparing the Rfd-values at 690 and 735 nm than by calculating only one Rfd-value. They therefore proposed to use a stress adaptation index Ap = 1–(1+Rfd735)/(1+Rfd690). This Ap-value has also been determined for other plants and stress constraints (Lichtenthaler and Rinderle, 1988). The Ap-value for the spruce needles shows higher values for the older needles compared to the younger needles, and lower values for one-year-old needles of the stressed tree as compared to one-year-old needles of the unstressed tree (Table 1). The stress adaptation index may vary between values of 0 and 0.4 indicating a stressed or damaged plant (low stress adaptation) and a healthy plant, respectively. Assuming that the chlorophyll fluorescence at 735 nm is partially emitted by photosystem I (Lombard and Strasser, 1984) a high Ap-value may be an expression for a healthy plant with a functioning state 1/state 2-transition of the photosynthetic apparatus. The leaf, which is in state 1 after a dark period of about 30 min, reaches the functional state 2 of photosynthesis after a light period of several minutes when the chlorophyll fluorescence is in its steady state. State 1 is characterized by an energy transfer predominantly towards photosystem II, whereas state 2 is characterized by a balanced energy transfer towards photosystem I and II (Bonaventura and Myers, 1969; Murata, 1969). During a light-induced induction kinetic, when state 1/state 2-transitions occur, more and more energy will be transferred towards photosystem I and thus, the fluorescence band at 735 nm containing a photosystem I component will decrease to a lower extent than the pure photosystem II fluorescence component at 690 nm, which, in addition, is reabsorbed by the leaf chlorophyll.

Conclusions

The shape of the emission spectrum of the chlorophyll fluorescence as well as the intensity changes (fluorescence induction kinetics after onset of illumination) can be taken as an indicator of chlorophyll content (ratio F690/F735) and physiological activity of a leaf and a plant (Rfd690 and Rfd735 values). These parameters clearly change during stress or damage to plants and can therefore be used for early stress detection in plants. The capability of the Karlsruhe CCD-OMA fluorosensor system to measure complete fluorescence emission spectra within milliseconds permits to monitor both long-term and short-term stress damage to the photosynthetic apparatus and plants. This instrument promises to be a valuable tool for laboratory and field measurements as well as for outdoor ground-truth control measurements during remote sensing of vegetation.

Acknowledgements

This joint study was carried out within the cooperation agreement between the Technical University of Budapest and the University of Karlsruhe. We wish to thank the Hungarian and German grant agencies for fincancial support. The supply of spruce needles from the Schöllkopf site by Dorothea Siefermann-Harms is greatfully acknowledged. We would like to thank Gabriele Jahnson for checking the English text.

References

D'Ambrosio, N., K. Szabó, and H. K. Lichtenthaler: Increase of the chlorophyll fluorescence ratio F690/F735 during autumnal breakdown. Radiat. Environ. Biophys. *31,* 51–62 (1992).

Babani, F., P. Richter, and H. K. Lichtenthaler: Changes of chlorophyll fluorescence signatures during greening of etiolated barley seedlings as measured with the CCD-OMA fluorometer. J. Plant Physiol. *148,* 471–477 (1996).

Bonaventura, C. and J. Myers: Fluorescence and oxygen evolution from *Chlorella pyrenoidosa.* Biochim Biophys. Acta *189,* 366–383 (1969).

Buschmann, C.: The characterization of the developing photosynthetic apparatus in greening barley leaves by means of (slow) fluorescence kinetic measurements. In: Akoyunoglou, G. (ed.): Photosynthesis, Vol. *5,* pp. 417–426. Balaban Intern. Sci. Serv., Philadelphia (1981).

Buschmann, C. and H. K. Lichtenthaler: Complete fluorescence emission spectra determined during the induction kinetic using a diode-array detector. In: Lichtenthaler, H. K. (ed.): Applications of Chlorophyll Fluorescence, pp. 77–84. Kluwer Academic Publishers, Dordrecht (1988).

Buschmann, C. and H. Schrey: Fluorescence induction kinetics of green and etiolated leaves by recording the complete *in vivo* emission spectra. Photosynth. Res. *1,* 233–241 (1981).

Buschmann, C., E. Nagel, S. Rang, and F. Stober: Reflexionsspektren von Blättern und Nadeln als Basis für die physiologische Beurteilung von Baumschäden – 3. Freilandmessungen an Laub- und Nadelbäumen. KFK-PEF-Report *61,* 323–333 (1990).

Hák, R., H. K. Lichtenthaler, and U. Rinderle: Decrease of the chlorophyll fluorescence ratio F690/F730 during greening and development of leaves. Radiat. Environ. Biophys. *29,* 329–336 (1990).

Kautsky, H. and A. Hirsch: Neue Versuche zur Kohlenstoffassimilation. Naturwiss. *19,* 964 (1931).

Krause, G. H. and E. Weiss: Chlorophyll fluorescence and photosynthesis: The basics. Annu. Rev. Plant Physiol. Plant Mol. Biol. *42,* 313–349 (1991).

Lichtenthaler, H. K.: Chlorophylls and carotenoids, the pigments of photosynthetic biomembranes. Methods in Enzymol. *148,* 350–382 (1987).

– The Kautsky Effect: 60 years of chlorophyll fluorescence induction kinetics. Photosynthetica *27,* 45–55 (1992).

– Vegetation stress: an introduction to the stress concepts of plants. J. Plant Physiol. *148,* 1–14 (1996).

Lichtenthaler, H. K. and C. Buschmann: Chlorophyll fluorescence spectra of green bean leaves. J. Plant Physiol. *129,* 137–147 (1987).

Lichtenthaler, H. K. and U. Rinderle: The role of chlorophyll fluorescence in the detection of stress conditions in plants. CRC Crit. Rev. Analyt. Chem. *19,* Suppl. I, 29–85 (1988).

Lichtenthaler, H. K., C. Buschmann, U. Rinderle, and G. Schmuck: Application of chlorophyll fluorescence in ecophysiology. Radiat. Environ. Biophys. *25,* 297–308 (1986).

Lichtenthaler, H. K., R. Hák, and U. Rinderle: The chlorophyll fluorescence ratio F690/F730 in leaves of different chlorophyll content. Photosynth. Res. *25,* 295–298 (1990).

Lombard, F. and R. J. Strasser: Evidence for spill-over changes during state-1 to state-2 transition in green leaves. In: Sybesma, C. (ed.): Advances in Photosynthesis Research, Vol. *3,* pp. 271–274. Nijhoff/Junk, The Hague (1984).

Murata, N.: Control of excitation transfer in photosynthesis. – 1. Light-induced change of chlorophyll *a* fluorescence in *Porphyridium cruentum.* Biochim. Biophys. Acta *172,* 242–251 (1969).

Rinderle, U. and H. K. Lichtenthaler: The chlorophyll fluorescence ratio F690/F735 as a possible stress indicator. In: Lichtenthaler, H. K. (ed.): Applications of Chlorophyll Fluorescence, pp. 189–196. Kluwer Academic Publishers, Dordrecht (1988).

Schreiber, U., W. Schliwa, and U. Bilger: Continuous recording of photochemical and non-photochemical chlorophyll fluorescence quenching with a new type of modulation fluorimeter. Photosynth. Res. *10,* 51–62 (1986).

Strasser, R. J., B. Schwarz, and P. Eggenberg: Fluorescence routine tests to describe the behaviour of a plant in its environment. In: Lichtenthaler, H. K. (ed.): Applications of Chlorophyll Fluorescence, pp. 181–187. Kluwer Academic Publishers, Dordrecht (1988).

Strasser, R. J., B. Schwarz, and J. B. Bucher: Simultane Messung der Chlorophyll-Fluoreszenz-Kinetik bei verschiedenen Wellenlängen als rasches Verfahren zur Frühdiagnose von Immissionsbelastungen an Waldbäumen: Ozoneinwirkungen auf Buchen und Pappeln. Eur. J. Forest Pathol. *17,* 149–157 (1987).

Szabó, K., H. K. Lichtenthaler, L. Kocsányi, and P. Richter: A CCD-OMA device for the measurement of complete chlorophyll fluorescence emission spectra of leaves during the fluorescence induction kinetics. Radiat. Environ. Biophys. *31,* 153–160 (1992).

Tuba, Z., H. K. Lichtenthaler, Z. Csintalan, Z. Nagy, and K. Szente: Reconstitution of chlorophylls and photosynthetic CO_2-assimilation in the desiccated poikilochlorophyllous plant *Xerophyta scabrida* upon rehydration. Planta *192,* 414–420 (1990).

J. Plant Physiol. Vol. 148. pp. 555–566 (1996)

Light-induced and Age-dependent Development of Chloroplasts in Etiolated Barley Leaves as Visualized by Determination of Photosynthetic Pigments, CO_2 Assimilation Rates and Different Kinds of Chlorophyll Fluorescence Ratios

FATBARDHA BABANI[1,2] and HARTMUT K. LICHTENTHALER[1]

[1] Botanisches Institut, Universität Karlsruhe, Kaiserstr. 12, D-76128 Karlsruhe, Germany

[2] Institute of Biological Research, Academy of Sciences, Tirana, Albania

Received June 24, 1995 · Accepted November 10, 1995

Summary

During the greening of etiolated barley seedlings the accumulation of chlorophylls and carotenoids was determined in primary leaves together with the chlorophyll fluorescence ratio F690/F735 (at maximum and steady-state fluorescence) as well as the variable fluorescence decrease ratio (Rfd-values in 690 nm and 735 nm) as calculated from the fluorescence induction kinetics recorded by the LITWaF fluorometer. The variable chlorophyll fluorescence parameters (Fv/Fm, Fv/Fo, ΔF/Fm′) were monitored using the pulse amplitude modulation chlorophyll fluorometer PAM, and photosynthetic net CO_2 fixation by using a CO_2/H_2O porometer. All parameters were used to characterize the development of photosynthetic activity under illumination with continuous white light in the upper (oldest), middle and lower (youngest) part of the primary leaf blade of etiolated barley seedlings. The time course of changes in the studied parameters provided information on the gradual and age-dependent development of photosynthetic activity in the three leaf parts of different age. During the greening process the chlorophyll fluorescence ratio F690/F735 at maximum and steady-state fluorescence, fm and fs, strongly correlated with the total chlorophyll content (in an inverse curvilinear relationship) in the upper, middle and lower part of etiolated leaves. Though the absolute values were different, there also existed a linear correlation between the Rfd-values measured at 690 nm and 735 nm.

The time dependence of the variable fluorescence parameters Fv/Fm, Fv/Fo, ΔF/Fm′ yielded a saturation curve, with maximum values at about 12 h of illumination. These variable fluorescence ratios characterized the changes of the efficiency of exciton capture by open PSII reaction centres and the quantum yield of non-cyclic electron transport during the greening of etiolated barley seedlings in the different parts of primary leaves. Fv/Fo exhibited a curvilinear relationship to Fv/Fm and ΔFv/Fm′, but a linear relationship to Rfd-values measured at 690 nm and 735 nm during the greening period. Fv/Fo and Rfd-values represent a similar pair of variable fluorescence ratios at the dark-adapted (state 1) and the light-adapted state (state 2) of the photosynthetic apparatus as the ratios Fv/Fm and ΔF/Fm′.

The variation of net CO_2 fixation in the middle part of the leaf blade demonstrated that during the first 6 h of greening the net CO_2 fixation remained near the zero line, but then rose very fast. In contrast, the variable chlorophyll fluorescence parameters had reached already after 6 h illumination 90 % (Fv/Fm), 88 % (ΔF/Fm′), 77 % (Fv/Fo) and 60 to 77 % (Rfd values at 690 and 735 nm, respectively) of the maximum value. A saturation level of these variable fluorescence ratios was obtained at or around an illumination time of 12 h. These results indicated that the photosynthetic quantum conversion and electron transport within the leaf proceeded at closed stomata and much earlier than measurable CO_2 fixation, apparently using internal respiratory CO_2 for photosynthesis. The greening process and development of photosynthetic quantum conversion was faster in the lower and the middle leaf blade parts, but was very much retarded in the upper, oldest part of the leaf blade.

Key words: Age dependence of greening, chloroplast development, chlorophyll fluorescence ratio F690/F735, CO_2 assimilation, photosynthetic function, chlorophyll formation, variable chlorophyll fluorescence parameters.

Abbreviations: a+b = chlorophyll *a* and *b;* c = carotenes; CCD = charge couple device; fd = fluoresence decrease from fm to fs; fm = chlorophyll fluorescence maximum; fs = steady state chlorophyll fluorescence; F690/F735 = ratio of the chlorophyll fluorescence bands at 690 and 735 nm; Fv = variable fluorescence; Fo = ground fluorescence; Fm or fm = maximum fluorescence; Fs = steady state fluorescence; ΔF/Fm′ = PS II photochemical efficiency in the light-adapted state; Fv/Fm = PS II photochemical efficiency in the dark-adapted state; LITWaF = laser induced two-wavelengths chlorophyll fluorometer; OMA = optical multichannel analyzer; Rfd = variable fluorescence decrease ratio; x = xanthophylls; x+c = total carotenoids.

Introduction

Light energy absorbed by the photosynthetic pigments (chlorophylls and carotenoids) in the leaf mesophyll cells is mainly used for photosynthesis. A small part of the absorbed light is, however, lost in the deexcitation process of excited chlorophyll *a* molecules as red chlorophyll fluorescence and infra-red radiation (heat emission) (Baker and Bradbury, 1981; Buschmann and Sironval, 1984; Lichtenthaler, 1986; Stein et al., 1986). Since the discovery of the «Kautsky effect», the measurement of chlorophyll fluorescence has been developed as one of the most frequently used measuring tools in basic photosynthesis research (Kautsky and Hirsch, 1931; Krause and Weis, 1991; Lichtenthaler, 1992; Lichtenthaler and Buschmann, 1987; Papageorgiou, 1975; Schreiber, 1983; Schreiber et al., 1986). Chlorophyll fluorescence induction kinetics and various parameters and ratios derived from these reveal valuable information on the physiological state of the photosynthetic apparatus in plants. Today various chlorophyll fluorescence parameters are applied to determine the photosynthetic capacity of leaves and the function of the photosynthetic apparatus (Baker and Bradbury, 1981; Krause and Weis, 1991; Lichtenthaler, 1986; Rinderle and Lichtenthaler, 1988; Schreiber, 1983; Schreiber et al., 1986).

Chlorophyll fluorescence emission spectra of a green leaf exhibit two emission maxima at room temperature, the first near 690 nm and the second near 735 nm (Buschmann and Schrey, 1981; Buschmann and Sironval, 1984; Lichtenthaler, 1986; Lichtenthaler and Buschmann, 1987; Lichtenthaler and Rinderle, 1988). The ratio of fluorescence intensity of the two maxima red/far-red (F690/F735) is strongly influenced by variations of photosynthetic activity, i.e. the rate of photosynthetic quantum conversion, and increases by 20 to 35 % when the photosynthetic electron transport is blocked, e.g. by the herbicide diuron (Lichtenthaler and Rinderle, 1988). The F690/F735 ratio has also been established as an indicator of the *in vivo* chlorophyll content (D'Ambrosio et al., 1992; Hák et al., 1990; Lichtenthaler and Rinderle, 1988; Lichtenthaler et al., 1990). With the increase of the chlorophyll content (per leaf area unit) during greening the decrease of the red fluorescence band F690 in the chlorophyll fluorescence emission spectra can be explained by a partial reabsorption of the emitted 690 nm region fluorescence by the *in vivo* chlorophylls. The increase of the chlorophyll content thus results in a decrease of the values of the chlorophyll fluorescence ratio F690/F735 (D'Ambrosio et al., 1992; Hák et al., 1990; Lichtenthaler and Buschmann, 1987; Lichtenthaler and Rinderle, 1988; Stober and Lichtenthaler, 1992; Szabó et al., 1992), and can be used as a non-invasive method for chlorophyll determination in intact leaves.

The study of the transient changes of fluorescence with time (Kautsky curve) measured on dark-adapted leaves provides a very essential information of understanding the relation between light harvesting, electron transport, thylakoid energetics and the processes of CO_2 fixation (Buschmann and Lichtenthaler, 1988; Buschmann and Lichtenthaler, 1988; Buschmann and Schrey, 1981; van Kooten and Snel, 1990; Lichtenthaler and Buschmann, 1987; Lichtenthaler and Rinderle, 1988; Schreiber and Bilger, 1993). Various variable chlorophyll fluorescence parameters (Fv/Fm, Fv/Fo, Fo/Fm, ΔF/Fm′, Rfd-values) are applied to study the organization and functioning of the photosynthetic apparatus. Any factor, which affects the efficiency to capture the exitation energy by open PSII reaction centers, also modifies the variable chlorophyll fluorescence ratios parameters Fv/Fm, Fv/Fo and ΔF/Fm′ and Rfd-values (Kitajima and Butler, 1975; Genty, 1989; Lawlor, 1993; Lichtenthaler and Rinderle, 1988; Schreiber and Bilger, 1993).

The greening process of leaves comprises the biosynthesis and accumulation of chlorophylls and carotenoids as well as the formation of redox carriers and proteins, which parallel the formation of thylakoids and the gradual appearence of photosynthetic activity (Lichtenthaler, 1967; Lichtenthaler, 1969; Sestak and Siffel, 1988; Sestak, 1985; Stein et al., 1986). The greening process can be considered an «inverse» model for chlorotic phenomena as caused by the damage of the photosynthetic apparatus by various natural and anthropogenic stress factors (Lichtenthaler, 1996). In the upper, oldest part of the leaf blade the velocity of greening is different from the middle and lower leaf parts, which are younger. By using different chlorophyll fluorometers we wanted to establish the changes and significance of the different variable chlorophyll fluorescence ratios during chloroplast development, increasing chlorophyll content and onset of photosynthetic quantum conversion and net CO_2 fixation in greening barley seedlings. The chlorophyll fluorescence parameters were separatley measured for upper and lower leaf parts in order to analyze, document and understand the developmental processes in the differentially old leaf parts.

Materials and Methods

Plants

Barley seedlings (*Hordeum vulgare* L.) were grown for 7 days in the dark on peat with a full complement of minerals for growth (TKS2). Barley seedlings were then exposed to moderate continuous white light (180 μmol $m^{-2}s^{-1}$ PAR) during a greening period of two days. Chlorophyll fluorescence measurements as well as pigment determination were performed in the upper (first 2 cm), middle (3 to 5 cm) and lower (6 to 8 cm from the tip) part of the leaf blade of primary leaves during greening of etiolated seedlings. The greening process was repeated four times.

Pigment determination

The pigments of the photosynthetic apparatus (chlorophyll *a*, chlorophyll *b* and total carotenoids x+c) were spectrophotometrically determined using the redetermined extinction coefficients and equations of Lichtenthaler in 100 % acetone extract (Lichtenthaler, 1987). At the beginning of the greening process (0–2 hour illumination) either no or only trace amounts of chlorophyll *b* (2 h light) were found in the etiolated barley leaves. The pigment content was determined from the absorption spectra in 100 % acetone using a Shimadzu Spectrophotometer UV-2101PC. The pigments were separately determined at the upper, middle and lower parts of the primary leaves of etiolated barley seedlings, which were of different age and responded differently in the velocity of greening. The values given are the mean values of 8 determinations.

Chlorophyll fluorescence emission spectra

Complete fluorescence emission spectra were measured of the middle part of the leaf blade of pre-darkened (20 min) etiolated barley leaves during the greening process using the new multichannel CCD-OMA chlorophyll spectrofluorometer as described in detail before (Szabó et al., 1992, Babani et al., 1996). A 10 mW He/Ne laser (λ 632.8 nm) was used as excitation light source (light intensity of ca. 40,000 μmol $cm^{-2}s^{-1}$). A charge coupled device (CCD) sensor (with 2048 elements) simultaneously detected the complete chlorophyll fluorescence emission spectrum in the 650–800 nm wavelength region within 100 ms. As compared to the diode-array OMA system applied so far (serial readout of signals), the CCD-OMA spectrofluorometer has the advantage of a parallel readout of signals, thus yielding more reliable spectra, in particular when the measuring times are short. Data acquisition and processing were accomplished by a PC.

Chlorophyll fluorescence induction kinetics

The chlorophyll fluorescence induction kinetics (Kautsky effect) of predarkened leaves (15 min) were measured simultaneously at 690 nm and 735 nm using the two-wavelength fluorometer LITWaF with a He/Ne laser excitation beam (Spectra Physics 5 mW; λ = 632.8 nm, light intensity ca. 500 μmol $m^{-2}s^{-1}$ at the sample). Measurements were carried out with circular leaf segments from the upper, middle and lower part of etiolated leaves. Chlorophyll fluorescence was excited and sensed from the upper leaf side. The chlorophyll fluorescence ratios F690/F735 at maximum (fm) and at steady-state conditions (fs) of fluorescence were calculated from the fluorescence induction kinetics at 690 nm and 735 nm, respectively. The Rfd-values as vitality index, i.e. the ratio of fluorescence decrease fd (from fm to fs) to the steady-state fluorescence fs (Rfd = (fm-fs)/fs = fd/fs) were determined at 690 nm and 735 nm (Rfd690 and Rfd735) (see Lichtenthaler and Rinderle, 1988; Rinderle and Lichtenthaler, 1988).

PAM fluorometer

The parameters of variable chlorophyll fluorescence: ground fluorescence Fo, variable fluorescence Fv (= Fm-Fo), maximum fluorescence Fm (in the dark), ΔF (= Fm′-Fs) and Fm′ (in the light) were measured using the pulse amplitude modulation chlorophyll fluorometer PAM (Walz, Effeltrich, Germany) as described in (van Kooten and Snel, 1990; Lawlor, 1993; Lichtenthaler and Rinderle, 1988; Schreiber and Bilger, 1993; Schreiber et al., 1986). Due to the wavelengths of the excitation light and the applied filter combinations, the PAM fluorometer senses the chlorophyll fluorescence only in the 720 nm region (Lichtenthaler and Rinderle, 1988), which shows a lower amplitude and sensitivity than the 690 nm fluorescence. The chlorophyll fluorescence was induced by the weak modulated measuring red light (1,6 kHz of 0.01 μmol $m^{-2}s^{-1}$), which did not induce photosynthetic activity. Maximum fluorescence Fm was obtained by a pulse of saturating white light (2500 μmol $m^{-2}s^{-1}$) of 1 s duration directly after dark adaptation. A second pulse (1 s duration) was applied at steady-state fluorescence to obtain the value of maximal fluorescence Fm′ at fully activated photosynthetic conditions. The light intensity of the red actinic light at the sample was 200 μmol $m^{-2}s^{-1}$ PAR and the measuring beam during the illumination with actinic light was modulated with 100 kHz. The value of Fo′ was obtained by switching off the actinic light, and giving a pulse (5 s duration) of weak far-red light to obtain the fully oxidized state of PSII. The chlorophyll fluorescence ratios Fv/Fm, Fv/Fo and ΔF/Fm′ (Genty et al., 1989) were determined from the PAM kinetics measured separately at the upper (oldest), middle and lower (youngest) part of primary leaves during the greening of 7 d old etiolated barley seedlings.

Photosynthetic CO_2 fixation

The light-induced CO_2 fixation (P_N) was measured in predarkened leaves (20 min) using a CO_2/H_2O porometer (Walz, D-91090 Effeltrich). The irradiance on the leaves was 700 μmol $m^{-2}s^{-1}$ PAR saturating with respect to P_N; the velocity of the air flow was 0.5 L min^{-1}. The calculation of the photosynthetic rates was carried out according to (van Caemmere and Farquhar, 1981).

Results

Pigment content and pigment ratio

The content of chlorophylls and total carotenoids increased during greening (Fig. 1). The pigments exhibited a different rate of accumulation in the upper, middle and lower part of the primary leaf blade of 7 d old etiolated seedlings. The chlorophyll content of the etiolated leaves, illuminated for 10 min, started from 0.2–0.3 μg cm^{-2} and after 30 hours reached a level of 12.9 μg cm^{-2}, 17.4 μg cm^{-2} and 12.6 μg cm^{-2} at the upper, middle and lower part, respectively. The total carotenoid content of the etiolated leaves (before illumination) were 2.5 μg cm^{-2}, 1.9 μg cm^{-2} and 0.8 μg cm^{-2} at the upper, middle and lower part. After a 30 h illumination, the level of carotenoids was the same in the upper and middle leaf parts (3.7–3.8 µg cm^{-2}), but was significantly lower in the lower, youngest part of the leaf blade (2.5 µg cm^{-2}). The rate of accumulation of chlorophylls was highest in the middle leaf part.

The ratio of chlorophyll *a/b* started from high values of 17.6, 14.9, and 10.8 in the upper, middle and lower leaf parts, respectively after about 10 min of illumination (Fig. 2). With increasing illumination time and chlorophyll content the ratio *a/b* decreased and reached constant values after 6 h of illumination in the middle and lower leaf part (3.3 to 3.1), whereas in the oldest upper leaf part the *a/b* ratio declined more slowly and remained even at a higher value (3.9) after 20 h of illumination as compared to the middle and lower leaf blade parts. The ratio of chlorophylls to carotenoids (a+b)/(x+c) increased continuously from 0.1 to 0.2 in etiolated leaves to final values of 5 in illuminated and green leaves exhibiting a saturation curve. In the upper leaf part the greening of leaves proceeded much more slowly, and the ratio of green to yellow pigments $(a+b)/(x+c)$ reached values of only 3.5 after 30 h of illumination, thus reflecting the retarded chlorophyll accumulation.

The ratio *a/b* was inversely correlated with the amount of chlorophyll $a+b$ (Fig. 3). The correlation could be expressed by a power function $y = ax^{-b}$: upper leaf part, a = 12.87, b = 0.44; middle leaf part, a = 1.55, b = 0.94; lower leaf part, a = 9.70, b = 0.532. The chlorophyll ratio *a/b* in the middle and lower leaf part reached the normal values (3.1–3.4) when the amount of chlorophyll $a+b$ was only 5 µg cm^{-2}. In contrast, in the oldest, upper leaf part the ratio *a/b* remained higher

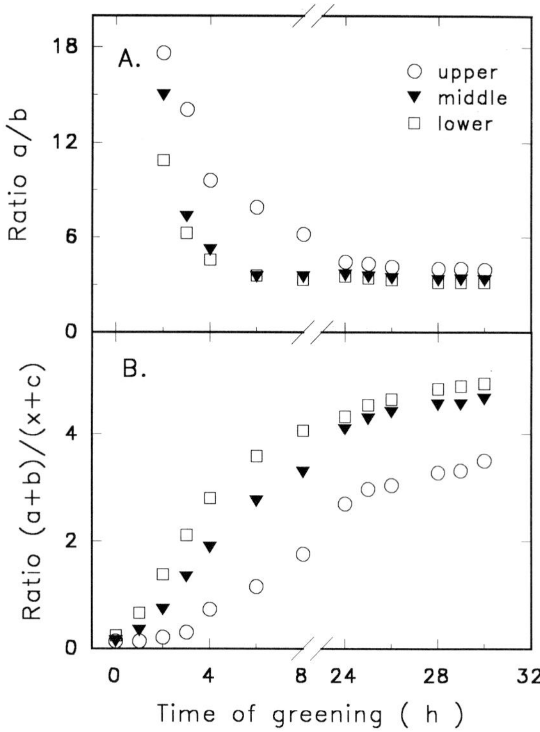

Fig. 2: Development of the ratios of chlorophyll *a/b* and green to yellow pigments $(a+b)/(x+c)$ during greening of primary leaf blades of barley seedlings.

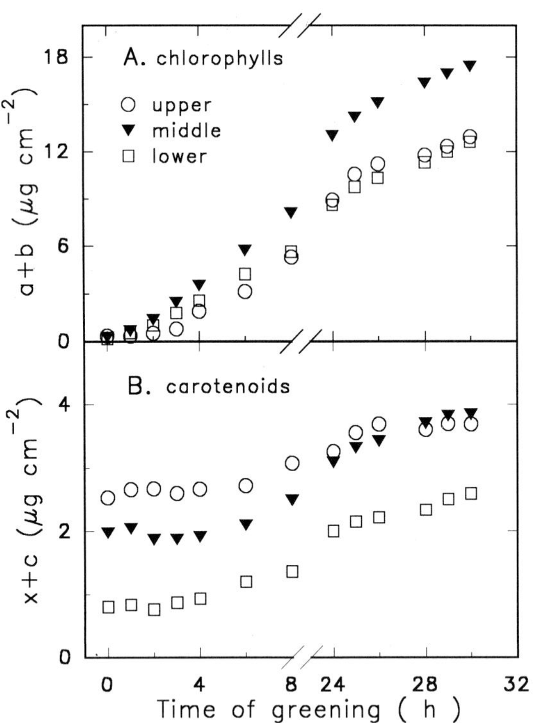

Fig. 1: Development of chlorophyll (**A.**) and carotenoid (**B.**) content in the upper, middle and lower part of primary leaves of the leaf blade of greening barley seedlings.

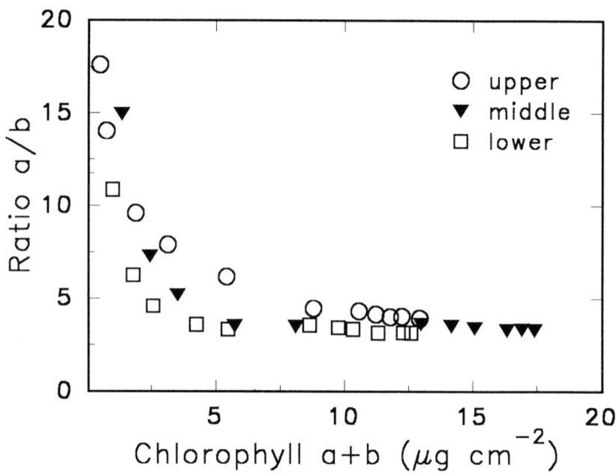

Fig. 3: Decrease of the pigment ratios chlorophyll *a/b* with increasing chlorophyll content of primary leaves during greening of etiolated barley seedlings. The relationship is given as a hyperbolic inverse curvilinear dependence: $y = ax^{-b}$; upper leaf part: a = 12.87, b = 0.464 ($r^2 = 0.967$); middle leaf part: a = 16.55, b = 0.694 ($r^2 = 0.818$); lower leaf part: a = 9.70, b = 0.532 ($r^2 = 0.822$).

(3.9) even after 30 h of illumination, although the level of chlorophyll had then reached 13 µg cm^{-2} leaf area already.

Chlorophyll fluorescence ratio F690/F735

The chlorophyll fluorescence induction kinetics were determined via the 690 nm and 735 nm fluorescence bands using the two-wavelength fluorometer LITWaF. At the beginning of greening the chlorophyll fluorescence ratios F690/F735 (Fig. 4) showed values of 2.6 to 3 at maximum and steady-state fluorescence fm and fs (Fig. 4). The values were higher in the upper part than in the middle and lower parts of the leaf blade. The values of the ratio F690/F735 decreased fast within 8 h of illumination parallel to the increasing chlorophyll content. The decrease of the F690/F735 ratio in the upper leaf part was very slow in the beginning of greening and correlated with the low rate of chlorophyll accumulation (Fig. 5). In the middle and lower parts of the etiolated leaf blades the F690/F735 decreased faster. After 24 h of illumination the values of the F690/F735 ratio reached nearly constant values of about 0.9 at fm and 0.6 at fs in the upper, middle and lower parts of the leaves. The chlorophyll fluorescence ratio F690/F735 was inversely correlated with the total chlorophyll content of the leaves. The correlation could be expressed by a power function $y = ax^{-b}$ at maximum and at steady-state of the fluorescence: upper leaf part, a = 2.087, b = 0.377 at fm and a = 2.278, b = 0.270 at fs; middle leaf part, a = 1.891, b = 0.372 at fm and a = 1.985, b = 0.307 at fs; lower leaf part, a = 1.598, b = 0.390 at fm and a = 1.774, b = 0.287 at fs. The values of the exponent were lower at fs than at fm.

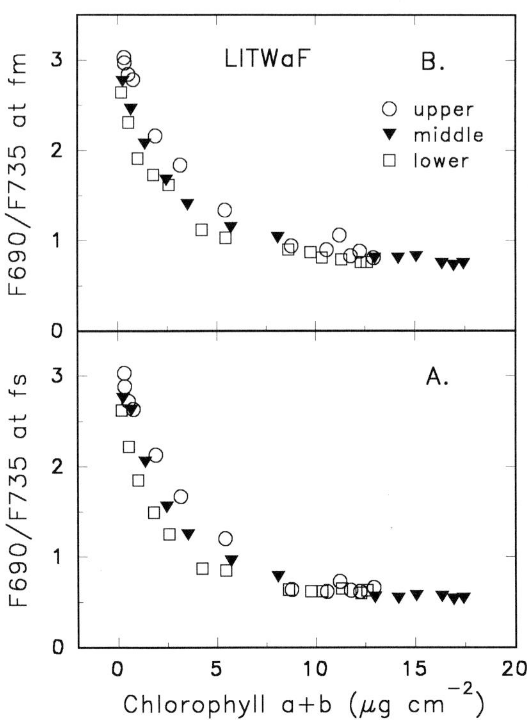

Fig. 4: Decrease of the values of the chlorophyll-fluorescence ratio F690/F735 with increasing chlorophyll *a*+ *b* content (curvilinear hyperbolic dependence $y = ax^{-b}$) during greening of etiolated barley seedlings measured at maximum and steady-state fluorescence, fm and fs, with the LITWaF fluorometer. The dependence exhibits correlation coefficients at fs of $r^2 = 0.960$, 0.975 and 0.957 for upper, middle and lower leaf part and at fm of $r^2 = 0.967$, 0.952 and 0.832 for upper, middle and lower leaf part, respectively.

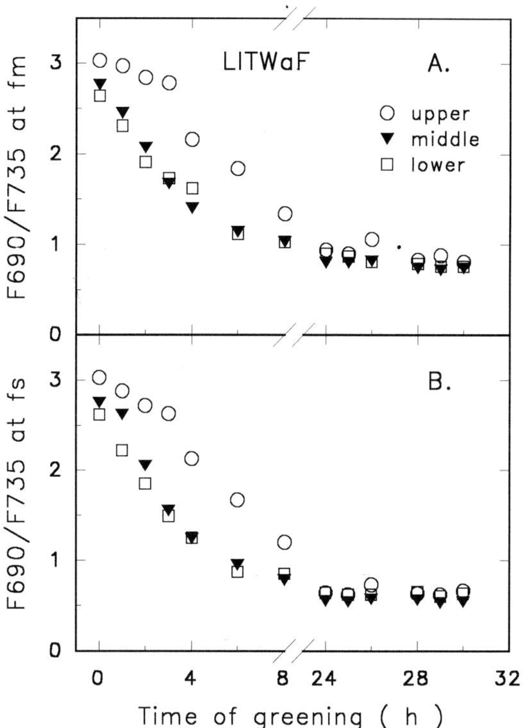

Fig. 5: Time dependence of the chlorophyll fluorescence ratio F690/F735 at fm and fs during greening of etiolated barley seedlings measured with the LITWaF fluorometer.

The fluorescence ratio F690/F735 as measured with the LITWaF provided slightly higher values at fm than at fs which had been described before (Kocsanyi et al., 1988). Higher values of the ratio F690/F735 are also found using other instrumentation such as the CCD-OMA. There existed, however, a linear relationship (y = ax+b) between the chlorophyll fluorescence ratios F690/F735 at fm and at fs (Fig. 6 A). This correlation was about the same in the upper, middle and lower part of leaves: upper: a = 0.938, b = 0.282; middle: a = 0.860, b = 0.310; lower: a = 0.920, b = 0.264.

There also existed a linear relationship between the fluorescencee chlorophyll ratios F690/F735 at fm and fs as determined from the chlorophyll fluorescence emission spectra which were measured during the fluorescence induction kinetics using the CCD-OMA spectrofluorometer (Fig. 6 B). The values of the parameters of the linear function (y = ax+b) were slightly different in comparison to the values determined with the LITWaF fluorometer (a = 1.099, b = 0.166). The first point measured at about 0.2 h (Fig. 6 B) was not included in the linear regression, since it reflects a too early step in the greening process in which photochlorophyllide transformations and shifts in physical chlorophyll *a* forms (e.g. Shibata shift) are still proceeding. The relationship between the chlorophyll fluorescence ratios F690/F735 determined by LITWaF and CCD-OMA measurements at fm (Fig. 7A.)and at fs (Fig. 7B.) was linear from 3 to 30 h of illumination and a chlorophyll content of leaves of more than 2.5 µg cm^{-2}. For the time of 0.2 to 3 h the LITWaF fluorometer was the more sensitive instrumentation for recording the early chlorophyll accumulation processes than the CCD-OMA spectrofluorometer.

The variable chlorophyll fluorescence ratios and vitality indices of the photosynthetic apparatus (Rfd-values) (Lichtenthaler and Rinderle, 1988) were determined from the fluorescence induction kinetics as measured by the LITWaF fluorometer. The Rfd690 values were always higher than the Rfd735 values which had been reported before (Lichtenthaler and Rinderle, 1988). The time dependence of Rfd690 and Rfd735 was different at three parts of the leaf blade (Fig. 8). In the middle and lower leaf parts of etiolated leaves the Rfd-values increased fast upon light exposure and exhibited normal values of photosynthetically active barley leaf tissue of 3 (Rfd690) and 2.2 (Rfd730) after 10 to 12 h of illumination showing the characteristics of a saturation curve (Fig. 8). In contrast, the Rfd values at 690 nm and 735 nm in the upper oldest leaf part remained at the same level near 0.3 for 6 h of illumination and increased only thereafter to the final values as found in the middle and lower leaf parts. This demonstrated again the much retarded greening in the upper and oldest part of the primary leaf blade. There also existed a linear relationship between the two Rfd-values measured at 690 and 735 nm (Fig. 9), which could be expressed by a linear function in the upper, middle and lower part of etiolated leaves ($y = ax+b$): upper a = 1.374, b = 0.078; middle a = 1.419, b = −0.211 and lower a = 1.416, b = −0.073.

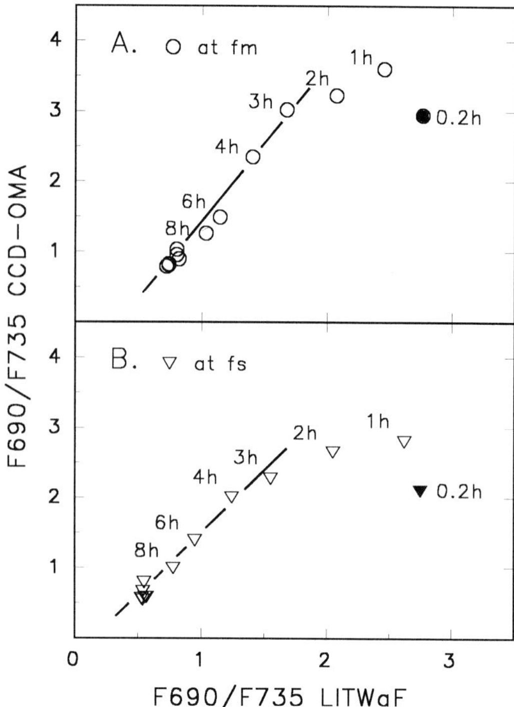

Fig. 7: Relationship between the fluorescence ratios F690/F735 measured by the CCD-OMA spectrofluorometer and by the LITWaF fluorometer at fm (○) and fs (▽). From 3 to 30 h of greening there exists a linear relationship with $r^2 = 0.942$ at fm and $r^2 = 0.890$ at fs.

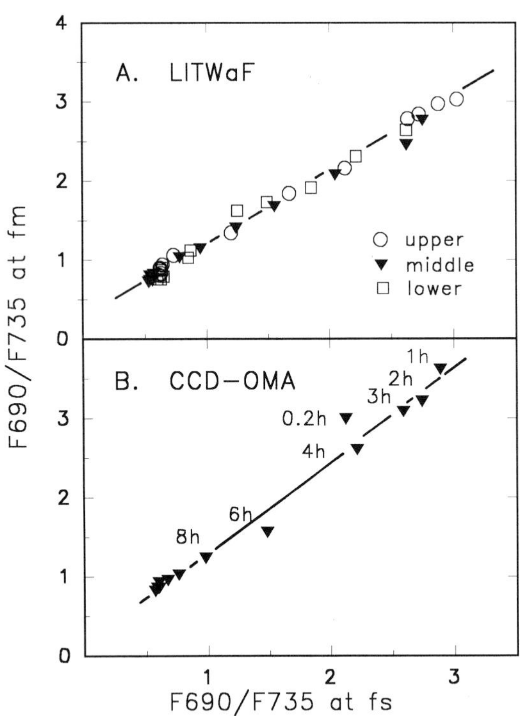

Fig. 6: Linear relationship ($y = ax+b$) between the fluorescence ratio F690/F735 at fm and fs during greening of etiolated barley seedlings measured by the LITWaF and CCD-OMA spectrofluorometers at a bandwith of 1 nm. The correlation coefficients for the LITWaF are $r^2 = 0.996$, 0.994 and 0.984 for upper, middle and lower leaf part, respectively and for the CCD-OMA $r^2 = 0.990$.

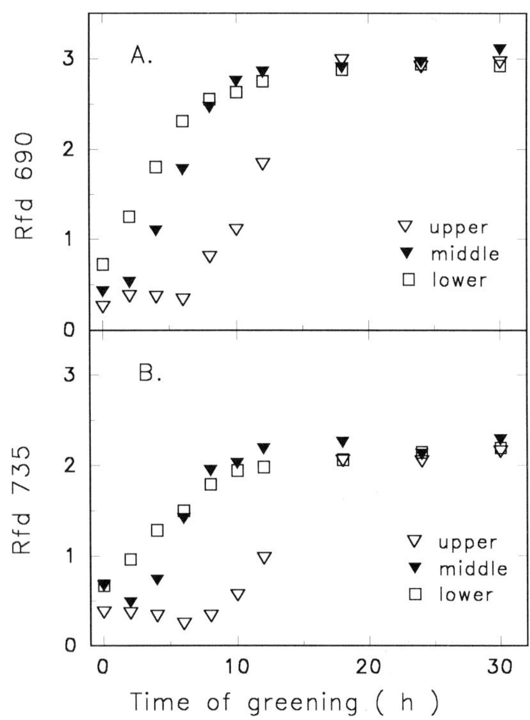

Fig. 8: Development of the fluorescence decrease ratios Rfd690 and Rfd735 in the upper (oldest) middle and lower (youngest) part of primary barley leaves during greening.

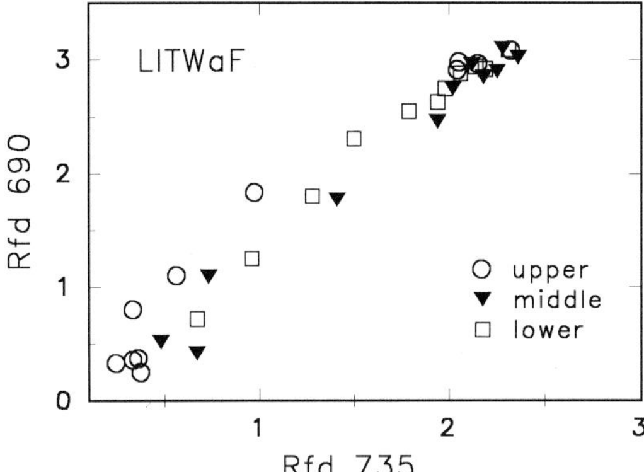

Fig. 9: Linear correlation between the values of the variable fluorescence ratios Rfd690 and Rfd735 measured in the upper, middle and lower part of the leaf blade of greening primary leaves of barley ($r^2 = 0.966$ for all values together).

PAM fluorometer and variable chlorophyll fluorescence parameters

The values of Fo (ground fluorescence) and Fm (maximum fluorescence) increased during greening of etiolated seedlings. The increase rate was high within the 18-hour illumination and thereafter the values of these parameters reached a fairly constant level. The values of Fo and Fm were significantly lower in the lower part than in the upper and middle part of leaves of greening seedlings (Fig. 10).

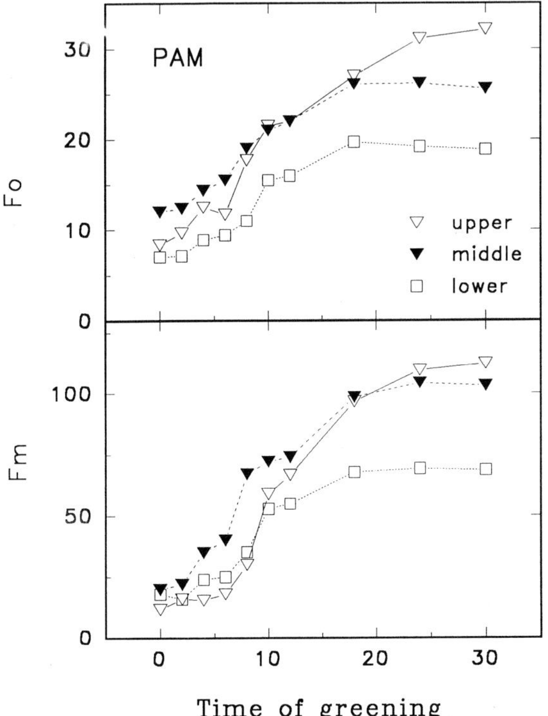

Fig. 10: Development of the two chlorophyll fluorescence parameters, ground and maximum fluorescence, Fo and Fm, in the upper, middle and lower part of the primary leaf blade of greening barley.

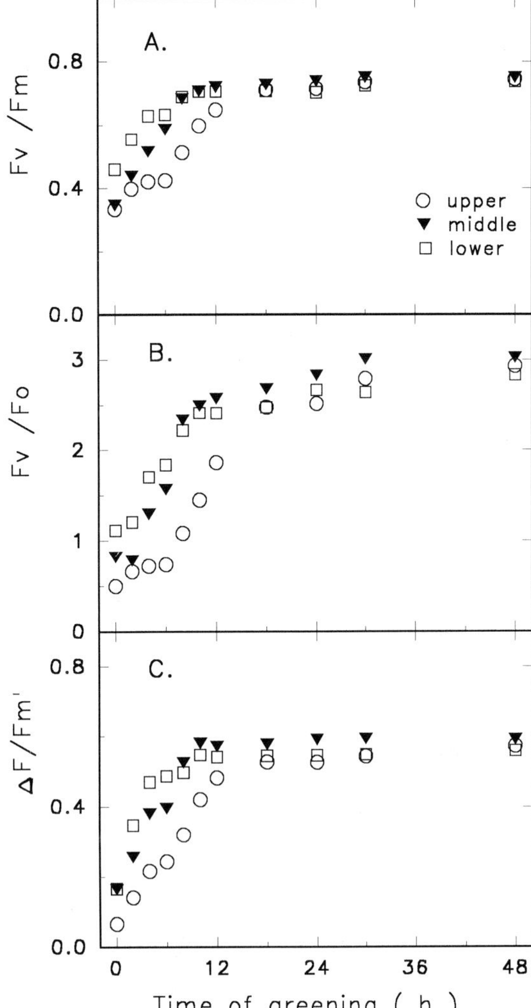

Fig. 11: Time dependence of the variable chlorophyll fluorescence parameters Fv/Fm (**A.**), Fv/Fo (**B.**) and ΔF/Fm′ (**C.**) during greening of the upper, middle and lower part of primary leaf blades of etiolated barley seedlings.

The time dependence of the variable chlorophyll fluorescence parameters Fv/Fo, Fv/Fm, ΔF/Fm′ (Fig. 11 A, B, C) determined from the induction kinetics measured by the PAM fluorometer showed the characteristics of a saturation curve (Fig. 11). These parameters reached the constant values (0.75, 2.4 and 0.55, respectively for Fv/Fm, Fv/Fo and ΔF/Fm′) within 10 h of illumination in the middle and lower part of the leaves and within 18 h in the upper oldest part of the leaf blade. The values of the Fv/Fm and Fv/Fo in the upper part of the leaves remained at low values in the beginning of the greening process (6 h). This characteristic was not observed in the case of ΔF/Fm′. The relationship between the Fv/Fo and Fv/Fm could be expressed by a hyperbolic function $y = ax/(1-bx)$ in the upper, middle and lower part of leaves (Fig. 12 A). The values of the coefficients a and b were about 1 (upper: $a = 0.999$, $b = 1.001$; middle: $a = 1.263$, $b = 0.902$; lower: $a = 1.131$, $b = 0.956$).

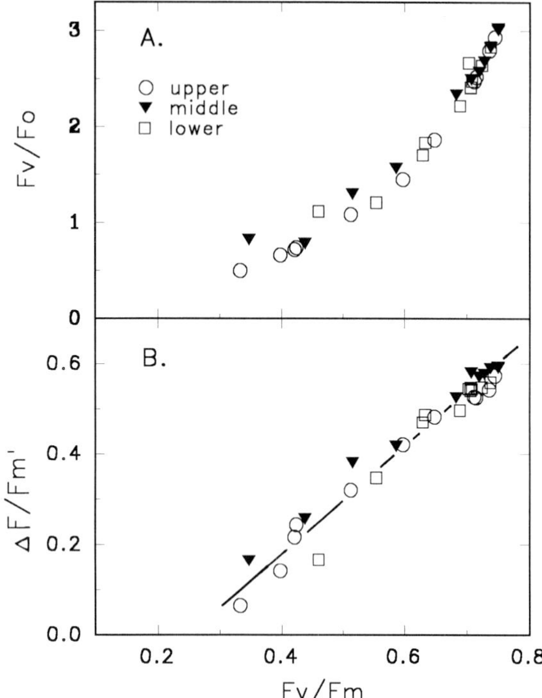

Fig. 12: Relationship between the variable chlorophyll fluorescence ratios Fv/Fo and ΔF/Fm′ to the ratio Fv/Fm during the greening of the upper, middle and lower part of the blade of etiolated barley seedlings. **A.** (hyperbolic dependence y = ax/(1-bx)) and **B.** linear relationship. The correlation coefficient r^2 in **A** is: 0.986, 0.940 and 0.919 for upper, middle and lower leaf part, respectively and in **B** r^2 = 0.982, 0.991 and 0.955 for upper, middle and lower leaf part.

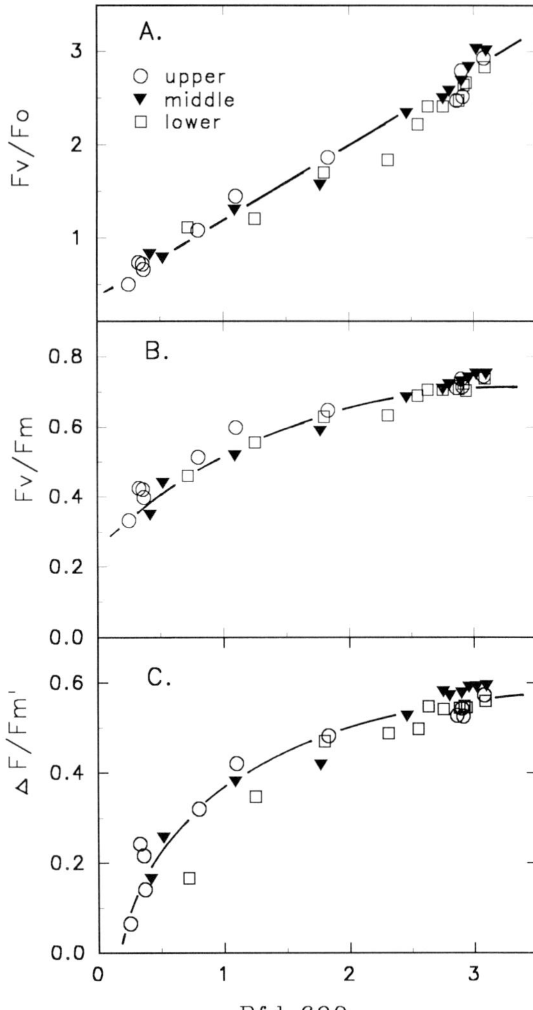

Fig. 13: Relationship between the chlorophyll fluorescence ratio Rfd690 (LITWaF) and the variable chlorophyll fluorescence ratios Fv/Fo (**A**), Fv/Fm (**B**) and ΔF/Fm′ (**C**) during greening of etiolated barley leaves. The relationships are linear in **A**, but hyperbolic in **B** and **C**. The correlation coefficients in **A** are r^2 = 0.984, 0.979 and 0.951 for upper, middle and lower leaf part, in **B**: r^2 = 0.962, 0.951 and 0.939 and in **C**: r^2 = 0.915, 0.976 and 0.959.

The relationship between the ΔF/Fm′ and Fv/Fm in the upper, middle and lower parts of the leaves could, however, be expressed by a linear function y = ax+b (Fig. 12 B). The values of coefficients were: upper: a = 1.153, b = 0.285; middle: a = 1.074, b = 0.203; lower: a = 1.279, b = 0.369. The dependence of the ratio Fv/Fo (measured by PAM) as plotted against the Rfd690 values (measured by LITWAF) also exhibited a linear relationship in the upper, middle and lower parts of the etiolated barley leaves (Fig. 13), which could be expressed by a linear function (y = ax+b), upper: a = 0.707, b = 0.436; middle: a = 0.816, b = 0.361; lower: a = 0.745, b = 0.386. The relationship between the Fv/Fm (Fig. 13 B), ΔF/Fm′ (Fig. 13 C) and Rfd690 can be expressed by a hyperbolic function y = ax/(b+x). These curvilinear relationships were expected from the relationship which exists between Fv/Fo and Fv/Fm and between ΔF/Fm′ and Fv/Fm (Fig. 12 B). The values of the coefficient in the case of Fv/Fm were: upper a = 0.802, b = 0.354; middle a = 0.884, b = 0,66; lower a = 0.882; b = 0,709 and in the case of F/Fm′: upper a = 0.741, b = 1.035; middle a = 0.897, b = 1.617; lower a = 1.109, b = 2.29.

Photosynthetic CO_2 fixation

A photosynthetic net CO_2 fixation was hardly detectable in the first 5 h of greening, but then it rose fast and reached a maximum value of 8.2 µmol $m^{-2} s^{-1}$ after 42 h of illumination (Fig. 14). The relationship between the rate of net CO_2 fixation P_N of primary leaves during greening of etiolated barley seedlings and the values of the ratios Rfd690 (Fig. 15 A), Fv/Fo (Fig. 15 B) and ΔF/Fm′ (Fig. 15 C) showed that at the beginning of the greening process (6 h illumination) the net CO_2 fixation remained very low (0–0.250 µmol $m^{-2} s^{-1}$), whereas the values of the fluorescence ratios Rfd690, Fv/Fo and ΔF/Fm′ steadily increased. After 8 h of illumination the values of net CO_2 fixation increased very fast parallel to the increase of the values of the analyzed variable fluorescence ratios. During this phase, which started after a minimum chlorophyll content of 8 to 9 µg cm^{-2} leaf area, the variable chlorophyll fluorescence ratios Rfd690 and Rfd735, Fv/Fo and Fv/Fm and also ΔF/Fm′ were linearily correlated to the

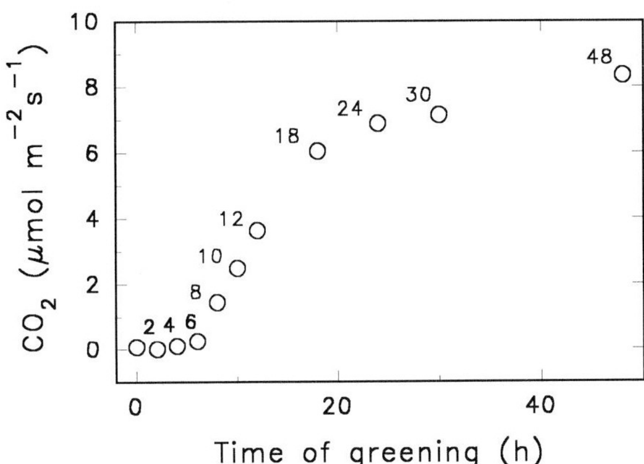

Fig. 14: Development of photosynthetic net CO_2 fixation rates P_N during greening of etiolated barley seedlings (middle and lower part of the primary leaf blade).

rising net CO_2 assimilation rate (Fig. 15). In this phase the variable fluorescence ratios Rfd690, Fv/Fo and ΔF/Fm′ represented approximate, non-invasive indicators of the *in vivo* net CO_2 fixation rates.

Discussion

The rate of chlorophyll $a+b$ accumulation during the illumination of 7 d etiolated barley seedlings was higher in the middle than in the lower part of the leaf blade and distinctively lower in the upper and oldest part of the leaf blade of the primary barley leaf (Fig. 1 A.). The value decrease of the ratio chlorophyll *a*/*b* in the first hours of illumination (0–8 h) demonstrated that the accumulation of chlorophyll *a* proceeded faster than that of chlorophyll *b*, and also faster in the middle and lower (youngest) part than in the upper (oldest) part of the etiolated primary leaf blades (Fig. 2 A.). The amount of carotenoids being formed in darkness and being present at the beginning of greening (10 min illumination) were higher in the oldest part and lower in the youngest parts of the etiolated leaf blades of barley seedlings (Fig. 1 B.). The increase of the ratio chlorophylls/carotenoids $(a+b)/(x+c)$ during greening demonstrated that the carotenoids were synthesized at a lower rate than the chlorophylls. Carotenoids exhibited a relatively linear accumulation curve (Fig. 1 B.), whereas the chlorophylls showed a saturation type accumulation curve (Fig. 1 A.). This has been described before for greening of etiolated wheat leaves (Stober and Lichtenthaler, 1992). The differences observed in the increase of the ratio chlorophylls/carotenoids were related to the different rate of accumulation of chlorophylls (higher in the middle leaf part) and carotenoids (higher in the lower leaf part).

The decrease of the fluorescence F690/F735 ratio paralleled the continuous increase of the chlorophyll $a+b$ level during the illumination of etiolated barley seedlings. This decrease can be explained with the reabsorption of the red F690 fluorescence by the *in vivo* chlorophyll *a* absorption bands, but this does not apply to the far-red chlorophyll fluorescence F735 (D'Ambrosio et al., 1992; Hák et al., 1990; Lichtenthaler et al., 1987; Lichtenthaler and Rinderle, 1988; Szabó et al., 1992). Reabsorption of the emitted chlorophyll *a* fluorescence at the beginning of greening (10 min light) can almost be excluded due to the very low chlorophyll content of the etiolated barley seedlings. At this stage the chlorophyll fluorescence ratio F690/F735 at maximum and steady-state fluorescence, fm and fs, exhibited high values near 3 that were highest in the upper part of the leaves (Fig. 4 A, B). The cur-

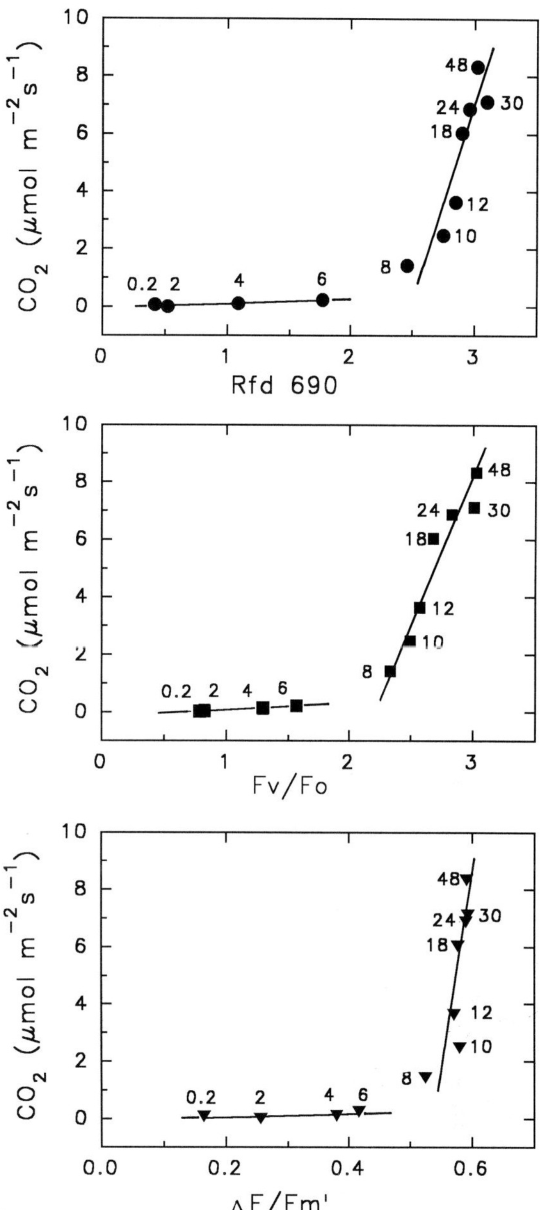

Fig. 15: Relationship between CO_2 fixation and the chlorophyll fluorescence decrease ratio Rfd690 (**A**) and the variable chlorophyll fluorescence Fv/Fo (**B**) and ΔF/Fm′ (**C**). The numbers in the diagram represent the duration of illumination and greening time in hours. From 8 to 48 h of greening the relationship is linear (with a higher correlation coefficient in **A** ($r^2 = 0.889$) and **B** ($r^2 = 0.895$) than in **C** ($r^2 = 0.629$)).

vilinear relationship between the chlorophyll fluorescence ratio F690/F735 and the total chlorophyll content (D'Ambrosio et al., 1992; Hák et al., 1990; Stober et al., 1992; Szabó et al., 1992) was also found in the upper, middle and lower part of etiolated barley seedlings and could be expressed by a power function $y = ax^{-b}$. The results demonstrate that the ratio F690/F735 is an excellent non-invasive indicator of the *in vivo* chlorophyll content.

The linear relationship ($y = ax+b$) between the two chlorophyll fluorescence ratios F690/F735, measured at maximum and steady-state of fluorescence, fm and fs, were the same in the upper, middle and lower part of leaves and independent of the rate of chlorophyll accumulation (Fig. 6). The values 20 to 30% higher at fm than fs are due to the fact that the F690 fluorescence yield declines much more during the chlorophyll fluorescence induction kinetics (Kautsky effect) than the far-red fluorescence band F735. A comparison of fluorescence ratios F690/F735 at fm and fs was made between those obtained by the two-wavelengths fluorometer LITWaF and those obtained by the CCD-OMA spectrofluorometer. At the beginning of greening (0.2 to 3 h illumination) the chlorophyll fluorescence ratios F690/F735 at fm and fs determined from the LITWaF kinetics were higher than those calculated from the fluorescence emission spectra of the CCD-OMA spectrofluorometer (Fig. 7). This connection was associated with a low chlorophyll content in the etiolated barley leaves of 0.2 to 2.5 $\mu g\,cm^{-2}$. Beyond 3 h of light exposure the relationship between the two F690/F735 ratios taken with the two instruments could, however, be expressed by a linear function.

The time dependence of the rising values of the fluorescence decrease ratios Rfd690 and Rfd735 exhibited the characteristics of a saturation curve in the middle and lower part of etiolated barley seedlings. In contrast, the upper oldest part of the primary leaf blade showed extremely low Rfd-values which remained constantly low up to 6 h (Rfd690) or 8 h (Rfd735) of greening. In this upper leaf part the greening was, however, much retarded, the photosynthetic quantum conversion started much later and the saturation level of the Rfd-values was reached only after 18 h of illumination (Fig. 8). The final saturation levels of the Rfd-values were, however, the same in the upper, middle and lower part of etiolated barley leaves. This indicates that the oldest upper leaf blade part of 7 d old barley seedlings can become photosynthetically functional, but requires a much longer time period than in young etiolated leaf tissue.

During the greening of etiolated barley seedlings we observed an increase of the values of ground fluorescence (Fo) and maximum fluorescence (Fm) as monitored by the PAM fluorometer kinetics. The increasing of Fo and Fm represented a saturation curve with different rates of increase and different saturation levels in the upper (oldest), middle and lower (youngest) part of etiolated leaves. These differences are related to the differentially rising chlorophyll content and appearence of photosynthetic activity in the three parts of the leaf blades during greening of etiolated barley seedlings.

The greening time dependence of the variable chlorophyll fluorescence ratios Fv/Fm, Fv/Fo and ΔF/Fm′ (Fig. 9), which can be used to monitor the changes of the efficiency of exiton capture by open PSII reaction centres and the quantum yield of non-cyclic electron transport (Genty et al., 1989; Lawlor, 1993), showed the characteristics of a saturation curve. The observed differences in the upper (reaching saturation level within 18 h illumination), middle and lower part (reaching saturation level within 10 h illumination) of the leaf blade were related to the different rates of chlorophyll accumulation. We found a curvilinear relationship between Fv/Fo and Fv/Fm which can be expressed by a power function (Fig. 10 A.) There existed, however, a linear correlation between the variable chlorophyll fluorescence parameters (ΔF/Fm′ and Fv/Fm) (Fig. 10 B), which can be explained by the fact that the changes in the quantum yield of non-cyclic electron transport are linearily related to the increasing efficiency of exiton capture by open PSII reaction centres. A correlation also showed up between the variable chlorophyll fluorescence parameters Fv/Fo, Fv/Fm and ΔF/Fm′ (determined from PAM measurements) on one hand, and the fluorescence decrease ratio Rfd690 (determined from LITWaF fluorescence kinetics) on the other hand. The correlation of Rfd690 is expressed by a linear function in the case of Fv/Fo (Fig. 13 A) and by a hyperbolic function in the case of Fv/Fm and ΔF/Fm′ (Fig. 13 B, C).

At the beginning of greening of etiolated barley seedlings up to 8 h of illumination the net CO_2 fixation rates remained at a very low level although the variable fluorescence decrease ratio Rfd690 and the variable chlorophyll fluorescence ratios Fv/Fo and ΔF/Fm′ increased rapidly (Fig. 14 A, B, C) indicating that photosynthetic quantum conversion was proceeding already at good rates. Thereafter, the values of net CO_2 uptake rose very fast, and the ratios Rfd690, Fv/Fm and ΔF/Fm′ reached the saturation level (Figs. 9 A and 11 B and C, respectively). During this stage (8 to 30 h of illumination) an four variable fluorescence ratios were linearily correlated to net CO_2 fixation rates.

The variable fluorescence ratio Fv/Fo is a much better indicator of changes in the rates of photosynthetic quantum conversion than the ratio Fv/Fm. In the fluorescence ratio Fv/Fo the changes of both components of the ratio, the variable fluorescence Fv and the ground fluorescence Fo, are considered at any time, and thus this ratio responds very sensitively to any changes in Fv and/or Fo. In contrast, the ratio Fv/Fm, which is the PSII photochemical efficiency in the dark-adapted state with fully open PSII reaction centers, is relatively inert and slow in response. It does not and cannot readily respond to small changes in Fv or Fo, since Fm is the sum of Fv+Fo. Thus, Fm often does not change at all, e.g. when Fv slightly decreases and Fo is slightly increased. Therefore changes in photosynthetic quantum conversion and photochemical efficiency of photosystem II are masked by forming the fluorescence ratio Fv/Fm. This also applies to the ratio ΔF/Fm′ which indicates the efficiency of energy capture by the yet open PSII reaction centers in the light-adapted photosynthetically active state. 2. Thus, Fv/Fm and ΔF/Fm′ represent a pair of related fluorescence ratios of the dark-adapted (state 1) and the light-adapted state of the photosynthetic apparatus (state 2), and this pair is linerarily correlated.

A similar pair of related variable chlorophyll fluorescence ratios describing the quantum conversion capacity at the dark-adapted and light-adapted state of the photosynthetic apparatus are Fv/Fo and the fluorescence decrease ratio Fd/Fs (or fd/fs), which is referred to as Rfd-values (measured at the chlorophyll fluorescence emission bands 690 nm and 735 nm). Both fluorescence ratios, Fv/Fo and the Rfd-values (ratio Fd/Fs), exhibit a much higher sensitivity to changes in photosynthetic quantum conversion than the fluorescence ratios Fv/Fm and ΔF/Fm′, which are fairly inert and exhibit only a low amplitude and sensitivity. Both ratios, Fv/Fo and Rfd-values, possess the same character and are linearily correlated as is shown in Fig. 13.

Due to their different mode of calculation and information on photosynthetic quantum conversion, which is somewhat different from the ratios Fv/Fm and ΔF/Fm′, the fluorescence ratios Fv/Fo anf Rfd possess a curvilinear relationship to the ratios Fv/Fm and ΔF/Fm′. From our comparative studies presented here and before (Lichtenthaler and Rinderle, 1988; Tuba et al., 1994) it appears that the fluorescence ratio Fv/Fo (dark-adapted state = state 1) and the Rfd-values (light-adapted state = state 2) are a good and fairly close measure of the potential photosynthetic capacity and the net CO_2 fixation of leaves.

In summary, the different chlorophyll fluorescence parameters and ratios studied here, provide valuable information on the different rates of accumulation of photosynthetic pigments, as well as on the appearance and increase of photosynthetic quantum conversion (photochemistry and electron transport) and the net CO_2 fixation rates during the illumination and greening of etiolated leaves.

Acknowledgements

This work was supported by a grant of the BMFT Bonn within the EUREKA research programme LASFLEUR (EU 380) to H. K. Lichtenthaler, and by a fellowship from DAAD, Bonn, and the European Science Foundation, Strasbourg to F. Babani, which is gratefully acknowledged. We wish to thank Dr. Claus. Buschmann for advice in measuring and calculating the net CO_2-fixation rates, Dr. Christiane Schindler for help with the application of the PAM fluorometer, Inge Jansche for help during the preparation of the manuscript, and Gabrielle Johnson for correcting the English text.

References

Babani, F., H. K. Lichtenthaler, and P. Richter: Changes of chlorophyll fluorescence signatures during greening of etiolated barley seedlings as measured with the CCD-OMA fluorometer. J. Plant Physiol. *148*, 471–477 (1996).

Baker, N. R. and M. Bradbury: Possible applications of chlorophyll fluorescence techniques for studying photosynthesis *in vivo*. In: Smith, H. (ed.): Plants and the Daylight Spectrum, pp. 355–373. Academic Press, London (1981).

Boardman, N. K. and J. M. Anderson: Composition, structure and photochemical activity of developing and mature chloroplasts. In: Akoyunoglou, G. and J. H. Argyroudi-Akoyunoglou (eds.): Chloroplast Development, pp. 1–14. Elsevier, Amsterdam (1978).

Buschmann, C. and H. K. Lichtenthaler: Complete fluorescence spectra determined during the induction kinetics using a diode-array detector. In: Lichtenthaler, H. K. (ed.): Applications of Chlorophyll Fluorescence, pp. 77–84. Kluwer Academic Publishers, Dordrecht (1988).

– – Reflectance and chlorophyll fluorescence signatures of leaves. In: Lichtenthaler, H. K. (ed.): Applications of Chlorophyll Fluorescence, pp. 325–332. Kluwer Academic Publishers, Dordrecht (1988).

Buschmann, C. and H. Schrey: Fluorescence induction kinetics of green and etiolated leaves by recording the complete *in vivo* emission spectra. Photosynth. Res. *1*, 133–241 (1981).

Buschmann, C. and C. Sironval: Fluorescence emission spectra of etiolated leaves measured at 296 and 77 K during the first second of continuous illumination. In: Sironval, C. and M. Brouers (eds.): Protochlorophyllide Reduction and Greening, pp. 139–148. Martinus Nijhoff/Dr. W. Junk Publishers, The Hague (1984).

Von Caemmere, S. and G. D. Farquhar: Some relationship between the biochemistry of photosynthesis and gas exchange of leaves. Planta *153*, 376–387 (1981).

D'Ambrosio, N., K. Szabó, and H. K. Lichtenthaler: Increase of the chlorophyll fluorescence ratio F690/F735 during the autumnal chlorophyll breakdown. Radiat. Environ. Biophys. *31*, 51–62 (1992).

Genty, B., J. M. Briantais, and N. R. Baker: The relationship between the quantum yield of photosynthetic electron transport and quenchers of chlorophyll fluorescence. Biochim. Biophys. Acta *990*, 87–92 (1989).

Hák, R., H. K. Lichtenthaler, and U. Rinderle: Decrease of the chlorophyll fluorescence ratio F690/F730 during greening and development of leaves. Radiat. Environ. Biophys. *29*, 329–336 (1990).

Kautsky, H. and A. Hirsch: Neue Versuche zur Kohlenstoffassimilation. Naturwissenschaften *19*, 964 (1931).

Kitajima, H. and W. L. Butler: Quenching of chlorophyll fluorescence and primary photochemistry in chloroplasts by dibromothymoquinone. Biochim. Biophys. Acta *376*, 105–115 (1975).

Kocsanyi, L., M. Haitz, and H. K. Lichtenthaler: Measurement of the laser-induced chlorophyll fluorescence kinetics using a fast acoustooptic device. In: Lichtenthaler, H. K. (ed.): Applications of Chlorophyll Fluorescence, pp. 99–107. Kluwer Academic Publishers, Dordrecht (1988).

Krause, G. H. and E. Weis: Chlorophyll fluorescence as a tool in plant physiology. II. Interpretation of fluorescence signal. Photosynth. Res. *5*, 139–157 (1984).

– – Chlorophyll fluorescence and photosynthesis: the basics. Ann. Rev. Plant. Pysiol. Mol. Biol. *42*, 313–349 (1991).

Lawlor, D. W.: Photosynthesis, pp. 99–104 and 286–290. Longman Scientific and Technical, Essex (1993).

Lichtenthaler, H. K.: Beziehungen zwischen Zusammensetzung und Struktur der Plastiden in grünen und etiolierten Keimlingen von *Hordeum vulgare* L. Z. Pflanzenphys. *56*, 273–281 (1967).

– Light-stimulated synthesis of plastid quinones and pigments in etiolated barley seedlings. Biochim. Biophys. Acta. *184*, 164–172 (1969).

– Laser-induced chlorophyll fluorescence of living plants. In: Proceedings of the International Geoscience and Remote Sensing Symposium (IGARSS), Zürich, Vol. III, pp. 1571–1576. ESA, Scientific and Technical Publications Branch, Noordwijk (1986).

– Chlorophylls and Carotenoids: pigments of photosynthetic biomembranes. Methods Enzymol. *148*, 350–382 (1987).

– : The Kautsky effect: 60 years of chlorophyll fluorescence induction kinetics. Photosynthetica *27*, 369–383 (1992).

Lichtenthaler, H. K. and C. Buschmann: Chlorophyll fluorescence spectra of green leaves. J. Plant Physiol. *129*, 137–147 (1987).

Lichtenthaler, H. K. and U. Rinderle: The role of chlorophyll fluorescence in the detection of stress conditions in plants. CRC Critical Reviews in Analytical Chemistry *19*, Suppl. 1, 29–85 (1988).

Lichtenthaler, H. K., R. Hák, and U. Rinderle: The chlorophyll fluorescence ratio F690/F730 in leaves of different chlorophyll content. Photosynth. Res. *25*, 295–298 (1990).

Papageorgiou, G.: Chlorophyll fluorescence, an intrinsic probe of phyotsynthesis. In: Govindjee (ed.): Bioenergetics of Photosynthesis, pp. 319–371. Academic Press, New York (1975).

Rinderle, U. and H. K. Lichtenthaler: The chlorophyll fluorescence ratio F690/F730 as a possible stress indicator. In: Lichtenthaler, H. K. (ed.): Applications of Chlorophyll Fluorescence, pp. 189–196. Kluwer Academic Publishers, Dordrecht (1988).

Schreiber, U.: Chlorophyll fluorescence yield changes as a tool in plant physiology. 1. The measuring system. Photosynth. Res. *4,* 361–372 (1983).

Schreiber, U. and W. Bilger: Progress in chlorophyll fluorescence research: Major developments during the past years in retrospect. Progress in Botany *54,* 150–175 (1993).

Schreiber, U., U. Schliwa, and W. Bilger: Continuous recording of photochemical and non-photochemical chlorophyll fluorescence quenching with a new type of modulation fluorometer. Photosynth. Res. *10,* 51–62 (1986).

Sestak, Z. and P. Siffel: Changes in chloroplast fluorescence during leaf development. In: Lichtenthaler, H. K. (ed.): Applications of Chlorophyll Fluorescence, pp. 85–91. Kluwer Academic Publishers, Dordrecht (1988).

Sestak, Z.: Photosynthesis During Leaf Development, pp. 76–107. Academia, Prague (1985).

Stein, U., R. Blaich, and C. Buschmann: Fluorscence kinetics of chloroplasts as indicator of disorders in the photosynthetic system. I. Comparative studies with greening leaves of *Vitis* and *Hordeum.* Vitis *25,* 129–141 (1986).

Stober, F. and H. K. Lichtenthaler: Changes of the laser-induced blue, green and red fluorescence signatures during greening of etiolated leaves of wheat. J. Plant Physiol. *140,* 673–680 (1992).

Szabó, K., H. K. Lichtenthaler, L. Kocsanyi, and P. Richter: A CCD-OMA device for the measurement of complete chlorophyll fluorescence emission spectra of leaves during the fluorescence induction kinetics. Radiat. Environ. Biophys. *31,* 153–160 (1992).

Van Kooten, O. and J. F. H. Snel: The use of chlorophyll fluorescence nomenclature in plant stress physiology. Photosynth. Res. *25,* 147–150 (1990).

J. Plant Physiol. Vol. 148. pp. 567–573 (1996)

Analysis of Reflectance and Fluorescence Spectra for Atypical Features: Fluorescence in the Yellow-green

William Philpot[1], Michael Duggin[2], Richard Raba[3], and Fu-An Tsai[1]

[1] Cornell University, Ithaca NY 14853, USA

[2] State University of New York at Syracuse, Syracuse NY, USA

[3] Boyce Thompson Institute for Plant Research, Ithaca NY 14853, USA

Received July 19, 1995 · Accepted October 12, 1995

Summary

An experiment designed to search for remotely detectable, identifying features of anthropogenic contaminants by remote sensing of vegetation is described. Reflectance and fluorescence measurements are made from single leaves, still attached to the plant. In order to maximize the opportunity for detection and identification of possibly subtle spectra features, all optical measurements are made under controlled conditions. Reflectance is measured using diffuse illumination and a well-defined illumination and viewing geometry over a wide range of wavelengths (0.35–2.5 μm). Fluorescence spectra are measured from live leaves at over 30 excitation wavelengths. Analysis procedures are designed to facilitate sorting through large amounts of data and allow interactive manipulation of the spectral data sets. Reflectance of single leaves from soybean plants subjected to water stress differed little from reflectance of the leaves from control plants. (This may not be indicative of canopy reflectance.) Blue and red fluorescence was somewhat greater for the water stressed plants. Unusual fluorescence features in the yellow-green part of the spectrum were also found in both control and the stressed plants. These features were exceptionally sensitive to stress.

Key words: Spectral reflectance, fluorescence, remote sensing, acute and chronic water stress, yellow-green fluorescence.

Introduction

If vegetation is subjected to man-made pollution and the pollutant produces an optical response that is atypical of natural stresses, the plant response could serve as an indicator of pollution and even serve to draw attention to the nature of the pollutant. If the optical response were detectable with remote sensing there would then be the potential of both detecting the and identifying the pollutant. This paper describes an experiment designed to explore the possibility of using vegetation as an integrating pollution detector. Any feature or set of reflectance and/or fluorescence features that would be diagnostic of a particular type of stress or a class of contaminants is sought. The emphasis is the use of vegetation as a collector; the health or condition of the plant is not a primary concern.

Vegetation responds and adapts to a wide range of variables in the environment, some natural, some induced by man. Some responses result in changes in the optical properties of the plant. When the optical changes are significant and consistent it is sometimes possible to distinguish stressed vegetation using remote sensing. The oldest, and still the most common remote sensing approach, is to exploit changes in reflectance in the red-infrared spectral range that coincide with stress (Bawden, 1933; Benedict and Swidler, 1961; Colwell, 1974; Richardson and Wiegand, 1977; Kauth and Thomas, 1976; Huete, 1988). Although some other reflectance features have also been noted (e.g., Baret et al., 1987; Demetriades-Shah, 1990; Adams, 1993) efforts to use more than simple ratios or differences of two spectral bands as vegetation indices have been rare. Vegetation response to stress is also apparent in leaf fluorescence emission spectra. Moreover, changes in fluorescence are more directly related to the pigment content and leaf physiology than reflectance spectra (Lichtenthaler and Rinderle, 1988). Fluorescence in the red originates from chlorophyll *a* in mesophyll cells (Lichtenthaler, 1987; Lichtenthaler and Buschmann, 1987), and changes in red fluorescence are indicative of stress (Lichtenthaler and Rinderle, 1988). Fluorescence in the blue and green is also sensitive to stress, but the source of the fluorescence is not so easily characterized (Chappelle et al., 1984, 1991; Lang et al., 1991; Stober and Lichtenthaler, 1993).

Clearly, there is a substantial literature documenting reflectance and fluorescence of vegetation and the typical changes with stress. The project described here is designed to search for atypical or unusual reflectance or fluorescence features of anthropogenic contaminants that can be detected by remote sensing using relatively high resolution fluorescence and reflectance spectra.

Materials and Methods

Experimental Design

Reflectance measurements are made under diffuse illumination in a specially designed hemispherical dome (Fig. 1) in order to avoid variations due to illumination geometry. The dome is 1.22 m in diameter, and the inside surface is coated with a titanium white paint. The target and a reflectance standard are placed at the center of the dome and diffusely illuminated by light reflected from the dome. Light is provided by four 3500° K color temperature lamps evenly spaced about the bottom of the dome and directed upward. The lamps are positioned below the level of the target and reference in order to avoid any direct illumination.

A spectroradiometer and a 4-lens, 35 mm camera are mounted on a moveable frame at the outer surface of the dome. The 4-lens camera is intended for polarization observations. A polarizing lens is

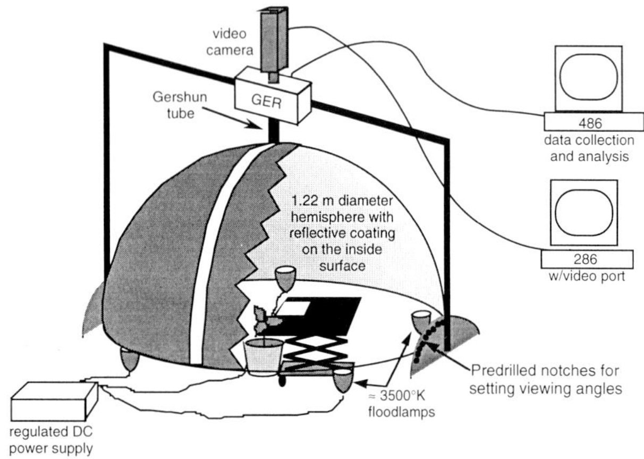

Fig. 1: The reflectance dome designed and constructed for this project is 1.22 m in diameter and is coated inside with a highly reflective titanium white paint. The target and a reflectance standard are placed at the center of the dome and diffusely illuminated by four 3200° K color temperature lamps evenly spaced about the bottom of the dome and directed upward. A spectroradiometer and a polarization camera are mounted on a moveable frame at the outer surface of the dome.

fixed to each of 3 of the 4 lenses with 45° offset relative to one another. The 4th lens is fitted with a neutral density filter to match the exposure for all lenses. Both the radiometer and the camera view the target plane at the center of the dome. The frame pivots about a line running through the center of the dome. Moving the frame allows the radiometer to be positioned at 10° intervals for viewing angles ranging from 0° (nadir) to 50°. The fixed positions help to insure repeatability. At the 0° position the radiometer is nadir-viewing and the 4-lens camera is offset by 5°. At all other viewing angles the camera viewing angle is 5° less than that of the radiometer.

The spectroradiometer is set back from the dome by a 30 cm-long tube. This serves to increase the effective field of view (FOV) of the instrument and minimizes the chance of direct illumination of the radiometer by one of the lamps. The radiometer is an IRIS Mark IV spectroradiometer (GER, Inc., Millbrook, New York, USA) which scans from 0.35 µm to 2.5 µm. This instrument uses two sets of two detectors, three different gratings and 4 different filters to scan the full wavelength range. Calibration of the instrument well beyond that which is standard with the instrument was essential in order to insure that the observations remained stable through all of the optical changes in the instrument as well as insuring cross calibration of the two optical channels.

The radiometer ia a «dual-beam» instrument, simultaneously measuring a target and a reference. The FOV is 2° × 3° for both the target and the reference. The two FOV's are separated by about 1° in the long axis. At the 90 cm viewing distance this translates into a 3 cm by 5 cm spot for nadir viewing of the target and reference at the target plane with a 1.5 cm separation between the two fields. The target FOV is large enough for a single soybean leaf (our target plant), but not large enough to view whole plants, much less canopies. Because of this, the data collection procedures were developed to optimize observation of single, attached leaves.

Fluorescence measurements were made using an SLM 6000 spectro-fluorometer. This is a double monochrometer instrument which can scan the excitation and emission independently under computer control. The exposure cell of the spectrofluorometer was modified to allow measurements to be made of whole, attached leaves. The standard components for the exposure cell were replaced with a mount that would hold a leaf at about 45° to the excitation and emission beams. The leaf remains attached to the plant at all times. Once a leaf is positioned in the beam, the access slit is sealed with black tape which serves both to hold the leaf in place and to reduce stray light.

Data are collected over a large range of excitation and emission wavelengths, $\lambda_{ex.}$ and λ_{em}, respectively. Fluorescence emission was measured at 2 nm intervals beginning at λ_{ex} + 10 nm and ending at 751 nm. Excitation ranged from 297 nm to 637 nm. 297 nm was expected to be below the major absorption range for plant pigments and within the absorption range for many important organic solvents, while 637 nm is within the strong chlorophyll absorption region. A single measurement sequence for one leaf takes over an hour and a half for this wavelength range.

A 2 mm slit width is used for the excitation beam and a 4 mm slit is used at the entrance slit of the emission monochrometer resulting in an effective resolution of about 4 nm for the emission spectra. For excitation wavelengths less than 387 nm ($\lambda_{ex} \leq 387$ nm) a low-pass blocking filter is required in order to reduce the 2nd order scatter from the diffraction grating in the excitation monochrometer. The blocking filter is designed to cut transmission by 10^5 for wavelengths longer than 600 nm in the excitation path. Since it also cuts transmission at the shorter wavelengths by about 20 %, the filter was removed for $\lambda_{ex} \geq 387$ nm. Variations due to changes in excitation intensity are corrected automatically by referencing the emission to a rhodamine dye standard.

An example of an excitation-emission (Ex-Em) matrix is shown in Fig. 2 a. Each continuous curve represents the set of emissions ob-

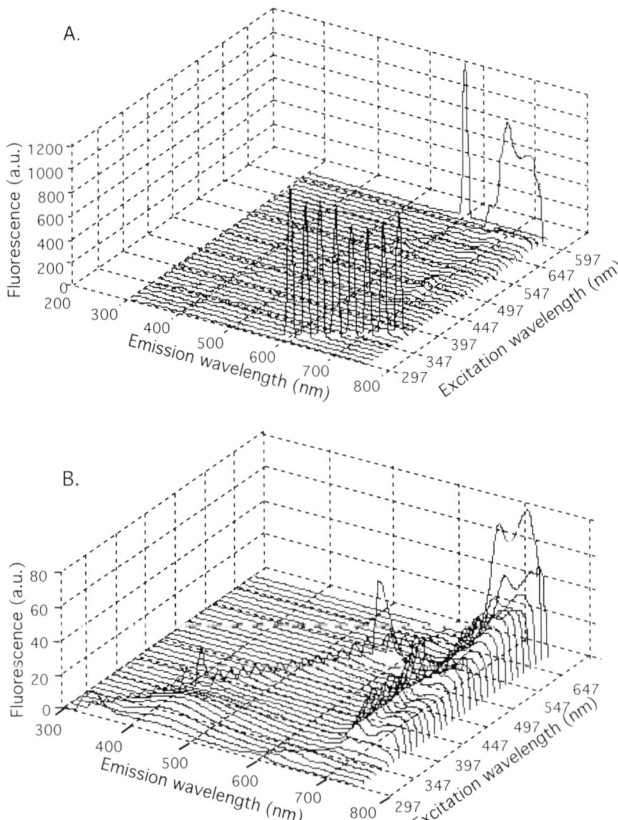

Fig. 2: An Ex-Em matrix. The x-axis is the fluorescence emission wavelength. The z-axis is the amplitude of the fluorescence emission. The y-axis is the spectrum number: spectrum # 1 = excitation at 297 nm, spectrum # 35 = excitation at 637 nm, and spectrum # 37 = excitation at 297 nm, a repeat of spectrum 1. A) Raw spectra. The large pulses in the red for spectra 1 through 9 represent the 2nd order scatter from the excitation wavelength. The diagonal line represents the 1st order scatter from the excitation wavelength. B) Spectra after the removal of the 2nd order scattering and rescaling.

served for a single excitation wavelength. The most obvious features in this figure are the strong 2nd order scatter peaks that appear in the red portion (590–750 nm) of the first nine scans and the last scan. This is the residual 2nd order scatter with the blocking filter in place. Obviously, the 5 orders of magnitude reduction provided by the blocking filter is insufficient to remove the 2nd order scatter. The last scan is simply a repetition of the first scan. Since the entire data collection procedure takes over 90 minutes to complete, we repeat the first scan at the end of the sequence to check for any changes that may have occurred during the data collection. Little difference has been noted between the first and last spectra suggesting that all the spectra should be directly comparable.

The second feature of note is the diagonal band running from λ_{em} = 307 nm for λ_{ex} = 297 nm to λ_{em} = 647 nm for λ_{ex} = 637. This is scatter from the excitation pulse and is strictly a noise feature. This represents the short wavelength limit of useful information for each emission spectrum. The last feature of note in Fig. 2 a is the next to last scan representing fluorescence emission from a leaf for excitation at 637 nm. The very strong fluorescence ($\approx 10^2$ over the background) indicates that 637 nm is within a very strong chlorophyll absorption region.

A better view of an Ex-Em matrix is shown in Fig. 2 b. The extraneous 2nd order peaks have been removed and the last three spectra have been removed from the plot. The rescaled plot shows most of the major fluorescence features in good detail.

Two preliminary experiments are described here. These experiments deal with two different forms of water stress – acute and chronic. Soybean plants were grown in Cornell mix[1] under greenhouse conditions. Acute stress was induced by withholding water for 24 hours prior to the measurement of reflectance. This caused the leaves to lose turgidity and droop although their color was visually unchanged. In a second experiment, the soybean plants were maintained at one of four levels of the maximum soil moisture capacity: 10–30 %, 30–55 %, 55–75 %, and 75–95 %. After determining the weight of a water-saturated soil, the water content was adjusted by adjusting the weight in each pot. Water levels were adjusted daily as necessary.

The general procedure was to make all reflectance measurements on one day and fluorescence measurements on the following day using the same plants and the same leaves. Reflectance measurements were made on the 3 different leaves on each of 2 different plants for each level of water stress tested. In order to insure that the leaf filled the instrument field-of-view, the measurement area on the leaf was away from the edge and included the central vein. Each reflectance spectrum took about 10 minutes including the time to position the plant in the dome.

Plants had to be transported to another building for the fluorescence measurements. Since the fluorescence measurement sequence took about 1.5 hours, only one plant was moved to the measurement facility at a time. The other plants remained in the greenhouse. The lengthy measurement time also meant that only 1 leaf was measured for each treatment.

Data Analysis

Three programs have been developed to facilitate the data analysis: HyperSpec and Correl, both written using MATLAB, and a third program written using MathCad. Hyperspec is designed to provide convenient and rapid manipulation of large spectral sets. The base data may be large sets of reflectance spectra or fluorescence Ex-Em matrices. HyperSpec allows the user to view the Ex-Em matrix (Fig. 2) from any angle and to select a single spectrum for examination as shown with spectrum # 10 (excitation at 397 nm) in Figs. 3 a and 3 b. One of the more important tools provided by HyperSpec is the ability to smooth the data and compute spectral derivatives. Fig. 3 b shows the display of the second derivative of spectrum # 10 computed for a range of bandwidths. The y-axis denotes the bandwidth and band separation used to compute the derivative (see Fig. 4). The higher values are representative of data collected with instruments having courser spectral resolution. In this case the bandwidth, bw, is equal to the band separation, $\Delta\lambda$, although the software allows the two to be varied independently.

In order to fully understand the spectral reflectance data and to maintain a continuous quality check on the accuracy of the data, a third set of routines were written in Mathcad 5.0 to plot the spectral reflectance of the standard reflectances, to manipulate the data and to plot the spectral reflectance and derivatives of various groupings of data sets from plants subjected to different stress levels, as well as controls. Derivative curves were plotted for various averages and after various degrees of smoothing were applied to improve signal-to-noise characteristics.

---

[1] Cornell mix is a 1 : 6 ratio v/v of peat moss and vermiculite with Peters Brand micronutrients and lime added. Peters Brand micronutrients is a 20–20–20 mix of N, P. K with Mg, B, Cu, Fe, Mn, Mo and Zn included as micronutrients.

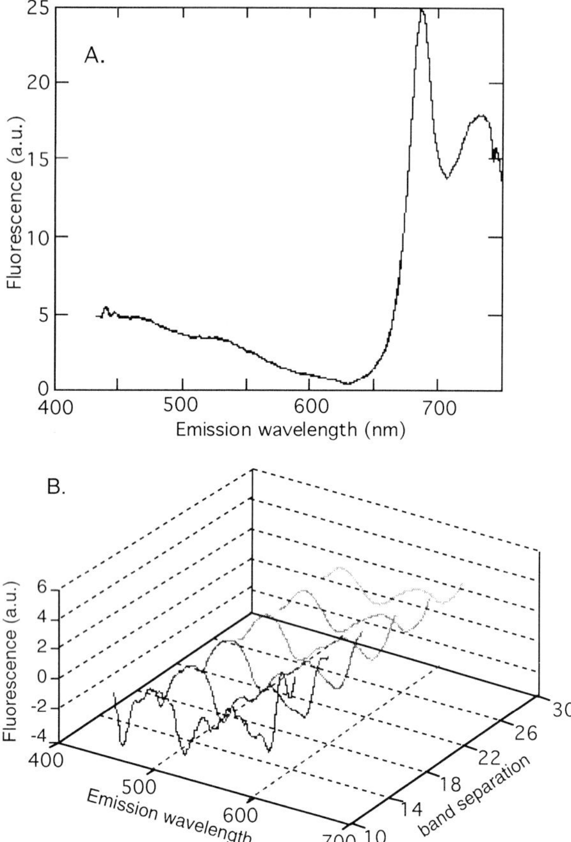

Fig. 3: A sequence of operations within HyperSpec using fluorescence emission spectrum # 10 from Fig. 4 (excitation at 397 nm). a) Fluorescence emission spectrum # 10. b) Second derivatives of spectrum # 10 from 450 nm to 650 nm at several different bandwidths.

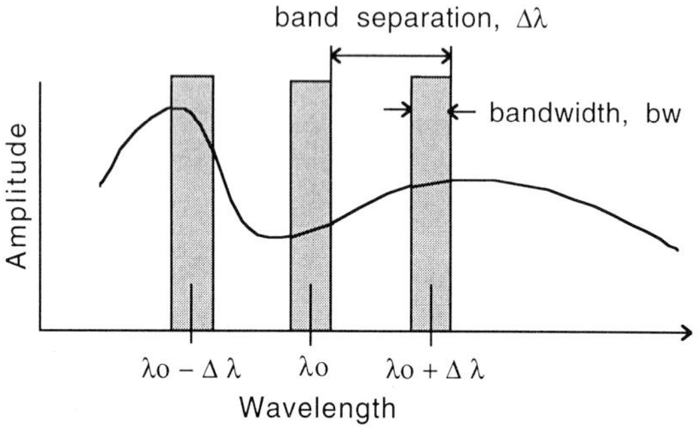

Fig. 4: Derivatives of reflectance or fluorescence emission spectra are computed using different bandwidths and different band separations to model the response of different detectors or a selection of bands from one detector. Bandwidth, bw, and band separation, $\Delta\lambda$. Samples are equally spaced at intervals of $\Delta\lambda$, with each sample having a bandwidth, bw. The bandwidth need not equal the band separation, although that would be typical of a hyperspectral instrument.

Derivatives are computed in HyperSpec using one of several smoothing options: 1) the Savitsky-Golay method (Savitsky and Golay, 1964; Madden, 1978), 2) the Kawata-Minami method (Kawata and Minami, 1984), and 3) a simple average over the effective bandwidth. Savitsky-Golay and Kawata-Minami both fit a polynomial to the spectral data in an attempt to preserve the fine structure of the spectrum while removing noise. The Savitsky-Golay method couples the derivative computation and the smoothing in a single operation while the Kawata-Minami method dissociates the two procedures.

The amplitude of the derivative decreases with the bandwidth, but that does not imply a decrease in information for the larger bandwidth data. Indeed, since many of the typical spectral reflectance and fluorescence features associated with vegetation are rather broad, not only can one improve the S/N ratio by using wider bandwidths, but the wider bandwidths are actually more appropriate to the spectral scale of broad reflectance and fluorescence features.

Some support for this idea is shown in Fig. 5, in which each spectrum represents the correlation at a series of wavelengths of the 2nd derivative of the reflectance to the chlorophyll concentration in an experiment to study Manganese deficiency in ‹Bragg› soybean. Again, the y-axis represents increasing bandwidth and band separation. For the smallest bandwidth considered, the correlations are very high at several places on the curve, indicating that a second derivative of the reflectance spectrum at these points correlates very well with the chlorophyll content. Note, however, that there are rapid changes between very high and very low correlations for the narrow bandwidth. In contrast, as the bandwidth/band separation increases, the correlation remains high and becomes more consistent over wavelength. It is worth noting that the reflectance features showing these high correlation are in the visible region. The high-correlation feature in the spectral range 550–650 nm probably represents the yellowing of the leaf (chlorosis).

Results

Reflectance spectra for soybean plants subjected to chronic water stress treatments are shown in Fig. 6. These measurements were made after the soybean plants had been held at the fixed soil water content for 14 days.

For both chronic (Fig. 6) and acute water stress (not shown) there was a slight decrease in reflectance in the infrared and a concurrent increase in reflectance in the visible. While these changes are consistent with expectations (Colwell, 1974), they may be too small (generally less than 4 % reflectance) to be seen with remote sensing. Each of the curves in Fig. 6 represents the average reflectance from several observations. The standard deviation for each curve is approximately 2 % reflectance in the visible and near IR increasing up to 5 % in the mid IR. The very small differences in reflectance between water-stressed soybeans and the control may be due in part to our data collection procedure which was to measure reflectance of single leaves. The apparent strategy for this species of soybean under chronic water stress is to allow the older leaves to wilt, yellow and die while maintaining the younger leaves in relatively good condition. A measurement of reflectance of the plant canopy would include both the healthy and dying leaves and might well yield much more noticeable differences. The optical properties of a canopy will also be substantially different than that of a single leaf. On the other hand, damage to leaves from the more severely stressed plants was visually apparent even though reflectance differences were not pronounced.

Fluorescence spectra for the water stress experiments are shown in Fig. 7. At the wavelengths normally monitored the changes are not too surprising. For acute stress there is a

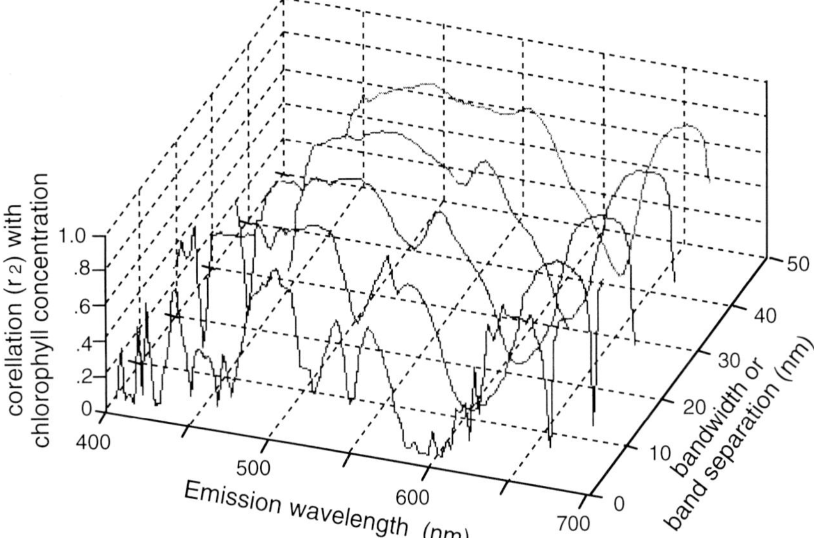

Fig. 5: Correlation of the 2nd derivative of reflectance with chlorophyll concentration in manganese deficient ‹Bragg› soybeans for increasingly large bandwidths/band separations. As the bandwidth increases the correlation remains high, but becomes more stable. The high-correlation feature in the spectral range 550–650 nm probably represents the yellowing of the leaf (chlorosis).

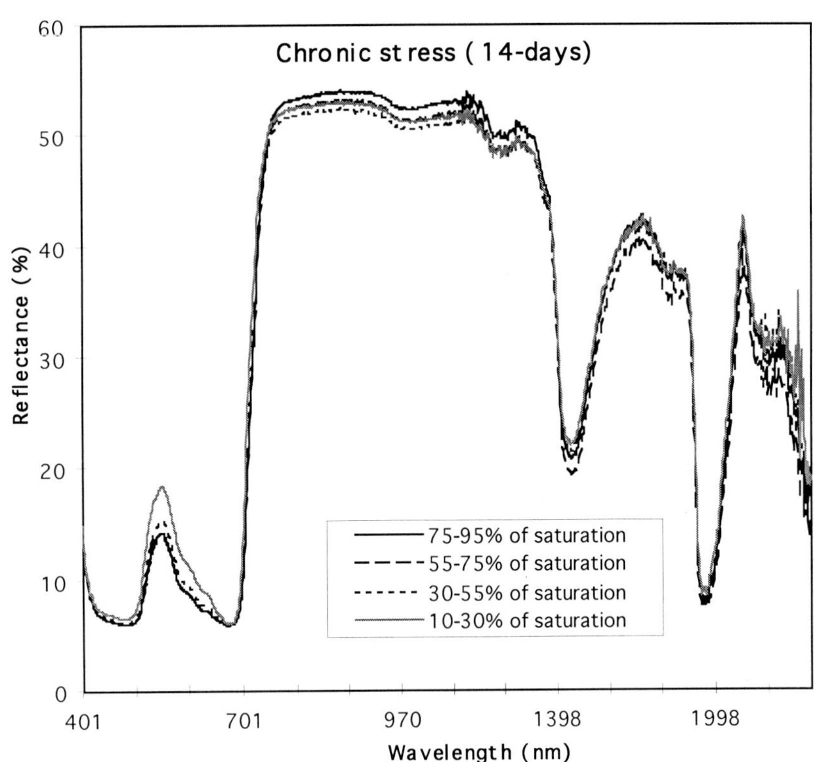

Fig. 6: Reflectance spectra for soybean plants subjected to chronic water stress. Plants were maintained at four different levels of water availability in terms of the percentage of water in the soil relative to water saturated soil: 10–30 %, 30–55 %, 55–75 % and 75–95 % (control).

slight increase in fluorescence in the blue and in the red/infrared chlorophyll emissions (Fig. 7a). There is also a substantial increase in fluorescence at these wavelengths for the chronic stress (Fig. 7b). It is worth noting that the increase in the blue and red/infrared fluorescence is more pronounced for the plants under chronic stress.

There are also several unusual features that appear in the yellow-green portions of these spectra (Table 1). These are remarkably strong fluorescence emissions and are apparently quite sensitive to stress. They are also unusual in other respects. The emission in the yellow-green is apparently related to a relatively narrow-band absorption. In each case emission only occurs for a very narrow range of excitation wavelengths (10–20 nm), and the excitations appear to be independent. Indeed, the shorter wavelength emission features result from excitation at longer wavelengths. This implies that there are

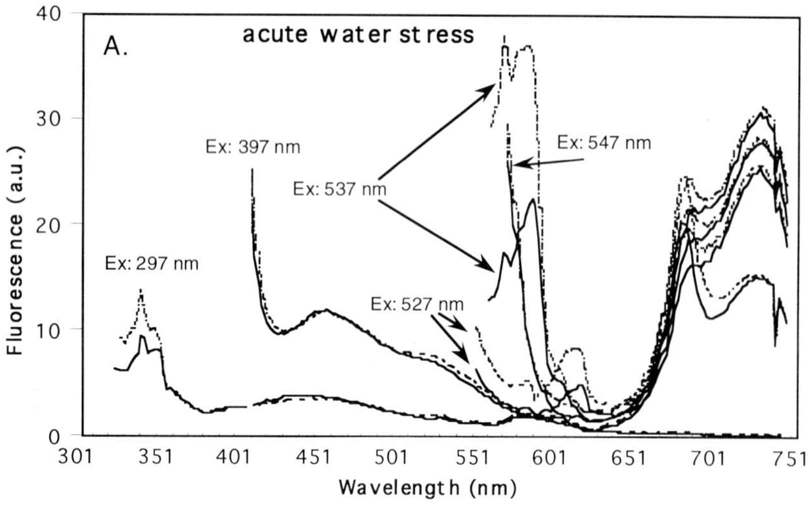

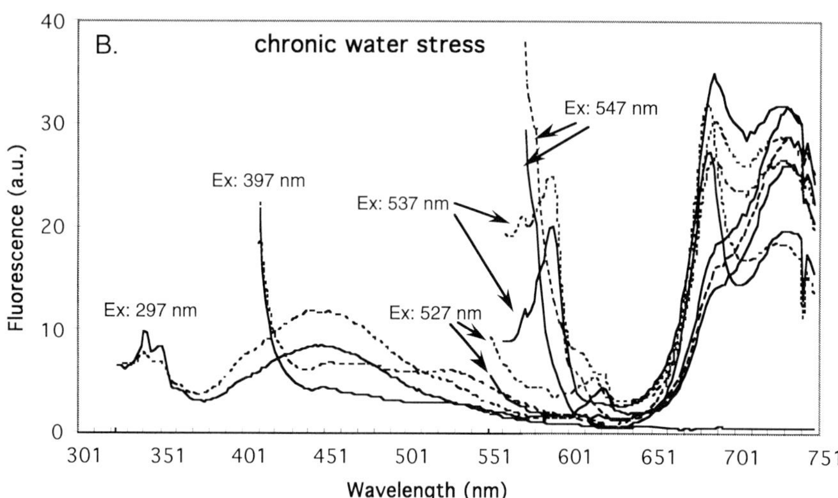

Fig. 7: Fluorescence emission spectra for pairs of stressed and control soybean plants at several excitation wavelengths: 297 nm, 397 nm, 527 nm, 537 nm and 547 nm. Solid lines represent control plants, dashes lines represent stressed plants: a) acute water stress vs. control, and b) chronic water stress vs. control.

Table 1: List of the unusual fluorescence peaks observed for the water stress experiment.

Excitation	Emission	Comments
297 nm	330–350 nm	Fluorescence nearly doubles for acute water stress and decreases slightly for chronic stress.
527 nm	620–637 nm	Relatively weak fluorescence that only occurs at this excitation wavelength. Apparently arises from a very narrow absorption band.
537 nm	570 and 600 nm	Strong emission that doubles for acute stress and nearly doubles for chronic stress. The shape of the emission curve is also rather different for the two.
547 nm	570–580 nm	Strong emission, but nearly lost in the 1st order scatter from the excitation.

relatively well-defined compounds responsible for the fluorescence and that the same molecule is not likely to be responsible for all of the fluorescence peaks.

At this time we have no hypothesis to explain these fluorescence features. They occur for both stressed and unstressed plants, and they change dramatically with stress conditions. For the unstressed plants the results are highly repeatable, having been observed in every fluorescence spectra examined in this project (Fig. 8). All this suggests that they are real features. On the other hand, we have no evidence yet that these are general features that will appear with all, or even most, vegetation.

Conclusion

An experiment designed to search for reflectance, fluorescence or polarization measurements that would be characteristic of anthropogenic contaminants has been described. The measurements are made from single, attached leaves, under controlled conditions. Early experiments were designed to characterize natural stresses since identification of a specific contaminant by the optical effects it induces would have to be distinguishable from similar effects induced by natural stresses.

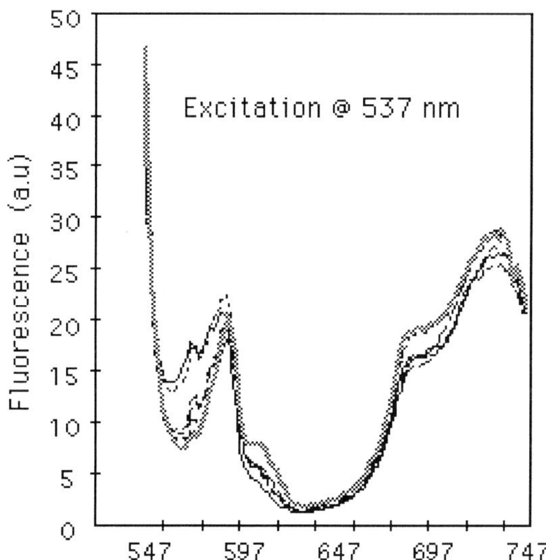

Fig. 8: Emission spectra (excitation at 537 nm) for the control soybean plants for six separate experiments, i.e., different nodes and different stages of development. Data were collected over a period of 2 months, and show very little variation compared to the changes in fluorescence seen in the stressed plants (Fig. 7).

Reflectance of soybean plants subjected to water stress (acute and chronic) differed little from reflectance of the control plants even though there was visible damage to the leaves. Blue and red fluorescence was somewhat greater for the water stressed plants. Unusual fluorescence features in the yellow-green part of the spectrum were also found in both control and stressed plants. These features were exceptionally sensitive to stress and appeared to be indicative of several distinct compounds in the soybean leaves.

Acknowledgements

The authors would like to recognize of Research Associates of Syracuse, Inc. for supporting this research, and especially for the assistance provided by Mr. Jack Monson and Mr. Joe Angilillo for the design and construction of the reflectance facility and by Ms. Betsy Chiesa for programming. We also wish to thank Dr. Jay Jacobson and Mr. Rudyard Edick for designing and conducted the water stress experiments.

References

Adams, M. L.: Use of leaf fluorescence and reflectance characteristics to detect Mn stress in ‹Bragg› soybeans. M.S. Thesis, Cornell University (1993).

Baret, F., I. Champion, G. Guyot, and A. Podaire: Monitoring wheat canopies with a high spectral resolution radiometer, Remote Sens. Environ. *22,* 367–378 (1987).

Bawden, F. C.: Infrared photography and plant virus diseases. Nature *132,* 168 (1933).

Benedict, H. M. and R. Swidler: Nondestructive method for estimating chlorophyll content of leaves. Science *133,* 2015–2016 (1961).

Chappelle, E. W., J. E. McMurtrey, and M. S. Kim: Identification of the pigment responsible for the blue fluorescence band in the laser induced fluorescence (LIF) spectra of green plants, and the potential use of this band in remotely estimating rates of photosynthesis, Remote Sens. Environ. *36,* 213–218 (1991).

Chappelle, E. W., F. M. Wood, J. E. McMurtrey, and W. W. Newcomb: Laser-induced fluorescence of green plants. 1: A technique for remote detection of plant stress and species differentiation. Applied Optics *23,* 134–138 (1984).

Colwell, J. E.: Vegetation canopy reflectance, Remote Sens. Environ. *3,* 175–183 (1974).

Demetriades-Shah, T. H., M. D. Steven, and J. A. Clark: High Resolution Derivative Spectra in Remote Sensing, Remote Sens. Environ. *33,* 55–64 (1990).

Huete, A. R.: Soil-adjusted vegetation index (SAVI), Remote Sens. Environ. *25,* 295–309 (1988).

Kawata, S. and S. Minami: Adaptive smoothing of spectroscopic data by a linear mean-square estimation. Applied Spectroscopy *38,* 49–58 (1984).

Kauth, R. J. and G. S. Thomas: The tasseled cap – a graphic description of the spectral-temporal development of agricultural crops as seen from Landsat. Paper presented at the Machine Processing of Remotely Sensed Data, Purdue University, West Lafayette, Indiana (1976).

Lang, M., F. Stober, and H. K. Lichtenthaler: Fluorescence emission spectra of plant leaves and plant constituents. Radiat. Environ. *30,* 333–347 (1991).

Lichtenthaler, H. K.: Chlorophylls and carotenoids: pigments of photosynthetic biomembranes. J. Plant Physiology *131,* 101–110 (1987).

Lichtenthaler, H. K. and C. Buschmann: Chlorophyll fluorescence spectra of green bean leaves. J. Plant Physiology *129,* 137–147 (1987).

Lichtenthaler, H. K. and U. Rinderle: The role of chlorophyll fluorescence in the detection of stress conditions in plants. CRC Crit. Rec. Anal. Chem. *19,* Suppl. 1, 29–85 (1988).

Madden, H. H.: Comments on the Savitsky-Golay convolution method for least-squares fit smoothing and differentiation of digital data. Analytical Chemistry *50,* 1383–1386 (1978).

Richardson, A. J. and C. L. Wiegand: Distinguishing vegetation from soil background information, Photogram. Eng. and Remote Sensing *43,* 1541–1552 (1977).

Savitsky, A. and M. H. E. Golay: Smoothing and differentiation of data by simplified least squares procedures. Analytical Chemistry *36,* 1627–1639 (1964).

Stober, F. and H. K. Lichtenthaler: Characterization of the laser-induced blue, green and red fluorescence signatures of leaves of wheat and soybean grown under different irradiance. Physiol. Plant. *88,* 696–704 (1993).

J. Plant Physiol. Vol. 148. pp. 574–578 (1996)

Fluorescence Emission Spectra of Paraquat Resistant *Conyza canadensis* During the Chlorophyll Fluorescence Induction as Determined by the CCD-OMA System

ZOLTÁN SZIGETI[1,2], PETER RICHTER[3], and HARTMUT K. LICHTENTHALER[2]

[1] Department of Plant Physiology, Eötvös University, P.O.B. 330, H-1445 Budapest, Hungary

[2] Botanical Institute, University of Karlsruhe, Kaiserstr. 12, D-76128 Karlsruhe, Germany

[3] Department of Atomic Physics, Technical University of Budapest, Budafoki ut 8, H-1111 Budapest, Hungary

Received September 25, 1995 · Accepted October 28, 1995

Summary

A multichannel chlorophyll spectrofluorometer based on a charge coupled device (CCD-OMA fluorosensor) was used for fast screening of the chlorophyll fluorescence characteristics of paraquat sensitive (S) and paraquat resistant (PqR) horseweed (*Conyza canadensis*) biotypes. It is demonstrated that S and PqR plants can easily be distinguished already from some leaf samples by only one series of fluorescence spectra measurements. The chlorophyll fluorescence parameters Fm, Fs, and the chlorophyll fluorescence ratios Rfd and F690/735, calculated from the spectra taken at different times during the fluorescence induction kinetics, clearly indicate the different paraquat sensitivity of the two horseweed biotypes. With the fluorosensor it was also shown that the paraquat resistant plants were, however, sensitive to the herbicide diuron. The results are in good agreement with earlier results achieved by more sophisticated and time consuming methods.

Key words: Conyza canadensis, diuron, chlorophyll fluorescence induction kinetics, paraquat resistance.

Abbreviations: CCD = charge coupled device; Fm = maximum chlorophyll fluorescence; Fs = steady-state fluorescence; OMA = optical multichannel analyzer; PqR = paraquat resistant; Rfd = variable chlorophyll fluorescence ratio (ratio of fluorescence decrease); S = paraquat sensitive.

Introduction

Among the various anthropogenic stress factors acting on plants frequent stressors are the herbicides used against weeds to protect crop plants. Paraquat (1,1′-dimethyl-4,4′-bipyridylium ion), normally applied in the form of the dichloride salt, is an active ingredient of several herbicide preparations and is extensively used in agriculture as a non-selective contact herbicide. The repeated use of paraquat for longer periods has led to the selection of paraquat resistant weed biotypes. Resistance to paraquat has so far appeared in 16 weed species following persistent selection pressure with this herbicide (Preston, 1994). Horseweed (*Conyza canadensis* (L.) Cronq.) is such a widespread weed which has developed high paraquat resistance in several countries (Hirata and Matsunaka, 1985; Pölös et al., 1988). The paraquat resistant *Conyza* biotypes found in Hungary simultaneously showed a relatively high resistance also against diquat (Szigeti et al., 1994).

In contrast to diuron or atrazine, paraquat does not inhibit photosynthetic electron transport at the Q_B-binding site of photosystem II, but deviates electrons on the reducing side of photosystem I. As a consequence the chlorophyll fluorescence of paraquat-treated plants is partially quenched. Therefore the use of different chlorophyll fluorescence spectroscopic methods are very useful in the study of paraquat effects in sensitive and resistant plants.

When illuminating a predarkened green leaf, the chlorophyll fluorescence shows characteristic fast and slow transients which reflect the functioning of the photosynthetic apparatus. This phenomenon is known as Kautsky effect (Kautsky and Hirsch, 1931; Lichtenthaler, 1992). The determination of the spectral composition of the emitted chlorophyll fluorescence during the fast rise and slow decline became only possible after the introduction of optical multichannel analyzer (OMA) systems (Buschmann and Sironval, 1984; Buschmann and Lichtenthaler, 1988). The charge coupled device CCD-OMA fluorosensor system, which permits a simultaneous parallel read-out of data, proved to be even more efficient to register the spectral changes in laser-induced chlorophyll fluorescence emission (Szabó et al., 1992; Buschmann et al., 1996) than OMA fluorosensors which only exhibit a serial read-out of data.

The aim of the present work was to demonstrate the possible application of the laser-equipped CCD-OMA fluorometric technique in the research of paraquat resistance and sensitivity of horseweed (*Conyza canadensis* (L.) Cronq.) plants.

Materials and Methods

Plants and treatment

Seeds of susceptible (S) and paraquat resistant (PqR) horseweed (*Conyza canadensis* (L.) Cronq.) plants were collected in a vineyard near Kecel (Hungary). Plants were grown in soil under laboratory conditions (illumination 130 μmol $m^{-2} s^{-1}$ PAR, 16 h light/8 h dark period, 22–25 °C) for 6 months. Then the plants in soil containers were transferred to field conditions. Approximately 8-month-old plants in the rosette stage were used in the experiments. The treatments were made by floating the detached leaves on Gramoxone (25 % paraquat) solution in the light. The concentration of paraquat was 0.5 mM. The herbicide diuron [3-(3,4-dichlorophenyl)-1,1-dimethylurea] was used in 0.1 mM concentration.

Pigment determination

Chlorophylls and carotenoids were extracted and quantified in 100 % acetone using the redetermined absorption coefficients and equations of Lichtenthaler (1987a).

Measurements of fluorescence spectra and kinetics

In order to follow the spectral changes during the fast rise and slow decline of the chlorophyll fluorescence induction kinetics the Karlsruhe/Budapest CCD-OMA fluorosensor system was applied. The CCD-OMA fluorometer consists of a 10 mW He/Ne laser (λ = 632.8 nm) as excitation light, a spectrograph and a CCD-line detector with parallel read-out as described earlier (Szabó et al., 1992; Buschmann et al., 1995). The light intensity on the sample surface was about 40,000 μmol $m^{-2} s^{-1}$ and saturating with respect to chlorophyll fluorescence excitation.

Results and Discussion

Chlorophyll fluorescence spectra, taken with the CCD-OMA spectrofluorometer at saturating light conditions with a rationally chosen timing during the chlorophyll induction kinetics, give all data for a correct determination of the actual chlorophyll fluorescence ratio F690/F735 as well as maximum (Fm) and steady state fluorescence (Fs) at 690 and 735 nm. From the Fm and Fs data the Rfd690 and Rfd735 values can be calculated (Fig. 1A., B.), which function as vitality index (Lichtenthaler, 1987b) and are related to the net CO_2-assimilation rate P_N (Tuba et al., 1994).

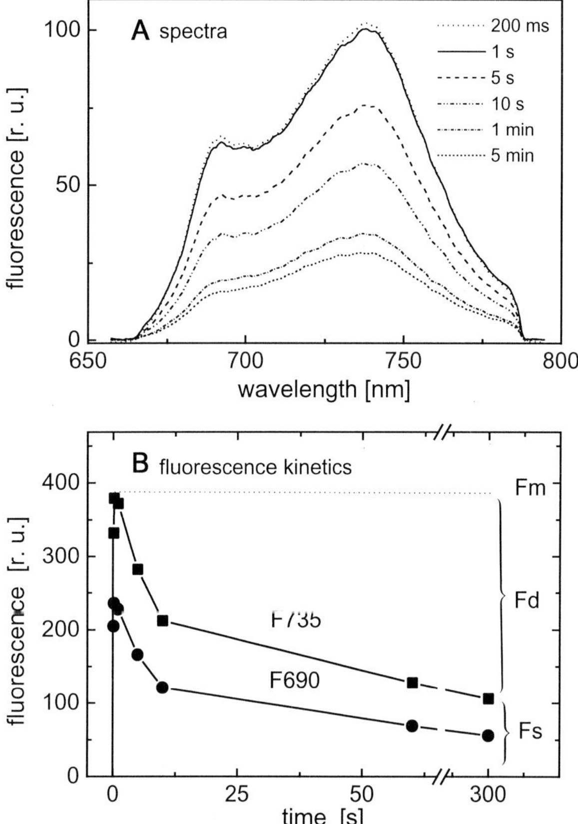

Fig. 1: Chlorophyll fluorescence spectra and kinetics in a leaf of a paraquat sensitive horseweed *Conyza canadensis* **A.)** Chlorophyll fluorescence emission spectra measured during the fluorescence induction kinetics at 200 ms, 1 s, 5 s, 10 s, 60 s and 5 min after onset of illumination. Integration time was 100 ms. **B.)** Fluorescence induction kinetics of the same leaf shown for the 690 and 735 nm region. The kinetics and Rfd values [Rfd = (Fm−Fs)/Fs or Fd/Fs] were drawn and calculated from the fluorescence emission spectra shown in Fig. 1 A. The Rfd-values calculated for the 690 nm and 735 nm kinetics are: Rfd690 = 3.25 and Rfd735 = 2.63.

It was earlier demonstrated that paraquat in horseweed quenched chlorophyll fluorescence in S-plants, but affected the fluorescence of PqR-plants only weakly (Pölös et al., 1988). The light induced fluorescence kinetics of dark adapted leaves of S and PqR-plants showed that Fm of paraquat treated S-plants was strongly quenched, while Fm in PqR-plants showed only a slight decrease (Szigeti et al., 1988). It is well known that the chlorophyll fluorescence emission spectra of green leaves at room temperature show maxima near 690 and 735 nm. At a higher chlorophyll content of the leaves the fluorescence maximum near 690 nm is decreased with respect to that near 735 nm (Szabó et al., 1992). The curvilinear relationship between chlorophyll content and the ratio F690/F735 had been demonstrated earlier (Lichtenthaler, 1987b; Hák et al., 1990). The lower intensity

Table 1: Pigment content and pigment ratios a/b and (a+b)/(x+c) in untreated paraquat sensitive (S) and paraquat resistant (PqR) horseweed (*Conyza canadensis*) biotypes. The pigment content is given in µg cm^{-2} leaf area. Mean of 8 determinations from 6 leaves. Standard deviation SD<5% for pigment content and <3% for pigment ratios.

	Sensitive (S)	Resistant (PqR)
Chlorophyll *a+b*	39.8	50.2
Total carotenoids, x+c	8.7	10.9
(a+b)/(x+c)	4.57	4.50

Table 2: Chlorophyll fluorescence ratio F690/F735 in leaves of paraquat sensitive (S) and paraquat resistant (PqR) horseweed (*Conyza canadensis*) biotypes at maximum fluorescence Fm and steady-state fluorescence Fs calculated from spectra taken during the laser-induced fluorescence induction kinetics before and after treatment with paraquat (0.5 mM) or diuron (0.1 mM). Mean of 10 determinations from 5 leaves, standard deviation SD<4%.

	at Fm	at Fs	% decrease from Fm to Fs
Sensitive (S)			
Control	0.64	0.52	18.9
+Paraquat	0.59	0.59	0
+Diuron	0.56	0.55	1.8
Resistant (PqR)			
Control	0.56	0.46	17.8
+Paraquat	0.56	0.46	17.8
+Diuron	0.57	0.56	1.8

of 690 nm fluorescence maximum in the leaves of the PqR, as compared to S-plants, was caused by a higher chlorophyll content (Table 1.) and the overlapping of the 690 nm chlorophyll fluorescence emission band with the red chlorophyll absorption band of the leaf. As a consequence the F690/F735 ratio showed lower values in the PqR-plants than in the S-plants (Table 2).

It is a general symptom that the decrease of the chlorophyll fluorescence maximum near 690 nm (F690) during the fluorescence decline under continuous illumination is stronger than that of the maximum near 735 nm (F735). Therefore the fluorescence ratio F690/F735 declines during the fluorescence induction kinetics (Buschmann and Lichtenthaler, 1988; Kocsányi et al., 1988). Fluorescence spectra of paraquat treated leaves of S-plants taken at Fm and Fs revealed that paraquat treatment eliminated the usual fluorescence decrease from Fm to Fs (Fig. 2 A, B). The lower fluorescence intensity of these leaves was accompanied by an unchanged F690/F735 ratio (Table 2). In contrast, the paraquat treated leaves of the PqR-plants showed the normal physiological response of an unaffected photosynthesizing system (Fig. 3 A, B). Correspondingly, the F690/F735 ratio decreased during the induction kinetics in the same way as in the untreated control leaves of PqR-plants (Table 2).

When the leaves of both S and PqR biotypes were treated with diuron in 0.1 mM concentration, the photosynthetic electron transport was blocked in both biotypes and the initial maximum chlorophyll fluorescence Fm remained high and did not show the usual decrease to a much lower Fs-value (Fig. 2 C, 3 C). Under these conditions of inhibited photosynthesis the chlorophyll fluorescence ratio F690/F735 amounted to 0.56–0.57 at Fm and remained nearly on the same level (0.55–0.56) at Fs i.e. 5 min after onset of illumination (Table 2). These data clearly indicate the diuron sensitivity of both *Conyza* biotypes.

The variable fluorescence ratio, the ratio of fluorescence decrease from Fm to the steady state fluorescence Fs, expressed as Rfd value, is a vitality and stress indicator of plants, which is proportional to the photosynthetic activity of the leaf (Lichtenthaler et al., 1987b; Tuba et al., 1994). The Rfd values presented in Table 3 show that the photosynthetic activity of leaves of S-plants was inhibited by both herbicides paraquat and diuron, whereas leaves of PqR-plants were affected only by diuron. The initially high Rfd values (controls)

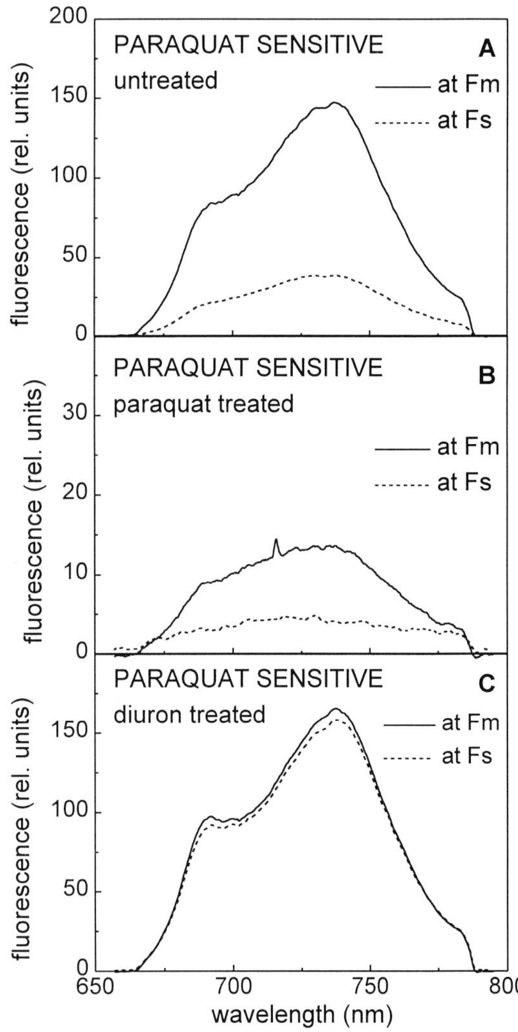

Fig. 2: Chlorophyll fluorescence emission spectra of leaves of sensitive (S) horseweed (*Conyza canadensis*) measured at Fm (after 500 ms) and at Fs (after 5 min). **A.)** untreated leaf, **B.)** leaf treated with 0.5 mM paraquat in light 8 h before and **C.)** leaf treated with 0.1 mM diuron in light 2 h before the measurement.

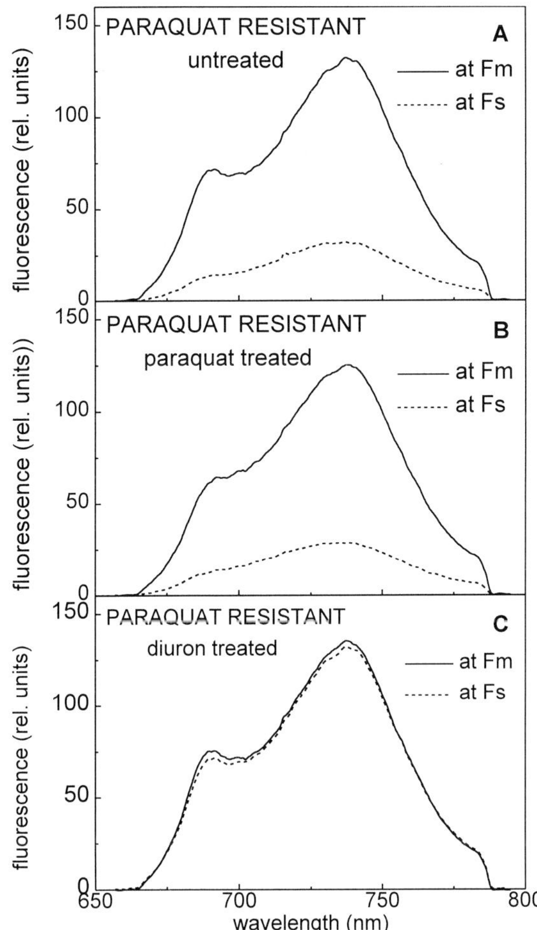

Fig. 3: Chlorophyll fluorescence emission spectra of leaves of paraquat resistant (PqR) horseweed (*Conyza canadensis*) measured at Fm (after 500 ms) and at Fs (after 5 min). **A.**) untreated leaf, **B.**) leaf treated with 0.5 mM paraquat in light 24 h before and **C.**) leaf treated with 0.1 mM diuron in light 2 h before the measurement.

Table 3: Ratio of the variable chlorophyll fluorescence expressed as Rfd-values [Rfd = (Fm-Fs)/Fs] in leaves of paraquat sensitive (S) and paraquat resistant (PqR) horseweed (*Conyza canadensis*) before and after treatment with paraquat (0.5 mM) or diuron (0.1 mM). The values of Rfd690 and Rfd735 were calculated from the chlorophyll fluorescence intensities measured at the 690 nm and 735 nm emission maxima, respectively. Mean of 10 determinations per condition from 5 leaves, standard deviation SD<4%.

	Rfd690	Rfd735
Sensitive (S)		
Control	3.11	2.96
+Paraquat	0.58	0.53
+Diuron	0.28	0.25
Resistant (PqR)		
Control	4.35	3.70
+Paraquat	4.05	3.38
+Diuron	0.31	0.20

declined in the diuron treated leaves to values of 0.2 to 0.3 which indicate a strongly inhibited photosynthesis (Table 3).

The results presented in this paper clearly show that chlorophyll fluorescence emission spectra, taken during the fluorescence induction kinetics, are rich in useful physiological information and represent valuable tools in characterization of paraquat resistance. The results are in good agreement with earlier observations as discussed above. The data presented here also demonstrate that responses of plants with varying sensitivity to different herbicide classes can easily be followed and distinguished by chlorophyll fluorescence measurements.

Acknowledgements

This work was supported by a grant of the Hungarian National Research Foundation (OTKA 013320) and by a fellowship of Deutscher Akademischer Austauschdienst, Bonn, for Z. Szigeti, which are gratefully acknowledged. Thanks are due to Dr. E. Lehoczki (Szeged, Hungary) for providing us the plant material and Dr. M. Lang (Karlsruhe) for assistance in the printing of the figures.

References

Buschmann, C. and C. Sironval: Fluorescence emission spectra of etiolated leaves measured at 296 and 77 K during the first seconds of continuous illumination. In: Sironval, C. and M. Brouers (eds.): Protochlorophyllide Reduction and Greening, pp. 139–148. M. Nijhoff/Dr. W. Junk Publ., The Hague (1984).

Buschmann, C. and H. K. Lichtenthaler: Complete fluorescence emission spectra determined during the induction kinetic using a diode-array detector. In: Lichtenthaler, H. K. (ed.): Applications of Chlorophyll Fluorescence, pp. 77–84. Kluwer Academic Publishers, Dordrecht, The Netherlands (1988).

Buschmann, C., J. Kisvári, K. Szabó, P. Richter, and H. K. Lichtenthaler: Detection of environmental stress to plants by measuring the laser-induced spectra of chlorophyll fluorescence with the CCD-OMA system. In: Richter, P. I. and R. C. Hendon (eds.): Proc. 2nd Internat. Symp. and Exhibition on Environ. Contamin. in Central and Eastern Europe, 1994 Budapest, pp. 154–156. Government Inst. Inc., Rockwell, Maryland, USA (1995).

Buschmann, C., P. Richter, and H. K. Lichtenthaler: Application of the Karlsruhe CCD-OMA LIDAR-Fluorosensor in stress detection of plants. J. Plant Physiol. *148,* 548–554 (1996).

Hák, R., H. K. Lichtenthaler, and U. Rinderle: Decrease of the chlorophyll fluorescence ratio F690/F730 during greening and development of leaves. Rad. Environ. Biophys. *29,* 329–336 (1990).

Hirata, T. and S. Matsunaka: Paraquat resistance in *Erigeron canadensis.* Weed Res. (Japan) 30, Suppl., 127–128 (1985).

Kautsky, H. and A. Hirsch: Neue Versuche zur Kohlenstoffassimilation. Naturwissenschaften *19,* 964 (1931).

Kocsányi, L., M. Haitz, and H. K. Lichtenthaler: Measurement of the laser-induced chlorophyll fluorescence kinetics using a fast acoustooptic device. In: Lichtenthaler, H. K. (ed.): pp. 99–107. Kluwer Academic Publishers, Dordrecht (1988).

Lichtenthaler, H. K.: Chlorophylls and carotenoids, the pigments of photosynthetic biomembranes. Methods in Enzymol. *148,* 350–382 (1987a).

– Chlorophyll fluorescence signatures of leaves during the autumnal breakdown. J. Plant Physiol. *131,* 101–110 (1987b).

– The Kautsky effect: 60 years of chlorophyll fluorescence induction kinetics. Photosynthetica *27,* 45–55 (1992).

Lichtenthaler, H. K., C. Buschmann, U. Rinderle, and G. Schmuck: Application of chlorophyll fluorescence in ecophysiology. Radiat. Environ. Biophys. *25,* 297–308 (1986).

Pölös, E., J. Mikulás, Z. Szigeti, B. Matkovics, Q. H. Do, Á. Párducz, and E. Lehoczki: Paraquat and atrazine coresistance in *Conyza canadensis* (L.) Cronq. Pestic. Biochem. Physiol. *30,* 142–154 (1988).

Preston, C.: Resistance to photosystem I disrupting herbicides. In: Powles, S. B. and J. A. M. Holtum (eds.): Herbicide Resistance: Biology and Biochemistry, pp. 61–82. Lewis Publ. Inc., Boca Raton, Florida (1994).

Szabó, K., H. K. Lichtenthaler, L. Kocsányi, and P. Richter: A CCD-OMA device for the measurement of complete chlorophyll emission spectra of leaves during the fluorescence induction kinetics. Radiat. Environ. Biophys. *31,* 153–160 (1992).

Szigeti, Z., É. Darkó, E. Nagy, and E. Lehoczki: Diquat resistance of different paraquat resistant *Conyza canadensis* (L.) Cronq. biotypes. J. Plant Physiol. *144,* 686–690 (1994).

Szigeti, Z., E. Pölös, and E. Lehoczki: Fluorescence properties of paraquat resistant *Conyza* leaves. In: Lichtenthaler, H. K. (ed.): Applications of Chlorophyll Fluorescence, pp. 109–114. Kluwer Academic Publishers, Dordrecht, The Netherlands (1988).

Tuba, Z., H. K. Lichtenthaler, Z. Csintalan, Z. Nagy, and K. Szente: Reconstitution of chlorophylls and photosynthetic CO_2 assimilation in the desiccated poikilochlorophyllous plant *Xerophyta scabrida* upon rehydration. Planta *192,* 414–420 (1994).

J. Plant Physiol. Vol. 148. pp. 579–585 (1996)

Detection of Fungal Infection of Plants by Laser-induced Fluorescence: An Attempt to Use Remote Sensing

WILHELM LÜDEKER, HANS-GÜNTER DAHN, and KURT P. GÜNTHER

Deutsche Forschungsanstalt für Luft- und Raumfahrt, Institut für Optoelektronik, Postfach 11 16, D-82230 Wessling, Germany

Received June 24, 1995 · Accepted October 20, 1995

Summary

Nowadays the detection and quantification of fungal infections of agricultural plants is performed by visual inspection. Farmers or scientists observe the leaves by eye and interpret the infection by estimating the amount of attacked leaf area. Trained and experienced interpreters yield good and mostly reliable results. But comparing the results of different interpreters may show ambiguous characterisation of the degree of infection of the same leaf ensemble. For industrial applications where e.g. for horticultural purposes thousands of growing seedlings must be analysed this method is very complicated, time consuming and thus very expensive. Therefore an automatic inspection and management system that supports the identification of fungal infection and selection process is highly desired.

Based on this task, the identification and quantification of fungal infections of different plant species were performed by fluorescence analysis to investigate the possibilities of the mobile fluorescence lidar of the German Aerospace Research Establishment (DLR). A systematic experiment with trained interpreters as reference was set up.

The results demonstrate that fungal infection is detectable by fluorescence remote sensing in spectral mode as well as in induction mode. It is shown that the infection can be monitored by the fluorescence signal earlier than by visual inspection. A comparison between visual inspection (bonitur) and the fluorescence data is presented.

Key words: Apple (Malus sylvestris), barley (Hordeum vulgare), cucumber (Cucumis sativa), wheat (Triticum aestivum), fungal infection, blue fluorescence, chlorophyll fluorescence, Kautsky effect.

Introduction

Fungal infections as mildew or rust (*Septoria apii*>) attack plants in many parts of the world. These infections can become severe on leaves of commercially grown plants and seedlings, especially when relatively cool, moist climatic conditions prevail.

Mildew is characterised by whitish to yellowish areas of irregular size and shape on infected leaves or seedlings stalks. Mostly mildew attacks the whole leaves by covering the leaf surface.

Rust is characterised by a spotted reddish to brownish discoloration of stems and leaves. The infected areas of rust are smaller than those of mildew because the rust fungi live in a parasitic way in the intercellular spaces of the leaves.

The fungal infections of mildew are produces by spores called sporangia on the surface of the leaves and seed stalks. The spores are disseminated through the air and can thus infect the leaves. The masses of mildew spores are at first transparent to greyish and normally not visible by eye, and then rapidly become white in colour while the rust spores become red to brown. Mildew infected leaves become girdled and may collapse.

Since long times fluorescence emission spectroscopy of plants either in the spectral or in the time-resolved mode has been a tool for the investigation of the physiology and photosynthesis of plants (Kautsky and Franck, 1943; Schreiber and Berry, 1977; Schreiber, 1983; Lichtenthaler, 1988; Krause and Weis, 1991). In order to understand the link between chlorophyll fluorescence and photosynthesis and plant physiology it is important to notice that photosynthesis and chlorophyll fluorescence occur in the same cellular organelles called chloroplasts. Within this structur, different pigment as e.g chlorophyll *a* and *b,* the biliproteins and the carotenoids are embedded and act as absorbers for the visible light energy. The major part of the absorbed light energy is used for photosynthesis while a minor part is emitted as fluorescence or dissipated by thermal processes as heat. Chlorophyll fluorescence is about 0.5–3 % of the total absorbed light energy.

Regarding plants, the main fluorescing pigment is chlorophyll *a,* which is either directly excited by blue light of 440 nm or by red light of 670 nm. The other so called accessory pigments. (i.e. carotenoides) transfer the absorbed energy to the reaction centers by a resonant dipol-dipol interaction. The amount of energy transfer varies widely among the pigments. Whereas e.g. chlorophyll *b* transfers its energy at nearly 100 %, the xanthophylls transfer only about 30 %. At the reaction centers which are highly specialised chlorophyll molecule containing electron donors and acceptors the transformation of the excitation energy into photochemical energy occurs by charge separation. The photochemical energy is then transported by redox reactions for carbon assimilation or oxygen evolution.

Steady state chlorophyll fluorescence is to a first approximation inversely related to the photosynthetic performance or carbon assimilation (Kautsky and Franck, 1943; Günther et al., 1994). Due to this close coupling of fluorescence and photosynthesis, fluorescence is a good remote indicator for the state of the electron transport chain of photosynthesis and the biochemical state of the plant. Several parameters can be deduced from the fluorescence induction curve describing the proportion of the excitation energy ‹trapped› by open reaction centres (photochemical quenching) or the non-photochemical quenching coefficient (for a review see: Krause and Weis, 1991). The non-photochemical quenching coefficient is related to the light-induced proton gradient across the thylakoid membrane, to the phosphorylation and when high light conditions occur to the photoinhibition. A full quantification of the relationship between the quantum yield of photosynthetic electron transport and the quenching of chlorophyll fluorescence was recently established (Weis and Berry, 1987; Genty et al., 1989; Seaton and Walker, 1990).

Even under steady state conditions (under continuous light) several techniques were developed in order to describe the behaviour of plants in its environment by fluorescence emission. The red fluorescence ratio (termed e.g. F690/F735) is a very suitable parameter for long term stress (Rinderle and Lichtenthaler, 1988). This parameter is related to the intrinsic chlorophyll concentration of the leaves. It could be shown that the red fluorescence ratio is nearly independent of changing light (Günther et al., 1994) and thus a good parameter for remote sensing.

With fungal infections plants may be stressed. For mildew, the fungi cover the leaf surface in form of a web. This web protects the leaf from the incoming light thus reducing the red fluorescence and increasing the blue fluorescence due to a strong blue fluorescence of mildew. During the development of the fungi, some parts of mildew penetrate the epidermal layer and partially grow towards the palisade parenchyma. Later in the development the fungi form on the surface of the leaves so called conidia which give the leaves the whitish colour. It is therefore expected that due to this parasitic growth of the fungi and the formation of conidia, the chloroplast are changed thus changing the chlorophyll fluorescence. For rusts, the situation is a little bit different. During the fungal infection, the spores enter the stomata and penetrate the intercellular spaces of the leaves where they grow. There the rusts are normally confined to the infected area not spreading out like mildew. During the development of the fungi rusts infect the neighbor cells growing towards the upper epidemis and penetrating the spongy parenchyma. On both sides of the leaves rust forms small grains. During the growth of the fungi the chloroplasts and thus the photosynthetic membranes are affected. Therefore, the photosynthetic activity may change which in turn should change the fluorescence. In addition, when the fungal infection is dominant on the leaf surface (as observed with mildew), one would expect an additional fluorescence due to the fungi. The goal of this investigation was to demonstrate spectral and temporal changes of the fluorescence of plants when attacked by mildew and rust and to show that fungal infection of plants can be monitored from remote by fluorescence emission.

Materials and Methods

For the investigation of the infected and non-infected plants three different fluorescence metering methods have been used. They allow measurements of fungal infections from punctual areas to complete ensembles.

The spectrofluorometer RF5001PC

For the control of the remote measurements (described below) a commercially available fluorometer RF5001PC (Shimadzu) was used. It allows the local investigation of an area of 1–2 mm^2. The leaves must be removed from the plants and are placed at an angle of about 50° to the excitation light and about 40° to the detector slit.

The excitation light is emitted by a 150W Xenon bulb. The excitation wavelength can be selected by a monochromator in the range from 350–700 nm. In this experiment only the wavelength 355 nm is used because this is the wavelength of the laser system used for the experiments with the other systems. Fluorescence is detected with a photomultiplier by scanning the spectral range of a monochromator from 400–800 nm. The detector-sensitivity above 600 nm has been enhanced by using a special type of photomultiplier (Hamamatsu R928).

Several measurements over one leaf have to be carried out systematically to get the spectral characteristic of fungal infection effects at a larger area.

A typical fluorescence emission spectrum of a plant leaf is shown in figure 1. The typical maxima at 685 nm and 730 nm are due to chlorophyll *a.* The origin of the blue fluorescence has not been discovered fully until today, but most probably it is the fluorescence of the epidermal cell walls.

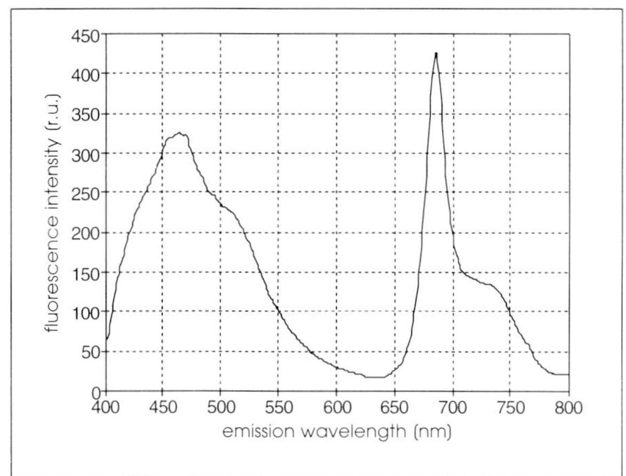

Fig. 1: Typical fluorescence spectrum of a plant leaf excited at 355 nm.

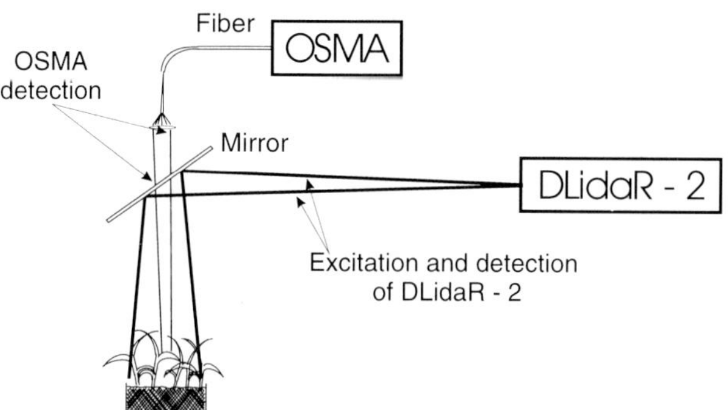

Fig. 2: Arrangement of the lidar systems for the fungal infection experiment.

Lidar with Optical Multichannel Analyser (OSMA)

For spectral high-resolution measurements of single plants or canopies a lidar system was operated which is described in detail in Günther et al. (1991). The excitation laser is a tripled Nd:YAG at 355 nm. The pulse length can be switched to 2.5 ns or 8 ns duration. Measurements were carried out at 12 Hz repetition rate, which is optimal for the pulse stability of laser output. The average energy is 35 mJ per pulse. After measuring the laser energy, the beam is expanded. The beam-divergence can be varied from 0.5 mrad to 20 mrad adjusting the energy density and the diameter of the beam according to the measuring task. For single plants the typical diameter is about 1 cm^2 which is about the size of a small plants leaf. The distance from the detector module to the sample can be varied from several centimetres to some meters. Figure 2 shows the lidar system set-ups with a multichannel analyser as detector and with photomultiplier as detector.

The detection is done in real-time by an optical multichannel analyser. The compact monochromator has a grating with 133 lines/mm yielding in a resolution of 0.6 nm/diode. With the slit width of 100 µm the spectral resolution is 3.5 nm. The spectral range of the system is 400–800 nm. The fluorescence light is coupled by a fiber into the monochromator. The fibers have 100 µm diameters, a change in the form of the beam from the circular image plane of the receiver to the slit form is done also by this fiber. The numerical aperture of the fiber is matched to the numerical aperture of the monochromator.

A cooled diode-array with 1024 pixels detects the fluorescence spectrum. For enhancing the sensibility of the detector a micro-channel plate as image intensifier is put in front of the diode array. In this arrangement only 700 pixels of the array are illuminated. A smaller micro-channel plate is much faster, thus it can be gated in less than 32 ns in order to achieve better reduction of the background light. The synchronisation of trigger and gate signals is done by a pulse from the laser. A complete spectrum can be transferred to disc within 4 ms. With direct memory transfer (DMA) the spectra can also be displayed in real-time with 12 Hz repetition rate. In this experiment spectra have been integrated for 20 seconds to achieve an optimal signal/noise ratio.

Vegetation lidar DLidaR-2

The vegetation lidar DLidaR-2 (Günther et al., 1993) was operated as a far-field sensor with an excitation spot of about 200 cm^2 (16 cm diameter). Thus the system is able to measure a whole plant or an ensemble of plants.

The transmitter is the same laser as used with the OSMA system. The receiver is a Cassegrain telescope with a Schmidt correction plate. The diameter of the entry mirror is 20 cm. The light is transmitted by a custom fiber optic system, which homogenises the intensities. After 80 cm mixing length the fiber is divided equally into four arms which are guided onto the cathode of multipliers. In front of each photocathode there is a combination of interference filter an cut-off filters, which yields the desired spectral response. In the two short wave channels there is an additional short-wave-pass filter, which filters the red part of the signal. The interference filters of those channels become transparent again for the red part of the spectrum. The wavelengths are: 440 nm ± 10 nm, 650 nm ± 5 nm, 685 nm ± 5 nm, 730 nm ± 5 nm. The channel at 440 nm has a larger bandwidth, because the peak of this spectral line is not very narrow.

For an effective discrimination of the ambient light, each multiplier is gateable. The signal is recorded either with a digital storage oscilloscope or by integrating analogue to digital converters. The DLidaR-2 is capable of single shot measurements.

The plants (barely, wheat, apple seedling, cucumber) were cultivated under controlled greenhouse condition for six weeks by BAYER AG, Monheim. They were infected with mildew and rust in the greenhouse and then transported to DLR for the measurements.

The visual inspection (bonitur) was performed by trained experts from BAYER AG, Monheim. The assessment of the bonitur is different for both plant types. In the case of monocots all leafs are taken into account. For dicots only the two lowest and oldest leafs are inspected, possible infections at younger leafs are disregarded. For judged leafs a bonitur of 0 % represents no infection and a bonitur of 100 % means a hundred percent coverage of the leaf area. Mildew infection is visible by a fraction of the leaf area which is homogeneously covered with the fungi. Rust shows a heterogeneous pattern of discoloration's and brownish spots, where the plantcells are more or less collapsed.

Results and Discussion

Fungal infections of monocots

Figure 3 shows the fluorescence ratio F685/F730 and F440/F730 of wheat in dependence of the degree of rust infection (*Septoria apii*). The fluorescence was detected from remote by the lidar system equipped with photomultipliers. For both parameter it is clearly seen that rust infection is hard to identify by the DLidaR-2 quantitatively. Within the individual groups (that means plants with the same bonitur) the

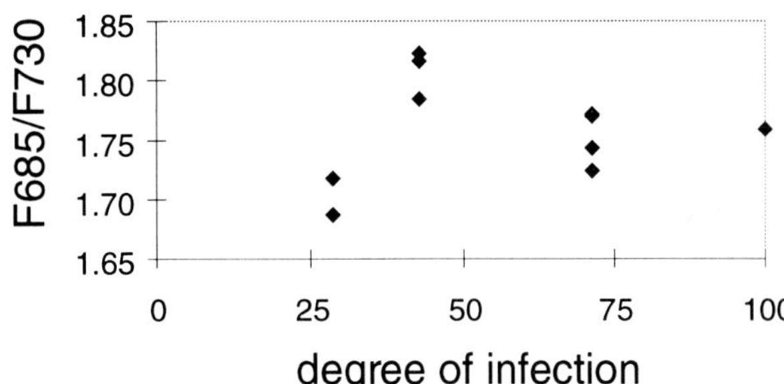

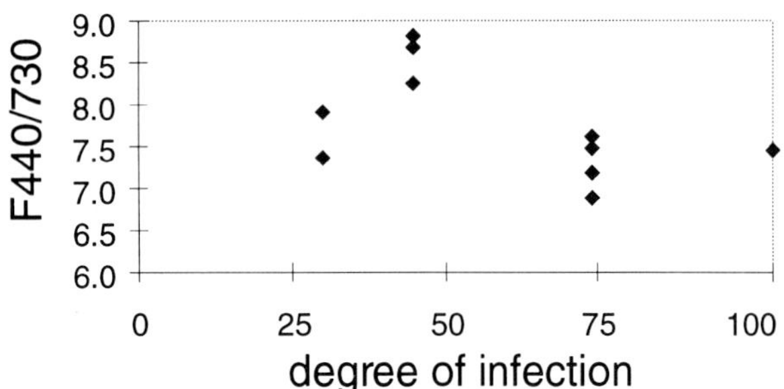

Fig. 3: Fluorescence ratio F685/F730 and F440/F730 of wheat in dependence of the bonitur. A bonitur of 0 means no *Septoria apii* infection.

mean fluorescence ratios show a small standard deviation. But a systematic trend of the fluorescence ratios (either F685/F730 or F440/F730) with the degree of infection is not obvious due to the statistic scatter of the mean values of the groups. The red fluorescence ratio F685/F730 is correlated to the chlorophyll concentration of the leaves (D'Ambrosio et al., 1991). Therefore one would argue from the data that the chlorophyll concentration has not changed with fungal infection. For the fluorescence ration F440/F730 one expects an increase with increasing fungal infection due to the pronounced blue fluorescence of the fungi. One reason for the variation between the individual groups may be the different canopy structure of the infected wheat plants which has a more pronounced effect on the fluorescence ratios than the fungal infection due to the larger excitation and observation spot in comparison to the infected areas of monocots. If theses plants are infected only partially healthy and infected plant leaves are measured simultaneously. Therefore the fluorescence of healthy and infected leaves is averaged over the spot size yielding to a damping of the characteristic signature of the fungal infection.

However fluorescence detection of fungal infection becomes more sensitive when the measurements are performed in a scanning mode with a small excitation spot size. In table 1, the fluorescence data of partially infected leaves are summarised. The measurements were performed with the fluorometer RF5001PC along the leaf surface by investigating

Table 1: Fluorescence ratios of wheat leaves infected with rust, measured with the spectrofluorometer RF5001PC.

Infection	non-infected area	beside infected area	infected area
Ratio			
F440/F520	1.75	1.49	1.34
F685/F730	1.21	1.49	2.14
F440/F730	0.77	1.91	1.80
F440/F685	0.64	1.29	0.85

small, point-like areas of typical size of 2 mm^2. The correlation with the bonitur is no more meaningful, because the bonitur is a parameter characterising the whole plant and not small areas on a leaf.

As mentioned earlier, the ratio F685/F730 is inversely correlated with the chlorophyll concentration. According to Dahn et al. (1992) under constant light conditions the ratio F685/F730 increases due to the lower reabsorption process of F685 when the concentration of the chlorophyll decreases. From table 1 one can therefore identify that the increase of the ratio F685/F730 near the infected areas and much more drastically in the infected areas is due to a local reduction of the chlorophyll concentration. This is in agreement with the plant pathological findings that the rust fungi penetrate the leaf and modify the chloroplasts.

The ratio F440/F730 is strongly increased with infection, which is due to the decreased chlorophyll concentration (indicated by the increase of F685/F730) and an increase of the blue fluorescence. Also a significant increase of the F520 signal in comparison to the F440 signal can be seen. There might be two reasons for this phenomenon. Either the blue/green fluorescence band of the fungi is responsible for the increase of F520 or intercellular substances are produced showing a fluorescence in the blue-green region. For a complete clarification of this aspect the chemistry of the tissue should be analysed.

Similar results were found for monocotyledonous plants (barley) infected with mildew. Lidar measurements show again no clear correlation with the degree of infection derived from visual inspection due to the large excitation area of the laser spot taking into account a whole ensemble of plants. Normally, a whole plant or an enseble of plants is not infected homogeneously even if the bonitur is classified as total infected. Therefore the fluorescence of healthy and infected leaves is averaged over the spot size yielding to a damping of the characteristic signature of the fungal infections. In accordance with the previous results point measurements with the fluorometer yield a clear spectral signature, comparable with the results described for rust infected monocots. A strong increase of nearly one order of magnitude of the F440/F730 ratio at the infected areas is detected. It can be explained by the increased blue fluorescence of the conidia primarily localised to the surface of the leaf.

Fungal infections of dicots

In figure 4 the fluorescence spectra of apple seedlings (healthy and infected by mildew) are shown. The spectra

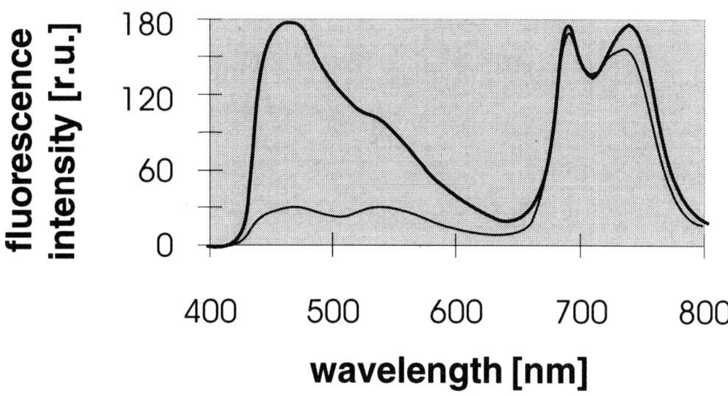

Fig. 4: Fluorescence spectra of apple seedlings (healthy – solid line and infected by mildew – dashed line) measured with the spectrofluorometer RF5001PC. The excitation wavelength was 355 nm.

were measured with the spectrofluorometer RF5001PC. The excitation wavelength was set to 355 nm for comparison with the lidar measurements. Mildew infected apple seedlings show again a strong increase of the blue fluorescence comparable to the results of the mildew infected monocots. This increase is due to the whitish mildew on the surface. The chlorophyll content of the apple leaves is only weakly influenced by the mildew as seen by the red fluorescence ratio. The ratio F440/F685 differs between healthy and infected leaves by about a factor of 6 for the dicots when determined by the point fluorometer. In general dicots have a significant lower leaf fluorescence in the blue region (F440) than monocots and a lower fluorescence ratio F440/F685 (Chappelle et al., 1985). This yields to a much higher contrast when dicots are infected by fungi.

The fluorescence ratio F685/F730 measured with the lidar equipment and presented in figure 5 shows a small decreasing tendency, which might lead to the assumption that the pigment content is increased with increasing infection. From these findings one could assume that the plant tries to compensate the stress by enhanced synthesis of chlorophyll. The ratio F440/F730 measured with the DLidaR-2 shows a linear correlation with the degree of infection. The regression coefficient is 0.82. It might be surprising that the increase of the fluorescence ratio F440/F730 is much lower than a factor of 3 when determined by lidar measurements than measured in the point mode. When measurements are performed with the OSMA system (thus increasing the specificity of the measurements by reducing the spot size) the increase of the ratio F440/F730 shows a higher correlation with a correlation coefficient of 0.95 (see Fig. 6) and the increase is factor of more than 3 between healthy and total infected plants. The reason for this differences is due to the extent of infected area inside the excitation spot and the inhomogeneity of the infection. If the spot size is reduced (as e.g. for the measurements with the fluorometer RF5001PC) an unambiguous results is observed because the investigated area is either infected or not.

Nevertheless, for the dicots the results obtained with the DLidaR-2 show a good correlation with the bonitur as well as the results of the other sensors. Therefore, it may be concluded that even with a large spot size the fungal infection of dicots can be quantified.

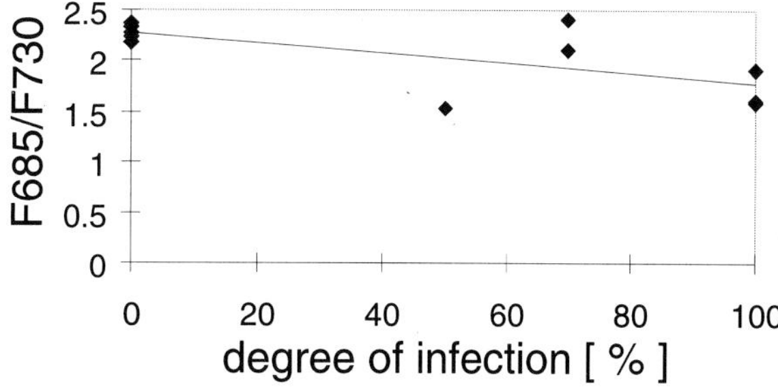

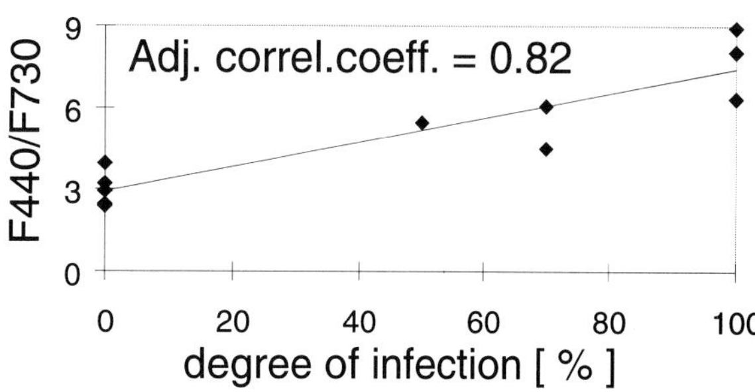

Fig. 5: F685/F730 and F440/F730 of apple seedlings dependent on the bonitur. A bonitur value of 0 % means no infection by mildew. The fluorescence was measured from remote with the DLidaR-2.

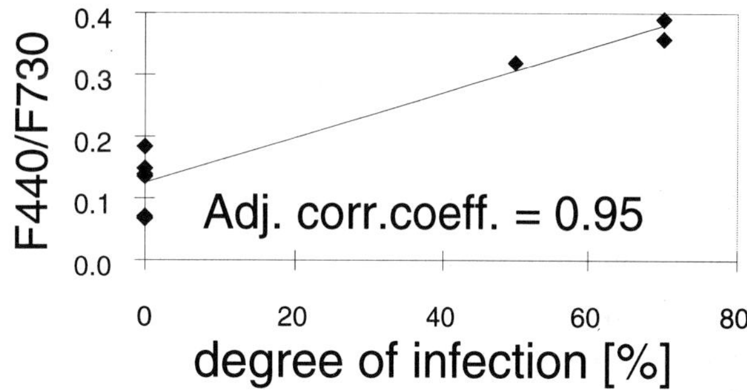

Fig. 6: Fluorescence ratio F440/F730 of apple seedlings measured by the OSMA system in relation to the bonitur. A degree of infection of 0 % means a healthy plant, not infected by mildew.

Measurements of fluorescence induction kinetics

In a third experiment the DLidaR-2 was used to record the chlorophyll fluorescence induction kinetics (Kautsky effect) of mildew infected cucumber plants. The plants were dark adapted for at least 30 minutes and then illuminated by a 500 W halogen lamp for 180 seconds. During dark adaptation and illumination the tripled Nd:YAG laser was operating, recording the ground fluorescence F_0 at 685 nm and during illumination the variable fluorescence. During the dark adaptation time the ground fluorescence F_0 was constant indicating that the laser pulses did not induce photosynthesis and that the reaction centre were open. The laser was operated at a rate of 12 Hz. Cucumber is well suitable for this experiment because the plants have relatively large leaves. Therefore the use of a large laser spot is adequate resulting in a low energy density of the excitation light. In this experiment not only the kinetics of the ratios, but also the kinetics of the absolute intensities can be used for the interpretation, because the fluorescence comes only from healthy or infected leaf surfaces.

The typical kinetics of a dark adapted healthy plant are given in figure 7. The reflection R355 shows no kinetics during illumination indicating that there was no movement or structural change of the plants. The chlorophyll fluorescence F685 shows the typical induction phenomenon with the maximum fluorescence F_{max} just after the begin of the illumination. The maximum fluorescence is due to the saturation of the reaction centres by the additional light (closed reaction centres). In this period the electron transfer is not working optimal. When the maximum of photosynthesis is reached after some minutes the fluorescence has gone to the steady state F_s. The time for reaching the steady state depends on the status of the plant.

In figure 8 the kinetics of the chlorophyll fluorescence signal F685 of mildew infected cucumber plants are shown. On left hand the induction of a 5 % infected plant and on the right hand the induction of a 30 % infected plant is presented.

For quantifying the temporal behaviour the so called «decay time» was calculated defined as the time when the fluorescence is reduced to $(F_{max} - F_s)/2 + F_s$. This time is characteristic for the vitality of a plant. Healthy plants are characterised by a short «decay time» of 21.5 seconds, the 5 % infected plant by a longer «decay time» of 43 seconds. The plant with the 30 % infection shows a long «decay time» of 45 seconds.

The Rfd-value successfully introduced by Lichtenthaler and Rinderle (1988) (defined as $(F_{max} - F_s)/F_s$) is used as a second parameter for characterising the physiological status of plants. This parameter is a good measure for the activity of the electron chain. If the electron transfer is slowed down, the number of closed reaction centres is higher, which means that F_s increases relatively to F_{max}. For healthy plants the Rfd value is typical 5.3, whereas the 5 % and the 30 % infected plants have a Rfd-value of 2.2. Both parameter show an early indication when fungal infection is present.

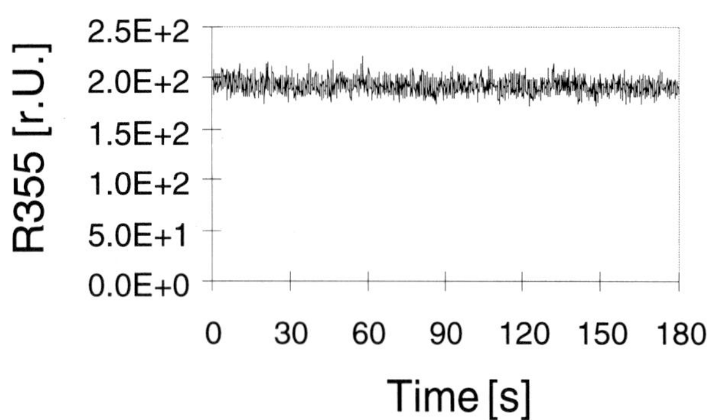

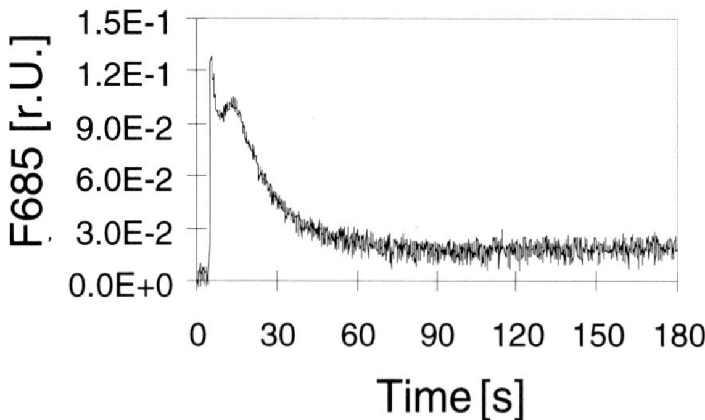

Fig. 7: Kautsky kinetics of the elastic reflected signal R355 and the chlorophyll fluorescence signal F685 of a healthy cucumber plant. The plant was dark adapted for 20 minutes and then illuminated by an additional actinic light (500 W).

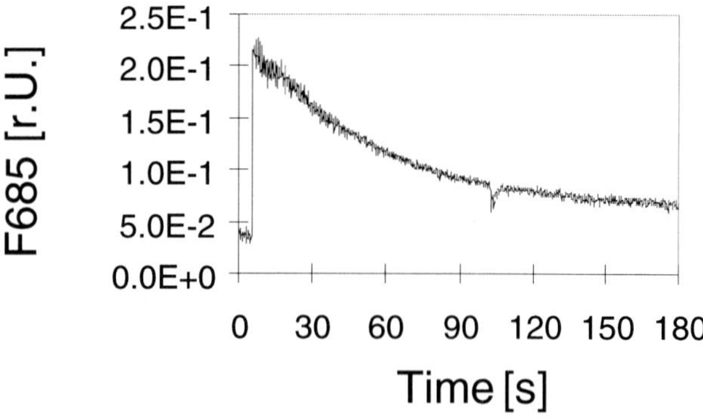

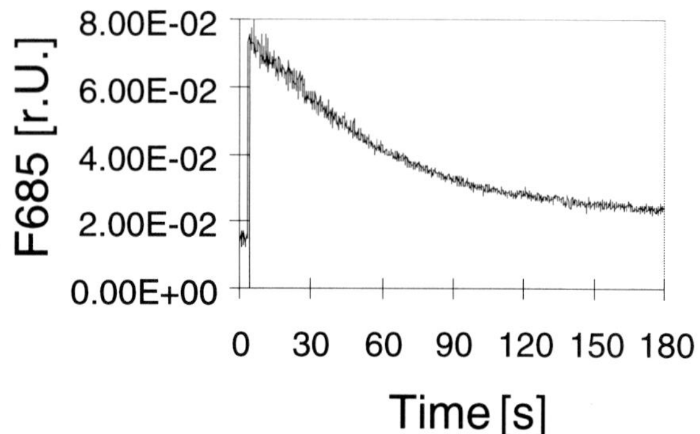

Fig. 8: Kautsky kinetics of the chlorophyll fluorescence signal F685 of mildew infected cucumber plants. The left curve is the induction curve of a 5 % infected plant and the right curve the induction kinetics of a 30 % infected plant.

Conclusion

The results of the investigations to identify and quantify fungal infection of plants by fluorescence spectroscopy can be summarised as follow:

- All infections investigated (mildew and rust) can be identified by the fluorescence signatures when a reference to healthy leaves is possible.

- The intensity of the chlorophyll signal, the reflection signal and the fluorescence ratios F685/F730, F440/F730, F440/F520 are the most important parameters for this type of investigation.
- F440/F730 clearly indicates a fungal infection and correlates quantitatively with the bonitur.
- Apple seedlings show a clear quantitative correlation of the fluorescence ratios with the degree of mildew infection.
- Mildew infected cucumber plants show a reduced photosynthetic activity observed by Kautsky kinetics and quantified by the Rfd-value and the «decay time».
- Remote detection of fungal infection might be possible with a modified lidar system, taking into account a proper, well adapted size of the excitation spot which should have a similar dimension as the leaf area of the plants.

The three used fluorescence systems were able to make a step towards automatic detection of fungal infections. Combining the results of all fluorometers the improvement for an operational instrument can be shown. For all applications the size of the excitation spot should be smaller than the size of the leaf. This finding is more important when monocots are investigated due to the lower contrast of the fluorescence ratio F440/F730 in comparison to dicots. Due to the small size of the excitation spot a scanning system with a high repetition rate is desirable for an automatic inspection system in order to map the plants and the infected areas. This mainly depends on the availability of high repetition rate lasers. On the data processing side an automatic statistical weighting of the data is absolutely necessary to gain enough throughput for real-time systems in the evaluation of plant infections.

Acknowledgements

We thank very much the Bayer AG, Plant Protection Centre Monheim, for preparing the plants with the different fungal infections. In addition we are very grateful to Dr. Schmitt and Dr. Dutzmann for contribution to the campaign.

References

Chappelle, E. W., E. M. Wood, W. W. Newcomb, and J. E. McMurtrey III: Laser-induced fluorescence of green plants. 3: LIF spectral signatures of five major plant types. Applied Optics *24*, 74–80 (1985).

D'Ambrosio, N., K. Szabó, and H. K. Lichtenthaler: Increase of the chlorophyll fluorescence ratio F690/F735 during the autumnal chlorophyll breakdown. Radiat. Environ. Biophys. *31*, 51–62 (1992).

Dahn, H.-G., K. P. Günther, and W. Lüdeker: Characterisation of drought stress of maize and wheat canopies by means of spectral resolved laser-induced fluorescence. EARSeL Advances in Remote Sensing *1*, 12–19 (1992).

Genty, B., J.-M. Briantais, and N. R. Baker: The relationship between the quantum yield of photosynthetic electron transport and quenching of chlorophyll fluorescence. Biochim. Biophys. Acta *990*, 87–92 (1989).

Günther, K. P., W. Lüdeker, and H.-G. Dahn: Design and testing of a spectral resolving fluorescence lidar system for remote sensing of vegetation. Proc. 5th Int. Coll. Physical measurements and Signatures in Remote Sensing, Courchevel, France, 14.–18. Jan. 1991, ESA SP-*319*, 723–726 (1991).

Günther, K. P., H.-G. Dahn, and W. Lüdeker: Remote sensing vegetation status by laser-induced fluorescence. Remote Sens. Environ. *47*, 10–17 (1994).

Kautsky, H. and U. Franck: Chlorophyllfluoreszenz und Kohlensäureassimilation. XI. Die Chlorophyllfluoreszenz von *Ulva lactuca* und ihre Abhängigkeit von Narkotika, Sauerstoff und Kohlendioxyd. Biochem. Z. *315*, 139–232 (1943).

Krause, G. H. and E. Weis: Chlorophyll fluorescence and photosynthesis: the basics. Annu. Rev. Plant Mol. Biol. *42*, 313–349 (1991).

Lichtenthaler, H. K.: *In vivo* chlorophyll fluorescence as a tool for stress detection in plants. In: Lichtenthaler, H. K. (ed.): Applications of Chlorophyll Fluorescence, pp. 129–142. Kluwer Academic Publishers, Dordrecht (1988).

Lichtenthaler, H. K. and U. Rinderle: Role of chlorophyll fluorescence in the detection of stress conditions in plants. CRC Critical Reviews in Analytical Chemistry *19*, Supp I, S29–S85 (1988).

Rinderle, U. and H. K. Lichtenthaler: The chlorophyll fluorescence ratio F690/F735 as a possible stress indicator. In: Lichtenthaler, H. K. (ed.): Applications of Chlorophyll Fluorescence, pp. 189–196. Kluwer Academic Publishers, Dordrecht (1988).

Schreiber, U. and J. A. Berry: Heat-induced changes of chlorophyll fluorescence in intact leaves correlated with damage of the photosynthetic apparatus. Planta *136*, 233–238 (1977).

Schreiber, U., U. Schliwa, and W. Bilger: Continuous recording of photochemical and non-photochemical chlorophyll fluorescence quenching with a new type of modulation fluorometer. Photosynt. Res. *10*, 51–62 (1986).

Seaton, G. G. R. and D. A. Walker: Chlorophyll fluorescence as a measure of photosynthetic carbon assimilation. Proc. R. Soc. Lond. B *242*, 29–35 (1990).

Weis, E. and J. A. Berry: Quantum efficiency of photosystem II in relation of energy dependent quenching of chlorophyll fluorescence. Biochem. Biophys. Acta *894*, 198–208 (1987).

J. Plant Physiol. Vol. 148. pp. 586–592 (1996)

Laser Induced Fluorescence Spectroscopy of Phytoplankton and Chemicals with Regard to an *in situ* Detection in Waters

ULRIKE UEBEL, JÖRG KUBITZ, and ANGELIKA ANDERS

Institut für Biophysik der Universität Hannover, Herrenhäuser Str. 2, D-30419 Hannover, Germany

Received June 24, 1995 · Accepted November 9, 1995

Summary

In order to be able to detect chemicals in waters the suitability of laser-induced fluorescence spectroscopy was evaluated as a tool for the *in situ* detection of water contaminants and possible interactions with phytoplankton and yellow-substances.

A frequency doubled dye laser was used as an excitation light-source. The substances were dissolved in water and excited in the range from 265 nm to 400 nm. The fluorescence signal was recorded in the range from 310 nm to 750 nm. The absorption-, fluorescence- and excitation-spectra of each contaminant were measured in distilled water, artificial seawater and natural seawater. Computer-simulations were applied to analyse the spectra of pollutant mixtures.

Firstly we investigated as to whether there are fluorescence-spectroscopically recognizable intermediate stages of phytoplankton and their breakdown-products (yellow-substances). Modifications of the fluorescence spectra during the ageing and dying of phytoplankton could be shown. Phytoplankton which is subject to sudden death display spectra differing from those of slowly dying organisms, e.g. the shift of the chlorophyll maximum to shorter wavelengths was measured. The spectra of mixtures containing phytoplankton and chemicals show differenes which depend on the characteristics of the chemicals concerned. PAHs and lead-acetate were used.

Key words: Scenedesmus subspicatus, phytoplankton, chemicals detection, dye laser, environmental monitoring, fluorescence spectra, yellow-substances.

Abbreviations: PAHs = polycyclic aromatic hydrocarbons; KDP = potassium-bi-phosphate.

Introduction

The deciding factor for an efficient damage limitation after for example an accident on sea which causes a contamination of water often is the time. Most detection methods for water pollutants need a laboratory, require sample preparation, are specific, time consuming or invasive, as e.g. chromatographical methods or Atomic Absorption Spectroscopy (AAS) (Rogge et al., 1992). Therefore, a period of time is generally necessary between the sampling process and the final contamination result. With these conventional methods it is not possible to trace the pollutants or the source of contamination. It will be also very useful to get some data about the spreading of the contaminants on the surface as well as in the water column and on the sea bed. This could be done with the help of laser induced fluorescence spectroscopy. Furthermore, this method will show results with high accuracy and selectivity and will moreover handle many different compounds, whilst being flexible and sufficient for on line detection. In the last years, different spectroscopic methods have already been proven (Reuter and Hengstermann, 1990; Reuter et al., 1990) as a method for reliable an *in situ* availability of data with high accuracy and with results obtained in a short time. They also provided a sensitivity comparable to that of non-spectroscopic methods. Several papers on investigations, using fluorescence and Raman spectroscopy for the detection of various compounds were published (Chudyk et al., 1985; Chudyk et al., 1989; Milanovich and Valey, 1986; Leugers and Lachlan, 1988).

The intensity of the fluorescence depends on the excitation wavelength for a given transition, and the resulting excitation-emission-matrix is characteristic for each compound (Kumke et al., 1995). Using the information from this matrix, by varying the excitation and the emission wavelength, therefore enhances the capability for the analysis of mixtures of contaminants, i.e. the detection of the various compounds, due to the change of the spectral shape when scanning both wavelengths. PAHs and some heavy metals are among the compounds, which can be detected by fluorescence spectroscopy. On the one hand investigations of the detection of PAH-mixtures were done for benzene, pyrene, toluene, xylene, naphtalene and anthracene (Rogge, 1992). On the other hand lead, thallium and cerium as heavy metal contaminants were investigated (Kubitz, 1992). Neither absorption and excitation spectra nor emission spectra of these compounds or their mixtures, being dissolved in pure water, are normally given in literature. Thus, the spectrum of each compound had to be measured to get information about their most efficient excitation wavelengths as well as for the ability to analyse the spectra of mixtures. To avoid background signals, the compounds were dissolved in bi-distilled water at first. Then the compounds were mixed, and also the bi-distilled water was replaced by a seawater-standard. Doing this, possible interactions of the different compounds and also dependencies on the spectra of the kind of water were investigated.

Furthermore, the contaminants were dissolved in filtered seawater from the Biological Institute of Helgoland (Germany). This water contains yellow-substances, which consists of dissolved organic substances and might cause problems, when detecting some heavy metals in natural samples. This is due to the fact, that the fluorescence of these compounds is influenced by yellow-substances. The detection of the PAHs does not seem to be affected by this problem. Nevertheless another aspect is the presence of phytoplankton in waters. There might be interactions between phytoplankton and the investigated chemicals. For these investigations we used *Scenedesmus subspicatus* and fresh water samples of the «Leine» (a river in North Germany) containing different phytoplankton. This alive phytoplankton reacts sensitively when influenced by some chemicals and could be an important indicator for water contaminants. However, it has to be considered, that the growing up of phytoplankton in natural waters is dependent on a lot of parameters.

Materials and Methods

In preliminary examinations a fluorimeter (Kontron, SFM 25) was used to test whether the pollutants can be determined by fluorescence spectroscopy at all. These measurements also yielded fundamental data about the excitation and emission range of the samples as well as about the shape of the spectra in water. A UV/VIS spectrophotometer (Perkin Elmer, Lambda 15) was used to record the absorption spectra.

For the main examination a laser system was built up. In order to get a tuneable laser radiation in the spectral range from 265 nm to 400 nm, the laser system consisted of a frequency doubled dye laser (Lambda Physik, FL 2000), which was pumped by an excimer laser (Lambda Physik, EMG 101). The frequency doubling was carried out with the help of an KDP-crystal. The excimer laser was filled with XeCl, the pulse energy is 100 mJ at the emission wavelength of 308 nm and the pulse duration 20 ns. Coumarin 153, Rhodamin B and 6G and RDC 480 (Radiant Dye Chem.) was used for the dye laser. Coumarin 153 dissolved in Methanol was used for all measurements which are shown in this paper. The concentration for the oscillator was 4.2 g/L and for the amplifier 2.1 g/L Coumarin 153. The pulse energy was about 10 mJ in front of and 1 mJ behind the frequency doubling crystals, with a duration of 10 ns.

Because of varying laser intensity a beam splitter reflected a part of the laser radiation towards a reference photodiode to correct the flourescence intensity by the PC.

The fluorescence light of the sample volume was imaged rectangular to the laser beam onto the entrance slit of a grating monochromator (Bausch & Lomb). Fig. 1 shows the experimental set-up. The halfwidth was mostly set to 5 nm to resolve the relative narrow Raman signal of water as well as to achieve enough intensity for the detection of weak fluorescing substances. The optical signal was detected by a photomultiplier (Hamamatsu and EMI, Typ 9558) and read out by a PC, which also controlled the set-up. Every spectrum was measured in three steps. First we measured the Raman signal in the range from 290 to 310 nm in 1 nm steps. Then the range from 310 to 500 nm was measured without a filter in 5 nm steps. After this we measured the last part of the spectrum with a GG 475 (SCHOTT) filter in front of the monochromator from 500 to 750 nm to avoid signals of the second order from the stimulation source. The three single spectra were connected with a special program to one spectrum (Kubitz, 1992).

In order to simulate the measured spectra and to get the contents of the specimen a program was used (Kubitz, 1992), which was based on a gradient method employing the detected spectra of the single compounds. The simulation was done by superimposition of the relevant spectra which yielded good results.

However, all spectra will be superimposed by the water Raman signal of the sample. Thus the spectra have to be corrected for the water Raman effect. Optical instabilities lead to changes in the absolute fluorescence intensity. To avoid these mistakes, the water Raman signal was determined of every spectrum (Kubitz, 1992; Rogge, 1992). As mentioned before the Raman signal of water, which was of course the main element of the sample, was measured in the range from 290 to 310 nm, corresponding to the excitation wavelength, e.g. about 292 nm with excitation wavelength of 265 nm and excited with 270 nm at about 298,1 nm.

For carrying out our investigations with phytoplankton, cultures of *Scenedesmus subspicatus SAG 86.81* from Pflanzenphysiologisches Institut der Universität Göttingen/Germany, were used. The algae were cultivated in a special culture medium for *Scenedesmus,* developed by Bohner, 1979. It was a discontinuous or «batch»-culture, which grew up in a simulated 16 hours day with a temperature of 26 °C. Fluorescent tubes of the firm Osram (Germany) (L 20 W/77 Fluora, L 20 W/15 daylight de Luxe, L 20 W/30 Warmton warm white) and the firm MazDAFLUOR (F5 TF 20 18 W blaue Industrie) delivered the simulated daylight. Fresh water phytoplankton of the river «Leine» was used. The samples included different phytoplankton, for example chlorophyta (e.g. *Scenedesmus* spec.) and diatoms (e.g. *Nitzschia spec.*). We also worked with natural seawater samples, in one period with filtered seawater from the Biological Institute of Helgoland (Germany/North Sea) and a natural, unfiltered sample from Borkum (Germany/North Sea).

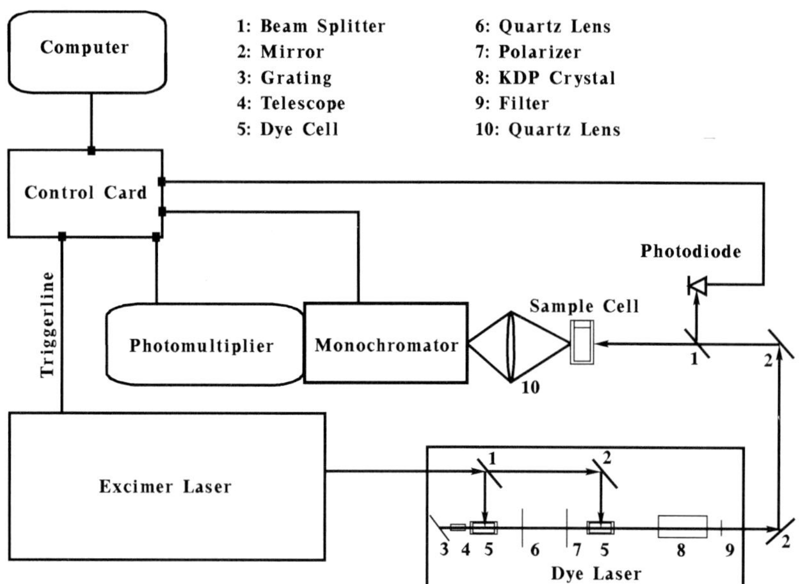

Fig. 1: Experimental set-up for an automatic record of laser induced fluorescence spectra.

Results

First of all we tried to find out as to whether there are fluorescence-spectroscopically recognizable intermediate stages of phytoplankton and their breakdown-products (yellow-substances). Modifications of the fluorescence spectra during the ageing and dying of phytoplankton and differences between phytoplankton which are subject to sudden death and those which were dying slowly could be shown. Then the organisms were mixed with different chemicals which were investigated before and from which the fluorescence spectra were well known.

Previously we found out that heavy metals can be excited by UV-light from 220 nm to 280 nm. There are some hints in the literature that thallium (Tl), lead (Pb), cerium (Ce), copper (Cu) and tin (Sn) are detectable by their fluorescence in aqueous solutions (Eisenbrand, 1966; Wünsch, 1976). In strong acids (3.3 mol/l HCl) they form complexes such as $(TlCl_4)^{3-}$ and $(PbCl_4)^{2-}$. These complexes show a characteristic fluorescence after excitation by UV-radiation.

The behaviour of these metals in mixtures, in seawater and in the presence of organic substances such as yellow-substances or alive phytoplankton is mostly unknown.

In general the following factors have an influence on the fluorescence of every fluorescing substance: the concentration of the fluorescent component, the solvent of the fluorescent component, the pH-value, the temperature, other fluorescent components and soils and last but not least the oxygen content.

The metals were dissolved in bi-distilled water and then in artificial seawater. Lead, for example, shows a small shift of the most efficient excitation wavelength from 250 nm in distilled water to 245 nm in artificial seawater, whereas the shape of the emission spectrum remains unchanged having a maximum at 480 nm. Remarkable is the quantum yield increase by a factor of 56. Subsequent examinations showed that the strong changes of the fluorescence properties of Tl and Pb result from the chlorine concentration of 0.5 mol/L in artificial seawater. The changes are presumably due to the formation of chlorine complexes instead of aqueous complexes (Avramenko and Belyi, 1959; Belyi and Okhrimenko, 1964 and 1964; Belyi and Rudko, 1960). Higher chlorine concentration did not lead to further spectral changes, but to a further rise of the quantum yield. Finally the metals were examined in a natural seawater sample from the Biological Institute of Helgoland (Germany), containing yellow-substances. There was a mutual quenching of the fluorescence of metals and the yellow-substances (Determann, 1990; Uebel, 1994).

In the case of the PAHs the wavelength region between 240 nm and 270 nm was chosen for excitation (Rogge, 1992). All the investigated compounds as well as most of other PAHs show absorption bands in this region; the transitions normally appear from the singlett ground state (S_0) to the first excited one (S_1), but in some cases higher excited states are affected. The distinct excitation spectra in combination with the emission spectra allowed an easy analysis of samples of any composition. For the PAH-measurements the use of excitation wavelengths between 240 nm and 270 nm was sufficient to identify the single compounds on their own or in mixtures. The application of different excitation and emission wavelengths, respectively, strongly displayed the enhanced selectivity.

Typical fluorescence spectra of *Scenedesmus subspicatus* at different time-intervals after inoculation and during ageing and dying of the culture are shown in Fig. 2. The sample was excited with 270 nm. There are typical fluorescence spectra which are similar to fluorescence spectra of natural water samples in the spring and autumn growing maxima of the phytoplankton in the temperate latitude. In the range from 300 to 400 nm with a maximum at about 330 nm we conjectured to have the fluorescence of proteins (probably mainly trypthophan). From 400 nm to about 585 nm we found the fluorescence of yellow-substances (probably mainly humus acid) with variable maxima and at 685 nm wavelength the

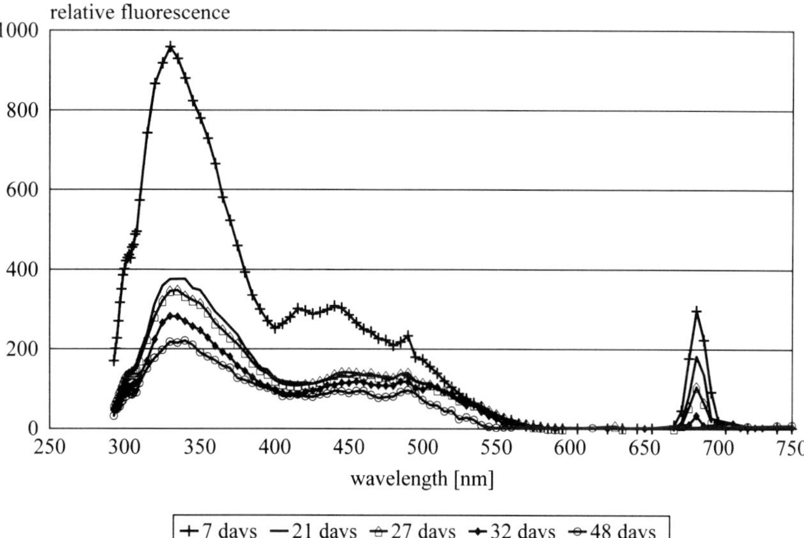

Fig. 2: Fluorescence spectra of *Scenedesmus subspicatus* at different time-intervals after inoculation and during ageing and dying of the culture. The excitation wavelength was 270 nm.

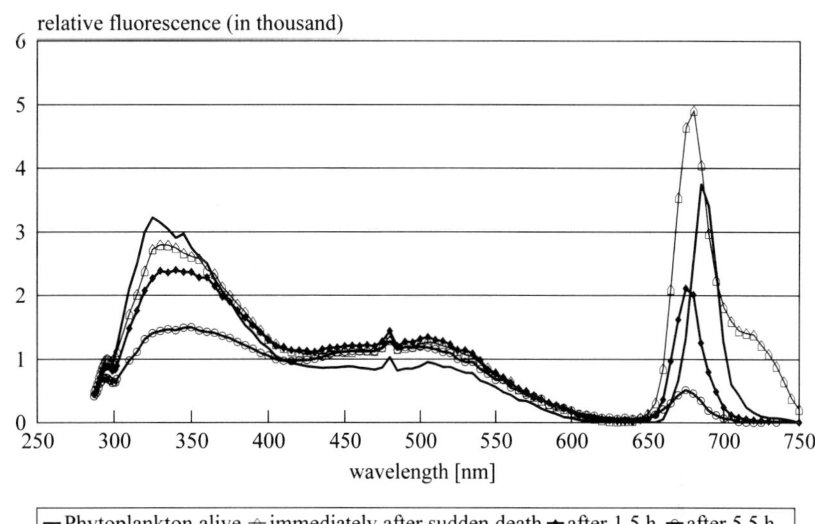

Fig. 3: Fluorescence spectra of phytoplankton of the river «Leine» at different time-intervals after subject to sudden death by boiling and in vitality stage. The excitation wavelength was 265 nm.

maximum of the chlorophyll. The pure medium was also measured before inoculation but the fluorescence signal intensity was not significant in contrast to the fluorescence signal intensity of the samples after inoculation. We subtracted this underground signal from our spectra but there was nearly no change.

Of course in waters there are other conditions. The easiest way to find out as to whether there are significant modifications of the fluorescence spectra of phytoplankton after sudden death, was to kill the phytoplankton by the influence of heat. Fig. 3 shows a phytoplankton sample of the river «Leine» alive and immediately after dying, then we measured every half hour and only show the spectra after 1.5 and 5.5 hours. A higher fluorescence intensity can be found directly after the dying (Kirk, 1983) and then a more or less rapid decrease of the signal, but we can also clearly find a shift of the chlorophyll maximum to shorter wavelengths. Directly after the dying the fluorescence intensity is higher but the maximum could be found at 680 nm, after half an hour we can recognize the maximum at 675 nm, this means a total shift of about 10 nm. Perhaps it will be possible to find a greater shift, if the fluorescence signal could be recorded in smaller steps, e.g. 1 nm, this will be done in the future.

By extremely lowered the pH-value from 7 to 2 by using 0.25 % HCl, similar results as in the example before, could be obtained. Then the increase of the fluorescence intensity of the chlorophyll maximum and also the shift to shorter wavelengths occurs more slowly. Step by step we also could find a shift of the chlorophyll maximum from 685 to 675 nm.

A similar investigation was done with *Scenedesmus subspicatus.* We killed the algae by boiling in a double boiler. Fig. 4 shows the fluorescence spectra after a few hours, one day, two days and 22 days. There were reproducible, similar intermediate stages of the fluorescence signals from the spectrum in the range from 300 to 400 nm with a maximum at about

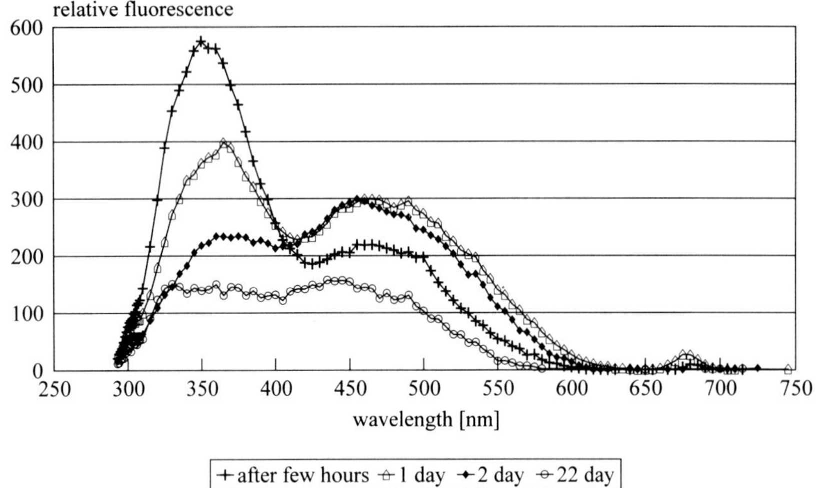

Fig. 4: Fluorescence spectra of *Scenedesmus subspicatus* at different time-intervals after subject to sudden death by boiling. The excitation wavelength was 270 nm.

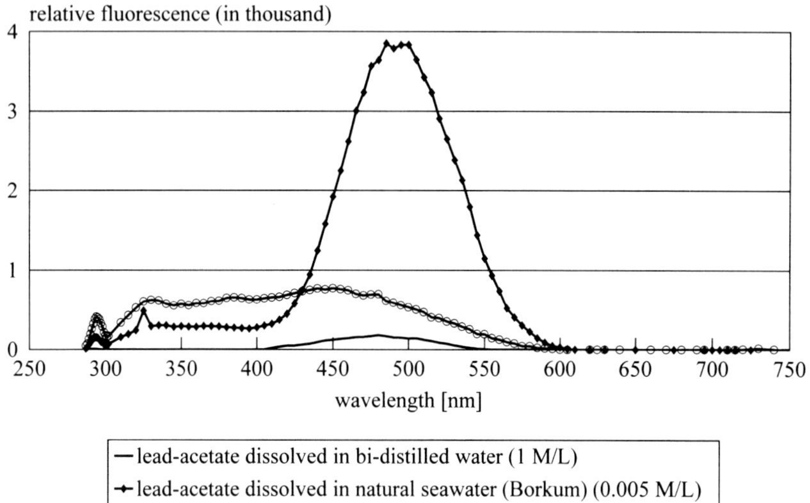

Fig. 5: Fluorescence spectra of lead-acetate in different solvents, excited with 265 nm. Fluorescence maxima in both cases at 480 nm. Recognizeable interactions with the yellow-substances of the seawater sample.

330 nm from the «proteinspectrum» and from 400 nm to about 585 nm from the fluorescence spectrum of yellow-substances. The fluorescence intensity of both main components varied, and were decreasing by the time.

We also used PAHs and mixed them with phytoplankton, but the concentration range which was used was not high enough for a recognizeable toxic effect, higher concentrations couldn't be measured, because the photomultiplier has a saturation point for the fluorescent signal. It was different, when heavy metals were used. Lead, as an example for heavy metals, has different fluorescence properties in different solvents (see above). Fig. 5 show one example for lead-acetate (1 M/L) in bi-distilled water and one in a natural seawater sample of Borkum (0.005 M/L), in both cases the fluorescence maxima are at 480 nm excited with 265 nm. Measurements with a concentration of 0.005 M/L lead-acetate in bi-distilled water were done but we couldn't show a recognizeable fluorescence signal with this concentration. The natural seawater sample which contains yellow-substances has recognizable interactions with the lead-acetate. We suppose this due to the fact that lead get in connection with the organic components. Also quenching effects were possible.

Fluorescence spectra from phytoplankton of the river «Leine» in vitality stage, with 0.005 M/L lead-acetate and with 1 M/L lead-acetate are shown in Fig. 6. In both cases we measured five minutes after the addition of the lead to the samples. The decrease of the fluorescence signal is visible. There was an interaction between the phytoplankton and the lead but no shift of a maximum could be recognized. Only the decrease of the fluorescence intensity is significant.

Discussion

The possibility to use the laser induced fluorescence spectroscopy for an *in situ* detection in waters, seems to be helpful. There are many scopes of work, where this method will bring considerable advances. Our present results could be only another indication in environmental monitoring. Thus there had to be more information about the natural variation

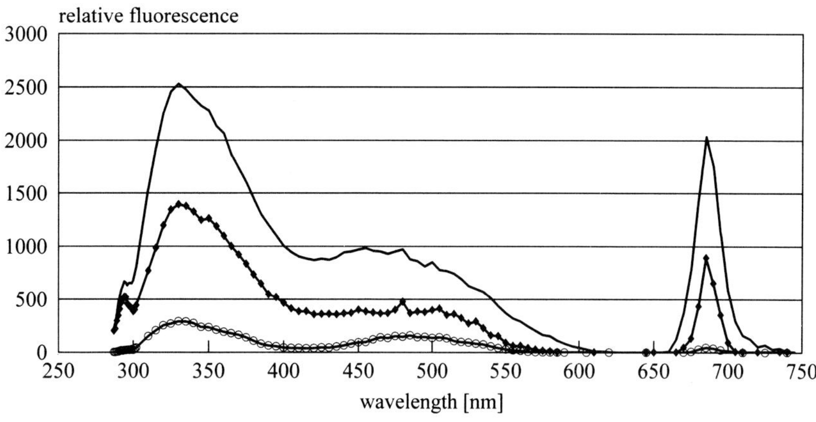

Fig. 6: Different fluorescence spectra excited with a wavelength of 265 nm. Phytoplankton of the river «Leine» in vitality stage, with 0.005 M/L and 1 M/L lead-acetate.

of the phytoplankton and the corresponding fluorescence spectra (Wilhelm and Manns, 1991; Yentsch, 1975). Furthermore, the sensitiveness to different chemicals in different phytoplankton groups had to be investigated, more information still will be needed (Bazzani et al., 1992; Babichenko et al., 1993). Last but not least the exact mixture of the yellow-substances or the breakdown-products of the phytoplankton should be known, because these differences in their fluorescence spectra could be an important indicator for water pollution.

Further investigations will be done by means of a dye laser, being pumped by a Nd:YAG laser. This system can be tuned by a PC. In combination with an intensified CCD camera, it is possible to get a fluorescence spectrum after about 10 laser pulses, reducing photochemical reactions. In addition, the fluorescence lifetime can be determined with a resolution of 5 ns. Beside the excitation and emission spectra, the lifetime is another parameter to separate compounds in a mixture (Bublitz and Schade, 1995).

References

Avramenko, V. G. and M. U. Belyi: Absorption and luminescence centers in thallium solutions. Bulletin of the academy of science of the USSR, Vol. *64,* Physical Series, Columbia Technical Translation, 5 Vermont Avenue, White Plains, New York (1959).

Babichenko, L., S. Kaitala, and H. Kuosa: Remote sensing of Phytoplankton using Laser-induced Fluorescence. Remote sens. Environ. *45,* 43–50 (1993).

Bazzani, M., B. Breschi, G. Cecchi, L. Panatoni, D. Tirelli, G. Valmori, P. Carlozzi, E. Pelosi, and G. Torzillo: Phytoplankton monitoring by laser induced fluorescence. EARSeL Advances in remote sensing, Vol. *1,* 106–118 (1992).

Belyi, M. U. and B. A. Okhrimenko: Effect of temperature on the luminescence and absorption spectra of heavy metal salt solutions, II. Inverstigations of tin salt solutions. Ukrainskij fiziceskij zurnal, 1059–1063 (1964).

– – Effect of temperature on the luminescence and absorption spectra of heavy metal salt solutions, III. Interpretation of the spectra of solutions containing Tl^{+}, Pb^{2+}, Sn^{2+} Ions. Ukrainskij fiziceskij zurnal, 1068–1073 (1964).

Belyi, M. U. and B. F. Rudko: Effect of temperature on the luminescence and absorption spectra of heavy metal salt solutions, I. Ukrainskij fiziceskij zurnal, 799–808 (1960).

Bublitz, J. and W. Schade: Laserspectroscopy for the Detection of Environmental Pollutants. GIT Fachz. Lab. *2/95,* 117–123 (1995).

Chudyk, W., K. Pohlig, L. Wolf, and R. Fordiani: Field determination of ground water contaminants using laser fluorescence and fiber optics. SPIE 1172, Chemical, biochemical and environmental sensors, 123–129 (1989).

Chudyk, W., M. M. Carrabba, and J. E. Kenny: Remote detection of groundwater contaminants using far-ultraviolet laser-induced fluorescence. Anal. Chem. *57,* 1237–1242 (1985).

Determann, S.: Wirkung von Metallionen auf das Gelbstoff-Fluoreszenz-Spektrum zur Bestimmung von Schwermetallbelastungen in Gewässern. Diplomarbeit, Uni. Oldenburg (1990).

Eisenbrand, J.: Fluorimetrie. Wissenschaftliche Verlagsgesellschaft Stuttgart (1966).

Kirk, J. T. O.: Light and Photosynthesis in aquatic ecosystems. Cambridge University Press, Cambridge (1983).

Kubitz, J.: Untersuchungen zum möglichen Einsatz der laserinduzierten Fluoreszenzspektroskopie bei der Erkennung von atomaren Schadstoffen im Meer. Diplomarbeit, Inst. für Biophysik, Uni. Hannover (1992).

Kumke, M. U., H.-G. Löhmannsröben, and T. H. Roch: Fluoreszenzspekroskopie in der Umweltanalytik. GIT Fachz. Lab. *2/95,* 112–116 (1995).

Leugers, W. A. and R. D. McLachlan: Remote analysis by fiber optic Raman spectroscopy. SPIE 990, Chemical, biochemical and environmental applications of fibers, 88–95 (1988).

Milanovich, F. P. and P. Daley: Remote detection of organochlorides with a fiber optic based sensor II. A dedicated portable fluorimeter. Analytical instrumentations *15* (4), 347–358 (1986).

Reuter, R. and T. Hengstermann: Lidar fluorosensing of mineral oil spills in the sea surface. Appl. Optics, Vol. *29,* No. 22 (1990).

Reuter, R., K. Grüner, and H. Smid: A new sensor system for airborne measurement of maritime pollution and of hydrographic parameter. GeoJournal *24.1,* 103–117 (1991).

Rogge, C., J. Kubitz, and A. Anders: Untersuchungen von Schadstoffen in Gewässern mit Hilfe der Laserspektroskopie. Verhdl. DPG(VI) *27,* 1416 (1992).

Rogge, C.: Untersuchungen zum möglichen Einsatz der laserinduzierten Fluoreszenzspektroskopie bei der Erkennung von atomaren Schadstoffen im Meer. Diplomarbeit, Inst. für Biophysik, Uni. Hannover (1992).

Uebel, U.: Aufbau und Anwendung einer Laser-Apparatur zur Messung der Fluoreszenz von Phytoplankton, Gelbstoff und Chemikalien, im Hinblick auf eine *in-situ* Detektion in Gewässern. Diplomarbeit, Inst. für Biophysik, Uni. Hannover (1994).

Wilhelm, C. and L. Manns: Changes in pigmentation of phytoplankton species during growth and stationary phase – consequences for reliability of pigment-based methods of biomass determination. Journal of Appl. Phycology *3,* 305–310 (1991).

Wünsch, G.: Optische Analysenmethoden zur Bestimmung anorganischer Stoffe. De Gruyter, Sammlung Göschen (1976).

Yentsch, C. S.: The fluorescence of chlorophyll and yellow substances in natural waters: a note on the problems of measurement and the importance of their remote sensing. NASA Conference on the Use of Lasers for Hydrographic Studies, NASA SP *375,* 137–145 (1975).

J. Plant Physiol. Vol. 148. pp. 593–598 (1996)

Time-Resolved Chlorophyll Fluorescence of Spruce Needles after Different Light Exposure

HERBERT SCHNECKENBURGER[1,2] and WERNER SCHMIDT[1,3]

[1] Fachhochschule Aalen, Optoelektronik, D-73428 Aalen

[2] Institut für Lasertechnologien in der Medizin und Meßtechnik an der Universität Ulm, Helmholtzstr. 12, D-89081 Ulm

[3] Universität Konstanz, Fachbereich Biologie, D-78465 Konstanz, Germany

Received June 24, 1995 · Accepted October 1, 1995

Summary

The impact of environmental light on the function of the photosystems was measured for a healthy and a declining spruce (*Picea abies* (L.) Karst) growing at an altitude of 840 m near Freudenstadt (Black Forest). Needles were collected from branches which had been exposed either to full sun light, or to reduced sun light (artificial shading by a wire mesh of 15–20 % transmission), or which had been growing under natural shadow (north-eastern side of the tree). More than two annual time courses of chlorophyll concentration, subnanosecond fluorescence decay kinetics and delayed luminescence were measured since April 1993. A long-lived component of (prompt) chlorophyll fluorescence (lifetime about 3 ns as compared with 100–600 ps for intact photosystems) indicated an obstruction of energy transfer from the antenna molecules to the reaction centres, whereas the integrals of delayed luminescence – measured within 0.5 s or 30 s after repetitive light pulses in the kHz range – appeared to be a measure of photosynthetic activity. All parameters showed a pronounced impact of high environmental light doses on the function of the photosystems – in particular in June–August – and a clear recovery after artificial shading. By measuring time-gated fluorescence spectra in the nanosecond range, the defect could be localized in Photosystem II of the needles from the damaged spruce.

Key words: Chlorophyll, delayed luminescence, fluorescence decay kinetics, forest decline, light stress, photosynthesis.

Introduction

Spruces growing at Schöllkopf near Freudenstadt (Black Forest, 840 m above sea level) are submitted to various stress factors, including Mg deficiency ($\leq$0.3 mg Mg/g needle dry matter), high ozone doses (>100 $\mu g/m^3$ monthly means during the summer period) and high global irradiance. At this site, apparently healthy spruces are growing among trees showing typical symptoms of yellowing and needle loss. In an early state of yellowing a simultaneous decrease in chlorophyll ($a+b$) and other pigments as well as a reduction of the photosynthetic capacity was measured (Siefermann-Harms et al., 1993).

In recent years we established various optical methods to quantify damages of the photosynthetic apparatus:

(1) Fluorescence decay kinetics of the chlorophyll antenna molecules probing the primary steps of energy transfer to the reaction centres of the photosystems I and II. A long-lived fluorescent component (I_3) with a decay time of about 3 ns (as compared with 100–600 ps for intact photosystems; Holzwarth, 1988; Evans and Brown, 1994) indicated that this energy transfer was partly obstructed (Schneckenburger and Schmidt, 1992 a);

(2) Delayed luminescence ranging from microseconds up to several seconds and reflecting the potential of the thylakoid membrane. According to Mitchell (1977), the «proton motive force» p (with the dimension «mV») can be expressed by the equation

$$p = \Delta\varphi - 59 \cdot \Delta pH \quad (1),$$

where $\Delta\varphi$ is the electrical gradient (up to 180 mV), and ΔpH the pH gradient (up to 3). In a previous article (Schmidt and Schneckenburger, 1995) the time course of delayed luminescence (induction kinetics) was correlated with the time course of p of dark adapted needles, when a rapid series of light pulses was applied – interrupted by dark periods for the measurement. Usually, delayed luminescence is explained (1) by a «back flow» of electron equivalents within the thylakoid membrane towards the Photosystem II and (2) by a recombination of the separated charges in the reaction centre.

In previous studies we found that the relative intensity I_3 of the long-lived component of prompt fluorescence increased (Schneckenburger and Schmidt, 1995), whereas the integral of delayed luminescence (measured during 30 s after dark adaptation; Schmidt and Schneckenburger, 1995) decreased, if needles of a declining spruce were exposed to high environmental light doses during the summer period. To study this effect in more detail, we carried out experiments over more than two years (April 1993–June 1995) with needles of a spruce of damage class 2 and an apparently healthy spruce (damage class 0) which had been growing either in full sun light, or in artificial shadow (using a wire mesh with 15–20 % transmission) or in natural shadow (north-eastern side of the tree). The measured parameters were I_3, two different integrals of delayed luminescence and the concentration of chlorophyll (a+ b).

Materials and Methods

Small twigs of a damaged spruce (*Picea abies* (L.) Karst; damage class 2 with 25 % needle loss and 40 % yellowing) and an apparently healthy spruce (damage class 0) were harvested in intervals of 4–7 weeks during the summer season (April–November), and needles of the second and third age classes of the 8th–10th whorl were measured. This means that in 1993 we measured needles of the years 1991 and 1992, in 1994 needles of the years 1992 and 1993, and in 1995 needles of the years 1993 and 1994. As mentioned above, these needles had been growing in full sun light, in artificial and in natural shadow.

Fluorescence decay kinetics of 8–10 densely packed needles were measured using a miniaturized equipment with a picosecond laser diode (Hamamatsu, PLP-01, emission wavelength 668 nm, pulse duration 90 ps, repetition rate adjusted to 100 kHz), two long pass filters for 690 nm, and time-correlated single photon counting (Hamamatsu R 928 photomultiplier, Tennelec NIM Electronics, IBH-199 M software by Edinburgh Instruments for reconvolution and 3-exponential curve fitting; Schneckenburger and Schmidt, 1992 a). The average power of the laser diode was limited by neutral glass filters to 0.035 µW, such that after 20 min of dark adaptation the reaction centres were kept open («F_0 condition», see Holzwarth, 1988; Krause and Weis, 1991). In addition, fluorescence spectra of individual needles were detected within short intervals of 5 ns at various delay times after the exciting pulses of a dye laser (430 nm, 2 ns, 10 Hz, pulse energy <10 µJ) pumped by the third harmonic of a Nd:YAG laser. A self-fabricated monochromator and an optical multichannel analyzer with a time-gated image intensifier (Hamamatsu, IMD 4562) were therefore adapted to a fluorescence microscope (Schneckenburger et al., 1994). This technique allowed to detect the short-lived and the long-lived (I_3) components of prompt chlorophyll fluorescence independently. All fluorescence spectra were integrated over 10 s, corresponding to 100 exciting laser pulses.

The *induction kinetics* of delayed luminescence were measured with a self-constructed phosphorimeter (for details see Schmidt and Schneckenburger, 1995) using a 3×3 array of pulsed super bright light emitting diodes with emission wavelengths of 655 ± 17 nm (RS components GmbH, Germany). The diodes were focused onto the sample consisting of an array of 15–20 spruce needles masked by a pattern of 10×20 mm^2 made from black cardboard (to guarantee a constant sample area). The delayed luminescence was measured within 200 µs following each light pulse using a red light sensitive photomultiplier (Hamamatsu, R 928), as well as an interference filter for 690–730 nm and a baffle for cutting off the excitation light. The average pawer density of illumination on the sample surface was 1 mW/cm^2. The detected signal was fed via a FET operational amplifier to an AD-converter card integrated into a personal computer (PC, MPU 486). A DA-converter on the same card was used to pulse the laser diodes via a Darlington transistor. Light pulses and detection of delayed luminescence were controlled exclusively by software, thus avoiding mechanical choppers as used in previous phosphoroscopes of the Becquerel type (Schneckenburger and Schmidt, 1992 b). In all cases, the induction kinetics of delayed luminescence were detected during 60 seconds after 20 minutes of dark adaptation; only a few seconds afterwards the curves were measured again in the range of 0–1000 milliseconds. In the latter case the needles could be regarded as light adapted. The sequence of light pulses followed a rectangular light/dark pattern with measurement of delayed light emission in darkness between two light pulses. For the present investigation we chose a light/dark ratio of 1:1 with a frequency of approximately 2000 Hz. This results in optimum signal to noise ratio for the current measuring situation.

For additional measurements of *chlorophyll (a+b) concentration,* the chlorophyll was extracted from the needles, dissolved in an acetone: water mixture (concentration ratio 80:20) and determined by absorption spectroscopy according to Lichtenthaler and Wellburn (1983). The mean values and standard deviations (related to the weight of fresh matter) were determined from 6 individual needles in each case.

Results

Prompt Fluorescence

Fig. 1 shows the fluorescence decay curves of light exposed needles from the declining (upper curve) and the healthy

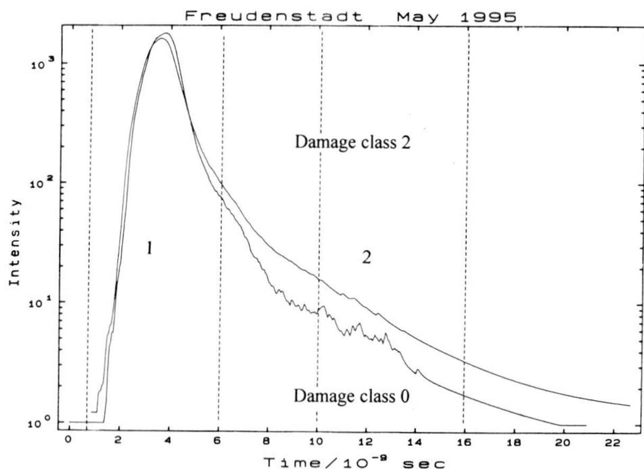

Fig. 1: Decay curves of chlorophyll fluorescence of dark adapted needles of a damaged (upper curve) and a healthy (lower curve) spruce after exposition to full sun light, as obtained on May 22, 1995. Age class 1993; excitation wavelength 668 nm; emission measured at 690–800 nm.

spruce (lower curve), as obtained on May 22, 1995, on a semi-logarithmic scale. Obviously, the needles of the damaged spruce exhibit some longer lasting fluorescence decay. From 3-exponential curve fitting, decay times of 100–200 ps, 400–600 ps and 2.5–3.5 ns were obtained. Within the given limits the lifetimes varied with the season and also between individual samples. In agreement with the literature (Holzwarth, 1988; Evans and Brown, 1994) the shortest component may be assigned to both photosystems (I+II), the middle component to the intact Photosystem II, and the long-lived component to those chlorophyll molecules whose energy transfer to the reaction centre is (partly) obstructed. The relative intensity I_3 of this long-lived component was calculated according to the equation

$$I_3 = A_3 T_3/(A_1 T_1 + A_2 T_2) \quad (2),$$

where T_i are the lifetimes and A_i the amplitudes of the individual components, as obtained from computer fitting.

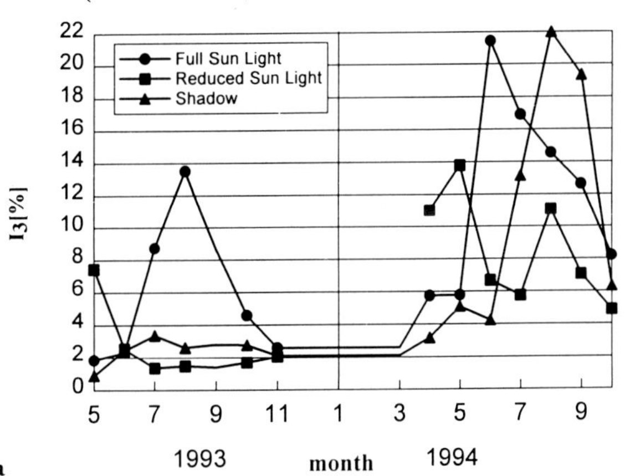

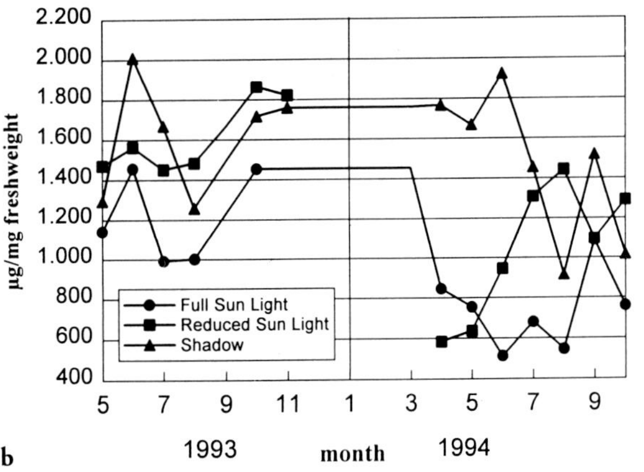

Fig. 2: Biannual course of long-lived chlorophyll fluorescence (I_3) (a) and chlorophyll concentration (b) of needles of the age class 1992 of the declining spruce after exposition to full sun light, artificial shading and natural shadow.

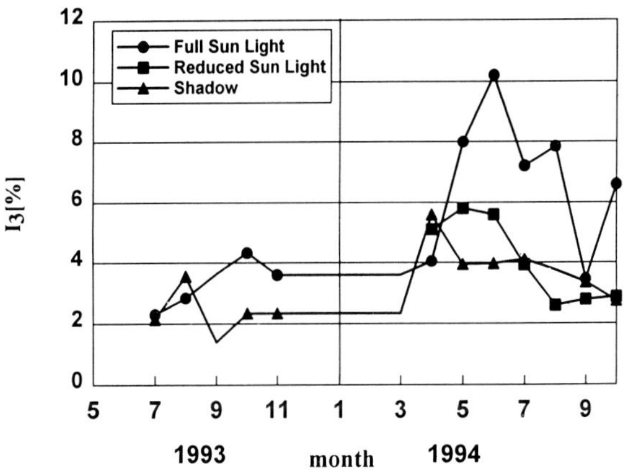

Fig. 3: Biannual course of long-lived chlorphyll fluorescence (I_3) of needles of the age class 1992 of the «healthy» spruce after exposition to full sun light, artificial shading and natural shadow.

Fig. 2 (a) shows the time course of I_3 of needles of the age class 1992 form the damaged spruce over two vegetation periods. As mentioned above, the needles had been growing either in full sun light, or in reduced sun light, or in natural shadow. For reducing the sun light, in 1993 and 1994 different branches were shaded by the wire mesh which was mounted in the period of May 20–25; therefore, the corresponding curve was interrupted between the two vegetation periods. According to Fig. 2 a, I_3 of the light exposed needles clearly shows an increase in summer and a decrease in automn during both years. The highest I_3 values were measured within the periods when solar irradiance (as well as temperature and ozone concentrations) reached maximum values. A pronounced decrease was detected immediately (i.e. after 2–3 weeks) after artificial shading. Needles which had been growing in natural shadow only showed a slight increase of I_3 in the first year (one year old needles), but a pronounced increase in the second year. This means that a partial obstruction of energy transfer in the photosystems occurred with a delay of one year (and also later in the season, since the maximum of I_3 was attained only in August as compared with fully light exposed needles which reached their maximum already in June).

As depicted in Fig. 2 (b), the chlorophyll concentration showed a negative correlation with the I_3 values. This means that chlorophyll concentration decreased in summer for the needles which had been growing in full sun light or in natural shadow. This decrease was more pronounced (1) for the light exposed than for the shadow needles and (2) for the second than for the first year. After artificial shading, an increase of the chlorophyll concentration was concomitant with a decrease of I_3.

In comparison with the damaged spruce, seasonal changes of I_3 were much less pronounced for the healthy spruce (Fig. 3). Nevertheless, similar time courses were found in Fig. 3 and Fig. 2 a. In particular, an increase of I_3 of the light exposed needles from 2–3 % to 7–10 % was found in the summer period of the second year, and a decrease of I_3 was detected after artficial shading. Again, the time course of chlorophyll concentration was somehow antiparallel to I_3, but the concentrations – with values ranging between 1100 and 1800 μg/g fresh matter for the sun light exposed needles and 1300–2200 μg/g for needles growing in natural shadow – were much higher than for the damaged spruce. These results clearly show that impairments of photosynthetic energy transfer can be deduced from fluorescence decay measurements, even if the needles visually appear green and healthy.

In Fig. 1 two time gates are indicated for the detection of the short-lived (0–5 ns) and the long-lived (10–15 ns) fluorescent components. The corresponding spectra are shown in Fig. 4 for the light exposed needles of the declining spruce and the healthy spruce. At 0–5 ns the spectra show a maximum at about 690 nm and a shoulder around 730 nm. The latter seems to reflect the concentration of chlorophyll, whereas the maximum around 690 nm is partly re-absorbed. At 10–15 ns the spectra show the long-lived component I_3 with a pronounced peak at 685 nm for the light exposed needles of the damaged spruce. In literature (see e.g. Buschmann, 1986) this peak was often attributed to Photosystem II.

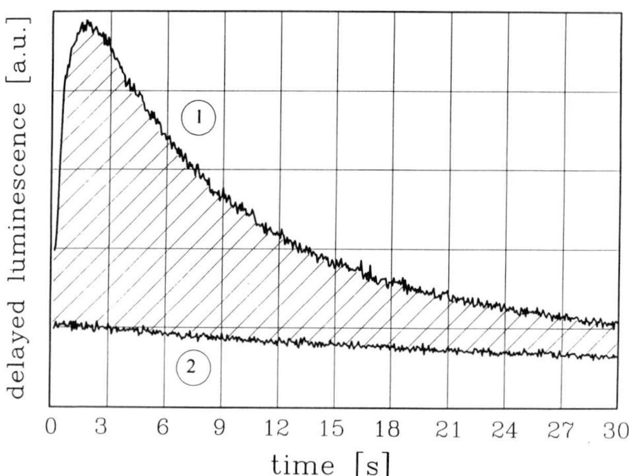

Fig. 5: Induction kinetics of delayed luminescence in dark adapted spruce needles at 0–30 s (curve 1) and 30–60 s (curve 2) (reproduced from Schmidt and Schneckenburger, 1995, with modifications). The shaded area between these two curves is easily integrated by a computer routine and serves as a quantitative measure of «vitality».

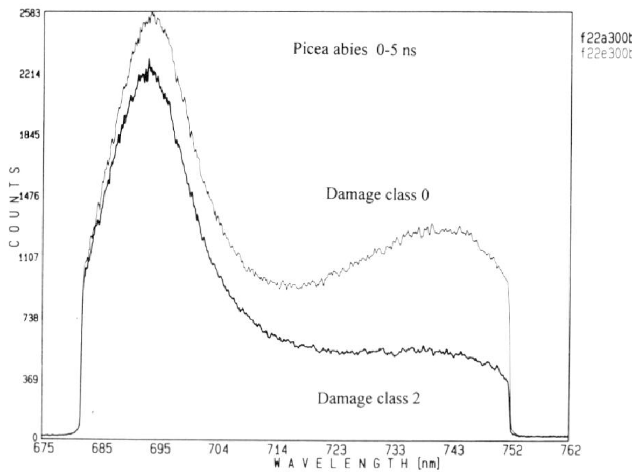

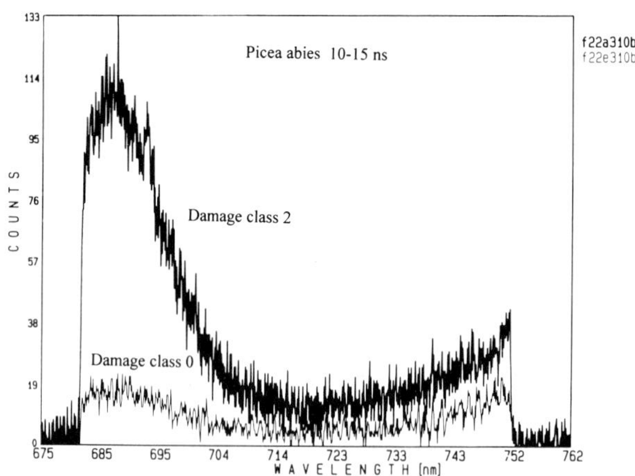

Fig. 4: Time-gated spectrum of chlorophyll fluorescence of light exposed spruce needles of the two damage classes at 0–5 ns (a) and 10–15 ns (b) after the exciting laser pulse (428 nm, 2 ns).

Delayed Luminescence

The induction kinetics of delayed luminescence after 20 min of dark adaptation are shown in Fig. 5 for light exposed needles of the healthy spruce. The lower curve (2) simply is the extension of the upper kinetics between 30 s and 60 s. The upper curve (1) between 0s and 30 s shows a strong increase within less than 3 seconds towards a maximum followed by a decline towards a low assymptotic level of equilibration. The lower curve (at 30–60 s) is close to this assymptotic level and represents some kind of «baseline». To quantify delayed light emission, we simply determined the area enveloped by the curves 1 and 2 in Fig. 5 («slow integral»). This is a more reliable procedure than to integrate the absolute area underneath the kinetics, since in kinetical emission experiments the exact offset is difficult to control. A few seconds after complete light adaptation (e.g. after measuring the kinetics shown in Fig. 5), the time course shown in Fig. 6 (curve 1) is detected in the short time range of 0–500 ms. Again, curve 2 is the extension of curve 1 at 500–1000 ms, and again the area between (1) and (2) was found to depend on the individual samples («fast integral»).

In Fig. 7 the seasonal courses of chlorophyll content, I_3 of prompt fluorescence and the two integrals of delayed luminescence are compared for light exposed needles of the age class 1993 of the declining (dc 2) and the healthy (dc 0) spruce. The numerical values of the integrals were corrected for variations in chlorophyll content (μg/g fresh weight) by division. Clearly, with increasing light exposure in the summer period, needles of the damaged spruce show lower chlorophyll concentrations, higher I_3 values, lower values of the «long» and higher values of the «short» integral of delayed luminescence, as compared with needles of the healthy spruce. In comparison with all other parameters, the increase of the «fast» integral appears later during the annual period.

Discussion

Various parameters of prompt and delayed chlorophyll luminescence were found to characterize the «vitality» of photosynthetic organisms such as spruce needles. The intensity of the long-lived component I_3 of *prompt fluorescence* seems to be a measure for the number of those chlorophyll molecules, whose energy transfer to the reaction centres is reduced or obstructed. They are probably located in Photosystem II. These primary steps in the photosynthetic pathway are progressively inhibited with light exposure – accompanied by heat and high ozone doses – during the summer period. Artificial shading leads to some recovery. The fact that the increase of I_3 in summer is more pronounced for damaged than for apparently healthy spruces may be due to photosensitization, i.e. formation of singlet oxygen ($^1O_2$) or oxygen radicals (O_2^-) by energy or electron transfer from the triplet state of chlorophyll molecules and subsequent photooxidation of cellular metabolites (see e.g. Bonnett, 1995). Indeed, the action spectrum of yellowing has recently been correlated with the absorption spectrum of chlorophyll (Siefermann-Harms, 1992). Some reduced protection against photosensitization may therefore account for light-induced yellowing of predamaged spruces.

Induction of *delayed luminescence* is commonly related to a «reverse electron flow» in the electron transport chain of photosynthetic organisms (Wermaas et al., 1984; Rutherford and Inoue, 1984). Even if its origin is largely attributed to Photosystem II, several convincing arguments suggest a small contribution of Photosystem I as well (Malkin, 1977; Jursenic, 1986). A functional thylakoid membrane is the indispensable prerequisite that induction kinetics of delayed luminescence can be detected. As soon as the membrane integrity is destroyed, delayed luminescence is completely lost – in contrast to prompt fluorescence (Schmidt and Senger, 1987). This finding makes delayed luminescence an excellent tool for monitoring vitality.

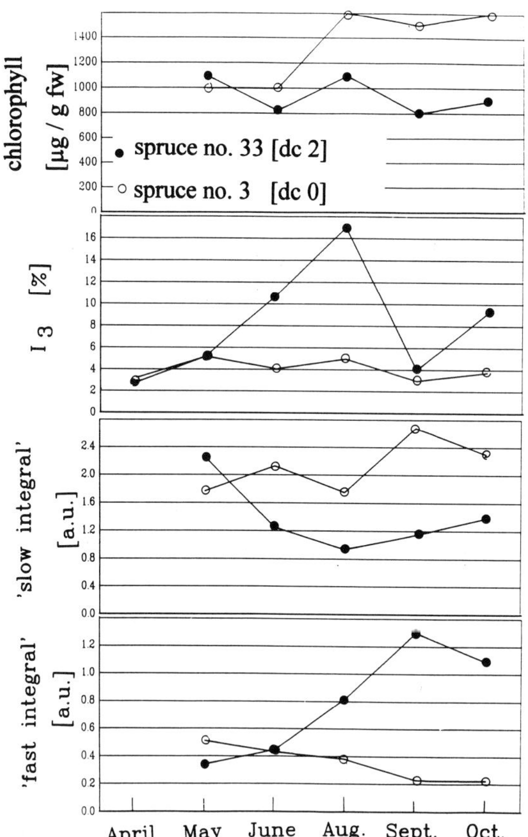

Fig. 7: Chlorophyll content, I_3 of prompt fluorescence, as well as «slow integral» (defined by Fig. 5) and «fast integral» (defined by Fig. 6) of delayed luminescence for light exposed needles of the declinig spruce (solid circles) and the «healthy» spruce (empty circles). The integrals of delayed luminescence were normalized by chlorophyll concentration. Standard measuring error appr. ±10%.

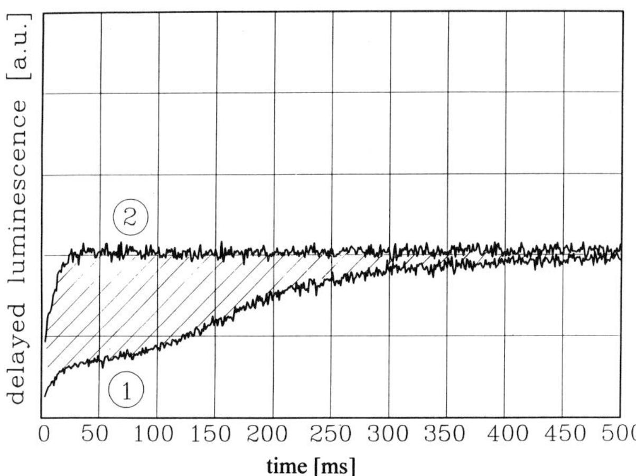

Fig. 6: Induction kinetics of delayed luminescence of light adapted needles at 0–500 ms (curve 1) and 500–1000 ms (curve 2) (reproduced from Schmidt and Schneckenburger, 1995). The shaded area between these two curves is integrated in analogy to Fig. 5. Additional induction kinetics are similar to curve 2; however, after a few seconds of darkness, curve 1 is reproduced.

Malkin et al. (1994) assume that the induction kinetics of delayed luminescence reflect the kinetics of formation of the proton motive force (p), whose main contribution is the electrical gradient in the millisecond range (Witt, 1979) and the pH gradient in the range of seconds (Davenport and McCarty, 1980), respectively. Therefore, the integral of the «slow» kinetics of delayed light emission (Fig. 5) is mainly a measure of the size of a pH gradient across the thylakoid membrane upon light excitation, i.e. of the vitality of the photosynthetic specimen. However, the «fast» kinetics (Fig. 6), which are essentially related to an electrical gradient, behave just in opposite: declining needles («less vitality») exhibit larger integrals. On this rapid change in the initial course (300 ms) we can only speculate: in analogy to the accepted interpretation of the so-called *Kautsky* kinetics of prompt fluorescence, the area enveloped by the kinetics 1 and 2 probably reflects a rapidly developing intermediate state of radical pairs of the reaction centre II.

Even, if the measured parameters of prompt and delayed luminescence are still not fully understood, we conclude that they are useful to characterize the function of the photosynthetic apparatus of intact biological specimen.

Acknowledgements

This work was supported by the Ministerium für Wissenschaft und Forschung (MWF) Baden-Württemberg. The authors thank D. Siefermann-Harms for stimulating discussions as well as R. Hahn, U. Bristle and M. Preiß for their cooperation in carrying out the experiments.

References

Bonnet, R.: Photosensitizers of the porphyrin and phthalocyanine series for photodynamic therapy. Chemical Society Reviews 1995, pp. 19–33.

Buschmann, C.: Fluoreszenz- und Wärmeabstrahlung bei Pflanzen – Anwendung in der Photosyntheseforschung. Naturwissenschaften *73,* 691–699 (1986).

Davenport, J. W. and R. E. McCarty: The onset of photophosphorylation correlates with the rise in transmembrane electrochemical proton gradients. Biochim. Biophys. Acta *589,* 353–357 (1980).

Evans, E. H. and R. G. Brown: An appraisal of photosynthetic fluorescence decay kinetics as a probe of plant function. J. Photochem. Photobiol. *B: Biol. 22,* 95–104 (1994).

Holtzwarth, A. R.: Time resolved chlorophyll fluorescence – what kind of information on photosynhetic systems does it provide? In: Lichtenthaler, H. K. (ed.): Applications of Chlorophyll Fluorescence, pp. 21–31. Kluwer Academic Publishers, Dordrecht (1988).

Jursinic, P. A.: Delayed fluorescence: Current concepts and status. In: Govindjee et al. (eds.): Light Emission by Plants and Bacteria, pp. 291–328. Academic Press Inc., Orlando (1986).

Krause, G. H. and E. Weis: Chlorophyll fluorescence and photosynthesis: The basis. Annu. Rev. Plant Physiol. Plant Mol. Biol. *42,* 313–349 (1991).

Lichtenthaler, H. K. and A. R. Wellburn: Determination of total carotenoids and chlorophyll *a* and *b* of leaf extracts in different solvents. Biochem. Soc. Transact. *603,* 591–592 (1983).

Malkin, S.: Induction pattern of delayed luminescence from isolated chloroplasts. (I) Response of delayed luminescence to changes of prompt fluorescence yield. In: Barber, J. (ed.): Primary Processes of Photosynthesis, pp. 349–431. Elsevier Biomedical Press, Amsterdam (1977).

Malkin, S., W. Bilger, and U. Schreiber: The relationship between millisecond luminescence and fluorescence in tobacco leaves during the induction period. Photosynthesis Research *39,* 57–66 (1994).

Mitchell, P.: Compartimentation and communication in living systems. Ligand conduction: a general catalytic principle in chemical, osmotic and chemiosmotic reaction systems. Eur. J. Biochem. *95,* 1–20 (1977).

Rutherford, W. and Y. Inoue: Oscillation of delayed luminescence from PS II: recombination of $(S_2Q_B^-)$ and $(S_3Q_B^-)$. FEBS Lett. *165,* 163–170 (1984).

Schmidt, W. and H. Senger: Long-term delayed luminescence in *Scenedesmus obliquus* (II) Influence of exogenous factors. Biochim. Biophys. Acta *890,* 22–27 (1987).

Schmidt, W. and H. Schneckenburger: Induction kinetics of delayed luminescence in photosynthetic organisms as measured by an LED-based phosphorimeter. Photochem. Photobiol. *62,* 745–750 (1995).

Schneckenburger, H. and W. Schmidt: Time-resolving luminescence techniques for possible detection of forest decline (II): Picosecond chlorophyll fluorescence. Radiat. Environ. Biophys. *31,* 73–81 (1992a).

– – Luminescence techniques for monitoring of forest decline. J. Photochem. Photobiol. *B: Biol. 13,* 187–193 (1992b).

Schneckenburger, H., K. König, T. Dienersberger, and R. Hahn: Time-gated microscopic imaging and spectroscopy in medical diagnosis and photobiology. Opt. Eng. *33,* 2600–2606 (1994).

Schneckenburger, H. and W. Schmidt: Time-resolved chlorophyll fluorescence of spruce needles under various light conditions. J. Fluoresc. *5,* 155–158 (1995).

Siefermann-Harms, D.: The yellowing of spruce in polluted atmosphere. Photosynthetica *27,* 323–341 (1992).

Siefermann-Hamrs, D., P. Bourgois, K. Baumann, G. Baumbach, C. Buschmann, W. Einig, S. Fink, A. Franke, R. Hampp, E. E. Hildebrand, E. Hochstein, H.-J. Jäger, M. Konnert, I. Kottke, H. K. Lichtenthaler, R. Manderscheid, W. Schmidt, H. Schneckenburger, D. Seemann, H. Spieker, W. Urfer, K. v. Vilpert, A. Wild, and U. Zimmer-Rinderle: Conclusions of coordinated research into the yellowing of spruce at higher elevations of the Black Forest: Research site Schöllkopf (Freudenstadt). KfK-PEF *104,* 79–126 (1993).

Wermaas, W. F. J., G. Renger, and G. Dohnt: The reduction of the oxygen-evolving system in chloroplasts by thylakoid components. Biochim. Biophys. Acta *764,* 194–202 (1984).

J. Plant Physiol. Vol. 148. pp. 599–612 (1996)

Detection of Vegetation Stress Via a New High Resolution Fluorescence Imaging System

H. K. Lichtenthaler[1], M. Lang[1], M. Sowinska[2], F. Heisel[2], and J. A. Miehé

[1] Botanisches Institut, Universität Karlsruhe, Kaiserstr. 12, D-76128 Karlsruhe, Germany

[2] Groupe d'Optique Appliquée, Centre de Recherches Nucléaires, 23, rue de Loess, F-67037 Strasbourg Cedex 2, France

Received August 15, 1995 · Accepted October 10, 1995

Summary

The UV-laser (λ 355 nm) induced fluorescence emission spectra of green leaves comprise the blue (F440) and green (F520) fluorescence bands as well as the red (F690) and far-red (F740) chlorophyll fluorescence emission bands. Based on the four UV-laser induced fluorescence bands blue, green, red and far-red a high resolution fluorescence imaging system was established, which allows a fast and large scale screening of fluorescence gradients and local disturbances in fluorescence emission over the whole leaf surface. The new imaging method not only permits to screen leaves by means of images in four fluorescence bands (*LIF images*) but, in addition, via images of the fluorescence ratios blue/red (F440/F690), blue/far-red (F440/F740), the chlorophyll fluorescence ratio red/far-red (F690/F740) and the ratio blue/green (F440/F520) (*LIF ratio images*). By fluorescence imaging we could prove that in aurea tobacco the major part of the leaves' blue and green fluorescence is emitted from the main and side leave veins, whereas the major part of the leaves' red and far-red chlorophyll fluorescence is emitted from the vein-free leaf regions, which also have the highest chlorophyll content. A smaller proportion of the aurea tobacco leaves' blue-green fluorescence emission is derived from the cell walls of epidermis cells. The fluorescence ratios blue/red and blue/far-red are very sensitive to environmental changes, and thus permit early stress and strain detection in plants, and the evaluation of damage to the photosynthetic apparatus. Via monitoring the increase in chlorophyll fluorescence LIF images allow to detect differences in the time-dependent uptake of diuron and the progressing inhibition of photosynthetic electron transport in the treated leaf part. The novel fluorescence imaging technique sets a new dimension for early stress detection in the photosynthetic apparatus and in plants. It has many advantages over the previously applied point-data measurements of selected leaf points using conventional spectrofluorometers. The new fluorescence imaging system proved to be very suitable for remote sensing of plants in the near distance, and can be further developed for far distance remote sensing of the state of health of terrestrial vegetation. Some examples in the many possible ways of computer-aided fluorescence data processing (formation of different fluorescence ratios, screening of fluorescence profiles, histogramme plotting) are indicated.

Key words: Fluorescence imaging, blue-green fluorescence, red and far-red chlorophyll fluorescence, inhibition of electron transport, diuron, fluorescence ratios blue/red, blue/far-red, red/far-red, stress detection, LIF images, LIF ratio images.

Abbreviations: LIF = laser-induced fluorescence; F440 = fluorescence emission band in the blue region (430 to 450 nm); F520 = green fluorescence band (near 520 to 530 nm); F690 = red chlorophyll fluorescence maximum (range: from 684 to 695 nm); F740 = far-red chlorophyll fluorescence emission maximum (range: 730 to 740 nm); F440/F690 = fluorescence ratio blue/red; F440/F740 = fluorescence ratio blue/far-red; F690/F740 = chlorophyll fluorescence ratio red/far-red; F440/F520 = fluorescence ratio blue/green.

Introduction

Field plants are exposed to a multitude of natural biotic and abiotic stressors as well as to particular anthropogenic stressors, many of which have shown up or had been established in the past decades only (Lichtenthaler, 1996). Early stress detection in plants, before visual damage symptoms are detectable, is thus today a substantial requirement to take suitable countermeasures against stress and damage in order to reactivate the plant's vitality.

In the past twenty years various types of chlorophyll signatures have been applied in order to describe and investigate the photosynthetic light processes and quantum conversion at physiological conditions, and to detect stress and strain in the photosynthetic apparatus (Lichtenthaler, 1986 and 1988; Lichtenthaler and Rinderle, 1988 a, b; Krause and Weis, 1988 and 1991; Schmuck and Lichtenthaler, 1986; Kocsanyi et al., 1988). These include the chlorophyll fluorescence induction kinetics (Kautsky effect) of pre-darkened leaves determining various ratios of the variable chlorophyll fluorescence, such as Fv/Fm, Fv/Fo, Fd/Fs (= Rfd-values at 690 and 735 nm as vitality indices) as well as the photochemical and non-photochemical quenching coefficients, qP, qN (see review Lichtenthaler and Rinderle, 1988 a), and various other chlorophyll fluorescence parameters, such as the chlorophyll fluorescence ratio red/far-red (F690/F735) which is an indicator of the *in vivo* chlorophyll content (Hák et al., 1990; D'Ambrosio et al., 1993). All of these parameters have also been applied as an efficient tool to detect disturbances and damage to the photosynthetic apparatus and its function. It had already been pointed out by Kautsky and co-workers that the chlorophyll fluorescence emission is inversely correlated to the performance of photosynthesis (Kautsky and Hirsch, 1931; see also the reviews of Lichtenthaler, 1992 and Govindjee, 1995). Screening of chlorophyll fluorescence parameters possesses the advantage of fluorescence measurements being non-invasive techniques. Therefore, they can be applied repeatedly to the same leaf, and thus permit to sense the development of damage as well as the regeneration of the photosynthetic function after removal of the stressor.

Within the European EUREKA research program LASFLEUR (EU 380) directed to remotely sense terrestrial vegetation by means of a fast screening technique, which can also be applied from airborne systems, we have chosen the red/far-red chlorophyll fluorescence ratio F690/F740 as a main parameter to judge the state of health of plants (Lichtenthaler, 1986 and 1988; Lichtenthaler and Lang, 1995; Lichtenthaler et al., 1995). During short-term stress, which impairs the photosynthetic quantum conversion, this ratio significantly increases by about 20 % to 35 % depending on the plant species and the functional state of the plant, and it rises by several times during long-term stress when the chlorophyll content declines (Hák et al., 1990; D'Ambrosio et al., 1993; Lichtenthaler et al., 1990; Buschmann et al., 1996).

Upon excitation with ultraviolet radiation green leaves not only emit red and far-red chlorophyll fluorescence, but show a fluorescence emission in the visible and near infra-red range of the spectrum from 400 nm up to 800 nm. The plants' fluorescence emission spectra possess a blue band ranging from 430 to 450 nm (F440), a green band between 520 and

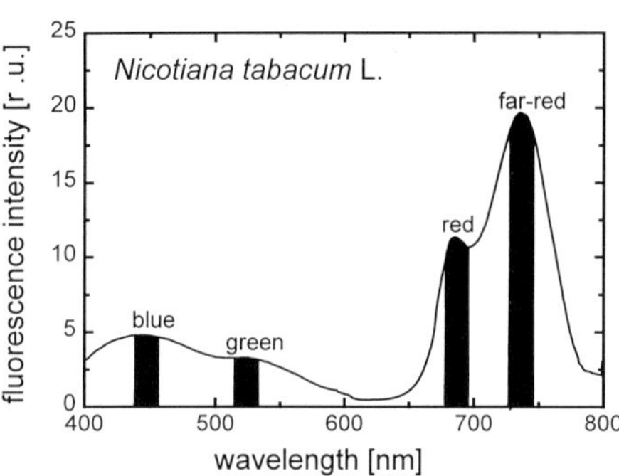

Fig. 1: UV-radiation induced fluorescence emission spectrum of a green tobacco leaf with maxima/shoulders in the blue (F440), green (F520), red (F690) and far-red (F740) spectral region. The four fluorescence bands, applied in the fluorescence imaging of plants, are indicated. (Excitation 355 nm; LS-50 Perkin-Elmer spectrofluorometer).

530 nm (F520), a red maximum near 684 to 695 nm (F690) and a far-red maximum between 730 and 740 nm (F740) (Chappelle et al., 1984; Lang and Lichtenthaler, 1991; Lang et al., 1991, 1994) as indicated in Fig. 1. The red and far-red fluorescences emanate from protein-bound chlorophyll *a* molecules in the pigment proteins of the photochemically active thylakoids of chloroplasts in the leaf's mesophyll cells. In contrast, the blue and green fluorescence emission is primarily emitted by various plant phenolics in the cell wall of the upper and lower epidermis (Harris and Hartley, 1976; Lang et al., 1991; Stober and Lichtenthaler, 1993 a, b, c), yet also the main and side leaf veins can contribute considerably to the blue-green fluorescence emission of leaves (Lang et al., 1994). With the re-detection of the UV-radiation induced blue and green fluorescence as regular fluorescence signatures of plants in the past 10 years, we have checked the possible significance of the blue and green fluorescence, together with the red/far-red chlorophyll fluorescence bands, on the near-distance detection of plant stress as well as on the potential remote sensing of the state of health of terrestrial vegetation. In this respect we have recognized that the fluorescence ratios blue/red (F440/F690), blue/far-red (F440/F740) are particularly suitable, since they are parameters sensitive to changes in the environment (Stober and Lichtenthaler, 1993 b, c; Stober et al., 1994). These ratios permit an early detection of stress and strain conditions in leaves, of damage to their photosynthetic apparatus, and thus allow an early stress diagnosis.

Although the various chlorophyll fluorescence parameters and ratios had been established for detection of the functional state of the photosynthetic apparatus and stress to plants, these fluorescence methods had a great disadvantage. Laser-induced fluorescence signatures (LIF-signatures) were only measured of single leaf spots, and thus only provided information on relatively small leaf points. Local differences in the fluorescence emission as well as fluorescence gradients over the leaf surface, which represent early stress symptoms (Lang et al., 1996), however, cannot be detected by single point data measurements. Even when several leaf points are measured, stress and damage can only be detected at a relatively progressed stage. Since punctuated fluorescence measurements yield only limited information on the state of health of plants and their photosynthetic apparatus, we were searching for a better technique which allows a large scale fluorescence imaging of whole leaves and even of several leaves of a plant. In a joint interfacultative cooperation between physicists from Strasbourg and plant biologists and biophysicists from Karlsruhe we developed for the first time a high resolution UV-laser-induced fluorescence imaging system, which allows a simultaneous large scale fluorescence screening of leaves on the basis of the blue, green, red and far-red fluorescence bands of plants and the formation of the corresponding fluorescence ratios. This imaging system is described here together with some examples of its applicability, including the computer-aided image processing.

Materials and Methods

Plants

Tobacco plants (*Nicotiana tabacum* L., green form su/su and aurea mutant Su/su; Schmidt, 1971) were cultivated for 4 months on peat containing mineral nutrients (TKS II) in the experimental greenhouse of the Botanical Garden (University of Karlsruhe). *Campelia zanonia* L. was grown on soil in the greenhouse. Plants were kept under natural light conditions at approximately 25 °C and 55 % relative humidity.

Determination of photosynthetic pigments

The photosynthetic pigments of tobacco leaves were determined spectrophotometrically from leaf extracts in 100 % acetone using the redetermined extinction coefficients and equations of Lichtenthaler (1987), which permit the determination of the chlorophylls *a* and *b* and of total carotenoids in the same extract solution.

Application of diuron

The photosynthetic herbicide diuron, which binds to the Q_B-protein, was applied in a 5×10^{-5} M aqueous solution by wetting the lower leaf-side (left part) 3 times with a thin hair pin. Diuron (DCMU) had been dissolved before in ethanol and was then diluted in water to yield a 5×10^{-5} M solution containing ca. 1 % of ethanol.

The Karlsruhe/Strasbourg fluorescence imaging system and image processing

The fluorescence imaging system (cf. Fig. 2), which is the first high resolution fluorescence imaging system for screening of leaves and whole plants, contains the following parts. The third harmonic of a cw Q-switched, mode locked and cavity dumped Nd:YAG laser, emitting at 355 nm, was used as excitation source. The UV-A pulses, which were emitted at an adjustable repetition rate, typically of 0.8 to 1 kHz, had an energy of 10 µJ and a FWHM of 100 ps. The laser beam was expanded in order to hit the complete leaf surface. Additional white light with a quanta fluence rate of 1000 µmol m^{-2} s^{-1} simulated day light conditions (steady state of chlorophyll fluorescence induction kinetics). Fluorescence emission was sensed from a distance of approximately 0.5 m via the fluorescence imaging system (RAGM6 & Animater V1; A.R.P., Strasbourg, France). The emission wavelength was selected by changeable interference filters (center wavelength 440 nm, 520 nm, 690 nm or 740 nm; full width at half maximum 10 nm; Oriel, France) and focused onto an image intensifier (Philips XX1414M/E image intensifier tube) with a lense.

The Karlsruhe/Strasbourg Fluorescence Imaging System

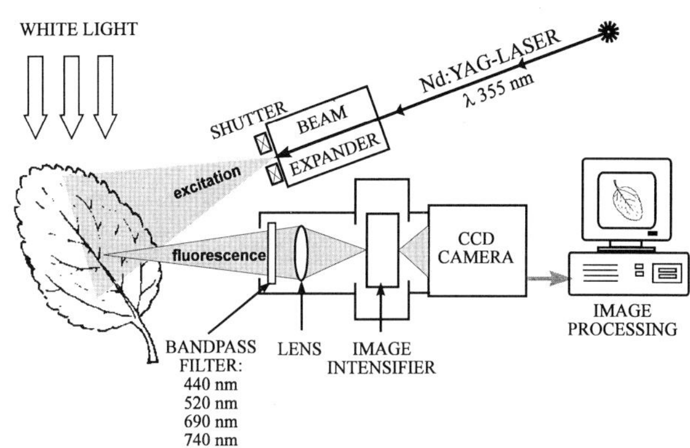

Fig. 2: Scheme of the Karlsruhe/Strasbourg laser-induced fluorescence imaging system LIFIS for near distance and remote sensing of leaves or whole plants. Due to its high resolution the LIFIS permits to screen and to detect in one measurement small local and stress-induced inhomogeneities in fluorescence emission as well as fluorescence gradients over the whole leaf surface.

The intensifier has an adjustable gain (up to 10^3) and can be gated synchronously with the laser pulses, the gate width can be varied from 20 ns up to 100 ms. For the fluorescence imaging performed in the presence of the ambient light, the gating is necessary to optimise the signal (fluorescence) to noise (ambient light) ratio, and to avoid a possible saturation of the camera by the ambient light. The gate width was adjusted to 20 ns, one time period longer than the mean duration of the fluorescence emission (a few ns). The digital CCD camera had a CCD array with 288×384 elements and operated with 50 frame transfers per second. Each readout was digitized on 8 bits and was transferred to the interface card of the microcomputer (e.g. PC486). On this card, an accumulation (up to 999) of successive readouts can be made up to 16 bits, followed or not by the subtraction (same number of accumulations) of the noise due to the ambient light and to the camera itself. Depending on the fluorescence intensities obtained the number of accumulations for one image was typically 200 to 400 at our experimental conditions. The resulting image was then automatically sent to the extended memory of the computer. The time for 200 accumulations including subtraction amounted to 8 seconds. The whole LIF images in the blue, green, red and far-red region can be recorded in less than 3 min.

Image processing, including image correction for a non-uniform excitation, the formation of images of the fluorescence ratios (LIF ratio images) as well as the calculation of fluorescence intensities were performed using the image processing software Animater V1 (A.R.P., Strasbourg, France), which also allows profiles and histogramme plotting. Within the 288×384 image, the software allows to select specific rectangular regions of adjustable sizes, and to determine their average fluorescence intensity and standard deviation. For quantitative values shown in figures or tables rectangular areas of 100 to 400 pixels were used. The resolution of the LIF imaging device (pixel size) depends on the distance of the leaf from the CCD camera, and amounts to 0.25 mm^2 at a distance of 50 cm.

The fluorescence intensities of the blue, green, red and far-red fluorescence images (LIF images) of the investigated leaves are presented in false colours with a colour scale from dark blue (zero fluorescence) via green and yellow to red (highest fluorescence intensity). Also, the images of the fluorescence ratios (LIF ratio images) blue/red, blue/far-red, red/far-red and blue/green are presented using the same false colour procedure from dark blue (zero point) to red as highest value of a fluorescence ratio. The different fluorescence intensities of LIF images and the different values of the fluorescence ratios of LIF ratio images can also be presented in grey scales from black (zero fluorescence or zero value) to white (highest fluorescence or highest value of fluorescence ratio). However, the false colour images provide a much better view of the gradients over the leaf surface as compared to the grey scales.

The LIF images and the LIF ratio images are processed and displayed on the computer screen in false colours. Then, colour slides are taken via computer-aided film exposure using the Polaroid digital palette film recorder system Cl-5000S and a daylight Agfachrome CT100i slide film (ISO 100/210). Colour images were also obtained with a Mitsubishi colour video copier (model CP200E). From these slides colour prints can be made in the desired size. At a black and white presentation (grey scales) of the images, they can be printed out directly with a high resolution laser printer.

This fluorescence imaging system allows to simultaneously screen large leaf areas of 25 to 30 cm transversely and at a longitudinal diameter, or several smaller leaves together. It cannot only be operated from a near distance, but also in a distance of several meters from the leaf and plant. With more sophisticated equipment (e.g. stronger lasers) this imaging system can easily be developed into a powerful field instrument for remote sensing of terrestrial vegetation. In fact, with a stronger laser and the use of a telescope, this imaging system has been used to record spatially well resolved images at a distance of 100 m.

Results and Discussion

Fluorescence imaging of plant leaves

With the new laser-induced fluorescence imaging system it was shown that the blue (F440) and green (F520) fluorescence are not evenly distributed over the leaf surface, but that they primarily emanate from the main and side leaf veins as is shown in Fig. 3 by false colour fluorescence images (LIF images) of a green aurea tobacco leaf. In contrast, the red (F690) and far-red chlorophyll fluorescence (F740) are derived predominantly from the vein-free leaf parts and intercostal fields, which also exhibit the higher chlorophyll content in this aurea leaf. The LIF images also demonstrated that the blue fluorescence F440 is stronga than the green fluorescence F520, and that the red chlorophyll fluorescence band F690 is stronger than the far-red chlorophyll fluorescence F740, as is visible from the image scales (Fig. 3). Furthermore, the intensities of the red and far-red chlorophyll fluorescence bands are higher than those of the blue and green fluorescence bands.

In any case, the LIF images indicate that the blue-green and the red and far-red fluorescence emissions exhibit a negative contrast in tobacco leaves. Such clear differences in the emission of the blue-green fluorescence and the red and far-red chlorophyll fluorescence were not detectable by point-data fluorescence measurements in the conventional spectrofluorometer applied so far. The Karlsruhe/Strasbourg fluorescence imaging system with its high resolution (pixel size 0.25 mm^2 at a distance of 50 cm of the sensor from the leaf), however, allows to detect and quantify even the smallest differences in fluorescence emission over the leaf surface. Spot-sized local differences in the blue fluorescence emission (red spots at the upper left and the lower left part of the leaf as seen in the F440 image of Fig. 3) are easily detectable, as well as the fact that these do not show up in red and far-red chlorophyll fluorescence F690 and F740. Also, gradients in the chlorophyll fluorescence emission found in leaves of water and heat-stressed plants (Lang et al., 1996) from the leaf edge to the center part of the leaf, can exactly be screened by means of UV-laser-induced fluorescence imaging. This emphasizes the high superiority of the fluorescence imaging over the previously applied point-data fluorescence measurements. The computer-aided image processing permits to present the differences in fluorescence emission over the leaf surface also in the form of a three-dimensional spatial relief histogramme plotting as shown for the blue and far-red fluorescence images in Fig. 4.

LIF ratio images

However, the fluorescence imaging of plants is not restricted to the LIF images in the four fluorescence bands blue, green, red and far-red. Via computer-aided image processing one can also print the fluorescence signals in the form of different fluorescence ratios, i.e. the LIF ratio images of leaves. This is shown for the fluorescence ratios blue/red (F440/F690), blue/far-red F440/F740), the chlorophyll fluorescence ratio red/far-red (F690/F740) and the ratio blue/green (F440/F520) in Fig. 5. However, these are not the only fluorescence ratio images that can be formed. Using image processing it is also possible to express images of the inverse fluorescence ratios green/blue and far-red/red (Fig. 6a and b), as well as the additional fluorescence ratios green/red and green/far-red (Fig. 6c and d). In the false colour presentations of figures 5 and 6 the LIF ratio images provide a wealth of information on the relative intensities of different fluorescence signatures. They distinctly show that the intensity of the four fluorescence bands, and consequently the values of the LIF ratios, are unequally distributed over the leaf surface. Although the fluorescence images are measured in the four fluorescence bands blue, green, red and far-red only, the LIF ratio images exhibit essential spectral information on the relative intensities of the individual fluorescence bands.

The chlorophyll fluorescence ratio and far-red

The values of the chlorophyll fluorescence ratio red/far-red F690/F740 are inversely correlated to the chlorophyll content of leaves (Hák et al., 1990; Dámbrosio et al., 1993; Buschmann et al., 1996; Babani et al., 1996), and they increase with decreasing chlorophyll content and *vice versa.* This is due to partial reabsorption of the emitted 690 nm fluorescence overlapping with the chlorophyll absorption bands, which is not the case for the 740 nm chlorophyll fluorescence band. In C3-plants with bifacial leaves this relationship particularly applies to the excitation and sensing of the chlorophyll fluorescence from the upper leaf surface. The fact is that the emitted fluorescence primarily comes from that leaf half which had been excited. In plants with bifacial leaves the upper leaf half contains the densely packed palisade parenchyma cells, which possess a great number of chloroplasts. For this reason, the upper leaf half contains a much higher chlorophyll content than the lower leaf half with its few spongy parenchyma cells and large aerial interspaces. As a

Fig. 3: UV-laser induced fluorescence images (LIF images) of the blue (F440) and green fluorescence (F520) as well as the red (F690) and far-red chlorophyll fluorescence (F740) of the lower leaf-side of a green leaf of the aurea tobacco mutant Su/su. (Excitation: Nd:YAG laser, 355 nm). The fluorescence intensity presented in false colours increases from dark blue via green and yellow to red. Note the difference in colour scale for the blue and green fluorescence and that of the red and far-red fluorescence.

Fig. 5: Fluorescence ratio images (LIF ratio images) of blue/red (F440/F690), blue/far-red (F440/F740), red/far-red (F690/F740) and blue/green (F440/F520) fluorescence ratios of the lower leaf side of a green leaf of the aurea tobacco mutant Su/su. The ratio images are based on the fluorescences shown in Fig. 3. False colour presentation of the fluorescence intensity increasing from dark blue via green and yellow to red.

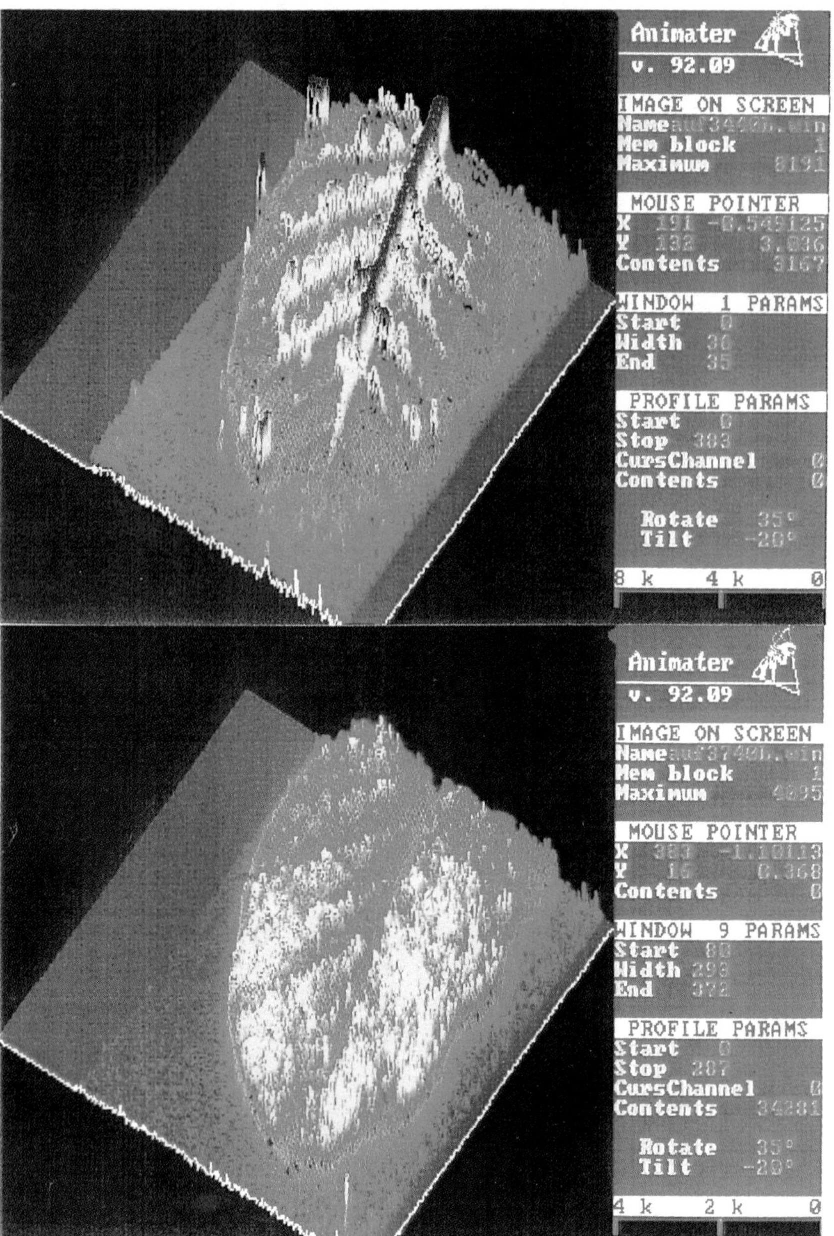

Fluoreszenzbildanalyse von Blättern (Fluorescence Imaging of Leaves):
Dreidimensionale Falschfarbendarstellung der Fluoreszenz-Intensität eines Tabakblattes.
(rot → gelb → grün → blau = abnehmende Fluoreszenz-Intensität)
(*Nicotiana tabacum* L., Aureamutante Su/su, Blattunterseite)

Fig. 4: Three-dimensional relief presentation of the inverse distribution of the blue fluorescence emission F440 (upper part) and the far-red chlorophyll fluorescence emission F740 (lower part) of the leaf surface (lower leaf-side) of green aurea tobacco Su/su. The fluorescence intensity given in false colours increases from blue via green and yellow to red as is also demonstrated by the height of the pixels.

consequence, the chlorophyll fluorescence ratio F690/F740 of the upper leaf half reflects in good approximation the total chlorophyll content of leaves and gradients within a leaf which is not the case for the fluorescence signals and ratios excited and sensed at the lower leaf half.

In the fluorescence ratio F690/F740 image (Fig. 5) of the lower leaf side of the aurea tabacco one sees values of 1.4 to 2.0 for this ratio which are unevenly distributed over the leaf surface. Higher values are found in the main and side leaf veins, which already visually seem to possess a lower chlorophyll content, but also near the left leaf edge, which in the lower leaf half of this C3-leaf of aurea tabacco apparently exhibits a lower chlorophyll content. Thus, chlorophyll fluorescence ratio images taken at the upper and the lower leaf side are a means to obtain information on the relative chlorophyll distribution in the upper and the lower leaf half, i. e. the palisade and spongy parenchyma cells. This is a further interesting application of the LIF images of plants.

Fluorescence ratio blue/red and blue/far-red

Due to the fact that a higher proportion of the blue-green fluorescence and lower amounts of chlorophyll fluorescence are emitted from the main and side leaf veins, the fluorescence ratios blue/red (F440/F690) and blue/far-red (F440/F740) exhibit significantly higher values in the leaf veins than in the vein-free leaf regions at the leaf edge (Fig. 5). This also applies to the ratios green/red and green/far-red as shown in Fig. 6 c and d. In the LIF ratio images the values of blue/red and green/red decline much more towards the leaf edge than the values of the ratios blue/far-red and green/far-red (Figs. 5 and 6). This indicates that the fluorescence ratios blue/red and green/red are more sensitive indicators of gradients or changes in fluorescence emission over the leaf surface than the fluorescence ratios blue/far-red and green/far-red.

Fluorescence ratios blue/green and green/blue

In the investigated aurea tobacco Su/su the blue fluorescence is generally higher (in particular in the leaf vein areas) than the green fluorescence emission as seen in Fig. 3 (compare F440 to F520). The LIF ratio image of blue/green (F440/F520) in Fig. 5 a shows that the blue fluorescence is particularly high in leaf vein regions compared to the green fluorescence. The question whether the intensity of the blue or the green fluorescence is higher in the vein-free leaf regions cannot sufficiently be answered, since the fluorescence images of Fig. 3 do not provide enough information. Both fluorescences showed up in the same range given in green colour of the colour scale. By forming the inverse LIF ratio image F520/F440 one realizes, however, that the vein-free leaf parts, which also possess the higher chlorophyll content of the leaf, exhibit a green fluorescence almost as high as the blue fluorescence, as presented in the LIF image in Fig. 6b by the red colour.

We have further checked the relative fluorescence intensities by separate determination of the fluorescence ratios blue/green and green/blue in different positions of the leaf surface, e.g. in the main and side leaf-veins as well as in the vein-free leaf parts (Fig. 7). The values of the fluorescence ratios F440/F520 and F520/F440 in the different leaf positions are given in Table 1. The highest value for F440/F520 of 2.4 was found in the isolated red spot «s» on the upper left part of the leaf in Fig. 7a, which coincided with the lowest value for the inverse ratio F520/F440 of 0.42 (green spot in the upper left part of the leaf in Fig. 7b). This spot visually exhibited the same chlorophyll content as compared to the other leaf parts, all of which were visually green. The values of the F440/520 ratio decreased from the main leaf veins (1.64 to 1.74) to the side leaf veins (1.62 to 1.40) via the inner predominantly vein-free leaf parts 9 and 10 (1.27 to 1.38) to the vein-free outer parts of the leaf (points 4 to 8 and 11) showing the lowest values (1.02 to 1.2).

This decrease in the F440/F520 values coincided with the increase in the value of the inverse ratio F520/F440 (Table 1), and reflected the gradient in the chlorophyll content per leaf area unit from the main leaf veins (with the lowest chlorophyll content: ca. 12 µg chl *a* + *b* per cm^2) to the vein-free outer parts of the leaf near the leaf edge with a higher chlorophyll content ranging from 15 to 18 µg chl *a* + *b* per cm^2. This gradient in the values of the fluorescence ratios blue/green and green/blue indicates that the intensity of the blue fluorescence decreases with increasing chlorophyll and carotenoid content of the different leaf positions. In fact, the emitted blue fluorescence (F430/F450) is increasingly re-absorbed by the protein-bound *in vivo* chlorophylls and carotenoids, which exhibit broad absorption maxima in the blue region between 430 to 470 nm (Lichtenthaler et al., 1981), as has been demonstrated before during the greening of etiolated wheat leaves (Stober and Lichtenthaler, 1993). The green fluorescence (F520 to F530) in turn, is much less affected by reabsorption of photosynthetic pigments. Therefore, its relative proportion increases from the leaf veins to the vein-free leaf parts with respect to the blue fluorescence, but never exceeds the blue fluorescence intensity.

However, one has to consider also another point, which is independent of differences in chlorophyll and carotenoid content. This is the fact that the main and side leaf veins of the lower leaf side of the aurea tobacco investigated here were standing out from the leaf surface. As a consequence, the blue and green fluorescences, which were reflected, refracted and scattered several or multiple times before they finally left the leaf epidermis, possess in leaf veins a much higher probability of leaving the leaf without much reabsorption by chlorophylls and carotenoids as compared to the blue-green fluorescence emitted in the predominantly vein-free intercostal leaf parts.

Table 1: Values of the fluorescence ratios blue/green (F440/F520) and green/blue (F520/F440) in different parts (as indicated by the numbers 1 to 14) of the lower leaf side of an aurea tobacco (Su/su) leaf. The ratio values are different in the main and side leaf veins from those in the vein-free leaf regions. The values are the mean (± standard deviation) of 160 pixels (positions 1, 2, 3, 6 and 7) or of 100 pixels (leaf positions 4, 5 and 8 to 14). The fluorescence ratios were taken in an elongated rectangle at each position. For the location of the leaf positions 1 to 14, see numbers in Fig. 7.

leaf position (numbered leaf points)	blue/green F440/F520	green/blue F520/F440
isolated spot «s» (upper left leaf part)	2.40 (±0.51)	0.42 (±0.15)
main leaf vein		
1	1.64 (±0.18)	0.59 (±0.05)
2	1.67 (±0.16)	0.60 (±0.06)
3	1.74 (±0.21)	0.56 (±0.07)
side leaf vein		
12	1.40 (±0.20)	0.71 (±0.09)
13	1.48 (±0.17)	0.69 (±0.08)
14	1.62 (±0.2)	0.61 (±0.07)
vein-free inner leaf parts		
7	1.26 (±0.13)	0.83 (±0.08)
9	1.27 (±0.14)	0.79 (±0.10)
10	1.38 (±0.15)	0.71 (±0.07)
vein-free outer leaf parts		
4	1.06 (±0.11)	0.93 (±0.10)
5	1.20 (±0.10)	0.82 (±0.08)
6	1.11 (±0.10)	0.91 (±0.07)
8	1.02 (±0.21)	0.98 (±0.19)
11	1.05 (±0.19)	0.96 (±0.19)

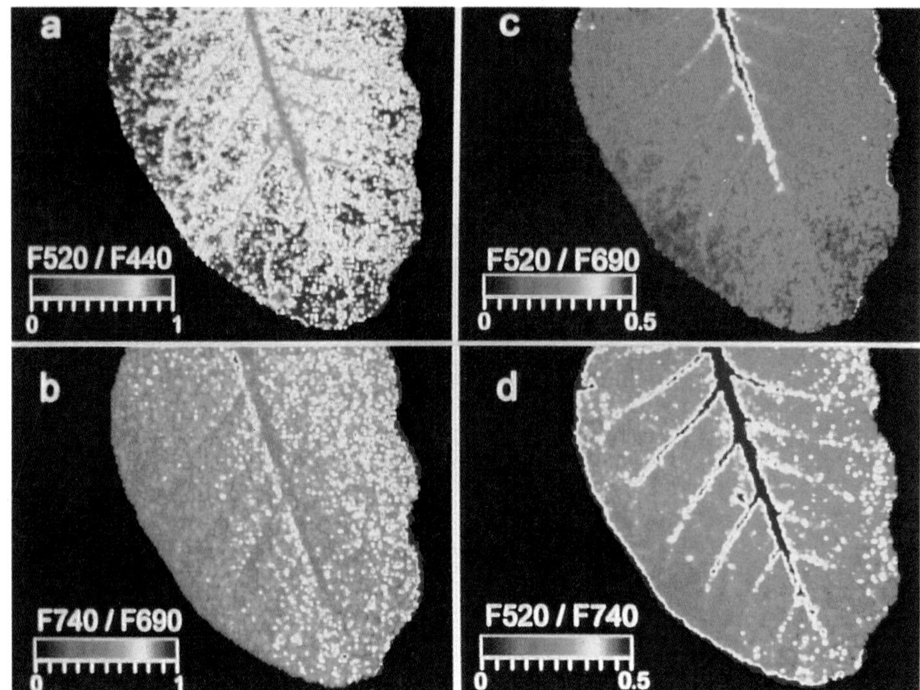

Fig. 6: Fluorescence ratio images (LIF ratio images) of the additional fluorescence ratios green/blue (F520/F440), green/red (F520/F690), green/far-red (F520/F740) and far-red/red (F740/F690) in an aurea tobacco Su/su leaf. The LIF ratio images are based on the fluorescences shown in Fig. 3. False colour presentation; the values increase from dark blue via green and yellow to red.

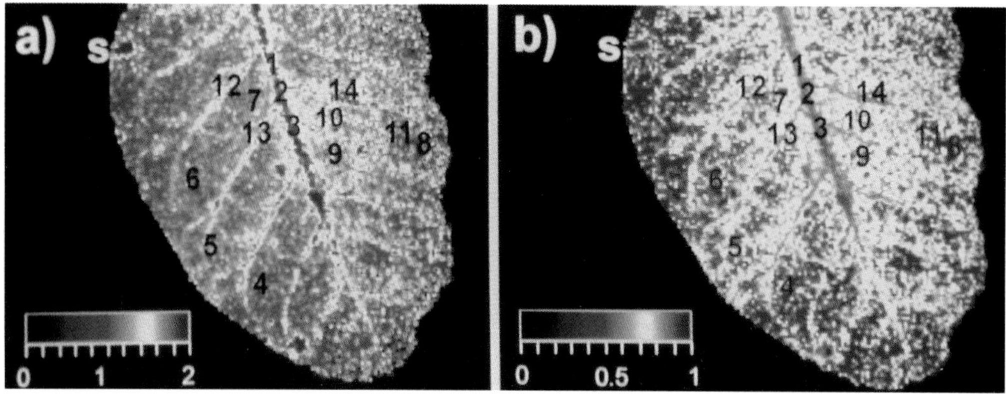

Fig. 7: Comparison of the inverse fluorescence ratio images of the blue/green (F440/F520) and the green/blue (F520/F440) fluorescence ratios in the lower leaf side of a green leaf of aurea tobacco Su/su. Presentation in false colours. The values increase from dark blue via green and yellow to red as highest values. The values of the fluorescence ratios, which differ considerably at the leaf positions 1 to 14, are shown in Table 1.

Thus, in the main and side leaf veins the blue and green fluorescence emission of the cell walls of the bundle sheet cells considerably contributes to the overall blue and green fluorescence emission of the leaf. In contrast, in the outer regions of the leaves, where the very thin bundle sheets are fully embedded in and surrounded by green mesophyll cells, the blue and green fluorescence emission is a lot lower. There, it mainly emanates from the cell walls of the leaf epidermis, whereas the blue and green fluorescence emitted by the cell walls of the thin bundle sheets and green mesophyll cells is reabsorbed by the chlorophylls and carotenoids, as has been shown before in wheat leaves by point data measurements (Stober and Lichtenthaler, 1993 c).

Evaluation by changing the sensitivity scale

It is possible that the false colour presentation of laser-induced fluorescence images (LIF images) or LIF ratio images may not show much contrast over the leaf surface within the lowest to highest value of either the fluorescence intensity or the fluorescence ratio. This is shown for the red/far-red fluorescence, which exhibits a fairly homogeneous green colour within the colour value scale from 0 to 4 (Fig. 8 a). By lowering the scale to the range of 0 to 2, the sensitivity of the false colour image of the ratio F690/F740 is increased for the ratio values in the range 0 to 2, whereas all values of 2 to 4 are no longer resolved and printed as red pixel points in the image Fig. 8 b). In a similar way, one can better visualize the gradient in the false colour LIF ratio image F440/F740 in Fig. 8 c by setting the colour scale from 0 to 0.5 instead of 0 to 1.0, which increases the recognition of the F440/F740 gradient in the leaf parts that are free of main and side leaf veins (Fig. 8 d as compared to Fig. 8 c).

This presentation demonstrates that also smaller differences in the gradient of either the fluorescence intensities or the LIF ratio images, which have not yet shown up in the false color image when using the full scale of values, can visually be resolved by changing the evaluation scale. This computer-aided fluorescence image processing provides various options for fine assessment of the fluorescence gradients over the leaf surface including the biomonitoring of small differences in spot-size leaf parts.

LIF images of diuron uptake

The herbicide diuron (DCMU) is known to efficiently block the photosystem II catalyzed photosynthetic electron transport by binding to the Q_B-binding site instead of the endogenous Q_B compound plastoquinone-9. As a consequence the photosynthetic quantum conversion of absorbed light energy declines giving rise to a considerable increase in the red and far-red chlorophyll fluorescence emission as has been demonstrated before by point-data fluorescence measurements (Lichtenthaler, 1988; Lichtenthaler and Pfister, 1978; Lichtenthaler and Rinderle, 1988 a; Szigeti et al., 1996). In order to obtain a more realistic view of the gradual uptake of the herbicide diuron and its progressive inhibition of photosynthetic quantum conversion, we took UV-laser-induced fluorescence images of diuron-treated and untreated leaf halves of the green tobacco variety su/su.

Diuron solution (5×10^{-5} M) in water with 1 % ethanol was applied via a hair pin to the lower leaf side of the left leaf-half of a ca. 20 cm long, attached, green tobacco leaf, whereas the right leaf-half represented the untreated control. The fluorescence was excited and screened on the upper leaf side. 10 min after diuron treatment the red (F690) and far-red chlorophyll bands (F740) had significantly increased in the diuron-treated leaf half and rose much further after 30 min of diuron treatment. In contrast, in the non-treated right leaf-half the red and far-red chlorophyll fluorescence remained very low as before the start of the treatment. The LIF images clearly demonstrated that the herbicide diuron was not homogeneously taken up by and distributed in the green mesophyll cells. In fact, several small leaf spots, e.g. on the left leaf edge (green areas in Fig. 9) did not exhibit any increase in chlorophyll fluorescence, whereas in other leaf areas the increase was lower (yellow regions in the left leaf half) than in those with the highest chlorophyll fluorescence (red regions in left leaf half of Fig. 9). Thus, LIF images show a certain leaf patchiness concerning the uptake of diuron and inhibition of the photosynthetic electron transport, which could not be detected by the point-data fluorescence measurement applied so far. In addition, the LIF images indicate that the diuron applied via the lower leaf side of the left leaf half had not been transported to the untreated right leaf half.

These LIF imaging data demonstrate that fluorescence images can routinely be applied in studies on the uptake and penetration speed as well as on transport and translocation of herbicides and other chemicals affecting photosynthesis. The blue and green fluorescence emission of the diuron-treated leaf half did not increase or hardly increased (up to 10 %) as compared to the untreated control. The blue-green fluorescence can therefore be used as internal standard for the rise in chlorophyll fluorescence. As a consequence the fluorescence ratios blue/red (F440/F690) and blue/far-red (F440/F740) decreased by four times, whereas the chlorophyll fluorescence ratio red/far-red (F690/F740) significantly rose only by ca. 30 % in the diuron-inhibited leaf regions as shown in Fig. 10. The LIF ratio blue/green (F440/F520) did not change after diuron treatment as would have been expected since the blue and green fluorescence did not change or hardly changed upon diuron treatment. This demonstrates that the changes in UV-laser induced blue, green, red and far-red fluorescence can be quantified by forming the LIF ratio images, whereby the LIF ratio images blue/red (F440/F690) and blue/far-red (F440/F740) were very sensitive even to small changes in chlorophyll fluorescence emission, and are thus very suitable indicators of changes in the photosynthetic quantum conversion and electron transport rates. These LIF ratios have also been found to be valid indicators of other stress factors affecting the photosynthetic apparatus (Lang et at., 1996). The LIF ratios green/red (F520/F690) and green/far-red (F520/F740) showed equivalent changes as did the LIF ratios blue/red and blue/far-red, and are also very sensitive and suitable indicators of herbicide treatment or stressors affecting photosynthetic function.

Fluorescence emission of variegated leaf tissues

There exist many plants with either white-green or yellow-green variegated leaves. These plants contain different plastid forms, some of which form chlorophylls and carotenoids and are functional chloroplasts, whereas others are only able to form some yellow carotenoids (as found in the yellow leaf parts of yellow-green variegated leaves), and again others are unable to accumulate carotenoids and chlorophylls as the colourless leucoplasts in the white leaf parts of white-green variegated plants. During the numerous cell divisions in the course of plant and leaf development the different plastid forms can become de-mixed. In the white-green variegated plants this eventually gives rise to white leaf cell lines and tissues (vertical stripes) which possess leucoplasts only and are unable to accumulate chlorophylls and carotenoids.

The epidermis cells of most higher plants contain leucoplasts and are free of chlorophylls and carotenoids except for the stomata cells possessing chloroplasts. This is shown in a photo taken by a fluorescence microscope of the epidermis strip isolated from *Commelina,* in which case the chloroplasts of stomata cells fluoresce red, whereas in the other leucoplast-containing epidermis cells one can only detect the blue-green

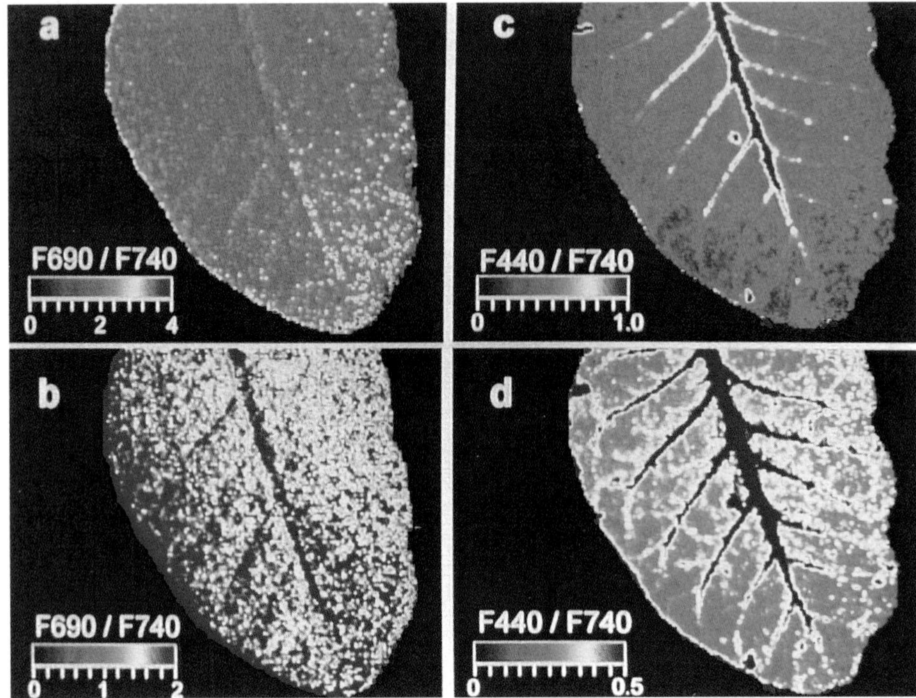

Fig. 8: Differences in the false colour presentation of the LIF ratio images red/far-red (F690/F740) and blue/far-red (F440/F740) shown in a) and b) by changing the sensitivity of the false colour scale by suppressing the higher ratio values from 2 to 4 for F690/F740 and from 0.5 to 1.0 for F440/F740.

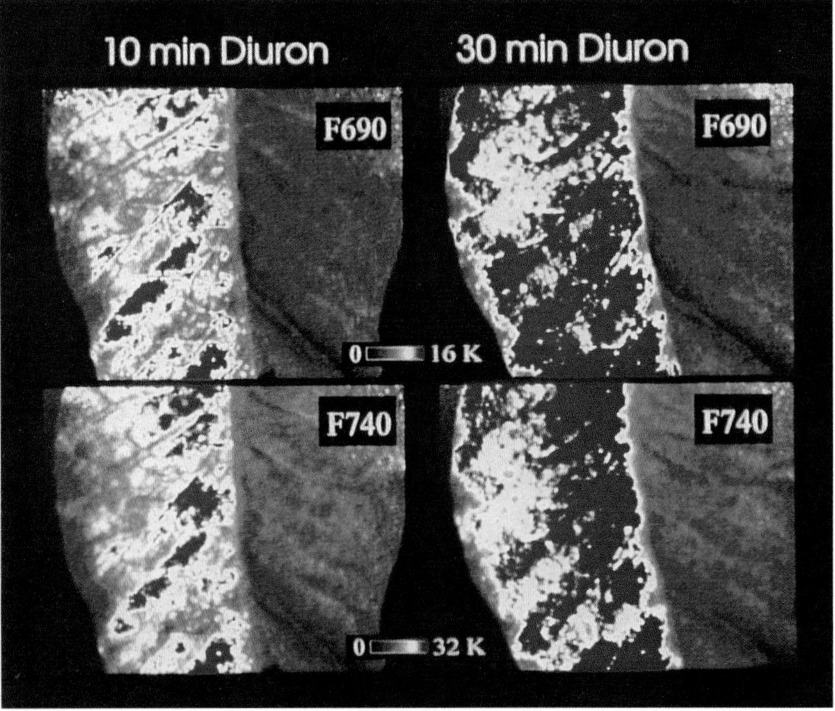

Fig. 9: Chlorophyll fluorescence imaging of the time-dependent uptake of the photosystem II inhibiting herbicide diuron (DCMU) in the left part of a leaf of green tobacco. False colour presentation. The intensity of the chlorophyll fluorescence increases from dark blue *via* green and yellow to red as highest intensity.

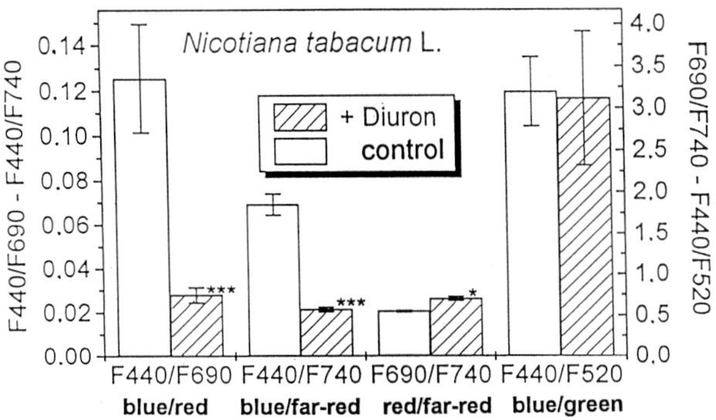

Fig. 10: Significant changes in the fluorescence ratios blue/red (F440/F690), blue/far-red (F440/F740) and the chlorophyll fluorescence ratio red/far-red (F690/F740) in a diuron-treated leaf half of green tobacco compared to an untreated leaf half. The ratios were directly read from LIF ratio images with 300 pixels each. The ratios marked by asterisks are significantly different from the controls: ***highly significant ($p<0.001$) and * significant ($p<0.05$).

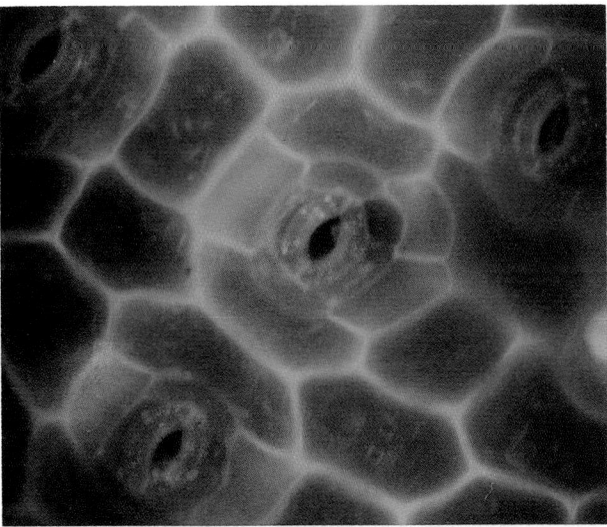

Fig. 11: UV-fluorescence microscope photo of blue cells of an epidermis stripped off from a leaf of *Commelina communis* L. The epidermal cell walls exhibit a blue-green fluorescence emission, and the chloroplasts in the stomatal guard cells exhibit a red chlorophyll fluorescence (excitation: UV-radiation) (Courtesy: Ya'acov Leshem, Ramat Gan, Israel).

fluorescence emission of the epidermis cell walls (Fig. 11). Also, the epidermal cells of the white leaf parts of the variegated *Campelia* leaves investigated here exhibited chloroplasts. In the fully green parts of *Campelia* leaves the blue-green fluorescence emission was significantly lower (−33 %), whereas the red and far-red chlorophyll fluorescence emission was expectedly much higher than in the white leaf parts as shown in Fig. 12. The maxima of the red and far-red chlorophyll fluorescence in green leaf parts were found at 686 and 740 nm, which correspond to the usual terminology of fluorescence bands red (F690) and far-red (F740). The fluorescence ratio red/far-red amounted to a value of 0.5, which is found in many green leaves. Although the white leaf parts did not visually contain chlorophylls, some very small amounts are found in the epidermal stomata cells containing chloroplasts. However, these extremely low chlorophyll amounts might not solely account for the low red- and far-red fluorescence emission in the white leaf parts. Since the red fluorescence emission maximum has been shifted to shorter wavelengths (676 nm) (see Fig. 11) in the white leaf parts as compared to green leaf parts (684 to 690 nm depending on the chlorophyll content of the leaf tissues), we assume that the mesophyll cells in the white leaf parts of variegated *Campelia* contain some traces of non-functionally bound chlorophyll *a*, which emits this 676 nm fluorescence.

Screening of fluorescence profiles from LIF images

From the white-green variegated *Campelia* leaves we took UV-laser induced fluorescence images using the Karlsruhe/Strasbourg fluorescence imaging system. Similar to the fluorescence emission spectra shown in Fig. 12 we found in the LIF images of the white leaf parts a blue fluorescence (red regions in the F440 image) and a green fluorescence yield (yellow regions in the F520 image) about two times higher than in the completely green leaf parts (Fig. 13). In contrast, the red chlorophyll fluorescence (green-yellow colour in the F690 image) and the far-red chlorophyll fluorescence (red-yellow colour in the F740 image) were high in the green leaf parts, but were extremely low and hardly to be detected in the fully white leaf parts of this variegated *Campelia* plant as seen in the F690 and F740 leaf image (Fig. 13).

The computer-aided fluorescence data processing permitted to measure horizontal profiles of the fluorescence emission across the leafs transverse axis. The profiles for the blue (F440), green (F520), red (F690) and far-red (F740) fluorescences of the variegated *Campelia* leaf are shown in Fig. 14. The negative contrast of blue (F440) and far-red fluorescence (F740) emissions (Fig. 14 A) as well as of green (F520) and red (F690) fluorescence emission (Fig. 14 B) is evident from these fluorescence profiles. The red and far-red chlorophyll fluorescences exhibit the same distribution pro-

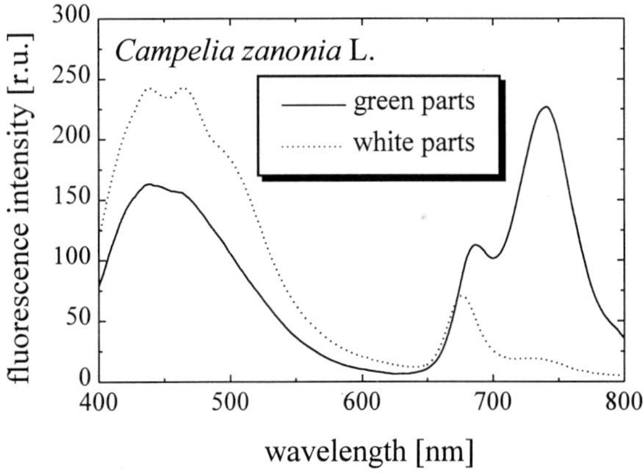

Fig. 12: Fluorescence emission spectra of a green and a white leaf region of the upper leaf side of a variegated green-white leaf of *Campelia zanonia* L. (Excitation 355 nm; measurements with the LS-50 Perkin-Elmer spectrofluorometer).

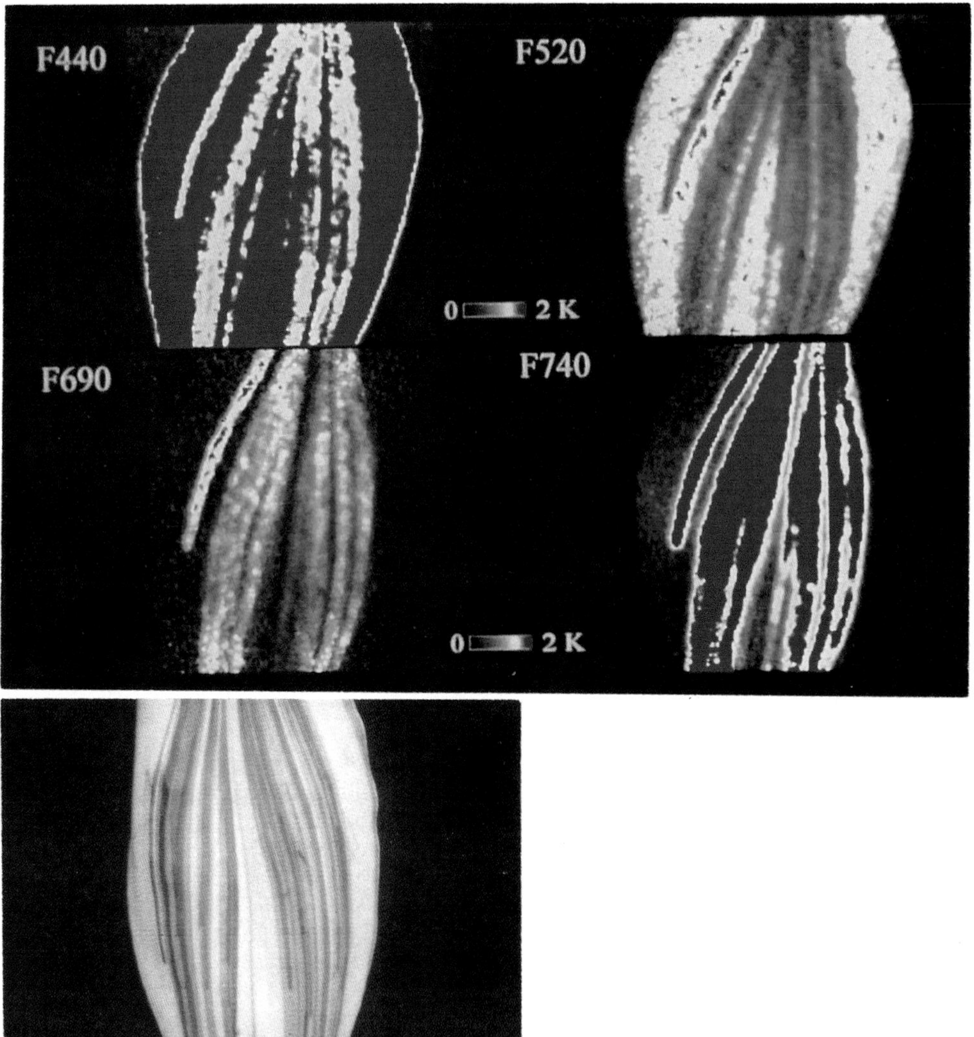

Fig. 13 A: UV-laser induced fluorescence images (LIF-images) of the blue (F440) and green fluorescence (F520) as well as the red (F690) and far-red chlorophyll fluorescence (F740) and **B:** a colour photo of the white-green leaf of the variegated plant *Campelia zonania* L. The LIF-images are presented in false colours from dark blue (very low intensity) via green and yellow to red (as highest fluorescence intensity). Note that the chlorophyll fluorescences F690 and F740 are only emitted from the green leaf stripes, whereas the broad white stripes on both leaf edges are not visible in the chlorophyll fluorescence images.

file. They are very low at white leaf parts, and increase at leaf parts which are either light green or completely green depending on the chlorophyll content. The chlorophyll fluorescence profiles thus fully reflected the chlorophyll distribution over the transverse leaf axis, and again demonstrated the uneven chlorophyll fluorescence emission of green leaf parts with a different chlorophyll content (compare colour photo of the leaf in Fig. 13 B).

The fluorescence profiles also document that the blue and green fluorescences, which leave the leaf epidermis, are much lower in the green leaf regions than in the white leaf regions. This is due to the partial reabsorption of the emitted blue-green fluorescence by the blue and green *in vivo* absorption bands of the different chlorophyll carotenoid pigment proteins of the thylakoids in the chloroplasts of the mesophyll cells (for absorption bands of these pigment proteins confer Lichtenthaler et al., 1981). Here, one has to take into account that the fluorescence is emitted into all directions and is reflected and refracted several times before it is finally emanated from the leaf epidermis. The reabsorption probability of the blue-green fluorescence in the green leaf parts is thus much higher than in the white leaf parts. A second point has to be considered in this respect, too: Since a major part of the applied UV excitation radiation (λ 355 nm) is absorbed by the photosynthetic pigments, less UV-quanta are available for the excitation of the blue-green fluorescence in the cells of the green leaf parts, and thus, this results in a lower blue-green fluorescence yield. In white leaf parts, however, the blue-green fluorescence not only comes from the epidermis cells but also from the cell walls of the mesophyll cells, which considerably contribute to the overall blue-green fluorescence emission of the white leaf parts. In contrast, in the greenish

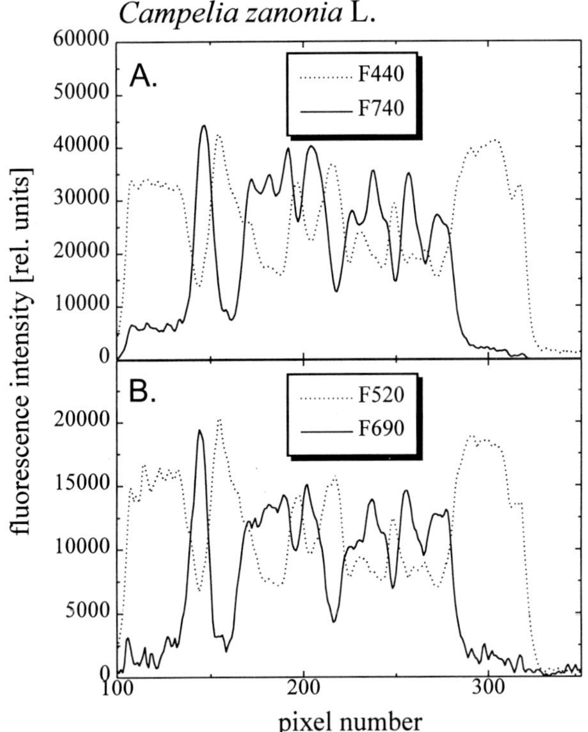

Fig. 14: Horizontal profile of the fluorescence intensities across the leaf's transverse axis from left to right of a white and green variegated leaf of *Campelia zanonia* L. Presentation of the inverse distribution of **A**) the blue fluorescence (F440) and far-red chlorophyll fluorescence (F740) as well as **B**) the green fluorescence (F520) and the red chlorophyll fluorescence (F690) over the transverse leaf axis. The fluorescences were measured of the upper leaf side from the left white leaf edge to the right white leaf edge over the light green, dark green and white leaf stripes (cf. Fig. 13 B). The measurements were performed with the Karlsruhe/Strasbourg fluorescence imaging system (Excitation: UV-laser Nd-YAG, 355 nm).

or completely green leaf parts of variegated *Campelia* the blue-green fluorescence emission primarily emanates from the cell walls of the leaf epidermis.

The fluorescence profile analysis of LIF images demonstrates the high suitability of this novel technique of leaf fluorescence imaging. Such an accuracy and resolution in the fluorescence profile distribution of leaves, as shown here for the first time, could not be determined so far, since the conventional point-data fluorescence measurements using a spectrofluorometer do not allow to monitor such subtle differences in fluorescence emission, even when various leaf points had been measured. This again shows the superiority of this novel fluorescence imaging technique.

Conclusion

UV-laser induced, high resolution fluorescence images of complete leaf surfaces provide for the first time precise and detailed information on the differential fluorescence distribution across the transversal and longitudinal leaf axis. LIF images are much superior to the point data measurements applied so far with conventional spectrofluorometers. The LIF images not only allow to screen the inverse emission and negative contrast of the blue-green fluorescence on one hand and the red and far-red chlorophyll fluorescence on the other hand, but also to monitor and detect small spot-sized differences and inhomogeneities in fluorescence emission as well as gradients from the leaf edge to the center part of the leaf. These differences in fluorescence emission can be quantified by forming the fluorescence ratios blue/red, blue/far-red, red/far-red and blue/green.

Impairment of photosynthetic quantum conversion by diuron application or short-term heat stress can easily be monitored and quantified via the LIF ratio images blue/red (F440/F690) and blue/far-red (F440/F740), which are very sensitive to changes or disturbances in the photosynthetic apparatus. Likewise also the LIF ratio images green/red and green/far-red can be applied, which provide similar information and are also very sensitive to environmental changes, or to a decline in photosynthetic quantum conversion like the LIF ratios blue/red and blue/far-red. With its high resolution, its large area screening possibilities and the multiple fluorescence data processing methods (LIF ratio images, LIF profile measurements, LIF histogramme plotting) as well as the detection of smalt spot-size differences in fluorescence emission or of fluorescence gradients over the leaf surface, the novel UV-A laser-induced fluorescence imaging technique provides a new dimension for stress detection in plants and terrestrial vegetation. LIF-images and LIF-ratio images can also be sensed from a far distance. Thus, the high resolution LIF-imaging technique presented here will further develop into a powerful technique in remote sensing of the state of health of the photosynthetic apparatus and of whole plants and canopies.

Acknowledgements

We wish to express our thanks to Gabrielle Johnson for checking the English text, and Dr. Ya'acov Leshem for providing Fig. 11.

References

Babani, F., H. K. Lichtenthaler, and P. Richter: Changes in different chlorophyll fluorescence signatures during greening of etiolated barley seedlings as measured with the CCD-OMA fluorometer. J. Plant Physiol. *148*, 471–477 (1996).

Buschmann, C. and H. K. Lichtenthaler: Reflectance and chlorophyll fluorescence signatures of leaves. In: Lichtenthaler, H. K. (ed.): Applications of Chlorophyll Fluorescence, pp. 325–332. Kluwer Academic Publishers, Dordrecht (1988).

Buschmann, C., P. Richter, and H. K. Lichtenthaler: Application of the Karlsruhe CCD-OMA LIDAR-fluorosensor in stress detection of plants. J. Plant Physiol. *148*, 548–554 (1996).

Chappelle, E. W., F. M. Wood, J. E. McMurtrey, and W. W. Newcomb: Laser-induced fluorescence of green plants. 1: A technique for remote detection of plant stress and species differentiation. Applied Optics *23*, 134–138 (1984).

D'Ambrosio, N., K. Szabó, and H. K. Lichtenthaler: Increase of the chlorophyll fluorescence ratio F690/F735 during the autumnal chlorophyll breakdown. Radiat. Environ. Biophys. *31*, 51–62 (1992).

Govindjee: Sixty-three years since Kautsky: Chlorophyll *a* fluorescence. Aust. J. Plant Physiol. *22*, 131–160 (1995).

Hák, R., H. K. Lichtenthaler, and U. Rinderle: Decrease of the chlorophyll fluorescence ratio F690/F730 during greening and develpoment of leaves. Radiat. Environ. Biophys. *29*, 329–336 (1990).

Harris, P. J. and R. D. Hartley: Detection of bound ferulic acid in cell walls of the gramineae by ultraviolet fluorescence microscopy. Nature *259*, 508–510 (1976).

Kautsky, H. and A. Hirsch: Neue Versuche zur Kohlenstoffassimilation. Naturwissenschaften *19*, 964 (1931).

Kocsanyi, L., M. Haitz, and H. K. Lichtenthaler: Measurement of laser-induced chlorophyll fluorescence kinetics using a fast acousto optic device. In: Lichtenthaler, H. K. (ed.): Applications of Chlorophyll Fluorescence, pp. 99–107. Kluwer Academic Publishers, Dordrecht (1988).

Krause, G. H. and E. Weis: The photosynthetic apparatus and chlorophyll fluorescence: an introduction. In: Lichtenthaler, H. K. (ed.): Applications of Chlorophyll Fluorescence, pp. 3–11. Kluwer Academic Publishers, Dordrecht (1988).

– – Chlorophyll fluorescence and photosynthesis: the basics. Ann. Rev. Plant Physiol. Plant Mol. Biol. *42*, 313–349 (1991).

Lang, M. and H. K. Lichtenthaler: Changes in the blue green and red fluorescence emission spectra of beech leaves during the autumnal chlorophyll breakdown. J. Plant Physiol. *138*, 550–553 (1991).

Lang, M., F. Stober, and H. K. Lichtenthaler: Fluorescence emission spectra of plant leaves and plant constituents. Rad. Environ. Biophysics *30*, 333–347 (1991).

Lang, M., H. K. Lichtenthaler, M. Sowinska, P. Summ, and F. Heisel: Blue, green and red fluorescence signatures and images of tobacco leaves. Botanica Acta *107*, 230–236 (1994).

Lang, M., H. K. Lichtenthaler, M. Sowinska, F. Heisel, J. A. Miehé, and F. Tomasini: Fluorescence imaging of water and temperature stress in plant leaves. J. Plant Physiol. *148*, 613–621 (1996).

Lichtenthaler, H. K.: Laser-induced Chlorophyll Fluorescence of Living Plants. In: Proceed. of the Remote Sensing Symposium (IGARSS) Zürich 1986, Vol. III. pp. 1571–1579. ESA Publications Division, Noordwijk (1986).

– Chlorophylls and carotenoids: pigments of photosynthetic biomembranes. Methods in Enzymology *148*, 350–382 (1987).

– *In vivo* chlorophyll fluorescence as a tool for stress detection in plants. In: Lichtenthaler, H. K. (ed.): Applications of Chlorophyll Fluorescence, pp. 143–149. Kluwer Academic Publishers, Dordrecht (1988).

– The Kautsky effect: 60 years of chlorophyll fluorescence induction kinetics. Photosynthetica *27*, 45–55 (1992).

– Vegetation stress: an introduction to the stress concept in plants. J. Plant Physiol. *148*, 4–14 (1996).

Lichtenthaler, H. K. and K. Pfister: Praktikum der Photosynthese. Quelle & Meyer, Heidelberg/Wiesbaden (1978).

Lichtenthaler, H. K. and U. Rinderle: The role of chlorophyll fluorescence in the detection of stress conditions in plants. CRC Critical Reviews in Analytical Chemistry *19*, Suppl. I, 29–85 (1988 a).

– – Chlorophyll fluorescence spectra of leaves as induced by blue light and red laser light. Proceed. 4th International Colloquium on Spectral Signatures of Objects in Remote Sensing, Aussois 1988, pp. 251–254. ESA Publications Division, Noordwijk (1988 b).

Lichtenthaler, H. K. and M. Lang: Biomonitoring of terrestrial vegetation by laser-induced fluorescence and reflectance measurements within the EUREKA Programme LASFLEUR (EU 380). In: Richter, P. I. and R. C. Herndon (eds.): Proceed. of the 2nd International Symposium and Exhibition on Environmental Contamination in Central and Eastern Europe, Budapest 1994, pp. 112–114. Government Institutes, Inc., Rockville, Maryland 20850, USA (1995).

Lichtenthaler, H. K., R. Hák, and U. Rinderle: The chlorophyll fluorescence ratio F690/F730 in leaves of different chlorophyll content. Photosynth. Research *25*, 295–298 (1990).

Lichtenthaler, H. K., G. Burkhard, G. Kuhn, and U. Prenzel: Light-induced accumulation and stability of chlorophylls and chlorophyll-proteins during chloroplast development in radish seedlings. Z. Naturforsch. *36c*, 421–430 (1981).

Schmid, G. H.: Origin and properties of mutant plants: Yellow tobacco. Methods in Enzymology *23*, 171–194 (1971).

Schmuck, G. and H. K. Lichtenthaler: Application of laser-induced chlorophyll *a* fluorescence in the forest decline research. In: Proceedings of the Remote Sensing Symposium (IGARSS) Zürich 1986, Vol. *III*, pp. 1587–1590. ESA Publications Division, Noordwijk (1986).

Stober, F. and H. K. Lichtenthaler: Studies on the localization and spectral characteristics of the fluorescence emission of differently pigmented wheat leaves. Botanica Acta *106*, 365–370 (1993 a).

– – Characterisation of the laser-induced blue, green and red fluorescence signatures of leaves of wheat and soybean leaves grown under different irradiance. Physiologia Plantarum *88*, 696–704 (1993 b).

– – Studies on the constancy of the blue and green fluorescence yield during the chlorophyll fluorescence induction kinetics (Kautsky effect). Radiat. Environ. Biophysics *32*, 357–365 (1993 c).

Stober, F., M. Lang, and H. K. Lichtenthaler: Studies on the Blue, Green and Red Fluorescence Signatures of Green, Etiolated and White Leaves. Remote Sensing of the Environment *47*, 65–71 (1994).

Szabó, K., H. K. Lichtenthaler, L. Kocsanyi, and P. Richter: A CCD-OMA device for the measurement of complete chlorophyll fluorescence emission spectra of leaves during the fluorescence induction kinetics. Radiat. Environm. Biophysics *31*, 153–160 (1992).

Szigeti, Z., P. Richter, and H. K. Lichtenthaler: Fluorescence emission spectra of paraquat resistant *Conyza canadensis* during the chlorophyll fluorescence induction as determined by the CCD-OMA system. J. Plant Physiol. *148*, 574–578 (1996).

J. Plant Physiol. Vol. 148. pp. 613–621 (1996)

Fluorescence Imaging of Water and Temperature Stress in Plant Leaves*

MICHAEL LANG[1], HARTMUT K. LICHTENTHALER[1], MALGORZATA SOWINSKA[2], FRANCINE HEISEL[2], and JOSEPH A. MIEHÉ[2]

[1] Botanisches Institut, Universität Karlsruhe, Kaiserstr. 12, D-76128 Karlsruhe, Germany

[2] Groupe d'Optique Appliquée, Centre de Recherches Nucléaires, 23 rue de Loess, F-67037 Strasbourg Cedex 2, France

Received July 12, 1995 · Accepted September 5, 1995

Summary

Fluorescence images of leaves from tobacco plants (green wild type and aurea mutant) were determined in the blue (F440), green (F520), red (F690) and far-red region (F740), and also expressed as fluorescence ratio images. Under long-term water stress tobacco plants initially showed constant ratios of the blue to red fluorescence (F440/F690) and the blue to far-red fluorescence (F440/F740). Below a distinct threshold in water content (84% in green and 88% in aurea tobacco), however, a linear increase of the fluorescence ratios blue/red and blue/far-red was observed. This was due to a distinct increase in the bluegreen fluorescence emission, whereas the red and far-red chlorophyll fluorescence increased to a lower proportion. These changes in fluorescence ratios could easily be monitored by high resolution fluorescence imaging of whole leaves. For each point of the leaf, the fluorescence ratio can be read from the fluorescence ratio images of the leaves. In contrast, a short-term heat plus water stress in green tobacco plants was very fast detected via fluorescence imaging as a significant increase of red and far-red chlorophyll fluorescence emission (F690 and F740) on the leaf rim, whereas the central part of the leaf still exhibited the regular fluorescence signatures of photosynthetically active leaves. A combined outdoor stress (light, heat and water stress) at a dry sunny summer period was detected in *Rhododendron* by fluorescence imaging due to a much reduced red and far-red chlorophyll fluorescence. The latter was caused by UV-absorbing substances (e.g. flavonols) which accumulated primarily in the epidermis of these stressed leaves. These compounds seemed to act as UV-radiation filter, thus reducing the amount of the UV-excitation radiation, which could penetrate the mesophyll and which resulted in a reduced chlorophyll fluorescence excitation and emission. These results demonstrate that fluorescence imaging of leaves in the blue, green, red and far-red emission bands is an excellent tool for an early stress detection in plants, which is much superior to the hitherto applied spectral point data measurements.

Key words: Blue-green fluorescence, chlorophyll fluorescence, fluorescence emission spectra, stress detection, fluorescence ratios blue/red and blue/far-red.

Abbreviations: F450 = blue fluorescence; F530 = green fluorescence; F690 = red chlorophyll fluorescence; F740 = far-red chlorophyll fluorescence; F440/F520 = ratio of blue to green fluorescence; F440/F690 = fluorescence ratio blue/red; F440/F740 = fluorescence ratio blue/far-red; F690/F740 = chlorophyll fluorescence ratio red/far-red.

* This paper is dedicated to Prof. Dr. Eberhard Schnepf, Heidelberg, on the occasion of his 65th birthday.

Introduction

In their natural environment plants are often exposed to stress conditions (cf. to Larcher, 1984 and 1987; Lichtenthaler, 1988 and 1996; Strasser, 1988). Examination of plant stress by applying techniques of chlorophyll fluorescence spectroscopy is well established (Lichtenthaler, 1987 a and 1988; Lichtenthaler and Rinderle, 1988; Strasser et al., 1988; Schreiber et al., 1988). When excited by UV-A or UV-B radiation, green leaves exhibit a fluorescence emission with maxima in the blue (F440), green (F520), red (F690) and far-red (F740) range (Chappelle et al., 1984; Goulas et al., 1990; Lang and Lichtenthaler, 1991; Lichtenthaler et al., 1992; Stober and Lichtenthaler, 1993). It has been estimated that more than 90 % of the ultraviolet radiation falling on a leaf is absorbed by the chlorophyll-free epidermis (Caldwell et al., 1983). Bluegreen fluorescing plant phenolics are covalently bound to the cell walls (Harris and Hartley, 1976), also in the epidermis walls, which are the main source of bluegreen fluorescence emission of leaves (Stober and Lichtenthaler, 1993 b). Removal of the epidermis results in an increase of the UV-induced green, red and far-red fluorescence emission (Lang et al., 1991; Lang 1995) indicating the UV filter effect of the epidermis. Without the epidermis the exciting UV-radiation can directly pass into the leaf mesophyll cells and excite the red and far-red chlorophyll fluorescence in the mesophyll. These results also demonstrated that the mesophyll essentially contributes to the green fluorescence of leaves as is also assumed by Cerovic et al. (1994). The contribution of the mesophyll cells to the blue fluorescence emission of the intact leaf, in turn, is negliable (Lang et al., 1992; Cerovic et al., 1993).

The conventional laser-induced fluorescence (LIF) emission spectra (LIF spectra) only allowed point measurements of randomly selected distinct leaf parts. The advantage of point measurements is that they provide information on the whole fluorescence spectra including position and intensity of fluorescence maxima. The great disadvatage of such punctuated measurements is the fact that neither local fluorescence differences nor fluorescence gradients over the whole leaf surface can easily be detected, since one leaf part only yields one spectral information. In contrast, the high resolution fluorescence imaging system, developed for near and far remote sensing of vegetation within the EUREKA project no. 380 LASFLEUR (Lang et al., 1994; Lichtenthaler and Lang, 1995 b), provides fluorescence information on the spatial resolution of the distinct fluorescence bands in the blue (F440), green (F520), red (F690) and far-red region (F740) over the whole leaf surface (Lang et al., 1994; Lichtenthaler et al., 1996). The information obtained by high resolution fluorescence imaging is of much higher quality and superior to the hitherto applied sensing of fluorescence emission spectra of particular leaf points.

There exists an inverse relationship between total chlorophyll content and the chlorophyll fluorescence ratio red/far-red (F690/F740) due to the partial overlapping of the red chlorophyll absorption band between 660 and 685 nm with the red chlorophyll fluorescence emission band in the 690 nm region. Thus, changes of the *in vivo* chlorophyll content can be monitored via the chlorophyll fluorescence ratio red/far-red F690/F740 (Lichtenthaler, 1987 a; Lichtenthaler and Rinderle, 1988; Hák et al., 1990; D'Ambrosio et al., 1992). The chlorophyll fluorescence ratio F690/F740 can also be sensed by high resolution fluorescence imaging over the whole leaf surface (Lang, 1995; Lichtenthaler et al., 1996).

Similar to the chlorophyll fluorescence, the bluegreen fluorescence emission of leaves is reduced by a partial reabsorption of the emitted fluorescence by the chlorophylls and carotenoids, which possess bluegreen absorption bands. Consequently, green tobacco leaves show lower values for the fluorescence ratios blue/red (F440/F690) and blue/far-red (F440/F740) than the chlorophyll-poor aurea-mutant of tobacco, which contains less photosynthetic pigments (Santrucek et al., 1992; Lang et al., 1994). The values of the fluorescence ratios blue/red and blue/far-red, which are increased under drought stress in maize or wheat (Dahn et al., 1992), are higher in field plants than greenhouse plants (Stober and Lichtenthaler, 1993). In contrast, treatment with the photosynthesis herbicide diuron (DCMU) decreased the fluorescence ratios blue/red and blue/far-red due to an increased chlorophyll fluorescence emission at the herbicide-inhibited photosynthetic quantum conversion (Edner et al., 1995). Testing the significance of the fluorescence ratios blue/red and blue/far-red under defined plant stress situations via the newly developed fluorescence imaging is the objective of the present work. Changes in the fluorescence ratios blue/red (F440/F690) and blue/far-red (F440/F740) of tobacco leaves in the course of long-term water stress and a short-term heat stress were studied in tobacco as well as a combined light, heat and water stress in *Rhododendron*.

Materials and Methods

Plants

Plants of a green tobacco and its aurea mutant (*Nicotiana tobacum* L.; Schmidt, 1971; Santrucek et al., 1992) were cultivated in the greenhouse of the Botanical Garden (University of Karlsruhe) at a quanta fluence rate of 600 µmol photons $m^{-2} s^{-1}$, a temperature of approximately 25 °C and 60 % rel. humidity.

Rhododendron spec. was grown in a private garden at the Turmberg-site of Karlsruhe.

Determination of pigments and water content

Photosynthetic pigments were determined spectrophotometrically using the redetermined pigment coefficients of Lichtenthaler (1987 b). The level of total flavonol of *Rhododendron* leaves was determined spectrophotometrically in methanol after a $AlCl_3$-induced bathochromic shift of the absorption maximum to 405 nm using a rutin (quercetin-3-rhamnoglucoside) standard.

Water content of leaves was determined by drying the leaves in an oven at 95 °C for 3 h. It was calculated as [(fresh weight − dry weight)/(fresh weight)] * 100 %.

Fluorescence spectroscopy

A Perkin-Elmer LS-50 Luminescence Spectrometer (Perkin-Elmer, Überlingen, Germany) was applied for measuring fluorescence emission spectra. Leaf samples (ca. 1 × 2 cm) were pre-illuminated for 5 min in the spectrofluorometer with red light (660 nm, ca.

500 µmol $m^{-2} s^{-1}$) in order to achieve the steady state of the chlorophyll fluorescence induction kinetic (Kautsky effect, as reviewed in Lichtenthaler, 1992). Fluorescence emission spectra of a leaf area of ca. 0.35 cm^2 were excited with UV-A radiation of 355 nm, which corresponded to the 3rd harmonic of the Nd:YAG-laser emission which has been applied as excitation source in the fluorescence imaging system (see below). The slits of excitation and emission of the monochromator of the Perkin-Elmer were adjusted to a spectral bandwidth of 15 nm. Excitation radiation was filtered using a UV-bandpass filter (center wavelength 340 nm, full-width at half maximum (FWHM) 70 nm; Schott, Mainz, Germany) in order to reduce stray light. The UV-A induced fluorescence emission spectra of leaves were recorded separately for the bluegreen fluorescence from 400 to 600 nm applying a 390 nm cut-off filter (Perkin-Elmer, Überlingen, Germany) and for the chlorophyll fluorescence from 600 to 800 nm using a 530 nm cut-off filter. Both spectra were combined yielding the complete fluorescence emission spectrum from 400 to 800 nm.

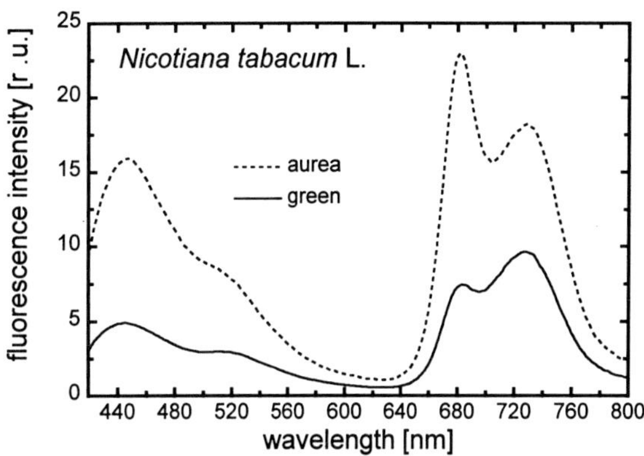

Fig. 1: UV-A (355 nm) excited fluorescence emission spectra of young tobacco leaves from the fully green wild type plant (su/su) and the yellowish-green aurea mutant (Su/su).

The Karlsruhe/Strasbourg fluorescence imaging system

Excitation source was a tripled Nd:YAG laser emitting 355 nm pulses at a 1 kHz repetition rate. Each pulse had the energy of 10 µJ and a FWHM of 150 ps. By expanding the laser beam the complete leaf surface even of large leaves was irradiated. Day light conditions (steady state of chlorophyll fluorescence induction kinetics) were simulated by illumination with white light of 1000 µmol $m^{-2} s^{-1}$. The fluorescence emission bands were sensed from a distance of approximately 0.2 to 0.6 m *via* a gated image intensifier (gate width 20 ns, Philips XX1414M/E image intensifier tube) and a digital CCD camera with a CCD array of 288 × 384 elements, operating with 50 frame transfers per second. Fluorescence images of the blue (F440), green (F520), red (F690) and far-red fluorescence bands (F740) were recorded separately using the appropriate changeable interference filters (center wavelength 440 nm, 520 nm, 690 nm or 740 nm; full width at half maximum 10 nm; Oriel, France). The resolution of the imaging system provided a high resolution of 0.2 to 0.5 mm^2 leaf area, depending on the distance of the CCD camera from the plant.

For image processing, the software Animater Vl (A.R.P., Strasbourg, France) was applied. The fluorescence images with 288 × 384 pixels were corrected for the inhomogenity of the excitation radiation and for the spectral sensitivity of the instrument. For a quantitative determination of the fluorescence ratios the fluorescence intensities were calculated from rectangular areas of 100 to 400 pixels (ca. 1–2 cm^2) leaf area. For quantitative data and fluorescence ratio determinations of each leaf, the fluorescence of at least six of such rectangles were averaged.

Results

Fluorescence emission spectra

The UV-A radiation (355 nm) induced fluorescence emission spectra of leaves of a normal green tobacco plant and a yellowish-green aurea mutant are shown in Fig. 1. The aurea mutant exhibits a considerably lower chlorophyll content, and also a slightly lower carotenoid content (Lang et al., 1994; Santrucek et al., 1992). The fluorescence emission spectra showed maxima/shoulders in the blue region near 440 nm (F440), the green region near 520 nm to 530 nm (F520) and the red and far-red chlorophyll fluorescence bands near 690 nm (F690) and between 730 to 740 nm (F740). The aurea mutant exhibited a higher yield in bluegreen fluorescence and red + far-red chlorophyll fluorescence than the fully green leaf of the wild-type tobacco. This is caused by the fact that in green tobacco the emitted fluorescence is reabsorbed by chlorophylls and carotenoids to a larger extent than in the aurea mutant which contains lower amounts of photosynthetic pigments. The lower content of chlorophyll in the aurea mutant was also documented by a higher value of the chlorophyll fluorescence ratio F690/F740 (value of 1.3) as compared to the green tobacco (value of 0.8).

Long-term water stress

Tobacco plants, green form and the aurea mutant, were kept unwatered for 3 weeks in the greenhouse, whereas control plants were regularly watered. The major effect of water stress and increasing water loss of the leaf tissue was a shrinking of the leaves. The decreasing percentage of water content of leaves was determined together with fluorescence emission spectra and fluorescence images. The UV-A induced (355 nm) fluorescence emission spectra of leaves of green tobacco did not yet change at a decreasing water content from 92 % (control) down to 84 % (Fig. 2). In contrast, at 82 % and lower percentages of water content the bluegreen fluorescence and the red + far-red chlorophyll fluorescence of green tobacco leaves increased. The relative fluorescence increase was larger for the bluegreen than for the chlorophyll fluorescence (Fig. 2).

The fluorescence ratios were also determined from fluorescence images. The ratios blue/red (F440/F690) and blue/far-red (F440/F740) exhibited constant values in the beginning of water loss (Fig. 3). Below a certain threshold in water content, both fluorescence ratios F440/F690 and F440/F740, however, linearly increased. This threshold value was 84 % water content for the green tobacco and 88 % water content for the aurea form. The chlorophyll fluorescence ratio F690/F740, in turn, remained more or less constant at values from about 0.4 to 0.6 (green tobacco), indicating that a distinct breakdown of chlorophyll had apparently not occurred. The fluorescence ratio F440/F520, in turn, proved to be very var-

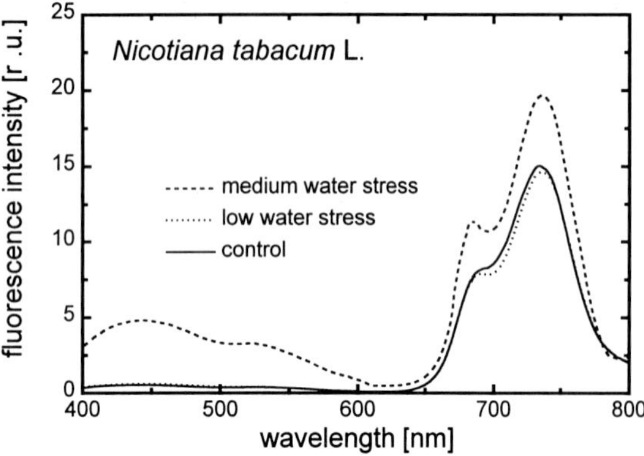

Fig. 2: UV-A (355 nm) induced fluorescence emission spectra of green tobacco leaves with different water content. Control: 88 % water content, low water stress (84 % water content) and medium water stress (82 % water content).

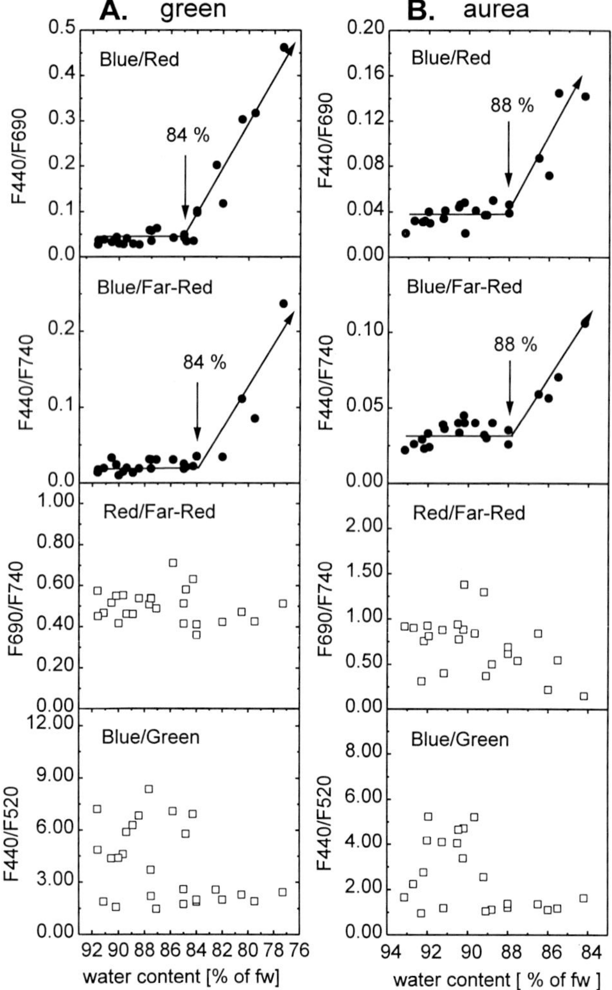

Fig. 3: Changes of the fluorescence ratios blue/red (F440/F690), blue/far-red (F440/F740), red/far-red (F690/F740) and blue/green (F440/F520) of *Nicotiana tabacum* L. during a 3 week water stress. **A** green tobacco, **B** aurea tobacco. A linear increase of the fluorescence ratios F440/F690 and F440/F740 started as soon as the water content dropped below 84 % and 88 % for green and aurea tobacco, respectively. The ratios were read from fluorescence ratio images of the leaves.

iable in green and aurea tobacco, not showing any clear tendency (Fig. 3).

Short term temperature stress

In one week unwatered tobacco plants, additionally treated for 6 h with 40 °C at an irradiance of 600 µmol quanta $m^{-2} s^{-1}$, no visual differences were detectable to leaves of control plants. Fluorescence images, however, showed a gradient in both chlorophyll fluorescence bands (F690 and F740) from the leaf rim (highest chlorophyll fluorescence) to the central part (lowest fluorescence yield) in the leaf of the heat treated plant (Fig. 4). The chlorophyll fluorescence ratio F690/F740 did, however, not increase at the leaf rim areas (Fig. 5b). In the lateral and leaf rim parts the red (F690) and far-red (F740) chlorophyll fluorescence were 2.5 times higher than in the central leaf parts. The latter still showed the same low chlorophyll fluorescence emission as the control leaf of the watered and non-heat treated plant (Fig. 4c and d). The green fluorescence emission at F520 slightly increased to a 1.3 times higher intensity by heat treatment, whereas the blue fluorescence (F440) remained constant. Due to the increase of chlorophyll fluorescence at the leaf rims, the fluorescence ratios blue/red (F440/F690) and blue/far-red (F440/740) significantly decreased at the leaf rim with a gradient from the center leaf part (highest values) to the leaf rim (lowest values) (Figs. 5 and 6).

Combined light, heat and water stress in Rhododendron

Leaves of *Rhododendron* were measured after a 4 week dry and sunny period in summer 1994 with midday temperatures of 34 to 36 °C and a high photon flux density of 2200 µmol $m^{-2} s^{-1}$. Sun-exposed leaves exhibited a yellowish-green colour on the upper leaf side, but a full green colour on the lower leaf side. Shaded leaves of the same plant showed a green colour on upper and lower leaf side. The yellowish-green colour of the upper leaf side of the sun-exposed leaf originated mainly from the epidermis, but the chlorophyll a+b content per leaf area was reduced by 25 % in the sun-exposed leaf (Table 1). The level of total flavonols was 2.4 times higher than in the shade leaf, and the largest proportion of these additional flavonols was found in the upper epidermis layer of the sun-exposed leaf.

Blue and green fluorescence were higher in sun-exposed stressed leaves as compared to shade leaves (control) (Fig. 7). The relative increase in green fluorescence was more pronounced than that of the blue fluorescence (Table 1). In contrast, the intensity of the emitted red and far-red chlorophyll fluorescence considerably decreased as shown in the fluorescence images (Fig. 7) and in Table 1. Apparently, the fluorescence exciting UV-radiation was absorbed to a large extent by the chemically modified epidermis (e.g. higher flavonol content of the sun-exposed stressed leaf) and this matter drasti-

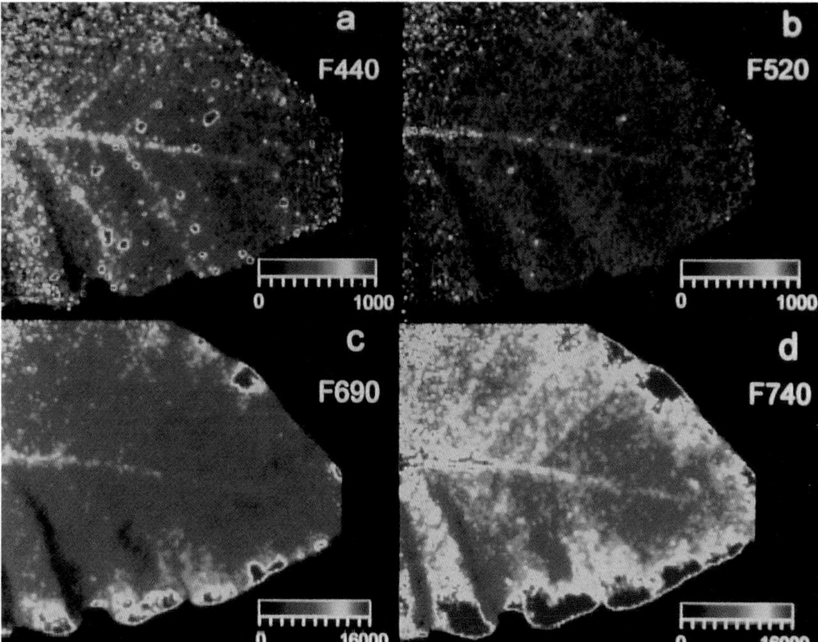

Fig. 4: False-colour images of the laser-induced blue (F440), green (F520, red (F690) and far-red (F740) fluorescence gradient in a green tobacco leaf after a 6 h heat plus water stress treatment at 40 °C. Fluorescence intensities increase from) blue via green and yellow to red (see colour scale in the figure).

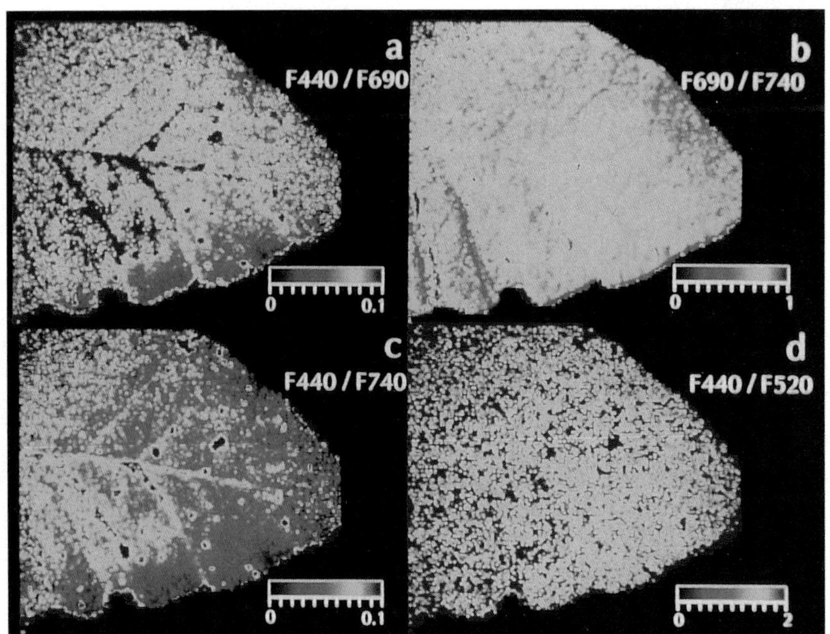

Fig. 5: False-colour images of the fluorescence ratios blue/red (F440/F690), blue/far-red (F440/F740), red/far-red (F690/F740) and blue/green (F440/F520) in a green tobacco leaf after a 6 h heat stress at 40 °C. The fluorescence ratios increase from blue via green and yellow to red (see colour scale).

cally reduced the yield of chlorophyll fluorescence in the subepidermal mesophyll cells (Figs. 7 and 8 A). Consequently, the fluorescence ratios blue/red and blue/far-red increased several fold due to the decreased chlorophyll fluorescence (Figs. 8 b and 9), whereas the fluorescence ratio blue/green decreased. The red/far-red chlorophyll fluorescence ratio F690/F740 also increased in the stressed sun-exposed leaf, thus reflecting the lower chlorophyll content (Table 1).

Discussion

Long-term water stress, as simulated in the greenhouse by slowly drying out tobacco plants, was monitored by fluorescence imaging and detected by an increase in the fluorescence emission in the blue (F440), green (F520), red (F690) and far-red (F740) spectral region. Since the bluegreen fluorescence increased to a higher extent than the red + far-red

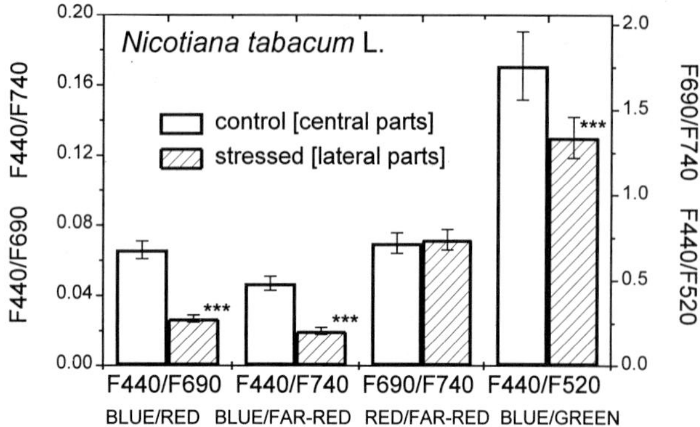

Fig. 6: Differences in the fluorescence ratios blue/red (F440/F690) and blue/far-red (F440/F740) [left scale], and in the fluorescence ratios red/far-red (F690/F740) and blue/green (F440/F520) [right scale] between stressed/damaged leaf parts of the leaf rim and central leaf parts of a tobacco leaf after a 6 h heat stress treatment at 40 °C. Mean values of 10 determinations (based on 180 pixels each) with standard deviation. ***These differences are highly significant ($p<0.001$).

Table 1: Fluorescence characteristics and pigment data (with standard deviation) of a shaded (control) and a sun-exposed leaf (stressed) of a *Rhododendron* hybrid. Intensities of the blue (F440), green (F520), red (F690) and far-red fluorescence (F740) were determined from fluorescence images together with the fluorescence ratios blue/red (F440/F690), blue/far-red (F440/F740), red/far-red (F690/F740) and blue/green (F440/F520). Mean values of 6 determinations. The values of chlorophylls, carotenoids and flavonols (given in $\mu g\ cm^{-2}$ leaf area) are based on 4 determinations. The arrows indicate increase or decrease in the sun-exposed leaf as compared to the shade leaf.

	shade leaf	sun-exposed leaf		
Fluorescence intensities:				
F440	275±18	419±13	***	↑
F520	223±11	645±38	***	↑
F690	1708±47	263±11	***	↓
F740	3268±76	370±12	***	↓
Fluorescence ratios:				
F440/F690	0.16±0.02	1.60±0.10	***	↑
F440/F740	0.09±0.01	1.16±0.19	***	↑
F690/F740	0.54±0.09	0.75±0.11	*	↑
F440/F520	1.27±0.17	0.66±0.11	***	↓
Flavonol content [$\mu g\ cm^{-2}$]	54±5	124±9	***	↑
Photosynthetic pigments				
Chlorophylls $a+b$ [$\mu g\ cm^{-2}$]	33.5±3.5	25.0±2.2	***	↓
Carotenoids x+c [$\mu g\ cm^{-2}$]	11.5±1.4	11.0±1.1		↔
Ratio Chl *a/b*	2.7±0.2	2.2±0.2	*	↓
Ratio (a+b)/(x+c)	2.9±0.2	2.3±0.2	*	↓

* and ***: significantly different to the control: *** $p<0.001$; * $p<0.05$.

chlorophyll fluorescence, the fluorescence ratios blue/red (F440/F690) and blue/far-red (F440/F740) changed, which could easily be monitored via images of the fluorescence ratios. Long-term water stress resulted in a particular behaviour of the fluorescence ratio blue/red and blue/far-red. These two fluorescence ratios remained constant for a longer period in the course of water loss, but below a distinct threshold in water content both fluorescence ratios increased linearily. Although the increase of the fluorescence ratios blue/red and blue/far-red in plants under drought stress was demonstrated before by spectral point data measurements (Dahn et al., 1992), the linear increase in the fluorescence ratios blue/red and blue/far-red below a threshold in water content, during drying of attached leaves, was observed and described here for the first time as well as its monitoring and detection by the newly developed fluorescence imaging technique. These results emphasize the superiority of fluorescence images of complete leaf areas over the hitherto applied fluorescence emission spectra of randomly selected small leaf parts. The advantage of the Karlsruhe/Strasbourg fluorescence imaging system is that it permits to directly image various fluorescence ratios and to recognize gradients and local disturbances in these ratios over the leaf area.

The increase in blue fluorescence during the induced water loss might have been caused, at least to some extent, by a transitory accumulation of such blue fluorescence emitting substances, which are known to be formed as intermediates in chlorophyll breakdown (Kräutler et al., 1992). Shrinking of cells as a consequence of turgor loss, however, also results in the change of optical leaf properties with an increased reflectance in the visible spectral range (Thomas et al., 1971).

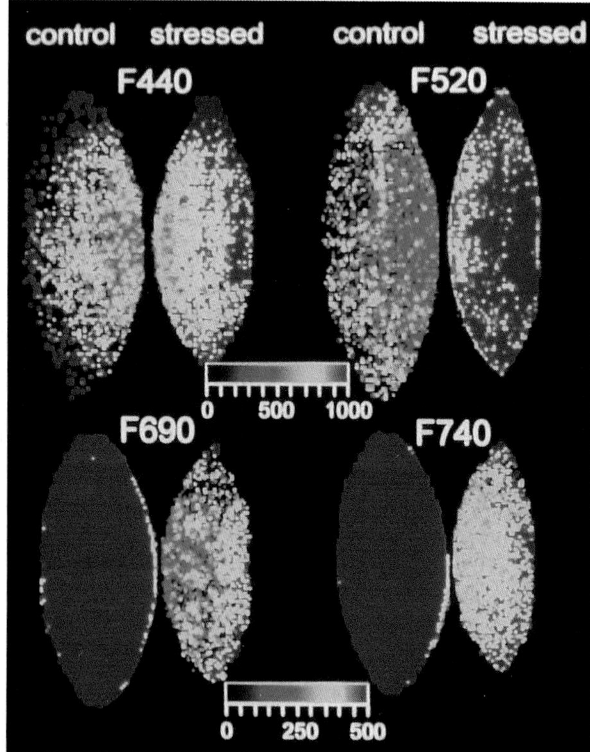

Fig. 7: False-colour images of the blue (F440), green (F520), red (F690) and far-red (F740) fluorescence distribution of a shaded leaf (left, control) and a sunexposed leaf (right, stressed) of *Rhododendron.* Fluorescence intensities increase from blue via green and yellow to red (see colour scale in the figure).

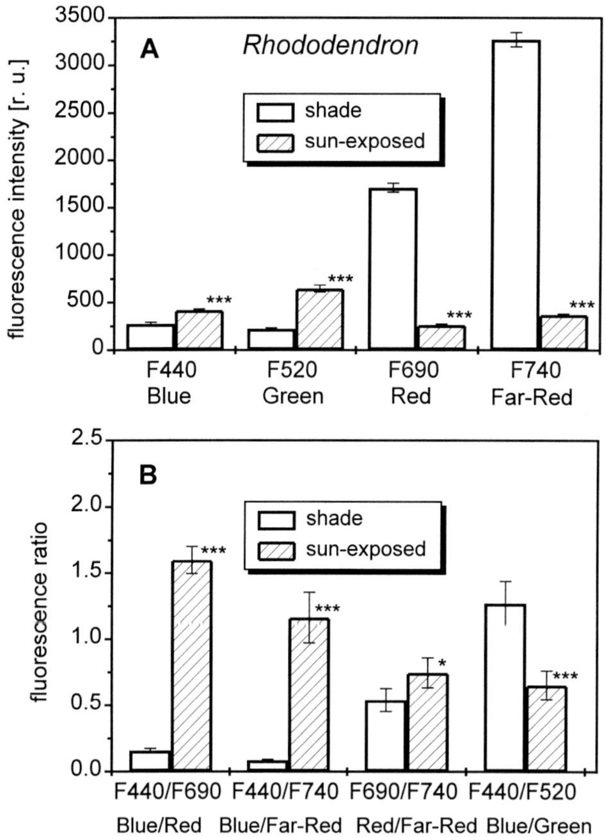

Fig. 8: Blue, green, red and far-red fluorescence yield (**A**) as well as fluorescence ratios blue/red (F440/F690), blue/far-red (F440/F740, red/far-red (F690/F740) and blue/green (F440/F520) (**B**) based on fluorescence images of sun-exposed and shaded leaves of *Rhododendron.* Mean values of 10 determinations (based on 200 pixels each). The differences are significant: ***p<0.001 and *p<0.05.

An enhanced reflection of the emitted blue fluorescence in the mesophyll and epidermis cells may have also contributed to the increase in blue fluorescence emission. Besides the changes in absolute fluorescence yield, the linear increase in the fluorescence ratios F440/F690 and F440/F740 proved to be good indicators of the progressing water stress in green and aurea tobacco leaves.

Short-term heat stress in tobacco plants was also detected by determination of fluorescence images in the four fluorescence bands blue, green, red and far-red. The fast detection of the increase of the chlorophyll fluorescence bands at the leaf rim indicated an impairment of photosynthetic quantum conversion at the lateral and leaf rim parts, whereas the photosynthetic processes in the leaf center part were still in full function. This demonstrates the excellent suitability of the fluorescence imaging system in early stress detection in plants. Since after 6 h of heat exposure (40 °C) virtually no changes were found in the blue fluorescence emission of the leaf, the latter can be taken as internal fluorescence standard. Thus the heat stress induced impairment of the photosynthetic quantum conversion yielding an increased chlorophyll fluorescence can be quantified in form of the fluorescence ratios blue/red (F440/F690) and blue/far-red (F440/F740), which decreased. Gradients in fluorescence emission and fluorescence ratios from the leaf rim to the central part as well as a certain patchiness of the leaf area concerning the emission of fluorescence signatures are easily to be screened by fluorescence imaging. This method is the only one to detect such gradients or local disturbances and changes in fluorescence emission and fluorescence ratios. The non-invasive fluorescence imaging procedure also allows to early detect whether a regeneration of a stress effect is still possible and will proceed when the stressor, e.g. heat stress or water stress, will be removed.

The strong decrease of the chlorophyll fluorescence emission together with the increase in the bluegreen fluorescence in light, heat and water stressed, sun-exposed *Rhododendron* leaves resulted again in increased values of the fluorescence ratios blue/red (F440/F690) and blue/far-red (F440/F740). The much stronger increase of the green fluorescence in the sun-exposed leaves with respect to the blue fluorescence indicated that UV-absorbing substances emitting green fluorescence were accumulated in the epidermis of the sun-exposed leaves, as compared to the control shade leaves. The much lower chlorophyll fluorescence emission in these sun-exposed leaves indicated that their photosynthetic apparatus in the subepidermal mesophyll cells was better protected against ultraviolet exposure than the non-stressed shade leaves of *Rhodo-*

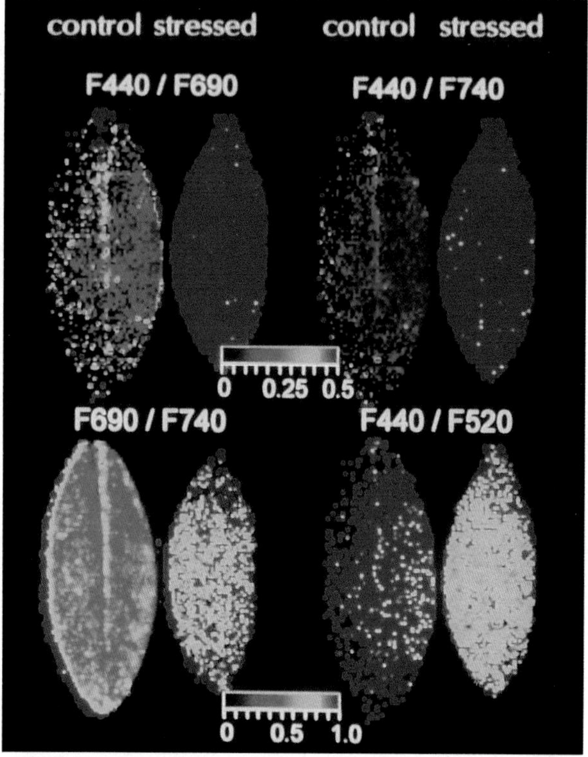

Fig. 9: False-colour fluorescence ratio-images blue/red (F440/F690), blue/far-red (F440/F740), red/far-red (F690/F740) and blue/green (F440/F520) of a shaded leaf (control) and a sun-exposed stressed leaf of *Rhododendron.* The values of the fluorescence ratios increase from blue to red as indicated in the scale.

dendron. The flavonol quercetin exhibits a green fluorescence, and other plant phenolics like catechin, coumarins and hydroxycinnamic acids possess broad bluegreen emission bands (Lang et al., 1991). Though an increase of the total flavonol content in sun-exposed as compared to shaded leaves was demonstrated in *Rhododendron* (Table 1), further investigations are required to prove whether these additionally accumulated flavonols caused the increase in bluegreen fluorescence. However, it is evident that the additional flavonols, accumulated in sun-exposed leaves (and predominantly in the leaves' upper epidermis), functioned as UV-absorbing filter and thus prevented the major part of the ultraviolet radiation (here: 355 nm) to penetrate through the epidermis into the mesophyll cells and there to excite chlorophyll fluorescence. This is the cause for the much lower chlorophyll fluorescence emission observed in sun-exposed *Rhododendron* leaves.

Conclusion

High resolution fluorescence images of complete leaf surfaces provide faster and more precise information on stress-induced changes in the fluorescence yields and the fluorescence ratios blue/red, blue/far-red, red/far-red and blue/green than conventional point measurements of leaves. This has been demonstrated here for long-term water stressed tobacco plants, in a short-term heat stress experiment with tobacco and also in a combined light, heat and water stress in *Rhododendron* plants. The high resolution fluorescence imaging, applied in this investigation, was shown to be an excellent tool for early stress detection, whereby gradients and a certain leaf patchiness in fluorescence emission over the leaf surface can easily be detected. Besides differences in the absolute fluorescence yields in the four fluorescence bands, changes in the fluorescence ratios blue/red and blue/far-red proved to be the best indicators for the detection of stress conditions in plants. Once a stress or damage has been recognized, secondary and complementary methods, such as fluorescence kinetics or chemical analysis including HPLC, can be applied to further define the stress effects in plants.

Acknowledgements

We thank Gabrielle Johnson for correcting the English text.

References

Caldwell, M. M., R. Robberecht, and S. D. Flint: Internal filters: Prospects for UV-acclimation in higher plants. Physiol. Plant. *58,* 445–450 (1983).

Cerovic, Z. G., M. Bergher, Y. Goulas, S. Tosti, and I. Moya: Simultaneous measurement of changes in red and blue fluorescence in illuminated isolated chloroplasts and leaf pieces. The contribution of NADPH to the blue fluorescence signal. Photosynth. Res. *36,* 193–204 (1993).

Cerovic, Z. G., F. Morales, and I. Moya: Time-resolved spectral studies of blue-green fluorescence of leaves, mesophyll and chloroplasts of sugar beet (*Beta vulgaris* L.) Biochim. Biophys. Acta *1188,* 58–68 (1994).

Chappelle, E. W., F. M. Wood, J. E. McMurtrey, and W. W. Newcomb: Laser-induced fluorescence of green plants. 1: A technique for remote detection of plant stress and species differentiation. Appl. Opt. *23,* 134–138 (1984).

D'Ambrosio, N., K. Szabó, and H. K. Lichtenthaler: Increase of the chlorophyll fluorescence ratio F690/F735 during the autumnal chlorophyll breakdown. Radiat. Environ. Biophys. *31,* 51–62 (1992).

Dahn, H. G., K. P. Günther, and W. Lüdecker: Characterisation of drought stress of maize and wheat canopies by means of spectral resolved laser induced fluorescence. EARSeL Advances in Remote Sensing *1,* 12–19 (1992).

Edner, H., J. Johansson, S. Svanberg, H. K. Lichtenthaler, M. Lang, F. Stober, C. Schindler, and L.-Ö. Björn: Remote multicolour fluorescence imaging of selected broad-leaf plants. EARSeL Advances in Remote Sensing *3,* No. 3, 2–14 (1995).

Goulas, Y., I. Moya, and G. Schmuck: Time resolved spectroscopy of the blue fluorescence of spinach leaves. Photosynth. Res. *25,* 299–307 (1990).

Hák, R., H. K. Lichtenthaler, and U. Rinderle: Decrease of the chlorophyll fluorescence ratio F690/F730 during greening and develpoment of leaves. Radiat. Environ. Biophys. *29,* 329–336 (1990).

Hák, R., U. Rinderle-Zimmer, H. K. Lichtenthaler, and L. Nátr: Chlorophyll *a* fluorescence signatures of nitrogen-deficient barley leaves. Photosynthetica *28,* 151–159 (1993).

Harris, P. J. and R. D. Hartley: Detection of bound ferulic acid in cell walls of the gramineae by ultraviolet fluorescence microscopy. Nature *259,* 508–510 (1976).

Kräutler, B., B. Jaun, W. Amrein, K. Bortlik, M. Schellenberg, and P. Matile: Breakdown of chlorophyll: Constitution of secoporphinoid chlorophyll catabolite isolated from senescent barley leaves. Plant Physiol. Biochem. *30,* 333–346 (1992).

Lang, M.: Studies on the bluegreen and chlorophyll fluorescence of plants and their application for fluorescence imaging of leaves. Karlsruhe Contrib. Plant Physiol. *29,* 1–110 (1995).

Lang, M. and H. K. Lichtenthaler: Changes in the blue green and red fluorescence emission spectra of beech leaves during the autumnal chlorophyll breakdown. J. Plant Physiol. *138,* 550–553 (1991).

Lang, M., F. Stober, and H. K. Lichtenthaler: Fluorescence emission spectra of plant leaves and plant constituents. Radiat. Environ. Biophys. *30,* 333–347 (1991).

Lang, M., P. Šiffel, Z. Braunova, and H. K. Lichtenthaler: Investigations on the blue-green fluorescence emission of plant leaves. Bot. Acta *105,* 435–440 (1992).

Lang, M., H. K. Lichtenthaler, M. Sowinska, P. Summ, and F. Heisel: Blue, green and red fluorescence signatures and images of tobacco leaves. Bot. Acta *107,* 230–236 (1994).

Larcher, W.: Streßkonzepte in der Biologie. Ber. Dtsch. Bot. Ges. *98,* 289–290 (1984).

– Streß bei Pflanzen. Naturwissenschaften *74,* 158–167 (1987).

Lichtenthaler, H. K.: Chlorophyll fluorescence signatures of leaves during autumnal chlorophyll breakdown. J. Plant Physiol. *131,* 101–110 (1987a).

– Chlorophylls and carotenoids: pigments of photosynthetic biomembranes. Methods Enzymol. *148,* 350–382 (1987b).

– *In vivo* chlorophyll fluorescence as a tool for stress detection in plants. In: Lichtenthaler, H. K. (ed.): Applications of Chlorophyll Fluorescence in Photosynthesis Research, Stress Physiology, Hydrobiology and Remote Sensing, pp. 129–142. Kluwer Academic Publishers, Dordrecht (1988).

– The Kautsky Effect: 60 years of chlorophyll fluorescence induction kinetics. Photosynthetica *27,* 45–55 (1992).

– Vegetation stress: an introduction to the stress concept in plants. J. Plant Physiol. *148,* 4–14 (1996).

LICHTENTHALER, H. K. and U. RINDERLE: The role of chlorophyll fluorescence in the detection of stress conditions in plants. CRC Crit. Rev. Anal. Chem. *19* (Suppl. 1), 29–85 (1988).

LICHTENTHALER, H. K. and M. LANG: Biomonitoring of terrestrial vegetation by laser-induced fluorescence and reflectance measurements within the EUREKA Programme LASFLEUR (EU380). In: RICHTER, P. I. and R. C. HERNDON (eds.): Proceed. 2nd. Internat. Symposium and Exhibition on Environmental Contamination in Central and Eastern Europe, Budapest 1994, pp. 112–114. Government Institutes, Inc., Rockville, Maryland (1995).

LICHTENTHALER, H. K., F. STOBER, and M. LANG: The nature of the different laser-induced fluorescence signatures of plants. EARSeL Advances in Remote Sensing *1,* 20–32 (1992) (**E**uropean **A**ssociation of **R**emote **S**ensing **L**aboratories).

LICHTENTHALER, H. K., M. LANG, M. SOWINSKA, F. HEISEL, and J. A. MIEHÉ: Detection of vegetation stress via a new high resolution fluorescence imaging system. J. Plant Physiol. *148,* 599–612 (1996).

SANTRUCEK, J., P. SIFFEL, M. LANG, H. K. LICHTENTHALER, C. SCHINDLER, H. SYNKOVÁ, V. KONECNÁ, and K. SZABÓ: Photosynthetic activity and chlorophyll fluorescence parameters in *aurea* and green forms of *Nicotiana tabacum.* Photosynthetica *27,* 529–543 (1992).

SCHMID, G. H.: Origin and properties of mutant plants: Yellow tobacco. Methods Enzymol. *23,* 171–194 (1971).

SCHREIBER, U., W. BILGER, C. KLUGHAMMER, and C. NEUBAUER: Application of the PAM fluorometer in stress detection. In: LICHTENTHALER, H. K. (ed.): Applications of Chlorophyll Fluorescence in Photosynthesis Research, Stress Physiology, Hydrobiology and Remote Sensing, pp. 151–155. Kluwer Academic Publishers, Dordrecht (1988).

STOBER, F. and H. K. LICHTENTHALER: Characterisation of the laser-induced blue, green and red fluorescence signatures of leaves of wheat and soybean leaves grown under different irradiance. Physiologia Plantarum *88,* 696–704 (1993 a).

STOBER, F. and H. K. LICHTENTHALER: Studies on the localization and spectral characteristics of the fluorescence emission of differently pigmented wheat leaves. Bot. Acta *106,* 365–370 (1993 b).

STRASSER, R. J., B. SCHWARZ, and J. B. BUCHER: Simultane Messung der Chlorophyll Fluoreszenz-Kinetik bei verschiedenen Wellenlängen als rasches Verfahren zur Frühdiagnose von Imissionsbelastungen an Waldbäumen. Ozoneinwirkungen auf Buchen und Pappeln. Eur. J. Forest Pathol. *17,* 149–157 (1987).

THOMAS, J. R., L. N. MANKEN, G. F. OERTHER, and R. G. BROWN: Estimating leaf water content by reflectance measurements. Agronom. J. *63,* 845-847 (1971).

J. Plant Physiol. Vol. 148. pp. 622–631 (1996)

Detection of Nutrient Deficiencies of Maize by Laser Induced Fluorescence Imaging

FRANCINE HEISEL[1], MALGORZATA SOWINSKA[1], JOSEPH ALBERT MIEHÉ[1], MICHAEL LANG[2], and HARMUT K. LICHTENTHALER[2]

[1] Groupe d'Optique Appliquée, Centre de Recherches Nucléaires, IN2P3, CNRS, 23, rue du Loess, F 67037 Strasbourg Cedex 2, France

[2] Botanisches Institut II der Universität Karlsruhe, Kaiserstr. 12, D-76128, Germany

Received August 3, 1995 · Accepted October 20, 1995

Summary

Laser-induced fluorescence is an active method of sensing the state of health of the plants and the photosynthetic apparatus, as it is related not only to the pigment concentrations but also to the physiological activity. A high gain and high spatial resolution fluorescence imaging set up, consisting of a pulsed Nd:YAG laser for the excitation (355 nm) and of an intensified gated CCD numerical camera, has been used for monitoring various nutrient deficiencies of maize (*Zea mays* L.) by recording fluorescence images of the leaves at 440, 520, 690 and 740 nm. The plant status was characterised by the fluorescence ratios F440/F520, F440/F690, F440/F740 and F690/F740. The experiments performed on field maize supplied with various amounts of nitrogen and on greenhouse maize with defined mineral deficiencies showed that all the deficiencies could be monitored by the fluorescence ratios and in some cases directly on the fluorescence images by considering the spatial distribution of the emission on the leaf surface. From this work it appeared that the efficiency of detection depended on the period of measurements and on the age of the leaves. The fluorescence ratios F440/F690 and F440/F740 were found more sensitive to the growth conditions than the most frequently used chlorophyll fluorescence ratio F690/F740.

Key words: Zea mays L., fluorescence ratios, laser-induced fluorescence imaging, leaf fluorescence, mineral deficiencies: Fe, Mg, Zn, nitrogen fertilisation.

Abbreviations: CCD = charge coupled device; F440 = blue fluorescence intensity at 440 nm; F520 = green fluorescence intensity at 520 nm; F690 = red chlorophyll fluorescence intensity at 690 nm; F740 = far-red chlorophyll fluorescence intensity at 740 nm.

Introduction

Remote sensing of the vegetation health status, based on the optical properties of the plants, is a very attractive tool since the method is non destructive and gives immediate results that allows, in the case of stress detection due to nutrient deficiencies, timely correction of the fertilisation. If far field reflectance measurements, which are at the present time technologically well developed (Slater, 1980; Brunel et al., 1991), have application for example in green biomass evaluation, they are insensitive to physiological variations of the plant status.

The fluorescence emission, which for plant leaves excited by UV-A radiation has a spectrum with maxima/shoulders near to 440, 520, 690 and 740 nm (Chappelle et al., 1984 a; Lang and Lichtenthaler, 1991; Lang et al., 1992, 1994 a), depends not only on the pigment concentration but also on the efficiency of the photosynthesis and other biochemical reactions. Much work has been performed to investigate the relation between the plant health and the fluorescence intensities (Kautsky and Hirsch, 1934; Buschmann et al., 1991; Lichtenthaler et al., 1991; Stober et Lichtenthaler, 1992, 1993; Günther et al., 1994; Morales et al., 1994) especially in what concerns the chlorophyll red emission (690 and 740 nm) (Krause and Weis, 1984; Lichtenthaler and Rinderle, 1988; Lichtenthaler, 1987, 1990; Snel and Van Kooten, 1990; Lippucci di Paola, 1992; Methy et al., 1994; Valentini et al., 1994). Indications came out that laser induced fluorescence signatures seem to be useful for remote detection of certain nutrient deficiencies (Chappelle et al., 1984 b; Kochubey et al., 1986; Hagg et al., 1992; Hák et al., 1993; McMurtrey et al., 1994; Subhash and Mohanan, 1994), in spite of the fact that the blue-green fluorescence is a very complex signal from various plant phenolics (Goulas et al., 1990; Lang et al., 1991; Lichtenthaler et al., 1991; Stober et al., 1994) and that the physiological significance of this blue-green fluorescence is not yet fully clear.

The aim of this work was to assess the possibility of nutrient deficiencies detection and characterisation by laser-induced fluorescence imaging (Edner et al., 1994; Balachandran et al., 1994; Lichtenthaler et al., 1995) with a newly developed high spatial resolution numerical camera used in conjunction with a high gain gated image intensifier that allows measurements in the presence of the ambient light (Lang et al., 1994 a, b, 1995). Experiments have been performed on maize (*Zea mays* L.) grown in the field with different nitrogen fertilisation degrees, or grown in a greenhouse with defined mineral deficiencies.

Since further investigations will be devoted to remote sensing at long distances in the fields, we were interested in characterising the plant status by various fluorescence ratios, namely F440/F520, F440/F690, F440/F740 and F690/F740, and not by the fluorescence intensities which strongly depend on external parameters like distance, geometry of the foliage subjected to the excitation beam, etc.

Material and Methods

Plant growth

Nitrogen deficiency has been studied on the Déa variety maize (*Zea mays* L.) planted at a density of 90,000 plants per hectare, on week 16 in 1994, in alluvial argillaceous land parcels cultivated by the Lycée Agricole d'Obernai (Bas-Rhin, France). These parcels were used since three years for testing the influence of the nitrification on the plant productivity. The ground contains all the other essential nutrients, in particular it is well furnished in K and P. The nitrogen was supplied at the 3–4 leaf stage in the ammonium-nitrate form at various amounts from 0 to 160 kg/ha, for a recommended supply of 120 kg/ha. For each N level, there were four randomly situated replicate areas: 6 m × 10 m, with 8 maize ranks spaced by 75 cm. The experiments which began at the 8-leaf stage after eight weeks of plant growth, were carried out during five weeks. Measurements were performed on the same leaf storeys for 0, 60 and 160 kg N/ha parcels. The leaves were cut from the maize plants of the four replicate parcels and the fluorescence of the upper leaf-side quickly measured in the laboratory (Strasbourg). The leaves were numbered from the bottom: as when becoming old the first leaves could fall down, a mark has been made on the stems at the beginning of the experiments to have a reference.

The Volga variety of maize has been used to study the effect of deficiency in Mg, Zn or Fe. The maize was grown in the greenhouse of the Centre de Recherches de la Société Commerciale des Potasses et de l'Azote (Aspach le Bas, Haut-Rhin, France), and planted in pots containing quartz sand, previously leached during four days with a HCl + oxalic acid solution and with doubly demineralised water during two days, to eliminate the endogenous elements. After three days of growth in a seed-bed, four seeds per pot were planted out and the pots were supplied with a complete nutritive solution of type Coïc-Lesaint (pH 6.2) (Lesaint and Coïc, 1983) for the control plants, and deficient in one element for the others: respectively Fe: 0, Mg: 1/10 of the full nutritive solution, Zn: ½ during 2 weeks and 0 after (a low amount of magnesium and zinc is necessary for the growth of maize). The temperature in the greenhouse was maintained between 20 °C and 28 °C with a relative humidity of 40–80 % and the plants were illuminated (≈ 30,000 lux) 15 hours per day, with 400 W high pressure natrium vapor lamps. There were two identical series of pots, one for measurements three weeks after the seed planting, the other for experiments two weeks later.

Fluorescence imaging

The laser induced fluorescence imaging arrangement is represented in Fig. 1. The excitation source was a cw Q-switched, mode locked and cavity dumped Nd:YAG laser emitting, after third harmonic generation, 355 nm light pulses at 1 kHz repetition rate, with an energy of 10 μJ and a width of 100 psec. The laser beam was directed onto the sample at right angle and shaped by a divergent lens to a 20 cm diameter spot on the leaf level.

The fluorescence imaging system (RAGM6 + Animater V1, ARP, Strasbourg, France) was composed of *i)* a gated intensified digital CCD camera operating at 50 frames per second and including all the electronic CCD's readout with 8 bits digitisation per pixel of the image due to one frame; the CCD array (Thomson TH 7863) is a 384 × 288 elements; the gated intensifier (Philips XX1414M/E image intensifier tube) has an adjustable gain (up to 10^3) and the width of the laser synchronous gating can be varied from 10 ns till to 100 ms *ii)* an interface card for the PC microcomputer where the images are stored *iii)* an image analysis software. After each digitisation, the result is transferred to the interface card where an addition of successive readouts can be made till to 16 bits (0–65,535), followed (or not) by the subtraction of the noise due to the camera itself and to the ambient light. The resulting image is then automatically sent in the extended memory of the PC. Besides the real-time visualisation and the storage of the images, the software allows fast recall and treatment of images, their number (maximum 99) de-

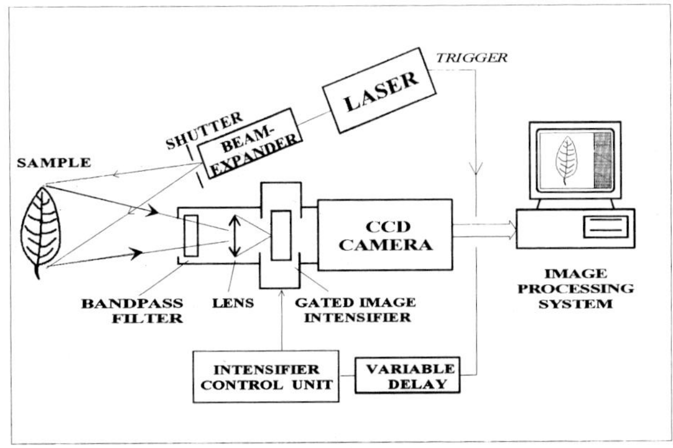

Fig. 1: Scheme of the laser induced fluorescence imaging set up.

pending on the size of the free extended memory (≈ 0.23 Mbyte per image).

Before and during the measurements, additional white light was applied on the leaves to simulate solar illumination so that the measured intensity of the chlorophyll fluorescence emission corresponded to the light adapted state of the chlorophyll fluorescence induction kinetic (Kautsky effect).

Fluorescence images were acquired, at an angle of ≈ 30°, through bandpass filters (bandwidth of 10 nm) in a distance of 50 cm via a focusing lens on the entrance of the intensifier. For this distance the height of the image corresponded to 20 cm at the leaf level. The intensifier was working with a 20 ns gate width, time long enough to collect the totality of the fluorescence and short enough to allow measurements in the presence of the white light. The filters were centred around the four characteristic emission wavelengths 440 nm (blue), 520 nm (green), 690 nm (red) and 740 nm (far-red). All images were corrected for the spectral sensitivity of the camera, including the attenuation factor of the bandpass filters and of the focusing lens.

Spectrofluorimetry

For having an overall view of the spectral distribution of the fluorescence, several additional emission spectra were recorded with a Spex Fluorolog 2 fluorimeter (Jobin Yvon, Longjumeau, France), which allows front face detection and provides spectra corrected for lamp intensity variations, spectral sensitivity of the monochromator-photomultiplier, etc. The excitation wavelength (355 nm) was the same as that used for imaging.

Results and Discussion

The first result to note is that the greenhouse Volga maize and the field Déa maize presented very different emission spectra, as illustrated in Fig. 2 where typical fluorescence curves are reported: the chlorophyll emission relative to the blue one was always more intense for the Volga maize than for the Déa maize. This behaviour is comparable to the enhancement of the red fluorescence observed by Stober et al. (1994) for various plants grown in the greenhouse with regard to those coming from the fields. Our results could be related to the visual observation that the epidermis is thicker for the outdoor maize plants: i) the epidermis is responsible for a major part of the blue-green fluorescence emission, ii) the excitation of the chloroplasts in the mesophyll cells is reduced by partial absorption of the excitation beam in the epidermis and in addition part of the emitted chlorophyll fluorescence is reabsorbed by the leaf chlorophyll.

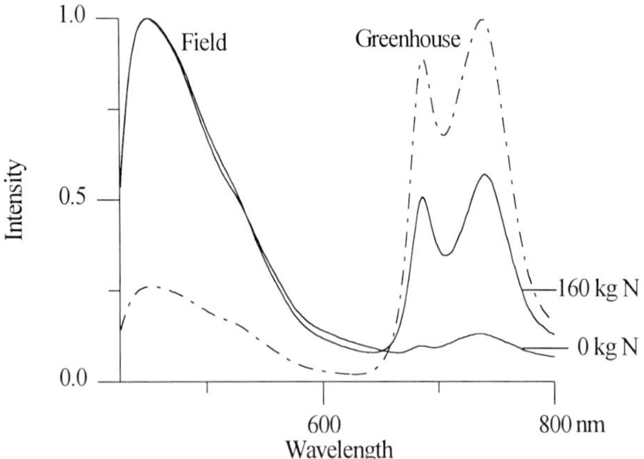

Fig. 2: Normalised fluorescence emission spectra of leaves of the Volga greenhouse maize and of the Déa field maize (excitation 355 nm, 400 nm long wave pass edge filter). The field maize was grown without (0 kg N per hectare) or with nitrogen fertilizer (160 kg N per hectare).

Table 1: Description of the leaves taken for the measurements on the field Déa maize planted on week 16 in 1994. The numbers (No.) of the leaves represent the leaf storeys labelled from the bottom of the stems. For each leaf number and each treatment the sample size was 8, except for the last week where it was 8 for the leaf No. 8 and 40 for the others.

Week No. in 1994	24	25	26	27	28
Leaves No.	4^a 5^a	6^a 7^a	8^a 9^a	10^a 11^a	8^a 11^a
(0-60-160 kg N/ha)	6^a 7^b	8^a 9^b	10^a 11^b	12^b 13^b	12^a
Number of leaves/plant					
0 kg N/ha	7–8	9–10	10–11	12–13	12–13
60 kg N/ha	8–9	10–11	11–12	13–14	13–14
160 kg N/ha	8–9	11–12	12–13	13–14	13–14

At a given week the leaves with different superscripts showed fluorescence ratio means significantly different ($p<0.05$) in the 0 kg N/há parcels.

Nitrogen fertilisation effect

A general feature of the laser induced fluorescence imaging results was that an increase of the soil nitrogen concentration did not significantly influence the intensity of the blue-green fluorescence emission, whereas, except for the first series of measurements, it enhanced the chlorophyll fluorescence (see also Fig. 2). The chlorophyll concentration has not been determined but for the 0 kg N/ha supply a chlorosis of some leaves was visible from the second week of measurements, and this chlorosis was associated with a weaker red fluorescence. It was also observed on all fluorescence images of maize leaves that the fluorescence intensity was roughly uniformly distributed on the leaf surface, with the main leaf vein fluorescence not markedly different from that of the other leaf parts. So, for each leaf the fluorescence ratio values were determined by averaging on the whole illuminated area.

In Table 1 are indicated, for each week of fluorescence imaging experiments, the storeys of the cut leaves as well as the number of leaves per plant for the studied nitrogen amounts. For each treatment and fluorescence ratio we have compared the leaves each to the others. As a general rule, less difference between the leaves of different ages was apparent for the 60 and 160 kg N/ha amounts than for the 0 kg N/ha parcels, that could be attributed to the fact that the latest cut leaves are younger in the absence of N. For the last week sampling (week 28), it has been observed that for the same excitation intensity, the fluorescence intensities of the leaves No. 11 and 12 were much higher than those of leaves No. 8, but that the fluorescence ratios were the same. For comparison between the different nitrogen fertilised plants, the mean

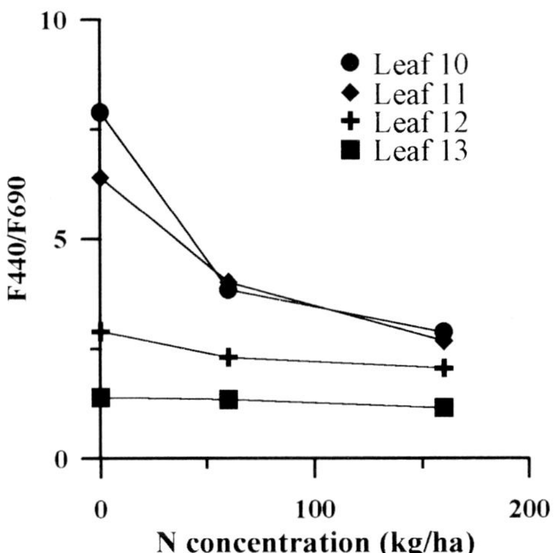

Fig. 3: Influence of the nitrogen supply on the fluorescence ratio F440/F690 measured for different leaves of the eleven-week-old field grown Déa maize (see Table 2).

Table 2: Mean values (with standard deviation) of the fluorescence ratios for the field Déa maize measured after eleven weeks of growth (week 27 in 1994). Each value is the mean of 8 leaves.

	F440/F520	F440/F690	F440/F740	F690/F740
0 kg N/ha				
Leaf 10	2.67^{a} (0.28)	7.88^{a} (2.87)	5.35^{a} (1.56)	0.70^{a} (0.08)
11	2.43^{a} (0.36)	6.40^{a} (2.16)	4.41^{a} (1.51)	0.69^{a} (0.04)
12	2.56^{a} (0.19)	2.90^{b} (1.26)	2.46^{b} (0.77)	0.90^{b} (0.16)
13	2.60^{a} (0.32)	1.39^{c} (0.40)	1.25^{c} (0.36)	0.90^{b} (0.06)
60 kg N/ha				
Leaf 10	2.86^{a} (0.30)	3.84^{a} (1.85)	2.88^{a} (1.41)	0.77^{a} (0.14)
11	2.70^{a} (0.35)	4.01^{a} (2.78)	2.68^{a} (1.21)	0.76^{a} (0.19)
12	2.83^{a} (0.31)	2.31^{ab} (0.97)	2.06^{a} (0.91)	0.94^{a} (0.34)
13	2.87^{a} (0.29)	1.34^{b} (0.76)	1.15^{b} (0.62)	0.87^{a} (0.06)
160 kg N/ha				
Leaf 10	2.58^{ab} (0.21)	2.94^{ab} (1.46)	2.12^{ab} (0.69)	0.79^{a} (0.19)
11	2.51^{a} (0.27)	3.36^{a} (1.23)	2.54^{a} (0.79)	0.79^{a} (0.20)
12	2.98^{b} (0.25)	2.12^{b} (0.75)	1.59^{b} (0.34)	0.80^{a} (0.24)
13	2.79^{b} (0.30)	1.34^{b} (0.83)	1.12^{c} (0.51)	0.89^{a} (0.12)

For one fluorescence ratio and one nitrogen amount the means followed by the same superscript were not significantly different ($p>0.05$).

values of the ratios have been calculated by taking into account only the leaves with the superscript a in Table 1.

The results of the experiments performed for eleven-week-old plants (Table 2 and Fig. 3), clearly show that the effect of nitrogen on the fluorescence ratios depended on the age of the leaves. The fluorescence properties of the youngest leaves had not yet been affected by the nitrogen shortage.

The overall results on the fluorescence ratios of maize leaves, are summarised in Table 3 which shows that the only fluorescence ratios which significantly varied with the nitrogen concentration were the blue/red F440/F690 and the blue/far-red F440/F740 (the blue emission was nearly constant). These ratios were increased (except for the first week of experiments when no nitrogen deficiency effect was detected) by a factor near to two when the nitrogen supply decreased from 160 kg to 0 kg/ha. This feature is promising for nitrogen lack detection by fluorescence imaging. It should be noted that inside a treatment, the standard deviations associated to the mean values were much more important for the fluorescence ratios F440/F690 and F440/F740 than for the fluorescence ratios F440/F520 and F690/F740. A consequence of these large standard deviations was that in some cases, a noticeable variation of the mean of the blue/red fluorescence ratio with the nitrogen amount was found not significant by using a robust statistical treatment.

In Fig. 4 all results on the dependence of the fluorescence ratios on the nitrogen supply are presented, upon which the following comments can be made:

1) The absence of a nitrogen deficiency effect for the first set of experiments can be understood by the fact that this period of growth corresponded to the beginning of the nitrogen consumption by the plants.

2) Since the blue fluorescence emission was constant, the increase of the ratios F440/F690 and F440/F740 for nitrogen deficient leaves corresponded to a decrease of the chlorophyll emission at 690 and 740 nm. This decrease is not surprising in view of the importance of the nitrogen nutrient for the chlorophyll production and accumulation (corroborated by the chlorosis of the 0 kg N/ha leaves). As the photosynthesis and the fluorescence are competitive reactions, and consequently have opposite effects on the red fluorescence emission intensities, the observed decrease of the chlorophyll fluorescence emission with decreasing nitrogen supply seems to indicate that the lack of nitrogen did not strongly inhibit the photosynthetic activity of the lower amounts of chlorophyll being present in these leaves.

3) In spite of the fact that the 690 and 740 nm intensities and hence the chlorophyll concentration increased with the

Table 3: Mean values (with standard deviation) at different times in the year, of the fluorescence ratios found for the field maize with different nitrogen fertilisation amounts.

Week in 1994	N/ha	F440/F520	F440/F690	F440/F740	F690/F740
24 (leaves 4+5+6)	0 kg	2.14^{a} (0.28)	7.48^{ab} (2.8)	5.20^{ab} (1.7)	0.72^{a} (0.14)
	60 kg	2.18^{a} (0.27)	8.12^{a} (3.2)	5.52^{a} (0.8)	0.73^{a} (0.17)
	160 kg	2.24^{a} (0.41)	6.48^{b} (1.8)	4.55^{b} (1.3)	0.71^{a} (0.14)
25 (leaves 6+7+8)	0 kg	2.54^{a} (0.38)	9.22^{a} (7.1)	5.71^{a} (2.4)	0.70^{a} (0.16)
	60 kg	2.72^{a} (0.34)	6.92^{a} (2.0)	4.97^{a} (2.1)	0.71^{a} (0.15)
	160 kg	2.62^{a} (0.35)	4.74^{b} (1.9)	3.33^{b} (1.0)	0.73^{a} (0.13)
26 (leaves 8+9+10)	0 kg	2.49^{a} (0.30)	11.87^{a} (3.4)	6.74^{a} (1.5)	0.58^{a} (0.10)
	60 kg	2.78^{b} (0.48)	8.46^{b} (3.4)	5.07^{b} (1.9)	0.61^{a} (0.61)
	160 kg	2.84^{b} (0.37)	5.62^{c} (2.4)	3.41^{c} (1.2)	0.63^{a} (0.11)
27 (leaves 10+11)	0 kg	2.55^{ab} (0.34)	7.14^{a} (2.6)	4.88^{a} (1.6)	0.69^{a} (0.06)
	60 kg	2.78^{a} (0.32)	3.93^{b} (2.3)	2.78^{b} (1.3)	0.76^{a} (0.16)
	160 kg	2.55^{b} (0.24)	3.15^{b} (1.3)	2.33^{b} (0.8)	0.79^{a} (0.19)
28 (leaves 8+11+12)	0 kg	2.54^{a} (0.34)	6.81^{a} (3.2)	4.78^{a} (2.0)	0.74^{a} (0.20)
	160 kg	2.51^{a} (0.34)	3.37^{b} (1.6)	2.47^{b} (1.1)	0.80^{a} (0.44)

Means for one ratio at a given week followed by the same superscript were not significantly different at the level of 5 % of probability ($p>0.05$). The sample size was 24 leaves for weeks 24, 25 and 26, 16 leaves for week 27 and 88 leaves for the last week 28.

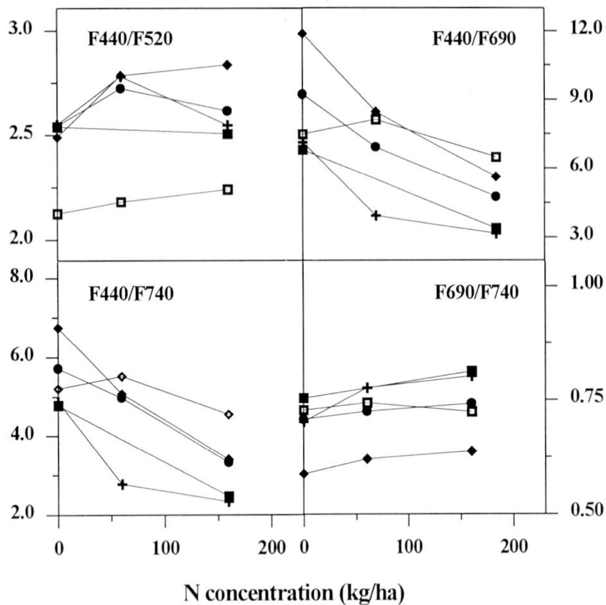

Fig. 4: Variation of the fluorescence ratios F440/F520, F440/F690, F440/F740 and F690/F740 versus the amount of nitrogen fertilisation at different periods of growth. Results for the groups of leaves indicated in Table 3. Week 24 □; 25 ●; 26 ◆; 27 +; and 28 ■ in 1994.

nitrogen supply, the chlorophyll fluorescence ratio F690/F740 of the maize leaves was almost constant. This is not necessarily in disagreement with the inverse relationship between the ratio F690/F740 and the chlorophyll concentration (Lichtenthaler, 1987; Lichtenthaler and Buschmann, 1987; Lichtenthaler and Rinderle, 1988; Hák et al., 1990; D'Ambrosio et al., 1992) since this fluorescence ratio does not much change at high chlorophyll concentrations and that it was established for bifacial or equifacial leaves, whereas maize is a C_4-plant with a particular arrangement of cells and leaf veins («Kranz anatomy») (cf. Lichtenthaler and Pfister, 1978). Taking into account that the red/far-red ratio was constant, the variations of F440/F690 and F440/F740 gave redundant results for these measurements.

4) Finally it should be remarked that in previous work, Chappelle et al. (1984b, 1994) have studied the influence of the nitrogen supply on the fluorescence spectra of maize. Our results are in good agreement with their finding of the absence of a nitrogen influence on the blue intensity, but for the red emission they found at once (1984b) a decrease in the intensity for nitrogen deficient maize and later for other growth conditions (1994) a higher intensity at 690 nm for the 0 kg N/ha samples.

Mineral nutrients deficiencies

As the maize plants were transferred in their pots from the greenhouse to the laboratory, images have also been recorded for some leaves still attached to their stem. Results obtained for cut leaves measured immediately after cutting or after one hour waiting showed in the case of maize practically no stress effect. Since, due to the ordering of the leaves, images could not be made on the stems for the oldest leaves, the data will be presented and discussed for cut leaves sensed directly after cutting.

The three-week-old maize plants had six leaves and two weeks later there were nine leaves for the control maize and only eight in an average for the deficient plants. Fluorescence images were recorded for all the leaves except the last week (in particular for magnesium lack) when the leaves number 1 and 2 were too much stunted or tumbled down (numbering from the bottom).

A first observation which has been made is that independently of the growth conditions, for the oldest leaves (in an average, number 1 to 4 in the first series and up to 5 in the second series) the blue and green emissions were always higher in the main leaf vein than in the neighbouring tissue, whereas for the youngest leaves the contrast was opposite (Fig. 5A). For the red emission, as expected from the chlorophyll localisation, the tissue aside from the main leaf vein was always more fluorescent (Fig. 5B). Consequently, the fluorescence ratios values were calculated only on the leaf tissue aside the main leaf vein.

– three-week-old maize

The overall results showed that within one treatment, the fluorescence ratios could be considered as identical for leaves 3, 4, 5 and 6, and different from the ratios for leaves 1 and 2, particularly for the magnesium deficiency and in a lesser extent in the other cases. So we have analysed the data in considering two groups of leaves. The results are summarised in Table 4 and Fig. 6 which demonstrate that the fluorescence ratios allowed the detection of differences between control and deficient plants. For the group of leaves 3, 4, 5 and 6 the fluorescence ratios F440/F690 and F440/F740 were found to be enhanced by each deficiency, but the fluorescence ratio F440/F520 did not present any significant difference between the control and the other plants. For the chlorophyll fluorescence ratio F690/F740, a small variation was observed only in the case of zinc deficiency. For the older magnesium deprived leaves (number 1 and 2), an important increase of all the fluorescence ratios was apparent: the values of F440/F690 and F440/F740 were ten times greater than for the other plants. For these senescent, magnesium deprived leaves we observed an increase of the blue emission and a strong decrease of the red fluorescence which could be related to a decrease of the chlorophyll concentration as suggested by the simultaneous increase of the value of F690/F740. That magnesium deficiency essentially appeared on the first leaves of the maize plants is not surprising since it is known that the upper leaves can import their magnesium from the lower and older leaves. On the contrary, the lack of the oligoelements zinc and iron affects primarily the health of the upper leaves. It has to be noted that the magnesium deficient old maize leaves presented visible symptoms, which was not the case for the other deficiencies.

– five-week-old plants

At this stage of growth, besides the central leaf vein, alternation between the leaf tissue and the lateral veins could be clearly seen by eye on all the leaves. In addition, the upper iron deficient leaves are clearer than the others. As before, the fluorescence ratio values were calculated in excluding only the main leaf vein.

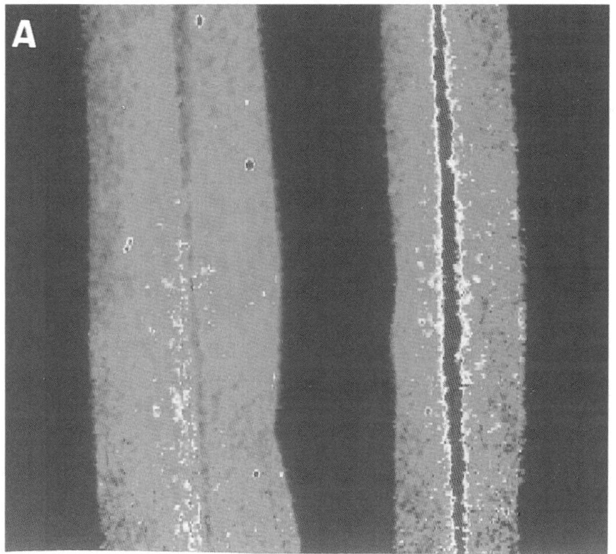

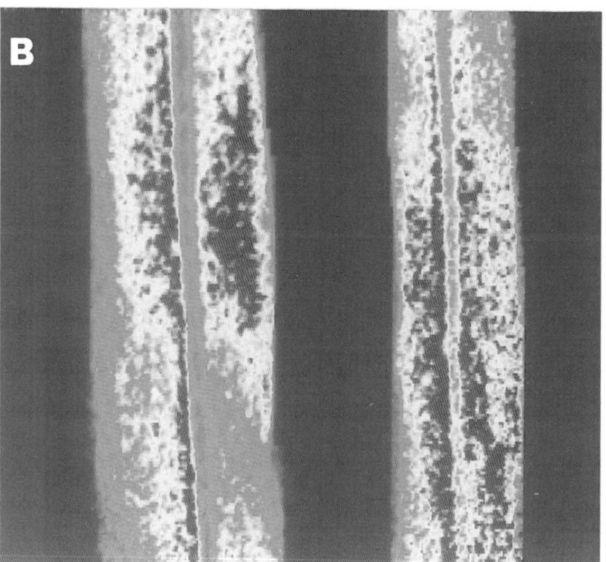

Fig. 5: Fluorescence images recorded in the blue region at 440 nm (Fig. 5 A) and in the chlorophyll fluorescence emission spectrum at 690 nm (Fig. 5 B) for leaf No. 4 (right) and leaf No. 6 (left) of the five-week-old greenhouse control maize. False-colour coding: increasing fluorescence intensity from blue via green, yellow to red.

For the analysis of the measurements, three groups of leaves were considered: the youngest (number 7 and 8, with also 9 for the control), the mature (number 5 and 6) and the senescent (number 3 and 4). Indeed, it was found that within each treatment the leaves of one group had the same fluorescence ratios, except for the first group in the case of zinc and iron deficiencies where all the ratios were higher in leaves 8 than in leaves 7, but with a difference much weaker than the difference to the other groups. When measured (excluded for magnesium deficient plants) leaves 1 and 2 were not different in their fluorescence ratios from leaves 3 and 4.

The results are collected in Table 5 and illustrated in Fig. 7. General findings were that (except for the magnesium deficient senescent maize leaves), *i)* both fluorescence ratios blue/red F440/F690 and blue/far-red F440/F740 strongly decreased from the top to the bottom of the plants, *ii)* in case of deficiency effect, these blue/red and blue/far-red fluorescence ratios were lower than for the control plants which is in contrast to what has been observed for the three-week-old plants, *iii)* a comparison with the data in Table 4 showed that for the control, magnesium and zinc deficient leaves 5 and 6, the fluorescence ratio F440/F690 markedly increased during growth while the ratio F690/F740 decreased. This could be due to the thickening of the epidermis or the accumulation of UV-absorbing compounds in the epidermis from which the major part of the blue emission originates and which hinders the penetration

Table 4: Greenhouse grown Volga maize. Mean values (with standard deviation) of the fluorescence ratios for two groups of leaves and for different treatments. Measurements were performed three weeks after planting.

	F440/F520	F440/F690	F440/F740	F690/F740
Leaves 3+4+5+6				
Control	1.95^{ab} (0.29)	0.21^{a} (0.05)	0.17^{a} (0.04)	0.82^{a} (0.05)
Minus Zn	2.12^{b} (0.16)	0.29^{b} (0.06)	0.23^{b} (0.04)	0.78^{b} (0.05)
Minus Mg	2.03^{bc} (0.46)	0.32^{b} (0.07)	0.26^{bc} (0.06)	0.81^{ab} (0.05)
Minus Fe	1.98^{ac} (0.13)	0.33^{b} (0.07)	0.29^{c} (0.07)	0.87^{a} (0.10)
Leaves 1+2				
Control	1.68^{a} (0.21)	0.24^{a} (0.15)	0.20^{a} (0.14)	0.79^{a} (0.05)
Minus Zn	2.17	0.26	0.18	0.70
Minus Mg	2.61^{b} (0.19)	2.09^{b} (1.06)	1.94^{b} (0.92)	0.94^{b} (0.09)
Minus Fe	2.08	0.15	0.11	0.77

For the first group the sample size was 16 and for the second 6 for control and 4 for magnesium deficient leaves. For zinc and iron lack only 2 leaves have been measured. Inside one leaves group and one ratio, the means with the same superscript were not significantly different at the 5 % level of probability ($p > 0.05$).

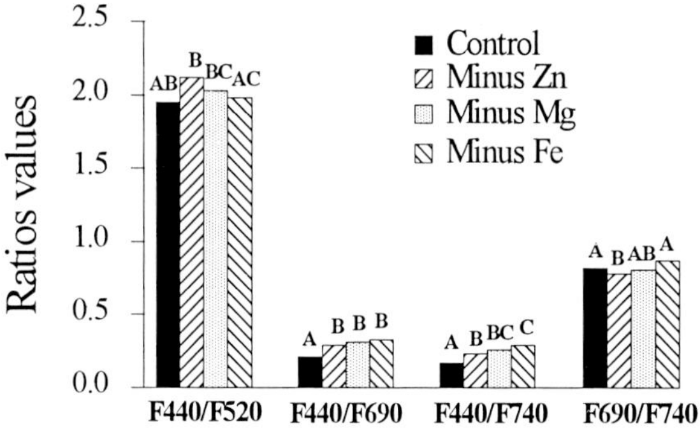

Fig. 6: Mineral nutrient deficiencies effect on the fluorescence ratios for the greenhouse grown Volga maize, investigated three weeks after planting. Mean values for the group of leaves 3, 4, 5 and 6 (see Table 4). For one fluorescence ratio, different letters indicate significant differences ($p < 0.05$).

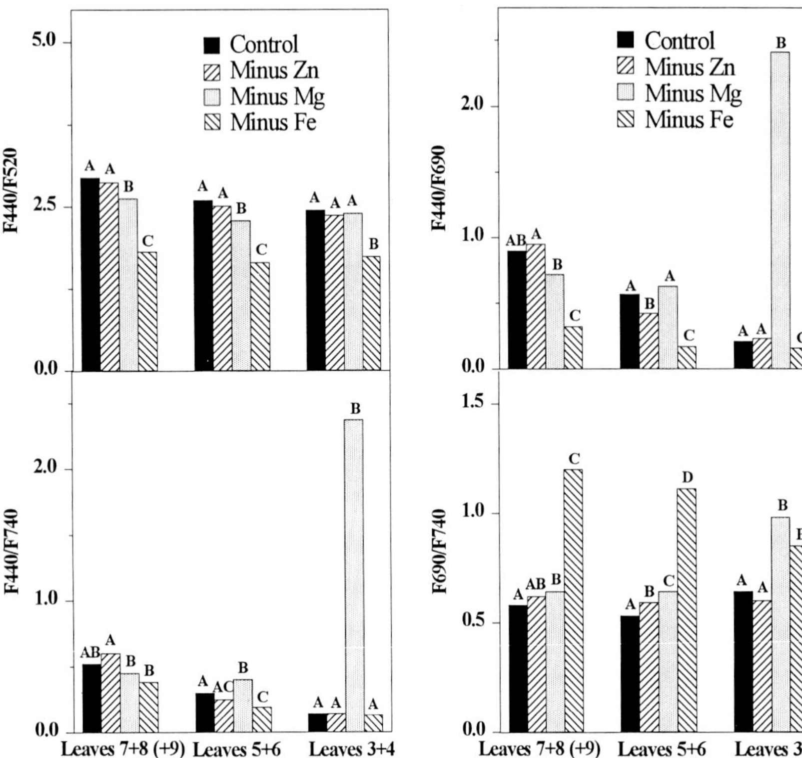

Fig. 7: Histogram of the different fluorescence ratios measured for different nutrient deficiencies of the five-week-old greenhouse Volga maize (see Table 5). The mean values of the fluorescence ratios have been determined for three groups of leaves: leaves 7+8 (+9 when available), leaves 5+6 and leaves 3+4. For one fluorescence ratio and one leaf group different letters indicate significant differences ($p<0.05$).

Table 5: Mean values (with standard deviation) of the fluorescence ratios for the greenhouse maize leaves measured after five weeks of growth at different nutritional conditions. For the first and second groups of leaves there were four repetitions of each leaf number, for the last group the sample size was 8 for control and minus Fe, and 7 and 6 for the zinc and magnesium deficiencies, respectively.

	F440/F520	F440/F690	F440/F740	F690/F740
Leaves 7+8 (+9)				
Control	2.94^{a} (0.18)	0.90^{ab} (0.19)	0.52^{ab} (0.09)	0.58^{a} (0.06)
Minus Zn	2.87^{a} (0.19)	0.95^{a} (0.18)	0.60^{a} (0.11)	0.62^{ab} (0.06)
Minus Mg	2.62^{b} (0.12)	0.72^{b} (0.22)	0.45^{b} (0.12)	0.64^{b} (0.04)
Minus Fe	1.81^{c} (0.06)	0.32^{c} (0.17)	0.38^{b} (0.19)	1.20^{c} (0.09)
Leaves 5+6				
Control	2.59^{a} (0.09)	0.57^{a} (0.07)	0.30^{a} (0.05)	0.53^{a} (0.05)
Minus Zn	2.51^{a} (0.12)	0.42^{b} (0.16)	0.25^{ac} (0.11)	0.59^{b} (0.03)
Minus Mg	2.28^{b} (0.19)	0.63^{a} (0.10)	0.40^{b} (0.05)	0.64^{c} (0.03)
Minus Fe	1.65^{c} (0.16)	0.17^{c} (0.03)	0.19^{c} (0.02)	1.11^{d} (0.10)
Leaves 3+4				
Control	2.44^{a} (0.20)	0.21^{a} (0.04)	0.14^{a} (0.02)	0.64^{a} (0.05)
Minus Zn	2.36^{a} (0.10)	0.23^{a} (0.07)	0.14^{a} (0.03)	0.60^{a} (0.04)
Minus Mg	2.39^{a} (0.23)	2.41^{b} (1.07)	2.37^{b} (1.11)	0.98^{b} (0.08)
Minus Fe	1.73^{b} (0.30)	0.16^{c} (0.04)	0.13^{a} (0.02)	0.85^{b} (0.18)

For one leaves group and one ratio the means followed by the same superscript were not significantly different at the 5 % level of probability ($p>0.05$).

of the UV-A excitation light so that less red and far-red chlorophyll fluorescence is excited in the leaf mesophyll cells.

For the two first groups of leaves, Table 5 and Fig. 7 display that the most striking changes in the fluorescence ratios resulted from the iron absence which led to variations of all the ratios: the F440/F690 ratios were about three times lower and F690/F740 twice higher than in the control maize. The intensities at 440 nm were roughly the same for both treatments and at 690 nm the fluorescence was strongly enhanced by the iron deficiency. The higher values of the chlorophyll fluorescence ratio F690/F740 indicate a lower concentration of the chlorophyll. These results would signify an important decrease in the photosynthetic capacity that is expected for iron absence during the plant growth. A small effect of zinc deficiency could be detected only for leaves 5 and 6 and as seen in the variations of the ratios F440/F690 and F690/F740. For the magnesium deficient plants, a decrease of the blue/green fluorescence ratio and an increase of F690/F740 with respect to the values measured for the control has been observed with an amount of only 10–15 %.

For the old leaves (3 and 4), the iron deprived leaves presented also a clear effect on the fluorescence ratios F440/F520 and F690/F740, but the most important variation was observed for the magnesium deficient plants for which an increase of the ratios F440/F690 and F440/F740 by a factor of more than ten was observed with practically the same values as those measured in the first series of experiments. The zinc deficiency, which was already difficult to detect in the earlier growth stage of maize plants did, in the senescent leaves, not lead to changes in the fluorescence ratios as compared to the control plants.

Table 5 and Fig. 7 also show that the deficiencies could always, with more difficulty for zinc, be distinguished by considering the amount of the variation of the ratios, which varied in a differential way.

Some of our results have to be compared to those obtained by Chappelle et al., 1984 b, who also studied some nutrient deficiencies on greenhouse maize, although their plants were seven weeks old. In case of iron absence they found, as demonstrated here for three-week-old maize plants, a considerable increase of the ratio F440/F690. For magnesium deficiency they did not observe any effect.

Finally, it has to be emphasized that the mean values of the fluorescence ratios, determined for practically the whole tissue surface, were only little sensitive to the zinc deficiency; the latter could only be detected by making use of the spatial distribution of the fluorescence intensities on the images. Indeed, the fluorescence was roughly uniformly distributed on the leaf tissue of the control and magnesium deficient leaves, which was not the case for the zinc deficient two penultimate leaves. For the latter the fluorescence images showed the presence of oblong area practically parallel to the main vein, with more blue and less red fluorescence emissions than on the average of the leaf surface (there was no visible symptom for the three-week-old plants). These fluorescence intensity inhomogeneities were enhanced in the images resulting from the fluorescence ratio pixel by pixel of the images recorded at 440 and 690 nm. An example is given in Fig. 8 A. Another outstanding feature to note is that for the five-week-old iron deprived plants, showing practically uniform blue emission, apparent streaks were observed on the images for the red fluorescence. These corresponded to the alternation between the lateral veins and the tissue with less intensity for the veins, while the blue emission was more uniform. This is illustrated by Fig. 8 B which represents the F440/F690 image for iron deficient leaves number 6 and 7.

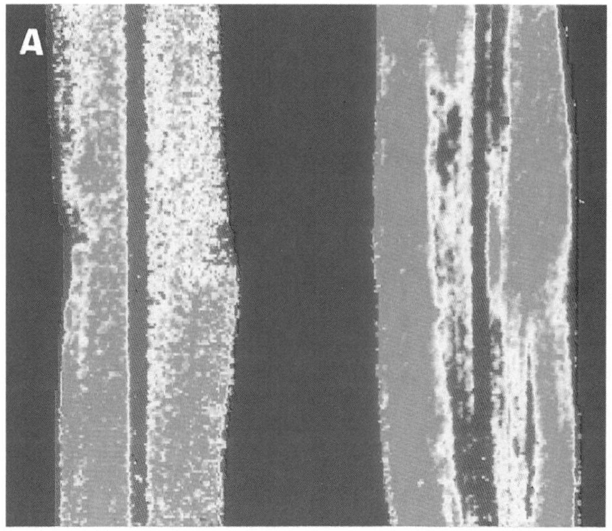

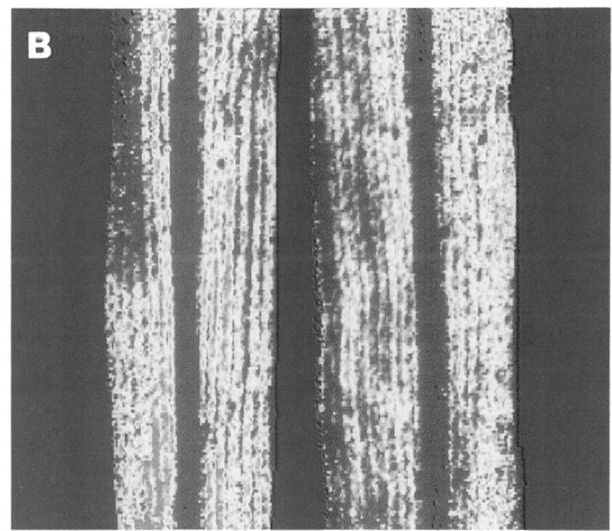

Fig. 8: Fluorescence ratio images F440/F690 of five-week-old greenhouse maize plants: A: control (left) and zinc deficient (right) leaves No. 6, B: iron deficient leaves No. 6 and 7 (from left to right). False-colour coding: increasing fluorescence intensity from blue via green, yellow to red.

Conclusion

This work has demonstrated in the case of maize plants, that the detection of laser-induced fluorescence of leaves with a very sensitive and high spatially resolved imaging system allows to make evident all the nutrient deficiencies we have investigated.

The difference in nitrogen fertilisation amounts was detected only by the blue/red and blue/far-red fluorescence ratios and our results emphasized the influence of the growth stage on the fluorescence ratios values and the importance of the age of the leaves.

Concerning the mineral deficiencies for the non-senescent leaves, the iron deficiency resulted in an important difference in the fluorescence ratio values between deficient and control plants. The magnesium and zinc deficiencies were also apparent in the change of the fluorescence ratios, but to a lesser extent and these could be discriminated only with difficulty.

Complementary results allowing discrimination between the deficiencies were obtained by taking advantage of the spatial resolution of the fluorescence imaging system. The control and magnesium deficient leaves showed uniform distribution of the emitted fluorescence over the leaf surface. For zinc deficiency the fluorescence images of mature leaves presented characteristic oblong area with fluorescence intensities different from those observed on a average of the leaf surface. In the case of iron absence, regularly distributed streaks appeared on the red fluorescence images of leaves of five-week-old plants.

These results are encouraging for future work on remote sensing of nutrient deficiencies by field measurements. It should be noted that with this new fluorescence imaging system, equipped with a telescope in front of the camera, well spatially resolved images can be and have been obtained at distance of up to 100 m applying a more powerful laser than the one described here.

Acknowledgements

The authors are grateful to Mrs. V. Lombaert of the Centre de Recherches de la Société Commerciale des Potasses et de l'Azote (SCPA, Aspach le Bas, France) for growth and care of the greenhouse Volga maize, and to Mrs. V. Gourbeau and Mr. F. Merkling of the Lycée Agricole d'Obernai (France) for placing at disposal the land parcels used for nitrogen supply testing, as well as for fruitful discussions and help in defining the experimental procedure.

References

Balachandran, S., C. B. Osmond, and P. F. Daley: Diagnosis of the earliest strain-specific interactions between tobacco mosaic virus and chloroplasts of tobacco leaves *in vivo* by means of chlorophyll fluorescence imaging. Plant Physiol. *104,* 1059–1063 (1994).

Brunel, P., M. Derrien, H. Legleau, and A. Marsouin: Routine mapping of vegetation index with the AVHRR of NOAA11. In: 5th Internat. Colloquium on Physical Measurements and Signature in Remote Sensing, Courchevel, France, 14–18 January 1991, ESA SP-319, Noordwijk, pp. 61–64 (1991).

Buschmann, C., U. Rinderle, and H. K. Lichtenthaler: Detection of stress of coniferous forest trees with the VIRAF spectrometer. IEEE Trans. Geosci. Remote Sens. *29,* 96–100 (1991).

Chappelle, E. W., J. E. McMurtrey, F. M. Wood, and W. W. Newcomb: Laser-induced fluorescence of green plants. 2: LIF caused by nutrient deficiencies in corn. Appl. Opt. *23,* 139–142 (1984 b).

Chappelle, E. W., F. M. Wood, J. E. McMurtrey, and W. W. Newcomb: Laser-induced fluorescence of green plants. 1: Technique for remote detection of plant stress and species differentiation. Appl. Opt. *23,* 134–138 (1984 a).

D'Ambrosio, N., K. Szabó, and H. K. Lichtenthaler: Increase of the chlorophyll fluorescence ratio F690/F735 during the autumnal chlorophyll breakdown. Radiat. Environ. Biophys. *31,* 51–62 (1992).

Edner, H., J. Johansson, S. Svanberg, H. K. Lichtenthaler, M. Lang, F. Stober, C. Schindler, and L. O. Björn: Remote Multi-Colour Fluorescence Imaging of Selected Broad-Leaf Plants. EARSeL Advances in Remote Sensing *3,* No. 3, 2–14 1995.

Edner, H., J. Johansson, S. Svanberg, and E. Wallinder: Fluorescence lidar multicolour imaging of vegetation. Appl. Opt. *33,* 2471–2478 (1994).

Goulas, Y., I. Moya, and G. Schmuck: Time resolved spectroscopy of the blue fluorescence of spinach leaves. Photosynth. Res. *25,* 299–307 (1990).

Günther, K. P., H.-G. Dahn, and W. Lüdeker: Remote sensing vegetation status by laser-induced fluorescence. Remote Sens. Environ. (Special Issue) *47,* 10–17 (1994).

Hagg, G., F. Stober, and H. K. Lichtenthaler: Pigment content, chlorophyll fluorescence and photosynthetic activity of spruce clones under normal and limited mineral nutrition. Photosynthetica *27,* 385–400 (1992).

Hák, R., H. K. Lichtenthaler, and U. Rinderle: Decrease of the chlorophyll fluorescence ratio F690/F730 during greening and development of leaves. Radiat. Environ. Biophys. *29,* 329–336 (1990).

Hák, R., U. Rinderle-Zimmer, H. K. Lichtenthaler, and L. Natr: Chlorophyll a fluorescence signatures of nitrogen deficient barley leaves. Photosynthetica *28,* 151–159 (1993).

Kautsky, H. and A. Hirsch: Chlorophyllfluoreszenz und Kohlensäureassimilation. 1. Mitteilung: das Fluoreszenzverhalten grüner Pflanzen, Biochem. Z. *247,* 423–434 (1934).

Kochubey, S. M., T. M. Shadchina, and N. S. Odinoky: Judgement of the nitrogen deficiency of crop plants by means of spectrofluorescence of leaves. Physiol. Biochem. Crop Plants (in Russian) *18,* 35–39 (1986).

Krause, G. H. and E. Weis: Chlorophyll fluorescence yield as a tool in plant physiology. II. Interpretation of fluorescence signal. Photosynth. Res. *5,* 139–157 (1984).

Lang, M. and H. K. Lichtenthaler: Changes in the blue-green and red fluorescence emission spectra of beech leaves during the autumnal chlorophyll breakdown. J. Plant Physiol. *138,* 550–553 (1991).

Lang, M., H. K. Lichtenthaler, M. Sowinska, F. Heisel, J. A. Miehé, P. Summ, and F. Tomasini: Application of laser-induced fluorescence imaging in the detection of plant stress. In: Richter, P. I. and R. C. Herndon (eds.): Proceed. 2nd Internat. Symposium and Exhibition on Environmental Contamination in Central and Eastern Europe, Budapest, pp. 88–90. Government Institutes, Inc., Rockville, Maryland 20850 (1995).

Lang, M., H. K. Lichtenthaler, M. Sowinska, P. Summ, and F. Heisel: Blue, green and red fluorescence signatures and images of tobacco leaves. Bot. Acta *107,* 230–236 (1994 a).

Lang, M., H. K. Lichtenthaler, M. Sowinska, P. Summ, F. Heisel, J. A. Miehé, and F. Tomasini: Sensing of plants using the laser-induced fluorescence imaging system. In: Proc. of the 6th Intern. Symposium on Physical Measurements and Signatures in Remote Sensing, Val d'Isère 1994, pp. 945–952. CNES, Toulouse (1994 b).

Lang, M., P. Siffel, Z. Braunova, and H. K. Lichtenthaler: Investigations on the blue-green fluorescence emission of plant leaves. Bot. Acta *105,* 435–440 (1992).

Lang, M., F. Stober, and H. K. Lichtenthaler: Fluorescence emission spectra of plant leaves and plant constituents. Radiat. Environ. Biophys. *30,* 333–347 (1991).

Lesaint, C. and Y. Coïc: Culture hydroponique. Techniques d'avenir. Maison rustique 1983.

Lichtenthaler, H. K.: Chlorophyll fluorescence signatures of leaves during the autumnal chlorophyll breakdown. J. Plant Physiol. *131,* 101–110 (1987).

– Applications of chlorophyll fluorescence in stress physiology and remote sensing. In: Steven, M. and J. A. Clark (eds.): Applications of Remote Sensing in Agriculture, pp. 287–305. Butterworths Scientific, London (1990).

Lichtenthaler, H. K. and C. Buschmann: Chlorophyll fluorescence spectra of green bean leaves. J. Plant Physiol. *129,* 137–147 (1987).

Lichtenthaler, H. K., M. Lang, and F. Stober: Nature and variation of the blue fluorescence spectra of terrestrial plants. In: Proc. Internat. Geoscience and Remote Sensing Symposium IGARSS '91, pp. 2283–2286. Helsinki University of Technology, Espoo (1991).

Lichtenthaler, H. K. and K. Pfister: Praktiken der Photosynthese. Quelle und Meyer Verlag, Heidelberg (1978).

Lichtenthaler, H. K. and U. Rinderle: The role of chlorophyll fluorescence in the detection of stress conditions of plants. CRC Crit. Rev. Anal. Chem. *19* (Suppl.), S29–S85 (1988).

Lipucci di Paola, M., P. Mazzinghi, A. Pardossi, and P. Vernieri: Vegetation monitoring of chilling stress by chlorophyll fluorescence ratio. EARSeL Adv. Remote Sens. *1,* 2–6 (1992).

McMurtrey, J. E., E. W. Chappelle, M. S. Kim, J. J. Meisinger, and L. A. Corp: Distinguish nitrogen fertilization levels in field corn (*Zea mays* L.) with actively induced fluorescence and passive reflectance measurements. Remote Sens. Environ. (Special Issue) *47,* 36–44 (1994).

Methy, M., A. Olioso, and L. Trabaud: Chlorophyll fluorescence as a tool for management of plant resources. Remote Sens. Environ. *47,* 2–9 (1994).

Morales, F., Z. G. Cerovic, and I. Moya: Characterization of blue-green fluorescence in the mesophyll of sugar beet (*Beta vulgaris* L.) leaves affected by iron deficiency. Plant Physiol. *106,* 127–133 (1994).

Slater, P.: Remote Sensing, pp. 438–515. Addison-Wesley, Reading, MA, 1980.

Snel, J. F. and O. van Kooten: The use of chlorophyll fluorescence and other non invasive spectroscopic techniques in plant stress physiology. Photosynt. Res. *25,* 146–332 (1990).

Stober, F., M. Lang, and H. K. Lichtenthaler: Blue, green and red fluorescence emission signatures of green, etiolated, and white leaves. Remote Sens. Environ. *47,* 65–71 (1994).

Stober, F. and H. K. Lichtenthaler: Changes of the laser-induced blue, green and red fluorescence signatures during greening of etiolated leaves of wheat. J. Plant Physiol. *140,* 673–680 (1992).

– – Characterisation of the laser-induced blue, green and red fluorescence signatures of leaves of wheat and soybean leaves grown under different irradiance. Physiol. Plant. *88,* 696–704 (1993).

Subhash, N. and C. N. Mohanan: Laser-induced red chlorophyll fluorescence signatures as nutrient stress indicator in rice plants. Remote Sens. Environ. *47,* 45–50 (1994).

Valentini, R., G. Cecchi, P. Mazzinghi, G. Scarascia Mugnozza, G. Agati, M. Bazzani, P. de Angelis, F. Fusi, G. Matteucci, and V. Raimondi: Remote sensing of chlorophyll *a* fluorescence of vegetation canopies: 2. Physiological significance of fluorescence signal in response to environmental stresses. Remote Sens. Environ. *47,* 29–35 (1994).

J. Plant Physiol. Vol. 148. pp. 632–637 (1996)

Remote Fluorescence Measurements of Vegetation Spectrally Resolved and by Multi-Colour Fluorescence Imaging

JONAS JOHANSSON, MATS ANDERSSON, HANS EDNER, JOHAN MATTSSON, and SUNE SVANBERG*

Department of Physics, Lund Institute of Technology, P.O. Box 118, S-221 00 Lund, Sweden

* Corresponding author

Received July 7, 1995 · Accepted October 2, 1995

Summary

A remote sensing system for combined fluorescence spectral recordings and multi-colour fluorescence imaging of vegetation will be presented. The system configuration will be explained and data from several field campaigns will be shown. The excitation light pulses for remote fluorescence detection are produced by a frequency tripled Nd:YAG laser emitting at 355 nm and shifted to 397 nm in a deuterium containing Raman cell. The receiving optics includes a Newtonian telescope and is shared by an optical fibre bundle connected to an OMA system and by a secondary Cassegrainian split-mirror telescope for fluorescence imaging at four wavelength bands simultaneously. The multi colour fluorescence images are computer processed to form images of a ratio at two wavelengths or if necessary a more complex mathematical function including the four wavelength bands is used. Examples of remote multi-colour imaging of vegetation from the past field campaigns will be given. Remote fluorescence spectra from several plant species were recorded at an excitation wavelength of 397 nm. Furthermore, the system was found capable of producing high-quality single-shot spectra at a range of 125 m. Lastly, a recently developed push-broom scanning technique for fluorescence imaging will be discussed and some preliminary data will be shown.

Key words: Laser-induced fluorescence, multi-colour fluorescence, vegetation, remote sensing, spectral analysis.

Abbreviations: LIDAR = light detection and ranging; CCD = charge coupled device; OMA = optical multi-channel analyser.

Introduction

The development of new techniques for remote monitoring of vegetation has recently become more important. The forests in Europe and elsewhere are endangered by environmental pollutants and it has become obvious that methods of early detection are needed. Satellite or air-borne multispectral imagery is useful for vegetation identification and also, to some extent, for monitoring of damage to vegetation (Rock et al., 1986). These techniques seem, however, to have limited usefulness. Laser-induced fluorescence has been shown to have a potential for remote sensing of plant health (See e.g. Svanberg, 1995). The red fluorescence from vegetation depends on both chlorophyll concentration and photosynthetic functioning and has been reported to change for plants exposed to different types of stresses such as water stress and nutrition deficiency (Chappelle et al., 1984; Krause and Weis, 1984; Lichtenthaler and Rinderle, 1988). Furthermore, laser-induced fluorescence is well suited for air-borne monitoring.

The first air-borne measurements utilising fluorescence were applied to the monitoring of marine constituents and oil spills at sea (Kim, 1973; Hoge and Swift, 1981). systems have also been tested for terrestrial monitoring of vegetation and have shown good potential also for this application (Hoge et al., 1983; Zimmermann and Günther, 1986; Cecchi et al., 1994). With the implementation of optical multichannel analysers (OMA) into remote fluorosensor systems, full fluorescence spectra have been monitored (Rosema et al., 1988; Edner et al., 1992 a). Since fluorescence spectra from green plants are very complex and depend on multiple factors, a broad spectral coverage would be desirable to be able to detect subtle damages to vegetation. So far, most measurements have been performed in a point monitoring mode. However, full imaging would be advantageous in terms of understanding and interpreting the results and would also yield better statistics than pointwise sampling. The first results on remote fluorescence imaging on vegetation were reported in 1992 (Edner et al., 1992 a). Here a multi-colour system facilitated simultaneous imaging at four wavelengths utilising a split mirror telescope (Edner et al., 1994, Edner et al., 1995). This system is very well suited for short range imaging, while air-borne application is limited by a low sensitivity. Hence, a different technique for remote fluorescence imaging was recently implemented into the system. This is a push-broom line scanning technique with simultaneous recordings at two wavelength bands. The laser beam is shaped to a linear streak of a length of about 2 m at 50 m distance and the fluorescence is imaged through the receiving Newtonian telescope and a secondary telescope consisting of two Fresnel lenses onto an image-intensified CCD detector. An important advantage of the push-broom system is that the images are constructed from single-shot data. In this paper we report on remote fluorescence imaging of vegetation and spectral point monitoring from areas within the images. Comparisons of the two imaging systems and their respective applications will be discussed here.

Material and Methods

One spectrally resolving point monitoring OMA system and two different multi-colour imaging systems, are described here. The first imaging system provides 2D imaging of a remote target, while the other is a one-dimensional line scanning (push-broom) system. The OMA system is utilised for pointwise spectral recordings at specific features within the target area, such as single leaves. The light source of the lidar system is a Nd:YAG laser with frequency doubling/tripling. In order to induce chlorophyll fluorescence efficiently while still staying eyesafe, the output from the frequency-tripled Nd:YAG laser (355 nm, 200 mJ pulse energy) was Raman-shifted in a high-pressure deuterium cell to generate radiation at 397 nm. An output pulse energy of 30 mJ was achieved at 20 Hz. The outgoing beam is directed coaxially with a vertically mounted 0.40 m diameter telescope and is transmitted towards the target via a large flat mirror in a retractable transmitting/receiving dome on the roof. Computer controlled stepper motors are used to turn the dome and to tilt the mirror. The fluorescence light was collected by the lidar telescope and directed to a fibre coupled OMA system (Spectroscopy Instruments model ICCD-576) or to a secondary telescope. The OMA based system has been described by Edner et al. (1994). In the case of 2-D imaging, the secondary telescope was a Cassegrainian telescope with its first mirror cut into four segments which can be individually adjusted. By tilting the mirror segments, each segment produces an image at the image intensifier; in total four identical images arranged as four quadrants on an image-intensified CCD detector (Spectroscopy Instruments model ICCD-576). The gate width of the image intensifier was set at 20 ns. Furthermore, different

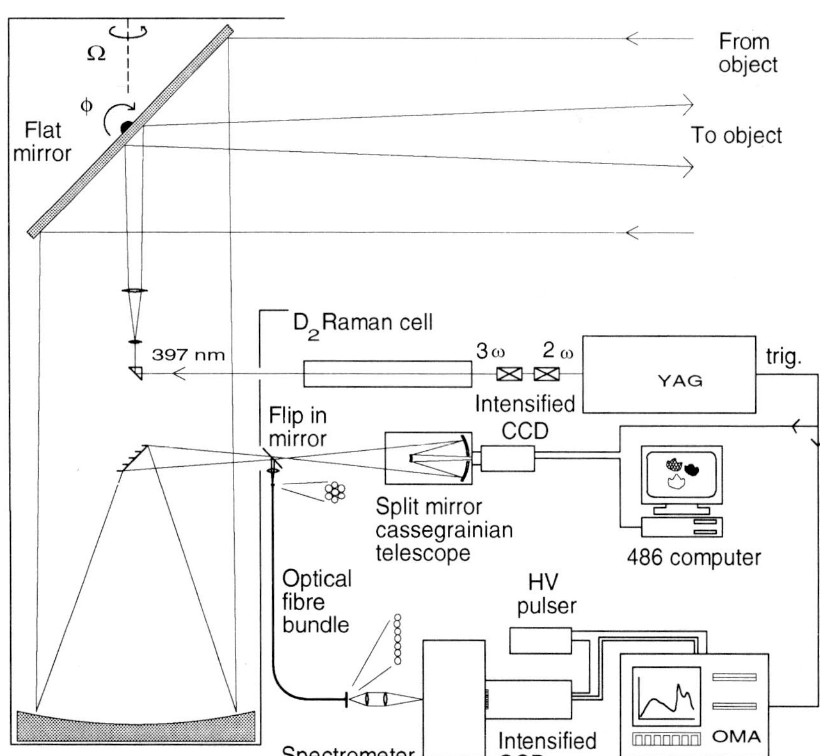

Fig. 1: Set-up of a fluorescence lidar system for point monitoring and multi-colour imaging.

interference filters or Schott coloured-glass filters were placed in front of each mirror segment matching specific features of the fluorescence spectra. Such features may include the broad-band blue-green fluorescence structures sampled e.g. at 450 nm and 520 nm and the chlorophyll fluorescence peaks at 685 nm and 740 nm. The fluorescence images are read out to a PC 486 computer. Computer processing makes it possible to generate a new image, pixel by pixel, from the four sub-images using a suitably designed spectral contrast function, which enhances features of interest. One example of such a spectral contrast function is the ratio I(685nm)/I(740nm), which yields an image that is related to the chlorophyll concentration of the target. The resulting image is shown on the screen in false colour. The system set-up is shown in Fig. 1 and was described by Edner et al. (1994).

In the case of the push-broom line scanning system, the laser beam was re-shaped to a linear streak. The length of the streak was about 2 m at 50 m distance (corresponding to 10 m at 250 m distance) and the width of the streak was about 2 cm. At the output of the Newtonian telescope, a secondary telescope was formed by two Fresnel lenses and an Al-coated plane mirror cut in two halves. Using different optical filters for each mirror segment, the image was split into two, identical images at different wavelengths. The pixels of the CCD were vertically binned into two groups forming two horizontal line scans of the fluorescence induced by the laser streak. These fluorescence intensity profiles were thus recorded simultaneously at the two selected wavelengths. Images of a size of about 2×2 m were scanned for several plant species including *Picea abies* at a distance of about 60 m. The excitation source and the receiving Newtonian telescope as well as the CCD and the electronics were the same as in Fig. 1. A close-up of the receiving optics for the push-broom system is shown in Fig. 2.

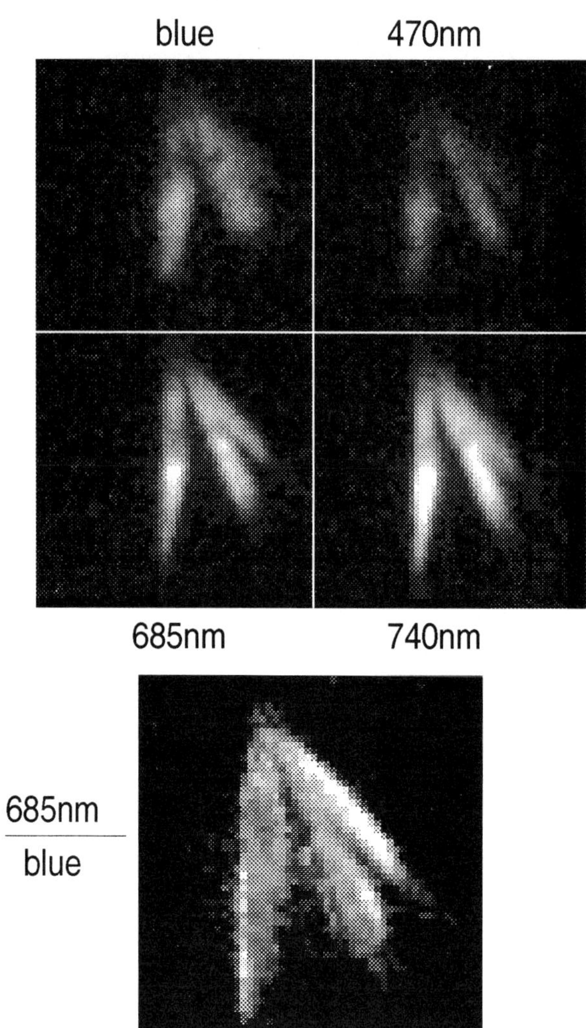

Fig. 3: Multi-colour fluorescence images (upper) and a ratio image (lower) of a maize plant. The filters used were a Schott BG7 (blue) filter and interference filters at 470, 685 and 740 nm, respectively. The ratio image was produced by division of the 685 nm and the blue image. The target distance was 40 m.

Results and Discussion

Multi-colour fluorescence images and fluorescence spectra were remotely recorded for test targets as well as for a number of different plant species. For images on a leaf level, the 2D imaging system was utilised, while the push-broom system was used for imaging on a canopy level.

A set of multi-colour fluorescence images from a maize plant at 40 m distance recorded with the 2D imaging system, is shown in Fig. 3. Fluorescence images were simultaneously recorded at four wavelength bands using a Schott BG7 (blue-green region) filter and interference filters at 470, 685 and 740 nm, respectively. The fluorescence was integrated over 100 laser shots. In the lower part of the figure, a computer processed ratio image displaying the 685 nm/"blue" ratio of the maize plant is shown. From the ratio image, a high value of the ratio is found for the smaller leaf to the right. This can also be understood by comparing the first and the third multi-colour sub-images (blue and 685 nm). In Fig. 4, remote fluorescence spectra from a part of the maize plant is shown. The first spectrum (left) is integrated over 100 laser shots, while the second (right) is a single-shot spectrum. It is interesting to note that high quality single-shot fluorescence spectra can be recorded at a distance of 40 m. Single-shot spectra at varying distance of up to 125 m were also successfully obtained. This suggests that it would be possible also to obtain remote single shot spectra from an air-borne platform,

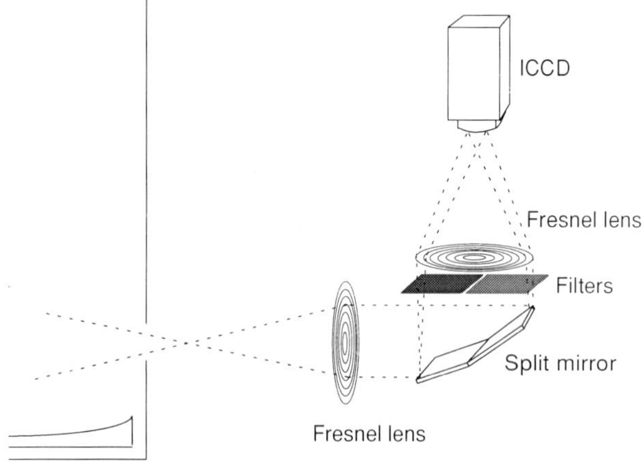

Fig. 2: Close-up of the telescope for the push-broom line scanning system for multi-colour imaging.

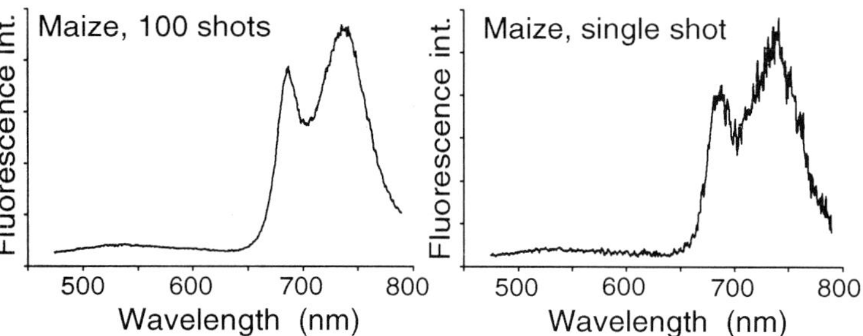

Fig. 4: Remotely recorded fluorescence spectra of a maize plant at 40 m distance. The fluorescence was integrated over 100 laser shots (left) and 1 shot (right), respectively.

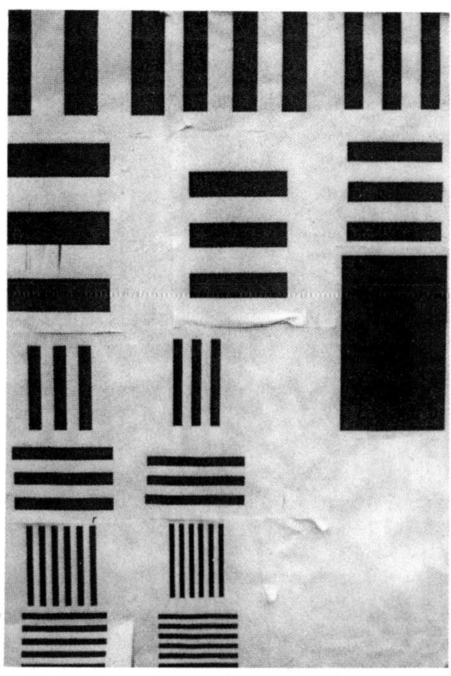

Fig. 5: (a) Photograph of a resolution test pattern made of black and white paper. (b) Scanned fluorescence image of a spatial resolution pattern. The size of the pattern was 70 × 100 cm. The target distance was 60 m.

in particular if a slight smoothing would be applied to the spectra, which was not the case here.

One limitation of the 2D imaging system is the poor signal-to-noise characteristics of single shot images (not shown here). This is partly due to a very low laser power density for larger target areas such as for whole trees. Therefore, a push-broom system utilising a line scanning technique, was developed. The first test object was a pattern of vertical and horizontal stripes on a white paper board that was utilised to determine the spatial resolution of the push-broom system. The width of the paper board was approximately 70 cm and the target distance was about 60 m. A photograph of the test pattern and the corresponding scanned fluorescence image is shown in Fig. 5. Modulation transfer functions (MTF) for the vertical and the horizontal directions were calculated and are shown in Fig. 6. As can be seen from the figure, the resolution was better for the vertical direction compared with the horizontal. An MTF value of 0.5 was reached for line pairs separated by 3 cm for the vertical direction (the scanning direction) and 5 cm for the horizontal direction. The superior vertical resolution can be explained in terms of the scanning direction. The fluorescence from the horizontal light streak is

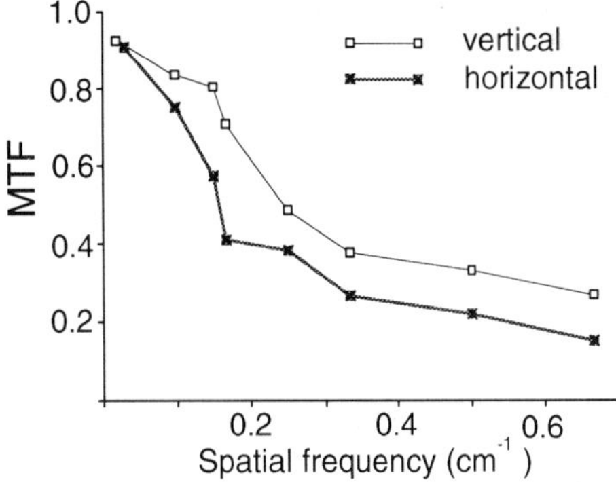

Fig. 6: Modular transfer functions (MTF) calculated from the fluorescence image in Fig. 5.

imaged on the CCD with some image aberrations. However, since the CCD is binned in the vertical direction, all available signal can be considered to originate from one and the same vertical position. In other words, we know that the fluorescence must originate from the laser illuminated streak and the resolution will thus be determined by the 2 cm height of the laser streak. In the horizontal direction, on the other hand, no binning is performed because this is the imaged direction. Thus, in the horizontal direction the resolution will be limited by image aberrations, which are fairly large for Fresnel lenses.

Remote fluorescence images were recorded for several plant species using the push-broom system. Initially, fluorescence images were scanned at one wavelengths at a time only, without the split mirror arrangement. In these cases sequential recordings at the interesting wavelengths were performed. The ratio image was formed by calculating the ratio, line by line, through the image. Such a set of sequential images were recorded for spruce trees (*Picea abies*) using different optical filters. In Fig. 7, an example of a set of fluorescence images recorded at 480, 685 and 740 nm, are shown. The target distance was also in this case 60 m and the height of the spruce was about 2 m. To the lower right in Fig. 7, a ratio image formed by division of the 685 and 740 nm images, is shown. As can be seen, the individual branches can easily be recognised in the fluorescence images as well as in the ratio image. In the middle part of the ratio image a white vertical streak indicating a high ratio, is seen. Measurements with the OMA system (data not shown) showed that this feature originated from the stem of the tree, which was confirmed to have a high value of the 685/740-nm ratio.

In the following measurements, the split mirror arrangement was included between the Fresnel lenses and fluorescence images were scanned at two wavelengths simultaneously. The CCD read out generated two separate scans corresponding to the two binned areas on the CCD chip and the ratio image was formed in the same way as above. In Fig. 8, fluorescence images recorded with Schott colour glass filters BG7 (blue) and RG645 (red), are shown. In this case the ratio image was formed by division of the "blue" and the "red" images. As can be seen, a clear ratio image of a 2 m spruce tree can be obtained also in this double scan mode of the push-broom system. It should be pointed out here, that there were difficulties in adjusting the two detection paths to obtain a sharp focus in both of them (see «red» fluorescence image in Fig. 8). This was due to the limitations in this early version of the set-up but will be replaced by a more ridgid set-up in the future.

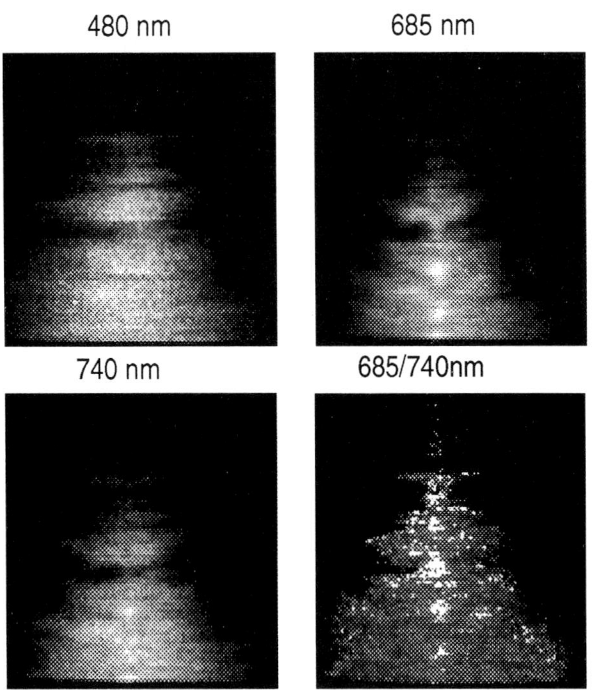

Fig. 7: Sequentially scanned fluorescence images of *Picea abies*. The selected wavelengths were 480, 685 (upper) and 740 nm (lower left). To the lower right a ratio of 685nm/740nm is shown. The spectral width of the filters were 10 nm.

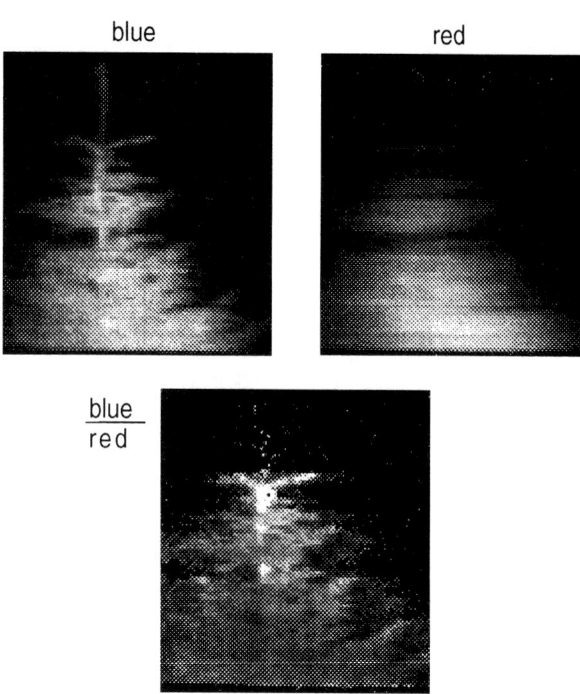

Fig. 8: Simultaneously scanned fluorescence images of *Picea abies* (upper) and a ratio image calculated by division of the «blue» by the «red» image. The optical filters were Schott BG7 (blue) and RG645 (red).

In this paper, a system for remote fluorescence recordings of vegetation, including an OMA system for spectral recordings and two different multi-colour imaging arrangements, has been presented. The OMA system is valuable for spectral recording within an image frame for a better understanding of the recorded images. The two different set-ups have their advantages and come to use in different situations. The 2-D imaging system provides a high resolution but is limited by the poor single-shot sensitivity at longer ranges and by a smaller field of view (0.5 m at 50 m distance). Thus, this system is best suited for short distance measurements where high quality images are required. The push-broom scanning system, on the other hand, provides high signal/noise characteristics at longer ranges and also a larger field of view. Therefore, the push-broom system is ideal for air-borne measurements, which are performed at long distance and where single-shot measurements are an absolute requirement due to the fast movement of an aircraft.

Acknowledgements

This work was supported by the Swedish Natural Science Research Council and by the Swedish Space Board.

References

Cecchi, G., M. Bazzani, V. Raimondi, and L. Pantani: Fluorescence lidar in vegetation remote sensing: system features and multiplatform operation. In: Proc. Internat. Geoscience Remote Sensing Symposium IGARSS '94, *1*, 637–639. California Institute of Technology, Pasadena, USA (1994).

Chappelle, E. W., F. M. Wood, and W. W. Newcomb: Laser induced fluorescence of green plants. 1: A technique for remote detection of plant stress and species differentiation. Appl. Opt. *23*, 134–138 (1984).

Edner, H., J. Johansson, S. Svanberg, E. Wallinder, M. Bazzani, B. Breschi, G. Cecchi, L. Pantani, B. Radicati, V. Raimondi, D. Tirelli, G. Valmori, and P. Mazzinghi: Laser-induced fluorescence monitoring of vegetation in Tuscany, EARSeL Advances in Remote Sensing *1*, 119–130 (1992 a).

Edner, H., J. Johansson, S. Svanberg, and E. Wallinder: Remote Multi Color Imaging of Vegetation Laser-Induced Fluorescence. Proc. CLEO'92, 432–433, Anaheim, Ca, USA (1992 b).

Edner, H., J. Johansson, S. Svanberg, and E. Wallinder: Fluorescence lidar multi-color imaging of vegetation, Appl. Opt. *33*, 2471–2479 (1994).

Edner, H., J. Johansson, S. Svanberg, H. K. Lichtenthaler, F. Stober, C. Schindler, and L.-O. Björn: Remote multi-colour fluorescence imaging of selected broad-leaf plants. EARSeL Advances in Remote Sensing *3*, No. 3, 2–14 (1995).

Hoge, F. E. and R.N. Swift: Airborne simultaneous spectroscopic detection of laser-induced water Raman backscatter and fluorescence from chlorophyll *a* and other naturally occurring pigments, Appl. Opt. *20*, 3197–3205 (1981).

Hoge, F. E., R. N. Swift, and J. K. Yungel: Feasibility of airborne detection of laser-induced fluorescence emission from green terrestrial plants. Appl. Opt. *22*, 2991–3000 (1983).

Kim, H. H.: New algea mapping technique by use of an airborne laser fluoresensor. Appl. Opt. *12*, 1454–1459 (1973).

Krause, G. H. and E. Weis: Chlorophyll fluorescence as a tool in plant physiology. II. Interpretation of fluorescence signals. Photosynth. Res. *5*, 139–157 (1984).

Lichtenthaler, H. K. and U. Rinderle: The role of chlorophyll fluorescence in the detection of stress conditions in plants. CRC Critical Reviews in Analytical Chemistry *19*, S29–S85 (1988).

Rock, B. N., J. E. Vogelmann, D. L. Williams, A. F. Vogelmann, and T. Hoshizaki: Remote detection of forest damage. BioScience *36*, 439–445 (1986).

Rosema, A., G. Cecchi, L. Pantani, B. Radicati, M. Romuli, P. Mazzinghi, O. van Kooten, and C. Kliffen: Results of the ‹LIFT› project: Air pollution effects on the fluorescence of Douglas fir and poplar, In: Lichtenthaler, H. K. (ed.): Applications of Chlorophyll Fluorescence, pp. 307–317. Kluwer Academic Publishers, Dordrecht (1988).

Svanberg, S.: Fluorescence Lidar Monitoring of Vegetation Status. Physica Scr. *T58*, 79–85 (1995).

Zimmermann, R. and K. P. Günther: Laser-induced chlorophyll *a* fluorescence of terrestrial plants, In: Proc. International Geoscience and Remote Sensing Symposium IGARSS '86, 1609–1613, Zürich. ESA Publications Division, Nordwijk (1986).

J. Plant Physiol. Vol. 148. pp. 638–644 (1996)

Remote Sensing of Plants by Streak Camera Lifetime Measurements of the Chlorophyll *a* Emission*

Malgorzata Sowinska[1], Francine Heisel[1], Joseph Albert Miehé[1], Michael Lang[2], Hartmut K. Lichtenthaler, and François Tomasini[3]

[1] Groupe d'Optique Appliquée, Centre de Recherches Nucléaires, IN2P3/CNRS, 23, rue du Loess, F-67037 Strasbourg Cedex 2, France

[2] Botanisches Institut II, Universität Karlsruhe, Kaiserstr. 12, D-76128 Karlsruhe, Germany

[3] Applications de la Recherche en Photonique, Centre de Transfert de Technologie, Route de Hausbergen, F-67309 Schiltigheim Cedex, France

Received August 10, 1995 · Accepted October 10, 1995

Summary

The possibility of time resolved chlorophyll *a* fluorescence measurements using an intensified and gated streak camera was demonstrated on different plants at a distance of 40–80 m. The investigations, carried out in various experimental conditions: different target geometries (one planar leaf, tilted leaves, sparce foliage, separated levels of leaves), variable slit widths (0.1 to 1 mm) of the camera and measurements at different weather conditions have shown a good reproducibility of the results. The decay profiles were analyzed in order to obtain the best fit between the experimental curves and those obtained by convoluting the experimental response function $R_{exp}(t)$ with the theoretical emission law $F(t)$ modelled by a sum of two or three exponentials. The mean value τ_m of the chlorophyll fluorescence lifetimes was determined for each experimental condition. For example for poplar leaves measured in October 1993, τ_m was equal to ≈ 0.8 ns in the morning, decreased strongly to ≈ 0.4 ns in midday and then slowly increased to 1.21 ns in the night.

Key words: Catalpa bignonioides WALT, Pinus sylvestris L., Populus spec., CCD camera, chlorophyll fluorescence, fluorescence lifetimes, streak camera.

Abbreviations: CCD = charge coupled device; FWHM = full width at half maximum; $R_{exp}(t)$ = experimental response function (reflectance); $F(t)$ = theoretical emission law; $I(t)$ = experimental fluorescence decay; τ_1, τ_2 and τ_3 = fast, medium and slow decay times; τ_m = mean value of the fluorescence lifetime.

Introduction

In the last years, non-invasive methods like spectroscopic measurements have been developed for the determination of the photosynthetic activity of intact leaves. Indeed, after light absorption by the various photosynthetic plant pigments (carotenoids, chlorophyll *a* and *b*), a minor part of the absorbed energy is reemitted as red chlorophyll fluorescence. For excitation by UV-A radiation the emitted fluorescence shows maxima/shoulders near to 440, 520, 690 and 740 nm, whereby only the last two peaks are emitted by the chlorophyll *a* (Chappelle et al., 1984; Lang et al., 1991, 1992, 1994). It had been shown that the chlorophyll fluorescence quantum yield varies in an antiparallel way to the photochemical conversion efficiency in photosynthesis (Kautsky and Hirsch, 1931; see review Lichtenthaler, 1992) and can be used to characterize the ecophysiology of plants (Lichtenthaler and Rinderle, 1988; Rinderle and Lichtenthaler, 1989; Genty et al., 1989; Lichtenthaler, 1988, 1990; Krause and Weis, 1991). More-

* This work has been performed in the frame of the EUREKA LASFLEUR Project (EU380).

over, in a general manner, there is a linear relationship between the mean fluorescence lifetimes and the quantum yields (Moya et al., 1986; Goulas, 1992; Schmuck et al., 1992). It results that both fluorescence characteristics can be used as a signature of the plant status and in particular for studying the physiological state of the stressed vegetation, and the usefulness of lifetime measurements for ecophysiological applications has been demonstrated (Schneckenburger and Frenz, 1986; Evans et al., 1992; Schmidt and Schneckenburger, 1992). In practice chlorophyll fluorescence quantum yield determination is very difficult especially by remote intensity measurements since the collected light strongly depends on uncontrollable parameters like the foliage surface reached by the excitation beam, the emission or absorption by the canopy, the distance etc. All these parameters do not influence the fluorescence decays and consequently experimental systems have been proposed for far field determination of the fluorescence decay curves (Moya et al., 1988; Schmuck et al., 1992; Goulas et al., 1993).

Here we describe and demonstrate the usefulness of a streak camera for fluorescence lifetime measurements at long distances.

Materials and Methods

Plant materials

All the measurements were performed on the trees growing on the Campus of the Centre de Recherches Nucléaires at Strasbourg: *Catalpa* (*Catalpa bignonioides* WALT) green leaves, poplar (*Populus spec.*) green and yellowish leaves, pine tree (*Pinus sylvestris* L.) green needles or branches. The plants were selected because of the different dimensions, geometries and densities of leaves and they were accessible for the measurements from our laboratory.

Experimental set-up

A schematic representation of the vegetation fluorescence lifetimes measuring system is given in Fig. 1. The excitation of the foliage is provided by a cw mode locked, Q-switched and cavity dumped Nd:YAG laser which after frequency doubling (tripling) delivered pulses at 532 nm (355 nm) with 100 ps duration, 100 μJ (10 μJ) energy and a repetition rate of 1 kHz. The laser beam was directed onto one or several leaves by means of flat mirrors free to turn and diffused by light expanders so that the size of the excitation spot could be changed. The back reflectance as well as the emitted fluorescence were collected with a Celestron 8 AZC 872 telescope (Schmidt Cassegrain conception): diameter = 203 mm, focal length = 2000 mm, aperture ratio *f*/10. The minimum focalisation distance is equal to 7.6 m with a visual field of 1.3 m at 100 m and a spatial resolution of 210 l/mm. The image at the telescope output was then focalized (*f*/50 lens) on the entrance slit of a streak camera (CBF-T1-ARP, Strasbourg, France), which allows real time measurements of the temporal profile of light phenomena. The streak camera is operating in the repetitive mode with a triggering jitter near to 10 ps (Geist et al., 1983) and is gated which allows fluorescence measurements even in the presence of ambient light (adjustable width >100 ns to 2 μs). The detailed description of the streak camera working is presented elsewhere (Heisel et al., 1979; Geist et al., 1984). The spatial streak appearing on the phosphor screen of the tube is then intensified (Photek image intensifier tube MCP-125 with adjustable gain up to 1000) and recorded on a 8 bit and 50 images per second numerical CCD camera of 384×288 pixels (one pixel is of 23×23 μm^2). The digitized images are stored in a computer using the ANIMATER-V1 software which allows realtime monitoring, addition and/or subtraction of successive readouts and the image treatment.

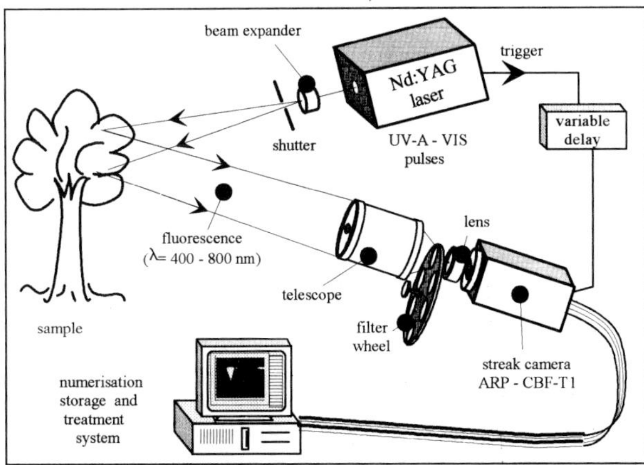

Fig. 1: Experimental set-up for the far-field fluorescence lifetime measurements.

The intrinsic temporal resolution of the camera (≈3 ps) is not a limiting factor for the experiments performed with the described Nd:YAG laser. With an entrance slit of 100 μm and for direct illumination of the camera by the attenuated laser beam, the full width at half maximum (FWHM) of the registered histogram was measured to be 110 ps, value corresponding to the FWHM of the laser pulses. The temporal definition of one pixel can be adjusted from 5 to 60 ps and the software allows to vary the acquisition duration.

It should be noted that the experimental system presented here can be successfully used for UV-A plant excitation and hence fluorescence detection at the four characteristic emission bands (440, 520, 690 and 740 nm) through 10 nm interference filters placed in a filter wheel. With the laser used for our experiments, measurements for 355 nm excitation were possible only at short distances (few meters) due to the laser excitation intensity, so all results here presented and discussed concern only chlorophyll fluorescence induced by 532 nm laser beam.

For the fluorescence recording, the emitted chlorophyll fluorescence was selected with a Corning CS 2-63 longpass filter ($\lambda > 585$ nm). The backscattered light, much more intense than the fluorescence light, was detected through a neutral filter, necessary to avoid the saturation of the streak camera. Both, the fluorescence and reflection were simultaneously registered owing to the optics particularly adapted for these measurements. For each measurement a subtraction of the background due to the ambient light was carried out by making use of the shutter situated at the front of the laser.

Decay curves convolution

The measured fluorescence decay curve $I(t)$ is the convolution of the temporal emission law via (response to a δ-excitation) with the experimental response function $R_{exp}(t)$:

$$I(t) = F(t) \otimes R_{exp}(t) \text{ where } R_{exp}(t) = L(t) \otimes R_i(t).$$

$L(t)$ is the excitation function of the plants by the laser spot and $R_i(t)$ the temporal instrumental response function of the camera. In practice, for each determination of $F(t)$ it is necessary to measure in the same experimental condition the fluorescence decay $I(t)$ and the response $R_{exp}(t)$ assumed to be given by the recording of the backscattered excitation pulses.

The first step of this work was to examine the effect of various experimental conditions on $R_{exp}(t)$ and on the fitted values of the parameters a_i, τ_i and τ_m characterizing the $F(t)$ law modelled by a sum of exponentials:

$$F(t) = \sum_i a_i \exp(-t/\tau_i)$$

with the mean lifetime τ_m defined as

$$\tau_m = \sum_i a_i \tau_i^2 / \sum_i a_i \tau_i$$

The values a_i and τ_i were extracted from the measured $I(t)$ decays by a numerical fitting method based on an iterative convolution procedure using the Marquardt algorithm (Marquardt, 1963).

Results and Discussion

Influence of the width of the camera entrance slit

The instrumental response function $R_i(t)$ depends not only on the intrinsic temporal resolution of the camera but strongly on the slit width. $R_i(t)$ will be deteriorated if the entrance slit width (equal to 100 µm in laboratory conditions) is increased for the detection of very weak fluorescence signals. Furthermore, long distance measurements conditions with the use of collecting optics (telescope-lens) can also lead to a broadening of $R_{exp}(t)$. Consequently, experiments have been performed in order to measure $R_{exp}(t)$ and $F(t)$ for a slit width varying from 0.1 to 1 mm by steps of 0.1 mm and that for one individual leaf (*Catalpa*) or a set of several leaves on a branch (poplar, pine) situated at the distance of 7 m from the telescope-camera system. On the leaf level, the diameter of the laser spot was equal to 20 cm diameter. Solar illumination was simulated by means of a 50 W halogen lamp. In Fig. 2 the $R_{exp}(t)$ and the chlorophyll fluorescence decay curves are represented for a *Catalpa* leaf (of about 20–30 cm diameter); these curves have been obtained with two different slit widths of the camera entrance. As can be seen in Table 1 (i) the FWHM of $R_{exp}(t)$ increased from 200 ps (Fig. 2 a) (value larger than for short distance punctuate excitation) to 820 ps (Fig. 2 b) for 0.1 and 1 mm slit, respectively; (ii) the fluorescence decays were clearly not monoexponential and (iii) a good agreement was obtained between the experimental and calculated curves. For example with 0.1 mm slit we have deduced, for the *Catalpa* leaf, three decay constants of 0.18, 0.6 and 1.0 ns giving a mean lifetime τ_m of 0.41 ns. Although with broader widths of $R_{exp}(t)$ the estimated value of the shortest decay constant had no meaning, the calculated τ_m = 0.47 ns value was only slighty affected by the deterioration of the resolution. For the other leaves, in all the cases τ_m could be evaluated with a good precision even for an entrance camera slit width of 1 mm (poplar: yellowish leaves τ_m = 1.25 ns, green leaves τ_m = 0.54 ns, pine needles τ_m = 0.54 ns). The accuracy of the measurements was better than 10 % (see Table 1).

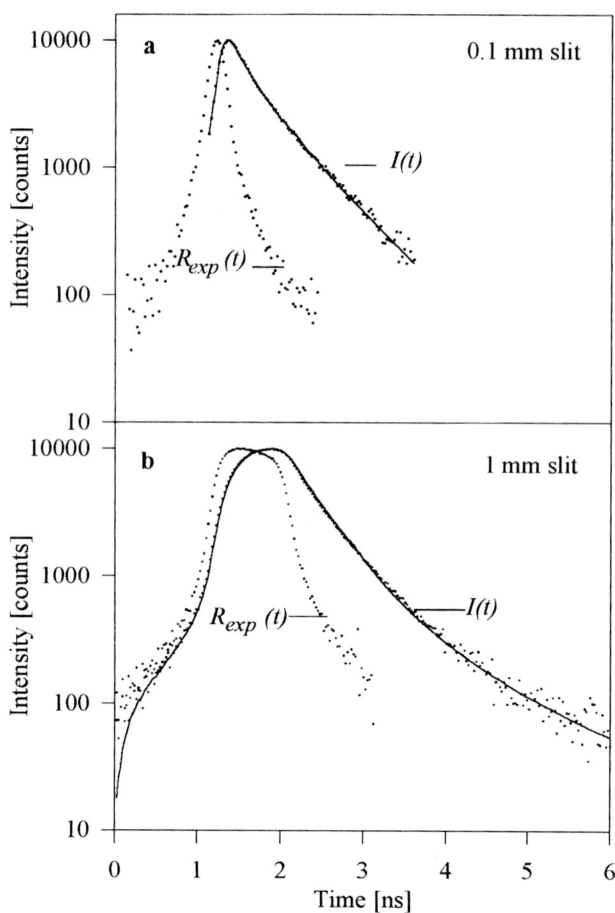

Fig. 2: Backscattered light profiles, $R_{exp}(t)$ and chlorophyll fluorescence decays, $I(t)$ of *Catalpa* leaves recorded at a distance of 7 m and for two slit widths of the camera: a) – 0.1 mm and b) – 1 mm (····· experimental curves; full line – fitted curves).

Table 1: Full widths at half maximum (FWHM) of the response $R_{exp}(t)$ and calculated values of mean fluorescence lifetime τ_m as a function of the camera slit width.

	Catalpa		Pine needles		Polar green leaves		Polar yellowish leaves	
slite [mm]	FWHM [ns]	τ_m [ns]	FWHM [ns]	τ_m [ns]	FWHM [ns]	τ_m [ns]	FWHM [ns]	τ_m [ns]
0.1	0.20	0.41	0.29	0.43	0.24	0.62	0.25	1.25
0.2	0.25	0.39	0.32	0.42	0.28	0.60	0.30	1.29
0.3	0.33	0.44	0.36	0.46	0.33	0.58	0.36	1.23
0.4	0.39	0.44	0.45	0.44	0.39	0.56	0.40	1.25
0.5	0.48	0.43	0.51	0.42	0.44	0.58	0.48	1.22
0.6	0.54	0.51	0.61	0.48	0.52	0.54	0.59	1.20
0.8	0.70	0.44	0.72	0.49	0.68	0.54	0.70	1.24
1.0	0.83	0.47	0.82	0.54	0.81	0.54	0.83	1.25

Influence of the spatial geometry of the illuminated plants

In the afore-mentioned experiments, the laser illuminated surface was roughly perpendicular to the excitation beam and was almost planar except for the pine branch for which the response function recorded with 0.1 mm slit showed a FWHM of ≈ 300 ps, significantly larger than the mean value

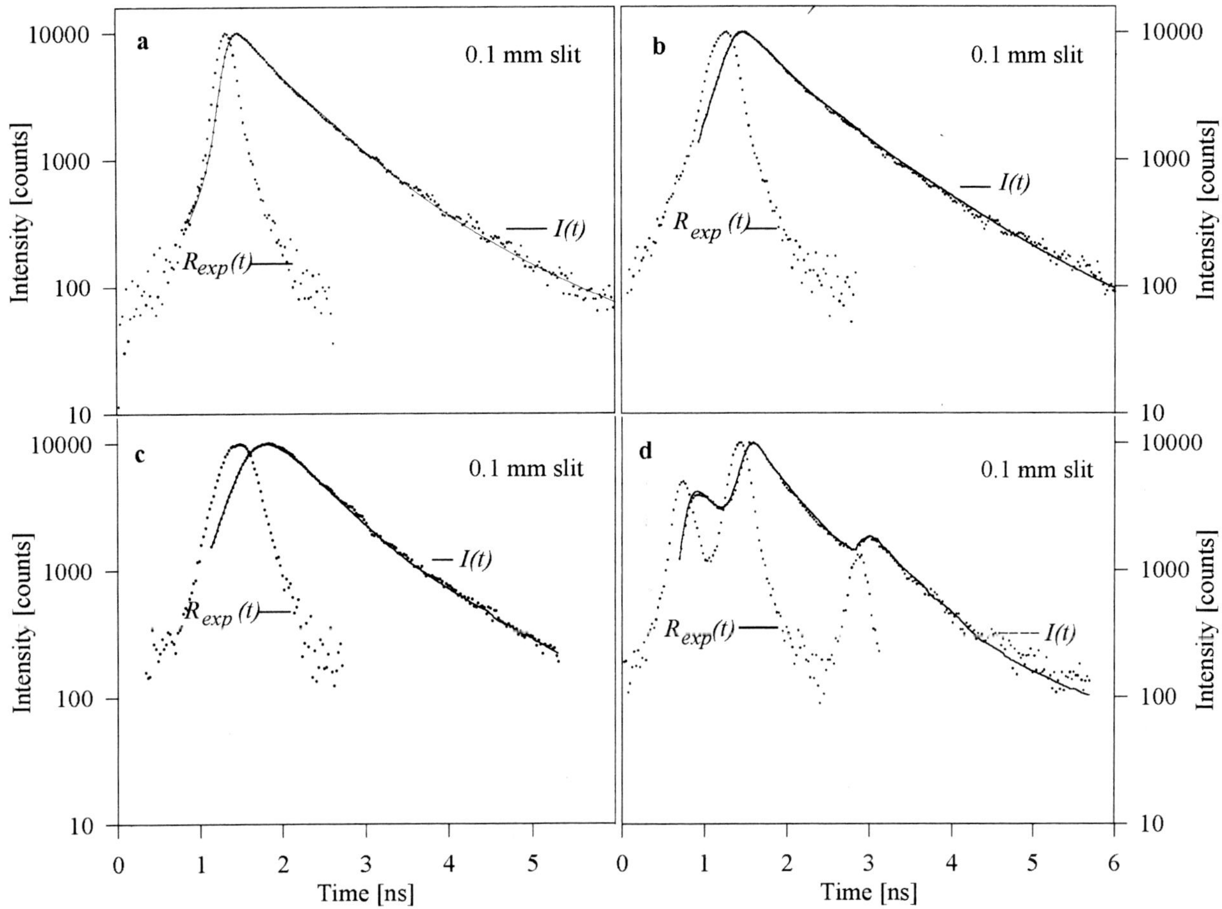

Fig. 3: Influence of the target geometry on $R_{exp}(t)$ and on $I(t)$ for poplar leaves (distance of 50 m, slit width of 0.1 mm) (see Table 2): a) planar leaf, b) tilted leaf (or leaves in the same plane), c) few leaves very close together and d) three separated leaves on a branch; (· · · · · experimental curves; full line – fitted curves).

(≈ 230 ps) obtained for the other leaves. This is due to the fact that the pine needles were not in the same plane leading to different propagation delays of the reflected light. As for far-field measurements, the laser beam can excite one leaf with any orientation or several leaves separated by a distance of few centimeters or more, we have tested if even in such complex situations, accurate determination of $F(t)$ can be made. Fig. 3 illustrates the results obtained with different typical conditions of illumination of the foliage of an outside poplar situated at 50 m from the telescope, with an entrance slit of 0.1 mm and a laser spot diameter of 20 cm on the leaves level. The excited surface could be clearly observed with the sighting system of the telescope. The response functions measured for the excitation of a single leaf whose plane is practically perpendicular (Fig. 3 a) to the excitation beam or strongly tilted (Fig. 3 b) had FWHM values respectively equal to 250 and 380 ps; when the laser spot reached several leaves very close together (Fig. 3 c) FWHM increased up to 500 ps. We have obtained the curves exhibiting for $R_{exp}(t)$ three peaks with FWHM ≈ 300 ps (Fig. 3 d) in a very sparce foliage for three leaves situated in different planes. If z_1 and z_2 are the distances between the nearest leaf and the second and third one, respectively, the response function can be simulated by

$$R_{exp}(t) = [b_1 L(t) + b_2 L(t - 2z_1/c) + b_3 L(t - 2z_2/c)] \otimes R_i(t)$$

where b_1, b_2 and b_3 represent the fraction of light falling on leaves 1, 2 and 3, respectively. The peak separations in Fig. 3 d corresponded to z_1 = 10 cm and z_2 = 33 cm. It was always possible to fit the recorded fluorescence decays $I(t)$ with $F(t)$ as a sum of two or three exponentials, as shown by the good agreement between the experimental and calculated curves of Fig. 3 a–d. For the single planar leaf, perpendicular to the excitation beam (Fig. 3 a), the experimental decay curve could be resolved into 3 components with τ_1 equal to about 100 ps, τ_2 = 0.35 ns and τ_3 = 1 ns giving τ_m = 0.647 ns. As already noted in the case of increasing slit widths, even if FWHM of $R_{exp}(t)$ response was deteriorated or presented several peaks, the τ_m values (≈ 0.64 ns) varied by less than 10 % (Table 2).

Remote measurements feasibility

Since the set-up is aimed to be used for far-field experiments, we have checked, on poplars at a distance of 40 to 80 m from the telescope, the reliability of the results for different weather conditions: in bright sunshine, under moving

Table 2: Chlorophyll fluorescence lifetimes (components τ_1, τ_2, τ_3 as well as mean value (τ_m) for different geometries of the leaf target of poplars measured at 50 m distance.

Poplar leaves	τ_1 [ns]	τ_2 [ns]	τ_3 [ns]	τ_m [ns]
single planar leaf (Fig. 3 a)	0.087	0.351	0.990	0.637
single tilted leaf (Fig. 3 b)	0.090	0.310	0.998	0.667
few leaves close together (Fig. 3 c)	0.116	0.490	1.209	0.702
three branches of leaves (Fig. 3 d)	0.047	0.433	1.663	0.619

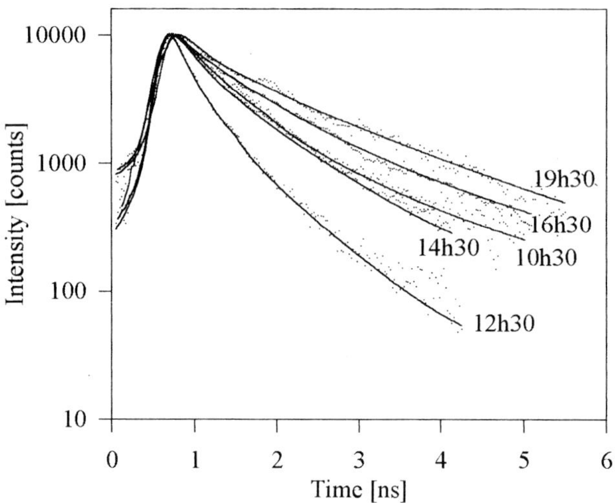

Fig. 4: Daily changes in the chlorophyll fluorescence decays of poplar leaves on the tree measured in October 1993.

clouds and also in the presence of wind or rain. A great number of measurements have been performed with, for each one, the simultaneous records of the response function $R_{exp}(t)$ and the chlorophyll fluorescence $I(t)$ curves. It is important to note that with our laser source and light collection system the detected chlorophyll fluorescence was high enough to allow the use of a small entrance slit of camera (0.1 mm) with illumination lasting only few seconds. For identical conditions a good reproducibility of the mean lifetime values has been found, without difficulties for numerical fitting.

Chlorophyll fluorescence lifetime variation during the day

It is very well known that the photosynthetic activity i.e. the efficiency and kinetics of the various different competiting chemical reactions depends not only on the state of plants (age, growing and physiological conditions, etc.) but can also vary during the daily cycle. Since the chemical activity and the chlorophyll fluorescence properties are correlated, we have performed lifetime measurements at different hours of the day in order to determine the τ_m variation. The experiments have been carried out during several days in October 1993 on the outside poplars at a distance of 40–60 m.

Typical chlorophyll fluorescence decays, recorded at different hours of the day are given in Fig. 4 which shows that the variation of the temporal properties of the chlorophyll *a* emission can be monitored by the streak camera measurements. The mean lifetime τ_m was about 0.8 ns in the morning, decreased to 0.4 ns in midday and then slowly increased to reach the value of 1.21 ns in the night (Table 3). In Fig. 5 are reported, for twenty consecutive measurements, carried out on the same leaves at 22 h, the τ_1 ($\approx$ 250 ps), τ_2 ($\approx$ 1.5 ns) and τ_m ($\approx$1.26 ns) values resulting from the numerical adjustement with a sum of two exponentials for $F(t)$ as well as τ_m = 1.19 ns obtained by fitting with three exponentials. One can observe, that the reproducibility was very good and that the average lifetimes τ_m deduced from two or three exponential decays were nearly identical. A comparison between the τ_1, τ_2 and τ_3 (Table 3) obtained from the fitting of the decays profiles represented in Fig. 4 seems to indicate that all of them depended on the illumination conditions. This is in agreement with some results obtained for modelling chloroplasts at low temperature where the authors preilluminated the samples before measurements (Hodges and Moya, 1987; Hwang-Schweitzer et al., 1992).

Fig. 6, where are collected all the results obtained in October 1993 (each point is the average value of at least 10 measurements), shows that the chlorophyll fluorescence lifetime

Table 3: Daily cycle variation of the poplars leaves chlorophyll fluorescence lifetimes given for the components τ_1, τ_2, τ_3 as well as the mean value τ_m.

Time of a day	τ_1 [ns]	τ_2 [ns]	τ_3 [ns]	τ_m [ns]
10 h 30	0.033	0.363	1.323	0.80
12 h	$\leq$ 0.010	0.138	0.618	0.42
14 h 30	0.063	0.208	0.964	0.78
16 h 30	0.070	0.446	1.688	1.15
19 h 30	0.170	0.619	1.662	1.21

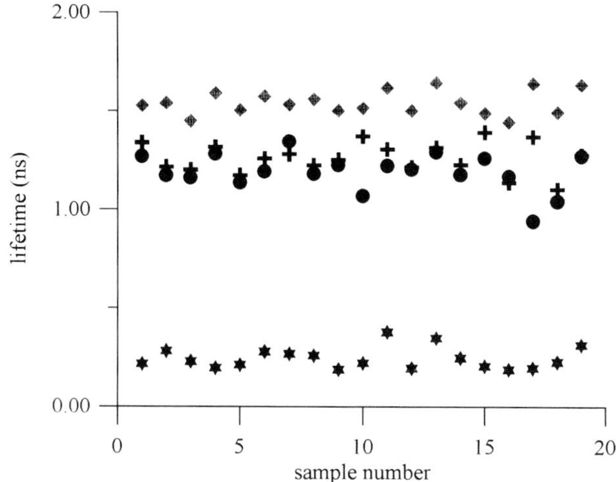

Fig. 5: Reproducibility of the chlorophyll fluorescence decay constants of poplar leaves at 50 m distance (series of measurements at 22 h), ★ – τ_1, ◆ – τ_2, ✚ – τ_m obtained after fitting with two exponentials, ● – τ_m after adjustment with three exponentials.

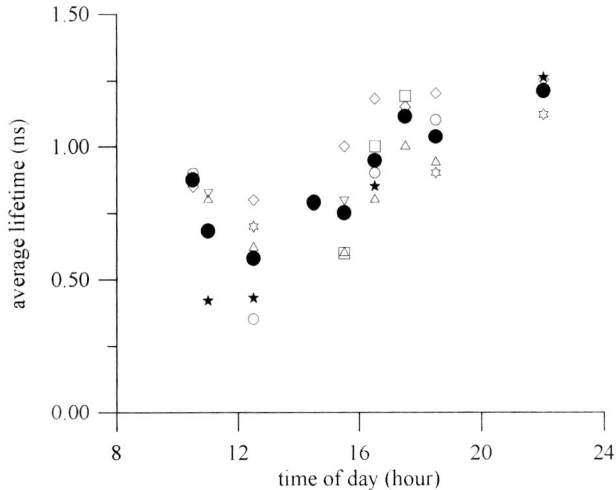

Fig. 6: Daily cycle of the chlorophyll fluorescence average lifetimes of poplar leaves: ◇ – October 15, 1993 (cloudy day); □ – October 16, 1993 (cloudy day); ○ – October 18, 1993 (sunny day); ☆ – October 19, 1993 (rainy); △ – October 20, 1993 (cloudy, rainy); ▽ – October 22, 1993 (rainy); ★ October 29, 1993 (sunny day) ● – average values over all the days.

was lowest around midday and exhibited a maximum during the night when the photosynthetic activity is very weak.

Conclusion

In this work we have demonstrated the usefulness of a streak camera for time resolved chlorophyll fluorescence measurements at long distance which can be applied for far-field detection of the health status of plants. For an excitation at 532 nm, the experimental set-up was sensitive enough to allow, for an illumination duration of a few seconds, the recording of the time resolved chlorophyll fluorescence emission of the leaves situated at 40–80 m from the telescope and applying an entrance slit width of the camera as small as 100 µm to which corresponded a temporal resolution of 200 ps.

Also shown is the possibility to easily restore the decay law *F(t)* by numerical fit by taking into account the response function $R_{exp}(t)$ and a two or three exponential law for the emission to determine with a good accuracy the average chlorophyll fluorescence lifetimes even in unfavorable conditions (large entrance slit width, not well defined excited surface, different weather conditions) for which $R_{exp}(t)$ is very broad or presents several peaks. The experiments have shown that although measured chlorophyll fluorescence decays *I(t)* depended on the canopy factors, the decay laws *F(t)* were found independent of the geometrical parameters after deconvolution.

The major advantage of the streak camera compared to other lifetime measurements techniques is that for high enough fluorescence signal it allows real time measurements and that when the accumulation is necessary it lasts always less time than for the experiments with most commonly used the single photon counting method.

References

Chappelle, E. W., F. M. Wood, J. E. McMutrey, and W. W. Newcomb: Laser induced fluorescence of green plants. I. A technique for remote detection of plant stress and species differentiation. Appl. Opt. *23,* 134–138 (1984).

Evans, E. H., R. G. Brown, and A. R. Wellburn: Chlorophyll fluorescence decay profiles of O_3-exposed spruce needles as measured by time-correlated single photon counting. New Phytologist *122,* 504–506 (1992).

Geist, P., F. Heisel, A. Martz, and J. A. Miehé: Temporal performances of a 400 kHz triggered streak camera used with mode-locked cavity dumped laser. Optics Comm. *45,* 17–20 (1983).

– – – Caméra à balayage de fente et métrologie picoseconde des impulsions de lasers fonctionnant à des taux de répétition élevés. Revue Phys. Appl. *19,* 619–629 (1984).

Genty, B. J., J. Briantais, and N. Baker: The relationship between the quantum yield of photosynthetic electron transport and quenching of chlorophyll fluorescence. Biochim. Biophys. Acta *990,* 87–92 (1989).

Goulas, Y.: Télédétection de la fluorescence des couverts végétaux: Temps de vie de la fluorescence chlorophyllienne et fluorescence bleue. Thesis, Université de Paris-Sud, (France) (1992).

Goulas, Y., L. Camenen, G. Schmuck, G. Guyot, F. Morales, and I. Moya: Picosecond fluorescence decay and backscattering measurements of vegetation over distances. Eureka LASFLEUR Rapport, 1993.

Heisel, F., J. A. Miehé, and B. Sipp: Détection et analyse de phénoménes lumineux brefs. Ann. Phys. Fr. *4,* 331–370 (1979).

Hodges, M. and I. Moya: Time resolved fluorescence studies on photosynthetic mutants of *Chlamydomonas reinhardtii:* origin of the kinetic decay components. Photosynth. Res. *13,* 125–141 (1987).

Hwang-Schweitzer, R. H., R. S. Knox, P. B. Gibbs, and J. Biggins: Fluorescence studies of photoregulation in the chrysophyte *Ochromonas danica.* J. Lumin. *51,* 99–109 (1992).

Kautsky, H. and A. Hirsch: Neue Versuche zur Kohlensäureassimilation. Naturwissenschaften *19,* 964 (1931).

Krause, G. H. and E. Weis: Chlorophyll fluorescence and photosynthesis: The basics. Annu. Rev. Plant Physiol. Plant Mol. Biol. *42,* 313–349 (1991).

Lang, M., H. K. Lichtenthaler, M. Sowinska, P. Summ, and F. Heisel: Blue, green and red fluorescence signatures and images of tobacco leaves. Botanica Acta *107,* 230–236 (1994).

Lang, M., P. Siffel, Z. Braunova, and H. K. Lichtenthaler: Investigations on the blue-green fluorescence emission of plant leaves. Bot. Acta *105,* 35–40 (1992).

Lang, M., F. Stober, and H. K. Lichtenthaler: Fluorescence emission spectra of plant leaves and plant constituents. Radiat. Environ. Biophys. *30,* 333–347 (1991).

Lichtenthaler, H. K.: *In vivo* chlorophyll fluorescence as a tool for stress detection in plants. In: Lichtenthaler, H. K. (ed.): Applications of Chlorophyll Fluorescence, pp. 143–149. Kluwer Academic Publishers, Dordrecht (1988).

– Application of chlorophyll fluorescence in stress physiology and remote sensing. In: Steven, M. and J. A. Clark (eds.): Application of Remote Sensing in Agriculture, pp. 287–305. Butterworths Scientific Ltd., London (1990).

– The Kautsky Effect; 60 years of chlorophyll fluorescence induction kinetics. Photosynthetica *27,* 45–55 (1992).

Lichtenthaler, H. K. and U. Rinderle: The role of chlorophyll fluorescence in the stress detection in plant. CRC Crit. Rev. in Anal. Chem. *19,* Suppl. I, 29–85 (1988).

Marquardt, D.: An algorithm for least-squares estimation of nonlinear parameters. J. SIAM *11,* 431–441 (1963).

Moya, I., Y. Goulas, and J. Briantais: Techniques pour la télédétection de la durée de vie et du rendement quantique de la fluorescence de la chlorophylle *in vivo,* Proc. 4th Int. Coll. on Spectral Signatures of Objects in Remote Sensing (Aussois (France)) (1988).

Moya, I., M. Hodges, and J. Barbet: Modification of room-temperature picosecond chlorophyll fluorescence kinetics in green algae by photosystem II trap closure. FEBS Lett. *198,* 256–262 (1986).

Rinderle, U. and H. K. Lichtenthaler: The various chlorophyll fluorescence signatures as a basis for physiological ground truth control in remote sensing of forest decline. In: Intern. Geosience and Remote Sensing Symposium, IGARSS'89, Vancouver, Vol. *2,* pp. 674–677 (1989).

Schmidt, W. and H. Schneckenburger: Time-resolving luminescence techniques for possible detection of forest decline. Radiat. Environ. Biophys. *31,* 63–72 (1992).

Schmuck, G., I. Moya, A. Pedrini, D. Van Der Linde, H. K. Lichtenthaler, F. Stober, C. Schindler, and Y. Goulas: Chlorophyll fluorescence lifetime determination of water stressed C_3 and C_4-plants. Radiat. Environ. Biophys. *31,* 141–151 (1992).

Schneckenburger, H. and M. Frenz: Time-resolved fluorescence of conifers exposed to environmental pollutants. Radiat. Environ. Biophys. *25,* 289–295 (1986).

Subject Index

The page numbers next to the individual keywords refer to the page where the paper begins dealing with that particular topic indicated by the keyword.

Author Index

The first author of a paper is indicated by bold print of the page number